PHANEROZOIC TIME

Era	Period	Epoch	Age (Millions of years ago)

Holocene (0.1)
Pleistocene (2)
Pliocene (5)
Quaternary

- *Homo sapiens:* 500,000 years ago
- Origin of *Homo:* 2 million years ago
- North and South America connected: 3.6 to 3.1 million years ago
- Formation of deserts: 4 million years ago to present
- Glaciation in arctic: 4 million years ago
- Origin of hominids: 5.5 million years ago
- Glaciation in Antarctica: 13 million years ago
- Worldwide forests recede: 15 million years ago to present

Cenozoic

Tertiary

Miocene — Earliest hominoids
Oligocene — South America separates from Antarctica
Eocene — Anthropoid primates
— 50 — Australia separates from Antarctica
Paleocene — Early insectivores and primates
— Fourth major extinction event: Dinosaurs and ammonites extinct
— 100 — Final separation of South America and Africa
— Angiosperms dominant worldwide

Mesozoic

Cretaceous

— First fossil angiosperm

— 150 — *Archaeopteryx:* first fossil bird

Jurassic

— 200 — Mammals appear
— Dinosaurs appear
— First fossil mammals

Triassic

— 250 — Disappearance of trilobites
— Appearance of therapsid reptiles

Permian

— Conifers appear

Paleozoic

Carboniferous

— 300 — Oldest fossil reptiles
— Oldest fossil insects
— Forests appear and dominate the land
— 350 — Amphibians appear on land
— Third major extinction event
— Age of fishes

Devonian

— 400 — Plants colonize the land
— First jawed fishes

Silurian

— First fossil plants
— 450 — Second major extinction event

Ordovician

— Diversification of mollusks
— Oldest fossil crustaceans and Onychophora
— 500 — First major extinction event
— Explosive evolution of phyla

Cambrian

— Evolution of lancelets and lampreys
— 550 —

— Evolution of external skeletons

— 600 —

UPDATED VERSION
BIOLOGY
Third Edition

UPDATED VERSION
BIOLOGY
Third Edition

Peter H. Raven
Director, Missouri Botanical Garden;
Engelmann Professor of Botany,
Washington University, St. Louis, Missouri

George B. Johnson
Professor of Biology,
Washington University, St. Louis, Missouri

Wm. C. Brown Publishers
Dubuque, Iowa•Melbourne, Australia•Oxford, England

Book Team

Editor *Carol J. Mills*
Developmental Editor *Diane Beausoleil*
Publishing Services Coordinator, Production *Julie Avery Kennedy*
Publishing Services Coordinator, Design *Barbara J. Hodgson*

Wm. C. Brown Publishers
A Division of Wm. C. Brown Communications, Inc.

Vice President and General Manager *Beverly Kolz*
Vice President, Publisher *Kevin Kane*
Vice President, Director of Sales and Marketing *Virginia S. Moffat*
Vice President, Director of Production *Colleen A. Yonda*
National Sales Manager *Douglas J. DiNardo*
Marketing Manager *Craig Johnson*
Advertising Manager *Janelle Keeffer*
Production Editorial Manager *Renée Menne*
Publishing Services Manager *Karen J. Slaght*
Permissions/Records Manager *Connie Allendorf*

Wm. C. Brown Communications, Inc.

President and Chief Executive Officer *G. Franklin Lewis*
Corporate Senior Vice President, President of WCB Manufacturing *Roger Meyer*
Corporate Senior Vice President and Chief Financial Officer *Robert Chesterman*

Stained glass cover designed and fabricated by St. Louis Stained Glass Studios
Stained glass part panels fabricated by St. Louis Glass Studios
Stained glass photographed by Leslie/Brauer Studios

The credits section for this book begins on page IA–1 and
is considered an extension of the copyright page.

A Times Mirror Company

ISBN 0–697–24251–X

Printed in the United States of America by Wm. C. Brown Communications, Inc.,
2460 Kerper Boulevard, Dubuque, IA 52001

10   9   8   7   6   5

Preface

With this third edition *Biology* comes of age. We began work on the first edition in 1982, when no full-color text had ever been published for biology majors (ours was the first). That first effort was very ambitious, attempting to recast the teaching of biology in terms of evolution—not just a chapter devoted to Darwin, but a restructuring of each chapter to reflect evolutionary principles. In some instances we over-reached. A chapter devoted to cell movement in the basic Cell Biology section was perhaps before its time. In large measure, however, the first edition succeeded very well in presenting biology from an evolutionary perspective, and was widely used.

The second edition of *Biology*, like the second edition of many texts, provided a chance to address the shortcomings of the first. In essence, the first edition was too much: too long, too detailed, too high level—the faults of enthusiasm, which now had to be tempered by experience. Using the input from hundreds of classrooms, we ruthlessly pared out the needless detail and rewrote and rewrote and rewrote until the reading level was the lowest of any majors text. We also reorganized the presentation within many chapters to more closely match that of teachers in classrooms across the country. The book that resulted was noticably trimmer than the first edition, and a far more powerful teaching tool. If the first edition, bulging out between its white covers, was the great white whale, then the second edition was orca, svelte and effective.

The second edition has been even more successful than the first, and is in use in classrooms all over the United States and Canada. The evolution of our text has not stopped there, however. Having addressed length and level in the second edition, we were now able to stand back and ask "How can we make a very good text into the best possible text?" While we were not the first authors to ask themselves this question, we have answered it in a novel way. After considerable reflection, a focus group, and a lot of discussion with our publisher, we have chosen to commit a considerable fraction of our royalties with matching funds from the publisher, to making *Biology*, third edition the most superbly illustrated majors text ever attempted. By now, all the competing majors texts have become full-color, but no art program, our own second edition included, begins to reach the potential we felt could be achieved. We felt it was time to attempt the next evolutionary stage. What has resulted is the book you hold in your hands. We believe that this third edition of *Biology* is the most effec-

tive teaching tool yet published to introduce the science of biology to students planning on majoring in the subject. While comprehensive and up-to-date, it is also easy and fun to read, and its illustrations support and guide a student's learning better than any book in the past. We are confident *Biology* will continue to evolve as new technologies change our classrooms. Today, however, we feel as its proud parents that *Biology*, third edition stands on the leading edge of textbook development, the very best that has ever been. Like any parent, we are probably not fully objective—who could be?—but look for yourself. We hope that what you see will spark your enthusiasm too.

Because of the importance we attach to this approach, we have continually revised and rethought the extent and placement of our evolutionary coverage in the second and third editions. *Biology*, third edition is the culmination of two editions worth of testing and perfecting our evolutionary framework of teaching basic biology, and we are confident that we have struck the right balance between conceptual presentation and our evolutionary perspective.

Science has not stood still, of course, as we perfected our text. Ten years is a lifetime in a rapidly advancing field like biology, and much of the content of the third edition of *Biology* has changed from those cold winter days of 1982 in Edinburgh and St. Louis when we wrote the first edition. Who would have dreamed that in this edition we would be describing the successful cure of an inherited disease—a healthy gene being transferred "piggy back" on a virus into a little girl unable to combat infection because neither of her own chromosomes had a workable copy of the gene encoding a key enzyme. Few nightmares in 1982 would have envisioned that the world's environment would deteriorate as rapidly as it in fact has, reeling from the stress of supporting 5.3 billion people. When we started to write, ten years ago, the first AIDS case had only recently been reported and fewer than 500 had died in the United States; as we write this, over 180,000 have died of AIDS in the United States and up to 2 million individuals are infected with the virus as researchers continue to feverishly seek a cure.

All of these changes, and many others, are reflected in this revision, as our attempt to keep *Biology* up-to-date is endless, a constant adjustment to the changing terrain of the science.

PEDAGOGICAL AIDS FOR THE STUDENT

One of the reasons why our text has been so popular is that we are committed to helping students use and understand our text. Our pedagogical program reinforces this commitment. Each chapter begins with a *"For Review"* list of terms that students should know before starting the chapter. Chapter numbers are given for each term for easy reference. A new element appears in this edition—the inclusion of an *outline* at the beginning of each chapter. An *overview*—a short paragraph describing the educational goal of the chapter—comes next. This overview provides a short synopsis of the chapter for the student, and can be used in reviewing chapters for exams. All key terms are set in **boldface** type. *Concept summaries* are short summaries that occur periodically throughout each chapter which "spot" summarize the preceeding material. Students have found these concept summaries invaluable, and they are an important component of our pedagogical program. Another feature popular with students are the *boxed essays*, short (usually no more than a page) presentations of a wide variety of topics that interest students today. We've kept our boxes on DNA fingerprinting and AIDS on the college campus, updating the information for the third edition. We've also added boxes on the first use of gene therapy to fight cancer; drugs and the central nervous system; dialysis; and the first "photograph" of DNA. These boxes provide reinforcement of the concepts presented and help to engage students in their study of biology by showing how biology applies to the world at large. An element hailed for its simplicity and practicality, the *mini-glossary*, has been carried over from the second edition. These mini-glossaries provide short lists with definitions of key terms in selected chapters notorious for their vocabulary, such as Chapter 15 (Genes and How They Work) and Chapter 20 (The Evidence for Evolution). Finally, the end-of-chapter pedagogy includes an enumerated *summary, review questions, thought questions,* and *For further reading*. These elements are designed to reiterate important themes, test knowledge, challenge comprehension, and provide additional resources for interested students. A few words in particular should be said about the *"For further readings"* sections: we include texts and articles specifically chosen for their readability and interest potential, and we are rigorous in updating. You'll notice *many* readings from 1990 and 1991; we've been known to insert "last minute" articles hot off the press in page proof, to the everlasting chagrin of our editors.

WHAT'S NEW AND WHY

Biology, third edition represents a major revision effort in both the illustration program and textual content. In today's biology texts, art and text are both critically important, and both facets must receive intense attention from the authors and editorial team. We have put much effort into revising art and text. Here's what we have accomplished:

SUCCESSFUL FEATURES CARRIED OVER FROM PREVIOUS EDITIONS
Organization

Biology, third edition, like the second edition, is organized into 11 parts, each with distinct "categories" of information. The first section is devoted to the principles of biology, which apply to all organisms—the origin of living things, cells, energetics, reproduction and heredity, and molecular genetics. We then devote the next four chapters to evolution, building upon the coverage of genes and heredity presented earlier. After ev-

olution, we proceed directly to ecology in order to provide the student with a sequential, cumulative flow of information. The second half of the text is devoted to the biology of particular organisms. It begins with a section on biological diversity, the product of the marriage of evolution and ecology that is the culmination of the previous section. We then treat the biology of plants. Finally, in the last 14 chapters, we detail vertebrate biology, examining animals from a structure/function perspective. We believe that this organization, one that builds on previously introduced themes and concepts, provides students with a pedagogically logical method of learning biology. However, for those professors who wish to use the chapters in a different order, our unique learning aid, *"For Review,"* lists terms at the beginning of each chapter that students should know before proceeding further.

Evolutionary Theme

Biological principles do not exist in a vaccuum; there is almost always an evolutionary explanation that explains why things are the way they are. From the outset of writing *Biology*, our goal has been to describe biology in its evolutionary context. For instance, the existence of two photosystems in the chloroplast is not just given to the student as a fact to memorize; rather, we seek to explain that two photosystems reflect the evolutionary history of photosynthesis. Similarly, evolution provides a framework for understanding why centrioles have DNA. Presenting biology in an evolutionary framework enables the student to *understand* principles, and so actually makes teaching, and learning, easier.

The Art Program

Using an art coordinator, a designer, and our developmental editor, we have developed illustrations that *teach* concepts. A few examples of these illustrations may be found in Chapter 47, (Hormones), such as Figures 47-6 and 47-7. Use of arrows, numbers, and sequential boxes giving students "direction" in understanding the illustrations are consistent themes we've employed in the art program. You'll also notice that colors are brighter and art sizes are larger—these two qualities contribute not only to the attractiveness of the text, but to the clarity of the illustrations. See, for example, the illustrations in Chapter 6 (Membranes). Larger, brighter art is easier to "read" than cramped, dingy illustrations. We have also added, in this edition, bold-faced titles to all of our figures. We feel that this small change increases the utility of the figures tremendously. Students now have a short title for the illustrated concept, making retention easier.

The Text

We've added, deleted, rewrote and clarified almost every chapter of the third edition. The changes are based on our own continuing research in biology and also on the reviews of professors. One consistent change that occurs throughout the text is the change in headings. We have organized information more effectively using headings, and have worded these headings in a more interesting way. For example, in Chapter 2, the heading "Electron Orbitals" has been changed to "Electrons Determine What Molecules are Like," a simple, more explanatory heading than the previous one. All of the new headings are gathered in the new chapter outline that appears on each chapter opening page. While we will not attempt here to detail each and every change that took place in preparing the third edition, the following are highlights (for easier reading, we've broken the following section down into parts).

Part I: The Origin of Life

In Chapter 1 (The Science of Biology) we've added more information on the scientific method, including sections on controls and the importance of prediction. In Chapter 2 (The Nature of Molecules) we're clarified information on atoms, elaborating a bit more on the nature of protons, neutrons and electrons. There is new information on oxidation reduction, energy levels, ionic bonds, and hydration shells. We've simplified the octet rule, and eliminated terms such as "quantum mechanics." Similarly, Chapter 3 (The Chemical Building Blocks of Life) provides more explanation and less extraneous detail about the chemical matters biology students need to know. There is more discussion of functional groups and important biological substances such as cellulose and lipids. An easier, lower level discussion of nucleic acids is featured in this chapter, with the structure of DNA introduced here for the first time, to be built upon later in the text. Chapter 4 (The Origin and Early History of Life) contains information on the latest theories concerning the origin of organic molecules and the earliest organisms. We've also kept our intriguing discussion about the possibility of life on other worlds. A new step by step progression through the fossil record in this chapter makes easier—and more interesting—reading.

Part II: Biology of the Cell

Chapter 5 (Cell Structure) uses the entire cell as an icon as each part of the cell is illustrated. For example, when the mitochondrion is illustrated, an icon of the animal cell presented in Figure 5-8 is also shown with its mitochondria highlighted. This method of illustration gives students "the big picture" while describing the details of cell structure. Also new to this chapter are clearer explanations of the nature of the cell and the structure and function of lysosomes. We've added a brief introduction to symbiosis and an overview of the plasma membrane structure. To our discussion of the nucleus, we've added brief descriptions of chromosomes and RNA, which will be elaborated upon later in the text. Finally, more information is given about centrioles and their interesting evolutionary origin. Chapter 6 (Membranes) features a stronger presentation of diffusion that clarifies movement of solutes, along with a clearer illustration (Figure 6-10). A new discussion of receptor-mediated endocytosis and all-new concept summaries complete the revision of this chapter.

Part III: Energetics

In Chapter 7 (Energy and Metabolism) we've added a discussion of enzyme inhibition and activation, as well as an all-new section on the evolution of metabolism, which details the evolution of anaerobic and aerobic processes. We've also reworked Figures 7-7 and 7-8, which illustrate endergonic and exergonic reactions and the role of enzymes in reactions. Chapter 8 (Cellular Respiration) contains a new, clearer overview of the cellular respiration, as well as a new, complete explanation of beta oxidation. Figure 8-18 has been modified to include the oxidation of fats and proteins. In Chapter 9 (Photosynthesis), we've rewritten our discussion of C_3 and C_4 photosynthesis, emphasizing the evolutionary aspects.

Part IV: Reproduction and Heredity

Mitosis and meiosis are very visual concepts and it is important to illustrate these concepts effectively; therefore new illustrations characterize the revision of the chapters in this part. Clearer diagrams of bacterial replication (Figure 10-2), levels of chromosomal organization (Figure 10-4), the cell cycle (Figure 10-7), the sexual life cycle (Figure 11-4), meiosis (Figure 11-9) and a comparison of meiosis and mitosis (Figure 11-11) make the concepts presented in Chapters 10 (The Cell Cycle: Mitosis) and 11 (Sexual Reproduction and Meiosis) easier to understand. Chapter 12 (Mendelian Genetics) has more concept summaries than in the previous edition, and it also features clearer discussions of multiple alleles, epistasis, continuous variation, pleitropy, incomplete dominance, and genetic effects. We've also added an increasingly popular genetic screening procedure, chorionic-villi sampling to Chapter 13 (Human Genetics).

Part V: Molecular Genetics

In Chapter 14 (DNA: The Genetic Material), you'll notice new headings that enhance the "scientific method" approach to this chapter. For example, the Messelson-Stahl experiment is introduced by its heading "Replication is Semiconservative," which briefly summarizes the results of the experiment for the student. New headings organize the discussion of genes, breaking material into discrete chunks for easier comprehension. Chapter 15 (Genes and How They Work)—a daunting chapter for most students—benefis from rewriting and reorganization. Many sections of this chapter have been revised for clarity and easier understanding, including the sections on the genetic code and activating enzymes. New illustrations depicting protein synthesis are also featured in this chapter. Chapter 18 (Gene Technology) clarifies the action of restriction enzymes for students in a new, clearer discussion, and the new illustrations provide solid graphic support. Many new findings are presented—this is indeed one of the fastest-changing areas in biology.

Part VI: Evolution

Expanded coverage of scientific creationism is featured in Chapter 20 (The Evidence for Evolution), explaining in unequivocal terms why scientific creationism is *not* scientific. A new illustration depicting punctuated equilibrium and gradualism (Figure 20-13) has been added to the chapter. Chapter 22 (Evolutionary History of Life on Earth) has been augmented by new illustrations showing C^{14} dating (Figure 22-4) and four possible human family trees (Figure 22-24).

Part VII: Ecology

Chapter 24 (Coevolution and Symbiosis—Interactions within Communities) clarifies the difference between predation and parasitism, a point of possible confusion for students. Chapter 25 (Dynamics of Ecosystems) has been enhanced by all new figures depicting the biogeochemical cycles (see, for instance, the nitrogen cycle, Figure 25-4). The Hubbard Brook experiment is used in this chapter as an example of cycling in a forested ecosystem. Not only are students acquainted with an elegant scientific design in this discussion, but they are also shown how scientific experiments relate to the "real" world and its potential problems. In Chapter 27 (The Future of the Biosphere), we supplement our discussion of world population growth and destruction of the rain forest with new information about nuclear power, pollution, and global warming.

We've also included a brief section on environmental science—its methods and goals, and how important this particular branch of science is likely to become in the future. We've also included "success" stories, such as the clean-up effort of the Rhine river, to show students that many of the environmental problems they will be confronting are *not* insurmountable, but can be solved.

Part VIII: Biology of Viruses and Simple Organisms

Chapter 28 (The Five Kingdoms of Life) features revised discussions of the binomial system and a clearer definition of species. We cut much detail out of Chapter 29 (Viruses), using boxed essays to cover hepatitis and flu viruses. Chapter 30 (Bacteria) features all-new coverage of Lyme disease, including two new photos (Figure 30-15). In Chapter 31 (Protists) we have yielded to the pleas of reviewers to provide a table organizing the protists into categories; our table is unique in that it does not use artificial ("animal-like," "plant-like") categories of classification, yet still provides students with a useful means of cataloging the different species. In Chapter 32 (Fungi), we've been careful to update the numbers for each phylum, which have changed since the last edition of *Biology* was published, carrying these changes into the classification scheme that appears in Appendix A.

Part IX: Biology and Plants

In the third edition of *Biology*, we've enhanced our excellent coverage of botany in several ways. First, in Chapter 33 (Diversity of Plants), we've simplified our explanation of the divisions of plants. We've also added more descriptive headings, such as "Mosses, Liverworts, and Hornworts." Our coverage of the seedless vascular plants has been completely rewritten by simplifying the narrative and providing comparisons to previously discussed plants. Likewise, our coverage of seed plants has been greatly revised, especially the discussion of gymnosperms. Also in this chapter, we've provided clearer diagrams of the life cycle of pine and the angiosperms. Finally, a new mini-glossary acquaints students with unfamiliar vocabulary. Chapter 36 (Nutrition and Transport in Plants) offers new figures showing pathways of mineral transport in roots (Figure 36-4), mass flow (Figure 36-8), and cavitation (Figure 36-3). Chapter 38 (Regulation of Plant Growth) is enhanced by new diagrams of the flowering response (Figure 38-12), the Darwin's experiment (Figure 38-1), and Fritz Went's experiment (Figure 31-7). In all the botany chapters, headings have been rewritten to be more descriptive, and the prose simplified for easier comprehension.

Part X: Biology of Animals

Continuing in the tradition of our highly popular coverage of diversity, the third edition of *Biology* actually improves upon the second edition with the addition of headings that contain both phylum and species name. New illustrations depicting the animal ancestral tree (Figure 39-2), anatomy of a sponge (Figure 39-4), anatomy of *Hydra* (Figure 39-9), life cycle of *Obelia* (Figure 39-11), structure of a tapeworm (Figure 39-20), patterns of embryonic development (Figure 40-2), locomotion in an earthworm (Figure 40-17), metamorphosis in a silkworm moth (Figure 41-34) and embryonic development of a vertebrate (Figure 42-15) are featured in this section.

Part XI: Vertebrate Biology

This last section of *Biology*, third edition has been extensively rewritten and revised. Physiology of vertebrate organisms is an extremely important part of an introductory biology course, and we have given much attention to our coverage of this important topic. Starting with Chapter 43 (Organization of the Vertebrate Body), our changes are evident. We've moved coverage of skin to Chapter 52 (The Immune System), since skin constitutes the body's first line of defense against invading microbes. We've completely rewritten Chapter 43 (Neurons), providing more introductory material on how nerve impulses are initiated before jumping into the details. We've simplified our explanation of voltage and how it works in the neuron and rewritten the sections on the resting potential and action potentials and their propagation. We've also added a new box on neurotoxins. Chapter 45 (The Nervous System) provides more coverage of the spinal cord, and gives more interesting information about the parts of the brain, including speech and language. We've also added information about the "right brain" and "left brain," as well as the most recent theories concerning memory. Two new boxes have been added: "Imaging the Brain" and "Neurotransmitters in the Central Nervous System." Chapter 46 (Sensory Systems) features entirely rewritten coverage of exteroception and enterception and receptor potentials. Clearer discussions of receptors and the special senses are also evident in this chapter. Chapter 47 (Hormones) is an entirely rewritten chapter that discusses each important hormone and its function in an organized, sequential fashion. We've added coverage of skeletons and bones to Chapter 48 (Locomotion), as well as a new box on the development of muscles in athletic training. Chapter 49 (Fueling Body Activities and Digestion) has been rewritten to include new sections on the agents of digestion and organs that support the digestive system, such as the pancreas, liver and gallbladder. In Chapter 50 (Respiration), we've rewritten sections on air flow in the lung, how the lungs work, forced inspiration and expiration, and the exchange process of oxygen and carbon dioxide. We've added information about the blood-brain barrier, and we've also included a box on hyaline membrane disease. Chapter 51 (Circulation) provides a clearer explanation of the regulation of the blood pressure. Chapter 52 (The Immune System) has been reorganized in a more linear fashion, so that students are acquainted with the roles of nonspecific and specific defenses before they are given detailed information about these defenses and how they work. Chapter 53 (Kidneys and Water Balance) has been extensively revised to include new sections on filtration, reabsorption and excretion. We've also added a new box on dialysis. In Chapter 54 (Sex and Reproduction), we've reorganized the chapter so that the anatomy of the male and female reproductive systems are grouped with their physiology—this type of organization works better, since structure and function are not separated, and students can understand how the various organs work. Information on birth control has also been updated. Chapter 55 (Development) offers new information on the maternal-fetal barrier and milk production. Finally, Chapters 56 (Animal Behavior) and 57 (Behavioral Ecology) are two entirely new chapters that cover all aspects of animal behavior and its ecology.

HELP FOR THE STUDENT:

Biology, third edition is accompanied by the following supplements that are designed to aid students and instructors:

- *Biology* Laboratory Manual by Darrell Vodopich of Baylor University and Randall Moore of Wright State University.
- *Biology* Study Guide by Margaret Gould Burke of the University of North Dakota and Ronald M. Taylor of Lansing Community College.
- Instructor's Resource Manual by Linda Van Thiel of Wayne State University. The Instructor's Resource Manual also includes 108 transparency masters.
- 102 full color transparency acetates. Test Bank by Richard Van Norman of the University of Utah. A Diploma II computerized version of the printed test bank is available in IBM PC or Apple II formats. The Diploma II test bank provides *Exam*, *Gradebook*, and *Proctor* capabilities. In both the printed and computerized versions, "Proctor Practice," a separate bank of questions for use in practice exams, is also provided.

Focus Group

Gerald Bergtrom, *University of Wisconsin–Milwaukee*
Norman Johnson, *Ohio State University*
Phil Laris, *University of California–Santa Barbara*
Darrell Moore, *East Tennessee State University*
Larry Sellers, *Louisiana Technological University*
Kemet Spence, *Washington State University*
Everett Wilson, *Sam Houston State University*

Reviewers

Robert Aldridge, *St. Louis University*
John and Florence Bernard, *Ithaca College*
Charles Biggers, *Memphis State University*
William Coffman, *University of Pittsburgh*
Steve Dina, *St. Louis University*
Robert Evans, *Rutgers University–Camden*
Judy Goodenough, *University of Massachusetts*
Norman Johnson, *Ohio State University*
Fred Kull, *State University of New York-Binghamton*
Waiston Lee, *Wayne Community College*
Tammy Levitt-Gilmour, *Penn State University*
James Liberatos, *Louisiana Technological University*
Owen Lind, *Baylor University*
Tom Lonergan, *University of New Orleans*
Pablo Mendoza, *El Paso Community College*
Mike Meighan, University of California-Berkeley,
Darrell Moore, *East Tennessee State University*
Randy Moore, *Wright State University*
Jeffrey Pommerville, *Glendale Community College*
Warren Porter, *University of Wisconsin–Madison*
Fred Rickson, *Oregon State University*
Keith Roberts, *University of Southwest Louisiana*
Gerry Schuman, *Valencia Community College*
Larry Sellers, *Louisiana Technological University*
Karen Stuedel, *University of Wisconsin-Madison*
Sue Thompson, *Auburn University*
James Traniello, *Boston University*
Fred Wasserman, *Boston University*
Carol Welsh, *Long Beach Community College*
Cherie Wetzel, *San Francisco City College*
Mark Wheelis, *University of California-Davis*
Gene Williams, *Indiana University*

ACKNOWLEDGEMENTS

In many ways, the acknowledgment is the hardest part of the book to write. It is impossible to be accurate and complete—too many people help in too many ways. Some people stand out to the authors like beacons: Charles Schauf, of the School of Science at Indiana University/Purdue University, Indianapolis dissected the animal physiology chapters, and James Traniello of Boston University the behavior chapters—no author could ask for better help. Kathleen Scogna, a developmental editor without equal, managed to goad us through a tight schedule despite our protestations and wails of agony—we rely on her efforts more than any acknowledgement can express. We have a visionary Irishman for a publisher and a company that is squarely behind our efforts. We thank all of you, and all the others we cannot name in a brief paragraph.

Our wives and children have, as always, had to put up with absent "daddies" more than is good for any family—Caitlin Johnson (age 5) announced in the process of this revision that she was going to go to "Mr. Mosby" and demand the return of her daddy, and that he pay a tax for using him up (one Barbie doll each week). We hope that when our children are older they will understand why we felt this book important enough to keep us away from them so often.

Peter H. Raven
George B. Johnson

About the Authors

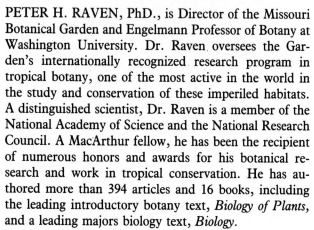

PETER H. RAVEN, PhD., is Director of the Missouri Botanical Garden and Engelmann Professor of Botany at Washington University. Dr. Raven oversees the Garden's internationally recognized research program in tropical botany, one of the most active in the world in the study and conservation of these imperiled habitats. A distinguished scientist, Dr. Raven is a member of the National Academy of Science and the National Research Council. A MacArthur fellow, he has been the recipient of numerous honors and awards for his botanical research and work in tropical conservation. He has authored more than 394 articles and 16 books, including the leading introductory botany text, *Biology of Plants*, and a leading majors biology text, *Biology*.

GEORGE B. JOHNSON, Ph.D., is Professor of Biology at Washington University, where he teaches one of the university's largest courses, freshman biology for nonmajors. Also Professor of Genetics at the Washington University School of Medicine, he has authored more than 50 research publications and 7 books. A Guggenheim and Carnegie fellow, Dr. Johnson is a recognized authority on population genetics and evolution and is renowned for his pioneering studies on genetic variability. He is the founding Director of the St. Louis Zoo's trend-setting educational center, The Living World. Dr. Johnson is coauthor of *Biology* with Peter Raven. Old friends, they have known each other and collaborated together for over 20 years.

Contents in Brief

Contents

I
The Origin of Living Things

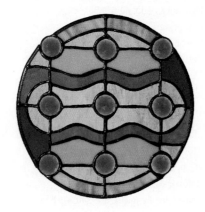

II
Biology of the Cell

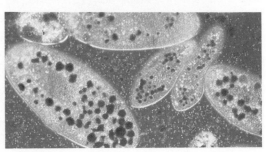

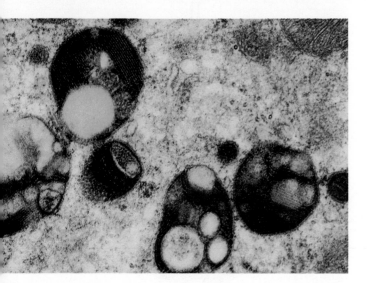

III
Energetics

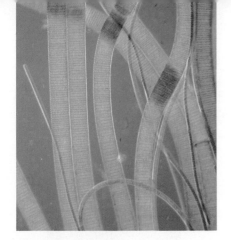

IV
Reproduction and Heredity

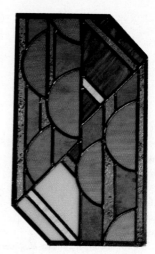

V
Molecular Genetics

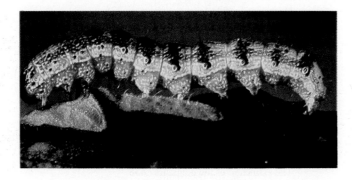

VI
Evolution

VII
Ecology

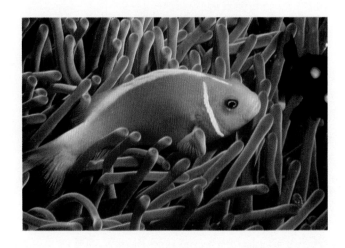

VIII
Biology of Viruses and Simple Organisms

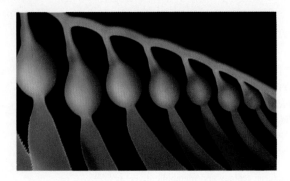

IX
Biology of Plants

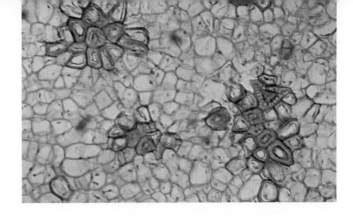

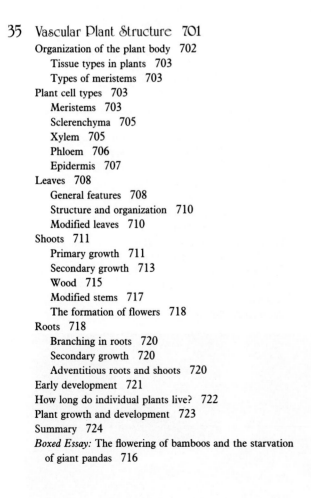

X
Biology of Animals

XI
Vertebrate Biology

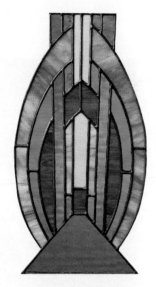

Appendices

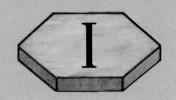

I

The Origin of Living Things

1

The Science of Biology

OVERVIEW

Scientists use the scientific method to infer general principles about how the world operates. Scientists first construct hypotheses, then test them with experiments, and finally eliminate hypotheses that are invalid. A hypothesis that is supported by a great deal of evidence is called a theory. Darwin's theory of evolution is an example of a scientific theory.

The ship in Figure 1-1 is a replica of HMS *Beagle*. It is shown traveling through a narrow passage at the southernmost tip of South America. In the background you can see the cold mountains of Tierra del Fuego. This was the view seen by Charles Darwin more than 150 years ago, when he made this same passage as a naturalist on HMS *Beagle*. What Darwin learned on his 5-year voyage led directly to his development of the theory of evolution by natural selection, a theory that has become the central core of the science of biology.

Darwin's voyage seems a fitting place to begin our consideration of biology, which is the scientific study of living organisms and how they have evolved. In this introductory chapter you will encounter information that Darwin never knew, from the structure of molecules to the number of toes of fossil horses. Before we begin, however, it is useful to take a moment to consider just what science is.

THE NATURE OF SCIENCE

In Egypt, more than 200 years before Christ was born, the Greek Eratosthenes correctly estimated the circumference of the earth. On the longest day of the year, when at high noon the sun's rays hit the water at the bottom of a deep well in the city of Syene, Eratosthenes measured the length of the shadow of a tall obelisk in the city of Alexandria, which was about 800 kilometers to the north. Because he knew the distance between the two cities and the height of the obelisk, he was able to employ the principles of Euclidian geometry to correctly deduce the circumference of the earth (Figure 1-2). This sort of analysis of specific cases using general principles is called **deductive reasoning.** It is the reasoning of mathematics and of philosophy, and the way in which the validity of general ideas is tested in all branches of knowledge. General principles are constructed and then used as the basis for examining specific cases to which they apply.

For the *construction* of scientific principles, a different logical process is used. Webster's Dictionary defines "science" as systematized knowledge derived from observation and experiment carried on to determine the principles underlying what is being studied. Said briefly, a scientist determines principles from observations. This method of discovering general principles by careful examination of specific cases is called **inductive reasoning.** It first became important to science in the 1600s in Eu-

FIGURE 1-1
A replica of the *Beagle* off the southern coast of South America. Charles Darwin, famous English naturalist, set forth on HMS *Beagle* in 1831, at the age of 22. The 5-year voyage took him to the coasts and coastal islands of South America, where Darwin gathered information that led him to formulate and test his hypothesis of evolution by means of natural selection.

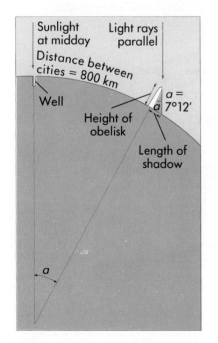

FIGURE 1-2
How Eratosthenes deduced the circumference of the earth.
1 On the day when sunlight shone directly down into a deep well at Syene, near the present location of the Aswan Dam in Egypt, he measured the length of the shadow cast by a tall obelisk in the city of Alexandria.
2 The shadow's length and the obelisk's height formed two sides of a triangle. From principles of Euclidian geometry recently developed at that time, he was able to use the height and length to deduce the angle *a*. It proved to be 7 degrees 12 minutes, exactly one fiftieth of a circle (360 degrees).
3 If angle *a* = one fiftieth of a circle, then the distance between the obelisk (Alexandria) and the well (Aswan) must equal one fiftieth of the circumference of the earth.
4 Eratosthenes had heard that it was a 50-day camel trip to the vicinity of the well, and so, figuring that a camel travels about 18.5 kilometers per day, he estimated the distance between obelisk and well as 925 kilometers. (Eratosthenes used different units of measure, of course)
5 Eratosthenes thus estimated the circumference of the earth to be 50 × 925 = 46,250 kilometers. In reality the distance from well to obelisk is just over 800 kilometers, 15% less than Eratosthenes' crude estimate, so his estimate of the earth's circumference was 15% too large.
Employing a value of 800 kilometers, Eratosthenes' estimate would have been 50 × 800 = 40,000 kilometers. The actual value is 40,075 kilometers.

rope, when Francis Bacon, Isaac Newton, and others began to use the results of particular experiments that they had carried out to infer general principles about how the world operates. If you release an apple from your hand, what happens? The apple falls to the ground. Although this result may not stun you, it is the sort of observation on which science is built. From a host of particular observations, each no more difficult to observe than the falling of an apple, Newton inferred a general principle—that all objects fall toward the center of the earth. What Newton did was to try to construct a mental model of how the world works, a family of working rules, general principles consistent with what he could see and learn. And, like Newton, that is what scientists do today—they are makers of models. They use observations to build models and then test the models to see how well they work.

Testing Hypotheses

How do scientists learn which general principles are actually true from among the many that might be true? They do this by systematically attempting to demonstrate that certain proposals are *not* valid—not consistent with what they learn from experimental observation. A great deal of careful and creative thinking is necessary for the construction of a hypothesis, which is a suggested explanation that accounts for the facts, observations, and experiments that are available concerning a particular area of science. Those hypotheses that have not yet been disproved are retained. They are useful because they fit the known facts, but they are always subject to future rejection if, in the light of new information, they are found to be incorrect.

A hypothesis is a proposition that might be true. We call the test of a hypothesis an **experiment** (Figure 1-3). An experiment evaluates alternative hypotheses. "There

FIGURE 1-3

The scientific method. This diagram illustrates the way in which scientific investigations proceed. First, a number of potential explanations (hypotheses) are suggested in answer to a question. Next, experiments are carried out in an attempt to eliminate one or more of these hypotheses, then predictions are made based on these hypotheses. Finally, further experiments are carried out to test these predictions. As a result of this process, the most likely hypothesis is selected. If it is validated by numerous experiments and stands the test of time, the hypothesis may eventually be considered a theory.

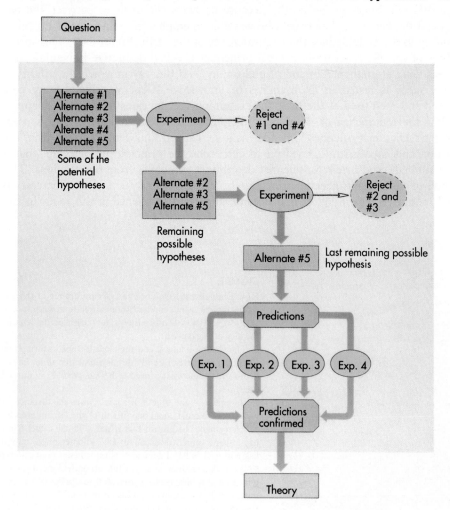

is no light in this dark room because the light switch is turned off" is a hypothesis; an alternative hypothesis is "There is no light in the room because the light bulb is burnt out"; yet another alternative might be "I am going blind." An experiment works by eliminating one of the hypotheses. For example, we might test these alternatives by reversing the position of the light switch. Let us say that when we do this the light does not come on. The result of our experiment is thus to disprove the first of the hypotheses—something other than the setting of the light switch must be at fault. Note that a test such as this does not prove that any one alternative is true, but rather demonstrates that one of them is not. In this instance, the fact that the light does not come on does not establish that the switch was in fact in the "on" position (it might have been either "on" or "off" if the light bulb was burnt out), but rather that the switch setting alone is not the sole reason for the darkness. A successful experiment is one in which one or more of the alternative hypotheses is demonstrated to be inconsistent with experimental or observational results and is thus rejected. Scientific progress is made the same way a marble statue is, by chipping away unwanted bits.

> **The scientific process involves the rejection of hypotheses that are not consistent with experimental results or observations. Hypotheses consistent with available data are conditionally accepted.**

As you proceed through this text, you will encounter a great deal of information, often coupled to explanations. These explanations are hypotheses that have withstood the test of experiment. Many will continue to do so; others will be revised as new observations are made by biologists. Biology, like all healthy science, is in a constant state of ferment, with new ideas bubbling up and replacing old ones.

Controls

Often we are interested in learning about processes that are influenced by many factors. We call each factor that influences a process a **variable**. To evaluate alternative hypotheses about one variable, all the other variables must be kept constant so that we do not get misled or confused by these other influences. This is done by carrying out two experiments in parallel: in the first "experimental" test we alter one variable in a known way to test a particular hypothesis; in the second **control** experiment we do not alter that variable. In all other respects the two experiments are the same. We then ask "What difference is there between the outcomes of the two experiments?" Any difference that we see must result from the influence of the variable that we changed, because all the other variables remained constant (the same in both experiments). Much of the challenge of experimental science lies in designing control experiments—in successfully isolating a particular variable from all other effects that might influence a process.

The Importance of Prediction

A successful scientific hypothesis needs to be not only valid but useful—it needs to tell you something that you want to know. When is a hypothesis most useful? When it makes predictions. The predictions that a hypothesis makes provide an important way to further test its validity. A hypothesis that your experiment does not reject, but which makes a prediction the experiment *does* reject, must itself be rejected. The more predictions a hypothesis makes that check out, the more demonstrably valid is the hypothesis. For example, Einstein's hypothesis of relativity was at first provisionally accepted because no one could think of an experiment that invalidates it. Acceptance soon became far stronger because the theory made a clear prediction, that the sun would bend the path of light passing by it. When this prediction was tested in a total eclipse, the light of background stars was indeed bent. Because the result was not known ahead of time, when Einstein's proposal was being formulated, it gave strong support to his hypothesis.

Theories

Hypotheses that stand the test of time, often tested and never rejected, are called **theories.** Thus one speaks of the general principle first noted by Newton as the theory of gravity. Theories are the solid ground of science, that of which we are most certain. There is no absolute truth in science, however—only varying degrees of uncertainty. The possibility always remains that future evidence will cause a theory to be revised. A scientist's acceptance of a theory is always provisional.

Thus scientists use the word "theory" in a very different sense than the general public. To a scientist, a theory represents that of which he or she is most certain, whereas to the general public, the word theory implies the *lack* of knowledge or a

---

HOW BIOLOGISTS DO THEIR WORK

The Consent

Late in November, on a single night
Not even near to freezing, the ginkgo trees
That stand along the walk drop all their leaves
In one consent, and neither to rain nor to wind
But as though to time alone: the golden and green
Leaves litter the lawn today, that yesterday
Had spread aloft their fluttering fans of light.

What signal from the stars? What senses took it in?
What in those wooden motives so decided
To strike their leaves, to down their leaves,
Rebellion or surrender? and if this
Can happen thus, what race shall be exempt?
What use to learn the lessons taught by time,
If a star at any time may tell us: Now.

Howard Nemerov

What is bothering the poet Howard Nemerov is that life is influenced by forces he cannot control, or even identify. It is the job of biologists to solve puzzles such as the one he poses—to identify and try to understand those things that influence life.

Nemerov asks "Why do ginkgo trees drop all their leaves at once?" To find an answer to questions such as this, biologists and other scientists pose *possible* answers and try to see which answers are false. Tests of alternative possibilities are called **experiments.** To learn why the ginkgo trees drop all their leaves simultaneously, a scientist would first formulate several alternative possible answers, called hypothetical answers, or **hypotheses:**

Hypothesis 1: Ginkgo trees possess an internal clock that times the release of leaves to match the season. On the day Nemerov describes, this clock sends a "drop" signal (perhaps a chemical) to all the leaves at the same time.

Hypothesis 2: The individual leaves of ginkgo trees are each able to sense day length, and when in the fall the days get short enough, each leaf responds independently by falling.

Hypothesis 3: A strong wind arose the night before Nemerov made his observation, blowing all the leaves off the ginkgo trees.

Next the scientist attempts to eliminate one or more of the hypotheses by conducting an experiment. In this case one might cover some of the leaves so that these leaves cannot use light to sense day length. If hypothesis 2 is true, then the covered leaves should not fall when the others do, since they are not receiving the same information. Suppose, however, that despite the covering of

some of the leaves, all the leaves fall together. This eliminates hypothesis 2 as a possibility. Either of the other hypotheses, and many others, remain possibilities.

This simple experiment with ginkgos serves to point out the essence of scientific progress: science does not prove that certain explanations are true, but rather that some possibilities are not. Hypotheses that are not consistent with experimental results are rejected. Hypotheses that are not proven false by an experiment or series of experiments are provisionally accepted. However, hypotheses may be rejected in the future when more information becomes available if they are not consistent with the new information. Just as a computer finds the correct path through a maze by trying and eliminating false paths, so a scientist gropes toward reality by eliminating false possibilities.

A hypothesis that stands the test of time, often tested and never rejected, is called a **theory.** In biology the hypothesis that tiny organelles, called mitochondria, within your cells are the descendants of bacteria is a theory. It is not certain that this idea is correct, but the overwhelming weight of evidence supports the hypothesis and most biologists accept it as "proven." There is no absolute truth in science, however, only varying degrees of uncertainty, and the possibility always remains that future evidence will cause a theory to be revised. A scientist's acceptance of a theory is always provisional.

guess—just the opposite! As you can imagine, confusion often results. In this text, the word "theory" will always be used in its scientific sense, in reference to a generally accepted scientific principle.

A theory is a hypothesis that is supported by a great deal of evidence.

Some theories are so strongly supported that the likelihood of their being rejected in the future is small. The theory of evolution, for example, is so broadly supported by different lines of inquiry that all but a few scientists accept it with as much certainty as they do the theory of gravity. We will examine this particular theory later in this chapter as an example of how science is carried out. It is a particularly important theory to biologists, since the theory of evolution provides the conceptual framework that unifies biology as a science.

The Scientific Method

It used to be fashionable to speak of the "scientific method" as consisting of an orderly sequence of logical "either/or" steps, each step rejecting one of two mutually incompatible alternatives, as if trial-and-error testing would inevitably lead one through the maze of uncertainty that always impedes scientific progress. If this were indeed so, a computer would make a good scientist. In fact, science is not done this way. As British philosopher Karl Popper has pointed out, if you ask successful scientists how they do their work, you would discover that without exception they design their experiments with a pretty fair idea of how the experiments are going to come out—what Popper calls an "imaginative preconception" of what the truth might be. A hypothesis that a successful scientist tests is not just any hypothesis, but a "hunch" or an educated guess in which the scientist integrates all that he or she knows and also allows his or her imagination full play, in an attempt to get a sense of what *might* be. It is because insight and imagination play such a large role in scientific progress that some scientists are so much better at science than others—for precisely the same reason that Beethoven and Mozart stand out above most other composers.

The scientific method is the experimental testing of a hypothesis formulated after the systematic, objective collection of data. Hypotheses are not usually formulated simply by rejecting a series of alternative possibilities, but rather often involve creative insight.

HISTORY OF A BIOLOGICAL THEORY: DARWIN'S THEORY OF EVOLUTION

The idea of evolution—the notion that living things change gradually from one form into another over the course of time—provides a good example of how an idea, an educated guess, is developed into a hypothesis, is tested, and is eventually accepted as a theory.

Charles Robert Darwin (1809-1882; Figure 1-4) was an English naturalist who, at the age of 50, after 30 years of study and observation, wrote one of the most famous and influential books of all time. The full title of this book, *On the Origin of Species by Means of Natural Selection, or the Preservation of Favoured Races in the Struggle for Life,* expressed both the nature of its subject and the way in which Darwin treated it. The book created a sensation when it was published in 1859, and the ideas expressed in it have played a central role in the development of human thought ever since.

In Darwin's time, it was traditional to believe that the various kinds of organisms and their individual structures resulted from direct actions of the Creator. Species were held to be specially created and unchangeable, or immutable, over the course of time. In contrast to these views, a number of earlier philosophers had presented the view that living things must have changed during the course of the history of life on earth. Darwin was the first to present a coherent, logical explanation for this process—natural selection—and the first to bring these ideas to wide public attention.

FIGURE 1-4
Charles Darwin. This portrait was painted when Darwin was 29 years old, 2 years after his return from the voyage on HMS *Beagle.* Darwin had just married his cousin Emma Wedgwood and was hard at work studying the materials he had gathered on the voyage.

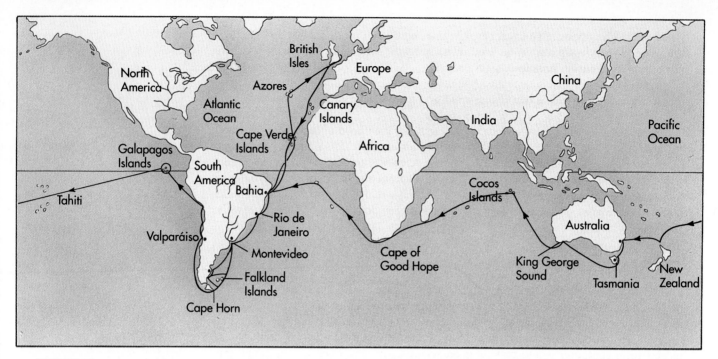

FIGURE 1-5
The 5-year voyage of HMS *Beagle*. Most of the time was spent exploring the coasts and coastal islands of South America, such as the Galapagos Islands. His studies of the animals of the Galapagos Islands played a key role in Darwin's eventual development of the theory of evolution by means of natural selection.

His book, as you can see from its title, presented a conclusion that differed sharply from conventional wisdom. Although his theory did not challenge the existence of a Divine Creator, Darwin argued that this Creator did not simply create things and then leave them forever unchanged. Darwin's God expressed Himself through the operation of natural laws that produced constant change and improvement. These views put Darwin at odds with most people of his time, who believed in a literal interpretation of the Bible and accepted the idea of a fixed and constant world. His theory was a revolutionary one that deeply troubled many of his contemporaries and Darwin himself.

The story of Darwin and his theory begins in 1831, when he was 22 years old. On the recommendation of one of his professors at Cambridge University, he was selected to serve as naturalist on a 5-year voyage around the coasts of South America, the voyage of HMS *Beagle* (1831-1836; Figure 1-5). During his long journey, Darwin had the chance to study plants and animals widely on continents and islands and in distant seas. He was able to experience at first hand the biological richness of the tropical forests, the extraordinary fossils of huge extinct mammals in Patagonia (Figure 1-6), and the remarkable series of related but distinct forms of life on the Galapagos Islands, off the west coast of South America. Such an opportunity clearly played an important role in the development of his thoughts about the nature of life on Earth.

FIGURE 1-6
Reconstruction of a glyptodont. This 2-ton fossil South American armadillo is much larger than a modern armadillo, which averages about 10 pounds. Finding the fossils of forms such as the glyptodonts and noting their similarity to living animals found in the same regions, Charles Darwin eventually concluded that evolution had taken place.

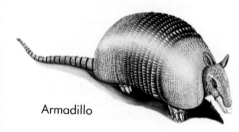

Armadillo

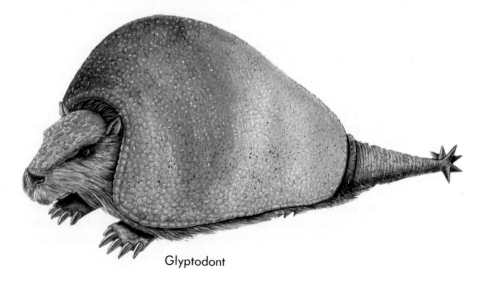

Glyptodont

When Darwin returned from the voyage at the age of 27, he began a long life of study and contemplation. During the next 10 years, he published important books on several different subjects, including the formation of oceanic islands from coral reefs and the geology of South America. He also devoted 8 years of study to barnacles, a group of small marine animals with shells that you may have seen on rocks and pilings; eventually, Darwin wrote a four-volume work on their classification and natural history. In 1842, he and his family moved a short distance out of London to a country home at Down, in the county of Kent. In these pleasant surroundings, Darwin lived, studied, and wrote for the next 40 years. During this period, he formulated for the first time a consistent theory of the process of evolution, an explanation that provided a mechanism for evolution. He presented his ideas in such convincing detail that they could be accepted as a logical explanation for the diversity of life on earth, the intricate adaptations of those living things, and the ways in which they are related to one another.

DARWIN'S EVIDENCE

So much information had accumulated by 1859 that the acceptance of the theory of evolution now seems, in retrospect, to have been inevitable. Darwin was able to arrive at a successful theory when many others had failed because he rejected supernatural explanations for the phenomena that he was studying. One of the obstacles that blocked the acceptance of any theory of evolution was the incorrect notion, still widely believed at that time, that the earth was only a few thousand years old. The discoveries of thick layers of rocks, evidences of extensive and prolonged erosion, and the increased numbers of diverse and unfamiliar fossils discovered during Darwin's time made this assertion seem less and less likely, however. For example, the great geologist Charles Lyell (1797-1875), whose works Darwin read eagerly while he was sailing on HMS *Beagle*, outlined for the first time the story of an ancient world of plants and animals in flux. In this world, species were constantly becoming extinct while others were emerging. It was this world that Darwin sought to explain.

TABLE 1-1   DARWIN'S EVIDENCE THAT EVOLUTION OCCURS

FOSSILS

1. Extinct species, such as the fossil armadillos shown in Figure 1-6, most closely resemble living ones in the same area, suggesting that one had given rise to the other.
2. In rock strata (layers), progressive changes in characteristics can be seen in fossils from earlier and earlier layers.

GEOGRAPHICAL DISTRIBUTION

3. Lands with similar climates, such as Australia, South Africa, California, and Chile, have unrelated plants and animals, indicating that differences in environment are not creating the diversity directly.
4. The plants and animals of each continent are distinctive, although there is no reason why special creation should invent this association: all South American rodents belong to a single group, structurally similar to the guinea pigs, for example, whereas most of the rodents found elsewhere belong to other groups.

OCEANIC ISLANDS

5. Although oceanic islands have few species, those they have are often unique ("endemic") and show relatedness to one another, such as the Galapagos tortoises (Figure 1-7). This suggests that the tortoises and other groups of endemic species developed after their mainland ancestors reached the islands, and are therefore more closely related to one another.
6. Species on oceanic islands show strong affinities to those on the nearest mainland. Thus the finches of the Galapagos Islands (Figure 1-8, *A*) closely resemble a finch seen on the western coast of South America (Figure 1-8, *B*). The Galapagos finches do *not* resemble the birds of the Cape Verde Islands, islands in the Atlantic Ocean off Africa that are similar to the Galapagos. Darwin visited the Cape Verde Islands and many other island groups personally and was able to make such comparisons on the basis of his own observations.

A

B

FIGURE 1-7

Galapagos tortoises. Tortoises with large, domed shells, **A**, are found in relatively moist habitats. Those with lower, saddleback-type shells, **B**, in which the front of the shell is bent up, exposing the head and part of the neck, are found among the tortoises of dry habitats. Taken together, differences of this kind make it possible to identify the races of tortoises that inhabit the different islands of the Galapagos.

What Darwin Saw

When HMS *Beagle* set sail, Darwin was fully convinced that species were unchanging and immutable. Indeed, he wrote that it was not until 2 or 3 years after his return that he began to consider seriously the possibility that they could change. Nevertheless, during his 5 years on the ship, Darwin observed a number of phenomena that were of central importance to him in reaching his ultimate conclusion (Table 1-1). For example, in the rich fossil beds of southern South America, he observed fossils of extinct armadillos that were directly related to the armadillos that still lived in the same area (Figure 1-6). Why would there be living and fossil organisms directly related to one another and in the same area unless one had given rise to the other?

Repeatedly, Darwin saw that the characteristics of different species varied from place to place. These patterns suggested to him that organisms change gradually as they migrate from one area into another. On the Galapagos Islands, off the coast of Ecuador, Darwin encountered giant land tortoises. Surprisingly, these tortoises were not all identical. Indeed, local residents and the sailors who captured the tortoises for food could tell which island a particular animal had come from just by looking at its shell (Figure 1-7). This pattern of physical variation suggested that all of the tortoises were related, but they had changed slightly in appearance after they had become isolated on the different islands.

In a more general sense, Darwin was struck by the fact that on these relatively young volcanic islands there was a profusion of living things, but that these plants and animals resembled those of the nearby coast of South America (Figure 1-8). If each one of these plants and animals had been created independently and simply placed on the Galapagos Islands, why did they not resemble the plants and animals of

FIGURE 1-8
Darwin's finches.
A One of Darwin's finches, the medium ground finch.
B The blue-black grassquit, which is found in grasslands along the Pacific Coast from Mexico to Chile. This species may be the ancestor of Darwin's finches.

A

B

faraway Africa, for example? Why did they resemble those of the adjacent South American coast instead?

> The patterns of distribution and the relationship of organisms that Darwin observed on the voyage of HMS *Beagle* made him certain that a process of evolution had been responsible for these patterns.

Darwin and Malthus

It is one thing to observe the results of evolution, but quite another to understand how it happens. Darwin's great achievement lies in his perception that evolution occurs because of natural selection. Of key importance to the development of Darwin's insight was his study of Thomas Malthus' *Essay on the Principles of Population*. In his book, Malthus pointed out that populations of plants and animals (including human beings) tend to increase geometrically, whereas, in the case of people, our ability to increase our food supply increases only arithmetically. A **geometric** progression is one in which the elements progress by a constant factor, as 2, 6, 18, 54, and so forth; in this example, each number is three times the preceding one. An **arithmetic** progression, in contrast, is one in which the elements increase by a constant difference, as 2, 6, 10, 14, and so forth. In this progression, each number is 4 greater than the preceding one.

Virtually any kind of animal or plant, if it could reproduce unchecked, would cover the entire surface of the world within a surprisingly short time. In fact, this does not occur; instead, populations of species remain more or less constant year after year, because death intervenes and limits population numbers. Malthus' conclusion was important in the development of Darwin's ideas, providing the key ingredient that was necessary for him to develop the hypothesis that evolution occurs by natural selection.

> A key contribution to Darwin's thinking was Malthus' concept of geometric population growth. The fact that real populations do not expand at this rate implies that nature acts to limit population numbers.

Natural Selection

Sparked by Malthus' ideas, Darwin saw that although every organism has the potential to produce more offspring than are able to survive, only a limited number actually survive and produce their own offspring. Combining this observation with what he had seen on the voyage of HMS *Beagle*, as well as with his own experiences in breeding domestic animals, Darwin made the key association (Figure 1-9): *those individuals that possess superior physical, behavioral, or other attributes are more likely to survive than those that are not so well endowed.* By surviving, they have the opportunity to pass on their favorable characteristics to their offspring. Because these characteristics will increase in the population, the nature of the population as a whole will gradually change. Darwin called this process **natural selection**. The driving force that he had identified has often been referred to as *survival of the fittest*.

> Natural selection results in the increase in succeeding generations of the traits of those organisms that leave more offspring. Its operation depends on the traits being inherited. The nature of the population gradually changes as more and more individuals with those traits appear.

Darwin was thoroughly familiar with variation in domesticated animals and began his *On the Origin of Species* with a detailed discussion of pigeon breeding. He knew that varieties of pigeons and other animals, such as dogs, could be selected by breeders to exhibit certain characteristics, a process called **artificial selection**. Once this had been done, the animals would breed true for the characteristics that had been concentrated in them. Darwin had also observed that the differences that could be

"Can we doubt . . . that individuals having any advantage, however slight, over others, would have the best chance of surviving and of procreating their kind? On the other hand, we may feel sure that any variation in the least degree injurious would be rigidly destroyed. This preservation of favorable variations, I call Natural Selection."

FIGURE 1-9
From Charles Darwin's *On the Origin of Species*.

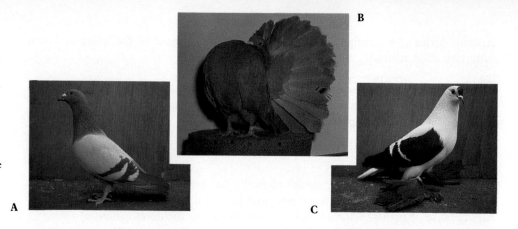

developed between domesticated races or breeds in this way were often greater than those that separated wild species. The breeds of the domestic pigeon shown in Figure 1-10, for example, show much greater variety than all of the hundreds of wild species of pigeons found throughout the world. Such relationships suggested to Darwin that evolutionary change could occur very rapidly under the right circumstances.

Darwin's theory provides a simple and direct explanation of biological diversity, of why animals are different in different places; because habitats differ in their requirements and opportunities, those kinds of organisms favored by natural selection will tend to vary in different places.

PUBLICATION OF DARWIN'S THEORY

Darwin drafted the overall argument for evolution by natural selection in 1842 and continued to enlarge and refine it for many years. The stimulus that finally brought it into print was an essay that he received in 1858. A young English naturalist named Alfred Russel Wallace (1823-1913; Figure 1-11) sent the essay to Darwin from Malaysia; it concisely set forth the theory of evolution by means of natural selection! Like Darwin, Wallace had been greatly influenced in his development of this theory by reading Malthus' 1798 essay. After receiving Wallace's essay, Darwin arranged for a joint presentation of their ideas at a seminar in London. He then proceeded to complete his own book, which he had been working on for so long, and submitted it for publication in what he considered an abbreviated version.

Darwin's book appeared in November, 1859, and caused an immediate sensation. Many people were deeply disturbed by the idea that human beings were descended from the same ancestor as apes, for example. This idea was not discussed by Darwin in his book, but it followed directly from the principles that he outlined. It had long been accepted that humans closely resembled the apes in all of their characteristics, but the possibility that there might be a direct evolutionary relationship between them was unacceptable to many people. Darwin's arguments for the theory of evolution by natural selection were so compelling, however, that his views were almost completely accepted within the intellectual community of Great Britain after the 1860s.

EVOLUTION AFTER DARWIN: TESTING THE THEORY

Darwin did more than propose a mechanism that accounts for how evolution has generated the diversity of life on earth. He also assembled masses of facts, otherwise seemingly without logic, that began to make sense when they were viewed in the light of his theory. After publication of his book other biologists continued this process, and it soon became evident that the theory of evolution was supported by a wide variety of biological information gathered by many investigators. These observations included the fact that members of different biological groups often share common features, the ways in which embryos resemble and differ from one another, and the increasing complexity that is observed to develop in the fossil record through time. All of these lines of evidence, and many more, are examined in Chapter 20. In the century since Darwin, evolution has become the main unifying theme of the biological sciences. It provides one of the most important insights that human beings have achieved into their own nature and that of the planet on which they have evolved.

More than a century has elapsed since Charles Darwin's death in 1882. During this period, the evidence supporting his theory has grown progressively stronger. There have also been many significant advances in our understanding of how evolution works. Although these advances have not altered the basic structure of Darwin's theory, they have taught us a great deal more about the mechanisms by which evolution occurs.

The Fossil Record

Darwin predicted that the fossil record would yield intermediate links (Figure 1-12) between the great groups of organisms, between fishes and the amphibians thought to have arisen from them, for example, and between reptiles and birds. The fossil record is now known to a degree that would have been unthinkable in the nineteenth century. Recent discoveries of microscopic fossils have extended the known history of life on earth back to about 3.5 billion years ago. The discovery of other fossils has shed light on the ways in which organisms have, over the course of this enormous time span, evolved from the simple to the complex. For vertebrate animals especially, the fossil record is rich and exhibits a graded series of changes in form, with the evolutionary parade visible for all to see.

The Age of the Earth

In Darwin's day, some physicists argued that the earth was only a few thousand years old. This bothered Darwin, as the evolution of all living things from some single original ancestor would have required a great deal more time. Using evidence obtained by studying rates of radioactive decay, we now know that the physicists of Darwin's time were wrong—the earth was formed about 4.5 billion years ago.

FIGURE 1-12

Fossil of an early bird, *Archaeopteryx*, a prehistoric ancestor of modern birds. A well-preserved fossil of this bird, about 150 million years old, was discovered within 2 years of the publication of *On The Origin of Species*. *Archaeopteryx* provides an indication of the evolutionary relationship that exists between birds and reptiles.

The Mechanism of Heredity

It was in the area of heredity that Darwin received some of his sharpest criticism. At that time no one had any concept of genes or of how heredity works, and so it was not possible for Darwin to explain completely how evolution occurs. Theories of heredity current in Darwin's day seemed to rule out the possibility of genetic variation in nature, a critical requirement of Darwin's theory. Genetics was established as a science only at the start of the twentieth century, 40 years after the publication of Darwin's *On The Origin of Species*. When the laws of inheritance became understood, the problem with Darwin's theory vanished, since the laws of inheritance (discussed in Chapter 12) account in a neat and orderly way for the production of new variations in nature that were required by Darwin's theory.

Comparative Anatomy

Comparative studies of animals have provided strong evidence for Darwin's theory. As vertebrates have evolved, for example, the same bones sometimes got put to different uses—and yet they are still present, betraying their evolutionary past. Thus the forelimbs seen in Figure 1-13 are all constructed from the same basic array of bones, modified in one way in the wing of a bat, in another way in the fin of a porpoise, and in yet another way in the leg of a horse. The bones are said to be **homologous** in the different vertebrates—that is, of the same evolutionary origin, although

Human

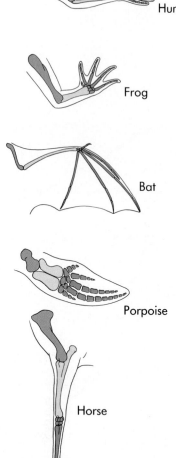
Frog

Bat

Porpoise

Horse

FIGURE 1-13
Homology among vertebrate limbs. Homologies among the forelimbs of four mammals and a frog, showing the ways in which the proportions of the bones have changed in relation to the particular way of life of the organism.

now differing in structure and function. This contrasts with **analogous** structures, which have similar structure and function, but different evolutionary origins. The eyes of vertebrates and octopuses, which evolved independently but are similar in design, are analogous structures.

Molecular Biology

Biochemical tools have become of major importance in our efforts to reach a better understanding of how evolution occurs. Within the last few years, for example, evolutionists have begun to "read" genes, much as you read this page. They have learned to recognize the order of the "letters" of the long DNA molecules, which are present in every living cell and which provide the genetic information or "instructions" for that organism. When the sequences of the "letters" in the DNA of different groups of animals or plants are compared, the degree of relationship among the groups can be specified more precisely than by any other means. In many cases, detailed family trees can then be constructed. When study of the DNA that encodes the structure of two different molecules leads to the same family tree, this provides strong evidence that we have indeed interpreted correctly the evolutionary history of the group. By measuring the degree of difference in the genetic coding, and by interpreting the information available from the fossil record (Figure 1-14), it is sometimes possible for us to estimate the rates at which evolution is occurring in different groups of organisms.

> In the century since Darwin, a large body of evidence contributed by many branches of science has supported his view that evolution occurs and that it takes place by means of the mechanism of natural selection that Darwin proposed.

WHY IS BIOLOGY IMPORTANT TO YOU?

Biologists do more than simply write books about evolution. They live with gorillas, collect fossils, and listen to whales. They isolate viruses, grow mushrooms, and grind up fruit flies. They read the message encoded in the long molecules of heredity and count how many times a hummingbird's wings beat each second. In its broadest sense, biology is the study of living organisms. Life, however, does not take the form of a uniform green slime covering the surface of the earth; rather, life consists of a diverse array of living forms. A biologist tries to understand the sources of this diversity, and in many cases attempts to harness particular life-forms to perform useful tasks. Even the narrowest study of a seemingly unimportant life-form is a study in biological diversity, one more brushstroke in the painting that biologists have labored over for centuries.

Biology is one of the most interesting of subjects, because of its great variety. But not only is it fun, it is also an important subject for you and for everyone, simply because biology will affect your future in many ways. The knowledge that biologists are gaining is fundamental to our ability to manage the world's resources in a suitable manner, to prevent or cure diseases, and to improve the quality of our lives and those of our children and grandchildren. Biologists are working on many problems that critically affect our lives, including dealing with the demands of the world's rapidly expanding population and attempting to find ways to prevent cancer and AIDS. Because the activities of biologists alter our lives in so many ways, an understanding of biology is becoming increasingly necessary for any educated person.

HOW THIS TEXT IS ORGANIZED TO HELP YOU LEARN BIOLOGY

In the century since Darwin's publication of *On The Origin of Species*, biology has exploded as a science, presenting today's student with a wealth of observation, hypothesis, information, and theory. There are many ways in which a beginner can be introduced to biology. In an introductory biology course you will encounter a wealth of experiment and observation that you *could* learn, and from this you must select a small body of information that you *will* learn. Your target is the basic body of principles that unite biology as a science.

From centuries of biological observation and inquiry, one organizing principle has emerged: biological diversity reflects *history*, a record of success, failure, and change extending back to a period soon after the formation of the earth. The weeding out of failures and the reward of success by increased reproduction is called **natural selection,** and the pattern of changes that result from this process is called **evolution,** the special subject of Chapters 19 to 22 of this book, but also the theme of all the chapters. The theory of evolution will form the backbone of your study of biological

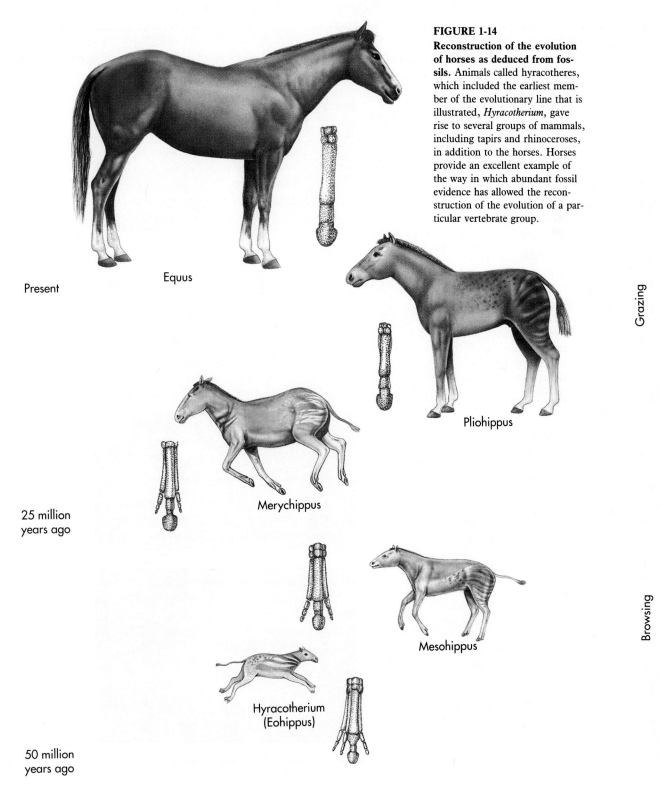

FIGURE 1-14
Reconstruction of the evolution of horses as deduced from fossils. Animals called hyracotheres, which included the earliest member of the evolutionary line that is illustrated, *Hyracotherium,* gave rise to several groups of mammals, including tapirs and rhinoceroses, in addition to the horses. Horses provide an excellent example of the way in which abundant fossil evidence has allowed the reconstruction of the evolution of a particular vertebrate group.

Present

Equus

Grazing

Pliohippus

25 million
years ago

Merychippus

Browsing

Mesohippus

Hyracotherium
(Eohippus)

50 million
years ago

science, just as the theory of the covalent bond is the backbone of the study of chemistry, or the theory of quantum mechanics is that of physics. It is a thread that runs through everything that you will learn in this text. Evolution is the essence of the science of biology.

A good way to begin your examination of basic biological principles is to focus on complexity, one of the essential features of life. Arranging the wealth of information about biology in terms of the level of complexity leads to what has been called a "levels-of-organization" approach to the field (Figure 1-15).

The first half of this text is devoted to a description of the basic principles of biology, using a levels-of-organization framework to introduce different principles, each

BIOLOGICAL ISSUES TODAY

Human Population Growth

There were about 10 million people alive in the world when people first spread across North America; today there are over 5 billion, straining the earth's ability to support them. And the population is growing fast—3 people are added to the world's population each second. At this rate, the number of people alive will double in less than 40 years.

Infectious Disease

AIDS is a serious and growing health problem worldwide. Other diseases are even more serious: malaria will kill over 3 million people this year, most in the underdeveloped countries of the world. Scientists are struggling to produce vaccines against these scourges.

Use of Addictive Drugs

The use of dangerous addictive drugs, particularly powerful heroin and cocaine derivatives, is creating a nightmare in many cities. Other dangerous but legal drugs are also in widespread use, such as cigarettes and other tobacco products that kill thousands by causing lung cancer.

Industrial Alteration of the Environment

The destruction of the atmosphere's ozone, the creation of acid rain, the greenhouse effect caused by increasing concentrations of CO_2 in the atmosphere, the pollution of rivers and underground water supplies, the dilemma of how to dispose of radioactive wastes—all of these problems require urgent attention.

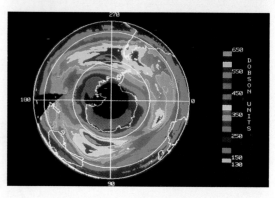

Levels of Biological Organization

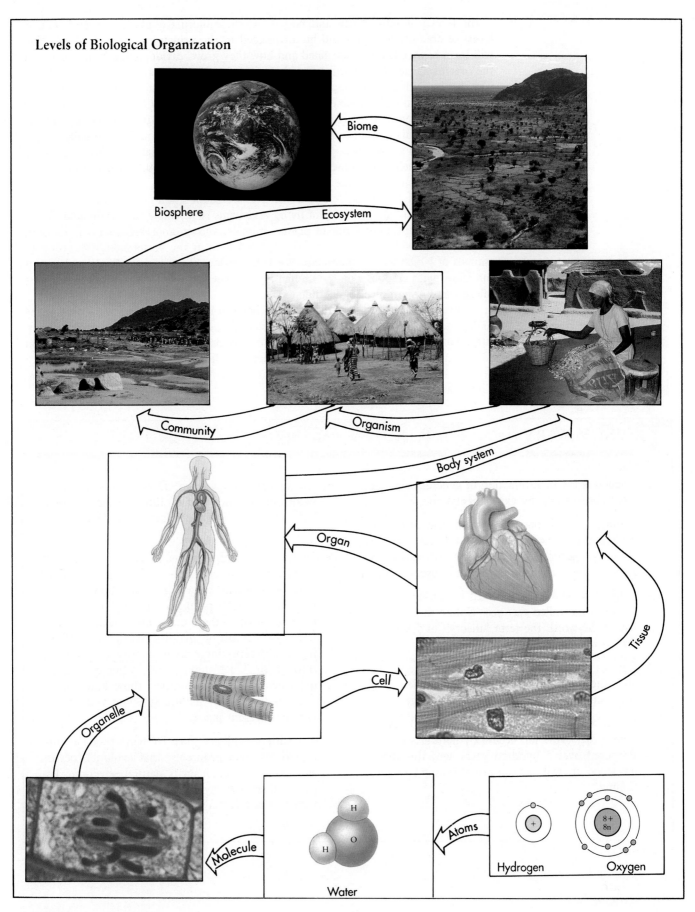

FIGURE 1-15
Levels of biological organization.

at the level it is most easily understood. At the molecular, subcellular, and cellular levels of organization, you will be introduced to the principles of **cell biology.** You will learn how cells are constructed and how they grow, divide, and communicate. At the organismal level, you will learn the principles of **genetics,** which deals with the way that individual traits are transmitted from one generation to the next. At the population level you will examine **evolution,** a field concerned with the nature of population changes from one generation to the next as a result of natural selection, and the way in which this has led to the biological diversity we see around us. Finally, at the community and global levels you will study **ecology,** which deals with how organisms interact with their environments and with one another to produce the complex communities characteristic of life on earth.

The second half of the text is devoted to an examination of organisms, the products of evolution. The diversity of living organisms is *incredible*. It is estimated that at least 5 million different kinds of plants, animals, and microorganisms exist. Later in the book, we will take a particularly detailed look at the vertebrates, the group of animals of which we are members. We will consider the vertebrate body and how it functions, with emphasis on the human body, since this is information that is of most interest and importance to all students.

As you proceed through this course, what you learn at one stage will give you the tools to tackle the next. In the following chapter we will examine some simple chemistry. You are not subjected to chemistry first to torture you, but rather to make what comes later easier to comprehend. To understand lions and tigers and bears, you first need to know the basic chemistry that makes them tick, for they are chemical machines, as are you.

SUMMARY

1. Science is the determination of general principles from observation and experimentation.

2. From among all possible hypothetical principles, or hypotheses, scientists select the best ones by using experiments to eliminate the alternative hypotheses that are inconsistent with observation.

3. Hypotheses that are supported by a large body of evidence are called theories. Unlike the everyday use of the word, the term "theory" in science refers to what we are most sure about.

4. Because even a theory is accepted only conditionally, there are no absolute truths in science, no propositions that are not subject to change.

5. One of the central theories of biology is Darwin's theory that evolution occurs by natural selection. Proposed over a hundred years ago, this theory has stood up well to a century of testing and questioning.

6. On an extended voyage, Darwin accumulated a wealth of evidence that evolution occurs, particularly while studying the animals and plants of oceanic islands.

7. Sparked by Malthus' ideas, Darwin proposed that evolution occurs as a result of natural selection. Some individuals have traits that allow them to produce more offspring in a given kind of environment than other individuals lacking these traits. As a result of this process, the traits allowing greater reproduction will increase in frequency through time. Because habitats differ in their requirements and opportunities, those kinds of organisms favored by natural selection will tend to vary in different places.

8. A wealth of evidence since Darwin's time has supported his twin proposals, that evolution occurs and that its mechanism is natural selection. Together these two hypotheses are now usually referred to as Darwin's theory of evolution.

REVIEW QUESTIONS

1. Define the term "hypothesis." How are inductive and deductive reasoning involved in the scientific process?

2. How does a hypothesis become a theory? Give several examples of scientific theories from your knowledge of biology, chemistry, and physics.

3. What was Darwin's evidence from his journey that evolution occurs?

4. What timely discoveries from other scientific disciplines lent support and direction to Darwin's theory of evolution?

5. What is natural selection and how did Darwin develop its concept from Malthus' writings?

6. What are the major points of Darwin's theory of natural selection?

7. What scientific discoveries since Darwin have furthered the support for evolution?

8. How does the examination of homologous structures support or refute evolution?

THOUGHT QUESTIONS

1. Imagine that you are assigned to investigate the cause of the Challenger disaster, in which a space shuttle exploded in flight. In particular, you are asked to assess the possibility that the temperature of the air at the time of take-off was too cold for proper operation of the equipment. How would you proceed to test this hypothesis?

2. It is sometimes argued that Darwin's reasoning is circular, that Darwin first defined the "fittest" individuals as those that leave the most offspring and then turned around and said that the fittest survive preferentially (i.e., leave the most offspring). Do you think this is a fair criticism of Darwin's theory as described in this chapter?

3. On the Galapagos Islands, Darwin saw a variety of finches, but few varieties of other small birds. Imagine that you are visiting another group of islands about as far away from the South American mainland as the Galapagos, but upwind, so that no birds travel between the two island groups. Do you expect that on your visit to this second island group you will find a variety of finches? Comment on how your knowledge of the birds of the Galapagos as discussed in this text aids you in predicting what you will find on the second island group, if indeed it does.

FOR FURTHER READING

BARTHOLOMEW, G.: "The Role of Natural History in Contemporary Biology," *Bioscience*, May 1986, pages 324-329. The study of natural history, which guided Darwin, can still pose key questions to modern molecular disciplines.

DARWIN, C.R.: *On The Origin of Species by Means of Natural Selection, or the Preservation of Favoured Races in the Struggle for Life*, Cambridge University Press, New York, 1975 reprint. One of the most important books of all time, Darwin's long essay is still comprehensible and interesting to modern readers.

DARWIN, C.R.: *The Voyage of the Beagle*, Natural History Press, Garden City, N.Y., 1962 reprint. Darwin's own account of his observations and adventures during the famous 5-year voyage he took in his twenties.

DESMOND, A. and J. MOORE: *Darwin*, Warner Books, N.Y., 1992. A marvelously detailed account of how Darwin developed his theory of natural selection, from a modern "social" approach that emphasizes Darwin the person.

GOULD, S.: "Darwinism Defined: The Difference Between Fact and Theory," *Discover*, January 1987, pages 64-70. A clear account of what biologists do and do not mean when they refer to the theory of evolution.

GOULD, S.: *Wonderful Life. The Burgess Shale and the Nature of History*, W.W. Norton & Company, Inc., New York, 1989. A marvelous book about the early evolution of animals and about evolution in general.

IRVINE, W.: *Apes, Angels, and Victorians*, McGraw-Hill Book Co, New York, 1954. The story of Darwin and the early years of the theory of evolution; beautifully written.

LEWIN, R.: *Thread of Life: The Smithsonian Looks at Evolution*, Smithsonian Books, Washington, D.C., 1982. A beautifully illustrated chronicle of the history of life on earth.

SMITH, J.M.: *The Problems of Biology*, Oxford University Press, London, 1986. A succinct and witty guide to what biology is all about, the deep problems on which all research elaborates.

2

The Nature of Molecules

CHAPTER OUTLINE

OVERVIEW

Organisms are chemical machines; to understand them, we must start by considering chemistry. Organisms are composed of molecules, which are collections of smaller units, called atoms, that are bound to one another. All of the atoms now in the universe are thought to have been formed long ago, when the universe itself was formed. The forces that hold atoms together in molecules are called chemical bonds, which often are created by two or more atoms sharing the electrons that orbit around each atom. Sometimes molecules reshuffle their atoms or trade atoms with one another. These processes are called chemical reactions. The molecule that is most important to the evolution of life is water. The electrons that orbit water molecules are distributed asymmetrically. This asymmetry leads directly to many of the properties that are characteristic of the chemistry of living organisms.

In a small hut in the Pine Barrens of New Jersey, in the predawn darkness of a summer morning in 1962, two research scientists from Bell Laboratories first listened to the beginnings of creation. These scientists were helping to design receiving stations for the early communications satellites, their job being to find ways to reduce background interference caused by cities and other sources of radio waves. To do this, they designed a special horn-shaped antenna that allowed them to pinpoint most sources of interference. However, they were unable to identify the source of one low-strength background emission at a frequency of 3 Hz (cycles per second). Wherever in the sky they pointed their antenna, the same 3 Hz signal arrived at the same low strength. The scientists dismantled and rebuilt their antenna, but to no avail. The signal that they were picking up, although it was very weak, was real. What they were hearing was the residual thermal energy left over from an enormous explosion that is thought to have marked the beginning of the universe about 20 billion years ago.

With this explosion began the process of evolution, which ultimately led to the origin and diversification of life on earth. When viewed from the perspective of 20 billion years, life within our solar system is a recent development. To understand the origin of life, we need to consider events that took place much earlier. The same processes that led to the evolution of life were responsible for the evolution of molecules. Thus our study of life on earth begins at a seemingly odd place, with physics and chemistry. As chemical machines ourselves, we must understand chemistry to begin to understand our origins.

Organisms are composed of molecules, which are collections of smaller units, called atoms, that are bound to one another. All atoms now in the universe are thought to have been formed long ago, as the universe itself evolved. Every carbon atom in your body was created in a star.

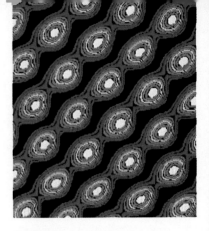

FIGURE 2-1
Individual atoms on the surface of a silicon crystal. This photograph was taken by means of the newly developed technique of tunneling microscopy.

ATOMS: THE STUFF OF LIFE

All matter is composed of small particles called **atoms.** Atoms are extremely small (Figure 2-1) and hard to study, and for a long time it was difficult for scientists to figure out their structure. It was only early in this century that experiments were carried out that suggested the first vague outlines of what an atom is like. We now know a great deal about the complexities of atomic structure, but the simple view proposed in 1913 by the Danish physicist Niels Bohr provides a good starting point. Bohr proposed that every atom possesses an orbiting cloud of tiny subatomic particles called **electrons** whizzing around the core like the planets of a miniature solar system. At the center of each atom is a small, very dense nucleus formed of two other kinds of subatomic particles, **protons** and **neutrons** (Figure 2-2).

Within the nucleus, the cluster of protons and neutrons is held together by a kind of force that works only over short subatomic distances. Each proton carries a positive (+) charge. The number of these charged protons (**atomic number**) determines the chemical character of the atom, because it dictates the number of electrons orbiting the nucleus and available for chemical activity. There is one electron for each proton. Neutrons are similar to protons in mass, but, as their name implies, they possess no charge. The **atomic mass** of an atom is equal to the sum of the masses of its protons and neutrons. Atoms that occur naturally on earth contain from 1 to 92 protons and up to 146 neutrons.

Isotopes

Different kinds of atoms are called elements. Formally speaking, an element is any substance that cannot be broken down to any other substance by ordinary chemical means. Each element is made up of one kind of atom, but may contain several versions of it.

Atoms with the same atomic number (that is, the same number of protons) have the same chemical properties and are said to belong to the same element. Not all atoms of an element, however, have the same number of neutrons. Two atoms of an element that possess different numbers of neutrons are said to be isotopes of the ele-

Hydrogen (H)

Deuterium (^{2}H)

Tritium (^{3}H)

FIGURE 2-2
Atoms. The smallest atom is hydrogen (atomic mass, 1), whose nucleus consists of a single proton. Hydrogen also has two naturally occurring isotopic forms, which possess neutrons as well as the single proton in the nucleus: deuterium (one neutron) and tritium (two neutrons).

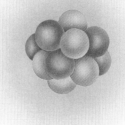

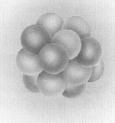

Carbon 14

6 Protons
8 Neutrons
6 Electrons

Carbon 12

6 Protons
6 Neutrons
6 Electrons

Carbon 13

6 Protons
7 Neutrons
6 Electrons

FIGURE 2-3

The three most abundant isotopes of carbon.

ment. Most elements in nature exist as mixtures of different isotopes. There are, for example, three **isotopes** of the element carbon, all of which possess 6 protons (Figure 2-3). The most common isotope of carbon has 6 neutrons. Because its total mass is 12 (6 protons plus 6 neutrons), it is referred to as carbon-12, and symbolized $^{12}_{6}C$. Over 99% of the carbon in nature is carbon-12. Most of the rest is carbon-13, an isotope with 7 neutrons.

The third isotope of carbon is carbon-14, with 8 neutrons. Carbon-14 is rare in nature. Unlike the other two isotopes of carbon, carbon-14 is unstable. Its nucleus tends to break up into elements with lower atomic numbers. This process is called radioactive decay, and it emits a significant amount of energy. Isotopes such as carbon-14 that decay in this fashion are said to be **radioactive.**

Some radioactive isotopes are more unstable than others, and so they decay more readily. For any given isotope, however, the rate of decay is constant. This rate is usually expressed as the half-life, the time it takes for 50% of the atoms in a sample to decay. Carbon-14, for example, has a half-life of about 5600 years. A sample of carbon containing 1 gram of the isotope carbon-14 today would contain 0.5 gram after 5600 years, 0.25 gram 11,200 years from now, 0.125 gram 16,800 years from now, and so on. By determining the ratios of the different isotopes of carbon and other elements in samples of biological origin and in rocks, scientists are able to make absolute determinations of the times when these materials formed.

> **Atoms that have the same number of protons but different numbers of neutrons are called isotopes. Isotopes of an atom differ in atomic mass but have similar chemical properties.**

Electrons

The positive charges in the nucleus of an atom are counterbalanced by negatively ($-$) charged electrons orbiting around the atomic nucleus at varying distances. The negative charge of one electron exactly balances the positive charge of one proton. Thus atoms with the same number of protons and electrons have no net charge. An atom in which the number of protons in the nucleus is the same as the number of orbiting electrons is known as a neutral atom.

Electrons have very little mass (only $\frac{1}{1840}$ of the mass of a proton). Of all the mass contributing to your weight, the portion that is contributed by electrons is less than the mass of your eyelashes. Electrons are maintained in their orbits by their attraction to the positively charged nucleus. Sometimes this attraction is overcome by other forces, and one or more electrons are lost by an atom. Sometimes, too, atoms may gain additional electrons. Atoms in which the number of electrons does not equal the number of protons are known as **ions,** which do carry an electrical charge. For example, an atom of sodium (Na) that has lost an electron becomes a positively charged sodium ion (Na^+), because the positive charge of one of the protons is not balanced by the negative charge of an electron.

> An atom is a core nucleus of protons and neutrons surrounded by a cloud of electrons. The number of its electrons largely determines the chemical properties of an atom.

Electrons Determine What Molecules Are Like

The key to the chemical behavior of atoms lies in the arrangement of the electrons in their orbits. It is convenient to visualize individual electrons as following discrete circular orbits around a central nucleus, as in the Bohr model of the atom. Such a simple picture is not realistic, however; it is not possible to locate the position of any individual electron precisely at any given time. Theory, in fact, says that a particular electron can be anywhere at a given instant, from close to the nucleus to infinitely far away from it.

However, a particular electron is not equally likely to be located at all positions. Some locations are much more probable than others. In other words, it is possible to say where an electron is *most likely* to be. The volume of space around a nucleus where an electron is most likely to be found is called the **orbital** of that electron (Figure 2-4). Some electron orbitals near the nucleus are spherical orbitals (*s* orbitals), whereas others are dumbbell-shaped orbitals (*p* orbitals). Still other orbitals, more distant from the nucleus, may have different shapes.

How far away from the nucleus are the orbiting electrons? Extremely far. Almost all the volume of an atom is empty space. If the nucleus of an atom were the size of an apple, the orbit of the nearest electron would be more than a mile out. It is for this reason that the electrons of an atom determine its chemical behavior—the nuclei of two atoms never come close enough to each other in nature to interact. That is why isotopes of an element, all of which have the same arrangement of electrons, behave the same way chemically.

Energy Within the Atom

Because electrons carry negative charges, they are attracted to the positively charged nucleus, and it takes work to keep them in orbit, just as it takes work to hold an apple in your hand when gravity is pulling the apple down toward the ground. The apple in your hand is said to possess energy, the ability to do work, because of its position—if you were to release it, the apple would fall. Similarly, electrons have energy of position, called potential energy. It takes work to oppose the attraction of the

FIGURE 2-4
Electron orbitals. The lowest energy level or electron shell, nearest the nucleus, is level *K*. It is occupied by a single *s* orbital, referred to as 1*s*. The next highest energy level, *L*, is occupied by four orbitals, one *s* orbital (referred to as the 2*s* orbital) and three *p* orbitals (each referred to as a 2*p* orbital). Viewed simultaneously, the four *L* level orbitals compactly fill the space around the nucleus, like two pyramids set base to base.

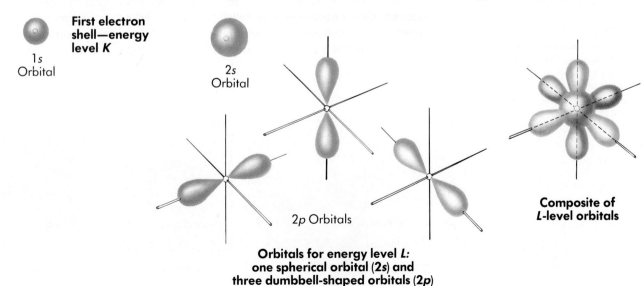

1*s*
Orbital

First electron shell—energy level *K*

2*s*
Orbital

2*p* Orbitals

Composite of *L*-level orbitals

**Orbitals for energy level *L*:
one spherical orbital (2*s*) and
three dumbbell-shaped orbitals (2*p*)**

FIGURE 2-5
Atomic energy levels. When an electron absorbs energy, it moves to higher energy levels further from the nucleus. When an electron releases energy, it falls inward to lower energy levels.

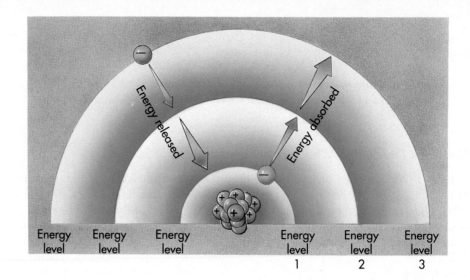

Energy level | Energy level | Energy level | Energy level 1 | Energy level 2 | Energy level 3

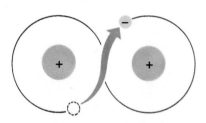

Oxidation   Reduction

FIGURE 2-6
Oxidation is the loss of an electron; reduction is the gain of one.

nucleus and move the electron further out, and so moving an electron out to a more distant orbit requires an input of energy and results in an electron with greater potential energy. For the same reason, a bowling ball released from the top of a building impacts with greater force than one dropped from a distance of a meter. Moving an electron in toward the nucleus has the opposite effect; energy is released, and the electron ends up with less potential energy (Figure 2-5).

Sometimes an electron is transferred during a chemical reaction from one atom to another. The loss of an electron is called **oxidation;** the gain of an electron is called **reduction** (Figure 2-6). It is important to realize that when an electron is transferred in this way, it keeps its energy of position. In living organisms, chemical energy is stored in high-energy electrons that are frequently transferred from one atom to another.

In a schematic drawing of an atom (Figure 2-7), the nucleus is represented as a small circle with the number of protons and neutrons indicated. The orbitals are depicted as concentric rings. In helium, for example, there is a single orbital containing two electrons. A given orbital may contain no more than two electrons. Such schematics are only diagrams; the actual orbitals have complex, three-dimensional shapes.

Most atoms have more than one orbital. When electrons from two different orbitals are the same distance from the nucleus, they are placed on the same ring by convention. When electrons in two orbitals are different distances from the nucleus, they are placed in separate concentric rings. These rings are called **energy levels,** or **shells.** The farther an electron is from the nucleus, the more energy it has.

Electrons orbit a nucleus in paths called orbitals. No orbital can contain more than two electrons, but many orbitals may be the same distance from the nucleus and thus possess electrons of the same energy.

FIGURE 2-7
Electron energy levels for helium and nitrogen. The electrons are indicated by orange dots. Each concentric circle represents a different distance from the nucleus and thus a different electron energy level.

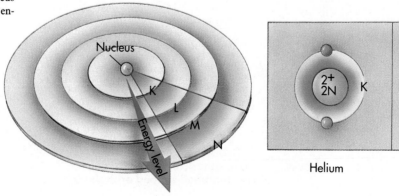

Helium   Nitrogen

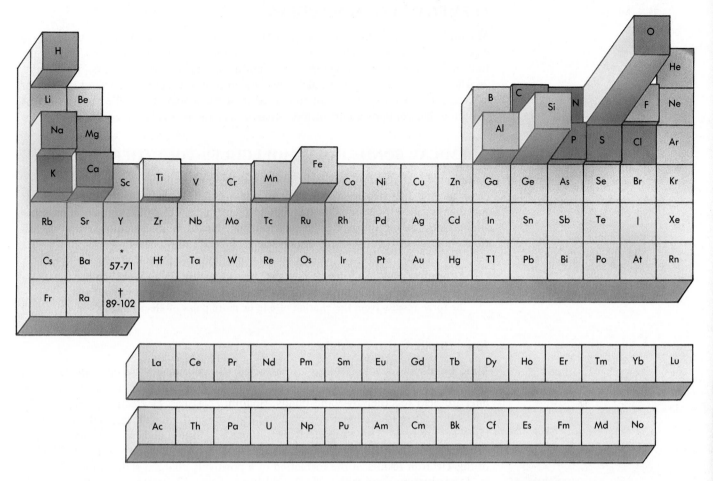

The Periodic Table

There are 92 naturally occurring kinds of atoms, called **elements.** Each element has a different number of protons and a different arrangement of electrons. When the nineteenth-century Russian chemist Dmitri Mendeleev arranged the known elements in a table according to their atomic mass (Figure 2-8), he discovered one of the great generalizations of all science. Mendeleev found that the entries in the table exhibited a repeating pattern of chemical properties, in recurring groups of eight elements.

The eight-element periodicity that Mendeleev found is based on the interactions that occur between the electrons of the outer energy levels of the different elements. These interactions, in turn, are the basis for the differing chemical properties of the elements. The maximum number of electrons that an outer energy level can possess is eight; the chemical behavior that we observe reflects how many of the eight positions are filled. Elements possessing all eight electrons at their outer energy level are not reactive (neon, argon, xenon, krypton). In sharp contrast, elements with one fewer than the maximum number of eight electrons at their outer energy level, such as chlorine, bromine, and fluorine, are highly reactive. Elements with only one electron in their outer energy level, such as sodium, lithium, or potassium are also reactive.

Mendeleev's **periodic table** thus leads to a useful generalization, the **octet rule** or **rule of eight** (Latin, *octo,* eight): atoms tend to establish completely full outer energy levels. For most of the atoms important to life, the number of electrons required to fill the outer shell is eight. Most chemical behavior can be predicted quite accurately on the basis of this simple rule and from the tendency of atoms to balance positive and negative charges.

All atoms tend to fill their outer energy levels with the maximum number of electrons.

FIGURE 2-8
Periodic table of the elements. In this representation, the frequency of elements that occur in the earth's crust in more than trace amounts is indicated in the vertical dimension. Elements found in significant amounts in living organisms are shaded in pink. Many elements that are common on the surface of the earth, such as silicon and iron, are not present in living organisms in significant concentrations.

ELEMENTS AND MOLECULES

Much of the core of the earth is thought to consist of iron, nickel, and other heavy elements. The crust of the earth is quite different in composition from that of the bulk of the planet, being composed primarily of the lighter elements. By weight, 74.3% of the earth's crust (its land, oceans, and atmosphere) consists of oxygen or silicon. Most of these atoms are combined in stable associations called **molecules**. A molecule that contains atoms of more than one element is called a **compound**.

CHEMICAL BONDS HOLD MOLECULES TOGETHER

A molecule is a group of atoms held together by energy. The energy acts as "glue," ensuring that the various atoms stick to one another. The force holding two atoms together is called a **chemical bond**. The force can result from the attraction of opposite charges, called an ionic bond, or from the sharing of one or more pairs of electrons, a covalent bond. Other, weaker kinds of bonds also occur.

> A chemical bond is a force holding two atoms together. In an ionic bond,
> the force results from the attraction of opposite charges. In a covalent bond,
> the force results from the sharing of one or more pairs of electrons.

Ionic Bonds Form Crystals

Ionic bonds form when atoms are attracted to one another by opposite electrical charges. Common table salt, sodium chloride (NaCl), is a lattice of ions in which atoms are held together by ionic bonds. Sodium (Na) has 11 electrons (Figure 2-9). Two of these are in the inner energy level, eight are at the next level, and one is at the outer energy level. The outer electron is unpaired ("free") and has a strong tendency to form a pair. A stable configuration is achieved if the outer electron is lost. The loss of this electron results in the formation of a positively charged sodium ion (Na$^+$).

The chlorine atom faces a similar dilemma. It has 17 electrons: 2 at the inner energy level, 8 at the next energy level, and 7 at the outer energy level. The outer energy level of the chlorine atom has an unpaired electron. The addition of an electron to the outer level causes the formation of a negatively charged chloride ion (Cl$^-$).

When placed together, metallic sodium and gaseous chlorine react swiftly and explosively, the sodium atoms donating electrons to chlorine atoms. The result of this

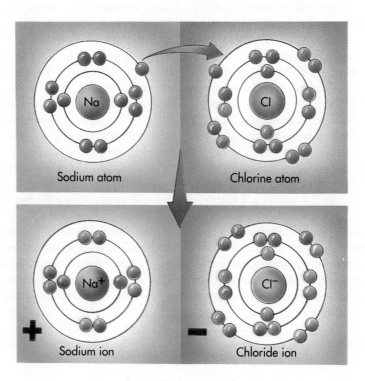

FIGURE 2-9

The formation of ionic bonds by sodium chloride. When a sodium atom donates an electron to a chlorine atom, the sodium atom, lacking that electron, becomes a positively charged sodium ion; the chlorine atom, having gained an extra electron, becomes a negatively charged chloride ion.

Sodium atom   Chlorine atom

Sodium ion   Chloride ion

chemical reaction between the two kinds of atoms is the production of Na^+ and Cl^- ions. Because opposite charges attract, electrical neutrality (the third chemical tendency) can be achieved by the association of these ions with each other. The chemical bond between Na^+ and Cl^- ions resulting from this electrical attraction is not directional, however, and discrete sodium chloride molecules do not form. Instead, the ions aggregate, or come together, into a crystal matrix, which has a precise geometry. Such aggregations are what we know as crystals of salt. If a salt such as NaCl is placed in water, the electrical attraction of the water molecules, for reasons we will point out later in this chapter, disrupts the forces holding the ions in their crystal matrix, causing the salt to dissolve into a roughly equal mixture of free Na^+ and free Cl^- ions. Approximately 0.06% of the atoms in your body are free Na^+ or Cl^- ions—about 40 grams, the weight of your fingernails.

> **An ionic bond is an attraction between ions of opposite charge. Such bonds are not directed between two specific ions but rather between a given ion and all of the oppositely charged ions in its immediate vicinity.**

True molecules do not result from ionic bonds, however, because the electrical attractive force is not directed between two specific ions with opposite charges; rather, it exists between any one ion and all neighboring ions of the opposite charge. True molecules are formed by much stronger bonds called covalent bonds, which involve attractive forces directed between specific atoms.

Covalent Bonds Build Stable Molecules

Covalent bonds form when two atoms share one or more pairs of the electrons of their outer energy levels. Consider hydrogen (H) as an example. Each hydrogen atom has an unpaired electron and an unfilled K energy level; for these reasons the hydrogen atom is chemically unstable. When two hydrogen atoms are close enough to one another, however, each electron can orbit both nuclei. In effect, nuclei in proximity are able to share their electrons (Figure 2-10). The result is a diatomic molecule (one with two atoms) of hydrogen gas (H_2).

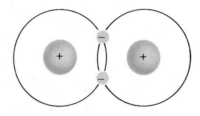

The diatomic hydrogen gas molecule formed as a result of this sharing of electrons is not charged, however, because it still contains two protons and two electrons. Each of the two hydrogen atoms can be considered to have two orbiting electrons at the K energy level. This relationship satisfies the octet rule because each shared K energy level electron orbits both nuclei and therefore is included in the outer energy level of *both* atoms. The relationship also results in the pairing of the two free electrons. For these reasons the two hydrogen atoms form a stable molecule. Note, however, that this stability is conferred by the electrons that orbit both nuclei; stability of this kind occurs only when the nuclei are close together. For this reason the strong chemical forces tending to pair electrons and fill the outer shell of electrons will act to keep the two nuclei near one another. The bond between the two hydrogen atoms of diatomic hydrogen gas is an example of a covalent bond.

> **A covalent bond is a chemical bond formed by the sharing of one or more pairs of electrons.**

The bonds that hold the molecules of your body together are covalent bonds. Covalent bonds are far better suited to this task than are ionic ones because covalent bonds are *directional*—a covalent bond is formed between two specific atoms, whereas an ionic bond is formed between a charged atom and the electric field contributed by all nearby atoms of opposite charge. Ionic bonds can form regular crystals, but such crystals dissolve in water. Complex stable shapes require the more specific associations made possible by covalent bonds.

Covalent bonds can be very strong. The strength of the bond depends on the degree to which breaking the bond violates the three central tendencies toward electrical neutrality, paired electrons, and full outer energy levels. Thus covalent bonds that

FIGURE 2-10
Hydrogen gas. Hydrogen gas is a diatomic molecule composed of two hydrogen atoms each sharing its electron with the other. The flash of fire that consumed the *Hindenburg* occurred when the hydrogen gas that was used to inflate the dirigible combined explosively with oxygen gas in the air to form water.

satisfy the octet rule by sharing *two* pairs of electrons, called **double bonds,** are stronger than covalent bonds that involve the sharing of only one electron pair, called single bonds. More chemical energy is required to disrupt a double bond than a single bond. Covalent bonds are represented in chemical formulations as lines connecting atomic symbols. Each line between two bonded atoms represents the sharing of one pair of electrons. Hydrogen gas is thus symbolized as H—H, and oxygen gas as O=O.

Molecules with several covalent bonds. Molecules are often composed of more than two atoms. One reason that larger molecules may be formed is that a given atom is able to share electrons with more than one other atom. An atom that requires two, three, or four additional electrons to fill its outer energy level completely may acquire them by sharing its electrons with two or more other atoms.

For example, carbon (C) atoms (atomic number, 6) contain six electrons, two at the K energy level and the other four at the L energy level. To satisfy the octet rule, a carbon atom must gain access to four additional electrons; it must form the equivalent of four covalent bonds. Because there are many ways in which four covalent bonds may form, carbon atoms are able to participate in many different kinds of molecules, which is why there are so many different kinds of carbon-containing molecules.

ATOMS IN LIVING ORGANISMS

Of the 92 elements that formed the crust of the cooling earth, only 11 are common in living organisms. Table 2-1 lists the frequency with which various elements occur in the human body; the frequencies that occur in the bodies of other organisms are similar. Inspection of this table suggests that the distribution of elements in living systems is by no means accidental. The life elements—those that make up 0.01% (1 in 10,000) or more of the atoms of organisms—are not the elements that are most abundant in the earth's crust. Unlike the elements that occur most abundantly in this crust, all of the elements common in living organisms are light, each having an atomic number less than 21 and thus a low mass.

The majority of atoms in living things—97.4% of the atoms in the human body, for example—are either nitrogen, oxygen, carbon, or hydrogen. Sulfur and phosphorus also play important roles, but are not as abundant. You can conveniently remember the four most abundant elements by their first initials, NOCH. Why are these four elements the most abundant?

1. They all form gases, either alone or in combination with one another. Life is thought to have evolved from complex molecules formed from the interaction of these gases in the primitive earth's atmosphere. Many of these molecules are water soluble. As water vapor in the atmosphere cooled and fell as rain, it brought these dissolved molecules to the primitive oceans where life began.
2. The NOCH elements all form molecules with one another by sharing electrons and so forming covalent bonds.
3. The molecules formed by the NOCH elements possess a variety of different chemical bonds that are not so strong that they cannot be broken.
4. Approximately 90% of the atoms common in living organisms are hydrogen and oxygen, reflecting the predominant role of water (H_2O) in living systems.

THE CRADLE OF LIFE: WATER

On the primitive earth many of the atomic elements formed complex molecules with silicon and oxygen. These molecules were bound to one another in strongly linked crystalline arrays, forming minerals, which in turn combined and formed rocks. In contrast to the others, the molecules of one compound that was released from these rocks as the primitive earth cooled did not form crystals: **water.** An oxide of hydrogen, water has the chemical formula H_2O. This seemingly simple molecule has many

TABLE 2-1   THE MOST COMMON ELEMENTS ON EARTH AND THEIR DISTRIBUTION IN THE HUMAN BODY

ELEMENT	SYMBOL	ATOMIC NUMBER	APPROXIMATE PERCENT OF EARTH'S CRUST BY WEIGHT	PERCENT OF HUMAN BODY BY WEIGHT	IMPORTANCE OR FUNCTION
Oxygen	O	8	46.6	65.0	Required for cellular respiration; component of water
Silicon	Si	14	27.7	Trace	
Aluminum	Al	13	6.5	Trace	
Iron	Fe	26	5.0	Trace	Critical component of hemoglobin in the blood
Calcium	Ca	20	3.6	1.5	Component of bones and teeth; triggers muscle contraction
Sodium	Na	11	2.8	0.2	Principal positive ion bathing cells; important in nerve function
Potassium	K	19	2.6	0.4	Principal positive ion in cells; important in nerve function
Magnesium	Mg	12	2.1	0.1	Critical component of many energy-transferring enzymes
Hydrogen	H	1	0.14	9.5	Electron carrier; component of water and most organic molecules
Manganese	Mn	25	0.1	Trace	
Fluorine	F	9	0.07	Trace	
Phosphorus	P	15	0.07	1.0	Backbone of nucleic acids; important in energy transfer
Carbon	C	6	0.03	18.5	Backbone of organic molecules
Sulfur	S	16	0.03	0.3	Component of most proteins
Chlorine	Cl	17	0.01	0.2	Principal negative ion bathing cells
Vanadium	V	23	0.01	Trace	
Chromium	Cr	24	0.01	Trace	
Copper	Cu	29	0.01	Trace	Key component of many enzymes
Nitrogen	N	7	Trace	3.3	Component of all proteins and nucleic acids
Boron	B	5	Trace	Trace	
Cobalt	Co	27	Trace	Trace	
Zinc	Zn	30	Trace	Trace	Key component of some enzymes
Selenium	Se	34	Trace	Trace	
Molybdenum	Mo	42	Trace	Trace	Key component of many enzymes
Tin	Sn	50	Trace	Trace	
Iodine	I	53	Trace	Trace	Component of thyroid hormone

surprising properties (Figure 2-11). For example, of all the molecules that are common on earth, only water exists as a liquid at the relatively cool temperatures that prevail on the earth's surface. When life was originating, water provided a medium in which other molecules could move around and interact without being bound by strong covalent or ionic bonds. Life evolved as a result of these interactions.

Life as it evolved on earth is thus inextricably tied to water. Three-fourths of the earth's surface is covered by liquid water. About two thirds of your body is composed of water, and you cannot exist long without it. All other organisms also require water. It is no accident that tropical rain forests are bursting with life, whereas dry deserts (Figure 2-12) are almost lifeless except when water becomes temporarily plentiful, such as after a rainstorm. Farming is possible only in those areas of the earth where rain is plentiful. No plant or animal can grow and reproduce in any but a water-rich environment.

The chemistry of life, then, is water chemistry. The way in which life first

FIGURE 2-11

Water takes many forms. As a liquid, it fills our rivers and runs down over the land to the sea, sometimes falling in great cascades, such as at Victoria Falls in Zimbabwe, Central Africa. The iceberg on which the penguins are holding their meeting was formed in Antarctica from a huge block of ice breaking away into the ocean water. When water cools below 0° C, it forms beautiful crystals, familiar to us as snow and ice. However, water is not always plentiful. At Badwater, Death Valley, there is no hint of water except for the broken patterns of dried mud.

FIGURE 2-12

Deserts. In deserts such as this one in Sonora, Mexico, little rain falls. Life cannot persist in such deserts for long. Only when rain falls is life plentiful in the desert. Those plants and animals which survive there do so by carefully managing the little water available to them.

evolved was determined in large part by the chemical properties of the liquid water in which that evolution occurred. The single most outstanding chemical property of water is its ability to form weak chemical associations with only 5% to 10% of the strength of covalent bonds. This one property of water, which as you will see derives directly from its structure, is responsible for much of the organization of living chemistry.

Water has a simple atomic structure. Just as a methane molecule results from the formation of four single covalent bonds among a carbon atom and four hydrogen atoms, thus satisfying the four vacancies in the outer energy level of the carbon atom, so water results from the formation of two single covalent bonds between an oxygen atom and two hydrogen atoms (Figure 2-13). The resulting molecule is stable: it satisfies the octet rule, has no unpaired electrons, and does not carry a net electrical charge.

Water Acts Like a Magnet

The remarkable story of the properties of water does not end there, however. Both the oxygen and the hydrogen atoms attract the electrons they share in the covalent bonds of the water molecule, an attraction called electronegativity. The oxygen atom is more electronegative than are hydrogen atoms, so that it pulls the electrons more strongly than do the two hydrogen atoms. As a result, the electron pair shared in each of the two single oxygen-hydrogen covalent bonds of a water molecule is far more likely, at a given moment, to be found near the oxygen nucleus than near one of the hydrogen nuclei. This relationship has a profoundly important result: the oxygen atom acquires a partial negative charge. It is as if the electron cloud were denser in the neighborhood of the oxygen atom and less dense around the hydrogen atoms. This charge separation within the water molecule creates negative and positive electrical charges on the ends of the molecule. These partial charges are much less than the unit charges of ions.

The water molecule thus has distinct "ends," each with a partial charge, like the two poles of a magnet. Molecules such as water that exhibit charge separation are called **polar molecules** because of these magnetlike poles. Water is one of the most polar molecules known. *The polarity of water underlies its chemistry and thus the chemistry of life.*

What would you expect the shape of a molecule such as water to be? Consider water's two covalent bonds. Each will have a partial charge at each end, because of the charge separation resulting from the unequal attraction of electrons by hydrogen and oxygen atoms. The most stable arrangement of these charges is one in which the two negative and two positive charges are equidistant from one another. Such an arrangement is called a **tetrahedron.** In water the oxygen molecule occupies the center of such a tetrahedron, with hydrogen atoms at two of the apexes and the partial negative charges at the other two apexes. This results in a bond angle between the two covalent oxygen-hydrogen bonds of 104.5 degrees (see Figure 2-13).

> Much of the biologically important behavior of water results because the
> oxygen atom attracts electrons more strongly than do the hydrogen atoms
> so that the water molecule has electron-rich (−) and electron-poor (+) regions,
> giving it positive and negative poles.

Polar molecules interact with one another. The partial negative charge at one end of a polar molecule is attracted to the partial positive charge of another polar molecule. This weak attraction, between a hydrogen atom and two electron-rich (electronegative) atoms, is called a **hydrogen bond** (Figure 2-14). Water forms a lattice of such hydrogen bonds. Each of these hydrogen bonds is individually very weak and transient. A given bond lasts only $\frac{1}{100,000,000,000}$ (10^{-11}) of a second. However, even though each bond is transient, a large number of such bonds can form, and the cumulative effects of large numbers of these bonds can be enormous. These cumulative effects are responsible for many of the important physical properties of water (Table 2-2).

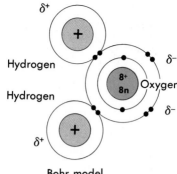

Bohr model

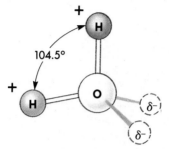

Ball and stick model

FIGURE 2-13
Water has a simple molecular structure. Each molecule is composed of one oxygen atom and two hydrogen atoms. The oxygen atom shares one electron with each hydrogen atom.

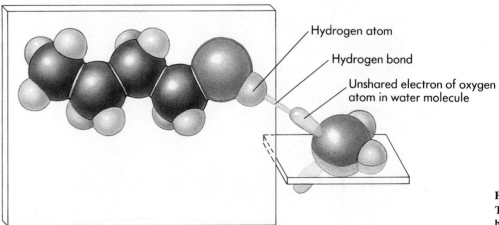

Hydrogen atom

Hydrogen bond

Unshared electron of oxygen atom in water molecule

FIGURE 2-14
The structure of a hydrogen bond.

TABLE 2-2   THE PROPERTIES OF WATER

PROPERTY	EXPLANATION	EXAMPLE OF BENEFIT TO LIFE
High polarity	Polar water molecules are attracted to ions and polar compounds, making them soluble	Many kinds of molecules can move freely in cells, permitting a diverse array of chemical reaction
High specific heat	Hydrogen bonds absorb heat when they break, and release heat when they form, minimizing temperature changes	Water stabilizes body temperature, as well as that of the environment
High heat of vaporization	Many hydrogen bonds must be broken for water to evaporate	Evaporation of water cools body surfaces
Lower density of ice	Water molecules in an ice crystal are spaced relatively far apart because of hydrogen bonding	Because ice is less dense than water, lakes do not freeze solid, and they overturn in spring
Cohesion	Hydrogen bonds hold molecules of water together	Leaves pull water upward from roots; seeds swell and germinate

Water Clings to Polar Molecules

Water molecules, being very polar themselves, are attracted to other polar molecules. When the other polar molecule is another water molecule, the attraction is referred to as **cohesion**. When the other polar molecule is a different substance, the attraction is called **adhesion.** It is because water is cohesive, forming a lattice of hydrogen bonds with itself, that it is a liquid, and not a gas, at moderate temperatures.

The cohesion of liquid water is also responsible for its surface tension. Insects, weighing little, can walk on water (Figure 2-15), even though they are denser than the water, because at the air-water interface all the hydrogen bonds face inward, causing the molecules of the water surface to cling together. Water is adhesive to any substance with which it can form hydrogen bonds. This is why things get "wet" when they are dipped in water—and why waxy substances do not (they are composed of nonpolar molecules).

The attraction of water to substances with surface electrical charges (called electrostatic attraction) is responsible for capillary action: if a glass tube with a narrow

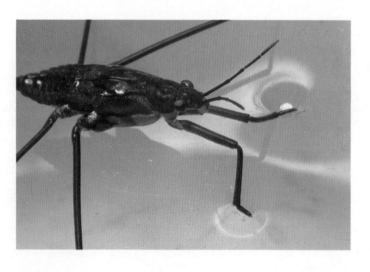

FIGURE 2-15

Some insects, such as the water strider, literally walk on water. In this photograph you can see the dimpling the insect's feet make on the water as its weight bears down on the surface. Because the surface tension of the water is greater than the force that one foot brings to bear, the strider does not sink, but rather glides along.

diameter is lowered into a beaker of water, the water will rise up in the tube above the level of the water in the beaker, held by adhesion of the water to the glass (Figure 2-16). The narrower the tube, the greater the electrostatic forces between the polar water molecules and the charged surface of the glass, and the higher the water rises.

Water Stores Heat

The temperature of any substance is a measure of how rapidly its individual molecules are moving. Because of the many hydrogen bonds that water molecules form with one another, a large input of thermal energy is required to disrupt the organization of water and raise its temperature. Water is said to have a high **specific heat.** Specific heat is defined as the amount of heat that must be absorbed or lost for one gram of a substance to change its temperature by 1° C. Specific heat measures the extent to which a substance resists changing its temperature when it absorbs or loses heat. The specific heat of water (1 calorie per gram per degree centigrade) is twice that of most carbon compounds and nine times that of iron; only ammonia, which forms very strong hydrogen bonds, has a higher specific heat than water (1.23 calories/gram/degree centigrade). Why does water have such a high specific heat? Because much of the heat energy water absorbs is used to break the many hydrogen bonds between water molecules, bonds that must be broken before the individual water molecules can begin moving about more freely and so produce a higher temperature. The more polar a substance, the higher its specific heat; ammonia has a higher specific heat than water because it is more polar than water.

Because of its high specific heat, water heats up more slowly than almost any other compound and will hold its temperature longer when heat is no longer applied. One consequence of organisms having evolved with a high water content is that even on land they are able to maintain a relatively constant internal temperature. The heat generated by your metabolism would cook your cells if it were not for the fact that the water within them has such a high specific heat and so absorbs the bulk of this energy.

If the temperature is low enough, only a few hydrogen bonds break and the water molecules maintain a crystal-like lattice of hydrogen bonds, forming a substance we call ice (Figure 2-17). Interestingly, ice is less dense than water—that is why icebergs float. Why is ice less dense? Because the hydrogen bonds space the water molecules relatively far apart.

A considerable amount of heat energy (586 calories) is required to change one gram of liquid water into vapor. Water is said to have a high **heat of vaporization.** Every gram of water that evaporates from the human body removes 586 calories of heat from the body. This amount is equal to the energy released by lowering the temperature of 586 grams of water 1° C. Thus the evaporation of water from a surface produces significant cooling. Many organisms dispose of excess heat by evaporative cooling. Human beings, for example, sweat.

Water Is a Powerful Solvent

Water molecules gather closely around any molecule that exhibits an electrical charge, whether the molecule carries a full charge (ion) or a charge separation (polar molecule). For example, sucrose (table sugar) is composed of molecules that contain slightly polar hydroxyl (—OH) groups. A sugar crystal dissolves rapidly when it is placed in water because water molecules can form hydrogen bonds with the polar hydroxyl groups of the sucrose molecules. Every time a sugar molecule dissociates or breaks away from the crystal, water molecules orient around it in a cloud. Such a **hydration shell,** formed by the water molecules, prevents every sucrose molecule from associating with other sucrose molecules. Similarly, hydration shells form around all polar molecules (Figure 2-18). Polar molecules that dissolve in water in this way are said to be **soluble** in water. Nonpolar molecules are not water-soluble. Oil is an example of a nonpolar molecule. Life originated in water not only because it is a liquid, but also because so many molecules are polar or ionized and thus are water soluble.

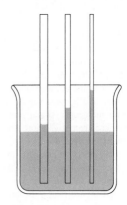

FIGURE 2-16
Capillary action. Capillary action will cause the water within a narrow tube to rise above the surrounding fluid; the adhesion of the water to the glass surface, drawing it upward, is stronger than the force of gravity, drawing it down. The narrower the tube, the greater the surface/volume ratio and the more the adhesion defies gravity.

FIGURE 2-17
The role of hydrogen bonds in an ice crystal. When water cools below 0° C, it forms a regular crystal structure in which the four partial charges of each atom in the water molecule interact with opposite charges of atoms in other water molecules. Because water forms a crystal latticework, ice is less dense than water and floats. If it did not, northern bodies of water in the United States might never thaw fully.

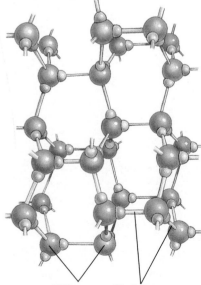

Water   Hydrogen
molecules   bonds

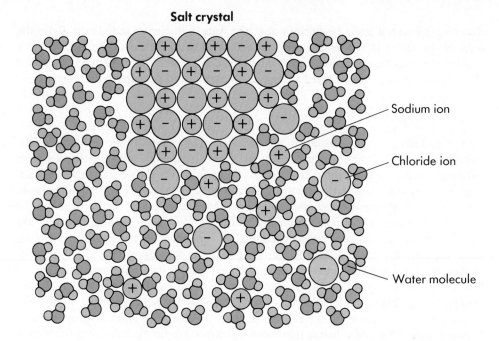

Salt crystal

Sodium ion

Chloride ion

Water molecule

FIGURE 2-18

Why salt dissolves in water. When a crystal of table salt dissolves in water, individual Na$^+$ and Cl$^-$ ions break away from the salt lattice and become surrounded by water molecules. Water molecules orient around Cl$^-$ ions so that their positive ends face in toward the negative chloride ion; water molecules surrounding Na$^+$ ions orient in the opposite way, the negative ends facing the positive sodium ion. Surrounded by water in this way, Na$^+$ and Cl$^-$ ions never re-enter the salt lattice.

Water Organizes Nonpolar Molecules

Water molecules in solution always tend to form the maximum number of hydrogen bonds possible. When nonpolar molecules, which do not form hydrogen bonds, are placed in water, the water molecules act in such a way as to exclude them, instead preferentially forming hydrogen bonds with other water molecules. The nonpolar molecules are forced into association with one another, thus minimizing their disruption of the hydrogen bonding of water. The result is a sort of molecular slum, in which all of the nonpolar molecules are crowded together. It seems almost as if the nonpolar compounds shrink from contact with the water, and for this reason they are called **hydrophobic** (Greek, *hydros*, water + *phobos*, fearing—although "feared by water" might be a more apt description). Polar molecules readily form hydrogen bonds with water. For that reason, polar compounds are freely soluble in water and are called **hydrophilic** (water-loving).

The tendency for nonpolar molecules to band together in water solution is referred to as **hydrophobic exclusion**. The hydrophobic forces induced by the hydrogen bonding of water were important in the evolution of life, because some of the exterior portions of many of the molecules on which life came to be based are nonpolar. By forcing these hydrophobic portions of molecules into proximity to one another, water causes such molecules to assume particular shapes in solution. Different molecular shapes have evolved by alteration of the location and strength of nonpolar regions. As you will see, much of the evolution of life reflects changes in the shapes of molecules, changes that can be induced in just this way.

Water Ionizes

The covalent bonds within a water molecule sometimes break spontaneously. When this happens, one of the protons (hydrogen atom nuclei) dissociates from the molecule. Because the dissociated proton lacks the negatively charged electron that it had been sharing in the covalent bond with oxygen, its own positive charge is not counterbalanced. Thus, a positively charged ion H$^+$ is produced. The proton is usually associated with another water molecule, forming a hydromium (H$_3$O$^+$) ion. The rest of the water molecule, which has retained the shared electron from the covalent bond, is negatively charged and forms the hydroxide ion (OH$^-$). This process of spontaneous ion formation is called **ionization**:

$$\text{H}_2\text{O} \rightarrow \text{OH}^- + \text{H}^+$$

WATER   HYDROXIDE   HYDROGEN

ION   ION

To quantify the concentration of H^+ ions in solution, a scale based on the slight degree of spontaneous ionization of water has been constructed. In a liter of water, roughly 1 molecule out of each 550 million is ionized at any instant in time, corresponding to $\frac{1}{10,000,000}$ (10^{-7}) of a mole of H^+ ions. (A **mole** is defined as the weight in grams that corresponds to the summed atomic weights of all the atoms of a molecule, that is, its molecular mass. In the case of H^+, the molecular mass equals 1, and a mole of H^+ ions would weigh 1 gram.) The molar concentration of hydrogen ions in pure water, $\frac{1}{10,000,000}$ moles/liter, can be written more easily by employing exponential notation (10^{-7}). This is done by counting the number of decimal places after the digit "1" in the denominator:

$$[H^+] = \frac{1}{10,000,000}$$

Brackets are used to indicate the chemical concentration of H^+. Since there are seven decimal places, the molar concentration is 10^{-7} moles per liter.

Any substance that dissociates to form H^+ ions when dissolved in water is called an **acid**. The exponents in the exponential notation of H^+ ion concentrations are used as a convenient indication of acid strength, called the pH scale. **pH** is formally defined as the negative logarithm of the hydrogen ion concentration:

$$pH = -\log [H^+]$$

What is a logarithm? In this case, simply the exponent of the molar concentration. pH is thus normally a positive number, the negative of the exponent of the molar concentration of a substance (it is because this exponent usually has a negative sign that the value of pH is positive). Thus pure water has a molar H^+ concentration of 10^{-7} and a pH of 7. The stronger an acid is, the more H^+ ions that it produces and the lower its pH. Hydrochloric acid (HCl), which is abundant in your stomach, ionizes completely, so the molar concentration of $[H^+]$ in water containing $\frac{1}{10}$ of a mole of hydrochloric acid (for example, 0.1 mole per liter) is 0.1, or 10^{-1} moles per liter, corresponding to a pH of 1. Some acids, such as nitric acid, are even stronger than this, although such very strong acids are rarely found in living systems (Figure 2-19).

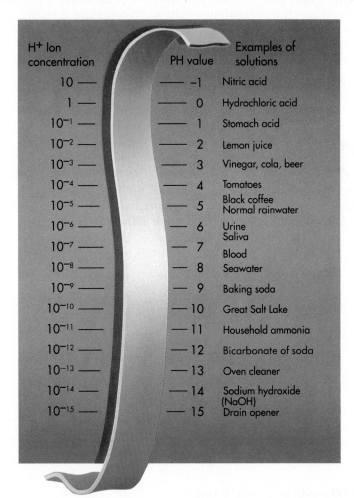

FIGURE 2-19

The pH scale. A fluid is assigned a value according to the number of hydrogen ions present in a liter of that fluid. The scale is logarithmic, so that a change of only 1 means a tenfold change in the concentration of hydrogen ions; thus lemon juice is 100 times more acidic than tomato juice, and seawater is 10 times more basic than pure water.

H⁺ Ion concentration	PH value	Examples of solutions
10	-1	Nitric acid
1	0	Hydrochloric acid
10^{-1}	1	Stomach acid
10^{-2}	2	Lemon juice
10^{-3}	3	Vinegar, cola, beer
10^{-4}	4	Tomatoes
10^{-5}	5	Black coffee Normal rainwater
10^{-6}	6	Urine Saliva
10^{-7}	7	Blood
10^{-8}	8	Seawater
10^{-9}	9	Baking soda
10^{-10}	10	Great Salt Lake
10^{-11}	11	Household ammonia
10^{-12}	12	Bicarbonate of soda
10^{-13}	13	Oven cleaner
10^{-14}	14	Sodium hydroxide (NaOH)
10^{-15}	15	Drain opener

The pH of champagne, which bubbles because of the carbonic acid dissolved in it, is about 2.

> pH refers to the relative concentration of H⁺ ions in solution. The numerical value of the pH is the negative of the hydrogen ion concentration. Thus low pH values indicate high concentrations of H⁺ ions (acids), and high pH values indicate low H⁺ concentrations.

Positively charged hydrogen ions are not the only types of ions that are produced when water ionizes. Negatively charged OH^- ions are also produced in equal concentration. Any substance that combines with H^+ ions, as OH^- ions do, is said to be a **base.** In pure water the concentrations of H^+ and OH^- ions are both 10^{-7} moles per liter, reflecting the spontaneous rate of dissociation of water. Any increase in base concentration has the effect of lowering the H^+ ion concentration, because base and H^+ ions join spontaneously. Bases, therefore, have pH values above 7. Strong bases, such as sodium hydroxide (NaOH), have pH values of 12 or more.

Buffers

The pH of the blood circulating through your body, and the pH inside almost all living cells, is about 7. It is quite important to your health that this value not change much because the many proteins that govern your metabolism are all extremely sensitive to pH—slight alterations in pH will change their shape and so disrupt their activities. The pH of your blood, for example, is 7.4, and you would survive only a few minutes if it were to fall to 7.0 or rise to 7.8. And yet we all eat substances that are acidic and others that are basic; Coca-Cola, for example, is a strong (although dilute) acid. In fact, our metabolism produces acids and bases within our cells all the time. What keeps our pH constant? Our cells contain substances called **buffers** that minimize changes in the concentrations of H^+ and OH^-.

A buffer is a substance that acts as a reservoir for hydrogen ions, donating them to the solution when their concentration falls and taking them from the solution when their concentration rises. What sort of substance will act in this way? Within organisms, most buffers consist of pairs of substances, one an acid and the other a base. The key buffer in human blood is the acid-base pair "carbonic acid (acid)-bicarbonate (base)." In water, carbonic acid (H_2CO_3) dissociates to yield bicarbonate ion and a hydrogen ion, which further dissociates to CO_2 and H_2O (Figure 2-20). Now if some other source adds H^+ ions to your blood (pH drops), the bicarbonate ion acts as a base and removes the excess H^+ ions from solution by forming H_2CO_3. Similarly, if some process removes H^+ ions from your blood (pH rises), the carbonic acid dissociates, releasing more hydrogen ions into the solution. pH is thus stabilized by the equilibrium between the forward and back reactions converting acid to base or base to acid.

The interaction of carbon dioxide and water has the important consequence that significant amounts of carbon enter into water solution from air in the form of carbonic acid. As we will discuss, biologists believe that life first evolved in the early oceans, which were rich in carbon because of this reaction.

FIGURE 2-20
Buffer formation. Carbon dioxide and water combine chemically to form carbonic acid (H_2CO_3), which dissociates in water, freeing H^+ ions. This reaction makes carbonated beverages acidic.

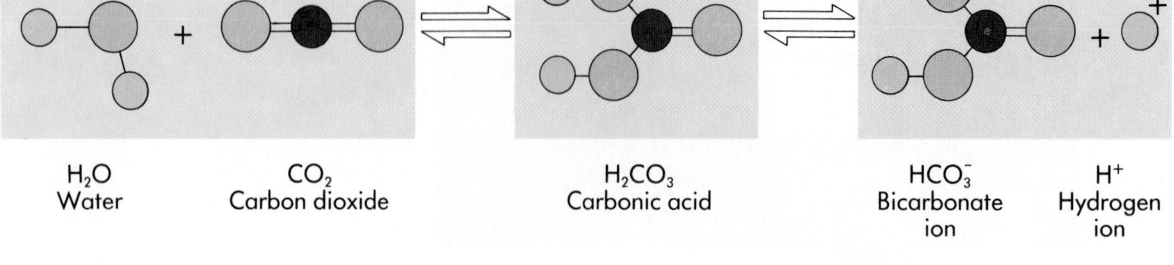

H_2O	CO_2	H_2CO_3	HCO_3^-	H^+
Water	Carbon dioxide	Carbonic acid	Bicarbonate ion	Hydrogen ion

1. The smallest stable particles of matter are protons, neutrons, and electrons, which associate to form atoms. The core, or nucleus, of an atom is composed of protons and neutrons; the electrons orbit around this core in a cloud. The farther out an electron orbits, the faster it goes and the more energy it possesses.

2. The chemical behavior of atoms is largely determined by the distribution of its electrons and in particular by the number of electrons in its outermost (highest) energy level. There is a strong tendency for atoms to have a fully populated outer level; electrons are lost, gained, or shared until this condition is reached.

3. A molecule is a stable collection of atoms. The forces holding the atoms together in a molecule are called chemical bonds. Ionic bonds involve the transfer of electrons and resultant attraction of oppositely charged ions. Covalent bonds involve the sharing of electrons between the two bonded nuclei. Covalent bonds are responsible for the formation of most biologically important molecules.

4. Over 99% of the atoms in the human body are nitrogen, oxygen, carbon, or hydrogen (NOCH), all of which form strong covalent bonds with one another.

5. The chemistry of life is the chemistry of water. The central oxygen atom in water tends to attract the electrons shared between it and the two hydrogen atoms. As a result, the oxygen atom is electron rich (partial negative charge) and the hydrogen atoms are electron poor (partial positive charge). This charge separation is like that of a magnet with positive and negative poles, and water is termed a polar molecule.

6. A hydrogen bond is formed by the weak attraction of the partial positive charge of a hydrogen atom of a molecule with the partial negative charge of an atom of another molecule, or another segment of the same molecule.

7. Water is cohesive, adhesive, has a great capacity for storing heat, is a good solvent for other polar molecules, and tends to exclude nonpolar molecules.

8. Water ionizes spontaneously to a slight degree, perhaps 1 molecule in 550 million at any one time. The molar concentration of H^+ ions that results is $1/10,000,000$, or 1×10^{-7}. This concentration is usually signified as the negative of the exponent of the molar concentration, termed pH. The pH of water is thus 7, because 7 is the negative of the exponent of 1×10^{-7}.

9. A buffer is a substance that acts as a reservoir for hydrogen ions, donating them to the solution when their concentration falls and taking them from the solution when the H^+ ion concentration rises.

10. One of the most biologically important buffers is formed when carbon dioxide dissolves in water to form carbonate and bicarbonate ions.

REVIEW QUESTIONS

1. What are the three subatomic particles? What is the charge of each? Which are found in the nucleus of an atom?

2. Which subatomic particle(s) is(are) associated with atomic number? Which is(are) associated with atomic mass?

3. How do isotopes of a single element differ from one another? Do they exhibit similar or different chemical properties?

4. How do ions of the same element differ from one another?

5. Of the three subatomic particles in an element, which one largely determines the chemical properties of that element?

6. Define the term "ionic bond." Define "covalent bond." Give examples of each.

7. Differentiate between reduction and oxidation.

8. What are the six life-giving properties of water? What one factor is most important in conferring these properties?

9. What is pH? What is the pH range for acids? . . . for bases? What pH value is assigned to a chemical that has a molar concentration of $1/10,000,000,000$ hydrogen ions? What is the pH if there are 0.001 hydrogen ions per liter?

10. How does a buffer work?

THOUGHT QUESTIONS

1. Carbon (atomic number, 6) and silicon (atomic number, 14) both have four vacancies in their outer energy levels. Ammonia is even more polar than water. Why do you suppose life evolved in organisms composed of carbon chains in water solution rather than ones of silicon in ammonia?

2. Champagne, a carbonic acid buffer, has a pH of about 2. How can we drink such a strong acid?

3. Carbon atoms can share four electron pairs when forming molecules. Why do you suppose that carbon does not form a bimolecular gas, as hydrogen (one pair of shared electrons), oxygen (two pairs of shared electrons), and nitrogen (three pairs of shared electrons) do?

FOR FURTHER READING

ATKINS, P.W.: *Molecules,* Scientific American Library, New York, 1987. A delightful journey among the molecules most familiar to us.

COX, T.: "Origin of the Chemical Elements," *New Scientist,* February 3, pages 1-4, 1990. The heavier elements making up most of the earth—and us—were created through the birth and death of generations of stars.

HOYLE, F.: *The Ten Faces of the Universe,* Freeman, Cooper, and Company, San Francisco, 1980. A readable account of how the universe is thought to have formed, written for the nonscientist.

MERTZ, W.: "The Essential Trace Elements," *Science,* vol. 213, pages 1332-1338, 1981. An account of the roles of rare elements in human metabolism and the effects of their absences.

MORRISON, P., and P. MORRISON: *Powers of Ten,* Scientific American Library, New York, 1982. A pictorial atlas explaining the macrocosm to microcosm.

WEINBERG, S.: *The Discovery of Subatomic Particles,* Scientific American Library, New York, 1983. Documents the discovery of subatomic particles and introduces the nontechnical reader to the fundamentals of classical physics.

3

The Chemical Building Blocks of Life

OVERVIEW

Living organisms are built from subassemblies, much as a wall is made of bricks. Some of these building blocks are long polymers, chains of similar units joined in a row. Among the polymers that make up the bodies of organisms are starches (molecules that store chemical energy), proteins (molecules that facilitate specific chemical reactions), and nucleic acids (molecules in which hereditary information is stored). Among the other building blocks of organisms are the phospholipids, which form biological membranes.

FOR REVIEW

Here are some important terms and concepts that you will encounter in this chapter. If you are not familiar with them, you should review them before proceeding.

Hydrophobic interactions (Chapter 2)

Reduced carbon compounds (Chapter 2)

Covalent bonds (Chapter 2)

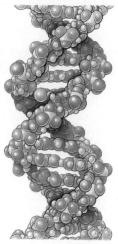

Nucleic acid

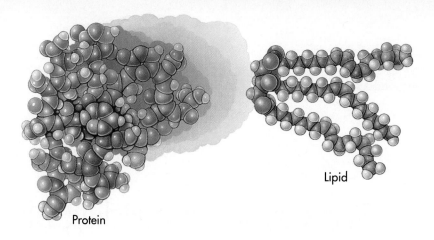

Protein

Lipid

Carbohydrate

FIGURE 3-1
The four fundamental kinds of biological macromolecules.

Atoms are extremely small when compared with the familiar world we see about us. A speck of sand contains many billions of atoms, more than there are grains of sand on all the beaches of California. Molecules that are composed of a few atoms, like water molecules, are also extremely small. There are more molecules of H_2O in a cup of water than there are stars in the sky. Not all molecules are small, however. The molecules in living organisms are often gigantic, by contrast, clusters of thousands of atoms assembled by linking together hundreds of smaller molecules into long chains. These enormous biological molecules are called **macromolecules.**

Your body contains four general types of macromolecules, which differ in the small molecules used to assemble the chains. The four general macromolecule types are: carbohydrates, lipids, proteins, and nucleic acids (Figure 3-1). They are the basic chemical building blocks of organisms, the mortar and bricks used to assemble an organism and its constituent parts. In this chapter we will discuss these building blocks. In following chapters we will discuss cells, the basic units of organization that are characteristic of all organisms; how cells evolved; and what their internal structure is like.

THE BUILDING BLOCKS OF ORGANISMS

The basic chemical building blocks of organisms, the mortar and bricks used to assemble a cell, are made of molecules. The molecules formed by living organisms, which contain carbon, are called **organic molecules.** Although many organic molecules are used in the construction of a cell, they need not confuse you. A good way to see through the tangle of different molecules is to focus on those bits of the molecules that are important, like focusing on who has the ball in a football game. Much of the complex structure of an organic molecule may have little to do with the biological process you are studying. It is often helpful to think of an organic molecule as a carbon-based core with special bits attached, groups of atoms with definite chemical properties. We refer to these groups of atoms as **functional groups.** For example, a hydrogen atom bonded to an oxygen atom (—OH) is a hydroxyl group. The most important functional groups are illustrated in Figure 3-2. Most chemical reactions that occur within organisms involve the transfer of a functional group from one molecule to another, or the breaking of a carbon-carbon bond. Proteins called kinases, for example, transfer phosphate groups from one kind of molecule to another.

Some of the molecules in organisms are simple organic molecules, often with a single reactive functional group protruding from a carbon chain. Other molecules are macromolecules, particularly those that play a structural role or store information in organisms. Most of these much larger macromolecules are composed of simpler components. Macromolecules fall into the following four classes: carbohydrates, lipids, proteins, and nucleic acids.

Many macromolecules are polymers. A **polymer** is a molecule built of a long chain of similar modules, like railroad cars coupled together to form a train. Complex carbohydrates, for example, are polymers of simple ring molecules called sugars. En-

Compound	Examples	
Hydroxyl group	— OH	
Carbonyl group	$-\overset{\displaystyle |}{\underset{\displaystyle \parallel O}{C}}-$	
Carboxyl group	$-C\!\!\begin{array}{l}\nearrow O\\ \searrow OH\end{array}$	
Amino group	$-N\!\!\begin{array}{l}\diagup H\\ \diagdown H\end{array}$	
Sulfhydryl	— S — H	
Phosphate	$-O-\overset{\displaystyle OH}{\underset{\displaystyle \parallel O}{P}}-OH$	

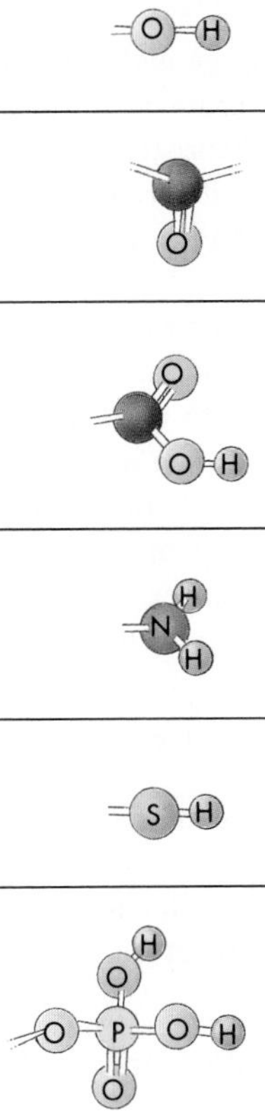

FIGURE 3-2
The primary functional chemical groups. These groups tend to act as units during chemical reactions and to confer specific chemical properties on the molecules that possess them. Hydroxyl groups, for example, make a molecule more basic, whereas carboxyl groups make a molecule more acidic.

zymes, membrane proteins, and other proteins are polymers of amino acids. DNA and RNA are two versions of a long-chain molecule called a nucleic acid, a polymer composed of a long series of complex molecules called nucleotides.

Building Macromolecules

The units that are linked together to form a macromolecule are called subunits. Although the four kinds of macromolecules are assembled from different kinds of subunits, they all put their subunits together in the same way: a covalent bond is formed between two subunit molecules in which a hydroxyl group (OH) is removed from one subunit and a hydrogen (H) is removed from the other (Figure 3-3). This process is called a **dehydration** (water-losing) **reaction,** because in effect the removal of the OH and H groups constitutes removal of a molecule of water. In the dehydration synthesis of a polymer, one water molecule is removed for every link in the chain of subunits. Energy is required to break the chemical bonds when water is extracted from the subunits, and so cells must supply energy to assemble polymers. The process also requires that the two subunits be held close together and that the correct chemical

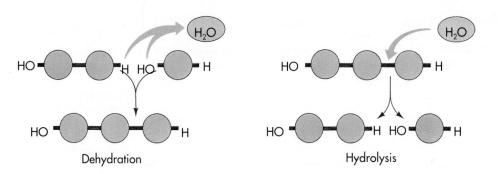

Dehydration   Hydrolysis

FIGURE 3-3
Making (and breaking) macromolecules. Biological macromolecules are formed by linking subunits together. The covalent bond between the subunits is formed in a dehydration reaction, in which a water molecule is eliminated. Breaking such a bond requires returning the water molecule, a hydrolysis reaction.

bonds be stressed and broken. This process of positioning and stressing is called **catalysis**. In cells, catalysis is carried out by a special class of proteins called **enzymes**.

> **Polymers are large molecules formed of long chains of similar molecules joined by dehydration, in which a hydroxyl (OH) group is removed from one subunit and a hydrogen (H) group is removed from the other. The process in which the subunits are held together and their bonds are stressed is called catalysis, which is carried out in organisms by enzymes.**

No molecule lasts forever. While cells are building up some macromolecules, they are disassembling others. When you eat a steak, the protein and fat that you consume are broken down into their subunit parts by enzymes of your digestive system. The process of tearing down a polymer is essentially the reverse of dehydration—instead of removing a molecule of water, one is added. A hydrogen is attached to one subunit and a hydroxyl to the other, breaking the covalent bond. The breaking up of a polymer in this way is an example of a **hydrolysis reaction** (literally, "to break with water"—*hydro*, water; *lyse*, break).

CARBOHYDRATES

In our discussion of macromolecules that make up the bodies of organisms, we will begin with carbohydrates (Table 3-1). Some carbohydrates are simple small molecules, whereas others form long polymers. Carbohydrates function as energy-storage vessels as well as structural elements.

Sugars are Simple Carbohydrates

The **carbohydrates** are a loosely defined group of molecules that contain carbon, hydrogen, and oxygen in the molar ratio $1:2:1$. A chemist would say that their empirical formula (a list of the atoms in a molecule with a subscript to indicate how many of each) is $(CH_2O)_n$, where n is the number of carbon atoms. Because they contain many carbon-hydrogen (C—H) bonds, which release energy when they are broken, carbohydrates are well suited for energy storage. Among the simplest of the carbohydrates are the simple sugars, or **monosaccharides** (Greek, *mono*, single + Latin, *saccharum*, sugar). Simple sugars may have as few as three carbon atoms, but those molecules that play the central role in energy storage have six. They have the following empirical formula:

$$C_6H_{12}O_6, \text{ or } (CH_2O)_6$$

TABLE 3-1	MACROMOLECULES		
MACROMOLECULE	SUBUNIT	FUNCTION	EXAMPLE
Carbohydrates			
Starch, glycogen	Glucose	Energy storage	Potatoes
Cellulose	Glucose	Cell walls	Paper
Chitin	Modified glucose	Structural support	Crab shells
Lipids			
Fats	Glycerol and 3 fatty acids	Energy storage	Butter; soap
Phospholipids	Glycerol, 2 fatty acids, and phosphate	Cell membranes	Lecithin
Steroids	Four-carbon rings	Membranes; hormones	Cholesterol; estrogen
Terpenes	Long carbon chains	Pigments; structural	Carotene; rubber
Proteins			
Globular	Amino acids	Catalysis; transport	Hemoglobin
Structural	Amino acids	Support	Hair; silk
Nucleic acids			
DNA	Nucleotides	Encodes genes	Chromosomes
RNA	Nucleotides	Needed for gene expression	Messenger RNA

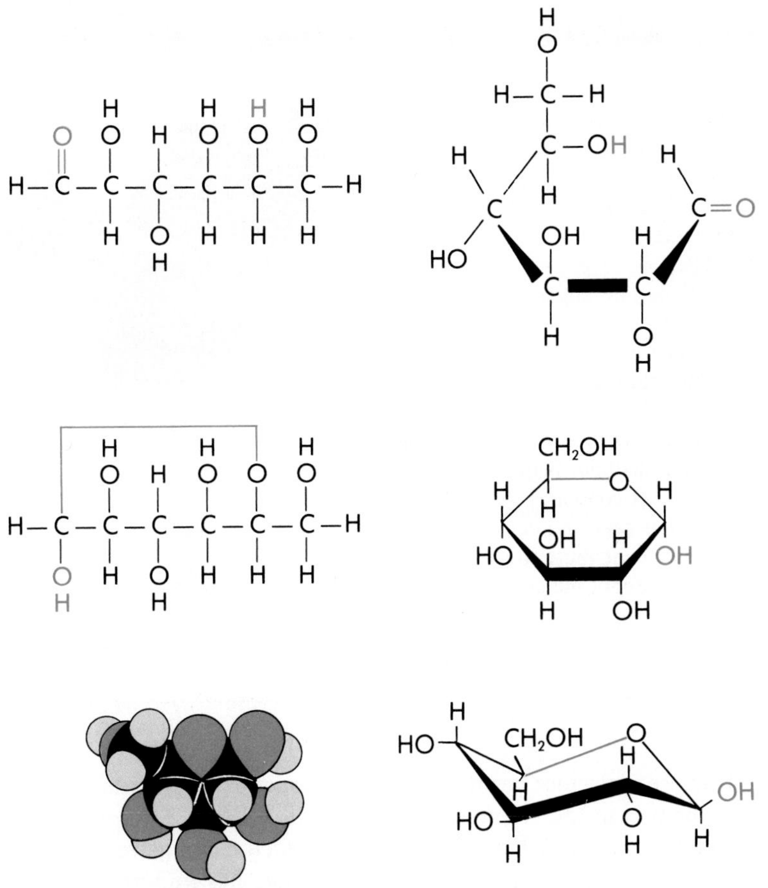

FIGURE 3-4
Structure of the glucose molecule. Glucose is a linear six-carbon molecule that forms a ring in solution. The structure of the ring can be represented in many ways; the ones shown here are the most common.

Sugars can exist in a straight-chain form, but in water solution they almost always form rings. The primary energy-storage molecule is **glucose** (Figure 3-4), a six-carbon sugar with seven energy-storing CH bonds.

> Sugars are among the most important energy-storage molecules in organisms. Simple sugars contain six carbon atoms and seven C—H bonds, which are energy-storing bonds.

Glucose is not the only sugar with the formula $C_6H_{12}O_6$. Among the other monosaccharides that have this same empirical formula are fructose and galactose (Figure 3-5). They are isomers, or alternative forms, of glucose. Glucose and fructose are **structural isomers.** In fructose the double-bonded oxygen is attached to an internal carbon rather than to a terminal one. Your taste buds can tell the difference: fructose tastes much sweeter than glucose. This structural difference also has an important chemical consequence: the two sugars form different polymers.

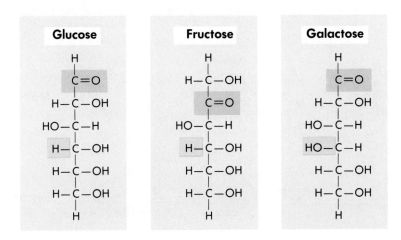

FIGURE 3-5
Isomers and stereoisomers. A structural isomer of glucose, such as fructose, has identical chemical groups bonded to different carbon atoms, whereas a stereoisomer of glucose, such as galactose, has identical chemical groups bonded to the same carbon atoms in a different orientation.

Imagine a kind of table sugar that looks, tastes, and cooks like the real thing, but has no calories or harmful side effects. You could eat mountains of candy made from such sweeteners without gaining weight. As Louis Pasteur discovered in the late 1800s, most sugars are "right-handed" molecules, the hydroxyl group that binds a critical carbon atom being on the right side. However, sugars can be made readily in the laboratory that are "left handed"—the critical ion is on the left side, producing a mirror-image chemical twin of the natural form. The enzymes that break down sugars in the human digestive system can tell the difference. To digest a sugar molecule, an enzyme must first grasp it, much like a shoe fitting onto a foot, and all the body's enzymes are right handed! A left-handed sugar just doesn't fit, any more than a right-handed shoe fits onto a left foot.

The Latin word for "left" is levo, and these new sugars are called levo- or l-sugars. These new laboratory-manufactured sweeteners do not occur in nature except for trace amounts in red algae, snail eggs, and seaweed. Because they pass through the body without being used, they can let diet-conscious sweet lovers have their cake and eat it, too. Nor will they contribute to tooth decay—bacteria cannot metabolize them either.

Unlike fructose, galactose has the same bond structure as glucose; the only difference between them is in the orientation of one hydroxyl (—OH) group. Because the two orientations are mirror images of one another, glucose and galactose are said to be **stereoisomers.** Again, this seemingly slight difference has important consequences, since this hydroxyl group is one that is often involved when bonds are formed to create polymers.

Transport Disaccharides

Most organisms transport sugars within their bodies. In humans, glucose circulates in the blood. The glucose that circulates in our bloodstream does so as a simple monosaccharide. However, in many other organisms, particularly plants, glucose is converted to a *transport form* before it is moved from place to place within the organism. In such a form it is less readily metabolized while it is being transported. Transport forms of sugars are commonly formed by linking two monosaccharide sugar molecules together to form a **disaccharide** (Greek, *di-*, two). Disaccharides serve as effective reservoirs of glucose that cannot be metabolized by the normal glucose-utilizing enzymes of the organism to obtain energy until the bond linking the two monosaccharide subunits is broken by a specific enzyme. These enzymes are typically present only in the tissue where the glucose is utilized.

Transport forms differ from one another in the identity of the monosaccharides

FIGURE 3-6

Disaccharides formed with glucose are often used to transport glucose from one part of an organism's body to another and to store it. Apples, for example, are rich in maltose, **A,** whereas sugar cane is rich in sucrose, **B,** Both are disaccharides.

A   CH₂OH + CH₂OH → CH₂OH + CH₂OH → **Maltose**

B   CH₂OH + CH₂OH → CH₂OH + CH₂OH → **Sucrose**

linked in the disaccharide. Glucose forms transport disaccharides with many other monosaccharides, including glucose itself, galactose, and fructose. When two glucose molecules form a bond between one glucose molecule and another, the resulting disaccharide is maltose (Figure 3-6). When glucose forms a disaccharide with its structural isomer fructose, the resulting disaccharide is sucrose. Commonly called cane sugar, sucrose is the form in which most plants transport glucose and the sugar that most people (and other animals) eat (Figure 3-7).

A different disaccharide results when glucose is linked to its stereoisomer galactose. The resulting disaccharide is lactose. Many mammals supply energy to their babies in the form of lactose. Channeling food energy into lactose production has the effect of reserving the energy for the child, since many adults, including virtually all non-white humans, lack the enzyme required to cleave the disaccharide into its two monosaccharide components. Since they lack this enzyme, adults cannot metabolize lactose.

Starches are Chains of Sugars

Organisms store the metabolic energy captured in glucose, first transporting the glucose as disaccharides, then converting them into an insoluble form and depositing it in specific storage areas in their bodies. The disaccharides are rendered insoluble when they are joined together by dehydration synthesis into long polymers called **polysaccharides.** Polysaccharides formed from glucose are called **starches.**

The starch with the simplest structure is amylose. Amylose is composed of many hundreds of glucose molecules linked together in long unbranched chains. Each linkage is between the carbon 1 of one glucose molecule and the carbon 4 of another so that amylose is, in effect, a longer form of maltose. The long chains of amylose tend to coil up in water, a property that renders amylose insoluble. Potato starch is about 20% amylose. When amylose is digested by a sprouting potato plant (or by you), enzymes first break it into fragments of random length, which are more soluble. Baking or boiling potatoes has the same effect. Another special enzyme then cuts these fragments into segments that are two glucose molecules long—the disaccharide maltose. Finally the maltose is cleaved into two glucose molecules, which the cell is able to metabolize.

Most plant starch, including 80% of potato starch, is a somewhat more complicated variant of amylose called amylopectin (Figure 3-8). **Pectins** are branched polysaccharides, and amylopectin is a form of amylose with short, linear amylose

FIGURE 3-7
This hungry butterfly has its proboscis extended down into a sucrose solution. Sensors in its footpads alert it to the presence of sugar and cause it to extend its proboscis. If the solution had been distilled water instead of sucrose, the reflex governing proboscis extension would not have been triggered.

FIGURE 3-8
Starch. Storage polymers of glucose are called starches. The simplest starches are long chains of maltose called amylose. Most plants contain more complex starches called amylopectins, which contain branches. The nucleus of the liver cell in the micrograph inset is surrounded by dense granules of animal starch called glycogen, which is even more highly branched.

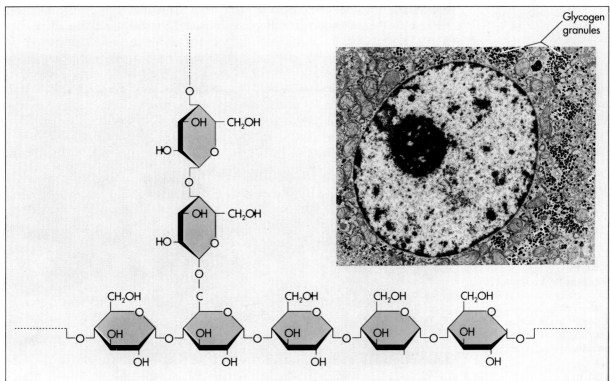

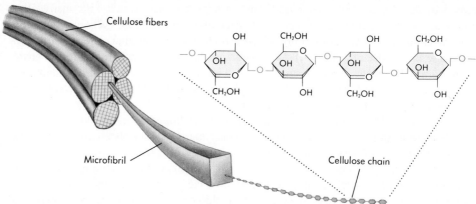

Cellulose fibers

CH₂OH

Microfibril

Cellulose chain

FIGURE 3-9
A journey into wood. The jumble of cellulose fibers shown above is from a yellow pine (*Pinus ponderosa*). Each fiber is composed of microfibrils (*above, right*), which are bundles of cellulose chains. Cellulose fibers can be very strong; they are quite resistant to metabolic breakdown, which is one reason that wood is such a good building material.

branches consisting of 20 to 30 glucose subunits. In some plants these chains are cross-linked. These cross-links create an insoluble mesh of glucose, which can be degraded only in the presence of yet another enzyme. The size of the mesh differs from one plant to another. In rice, for example, it is made up of about 100 amylose chains, each with one or two cross-links.

Animals also store glucose in branched amylose chains. In animals, however, the average chain length is much longer and there are more branches than in plant starch, producing a highly branched, insoluble structure called glycogen (see Figure 3-8, *inset*).

> **Starches are glucose polymers. Most starches are branched and some are cross-linked. The branching and cross-linking render the polymer insoluble as well as protecting it from degradation.**

Cellulose is a Starch that is Hard to Digest

Imagine that you could draw a line down the central axis of a starch molecule, like threading a rope through a pipe. Because all the glucose subunits of the starch chain are joined in the same orientation, all the CH_2OH groups would fall on the same side of the line (see Figure 3-8). There is another way to build a chain of glucose molecules, however, in which the glucose subunit orientations switch back and forth (the CH_2OH groups alternate on opposite sides of the line). The resulting polysaccharide is **cellulose,** the chief component of plant cell walls. Cellulose is chemically similar to amylose, with one important difference (Figure 3-9): the starch-degrading enzymes

FIGURE 3-10
Chitin. Chitin, which might be considered to be a modified form of cellulose, is the principal structural element in the external skeletons of many invertebrates, such as this crab, and in the cell walls of fungi.

that occur in most organisms cannot break the bond between two sugars in opposite orientation. It is not that the bond is stronger, but rather that its cleavage requires the aid of a different protein, one not usually present. Because cellulose cannot readily be broken down, it works well as a biological structural material and occurs widely in this role in plants. For those few animals able to break down cellulose, it provides a rich source of energy. Certain vertebrates, such as cows, which do not themselves produce the enzymes necessary to digest cellulose, do so by means of the bacteria and protists they harbor in their intestines. In humans, cellulose is a major component of dietary fiber, necessary for the proper functioning of the digestive system.

The structural material in insects, many fungi, and certain other organisms is **chitin** (Figure 3-10). Chitin is a modified form of cellulose in which a nitrogen group has been added to the glucose units. Chitin is a tough, resistant surface material. Few organisms are able to digest chitin.

LIPIDS
Fats

When organisms store glucose molecules for long periods, they usually convert the glucose into another kind of insoluble molecule that contains more C—H bonds than do carbohydrates. These storage molecules are called **fats.** The ratio of H to O in carbohydrates is $2:1$, but in fat molecules it is much higher. Like starches, fats are insoluble and can therefore be deposited at specific storage locations within the organism. Starches are insoluble because they are long polymers; fats, in contrast, are insoluble because they are nonpolar. Unlike the H—O bonds of water, the C—H bonds of carbohydrates and fats are nonpolar and cannot form hydrogen bonds. Because fat molecules contain a large number of C—H bonds, they are hydrophobically excluded by water because water molecules tend to form hydrogen bonds with other water molecules. The result is that the fat molecules cluster together, insoluble in water.

Fats are one kind of **lipid,** a loosely defined group of molecules that are insoluble in water but soluble in oil. Oils such as olive oil, corn oil, and coconut oil are also lipids, as are waxes such as bee's wax and ear wax (see Table 3-1). Fats are composite molecules, each made up of two kinds of subunits:

1. *Glycerol.* A three-carbon alcohol, each of whose carbons bears a hydroxyl group. The three carbons form the backbone of the fat molecule, to which three fatty acids are attached.
2. *Fatty acids.* Long hydrocarbon chains ending in a carboxyl (—COOH) group. Three fatty acids are attached to each glycerol backbone.

FIGURE 3-11

Triglycerides are made from simple pieces. Triglycerides are composite molecules, made up of three fatty acid molecules coupled in a dehydration reaction to a single glycerol backbone.

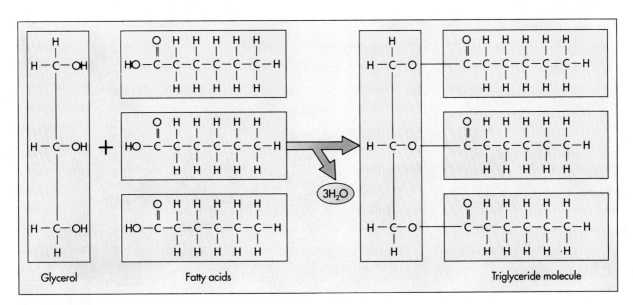

Glycerol   Fatty acids   Triglyceride molecule

The structure of an individual fat molecule, such as the one diagrammed in Figure 3-11, consists simply of a glycerol molecule with a fatty acid joined to each of its three carbon atoms:

$$
\begin{array}{c}
\text{H} \\
| \\
\text{H—C—Fatty acid} \\
| \\
\text{H—C—Fatty acid} \\
| \\
\text{H—C—Fatty acid} \\
| \\
\text{H}
\end{array}
$$

Because there are three fatty acids, the resulting fat molecule is called a **triglyceride.** The three fatty acids of a triglyceride do not have to be identical, and often differ markedly from one another.

Fatty acids vary in length. The most common are even-numbered chains of 14 to 20 carbons. Fatty acids in which all of the internal carbon atoms possess hydrogen side groups are said to be **saturated** because they contain the maximum number of hydrogen atoms that are possible. Some fatty acids have double bonds between one or more pairs of successive carbon atoms (Figure 3-12). Fats that are composed of such fatty acids are said to be **unsaturated,** since the double bonds replace some of the hydrogen atoms and they therefore contain fewer than the maximum number of such atoms. If a given fat has more than one double bond, it is said to be **polyunsaturated.** Polyunsaturated fats have low melting points, because their chains bend at the double bonds so that the fat molecules cannot be aligned closely with one another. Consequently, the fat may be fluid. A liquid fat is called an **oil.** Many plant fatty acids, such as oleic acid (a vegetable oil) and linolenic acid (a linseed oil) are unsaturated. Animal fats, in contrast, are often saturated and occur as hard fats.

It is possible to convert an oil into a hard fat by adding hydrogen. The peanut butter that you buy in the store has usually been hydrogenated to convert the peanut fatty acids to hard fat and thus to prevent them from separating out as oils while the jar sits on the store shelf.

Fats are efficient energy-storage molecules because of their high concentration of C—H bonds. Most fats contain over 40 carbon atoms. The ratio of energy-storing C—H bonds to carbon atoms is more than twice that of carbohydrates, making fats much more efficient vehicles for storing chemical energy. On the average, fats yield about 9 kilocalories (kcal) of chemical energy per gram of fat, as compared with somewhat less than 4 kcal per gram obtainable from carbohydrates. As you might expect, the more highly saturated fats are richer in energy than less saturated ones. Animal

FIGURE 3-12

Saturated and unsaturated fats.
A Palmitic acid, with no double bonds and thus a maximum number of hydrogen atoms bonded to the carbon chain, is a saturated fatty acid. Linolenic acid, with three double bonds and thus fewer than the maximum number of hydrogen atoms bonded to the carbon chain, is an unsaturated fatty acid.
B Many animal triglyceride fats are saturated. Because their fatty acid chains can fit closely together, these triglycerides form immobile arrays called *hard fat.* Plant fats, by contrast, are typically unsaturated, and the many kinks that the double bonds introduce into the fatty acid chains prevent close association of the triglycerides and produce oils such as linseed oil, which is obtained from flax (*Linum*) seed.

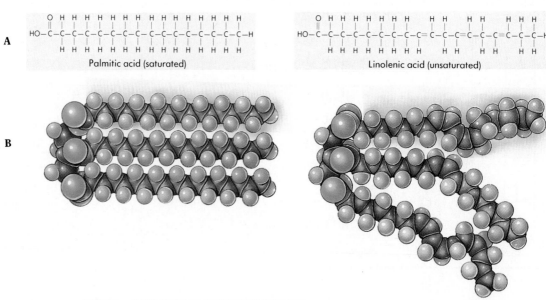

A Palmitic acid (saturated)

Linolenic acid (unsaturated)

B

Chemical structures (Figure 3-13)

CH₃—N—CH₂—CH₂—O—P—O—CH₂ (with CH₃ groups on N and O, P double bond O)

Phospholipid
(phosphatidyl choline)

Terpene
(citronellol)

Steroid
(cholesterol)

Prostaglandin
(PGE)

FIGURE 3-13
Four major classes of biologically important lipids.

fats contain more calories than do vegetable fats. Human diets that contain relatively large amounts of saturated fats appear to upset the normal balance of fatty acids in the body, a situation that may lead to heart disease.

The total amount of carbohydrate that organisms consume or produce is allocated three different ways: (1) some is maintained as glucose and is available for immediate use; (2) a second portion is converted to transport disaccharides and can then be carried with minimum loss to other parts of the organism; and (3) some is converted to starches or fats and reserved for future use. The reason that many people gain weight as they grow older is that the amount of carbohydrate needed daily decreases with age, whereas their intake of food does not. A greater and greater portion of the ingested carbohydrate is available to be converted to fat.

There Are Many Other Kinds of Lipids

Fats are but one example of the oily or waxy class of macromolecules called lipids. Your body also contains other kinds of lipids (Figure 3-13). The membranes of your cells are composed of a kind of modified fat called phospholipid. Phospholipids have a polar group at one end and a long nonpolar tail. In water, the nonpolar ends of phospholipids aggregate, forming two layers of molecules with the nonpolar tails of each layer pointed inside—a lipid bilayer (Figure 3-14). Lipid bilayers are the basic framework of biological membranes, discussed in detail in Chapter 6. Membranes often also contain a different kind of lipid called steroids, composed of four carbon rings; most of your membranes, for example, contain the steroid cholesterol. Male and female sex hormones are also steroids.

A different kind of lipid forms many of the biologically important pigments, such as the photosynthetic pigment chlorophyll in plants and the pigment retinal, which

FIGURE 3-14
Phospholipids.
A At an oil-water interface, phospholipid molecules will orient so that their polar heads are in the polar medium, water, and their nonpolar tails are in the nonpolar medium, oil.
B When surrounded by water, phospholipid molecules arrange themselves into two layers with their polar heads extending outward and their nonpolar tails inward.

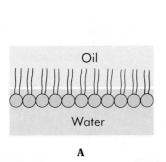

A

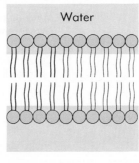

B

absorbs light in your eyes. These pigments are examples of long-chain lipids called terpenes. Rubber is also a terpene.

Prostaglandins are a group of about 20 lipids that are modified fatty acids, with two nonpolar "tails" attached to a five-carbon ring. Prostaglandins occur in many vertebrate tissues, where they appear to act as local chemical messengers. Some stimulate smooth muscle to contract or to relax; others constrict or expand the diameter of small blood vessels. Prostaglandins have been shown to be involved in many aspects of reproduction, and in the inflammatory response to infection; it is because aspirin inhibits prostglandin synthesis that it reduces pain, inflammation, and fever.

PROTEINS

Proteins are the third major group of macromolecules that make up the bodies of organisms (see Table 3-1). We have already encountered one class of proteins, the **enzymes,** which are biological catalysts able to facilitate specific chemical reactions. Because of this property, the appearance of enzymes was one of the most important events in the evolution of life. Other kinds of proteins also have important functions (Table 3-2 and Figure 3-15). Short proteins called **peptides** are used as chemical messengers within your brain and throughout your body. Despite their diverse functions, however, all proteins have the same basic structure: a long polymer chain of amino acid subunits linked end to end.

TABLE 3-2   THE MANY FUNCTIONS OF PROTEINS

FUNCTION	CLASS OF PROTEIN	EXAMPLES	USE
Structure	Fibers	Collagen	Cartilage
		Keratin	Hair, nails
		Fibrin	Blood clot
Metabolism	Enzymes	Lysosomes	Cleave polysaccharide
		Proteases	Break down proteins
		Polymerases	Produce nucleic acids
		Kinases	Phosphorylate sugars and proteins
Membrane transport	Channels	Sodium-potassium pump	Excitable membranes
		Proton pump	Chemiosmosis
		Anion channels	Transports Cl^- ions
Cell recognition	Cell surface antigens	MHC proteins	"Self" recognition
Osmotic regulation	Albumin	Serum albumin	Maintains osmotic concentration of blood
Regulation of gene action	Repressors	*lac* repressor	Regulates transcription
Regulation of body functions	Hormones	Insulin	Controls blood glucose levels
		Vasopressin	Increases water retention by kidney
		Oxytocin	Regulates mild production
Transport throughout body	Globins	Hemoglobin	Carries O_2 and CO_2 in blood
		Myoglobin	Carries O_2 and CO_2 in muscle
		Cytochromes	Electron transport
Storage	Ion-binding	Ferritin	Stores iron, especially in spleen
		Casein	Stores ions in milk
		Calmodulin	Binds calcium ions
Contraction	Muscle	Actin	Contraction of muscle fibers
		Myosin	Contraction of muscle fibers
Defense	Immunoglobulins	Antibodies	Mark foreign proteins for elimination
	Toxins	Snake venom	Blocks nerve function

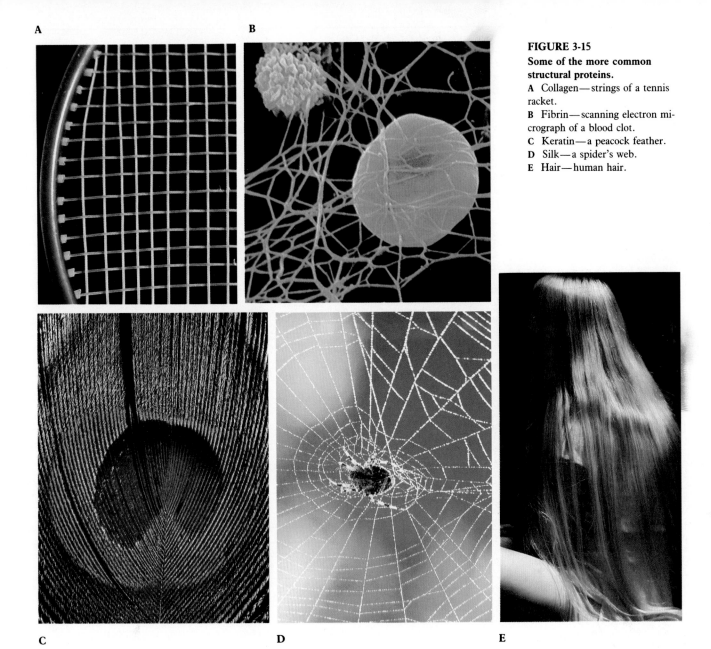

A

B

C

D

E

Amino Acids are the Building Blocks of Protein

Amino acids are thought to have been among the first molecules formed in the atmosphere of the early earth. The next chapter describes an experiment in which an attempt was made to reproduce such an atmosphere; when sparks were passed through a mixture of methane, ammonia, hydrogen, and water, chemical reactions occurred that produced amino acids, along with many other organic molecules. It seems highly likely that the oceans that existed early in the history of the earth contained a wide variety of amino acids.

An amino acid can be defined as a molecule containing an **amino group** ($-NH_2$), a carboxyl group ($-COOH$), and a hydrogen atom, all bonded to a central carbon atom:

$$H_2N-\underset{\underset{H}{|}}{\overset{\overset{R}{|}}{C}}-COOH$$

The identity and unique chemical properties of each amino acid are determined by the nature of the side group (indicated by R) covalently bonded to the central carbon

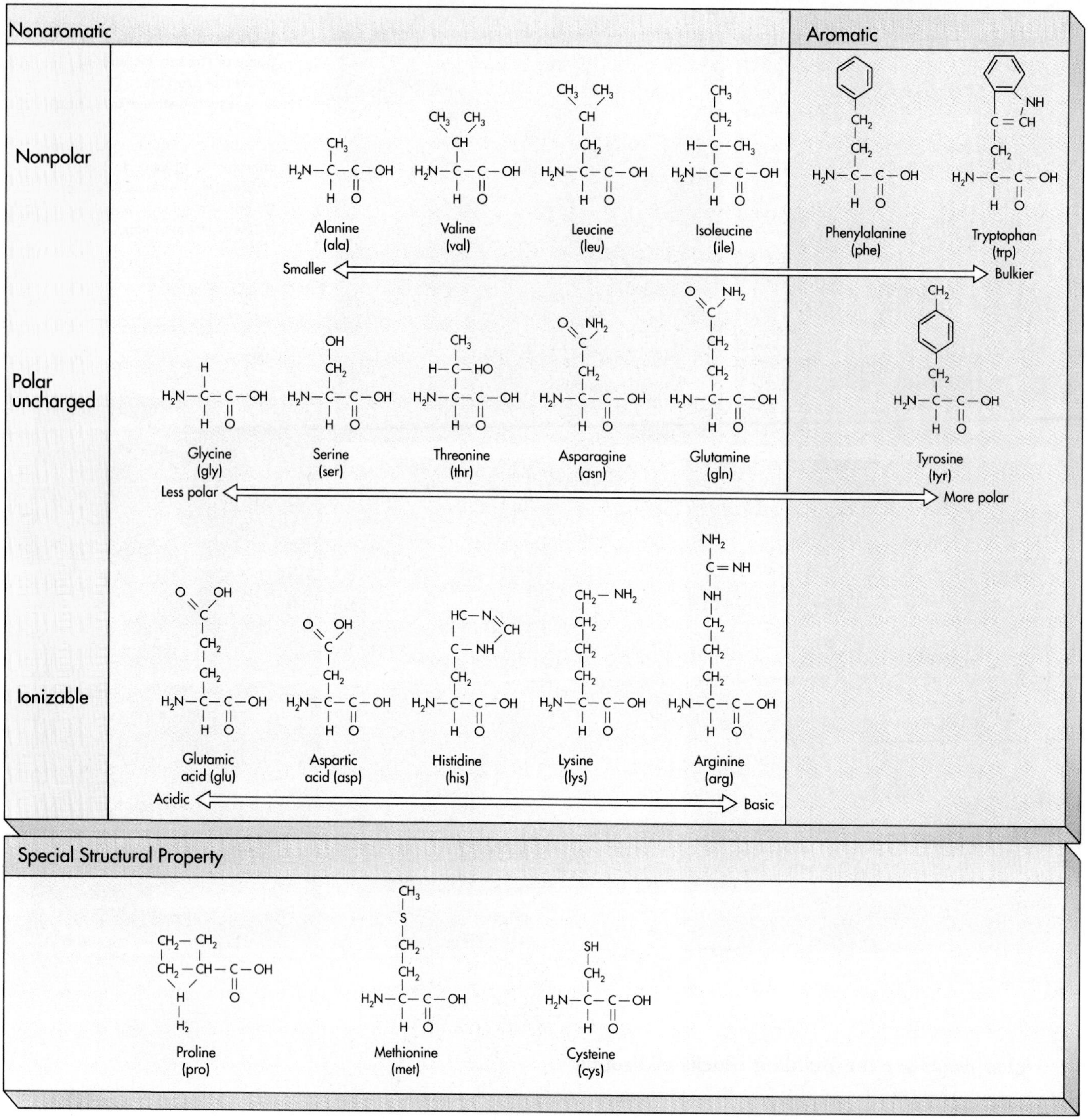

FIGURE 3-16

The 20 common amino acids. Each amino acid has the same chemical backbone, but differs from the others in the side or R group that it possesses. Six of the amino acid R groups are nonpolar because they have CH_2 and CH_3 groups in their R groups. Some are more bulky than others (particularly the ones containing ring structures, which are called *aromatic*). Another six are polar because they have oxygen in their R groups, but are uncharged; these differ from one another in how polar they are. Five are polar and, because they have a terminal acid or base in their R group, are capable of ionizing to a charged form; under typical cell conditions some of these are acids, others bases. The remaining three have special chemical properties that play important roles in forming links between protein chains or kinks in their shape.

atom. Although many different amino acids occur in nature, only 20 are used commonly in proteins. These 20 "common" amino acids and their side groups are illustrated in Figure 3-16. The different side groups of these 20 amino acids give each amino acid distinctive chemical properties. For example, when the side group is —H, the amino acid (glycine) is polar, whereas when the side group is —CH$_3$, the amino acid (alanine) is nonpolar. The 20 amino acids that occur in proteins are commonly grouped into five chemical classes, based on their side groups:

1. *Nonpolar* amino acids, such as leucine, whose R groups contain —CH$_2$ and —CH$_3$ groups
2. *Polar uncharged* amino acids, such as threonine, whose R groups contain oxygen
3. *Ionizable* amino acids, such as glutamic acid, whose R groups contain acids or bases
4. *Aromatic* amino acids, such as phenylalanine, whose R groups contain an organic ring with alternating single and double bonds
5. *Special function* amino acids, such as methionine (which often initiates polypeptide chains), proline (which causes kinks in chains of amino acids), and cysteine (which links chains together)

The way in which each amino acid affects the shape of a protein depends on the chemical nature of its side group. Portions of a protein chain with numerous nonpolar amino acids, for example, tend to be shoved into the interior of the protein by hydrophobic interactions, with polar water molecules tending to exclude nonpolar amino acid side groups.

Proteins can contain up to 20 different kinds of amino acids. These amino acids fall into five chemical classes, which have properties that are quite different from one another. These differences determine what the polymers of amino acids, known as proteins, are like.

Note that in addition to its R group, each amino acid, when ionized, has a positive (amino, or NH$_4^+$) group at one end and a negative (carboxyl, or COO$^-$) group at the other end. These two groups can undergo a chemical (dehydration) reaction, losing a molecule of water and forming a covalent bond between two amino acids. A covalent bond that links two amino acids is called a **peptide bond**. As illustrated in Figure 3-17, the two amino acids linked by such a bond are not free to rotate around

FIGURE 3-17

The peptide bond. Because of the partial double-bond nature of the C—N peptide bond, which forms when the —NH$_2$ end of one amino acid joins to the —COOH end of another, the resulting peptide chain cannot rotate freely around the peptide bond.

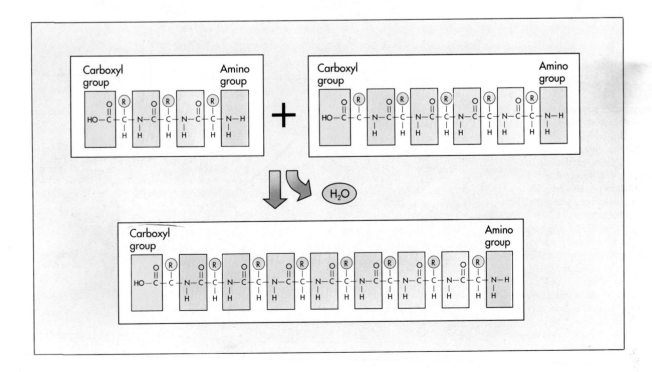

the N—C linkage, because the peptide bond has a partial double-bond character. The stiffness of the peptide bond is one factor that leads chains of amino acids linked in this way to form spirals and other regular shapes (referred to as secondary structures).

Polypeptides are Chains of Amino Acids

As just mentioned, a protein is composed of one or more long chains of amino acids. These chains are called polypeptides. Within a polypeptide the amino acids are linked end to end by peptide bonds. The linear sequence of amino acids that makes up a particular polypeptide chain is termed its **primary structure** (Figure 3-18). Because the R groups that distinguish the various amino acids play no role in the peptide

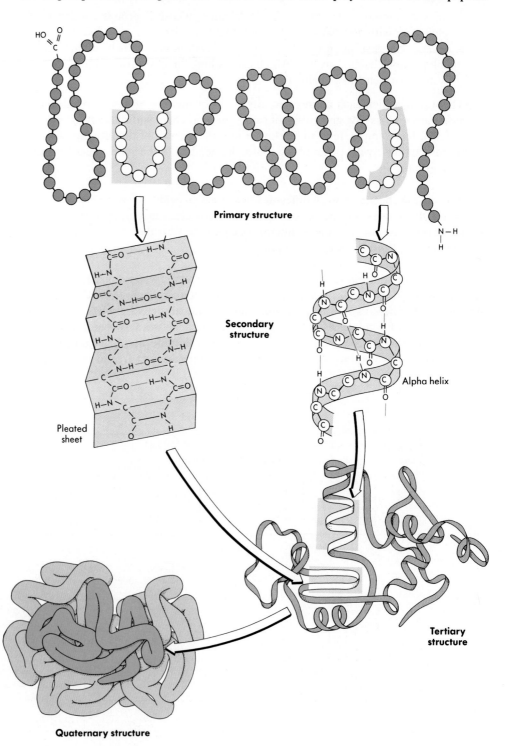

FIGURE 3-18
How primary structure determines a protein's shape. The amino acid sequence of the enzyme protein lysozyme, called its *primary structure*, encourages the formation of hydrogen bonds between nearby amino acids, producing coils called *alpha helixes* and fold-backs called *pleated sheets;* these coils and fold-backs are called the *secondary structure.* The lysozyme protein assumes a three-dimensional shape like a cupped hand; this is called its *tertiary structure.* Many proteins (not lysozyme) aggregate in clusters like the one illustrated here; such clustering is called the *quaternary structure* of the protein.

Primary structure

Secondary structure

Pleated sheet

Alpha helix

Tertiary structure

Quaternary structure

backbone of proteins, a protein can be composed of any sequence of amino acids. A protein composed of 100 amino acids linked together in a chain might have any of 20^{100} different amino acid sequences. This is perhaps the most important property of proteins, permitting great diversity in the kinds of proteins that are possible.

Each of the amino acids of a polypeptide interacts with its nearby neighbors, forming hydrogen bonds. Because of these near-neighbor interactions, polypeptide chains tend to fold spontaneously into sheets or wrap into coils. The form that a region of a polypeptide assumes is called its local **secondary structure** (see Figure 3-18).

Amino acids in a protein chain also interact with water. Because the nonpolar side chains of a protein tend to aggregate in such a way as to minimize disruption of the hydrogen bonding of water molecules, the secondary structure of polypeptide chains tends to further fold up in water into a complicated globular shape called the protein's **tertiary structure** (see Figure 3-18). The shape, or conformation, of a protein's tertiary structure depends greatly on the order and nature of amino acids in the sequence. A change in the identity of a single amino acid can have extremely subtle, or profound, effects on protein shape. It is because proteins can assume so many different specific shapes that they are such effective biological catalysts.

When two or more polypeptide chains associate in forming a functional protein, the chains are referred to as subunits. Hemoglobin, for example, is a protein composed of four subunits. The subunits need not all be the same. Hemoglobin possesses two identical alpha-chain subunits and two beta-chain subunits. A protein's subunit structure is often referred to as its **quaternary structure.**

In some proteins, sulfur-containing amino acid R groups form cross-links, called disulfide bridges, that stabilize the structure. It is important to realize that, except for such stabilizing links, the shape of a protein is largely the result of the interaction of R groups with water, which tends to shove nonpolar portions of the polypeptide into the protein's interior. If the polar nature of the protein's environment is altered, the protein may change its shape, or even unfold (a process called denaturation); when the polar nature of the solvent is reestablished, a small protein may spontaneously refold into its natural shape (Figure 3-19), driven to assume the correct shape by the chemical nature of its R groups. Larger proteins can rarely refold spontaneously because of the complex nature of their final shape—such proteins can sometimes reachieve their native configuration with the help of special protein cofactors called *chaperones.*

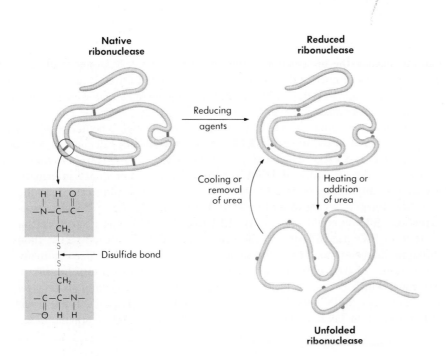

FIGURE 3-19
Primary structure determines tertiary structure. When the protein ribonuclease is treated to break the disulfide covalent bonds that cross-link its chains, and then is placed in urea or heated, the protein unfolds and loses its enzyme activity. Upon cooling or removal of urea, it refolds—and regains its enzyme activity. This demonstrates that no information but the amino acid sequence of the protein is required for proper folding: primary structure determines tertiary structure.

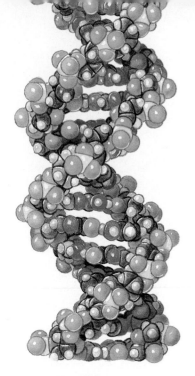

FIGURE 3-20
The DNA double helix.

Proteins perform many functions in your body. The thousands of different enzymes that carry out your body's chemical reactions are **globular proteins.** So are the antibodies that protect you from infection and cancer. **Fibrous proteins** play structural roles. The keratin in your hair is a fibrous protein, and so are the actin and myosin that make up your muscles. Indeed, the most abundant protein in your body is a fibrous protein—collagen, the protein that forms the matrix of your skin, ligaments, tendons, and bones.

NUCLEIC ACIDS

The fourth major kind of macromolecule within cells is nucleic acids (see Table 3-1). Nucleic acids serve as the information storage devices of cells, just as disks or tapes store the information that computers use, blueprints store the information that builders use, and roadmaps store the information that tourists use (Figure 3-20). Nucleic acids come in two varieties: deoxyribonucleic acid (DNA) and ribonucleic acid (RNA). The way organisms store the information used to assemble proteins as a coded sequence of subunits in DNA is similar to the way in which the sequence of letters on this page encodes information. Unique among macromolecules, nucleic acids are able to produce precise copies of themselves, so that the information that specifies what an organism is like can be copied and passed down to its descendants. For this reason, DNA is often referred to as the hereditary material (see Chapter 14).

The alternative form of nucleic acid, RNA, is used by organisms to read the cell's DNA-encoded information and use the information to make proteins. This process will be described in detail in Chapter 15.

Nucleic acids are long polymers of repeating subunits called **nucleotides.** Each nucleotide is a molecule composed of the following three smaller building blocks (Figure 3-21):

1. A five-carbon sugar
2. A phosphate group (PO_4)
3. An organic nitrogen-containing base

In the formation of a nucleic acid chain, the individual nucleotide subunits are linked together in a line by the phosphate groups. Thus the phosphate group of one nucleotide binds to the hydroxyl group of another, forming what is called a **phosphodiester bond.** A nucleic acid, then, is simply a chain of ribose or deoxyribose sugars linked together by phosphodiester bonds with an organic base protruding from each sugar (Figure 3-22).

DNA

Organisms encode the information specifying the amino acid sequence of their proteins as sequences of nucleotides in the nucleic acid called DNA. This encoded information is used in the everyday metabolism of the organism as well as being stored and passed on to the organism's descendants.

Nucleotides play other critical roles in the life of the cell. The nucleotide adenine is a key component of the molecule **ATP,** the energy currency of the cell, and of the molecules **NAD** and **FAD,** which carry electrons whose energy is used to make ATP.

How does the structure of DNA permit it to store hereditary information? If DNA were a simple, monotonously repeating polymer, it could not encode the message of life. Imagine trying to write a story using only the letter E and no spaces or punctuation. All you could ever say is "EEEEEEE. . . ." You need more than one letter to write. We use 26 letters in the English alphabet. The Chinese use thousands of different characters to convey the same messages. You do not need so many individual symbols, of course, if the individual "letters" are grouped together into words. Morse code, which is used to transmit messages by telegraph, employs only two elements ("dot" and "dash"), as do most modern computers (0 and 1). Nucleic acids can encode information because they contain more than one kind of organic base. Each

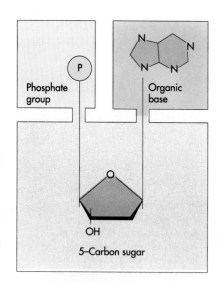

FIGURE 3-21
The structure of DNA and RNA.
The nucleotide subunits of DNA and RNA have a composite structure; each is made up of three elements: a five-carbon sugar, an organic base, and a phosphate group.

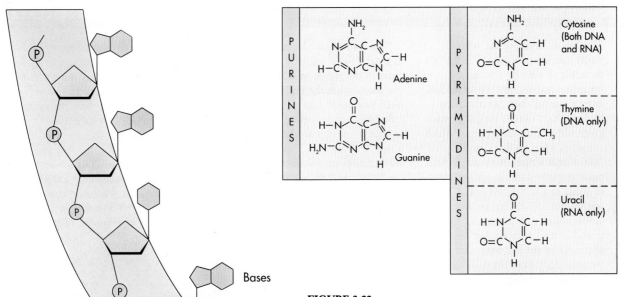

Bases

Backbone

FIGURE 3-22
The five nitrogenous bases of nucleic acids. In DNA, thymine replaces the uracil found in RNA. In a nucleotide chain, nucleotides are linked to one another via phosphodiester bonds, as shown on the left.

sugar link in a nucleic acid chain can have any one of four different organic bases attached to it. Just as in the English language, the sequence of letters encodes the information. In nucleic acids there are not 26 letters, as in English, but only four letters, the four organic bases that occur in nucleic acids. (As you will see shortly, one of the four bases is present in different versions in the two principal forms of nucleic acid.)

Organisms store and use hereditary information by encoding the sequence of the amino acids of each of their proteins as a sequence of nucleotides in nucleic acids.

Organisms store hereditary information in two forms of nucleic acid. One form, DNA, provides the basic storage vehicle, or master plan. The other form, RNA, is similar in structure and is made as a template copy of portions of the DNA. This copy passes out into the rest of the cell, where it provides a blueprint specifying the amino acid sequence of proteins.

Two of the four organic bases that make up DNA and RNA (see Figure 3-22), adenine and guanine, are large, double-ring compounds called **purines.** The other organic bases that occur in these molecules, cytosine (in both DNA and RNA), thymine (in DNA only), and uracil (in RNA only), are smaller, single-ring compounds called **pyrimidines.** In discussing the sequence of bases in RNA and DNA, the organic bases are usually referred to by their first initials: A, G, C, T, and U.

DNA chains in organisms exist not as single chains folded into complex shapes, as in proteins, but rather as double chains (Figure 3-23). Two of the polymers wind around each other like the outside and inside rails of a circular staircase. Such a winding shape is called a **helix,** and one composed of two molecules winding about one another, as in DNA, is called a **double helix.** The steps of the helical staircase are hydrogen bonds between the bases in one polymer chain and those opposite them in the other chain. These hydrogen bonds hold the two chains together as a duplex. Details of the structure of DNA and how it interacts with RNA in the production of proteins are presented in Chapter 14 and 15.

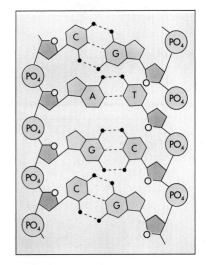

FIGURE 3-23
Hydrogen bonds hold the DNA double helix together. Hydrogen bond formation between the organic bases, called base pairing, causes the two chains of a DNA duplex to bind to one another.

The beautiful pattern of yellow peaks you see in Figure A is a DNA molecule as seen by a scanning-tunneling microscope, the bases following one another like marching soldiers. You cannot see DNA molecules with a light microscope, which cannot resolve anything smaller than 1000 atoms across. With an electron microscope you can image structures as small as a few dozen atoms across, but still cannot see the individual atoms that make up a DNA strand. This limitation has been overcome in the last few years with the introduction of the scanning-tunneling microscope.

What is so new? Imagine you are in a dark room and that there is a chair in the room, too. How would you know what the chair looked like? You could find out by shining a flashlight on it, letting the light bounce off the chair and form an image on your eye. That's what optical and electron microscopes do (in the latter, the "flashlight" shoots out electrons instead of light). Another thing you could do would be to reach over and feel the chair's outline with your hands. You are "see-ing" the chair by putting a probe (your hands) near its surface and measuring how far away the surface is. In a scanning-tunneling microscope, computers advance the probe over the surface of a molecule in tiny steps less than the diameter of an atom. Figure B is the first photograph ever made of DNA. It shows the two strands magnified a million times! The strand is so slender it would take 50,000 of them to reach the diameter of a human hair.

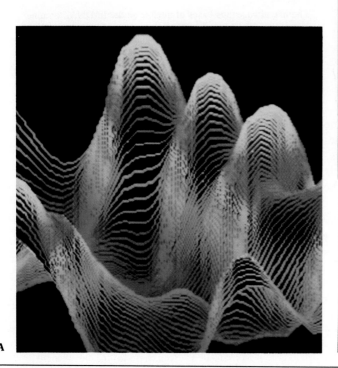

A

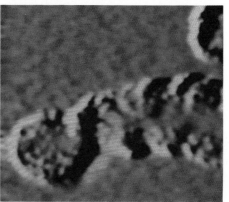

B

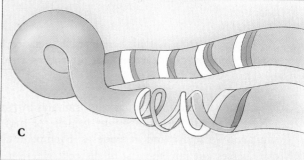

C

RNA

RNA is similar in structure to DNA, but with the following two major chemical differences:

1. RNA molecules contain ribose sugars in which carbon 2 has a hydroxyl group bound to it. In DNA molecules, in contrast, this hydroxyl group is replaced by a hydrogen atom.
2. One of the four organic bases is changed slightly. In RNA the base corresponding to thymine has the same structure as thymine except that one of the carbons lacks a CH_3 group. This new RNA base is called **uracil.** RNA molecules utilize uracil in place of thymine.

Copying the DNA message onto a chemically different molecule such as RNA allows the cell to tell which is the original and which is the copy. DNA molecules are always double stranded (except for a few single-strand DNA viruses—p. 584), whereas the RNA molecules copied from DNA (called messenger RNA) are typically single stranded. Although there is no chemical reason why RNA cannot form double helices as DNA does, cells do not in fact possess the enzymes necessary to assemble double strands of messenger RNA, as they do with DNA. Using two different molecules, one single stranded and the other double stranded, allows the separation of the role of DNA in storing hereditary information from the role of RNA in using this information to specify proteins.

Which Came First, DNA or RNA?

DNA probably evolved from RNA as a means of preserving the genetic information, protecting it from the ongoing wear and tear associated with cellular activity. By storing the information in DNA and then using a complementary RNA sequence to direct protein synthesis, the information-encoding DNA chain is not exposed to the dangers of single-strand cleavage every time the information is used. In all organisms the information encoded in the RNA base sequence is derived from a complementary "master" DNA sequence. (In some viruses other systems occur.) Cells have evolved enzymes, called DNA and RNA polymerases, that synthesize strands that are complementary to single-stranded portions of a DNA molecule. Other enzymes inspect the two strands, ensuring that the polymerase has made no mistakes and that no damage has occurred subsequently. Cells employ a large battery of devices that ensure the accurate replication of the genetic message.

This genetic system is the one that has come down to us from the very beginnings of life. The information necessary for the synthesis of all the protein molecules required by the cell is stored in a double-stranded DNA base sequence. The cell uses this information by making a working copy of it: RNA nucleotides are paired to complementary DNA ones at positions where the DNA is temporarily opened up into single strands. The resulting single-stranded, short-lived RNA copy is used to direct the synthesis of a protein that has a specified sequence of amino acids. In this way the information flows from DNA to RNA to protein. In bacterial cells this process, which has been termed the "central dogma" of heredity, can be visualized directly: RNA dangles from the DNA as it is synthesized, with protein chains already beginning their synthesis on each of the RNA strands while the RNAs are still being completed.

Proteins certainly evolved in synchrony with the evolution of the genetic system as a whole. They are in fact two aspects of a single system, neither part of which is functional without the other.

SUMMARY

1. Organisms store energy in carbon-hydrogen (C—H) bonds. Short-term storage usually occurs in carbohydrates. The most important of these is glucose, a six-carbon sugar.

2. Organisms often transport sugars as disaccharides, two simple sugars (monosaccharides) that are linked together and cannot therefore be utilized while they are being transported.

3. Excess energy resources may be stored in complex sugar polymers called starches, especially in plants. Glycogen, which is a comparable storage polymer that is common in animals and fungi, is characterized by complex branching.

4. Fats are molecules that contain many more C—H bonds than carbohydrates and thus provide more efficient long-term energy storage. Fats are one form of water-insoluble molecules called lipids; others are steroids and prostaglandins.

5. Proteins are linear polymers of amino acids. Because the 20 amino acids that occur in proteins have side groups with very different chemical properties, the function and shape of a protein are critically affected by its particular sequence of amino acids.

6. Hereditary information is stored as a sequence of nucleotides in a linear nucleotide polymer called DNA. DNA is a double helix.

7. The cell uses the information within DNA by transcribing a complementary single strand of RNA from the DNA duplex. Elsewhere in the cell, this single RNA strand directs the synthesis of a protein with an amino acid sequence corresponding to the nucleotide sequence of DNA from which it was copied.

REVIEW QUESTIONS

1. What are the four macromolecules that are the chemical building blocks of all organisms? What elements predominate in each?

2. What kinds of reactions are dehydration synthesis and hydrolysis?

3. What are the three most common monosaccharides? Which are structural isomers? Which are stereoisomers?

4. How do cellulose and chitin differ from starch? How does chitin differ structurally from cellulose?

5. What is the subunit composition of a typical fat?

6. Differentiate between saturated and unsaturated fats. What is a polyunsaturated fat?

7. Which are more efficient in terms of storing chemical energy, fats or carbohydrates? Why?

8. What is the general class of subunits that make up proteins? What is the characteristic chemical structure of the subunit?

9. Does the nature of an amino acid R group most directly affect the protein's primary, secondary, tertiary, or quaternary structure? Why?

10. What are the three components of a nucleic acid?

11. How do DNA and RNA differ from one another?

THOUGHT QUESTIONS

1. DNA exists in cells as a double-stranded duplex molecule, whereas RNA, which is composed of similar nucleotides linked together in the same way, does not form a double-stranded duplex within cells. Why not? If you were to place complementary single strands of RNA in a test tube, would they spontaneously form duplex molecules?

2. Of all possible DNA nucleotide sequences, what sequence of base pairs would most easily dissociate into single strands on gentle heating of the DNA duplex? Why did you choose this sequence?

3. Why do you suppose humans circulate free glucose in their blood, rather than employing a disaccharide such as sucrose as a transport sugar, like plants?

FOR FURTHER READING

DOOLITTLE, R.: "Proteins," *Scientific American*, October 1985, pages 88-99. Proteins, by virtue of their flexible structure, are able to bind to many other molecules, making the machinery of life possible.

KANTROWITZ, E., and W. LIPSCOMB: "Aspartate Transcarbamylase—The Relationship between Structure and Function," *Science*, vol. 241, pages 669-674, 1988. Insights into how protein shape dictates enzyme function.

KARPUS, M., and J. McCAMMON: "The Dynamics of Proteins," *Scientific American*, April 1986 pages 42-51. A description of how computer simulation is aiding our understanding of how proteins flex and change their shape.

OSTRO, M.: "Liposomes," *Scientific American*, January 1987, pages 103-111. Spontaneous aggregation of lipid spheres offers a novel way to deliver medicine to tissues without diluting it in the blood.

SHARON, N.: "Carbohydrates," *Scientific American*, November 1980, pages 90-116. An overview of the structures of carbohydrates and the diverse roles they assume in organisms.

"The Molecules of Life," *Scientific American*, October 1985. An issue devoted entirely to presenting a comprehensive view of what is known about biologically important molecules, with individual articles on DNA, RNA, and many kinds of proteins.

4

The Origin and Early History of Life

OVERVIEW

Life originated on earth more than 3.5 billion years ago, within 1 billion years of our planet's formation. We do not know how life formed, although the evidence is consistent with the hypothesis that it evolved spontaneously from chemicals. There is considerable discussion among biologists about what the early stages of such an evolutionary process might have been like.

FOR REVIEW

Here are some important terms and concepts that you will encounter in this chapter. If you are not familiar with them, you should review them before proceeding.

Oxidation and reduction (Chapter 2)

Covalent bonds (Chapter 2)

Proteins (Chapter 3)

Nucleic acids (Chapter 3)

FIGURE 4-1

Lightning. Before life evolved, the simple molecules in the earth's atmosphere combined to form more complex molecules. The energy that drove these chemical reactions came from lightning and forms of geothermal energy.

The earth was formed about 4.5 billion years ago. We know nothing directly of these very early times, which are called Hadean times, since no rocks older than 3.9 billion years have been found. The earth's crust, on which we live, formed in this early period, solidifying over a hot and very thick mantle. You would not have recognized the early earth, a land of molten rock and violent volcanic activity. The oldest terrestrial rocks that have survived contain no definite traces of life, or at least none that can be recognized with our current level of technology. Today, however, the earth is teeming with life; except within a blast furnace, you would be hard pressed to think of a place anywhere on the surface of the earth where life does not exist in profusion.

The question of the origin of life is not simple. It is not possible to go back in time and watch how life originated; nor are there any witnesses. There is testimony, in the rocks of the earth, but it is not easily read, and often this record is silent on issues crying out for answers. Perhaps the most fundamental of these issues is the nature of the agency or force that led to the appearance of the first living organisms on earth—the creation of life. There are, in principle, at least three possibilities:

1. *Extraterrestrial origin.* Life may not have originated on earth at all but instead may have been carried to it, perhaps as an extraterrestrial infection of spores originating on a planet of a distant star. How life came to exist on *that* planet is a question we cannot hope to answer soon.
2. *Special creation.* Life-forms may have been put on earth by supernatural or divine forces. This viewpoint, common to most Western religions, is the oldest hypothesis and is widely accepted by non-scientists. It forms the basis of the very unscientific "scientific creationism" viewpoint discussed on page 400.
3. *Evolution.* Life may have evolved from inanimate matter, with associations among molecules becoming more and more complex. In this view, the force leading to life was selection; changes in molecules that increased their stability caused the molecules to persist longer.

In this book we deal only with the third possibility, attempting to understand whether the forces of evolution could have led to the origin of life and, if so, how the process might have occurred. This is not to say that the third possibility is definitely the correct one. Any one of the three possibilities might be true. Nor does the third possibility preclude religion: a divine agency might have acted via evolution. Rather, we are limiting the scope of our inquiry to scientific matters. Of the three possibilities, only the third permits testable hypotheses to be constructed and so provides the only *scientific* explanation, that is, one that could potentially be disproven by experiment, by obtaining and analyzing actual information.

In our search for this understanding, we must look back to the early times before life appeared, when the Earth was just starting to cool (Figure 4-1). We must go back at least that far because there are fossils of simple living things, bacteria, in rocks from 3.0 to 3.5 billion years old, some of the oldest rocks that have persisted on earth. Because of the existence of these fossils, we know that life originated during the first billion years of the history of our planet. As we attempt to determine how this process took place, we will first consider how organic molecules may have originated. Then we will consider how organic molecules might have become organized into living cells.

THE ORIGIN OF ORGANIC MOLECULES: CARBON POLYMERS

Scientists who study the conditions of the primitive earth are called geochemists. Geochemists believe that as the primitive earth cooled and its rocky crust formed, many gases were released from the molten core. It was a time of volcanoes, blasting enormous amounts of material skyward. These gases formed a cloud around the earth and were held as an atmosphere by the earth's gravity. The atmosphere we breathe now is quite different from what it used to be; it has been changed by the activities of organisms, as we shall see later. Despite the changes, geochemists have been able to learn what the early atmosphere must have been like by studying the gases released by volcanoes and by deep sea vents in the earth's crust. Geochemists do not all agree on the exact composition of this original atmosphere, but they do agree that it was principally composed of nitrogen gas. It also contained significant amounts of carbon dioxide and water. It is probable, although not certain, that compounds in which hydrogen atoms were bonded to other light elements such as sulfur, nitrogen, and carbon were also present in the atmosphere of early earth. These compounds would have been hydrogen sulfide (H_2S), ammonia (NH_3), and methane (CH_4).

The atmosphere of the early earth is also thought by most investigators to have been rich in hydrogen, although there is debate on this point. We refer to such an atmosphere as a reducing one, because of the ample availability of hydrogen atoms and associated electrons. There was little if any oxygen gas present. In such a reducing atmosphere, it did not take as much energy as it would today to form the carbon-rich molecules from which life evolved. Our atmosphere has since changed as living organisms began to harness the energy in sunlight to split water molecules and form complex carbon molecules, giving off gaseous oxygen molecules in the process. Our atmosphere is now approximately 21% oxygen. In the oxidizing atmosphere that exists today the spontaneous formation of complex carbon molecules cannot occur.

The first step in the evolution of life, therefore, probably occurred in a reducing atmosphere, devoid of gaseous oxygen and different from the atmosphere that exists now. Those were violent times, and the earth was awash with energy: solar radiation, lightning from intense electrical storms, violent volcanic eruptions, and heat from radioactive decay. On earth today, we are shielded from the effects of solar ultraviolet radiation by a layer of ozone gas (O_3) in the upper atmosphere, and it is particularly difficult to imagine the enormous flux of ultraviolet energy to which the early earth's surface must have been exposed. Subjected to ultraviolet energy and to the other sources of energy as well (Figure 4-2), the gases of the early earth's atmosphere un-

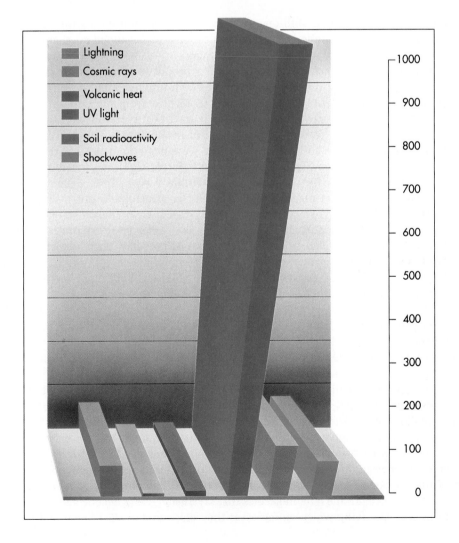

FIGURE 4-2
Sources of energy for the synthesis of complex molecules in the atmosphere of the primitive earth. Of the ultraviolet radiation, only the very short wavelengths (less than 200 nanometers) would have been effective in promoting chemical reactions. Electrical discharges are thought to have been more common on the primitive earth than they are now.

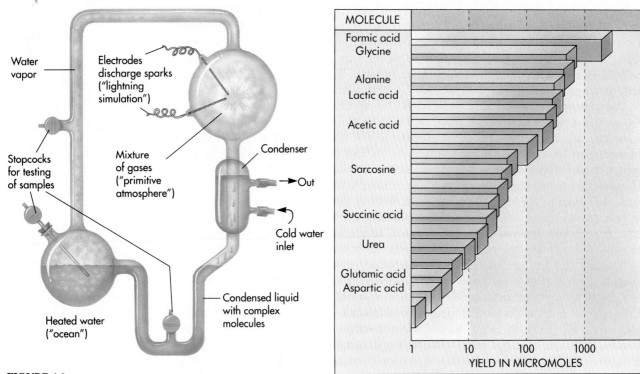

FIGURE 4-3

The Miller-Urey experiment.
A The apparatus consisted of a closed tube connecting two chambers. The upper chamber contained a mixture of gases thought to resemble the primitive earth's atmosphere. Through this mixture electrodes discharged sparks, simulating lightning. Condensers then cooled the gases, causing water droplets to form, which passed into the second heated chamber, the "ocean." Any complex molecules formed in the atmosphere chamber would be dissolved in these droplets and carried to the ocean chamber, from which samples were withdrawn for analysis. **B** Some of the 20 most common complex molecules detected in the original Miller-Urey experiments are indicated. Among these 20 molecules are 4 amino acids.

derwent chemical reactions with one another, forming a complex assemblage of molecules. In the covalent bonds of these molecules some of the abundant energy present in the atmosphere was captured as chemical energy.

What kinds of molecules might have been produced? One way to answer this question is to repeat the process: (1) assemble an atmosphere similar to that thought to exist on the early earth; (2) exclude gaseous oxygen from the atmosphere, since none was present in the atmosphere of the early earth; (3) place this atmosphere over liquid water, which was present on the surface of the cooling earth; (4) maintain this mixture at a temperature somewhat below 100° C; and (5) bombard it with energy in the form of sparks. When Stanley L. Miller and Harold C. Urey carried out this experiment in 1953 (Figure 4-3), they found that within a week, 15% of the carbon that was originally present as methane gas had been converted into other simple compounds of carbon.

Among the first new substances produced in the Miller-Urey experiments (Figure 4-4) were molecules derived from the breakdown of methane, such as formaldehyde (CH_2O) and hydrogen cyanide (HCN). These then combined to form simple molecules, such as formic acid (HCOOH) and urea (NH_2CONH_2), and more complex molecules containing carbon–carbon bonds, including the amino acids glycine and alanine.

As you remember from Chapter 3, amino acids are important because they are the basic building blocks of proteins, which are one of the major kinds of molecules of which organisms are composed. About 50% of the dry weight of each cell in your body consists of amino acids, alone or linked together into protein chains.

In later experiments by other scientists, more than 30 different carbon compounds were identified, including the amino acids glycine, alanine, glutamic acid, valine, proline, and aspartic acid. The production of amino acids indicates that proteins could have formed under conditions similar to those thought to have existed on the early earth. Other biologically important molecules were also formed in these experiments: the presence of hydrogen cyanide was shown to lead to the production of a complex ring-shaped molecule called adenine. Adenine is one of the bases found in DNA and RNA, which are the long molecules that organisms use to encode and transfer the information specifying which amino acids they place in a given protein and in what

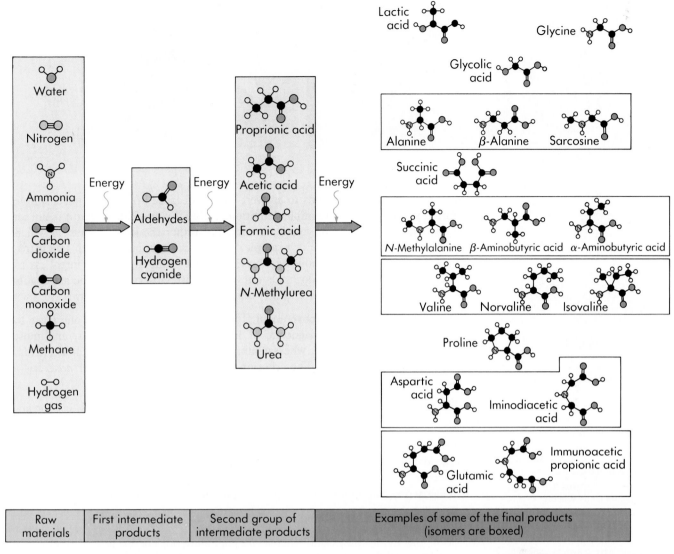

Raw materials	First intermediate products	Second group of intermediate products	Examples of some of the final products (isomers are boxed)

FIGURE 4-4

Results of the Miller-Urey experiment. Seven simple molecules, all gases, were included in the original mixture; note that oxygen was not among them, the atmosphere instead being rich in hydrogen. At each stage of the experiment more complex molecules were formed: first aldehydes, then simple acids, then more complex acids. The molecules that are structural isomers of one another are grouped together in boxes. In most cases only one isomer of a compound is found in living systems today, although many may have been produced in the Miller-Urey experiment.

order. Thus the key molecules from which life evolved were created in the atmosphere of the early earth as a by-product of its birth.

> **Among the molecules that form spontaneously under conditions thought to be similar to those of the primitive earth are those that form the building blocks of organisms.**

NATURE OF LIFE PROCESSES

As the earth cooled, much of the water vapor present in its atmosphere condensed into liquid water, which accumulated in the ever-expanding oceans. Judging from the results of the Miller-Urey experiments, we may presume that the water droplets carried the precursors of amino acids, nucleotides, and other compounds produced by chemical reactions in the atmosphere. The primitive oceans must not have been pleasant places. It is odd to think of life originating from a dilute, hot, smelly soup of ammonia, formaldehyde, formic acid, cyanide, methane, hydrogen sulfide, and organic hydrocarbons. Yet from such an ocean emerged the organisms from which all subsequent life-forms are derived. The way in which this happened is a puzzle and may forever remain so. Nevertheless, one cannot escape a certain curiosity about the earliest steps that eventually led to the origin of all living things on earth, including ourselves. How did organisms evolve from complex molecules? What is the "origin of life"?

FIGURE 4-5
Movement. A graceful gull gliding through the air, a killer whale exploding from the sea—animals have evolved mechanisms that allow them to move about in any medium. Land-dwellers ourselves, we move on land. Whether it is the awkward galumphing of a giraffe, the laborious crawling of a weevil, or a human's first faltering steps, we have grown to expect some kind of movement from all land-dwelling animals.

These are difficult questions to answer, largely because life itself is not a simple concept. If you try to write a simple definition of "life," you will find that it is not an easy task. The problem is not your ignorance, but rather the loose manner in which the concept of "life" is used. For example, imagine a situation in which two astronauts encounter a large amorphous blob on the surface of the moon. One might say to the other, "Is it alive?" What does this question mean? Observe what the astronauts do to find out. Probably they would first observe the blob to see if it moves.

Movement. Most animals move about (Figure 4-5). Movement from one place to another is not in itself diagnostic of life, however. Some animals, and most plants, do not move about, whereas numerous nonliving objects, such as clouds, can be observed to move. The criterion of movement is thus neither *necessary*—possessed by all life—nor *sufficient*—possessed only by life.

The astronauts might prod the blob, to see if it responds, and thus test for another criterion, sensitivity.

Sensitivity. Almost all living things respond to stimuli (Figure 4-6). Plants grow toward light, and animals retreat from fire. Not all stimuli produce responses, however. Imagine kicking a redwood tree or singing to a hibernating bear. Thus this criterion, although superior to the first one, is still inadequate.

The astronauts might attempt to kill the blob.

Death. All living things die, whereas no inanimate objects do. Death is not easily discriminated from disorder, however; a car that breaks down does not die—it was never alive. Death is simply the loss of life. Unless one can detect life, death is a meaningless concept. This is a terribly inadequate criterion.

FIGURE 4-6
Sensitivity. This father lion is responding to a stimulus—he has just been bitten on the rump by his cub. As far as we know, all organisms respond to stimuli, although not always to the same ones. Had the cub bitten a tree instead of its father, the response would not have been as dramatic.

Finally, the astronauts might cut up the blob, to see if it is complexly organized.

Complexity. All living things are complex. Even the simplest bacteria contain a bewildering array of molecules, organized into many complex structures. Complexity is not diagnostic of life, however. A computer is also complex, but it is not alive. Complexity is a necessary criterion of life, but it is not sufficient in itself to identify living things, since many complex things are not alive.

To determine whether the blob is alive, the astronauts must learn much more about it. Probably the best thing they could do would be to examine it more carefully and determine the ways in which it resembles living organisms. All known organisms share certain general properties, ones that we think must ultimately have derived from the first organisms that evolved on earth. It is by these properties that we recognize other living things, and to a large degree these properties define what we mean by the process of life. The following three fundamental properties are shared by all organisms on earth:

Cellular organization. All organisms are composed of one or more cells, which are complex organized assemblages of molecules enclosed within membranes (Figure 4-7).

Growth and metabolism. All living things assimilate energy and use it to grow. This process is called **metabolism.** Plants, algae, and some bacteria use sunlight to create carbon-carbon covalent bonds from CO_2 and H_2O by the process known as photosynthesis. Nearly all organisms obtain energy by metabolizing these bonds (Figure 4-8). The conversion of energy is essential to all life on earth. All known living organisms are similar in another respect: metabolic energy captured from carbon compounds is transferred from one molecule to another via special energy-storing phosphate bonds.

Reproduction. All living things reproduce, passing on traits from one generation to the next (Figure 4-9). Some organisms live for a very long time; for example, some of the bristlecone pines *(Pinus longaeva)* that grow near timberline in the White Mountains of the western Great Basin have been alive for nearly 5000 years. But no organism lives forever, as far as we know. Because all organisms die, the ongoing process of life is impossible without reproduction.

Are these properties adequate to define life? Is a membrane-enclosed entity that grows and reproduces alive? Not necessarily. Phospholipid molecules (long, linear carbon polymers with polar and nonpolar ends) in a water solution spontaneously form hollow spheres, membranes that enclose a small volume of fluid (Figure 4-10). These spheres, called **coacervates,** may contain energy-processing molecules, and they may also grow and subdivide. Despite these features, they are certainly not alive, any more than soap bubbles are alive. Therefore the three criteria just listed, although necessary for life, are not sufficient to define life. One ingredient is missing—heredity, a mechanism for the preservation of improvement:

Heredity. All organisms on earth possess a genetic system that is based on the replication of a long, complex molecule called DNA. The order of DNA subunits encodes the information that determines the individual organism's characteristics.

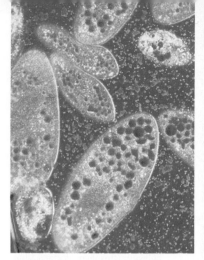

FIGURE 4-7
Cellular organization. These *Paramecia,* complex protists, have just ingested several yeast cells, which are stained red in this photograph. *Paramecia* are single-cell organisms. The yeasts are enclosed within membrane vesicles called digestive vacuoles. A variety of other organelles are also visible.

FIGURE 4-9
Reproduction. The reptile crawling out of this egg represents successful reproduction. All species reproduce, although not all hatch from eggs. Some organisms reproduce several generations each hour; others, once in a thousand years.

FIGURE 4-8
Metabolism. The energy that this robin uses to grow is obtained from the food it eats, such as the berry it is choking down. The plant made the berry with energy it obtained through the process of photosynthesis. In this process the energy of light is captured by chlorophyll and used to build carbon-containing molecules from atmospheric carbon dioxide gas.

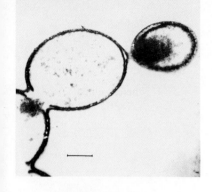

FIGURE 4-10
Coacervates. These protocells, hollow microspheres formed from protein molecules, possess many of the characteristics of living cells. If you look carefully, for example, you can see that each is bounded by a bilayer membrane. The first cells are thought to have evolved from protein or lipid microspheres.

To understand the role of heredity in our definition of life, we will return for a moment to the coacervates. When examining an individual coacervate globule, we see it at that precise moment in time but learn nothing of its predecessors. It is likewise impossible to guess what future globules will be like. The globules are the passive prisoners of a changing environment, and it is in this sense that they are not alive. The essence of being alive is the ability to encompass change and to reproduce the results of change permanently. Heredity, therefore, provides the basis for the great division between the living and the nonliving. Change does not become evolution unless it is passed on to a new generation. A genetic system is the sufficient condition of life. When we look at any living organism, we are seeing its history, carried through its genes from the earliest time. Some changes are preserved because they improve chances of survival in a hostile world, whereas others are lost. Not only did life evolve, evolution is the very essence of life.

All living things on earth are characterized by cellular organization, growth, reproduction, and heredity. These characteristics serve to define the term *life*. Other properties that are commonly exhibited by living organisms include movement and sensitivity to stimuli.

ORIGIN OF THE FIRST CELLS

Many different kinds of molecules will aggregate with one another in water, much as people from the same foreign country tend to aggregate within a large city. Sometimes the aggregation of molecules of one kind will form a cluster big enough to see, spherical protocells 1 to 2 micrometers (one millionth of a meter) in diameter. Since microspheres will form spontaneously from fat molecules when suspended in water, and can also form from amino acids, simple protocells could have formed in the primeval soup, and thus represent the first step in the evolution of cellular organization.* Coacervates, for example, have the following remarkably cell-like properties:

1. Coacervates form an outer boundary that resembles a biological membrane in that it has two layers.
2. Coacervates grow, accumulating more subunit lipid molecules from the surrounding medium.
3. Coacervates form budlike projections and divide by pinching in two, like bacteria.
4. Coacervates can contain amino acids and use them to facilitate the occurrence of a variety of acid-base reactions, including the decomposition of glucose.

It is not difficult to imagine a process of chemical evolution, involving such coacervate microdrops, that may have taken place before the origin of life. The early oceans must have contained untold numbers of these microdrops, billions in a spoonful, each one forming spontaneously, persisting for a while, and then dispersing. Some of the droplets would by chance have contained amino acids with side groups that were better able to catalyze growth-promoting reactions than the other droplets. These droplets would have survived longer than the others, because the persistence of both protein and lipid coacervates is greatly increased when they carry out metabolic reactions such as glucose degradation and when they are growing actively.

Over millions of years, those complex microdrops that were better able to incorporate molecules and energy from the lifeless oceans of the early earth would have tended to persist longer than the others. Also favored would have been those micro-

*There is considerable controversy as to the nature of the first primeval protocells. Lipid coacervate droplets were suggested by the Russian Oparin, one of the first to suggest that life evolved spontaneously in the oceans of the early earth, but the idea has fallen into disfavor, as it is difficult to see how such droplets could evolve. Fox and others have more recently proposed that the first protocells were proteinoid microspheres, created by the spontaneous aggregation of amino acids; this aggregation would have had to have taken place away from water, as the formation of amino acid polymers is a dehydration reaction, and excess water would tend to push the aggregation reaction in the wrong direction.

It is not clear how in the dilute, nonliving (**prevital**) soup of the early earth the more complex molecules characteristic of life were formed. Among the many questions that arise, one is particularly significant: *How did amino acids aggregate spontaneously to form the first proteins?* The question is a puzzle because it seems to defy what we know of the laws of chemistry. Biological proteins form by the joining of subunits into chains. The addition of each element to the growing chain requires the input of energy and the removal of a water molecule. This addition is accomplished by the removal of an —OH group from the end of the chain and an —H from the incoming element. Because this reaction is chemically reversible, an excess of water should in principle drive the reaction in the direction of breakdown of the molecule rather than synthesis. The puzzle is that these reactions critical to the evolution of life are thought to have taken place within the oceans and therefore in a high excess of water. It is difficult to imagine that the spontaneous formation of proteins could have occurred under these conditions.

It has been suggested that the first proteins may have formed on the surfaces present within silicate clays. The interior of clays such as kaolinite is made up of thin layers, only 0.71 nanometer apart, separated by water. A nanometer is a billionth of a meter (10^{-9} meter). Thus a cube of kaolinite 1 centimeter on a side has a surface area of 2800 square meters! A one-pound lump would have a surface greater than 50 football fields. Clay would thus have provided ample surfaces for protein formation to occur. In addition there are many positive and negative charges on the surfaces of these layers, which would have facilitated the process. Supporting the hypothesis that clays may have played an important role in the origin of life, clays such as kaolinite have been shown to catalyze the formation of long polypeptide chains from amino acids when the amino acids are first joined to adenosine monophosphate (AMP), a molecule that produces the energy that makes the reaction possible. This same initial joining of amino acids to AMP is universally employed by all living systems in their synthesis of proteins.

FIGURE 4-A

Magnified nearly 10,000 times, the layers of clay are stacked, one on another, in chunks. At the molecular level, the layers provide enormous surface area and spaces for water and other substances.

drops that could use these molecules to expand in size, growing large enough to divide into daughter microdrops with features similar to those of their "parent" microdrop. The daughter microdrops would have been able to use the same favorable combination of characteristics as their parent, and grow and divide, too. When a means occurred to facilitate this transfer of new ability from parent to offspring, heredity—and life—began.

There is considerable discussion among biologists as to how the first cells evolved. The recent discovery by Thomas Cech and his colleagues at the University of Colorado that RNA can act like an enzyme to assemble new RNA molecules on an RNA template has raised the interesting possibility that coacervates may not have been the first step in the evolution of life. Nucleotides were also produced in the Miller-Urey experiments. Perhaps the first macromolecules were RNA molecules, and the initial steps on the evolutionary journey were ones leading to the more complex and stable RNA molecules. Later, stability might have been improved by surrounding the RNA within a coacervate. There is as yet no consensus among those studying this problem as to whether RNA evolved before a coacervate-like structure, or after.

THE EARLIEST CELLS

The fossils found in ancient rocks (Figure 4-11) show an obvious progression from simple to complex organisms during the vast period of time that began no more than 1 billion years after the origin of the earth. Life-forms may have been present earlier, but rocks of such great antiquity are rare, and fossils have not yet been found in them.

What do we know about these early life-forms? Early microfossils show that, for most of the history of life, all organisms resembled living bacteria in their physical characteristics, although some ancient forms cannot be matched exactly. The ancient bacteria were small (1 to 2 micrometers in diameter) and single-celled, lacked external appendages, and had little evidence of internal structure.

We call these simple organisms with a body plan of this sort **prokaryotes,** from the Greek words for "before" and "kernel," or nucleus. The name reflects their lack of a **nucleus,** which is a spherical organelle (structure) characteristic of the more complex **eukaryotic** cells that evolved much later. We refer to the prokaryotes collectively as **bacteria.** Judging from the fossil record, eukaryotes did not appear until about 1.5 billion years ago. Therefore for at least 2 billion years—nearly half the age of the earth—bacteria were the only organisms that existed.

Living Fossils

Most organisms living today resemble one another fundamentally, having the same kinds of membranes and hereditary systems and many similar aspects of metabolism. However, not all living organisms are exactly the same in these respects. If we look carefully in uncommon environments, we occasionally encounter organisms that are quite unusual, differing in form and metabolism from most other living things. Sheltered from evolutionary alteration in unchanging habitats that resemble those of earlier times, these living relics are the surviving representatives of the first ages of life on earth. In those ancient times, biochemical diversity was the rule, and living things did not resemble each other in their metabolic features as closely as they do today. In places such as the oxygenless depths of the Black Sea or the boiling waters of hot springs, we can still find bacteria living without oxygen and displaying a bewildering array of metabolic strategies. Some of these bacteria have shapes similar to those of the fossils of bacteria that lived 2 or 3 billion years ago.

Methane-Producing Bacteria

What were these early bacteria like? Among the most primitive ones that still exist today are the methane-producing bacteria. These organisms are typically simple in form and are able to grow only in an oxygen-free environment. For this reason they are said to grow "without air," or **anaerobically** (Greek *an,* without + *aer,* air + *bios,* life), and are poisoned by oxygen. The methane-producing bacteria convert CO_2 and H_2 into methane gas (CH_4). They resemble all other bacteria in that they possess hereditary machinery based on DNA, a cell membrane composed of lipid molecules, an exterior cell wall, and a metabolism based on an energy-carrying molecule called ATP. However, the resemblance ends at that point.

When the details of membrane and cell wall structure of the methane-producing bacteria are examined, they prove to be different from those of all other bacteria. There are also major differences in some of the fundamental biochemical processes of metabolism that are the same in all other bacteria. The methane-producing bacteria are survivors from an earlier time when there was considerable variation in the mechanisms of cell wall and membrane synthesis, in reading hereditary information, and in energy metabolism. They represent a road not taken, a side branch of evolution.

Photosynthetic Bacteria

One additional kind of bacteria deserves mention here: that which has the ability to capture the energy of light and transform it into the energy of chemical bonds within cells. These bacteria are photosynthetic, like plants and algae. The pigments used to

FIGURE 4-11
A fossil of a fish. The rocks of Colorado and Utah are rich with such fossils, since the western United States was submerged beneath an inland sea in prehistoric times.

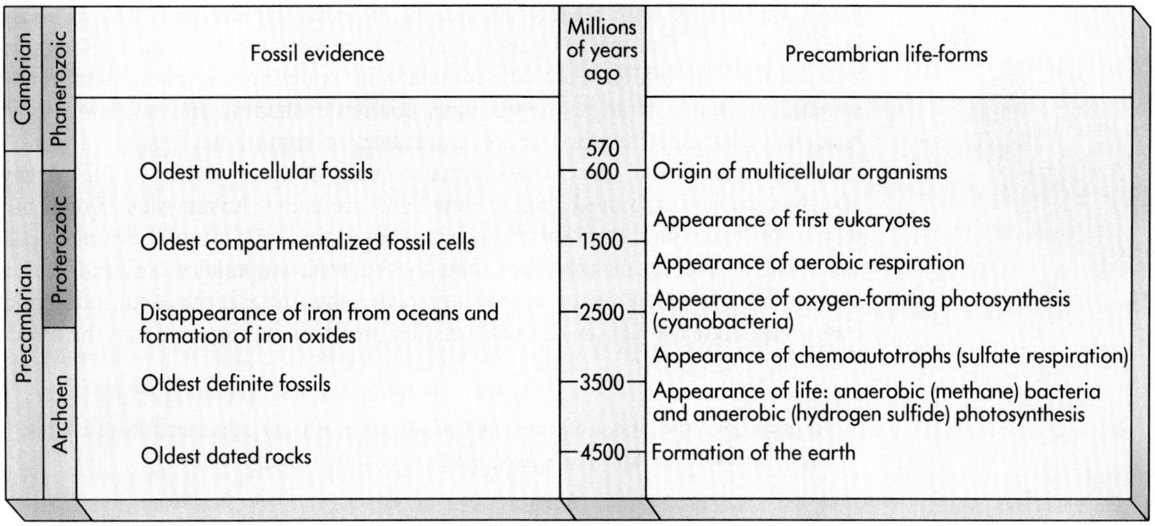

	Fossil evidence	Millions of years ago	Precambrian life-forms
Cambrian / Phanerozoic		570	
	Oldest multicellular fossils	600	Origin of multicellular organisms
Precambrian / Proterozoic			Appearance of first eukaryotes
	Oldest compartmentalized fossil cells	1500	Appearance of aerobic respiration
	Disappearance of iron from oceans and formation of iron oxides	2500	Appearance of oxygen-forming photosynthesis (cyanobacteria)
			Appearance of chemoautotrophs (sulfate respiration)
Archean	Oldest definite fossils	3500	Appearance of life: anaerobic (methane) bacteria and anaerobic (hydrogen sulfide) photosynthesis
	Oldest dated rocks	4500	Formation of the earth

capture light energy vary in different groups; when these bacteria are massed, they often color the earth, water, or other areas where they grow with characteristic hues.

One of the groups of photosynthetic bacteria important in the history of life on earth is the **cyanobacteria,** sometimes called "blue-green algae." They have the same kind of chlorophyll pigment that is most abundant in plants and algae, plus other pigments that are blue or red. Cyanobacteria produce oxygen as a result of their photosynthetic activities, and when they appeared at least 3 billion years ago, they played the decisive role in increasing the concentration of free oxygen in the earth's atmosphere from below 1% to the current level of 21%. As the concentration of oxygen increased, so did the amount of ozone in the upper layers of the atmosphere, thus affording protection from most of the ultraviolet radiation from the sun—radiation that is highly destructive to proteins and nucleic acids. Certain cyanobacteria are also responsible for the accumulation of massive limestone deposits.

The Origin of Modern Bacteria

The early stages of the history of life on earth seem to have been rife with evolutionary metabolic experimentation. Novelty abounded, and many biochemical possibilities were apparently represented among the organisms alive at that time. From the array of different early living forms, representing a variety of biochemical strategies, only a few became the ancestors of the great majority of organisms that are alive today. A few of the other "evolutionary experiments," such as the methane-producing bacteria, have survived locally or in unusual habitats, but most others became extinct millions or even billions of years ago.

Most organisms now living are descendants of a few lines of early bacteria. Many other diverse forms have not survived.

Modern bacteria, for the most part, seem to have stemmed from a tough, simple little cell; its hallmark was adaptability. For at least 2 billion years, bacteria were the only form of life on earth (Figure 4-12). All of the eukaryotes, including animals, plants, fungi, and protists, are their descendants.

THE APPEARANCE OF EUKARYOTIC CELLS

All fossils that are more than 1.5 billion years old are generally similar to one another structurally. They are small, simple cells (Figure 4-13); most measure 0.5 to 2 micrometers in diameter, and none are more than about 6 micrometers thick.

In rocks about 1.5 billion years old we begin to see for the first time microfossils that are noticeably different in appearance from the earlier, simpler forms. These cells are much larger than bacteria and have internal membranes and thicker walls (Figure

FIGURE 4-12

The geological time scale. The periods refer to different stages in the evolution of life on earth. Certain fossils date back to the Archean Period, and very different ones are found in samples from the Proterozoic Era. The time scale is calibrated by examining rocks containing particular kinds of fossils; the fossils are dated by determining the degree of spontaneous decay of radioactive isotopes locked within rock when it was formed.

— 10 μm —

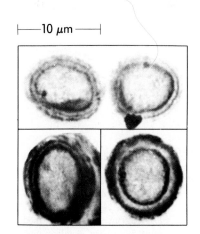

FIGURE 4-13

Fossil bacteria from the Bitter Springs formation of Australia. These fossils, far too small to be seen except with an electron microscope, are about 850 million years old. Similar fossil bacteria as much as 3.5 billion years old have been detected.

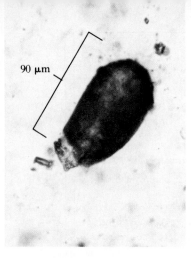

90 μm

FIGURE 4-14
Fossil of unicellular eukaryote about 800 million years old. All life was unicellular until about the past 700 million years.

4-14). Cells more than 10 micrometers in diameter rapidly increased in abundance. Some fossil cells that are 1.4 billion years old are as much as 60 micrometers in diameter. Others, 1.5 billion years old, contain what appear to be small, membrane-bound structures. Many of these fossils have elaborate shapes, and some exhibit highly branched filaments, tetrahedral configurations, or spines.

These early fossil traces mark a major event in the evolution of life. A new kind of organism had appeared. These new cells are called **eukaryotes,** from the Greek words for "true" and "nucleus," because they possess an internal chamber called the cell nucleus. All organisms other than the bacteria are eukaryotes, and they rapidly evolved to produce all of the diverse organisms that inhabit the earth today, including ourselves (Figure 4-15). In the next chapter we shall explore in detail the structure of eukaryotes in relation to the factors involved in their origin.

For at least the first 2 billion years of life on earth, all organisms were bacteria. About 1.5 billion years ago, the first eukaryotes appeared.

IS THERE LIFE ON OTHER WORLDS?

The life-forms that evolved on earth closely reflect the nature of this planet and its history. If the earth were farther from the sun, it would be colder, and chemical processes would be greatly slowed down. Water, for example, would be a solid, and many carbon compounds would be brittle. If the earth were closer to the sun, it would be warmer, chemical bonds would be less stable, and few carbon compounds would be stable and persist. The evolution of a carbon-based life-form is probably possible only within the narrow range of temperatures that exist on earth, which is directly related to the distance from the sun.

The size of the earth has also played an important role because it has permitted a gaseous atmosphere. If the earth were smaller, it would not have a sufficient gravitational pull to hold an atmosphere. If it were larger, it might hold such a dense atmosphere that all solar radiation would be absorbed before it reached the surface of the earth.

Has life evolved on other worlds? In the universe there are undoubtedly many worlds with physical characteristics that resemble those of our planet (Figure 4-16).

FIGURE 4-15
A clock of biological time. A billion seconds ago it was 1953, and most students using this text had not yet been born. A billion minutes ago Jesus was alive and walking in Galilee. A billion hours ago the first human had not been born. A billion days ago no biped walked on earth. A billion months ago the first dinosaurs had not yet been born. A billion years ago no creature had ever walked on the surface of the earth.

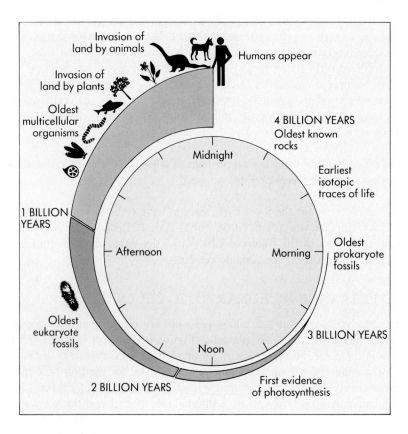

PART 1 THE ORIGIN OF LIVING THINGS

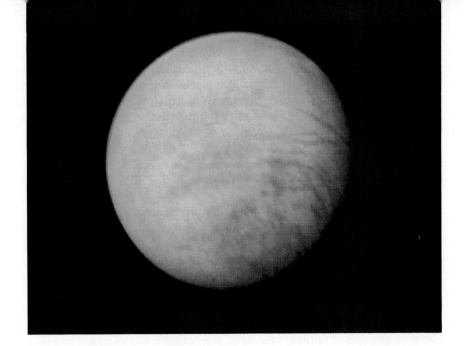

IS ANYBODY OUT THERE?

In a heavily forested mountain valley outside the small town of Arecibo in Puerto Rico rests a huge white dish. One thousand feet across, it is the world's largest radio telescope, and it listens to the stars. In 1992 NASA is proposing to aim it, and other smaller radio telescopes around the world, at one thousand of the stars nearest earth to listen for signals from other worlds. The Search for Extra Terrestrial Intelligence (SETI) project will be the most comprehensive search ever made for signs that intelligent life is out there somewhere in the stars, trying to get in touch with us.

We have made earlier attempts to send messages telling of our presence. Every radio and television program ever broadcast travels outward, although the signals are very weak. Elvis Presley's appearance on the Ed Sullivan show in 1957 will arrive at the star Zeta Herculis this year, having travelled 31 light years (1 light year = 5,878,000,000,000 miles). In 1974 a group of Cornell scientists led by Carl Sagan beamed a signal (much of it in image form) toward the Great Cluster, a massive aggregation of stars 25,000 light years away. We cannot expect a reply soon. If they respond the same year they receive our message, we will hear from them 50,000 years from now.

Rather than transmitting messages, the SETI project is designed to listen. For several years an ongoing project called META (for Megachannel Extra Terrestrial Assay) has already been listening in a narrow channel surrounding frequencies deemed likely for alien messages, with no success yet. The new NASA project will dwarf these attempts. Using new computer technology called *multi-channel spectrum analysis*, it will simultaneously monitor millions of channels over a frequency range of 9 billion hertz (Hz), the entire range of clear radio frequencies that reach the earth. Static noise in our galaxy limits useful transmission by anybody to a "quiet window" of some 60 billion Hz, and all but frequencies between 1 billion and 10 billion Hz are obscured by water in the earth's atmosphere. Any future look at the rest of the 60 billion Hz window will have to be done from space. What does NASA intend to listen for? How will we know when we are being spoken to? The only approach that makes sense is to assume that the signals are designed to be easily detected and understood. What do you do when you want your dog's attention? You shout or clap your hands loudly—the sound you make is loud enough and different enough that it doesn't blend in with the rest of the background noise and so it gets the dog's attention. An alien signal ought to be something similar, a patterned pulse, perhaps. Or maybe, like Carl Sagan, they will send a picture of themselves.

The universe contains some 10^{20} (100,000,000,000,000,000,000) stars with physical characteristics that resemble those of our sun; at least 10% of these stars are thought to have planetary systems. If only 1 in 10,000 planets is the right size and at the right distance from its star to duplicate the conditions in which life originated on earth, the "life experiment" will have been repeated 10^{15} times (that is, a million billion times). It does not seem likely that we are alone.

It seems very possible that life has evolved on other worlds in addition to our own.

Nor should we overlook the possibility that life processes might have evolved in different ways on other planets. A functional genetic system, capable of the accumulation and replication of changes and thus of adaptation and evolution, could theoretically evolve from molecules other than carbon, hydrogen, nitrogen, and oxygen in a different environment. Silicon, like carbon, needs four electrons to fill its outer energy level, and ammonia is even more polar than water. Perhaps under radically different temperatures and pressures, these elements might have formed molecules as diverse and flexible as those that carbon forms on earth.

The final question about life, of course, is whether it has an end. Does evolution tend toward any fixed form or state? It has been suggested that with 10^{15} worlds on which life might have emerged, we should have heard from someone by now—unless the life experiment does not work, with life always tending to destroy itself. We can only hope that this is not so and that peace and progress are perpetual possibilities.

SUMMARY

1. Of the many explanations of how life might have originated, only the theory of evolution provides a scientifically testable explanation.

2. The experimental re-creation of the atmosphere of primitive earth, with the energy sources and temperatures thought to be prevalent, led to the spontaneous formation of amino acids and other biologically significant molecules.

3. Life cannot be defined simply by movement or sensitivity. Its key characteristics are cellular organization, growth, reproduction, and especially heredity.

4. The first cells are thought to have arisen by a process of selecting for aggregations of molecules that were more stable and therefore persisted longer.

5. Before the Phanerozoic Era, which began about 630 million years ago, organisms were too small to be seen with the naked eye. Microscopic fossils—bacteria—are found continuously in the fossil record as far back as 3.5 billion years ago, in the oldest rocks suitable for the preservation of organisms.

6. Bacteria were the only life-forms on earth for 2 billion years or more. At least three kinds of bacteria were present in ancient times: methane utilizers, anaerobic photosynthesizers, and eventually O_2-forming photosynthesizers.

7. The first eukaryotes can be seen in the fossil record 1.3 to 1.5 billion years ago. All organisms other than bacteria are their descendants.

8. The number of stars in the universe similar to our sun exceeds 10^{20} (100,000,000,000,000,000,000). It is almost certain that life has evolved on planets circling some of them.

REVIEW QUESTIONS

1. What compounds were present in the atmosphere of the early earth? Which compound in our present atmosphere was notably absent? What type of atmosphere existed then?

2. What is the significance of the Urey-Miller experiments in the 1950s?

3. What are the necessary characteristics of life? What are the secondary characteristics exhibited by most living things?

4. What are coacervates, and what characteristics do they have in common with living organisms? Are they alive? Why or why not?

5. What evidence supports the reasoning that the earliest known organisms were photosynthesizers? To which present group of organisms are they most closely related?

6. How long ago did the atmosphere acquire plentiful amounts of oxygen? How is this determined?

7. In contemporary photosynthesizers, what molecule(s) is (are) the source(s) of hydrogen and carbon? What is the by-product of this reaction? How did their ancestors change the face of the earth?

8. What is the primary metabolic difference between autotrophs and heterotrophs?

9. How old are the earliest known eukaryotes? How do they differ from the earlier prokaryotes?

10. From a scientific viewpoint, is it likely that there is some form of life on other planets? Why or why not?

THOUGHT QUESTIONS

1. In Fred Hoyle's science fiction novel, *The Black Cloud*, the earth is approached by a large interstellar cloud of gas. The cloud orients itself around the sun. Scientists soon discover that the cloud is feeding, absorbing the sun's energy through the excitation of electrons in the outer energy levels of cloud molecules, a process similar to the photosynthesis that occurs on earth. Different portions of the cloud are isolated from one another by associations of ions created by this excitation. Electron currents pass between these sectors, much as they do on the surface of the human brain, endowing the cloud with self-awareness, memory, and the ability to think. Using static electricity produced by static discharges, the cloud is able to communicate with humans and describe its history, as well as to maintain a protective barrier around itself. It tells human scientists that it once was smaller, having originated as a small extrusion from an ancestral cloud, but has grown by the adsorption of molecules and energy from stars like our sun, on which it has been grazing. Soon the cloud moves off in search of other stars. Is it alive?

2. The nearest galaxy to ours is the spiral galaxy Andromeda. It contains millions of stars, many of which resemble our sun. The universe contains more than a billion galaxies. Each galaxy, like Andromeda, contains countless thousands of stars. It is interesting to speculate: on planets orbiting these stars, are there students speculating on *our* existence? If 1 in 10 of the stars that are like our sun has planets, if 1 in 10,000 of these planets is capable of supporting life, and if 1 in each million life-supporting planets evolves an intelligent life-form, how many planets in the universe support intelligent life? Can you think of any objections to this estimate?

FOR FURTHER READING

BEEN, M., and T. CECH: "RNA as an RNA Polymerase," *Science*, vol. 239, pages 1412-1416, 1988. The classic "chicken or egg" problem of biology—which came first, DNA or RNA?—is solved in favor of RNA, in work for which the Nobel prize was awarded to Cech in 1989.

CAIRNS-SMITH, A.G.: "The First Organisms," *Scientific American*, June 1985, pages 90-101. Interesting article arguing that clay, not primordial soup, provided the fundamental materials from which life arose.

DAWKINS, R.: *The Blind Watchmaker*, W.W. Norton, New York, 1986. Modern evolutionary theory is put into clear and vivid language by this enthusiastic popular paperback writer.

HORGAN, J.: "In the Beginning. . .," *Scientific American*, February 1991, pages 116-125. A review of the very active controversy among scientists about how life first emerged on earth.

JOYCE, G.: "RNA Evolution and the Origin of Life," *Nature*, March 1989, vol. 338, pages 217-224. A review of what scientists now think about the role of RNA in early life forms, in light of new findings from molecular biology.

KASTING, J., O. TOON, and J. POLLACK: "How Climate Evolved on the Terrestrial Planets," *Scientific American*, February 1988, pages 90-97. There may be many habitable plants outside our solar system.

MARGULIS, L., and D. SAGAN: *Microcosmos: Four Billion Years of Evolution from Our Microbial Ancestors*, Summit Books, Simon & Schuster, Inc., New York, 1986. In a beautifully written essay, this mother-son team outlines the evolution of life on earth, showing how all the features that we see today are derived from the early evolution of bacteria. Highly recommended.

MILLER, S.L.: "A Production of Amino Acids Under Possible Primitive Earth Conditions," *Science*, 1953, vol. 117, pages 528-529. The original paper describing the experiment that first demonstrated that biological molecules might have been generated by natural forces on the primitive earth.

NATIONAL RESEARCH COUNCIL: *The Search for Life's Origins*, National Academy Press, Washington DC, 1990. A report on current progress and future directions in the study of the way in which life started on earth.

SCHOPE, J.: "The Evolution of the Earliest Cells," *Scientific American*, September 1978, pages 110-139. How the earliest cells gave rise to the oxygen present in the earth's atmosphere today.

SCOTT, J.: "How Did 'Biopolymers' Evolve Before Life Began?" *Tree*, vol 3, pages 340-342, 1988. A thoughtful consideration of the chemical processes, particularly hydrated electrons, that might have carried out this key step in the evolution of life.

SHAPIRO, R.: *Origins: A Skeptic's Guide to the Creation of Life on Earth*, Simon & Schuster, New York, 1986. A witty, well-informed look at the controversial issues involved in trying to understand how life originated.

TENNESEN, M.: "Mars: Remembrance of Life Past," *Discover*, July 1989, pages 82-88. A lively discussion of the way life may have originated, then disappeared, on another planet of our solar system.

Biology of the Cell

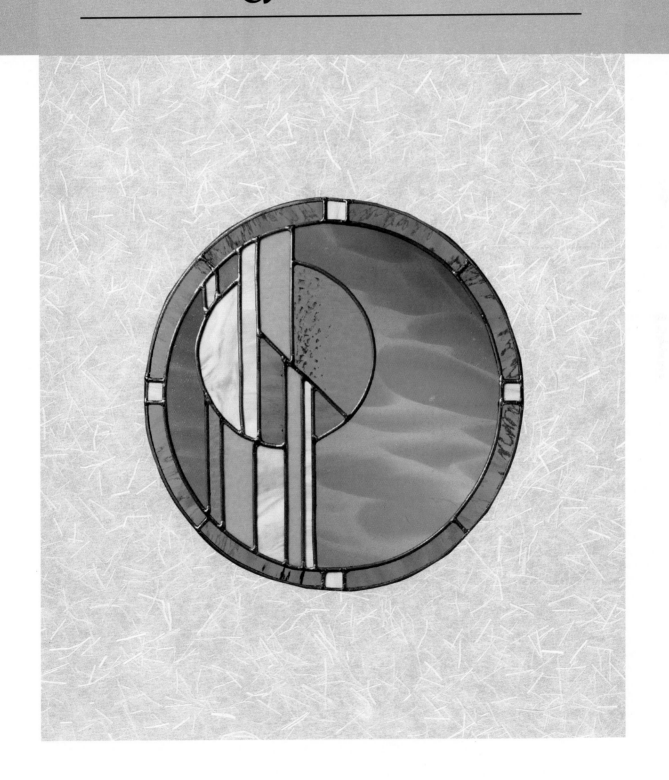

Cell Structure

OVERVIEW

Unlike the simple cells of bacteria, eukaryotic cells exhibit a considerable degree of internal organization, with a dynamic system of membranes forming internal compartments. Some of these compartments are relatively permanent, such as the nucleus that isolates the hereditary apparatus from the rest of the cell. Others are more transient, such as the lysosomes that contain digestive enzymes. The partitioning of the cytoplasm into functional compartments is the most distinctive feature of the eukaryotic cell.

FOR REVIEW

Here are some important terms and concepts that you will encounter in this chapter. If you are not familiar with them, you should review them before proceeding.

Proteins (Chapter 3)

Formation of lipid bilayers (Chapter 3)

Distinction between prokaryotic and eukaryotic cells (Chapter 4)

Evolution of eukaryotes (Chapter 4)

All organisms are composed of cells. Some are composed of a single cell (Figure 5-1), and some, like us, are composed of many cells. The gossamer wing of a butterfly is a thin sheet of cells, and so is the glistening layer covering your eyes. The hamburger you eat is composed of cells, and its contents soon become part of your cells. Your eyelashes and fingernails, orange juice, the wood in your pencil—all were produced by cells. Cells are so much a part of life as we know it that we cannot imagine an organism that is not cellular in nature. In this chapter we will take a close look at cells and learn something of their internal structure. In the following chapters we will focus on cells in action, on how they communicate with their environment, grow, and reproduce.

WHAT CELLS ARE LIKE

Before launching into a detailed examination of cell structure, it is useful first to gain an overview of what to expect. What would we find on the inside of a cell? What is a typical cell like?

A bacterial cell, with its prokaryotic organization, is enclosed by an outer cell wall that surrounds a **plasma membrane.** In eubacteria the plasma membrane is a lipid bilayer with embedded proteins that controls the permeability of the cell to water and dissolved substances. In archaebacteria, by contrast, many lipid membranes are not bilayers. Within, bacterial cells are relatively uniform in appearance. Their **flagella,** if present, are uniform, threadlike protein structures that move the bacteria by rotating. Eukaryotic cells are far more complex; those of animals and a few protists lack cell walls. Within every eukaryotic cell is a **nucleus,** which is its control center. The power that drives a eukaryotic cell comes from internal bacteria-like inclusions called **mitochondria.** A eukaryotic cell is further subdivided into separate compartments by a winding membrane system called the **endoplasmic reticulum.** The flagella that propel motile eukaryotic cells are much more complex than those of bacteria and are not related to bacterial flagella in an evolutionary sense. They propel the cell through its medium by undulating rapidly.

A typical cell, then, is composed of the following three elements:

1. A membrane surrounds the cell, isolating it from the outside world. Chapter 6 describes the many passageways and communications channels that span these membranes. They provide the only connection between the cell and the outside world.
2. The nuclear region directs the activities of the cell. In bacteria, the genetic material is mostly included in a single, closed, circular molecule of DNA, which resides in a central portion of the cell, unbounded by membranes. In eukaryotes, by contrast, a double membrane—the **nuclear membrane,** surrounds the nucleus, which contains the DNA.
3. A semifluid matrix called the **cytoplasm** occupies the volume between the nuclear region and the cell membrane. In bacteria, the cytoplasm contains the chemical wealth of the cell: the sugars, amino acids, and proteins that the cell uses to carry out its everyday activities of growth and reproduction. In addition to these elements, the cytoplasm of a eukaryotic cell contains numerous organized structures, called **organelles.** Many of these organelles are created by the membranes of the endoplasmic reticulum, which close off compartments within which different activities take place. The cytoplasm of eukaryotic cells also contains bacteria-like organelles, called **mitochondria,** that provide power for the cell.

All cells share this architecture. In different broad classes of cells, however, the general plan is modified in various ways. For example, the cells of most kinds of organisms—plants, bacteria, fungi, some protists—possess an outer cell wall that provides structural strength, whereas animal cells do not. The cells of plants frequently contain large membrane-bound sacs called **central vacuoles,** which are used for storage of proteins and waste chemicals; vacuoles in animal cells, called **vesicles,** are

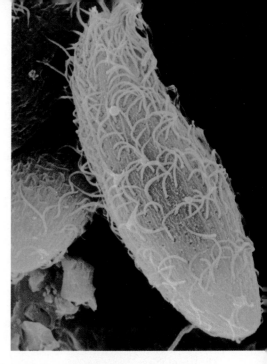

FIGURE 5-1
Two individuals of the single-celled protist *Dileptus*. The hair-like projections that cover their surfaces are cilia, which they undulate to propel themselves through the water.

much smaller. The cells of the majority of organisms possess a single nucleus, although the cells of fungi and some other groups have several to many nuclei. Most cells derive all of their power from mitochondria, whereas plant cells contain a second kind of bacteria-like powerhouse, **chloroplasts,** in addition to their mitochondria. As we shall see, these differences are relatively minor compared with the many ways in which all cells resemble one another.

> A cell is a membrane-bound unit that contains the DNA hereditary machinery as well as membranes, organelles, and cytoplasm.

MOST CELLS ARE VERY SMALL

Sometimes important things seem so obvious that they are overlooked; when we study cells, for example, it is important that we not overlook one of their most remarkable traits—their very small size. Cells are not like shoeboxes, big and easy to study. Instead, they are much smaller. Most kinds are so small that you cannot see a single one with the naked eye (Figure 5-2). Your body contains over 100 trillion cells. If each cell were the size of a shoebox, and lined up end to end, the line would extend to Mars and back, over 500 thousand million kilometers!

FIGURE 5-2

The size of things. Bacteria are generally 1 to 2 micrometers (μm) thick, and human cells are typically orders of magnitude larger. The scale goes from nanometers (nm) to micrometers (μm) to millimeters (mm) to centimeters (cm) and finally meters (m).

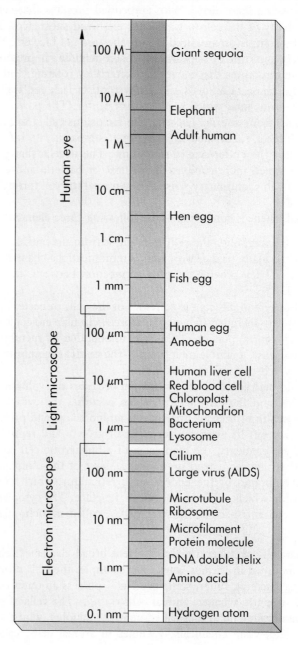

PART II   BIOLOGY OF THE CELL

The Cell Theory

It is because cells are so small that they were not observed until microscopes were invented in the mid-seventeenth century. Cells were first described by Robert Hooke in 1665, when he used a microscope he had built to examine a thin slice of cork. Hooke observed a honeycomb of tiny empty compartments, similar to that shown in a photograph taken through a microscope of his time (see Figure 5-B, 2). He called the compartments in the cork *cellulae*, using the Latin word for a small room. His term has come down to us as **cells**. The first living cells were observed a few years later by the Dutch naturalist Antonie van Leeuwenhoek. van Leeuwenhoek called the tiny organisms that he observed "animalicules"—little animals (Figure 5-3). For another century and a half, however, the general importance of cells was not appreciated by biologists. In 1838 Matthias Schleiden, after a careful study of plant tissues, made the first statement of the cell theory. He stated that all plants "are aggregates of fully individualized, independent, separate beings, namely the cells themselves." In 1839, Theodor Schwann reported that all animal tissues are also composed of individual cells.

The cell theory, in its modern form, includes the following three principles:

1. All organisms are composed of one or more cells, within which the life processes of metabolism and heredity occur.
2. Cells are the smallest living things, the basic unit of organization of all organisms.
3. Cells arise only by division of a previously existing cell. Although life evolved spontaneously in the hydrogen-rich environment of the early earth, biologists have concluded that additional cells are not originating at present. Rather, life on earth represents a continuous line of descent from those early cells.

All the organisms on earth are cells or aggregates of cells, and all of us are descendants of the first cells.

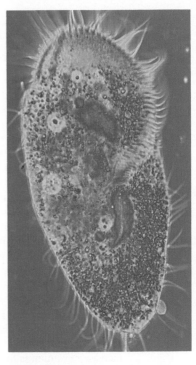

FIGURE 5-3
Cells. Among the "animalicules" that can be seen with a microscope are individuals of *Paramecium*, which dash this way and that, engulfing smaller organisms, and are clearly vibrantly alive.

Why Aren't Cells Larger?

Not all cells are the same size. Individual cells of the marine alga *Acetabularia*, for example, are up to 5 centimeters long (Figure 5-4), whereas the cells of your body are typically 5 to 20 micrometers (one millionth of a meter) thick. If a typical cell in your body were the size of a shoebox, then an *Acetabularia* cell to the same scale would be about 2 kilometers high!

Why are our bodies made up of so many tiny cells? Each cell must maintain centralized control to function efficiently. The nucleus sends commands to all parts of the cell, using molecules that direct the synthesis of certain enzymes, the entry of ions from the exterior, and the assembly of organelles. These molecules must pass by diffusion from the nucleus to all parts of the cell, and it takes them a long time to reach the periphery of a large cell. For this reason, an organism made up of relatively small cells has an advantage over one composed of larger cells.

The advantage of small cell size is also seen in terms of what is called the surface-to-volume ratio. As cell size increases, the volume grows much more rapidly than does the surface area. For a round cell, the increase of the surface area is equal to the square of the increase in the diameter, whereas the increase of the volume is equal to the cube of the increase in the diameter. Thus a cell with 10 times greater diameter would have 10^2 (squared) or 100 times the surface area but 10^3 (cubed) or 1000 times the volume. A cell's surface provides its only opportunity for interaction with the environment, and large cells have far less surface area per unit of volume than do small ones. All substances enter and exit from a cell via the **plasma membrane,** a structure of fundamental importance that will be discussed later in this chapter and in more detail in Chapter 6. This membrane plays a key role in controlling cell function, something that is more effectively done when cells are relatively small.

We have many small cells rather than a few large ones because small cells can be commanded more efficiently and, because of their greater relative surface area, enables better communication with their environment.

FIGURE 5-4
Acetabularia. Although the marine green alga *Acetabularia*, a protist, is a large organism with clearly differentiated parts, such as the stalks and elaborate "hats" visible here, individuals are actually single cells (each with one nucleus), several centimeters tall.

Cells are so small that you cannot see them with the naked eye. Most eukaryotic cells are between 10 and 30 micrometers in diameter. Why can't we see such small objects? Because when two objects are closer together than about 100 micrometers, the two light beams fall on the same "detector" cell at the rear of the eye. Only when two dots are farther apart than 100 micrometers will the beams fall on different cells, and only then can your eye resolve them—tell that they are two objects and not one.

Resolution is defined as the minimum distance that two points can be separated and still be distinguished as two separate points. One way to increase resolution is to increase magnification—to make small objects seem larger. Robert Hooke and Antonie van Leeuwenhoek were able to see small cells by magnifying their size, so that the cells appeared larger than the 100 micrometer limit imposed by the structure of the hu-

man eye. Hooke and Leeuwenhoek accomplished this with microscopes that magnified images of cells by bending light through a glass lens. To understand how such a single-lens microscope is able to magnify an image, examine Figure 5-A, *1*. The size of the image that falls on the picture screen of detector cells lining the back of your eye depends on how close the object is to your eye—the closer the object, the bigger the picture. Your eye, however, is not able to focus comfortably on an object closer than about 25 centimeters (Figure 5-A, *2* [top]), because it is limited by the size and thickness of its lens. What Hooke and Leeuwenhoek did was assist the eye by interposing a glass lens between object and eye (Figure 5-A, *2* [bottom]). The glass lens added additional focusing power, producing an image of the close-up object on the back of the eye. Because the object is closer, however, the image on the back of the eye is bigger than

it would have been, had the object been 25 centimeters away from the eye—as big as a much larger object 25 centimeters away would have produced without the lens. The object is perceived as magnified, bigger.

The microscope employed by Leeuwenhoek is shown in Figure 5-B, *1*. The microscope consists of (*1*) a plate with a single lens, (*2*) a mounting pin that holds the specimen to be observed, (*3*) a focusing screw that moves the specimen nearer to or farther from the eye, and (*4*) a specimen-centering screw.

Leeuwenhoek's microscope, although simple in construction, is very powerful. One of Leeuwenhoek's original specimens, a thin slice of cork, was recently discovered among his papers. Figure 5-B, *2* is the image of that section obtained with Leeuwenhoek's own microscope. The magnification is 266 times, about half as good as a modern student microscope. The finest structures visi-

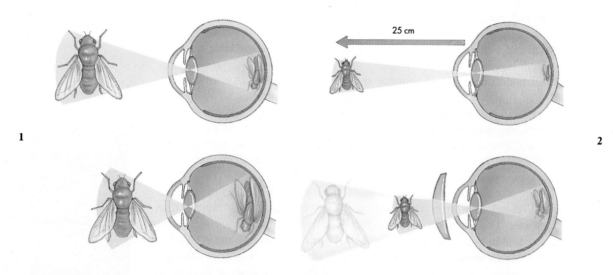

FIGURE 5-A
How does a microscope magnify?
1 The closer an object is to the eye, the larger the image that falls on the back of that eye. The rear surface of the eye is covered with cells called photoreceptors.
2 The eye will not comfortably focus an object closer than 25 centimeters, because the lens of the eye must change shape to focus, and cannot exceed this limit. The glass lens aids the eye in focusing the close object. Because the object is closer, it produces a larger image on the back of the eye, and so appears "larger." A much larger object would have been required to produce an image of the same size without the lens.

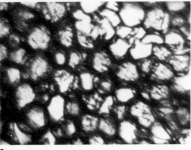

FIGURE 5-B
Early microscopy.
1 Antonie van Leeuwenhoek's microscope.
2 An image of van Leeuwenhoek's sample of cork, obtained with his microscope, which is preserved at Utrecht, in The Netherlands. In this image, the cork is modified ×266; it compares well with modern images of thin sections of cork. The finest structures that are visible are less than one micrometer across.

ble are less than 1 micrometer (1000 nanometers) in thickness.

Modern microscopes use two magnifying lenses (and a variety of correcting lenses) that act like back-to-back eyes, the first lens focusing the image of the object on the second lens; the image is then magnified again by the second lens, which focuses it on the back of the eye. Microscopes that magnify in stages by using several lenses are called compound microscopes. The finest structures visible with modern compound microscopes are about 200 nanometers in thickness.

A contemporary light micrograph is shown in Figure 5-C, *1*. Compound light microscopes are not powerful enough to resolve many structures within cells. A membrane, for example, is only 5 nanometers thick. Why not just add another magnifying stage to the microscope, and so increase the resolving power? This approach doesn't work, because

when two objects are closer than a few hundred nanometers the light-beams of the two images start to overlap. A light beam vibrates like a vibrating string, and the only way two beams can get closer together and still be resolved is if the "wavelength" is shorter.

One way to do this is by using a beam of electrons rather than a light beam. Electrons have a much shorter wavelength, and a microscope employing electron beams has 400 times the resolving power of a light microscope. Transmission electron microscopes today are capable of resolving objects only 0.2 nanometer apart—just five times the diameter of a hydrogen atom! The specimen is prepared as a very thin section, and those areas that transmit more electrons (are less dense) show up as bright areas in the micrographs (Figure 5-C, *2*).

Transmission electron micrographs are so called because the elec-

rons used to visualize the specimens are *transmitted* by the material. A second kind of electron microscope, the scanning electron microscope, beams the electrons onto the surface of the specimen in the form of a fine probe that passes back and forth rapidly. In images made with a scanning electron microscope, depressed areas and cracks in the specimen appear dark, whereas elevated areas such as ridges appear light. The electrons reflected back from the surface of the specimen, together with other electrons that the specimen itself emits as a result of the bombardment, are amplified and transmitted to a television screen, where the image can be viewed and photographed. Scanning electron microscopy yields striking three-dimensional images and has proved useful in understanding many biological and physical phenomena (Figure 5-C, *3*).

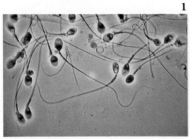

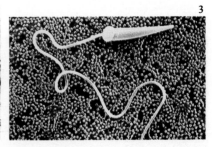

FIGURE 5-C
Three views of sperm cells from three different microscopes.
1 Image of sperm cell taken with a light microscope (×400).

2 Transmission electron micrograph of sperm cell (×15,000).

3 Scanning electron micrograph of sperm cell (×8500).

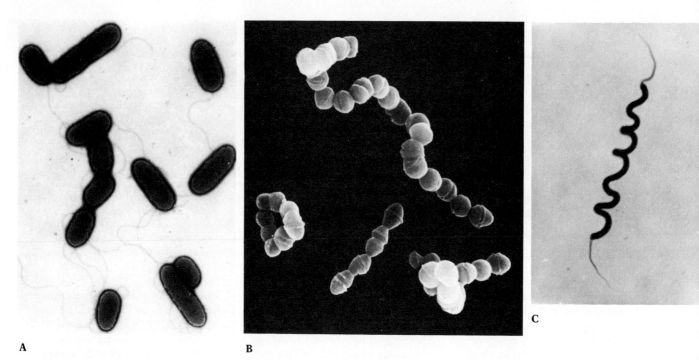

A   **B**   **C**

FIGURE 5-5

Bacterial cells have several different shapes.

A A rod-shaped bacterium, *Pseudomonas*, is a type associated with many plant diseases.
B *Streptococcus* is a more or less spherical bacterium in which the individuals adhere in chains.
C *Spirillum* is a spiral bacterium; this large bacterium has a tuft of flagella at each end.

THE STRUCTURE OF SIMPLE CELLS: BACTERIA

Bacteria are the simplest cellular organisms. Over 2500 species that have been given names are considered to be distinct, but doubtless many times that number actually exist and have not yet been described properly. Although these species are diverse in form (Figure 5-5), their organization is fundamentally similar: small cells about 1 to 10 micrometers thick, enclosed within a membrane and encased within a rigid cell wall, with no distinct interior compartments (Figure 5-6). Sometimes the cells of bacteria adhere in chains or masses, but fundamentally the individual cells are separate from one another.

> **Compared with the other kinds of cells that have evolved from them, bacteria are smaller and lack interior organization.**

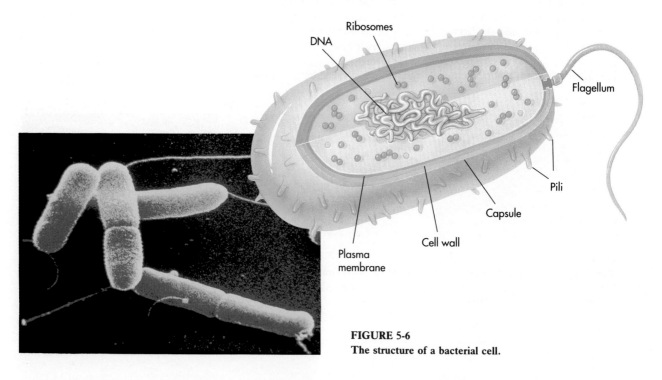

FIGURE 5-6
The structure of a bacterial cell.

Strong Cell Walls

Bacteria are encased by a strong **cell wall,** in which a carbohydrate matrix (a polymer of sugars) is cross-linked by short polypeptide units. No eukaryotes possess cell walls with a chemical composition of this kind. Bacteria are commonly classified as gram-positive and gram-negative by differences in their cell walls. The name refers to the Danish microbiologist Hans Christian Gram, who developed a staining process that distinguishes the two classes of bacteria as a way to detect the presence of certain disease-causing bacteria. **Gram-positive** bacteria have a single, thick cell wall that retains the Gram stain within the cell, causing the stained cells to appear purple under the microscope. More complex cell walls have evolved in other groups of bacteria. In them, the cell wall is thinner and it does not retain the Gram stain; such bacteria are called **gram-negative.** Bacteria are often susceptible to different kinds of antibiotics, depending on the structure of their cell walls.

Simple Interior Organization

If you were to look at an electron micrograph of a thin section of a bacterial cell, you would be struck by its simple organization. There are few if any internal compartments bounded by membranes and no membrane-bounded organelles—the kinds of distinct structures so characteristic of eukaryotic cells. The entire cytoplasm of a bacterial cell is one unit with no internal support structure, thus the strength of the cell comes primarily from its rigid wall (see Figure 5-6).

The plasma membrane of bacterial cells often intrudes into the interior of the cell, where it may play an important role (Figure 5-7). When the cells of bacteria divide, for example, the circular, closed DNA molecule replicates first and the two DNA molecules that result attach to the cell membrane at different points. Their attachment at different points ensures that the resulting daughter cells will each contain one of the identical units of DNA. In some photosynthetic bacteria, the cell membrane is often extensively folded, with the folds extending into the cell's interior. These folded membranes are where the bacterial pigments connected with photosynthesis are located.

Since there are no membrane-bounded compartments within a bacterial cell, however, both the DNA and the enzymes have access to all parts of the cell. Reactions are not compartmentalized as they are in eukaryotic cells, and the whole bacterium operates as a single unit.

> **Bacteria are encased by an exterior wall composed of carbohydrates cross-linked by short polypeptides. They lack interior compartments.**

SYMBIOSIS AND THE ORIGIN OF EUKARYOTES

The first eukaryotes had a cell structure that was radically different from that of all organisms that had existed earlier. Eukaryotic cells are far more complex internally than their bacterial ancestors. We know little about how this increase in internal organization first evolved, 1.5 billion years ago—with one exception. Within the cells of virtually all eukaryotes are organelles that resemble bacterial cells. Most biologists interpret a wealth of evidence as indicating that that is just what they are, ancient bacteria that lived within ancestral eukaryote cells, where they provided (and still provide) their host the advantages associated with their special metabolic abilities. Mitochondria, the energy factories of eukaryotic cells, are organelles of this sort, and so are chloroplasts, the organelles in which photosynthesis takes place in eukaryotic cells. These critically important metabolic functions evolved in different groups of bacteria, but are combined in individual eukaryotic cells through the process of symbiosis, the living together in close association of two or more dissimilar organisms.

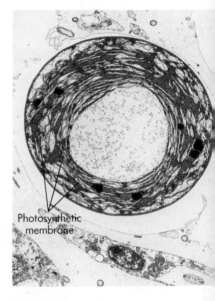

Photosynthetic membrane

FIGURE 5-7

Electron micrograph of a photosynthetic bacterial cell, *Prochloron*. Extensive folded photosynthetic membranes are visible. The single, circular DNA molecule is located in the clear area in the central region of the cell.

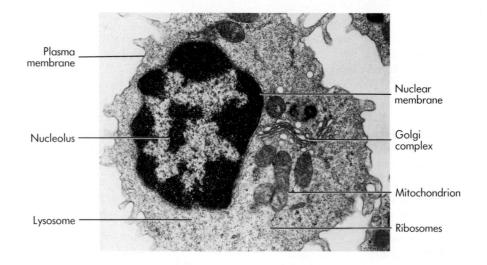

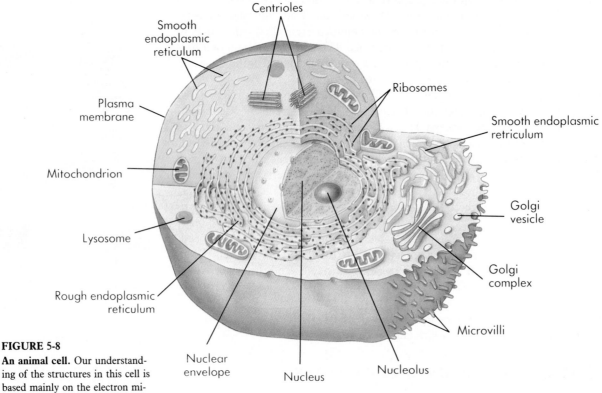

FIGURE 5-8

An animal cell. Our understanding of the structures in this cell is based mainly on the electron micrographs ($\times$40,500) shown with the cell diagram.

A COMPARISON OF BACTERIA AND EUKARYOTIC CELLS

Eukaryotic cells (Figure 5-8) are far more complex than prokaryotic ones. Compared with their bacterial ancestors, eukaryotic cells exhibit the following significant morphological differences:

1. Their DNA is packaged tightly into compact units called **chromosomes** that are located within a separate organelle, the nucleus.
2. The interiors of eukaryotic cells are subdivided into membrane-bounded compartments, which permit one biochemical process to proceed independently of others that may be going on at the same time. The compartmentalization of biochemical activities in eukaryotes increases the efficiency of the various processes.
3. The cells of animals and some protists lack cell walls. Plants possess cell walls, but they are totally different in character from bacterial cell walls: cellulose fi-

bers are embedded in a matrix of other polysaccharides and protein, producing a strong fiberglass-like structure. Plant cell walls are discussed in detail in Chapter 35.

4. Only rarely do prokaryotes contain internal sacs. By contrast, the mature cells of plants often contain large fluid-filled internal sacs called central vacuoles (Figure 5-9). Animal cells do not possess such large vacuoles.

The interiors of eukaryotic cells are subdivided by membranes. The DNA of the cell is packaged with protein and contained in the nucleus in units called chromosomes. Most eukaryotes possess cell walls, although cell walls are lacking in animals and some single-celled organisms.

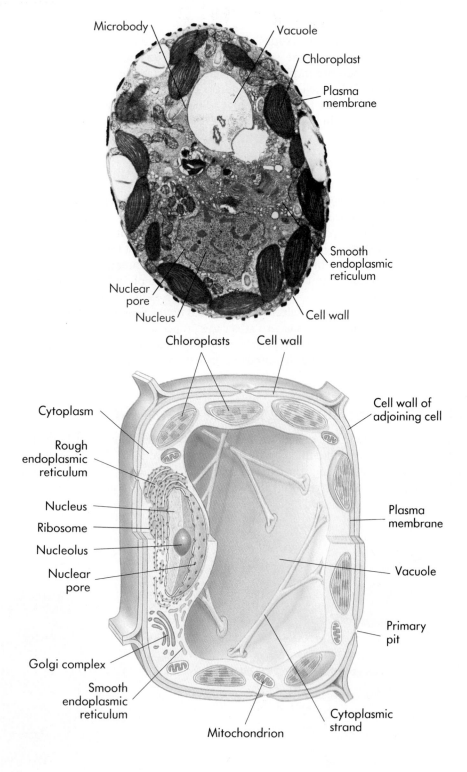

FIGURE 5-9

A plant cell. Most mature plant cells contain large central vacuoles, which occupy a major portion of the internal volume of the cell. The cytoplasm occupies a thin layer between vacuole and cell membrane. Within it are all of the cell's mitochondria and other organelles.

THE INTERIOR OF EUKARYOTIC CELLS: AN OVERVIEW

Although eukaryotic cells are diverse in form and function, they share a basic architecture. They all are bounded by a membrane called the **plasma membrane,** they all contain a supporting matrix of protein called a **cytoskeleton,** and they all possess numerous organelles. The major organelles are of three general kinds: *class 1*, membrane structures or organelles derived from membranes; *class 2*, bacteria-like organelles involved in energy production; and *class 3*, organelles involved in gene expression.

CLASS 1	CLASS 2	CLASS 3
nucleus	mitochondria	chromosomes
endoplasmic reticulum	chloroplasts	ribosomes
Golgi bodies		nucleolus
lysosomes		
microbodies		

The membranes of the eukaryotic cell and the organelles derived from them (class 1) interact as an **endomembrane system.** All of these endomembranous structures give rise to one another, are in physical contact with one another, or pass tiny membrane-bound sacs called **vesicles** to one another. The endomembrane system includes the endoplasmic reticulum, nuclear envelope, Golgi complex, lysosomes, microbodies, and the plasma membrane (which is not really an endomembrane, but is continuous with it).

MEMBRANES

As we have seen, all biological membranes are phospholipid bilayers about 7 nanometers thick, with embedded proteins; viewed with the electron microscope in cross section, such membranes appear as two dark lines separated by a lighter area (Figure 5-10). This distinctive appearance arises from the tail-to-tail packing of the phospholipid molecules that make up a membrane. The major proteins of a membrane are hydrophobic, therefore they associate with and become embedded in the phospholipid matrix. Few molecules—water is one, as we shall see—are able to cross the lipid bilayer of a membrane; instead, their passage is selectively controlled by the proteins that lie embedded in this matrix, and these proteins also affect the properties of the individual kind of membrane. The properties of the plasma membrane and the various internal membranes of a eukaryotic cell will be discussed in relation to the various structures involved.

A CELL'S BOUNDARY: THE PLASMA MEMBRANE

The plasma membrane of a eukaryotic cell is a double layer (bilayer) of lipid about 9 nanometers (that is, 9 billionths of a meter) thick (Figure 5-10). This membrane envelops the cell, and nothing can enter or leave the cell without crossing it. In fact, little does cross the lipid layer itself. However, traversing the lipid layer are a variety

FIGURE 5-10

The plasma membrane. The electron micrograph of a thin section of a red blood cell clearly shows the double nature of the plasma membrane. The membrane, indicated by *arrows*, consists of two layers of phospholipid molecules, with their hydrophobic tails pointing inward. The phospholipid molecules themselves are electron-dense, and therefore dark in this micrograph, whereas the area where the tails are concentrated is electron-transparent.

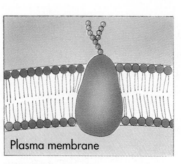

Plasma membrane

of proteins that control the interactions of the cell with its environment, including the following:

1. *Channel-forming proteins.* Some membrane proteins form channels through the plasma membrane that admit specific molecules to the cell; for example, some membranes possess specific channels for sodium ions, and other membranes have specific channels for glucose.

2. *Receptors.* Other proteins that cross the membrane transmit information rather than molecules. These proteins, called receptors, induce changes within the cell when they come in contact with particular molecules on the cell surface. Many hormones induce changes in cells by first binding to such surface receptors.

3. *Markers.* A third class of proteins embedded within the membrane identify the cell as being of a particular type. This is important in a multicellular individual, since cells must be able to recognize one another for tissues to form and function correctly.

Chapter 6 presents a more detailed look at the structure of cell membranes and how they function.

THE ENDOPLASMIC RETICULUM: WALLS WITHIN THE CELL

When viewed with a light microscope, the interiors of eukaryotic cells exhibit a relatively featureless matrix, within which various organelles are embedded. When viewed with an electron microscope, however, a striking difference becomes evident—the interiors of eukaryotic cells are seen to be packed with membranes. So thin that they are not visible with the relatively low resolving power of light microscopes, these membranes fill the cell, dividing it into compartments, channeling the transport of molecules through the interior of the cell, and providing the surfaces on which enzymes act. This system of internal compartments created by these membranes in eukaryotic cells constitutes the most fundamental distinction between eukaryotes and prokaryotes.

The extensive system of internal membranes that exists within the cells of eukaryotic organisms is called the **endoplasmic reticulum,** often abbreviated ER (Figure 5-11). The term *endoplasmic* means "within the cytoplasm," and the term *reticulum*

FIGURE 5-11
Rough endoplasmic reticulum.
The electron micrograph is of a rat liver cell, rich in ER-associated ribosomes. In the drawing you can see that the ribosomes are associated with only one side of the rough ER; the other side is the boundary of a separate compartment within the cell into which the ribosomes extrude newly made proteins destined for secretion.

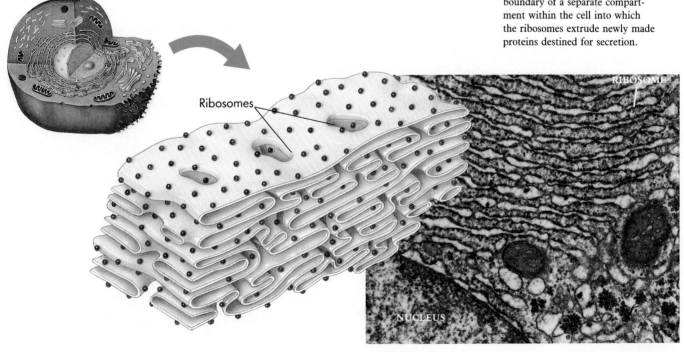

Ribosomes

RIBOSOME

NUCLEUS

comes from a Latin word that means "a little net." Like the plasma membrane, the endoplasmic reticulum is composed of a bilayer of lipid with various enzymes attached to its surface. The endoplasmic reticulum, weaving in sheets through the interior of the cell, creates a series of channels and interconnections between its membranes that isolates some spaces as membrane-enclosed sacs called vesicles.

The surfaces of those regions of the ER devoted to the synthesis of transported proteins are heavily studded with ribosomes (see Figure 5-11). When viewed with an electron microscope, their membrane surfaces appear pebbly, like the surface of sandpaper. Because of this "rocky beach" appearance, the regions of ER rich in bound ribosomes are often termed **rough ER**. Regions of the endoplasmic reticulum with relatively few bound ribosomes are correspondingly called **smooth ER.**

Rough ER: Manufacturing for Export

The surface of the rough ER is where the cell manufactures proteins intended for export. Enzymes and protein hormones like insulin are secreted from the cell surface. The manufacture of proteins is carried out by ribosomes, which are large molecular aggregates of protein and ribonucleic acid (RNA) that translate RNA copies of genes into protein.

Proteins intended for export contain special amino acid sequences called *signal sequences* (Figure 5-12). As a new protein is made by a free ribosome (one not attached to a membrane), the signal portion of the growing polypeptide attaches a recognition factor that carries the aggregate of genetic message, ribosome, and partially completed protein to a "docking site" on the surface of the endoplasmic reticulum. When the protein is complete, it passes through the ER membrane into the vesicle-forming system called the Golgi complex (discussed later). It then travels within vesicles to the inner surface of the cell, where it is released to the outside. From the time the protein is first attached to the ER, it is, in a sense, already located outside the cell.

Smooth ER: Organizing Internal Activities

Many of the cell's enzymes cannot function when floating free in the cytoplasm; they are active only when they are associated with a membrane. ER membranes contain many such enzymes embedded within them. Enzymes anchored within the ER, for example, catalyze the synthesis of a variety of carbohydrates and lipids. In cells that carry out extensive lipid synthesis, such as the cells of the testicles, smooth ER is particularly abundant. Intestinal cells, which synthesize triglycerides, and brain cells are also rich in smooth ER. In the liver, enzymes embedded within the smooth ER are involved in a variety of detoxification processes. Drugs such as amphetamines, mor-

FIGURE 5-12

The signal hypothesis. Many biologists now believe that short 20–amino acid sequences at the tips of proteins direct their destinations in the cell. In this example, a sequence of hydrophobic amino acids on a secretory protein attaches them (and the ribosomes making them) to ER, injecting the leading end of the newly made protein across the ER into the lumen. As the synthesis of the protein continues, it passes across the ER, "exported" into the lumen. The signal sequence is clipped off after the leading edge of the protein enters the lumen.

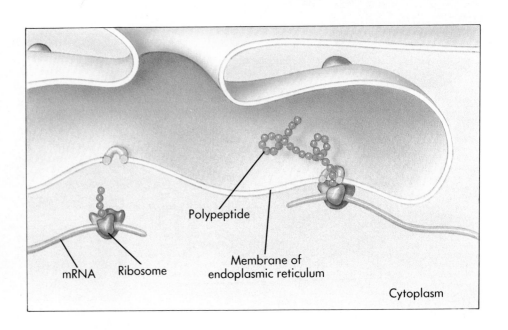

Polypeptide

mRNA   Ribosome

Membrane of
endoplasmic reticulum

Cytoplasm

phine, codeine, and phenobarbital are detoxified in the liver by components of the smooth ER.

> The endoplasmic reticulum (ER) is an extensive system of membranes that divides the interior of eukaryotic cells into compartments and channels. Rough ER synthesizes proteins to be exported, whereas smooth ER organizes the synthesis of lipids and other biosynthetic activities.

THE NUCLEUS: CONTROL CENTER OF THE CELL

The largest and most easily seen of the organelles within a eukaryotic cell is the **nucleus,** which was first described by the English botanist Robert Brown in 1831. The word "nucleus" is derived from the Latin word meaning kernel or nut. In fact, nuclei are more or less spherical in shape and do bear some resemblance to nuts (Figure 5-13). In animal cells the nucleus is typically located in the central region. In some cells the nucleus seems to be cradled in this position by a network of gossamer-fine cytoplasmic filaments. The nucleus is the repository of the genetic information that directs all of the activities of a living cell. Some kinds of cells, such as mature red blood cells, discard their nuclei during the course of development. The development of these cells terminates when their nuclei are lost. They lose all ability to grow, change, and divide, and become merely passive vessels for the transport of hemoglobin.

> The nucleus of a eukaryotic cell contains the cell's hereditary apparatus and isolates it from the rest of the cell.

FIGURE 5-13
The nucleus. The nucleus is composed of a double membrane, called a nuclear envelope, enclosing a fluid-filled interior containing the chromosomes. In cross section, the individual nuclear pores are seen to extend through the two membrane layers of the envelope; the dark material within the pore is protein, which acts to control access through the pore.

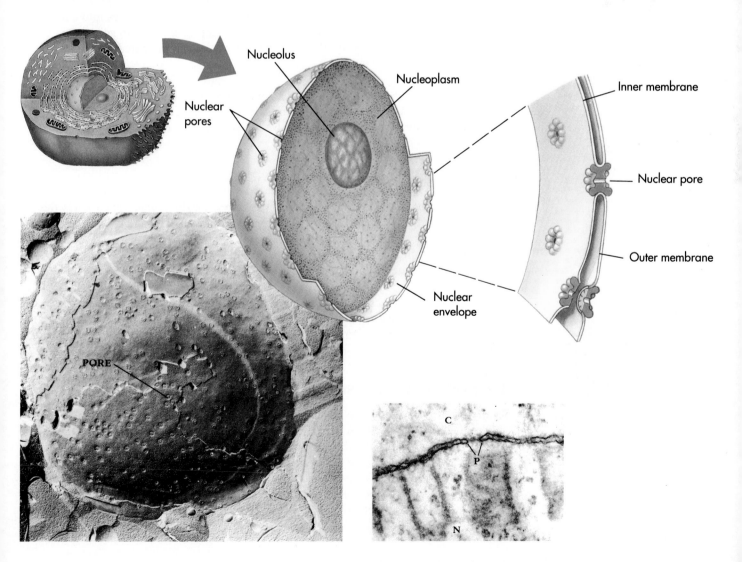

Nucleolus

Nucleoplasm

Nuclear pores

Inner membrane

Nuclear pore

Outer membrane

Nuclear envelope

PORE

C

P

N

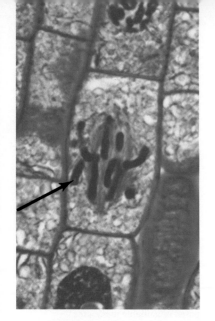

FIGURE 5-14
Eukaryotic chromosomes (*arrow*)
within an onion root tip.

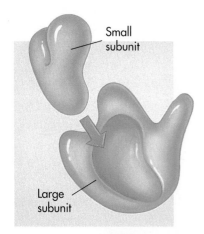

Small
subunit

Large
subunit

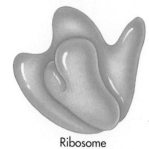

Ribosome

FIGURE 5-15
A ribosome. To make proteins,
the cell uses the ribosome to read
the RNA copy of DNA and uses
the information it finds there to
direct the synthesis of a protein.

Getting In and Out: The Nuclear Envelope

The surface of the nucleus is bounded by *two* layers of membrane, the **nuclear envelope** (see Figure 5-13). Many researchers believe that endoplasmic reticulum is the source of the outer membrane of the nuclear envelope, which seems to be continuous with it. Scattered over the surface of the nuclear envelope, like craters on the moon, are shallow depressions called **nuclear pores**. These pores, 50 to 80 nanometers apart, form at locations where the two membrane layers of the nuclear envelope pinch together. The nuclear pores are not empty openings in the same sense that the hole in a doughnut is, with nothing filling the actual hole. Rather, nuclear pores are embedded with many proteins that act as molecular channels, permitting certain molecules to pass into and out of the nucleus. Passage is restricted primarily to two kinds of molecules: (1) Proteins moving into the nucleus, where they will be incorporated into nuclear structures or serve to catalyze nuclear activities; and (2) RNA and protein-RNA complexes formed in the nucleus and subsequently exported to the cytoplasm.

The Chromosomes of Eukaryotes Are Complex

Both in bacteria and in eukaryotes, all the hereditary information specifying cell structure and function is encoded in DNA with proteins bound to it that aid coiling. However, unlike the DNA of bacteria, the DNA of eukaryotes is divided into several segments and is associated with packaging proteins and RNA to form **chromosomes** (Figure 5-14). Association with these packaging protein enables eukaryotic DNA to wind up into a highly condensed form during cell division. Under a light microscope, these condensed chromosomes are readily seen in dividing cells as densely staining rods. After cell division, eukaryotic chromosomes uncoil and can no longer be distinguished individually with a light microscope. Uncoiling the chromosomes into a more extended form permits the enzymes that make RNA copies of DNA to gain access to the DNA molecule. Only by means of these RNA copies can the hereditary information be used to direct the synthesis of enzymes.

> A distinctive feature of eukaryotes is the organization of their DNA into chromosomes. Chromosomes can be condensed into compact structures when a eukaryotic cell divides, and later they can be unraveled so that the information the chromosomes contain can be used.

Ribosomes Construct Proteins

To make proteins, a cell employs a special structure, the **ribosome,** which reads the RNA copy of a DNA gene and uses the information it finds there to direct the synthesis of a protein (Figure 5-15). Ribosomes are made up of several special forms of RNA bound within a complex of several dozen different proteins. When a lot of proteins are being made, a cell needs a lot of ribosomes to handle the workload and therefore needs to be able to make large numbers of ribosomes quickly. To do this, many thousands of copies of the portion of the DNA encoding the RNA components of ribosomes, called **ribosomal RNA** (rRNA), cluster together on the chromosome. Copying RNA molecules from this cluster rapidly generates large numbers of the molecules needed to produce ribosomes.

The Nucleolus Is Not a Structure

At any given moment, many rRNA molecules dangle from the chromosome at the sites of these clusters of RNA genes. The proteins that will later form part of the ribosome complex bind to the dangling rRNA molecules. These areas where ribosomes are being assembled are easily visible within the nucleus as one or more dark-staining regions, called **nucleoli** (singular, *nucleolus;* Figure 5-16). The nucleoli can be seen under the light microscope even when the chromosomes are extended, unlike the rest of the chromosomes, which are visible only when condensed. Because they are visible in nondividing cells, early scientists thought that the nucleoli were distinct cellular

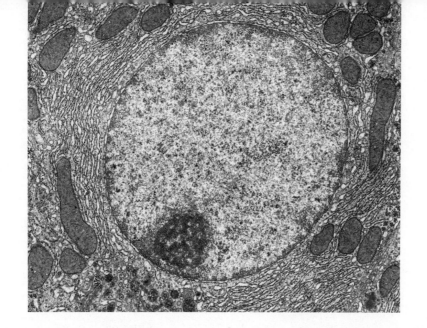

FIGURE 5-16
The interior of a rat liver cell, magnified about 6000 times. A single large nucleus occupies the center of the micrograph. The electron-dense area in the lower center of the nucleus is the nucleolus, the area where the major components of the ribosomes are produced. In the nucleoplasm around the nucleolus can be seen partly formed ribosomes.

structures. We now know that this is not true. Nucleoli are, in fact, aggregations of rRNA and some ribosomal proteins that are transported into the nucleus from the rough ER and accumulate at those regions on the chromosomes where active synthesis of rRNA is taking place.

THE GOLGI COMPLEX: THE DELIVERY SYSTEM OF THE CELL

At various locations within the cytoplasm, flattened stacks of membranes called **Golgi bodies** occur (Figure 5-17). These structures are named for Camillo Golgi, the nineteenth century Italian physician who first called attention to them. Animal cells contain 10 to 20 Golgi bodies each (they are especially abundant in glandular cells, which manufacture the substances that they secrete), whereas plant cells may contain several hundred. Protists typically contain a single or very few Golgi. Collectively the Golgi bodies are referred to as the **Golgi complex.**

Golgi bodies function in the collection, packaging, and distribution of molecules synthesized in the cell (Figure 5-18). The proteins and lipids manufactured on the rough and smooth ER membranes are transported through the channels of the endoplasmic reticulum, or as vesicles budded off from it, into the Golgi bodies. Within the Golgi bodies, many of these molecules are bound to polysaccharides, forming compound molecules. Among these are **glycoproteins,** which consist of a polysaccharide

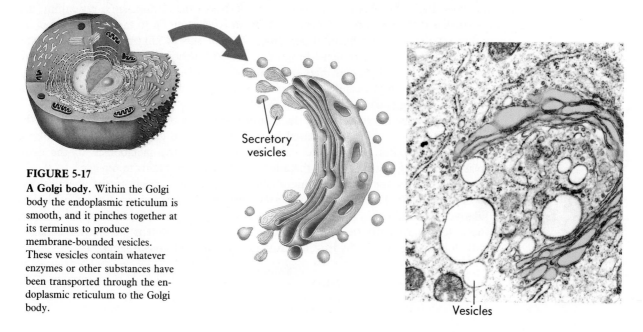

FIGURE 5-17
A Golgi body. Within the Golgi body the endoplasmic reticulum is smooth, and it pinches together at its terminus to produce membrane-bounded vesicles. These vesicles contain whatever enzymes or other substances have been transported through the endoplasmic reticulum to the Golgi body.

Secretory vesicles

Vesicles

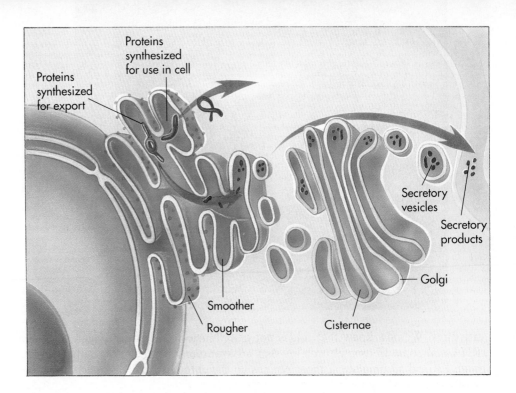

FIGURE 5-18
How proteins are secreted across membranes.

complexed to a protein, and **glycolipids,** consisting of a polysaccharide bound to a lipid. The newly formed glycoproteins and glycolipids collect at the ends of the membranous folds of the Golgi bodies; these folds are given the special name **cisternae** (Latin, "collecting vessels"). Intermittently in such regions, the membranes of the cisternae push together, pinching off small membrane-bound vesicles containing the glycoprotein and glycolipid molecules. These vesicles then move to other locations in the cell, distributing the newly synthesized molecules within them to their appropriate destinations.

> The Golgi complex is the delivery system of the eukaryotic cell. It collects, packages, modifies, and distributes molecules that are synthesized at one location within the cell and used at another.

PEROXISOMES: CHEMICAL SPECIALTY SHOPS

Eukaryotic cells contain a variety of enzyme-bearing, membrane-bound vesicles called **microbodies.** Microbodies, thought to be derived from endoplasmic reticulum, are organelles that carry a set of enzymes active in converting fat to carbohydrate and another set that destroys harmful peroxides. Similar sets of enzymes are found in the microbodies of plants, animals, fungi, and protists, although traditionally animal microbodies are called **peroxisomes** and plant microbodies are called **glyoxysomes.** The distribution of enzymes into microbodies is one of the principal ways in which eukaryotic cells organize their metabolism.

Many of the enzymes within microbodies are oxidative enzymes that catalyze the removal of electrons and associated hydrogens. If these enzymes were not isolated within microbodies, they would tend to short-circuit the metabolism of the cytoplasm, much of which involves the addition of hydrogen atoms to oxygen.

The name given to animal microbodies, peroxisomes, refers to the chemical hydrogen peroxide (H_2O_2) produced as a by-product of the activities of many of the oxidative enzymes within the microbody. Hydrogen peroxide is a dangerous by-product because it is violently reactive chemically. Microbodies contain the enzyme catalase, which breaks down the hydrogen peroxide into harmless constituents:

$$2H_2O_2 \xrightarrow{\text{[CATALASE]}} 2H_2O + O_2$$

HYDROGEN   WATER   OXYGEN
PEROXIDE   GAS

LYSOSOMES: RECYCLING CENTERS OF THE CELL

Lysosomes (Figure 5-19), another class of membrane-bound organelles, are about the same size as peroxisomes. They provide an impressive example of the metabolic compartmentalization achieved by the activity of the Golgi complex. They contain in a concentrated mix the digestive enzymes of the cell, enzymes that catalyze the rapid breakdown of proteins, nucleic acids, lipids, and carbohydrates. Lysosomes contain the enzymes responsible for the breakdown of macromolecules.

Lysosomes digest worn-out cellular components, making way for newly formed ones while recycling the materials locked up in the old ones. Cells can persist for a long time only if their components are constantly renewed. Otherwise the ravages of use and accident chip away at their metabolic capabilities and slowly degrade the cell's ability to survive. Cells age for the same reason that people do, because of a failure to renew themselves. Throughout the lives of eukaryotic cells, lysosomes break down the organelles and recycle their component proteins and other molecules at a fairly constant rate. As an example, mitochondria are replaced in some tissues every 10 days, with lysosomes digesting the old ones as new ones are produced.

Lysosomes that are actively engaged in digestive activities keep their battery of hydrolytic enzymes (those that catalyze the hydrolysis of molecules) fully active by maintaining a low internal pH—they pump protons into their interiors. Only at such acid pH values are the hydrolytic enzymes maximally active. Lysosomes that are not functioning actively do not maintain such an acid internal pH. A lysosome in such a "holding pattern" is called a **primary lysosome.** It is not until a primary lysosome fuses with a food vacuole or other organelle that its pH falls and the arsenal of hydrolytic enzymes is activated. When it becomes active, it is called a **secondary lysosome.**

It is not known what prevents lysosomes from digesting *themselves,* but the process requires energy, which is the reason why metabolically inactive eukaryotic cells die. Without a constant input of energy, the hydrolytic enzymes of primary lysosomes digest the lysosomal membrane from within. When these membranes disintegrate, the digestive enzymes of the lysosomes pour out into the cytoplasm of the cell and destroy it. Bacteria, in contrast, do not possess lysosomes and do not die when they are metabolically inactive. They are simply able to remain quiescent until altered conditions restore their metabolic activity, a property that greatly heightens their ability to persist under unfavorable environmental conditions. For us, the very process that repairs the ravages of time to our cells may also lead to their destruction. An absolute dependency on a constant supply of energy is the price that we pay for our long lives.

In addition to their role in eliminating organelles and other structures within cells, lysosomes also eliminate whole cells. Selective cell death is one of the principal mechanisms used by multicellular organisms in their achievement of complex patterns of development. When a tadpole develops into a frog, the cells of the tail are destroyed by the enzymes from lysosomes. Many cells in your brain die during development, as do the cells of the "tail" you had as an embryo. This directed cellular suicide is accomplished by the rupture of the lysosomes within the cells that are being eliminated. Once released from ruptured lysosomes, the hydrolytic enzymes proceed to digest the entire cell, a process that is irreversible and quickly leads to cell death.

> **Lysosomes are vesicles, formed by the Golgi complex, that contain digestive enzymes. The isolation of these enzymes in lysosomes protects the rest of the cell from inappropriate digestive activity.**

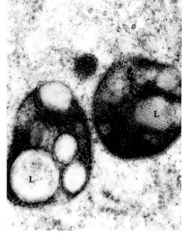

FIGURE 5-19

Lysosomes. Lysosomes are digestive organelles found in many eukaryotic cells. These lysosomes are within the cytoplasm of mouse kidney cells.

SOME DNA-CONTAINING ORGANELLES RESEMBLE BACTERIA
Mitochondria: The Cell's Chemical Furnace

Mitochondria are thought by biologists to have originated as symbiotic, **aerobic** (oxygen-requiring) bacteria. The theory of the symbiotic origin of mitochondria has had a controversial history, and a few biologists still do not accept it. The evidence supporting the theory is so extensive, however, that in this book we will treat it as established. We will present the evidence as we proceed.

According to this theory, the bacteria that became mitochondria were engulfed by

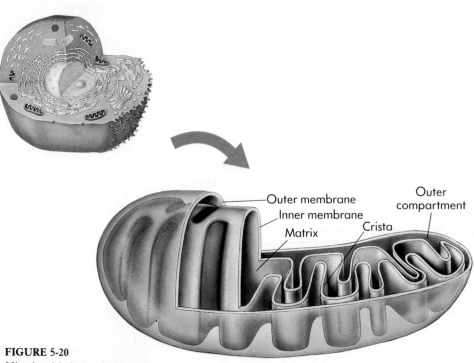

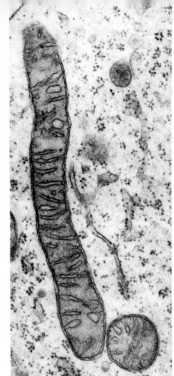

Outer membrane
Inner membrane
Matrix
Crista
Outer compartment

FIGURE 5-20

Mitochondria in longitudinal and cross section. These organelles are thought to have evolved from bacteria that long ago took up residence within the ancestors of present-day eukaryotes. Plant mitochondria tend to be shorter and thicker than the animal mitochondrion shown here.

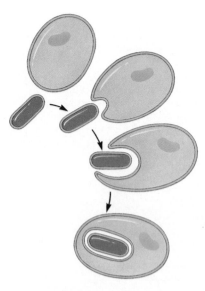

FIGURE 5-21

Endosymbiosis. This figure shows how a double membrane may have been created during the symbiotic origin of mitochondria.

ancestral eukaryotic cells early in their history. Before they had acquired these bacteria, the host cells were unable to carry out the metabolic reactions necessary for living in an atmosphere that contained increasing amounts of oxygen. Reactions requiring oxygen are collectively called **oxidative metabolism,** a process that the engulfed bacteria were able to carry out. The engulfed bacteria became the inner part of mitochondria.

Mitochondria (singular, *mitochondrion*) are tubular or sausage-shaped organelles 1 to 3 micrometers long (Figure 5-20); thus they are about the same size as most bacteria. Mitochondria are bounded by two membranes. The outer membrane is smooth and was apparently derived from the endoplasmic reticulum of the host cell (Figure 5-21), whereas the inner one is folded into numerous contiguous layers called **cristae.** These cristae resemble the folded membranes that occur in various groups of bacteria. The cristae partition the mitochondrion into two compartments, an inner **matrix** and an **outer compartment.** On the surfaces of the inner membrane, and also submerged within it, are the proteins that carry out oxidative metabolism.

During the billion and a half years in which mitochondria have existed as endosymbionts in eukaryotic cells, most of their genes have been transferred to the chromosomes of their host cells. For example, the genes that produce the enzymes involved with oxidative metabolism, the process characteristic of mitochondria, are located in the nucleus. Mitochondria still have their own genome, however, contained within a circular, closed molecule of DNA similar to that found in bacteria. On this mitochondrial DNA are located several genes that produce some of the proteins essential for the role of the mitochondria as the site of oxidative metabolism. All of these genes are copied into RNA within the mitochondrion and used there to make proteins. In this process the mitochondria use small RNA molecules and ribosomal components that are also encoded within the mitochondrial DNA. These ribosomes are smaller than those of eukaryotes in general, resembling bacterial ribosomes in size and structure.

A eukaryotic cell does not produce brand new mitochondria each time the cell itself divides. Instead, the ones it has divide in two, doubling the number, and these are partitioned between the new cells. Thus all mitochondria within a eukaryotic cell are produced by the division of existing mitochondria, just as all bacteria are produced from existing bacteria by cell division. Mitochondria divide by simple fission, splitting in two just as bacterial cells do, and apparently replicate and partition their circular DNA molecule in much the same way as do bacteria. Mitochondrial reproduction is not autonomous (self-governed), however, as is bacterial reproduction. Most of the components required for mitochondrial division are encoded as genes within the eukaryotic nucleus and translated into proteins by the cytoplasmic ribosomes of the cell itself. Mitochondrial replication is thus impossible without nuclear participation, and mitochondria cannot be grown in a cell-free culture.

Chloroplasts: Where Photosynthesis Takes Place

Symbiotic events similar to those postulated for the origin of mitochondria also seem to have been involved in the origin of chloroplasts, which are characteristic of photosynthetic eukaryotes (algae and plants). Chloroplasts, which apparently were derived from symbiotic photosynthetic bacteria, give these eukaryotes the ability to perform photosynthesis. The advantage that chloroplasts bring to the organisms that possess them is therefore obvious: these organisms can manufacture their own food.

Mitochondria apparently originated as endosymbiotic aerobic bacteria, whereas chloroplasts seem to have originated as endosymbiotic aerobic photosynthetic bacteria.

The chloroplast body is bounded, like the mitochondrion, by two membranes that resemble those of mitochondria (Figure 5-22) and that apparently were derived in a similar fashion. Chloroplasts are larger than mitochondria, and their inner membranes have a more complex organization. Instead of forming a single isolated compartment within the organelle, as do mitochondrial cristae, the internal chloroplast membranes lie in close association with one another; by fusing along their peripheries, two adjacent membranes form a disk-shaped closed compartment called a **thylakoid**. On the surface of the thylakoids are the light-capturing photosynthetic pigments, discussed in depth in Chapter 9. Chloroplasts contain stacks of such thylakoids, which, when viewed with a microscope, resemble stacks of coins. Each stack, called a **granum** (plural, *grana*), may contain from a few to several dozen thylakoids, and a chloroplast may contain a hundred or more grana.

The circular DNA molecule of chloroplasts is larger than that of mitochondria, but many of the genes that specify chloroplast components are located in the nucleus, so that the transfer of genetic material has been a part of their history also. Some components of the photosynthetic process are synthesized entirely within the chloroplast, which includes the specific RNA and protein components necessary to accomplish this. Photosynthetic cells typically contain from one to several hundred chloroplasts, depending on the organism involved or, in the case of multicellular photosynthetic organisms, the kind of cell.

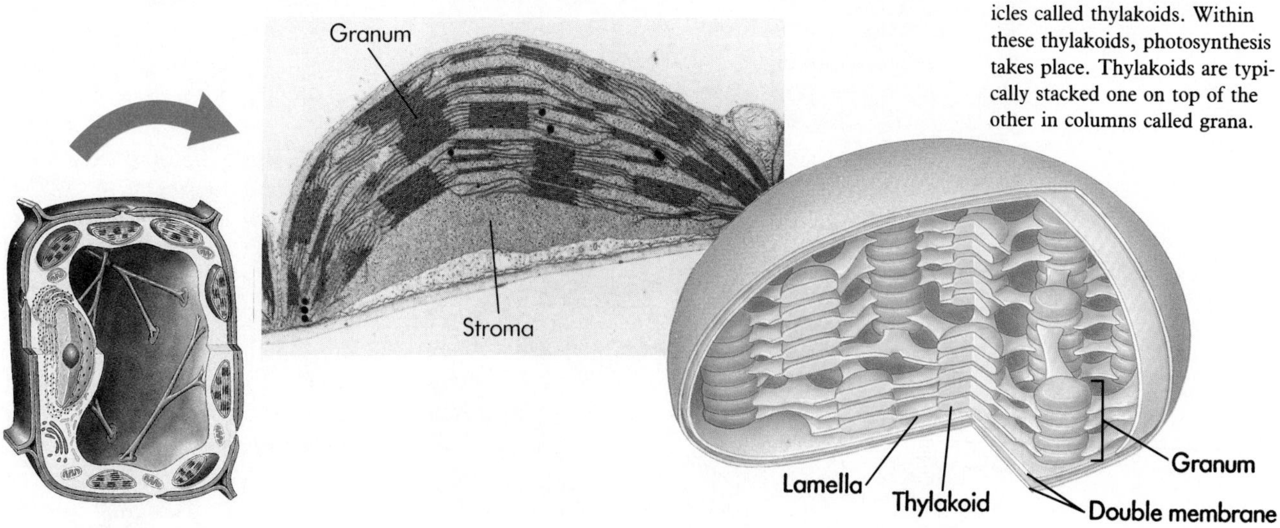

FIGURE 5-22
Chloroplast structure. The inner membrane of a chloroplast is fused to form stacks of closed vesicles called thylakoids. Within these thylakoids, photosynthesis takes place. Thylakoids are typically stacked one on top of the other in columns called grana.

Most green plants can synthesize chlorophyll only in the presence of light. In the dark the production of chlorophyll ceases, and in many plants most of the lamellae are reabsorbed. When reabsorption occurs, the chloroplast, largely devoid of internal membrane invaginations, is called a **leucoplast.** In the root cells and various storage cells of plants, leucoplasts may serve as sites of storage for starch. A leucoplast that stores starch is sometimes termed an **amyloplast.** Many plant pigments other than chlorophyll likewise occur in chloroplasts. A collective term for these different kinds of organelles, all derived from chloroplasts, is **plastid.** Like mitochondria, all plastids come from the division of existing plastids.

> Both mitochondria and chloroplasts have lost the bulk of their genomes to the host chromosomes but retain certain specific genes related to their functions. Neither kind of organelle can be maintained in a cell-free culture.

Centrioles: Producing the Cytoskeleton

Centrioles are organelles associated with the assembly and organization of **microtubules** in the cells of animals and most protists. Microtubules are long, hollow cylinders about 25 nanometers in diameter, composed of the protein tubulin. They influence cell shape, move the chromosomes in cell division, and provide the functional internal structure of cilia and flagella, as we discuss below.

Centrioles occur in pairs within the cytoplasm of eukaryotic cells, usually located at right angles to one another near the nuclear envelope (Figure 5-23). They resemble tubes and are among the most structurally complex organelles of the cell. In cells that contain flagella or cilia, each flagellum is anchored by a form of centriole called a **basal body.** Most animal and protist cells have both centrioles and basal bodies. The cells of plants and fungi lack centrioles and basal bodies, and their microtubules are organized by amorphous structures.

In many respects, centrioles resemble spirochaete bacteria. For this reason it has been hypothesized, notably by Lynn Margulis of the University of Massachusetts,

FIGURE 5-23

Centrioles. The centriole is composed of nine triplets of microtubules. The electron micrograph shows a pair of centrioles. The round shape is a centriole in cross-section. The rectangular shape is a centriole in longitudinal section.

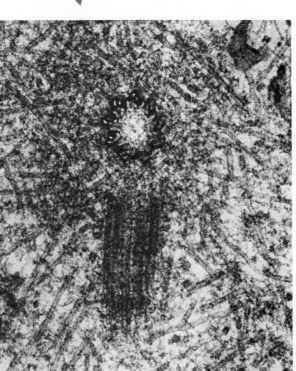

Microtubule triplet

PART II   BIOLOGY OF THE CELL

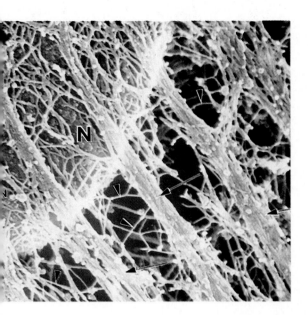

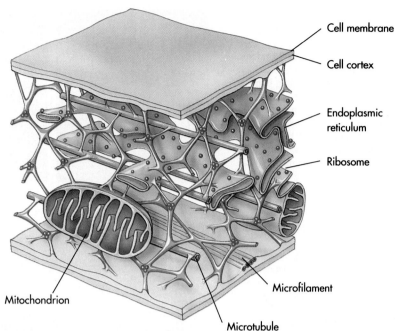

Cell membrane

Cell cortex

Endoplasmic reticulum

Ribosome

Microfilament

Microtubule

Mitochondrion

that centrioles actually originated as endosymbiotic spirochaetes. In 1989, David Luck and his colleagues at the Rockefeller University demonstrated that at least some centrioles contain DNA, which apparently is involved in the production of their structural proteins. This discovery appears to lend support to Lynn Margulis' theory, which is controversial and will doubtless become an area of even more active research in the future.

FIGURE 5-24

The cytoskeleton. In this diagrammatic cross section of a eukaryotic cell, the mitochondria, ribosomes, and endoplasmic reticulum are all supported by a fine network of filaments, through which pass microtubules linking various portions of the cell.

THE CYTOSKELETON: INTERIOR FRAMEWORK OF THE CELL

The cytoplasm of all eukaryotic cells is crisscrossed by a network of protein fibers that supports the shape of the cell and anchors organelles such as the nucleus to fixed locations (Figure 5-24). This network, called the **cytoskeleton,** cannot be seen with an ordinary microscope because the fibers are single chains of protein much too fine for microscopes to resolve. The fibers of the cytoskeleton are a dynamic system, constantly being formed and disassembled. Individual fibers form by **polymerization,** a process in which identical protein subunits are attracted to one another chemically and spontaneously assemble into long chains. Fibers are disassembled in the same way, by the removal of first one subunit, then another from one end of the chain.

Cells from plants and animals contain the following three different types of cytoskeleton fibers, each formed from a different kind of subunit (Figure 5-25):

1. *Actin filaments.* Actin filaments (also called microfilaments) are long protein fibers about 7 nanometers in diameter. Each fiber is composed of two chains of protein loosely twined around one another like two strands of pearls (Figure 5-25, *A*). Each "pearl" of a filament is a ball-shaped molecule of a protein called **actin,** the size of a small enzyme. Actin molecules left alone will spontaneously form these filaments, even in a test tube; a cell regulates the rate of their formation by means of other proteins that act as switches, turning on polymerization only when appropriate.

2. *Microtubules.* Microtubules are hollow tubes about 25 nanometers in diameter, each a ring of 13 protein *protofilaments.* Each a stack of tubulin molecules, the 13 protofilaments are arrayed side by side around a central core (Figure 5-25, *B*). The basic protein subunit of a microtubule is a molecule a little larger than actin, called **tubulin.** Like actin filaments, microtubules form spontaneously, but in a cell microtubules form only around specialized structures called **organizing centers,** which provide a base from which they can grow.

3. *Intermediate filaments.* The most durable element of the cytoskeleton is a system of tough protein fibers, each a rope of threadlike protein molecules

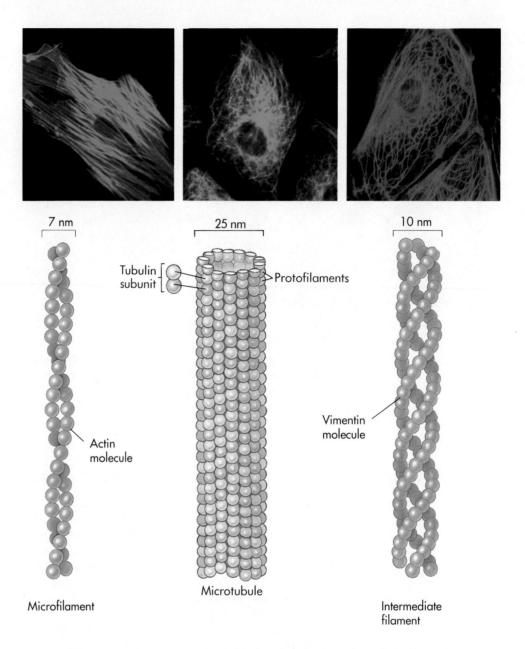

wrapped around one another like the strands of a cable (Figure 5-25, *C*). These fibers are characteristically 8 to 10 nanometers in diameter, intermediate in size between actin filaments and microtubules; this is why they are called intermediate filaments. Once formed, intermediate filaments are stable and do not usually break down. The most common basic protein subunit of an intermediate filament is called **vimentin,** although some cells employ other fibrous proteins instead. Skin cells, for example, form intermediate filaments from a protein called **keratin.** When skin cells die, the intermediate filaments of their cytoskeleton persist—hair and nails are formed in this way.

Both actin filaments and intermediate filaments are anchored to proteins embedded within the plasma membrane and provide the cell with mechanical support. Intermediate filaments act as intracellular tendons, preventing excessive stretching of cells, and actin filaments play a major role in determining the shape of cells. Because actin filaments can form and dissolve so readily, the shape of an animal cell can change quickly. If you look at the surface of an animal cell under a microscope, you will find it alive with motion, projections shooting outward from the surface and then retracting, only to shoot out elsewhere moments later (Figure 5-26).

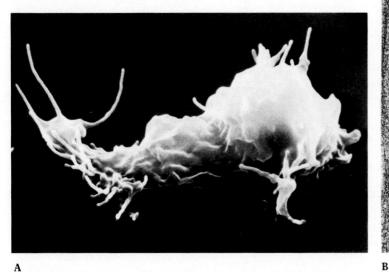

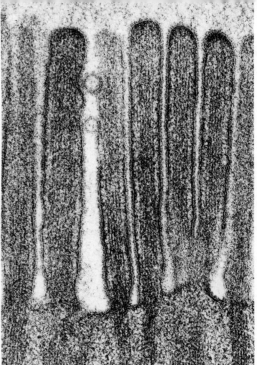

A

B

FIGURE 5-26
The surfaces of animal cells are in constant motion.
A This amoeba, a single-celled protist, is advancing toward you, its advancing edges extending projections outward. The moving edges have been said to resemble the ruffled edges of a skirt.
B Animal cells often produce projections. This figure shows finger-like projections, called microvilli, in the cells lining the human intestine. Microvilli can change their length quickly. They often appear to pop up almost instantaneously and to disappear just as quickly.

Not only is the cytoskeleton responsible for the cell's shape, but it also provides a scaffold on which the enzymes and other macromolecules are located in defined areas of the cytoplasm. Many of the enzymes involved in cell metabolism, for example, bind to actin filaments, as do ribosomes that carry out protein synthesis. By anchoring particular enzymes near one another, the cytoskeleton serves, like the endoplasmic reticulum, to organize the cell's activities (Table 5-1).

Actin filaments and microtubules also play important roles in cell movement. Pairs of microtubules are cross-linked at numerous positions by molecules of protein. The shifting positions of these cross-links determine the relative motion of the microtubules. As an example of the results of microtubule movement, when we study cell reproduction in Chapter 10, we will see that chromosomes move to opposite sides of dividing cells because they are attached to shortening microtubules, and that the cell pinches into two because a belt of actin filaments contracts like a purse-string. Your own muscle cells use actin filaments to contract their cytoskeletons. Indeed, all cell motion is tied to these same processes. The fluttering of an eyelash, the flight of an eagle, the awkward crawling of a baby, all depend on the movements of actin filaments in the cytoskeletons of muscle cells.

FLAGELLA: SLENDER WAVING THREADS

Flagella (singular, *flagellum*) are fine, long, threadlike organelles protruding from the surface of cells; they are used in locomotion and feeding. The flagella of bacteria are long protein fibers, which are so efficient that the bacteria that possess them can move about 20 cell diameters per second. Imagine trying to run 20 body lengths per second! The bacteria swim by rotating their flagella (Figure 5-27). One or more flagella trail behind each swimming bacterial cell, depending on the species of bacterium. Each has a motion like a propeller, caused by a complex rotary "motor" embedded within the cell wall and membrane. This rotary motion is virtually unique to bacteria; only a few eukaryotes have organs that truly rotate.

Eukaryotic cells have a completely different kind of flagellum, based on a kind of cable made up of microtubules. Their flagella consist of a circle of nine microtubule pairs surrounding two central ones; this arrangement is referred to as the **9 + 2 structure** (Figure 5-28). Completely different from bacterial flagella, this complex mi-

TABLE 5-1 EUKARYOTIC CELL STRUCTURES AND THEIR FUNCTIONS

STRUCTURE	DESCRIPTION	FUNCTION
STRUCTURAL ELEMENTS		
Cell wall	Outer layer of cellulose or chitin, or absent	Protection; support
Cytoskeleton	Network of protein filaments	Structural support; cell movement
Flagella (cilia)	Cellular extensions with 9 + 2 arrangement of pairs of microtubules	Motility or moving fluids over surfaces
ENDOMEMBRANE SYSTEM		
Plasma membrane	Lipid bilayer in which proteins are embedded	Regulates what passes into and out of cell; cell-to-cell recognition
Endoplasmic reticulum	Network of internal membranes	Forms compartments and vesicles
Microbodies	Vesicles derived from ER, containing oxidative and other enzymes	Isolate particular chemical activities from rest of cell
Nucleus	Spherical structure bounded by double membrane; contains chromosomes	Control center of cell; directs protein synthesis and cell reproduction
Golgi complex	Stacks of flattened vesicles	Packages proteins for export from cell; forms secretory vesicles
Lysosomes	Vesicles derived from Golgi complex that contain hydrolytic digestive enzymes	Digest worn-out mitochondria and cell debris; play role in cell death
ENERGY-PRODUCING ORGANELLES		
Mitochondria	Bacteria-like elements with inner membrane	Power plant of the cell; site of oxidative metabolism
Chloroplasts	Bacteria-like elements with vesicles containing chlorophyll	Site of photosynthesis in plant cells
ORGANELLES OF GENE EXPRESSION		
Chromosomes	Long threads of DNA that form a complex with protein	Contain hereditary information
Nucleolus	Site of chromosomes of rRNA synthesis	Assembles ribosomes
Ribosomes	Small, complex assemblies of protein and RNA, often bound to ER	Sites of protein synthesis

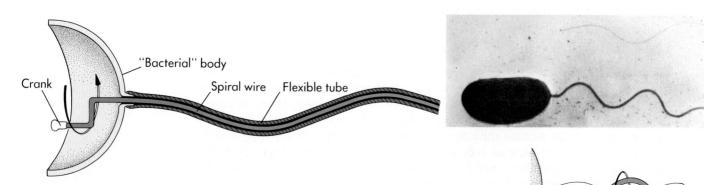

FIGURE 5-27

Bacteria swim by rotating their flagella. The photograph is of *Vibrio cholerae*, the microbe that causes the serious disease cholera. The unsheathed core visible at the top of the photograph is composed of a single crystal of the protein **flagellin**. In intact flagella, this core is surrounded by a flexible sheath. Imagine that you are standing inside the *Vibrio* cell, turning the flagellum like a crank. You would create a spiral wave that travels down the flagellum, just as if you were turning a wire within a flexible tube. The bacterium employs this kind of rotary motion when it swims.

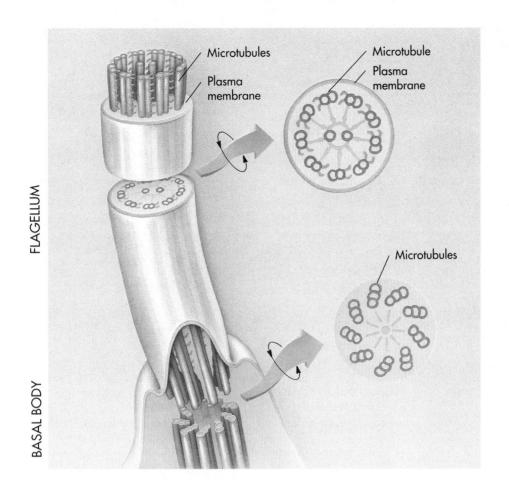

FIGURE 5-28

Figure 5-28

Structure of a flagellum. A eukaryotic flagellum springs directly from a basal body, is composed of nine pairs of microtubules, and has two microtubules in its core connected to the outer ring of paired microtubules by radial spokes.

crotubular apparatus evolved early in the history of the eukaryotes. Although the cells of many multicellular and some unicellular eukaryotes today no longer exhibit 9 + 2 flagella and are nonmotile, the 9 + 2 arrangement of microtubules can still be found within them, the apparatus having been put to other uses. Like mitochondria, the 9 + 2 flagella structure appears to be a fundamental component of the eukaryotic endowment.

Eukaryotic flagella are not difficult to see. If you were to examine a motile, single-celled eukaryote under a microscope, you would immediately notice the flagella protruding from its cell surface. The constant undulating motion of these flagella, caused by the movement of pairs of microtubules past one another, propels the cell through its environment and requires a constant input of energy into the flagellum. When examined carefully, each flagellum proves to be an outward projection of the interior of the cell, containing cytoplasm and enclosed by the cell membrane. The microtubules of the flagella are derived from a basal body, situated just below the point at which the flagellum protrudes from the surface of the cell membrane. The arrangement of flagella differs greatly in different eukaryotes. If there are many, short flagella organized in dense rows (see Figure 5-1), they are called **cilia** (singular, *cilium*), but cilia do not differ at all from flagella in structure.

In many multicellular organisms, cilia carry out tasks far removed from their original function of propelling cells through water. In several kinds of human tissues, for example, the beating of rows of cilia move water over the tissue surface. The 9 + 2 arrangement of microtubules occurs in the sensory hairs of the human ear, where the bending of these hairs by pressure constitutes the initial sensory input of hearing. Throughout the evolution of eukaryotes, 9 + 2 flagella have been an element of central importance.

TABLE 5-2   A COMPARISON OF BACTERIAL, ANIMAL, AND PLANT CELLS

	BACTERIUM	ANIMAL	PLANT
EXTERIOR STRUCTURE			
Cell wall	Present (protein-polysaccharide)	Absent	Present (cellulose)
Cell membrane	Present	Present	Present
Flagella	May be present (1 strand)	May be present	Absent except in sperm of a few species
INTERIOR STRUCTURE			
ER	Absent	Present	Usually present
Microtubules	Absent	Present	Present
Centrioles	Absent	Present	Absent
Golgi bodies	Absent	Present	Present
ORGANELLES			
Nucleus	Absent	Present	Present
Mitochondria	Absent	Present	Present
Chloroplasts	Absent	Absent	Present
Chromosomes	A single circle of naked DNA	Multiple units, DNA associated with protein	Multiple units, DNA associated with protein
Ribosomes	Present	Present	Present
Lyosomes	Absent	Usually present	Equivalent structures called "spherosomes"
Vacuoles	Absent	Absent or small	Usually a large single vacuole in mature cell

*The structures found in the cells of the other two kingdoms of eukaryotic organisms, Protista and Fungi, are diverse, but share the basic features of eukaryotic cells; most have cell walls and centrioles, and chloroplasts are present only in the photosynthetic protists known as algae.

AN OVERVIEW OF CELL STRUCTURE

The structure of eukaryotic cells is much more complicated and diverse than the structure of bacterial cells (Table 5-2). The most distinctive difference between these two fundamentally different cell types is the extensive subdivision of the interior of eukaryotic cells by membranes. The most visible of these membrane-bound compartments, the nucleus, gives eukaryotes their name. There are no equivalent membrane-bound compartments within prokaryotic cells. The membranes of some photosynthetic bacteria are extensively folded inwardly, but they do not isolate any one portion of the cell from any other portion. A molecule can travel unimpeded from any location in a bacterial cell to any other location.

In the following chapters we will consider the consequences of these structural differences, and how they influence the metabolism and biochemistry of eukaryotes. We shall see that the metabolic processes that go on within eukaryotic cells differ from those of bacteria and that the differences, like the structural ones discussed in this chapter, are substantial.

1. The cell is the smallest unit of life. It is composed of a nuclear region containing the hereditary apparatus that is enclosed within a larger volume called the cytoplasm that performs the day-to-day functions of the cell under the supervision of the nuclear region. In all cells the cytoplasm is bounded by a lipid membrane.

2. Bacteria, which have prokaryotic cell structure, do not have membrane-bound organelles within their cells. Their DNA is mostly included in a single, closed, circular molecule.

3. The endoplasmic reticulum is a dynamic series of internal endomembranes that subdivides the interior of eukaryotic cells into separate compartments. It is the most distinctive feature of eukaryotic cells and one of the most important, since it enables the cells to carry out different metabolic functions separately from one another.

4. The nucleus is the largest of the compartments created by the endomembranes of the cell. Within it, the cell's DNA is separated from the rest of the cytoplasm.

5. A third key organelle associated with endomembranes is the Golgi complex, which serves as a cellular "express package service," packaging molecules within special membrane vesicles and transporting them to various locations in the cell.

6. One class of vesicle created by the Golgi complex consists of the lysosomes, vesicles containing high concentrations of digestive enzymes. A cell constantly expends energy to prevent digestive damage to these vesicles, and when this preventive activity ceases, the vesicles soon burst, digesting and killing the cell.

7. Not all of the eukaryotic cell's internal organelles are created by internal membrane systems. Others appear to have been derived from symbiotic bacteria. By far the most widespread of these are mitochondria and chloroplasts.

8. Many, but not all, of the genes originally present in mitochondrial and chloroplast DNA seem to have been transferred to, or had their functions taken over by, the DNA in the chromosomes of the host cell. Both classes of organelles, however, have retained the genes necessary to create their particular distinctive structures. Without these structures, mitochondria would be unable to function in aerobic respiration, and chloroplasts in photosynthesis.

9. Eukaryotic cells possess a cytoskeleton of protein fibers that support the shape of the cell and anchor its organelles.

10. Many eukaryotic cells possess flagella consisting of a 9 + 2 arrangement of microtubules; the sliding of these microtubules past one another bends the flagellum.

REVIEW QUESTIONS

1. What are the three principles of the cell theory?

2. What are the limitations controlling the size that cells can attain?

3. What is the composition of the plasma membrane? What classes of proteins are embedded in it?

4. What is endoplasmic reticulum? What is its function? How does rough ER differ from smooth ER?

5. What are microbodies? What types of molecules do they contain?

6. How is the nuclear membrane different from that of other organelles of the endomembrane system? What types of molecules move through the nuclear pores?

7. What is the function of the nucleolus? Where is it located? Is it a permanent structure?

8. What is the function of the Golgi bodies? With what other organelle are they closely associated?

9. What two organelles are involved in cellular energetics? Which of the two are found in plants? in animals? What is the probable origin of these organelles?

10. Describe the basic structure of a mitochondrion, from the outside inward. What function does it serve?

11. What unique metabolic activity occurs in chloroplasts? What is the functional unit of the chloroplast?

12. What is the composition of the cytoskeleton? Which of the components are stable and which are changeable?

13. What is the function of the cytoskeleton?

14. What is meant by the term "9 + 2 flagella"? Are they characteristic of prokaryotes or eukaryotes? How do cilia compare with flagella?

THOUGHT QUESTIONS

1. Mitochondria are thought to be the evolutionary descendants of living cells, probably of symbiotic aerobic bacteria. Are mitochondria alive?

2. Some cells are much larger than others. What would you expect the relationship to be between cell size and level of cell activity?

3. Trace a synthesizing protein through the endomembrane system, culminating in release of the protein outside the cell.

FOR FURTHER READING

ALBERTS, B., D. BRAY, J. LEWIS, M. RAFF, K. ROBERTS, and J. WATSON: *Molecular Biology of the Cell*, ed. 2, Garland Press, New York, 1989. A very comprehensive text on modern cell biology.

ALLEN, R.: "The Microtubule as an Intracellular Engine," *Scientific American*, February 1987, pages 42-49. The use of video-enhancement techniques provides our first look at microtubules in action within cells.

BERSHADSKY, A., and J. VASILIEV: *Cytoskeleton*, Plenum Press, New York, 1988. A broad and up-to-date synthesis of our knowledge of the structure and function of the cell cytoskeleton.

DARNELL, J., H. LODISH, and D. BALTIMORE: *Molecular Cell Biology*, ed. 2, Scientific American Books, New York, 1990. One of the best available cell biology texts today.

HALL, J., Z. RAMANIS, and D. LUCK: "Basal Body/Centriolar DNA," *Cell*, vol 59, 1989, pages 121-132. The exciting discovery that centrioles contain DNA, a key prediction of the Margulis endosymbiosis theory.

HYNES, R.: "Fibronectins," *Scientific American*, June 1986, pages 42-51. A detailed description of the adhesive proteins that hold cells in position and guide their migration.

MARGULIS, L.: *Symbiosis in Cell Evolution*, W.H. Freeman Company, San Francisco, 1980. A broad-ranging exposition of the theory that mitochondria, chloroplasts, and other organelles were acquired by eukaryotes from symbiotes, written by the chief proponent of the idea.

McDERMOTT, J.: "A Biologist Whose Heresy Redraws Earth's Tree of Life," *Smithsonian*, August 1989, pages 71-81. Engaging account of the scientific approach taken by Lynn Margulis, leading contemporary advocate of the endosymbiotic theory of the origin of some organelles.

MURRAY, M.: "Life on the Move," *Discover*, March 1991, pages 72-76. The Living cell is a world of motion, powered by delicate protein filaments and tiny protein motors.

ROTHMAN, J.: "The Compartmental Organization of the Golgi Apparatus," *Scientific American*, February 1987, pages 74-89. A description of current ideas about how Golgi bodies are functionally divided into three zones, each specialized in a different way.

STOSSEL, T.: "How Cells Crawl," *American Scientist*, vol 78, 1990, pages 408-423. With the discovery that the cellular "motor" contains muscle proteins, biologists are beginning to understand cell motility in molecular terms.

SYMMONS, M., A. PRESCOTT, and R. WARN: "The Shifting Scaffolds of the Cell," *New Scientist*, February 1989, pages 44-47. New findings show that the microtubules of which our cells are built are constantly appearing and disappearing.

WEBER, K., and OSBORN, M.: "The Molecules of the Cell Matrix," *Scientific American*, October 1985, pages 110-120. An up-to-date description of the many different molecules that make up the cell cytoskeleton.

6

Membranes

OVERVIEW

Every cell is enveloped within a liquid layer of lipid, a fluid shell that isolates its interior. Anchored within this liquid are a host of proteins that move about on the surface like ships on a lake. These proteins provide passages across the lipid layer for molecules and information, and they are the cell's only connection with the outer world. The combination of lipid shell and embedded proteins is called a biological membrane.

FOR REVIEW

Here are some important terms and concepts that you will encounter in this chapter. If you are not familiar with them, you should review them before proceeding.

Polar nature of water (Chapter 2)

Hydrogen bonds (Chapter 2)

Structure of fat molecules (Chapter 3)

Hydrophobic (Chapter 3)

Types of membrane proteins (Chapter 5)

FIGURE 6-1

A foraminiferan. This beautiful foraminiferan carries out the same essential life processes that you do, and, as in your body, all these activities require that substances enter and leave the cell.

Among a cell's most important activities are its transactions with the environment, a give and take that never ceases. Cells are constantly feasting on food they encounter, ingesting molecules and sometimes entire cells. They dump their wastes, together with many other kinds of molecules, back into the environment. Cells continuously garner information about the world around them, responding to a host of chemical clues and often passing on messages to other cells. This incessant interplay with the environment is a fundamental characteristic of all cells. Without it, life could not persist (Figure 6-1).

Imagine, now, that you were to coat a living cell with plastic, giving it a rock-hard, impermeable shell. All of the cell's transactions with the environment would stop. No molecules could pass in or out. Nor could the cell learn anything about the molecules around it. The cell might as well be a rock. Life in any meaningful sense would cease—unless you allowed for doors and windows in the shell.

That is, in fact, what living cells do. Every cell is encased within a lipid membrane, an impermeable shell through which few water-soluble molecules and little information (data about its surroundings) can pass—but the shell contains doors and windows made of protein. Molecules pass in and out of a cell through these passageways, and information passes in and out through the windows. A cell interacts with the world through a delicate skin of protein molecules embedded in a thin sheet of lipid. We call this assembly of lipid and protein a **plasmalemma** or **plasma membrane.** The structure and function of this membrane are the subject of this chapter.

THE LIPID FOUNDATION OF MEMBRANES

The membranes that encase all living cells are sheets only a few molecules thick; it would take more than 10,000 of these sheets piled on one another to equal the thickness of this sheet of paper. These thin sheets are not simple in structure, however. Rather, they are made up of diverse collections of proteins enmeshed in a lipid framework.

Phospholipids

The lipid layer that forms the foundation of cell membranes is composed of molecules called **phospholipids** (Figure 6-2). Like the fat molecules you studied in Chapter 3, a phospholipid has a backbone derived from a three-carbon molecule called glycerol, with long chains of carbon atoms called fatty acids attached to this backbone. A fat molecule has three such chains, one attached to each carbon of the backbone; because these chains are nonpolar (do not form hydrogen bonds with water), the fat molecule

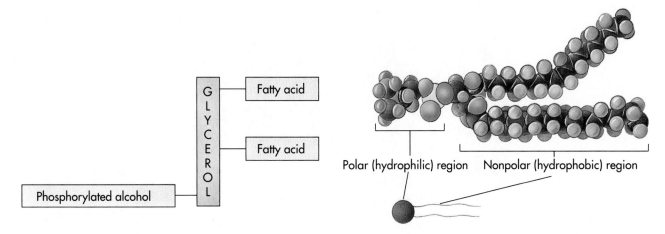

Polar (hydrophilic) region   Nonpolar (hydrophobic) region

FIGURE 6-2
Phospholipid structure. A phospholipid is a composite molecule similar to a triglyceride, except in this case only two fatty acids are bound to the glycerol backbone, the third position being occupied by another kind of molecule called a phosphorylated alcohol. Because the phosphorylated alcohol usually extends from one end of the molecule, and the two fatty acid chains from the other, phospholipids are often diagrammed as a polar head with two nonpolar hydrophobic tails.

is not soluble in water. A phospholipid, by contrast, has only two such chains attached to its backbone (see Figure 6-2). The third position is occupied instead by a highly polar organic alcohol that readily forms hydrogen bonds with water. Because this alcohol is attached by a phosphate group, the molecule is called a *phospho*lipid.

One end of a phospholipid molecule is therefore strongly nonpolar (water insoluble), and the other end is extremely polar (water soluble). The two nonpolar fatty acids extend in one direction, roughly parallel to each other, and the polar alcohol group points in the other direction. Because of this structure, phospholipids are often diagrammed as a (polar) head with two dangling (nonpolar) tails.

Phospholipids Form Bilayer Sheets

Imagine what happens when a collection of phospholipid molecules is placed in water. The long nonpolar tails of phospholipid molecules are pushed away by the water molecules that surround them, shouldered aside as the water molecules seek partners able to form hydrogen bonds. Water molecules always tend to form the maximum number of such bonds; the long nonpolar chains that cannot form hydrogen bonds get in the way, like too many chaperones at a party. The best way to rescue the party is to put all the chaperones together in a separate room, and that is what water molecules do—they shove all the long nonpolar tails of the lipid molecules together, out of the way. The polar heads of the phospholipids are "welcomed," however, because they form good hydrogen bonds with water. What happens is that every phospholipid molecule orients so that its polar head faces water and its nonpolar tails face away. By forming *two* layers with the tails facing each other, no tails are ever in contact with water! The structure that results is called a **lipid bilayer** (Figure 6-3). Lipid bilayers form spontaneously, driven by the forceful way in which water tends to form hydrogen bonds.

> **The basic foundation of biological membranes is a lipid bilayer, which forms spontaneously. In such a layer, the nonpolar hydrophobic tails of phospholipid molecules point inward, forming a nonpolar zone in the interior of the bilayer.**

Lipid bilayer sheets of this sort are the foundation of all biological membranes. Because the interior of the bilayer is completely nonpolar, it repels any water-soluble molecules that attempt to pass through it, just as a layer of oil stops the passage of a drop of water (which is why ducks do not get wet). This barrier to the passage of water-soluble molecules is the key biological property of the lipid bilayer. It means that a cell, if fully encased within a pure lipid bilayer, would be completely impermeable to water-soluble molecules such as sugars, polar amino acids, and proteins. No cell, however, is so imprisoned. In addition to the phospholipid molecules that make up the lipid bilayer, the membranes of every cell also contain proteins, which extend across the lipid bilayer, providing passageways across the membrane.

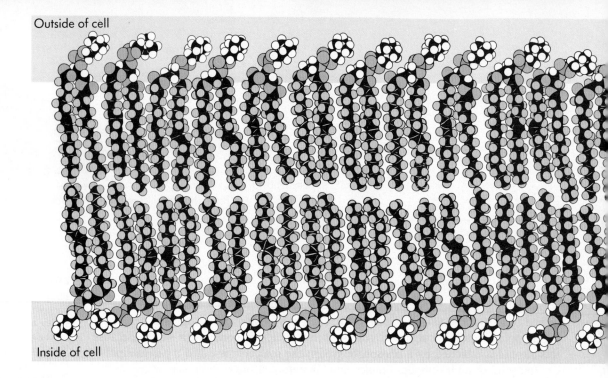

Outside of cell

Inside of cell

FIGURE 6-3

A phospholipid bilayer. The diagram above illustrates how the long nonpolar tails of the phospholipids orient toward one another. Because some of the tails contain double bonds, which introduce kinks in their shape, the tails do not align perfectly and the membrane is "fluid"—individual phospholipid molecules can move from one place to another in the membrane.

The Lipid Bilayer is a Fluid

A lipid bilayer is stable because the hunger of water for hydrogen bonding never stops. But while water continually urges phospholipid molecules into this orientation, it is indifferent to where individual phospholipid molecules are located. Water forms just as many hydrogen bonds when a particular phospholipid molecule is located here as there. As a result, individual lipid molecules are free to move about within the membrane. Because the individual molecules are free to move about, the lipid bilayer is not a solid like a rubber balloon, but rather a liquid like the "shell" of a soap bubble. The bilayer itself is a fluid, with the viscosity of olive oil. Just as the surface tension holds the soap bubble together, even though it is made of a liquid, so the hydrogen bonding of water holds the membrane together.

Some membranes are more fluid than others, however. The tails of individual phospholipid molecules are attracted to one another when they line up close together. This causes the membrane to stiffen, because aligned molecules must pull apart from one another before they can move about in the membrane. The more alignment, the less fluid the membrane is. Some phospholipids have tails that do not align well because they contain one or more double CC bonds, which introduce kinks in the tail. Membranes containing these types of phospholipids are more fluid than those that lack them. Sometimes membranes contain other short lipids like cholesterol, which prevent the phospholipid tails from coming into contact with one another and stiffening the membrane.

ARCHITECTURE OF THE PLASMA MEMBRANE

A eukaryotic cell contains many membranes. They are not all identical, but they share the same fundamental architecture. Cell membranes are assembled from four components:

1. *Lipid bilayer foundation.* Every cell membrane has a phospholipid bilayer as its basic foundation. The other components of the membrane are enmeshed within the bilayer, which provides a flexible matrix and, at the same time, imposes a barrier to permeability (Figure 6-4).
2. *Transmembrane proteins.* A major component of every membrane is a collection of proteins (Figure 6-5) that float within the lipid bilayer like boats on a lake. These proteins provide channels into the cell through which molecules and in-

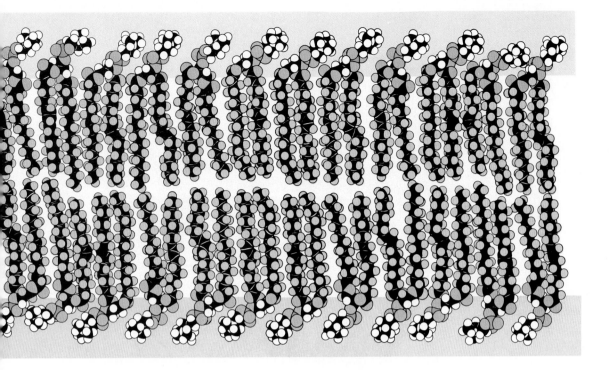

formation pass. Membrane proteins are not fixed in position; instead they move about. Some membranes are crowded with proteins, side by side, just as some lakes are so crowded with boats that you can hardly see the water. In other membranes, the proteins are more sparsely distributed.

3. *Network of supporting fibers.* Membranes are structurally supported by proteins that reinforce the membrane's shape. This is why a red blood cell is shaped like a doughnut rather than being irregular—the plasma membrane is held in that shape by a scaffold of protein on its inner surface that tethers proteins within the membrane to actin filaments in the cell's cytoskeleton. Membranes use networks of other proteins to control the lateral movements of some key membrane proteins, anchoring them to specific sites so that they do not simply

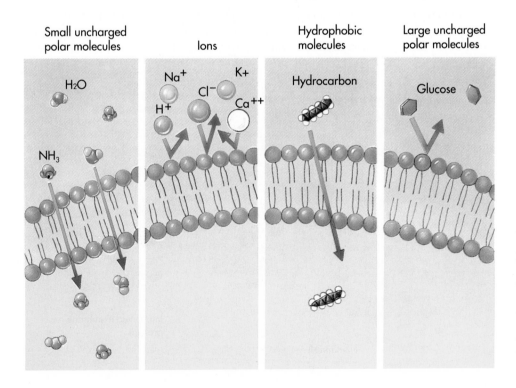

FIGURE 6-4

Some molecules pass across bilayer membranes, others do not. Lipid bilayer membranes are permeable to oxygen, to lipids, and to small uncharged molecules even if they are polar (such as water); they are not permeable to large molecules if they are polar, or to anything that is charged, such as ions or proteins.

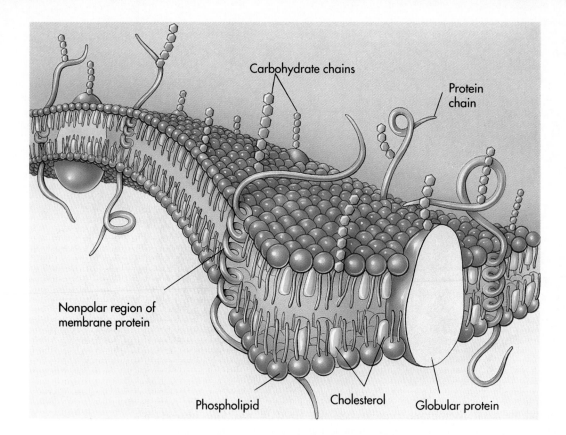

Carbohydrate chains

Protein chain

Nonpolar region of membrane protein

Phospholipid

Cholesterol

Globular protein

FIGURE 6-5

Cells have complex surfaces. A variety of proteins protrude through the lipid membrane of animal cells, and nonpolar regions of the proteins tether them to the membrane's nonpolar interior. The three principal classes of membrane protein are channels, receptors, and cell surface markers. Carbohydrate chains (strings of sugar molecules) are often bound to these proteins, and to lipids in the membrane itself as well. These chains serve as distinctive identification tags, unique to particular types of cells.

drift away. Unanchored proteins have been observed to move as much as 10 micrometers in 1 minute (Figure 6-6).

4. *Exterior proteins and glycolipids.* Membrane sections are assembled in the endoplasmic reticulum, transferred to the Golgi complex, and then transported to the cell plasma membrane. The Golgi complex adds chains of sugar molecules to membrane proteins and lipids, creating a thicket of protein and carbohydrate that extends from the cell surface. These act as cell identity markers, and different cell types exhibit various kinds of protein and carbohydrate chains on their surfaces.

A membrane, then, is a sheet of lipid and protein that is supported by other proteins and to which carbohydrates are attached (Table 6-1). The key functional proteins act as passages through the membrane, extending all the way across the bilayer.

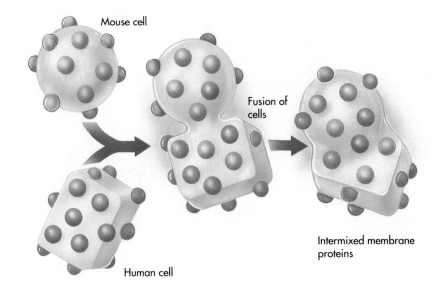

Mouse cell

Fusion of cells

Intermixed membrane proteins

Human cell

FIGURE 6-6

Proteins in membranes move around. Protein movement within membranes can be easily demonstrated by labeling the proteins of a mouse cell with fluorescent antibodies, and then fusing that cell with a human cell. Within 1 hour, the labeled and unlabeled proteins are intermixed throughout the hybrid cell's membranes.

TABLE 6-1 COMPONENTS OF THE CELL MEMBRANE

COMPONENT	COMPOSITION	FUNCTION	HOW IT WORKS	EXAMPLE
Lipid foundation	Phospholipid bilayer	Permeability barrier matrix for proteins	Water-soluble molecules excluded from nonpolar interior of bilayer	Bilayer of cell is impermeable to water-soluble molecules
Transmembrane proteins	Carriers	Transport of molecules across membrane against gradient	Carrier "flip-flops"	Glycophorin channel for sugar transport
	Channels	Passive transport of molecules across membrane	Create a tunnel that acts as a passage	Photoreceptor
	Receptors	Transmit information into cell	Bind to cell surface portion of protein; alter portion within cell, inducing activity	Peptide hormones, neurotransmitters bind to specific receptors
Cell surface markers	Glycoprotein	"Self"-recognition	Shape of protein/carbohydrate chain is characteristic of individual	Major histocompatibility complex protein recognized by immune system
	Glycolipid	Tissue recognition	Shape of carbohydrate chain is characteristic of tissue	A, B, O blood group markers
Interior protein network	Spectrin	Determines shape of cell	Forms supporting scaffold beneath membrane, anchored to both membrane and cytoskeleton	Red blood cell
	Clathrins	Anchor certain proteins to specific sites	Form network above membrane to which proteins are anchored	Localization of low-density lipoprotein receptor within coated pits

How do these transmembrane proteins manage to span the membrane, rather than just floating on the surface, in the way that a drop of water floats on oil? Many surface proteins are anchored into the cell membrane by a glycolipid* that extends down into the lipid bilayer. Other surface proteins actually traverse the lipid bilayer. The part of the protein that extends through the lipid bilayer is specially constructed, a spiral helix of nonpolar amino acids (Figure 6-7). Because water responds to nonpolar amino acids much as it does to nonpolar lipid chains, the nonpolar spiral is held within the interior of the lipid bilayer by the strong tendency of water to avoid contact with these amino acids. Although the polar ends of the protein protrude from both sides of the membrane, the protein itself is locked into the membrane by its nonpolar helical segment.

A single nonpolar helical segment is adequate to anchor a protein into the membrane. Many receptor proteins, proteins that act as windows for information, are of this sort. The portion of a receptor protein that extends out from the cell surface is able to bind to specific hormones or other molecules when the cell encounters them; the binding induces changes at the other end of the protein, in the cell's interior. In this way, information on the cell's surface is translated into action within the cell.

Other proteins thread their way back and forth through the membrane many times, creating a passage through the bilayer like the hole in a doughnut. These pro-

*The name of the glycolipid is phosphatidylinositol (PTI). Its function in mediating the protein composition of the cell surface is one of the most active areas of current membrane research.

FIGURE 6-7

How proteins are anchored to membranes. Many membrane proteins are anchored within the lipid bilayer by nonpolar segments. In all cases studied to date, these segments have proved to be helical in secondary structure. Two general classes of membrane protein occur: proteins that traverse the membrane only once (receptors and some channels are of this sort), and proteins that traverse the membrane many times, creating a hollow "pipe" through the bilayer, as illustrated here. Many channels are of this sort.

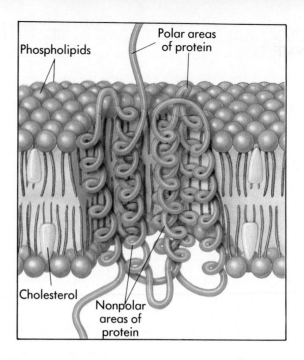

FIGURE 6-8

Functions of plasma membrane proteins.

teins are carriers and channels. They are locked into pipe-like shapes because they possess several nonpolar helical segments, each of which is driven by water into the interior of the bilayer. For example, one of the key transmembrane proteins that carries out photosynthesis, is composed of two different subunits, each of which has five nonpolar helical segments. The protein thus passes across the membrane 10 times, creating a crescent-shaped channel through the membrane (see Figure 6-7). All water-soluble molecules that enter or leave the cell are carried by carriers or pass through channels, each of which, like a guarded entrance, admits only certain ones.

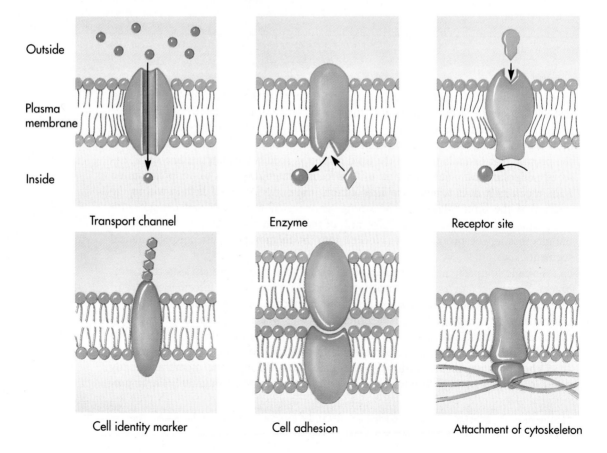

Outside

Plasma membrane

Inside

Transport channel   Enzyme   Receptor site

Cell identity marker   Cell adhesion   Attachment of cytoskeleton

HOW A CELL'S MEMBRANES REGULATE INTERACTIONS WITH ITS ENVIRONMENT

The plasma membrane, then, is a complex assembly of proteins floating like loosely anchored ships on a lipid sea. This is probably not the sort of cell surface you or I would have designed. It is not strong, and its surface components are not precisely located. It is, however, a design of enormous flexibility, permitting a broad range of interactions with the environment. A cell membrane is like a rack that can hold many different tools. A list of all the different kinds of proteins in one cell's plasma membrane would run to many pages. With these tools, the cell can interact with its environment in many ways (Figure 6-8), admitting a particular molecule here, sensing the presence of a hormonal signal there—like the busy floor of a factory, a cell's membrane proteins are in a constant state of high activity. If the proteins of a membrane made noise when they worked, as a factory's tools do, a cell would be in a state of constant uproar.

Among the many ways in which a cell's membranes regulate its interactions with the environment, we shall mention only six:

1. *Passage of water.* Membranes are freely permeable to water, but the spontaneous movement of water into and out of cells sometimes presents problems.
2. *Passage of bulk material.* Cells sometimes engulf large hunks of other cells, or gulp liquids.
3. *Selective transport of molecules.* Membranes are very specific about which molecules they allow to enter or leave the cell.
4. *Reception of information.* Membranes can sense chemical messages with exquisite sensitivity.
5. *Expression of cell identity.* Membranes carry name tags that tell other cells who they are.
6. *Physical connection with other cells.* When forming tissues, membranes make special connections with each other.

THE PASSAGE OF WATER INTO AND OUT OF CELLS

Molecules dissolved in liquid are in constant motion, moving about randomly. This random motion causes a net movement of molecules toward zones where the concentration of the molecules is lower, a process called **diffusion** (Figure 6-9). Driven by random motion, molecules always "explore" the space around them, diffusing out until they fill it uniformly. A simple experiment will demonstrate this. Take a jar, fill it with ink to the brim, cap it, place it at the bottom of a bucket of water, and then remove the cap. The ink molecules will slowly diffuse out until there is a uniform concentration in the bucket and the jar.

> **Diffusion is the net movement of molecules to regions of lower concentration as a result of random spontaneous molecular motions. Diffusion tends to distribute molecules uniformly.**

The cytoplasm of a cell consists of molecules such as sugars, amino acids, and ions dissolved in water. The mixture of these molecules and water is called a **solution.** Water, the most common of the molecules in the mixture, is called the **solvent,** and the other kinds of molecules dissolved in water are called **solutes.**

Because of diffusion, both solvent and solute molecules in a cell will move from regions where their concentration is greater to a region where their concentration is less. When two regions are separated by a membrane, what happens depends on whether or not the molecule can pass freely through that membrane; most kinds of solutes that occur in cells cannot do so. Sugars, amino acids, and other solutes are water soluble and not lipid soluble, and so are imprisoned within the cell: they are unable to cross the lipid bilayer of the membrane. Water molecules, in contrast, can pass through slight imperfections in the sheet of lipid molecules and so diffuse across the membrane into the cell. Water molecules stream into the cell across the mem-

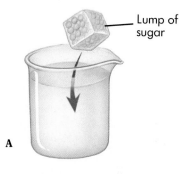

Lump of sugar

A

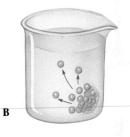

B

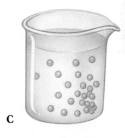

C

D

FIGURE 6-9
Diffusion. If a lump of sugar is dropped into a beaker of water, its molecules dissolve (**A**) and diffuse (**B** and **C**). Eventually, diffusion results in an even distribution of sugar molecules throughout the water (**D**).

FIGURE 6-10

An experiment demonstrating osmosis.

A The end of a tube containing a 3% salt solution is closed by stretching a differentially permeable membrane across its face that will pass water molecules but not salt molecules.

B When this tube is immersed in a beaker of distilled water, the salt cannot cross the membrane; however, water can. The added water causes the salt solution to rise in the tube.

C Water will continue to enter the tube from the beaker until the weight of the column of water in the tube exerts a downward force equal to the force drawing water molecules upward into the tube. This force is referred to as osmotic pressure.

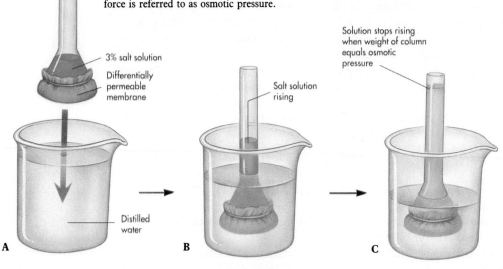

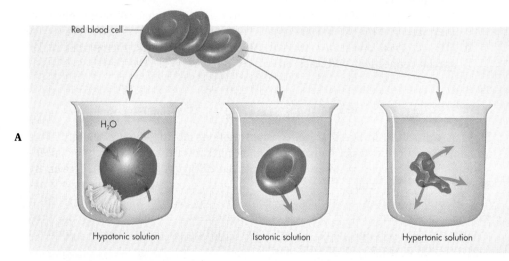

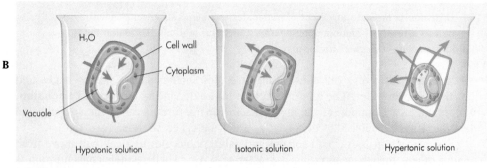

FIGURE 6-11

Osmosis.

A Animal cells. When the outer solution is hypotonic with respect to the cell, water will move in; when it is hypertonic, water will move out.

B Plant cells. In plant cells, the large central vacuole contains a high concentration of solutes, so water tends to diffuse inward causing the cells to swell outward against their rigid cell walls. However, if a plant cell is immersed in a high-solute (hypertonic) solution, water will leave the cell, causing the cytoplasm to shrink and pull in from the cell wall.

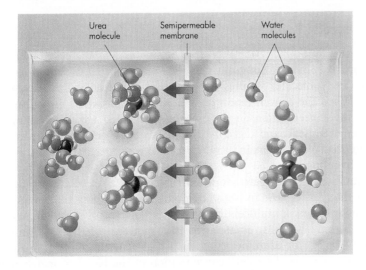

Urea molecule   Semipermeable membrane   Water molecules

FIGURE 6-12
How solutes create osmotic pressure. Charged or polar molecules are soluble in water because they form hydrogen bonds with water molecules clustered around them. When such a polar solute (illustrated here with urea) is added to one side of a membrane, the water molecules that gather around each urea molecule are no longer free to diffuse across the membrane—in effect, the polar solute has reduced the number of free water molecules on that side of the membrane. Because the hypotonic side of the membrane (on right, with less solute) has more unbound water molecules than the hypertonic side with more solute, water moves by diffusion from the right to the left.

brane, thus diluting the high concentration of solutes within the cell so that it matches more and more closely the lower concentration in the outside solution. This form of net water movement into or out of a cell is called **osmosis** (Figure 6-10).

> **Osmosis is the diffusion of water across a membrane that permits the free passage of water but not that of one or more solutes.**

The fluid content of a cell suspended in pure water is said to be **hypertonic** (Greek *hyper*, more than) with respect to its surrounding solution, because it has a higher concentration of solutes than the water does (Figure 6-11). The surrounding solution, which has a lower concentration of solutes than the cell, is said to be **hypotonic** (Greek *hypo*, less than) with respect to the cell. A cell with the same concentration of solutes as its environment is said to be **isotonic** (Greek *iso*, the same) with the environment.

Intuitively you might think that as new water molecules diffuse inward, the pressure of the cytoplasm pushing out against the cell membrane, called **hydrostatic pressure,** would build up, and this is indeed what happens. As water molecules continue to diffuse inward toward the area of lower concentration of unbound water molecules, the hydrostatic water pressure within the cell increases. We refer to pressure of this kind as **osmotic pressure** (Figure 6-12). Osmotic pressure is defined as the force that must be applied to stop the osmotic movement of water across a membrane. Because osmotic pressure opposes the movement of water inward, such diffusion will not continue indefinitely. The cell will eventually reach an equilibrium, at which point the osmotic force driving water inward is counterbalanced exactly by the hydrostatic pressure driving water out. In practice, the hydrostatic pressure at equilibrium is typically so high that an unsupported cell membrane cannot withstand it, and such an unsupported cell suspended in water will burst like an overinflated balloon. In contrast, cells whose membranes are surrounded by cell walls can withstand high internal hydrostatic pressures (Figure 6-13).

> **Within the closed volume of a cell that is hypertonic with respect to its surroundings, the movement of water inward, which tends to lower the relative concentration difference of water, will also increase the internal hydrostatic pressure. The net movement of water stops when an equilibrium condition is reached or the cell bursts.**

Stoma

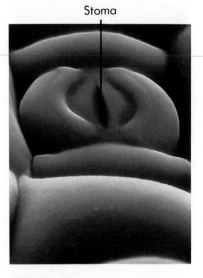

FIGURE 6-13
Cells use osmotic pressure to control water loss. The cells surrounding this opening in a *Tradescantia* leaf are swollen by osmosis. The opening is called a stoma, and the swollen cells are called guard cells. When relaxed, these guard cells rest against one another, closing the opening. When swollen by osmosis from the surrounding cells, as they are here, the cells assume a rigid bowed shape, creating an opening between them.

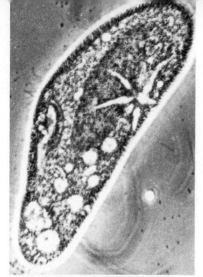

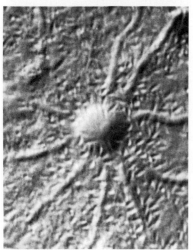

FIGURE 6-14
How a *Paramecium* removes water. Protists such as this *Paramecium* contain star-shaped contractile vacuoles. The long arms act as collecting ducts into which water diffuses by osmosis. The water is channeled to a central vacuole near the cell surface, which somewhat resembles a balloon. By contracting microfilaments surrounding this vacuole, the cell expels its water to the exterior.

Avoiding Osmotic Pressure

Early eukaryotes found many solutions to the osmotic dilemma posed by being hypertonic to their environment. Some, living in the sea, adjusted their internal concentration of nonwater molecules (solutes) to be the same as that of the seawater in which they swam. When the concentration of solutes on the inside of the cell membrane is the same as the concentration on the outside, the cell is said to be isotonic with respect to its environment, and there is no tendency for a net flow of water into or out of the cell to occur. The cell is in osmotic balance with its environment. The problem is solved by avoiding it.

Many multicellular animals adopt similar solutions, circulating a fluid through their bodies that bathes cells in isotonic liquid. By controlling the composition of its circulating body fluids, a multicellular organism can control the solute concentration of the fluid bathing its cells, adjusting it to match that of the cells' interiors. The blood in your body, for example, contains a high concentration of a protein called albumin, which elevates the solute concentration of the blood to match that of your tissue so that osmosis does not occur.

Water Removal. When early eukaryotes colonized fresh water, they had to face the osmotic dilemma again. Eukaryotes that had adjusted their internal environment to match that of seawater found that in fresh water they were again hypertonic with respect to their environment. The solution adopted by many single-celled eukaryotes was **extrusion.** The water that moved into the cell was dumped back out by one means or another.

In *Paramecium*, for example, the cell contains one or more special organelles known as **contractile vacuoles.** In the photographs of *Paramecium* in Figure 6-14 the contractile vacuole resembles a spider, with many legs that collect osmotic water from various parts of the cell's interior and transport it to the central body, which is a vacuole near the cell surface. This vacuole is bounded by microfilaments and possesses a small pore that opens to the outside of the cell. By rhythmic contractions, it pumps the accumulating osmotic water out through the pore. Because contraction of microfilaments involves the use of ATP, the cell must constantly expend energy to keep the contractile vacuole functioning and so survive in a hypotonic environment.

Plant Cell Walls. Unlike animals, plants do not circulate an isotonic solution. Most plant cells are hypertonic with respect to their immediate environment, with a high concentration of solutes in the central vacuole. The resulting osmotic pressure presses the cytoplasm firmly against the interior of the plant cell wall, making the individual plant cells rigid. The internal pressure in these cells is referred to as **turgor pressure.** The shapes of plants that lack wood and of the newer, soft portions of trees and shrubs depend to a large extent on the turgor of their individual cells. Because plants depend on the cell rigidity imparted by turgor pressure to maintain their shape, they wilt when they lack sufficient water.

These indirect solutions are the best that evolution has achieved, each a compromise. They all satisfy one important criterion—they all work.

BULK PASSAGE INTO THE CELL

The lipid nature of biological membranes raises a second problem for growing cells. The metabolites required by cells as food are for the most part polar molecules, which will not pass across the hydrophobic barrier interposed by a lipid bilayer. How then are organisms able to get food molecules into their cells? Particularly among single-celled eukaryotes, the dynamic cytoskeleton is employed to extend the cell membrane outward toward food particles, such as bacteria. The membrane encircles and engulfs a food particle. Its edges eventually meet on the other side of the particle where, because of the fluid nature of the lipid bilayer, the membranes fuse together, forming a vesicle around the fluid. This process is called **generalized endocytosis** (Figure 6-15, *A* to *D*). Generalized endocytosis involves the incorporation of a portion of the exterior medium into the cytoplasm of the cell by capturing it within a vesicle.

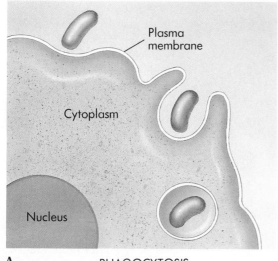

Plasma membrane

Cytoplasm

Nucleus

A   PHAGOCYTOSIS

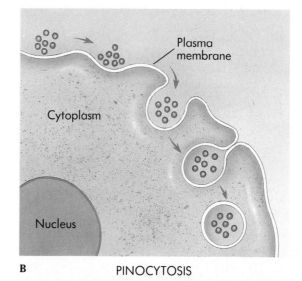

Plasma membrane

Cytoplasm

Nucleus

B   PINOCYTOSIS

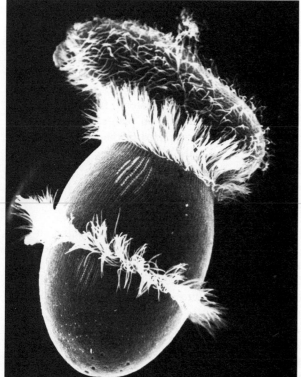

C

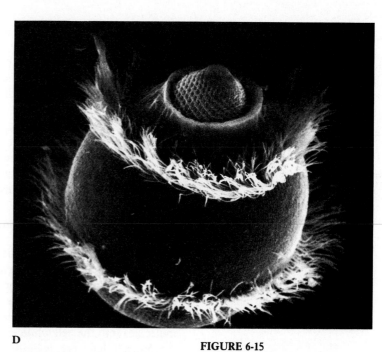

D

FIGURE 6-15
Endocytosis. Both phagocytosis
(**A**) and pinocytosis (**B**) are forms
of endocytosis. The large egg-
shaped protist *Didinium nasutum*
(**C**) has just begun eating the
smaller protist *Paramecium;* (**D**)
its meal is practically over.

If the material brought into the cell is particulate, such as an organism (Figure
6-15, *C* and *D*) or some other fragment of organic matter, that particular kind of endocytosis is called **phagocytosis** (Greek *phagein,* to eat + *cytos,* cell). If the material
brought into the cell is liquid and contains dissolved molecules, the endocytosis is referred to as **pinocytosis** (Greek *pinein,* to drink; Figure 6-15, *B*). Pinocytosis is common among the cells of multicellular animals. Human egg cells, for example, are
"nursed" by surrounding cells, which secrete nutrients that the maturing egg cell
takes up by pinocytosis.

It is important to understand that endocytosis does not by itself internalize things
into the cell—the material taken in is still separated from the cytoplasm by a membrane, as can be clearly seen in Figure 6-15, *A* & *B*.

Virtually all eukaryotic cells are constantly carrying out generalized endocytosis,

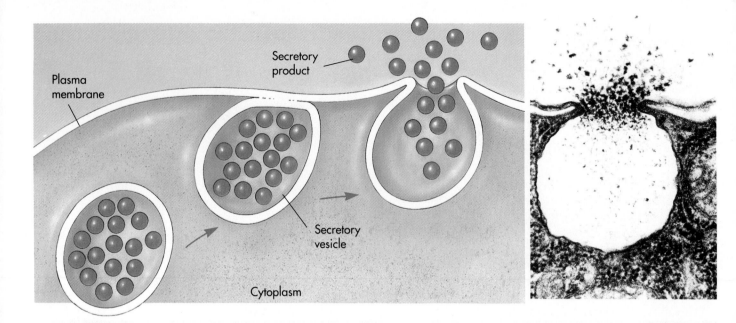

FIGURE 6-16
Exocytosis. Proteins and other molecules are secreted from cells in small pockets called vesicles, whose membranes fuse with the cell membrane, releasing their contents to the cell surface.

(Labels in figure: Plasma membrane; Secretory product; Secretory vesicle; Cytoplasm)

trapping particles and extracellular fluid in vesicles and ingesting it. Rates of generalized endocytosis vary from one cell type to another, but can be surprisingly large. Some types of white blood cells, for example, ingest 25% of their cell volume each hour!

The reverse of endocytosis is **exocytosis,** the extrusion of material from a cell by discharging it from vesicles at the cell surface (Figure 6-16). In plants, vesicle discharge constitutes a major means of exporting the materials used in the construction of the cell wall through the plasma membrane. In animals, many cells are specialized for secretion using the mechanism of exocytosis.

SELECTIVE TRANSPORT OF MOLECULES

The generalized endocytosis discussed above is an awkward means of governing entrance to the interior of the cell. It is expensive from an energy standpoint, since a considerable amount of membrane movement is involved. More important, generalized endocytosis is not selective. Particularly when the process is pinocytotic, it is difficult for the cell to discriminate between different solutes, admitting some kinds of molecules into the cell while excluding others.

No cell, however, is limited to endocytosis. As we have seen, cell membranes are studded with proteins, which act as **carriers** or provide **channels** across the membrane. Because a given carrier or channel will transport only certain kinds of molecules, the cell membrane is **selectively permeable**—that is, it is permeable to some molecules and not to others.

> Selective permeability allows the passage across a membrane of some solutes but not others. Selective permeability of cell membranes is the result of specific protein passages extending across the membrane; some molecules can pass through a specific kind of carrier or channel, whereas others cannot do so.

The Importance of Selective Permeability

The most important property of any cell, in general, is that the cell constitutes an isolated compartment within which certain molecules can be concentrated and brought together in particular combinations. This essential isolation depends on allowing molecules to enter the cell selectively. Your home is private in the same sense—unusable if no one, including you, can enter, but not private if everyone in the neighborhood is free to wander through at all times. The solution adopted by cells is the same one that most homeowners adopt: there are doors with keys, and only those possessing the proper keys can enter or leave. The protein carriers and channels through cell membranes are the doors to cells. These doors are not open to any molecule that presents itself; only particular molecules can pass through a given kind of door. A cell is able to control the entry and exit of many kinds of molecules by possessing many different kinds of doors. These channels through its membrane are among the most important functional features of any cell.

Facilitated Diffusion

Channels in the cell membrane are highly selective open passageways across the membrane, facilitating *only* the passage of specific molecules—but in either direction. An example is provided by the channel of vertebrate red blood cell membranes that transports negatively charged ions, or **anions**. As you will learn in Chapter 50, this channel plays a key role in the oxygen-transporting function of these cells. The anion channels of the red blood cell membrane readily pass chloride ions (Cl^-) or carbonate ions (HCO_3^-) across the red blood cell membrane. We can easily demonstrate that the ions are moving by diffusion through the channels in the red blood cell membrane: if there are more Cl^- ions within the cell, we will see that the net movement is outward, whereas if there are more Cl^- ions outside the cell, then the net movement will be into the cell. Since the ion movement is always toward the direction of lower ion concentration, the transport process is one of diffusion. A channel can never be used to go against a gradient.

Are these channels simply holes in the membrane, somehow specific for these anions? Not at all. If we repeat our hypothetical experiment, progressively increasing the concentration of Cl^- ions outside the cell above that inside the cell, the rate of movement of Cl^- ions into the cell increases only up to a certain point, after which it levels off and will proceed no faster despite increases in the concentration of exterior Cl^- ions. Whatever is passing the Cl^- ions across the membrane, there must be a limited number of them. In fact, the reason that the diffusion rate will increase no further is that the Cl^- ions are being transported across the membrane by channels, and all available channels are in use: we have saturated the capacity of the transport system. The transport of these anions, then, is a diffusion process facilitated by specific channels. Transport processes of this kind are called **facilitated diffusion** (Figure 6-17).

> Facilitated diffusion is the transport of molecules across a membrane by specific channels toward the direction of lowest concentration.

Other molecules, particularly sugars, also pass in and out of cells by facilitated diffusion (Figure 6-18). The particular kind of channel responsible for the passage of sugars passes through the plasma membrane only once; it is anchored by a central coil of nonpolar amino acids.

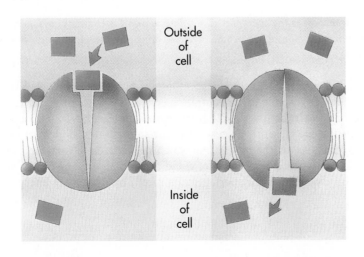

FIGURE 6-17
Facilitated diffusion is a carrier-mediated transport process.

Outside of cell

Inside of cell

FIGURE 6-18

The molecular structure of the glucose sugar transport carrier is known in considerable detail. The protein's 492 amino acids form a folded chain that traverses the lipid membrane in 12 segments. Amino acids with charged groups are less stable in lipid, and are located outside the membrane. Researchers think that the pore consists of five helical segments with glucose-binding sides (red) facing inward. A conformational shift in the protein transports glucose through the pore by shifting the position at the glucose-binding sites.

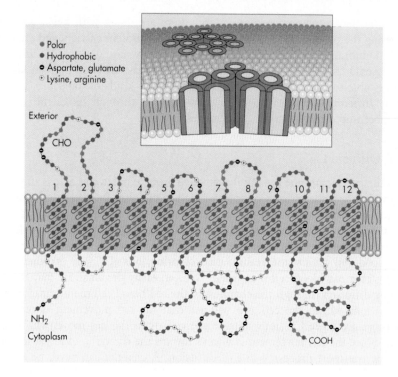

Facilitated diffusion provides the cell with a ready means of preventing the buildup of unwanted molecules within the cell or of gleaning from the external medium molecules that are present there in high concentration. Facilitated diffusion has three essential characteristics:

1. It is *specific*, with only certain molecules able to traverse a given channel.
2. It is *passive*, the direction of net movement determined by the relative concentrations of the transported molecule inside and outside the membrane.
3. It may become *saturated* if all of the protein channels are in use.

Active Transport

The cell admits many molecules across its membrane that are maintained within the cell at a concentration different from that of the surrounding medium. In all such cases, the cell employs highly selective protein carriers within the membrane that permit active transport against a concentration gradient. Because the carriers require energy to operate, the cell must expend energy to maintain the concentration difference. The kind of transport that requires the expenditure of energy is called **active transport.**

> **Active transport is the transport of a solute across a membrane, independent of solute concentration, using protein carriers driven by the expenditure of chemical energy.**

Active transport is one of the most important functions of any cell. It is by active transport that a cell is able to concentrate metabolites. Without it, for example, your red blood cells would be unable to harvest glucose molecules—a major source of energy—from the blood, since glucose concentration is often higher in the red blood cells already than it is in the blood from which it is extracted. Imagine how difficult it would be to survive as a beggar if you could only obtain money from those who had less than you did! Active transport permits a cell, by expending energy, to take up

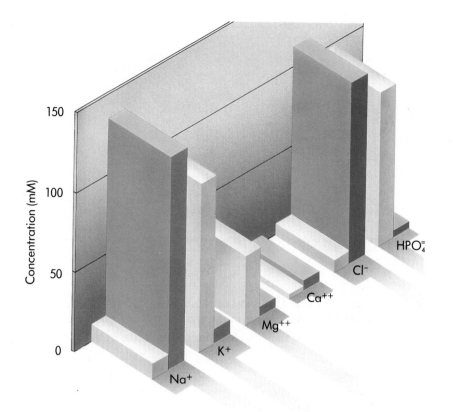

FIGURE 6-19

The ionic composition of animal cells. Many ions are present at far higher concentrations within cells (*gold*), than in the fluid surrounding the cells (*blue*). These differences, maintained by active transport of ions, are responsible for many of the most important properties of cells.

additional molecules of a substance that is already present in its cytoplasm in concentrations higher than in the cell's environment.

There are many molecules that a cell takes up or eliminates against a concentration gradient. Some, like sugars and amino acids, are simple metabolites Others are ions (Figure 6-19). Still others are the nucleotides that the cell uses to synthesize DNA. These various kinds of molecules enter and leave cells by way of a wide variety of selectively permeable transport channels. Some of them are permeable to one or a few sugars, others to a certain size of amino acid, and still others to a specific ion or nucleotide. You might suspect that active transport occurs at each of these channels, but you would be wrong. In animals there is *one* major active transport channel in cell membranes that transports sodium and potassium ions; all the others work by tying their activity to this all-important channel. The many channels in the membrane that the cell uses to concentrate metabolites and ions are called **coupled channels.** We will first discuss the sodium-potassium channel and then these coupled channels.

The Sodium-Potassium Pump.

More than one third of all the energy expended by a cell that is not actively dividing is used to actively transport sodium (Na^+) and potassium (K^+) ions. The remarkable carrier that transports these two ions across the cell membrane is referred to as the **sodium-potassium pump.** Most animal cells have a low internal concentration of Na^+ ions, relative to their surroundings, and a high internal concentration of K^+ ions. They maintain these concentration differences by actively pumping Na^+ ions out of the cell and K^+ ions in.

The transport of these ions is carried out by a highly specific transmembrane protein carrier, the sodium-potassium pump (Figure 6-20). This ATP-driven active transport carrier plays an important role in vertebrate biology. It is responsible for establishing the charge difference between nerve cell interior and exterior on which all nerve conduction depends. In addition, many other transport processes are driven by the pumping of Na^+ and K^+ ions by this carrier. The pump works by a series of conformational changes in the transmembrane protein:

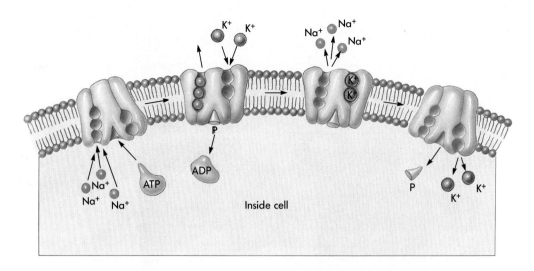

FIGURE 6-20
The sodium-potassium pump.

Step 1. Three molecules of Na^+ bind to the interior ends of the two *a* subunits, producing a conformational change in the transmembrane protein complex.

Step 2. The new shape of the complex binds a molecule of ATP and cleaves it into ADP + P_i. ADP is released, but the terminal phosphate group remains bound to the complex.

Step 3. The binding of the phosphate group to the complex induces a second conformational change in the complex. This change results in the passage of the three Na^+ ions across the membrane, where they are now positioned facing the exterior. In this new conformation the complex has a low affinity for the Na^+ ions, which dissociate and diffuse away. The new conformation does, however, have a high affinity for K^+ ions and binds two of them as soon as it is free of the Na^+ ions.

Step 4. The binding of the K^+ ions leads to another conformational change in the transmembrane complex, this time resulting in the dissociation of the bound phosphate group. Freed of the phosphate group, the complex reverts to its original conformation, with the two K^+ ions exposed on the interior side of the membrane. The new conformation has a low affinity for K^+ ions, which dissociate from it and diffuse away into the interior of the cell. This conformation, which is the one we started with, does have a high affinity for Na^+ ions. When these ions bind, they initiate another pump cycle. In every cycle, three Na^+ leave and two K^+ enter.

The unique characteristic of the sodium-potassium pump is that it is an active transport process, transporting Na^+ and K^+ ions from areas of low concentration to areas of high concentration. This transport into a zone of higher concentration is just the opposite of that which occurs spontaneously in diffusion; it is achieved only by the constant expenditure of metabolic energy. The energy used in the process is obtained from a molecule called **adenosine triphosphate** (ATP), the functioning of which will be explained in Chapter 8. Some membranes contain large numbers of $Na^+ - K^+$ carriers, whereas others have few. The changes in protein shape that go on within an individual carrier are rapid. Each carrier is capable of transporting as many as 300 Na^+ ions per second, when working at top speed.

Coupled Channels. The accumulation of many amino acids and sugars by cells is also driven against a concentration gradient: the molecules are harvested from a surrounding medium in which their concentration is much lower than it is inside the cell. The active transport of these molecules across the cell membrane takes place by coupling them with Na^+ ions, which pass simultaneously through a channel by facilitated diffusion. This process is driven by the low concentration of Na^+ ions within the cell, maintained by active transport of these ions outward by the sodium-potas-

Cystic fibrosis is a fatal disease of human beings in which affected individuals secrete a thick mucus that clogs the airways of the lungs. These same secretions block the ducts of the pancreas and liver so that the few patients who do not die of lung disease die of liver failure. Cystic fibrosis is usually thought of as a children's disease, because few affected individuals live long enough to become adults. There is no known cure.

Cystic fibrosis is a hereditary disease, resulting from a defect in a single gene passed down from parent to child. It is the most common fatal genetic disease of Caucasians. One in 20 individuals possesses at least one copy of the defective gene. Most carriers are not afflicted with the disease; only those children who inherit two copies of the defective gene, one from each parent, succumb to cystic fibrosis—about 1 in 1800 Caucasian children.

Cystic fibrosis has proved to be difficult to study. Many organs are affected, and until recently it was impossible to identify the nature of the defective gene responsible for the disease. In 1985 the first clear clue was obtained. An investigator, Paul Quinton, seized on a commonly observed characteristic of cystic fibrosis patients, that their sweat is abnormally salty, and performed the following experiment. He isolated a sweat duct from a small piece of skin and placed it in a solution of salt (NaCl) that was three times as concentrated as the NaCl inside the duct. He then monitored the movement of ions. Diffusion tends to drive both the Na^+ and Cl^- ions into the duct because of the higher outer ion concentrations. In skin isolated from normal individuals, Na^+ ions indeed entered the duct, transported by the sodium-potassium pump; Cl^- ions followed, passing through a passive channel. Both ions crossed the membrane easily. In skin isolated from individuals with cystic fibrosis, the sodium-potassium pump transported Na^+ ions into the ducts, but no Cl^- ions entered. The passive chloride channels were not functioning in these individuals.

We now know that cystic fibrosis results from a defective channel within plasma membranes, one that transports Cl^- ions across the membranes of normal individuals but not across those of affected persons. The genetic defect is the result of an alteration in a protein that both serves as a Cl^- ion channel and acts to regulate the activity of the channel. The defective gene was isolated in 1987, and its position on a particular human chromosome was pinpointed in 1989. Now scientists are searching for ways to transfer working copies of the gene into cystic fibrosis patients. Initial attempts in 1992, using common cold viruses as carriers and animals as patients, have been successful—for the first time, there is hope for a cure.

sium pump. The special transmembrane protein of the **coupled channel** acts as a carrier of Na^+ only when the sugar or amino acid that is transported at the same time is also bound to the protein's exterior surface (Figure 6-21). As a result, the facilitated diffusion inward of the Na^+ ion also results in the importation into the cell of the sugar or amino acid. In this process, the sugar or amino acid is literally dragged along osmotically, because the Na^+ ions are so much less concentrated within the cell than outside. Thus the transport of the sugar or amino acid inward to an area of its higher concentration—against its concentration gradient—occurs as a direct consequence of the activity of the sodium-potassium pump.

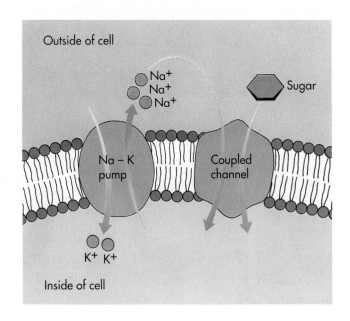

FIGURE 6-21

A coupled channel. The sodium-potassium pump keeps the Na^+ ion concentration higher outside the cell than inside. There is thus a strong tendency for Na^+ ions to diffuse back in through the coupled channel, a co-transport carrier, but their passage requires the simultaneous transport of a sugar molecule as well. The diffusion gradient driving Na^+ entry is so great that sugar molecules are pulled in two, even against a sugar concentration gradient.

The Proton Pump.

The sodium-potassium pump is of central importance because it drives the uptake of so many different molecules. There is a second carrier of equal importance in the life of the cell. It is called the **proton pump.**

The proton pump involves two special transmembrane proteins: the first pumps protons (H^+ ions) out of the cell (or into an organelle), using energy derived from energy-rich molecules or from photosynthesis to power the active transport. This creates a proton gradient, in which the concentration of protons is higher outside the membrane (or inside the organelle) than inside. As a result, diffusion acts as a force to drive protons back across the membrane toward a zone of lower proton concentration. The plasma membrane, however, is impermeable to protons, so the only way protons can diffuse back into the cell is through a special channel, which couples the transport of protons to the production of ATP, the energy-storing molecule we mentioned above. This process occurs within the cell, and not primarily at the cell membrane. The net result is the expenditure of energy to produce ATP, which provides the cell with a convenient energy source that it can employ in its many activities.

This coupling of a proton pump to ATP synthesis, called **chemiosmosis,** is responsible for the production of almost all the ATP that you harvest from food that you eat—and for all of the ATP produced by photosynthesis. We discuss how you obtain ATP from food in Chapter 8, and photosynthesis in Chapter 9. We know that proton pump channels are ancient, since they are present in bacteria as well as in eukaryotes. In fact, they appear to have evolved first as a mechanism to drive bacterial transport.

> The proton pump creates a proton gradient across a membrane. When protons diffuse back across the membrane through special channels, their transport is coupled to the production of ATP within the cell.

Transfer Between Cell Compartments

Relatively little is known about how molecules pass from one compartment of a cell to another. One process thought to be important is receptor-mediated endocytosis.

Receptor-Mediated Endocytosis.

Within most eukaryotic cells there are vesicles that transport molecules from one place to another. In electron micrographs, some of these vesicles appear like membrane-bounded balloons coated with bristles on their interior (cytoplasmic) surfaces, as if they had a 2-day growth of beard. These coated vesicles appear to form by the pinching in of local regions of plasma membrane that are coated with a network of proteins called coated pits (Figure 6-22). Coated pits act like molecular mousetraps, closing over to form an internal vesicle when the right molecule enters the pit. The trigger that releases the mousetrap is a protein called a receptor, embedded within the pit, that detects the presence of a particular target molecule and reacts by initiating endocytosis of the coated pit. By doing so, it traps the target molecule within the new vesicle. This process is called **receptor-mediated endocytosis.** It is quite specific, so that transport across the membrane occurs *only* when the correct molecule is present and positioned for transport.

FIGURE 6-22

How coated vesicles form. A coated pit appears in the plasma membrane of a developing egg cell, covered with a layer of proteins (**A**). A coating of protein molecules can be seen just beneath the pit, on the interior side of the membrane. When an appropriate collection of molecules gathers in the coated pit, the pit deepens (**B**) as the outer membrane of the cell closes in behind the pit (**C**), and the pit buds off to form a coated vesicle, which carries the molecules into the cell (**D**).

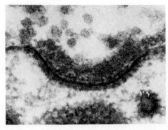

A

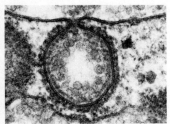

B   C

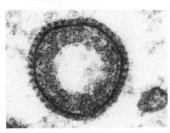

D

TABLE 6-2   MECHANISMS FOR TRANSPORT ACROSS CELL MEMBRANES

PROCESS	PASSAGE THROUGH MEMBRANE	HOW IT WORKS	EXAMPLE
NONSPECIFIC PROCESS			
Diffusion	Direct	Random molecular motion produces net migration of molecules toward region of lower concentration	Movement of oxygen into cells
Osmosis	Direct	Diffusion of water across differentially permeable membrane	Movement of water into cells placed in distilled water
Endocytosis			
Phagocytosis	Membrane vesicle	Particle is engulfed by membrane, which folds around it and forms a vesicle	Ingestion of bacteria by white blood cells
Pinocytosis	Membrane vesicle	Fluid droplets are engulfed by membrane, which forms vesicles around them	Nursing of human egg cells
Exocytosis	Membrane vesicle	Vesicles fuse with plasma membrane and eject contents	Secretion of mucus
SPECIFIC PROCESS			
Carrier-mediated endocytosis	Membrane vesicle	Endocytosis triggered by a specific receptor	Cholesterol uptake
Facilitated diffusion	Protein channel	Molecule binds to carrier protein in membrane and is transported across; net movement is in direction of lowest concentration	Movement of glucose into cells
Active transport			
Na^+-K^+ pump	Protein carrier	Carrier expends consumed energy to export Na^+ ions against a concentration gradient	Coupled uptake of many molecules into cells, against a concentration gradient
Proton pump	Protein carrier	Carrier expends energy to export protons against a concentration gradient	Chemiosmotic generation of ATP

The mechanisms for transport across plasma membranes that we have considered in this chapter are summarized in Table 6-2.

RECEPTION OF INFORMATION

So far in this chapter we have focused on membrane proteins that act as channels—doors across a membrane through which only particular molecules may pass. But cells also interact with their environments in many ways that do not involve the passage of molecules across membranes. A second major class of transaction that cells carry out with their environments involves information. In examining receptor-mediated endocytosis, we encounter for the first time a second general class of membrane protein, **cell surface receptors.** The LDL receptor, a protein embedded within the membrane, binds specifically to LDL particles but does *not* itself provide a transport channel for the particle; what the receptor transmits into the cell is information.

In general, a cell surface receptor is an information-transmitting protein that extends across a cell membrane. The end of the receptor protein exposed on the cell surface has a shape that fits to specific hormones or other "signal" molecules, and when such molecules encounter the receptor on the cell surface, they bind to it. This binding produces a change in the shape of the other end of the receptor protein, the end protruding into the interior of the cell, and this change in shape in turn causes a change in cell activity in one of several ways.

Cell surface receptors play an important role in the lives of multicellular animals. Among these receptors are, for example, the receptors for the signals that pass from one nerve to another; the receptors for protein hormones such as adrenaline and insulin, which your body uses to regulate its metabolic level; and the receptors for growth factors such as epidermal growth factor, which regulate development. All of these substances bind to specific cell surface receptors. The antibodies that your body uses to defend itself against infection are themselves free forms of receptor proteins. Without receptor proteins, the cells of your body would be "blind," unable to detect the wealth of chemical signals that the tissues of your body use to communicate with one another.

> **Cell surface receptors in membranes bind specifically to particles and transmit information about them into the cells that they surround. Among the information that they transmit are the presence of hormones and the signals that pass from one nerve to another. Antibodies are free forms of receptor proteins.**

EXPRESSION OF CELL IDENTITY

In addition to passing molecules across its membranes and acquiring information about its surroundings, a third major class of cell interaction with the environment involves conveying information *to* the environment. To understand why this is important, let us consider multicellular animals such as ourselves. One of their fundamental properties is the development and maintenance of highly specialized groups of cells called **tissues.** Your blood is a tissue and so is your muscle. What is remarkable about having tissues is that each cell within a tissue performs the functions of a member of that tissue and not some other one, even though all the cells of the body have the same genetic complement of DNA and are derived from a single cell at conception. How does a cell "know" to what tissue it belongs? In the course of development in human beings and other vertebrates, some cells move over others, as if they were seeking particular collections of cells with which to develop. How do they sense where they are? Every day of your adult life, your immune system inspects the cells of your body, looking for cells infected by viruses. How does it recognize them? When foreign tissue is transplanted into your body, your immune system rejects it. How does it know that the transplanted tissue is foreign?

The answer to all of these questions is the same: during development, every cell type in your body is given a banner proclaiming its identity, a set of proteins (cell surface markers) that are unique to it alone. Cell surface markers are the tools a cell uses to signal to the environment what kind of cell it is.

Some cell surface markers are proteins anchored in plasma membranes. The immune system uses such marker proteins to identify "self," for example. All the cells of a given individual have the same "self" markers, called the major histocompatibility complex (MHC) proteins. Because practically every individual makes a different set of the MHC marker proteins, they serve as distinctive markers for that individual.

Other cell surface markers are glycolipids, lipids with carbohydrate tails. These are the cell surface markers that differentiate the different organs and tissues of the vertebrate body. The markers on the surfaces of red blood cells that distinguish different blood types, such as A, B, and O, are glycolipids. During the course of development, the cell population of glycolipids changes dramatically as the cells divide and differentiate.

> **The cells that make up specific kinds of tissues are marked by proteins called cell surface markers. These are either proteins anchored in plasma membranes or glycolipids.**

PHYSICAL CONNECTIONS BETWEEN CELLS

So far we have seen how cells pass molecules to and from the environment, how they acquire information from the environment, and how they convey information about their identity to the environment. A fourth major class of interaction with the environment concerns physical interactions with other cells. Most of the cells of multicellular organisms are in contact with other cells, usually as members of organized tissues such as lungs, heart, or gut. The immediate environment of many of these cells is the mass of other cells clustered around it. The nature of the physical connections between a cell and the other cells of a tissue in large measure determines that cell's contribution to what the tissue will be like.

The places on the cell surface where cells of a tissue adhere to one another are called cell junctions. There are three general classes in animals:

1. *Adhering junctions.* These hold cells together like welds constructed of protein. They are called **desmosomes** (Figure 6-23), and are often especially common in sheets of tissue that are subject to severe stress, such as skin and heart muscle.

2. *Organizing junctions.* These cell junctions partition the cell membrane into separate compartments. Many tissues, such as the cells that line the gut, form sheets only one cell thick. In the gut, one surface of this sheet of cells faces the digestive tract and the other faces the blood. For the tissue to function in food absorption from gut to bloodstream, the membrane channels on the cell membranes facing the digestive tract must pump food molecules inward and those facing the bloodstream must pump food molecules outward. It is important that the two kinds of protein channels used for these two tasks be prevented from mixing within the fluid membrane. To prevent this, the cells of the gut wall are joined with what are known as **tight junctions.** Tight junctions are belts of protein that girdle each cell like the belt around a pair of bluejeans. These protein belts act like fences, preventing any membrane proteins afloat in the lipid bilayer from drifting across the boundary from one side of the cell to the other. The belts of the different cells in the gut wall are all aligned the same way. In the gut, the belts of adjacent cells are shoved tightly together, and there is no space between them through which leakage could occur. Therefore, the two sides of the gut wall remain functionally distinct.

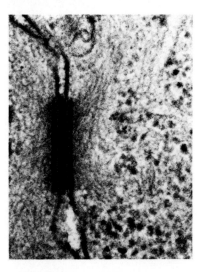

FIGURE 6-23
Desmosomes are simple points of attachment between animal cells. They do not contain channels connecting the interiors of the two attached cells.

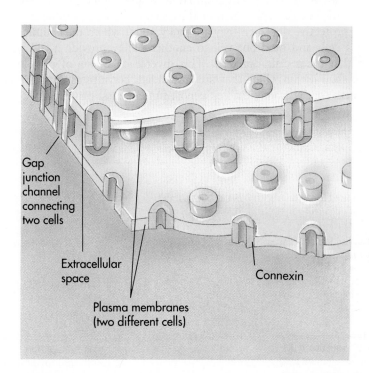

Gap junction channel connecting two cells

Extracellular space

Plasma membranes (two different cells)

Connexin

FIGURE 6-24
Gap junctions. Gap junctions are open channels connecting animal cells, like so many pipes between two rooms. In this rat liver cell many gap junctions are clustered together. Each junction is made of proteins known as connexins that align to form a continuous channel. In normal cells regulatory messages are constantly passed through these communicating tunnels. Connexin defects that block the passage at growth-inhibiting signals appear to be at the root of some cancers—and, in test tubes, correcting the connexin defect restores the cancer cells to healthy growth.

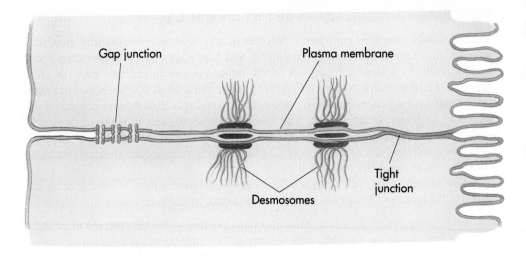

FIGURE 6-25
The three principal types of intercellular connection.

3. *Communicating junctions.* Cell junctions of this kind pass small molecules from one cell to another. Sometimes the cells of tissue sheets are connected beneath their tight junction belts by open channels, called **gap junctions** (Figure 6-24). Such junctions are passageways large enough to permit small molecules, such as sugar molecules and amino acids, to pass from one cell to another, while preventing the passage of larger molecules, such as proteins.

In plants, the plasma membranes of adjacent cells come together through pairs of holes in the walls; the cytoplasmic connections that extend through such holes are called **plasmodesmata**. Animal cells, in contrast, are not enclosed in cell walls. A summary of the types of intercellular connections in animals is illustrated in Figure 6-25, and their characteristics are listed in Table 6-3.

HOW A CELL COMMUNICATES WITH THE OUTSIDE WORLD

Every cell is a prisoner of its lipid envelope, unable to communicate with the outside world except via the proteins that traverse its lipid shell. Anchored within the lipid layer by nonpolar segments, these proteins are the "senses" of the cell. They detect the presence of other molecules, often initiating responses within the cell. They provide doors into the cell through which food molecules, ions, and other molecules may pass, but like protective doormen they are picky about whom they admit. They form the shape of a cell, and bind one cell to another. They provide a cell with surface identity tags that other cells can read. This diverse collection of proteins, together with the lipid shell within which they are embedded, constitutes the cell's membrane system, a system that is among the cell's most fundamental properties.

		CHARACTERISTIC SPECIALIZATIONS IN	WIDTH OF INTERCELLULAR	
TYPE	NAME	CELL MEMBRANE	SPACE	FUNCTION
Adhering junctions	Desmosomes	Buttonlike welds joining opposing cell membranes	Normal size 24 nm	Hold cells tightly together
Organizing junctions	Tight junctions	Belts of protein that isolate parts of plasma membrane	Intercellular space disappears as the two adjacent membranes come together	Form a barrier separating surfaces of cell
Communicating junctions	Gap junctions	Channels or pores through the two cell membranes and across the intercellular space	Intercellular space greatly narrowed to 2 nm	Provide for electrical communication between cells and for flow of ions and small molecules

TABLE 6-3   TYPES OF INTERCELLULAR CONNECTIONS

1. Every cell is encased within a bilayer sheet of phospholipid, which exists as a liquid.

2. Because cells contain significant concentrations of sugars, amino acids, and other solutes, water tends to diffuse into them. As it does so, a hydrostatic pressure builds that will rupture cells lacking a wall or other means of support.

3. Lipid bilayers are selectively permeable and do not permit the diffusion of water-soluble molecules into the cell. These molecules gain entry by crossing one of a variety of transmembrane proteins embedded within the membrane, the proteins acting as transport channels.

4. Passage across membranes from one cell compartment to another takes place when a particular molecule triggers a receptor in the membrane, initiating a form of endocytosis. Other molecules are synthesized across special secretory channels in the membrane; these molecules have special ends that determine which compartment membranes they can traverse.

5. Some channels involve carriers that transport molecules across the membrane much like cars of a train carry passengers across a bridge. This process is called facilitated diffusion, because net movement is always in the direction of lowest concentration but cannot proceed any faster than the capacity of the number of carriers.

6. Some channels transport molecules against a concentration gradient by expending energy. The two most important ones transport sodium ions and protons. These channels create low concentrations of sodium ions and protons (hydrogen ions) within the cell. These ions can diffuse back into the cell only through a second set of special channels, where their passage is often coupled to transport of another molecule inward or to the synthesis of ATP.

7. Many proteins embedded within the plasma membrane transmit information into the cell rather than transporting molecules. These proteins, called receptors, initiate chemical activity inside the cell in response to the binding of specific molecules on the cell surface.

8. Both proteins and glycolipids are used by cells as identification markers. These cell surface recognition markers permit cells of a given tissue to identify one another and also provide your body with a means of identifying foreign cells.

9. There are three general classes of cell junctions—connections between cells—in animal cells: (1) adhering junctions, or desmosomes, which hold cells together; (2) organizing junctions, which partition the plasma membrane into separate compartments; and (3) communicating junctions, which pass small molecules from one cell to another. In plants, the cell walls have openings that allow cytoplasmic connections, called plasmodesmata, to connect adjacent cells.

 ## REVIEW QUESTIONS

1. What type of lipid composes biological membranes? How does it differ from a typical fat in both structure and polarity?

2. Why do phospholipids spontaneously form bilayers when placed in water? Why is this structure such a good waterproofing agent for the exterior of a cell?

3. What are the four components of cell membranes? What is the function of each?

4. What are seven major ways in which a cell membrane controls interactions with the environment?

5. Define diffusion. What is the driving force behind diffusion?

6. How would the amount of solute in a solution compare with the amount of solute in a cell for the solution to be hypertonic to the cell? hypotonic to the cell? isotonic to the cell?

7. How does osmosis differ from diffusion? What limits the passage of water into a cell immersed in a hypertonic solution?

8. What is the general term that describes the bulk passage of materials into the cell? What term reflects the opposite effect? How do phagocytosis and pinocytosis differ from one another?

9. What is selective permeability? Why is it important to a living cell?

10. Describe the mechanism of receptor-mediated endocytosis.

11. How does facilitated diffusion differ from ordinary transmembrane diffusion? What are its three essential characteristics?

12. How does active transport differ from facilitated diffusion?

13. In basic terms, how does the sodium-potassium pump work? What is the net change in terms of sodium, potassium, and energy?

14. To which process is a coupled channel related? What is the biological consequence?

15. What is the proton pump? What is another name given to this process? Does this occur at the cell membrane or within membranes inside the cell?

16. What is a cell surface receptor? How does it work?

17. What are cell surface markers? What function do they serve?

18. What are the classes of physical connections between animal cells? In contrast, how do plant cells communicate with one another?

 THOUGHT QUESTIONS

1. When a hypertonic cell is placed in water solution, water molecules move rapidly into the cell. If the introduced cell is hypotonic, however, water molecules leave the cell and enter the surrounding solution. How does an individual water molecule *know* what solutes are on the other side of the membrane?

2. Cells maintain many internal metabolite molecules at high concentrations by coupling their transport into the cell to the transport of sodium and potassium ions by the sodium-potas-sium pump. What happens to all the potassium ions that are constantly being pumped into the cell?

3. Why is a lipid bilayer membrane freely permeable to water, which is quite polar, when it is not freely permeable to ammonia, which is also polar and about the same size?

 FOR FURTHER READING

BRETSCHER, M.: "The Molecules of the Cell Membrane," *Scientific American*, October 1985, pages 100-108. A good description of the structure of the cell membrane and of how transmembrane proteins are anchored within the lipid bilayer.

BRETSCHER, M.: "How Animal Cells Move," *Scientific American*, December 1987, pages 72-90. An exciting suggestion that receptor-mediated endocytosis may prove to be the basic mechanism underlying animal cell movement.

FRANCIOLINI, F. and A. PETRIS: "Evolution of Ionic Channels of Biological Membranes," *Molecular Biological Evolution* vol. 6, 1989, pages 503-513. A broad view of how membrane channels evolved, emphasizing the key role of calcium channels.

GENNIS, R.: *Biomembranes: Molecular Structure and Function*, Springer-Verlag, New York, 1989. A text devoted to the biochemistry and dynamics of membranes.

HAKOMORI, S.: "Glycosphingolipids,"*Scientific American*, May 1986, pages 44-53. Carbohydrate chains attached to lipid molecules play an important role in cell-to-cell recognition, an area of intensive present-day research.

LIENHARD, G. and others: "How Cells Absorb Glucose," *Scientific American*, January 1992, pages 86-91. A description of recent research on the structure of the glucose transporter.

SLAYMAN, C.L.: "Proton Chemistry and the Ubiquity of Proton Pumps," *BioScience*, January 1985, pages 16-47. This entire issue is devoted to a collection of seven articles on proton pumps; together they illustrate the central role these transmembrane channels play in the biology of both bacteria and eukaryotes.

TOSTESON, D.: *Membrane Transport: People and Ideas*, Oxford University Press, London, 1989. The origins and development of the principles of membrane biology, from the recognition of the lipid bilayer 57 years ago to the suggestion 18 years ago that proteins are embedded in the bilayer.

UNWIN, N., and R. HENDERSON: "The Structure of Proteins in Biological Membranes," *Scientific American*, February 1984, pages 78-94. A lucid account of how proteins are anchored within membranes via nonpolar segments.

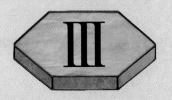

Energetics

7

Energy and Metabolism

CHAPTER OUTLINE

OVERVIEW

The life processes of every organism are driven by energy. Cell growth, movement, the transport of molecules across cell membranes—all of a cell's activities require energy. Just as a car is powered by burning gasoline, so our bodies burn chemical fuel to obtain energy. The energy in a chocolate bar exists in electrons, which spin in energetic orbitals about their atomic nuclei. Your cells strip these electrons away and use them to power their lives. Your every thought is fueled by the energy of electrons.

FOR REVIEW

Here are some important terms and concepts that you will encounter in this chapter. If you are not familiar with them, you should review them before proceeding.

Nature of chemical bond (Chapter 2)

Protein structure (Chapter 3)

Nucleotides (Chapter 3)

Proton pump (Chapter 6)

Once, a famous story goes, there was a wealthy man who purchased an expensive car. He drove it for several hours until it sputtered and stopped—and then went and bought another! You and I and every other living creature are like this man's car—we cannot run without fuel. Fuel is a source of energy, and energy drives everything that we do. When you whistle or walk down the street, you use energy. When you plan or hope or dream, whether standing in the shower or sleeping in bed, you use energy. In fact, it takes more energy to power your brain when you are sound asleep in bed than it does to power the 60-watt bulb that lights your bedroom.

You obtain your energy from the food you eat. If you stopped eating, you would soon begin to lose weight as your body used up its stored energy. Eventually, if you were not supplied with some outside source of energy, you would die. The same is true of all other living organisms. Deprived of a source of energy, life stops. Why does this happen? Why cannot life simply continue? The reason is that each of the significant properties by which we define life—growth, reproduction, and heredity—uses up energy. And once this energy has been used, it is gone, dissipated as heat, and cannot be used again. To keep life going, more energy must be supplied, like putting more logs on a fire.

Life can be viewed as a constant flow of energy, channeled by organisms to do the work of living. In this chapter we will focus on energy, what it is and how organisms capture, store, and use it. In the following two chapters, we will explore the energy-capturing and energy-using engines of cells, a network of chemical reactions that are the highway system for the energy of your body. This living chemistry, the total of all chemical reactions carried out by an organism, is called **metabolism.**

WHAT IS ENERGY?

Energy is defined as the ability to bring about change, or, more generally, as the capacity to do work. Instinctively we all know something about energy. It is "work": the force of a falling boulder, the pull of a locomotive, the swift dash of a horse; it is also "heat": the blast from an explosion, a warming fire. Energy can exist in many forms (Figure 7-1): as mechanical force, as heat, as sound, as an electrical current, as light, as radioactive radiation, as the pull of a magnet—all are able to create change, to do work.

Energy can be considered to exist in two states. Some energy is actively engaged in doing work, driving a speeding bullet or lifting a brick. This form of energy is called **kinetic energy,** or energy of motion. Other energy is not actively doing work, but has the capacity to do so, just as a boulder perched on a hilltop has the capacity to

FIGURE 7-1
Energy comes in many forms. Solar flares, raging torrents of water, fiercely burning fire—and the push of a baby's hand—all are forms of energy.

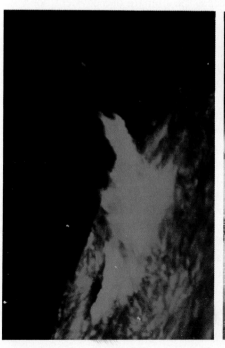

FIGURE 7-2
Transforming potential energy into kinetic energy. This hare has just sprung into motion. Much of the work carried out by organisms involves this transformation.

roll downhill. This form of energy is called **potential energy,** or stored energy. Much of the work carried out by living organisms involves the transformation of potential energy to kinetic energy (Figure 7-2).

> **Energy is the capacity to do work. It can exist in many forms, some of which are actively engaged in doing work (kinetic energy) and others that store the energy for later use (potential energy).**

Because energy can exist in so many forms, there are many ways to measure it. The most convenient way is in terms of heat, because all other forms of energy can be converted into heat. Indeed, the study of energy is called **thermodynamics**—that is, heat changes. The unit of heat most commonly employed in biology is the **kilocalorie (kcal)**. A kilocalorie is equal to 1000 calories. A calorie is the heat required to raise the temperature of 1 gram of water 1 degree Celsius. It is important not to confuse calories with a term often encountered in diets and discussions of nutrition, the Calorie (with a capital C), which is actually another term for kilocalorie.

THE LAWS OF THERMODYNAMICS DESCRIBE HOW ENERGY CHANGES

All the changes in energy that take place in the universe, from nuclear explosions to the buzzing of a bee, are governed by two laws called the Laws of Thermodynamics. The **First Law of Thermodynamics** concerns amounts of energy. It states that energy can change from one form to another and can be transformed from potential to kinetic energy, but it can never be destroyed. Nor can any new energy be made. The total amount of energy in the universe remains constant.

> **The First Law of Thermodynamics states that energy cannot be created or destroyed; it can only undergo conversion from one form to another.**

The lion that you see gnawing on a giraffe bone in Figure 7-3 is busy acquiring energy. He is not creating new energy, but rather is transferring potential energy stored in the giraffe's tissues to his own body, where that potential energy will be converted to kinetic energy to fuel running, growling, and all his other lionly activities. Where is the potential energy in giraffe tissue? It is stored in chemical bonds. Recall the nature of covalent chemical bonds, which were discussed in Chapter 2. A covalent bond is created when two atomic nuclei share electrons, and to break such a bond requires energy to pull the nuclei apart. Indeed, the strength of a covalent bond is measured by the amount of energy required to break it. For example, it takes 98.8 kcal of energy to break a mole (the atomic weight of a substance, expressed in grams) of carbon-hydrogen (C—H) bonds.

What happens to the energy after the lion uses it? Some of it is transferred to other forms of potential energy—stored as fat, for example. Another portion accomplishes mechanical work, bending blades of grass, breaking bones. Almost half is dissipated to the environment as heat, where it speeds up the random motions of molecules. This energy is not lost, but rather is converted to a nonuseful form, random molecular motion.

The **Second Law of Thermodynamics** concerns this transformation of potential energy to a form of kinetic energy, the random molecular motion of molecules—to heat. It states that all objects in the universe tend to become more-disordered, and that the disorder in the universe is continuously increasing. The idea behind this law is easy to understand and is part of everyone's experience. It is much more likely that a stack of six soft drink cans will tumble over than that six cans will spontaneously leap one onto another and form a stack. Stated simply, disorder is more likely than order. This is true of a child's room, of the desk you study or work at, of a waiting crowd of people—and of molecules.

> **The Second Law of Thermodynamics states that disorder in the universe constantly increases. Energy spontaneously converts to less organized forms.**

FIGURE 7-3
Transferring potential energy. This lion is eating a giraffe. The tissue the lion is consuming will undergo many changes on its metabolic journey, some eventually becoming part of the lion and of the gases that the lion exhales, as well as solid waste products.

As energy is transferred from one molecule to another, some always leaks away as kinetic energy of motion, increasing the random movements of molecules. At normal temperatures all molecules dance about randomly, and this added energy increases the pace of that dance. We refer to this form of kinetic energy as **heat energy.**

Heat is the energy of random molecular motion.

With every transfer of energy some potential energy is always dissipated as heat. Although heat can be made to do work when there is a gradient (that is how a steam engine works), it is generally not a useful source of energy for biological systems. Thus, from a biological point of view, the amount of useful energy in a system dissipates progressively. Although the total amount of energy does not change, the amount of useful energy available to do work does decrease, as progressively more is degraded to heat.

When energy becomes so randomized and uniform in a system that it is no longer available to do work, it is referred to as **entropy.** Entropy is a measure of the disorder of a system. Sometimes, the Second Law of Thermodynamics is stated simply as "Entropy increases." When the universe was formed 14 billion years ago, it had all the potential energy it ever will have. It has been becoming progressively more disordered ever since, with every energy exchange frittering away useful energy and increasing entropy. Someday all the energy will be random and uniform in distribution, the universe having wound down like an abandoned clock. No stars will shine, no waves will break upon a beach. But this is not going to happen soon. Perhaps it will occur 100 billion years from now.* Our species has been around for less than 1 million years, and life on earth for at least 4 billion, we think—clearly many more immediate problems face us than the final end predicted by the Second Law of Thermodynamics.

The First Law of Thermodynamics tells us that the universe as a whole is a closed system: energy does not come in and it does not go out. The earth, however, is not a closed system; it is constantly receiving energy from the sun. It has been estimated that every year the earth receives in excess of 13×10^{23} calories of energy from the sun. That is 2 million trillion calories per second! A lot of this energy heats up the oceans and continents, but some of it is captured by photosynthetic organisms: plants (Figure 7-4), algae (photosynthetic protists), and photosynthetic bacteria. In photosynthesis, energy garnered from sunlight is converted to chemical energy, combining small molecules (water and carbon dioxide) into ones that are more complex (sugars). The energy is stored as potential energy in the bonds of the sugar molecules. This energy can then be shifted to other molecules by forming different chemical bonds, or converted into motion, light, electricity—and heat. As each shift and conversion takes place, more energy is dissipated as heat. Energy continuously flows through the biological world, new energy from the sun constantly flowing in to replace the energy that is dissipated as heat.

Life converts energy from the sun to other forms of energy that drive life processes; the energy is never lost, but as it is used, more and more is converted to heat energy, a form of energy that is not useful in performing work.

FREE ENERGY

Because it takes energy to break chemical bonds, the bonds between the atoms of a molecule tend to hold the atoms together, as if linked by chains. Acting in the opposite way, thermal energy increases atomic motion and so encourages atoms to move apart. Both thermal effects and bonding have a significant influence on molecules, the first tending to promote disorder, the second to reduce it. The net effect, the amount of energy actually available to form other chemical bonds, is called the **free energy** of

FIGURE 7-4
Practically all the energy that powers life is captured from sunlight. These trees in a Michigan forest are the first of many plants to snare sunlight as it falls toward the forest floor.

*There is a lively discussion among cosmologists about this possibility, called the Warmetod Hypothesis. Its validity depends on the assumption that our universe is closed, a point that has proven very difficult to settle.

that molecule, denoted by the symbol **G.** Free energy in a more general sense is defined as the energy in any system available to do work. In a molecule within a cell, where pressure and volume usually do not change, the energy available to do work is equal to the energy contained in the chemical bonds of the molecule (the technical name for this energy is **enthalpy;** it is designated **H**) minus the total thermal energy at that temperature that is unavailable because of disorder (the entropy, designated **S,** times the temperature **T** measured on a scale of Celsius degrees above absolute zero):

$$G \quad = \quad H \quad - \quad TS$$

FREE ENERGY   ORDERING INFLUENCES   DISORDERING INFLUENCES

Because chemical reactions can involve changes in bond energies (H) and can also in many cases promote thermal motion among molecules and so increase disorder (S), chemical reactions can produce changes in free energy. The change (Δ) in available energy that a chemical reaction produces is referred to as the **change in free energy,** symbolized $\Delta\mathbf{G}$. When a chemical reaction occurs under conditions of constant temperature, pressure, and volume, as do most biological reactions, the change in free energy is simply

$$\Delta G = \Delta H - T\Delta S$$

Free energy is the energy available to do work. In a chemical reaction carried out within a cell, the change in free energy is the difference in bond energies between reactants and products, corrected for changes in the degree of disorder of the system.

The change in free energy, ΔG, is the most fundamental property of any chemical reaction. Any reaction in which ΔG is negative will proceed spontaneously. Consider for example the combustion of gasoline. In the equation above, ΔG for the combustion of gasoline would be negative because the term ΔH has a negative value (the products of the reaction have lower bond energies) and also because the term ΔS has a positive value (higher disorder) coming after a negative sign in the equation. The chemical reactions that take place in all living things obey this fundamental rule: the balance between bond energies and disorder must be negative for a reaction to occur spontaneously.

Any reaction that produces products containing less free energy than the original reactants contained will tend to proceed spontaneously.

OXIDATION-REDUCTION: THE FLOW OF ENERGY IN LIVING THINGS

The flow of energy into the biological world is from the sun, which shines a constant beam of light on our earth. Life exists on earth because it is possible to capture some of that continual flow of energy, transforming it to chemical energy that can be transferred from one organism to another and used to create cattle, and fleas, and you. Where is the energy in sunlight, and how is it captured? To understand the answers to these questions, we need to look more closely at the atoms on which the sunlight shines. As described in Chapter 2, an atom is composed of a central nucleus surrounded by one or more orbiting electrons and different electrons that possess different amounts of energy, depending on how far from the nucleus they are and how strongly they are attracted to it. Light (and other forms of energy) can boost an electron to a higher energy level. The effect of this boost is to store the added energy as potential chemical energy that the atom can later release when the electron drops back to its original energy level.

Energy stored in chemical bonds may be transferred to other new chemical bonds, with the electrons shifting from one energy level to another. In some, although not all, of these chemical reactions, electrons actually pass from one atom or molecule to another. This class of chemical reaction is called an **oxidation-reduction reaction.** Oxidation-reduction (or redox) reactions are of critical importance to the flow of energy through living systems.

When an atom or molecule loses an electron, it is said to be oxidized, and the process by which this occurs is called **oxidation** (Figure 7-5). The name reflects the fact that in biological systems oxygen, which attracts electrons strongly, is the most common electron acceptor.

When an atom or molecule gains an electron, it is said to be reduced, and the process is called **reduction.** Oxidation and reduction always take place together, because every electron that is lost by one atom (oxidation) is gained by some other atom (reduction) (Figure 7-6).

Oxidation is the loss of an electron; reduction is the gain of an electron.

Oxidation-reduction reactions play a key role in energy flow through biological systems, because the electrons that pass from one atom to another carry with them their potential energy of position (they maintain their distance from the nucleus). Energy is passed from one molecule to another via these electrons. The energy originally entered the system when light boosted electrons in a plant pigment to higher energy levels. Because the electrons have not yet returned to their original low energy level, they still store potential energy.

Atoms can store potential energy by means of electrons that orbit at higher than usual energy levels. When such an energetic electron is removed from one atom (oxidation) and donated to another (reduction), it carries the energy with it, orbiting the second atom's nucleus at the higher energy level.

Very often in biological systems electrons do not travel alone from one atom to another, but rather in the company of a proton. Recall that a proton and an electron together make up a hydrogen atom. Thus oxidation in a chemical reaction usually involves the removal of hydrogen atoms from one molecule; reduction involves the gain of hydrogen atoms by another molecule. In photosynthesis, hydrogen atoms are transferred from water to carbon dioxide, reducing the carbon dioxide to form glucose:

$$6CO_2 + 12H_2O + Energy \xrightarrow{\text{LIGHT}} C_6H_{12}O_6 + 6O_2 + 6H_2O$$

CARBON DIOXIDE   WATER   GLUCOSE   OXYGEN   WATER

This is a redox reaction—an example of one in which electrons in water move to higher energy levels in glucose. The reduction of carbon dioxide to form a mole of glucose stores 686 kcal of energy in the chemical bonds of the glucose.

The energy stored in a glucose molecule is released in a process called **cellular respiration,** in which the glucose is oxidized. Hydrogen atoms are lost by glucose and gained by oxygen:

$$C_6H_{12}O_6 + 6O_2 \longrightarrow 6CO_2 + 6H_2O + Energy$$

The oxidation of a mole of glucose releases 686 kcal of energy, the same amount that was stored in making it.

ACTIVATION ENERGY: PREPARING MOLECULES FOR ACTION

As the Laws of Thermodynamics predict, all chemical systems tend to proceed toward a state of maximum disorder and minimum free energy. Those reactions in which the products contain less free energy or more disorder than the reactants release the excess free (usable) energy. These reactions occur spontaneously and are called **exergonic.** In contrast, those reactions in which the products contain more free energy than the reactants require an input of usable energy from an outside source before they will proceed. These reactions do not occur spontaneously, and are called **endergonic** (Figure 7-7).

Exergonic reactions generate products containing less free energy than the original reactants and tend to proceed spontaneously. Endergonic reactions do not proceed without the addition of energy, because their products contain more energy than the initial reactants.

FIGURE 7-5
An animal's body is an efficient furnace. The oxidation of foodstuffs that takes place during metabolism leads to the same chemical end-point that burning it would have done—it is converted to CO_2 and H_2O. Organisms, however, siphon off much of the energy, whereas burning disperses the chemical bond energy as heat.

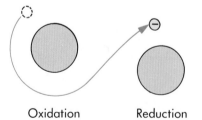

Oxidation   Reduction

FIGURE 7-6
Oxidation is the loss of an electron; reduction is the gain of one.

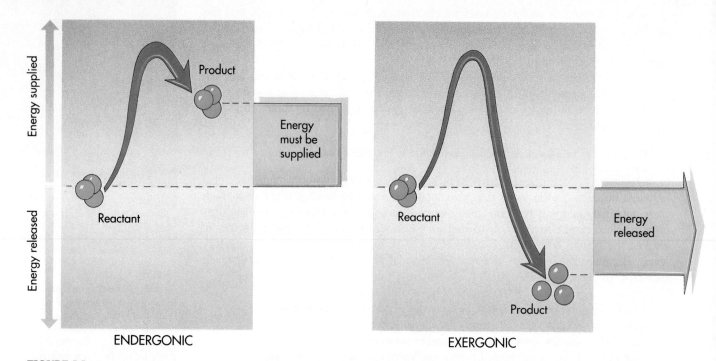

ENDERGONIC

EXERGONIC

FIGURE 7-7

Energy in chemical reactions. In an *endergonic* reaction, the products of the reaction contain more energy than the reactants, so the extra energy must be supplied for the reaction to proceed. In an *exergonic* reaction, the products contain less energy than the reactants, and the excess energy is released.

If all chemical reactions that release free energy tend to occur spontaneously, it is fair to ask, "Why haven't all such reactions already occurred?" Clearly they have not. If you ignite gasoline, for example, the resulting chemical reaction proceeds with a net release of free energy. So why doesn't all the gasoline within all the automobiles of the world and beneath all filling stations, just burn up right now? It doesn't because most reactions require an input of energy to get started. Before it is possible to form new chemical bonds with less energy, it is first necessary to break the existing bonds, and this takes energy. The extra energy required to destabilize existing chemical bonds and initiate a chemical reaction is called **activation energy** (Figure 7-8, *A*). It is because the burning of gasoline involves appreciable activation energy that all the world's gasoline has not burned up yet. An initial input of activation energy is required, such as the heat from the flame of a match.

The speed of an exergonic reaction, its reaction rate, does not depend on how much energy the reaction releases but rather on the amount of activation energy required for the reaction to begin. Reactions that involve larger activation energies tend to proceed more slowly, since fewer molecules succeed in overcoming the initial energy hurdle. Activation energies, however, are not fixed constants. Stressing particular chemical bonds can make them easier to break. The process of influencing chemical bonds in a way that lowers the amount of activation energy needed to initiate a reaction is called **catalysis.** Substances that carry out catalysis are known as **catalysts** (Figure 7-8, *B*).

Catalysts cannot violate the basic laws of thermodynamics; they cannot, for example, make an endergonic reaction proceed spontaneously. A reduction in activation energy accelerates both the forward and the reverse reactions by exactly the same amount and so cannot alter the final proportion of substrate converted to product. Catalysts only hasten the inevitable. A catalyst can never by itself make an energetically unfavorable reaction proceed; in fact it would simply accelerate the reverse process. Imagine lowering a fence between two identical pastures, one pasture containing many more sheep than the other. There is no way to lower the barrier between the two pastures in such a way that more sheep will jump from the sparsely populated pasture than from the full one; lowering the barrier simply makes movement in either direction easier. The lowering of the fence will promote a faster *net* movement of sheep into the empty pasture, no matter how it is lowered. The direction in which a chemical reaction proceeds is determined solely by the difference in free energy. Only

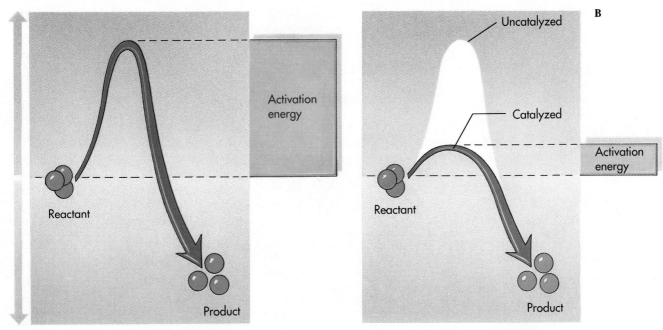

exergonic reactions proceed spontaneously, and catalysis cannot change that. What catalysts *can* do is make a reaction rate much faster.

The speed of a reaction depends on the activation energy necessary to initiate it. Catalysts reduce the amount of required activation energy and so speed the rate of reactions, although they do not change their final proportions.

ENZYMES: THE WORKERS OF THE CELL

The chemistry of living things, metabolism, is organized by controlling the points at which catalysis takes place: life itself is therefore a process regulated by catalysts. The agents that carry out most catalysis in living organisms are **enzymes,** globular proteins whose shapes are specialized to permit formation of temporary associations with the molecules that are reacting. By bringing two substrates together in correct orientation or by stressing particular chemical bonds of a substrate, an enzyme lowers the activation energy required for new bonds to form. The reaction thus proceeds much faster than it otherwise would. Because the enzyme itself is not changed, it can be used over and over.

As an example of how an enzyme works, consider the joining of carbon dioxide and water to form carbonic acid:

$$CO_2 + H_2O \longleftrightarrow H_2CO_3$$
$$\text{CARBON DIOXIDE} \quad \text{WATER} \quad \text{CARBONIC ACID}$$

This reaction may proceed in either direction, but in the absence of an enzyme it is very slow, because there is an appreciable activation energy. Perhaps 200 molecules of carbonic acid form in an hour. Given the speed at which events occur within cells, this reaction rate is like a snail racing in the Indianapolis 500. Cells overcome this problem by employing an enzyme within their cytoplasm called **carbonic anhydrase,** which speeds the reaction dramatically (enzymes are usually given names that end in *-ase*). In the presence of the enzyme carbonic anhydrase, an estimated 600,000 molecules of carbonic acid form every second! The enzyme has speeded the reaction rate about 10 million times.

Cells employ proteins called enzymes as catalysts to lower activation energies.

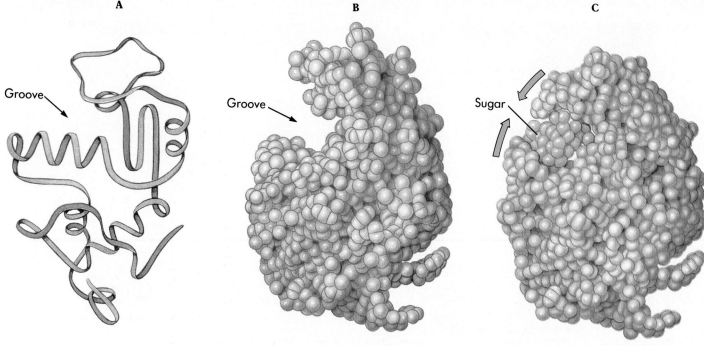

A   **B**   **C**

Groove   Groove   Sugar

FIGURE 7-9

How the enzyme lysozyme works. The tertiary structure of the lysozyme, diagrammed as a ribbon in **A** and shown in a three-dimensional model in **B** and **C,** forms a groove through the middle of the protein. This groove fits the shape of the chains of sugars that make up bacterial cell walls. When such a chain of sugars, indicated in yellow in **C,** slides into the groove, its entry induces the protein to alter its shape slightly to embrace the substrate more intimately. This *induced fit* positions a glutamic acid in the protein right next to the bond between two adjacent sugars, and the glutamic acid "steals" an electron from the bond, causing it to break.

Thousands of different kinds of enzymes are now known, each catalyzing a different chemical reaction. The enzymes in a cell, by facilitating particular chemical reactions, determine the course of metabolism in that cell, much as traffic lights determine the flow of traffic in a city. Not all cells contain the same set of enzymes, which is one reason why there is more than one type of cell. The chemical reactions going on within a red blood cell are very different from those going on within a nerve cell because the cytoplasm and membranes of red blood cells contain a different array of enzymes.

How Enzymes Work

Enzymes are globular proteins with one or more pockets or clefts on their surface, like a deep crease in a prune (Figure 7-9). These surface depressions are called **active sites,** and are the locations at which catalysis occurs. For catalysis to occur, the molecule on which the enzyme acts, called a **substrate,** must fit precisely into the active site so that many of its atoms nudge up against atoms of the enzyme. The substrate fits into its active site like a person into a tight-fitting pair of jeans, in very close contact. Proteins are not rigid, however, and in some cases the binding of substrate may induce the protein to adjust its shape slightly, leading to a better **induced fit** (Figure 7-10).

When a substrate molecule has bound to the active site of an enzyme, amino acid side groups of the enzyme are placed in close proximity to certain bonds of the substrate, just as when you sit in a chair certain parts of your body press against the seat. These amino acid side groups chemically interact with the substrate, usually stressing or distorting a particular bond and consequently lowering the activation energy for the bond's cleavage.

Enzymes typically catalyze only one or a few similar chemical reactions because they are specific in their choice of substrate. The active site of each kind of enzyme is shaped so that only a certain substrate molecule will fit into it.

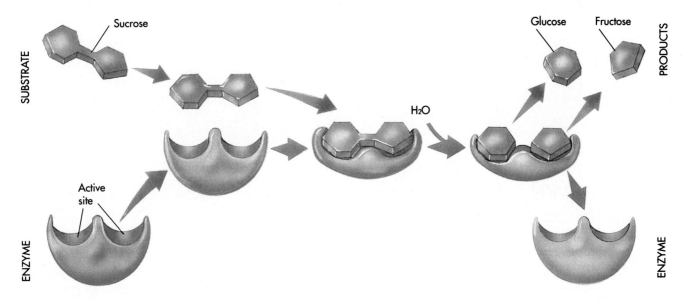

FIGURE 7-10

The catalytic cycle of an enzyme. Enzymes increase the speed with which chemical reactions occur, but they are not altered themselves as they do this. In the reaction illustrated here, the enzyme sucrase is splitting the sugar sucrose (the sugar present in most candy) into its two parts, the simpler sugars glucose and fructose.

A First, the sucrose substrate binds to the active site of the enzyme, fitting into a depression in the enzyme surface.

B The binding of sucrose induces the sucrase molecule to alter its shape, fitting more tightly around the sucrose molecule.

C Amino acids in the active site, now in close proximity to the bond between the glucose and fructose parts of sucrose, break the bond.

D The enzyme releases the resulting glucose and fructose fragments, the products of the reaction, and is then ready to bind another molecule of sucrose and run through the catalytic cycle once again.

Factors Affecting Enzyme Activity

The activity of an enzyme is affected by any change in conditions that alters its three-dimensional shape, such as temperature, pH, or the binding to the protein of specific chemicals that regulate its activity.

Temperature.   The shape of a protein is determined by hydrogen bonds that hold its polypeptide arms in particular positions and also by the tendency of noncharged ("nonpolar") segments of the protein to avoid water (hydrophobic interactions). Both hydrogen bonds and hydrophobic interactions are easily disrupted by slight changes in temperature. Most human enzymes function best within a relatively narrow temperature range, between 35° and 40° C (close to our body temperature). Below this temperature level, the bonds that determine protein shape are not flexible enough to permit the induced change in shape that is necessary for catalysis; above it, the bonds are too weak to hold the protein's polypeptide arms in the proper position. Bacteria that live in hot springs have proteins with stronger bonding between their polypeptide arms and can function at temperatures of 70° C or higher (Figure 7-11, *A*).

pH.   A third kind of bond that acts to hold the polypeptide arms of proteins in position is the bond between oppositely charged amino acids, such as glutamic acid (−) and lysine (+). These bonds are sensitive to hydrogen ion concentration. The more hydrogen ions available in the solution, the fewer negative charges and the more positive charges that occur. For this reason, most enzymes have a **pH optimum** just as they do a temperature optimum; it usually lies in the range of pH 6 to 8. Those proteins able to function in very acid environments have amino acid sequences that main-

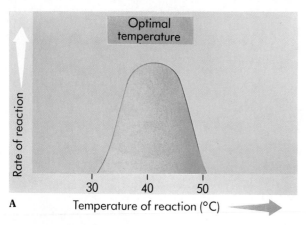

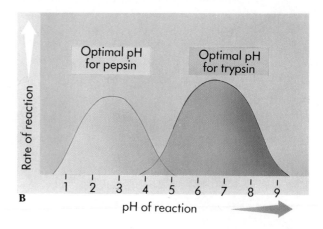

FIGURE 7-11

Enzymes are sensitive to their environment. The activity of an enzyme is influenced both by temperature (**A**) and by pH (**B**). Most human enzymes, such as the protein-degrading enzyme trypsin, tend to work best at temperatures of about 40° C, and within a pH range of 6 to 8.

tain their ionic and hydrogen bonds even in the presence of high levels of hydrogen ion (Figure 7-11, *B*). The enzyme pepsin, for example, digests proteins in your stomach at pH 2, a very acidic level.

How Enzyme Activity is Regulated

The activity of an enzyme is sensitive not only to temperature and pH, but also to the presence of specific chemicals that bind to the enzyme and cause changes in its shape. By means of these specific chemicals, a cell is able to regulate which enzymes are active and which are inactive at a particular time. When the binding of the chemical alters the shape of the protein and thus shuts off enzyme activity, the chemical is called an **inhibitor;** when the change in the enzyme's shape is necessary for catalysis to occur, the chemical is called an **activator.**

The change in shape that occurs when an activator or inhibitor binds to an enzyme is called an **allosteric** change (Greek *allos,* other + *steros,* shape). Enzymes usually have special binding sites for the activator and inhibitor molecules that affect them, and these binding sites are different from their active sites. The enzyme catalyzing the first step in a series of chemical reactions, or **biochemical pathway,** often has an inhibitor-binding site to which the molecule produced by the last step in the series binds. As the amount of this molecule builds up in the cell, it begins to bind to the initial enzyme in the biochemical pathway, thus inhibiting the activity of that en-

NOT ALL ENZYMES ARE PROTEINS

Until a few years ago, most biology textbooks contained statements such as, "Enzymes are the catalysts of biological systems." We can no longer make that statement without qualification. Cech and colleagues at the University of Colorado first reported in 1981 that certain reactions in which bits are cut out from RNA molecules appeared to be catalyzed in cells by RNA rather than by enzymes. Evidence has accumulated in the last few years that other RNA molecules are also capable of acting as catalysts. Like enzymes, these RNA catalysts, loosely called "ribozymes," greatly accelerate the rate

of a biochemical reaction and show extraordinary specificity with respect to the substrates on which they act.

There appear to be at least two sorts of ribozymes. Those that carry out intramolecular catalysis have a folded structure that mediates a reaction on another part of itself. Those that carry out intermolecular catalysis act on other molecules without themselves being changed in the process. Many important cellular reactions involve small RNA molecules, including the reactions that clip out unnecessary bits from RNA copies of genes, other reactions that prepare ribosomes for protein syn-

thesis, and still others that facilitate the replication of DNA within mitochondria. In all of these cases, the possibility of RNA catalysis is being actively investigated. It seems likely, particularly in the complex process of protein synthesis, that both enzymatic and RNA catalysis will play important roles.

The ability of RNA, an informational molecule, to act as a catalyst has created great excitement among biologists, as it appears to answer the chicken-and-egg problem of the origin of life—which came first, the protein or the nucleic acid?—in favor of RNA.

zyme. By this process the pathway is shut down when it is no longer needed. Such **feedback inhibition** by end products is a good example of the way many enzyme-catalyzed processes within cells are self-regulating.

> **The activity of enzymes is regulated by allosteric changes in enzyme shape; these changes result when specific, small molecules bind to the enzyme, molecules that are not substrates of that enzyme.**

Coenzymes are Tools Enzymes Use to Aid Catalysis

Often enzymes use additional chemical components as tools to aid catalysis. These additional components are called **cofactors.** Many enzymes, for example, have metal ions locked into their active sites, ions that help draw electrons from substrate molecules. The enzyme carboxypeptidase chops up proteins by employing a zinc ion to help draw electrons away from the bonds joining amino acids. Many trace elements

---

A VOCABULARY OF METABOLISM

catalysis   Accelerating the rate of a chemical reaction by lowering the activation energy, without being used up in the reaction. Enzymes are the catalysts of cells.

coenzyme   A nonprotein organic molecule that plays an accessory role in enzyme-catalyzed processes, often by acting as a donor or acceptor of electrons. NAD^+ is a coenzyme.

endergonic   A chemical reaction to which energy from an outside source must be added before the reaction proceeds; opposite of exergonic.

energy of activation   The energy required to destabilize chemical bonds and so initiate a chemical reaction.

entropy   A measure of the randomness or disorder of a system. In cells, a measure of how much energy has become so dispersed (usually as evenly distributed heat) that it is no longer available to do work.

enzyme   A protein that is capable of speeding up specific chemical reactions by lowering the required activation energy while being itself unaltered by the process; a biological catalyst.

exergonic   An energy-yielding chemical reaction. Exergonic reactions tend to proceed spontaneously, although activation energy is required to initiate them in living systems.

feedback inhibition   The regulation of the level of a factor in response to its own magnitude. In cells, a control mechanism whereby an increase in the concentration of some molecule inhibits the synthesis or activity of that molecule.

free energy   Energy available to do work.

kilocalorie   1000 calories. A calorie is the heat required to raise the temperature of 1 gram of water 1° C.

metabolism   The sum of all chemical processes occurring within a living cell or organism.

oxidation   The loss of an electron by an atom or molecule. Takes place simultaneously with reduction (gain of an electron) of some other atom, because an electron that is lost by one atom is gained by another.

reduction   The gain of an electron by an atom. Oxidation-reduction reactions are an important means of energy transfer within living systems.

respiration   An ATP-generating process in which an inorganic compound such as O_2 serves as the ultimate electron acceptor. In animal cells, food molecules are oxidized to obtain the high-energy electrons.

substrate   A molecule on which an enzyme acts; the initial reactant in an enzyme-catalyzed reaction.

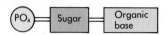

Basic structure of a nucleotide

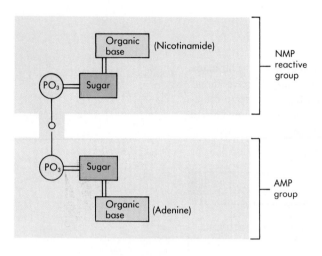

Basic structure of nicotinamide
adenine dinucleotide (NAD⁺)

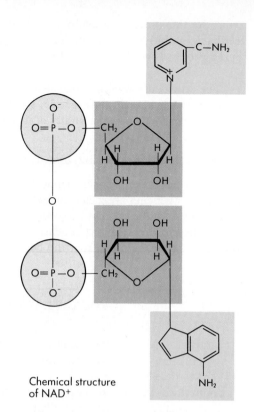

Chemical structure
of NAD⁺

FIGURE 7-12

The chemical structure of NAD⁺. Nicotinamide adenine dinucleotide (NAD⁺) is a composite molecule composed of two nucleotides bound together. In a nucleotide, one or more phosphate group(s) is attached to a five-carbon sugar, and an organic base is attached to the sugar's other end. The two nucleotides that make up NAD⁺ are nicotinamide monophosphate (NMP) and adenine monophosphate (AMP), joined head to head by their phosphate groups. The two nucleotides serve different functions in the NAD⁺ molecule: AMP acts as a core, providing a shape recognized by many enzymes; NMP is the active part of the molecule, contributing a site that is readily reduced.

such as molybdenum and manganese, which are necessary for your health, are used by enzymes as cofactors in this way. When the cofactor is a nonprotein organic molecule, it is called a **coenzyme.** Many of the vitamins that your body requires are parts of coenzymes.

> **Many enzymes employ metal ions or organic molecules to facilitate their activity; these are called cofactors. Cofactors that are nonprotein organic molecules are called coenzymes.**

In many enzyme-catalyzed oxidation-reduction reactions that harvest energy-bearing electrons, the electrons are passed in pairs from the active site of the enzyme to a coenzyme that serves as the electron acceptor. The coenzyme then carries the electrons away to a different enzyme, catalyzing another reaction, releasing them (and the energy they bear) to that reaction, and then returning to the original enzyme for another load of electrons. In most cases, the electrons are paired to protons as hydrogen atoms. Just as armored cars transport cash around a city, so coenzymes shuttle energy, as hydrogen atoms, from one place to another in a cell.

One of the most important coenzymes is the hydrogen acceptor **nicotinamide adenine dinucleotide,** usually referred to by the abbreviation **NAD⁺** (Figure 7-12). When NAD⁺ acquires an electron and a hydrogen atom (actually, two electrons and a proton) from the active site of an enzyme, it becomes reduced as NADH. The two energetic electrons and the proton are now carried by the NADH molecule, like money in your wallet. The oxidation of foodstuffs in your body, from which you get the energy to drive your life, takes place by stripping electrons from food molecules and donating them to NAD⁺, forming a wealth of NADH. This wealth is the principal energy income of your cells—although, as we shall see, much of it is converted to another currency.

ATP: THE ENERGY CURRENCY OF LIFE

The chief energy currency of all cells is a molecule called **adenosine triphosphate,** or **ATP.** The bulk of the energy that plants harvest during photosynthesis is channeled into production of ATP, and so is most of the NADH that soaks up the energy harvested from the oxidation of your food. The energy stored in the chemical bonds of fat and starch is converted to ATP, and so is the energy carried by the sugars circu-

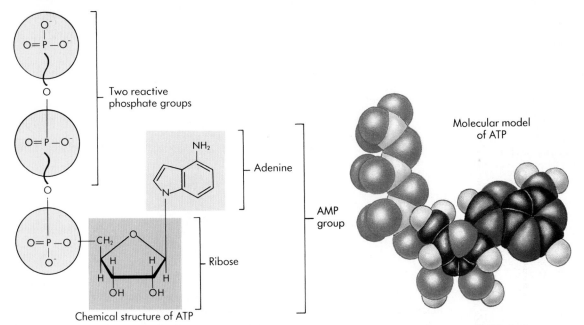

Two reactive phosphate groups

Adenine

AMP group

Ribose

Chemical structure of ATP

Molecular model of ATP

FIGURE 7-13

ATP is the primary energy currency of all cells. Like NAD^+, it has a core of AMP, a shape recognized by many enzymes. In this case, however, the reactive group added to the end of the AMP phosphate group is not another nucleotide but rather a chain of two additional phosphate groups. The bonds connecting these two phosphate groups to each other and to AMP are high-energy ($\sim$) bonds. When the outer bond of ATP is cleaved, yielding adenosine diphosphate (ADP) and phosphate (P), 7.3 kcal/mole of energy is released. Similarly, when the second bond is cleaved, yielding AMP and either P or P/P, 7.3 kcal/mole of energy is released. Most energy exchanges in cells involve cleavage of only the outermost of the two bonds, converting ATP into ADP + P.

lating in your blood. Cells then use their supply of ATP to drive active transport across membranes, to power movement, to provide activation energy for chemical reactions, to grow—almost every energy-requiring process that cells carry out is powered by ATP. Because ATP plays this central role in *all* organisms, it is clear that its role as the major energy currency of cells evolved early in the history of life.

Each ATP molecule (Figure 7-13) is composed of three subunits. The first subunit is a five-carbon sugar called ribose, which serves as the backbone to which the other two subunits are attached. The second subunit is adenine, an organic molecule composed of two carbon-nitrogen rings. Each of the nitrogen atoms in the ring has an unshared pair of electrons and weakly attracts hydrogen ions. Adenine therefore acts as a chemical base and is usually referred to as a "nitrogenous base." As we learned in Chapter 3, adenine has another major role in the cell—it is one of the four nitrogenous bases that are the principal components of DNA, the genetic material.

The third subunit is a triphosphate group (three phosphate groups linked in a chain). The two covalent bonds linking these three phosphates together are usually indicated by a squiggle ($\sim$) and are called **high-energy** bonds. Due to electrostatic repulsion between the phosphate groups these bonds release much energy when hydrolyzed, due to the increased stability of the products relative to the substrates. When one is broken, about 7 kcal/mole of energy is released; this is about twice the activation energy of an average chemical reaction in a cell. These high-energy phosphate bonds possess what a chemist would call "high transfer potential": they are bonds that carry considerable energy but have a low activation energy and are broken easily, releasing their energy. In a typical energy transaction, only the outermost of the two high-energy bonds is broken, cleaving off the phosphate group on the end. When this happens, ATP becomes ADP, adenosine diphosphate, and energy equal to 7.3 kcal/mole is released.

Cells use ATP to drive endergonic reactions, reactions whose products possess more energy than their substrates. Such reactions will not proceed, any more than a boulder will roll uphill, unless the reactants are supplied with the necessary energy. As long as the cleavage of ATP's terminal high-energy bond is more exergonic than the other reaction is endergonic, however, the overall energy change of the two "coupled" reactions is exergonic and the reaction will proceed. Because almost all endergonic reactions in the cell require less than 7.3 kcal/mole of activation energy, ATP is able to power all of the cell's activities. Thus you see that although the high-energy bond of ATP is not highly energetic in an absolute sense—it is not nearly as strong as a carbon-hydrogen bond—it is more energetic than the activation energies of almost all energy-requiring cell activities, and that is why it is able to serve as a universal energy donor.

FIGURE 7-14

The ATP-ADP cycle. In mitochondria and chloroplasts, chemical or photosynthetic energy is harnessed to form ATP from ADP and inorganic phosphate. When ATP is used to drive the living activities of cells, the molecule is cleaved back to ADP and inorganic phosphate, which are then available to form new ATP molecules.

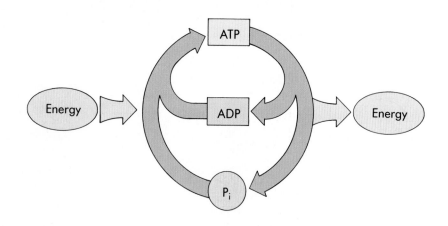

Cells contain a pool of ATP, ADP, and phosphate. ATP is constantly being cleaved to ADP plus phosphate to drive the endergonic, energy-requiring processes of the cell. Cells, however, do not maintain large stockpiles of ATP, any more than most people carry large amounts of cash with them. Instead, cells are constantly withdrawing from their energy reserves to rebuild more ATP. Using the energy and phosphorus derived from foodstuffs, from stored fats or starches, or (in the case of plants) from photosynthesis, ADP and phosphate are joined back together to form ATP, with 7.3 kcal/mole of energy being contributed to each newly formed high-energy bond. If you were able to mark every ATP molecule in your body at one instant in time, and then watch them, they would be gone in a flash. Most cells maintain only a few seconds' supply of ATP, which they constantly recycle (Figure 7-14).

BIOCHEMICAL PATHWAYS: THE ORGANIZATIONAL UNITS OF METABOLISM

Your body contains over a thousand different kinds of enzymes. These enzymes catalyze a bewildering variety of reactions. Many of the reactions occur in sequences, called **biochemical pathways,** in which the product of one reaction becomes the substrate for another. For example, the amino acid proline is synthesized from atmospheric nitrogen (N_2) and a simple carboxylic acid, alpha-ketoglutarate, in a series of enzyme-catalyzed steps (Figure 7-15). Biochemical pathways are the organizational units of metabolism, the elements that an organism controls to achieve coherent metabolic activity, just as the many metal parts of an automobile are organized into distinct subassemblies such as carburetor, transmission, and brakes.

How Biochemical Pathways Evolved

In primitive organisms the first biochemical processes were probably **heterotrophic processes,** in which energy-rich molecules were scavenged from the environment. Most of the molecules necessary for these processes are thought to have existed in the organic soup of the early oceans. The first catalyzed reactions were probably simple one-step processes that brought these energy-rich molecules together in various combinations. Eventually, of course, the original energy-rich molecules became depleted in the external environment, and only organisms that evolved some means of making the energy-rich molecule from other raw materials in the organic soup could survive. Thus a hypothetical process

where two hypothetical energy-rich molecules (*G* and *F*) react to produce compound *H* and release energy, becomes more complex when the supply of *F* in the environ-

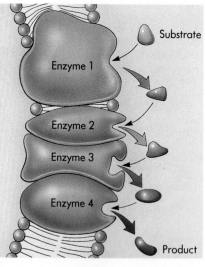

FIGURE 7-15

A biochemical pathway. The original substrate is acted on by enzyme 1, changing the substrate to a new form recognized by enzyme 2. Each enzyme in the pathway acts on the product of the previous stage.

ment runs out. A new reaction is added in which the depleted molecule, F, is made from another molecule, E, present in the environment:

$$E \longrightarrow F \overset{\overset{\textstyle G}{\searrow}}{\longrightarrow} H$$

When the supply of compound E in turn becomes depleted, organisms that are able to make it from some other available precursor, D, survive. When D becomes depleted, these organisms in turn are replaced by ones able to synthesize D from another molecule, C:

$$C \longrightarrow D \longrightarrow E \longrightarrow F \overset{\overset{\textstyle G}{\searrow}}{\longrightarrow} H$$

Thus our hypothetical biochemical pathway slowly evolves through time, with the final reactions evolving first and earlier reactions evolving only later. Looking at this pathway now, we would say that the organism, starting with compound C, is able to synthesize H by means of a complicated series of steps. This is how the biochemical pathways within organisms are thought to have evolved—not all at once, but one step at a time, backward.

How Biochemical Pathways Are Regulated

Once organisms evolved the means to catalyze ordered sequences of biochemical pathways, it became necessary to develop ways of controlling the output of these pathways. Not only does it make little sense to synthesize compound H from compound C when there is already plenty of H present, but also energy and raw materials that could better be put to use elsewhere are used up. The problem, then, was to find a way to shut down biochemical pathways at times when their products were not needed.

Primitive organisms evolved an ingenious mechanism, the **feedback loop**, to solve the problem of shutting down biochemical pathways. To understand how the mechanism works, we need to briefly reconsider protein structure. Recall from Chapter 3 that proteins maintain their shape largely as a result of hydrophobic interactions and internal hydrogen bonds. Such bonding forces are weak, and relatively small changes in the surface character of an enzyme can lead to an alteration of its shape. Thus the binding of substrate to enzyme often induces a change in the shape of the enzyme, which enhances its catalytic activity.

The opposite action can also occur. The binding of a molecule to an enzyme can alter the shape of the enzyme in such a way as to make it less active (Figure 7-16). This effect underlies the simplest, and probably most primitive, means of regulating enzyme activity: enzymes evolved secondary binding sites to which nonsubstrate molecules could bind. These sites served as chemical on/off switches, because the binding of the nonsubstrate molecule changed the shape of the enzyme and thus its activity. If the nonsubstrate binding induced a change in enzyme shape that improved catalytic activity, then the binding of the nonsubstrate molecule acted as an "on" command; if the binding led to an altered enzyme shape that was not catalytically active, then the nonsubstrate molecule acted as an "off" signal. Enzymes controlled in this way are said to be **allosteric** (Greek, *allos*, other + *steros*, form).

Enzymes can assume different allosteric forms, with various levels of catalytic activity, when bound by specific nonsubstrate molecules.

The regulation of simple biochemical pathways often takes place by means of an elegant feedback mechanism, in which the enzyme catalyzing the first step in a biochemical pathway possesses a second binding site, one that binds not the substrate of that enzyme, but rather the molecule that is the end product of the whole biochemical pathway. In our hypothetical example, the enzyme catalyzing the reaction $C \rightarrow D$ would possess a secondary binding site for H, the end product of the pathway. As the

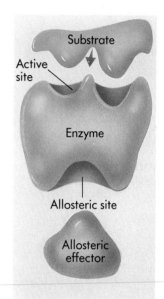

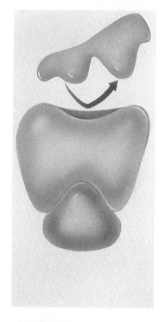

FIGURE 7-16
Allosteric inhibition. The binding to the enzyme of a particular metabolite, the "allosteric effector," causes the enzyme to alter its shape in such a way that the substrate no longer fits into the active site.

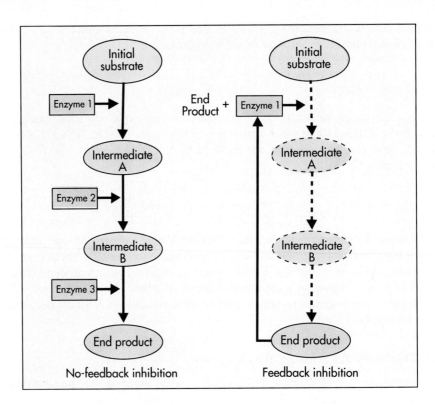

pathway proceeds and the end product H becomes present in the cell in ever-greater quantities, it becomes increasingly likely that one of the H molecules will encounter the secondary binding site of the $C \rightarrow D$ enzyme and bind to it. When this occurs, the binding induces an allosteric change in the shape of the enzyme that causes it to stop catalyzing the reaction $C \rightarrow D$. Because shutting down this reaction stops the first reaction in the sequence, the effect is to shut down the whole pathway. Thus when the cell has made the end product H in sufficient quantity, it is able to stop making more. Such a mode of regulation is called **feedback inhibition** (Figure 7-17).

THE EVOLUTION OF METABOLISM

The way in which contemporary organisms harvest energy and employ it to drive chemical reactions is different from the way in which the earliest organisms probably operated. For most of the period in which life has existed on earth, metabolic processes took place in what was essentially an oxygen-free, or **anaerobic,** environment. In fact, you will learn in the next chapter that, not only does anaerobic metabolism still exist, it remains the fundamental chemistry of all life processes. In this process, which is called cellular respiration, organic compounds are oxidized; it is the subject of Chapter 7. Respiration converts carbon compounds and O_2 gas into CO_2 and chemical energy (ATP).

The other half of the cycle, which is characteristic of our world today, is oxidative, or **aerobic,** metabolism, in which carbon atoms are cycled between organisms and the atmosphere (Figure 7-18). The portion of the cycle that acquires carbon atoms from atmospheric carbon dioxide and incorporates them into organisms is photosynthesis, the subject of Chapter 8. The photosynthesis carried out by today's plants, algae, and photosynthetic bacteria converts atmospheric CO_2 and the energy of sunlight into organic compounds with the release of oxygen gas (O_2).

This kind of chemical machinery, which is so characteristic of biological systems, is nevertheless a relatively recent development in life's evolutionary history. The more modern biochemical machinery of aerobic metabolism has simply been added on. In the next two chapters we examine respiration and photosynthesis and consider how evolution has shaped these two most fundamental metabolic processes.

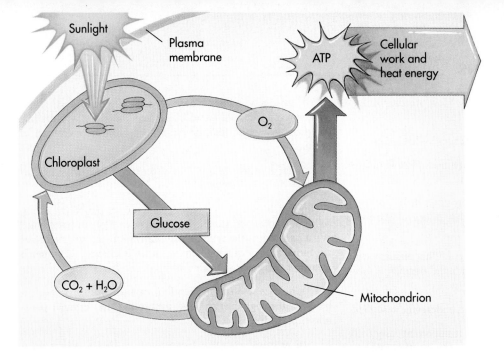

FIGURE 7-18
An overview of aerobic metabolism.

SUMMARY

1. Energy is the capacity to bring about change, to do work. Kinetic energy is actively engaged in doing work, whereas potential energy has the capacity to do so. Most energy transformations of living things involve transformation of potential to kinetic energy.

2. The First Law of Thermodynamics states that the amount of energy in the universe is constant, no energy being lost or created. Energy can, however, be converted from one form to another.

3. The Second Law of Thermodynamics says that disorder in the universe tends to increase. As a result, energy spontaneously converts to less organized forms.

4. The least organized form of energy is heat, random molecular motion. Heat energy may be used to do work when heat gradients exist, as in a steam engine, but it cannot accomplish work in cells.

5. A redox reaction is one in which an electron is taken from one atom (oxidation) and donated to another (reduction). The electrons in redox reactions often travel in association with protons, as hydrogen atoms.

6. If the electron transferred in a redox reaction is an excited electron, then the extra energy is transferred when the electron is, with the electron moving to the same higher energy orbital in the reduced atom as it had achieved in the original atom. Many energy transfers in cells take place by means of redox reactions.

7. Any chemical reaction whose products contain less free energy than the original reactants will tend to proceed spontaneously. The difference in free energy does not, however, determine the rate of the reaction.

8. The speed of a reaction depends on the amount of activation energy required to break existing bonds. Catalysis is the process of lowering activation energies by stressing chemical bonds. Enzymes are the catalysts of cells. Enzymes only affect the rate of a reaction, not the chemical equilibrium.

9. Cells contain many different enzymes, each of which catalyzes a different reaction. A given enzyme is specific because its active site fits only one or a few potential substrate molecules.

10. Cells focus all of their energy resources on the manufacture of ATP from ADP and phosphate, which requires them to supply about 7 kcal/mole of energy obtained from photosynthesis or from electrons stripped from foodstuffs. Cells then use this ATP to drive endergonic reactions.

11. Generally, the final reactions of a biochemical pathway evolve first and earlier reactions are added later, one step at a time.

REVIEW QUESTIONS

1. What are the two forms of energy? How are they different?

2. State the First Law of Thermodynamics. What is heat?

3. State the Second Law of Thermodynamics. How is this related to entropy?

4. Define oxidation and reduction. Why must these two reactions always occur in concert? What atom is one of the most common electron acceptors?

5. How do atoms store potential energy? What element with a very low atomic mass is associated with the transfer of electrons and redox reactions in biological systems?

6. Differentiate between exergonic and endergonic reactions.

7. What is activation energy? Why is it important, especially to biological systems? How is it related to the reaction rate?

8. How can a reaction's activation energy be altered? What term is given to a compound that effects this change? Does this compound alter the final proportion of substrate converted to product?

9. What are three factors that affect enzyme activity? In general how does this happen?

10. What is the chief energy molecule employed by living cells? What are the three main components of this molecule? What part of it is the actual energy carrier? What two molecules are formed when this high-energy bond is cleaved? How much energy is liberated?

11. Which is a stronger bond, a carbon-hydrogen covalent bond or the high-energy bond connecting the terminal phosphates of ATP?

12. In terms of complex biochemical pathways, which reactions generally evolved first: the initial reactions in the series or the final reactions?

 THOUGHT QUESTIONS

1. Oxidation-reduction reactions occur between a wide variety of molecules. Why do you suppose those concerning hydrogen and oxygen are the ones of paramount importance in biological systems?

2. In "The Wizard of Oz," Dorothy was afraid of lions and tigers and bears. On earth, there are many more bears than there are lions and tigers. Why do you suppose this is so?

3. In the deep ocean it is very cold and almost no light penetrates. Many fish that live there attract prey and potential mates by producing their own light. These luminescent fish glow like fireflies. Where do you suppose the light comes from? Does its generation require energy?

 FOR FURTHER READING

HAROLD, F.: *The Vital Force: A Study of Bioenergetics*, San Francisco, W.H. Freeman and Company, 1986. A comprehensive review of the advances that have been made in the last 10 years in our understanding of how organisms process energy.

HINKLE, P.C., and R.E. McCARTY: "How Cells Make ATP," *Scientific American*, March 1978, pages 104-123. Describes how cells use electrons stripped from foodstuffs to make ATP.

KRAUT, J.: "How Do Enzymes Work?," *Science*, 1988, vol. 242, pages 533-540. A clear presentation of the idea that enzymes work by stabilizing transitory transition states.

WACHTERSHAUSER, G.: "Evolution of the First Metabolic Cycles," *Proceedings of the National Academy of Science USA*, vol. 87, pages 200-204, 1990. A suggestion that metabolic cycles played a key role in the evolution of life.

WESTHEIMER, F.: "Why Nature Chose Phosphates," *Science*, March 6, 1987, vol. 235, pages 1173-1177. A wonderfully thoughtful analysis of the appropriateness of phosphates for biological but not laboratory chemistry.

8

Cellular Respiration

OVERVIEW

All organisms drive their metabolism with energy from ATP. The simplest processes for generating ATP, and the ones that evolved first, involve the rearrangement of chemical bonds; later, organisms evolved far more efficient means of carrying out this process. Plants, algae, and some bacteria obtain ATP by using energetic electrons to drive proton pumps, obtaining the electrons both by photosynthesis and by oxidizing sugars and fats. Animals, fungi, most protists, and most bacteria are heterotrophs; they use only the second oxidative process, which occurs within the symbiotically derived mitochondria that occur in all but a very few eukaryotes.

FOR REVIEW

Here are some important terms and concepts that you will encounter in this chapter. If you are not familiar with them, you should review them before proceeding.

Glucose (Chapter 3)

Chemiosmosis (Chapter 6)

Oxidation-reduction (Chapter 7)

ATP (Chapter 7)

Exergonic and endergonic reactions (Chapter 7)

Life is driven by energy. All the activities that organisms carry out—the swimming of bacteria, the purring of a cat, our reading these words—use energy. Even though the ways that organisms use energy are many and varied, all of life's energy ultimately has the same beginning: the sun. Plants, algae, and some bacteria harvest the energy of sunlight by the process of photosynthesis, thus converting radiant energy to chemical energy. These organisms, along with a few others that use chemical energy in a similar way, are called **autotrophs** (self-feeders). All organisms live on the energy produced by these autotrophs. Those that do not have the ability to produce their own food are called **heterotrophs** (fed by others). At least 95% of the kinds of organisms on earth—all animals, all fungi, and most protists and bacteria—are heterotrophs; most of them live by feeding on the chemical energy fixed by photosynthesis. All of us—plants and bacteria, you and I—share the same ultimate dependency on the sun. We are all children of light.

In this chapter, we will discuss the processes that all cells use to derive energy from organic molecules and to convert that energy to ATP. We will consider photosynthesis in detail in Chapter 9. We treat the conversion of chemical energy to ATP first because all organisms, both photosynthesizers and the heterotrophs that feed on them, are capable of transforming energy in this way. In contrast, fewer than 1 in 20 of the kinds of organisms on earth is capable of carrying out photosynthesis. As you will see, though, the two processes have much in common.

USING CHEMICAL ENERGY TO DRIVE METABOLISM

Your body contains over 1000 different kinds of enzymes. These enzymes catalyze a bewildering variety of reactions. Many of the reactions occur in sequences called **biochemical pathways.** We will encounter several biochemical pathways in this chapter. Biochemical pathways are the organizational units of metabolism, the pathways that energy and materials follow in the cell. Just as the many parts of an automobile are organized into distinct subassemblies such as fuel pump, transmission, and brakes, so the many enzymes of an organism are organized into biochemical pathways.

In biochemical pathways, **exergonic reactions**—those that involve a release of free energy—occur without the net input of energy; **endergonic reactions**—those that require the addition of energy—do not. To drive endergonic reactions, all organisms use the same mechanism—they *couple* them to the energy-yielding splitting of an energy-carrying molecule such as ATP (Figure 8-1).

Chemical energy powers metabolism by driving endergonic reactions through the use of ATP. Chemical energy is used to create ATP, and the splitting of ATP is coupled to endergonic reactions, providing the necessary energy.

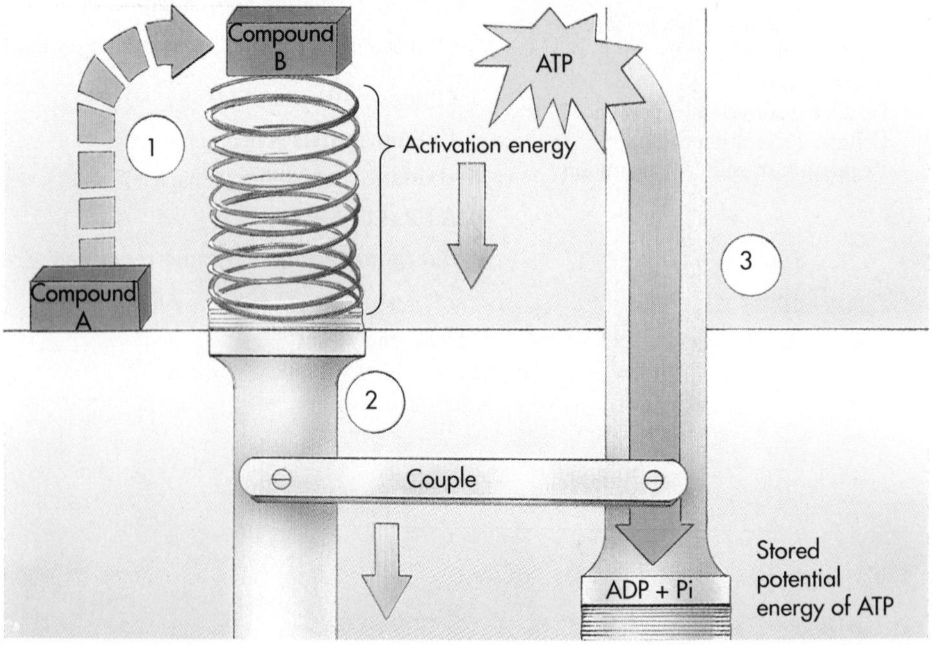

FIGURE 8-1

How coupled reactions work. The conversion of compound A to compound B (1) requires the net input of 4 kcal/mole of energy—in this diagram, the left piston must be driven down to the same level as compound A (2), and to do this, the left spring must be pulled, extending it. In cells, this extra energy is acquired from ATP. In this diagram, the cleavage of ATP to ADP drives the right piston down, compressing the right spring. *Because the two pistons are linked by a crossbar ("coupled"), the movement of the right piston pulls the left one down, too.* In cells, enzymes provide the physical link.

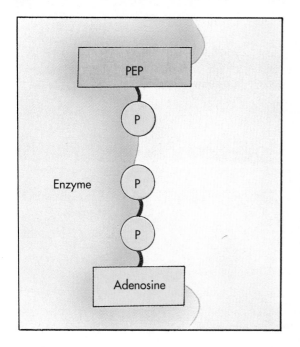 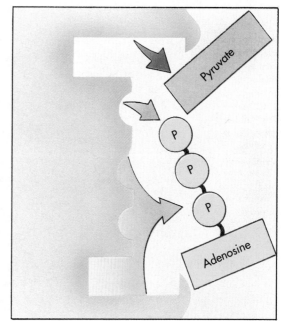

FIGURE 8-2
Substrate-level phosphorylation.
Some molecules such as phospho-
enol pyruvic acid (PEP) possess a
high-energy phosphate bond similar to those of ATP. When the
phosphate group of PEP is transferred enzymatically to ADP, the
energy in the bond is conserved
and ATP is created. The reaction
takes place because the phosphate
bond of PEP has higher energy
than the terminal phosphate bond
of ATP.

HOW CELLS MAKE ATP

ATP is the energy currency of all living organisms. How do organisms make ATP? They do so in one of two ways:

1. *Substrate-level phosphorylation.* Because the formation of ATP from ADP + inorganic phosphate (P_i) requires an input of free energy, ATP formation is endergonic—it does not occur spontaneously. When coupled to an exergonic reaction that has a strong tendency to occur, however, the synthesis of ATP from ADP + P_i does take place. The reaction occurs because the release of energy from the exergonic reaction is greater than the input of energy required to drive the synthesis of ATP. The generation of ATP by coupling strongly exergonic reactions with the synthesis of ATP from ADP and P_i is called **substrate level phosphorylation** (Figure 8-2). Many bacteria subsist entirely on ATP generated in this way.

2. *Chemiosmotic generation of ATP.* Almost all organisms possess transmembrane channels that function in pumping protons out of cells. We encountered these in Chapter 6. Proton-pumping channels use a flow of excited electrons to induce a change in the shape of a transmembrane protein, which in turn causes protons to pass outward. As the proton concentration outside the membrane rises higher than that inside, the outer protons are driven inward by diffusion, passing *backward* through special proton channels that use their passage to induce the formation of ATP from ADP + P_i. Because the chemical formation of ATP is driven by a diffusion force similar to osmosis, this process is referred to as a **chemiosmotic** one (Figure 8-3).

The harvesting of chemical energy can be considered to take place in one or more of two stages:

1. **A reshuffling of chemical bonds to couple ATP formation to a highly exergonic reaction, called substrate-level phosphorylation.**

2. **The transport of high-energy electrons to a membrane where they drive a proton pump and so power the chemiosmotic synthesis of ATP.**

Substrate-level phosphorylation was probably the first of the two ATP-forming mechanisms to evolve. The process of glycolysis, which is the most basic of all ATP-generating processes and is present in almost every living cell, employs this mechanism.

However, most of the ATP that organisms make is produced by chemiosmosis.

FIGURE 8-3

Chemiosmosis. High-energy electrons harvested from food molecules during metabolism are transported by NADH carrier molecules to "proton pumps," which use the energy to pump protons out across the membrane. As a result, the concentration of protons on the outside of the membrane rises, encouraging protons to diffuse back. They pass through the only channel open to them, a special ATP-forming protein complex. As each proton passes, an ATP molecule is formed.

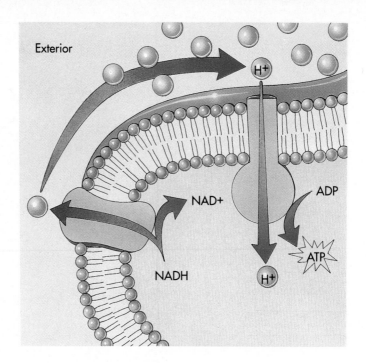

The electrons that drive proton-pumping channels in chemiosmosis are obtained by organisms from two sources:

1. In photosynthetic organisms, light energy boosts electrons to higher energy levels, and these electrons are channeled to proton pumps.
2. In all organisms, photosynthetic and nonphotosynthetic alike, high-energy electrons are extracted from chemical bonds and carried by coenzymes to proton pumps.

The extraction of electrons from chemical bonds is an oxidation-reduction process. Recall that oxidation is defined as the removal of electrons. When electrons are taken away from the chemical bonds of food molecules to drive proton pumps, the food molecules are being oxidized chemically. This electron-harvesting process is given a special name: **cellular respiration.** Cellular respiration is the oxidation of food molecules to obtain energy. Do not confuse this with the breathing of oxygen gas that your body carries out, which is also called respiration.

CELLULAR RESPIRATION

Animals can use many different kinds of molecules as metabolic food, oxidizing carbohydrates, fats, and proteins. To learn how the oxidizing machinery works, we will first focus on carbohydrates, looking to see how the energy-rich electrons are stripped from the sugar glucose and put to work to make ATP. We will consider fats and proteins later.

Cellular respiration of glucose is carried out in three stages in most organisms (Figure 8-4). The first stage is a biochemical pathway called **glycolysis.** The enzymes that carry out glycolysis are present in the cytoplasm of the cell, not bound to any membrane or organelle. Some of these enzymes form ATP by substrate-level phosphorylation, whereas others strip away a few energetic electrons from food molecules and donate them to NAD^+ molecules to form NADH molecules. Because electrons are removed from them, the glucose molecules are said to be oxidized. In this process, the electrons are accompanied by protons and thus travel as hydrogen atoms. These NADH molecules then donate the energetic electrons to proton-pumping channels to drive the chemiosmotic synthesis of ATP.

Glycolysis will produce ATP by substrate-level phosphorylation whether or not oxygen is present, but in animals the use of energetic electrons to drive the chemios-

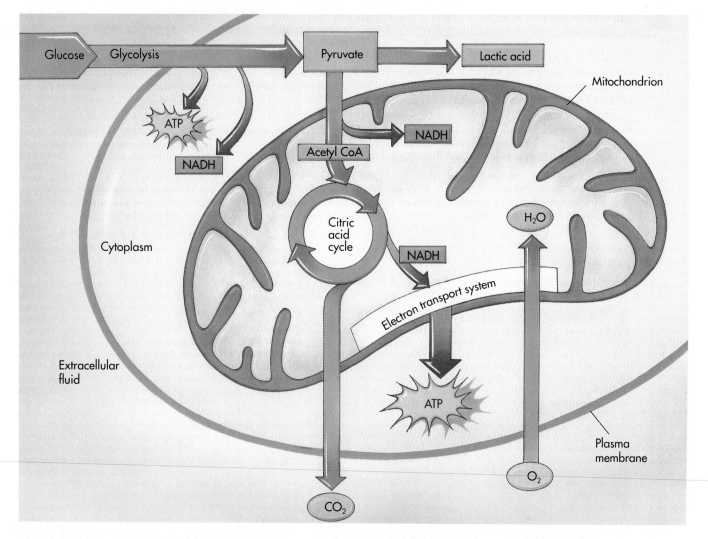

FIGURE 8-4
An overview of cellular respiration.

motic synthesis of ATP does not take place in the absence of oxygen, because oxygen
serves as the final acceptor of the electrons harvested from the glucose molecules:

$$C_6H_{12}O_6 + 6H_2O + 6O_2 \longrightarrow 6CO_2 + 12H_2O + Energy$$

In the absence of oxygen, animal cells are restricted to the substrate-level phosphorylation reactions of glycolysis to obtain ATP. Other organisms respire using different electron acceptors. Many bacteria, for example, use sulfur, nitrate or other inorganic compounds as the electron acceptor in place of oxygen.

All cells carry out glycolysis, but glycolysis is not efficient at extracting energy
from glucose molecules—only a small percentage of the energy in the chemical bonds
of food molecules is converted by glycolysis to ATP. In the presence of oxygen, however, two other pathways of cellular respiration also occur in your body, ones that
extract far more ATP. The second pathway is the **oxidation of pyruvate.** The third
pathway is called the **citric acid cycle,** after the six-carbon citric acid molecule
formed in its first step. Alternatively, it is called the Krebs cycle after the British biochemist, Sir Hans Krebs, who discovered it. (Less commonly, it is called the tricarboxylic acid cycle, because citric acid has three carboxyl groups.) In eukaryotes, the
oxidation of pyruvate and the reactions of the citric acid cycle take place only within
the cell organelles called mitochondria.

Pyruvate oxidation and the reactions of the citric acid cycle take place in all organisms that possess mitochondria, as well as in many forms of bacteria. (Chapter 5
recounts how mitochondria are thought to have evolved from bacteria.) Thus plants
as well as animals carry out oxidative respiration in this way. Because plants also produce ATP by photosynthesis, it is sometimes easy to lose sight of the fact that they
also produce ATP in this other oxidative fashion, just as you do.

If a heterotroph preserves only 2% of the energy of a photosynthetic organism that it consumes, then any other population of heterotrophs that consumes this kind of heterotroph has only 2% of the original energy available to it, and can garner from this by glycolysis only 2% of that, or 0.04% of the original amount of available energy. A very large base of autotrophs would thus be needed to support a small number of heterotrophs.

When organisms became able to extract energy from organic molecules by oxidative metabolism, this constraint became far less severe, because oxidative processes are far more efficient. The efficiency of oxidative respiration is 38%, so that about two-thirds of the available energy is lost at each trophic step. Thus animals that eat plants can obtain no more than approximately a third of the energy in the plants, and other animals that eat them no more than a third in the original plant-eaters, or **herbivores**. Losses of this magnitude, however, result in the transmission of far more energy from one **trophic level** (defined as a step in the movement of energy through an ecosystem) to another than does anaerobic glycolysis, where more than 98% of the energy is lost at each step. Because of the improved efficiency of oxidative metabolism, it has been possible for **food chains**, in which some kinds of

FIGURE 8-A
A food chain in the savannas, or open grasslands, of East Africa.

Stage 1 *Photosynthesizer*. The grass growing under these palms grows actively during the hot, rainy season and, capturing the energy of the sun abundantly, converts it into molecules of glucose that are stored in the grass plants as starch.

Stage 2 *Herbivore*. These large antelopes, known as wildebeests, consume the grass and convert some of its stored energy into their own bodies.

THE FATE OF A CANDY BAR: CELLULAR RESPIRATION IN ACTION

When you metabolize food and thus obtain the ATP that powers your life, you employ both substrate-level phosphorylation and oxidative respiration. To get an overview of what goes on, it is instructive to follow the fate of something you eat and see what happens to it. Let us eat, then, a chocolate bar (Figure 8-5).

A chocolate bar, like many of the things that we consume, is a complex mixture of sugars, lipids, proteins, and other molecules. The first thing that happens in its heterotrophs consume plants, others consume them, and so forth, to evolve. You will read more about food chains in Chapter 25.

There is a limit to how long a food chain can be, however, and that limit is dictated by the efficiency of oxidative metabolism; most are like the one illustrated in Figure 8-A, *1* to *5*, and involve only three, and rarely four levels. Too much energy is lost at each transfer point to allow the chains to become much longer than that. You could not support a large human population by subsisting on lions captured from the Serengeti plain of Africa—the amount of grass will support only so many zebras, not enough to maintain a large population of lions. Thus the ecological complexity of our world is fixed in a fundamental way by the chemistry of oxidative respiration.

Stage 3 *Carnivore.* The lion feeds on wildebeests and other animals, converting part of their stored energy into its own body.

Stage 4 *Scavenger.* This hyena and the vulture occupy the same stage in the food chain as the lion. They are also consuming the body of the dead wildebeest, which has been abandoned by the lion.

Stage 5 *Refuse utilizer.* These butterflies, mostly *Precis octavia,* are feeding on the material left in the hyena's dung after the material the hyena consumed has passed through its digestive tract. At each of these four levels, only about a third or less of the energy present is used by the recipient.

journey toward ATP production is that the complex molecules are degraded to simple ones. Disaccharides, such as sucrose, are split into simple sugars, which in turn are converted to glucose; proteins are split up into amino acids; and complex lipids are broken into smaller bits called fatty acids. These initial steps usually yield no usable energy, but they serve to marshal the energy wealth of a diverse array of complex molecules into a small number of simple molecules, such as glucose.

For simplicity, let us assume that our chocolate bar is entirely degraded to molecules of the six-carbon sugar glucose. Glucose occupies a central place in metabolism,

FIGURE 8-5

The fate of a chocolate bar. The bar is composed of sugar, chocolate, and other lipids and fats, protein, and many other molecules. This diverse collection is broken down into simple molecules, the sugars to glucose molecules, and the proteins, fats, and other lipids to two-carbon molecules called acetyl-CoA. These breakdowns produce little or no energy but prepare the way for three major energy-producing processes. The first process to occur is glycolysis, in which glucose is converted into two molecules of pyruvate. The second process to occur is the oxidation of this pyruvate to two molecules of acetyl-CoA. The third process is the oxidation of acetyl-CoA molecules in the citric acid cycle.

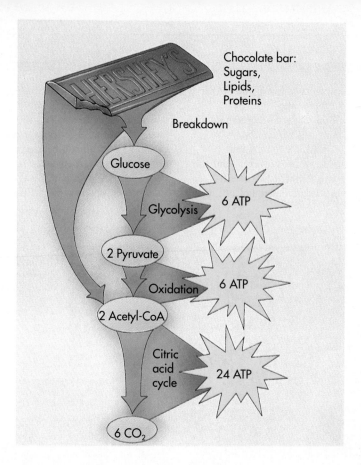

since many different foodstuffs are converted to it. Glucose is where the making of ATP begins.

The first stage of extracting energy from glucose is a 10-reaction biochemical pathway called **glycolysis.** In glycolysis, ATP is generated in two ways. For each glucose molecule, two ATP molecules are used up in preparing the glucose molecule, and four ATP molecules are formed by substrate-level phosphorylation, for a net yield of two ATP molecules. In addition, four electrons are harvested and used to form ATP molecules by oxidative respiration. But the total yield of ATP molecules is small. When the glycolytic process is completed, the two molecules of **pyruvate** that are left still contain most of the energy that was present in the original glucose molecule.

The second stage of extracting energy from glucose, after glycolysis, is a cycle of nine reactions, the citric acid cycle. The pyruvate left over from glycolysis is first converted to a two-carbon molecule, which feeds into the cycle. In the cycle, two more ATP molecules are extracted by substrate-level phosphorylation, and a large number of electrons are removed and donated to electron carriers. It is the harvesting of these electrons that leads to the greatest amount of ATP formation.

When this whole series of steps has been completed, the six-carbon glucose molecule has been divided into 6 molecules of CO_2, and 38 ATP molecules have been generated, 6 by substrate-level phosphorylation and 32 by oxidative respiration. Even though 2 of the ATP molecules must be expended to transport NADH out into the cytoplasm, the net yield of 36 ATP molecules is still very good. The overall process for the catabolism of glucose can be summarized as:

Substrate-level phosphorylation	+	Oxidative respiration	−	Cost of NADH transport	=	Net ATP production
6		32		2		36

Each ATP molecule represents the capture of 7.3 kilocalories of energy per mole of glucose; therefore 36 ATP molecules represent a total capture of $7.3 \times 36 = 263$ kilocalories per mole. The total energy content of the chemical bonds of glucose is only 686 kilocalories per mole; so we have succeeded in harvesting 38% of the available energy. By contrast, a car converts only about 25% of the energy in gasoline into useful energy.

This brief overview gives some sense of how cells organize their production of ATP from a food source such as a candy bar. We shall now examine glycolysis and the citric acid cycle as processes and study the ways in which they direct the flow of energy.

GLYCOLYSIS

Among the many simple molecules that were available as a consequence of degradation, the metabolism of primitive organisms focused on the simple six-carbon sugar glucose, undoubtedly in part because glucose was a major constituent of the carbohydrates of the cell. Glucose molecules can be dismantled in many ways, but primitive organisms evolved the ability to do it in one way that involved reactions releasing enough free energy to drive the synthesis of ATP in coupled reactions. The process involves a sequence of 10 reactions that convert glucose into two three-carbon molecules of pyruvate. For each molecule of glucose that passes through this transformation, the cell acquires two ATP molecules by substrate phosphorylation. The overall process, as we have seen, is called glycolysis.

An Overview of Glycolysis

Glycolysis consists of two different sorts of processes, one wedded to the other:

1. Glucose is converted to two molecules of the three-carbon compound **glyceraldehyde-3-phosphate (G3P)**, with the expenditure of ATP.
2. ATP is generated from the conversion of G3P to pyruvate.

The 10 reactions of glycolysis proceed in four stages (Figure 8-6):

Stage A: Three reactions change glucose into a compound that can readily be cleaved into three-carbon phosphorylated units. Two of these reactions require the cleavage of an ATP molecule, so that this stage, **glucose priming,** requires the investment by the cell of two ATP molecules.

Stage B: The second stage is **cleavage and rearrangement,** in which the six-carbon product of the first stage is split into two three-carbon molecules. One is G3P, and the other is converted to G3P by another reaction.

Stage C: The third stage is **oxidation,** in which a pair of electrons is removed from G3P and donated to NAD^+. NAD^+ is a coenzyme that acts as an electron carrier in the cell, in this case accepting the two electrons from G3P to form NADH. Note that NAD^+ is an ion, and that *both* electrons in the new covalent bond come from G3P.

Stage D: The final stage, **ATP generation,** is composed of a series of four reactions that convert G3P into another three-carbon molecule, pyruvate, and in the process generate two ATP molecules.

Because each glucose molecule is split into *two* G3P molecules, the overall net reaction sequence yields two ATP molecules, as well as two NADH molecules and two molecules of pyruvate:

$$\begin{array}{ll} -2\text{ATP} & \text{Stage A: priming} \\ \underline{2\,(+2\text{ATP})} & \text{Stage D: 2 ATP for each of the 2 G3P molecules} \\ +2\text{ATP} & \end{array}$$

This is not a great deal of energy. When you consider that the free energy of the total oxidation of glucose to CO_2 and H_2O is -686 kcal/mole (the sign is negative to indicate that the products of the oxidation have less energy) and that ATP's high-energy bonds each have an energy content of -7.3 kcal/mole, then the efficiency with which glycolysis harvests the chemical energy of glucose is only $14.6/686$, or 2.1%. Although far from ideal in terms of the amount of energy that it releases, however, gly-

FIGURE 8-6

The glycolytic pathway. The first five reactions convert a molecule of glucose into two molecules of G3P. This process is endergonic, and requires the expenditure of two ATP molecules to drive it. The second five reactions convert G3P molecules into pyruvate molecules, and generate four molecules of ATP for each two molecules of G3P. They also generate two molecules of NADH, which as we shall see is further oxidized to produce four more molecules of ATP. Subtracting the two ATP molecules expended in driving the initial endergonic reactions, the body's net yield, including oxidation of NADH, is six ATP molecules for each molecule of glucose. When cells are forced to operate without oxygen, the two molecules of NADH cannot be used to produce ATP, and the net yield is then only two molecules of ATP per molecule of glucose.

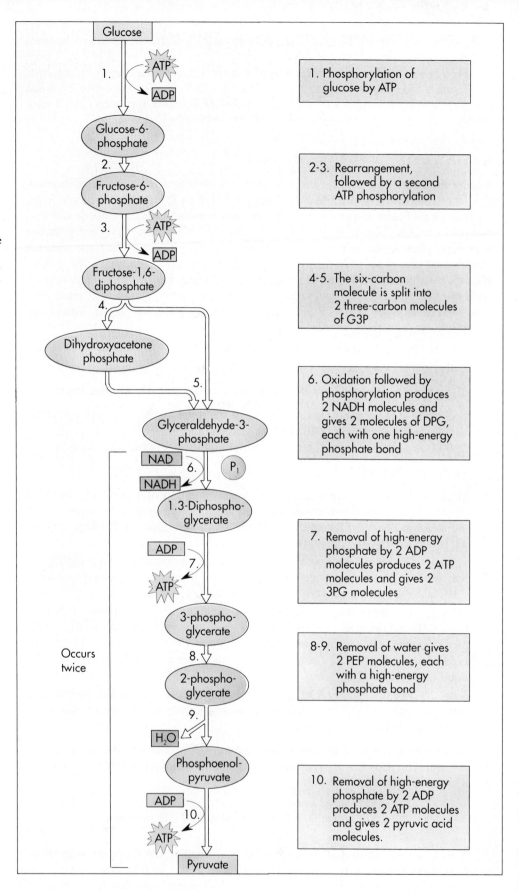

colysis does generate ATP, and for more than a billion years during the long, anaerobic first stages of the history of life on earth, this reaction sequence was the primary way for heterotrophic organisms to generate ATP from organic molecules.

The glycolytic reaction sequence generates a small amount of ATP by reshuffling the bonds of glucose molecules. Glycolysis is a very inefficient process, capturing only about 2% of the available chemical energy of glucose. Most of the remaining energy is unrecovered in the molecules that glycolysis produces, particularly pyruvate.

A **catabolic process** is one in which complex molecules are broken down into simpler ones; in contrast, an **anabolic process** is one in which more complex molecules are built up. Perhaps the original catabolic process involved only the ATP-yielding breakdown of G3P; the generation of G3P from glucose may have evolved later when alternative sources of G3P were depleted. Like many biochemical pathways, glycolysis may have evolved backwards, the last stages in the process being the most ancient.

All Cells Use Glycolysis

The glycolytic reaction sequence is thought to have been among the earliest of all biochemical processes to evolve. It uses no molecular oxygen and occurs readily in an anaerobic environment. All of its reactions occur free in the cytoplasm; none is associated with any organelle or membrane structure. Except for a few primitive bacteria, every living creature is capable of carrying out the glycolytic sequence. Most present-day organisms, however, are able to extract considerably more energy from glucose molecules than glycolysis does. You, for example, obtain 36 ATP molecules from each glucose molecule that you metabolize, and only two of these are obtained by substrate phosphorylation in glycolysis. Why is glycolysis maintained even now, considering that its energy yield is comparatively so paltry?

This simple question has an important answer: evolution is an incremental process. Change occurs during evolution by improving on past successes. In catabolic metabolism, glycolysis satisfied the one essential evolutionary criterion: it was an improvement. Cells that could not carry out glycolysis were at a competitive disadvantage. By studying the metabolism of contemporary organisms, we can see that only those cells that were capable of glycolysis survived the early competition of life. Later improvements in catabolic metabolism built on this success. Glycolysis was not discarded during the course of evolution, but rather used as the starting point for the further extraction of chemical energy. Nature did not, so to speak, go back to the drawing board and design a different and better metabolism from scratch. Rather, metabolism evolved as one layer of reactions was added to another, just as successive layers of paint can be found on the walls of an old apartment. We all carry glycolysis with us, a metabolic memory of our evolutionary past.

THE NEED TO CLOSE THE METABOLIC CIRCLE: REGENERATING NAD$^+$

Inspect for a moment the net reaction of the glycolytic sequence:

Glucose + 2ADP + 2P$_i$ + 2NAD$^+$ $\longrightarrow$ 2 Pyruvate + 2ATP + 2NADH + 2H$^+$ + 2H$_2$O

You can see that three changes occur in glycolysis: (1) glucose is converted to two molecules of pyruvate, (2) two molecules of ADP are converted to ATP, and (3) two molecules of NAD$^+$ are converted to NADH.

As long as foodstuffs are available that can be converted to glucose, a cell can continually churn out ATP to drive its activities—except for one problem: the NAD$^+$. The cell accumulates NADH molecules at the expense of the pool of NAD$^+$ molecules. A cell does not contain a large amount of NAD$^+$, and for glycolysis to continue, it is necessary to recycle the NADH that it produces back to NAD$^+$. Some

FIGURE 8-7
Fermentation. The conversion of pyruvate to ethanol takes place naturally in grapes left to ferment on vines, as well as in fermentation vats of crushed grapes. Often wine will be aged in casks to allow time for more complex chemical reactions to occur that subtly alter the flavor of the wine. The casks must be airtight to prevent bacterial contamination, because bacteria further metabolize the ethanol to acetic acid (vinegar).

other home must be found for the hydrogen atom taken from G3P—some other molecule that will accept the hydrogen and be reduced.

What happens after glycolysis depends critically on the fate of this hydrogen atom. One of two things generally happens:

1. *Oxidative respiration.* Oxygen is an excellent electron acceptor, and in the presence of oxygen gas the hydrogen atom taken from G3P can (through a series of electron transfers) be donated to oxygen, forming water. This is what happens in your body. Because air is rich in oxygen, this is referred to as aerobic metabolism.

2. *Fermentation.* When oxygen is not available, another organic molecule can accept the hydrogen atom instead. Such a process is called **fermentation** (Figure 8-7). This is what happens, for example, when bacteria grow without oxygen, and is referred to as anaerobic metabolism, metabolism without oxygen.

FERMENTATION

In the absence of oxygen, oxidative respiration cannot occur in animals for lack of a final electron acceptor. In such a situation a cell must rely exclusively on glycolysis to produce its ATP. Although it is much less efficient than oxidative respiration, glycolysis has one great advantage in this situation—it does work. In the absence of oxygen, cells donate the hydrogen atoms generated by glycolysis to organic molecules in a process called **fermentation.**

Bacteria carry out many kinds of fermentations, all using some form of organic molecule to accept the hydrogen atom from NADH and thus reform NAD^+:

$$\text{Organic molecule} + \text{NADH} \longrightarrow \text{Reduced organic molecule} + NAD^+$$

More than a dozen fermentation processes evolved among the bacteria, each using a different organic molecule as the hydrogen acceptor. Often the resulting reduced compound is an organic acid, such as acetic acid, butyric acid, propionic acid, or lactic acid. In other organisms, such as yeasts, the reduced compound is an alcohol.

Of these various bacterial fermentations, only a few occur among eukaryotes (Figure 8-8). In one fermentation process, which occurs, for example, in the single-celled

FIGURE 8-8
What happens to pyruvate, the product of glycolysis? In the presence of oxygen, pyruvate is oxidized to acetyl-CoA and enters the citric acid cycle. In the absence of oxygen, pyruvate instead is reduced, accepting the electrons extracted during glycolysis and carried by NADH. When pyruvate is reduced directly, as it is in your muscles, the product is lactic acid; when CO_2 is first removed from pyruvate and the remainder reduced, as it is in yeasts, the product is ethanol.

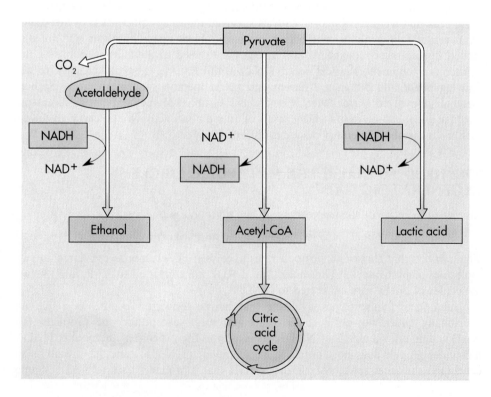

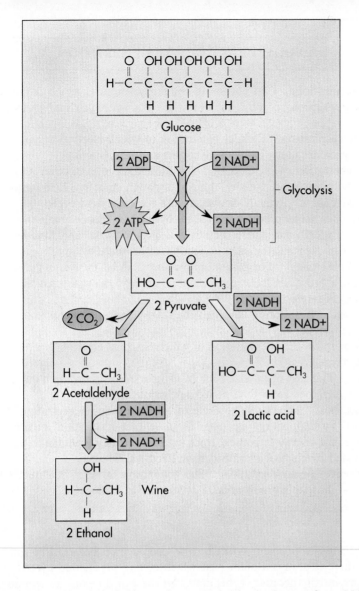

FIGURE 8-9

How wine is made. The conversion of pyruvate to ethanol is carried out by yeasts. It takes place naturally in grapes left to ferment on vines, as well as in fermentation vats of crushed grapes. When ethanol concentration reaches about 12%, the toxic effects of the alcohol kill the yeasts. What is left is wine. In your muscles, pyruvate is instead converted to lactic acid, which is less toxic than ethanol.

fungi called yeasts, the carbohydrate that accepts the hydrogen from NADH is the end product of the glycolytic process itself, pyruvate. Yeast enzymes remove a terminal CO_2 group from pyruvate (a process called decarboxylation), producing a toxic two-carbon molecule, acetaldehyde, and CO_2. Because of this CO_2, bread made with yeast rises and unleavened bread—that is, bread made without yeast—does not. Acetaldehyde then accepts the hydrogen from NADH, producing NAD^+ and ethyl alcohol, also called **ethanol.**

This particular type of fermentation has been of great interest to human beings, since it is the source of the ethyl alcohol in wine and beer (Figure 8-9). This alcohol is an undesirable by-product from the yeast's standpoint, since it becomes toxic to the yeast when it reaches high levels. That is why natural wine contains only about 12% alcohol, approximately the amount of alcohol that it takes to kill the yeast fermenting the sugars.

Most multicellular animals regenerate NAD^+ without decarboxylation, and the processes they use involve the production of by-products that are less toxic than alcohol. Your muscle cells, for example, use an enzyme called lactate dehydrogenase to add the hydrogen of the NADH produced by glycolysis back to the pyruvate that is the end product of glycolysis, converting pyruvate + NADH into lactic acid + NAD^+.

This process closes the metabolic circle, allowing glycolysis to continue for as long as the glucose holds out. Blood circulation removes excess lactic acid from muscles. When the lactic acid cannot be removed as fast as it is produced, our muscles cease to work well. Subjectively, they feel tired or leaden. Try raising and lowering your arm rapidly a hundred times and you will soon experience this sensation. A more efficient circulatory system will let you run longer before the accumulation of

aerobic respiration   That portion of cellular respiration which requires oxygen as an electron acceptor; includes the citric acid cycle and pyruvate oxidation.

anaerobic respiration   Cellular respiration in which electron acceptors other than oxygen are used; includes glycolysis and fermentation.

cellular respiration   The oxidation of organic molecules in cells; ATP is produced both by substrate-level phosphorylation and by chemiosmosis; includes glycolysis, pyruvate oxidation, and the citric acid cycle.

chemiosmosis   The production of ATP by the transport of high-energy electrons to a membrane where they drive a proton pump and thus power the synthesis of ATP via the osmotic reentry of protons.

maximum efficiency   The maximum number of ATP molecules generated by oxidizing a substance, relative to its bond energy; in organisms, the realized efficiency is rarely maximal.

fermentation   Respiration in which the final electron acceptor is an organic molecule.

oxidative respiration   Respiration in which the final electron acceptor is molecular oxygen.

oxidation   The loss of an electron. In cellular respiration, high-energy electrons are stripped from food molecules, oxidizing them.

photosynthesis   The energy to drive the chemiosmotic generation of ATP is provided by high-energy electrons as in cellular respiration, but in photosynthesis the energy is derived from light, whereas in cellular respiration it is provided by electrons stripped from chemical bonds.

substrate-level phosphorylation   The generation of ATP by directly transferring a phosphate group to ADP from another molecule.

---

lactic acid becomes a problem—this is why people train before they run marathon races—but there is always a point past which lactate production exceeds lactate removal. It is essentially because of this limit that the world record for running a mile is just under 4 minutes and not significantly less.

> **In fermentations, which are anaerobic processes, the electrons generated in the glycolytic breakdown of glucose are donated to an oxidized organic molecule.**
> **In aerobic metabolism, in sharp contrast, such electrons are transferred to oxygen, generating ATP in the process.**

OXIDATIVE RESPIRATION

Even in aerobic organisms, not all cellular respiration is aerobic; glycolysis and fermentation play important roles in the metabolism of most organisms. In all aerobic organisms, however, the oxidation of glucose that began in stage C of glycolysis is continued where glycolysis leaves off—with pyruvate. The evolution of this new biochemical process was conservative, as is almost always the case in evolution: the new process was simply tacked onto the old one. In eukaryotic organisms, oxidative respiration takes place exclusively in the mitochondria, which apparently originated as symbiotic bacteria with these characteristics. The oxidation of pyruvate takes place in two stages: the oxidation of pyruvate to form acetyl-CoA, and the subsequent oxidation of the acetyl-CoA. We will consider them in turn.

The Oxidation of Pyruvate

The first stage of oxidative respiration is a single oxidative reaction in which one of the three carbons of pyruvate is cleaved off, departing as CO_2 (chemists call a reaction

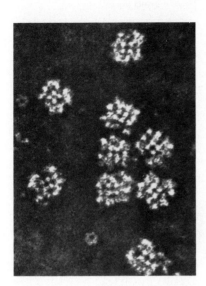

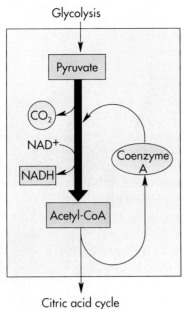

Glycolysis

Pyruvate

CO_2

NAD^+

NADH

Coenzyme A

Acetyl-CoA

Citric acid cycle

FIGURE 8-10

The oxidation of pyruvate. This complex reaction involves the reduction of NAD^+ to NADH and is thus a significant source of metabolic energy. Its product, acetyl-CoA, is the starting material for the citric acid cycle. The enzyme that carries out the reaction, pyruvate dehydrogenase, is one of the most complex enzymes known—it has 72 subunits, many of which can be seen in the electron micrograph.

of this kind a **decarboxylation**) and leaving behind two remnants (Figure 8-10): (1) a pair of electrons and their associated hydrogen, which reduce NAD^+ to NADH; and (2) a two-carbon fragment called an **acetyl group.**

This reaction is complex, involving three intermediate stages, and is catalyzed within mitochondria by an assembly of enzymes, a **multienzyme complex.** Such multienzyme complexes organize a series of reactions so that the chemical intermediates do not diffuse away or undergo other reactions. Within such a complex, component polypeptides pass the reacting substrate molecule from one enzyme to the next in line, without ever letting go of it. The complex of enzymes that removes the CO_2 from pyruvate, called **pyruvate dehydrogenase,** is one of the largest enzymes known—it contains some 72 subunits! In the course of the reaction, the two-carbon acetyl fragment removed from pyruvate is added to a cofactor, a carrier molecule called coenzyme A, forming a compound called **acetyl-CoA:**

$$\text{Pyruvate} + NAD^+ + \text{CoA} \longrightarrow \text{Acetyl CoA} + \text{NADH} + CO_2$$

This reaction produces a molecule of NADH, which is later used to produce ATP. Of far greater significance than the reduction of NAD^+ to NADH, however, is the residual fragment, acetyl-CoA (Figure 8-11). Acetyl-CoA is important because it is the end product of many different metabolic processes. It is produced not only by the oxidation of the product of glycolysis as outlined above, but also by the metabolic breakdown of proteins and of fats and other lipids. Acetyl-CoA thus provides a single focus for the many catabolic processes of the eukaryotic cell, all of whose resources are channeled into this single metabolite. Acetyl-CoA is the "currency" of oxidative metabolism.

Although acetyl-CoA is formed by many catabolic processes in the cell, there are only a limited number of processes that use acetyl-CoA, ones that expend this metabolic currency. Most acetyl-CoA either is directed toward energy storage (it is used in lipid synthesis) or is oxidized to produce ATP. Which of these two processes occurs depends on the level of ATP in the cell. When ATP levels are high, the oxidative pathway is inhibited and acetyl-CoA is channeled into fatty acid biosynthesis. This is why people get fat when they eat too much. When ATP levels are low, the oxidative pathway is stimulated and acetyl-CoA flows into energy-producing oxidative metabolism.

The Oxidation of Acetyl-CoA

The acetyl group of acetyl-CoA is oxidized by first binding it to a four-carbon molecule. The resulting six-carbon molecule then goes through a series of electron-yielding

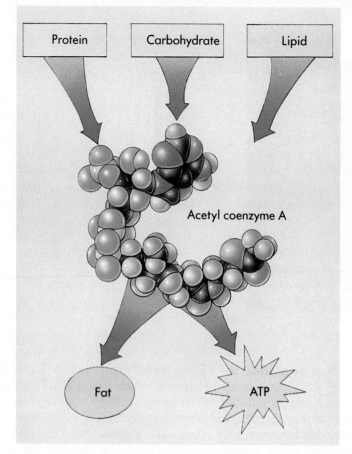

FIGURE 8-11
Acetyl-CoA is the central molecule of energy metabolism. Almost all the molecules that you use as foodstuffs are converted to acetyl-CoA when you metabolize them; the acetyl-CoA is then channeled into fat synthesis or into ATP production, depending on your body's energy requirements.

oxidation reactions, during which two CO_2 molecules are split off, restoring the four-carbon OAA molecule, which is then free to bind another acetyl group. The process is thus a cycle, a constant flow of carbon. In each turn of the cycle, a new acetyl group replaces the two CO_2 molecules lost, and more electrons are extracted.

The Reactions of the Citric Acid Cycle

The citric acid cycle consists of nine reactions (numbered 1 to 9 in the boxed essay) diagrammed in Figure 8-12. The cycle has two stages:

Stage A: Three **priming** reactions (reactions 1 to 3) set the scene. In the first reaction, acetyl-CoA joins the cycle, and then chemical groups are rearranged.

Stage B: It is in the second stage (reactions 4 to 9) that **energy extraction** occurs. Four of the six reactions are oxidations in which electrons are removed, and one generates an ATP equivalent directly by substrate-level phosphorylation.

The nine reactions together constitute a cycle, which begins and ends with oxaloacetate. At every turn of the cycle, acetyl-CoA enters and is oxidized, the electrons being channeled off to drive proton pumps that generate ATP. Each glucose molecule requires *two* turns of the cycle—one for each pyruvate molecule generated by glycolysis.

The Products of the Citric Acid Cycle

In the oxidation of glucose described in this chapter, we have removed high-energy electrons and have generated some ATP. The extracted electrons are temporarily housed within NADH molecules. In one reaction, the extracted electrons are not energetic enough to reduce NAD^+, and a different coenzyme, **flavin adenine dinucleotide (FAD),** is used to carry these less energetic electrons, and is reduced to $FADH_2$. Let us count the number of molecules of ATP and electron carriers that we have generated, starting from glucose (Table 8-1).

In the process of aerobic respiration the glucose molecule has been consumed entirely. Its six carbons were first cleaved into three-carbon units during glycolysis. One

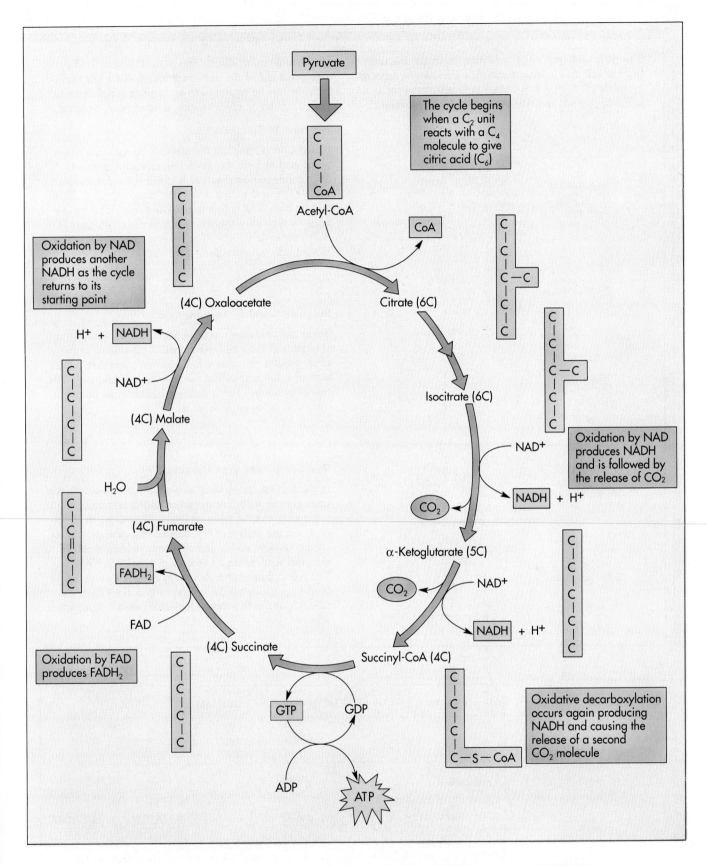

FIGURE 8-12
The citric acid cycle.

The citric acid cycle consists of nine sequential reactions that the cell uses to extract energetic electrons to drive the synthesis of ATP. A two-carbon molecule enters the cycle at the beginning, and two CO_2 molecules and several elec-

trons are given off during the cycle. The citric acid cycle represents one of the best examples of how a biochemical pathway can be organized to accomplish a sophisticated goal. It is clever, efficient, and, as chemistry, beautiful.

Reaction 1   Condensation

Acetyl-CoA is joined to the four-carbon molecule oxaloacetic acid to form the six-carbon molecule citric acid. This **condensation reaction** is irreversible, committing acetyl-CoA to the citric acid cycle. The reaction is inhibited by high ATP concentrations and stimulated by low ones, so that when the cell possesses ample amounts of ATP, the cycle is shut off and acetyl-CoA is channeled instead into fat synthesis.

Reactions 2 and 3   Isomerization

Before the oxidation reactions begin, the hydroxyl group of citric acid must be repositioned. This is done in two steps: a water molecule is first removed, and then added back to a different carbon, so that an —H and an —OH change positions. The resulting molecule is an isomer of citric acid called **isocitric acid.**

Reaction 4   The First Oxidation

In the first energy-yielding step of the cycle, isocitric acid undergoes an **oxidative decarboxylation** reaction, such as you encountered in the conversion of pyruvate to acetyl-CoA. On the surface of the enzyme that catalyzes this reaction, isocitric acid is first oxidized, yielding a pair of electrons that reduce a molecule of NAD^+ to NADH. The reduced carbohydrate intermediate is then decarboxylated by splitting off the central carbon as a CO_2 molecule, yielding a five-carbon molecule called α-ketoglutaric acid.

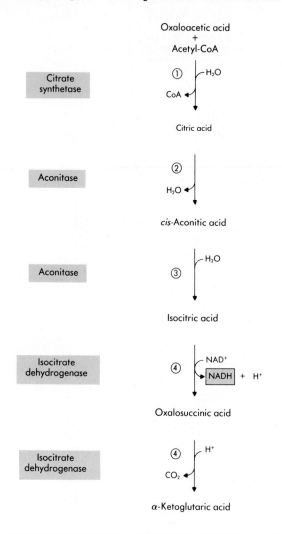

TABLE 8-1	THE OUTPUT OF AEROBIC METABOLISM		
	SUBSTRATE LEVEL PHOSPHORYLATION	OXIDATION	
Glycolysis	2 ATP	2 NADH	
Oxidation of pyruvate (×2)		2 NADH	
Citric acid cycle (×2)	2 ATP	6 NADH	2 $FADH_2$
	4 ATP	10 NADH	2 $FADH_2$

The oxidative metabolism of one molecule of glucose proceeds in three stages: glycolysis (which can be anaerobic), the oxidation of pyruvate, and the citric acid cycle. Both glycolysis and the citric acid cycle produce two ATP molecules by substrate level phosphorylation, coupling ATP production to a reaction that involves a large release of free energy. All three processes involve oxidation reactions, that is, reactions that produce electrons. The electrons are donated to a carrier molecule, NAD^+ (or in one instance FAD^+).

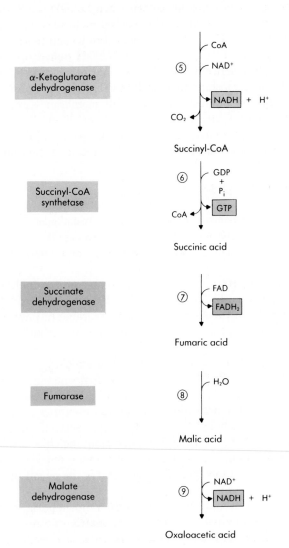

α-Ketoglutarate dehydrogenase

⑤ CoA / NAD⁺

NADH + H⁺

CO_2

Succinyl-CoA

Succinyl-CoA synthetase

⑥ GDP + P_i

CoA

GTP

Succinic acid

Succinate dehydrogenase

⑦ FAD

$FADH_2$

Fumaric acid

Fumarase

⑧ H_2O

Malic acid

Malate dehydrogenase

⑨ NAD⁺

NADH + H⁺

Oxaloacetic acid

Reaction 5   The Second Oxidation

α-Ketoglutaric acid is then oxidatively decarboxylated itself by a multienzyme complex much like pyruvate dehydrogenase. As in that reaction, the fragment left after removal of CO_2 is joined to a molecule of coenzyme A, in this case forming succinyl-CoA, and a pair of electrons is extracted, which reduces a molecule of NAD⁺ to NADH.

Reaction 6   Substrate Level Phosphorylation

The bond linking the four-carbon succinyl group to the CoA carrier molecule is a high-energy ester linkage, similar in character to the high-energy phosphoanhydride linkage of ATP. Such a bond represents metabolic energy that is "ripe for the picking." In a coupled reaction similar to those encountered in glycolysis, this bond is cleaved, its energy driving the phosphorylation of guanosine diphosphate (GDP). The GTP that results is readily converted to ATP. The four-carbon fragment that remains at the completion of this reaction is called succinic acid.

Reaction 7   The Third Oxidation Reaction

In the third of the four oxidation-reduction reactions of the cycle, succinic acid is oxidized to fumaric acid. Because the free energy change in this reaction is not large enough to reduce NAD⁺, a different electron acceptor, **flavin adenine dinucleotide (FAD⁺)**, is used. Unlike NAD⁺, FAD⁺ is not free to diffuse into the surroundings; rather, it is an integral part of the membrane. Its reduced form, $FADH_2$, contributes electrons directly to the electron transport network of the membrane.

Reactions 8 and 9   Regenerations of Oxaloacetic Acid

In the final two reactions of the cycle, a water molecule is added to fumaric acid and the resulting malic acid is oxidized. The oxidation yields a four-carbon molecule of oxaloacetic acid and two electrons that reduce a molecule of NAD⁺ to NADH. Oxaloacetic acid, the molecule with which the cycle began, is now free to bind another molecule of acetyl-CoA and reinitiate the cycle.

of the carbons of each three-carbon unit was then lost as CO_2 in the conversion of pyruvate to acetyl-CoA, and the other two carbons were lost during the oxidations of the citric acid cycle. All that is left to mark the passing of the glucose molecule is its energy, which is preserved in four ATP molecules and in the reduced state of 12 electron carriers.

The systematic oxidation of the two pyruvate molecules left after glycolysis generates two ATP molecules, which is as many as glycolysis produced. More importantly, this process harvests many energized electrons, which can then be directed to the chemiosmotic synthesis of ATP.

USING ELECTRONS GENERATED BY THE CITRIC ACID CYCLE TO MAKE ATP

The NADH and $FADH_2$ molecules formed during glycolysis and the subsequent oxidation of pyruvate each contain a pair of electrons gained when NADH was formed from NAD^+ and $FADH_2$ was formed from FAD^+. The NADH molecules carry their electrons to the cell membrane (the $FADH_2$ is already attached to it) and there transfer them to a complex membrane-embedded protein called **NADH dehydrogenase.** The electrons are then passed on to a series of cytochromes and other carrier molecules (Figure 8-13), one after the other, losing much of their energy in the process by driving several transmembrane proton pumps. This series of membrane-associated electron carriers is collectively called the **electron transport chain** (Figure 8-14). At the terminal step of the electron transport chain, the electrons are passed to the **cytochrome c oxidase complex,** which uses four of the electrons to reduce a molecule of oxygen gas and form water:

$$O_2 + 4H^+ + 4e^- \rightarrow 2H_2O$$

> The electron transport chain puts the electrons garnered from the oxidation of glucose to work driving proton-pumping channels. The ultimate acceptor of the electrons harvested from pyruvate is oxygen gas, which is reduced to form water.

It is the availability of a plentiful electron acceptor (that is, an oxidized molecule) that makes oxidative respiration possible. In the absence of such a molecule, oxidative respiration is not possible. The electron transport chain used in aerobic respiration is similar to, and is thought to have evolved from, the one employed in aerobic photosynthesis.

Thus we see that the final products of oxidative metabolism are CO_2 and water. This is not an impressive result in itself. Recall, however, what happened in the process of forming that CO_2 and water. The electrons contributed by NADH molecules passed down the electron transport chain, activating three proton-pumping channels. Similarly, the passage of electrons contributed by $FADH_2$, which entered the chain later, activated two of these channels. The electrons harvested from the citric acid cycle have in this way been used to pump a large number of protons out across the membrane, driving the chemiosmotic synthesis of ATP, and *that* is the payoff of oxidative respiration.

In eukaryotes, oxidative metabolism takes place within the mitochondria, which are present in virtually all cells. The internal compartment or **matrix** of a mitochondrion contains the enzymes that carry out the reactions of the citric acid cycle. Electrons harvested there by oxidative respiration are used (via energy changes along the electron transport chain) to pump protons out of the matrix into the **outer compartment,** sometimes called the intermembrane space. As a high concentration of protons builds up, protons recross the membrane back into the matrix, driven by diffusion. The only way they can get in is through special channels, which are visible in electron micrographs (see Figure 8-14) as projections on the inner surface of the inner membrane. This protein complex synthesizes ATP within the matrix when protons travel inward through the channel (Figure 8-15). The ATP then passes by facilitated diffusion out of the mitochondrion and into the cell's cytoplasm.

> The electrons harvested from glucose and carried to the membrane by NADH drive protons out across membranes. The return of the protons into the mitochondrion by diffusion generates ATP.

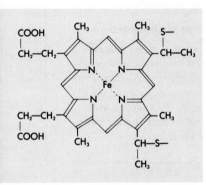

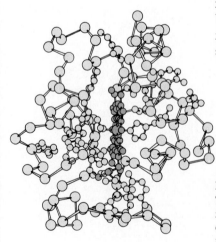

FIGURE 8-13

The structure of cytochrome c. Cytochromes are respiratory proteins that contain "heme" groups, complex iron-containing carbon rings often found in molecules that transport electrons. In the molecular model of cytochrome c, the heme group is indicated in red. Heme is an iron-containing pigment that has a complex ring structure with many alternating single and double bonds.

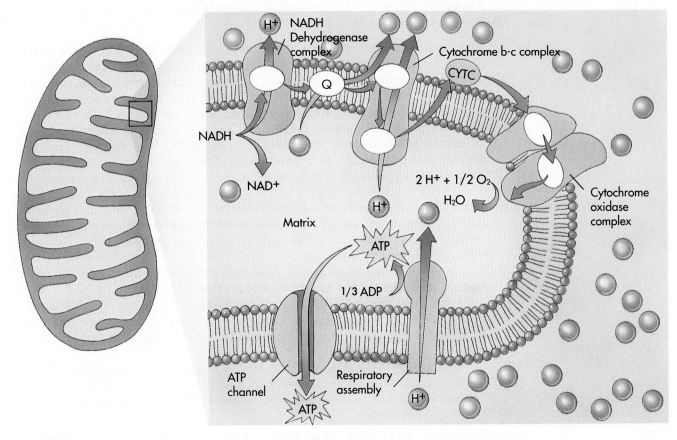

FIGURE 8-14

The electron transport chain: how mitochondria use electrons to make ATP. High-energy electrons harvested from food molecules *(red arrows)* are transported along a chain of membrane proteins, three of which use portions of the electron's energy to pump protons *(blue arrows)* out of the matrix into the other compartment. The electrons are finally donated to oxygen to form water. ATP is formed by chemiosmosis as protons re-enter the matrix, driven by the high outside concentration.

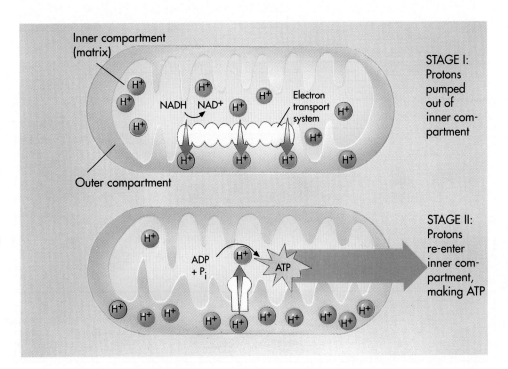

FIGURE 8-15

How proton movement in the mitochondrion drives ATP production. A summary of proton movement in mitochondria during chemiosmosis.

Stage I The electron transport system uses energetic electrons from NADH to drive the pumping of electrons out of the inner matrix compartment. The result is a deficit of protons in the matrix and an excess in the outer compartment.

Stage II Protons re-enter the matrix, driven by diffusion, through special channels that produce ATP. These ATP molecules diffuse out of the mitochondrion through passive channels *(red arrow)*.

AN OVERVIEW OF GLUCOSE CATABOLISM: THE BALANCE SHEET

How much metabolic energy does the chemiosmotic synthesis of ATP produce? One ATP molecule is generated chemiosmotically for each activation of a proton pump by the electron transport chain. Thus each NADH molecule that the citric acid cycle generates ultimately causes the production of three ATP molecules, as its electrons activate three pumps. Each $FADH_2$, activating two of the pumps, leads to the production of two ATP molecules. Because of a later evolutionary development that we will discuss below, however, eukaryotes carry out glycolysis in their cytoplasm and the citric acid cycle within their mitochondria. This separation of the two processes into separate compartments within the cell leads to the requirement that the electrons of the NADH created during glycolysis must be transported across the mitochondrial membrane, an event that costs one ATP molecule per NADH. Thus each glycolytic NADH produces only two ATP molecules in the final balance sheet instead of three. Overall, 32 ATP molecules result from chemiosmotic phosphorylation, compared with four produced by substrate-level phosphorylation.

The overall efficiency of glucose catabolism is high: the aerobic oxidation (combustion) of glucose yields a maximum of 36 ATP molecules (a total of $-7.3 \times 36 = -263$ kcal/mole) (Figure 8-16). The aerobic oxidation of glucose thus has an efficiency of $263/686$, or 38%. Compared with the two ATP molecules generated by glycolysis, aerobic oxidation is 18 times more efficient, an enormous improvement.

Oxidative respiration is 18 times more efficient than glycolysis at converting the chemical energy of glucose into ATP. It produces 36 molecules of ATP from each glucose molecule consumed, compared with the two that are produced by glycolysis.

FIGURE 8-16
An overview of the energy extracted from the oxidation of glucose.

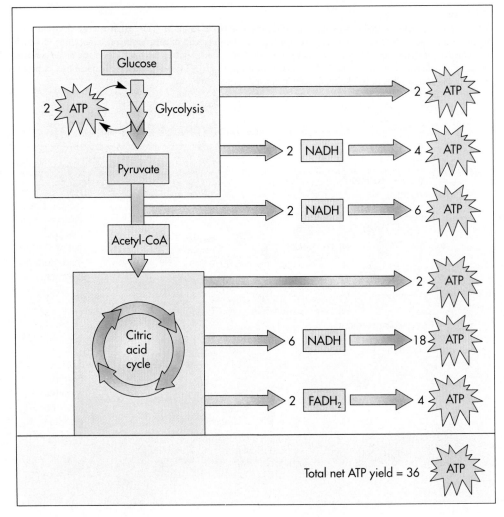

EATING OTHER FOODS

Thus far we have discussed oxidative respiration in terms of glucose, which organisms obtain from the digestion of carbohydrates. The other building blocks of cells also serve as important sources of food energy, particularly fats and protein. Humans usually obtain more of their energy from oxidizing fatty acids (from fats) than from oxidizing glucose.

Cellular Respiration of Protein

To use the protein within cells as food, organisms first digest the protein molecules, breaking down the amino acid chains into individual amino acids. The nitrogen side group (the "amino group") is then removed from each amino acid, and the carbon chain that remains is converted, through a series of reactions, to a substance that is part of the citric acid cycle. Alanine, for example, is converted to pyruvate, glutamate to α-ketoglutarate, aspartate to oxaloacetate. The machinery of the citric acid cycle is then used to extract the high energy electrons, which are put to work making ATP.

Cellular Respiration of Fat

To use fats as food, organisms first digest the fat, breaking it down into individual fatty acids. The long carbon chains of these molecules have many hydrogen atoms and so provide a rich harvest of energy—one gram contains more than twice as many kilocalories as one gram of glucose or amino acids. When completely oxidized by cellular respiration, a molecule of a six-carbon fatty acid can generate up to 44 ATPs, compared with 36 for glucose.

Fatty acids are oxidized in the matrix of the mitochondrion. There an assembly line of four enzymes working in succession attacks the ends of the long fatty acid

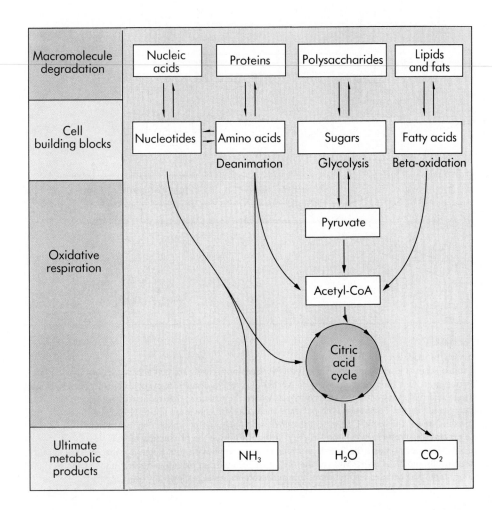

FIGURE 8-17

How cells extract chemical energy. All eukaryotes and many prokaryotes extract energy from organic molecules by oxidizing them. The first stage of this process, tearing down macromolecules into their constituent parts, yields little energy. The second stage, oxidative respiration, extracts energy, primarily in the form of high-energy electrons, and leaves simple inorganic molecules as final end products.

chains, splitting off two-carbon chunks (acetyl groups), which they stick onto coenzyme A—that is, each chunk is converted to acetyl-CoA. Like nibbling down the end of a candy cane, the long fatty acid chain (which might typically have 16 or more CH_2 links) is progressively shortened, until the entire fatty acid is converted to acetyl-CoA. This process is known as **beta-oxidation**.

Each cycle of beta-oxidation requires one molecule of ATP to prime the process, and produces not only an acetyl CoA molecule, but also one NADH and one $FADH_2$. When their electrons are passed through the electron transport system, the NADH produces three and the $FADH_2$ produces two ATPs, for a total of five ATPs. Each acetyl CoA produces an additional 12 ATPs as it is oxidized in the citric acid cycle. Thus for a six-carbon fatty acid, the energy yield of beta-oxidation is 10 ATPs from two rounds of splitting, minus 2 ATPs required to prime the two rounds of splitting, plus 36 ATPs from oxidizing the three acetyl CoAs: $10 - 2 + 36 = 44$ ATPs. Obviously, the cellular respiration of fat yields a great deal of energy. That is why your body stores excess energy as fat—it is a very efficient storage vehicle. If you stored excess energy as carbohydrate instead, as a potato does, you would be a lot bulkier!

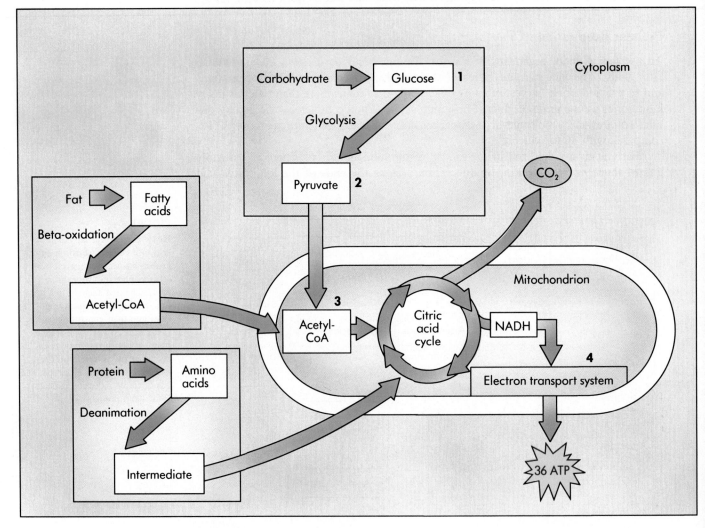

FIGURE 8-18

An overview of oxidative respiration.

1 The initial stage of glucose metabolism, glycolysis, does not require oxygen and occurs in the cytoplasm of cells; during it, each glucose molecule is converted into two molecules of pyruvic acid. In eukaryotic cells, the pyruvic acid molecules enter the mitochondria, where the reactions that require oxygen occur.

2 In the second stage of respiration, the pyruvate molecules are converted to acetyl-CoA.

3 In the third stage, the acetyl-CoA enters the citric acid cycle and is broken down to carbon dioxide. Both of these processes harvest chemical bond energy, using it to drive the production of NADH.

4 The NADH in turn powers the fourth and final stage of respiration, the electron transport system that chemiosmotically produces ATP. Most of the ATP that results from oxidative respiration is at the final chemiosmotic stage. The cellular respiration of fats and proteins are also shown in this diagram.

Oxidative metabolism is so efficient at extracting energy from organic molecules that its development opened up an entirely new avenue of evolution. For the first time it became feasible for heterotrophic organisms to evolve that derived their metabolic energy exclusively from the oxidative breakdown of other organisms. Sugars, lipids, fats, proteins, nucleic acids—all provide potential sources of energy, since all can be degraded to compounds that feed into the citric acid cycle (Figure 8-17). As long as some organisms produce energy by photosynthesis, heterotrophs can exist solely by feeding on these autotrophs. The high efficiency of oxidative respiration was one of the key conditions that fostered the evolution of nonphotosynthetic heterotrophic organisms. An overview of the process is presented in Figure 8-18.

Regulation of Cellular Respiration

When cells possess plentiful amounts of ATP, negative feedback mechanisms shut down key reactions of glycolysis, the citric acid cycle, and fatty acid breakdown, and thus slow down ATP production. Conversely, when ATP levels in the cell are low, the ADP will activate enzymes in the pathways of glucose, protein, and fat catabolism to stimulate production of more ATP. The regulation of these biochemical pathways by levels of ATP is an example of **feedback inhibition.** Almost all of the biochemical pathways in cells are regulated by the sensitivity of key enzymes to particular metabolites. The cell's metabolic machinery operates like a well-designed automobile engine, every part working in concert.

SUMMARY

1. Metabolism is driven by the application of energy to endergonic reactions. Those reactions that require a net input of free energy are coupled to the cleavage of ATP, which involves a large release of free energy.

2. Organisms acquire ATP from photosynthesis or by harvesting the chemical energy in the bonds of organic molecules.

3. Harvesting energy can be done in two ways: substrate-level phosphorylation, in which some reactions involving a large decrease in free energy are coupled to ATP formation; or oxidative respiration, in which electrons are used to drive proton pumps, and thus the chemiosmotic synthesis of ATP.

4. Glycolysis harvests chemical energy in the first of these two ways, by rearranging the chemical bonds of glucose to form two molecules of pyruvate and two molecules of ATP.

5. Organisms living in anaerobic environments require a mechanism to dispose of the electron and associated hydrogen produced in the oxidation step of glycolysis. They are donated to one of a number of carbohydrates in a process called fermentation.

6. Organisms living in aerobic environments dispose of the electrons produced in glycolysis by using them in chemiosmosis to generate ATP; the hydrogen and associated electron are then added to oxygen to form water. Further electrons are stripped from the end-product of glycolysis (pyruvate) in two stages: (1) pyruvate is oxidized to acetyl-CoA, and (2) acetyl-CoA is oxidized further in the citric acid cycle. Both oxidations occur within mitochondria, generating a stream of high-energy electrons.

7. Aerobic eukaryotic organisms direct these electrons to a mitochondrial membrane, where they drive proton pumps.

8. The excess of outside protons created by the proton pumps diffuses back across the membrane through special channels, and the passage of these protons drives the production of ATP.

9. The aerobic oxidation of glucose results in a maximal net production of 36 ATP molecules, all but four of them produced by chemiosmosis.

10. Within all eukaryotic cells, oxidative respiration of pyruvate takes place within the matrix of mitochondria, which act as closed osmotic compartments from which protons are pumped out into the outer compartment. They diffuse back in to create ATP. The ATP then leaves the mitochondrion by facilitated diffusion.

11. The use of fat to fuel ATP synthesis is called beta-oxidation. Humans obtain more energy from oxidizing fats than from oxidizing glucose.

REVIEW QUESTIONS

1. Living organisms generate ATP in two ways; what are they, and how does each work? Which reaction was the first to evolve? Which generates the greatest amount of ATP?

2. Where in the cell does glycolysis occur? How does this reaction differ when oxygen is absent as compared with when it is present?

3. What molecule starts glycolysis? What molecule remains at the completion of glycolysis? How many ATPs are produced by glycolytic substrate level phosphorylation? What is its net production of ATP, and why is it different from the total production? How many additional ATPs are produced when oxygen is present?

4. Given an unlimited source of glucose, what is the limiting factor in the glycolytic reaction?

5. What two mechanisms exist to recycle the postglycolytic NADH? How do they differ from one another? Which is an aerobic process, and which is anaerobic?

6. How is fermentation different from oxidative respiration in terms of electron acceptors and energy production?

7. What are the end-products of alcoholic fermentation? Which products are toxic to a living cell?

8. Why is anaerobic respiration important to the human muscular system?

9. How is the exchange of electrons for ATP different among the NADH produced in glycolysis, the NADH produced in the citric acid cycle, and the $FADH_2$ produced in the citric acid cycle?

10. What is the overall efficiency of the complete aerobic degradation of one molecule of glucose? How does this compare with the efficiency of glycolysis alone?

 THOUGHT QUESTIONS

1. If you poke a hole in a mitochondrion, can it still carry out oxidative respiration? Can fragments of mitochondria carry out oxidative respiration? Explain.

2. Why have eukaryotic cells not dispensed with mitochondria, placing the mitochondrial genes in the nucleus and incorporating the mitochondrial ATPase and proton pump into the cell's smooth ER?

 FOR FURTHER READING

DICKERSON, R.: "Cytochrome *c* and the Evolution of Energy Metabolism," *Scientific American*, March 1980, pages 136-154. A superb description of how the metabolism of modern organisms evolved.

HINKLE, P., and R. McCARTY: "How Cells Make ATP," *Scientific American*, March 1978, pages 104-125. A good summary of oxidative respiration, with a clear account of the events that happen at the mitochondrial membrane.

LEVINE, M., AND OTHERS: "Structure of Pyruvate Kinase and Similarities with Other Enzymes: Possible Implications for Protein Taxonomy and Evolution," *Nature*, vol. 271, pages 626-630, 1978. An advanced article, well worth the effort, that recounts how the enzymes of oxidative metabolism may have evolved.

McCARTY, R.: "H^+-ATPases in Oxidative and Photosynthetic Phosphorylation," *BioScience*, January 1985, pages 27-33. An excellent overview of the key protein channels that carry out chemiosmosis.

9

Photosynthesis

OVERVIEW

We all depend on the process of photosynthesis, by means of which the energy that builds our bodies is captured from sunlight. Photosynthesis is one of the oldest and most fundamental of life processes, and the major types first evolved billions of years ago among the bacteria. The first of these to evolve apparently used the energy obtained from sunlight to split hydrogen sulfide, generating sulfur as a by-product. Later, in the cyanobacteria, a system evolved in which water is split in a similar way, yielding oxygen gas. The outcome of this form of photosynthesis, carried out over nearly 3 billion years, has been the production of an atmosphere rich in oxygen; this atmosphere has set the stage for the evolution of all complex forms of life on earth, including ourselves.

FOR REVIEW

Here are some important terms and concepts that you will encounter in this chapter. If you are not familiar with them, you should review them before proceeding.

Electron energy levels (Chapter 2)

Chloroplasts (Chapter 5)

Chemiosmosis (Chapters 6 and 8)

Oxidation-reduction (Chapter 7)

Glycolysis (Chapter 8)

FIGURE 9-1
Capturing energy. These sunflowers, growing vigorously in the August sun, capture enough energy to power your body for hours.

For all its size and diversity, our universe might never have spawned life, except for one characteristic of overriding importance: it is awash in energy. Everywhere in the universe, matter is continually being converted to energy by thermonuclear processes. The energy from those processes streams from the stars, which include our sun, in all directions. The amount of radiant energy that reaches the earth from the sun each day is the equivalent of about 1 million Hiroshima-sized atomic bombs. About a third of this energy is radiated back into space immediately, and most of the remainder is absorbed by the earth and converted to heat. Less than 1% of the energy that reaches the earth is captured as a result of the process of photosynthesis and provides the energy that drives all of the activities of life on earth (Figure 9-1).

In this chapter we will provide a detailed account of photosynthesis, both in an evolutionary and in a functional context. We will review the way in which photosynthesis is thought to have evolved among the photosynthetic bacteria, and we will then conduct a more detailed examination of the particular pattern of photosynthesis that is characteristic of the photosynthetic eukaryotes, which are the plants and algae.

THE BIOPHYSICS OF LIGHT
The Photoelectric Effect

Where is the energy in light? What is there about sunlight that a plant can use to create chemical bonds? To answer these questions, we need to begin by considering the physical nature of light itself. Perhaps the best place to start is with a curious experiment carried out in a laboratory in Germany in 1887. A young German physicist, Heinrich Hertz, was attempting to verify a highly mathematical theory that predicted the existence of electromagnetic waves. To see whether such waves existed, Hertz constructed a spark generator in his laboratory—a machine that looked like two shiny metal spheres standing near each another on slender rods. When a very high static electrical charge built up on one sphere, sparks would jump across to the other sphere.

After Hertz had constructed this system, he set out to investigate whether the sparking would create invisible electromagnetic waves, as predicted by the mathematical theory. On the other side of the room, he placed a metal hoop—a thin circle that was almost, but not quite, closed—on an insulating stand. When he turned on the spark generator across the room, tiny sparks could be seen to pass across the gap in the hoop! This was the first demonstration of radio waves. But Hertz noted a curious side-effect as well. When light shone on the ends of the hoop, the sparks came more readily. This unexpected facilitation, called the **photoelectric effect,** puzzled investigators for many years.

Especially perplexing was the fact that the strength of the photoelectric effect depended not only on the brightness of the light shining on the gap in the hoop, but also on its wavelength. Short wavelengths were much more effective than long ones in producing the photoelectric effect. This effect was finally explained by Albert Einstein as a natural consequence of the physical nature of light. It turns out that light was literally blasting electrons from the metal surface at the ends of the hoop, creating ions and thus facilitating the passage of the electronic spark induced by the radio waves. Light consists of units of energy called **photons,** and some of these photons were being absorbed by the metal atoms of the hoop. In this process, some of the electrons of the metal atoms were being boosted into higher energy levels, and so ejected from the atoms.

The Role of Photons

Photons do not all possess the same amount of energy. Some possess a great deal of energy, others far less. The energy content of light is inversely proportional to the wavelength of the light: short-wavelength light contains photons of higher energy than those of long-wavelength light. For this reason, the photoelectric effect was more pronounced with light of shorter wavelength: the metal atoms of the hoop were being bombarded with higher-energy photons. Sunlight contains photons of many different energy levels, only some of which our eyes perceive as visible light. The highest-energy photons, at the short-wavelength end of the electromagnetic spectrum (Figure

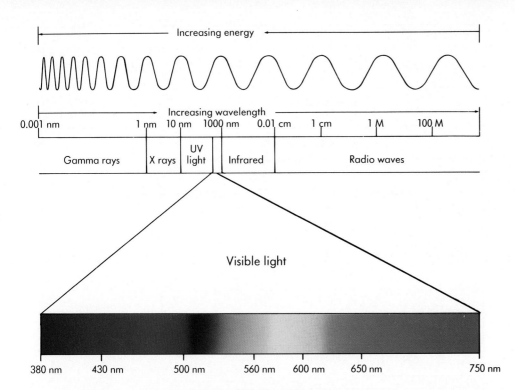

FIGURE 9-2
The electromagnetic spectrum.
Light is a form of electromagnetic
energy and is conveniently
thought of as a wave. The shorter
the wavelength of light, the
greater the energy. Visible light
represents only a small part of the
electromagnetic spectrum, that
between 380 and 750 nanometers.

9-2), are gamma rays, with wavelengths of less than 1 nanometer; the lowest-energy
photons, with wavelengths of thousands of meters, are radio waves. Within the visible
spectrum, blue light has the shortest wavelength and the most energetic photons, and
red light has the longest wavelength and the least energetic photons.

Ultraviolet Light

Sunlight contains a significant amount of ultraviolet light, which because it has a
shorter wavelength than visible light possesses considerably more energy. Ultraviolet
light is thought to have been an important source of energy on the primitive earth
when life originated. Today's atmosphere contains ozone (derived from oxygen gas),
which absorbs most of the ultraviolet-energy photons in sunlight—but anyone who
has been sunburned knows that considerable quantities of ultraviolet-energy photons
still manage to penetrate the atmosphere. As we will learn in a later chapter, mysterious "holes" have recently been discovered in the ozone layer; these holes, which apparently have been caused by human activities, threaten to cause an enormous jump
in the incidence of skin cancers in human beings throughout the world.

As we saw in Chapter 2, electrons occupy discrete energy levels in their orbits
around the atomic nucleus. To boost an electron to a different energy level requires
just the right amount of energy, no more and no less; similarly, if you are climbing a
ladder, you must raise your foot just so much to climb to a higher rung, not a centimeter more or less. Specific atoms can therefore absorb only certain photons of
light—those that correspond to available energy levels. A given atom or molecule has
a characteristic range, or **absorption spectrum,** of photons that it is capable of absorbing. The absorption spectrum of a given atom or molecule depends on the electron energy levels that are available in it.

CAPTURING LIGHT ENERGY IN CHEMICAL BONDS

The energy of light is "captured" by a molecule that absorbs it, in that the photon of
energy is used to boost an electron of the molecule to a higher energy level. Molecules
that absorb light are called **pigments,** and organisms have evolved a variety of such
pigments. There are two general sorts of photosynthetic pigments:

1. *Carotenoids.* Carotenoids consist of carbon rings linked to chains in which single and double bonds alternate. They can absorb photons of a wide range of
   different energies, although they are not always highly efficient. A typical car-

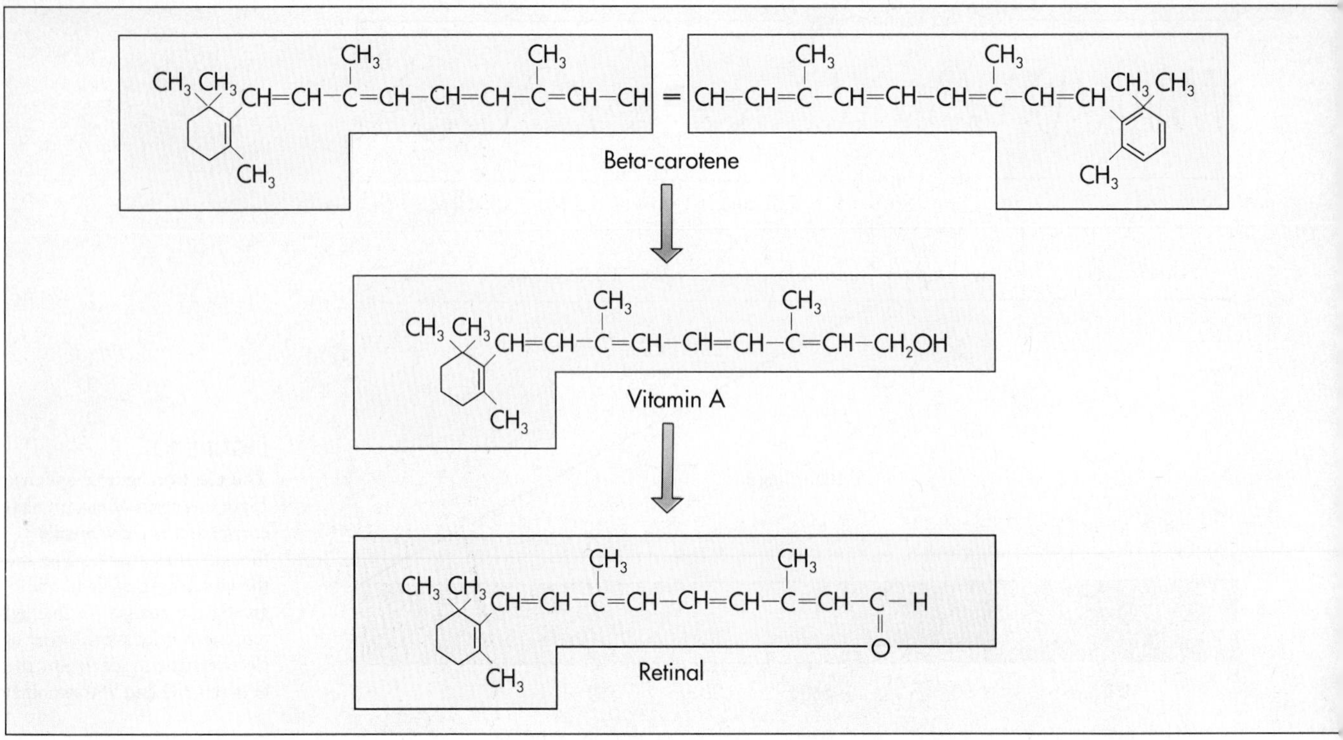

FIGURE 9-3

What do carrots have to do with vision? The pigment beta-carotene is what makes carrots look orange. When the double bond that links the two halves of beta-carotene is broken, two molecules of vitamin A are produced. Retinal, which functions as the key visual pigment in your eyes, is produced from vitamin A.

otenoid is **beta-carotene,** which is shown in Figure 9-3; in it, two carbon rings are connected by a chain of 18 carbon atoms connected alternately by single and double bonds. Splitting a molecule of beta-carotene into equal halves produces two molecules of **vitamin A.** When vitamin A is oxidized in turn, **retinal,** the pigment used in human vision, is produced. That explains the connection between carrots (rich in beta-carotene), vitamin A, and vision.

2. *Chlorophylls.* Other biological pigments, called chlorophylls, absorb photons by means of an excitation process analogous to the photoelectric effect. These pigments use a metal atom (magnesium). Photons absorbed by the pigment molecule excite electrons which are then channeled away through the carbon-bond system. The magnesium lies at the center of a complex ring structure of alternating single and double bonds, called a **porphyrin ring.** This ring structure consists of alternating single and double carbon bonds. Photons absorbed by the pigment molecule excite electrons of the porphyrin ring, which are then channeled away through the carbon-bond system. Several small side groups are attached to the outside of the porphyrin ring. These side groups alter the absorption properties of the pigment in different kinds of chlorophyll (Figure 9-4).

The different kinds of chlorophyll absorb photons only of a narrow energy range. As you can see from their absorption spectra, the two kinds of chlorophyll that occur in plants, chlorophylls *a* and *b*, absorb primarily violet-blue and red light, respectively. Light between wavelengths of 500 and 600 nanometers is not absorbed by chlorophyll pigments and is therefore reflected by plants. The light reflected from a chlorophyll-containing plant has had all of its middle-energy photons absorbed by the chlorophyll except those in the 500 to 600 nanometer range. When these photons are subsequently absorbed by the retinal in our eyes, we perceive them as green.

Chlorophyll *b* has an absorption spectrum shifted toward the green wavelengths. It acts as an **accessory pigment** within the photocenter of plants, one that is able to absorb photons that chlorophyll *a* cannot. By doing this, chlorophyll *b* greatly increases the proportion of the photons of sunlight that a plant can harvest.

A pigment is a molecule that absorbs light. The wavelengths absorbed by a particular pigment depend on the available energy levels to which light-excited electrons can be boosted.

A

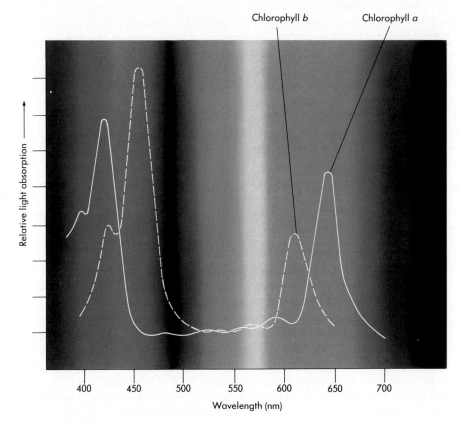

Chlorophyll *b*   Chlorophyll *a*

B

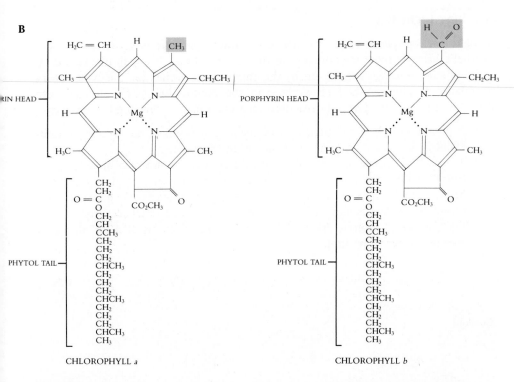

CHLOROPHYLL *a*   CHLOROPHYLL *b*

FIGURE 9-4
Chlorophyll. The chlorophylls absorb violet-blue and red light in two narrow bands of the spectrum.
A Absorption spectra for chlorophyll *a* and chlorophyll *b*. These spectra were obtained from solvent-
extracted pigments in vitro (values are higher in vivo when the pigments are associated with proteins).
B The structures of chlorophyll *a* and chlorophyll *b*. The only difference between the two chlorophyll mole-
cules is the substitution of a —CHO group in chlorophyll *b* for a —CH₃ group in chlorophyll *a*.

All plants and algae, and all but one primitive group of photosynthetic bacteria, use chlorophylls as their primary light-gatherers. It is reasonable to ask why these photosynthetic organisms do not use a pigment like retinal, which has a broad absorption spectrum and can harvest light in the 500 to 600 nanometer wavelength range as well as at other wavelengths. The most likely hypothesis involves photoefficiency. Retinal absorbs a broad spectrum of light wavelengths, but does so with relatively low efficiency; chlorophyll, in contrast, absorbs in only two narrow bands, blue and red, but does so with high efficiency. By using chlorophyll, therefore, plants and most other photosynthetic organisms achieve far higher overall photon capture rates than if they used a pigment with a broader but less efficient spectrum of absorption.

AN OVERVIEW OF PHOTOSYNTHESIS

Photosynthesis is a complex series of events that involves three kinds of chemical processes (Figure 9-5):

1. The first process to occur is the chemiosmotic generation of ATP by electrons harvested by using sunlight. The reactions involved in this process are called the **light reactions** of photosynthesis, since the photosynthetic synthesis of ATP takes place only in the presence of light.
2. The light reactions are followed by a series of enzyme-catalyzed reactions that use this newly generated ATP to drive the formation of organic molecules from atmospheric carbon dioxide (CO_2). These are called the "light independent" or **dark reactions** of photosynthesis, since they will occur as readily in the absence of light as in its presence, as long as ATP is available.
3. Finally, the pigment that absorbed the light in the first place is rejuvenated, ready to initiate another light reaction by absorbing another photon.

Absorbing Light Energy

The light reactions occur on membranes, called **photosynthetic membranes.** In photosynthetic bacteria these membranes are the cell membrane itself; in plants and algae, all photosynthesis is carried out by the **chloroplasts**, which are the evolutionary descendants of photosynthetic bacteria, and the photosynthetic membranes exist within the chloroplasts. The light reactions take place in the following three stages:

1. A photon of light is captured by a pigment. The result of this **primary photoevent** is the excitation of an electron within the pigment.
2. The excited electron is shuttled along a series of electron-carrier molecules embedded within the photosynthetic membrane until it arrives at a transmembrane proton-pumping channel, where its arrival induces the transport of a proton in across the membrane. The electron is passed to an acceptor.
3. The passage of protons drives the chemiosmotic synthesis of ATP, just as it does in aerobic respiration.

FIGURE 9-5
Overview of the three processes of photosynthesis.

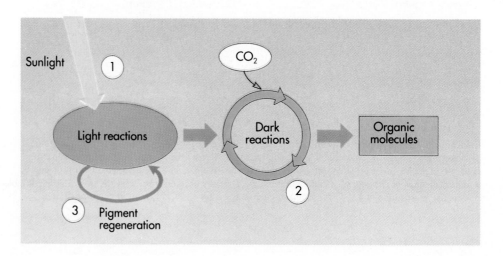

Fixing Carbon

The chemical events that use the ATP generated by the light reactions take place readily in the dark. Building an organism requires not only ATP, however, but also a way of making organic molecules from carbon dioxide (CO_2). Organic molecules are said to be reduced, compared with CO_2, since they contain many C—H bonds. To build organic molecules, one needs a source of hydrogens and the electrons to bind them to carbon. Photosynthesis uses the energy of light to extract the **reducing power** (hydrogen atoms) from water in a way that we will explain later in this chapter. In the dark reactions overall, atmospheric CO_2 is incorporated into carbon-containing molecules, a process called **carbon fixation.**

Replenishing the Pigment

The third kind of chemical event that occurs during continuous photosynthesis involves the electron that was stripped from the chlorophyll at the beginning of the light reactions. This electron must be returned to the pigment or another source of electrons must be used to replenish the supply of electrons in the pigment. Otherwise, continual electron removal would cause the pigment to become deficient in electrons (bleached), and then it could no longer trap photon energy by electron excitation. As we will see, various organisms have evolved different approaches to this problem during the course of their evolutionary history.

> **Photosynthesis involves three processes: the use of light-ejected electrons to drive the chemiosmotic synthesis of ATP; the use of the ATP to fix carbon; and the replenishment of the electrons in the photosynthetic pigment.**

The overall process of photosynthesis may be summarized by a simple oxidation-reduction equation:

$$6CO_2 \; + \; 12H_2 \star O \xrightarrow{\text{LIGHT}} C_6H_{12}O_6 + 6H_2O \; + \; 6 \star O_2$$

ATMOSPHERIC CARBON DIOXIDE   WATER VAPOR   SUGAR   WATER   OXYGEN GAS FROM THE ORIGINAL WATER MOLECULES

H_2O appears on both sides of the equation, but these are *not* the same water molecules. This can be demonstrated by carrying out photosynthesis with water vapor in which the oxygen atom is a heavy isotope. The majority of the heavy oxygen atoms (indicated above by the symbol $\star O$) end up in oxygen gas and not in water.

HOW LIGHT DRIVES CHEMISTRY: THE LIGHT REACTIONS

Photosynthesis in plants, algae, and those bacteria in which it occurs is the result of a long evolutionary process. As this process took place, new reactions were added to older ones, thus making the overall series of reactions that are involved more complex. Much of the evolution has centered on the light reactions, which apparently have changed considerably since they first evolved.

Evolution of the Photosystem

In chloroplasts and all but the most primitive bacteria, light is captured by a network of chlorophyll pigments working together on a membrane surface (see Chapter 5). The battery of chlorophyll molecules, called a **photosystem,** is capable of efficiently capturing photons. Pigment molecules are held on a protein matrix within the photosystem, and their arrangement permits the channeling of excitation energy from anywhere in the array to a central point. The assembly of chlorophyll molecules thus acts collectively as a sensitive "antenna" to capture and focus photon energy (Figure 9-6). Its mode of operation can be thought of as similar to the way that a magnifying glass, by focusing light, can generate enough heat energy at the point of focus to burn paper. Similarly, the photosystem channels the excitation energy gathered by any one of the pigment molecules to the **reaction center chlorophyll** (in plants P_{700}), which in

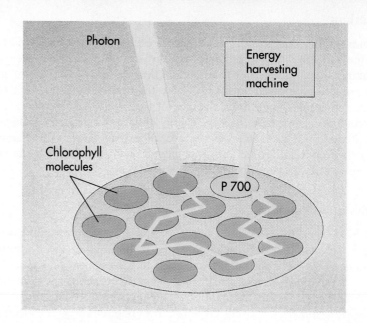

turn passes the energy on to a **primary electron acceptor** which in plants was thought to be **ferredoxin**, a kind of membrane bound protein*. The photosystem thus funnels to the primary acceptor many more electrons than would otherwise be possible.

When light of the proper wavelength strikes any pigment molecule of the photosystem, the resulting excitation passes from one chlorophyll molecule to another. The excited electron does not transfer physically from one pigment molecule to the next. Instead, the pigment passes along the energy to an adjacent molecule of the photosynthetic unit, after which its electron returns to the low energy level it had before the photon was absorbed.

A crude analogy to this form of energy transfer exists in the initial "break" in a game of pool. If the cue ball squarely hits the point of the triangular array of 15 pool balls, the two balls at the far corners of the triangle fly off, and none of the central balls move at all. The energy is transferred through these balls to the most distant ones. The protein matrix of the photosystems, in which the molecules of chlorophyll are embedded, serves as a sort of scaffold, holding individual pigment molecules in orientations that are optimal for energy transfer. In this way, the process channels excitation energy to the membrane-bound ferredoxin in the form of electrons.

Where Does the Electron Go?

Photosynthetic units are thought to have evolved more than 3 billion years ago in bacteria similar to the groups called green and purple sulfur bacteria. In these bacteria, the absorption of a photon of light by the photosynthetic unit results in the transmission of an electron from the pigment P(P_{840} or P_{870}) to ferredoxin. It leaves ferredoxin, combining with a proton to form a hydrogen atom. Where does the proton come from? In green sulfur bacteria the proton is extracted from hydrogen sulfide (H_2S), a process that produces elemental sulfur as a residual by-product.

The ejection of an electron from P and its donation to ferredoxin leaves P short one electron. Before the photosynthetic unit of the sulfur bacteria can function again, the electron must be returned. These bacteria channel the electron back to the pigment through an electron-transport system similar to the one we encountered in Chapter 8, in which the electron's passage drives a proton pump and thus promotes the chemiosmotic synthesis of ATP. Viewed overall (Figure 9-7), the path of the electron originally extracted from P is thus a circle. Chemists call this process **cyclic photophosphorylation.** Note carefully, however, that the process is not a true circle. The

---

*Recent research suggests that the primary electron acceptor may not be ferredoxin, as has been commonly thought, but rather molecules of chlorophyll (photosystem I) and pheophytin (photosystem II). The matter is a subject of intensive research.

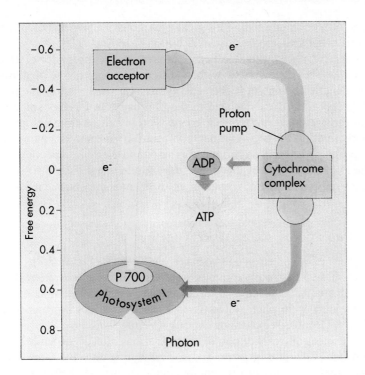

FIGURE 9-7
The path of an electron in bacterial photosynthesis. When an electron is ejected from this photocenter by a photon of light, its passage is a circle, the electron returning to the photosystem from which it was initially ejected. The photosystem is P_{700} in green plants; other photosystems occur in different photosynthetic bacteria and algae.

electron that left P was a high-energy-level electron, boosted to its high energy level by the absorption of a photon of energy; but the electron that returns has only as much energy left as it had before the photon absorption. The difference in the energy of that electron is the photosynthetic payoff, the energy that drives the proton pump.

> **In bacteria, the electron ejected from the pigment by light travels a circular path, in which light-excited electrons pass a proton pump and then return to the photocenter where they originated. For every three electrons, the energy yield is one ATP molecule.**

The light reactions of all the more complex photosynthetic systems that evolved later are built up from the simple cyclic photophosphorylation process of the green sulfur bacteria. Just as glycolysis is retained as a fundamental component of the respiratory metabolism of all organisms, so cyclic photophosphorylation remains a fundamental component of photosynthesis in the chloroplasts of all plants, algae, and most photosynthetic bacteria.

LIGHT REACTIONS OF PLANTS

For more than a billion years, cyclic photophosphorylation was the only form of photosynthetic light reaction that organisms used. It has, however, a fundamental limitation: it is geared only toward energy production, not toward biosynthesis. To understand this point, consider for a moment that the ultimate point of photosynthesis is *not* to generate ATP, but rather to fix carbon—to incorporate atmospheric carbon dioxide into new carbon compounds. Because the molecules produced during carbon fixation (sugars) are more reduced (have more hydrogen atoms) than their precursor (CO_2), it is necessary to provide a source of hydrogens. Cyclic photophosphorylation does not do this, and bacteria that use this process must scavenge hydrogens from other sources, an inefficient undertaking.

The Advent of Photosystem II

After the sulfur bacteria appeared, other kinds of bacteria evolved an improved version of the photocenter, one that solved the reducing power problem in a neat and simple way. To the original photosystem was grafted a second more powerful one, using a new form of chlorophyll. This great evolutionary advance took place when the cyanobacteria originated, no less than 2.8 billion years ago.

In this second photosystem, called **photosystem II,** molecules of chlorophyll *a* are arranged in the photosystem with a different geometry, so that more of the shorter-wavelength photons of higher energy are absorbed than in the more ancient ancestral photosystem (closely related to the **photosystem I** plants still employed today). As in the bacterial photosystem, energy is transmitted from one pigment molecule to another within the photosynthetic unit until it encounters a particular pigment molecule, positioned near a strong membrane-bound electron acceptor. In photosystem II, the absorption peak of this pigment molecule is not substantially shifted as it is in the bacterial photosystem, remaining near 680 nanometers; the molecule is called P_{680}.

How the Two Photosystems Work Together

Plants use both photosystems—a two-stage photocenter (Figure 9-8). The new photosystem acts first. When a photon jolts an electron from photosystem II, the excited electron is donated to an electron transport chain, which passes it along to photosystem I. In its journey to photosystem I, each electron drives a proton pump and so generates an ATP molecule chemiosmotically (Figure 9-9).

When the electron reaches photosystem I, it has already expended its excitation energy in driving this proton pump and has remaining only the same amount of energy as the other electrons of this photosystem. Its arrival, however, does give the photosystem an electron that it can afford to lose. Photosystem I now absorbs a photon, boosting one of its pigment electrons to a high energy level. The electron is now channeled to ferredoxin and used to generate reducing power. In plants and algae, ferredoxins contribute two electrons to reduce **nicotine adenine dinucleotide phosphate, NADP$^+$,** generating NADPH. By using this molecule instead of the NAD$^+$ used in oxidative respiration, plants and algae keep the flow of electrons in the two processes separate.

Plants and algae use a two-stage photocenter. First a photon is absorbed by photosystem II, which passes an electron to photosystem I, using its photon-contributed energy to drive a proton pump and so generate a molecule of ATP. Then another photon is absorbed, this time by photosystem I, which also passes on a photon-energized electron. This second electron is channeled away to provide reducing power in the form of NADPH.

FIGURE 9-8

The path of an electron in a chloroplast. There are two photosystems that work one after the other. First, a photon of light ejects a high-energy electron from photosystem II that is used to pump a proton across the membrane, contributing chemiosmotically to the production of a molecule of ATP. The ejected electron passes along a chain of cytochromes to photosystem I. When this photosystem in turn absorbs a photon of light, it ejects a high-energy electron. The electron's energy is used to drive the formation of the electron carrier NADPH.

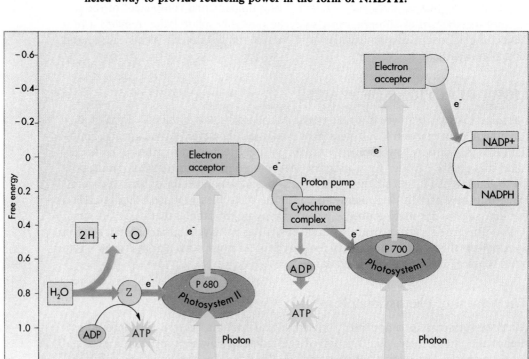

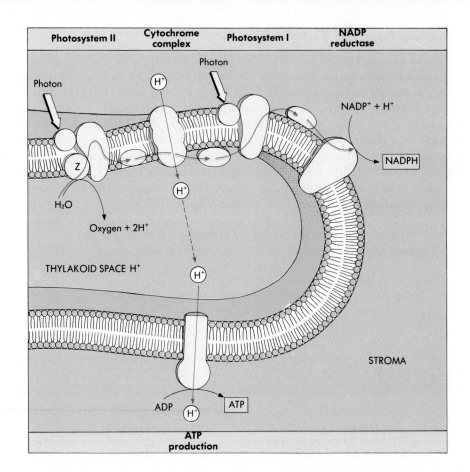

FIGURE 9-9
The photosynthetic electron transport system. When a photon of light strikes a pigment molecule (PO in photosystem II, it excites an electron. This electron is coupled to a proton stripped from water by a Z protein and passes along a chain of membrane-bound cytochrome electron carriers (*red arrow*) to a proton pump. There the energy supplied by the photon is used to transport a proton across the membrane into the thylakoid. The resulting proton gradient drives the chemiosmotic synthesis of ATP. The spent electron then passes to photosystem I. When photosystem I absorbs a new photon of light, P_{700} passes a second high-energy electron to a reduction complex, which drives NADPH generation.

Thus, the energy produced in the first photo event is spent in ATP synthesis; the energy of the second creates reducing power.

The Formation of Oxygen Gas

By now you should have begun to wonder about the fate of P_{680}, the pigment of photosystem II. It started the photosynthetic process by donating an electron. How does it make up for this loss if the electron is not returned but is instead expended to synthesize NADPH? As you might expect, an electron is obtained from another source. The loss of the excited electron from photosystem II converts P_{680} into a powerful oxidant (electron-seeker), and it proceeds to obtain the required electron from a protein called **Z.** The removal of this electron from Z renders it a strong electron-acceptor in turn, and Z obtains electrons to funnel to P_{680} from water. Z catalyzes a complex series of reactions in which water is split into electrons (passed to P_{680}), H^+ ions, and OH radicals. The OH radicals are collected and reassembled as water and oxygen gas. The hydrogen ions (protons) augment the gradient in proton concentration established during the passage of electrons to photosystem I.

> **The electrons and associated protons that oxygen-forming photosynthesis employs to form energy-rich reduced organic molecules are obtained from water, the residual oxygen atoms of the water molecules combining to form oxygen gas.**

The use of a two-stage photocenter containing both photosystem I and photosystem II thus solves in a simple way the evolutionary problem of how to obtain reducing power for biosynthetic reactions. The cyclic photophosphorylation of sulfur bacteria, even though it provides ATP by cyclic photophosphorylation, does not provide a ready means of generating NADPH. For this reason, organisms that use this form of photosynthesis must make NADPH in a roundabout way and expend a lot of ATP to do so.

Comparing Plant and Bacterial Light Reactions

It is useful to compare the two-stage P_{680}/P_{700} photocenter with the P_{700} photocenter from which it evolved. The removal of an electron from P_{700} creates an oxidant strong enough to extract hydrogen from H_2S (78 kcal), but not from H_2O (118 kcal). The removal of an electron from P_{680}, by contrast, yields a considerably stronger oxidant, adequate to cleave water, producing gaseous oxygen as a by-product. In cyanobacteria, algae, and plants, which use this double photocenter, electrons and associated hydrogen atoms are extracted continually from water, eventually being used to reduce $NADP^+$ to NADPH. As a result of stripping hydrogens from water, oxygen gas is continuously generated as a product of the reaction. This photosynthetic process generates all of the oxygen in the air we breathe (Figure 9-10).

Every oxygen molecule in the air you are breathing was once split from a water molecule by an organism carrying out oxygen-forming photosynthesis.

FIGURE 9-10
Our every breath. All of the oxygen we breathe has been generated by photosynthesis.

THE EVOLUTION OF METABOLISM

The evolution of metabolism has involved many changes, but six events were of crucial importance:

Abiotic Synthesis

The first organic molecules were created by natural forces before life evolved. The creation of organic molecules by nonliving forces is called **abiotic synthesis.** Among these molecules were amino acids, carboxylic acids, and hydrocarbons. Produced in an atmosphere rich in hydrogen gas and free of oxygen gas, these early organic molecules were highly reduced, with hydrogen atoms bonded to many of their carbon atoms.

Degradation

The most primitive forms of life are thought to have obtained chemical energy by degrading, or breaking down, organic molecules that were abiotically produced.

The first major event in the evolution of metabolism was the origin of the ability to obtain energy from chemical bond energy. At an early stage, organisms began to convert this energy into ATP, an energy carrier used by all organisms today.

Glycolysis

As proteins evolved diverse catalytic functions, it became possible to capture a larger fraction of the chemical bond energy in abiotic organic molecules by breaking a series of chemical bonds in successive steps. For example, the progressive breakdown of the six-carbon sugar glucose into smaller two-carbon molecules was performed in a series of 10 steps that resulted in the net production of two ATP molecules. The energy for the synthesis of ATP was obtained by breaking a series of chemical bonds and forming new ones with less bond energy, the energy difference being channeled into ATP production. This biochemical pathway is called **glycolysis.**

The glycolytic sequence of reactions undoubtedly evolved early in the history of life on earth, since this biochemical pathway has been retained by all modern organisms. It is a chemical process that does not appear to have changed for well over 3 billion years.

The second major event in the evolution of metabolism was the development of organized sequences of catalyzed degradation reactions.

Anaerobic Photosynthesis

Early in the history of life, organisms evolved a different way of generating ATP, called photosynthesis. Instead of obtaining energy for ATP synthesis by reshuffling chemical bonds, as in glycolysis, organisms

developed the ability to use light to pump protons out of their cells. ATP was then produced chemiosmotically. The photosynthetic process evolved in the absence of oxygen and works well without it. Dissolved H_2S, present in the oceans beneath an atmosphere free of oxygen gas, served as a ready source of hydrogen atoms for photosynthesis.

When photosynthesis used H_2S as a source of hydrogen atoms, free sulfur was produced as a by-product.

The third major event in the evolution of metabolism was the advent of photosynthesis using H_2S.

Nitrogen Fixation

Proteins and nucleic acids cannot be synthesized from the products of photosynthesis, since both of these biologically critical molecules contain nitrogen. Obtaining nitrogen atoms from N_2 gas, a process called **nitrogen fixation,** requires the breaking of an $N\equiv N$ triple bond. This important reaction evolved in the hydrogen-rich atmosphere of the early earth, an atmosphere in which no oxygen was present. Oxygen acts as a poison to nitrogen fixation, since the reaction mechanism will not work in its presence.

One consequence of the appearance of plentiful oxygen in the atmosphere was the ability to fix nitrogen

HOW THE PRODUCTS OF PHOTOSYNTHESIS ARE USED TO BUILD ORGANIC MOLECULES FROM CO₂

The preceding section concerns the light reactions of photosynthesis. These reactions use light energy to produce metabolic energy in the form of ATP and to produce reducing power in the form of NADPH. But this is only half of the story. Photosynthetic organisms employ the ATP and NADPH produced by the light reactions to build organic molecules from atmospheric carbon dioxide. This later phase of photosynthesis, which comprises the so-called dark reactions (light independent), is carried out by a series of enzymes present in the chloroplasts of plants and algae and in the cells of many photosynthetic bacteria.

Carbon fixation depends on the presence of a molecule to which CO_2 can be attached. The cell produces such a molecule by reassembling the bonds of two of the intermediates of glycolysis, fructose-6-phosphate (F6P) and glyceraldehyde-3-phosphate (G3P), to form a five-carbon sugar, **ribulose 1,5 bisphosphate (RuBP).**

The dark reactions of photosynthesis form a cycle, a circle of enzyme-catalyzed steps, just as the citric acid cycle is a circular biochemical pathway. The cycle begins when carbon dioxide is bound to the five-carbon molecule RuBP (Figure 9-11), which then splits to form two molecules of **phosphoglycerate (PGA),** a 3-carbon molecule.

became restricted to organisms that lived in strictly anaerobic habitats and to the cyanobacteria. In the nitrogen-fixing cyanobacteria there are specialized inclusions called **heterocysts,** which contain the nitrogenase complex and are relatively impermeable to oxygen. In the nodules on the roots of legume plants, other groups of nitrogen-fixing bacteria called *Rhizobium* live symbiotically.

The fourth major event in the evolution of metabolism was the evolution of a mechanism for fixing the atmospheric nitrogen.

Oxygen-Forming Photosynthesis

Oxygen-forming photosynthesis employs H_2O rather than H_2S as a source of hydrogen atoms and their associated electrons. Because it garners its hydrogen atoms from reduced oxygen rather than from reduced sulfur, oxygen-forming photosynthesis generates oxygen gas rather than free sulfur.

More than 2 billion years ago, small cells carrying out the new oxygen-forming photosynthesis, such as the cyanobacteria, became the dominant forms of life on earth. Oxygen gas began to disperse into the atmosphere. This was the beginning of the great transition that changed conditions on earth permanently.

Our atmosphere is now 20.9% oxygen, every molecule of which is derived from an oxygen-forming photosynthetic reaction.

The fifth and pivotal event in the evolution of metabolism was the substitution of H_2O for H_2S in photosynthesis.

Chemoautotrophs

When oxygen-forming photosynthesis first began to enrich the oceans (and eventually the atmosphere) with oxygen, these oceans were already rich in elemental sulfur, the residue of eons of anaerobic photosynthesis. Organisms soon evolved that could combine this sulfur with oxygen and thus generate reducing power and energy. An organism that obtains its energy in this way is called a **chemoautotroph** (Greek, feeding itself by chemicals). Such an organism requires no additional source of energy other than the chemical to which it adds oxygen, and so it is, in a sense, "feeding" on inorganic compounds. Other chemoautotrophic organisms have evolved that carry out the oxidation of other inorganic molecules.

Aerobic Respiration

Aerobic respiration probably evolved from photosynthesis, as a modification of the basic photosynthetic machinery, and employs the same kinds of proton pumps. However, the hydrogens and their associated electrons are not obtained from H_2S or H_2O, as in photosynthesis, but rather from the breakdown of organic matter. We think that the ability to carry out photosynthesis without H_2S first evolved among purple nonsulfur bacteria. These bacteria carry out photosynthesis in the absence of H_2S, obtaining their hydrogens from organic compounds instead.

It was perhaps inevitable that among the descendants of these respiring photosynthetic cells, some would eventually do without photosynthesis entirely, subsisting only on the energy and reducing power derived from the breakdown of organic molecules. The mitochondria within all eukaryotic cells are thought to be their descendants.

The sixth and final stage in the evolution of metabolism was the development of a mechanism, aerobic respiration, in which electrons were stripped from the glucose molecules manufactured by plants and employed to drive ATP-generating proton pumps.

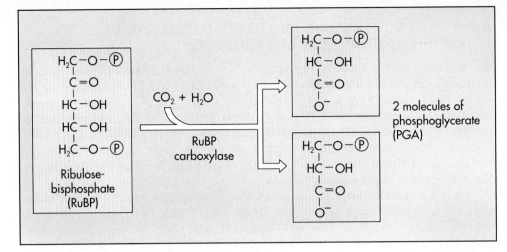

FIGURE 9-11

The key step in carbon fixation. Melvin Calvin and his coworkers at the University of California worked out the first step of what later became known as the Calvin cycle by exposing photosynthesizing algae to radioactive carbon dioxide ($^{14}CO_2$). Following the fate of the radioactive carbon atom (shown in color in these diagrams), they found that it is first bound to a molecule of ribulose bisphosphate (RuBP), which splits immediately, forming two molecules of phosphoglycerate (PGA), one of which contains the radioactive carbon atom.

It is because PGA contains three carbon atoms that this process is called **C₃ photosynthesis.** A series of reactions then takes place (Figure 9-12), which convert PGA to glyceraldehyde phosphate molecules, some of which are in turn used to reconstitute RuBP and others of which are assembled into sugars.

This cycle of reactions is called the **Calvin cycle,** after its discoverer, Melvin Calvin of the University of California, Berkeley. At each full turn of the cycle, a molecule of carbon dioxide enters the cycle and a molecule of RuBP is regenerated. Six revolutions of the cycle, with the introduction of six molecules of carbon dioxide, are

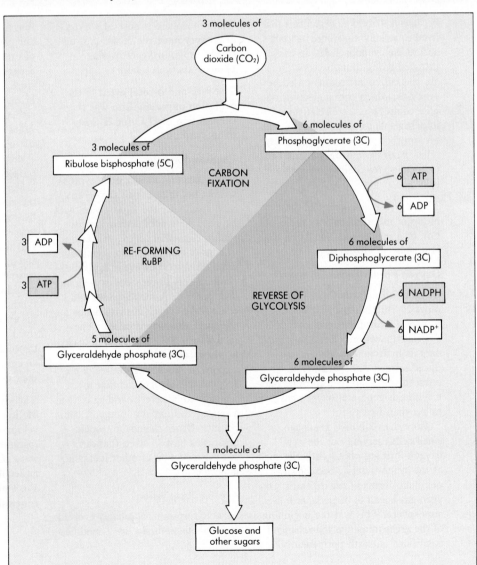

FIGURE 9-12

The Calvin cycle. For every three molecules of CO₂ that enter the cycle, one molecule of the three-carbon compound glyceraldehyde phosphate is produced. Notice that the process requires energy, supplied as ATP and NADPH. This energy is generated by the light reactions, which generate ATP and NADPH.

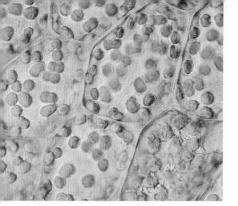

FIGURE 9-13
Chloroplasts make plants green. These moss cells are densely packed with bright green chloroplasts.

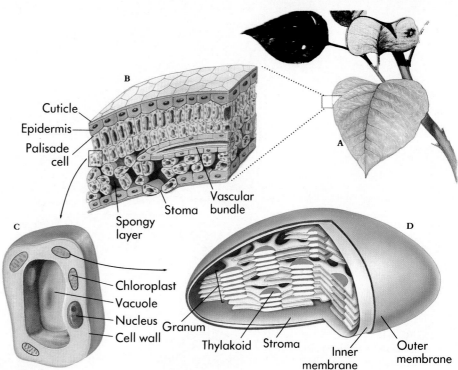

B
Cuticle
Epidermis
Palisade cell
Spongy layer
Stoma
Vascular bundle
A
C
Chloroplast
Vacuole
Nucleus
Cell wall
Granum
Thylakoid
Stroma
Inner membrane
Outer membrane
D

FIGURE 9-14
Journey into a leaf.
A The leaf shown is from a plant that carries out the Calvin cycle.
B In cross section the leaf possesses a thick palisade layer whose cells, **C**, are rich in chloroplasts, **D**.

necessary to produce the equivalent of a six-carbon sugar, such as glucose. All the carbon-containing organic molecules are derived from sugars made in this way.

> **Most plants incorporate carbon dioxide into sugars by means of a cycle of reactions called the Calvin cycle, which is driven by the ATP and NADPH produced in the light reactions but which can itself take place in the dark.**

THE CHLOROPLAST AS A PHOTOSYNTHETIC MACHINE

In eukaryotes, all photosynthesis takes place in chloroplasts (Figure 9-13). As you will recall from Chapter 5, the internal membranes of chloroplasts are organized into flattened sacs called **thylakoids,** and numerous thylakoids are stacked on top of one another in arrangements called **grana** (Figure 9-14). The photosynthetic pigments are associated with proteins embedded within the membrane of the thylakoids.

Each thylakoid is a closed compartment into which protons can be pumped. The thylakoid membrane is impermeable to protons and to most molecules, so transit across it occurs almost exclusively via transmembrane channels. As in mitochondria, the exit of protons from the thylakoid interior, driven by diffusion, takes place at distinctive ATP-synthesizing proton channels. These channels protrude as knobs on the external surface of the thylakoid membrane, from which ATP is released into the surrounding fluid. This fluid matrix inside the chloroplast, within which the thylakoids are embedded, is called the **stroma.** It contains the Calvin-cycle enzymes, which catalyze the dark reactions of carbon fixation, using the ATP and NADPH that the photosynthetic activity of the thylakoids produces (Figure 9-15). Within the chloroplast,

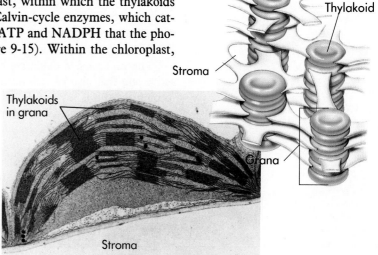

Thylakoid
Stroma
Thylakoids in grana
Grana
Stroma

FIGURE 9-15
Internal structure of a chloroplast. The basic photosynthetic unit is the thylakoid, a flattened membranous sac. These are stacked into columns called grana (singular, *granum*). In the micrograph you can see that a single chloroplast may contain many such grana. The interior matrix of a chloroplast, which is fluid, is called the stroma. The stroma contains the Calvin cycle enzymes that carry out the carbon-fixing "dark reactions." Thylakoid membranes are the sites of the light reactions, generating the ATP and NADPH that fuel the dark reactions.

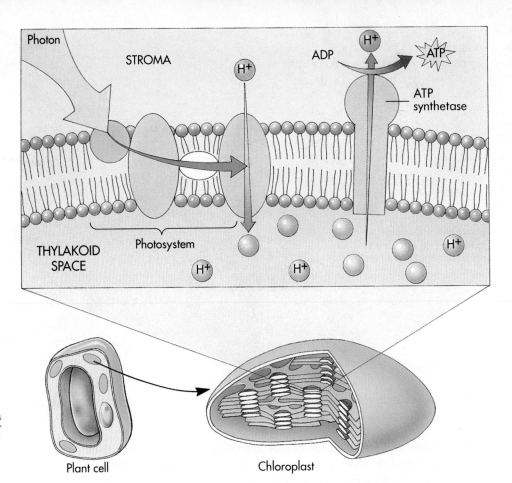

FIGURE 9-16
Chemosmosis in a chloroplast.
Chloroplasts contain sacs called thylakoids. Proteins in the thylakoid membrane pump protons into the interior of the sac. ATP is produced on the outside surface of the membrane (stroma side) as protons pass back out of the sac through ATPase channels.

the thylakoid membrane pumps protons from the stroma into the thylakoid compartment. As hydrogen ions pass back through the membrane, phosphorylation of ADP occurs on the stroma side of the membrane (Figure 9-16).

PHOTOSYNTHESIS IS NOT PERFECT

One of the ironies of evolution is that processes evolve to be only as good as they need to be, rather than as good as they potentially might be. Evolution favors not optimum solutions, but rather workable ones that can be derived from others that already exist. Photosynthesis is no exception. RuBP carboxylase, the enzyme that catalyzes the key carbon-fixing reaction of photosynthesis, provides a decidedly suboptimal solution. This enzyme has another activity, and its second activity interferes with the successful performance of the Calvin cycle: RuBP carboxylase also initiates the *oxidation* of ribulose 1,5 bisphosphate. In this process, called **photorespiration,** CO_2 is released without the production of ATP or NADPH. Because it produces neither ATP nor NADPH, photorespiration undoes the work of photosynthesis.

The undesirable oxidative reaction is initiated by RuBP carboxylase at the same active site that also carries out the carbon-fixing carboxylation reaction so important to photosynthesis. When photosynthesis first evolved, there was little oxygen in the atmosphere and, because the decarboxylation reaction requires oxygen, there was thus little or no photorespiration. Under these conditions, the fact that the active site of RuBP carboxylase was capable of carrying out both reactions presented no problem. Only after millions of years of oxygen buildup in the atmosphere did the competition of CO_2 and O_2 for the same site lead to the problem that photorespiration now poses.

The loss of fixed carbon as a result of photorespiration is not trivial. Plants that use the Calvin cycle to fix carbon, which are called **C_3 plants,** lose between a fourth and a half of their photosynthetically fixed carbon in this way. The extent of carbon loss depends a great deal on temperature, since the oxidative activity of the RuBP carboxylase enzyme increases far more rapidly with temperature than does its carbon-fixing activity. In tropical climates, especially those in which the temperature is often above 28° C, the problem is a severe one, and it has a major impact on tropical agriculture.

Plants that adapted to these warmer environments evolved two principal ways to deal with this problem. One approach, taken by a number of grasses, including corn, sugarcane, and sorghum, as well as members of about two dozen other plant groups, is called **C_4 photosynthesis.** In them, the first product of CO_2 fixation to be detected is not a three-carbon molecule, as in the Calvin cycle, but rather a four-carbon molecule, oxaloacetate. Such plants concentrate CO_2 by carboxylating a three-carbon metabolite called **phosphoenolpyruvate.** The resulting four-carbon molecule, oxaloacetate, is in turn converted to the citric acid cycle intermediate *malate* and transported to an adjacent **bundle-sheath cell.** Such cells are impermeable to CO_2 and therefore hold CO_2 within them. The malate is oxidatively decarboxylated to pyruvate, releasing a CO_2 molecule into the interior of the bundle-sheath cell. Pyruvate returns to the leaf cell, where an ATP is converted to adenosine monophosphate, or AMP (*two* high-energy bonds are split), in converting the pyruvate back to phosphoenolpyruvate, thus completing the cycle.

C_4 plants expend a considerable amount of ATP to concentrate CO_2 within the cells that carry out the Calvin cycle. Because CO_2 binds to the same place within the RuBP carboxylase active site that O_2 does, high concentrations of CO_2 commandeer the available enzyme for the CO_2-fixing reaction.

The path of carbon fixation in C_4 plants is diagrammed in Figure 9-17. The en-

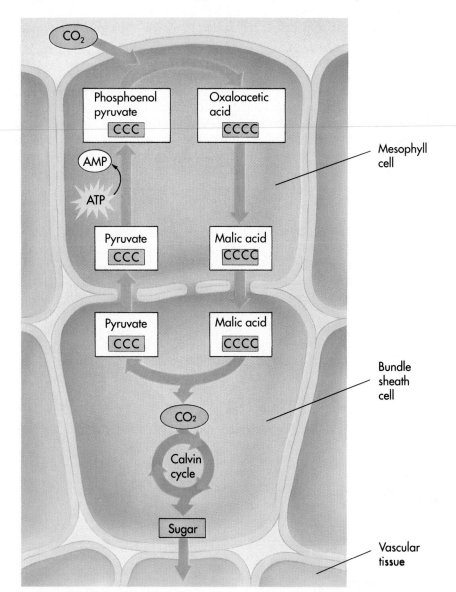

FIGURE 9-17
A common path of carbon fixation in C_4 plants. The process is called the C_4 pathway because the starting material, OAA, is a molecule with four carbons.

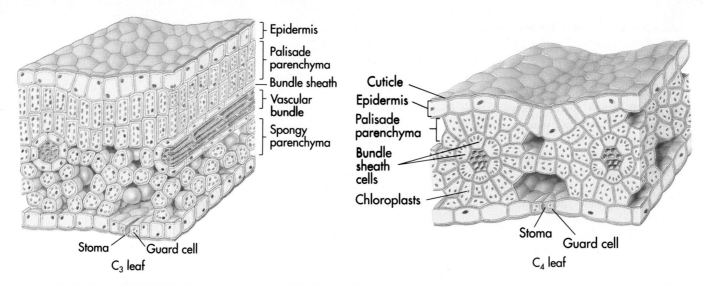

Epidermis
Palisade parenchyma
Bundle sheath
Vascular bundle
Spongy parenchyma

Stoma Guard cell

C_3 leaf

Cuticle
Epidermis
Palisade parenchyma
Bundle sheath cells
Chloroplasts

Stoma Guard cell

C_4 leaf

FIGURE 9-18

The leaf of a C_4 plant is very different from that of a C_3 leaf. In a C_3 leaf, carbon dioxide is taken up by widely spaced mesophyll cells (the spongy parenchyma), which pass it to the vascular bundles (veins). In a C_4 leaf, however, CO_2 is taken up by similar mesophyll cells that incorporate it into C_4 compounds. These C_4 compounds then pass inward to the bundle-sheath cells, which surround the veins, and release carbon dioxide there. Because these cells are relatively impermeable to carbon dioxide, this process can build up high carbon-dioxide concentrations in the bundle-sheath cells, which carry out the Calvin cycle.

zymes that carry out the Calvin cycle are located within the bundle-sheath cells, where the increased CO_2 concentration inhibits photorespiration. Because each CO_2 molecule is transported into the bundle-sheath cells at a cost of two high-energy ATP bonds, and because six carbons must be fixed to form a molecule of glucose, 12 additional molecules of ATP are required to form a molecule of glucose. The unique kind of leaf structure in which these processes occur is illustrated in Figure 9-18. In C_4 photosynthesis the energetic cost of forming glucose is almost doubled—from 18 to 30 molecules of ATP. However, in a hot climate in which photorespiration would otherwise remove more than half of the carbon fixed, it is the best compromise available. For this reason, C_4 plants are more abundant in warm regions than in cooler ones (Figure 9-19).

A second strategy to facilitate photosynthesis in hot regions has been adopted by many succulent (water-storing) plants such as cacti, pineapples (Figure 9-20), and some members of about two dozen other plant groups. This mode of carbon fixation is called crassulacean acid metabolism, or CAM, after the plant family Crassulaceae (the stonecrops or hens-and-chickens) in which it was first discovered. In them, the **stomata** (singular, *stoma*), specialized openings that occur in the leaves of all plants

FIGURE 9-19

The distribution of species of C_4 grasses in North America. Many more C_4 species occur toward the south, where the average temperatures during the growing season are higher; at higher temperatures, photorespiration wastes more of the products of photosynthesis, and the ability of C_4 plants to counteract photorespiration is more of an advantage than in cooler regions, where C_3 grasses predominate.

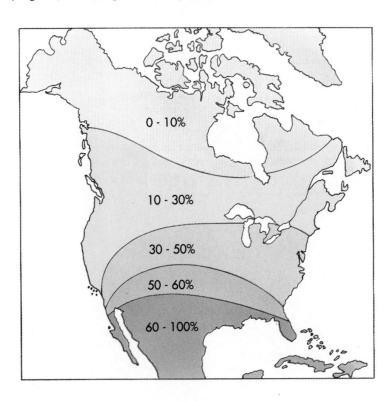

0 - 10%

10 - 30%

30 - 50%

50 - 60%

60 - 100%

and through which CO_2 enters and water vapor is lost, open during the night and close during the day (the reverse of how plants normally behave). Closing stomata during the day prevents water loss and reduces photorespiration by preventing CO_2 from leaving the leaves, so that the ratio of CO_2 to O_2 within the leaves is high. The CO_2 necessary for producing sugars is provided from organic molecules made the night before. Like C_4 plants, these plants use both C_4 and C_3 pathways. They differ from C_4 plants in that the C_4 pathway operates at night and the time of operation of the two pathways differs, with the C_3 pathway operating *within the same cells* during the day. In C_4 plants, on the other hand, the two cycles take place in different, specialized cells.

Photorespiration releases CO_2 without the production of ATP and so short-circuits photosynthesis. C_4 plants and CAM plants circumvent this waste by modifications of leaf architecture.

A LOOK BACK

Perhaps more than any other phase of a cell's life, its metabolism betrays its evolutionary past. This is particularly true of photosynthesis. The two-stage photocenter of modern plants (Figure 9-21) has as its second stage a photosystem that first evolved

FIGURE 9-20
CAM photosynthesis. Many CAM plants, including this pineapple, grow in warm tropical climates; far fewer grow in cooler northern temperate climates.

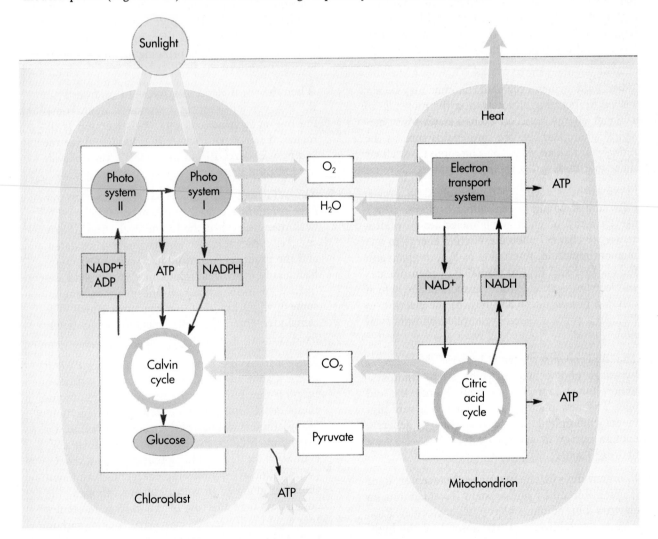

FIGURE 9-21
The metabolic machine. Within the chloroplast, sunlight drives the production of glucose, using up CO_2 and H_2O while generating O_2. This O_2 is converted back into water in the mitochondrion by accepting electrons harvested from glucose molecules after the electrons have been used by the electron transport system to drive the synthesis of ATP. Water and oxygen gas thus cycle between chloroplasts and mitochondria within a plant cell, as do glucose and CO_2. Cells without chloroplasts, such as your own cells, require an outside source of glucose and oxygen, and generate CO_2 and water.

millions of years earlier in anaerobic bacteria that used hydrogen sulfide rather than water as a source of reducing hydrogen. The Calvin cycle uses part of the ancient glycolytic pathway, run in reverse, to produce glucose. The principal chlorophyll pigments of plants are but simple modifications of bacterial chlorophylls employed by anaerobic photosynthetic bacteria for billions of years before there were any plants. Thus by looking at a plant today, we see as well its history.

In Chapters 33 through 38, we will examine plants in detail; photosynthesis is but one aspect of plant biology, although an important one. We have treated photosynthesis here, in a section devoted to cell biology, because photosynthesis arose long before plants did, and all organisms depend directly or indirectly on photosynthesis for the energy that powers their lives.

SUMMARY

1. Light consists of energy packets called photons; the shorter the wavelength of light, the more energy in its photons. Much of short and long wavelengths are absorbed in the earth's atmosphere; largely middle-range wavelengths reach the earth's surface.

2. When light strikes a pigment, photons may be absorbed by boosting an electron in the molecule to a higher energy level, an excited state. Most biological photopigments are carotenoids such as the retinal of human eyes, or chlorophylls such as those that make grass green.

3. Photosynthesis seems to have evolved in organisms similar to the green sulfur bacteria, which use an array of chlorophyll molecules (a photocenter) to channel photon excitation energy to one pigment molecule, referred to as P_{840} in green sulfur bacteria and P_{700} in plants. P_{700} then donates an electron to an electron transport chain, which drives a proton pump and returns the electron to P_{700}, in a process called cyclic photophosphorylation.

4. The descendants of these bacteria developed a two-stage photocenter in which a new photosystem, photosystem II, was grafted onto the old one. The new photosystem employed a new pigment, chlorophyll a, and was able to generate enough energy to use H_2O rather than H_2S as a hydrogen source.

5. In organisms with the two-stage photocenter, light is first absorbed by photosystem II, which jolts an electron out of one chlorophyll molecule, P_{680}. This has two effects: (1) the absence of the high-energy electron causes photosystem II to seek another one actively, which results in the eventual splitting of water to obtain it, with O_2 as the by-product; and (2) the high-energy electron is passed to photosystem I, driving a proton pump in the process and thus bringing about the chemiosmotic synthesis of ATP.

6. When the spent electron arrives at P_{700} from P_{680}, the P_{700} pigment absorbs a photon of light, boosting one of its electrons out. This electron is directed to ferredoxin, where it is used to drive the synthesis of NADPH from $NADP^+$, thus providing reducing power.

7. The ATP and reducing power produced by the light reactions are used to fix carbon in a series of dark reactions called the Calvin cycle. In this process, ribulose 1,5 bisphosphate is carboxylated, and the products run through a series of reactions backwards from that found in the glycolytic sequence to form fructose-6-phosphate molecules, some of which are used to reconstitute RuBP. The remainder enter the cell's metabolism as newly fixed carbon.

8. RuBP carboxylase, the enzyme that fixes carbon in the Calvin cycle, also carries out an oxidative reaction that burns up the products of photosynthesis, a wasteful process called photorespiration. Many tropical plants inhibit photorespiration by expending ATP to increase the intracellular concentration of CO_2. This process, called the C_4 pathway, nearly doubles the energetic cost of synthesizing glucose. In warm climates, though, the cost is less than the cost of the glucose molecules that would otherwise be converted back to carbon dioxide by photorespiration and lost.

REVIEW QUESTIONS

1. What is the basic unit of light energy? In terms of color and length, which wavelength of visible light is most energetic, blue or red? Which is least energetic?

2. Why do plants appear green?

3. What three chemical processes occur in photosynthesis? Which of these occur only in the presence of light and which can occur in the absence of light?

4. What is the overall equation for photosynthesis? From which reactant molecule is the final oxygen derived?

5. What molecules are involved in the capture and transfer of light energy through a plant's photocenter? Are electrons physically passed from one molecule to the next? Explain.

6. How is bacterial photosynthesis a cyclic process? Why is it not a precise circle? What is the energy yield in electrons to ATPs?

7. How does the photocenter in plants work? Which system generates ATP and which generates NADPH?

8. How is chlorophyll *b* related to the PS II/PS I photosystem?

9. What occurs in the dark reaction? How is this related to the ATP and NADPH produced in the light reaction? What is another name for this reaction?

10. What happens to the RuBP in the Calvin cycle? How does glyceraldehyde-3-phosphate (G3P) fit into the Calvin cycle?

11. Where in the chloroplast does the light reaction occur? Where does the dark reaction occur?

12. What is photorespiration? What is the advantage of C_4 photosynthesis in those plants that have it? Why don't all plants use C_4 photosynthesis?

 THOUGHT QUESTIONS

1. What is the advantage of having many pigment molecules in each photocenter for every P_{700}? Why not couple *every* pigment molecule directly to an electron acceptor?

2. Why are plants that consume 30 ATP molecules to produce one molecule of glucose (rather than the usual 18 molecules of ATP per glucose) favored in hot climates but not in cold ones? What role does temperature play?

 FOR FURTHER READING

BARBER, J.: "A Quantum Step Forward in Understanding Photosynthesis," *Plants Today*, September 1989, pages 165-169. An engaging account of the discovery of photosystem II, for which the Nobel prize was given in 1988.

BJORKMAN, O., and J. BERRY: "High Efficiency Photosynthesis," *Scientific American*, October 1973, pages 80-93. A description of C_4 photosynthesis and other strategies carried out by plants that live in Death Valley, where the problem posed by photorespiration is acute.

DEISENHOFER, J. and H. MICHEL: "The Photosynthetic Reaction Center of the Purple Bacterium *Rhodopseudomonas*," *Science*, vol. 245, pages 1463-1473, 1988. The Nobel Prize address for the first description at the molecular level of a photosynthetic reaction center.

GOVINDJEE and W. COLEMAN: "How Plants Make Oxygen," *Scientific American*, February 1990, pages 50-58. How plants and some bacteria exploit solar energy to split water molecules into oxygen gas.

HALL, D., and K. RAO: *Photosynthesis*, ed. 4, Arnold, Baltimore, 1987. A great overview of the photosynthetic process, and of how our understanding of it has developed.

KEELEY, J.: "Photosynthesis in Quillworts, or Why Are Some Aquatic Plants Similar to Cacti?" *Plants Today*, July 1988, pages 127-132. An engaging account of CAM photosynthesis.

KUHLBRANDT, W., and D. WANG: "Three-dimensional structure of Plant Light-Harvesting Complex," *Nature*, March 1991, vol. 350, pages 130-134. For the first time, the plant photosystem's molecular architecture is known.

O'LEARY, M.: "Carbon Isotopes in Photosynthesis," *Bioscience*, vol. 38, pages 328-336, 1988. Plants discriminate against the heavy isotope of carbon in photosynthesis—this article tells you how they manage this feat.

SCHNAPF, J., and D. BAYLOR: "How Photoreceptor Cells Respond to Light," *Scientific American*, April 1987, pages 40-47. The photoreceptor of the eye has many parallels with the photocenter of plants, when viewed at a molecular level.

TING, I.P.: "Photosynthesis of Arid and Subtropical Succulent Plants," *Aliso*, vol. 12, 1989, pages 387-406. An excellent review of the occurrence and significance of CAM metabolism.

YOUVAN, D., and B. MARRS: "Molecular Mechanisms of Photosynthesis," *Scientific American*, June 1987, pages 42-48. Molecular genetics and x-ray crystallography are used to study how the photosynthetic reaction center works.

Reproduction and Heredity

10

The Cell Cycle: Mitosis

CHAPTER OUTLINE

Cell division in bacteria
Cell division among eukaryotes
 The structure of eukaryotic chromosomes
 How many chromosomes are in a cell?
 The cell cycle
Mitosis
 Preparing the scene: interphase
 Formation of the mitotic apparatus: prophase
 Division of the centromeres: metaphase
 Separation of the chromatids: anaphase
 Reformation of the nuclei: telophase
Cytokinesis
Comparing cell division in eukaryotes and prokaryotes

OVERVIEW

Your body consists of some two hundred trillion cells, all of them derived from a single cell at the start of your life as a fertilized egg. Many millions of successful cell divisions occurred while your body was reaching its present form. In each of these, the genetic material of the dividing cells was equally partitioned between the two resulting cells. To accomplish this, growing eukaryotic cells attached microtubules to each replicated chromosome and segregated daughter chromosomes to opposite ends of the cell in a process called mitosis.

FOR REVIEW

Here are some important terms and concepts that you will encounter in this chapter. If you are not familiar with them, you should review them before proceeding.

DNA (Chapter 5)

Chromosomes (Chapter 5)

Nuclear membrane (Chapter 5)

Microtubules (Chapter 5)

Microtubular organizing center (Chapter 5)

FIGURE 10-1
All reproduction of organisms depends on the reproduction of cells. Like you, this turtle started life as a fertilized egg.

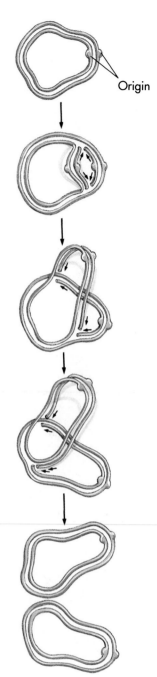

Origin

All living organisms grow and reproduce. Bacteria too small to see, alligators, the weeds growing on your lawn—from the smallest of creatures to the largest, all species produce offspring like themselves and pass on to them the hereditary information that makes them what they are (Figure 10-1). In this chapter, we begin our consideration of heredity with an examination of how cells reproduce themselves. The ways in which cell reproduction is achieved and their biological consequences have changed significantly during the course of the evolution of life on earth.

CELL DIVISION IN BACTERIA

Among the bacteria the process of cell division is simple. In them the genetic information, or **genome,** exists as a single, circular, double-stranded DNA molecule attached at one point to the interior surface of the cell membrane. This genome is replicated early in the life of the cell. At a special site on the chromosome called the **replication origin,** a battery of more than 22 different enzymes goes to work and starts to make a complete copy of the DNA molecule (Figure 10-2). When these enzymes have proceeded all the way around the circle of DNA, the cell then possesses two copies of the genome, attached side-by-side to the interior cell membrane.

The growth of a bacterial cell to an appropriate size induces the onset of cell division. First, new plasma membrane and cell wall materials are laid down in the zone between the attachment sites of the two "daughter" DNA genomes. This begins the process of binary fission. **Binary fission** is the division of a cell into two equal or nearly equal halves. As new material is added in the zone between the attachment sites, the growing plasma membrane pinches inward (invaginates) and the cell is progressively constricted in two (Figure 10-3). Initiating the constriction at a position be-

FIGURE 10-2
How bacterial DNA replicates. The circular DNA molecule that constitutes the genome of a bacterium initiates replication at a single site, moving out in both directions. When the two moving replication points meet on the far side of the molecule, its replication has been completed.

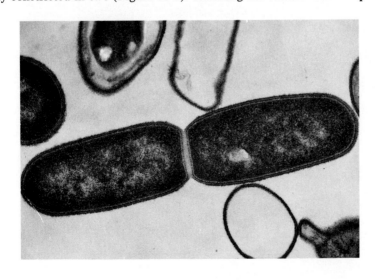

FIGURE 10-3
Fission. Bacteria divide by a process of simple cell fission. Note here the septum between the two daughter cells.

tween the two daughter DNA genomes ensures that each of the two new cells will contain one of the two identical genomes. Eventually the invaginating circle of membrane reaches all the way into the cell center, pinching the cell in two. A new cell wall forms around the new membrane, and what was originally one cell is now two.

> Bacteria divide by binary fission, a process in which a cell pinches in two. The point where the constriction begins is located between where the two replicas of the chromosome are bound to the cell membrane, ensuring that one copy will end up in each daughter cell.

CELL DIVISION AMONG EUKARYOTES

The evolution of the eukaryotes introduced several additional factors into the process of cell division. Eukaryotic cells are much larger than bacteria, and they contain genomes with much larger quantities of DNA. This DNA, however, is located in a number of individual linear chromosomes, rather than in one single circular molecule. In these chromosomes, the DNA forms a complex with packaging proteins and is wound into tightly condensed coils (Figure 10-4). The eukaryotic chromosome is a structure with complex organization, one that contrasts strongly with the single, circular DNA molecule that plays the same role in bacteria.

FIGURE 10-4
Levels of chromosomal organization.

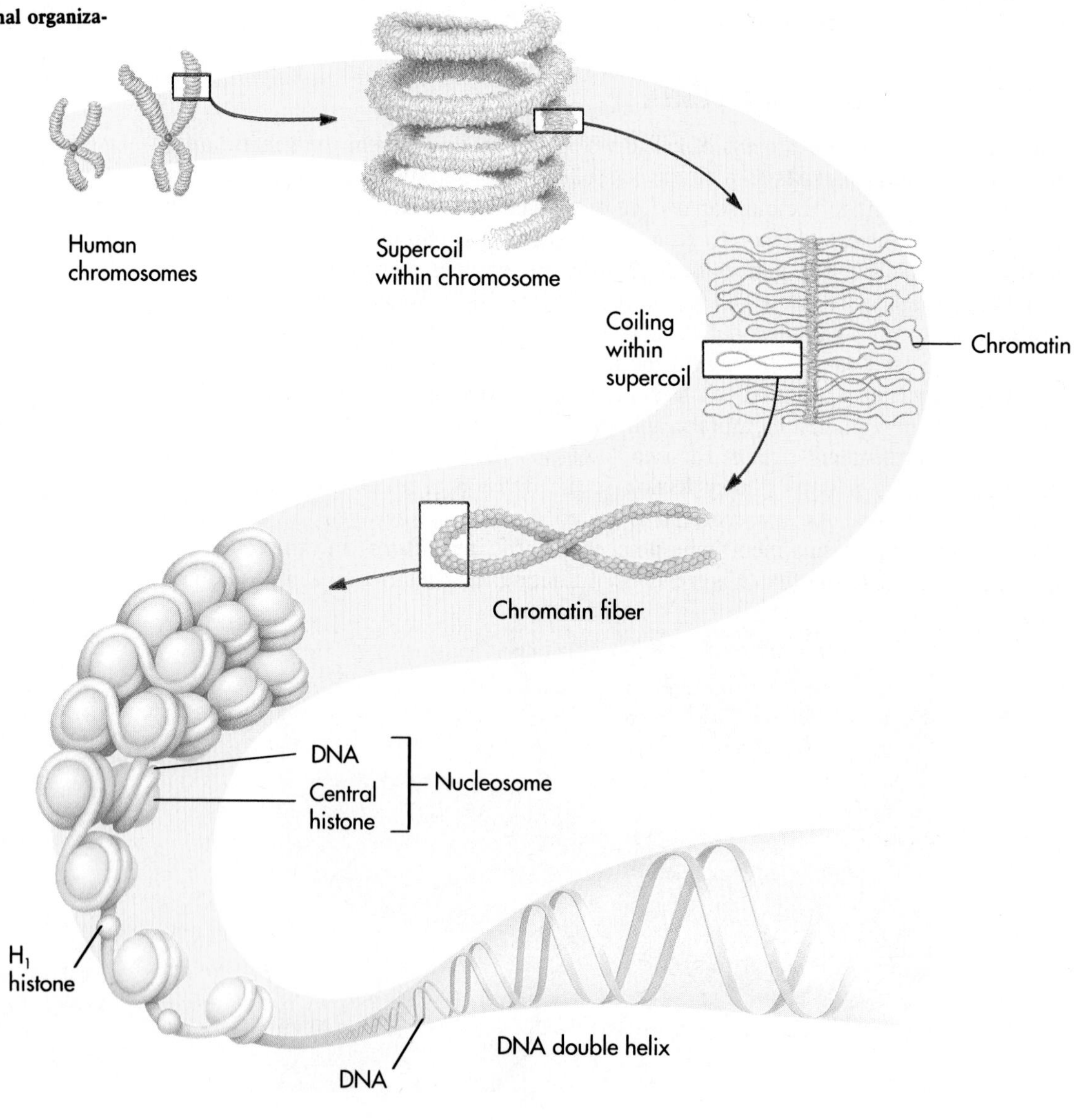

Human chromosomes

Supercoil within chromosome

Coiling within supercoil

Chromatin

Chromatin fiber

DNA
Central histone
— Nucleosome

H₁ histone

DNA

DNA double helix

TABLE 10-1 CHROMOSOME NUMBER IN SELECTED EUKARYOTES

DIVISION	NUMBER OF CHROMOSOMES	DIVISION	NUMBER OF CHROMOSOMES
FUNGI (haploid)		**VERTEBRATES (diploid)**	
Penicillium	1-4	Opossum	22
Neurospora	7	Toad	22
Dictyostelium	7	Salamander	24
Saccharomyces (a yeast)	18	Frog	26
		Vampire bat	28
INSECTS (diploid)		Lungfish	38
Mosquito	6	Mouse	40
Drosophila	8	Rat	42
Housefly	12	Human	46
Honeybee	32	Chimpanzee	48
Silkworm	56	Orangutan	48
		Gorilla	48
PLANTS (ploidy varies)		Cow	60
Haplopappus gracilis	2	Horse	64
Barley	14	Black bear	76
Garden pea	14	Chicken	78
Corn	20	Dog	78
Bread wheat	42	Duck	80
Tobacco	48		
Sugarcane	80		
Horsetail	216		
Adder's tongue fern	1262		

A wide range of organisms is presented to demonstrate the very different values that are possible. The number of chromosomes in the body cells of most eukaryotes falls between 10 and 50.

The Structure of Eukaryotic Chromosomes

Chromosomes were first observed by the German embryologist Walther Fleming in 1882, while he was examining the rapidly dividing cells of salamander larvae. When Fleming looked at these cells through what we would now regard as a rather primitive light microscope, he saw minute threads within the nuclei of the cells. These threads appeared to be dividing lengthwise, and Fleming called their division mitosis, basing his term on the Greek word *mitos*, meaning "thread."

Since their initial discovery, chromosomes have proved to be present in the cells of all eukaryotes. Their number may vary enormously from one species to another. A few kinds of plants and animals—such as *Haplopappus gracilis*, a relative of the sunflower that grows in North American deserts, or the fungus *Penicillium*—have as few as 1 or 2 pairs of chromosomes, whereas some ferns have more than 500 pairs (Table 10-1).

In the century since their discovery, we have learned a great deal about the structure and function of chromosomes. Eukaryotic chromosomes are composed of **chromatin,** a complex of DNA, and protein. Most eukaryotic chromosomes are about 60% protein and 40% DNA. A significant amount of RNA is also associated with chromosomes, because they are the sites of RNA synthesis. The DNA of a chromosome exists as one very long double-stranded fiber, a **duplex,** which extends unbroken through the entire length of the chromosome. A typical human chromosome contains about half a billion (5×10^8) nucleotides in its DNA fiber. The amount of information one chromosome contains, therefore, would fill about 2000 printed books of 1000 pages each, assuming that the nucleotides were "words" and that each page had about 500 of them on it. If the strand of DNA from a single chromosome were laid out in a straight line, it would be about 5 centimeters (2 inches) long. This is much too long to fit into a cell. In the cell, however, the DNA is coiled, thus fitting into a much smaller space than would be possible if it were not.

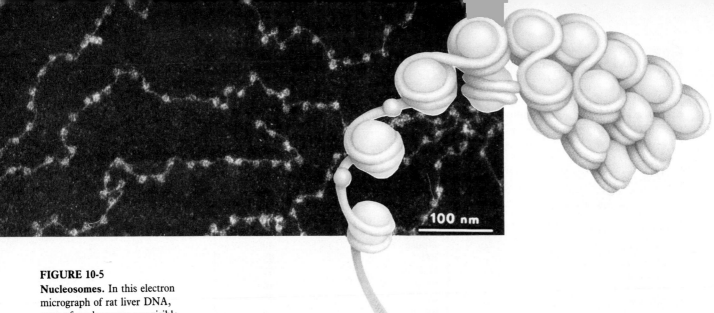

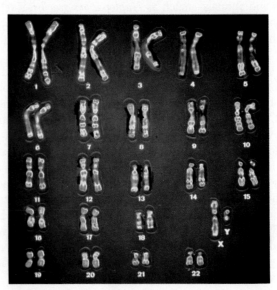

FIGURE 10-5

Nucleosomes. In this electron micrograph of rat liver DNA, rows of nucleosomes are visible, like beads on a string. Each nucleosome is a region in which the DNA duplex is wound tightly around a cluster of histone proteins.

How is the coiling of this long DNA fiber achieved? If we gently disrupt a eukaryotic nucleus and examine the DNA with an electron microscope, we find that it resembles a string of beads (Figure 10-5). Every 200 nucleotides, the DNA duplex is coiled about a complex of **histones,** which are small, very basic polypeptides, rich in the amino acids arginine and lysine. Eight of these histones form the core of an assembly called a **nucleosome.** Because so many of their amino acids are basic, histones are very positively charged. The DNA duplex, which is negatively charged, is strongly attracted to the histones and wraps tightly around the histone core of each nucleosome. The core thus acts as a "form" that promotes and guides the coiling of the DNA. Further coiling of the DNA occurs when the string of nucleosomes wraps up into higher-order coils called **supercoils** (see Figure 10-4).

Highly condensed portions of the chromatin are called **heterochromatin.** Some remain condensed permanently, so that their DNA is never expressed. The remainder of the chromosome, called **euchromatin,** is not condensed except during cell division, when the movement of the chromosomes is facilitated by the compact packaging that occurs at that stage. At all other times, the euchromatin is present in an open configuration and its genes can be activated.

Chromosomes may differ widely from one another in appearance. They vary in such features as the location of a region called a **centromere** present on all chromosomes, the relative length of the two arms (regions on either side of the centromere), size, staining properties, and the position of constricted regions along the arms. The particular array of chromosomes that an individual possesses, called its **karyotype**

FIGURE 10-6

A human karyotype. The individual chromosomes that make up the 23 pairs can be seen to differ widely in size and the position of the centromere, which is indicated in each case. In this preparation, they have been specially stained to indicate further differences in their composition and to distinguish them clearly from one another.

(Figure 10-6), may differ greatly between different species, or sometimes even between particular individuals.

To examine human chromosomes, investigators collect a blood sample, add chemicals that induce the cells in the sample to divide, and then add other chemicals that stop cell division at metaphase. Metaphase is the stage of mitosis at which the chromosomes are most condensed and thus most easily distinguished from one another. After arresting the process of cell division in their sample at this stage, the biologists break the cells to spread out their contents, including the chromosomes, and then stain and examine the chromosomes. To facilitate the examination of the karyotype, the chromosomes are usually photographed and then the outlines of each chromosome are cut out, like paper dolls, and arranged in order (see Figure 10-6).

Karyotypes of individuals are often examined to detect genetic abnormalities, such as those arising from extra or lost chromosomes. The human congenital defect known as Down syndrome (or trisomy 21), for example, is associated with the presence of an extra copy of a particular segment of chromosome 21, usually the result of an extra chromosome 21, and can be recognized in photographs of the chromosomes.

How Many Chromosomes Are in a Cell?

Each of the cells in your body, except your **gametes** (sex cells), is **diploid** and contains 46 chromosomes, consisting of two nearly identical copies of each of the basic set of 23 chromosomes. This basic set of 23 chromosomes, called the **haploid complement,** is present in all of your gametes, in eggs or sperm. The two nearly identical copies of each of the 23 different kinds of chromosomes are called homologous chromosomes or **homologues** (Greek, *homologia,* agreement). Before cell division, each of the two homologues replicates, producing in each case two identical copies called sister chromatids that remain joined together at the centromere. Thus at the beginning of cell division a body cell contains a total of 46 replicated chromosomes, each composed of two sister chromatids joined by one centromere. How many centromeres does the cell contain? 46. How many copies of the basic set of 23 chromosomes does it contain? Four, with a total of 92 chromosome copies (23 basic set × 2 homologues × 2 sister chromatids). The cell is said to contain 46 chromosomes and not 92 because by convention, the number of chromosomes is obtained by counting centromeres.

The Cell Cycle

The profound change in genome organization from bacteria (a single circle of naked DNA) to eukaryotes (several segments of DNA packaged with protein) required radical changes in the way cells partition two replicas of the genome accurately, one into each of two daughter cells. The processes that occur during the division of eukaryotic cells are conveniently diagrammed as a cell cycle:

$$G_1 \rightarrow S \rightarrow G_2 \rightarrow M \rightarrow C$$

G₁ phase: The primary growth phase of the cell; for many organisms, this occupies the major portion of the cell's life span

S phase: The phase in which a replica of the genome is synthesized

G₂ phase: The stage in which preparations are made for genomic separation; this includes the replication of mitochondria and other organelles, chromosome condensation, and the synthesis of microtubules

M phase: The phase in which the microtubular apparatus is assembled, binds to the chromosomes, and moves the sister chromosomes apart; this stage, called **mitosis,** is the essential step in the separation of the two daughter genomes

C phase: The phase in which the cell itself divides, creating two daughter cells; this phase is called **cytokinesis**

Growth occurs throughout the G_1, S, and G_2 phases. The two G phases are sometimes referred to as "gap" phases, as they separate the S phase from the M phase.

In fungi and some groups of protists, the nuclear membrane does not dissolve and mitosis is confined to the nucleus. When mitosis is complete in these organisms,

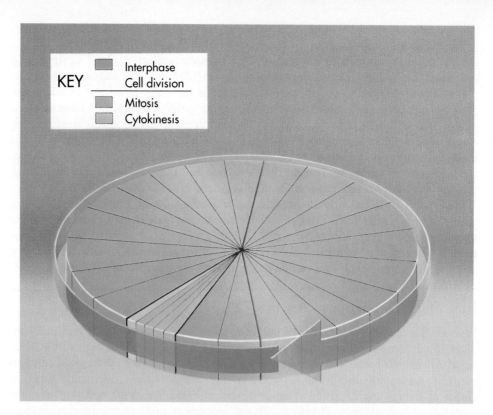

the nucleus divides into two daughter nuclei. One goes to each daughter cell during cytokinesis. This separate nuclear division phase of the cell cycle does not occur in most protists, or in plants and animals. The absence of nuclear division as a separate step is a simplification of the overall process, one that allows less room for error in chromosome separation and, by combining several steps into one, is less expensive in terms of energy.

MITOSIS

Two phases of the eukaryotic cell division cycle that have received close attention from biologists accomplish the physical division of the cell and its contents: phase M (mitosis) and phase C (cytokinesis). The time devoted to these two phases often represents only a small part of the cell cycle (Figure 10-7). We will discuss them in turn. The M phase of cell division, mitosis, has received more attention than any other aspect of the eukaryotic cell cycle. Biologists have long been fascinated by the intricate movements of the chromosomes as they separate (Figure 10-8). We will describe the process as it occurs in animals and plants. Among fungi and some protists, different forms of mitosis occur, but the process varies little among different kinds of animals and plants.

FIGURE 10-8
Sister chromatids. The photograph and companion drawing show human chromosomes as they appear immediately before nuclear division. Each DNA strand has already been replicated, forming sister chromatids that are identical to each other, held together by the centromere.

Traditionally, mitosis is subdivided into four stages: prophase, metaphase, anaphase, and telophase. Such a subdivision is convenient, but the process is actually continuous, the stages flowing smoothly into one another.

The stages of mitosis are illustrated in Figure 10-9.

Preparing the Scene: Interphase

Before the initiation of mitosis, the following events have occurred in the preceding **interphase** (that is, the G_1, S, and G_2 phases) that are of great importance for the successful completion of the mitotic process:

1. During S phase, each chromosome replicates to produce two daughter copies, called **sister chromatids** (see Figure 10-8). The two copies remain attached to each other at a point of constriction, the centromere. The **centromere** is a specific DNA sequence of about 220 nucleotides, to which is bound a disk of protein called a **kinetochore.** The centromere occurs at a specific site on any given chromosome. At the completion of S phase, each replicated chromosome consists of two sister chromatids joined at the centromere sequences.

2. After the chromosomes are replicated in phase S, they remain fully extended and uncoiled and are not visible under the light microscope. In G_2 phase, the chromosomes begin the long process of **condensation,** coiling into more and more tightly compacted bodies.

3. During G_2 phase, the cells begin to assemble the machinery that they will later use to move the chromosomes to opposite poles of the cell. In animal cells, a class of nuclear microtubule-organizing centers called **centrioles** replicate; plants and fungi lack visible centrioles. All cells undertake an extensive synthesis of **tubulin,** the protein of which microtubules are formed.

Interphase is that portion of the cell cycle in which the condensed chromosomes are not visible under the light microscope. It includes the G_1, S, and G_2 phases. In the G_2 phase, the cell mobilizes its resources for cell division.

Formation of the Mitotic Apparatus: Prophase

When the chromosome condensation initiated in G_2 phase reaches the point at which individual condensed chromosomes first become visible with the light microscope, the first stage of mitosis, **prophase,** is said to have begun. This condensation process continues throughout prophase, so that chromosomes that start prophase as minute threads may appear quite bulky before its conclusion. Ribosomal RNA synthesis ceases when that portion of the chromosome bearing the rRNA genes is condensed, with the result that the nucleolus, which was previously conspicuous, disappears.

Prophase is the stage of mitosis characterized by the condensation of the chromosomes.

Although the chromosomes are beginning to condense in prophase, another series of equally important events is also occurring: the assembly of the microtubular apparatus, which will be employed to separate the sister chromatids. Early in prophase in animal cells, the two centriole pairs start to move apart, forming between them an axis of microtubules referred to as **spindle fibers.** The centrioles continue to move apart until they reach the opposite poles of the cell, with a bridge of microtubules called the **spindle apparatus** extending between them. In plant cells a similar bridge of microtubular spindle fibers forms between opposite poles of the cell, although in this case the microtubular-organizing center is not visible with the light microscope, and centrioles are absent.

During the formation of the spindle apparatus, the nuclear envelope breaks down and its components are reabsorbed into the endoplasmic reticulum. The microtubular spindle thus extends right across the cell from one pole to the other. Its position de-

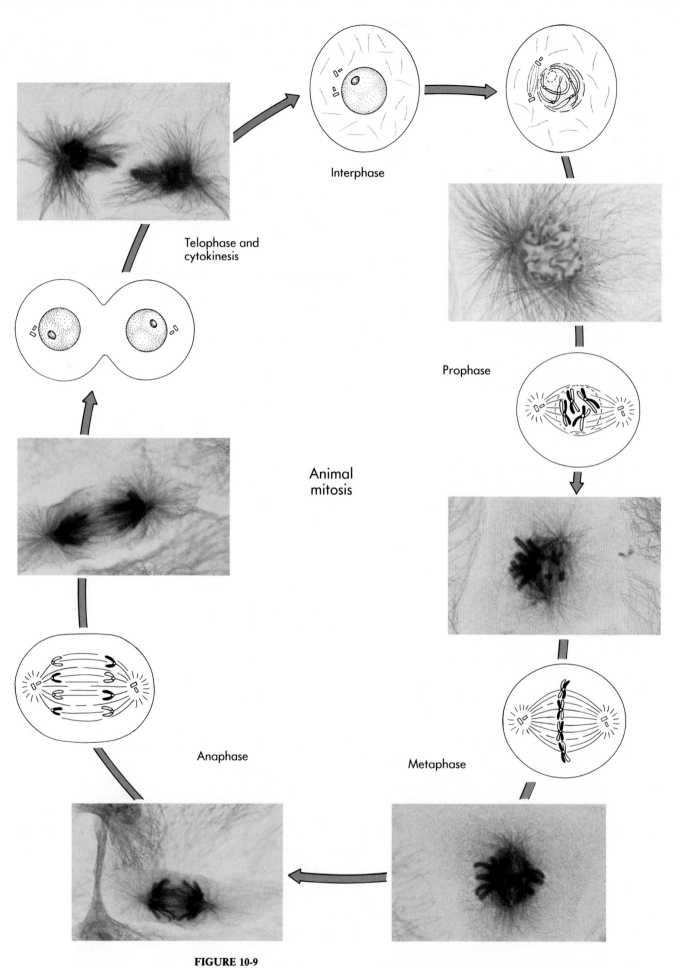

Interphase

Prophase

Telophase and
cytokinesis

Animal
mitosis

Anaphase

Metaphase

FIGURE 10-9

Animal mitosis. In this remarkable series of photographs, the chromosomes of a kangaroo have been stained blue and the spindle fibers stained brown.

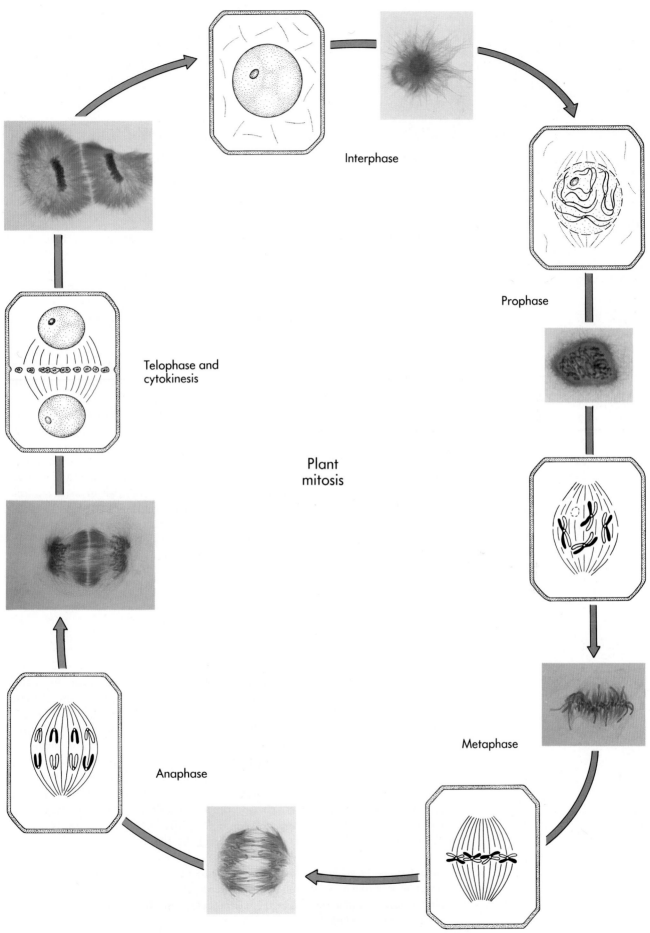

Interphase

Prophase

Telophase and
cytokinesis

Plant
mitosis

Metaphase

Anaphase

FIGURE 10-9, CONT'D
Plant mitosis. The chromosomes of the African blood lily, *Haemanthus katharinae*, are stained blue, and microtubules are stained red. In both animals and plants, the importance of microtubules in the process of mitosis is clearly illustrated.

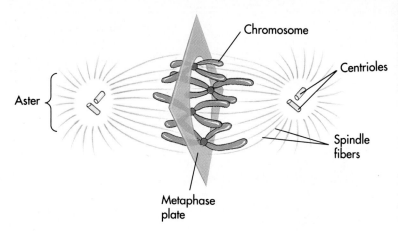

FIGURE 10-10
The metaphase plate. In metaphase the chromosomes array themselves in a circle around the spindle midpoint.

termines the plane in which the cell will subsequently divide, a plane that passes through the center of the cell at right angles to the spindle.

> During prophase the nuclear envelope breaks down, and a network of microtubules called the spindle apparatus forms between opposite poles of the cell. The position of the spindle apparatus determines the plane in which the cell will divide.

In animal mitosis, when the centrioles reach the poles of the cell, they extend a radial array of microtubules outward, thus bracing the centrioles against the cell membrane. This arrangement of microtubules is called an **aster.** The function of the aster is not well known, but is probably mechanical, acting to stiffen the point of microtubular attachment during the subsequent retraction of the spindle. Plant cells, which have rigid cell walls lack centrioles and do not form asters.

As prophase continues, a second group of microtubules appears to grow out from the individual centromeres to the poles. Two sets of microtubules extend from each chromosome (Figure 10-10), connecting the kinetochore of each pair of sister chromatids to the two poles of the spindle. The two sets of microtubules attached to a centromere both continue to grow until each has made contact with one of the poles of the cell. Because microtubules extending from the two poles attach to opposite sides of the centromere, the effect is to attach one sister chromatid to one pole, and the other sister chromatid to the other.

> At the end of prophase, the centromere joining each pair of sister chromatids is attached by microtubules to opposite poles of the spindle.

Division of the Centromeres: Metaphase

The second phase of mitosis, **metaphase,** begins when the pairs of sister chromatids align in the center of the cell. When viewed with a light microscope, the chromosomes appear to be lined up along the inner circumference of the cell in a circle perpendicular to the axis of the spindle. The three-dimensional configuration is that of a line inscribing a great circle on the periphery of a sphere, as the equator girdles the earth (see Figure 10-10). An imaginary plane passing through this circle is called the **metaphase plate.** The metaphase plate is not a structure in the physical sense of the word but rather an indication about where the future axis of cell division will occur. Positioned by the microtubules attached to the kinetochores of their centromeres, all of the chromosomes line up on the metaphase plate, their centromeres neatly arrayed in a circle, each equidistant from the two poles of the cell.

> Metaphase is the stage of mitosis characterized by the alignment of the chromosomes in a ring along the inner circumference of the cell. Each chromosome is drawn to that position by the microtubules extending from it to the two poles of the spindle.

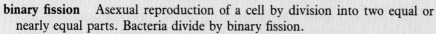

binary fission   Asexual reproduction of a cell by division into two equal or nearly equal parts. Bacteria divide by binary fission.

centromere   A constricted region of the chromosome joining two sister chromatids, to which the kinetochore is attached. About 220 nucleotides in length, it is composed of highly repeated DNA sequences (satellite DNA).

chromatid   One of the two replicated strands of a duplicated chromosome, joined by a single centromere to the other replicated strand.

chromatin   The complex of DNA and proteins of which eukaryotic chromosomes are composed.

chromosome   The organelle within cells that carries the genes. In bacteria, the chromosome consists of a single naked circle of DNA; in eukaryotes, a chromosome consists of a single linear DNA molecule associated with proteins.

cytokinesis   Division of the cytoplasm of a cell after nuclear division.

euchromatin   That portion of chromatin that is extended except during cell division, and from which RNA is transcribed.

heterochromatin   That portion of a eukaryotic chromosome that remains permanently condensed and therefore is not transcribed into RNA. Most centromere regions are heterochromatic.

homologues   Homologous chromosomes. In diploid cells, one chromosome of a pair that carry equivalent genes.

kinetochore   A disk of protein bound to the centromere to which microtubules attach during mitosis, linking each chromatid to the spindle.

microtubule   A hollow protein cylinder, about 25 nanometers in diameter, composed of subunits of the protein tubulin. Microtubules grow in length by the addition of tubulin subunits to the end(s), and shorten by their removal.

mitosis   Nuclear division in which duplicated chromosomes separate to form two genetically identical daughter nuclei; usually accompanied by cytokinesis, producing two identical daughter cells.

nucleosome   The basic packaging unit of eukaryotic chromosomes, in which the DNA molecule is wound around a ball of histone proteins. Chromatin is composed of long strings of nucleosomes, like beads on a string.

Each chromosome possesses two kinetochores, one attached to the centromere region of each sister chromatid. One or more microtubules are attached to each kinetochore, extending to the opposite pole. This arrangement is absolutely critical to the process of mitosis. Any mistakes in this positioning of the microtubules are disastrous. The attachment of the two microtubules to the same pole, for example, leads to a failure of the sister chromatids to separate, so that they end up in the same daughter cell.

At the end of metaphase the centromeres divide. Each centromere splits in two, freeing the two sister chromatids from their attachment to one another. Centromere separation is simultaneous for all the chromosomes; the mechanism that achieves this synchrony is not known.

At the end of metaphase the centromeres divide, thus freeing the sister chromatids so that in the next phase they can be drawn to opposite poles of the spindle by the microtubules attached to their kinetochore.

Separation of the Chromatids: Anaphase

Of all the stages of mitosis, anaphase is the most beautiful to watch and the shortest. No longer tugged in two directions at once by opposing microtubules, like a rope in a tug of war, each sister chromatid now rapidly moves toward the pole to which its microtubule is attached (see Figure 10-10). The following two forms of movement take place simultaneously, each driven by microtubules:

1. *The poles move apart.* The microtubular spindle fibers slide past one another.* Because the two members of each microtubule pair are physically anchored to opposite poles, their sliding past one another pushes the poles apart. Because the chromosomes are attached to these poles, they move apart, too. In this process the cell, if it is bounded by a flexible membrane, becomes visibly elongated.

2. *The centromeres move toward the poles.* The microtubules shorten as tubulin subunits are continuously removed from their polar ends. This shortening process is not a contraction, since the microtubules do not get any thicker. Instead, tubulin subunits are removed from the polar ends of the microtubules by the organizing center. As more and more subunits are removed, the progressive disassembly of the chromosome-bearing microtubule renders it shorter and shorter, pulling the chromosome ever closer to the pole of the cell.

Anaphase is the stage of mitosis characterized by the physical separation of sister chromatids. The poles of the cell are pushed apart by microtubular sliding, and the sister chromatids are drawn to opposite poles by the shortening of the microtubules attached to them.

Reformation of Nuclei: Telophase

The separation of sister chromatids achieved in anaphase completes the accurate partitioning of the replicated genome, the essential element of mitosis. With the play complete, the only tasks that remain in telophase are to dismantle the stage and remove the props. The spindle apparatus is disassembled, with the microtubules being disassembled into tubulin monomers ready for use in constructing the cytoskeleton of the new cell. The nuclear envelope re-forms around each set of sister chromatids, now chromosomes in their own right because each now has its own centromere. The chromosomes soon begin to uncoil into the more extended form that permits gene expression. An early group of genes to regain expression are the rRNA genes, resulting in the reappearance of the nucleolus.

Telophase is the stage of mitosis during which the mitotic apparatus assembled during prophase is disassembled, the nuclear envelope is reestablished, and the normal use of the genes present in the chromosomes is reinitiated.

CYTOKINESIS

At the end of telophase mitosis is complete. The eukaryotic cell has partitioned its replicated genome into two nuclei, which are positioned at opposite ends of the cell. While this process has been going on, the cytoplasmic organelles, such as the mitochondria and (if they are present) the chloroplasts, have also been reassorted to the areas that will separate and become the daughter cells. Their replication takes place before cytokinesis, often in the S or G_2 stage. At this point, the process of cell division is still not complete, however. Indeed, the division of the cell proper has not yet even begun. The stage of the cell cycle at which cell division occurs is called **cytokinesis.** Cytokinesis generally involves the cleavage of the cell into roughly equal halves.

Cytokinesis is the physical division of the cytoplasm of a eukaryotic cell into two daughter cells.

---

*The details of how microtubules slide and shorten are active areas of research. The sliding of spindle-fiber microtubules past one another is thought to involve ATP-dependent movement of crossbridges made of the protein dynein, and can be inhibited with the same chemicals that block motion of flagella. See for example the excellent work of Z. Cande and McDonald J, *Nature*, 316:168-170, 1985.

In the cells of animals and all other eukaryotes that lack cell walls, cytokinesis is achieved by pinching the cell in two with a constricting belt of microfilaments. As the actin filaments within the belt slide further and further past one another, a **cleavage furrow** becomes evident around the circumference of the cell (Figure 10-11) where the cytoplasm is being progressively pinched inward by the decreasing diameter of the microfilament belt. As constriction proceeds, the furrow deepens until it eventually extends all the way into the residual spindle, and the cell is literally pinched into two.

Plant cells possess a rigid cell wall, one far too strong to be deformed by microfilament contraction. Plants have evolved a different strategy of cell division. Plants assemble membrane components in their interior, at right angles to the spindle (Figure 10-12). This expanding partition is called a **cell plate.** It continues to grow outward until it reaches the interior surface of the cell membrane and fuses with it, at which point it has effectively divided the cell into two. Cellulose is then laid down on the new membranes, creating two new cells. The space between the two new cells becomes impregnated with pectins and is called a **middle lamella.**

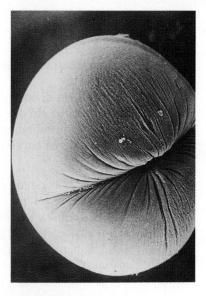

FIGURE 10-11
A cleavage furrow around a dividing sea urchin egg.

COMPARING CELL DIVISION IN EUKARYOTES AND PROKARYOTES

When we consider the process of cell division, it is impossible not to be struck by the sophisticated way in which eukaryotic cells partition their chromosomes among daughter cells, relative to the simple way this is achieved in prokaryotes. This complex mechanism of moving chromosomes to opposite ends of a dividing cell by attaching them to microtubules permits a rapid and accurate separation of sister chromatids into daughter cells. Separation of daughter chromosomes within a dividing bacterium, by contrast, is a slower process whose success depends on uninterrupted membrane growth. The process of binary fission that occurs in bacteria works well, but mitosis is a surer, more efficient process.

Most eukaryotes possess much more DNA than do prokaryotes. If human cells divided by binary fission, like those of bacteria, all 46 chromosomes would be stitched together in one enormous circle some 3 billion nucleotides around. The circle would be so large (about 3 meters in circumference) that it would be very difficult to partition it equally between the two daughter cells. Eukaryotes achieve this difficult feat by the process of mitosis.

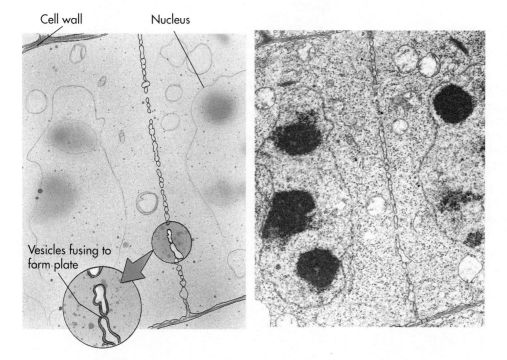

Cell wall   Nucleus

Vesicles fusing to form plate

FIGURE 10-12
Cytokinesis in plant cells. In this photograph and companion drawing, a cell plate is forming between daughter nuclei. Once the plate is complete there will be two cells.

SUMMARY

1. Bacterial cells divide by simple binary fission. The two replicated circular DNA molecules attach to the plasma membrane at different points, and fission is initiated between the points to which they are attached.

2. The cells of eukaryotes contain much more DNA than do those of prokaryotes. Eukaryotic DNA forms a complex with histones and other proteins and is divided among chromosomes. Some of the DNA is permanently condensed into heterochromatin, whereas the rest is condensed only during cell division.

3. DNA replication is completed during the S phase before mitosis begins. Immediately before the onset of mitosis, condensation of the chromosomes and synthesis of microtubules begins. This period preceding mitosis is called interphase.

4. The first stage of mitosis proper is prophase, during which the mitotic apparatus forms. At the end of prophase the nuclear envelope disassembles and microtubules attach each pair of sister chromatids to the two poles of the cell.

5. The second stage of mitosis is metaphase, during which the chromosomes align along the periphery of a plane cutting through the center of the cell at right angles to the spindle axis. At the end of metaphase the centromeres joining each pair of sister chromatids separate, freeing each chromatid to be pulled to one of the poles of the cell by the microtubules attached to it.

6. The third stage of mitosis is anaphase, during which the chromatids physically separate, moving to opposite poles of the cell.

7. The fourth and final stage of mitosis is telophase, during which the mitotic apparatus is disassembled, the nuclear envelope reforms, and the chromosomes uncoil.

8. Following mitosis, most cells undergo cytoplasmic cleavage, or cytokinesis. In cells without a cell wall, the cell body is pinched in two by a belt of microfilaments drawing inward around its midsection. In plant cells and other cells with a cell wall, an expanding cell plate forms along the spindle midline.

REVIEW QUESTIONS

1. What is the molecular composition of eukaryotic chromosomes? Why is there RNA associated with the chromosomes? What are nucleosomes, and how are they associated with DNA coiling? What is supercoiling?

2. Compare heterochromatin and euchromatin.

3. What is a karyotype? How are chromosomes differentiated from one another?

4. A normal human possesses how many different kinds of chromosomes (consider the sex chromosomes as a single kind)? What are the chromosome pairs called? How many individual chromosomes are present? . . . chromatids? The number of individual chromosomes is best determined by counting the number of what chromosomal structure?

5. What are the five stages of the cell cycle? Concisely state what occurs in each stage. Which two of these stages are specifically associated with the process of cell division? Which is generally the lengthiest stage in terms of an individual cell's life span?

6. In chronological order, what are the stages of mitosis? Indicate a key characteristic of each stage.

7. How is a pre-S chromosome different from a post-S chromosome? What is the structure of an individual chromosome? Is the genetic material on each sister chromatid identical or different?

8. What events occur near the end of interphase, in G2, that are associated with cell division? How do these events differ between animal cells and plant and fungus cells?

9. What happens to ribosomal synthesis at the beginning of prophase? What characteristic structure of the nucleus does this affect?

10. What observation signals the initiation of metaphase? What event indicates the end of metaphase?

11. What events characterize anaphase? What are the two types of microtubular movements and how do they affect the sister chromatids?

12. What events characterize telophase?

13. How is cytokinesis different between animals and plants?

THOUGHT QUESTIONS

1. Chromosomes separate from one another in anaphase because of microtubular sliding and shortening. Why do you suppose mitosis is organized in this fashion, rather than being powered by microfilament movements?

2. The plant *Haplopappus gracilis* has only 2 chromosomes, whereas the adder's tongue fern has 1,262. Can you suggest a reason for this wide variation in chromosome number? There is much less variation among mammals. Can you propose an explanation?

3. Imagine you were constructing an artificial chromosome. What elements would you introduce into it, at a *minimum*, so that it could function normally in mitosis?

FOR FURTHER READING

BASERGA, R.: *The Biology of Cell Reproduction*, Harvard University Press, Cambridge, Mass., 1985. A very stimulating and insightful look at mitosis, in a chatty style that is fun to read.

KORNBERG, R.D., and A. KLUG: "The Nucleosome," *Scientific American*, Februray 1981, pages 52-64. The DNA superhelix is wound on a series of protein spools.

MANUELIDIS, L.: "A View of Interphase Chromosomes." *Science*, vol. 250, December, 1990, pages 1533-1540. A review of how chromosomes are organized to express their genes.

MARX, J.: "The Cell Cycle Coming Under Control," *Science*, vol 245, July 1989, pages 252-255. A highly readable account of how researchers uncovered the biochemical machinery that controls cell division in all species of eukaryotes, from yeast to man.

MAZIA, D.: "The Cell Cycle," *Scientific American*, January 1974, pages 54-64. An account by one of those responsible for the development of the concept.

McINTOSH, R., and K. McDONALD: "The Mitotic Spindle," *Scientific American*, October 1989, pages 48-57. New information on how microtubules parted the DNA of dividing cells into two equal clusters.

MOYZIS, R.: "The Human Telomere," *Scientific American*, August 1991, pages 48-55. A unique nucleotide sequence repeated thousands of times forms a protective cap on the ends of chromosomes that protects them from shortening during DNA replication.

MURRAY, A., and KIRSCHNER, M.: "What Controls the Cell Cycle," *Scientific American*, March 1991, pages 56-63. A modern look at how cell division is controlled. One protein plays a key role in virtually all organisms.

MURRAY, A., and J. SZOSTAK: "Artificial Chromosomes," *Scientific American*, November 1987, pages 62-68. Artificial chromosomes made from the DNA of yeast cells and other organisms allow investigators to define the functional elements of eukaryotic chromosomes.

SCIENCE: "The Cell Cycle," *Science*, vol. 246, no. 4930, November 3, 1989, pages 545-640. A comprehensive account of all aspects of modern research into the cell cycle.

SLOBODA, R.D.: "The Role of Microtubules in Cell Structure and Cell Division," *American Scientist*, vol. 68, pages 290-298, 1980. A good description of how the spindle apparatus works.

Sexual Reproduction and Meiosis

OVERVIEW

Hereditary traits are specified by genes, which are integral parts of chromosomes. Humans and many other organisms reproduce sexually. When diploid organisms such as humans produce haploid gametes, the particular chromosomal homologue that is included in a particular gamete is chosen at random. This random assortment of chromosomes among gametes is one of the mechanisms that leads to much of the genetic variation among individuals. The process of gamete formation in sexual reproduction occurs by a special form of cell division called meiosis.

FOR REVIEW

Here are some important terms and concepts that you will encounter in this chapter. If you are not familiar with them, you should review them before proceeding.

Microtubules (Chapter 5)

Centromere (Chapter 10)

Mitosis (Chapter 10)

Homologous chromosomes (Chapter 10)

THE DISCOVERY OF MEIOSIS

Only a year after Fleming's discovery of chromosomes, the Belgian cytologist Pierre-Joseph van Beneden was studying the chromosomes of *Ascaris* and found, to his surprise, that the number of chromosomes was different in different types of cells. Specifically, he observed that eggs and sperm each contained two chromosomes, whereas the cells of the young embryo contained four chromosomes, as do all the cells in the body of a mature individual of *Ascaris*. From his observations, van Beneden was able to outline the basic process of what we now call sexual reproduction:

1. The gametes (eggs and sperm) of an organism each contain a single basic complement of chromosomes.
2. The zygote produced by fusion of egg and sperm (and the adult that the zygote becomes) contains in each of its body cells two copies of each chromosome.

The fusion of haploid gametes to form a new diploid cell is called **fertilization** or **syngamy.** As you can readily imagine, cell fusion is not a process that can continue to occur over and over. If it did, the number of chromosomes in each cell would become impossibly large. For example, in 10 generations the 46 chromosomes present in each of your cells would have increased to over 47,000 (46×2^{10}).

It was clear even to early investigators that there must be some mechanism during the course of gamete formation to reduce the number of chromosomes to half the number that is characteristic of that species. If a special reduction division occurs, in which cells are formed with half the number of chromosomes characteristic of most cells of that species, then the subsequent fusion of these cells would make possible a stable chromosome number. The mature individuals of each successive generation would have the same number of chromosomes. The expected reduction division process was soon observed (Figure 11-1). It is called **meiosis.**

Meiosis is a process of nuclear division in which the number of chromosomes in certain cells is halved during gamete formation.

SEXUAL VERSUS ASEXUAL REPRODUCTION

Fertilization and meiosis together constitute a cycle of reproduction in which there are two sets of chromosomes present in the body cells of adult individuals, which are said to be **diploid** (Greek *di-*, two), and one set present in the gametes, which are said to be **haploid** (Greek *haploos,* one). Reproduction that involves this alternation of fertilization and meiosis is called **sexual reproduction.** Its outstanding characteristic is that an individual offspring inherits genes from two different parent individuals. You, for example, inherited genes from your mother and your father, your mother's genes being contributed by the egg fertilized at your conception, and your father's by the sperm that fertilized that egg cell (Figure 11-2).

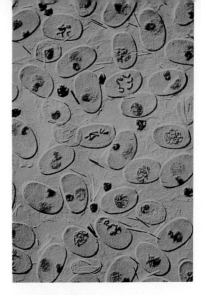

FIGURE 11-1
The stages of meiosis. This preparation of pollen cells of a spiderwort, *Tradescantia,* was made by freezing the cells and then fracturing them. It shows several stages of meiosis.

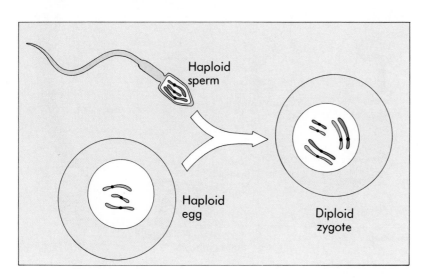

FIGURE 11-2
Diploid cells carry their parents' chromosomes. A diploid cell contains two versions of each chromosome, one contributed by the haploid egg of the mother, the other by the haploid sperm of the father.

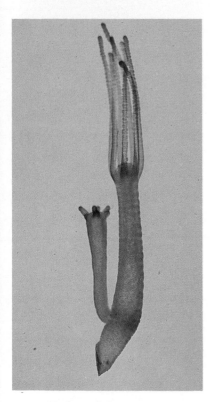

FIGURE 11-3
Asexual reproduction.

Not all reproduction is sexual. In asexual reproduction, an individual inherits all of its genes from a single parent. Bacteria reproduce in this way, undergoing binary fission of a single cell and producing two daughter cells with the same genetic information. Most protists reproduce asexually except under conditions of stress, when they switch to sexual reproduction. Many multicellular organisms are also capable of reproducing asexually. In animals, asexual reproduction often involves the budding off of a localized mass of cells, which grows by mitosis to form a new individual (Figure 11-3). Because these cell divisions are mitotic, every cell of the new individual has the same genetic makeup as the individual from which the bud arose. Asexual reproduction is common among plants, as we shall see in Chapter 33. The outstanding characteristic of asexual reproduction is that an individual offspring is genetically identical to its parent, like a Xerox copy.

THE SEXUAL LIFE CYCLE

The life cycles of all sexually reproducing organisms follow the same basic pattern of alternation between the diploid and the haploid chromosome number (Figure 11-4). Each of the somatic (body) cells in your body possesses 46 chromosomes, the diploid number. When your body was first formed as a zygote as a result of the fertilization of an egg from your mother by a sperm from your father, 23 chromosomes, the haploid number, were contributed by your mother's egg, one of each of the 23 different kinds; a second version of each of these 23 was contributed by your father's sperm.

After syngamy—the fusion of egg and sperm—the resulting single diploid cell (the zygote) begins to divide by mitosis. This single initial cell eventually gives rise to an adult body with some 100 trillion cells! As you read this line, you are the result of a long journey of cell division, all carried out by mitosis. Except for rare accidents along the way, every one of your 100 trillion cells is genetically identical to that first fertilized zygote cell.

In plants, the haploid cells that result from meiosis divide by mitosis, forming a multicellular haploid phase (see Figure 28-13, *C*). Certain cells of this haploid phase

FIGURE 11-4
The sexual life cycle. In animals the vast majority of the life cycle is spent as a diploid organism, the completion of meiosis being followed soon by fertilization.

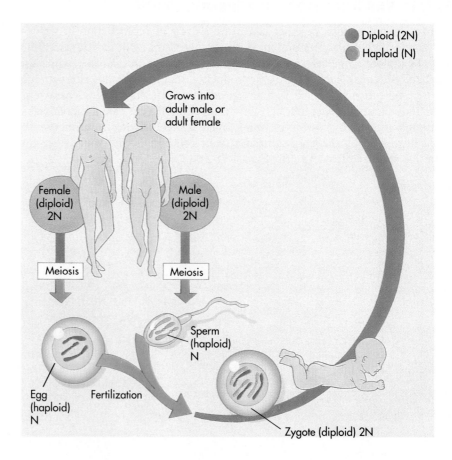

eventually differentiate into eggs or sperm. In unicellular organisms, such as most protists, individual cells function directly as gametes, and the zygote may either divide immediately by meiosis to give rise to haploid individuals or by mitosis (see Figure 28-13, A).

It is in the sex organs of your body that meiosis occurs and gametes are produced (this process is described in detail in Chapter 54). Meiosis should be distinguished clearly from mitosis, the subject of Chapter 10. In humans, as in most animals and many plants, both the somatic or body cells and the gamete-producing cells are diploid. When a somatic cell undergoes mitosis, it divides to form two diploid daughter cells exactly like it; whereas when a gamete-producing cell undergoes meiosis, it produces haploid gamete cells with half the diploid number of chromosomes. In animals the cells that will eventually undergo meiosis to produce the gametes are set aside from somatic cells early in the course of development and are often referred to as germ line cells.

THE STAGES OF MEIOSIS

Although meiosis is a continuous process, it is most easily studied when it is divided into arbitrary stages, just as was done in describing the process of mitosis in Chapter 10. Indeed, the two forms of nuclear division have much in common. Meiosis in a diploid organism consists of two rounds of nuclear division, during the course of which two unique events occur:

1. In an early stage of the first nuclear division, the two versions of each chromosome, called **homologous chromosomes,** or **homologues,** pair all along their length. While they are thus paired, genetic exchange occurs between them, physically joining them. This exchange process is called **crossing-over.** The homologous chromosomes are then drawn together to the equatorial plane of the dividing cell; subsequently, each homologue is pulled by microtubules toward the opposite poles of the cell. When this process is complete, the group of chromosomes at each pole contains one of the two homologues of each chromosome. The clusters of chromosomes at each pole are thus haploid, since each chromosome within a cluster *has no homologue* in that cluster. Each pole contains half the number of chromosomes present in the original diploid cell. At the same time, each chromosome is still composed of two chromatids.
2. The second meiotic division is identical to a normal mitotic division, except that *the chromosomes do not replicate between the two divisions.* Because of the crossing-over that occurred earlier, the sister chromatids that separate in this second division are not identical to one another.

Two important properties distinguish meiosis from mitosis:
1. **In meiosis the homologous chromosomes pair lengthwise, and their chromatids exchange genetic material by crossing-over.**
2. **The sister chromatids, which are not identical after crossing-over, do not separate from one another in the first nuclear division, and the chromosomes do not replicate between the two nuclear divisions.**

The two stages of meiosis are traditionally called meiosis I and meiosis II, each stage being subdivided further into prophase, metaphase, anaphase, and telophase, just as in mitosis. In meiosis, however, prophase is more complex than it is in mitosis.

The First Meiotic Division

Prophase I. In prophase I, individual chromosomes first become visible under the light microscope as their DNA coils more and more tightly during condensation and a matrix of fine threads becomes apparent. Since the DNA has already replicated before the onset of meiosis, each of these threads actually consists of two sister chromatids joined at their centromeres.

FIGURE 11-5
Synapsis. A portion of a synaptonemal complex of the ascomycete *Neotiella rutilans*, a cup fungus. Similar complexes can be observed in a wide variety of eukaryotes during prophase I of meiosis.

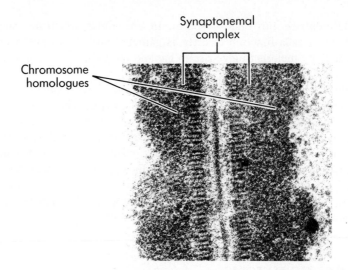

The ends of these chromatids attach to the nuclear envelope at specific sites. The membrane sites to which homologous chromosomes attach are near one another so that the members of each homologous pair of chromosomes are brought close together. They then line up side by side, a process called **synapsis.** A lattice of protein is laid down between the homologous chromosomes (Figure 11-5). This lattice holds the two replicated chromosomes in precise register with one another, each gene located directly across from its corresponding sister on the homologous chromosome. The effect is much like that of zipping up a zipper. The resulting complex is called a **synaptonemal complex.** Within it the DNA duplexes unwind and single strands of DNA pair with their opposite number *from the other homologue.*

> **Synapsis is the close pairing of homologous chromosomes that takes place early in prophase I of meiosis. During synapsis, a molecular scaffold called the synaptonemal complex aligns the DNA molecules of the two homologous chromosomes side by side. As a result, a DNA strand of one homologue can pair with the corresponding DNA strand of the other.**

The process of synapsis initiates a complex series of events called crossing-over, in which DNA is exchanged between the two paired DNA strands. Once the process of crossing-over is complete, the synaptonemal complex breaks down. At that point the homologous chromosomes are released from the nuclear envelope and the chromatids begin to move apart from one another. There is a total of four chromatids for each type of chromosome at this point: two homologous chromosomes, each replicated and so present twice (as sister chromatids). The four chromatids do not separate completely, however, because they are held together in the following two ways:

1. The two sister chromatids of each homologue, recently created by DNA replication, are held together by their common centromere.
2. The paired homologues are held together at the points where crossing-over occurred within the synaptonemal complex.

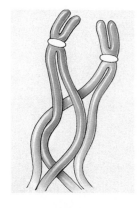

The points at which portions of chromosomes have been exchanged can often be seen under the light microscope as an X-shaped structure known as a **chiasma** (Greek, cross; plural, **chiasmata**) (Figure 11-6). The presence of a chiasma indicates that two of the four chromatids of paired homologous chromosomes have exchanged parts, one participant from each homologue. The X-shaped chiasma figures soon begin to move out to the end of the respective chromosome arm as the chromosomes separate (imagine moving a small ring down two strands of rope).

> **In prophase I, individual DNA strands of the two homologues pair with one another. Crossing-over occurs between the paired DNA strands, creating the chromosomal configurations known as chiasmata. These exchanges lock the two homologues together, and they do not disengage readily.**

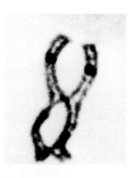

FIGURE 11-6
Chiasmata.

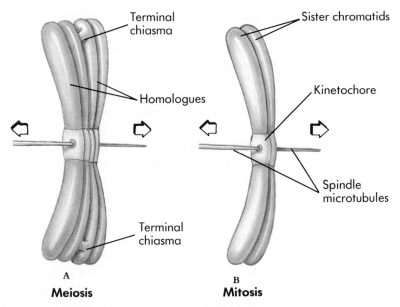

FIGURE 11-7

A key mechanical element necessary for reduction division in meiosis is the presence of chiasmata.
A These chiasmata are created by crossing-over during prophase I, when the chromosome homologues are close to one another within the synaptonemal complex. The chiasmata hold the two pairs of sister chromatids together; consequently, the spindle microtubules are able to bind only one face of each kinetochore. Later, when these microtubules shorten by sliding past one another, the terminal chiasmata break, and the two pairs of sister chromatids are drawn to opposite poles. However, no microtubule-driven separation of the individual chromatids occurs in the pairs of sister chromatids.
B In mitosis, by contrast, microtubules attach to both faces of each kinetochore; when they shorten, the pair of sister chromatids is split and the two chromatids are drawn to opposite poles.

Metaphase I. By the second stage of meiosis I, the nuclear envelope has dispersed and the microtubules form a spindle, just as in mitosis. There is, however, a crucial difference between this metaphase and that of mitosis: in meiosis one face of each centromere is inaccessible to spindle microtubules. To understand why, consider again the chiasmata produced by crossing-over in prophase. They continue their movement down the paired chromosomes from the original point of crossover, eventually reaching the ends of the chromosomes. At this point they are called **terminal chiasmata.** Their presence has an important consequence: a terminal chiasma holds the two homologous chromosomes together, like a couple dancing closely, so that only one side of each centromere faces outward from the complex, just as only one side of each dancer's body does. The other face of each centromere is directed inward toward the homologue, like the belt buckles of the dancers. Microtubules from the spindle are able to attach to kinetochore proteins of only the outside centromere faces (Figure 11-7). This one-sided attachment is in marked contrast to what occurs in mitosis, when kinetochores of *both* faces of a centromere are bound by microtubules.

As a result of microtubules binding only to kinetochores on the outside of each centromere, the centromere of one homologue becomes attached to microtubules originating from one pole, whereas that of the other homologue becomes attached to microtubules originating from the other pole. Each joined pair of homologues then lines up on the metaphase plate. For each pair, the orientation on the spindle axis is random; which homologue is oriented toward which pole is a matter of chance (Figure 11-8).

Anaphase I. After spindle attachment is complete, the microtubules of the spindle fibers begin to shorten. As they do, they break the chiasmata apart and pull the centromeres toward the two poles, dragging the chromosomes along with them. Because the microtubules are attached to kinetochores on only one side of each centromere,

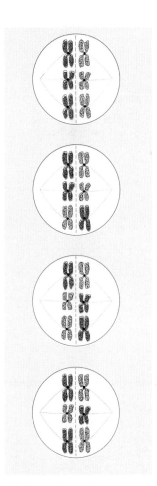

FIGURE 11-8

Independent assortment occurs because the orientation of chromosomes on the metaphase plate is random. Many combinations are possible—in fact, 2 raised to a power equal to the number of chromosome pairs. In this hypothetical cell with three chromosome pairs, four of the eight possible orientations (2^3) are illustrated. Each orientation results in gametes with different combinations of parental chromosomes.

the individual centromeres are not pulled apart to form two daughter centromeres, as they are in mitosis. Instead, the entire centromere proceeds to one pole, taking both sister chromatids with it. When the contraction of the spindle fibers is complete, each pole has a complete haploid set of chromosomes, consisting of one member of each homologous pair. Because the orientation of each pair of homologous chromosomes on the metaphase plate is random, the chromosome that a pole receives from each pair of homologues is random with respect to all other chromosome pairs.

> **The random orientation of homologous chromosome pairs on the metaphase plate and the subsequent separation of homologues from one another in anaphase are responsible for the independent assortment of traits located on different chromosomes.**

Telophase I.   At the completion of anaphase I, each pole has a complete complement of chromosomes, one member of each pair of homologues. Each of these chromosomes replicated before meiosis began and thus contains two chromatids attached by a common centromere. These chromatids, however, are not identical, because of the crossing-over that occurred in prophase I. The stage at which the two complements gather together at their respective poles to form two chromosome clusters is called telophase I. Cytokinesis may or may not occur after telophase I. After an interval of variable length the second meiotic division, meiosis II, occurs.

> **The first meiotic division is traditionally divided into four stages:**
> **Prophase I—homologous chromosomes pair and exchange segments**
> **Metaphase I—homologous chromosomes align on a central plane**
> **Anaphase I—homologous chromosomes move toward opposite poles**
> **Telophase I—individual chromosomes gather together at the two poles**

The Second Meiotic Division

Meiosis II is simply a mitotic division, involving the products of meiosis I. At the completion of anaphase I, each pole has a full haploid complement of chromosomes, each of which is composed of two sister chromatids attached by a common centromere. Because of crossing-over in the first phase of meiosis, however, these sister chromatids are not identical to one another genetically. At both poles of the cell, these two complements of chromosomes now divide mitotically, spindle fibers binding each side of the centromeres, which divide and move to opposite poles. The end result of this mitotic division is four haploid complements of chromosomes. At this point the nuclei are reorganized, nuclear envelopes forming around each of the four haploid complements of chromosomes (Figure 11-9). The cells that contain these haploid nuclei may function directly as gametes, as they do in animals, or they may themselves divide mitotically, as they do in plants, fungi, and many protists, then eventually producing greater numbers of gametes.

> **Each of the four haploid products of meiosis contains a full complement of chromosomes. These haploid cells may function directly as gametes, as they do in animals, or may divide by mitosis, as they do in plants, fungi, and many protists.**

WHY SEX?

Sexual reproduction evolved long before the advent of the vertebrates. However, it is not the only way that reproduction can occur among animals. Consider, for example, sponges. They can reproduce by simply fragmenting their bodies. In such a process of asexual reproduction, a small portion of the animal divides from it and gives rise to a new organism. In asexual reproduction there is no alternation of haploid and diploid cells, no meiosis, and no gametes. Instead, differentiated diploid tissue simply undergoes a new cycle of developmental events.

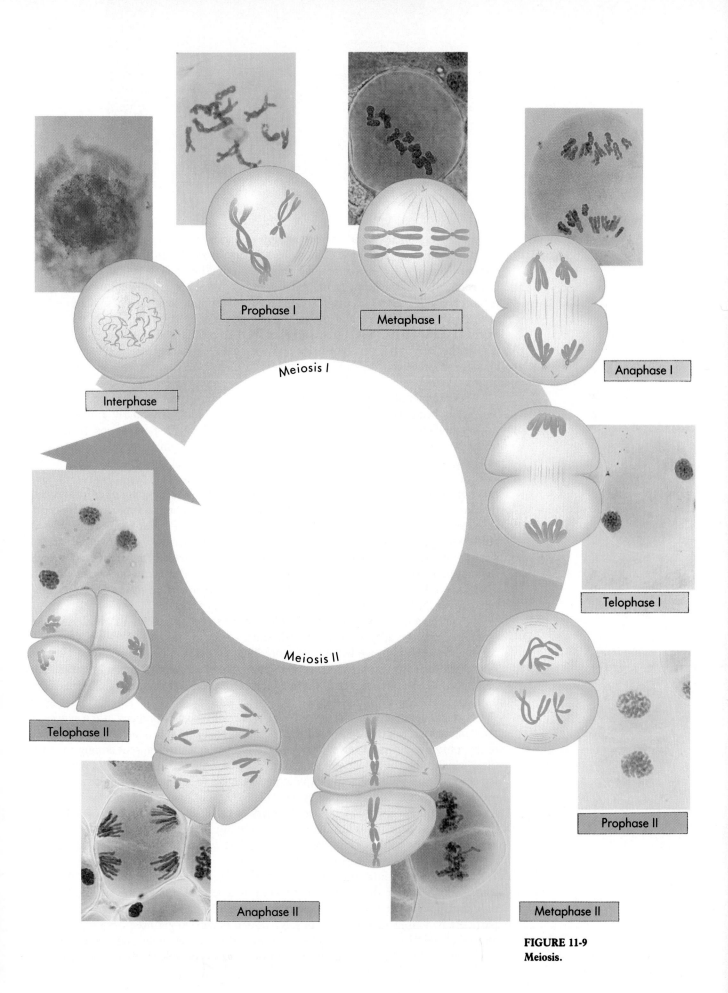

Prophase I

Metaphase I

Meiosis I

Interphase

Anaphase I

Telophase I

Meiosis II

Telophase II

Prophase II

Anaphase II

Metaphase II

FIGURE 11-9
Meiosis.

FIGURE 11-10
A parthenogenetic lizard.

Even when meiosis and the production of gametes occur, reproduction may still occur without sex. The development of an adult from an unfertilized egg is called **parthenogenesis,** and is a common form of reproduction in arthropods. Among bees, for example, the development of eggs into adults does not require fertilization. The fertilized eggs develop into diploid females, but the unfertilized eggs also undergo development, becoming haploid males. Parthenogenesis even occurs among the vertebrates (Figure 11-10). Some lizards, fishes, and amphibians are capable of reproducing in this way, their unfertilized eggs undergoing a mitotic division without cell cleavage to produce a diploid cell, which then undergoes development as if it were a diploid zygote.

If reproduction can occur without sex, it is fair to ask why sex occurs at all. There has been considerable discussion of this question, particularly among evolutionary biologists. Sex is of great evolutionary advantage for populations or species, which benefit from the variability generated by meiotic recombination (crossing-over) and gene segregation. However, evolution occurs because of changes that occur at the level of *individual* survival and reproduction, rather than at the population level, and it is not immediately obvious what advantage accrues to the progeny of an individual that engages in sexual reproduction. The segregation of chromosomes, which occurs in meiosis, tends to disrupt advantageous combinations of genes more often than it creates new, better-adapted ones; as a result, some of the diverse progeny produced as a result of any sexual mating will be less well adapted than their parents were. It is therefore a puzzle to know what a well-adapted individual gains from participating in sexual reproduction, since *all* of its progeny could maintain its successful gene combinations if that individual were simply to reproduce asexually.

Recombination is both a destructive and a constructive process in evolution. The more complex the adaptation of an individual organism, the less likely it is that recombination will improve it, and the more likely it is that recombination will disrupt it. It is no accident that asexually reproducing plants are more common in harsh habitats, such as the Arctic, where there is a premium on large numbers of genetically uniform, well-adapted individuals.

If recombination is often detrimental to an individual's progeny, then where is the benefit from sex that promoted the evolution of sexual reproduction? The answer to this puzzle is not yet known. Meiotic recombination is often absent among the protists, which typically undergo sexual reproduction only occasionally. The flagellates, for example, are predominantly haploid and asexual. Many other protists, such as the green algae, exhibit a true sexual cycle, but the diploid phase of such a cycle may be transient. In many protists the great majority of the cells are haploid almost all of the time, and such cells reproduce themselves asexually. Often the fusion of two haploid cells occurs only under stress, creating a diploid zygote, and the diploid stage that results from such a fusion event may not persist.

To understand how sex evolved, consider the protists more carefully. Why do some protists form a diploid cell in response to stress? We think this occurs because only in a diploid cell can certain kinds of chromosome damage be repaired effectively, particularly double-strand breaks in DNA. Such breaks are induced both by radiation and by chemical events within cells. As organisms became larger and longer lived, it must have become increasingly important for them to be able to repair such damage.

The synaptonemal complex, which in early stages of meiosis precisely aligns pairs of homologous chromosomes, may well have evolved originally as a mechanism for repairing double-strand damage to DNA by using the homologous chromosome as a template. A transient diploid phase would have provided an opportunity for such repair.

The origins of crossing-over in meiosis appear to have been in the gene repair processes. The current view of the molecular mechanism of crossing-over is that it evolved from a preexisting synaptonemal mechanism for repairing accidental double-strand breaks in DNA. In yeast, mutations that inactivate the repair system for double-stranded breaks of the chromosomes also prevent crossing-over, suggesting a common mechanism.

> **Sexual reproduction and the close association between homologous chromosomes that occurs during meiosis probably evolved as mechanisms to repair chromosomal damage by using the homologous chromosome as a template. This repair mechanism then provided a ready means for generating genetic recombination by crossing-over.**

THE EVOLUTIONARY CONSEQUENCES OF SEX

The reassortment of genetic material that takes place during meiosis is the principal factor that has made possible the evolution of eukaryotic organisms, in all their bewildering diversity, over the past 1.5 billion years. Sexual reproduction represents an enormous advance in the ability of organisms to generate genetic variability. To understand why this is so, recall that most organisms have more than one chromosome. Human beings, for example, have 23 different pairs of homologous chromosomes. Each human gamete receives one of the two copies of each of the 23 chromosomes, but which copy of a particular chromosome it receives is random. For example, the copy of chromosome number 14 that a particular human gamete receives in no way influences which copy of chromosome number 5 it will receive. Each of the 23 pairs of chromosomes segregates independently of all the others, so there are 2^{23} (more than 8 million) different possibilities for the kinds of gametes that can be produced, no two of them alike.

Because the zygote that forms a new individual is created by the fusion of two gametes, each produced independently, fertilization squares the number of possible outcomes (8 million × 8 million = 64 trillion).

The exchange that occurs as a result of crossing-over between the arms of homologous chromosomes adds even more recombination to the random assortment of chromosomes that occurs subsequently in meiosis (Figure 11-11). Thus the number of possible genetic combinations that can occur among gametes is virtually unlimited.

Sexual reproduction increases genetic variability in three ways:

1. **Independent assortment in meiosis**
2. **Fertilization**
3. **Crossing-over in prophase I of meiosis**

Whatever the forces that led to sexual reproduction, its evolutionary consequences have been profound. No genetic process generates diversity more quickly, and, as you have seen, genetic diversity is the raw material of evolution, the fuel that drives it and determines its potential directions. In many cases the pace of evolution appears to be geared to the level of genetic diversity: the greater the genetic diversity, the greater the evolutionary pace. Programs for selecting larger stature in domesticated animals such as cattle and sheep, for example, proceed rapidly at first, but then slow as all of the existing genetic combinations are exhausted; further progress must then await the generation of new gene combinations.

> **The genetic recombination associated with sexual reproduction has had an enormous evolutionary impact because of the extensive variability that it can rapidly generate within the genome.**

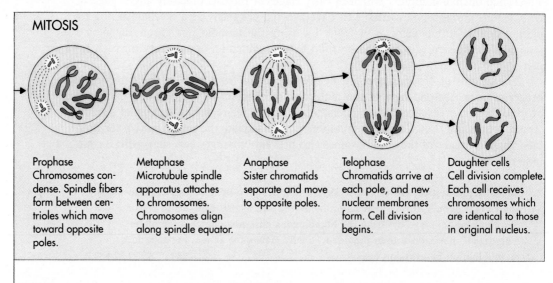

MITOSIS

Prophase
Chromosomes condense. Spindle fibers form between centrioles which move toward opposite poles.

Metaphase
Microtubule spindle apparatus attaches to chromosomes. Chromosomes align along spindle equator.

Anaphase
Sister chromatids separate and move to opposite poles.

Telophase
Chromatids arrive at each pole, and new nuclear membranes form. Cell division begins.

Daughter cells
Cell division complete. Each cell receives chromosomes which are identical to those in original nucleus.

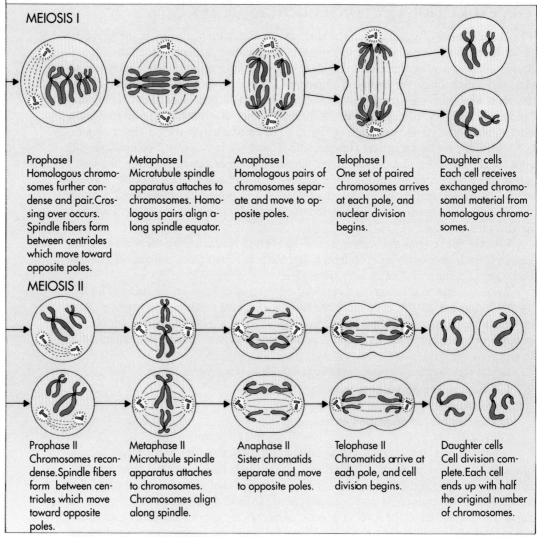

MEIOSIS I

Prophase I
Homologous chromosomes further condense and pair. Crossing over occurs. Spindle fibers form between centrioles which move toward opposite poles.

Metaphase I
Microtubule spindle apparatus attaches to chromosomes. Homologous pairs align along spindle equator.

Anaphase I
Homologous pairs of chromosomes separate and move to opposite poles.

Telophase I
One set of paired chromosomes arrives at each pole, and nuclear division begins.

Daughter cells
Each cell receives exchanged chromosomal material from homologous chromosomes.

MEIOSIS II

Prophase II
Chromosomes recondense. Spindle fibers form between centrioles which move toward opposite poles.

Metaphase II
Microtubule spindle apparatus attaches to chromosomes. Chromosomes align along spindle.

Anaphase II
Sister chromatids separate and move to opposite poles.

Telophase II
Chromatids arrive at each pole, and cell division begins.

Daughter cells
Cell division complete. Each cell ends up with half the original number of chromosomes.

FIGURE 11-11
A comparison of meiosis with mitosis. Meiosis involves two serial nuclear divisions with no DNA replication between them, and so produces four daughter cells, each with half the original amount of DNA. Crossing-over occurs in prophase I of meiosis.

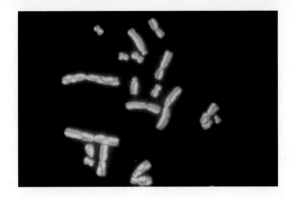

FIGURE 11-12
"Harlequin" chromosomes. So-called harlequin chromosomes provide vivid testimony of the reciprocal exchange of genetic material, in this case between sister chromatids.

Paradoxically, the evolutionary process is thus both revolutionary and conservative (Figure 11-12). It is revolutionary in that the pace of evolutionary change is quickened by genetic recombination, much of which results from sexual reproduction. It is conservative in that evolutionary change is not always favored by selection, which may instead preserve existing combinations of genes. These conservative pressures appear to be greatest in some asexually reproducing organisms that do not move around freely and that live in habitats that are especially demanding. In vertebrates, on the other hand, the evolutionary premium appears to have been on versatility, and sexual reproduction is, by an overwhelming margin, the predominant mode of reproduction.

SUMMARY

1. Sexual reproduction is a form of reproduction in which meiosis and syngamy alternate, producing haploid and diploid states respectively. Each diploid individual is produced by the union of two haploid gametes originating from different parents.

2. In asexual reproduction, mitosis produces offspring genetically identical to the parent.

3. Meiosis is a special form of nuclear division that produces the gametes of the sexual cycle. It involves two chromosome separations and only one chromosome replication.

4. Meiosis consists of a pair of serial nuclear divisions. Three unique events that occur during meiosis are synapsis—the intimate pairing of homologous chromosomes, crossing-over, and a lack of chromosome replication before the second nuclear division.

5. Crossing-over is an essential element of meiosis. The crossing-over between homologues that occurs during synapsis binds the two homologous chromosomes together, like dancers whose belt buckles are linked together. As a result of this, the spindle fibers are able to bind to only one face of each homologous chromosome, since the other face is covered by the opposite homologue. For this reason, the spindle fibers do not pull the sister chromatids apart. Rather, their shortening pulls the paired homologous chromosomes apart, ultimately breaking the terminal chiasmata that link them.

6. At the end of meiosis I, one paired homologue is present at each of the two poles of the dividing nucleus. These chromosomes still consist of two chromatids, which differ from one another as a result of the crossing-over that occurred when the chromosomes were paired with their homologues.

7. No further replication occurs before the next nuclear division, which is a normal mitosis that occurs at each of the two poles. The sister chromatids simply separate from one another. This process results in the formation of four clusters of chromosomes, each with a complement of half the number of chromosomes that were present initially. Because each daughter nucleus has one copy of each chromosome, any or all of the clusters can function as a gamete. Cytokinesis may or may not occur.

8. Meiosis is thought to have evolved initially as a mechanism to repair double-strand breaks in DNA, by pairing the broken chromosome to a homologous one while it is being repaired.

9. The evolutionary significance of meiosis is that it generates large amounts of recombination, rapidly reshuffling gene combinations and so producing variability on which evolutionary processes can act.

REVIEW QUESTIONS

1. Define meiosis. Why are both meiosis and syngamy necessary components of sexual reproduction?

2. What are some examples of asexual reproduction? How do the progeny of asexual reproduction differ from those produced via sexual reproduction with respect to their genetic material?

3. Are human somatic cells generally haploid or diploid? Are gamete-producing cells haploid or diploid?

4. What three important properties differentiate meiosis from mitosis?

5. What two mechanisms hold the homologues together in prophase I?

6. What events occur at the level of the DNA during synaptonemal crossing-over?

7. How is the spindle fiber–centromere attachment at metaphase I different from that which occurs at metaphase in mitosis?

8. What is meant by chromosomal independent assortment?

9. What events occur during anaphase I? In what three ways is this stage different from mitotic anaphase?

10. Define parthenogenesis.

11. What is the current scientific explanation for the evolution of sexual reproduction?

12. In just two words, what is the ultimate evolutionary consequence of sexual reproduction? What three events strengthen this consequence?

 THOUGHT QUESTIONS

1. Humans have 23 pairs of chromosomes. Ignoring the effects of crossing-over, what proportion of a woman's eggs contain only chromosomes she received from her mother?

2. Many sexually reproducing lizard species are able to generate local populations that reproduce asexually. What do you imagine the sex of the local asexual populations would be, male or female or neuter? Explain your reasoning.

3. Can you suggest any mechanism that might permit humans to reproduce asexually? If so, why doesn't this happen?

 FOR FURTHER READING

BELL, G.: *The Masterpiece of Nature: The Evolution and Genetics of Sexuality,* University of California Press, Berkeley, 1982. A somewhat advanced but very rewarding look at the various theories of why sexual reproduction evolved the way it did.

HAMILTON, W., R. AXELROD, and R. TANESE: "Sexual Reproduction as an Adaptation to Resist Parasites," *Proceedings of the National Acadamy of Science,* vol. 87, pages 3566-3573, 1990. A new answer to the still-puzzling question of why sex evolved.

JOHN, B.: "Myths and Mechanisms of Meiosis," *Chromosoma,* vol. 54, pages 295-325, 1976. A fine description of meiosis as a physical process, with focus on mechanisms.

JOHN, B.: *Meiosis,* Cambridge University Press, New York, 1990. A broad survey of meiosis in the plant and animal kingdom, with emphasis on the surprising diversity in how this key form of cell division is carried out.

MARGULIS, L., and D. SAGAN: *Origins of Sex,* Yale University Press, New Haven, Conn., 1986. A controversial and highly original viewpoint of how sex first evolved.

MAYNARD-SMITH, J.: *The Evolution of Sex,* Cambridge University Press, New York, 1978. An important viewpoint on the origin of sex that outlines clearly the issues currently being argued.

PICKETT-HEAPS, J., D. TIPPIT, and K. PORTER: "Rethinking Mitosis," *Cell,* vol. 29, pages 729-744, 1982. An advanced but very rewarding assessment of the evidence concerning the evolution of mitosis and meiosis.

STAHL, F: "Genetic Recombination," *Scientific American,* February 1987, pages 90-101. A molecular analysis of the events that take place during crossing-over.

12

Patterns of Inheritance

OVERVIEW

Our understanding of how heredity works dates back to the studies of Gregor Mendel over a century ago. Although Mendel was not the first person to observe patterns of inheritance, he was the first to quantify them. By doing so, he was able to detect the underlying principle of inheritance: that inherited traits are specified by discrete factors that assort independently during gamete formation and then form new combinations during fertilization. We now know that traits such as those studied by Mendel are specified by genes, which are integral parts of chromosomes. Mendel proposed his model before the existence of chromosomes was known—one of the greatest intellectual accomplishments in the history of science.

FOR REVIEW

Here are some important terms and concepts that you will encounter in this chapter. If you are not familiar with them, you should review them before proceeding.

Scientific method (Chapter 1)

Chromosomes (Chapters 5 and 10)

Meiosis (Chapter 11)

FIGURE 12-1
Human beings are extremely diverse in appearance. The differences between us are partly inherited and partly arise as a result of the environmental factors that we encounter during the course of our lives.

Every creature now living is a product of the long evolutionary history of life on earth. You, your dog or cat, the smallest bacterium—all of us share this history. Only humans, however, puzzle about the processes that led to their origin. Attempts to understand where we come from have been made throughout the course of history. We are still far from understanding everything about our origins, but we have learned a great deal. Like a partially completed jigsaw puzzle, the boundaries are now known, and much of the internal structure has become apparent. In this chapter we will discuss one piece of that puzzle, the enigma of heredity. Why do the members of a family tend to resemble one another more than they do members of other families?

THE FACES OF VARIATION

Some variation is evident to all of us (Figure 12-1). Groups of people from different parts of the world are often quite different in appearance. Within any of these groups, individuals of one family often differ greatly from those of another. Look at your classmates. Rarely will any two of you resemble one another closely. Even your brothers and sisters do not resemble you exactly, unless of course you have an identical twin. Nor are we human beings unique in this respect. Great differences in appearance often exist within other species. For example, there is a bewildering array of varieties and breeds of dogs, of every size and form imaginable, and yet all are still dogs, able to breed with one another and produce puppies (Figure 12-2).

Variation is not surprising in itself. Differences in diet during development can have marked effects on adult appearance, as can variation in the environments that different individuals experience. Many arctic mammals, for example, develop white fur when they are exposed to the cold of winter and dark fur during the warm summer months (Figure 12-3). The remarkable property of some of the patterns of variation that we can observe—the property that has always fascinated and puzzled us— is that some of the differences that we observe between individuals are inherited, passed down from parent to offspring.

As far back as there is a written record, such patterns of resemblance among the members of particular families have been noted and commented on. Some features by which the members of families resemble one another are unusual, such as the protruding lower lip of the European royal family Hapsburg, which is evident in pictures and descriptions of family members from the thirteenth century onward. Other characteristics are more familiar, such as the common occurrence of redheaded children

FIGURE 12-2
Differences in appearance often exist within species. All of the different breeds of dogs are members of the same species, *Canis familiaris*, and all can breed successfully with one another. The great differences among breeds of dogs were produced by artificial selection.

within families of redheaded parents (Figure 12-4). Such inherited features, the building blocks of evolution, will be our concern in this chapter.

EARLY IDEAS ABOUT HEREDITY: THE ROAD TO MENDEL

Like many great puzzles, the riddle of heredity seems simple now that it has been solved. The solution was not an easy one to find, however. Our present understanding is the result of a long history of thought, surmise, and investigation. At every stage we have learned more, and as we have done so, the models used to describe the mechanisms of heredity have been changed to encompass new facts.

Two concepts provided the basis for most of the thinking about heredity before the twentieth century:

FIGURE 12-4
Many human traits are inherited within families. This little redhead is exhibiting such a trait.

1. *Heredity occurs within species.* For a very long time people believed that it was possible to obtain bizarre composite animals by breeding (crossing) widely different species. The minotaur of Cretan mythology, a creature with the body of a bull and the torso and head of a man, is one example. The giraffe was thought to be another; its scientific name, *Giraffa camelopardalis*, suggests that it was believed to be the result of a cross between a camel and a leopard. From the Middle Ages onward, however, people discovered that such extreme crosses were not possible and that variation and heredity occur mainly within the boundaries of a particular species. Species were thought to have been maintained without significant change from the time of their creation.

2. *Traits are transmitted directly.* When variation is inherited by offspring from their parents, *what* is transmitted? The ancient Greeks suggested that parts of the bodies of parents were transmitted directly to their offspring. Reproductive material, which Hippocrates called *gonos*, meaning "seed," was thought to be contributed by all parts of the body; hence a characteristic such as a misshapen limb was transmitted directly to the offspring by elements that came from the misshapen limb of the parent. Information from each part of the body was thought to be passed along independently of the information from the other parts, and the child was formed after hereditary material from all parts of the parents' bodies had come together.

This idea was predominant until fairly recently. For example, in 1868 Charles Darwin proposed that all cells and tissues excrete microscopic granules, or "gemmules," that are passed along to offspring, guiding the growth of the corresponding part in the developing embryo. Most similar theories of the direct transmission of hereditary material assumed that the male and female contributions *blended* in the offspring. Thus parents with red and brown hair would be expected to produce children with reddish brown hair, and tall and short parents would produce children of intermediate height.

Taken together, however, these two concepts lead to a paradox. If no variation enters a species from outside, and if the variation within each species is blended in every generation, then all members within a species should soon resemble one another exactly. This does not happen, however. Individuals within most species differ widely

FIGURE 12-5

The garden pea, *Pisum sativum*. Easy to cultivate and with many distinctive varieties, the garden pea had been a popular choice as an experimental subject in investigations of heredity as much as a century before Gregor Mendel's investigations.

from one another, and they differ in characteristics that are transmitted from generation to generation.

How can this paradox be resolved? Actually, the resolution had been provided long before Darwin, in the work of the German botanist Josef Koelreuter. In 1760 Koelreuter carried out the first successful **hybridizations** of plant species. He was able to cross different species of tobacco and obtain fertile offspring. The hybrids differed in appearance from both of their parent strains. When crosses were made within the hybrid generation, the offspring were highly variable. Some of these offspring resembled plants of the hybrid generation, and a few resembled not the hybrid generation, but rather the original parent strains (that is, the grandparents of these individuals).

Koelreuter's work provided an important clue about how heredity works: the traits that he was studying were capable of being masked in one generation, only to reappear in the next. This pattern is not predicted by the theory of direct transmission. How could a characteristic that is transmitted directly be latent and then reappear? Nor were Koelreuter's traits "blended." A contemporary account records that they reappeared in the next generation "fully restored to all their original powers and properties."

It is important to note that the offspring of Koelreuter's crosses were not identical to one another. Some resembled the parents of the crosses, whereas others did not; the alternative forms of the traits Koelreuter was studying were distributing themselves among the offspring. A modern geneticist would say the alternative forms of a trait were **segregating** among the progeny of a single mating, meaning that some offspring exhibit one alternative form of a trait (for example, hairy leaves), whereas other offspring from the same mating exhibit a different alternative (smooth leaves). This segregation of alternative forms of a trait provided the clue that led Mendel to his understanding of the nature of heredity.

Over the next hundred years, Koelreuter's work was elaborated on by other investigators. Prominent among them were English gentleman farmers who were trying to improve varieties of agricultural plants. In one such series of experiments, carried out in the 1790s, T.A. Knight crossed two **true-breeding** varieties (varieties that were uniform from one generation to the next) of the garden pea, *Pisum sativum* (Figure 12-5). One of these varieties had purple flowers; the other, white flowers. All of the progeny of the cross had purple flowers. Among the offspring of these hybrids, however, were some plants with purple flowers and others, less common, with white ones. Just as in Koelreuter's earlier studies, a character trait from one of the parents was hidden in one generation, only to reappear in the next.

> Early geneticists demonstrated that (1) some forms of an inherited trait can be masked in some generations but may subsequently reappear unchanged in future generations, (2) forms of a trait segregate among the offspring of a cross, and (3) some forms of a trait are more likely to be represented than their alternatives.

In these deceptively simple results were the makings of a scientific revolution. Another century passed, however, before the process of segregation of genes was appreciated properly. Why did it take so long? One reason was that early workers did not quantify their results. A numerical record of results proved to be crucial to the understanding of this process. Knight and later experimenters who carried out other crosses with pea plants noted that some traits had a "stronger tendency" to appear than others, but they did not record the numbers of the different classes of progeny. Science was young then, and it was not obvious that the numbers were important.

MENDEL AND THE GARDEN PEA

The first quantitative studies of inheritance were carried out by Gregor Mendel (Figure 12-6), an Austrian monk. Born in 1822 to peasant parents, Mendel was educated in a monastery and went on to study science and mathematics at the University of Vienna, where he failed his examinations for a teaching certificate. Returning to the

FIGURE 12-6
Gregor Johann Mendel. Cultivating his plants in the garden of his monastery in Brunn, Austria (now Brno, Czechoslovakia), Mendel studied how differences among varieties of peas were inherited when the varieties were crossed with one another. Such experiments with peas had been done often before, but Mendel was the first to appreciate the significance of the results of such crosses.

monastery (where he spent the rest of his life, eventually becoming abbot), Mendel initiated a series of experiments on plant hybridization in its garden. The results of these experiments would ultimately change our views of heredity irrevocably.

For his experiments Mendel chose the garden pea, the same plant that Knight and many others had studied earlier. The choice was a good one for several reasons:

1. Many earlier investigators had produced hybrid peas by crossing different varieties. Mendel knew that he could expect to observe segregation among the offspring.
2. A large number of true-breeding varieties of peas were available. Mendel initially examined 32. Then, for further study, he selected lines that differed with respect to seven easily distinguishable traits, such as smooth versus wrinkled seeds (Figure 12-7) and purple versus white flowers (a characteristic that Knight had studied 60 years earlier).
3. Pea plants are small, are easy to grow, and have a short generation time. Thus one can conduct experiments involving numerous plants, grow several generations in a single year, and obtain results relatively quickly.
4. The sexual organs of the pea are enclosed within the flower (Figure 12-8). The flowers of peas, like those of most flowering plants, contain both male and female sex organs. Furthermore, the gametes produced by the male and female parts of the same flower, unlike those of many flowering plants, can fuse to form a viable offspring. Fertilization takes place automatically within an individual flower if it is not disturbed. As a result of this process, the offspring of

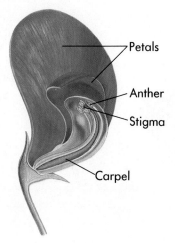

Petals
Anther
Stigma
Carpel

FIGURE 12-7
A Mendelian trait. One of the differences among varieties of pea plants that Mendel studied affected the shape of the seed. In some varieties the seeds were round, whereas in others they were wrinkled. As you can see, wrinkled seeds look like dried-out, shrunken versions of the round ones.

FIGURE 12-8
Structure of the pea plant. In a pea plant flower the petals enclose the male (anther) and female (stigma and carpel) parts, ensuring that self-fertilization will take place unless the flower is disturbed. The anther contains the pollen grains, which will give rise to haploid male gametes. The stigma and carpel contain ovules, which will give rise to haploid female eggs.

garden peas are the progeny of a single individual. Therefore one can either let **self-fertilization** take place within an individual flower, or remove its male parts before fertilization and introduce pollen from a strain with alternative characteristics, thus performing an **experimental cross.**

MENDEL'S EXPERIMENTAL DESIGN

Mendel usually conducted his experiments in three stages:

1. He first allowed pea plants of a given variety to produce progeny by self-fertilization for several generations. Mendel was thus able to assure himself that the forms of traits that he was studying were indeed constant, transmitted unchanged from generation to generation. Pea plants with white flowers, for example, when crossed with each other, produced only offspring with white flowers, regardless of the number of generations for which the experiment was continued.

2. Mendel then conducted crosses between varieties exhibiting alternative forms of traits (Figure 12-9). For example, he removed the male parts from a flower of a plant that produced white flowers and fertilized it with pollen from a purple-flowered plant. He also carried out the reciprocal cross, by reversing the pollen donor, using pollen from a white-flowered individual to fertilize a flower on a pea plant that produced purple flowers.

3. Finally, Mendel permitted the "hybrid" offspring produced by these crosses to self-pollinate for several generations. By doing so, he allowed the alternative forms of a trait to segregate among the progeny. This was the same experimental design that Knight and others had used much earlier. But Mendel added a new element: he counted the numbers of offspring of each type and in each succeeding generation. No one had ever done that before. The quantitative results that Mendel obtained proved to be of supreme importance in helping him (and us) understand the process of heredity.

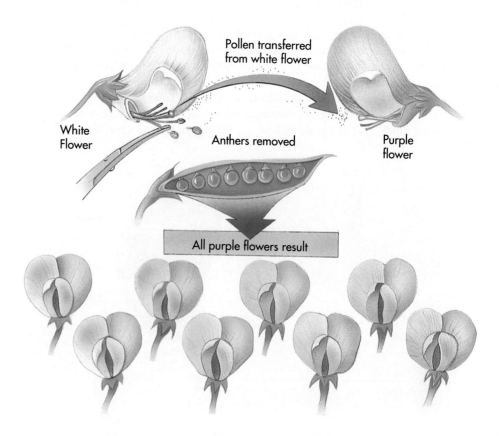

FIGURE 12-9
How Mendel conducted his experiments. Mendel pushed aside the petals of a purple flower and cut off the anthers, where the male gametes are produced and enclosed in pollen grains. He then placed that pollen onto the female parts of a similarly castrated purple flower, where cross-fertilization took place. All of the seeds in the pod that resulted from this pollination were hybrids with a white-flowered female parent and a purple-flowered male parent. Planting these seeds, Mendel observed what kinds of plants they produced. In this instance all of them had purple flowers.

WHAT MENDEL FOUND

When Mendel crossed two contrasting varieties, such as purple-flowered plants with white-flowered plants, the hybrid offspring that he obtained were not intermediate in flower color, as the theory of blending inheritance would have predicted. Instead, the hybrid offspring in every case resembled one of their parents. It is customary to refer to these hybrid offspring as the **first filial**, or **F₁**, generation. Thus in a cross of white-flowered with purple-flowered plants, the F_1 offspring all had purple flowers, just as Knight and others had reported earlier. Mendel referred to the trait that was expressed in the F_1 plants as **dominant** and to the alternative form that was not expressed in the F_1 plants as **recessive**. For each of the seven pairs of contrasting forms of traits that Mendel examined, one of the pair proved to be dominant; the other was recessive.

After individual F_1 plants had been allowed to mature and self-pollinate, Mendel collected and planted the seed from each plant to see what the offspring in this **second filial**, or **F₂**, generation would look like. He found, just as Knight had earlier, that some F_2 plants exhibited the recessive form of the trait. Latent in the F_1 generation, the recessive alternative reappeared among some F_2 individuals.

At this stage Mendel instituted his radical change in experimental design. He counted the numbers of each type among the F_2 progeny. Mendel was investigating whether the proportions of the F_2 types would provide some clue about the mecha-

FIGURE 12-10

Mendel's experimental results. The table illustrates the seven pairs of contrasting traits studied by Mendel in the garden pea and presents the data he obtained in crosses of these traits. Every one of the seven traits that Mendel studied yielded results very close to a theoretical $3:1$ ratio.

Trait	Dominant vs recessive		F_2 generation results Dominant form	Recessive form	Ratio
Flower color	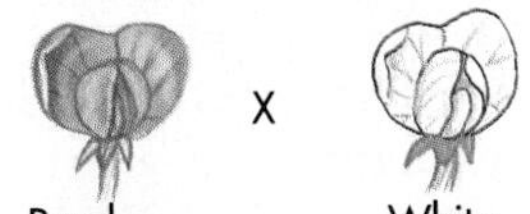 Purple X	White	705	224	3.15:1
Seed color	 Yellow X	Green	6022	2001	3.01:1
Seed shape	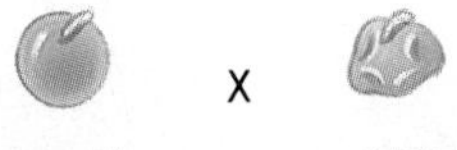 Round X	Wrinkled	5474	1850	2.96:1
Pod color	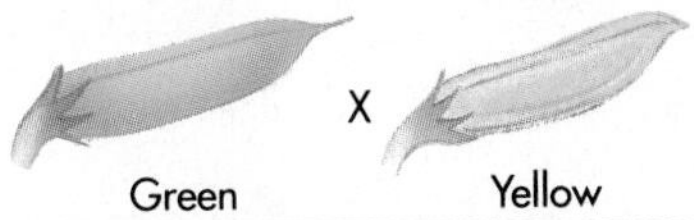 Green X	Yellow	428	152	2.82:1
Pod shape	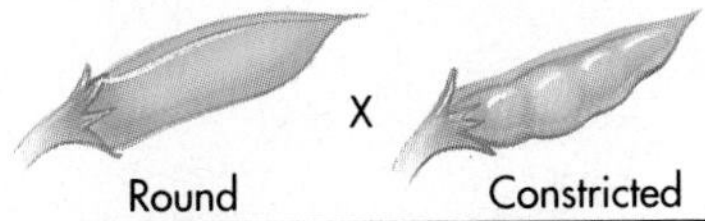 Round X	Constricted	882	299	2.95:1
Flower position	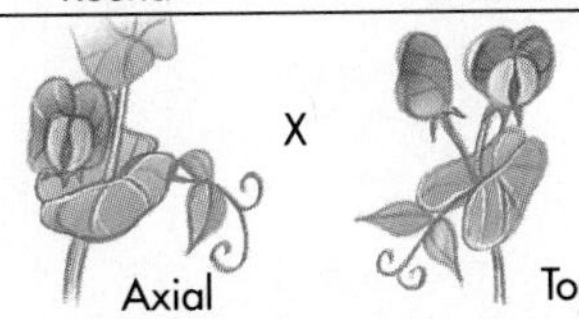 Axial X	Top	651	207	3.14:1
Plant height	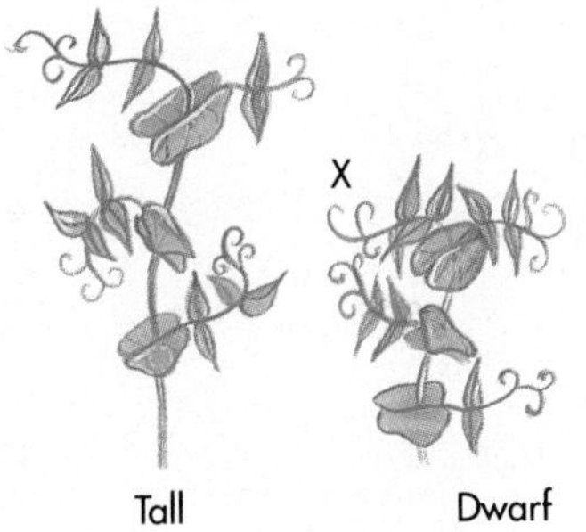 Tall X	Dwarf	787	277	2.84:1

nism of heredity. For example, he scored a total of 929 F_2 individuals in the cross between the purple-flowered F_1 plants described above. Of these F_2 plants, 705 had purple flowers and 224 had white flowers. Almost precisely one fourth of the F_2 individuals (24.1%) exhibited white flowers, the recessive trait.

Mendel examined seven traits with contrasting alternative forms (Figure 12-10), and the numerical result was always the same: three fourths of the F_2 individuals exhibited the dominant form of the trait, and one fourth displayed the recessive form of the trait. The dominant/recessive ratio among the F_2 plants was always $3:1$.

Mendel went on to examine how the F_2 plants behaved in subsequent generations. He found that the recessive one-fourth were always true-breeding. In the cross of white-flowered with purple-flowered plants, for example, the white-flowered F_2 individuals reliably produced white-flowered offspring when they were allowed to self-fertilize. By contrast, only one third of the dominant purple-flowered F_2 individuals (one fourth of the total offspring) proved true-breeding, whereas two thirds were not. This last class of plants produced dominant and recessive F_3 individuals in a ratio of $3:1$. This result suggested that, for the entire sample, the $3:1$ ratio that Mendel observed in the F_2 generation was really a disguised $1:2:1$ ratio: one fourth pure-breeding dominant individuals to one half not-pure-breeding dominant individuals to one fourth pure-breeding recessive individuals.

HOW MENDEL INTERPRETED HIS RESULTS

From these experiments Mendel was able to understand four things about the nature of heredity. First, plants exhibiting the traits he studied did not produce progeny of intermediate appearance when crossed, as a theory of blending inheritance would have predicted. Instead, alternatives were inherited intact, as discrete characteristics that either were or were not seen in a particular generation. Second, for each pair of alternative forms of a trait that Mendel examined, one alternative was not expressed in the F_1 hybrids, although it reappeared in some F_2 individuals. *The "invisible" trait must therefore have been latent (present but not expressed) in the F_1 individuals.* Third, the pairs of alternative forms of the traits that Mendel examined segregated among the

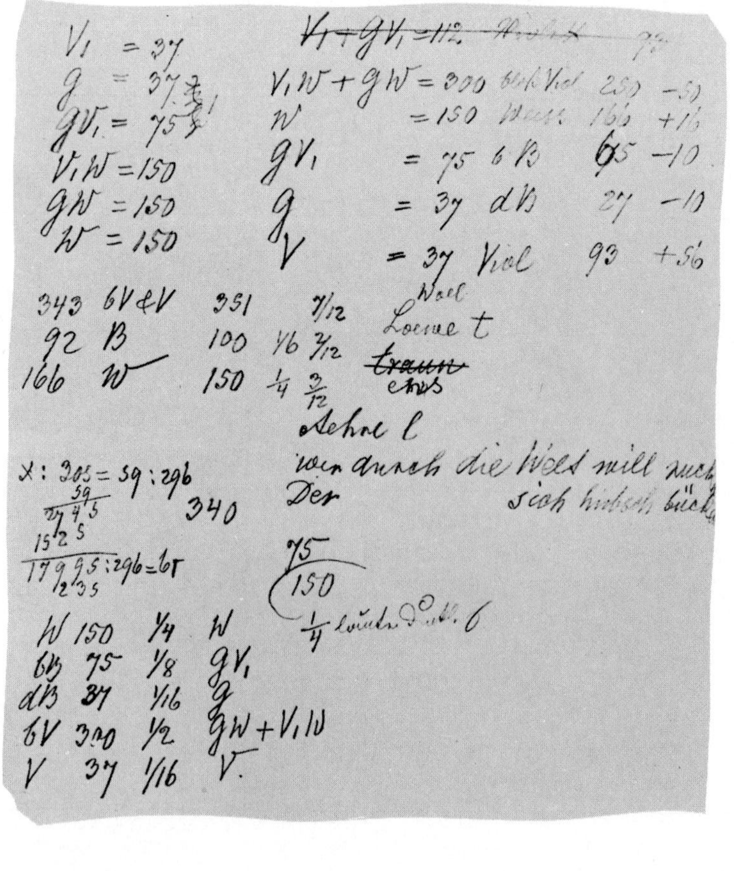

FIGURE 12-11

A page from Mendel's experiment notebook. Here he is trying various ratios in an unsuccessful attempt to explain a segregation ratio disguised by phenotypes that are so similar he cannot distinguish them from one another.

progeny of a particular cross, some individuals exhibiting one form of a trait, some the other. Fourth, pairs of alternatives were expressed in the F_2 generation in the ratio of three-fourths dominant to one-fourth recessive (Figure 12-11). This characteristic 3:1 segregation is often referred to as the **Mendelian ratio.**

To explain these results, Mendel proposed a simple model. It has become one of the most famous models in the history of science, containing simple assumptions and making clear predictions. The model has five elements. For each, we will first state Mendel's assumption and then rephrase it in modern terms.

1. Parents do not transmit their physiological traits or form directly to their offspring. Rather, they transmit discrete information about the traits, what Mendel called "factors." These factors later act in the offspring to produce the trait. In modern terms we would say that the forms of traits that an individual will express are *encoded* by the factors (genes) that it receives from its parents.

2. Each individual, with respect to each trait, contains two factors that may code for the same form of the trait or that may code for two alternative forms of the trait. We now know that there are two factors for each trait present in each individual because these factors are carried on chromosomes, and each adult individual is *diploid*. When the individual forms gametes (eggs or sperm), only one of each kind of chromosome is included in each gamete: it is *haploid*. Therefore only one factor for each trait of the adult organism is included in the gamete. Which of the two factors for each trait is included in a particular gamete is random.

3. Not all copies of a factor are identical. The alternative forms of a factor, leading to alternative forms of a trait, are called **alleles.** When two haploid gametes containing exactly the same allele of a factor fuse during fertilization, the offspring that develops from that zygote is said to be **homozygous;** when the two haploid gametes contain different alleles, the individual offspring is **heterozygous.**

In modern terminology Mendel's factors are called **genes.** We now know that one of Mendel's "factors," a gene, is composed of a DNA nucleotide sequence (Chapter 3). The position on a chromosome where a gene is located is often referred to as a **locus.** Most genes exist in alternative versions, or alleles, with differences at one or more nucleotide positions in the DNA. Different alleles of a gene are usually recognized by the change in appearance or function that results from the nucleotide differences.

4. The two alleles, one each contributed by the male and female gametes, do not influence each other in any way. In the cells that develop within the new individual these alleles remain discrete (Mendel referred to them as "uncontaminated"). They neither blend with one another nor become altered in any other way. Thus when this individual matures and produces its own gametes, the alleles for each gene are segregated randomly into these gametes, just as described in point 2.

5. The presence of a particular element does not ensure that the form of the trait encoded by it will actually be expressed in the individual carrying that allele. In heterozygous individuals only one (dominant) allele achieves expression, the other (recessive) allele being present but unexpressed. In modern terms each element *encodes* the information that specifies an alternative form of a trait, rather than containing the trait itself. The presence of information does not guarantee its expression, as any undergraduate student taking an examination appreciates. To distinguish between the presence of an element and its expression, modern geneticists refer to the totality of alleles that an individual contains as the **genotype** and to the physical appearance of an individual as the **phenotype.** The phenotype of an individual is the observable outward manifestation of the genes that it carries. The phenotype is the end result of the functioning of the enzymes and proteins encoded by the genes of the individual, its genotype. The genotype is the blueprint; the phenotype is the realized outcome.

TABLE 12-1   SOME DOMINANT AND RECESSIVE TRAITS IN HUMANS

TRAIT	PHENOTYPE	TRAIT	PHENOTYPE
RECESSIVE TRAITS		**DOMINANT TRAITS**	
Common baldness	M-shaped hairline receding with age	Mid-digital hair	Presence of hair on middle segment of fingers
Albinism	Lack of melanin pigmentation	Brachydactyly	Short fingers
Alkaptonuria	Inability to metabolize homogentisic acid	Huntington's disease	Degeneration of nervous system, starting in middle age
Red-green color blindness	Inability to distinguish red or green wavelengths of light	Phenylthiocarbamide (PTC)	Ability to taste PTC as bitter
Cystic fibrosis	Abnormal gland secretion, leading to liver degeneration and lung failure	Camptodactyly	Inability to straighten the little finger
Duchenne muscular dystrophy	Wasting away of muscles during childhood	Hypercholesterolemia (the most common human Mendelian disorder—1:500)	Elevated levels of blood cholesterol and risk of heart attack
Hemophilia	Inability of blood to clot	Polydactyly	Extra fingers and toes
Sickle cell anemia	Defective hemoglobin that collapses red blood cells		

FIGURE 12-12
Blue eyes are considered a recessive trait in humans, although many genes influence the final color.

Mendel's results were clear because he was studying alternatives that exhibited complete dominance. Many traits in humans also exhibit dominant or recessive inheritance, in a manner similar to the traits Mendel studied in peas (Figure 12-12). Table 12-1 lists a few of the many human traits known to be caused by recessive or dominant alleles.

> **The genes that an individual has are referred to as its genotype; the outward appearance of the individual is referred to as its phenotype.**

These five elements, taken together, constitute Mendel's "model" of the hereditary process. Does Mendel's model predict the result that he actually obtained?

The F₁ Generation

Consider again Mendel's cross of purple-flowered with white-flowered plants. We will assign the symbol w to the recessive allele, associated with the production of white flowers, and the symbol W to the dominant allele, associated with the production of purple flowers. By convention, genetic traits are usually assigned a letter symbol referring to their less common state, in this case the letter "W" for white flower color. The recessive allele (white flower color) is written in lower case, as w; the alternative dominant allele (purple flower color) is assigned the same symbol in upper case, W.

In this system the genotype of an individual that is true-breeding for the recessive white-flowered trait would be designated ww. In such an individual both of the copies of the allele specify the white flower phenotype. Similarly, the genotype of a true-breeding purple-flowered individual would be labeled WW, and a heterozygote would be designated Ww (the dominant allele is usually written first). Using these conventions, and denoting a cross between two lines with X, we can symbolize Mendel's original cross as $ww \times WW$ (Figure 12-13). Since the white-flowered parent can produce only w gametes and the purple-flowered parent can produce only W gametes, the union of an egg and a sperm from these parents can produce only heterozygous Ww offspring in the F₁ generation. Because the W allele is dominant, all of these F₁ individuals are expected to have purple flowers. The w allele is present in these heterozygous individuals, but it is not phenotypically expressed.

> **In a cross between homozygous-dominant and homozygous-recessive individuals, all of the F₁ progeny will be heterozygous; they will all resemble the homozygous-dominant parent in their phenotype.**

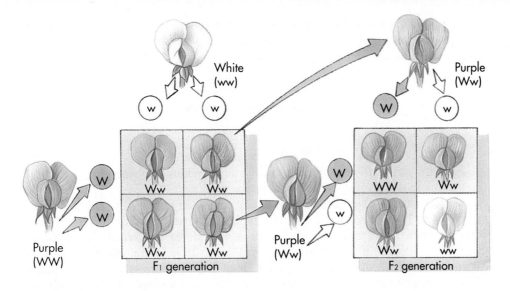

FIGURE 12-13
Mendel's cross of peas differing in flower color. The only possible offspring of the first cross are *Ww* heterozygotes, purple in color. These individuals are known as the F₁ generation. When two heterozygous F₁ individuals cross, three kinds of offspring are possible: *WW* homozygotes (purple); *Ww* heterozygotes (purple), which may form two ways; and *ww* homozygotes (white). Among these individuals, known as the F₂ generation, the ratio of dominant type to recessive type is 3:1.

The F₂ Generation

When F₁ individuals are allowed to self-fertilize, the *W* and *w* alleles segregate at random during gamete formation. Their subsequent union at fertilization to form F₂ individuals is also random, not being influenced by which alternative alleles the individual gametes carry. What will the F₂ individuals look like? The possibilities may be visualized in a simple diagram called a **Punnett square,** named after its originator, the English geneticist Reginald Crundall Punnett (Figure 12-14). Mendel's model, analyzed in terms of a Punnett square, clearly predicts that in the F₂ generation one should theoretically observe phenotypes that are three-fourths purple-flowered plants and one-fourth white-flowered plants, a phenotypic ratio of 3:1.

> In an F₂ generation derived from a heterozygous individual, any gamete can receive either allele. Their fusion following fertilization is random and is not influenced by the particular allele contained in a particular gamete. The outcome of such a cross can be illustrated graphically by a Punnett square.

	W (Male) w Pollen	
W (Female) Eggs	WW (Purple)	Ww (Purple)
w	Ww (Purple)	ww (White)

FIGURE 12-14
A Punnett square. To make a Punnett square, place the different possible types of female gametes (eggs) along one axis of a square and the different possible types of male gametes (sperm or pollen) along the other. Each potential zygote can then be represented as the intersection of a vertical (male gamete) and horizontal (female gamete) line.

Further Generations

As you can see in Figure 12-13, there are really three kinds of F₂ individuals: one fourth are pure-breeding *ww* white-flowered individuals, one half are heterozygous *Ww* purple-flowered individuals, and one fourth are pure-breeding *WW* purple-flowered individuals. The 3:1 phenotypic ratio is really a disguised 1:2:1 genotypic ratio.

THE TESTCROSS

To test his model further, Mendel devised a simple and powerful procedure called the **testcross.** Consider a purple-flowered individual: is it homozygous or heterozygous? It is impossible to tell simply by looking at its phenotype. To learn its genotype, you must cross it with some other plant. With what kind of plant? If you cross it with a homozygous dominant individual, all of the progeny will show the dominant phenotype whether the test plant is homozygous or heterozygous. It is also difficult (but not impossible) to distinguish between the two possible test plant genotypes by crossing with a heterozygous individual. If you cross the test individual with a homozygous recessive individual, however, the two possible test plant genotypes give totally different results:

Alternative 1: unknown individual homozygous
WW × *ww:* all offspring have purple flowers (*Ww*)
Alternative 2: unknown individual heterozygous
Ww × *ww:* one half of offspring have white flowers (*ww*) and one half have purple flowers (*Ww*)

To perform his test cross, Mendel crossed heterozygous F_1 individuals back to the parent homozygous for the recessive trait. He predicted that the dominant and recessive traits would appear in a 1:1 ratio.

		Gametes of homozygous recessive parent
		w
Gametes of F_1 individual	W	Ww
	w	ww

For each pair of alleles he investigated, Mendel observed phenotypic testcross ratios very close to 1:1, just as his model predicted.

Mendel's model thus accounted in a neat and satisfying way for the segregation ratios that he had observed. Its central assumption—that alternative alleles of a trait segregate from one another in heterozygous individuals and remain distinct—has since been verified in countless other organisms. It is commonly referred to as **Mendel's First Law of Heredity,** or the **Law of Segregation.** As you saw in Chapter 11, the segregational behavior of alternative alleles has a simple physical basis, one that was unknown to Mendel. It is a tribute to the intellectual power of Mendel's analysis that he arrived at the correct scheme with no knowledge of the cellular mechanisms of inheritance: neither chromosomes nor meiosis had yet been described (Figure 12-15).

Mendel's original paper describing his experiments, published in 1866, remains charming and interesting to read. His explanations are clear, and the logic of his arguments is presented in lucid detail. Unfortunately, Mendel failed to arouse much interest in his findings, which were published in the journal of the local natural history society. Only 115 copies of the journal were sent out, in addition to 40 reprints, which Mendel distributed himself. Only the German botanist Carl Naegeli was interested enough to correspond with Mendel about his findings. Naegeli believed that Mendel was wrong; he was convinced that the offspring of all hybrids must be variable. Although Mendel's results did not receive much notice during his lifetime, in 1900, 16 years after his death, three different investigators independently rediscovered his pioneering paper. They came across it while searching the literature in preparation for publishing their own findings, which were very similar to those Mendel had quietly presented more than three decades earlier.

Mendel's First Law states that (1) the alternative forms of a trait encoded by a gene are specified by alternative alleles of that gene and are discrete (do not blend in heterozygotes); (2) when gametes are formed in heterozygous diploid individuals, the two alternative alleles segregate from one another; and (3) each gamete has an equal probability of possessing either member of an allele pair.

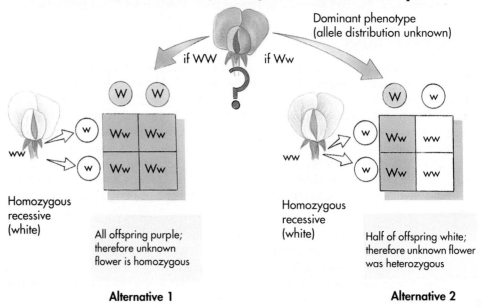

FIGURE 12-15

A testcross. To determine whether an individual exhibiting a dominant phenotype, such as purple flowers, is homozygous or heterozygous for the dominant allele, Mendel crossed the individual in question to a plant that he knew to be homozygous recessive, in this case a plant with white flowers.

INDEPENDENT ASSORTMENT

After Mendel had demonstrated that different alleles of a given gene segregate independently of one another in crosses, he asked whether different genes also segregated independently of one another. Would the possession of a particular allele for one trait (say seed shape) influence which allele the gamete had for another trait (say color)?

Mendel set out to answer this question in a straightforward way. He first established a series of pure-breeding lines of peas that differed from one another with respect to two of the seven pairs of characteristics that he had studied. His second step was to cross contrasting pairs of the pure-breeding lines. In a cross involving different seed shape alleles (round, W, and wrinkled, w) and different seed color alleles (yellow, G, and green, g), all the F_1 individuals were identical, each being heterozygous for both seed shape (Ww) and seed color (Gg). The F_1 individuals of such a cross are dihybrid individuals. A **dihybrid** is an individual heterozygous for two genes.

The third step in Mendel's analysis was to allow the dihybrid individuals to self-fertilize. If the segregation of alleles affecting seed shape were independent of the segregation of those affecting seed color, then the probability that a particular pair of seed shape alleles would occur together with a particular pair of seed color alleles would be simply the product of the individual probabilities that each pair would occur separately. Thus the probability that an individual with wrinkled, green seeds would appear in the F_2 generation would be equal to the probability of observing an individual with wrinkled seeds (one fourth) times the probability of observing an individual with green seeds (one fourth), or one sixteenth.

Since the genes concerned with seed shape and those concerned with seed color are each represented by a pair of alternative alleles in the dihybrid individuals, four types of gametes are expected: WG, Wg, wG, wg. Thus in the F_2 generation there are 16 possible combinations of alleles, each of them equally probable (Figure 12-16). Of the 16 combinations, 9 possess at least one dominant allele for each gene (usually signified $W—G—$, where the dash indicates the presence of either allele) and thus

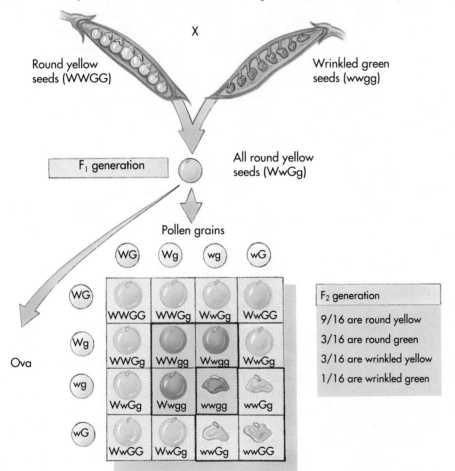

FIGURE 12-16
A Punnett square explaining the results of Mendel's dihybrid cross of round vs wrinkled seeds and yellow vs green seeds. The ratio of the four possible combinations of phenotypes is predicted to be 9:3:3:1, the ratio that Mendel found.

allele One of two or more alternative states of a gene.

diploid Having two sets of chromosomes, referred to as *homologues*. Animals, the dominant phase in the life cycle of most plants, and some protists are diploid.

dominant allele An allele that dictates the appearance of heterozygotes. One allele is said to be dominant over another if an individual who is heterozygous for that allele has the same appearance as an individual who is homozygous for it.

gene The basic unit of heredity. A sequence of DNA nucleotides on a chromosome that encodes a polypeptide or RNA molecule and so determines the nature of an individual's inherited traits.

genotype The total set of genes present in the cells of an organism. This term is often also used to refer to the set of alleles at a single gene locus.

haploid Having only one set of chromosomes. Gametes, certain protists and fungi, and certain stages in the life cycle of plants are haploid.

heterozygote A diploid individual carrying two different alleles of a gene on its two homologous chromosomes. Most human beings are heterozygous for many genes.

homozygote A diploid individual whose two copies of a gene are the same. An individual carrying identical alleles on both homologous chromosomes is said to be *homozygous* for that gene.

locus The location of a gene on a chromosome.

phenotype The realized expression of the genotype. The phenotype is the observable expression of a trait (affecting an individual's structure, physiology, or behavior) that results from the biological activity of the DNA molecules.

recessive allele An allele whose phenotypic effect is masked in heterozygotes by the presence of a dominant allele.

should have round, yellow seeds. Three possess at least one dominant W allele but are homozygous recessive for color (W—gg), three others possess at least one dominant G allele but are homozygous recessive for shape (wwG—), and one combination among the 16 is homozygous recessive for both genes ($wwgg$). The hypothesis that color and shape genes assort independently thus predicts that the F_2 generation of this dihybrid cross will display a ratio of 9 individuals with round, yellow seeds to 3 individuals with round, green seeds to 3 individuals with wrinkled, yellow seeds to 1 with wrinkled, green seeds: a 9:3:3:1 ratio (see Figure 12-16).

What did Mendel actually observe? He examined a total of 556 seeds from dihybrid plants that had been allowed to self-fertilize, and he obtained the following results:

315	Round, yellow	W—G—
108	Round, green	W—gg
101	Wrinkled, yellow	wwG—
32	Wrinkled, green	wwgg

This is very close to a perfect 9:3:3:1 ratio (313:104:104:35). Thus the two genes appeared to assort completely independently of one another. Note that this independent assortment of different genes in no way alters the independent segregation of individual pairs of alleles. Round and wrinkled seeds occur approximately in a ratio of 3:1 (423:133), as do yellow and green seeds (416:140). Mendel obtained similar results for other pairs of traits.

Mendel's observation is often referred to as **Mendel's Second Law of Heredity,** or the **Law of Independent Assortment.** Genes that assort independently of one another, as did Mendel's seven genes, usually do so because the genes are located on different chromosomes, which segregate from one another during the meiotic process of gamete formation. A modern restatement of Mendel's Second Law would be that *genes that are located on different chromosomes assort independently during meiosis.*

> Mendel's Second Law of Heredity states that genes located on different chromosomes assort independently of one another.

CHROMOSOMES: THE VEHICLES OF MENDELIAN INHERITANCE

Chromosomes are not the only kinds of organelles that segregate regularly when eukaryotic cells divide. Centrioles also divide and segregate in a regular fashion, as do the mitochondria and chloroplasts in the cytoplasm. Thus in the early twentieth century it was by no means obvious that chromosomes were the vehicles for the information of heredity. A central role for them was first suggested in 1900 by the German geneticist Karl Correns, in one of the papers announcing the rediscovery of Mendel's work. Soon after, observations that similar chromosomes paired with one another in the process of meiosis led directly to the chromosomal theory of inheritance, first formulated by the American Walter Sutton in 1902. Sutton's argument was as follows:

1. Reproduction involves the initial union of only two cells, egg and sperm. If Mendel's model is correct, then these two gametes must make equal hereditary contributions. Sperm, however, contain little cytoplasm. Therefore the hereditary material must reside within the nuclei of the gametes.

PROBABILITY AND ALLELE DISTRIBUTION

Many, although not all, alternative alleles produce discretely different phenotypes. Mendel's pea plants were either tall or dwarf, were green or white, and had wrinkled or smooth seeds. The eye color of a fruit fly may be red or white; the skin color of a human, pigmented or albino. When the number of alternative phenotypes is two, rather than three or some other number, the distribution of phenotypic types seen among the progeny of a cross is referred to as a **binomial distribution.**

To illustrate the distribution of phenotypes that will occur in a cross as a result of the segregation of two alternative alleles, consider the distribution of sexes in human families that results from the segregation of particular sex-determining chromosomes. Imagine that you choose to have three children. Let the probability of having a boy at any given birth be symbolized p and the probability of having a girl be symbolized q. Table 12-A describes all the possibilities for this family of three.

Because there are only eight ways in which a family of three can occur, the combinations listed in Table 12-A are all that are possible. The frequency with which any particular possibility occurs is referred to as its **probability** of occurrence, and the sum of the probabilities of the eight different possibilities must equal one:

$$p^3 + 3p^2q + 3pq^2 + q^3 = 1$$

To calculate the probability that the three children will be two boys and one girl, with $p = \frac{1}{2}$ and $q = \frac{1}{2}$, one calculates that $3p^2q = 3 \times [\frac{1}{2}] \times [\frac{1}{2}] \times [\frac{1}{2}] = \frac{3}{8}$. To test your understanding, try to estimate the probability that two parents heterozygous for the recessive allele producing albinism will have one albino child in a family of three:

FATHER'S GAMETES

		A	a
MOTHER'S	A	AA	Aa
GAMETES	a	Aa	aa

You can see that one fourth of the children are expected to be aa, albino. Thus for any given birth the probability of an albino child is one in four. This probability can be symbolized as q. The probability of a nonalbino child is three out of four, symbolized as p. The probability that among the three children there will be one albino child is $3p^2q = 27/64$, or 42%.

TABLE 12-A	THE SEXES OF CHILDREN IN HUMAN FAMILIES EXHIBIT A BINOMIAL DISTRIBUTION		
COMPOSITION OF FAMILY	ORDER OF BIRTH	CALCULATION	PROBABILITY
3 boys	♂ ♂ ♂	$p \times p \times p$	p^3
2 boys and 1 girl	♂ ♂ ♀ ♂ ♀ ♂ ♀ ♂ ♂	$p \times p \times q$ $p \times q \times p$ $q \times p \times p$	p^2q p^2q } $3p^2q$ p^2q
1 boy and 2 girls	♀ ♀ ♂ ♀ ♂ ♀ ♂ ♀ ♀	$q \times q \times p$ $q \times p \times q$ $p \times q \times q$	pq^2 pq^2 } $3pq^2$ pq^2
3 girls	♀ ♀ ♀	$q \times q \times q$	q^3

2. Chromosomes segregate during meiosis in a manner similar to that exhibited by the elements of Mendel's model.

3. Gametes have one copy of each pair of homologous chromosomes; diploid individuals have two copies. In Mendel's model gametes have one copy of each element; diploid individuals have two copies.

4. During meiosis, each pair of homologous chromosomes orients on the metaphase plate independently of any other pair. This independent assortment of chromosomes is a process reminiscent of the independent assortment of factors postulated by Mendel.

There was one problem with this theory, as many investigators soon pointed out. If Mendelian traits are determined by factors located on the chromosomes, and if the independent assortment of Mendelian traits reflects the independent assortment of these chromosomes in meiosis, why is it that the number of genes that assort independently of one another in a given kind of organism is often much greater than the number of chromosome pairs that the organism possesses? This seemed a fatal objection, and it led many early researchers to have serious reservations about Sutton's theory.

SEX LINKAGE

The essential correctness of the chromosomal theory of heredity was demonstrated long before this paradox was resolved. The proof was provided by a single, small fly. In 1910 Thomas Hunt Morgan, studying the fruit fly *Drosophila melanogaster*, detected a **mutant** fly, a male fly that differed strikingly from normal flies of the same species. In this fly the eyes were white (Figure 12-17) instead of the normal red.

Morgan immediately set out to determine if this new trait would be inherited in a Mendelian fashion. He first crossed the mutant male to a normal female to see if red or white eyes were dominant. All F_1 progeny had red eyes, and Morgan therefore concluded that red eye color was dominant over white. Following the experimental procedure that Mendel had established long ago, Morgan then crossed flies from the F_1 generation with each other. Eye color did indeed segregate among the F_2 progeny, as predicted by Mendel's theory. Of 4252 F_2 progeny that Morgan examined, 782 had white eyes—an imperfect 3:1 ratio, but one that nevertheless provided clear evidence of segregation. Something was strange about Morgan's result, however, something totally unpredicted by Mendel's theory: *all of the white-eyed F_2 flies were males!*

How could this strange result be explained? Perhaps it was not possible to be a white-eyed female fly; such individuals might not be viable for some unknown reason. To test this idea, Morgan testcrossed the F_1 progeny back to the original white-eyed male and obtained white-eyed and red-eyed males and females in a 1:1:1:1 ratio, just as Mendelian theory predicted. So a female could have white eyes. Why then were there no white-eyed females among the progeny of the original cross?

The solution to this puzzle involved sex. In *Drosophila* the sex of an individual is influenced by the number of copies of a particular chromosome, the **X chromosome,** that an individual possesses. An individual with two X chromosomes is a female, and an individual with only one X chromosome—which pairs in meiosis with a large, dissimilar partner called the **Y chromosome**—is a male. The female thus produces only X gametes, whereas the male produces both X and Y gametes. When fertilization involves an X sperm, the result is an XX zygote, which develops into a female; when fertilization involves a Y sperm, the result is an XY zygote, which develops into a male.

The solution to Morgan's puzzle lies in the fact that in *Drosophila* the white-eye trait resides on the X chromosome and is absent from the Y chromosome. (We now know that the Y chromosome carries almost no functional genes.) A trait that is determined by a factor on the X chromosome is said to be **sex linked.** Knowing the white-eye trait to be recessive to the red-eye trait, we can now see that Morgan's result was a natural consequence of the Mendelian assortment of chromosomes (Figure 12-18).

Morgan's experiment is one of the most important in the history of genetics because it presented the first clear evidence that Sutton was right and that the factors determining Mendelian traits do indeed reside on the chromosomes. The segregation

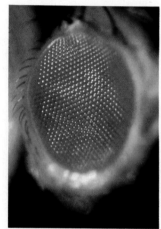

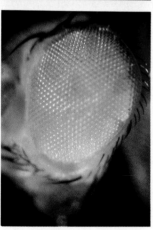

FIGURE 12-17

White-eyed (mutant) and red-eyed (normal) *Drosophila.* The white-eyed defect in eye color is hereditary, the result of a mutation in a gene located on the sex-determining X chromosome. By studying this mutation, Morgan first demonstrated that genes are on chromosomes.

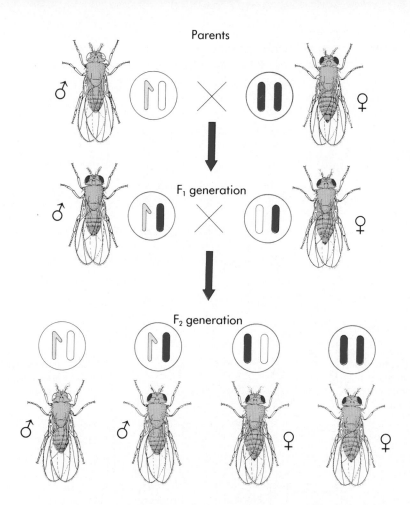

Parents

F₁ generation

F₂ generation

FIGURE 12-18
Morgan's experiment demonstrating the chromosomal basis of sex linkage in ***Drosophila.*** The white-eyed mutant male fly was crossed to a normal female. The F₁ generation flies all exhibited red eyes, as expected for flies heterozygous for a recessive white-eye allele. In the F₂ generation, all the white-eyed F₂-generation flies were male.

of the white-eye trait, evident in the eye color of the flies, has a one-to-one correspondence with the segregation of the X chromosome, evident from the sexes of the flies.

The white-eye trait behaves exactly as if it were located on an X chromosome, and this is indeed the case. The eye color gene, which specifies eye color in *Drosophila*, is carried through meiosis as part of an X chromosome. In other words, Mendelian traits such as eye color in *Drosophila* assort independently because chromosomes do. When Mendel observed the segregation of alternative traits in pea plants, he was observing a reflection of the meiotic segregation of chromosomes.

> **Mendelian traits assort independently because they are determined by genes located on chromosomes that assort independently in meiosis.**

CROSSING-OVER

Morgan's results led to the general acceptance of Sutton's chromosomal theory of inheritance. Scientists then attempted to resolve the paradox posed by the fact that there are more independently assorting Mendelian factors than there are chromosomes. In 1903 the Dutch geneticist Hugo de Vries had suggested that this paradox could be resolved only by assuming that homologous chromosomes exchange elements during meiosis. In 1909 the cytologist F.A. Janssens provided evidence for this suggestion. Investigating chiasmata produced during amphibian meiosis, Janssens noticed that of the four filaments involved in the X configuration, two crossed each other and two did not. He suggested that this crossing of chromatids reflected a switch in chromosomal arms between paternal and maternal chromatids, that the two homologues were exchanging arms with one another. His suggestion was not accepted widely, primarily because it was difficult to see how the two chromatids could break and rejoin at exactly the same position.

Janssens was right, however. Later experiments clearly established the correctness of his hypothesis, just as they had for the earlier hypotheses of Mendel and Sut-

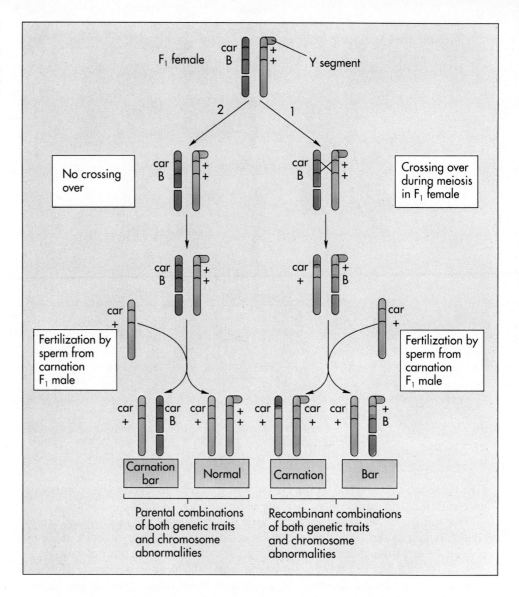

ton. One of these experiments, performed in 1931 by American geneticist Curt Stern, is described in Figure 12-19. Stern studied two sex-linked traits in strains of *Drosophila* whose X chromosomes were visibly abnormal at both ends. He first examined many flies and identified those in which an exchange had occurred with respect to the two eye traits. Stern then studied the chromosomes of these flies to see if their X chromosomes had exchanged arms. He found that all of the individuals that had exchanged eye traits also possessed chromosomes whose abnormal ends could be seen to have exchanged. The conclusion was inescapable: genetic exchanges of traits on a chromosome, such as eye color, involve **crossing-over,** the physical exchange of chromosome arms (see Figure 11-6).

The chromosomal exchanges demonstrated by Stern provide the solution to our paradox. Crossing-over can occur between homologues anywhere along the length of the chromosome; where it actually occurs seems to be random. Thus if two different genes are located relatively far apart from one another on the chromosome, crossing-over is more likely to occur somewhere in the long interval between them than it is if they are located closer together and the interval between them is short. Two genes can be on the same chromosome and still show independent assortment if they are located so far apart on the chromosome that crossing-over occurs regularly between them (Figure 12-20).

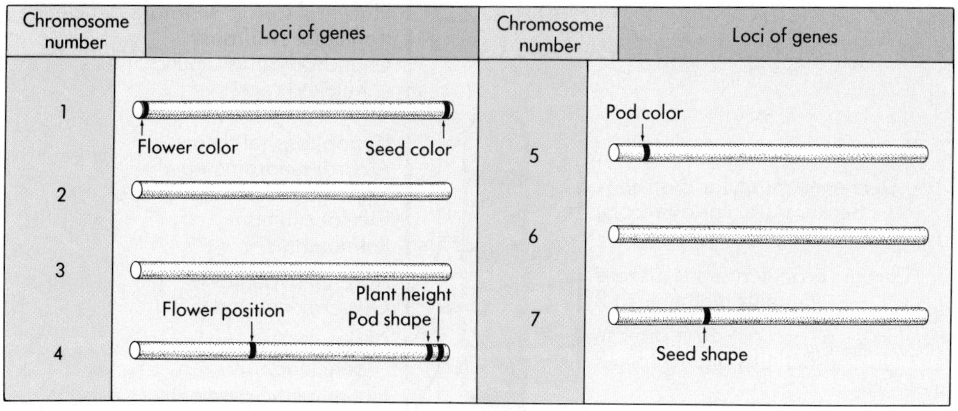
Chromosome number
Loci of genes
Chromosome number
Loci of genes
1
Flower color
Seed color
2
3
Plant height
Flower position
Pod shape
4
Pod color
5
6
7
Seed shape

FIGURE 12-20
The chromosomal locations of the seven genes studied by Mendel in the garden pea. The genes for plant height and pod shape are very close to one another and do not recombine freely. Pod shape and plant height were not among the pairs of traits that Mendel examined in dihybrid crosses. One wonders what Mendel would have made of the linkage he would surely have detected had he tested this pair of traits.

Genetic Maps

Because crossing-over is more frequent between two genes that are relatively far apart than between another set of two genes that are relatively close to each other, the frequency with which crossing-over occurs can be used to map the relative positions of genes on chromosomes. In a cross, the proportion of progeny in which an exchange has occurred between two genes is a measure of the frequency with which crossover events occur between them, and thus of the distance separating them.

Thus the frequencies with which crossing-over occurs in crosses can be used to construct a genetic map, in which distance is measured in terms of the frequency of recombination. One "map unit" is defined as the distance within which a crossover event is expected to occur, on the average, in 1 out of 100 gametes. This unit, 1% recombination, is now called a **centimorgan,** after Thomas Hunt Morgan.

In constructing a genetic map, one simultaneously monitors recombination among three or more genes located on the same chromosome. By convention, the most common allele of a locus is often designated on the map with the symbol + and is referred to as **wild type.** All other alleles are assigned specific symbols. Genes located on the same chromosome are said to be **syntenic.** When genes are close enough together on a chromosome that they do not assort independently, they are said to be **linked** to one another. A cross involving three linked genes, is called a **three-point cross.** Morgan studied three sex-linked traits, and these data of Morgan were used by his student A.H. Sturtevant to draw the first genetic map (Figure 12-21).

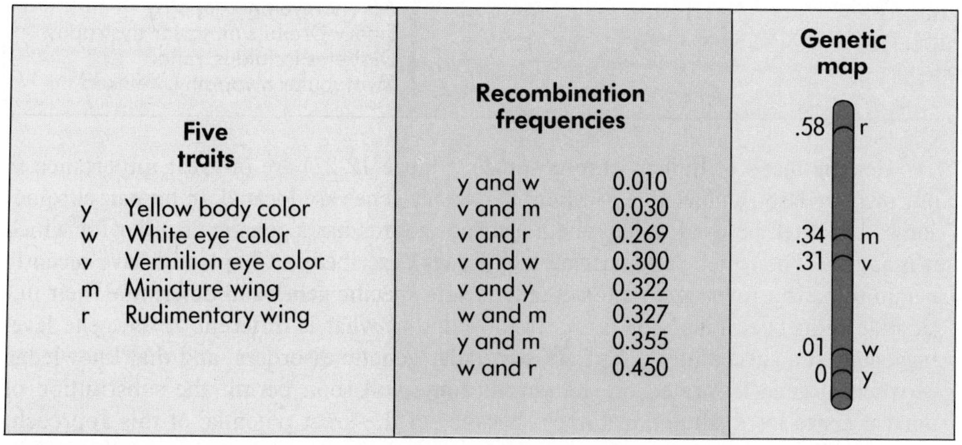
Five traits
y Yellow body color
w White eye color
v Vermilion eye color
m Miniature wing
r Rudimentary wing
Recombination frequencies
y and w 0.010
v and m 0.030
v and r 0.269
v and w 0.300
v and y 0.322
w and m 0.327
y and m 0.355
w and r 0.450
Genetic map
.58 r
.34 m
.31 v
.01 w
0 y

FIGURE 12-21
The first genetic map. This map of the X chromosome of *Drosophila* was prepared in 1913 by A.H. Sturtevant, a student of Morgan. On it he located the relative positions of five recessive traits that exhibited sex linkage, by estimating their relative recombination frequencies in genetic crosses. Sturtevant arbitrarily chose the position of the *yellow* gene as zero on his map to provide a frame of reference.

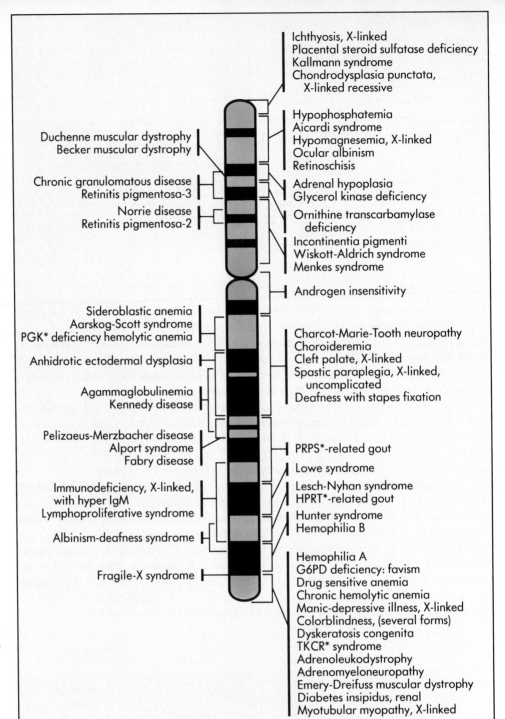

FIGURE 12-22

The human X-chromosome gene map. Over 59 diseases have now been traced to specific segments of the X-chromosome. Many of these disorders are also influenced by genes on other chromosomes.*KEY: *PGK*, phosphoglycerate kinase; *PRPS*, phosphoribosyl pyrophosphate synthetase; *HPRT*, hypoxanthine phosphoribosyl transferase; *TKCR*, torticollis, keloids, cryptorchidism, and renal dysplasia.

Ichthyosis, X-linked
Placental steroid sulfatase deficiency
Kallmann syndrome
Chondrodysplasia punctata,
   X-linked recessive

Hypophosphatemia
Aicardi syndrome
Hypomagnesemia, X-linked
Ocular albinism
Retinoschisis

Adrenal hypoplasia
Glycerol kinase deficiency

Ornithine transcarbamylase
   deficiency
Incontinentia pigmenti
Wiskott-Aldrich syndrome
Menkes syndrome

Androgen insensitivity

Charcot-Marie-Tooth neuropathy
Choroideremia
Cleft palate, X-linked
Spastic paraplegia, X-linked,
   uncomplicated
Deafness with stapes fixation

PRPS*-related gout

Lowe syndrome

Lesch-Nyhan syndrome
HPRT*-related gout

Hunter syndrome
Hemophilia B

Hemophilia A
G6PD deficiency: favism
Drug sensitive anemia
Chronic hemolytic anemia
Manic-depressive illness, X-linked
Colorblindness, (several forms)
Dyskeratosis congenita
TKCR* syndrome
Adrenoleukodystrophy
Adrenomyeloneuropathy
Emery-Dreifuss muscular dystrophy
Diabetes insipidus, renal
Myotubular myopathy, X-linked

Duchenne muscular dystrophy
Becker muscular dystrophy

Chronic granulomatous disease
Retinitis pigmentosa-3

Norrie disease
Retinitis pigmentosa-2

Sideroblastic anemia
Aarskog-Scott syndrome
PGK* deficiency hemolytic anemia

Anhidrotic ectodermal dysplasia

Agammaglobulinemia
Kennedy disease

Pelizaeus-Merzbacher disease
Alport syndrome
Fabry disease

Immunodeficiency, X-linked,
   with hyper IgM
Lymphoproliferative syndrome

Albinism-deafness syndrome

Fragile-X syndrome

Genetic maps of human chromosomes (Figure 12-22) are of great importance to our own welfare. Knowledge of where particular genes are located on human chromosomes can often be used to tell whether or not a fetus has a genetic disorder for which it is at risk. The genetic engineering techniques described in Chapter 18 have recently begun to permit investigators to actually isolate specific genes and determine their nucleotide sequence. The hope is that knowledge of what is different at the gene level may suggest a successful therapy for particular genetic disorders, and that knowledge of where a gene is located on the chromosome will soon permit the substitution of normal genes for dysfunctional ones. Because of the great potential of this approach, investigators are working hard to assemble a detailed map of the human genome. Initially, this map will consist of thousands of small lengths of DNA whose position is known. Investigators wishing to study a particular gene will first use techniques described in Chapter 18 to screen this "library" of fragments to determine which one carries the gene of interest to them. They can then analyze that fragment in detail.

FROM GENOTYPE TO PHENOTYPE:
HOW GENES INTERACT

It is important to keep in mind that the "gene" of Mendelian genetics is an abstract concept that Mendel used to explain the results of crosses. Later investigators determined that these elements are located on the chromosomes. In fact, as we will learn in Chapter 14, the Mendelian gene is no more than a segment of a DNA molecule. Genes act in ways Mendel did not understand to produce differences among progeny of crosses. How genes work is the subject of Chapter 15.

The relationship between chromosomal genes and the phenotype that Mendelian traits exhibit is not always a simple one to understand, even with the extensive knowledge of DNA that we have today. Mendel was lucky in his choice of traits. Often genes reveal more complex patterns of inheritance than simple 3:1 ratios, including the following:

1. *Multiple alleles.* Although a diploid individual may possess no more than two alleles at one time, this does not mean that only two allele alternatives are possible for a given gene in the entire population. On the contrary, almost all genes that have been studied exhibit several different alleles. The gene that determines the human ABO blood group, for example, has three common alleles (Chapter 11).

2. *Gene interaction.* Few phenotypes are the result of the action of only one gene. Most traits reflect the action of many genes that act sequentially or jointly. When genes act sequentially, as in a biochemical pathway, an allele expressed as a defective enzyme early in the pathway blocks the flow of material through the pathway and thus makes it impossible to judge whether the later steps of the pathway are functioning properly. Such interactions between genes are the basis of the phenomenon called **epistasis.**

3. *Epistasis* is an interaction between the products of two genes in which one of them modifies the phenotypic expression produced by the other. For example, in a two-step biochemical pathway where the gene governing the second step has two alleles yielding black (dominant) or blonde (recessive) hair color, it is impossible to deduce which of these two alleles is present in individuals whose alleles in the first step are nonfunctional—the hair is white, no matter which allele governs step two. Epistatic interactions between genes make the interpretation of particular patterns of inheritance very difficult.

4. *Continuous variation.* When multiple genes act jointly to influence a trait such as height or weight, the contribution caused by the segregation of the alleles of one particular gene is difficult to monitor, just as it is difficult to follow the flight of one bee within a swarm. Because all of the genes that play a role in determining phenotypes such as height or weight are segregating independently of one another, one sees a gradation in degree of difference when many individuals are examined (Figure 12-23).

5. *Pleiotropy.* Often an individual allele will have more than one effect on the phenotype. Such an allele is said to be **pleiotropic.** Thus, when the pioneering French geneticist Lucien Cuenot studied yellow fur in mice, a dominant trait, he was unable to obtain a true-breeding yellow strain by crossing individual yellow mice with one another—individuals that were homozygous for the yellow allele died. The yellow allele was pleiotropic: one effect was yellow color, another effect was a lethal developmental defect. A pleiotropic gene alteration may be dominant with respect to one phenotypic consequence (yellow fur) and recessive with respect to another (lethal developmental defect). Pleiotropic relationships occur because the characteristics of organisms result from the interactions of products made by genes. These products often also perform other functions about which we may be ignorant.

6. *Incomplete dominance.* Not all alternative alleles are fully dominant or recessive in heterozygotes. Sometimes heterozygous individuals do not resemble one parent precisely. Some pairs of alleles produce instead a heterozygous phenotype that (1) is intermediate between the parents (intermediate or incomplete

FIGURE 12-23
Height is a continuously varying trait.
A Variation in height among students of the 1914 class of the Connecticut Agricultural College. Because many genes contribute to height and tend to segregate independently of one another, there are many possible combinations.
B The cumulative contribution of different combinations of alleles to height forms a continuous spectrum of possible heights—a random distribution, in which the extremes are much rarer than the intermediate values.

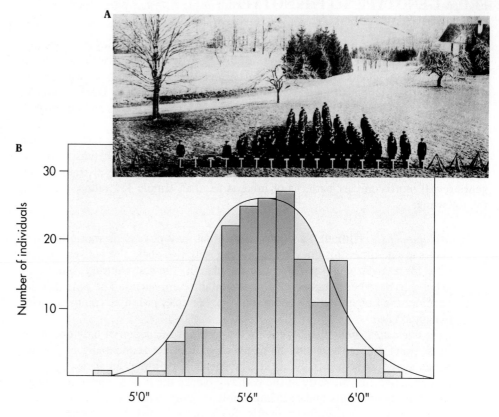

dominance, Figure 12-24), (2) resembles one allele closely but can be distinguished from it (partial dominance), or (3) is one in which both parental phenotypes can be distinguished in the heterozygote (co-dominance).

7. *Environmental effects.* The degree to which many alleles are expressed depends on the environment. Some alleles encode an enzyme whose activity is more sensitive to conditions such as heat (see Figure 12-3) or light than are other alleles.

Modified Mendelian Ratios

When individuals heterozygous for two different genes mate (a dihybrid cross), four different phenotypes are possible among the progeny: the dominant phenotype of both genes is displayed, one of the dominant phenotypes is displayed, or neither dom-

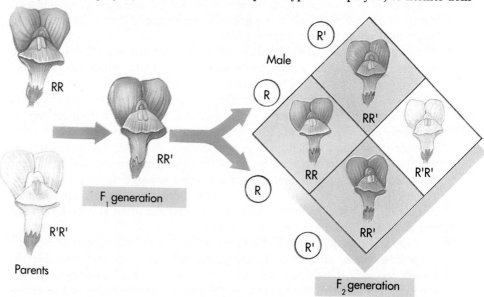

FIGURE 12-24

Incomplete dominance. A cross between a pink-flowered snapdragon, which has the genotype RR, and a white-flowered one (R'R'). Neither allele is dominant, and the heterozygotes have purple flowers and the genotype RR'.

inant phenotype is displayed. Mendelian assortment predicts that these four possibilities will occur in the proportions 9:3:3:1. Sometimes, however, it is not possible for an investigator to successfully identify each of the four possible phenotypic classes, because two or more of the classes look alike.

One example of such difficulty in identification is seen in analysis of particular varieties of corn, *Zea mays*. Some commercial varieties exhibit a purple pigment called anthocyanin in their seed coats, whereas others do not. When in 1918 the geneticist R.A. Emerson crossed two pure-breeding corn varieties, neither of which typically exhibits any anthocyanin pigment, he obtained a surprising result: all of the F_1 plants produced purple seeds. The two white varieties, which had never been observed to make pigment, would, when crossed, produce progeny that uniformly make the pigment.

When two of these pigment-producing F_1 plants were crossed to produce an F_2 generation, 56% were pigment producers and 44% were not. What was happening? Emerson correctly deduced that two genes were involved in the pigment-producing process and that the second cross had thus been a dihybrid cross such as described by Mendel. Mendel predicted 16 possible genotypes in equal proportions ($9 + 3 + 3 + 1 = 16$) suggesting to Emerson that the total number of genotypes in his experiment was also 16. How many of these were in each of the two types Emerson obtained? He multiplied the fraction that were pigment producers ($.56) \times 16 = 9$, and multiplied the fraction that were not ($.44) \times 16 = 7$. Thus Emerson had in fact a **modified ratio** of 9:7 instead of the usual 9:3:3:1 ratio.

In this case the pigment anthocyanin is produced from a colorless molecule by two enzymes that work one after the other. In other words the pigment is the product of a two-step biochemical pathway:

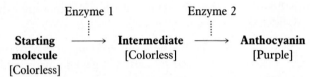

For pigment to be produced, a plant must possess at least one good copy of each enzyme. The dominant alleles encode functional enzymes; the recessive alleles, defective nonfunctional ones. Of the 16 genotypes predicted by random assortment, 9 contain at least one dominant allele of both genes—these are the purple progeny. The 9:7 ratio that Emerson observed resulted from the pooling of the three phenotypic classes that lack dominant alleles at either or both loci ($3 + 3 + 1 = 7$) and so all looked the same, nonpigmented.

THE SCIENCE OF GENETICS

After many centuries of speculation about heredity, the puzzle was finally solved in the space of a few generations. Guided by the work of Mendel and a generation of investigators determined to explain his results, the basic outline of how heredity works soon became clear. Hereditary traits are specified by genes, which are integral parts of chromosomes. The movements of chromosomes during meiosis produce the patterns of segregation and independent assortment that Mendel reported. Two of the most important discoveries made after Mendel were that (1) chromosomes exchange genes during meiosis and (2) genes located far apart on chromosomes are more likely to have an exchange occur between them. These findings allowed investigators to learn how genes were distributed on chromosomes long before we knew how to isolate them.

This core of knowledge, this basic outline of heredity, has led to a long chain of investigation and questions. Is human heredity like that of a garden pea? What is the physical nature of a gene? How do genes change, and why? How does a difference in a gene produce a difference in a phenotype? The people who ask such questions are called geneticists, and the body of what they have learned and are learning is called genetics. Genetics is one of the most active subdisciplines of biology. The next six chapters will answer the questions just posed and many others.

1. Koelreuter noted the basic facts of heredity a century before Mendel. He found that alternative traits segregate in crosses and may mask each other's appearance. Mendel, however, was the first to quantify his data, counting the numbers of each alternative type among the progeny of crosses.

2. From counting progeny types, Mendel learned that the alternatives that were masked in hybrids appeared only 25% of the time when they subsequently segregated in the F_2 generation. This finding, which led directly to Mendel's model of heredity, is usually referred to as the Mendelian ratio of $3:1$ dominant to recessive traits.

3. Mendel deduced from the $3:1$ ratio that traits are specified by discrete "factors," which do not blend. We refer to Mendel's factors as genes and to alternative forms of his factors as alleles. Mendel deduced that pea plants contain two factors for each feature that he studied (we now know this is because they are diploid). When the two copies of a factor are not the same—when the plant is heterozygous for a trait—one factor, which Mendel described as dominant, determines the appearance, or phenotype, of the individual.

4. When two heterozygous individuals mate, an individual offspring has a 50% (that is, random) chance of obtaining the dominant allele from the father and a 50% chance of obtaining the dominant allele from the mother, so the probability of obtaining two dominant alleles—of being homozygous dominant—is .5 × .5 = .25, or 25%. Similarly, the probability of being homozygous recessive is 25%. The rest of the progeny, one half, are heterozygotes. Because the appearance of heterozygotes is specified by the dominant allele, the progeny thus appear as three-fourths dominant and one-fourth recessive, a ratio of $3:1$ dominant to recessive.

5. When two genes are located on different chromosomes, the alleles included in an individual gamete are distributed at random. The allele for one gene included in the gamete has no influence on which allele of the other gene is included in the gamete. Such genes are said to assort independently.

6. The first clear evidence that genes reside on chromosomes was provided by Thomas Hunt Morgan. Morgan demonstrated that the segregation of the white-eye trait in *Drosophila* was associated with the segregation of the X chromosome, the one responsible for sex determination.

7. The first evidence that crossing-over occurs between chromosomes was provided by Curt Stern. Stern showed that when two Mendelian traits exchange during a cross, so do visible abnormalities on the ends of the chromosomes bearing the traits.

8. The frequency of crossing-over between genes can be used to construct genetic maps, which are representations of the physical locations of genes on chromosomes, inferred from the degree of crossing-over between particular pairs of genes.

9. Because phenotypes are often influenced by more than one gene, the ratios of alternative phenotypes observed in crosses sometimes deviate from the simple ratios predicted by Mendel. This is particularly true in epistatic situations, where alleles of one locus mask the ability of an investigator to identify another.

REVIEW QUESTIONS

1. How did Koelreuter's hybridization of tobacco plants alter the then current theories regarding heredity? How did the progeny of his hybridizations appear compared with one another?

2. Why did Mendel's research succeed in clarifying gene segregation among offspring where others had failed?

3. Mendel's choice of the garden pea for his breeding experiments was fortunate for what four reasons?

4. What were the three general stages of Mendel's experiments?

5. What four conclusions did Mendel make about his experiments with pea plants?

6. Briefly state the five elements of Mendel's model of heredity.

7. Define genotype and phenotype. Define homozygous and heterozygous.

8. A testcross is performed to determine whether a purple-flowered plant is homozygous or heterozygous. With what genotype plant is the unknown plant crossed? Why?

9. State Mendel's First Law of Heredity.

10. What four points were made by Sutton to explain that chromosomes were the vehicles for Mendelian inheritance?

11. What experiments enabled scientists to conclude that genes were actually located on the chromosomes?

12. Mendel studied flower color and seed color in pea plants. We now know that these genes are located on the same chromosome. How do you explain Mendel's results showing that these traits assort independently?

13. List and briefly describe six gene interactions that can alter Mendelian ratios.

THOUGHT QUESTIONS

1. Why did Mendel observe only two alleles of any given trait in the crosses that he carried out?

2. There once was a lonely and rather sour buzzard named Clyde. It came as no surprise to any who knew him that Clyde had no offspring—no female buzzard would come anywhere near a buzzard with his personality. Clyde, however, nursed a secret desire to pass on his genes, and one day he hit upon a plan. He had heard from his boss, Professor Johnson, that the St. Louis Zoo practices birth control among its captive birds by the simple expedient of keeping only female birds. In the dark of night he invaded the zoo and there wooed numerous female buzzards, none of whom knew the meanness of his nature. Soon buzzard eggs were hatching all over the zoo. Now a new issue arose to give Clyde pain. In reading his boss's genetics notes he made a dread discovery: genes recombine during meiosis. This meant that there was a chance that his wonderfully horrible combination of characteristics might be diluted out by other more "normal" alternatives in subsequent generations. Clyde brooded on this for quite a while. But he finally decided that he need not worry, since he remembered his scrawny mother telling him on her knee that the two traits he most cared about, *small mind* and *hard heart*, were closely linked to one another; his mother was in fact homozygous for these traits, as well as for *scrawniness of frame*. Because Clyde shows all three traits,

even though his mother was normal and did not, all three traits are dominant.

Clyde asked his boss to look into this matter for him by examining the baby buzzards at the zoo. Dr. Johnson located 1000 baby buzzards apparently sired by Clyde in his nightly visits and checked out the smallness of their minds, the hardness of their hearts, and the scrawniness of their frames. Here is what he found:

NUMBER	MIND	FRAME	HEART
235	Normal	Normal	Normal
230	Small	Normal	Hard
226	Small	Scrawny	Hard
221	Normal	Scrawny	Normal
24	Small	Normal	Normal
23	Normal	Normal	Hard
21	Normal	Scrawny	Hard
20	Small	Scrawny	Normal

Dr. Johnson went home and told Clyde he had nothing to worry about, that hardness of heart and smallness of mind are very closely linked, relative to other Clyde traits such as scrawniness of frame. Was he right? Explain, backing up your argument with an appropriate genetic map.

FOR FURTHER READING

BLIXT, S.: "Why Didn't Gregor Mendel Find Linkage?" *Nature*, vol. 256, page 206, 1975. Modern information on the location on chromosomes of the genes that Mendel studied.

CORCOS, A., and F. MONAGHAN: "Mendel's Work and Its Rediscovery: A New Perspective," *Critical Reviews in Plant Sciences*, May 1990, vol. 9(3), pages 197-212. An evaluation of the many myths surrounding Mendel's work.

HODGKIN, J.: "Sex determination compared in *Drosophila* and nematodes," *Nature*, April 1990, vol. 344, pages 721-728. Molecular mechanisms of sex determination have not been strongly conserved in evolution.

MENDEL, G.: "Experiments on Plant Hybridization," (1866). Translation, reprinted in C. Stern and E. Sherwood (eds.): *The Origins of Genetics: A Mendel Source Book*, W.H. Freeman and Co., San Francisco, 1966.

MORGAN, T.H.: "Sex-Limited Inheritance in *Drosophila*," *Science*, vol. 32, pages 120-122, 1910. Morgan's original account of his famous analysis of the inheritance of the white-eye trait.

MURRAY, J.: "How the Leopard Gets Its Spots," *Scientific American*, March 1988, pages 80-87. The wide variety of animal coat markings found in nature can be understood as a single, simply inherited, pattern-forming mechanism.

PLOMIN, R.: "The Role of Inheritance in Behavior," *Science*, April 1990, vol. 185, pages 183-188. Genes affect behavior, but not in a simple Mendelian fashion.

SAPIENZA, C.: "Parental Imprinting of Genes," *Scientific American*, October 1990, pages 52-60. Sometimes cells alter the genes they carry; as a result, even when fathers and mothers contribute identical genes to their offspring, the genes may have different effects.

SUTTON, W.S.: "The Chromosomes of Heredity," *Biological Bulletin*, vol. 4, pages 213-251, 1903. The original statement of the chromosomal theory of heredity.

SUZUKI, D., A. GRIFFITHS, and R. LEWONTIN: *An Introduction to Genetic Analysis*, ed. 4, W.H. Freeman and Co., San Francisco, 1989. The most widely used undergraduate genetics text, with a classical approach.

MENDELIAN GENETICS PROBLEMS

1. Why did Mendel observe only two alleles of any given trait in the crosses that he carried out?

2. The illustration at right describes Mendel's cross of *wrinkled* and *round* seed characters. What is wrong with this diagram?

3. The annual plant *Haplopappus gracilis* has two pairs of chromosomes 1 and 2. In this species, the probability that two traits *a* and *b* selected at random will be on the same chromosome of *Haplopappus* is the probability that they will both be on chromosome 1 (½), times the probability that they will both be on chromosome 2 (also ½): $\frac{1}{2} \times \frac{1}{2} = \frac{1}{4}$, or 25%. This is often symbolized

$$\frac{1}{2}^{(\text{# of pairs of chromosomes})}$$

Human beings have 23 pairs of chromosomes. What is the probability that any two human traits selected at random will be on the same chromosome?

4. Among Hereford cattle there is a dominant allele called *polled;* the individuals that have this allele lack horns. After college, you become a cattle baron and stock your spread entirely with polled cattle. You personally make sure that each cow has no horns, and none does. Among the calves that year, however, some grow horns. Angrily you dispose of them, and make certain that no horned adult has gotten into your pasture. None has. The next year, however, more horned calves are born. What is the source of your problem? What should you do to rectify it?

5. An inherited trait among human beings in Norway causes affected individuals to have very wavy hair, not unlike that of a sheep. The trait is called *woolly*. The trait is very evident when it occurs in families; no child possesses woolly hair unless at least one parent does. Imagine you are a Norwegian judge, and that you have before you a woolly haired man suing his normal-haired wife for divorce because their first child has woolly hair but their second child has normal long, blonde hair. The husband claims this constitutes evidence of infidelity on the part of his wife. Do you accept his claim? Justify your decision.

6. In human beings, Down syndrome, a serious developmental abnormality, results from the presence of three copies of chromosome 21 rather than the usual two copies. If a female exhibiting Down syndrome mates with a normal male, what proportion of her offspring would be expected to be affected?

7. Many animals and plants bear recessive alleles for *albinism*, a condition in which homozygous individuals completely lack any pigments. An albino plant, for example, lacks chlorophyll and is white. An albino person lacks any melanin pigment. If two normally pigmented persons heterozygous for the same albinism allele marry, what proportion of their children would be expected to be albino?

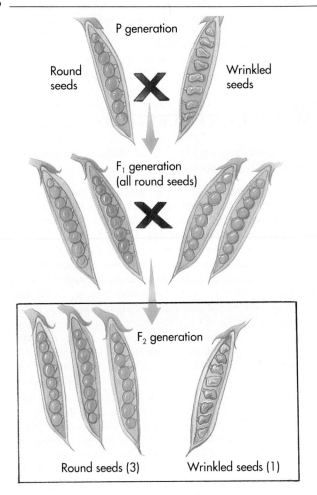

Round seeds (3)   Wrinkled seeds (1)

8. Your uncle dies and leaves you his racehorse, Dingleberry. To obtain some money from your inheritance, you decide to put the horse out to stud. In looking over the stud book, however, you discover that Dingleberry's grandfather exhibited a rare clinical disorder that leads to brittle bones. The disorder is hereditary and results from homozygosity for a recessive allele. If Dingleberry is heterozygous for the allele, it will not be possible to use him for stud, since the genetic defect may be passed on. How would you go about determining whether Dingleberry carries this allele?

9. In the fly *Drosophila*, the allele for dumpy wings (symbolized *d*) is recessive to the normal long-wing allele (symbolized *D*). The allele for white eye (symbolized *w*) is recessive to the normal red-eye allele (symbolized *W*). In a cross of *DDWw* × *Ddww*, what proportion of the offspring are expected to be "normal" (long wing, red eye)? What proportion "dumpy, white"?

10. As a reward for being a good student, your instructor presents you with a *Drosophila* named "Oscar." Oscar has red eyes, the same color that normal flies do. You add Oscar to your fly collection, which also contains "Heidi" and "Siegfried," flies with white eyes, and "Dominique" and "Ronald," which are from a long line of red-eyed flies. Your previous work has shown that the white-eye trait exhibited by Heidi and Siegfried is caused by their being homozygous for a recessive allele. How would you determine whether or not Oscar was heterozygous for this allele?

11. In some families, children are born that exhibit recessive traits (and therefore must be homozygous for the recessive allele specifying the trait), even though one or both of the parents do not exhibit the trait. What can account for this?

12. You collect in your backyard two individuals of *Drosophila melanogaster*, one a young male and the other a young, unmated female. Both are normal in appearance, with the typical vivid red eyes of *Drosophila*. You keep the two flies in the same vial, where they mate. Two weeks later, hundreds of little offspring are flying around in the vial. They all have normal red eyes. From among them, you select 100 individuals, some male and some female. You cross each individually to a fly you know to be homozygous for a recessive allele called "sepia," which leads to black eyes when homozygous (these flies thus have the genotype *se/se*). Examining the results of your 100 crosses, you observe that in about half of them, only normal red-eyed progeny flies are produced. In the other half, however, the progeny are about 50% red eyed and 50% black eyed. What must have been the genotypes of your original backyard flies?

13. Hemophilia is a recessive sex-linked human blood disease that leads to failure of blood to clot normally. One form of hemophilia has been traced to the royal family of England, from which it spread throughout the royal families of Europe. For the purposes of this problem, assume that it originated as a mutation either in Prince Albert or in his wife, Queen Victoria (the actual explanation is in Chapter 13).

a. Prince Albert did not himself have hemophilia. If the disease is a sex-linked recessive abnormality, how can it have originated in Prince Albert, who is a male and therefore is expected to exhibit recessive sex-linked traits, since he did not suffer from hemophilia?

b. Alexis, the son of Czar Nicholas II of Russia and Empress Alexandra (a granddaughter of Victoria), had hemophilia, but their daughter Anastasia did not. Anastasia died, a victim of the Russian revolution, before she had any children. Can we assume that Anastasia would have been a carrier of the disease? How is your answer influenced if the disease originated in Nicholas II or in Alexandra?

13

Human Genetics

OVERVIEW

Humans possess 46 chromosomes, which are inherited in Mendelian fashion. The degree to which you resemble your father or mother was largely established before your birth by the chromosomes that you received from them, just as the course of meiosis in peas determined the segregation of Mendel's traits. The inheritance of human traits is more difficult to study, however. It usually involves analysis of family pedigrees in an attempt to deduce how particular alleles are segregating. Some of the most serious of human disorders are inherited, the result of alleles that specify defective forms of proteins that have important functions in our bodies. By studying human heredity, we are beginning to learn how to limit the misery that these disorders bring to so many human families.

FOR REVIEW

Here are some important terms and concepts that you will encounter in this chapter. If you are not familiar with them, you should review them before proceeding.

Karyotypes (Chapter 10)

Dominant and recessive traits (Chapter 12)

Sex linkage (Chapter 12)

FIGURE 13-1
End of the line.
Czar Nicholas II of Russia, the
last of the czars, and his wife Alexandra with their children: Olga,
Tatiana, Maria, Anastasia, and
Alexis. Alexandria was a carrier of
the genetic disease hemophilia, a
disorder in which the blood does
not clot properly, and she passed
it on to her son Alexis. Since the
kind of hemophilia they had is
caused by a recessive mutant allele
located on the X chromosome, it
is not expressed in the heterozygous condition in women, who
have two such chromosomes. It is
not known which of her four
daughters might have received the
hemophilia allele from her, as
none of the four lived to bear children.

We devote a special chapter to human heredity because we are interested in ourselves. Although we humans pass our genes on to the next generation (Figure 13-1) in much the same way that other organisms do, we naturally have a special curiosity about ourselves. Practically every one of us knows of someone, a relative or a friend's relative, who suffers from a condition that might be hereditary. If someone in your family has had a stroke, for example, it is difficult not to worry about your own future health, knowing that the propensity to suffer strokes can be hereditary. Few women have babies without worrying to some extent about the possibility that their child may have some defects. In this sense, we are all human geneticists, interested in what the laws of genetics tell us about ourselves, our families, our friends, and our future children.

HUMAN CHROMOSOMES

The exact number of chromosomes that humans possess was not established accurately until 1956. At that time, appropriate techniques were developed that allowed investigators to determine accurately the number, shape, and form of human chromosomes and those of other mammals. Earlier, the number of chromosomes characteristic of human beings had been known only approximately. We now know that each typical human cell has 46 chromosomes, which come together in meiosis to form 23 pairs.

By convention, the 23 different kinds of human chromosomes are arranged into seven groups, each characterized by a different size, shape, and appearance. These groups are designated A through G. Of the 23 pairs, 22 are perfectly matched in both males and females and are called **autosomes**. The remaining pair consists of two unlike members in males; in females, it consists of two similar members. The chromosomes that constitute this pair are called the **sex chromosomes**. Just as in *Drosophila*, females are designated XX and males XY, to indicate that the male pair of sex chromosomes contains one member (the Y chromosome) that bears few functional genes and differs in this respect from all other chromosomes.

Down Syndrome

The 46 chromosomes seen in Figure 10-6 represent a karyotype that is characteristic of most cells of a human male. The reason that almost everyone has the same karyotype is the same reason that almost all cars have engines, transmissions, and wheels—other arrangements don't work well. Humans who have lost even one copy of an autosome (called **monosomics**) do not survive development. In all but a few cases, humans who have gained an extra autosome (called **trisomics**) do not survive either. Five of the smallest chromosomes—those numbered 13, 15, 18, 21, and 22—can be present in humans as three copies, and still allow the individual to survive for a time. Individuals with an extra chromosome 13, 15, or 18, however, undergo severe developmental defects, and infants with such a genetic makeup die within a few months. In contrast, individuals who have an extra copy of chromosome 21 (Figure 13-2), and more rarely those who have an extra copy of chromosome 22, usually do survive to adulthood. In such individuals the maturation of the skeletal system is delayed, so that they generally are short and have poor muscle tone. Their mental development is also affected, and children with trisomy 21 or trisomy 22 are always mentally retarded. The developmental defect produced by trisomy 21 was first described in 1866 by J. Langdon Down; for this reason, it is called **Down syndrome** (formerly "Down's syndrome").

About 1 in every 750 children exhibits Down syndrome. The frequency is similar in all human racial groups; similar conditions also occur in chimpanzees and other related primates. In humans the defect is associated with a particular small portion of chromosome 21. When this chromosomal segment is present in three copies instead of two, Down syndrome results. In 97% of the human cases examined, all of chromosome 21 is present in three copies (trisomy 21). In the other 3%, a small portion containing the critical segment has been added to another chromosome by translocation, in addition to the normal two copies of chromosome 21, a condition known as "translocation Down syndrome" (see Chapter 17).

We do not yet know a great deal about the developmental role of the genes whose duplication produces Down syndrome, although clues are beginning to emerge from current research. There is a suspicion that the gene or genes that produce Down syndrome are similar to, if not identical with, some of the genes that are associated with cancer and with Alzheimer's disease. When human cancer-causing genes (to be described in Chapter 16) were located on human chromosomes, one of them turned out, like the gene causing Alzheimer's disease, to be on chromosome 21, on the segment associated with Down syndrome. Is cancer more common in children with Down syndrome? Yes. The incidence of leukemia, for example, is 11 times higher in children with Down syndrome than in unaffected children of the same age.

FIGURE 13-2

Down syndrome.

A As shown in this karyotype of a male, Down syndrome is usually associated with trisomy of chromosome 21.

B Child with Down syndrome sitting on his father's knee.

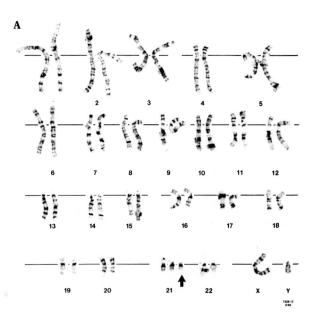

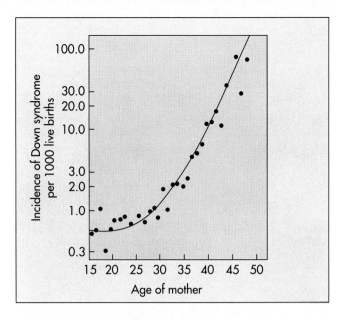

FIGURE 13-3
Correlation between maternal age and the incidence of Down syndrome. As women age, the chance they will bear a child with Down syndrome increases. After age 35, the frequency of nondisjunction of chromosome 21 increases rapidly.

How does Down syndrome arise? In humans, it comes about almost exclusively as a result of **primary nondisjunction** (failure of chromosomes to separate in meiosis) of chromosome 21 in the meiotic event that leads to the formation of the egg. The cause of these primary nondisjunctions is not known, but just as in the case of cancer, the incidence of primary nondisjunctions increases with age (Figure 13-3). In mothers less than 20 years of age, the incidence of Down syndrome children is only about 1 per 1700 births; in mothers 20 to 30 years old, the risk is only slightly greater, about 1 per 1400. In mothers 30 to 35 years old, however, the risk doubles, to 1 per 750. In mothers over 45, the risk is as high as 1 in 16 births!

Primary nondisjunctions are far more common in women than in men because all the eggs that a woman will ever produce have begun their development, to the point of prophase of the first meiotic division, by the time she is born; by the time she has children, her eggs are many years old—as old as she is. In men, by contrast, development of new sperm occurs daily. There is, therefore, a much greater chance for problems of various kinds, including primary nondisjunction, to accumulate over time in the gametes of women than those of men. For this reason, the age of the mother is more critical than that of the father in couples contemplating childbearing.

Sex Chromosomes

As you saw in Chapter 10, humans possess 22 pairs of autosomes that are the same in both sexes, and one pair of sex chromosomes that differs between males and females. Females are designated XX, and males are designated XY. The Y chromosome is highly condensed, and few of its genes are expressed; for this reason, recessive alleles that are present on the single X chromosome of males have no counterpart on the Y chromosome, at least no *active* counterpart.

The Y chromosome is not completely inert genetically, however; it does possess some active genes. For example, the genes determining the features associated with "maleness" are located on the Y chromosome in humans: any individual with at least one Y chromosome is a male, and any individual without a Y chromosome is a female.

Because a male has only one copy of the X chromosome and a female has two copies, you might think that female cells would produce twice as much of the proteins encoded by genes on the X chromosome. This does not in fact happen, because in females one of the X chromosomes is inactivated shortly after sex determination, early in embryonic development.* The inactivated chromosome can be seen as a deeply staining body, the **Barr body,** which remains attached to the nuclear membrane.

---

*In different embryonic cells, different X chromosomes are inactivated. Apparently the choice of *which* of the two to inactivate is made completely at random. If a woman is heterozygous for a sex-linked trait, some of her cells will express one allele, some the other.

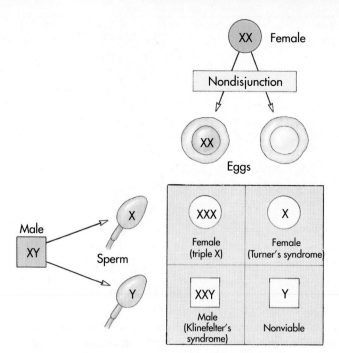

FIGURE 13-4
How nondisjunction can result in
abnormalities in the number of
sex chromosomes.

Individuals that gain an extra copy of the X chromosome, or lose one, are not subject to the severe developmental abnormalities usually associated with similar changes in autosomes. When primary nondisjunction results in the acquisition of an extra X chromosome in a particular individual, or results in a female with only one X chromosome, the affected individuals may become mature, but they have somewhat abnormal features.

The X Chromosome. When X chromosomes fail to separate in meiosis, some gametes are produced that possess both of the X chromosomes and so are XX gametes; the other gametes that result from such an event have no sex chromosome (designated O). If the XX gamete joins an X gamete to form an XXX zygote, the zygote develops into a female individual with one functional X chromosome and two Barr bodies. She is sterile but usually quite normal in other respects. If the XX gamete instead joins a Y gamete, the effects are more serious. The resulting XXY zygote develops into a sterile male who has many female body characteristics and, in some cases, diminished mental capacity. This condition, called **Klinefelter's syndrome,** occurs in about 1 out of every 500 male births.

The other gamete produced when the X chromosomes fail to separate, O, lacks any X chromosome. If this O gamete fuses with a Y gamete, the resulting OY zygote is inviable and fails to develop further. Humans cannot survive without any of the genes on the X chromosome. If, on the other hand, the O gamete fuses with an X gamete to form an XO zygote, the result is a sterile female of short stature, a webbed neck, and immature sex organs that do not undergo puberty changes. The mental abilities of an XO individual are in the low normal range. This condition, called **Turner's syndrome,** occurs roughly once in every 5000 female births. The ways in which nondisjunction can result in abnormal numbers of sex chromosomes are shown in Figure 13-4.

The Y Chromosome. Like the X chromosome, the Y chromosome occasionally fails to separate in meiosis. Failure of the Y chromosome to separate leads to the formation of YY gametes and viable XYY zygotes, which develop into fertile males of normal appearance. The frequency of XYY among newborn males is about 1 per 1000. Interestingly, the frequency of XYY males in penal and mental institutions has been reported to be approximately 2% (that is, 20 per 1000), which is twenty times as many such individuals as exist in the population at large. This observation has led to the suggestion that XYY males are inherently antisocial, a suggestion that has proven highly controversial. The observation is confirmed in some studies, but not in others. In any case, most of XYY males do not develop patterns of antisocial behavior.

PATTERNS OF INHERITANCE

Many human traits are inherited. Hair and eye color, facial characteristics, aspects of personality—all are passed from parents to children. Some major clinical disorders are also inherited, and it is important for investigators to be able to study them, both to assess the likelihood that parents at risk will produce affected children, and to seek future cures. Imagine that you are trying to learn about an inherited disorder present in the history of your family. How would you find out if the trait is dominant or recessive, how many genes contribute to it, and how likely you might be to transmit it to your future children? If you were studying such a trait in *Drosophila,* you could conduct crosses, occasionally squashing a fly to examine its chromosomes. Studying your own heredity requires a more indirect approach.

To study how a human trait is inherited, investigators look at the results of crosses that have already been made—they study family histories, called **pedigrees.** By studying which relatives exhibit a trait, it is often possible to say if the gene producing the trait is sex linked or autosomal and to determine if the trait's phenotype is dominant or recessive. In many cases it is possible to infer which individuals are homozygous and which are heterozygous for the allele specifying the trait.

A Pedigree: Albinism

An albino is an individual that lacks all pigmentation, so that his or her hair and skin are white. In the United States, about one caucasian person in 38,000 and one African-American person in 22,000 are albinos. In the pedigree of albinism presented in Figure 13-5, each symbol represents one individual in the family history, with the circles representing females and the squares males. In such a pedigree, individuals that exhibit a trait being studied, in this case albinism, are indicated by blue solid symbols. Marriages are represented by horizontal lines connecting a circle and a square, from which a cluster of vertical lines indicate the progeny, arranged in order of birth.

How does one analyze the pedigree in Figure 13-5?

1. First, let us inquire whether albinism is sex linked or autosomal. If the trait is sex linked, it will be expressed more frequently in males, whereas if it is autosomal it will appear in both sexes equally. In Figure 13-5 the proportion of affected males (5/13, or 39%) is similar to the proportion of affected females (8/12, or 37%), and so we can conclude that the trait is autosomal rather than sex linked.
2. Second, let us ask whether albinism is dominant or recessive. If the trait is dominant, then every albino child will have an albino parent; if it is recessive,

FIGURE 13-5

A pedigree typical of albinism. In pedigrees, males are conventionally shown as squares, females as circles, and marriages as horizontal lines connecting them, with the offspring shown below. The individuals who exhibit the trait being considered are indicated by solid symbols.

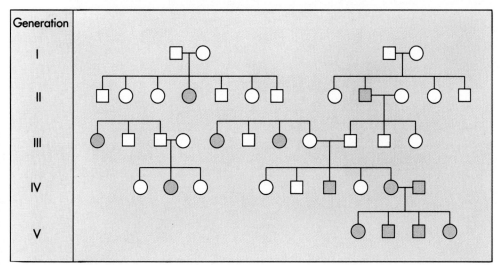

an albino child's parents can appear normal (both parents may be heterozygous). In Figure 13-5 the parents of most of the albino children do not exhibit the trait, which indicates that albinism is recessive. The reason that the four children in one family *do* have albino parents is that the allele is very common among the Hopi Indians, from which this pedigree was derived, so that among them homozygous individuals (such as these albino parents) are common enough that they sometimes marry. Note that in this marriage *both* parents were albino, and that all four of their children are albino, as you would expect if the trait were recessive and both of these parents, therefore, are homozygous for this allele.

3. Third, let us see if the trait is determined by a single gene, or by several. If the trait is determined by a single gene, then albinos born to heterozygous parents should occur in families in 3:1 proportions, reflecting Mendelian segregation. Thus you would expect that about 25% of the children should be albinos. If the trait were determined by several genes, on the other hand, then the proportion of albinos would be much lower, only a few percent. In this case, 9/34 (you don't count the four children of the marriage between two homozygous individuals, as this is not a cross between two heterozygotes) or 27% of the children are albinos, strongly suggesting that only one gene is segregating in these crosses.

Thus, looking at the pattern of inheritance of albinism, we were able to learn that albinism is an autosomal recessive trait controlled by a single gene. This is how pedigree analysis is done. Other inherited human traits are studied in a similar way. As a second example, consider red-green color blindness, a rare inherited trait in humans. In the pedigree shown in Figure 13-6, a color-blind man has five children with a woman who is heterozygous for the allele.

1. Is red-green color blindness sex linked or autosomal? Of the seven affected individuals, all are male. The trait is clearly sex linked.
2. Is red-green color blindness dominant or recessive? If the trait is dominant, then every color-blind child will have a color-blind parent—in this pedigree, that is not true in any family after that of the original male. The trait must be recessive.
3. Is the trait determined by a single gene? If so, then children born to heterozygous parents should be color-blind in about 25% of cases, reflecting a 3:1 Mendelian segregation of the trait. In this case 4/14 or 28% of the children of heterozygous mothers are color-blind, indicating that a single gene is segregating.

Pedigree analysis is the acquiring of information about the phenotypes of family members to infer the gentic nature of a trait from the pattern of its inheritance.

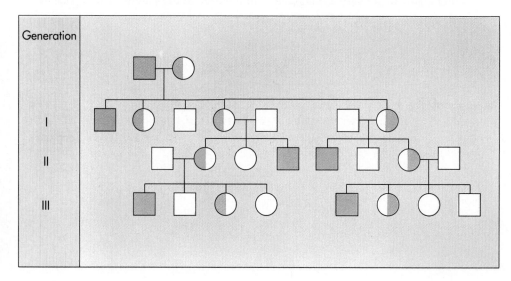

FIGURE 13-6
A pedigree showing color blindness. Squares indicate males, circles indicate females. Solid symbols indicate a color-blind individual; half-filled symbols indicate a heterozygous individual who carries the trait but does not express it.

MULTIPLE ALLELES

It is important to understand that a gene may exhibit more than two alleles in a population. Mendel deliberately limited his study to pairs of contrasting traits in pea plants; his plants were either tall or short, their flowers purple or white. Similarly, Morgan's *Drosophila* had eyes that were either white or red (the normal color). Many human genes also exhibit two alternative alleles. An individual, for example, is either albino or pigmented. It is important to remember, however, that most genes possess more than two possible alleles. Any change in the long sequence of nucleotides that make up a gene is potentially a new allele.

A human gene that typically exhibits more than one allele is the gene that encodes an enzyme that adds sugar molecules to lipids on the surface of red blood cells. These sugars act as recognition markers for our immune system and are called **cell surface antigens.** The gene encoding the enzyme is designated I and possesses three common alleles: (1) allele B, which adds the sugar galactose; (2) allele A, which adds a modified form of the sugar, galactosamine; and (3) allele O, which does not add a sugar.

Many genes possess multiple alleles, several of which may be common within populations.

When an individual is heterozygous for a gene with many possible alleles, which allele is dominant? Often, no one allele is; instead, each allele has its own effect. Thus an individual heterozygous for the A and B alleles of the I gene produces both forms of the enzyme and adds both galactose and galactosamine to the surfaces of this individual's blood cells; the cells thus possess antigens with both kinds of sugar attached to them. Because both alleles are expressed simultaneously in heterozygotes, the A and B alleles are said to be **co-dominant.** Either A or B alleles are dominant over the O allele, because the A or B allele leads to sugar addition and the O allele does not.

ABO Blood Groups

Different combinations of the three possible I gene alleles occur in different individuals, because each person possesses two copies of the chromosome bearing the I gene and may be homozygous for any allele or heterozygous for any two. The different combinations of the three alleles produce four different phenotypes:

1. Persons who add only galactosamine are called **type A** individuals. They are either AA homozygotes or AO heterozygotes.
2. Persons who add only galactose are called **type B** individuals. They are either BB homozygotes or BO heterozygotes.
3. Persons who add both sugars are called **type AB** individuals. They are, as we have seen, AB heterozygotes.
4. Persons who add neither sugar are called **type O** individuals. They are OO homozygotes.

The four different cell-surface phenotypes listed above are called the **ABO blood groups,** or, less commonly, the Landsteiner blood groups, after the man who first described them. As Landsteiner first noted, your immune system can tell the difference between these four phenotypes. If a type A individual receives a transfusion of type B blood, the recipient's immune system will recognize that the type B blood cells possess a "foreign" antigen (galactose) and attack the donated blood cells. If the donated blood is type AB, this will also happen. However, if the donated blood is type O, no attack will occur, as there are no foreign galactose antigens on the surfaces of blood cells produced by the type O donor. In general, any individual's immune system will tolerate a transfusion of type O blood. Because neither galactose nor galactosamine is foreign to type AB individuals (they add both to their red blood cells), AB individuals may receive any type of blood. The combinations in which agglutination of blood cells will occur are shown in Figure 13-7.

Some of the ABO blood group phenotypes are more common than others in human populations. In general, type O individuals are the most common, and type AB

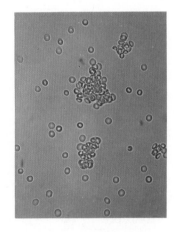

Blood types and
agglutination

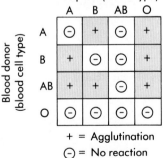

		Recipient (serum type)			
		A	B	AB	O
Blood donor (blood cell type)	A	⊖	+	⊖	+
	B	+	⊖	⊖	+
	AB	+	+	⊖	+
	O	⊖	⊖	⊖	⊖

+ = Agglutination
⊖ = No reaction

FIGURE 13-7

The agglutination reaction. Type A red blood cells agglutinating in type O serum. The unclumped red blood cells are type O cells. The diagram shows the combinations in which agglutination will occur.

individuals the least common. Human populations differ from one another a great deal, however. Among North American Indians, for example, the frequency of type A individuals is 31%, whereas among South American Indians, it is only 4%.

The Rh Blood Group

Another set of cell surface markers on human red blood cells are the **Rh blood group** antigens, named for the rhesus monkey in which they were first described. About 85% of adult humans have the Rh cell surface marker on their red blood cells, and are called Rh positive (Rh+). Rh negative persons lack this cell surface marker because they are homozygous recessive for the gene encoding it.

If an Rh negative person is exposed to Rh positive blood, the Rh surface antigens of that blood are treated like foreign invaders by the Rh negative person's immune system, which proceeds to make antibodies directed against the Rh antigens. This most commonly happens when an Rh negative woman gives birth to an Rh positive child (the father being Rh+). Some fetal red blood cells cross the placental barrier and enter the mother's bloodstream, where they induce the production of "anti-Rh" antibodies, which in later pregnancies can cross back to another fetus and cause its red blood cells to clump, leading to a potentially fatal condition called **erythroblastosis fetalis.**

GENETIC DISORDERS

There are few human genes that do not vary from one person to the next. That is to say, among all the people living, there are almost always some individuals that possess a different allele of a particular gene than others. However, most people possess the same allele as most of their genes; variant alleles that influence the phenotype are usually rare, and only a few occur commonly in human populations. In Chapter 16 we will discuss the process, called **mutation,** that is responsible for the production of such variant alleles. Here we need only note that the mutation process involves making random changes in genes, and that changing a gene randomly rarely improves the

TABLE 13-1   SOME IMPORTANT GENETIC DISORDERS

DISORDER	SYMPTOM	DEFECT	DOMINANT/ RECESSIVE	FREQUENCY AMONG HUMAN BIRTHS
Cystic fibrosis	Mucus clogging lungs, liver, and pancreas	Failure of chloride ion transport mechanism	Recessive	$1/1800$ (caucasian)
Sickle cell anemia	Poor blood circulation	Abnormal hemoglobin molecules	Recessive	$1/1600$ (African-Americans)
Tay-Sachs disease	Deterioration of central nervous system in infancy	Defective form of enzyme hexosaminidase A	Recessive	$1/1600$ (Ashkenazim Jews)
Phenylketonuria	Failure of brain to develop in infancy	Defective form of enzyme phenylalanine hydroxylase	Recessive	$1/18,000$
Hemophilia (Royal)	Failure of blood to clot	Defective form of blood clotting factor IX	Sex-linked recessive	$1/7000$
Huntington's disease	Gradual deterioration of brain tissue in middle age	Production of an inhibitor of brain cell metabolism	Dominant	$1/10,000$
Muscular dystrophy (Duchenne)	Wasting away of muscles	Degradation of myelin coating of nerves stimulating muscles	Sex-linked recessive	$1/10,000$
Hypercholesterolemia	Excessive cholesterol levels in blood, leading to heart disease	Abnormal form of cholesterol cell-surface receptor	Dominant	$1/500$

functioning of its encoded protein, any more than your randomly changing a wire in a computer is likely to improve the computer's functioning.

Usually, but not always, alternative alleles with detrimental effects are rare in human populations. Sometimes a detrimental allele is common. A common allele that results in unfavorable characteristics can have disastrous effects on the group of humans in which it occurs. When such an allele is recessive, as they usually are, two seemingly normal people who are heterozygous can produce children homozygous for the recessive allele who cannot avoid the detrimental effect of the mutant allele. This tragedy strikes many families. Learning how to avoid it is one of the principal goals of human genetics.

In those instances in which a detrimental allele occurs at a significant frequency in human populations, the harmful effect that it produces is called a **genetic disorder.** Table 13-1 (see p. 266) lists some of the most important human genetic disorders. We know a great deal about some of them, and much less about many others.

Cystic Fibrosis

The most common fatal genetic disorder of caucasian persons is **cystic fibrosis** (Figure 13-8). As we saw in the boxed essay in Chapter 6 (p. 125), the affected individuals secrete a thick mucus that clogs the airways of their lungs and the passages of their pancreas and liver. Among white persons, about 1 in 20 individuals has a copy of the defective gene but shows no symptoms; the double recessive individuals make up about 1 in 1800 white children. These individuals inevitably die from complications that result from their disease.

We have learned only recently that the cause of cystic fibrosis is a defect in the way that cells regulate the transport of chloride ions across their membranes. Cystic fibrosis occurs when an individual is homozygous for an allele encoding a defective version of the protein regulating the chloride-transport channel. This allele is recessive to the normal-functioning version of the regulating protein, so that the chloride channels of heterozygous individuals function normally and such individuals do not develop cystic fibrosis.

Sickle Cell Anemia

Sickle cell anemia is a heritable disorder in which the afflicted individuals are unable to transport oxygen to their tissues properly because the molecules within red blood cells that carry oxygen, molecules of the protein **hemoglobin,** are defective. When oxygen is scarce, these defective hemoglobin molecules become insoluble and combine with one another, forming stiff, rodlike structures. This results in the formation of sickle shaped red blood cells.

Surprisingly, the hemoglobin that occurs in such defective red blood cells differs from that which occurs in normal red blood cells in only one out of a total of about 300 amino acid molecules. In the defective hemoglobin, one molecule of valine occurs in place of the glutamic acid that occurs in the same position in normal hemoglobin. Red blood cells that contain large proportions of such defective molecules become sickle-shaped and stiff; normal red blood cells are disk-shaped and much more flexible (Figure 13-9). As a result of their stiffness and irregular shape, the sickle-shaped red blood cells are able to move through the smallest blood vessels only with great difficulty. For the same reason, they also tend to accumulate in the blood vessels, forming clots. As a result, people who have large proportions of sickle-shaped red blood cells tend to have intermittent illness and a shortened life span.

Individuals who are heterozygous for the sickle cell allele are generally indistinguishable from normal persons. In the blood of people who are heterozygous for this trait, however, some of the red cells show the sickling characteristic when they are exposed to low levels of oxygen. The allele responsible for the sickle cell characteristic is particularly common among people of African descent. For example, about 9% of African-Americans are heterozygous for this allele, and about 0.2% are homozygous and therefore have sickle cell anemia. In some groups of people in Africa, up to 45% of the individuals are heterozygous for this allele. The factors determining the frequency of sickle cell anemia are discussed in Chapter 20.

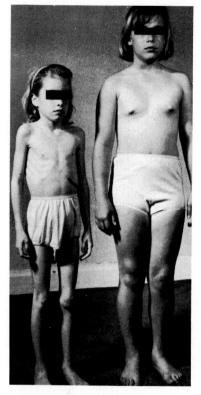

FIGURE 13-8
A child with cystic fibrosis. In affected individuals, such as the child on the left the mucus that normally lines the insides of the lungs thickens, making breathing difficult. Few affected children live to be adults.

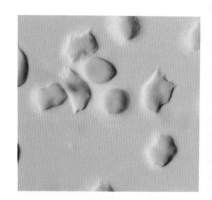

FIGURE 13-9
"Sickled" red blood cells. In individuals who are homozygous for the sickle cell trait, many of the red blood cells have such shapes.

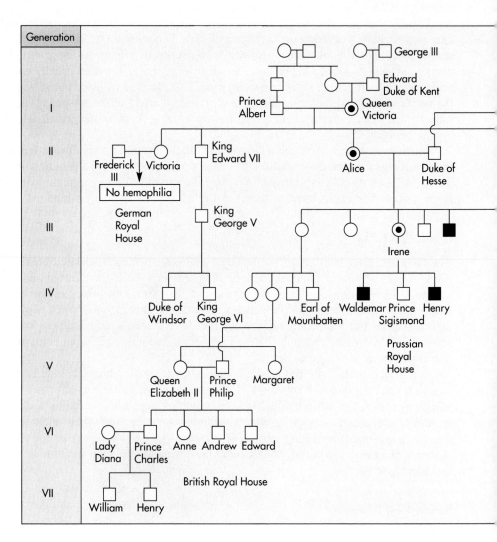

FIGURE 13-10

The Royal hemophilia pedigree. From Queen Victoria's daughter Alice, the disorder was introduced into the Russian and Austrian royal houses, and from her daughter Beatrice, it was introduced into the Spanish royal house. Victoria's son Leopold, himself a victim, also transmitted the disorder in a third line of descent.

Hemophilia

The reason that you do not bleed to death when you cut your finger is that the blood in the immediate area of the cut is converted to a solid gel that seals the cut in a process similar to that which occurs in a puncture-proof tire. Such a blood clot forms from the polymerization of protein fibers circulating in the blood, in a way similar to how gelatin hardens. A variety of proteins are involved in this process, and all of them must function properly for a blood clot to form. A mutation causing the loss of activity of any of these factors leads to a form of **hemophilia.** Hemophilia is a hereditary condition in which the blood is slow to clot or does not clot at all.

Hemophilias are recessive disorders, expressed only when an individual does not possess at least one copy of the gene that is normal and so cannot produce one of the proteins necessary for clotting. Individuals homozygous for a mutant allele do not produce any active version of the affected clotting protein, and so cannot clot blood. Most of the dozen protein-clotting genes are on autosomes, but two (designated VIII and IX) are known to be located on the X chromosome. In the case of these particular protein-clotting genes, any male who inherits a mutant allele will develop hemophilia because his other sex chromosome is the inactive Y and thus he lacks a normal allele of the protein-clotting gene.

The most famous form of hemophilia, often called the Royal hemophilia (Figure 13-10), is a mutation in factor IX that occurred in one of the parents of Queen Victoria of England (1819-1901) (Figure 13-11). In the five generations since Queen Victo-

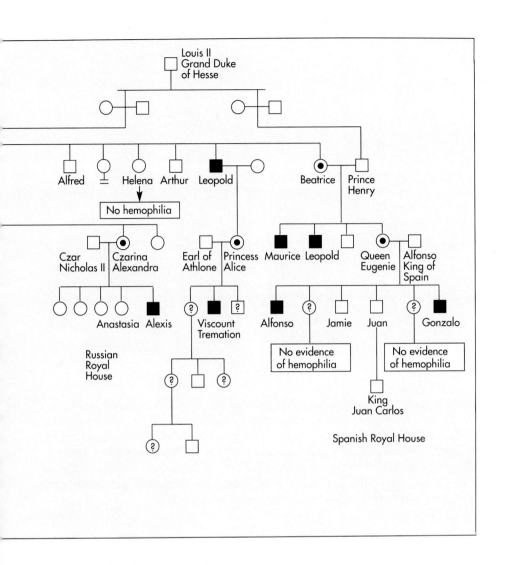

FIGURE 13-11

Queen Victoria of England in 1894, surrounded by some of her descendents. Of Victoria's four daughters who lived to bear children, two, Alice and Beatrice, were carriers of Royal hemophilia. Two of Alice's daughters are standing behind Victoria (wearing feathered boas): to Victoria's right is Princess Irene of Prussia; to her left, Alexandra, who would soon become Czarina of Russia. Both Irene and Alexandra were also carriers of hemophilia.

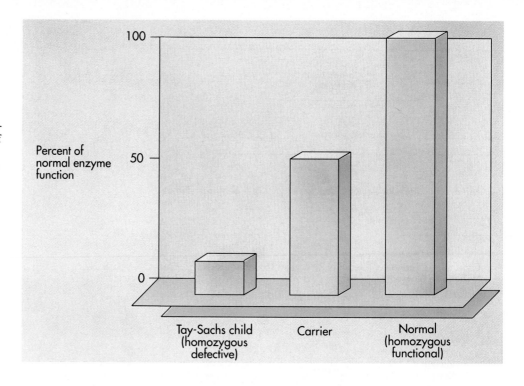

FIGURE 13-12
Tay-Sachs disease. Tay-Sachs disease is a genetic disorder in which an enzyme critical to lipid metabolism does not function, leading to harmful accumulations of fatty acids in the lysosomes of brain cells. Homozygous individuals typically have less than 10% of normal levels of the enzyme hexosaminidase A *(left)*, whereas heterozygous individuals have about 50% of normal levels, enough to prevent deterioration of the central nervous system.

ria, ten of her male descendants have had hemophilia. The British royal family escaped the disorder because Queen Victoria's son King Edward VII did not inherit the defective factor IX allele. Three of Victoria's nine children did receive the defective allele, however, and carried it by marriage into many of the royal families of Europe. It is still being transmitted to future generations among these family lines, except in the Soviet Union, where the five children of Alexandra, Victoria's granddaughter, were killed soon after the Russian revolution (see Figure 13-1).

Tay-Sachs Disease

Tay-Sachs disease is an incurable hereditary disorder in which the brain deteriorates. Affected children appear normal at birth and do not usually develop symptoms until about the eighth month, at which time signs of mental deterioration become evident. Within a year of birth, affected children are blind; they rarely live past their fifth year.

Tay-Sachs disease is rare in most human populations. In the United States it occurs in 1 in 300,000 births. However, Tay-Sachs disease has a high incidence among Jews of Eastern and Central Europe (Ashkenazim), and among American Jews (90% of whom are descendants of Eastern and Central European ancestors). Approximately 1 in 3600 such Jewish infants has this genetic disorder. Because it is a recessive condition, most of the people that carry the defective allele do not themselves develop the characteristic symptoms. It is estimated that 1 in 28 individuals in these Jewish populations is a heterozygous carrier of the allele.

How does the Tay-Sachs allele produce the symptoms of the disease? The Tay-Sachs allele encodes a nonfunctional form of a critical enzyme (Figure 13-12). Individuals homozygous for the allele lack an enzyme necessary to break down a special class of lipid called **gangliosides,** which occur within the lysosomes of brain cells. As a result, the lysosomes fill with gangliosides, swell, and eventually burst, releasing oxidative enzymes that kill the brain cell. There is no known cure for this condition.

Phenylketonuria

Although less common than cystic fibrosis, sickle cell anemia, or Tay-Sachs disease, some additional genetic disorders do affect significant numbers of people. A good ex-

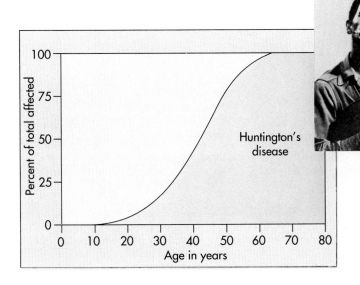

FIGURE 13-13
Huntington's disease is a dominant genetic disorder. The folksinger Woody Guthrie was a victim of Huntington's disease. It is because of this late age of onset that the disorder persists despite the fact that it is dominant and fatal.

ample of a relatively infrequent genetic disorder is **phenylketonuria (PKU),** a hereditary condition in which the affected individuals are unable to break down the amino acid phenylalanine. In such individuals, phenylalanine is instead converted to other chemicals that accumulate in the bloodstream. Although not harmful to an adult, these abnormal derivatives of phenylalanine are harmful to infants because they interfere with the development of brain cells. An infant with this disorder suffers severe mental retardation, and affected individuals rarely live more than 30 years. When it is detected early enough, however, PKU can be treated nutritionally, and individuals with this genetic constitution can then develop and mature normally.

Phenylketonuria is a recessive disorder caused by a mutant allele of the gene encoding the enzyme that normally breaks down phenylalanine. Only individuals homozygous for the mutant allele (in the United States, about 1 in every 15,000 infants) develop the disorder.

Huntington's Disease

Not all hereditary disorders are recessive. **Huntington's disease** is a hereditary condition caused by a dominant allele that causes progressive deterioration of brain cells. It is the disorder that killed folksinger and songwriter Woody Guthrie (Figure 13-13). Perhaps 1 in 10,000 individuals develops the disorder. Because Huntington's disease is a dominant condition, every individual that carries an allele expresses it. You might wonder why in this case the genetic disorder doesn't die out. The answer is that symptoms of Huntington's disease do not usually develop until the individuals are more than 30 years old, by which time most of them have already had children. For this reason, the allele is transmitted before the lethal condition develops.

GENETIC COUNSELING

Although most genetic disorders cannot yet be cured, we are learning a great deal about them, and progress toward successful therapy is being made in many cases. Table 13-1 presents a list of the most significant genetic disorders. In the absence of a cure, however, the only recourse is to try to avoid producing children subject to these conditions. The process of identifying parents at risk of producing children with genetic defects and of assessing the genetic state of early embryos is called genetic counseling.

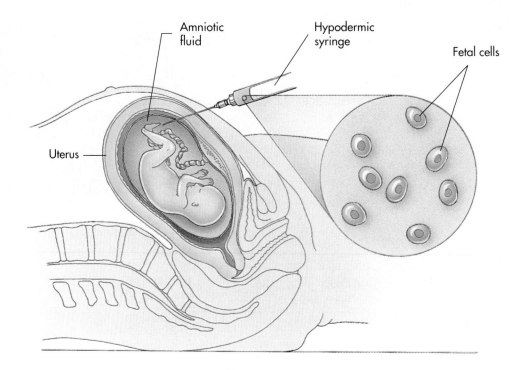

FIGURE 13-14
Amniocentesis. A needle is inserted into the amniotic cavity, and a sample of amniotic fluid, containing some free cells derived from the embryo, is withdrawn into a syringe. The fetal cells are then grown in tissue culture, so that their karyotype and many of their metabolic functions can be examined.

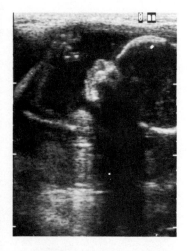

FIGURE 13-15
The appearance of a fetus as detected with ultrasound. During the fourth month of pregnancy, when amniocentesis is normally performed, the fetus usually moves about actively.

If the genetic defect is a recessive allele, how can potential parents determine if it is likely that they carry the allele? Pedigree analysis is often employed as an aid in genetic counseling. If one of your relatives has been afflicted with a recessive genetic disorder such as cystic fibrosis, there is a possibility that you are a carrier of the trait; in other words, that you carry the recessive allele in the heterozygous state. By analyzing your pedigree, it is often possible to estimate the likelihood that you are a carrier. When a couple is expecting a child and pedigree analysis indicates that both parents have a significant probability of being heterozygous carriers of a recessive allele that is responsible for a serious genetic disorder, the pregnancy is said to be a **high-risk pregnancy.** In such a pregnancy, there is a significant probability that the child will exhibit the clinical disorder.

Another class of high-risk pregnancies are those in which mothers are more than 35 years old. As we have seen, the frequency of birth of infants with Down syndrome increases dramatically among the pregnancies of older women (see Figure 13-3).

When a pregnancy is diagnosed as being high risk, many women elect to undergo **amniocentesis,** a procedure that permits the prenatal diagnosis of many genetic disorders (Figure 13-14). In the fourth month of pregnancy, a small sample of amniotic fluid is removed by means of a sterile hypodermic needle. When the needle is inserted into the expanded uterus of the mother, its position and that of the fetus are usually observed simultaneously by means of a technique called **ultrasound.** Sound waves allow an image of the fetus to be seen (Figure 13-15), as do x rays, but the sound waves are not damaging to the mother or to the fetus. Since ultrasound allows the position of the fetus to be determined, the person withdrawing the amniotic fluid can do so in such a way as to avoid damaging the fetus. In addition, the fetus can be examined for the presence of major abnormalities. The amniotic fluid, which bathes the fetus, contains within it free-floating cells derived from the fetus; once removed, these cells can be grown as tissue cultures in the laboratory.

In the last few years, physicians have increasingly turned to a new, less invasive procedure, **chorionic villi sampling,** for genetic screening. In this procedure the physician removes cells from the chorion, a membrane part of the placenta that nourishes the fetus. This procedure can be used earlier in pregnancy (by the eighth week) and yields results much more rapidly than does amniocentesis.

By studying tissue cultures from amniocentesis or tissue from chorionic villi sampling, genetic counselors can test for many of the most common genetic disorders.

1. *Enzyme activity tests.* In many cases it is possible to test directly for the proper functioning of the enzymes involved in genetic disorders; the lack of proper acitivy signals the presence of the disorder. Thus the lack of the enzyme responsible for breaking down phenylalanine signals PKU, the absence of the enzyme reponsible for the breakdown of gangliosides indicates Tay-Sachs disease, and so forth.

2. *Association with genetic markers.* For sickle cell hemoglobin, Huntington's disease, and one form of muscular dystrophy (a condition characterized by weakened muscles that do not function normally), investigators have found other mutations on the same chromosome that, by chance, occurred at about the same place as the disorder-causing mutation. By testing for the presence of the second mutation, an investigator can identify individuals with a high probability of possessing the disorder-causing defect. Identifying such mutations in the first place is a little like searching for a needle in a haystack, but persistent efforts have proved successful in these cases. The associated mutations are detected because they alter the length of DNA segments produced by enzymes that cut strands of DNA at particular places. Such enzymes, called **restriction enzymes,** are discussed in Chapter 17. The mutations produce restriction fragment-length polymorphisms, or **RFLPs.**

3. *Identification of heterozygotes.* As demonstrated in Figure 13-16, heterozygous individuals can often be detected during genetic counseling and can be given the opportunity to plan accordingly.

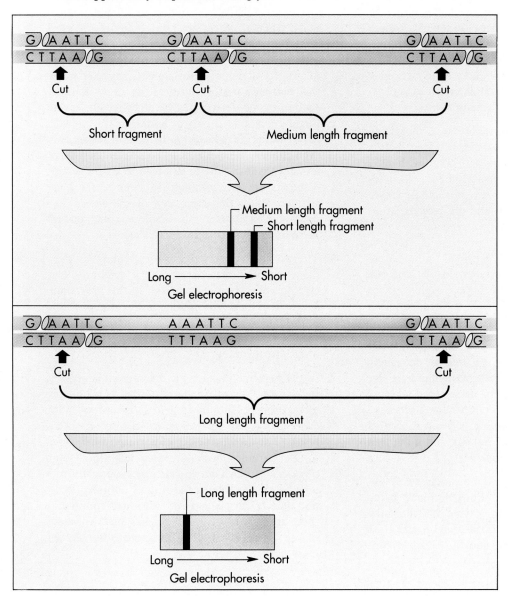

FIGURE 13-16

RFLPs. Restriction fragment length polymorphisms (RFLPs) are playing an increasingly important role in genetic identification. Here you can see how the mutation of a single nucleotide from G to A alters a restriction endonuclease cutting site so that the enzyme no longer cuts there. As a result, a single long fragment is obtained, rather than two shorter ones. Such a change is easy to detect by gel electrophoresis of the fragments.

Genetic Therapy

When the analysis of amniotic fluid indicates a severe disorder in the fetus, the parents may consider terminating the pregnancy by therapeutic abortion. In some instances, other options are available. If PKU is diagnosed, for example, it is possible to avoid the effects of the disorder by placing the mother on a low-phenylalanine diet. This provides her (and her unborn baby) with enough phenylalanine to make proteins, but not enough to lead to buildup. After birth, the child is maintained on a low phenylalanine diet until age 6. At that age, the child's brain is fully developed and PKU is no longer a potential health problem. We do not yet have the knowledge that would allow us to correct undesirable genetic conditions directly, but as we learn more, it seems increasingly likely that such techniques will eventually exist.

SUMMARY

1. Human somatic cells contain 46 chromosomes: 44 autosomes and 2 sex chromosomes. The autosomes form 22 pairs of homologous chromosomes in meiosis. The 2 sex chromosomes may either be similar in size and appearance—in females, where they are designated XX—or one, the Y chromosome, may be much smaller than the other, which is the X chromosome. XY individuals are male because the genes that determine male characteristics are located on the Y chromosome.

2. In humans, the loss of an autosome is invariably fatal. Gaining an extra autosome, which leads to a condition called trisomy, is also fatal, with only two exceptions: chromosomes 21 and 22. Individuals with an extra copy of chromosome 21 (three copies in all) are retarded, and are referred to as having Down syndrome. Down syndrome is much more frequent among the children of mothers more than 35 years old. The primary nondisjunction results when chromosomes do not separate during meiosis. This occurs almost exclusively in the meiotic division that leads to the production of the egg.

3. Patterns of inheritance observed in family histories, called pedigrees, can be used to determine the mode of inheritance of a particular trait. By such analysis, it can often be determined whether a trait is associated with a dominant or a recessive allele, if the gene determining the trait is located on the X chromosome (sex-linked), and if the trait is specified by more than one gene.

4. A few human genes possess more than two common alleles. An example is the ABO blood group gene. This gene encodes an enzyme that adds sugars to the surfaces of blood cells. The A and B alleles add different sugars, and the O allele adds none.

5. Genetic disorders are caused by alleles that encode abnormal proteins; the effects of these proteins lead to serious health problems. Some genetic disorders are relatively common in human populations, whereas others are rare. Many of the most important genetic disorders are associated with recessive alleles, the functioning of which may lead to the production of defective versions of enzymes that normally perform critical functions. Because such traits are determined by recessive alleles and therefore expressed only in homozygotes, the alleles are not eliminated from the human population, even though their effects in homozygotes may be lethal. Dominant alleles that lead to severe genetic disorders are less common; in some of the more frequent ones, the expression of the alleles does not occur until after the individuals that possess them have reached reproductive age.

6. Women who suspect that their children may express a genetic disorder or Down syndrome may elect to undergo amniocentesis. In this procedure, a sample of fetal cells obtained from amniotic fluid is used to establish a tissue culture, which can then be checked for the presence of a wide variety of genetic disorders.

7. In the case of a few genetic disorders it is possible to initiate therapy if the disorder is diagnosed early enough. For most genetic disorders, however, we do not know how to accomplish such a result. The direct transfer of normal human genes to the chromosomes of individuals who would otherwise suffer from a particular genetic disorder may prove possible in the future, although it has not been accomplished yet. Meanwhile, although there are no cures for any genetic disorder, many of the conditions can be managed with increasingly positive results.

REVIEW QUESTIONS

1. How do monosomic individuals differ from normal individuals? How do trisomic individuals differ from normal individuals? Of monosomy and trisomy, which condition is always lethal in humans?

2. Define primary nondisjunction. How is it related to Down syndrome? What age and sex of an individual has a greater propensity to produce gametes damaged by nondisjunction? Why?

3. What is the constitution of the sex chromosomes in a normal female? a normal male? Which sex chromosome possesses fewer active alleles? What kinds of genes do exist on this chromosome?

4. What is the sex chromosome genotype of an individual that exhibits Klinefelter's syndrome? Is this individual genetically male or female? Why? Does this individual generally appear male or female?

5. What is the sex chromosome genotype of an individual that exhibits Turner's syndrome? Is this individual genetically male or female? Why? Does this individual generally appear male or female?

6. In the ABO blood group system, what are the phenotypes and genotypes of the four blood groups that result from the combinations of the three antigen alleles? What sugars are present on the surface of each type of red blood cell?

7. What physiological defects occur in individuals homozygous for the cystic fibrosis gene? What is the biochemical basis of this disorder?

8. What is the physiological basis of hemophilia? How do type VIII and IX differ from the other forms of Royal hemophilia? Can an otherwise normal human male be a carrier for Royal hemophilia? Why or why not?

9. Is Huntington's disease a dominant or recessive genetic disorder? Why is it maintained at its current frequency in human populations?

THOUGHT QUESTIONS

1. If a mother is Rh positive and her child is Rh negative (she is heterozygous, the father homozygous Rh negative), will the fetus develop anti-Rh antibodies and kill the mother? Why not?

2. Schizophrenia is a mental disorder in which a split occurs between thoughts and feelings and contact is lost with the environment. To assess whether schizophrenia is hereditary, twins who had been reared apart were studied; all they had in common was the fact that they were twins. Two kinds of twins were compared: those that were identical (monozygotic) and those that were not (dizygotic). Monozygotic twins are genetically identical, whereas dizygotic twins are brothers and/or sisters who happen to be born at the same time. Investigators asked if twins reared apart develop schizophrenia more commonly if they are genetically identical. Here is what they found, using data collected over 40 years: of 289 monozygotic twins studied, in 51% of the cases both twins develop the disorder if one does; of 398 dizygotic twins studied, in 10% of the cases both develop the disorder if one does. Do you think that these results suggest that schizophrenia is a hereditary disorder?

DIAMOND, J.: "Blood, Genes, and Malaria," *Natural History*, February 1989. A lucid account of the evolutionary history of sickle cell anemia.

DIAMOND, J.: "The Cruel Logic of Our Genes," *Discover*, November 1989, pages 72-79. A discussion of why evolution has not eliminated genetic diseases.

EDELSTEIN, S.: *The Sickled Cell: From Myths to Molecules*, Harvard University Press, Cambridge, Mass., 1986. An engaging account of sickle cell anemia and how it is currently being diagnosed and treated. Current efforts are focused on developing antisickling drugs to modify hemoglobin molecules.

FRIEDMANN, T.: "Progress Toward Human Gene Therapy," *Scientific American*, June 1989. A discussion of the new gene therapy—approaches to genetic disease treatment by attacks directly on mutant genes.

GOODFIELD, J.: *Quest for the Killers*, Birkhauser Press, Boston, 1985. An account of how biomedical scientists do their work, out in the field where diseases are. The diseases kuru, hepatitis B, smallpox, leprosy, and schistosomiasis are each examined as social as well as biomedical problems.

GOULD, S.J.: "Dr. Down's Syndrome," *Natural History*, vol. 89, pages 142-148, 1980. An account of the history of Down syndrome, a relatively common chromosomal abnormality that results in severe mental retardation.

HOOK, E.B.: "Behavioral Implications of the Human XYY Genotype," *Science*, vol. 179, pages 139-150, 1973. A review of some of the conflicting data on this controversial subject.

LAWN, R. and G. VEHAR: "The Molecular Genetics of Hemophilia," *Scientific American*, March 1986, pages 48-65. A review of the complex way in which blood clots, and the many genes that affect this process.

PATTERSON, D.: "The Causes of Down Syndrome," *Scientific American*, August 1987, pages 52-60. A cluster of genes on chromosome 21 associated with Down syndrome are being identified and studied.

RENSBERGER, B.: "Creating the Ultimate Map of Our Genes," *Science Year 1990*, pages 159-171. A very readable account of the human genome project.

SCIENCE: *Cystic Fibrosis: Cloning and Genetics*. An entire issue of *Science*, September 1989, devoted to the finding of the gene responsible for this widespread genetic defect.

VERMA, I.: "Gene Therapy," *Scientific American*, November 1990, pages 68-84. Treatment of genetic disorders by introducing healthy genes into the body of an affected individual is producing exciting results.

WATSON, J.D.: "The Human Genome Project: Past, Present, and Future," *Science*, vol. 248, April 1990, pages 44-51. An overview of how the project has evolved, by its Nobel-prize-winning director.

WHITE, R. and J. LALOUEL: "Chromosome Mapping with DNA Markers," *Scientific American*, February, 1988, pages 40-48. An account of how restriction fragment polymorphisms are being used as signposts in developing a detailed map of the human genome. This advance will revolutionize human genetics, as it will for the first time place the 3000 known genetic diseases within reach of the molecular tools for cloning DNA.

HUMAN GENETICS PROBLEMS

1. This pedigree is of a rare trait in which children have extra fingers and toes. Which if any of the following patterns of inheritance is consistent with this pedigree?

    a. Autosomal recessive
    b. Autosomal dominant
    c. Sex-linked recessive
    d. Sex-linked dominant
    e. Y-linkage

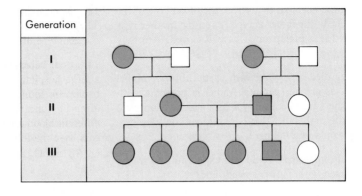

2. This pedigree is for a common form of inherited color blindness. What pattern of inheritance best accounts for this pedigree?

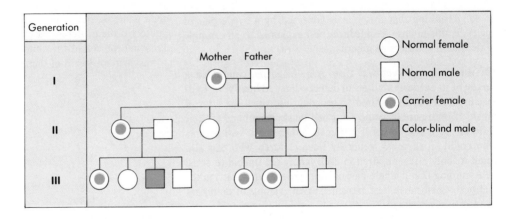

3. Of the 43 people in the five generations of this family, over one third exhibit an inherited mental disorder. Is the trait dominant or recessive?

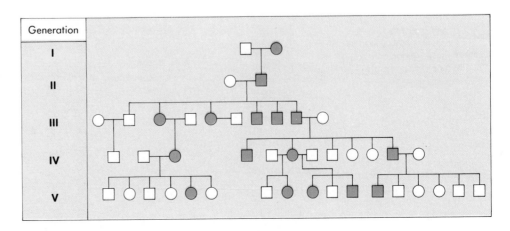

4. A man, George, with Royal hemophilia, marries his mother's sister's daughter Patricia. His maternal grandfather also had hemophilia. George and Patricia have five children: two daughters are normal, whereas two sons and one daughter develop hemophilia. Draw the pedigree.

5. A couple with a newborn baby are troubled that the child does not appear to resemble either of them. Suspecting that a mixup occurred at the hospital, they check the blood type of the infant. It is type O. As the father is type A and the mother type B, they conclude that a mixup must have occurred. Are they correct?

6. Mabel's sister died as a child with cystic fibrosis. Mabel herself is healthy, as are her parents. Mabel is pregnant with her first child. If she were to consult you as a genetic counselor, wishing to know the probability that her child will develop cystic fibrosis, what would you tell her?

7. How many chromosomes would you expect to find in the karyotype of a person with Turner's syndrome?

8. A woman is married for the second time. Her first husband was ABO blood type A and her child by that marriage was type O. Her new husband is type B, and when they have a child it is type AB. What is the woman's ABO genotype and blood type?

9. Two intensely freckled parents have five children, three of whom eventually become intensely freckled and two that do not do so. Assuming that this trait is governed by a single pair of alleles, is this intense freckledness best explained as an example of dominant or recessive inheritance?

10. In 1986, the *National Geographic* magazine conducted a survey of its readers' abilities to detect odors. About 7% of Caucasian persons in the United States could not smell the odor of musk. If both parents could not smell musk, then none of their children were able to smell it. On the other hand, two parents who could smell musk generally have children who can also smell it; only a few children in each family are unable to smell it. Assuming that a single pair of traits governs this trait, is the ability to smell musk best explained as an example of dominant or recessive inheritance?

11. Total color blindness is a rare hereditary disorder among humans in which no color is seen, only shades of grey. It occurs in individuals homozygous for a recessive allele, and it is not sex linked. A man whose father is totally color blind intends to marry a woman whose mother was totally color blind. What are the chances that they will produce offspring that are totally color blind?

12. A normally pigmented man married an albino woman. They have three children, one of whom is an albino. What is the genotype of the father?

13. Four babies are born within a few minutes of each other in a large hospital, when suddenly an explosion occurs. All four babies are found alive among the rubble. None had yet been given identification bracelets. The babies proved to be of four different blood groups: A, B, AB, and O. The four pairs of parents have the following pairs of blood groups: A and B, O and O, AB and O, and B and B. Which baby belongs to which parents?

14. An American couple both work in an atomic energy plant, and both are exposed daily to low-level background radiation. After several years the couple have a child, and this child proves to be affected by Duchenne muscular dystrophy, a sex-linked recessive genetic defect in which the mutant locus is on the X chromosome. Both parents are normal, as are the grandparents. The couple sue the plant, claiming that the abnormality in their child was the direct result of radiation-induced mutation of their gametes, radiation against which the company should have protected them. Before reaching a decision, the judge insists on knowing the sex of the child. Which sex would be more likely to result in an award of damages, and why?

Molecular Genetics

14

DNA: The Genetic Material

CHAPTER OUTLINE

OVERVIEW

Mendel's laws of heredity and the chromosomal theory of inheritance that explained them answered the age-old question of how traits are inherited only to raise another question: how does it work? What is the mechanism that a cell uses to specify what its descendants will be like? We now know that cells store this information in DNA, copies of which are faithfully made and passed down from generation to generation. The experiments that uncovered this mechanism are among the most elegant in biology.

FOR REVIEW

Here are some important terms and concepts that you will encounter in this chapter. If you are not familiar with them, you should review them before proceeding.

Nature of scientific experiments (Chapter 1)

Structure of DNA (Chapter 3)

Nucleotides (Chapter 3)

Structure of proteins (Chapter 3)

Structure of chromosomes (Chapter 10)

The realization, reached early in this century, that patterns of heredity can be explained by the segregation of chromosomes in meiosis was one of the most important advances in human thought. Not only did it lead directly to formulation of the science of genetics and thus to great progress in agriculture and medicine, but also it profoundly influenced the way in which we view ourselves. By removing the mystery from the process of heredity, it made the biological nature of humans more approachable (Figure 14-1). It also raised the question that occupied biologists for over 50 years: what is the exact nature of the connection between hereditary traits and the chromosomes?

We are now able to answer this question. We understand in considerable detail the mechanism by which the information on the chromosomes is converted into organisms with eyes, arms, and inquiring minds. Our understanding was not acquired all at once—deduced in a single flash of insight—but rather was developed slowly over many years by a succession of investigators. How they did this and what they learned are the subjects of this chapter.

WHERE DO CELLS STORE HEREDITARY INFORMATION?

Perhaps the most basic question that one can ask about hereditary information is where it is stored in the cell. Of the many approaches that one might take to answer this question, let us start with a simple one: cut a cell into pieces and see which of the pieces are able to express hereditary information. For this experiment we will need a single-celled organism that is large enough to operate on conveniently and differentiated enough that the pieces can be distinguished.

An elegant experiment of this kind was performed by the Danish biologist Joachim Hammerling in the 1930s. As an experimental subject, Hammerling chose the large unicellular green alga *Acetabularia* (Figure 14-2). Individuals of this genus have distinct foot, stalk, and cap regions, all of which are differentiated parts of a single cell. The nucleus of this cell is located in the foot. As a preliminary experiment, Hammerling tried amputating the caps or feet of individual cells. He found that when the cap is amputated, a new cap regenerates from the remaining portions (foot and stalk) of the cell. When the foot is amputated and discarded, however, no new foot is regenerated from the cap or the stalk. Hammerling concluded that the hereditary information resided within the foot, or basal portion, of *Acetabularia*.

To test this hypothesis, Hammerling selected individuals from two species of the genus in which the caps looked very different from one another: *Acetabularia mediterranea*, which has a disk-shaped cap, and *Acetabularia crenulata*, which has a branched, flowerlike cap. Hammerling cut the stalk and cap away from an individual of *A. mediterranea;* to the remaining foot he grafted a stalk cut from a cell of *A. crenulata* (Figure 14-3). The cap that formed looked something like the flower-shaped one characteristic of *A. crenulata*, although it was not exactly the same.

FIGURE 14-2
The marine green alga *Acetabularia* has been the subject of many elegant experiments in developmental biology. Although *Acetabularia* is a large organism with clearly differentiated parts, such as the stalks and elaborate caps visible here, individuals are actually single cells.

FIGURE 14-3
Hammerling's *Acetabularia* **reciprocal graft experiment.** To the base of each species he grafted a stalk of the other. In each case, the cap that eventually developed was dictated by the foot and not the stalk.

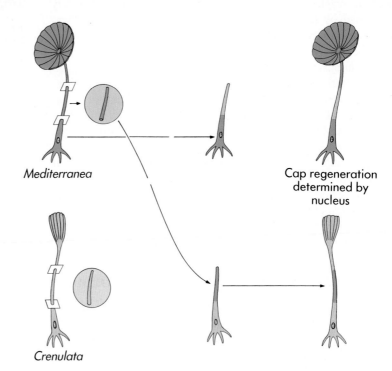

Mediterranea

Cap regeneration determined by nucleus

Crenulata

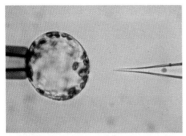

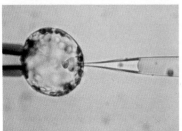

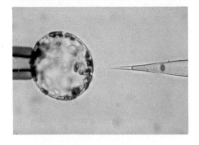

FIGURE 14-4

Microinjection. Working with a microscope, it is possible to pierce a cell with the fine tip of a glass micropipette without rupturing the cell. These micrographs show microinjection into a cell from the leaf of a tobacco plant (the cell, which lacks a cell wall, is about 50 micrometers in diameter).

Hammerling then cut off this regenerated cap and found that a disk-shaped one exactly like that of *A. mediterranea* formed in the second regeneration and in every regeneration thereafter. This experiment strengthened Hammerling's earlier conclusion that the instructions that specify the kind of cap that is produced are stored in the foot of the cell—and probably therefore in the nucleus—and that these instructions must pass from the foot through the stalk to the cap. In his regeneration experiment the initial flower-shaped cap was formed as a result of the instructions that were already present in the transplanted stalk when it was excised from the original *A. crenulata* cell. In contrast, all subsequent caps used new information, derived from the foot of the *A. mediterranea* cell onto which the stalk had been grafted. In some unknown way the original instructions that had been present in the stalk were eventually "used up."

Hammerling's experiments identified the nucleus as the likely repository of the hereditary information but did not prove definitely that this was the case. To do that, isolated nuclei had to be transplanted. Such an experiment was carried out in 1952 by American embryologists Robert Briggs and Thomas King. Using a glass pipette drawn to a fine tip and working with a microscope (Figure 14-4), Briggs and King removed the nucleus from a frog egg; without the nucleus, the egg would not develop. They then replaced the absent nucleus with one that they had isolated from a cell of a more advanced frog embryo (Figure 14-5). The implant of this nucleus ultimately caused an adult frog to develop from the egg. Clearly the nucleus was directing the frog's development.

Hereditary information is stored in the nucleus of eukaryotic cells.

Can each and every nucleus of an organism direct the development of an entire adult individual? The Briggs and King experiment did not answer this question definitively, since the nuclei that they took from more advanced frog embryos often caused the eggs into which the nuclei were transplanted to develop abnormally. But at Oxford and Yale, John Gurdon, working with another amphibian, was able to transplant nuclei isolated from developed tadpole tissue into eggs from which the nuclei had been removed and obtain normal development.

Have the nuclei in the cells of adult animals lost their hereditary information? This question has proved difficult to answer, since animal development is so complex. In plants, on the other hand, a simple experiment did yield a clear-cut answer. At

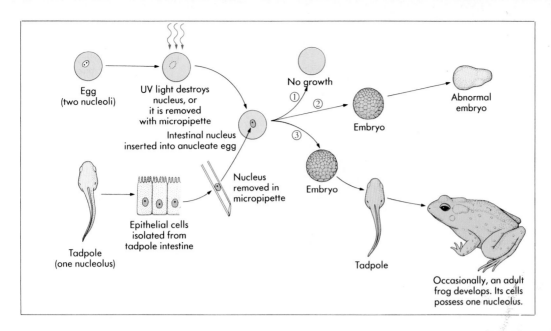

FIGURE 14-5

Briggs and King's nuclear transplant experiment. Two strains of frogs were used that differed from each other in the number of nucleoli their cells possessed. The nucleus was removed from an egg of one strain, either by sucking the egg nucleus into a micropipette or, more simply, by destroying it with ultraviolet light. Briggs and King then injected into this anucleate egg a nucleus obtained from a differentiated cell of the other species—in this case a cell isolated from the intestine of a tadpole of that species. The hybrid egg was then allowed to develop. One of three results was obtained in individual experiments: (1) no growth occurred, perhaps reflecting damage to the egg cell during the nuclear transplant operation; (2) normal growth and development occurred up to an early embryo stage, but subsequent development was not normal and the embryo did not survive; or (3) normal growth and development occurred, eventually leading to the development of an adult frog. That frog was of the species that contributed the nucleus and not of the species that contributed the egg. Only a few experiments gave this third result, but they serve to clearly establish that the nucleus directs frog development.

Cornell University in 1958, plant physiologist F.C. Steward let fragments of fully developed carrot tissue (bits of conducting tissue called phloem) swirl around in a rotating flask containing liquid growth medium. Individual cells broke away and tumbled through the liquid. Steward observed that these cells often divided and differentiated into multicellular roots. If these roots were then immobilized by placing them in a gel, they would go on to develop into entire plants that could be transplanted to soil and develop normally into maturity. Figure 37-2 presents a fuller description of this experiment. Steward's experiment makes clear that in plants at least some of the cells present in adult individuals do contain a full complement of hereditary information.

> **With rare exceptions, the nuclei of all cells of multicellular eukaryotes contain a full complement of genetic information. In many of the tissues of adult animals, however, the expression of much of this information is blocked.**

WHICH COMPONENT OF THE CHROMOSOMES CONTAINS THE HEREDITARY INFORMATION?

The identification of the nucleus as the source of hereditary information focused attention on the chromosomes, which were already suspected to be the vehicles of Mendelian inheritance. Specifically, biologists wondered how the actual hereditary information was arranged in the chromosomes. It was known that chromosomes contain both protein and DNA. On which of these was the hereditary information written?

Over a period of about 30 years, starting in the late 1920s, a series of investigators addressed this issue, resolving it clearly. We will describe three different kinds of experiments, each of which yields a clear answer in a simple and elegant manner.

The Griffith-Avery Experiments: Transforming Principle is DNA

As early as 1928, British microbiologist Frederick Griffith made a series of unexpected observations while experimenting with pathogenic (disease-causing) bacteria. When Griffith infected mice with a virulent strain of *Pneumococcus* bacteria, the mice died of blood poisoning, but when he infected similar mice with a strain of *Pneumococcus* that lacked a polysaccharide coat like that possessed by the virulent strain, the mice showed no ill effects. The coat was apparently necessary for successful infection.

As a control, Griffith injected normal but heat-killed bacteria into the mice to see if the polysaccharide coat itself had a toxic effect. The mice remained perfectly healthy. As a final control, he blended his two ineffective preparations—living bacteria whose coats had been removed and dead bacteria with intact coats—and injected

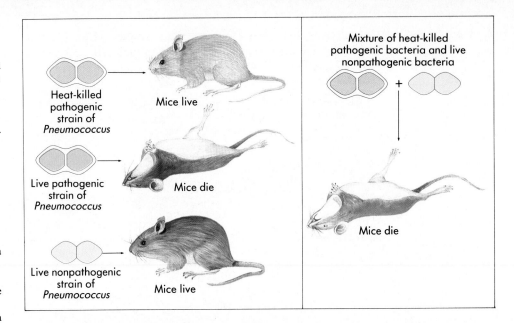

FIGURE 14-6

Griffith's discovery of transformation. The pathogenic bacterium *Pneumococcus* will kill many of the mice into which it is injected, but only if the bacterial cells are covered with a polysaccharide coat. Living bacteria without such polysaccharide coats do not harm the mice, since the coat is necessary for successful infection. However, it is not the coat itself that is the agent of disease. When Griffith injected dead bacteria possessing polysaccharide coats, the mice were not harmed. However, if he injected a mixture of dead bacteria with polysaccharide coats (harmless) and live bacteria without such coats (harmless), many of the mice died and virulent bacteria with coats were recovered. Griffith concluded that the live cells had been "transformed" by the dead ones—that the genetic information specifying the polysaccharide coat had passed from the dead cells to the living ones.

the mixture into healthy mice (Figure 14-6). Unexpectedly, the injected mice developed disease symptoms and many of them died. The blood of the dead mice was found to contain high levels of normal virulent *Pneumococcus* bacteria. Somehow the information specifying the polysaccharide coat had passed from the dead bacteria to the live but coatless ones in the control mixture, transforming them into normal virulent bacteria that infected and killed the mice.

The agent responsible for transforming *Pneumococcus* was not discovered until 1944. In an elegant series of experiments Oswald Avery and his coworkers characterized what they referred to as the **"transforming principle."** Its properties resembled those of DNA rather than protein: the activity of the transforming principle was not affected by protein-destroying enzymes but was lost completely in the presence of the DNA-destroying enzyme DNase.

When the transforming principle was purified, it indeed consisted predominantly of DNA. Subsequently, it was shown that all but trace amounts of protein (0.02%) could be removed without reducing the transforming activity. The conclusion was inescapable: DNA is the hereditary material in bacteria. It has since proved possible to use purified DNA to change the genetic characteristics of eukaryotic cells in tissue culture and even possible to inject pure DNA into fertilized *Drosophila* eggs and thereby alter the genetic characteristics of the resulting adult.

The Hershey-Chase Experiment: Bacterial Viruses Direct Their Heredity with DNA

Avery's results were not widely appreciated at first, many biologists preferring to believe that proteins were the depository of the hereditary information. Another convincing experiment was soon performed, however, that was difficult to ignore. It was done in a simple system—viruses—so that a direct experimental question could be asked. Viruses consist of either RNA or DNA with a protein coat; they are described in more detail in Chapter 29. These investigators focused on **bacteriophages,** viruses that attack bacteria, and carried out an experiment analogous to the transplant experiments described before. When a lytic bacteriophage infects a bacterial cell, it first binds to the cell's outer surface and then injects its hereditary information into the cell. There the hereditary information directs the production of thousands of new virus particles within the cell. The host bacterial cell eventually falls apart, or lyses, releasing the newly made viruses.

In 1952 Alfred Hershey and Martha Chase set out to identify the material injected into the bacterial cell at the start of an infection. They used a strain of bacteriophage known as T2, which contains DNA rather than RNA, and designed an experiment to distinguish whether the genetic material was DNA or protein (Figure 14-7). Hershey and Chase labeled the DNA of these T2 bacteriophages with a radioactive isotope of phosphorus, ^{32}P, and also labeled their protein coats with a radioactive isotope of sulfur, ^{35}S. Since the radioactive ^{32}P and ^{35}S isotopes emit particles of different energies when they decay, they are easily distinguished. The labeled viruses were permitted to infect bacteria. The bacterial cells were then agitated violently to shake the protein coats of the infecting viruses loose from the bacterial surfaces to which they were attached. Spinning the cells at high speed so that they were pulled from solution by the centrifugal forces (in this procedure, called centrifugation, the cells experience the

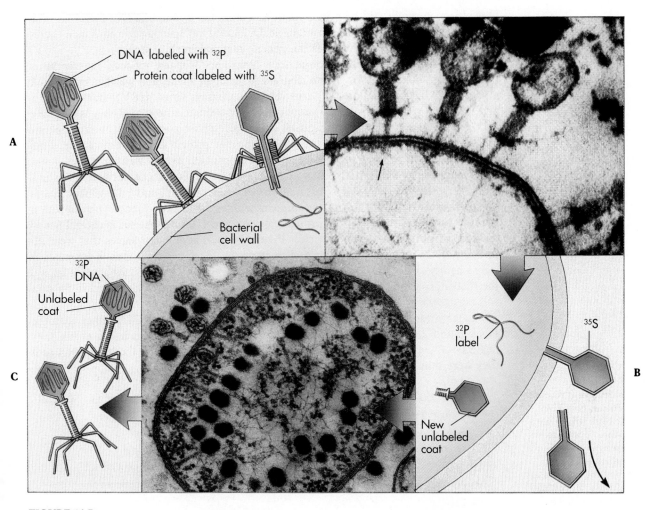

FIGURE 14-7

The Hershey-Chase experiment. The T2 bacterial viruses that they employed have a simple structure: they are composed of a protein envelope within which DNA is packaged.

A DNA released from a single virus particle. Hershey and Chase labeled the T2 virus particles with either of two radioisotopes: in one experiment the protein coats were labeled with ^{35}S (sulfur occurs in protein in the amino acids cysteine and methionine but does not occur in DNA), and in the other experiment the DNA molecules were labeled with ^{32}P (phosphorus occurs in the phosphate groups of DNA but does not occur in proteins). These T2 particles were then allowed to infect bacterial cells. Each virus binds to the outside of the cell but, instead of entering, injects its DNA into the cell.

B Individual DNA strands entering the cell from virus particles bound to its surface. Within the cell, the injected DNA commandeers the machinery of the cell and directs the synthesis of all the parts necessary to make new viruses.

C New virus particles are being assembled from parts within an infected cell. Eventually these new viruses will rupture the cell and be released into the surroundings. When Hershey and Chase used a Waring blender to knock the virus particles off the bacterial cells after the initial injection and separated the bacterial cells (with the injected viral DNA) from the liquid medium (with the dislodged viral protein coats in it), they found that, when ^{35}S–labeled virus was used, the bulk of the radioactivity (and thus the virus protein) was now in the medium; when ^{32}P–labeled virus was used, the radioactivity (and thus the virus DNA) was present in the *interiors* of the bacterial cells. It follows that the virus DNA, not the virus protein, was responsible for directing the production of new viruses.

same forces you do when pressed against the floor of a rapidly-rising elevator), Hershey and Chase found that the ^{35}S label (and thus the virus protein) was now predominantly in solution with the dissociated virus particles, whereas the ^{32}P label (and thus the DNA) had transferred to the interior of the cells. The viruses subsequently released from the infected bacteria contained the ^{32}P label. The hereditary information injected into the bacteria that specified the new generation of virus particles was DNA and not protein.

The Fraenkel-Conrat Experiment: Some Viruses Direct Their Heredity with RNA

One objection might still be raised about accepting the hypothesis that DNA was the genetic material. Some viruses contain *no* DNA and yet manage to reproduce themselves quite satisfactorily. What is the genetic material in this case?

In 1957 Heinz Fraenkel-Conrat and coworkers isolated tobacco mosaic virus (TMV) from tobacco leaves. From ribgrass (*Plantago*), a common weed, they isolated a second, rather similar kind of virus, Holmes ribgrass virus (HRV). In both TMV and HRV the viruses consist of a protein coat and a single strand of RNA. When they had isolated these viruses, the scientists chemically dissociated each of them, separating their protein from their RNA. By putting the protein component of one virus with the RNA of another, they were able to reconstitute hybrid virus particles.

The payoff of the experiment was in its next step. To choose between the alternative hypotheses "the genetic material of viruses is protein" and "the genetic material of viruses is RNA," Fraenkel-Conrat now infected healthy tobacco plants with a hybrid virus composed of TMV protein capsules and HRV RNA, being careful not to include any nonhybrid virus particles (Figure 14-8). The tobacco leaves that were infected with the reconstituted hybrid virus particles developed the kinds of lesions that were characteristic of HRV and that normally formed on infected ribgrass. Clearly, the hereditary properties of the virus were determined by the nucleic acid in its core and not by the protein in its coat.

Later studies have shown that many virus particles contain RNA rather than the DNA found universally in cellular organisms. When these viruses infect a cell, they make DNA copies of themselves, which can then be inserted into the host DNA as if they were cellular genes.

DNA is the genetic material for all cellular organisms and most viruses, although some viruses use RNA.

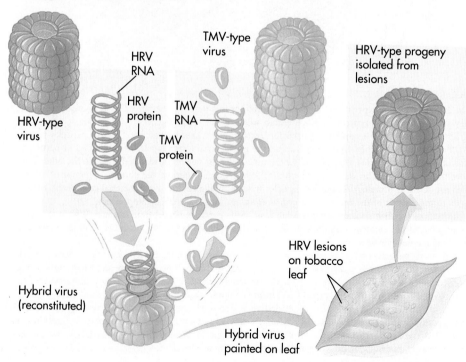

FIGURE 14-8

Fraenkel-Conrat's virus reconstitution experiment. Both TMV and HRV are plant RNA viruses that infect tobacco plants, causing lesions on the leaves. Because the two viruses produce different kinds of lesions, the source of any particular infection can be identified. In this experiment, TMV and HRV were both dissociated into protein and RNA and the protein and RNA were separated from one another. Then hybrid virus particles were produced by mixing the HRV RNA and the TMV protein and allowing virus particles to form from these ingredients. When the reconstituted virus particles were painted onto tobacco leaves, lesions developed of the HRV type. From these lesions, normal HRV virus particles could be isolated in great numbers; no TMV viruses could be isolated from the lesions. Thus the RNA (HRV) and not the protein (TMV) contains the information necessary to specify the production of the viruses.

HRV-type virus

HRV RNA

HRV protein

TMV RNA

TMV protein

TMV-type virus

HRV-type progeny isolated from lesions

Hybrid virus (reconstituted)

HRV lesions on tobacco leaf

Hybrid virus painted on leaf

THE CHEMICAL NATURE OF NUCLEIC ACIDS

DNA was first discovered only 4 years after the publication of Mendel's work. In 1869 a German chemist, Friedrich Miescher, extracted a white substance from the cell nuclei of human pus and from fish sperm nuclei. The proportion of nitrogen and phosphorus was different from any other known constituent of cells, which convinced Miescher he had discovered a new biological substance. He called this substance "nuclein," since it seemed to be specifically associated with the cell nucleus.

Because Miescher's nuclein was slightly acidic, it came to be called **nucleic acid.** For 50 years little work was done on it by biologists because nothing was known of its function in cells and there seemed little to recommend it to investigators. In the 1920s the basic chemistry of nucleic acids was determined by the biochemist P.A. Levine. There were two sorts of nucleic acid: ribonucleic acid or RNA, which had a hydroxyl group attached to a particular carbon atom, and deoxyribonucleic acid or DNA, which did not. Levine found that DNA contained three basic elements (Figure 14-9):

1. Phosphate (PO_4) groups
2. Five-carbon sugars
3. Four nitrogen-containing bases: adenine and guanine (double-ring compounds called purines) and thymine and cytosine (single-ring compounds called pyrimidines); RNA contained the pyrimidine uracil in place of thymine

From the roughly equal proportions of the three elements, Levine concluded correctly that DNA and RNA molecules are composed of units of these three elements, strung one after another in a long chain. Each element, a five-carbon sugar to which is attached a phosphate group and a nitrogen-containing base, is called a **nucleotide.** The identity of the base distinguishes one nucleotide from another.

To identify the various chemical groups in DNA and RNA, it is customary to number the carbon atoms of the organic base and the ribose sugar and then to refer to any chemical group attached to that carbon by that number. In the ribose sugar, four of the carbon atoms together with an oxygen atom form a five-membered ring. As illustrated in Figure 14-10, we number the different carbon atoms as 1' to 5', starting just to the right of the oxygen atom (the "prime" symbol ['] indicates that the number refers to a carbon of the sugar, as opposed to the organic base). Now look again at the diagram of a nucleotide in Figure 14-9. The phosphate is attached to the 5' carbon atom of the ribose sugar, and the organic base to the 1' carbon atom. There is in addition a free hydroxyl (OH) group attached to each 3' carbon.

The presence of the 5'-phosphate and the 3'-hydroxyl groups is what allows DNA and RNA to form a long chain of nucleotides: these two groups can chemically react with one another. The chemical reaction between the phosphate group of one unit and the OH group of another causes the elimination of a water molecule and the formation of a covalent bond linking the two groups (Figure 14-11). The linkage is called a **phosphodiester bond** because the phosphate group is now linked to the two sugars by means of two (—O—) bonds. Further linking can occur in the same way, since the two-unit polymer resulting from the condensation reaction we have just described still has a free 5'-phosphate group at one end and a free 3'-hydroxyl group at the other. In this way, many thousands of nucleotides can be linked together in long chains.

A single strand of DNA or RNA is a long chain of nucleotide subunits joined together like cars in a train.

Any linear strand of DNA or RNA, no matter how long, will always have a free 5'-phosphate group at one end and a free 3'-hydroxyl group at the other. This difference between the two ends of the nucleic acid molecule allows us to refer unambiguously to a particular end of the molecule; we can talk about the 5' end or the 3' end with no possibility of confusion. It is important to recognize that a DNA or RNA molecule has this intrinsic directionality. By convention, the sequence of DNA bases

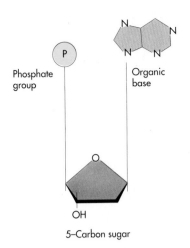

FIGURE 14-9
Nucleotide subunit of DNA. The nucleotide subunits of DNA are composed of three elements: an organic base, a phosphate group, and a 5-carbon sugar.

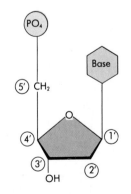

FIGURE 14-10
Numbering the carbon atoms in a nucleotide. The carbon atoms of the sugar of the nucleotide are numbered counting clockwise past the oxygen atom, as 1', 2' . . . 5'. The "prime" symbol indicates that the carbon belongs to the sugar, and not to the organic base.

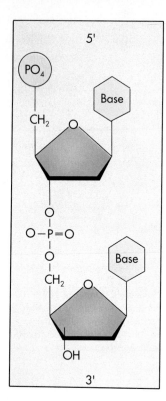

FIGURE 14-11
A phosphodiester bond.

is usually expressed in a 5′ to 3′ direction. Thus the base sequence "GTCCAT" refers to the sequence

$$5' \text{ pGpTpCpCpApT-OH } 3'$$

with phosphodiester bonds indicated by "p." Note that this is NOT the same molecule as that represented by the reverse sequence,

$$5' \text{ pTpApCpCpTpG-OH } 3'$$

Levine's early studies indicated that all four types of DNA nucleotides were present in roughly equal amounts. This result, which later proved an error, led to the mistaken belief that DNA was a simple repeating polymer in which the four nucleotides occurred together in simple repeating units (for instance . . . GCAT . . . GCAT . . . GCAT . . . GCAT). In the absence of sequence variation in such a repeating chain, it was difficult to see how DNA might contain the hereditary information necessary to specify even the most simple of organisms. For this reason, Avery's experiments on transforming principle, although crystal clear, were not readily accepted at first. It seemed more plausible that DNA was no more than a structural element of the chromosomes, with proteins playing the central genetic role.

The key advance came after World War II, when Levine's chemical analysis of DNA was repeated using more accurate techniques than had previously been available. Quite a different result was obtained. The four nucleotide bases were NOT present in equal proportions in DNA molecules after all. A careful study carried out by Erwin Chargaff showed that differences did exist in DNA nucleotide composition. Chargaff found that the nucleotide composition of DNA molecules varied in complex ways, depending on the source of the DNA (Table 14-1). This strongly suggested that DNA was not a simple repeating polymer and might after all have the information-encoding properties required of genetic material. Despite DNA's complexity, however, Chargaff observed an important underlying regularity: *the amount of adenine present in DNA molecules was always equal to the amount of thymine, and the amount of guanine was always equal to the amount of cytosine.* Chargaff's results are commonly referred to as **Chargaff's rules:**

1. The proportion of A always equals that of T, and G is similarly equal to C:

$$A = T, G = C$$

2. From the above rule, it follows that there is always an equal proportion of purines (A and G) and pyrimidines (C and T).

Chargaff pointed out that in all natural DNA molecules, the amount of A = T and the amount of G = C.

TABLE 14-1	CHARGAFF'S ANALYSIS OF DNA NUCLEOTIDE BASE COMPOSITIONS			
	BASE COMPOSITION (MOLE PERCENT)			
ORGANISM	A	T	G	C
Escherichia coli (K12)	26.0	23.9	24.9	25.2
Streptococcus pneumoniae	29.8	31.6	20.5	18.0
Mycobacterium tuberculosis	15.1	14.6	34.9	35.4
Yeast	31.3	32.9	18.7	17.1
Sea urchin	32.8	32.1	17.7	18.4
Herring	27.8	27.5	22.2	22.6
Rat	28.6	28.4	21.4	21.5
Human	30.9	29.4	19.9	19.8

SOURCE: Chargaff, E., and J. Davidson, (editors): *The Nucleic Acids*, New York, Academic Press, 1955.

THE THREE-DIMENSIONAL STRUCTURE OF DNA

As it became clear that the DNA molecule was the molecule in which the hereditary information was stored, investigators began to puzzle over how such a seemingly simple molecule could carry out such a complex function. The significance of the regularities pointed out by Chargaff, although not immediately obvious, soon became clear. A British chemist, Rosalind Franklin, had carried out x-ray crystallographic analysis of fibers of DNA. In this process the DNA molecule is bombarded with a beam of x rays. When individual rays encounter atoms, their path is bent or diffracted; the pattern created by the total of all these diffractions can be captured on a piece of photographic film. Such a pattern resembles the ripples created by tossing a rock into a smooth lake. By carefully analyzing the diffraction pattern, it is possible to develop a three-dimensional image of the molecule.

Franklin's studies were severely handicapped by the fact that it had not proven possible to obtain true crystals of natural DNA, so she had to work with DNA in the form of fibers. Although the DNA molecules in a fiber are all aligned with one another, they do not form the perfectly regular crystalline array required to take full advantage of x-ray diffraction. Franklin worked in the laboratory of British biochemist Maurice Wilkins, who was able to prepare more uniformly oriented DNA fibers than had been possible previously; using these fibers, Franklin was able, by working carefully, to obtain crude diffraction information on natural DNA. The diffraction patterns that Franklin obtained (Figure 14-12) suggested that the DNA molecule was a helical coil, a springlike spiral. It was also possible from her photographs of the diffraction pattern to determine some of the basic structural parameters of the molecule; the pattern indicated that the helix had a diameter of about 2 nanometers and made a complete spiral turn every 3.4 nanometers.

Learning informally of Franklin's results before they were published in 1953, James Watson and Francis Crick, two young investigators at Cambridge University, quickly worked out a likely structure of the DNA molecule (Figure 14-13), which we now know to be substantially correct. They analyzed the problem deductively: first they built models of the nucleotides, and then they tested how these could be assem-

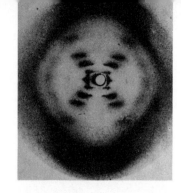

FIGURE 14-12

The essential clue. This x-ray diffraction photograph of fibers of DNA was made in 1953 by Rosalind Franklin in the laboratory of Maurice Wilkins. It suggested to Watson and Crick that the DNA molecule was a helix, like a winding staircase.

FIGURE 14-13

The discovery of DNA's structure. A key breakthrough in our understanding of heredity occurred in 1953, when James Watson, a young American postdoctoral student (he is the one peering up as if afraid their homemade model of the DNA molecule will topple over), and the English scientist Francis Crick (pointing) deduced the structure of DNA, the molecule that stores the hereditary information.

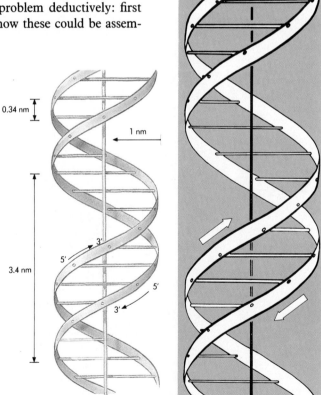

0.34 nm

1 nm

3.4 nm

3'

5'

5'

3'

FIGURE 14-14

DNA is a double helix. On the right is a reproduction of the original diagram of the double-helical structure of DNA, presented in a 1953 paper by Watson and Crick. The diagram on the left presents the exact dimensions of the double helix.

bled into a molecule that fit what they knew from Chargaff's and Franklin's work about the structure of DNA. They tried various possibilities, first assembling molecules with three strands of nucleotides wound around one another to stabilize the helical shape. None of these early efforts proved satisfactory. They finally hit on the idea that the molecule might be a simple **double helix,** one in which the bases of two strands pointed inward toward one another (Figure 14-14). By always pairing a purine, which is large, with a pyrimidine, which is small, the diameter of the duplex stays the same, 2 nanometers. Because hydrogen bonds can form between the two strands, the helical form (Figure 14-15) is stabilized.

It immediately became apparent why Chargaff had obtained the results that he had (Figure 14-16)—because the purine adenine (A) will not form proper hydrogen bonds in this structure with cytosine (C) but will with thymine (T), every A is paired to a T. Similarly the purine guanine (G) will not form proper hydrogen bonds with thymine but will with cytosine (C), so that every G is paired with C.

The Watson and Crick structure of DNA is a "double helix," a spiral staircase composed of two polynucleotide chains hydrogen-bonded to each other, wrapped around a central axis.

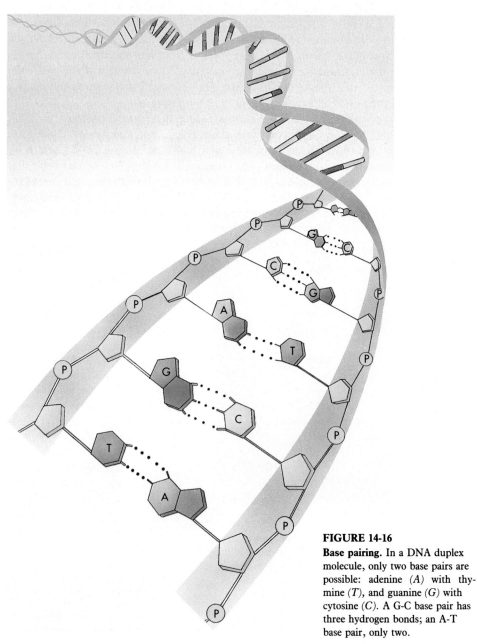

FIGURE 14-15
A three-dimensional model of DNA.

FIGURE 14-16
Base pairing. In a DNA duplex molecule, only two base pairs are possible: adenine (A) with thymine (T), and guanine (G) with cytosine (C). A G-C base pair has three hydrogen bonds; an A-T base pair, only two.

HOW DNA REPLICATES

The Watson-Crick model immediately suggested that the basis for copying the genetic information is **complementarity**. One chain of the DNA molecule may have any conceivable base sequence, but this sequence completely determines that of its partner in the duplex. If the sequence of one chain is ATTGCAT, the sequence of its partner in the duplex *must* be TAACGTA. Each chain in the duplex is a complementary mirror image of the other. To copy the DNA molecule, one need only "unzip" it and construct a new complementary chain along each naked single strand.

Replication is Semiconservative

The form of DNA replication suggested by the Watson-Crick model is called **semiconservative** because, after one round of replication, the original duplex is not conserved; instead, each strand of the duplex becomes part of another duplex.

The complementary nature of the DNA duplex provides a ready means of duplicating the molecule. If one were to unzip the molecule, one would need only to assemble the appropriate complementary nucleotides on the exposed single strands to form two daughter duplexes of the same sequence. This prediction of the Watson-Crick model was tested in 1958 by Matthew Meselson and Frank Stahl of the California Institute of Technology (see Figure 14-17). These two scientists grew bacteria for

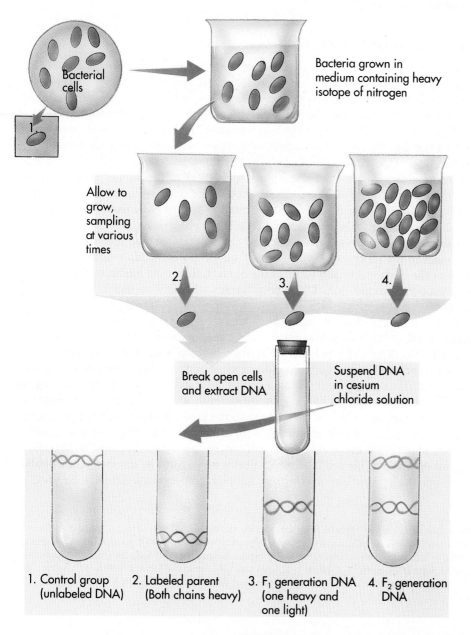

Bacteria grown in medium containing heavy isotope of nitrogen

Allow to grow, sampling at various times

Break open cells and extract DNA

Suspend DNA in cesium chloride solution

1. Control group (unlabeled DNA)
2. Labeled parent (Both chains heavy)
3. F_1 generation DNA (one heavy and one light)
4. F_2 generation DNA

FIGURE 14-17

The Meselson and Stahl experiment. Bacterial cells were grown for several generations in a medium containing heavy nitrogen isotopes and then were transferred to a new medium containing only the normal lighter nitrogen isotope. At various times thereafter, samples were taken, and the DNA was centrifuged in a cesium chloride solution. Because the cesium ion is so massive, the cesium tends to settle in the rapidly spinning tube, establishing a gradient of cesium concentration. DNA molecules sink in the gradient until they reach a place where the cesium concentration has the same density as the density of the DNA; it then "floats" at that position. Because DNA containing heavy nitrogen isotopes (^{15}N) is denser than DNA containing the normal lighter isotope of nitrogen (^{14}N), it sinks to a lower position on the cesium gradient. What Meselson and Stahl found was that after one generation in "light" medium, a single band of intermediate density halfway between heavy and light was obtained (one strand of each duplex was labeled; the other was not). After a second cell division, two bands were obtained, one intermediate (one of the two strands was labeled) and one light (neither strand was labeled). Meselson and Stahl concluded that replication of the DNA duplex involves building new molecules by separating strands and assembling new partners on each of these templates.

several generations in a medium containing the heavy isotope of nitrogen ^{15}N, so the DNA of their bacteria was eventually denser than normal. They then transferred the growing cells to a new medium containing the lighter isotope ^{14}N and harvested the DNA at various intervals.

At first the DNA that the bacteria manufactured was all heavy. But as the new DNA that was being formed incorporated the lighter nitrogen isotope, DNA density fell. After one round of DNA replication was complete, the density of the bacterial DNA had decreased to a value intermediate between all-light isotope and all-heavy isotope DNA. After another round of replication, two density classes were observed, one intermediate and the other light, corresponding to DNA that included none of the heavy isotope. These results indicate that after one round of replication, each daughter DNA duplex possessed one of the labeled heavy strands of the parent molecule. When this hybrid duplex replicated, it contributed one heavy strand to form another hybrid duplex and one light strand to form a light duplex. Meselson and Stahl's experiment thus clearly confirmed the prediction of the Watson-Crick model that DNA replicates in a semiconservative manner.

> **The basis for the great accuracy of DNA replication is complementarity. A DNA molecule is a duplex, containing two strands that are complementary mirror images of each other, so either one can be used as a template to reconstruct the other.**

How DNA Unzips to Copy Itself

When a DNA molecule replicates, the double-stranded DNA molecule separates at one end, forming a **replication fork,** and each separated strand serves as a template for the synthesis of a new complementary strand. Indeed, electron micrographs reveal Y-shaped DNA molecules at the point of replication, just as the model predicts.

To the surprise of those investigating the way in which DNA molecules replicate, it turned out that the two new daughter strands are synthesized on their templates in different ways. One strand is built by simply adding nucleotides to its growing end. This strand grows inward toward the junction of the Y as the duplex unzips. Because this strand ends with an —OH group attached to the third carbon of the ribose sugar (called the "3-prime," or 3′ carbon), the strand is said to grow from its 3′ end. The enzyme that catalyzes this process is called a **DNA polymerase.**

However, when investigators searched for a corresponding enzyme that added nucleotides to the other strand (which ends with a —PO$_4$ group attached to the fifth carbon of the ribose sugar and is called the 5′ end), they were unable to find one. Nor has anyone ever found one. DNA polymerases add only to the 3′ ends of DNA strands.

How does the polymerase build the 5′ strand? Along this strand, the chain is also formed in the 3′ direction, the polymerase jumping ahead and filling in backward (Figure 14-18). The DNA polymerase starts a burst of synthesis at the point of the replication fork and moves outward, adding nucleotides to the 3′ end of a short new chain* until this new segment fills in a gap of 1000 to 2000 nucleotides between the replication fork and the end of the growing chain to which the previous segment was added. The short new chain is then added to the growing chain, and the polymerase jumps ahead again to fill in another gap. In effect, it copies the template strand in segments about 1000 nucleotides long and stitches each new fragment to the end of the growing chain. This mode of replication is referred to as **discontinuous synthesis.** If one looks carefully at electron micrographs showing DNA replication in

---

*The fragment is called an Okazaki fragment after the discoverer, who performed a "pulse-chase" experiment: a brief addition of radioactive nucleotides appears first in 1000 to 2000 nucleotide DNA fragments. Sampled a little later, such fragments no longer exhibit radioactivity; instead, the radioactive nucleotides are in the main DNA molecule. Okazaki concluded that the 5′ strand of DNA is built by first making short fragments and then stitching them to the growing end of the DNA chain with DNA-joining enzymes called ligases.

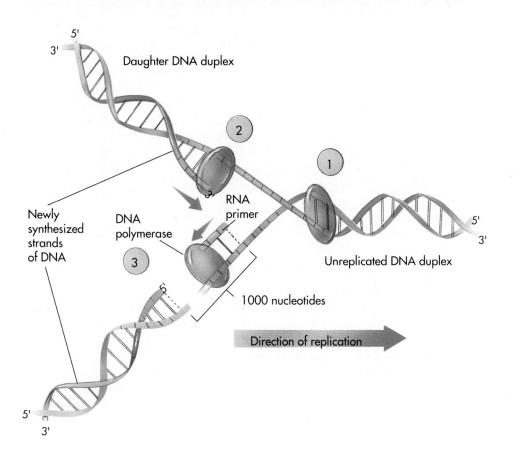

5'
3'

Daughter DNA duplex

2

3'
RNA
primer

Newly
synthesized
strands
of DNA

DNA
polymerase

3

5'

1000 nucleotides

1

Unreplicated DNA duplex

5'
3'

Direction of replication

5'
3'

FIGURE 14-18
A DNA replication fork. Replication occurs in three stages:
1 *Unwinding.* Special proteins separate and stabilize the strands of the double helix.
2 *Continuous synthesis.* Nucleotides are added by DNA polymerase to the end of the "leading strand" (signified by 3′).
3 *Discontinuous synthesis.* A short RNA primer is added 1000 nucleotides ahead of the end of the "lagging" strand (signified by 5′). DNA polymerase then adds nucleotides to the primer until the gap is filled in.

progress, one of the daughter strands behind the polymerase appears single-stranded for about 1000 nucleotides.

> A DNA molecule copies (replicates) itself by separating its two strands and using each as a template to assemble a new complementary strand, thus forming two daughter duplexes. The two strands are assembled in different directions.

How Bacteria Replicate Their DNA

The genetic material of bacteria is organized as a single, circular molecule of DNA. Such a structure cannot be compared directly with the complex chromosomes that are characteristic of eukaryotes. The replication of the DNA in bacteria is a complex process involving many enzymes; it is not, however, difficult to visualize the overall process. The entire genome can be replicated simply by nicking the duplex at one site and displacing the strand on one or both sides of the nick, creating one or two replication forks. These replication forks then proceed around the circle (Figure 14-19), creating a new daughter duplex as they go. The mitochondria and chloroplasts of eukaryotic cells contain similar circular molecules of DNA, which replicate in the same way as they do in the bacteria from which they evolved.

FIGURE 14-19

How a bacterial chromosome replicates. The circular chromosome of a bacterium initiates replication at a single site, moving out in both directions. When the two moving replication forks meet on the far side of the chromosome, the circular chromosome has been duplicated.

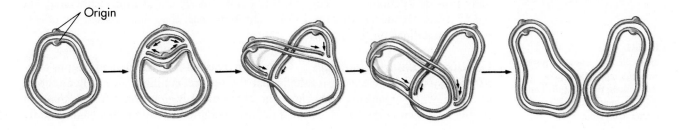

Origin

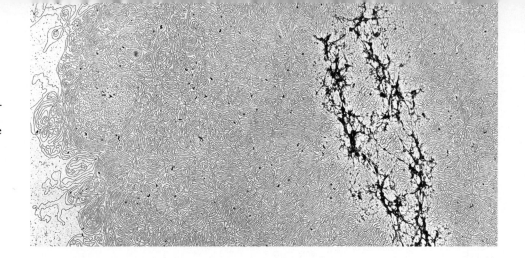

FIGURE 14-20
A human chromosome contains an enormous amount of DNA. The dark element at the right in the photograph is the protein matrix of a single chromosome. All of the surrounding material is the DNA of that chromosome.

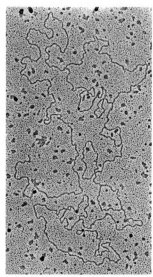

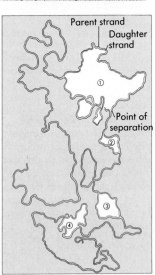

Parent strand
Daughter strand

①

Point of separation

②

③

④

FIGURE 14-21
Eukaryotic chromosomes possess numerous replication forks spaced along their length. Four replication units are visible in this electron micrograph.

How a Eukaryote Replicates its Chromosomes

The DNA within a eukaryotic chromosome is not circular (Figure 14-20), and, when examined under the electron microscope (Figure 14-21), it proves to have numerous replication forks spaced along the chromosome, rather than a single replication fork as prokaryotic chromosomes do. Each individual zone of the chromosome replicates as a discrete unit, called a **replication unit.** Replication units vary in length from 10,000 to 1 million base pairs; most are about 100,000 base pairs in length. They have been described for many different eukaryotes. Because each chromosome of a eukaryote possesses so much DNA, the orderly replication of DNA in eukaryotes undoubtedly requires sophisticated controls, which are as yet largely unknown.

GENES: THE UNITS OF HEREDITARY INFORMATION

In 1902 a British physician, Archibald Garrod, who was working with one of the early Mendelian geneticists, his countryman William Bateson, noted that certain diseases among his patients were prevalent in particular families. Indeed, if one examined several generations within such families, some of these disorders seemed to behave as if they were controlled by simple recessive alleles. Garrod concluded that these disorders were Mendelian traits and that they had resulted from changes in the hereditary information that had occurred in the past to an ancestor of the affected families.

Garrod examined several of these disorders in detail. In one, alkaptonuria, the patients passed urine that rapidly turned black on exposure to air. Such urine contained homogentisic acid (alkapton), which air oxidized. In normal individuals homogentisic acid is broken down into simpler substances, but the affected patients were unable to carry out that breakdown. With considerable insight, Garrod concluded that the patients suffering from alkaptonuria lacked the enzyme necessary to catalyze this breakdown and, more generally, that many inherited disorders might reflect enzyme deficiencies.

The One Gene–One Enzyme Hypothesis

From Garrod's finding it is but a short leap of intuition to surmise that the information encoded within the DNA of chromosomes is used to specify particular enzymes. This point was not actually established, however, until 1941, when a series of experiments by the Stanford University geneticists George Beadle and Edward Tatum finally provided definitive evidence on this point. Beadle and Tatum deliberately set out to create Mendelian mutations in the chromosomes; they then studied the effects of these mutations on the organism.

Creating Genetic Differences. One of the reasons that Beadle and Tatum's experiments produced clear-cut results is that the researchers made an excellent choice of experimental organism. They chose the bread mold *Neurospora,* a fungus that can readily be grown in the laboratory on a defined medium (a medium that contains only known substances such as glucose and sodium chloride, rather than some uncharacterized cell extract such as ground-up yeasts). Beadle and Tatum induced mutations by exposing *Neurospora* spores to x-rays. They then allowed the progeny to grow on complete medium (a medium that contained all necessary metabolites). In this way the investigators were able to keep alive strains that, as a result of the earlier irradiation, had experienced damage to their DNA in a region encoding the ability to make one or more of the compounds that the fungus needed for normal growth. Change of this kind in the DNA is called **mutation,** and strains or organisms that have undergone such change (in this case losing the ability to use one or more compounds) are called **mutants.**

Identifying Mutant Strains. The next step was to test the progeny of the irradiated spores to see if any mutations leading to metabolic deficiency actually had been created by the x-ray treatment. Beadle and Tatum did this by attempting to grow subdivisions of individual fungal strains on minimal medium, which contained only sugar, ammonia, salts, a few vitamins, and water. A cell that had lost the ability to make a necessary metabolite would not grow on such a medium. Using this approach, Beadle and Tatum succeeded in identifying and isolating many deficient mutants.

Pinpointing the Problem. To determine the nature of each deficiency, Beadle and Tatum tried adding various chemicals to the minimal medium in an attempt to find one that would make it possible for a given strain to grow (Figure 14-22). In this way they were able to pinpoint the nature of the biochemical problems that many of their mutants had. Many of the mutants proved unable to synthesize a particular vitamin or amino acid. The addition of arginine, for example, permitted the growth of a group of mutant strains, dubbed *arg* mutants. When the chromosomal position of

FIGURE 14-22
Beadle and Tatum's procedure for isolating nutritional mutations in *Neurospora*. This fungus grows easily on an artificial medium in test tubes. In this experiment spores were first irradiated to increase the frequency of mutation and then were placed on complete medium and allowed to grow. Any mutation that might have occurred in genes that were normally used by the fungus to produce its necessary amino acids or vitamins would not prevent growth, since all of these substances were present in the complete medium. Once the colonies were established, individual spores were taken and tested to see whether they would grow on minimal medium, which lacks the amino acids and vitamins that the fungus normally manufactures. Any strains that will not grow on minimal medium but will grow on complete medium contain one or more mutations in the genes that are necessary to produce one of the substances in the complete but not the minimal medium. To find out which one, the line is tested for its ability to grow on minimal medium supplemented with particular substances. The mutation illustrated here is an arginine mutant, a cell line that has lost the ability to produce arginine. It will not grow on minimal medium but will grow on minimal medium to which only arginine has been added.

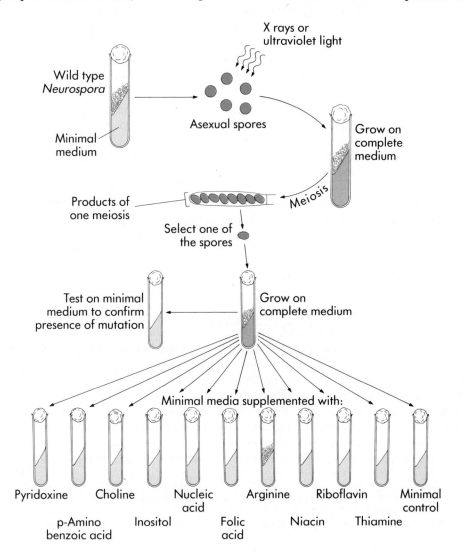

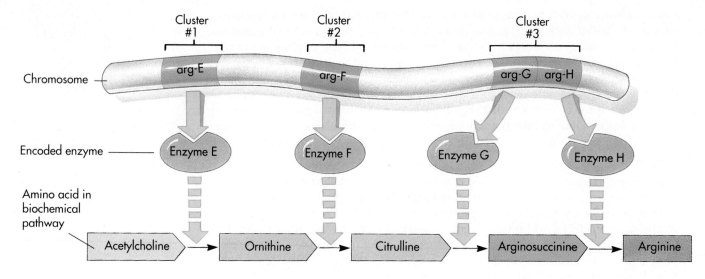

Cluster #1   Cluster #2   Cluster #3

Chromosome — arg-E   arg-F   arg-G arg-H

Encoded enzyme — Enzyme E   Enzyme F   Enzyme G   Enzyme H

Amino acid in biochemical pathway

Acetylcholine → Ornithine → Citrulline → Arginosuccinine → Arginine

FIGURE 14-23
Evidence for the "one gene–one enzyme" hypothesis. The chromosomal locations of the many arginine mutations isolated by Beadle and Tatum cluster around three locations, corresponding to the locations of the genes encoding the enzymes that carry out arginine biosynthesis.

each mutant *arg* gene was located, they were found to cluster in three areas (Figure 14-23).

For each enzyme in the arginine biosynthetic pathway, Beadle and Tatum were able to isolate a mutant strain with a defective form of that enzyme, and the mutation always proved to be located at *one* of a few specific chromosomal sites, a different site for each enzyme. Thus each of the mutants that Beadle and Tatum examined could be explained in terms of a defect in one (and only one) enzyme, which could be localized at a single site on one chromosome. The geneticists concluded that genes produce their effects by specifying the structure of enzymes, and that each gene encodes the structure of a single enzyme. They called this relationship the **one gene–one enzyme hypothesis.★**

Enzymes are responsible for catalyzing the synthesis of all the parts of the cell. They mediate the assembly of nucleic acids, the synthesis of proteins, carbohydrates, fats, and lipids. From the hair on your head to the toenails of your feet, all of you represents the product of enzyme-directed chemical reactions. By specifying your enzymes, DNA specifies you.

> **Genetic traits are expressed largely as a result of the activities of enzymes. Organisms store hereditary information by encoding the structures of enzymes in the DNA of their chromosomes.**

How DNA Encodes Proteins

What kind of information must a gene encode to specify a protein? For some time, the answer was not clear, since protein structure seemed to be impossibly complex. It was not evident, for example, whether or not a particular kind of protein had a consistent sequence of amino acids that would be the same in one individual molecule as it would be in another of that same kind of protein. The picture changed in 1953, the same year in which Watson and Crick unraveled the structure of DNA. The great English biochemist Frederick Sanger, after many years of work, announced the complete sequence of amino acids in the protein insulin. Sanger's achievement was extremely significant because it demonstrated for the first time that proteins consisted of definable sequences of amino acids. For any given form of insulin, each molecule has the same amino acid sequence as every other, and this sequence can be learned and

★Their idea has since been modified somewhat, as you will see in later chapters. We now know that many proteins are composed of several kinds of polypeptide chains, each specified by a separate gene. A modern restatement of the Beadle and Tatum proposal would be that one gene specifies one polypeptide.

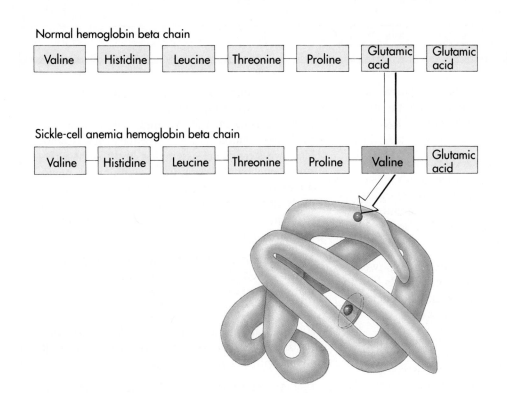

Normal hemoglobin beta chain

| Valine | Histidine | Leucine | Threonine | Proline | Glutamic acid | Glutamic acid |

Sickle-cell anemia hemoglobin beta chain

| Valine | Histidine | Leucine | Threonine | Proline | Valine | Glutamic acid |

FIGURE 14-24

The molecular basis of a hereditary disease. Sickle cell anemia is produced by a recessive allele of the gene encoding the beta chain of hemoglobin. It represents a single amino acid change from glutamic acid to valine at the sixth position in the chain.

written down. All enzymes and other proteins are strings of amino acids arranged in a certain definite order. The information necessary to specify an enzyme, therefore, is an ordered list of amino acids.

In 1956 Sanger's pioneering work was followed by Vernon Ingram's analysis of the molecular basis of sickle cell anemia, a protein defect inherited as a Mendelian disorder. By analyzing the structure of normal and sickle-cell hemoglobin (Figure 14-24), Ingram, working at Cambridge University, showed that sickle cell anemia was caused by the change of the amino acid at a single position in the protein from glutamic acid to valine. The alleles of the gene encoding hemoglobin differed only in their specification of this one amino acid in the hemoglobin amino acid chain.

These experiments, and other related ones, have finally brought us to a clear understanding of what the unit of heredity is. It is the information that encodes the amino acid sequence of an enzyme or other protein. The sequence of nucleotides that encodes this information is called a **gene.** Although most genes encode proteins, there are also genes devoted to the production of special forms of RNA, many of which play important roles in protein synthesis.

> **The amino acid sequence of a particular protein is specified by a corresponding sequence of nucleotides in the DNA. This nucleotide sequence is called a gene.**

In the next four chapters we will focus on the molecular nature of genes, exploring how they work and what happens when one is changed. As we will see, cancer is one example of what can happen when a gene is altered. Other deliberate changes induced by genetic engineers have proved very beneficial; most of the insulin used today by diabetics, for example, is the product of a human gene introduced into bacterial cells. Our understanding of genes as the units of heredity, the foundations of which you have encountered in this chapter, represents one of the high-water marks of biology as a science. The intellectual path that biologists have followed in their pursuit of this understanding has not always been a straight one, the best questions not always obvious. But however erratic and lurching the experimental journey, our picture of heredity has become progressively clearer, the image more sharply defined.

1. Eukaryotic cells store hereditary information within the nucleus. When one transplants the nucleus, one also transplants the hereditary specifications of an organism.

2. In viruses, bacteria, and eukaryotes, the hereditary information resides in nucleic acids. The transfer of pure nucleic acid can lead to the transfer of hereditary traits. In all cellular organisms the genetic material is DNA, although in some viruses the genetic material is RNA.

3. DNA has the structure of a double helix, two chains held to each other by hydrogen bonds between nucleotides on different chains. The nucleotide *A* bonds with *T,* and *G* bonds with *C.*

4. During cell division, the hereditary message is duplicated with great accuracy. The mechanism that achieves this high degree of accuracy is complementarity: because the DNA molecule is organized as a double helix with two strands that are complementary mirror images of one another, either one of the strands can be used to re-form the helix by adding complementary bases.

5. DNA is replicated by a battery of enzymes, which include a DNA polymerase and a variety of proteins that unzip the DNA in front of the polymerase. DNA is replicated by unwinding the duplex and using each of the single strands as a template to assemble a complementary new strand. The two strands are assembled in different directions, one by the continuous addition of nucleotides to the growing end, the other discontinuously in the opposite direction in 1000-nucleotide segments, which are then joined to the end of the growing strand.

6. Most hereditary traits reflect the actions of enzymes. The traits are hereditary because the information necessary to specify these enzymes is encoded within the DNA.

7. Enzymes are encoded within DNA by specifying the amino acid sequence of individual polypeptide chains with a sequence of nucleotides.

8. Changes in individual nucleotides can alter the identity of a particular amino acid in a protein. When such an alteration in DNA, called a mutation, results in the production of a protein with altered biological activity, or no activity at all, the mutation may alter the phenotype.

REVIEW QUESTIONS

1. Of the four sets of mice in Griffith's *Pneumococcus* bacteria experiment, which survived and which did not? What did Griffith conclude from these data?

2. What did Avery determine was the primary component of the transforming principle? What was his evidence? Was his conclusion accepted by other scientists of that time?

3. Are there equal amounts of all four nucleotides in a molecule of DNA? What does this indicate? What proportion of nucleotides did Chargraff's research indicate?

4. Based on Franklin's x-ray diffraction photographs, what did Watson and Crick deduce was the three-dimensional shape of DNA? What was their explanation for maintaining the diameter of the helix at 2 nanometers throughout the length of the strand? How did this fit with Chargraff's research on the proportions of purines and pyrimidines in DNA?

5. What is the importance of Watson and Crick's explanation of DNA complementarity? What term is given to this form of replication?

6. Are the two strands of a DNA molecule replicated in the same way? Explain.

7. What enzyme catalyzes DNA replication from the 3' to 5' direction? What enzyme catalyzes DNA replication from the 5' to 3' direction? What name is given to the type of replication shown by the growing 3' strand? What name is given to the type of replication shown by the growing 5' strand?

8. How does eukaryotic DNA replication differ from bacterial replication?

9. What intuitive hypothesis did Garrod propose concerning inheritance of certain genetic disorders? What organism did Beadle and Tatum use to support Garrod's hypothesis? What did they do to change the organism?

10. Why did Beadle and Tatum initially grow their fungi on a complete medium? Why did they grow the progeny of the irradiated spores on minimal medium?

THOUGHT QUESTIONS

1. In Hammerling's experiments with *Acetabularia*, the grafting of a nucleus-containing basal portion of an *A. mediterranea* individual (with a disk-shaped cap) to the stalk of an *A. crenulata* individual (with a flower-shaped cap) resulted in the formation of a cap above the grafted stalk. The cap that formed was flower-shaped, like that of *A. crenulata*. When that cap was removed, another formed, only this one was disk-shaped, like that of *A. mediterranea*. What is responsible for the change in morphology of these two regenerated caps?

2. From an extract of human cells growing in tissue culture, you obtain a white fibrous substance. How would you distinguish whether it was DNA, RNA, or protein?

3. In analyzing DNA obtained from your own cells, you are able to determine that 15% of the nucleotide bases that it contains are thymine. What percentage of the bases are cytosine?

4. From a hospital patient afflicted with a mysterious illness, you isolate and culture cells, and from the culture purify DNA. You find that the DNA sample obtained from the culture contains two quite different kinds of DNA, one double-stranded human DNA and the other single-stranded virus DNA. You analyze the base composition of the two purified DNA preparations, with the following results:

Tube 1:22.1%A:27.9%C:27.9%G:22.1%T
Tube 2::31.3%A:31.3%C:18.7%G:18.7%T

Which of the two tubes contains single-stranded virus DNA?

5. The human genome contains more than 3 billion (3×10^9) nucleotide base pairs; each nucleotide of the sugar phosphate backbone of DNA takes up about 0.34 nanometer (1000 nanometers = 1 micrometer; 1000 micrometers = 1 millimeter; 10 millimeters = 1 centimeter). If a human genome were fully extended, how long would it be?

FOR FURTHER READING

CRICK, F.H.C.: "The Discovery of the Double Helix Was a Matter of Selecting the Right Problem and Sticking to It," *The Chronicle of Higher Education*, October 5, 1988. Francis Crick's own recollections of the hectic days when he and James Watson deduced that the structure of DNA is a double helix.

FELSENFELD, G.: "DNA," *Scientific American*, October 1985, pages 58-67. A good overview of how the shape of DNA determines its activity.

HALL, S.S.: "James Watson and the Search for Biology's 'Holy Grail'," *Smithsonian*, February 1990, pages 41-49. The effort to provide a complete sequence of base pairs for the human genome is now underway.

JUDSON, H.F.: *The Eighth Day of Creation*, Simon and Schuster, New York, 1979. The definitive historical account of the experimental unraveling of the mechanisms of heredity, based on personal interviews with the participants. This book is full of the feel of how science is really conducted.

KORNBERG, R.D., and A. KLUG: "The Nucleosome," *Scientific American*, February 1981, pages 52-64. This article describes how the DNA helix is wound on a series of protein spools.

MULLIS, K.: "The Unusual Origin of the Polymerase Chain Reaction," *Scientific American*, April 1990, pages 56-65. A charming account of how a young molecular biologist thought up a key procedure for making unlimited copies of DNA fragments.

RADMAN, M., and R. WAGNER: "The High Fidelity of DNA Duplication," *Scientific American*, August 1988, pages 40-46. A brief account of how "proofreading" ensures that few mistakes are made when DNA is replicated.

SCHLIEF, R.: "DNA Binding by Proteins," *Science*, vol. 241, pages 1182-1187, 1988. An account of the different ways in which proteins bind to the DNA double helix.

WATSON, J.D.: *The Double Helix*, Athenaeum Publishing Company, Inc., New York, 1968. A lively, often irreverent account of what it was like to discover the structure of DNA, recounted by someone in a position to know.

WATSON, J.D., and F.H.C. CRICK: "A Structure for Deoxyribose Nucleic Acid," *Nature*, vol. 171, page 737, 1953. The original report of the double helical structure of DNA. Only one page long, this paper marks the birth of molecular genetics.

WATSON, J., AND OTHERS: *Molecular Biology of the Gene*, ed. 4, Benjamin/Cummings Publishing Co., Menlo Park, California, 1987. The latest edition of a classic text on DNA structure and function.

Genes and How They Work

OVERVIEW

The discovery that Mendel's genes were composed of DNA within chromosomes and that genes acted by directing the production of specific proteins still left unanswered the question of how this is accomplished. We now know that special proteins bind to the DNA at the beginning of genes and then move along them, making an RNA copy of the gene as they go. It is this RNA copy that cells use to produce the polypeptides specified by the genetic information. Cells often control when a gene is "turned on" by controlling production of these RNA copies.

FOR REVIEW

Here are some important terms and concepts that you will encounter in this chapter. If you are not familiar with them, you should review them before proceeding.

Structure of DNA and RNA (Chapters 3 and 14)

Ribosomes (Chapter 5)

Enzymes and enzyme activity (Chapter 7)

Structure of eukaryotic chromosomes (Chapter 10)

Every cell in your body contains more information than is stored in this book. This information is the hereditary instructions that specify that you will have arms and not fins, hair and not feathers, two eyes and not one. The color of your eyes, the texture of your fingernails, whether you dream in color—all of the many traits that you receive from your parents are recorded in the cells of your body. As we have seen, biologists learned by experiment that long DNA molecules, which in eukaryotes complex with proteins to form chromosomes, contain this information. The information itself is arrayed in little blocks like entries in a dictionary, each block a gene specifying a particular polypeptide. Some of these polypeptides are entire proteins, while many other proteins are formed of two or more gene products. Proteins are the tools of heredity. Many of them are enzymes that carry out reactions within cells: what you are is a result of what they do. The essence of heredity is the ability of a cell to use information in its DNA to bring about the production of particular polypeptides, and so affect what that cell will be like. In this chapter we will examine how this happens.

CELLS USE RNA TO MAKE PROTEIN

To find out how a cell uses its DNA to direct the production of particular proteins, perhaps the simplest question you might ask is "Where in the cell are proteins made?" You can answer this question by placing cells for a short time in a medium containing radioactive amino acids; the cells will take up the radioactively labeled amino acids for the short time that they are exposed to them. This is known as pulse labeling as the cells are exposed to a "pulse" of radioactive label. When investigators looked to see where in the cells radioactive proteins first appeared, they found that proteins were assembled not in the nucleus, where the DNA is, but rather in the cytoplasm, on large protein aggregates called **ribosomes** (Figure 15-1). These little polypeptide-making factories proved to be very complex, containing over 50 different proteins. They also contain a different sort of molecule, RNA. As you will recall from Chapter 3, RNA is similar to DNA (Figure 15-2), and its presence in ribosomes hints that RNA molecules play an important role in polypeptide synthesis.

A cell contains many kinds of RNA. There are three major classes:

Ribosomal RNA. The class of RNA found in ribosomes is called **ribosomal RNA,** or **rRNA.** During polypeptide synthesis, rRNA molecules provide the site on the ribosome where the polypeptide is assembled.

Transfer RNA. A second class of RNA, called **transfer RNA,** or **tRNA,** is much smaller. Human cells contain more than 40 different kinds of tRNA molecules, which float free in the cytoplasm. During polypeptide synthesis, tRNA molecules

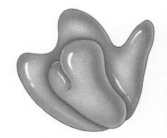

Ribosome

FIGURE 15-1

A ribosome is composed of two subunits. The smaller subunit fits into a depression on the surface of the larger one.

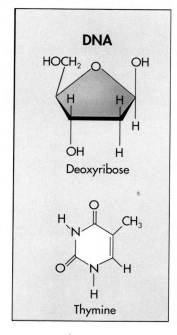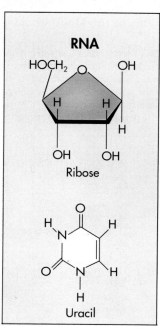

FIGURE 15-2

How RNA is different from DNA. There are two important differences: first, in place of the sugar deoxyribose, RNA contains ribose, which has an additional oxygen atom. Second, RNA contains the pyrimidine uracil (U) instead of thymine (T). Another difference is that RNA does not have a regular helical structure and is usually single-stranded.

transport the amino acids to the ribosome for use in building the polypeptide, and position each amino acid at the correct place on the elongating polypeptide chain.

Messenger RNA. A third class of RNA is **messenger RNA,** or **mRNA.** Each mRNA molecule is a long, single strand of RNA that passes from the nucleus to the cytoplasm. During polypeptide synthesis, mRNA molecules bring information from the chromosomes to the ribosomes to direct the assembly of amino acids into a polypeptide.

These molecules, together with ribosomal proteins and certain enzymes, constitute a system that carries out the task of reading the genetic message and producing the polypeptide that the particular message specifies. They are the principal components of the apparatus that a cell uses to translate its hereditary information. You can think of this information as a message written in the code specified by the sequence of nucleotides in the DNA. The cell's polypeptide-producing apparatus reads this message one gene after another, translating the genetic code of each gene into a particular polypeptide. As we will see, biologists have also learned to read this code, and in so doing have learned a great deal about what genes are and how they work in dictating what a protein will be like and when it will be made.

AN OVERVIEW OF GENE EXPRESSION

The hereditary apparatus of your body works in much the same way as that of the most primitive bacteria—all organisms use the same basic mechanism. An RNA copy of each active gene is made, and at a ribosome the RNA copy directs the sequential assembly of a chain of amino acids. There are many minor differences in the details of gene expression between bacteria and eukaryotes, and a single major difference that we will discuss later in this chapter. The basic apparatus used in gene expression, however, appears to be the same in all organisms (Figure 15-3); it apparently has persisted virtually unchanged since early in the history of life. The process of gene expression occurs in two phases, which are called transcription and translation.

Transcription

The first stage of gene expression is the production of an RNA copy of the gene, called messenger RNA or mRNA (Figure 15-4). Like all classes of RNA that occur in cells, mRNA is formed on a DNA template. The production of RNA is called **transcription**; the messenger RNA molecule (as well as other RNA classes such as rRNA and tRNA) is said to have been transcribed from the DNA. Transcription is initiated when a special enzyme, called an **RNA polymerase,** binds to a particular sequence of nucleotides on one of the DNA strands,* a sequence located at the edge of a gene.

---

*How does the RNA polymerase know which strand is the "sense" strand? Because it searches for a particular nucleotide sequence called a promoter site. Analysis of over 100 such sites shows that two 6-nucleotide sequences occur, with minor variations, in all promoter sites: TTGACA and TATAAT. It is to these two sequences on the sense strand that the RNA polymerase binds. RNA polymerase does not bind the other strand because that strand contains the complementary sequences AACTGT and ATATTA, sequences which RNA polymerase does not recognize.

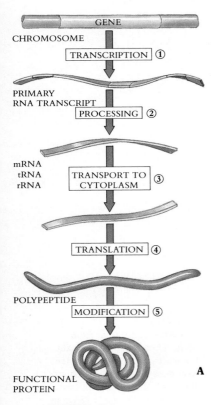

FIGURE 15-3

Gene expression.

A The 5 stages of gene expression.

B The central dogma.

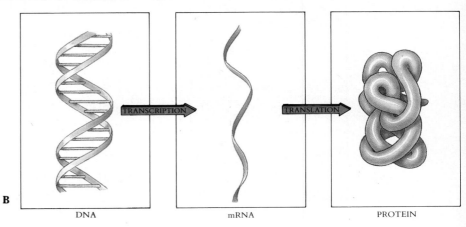

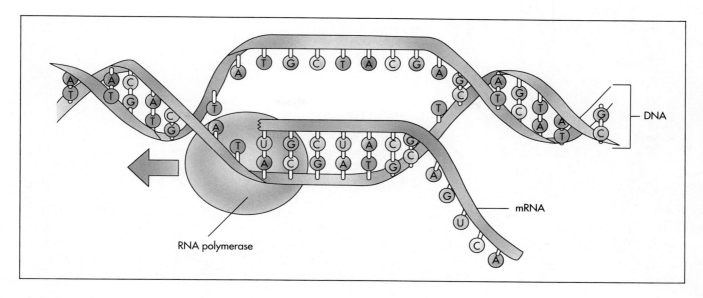

RNA polymerase

DNA

mRNA

Starting at that end of the gene, the RNA polymerase proceeds to assemble a single strand of mRNA with a nucleotide sequence complementary to that of the DNA strand it has bound. **Complementarity** refers to the way in which the two single strands of DNA that form a double helix relate to one another, with A (adenine) pairing with T (thymine) and G (guanine) pairing with C (cytosine). An RNA strand complementary to a DNA strand has the same relationships, but with U (uracil) in place of thymine.

As the RNA polymerase moves along the strand into the gene, encountering each DNA nucleotide in turn, it adds the corresponding complementary RNA nucleotide to the growing RNA strand. When the enzyme arrives at a special "stop" signal at the far edge of the gene, it disengages from the DNA and releases the newly assembled RNA chain. This chain is complementary to the DNA strand from which the polymerase assembled it; thus, it is an RNA transcript (copy), called the **primary RNA transcript,** of the DNA nucleotide sequence of the gene.

Translation

The second stage of gene expression is the synthesis of a polypeptide by ribosomes, which use the information contained in an mRNA molecule to direct the choice of amino acids. This process of mRNA-directed polypeptide synthesis by ribosomes is called **translation** because nucleotide-sequence information is translated into amino acid–sequence information. Translation begins when an rRNA molecule within the ribosome binds to one end of an mRNA transcript. Once it has bound to the mRNA molecule, a ribosome proceeds to move along the mRNA molecule in increments of three nucleotides. At each step, it adds an amino acid to a growing polypeptide chain. It continues to do this until it encounters a "stop" signal that indicates the end of the polypeptide. It then disengages from the mRNA and releases the newly assembled polypeptide. An overview of protein synthesis is presented in Figure 15-5.

> The information encoded in genes is expressed in two stages: transcription, in which an RNA polymerase enzyme assembles an mRNA molecule whose nucleotide sequence is complementary to the gene's template DNA strand; and translation, in which a ribosome assembles a polypeptide, using the mRNA to specify the amino acids.

HOW ARE GENES ENCODED?

The essential question of gene expression is: How does the *order* of nucleotides in a DNA molecule encode the information that specifies the order of amino acids in a

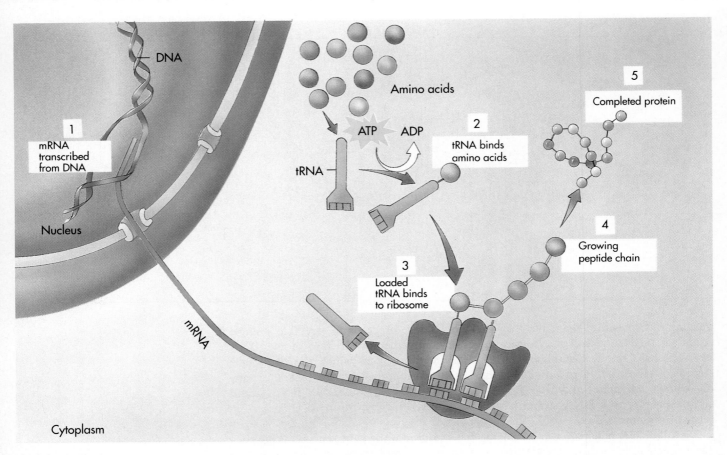

FIGURE 15-5
An overview of protein synthesis.

protein? What, in other words, is the nature of the genetic code? The answer came in 1961, as the result of an experiment led by Francis Crick. Crick's experiment is so elegant and the result so critical to our understanding of genes that we will describe it in detail.

Crick and his colleagues reasoned that the "genetic code" most likely consisted of a series of blocks of information, each block corresponding to an amino acid in the encoded protein. They further hypothesized that the information within one block was probably a sequence of three nucleotides specifying a particular amino acid. They arrived at the number three because a two-nucleotide block would not yield enough different combinations to code for the 20 different kinds of amino acids that commonly occur in proteins. Imagine that you have some fruit: apples, oranges, pears, and grapefruits. How many different pairs of fruit can you construct? Only 4^2, or 16 different pairs. How many groups of three can you construct? A lot more—4^3, or 64 different combinations, more than enough to code for the 20 amino acids.

But are these three-nucleotide blocks, called **codons,** read in simple sequence, one after the other? Or is there punctuation (a silent nucleotide) between each of the three-nucleotide blocks? The sentences of this book, for example, use spaces to punctuate between words. Without the spaces, this sentence would read "withoutthespacesthissentencewouldread." In principle, either way of reading will work. However, it is important to know which method is employed by cells, since the two ways of reading DNA imply different translating mechanisms. To choose between these two alternative hypotheses, Crick and his colleagues used a chemical that adds or deletes a single nucleotide to a DNA strand. Such an addition or deletion **changes the reading frame** of a genetic message, whether or not it is punctuated, but the two alternative hypotheses differ in what is required to restore the correct reading frame register (Figure 15-6).

From the examples in Figure 15-6, you can see that the deletion (or addition) of three nucleotides restores a three-digit unpunctuated code to the correct reading frame but does not restore a punctuated code. What Crick and his coworkers did was

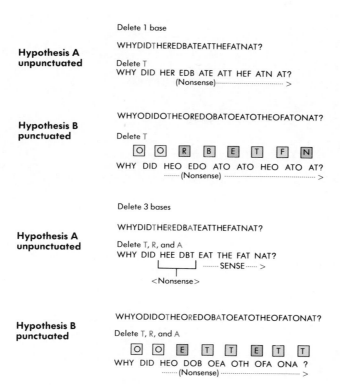

Delete 1 base

Hypothesis A
unpunctuated

WHYDIDTHEREDBATEATTHEFATNAT?

Delete T
WHY DID HER EDB ATE ATT HEF ATN AT?
          (Nonsense)⋯⋯⋯⋯⋯⋯⋯ >

Hypothesis B
punctuated

WHYODIDOTHEOREDOBATOEATOTHEOFATONAT?

Delete T
☐O☐ ☐O☐ ☐R☐ ☐B☐ ☐E☐ ☐T☐ ☐F☐ ☐N☐

WHY DID HEO EDO ATO ATO HEO ATO AT?
      ⋯⋯ (Nonsense) ⋯⋯⋯⋯⋯⋯⋯⋯ >

Delete 3 bases

Hypothesis A
unpunctuated

WHYDIDTHEREDBATEATTHEFATNAT?

Delete T, R, and A
WHY DID HEE DBT EAT THE FAT NAT?
            └⎯⎯⎯⎯┘ ⋯⋯ SENSE ⋯⋯ >
        <Nonsense>

Hypothesis B
punctuated

WHYODIDOTHEOREDOBATOEATOTHEOFATONAT?

Delete T, R, and A
☐O☐ ☐O☐ ☐E☐ ☐T☐ ☐T☐ ☐E☐ ☐T☐ ☐T☐

WHY DID HEO DOB OEA OTH OFA ONA ?
      ⋯⋯ (Nonsense) ⋯⋯⋯⋯⋯⋯⋯⋯ >

FIGURE 15-6
Using the correction of frame-shift alterations of DNA to determine if the genetic code is punctuated. The hypothetical genetic message presented here is "Why did the red bat eat the fat nat?" Under hypothesis B, that the message is punctuated, the three-letter words are separated by nucleotides that are not read (here indicated by the letter "O").

to use genetic recombination to put three deletions together near one another on a virus DNA and then look to see if the genes downstream were read correctly or as nonsense. When one or two deletions were placed at the beginning of the region, the downstream gene was translated as nonsense, but three deletions (or three additions) restored the correct reading frame so that sequences downstream were translated correctly. Thus the genetic code was read in increments consisting of three nucleotides, and reading occurs without punctuation between the three-nucleotide units.

> Within genes that encode proteins, the nucleotide sequence of DNA is read in increments of three consecutive nucleotides, without punctuation between increments. Each block of three nucleotides codes for one amino acid.

Just what the code words are was soon worked out, the first results coming within a year of Crick's experiment. Researchers had developed mixtures of RNA and protein isolated from ruptured cells ("cell-free systems") that would synthesize proteins in a test tube. To determine which three-nucleotide sequences specify which amino acids, researchers added artificial RNA molecules to these cell-free systems and then looked to see what proteins were made. For example, when Marshall Nierenberg, of the National Institutes of Health, added poly-U (an RNA molecule consisting of a string of uracil nucleotides) to such a system, it proceeded to synthesize polyphenylalanine (a protein consisting of a string of phenylalanine amino acids). This result indicated that the three-nucleotide sequence (or **triplet**) specifying phenylalanine was UUU. In this and other ways all 64 possible triplets were examined, and the full genetic code was determined.

THE GENETIC CODE

Working out the genetic code was a great step forward in removing the mystery from the process of gene expression. However, it leaves a great question unanswered. How is the information stored in a sequence of nucleotides like UUU used to identify a specific amino acid such as phenylalanine? How is the genetic code deciphered?

To answer this question, we must look more carefully at the first events of the

FIGURE 15-7

Translation in action. Bacteria have no nucleus and hence no membrane barrier between DNA and cytoplasm. In the electron micrograph of genes being transcribed in the common colon bacterium, *Escherichia coli*, you can see every stage of the process. The arrows point to RNA polymerase molecules. From each mRNA molecule dangling from the DNA, a series of ribosomes is assembling polypeptides, one seeming to follow the next down the mRNA. These clumps of ribosomes are sometimes called "polyribosomes."

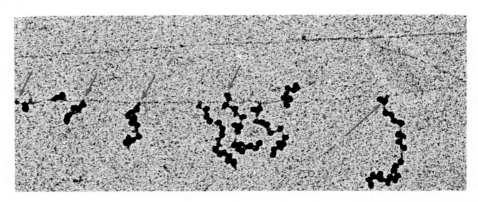

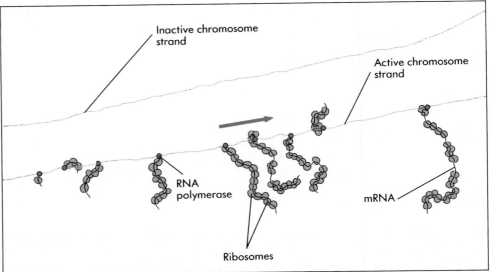

translation process. Translation occurs on the ribosomes. First, the initial portion of the mRNA transcribed from a gene binds to an rRNA molecule interwoven in the ribosome (Figure 15-7). The mRNA lies on the ribosome in such a way that only the three-nucleotide portion of the mRNA molecule—the codon—is exposed at the polypeptide-making site. As each bit of the mRNA message is exposed in turn, a molecule of tRNA with the complementary three-nucleotide sequence, or **anticodon**, binds to the mRNA (Figure 15-8). Because this tRNA molecule carries a particular amino acid, that amino acid and no other is added to the polypeptide in that position. Protein synthesis occurs as a series of tRNA molecules bind one after another to the exposed portion of the mRNA molecule as it moves through the ribosome. Each of these tRNA molecules has attached to it an amino acid, and the amino acids which they bring to the ribosome are added, one after another, to the end of a growing polypeptide chain.

How does a particular tRNA molecule come to possess the amino acid that it does, and not just any amino acid? The correct amino acid is placed on each tRNA molecule by a collection of enzymes called **activating enzymes**. There are activating enzymes for each of the 20 common amino acids. An activating enzyme binds the amino acid that it recognizes (Figure 15-9) to a tRNA molecule. If one considers the nucleotide sequence of mRNA to be a coded message, then the 20 activating enzymes are the code books of the cell—the instructions for decoding the message. An activating enzyme recognizes both nucleotide-sequence information (a specific anticodon sequence of a tRNA molecule) and protein-sequence information (a particular amino acid).

The code word recognized by an activating enzyme is three nucleotides long. Because there are four kinds of nucleotides in mRNA (cytosine, guanine, adenine, and uracil instead of thymine), there are 4^3 or 64 different three-letter code words, or codons, possible. Some of the activating enzymes recognize only one tRNA molecule,

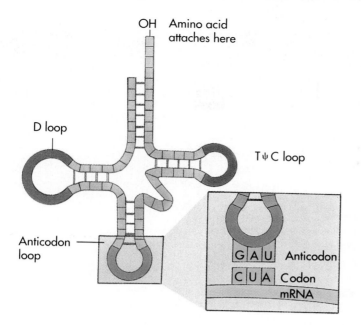

OH Amino acid
attaches here

D loop

TΨC loop

Anticodon
loop

G A U Anticodon
C U A Codon
mRNA

FIGURE 15-8
Structure of a tRNA molecule. The T, C loop, and the D loop function in binding to the ribosomes during polypeptide synthesis. The third loop contains the anticodon sequence. The activating enzyme adds an amino acid to the free single-stranded —OH end.

corresponding to one of these code words; others recognize two, three, four, or six different tRNA molecules, each containing a different anticodon. The base sequences of the tRNA anticodons are complementary to the associated sequences of mRNA and relate to the same amino acid as do those of their partner. The list of different mRNA codons specific for each of the 20 amino acids—the genetic code—is presented in Table 15-1.

The genetic code is the same in all organisms, with only a few exceptions. A particular codon such as AGA corresponds to the same amino acid (arginine) in bacteria as in humans. The universality of the genetic code is among the strongest evidence that all living things share a common evolutionary heritage.

Three of the 64 codons (UAA, UAG, and UGA), which are called **nonsense codons,** are not recognized by any activating enzyme. These codons serve as "stop" signals in the mRNA message, marking the end of a polypeptide. The "start" signal, which marks the beginning of a polypeptide amino-acid sequence within a mRNA message, is the codon AUG, a three-base sequence that also encodes the amino acid methionine. The ribosome uses the first AUG that it encounters in the mRNA message to signal the start of its translation.

The mRNA codons specific for the 20 common amino acids constitute the genetic code. All organisms possess a battery of enzymes, called activating enzymes, each of which recognizes a particular anticodon, and a particular amino acid.

FIGURE 15-9
Activating enzymes "read" the genetic code. Each kind of activating enzyme recognizes and binds a specific amino acid such as tryptophan, on the one hand, and also recognizes and binds the tRNA molecules with anticodons specifying that amino acid, such as ACC for tryptophan.

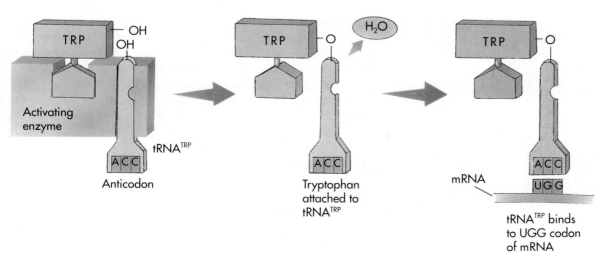

TRP — OH
OH

Activating
enzyme

A C C tRNA^TRP
Anticodon

TRP — O

H_2O

Tryptophan
attached to
tRNA^TRP

A C C

TRP — O

mRNA

A C C
U G G

tRNA^TRP binds
to UGG codon
of mRNA

TABLE 15-1	**THE GENETIC CODE**		

FIRST LETTER	SECOND LETTER				THIRD LETTER
	U	**C**	**A**	**G**	
U	Phenylalanine	Serine	Tyrosine	Cysteine	U
	Phenylalanine	Serine	Tyrosine	Cysteine	C
	Leucine	Serine	Stop	Stop	A
	Leucine	Serine	Stop	Tryptophan	G
C	Leucine	Proline	Histidine	Arginine	U
	Leucine	Proline	Histidine	Arginine	C
	Leucine	Proline	Glutamine	Arginine	A
	Leucine	Proline	Glutamine	Arginine	G
A	Isoleucine	Threonine	Asparagine	Serine	U
	Isoleucine	Threonine	Asparagine	Serine	C
	Isoleucine	Threonine	Lysine	Arginine	A
	(Start); Methionine	Threonine	Lysine	Arginine	G
G	Valine	Alanine	Aspartate	Glycine	U
	Valine	Alanine	Aspartate	Glycine	C
	Valine	Alanine	Glutamate	Glycine	A
	Valine	Alanine	Glutamate	Glycine	G

A, codon consists of three nucleotides read in the sequence shown above. For example ACU codes threonine. The first letter, A, is read in the first column; the second letter mRNA, C, from the second letter column; and the third letter, U, from the third letter column. Each of the mRNA codons is recognized by a corresponding anticodon sequence on a tRNA molecule. Some tRNA molecules recognize more than one codon sequence mRNA but always for the same amino acid. Most amino acids are encoded by more than one codon. For example, threonine is encoded by four codons (ACU, ACC, ACA, and ACG), which differ from one another only in the third position.

Not All Organisms Use the Same Genetic Code

Starting in 1979, investigators began to determine the complete nucleotide sequences of mitochondrial genomes. The sequences for humans, cattle, and mice were all reported within a short period. It came as something of a shock when these investigators learned that the genetic code that these mammalian mitochondria were using was not quite the same as the "universal code" that had by then become so familiar to biologists. Most of the code words were the same—but not all. In mammalian mitochondria, what should have been a "stop" codon, UGA, was instead read as an amino acid, tryptophan; AUA was read as methionine rather than isoleucine; and AGA and AGG were read as "stop" rather than arginine. Also, when the first chloroplasts were sequenced, they too contained minor differences from the universal code.

Thus it appears that the genetic code is not quite universal. Sometime, presumably after beginning their symbiotic existence, mitochondria and chloroplasts began to read the code differently, particularly that portion of the code associated with "stop" signals. Nor is this phenomenon limited to subcellular organelles. At some point early in the evolution of ciliates (one of the phyla of the kingdom Protista), they also changed the way in which they read chain-terminating "stop" codons. Under most conditions, any genetic change involving termination signals would be expected to be lethal to an organism. How this change could have arisen in organelles, much less in eukaryotes such as ciliates, is a puzzle to which we as yet have no answer.

THE MECHANISM OF PROTEIN SYNTHESIS

Polypeptide synthesis begins with the formation of an **initiation complex.** First, a methionine-carrying tRNA—met-tRNA—molecule binds to the small ribosomal subunit. Special proteins called **initiation factors** position the met-tRNA on the ribo-

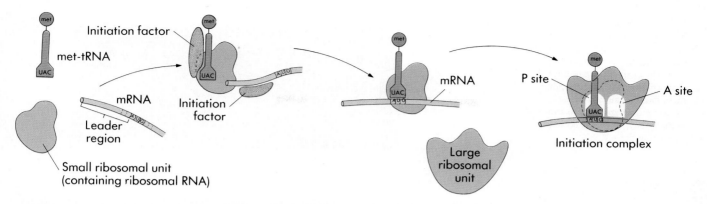

FIGURE 15-10

Formation of the initiation complex. Proteins called initiation factors play key roles in positioning the small ribosomal subunit and the met-tRNA molecule at the beginning of the mRNA message. When the met-tRNA is positioned over the first AUG codon sequence of the mRNA, the large ribosomal subunit binds, forming the A and P sites, and polypeptide synthesis begins.

somal surface (Figure 15-10). The proper positioning of this first amino acid is critical, since it determines the reading frame (the groups of three) with which the nucleotide sequence will be translated into a polypeptide. This initiation complex, guided by another initiation factor, then binds to mRNA. It is important that the complex bind to the beginning of a gene, so that all of the gene will be translated. In bacteria, the beginning of each gene is marked by a sequence that is complementary to one of the rRNA molecules on the ribosome. This ensures that genes are read from the beginning; each mRNA binds to the ribosomes that read it by base-pairing between the sequence at its beginning and the complementary sequence on the rRNA which is a part of the ribosome.

> **An initiation complex consists of a small ribosomal subunit, mRNA, and a tRNA molecule carrying methionine.**

After the initiation complex has been formed, the synthesis of the polypeptide proceeds as follows (Figure 15-11):

1. The ribosome exposes the codon on the mRNA immediately adjacent to the initiating AUG codon, positioning it for interaction with another incoming tRNA molecule. When a tRNA molecule with the appropriate anticodon appears, this new incoming tRNA briefly binds to the mRNA molecule at its exposed codon position. Special proteins called **elongation factors** (because they aid in making the polypeptide longer) help to position the incoming tRNA. Binding the incoming tRNA to the ribosome in this fashion places the amino acid at the other end of the incoming tRNA molecule directly adjacent to the initial methionine, which is dangling from the initiating tRNA molecule still bound to the ribosome.

2. The two amino acids undergo a chemical reaction, in which the initial methionine is released from its tRNA and is attached instead by a peptide bond to the adjacent incoming amino acid. The abandoned tRNA falls from its site on the ribosome, leaving that site vacant.

FIGURE 15-11

How polypeptide synthesis proceeds. The A site is occupied by the tRNA with an anticodon complementary to the mRNA codon exposed in the A site. The growing polypeptide chain (here, met) is transferred to the incoming amino acid (here, leucine), and the ribosome moves three nucleotides to the right.

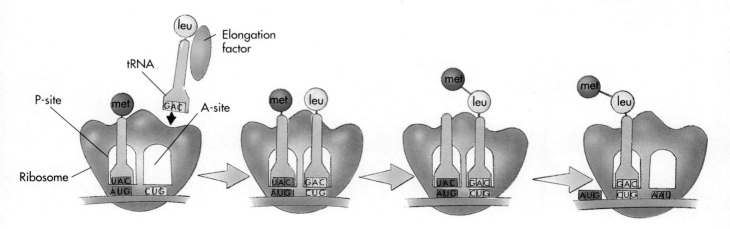

FIGURE 15-12

Translocation. These ribosomes are reading along an mRNA molecule of the fly *Chironomus tentans* from left to right, assembling polypeptides that dangle behind them like the tail of a tadpole. Clearly visible are the two subunits (*arrows*) of each ribosome translating the mRNA.

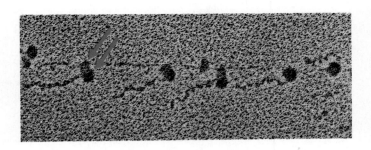

3. In a process called **translocation** (Figure 15-12), the ribosome now moves along the mRNA molecule ("translocates") a distance corresponding to three nucleotides, guided by other elongation factors. This movement repositions the growing chain, at this point containing two amino acids, and exposes the next codon of the mRNA. This is the same situation that existed in step 1. When a tRNA molecule that recognizes this next codon appears, the anticodon of the incoming tRNA binds this codon, placing a new amino acid adjacent to the growing chain. The growing chain transfers to the incoming amino acid, as in step 2, and the elongation process continues.

4. When a chain-terminating nonsense codon is encountered (Figure 15-13), no tRNA exists to bind to it. Instead, it is recognized by special **release factors,** proteins that bring about the release of the newly made polypeptide from the ribosome.

Protein synthesis is carried out on the ribosomes, which bind to sites at one end of the mRNA and then move down the mRNA in increments of three nucleotides. At each step of the ribosome's progress, it exposes a three-base sequence to binding by a tRNA molecule with the complementary nucleotide sequence. Ultimately, the amino acid carried by that particular tRNA molecule is added to the end of the growing polypeptide chain.

PROTEIN SYNTHESIS IN EUKARYOTES

Protein synthesis occurs in a similar way in both bacteria and eukaryotes, although there are differences (Table 15-2). One difference is of particular importance. Unlike bacterial genes, most eukaryotic genes are much larger than they need to be, containing long stretches of nucleotides that are cut out of the mRNA transcript before it is used in polypeptide synthesis. Because these sequences are removed from the mRNA transcript before it is used, they are not translated and do not correspond to any portion of the polypeptide. The sequences that intervene between the polypeptide-speci-

FIGURE 15-13

Termination of protein synthesis. There is no tRNA with an anticodon complementary to any of the three termination signal codons such as the UAA nonsense codon illustrated here. For this reason, when a ribosome encounters a termination codon, it stops translocating. Release of the polypeptide chain involves the breaking of the covalent bond linking the polypeptide to the P-site tRNA, a process facilitated by a specific release factor.

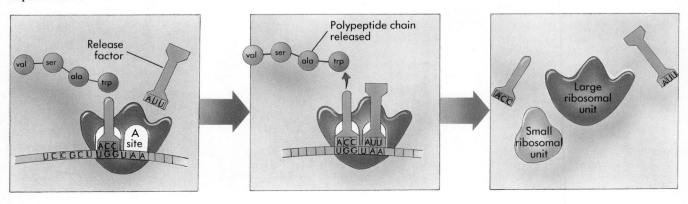

1. Because eukaryotes possess a nucleus, their mRNA molecules must pass across a membrane before they are used, whereas in bacteria this is not the case. Bacteria often begin the translation of an mRNA molecule before its transcription is completed.

2. Individual bacterial mRNA molecules often contain transcripts of several genes. By placing genes with related functions on the same mRNA, bacteria coordinate the regulation of the function. Eukaryotic mRNA molecules, by contrast, rarely contain transcripts of more than one gene. Regulation of eukaryotic gene expression is achieved in other ways.

3. The ribosomes of eukaryotes are a little larger than those of bacteria.

4. Rather than starting translation with those AUG codons preceeded by a special nucleotide sequence, as bacterial ribosomes do, eukaryotic mRNA molecules are modified after transcription at the 5′ leading end, where a methylated guanine triphosphate, called the CAP, is added. This CAP binds the mRNA to the small ribosomal subunit to initiate translation at the first AUG encountered.

5. Bacterial mRNA molecules are translated by ribosomes as they are transcribed. Eukaryotic mRNA molecules, which are released after transcription, are modified before being translated in order to protect them, since free mRNA molecules are subject to attack by cellular enzymes: to the 3′ end of each eukaryotic mRNA transcript is added a poly-A tail of some 200 adenine nucleotides, which appears to delay the destruction of the mRNA by cellular enzymes.

6. Almost all eukaryotic genes possess introns, whereas prokaryotic genes generally do not, except a few genes of primitive archaebacteria.

fying portions of the gene (Figure 15-14) are called **introns.** The remaining segments of the gene—the nucleotide sequences that encode the amino-acid sequence of the polypeptide—are called **exons.** Exons are typically much shorter than introns, and are scattered among the larger noncoding sequences. In a typical human gene the nontranslated intron portion of a gene can be 10 to 30 times larger than the coding exon portion. For example, even though only 432 nucleotides are required to encode the 144 amino acids of hemoglobin, there are actually 1356 nucleotides in the primary mRNA transcript of the hemoglobin gene.

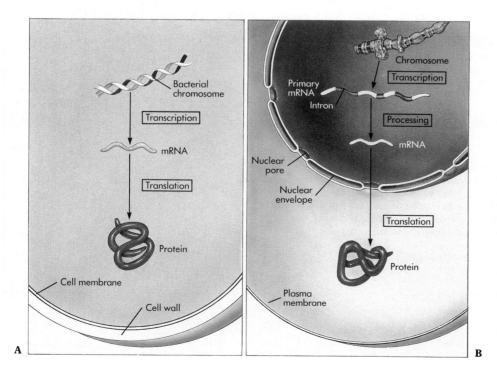

FIGURE 15-14

Genes are used differently in prokaryotes and eukaryotes.
A The genes on a bacterial chromosome are transcribed into mRNA, which is translated directly. In other words, the sequence of DNA nucleotides in bacteria corresponds to the sequence of amino acids in the encoded polypeptide.
B The genes of a eukaryote are typically different, containing long stretches of nucleotides called introns, which do not correspond to amino acids within the encoded polypeptide but rather serve some other function. These introns are removed from the mRNA transcript of the gene before the mRNA is used to direct the synthesis of the encoded polypeptide.

Virtually every nucleotide within the transcribed portion of a bacterial gene participates in an amino acid—specifying codon, and the order of amino acids in the protein is the same as the order of the codons in the gene. It was assumed for many years that all organisms would naturally behave in this logical way. In the late 1970s, however, biologists were amazed to discover that this relationship, one with which they had become completely familiar, did not in fact apply to eukaryotes. Instead, eukaryotic genes are encoded in segments that are excised from several locations along the transcribed mRNA and subsequently stitched together to form the mRNA that is eventually translated in the cytoplasm. With the benefit of hindsight, it is not difficult to design an experiment that reveals this unexpected mode of gene organization:

1. Isolate the mRNA corresponding to a particular gene. Much of the mRNA of red blood cells, for example, is related to the production of the proteins hemoglobin and ovalbumin, making it easy to purify the mRNAs from the genes related to these proteins.
2. Using an enzyme called reverse transcriptase, it is possible to make a DNA version of the mRNA that has been isolated. Such a version of a gene is called "copy" DNA (cDNA).
3. Using genetic engineering techniques (Chapter 18), isolate from the nuclear DNA the portion that corresponds to one of the actual hemoglobin genes. This procedure is referred to as "cloning" of the gene in question.
4. Mix single-strand forms of this hemoglobin cDNA and nuclear DNA and permit them to pair with each other ("hybridize") and form a duplex.

When this experiment was actually carried out and the resulting duplex DNA molecules were examined with the electron microscope, the hybridized DNA did not appear as a single duplex. Instead, unpaired loops were observed (Figure 15-A). In a related example, there are seven different sites within the ovalbumin gene at which the nuclear version contains long nucleotide sequences that are not present in the cytoplasmic cDNA version. The conclusion is inescapable: nucleotide sequences are removed from within the gene transcript before the cytoplasmic mRNA is translated into protein. As we noted earlier, these internal noncoding sequences are called introns, and the coding segments are called exons. Because introns are removed from the mRNA transcript before it is translated into tRNA, they do not affect the structure of the protein that is encoded by the gene in which they occur.

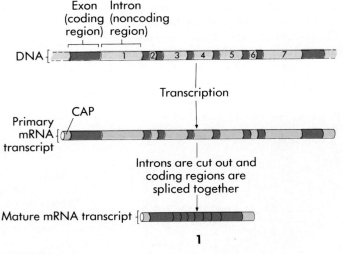

FIGURE 15-A
1 The ovalbumin gene and its primary transcript contain seven segments not present in the mRNA version, which the ribosomes use to direct the synthesis of protein. These intron segments are removed by enzymes that cut out the introns and splice together the exons.
2 The seven loops are the seven introns represented in the schematic drawing,
3, of the DNA and primary mRNA transcript.

REGULATING GENE EXPRESSION

A cell must know not only how to make a particular protein, but also when to make it. It is important for an organism to be able to control which of its genes are being transcribed and when. There is, for example, little point for a cell to produce an enzyme when the enzyme's substrate, the target of its activity, is not present in the cell. Much energy can be saved if the enzyme is not produced until the appropriate substrate is encountered and the enzyme's activity will be of use to the cell.

From a broader perspective, the growth and development of multicellular organisms entails a long series of biochemical reactions, each delicately tuned to achieve a precise effect. Specific enzyme activities are called into play and bring about a particular change. Once this change has occurred, those particular enzyme activities cease, lest they disrupt other activities that follow. During development, genes are transcribed in a carefully prescribed order, each gene for a specified period of time. The hereditary message is played like a piece of music on a grand organ, in which particular proteins are the notes and the hereditary information that regulates their expression is the score.

Organisms control the expression of their genes largely by controlling when the transcription of individual genes begins (Figure 15-15). Most genes possess special nucleotide sequences called **regulatory sites,** that act as points of control. These nucleotide sequences are recognized by specific regulatory proteins within the cell that bind to the sites.

Negative Control

Regulatory sites often function to shut off transcription, a process called **negative control.** In these cases, the site at which the regulatory protein binds to the DNA is located between the site where the polymerase binds and the beginning edge of the gene that the polymerase is to transcribe. When the regulatory protein is bound to its regulatory site, its presence there blocks the movement of the polymerase toward the gene. To understand this more clearly, imagine that you are shooting a cue ball at the eight ball on a pool table and someone places a brick on the table between the cue ball and the eight ball. Functionally, this brick is like the regulatory protein that binds to a regulatory site on the DNA: its placement blocks movement of the cue ball to the eight ball, just as placement of the regulatory protein between the polymerase and the gene blocks movement of the polymerase to the gene. The process of blocking transcription in this way is called **repression,** and the regulatory protein that is responsible for the blockage is called a **repressor protein** (Figure 15-16).

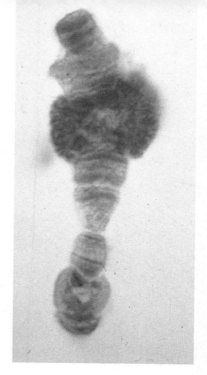

FIGURE 15-15
Chromosome puffs. In the giant chromosomes of the fly *Drosophila melanogaster*, the individual strands of DNA are replicated many times in parallel. For this reason, the activity of individual genes can readily be detected by special techniques as "puffs." The RNA that is being transcribed from the DNA template has been radioactively labeled. Its position on the chromosome can be detected by the dark specks.

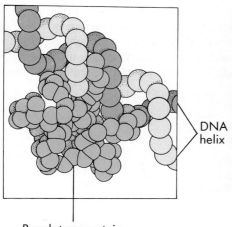

DNA helix

Regulatory protein (repressor)

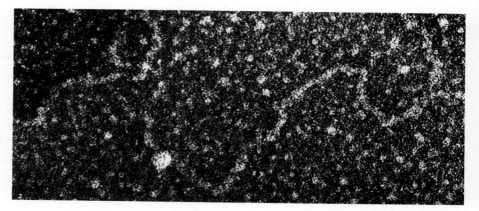

FIGURE 15-16
Repression. The regulatory protein controlling transcription of the lactose enzyme is the large whitish sphere bound to the DNA strand at the lower left. Because the repressor protein fills the groove of the DNA double helix, the RNA polymerase cannot attach there and the repressor blocks transcription.

Positive Control

Regulatory sites may also serve to turn on transcription, a process called **positive control.** In these situations, the binding of a regulatory protein to the DNA is necessary before the transcription of a particular gene may begin. The regulatory protein whose binding turns on transcription is called an **activator protein,** and the "turning on" of the transcription of specific genes in this way is called **activation.** Activation can be achieved by a variety of mechanisms. In some cases, the activator protein's binding promotes the unwinding of the DNA duplex. This facilitates the production of an mRNA transcript of a gene, because the polymerase, although it can bind to a double-stranded DNA duplex, cannot produce an mRNA transcript from such a duplex: mRNA is transcribed from a single strand of the duplex.

How Regulatory Proteins Work

How does the cell use regulatory sites to control which genes are transcribed? It does it by influencing the shape of the regulatory proteins. Regulatory proteins possess binding sites not only for DNA but also for specific small molecules. The binding of one of these small molecules can change the shape of a regulatory protein and thus destroy or enhance its ability to bind DNA. In some cases, the protein in its new shape may no longer recognize the regulatory site on the gene. In other cases, the recontoured regulatory protein may begin to recognize a regulatory site that it had previously ignored.

The cell thus uses the presence of particular "signal" molecules within the cell to incapacitate particular regulatory proteins or to mobilize them for action. These regulatory proteins, in turn, repress or activate the transcription of particular genes. The pattern of metabolites in the cell sets "on/off" protein regulatory switches, and, by doing so, achieves a proper configuration of gene expression.

> **Organisms control the expression of their hereditary information by selectively inhibiting the transcription of some genes and facilitating the transcription of others. Control over transcription is exercised by modifying the shape of regulatory proteins, thus influencing their tendency to bind to sites on the DNA that influence the initiation of transcription.**

THE ARCHITECTURE OF A GENE

Now that we have surveyed in general terms how specific polypeptides are assembled by ribosomes from mRNA copies of genes, and how the production of these mRNA copies is regulated, we will examine in more detail a specific example and trace how the structure of a particular set of genes achieves the precise and timed production of the proteins it encodes. The set of genes we will examine is the *lac* **system** (Figure 15-17), a cluster of genes encoding three proteins that bacteria use to obtain energy from the sugar lactose. These proteins include two enzymes and a membrane-bound

FIGURE 15-17
The *lac* region of the *Escherichia coli* chromosome. The left-most segment (P*i*) is a site where RNA polymerase binds, called a promoter site. Immediately downstream of it (that is, in the direction that the polymerase moves after it binds) is the gene *i*, which encodes a regulatory protein called the repressor protein. The RNA polymerase reads this sequence, producing the corresponding mRNA, and then dissociates from the DNA. Just to the right of the *i* gene is a second RNA polymerase binding site, called P*lac*, and a site called the operator, to which the repressor protein can bind. Just downstream of these two sites—farther along the DNA molecule in the direction in which mRNA transcription is taking place—are three genes encoding the two enzymes and the permease that are involved in the metabolism of lactose. Because they encode the primary structure (amino-acid sequence) of enzymes, they are called structural genes.

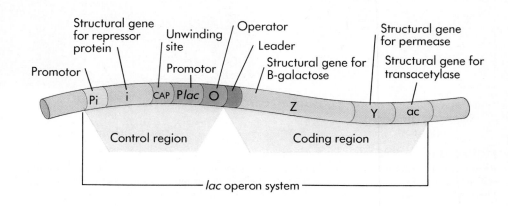

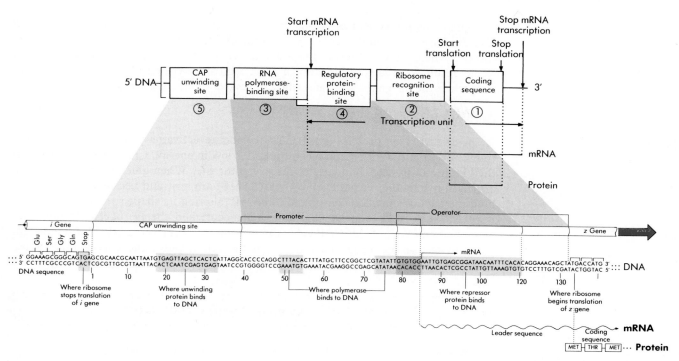

transport protein (a permease). Researchers have found this cluster to be typical of how genes are organized in bacteria. Within the cluster there are five different regions (Figure 15-18):

1. **Coding sequence.** Three coding sequences specify the three lactose-using enzymes. All three sequences are transcribed onto the same piece of mRNA and constitute part of an operational unit called an operon. An **operon** consists of one or more structural genes and the associated regulatory elements, the **operator** and the **promoter,** which are discussed below. This particular operon is called the *lac* operon, because the three genes that it includes are all involved in lactose use. Such a pattern of clustering of coding sequences onto single transcription units is common among bacteria but not in eukaryotes.

2. **Ribosome recognition site.** Upstream from the three coding sequences is the binding site for the bacterial ribosome. This series of nucleotides is within an initial untranslated portion of the mRNA sometimes called a **leader region.** Each mRNA molecule transcribed from the cluster is composed of the leader region and the three coding sequences, transcribed in order. The cluster and the leader region are often referred to as a **transcription unit.**

3. **RNA polymerase binding site.** Upstream from the transcription unit is a specific DNA nucleotide sequence that the polymerase recognizes and to which it binds. Such polymerase-recognition sites are called **promoters** because they promote transcription.

4. **Regulatory protein binding site.** Between the promoter site and the transcription unit is a regulatory site, the **operator,** where a repressor protein binds to block transcription.

5. **CAP site.** Upstream from the promoter is another regulatory site, **CAP,** where an activator protein binds. This in turn facilitates the unwinding of the DNA duplex and so enables the polymerase to bind to the nearby promoter.

Genes encoding enzymes possess regulatory regions. The segment that is transcribed into mRNA is called a transcription unit and consists of the elements that are involved in the translation of the mRNA: the ribosome-binding site and the coding sequences. In front of the transcription unit on the DNA are the elements involved in regulating its transcription: binding sites for the polymerase and for regulatory proteins.

FIGURE 15-18

The *lac* system of *Escherichia coli* is composed of five regions. *(1)* sequences encoding the lactose-metabolizing enzymes; *(2)* a ribosome recognition site; *(3)* a negative control site, the operator, where a repressor protein binds; *(4)* a promoter site, where RNA polymerase binds; and *(5)* a positive control site, where CAP binds to facilitate unwinding.

This complex system of regulatory sites works to ensure that mRNA is copied from the three structural genes only when the cell can effectively use the proteins they encode: when there is lactose present and when the cell requires the energy that would result from lactose breakdown. The regulatory region of the *lac* operon controls when the *lac* genes are transcribed in several interacting ways.

Activation

A special activator protein called CAP (catabolite activator protein) stimulates the transcription of the *lac* operon when the cell is low in energy. CAP appears to assist in the binding of RNA polymerase to the promotor site. When glucose levels in the cell are high, metabolites called **cyclic AMP (cAMP)** are low, and because CAP can only bind the *lac* operon when complexed to cAMP (altering its shape so that it recognizes the CAP site), activation by CAP does not occur, and the *lac* operon is not transcribed. As a result of this CAP activation system, the enzymes needed for the metabolism of lactose are produced only when the cell requires the energy that lactose would provide.

A VOCABULARY OF GENE EXPRESSION

anticodon   The three-nucleotide sequence at the tip of a transfer RNA molecule that is complementary to and base pairs with an amino acid–specifying codon in messenger RNA.

codon   The basic unit of the genetic code; a sequence of three adjacent nucleotides in DNA or mRNA that code for one amino acid or for polypeptide termination.

exon   A segment of DNA that is both transcribed into mRNA and translated into protein; contrasts with introns. In eukaryotic genes, exons are typically scattered within much longer stretches of nontranslated intron sequences.

intron   A segment of DNA that is transcribed into mRNA but removed before translation. These untranslated regions make up the bulk of most eukaryotic genes.

nonsense codon   A chain-terminating codon; a codon for which there is no tRNA with a complementary anticodon. There are three: UAA, UAG, and UGA.

operator   A site of negative gene regulation; a sequence of nucleotides that may overlap the promoter, which is recognized by a repressor protein. Binding of the repressor protein to the operator prevents binding of the polymerase to the promoter (just as two people cannot sit in one chair) and so blocks transcription of the structural genes of an operon.

operon   A cluster of functionally related genes transcribed onto a single mRNA molecule. A common mode of gene regulation in prokaryotes, but it is rare in eukaryotes other than fungi.

promoter   An RNA polymerase binding site; the nucleotide sequence at the 5′ end of a gene to which RNA polymerase attaches to initiate transcription of mRNA.

RNA polymerase   The enzyme that transcribes RNA from DNA.

repressor   A protein that regulates transcription of mRNA from DNA by binding to the operator and so preventing RNA polymerase from attaching to the promoter.

transcription   The polymerase-catalyzed assembly of an RNA molecule complementary to a strand of DNA.

translation   The assembly of a protein on the ribosomes, using mRNA to direct the order of amino acids.

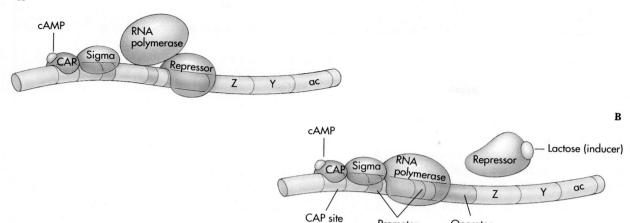

Repression

The sugar lactose is encountered by bacteria only occasionally, and so the enzymes that metabolize lactose are not usually able to function for lack of a substrate on which to act. Bacteria do not produce the enzymes under these conditions. The *lac* repressor protein blocks the binding of RNA polymerase to the *lac* promoter region under most circumstances; cells in this condition are said to be repressed with respect to *lac* operon transcription (Figure 15-19, *A*).

Like the activator protein, the *lac* repressor protein is capable of changes in shape. When lactose binds to the repressor protein, the protein assumes a different shape, one that does not recognize the operator sequence. If the cell contains much lactose, the *lac* repressor proteins fail to bind to the operator. This removes the block from in front of the polymerase (Figure 15-19, *B*), and so permits transcription of the *lac* genes to begin. It is for this reason that addition of lactose to a growing bacterial culture causes a burst of synthesis of the lactose-using enzymes (Figure 15-20). The transcription of the enzymes is said to have been **induced** by the lactose. This element of the control system ensures that the *lac* operon is transcribed only in the presence of lactose.

The *lac* operon is thus controlled at two levels: the lactose-using enzymes are not produced unless the sugar lactose is available; even if lactose is available, these enzymes are not produced unless the cell has need of the energy. High levels of cAMP reflect the cell's need for energy. Other similarly precise control mechanisms are known in eukaryotes, but the example provided here illustrates the complex nature of cellular control of protein synthesis.

FIGURE 15-19

How *lac* operon works.

A The *lac* operon is shut down ("repressed") when the repressor protein is bound to the operator site. Because promotor and operator sites overlap, polymerase and repressor cannot bind at the same time, any more than two people can sit in one chair.

B The *lac* operon is transcribed ("induced") when lactose binding to the repressor protein changes its shape so that it can no longer site on the operator site and block polymerase binding.

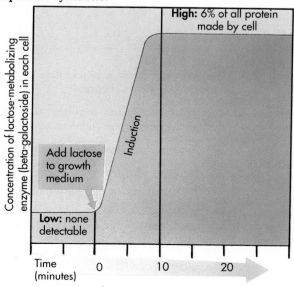

FIGURE 15-20

Induction of enzyme synthesis. Bacterial cells growing in the absence of lactose do not contain measurable amounts of the enzyme beta-galactosidase, which splits lactose into glucose and galactose, which bacteria can use as food. When lactose is added to a solution containing growing bacteria, the bacteria begin to produce large amounts of the enzyme. Within 10 minutes after lactose is added, 6% of all the protein made by a typical cell will be the enzyme beta-galactosidase.

SUMMARY

1. The expression of hereditary information in all organisms takes place in two stages. First, an mRNA molecule with a nucleotide sequence that is complementary to a particular segment of the DNA is synthesized by the enzyme RNA polymerase. Second, an amino-acid chain is assembled by a ribosome, using the mRNA sequence to direct its choice of amino acids. The first process is called transcription, the second translation.

2. A single mRNA molecule is called a transcription unit. It contains both the coding sequence, which directs the ribosome's choice of amino acids, and a leader region to which the ribosome binds.

3. The coding sequence of a gene is read in increments of three nucleotides, called a codon, from the mRNA by a ribosome. The ribosome positions the three-nucleotide segments of the message so that a tRNA molecule with the complementary base sequence can bind to it. Attached to the other end of the tRNA molecule is an amino acid, which is added to the end of the growing polypeptide chain.

4. The actual decoding of the genetic message is carried out by a family of activating enzymes, each of which binds the amino acid that it recognizes to a specific tRNA molecule. The three-base sequence, or anticodon, on the tRNA molecule is complementary to the corresponding codon of an mRNA molecule.

5. In eukaryotes, the mRNA transcript is processed in the nucleus and transported from the nucleus to the cytoplasm before it is translated.

6. Most eukaryotic genes contain additional sequences called introns embedded within the coding sequences of their transcription units. These introns are removed from the transcript before it is translated.

7. The control of gene expression is exercised largely by regulating the initiation of transcription. Control is both positive, in which cellular signals are required before transcription can start, and negative, in which other cellular signals prevent transcription except under appropriate circumstances.

8. The control of transcription is achieved largely by controlling the ability of RNA polymerase molecules to initiate transcription. In some instances, a nucleotide sequence located between the promoter and the transcription unit acts as a binding site for a repressor protein. When bound to the site, this repressor protein prevents the read-through of the RNA polymerase, just as a tree fallen across a road blocks traffic.

9. In other instances, nucleotide sequences located near the promoter site act as binding sites for activator proteins. The binding of the activator protein to the promoter site causes the unwinding of the DNA duplex. Since RNA polymerase can translate only a single strand of DNA, this forced unwinding greatly facilitates the transcription process.

REVIEW QUESTIONS

1. What are the three major classes of RNA? What is the abbreviation for each? What is the function of each type?

2. What types of RNA are involved in translation? Briefly describe the process of translation.

3. Are the nucleotide sequences read in a simple or a punctuated sequence? How was this confirmed by Crick and his colleagues? What were their results?

4. How did scientists determine the nucleic acid triplet that coded for each amino acid?

5. What is a codon? What is an anticodon? What confers specificity to the tRNA—amino acid complex?

6. What is the first step in protein synthesis? What occurs in this process? Why is it critical?

7. Which is the most important difference between protein synthesis in prokaryotes and eukaryotes?

8. How does repression of transcription work?

9. How are regulatory proteins themselves controlled within the cell?

10. What is the *lac* operon CAP site? When does CAP stimulate transcription of the *lac* operon? How is it incapacitated at other times?

THOUGHT QUESTIONS

1. You are provided with a sample of aardvark DNA, obtained at great risk of life and limb. As part of your investigation of this DNA, you transcribe messenger RNA from the DNA and purify it. You then separate the two strands of the DNA and analyze the base composition of each strand and of the mRNA. You obtain the following results:

	A	G	C	T	U
DNA strand #1	19.1	26.0	31.0	23.9	0
DNA strand #2	24.2	30.8	25.7	19.3	0
mRNA	19.0	25.9	30.8	0	24.3

Which strand of the DNA is the "sense" strand, serving as the template for mRNA synthesis?

2. In yeasts, investigators have found that 96% of the amino acids in yeast proteins are coded by only 25 of the 61 coding triplets; the other 36 coding triplets are rarely used, and their corresponding tRNAs are at much lower concentrations in cells. Can you suggest a reason why so many tRNAs are rarely used?

FOR FURTHER READING

DARNELL, J.: "The Processing of RNA," *Scientific American*, October 1983, pages 90-102. A description of how intron sequences are removed from messenger RNA transcripts before they are employed to direct the assembly of polypeptides.

DICKERSON, R.E.: "The DNA Helix and How It is Read," *Scientific American*, December 1983, pages 94-112. X-ray analysis of DNA molecules of different sequence shows that sequence differences create subtle differences in shape that may influence DNA site recognition by proteins. A clear exposition of a complex subject.

DOOLITTLE, W.F.: "The Origin and Function of Intervening Sequences in DNA—A Review," *American Naturalist*, vol. 130, December 1987, pages 915-928. A thoughtful and comprehensive review of the many ideas concerning why introns exist and what they do.

DORIT, R., L. SCHOENBACH, and W. GILBERT: "How Big is the Universe of Exons?", *Science*, December 1990, vol. 250, pages 1377-1382. A controversial attempt to guess how many exons are needed to construct all known proteins. A very important first step in addressing this key evolutionary question.

HOLLIDAY, R.: "A Different Kind of Inheritance," *Scientific American*, June 1989, pages 60-73. A speculative but very interesting account of the role that the methylation of DNA may play in determining patterns of gene activity.

MANIATIS, T., S. GOODBOURN, and J. FISCHER: "Regulation of Inducible and Tissue-Specific Gene Expression," *Science*, vol. 236, June 1987, pages 1237-1245. A comprehensive evaluation of the ongoing research into mechanisms of gene regulation in eukaryotes.

McKNIGHT, S.: "Molecular Zippers in Gene Regulation," *Scientific American*, April 1991, pages 54-64. Proteins that bind to DNA often do so in pairs "zipped" together by strings of leucine amino acids.

NIERHAUS, K.: "The Three-site Elongation Model for the Ribosome Elongation Cycle," *Biochemistry*, May 1990, vol. 29, pages 4997-5007. An exciting suggestion that the ribosome contains an E (exit) site as well as S and P sites.

PTASHNE, M.: "How Gene Activators Work," *Scientific American*, January 1989, pages 41-47. A lucid overview of how genes are turned on and off.

ROSS, J.: "The Turnover of Messenger RNA," *Scientific American*, April 1989, pages 48-55. The level of many proteins in the body is determined by how fast the messenger RNA encoding them is broken down, rather than by how speedily new transcripts are churned out.

STEITZ, J.: "Snurps," *Scientific American*, June 1988, pages 56-63. Small nuclear ribonuclear proteins, or "snurps," help to remove introns from mRNA transcripts. They were discovered less than 10 years ago, and are now an active subject of research.

TODOROV, I.: "How Cells Maintain Stability," *Scientific American*, December 1990, pages 66-75. An account of how cells shut down their protein-making factories in times of stress, and start them back up afterward.

WANG, J. and G. GIAEVER: "Action at a Distance Along a DNA," *Science*, vol. 240, April 1988, pages 300-304. How does an enhancer sequence affect the transcription of a gene thousands of base pairs away?

16

Mutation

OVERVIEW

The evolution of life on earth depends critically on the existence of genetic variation in nature. Only when heritable differences exist can one life-form replace another. Broadly speaking, there are two factors responsible for creating genetic change, mutation, and recombination, each occurring in many guises. In this chapter we consider the first of these, mutation. Mutations arise from many causes and can have serious consequences. Cancer is caused by mutation.

FOR REVIEW

Here are some important terms and concepts that you will encounter in this chapter. If you are not familiar with them, you should review them before proceeding.

Synaptonemal complex (Chapter 11)

Triplet reading frames (Chapter 15)

Repression of gene transcription (Chapter 15)

Codon (Chapter 15)

There is an enormous amount of DNA within the cells of your body. This DNA represents a long series of DNA replications, starting with the DNA of a single cell, the fertilized egg. The replication of DNA that took place in producing the cells of your body is equivalent to producing a length of DNA nearly 97×10^9 kilometers long from an original 0.9-meter piece. A rope that long would stretch from Earth to the planet Jupiter—60 times! Living cells have evolved many different kinds of mechanisms to avoid errors during DNA replication and to preserve the DNA from damage. These mechanisms ensure the accurate replication of DNA duplexes by proofreading the strands of each daughter cell against one another for accuracy and correcting any mistakes. The proofreading is not perfect, however. If it were, no mistakes would occur, no variation in gene sequence would result, and evolution would come to a halt.

In fact, living cells do make mistakes, and changes in the genetic message do occur (Figure 16-1), though only rarely. If changes were common, the genetic instructions encoded in DNA would soon degrade into meaningless noise. Typically, a particular gene is altered in only one out of a million gametes. Limited as it might seem, this steady trickle of change is the very stuff of evolution. Every single difference between the genetic message specifying you and the one specifying your cat, or the fleas on your cat, arose as the result of genetic change.

FIGURE 16-1
Mutation. Fruit flies normally have one pair of wings, extending from the thorax. This fly is a mutant, *bithorax*. Because of a mutation in a gene regulating a critical stage at development, it possesses *two* thorax segments and thus two sets of wings.

GENE MUTATION

A change in the genetic message of a cell is referred to as a **mutation.** This chapter will discuss mutational changes that affect the message itself, producing alterations in the sequence of DNA nucleotides. These alterations involving only one or a few nucleotides in the coding sequence are called **point mutations,** since they usually involve only one or a few nucleotides. Other classes of mutation, to be discussed in Chapter 17, involve changes in the way in which the genetic message is organized. In both bacteria and eukaryotes individual genes may move from one place on the chromo-

TABLE 16-1	SOURCES AND TYPES OF MUTATION	
SOURCE	PRIMARY EFFECT	TYPE OF MUTATION
MUTATIONAL		
Ionizing radiation	Two-strand breaks in DNA	Deletions, translocations
Ultraviolet radiation	Pyrimidine dimers	Errors in nucleotide choice during repair
Chemical mutagens	Base analogue mispairing	Single nucleotide substitution
	Modification of a base leads to mispairing	Single nucleotide substitution
Spontaneous	Isomerization of a base	Single nucleotide substitution
	Slipped mispairing	Frameshift, short deletion
RECOMBINATIONAL		
Transposition	Insertion of transposon into gene	Insertional inactivation
Mispairing of repeated sequences	Unequal crossing-over	Deletions, addition, inversions
Homologue pairing	Gene conversion	Single nucleotide substitution

some to another by a process called **transposition.** When a particular gene moves to a different location, there is often an alteration in its expression or in that of the neighboring genes. In eukaryotes large segments of chromosomes may change their relative location or undergo duplication. Such **chromosomal rearrangement** often has drastic effects on the expression of the genetic message. Sources and types of mutations are summarized in Table 16-1.

Point mutations, involving only one or a few nucleotides, result either from chemical or physical damage to the DNA or from spontaneous pairing errors that occur during DNA replication. The first class of mutation is of particular practical importance because modern industrial societies produce and release into the environment many chemicals capable of damaging DNA, chemicals called **mutagens.**

> **Point mutations are changes in the hereditary message of an organism. They may result from physical or chemical damage to the DNA or from spontaneous errors during DNA replication.**

DNA Damage

Although there are many different ways in which a DNA duplex can be damaged, three are of major importance: (1) ionizing radiation, (2) ultraviolet radiation, and (3) chemical mutagens.

Ionizing Radiation.   High-energy radiation, such as x rays and gamma rays, is highly mutagenic. Nuclear radiation is of this sort. When such radiation reaches a cell, it is absorbed by the atoms that it encounters, imparting energy to the electrons of their outer shells and causing these electrons to be ejected from the atoms. The ejected electrons leave behind ionized atoms with unpaired electrons, each called a **free radical.** Because most of a cell's atoms reside in water molecules and not in DNA, the great majority of free radicals created by **ionizing radiation** are produced from water molecules and not DNA.

Most of the damage that the DNA suffers is thus indirect. It occurs because free radicals are highly reactive chemically, reacting violently with the other molecules of the cell, including DNA. The action of free radicals on a chromosome is like that of shrapnel from a grenade blast tearing into a human body.

When a free radical breaks *both* phosphodiester bonds of a DNA helix—a **double-strand break**—the cell's usual mutational repair enzymes cannot fix the damage. To repair a double-strand break, the two fragments created by the break must be immobilized end to end so that the phosphodiester bond can be re-formed. Bacteria have no mechanism that can achieve this alignment, and double-strand breaks are lethal to their descendants. Eukaryotes, almost all of which possess multiple copies of their chromosomes, are able to position the two fragments created by a double-strand break end to end by using the synaptonemal complex that is assembled in meiosis to pair the fragmented duplex with another copy of the chromosome. In fact, it is thought that meiosis may have initially evolved as a mechanism to repair double-strand breaks in DNA.

Ultraviolet Radiation.   Ultraviolet (UV) radiation, the component of sunlight that leads to suntan (and sunburn), is much lower in energy than are x rays. When molecules absorb UV radiation, the radiation does not impart enough energy to cause the molecules to eject electrons; consequently, free radicals are not formed. The only molecules that are capable of absorbing UV radiation, in fact, are certain organic ring compounds.

In DNA, the principal absorbers of UV radiation are the pyrimidine bases thymine and cytosine. When these bases absorb UV energy, the electrons in their outer shells become reactive. If one of the nucleotides on either side of the absorbing pyrimidine is also a pyrimidine, a double covalent bond is formed between them. The resulting cross-link between adjacent bases of the DNA strand is called a **pyrimidine dimer** (Figure 16-2). If such a cross-link is left unrepaired, it can potentially block DNA replication, in which case the damage would be lethal. What actually happens

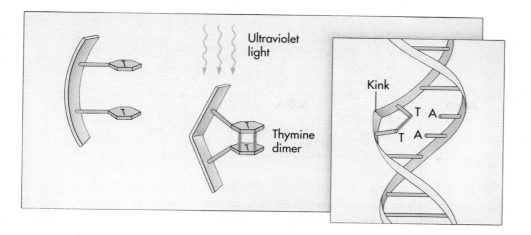

FIGURE 16-2

Making a thymine dimer. When two thymines are adjacent to one another in a DNA strand, the absorption of UV radiation can cause the formation of a covalent bond between them—a thymine dimer. Such a dimer introduces a "kink" into the double helix, which prevents replication of the duplex by DNA polymerase.

in most cases, however (Figure 16-3), is that cellular UV repair systems either (1) cleave the bond that links the adjacent pyrimidines or (2) excise the entire pyrimidine dimer from the strand and fill in the gap by using the other strand as a template. In those rare instances in which a pyrimidine dimer does escape such cleavage or excision, the replicating polymerase simply fails to replicate the portion of the strand that includes the pyrimidine dimer, skipping ahead and leaving the problem area to be filled in later. This filling-in process is often error prone, however, and it may create mutational changes in the base sequence of the gap region.

Without a mechanism for repairing the damage to DNA caused by UV radiation, sunlight would wreak havoc on the cells of your skin. There is a rare hereditary disorder among humans called **xeroderma pigmentosum** (Figure 16-4) in which just this problem occurs. Individuals homozygous for a mutation that destroys the ability of the body's cells to repair UV damage develop extensive skin tumors after exposure to sunlight. Because of the many different proteins involved in excision and repair of pyrimidine dimers, mutations in as many as six different genes can cause the disease.

Chemical Mutagens.

Many mutations result from the direct chemical modification of the DNA bases. The chemicals that act on DNA fall into three classes: (1) chemicals that look like DNA nucleotides but pair incorrectly when they are incorporated into DNA (Figure 16-5); (2) chemicals that remove the amino group from adenine or cytosine, causing them to mispair; and (3) chemicals that add hydrocarbon groups to nucleotide bases, also causing them to mispair. This last group includes

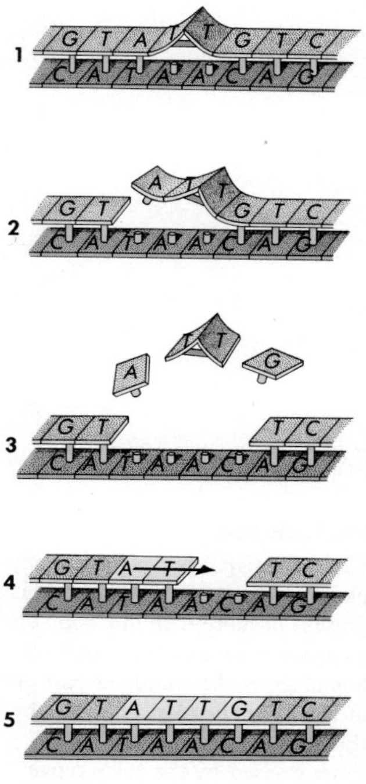

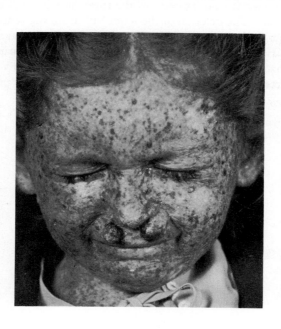

FIGURE 16-3

Repair of a thymine dimer. Thymine dimers are repaired by excising the troublesome dimer, as well as a short run of nucleotides on either side of it, and then filling in the gap using the other strand as a template.

FIGURE 16-4

The inherited disorder xeroderma pigmentosum. Those who have the disease develop extensive malignant skin tumors after exposure to sunlight. The tumors develop because affected individuals are less efficient in the repair of DNA damage caused by exposure to sunlight or other sources of ultraviolet radiation.

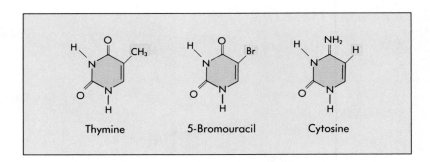

many particularly potent mutagens that are commonly used in the laboratory, as well as compounds that are sometimes released into the environment, such as mustard gas.

The three major sources of mutational damage to DNA are (1) high-energy radiation, such as x rays, which physically break the DNA strands; (2) low-energy radiation, such as UV light, which creates DNA cross-links whose removal often leads to errors in base selection; and (3) chemicals that modify DNA bases and thus alter their base-pairing behavior.

Spontaneous Mispairing

Many point mutations occur spontaneously, without exposure to mutagenic chemicals or radiation. Sometimes nucleotide bases spontaneously shift to alternative conformations, or **isomers.** These isomers form different kinds of hydrogen bonds than do the normal conformations. During the replication of a DNA strand, the polymerase selects a different nucleotide to pair with an isomer than the one it would have otherwise selected. These spontaneous errors occur in fewer than one in a billion nucleotides per generation, but they are still an important source of mutation.

When chromosomes pair, sequences may sometimes misalign, looping out a portion of one strand. Usually such misaligned pairing, called **slipped mispairing,** is only transitory, and the strands revert quickly to the normal arrangement (Figure 16-6). If the error-correcting system of the cell encounters such a mispairing before it reverts, however, it will attempt to "correct" it, usually by excising the loop. This results in a **deletion** of several hundred nucleotides from one of the strands. Many of the deletions created in this way start or stop in the middle of a codon. When they occur, therefore, they lead to the creation of a genetic message that is out of synchrony with the normal reading pattern, the three-base increments being displaced one or two positions. The deletion of the letter "F" from the sentence "THE FAT CAT ATE THE RAT" similarly shifts the reading frame of that message, producing the message "THE ATC ATA TET HER AT," which is meaningless. Such an event results in what is called a **frameshift mutation** and leads to a message that is read in the wrong three-base groupings. Some chemicals specifically promote the occurrence of deletions

FIGURE 16-6
Slipped mispairing. Slipped mispairing occurs when a sequence is present in more than one copy on a chromosome and the copies on homologous chromosomes pair out of register. The loop produced by this mistake in pairing is sometimes excised by the cell's repair enzymes, producing a short deletion and often altering the reading frame. Any chemical that tends to stabilize the loop increases the chance that this will happen.

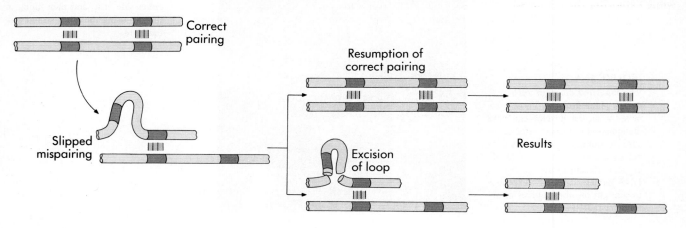

and frameshift mutations by stabilizing the loops that are produced by slipped mispairing, thus increasing the length of time during which the loops are vulnerable to excision.

Spontaneous errors in DNA replication occur very rarely. They result from transient changes in the conformation of nucleotides and also from the accidental mispairing of nonidentical but similar sequences.

THE BIOLOGICAL SIGNIFICANCE OF MUTATION

Because mutations are random events and can occur anywhere in a cell's DNA, most mutations are detrimental—for the same reason that making a random change in a computer program or altering a musical score at random usually worsens performance. The consequences of a detrimental mutation may be minor or catastrophic, depending on the function of the gene that is altered.

The consequences of a mutation also critically depend on the identity of the cell in which the mutation occurs. In the course of the differentiation of all multicellular organisms, there comes a point early in animal development, when cells destined to form gametes (**germline cells**) are segregated from those destined to differentiate into other cells of the body (**somatic cells**). In other organisms, such as plants and fungi, this developmental decision comes much later. In plants, a mutation of any nucleate cell can potentially be passed to the next generation, since any cell can potentially develop into an adult individual. In humans and most other animals, by contrast, only when a mutation occurs within a germline cell is the mutational change passed along to subsequent generations as part of the hereditary endowment contributed by the gametes derived from that cell.

Mutations in germline tissue are of enormous biological importance, since they provide the raw material from which natural selection produces evolutionary change. Evolutionary change can only occur if new alternatives are available, different allele combinations to replace the old. Two processes are critical to generating an array of allele combinations upon which selection can act: mutation, which produces new alleles, and recombination (Chapter 17), which puts alleles together in different combinations. In animals, it is the mutation and recombination that occur in germline tissue that are of importance to evolution, since somatic mutation and recombination are not passed from one generation to the next in animals.

When a mutation occurs within a somatic cell of an animal, such as in one of the cells lining the lung, the mutation is not passed along to subsequent generations of organisms, since the mutation does not affect the gametes. Mutation of a somatic cell (**somatic mutation**) may, however, have drastic effects on the individual organism in which it occurs, since it *is* passed on to all the cells of the organism that are descended from that cell. Thus if a mutant lung cell divides, all the cells that are derived from it will carry the mutation. Somatic mutations of lung cells are, as we shall now see, the principle cause of lung cancer.

In Chapter 17 we shall consider further the consequences of germline mutations while exploring the evolutionary consequences of mutation. Now we shall examine in detail the consequences of somatic mutation, focusing on cancer, the most important consequence of somatic mutation.

THE RIDDLE OF CANCER

Cancer is a growth disorder of cells. It starts when an apparently normal cell begins to grow in an uncontrolled and invasive way (Figure 16-7). The result is a ball of cells, called a **tumor,** that constantly expands in size. When this ball remains a hard mass, it is called a **sarcoma** if connective tissue such as bone is involved, or a **carcinoma** if epithelial tissue such as skin is involved. If its cells leave the mass and spread throughout the body (Figure 16-8), forming new tumors at distant sites, the spreading cells are called **metastases.** Cancer is perhaps the most pernicious human disease. Of the children born in 1985, one third will contract cancer; one fourth of the male

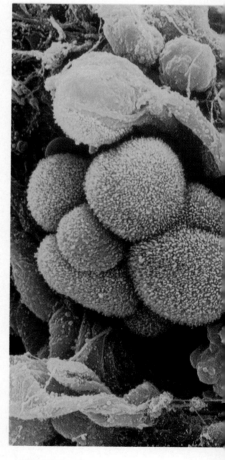

FIGURE 16-7
Lung cancer cells. These cells are from a tumor located in the alveolus of a lung.

FIGURE 16-8
Portrait of a cancer. The ball of cells is a carcinoma, developing from epithelial cells lining the interior surface of a human lung. As the mass of cells grows, it invades surrounding tissues, eventually penetrating into lymphatic vessels and blood vessels, both of which are plentiful within the lung. These vessels carry metastatic cancer cells throughout the body, where they lodge and grow, forming new foci of cancerous tissue.

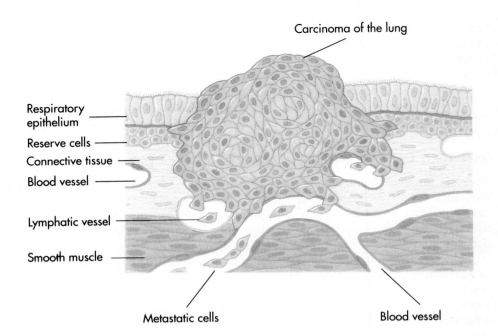

Carcinoma of the lung

Respiratory epithelium

Reserve cells

Connective tissue

Blood vessel

Lymphatic vessel

Smooth muscle

Metastatic cells

Blood vessel

children and fully one third of the female children will someday die of cancer. Each of us has had family or friends affected by the disease. Many of us will die of cancer. Not surprisingly, a great deal of effort has been expended to learn the cause of this disease. Using the new techniques of molecular biology a great deal of progress has been made, and the rough outlines of understanding are now emerging.

The search for a cause of cancer has focused in part on those environmental factors that were considered potential agents in causing the disease (Figure 16-9). Many were found, including ionizing radiation such as x rays and a variety of chemicals. Many of these cancer-causing agents, or carcinogens, had in common the property of being potent mutagens. This observation led to the suspicion that cancer might be caused, at least in part, by the creation of mutations.

Mutagens were not the only cancer-causing agents that were found, however. Some tumors seemed almost certainly to have resulted from viral infection. As early as 1910, American medical researcher Peyton Rous demonstrated that a virus was associated with a cancerous sarcoma in chickens. Over the next 50 years, a number of experiments demonstrated that viruses could be isolated from certain tumors and that these viruses would cause virus-containing tumors to develop in other individuals.

What do mutagens and viruses have in common, that they can both induce cancer? At first it seemed that they had little in common: mutagens alter genes directly, whereas viruses introduce foreign genes into cells. However, this seemingly fundamental difference is less significant than it might first appear, and the two causes of cancer have in fact proved to be one and the same.

THE STORY OF CHICKEN SARCOMA

As early as 1910, Peyton Rous reported the presence of a virus, subsequently named **Rous avian sarcoma virus (RSV),** that was associated with chicken sarcomas; he was awarded the 1966 Nobel prize in physiology and medicine for his discovery. The RSV virus proved to be an RNA virus, or **retrovirus.** Retroviruses are unusual in that their genomes are composed of RNA rather than DNA. When they infect a cell (Figure 16-10) they make a DNA copy of their RNA—a copy that can be inserted into the animal DNA! RSV was able to initiate cancer in chicken fibroblast (connective tissue) cells growing in culture when they were infected with the virus; from these transformed cells, more virus could be isolated.

How does RSV initiate cancer? When RSV was compared with a closely related virus, RAV-O, which was not able to transform chicken cells into cancer cells, the

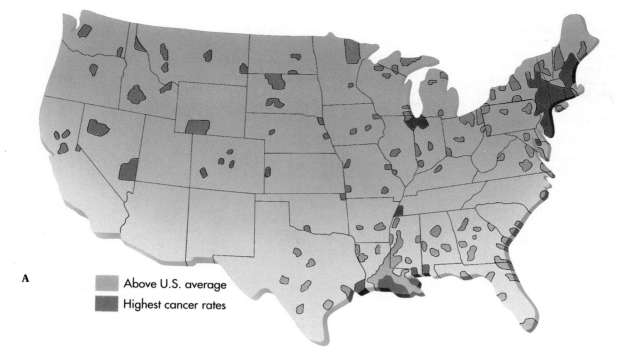

A

Above U.S. average

Highest cancer rates

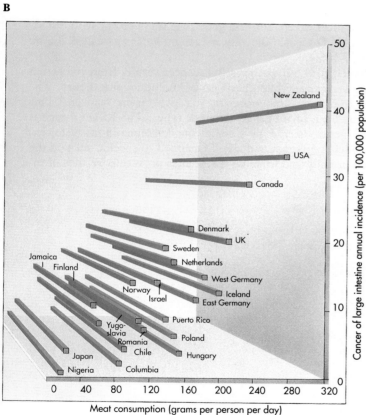

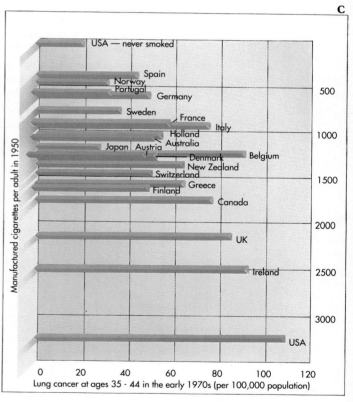

FIGURE 16-9

Potential cancer-causing agents.

A The incidence of cancer per 1000 people is not uniform throughout the United States. Rather, it is centered in cities and in the Mississippi Delta. This suggests that pollution and pesticide runoff may contribute to the development of cancer.

B One of the most deadly cancers in the United States, cancer of the large intestine, is not at all common in many other countries, such as Japan. Its incidence appears to be related to the amount of meat an average person consumes—a high meat diet slows passage of food through the intestine, prolonging exposure of the intestinal wall to digestion waste.

C The biggest killer among cancers is lung cancer, and the most important environmental agent producing lung cancer is cigarette smoking. When levels of lung cancer in many countries are compared, the incidence of lung cancer among men between 40 and 50 years of age is strongly correlated with the cigarette consumption in that country 20 years earlier.

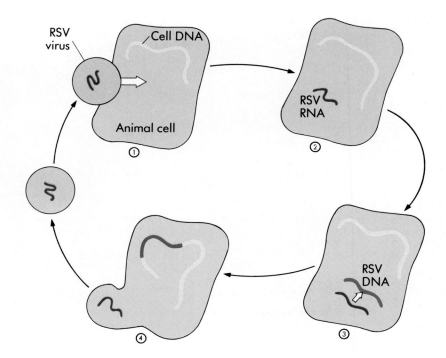

two viruses proved to be identical except for one gene that was present in the RSV virus but absent from the RAV-O virus. The cancer-causing gene was called the *src* gene (for sarcoma) (Figure 16-11).

These and other results led to the hypothesis that cancer results from the action of a specific tumor-inducing *onc* gene, a hypothesis called the **oncogene theory.** The term oncogene was derived from the Greek word *oncos*, which means "cancer." This theory soon proved to be true for a broad range of tumor types. The RSV *src* gene is but one example of such *onc* genes. Seemingly spontaneous leukemia in mice, for example, is inherited in a Mendelian manner and can be mapped to specific sites on the mouse chromosomes. When investigated in detail, these sites proved to be sites of integrated virus genomes.

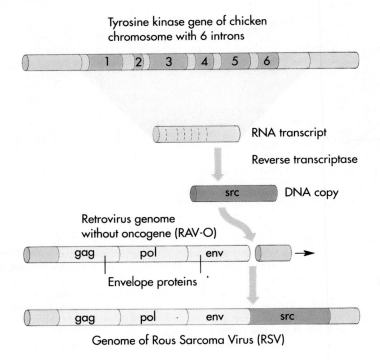

FIGURE 16-11

Structure of the Rous sarcoma virus (RSV). The virus contains only a few virus genes, encoding the virus protein envelope (called the *env* and *gag* genes) and reverse transcriptase, which produces a DNA copy of its RNA genome (called *pol*). It also contains a gene, *src* (for sarcoma). The RAV-O virus shown here lacks this *src* gene, but otherwise is identical to the RSV retrovirus. RSV causes cancer in chickens, RAV-O does not.

What might be the nature of a virus gene that causes cancer? An essential clue came in 1970, when RSV mutants were isolated that were temperature-sensitive. These mutants would transform tissue culture cells into cancer cells at 35° C, but not at 41° C. Temperature sensitivity of this kind is almost always associated with proteins. It seemed very likely, therefore, that the *src* gene was actively transcribed by the cell, rather than serving as a recognition site for some unknown regulatory protein. This was an exciting result because it suggested that the protein specified by this *onc* gene could be isolated and its properties studied.

The *src* protein was first isolated in 1977. It proved to be an enzyme of moderate size that phosphorylates (adds a phosphate group to) the tyrosine amino acids of proteins; such an enzyme is called a **tyrosine kinase.** Tyrosine kinases are not common in animal cells. Among the few that are known is **epidermal growth factor,** a protein that signals the initiation of cell division by binding to a special receptor site on the plasma membrane and phosphorylating key protein components. This fact raised the exciting possibility that RSV perhaps causes cancer by introducing into cells an active form of a normally quiescent enzyme. As you will see, this indeed proved to be the case.

Does the *src* gene actually integrate into the host chromosome with the RSV virus? One way of investigating this question is to prepare a radioactive version of the *src* gene. We can then permit this radioactive *src* DNA to bind to complementary sequences on the chicken genome and examine where the chicken chromosomes become radioactive. Sites of radioactivity are sites where a sequence complementary to *src* occurs. As expected, radioactive *src* DNA binds to the site where RSV is inserted into the chicken genome—but unexpectedly it also binds to a second site where there is no RSV.

From such experiments investigators learned that the *src* gene is not a virus gene at all, but rather a growth-promoting gene that evolved in and occurs normally in chickens. This normal chicken gene is the second site where *src* binds to chicken DNA. Somehow a copy of the normal chicken gene was picked up by an ancestor of the RSV virus in some past infection. Now part of the virus, the gene is active in a pernicious new way, escaping the normal regulatory controls of the chicken genome; its transcription is governed instead by virus promoters, which are actively transcribed during infection.

Thus the picture of cancer that emerges from the study of RSV is of a malady that results from the inappropriate activity of growth-promoting cellular genes that are normally less active or completely inactive.

> Sarcoma in chickens results from the inappropriate activity of a normal chicken gene. Taken up by the RSV virus, it escapes the genetic regulation that is imposed by the chicken genome and induces cancer by being active when it should not be.

THE MOLECULAR BASIS OF CANCER

Not all cancer is associated with viruses. Other kinds of tumors have been studied by using techniques different from those we have just been exploring. The most important of these techniques, called **transfection,** consists of (1) the isolation of the nuclear DNA from human cells that have been isolated from tumors, (2) its cleavage into random fragments, and (3) the testing of the fragments individually for the ability of any particular fragment to induce cancer in the cells that assimilate it.

By using transfection techniques, researchers found that a single gene isolated from a cancer cell is all that is needed to transform normally dividing cells in tissue culture into cancerous ones. The cancerous cells differed from the normal ones only with respect to this one gene. In some cases the cancer-inducing gene identified by transfection proved to be the same as one of the *onc* genes that had previously been identified as having been included in a cancer-causing virus.

Onc genes are normal genes gone wrong. By identifying *onc* genes by transfection and then isolating them, investigators have been able to compare them with their nor-

FIGURE 16-12

How a mutation can cause cancer. Mutations that lead to cancer often involve proteins of the plasma membrane associated with cell division. In a normal cell, cell division is triggered by a protein called epidermal growth factor (EGF), which binds to an EGF receptor protein on the exterior surface of the cell. This alters the shape of the portion of the receptor protruding into the cell, initiating a signal that passes to the cell nucleus and initiates cell division. The level of EGF necessary to start this process is affected by another protein, *ras*. A mutation increasing receptor efficiency or *ras* signal facilitation may trigger more frequent cell division—cancer.

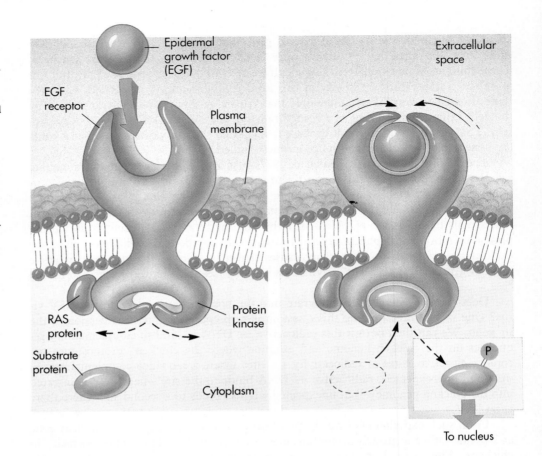

mal counterparts. In this way researchers have been able to study why normal genes were converted into cancer-causing ones. The analysis of a number of *onc* genes associated with cancers of various tissues has led to the following conclusions:

1. The induction of many cancers involves changes in cellular activities that occur at the inner surface of the plasma membrane. In a normal cell these activities are associated with the initiation of cell division. In one well-studied example (Figure 16-12), a cellular regulatory protein, epidermal growth factor (EGF), binds to a specific receptor on the inner plasma membrane, acts as a tyrosine kinase, and triggers cell division. In this process a protein encoded by a gene called *ras* is associated with the membrane EGF receptor site and determines what cellular levels of EGF are adequate to initiate cell division. In several forms of human cancer the cancerous, or *onc*, version of the *ras*-encoded protein activates the receptor site in response to much lower levels of EGF than does the normal version of the protein (Figure 16-13).

2. The difference between a normal gene encoding the proteins that carry out cell division and a cancer-inducing *onc* version need only be a single-point mutation in the DNA. In several human carcinomas induced by *ras*, a single nucleotide alteration is the only difference between the normal and the cancer-caus-

FIGURE 16-13

How the *ONC* version of *RAS* initiates cell division. The amount of epidermal growth factor (EGF) in most cells is not sufficient to trigger cell division. The *onc* version of *RAS* increases the effectiveness of encounters between EGF molecules and receptor sites, so that the amount of EGF in most cells is now sufficient to induce cell division.

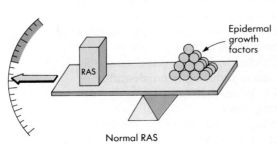

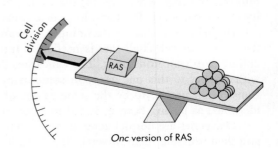

ing version of the gene; in one case of *ras*-induced bladder cancer, for example, a single DNA base change from G to T converts a glycine in the normal *ras* protein into a valine in the cancer-causing protein. There is no other difference between the normal and cancer-inducing forms of the *ras* gene.

3. The mutation of a gene such as *ras* to an *onc* form in different tissues can lead to different forms of cancer. There are probably no more than a few dozen different genes whose mutation can lead to cancer, and there may be far fewer than that.

4. The induction of many cancers involves the action of two or more different *onc* genes. The initiation of cancer may require changes at both the plasma membrane and in the nucleus. This may be the reason why most cancers occur in people over 40 years old (Figure 16-14). It is as if human cells accumulate mutational changes, and time is required for several such mutations to occur in the same cells.

The emerging picture of cancer is one that involves aborted regulation of the genes that normally signal the onset of cell proliferation (Figure 16-15). Cancer seems to occur when several of the controls that cells normally impose on their own growth and division become inoperative. Among the examples that have been studied in detail, the specific means whereby these controls are evaded vary, but many of them involve one of two general causes. In the first case, cancer may result from the mutation of a cellular gene with a regulatory function, as in the case of the *ras* gene, whose mutant *onc* form induces human bladder carcinomas. In the second case, cancer may result from the introduction of a normal cellular gene with a regulatory function into a cell as part of a virus, so that the gene escapes the cell's regulation.

Cancer is a growth disease of cells, in which the controls that normally restrict cell proliferation do not operate. The critical factor in inducing cancer seems to be the inappropriate activation of one or more proteins that regulate cell division, transforming cells to a state of cancerous growth.

FIGURE 16-14
The annual death rate from cancer is a function of age. A logarithm plot of age versus death rate is linear, suggesting that several independent events are required to give rise to cancer.

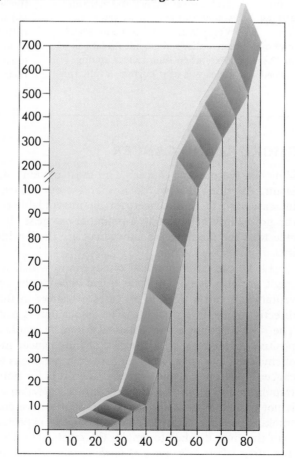

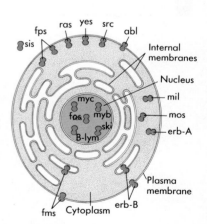

FIGURE 16-15
A representation of a eukaryotic cell, showing sites of action of the proteins encoded by 16 of the 30 known cancer-causing genes. The activities of some genes are associated with plasma membranes; others are associated with internal membranes, the nucleus, or the cytoplasm of cells.

Of the 2.2 million people who died in the United States last year, one fourth died of cancer and almost one-third of these—106,000 people—died of lung cancer. About 140,000 cases of lung cancer were diagnosed each year in the 1980s, and 90% of these persons died within 3 years. Of those who died, 96% were cigarette smokers.

Smoking is a popular pastime in the United States. Twenty-nine percent of the U.S. population smokes. American smokers consumed 520 billion cigarettes in 1989. These cigarettes emit in their tobacco smoke some 3000 chemical components, among them vinyl chloride, benzo (a) pyrenes, and nitroso-nor-nicotine, all of them potent mutagens. Smoking introduces these mutagens to the tissues of your lungs. Figure 16-A, a photograph of lung cancer in an adult human, illustrates the result. The bottom half of the lung is normal; the top half has been taken over completely by a cancerous growth. As you might imagine, a lung in this condition does not function well, but difficulty in breathing is rarely the cause of death from lung cancer. As the cancer grows within the lung, its cells invade the surrounding tissues, as diagrammed in Figure 16-8, and

eventually break through into the lymph and blood vessels. Once the cancer cells have done this, they spread rapidly through the body, lodging and growing at many locations, particularly in the brain. Death soon follows.

Among cigarette manufacturers, it has been popular to argue that the causal connection between smoking and cancer has not been proved, that somehow the relationship is coincidental. Look carefully at the data presented in Figure 16-B and see if you agree. The upper graph presents data collected for American men, including the incidence of smoking from the turn of the century until now and the incidence of lung cancer over the same period. Note that as late as 1920, lung cancer was a rare disease. With a lag of some 20 years behind the increase in smoking, it became progressively more common.

Now look at the lower graph, which presents data on American women. Because of social mores, significant numbers of American women did not smoke until after World War II, when many social conventions changed. As late as 1963, when lung cancer among males was near current levels, this

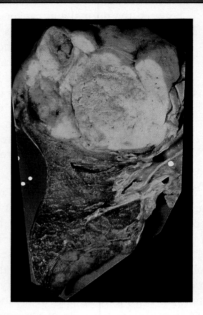

FIGURE 16-A

disease was still rare in women. In the United States that year, only 6588 women died of lung cancer. But as their smoking increased, so did their incidence of lung cancer, with the same inexorable lag of about 20 years. American women today have achieved equality with their male counterparts in the

SMOKING AND CANCER

How can we prevent cancer? The most obvious strategy is to minimize mutational insult. Anything that we do to increase our exposure to mutagens will result in an increased incidence of cancer, for the unavoidable reason that such exposure increases the probability of mutating a potential *onc* gene. It is no accident that the most reliable tests for carcinogenic substances are those which measure their mutagenic ability.

Of all the environmental mutagens to which we are exposed, perhaps the most tragic are cigarettes. They are tragic because the cancers they cause are largely preventable. About a third of all cases of cancer in the United States can be attributed directly to cigarette smoking. The association is particularly striking for lung cancer. The dose-response curve obtained for male smokers (Figure 16-16) shows a highly positive correlation, with the risk of lung cancer increasing with increasing amounts of smoking. For those smoking two or more packs a day, the risk of contracting lung cancer is 40 times or more greater than it is for nonsmokers. Note that the curve extrapolates back approximately to zero. Clearly, an effective way to avoid lung cancer is not to smoke. Life insurance companies have computed that on a statistical basis smoking a single cigarette lowers your life expectancy 10.7 minutes. (That is more than the time it takes to smoke the cigarette!) Every pack of 20 cigarettes bears an unwritten label: *"The price of smoking this pack of cigarettes is 3½ hours of your life."*

numbers of cigarettes that they smoke—and their lung cancer death rates are now approaching those for men. In 1990 more than 49,000 women died of lung cancer in the United States.

Among smokers, the current rate of deaths resulting from lung cancer is 180 per 100,000, or about 2 of each 1000 smokers *each year*. Smoking is very much like going into a totally dark room, standing still, and then calling in someone with a gun and closing the door. The person with the gun cannot see you, does not know where you are, and so just shoots once in a random direction and leaves the room. Every time an individual smokes a cigarette, he or she is "shooting" mutagens at their genes. Just as in the dark room, a hit is unlikely, and most shots will miss potential *onc* genes. As one keeps shooting, however, the chance of eventually scoring a hit becomes more likely. Nor do statistics protect any one individual: nothing says the first shot will not hit. Older people are not the only ones to die of lung cancer.

Except for eating powerful radio-isotopes, there is probably no more certain way to develop cancer than to smoke cigarettes.

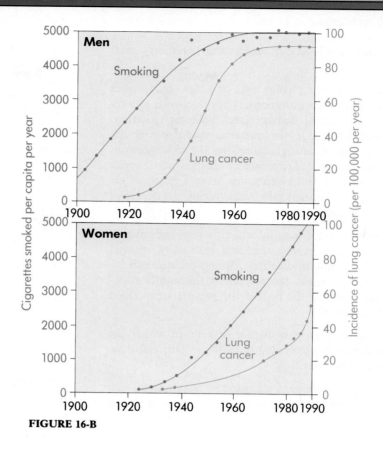

FIGURE 16-B

Do we have any idea how to cure cancer once it has started? Although surgery, radiotherapy, and chemotherapy can in some instances lead to remission, there is still no general "cure" for cancer. The path to a general cure lies in a better understanding of the thwarted cellular controls that are responsible for cancerous growth so that therapies can be devised to reinstitute appropriate cellular regulation within cancerous tissue. As you can imagine, this is an area of intense research activity.

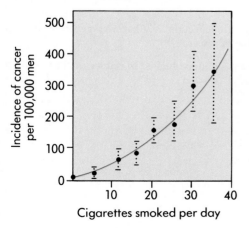

FIGURE 16-16

Smoking causes cancer. The annual incidence of lung cancer per 100,000 men is clearly related to the number of cigarettes smoked per day. Of every 100 American college students who smoke cigarettes regularly, 1 will be murdered, about 2 will be killed on the roads, and about 25 will be killed by tobacco.

SUMMARY

1. A mutation is any change in the hereditary message.

2. Mutations that change one or a few nucleotides are called point mutations. They may arise as a result of physical damage from ionizing radiation, chemical damage, or mistakes made during the correction of damage caused by the absorption of ultraviolet light. Occasionally changes occur spontaneously as DNA replicates or as a result of errors in pairing that take place during the course of DNA replication.

3. Slipped mispairing, in which a gene mistakenly pairs with a nearby sequence instead of its true partner, leads to deletions. When the deletion interrupts a codon, the result is a frameshift mutation, in which the rest of the gene is transcribed out of synch.

4. Cancer is a growth disease of cells in which the regulatory controls that usually restrain cell division do not work.

5. Sarcoma in chickens is caused by the presence in a virus of an otherwise normal chicken gene whose activity mimics that of epidermal growth factor, a regulator of cell division. Because it is located in the virus genome, the gene is not under the control of the chicken regulator and is transcribed far more often than it otherwise would be, triggering cell division.

6. The testing of DNA fragments from tumor cells to identify those fragments capable of inducing cancer has led to the isolation of a variety of cancer-causing genes. In every case the cancer-causing gene is a normal gene whose product has a role in cell proliferation but that has become active when it should not.

7. In some cases cancer induction results from a single nucleotide mutation in a normal gene. In other cases a normal gene is acquired by a virus and is transcribed at the high rate characteristic of the virus.

8. The best way to avoid cancer is to avoid those things that cause mutation, notably cigarette smoking.

REVIEW QUESTIONS

1. What is a free radical? What is the source of most of a cell's free radicals? Why are they so damaging?

2. What is a pyrimidine dimer? Why is it harmful to a cell? How does a cell repair this damage? What may happen if the dimer is not repaired?

3. What is a slipped mispairing? What type of mutation may result from this repair?

4. Are most mutations beneficial or detrimental? Why? What determines whether the consequences of a mutation are major or minor?

5. Is it more likely that a mutation will affect future progeny in an animal cell or a plant cell? Why?

6. How does cancer affect the cells with which it is associated? What is a tumor? What type of tissue forms a sarcoma? What type of tissue forms a carcinoma? What term is given to cancer cells that leave the tumorous mass and spread throughout the body, forming new tumors at distant locations?

7. What is the nature of the protein transcribed from the *src* gene? What is an important function of this protein?

8. How did scientists determine whether or not the *src* gene was integrated into the host chromosome? What were the experimental results? What did this indicate about the origin of the *src* gene?

9. What is "transfection"? What was generally determined from this type of research?

10. What can any individual do to reduce the chance of developing cancer?

THOUGHT QUESTIONS

1. In medical research, mice are often used as model systems in which to study the immune system and other physiological systems that are important to human health. Many medical centers maintain large colonies of mice for these studies. In one such colony under your supervision, a hairless mouse is born. What minimum evidence would you accept that this variant represents a genetic mutation?

2. The evidence associating lung cancer with smoking is overwhelming, and, as you have learned in this chapter, we now know in considerable detail the mechanism whereby smoking induces cancer. It introduces powerful mutagens into the lungs, which cause mutations to occur; when a growth-regulating gene is mutated by chance, cancer results. In light of this, why do you think that cigarette smoking is not illegal?

3. Analysis of the intake of dietary fat in 40 countries is presented in Figure 16-C, compared with rates of cancer deaths. What would you conclude from these data?

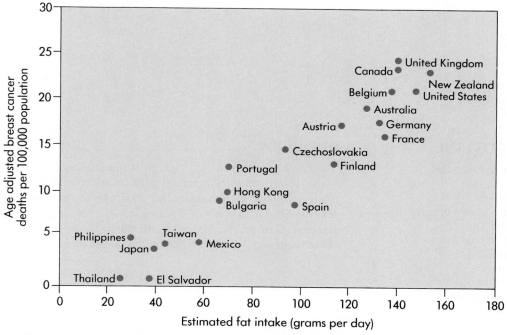

FOR FURTHER READING

BISHOP, J.M.: "Oncogenes," *Scientific American*, March 1982, pages 82-92. The story of *src* and how it causes cancer, by the man who first identified the protein product of the *src* gene. This article is particularly good at showing the chain of reasoning that underlies a major scientific advance.

CAIRNS, J.: "The Treatment of Diseases and the War Against Cancer," *Scientific American*, November 1985, pages 51-59. There are many different kinds of cancer, and this article does an unusually good job of describing the different forms and the progress that has been made in treating some of them.

COHEN, L.: "Diet and Cancer," *Scientific American*, November 1987, pages 42-48. Diet appears to play an important role in cancer, particularly breast cancer, which is associated with high levels of dietary fat.

KARTNER, N., and V. LING: "Multidrug Resistance in Cancer," *Scientific American*, March 1989, pages 44-51. Cancers resistant to many forms of chemotherapy seem to result from defects in a pump that flushes toxins out of cells.

LIOTTA, L.: "Cancer Cell Invasion and Metastasis," *Scientific American*, February 1992, pages 54-63. An up-to-date account of what is known about what makes a tumor cell metastatic.

OLD, L.: "Tumor Necrosis Factor," *Scientific American*, May 1988, pages 59-75. One of the few cancer cures that (sometimes) works, this small protein plays an important role in your body's defense against cancer.

SAGER, R.: "Tumor Suppressor Genes," *Science*, vol. 246, December 1989, pages 1406-1416. Some genes protect from cancer by preventing tumors from arising. These genes may play a key role in future gene-transfer therapies.

WEINBERG, R.: "The Molecular Basis for Cancer," *Scientific American*, November 1983, pages 126-144. This important paper by a developer of the transfection approach describes how transfection experiments yielded our first look at a cancer-causing gene.

WEINBERG, R.: "Finding the Anti-Oncogene," *Scientific American*, September 1988, pages 44-51. Mutation of certain growth-suppressing genes confers susceptibility to cancer. This article describes the first isolation of such a gene.

WILLETT, W.: "The Search for the Causes of Breast and Colon Cancer," *Nature*, vol. 338, March 1989, pages 389-394. The evidence is mounting that diet plays a key role in these two major killers.

17

Recombination

OVERVIEW

Genetic change, which has such great implications for evolution and human health, is the product of mutation, which creates the changes and recombination, which rearranges it. Recombination is responsible for most of the differences between individuals. It has also played a major role in organizing the eukaryotic genome and in determining the number of copies of genes and which genes are located near one another. Recombination is the architect of the genome.

FOR REVIEW

Here are some important terms and concepts that you will encounter in this chapter. If you are not familiar with them, you should review them before proceeding.

Heterochromatin (Chapter 10)

Crossing-over (Chapter 11)

Double-strand break repair theory (Chapter 11)

Independent assortment (Chapters 11 and 12)

All genetic change is ultimately produced by mutation. If we were genetically identical, recombination would not make us any more different, any more than exchanging numbers between 777 and 777 would alter either number. However, once mutation has created alternative versions of genes within the individuals of a population, then recombination offers a powerful means of reshuffling the combinations of alleles that occur in particular individuals. In Chapter 11 we discussed how meiosis and fertilization result in extensive recombination during sexual reproduction, a form of recombination that involves reassorting the array of chromosomes every generation, like shuffling a deck of cards between deals of a poker game. In this chapter we will focus on recombination occurring *within* chromosomes. We have met such within-chromosome recombination once already—the crossing-over described in Chapter 11. Many other forms also occur.

AN OVERVIEW OF RECOMBINATION

If we define mutation as any change in an organism's genetic message, then you might think that mutation is the only source of genetic diversity and that all change must arise as a result of it. Change, however, can also result from moving around existing elements of the genetic message on the chromosomes. As an analogy, consider the pages of this book. A mutation would correspond to a change in one or more of the letters on the pages. For example, ". . . in one or more of the letters *of* the pages" is a mutation of the previous sentence, in which a letter "n" is changed to an "f." A significant alteration is also achieved, however, by moving the position of words; ". . . in one or more of the *pages* on the *letters*," for example, alters the meaning of the sentence (and destroys it) by exchanging the position of the words "*letters*" and "*pages*." This second kind of genetic change, which is an alteration in the chromosomal position of a gene or a fragment of a gene, is a form of **genetic recombination.**

Viewed broadly, three kinds of genetic recombination occur in organisms: (1) **gene transfer,** in which one chromosome donates a segment to another; (2) **reciprocal recombination,** in which two chromosomes trade segments; and (3) **chromosome assortment,** discussed in Chapter 12, in which chromosomes assort during meiosis. The acquisition of an AIDS-bearing virus by a human chromosome is an example of gene transfer, the crossing-over that occurs between the homologous chromosomes of eukaryotes during meiosis is an example of reciprocal recombination, and the 9:3:3:1 dihybrid Mendelian ratio is an example of independent assortment. Gene transfer processes occur among both bacteria and eukaryotes; reciprocal recombination between chromosomes and independent assortment are strictly eukaryotic phenomena. For this reason, gene transfer is thought to be the more primitive process.

> Genetic recombination is a change in the chromosomal association among genes. It often involves a change in the position of a gene or portion of a gene. Recombination of this sort may result from one-way gene transfer or from reciprocal gene exchange.

GENE TRANSFER

The first of the major kinds of recombination that we will discuss is gene transfer, the unidirectional transfer of genes from one chromosome to another. Genes are not fixed in their locations on chromosomes, like words engraved in granite; they move around. Some genes move about because they are part of small auxiliary chromosomes called **plasmids,** which are able to enter and leave the main chromosome. Plasmids are able to do so, however, only at specific places on the chromosome where a nucleotide sequence occurs that is also present on the plasmid DNA. Other genes move as part of small fragments of the chromosome called **transposons,** which migrate from one chromosomal position to another at random, like fleas jumping along a string. Plasmids occur primarily in bacteria, in which the DNA can interact readily with other DNA fragments; transposons occur in both bacteria and eukaryotes. Both of these gene transfer processes, plasmid movement and transposon movement, probably together represent the earliest form of genetic recombination to evolve, one that is still responsible for much of the genetic recombination that occurs today.

These two forms of gene movement were discovered within 3 years of one another, the first (plasmid movement) by Joshua Lederberg and Edward Tatum in 1947, the second (transposon movement) by Barbara McClintock in 1950. Because the second mode of gene movement is random and implies that gene position on chromosomes is not constant, its discovery was not readily accepted by researchers accustomed to viewing chromosomes as composed of genes in fixed positions, like beads on a string. Lederberg and Tatum received a Nobel prize for their discovery in 1958. McClintock was awarded a Nobel prize for hers 25 years later, in 1983 (Figure 17-1).

Plasmids

Most, but not all, of the DNA of bacteria exists in the form of a single, circular, duplex molecule. About 5% of the total DNA, however, is found outside the main chromosome in small, circular DNA plasmids. Some plasmids are very small, containing only one or a few genes; others are quite complex and contain many genes.

To understand how plasmids arise, consider a hypothetical stretch of DNA along a bacterial chromosome that contains two copies of a repeated sequence. Because this sequence occurs twice on the chromosome, it is possible for the chromosome to form a transient "loop" in such a way that the two sequences base-pair with each other and create a transient double duplex. All cells have enzymes, called recombination enzymes, that can recognize such double duplexes. These enzymes can cause the two duplexes to exchange strands, to undergo a **reciprocal exchange.** The result of such an exchange, as you can see in Figure 17-2 (steps 1 to 3), is to free the loop from the rest of the chromosome. The resulting free loop is a plasmid. The DNA that makes up the plasmid corresponds to the genes, such as gene A between the two duplicated sequences in Figure 17-2. These genes are no longer present in the main chromosome and reside only on the plasmid.

Once a plasmid has been created by this kind of reciprocal exchange of DNA strands, it is a free element, able to move about within the cell. The plasmid may contain a replication origin, a site that the cell's DNA-replicating enzyme, DNA polymerase, recognizes. If it does, the plasmid will be replicated by the DNA polymerase, often without the controls that restrict replication of the chromosome to only once per cell division. Because their replication is independent from that of the cell, some plasmids may be represented by many copies and others by few copies in a given cell.

A plasmid that was created by recombination can reenter a chromosome the same way that it left it. Sometimes the region of the plasmid DNA involved in the original

FIGURE 17-2

Integration and excision of a plasmid. Because the ends of the two sequences on the bacterial chromosome are the same—D′, C′, B′, and D, C, B—it is possible for the two ends to pair. Steps 1 to 3 show the sequence of events if the strands exchange during the pairing. The result is excision of the loop and a free circle of DNA—a plasmid. Steps 4 to 6 show the sequence followed when a plasmid integrates itself into a bacterial chromosome.

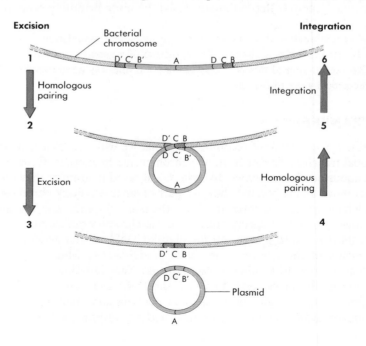

exchange, called the **recognition site,** aligns with its mate on the main chromosome; if a recombination event subsequently occurs anywhere in the common sequence while they are aligned, it will result in the reincorporation of the plasmid sequence into the chromosome. The plasmid is then said to have been **integrated** into the chromosome (Figure 17-2, steps 4 to 6). This integration can occur wherever the shared sequence (or any other shared sequence) exists, so that plasmids sometimes integrate at positions different from those at which they arose. If they are reincorporated into the chromosome at a new position, they transport their genes with them to the new location.

> **The locations where plasmids excise and integrate genes correspond to the positions of DNA sequences on the main chromosome that also occur in the plasmids.**

Gene Transfer Among Bacteria

A startling discovery that Lederberg and Tatum made in the 1950s was that bacteria can pass plasmids from one cell to another. The plasmid that was studied by Lederberg and Tatum was a fragment of the chromosome of *Escherichia coli.* It was given the name "F" for fertility factor, since only cells containing that plasmid could act as plasmid donors. The F plasmid contains a DNA replication origin and several special genes that promote its transfer to other cells. These genes encode protein subunits that assemble on the surface of the bacterial cell, forming a hollow tube called a **pilus.**

When the pilus of one cell contacts the surface of another cell that lacks pili, and therefore does not contain the F plasmid, the pilus forms a **conjugation bridge** between the two cells and mobilizes the plasmid within the first cell for transfer. First, the F plasmid binds to a site on the interior of the bacterial cell just beneath the pilus. The F plasmid then begins to replicate its DNA at that point of binding, passing the replicated copy out across the pilus and into the other cell (Figure 17-3). This **rolling-circle replication** can eventually place an entire single-strand copy of the plasmid into the recipient cell, where a complementary strand is added, creating a new stable F plasmid.

The F plasmid does not have to be free to carry out transfers of this kind. It will transfer just as readily if it is integrated into the bacterial chromosome. The F region binds beneath the pilus and initiates synthesis across the bridge between the two cells as just described. In this case, however, the cell begins to replicate a copy of *the entire bacterial chromosome* across to the recipient cell. Transfer proceeds just as if the bacterial chromosome were simply part of the small plasmid. Such transfer of bacterial genes is called **conjugation.** Researchers have used knowledge of this process to locate the positions of different genes on bacterial chromosomes (Figure 17-4).

> **Plasmids, such as the F plasmid, transfer bacterial genes from one individual bacterium to another by integrating the entire circular bacterial chromosome into the small circular plasmid DNA molecule. When a copy of the plasmid DNA molecule is replicated across to another cell, the integrated bacterial DNA is replicated across as if it were part of the plasmid.**

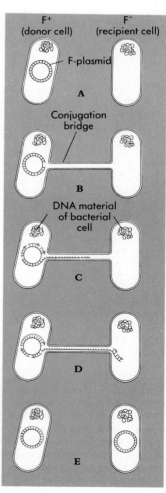

FIGURE 17-3

Gene transfer between bacteria.

A Donor cells contain an F plasmid that recipient cells lack.

B The long pilus connecting the two cells is called a conjugation bridge.

C Across the bridge the F plasmid replicates a copy of its chromosome. One strand is passed across, the other serving as a template to build a replacement.

D When the single strand enters the recipient cell, it serves as a template to assemble a double-stranded duplex.

E When the process is complete, both cells contain a complete copy of the circular plasmid DNA molecule.

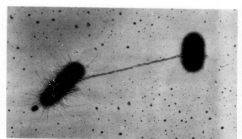

FIGURE 17-4
Scientists are now able to locate the position of genes on bacterial chromosomes. For the entire bacterial chromosome to pass across the conjugation bridge from donor to recipient cell requires over an hour. Because this bridge is fragile, it is often broken before all the chromosome has passed across. It is possible for an investigator to break all the conjugation bridges in a cell suspension by agitating the suspension rapidly in a blender. By conducting parallel experiments in which a blender is turned on at different times after the start of conjugation, investigators have been able to locate the positions of various genes on the bacterial chromosome. The closer the genes are to the origin of replication, the sooner one has to turn on the blender to block their transfer.

A The time elapsed from the beginning of conjugation before turning on the blender no longer blocks the genes indicated.

B A map of the *Escherichia coli* chromosome developed using this method.

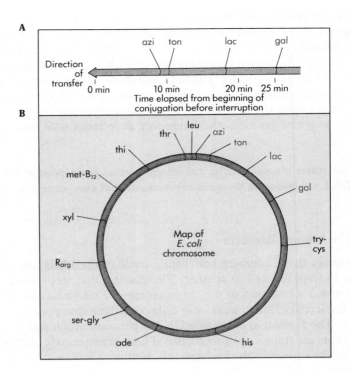

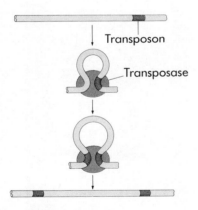

FIGURE 17-5
Transposition. The transposase does not recognize any particular DNA sequence, but rather selects one at random, with the result that the movement is to a random location. Some, but not all, transposons leave a copy of themselves behind when they move.

Transposition

Transposition is a form of gene transfer that occurs both in bacteria and in eukaryotes. Transposing genes do not stay put on chromosomes. Every once in a while, after many generations in one location, a transposing gene will abruptly move to a new position on the chromosomes, the location of its new residence apparently chosen at random. Transposing genes move about the chromosomes like so many Mexican jumping beans. These nomadic genes behave in this unusual way because they are carried along as parts of randomly moving genetic elements called transposons, passengers in a car with no driver.

Like a plasmid, a transposon may move from one chromosomal location to another. Rather than pairing a sequence shared in common with the chromosome, as plasmids do, transposons encode an enzyme called a **transposase,** which inserts the transposon into the chromosome. However, the transposase does not recognize any particular sequence on the host chromosome. Instead, it selects a site more or less randomly (Figure 17-5). Therefore the genetic locations of transposable elements change in a random manner.

> **Transposition is a one-way gene transfer to a random location on the chromosome. Genes move because they are associated with mobile genetic elements called transposons.**

Many transposons exist in multiple copies scattered about on the chromosomes. In *Drosophila,* for example, more than 30 different transposons are known, most of them present in numerous copies located at some 20 to 40 different sites throughout the genome. In all, the known transposons in a *Drosophila* cell account for perhaps 5% of its total DNA. Fewer different transposons are thought to occur in humans, but some of them are repeated many thousands of times.

The Impact of Transposition

The transposition of a given transposon element (Figure 17-6) is relatively rare, although an element is perhaps 10 times more likely to move than to undergo a random mutational change. The transposition of a particular mobile element occurs perhaps once in every 100,000 cell generations. There are many elements in most cells, however, and many generations to consider. Viewed over long periods of time, transposition has had enormous evolutionary impact, transposition causes both mutation and changes in gene location:

1. *Mutation.* The insertion of a mobile element within a gene often destroys the gene's function. Such a loss of gene function is termed **insertional inactivation.** It is thought that a significant fraction of the spontaneous mutations observed in nature has in fact resulted from this effect.

2. *Gene mobilization.* Natural selection sometimes favors **gene mobilization,** the bringing together in one place of genes that are usually located at different positions on the chromosomes. In bacteria, for example, a number of different genes encode enzymes that make the bacteria resistant to one or more antibiotics, such as penicillin. Many of these genes are encoded within small plasmids, different genes on different plasmids. The administration of several antibiotic drugs simultaneously to sick people was a common medical practice some years ago. Unfortunately, this simultaneous exposure to many antibiotics favors the persistence of plasmids that have managed to acquire several resistance genes. Transposition can rapidly generate such composite plasmids, with the antibiotic resistance genes moving by transposition from one plasmid to another. Called **resistance transfer factors,** these composite plasmids each possess many antibiotic resistance genes. The bacteria in which these plasmids occur are able to resist or destroy a variety of antibiotic drugs and are thus resistant to all of them simultaneously.

RECIPROCAL RECOMBINATION

The second major class of recombination that we will discuss is reciprocal recombination, in which two chromosomes trade segments. Reciprocal recombination is important primarily in eukaryotes, which typically possess two or more copies of their chromosomes at some stage of their life cycle. Human beings, for example, like most other animals, are diploid; they have two copies of each of their chromosomes in each of their cells throughout their entire life cycle, except when they form gametes. The possession of duplicate copies of chromosomes by eukaryotes has a profound consequence: because the two copies can be paired together, it is possible for them to exchange segments with each other. You encountered this process in Chapter 11 as crossing-over.

Crossing-Over

Crossing-over occurs in the first prophase of meiosis, when two homologous chromosomes are lined up side by side within the synaptonemal complex; an exchange of strands occurs between the two homologues at one or more locations. Such an exchange may result in the physical exchange of one of the paired chromosome strands at that point (see Figure 11-6).

When a reciprocal crossover event has concluded, the participating chromosomes often have physically exchanged chromosome arms. If the two chromosomes contain different mutations on each side of the crossover event, the physical exchange of chromosomal arms produces chromosomes that have different combinations of mutations. When such chromosomes continue on to segregate in meiosis, they form gametes that have new combinations of alleles.

Imagine, for example, that a giraffe has a gene encoding neck length on one of its chromosomes and a gene encoding leg length that is located elsewhere on that same chromosome. Now imagine further that a recessive mutation occurs within the giraffe population at the neck length locus, leading after several rounds of independent assortment to some individuals that are homozygous for a variant "long-neck" allele. Similarly, a recessive mutation occurs at the leg length locus, leading to individuals homozygous for a "long-leg" allele.

That the two independent mutations would occur at the same time in the same individual is very unlikely, since the probability of two independent events occurring

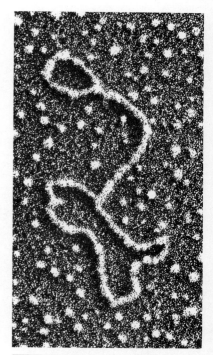

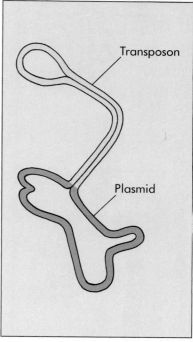

FIGURE 17-6
Electron micrograph of a transposon. Transposons form characteristic stem-and-loop structures, called "lollipops," because their two ends have the same nucleotide sequence, as inverted repeats, and so pair together to form the stem of the lollipop.

together is the product of their individual probabilities. If the only way to produce a giraffe that had both a long neck and long legs would be to await the spontaneous occurrence of both of these mutations in a single individual, it would be extremely unlikely that such an individual would ever occur. By recombination, however, a cross-over in the interval between the two genes could lead in one meiosis to the production of a chromosome bearing both variant alleles. It is because of this ability to reshuffle gene combinations rapidly that recombination is so important in the production of natural variation and in the process of evolution by natural selection in eukaryotes.

Gene Conversion

Because the two homologues that pair within a synaptonemal complex are not identical, places exist along the length of the paired chromosomes where one or more nucleotides of one homologue are not complementary to the strand with which it is paired. These occasional nonmatching pairs of nucleotides are called **mismatch pairs.**

As you might expect, the error-correcting machinery of the cell is able to detect mismatch pairing. Error correction is achieved during meiosis by a search for mismatches between the strands of a duplex within the syaptonemal complex. If a mismatch is detected, it is "corrected" by the enzymes that "proofread" new DNA strands during DNA replication (Figure 17-7): one of the mismatched strands is excised and the gap is filled in a way that is complementary to the other strand, producing two chromosomes with the same sequence. One of the two mismatched sequences has been lost. The conversion of one nucleotide sequence to another is called **gene conversion.**

> Gene conversion is the alteration of one homologous chromosome by the cell's error detection and repair system to make it resemble the other homologue.

Unequal Crossing-Over

Reciprocal recombination can occur between two chromosomes in any region where the sequences are similar enough to permit close pairing. Mistakes in pairing occasionally occur when identical sequences exist at several different locations on a chromosome. In such cases one copy of a sequence may line up not with its homologous segment on the other chromosome, but rather with one of the other duplicate copies on that chromosome. Such misalignment is responsible for the slipped mispairing that leads to the small deletions and frameshift mutations that were discussed in the previous chapter. If a crossover occurs in the region of pairing, this process is termed **unequal crossing-over,** because, as you can see in Figure 17-8, the two chromosomes exchange segments of unequal length.

In unequal crossing-over, one chromosome gains extra copies of the multicopy sequences and the other chromosome loses these copies. This process is capable of generating a chromosome that has hundreds of copies of a particular gene, lined up side by side along the chromosome in tandem array.

> Unequal crossing-over is the occurrence of a crossover event between chromosomal regions that are similar in nucleotide sequence but are not homologous. Unequal crossing-over can rapidly expand the number of copies of genes.

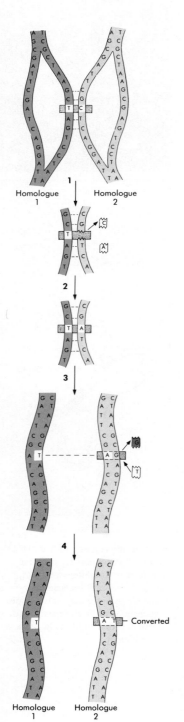

Homologue 1   Homologue 2

FIGURE 17-7

Gene conversion. When a cell's DNA repair enzymes detect an improper base pair (*1*), they set about "repairing" the problem base pair by excising one of the two bases, chosen at random, and replacing it with a base that pairs properly (*2*). When the two DNA strands that are pairing are not members of the same duplex, however, but rather homologues paired during meiosis, this repair process will tend to make one homologue a hybrid (*3*). This hybrid base, an AG pair in this example, will also be repaired, and half the time, on average, the G will be removed and replaced with a T (*4*). In these instances one strand (left) has actually been converted to the sequence of the other.

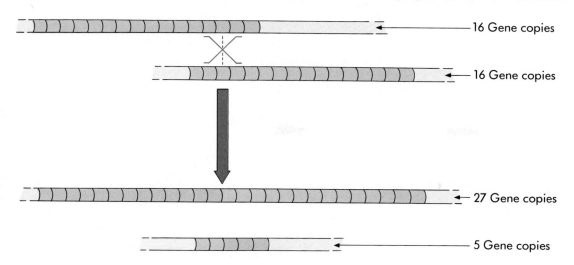

16 Gene copies

16 Gene copies

27 Gene copies

5 Gene copies

FIGURE 17-8

Unequal crossing-over. When a repeated sequence pairs out of register, a crossover within the region will produce one chromosome with fewer gene copies and one with more. It is thought that much of the gene duplication that has occurred during eukaryotic evolution is the result of unequal crossing-over.

Unequal crossing-over can occur between any two regions of the chromosome that are similar in sequence. Because the genomes of most eukaryotes possess multiple copies of transposons scattered throughout the chromosomes, unequal crossing-over between copies of transposons located at different positions on chromosomes has had a profound influence on gene organization in eukaryotes, as we shall discuss below. Most of the genes of eukaryotes appear to have been subject to duplication at one or more times during the course of their evolution.

A summary of the different kinds of genetic recombination that we have discussed is presented in Table 17-1.

THE EVOLUTION OF GENE ORGANIZATION

The two recombination processes we have discussed in this chapter, gene transfer and reciprocal recombination, are directly responsible for the architecture of the eukaryotic chromosome—what genes occur where, and in how many copies. To understand how recombination shapes the genome, it is instructive to compare the effects of recombination in prokaryotes and eukaryotes.

Bacterial and viral genomes are relatively simple compared with those of eukaryotes; genes almost always occur in them as single copies. Unequal crossing-over be-

TABLE 17-1	CLASSES OF GENETIC RECOMBINATION
CLASS	OCCURRENCE
GENE TRANSFERS	
Plasmid transfer	Occurs predominantly, if not exclusively, in bacteria and is targeted to specific locations on their chromosomes
Transposition	Common in both bacteria and eukaryotes; genes move from one chromosomal location to another at random
MEIOTIC RECOMBINATIONS	
Crossing-over	Occurs only in eukaryotes; requires the pairing of homologous chromosomes and may occur anywhere along their length
Unequal crossing-over	The result of crossing-over between mismatched segments; leads to gene duplication
Gene conversion	Occurs when homologous chromosomes pair and one is "corrected" to resemble the other
Independent assortment	Haploid cells produced by meiosis contain only one member of each pair of homologous chromosomes, selected at random

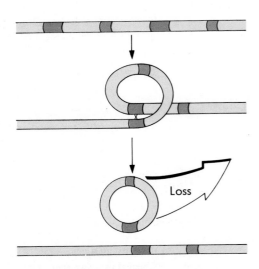

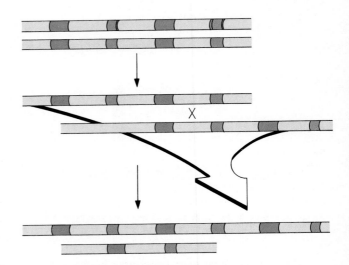

A Unequal crossing-over within a chromosome leads to loss of material

B Unequal crossing-over between chromosomes leads to addition of material

FIGURE 17-9

Unequal crossing-over has different consequences in bacteria and eukaryotes. Bacteria have only one (circular) chromosome, and a crossover WITHIN the chromosome between duplicate regions leads to loss of the intervening material. In eukaryotes, with two versions of each chromosome, crossover BETWEEN two versions leads to one chromosome with added material. As this chromosome goes on to form a gamete, gene amplification results.

tween repeated transposition elements on a circular chromosome such as that of a bacteria or virus tends to delete material (Figure 17-9, *A*), fostering the maintenance of a minimum genome size. For this reason bacterial and viral genomes are very tightly packed, with little or no wasted space. Note the efficient use of space in the organization of the *lac* genes described in Chapter 15 (see Figure 15-18)—hardly a nucleotide is wasted. The use of overlapping reading frames in viruses provides another example of the efficient use of space within circular chromosomes.

In eukaryotes, by contrast, the introduction of *pairs* of homologous chromosomes (presumably because of their importance in repairing breaks in two-strand DNA) has led to a radically different situation. *Unequal crossing-over between homologous chromosomes tends to promote the **duplication** of material rather than its reduction* (Figure 17-9, *B*). Consequently, the eukaryotic genome has during the course of its evolution been in a constant state of flux, with genes evolving multiple copies, some of which subsequently diverge in sequence to become new and different genes, which in their turn duplicate and diverge. Six different classes of eukaryotic DNA sequence are commonly recognized, based on copy number.

Satellite DNA

Some short nucleotide sequences are repeated several million times in the genomes of eukaryotes. About 4% of the human genome consists of repeated sequences of this kind. They are collectively called **satellite DNA.**

Why should such sequences be repeated so many times? Almost all of the copies of satellite DNA are either clustered together around the centromere (Figure 17-10) or are located near the ends of the chromosomes. These regions of the chromosomes remain highly condensed, tightly coiled, and untranscribed throughout the cell cycle; the satellite DNA sequences thus seem to be serving structural functions.

Transposition Elements

Other sequences in eukaryotic genomes are repeated thousands of times but are much longer than satellite sequences and are scattered randomly throughout the genome. These segments are transposons. From one generation to the next, a few transposons in each cell jump sporadically to new, apparently random locations. Fewer different transposons appear to be found in mammals than in many other organisms, but the ones that are present in mammals are repeated more often. The family of human transposons called *ALU* elements, for example, typically occurs about 300,000 times in each cell. Transposons are transcribed but appear to play no functional role in the life of the cell that bears them.

Tandem Clusters

A third class of repeated sequences encodes products the cell requires in large amounts. These sequences are repeated many times, one copy following another in tandem array. The cell is able to obtain large amounts of the encoded product rapidly by transcribing all of the numerous copies simultaneously. The genes encoding ribosomal RNA (rRNA), for example, are present in several hundred copies in most eukaryotes (Figure 17-11). Because these clusters are active sites of rRNA synthesis, they are readily visible in cytological preparations; they are called **nucleolar organizer regions.** When transcription of these rRNA gene clusters ceases during cell division, the nucleolus "disappears" from view under the microscope as the localized expansion of the chromosome subsides, only to reappear when transcription is reinitiated.

All of the many transcribed genes that are present in a **tandem cluster** are similar in sequence. Not all of the copies are identical—some differ by one or a few nucleotides—but their overall similarity is great. Each of the genes in a cluster is separated from its neighboring copies by a short "spacer" sequence, which is not transcribed. Unlike the transcribed genes of a cluster, spacers are not similar to one another. They vary considerably within a cluster, both in sequence and in length.

Multigene Families

As we learn more about the nucleotide sequence of the human genome, it has become apparent that many of our genes exist not in single copies, but rather as **multigene families,** groups of related but distinctly different genes often occurring together in a

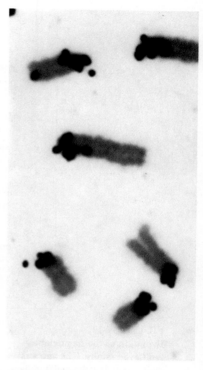

FIGURE 17-10
Satellite DNA. These mouse chromosomes have been exposed to a solution containing radioactive RNA complementary to mouse satellite DNA. The labeled RNA (dark spots) has bound to its complementary sequences on the DNA, showing that the satellite sequences are localized in centromeres.

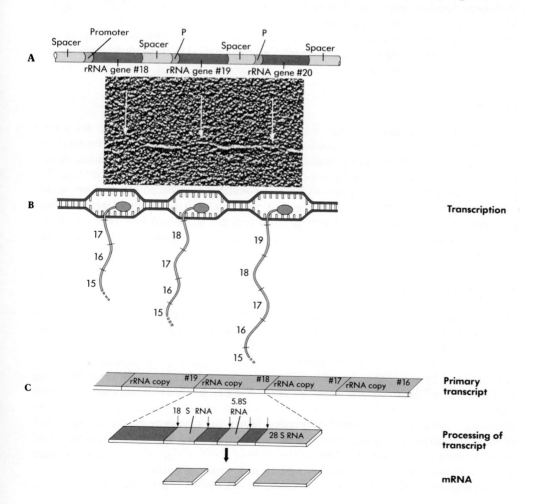

FIGURE 17-11
A tandem cluster. The genes encoding rRNA are repeated several hundred times, each copy separated from the next by a spacer.
A In a region of active rRNA synthesis, progressively longer segments of rRNA primary transcript can be seen as more and more of the repeated gene copies are passed by the RNA polymerase.
B A close-up view illustrates how the DNA duplex is opened in the zones corresponding to rRNA-encoding genes.
C Once a full transcript of the repeat region is produced, it is processed in two stages: first it is cut into lengths corresponding to the repeat unit; then each unit is cut into segments corresponding to the three different rRNA molecules.

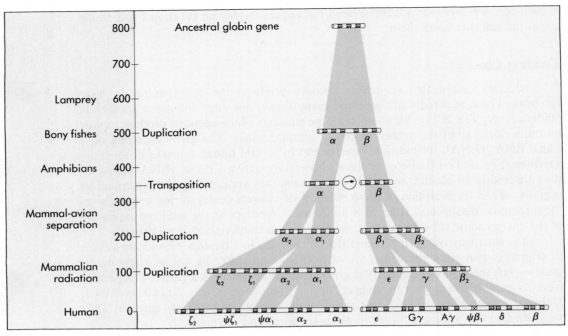

FIGURE 17-12

Evolution of the hemoglobin multigene family. The ancestral globin gene is at least 800 million years old. By the time that modern fishes had evolved, this ancestral gene had already duplicated, forming the α and β forms. Later, after the evolutionary divergence of amphibians and reptiles, these two forms moved apart on the chromosome *(arrow);* the mechanism is not known, but may have involved transposition. Each of these two sequences duplicated in mammals by the time their ancestors and those of the birds had differentiated as distinct lines of reptiles. In mammals two more waves of duplication occurred to produce the array of genes that occurs in humans. There are 11 globin genes in the human genome. Three of them (ψ) are silent, encoding nonfunctional proteins. Other forms are produced only during embryonic (ζ and ε) or fetal (γ) development. Of the 11 genes, only four (Δ, β, $α_1$, $α_2$) encode the chains of adult human hemoglobin.

cluster. They differ from tandem clusters in that they contain far fewer genes, and these genes are far more different from one another than those in the tandem clusters. Multigene families may contain as few as three or as many as several hundred genes. Although they do differ from one another, the members of a multigene family are clearly related in their sequences. All members of multigene families are thought to have arisen from a single ancestral sequence (Figure 17-12) through a series of unequal crossing-over events.

Dispersed Pseudogenes

Silent copies of a gene, which have been inactivated by mutation, are called **pseudogenes.** In some cases pseudogenes result from mutations in promoters; in other cases they result from frameshift mutations or small deletions. Pseudogenes may occur within a multigene family cluster; they may also occur at a distant site. We call distantly located pseudogenes **dispersed pseudogenes,** because they have somehow moved far from their original position within a multigene family gene cluster. The existence of such dispersed pseudogenes was not suspected as recently as a few years ago, but they are now thought to be of major evolutionary significance in eukaryotes.

Single-Copy Genes

Over the long time since the appearance of the eukaryotes, processes such as unequal crossing-over between different copies of transposons have time and time again caused segments of the eukaryotic chromosome to duplicate; it appears that no portion of the genome has been immune to this process. Indeed, this process of duplication followed by the conversion of copies to pseudogenes has probably been the major source of "new" genes during evolution; pseudogenes accumulate mutational changes until a fortuitous combination of changes in one pseudogene results in an active gene encoding a protein with different properties. When the new gene encoding this useful protein first arises, it is a single-copy gene but it in its turn will undergo duplication. A single-copy gene is but one stage in the cycle of duplication and divergence (Figure 17-13) that characterizes the evolution of the eukaryotic genome.

A summary of the classes of genes found in eukaryotes is presented in Table 17-2.

TABLE 17-2	THE CLASSES OF GENES FOUND IN EUKARYOTES
CLASS	COPY NUMBER
Satellite DNA	Short sequences present in millions of copies per genome
Transposition elements	Thousands of copies of transposons scattered around the genome
Tandem clusters	Clusters containing hundreds of nearly identical copies of a gene
Multigene families	Clusters of a few copies of related but distinctly different genes
Dispersed pseudogenes	Nonactive members of a multigene family located elsewhere
Single-copy genes	Genes that exist in only one copy

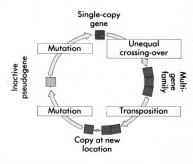

FIGURE 17-13
The cycle of gene duplication and divergence in eukaryotes.

CHROMOSOMAL REARRANGEMENTS

Although transposable sequences move around within the genome, most genes appear to maintain relatively stable positions even over long evolutionary periods. Why are genes not scattered to different locations by recombination and transposition? For most genes, chromosome location plays an important role in the cell's control of gene transcription, concerning when particular genes are used and when they are not. For example, some eukaryotic genes are not capable of being transcribed if they are adjacent to a tightly coiled heterochromatic sequence, even though the same genes are transcribed normally in any nonheterochromatic location. The transcription of many chromosomal regions appears to be regulated in this manner, with the binding of specific chromosomal proteins regulating the degree of coiling in local regions of the chromosome and so decreasing the accessibility of genes located within the region to RNA polymerase.

Chromosomes undergo several different kinds of gross physical alterations that have significant effects on the locations of their genes. The two most important are **translocations,** in which a segment of one chromosome becomes a part of another chromosome, and **inversions,** in which the orientation of a portion of a chromosome becomes reversed. Translocations, like transpositions, often have important effects on gene expression. Inversions, in contrast, usually do not alter gene expression but are important because of their effect on recombination. Recombination within a region of the chromosome that is inverted on one chromosome and not on the other (Figure 17-14) leads to serious problems during meiosis, because none of the gametes produced following such a crossover event will have a complete set of genes.

Other chromosomal alterations change the number of gene copies that an individual possesses. Individual genes may be deleted or duplicated. Whole chromosomes may be gained or lost. An individual that has gained or lost a whole chromosome is

FIGURE 17-14
The consequence of inversion.
When a segment of a chromosome is inverted (*1*), it can pair in meiosis only by forming an internal loop (*2*). Any crossing-over (*3*) that occurs within the inverted segment during meiosis will result in nonviable gametes; as some genes are lost from each chromosome, others are duplicated (*4* and *5*). The pairing that occurs between inverted segments can sometimes be seen under the microscope, producing characteristic loops such as those illustrated in the inset. For clarity, only two strands are shown although crossing over occurs in the four strand stage.

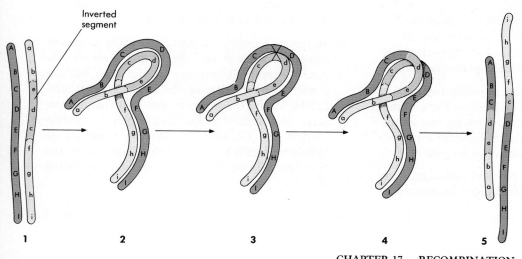

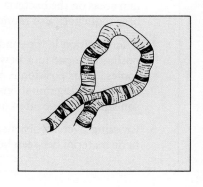

Inverted segment

1   2   3   4   5

said to be **aneuploid.** Whole diploid sets of chromosomes may be multiplied; an individual, tissue, or cell that has three or more sets of chromosomes is said to be **polyploid** (see Chapter 21). An individual chromosome segment may be deleted from the genome. Most deletions are harmful; they halve the number of gene copies within a diploid genome and thus seriously affect the level of transcription. A part of a chromosome may be duplicated, causing gene imbalance, and is usually harmful.

THE IMPORTANCE OF GENETIC CHANGE

All evolution begins with genetic change, with the processes described in this and the previous chapter: mutation, which creates new alleles; gene transfer and transposition, which alter gene location; reciprocal recombination, which shuffles and sorts the changes; and chromosomal rearrangement, which alters the arrangement of entire chromosomes. Some changes produce alterations in an organism that enable it to leave more offspring, and these changes are preserved as the genetic endowment of future generations. Other changes reduce the ability of an organism to leave offspring. These changes tend to be lost, as the organisms that carry them contribute fewer members to future generations than those which lack them.

Evolution can be viewed as the selection of particular combinations of alleles from a pool of different alternatives. The rate of evolution is ultimately limited by the rate at which these different alternatives are generated. Genetic change—mutation and recombination—provides the raw material for evolution.

SUMMARY

1. Recombination is defined as the creation of new gene combinations. It includes both the exchange of entire chromosomes that occurs during meiosis and changes in the position of a gene or a fragment of a gene on a chromosome.

2. Recombination within a chromosome occurs when the position on the chromosome of a gene or a fragment of a gene changes. Recombination encompasses the transfer of a gene to the same position on a homologous chromosome or to a different position on the same or a different chromosome. It also includes the reciprocal exchange of genes between two chromosomes.

3. Bacterial genes may be transferred by association with plasmids, which are small circles of DNA. The plasmid enters the bacterial genome by a reciprocal exchange of strands, a process that can only occur in a region of sequence similarity. For this reason, a plasmid may carry genes only to certain areas on the bacterial chromosome—sites with base sequences also represented on the plasmid.

4. Transposition is the random movement of a transposing element to a new location on the chromosomes. Transposition is responsible for many naturally-occurring mutations, as insertion of a transposon within a gene often inactivates the gene.

5. Crossing-over involves a physical exchange of genetic material between homologous chromosomes

during the close pairing that occurs in meiosis. It may produce chromosomes that have different combinations of mutations.

6. Satellite sequences are short sequences repeated millions of times. They are commonly associated with nontranscribed regions of chromosomes and probably play a structural role.

7. Transposition elements (transposons) are scattered throughout the genomes of prokaryotes and eukaryotes. In humans and other mammals only a few different sequences exist, but they are repeated hundreds of thousands of times. Transposons are transcribed, but they are not thought to any regular function in the life of the cell.

8. Tandem clusters are genes that occur in thousands of copies clustered together at one or a few sites on the chromosome. These encode for products the cell requires in large amounts.

9. Multigene families are genes that occur in tens of copies clustered at one site on the chromosome. Because these clusters are smaller than tandem clusters, less pairing occurs between members. Consequently, the copies in multigene families diverge in sequence more than copies in tandem clusters do. Occasionally, inactive members of multigene families move to other locations on the chromosome. Such "distant cousins" are called dispersed pseudogenes.

REVIEW QUESTIONS

1. Define the term genetic recombination. What are three kinds of genetic recombination that occur naturally in living organisms? Which of these occur in eukaryotes? Which occur in prokaryotes? Which is most primitive? Why?

2. What is a plasmid? At what point can it insert on the main chromosome? What is a transposon? In which kinds of organisms are plasmids found? In which are transposons found?

3. Can bacteria pass plasmids from one cell to another? What occurs if the bacterial F plasmid is inserted in the main chromosome at the time of transfer? What is this process called?

4. Define insertional inactivation.

5. When does crossing-over between homologous chromosomes occur? Is it more likely that variation will result from two independent mutations in a single individual or that it will result from exchange of a mutation in crossing-over?

6. What is a mismatched pair? How does the cell react to these mismatchings? What is this process called?

7. What is unequal crossing-over? When is this most likely to occur? How does this ultimately affect the resultant gametes?

8. What are the six classes of eukaryotic DNA sequences?

9. What is satellite DNA? Where is it primarily located, and what form does it take? What is the probable function of satellite DNA?

10. What kind of information do tandem clusters encode? How does this occur? What is an important example of the information encoded by a tandem cluster?

11. What kinds of genes exist in multigene families? How do they differ from tandem clusters? How are multigene families thought to have evolved?

12. What are pseudogenes? How do dispersed pseudogenes differ from those located within a multigene family cluster?

13. How are genes protected from being scattered all over the genome? What is a translocation? What is an inversion?

 THOUGHT QUESTIONS

1. Many eukaryotic genes exist in multiple copies. Most bacterial genes, in contrast, exist in single copies. As a result, a newly arisen mutation usually alters the phenotype of the bacterium in which it occurs. In contrast, when a mutation arises in a eukaryotic gene, it alters only one member of a multigene family. Does this mean that mutations in eukaryotes do not usually alter the phenotype, and if so, how does evolution occur in eukaryotes?

2. The analysis of the structure of vertebrate-dispersed-pseudogenes suggests that RNA transcripts of genes, after being processed in the cytoplasm, may gain reentry into the genome as DNA copies of the transcripts. Since many processes that go on within the cytoplasm alter RNA molecules, does this passage of RNA back into the genome imply that experience may act to modify the genetic message?

 FOR FURTHER READING

BROSIUS, J.: "Retroposons—Seeds of Evolution," *Science*, February 1991, vol. 251, page 753. A brief description of an exciting new development–the possibility that retrovirus-like transposons are responsible for many introns.

DONELSON, J., and M. TURNER: "How the Trypanosome Changes Its Coat," *Scientific American*, February 1985, pages 44-51. An easily understood discussion of how parasites responsible for sleeping sickness use transposition.

FEDEROFF, N.: "Transposable Genetic Elements in Maize," *Scientific American*, June 1984, pages 85-98. An excellent introduction to the studies for which Barbara McClintock was awarded the 1983 Nobel prize in physiology or medicine, rephrased in molecular terms that were not available when she made her important discoveries.

NICOLAS, A., AND OTHERS: "An Initiation Site for Meiotic Gene Conversion in Yeast," *Nature*, vol. 338, March 1989, pages 35-39. The powerful techniques of molecular biology solve a key puzzle, the molecular nature of meiotic crossing over. The evidence favors a double-strand-break and repair mechanism.

PATRUSKY, B.: "DNA on Target: Homologous Recombination," *Mosaic*, vol. 21, May 1990, pages 44-52. Recombination can be used to replace damaged genes.

SHEN, S., J. SLIGHTOM, and O. SMITHIES: "A History of the Human Fetal Globin Gene Duplication," *Cell*, vol. 26, pages 191-203, 1981. A concise summary of the evolution of perhaps the best understood multigene family.

STAHL, F.: "Genetic Recombination," *Scientific American*, February 1987, pages 91-101. A detailed look at the molecular events underlying recombination, using viruses as a model system.

18

Gene Technology

CHAPTER OUTLINE

OVERVIEW

Recent advances in molecular biology have given us powerful new tools with which to investigate genetics at the molecular level. Particularly important new techniques permit geneticists to isolate individual genes and to transfer them from one kind of organism to another. These techniques are revolutionizing medicine and agriculture.

FOR REVIEW

Here are some important terms and concepts that you will encounter in this chapter. If you are not familiar with them, you should review them before proceeding.

Enzymes (Chapter 7)

DNA (Chapters 3 and 14)

Point mutation (Chapter 16)

Plasmids (Chapter 17)

During the last decade, a revolution in genetics has taken place since new and powerful techniques have been developed to study and manipulate DNA. These techniques have allowed biologists for the first time to intervene directly in the genetic fate of organisms (Figure 18-1). In this chapter we will discuss these techniques and consider their application to specific problems of great practical importance. Few areas of biology will have as great an impact on our future lives.

PLASMIDS AND THE NEW GENETICS

In 1980 geneticists first succeeded in introducing a human gene, the one that encodes the protein **interferon,** into a bacterial cell. Interferon is a rare protein that increases human resistance to viral infection and is difficult to purify in any appreciable amount. It may prove to be the basis of a useful therapy against cancer. This possibility has been difficult to explore, however, since the purification of substantial amounts of interferon required for large-scale clinical testing would, until recently, have been prohibitively expensive. An inexpensive way to produce interferon was needed, and introducing the gene responsible for its production into a bacterial cell made this possible.

The bacterial cell that had acquired the human interferon gene proceeded to produce interferon at a high rate, and to grow and divide. Soon millions of bacterial cells were in the culture, all of them descendants of the original bacterial cell that had the human interferon gene, and all of them furiously producing interferon. This procedure of producing a line of genetically identical cells from a single "altered" cell, called **cloning,** had succeeded in making every cell in the culture a miniature factory for the production of human interferon. In a similar way, the successful cloning of insulin was commercially significant, providing the basis for producing large amounts of a clinically important drug for relatively little expense. Cloning also has considerable theoretical significance. Such molecular techniques can and will be used increasingly to manipulate genes, and by doing so, enable us to learn more about them. This interferon experiment and others like it mark the beginning of a new genetics, the birth of **genetic engineering.**

Genetic engineering is based on the ability to cut up DNA into recognizable pieces and to rearrange these pieces in different ways. In the experiment just described the gene segment carrying the interferon gene was inserted into a plasmid, which brought the inserted gene in with it when it infected the bacterial cell. Most other genetic engineering approaches have used the same general strategy of carrying the gene of interest into the target cell by first incorporating the gene into an infective plasmid or virus.

The success of the initial step in a genetic engineering experiment is the key to the whole procedure. As you might expect, success depends on being able to cut up the source DNA (human DNA in the interferon experiment, for example) and the plasmid DNA in such a way that the desired fragment of source DNA can be spliced permanently into the plasmid genome. This cutting is performed by a special kind of enzyme called a **restriction endonuclease.*** These restriction enzymes are able to recognize and cleave specific sequences of nucleotides in a DNA molecule. They are the basic tools of genetic engineering.

RESTRICTION ENZYMES

Scientific discoveries often have their origins in odd little crannies—seemingly unimportant areas that receive little attention by researchers before their general significance is appreciated. In the example we are considering now, the particular obscure topic was the warfare that takes place between bacteria and viruses.

FIGURE 18-1
Larger than life—a product of genetic engineering. These two mice are genetically identical, except that the large one has one extra gene—the gene encoding a potent growth hormone not normally present in mice. The gene was added to the mouse genome by human genetic engineers and is now a stable part of the mouse's genetic endowment.

---

*There are two classes of restriction endonucleases, called Type I and Type II. Type I nucleases make cuts at random locations, and are not useful as tools in genetic engineering. Type II nucleases recognize specific sequences. It is this second class that are commonly referred to as restriction enzymes and used in genetic engineering procedures.

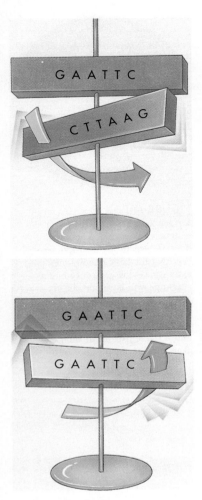

FIGURE 18-2
Restriction sites have internal symmetry. The nucleotide sequences recognized by restriction enzymes have twofold rotational symmetry—the second half of the sequence is the complementary mirror of the first. As a result, if the sequence CTTAAG is rotated through the plane of the paper as illustrated here, it becomes its complementary sequence, GAATTC. The importance of twofold rotational symmetry is that both DNA strands have the same sequence, read in opposite directions.

Most organisms in nature eventually evolve means of defending themselves from predators, and bacteria are no exception to this rule. Among the natural enemies of bacteria are the viruses called bacteriophages, which infect bacteria, multiply within them, and eventually burst the bacterial cell, releasing thousands of viral particles. For a bacterium, a virus is a potentially lethal adversary. As a result of natural selection, those bacterial individuals that can somehow resist viral infection will be favored and leave more progeny, on the average, than those that lack such means. Some bacteria have powerful weapons against viruses: enzymes that chop up the foreign viral DNA as soon as it enters the bacterial cell but without harming the bacterial DNA at all. When viruses insert their DNA into a bacterial cell that is protected in this way, the viral DNA is immediately attacked by these enzymes, called restriction endonucleases, and degraded. Why is the DNA of the bacteria not degraded by the restriction enzymes? Because the bacterial cell has modified its own DNA in such a way that the restriction enzymes do not recognize it as DNA.

Restriction endonucleases recognize specific nucleotide sequences within a DNA strand, bind to DNA strands at sites where these sequences occur, and cleave the bound strand of DNA at a specific place in the recognition sequence. In this way they cut up DNA. Other bacterial enzymes called **methylases** recognize the same bacterial DNA sequences, bind to them, and add methyl (—CH$_3$) groups to the nucleotides. When the recognition sites of bacterial DNA have been modified with methyl groups in this way, they are no longer recognized by the restriction enzymes. Consequently the bacterial DNA is protected from being degraded. Viral DNA, on the other hand, is not protected because it has not been methylated.

The sequences that restriction enzymes recognize are typically four to six nucleotides long and symmetrical. Their symmetry is of a special kind, called **twofold rotational symmetry** (Figure 18-2). The nucleotides at one end of the recognition sequence are complementary to those at the other end, so that the two strands of the duplex have the same nucleotide sequence running in opposite directions for the length of the recognition sequence. This arrangement has two consequences. The first is of great importance to the bacteria; the second is of little significance in the bacterial systems, but of paramount importance to us:

1. Because the same recognition sequence occurs on both strands of the DNA duplex (running in opposite directions), the restriction enzyme is able to recognize and cleave both strands of the duplex, effectively cutting the DNA duplex in half. This ability to chop across both strands is almost certainly the reason that restriction enzymes have evolved in such a way that they specifically recognize nucleotide sequences with twofold rotational symmetry—it lets them use one sequence to bind both strands.
2. Because the position of the bond cleaved by a particular restriction enzyme is typically not in the center of the recognition sequence to which it binds* and because the sequence is running in opposite directions on the two strands, the sites at which the two strands of a duplex are cut are offset from one another. An example of being offset can be illustrated by taking your two hands, one palm up and the other palm down, and fitting the little fingers and ring fingers together. After cleavage, the two fragments of DNA duplex each a short single strand a few nucleotides long dangling from the end. Look carefully at Figure 18-3: the two single-stranded tails are complementary to one another.

There are hundreds of bacterial restriction enzymes recognizing a variety of four- to six-nucleotide sequences. Six-nucleotide sequences are the most common. Every cleavage by a given kind of restriction enzyme takes place at the same recognition sequence. By chance, this sequence will probably occur somewhere in any given sample of DNA, so that a restriction endonuclease will cut DNA from any source into frag-

---

*A few restriction enzymes are known in which the cleavage position is in the center of a four- or six-nucleotide sequence. This results in fragments without dangling single-stranded ends—"blunt" ends. Such restriction enzymes do not generate fragments that spontaneously reassociate, and are used in genetic engineering procedures to prevent such reassociation from occurring.

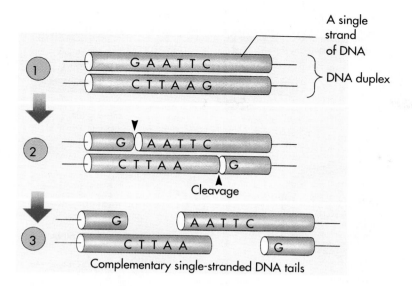

A single strand of DNA

DNA duplex

Cleavage

Complementary single-stranded DNA tails

FIGURE 18-3
The method in which restriction enzymes produce DNA fragments with "sticky ends." The restriction enzyme Eco R1 always cleaves the sequence GAATTC at the same spot, after the first G. Because the same sequence occurs on both strands, both are cut. But the position of G is not the same on the two strands; the sequence runs the opposite way on the other strand. As a result, single-stranded tails are produced. Because of the twofold rotational symmetry of the sequence, the single-stranded tails are complementary to each other, or "sticky."

ments. The shorter the sequence, the more often it will arise by chance within a genome. Each of these fragments will have the dangling sets of complementary nucleotides (sometimes called "sticky ends") characteristic of that endonuclease. Because the two single-stranded ends produced at a cleavage site are complementary, they can pair with each other. Once they have done so, the two strands can then be joined back together with the aid of a sealing enzyme called a **ligase,** which reforms the phosphodiester bonds—the bonds between the sugars and phosphates of DNA. This latter property makes restriction endonucleases the invaluable tools of the genetic engineer: *any* two fragments produced by the same restriction enzyme can be joined together. Fragments of elephant and ostrich DNA cleaved by the same bacterial restriction enzyme can be joined to one another just as readily as can two bacterial fragments because they have the same complementary sequences at their ends.

A restriction enzyme cleaves DNA at specific sites, generating in each case two fragments whose ends have one strand of the duplex longer than the other. Because the tailing strands of the two cleavage fragments are complementary in nucleotide sequence, *any* pair of fragments produced by the same enzyme, from any DNA source, can be joined together.

CONSTRUCTING CHIMERIC GENOMES

A **chimera** is a mythical creature with the head of a lion, the body of a goat, and the tail of a serpent. No such chimera ever existed in nature. Human beings, however, have made them—not the lion-goat-snake variety, but chimeras of a more modest kind.

The first actual chimera was a bacterial plasmid that American geneticists Stanley Cohen and Herbert Boyer made in 1973. Cohen and Boyer used a restriction endonuclease to cut up a large bacterial plasmid called a resistance transfer factor. From the resulting fragments, they isolated one fragment 9000 nucleotides long, which contained both the sequence necessary for replicating the plasmid—the replication origin—and a gene that conferred resistance to an antibiotic—tetracycline.

Because both ends of this fragment were cut by the same restriction enzyme (called *Escherichia coli* restriction endonuclease 1, or Eco R1), they could be joined together to form a circle, a small plasmid that Cohen dubbed **pSC101** (Figure 18-4). Cohen and Boyer used the same restriction enzyme, Eco R1, to cut up DNA that they had isolated from an adult amphibian, the African clawed toad, *Xenopus laevis.* They then mixed the toad DNA fragments with opened-circle molecules of pSC101, allowed bacterial cells to take up DNA from the mixture, and selected for bacterial cells that had become resistant to tetracycline (Figure 18-5). From among these pSC101-

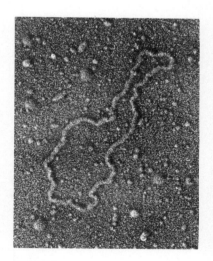

FIGURE 18-4
A famous plasmid. The circular molecule in the electron micrograph was the first plasmid used successfully to clone a vertebrate gene, pSC101. Its name refers to it being the one hundred-and-first plasmid isolated by Stanley Cohen.

FIGURE 18-5

FIGURE 18-5

The first genetic engineering experiment. This diagram illustrates how Cohen and Boyer inserted an amphibian gene encoding rRNA into a bacterial plasmid. pSC101 contains a single site cleaved by the restriction enzyme Eco R1. The rRNA-encoding region was inserted into pSC101 at that site by cleaving the rRNA region with Eco R1 and allowing the complementary sequences to pair.

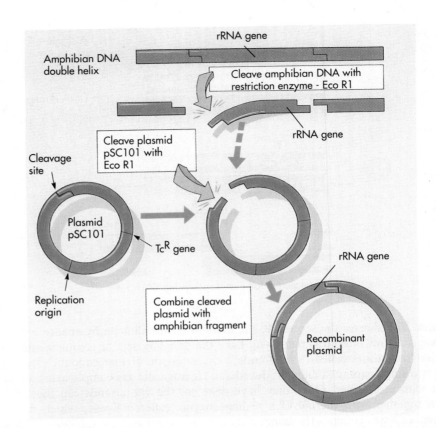

containing cells they were able to isolate ones containing the toad ribosomal RNA gene. These versions of pSC101 had the toad gene spliced in at the Eco R1 site. Instead of joining to one another, the two ends of the pSC101 plasmid had joined to the two ends of the toad DNA fragment that contained the ribosomal RNA gene.

The pSC101 containing the toad ribosomal RNA gene is a true chimera. It is an entirely new creature that never existed in nature and would never have evolved there. It is a form of **recombinant DNA,** a DNA molecule created in the laboratory by molecular geneticists who joined together bits of several genomes into a novel combination.

> **The first recombinant genome produced by human genetic engineering was a bacterial plasmid into which an amphibian ribosomal RNA gene was inserted in 1973.**

The insertion of fragments of foreign DNA into bacterial cells by carrying them into the cells piggyback in plasmids or viruses has become common in molecular genetics. Newer-model plasmids with exotic names, such as pBR322, can be induced to make hundreds of copies of themselves and thus of the foreign genes that are included in them within such bacterial cells. Even easier entry into bacterial cells can be achieved by inserting the foreign DNA fragment into the genome of a bacterial virus, such as lambda virus, instead of into a plasmid. The infective genome that harbors the foreign DNA and carries it into the target cell is called a **vector,** or **vehicle.** Not all vectors have bacterial targets. Animal viruses, for example, have been used as vectors to carry bacterial genes into monkey cells. Animal genes have even been carried into plant cells by using methods of this kind.

There has been considerable discussion about the potential danger of inadvertently creating an undesirable life-form in the course of a recombinant DNA experiment. What if one fragmented the DNA of a cancer cell and then incorporated the fragments at random into viruses that were propagated within bacterial cells? Might there not be a danger that among the resulting bacteria there could be one capable of constituting an infective form of cancer?

Even though most recombinant DNA experiments are not dangerous, such concerns are real and need to be taken seriously. Both scientists and individual governments monitor these experiments to detect and forestall any hazard of this sort. Experimenters have gone to considerable lengths to establish appropriate experimental safeguards. The bacteria used in many recombinant DNA experiments, for example, are unable to live outside of laboratory conditions; many of them are obligate anaerobes, poisoned by oxygen. Decidedly dangerous experiments such as cancer cell shotgun experiments (in which genomes of cancer cells are cleaved randomly and the fragments inserted into plasmids and screened) are prohibited.

GENETIC ENGINEERING

The movement of genes from one organism to another is often referred to as recombinant DNA technology or genetic engineering. Although each experiment presents unique problems, all genetic engineering approaches share in common four distinct stages:

Stage 1: Cleavage. The first stage of any genetic engineering experiment is the generation of specific DNA fragments by cleavage of a genome with restriction endonuclease enzymes. Because a given six-nucleotide sequence will occur many times within a genome, restriction endonuclease cleavage will produce a large number of specific fragments, called a **library.** Different "libraries" of fragments may be obtained by employing enzymes that recognize different sequences. Fragments are usually compared by electrophoresis (Figure 18-6), which permits estimation of relative size.

Stage 2: Producing recombinant DNA. The fragments of a restriction fragment library are put into plasmids or virus "vehicles," called **vectors,** which will later carry the fragments into other cells. The key property of the recombinant vector is that the fragment is now replicated as part of the plasmid or virus genome. At this stage or later it is necessary to eliminate those vectors that do NOT contain a fragment.

Stage 3: Cloning. The fragment-containing plasmid or virus vectors are introduced into bacterial cells. Each such cell then reproduces, forming a clone of cells that all contain the fragment-bearing plasmid. Each of the cell lines is maintained separately; together they constitute a clone library of the original genome.

Stage 4: Screening. From the many clonal lines that contain fragments of the original library, it is necessary to identify those containing the specific fragment of interest, often a fragment containing a particular gene.

FIGURE 18-6
Gel electrophoresis. This process separates DNA or protein fragments (**A**), causing them to migrate within a gel in response to an electric field. The fragments migrate according to size and can be visualized easily (**B**), as the migrating bands glow in fluorescent light.

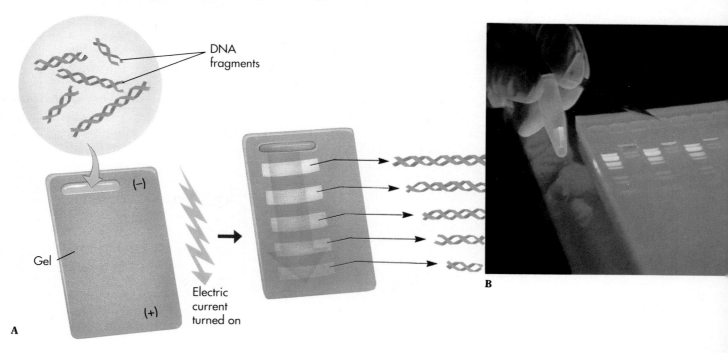

The fourth stage, screening, is the most critical—and difficult—experimental step.

Initial Screening of Clones

The key to a successful genetic engineering experiment lies in the strategy adopted to identify and select the desired fragment. To make this job easier, investigators usually try to eliminate from the final mix any bacterial cells that do not contain a plasmid or virus vector and any vectors that do not contain a fragment from the original library. They do this by using genes that make bacteria resistant to antibiotics (Figure 18-7), chemicals such as penicillin, tetracycline, or ampicillin, which would otherwise block bacterium growth.

1. To eliminate bacteria without a vector, a vector is employed with an antibiotic resistance gene, such as one conferring tetracycline resistance. By culturing the clones of stage 4 on a medium containing tetracycline, the investigator ensures that only bacteria resistant to this antibiotic (because they contain the vehicle) will be able to grow.

2. To eliminate bacteria with a vector that does not contain a fragment, a vector is employed that has only one restriction site for the cleavage enzyme used, a site located within a second antibiotic resistance gene, such as one conferring ampicillin resistance; by testing the clones in stage 4 individually for ampicillin resistance, it becomes possible to identify directly those clones derived from cells that have successfully taken up a fragment (they have lost their resistance to this antibiotic because one of the library fragments is now sitting within the resistance gene).

Finding the Gene You Want

A library of restriction fragments, such as results from stage 3 above, may contain anywhere from a few dozen to many thousand individual members, each representing a different fragment of genomic DNA. A complete *Drosophila* (fruit fly) library, for example, contains more than 40,000 different clones; a complete human library of fragments 20 kilobases long would contain 150,000 clones. To identify a single fragment containing a particular gene from within that immense clone library often requires ingenuity, but many different approaches have been used successfully.

FIGURE 18-7
Screening restriction fragment clones.

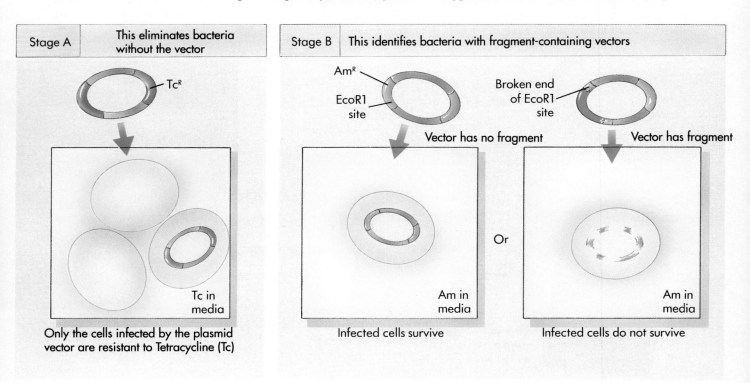

Stage A	This eliminates bacteria without the vector

Tc^R

Tc in media

Only the cells infected by the plasmid vector are resistant to Tetracycline (Tc)

Stage B	This identifies bacteria with fragment-containing vectors

Am^R

EcoR1 site

Vector has no fragment

Am in media

Infected cells survive

Broken end of EcoR1 site

Vector has fragment

Or

Am in media

Infected cells do not survive

One of the most useful procedures for identifying a specific gene has been the **Southern blot,** a procedure that employs purified mRNA as a "probe"; among thousands of fragments, only that fragment containing the proper gene will **hybridize** (bind by complementary base pairing) with the probe, since only that fragment has a nucleotide sequence complementary to the mRNA's sequence. In such a screening procedure, the fragments are spread apart by electrophoresis, and a radioactive probe (synthesized in the presence of ^{32}P) is "blotted" onto the resulting gel pattern, incubated, and then the excess washed off; because only the gene-containing fragment will hybridize with the probe, only that fragment produces a radioactive band on film left beneath the gel.

Getting Enough DNA to Work With: PCR

Once a particular gene is identified within the library of DNA fragments, it only remains to make multiple identical copies of it. This can be done by inserting the identified fragment into a bacterium--after repeated cell replications and divisions, millions of cells will contain identical copies of the fragment. A far more direct approach, however, is to use DNA polymerase to make millions of identical copies of the gene of interest, an approach called **polymerase chain reaction,** or **PCR** (Figure 18-8):

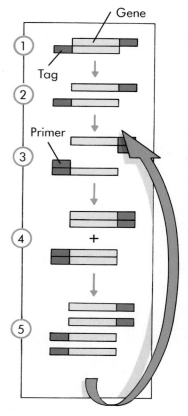

FIGURE 18-8
The Polymerase Chain Reaction.

1. **TAGGING.** First, the experimenter adds to each end of the fragment a different short single-stranded nucleotide sequence. Called **oligonucleotides,** such short 20-30 nucleotide segments can be manufactured in large quantities by machine. When this step is completed, the fragment is a little longer than it was, and each of its ends has a unique short tag.
2. **HEATING.** Now a solution of the tagged fragment is heated to about 98° C, just below boiling, and a large excess of the oligonucleotide duplexes is added to the solution. At this high temperature, the DNA duplex of the fragment and oligonucleotides dissociate into single strands.
3. **PRIMING.** Next the solution is allowed to cool to about 60° C. As it does, the single strands of DNA reassociate into double strands—*but because there is a large excess of oligonucleotide, each fragment strand picks an oligonucleotide as a double-strand partner.* The oligonucleotide binds at the end, to its complementary sequence tag.
4. **COPYING.** Now DNA polymerase is added, a very heat-stable type extracted from the bacterium *Thermus aquaticus,* which lives in hot springs. Using the oligonucleotide as a starting point, or primer, the polymerase proceeds to copy the rest of the fragment as if it were replicating DNA. When it is done, what used to be the oligonucleotide primer is now lengthened into a complementary copy of the entire single-stranded fragment--and because BOTH single strands behave this way, we are left with two copies of the fragment we started with!
5. **REPEATING THE CYCLE.** Steps 2-4 are now repeated, and the two copies become four. It is not necessary to add any more polymerase, as the heating step does not harm this *T. aquaticus* enzyme. Each heating and cooling cycle can be as short as one or two minutes, and each cycle doubles the number of DNA target molecules. After 20 cycles, a single fragment produces 2^{20} number of copies--over a million! In a few hours, 100 billion copies of the fragment can be easily manufactured.

PCR has revolutionized many aspects of molecular biology, because it allows the investigation of minute samples of DNA. In criminal investigations DNA fingerprints can be prepared from the cells in a tiny speck of dried blood or at the base of a single human hair. Doctors can detect genetic defects in very early embryos by collecting a few sloughed-off cells and amplifying their DNA with PCR, or use PCR to examine the DNA of Abraham Lincoln.

George Poinar Jr. of the University of California at Berkeley has suggested a novel use, popularized in the novel *Jurassic Park: 1. Find a bead of amber that contains a blood-sucking insect from the age of dinosaurs; 2. extract DNA from blood cells of a bitten dinosaur and amplify with PCR; 3. process and inject into the embryo of an alligator; 4. wait until it hatches!* This recipe makes many as-yet-untested assumptions, but points out the great impact this technique is having on all aspects of biology.

A SCIENTIFIC REVOLUTION

The 1980s saw an explosion of interest in applying genetic engineering techniques to solving practical human problems.

Pharmaceuticals

Perhaps the most obvious commercial application and the one seized first was to introduce human genes that encode clinically important proteins into bacteria. Because bacterial cells can be grown cheaply in bulk, bacteria that incorporate human genes can produce large amounts of the human proteins that they specify. This method has been used to produce several forms of human insulin and interferon, to place rat and human growth hormone genes into mice (see Figure 18-1), and to manufacture many commercially valuable nonhuman enzymes.

Among the many medically important proteins now being produced by these approaches is **tissue plasminogen activator,** a protein produced in minute amounts by the body that causes blood clots to dissolve and may be effective in preventing heart attacks and strokes. Also being produced are **atrial peptides,** small proteins being tested as possible new ways to treat high blood pressure and kidney failure.

Agriculture

A second major area of genetic engineering activity is the manipulation of the genes of key crop plants. In plants the primary experimental difficulty has been identifying a suitable vector for introducing genes into target organisms. Plants do not possess the many plasmids that bacteria do, so the choice of potential vectors is therefore limited. The most successful results obtained thus far with plant systems have involved a plasmid of the plant bacterium *Agrobacterium tumefaciens*, called the **T_i plasmid,** which infects broadleaf plants such as tomato, tobacco, and soybeans. Part of this T_i plasmid integrates into the plant DNA; it has proved possible to attach other genes to this portion of the plasmid (Figure 18-9). The characteristics of a number of plants have

FIGURE 18-9
The use of the T_i plasmid of *Agrobacterium tumefaciens* in plant genetic engineering.

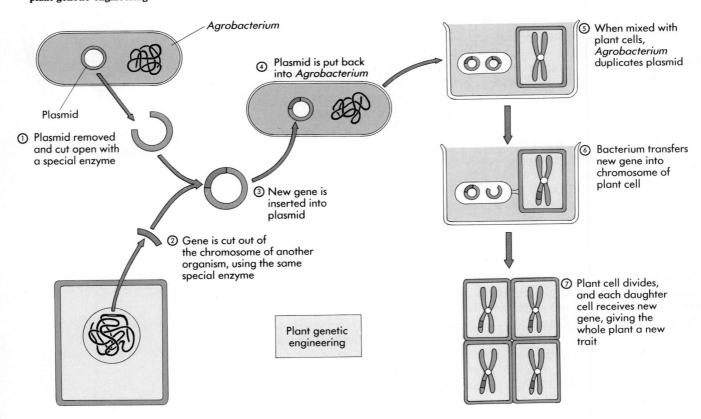

Agrobacterium

Plasmid

① Plasmid removed and cut open with a special enzyme

② Gene is cut out of the chromosome of another organism, using the same special enzyme

③ New gene is inserted into plasmid

④ Plasmid is put back into *Agrobacterium*

Plant genetic engineering

⑤ When mixed with plant cells, *Agrobacterium* duplicates plasmid

⑥ Bacterium transfers new gene into chromosome of plant cell

⑦ Plant cell divides, and each daughter cell receives new gene, giving the whole plant a new trait

FIGURE 18-10
Genetically engineered herbicide resistance. All four of these petunia plants were exposed to equal doses of the herbicide Roundup. The two on top were genetically engineered to be resistant to glyphosate, the active ingredient of Roundup, whereas the dead ones were not.

been altered through the use of this technique, which will be valuable in improving crops and forest trees. Among the features that scientists would like to effect are resistance to disease, frost, and other forms of stress; nutritional balance and protein content; and herbicide resistance. Unfortunately, *Agrobacterium* generally does not infect the cereals such as corn, rice, and wheat, and alternative methods are being developed to introduce new genes into them.

Herbicide Resistance. Recently, broadleaf plants have been genetically engineered to be resistant to the herbicide **glyphosate,** the active ingredient in Roundup, a powerful and broad-spectrum biodegradable chemical that kills most green, actively growing plants (Figure 18-10). Glyphosate kills plants by inhibiting an enzyme called EPSP synthetase, which is required by plants to produce aromatic amino acids. Humans do not make aromatic amino acids but get them from their diet, and so are unaffected by glyphosate. To obtain Roundup-resistant plants, agricultural scientists first inserted extra copies of the EPSP genes into plants, carrying them in "piggyback" on a T_i plasmid. The engineered plants "overproduce" the enzyme, producing 20 times normal levels, and so are able to withstand EPSP suppression caused by glyphosate; the extra amounts of the EPSP enzyme still permit them to carry out protein production and growth. In later experiments, a bacterial form of the EPSP synthetase gene that differs from the plant form by a single nucleotide was introduced into plants via T_i plasmids with the same result—the plants became resistant to Roundup.

This advance is of great interest to farmers, since a crop resistant to Roundup would never have to be weeded if the field were simply treated with the herbicide. Because Roundup is a broad-spectrum herbicide, farmers would no longer need to employ a variety of different herbicides on a single crop (most herbicides kill only a few kinds of weeds). Finally, glyphosate is readily broken down in the environment, a great improvement over many commercial herbicides commonly used in agriculture. A plasmid that will permit introduction of EPSP genes into cereal plants making them Roundup-resistant is being actively sought.

Virus Resistance. T_i plasmids have been used to introduce a variety of other genes into broadleaf crop plants. In one of the most interesting experiments, Roger Beachy of Washington University developed plants immune to a virus (Figure 18-11). Unlike humans, plants have no immune system of their own. What Beachy did was to introduce the tobacco mosaic virus (TMV) gene encoding the virus protein coat into a tobacco cell chromosome, using the T_i plasmid-transfer system; he then grew the plasmid-infected plant cell in tissue culture, and from the culture grew a whole tobacco plant. Every cell of this new plant was derived from the T_i-infected cell and contained the TMV gene. When the genetically engineered tobacco plants are reinfected with the virus, they do not contract the disease!★ This exciting avenue is being

---

★For reasons that are not clear, the TMV virus does not infect a cell that is already TMV-infected, and Beachy's experiment tricks the TMV viruses into behaving as if all the cells of the tobacco plants were already infected.

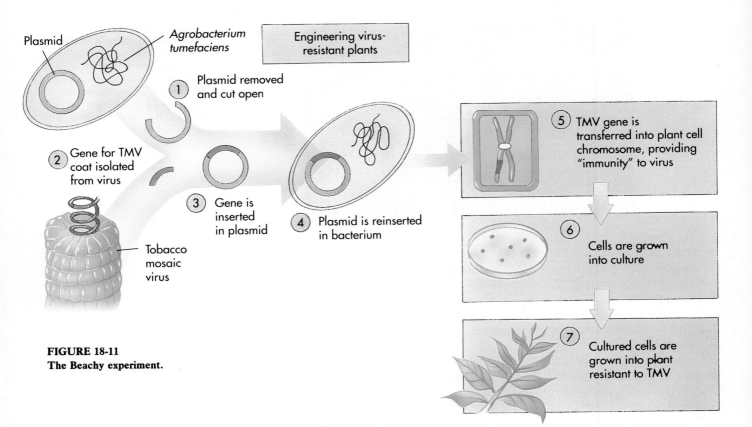

FIGURE 18-11
The Beachy experiment.

Plasmid

Agrobacterium tumefaciens

Engineering virus-resistant plants

① Plasmid removed and cut open

② Gene for TMV coat isolated from virus

Tobacco mosaic virus

③ Gene is inserted in plasmid

④ Plasmid is reinserted in bacterium

⑤ TMV gene is transferred into plant cell chromosome, providing "immunity" to virus

⑥ Cells are grown into culture

⑦ Cultured cells are grown into plant resistant to TMV

vigorously pursued, since it offers hope of engineering resistance to many viral diseases of commercial crops.

Immunity to Insects. Many commercially important plants are attacked by insects. Over 40% of the chemical insecticides used today, for example, are used to kill boll weevil, boll worm, and other insects that eat cotton plants (Figure 18-12). Using genetic engineering, we may be able to produce cotton plants that are naturally resistant to these insect pests, thereby removing the need for much chemical insecticide. What is being attempted is to engineer an insecticide toxic to boll weevils and boll worms but harmless to all other animals and plants.

The approach is to place into crop plants genes harmful to insects that feed on the plant but not harmful to other creatures. In one approach, the gene for a suitable insecticidal protein has been identified in *Bacillus thuringiensis,* a bacterium that lives in the soil (Figure 18-13). When tomato pests such as the caterpillar-like tomato horn-

FIGURE 18-12
An agricultural use of genetic engineering. Over 40% of the chemical insecticides used in the world in 1987 were used to control insects on cotton. Using genetic engineering, scientists may be able to produce cotton plants that are naturally resistant to insect pests.

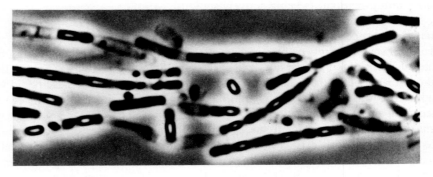

FIGURE 18-13
Bacillus thuringiensis, **a bacterium which produces a protein toxic to many caterpillars.** The white spots within the bacteria are protein crystals that are transformed into a toxin by enzymes in the caterpillar's stomach.

FIGURE 18-14
A target for future genetic engineering. Crop plants have been engineered to produce a naturally occurring protein that kills certain insect pests. Advances like this could reduce the farmer's dependence on chemical insecticides.

worm (Figure 18-14) ingest this protein, enzymes in the caterpillar's stomach convert it into an insect-specific toxin, causing paralysis and death. Since the necessary enzymes are not found in other insects or animals, the protein is harmless to them. Using the T_i plasmid system, the gene encoding for the insecticidal protein has been transferred to tomato and tobacco plants, and the plants have been tested in the open field. The plants containing the gene are protected from attack by insects that feed on them (Figure 18-15).

Many important plant pests attack roots, and to counter this threat biologists have introduced the *Bacillus thuringiensis* insecticidal protein gene into root-colonizing bacteria, especially strains of *Pseudomonas*. *Bacillus thuringiensis* does not normally colonize plant roots as *Pseudomonas* does, so this gene transfer affords a novel form of protection. The field testing of this promising approach is currently awaiting approval by the Environmental Protection Agency, which is assessing the possibility that other soil bacteria might be affected by the altered form of *Pseudomonas*. A variety of novel approaches is being employed to check this unlikely but important possibility (Figure 18-16).

FIGURE 18-15
A successful experiment in pest resistance. Both of these tomato plants were exposed to destructive caterpillars under laboratory conditions. The non-engineered plant on the left has been completely eaten, while the engineered plant shows virtually no signs of damage.

Nitrogen Fixation.

A more long-range goal of agricultural genetic engineering is to increase yield and plant size in crop plants; however, we do not yet know, for the most part, which genes are responsible for these complex characteristics. An important long-range goal is to introduce into key crop plants the genes that enable soybeans and other legume plants to "fix" nitrogen. These genes belong to certain symbiotic bacteria living in the plant's roots; they enable the legume plant to convert the nitrogen in the air to ammonia, a form of nitrogen that the plant can use to make amino acids and other molecules. Other plants, which lack these bacteria and so cannot fix nitrogen themselves, must obtain their nitrogen from the soil; farmland where crops are grown repeatedly soon becomes depleted of nitrogen unless nitrogen fertilizers are applied. Worldwide, farmers applied over 60 million metric tons of nitrogen fertilizers in 1987, an expensive undertaking. Farming costs would be much cheaper if major crops like wheat or corn could be engineered to carry out biological nitrogen fixation. However, introducing the nitrogen-fixing genes from bacteria into plants has thus far proved difficult, because these genes do not seem to function properly in their new eukaryotic environment. Experiments are being pursued actively, particularly with other species of nitrogen-fixing bacteria whose genes might function better in non-legume plants.

Farm Animals.

One of the first genes to be cloned successfully, that encoding the growth hormone somatotropin, is now undergoing final USDA testing before becoming commercially available to dairy farmers and beef growers. Instead of extracting the hormone from cow pituitaries at great expense, bovine somatotropin is produced

FIGURE 18-16
Keeping track of genetically altered bacteria. To monitor the location of genetically engineered *Pseudomonas* during and after a field test, scientists have introduced into them the bacterial *lac* genes; in the presence of the sugar lactose, which natural *Pseudomonas* ignore, the genetically engineered *Pseudomonas* break down the lactose and, as a result, turn a detector dye blue.

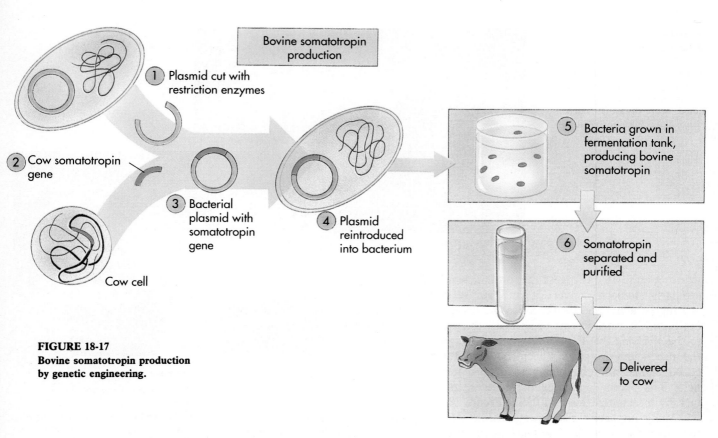

FIGURE 18-17
Bovine somatotropin production by genetic engineering.

Bovine somatotropin production

① Plasmid cut with restriction enzymes

② Cow somatotropin gene

③ Bacterial plasmid with somatotropin gene

④ Plasmid reintroduced into bacterium

Cow cell

⑤ Bacteria grown in fermentation tank, producing bovine somatotropin

⑥ Somatotropin separated and purified

⑦ Delivered to cow

FIGURE 18-18
A fatter pig. Research is currently being conducted that fattens pigs by the introduction of human growth hormone.

in large amounts by a gene introduced into bacteria. The hormone, used as a supplement to a dairy cow's diet, improves the animal's milk production efficiency (Figure 18-17). It has no effect on humans because it is a protein and is digested in the stomach like all the other proteins we ingest. Other experiments using somatotropin to increase the weight of cattle and pigs are underway (Figure 18-18). The human version of this same growth hormone is currently being used as a potential treatment for disorders in which the pituitary gland fails to make adequate levels of somatotropin, producing dwarfism.

Probing the Human Genome

Genetic engineering techniques are enabling us to learn a great deal more about the human genome. Any cloned gene can now be localized to a specific chromosomal location by using radioactive probes to detect *in situ* hybridization—only at the site where the probe binds to the chromosome DNA does the radioactive label accumulate. Several clonal libraries of the human genome have been assembled, using large-size restriction fragments. Maps of these fragments' locations on the human chromosomes are being constructed, and a very detailed map of the human genome will soon result. These attempts are most important, since many genetic diseases can be shown to be associated with specific restriction fragments. When the disease-associated genes have been cloned, they can be used as probes to screen the human clone library and the appropriate restriction fragment identified. Mutations often affect the way in which the fragment moves on an electrophoresis gel and can always be detected as a change in nucleotide sequence of the fragment. Genetic screening for potential birth defects by the use of restriction fragment analysis is a very real possibility in the near future.

A major effort has been mounted in the United States to sequence the entire human genome, one restriction fragment at a time (Figure 18-19). Since the human genome contains some 3×10^9 nucleotide base pairs, this task presents no

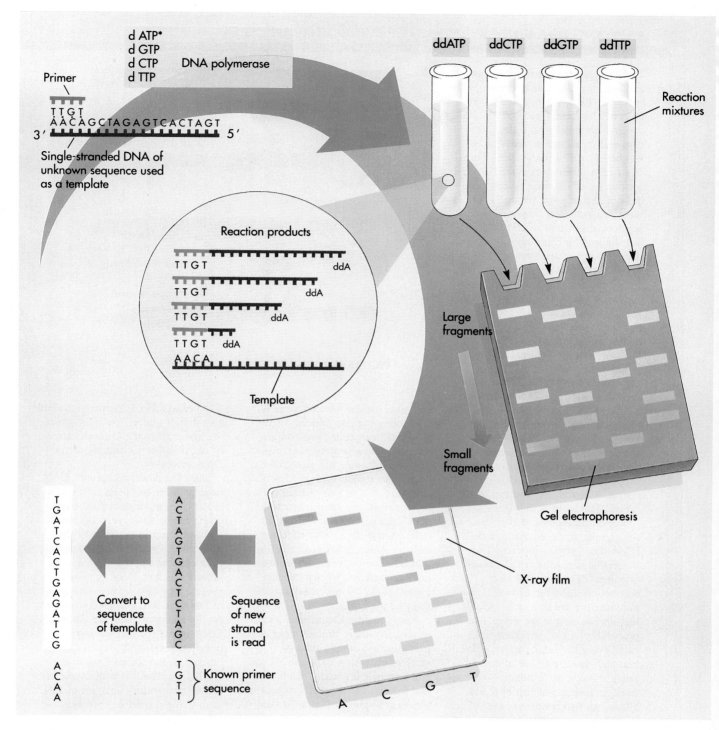

FIGURE 18-19

Steps in sequencing DNA. Most DNA sequencing is currently done using the technique of "primer extension." A short two-stranded primer is added to the end of a single-stranded fragment to be sequenced. This provides a 3′ end for DNA polymerase to add to. The primed fragment is placed in four in vitro synthesis tubes, each containing a different synthesis-stopping dd-nucleotide. The first tube, for example, contains ddATP and synthesis stops whenever ddATP is incorporated instead of dATP. Thus this tube will come to contain a series of fragments, corresponding to the different lengths the polymerase can travel from the primer before A nucleotides are encountered (ddATP is added at random, and will not always hit every site). Electrophoresis separated these fragments according to size. A radioactive label (here dATP*) allows the fragments to be visualized on x-ray film, and the newly-made sequence can be read directly. Try it. The fragment you started with has the complementary sequence.

The 27-year old woman never saw her assailant. She was asleep on February 22, 1987, when a man broke in to her house in Orlando, Florida, covered her head with a sleeping bag, and raped her in bed. On November 3, 1987 a man named Tommie Lee Andrews was placed on trial for this rape. The case against him did not at first seem strong. There were two indistinct fingerprints on a window screen that resembled his, but the woman could not identify Andrews as her attacker—her head covered, she had never seen the man who raped her. Andrew's girlfriend and sister swore he never left home that night. Like many rape cases, this one presented the jury with conflicting testimony and evidence open to more than one interpretation.

Then the prosecuting attorney introduced a new line of evidence: there was a witness. The man who raped this woman could be clearly identified, without ambiguity and beyond the shadow of a doubt. The witness was not human, someone who saw what happened. Such a witness could always be doubted, as people do make mistakes in identification. The witness was DNA.

The chromosomes of every human cell contain scattered through their DNA short, highly-repeated 15-nucleotide segments called "mini-satellites." The locations and number-of-repeats of any particular mini-satellite are so highly variable that no two people are alike. The probability of two unrelated individuals having the same pattern of location and repeat-number of mini-satellite is one in 10 billion—with a world population of just over 5.4 billion, no two people are ever alike. If one analyzes two mini-satellites to be sure, the probability is a minuscule 5×10^{-19}, which for all practical purposes is zero.

In Figure 18-A you can see the evidence presented by the prosecuting attorney. It consisted of autorad-

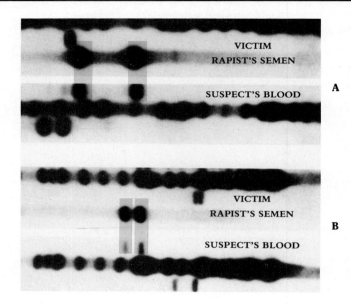

FIGURE 18-A

iographs, parallel bars on x-ray film resembling the line patterns of the universal price code found on groceries. Each bar represents the position of a mini-satellite, detected by techniques similar to those described in Figure 18-6 of this chapter. A vaginal swab had been taken from the victim within hours of her attack, from it semen was collected, and the semen DNA analyzed for its mini-satellite patterns. These mini-satellite patterns, labeled "semen" in Figure 18-A, are beyond question those of the rapist.

Now compare a mini-satellite pattern of the rapist to that of the suspect Andrews, lining the two samples up with a standardized control lane that has many bands: you can see that the suspect's pattern is identical to that of the rapist (and not at all like that of the victim). And all the patterns are similarly the same. Clearly the semen collected from the rape victim and the blood sampled from Tommie Lee Andrews came from the same person.

On November 6 the jury returned a verdict of guilty. Andrews became the first man in the United States to be convicted of a crime based on DNA evidence.

Since the Andrews verdict, DNA fingerprinting has been admitted as evidence in more than 2000 court cases. Just as fingerprinting revolutionized forensic evidence in the early 1900s, so DNA fingerprinting is revolutionizing it today. A hair, a minute speck of blood, a drop of semen, they all can serve as sources of DNA, to damn or clear a suspect. As the man who analyzed Andrew's DNA says: "It's like leaving your name, address, and social security number at the scene of the crime. It's that precise." It is of course critical that laboratory analyses of DNA samples be carried out properly—sloppy procedures could lead to a wrongful conviction. After several widely publicized instances of questionable laboratory procedures, national standards are now being developed.

small challenge. Other simpler organisms are also being sequenced. In 1991 the first eukaryotic chromosome to be entirely sequenced, yeast chromosome III, yielded a surprising finding: over half the genes on the chromosome are "new," previously undetected in attempts to produce mutations. These genes apparently act to modulate cellular functions and may play key roles in evolutionary adaptation.

Piggyback Vaccines

Another area of potential significance involves the use of genetic engineering techniques to help produce **sub-unit vaccines.** For example, vaccines against herpes virus and hepatitis viruses have recently been manufactured by using gene-splicing techniques. Genes specifying the protein-polysaccharide coat of the herpes simplex virus or hepatitis B virus were spliced into a fragment of the DNA genome of the cowpox, or vaccinia, virus—the same virus that British physician Edward Jenner used almost 200 years ago in his pioneering vaccinations against smallpox.

Live vaccinia virus is introduced into a mammalian cell culture (Figure 18-20) along with spliced fragments of vaccinia virus DNA, into which have been incorporated genes encoding the protein-polysaccharide coat of the herpes simplex virus genome (or, in other experiments, of hepatitis B virus). Under such circumstances the spliced herpes simplex gene is able to recombine into the vaccinia virus genome, producing a recombinant vaccinia virus. When this recombinant virus is injected into a mouse or rabbit, it dictates the production not only of the proteins that the vaccinia virus genome specifies, but also of the protein-polysaccharide coat specified by the herpes genes that it carries. The viruses that it produces are thus somewhat like sheep wearing wolf's clothing—they have the interior of a benign vaccinia virus and the exterior surface of a herpes virus. Infected individuals contract cowpox and not herpes, and cowpox is relatively harmless. At the same time infected individuals make antibodies against herpes and so become immune to it. The antibody-producing cells of the infected animal react to the **outer surface** of the recombinant virus particles, thereby causing the animal to develop a high level of antibodies directed against the

FIGURE 18-20
Constructing a "subunit" vaccine.

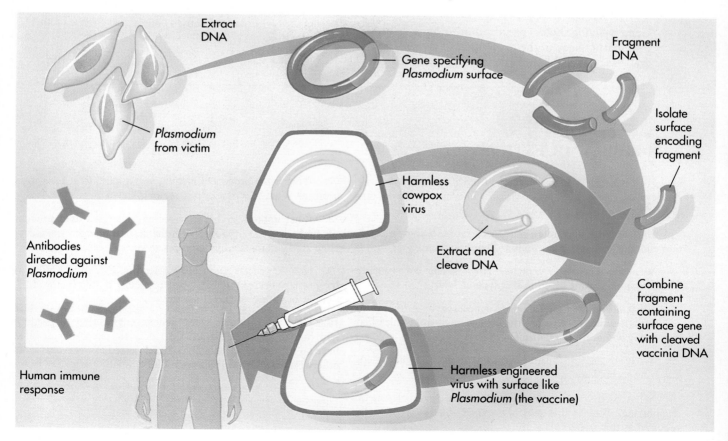

Extract DNA

Gene specifying *Plasmodium* surface

Fragment DNA

Plasmodium from victim

Isolate surface encoding fragment

Harmless cowpox virus

Extract and cleave DNA

Combine fragment containing surface gene with cleaved vaccinia DNA

Antibodies directed against *Plasmodium*

Human immune response

Harmless engineered virus with surface like *Plasmodium* (the vaccine)

cell surface of the herpes virus. Vaccines produced in this way are harmless, since only a small fragment of the DNA of the disease-associated virus is introduced via the recombinant virus.

The great attraction of this approach is that it does not depend on the nature of the virus-caused disease. In the future similar recombinant viruses injected into humans may prove able to confer resistance to a wide variety of viral diseases.

SUMMARY

1. Genetic engineering involves the isolation of specific genes and their transfer to new genomes.

2. The key to genetic engineering technology is a special class of enzymes called restriction endonucleases, which cleave DNA molecules into fragments. These fragments have short, one-strand tails that are complementary in nucleotide sequence to each other.

3. Because the sequences of the tails of the two cleavage products are complementary, they can pair with one another and rejoin together, whatever their source. For this reason DNA fragments from very different genomes can be combined.

4. The first combination of fragments from different genomes (recombinant DNA) was achieved by Stanley Cohen and Herbert Boyer in 1973. They inserted an amphibian ribosomal RNA gene into a bacterial plasmid.

5. Gene splicing holds great promise as a clinical tool, particularly in the prevention of disease. Several vaccines are being prepared by stitching the coat-specifying genes of disease-causing viruses or cells into the genomes of harmless viruses and using the recombinant form, which is harmless and has the outside coat of the disease form, as a vaccine to induce antibody formation against the disease.

6. Genetic engineering has fostered many advances in the production of affordable drugs, including insulin, interferon, tissue plasminogen activator, and atrial peptides.

7. Agriculture has been a major focus of genetic engineering activity. Successes have included incorporation of genes for herbicide resistance into crop plants, and, in parallel experiments, incorporation of genes conferring natural resistance to insect pests.

8. One of the great potential benefits of genetic engineering is the production of vaccines directed against human diseases such as herpes and hepatitis.

REVIEW QUESTIONS

1. How is bacterial DNA protected from attack by its own virus-degrading enzymes? What class of enzymes performs this function and how do they work?

2. How are the nucleotide sequences recognized by restriction endonuclease enzymes unique? Describe them.

3. What is a chimeric genome? How is one constructed? What is another name given to this type of DNA?

4. What is the primary safeguard employed by scientists to reduce the possibility of risk to the environment from unleashed genetically engineered organisms?

5. What are the four distinct steps common to all genetic engineering technology?

6. What occurs during the recombination phase of genetic engineering? Why is this step necessary? What is additionally required at this step?

7. What is the process of cloning as it applies to genetic engineering? What function does screening serve?

8. What important characteristic of bacteria is used to identify and select for a plasmid with the desired inserted fragment? How are bacteria lacking appropriate vectors selected against?

9. How is genetic engineering affecting agriculture? What is the primary vector for such plant research? Is this vector effective for broadleaf or cereal plants?

10. How are scientists using genetic engineering to produce new human vaccines? What effect does the recombinant virus have on the organism into which it is injected? Why is this virus otherwise harmless to the injected individual?

THOUGHT QUESTIONS

1. You are having a genetic engineering nightmare. In your dream, a well-meaning student digests the DNA of reticulocytes (red blood cells) isolated from a man dying of leukemia with the restriction enzyme Eco R1. The student mixes the resulting fragments with the Eco R1-treated plasmid pSC101, infects a growing culture of *E. coli* bacteria with the resulting mix of normal and chimeric plasmids, and selects for cells containing plasmids by adding the antibiotic tetracycline to the liquid culture of growing bacteria. Many cells continue to grow. In an excess of joy at this intimation of success, the student accidently drops the tube in the sink. It shatters, and the liquid flows down the drain. Knowing that *E. coli* are common inhabitants of the human intestinal tract, the student breaks out in a cold sweat—and you wake up. Was the student right to be scared? Why?

2. A major focus of genetic engineering has been the attempt to produce large quantities of scarce human metabolites by plac-ing the appropriate human gene into bacteria; since prodigious numbers of bacteria can be readily and cheaply produced, large amounts of the metabolite can be made. Human insulin is now manufactured in this way. However, if one attempts to use this approach to produce human hemoglobin (beta-globin), the experiment does not work. If you first identify the proper clone from a clone library by using an appropriate radioactive probe, insert the fragment containing the beta-globin gene into a plasmid, and then infect bacterial cells with the chimeric plasmid, no human hemoglobin is produced by the infected cells. This negative result is obtained despite the fact that the proper fragment clone was chosen, the beta-globin gene was successfully incorporated into the plasmid, and the chimeric plasmid did successfully infect the *E. coli* cells. Why isn't this experiment working?

FOR FURTHER READING

ANGIER, N.: *Natural Obsessions—The Search for the Oncogene,* Houghton Mifflin Co., Boston, 1988. An exciting, very readable account of cancer research as it is done today. Highly recommended. The author's science writing won the Pulitzer Prize in 1991.

CASKEY, C.T.: "Disease Diagnosis by Recombinant DNA Methods," *Science,* June 1987, pages 1223-1228. A review of how restriction fragment length variation and probes constructed from cloned genes are being used to aid clinical diagnosis of many disease, some in the fetus.

CHILTON, M.: "A Vector for Introducing New Genes into Plants," *Scientific American,* June 1983, pages 50-60. A description of the only successful plant genetic engineering vector developed to date, important because of its agricultural implications.

DiLISI, C.: "The Human Genome Project," *American Scientist,* vol. 76, 1988, pages 438-439. Mapping and deciphering the complete sequence of human DNA will stimulate research in fields ranging from computer technology to theoretical chemistry.

ERLICH, H. and others: "Recent Advances in the Polymerase Chain Reaction," *Science,* June 1991, vol. 252, pages 1643-1650. An up-to-date overview of how this exciting new technique is being used in genetic engineering.

GILBERT, W., and L. VILLA-KAMAROFF: "Useful Proteins from Recombinant Bacteria," *Scientific American,* April 1980, pages 74-94. An important byproduct of the new gene technology is the movement of human genes into bacteria, which produce their products in large amounts. This article, by a Nobel prize winner, provides a clear description of how this is done.

GODSON, G.: "Molecular Approaches to Malaria Vaccines," *Scientific American,* May 1985, pages 52-59. A revealing look at how the new techniques of molecular engineering are being used to defeat an ancient scourge.

HALL, S.: "James Watson and the Search for Biology's 'Holy Grail'," *Smithsonian,* February 1990, pages 41-47. A fascinating and well-written account of the human genome project and the man who heads it up.

HOFFMAN, P.: "The Human Mouse," *Discover,* August 1989, pages 4, 48-55. To defeat AIDS, an innovative young researcher, using the methods discussed in this chapter, has created remarkable new hybrids of mice and men.

MacKENZIE, D.: "How to Build a Better Fish," *New Scientist,* April 22, 1989, pages 52-56. To apply genetic engineering principles to the improvement of fishes, we still have to learn a great deal about fish genetics.

MONTGOMERY, G.: "The Ultimate Medicine," *Discover,* March 1990, pages 60-68. An account of how virus vectors are being used in human gene transfer therapy.

MULLIS, K.: "The Unusual Origin of the Polymerase Chain Reaction," *Scientific American,* April 1990, pages 56-65. How a critical advance in gene engineering technology was made.

NEUFELD, P.J. and N. COLMAN: "When Science Takes the Stand," *Scientific American,* May 1990, pages 46-53. DNA and other evidence is increasingly being applied to the solution of criminal cases but must be used with caution.

RAFFA, K.: "Genetic Engineering of Trees to Enhance Resistance to Insects," *BioScience*, September 1989, pages 524-534. An evaluation of the promise of this new approach, and of the risks.

VERMA, I.: "Gene Therapy," *Scientific American*, November 1990, pages 68-84. The first attempts to treat inherited genetic disorders by inserting healthy genes into patients are underway.

WEAVER, R.: "Beyond Supermouse," *National Geographic*, December 1984, pages 818-847. A well-illustrated and entertaining account of current progress in genetic engineering.

WHITE, R., and J. LALOUEL: "Chromosome Mapping with DNA Markers," *Scientific American*, February 1988, pages 40-48. Using variable sequences in the DNA of human chromosomes, investigators have begun to compile a chromosomal map on which the positions of many human genetic disorders are being located.

Evolution

19

Genes within Populations

OVERVIEW

Evolution occurs as a result of the progressive change in the heritable characteristics of populations of organisms in response to their environments; Darwin defined this process as natural selection. Such a change is defined as adaptation, a microevolutionary process. In contrast, the grand outline of the history of life on earth, a process in which the origin of species plays a central role, is called macroevolution. In nature, most populations of organisms consist of individuals that differ from one another, chiefly as a result of mutation and recombination. Such variation is the raw material of evolution, since these features are often directly correlated with the probability of individuals producing offspring, and with the reproductive success of those offspring once produced. The reasons that different alleles are maintained in natural populations include natural selection, adaptation to different environmental factors by different individuals, and intrinsic genetic features of a population.

FOR REVIEW

Here are some important terms and concepts that you will encounter in this chapter. If you are not familiar with them, you should review them before proceeding.

Natural selection (Chapter 1)

Mechanism of hereditary (Chapter 12)

Alleles (Chapter 12)

Heterozygosity and homozygosity (Chapter 12)

Mutation and recombination (Chapters 16 and 17)

FIGURE 19-1
We humans are used to detecting subtle variations in each other's appearance. No two of the individuals here look alike to us. Yet can you tell two fruit flies apart, or two fish?

The genetic machinery of cells ensures that the hereditary instructions that specify the characteristics of each organism will be followed faithfully. If you plant a radish seed in your garden, you get a radish and not a turnip. However, as we have seen in Chapters 16 and 17, mutation and recombination alter genetic instructions over the course of time. Because of the operation of these factors, few radish plants are identical—and no other human being is exactly like you (unless you have an identical twin) (Figure 19-1). Often the particular characteristics of individual organisms have an important bearing on their chances to bear offspring, and in turn on the success of those offspring.

It was insights of this kind that led Darwin to formulate his theory that natural selection provided the explanation for evolution, which you studied in Chapter 1. Darwin agreed with many earlier philosophers and naturalists, who deduced that the many kinds of organisms around us in the world were produced by a process of evolution. Unlike his predecessors, however, Darwin proposed a mechanism capable of accounting for the process of evolution—natural selection. He proposed that new kinds of organisms evolve from existing ones because they occur in different environments, in which natural selection favors those heritable features that better suit an organism to survive and reproduce (Figure 19-2).

When the word "evolution" is mentioned, it is difficult not to conjure up images of dinosaurs, of woolly mammoths frozen in blocks of ice, or of Darwin confronting a monkey. Traces of ancient forms of life, now extinct, survive as fossils that help us piece together the evolutionary story. With such a background, we usually think of evolution as meaning changes in the kinds of animals and plants on earth, changes that take place over long periods of time, with new forms replacing old ones. This kind of evolution is called **macroevolution.** Macroevolution is evolutionary change on a grand scale, encompassing the origin of novel designs, evolutionary trends, new kinds of organisms penetrating new habitats, and major episodes of extinction.

Much of the focus of Darwin's theory, however, is directed not at the way in which new species are formed from old ones, but rather at the way that changes occur within species. Natural selection, the explanation that Darwin proposed for evolutionary change, is the process whereby some individuals in a population produce more surviving offspring than others lacking these characteristics. As a result, the population gradually will come to include more and more individuals with the characteristics that provide an advantage to the individuals in which they occur, generation after generation, assuming that the characteristics have a genetic basis. In this way the population evolves. Change of this sort within populations—progressive change in gene frequencies—is called **microevolution.** Natural selection is the process by which microevolutionary change occurs; the result of the process, the features that promote the likelihood of survival and reproduction by an organism in a particular environment, is called **adaptation.** In essence, Darwin's explanation of evolution is that progressive adaptation by natural selection is responsible for evolutionary changes *within* a spe-

FIGURE 19-2
Two peppered moths, *Biston betularia*, resting on lichens growing on a piece of bark. The dark moth is more easily seen on this light bark, which shows no evidence of industrial pollution. This chapter and the next deal with the ways in which such differences arise, are maintained, and change in frequency.

cies, changes which when they accumulate lead to the creation of new kinds of organisms, new species. We will return to the ways in which new species originate in Chapter 21.

GENE FREQUENCIES IN NATURE

When considering Darwin's theory that evolution is a progressive series of adaptive changes brought about by natural selection, it is best to start by looking at the raw material available for the selective process—the genetic variation present among individuals within a species from which natural selection chooses the best-suited alleles.

As we saw in Chapter 13, a group of individuals that live together, a natural population, can contain among its members a great deal of genetic variation. This is true not only of humans but of all organisms. How much variation? Biologists have looked at many different genes in an effort to answer this question:

1. *Blood groups.* Chemical analysis has revealed the existence of more than 30 blood group genes in humans, in addition to the ABO locus. At least a third of these genes are routinely found to be present in several alternative alleleic forms in human populations. In addition to these, more than 45 additional variable genes are also known that encode proteins in human blood cells and plasma, but which are not considered to define blood groups. Thus there are more than 75 genetically variable genes in this one system alone.

2. *Enzymes.* Alternative alleles of the genes specifying particular enzymes are easy to distinguish. The differences in their nucelotide sequences alter the ways in which the proteins specified by these alleles behave in simple physical tests. One of the most popular of these is to measure how fast the alternative proteins migrate in an electric field (a process called **electrophoresis**). A great deal of variation is found at enzyme-specifying loci (Table 19-1). About 5% of the loci of a typical human are heterozygous. That is, if you picked an individual at random, and in turn selected one of the genes of that individual at random, the chances are 1 in 20 (5%) that the gene you selected would be heterozygous in that individual.

Considering the entire human genome, it is fair to say that almost all people differ from one another; this is also true of other organisms as well, except for those that reproduce without genetic recombination. In nature, genetic variation is the rule.

TABLE 19-1 LEVELS OF GENETIC VARIATION AT ENZYME-ENCODING GENES	
GROUP	PERCENTAGE OF GENES HETEROZYGOUS IN AN AVERAGE INDIVIDUAL
INVERTEBRATES	
Drosophila	15%
Other insects	15%
Marine invertebrates	12%
Land snails	15%
VERTEBRATES	
Fishes	8%
Amphibians	8%
Reptiles	5%
Birds	4%
Mammals	5%
PLANTS	
Self-pollinating	3%
Outcrossing	8%

How many of the loci in a given population will include more than one allele at a frequency greater than that which would be associated with mutation alone? In other words, what proportion of the loci exhibit **genetic polymorphism?** The extent of such variation was not even suspected more than 20 years ago, until modern techniques made it possible to discover many more polymorphic loci than could be detected on the basis of their external, physical effects (phenotype). As we now know, most populations of insects and plants are polymorphic at more than half of their loci, although vertebrates are somewhat less polymorphic. Such high levels of genetic variability provide ample supplies of raw material for evolution.

POPULATION GENETICS

When considering Darwin's theory that evolution is a progressive series of adaptive changes brought about by natural selection, it is best to start by looking at the raw material available for the selective process—the genetic variation present among individuals within a species from which natural selection chooses the best-suited alleles.

In the early part of this century the science of genetics contributed relatively little to our understanding of the process of evolution. At first, geneticists were involved primarily with understanding the actions of individual genes. Evolutionists in turn could not understand how such observations would have a bearing on the evolution of a complex structure, such as an eye.

The gap between the geneticists and the evolutionists finally started to close in the 1920s, with the development of the field of **population genetics,** simply defined as the study of the properties of genes in populations. At that time, scientists began to formulate a comprehensive theory of how alleles behave in populations and the ways in which changes in allele frequencies lead to evolutionary change. The most fundamental model in the field of population genetics, the Hardy-Weinberg principle, was developed in the early years of this century; all other aspects of this synthetic field can be viewed in relation to it.

The Hardy-Weinberg Principle

Genetic variation within natural populations was a puzzle to Darwin and his contemporaries. The way in which meiosis produces genetic segregation among the progeny of a hybrid had not yet been discovered. The theories of the time predicted that dominant alleles should eventually drive recessive ones out of a population, thus eliminating any genetic variation. Selection, they thought, should favor an optimal form.

The solution to the puzzle of why genetic variation persists was developed independently and published almost simultaneously in 1908 by G.H. Hardy, an English mathematician, and G. Weinberg, a German physician. They pointed out that in a large population in which there is random mating and in the absence of forces that change the proportions of the alleles at a given locus (these forces will be discussed below), the original proportions of the genotypes will remain constant from generation to generation. Dominant alleles do not in fact replace recessive ones. Because their proportions do not change, the genotypes are said to be in **Hardy-Weinberg equilibrium.**

In algebraic terms, the Hardy-Weinberg principle is written as an equation. Its form is what is known as a binomial expansion. For a gene with two alternative alleles, which we will call A and a, the equation looks like this:

$$(p + q)^2 = \underset{\substack{\text{(INDIVIDUALS} \\ \text{HOMOZYGOUS} \\ \text{FOR ALLELE A)}}}{p^2} + \underset{\substack{\text{(INDIVIDUALS} \\ \text{HETEROZYGOUS} \\ \text{WITH ALLELES} \\ \text{A + a)}}}{2pq} + \underset{\substack{\text{(INDIVIDUALS} \\ \text{HOMOZYGOUS} \\ \text{FOR ALLELE a)}}}{q^2}$$

In statistics, **frequency** is defined as the proportion of individuals falling within a certain category in relation to the total number of individuals being considered. Thus, in a population of 100 cats, with 84 black and 16 white cats, the respective frequencies would be 0.84 (or 84%) and 0.16 (or 16%). In the algebraic terms of this equation, the letter p designates the frequency of one allele, the letter q the frequency of

Phenotypes	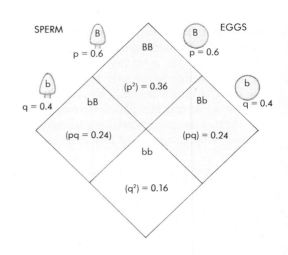		
Genotypes	BB	Bb	bb
Frequency of genotype in population	0.36	0.48	0.16
Frequency of gametes	$\dfrac{0.36 + 0.24}{0.6B}$		$\dfrac{0.24 + 0.16}{0.4b}$

SPERM · EGGS

B p = 0.6   BB   B p = 0.6

b q = 0.4   bB $(p^2) = 0.36$ Bb   b q = 0.4

$(pq = 0.24)$ $(pq) = 0.24$

bb

$(q^2) = 0.16$

FIGURE 19-3
The Hardy-Weinberg equilibrium. In the absence of factors that alter them, the proportions of alleles, genotypes, and phenotypes remain constant generation after generation.

the alternative allele. By convention, the more common of the two alleles is designated p, the rarer allele q. Because there are only two alleles, p and q must always equal 1.

If we assume that the white cats were homozygous recessive for a gene that we might designate b, and the black cats were therefore either homozygous dominant BB or heterozygous Bb, we can calculate the frequencies of the two alleles in the population from the proportion of black and white individuals. If $q^2 = 0.16$, p, the frequency of the allele B, therefore equals 0.6. In addition to the 16 white cats, which have a bb genotype, there are $2pq$, or $2 \times 0.6 \times 0.4 \times 100$ (the number of individuals in the total population), or 48 heterozygous individuals. The heterozygous individuals have the Bb genotype. We can also calculate easily that there are $p^2 = (0.6)^2$, or 36 homozygous dominant BB individuals.

Figure 19-3 allows you to trace genetic reassortment during sexual reproduction, and to see how it affects the frequencies of the B and b alleles during the next generation. In constructing this diagram, we have assumed that the union of sperm and egg in these cats is random, so that all combinations of b and B alleles are equally likely. For this reason, the alleles are, in effect, mixed randomly and represented in the next generation in proportion to their original representation; there is no inherent reason for them to change in frequency from one generation to the next. Each individual in each generation has a 0.6 chance of receiving a B allele, and a 0.4 chance of receiving a b allele.

In the next generation, therefore, the chance of combining two B alleles is 0.36 (that is, 0.6×0.6), and approximately 36% of the individuals in the population will continue to have the BB genotype. The frequency of bb individuals (0.4×0.4) will continue to be about 16%, and the frequency of Bb individuals will be $2 \times 0.6 \times 0.4$, or approximately 48%. Phenotypically, there will still be approximately 84 black individuals (with either BB or Bb genotypes) and 16 white individuals (with the bb genotype) in the population.

This simple relationship has proved extraordinarily useful in assessing actual situations. As an example, consider the recessive allele responsible for the serious human disease cystic fibrosis. This allele is present in white North Americans at a frequency of about 22 per 1000 individuals, or 0.022. What proportion of white North Americans, therefore, is expected to express this trait? The frequency of double recessive individuals (q^2) is expected to be 0.022×0.022, or 1 in every 2000 individuals. What proportion is expected to be heterozygous carriers? If the frequency of the recessive allele q is 0.022, then the frequency of the dominant allele p must be $1 - 0.022$, or 0.978. The frequency of heterozygous individuals ($2pq$) is thus expected to be $2 \times 0.978 \times 0.022$, or 43 in every 1000 individuals.

How valid are these calculated predictions? For many genes, they prove to be very accurate. Most human populations, for example, are large and effectively random-mating, and so are similar to the "ideal" population envisioned by Hardy and

Weinberg. As we will see, however, for some genes the calculated predictions do *not* match the actual values. The reasons they do not do so tell us a great deal about evolution.

> The Hardy-Weinberg principle states that in a large population mating at random and in the absence of other forces that would change the proportions of the different alleles at a given locus, the process of sexual reproduction (meiosis and fertilization) alone will not change these proportions.

WHY DO ALLELE FREQUENCIES CHANGE?

According to the Hardy-Weinberg principle, both the allele and genotype frequencies in a large, random-mating population will remain constant from generation to generation if there is no mutation, no migration, and no selection. The reservations tacked onto the end of the statement are important and will be explained further. In fact, they are the key to the importance of the Hardy-Weinberg principle for biology, because individual allele frequencies *are* changing all the time in natural populations, with some alleles becoming more common than others. The Hardy-Weinberg principle establishes a convenient baseline against which to measure such changes. By looking at the ways in which various factors alter the proportions of homozygotes and heterozygotes in populations, we can identify those forces that are affecting particular situations that we observe.

Many factors can alter allele frequencies. Only five, however, alter the proportions of homozygotes and heterozygotes enough to produce significant deviations from the proportions predicted by the Hardy-Weinberg principle. These are **mutation; migration** (including both immigration into and emigration out of a given population); **genetic drift** (random loss of alleles, which is more likely in small populations), **nonrandom mating;** and **selection.** Of these, only the last factor, selection, produces adaptive evolutionary change, because only in the case of selection does the result depend on the nature of the environment. The other factors operate relatively independently of the environment, so that the changes they produce are not shaped by environmental demands.

> Five factors can bring about a deviation from the proportions of homozygotes and heterozygotes predicted by the Hardy-Weinberg principle: mutation, migration, genetic drift (random loss of alleles), nonrandom mating, and selection.

Mutation

Mutation from one allele to another obviously can change the proportions of particular alleles in a population. Mutation rates are generally so low, however, that they cannot alter Hardy-Weinberg proportions of common alleles very much. Many genes mutate about 1 to 10 times per 100,000 cell divisions, although of course a mutation in *some* gene of an individual will occur much more frequently than that. Since most environments are constantly changing, it is rare for a population to be stable enough to accumulate differences in allele frequency produced this slowly. Nonetheless, as we have seen, mutation is the ultimate source of genetic variation, and thus makes evolution possible.

Migration

Migration is defined in genetic terms as the movement of individuals from one population into another. It can be a powerful force in upsetting the genetic stability of natural populations. Sometimes migration is obvious, as when an animal moves from one place to another. If the characteristics of the newly arrived animal differ from those already there, the genetic composition of the receiving population may be altered if the newly arrived individual or individuals are well enough adapted to survive

FIGURE 19-4
The yellowish-green cloud around these Monterey pines, *Pinus radiata,* **is pollen being dispersed by the wind.** The male gametes within the pollen reach the egg cells of the pine passively in this way. In genetic terms, such dispersal is a form of migration.

and mate successfully in the new area. Other important kinds of migration are not as obvious to the observer. These subtler movements include, for example, the drifting of gametes or immature stages of marine animals or plants from one place to another (Figure 19-4). The male gametes of flowering plants are often carried great distances by insects and other animals that visit their flowers. Seeds may also be blown in the wind or carried by animals or other agents to new populations far from their place of origin. However it occurs, migration can alter the genetic characteristics of populations and prevent them from maintaining Hardy-Weinberg equilibrium. Its evolutionary role is more difficult to assess, however, and depends heavily on the prevailing selective forces.

One way of imagining the genes in a population or species is to picture them as a **gene pool,** consisting of all the alleles present in that population. Sometimes the concept of a gene pool is extended to an entire species, but the local nature of most populations makes such an extension unwarranted for many species. Microevolution— changes in the allele frequencies of a population—can be visualized as affecting the frequencies of alleles in the gene pool. A related concept is that of **gene flow,** defined as the movement of genes from one population to another. Such movement can take place by migration, of course, but it can also come about by means of hybridization between individuals belonging to adjacent populations. The degree to which gene flow takes place depends on the size of the gaps that separate the populations of a particular species in a particular area, the degree to which individuals move during or before they reach reproductive age, and on the distance over which mating usually takes place.

Genetic Drift

In small populations, the frequencies of particular alleles may be changed drastically by chance alone. The individual alleles of a given gene are all represented in few individuals, and some of them may even be lost from the population if those individuals fail to reproduce. Since these changes in allele frequency appear to occur randomly, as if the frequencies were drifting, it is known as **genetic drift.** A series of small populations that are isolated from one another may come to differ strongly as a result of genetic drift. In this connection, it is interesting to realize that humans have lived in small groups for much of the course of their evolution; genetic drift, consequently, may have been a particularly important factor in the evolution of our species.

> **Genetic drift is random change in the frequency of alleles at a locus. In small populations, such fluctuations may lead to the loss of particular alleles.**

Sometimes one or a few individuals are dispersed and become the founders of a new, isolated population at some distance from their place of origin. When this occurs, the alleles that they carry are of special significance. Even if these alleles are rare in the source population, in their new area they will be a significant fraction of the whole population's genetic endowment. This effect—by which rare alleles and combinations of alleles may be enhanced in the new populations—is called the **founder principle.** It is a particularly important factor in the evolution of organisms on distant oceanic islands, such as the Hawaiian Islands and the Galapagos Islands visited by Darwin. Most of the kinds of organisms that occur in such areas have probably been derived from one or a few initial "founders" (Figure 19-5). In a similar way, isolated human populations are often dominated by the genetic features that were characteristic of their founders, if only a few individuals were involved initially.

Even if organisms do not move from place to place, their populations may sometimes be drastically reduced in size. This may result from flooding, drought, earthquakes, or other natural forces, or by progressive changes in the environment. If the factors operate strongly enough, the surviving individuals may constitute a random genetic sample of the original population. Such a restriction in genetic variability has been termed the **bottleneck effect.**

Some living species appear to be severely depleted genetically, and have probably suffered from a bottleneck effect in the past. As shown by Steve O'Brien of the National Institutes of Health, for example, all living cheetahs are practically identical genetically, and therefore very susceptible to disease. The simplest explanation of this similarity would be that a drastic reduction in the size of cheetah populations occurred in the past, and that all cheetahs living today have descended from a very few individuals—they have passed through a genetic bottleneck.

Nonrandom Mating

Individuals with certain genotypes sometimes mate with one another more commonly than would be expected on a random basis, a phenomenon known as nonrandom mating. **Inbreeding** (mating with relatives), a type of nonrandom mating characteristic of many groups of organisms, will cause the frequencies of particular genotypes to differ

A

B

FIGURE 19-5
Evolution in the tarweeds.
A *Adenothamnus validus* is a primitive member of the tarweeds (subtribe Madiinae), a group of more than 100 western North American species of the sunflower family, Asteraceae.
B In the Hawaiian Islands, a group of unusual species of tarweeds, including this silversword, *Argyroxiphium sandwicense* subspecies *macrocephalum*, have evolved from ancestors similar to *Adenothamnus*, which reached Hawaii by chance dispersal over long distances. The particular genetic constitution of the founding individuals played key roles in the evolution of Hawaiian plants and animals.

FIGURE 19-6
Bees help flowers find distant mates. This wild bee *(Ptilothrix)*, loading bristly areas on its hind legs with pollen from the desert "poppy" *(Kalstroemia)* in eastern Arizona, illustrates the way that flowering plants have co-opted animals to the task of dispersing their pollen precisely from plant to plant. The pollen that the bee has gathered will be used to provision the cell in which its larva will reach maturity. Other pollen grains, adhering to the hairy body of the bee, will have a good chance of reaching the female parts of other flowers of this same species.

FIGURE 19-7

Inbreeding often has detrimental effects. In Japan it is common for first cousins to marry, and the Japanese population as a result is more homozygous than that of the United States, where such marriages are generally forbidden. Because of this increased homozygosity, recessive alleles tend to be expressed more often in Japan. For each of the five genetic disorders illustrated here, affected children in Japan are far more likely to have parents who are first cousins.

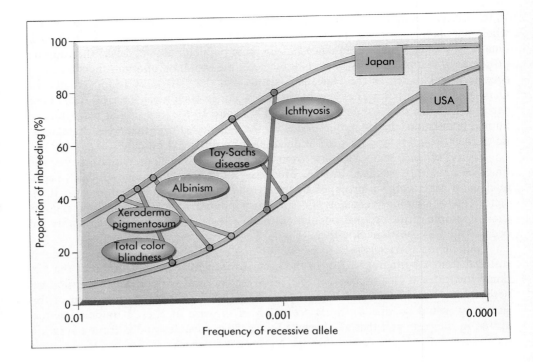

greatly from those predicted by the Hardy-Weinberg principle. Inbreeding does not change the frequency of the alleles, but rather the proportion of individuals that are homozygous. In inbred populations there are more homozygous individuals than predicted by the Hardy-Weinberg principle. It is for this reason that populations of self-fertilizing plants consist primarily of homozygous individuals, whereas **outcrossing** plants, which interbreed with individuals different from themselves, have a higher proportion of heterozygous individuals (Figure 19-6).

Because inbreeding increases the proportion of homozygous individuals in a population, it tends to promote the occurrence of double recessive combinations. It is for this reason that marriage between relatives is discouraged—it greatly increases the possibility of producing children homozygous for an allele associated with one or more of the genetic disorders discussed in Chapter 13. A dramatic example of this can be seen in Figure 19-7, which compares the incidence of five important genetic diseases in the United States, where inbreeding is rare, with the incidence in Japan, where it is more common. Even though alleles specifying some disorders, such as Tay-Sachs disease, are rarer in Japan, there is a much greater likelihood that a marriage between first cousins will produce affected children there, because the Japanese population is more homozygous in general.

Selection

As Darwin pointed out, some individuals leave behind more progeny than others, and the rate at which they do so is affected by their inherited characteristics. We describe the results of this process as **selection,** and speak both of **artificial selection** and of **natural selection.** In artificial selection, the breeder selects for the desired characteristics, such as those of the pigeons shown in Figure 1-10. In natural selection, the environment plays this role, conditions in nature determining which kinds of individuals in a population are the most fit and so affecting the proportions of genes among individuals of future populations. This is the key point in Darwin's proposal, that evolution occurs because of natural selection: the environment imposes the conditions that determine the results of selection and thus the direction of evolution.

Like mutation, migration, nonrandom mating, and genetic drift, selection causes deviations from Hardy-Weinberg proportions by directly altering the frequencies of the alleles. Darwin argued that the more successful reproduction of particular genotypes, which is how he defined selection, is the primary force that shapes the pattern of life on earth. The selection of these genotypes, however, is indirect: selection acts directly on the phenotype. The phenotype is determined by the interaction of the genotype and the environment, and the linkage between particular alleles and particular characteristics of the phenotype is less direct for some features than for others.

Although selection is perhaps the most powerful of the five principal agents of genetic change, there are limits to what it can accomplish. These limits arise because alternative alleles may interact in different ways with other genes. These interactions tend to set limits on how much a phenotype can be altered. For example, selecting for large clutch size in barnyard chickens eventually leads to eggs with thinner shells that break more easily. Because of the limits imposed by gene interactions, strong selection is apt to result in rapid change initially, but the change soon comes to a halt as the interactions between genes increase. For this reason, we do not have gigantic cattle that yield twice as much meat as our leading strains, chickens that lay twice as many eggs as the best layers do now, or corn with an ear at the base of every leaf, instead of just one or a few.

In 1988, two investigators at Trinity College, Dublin, Ireland, carried out a fascinating analysis of the performances of thoroughbred horses that underscores this point. Over 80% of the gene pool of the thoroughbred horses racing today goes back to 31 known ancestors from the late eighteenth century; the first Stud Book was established in 1791. Despite intense directional selection on thoroughbreds, especially on males, their performances have not improved for the last 50 years. Although the

A VOCABULARY OF GENETIC CHANGE

adaptation A change in structure, physiology, or behavior that promotes the likelihood of an organism's survival and reproduction in a particular environment.

allele One of two or more alternative versions of a gene.

allele frequency The relative proportion of a particular allele among individuals of a population. Not equivalent to "gene frequency," although the two terms are sometimes confused.

evolution Genetic change in a population of organisms over time (generations). Darwin proposed that natural selection was the mechanism of evolution. Microevolution occurs within a species, whereas macroevolution is the creation of new species and the extinction of old ones.

fitness The genetic contribution of an individual to succeeding generations, relative to the contributions of other individuals in the population.

gene The basic unit of heredity; a sequence of DNA nucleotides on a chromosome that encodes a protein or RNA molecule, or regulates the transcription of such a sequence.

gene frequency The frequency with which individuals in a population possess a particular gene. Often confused with allele frequency.

genetic drift Random fluctuations in allele frequencies over time.

mutation A permanent change in a gene, such as an alteration of its nucleotide sequence.

natural selection The differential reproduction of genotypes in response to factors in the environment; artificial selection, by contrast, is in response to demands imposed by human intervention.

polymorphism The presence in a population of more than one allele of a gene at a frequency greater than that of newly arising mutations.

population Any group of individuals, usually of the same species, occupying a given area at the same time.

species A kind of organism. In nature, individuals of one species usually do not interbreed with individuals of other species.

investigators concluded that the lack of improvement was not likely to be caused by a depletion of the genetic variation of the thoroughbred stock, it is difficult to devise alternative explanations. During the same period of time, human performance for the 1500 meter race improved by some 15 seconds (1936 to 1984), or about 7%.

There is a second factor that limits what selection can accomplish: selection acts only on phenotypes. Only those characteristics that are expressed in the phenotype of an organism can affect the ability of that organism to produce progeny. For this reason, selection does not operate efficiently on rare recessive alleles, simply because they do not often come together as homozygotes and there is no way of selecting them unless they do come together. For example, when a recessive allele a is present at a frequency q equal to 0.1, 10% of the alleles for that particular gene will be a, but only one out of a hundred individuals (q^2) will be double recessive and so display the phenotype associated with this allele. For lower allele frequencies, the effect is even more dramatic: if the frequency in the population of the recessive allele $q = 0.01$, the frequency of homozygotes in that population will be only 1 in 10,000.

> **Selection, the differential reproduction of genotypes as determined by phenotypic characteristics that lead to reproductive success in the environment, is a powerful mechanism for producing deviations from Hardy-Weinberg equilibrium.**

What this means is that selection against undesirable genetic traits in humans or domesticated animals is difficult unless the heterozygotes can also be detected. For example, if a particular recessive allele r, ($q = 0.01$) was considered undesirable, and none of the homozygotes for this allele were allowed to breed, it would take 1000 generations, or about 25,000 years in humans, to lower the allelic frequency by half to 0.005. At this point, after 25,000 years of work, the frequency of homozygotes would still be one in 40,000, or 25% of what it was initially. This is the basic reason that few geneticists advocate **eugenics,** the field that deals with efforts to change the genetic characteristics of human beings by artificial selection. Aside from the moral implications, such efforts are essentially doomed to failure by the sheer difficulty of producing the desired results within any plausible human timeframe.

SELECTION IN ACTION

Selection operates in natural populations of a species something like skill does in football games; in any individual game, it is difficult to predict the winner, since chance can play an important role in the outcome; but over a long season the teams with the most skillful players usually win the most games. In nature, too, those individuals best suited to their environment tend to win the evolutionary game by leaving the most offspring, although chance can play a major role in the life of any one individual. Selection is a statistical concept, just as betting is. Although you cannot predict the fate of any one individual, or any one coin toss, it *is* possible to predict which kind of individual will tend to become more common in populations of a species.

Forms of Selection

In nature many traits, perhaps most, are affected by more than one gene. The interactions between genes are typically complex, as you saw in Chapter 12. For example, alleles of many different genes play a role in determining human height (see Figure 12-23). In such cases, selection operates on all the genes, influencing most strongly those that make the greatest contribution to the phenotype. How selection changes the population depends on which genotypes are favored.

Directional Selection. When selection acts to eliminate one extreme from an array of phenotypes (Figure 19-8, A), the genes promoting this extreme become less frequent in the population. Thus in the *Drosophila* population illustrated in Figure 19-9, the elimination of flies that move toward light causes the population to contain fewer individuals with alleles promoting such behavior. The result is that if you were to pick an individual at random from the new fly population, there is a lesser chance

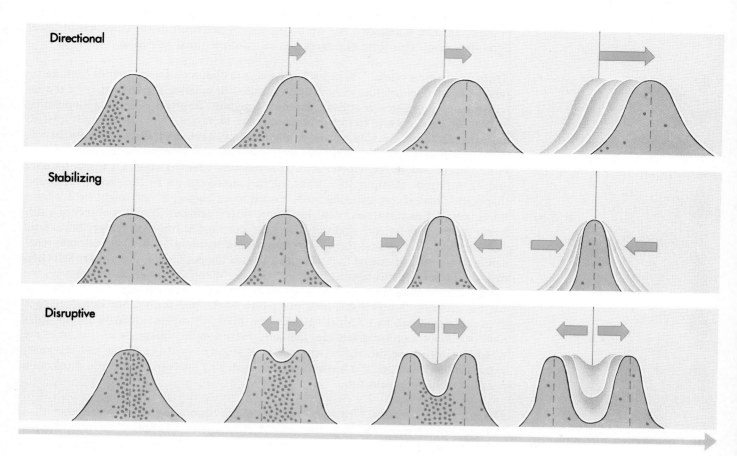

Directional

Stabilizing

Disruptive

FIGURE 19-8

Schematic representation of the three different kinds of natural selection acting on a trait, such as height, that varies in a population. In this diagram, dots represent individuals that do not contribute to the next generation. The curves represent measurements from the trait taken on each individual in the population. All three kinds of natural selection have the same starting point, and the three series show the way that selection alters the distribution of the characteristics as time passes, moving to the right.

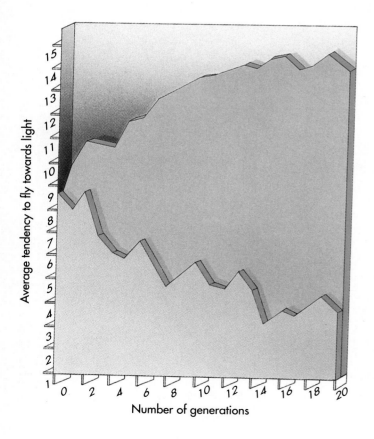

FIGURE 19-9

Microevolution in action. In generation after generation (expressed along the bottom of the graph), individuals of the fly *Drosophila* were selected for their tendency to fly toward light more strongly than usual *(top curve);* the bottom curve represents flies selected *against* the tendency to fly toward light. When flies with a strong tendency to fly toward light were used as the parents for the next generation, their offspring had a greater tendency to fly toward light.

it would spontaneously move toward light than if you had selected a fly from the old population. The population has been changed by selection in the direction of lower light attraction. This form of selection is called **directional selection.**

Stabilizing Selection.   When selection acts to eliminate *both* extremes from an array of phenotypes (Figure 19-8, *B*), the result is to increase the frequency of the intermediate type, which is already the most common. In effect, selection is operating to prevent change away from this middle range of values. In a classic study carried out after an "uncommonly severe storm of snow, rain, and sleet" on February 1, 1898, 136 starving English sparrows were collected and brought to the laboratory of H.C. Bumpus in Brown University at Providence, Rhode Island. Of these, 64 died and 72 survived. Bumpus took standard measurements on all the birds. He found that among males, the surviving birds tended to be bigger, as one might expect from the action of directional selection. However, among females, Bumpus observed a different result. The females that perished were not smaller, on the average, than those that survived, but among them were many more individuals that had extreme measurements—measurements that were unusual for the population as a whole. Selection had acted most strongly against these individuals. When selection acts in this way, the population contains fewer individuals with alleles promoting extreme types. Selection has not changed the most common phenotype of the population, but rather made it even more common by eliminating extremes. Many examples similar to Bumpus' female sparrows can be given. In humans, for example, infants with intermediate weight at birth have the highest survival rate (Figure 19-10). In ducks and chickens, eggs of intermediate weight have the highest hatching success. This form of selection is called **stabilizing selection.**

Disruptive Selection.   In some situations, selection acts to eliminate rather than to favor the intermediate type (Figure 19-8, *C*). A clear example is provided by the different color patterns of the African butterfly *Papilio dardanus*. In different parts of Africa, the color pattern of this butterfly is dramatically different, in each instance being a close copy of some other species that birds do not like to eat (*Papilo dardanus* is said to be a "mimic"). Any patterns that do not look like distasteful butterflies would be detected readily and eaten by birds, and so any intermediate patterns that occur are selected against. In this case, selection is acting to eliminate the intermedi-

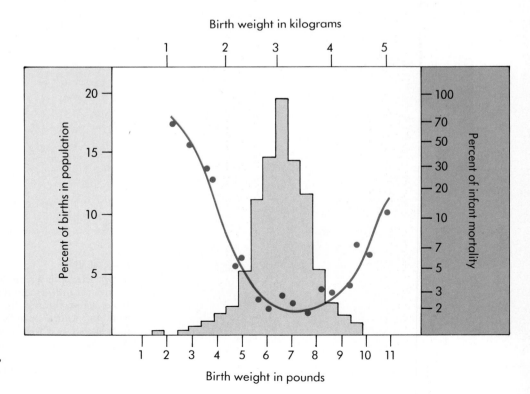

FIGURE 19-10
Stabilizing selection for birth weight in human beings. As you can see, the death rate among babies is lowest at an intermediate birth weight; both smaller and larger babies have a greater tendency to die at or near birth than those around the optimum weight, which is between 7 and 8 pounds.

ate phenotypes, in effect partitioning the population into homozygous groups. This form of selection is called **disruptive selection.**

Which Force Is The Most Important in Evolution?

Any of the five forces discussed above can cause genetic variation to occur in a population. Individual alleles may make different contributions to the **fitness** of the organisms in which they occur (selection); organisms that are more fit make greater contributions to the genetic endowments of succeeding generations, relative to the contributions of other individuals. Alternatively, alleles may be maintained in different frequencies by genetic drift or by migration from other populations. Mutation may affect the frequencies of very rare alleles. In practice, which of these effects are the most important in evolution? Although many alleles in natural populations have been examined in an effort to evaluate these relationships precisely in individual cases, exact answers have been difficult to obtain. This is because assigning a precise role to the contribution that a particular allele makes to the fitness of an organism in nature is so difficult. A population may maintain particular alleles at high frequencies in natural populations because the enzymes they produce make different contributions to individual fitness or simply because of the structure and dynamics of the population.

Only selection regularly produces adaptive evolutionary change, whereas the other effects are essentially random in direction, and so "neutral" in their evolutionary effect. Natural selection is clearly the primary force that leads to evolution, as Darwin pointed out, but the genetic constitution of individual populations, and thus the course of evolution, can also be affected by mutation, migration, genetic drift, and nonrandom mating. In the next chapter, we will focus on the role of selection and the evidence for the process of evolution as a whole.

SUMMARY

1. Macroevolution describes the grand outlines of evolution, which is based on the evolution of species. Microevolution, also called adaptation, refers to the evolutionary process itself. Adaptation leads to species formation, and thus ultimately to macroevolution.

2. By the 1860s, natural selection was widely accepted as the correct explanation for the process of evolution. The field of evolution did not progress much further, however, until the 1920s because of the lack of a suitable explanation of how hereditary traits are transmitted.

3. Invertebrates and outcrossing plants are often heterozygous at about 12% to 15% of their loci; the corresponding value for vertebrates is about 4% to 8%. These levels of genetic variation are much higher than any imagined before methods suitable to detect them were introduced in the 1960s.

4. Population genetics is the branch of genetics that deals with the behavior of genes in populations. Understanding population genetics is essential to understanding evolution.

5. The Hardy-Weinberg principle provides the baseline for all population genetic theory. It illustrates the fact that in large populations with random mating, allele and genotype frequencies—and consequently phenotype frequencies—will remain constant indefinitely, provided that selection, net mutation or migration in one direction, nonrandom mating, and genetic drift do not occur.

6. Fitness is a measure of the tendency of some organisms to leave more offspring than competing members of the same population. The genetic traits possessed by the fit individuals will appear in greater proportions among members of succeeding generations. This process is called selection.

7. Selection acts only indirectly, on the phenotype, and not directly on the genotype; it also acts on entire individuals, and thus assemblages of genes, rather than on individual genes. Both of these points are important in determining the consequences of selection.

8. There are three principal kinds of selection. Directional selection acts to eliminate one extreme from an array of phenotypes; stabilizing selection acts to eliminate *both* extremes; and disruptive selection acts to eliminate rather than to favor the intermediate type.

REVIEW QUESTIONS

1. Define macroevolution and microevolution.

2. What is adaptation? How does it fit into Darwin's concept of evolution?

3. State the Hardy-Weinberg principle in mathematical terms and define all variables.

4. Given that allele A is present in a large random mating population at a frequency of 54 per 100 individuals, what is the proportion of individuals in that population expected to be heterozygous for the allele? __ homozygous dominant? __ homozygous recessive?

5. What are the five factors that can alter the proportions of homozygotes and heterozygotes from the predicted Hardy-Weinberg values?

6. What is genetic drift? Why is it dependent on the size of the population?

7. Why does the founder principle have such a profound effect on a population's genetic makeup?

8. Why does nonrandom mating adversely affect the Hardy-Weinberg prediction?

9. How do artificial and natural selection differ from one another?

10. Why are there limitations to the success of selection?

THOUGHT QUESTIONS

1. The North American human population is similar to the ideal Hardy-Weinberg population in that it is very large (over 270 million people in the United States and Canada alone) and generally random-mating. Although mutation occurs, it alone does not lead to great changes in allele frequencies. Migration from Latin American and Asian countries occurs at relatively high levels—perhaps 1% per year—however. The following data were obtained in 1976 by the geneticist A.E. Mourant about relative numbers of individuals bearing the two alleles of what is known as the MN blood group:

	MM	MN	NN	Total
Observed individuals	1787	3037	1305	6129

Do these data suggest that migration, selection, or some other factor is acting to perturb the Hardy-Weinberg proportions of the three genotypes?

2. Will a dominant allele that is lethal be removed from a large population as a result of natural selection? What factors might prevent this from happening?

3. In a large, random-mating population with no forces acting to change gene frequencies, the frequency of homozygous recessive individuals for the characteristic of extra-long eyelashes is 90 per 1000 or .09. What percent of the population carries this desirable trait but displays the dominant phenotype, short eyelashes? Would the frequency of the extra-long eyelash allele increase, decrease, or remain the same if long-lashed individuals preferentially mated with each other and no one else?

FOR FURTHER READING

CHRISTIANSEN, F.B., and M.W. FELDMAN: *Population Genetics*, Blackwell Scientific Publications, Oxford, England, 1986. An excellent outline of the field, with an emphasis on human population genetics.

COHN, J.: "Genetics for Wildlife Conservation," *BioScience*, March 1990, pages 167-171. DNA analysis provides valuable information for managing both natural and captive populations of endangered species.

CROWN, J.F.: *Basic Concepts in Population, Quantitative, and Evolutionary Genetics*, W.H. Freeman and Co., New York, 1986. An outstanding treatment of the entire field.

CUNNINGHAM, P.: "The Genetics of Thoroughbred Horses," *Scientific American*, May 1991, pages 92-98. An enjoyable account of how breeding has influenced thoroughbred racehorses.

ENDLER, J.: *Natural Selection in the Wild*, Princeton University Press, Princeton, N.J. 1986. An overview of what we know about how natural selection operates in wild populations.

KIMURA, M.: "The Neutral Theory of Molecular Evolution," *New Scientist*, vol. 107, pages 41-43, 1985. Professor Kimura has demonstrated that many of the alleles that occur in natural populations are neutral selectively.

KOEHN, R.K., and T.J. HILBISH: "The Adaptive Importance of Genetic Variation," *American Scientist*, vol. 75, pages 134-140, 1987. Interesting new information about the role of variation in allozymes in nature.

O'BRIEN, S., D.E. WILDT, and M. BUSH: "The Cheetah in Genetic Peril," *Scientific American*, May 1986, pages 84-92. Evidence for an ancient population bottleneck in cheetahs, which has put the survival of the species in doubt.

READ, A.F., and P.H. HARVEY: "Genetic Management in Zoos," *Nature*, vol. 322, pages 408-410, 1986. Good discussion of the ways in which some of the principles discussed in this chapter can be applied to conservation.

SLATKIN, M.: "Gene Flow and the Geographic Structure of Natural Populations," *Science*, vol. 236, pages 787-792, 1987. Although somewhat technical, this article provides an example of modern reasoning about populations and the origin of species.

20

The Evidence of Evolution

OVERVIEW

Most biologists accept Charles Darwin's conclusion that natural selection is the primary explanation for evolution. During the time since Darwin first suggested this explanation, we have learned a great deal about the mechanisms that bring about evolution. In this chapter we will consider natural selection and other reasons for change in the heritable characteristics of populations, and also some of the features of major evolutionary change. Even though all available evidence points to the fact that evolution has occurred, and this has been accepted generally since well before the time of Darwin, the validity of this fact has been challenged recently, mainly in the United States, on the basis of a set of religious beliefs labeled scientific creationism.

FOR REVIEW

Here are some important terms and concepts that you will encounter in this chapter. If you are not familiar with them, you should review them before proceeding.

Radioactive carbon dating (Chapter 1)

Natural selection (Chapters 1 and 19)

Mutation and recombination (Chapters 16 and 17)

Hardy-Weinberg principle (Chapter 19)

FIGURE 20-1
Evolution in action. The Kaibab squirrel, *Sciurus aberti kaibabensis* (left), became isolated on the North Rim of the Grand Canyon in Arizona about 10,000 years ago, as the forests of the southwestern United States contracted after the most recent retreat of the glaciers. During this period, the features such as the black belly that distinguishes the Kaibab squirrel from its closest relative, the Abert squirrel, *S. aberti aberti* (right), evolved. The Abert squirrel occurs throughout a wide area in the desert woodlands of the southwestern United States. There are only about 22,000 individuals of the Kaibab squirrel in existence.

Of all the major ideas of biology, evolution is perhaps the best known to the general public. This is not because the basic facts of evolution are well understood by the average person, but rather because many people mistakenly believe that it represents a challenge to their religious beliefs (Figure 20-1). Within the last few years, controversial attempts have been made to require the teaching of a set of religious dogmas known as "scientific creationism" in public school science classes. Similar highly publicized criticisms of evolution have occurred ever since the time of Darwin, and it is likely that others will occur in the future. For this reason it is important that, during the course of your study of biology, you address the issue squarely. Just what *is* the evidence for evolution?

In the previous chapter we examined how biologists currently view the evolutionary process, and asked "What factors are thought to bring about evolutionary change?" In this chapter we will look at individual cases in which evolutionary change in natural populations can be seen to be adaptive, a key tenet of Darwin's theory of evolution by means of natural selection. Finally, we will review the evidence that supports the general validity of Darwin's proposal.

THE EVIDENCE THAT NATURAL SELECTION EXPLAINS MICROEVOLUTION

Although we do not know how much of the wealth of genetic variation that we see in nature is present because it is being maintained by natural selection, it is abundantly clear that at least some of it is (Table 20-1). Much of the genetic variation that occurs in nature is adaptive, as can be documented easily for some of the better-studied examples.

Sickle Cell Anemia

Sickle cell anemia is a hereditary disease affecting hemoglobin molecules in the blood. Sickle cell anemia was first reported in 1910 by a Chicago physician, James B. Herrick. A West Indian black student exhibited symptoms of severe anemia that appeared related to abnormal red blood cells: "The shape of the red cells was very irregular, but what especially attracted attention was the large number of thin, elongated, sickle-shaped and crescent-shaped forms." The disease was soon found to be common among African-Americans. In Chapter 13 we noted that this disorder, which affects roughly two African-Americans out of every 1000, is associated with a particular recessive allele. Using the Hardy-Weinberg equation, you can calculate that the frequency with which the sickle cell allele occurs in the African-American population; this frequency is the square root of 0.002, or approximately 0.045. In contrast, the frequency of the allele among white Americans is only about 0.001.

Sickle cell anemia is usually fatal. The disease occurs because of a single amino acid change, repeated in the two beta chains of the hemoglobin molecule. In this change, a valine is substituted for the usual glutamic acid at a location on the surface of the protein near the oxygen-binding site. Unlike glutamic acid, valine is nonpolar (hydrophobic), and its presence on the surface of the molecule creates a "sticky"

TABLE 20-1 THE LOGICAL ARGUMENT FOR NATURAL SELECTION

Fact 1: All species have such great potential fertility that their population size would increase exponentially if all individuals born would again reproduce successfully.

Fact 2: Except for minor annual fluctuations and occasional major fluctuations, populations normally display stability.

Fact 3: Natural resources are limited. In a stable environment they remain relatively constant.

Inference 1: Since more individuals are produced than can be supported by the available resources but population size remains stable, it means that there must be a fierce struggle for existence among individuals of a population, resulting in the survival of a part, often a very small part, of the progeny of each generation.

Fact 4: No two individuals are exactly the same; rather, every population displays enormous variability.

Fact 5: Much of this variation is heritable.

Inference 2: Survival in the struggle for existence is not random but depends in part on the hereditary constitution of the surviving individuals. This unequal survival constitutes the process of NATURAL SELECTION.

Inference 3: Over the generations this process of natural selection will lead to a continuing gradual change of populations, that is, EVOLUTION, and to the production of new species.

From Mayr, E.: *The growth of biological thought,* Harvard University Press, Cambridge, 1982. (Emphasis added.)

patch that will attempt to escape from the polar water environment by binding to another similar patch. As long as oxygen is bound to the hemoglobin molecule there is no problem, since the oxygen atoms shield the critical area of the surface. When oxygen levels fall, however, such as after exercise or at high altitudes, then oxygen is not so readily bound to hemoglobin, and the exposed sticky patch binds to similar patches on other molecules, eventually producing long fibrous clumps. The result is a deformed, "sickle-shaped" red blood cell.

Individuals who are heterozygous for the dominant valine-specifying allele (designated allele *S*) are said to possess the sickle cell trait. They produce some sickle-shaped red blood cells, but only 2% of the level seen in homozygous individuals.

The average incidence of the *S* allele in the Central African population is about 0.12, a far higher value than that found among African-Americans. From the Hardy-Weinberg principle you can calculate that one in five Central African individuals are heterozygous at the *S* allele, and one in a hundred develops the fatal form of the disorder. People who are homozygous for the sickle cell allele almost never reproduce, because they usually die before they reach reproductive age. Why is the *S* allele not eliminated from the Central African population by selection, rather than being maintained at such high levels? People who are heterozygous for the sickle cell allele are also much less susceptible to malaria—one of the leading causes of illness and death, especially among young children—in the areas where the allele is common. In addition, for reasons that are not understood, women who are heterozygous for this allele are more fertile than are those who lack it. Consequently, even though most people who are homozygous recessive die before they have children, the sickle cell allele is maintained at high levels in these populations because of its role in resistance to malaria in heterozygotes and its association with increased fertility in female heterozygotes.

As Darwin's theory predicts, it is the environment that acts to maintain the sickle-cell allele at high frequency. In this case the characteristic of the environment of Central Africa that is exercising selection is the presence of malaria. For the people living in areas where malaria is frequent, maintaining a certain level of the sickle cell allele in the populations has adaptive value (Figure 20-2). Among African-Americans, many of whom have lived for some 15 generations in a country where malaria has been relatively rare in most areas and is now essentially absent, the environment does

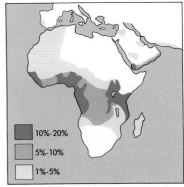

Frequency of sickle cell gene

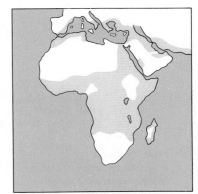

Distribution of Falciparum malaria

FIGURE 20-2
Frequency of sickle cell allele (*top*), and distribution of *Falciparum* malaria (*bottom*).

not place a premium on resistance to malaria, so that there is no adaptive value to counterbalance the ill effects of the disease; in this nonmalaria environment, selection is acting to eliminate the S allele.

Peppered Moths and Industrial Melanism

The peppered moth, *Biston betularia*, is a European moth that rests on the trunks of trees during the day (Figure 20-3). Until the mid-nineteenth century, almost every individual of this species that was captured had light-colored wings. From that time on, individuals with dark-colored wings increased in frequency in the moth populations near industrialized centers until they made up almost 100% of these populations. The black individuals have a dominant allele that was present in populations before 1850, although it was very rare then. Biologists soon noticed that in industrialized regions where the dark moths were common, the tree trunks were darkened almost black by the soot of pollution, and the dark moths were much less conspicuous when resting on them than were the light moths. In addition, the air pollution that was spreading in the industrialized regions had killed many of the light-colored lichens that had occurred on the tree trunks before, thus making these trunks even darker than they would have been otherwise.

Can Darwin's theory explain the increase in frequency of the dark allele? Why was it an advantage for the dark moths to be less conspicuous? Although initially there was no evidence, the ecologist H.B.D. Kettlewell hypothesized that the moths were eaten by birds while they were resting on the trunks of the trees during the day. He tested the hypothesis by rearing populations of peppered moths in which dark and light individuals were evenly mixed. Kettlewell then released these populations into two sets of woods: one, near Birmingham, was heavily polluted; the other, in Dorset, was unpolluted. Kettlewell set up rings of traps around the woods to see how many of both kinds of moths survived. To be able to evaluate his results, he had marked the moths that he had released with a dot of paint on the underside of their wings, where it could not be seen by the birds.

In the polluted area near Birmingham, Kettlewell trapped 19% of the light moths, but 40% of the dark ones. This indicated that the dark moths had had a far better chance of surviving in these polluted woods, where the tree trunks were dark. In the relatively unpolluted Dorset woods, Kettlewell recovered 12.5% of the light moths but only 6% of the dark ones. These results indicated that where the trunks of the trees were still light-colored, the light moths had a much better chance of survival than the dark ones. He later solidified his argument by placing hidden blinds in the woods and actually filming birds eating the moths. The birds that Kettlewell observed sometimes actually passed right over or next to a moth that was of the "correct" color and thus well concealed.

Industrial melanism is a term used to describe the evolutionary process in which initially light-colored organisms become dark as a result of natural selection. The process, which is common among moths that rest on tree trunks, takes place because the dark organisms are better concealed from their predators in habitats that have been darkened by soot and other forms of industrial pollution.

FIGURE 20-3
Industrial melanism. These photographs show color variants of the peppered moth, *Biston betularia*. The dark moth is more easily visible in the photo on the left. The light moth in the photo on the right is more easily visible on the darker bark, which has been affected by industrial pollution.

Dozens of other species of moths have changed in the same way as the peppered moth in industrialized areas throughout Eurasia and North America, with dark forms becoming more common from the mid-nineteenth century onward as industrialization spread. In the second half of the twentieth century, with the widespread implementation of pollution controls, the trends are being reversed, not only for the peppered moth in many areas in England, but also for many other species of moths throughout the northern continents. Such examples provide some of the best documented instances of changes in allelic frequencies of natural populations because of natural selection in relation to specific factors in the environment.

Lead Tolerance

A.D. Bradshaw and other investigators have studied the grasses that grow on the tailings, or refuse, around lead mines in Wales (Figure 20-4). These tailings are areas in which the soil is rich in lead, copper, and other unusual and potentially toxic materials and are almost bare of plants, but a few kinds, including bent grass, *Agrostis tenuis,* are found there. Bradshaw compared the growth patterns of *Agrostis* plants taken from nearby pastures and areas where lead is not abundant in the soil with those of the plants from the mine tailings. He grew plants taken from the two different kinds of soils side by side in samples of both soil types.

In normal pasture soil, the plants from the mine area were smaller and grew more slowly than the ordinary pasture plants, but they did survive. In the altered environment of lead-rich mine soil, the plants that Bradshaw had collected from the mine area grew well. In complete contrast, the pasture plants were, with a few exceptions, unable to grow in the mine soil, most of them dying within a few months. The exceptions, however, were significant: in one sample of 60 plants from a pasture, 3 showed some ability to grow in the lead-rich soil. Such plants were undoubtedly the kind from which those that could grow in the mine soil had been selected originally. In this way, a race of bent grass that was able to grow well in lead-rich mine soil evolved, the altered environment selecting in favor of individuals tolerant of high levels of lead.

Because plants able to grow on lead-rich soil are found in association with mines less than a century old, the populations of *Agrostis tenuis* clearly are able to evolve— change their genotypic frequencies—quickly when the environment demands it. Similar rapid changes have now been documented for other populations of organisms; they tell us a great deal about the process of evolution.

An Overview of Adaptation

These three case histories, of sickle cell anemia, industrial melanism, and lead tolerance, are among the best documented cases of adaptation, and provide clear evidence

FIGURE 20-4
Lead tolerance.
A Mine at Drws-y-Coed, Wales, showing soil contaminated by lead-rich tailings.
B Bentgrass, *Agrostis tenuis,* tolerant and nontolerant strains growing in 0.5 millimolar copper, one of the metals that contaminates the mine tailings.

A

B

that microevolutionary changes can be produced by natural selection. They are typical of many other situations, all of which share the same fundamental characteristic: changes occur in the frequencies of alleles in populations, which alter the characteristics of the population to make it better adapted to its environment, in which it is living. In every case, it is the *environment* that dictates the direction and extent of the change. Just as Darwin's theory demands, it is the nature of the environment that leads to natural selection and so determines the direction of evolutionary change.

THE EVIDENCE FOR MACROEVOLUTION

Adaptation within natural populations such as those we have just described constitutes strong evidence that Darwin was right in arguing that selection could bring about genetic change within populations (Table 20-2). The examples we reviewed offer direct and compelling evidence of microevolutionary change. Because we can see the evolution as it occurs, we know that Darwin was correct. What of macroevolution, however? What is the evidence that macroevolution has led to the diversity of life on earth?

In Chapter 1 we outlined the evidence that Darwin presented in favor of the hypothesis that macroevolution had occurred, a view that had become generally accepted by the time he carried out his studies, but for which he was the first to provide an explanation of the mechanism. It is summarized in Table 1-1. A great deal of additional evidence has accumulated since then, much of it far stronger than that available to Darwin and his contemporaries. Among the many lines of available evidence, we will review seven.

The Fossil Record

The most direct evidence of macroevolution is to be found in the fossil record; we now have a far more complete understanding of this record than was available in Darwin's time. Fossils are created when organisms become buried in sediment, the calcium in bone and other hard tissue is mineralized, and the sediment eventually is converted to rock. The fossils contained in sedimentary layers of rock reveal a history of life on earth.

By dating the rocks in which fossils occur, we can get an accurate idea of how old the fossils are. In Darwin's day rocks were dated by their position with respect to one

TABLE 20-2	**EXAMPLES OF THE EVIDENCE FOR EVOLUTION**
The fossil record	When fossils are arrayed in the order of their age, a progressive series of changes are seen.
The molecular record	The longer organisms have been separated according to the fossil record, the more differences are seen in the structure of their DNAs and proteins.
Homology	All vertebrates contain a similar pattern of organs, arguing that they are related to one another.
Development	During intrauterine development, human embryos and fetuses exhibit characteristics of other vertebrates, which suggests that humans are related to the other forms.
Vestigial structures	Many vertebrates contain structures with no function, but which resemble functional structures of other vertebrates, suggesting that the structures are inherited from a common ancestor.
Parallel adaptation	The marsupials in Australia closely resemble the placental mammals of the rest of the world, which argues that parallel selection has occurred.
Patterns of distribution	Inhabitants of oceanic islands resemble forms of the nearest mainland but show some differences, which suggests that they have evolved from mainland migrants.

another; rocks in deeper strata are generally older. Knowing the relative positions of sedimentary rocks and the rates of erosion of different kinds of sedimentary rocks in different environments, geologists of the nineteenth century had derived a fairly accurate idea of the relative ages of rocks.

More recently, much more accurate ways of dating rocks have been derived, and these provide dates that are absolute, rather than relative. Today, rocks are dated by measuring the degree of decay of certain radioisotopes contained in the rock; the older the rock, the more its isotopes have decayed. Because radioactive isotopes decay at a constant rate that is not altered by temperature or pressure, the isotopes in a rock act as an internal clock, measuring the time since the rock was formed. We will discuss these dating methods further in Chapter 22.

When fossils are arrayed according to their age, from oldest to youngest, they often provide evidence of progressive evolutionary change in the direction of greater complexity. Among the hoofed mammals illustrated in Figure 20-5, for example, small bony bumps on the nose can be seen to change progressively, until they become large blunt horns. In the evolution of horses (see Figure 1-15), the number of toes on the front foot is gradually reduced from 4 to 1. About 200 million years ago, oysters underwent a change from small curved shells to larger flatter ones, progressively flatter fossils being seen in the fossil record over a period of 12 million years (Figure 20-6). A host of other examples are known, all illustrating a record of progressive change. The demonstration of this progressive change is one of the strongest lines of evidence that evolution has occurred.

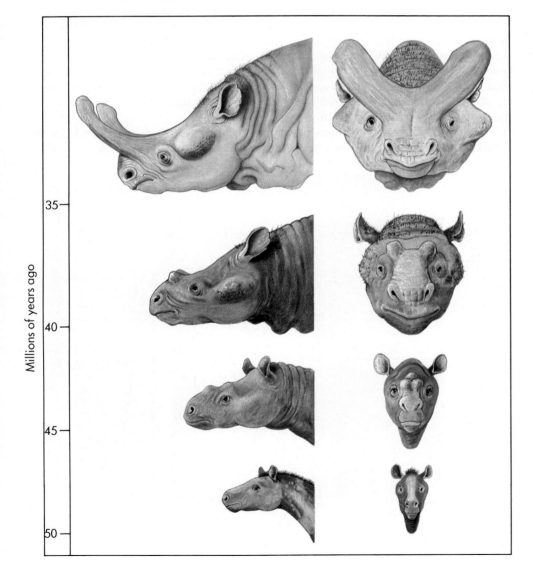

FIGURE 20-5
Macroevolution. Evolution in a group of hoofed mammals known as titanotheres between the Early Eocene Epoch (about 50 million years ago) and the Early Oligocene Epoch (about 36 million years ago). During this period of time, the small, bony protuberances that began to appear by the Middle Eocene (about 45 million years ago) evolved into relatively large, blunt horns.

Millions of years ago

35 —

40 —

45 —

50 —

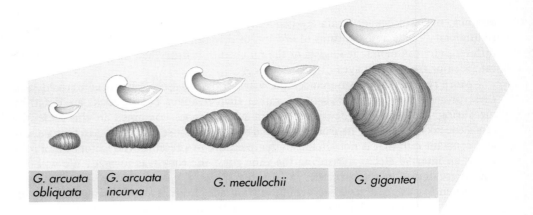

FIGURE 20-6

Evolution of shell shape in oysters. The apparently gradual evolution of a group of coiled oysters during a portion of the Early Jurassic Period that lasted about 12 million years. During this interval, the shell became larger, thinner, and flatter. These animals rested on the ocean floor in a special position called the "life position," and it may be that the larger, flatter shells were more stable against potentially disruptive water movements.

G. arcuata obliquata *G. arcuata incurva* *G. mecullochii* *G. gigantea*

The Molecular Record

If you think about it, the fact that organisms have evolved progressively from relatively simple ancestors implies that a record of evolutionary change is present in the cells of each of us, in our DNA. According to evolutionary theory, every evolutionary change involves the substitution of new versions of genes for old ones, the new arising from the old by mutation and coming to predominance because of favorable selection. Thus a series of evolutionary changes involves a progressive accumulation of genetic change in the DNA. Organisms that are more distantly related will accumulate a greater number of evolutionary differences. This is indeed what is seen when DNA sequences are compared among various organisms: for example, the longer the time since the organisms diverged, the greater the number of differences in the nucleotide sequence of the gene for cytochrome c, a protein that you will recall from Chapter 8 that plays a key role in oxidative metabolism (Figure 20-7). The same regular pattern of change is seen in hemoglobins and many other proteins. Again we see that evolutionary history involves a pattern of progressive change.

Some genes, such as the ones specifying the protein hemoglobin, have been particularly well studied, and the entire time course of their evolution can be laid out

FIGURE 20-7

The evolution of cytochrome c. Comparing various pairs of organisms, investigators have counted the numbers of nucleotides in the cytochrome c genes that are not the same. When the number of such "substitutions" is plotted against the time evolutionists believe has elapsed since the pair of organisms diverged, a straight line is obtained. This constant rate of nucleotide substitution suggests that the cytochrome c gene is evolving at a constant rate.

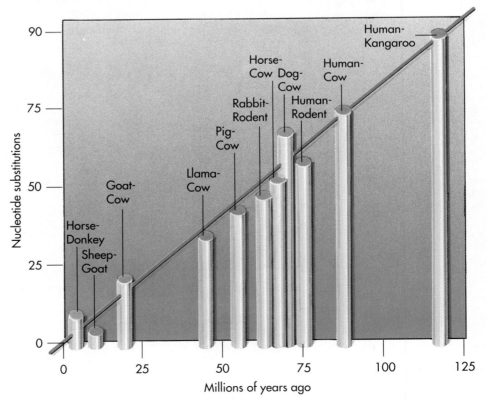

with confidence by tracing the origin of particular substitutions in their nucleotide sequences (Figure 20-8). The pattern of descent that is obtained is called a **phylogenetic tree.** It represents the evolutionary history of the gene. You should note that the progressive changes seen in the hemoglobin molecule produce a tree that reflects precisely the evolutionary relationships predicted by a study of anatomy. Whales, dolphins, and porpoises cluster together, as do the primates and the hoofed animals. The pattern of progressive change seen in the molecular record constitutes strong direct evidence for macroevolution.

FIGURE 20-8
Evolution of the globin gene.
The length of the various lines is relative to the number of nucleotide substitutions in the gene.

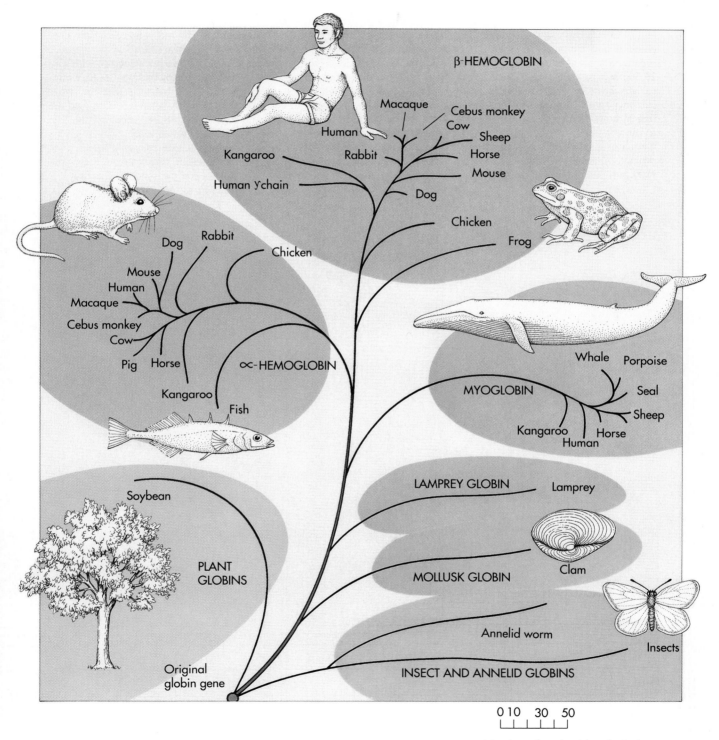

Ever since the giant panda, a unique mammal of the mountain forests of southwestern China, was brought to the attention of European scientists, scientists worldwide have speculated about its true relationships. Although it was widespread throughout most of central China as recently as 150 to 200 years ago, the giant panda is now confined to six widely separated areas amounting in total to 29,500 square kilometers. These areas are mostly in central Sichuan Province, with an outlier in the Qinling Mountains of Shaanxi Province, to the northeast. The total wild population of giant pandas at present probably numbers as few as 1150 to 1200 individuals. In addition,

13 giant pandas are still alive in zoos outside of China, and 60 to 70 are in zoos within China.

Although Pere David, who provided the first scientific description of the giant panda for European scientists (in 1869), considered it to be a bear, others pointed out that its bones and teeth resembled those of the raccoons, a family of mammals from the New World. The argument has raged ever since and although there is general acceptance of the fact that pandas resemble both bears and raccoons, it has not been possible to reach general agreement about which group of mammals they resemble more closely.

The lesser panda, or red panda

(Ailurus fulgens), which is another fascinating Asian mammal, seems similar to giant pandas in many respects, yet it is clearly similar to the raccoons. Partly because of these relationships, the red panda has often been grouped with the giant panda in a subfamily of the raccoon family. The red panda lacks the remarkable false thumb of the giant panda, an "extra" structure with which it grasps bamboo stalks while eating them; but otherwise the two are similar in behavior; both feed on bamboos, and they occur in the same general area.

By matching reactions between blood serum samples from these animals, a comparison that was first made in the 1950s, scientists soon

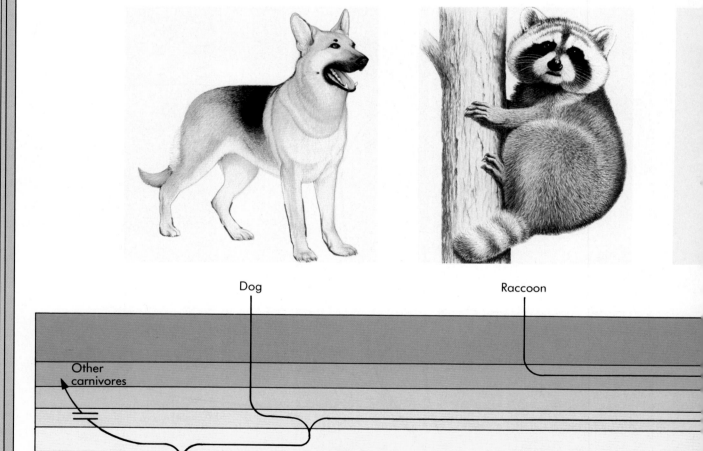

Dog

Raccoon

Other carnivores

FIGURE 20-A

came to the conclusion that the giant panda was indeed related to the bears. By the same methods, the red panda showed up as a separate evolutionary line, independent from, but related to, both bears and raccoons. Zoologist George Schaller, of the New York Zoological Society's Wildlife Conservation International Division, who has probably spent more time studying the behavior of pandas in the wild than anyone else, has concluded on the basis of their behavioral traits that red pandas and giant pandas are closely related to one another, and that they are both more closely related to bears than to raccoons.

Recently, Stephen O'Brien of the U.S. National Cancer Institute and his coworkers at the U.S. National Zoological Park have neatly solved this problem with molecular methods of analysis—DNA hybridization, isozyme similarities, immunological comparisons, and the study of chromosomal morphology after the application of special staining techniques that reveal satellite DNA and other differentiated regions. Their conclusions are illustrated in the diagram (Figure 20-A). They clearly indicated that the red panda represents an evolutionary line that diverged from the raccoons fairly soon after they had separated from the bears, and that the giant panda diverged from the bears

much more recently. As to timing, the original split between raccoons and bears probably occurred 30 to 50 million years ago; that between the red panda and the raccoons about 10 million years later; and that between the giant panda and the bears about 10 million years later than that, or about 10 to 20 million years ago. Although the red panda and the giant panda share a number of structural and behavior features, they do not share a common ancestor closer than the common ancestor of the bears and the raccoons. In these studies, modern molecular information has led to the solution of the longstanding evolutionary problem.

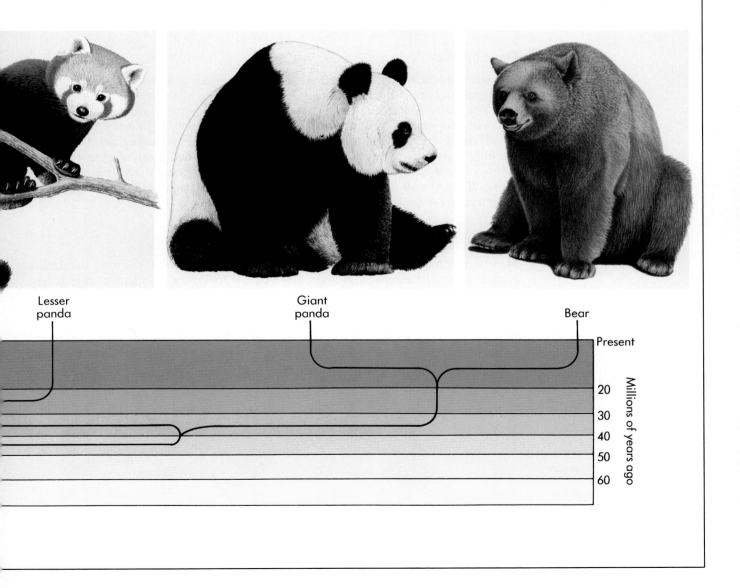

Lesser panda

Giant panda

Bear

Present

Millions of years ago

20

30

40

50

60

FIGURE 20-9
Our embryos show our evolutionary history. The embryos of various groups of vertebrate animals showing the primitive features that all share early in their development, such as gills and a tail.

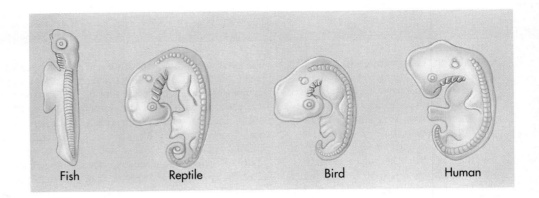

Fish   Reptile   Bird   Human

Homology

A third demonstration of the process of macroevolution lies in the fact that many organisms exhibit structures that appear to have been derived from a common ancestral form. The forelimbs of all mammals, for example, contain the same pattern of bones, although they now carry out a variety of different functions (see Figure 1-14). All vertebrates have the same pattern of bones, muscles, nerves, blood circulation and organs, the pattern becoming gradually more complex as one moves from the fishes to amphibians to reptiles to mammals. It is difficult to avoid the conclusion that progressive change is taking place.

Development

In many cases the evolutionary history of an organism can be seen to unfold during its development, with the embryo exhibiting characteristics of the embryos of its ancestors. For example, early in their development, human embryos possess gill slits like a fish, and later exhibit a tail, the vestige of which we carry to adulthood as the coccyx at the end of our spine. Human fetuses even possess a fine fur (called lanugo) during the fifth month of development. These relict developmental forms suggest strongly that our development has evolved, with new instructions being layered on top of old ones and the overall developmental program getting progressively longer. Some vertebrate embryos are shown in Figure 20-9.

Vestigial Structures

Many organisms possess structures with no apparent function that resemble the structures of presumed ancestors. Humans, for example, possess a complete set of muscles for wiggling their ears, just as a coyote does. Figure 20-10 illustrates the skeleton of a

FIGURE 20-10

Vestigial features. The skeleton of a baleen whale, a representative of the group of mammals that contains the largest living species. The enlargement shows the pelvic bones, which resemble those of other mammals, but are only weakly developed in the whale and have no apparent function.

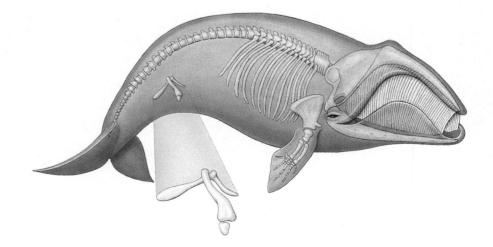

baleen whale, a representative of the group that contains the largest living mammals. It contains pelvic bones just as other mammals do, even though such bones serve no known function in the whale. The human vermiform appendix is a hollow, wormlike appendage of the caecum, or sac in which the large intestine begins. The vermiform appendix apparently is vestigial, and represents the degenerate terminal part of the caecum, the blind pouch or sac in which the large intestine begins. Although some suggestions have been made, it is difficult to assign any current function to the vermiform appendix. In many respects, it is a dangerous organ: quite often it becomes infected, leading to an inflammation called appendicitis. If it is not removed surgically when inflamed, the vermiform appendix may burst, allowing the contents of the gut to come in contact with the lining of the body cavity. This condition can be fatal if unchecked. It is difficult to understand vestigial structures such as these in any way other than as evolutionary relicts, holdovers from the evolutionary past. They argue strongly for the common ancestry of the members of the groups that share them, regardless of how different they have become subsequently.

Parallel Adaptation

Different geographical areas sometimes exhibit plant and animal communities of similar appearance, even though the individual plants and animals that make up these communities may be only distantly related to one other. It is difficult to explain these similarities as resulting from coincidence. In the best-known case, the continent of Australia separated from the other continents more than 50 million years ago, before placental mammals—the group that dominates throughout most of the world—are thought to have arrived in the area. Today only a few mammals in Australia are placental and these were all recently introduced. The bulk of Australian mammals are marsupials, members of a group in which the young are born in a very immature condition and held in a pouch until they are ready to emerge into the outside world. Marsupials may have evolved earlier than placental mammals and probably arrived in Australia before its separation from Antarctica. What are the Australian marsupials like? To an astonishing degree, they resemble the placental mammals present on the other continents (Figure 20-11). The similarity of some of the individual members of these two sets of mammals, in which specific kinds have similar habits and find their food in similar ways, argues strongly that they have evolved in different, isolated areas as a result of natural selection in relation to similar environments.

Patterns of Distribution

Darwin was the first to present evidence that the animals and plants living on oceanic islands resemble most closely the forms of the nearest continent—a relationship that would not make sense if they were all specially created. This kind of relationship, which has been observed many times since Darwin with the increasing exploration of the earth's surface, strongly suggests that the island forms evolved from individuals that came to the islands from the adjacent mainland at some time in the past. In many cases, the island forms are *not* identical with those that still occur on the nearby continents. The Galapagos finch of Figure 1-8 has a different beak than its South American relative, for example. In the absence of evolution, there seems to be no logical explanation of why individual kinds of plants and animals would clearly be related to, but have diverged in their features from, other kinds of plants and animals that occur on the adjacent mainlands. As Darwin pointed out, this relationship provides strong evidence that macroevolution has occurred.

In sum total, the evidence for macroevolution is overwhelming. Almost all biologists would agree both that: (1) macroevolutionary changes have occurred, and (2) microevolutionary changes result from natural selection. In the next chapter we will consider Darwin's proposal that microevolutionary changes have led directly to macroevolutionary ones, the key argument in his theory that evolution occurs by natural selection.

Mole

Marsupial mole

Anteater

Numbat (anteater)

Mouse

Marsupial mouse

Lemur

Spotted cuscus

Flying squirrel

Flying phalanger

Bobcat

Tasmanian "tiger cat"

Wolf

Tasmanian wolf

FIGURE 20-11
The parallel adaptation of mar-
supials in Australia and placental
mammals in the rest of the
world.

FIGURE 20-12
A lungfish. The lungfish is a member of a group of vertebrates that has changed very little during the past 150 million years.

THE TEMPO AND MODE OF EVOLUTION

On the basis of a relatively complete fossil record, it has been estimated that most living species of mammals go back at least 100,000 years, almost none a million years. A good average value for the life of a mammal species might be about 200,000 years. Mammal genera, on the other hand, seem to persist several million years on the average. The American paleontologist George Gaylord Simpson has pointed out that certain other groups, such as the lungfishes (Figure 20-12) are apparently evolving much more slowly than the mammals. In fact, Simpson estimated that there has been little evolutionary change among the lungfishes over the past 150 million years, and even slower rates of evolution are known in other groups.

Not only does the tempo, or rate, of evolution differ greatly from group to group, but groups apparently have rapid and relatively slow periods in their evolution. The fossil record provides evidence for such variability in evolutionary rates, and evolutionists are anxious to understand the environmental and other factors that account for it. In 1972 the paleontologists Niles Eldredge of the American Museum of Natural History in New York and Stephen Jay Gould of Harvard University proposed that it was the norm of evolution to proceed in spurts. They claimed that the process of evolution includes a series of **punctuated equilibria.** Evolutionary innovations would occur and give rise to new lines; then these lines might persist unchanged for a very long time. Eventually there would be a new spurt of evolution, creating a "punctuation" in the fossil record (Figure 20-13).

Eldredge and Gould proposed that rapid evolution would usually occur when populations were small, possibly different from their parental populations as a result of the founder effect, and may still have been local enough for rapid adaptation to novel ecological circumstances. In contrast, **stasis,** or lack of evolutionary change, would be expected for large populations under diverse and conflicting selective pressures.

Unfortunately, the distinctions are not as clearcut as implied by this discussion. The fossil record is incomplete because of changes in the conditions under which fossils are deposited, and the interpretation of many of its "gaps" is problematical. Notwithstanding these difficulties, the punctuated equilibrium model has provided a useful perspective for considering the mode of evolution and will continue to do so in the future. Eldredge and Gould contrasted their theory of punctuated equilibrium with that of **gradualism,** or gradual evolutionary change, which they claimed was what Darwin and most earlier students of evolution had considered normal. Whether they did so or not is debatable.

> The punctuated equilibrium model assumes that evolution occurs in spurts, between which there are long periods in which there is little evolutionary change. The gradualism model assumes that evolution proceeds gradually, with progressive change in a given evolutionary line.

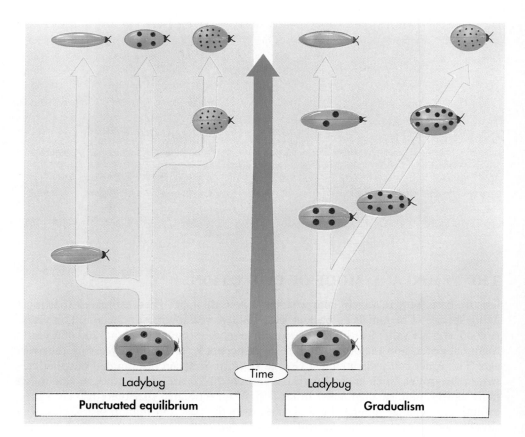

FIGURE 20-13

Two views of the pace of macro-evolution. Punctuated equilibrium surmises that species formation occurs in bursts, separated by long periods of quiet, whereas gradualism surmises that species formation is constantly occurring.

Ladybug

Punctuated equilibrium

Time

Ladybug

Gradualism

SCIENTIFIC CREATIONISM

In the century since he proposed it, Darwin's theory of evolution by natural selection has become nearly universally accepted by biologists as the best available explanation for biological diversity. Its predictions have been supported by the experiments and observations of generations of scientists. The fact that evolution has occurred and is occurring is as well accepted as is the existence of gravity, and its operation is no more doubted than predictions that a dropped apple will fall or that the sun will rise tomorrow morning.

Evolution is not, however, the only way in which the diversity of life on earth can be explained. The argument advanced in Darwin's day, that the Bible provides the correct and literal explanation of biological diversity, is still widely accepted by many people today, those who prefer a religious belief to a scientific explanation. Now, as in Darwin's time, the scientific perspective is only one of many world views. It is important in this regard to understand the limits of science. What science provides is a coherent means of organizing observations, of making predictions about how the world is going to behave. It is not a substitute for religion, which addresses a different arena of human concerns: questions of ethics and of ultimate causes. Religion and science do not preclude one another, but are regarded by many as complementary ways of viewing the world.

The clear distinction between science and religion sometimes gets muddled. Thus a number of individuals, mainly in the United States and starting largely in the 1970s, have put forward a view they title **scientific creationism**, which holds that the Biblical account of the origin of the earth is literally true, that the earth is much younger than most scientists believe, and that all species of organisms were individually created and appeared at their creation the same as they appear today. Scientific creationists are arguing in the courts that their views should be taught alongside evolution in classrooms. If both evolution and scientific creationism provide scientific explanations of biological diversity, they argue, then teachers have an obligation to present *both* views, taught side by side, so that students can choose knowledgeably between them.

This does not seem to be a bad argument if you accept the premise, which is that the view of the "scientific creationists" is indeed scientific. The confusion is not in the beliefs of the scientific creationists, which are religious beliefs that many people hold, but rather in their labeling of these beliefs as scientific. There is no scientific evidence to support the hypothesis that the earth is only a few thousand years old, and none that indicates that every species of organism was created separately. These conclusions can be reached only on the basis of arbitrary faith; they are untestable, and as such, they lie outside the realm of science.

Science, as represented by the observations of scientists, has come to different conclusions. There are virtually no differences of opinion among biologists, and indeed nearly all scientists, on the following major points: (1) the earth is about 4.5 billion years old; (2) organisms have inhabited it for the greater part of that time; and (3) all living things, including human beings, have evolved from earlier, simpler living things. The antiquity of the earth and the role the process of evolution played in the production of all organisms, living and extinct, is accepted by virtually all scientists.

In addition to the fact that it contradicts the considered judgment of almost all scientists, there is an even more fundamental problem associated with labeling "scientific creationism" as science: scientific creationism implicitly denies the intellectual basis of science itself, the reasoning on which the operation of science depends. Science insists on acceptance of the most predictive explanation of biological diversity, which is evolution, whereas scientific creationism says "Yes, but God just made it look that way." Perhaps so, but this is simply substituting religious dogma for a scientific explanation. Science consists of inferring principle from observation, and when faith is substituted for observation, the conclusion is not science. It is in just this sense that scientific creationism is not science.

Scientific creationism should not be labeled science for three reasons:
1. **It is not supported by any empirical observations**
2. **It does not infer its principles from observation, as does all science**
3. **Its assumptions lead to no testable and falsifiable hypotheses**

Scientific creationism implicitly denies the whole intellectual basis for the set of facts than human beings have assembled over the centuries about the nature of life on earth. It implies that a creator has made a world in which all phenomena are deceptively consistent with a great age for the earth and an evolutionary relationship between organisms. Evolutionary theory, in contrast, provides a coherent scientific explanation for the nature of the world in which we live and enables us to make predictions about that world. Certainly there is ample controversy among serious students as to the details of how evolution has occurred, just as there is controversy in every active scientific field. But there is no controversy about Darwin's basic finding that natural selection has played and is continuing to play the central role in the process of evolution.

The future of the human race depends largely on our collective ability to deal with the science of biology and all the phenomena that it comprises. We need the information that we have gained to deal with the problems, challenges, and uncertainties of the world in an appropriate way. We cannot afford to discard the advantages that this knowledge gives us because some of us wish to do so as an act of what we construe as religious faith. Instead, we must use all of the knowledge that we are able to gain for our common benefit. With its help, we can come to understand ourselves and our potentialities better. In no way should such rational behavior be taken as denial of the existence of a Supreme Being; it should rather be considered by those who do have religious faith as a sign that they are using their God-given gifts to reason and to understand.

SUMMARY

1. There is clear evidence of microevolutionary change in natural populations. For example, the disease called sickle cell anemia occurs when an altered hemoglobin molecule, associated with a particular allele, is well represented in the red blood cells. If this allele is present in homozygous form, it is almost invariably lethal; if it is present in heterozygous form, it not only does not produce anemia, but it also confers resistance to malaria and increased female fertility on heterozygous persons. For these reasons the allele has reached high frequencies in certain African populations in areas where malaria occurs frequently.

2. The British populations of the peppered moth, *Biston betularia*, consisted mostly of light-colored individuals before the Industrial Revolution. Over the last two centuries, however, populations that occur in heavily polluted areas, where the tree trunks are darkened with soot, have come to consist mainly of dark-colored (melanic) individuals—a result of rapid natural selection.

3. The bent grass *Agrostis tenuis* has colonized mine tailings in Wales as a result of the selection of individuals tolerant of high concentrations of lead in the soil. Such individuals are present in populations in nonpolluted areas, but in low numbers. Very strong selection has changed the nature of the populations on the tailings rapidly.

4. Two direct lines of evidence argue that macroevolution has occurred: (1) the fossil record, which exhibits a record of progressive change correlated with age; and (2) the molecular record, which exhibits a record of accumulated changes, the amount of change correlated with age as determined in the fossil record.

5. Several indirect lines of evidence argue that macroevolution has occurred, including progressive changes in homologous structures, vestigial developmental structures, parallel patterns of evolution, and patterns of biogeographical distribution.

6. The rate of macroevolution has not been constant, although there is considerable disagreement among biologists about the reason why.

7. A nonscientific theory is one whose principles are not derived from observation or supported by observation. Scientific creationism is an example of a nonscientific theory.

REVIEW QUESTIONS

1. Give several examples of how natural selection explains evolution.

2. In their natural environment, why are sickle cell heterozygotes more fit than either homozygote?

3. Why did the frequency of light-colored moths decrease whereas that of dark-colored moths increase with the advent of industrialism?

4. What evidence exists to support macroevolution?

5. Why are scientists today able to date rocks more accurately than in Darwin's day?

6. How does the molecular record indicate evolutionary change?

7. What are vestigial structures? How are they important to the study of evolution?

8. What is parallel adaptation? Give several examples.

9. How does the concept of punctuated equilibria differ from gradualism as they both concern the process of evolution?

10. Based on the explanation of the scientific process outlined in Chapter 1, is scientific creationism truly scientific? Why or why not?

THOUGHT QUESTIONS

1. In Central Africa there is a low frequency of a third hemoglobin allele, called *C*, in addition to the *A* and *S* alleles discussed in this chapter. Individuals that are heterozygous for *C* and the normal allele *A* are susceptible to malaria just as *AA* homozygotes are, but *CC* individuals are resistant to malaria—and do *not* develop anemia! Assuming that the Bantu people entered Central Africa relatively recently from a land where malaria is not common (we think this is what happened), and that among the original settlers both *C* and *S* alleles were rare, can you suggest a reason that *CC* individuals have not become predominant?

2. Imagine that you sat on the Supreme Court in the fall of 1986, hearing a case in which it was argued that creation science be taught in public schools alongside evolution as a legitimate alternative scientific explanation of biological diversity. What is the best case that lawyers might have made for and against this proposition? A decision of the Supreme Court was announced in June, 1987. How would you have voted and why?

FOR FURTHER READING

CARSON, H.L.: "The Process Whereby Species Originate," *Bioscience*, vol. 37, pages 715-720, 1987. Carson believes that geographical separation is important in the process of speciation, and presents interesting, contemporary views on the subject.

DAWKINS, R.: *The Blind Watchmaker*, W.W. Norton & Company, New York, 1986. A brilliant exposition of the factors involved in evolution by natural selection, and of contemporary reasoning concerning them.

FRENCH, M.: *Invention and Evolution*, Cambridge University Press, NY, 1990. Evolutionary adaptation as seen through the eyes of an engineer.

FUTUYMA, D.: *Science on Trial: The Case for Evolution*, Pantheon Books, New York, 1983. An excellent exposition of the serious errors in the creationist argument.

GILLIS, A.: "Can Organisms Direct their Evolution?," *Bioscience*, April 1991, pages 202-205. Biologists are rethinking this question in light of recent findings that challenge the randomness of bacterial mutations.

GILKEY, L.: *Creationism on Trial: Evolution and God at Little Rock*, Winston Press, Minneapolis, Minn., 1985. Excellent book, written by a theologian, outlining the case for creationism as argued in the courts at Little Rock.

GOULD, S.J.: "Darwinism Defined: The Difference Between Fact and Theory," *Discover*, January 1987, pages 64-70. A clear account of what biologists do and do not mean when they refer to the theory of evolution by natural selection.

GOULD, S.J.: "Through a Lens Darkly," *Natural History*, September 1989, pages 16-24. Do species change by random molecular shifts or natural selection?

HITCHING, F.H.: *The Neck of the Giraffe or Where Darwin Went Wrong*, Chaucer Press (Pan), London, 1982. An entertaining and informal presentation of all the arguments currently being advanced *against* Darwin's theory.

MAYNARD SMITH, J.: *Evolutionary Genetics*, Oxford University Press, Oxford, 1989. A first rate text of the "new" population genetics, emphasizing long term evolutionary change.

McDONALD, J.: "Macroevolution and Retroviral Elements," *Bioscience*, March 1990, pages 183-191. Insertion of DNA segments may quicken the pace of gene evolution.

O'BRIEN, S.J.: "The Ancestry of the Giant Panda," *Scientific American*, November 1987, pages 102-107. Modern molecular research has illuminated the relationship of pandas to raccoons and bears.

SEELEY, T.: "The Honey Bee Colony as a Superorganism," *American Scientist*, vol. 77, December 1989, pages 546-553. In these insects, evolution has made the colony, not the individual, the vehicle for survival of genes.

SIBLEY, C., and J. ALQUIST: "Reconstructing Bird Phylogenies by Comparing DNAs," *Scientific American*, February 1986, pages 82-92. A good introduction to the way in which biologists are beginning to use the tools of molecular biology to answer questions about evolution.

SIMPSON, G.G.: *Fossils and the History of Life*, Scientific American Library, New York, 1983. A short and beautifully illustrated account of how fossils are used to learn about life's evolutionary past by a master of the field.

THOMSON, K.: *Morphogenesis and Evolution*, Oxford University Press, Oxford, 1989. A fascinating analysis of the role of development in evolution, focusing on the way in which developmental mechanisms constrain the range of possible body forms.

21

The Origin of Species

OVERVIEW

As a result of adaptive change, populations that are originally identical eventually become more and more dissimilar. No two populations exist under exactly the same circumstances, and none has exactly the same genetic material combined in the same way. When selective forces are strong, for example, when conditions are changing rapidly or organisms are moving into a new habitat, either the characteristics of the population change rapidly or the population becomes extinct. When conditions are stable, these characteristics may remain constant for long periods even if the populations are widely separated. As a result of the process of adaptive change, some populations eventually become different enough that they are recognized as species. On islands, or in places where there are many available habitats and little competition from organisms that are already established, large numbers of distinct species may originate rapidly. For this reason, Darwin learned a lot about the process of evolution during his stay on the Galapagos Islands.

FOR REVIEW

Here are some important terms and concepts that you will encounter in this chapter. If you are not familiar with them, you should review them before proceeding.

Natural Selection (Chapter 19)

Fitness (Chapter 19)

Microevolution (Chapter 20)

Change in populations (Chapter 20)

The kinds of changes that have occurred over the last few hundred years in populations of peppered moths and bent grass (Chapter 20) occur continuously in nature. Not surprisingly, many of the changes in natural populations that have been documented best are those that have occurred in response to alterations in the environment caused by people. Human activities have dominated habitats all over the world for centuries, and populations of moths, grasses, bacteria, and other organisms have had to react to them or have become extinct. Changes similar to those that have occurred in peppered moths and bent grass also occur in populations of organisms undisturbed by human activities, however. These changes allow the organisms in which they occur to live in diverse habitats, by changing their characteristics so that they can live successfully in the new areas or under new conditions. All such changes come about as a result of changes in allele frequencies.

Sickle cell anemia in human populations provides an illustration of the way in which such a process of adaptation works. As people migrated for the first time into regions where malaria was prevalent, the frequency of the sickle cell allele would have increased. In populations of people who lived in other regions the allele would have remained rare (see Figure 20-2).

Many natural changes are occurring that have effects on organisms just as profound as, or more profound than, those that result from the impact of human activities. As successive periods of expansion and contraction of continental glaciers have occurred over the past few million years, for example, climates have gotten warmer and cooler, wetter and drier. In response to these changes, whole communities of organisms have had to adjust in their characteristics, primarily by selective shifts in the frequencies of those alleles that are related to survival under the new circumstances. Species of plants and animals that could not adjust have become extinct, at least locally, while others have taken their places.

THE NATURE OF SPECIES

How do such adaptive changes in natural populations lead to the origin of species? Darwin was extremely interested in this question because he considered species to be the most important evolutionary units. As we begin to think about the problem, we must first examine what the concept **species** means and how this concept has changed through the years.

John Ray (1627-1705), an English clergyman and scientist, was one of the first to propose a definition of the category species. In about 1700 he pointed out how a species could be recognized: all of the individuals that belonged to it could breed with one another and produce progeny that were still of that species. Even if two different-looking individuals appeared among the progeny of a single mating, they were still considered to belong to the same species. All dogs were one species, all pigeons, and so on; carp, however, were not the same species as goldfish, nor mallards the same species as teal, and so forth (Figure 21-1).

In an informal way, of course, people had always recognized species. Indeed, the word "species" is simply Latin for "kind." With Ray, however, the species began to be regarded as an important biological unit that could be catalogued and understood. Along with other scientists of his time, Ray believed that species were individually created by the Supreme Being and did not change, a view that was widely held until it was challenged by Darwin in 1859. Darwin was interested in those situations in which it was not clear whether he was dealing with distinct species or not. He considered the fact that some species intergraded with one another in their features to be important evidence in support of his theory of evolution. Darwin explained the relative constancy of species by saying that each had its own distinctive role in nature, a role that we would call a **niche**. Thus each species occurs in a particular kind of place, displays different activities at different times of the year, has different habitats, and so forth.

From the 1920s onward, with the emergence of population genetics, there was a desire to define the category species more precisely. The definition that began to emerge was stated by the American evolutionist Ernst Mayr as follows: species were "groups of actually or potentially interbreeding natural populations which are reproductively isolated from other such groups." In other words, hybrids between species

FIGURE 21-1
The mule, a sterile hybrid between a female horse and a male donkey. By any standard, the horse and donkey are distinct species; mules, of course, cannot interbreed with either.

A   B   C

D   E   F

FIGURE 21-2
Butterflies and annual flowering plants: distinct species in one genus.
Butterflies: (**A**) American painted lady, *Vanessa virginiensis*. (**B**) Western painted lady, *Vanessa annabela*, (**C**) Red admiral, *Vanessa atalanta*.
Annual flowering plants: (**D**) *Clarkia concinna*. (**E**) *Clarkia speciosa*. (**F**) *Clarkia rubicunda*.

occur rarely in nature. On the other hand, individuals that belong to the same species are able to interbreed freely. In practice, however, scientists usually have little or no experimental evidence on which to base their decisions about what constitutes a species in most particular groups. They recognize species primarily because they differ from one another in their features (Figure 21-2). In fact, there are essentially no barriers to hybridization between the species in some groups of organisms, and strong barriers to hybridization between the species of other groups. That is why we have not used so-called definitions of species that claim that all of the individuals of a species can interbreed with one another, but not with the individuals of other species; such "definitions" are simply not valid reflections of what we know.

Within the units that are classified as species, the populations that occur in different places may be more or less distinct from one another (Figure 21-3). When such

FIGURE 21-3
Extinction in our own backyard.
Subspecies of the seaside sparrow, *Ammodramus maritimus*, are quite local in distribution, and some of them are in danger of extinction because of the alteration of their habitats. The widespread subspecies *Ammodramus maritimus maritimus* (adult, *1*, and juvenile, *2*) is the most common; *A. m. fisheri* (*3*) occurs along the Gulf Coast. The Cape sable seaside sparrow, *A. m. mirabilis* (*4*), occurs in a small area of southwestern Florida. The last subspecies, the dusky seaside sparrow (*5*), *A. m. nigrescens*, occurred only near Titusville, Florida. The last individual, a male, died in captivity in 1987.

populations occur in the same areas, however, individuals with intermediate or mixed characteristics usually are frequent; in other words, the distinct-appearing populations within a species usually **intergrade** with one another when they occur together. In areas where these different-looking races, which may be classified taxonomically as **subspecies** or **varieties,** approach one another, many individuals may occur in which the distinctive features that are characteristic of each of the intergrading races are combined. Many of these individuals may not match either race in their characteristics. In contrast, when *species* occur together, they usually do not intergrade, although they may hybridize occasionally.

In some groups of organisms even local races are not capable of interbreeding with one another. This pattern occurs, for example, in many annual plants. In contrast, species of trees, some groups of mammals, and fishes generally *are* able to form fertile hybrids with one another, even though they may not do so in nature. For still other kinds of plants and animals, we do not know whether the species can form hybrids. We define a species, therefore, as a group of organisms that is unlike other such groups of organisms and does not intergrade extensively with them in nature.

> **Species are groups of organisms that differ in one or more characteristics and do not intergrade extensively if they occur together in nature.**

THE DIVERGENCE OF POPULATIONS

Local populations are usually more or less separated geographically, and the conditions in which they occur are dissimilar. Populations of interbreeding individuals are often extremely small (Figure 21-4), and the exchange of individuals, and thus of ge-

FIGURE 21-4
Some animal populations are very localized. The butterfly *Euphydryas editha bayensis* (**A**) occurs on Jasper Ridge (**B**), a biological preserve in the foothills above Stanford University, just south of San Francisco, California (see also Figure 25-10). The butterflies apparently constitute one continuous population throughout the open grassland, which occurs as an "island" surrounded by oak forest and chaparral. Extensive studies by Paul Ehrlich and his colleagues over more than 30 years, however, involving marking the butterflies with dots of ink on their wings, releasing them, and recapturing them, have revealed that this species of butterfly actually exists as a series of discontinuous populations between which there is little movement of individuals. Each of these local populations, separated by dashed lines on the maps (C), is free to respond to the selective forces that are characteristic of its particular part of the ridge. Changes in the density of these populations, which have remained essentially constant in overall distribution since the late 1950s at least, are shown in the maps.

netic material, between such populations may be limited. For these reasons local populations often are able to adjust individually and effectively to the demands of their particular environment. As they do so, their characteristics change. The rate at which they change will depend primarily on the selective forces to which they are responding. If these forces are strong, the populations will change rapidly; populations certainly do not need to be isolated on islands to be able to become distinct species.*

> The characteristics of populations tend to diverge more or less rapidly, depending on the features of their particular environment. Since there is only a limited exchange of genetic material between populations, even if they are geographically close, all populations tend to become increasingly divergent from one another in their characteristics over time.

Many genes interact in the production of most phenotypic characteristics. Furthermore, all of the developmental processes that occur within a single organism are closely integrated with one another. In addition, every population begins with a different endowment of alleles, as we saw in our discussion of the founder effect in Chapter 19. Because of these three factors, the response of a given population of organisms to a combination of selective factors tends to be unpredictable. Even if two populations respond to similar selective forces, and even though they may be geographically close to one another, they still tend to change and to differ more and more from one another over time. If their environments are dissimilar enough or change rapidly enough, the populations may diverge rapidly and strikingly, and soon come to look different from one another.

ECOLOGICAL RACES

What kinds of patterns can be expected as a result of the differentiation of populations? One consequence is that the individuals of a species that occur in one part of its range often look different from those that occur elsewhere (see Figure 21-4). Such groups of distinctive individuals are informally called races, and they may, as we have mentioned, be classified taxonomically as subspecies or varieties. The existence of such races proved fascinating to Darwin, because he considered them to be an intermediate stage in the evolution of species. The kinds of ecologically defined races that we will discuss first may change over time to the clusters of species that we will consider next. Both provide important examples of the evolution of populations in nature.

---

*The term "allopatric speciation" has been used to describe the differentiation of geographically isolated populations into distinct species, whereas the term "sympatric speciation" has been used to describe the splitting of populations within a common area into species. We have not used these terms because most populations of any species are largely isolated from one another in a genetic sense, and may therefore diverge if they are subject to different combinations of selective forces. The distinction implied by the terms mentioned, therefore, is misleading; geographic separation, in the normal sense of the word, is not necessary for populations to diverge and become distinct species.

FIGURE 21-5
Two forms of Bishop pine, *Pinus muricata*, in Marin County, just north of San Francisco, California.
A This tree is growing in a flat field behind Inverness Ridge, where it is protected from the ocean winds.
B This tree, growing in an exposed place, has been dwarfed and shaped by the winds. Without experimental study, it is not possible to determine whether these two individuals differ genetically or not.

A

B

Ecological races were first studied in detail in plants. As every gardener knows, the same plants may differ greatly in appearance, depending on the places where they are grown (Figure 21-5). This is true even for genetically identical divisions of the same plant—**clones.** The part of an individual plant that is usually in the sun often produces leaves that are unlike the leaves the same plant produces in the shade. For example, shade leaves are usually thinner and broader and have more internal air spaces than do sun leaves. Thus it seemed to many botanists in the nineteenth century that environmental factors, rather than genetic differences, might account for many of the differences between races and even between species of plants.

Ecotypes in Plants

In the 1920s and 1930s the Swedish botanist Gôte Turesson performed a series of experiments that were designed to test whether differences between the races of plants were largely genetically determined or mainly caused by environmental factors. Turesson observed that many plants had distinctive races that grew in different habitats. These races differed from one another in characteristics such as height, leaf size and shape, degree of hairiness, flowering time, and branching pattern. Turesson dug up individuals representing these races and cultivated them together in his experimental garden at Lund, Sweden. In nearly every case he found that the unique features of individual races were maintained when the plants were grown in a common environment. Most of the characteristics that he observed, therefore, had a genetic basis; a few of them were environmental. Turesson called the ecological races that he studied and showed to have a genetic basis **ecotypes.**

The important studies initiated by Turesson were continued in California by a group of scientists from the Department of Plant Biology of the Carnegie Institution of Washington, located on the campus of Stanford University. The Carnegie investigators used three major transplant stations for their investigations, one near Stanford at sea level, the second at 1400 meters elevation on the forested, western slope of the

HUMAN RACES

Human beings, like all other species, have differentiated in their characteristics as they have spread throughout the world. The local populations in one area often look impressively different from those that live elsewhere. For example, northern Europeans often have blond hair, fair skin, and blue eyes, whereas Africans often have black hair, dark skin, and brown eyes. In addition, other features, such as the ABO blood groups discussed in Chapter 13, differ in proportion from area to area. These traits may play a role in adapting the particular populations to their environments. Thus, the blood groups may be associated with immunity to diseases characteristic of certain geographical areas, and dark skin shields the body from the damaging effects of ultraviolet radiation, which is much stronger in the tropics than in temperate regions.

All human beings, however, are capable of mating and producing fertile offspring. The reasons that they do or do not choose to associate with one another are purely psychological and behavioral (cultural). The number of races into which the human species might logically be divided, as well as which should be given names, has long been a point of contention. Some contemporary anthropologists divide people into as many as 30 races, others as few as three: Caucasoid, Negroid, and Oriental. American Indians, Bushmen, and Aborigines are particularly distinctive units that are sometimes regarded as distinct races.

The problem with classifying people or other organisms into races is that the characteristics used to define the races are usually not well correlated with one another. The way one determines the race to which a given group of people should be assigned is therefore always somewhat arbitrary. In human beings, it is simply not possible to delimit clearly defined races that can be recognized by a particular combination of characteristics. Variation patterns are somewhat correlated with geographical distribution, but they do not form clearly defined units that we can classify. Different groups of people have constantly intermingled and interbred with one another during the entire course of history. Today, the differences between human "races" are tending to break down rapidly as large numbers of individuals are constantly moving over the face of the globe, mixing with one another and recombining their characteristics.

In any event, individual differences within what might be considered races are greater than the differences between such units. This is a sound biological basis for dealing with each human being on his or her own merits, and not as a member of a particular "race."

Sierra Nevada, and their third at 3050 meters elevation near the crest of that range. Planting genetically identical individuals—divisions of the same plant—at the three stations, the scientists were able to demonstrate the existence of ecotypes in a number of different plant species, and also to confirm the physiological differences between them. These differences made the respective ecotypes better suited to grow in particular areas; for example, those that occurred naturally in areas of summer drought exhibited dormancy during that period.

Ultimately, these studies and others led to the conclusion that most of the differences between individuals, populations, races, and species of plants have a genetic basis. Despite the fact that plants change their characteristics in relation to the environments in which they grow, the differences between them are usually fixed genetically in the course of their evolution.

Ecological Races in Animals

Similar patterns of variation are, of course, also found in animals (see Figure 21-3). The differences may be morphological or physiological, and they have the same basis as do ecotypes in plants. The differences between subspecies may be striking. For example, the larger races of some species of birds may often consist of individuals that weigh three or four times as much as individuals of the smaller races. Races may differ from one another in their tolerance to different temperatures, in the speed of their larval development, in their behavioral characteristics—in short, in any feature that can be measured and studied. Their features are almost always genetically determined.

BARRIERS TO HYBRIDIZATION

As isolated populations become increasingly different from one another in their overall characteristics, they may eventually occupy different niches; in other words, they may exploit different resources in different ways. If such differentiated populations do ever migrate back into contact with one another, they may still remain distinct in their characteristics. The populations may occur in different habitats, have different feeding habits, or otherwise be separated by differences that arose in them while they were apart. For these or other reasons the individuals of the differentiated populations may not hybridize with each other; if they do, functional hybrid individuals may not be formed, or the hybrids that are formed may be sterile. For animals that choose their mates, the differences between the populations that have originated in isolation may be so great that they may choose mates of their own kind, rather than those of the other, formerly isolated population. In other words, and for a variety of possible reasons, the populations may have become species.

> **Populations of organisms tend to become increasingly different from one another in all of their characteristics. If the process continues long enough, or the selective forces are strong enough, the populations may become so different that they are considered distinct species.**

Once species have formed, how do they keep their identity? The reasons that they do retain their identity may be grouped into two categories: **prezygotic isolating mechanisms,** those preventing the formation of zygotes; and **postzygotic isolating mechanisms,** those preventing the proper functioning of zygotes after they are formed. Some of these isolating mechanisms also occur, although often in a less marked form, within species. When they do, they illustrate stages in the evolution of new species. In the following sections we will discuss various isolating mechanisms in these two categories and offer examples that illustrate how the isolating mechanisms operate to help species retain their identity.

Prezygotic Isolating Mechanisms

Geographical Isolation.   Most species simply do not exist together in the same places. Species are generally adapted to different kinds of climates, or to different

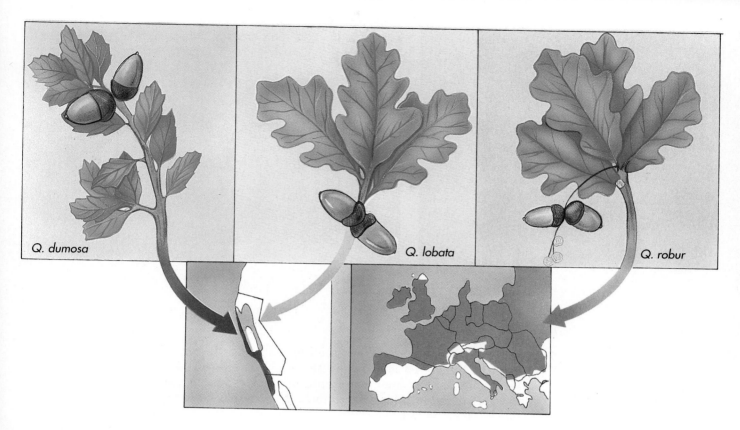

FIGURE 21-6

Distance is an effective barrier for keeping species apart. There are many differences in leaf and acorn characteristics of three species of oaks, *Quercus robur*, *Q. lobata*, and *Q. dumosa*. The two Californian species occur in different habitats and so remain different.

habitats. If this is the case, there is no possibility of natural hybridization between them. They may, however, hybridize if they are brought together in zoos, parks, or botanical gardens.

As an example of geographical isolation, the English oak, *Quercus robur*, occurs throughout those areas of Europe that have a relatively mild, oceanic climate. In its characteristics it is quite similar to the valley oak, *Quercus lobata*, of California, and quite different from the scrub oak, *Quercus dumosa*, also of California and adjacent Baja California (Figure 21-6). All of these species can hybridize with one another and form fertile hybrids. The English oak does not hybridize with the others in nature, however, simply because its geographical range does not overlap with theirs. Similarly, although lions, *Panthera leo*, and tigers, *P. tigris*, do not now occur together in nature, they do mate and produce hybrids in zoos. The hybrids in which the tiger is the father, called "tiglons," are viable and fertile; less is known about "ligers," hybrids in which the lion is the father.

Ecological Isolation. Even if two species occur in the same area, they may occur in different habitats and thus may not hybridize with one another. If, on the other hand, they do hybridize with one another, the hybrids may not be well represented in the overall population, since they may not be as fit in the habitat of either of their parents. In the latter case, one would speak of a postzygotic isolating mechanism of the kind that will be discussed in the next section.

For example, in India the ranges of lions and tigers overlapped until about 150 years ago. Even when they did, however, there were no records of natural hybrids. Lions stayed mainly in the open grassland and hunted in groups called *prides;* tigers tended to be solitary creatures of the forest. Because of the ecological and behavioral differences between them, lions and tigers rarely came into direct contact with one another, even though their ranges overlapped over thousands of square kilometers.

Similar situations occur among plants. We have already mentioned two species of oaks that occur in California: the valley oak, *Quercus lobata*, and the scrub oak, *Q. dumosa* (Figures 21-6 and 21-7). Valley oak, a graceful deciduous tree, which can be as much as 35 meters tall, occurs in the fertile soils of open grassland on gentle slopes

A
B

FIGURE 21-7
Closely related species may look very different.
A *Quercus lobata* in Yosemite Valley.
B *Q. dumosa*, a shrub, in the Coast Ranges south of San Francisco.

and valley floors in central and southern California. In contrast, scrub oak, *Quercus dumosa*, is an evergreen shrub, usually only 1 to 3 meters tall, which often forms the kind of dense scrub known as chaparral. The scrub oak is found on steep slopes, in less fertile soils. Hybrids between these different oaks do occur and they are fully fertile, but they are rare. The sharply distinct habitats of their parents limit their occurrence together, and there is no intermediate habitat where the hybrids might flourish.

Temporal Isolation. *Lactuca graminifolia* and *L. canadensis*, two species of wild lettuce, grow together along roadsides throughout the southeastern United States. Hybrids between these two species can easily be made experimentally and are completely fertile. Such hybrids are rare in nature, however, because *L. graminifolia* flowers in early spring and *L. canadensis* flowers in summer. When their blooming periods overlap, as they do occasionally, the two species do form hybrids, which may even be abundant locally.

Many species of amphibians that are closely related have different breeding seasons. Differences of this kind prevent hybridization between such species. For example, five species of frogs of the genus *Rana* occur together in most of the eastern United States. The peak time of breeding is different for each of them; because of this difference, hybrids are rare. In insects such as termites and ants, mating occurs when winged, reproductive individuals swarm from the nest. Species often differ in their swarming times, which eliminates the possibility of hybridization between them.

Behavioral Isolation. In Chapter 56 we will consider the often elaborate courtship and mating rituals of some groups of animals. Related species of organisms such as birds often differ in their mating rituals, which tend to keep these species distinct in nature even if they do occur in the same places. Indeed, much animal communication is related to the selection of mates.

In the Hawaiian Islands there are more than 500 species of flies of the genus *Drosophila*. This is one of the most remarkable concentrations of species in a single animal genus found anywhere. Many of these flies differ greatly from other species of *Drosophila*, exhibiting characteristics that can only be described as bizarre (Figure 21-8). The genus occurs throughout the world, but nowhere are the flies more diverse in their external appearance or behavior than in Hawaii.

The Hawaiian species of *Drosophila* are long-lived and often very large as compared with their relatives on the mainland. The females are more uniform than the males, which are often bizarre and highly distinctive. The males display complex territorial behavior and elaborate courtship rituals. Some of these are shown in Figure 21-8, *A* to *C;* another kind of courtship behavior is exhibited by *D. clavisetae* (Figure 21-8, *D*).

A

B

C

D

The patterns of mating behavior among the Hawaiian species of *Drosophila* are of great importance in maintaining the distinctiveness of the individual species. Despite the great differences between them, which are so evident in Figure 21-8, for example, *D. heteroneura* and *D. silvestris* are very closely related. Hybrids between them are fully fertile. The two species occur together over a wide area on the island of Hawaii, and yet hybridization has been observed at only one locality. The very different and complex behavioral characteristics of these flies obviously play the major role in maintaining their distinctiveness.

Mechanical Isolation.

Structural differences between some related species of animals prevent mating. Aside from such obvious features as size, the structure of the male and female copulatory organs may be so incompatible that mating cannot occur. In many insect and other arthropod groups the sexual organs, particularly those of the male, are so diverse that they are used as a primary basis for classification. This diversity in structure is generally presumed to have some importance in maintaining differences between species.

Similarly, the flowers of related species of plants often differ significantly in their proportions and structures. Some of these differences limit the transfer of pollen from one plant species to another. For example, bees may pick up the pollen of one species on one place on their bodies, and this area may not come into contact with the receptive structures of the flowers of another plant species, so that the pollen is not transferred. This difference would then decrease the frequency of hybridization between them.

Prevention of Gamete Fusion.

In animals that simply shed their gametes into water, the eggs and sperm derived from different species may not attract one another. Many hybrid combinations involving land animals are not realized because the sperm of one species may function so poorly within the reproductive tract of another that fertilization never takes place. The growth of the pollen tubes may be impeded in hybrids between different species of plants. In both plants and animals the operation of such isolating mechanisms may prevent the union of gametes even following successful mating. The prevention of fusion between gametes is the last kind of prezygotic isolating mechanism possible before hybrids are formed.

Prezygotic isolating mechanisms lead to reproductive isolation by preventing the formation of hybrid zygotes. The principal isolating mechanisms are geographical, ecological, temporal, behavioral, and mechanical isolation and the prevention of gamete fusion.

FIGURE 21-8
Evolution among the Hawaiian *Drosophila*. Closely related species of Hawaiian flies of the genus *Drosophila* engaging in territorial defense and courtship.
A *Drosophila silvestris*. After approaching the female from the rear, the male has lunged forward while vibrating his wings with his head under the wings of the female, and raised his forelegs up and over the female's abdomen. Specialized hairs on the dorsal surface of one of the leg segments are then "drummed" over the dorsal surface of the female's abdomen.
B *Drosophila heteroneura*. A male with extended wings approaching and displaying toward a female in typical courtship posture.
C *Drosophila heteroneura*. Two males have locked antennae as part of the aggressive behavior involved in the defense of their territories.
D *Drosophila davisetae*. In this species, the males spray a chemical signal over the female.

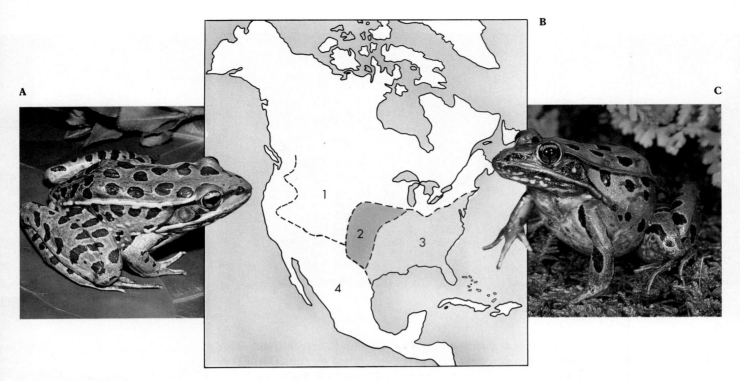

A

B

C

FIGURE 21-9
Sibling species of leopard frogs.
A The southern leopard frog, *Rana berlandieri*, in California.
B The leopard frog, *Rana pipiens*, in Wisconsin.
C Numbers indicate the following species in the geographical ranges shown: *1, Rana pipiens; 2, Rana blairi; 3, Rana sphenocephala; 4, Rana berlandieri.*
These four species resemble one another closely in their external features. The fact that they are separate species was first suspected when hybrids between them produced defective embryos in some combinations. At first it was thought that this occurrence indicated differentiation within a species along a geographical gradient, but when it was demonstrated that the mating calls of these four species differed substantially, their distinctiveness was recognized. These species usually do not hybridize where they occur together, although hybrids do occur locally. Such closely related species, often distinguished primarily by behavioral or other nonevident characteristics, are called **sibling species.**

Postzygotic Isolating Mechanisms

All of the factors that we have discussed up to this point tend to prevent hybridization. If hybridization does occur, however, and zygotes are produced, there are still many factors that may prevent those zygotes from developing into normal, functional, fertile F_1 individuals. Development in any species is a complex process. In hybrids the genetic complements of two species may be so different that they cannot function together normally in embryonic development. For example, hybridization between sheep and goats usually produces embryos that die in the earliest developmental stages.

The leopard frogs (*Rana pipiens* complex) of the eastern United States are a series of very similar species, which were assumed for a long time to constitute a single species. Before the nature of the species of this complex was elucidated, it was assumed that there were simply difficulties in producing hybrids between some of these frogs because they came from different regions. Although the species are similar, hybrids between them are usually rare; a number of the hybrid combinations cannot be produced, even in the laboratory (Figure 21-9).

Many examples of this kind, in which similar species initially have been distinguished only as a result of hybridization experiments, are known in plants. Sometimes the hybrid embryos can be removed at an early stage and grown in an appropriate medium. When these hybrids are supplied with extra nutrients or other growth requirements that compensate for their weakness or inviability, they may complete their development normally.

Even if the hybrids survive the embryo stage, they may not develop normally. If the hybrids are weaker than their parents, they will almost certainly be eliminated in nature. Even if they are vigorous and strong, as in the case of the mule, which is a hybrid of the horse and the donkey (see Figure 21-1), they may still be sterile and thus incapable of contributing to succeeding generations. The development of sex organs in hybrids may be abnormal, the chromosomes derived from the respective parents may not pair properly, or their fertility may simply be lower than normal for other reasons.

> Postzygotic isolating mechanisms are those in which hybrid zygotes develop abnormally or fail to develop entirely, or in which hybrids cannot become established in nature.

TABLE 21-1 REPRODUCTIVE ISOLATING MECHANISMS

1. Prezygotic mechanisms: prevent the formation of zygotes
   a. Geographical isolation: the species occur in different areas
   b. Ecological isolation: the species live in different habitats and do not meet; there may be no habitat suitable for their hybrids
   c. Temporal isolation: the species reproduce at different seasons, or different times of day
   d. Behavioral isolation: the behavior of the species may differ, so that there is little or no attraction between them
   e. Mechanical isolation: structural differences between species may prevent mating
   f. Prevention of gamete fusion: the gametes may not fuse, usually because of chemical factors
2. Postzygotic mechanisms: prevent the proper functioning of zygotes once they are formed; these include the inviability or sterility of the hybrids or their irregular development

Reproductive Isolation: An Overview

All of the kinds of reproductive isolation that we have discussed can arise in populations that are diverging from one other (Table 21-1). These kinds of reproductive isolation all occur within certain species, as would be expected from the way in which such differences evolve. Some populations of particular species, therefore, cannot hybridize with one another, just as if they were distinct species as judged by this criterion.

The formation of species is a continuous process, one that we can understand because of the existence of intermediate stages at all levels of differentiation. If populations that are partly differentiated come into contact with one another, they may still be able to interbreed freely, and the differences between them may then disappear over the course of time. If their hybrids are partly sterile, or not as well adapted to the existing habitats as their parents, these hybrids will be at a disadvantage. As a result, there may be selection for factors that limit the ability of the differentiated populations to hybridize. If hybrids are sterile or not as successful as their parents, individual plants or animals that do not hybridize may be more fit, in the sense explained in Chapter 19, than those that do so.

Most species are separated by combinations of the isolating mechanisms that we have discussed. For example, two related species may occur in different habitats, produce their gametes at different times of the year, have different behavioral patterns, and produce inviable embryos even if hybridization does take place. Such patterns, in which more than one factor functions in limiting the frequency of hybrids between two species, presumably arise for two reasons:

1. The factors that limit hybridization arise primarily as by-products of adaptive change in populations. Consequently, several different kinds of factors that limit hybridization often emerge simultaneously and may characterize the differentiated populations.
2. If differentiated populations do come into contact with one another, natural selection may strengthen the isolating mechanisms that are already present. If, for example, the hybrids do not complete their development beyond the embryo stage, then any factor that limits the hybridization that produces them (prezygotic isolating mechanisms) would be an advantage. Individuals that form hybrids that do not function well in nature waste reproductive energy by doing so and are less fit than individuals that do not form such hybrids.

Both kinds of forces that limit hybridization are built up as a part of the overall process of change in populations that are isolated from one another, although they may sometimes be strengthened when and if the differentiated populations migrate into contact with one another.

> As a result of the way in which they originate, species are often separated by more than one factor. Some of these factors prevent the species from hybridizing at all; others limit the success of the interspecific hybrids once they are formed.

CLUSTERS OF SPECIES

One of the most visible manifestations of evolution is the existence of groups of closely related species in various genera. These species often have evolved relatively recently from a common ancestor. The phenomenon by which they change in occupying a series of different habitats within a region is called **adaptive radiation.** Such clusters are often particularly impressive on groups of islands, in series of lakes, or in other sharply discontinuous habitats. We will discuss two examples from islands to illustrate these species clusters.

Darwin's Finches

Thirteen species of Darwin's finches occur on the Galapagos Islands, and one lives on Cocos Island, which lies about 1000 kilometers to the north. This group of birds provides one of the most striking and best-studied examples of evolution on islands. Although Darwin was, as we have seen, most impressed by the differences in the land tortoises from the different islands, these finches have excited subsequent students of evolution.

Just as Darwin found them on his voyage, the Galapagos Islands are a particularly striking natural laboratory of evolution. The islands are all relatively young in geological terms (several million years), and they have never been connected with the adjacent mainland of South America or with any other source area. The lowlands of the Galapagos Islands are covered with thorn scrub; at higher elevations, which are attained only on the larger islands, there are moist, dense forests. All of the organisms that occur on these islands have reached them by crossing the sea, as a result of chance dispersal in the water, by wind, or by transport via another organism.

On islands there is often a disproportionate representation of certain groups of organisms. For example, in addition to the 13 species of Darwin's finches on the Galapagos, there are only 7 other species of land birds. Presumably the ancestor of Darwin's finches reached these islands earlier than the ancestors of the other birds. All of the kinds of habitats where birds occur on the mainland were unoccupied, and the ancestor of Darwin's finches was able to take advantage of them all. As the new arrivals moved into these vacant niches, adopting new life-styles in doing so, they were subjected to diverse sets of selective pressures. Under these circumstances, the ancestral finches rapidly split into a series of diverse populations, and some of these became species.

In addition, since the Galapagos are a group of islands, small flocks of birds could occasionally reach new islands where no other land birds were present. On these new islands the immigrants would become adjusted to the local conditions through the process of natural selection. In doing so, they would have become progressively more different from the original populations.

The descendants of the original finches that reached the Galapagos Islands now occupy many different kinds of habitats on the islands (Figure 21-10). These habitats encompass a variety of niches comparable to those occupied by several distinct groups of birds on the mainland. Among the 13 species of Darwin's finches that inhabit the Galapagos, there are three main groups. Some scientists consider these birds to belong to several different genera, but most now assign all of them to the genus *Geospiza.* The three groups are:

1. *Ground finches.* There are six species of ground finches (Figure 1-8, *A*). Most of the ground finches feed on seeds of different sizes. The size of their bills is related to the size of the seeds on which the birds feed. One of the ground

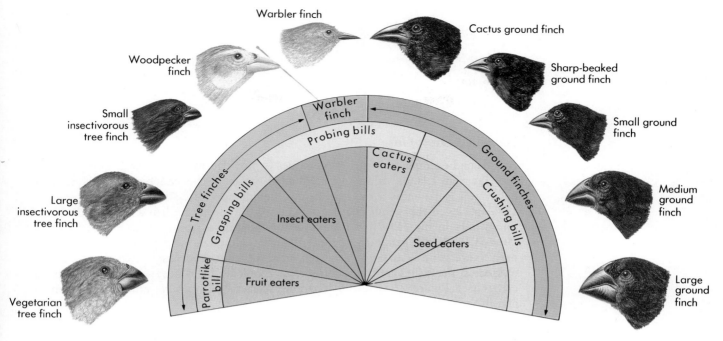

Warbler finch

Woodpecker finch

Small insectivorous tree finch

Large insectivorous tree finch

Vegetarian tree finch

Cactus ground finch

Sharp-beaked ground finch

Small ground finch

Medium ground finch

Large ground finch

Warbler finch

Probing bills

Cactus eaters

Tree finches

Grasping bills

Ground finches

Crushing bills

Insect eaters

Seed eaters

Parrotlike bill

Fruit eaters

finches is the cactus ground finch, *Geospiza conirostris*. It feeds primarily on cactus flowers and fruits and has a longer, larger, more pointed bill than the others.

2. *Tree finches*. There are six species of tree finches, belonging to several diverse groups. These species also differ from one another primarily in bill size and shape, which reflect adaptations to their food. Four species have bills that are suitable for feeding on insects: the ones with larger bills feed on larger insects, and those with smaller bills feed on smaller insects. One of the tree finches has a parrotlike beak; it feeds on buds and fruit in trees. The sixth species of tree finch, the woodpecker finch, has a chisel-like beak. This unusual bird carries around a twig or a cactus spine, which it uses to probe for insects in deep crevices. It is an extraordinary example of a bird that uses a tool.

3. *Warbler finch*. This unusual bird differs from the ground finches and the tree finches in its habits and morphological adaptations. The warbler finch plays the same ecological role in the Galapagos woods that warblers play on the mainland, searching continually over the leaves and branches for insects. It has a slender, warblerlike beak.

The warbler finch is the most distinctive of all of Darwin's finches, and the question has been raised as to whether it is directly related to the other species. Detailed studies, however, have shown that it is derived from the same common ancestor as the others: presumably a pair of windblown finches, or a pregnant female, that reached the Galapagos Islands perhaps 100,000 or more years ago.

David Steadman, of the New York State Museum, believes that this ancestor was similar to the blue-black grassquit, *Volatinia jacarina* (see Figure 1-8, *B*). This species of bird is abundant throughout the lowlands of Latin America and occurs from Mexico to Chile along the Pacific Coast. Steadman bases his conclusion on the characteristics of this widespread bird, including its short, stout, cone-shaped bill, adapted for seed crushing. He believes that the blue-black grassquit colonized the Galapagos Islands and Cocos Island independently and gave rise to similar derivatives in both areas. On the basis of his morphological and behavioral studies, Steadman places all of Darwin's finches together with the blue-black grassquit in the genus *Geospiza*.

The evolution of Darwin's finches on the Galapagos Islands and Cocos Island provides one of the classic examples of species formation. Although spectacular, it illustrates the same kinds of processes by which species are originating continuously in all groups of organisms. Isolated populations subjected to unique combinations of selec-

FIGURE 21-10

Darwin's finches. Ten species of Darwin's finches from Isla Santa Cruz, one of the Galapagos Islands, showing differences in bills and feeding habits. The bills of several of these species resemble those of different distinct families of birds on the mainland, a condition that presumably arose when the finches evolved new species in habitats where other kinds of birds normally occur on the mainland and which are not available to finches there. The woodpecker finch uses cactus spines to probe in crevices of bark and rotten wood for food. All of these birds are thought to have been derived from a single common ancestor, a finch like that shown in Figure 1-8, *A*.

tive pressures diverge from one another and may ultimately become so different that they are distinct species.

> Darwin's finches are a group of 14 species of small birds, 13 of which occur on the Galapagos Islands. They have radiated widely in the absence of competition, and all seem to have been derived from one closely similar species from the mainland.

Hawaiian *Drosophila*

Our second example of a cluster of species is provided by the fly genus *Drosophila* on the Hawaiian Islands, which we mentioned earlier in this chapter as an example of behavioral isolation as a mechanism for keeping species distinct. There are at least 1250 species of this genus throughout the world, and more than a third of them are found only in the Hawaiian Islands, a tiny area on a world scale. More species of *Drosophila* are still being discovered in Hawaii, although the rapid destruction of the native vegetation is making the search more difficult. Aside from their sheer number of species, these flies are unusual in the morphological and behavioral traits discussed earlier (see Figure 21-8). No comparable species of *Drosophila* are found anywhere else in the world.

A second, closely related genus of flies, *Scaptomyza*, is also found in Hawaii, where it may be represented by as many as 300 species. A few species of *Scaptomyza* are found outside of Hawaii, but the genus is better represented there than elsewhere. In addition, species intermediate between *Scaptomyza* and *Drosophila* exist in Hawaii, but nowhere else. The genera are so closely related that scientists have suggested that all of the estimated 800 species of these two genera that occur in Hawaii may have been derived from a single common ancestor.

The native Hawaiian flies are closely associated with the remarkable native plants of these islands and are often abundant in the native vegetation. Evidently, when their ancestors first reached these islands, they encountered many "empty" habitats, ones in which other kinds of flies had not yet become established. The evolutionary opportunities that the ancestral *Drosophila* flies found were similar to those encountered by the ancestors of Darwin's finches in the Galapagos Islands, and both groups evolved in a similar way. Many of the Hawaiian *Drosophila* species are highly selective in their choice of host plants for their larvae and in the part of the plant that they use. The larvae of various species live in rotting stems, fruits, bark, leaves, or roots or feed on sap.

New islands have continually arisen from the sea in the region of the Hawaiian Islands. As they have done so, they appear to have been invaded successively by the various groups of *Drosophila* that were present on the older islands. New species have evolved as new islands have been colonized. The Hawaiian species of *Drosophila* have had even greater evolutionary opportunities than have Darwin's finches, because of their restricted ecological niches and the variable ages of the islands. They clearly present one of the most unusual evolutionary stories found anywhere in the world.

As a result of the geographical position of the Hawaiian Islands, which emerged from the sea in isolation, the individuals of *Drosophila* that reached them first were able to move into many habitats that were occupied elsewhere by other kinds of insects and other animals. The extraordinary adaptive radiation—the emergence of numerous species from a common ancestor in a single area—that followed resulted in the evolution of at least 800 species of this group in a relatively small area and provides an excellent model for similar processes, usually taking place on a smaller scale, that have resulted in the evolution of species generally.

> The adaptive radiation of about 800 species of the flies *Drosophila* and *Scaptomyza* on the Hawaiian Islands, probably from a single common ancestor, is one of the most remarkable examples of active evolution found anywhere on earth.

Sexual Selection and the Origin of Species

In addition to natural selection, Charles Darwin considered sexual selection a major feature of evolution; he devoted an entire book—*Sexual Selection and the Descent of Man*—to this topic. Most subsequent writers on evolution, however, have denied both the major significance and the distinctiveness of sexual selection, a topic that is now being considered actively once more.

Sexual selection is defined as differential reproduction that results from variable success in obtaining mates, because of combat or courtship; it is contrasted with natural selection, which Darwin defined as differential reproduction related to the "struggle for existence." Sexual selection, then, results from a contest between rivals of the same sex.

The theory of sexual selection argues that evolutionary change can originate as a result of the divergence of characteristics used in sexual displays; social factors may sometimes be more important than ecological factors in determining the evolution of distinctive features. They may lead to the incompatibility of populations, because of social and courtship incompatibility. Even a slight divergence in key aspects of courtship behavior may be sufficient to put "misfit" males at a severe disadvantage under courtship selection, with female discrimination against them a strong barrier to interbreeding. Such sexual selection is clearly a distinctive factor in the evolution of life, and it is an important subject of research at present.

Rapid Evolution

As you have seen, the evolution of organisms on islands differs from that which occurs in mainland areas only in the degree of opportunity afforded for *rapid* evolution. Because evolution on islands is often rapid, and therefore recent, Darwin and other investigators have found the study of island plants and animals especially informative.

In continental areas, too, the process of evolution sometimes results in the production of clusters of species. These clusters are more frequent in certain regions. For example, more than a third of the total number of plant species in the United States and Canada, about 6000 out of perhaps 18,000 species, occur in California, and nearly half of these are found nowhere else. In the dry, semiarid, and variable habitats of that state, there are opportunities for evolution comparable to those found on islands throughout the world. Climatic changes in relatively recent times, together with a diversity of habitats, have provided conditions in California that are especially suitable for the rapid evolution of new species. Similarly, there are opportunities for rapid evolution in areas of active vulcanism or other ecological catastrophe, or when dominant groups of organisms become extinct—a situation that we shall discuss further in Chapter 22.

> In general, rapid evolution takes place in areas where the habitats are varied and located near one another. The pace of evolution is accelerated further by rapid climatic change.

THE ROLE OF HYBRIDIZATION IN EVOLUTION

Among perennial plants, especially trees and shrubs, there are often few postzygotic barriers to hybridization among the species of a given genus. For example, all white oaks are capable of forming fertile hybrids with one another. So are all of the members of the major groups of *Eucalyptus*, many of the cottonwoods and aspens, and so forth. Such hybrids recombine the characteristics of their parents in unusual ways, and these genetic combinations may be particularly suited to some habitats, where they may flourish to a greater degree than either of their parents. If so, the hybrids may give rise to distinctive populations that may replace their parental species in the appropriate habitats. The opportunity to recombine alleles from different species may be important in helping to bring about rapid changes in the overall populations. If it is, hybridization can be considered an important force in the production of new species in some groups, especially among plants.

FIGURE 21-11
A parthenogenic species of desert whiptail lizard, *Cnemidophorus*, from Arizona. All individuals of this species, *C. sonorae*, are triploid females, containing three sets of chromosomes; the species is of hybrid origin. All of the members of this species are genetically identical (except when they have been changed by somatic mutation) because of the way in which they reproduce.

Even if such hybrids are sterile, they may have a future. Sometimes hybrid plants simply reproduce vegetatively; for example, broken-off pieces of stems or roots may become established away from the original plant. If they are able to do so, and if such a clone has a genotype that is particularly well adapted to a certain habitat, the sterile hybrid individuals may become quite common. Among some groups of animals and plants, the corresponding phenomenon of **parthenogenesis** is of importance. In it, egg cells may give rise directly to new individuals or other somatic cells may function to produce embryos (Figure 21-11). Depending on the way in which they are produced, these individuals may either be diploid or haploid. Most races of dandelions, including the ones you see in lawns, reproduce in this same way; their seeds contain "embryos" that are produced asexually and that have the same genetic constitution as the parent plant.

Fertile individuals may also arise from sterile ones by the process of **polyploidy**, by which the chromosome number of the original sterile hybrid individual is doubled. A **polyploid** cell, tissue, or individual has more than two sets of chromosomes. Polyploid cells and tissues derived from them occur spontaneously and reasonably often in all organisms, although in many they are soon eliminated. A hybrid may be sterile simply because the sets of chromosomes derived from its male and female parents do not pair with one another, having come from different species. If the chromosome number of such a hybrid doubles, the hybrid, as a result of the doubling, will have a duplicate of each chromosome. In that case the chromosomes will pair, and the fertility of the polyploid hybrid individual may be restored (Figure 21-12). It is estimated that about half of the approximately 260,000 species of plants are of polyploid origin, including many that are of great commercial importance, such as bread wheat, cotton, tobacco, sugarcane, bananas, and potatoes.

> **Hybrids between species may be wholly or partly sterile but still have genotypes with high fitness in a particular environment. Such hybrids may perpetuate themselves through selection for fertility, through polyploidy, or through parthenogenesis, among other mechanisms.**

In this chapter we have traced the way in which adaptive changes in populations—changes that are occurring at all times and in all populations—lead to the origin of diverse races of organisms. Factors of the kind that make it difficult or impossible for species to hybridize arise as a consequence of such adaptive change but are sometimes strengthened if the unlike populations come back into contact with one another. Even when differentiation is relatively complete, however, hybridization between species can often have important evolutionary consequences, particularly among plants. Processes of the kinds described in this chapter have ultimately been responsible for the origin of at least several million species of plants, animals, and microorganisms. In Chapter 22 we will discuss the way in which these changes, occurring over the billions of years of the history of life on earth, have led to the overall patterns that we observe today.

FIGURE 21-12
Bread wheat, *Triticum aestivum*. Known only in cultivation, bread wheat has 42 chromosomes in its somatic cells and 21 chromosomes in its eggs and sperm. It appeared at least 8000 years ago, probably in central Europe, following spontaneous hybridization between a species of cultivated wheat with 28 chromosomes and a wild relative with 14 chromosomes. The chromosomes in such a hybrid doubled by chance, thus giving rise to one of our most important crop plants.

SUMMARY

1. Species differ from one another in one or more characteristics and do not normally hybridize freely with other species when they come into contact in nature. They often cannot hybridize with one another at all. Individuals within a given species, on the other hand, usually are able to interbreed with one another freely.

2. Populations change as they adapt to the demands of the environments where they occur. Even populations of a given species that are close to one another geographically are normally effectively isolated. Such populations are free to diverge in ways that are responsive to the needs of their particular environment.

3. In general, if the selective forces bearing on populations differ greatly, the populations will diverge rapidly; if these selective forces are similar, the populations will diverge slowly.

4. Among the factors that separate populations, and species, from one another are geographical, ecological, temporal, behavioral, and mechanical isolation, as well as factors that inhibit the fusion of gametes or the normal development of the hybrid organisms. Reproductive isolation between species arises as a normal by-product of the progressive differentiation of populations.

5. Ecological races and subspecies differentiate within species but often still intergrade with one another. The differences between them, in both plants and animals, are mostly genetically fixed.

6. Clusters of species arise when the differentiation of a series of populations proceeds further. On islands such differentiation often is rapid, because of the numerous open habitats that are available. In many continental areas, differentiation is not as rapid, but there are local situations, such as those where many different kinds of habitats are developing close to one another, where differentiation may be rapid.

7. Hybridization between differentiated species, especially in trees and shrubs, affords a source of new genotypes on which selection can act. If the chromosomes of the parents cannot pair in a hybrid, spontaneous doubling may produce a polyploid individual, one with multiple complements of chromosomes, in which each chromosome has a partner with which it can pair. Alternatively, sterile hybrids may propagate by means of parthenogenesis.

REVIEW QUESTIONS

1. What is the definition of a species?

2. Define the term niche.

3. Why do populations of organisms change?

4. How does the force of selection correlate with population divergence?

5. What barriers exist to the formation of hybrids? Which are prezygotic and which are postzygotic isolating mechanisms?

6. When are species hybrids at a fitness disadvantage?

7. What is adaptive radiation? Give an example.

8. What is the primary mechanism that maintains the integrity of the species clusters of *Drosophila* in the Hawaiian Islands?

9. What environmental conditions foster rapid evolution?

10. Define the terms polyploidy and parthenogenesis.

THOUGHT QUESTIONS

1. Under what circumstances is hybridization between species selected against? When is it selected for?

2. There are polyploid animals in some groups, but they are far less frequent than they are among plants, particularly perennial plants. Why do you think this might be so?

3. Could artificial selection by human beings (as for dogs) eventually lead to the production of new species? Outline the basis for your answer.

4. Can species ever originate in a single step? If so, what are some of the ways in which this could occur.

Biological Journal of the Linnaean Society, vol. 21, pages 1-258, 1984. Special issue on Charles Darwin and the Galapagos Islands. A fine collection of papers, including historical accounts and details of modern scientific work. They indicate that the Galapagos Islands are still of great interest as a natural laboratory today, a century and a half after Darwin's 6-week visit in 1835.

COLE, C.J.: "Unisexual Lizards," *Scientific American,* January 1984, pages 94-100. An extraordinary account of the populations of whiptail lizards that consist of females only and reproduce by parthenogenesis.

CREWS, D.: "Courtship in Unisexual Lizards: A Model for Brain Evolution," *Scientific American,* December 1987, pages 116-121. An all-female species of whiptail lizard presents a unique opportunity to test hypotheses regarding the nature and evolution of sexual behavior.

FUTUYMA, D.: *Evolutionary Biology,* ed. 2, Sinauer Associates, Inc., Sunderland, Mass., 1986. This text presents a good discussion of the development of evolutionary thought, combined with a process-oriented treatment of the whole field.

GOULD, S.J.: *The Panda's Thumb: More Reflections in Natural History,* W.W. Norton & Co., New York, 1980. Outstanding essays that illustrate some of the kinds of problems that are of interest to contemporary students of evolution.

GOULD, J., and C. GOULD: *Sexual Selection,* W.H. Freeman, San Francisco, 1989. A broad survey of how sex works in nature, and why related species sometimes take quite different approaches.

GRANT, P.: "Natural Selection and Darwin's Finches," *Scientific American,* October 1991, pages 82-87. In the Galápagos, finches evolve very rapidly—populations are altered significantly by a single season of drought.

MARANTO, G.: "Will Guenons Make a Monkey of Darwin?" *Discover,* vol. 7(11), pages 87-101, 1986. A fascinating account of a large genus of African monkeys; the species differ greatly in markings, behavior, and chromosome number and morphology.

RYAN, M.: "Signals, Species, and Sexual Selection," *American Scientist,* January 1990, pages 46-52. The important role of species recognition is emphasized in this overview of why sexual selection occurs so often in nature.

WATERS, A. and others: "*Plasmodium falciparum* appears to have arisen as a result of lateral transfer between avian and human hosts," *Proc. Nat. Acad. Sci. USA,* April 1991, vol. 88, pages 3140-3144. The microbe which causes malaria kills over 3 million people each year. It now appears that humans caught the parasite from birds in the recent past, with the advent of agricultural society.

Evolutionary History of the Earth

OVERVIEW

No more than a billion years after the condensation of the earth, which was complete by about 4.5 billion years ago, bacteria were abundant. After another 2 billion years, eukaryotic cells appeared—about 1.5 billion years ago. Half that long ago, the eukaryotic cells began to aggregate as multicellular organisms that were more efficient than their unicellular ancestors, thus extending the mastery of living things over their environment. More than 410 million years ago the ancestors of the groups that are so successful on land today, the insects and the plants, appeared for the first time; 50 million years later they were followed by the first terrestrial vertebrates. One of the groups of mammals, the primates, gave rise to our genus, Homo, *about 2 million years ago, and to our species perhaps 500,000 years ago.*

FOR REVIEW

Here are some important terms and concepts that you will encounter in this chapter. If you are not familiar with them, you should review them before proceeding.

Isotopes (Chapter 2)

The origin of life (Chapter 4)

Punctuated equilibrium (Chapter 20)

Origins of species (Chapter 21)

FIGURE 22-1

Trilobites. This notable assemblage of trilobites, predaceous arthropods that flourished in the seas of the Cambrian Period, 500 to 550 million years ago, was found in an abandoned Ohio cement quarry by collector Thomas T. Johnson. Trilobites disappeared about 245 million years ago.

Plants, animals, amoebas—individuals that belong to distinct major groups of organisms—differ so greatly from one another that it is difficult at first to identify the features they have in common. Nevertheless, the biochemical processes that occur within an oak tree, a bear, and a bacterium in your intestine resemble one another so closely that we are justified in concluding that all of these kinds of organisms are related by descent. The purpose of this chapter is to provide an overview of the major features of the evolution of life on earth, relating this overview both to time and to geography. It sketches the way in which the different kinds of organisms have diverged from one another over time, and tells something about their abundance in the past. Further details will be given in Chapters 28 to 42, in the course of our discussion of individual groups.

FOSSILS AND FOSSILIZATION

Major groups of organisms evolve in the same way as species. If we thoroughly understand the homologies of structures in members of diverse groups, we will be able to outline the relationships between them. Comparisons of their nucleic acids and proteins likewise provide important evidence on their relationships. Also very important, however, is the fossil record, which is studied and interpreted by paleontologists (Figure 22-1). A **fossil** is any record of a dead organism; fossils may be nearly complete impressions of organisms, or merely burrows, tracks, molecules, or traces of their existence.

Only fossils provide definite evidence about what extinct organisms looked like, including the ancestors of those that are living now. On the basis of the characteristics of living organisms, we can make deductions about their ancestors, but fossils provide the means whereby these deductions can be checked. Fossils provide an actual record of organisms that once lived, an accurate understanding of where and when they lived, and some appreciation of the environment in which they lived. Fossils also enable us to trace lines of evolutionary progression among individual groups of organisms, provided that enough fossils of various ages are available. For many reasons, however, the fossil record is not as complete as we might wish, and the samples that we have been able to obtain are not as extensive as would be ideal. Both mechanical and theoretical problems have contributed to this lack of completeness.

Mechanical Problems in Preservation

Only a minute fraction of the organisms living at any one time are preserved as fossils. Most of the fossils that we do have are preserved in **sedimentary** rocks. Sedimentary rocks are formed of particles of other rocks that are weathered off, deposited at the bottom of bodies of water or accumulated by wind, and then hardened into strata of the sort that Darwin found so filled with intriguing fossils in South America. During the formation of sedimentary rocks, dead organisms are sometimes washed down along with inorganic debris, such as mud or sand, and eventually reach the bottom of some pond, lake, or the ocean itself. Occasionally, dead organisms are covered by debris accumulated by wind. In exceptional circumstances, fossils may be preserved in an organic substance, such as tar: the La Brea Tar Pits in Los Angeles (Figure 22-2) illustrate this phenomenon. In other instances, organisms themselves may form the whole deposit, as in the formation of coal or oil deposits. However, many organisms do not live in places where their remains are likely to be preserved in sedimentary rocks; for this reason they are unlikely to be preserved as fossils.

Sedimentation proceeds steadily in the sea, but the seabed is constantly being renewed, as you will see in the next section. Most of the existing seabed is less than 100 million years old, but despite that limitation, we do find a more complete record of life in the sea than anywhere else. For the older record to reach us, however, the seafloor must be located in strata that are eventually uplifted and become land.

For specific organisms to become preserved as fossils, they must usually be buried before their decay is complete, and before they are scattered by scavengers. In addition, the process of decay, for at least some portions of the organism, must stop after they are buried. Since the soft parts usually decay rapidly, normally only structures such as bones, teeth, shells, scales, leaves, and wood are available in the fossil record. If fossils are exposed, they often disintegrate quickly; most of those we find, therefore, were exposed recently. When fossils are formed, the actual parts of the organisms are almost always replaced by minerals; a fossil rarely contains any of the material that made up the body of the organism originally, but rather is a mineralized replica of that body part.

When interpreting the structure of fossil organisms, we must normally use the hard parts to try to interpret what the soft parts looked like. For organisms such as worms, which have no hard parts, fossils are understandably rare. Even though soft-

FIGURE 22-2
Life in what is now downtown Los Angeles about 15,000 years ago. The skeletons of these animals—sabretooth cats, dire wolves, giant condors, sloths, associated grazing mammals—which were similar in size and abundance to those that have survived from the Pleistocene Epoch on the plains of East Africa, were preserved in large pools of tar (asphalt) into which they fell.

bodied animals undoubtedly evolved before their hard-bodied counterparts, we have little evidence of their history in the fossil record. Sometimes soft-bodied animals are preserved in exceptionally fine-grained muds, in conditions under which the supply of oxygen was poor while the muds were being deposited and deterioration was therefore slowed (Figure 22-3).

> Fossils provide the concrete means whereby we can judge our deductions about the history of particular groups of organisms. Fossils are found mainly in sedimentary rocks, and those organisms with hard body parts or which lived in areas where sedimentary rocks are being formed are the most likely to be preserved. The fossil record is rarely complete enough to allow whole evolutionary sequences to be studied for a particular group.

Dating Fossils

Correlations of Strata.

One of the ways of determining the age of particular fossils is to compare the sequences in which they appear in different strata. Such sequences demonstrate the evolutionary progression of life on earth. Sedimentation deposits new layers on older ones, so the fossils in upper layers represent younger species, for the most part, than the fossils in lower layers. When you go to different places and find similar fossils in the same strata, the fossils are assumed to be of the same age. Such correlations were used to establish the broad patterns of progression of life on earth and the relative ages of different sequences of fossils well before Darwin completed his formation of the theory of evolution.

Many of the geological periods were recognized and given names before methods were available that would allow their actual dates to be established. An outline of the eras, periods, and epochs, including their dates (which were assigned later) and major features, is presented on the inside cover of this book. The past periods were usually named for the areas where they were especially well represented or first studied, or for other outstanding characteristics.

> The geological eras, periods, and epochs, summarized on the inside cover of this book, provide a means for organizing data about the history of life on earth. They were named for the characteristic strata associated with them long before their actual dates had even been estimated.

Direct Age Determination.

Direct methods of dating rocks and fossils, mentioned in Chapter 20, first became available in the late 1940s. Naturally occurring radioactive isotopes of certain elements are employed in this process. Such isotopes are unstable and decay over the course of time at a steady rate, producing other isotopes. One of the most widely used methods of dating, the **carbon-14** (^{14}C) method, employs estimates of the different isotopes present in samples of carbon.

Most carbon atoms have an atomic weight of 12; the symbol of this particular isotope of carbon is ^{12}C. A fixed proportion of the atoms in a given sample of carbon, however, consists of carbon with an atomic weight of 14 (^{14}C), an isotope that has two more neutrons than ^{12}C. ^{14}C is produced from ^{12}C as a result of bombardment by particles from space. The carbon incorporated into the bodies of living organisms consists of the same fixed proportion of ^{14}C and ^{12}C that occurs generally. After an organism dies, however, and is no longer incorporating carbon, the ^{14}C in it gradually decays into nitrogen over time. It takes 5730 years for half of the ^{14}C present in a sample to be converted by this process; this length of time is called the **half-life** of the ^{14}C isotope.

It is possible, therefore, to accurately date fossil material that contains carbon, as all organic material does, provided that the fossil material is less than about 50,000 years old. This can be done by estimating as accurately as possible the proportion of ^{14}C that is still present in the carbon (Figure 22-4). For older fossils, the amount of ^{14}C remaining is so small that it is not possible to measure it precisely enough to provide accurate estimates of age.

To determine the ages of older samples, the decay of other radioactive isotopes can sometimes be studied. For example, ^{40}K (potassium-40) decays very slowly into

FIGURE 22-3
Soft-bodied animals are sometimes well preserved in mud.
A *Pteridinum carolinaensis*, a Precambrian fossil from the Piedmont of south-central North Carolina. This fossil, perhaps about 630 million years old, represents a segmented animal about 9 centimeters long and about 2.5 centimeters wide; it had 26 segments. Its relationships, and those of most other animals that were contemporary with it, are unknown. It is the best-preserved multicellular fossil of its age that has yet been reported from the United States.
B *Mawsonites spriggi*, a fossil jellyfish (phylum Cnidaria) from the Ediacara formation in South Australia, illustrating the remarkable preservation of soft-bodied fossils in these fine-grained sedimentary rocks. This jellyfish was originally conical, with distinct lobes on its upper surface, and about 11 centimeters across. During Precambrian time, nearly all of the major phyla and most of the classes and orders of animals (other than chordates) first appeared.

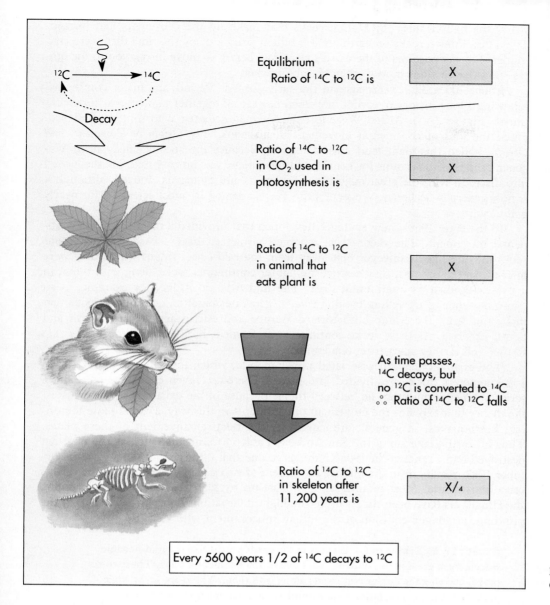

Equilibrium
Ratio of ^{14}C to ^{12}C is

$^{12}C \longrightarrow \: ^{14}C$

Decay

Ratio of ^{14}C to ^{12}C
in CO_2 used in
photosynthesis is

Ratio of ^{14}C to ^{12}C
in animal that
eats plant is

As time passes,
^{14}C decays, but
no ^{12}C is converted to ^{14}C
∴ Ratio of ^{14}C to ^{12}C falls

Ratio of ^{14}C to ^{12}C
in skeleton after
11,200 years is

| X |
| X |
| X |
| X/₄ |

Every 5600 years 1/2 of ^{14}C decays to ^{12}C

FIGURE 22-4
C-14 dating.

^{40}Ar (argon-40) and can be used to date much older rocks. The half-life of ^{40}K is 1.3 billion years; other isotopes have even longer half-lives. With the use of such dating methods, our knowledge of the ages of various rocks has become more precise. Unfortunately, such dating methods can establish only the ages of the materials in sedimentary rocks and not the time when they were incorporated into those rocks. They do, however, allow us to estimate the ages of all of the rocks on earth, regardless of how old they may be.

> **The ages of fossils may be estimated by correlating the strata in which they occur. A more accurate method is to determine the proportions of different isotopes of carbon in the organic material preserved in them, or the proportions of different isotopes in the rocks where they are found. Radioactive isotopes change from one form to another over time, so that the proportions in a given sample will provide an absolute date for the age of that sample.**

CONTINENTAL DRIFT AND BIOGEOGRAPHY

As you saw in Chapter 4, the planets of our solar system, including the earth, began to coalesce approximately 4.6 billion years ago. The accretion of the early earth seems to have been completed about 4.5 billion years ago. After several hundred million

A 200 million years ago

B 135 million years ago

C 65 million years ago

D Present

FIGURE 22-5
How the continents have moved.
The positions of the continents at 200, 135, and 65 million years ago, and at present.

years, the lighter silica (SiO_2)-rich rocks that make up the continents had formed. The oldest known rocks on earth are 3.8 billion years old; by the time they were consolidated, some portions of the earth's crust had begun to move in relation to the others. As these sections moved, they became thicker.

About 200 million years ago, in the early Jurassic Period, the major continents, following a long history of earlier movements, were all together in one great supercontinent (Figure 22-5). Alfred Wegener, the German scientist who in 1915 first proposed the idea of continental movement in his book, *The Origin of Continents and Oceans*, called this giant land mass **Pangaea.** Because the specific mechanism Wegener proposed to account for continental movement was not feasible, his theory fell into discredit with the great majority of geologists and biologists. Indeed, although it is now generally accepted, Wegener's theory was denied by most scientists for nearly half a century.

In the early 1960s, new evidence developed that provided a mechanism for continental movement. The theory that emerged visualized heavy, basaltic, ocean-floor rocks moving like a conveyor belt away from the mid-ocean ridges where they were formed. In the process, they carried the lighter continental rocks along with them, in a process known as **continental drift.** The mid-Atlantic Ridge, for example, is an enormous zone of upwelling basaltic lava. It generates basaltic, ocean-floor rocks that move out from the Ridge both toward Europe and Africa and toward North and South America. As these rocks continue to be formed, both pairs of continents ride farther and farther apart from one another.

The earth's crust and associated upper mantle, which together are about 100 to 150 kilometers thick, are divided into plates. There are seven enormous ones and a series of smaller ones that lie between them. Because of the existence of these plates, the theory that explains the movement of continents in this way is called **plate tectonics.** Earthquakes, in general, are caused by the relative movement of these plates. Thus an earthquake along the San Andreas Fault in California, such as the one that destroyed San Francisco in 1906 or the large one that occurred in Santa Cruz in October 1987, results from the relative movement of two gigantic segments of the earth's crust and mantle. Most mountains are thrust up by plate movements; for example, the Himalayas have been thrust up to the highest elevations on earth as a result of the grinding, prolonged collision of the Indian subcontinent with Asia.

> **Basaltic lavas, flowing out of fissures in the earth's crust like the mid-oceanic ridges, form great plates of heavy rock 100 to 150 kilometers thick. These rocks move, carrying the lighter continents along with them. There are seven huge plates, all moving in relation to one another, and a number of smaller ones.**

History of Continental Movements

The continents have gradually moved apart from their positions as parts of Pangaea 200 million years ago. For example, the opening of the South Atlantic Ocean began about 125 to 130 million years ago, with Africa and South America connected directly earlier. Subsequently South America has moved slowly toward North America, with which it eventually became connected by a land bridge, the Isthmus of Panama, between 3.1 and 3.6 million years ago. When the bridge was complete, many South American plants and animals, such as the opossum (Figure 22-6) and the armadillo (see Figure 1-6), migrated overland into North America. At the same time, numerous North American plants and animals such as oaks, deer, and bears moved into South America for the first time. Changes on a greater or lesser scale occurred in the positions of all the continents, playing a major role in the patterns of distribution of organisms that we see today. For example, ratite birds—ostriches (see Figure 42-27), emus, rheas, casowaries, and kiwis—seem to have diverged early in the course of their evolution from all other living groups of birds. The members of this group undoubtedly migrated between southern continents and islands, which are now widely separated from one another, at a time when direct overland migration was possible: the Mesozoic Era. Collectively, **biogeography** is the study of the past and present distributions of organisms, and attempts to explain how these distributions originated.

     PART VI   EVOLUTION

FIGURE 22-6
The North American opossum,
***Didelphis virginiana* with young.**
Marsupials, which once lived in
Europe, Africa, and Antarctica,
are now best represented in Australia and South America. They
formerly occurred in North America but became extinct there about
40 million years ago. The opossum, like the armadillo, is a mammal that migrated into North
America from South America once
the Isthmus of Panama had been
uplifted above the sea, an event
that occurred several million
years ago.

Australia provides a striking example of the way in which continental movements have affected the nature and distribution of organisms. Until about 53 million years ago, Australia was joined with a much warmer Antarctica, as was South America. Marsupials, which are best represented in Australia and South America at present, seem clearly to have moved overland between these continents via Antarctica when this was still possible. Isolated from the placental mammals that had become abundant and diverse in other parts of the world, marsupials underwent a major episode of adaptive radiation in Australia (see Figure 20-11). As Australia moved northward toward the tropical islands fringing Southeast Asia, other groups of plants and animals also radiated in a similar manner.

On the other side of the world, the connection of southern South America with West Antarctica lasted until about 23 million years ago. With the warmer climates of the early Cenozoic Era, therefore, more or less direct overland migration between Australia and South America, via Antarctica, was possible until about 40 million years ago. Eventually the formation of gigantic ice sheets, first in the south and then in the north, accompanied the deterioration and diversification of world climates.

> **The movements of continents over the past 200 million years have pro-
> foundly affected the distribution of organisms on earth. Some of the major
> events include the linkage of South America with North America about 3.1
> to 3.6 million years ago, and the separation of Australia and South America
> from Antarctica.**

THE EARLY HISTORY OF LIFE ON EARTH

The main divisions of earth history, as outlined by contemporary authors, are Hadean, Archean, Proterozoic, and Phanerozoic. Archean time, during which the oldest recognizable rocks were formed, extended from 3.8 billion years ago until an oxygen-containing atmosphere was stabilized, an event that marked the start of Proterozoic time about 2.5 billion years ago. The oldest known fossils, all bacteria are from Archean time, about 3.5 billion years ago. As far as we can judge from their appearance, some of these bacteria may have been cyanobacteria, and therefore capable of producing oxygen by photosynthesis.

These cyanobacteria were associated with the earliest known **stromatolites** (Figure 22-7), massive limestone deposits that became frequent in the fossil record by about 2.8 billion years ago. Cyanobacteria produced stromatolites abundantly in virtually all freshwater and marine communities until about 1.6 billion years ago. Today stromatolites are still being formed, but usually under conditions of high salinity, aridity, and high light intensities, as in lagoons along the edge of warm seas.

During Proterozoic time, 2.5 billion to 630 million years ago, oxygen was stabilized as a component of the atmosphere, and large continents coalesced from the light

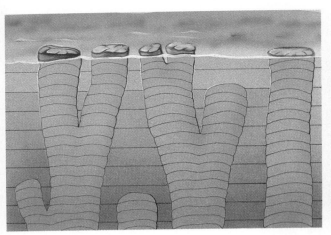

A **B**

FIGURE 22-7
Stromatolites.

A This diagram shows how aggregations of calcium carbonate build up around massive colonies of cyanobacteria, layer after layer, in the water and are progressively buried. Such deposits are known as stromatolites.

B Stromatolites in the intertidal zone at Shark Bay, Western Australia. The largest structures are about 1.5 meters across. These stromatolites formed during a time of slightly higher sea level, perhaps 1000 to 2000 years ago.

crustal rocks. During all of Archean and Proterozoic time, the only kinds of organisms in existence were bacteria and protists. Microfossils called **acritarchs** (Figure 22-8) first appeared in the fossil record about 1.5 billion years ago; they seem to represent the earliest eukaryotes.

The oldest probable eukaryotic fossils, the acritarchs, appeared about 1.5 billion years ago.

The onset of Phanerozoic time is marked by the appearance of easily visible, multicellular fossils. The earliest such fossils are found in rocks at least 630 million years old, in the strata of the Ediacara series of southern Australia. Of the roughly 250,000 different kinds of fossils that have been identified, described, and named, only a few dozen are more than 630 million years old.

Within Phanerozoic time, we begin to speak in terms of geological eras, periods, epochs, and ages, as summarized inside the front cover of this book. For many years, the oldest known fossils were those from the Cambrian Period (590 to 505 million years ago), the first period of the Paleozoic Era. The geological eras, with their dates in millions of years before the present, are as follows:

1. **Paleozoic Era**, 590 to 248 million years ago. The name of this era is derived from the Greek words *paleos*, "old," and *zoos*, "life." Until the discoveries outlined above were made, this was the oldest period from which fossils were known.
2. **Mesozoic Era**, 248 to 65 million years ago. The name is from the Greek *mesos*, "middle."
3. **Cenozoic Era**, 65 million years ago to the present. This name is from the Greek *coenos*, "recent."

All of the strata older than the Cambrian Period, and thus older than the Paleozoic Era, are classified as Precambrian. Earlier scientists found no fossils in Precambrian rocks, and their absence was regarded as a great mystery. Now, however, we know that many of these fossil organisms were not detected simply because they were unicellular and therefore so small that they were difficult to observe. Others were soft-bodied, and found only under special conditions of preservation.

We now have abundant and diverse fossils of multicellular animals from the Ediacaran Period, comprising the last 40 million years of Precambrian time, from 630 to 590 million years ago. Members of the phylum Cnidaria (Chapter 39) were abundant and diverse worldwide during this period. Among the contemporary members of this phylum are corals, sea anemones, and hydroids. Other animal fossils from the Ediacaran Period include some segmented worms (annelids), as well as early arthropods and some animals that cannot be assigned clearly to any phylum that we know. Annelida (Chapter 40) is a large phylum of worms, and Arthropoda (Chapter 41) is an enormous phylum that includes insects, crustaceans, and spiders.

Even older than the Ediacaran fossils are some traces of soft-bodied, multicellular

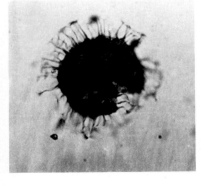

FIGURE 22-8
An acritarch from the early Cambrian Period of Spitsbergen, Norway. This fossil is about 550 million years old, but acritarchs are the earliest fossils that are considered probable eukaryotes; some fossils of this group are as much as 1.5 billion years old.

animals. These include what are apparently worm burrows that have been found in Australia, China, and the USSR, suggesting that multicellular organisms actually may have evolved about 700 million years ago. Such fossils are preserved only under unusually favorable conditions, such as when they were buried in fine mud.

The oldest known multicellular fossils are from Ediacaran time, 630 to 590 million years ago; they include members of the phyla Cnidaria, Annelida, and Arthropoda. Older traces of multicellular life, including what are apparently worm burrows, go back to about 700 million years ago.

With the evolution of hard external skeletons, shells, and other structures that were easily preserved, the nature of the fossil record changed dramatically. The evolution of nearly all major groups of organisms of which we are aware took place within a few tens of millions of years after the appearance of skeletons in their ancestors about 570 million years ago. The ability of organisms to manufacture such skeletons seems to have been a major evolutionary advance that made the further evolutionary radiation of these groups possible.

THE PALEOZOIC ERA

Except for plants, all of the main phyla and divisions of organisms that exist today had evolved by the end of the first period of the Paleozoic Era, the Cambrian, more than 505 million years ago. Because so many new kinds appeared in such a short time, paleontologists speak of a Cambrian "explosion" of living forms. The evolution of all of these different kinds of organisms took place in the sea, with plants and other groups of organisms colonizing the land about 400 million years ago, during the middle part of the Paleozoic Era. The Cambrian Period was a time of intense evolutionary radiation, the greatest period of diversification for multicellular animals that has occurred during the entire history of life on earth. These animals were entering new **adaptive zones,** which might be thought of as major niches that differ greatly from one another, and organisms were changing their characteristics in relation to those of their new ways of life.

Many kinds of multicellular organisms that were present in the early Paleozoic Era have no living relatives. When studying their remains in the rocks, it becomes evident that this was a period of "experimentation" with different body forms and ways of life (Figures 22-9 and 22-10). Multicellular organisms were stronger and more

FIGURE 22-9
Reconstruction of a shallow seafloor during the Cambrian Period, about 530 million years ago. Fossils of the organisms shown here, together with well over 100 others, are found together in the Burgess Shale, a formation that has been uplifted to high elevations in the Rocky Mountains of British Columbia, Canada. *Opabinia*, which was about 12 to 15 centimeters long, had a jointed grasping organ unlike anything seen in living animals. *Hallucigenia* is a very peculiar organism that also has no known close relatives among living organisms. *Eldonia*, a cnidarian, and *Vauxia*, a sponge, are shown with multicellular alga. These three organisms represent living phyla. These organisms are beautifully preserved in the Burgess Shale, which was formed from a kind of fine-grained mud.

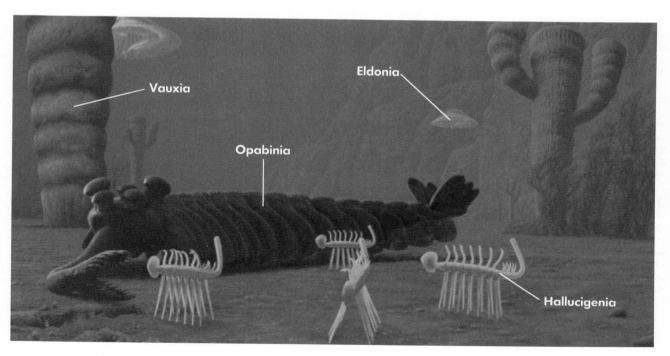

A

C

D

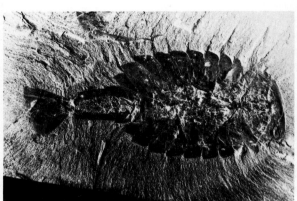

B

E

FIGURE 22-10
Cambrian Period fossils from the Burgess Shale, British Columbia, Canada, about 530 million years old.

A and **B** *Sidneyia inexpectans*, an ancient arthropod; shown are a specimen and a model.

C *Hallucigenia sparsa*, shown in Figure 22-9. The relationships of this bizarre animal are obscure; this specimen is 12.5 millimeters long. Fossils found recently in China suggest that *Hallucigenia* may be an early representative of *Onychophora*, an obsure phylla of velvet worms. *Hallucigenia* may have been a scavenger on the bottom of the sea.

D *Wiwaxia corrugata.* The body of this animal, also of unknown relationships, was largely covered with scales and also bore elongate spines; this specimen is 30.5 millimeters across, excluding the spines. It may have been a distant relative of the mollusks.

E *Burgessochaeta setigera.* A polychaete worm, a member of the phylum Annelida. The photograph shows the anterior end of the worm, which had a pair of tentacles. Parapodia (footlike structures) are borne in pairs along the sides of the body. This specimen is 16.5 millimeters long.

mobile than their unicellular ancestors. Consequently, they could obtain their food and satisfy other requirements in more diverse, active ways. These possibilities led them to diversify along new evolutionary pathways, some of which ultimately led to the contemporary divisions and phyla of animals (see Figure 22-9). As another example, the trilobites (see Figure 22-1) appear to be derived from the same evolutionary line that gave rise to one living group of arthropods, the horseshoe crabs (Chapter 41). Trilobites, with their efficient jaws and tough external skeletons, were efficient predators. One arthropod that inhabited these ancient seas, *Anomalocaris*, was shown in 1985 to be a predator on trilobites; these animals were half a meter long when mature, and probably the largest animals in existence at the time.

> The early Paleozoic Era was a time of extensive diversification for marine animals. Many new kinds of animals appeared, and some of them have persisted to the present.

The Invasion of the Land

The first organisms that colonized the land were plants, and they did so more than 410 million years ago. The features of plants—including multicellular bodies and efficient mechanisms for conserving water and transporting it within their bodies—evolved in relation to their colonization of the land and were those that made them best suited for such colonization. The ancestors of plants were specialized green algae (division Chlorophyta; see Chapter 31). Although most of the members of this division are aquatic, the immediate ancestors of plants might themselves have been semiterrestrial. It was, however, with the plants themselves that the occupation of the land truly began.

Several major groups of animals also originated in connection with their invasion of the land; they had the same problems of conserving water as did ancestral plants, but they solved them in different ways. Among the arthropods, the insects originated on land from an evolutionary line derived from annelid worms (Chapter 40). This occurred at about the same time as the evolution of the plants, about 410 million years ago. The body plan of arthropods has proved so well adapted to life on land—for example, by virtue of the drought-resistant outer cuticle that covers their bodies—that insects and other classes of this phylum have radiated extensively, producing a

FIGURE 22-11
Amphibians were the first verte-brates to walk on land. Recon-struction of *Ichthyostega*, one of the first amphibians with efficient limbs for crawling on land, an improved olfactory sense associ-ated with a lengthened snout, and a relatively advanced ear structure for picking up airborne sounds. Despite these features, *Ichthy-ostega*, which lived about 350 mil-lion years ago, was still quite fishlike in overall appearance and represents a very early amphibian.

probable majority of all species of organisms. During the subsequent history of the group, only a few kinds of insects have moved back into freshwater habitats, and a handful have returned to marine habitats. It seems certain that the plants colonized the land before animals did, since they would have been essential sources of food and shelter for the first arthropods that emerged from the sea.

Among the vertebrates, the ancestors of amphibians were among the first colo-nists of the land. The earliest amphibians known are about 360 million years old (Fig-ure 22-11). Among their descendants on land are the reptiles. Different groups of rep-tiles, in turn, ultimately became the ancestors of the birds and the mammals, as we will discuss further in Chapter 42.

The fact that all three of these major groups of organisms—plants, arthropods, and vertebrates—colonized the land within a few tens of millions of years of one an-other is probably related to the development of suitable environmental conditions, for example, the formation of a layer of ozone in the atmosphere that blocked ultraviolet radiation. These conditions have allowed the existence of multicellular organisms in terrestrial habitats for over 400 million years, about a tenth of the age of the earth.

A fourth major group of organisms that invaded the land successfully is the fungi, which constitute a distinct kingdom of organisms. The success of the fungi on land, where they probably have been present for as long as any group of organisms, might be related to the structure of their cell walls, which are rich in chitin, a drought-resis-tant substance that also characterizes the outer skeletons of the arthropods. The mem-bers of one division of fungi, the Zygomycetes, were associated with the roots of the first terrestrial plants.

Plants and arthropods colonized the land about 410 million years ago; am-phibians did so about 50 million years later. Fungi may also have colonized the land at about the same time as plants. Earlier, there were no terrestrial, multicellular organisms.

Mass Extinctions

One of the most prominent features of the history of life on earth has been the peri-odic occurrence of major episodes of extinction. During the course of geological time there have been five such events, in each of which a large proportion of the organisms on earth at that time became extinct. Four of these events occurred during the Paleo-zoic Era, the first of them near the end of the Cambrian Period (about 505 million years ago). At that time most of the existing families of trilobites, a well-known group of marine arthropods (see Figure 22-1), became extinct. Additional major extinction events occurred about 438 and about 360 million years ago.

The fourth and most drastic extinction event in the history of life on earth hap-pened during the last 10 million years of the Permian Period, which was the final por-tion of the Paleozoic Era, about 248 to 258 million years ago. It is estimated that ap-proximately 96% of all species of marine animals that were living at that time may have become extinct! All of the trilobites, and many other groups of organisms as well, disappeared forever. The fifth major extinction event occurred at the close of the

Mesozoic Era (and therefore of the Cretaceous Period, the last part of the Mesozoic), 65 million years ago. That last event was the famous one when dinosaurs became extinct; we will discuss it further below.

Major episodes of extinction clearly produce conditions appropriate for rapid evolution for those relatively few plants, animals, and microorganisms that survive the extinction episode. Many paleontologists believe that the great episodes of extinction that have occurred periodically have had much to do with the pattern of evolution of life on earth. Virtually unlimited opportunities for evolutionary radiation among multicellular organisms, however, occurred only before and during the Cambrian Period, when the first organisms of this kind evolved.

Large-scale extinction events other than the five major ones summarized above also have occurred, and scientists are searching for extraterrestrial or other factors that might have been involved with all of these. It has been hypothesized that the major extinction event that ended the Mesozoic Era is correlated with the impact of a large meteorite, as we will discuss.

THE MESOZOIC ERA

The Mesozoic Era, which began about 248 million years ago and ended about 65 million years ago, was a time very different from the present and one of intense evolution of terrestrial plants and animals. The major evolutionary lines on land had been established during the mid-Paleozoic Era, but the evolutionary radiation of these lines—a radiation that led to the establishment of the major groups of organisms living today—took place in the Mesozoic Era (Figure 22-12). When tracing the evolution of these lines, we need to refer to events in the Permian Period (286 to 248 million years ago), a period of drought and extensive glaciation that concluded the Paleozoic Era. Many of the evolutionary events that occurred then had important consequences during the Mesozoic Era, influences that have continued to the present.

The Mesozoic Era has traditionally been divided into three periods, the Triassic, Jurassic, and Cretaceous. Because of the major extinction event ending the Paleozoic Era, only about 4% of the species that were present earlier persisted into the Mesozoic Era. The few kinds of marine organisms that survived into the Mesozoic Era—including gastropod and bivalve mollusks, crustaceans, fishes, and echinoderms, among others—began to evolve into many new species, genera, and families. In addition, a number of new life-forms evolved in the sea during the Mesozoic Era. For example, the first efficient burrowers appeared among the echinoderms (Chapter 42).

Both on land and in the sea, the number of species of almost all groups of organisms has been climbing steadily for the past 250 million years and is now at an all-time high. Even though the evolutionary radiation of marine organisms during this period has been spectacular, the story of the evolution of life on land during the Mesozoic Era is of even greater interest for us, who are products of that history.

The History of the Vertebrates

We trace the history of vertebrates here because so many of its significant events took place during the Mesozoic Era (Figure 22-13). Although we think of the vertebrates as advanced organisms, the phylum that includes them, Chordata, is nearly as old as any other. We will consider this phylum in more detail in Chapter 42. The first chordates for which we have a fossil record, the lancelets and lampreys, appeared about 540 to 550 million years ago, in the middle Cambrian Period. Over the next 100 million years, during the early Paleozoic Era, fishes were becoming abundant and diverse. Amphibians, the descendants of fishes, appear in the fossil record by the end of this period, and subsequently became abundant in the great swamps that formed at that time.

The amphibians gave rise to the first reptiles about 300 million years ago. The reptiles, being better suited for a terrestrial existence than their amphibian ancestors, replaced them as dominants over the next 50 million years, becoming abundant during the Permian Period. During the course of the Mesozoic Era, there were extensive

FIGURE 22-12
Basic body patterns were established long ago. The fossil dragonfly, which is about 170 million years old (Jurassic Period), closely resembles its modern counterpart. It dramatically illustrates the establishment of modern groups of organisms in the Mesozoic Era; many of these have persisted to the present day.

evolutionary radiations of flying reptiles, the pterosaurs, and swimming reptiles, including the plesiosaurs, in addition to the well-known radiation of terrestrial ones.

Dinosaurs, the major group of reptiles that was dominant throughout the Mesozoic Era, first appeared at least 220 million years ago, and persisted to the end of the Mesozoic Era—in other words, for more than 150 million years. They gradually became more diverse (Figure 22-14), and finally disappeared about 65 million years ago for reasons that are still much debated. However, one evolutionary line related to the dinosaurs did not become extinct: the birds. The earliest known fossils that may be a

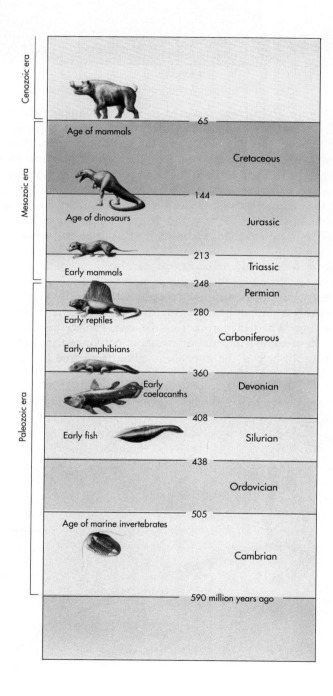

FIGURE 22-13
Evolutionary history of the vertebrates. Successive groups of vertebrates have flourished for more than 400 million years of earth history.

FIGURE 22-14

Dinosaurs. Some of the remarkable diversity of dinosaurs, as shown in a reconstruction from the Peabody Museum, Yale University. This painting covers a span of approximately 320 million years (reading right to left) during the later Paleozoic and Mesozoic Eras, ending 65 million years ago at the end of the Cretaceous Period (far left). Throughout this vast period of time, the remarkable increase in structural complexity and overall diversity of the dinosaurs can be seen, until they abruptly became extinct, giving way to the dominant mammals of the Cenozoic Era. Flowering plants can be seen for the first time at the left-hand side of the illustration. The names of the geological periods are given along the bottom of the painting.

bird, *Protoavis* (Figure 22-15), was reported in 1986 by Sankar Chaterjee of Texas Tech University. *Protoavis,* which lived in the tropical forests of what is now west Texas about 225 million years ago, was about the size of a chicken. It had distinctly dinosaurian features, but was contemporary with the earliest dinosaurs known. It possessed clawed fingers, a tail, and teeth in the front part of its jaws; on the other hand, it had a wishbone and a narrow shoulder girdle like that of modern birds, both probably adaptations for flight. Although no feathers were found associated with the fossils of *Protoavis,* bumps on the forearm and "hands" of the fossil probably indicate places where feathers were attached, according to Chaterjee. It probably was a weak flier.

A better known fossil bird, *Archaeopteryx,* is about 150 million years old; it is structurally little more than a bipedal dinosaur with feathers (Figure 22-16). *Archaeopteryx* had feathers, but apparently not enough of them to fly very effectively. Its ancestor was probably a dinosaur-like reptile that ran on its hind legs and used its front legs for grasping. Whether it arose from the same evolutionary line as *Protoavis,* or independently from another bipedal dinosaur, is not known. In early birds, however, feathers probably helped the animals to glide from tree to tree. Ultimately,

FIGURE 22-15

Protoavis, **a 225 million year old bird about the size of a chicken, from west Texas.** This controversial find, which if valid will extend the history of birds by 75 million years, was announced by Sarkar Chaterjee of Texas Tech University in 1986.

TRIASSIC   PERMIAN   CARBONIFEROUS

feathers made possible the subsequent evolution of birds that *could* fly efficiently, and the group began to become large and diverse about 90 to 65 million years ago, toward the end of the Mesozoic Era. The ability of birds to fly, and thus to exploit an unoccupied adaptive zone, made possible the evolution of the entire class Aves, the birds, a group that comprises about 9000 living species.

Mammals originated during the first part of the Mesozoic Era from a different group of reptiles and represent a line parallel to the dinosaurs. Mammals became dominant during the Cenozoic Era, replacing the dinosaurs as ecological dominants. Dinosaurs and mammals appeared at about the same time, but mammals did not become abundant and diverse until after the dinosaurs had become extinct. They seem to present an example of adaptive radiation into an adaptive zone, or rather a series of adaptive zones, left vacant by the extinction of the dinosaurs. For example, many of the dinosaurs weighed thousands of kilograms, but no land animal weighing more than 20 kilograms survived into the Cenozoic Era. The ancestral small mammals clearly had ample opportunity to take up ways of life similar to those of the extinct larger dinosaurs.

> During the Mesozoic Era, the reptiles, which had evolved earlier from amphibians, became dominant and in turn gave rise to the mammals (about 200 million years ago) and the birds (at least 150 million years ago).

FIGURE 22-16
Archaeopteryx, which was for more than a century the earliest known bird, was about the size of a crow. *Archaeopteryx* lived in the forests of central Europe 150 million years ago. The teeth and long, jointed tail are features not found in any modern birds, but they are shared with *Protoavis*. The colors of this reconstruction are as imagined by the artist. Discovered in 1862, *Archaeopteryx* was cited by Darwin in support of his theory of evolution.

The History of Plants

The earliest known fossil plants are about 410 million years old. Within the next 100 million years, plants became abundant and diverse, eventually forming extensive forests. During the Carboniferous Period (360 to 286 million years ago), these forests formed many of the great coal deposits that we are consuming now. Much of the land was low and swampy, providing excellent conditions for the preservation of plant remains. In these coal deposits, we have a relatively complete record of the horsetails, ferns, and primitive seed-bearing plants that made up these ancient forests (Figure 22-17).

The Permian Period (286 to 248 million years ago) was, as we have seen, cool and dry. The swamps of the preceding Carboniferous Period, which existed at a time of

Figure 22-17

Medullosa. An example of a plant that was very prevalent in the Carboniferous period. Their fossilized remains are found frequently in sediments from this period. Their bizzare appearance resembles that of ferns. They were the size of shrubs or small trees.

FIGURE 22-18
Fossils from the shore of Fossil Lake, a locality on the border of Wyoming and Utah, which was subtropical when this palm and these fish flourished there 50 million years ago. Most families of organisms that exist now had appeared by the first half of the Cenozoic Era, and the worldwide climate was much milder than it is now.

worldwide moist and warm climates, largely disappeared. The Permian Period seems to have been one of ecological stress and evolutionary innovation; for example, the conifers—a group of seed-bearing plants represented today by pines, spruces, firs, and similar trees and shrubs—originated then. Seed-bearing plants with featherlike leaves, similar to the living cycads, were abundant in the Mesozoic Era and helped give that period its nickname, "the age of dinosaurs and cycads."

The oldest fossils definitely known to be flowering plants, or angiosperms, are from the early Cretaceous Period, about 127 million years ago. It seems certain that the group originated earlier, but no one is certain how much earlier. At any rate, they became dominant on land about 100 million years ago. Today there are about 240,000 species of flowering plants, which greatly outnumber all other kinds of plants. As the flowering plants became more diverse, so did the insects, which had feeding habits that were closely linked with the characteristics of the flowering plants. Insects and flowering plants have **coevolved** with one another, a relationship that we will explore further in Chapters 24 and 34. Indeed, all groups of terrestrial organisms, including mammals, birds, and fungi, have evolved their characteristics largely in relation to those of the flowering plants. These groups now dominate life on the land, literally making our world look the way it does (Figure 22-18).

> **The earliest plants, about 410 million years ago, evolved into others that formed extensive forests within 50 million years of their appearance. The angiosperms (flowering plants) have been dominant in terrestrial habitats for about 100 million years.**

The Extinction of the Dinosaurs

Everyone is generally familiar with the disappearance of the dinosaurs, an event of global importance that took place at the end of the Cretaceous Period, the closing period of the Mesozoic Era. Less discussed, but actually of more fundamental importance, was the disappearance of many other kinds of organisms that took place at about the same time. Recently, important strides have been made in understanding the nature of this, the most recent of the major extinction events.

The Mesozoic Era has traditionally been separated from the Cenozoic Era by abrupt shifts in the kinds of marine organisms that occur in strata exposed in Europe. Among the plankton, which are free-drifting protists, and members of other groups of organisms that are still abundant in the sea, many of the larger (but still microscopic) forms suddenly disappeared about 65 million years ago and a much lower number of smaller ones took their place. The same rapid changes occurred in at least some nonplanktonic marine animal groups, such as the bivalve mollusks (clams and their relatives). The ammonites, a large and diverse group related to octopuses, but with shells, abruptly disappeared. In 1980 a group of distinguished scientists headed

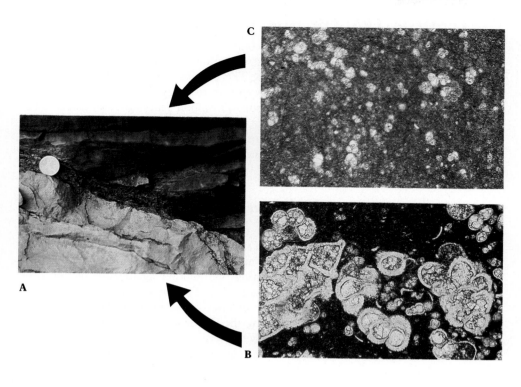

FIGURE 22-19
The end of the Cretaceous Period can be seen in rocks.
A The white limestone was deposited under the sea in the closing years of the Cretaceous Period, and the red limestone above was deposited during the first years of the Tertiary Period. They are separated by a layer of clay about 1 centimeter thick, in which iridium is abundant.
B Large, ornamented, diverse planktonic forams (members of the phylum Foraminifera, a group of marine protists) in sediments laid down during the late Cretaceous Period.
C Small, relatively unornamented, much less diverse forams deposited a few millions years later during the early years of the Tertiary Period. They are about one sixth the diameter of those that existed before the formation of the iridium layer.

by physicist Luis W. Alvarez of the University of California, Berkeley, presented a dramatic hypothesis about the reasons for these changes.

Alvarez and his associates discovered that the usually rare element iridium was abundant in a thin layer that marked the end of the Cretaceous Period, not only in the strata where the period had first been defined, in Italy, but also in many other parts of the world at the same time (Figure 22-19). Iridium is rare in the earth's outer crust but common in meteorites. Alvarez and his colleagues proposed that if a large meteorite or asteroid had struck the surface of the earth, a dense cloud would have been thrown up that would for a time have darkened the earth. The cloud would have been rich in iridium, and as its particles settled, the iridium would have been incorporated in the layers of sedimentary rock that were being deposited at that time. By darkening the world, the cloud would have greatly slowed or temporarily halted photosynthesis and driven many kinds of organisms to extinction.

Subsequent calculations have shown that the cloud produced by a meteorite about 20 kilometers in diameter would certainly have caused these effects to occur. Daytime conditions would have resembled those on a moonless night, with less than the amount of light required for photosynthesis, and the condition would have persisted for several months. In biological communities such as the marine plankton, which are based directly on the continuous production of food by photosynthesis, such darkness would have had seriously disruptive effects and could have produced the sudden change seen in the fossil record. Disruption of photosynthesis may also have been responsible for the extinction of certain other kinds of organisms, but is it reasonable to assume that it would have done away with the dinosaurs? A number of groups of organisms do not display abrupt changes at the Cretaceous-Tertiary boundary, and many factors are being evaluated in relation to their possible contribution to the demise of the dinosaurs.

A worldwide layer rich in iridium, an element that is much more common in meteorites than it is in general on earth, marks the end of the Cretaceous Period and therefore of the Mesozoic Era. It appears likely that this iridium settled out of an enormous, worldwide cloud cover. Such a cloud cover would have darkened the earth greatly and seems to have been responsible for at least some of the extinction events that occurred at this time.

THE CENOZOIC ERA: THE WORLD WE KNOW

We conclude this chapter with a brief account of some of the major evolutionary changes that have occurred during the past 65 million years, changes that have resulted in the conditions we now experience. The relatively warm and moist climates of the early Cenozoic Era have gradually given way to today's climates. As glaciation in Antarctica became fully established by about 13 million years ago, and glaciation in the north over the past several million years, regional climates changed dramatically. The ice mass that formed as a result of this glaciation has made the climate cooler near the poles, warmer near the equator, and drier in the middle latitudes than ever before.

The Origin of Contemporary Biogeography

In general, forests covered most of the land area of the continents, except for Antarctica, until about 15 million years ago, when they began to recede rapidly. During the time when these forests were receding and modern plant communities were appearing, some of the continents were approaching one another again after having been widely separated during most of the Cenozoic Era. The organisms in Australia and South America, particularly, evolved their distinctive features in isolation from all those in the rest of the world, mainly during the Cenozoic Era. During the past several million years, the formation of extensive deserts in northern Africa, the Middle East, and India has made migration between Africa and Asia difficult for the organisms of tropical forests. This formation of desert barriers, in turn, has provided further opportunities for evolution in isolation. In general, the overall character of the Cenozoic Era has been marked by a deteriorating climate, sharp differences in habitats even within small areas, and the regional evolution of distinct groups of plants and animals.

> Throughout the 65 million years of the Cenozoic Era, the world climate has deteriorated steadily and the distributions of organisms have become more and more regional in character.

THE EVOLUTION OF PRIMATES

The small Mesozoic mammals that ultimately gave rise to the primates, our ancestors, fed largely on insects and were generalized in their features (Figure 22-20). These mammals were rare during the first 10 million years of the Cenozoic Era, when other kinds of mammals appeared and became diverse very rapidly. Monkeylike primates first appear as fossils about 36 million years ago, in the early Oligocene Epoch. These **anthropoid** primates—monkeys, apes, and humans and their direct ancestors—seem to have replaced earlier forms of primates rather rapidly.

Whereas these earlier primates were mostly **nocturnal** (active primarily at night, like most kinds of mammals), monkeys and apes are almost all **diurnal** (active during the day). They are agile animals that feed mainly on fruits and leaves, and most of them live in trees. The well-developed **binocular** vision (vision in which both eyes are located at the front of the head and function together to provide a three-dimensional view of the object being examined) evolved in relation to a reduction in their snouts and is a great advantage in leaping through the treetops. This evolutionary trend, in turn, resulted in the flat faces that are characteristic of most living primates.

Anthropoid primates have larger brains than their ancestors, and their braincase forms a larger portion of the head than it does in other mammals. The thumb of anthropoids is opposable: it stands out at an angle to the other digits and can be bent back against them, as when the animal grasps an object. Some of the advanced anthropoids have developed an extraordinary ability to manipulate objects by using this opposable thumb. Monkeys and apes live in socially complex groups, and they tend to care for their young for prolonged periods, thus enabling the prolonged period of learning characteristic of these animals. The maternal care given the young of anthro-

FIGURE 22-20
Where primates come from. Reconstruction of an insect-eating mammal from the early Cenozoic Era, a mouse-sized animal that probably resembled the common ancestor of the primates and insectivores, two contemporary orders of mammals.

poids seems somehow related to their large brains and the long periods necessary for their development.

Contemporary primates are characterized by the expansion and elaboration of their brains; the shortening of their snouts, together with well-developed binocular vision; the free mobility of the fingers and toes, with the forefinger and thumb opposable; and an increased complexity and quantity of social behavior.

The Evolution of Hominoids

The apes, together with the **hominids** (humans and their direct ancestors), make up a group called the **hominoids.** The earliest known fossils of hominoids are from the Early Miocene Epoch (20 to 25 million years ago) and occur in the Old World. The living apes are the chimpanzees, gorillas, orangutans, and gibbons (Figure 22-21). The living apes are confined to relatively small areas in Africa and Asia; no apes ever occurred in the New World (North or South America).

Apes have larger brains than monkeys. Apes exhibit the most adaptable behavior of all mammals except for human beings. All apes lack tails. With the exception of the gibbons, which are relatively small, all apes are larger than any surviving monkey.

Although the fossil evidence is still being accumulated, biochemical studies have told us a great deal about our relationships to the apes and their relationships to one another (Figure 22-22). Differences in amino acid residues seem to accumulate at a relative rate in proteins, a relationship that forms the basis of what has been called a **molecular clock.** Based on these differences, investigators have calculated that the evolutionary line leading to gibbons diverged from that leading to the other apes about 10 million years ago; the line leading to orangutans split off about 8 million years ago; and the split between hominids and the line leading to chimpanzees and gorillas may

FIGURE 22-21
The living apes.
A Gorilla, *Gorilla gorilla.*
B Mueller gibbon, *Hylobates muelleri.*
C Chimpanzee, *Pan troglodytus.*
D Orangutan, *Pongo pygmaeus.*

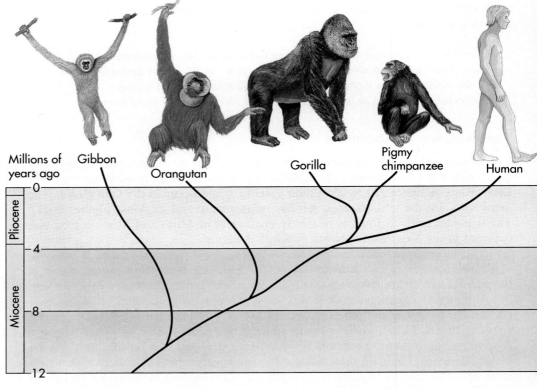

FIGURE 22-22
The evolution of living hominoids, the apes and human beings.

Millions of years ago

Gibbon
Orangutan
Gorilla
Pigmy chimpanzee
Human

Pliocene
Miocene

0
4
8
12

have occurred about 4 million years ago. The time of divergence between chimpanzees and gorillas appears to have been a little less than 3.2 million years ago, as determined by the same methods. All three hominoids are so close to one another that still more recent evidence, based on the analysis of the sequence of 7100 nucleotides in a segment of DNA from the hemoglobin gene, suggests that humans and chimpanzees are closer to one another than chimpanzees and gorillas; they matched at 98.4% of the bases, with chimpanzees and gorillas matching at 97.9% of them. Indeed, if chimpanzees, gorillas, and human beings were members of most other groups of organisms, they would almost certainly be considered to belong to a single genus.

Molecular evidence suggests that chimpanzees and gorillas diverged from one another a little less than 3.2 million years ago, hominids from that line about 5 million years ago, orangutans from all of them about 8 million years ago, and gibbons from the rest about 10 million years ago. The absolute timing of these events depends on the interpretation of the fossil record and on the relationship of molecular patterns to time.

FIGURE 22-23
"Lucy," from Ethiopia, the most complete skeleton of *Australopithecus* discovered so far. The reconstruction was made by a careful study of muscle attachments on the skull and skeleton.

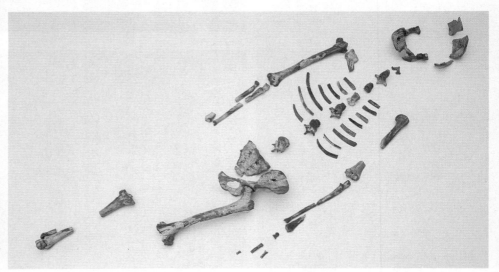

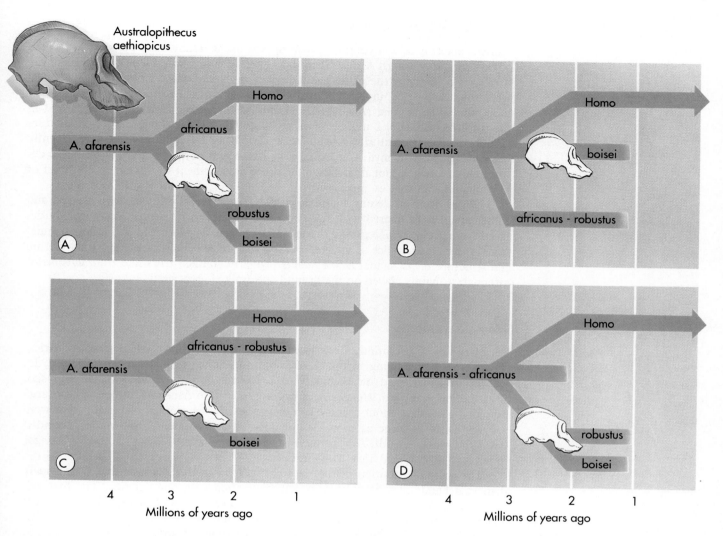

Australopithecus
aethiopicus

(A) A. afarensis — africanus — Homo; robustus; boisei

(B) A. afarensis — Homo; boisei; africanus - robustus

(C) A. afarensis — africanus - robustus — Homo; boisei

(D) A. afarensis - africanus — Homo; robustus; boisei

4   3   2   1
Millions of years ago

4   3   2   1
Millions of years ago

THE APPEARANCE OF HOMINIDS
The Australopithecines

The two critical steps in the evolution of human beings were the evolution of bipedalism and the enlargement of the brain. The earliest definite hominids, which share many of the characteristics that we regard as distinctively human, are assigned to the genus *Australopithecus* (Figure 22-23). Among the human features were bipedal locomotion and rounded jaws. The australopithecines lived on the ground in the open grasslands of eastern and southern Africa, weighed up to 20 kilograms, and were up to 1.3 meters tall. Their brains were about the size of those of present-day gorillas, about 350 to 450 cubic centimeters in volume, as compared with an average volume of about 1450 cubic centimeters in modern humans. Individuals of *Australopithecus* appear to have eaten many different kinds of plant and animal food, judging from the characteristics of their teeth and jaws. Although there is no general agreement on this point, there may have been four or more species of the genus (Figure 22-24). Their characteristics indicate clearly that *Australopithecus* was ancestral to our genus, *Homo*.

The oldest known fossils that may represent *Australopithecus* are 5 to 5.5 million years old. They thus date from close to the postulated time of divergence of the hominid and chimpanzee-gorilla evolutionary lines. Other fossils of *Australopithecus* are commonly found from 4 million years ago onward at a number of localities in Africa, disappearing about 1.3 million years ago.

The first definite hominids, *Australopithecus*, appeared about 5 million years ago in Africa. They lived on the ground, weighed less than 20 kilograms, and were less than 1.3 meters tall. They disappeared about 1.3 million years ago.

FIGURE 22-24
Four possible human family trees. Version B is the most popular among investigators of human fossils, but the evidence now available does not clearly eliminate any of the other three. The illustrated skull is of a new find, tentatively called *Australopithecus aethiopicus*, which appears to be an ancestor of *A. boisei*.

The Use of Tools: *Homo habilis*

Starting about 2 million years ago, tools appear in the sedimentary beds of a wide area of eastern and southern Africa. Such tools have been associated with the fossils of the first known humans, the earliest members of the genus *Homo*. The fossils and tools were first discovered in the Olduvai Gorge, Tanzania. Because these people used tools, they were named *Homo habilis* (Latin *habilis*, able). *Homo habilis* appeared in the fossil record about 2 million years ago and persisted for about half a million years (Figure 22-25). Partial *Homo* skull bones as old as 2.4 million years have recently been identified in Kenya, suggesting that *Homo* was born at a time of dramatic climate change, when global cooling altered the environment drastically and a burst of new mammalian species appeared.

Judged from skeletons found in 1987, *Homo habilis* was small in stature and rather apelike, compared with the two other species of its genus, and certainly derived from *Australopithecus*. As the use of tools was developed in its earliest populations, the characteristics of the genus *Homo* may have begun to emerge. Judged from the structure of its hands, *Homo habilis* regularly climbed trees like *Australopithecus*, although it mainly stayed on the ground and walked erect on its two legs.

Homo Erectus

All of the early evolution of the genus *Homo* seems to have taken place in Africa. There, fossils belonging to the second and only other extinct species of *Homo*, *H. erectus*, are widespread and abundant from 1.7 million until about 500,000 years ago. By 1 million years ago, *Homo erectus* had migrated into Asia and Europe. These large humans, who were about the size of our species, *Homo sapiens*, had brains that were roughly twice as large as those of our ancestors, prominent brow ridges, rounded jaws, and massive teeth. *Homo erectus* survived in Asia until about 250,000 years ago, much longer than in Africa or in Europe.

The wide distribution of tools made by *Homo erectus* suggests that their widely dispersed bands may have been in communication with one another. The first evidence of the use of fire by humans occurs at the campsites of this species in the Rift Valley of Kenya at least 1.4 million years ago, and fire was characteristically associated with populations of *Homo erectus* from that time onward.

Homo erectus, **the second species of our genus to evolve, appeared in Africa about 1.7 million years ago and migrated throughout Eurasia by 1 million years ago. It disappeared in Africa about 500,000 years ago and in Asia about 250,000 years ago.** *Homo erectus* **used fire, starting at least 1.4 million years ago, and made characteristic stone tools.**

Modern Humans: *Homo sapiens*

Many scientists place the first appearance of our own species at about 500,000 years ago, because of the larger brains, reduced teeth, and enlargement of the rear of some skulls that date from that time, as compared with those of *Homo erectus*. *Homo sapiens* appeared in Europe as *H. erectus* was becoming rarer. It appears likely, in fact, that *Homo sapiens* represents a derivative of *H. erectus* with certain improved, "modern" features.

Recently, scientists analyzing mitochondrial DNA have calculated that all the races of living human beings might have originated in Africa perhaps 200,000 years ago. Because the DNA within mitochondria is transmitted only from the egg and not the sperm, lineages do not get rearranged by Mendelian assortment and so it is possible to trace the parentage of humans back from mother to grandmother to great-grandmother. When this is done, scientists find that all humans alive today share mitochondrial DNA derived from a single ancestral female who lived about 200,000 years ago in Central Africa. The popular press quickly dubbed her "Eve."

FIGURE 22-25
The path of human evolution, illustrated here, shows the many early branches that are known from fossil evidence to have existed. *Australopithecus robustus* and *A. boisei* seem to represent evolutionary dead ends, with no living descendants. Whether *A. africanus* was ancestral to *Homo habilis* or to *Australopithecus robustus* is currently in dispute, as is whether Neanderthal man is an ancestor of modern man or a parallel line that did not survive.

In 1992 this analysis was shown to be inadequate because of statistical problems—the data simply aren't clear about where Eve originated. A variety of other lines of evidence, however, still supports a theory of African origin. The diversity of mitochondrial DNA is much greater in Africa, for example, implying a longer history there to accumulate the greater number of differences.

Most researchers now believe that the human races did not evolve independently in Asia, Europe, and Africa from different *H. erectus* lineages as some have argued, but rather that modern *H. sapiens* evolved only once, in Africa, and from there spread to all parts of the world, supplanting everywhere the *H. erectus* which had itself come out from Africa far earlier.

Neanderthals. Populations of *Homo sapiens* of the sort that are called Neanderthals were abundant in Europe and western Asia between about 70,000 and 32,000 years ago, judging from the fossil record. Such fossils were first discovered in the valley of the Neander River in Germany in 1856, and the name given to humans of this sort is derived from that of the valley.

Compared with ourselves, Neanderthals were powerfully built, short, and stocky. Their skulls were massive, with protruding faces, projecting noses, and rather heavy bony ridges over their brows. Their brains were even larger than those of modern humans, a fact that may have been related to their heavy, large bodies. The Neanderthals made diverse tools, including scrapers, borers, spearheads, and hand axes. Some of the Neanderthal tools were used for scraping hides, which they used for clothing. They lived in huts or caves. Neanderthals took care of their injured and sick and commonly buried their dead, often placing food and weapons, and perhaps even flowers, with the bodies. Such attention to the dead suggests strongly that they believed in a life after death. For the first time, the kinds of thought processes that are characteristic of modern *Homo sapiens*, including symbolic thought, are evident in these acts.

Our species, *Homo sapiens*, may be as much as 500,000 years old; it has been frequent for the past 150,000 years. Neanderthals were abundant in Europe and western Asia from about 70,000 to about 32,000 years ago.

Cro-Magnons. About 34,000 years ago the European Neanderthals were abruptly replaced by people of essentially modern character, the so-called Cro-Magnons. We can only speculate about why this sudden replacement occurred, but it was complete all over Europe in a short period of time. There is some evidence that the Cro-Magnons came from Africa, where human fossils of essentially modern aspect, but as much as 100,000 years old, have been found. They seem to have replaced the Neanderthals completely in southwest Asia by 40,000 years ago, and then spread across Europe during the next few thousand years, coexisting and possibly even interbreeding with the Neanderthals for several thousand years.

The Cro-Magnons used sophisticated stone tools that rapidly became more diverse and thus more useful for various purposes. They also made a wide variety of tools out of bone, ivory, and antler, materials that had not been used earlier. The Cro-Magnons were hunters who killed game by using complex tools. In addition, their evident social organization suggests that they may have been the first people to have fully modern language capabilities. Soon after Cro-Magnon people appeared, they began to make what were apparently ritual paintings in caves. The culture associated with these impressive and mysterious works of art existed during a period that was much cooler than the present. Such conditions prevailed at the time of the last great expansion of continental ice. During that period, grasslands inhabited by large herds of grazing mammals occurred across Europe. The animals that occurred in such habitats are often depicted in the cave paintings.

Humans of modern appearance eventually spread across Siberia to the New World, which they reached at least 12,000 to 13,000 years ago, after the ice had begun to retreat at the end of the last glacial period. As people spread throughout the world, they were apparently responsible for the extinction of many populations of animals and plants, a process that has accelerated recently. Human remains are often associated with the bones of the large animals they hunted, and these animals often disappeared promptly from different parts of the world shortly after humans arrived for the first time. By the end of the Pleistocene Epoch, about 10,000 years ago, there were only about 10 million people throughout the entire world (compared with well over 5 billion now). The subsequent history of our species and its impact on the global ecosystem will be considered in Chapter 27.

SUMMARY

1. The fossil record is relatively complete for marine organisms, since there is a steady rate of sedimentation in the sea. Normally only structures such as shells, teeth, bones, and wood are preserved, and we generally have a very poor record of soft-bodied organisms.

2. The earth originated about 4 to 5 billion years ago, and distinct crustal plates had been organized by about 8 billion years ago, when the oldest rocks were formed. About 200 million years ago, the continents were clustered. They have been separating from one another ever since, resulting in realignments in their positions.

3. The oldest known fossils, which are bacteria, are about 3.5 billion years old, with stromatolites, deposits of limestone formed by cyanobacteria, appearing at about that same time.

4. Multicellularity appeared about 700 million years ago, but the first multicellular animals were soft-bodied and are rare in the fossil record.

5. All of the phyla and divisions, as well as most of the classes, of organisms that we know about evolved during the Cambrian Period, with the exception of the plants. The evolution of hard skeletons, including shells and similar structures, about 570 million years ago seems to have been a fundamental evolutionary advance that made this evolutionary radiation possible.

6. During the Paleozoic Era, plants and terrestrial arthropods appeared about 410 million years ago and terrestrial vertebrates (amphibians) appeared about 360 million years ago. These groups of organisms, together with the fungi, have dominated life on the land since then.

7. The Mesozoic Era, the "age of dinosaurs and cycads," was a time when the outlines of life on earth as we know it were established. Flowering plants were dominant by the end of the era, and insects, mammals, birds, and other groups had begun to evolve in relation to the diversity of these plants.

8. There have been five major mass extinctions during the history of life on earth, with the most drastic at the end of the Permian Period, when about 96% of marine animals became extinct. At the end of the Cretaceous Period, 65 million years ago, the dinosaurs and many other kinds of organisms disappeared.

9. Primates first appeared 38 to 55 million years ago, with the first anthropoid primates (now including monkeys, apes, and humans) appearing about 36 million years ago. Primates have proportionately large brains, binocular vision, and five digits with an opposable thumb.

10. The apes, which appeared 20 to 25 million years ago, gave rise to the gibbons, orangutans, and then to the African apes (chimpanzees and gorillas) and hominids (humans and their direct ancestors), starting more than 10 million years ago. Apes and hominids collectively are termed hominoids.

11. The earliest hominids belong to the genus *Australopithecus*. They were the direct ancestors of humans. They appeared in Africa about 5 million years ago and were up to about 3 meters tall and about 20 kilograms in weight, with brains about 450 cubic centimeters in volume.

12. About 2 million years ago, with an enlargement in brain size that was perhaps associated with the increased use of tools, the genus *Australopithecus* gave rise to humans belonging to the genus *Homo*. Our species, *Homo sapiens*, seems to have originated in Africa about 500,000 years ago.

REVIEW QUESTIONS

1. Why are fossil records primarily found in sedimentary rocks? At which location on the earth is this sedimentary record most complete?

2. What do gaps in the fossil record indicate?

3. How old is the planet earth? How old are the oldest known rocks? Why are these ages not the same?

4. What is continental drift? How is it associated with plate tectonics?

5. Why is the Cambrian period so important to the evolutionary biologists?

6. What types of organisms were the first to successfully colonize the land? When did this occur? What special features needed to be developed before it could occur?

7. Which were the first vertebrates to become terrestrial? When did this occur? Into which forms did these evolve?

8. How many mass extinctions have occurred? During which eras did they occur? When did the most drastic event in terms of overall loss of species occur? Which is the most famous extinction?

9. What are hominids and how do they differ from other groups of animals?

10. What are the species of the genus *Homo,* and when did they live?

THOUGHT QUESTIONS

1. How can you tell whether a new species will give rise to a whole new group of organisms? Under what circumstances is it most likely to do so?

2. Why do you think that at least 2 billion years of evolution preceded the origin of multicellularity, and more than 3 billion years preceded the invasion of the land? What kinds of information would you need to test your ideas?

3. There seems to be no iridium layer preserved in the rocks from four of the five major periods when mass extinction occurred. How could one account for these four events?

FOR FURTHER READING

BAKKER, R.T.: *The Dinosaur Heresies,* William Morrow & Co., Inc., New York, 1986. A fine discussion of the controversy about whether dinosaurs could control their own temperatures internally; well-written and beautifully illustrated.

BONATTI, E.: "The Rifting of Continents," *Scientific American,* March 1987, pages 97-103. Excellent presentation of modern research into the ways that continents move.

CLOUD, P.: *Oasis in Space,* W.W. Norton & Company, New York, 1988. Subtitled "Earth History from the Beginning," this well-illustrated book outlines approximately the same material as the present chapter, but in greater detail; well-written and readable.

CZERKAS, S.J., and E.C. OLSON, editors: *Dinosaurs Past and Present,* vols. 1 and 2, University of Washington Press, Seattle, 1987. Outstanding and beautifully illustrated work on the dinosaurs, based on an exhibition at the Los Angeles County Museum of Natural History.

GLAESSNER, M.F.: *The Dawn of Animal Life,* Cambridge University Press, Cambridge, England, 1984. Outstanding account of the earliest history of animals.

GOULD, S.J.: *Wonderful Life,* W.W. Norton & Company, New York, 1989. A beautifully written account of some of the earliest animal fossils, and the ways in which they have been interpreted.

KNOLL, A.: "End of the Proterozoic Eon," *Scientific American,* October 1991, pages 64-73. A marvelous account of how multicellular animals big enough to see first evolved 800 million years ago.

LAMBERT, D. and THE DIAGRAM GROUP: *The Field Guide to Early Man,* Facts on File Publications, New York, 1987. A richly-illustrated guide to the history of our species and its relatives.

LITTLE, C.: *Terrestrial Invasion. An Ecophysiological Approach to the Origins of Land Animals.* Cambridge University Press, Cambridge, 1990. An excellent treatment of the factors involved in the evolution of land animals.

MACDONALD, K.C. and P.J. FOX: "The Mid-Ocean Ridge," *Scientific American,* June 1990, pages 72-79. The formation and evolution of the longest mountain chain on earth, the mid-ocean ridge, provides insight into modern concepts of plate tectonics.

McMENAMIN, M.A.S.: "The Emergence of Animals," *Scientific American,* April 1987, pages 94-101. During the major period of diversification of animals starting about 570 million years ago, new ways of living emerged rapidly.

O'BRIEN, J.: "The Ancestry of the Giant Panda," *Scientific American,* November 1987, pages 102-107. A clear account of how modern molecular techniques are being used to solve a famous evolutionary puzzle: is the panda a raccoon or a bear?

RAUP, D.M.: *The Nemesis Affair,* W.W. Norton and Co., New York, 1986. Subtitled "A story of the death of dinosaurs and the ways of science," this small book provides an excellent account of the theories and controversies surrounding its subject.

SCIENTIFIC AMERICAN DEBATE: *What caused the mass extinction,* Extraterrestrial Impact (W. Alverez and F. Asaro) or Volcanic Eruption (V. Courtillot)? *Scientific American,* October 1990, pages 80-92. A spirited debate by proponents of two very different answers to the question "What happened to the dinosaurs?".

STRINGER, C.B.: "The Emergence of Modern Humans," *Scientific American,* December 1990, pages 98-104. The fossil record agrees with biochemical evidence that all humans are descended from a recent African ancestor.

TUTTLE, R.H.: "Apes of the World," *American Scientist,* vol. 78, pages 115-126, 1990. A survey of our closest living relatives reveals rich prospects for further study and an urgent need for conservation.

WALKER, A. and M. TEAFORD: "The Hunt for Proconsul," *Scientific American,* January 1989, pages 76-82. An exciting account of the last common ancestor of great apes and human beings.

WALLACE, J.: "New Discoveries About Dinosaurs," *Science Year,* 1989, pages 42-76. Annual supplement to World Book. New evidence has convinced most scientists that many dinosaurs once considered dull and plodding were actually agile, fast-moving, and even social.

WELLNHOFER, P.: "Archaeopteryx," *Scientific American,* May 1990, pages 70-77. New information is continuing to provide insight into the evolution of flight in birds.

WOOD, B.: "Origin and Evolution of the Genus *Homo,*" Nature, February 1992, pages 783-790. A current review of the argument that rages today about how humans evolved.

Ecology

Population Dynamics

OVERVIEW

Ecology is the study of the relationships of organisms with one another and with their environment; as such, ecology is concerned with populations of organisms. Natural populations are affected by the ways in which they interact with other populations and with nonliving factors as well. Each population grows in size until it eventually reaches the limits of its habitat to support it. Some of the limits to the growth of a population are related to the density of that population, but others are not. Competition between populations of any two species may limit their coexistence. Predation and other interactions among populations also play an important role in limiting population size.

FOR REVIEW

Here are some important terms and concepts that you will encounter in this chapter. If you are not familiar with them, you should review them before proceeding.

Adaptation (Chapter 20)

How species originate (Chapter 21)

Major features of evolution (Chapter 22)

Ecology, the study of the relationships of organisms with one another and with their environment, is a complex but fascinating area of biology that has many important implications for each of us. The human population is climbing rapidly, and has now attained the unprecedented level of more than 5 billion people. This severely strains the earth's capacity to sustain us all. In the face of this situation the principles of ecology assume a crucial importance and may enable us to chart a sound future.

Ecology, however, is not intrinsically an action-oriented field; it is an area of scientific knowledge that is concerned with the most complex level of biological integration. Ecology attempts to tell us why particular kinds of organisms can be found living in one place and not another—the physical and biological variables that govern their distribution; the factors that control the numbers of particular kinds of organisms and maintain them at certain levels; and the principles that may allow us to predict the future behavior of assemblages of organisms.

Ecologists consider groups of different organisms at three levels of organization that become progressively more inclusive. Thus populations of different organisms that live together are called **communities.** A community, together with the nonliving factors with which it interacts, is called an **ecosystem.** An ecosystem regulates the flow of energy, ultimately derived from the sun, and the cycling of the essential elements on which the lives of its constituent plants, animals, and other organisms depend. **Biomes** are major terrestrial assemblages of plants, animals, and microorganisms that occur over wide geographical areas and have definite characteristics that identify them as distinct from other such assemblages. They include deserts, tropical forests, and grasslands. Similar major groupings can be distinguished in marine and freshwater habitats.

> **A community is the interacting set of different kinds of organisms that occur together at a particular place. An ecosystem includes that set of organisms, together with the nonliving factors with which it interacts. A biome is an assemblage of organisms that has a characteristic appearance and that occurs over a wide geographical area on land.**

In our exploration of ecological principles we will first consider in this chapter the properties of populations, emphasizing their dynamics; discuss communities and the sorts of interactions that occur in them in Chapter 24; move on to the dynamics of ecosystems in Chapter 25; and consider the properties of biomes and similar major aggregations in the seas in Chapter 26. Chapter 27 will deal with the future of the biosphere, a future that will depend to an ever-increasing extent on the ecological principles presented in this section.

POPULATIONS

A **population** consists of the individuals of a given species that occur together at one place and at one time. This is a flexible definition, which allows us to speak in similar terms of the world's human population, the population of protozoa in the gut of an individual termite, or the deer that inhabit a forest.

All populations have characteristic features, such as size, density, dispersion, and demography. Each occupies a particular place and plays a particular role in its ecosystem; that role is defined as the **niche** of that population. Understanding these properties of populations is crucial for understanding the nature of life on earth.

POPULATION SIZE AND DISPERSION

One of the important features of any population is its size. Population size has a direct bearing on the ability of a given population to survive. Very small populations are the most likely to become extinct. Random events or natural disturbances may endanger a few individuals more than they endanger many, and inbreeding can also be a negative factor in population survival if few potential mates are available. Not only does inbreeding often lead directly to a lowering of vigor by direct genetic effects, but the

A

B

C

FIGURE 23-1

At risk. These animals exist in small populations and therefore are in danger of extinction unless the activities that are threatening them are brought under control.

A The Sumatran rhinoceros, *Didermoceros sumatrensis*, reduced to a few hundred individuals on the island of Sumatra and on the adjacent Asian mainland. Poaching to obtain the horns, which are in great demand for medicinal uses, and destruction of habitat are responsible for the decline of this species.

B The kagu, *Rhynochetus jubatus*, the only member of a family of birds that is restricted to the island of New Caledonia, in the southwestern Pacific Ocean. It has been brought to the brink of extinction by wild dogs and habitat destruction.

C Mediterranean monk seal, *Monachus monachus*. Fewer than 500 individuals of this species are believed to exist; they are confined to remote cliffbound coasts and islands in the Mediterranean. The seals have low birth and survival rates, and are killed by fishermen.

reduced levels of variability that result from inbreeding are likely to detract from the population's ability to adjust to changing conditions. If an entire species consists of only one or a few small populations, that species is likely to become extinct, especially if it occurs in areas that have been or are being changed radically (Figure 23-1).

In addition to population size, population **density** is also extremely important. If the individuals that make up the population are widely spaced, they may rarely, if ever, encounter one another, and the reproductive capabilities—and therefore the future—of the population may be very limited, even if the absolute numbers of individuals over a wide area are relatively high. A related measure is **dispersion,** the way in which the individuals of a population are arranged. They may be **randomly spaced, evenly spaced,** or **clumped.** Each of these patterns reflects the interactions between a given population and its environment, including the other species that are present.

Individuals may be evenly spaced or clumped regardless of how abundant they are in a given habitat, and the patterns may not be at all obvious. For example, the creosote bush *(Larrea divaricata)* is often dominant over wide areas in the deserts of Mexico and the southwestern United States. Within these populations, the individuals are evenly spaced, probably because the shrubs secrete chemicals that retard the establishment of other individuals near established ones; however, they still form more or less continuous populations. On the other hand, in tropical rain forests, any given plant species may be represented by only a few individuals per hectare; when

The number of species that live on an island is related to the size of that island, an important finding that ws first expressed in quantitative terms in the 1950s. In 1967, Robert Mac-Arthur of Princeton University and E.O. Wilson of Harvard published an important short book on this relationship, *The Theory of Island Biogeography*. They explained that species are constantly being dispersed to islands, so islands have a tendency to accumulate more and more species. At the same time that new species are being added, however, other species are being lost. Once the number of species in a particular group is high enough that it completely fills the capacity of that island, no more species can be established, on the average, unless one of the species that is already there becomes extinct. Every island of a given size, then, has a characteristic number of species that tends to be maintained through time, even though many of the individual species might change.

The initial stages in this process resemble those that occurred on Surtsey (see p. 502). After such a new island is formed, if it is relatively close to a source area, colonization may be rapid. Eventually, however, an equilibrium number of species will be reached and subsequently maintained for each group of organisms.

The relationship, if expressed on a log/log basis, approximates a straight line. Thus Figure 23-A, modified from MacArthur and Wilson's book, expresses the relationship between the number of species of reptiles and amphibians on certain islands in the West Indies and the area of the islands.

The analysis of a great deal of data has shown that relationships such as this are best approximated by the formula

$$S = CA^z$$

In this formula S represents the number of species; C, the density of species, a factor that varies according to the group of organisms being considered and the region; A, the area of the island; and z the slope of

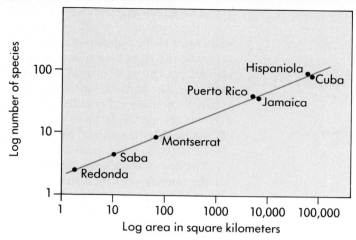

FIGURE 23-A

the line in a graph like the one presented here. The number of species present on an individual island or set of islands varies somewhat with the height of the islands and their distance from the mainland or other source areas; more distant islands have fewer species, and higher ones have more species. Despite these variables, the relationship between species number and area is strikingly constant in a given region anywhere in the world. Thus useful predictions can be made about the expected numbers of species on island or other areas when little or no information is available.

An experimental approach to understanding this relationship was carried out in small mangrove islands off the southern tip of Florida. In this region, islands about 12 meters across are inhabited by about 20 to 45 species of arthropods. Larger islands had more species; smaller ones had fewer species. When some of the islands were fumigated with an insecticide that killed all arthropods, they were recolonized within a year and rapidly attained a total number of species that roughly equaled the number before fumigation. Most interesting, however, was the fact that the new inhabitants of a given island mainly represented different species from those that lived there initially. The numbers of species that were able to coexist on a particular island tended

to be set by the capacity of that island, which was presumably affected by the principles of competition, availability of habitat, and other factors we have discussed in this chapter. The exact species on an island, however, were determined largely by chance.

On islands far from source areas, like those in the central Pacific, the replacement of extinct species tends to be slow. This obvservation has important implications for conservation biology. The introduction of a foreign species is apt to be especially disruptive in such ecologically fragile communities. The members of those groups that do disperse well, such as small, light insects or plants whose seeds drift in the sea or blow through the air, are apt to be better represented than organisms with more limited powers of dispersal. Furthermore, if there is no source area or if the source area is very distant (patterns that arise when the ecosystems of former source areas are destroyed by human activities or other activities), recolonization may not occur at all. As tropical vegetation is being cut up into small "islands," for example, the nature and size of these islands and the distance that separates them from the areas from which they might be recolonized become critical to species survival.

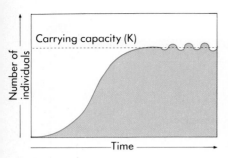

FIGURE 23-4
The sigmoid growth curve.

In other words, the growth rate of the population under consideration (dN/dt) equals its rate of increase (r) multiplied by N, the number of individuals present at any one time, and then multiplied by an expression equal to K, the carrying capacity of the environment, minus N divided by K. As N increases (the population grows in size), the fraction by which r is multiplied becomes smaller and smaller, and the rate of increase of the population declines. In practical terms, this amounts to such factors as increasing competition among more and more individuals for a given set of resources present in a particular system, the buildup of waste, or an increased rate of predation. Graphically, this relationship is the S-shaped sigmoid growth curve characteristic of biological populations (Figure 23-4). The curve is called **sigmoid** because it resembles the Greek letter sigma (σ). As the size of a population stabilizes, its rate of growth slows down, and eventually it does not increase further.

> **The size at which a population stabilizes in a particular place is defined as the carrying capacity of that place for that species. Populations grow to the carrying capacity of their environment.**

Processes such as competition for resources, emigration, and accumulation of toxic waste products all tend to increase as a population approaches its carrying capacity for a particular habitat. The resources for which the members of the population are competing may be food, shelter, light, mating sites, mates, or any other factor necessary for the species to carry out its life cycle and reproduce itself. Among animals, territoriality, which is the result of competition for food and other resources between the members of the species in which it occurs, limits the size of the population.

Density-Dependent and Density-Independent Effects

Effects such as those we have just discussed regulate the growth of populations and are called **density-dependent effects.** Among animals they may be accompanied by hormonal changes that may bring about alterations in behavior that directly affect the ultimate size of the population. One striking example occurs in migratory locusts ("short-horned" grasshoppers), which, when they are crowded, produce hormones that cause them to enter a migratory phase, in which the locusts take off as a swarm and fly long distances to new habitats (Figure 23-5). Density-dependent effects, in general, are those that have an increasing effect as population size increases; that is, the individuals in a population compete with increasing intensity for limited resources as the population grows. They were described by Charles Darwin as resulting in natural selection and the improved adaptation of individuals as they compete for these limiting factors. In contrast, **density-independent effects** are caused by factors such

FIGURE 23-5
Density-dependent effects. Migratory locusts, *Locusta migratoria,* a legendary plague of large areas of Africa and Eurasia. When the population density reaches a certain level, the locusts change their characteristics and take off as a swarm. The most serious infestation of locusts in 30 years occurred in North Africa in 1988. These Moroccan villagers are gathering locusts killed in 1988 spraying operations, to burn them.

as the weather and physical disruption of the habitat which operate regardless of population size.

Density-dependent effects are caused by factors that come into play particularly when the population size is larger; density-independent effects are controlled by factors that operate regardless of population size.

Agriculture depends in part on the characteristics of the sigmoid growth curve, which we can apply to ecosystems in their early stages of development; for example, after an area has been cleared. At such times, populations and individuals are growing rapidly, and net productivity—in terms of the amount of material incorporated into the bodies of these organisms—is highest. Commercial fisheries exploit populations during their rapid growth phase, and they attempt to operate so that they are always harvesting populations in the steep, rapidly growing parts of the curve. The point of **optimal yield**—maximum sustainable catch from the population—lies partway up the sigmoid curve. Harvesting the population of an economically desirable species near this point will result in much better sustained yields than can be obtained either when the population approaches the carrying capacity of its habitat or when it is small (Figure 23-6). Overharvesting a population that is smaller than this critical size can destroy its productiveness for many years or even drive it to extinction. This evidently happened to the Peruvian anchovy fishery after the populations had been depressed by the 1972 El Niño (see Chapter 26). It is difficult to determine the population levels of commercially valuable species that are most suitable for long-term, productive harvesting, but this is now the subject of much study.

In natural systems that are exploited by humans, such as agricultural systems and fisheries, the aim is to exploit the population at the early, most productive part of the rising portion of the sigmoid growth curve.

FIGURE 23-6
Fishing in the ocean. The numbers and ages of the individuals harvested by the fisheries industry will determine in large measure the future of the world's fish populations.

r Strategists and K Strategists

Many species, such as annual plants, a number of insects, and bacteria, have very fast rates of population growth. This growth cannot be controlled effectively by reducing their population sizes. In such species, small surviving populations will soon enter an exponential pattern of growth and regain their original sizes. In contrast, a comparable reduction in population size among relatively slow-breeding organisms, such as whales, rhinoceroses, California redwoods, or most tropical rain forest trees, can lead directly to extinction.

The populations of organisms that have sigmoid population growth curves are limited in number by the carrying capacity of the environment, or K. Such organisms, including the relatively slow-breeding ones just mentioned, tend to live in habitats that are fairly stable and predictable. Organisms for which an S-shaped growth curve is characteristic are called K strategists. By contrast, species whose populations are characterized by exponential growth followed by sudden crashes in population size tend to live in unpredictable and rapidly changing environments. They have a high intrinsic rate of increase, or r, and are called r strategists. Many organisms are neither "pure" r strategists nor "pure" K strategists. Rather, their reproductive strategies lie somewhere between these two extremes or change from one extreme to the other under certain environmental circumstances.

In general, r strategists reproduce early and have many offspring (Figure 23-7). These offspring are small, mature rapidly, and receive little or no parental care; their generations are relatively short. Many offspring are produced as a result of each reproductive event. Examples of such organisms include dandelions, aphids, mice, and cockroaches.

In contrast, the K strategists tend to reproduce late and to have few offspring. These offspring are large, mature slowly, and often receive intensive parental care; their generations are relatively long. This group consists of such organisms as coconut

FIGURE 23-7

r-Strategists. Different kinds of organisms produce various numbers of offspring, but all have the potential to produce populations larger than those that actually occur in nature. The German cockroach (*Blatella germanica*), a major household pest, produces 80 young every 6 months. If every cockroach that hatched survived, kitchens might look like this.

palms, whooping cranes, and whales (Figure 23-8). Many of the plants and animals that are in danger of extinction are *K* strategists (see Figure 23-1).

> The *r* strategists are characterized by an early age of first reproduction, large brood size, numerous offspring, no parental care, and short generations. The *K* strategists, in contrast, have delayed reproduction, small broods, few offspring, parental care, and long generations. Many *K* strategists are highly prone to extinction.

Human Populations

The size of human populations, like those of other organisms, is controlled by their environment. Throughout history, however, humans have expanded the carrying capacity of the habitats in which they lived because of their ability to develop technical innovations. Earlier in our history, our populations were regulated by both density-dependent and density-independent effects, including food supply, disease, and predators; there was also ample room on earth for migration to new areas to relieve overcrowding in specific regions. To a lesser extent, migration still plays a role in the adjustment of human populations to particular areas. Unusual disturbances, including floods, extreme temperatures, and droughts have also affected the pattern of human population growth.

Gradually, changes in technology have given humans more control over their food supply and enabled them to develop superior weapons to ward off predators, as well as the means to cure diseases. Improvements in transportation and housing have increased the efficiency of migration. At the same time, improvements in shelter and storage capabilities have made humans less vulnerable to climatic uncertainties.

As a result of our ability to manipulate these factors, the human population has grown explosively to its present level of more than 5.4 billion people. It is continuing to grow at the rate of approximately 1.8% per year, so that nearly 95 million people are added to the world population annually. At this rate, the human population will double in 40 years. As we will discuss in Chapter 27, both the current human population level and the projected rate of growth have potential consequences for our future that are extremely grave.

FIGURE 23-8

A female humpbacked whale (*Megaptera*), a *K* strategist, swimming with her calf in the waters off Hawaii. Most whales have one calf at a time. Roger Payne, who has contributed a great deal to our understanding of whales, has written: "In our long sad history, we have brought hundreds of species to extinction. Unless we change our priorities, we will soon eradicate hundreds of thousands more. There is a curious fact, however: among all of the species we have destroyed, there is not one that occurred worldwide . . . the closest that we have ever come to taking that fatally insane step is with the right whale—a species that once occurred off the western and eastern shores of every continent. We came so close to destroying that species that it really is something of a miracle that it survived."

MORTALITY AND SURVIVORSHIP

A population's intrinsic rate of increase depends on the ages of the organisms in it and the reproductive performance of the individuals in the various age groups. When a

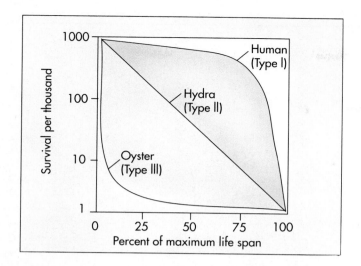

FIGURE 23-9
Survivorship curves for the oyster, for a microscopic freshwater animal called the hydra, and for humans. Each population is assumed to have started with 1000 individuals. Humans have a type I life cycle, the hydra type II, and oysters type III. The shapes of the respective curves are determined by the percentages of individuals in populations that die at different ages.

population lives in a constant environment for a few generations, its **age distribution**—the proportion of individuals in the different age categories—tends to become stable. This distribution, however, differs greatly from species to species and even, to some extent, from place to place within a given species. Depending on the mating system of the species, sex distribution can likewise have an important effect on population growth statistics. In addition to age and sex distribution, another important factor in determining a population's rate of growth is **generation time**—the length of time that separates successive generations.

One way to express the characteristics of populations with respect to age distribution is the survivorship curve. **Survivorship** is defined as the percentage of an original population that survives to a given age. Samples of different kinds of survivorship curves are shown in Figure 23-9. In hydra, individuals are equally likely to die at any age, as indicated by the straight survivorship curve (type II). Oysters, on the other hand, produce vast numbers of offspring, only a few of which live to reproduce. Once they become established and grow into reproductive individuals, their rate of death, or **mortality,** is extremely low (type III survivorship curve). Even though human babies are susceptible to death at relatively high rates, the highest mortality in people occurs later in life, in their postreproductive years (type I survivorship curve).

Many animal and protist populations in nature probably have survivorship curves that lie somewhere between those characteristic of type II and type III, and many plant populations, with high mortality at the seed and seedling stages, are probably closer to type III. Humans have probably approached type I more and more closely through the years, with the birth rate remaining relatively constant or declining somewhat, but the death rate dropping markedly.

In type I survivorship curves, a large number of individuals reach their theoretical maximum age. In type II curves, mortality is largely independent of age. In type III curves, mortality is highest during the young stages.

DEMOGRAPHY

Demography is the statistical study of populations. The term comes from two Greek words, *demos*, "the people" (the same root we see in the word "democracy") and *graphos*, "measurement." It therefore means measurement of people, or, by extension, of the characteristics of populations. Demography is the science that helps us to predict the ways in which the sizes of populations will alter in the future. It takes into account the age distribution of the population and its changing size through time.

A population whose size remains the same through time is called a **stable population.** In such a population, births plus immigration must exactly balance deaths plus emigration. In such a population, not only does the size remain constant, but so does the age structure.

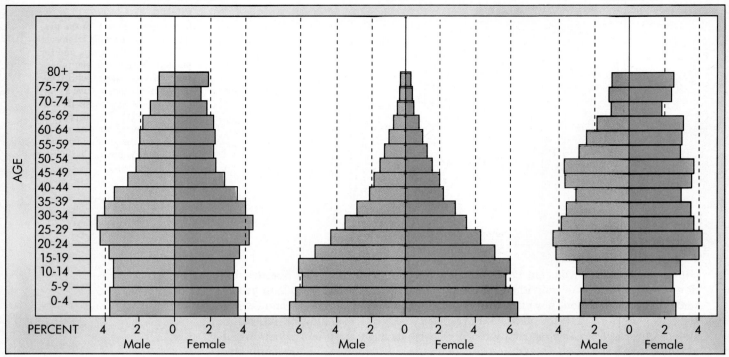

FIGURE 23-10
Population pyramids. The age
distribution of human males and
females in 1990 in the populations
of Mexico, representing rapid
growth; the United States, slow
growth; and Germany, negative
growth.

The characteristics of a population can be illustrated graphically by means of a **population pyramid**—a bar graph using single-year or 5-year age categories (Figure 23-10). In a population pyramid, males are conventionally enumerated to the left of the vertical age axis, and females to the right. A population pyramid shows the composition of a population by age and sex. By viewing such a pyramid, one can see the historical trends of demographic events such as births and deaths.

Examples of population pyramids for the United States, Mexico, and Germany in 1990 are shown in Figure 23-10. In the population pyramid for the United States, the cohorts (group of individuals) 45 to 49 years old is smaller in size than those preceding and following (these people were born during the depression years). The cohorts 20 to 24 and 25 to 29 years old represent the "baby boom." In most human population pyramids, the number of females will be disproportionately large as compared with the number of males: females in most regions have a longer life expectancy than males.

INTERSPECIFIC INTERACTIONS THAT LIMIT POPULATION SIZE

Competition between individuals of two or more species for the same limiting resources, as well as the direct effects of predation or parasitism of one species on another, are among the kinds of interspecific interaction that may limit population size. We will discuss them in this context here, and in Chapter 24 we will study their role in the structure of the community.

Competition: General

Two kinds of organisms that are each using a resource that is in short supply *compete* with one another. Interspecific competition is often greatest between organisms that obtain their food in similar ways; thus green plants compete mainly with other green plants, herbivores with other herbivores, and carnivores with carnivores. In addition, competition is more acute between closely similar kinds of organisms than between ones that are less similar. **Interspecific competition** is to be distinguished from competition between individuals of a single species, which is called **intraspecific competition.**

More than 50 years ago, the Soviet ecologist G.F. Gause formulated what is called the principle of **competitive exclusion.** This principle states that if two species are competing with one another for the same limited resource, then one of the species will be able to use that resource more efficiently than the other, and the former will therefore eventually eliminate the latter locally.

In many experiments, the results of growing individuals of two species together in the laboratory have not always been easily predictable. For example, when Thomas Park and his colleagues at the University of Chicago grew two species of flour beetle, *Tribolium* (Figure 23-11), together in the same container of flour, one always became extinct. The species that survived, however, varied. *Tribolium castaneum* usually won under relatively hot and damp conditions, whereas *Tribolium confusum* won under cooler, drier conditions. Subsequent experiments with these species demonstrated that a genetic component was also involved in the unpredictability of the outcome. Some strains of one species would win over some—but not all—strains of the other, under a given set of conditions.

Similarly, John Harper and his colleagues at the University College of North Wales grew two species of clover together, *Trifolium repens* (white clover) and *Trifolium fragiferum* (strawberry clover). Each was sown at two densities, 36 and 64 plants per square foot (930 square centimeters), using all the possible combinations. Although white clover initially formed a dense canopy of leaves, the slower-growing strawberry clover, whose leaf stalks are longer, eventually produced enough leaves that overtopped the lower white clover and overcame it. Strawberry clover did this by competing more effectively for light; the outcome was the same regardless of the initial densities at which the plants were sown.

Among plants generally, competition between root systems for the nutrients that the plants require and which they obtain from soil is of central importance. The roots of one species not only may deplete essential minerals and thus outcompete another, but may also excrete toxic substances that depress the growth of the other species. Such interactions are poorly understood, but experimental studies are beginning to resolve their complexity.

FIGURE 23-11
Competitive exclusion. The flour beetle *Tribolium confusum,* and the closely similar *T. castaneum,* were used in experiments that demonstrated the unpredictability of results when two species of organisms are competing with one another.

Competition: Examples from Nature

Examples that demonstrate the principle of competitive exclusion can be found in nature and studied, both by observation and by experiment. Such examples are directly related to the structure of the communities in which they occur, and they will be discussed more extensively in Chapter 24. However, we include a few examples here, as we consider the contribution of competitive interactions to the structure of natural communities.

The competitive interactions between two species of barnacles that grew together on the same rocks along the coast of Scotland have been investigated by J.H. Connell of the University of California, Santa Barbara. Barnacles are marine animals (crustaceans) that have free-swimming larvae that settle down, cement themselves to rocks, and then remain permanently attached to that point. Of the two species Connell studied, *Chthamalus stellatus* lives in shallower water, where it is often exposed to air by tidal action, and *Balanus balanoides* occurs lower down, where it is rarely exposed to the atmosphere (Figure 23-12). In this deeper zone, *Balanus* could always outcompete *Chthamalus* by crowding it off the rocks, undercutting it, and replacing it even where it had begun to grow. When Connell removed *Balanus* from the area, however, *Chthamalus* was easily able to occupy the deeper zone, indicating that no physiological or other general obstacles prevented it from becoming established there. In contrast, *Balanus* cannot survive in the shallow-water habitats where *Chthamalus* normally occurs. It evidently does not have the special physiological and morphological adaptations that allow *Chthamalus* to occupy this zone.

Viewing the same situation in evolutionary terms, one might suppose that *Chthamalus* evolved later than *Balanus* or that it reached the Scottish coast later than its rival. Unable to compete with *Balanus* in its main area of distribution, it nevertheless could survive in the relatively less favorable shallow-water habitats where *Balanus* did

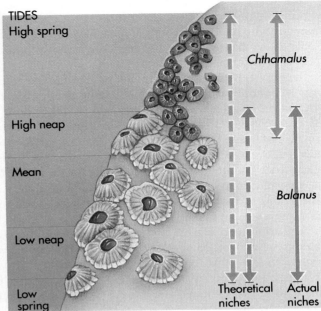

TIDES
High spring

High neap

Mean

Low neap

Low spring

Chthamalus

Balanus

Theoretical niches

Actual niches

FIGURE 23-12
Competition among barnacles.
A *Chthamalus* (the smaller, smooth barnacles) and *Balanus* (the larger, ridged barnacles) growing together on a rock. Competing species belonging to these two genera were studied by J.H. Connell along the coast of Scotland. The microscopic larvae of barnacles are dispersed widely, but they settle down as adults.
B The distribution of the two species with respect to different water levels and tidal effects is shown at the right. The neap tide is the extreme tide that occurs in each lunar month, whereas the spring tide is greater than usual, and is produced when the sun and the moon are on opposite sides of the earth or lined up with one another.

not occur. Conversely, it is possible that *Balanus* evolved more recently or reached the area more recently than *Chthamalus* and possibly eliminated *Chthamalus* from most of its former range. In doing so it would have restricted *Chthamalus* to the less favorable sites where it was subject to periodic drying.

In another set of field observations, the late Princeton ecologist Robert MacArthur studied five species of warblers, which are small, insect-eating birds that coexist part of the year in the forests of the northeastern United States and adjacent Canada. Although they all appeared to be competing for the same resources, MacArthur found that each species actually spent most of its time feeding in different parts of the trees, and so each ate different subsets of insects in those trees. Some of the species of warblers fed on insects near the ends of the branches, whereas others regularly penetrated well into the foliage; some stayed high on the trees, others fed on the lower branches; and these patterns were recombined in different ways characteristic of each of the warbler species (Figure 23-13). As a result of these different feeding habits, each species of warbler actually occupied a different niche; in other words, they had different ways of using the resources of the environment and thus were not really in direct competition with one another for limited resources.

> **Species of barnacles, warblers, and many other kinds of organisms coexist because the niches of the species involved differ.**

Predator-Prey Interactions

Predation is another factor that may limit the size of populations. In this sense, predation includes everything from one kind of animal capturing and eating another to parasitism (the condition of an organism living in or on another organism, at whose expense the parasite is maintained). Predation and parasitism are two ends of a biological spectrum, between which there is no clearly marked distinction. They are governed by similar principles.

When experimental populations are set up under simple conditions in the laboratory, the predator often exterminates its prey and then becomes extinct itself, having nothing to eat. If refuges are provided for the prey, however, its population will be driven to low levels but can recover; one would then expect the populations of predators and prey to follow a cyclical pattern. The low population levels of the prey species provide scant food for the predators; then the predators in turn become scarce. As this occurs, the prey recover and again become abundant. Population cycles are

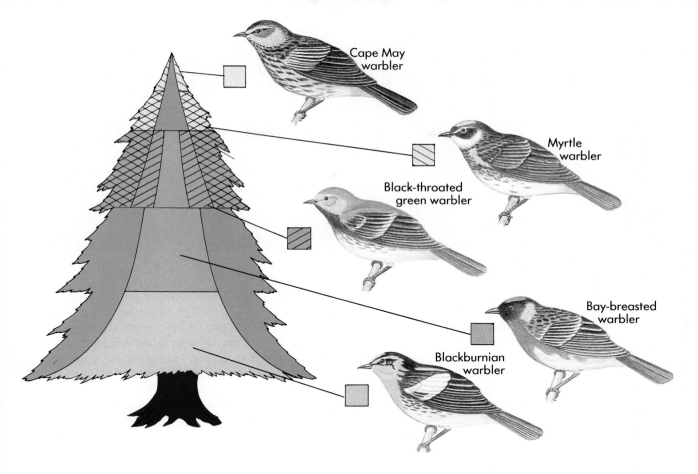

characteristic of some species of small mammals, such as lemmings, and may be stimulated, at least in some situations, by their predators. An example of such fluctuating populations is expressed diagrammatically in Figure 23-14.

Relationships of this sort are of the greatest importance for biological control. If the prey becomes so rare that it is an infrequent event for the predator species to encounter it, the predator itself may become extinct. Ideally, in the case of deliberate biological control, the prey will survive in small numbers, thus making it possible for small populations of the predator to survive also and for the prey species to be controlled indefinitely.

In Australia, for example, prickly pear cactus (*Opuntia*), introduced from America, once overran the ranges and became so abundant that vast areas were effectively

FIGURE 23-13
Competition among warblers. Although all five species of warbler shown here fed on insects in the same spruce trees at the same time, Robert MacArthur was able to demonstrate that they mainly feed in different parts of the tree and in different ways. The shaded areas indicate areas where the given species spends at least half of its feeding time. In this way, competition between them is reduced.

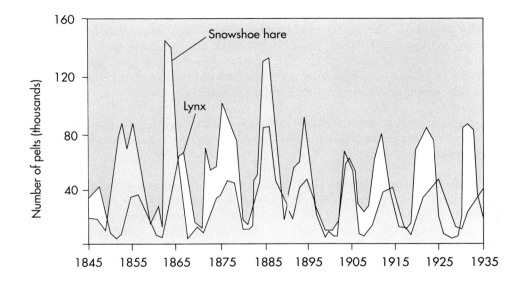

FIGURE 23-14
A predatory-prey cycle. The data are based on numbers of animal pelts from 1845 to 1935. Recent evidence suggests that as the numbers of hares grow, so does the number of lynxes; the cycle repeats about every 9 years. The numbers of hares is controlled not by predators, but by available resources; the number of lynxes is controlled by the availability of prey, hares.

FIGURE 23-15

A B

Biological control of pests. After an initial period in which prickly pear cacti *(Opuntia)*, introduced from Latin America, choked many of the pastures of Australia with their rampant growth, they were controlled by the introduction of a cactus-feeding moth, *Cactoblastis*, from the areas where the cacti were native.

A An infestation of prickly pear cacti in scrub in Queensland, Australia, in October, 1926.

B The same view in October, 1929, after the introduction of *Cactoblastis*.

closed to cattle grazing (Figure 23-15). The situation was changed dramatically with the introduction of the moth *Cactoblastis* (Figure 23-16). The larvae of the moth feed on the pads of the cactus and rapidly destroy the plants. Within relatively few years, the moth had reduced the cactus to the status of a rare species in many regions where it was formerly abundant. It is now exceptional to find an individual of *Cactoblastis*, but the moth is still present and evidently keeps the cactus in check.

The future is more problematic for the American chestnut *(Castanea americana)*, a species of tree that has virtually been driven to extinction by the accidental introduction of an Asiatic fungus, *Cryptonectria (Endothia) parasitica*. The chestnut used to be a dominant or codominant tree throughout much of the forests of eastern and central temperate North America. Chestnut blight, the disease caused by the fungus, was first seen in North America in New York State in 1904 and killed most of the chestnut trees in North America during the following 30 years. Today the American chestnut survives largely as sprouts that grow every year from the trunks of trees that were killed decades ago. A few populations also seem to be isolated from the chestnut blight in certain remote areas.

Many additional examples of the ways in which predator-prey relationships operate are familiar. Organisms that cause diseases that completely kill their host species are not "successful," because once they have done so they have also eliminated their own source of food. Thus those strains of the disease-causing organism that are less virulent will be favored by natural selection and will survive.

The history of the viral disease myxomatosis, which is caused by a virus introduced into Australia and New Zealand to control rabbits, provides an instructive example of this principle. The rabbits were brought to these countries as a convenient source of meat, but they soon ran wild, with devastating effects on the countryside (Figure 23-17). When the virus causing myxomatosis was introduced, most of the rabbits soon died. The most virulent strains of the virus disappeared with the dead rabbits, and less lethal strains became apparent in the remaining rabbit populations. At the same time, strains of rabbits that were resistant to the disease began to appear. Now the populations of both kinds of organisms have achieved the kind of equilibrium relationship in which they can coexist indefinitely.

An organism that causes a disease that always kills its host will die with it; one that produces sublethal effects will have the opportunity to spread to another host.

The relationships between large carnivores and grazing mammals are a subject of great interest and importance in many parts of the world. Appearances are sometimes deceiving. On Isle Royale in Lake Superior, for example, moose reached the island and multiplied freely there in isolation. When wolves later reached the island by crossing over the ice in winter as the moose had done earlier, it was widely assumed at first that they were playing the determining role in controlling the moose populations

A

B

FIGURE 23-16

Cactoblastis, a moth introduced from Latin America, controls prickly pear cacti in Australia effectively.

A Larvae of Cactoblastis feeding on a cactus pad.

B Adult *Cactoblastis* on a cactus spine.

(Figure 23-18). More careful studies, however, have demonstrated that this is not the case. The moose that the wolves eat are old and diseased, for the most part, and would not survive long anyway. In general the moose are controlled by the amount of food available to them, their diseases, and many factors other than the wolves.

The intricate interactions between predators and prey are an essential factor in the maintenance of groups of organisms occurring together that are rich and diverse in species. By controlling the levels of some species, the predators make possible the continued existence of others in that same community. In other words, by keeping the numbers of individuals of some of the competing species low, the predators prevent or greatly reduce competitive exclusion. Such patterns are particularly characteristic of biological communities in intertidal, marine habitats. For example, in preying selectively on bivalves, sea stars prevent them from monopolizing all the space in such habitats, opening them up to many other kinds of organisms.

A given predator may very often feed on two, three, or more kinds of plants or

 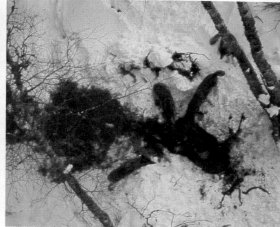

FIGURE 23-18   B
Predation often hits the old and infirm.
A On Isle Royale, Michigan, a large pack of wolves in unsuccessful pursuit of a moose. This moose was chased for almost 2 kilometers; it then turned and faced the wolves, who by that time were exhausted from running through chest-deep snow. The wolves lay down and the moose walked away.
B Wolves beginning to feed on a freshly killed moose. A dominant female begins to open up the chest cavity, where the internal organs will be consumed first. Such dominant females, called alpha wolves, control the activities of the pack and generally produce the only offspring. Wolf predation selectively removes calves and old moose from the population—usually those more than 8 years old, and often with skeletal pathology such as arthritis. Isle Royale National Park provides an exceptional outdoor laboratory, the only site on earth where naturally regulated wolves and prey have been subjected to long-term study.

animals in a given community in a way that partly depends on their relative abundance. In other words, a predator may feed on species *A* when it is abundant and then switch to species *B* when *A* is rare. Similarly, a given prey species may be a primary source of food for increasing numbers of species as it becomes more abundant, which tends to limit the size of its population automatically. Such feedback systems are key factors in determining the structure of many natural communities.

THE NICHE

Studies of the principle of competitive exclusion led to the development of the concept of the niche, mentioned earlier. A niche may be described in terms of space, food, temperature, appropriate conditions for mating, requirements for moisture, and so on. A full portrait of an organism's niche would also take into account its behavior and the ways in which this behavior changes at different seasons and different times of the day. In other words, *niche* is not synonymous with **habitat**: the habitat of an organism, the place where it lives, is defined by some, but by no means all, of the factors that make up its niche. The factors that make up an organism's niche determine whether it can exist in a given ecosystem, and also how many species can exist there together.

Niche, in the sense we have just been discussing, is sometimes called the **realized niche** (actual niche) of an organism—the role the organism actually plays in a particular ecosystem. It is thus distinguished from the niche that it might occupy if competitors were not present, called the **fundamental niche** (theoretical niche). Thus the fundamental niche of the barnacle *Chthamalus* in Connell's experiments in Scotland included that of *Balanus,* but its realized niche was much narrower because *Chthamalus* was outcompeted by *Balanus* in its theoretical niche.

> **A niche may be defined as the way in which an organism interacts with its environment. The fundamental niche of an organism is the niche that it would occupy if competitors were not present. The realized niche is the niche that it actually occupies under natural circumstances.**

Another interesting example of some of the properties of a niche involves the flour beetles of the genus *Tribolium* that we considered earlier. If *Tribolium* is grown in pure flour along with beetles of a second genus, *Oryzaephilus,* it will drive *Oryzaephilus* to extinction by means that are only partly understood. If, however, small pieces of glass tubing are added to the flour, providing refuges for *Oryzaephilus,* then both kinds of flour beetle will coexist indefinitely. In a sense this experiment suggests why so many kinds of organisms can coexist in a really complex ecosystem, such as a tropical rain forest, and why competition is more direct in an ecosystem with fewer species, such as the tundra.

When organisms reach habitats where no other plants, animals, or microorganisms have yet arrived, such as islands that have been raised above sea level or newly formed ponds, they are often able to occupy a much wider niche, that is, play a wider variety of ecological roles, than their relatives in the communities from which they were derived. We considered this relationship in Chapter 21, when discussing the evolution of new species on islands. For example, the species of Darwin's finches on the Galapagos Islands apparently occupy wider niches than do their relatives on the continent of South America, and the group of species as a whole occupies many niches occupied by other species of birds on the mainland.

As species evolve, they become increasingly different from one another; eventually, they may be able to coexist, not then being subject to competitive exclusion. If two species become so different in isolation from one another that their niches differ sufficiently when they come into contact with each other, there will be no problem; if the niches do not, there will be selection so that they become more different, or the species cannot continue to coexist. In other words, the differences may evolve either before the species come into contact or after they do so.

Gause's principle of competitive exclusion can be restated, in terms of niches, as follows: *no two species can occupy the same niche indefinitely.* Certainly species can and do coexist while competing for the same resources, and we have just seen some examples of such relationships. Nevertheless, Gause's theory predicts that when two species do coexist on a long-term basis, one or more features of their niches will always differ; otherwise the extinction of one species will inevitably result.

Niche is, of course, a complex concept, one that involves all facets of the environment that are important to individual species. No two species can coexist indefinitely if their niches overlap in too many ways. Within the past decade there has been a vigorous debate concerning the role of competitive exclusion, not only in determining the structure of communities, as we mentioned earlier, but also in setting the course of evolution. Where one or more resources are obviously limiting, as in periods of drought, the role of competition becomes much more obvious than when they are not. On the other hand, especially for plants, the factors that are important in defining a niche are often difficult to determine, and alternative explanations are being sought for the coexistence of large numbers of species.

SUMMARY

1. Populations of different organisms that live together in a particular place are called communities. A community together with the nonliving components of its environment is called an ecosystem. Biomes are major terrestrial assemblages of plants, animals, and microorganisms that occur together over wide geographical areas and have definite characteristics that distinguish them from other such assemblages.

2. Populations consist of the individuals of a given species that occur together at one place and at one time. They may be dispersed in an evenly spaced, clumped, or random manner. Clumped patterns are the most frequent.

3. The rate of growth of any population is defined as the difference between the birth rate and the death rate per individual per unit of time. The actual rate is affected both by emigration from the population and by immigration into it.

4. The intrinsic rate of increase of a population is defined as its biotic potential.

5. Most populations exhibit a sigmoid growth curve, which implies a relatively slow start in growth, a rapid increase, and then a leveling off when the carrying capacity of the species' environment is reached. Populations can be harvested most effectively when they are in the rapid growth phase.

6. Density-dependent effects are those that have an increasing effect as population size increases; density-independent effects, such as the weather, have the same effect regardless of population size.

7. *r* Strategists have large broods and rapid rates of population growth. *K* strategists are limited in population size by the carrying capacity of their environments; they tend to have fewer offspring and slower rates of population growth.

8. Survivorship curves are used to describe the characteristics of mortality in different kinds of populations. Type I populations are those in which a large proportion of the individuals live out their physiologically determined life span. Type II populations have a constant mortality rate throughout their life span. Type III populations have high mortality in their early stages of growth, but an individual surviving beyond that point is likely to live a long time.

9. Gause's principle of competitive exclusion states that if two species compete with one another for the same limited resources, then one will be able to use them more efficiently than the other and will drive its less efficient competitor to extinction. This principle can be restated in terms of the niche concept: no two species can occupy the same niche indefinitely. Organisms studied under field conditions display many different ways of minimizing competition for the same scarce resources.

10. Predator-prey relationships are of crucial importance in limiting population size in nature. They appear to maintain a kind of balance, because if the prey are completely eliminated, the predator will die also.

11. Each species plays a specific role in its ecosystem; this role is called its niche.

12. An organism's fundamental niche is the total niche that the organism would occupy in the absence of competition. The realized niche of an organism is the actual niche that it occupies in nature.

REVIEW QUESTIONS

1. Which is more likely to become extinct: a small population or a large population? Why? How does inbreeding affect the possibility of extinction?

2. What are the three types of dispersion in a population? Which type is most frequent in nature? Why?

3. Define the biotic potential of a population. What is the definition for the actual rate of population increase? What other two factors affect it? What is meant by an exponential capacity for growth?

4. In general, what types of growth-regulating effects are density-dependent? What types of effects are density-independent?

5. Why is it best to harvest individuals of a productive population partway up the sigmoid growth curve rather than when the population is at its carrying capacity? What is the result of harvesting small populations?

6. What type of growth curve is exhibited by a K strategist? What limits the growth of these populations? When in their life span do these types of organisms reproduce? What is the relative generation time of these organisms? What is the relative number and size of their offspring? To what degree do the par-

ents provide care for their young? How rapidly do the offspring mature?

7. What type of growth curve is exhibited by an *r* strategist? What limits the growth of these populations? When in their life span do these types of organisms reproduce? What is the relative generation time of these organisms? What is the relative number and size of their offspring? To what degree do the parents provide care for their young? How rapidly do the offspring mature?

8. Define survivorship. Describe the three types of survivorship curves and give examples of each.

9. What is demography? What two factors are taken into account in demographic studies? What are the characteristics of a stable population?

10. Is the term niche synonymous with the term habitat? Why or why not? How does an organism's theoretical niche differ from its actual niche?

11. What occurs to the niche of an organism when it newly occupies a habitat lacking many other organisms? What happens to the niches of evolving species in terms of competitive exclusion?

THOUGHT QUESTIONS

1. How is it possible for ecological generalists—those that, for example, feed on a wide range of foods—to coexist with specialists in natural communities, as they do throughout the world? Why do all species not simply become generalists or specialists? Which of these habits is a better strategy for survival? Why?

2. How can the principle of competitive exclusion be applied to the origin of large numbers of species of certain genera in the Hawaiian Islands? What would you expect to happen in the fu-

ture to the hundreds of species of *Drosophila* that occur there, leaving aside the possibility of their extinction through the destruction of their habitats by humans?

3. How do you think the intensity of competition would differ between an Arctic tundra and a tropical rain forest? How would you test your answer?

4. Which kinds of natural communities are most resistant to introduced species? Why?

FOR FURTHER READING

BEDDINGTON, J.R., and R.M. MAY: "The Harvesting of Interacting Species in a Natural Ecosystem," *Scientific American*, November 1982, pages 62-69. An analysis of the changing populations of whales and other animals that feed on the krill (shrimp) populations in the Antarctic Ocean as an example of the problems of using a biological resource without destroying it.

BEGON, M., J.L. HARPER, and C.R. TOWNSEND: *Ecology: Individuals, Populations, and Communities*, Sinauer Associates, Sunderland, Massachusetts, 1986. This book provides a superb treatment of the population aspects of ecology in a thoroughly modern context.

BEGON, M., and M. MORTIMER: *Population Ecology: A Unified Study of Animals and Plants*, ed. 2, Sinauer Associates, Sunderland, Massachusetts, 1987. An outstanding short treatment of all aspects of population ecology.

CASE, T.J., and M.L. CODY: "Testing Theories of Island Biogeography," *American Scientist*, vol. 75, pages 402-411, 1987. On islands in the Gulf of California, a natural laboratory for testing theories of island biogeography exists.

COLINVAUX, P.A.: *Ecology*, John Wiley & Sons, Inc., New York, 1986. Excellent book that covers all aspects of ecology, with a particularly good treatment of ecosystems.

CONNOR, E.F., and D. SIMBERLOFF: "Competition, Scientific Method, and Null Models in Ecology," *American Scientist*, vol. 74, pages 155-162, 1986. A central article in the continuing debate about the role of competition in structuring natural communities.

EHRLICH, P.R., and J. ROUGHGARDEN: *The Science of Ecology*, Macmillan Publishing Company, New York, 1987. Excellent, up-to-date account of the entire field.

MacARTHUR, R.H., and E.O. WILSON: *The Theory of Island Biogeography*, Princeton University Press, Princeton, N.J., 1967. This short book introduced one of the more productive fields of modern ecology.

WOLFF, J.O.: "Getting Along in Appalachia," *Natural History*, vol. 95, number 9, pages 44-49, 1986. Fine account of the author's investigations of two species of mice that occur together in the forests of the southern Appalachians, and of the different ways in which they relate to the habitat.

24

Coevolution and Symbiosis: Interactions within Communities

OVERVIEW

Ecosystems function because of the interactions of the organisms that make them up; such a set of interacting organisms is called a community. Through time, the features of these organisms coevolve in relation to one another. Within a community, the herbivores avoid feeding on certain plants and selectively consume others; their actions are mediated by various characteristics of the plants. As a result, the herbivores themselves become specialized. In some cases, herbivores use chemicals obtained from the plants that they eat to protect themselves from their own predators; in others, they break down the chemicals and thus avoid their toxic effects. Many other phenomena, such as warning or protective coloration and mimicry, are related to the nature of the chemicals that animals obtain from their food or that they manufacture. Symbiotic relationships generally act to order the flow of energy through ecosystems and to regulate the cycling of nutrients. Learning about these relationships individually is the key to learning about the functioning of communities in nature.

FOR REVIEW

Here are some important terms and concepts that you will encounter in this chapter. If you are not familiar with them, you should review them before proceeding.

Natural selection (Chapter 19)
Adaptation (Chapter 20)
Demography (Chapter 23)
Competition (Chapter 23)

FIGURE 24-1
The redwood community. The redwood forest of coastal California and southwestern Oregon is dominated by the redwoods (*Sequoia sempervirens*) themselves.

A

B

C

FIGURE 24-2

Plants and animals that occur regularly in the redwood community.

A Swordfern (*Polystichum munitum*)

B Redwood sorrel (*Oxalis oregana*).

C One of the many characteristic animals is this ground beetle (family Carabidae), *Scaphinotus velutinus*, shown feeding on a slug on a leaf of sword fern.

The ecological requirements of each of the organisms associated with the redwoods differ, but they overlap enough that they are able to occur together in what we perceive as the redwood community.

The magnificent redwood forest that extends along the coast of central and northern California and into the southwestern corner of Oregon is an example of a community. Within it, the most obvious organisms are redwood trees, *Sequoia sempervirens* (Figure 24-1). These trees are the sole survivors of a genus that once was distributed throughout much of the Northern Hemisphere. A number of other plants and animals are regularly associated with them, including the sword fern and the beetle illustrated in Figure 24-2; their coexistence is in part made possible because of the special conditions created by the redwood trees themselves: shade, water (dripping from the branches), and relatively cool temperatures, for example. This particular distinctive assemblage of organisms is called the redwood community.

We recognize this community, however, largely because of the redwood trees themselves. The overall distributions of the other kinds of organisms that occur together with the redwoods differ greatly from one another—some are smaller than that of the redwoods, and some are larger. In the redwood community or any other community, their ranges overlap; that is why they occur together.

The redwood community also has a historical dimension; the organisms characteristic of this community have each had a complex and unique evolutionary history. At different times in the past, they evolved and then came to be associated with the redwoods. Certainly the present-day redwood community has come into existence only during the past few million years. During this period, the range of the redwood was reduced as the tree was eliminated by climatic changes that occurred in many of the areas in which it grew previously. In a historical sense, then, the redwood community that we recognize today represents the concordance of the individual histories of the plants, animals, and microorganisms that comprise it.

We recognize a redwood community largely because of the presence of the redwood trees themselves, but many other kinds of organisms are characteristic of this community. The community exists because the ranges of these species overlap in it.

Many communities are very similar in their species composition and their appearance over wide areas. For example, the open savanna grassland that stretches across much of Africa includes many plant and animal species that coexist over thousands of square kilometers. Similar interactions between these organisms, some of which have evolved over millions of years, occur throughout these communities. We will now consider some of these interactions, and then expand our discussion to consider the ways in which the distributions of organisms are limited—the reasons that particular kinds of organisms occur in particular places, and not in others.

COEVOLUTION

Interactions between organisms that characterize particular communities have arisen as a result of their evolutionary history. Thus the plants, animals, protists, fungi, and bacteria that live together in communities have changed and adjusted to one another continually over millions of years. For example, many of the features of flowering plants (as we will discuss in Chapter 34) have evolved in relation to the dispersal of the plant's gametes to other members of the same species by animals, especially insects. These animals, in turn, have evolved a number of special traits that enable them to obtain food or other resources efficiently from the plants they visit, often from their flowers. In addition, the seeds of many flowering plants have features that make them more likely to be dispersed to new areas of favorable habitat.

Such interactions, which involve the long-term, mutual evolutionary adjustment of the characteristics of the members of biological communities in a reciprocal relation to one another, are examples of **coevolution.** Many of the interspecific interactions we have covered earlier (for example, in Chapter 23) fall under this heading. Here we will consider some additional examples, grouping them under the general headings of predator-prey interactions and symbiosis.

> **Coevolution is a term that describes the long-term evolutionary adjustment of one group of organisms to another and vice versa.**

PREDATOR-PREY INTERACTIONS

Predator-prey interactions are competitive interactions between organisms. Although most of us think of a typical predator-prey interaction as the hunting strategies of a lion for a gazelle, there are other such interactions in nature that are not as obvious. Plants, for example, can contain chemical substances that are poisonous to other organisms. Other defenses include structures such as spikes and thorns that effectively shield plants from attack. Some animals also employ chemical defenses to ward off prey and herald their poisonous nature by being dramatically colored. Still other animals mimic poisonous species by adopting their coloration. All of these strategies are examples of predator-prey interactions, which we will now consider in detail.

Plant Defenses

The most obvious ways in which plants limit the activities of herbivores are **morphological defenses.** Thorns, spines, and prickles, for example, play an important role in discouraging browsers, although some of them are able to overcome these obstacles (Figure 24-3). The defensive role of plant hairs, especially those that have a glandular, sticky tip (Figure 24-4), is less obvious. The enormous abundance of insects and

FIGURE 24-3
Some animals can overcome plant defenses. Gerenuk, an antelope that grazes successfully on some of the spiny shrubs and trees of the African savanna that other browsers do not use as food.

FIGURE 24-4
Glandular hairs of the wild potato *Solanum berthaultii*. The hairs are defenses against aphids, which are the most important herbivores on these plants. The hairs are of two types, type A and type B, both seen in this scanning electron micrograph. The shorter type A hairs have a four-lobed head that ruptures on contact, entrapping the aphid or other insect in a quick-setting sticky liquid. The type B hairs play an even more fascinating role in the plant's defenses. The sticky exudate that they secrete contains a chemical that is the main ingredient of the alarm pheromone in aphids. An extract prepared from these hairs greatly disturbed aphids, driving many of them away when it was tested experimentally by scientists at the Rothamsted Experimental Station in England. Together, the two types of hairs constitute a formidable barrier against the aphids. Breeding experiments have now succeeded in transferring these hairs to the cultivated tomato and potato; the new strains are much more resistant to pests than the nonhybrid ones.

the scale at which they operate as herbivores provide some indication of why hairs of various kinds are so important to the plants that possess them. Simple strengthening of the plant parts, too, by depositing silica in the leaves, is an element in the protective system of some plants, such as grasses (see Figure 36-10). If enough silica is present in their cells, the plants may simply be too tough to eat.

Significant as these morphological adaptations are, the chemical defenses that occur so widely in plants are even more crucial. As research in this area proceeds, we continue to be astonished by the number and diversity of these features. Most animals can manufacture only 8 of the 20 required amino acids, although the exact number varies from species to species and even during the life span of an individual in some species; they must obtain all of the others from plants. By restricting the amount of one or more of the amino acids that animals require from them, plants can limit their nutritional suitability for animals. As a result, they may be protected from herbivores.

Best known and perhaps most important in the defenses of plants against herbivores are the **secondary chemical compounds,** so called to distinguish them from the **primary compounds,** or compounds that are regularly formed as components of the major metabolic pathways, such as respiration. Virtually all plants, and apparently many algae as well, contain secondary compounds that play a defensive role, and these are structurally very diverse. Estimates suggest that perhaps 50,000 to 100,000 different secondary plant compounds may exist; some 15,000 of these have been characterized chemically.

Secondary compounds, chemicals that are not involved in primary metabolic processes, play the dominant role in protecting plants from being eaten by herbivores or predators.

We now know a great deal about the roles of secondary compounds in nature. As a rule, different kinds of compounds are characteristic of particular groups of plants. For example, the mustard family (Brassicaceae) is characterized by a group of chemicals known as mustard oils, the substances that give the pungent aromas and tastes to such plants as mustard, cabbage, watercress, radish, and horseradish. The same tastes that we enjoy signal the presence of chemicals that are toxic to many groups of insects.

Another group of plants, the potato and tomato family (Solanaceae), is rich in alkaloids and steroids, complex ring-forming molecules that exhibit an enormous range of structural diversity. Such compounds occur widely among the flowering plants, but the particular kinds found in various groups differ greatly. Sometimes they are involved in more complex defensive systems; for example, plants of the milkweed family (Asclepiadaceae) and the closely related dogbane family (Apocynaceae) tend to produce a milky sap that deters herbivores from eating them. In addition, these plants usually contain cardiac glycosides, molecules named for their drastic effect on heart function in vertebrates.

One of the best known plant groups with toxic effects consists of the related poison ivy, poison oak, and poison sumac (*Toxicodendron* species). All contain a gummy oil called urushiol, which causes a severe rash in susceptible people; urushiol can persist for years on clothes or other objects with which it has come into contact. At least 140,000 cases that cause absence from work occur in the United States annually, and many more are not reported. Approximately one out of two people is at least moderately sensitive to these plants. Immediate washing with soap will remove the excess urushiol and prevent its further spread, but the reaction of exposed body parts seems to be immediate. Although this toxic substance almost certainly functions to protect the plants in which it occurs, evidence for its other roles in nature is absent.

The seeds of castor bean (*Ricinus communis*) produce a protein, ricin, that attacks ribosomes and thus blocks protein synthesis. Ricin is an enzyme similar to a toxin produced by the bacterium that causes bacterial dysentery (*Shigella dysenteriae*). It removes an adenine residue from a specific position in the RNA chain within a ribosome; this area of the ribosome is the same in all animals, which accounts for the wide toxicity of castor bean seeds. Why such an unusual enzyme in a bacterium should

When plants are strongly protected from most herbivores by their secondary compounds or other means, they may be difficult to control biologically. Prickly pear cactus spread in Australia, as we saw in Chapter 23, partly because its spines made it resistant to grazers; ultimately a moth was introduced from Latin America that brought it under control.

In California and many other regions of the world, the Australian tree *Eucalyptus* is planted widely; it sometimes becomes naturalized, reproducing itself in native habitats by producing seedlings. *Eucalyptus* was first introduced into California in 1853 as a potential source of fuel wood and timber, but it has not proved particularly useful for these purposes and has become well established in many areas. The aromatic oils and other chemicals that *Eucalyptus* produces apparently make it immune to attacks from most herbivores and other biological control agents. It also appears to suppress the growth of native plants, judging from the bare ground that is characteristic of eucalyptus groves. It certainly lends beauty to many areas in California, and may yet become a commercial source of paper and pulp in the state.

In October, 1984, an Australian longhorn wood-boring beetle (*Phoracantha semipunctata*) was found infesting eucalyptus trees near El Toro, California. By the late 1980s, the beetle (Figure 24-A, *1* and *2*) had spread throughout the coastal regions of southern California, laying eggs on diseased or moisture-stressed trees and freshly cut logs. The trees of certain common species of eucalyptus are highly susceptible to massive beetle attacks (Figure 24-A, *3*); therefore, the introduction of the beetle threatens to alter the California landscape drastically unless some of the predators and diseases that control it in its native habitat in Australia are introduced soon (Figure 24-A, *4*).

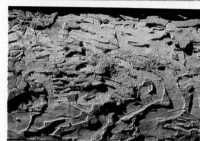

Another plant with strong chemical defenses is braken fern (*Pteridium aquilinum*), a species found throughout the world in relatively cool habitats (Figure 24-A, *5*). Braken spreads aggressively and forms large patches particularly in areas where people have disturbed the native woods. It causes many diseases in animals that graze on it and, despite the fact that it is apparently highly carcinogenic, it is eaten by humans in certain areas as a salad delicacy. Although it is a native plant, braken is a serious pest in the British Isles; discussion about whether to try to control it by introducing two species of South African moths whose larvae devour it voraciously are under way. There is some reluctance to take this step, however, because braken is a native species, of value in certain areas, and it probably would be impossible to control the moths once they were introduced.

FIGURE 24-A

The long-horned beetle *Phoracantha semipunctata*, recently introduced fron Australia, is causing widespread damage on eucalyptus trees in southern California.

1 An adult beetle visiting eucalyptus flowers.

2 Larva of the beetle.

3 Galleries such as these penetrate the living tissue between the bark and the wood of the tree, eventually girdling it and leading to death.

4 A eucalyptus tree, damaged by *Phoracantha*, growing next to a healthy one.

5 Bracken fern (*Pteridium aquilinum*), which is highly protected from most herbivores by the secondary compounds it produces, forms large masses, spreading underground, in disturbed woods and pastures.

resemble so closely one found in the seeds of a plant is unclear; perhaps the bacterium acquired the gene from the plant in the course of its evolution.

Many plant parts are toxic to humans. One familiar example is the houseplant called dumb cane, *Dieffenbachia*, a member of the same plant family as philodendrons. Touching a leaf of this plant with the tongue will result in intense pain and eventually produce ulcers and corrosive burns. These effects occur because the leaves of dumb cane have special ejector cells that expel needlelike crystals of calcium oxalate that penetrate the skin and cause the release of histamines. Particularly in families with children, people should take care to know the properties of the plants they grow.

These few examples demonstrate that many angiosperms are protected by a rich and varied chemical arsenal. Mustard oils, alkaloids, steroids, and other classes of secondary compounds either are toxic to most herbivores or disturb their metabolism so greatly that they are unable to complete their development normally. As a consequence, most herbivores tend to avoid the plants that possess these compounds. The pattern of occurrence of such chemicals, therefore, has had major consequences on the evolution both of the plants themselves and of the herbivores, especially the insects, that feed on them.

Producing Defenses When They Are Needed

Some of the secondary compounds that plants use to defend themselves from herbivores are not normally present in their tissues. Rather, the plants produce them only when they are needed. For example, when a leaf of tomato, potato, or alfalfa, is injured, a chemical message derived from fragments of ruptured cell walls travels rapidly through the plant. That induces the synthesis and accumulation of proteins that inhibit digestion in an animal's gut. These inhibitory enzymes are being identified in a number of laboratories at present. Similarly, the infection of a plant by a fungus or bacterium may cause the plant to synthesize a molecule that retards the spread of the infection. A familiar example of such a process is provided by the chemicals that color apples brown when they are cut; interacting with proteins, these chemicals make the apple less attractive to caterpillars, fungi, and other organisms. Such defensive strategies are metabolically efficient: the plants are able to avoid the necessity of producing the chemical until it is actually needed. At other times, they are able to use the energy available to them for their growth and maintenance. It may not surprise you to learn that parasites, such as fungi, that regularly infect particular kinds of plants are often tolerant of their particular defensive substances.

The Evolution of Herbivores

Associated with each family or other group of plants protected by a particular kind of secondary compound are certain groups of herbivores that are able to feed on these plants, often as their exclusive food source. How do these animals manage to avoid the chemical defenses of the plants, and what are the evolutionary antecedents and ecological consequences of such patterns of specialization?

As a starting point, we can offer the observation that some herbivore groups feed on plants of many different families, others on plants of just a few families. In general, those herbivores that feed on a restricted array of plant families, perhaps only one, feed on plant groups in which certain kinds of secondary compounds are well represented. For example, the larvae of cabbage butterflies (subfamily Pierinae) feed almost exclusively on plants of the mustard and caper families, as well as on a few other small families of plants that are characterized by the presence of mustard oils (Figure 24-5, *A* and *B*). Similarly, the caterpillars of the monarch butterflies and their relatives (subfamily Danainae) feed on plants of the milkweed and dogbane families (Figure 24-5, *C* and *D*). No other groups of butterflies have larvae that feed on plants of the families that are particularly well-protected by their secondary plant compounds, indicating the efficiency of the chemicals in retarding feeding by most kinds of butterflies. Other kinds of insects that feed on them are usually specialized within their groups, as are the cabbage butterflies and monarch butterflies.

A **B**

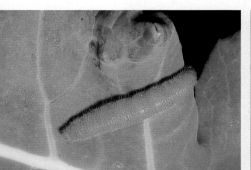

D

FIGURE 24-5

Insect herbivores are well-suited to their hosts.

A The green caterpillars of the cabbage butterfly, *Pieris rapae,* are camouflaged on the leaves of cabbage and other plants on which they feed. Although these plants are protected against most potential herbivores by mustard oils, the cabbage butterfly caterpillars break down the compounds; thus they are not protected by them.
B An adult cabbage butterfly. All stages of the life cycle of the monarch butterfly (*Danaus plexippus*) are protected from birds and other predators by the cardiac glycosides that they incorporate from the milkweeds and dogbanes on which they feed as larvae. Both the caterpillars (**C**) and the adult butterflies (**D**) "advertise" their poisonous nature with warning coloration.

We can explain the evolution of these particular patterns as follows. Once the ancestors of the caper and mustard families acquired the ability to manufacture mustard oils, they were protected for a time against most or all herbivores that were feeding on other plants in their area. The plants that contained the mustard oils apparently succeeded very well as a result of this particular defense, and they and their descendants thus evolved and migrated, eventually giving rise to the thousands of species of mustards and capers that now grow worldwide.

At some point, certain groups of insects—for example, the cabbage butterflies—developed the ability to break down the mustard oils and thus feed on these mustards and capers without harming themselves. Having acquired this ability, the butterflies themselves had registered an important evolutionary "breakthrough." They were able to use a new resource without having to face competition for it from other herbivores. Often, in groups of insects such as the cabbage butterflies, sense organs have evolved that are particularly sensitive to the secondary compounds that their food plants produce. Clearly, the relationship that has formed between the cabbage butterflies and the plants of the mustard and caper families is an example of coevolution.

> **The members of many groups of plants are protected from most herbivores by their secondary compounds. Once the members of a particular herbivore group acquire the ability to feed on them, however, these herbivores have gained access to a new resource, which they can exploit without competition from other herbivores.**

Although the role of secondary compounds in protecting plants from herbivores was first discovered in flowering plants, other groups of photosynthetic organisms have similar defenses. Many red, brown, and green algae produce terpenoids and other compounds that deter feeding by herbivores such as fishes, and also inhibit the growth of bacteria in culture. Some evidence suggests that the group of shrimplike crustaceans known as amphipods, which are often abundant on algae, may be as important in feeding on algae in marine environments as insects are in feeding on plants on land. Therefore, similar systems seem to operate both in the sea and on land that deter the dominant herbivores from consuming the primary producers of each community, and that deter carnivores from consuming herbivores.

A   B

FIGURE 24-6

A bluejay learns that monarch butterflies taste bad.

A A cage-reared jay that had never seen a monarch butterfly before, eating one.

B The same jay a few minutes later regurgitating the butterfly. Such a bird is not likely to attempt to eat an orange-and-black insect again.

Chemical Defenses in Animals

Some groups of animals that feed on plants rich in secondary compounds receive an extra benefit, one of great ecological importance. When the caterpillars of monarch butterflies feed on plants of the milkweed family, for example, they do not break down the cardiac glycosides that protect these plants from most herbivores. Instead, they store them in fat bodies. As a result, these butterflies are themselves protected from predators! The cardiac glycosides in the leaves that the caterpillars eat are concentrated, stored, and passed through the chrysalis stage to the adult and even to the eggs; all stages are protected from predators. A bird that eats a monarch butterfly quickly regurgitates it and thenceforth avoids the kind of conspicuous orange-and-black pattern that characterizes the adult monarch (Figure 24-6). Locally, however, some birds have acquired the ability to tolerate the protective chemicals and eat the monarchs.

Insects that do feed regularly on plants of the milkweed family are, in general, brightly colored (Figure 24-7). Among them are brightly colored cerambycid beetles, whose larvae feed on the roots of the milkweed plants; bright blue or green chrysomelid beetles; and bright red true bugs of the order Hemiptera. In some parts of the world, there are also bright red grasshoppers and other very obvious insects. These herbivores clearly are "advertising" their poisonous nature by their bright colors, using an ecological strategy known as **warning coloration.** Similar relationships occur in marine communities; a recent investigation revealed that of the exposed common coral reef invertebrates at Lizard Island, on the Great Barrier Reef off the northeast coast of Australia, three fourths were toxic to fish; of the protectively camouflaged ones, only one fourth were toxic.

Insects that eat plants that lack specific chemical defenses are seldom brightly colored. In fact, many of these insects are **cryptically** colored—colored so as to blend in with their surroundings and thus to be hidden from predators (Figure 24-8). This is also true of insects such as the larvae of cabbage butterflies, which, although they feed on plants with well-marked chemical defenses, are able to do so because of their ability to break down the molecules involved, rather than store them.

Some marine animals—such as certain nudibranchs (sea slugs; see Chapter 40)—acquire defensive chemicals or defensive cells from their prey. Hydroids often provide such stinging cells to animals that graze on them, for example. Off the coast of Panama, the large nudibranch *Aplysia* grazes selectively on red algae of the genus *Laurencia*, which is protected by elatol, a powerful inhibitor of cell division. Perhaps this is the reason that few fish feed on *Aplysia*. An intensive investigation of marine animals, algae, and flowering plants for new drugs against cancer and other diseases, or as sources of antibiotics, is now being carried out. It holds great promise, because of the enormous diversity of chemical compounds that occur in these organisms.

Animals also manufacture many chemicals that they use in their defense. In fact, animals manufacture and use a startling array of substances to perform a wide variety of defensive functions. Venomous snakes, lizards, and fishes are well known; in addition, bees, wasps, predatory bugs, scorpions, spiders, and many other arthropods have chemicals that they use to defend themselves and to kill their prey.

FIGURE 24-7

The bright colors of poisonous herbivores. The herbivores feed on milkweeds that obtain toxic cardiac glycosides, and advertise their poisonous nature by displaying warning coloration to their potential predators. Examples of such herbivores are the longhorn beetle *Tetraopes* (**A**) and the milkweed bug, *Oncopeltus fasciatus* (**B**). The larvae of *Tetraopes* feed on the roots of the milkweed plants, and the adults mate on the flower clusters. All stages of the life cycle of *Oncopeltus* take place on the leaves and stems.

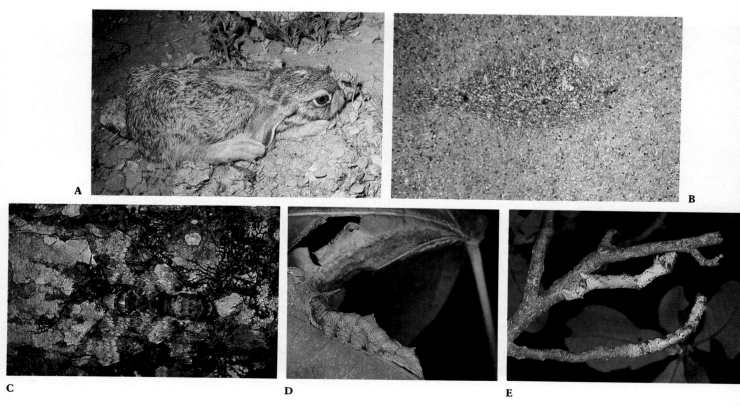

FIGURE 24-8
A few striking examples of cryptic coloration.
A A young jackrabbit in the desert near Tucson displays both adaptive stillness and camouflage.
B Protective coloration in a flatfish on the sea floor.
C A tropical hawkmoth at rest, blending in perfectly with the moss- and lichen-covered tree bark.
D The brown-green-brown coloring of the unicorn caterpillar gives it the appearance of a dried leaf edge.
E An inchworm caterpillar (*Necophora quernaria*), which closely resembles a twig.

Various chemical defenses are found among the vertebrates, too. The dart-poison frogs of the family Dendrobatidae, for example, produce toxic alkaloids in the mucus that covers their skin (Figure 24-9). Some of these toxins are so powerful that a few micrograms will kill a person if injected into the bloodstream. Most species of these frogs, which number more than 100 species, are brightly colored; they are favorites in zoos and aquaria for that reason.

Aposematic Coloration

Warning coloration, as discussed previously, is also known as **aposematic coloration**. Such coloration is characteristic of animals that have effective defense systems, including not only poisons, but also stings, bites, and other means of repelling predators. An organism possessing such a system will benefit by advertising the fact clearly—for example, by showy colors that are not normally found in that particular habitat. Otherwise, the distasteful or poisonous individual runs the risk of being killed while protecting itself. An individual protected in this way has a selective advantage over others that are not protected; and other individuals of the same species will also benefit from similar coloration.

> **Aposematic, or warning, coloration serves to keep potential predators away from poisonous or otherwise dangerous prey.**

Of course, the animals that display aposematic coloration must occur at relatively high densities, too, if the system is to be effective. If genetically related individuals are similarly colored and live in the same vicinity, the selective advantage is obvious. Predators will tend to avoid such individuals.

FIGURE 24-9
Vertebrate chemical defenses.
Frogs of the family Dendrobatidae, like this individual of *Dendrobates reticulatus*, are abundant in the forests of Latin America and are extremely poisonous to vertebrates. More than 200 different alkaloids have been isolated and identified from these frogs, and some of them are playing important roles in neuromuscular research. Two groups of Indians in western Colombia obtain a potent poison for blowgun darts from extremely toxic species of these frogs that occur in their region.

FIGURE 24-10

Warning coloration. Warning coloration is displayed by the spotted skunk, *Spilogale putorius* (**A**); the poisonous gila monster, *Heleoderma suspectum*, a member of the only genus of poisonous lizards in the world (**B**); and the red-and-black African grasshopper *Phymateus morbillosus* (**C**), which feeds on succulent spurges, such as *Euphorbia*; plants that are highly poisonous. Grasshoppers of this genus make no effort to conceal themselves from predators.

Some examples of animals with aposematic coloration are shown in Figure 24-10. Such animals tend to live together in family groups, unlike those that are cryptically colored. If camouflaged animals lived together in groups, one might be discovered by a potential predator, offering a valuable clue to the presence of others.

Mimicry

During the course of their evolution, many unprotected species have come to resemble distasteful ones that exhibit aposematic coloration. Two types of mimicry have been identified: Batesian mimicry and Muellerian mimicry.

Batesian Mimicry. Batesian mimicry is named for W.H. Bates, a British naturalist who first brought it to general attention in 1857. Bates discovered that if the unprotected (nonpoisonous) animals are present in numbers that are low relative to those of the species that they resemble, they will be avoided by predators. If the unprotected animals are too common, of course, many of them will be eaten by predators that have not yet learned to avoid individuals with particular characteristics.

> **In Batesian mimicry, unprotected species resemble others that are distasteful. Both species exhibit aposematic coloration. The unprotected mimics will be avoided by predators if they are relatively scarce.**

Many of the best-known examples of Batesian mimicry occur among butterflies and moths. Obviously, predators in systems of this kind must use visual cues to hunt for their prey, otherwise similar patterns of coloration would not matter to potential predators. Evidence is also increasing that Batesian mimicry can also involve nonvisual cues, such as olfaction, although such examples are less obvious to humans.

The groups of butterflies that provide the **models** in Batesian mimicry are, not surprisingly, members of groups whose larvae feed on only one or a few closely related plant families; the plants on which they feed are strongly protected chemically. The model butterflies take poisonous molecules from these plants and retain them in their own bodies. The mimic butterflies, in contrast, belong to groups in which the feeding habits of the larvae are not so restricted. As caterpillars, these butterflies feed on a number of different plant families, but not those protected by toxic chemicals.

One often-studied mimic among North American butterflies is the viceroy, *Limentitis archippus* (Figure 24-11). This butterfly, which resembles the poisonous monarch, ranges from central Canada south through much of the United States east of the Sierra Nevada and Cascade Range into Mexico. The larvae feed on willows and cottonwoods, and neither the larvae nor the adults were thought to be distasteful to birds, although recent findings dispute this. Interestingly, the larvae of the viceroy are camouflaged on leaves because they resemble bird droppings, whereas the distasteful larvae of the monarch are very conspicuous.

Muellerian Mimicry. Another kind of mimicry, **Muellerian mimicry,** was named for the German biologist Fritz Mueller, who first described it in 1878. In Muellerian mimicry, several unrelated but protected animal species come to resemble one another. Thus different kinds of stinging wasps have yellow-and-black striped abdomens, but they may not all be descended from a common ancestor that had similar

A

B

C

D

FIGURE 24-11
Monarch mimics.
A and **B** The viceroy butterfly, *Limenitis archippus,* is a North American mimic of the poisonous monarch (see Figure 24-7). Although the viceroy is not related to the monarch, it looks a lot like it, and so predators that have learned how distasteful monarchs are avoid viceroys, too. New evidence suggests that the viceroy may be a Muellerian rather than a Batesian mimic of the monarch. **C** and **D** The red-spotted purple, *Limenitis arthemis astyanax,* is another member of the same genus as the viceroy, and thus much more closely related to it than is the monarch. However, it does not look at all like the viceroy. Instead, the red-spotted purple is a Batesian mimic of another poisonous butterfly, the pipevine swallowtail, *Battus philenor.*

coloration. In general, yellow-and-black and bright red tend to be common color patterns that presumably warn predators relying on vision that animals with such coloration are to be avoided. It is more difficult to prove that a resemblance between two protected animals actually represents Muellerian mimicry than it is to demonstrate Batesian mimicry experimentally.

In both Batesian and Muellerian mimicry, mimic and model must not only look alike but also act alike if predators are to be deceived. For example, the members of several families of insects that resemble wasps (Figure 24-12) behave surprisingly like the wasps they mimic, flying often and actively from place to place. Mimics must also spend most of their time in the same habitats as do their models; such collective living provides a greater learning signal for potential predators. If they did not, predators would discover that all of those conspicuous animals are not only easily seen but also quite tasty! If animals that resemble one another are all poisonous or dangerous, they still gain an advantage by resembling one another, thus achieving collective protection.

In Muellerian mimicry, two or more unrelated but protected species resemble one another, thus achieving a kind of group defense.

A

B

C

D

FIGURE 24-12
Yellow jacket mimics. The familiar yellow-and-black stripes of the yellow jackets and other wasps, such as *Vespula arenaria* (**A**), form the basis for large Batesian and Muellerian mimicry complexes. Here we see Batesian mimics representing three separate orders of insects, all rarer than yellow jackets and with patterns of behavior similar to those of the dangerous wasps that they resemble: (**B**) a flower fly, *Chrysotoxum;* (**C**) a longhorn beetle; and (**D**) a sesiid moth, *Aegeria rutilans,* whose caterpillars are a pest on strawberries. None of these mimics stings or is poisonous, yet all are conspicuous members of the communities where they occur, flying about actively and remaining in full view at all times. They would be easy prey if they were not protected by their resemblance to yellow jacket wasps. Many stinging insects also resemble yellow jackets in their color patterns and thus form Muellerian complexes.

SYMBIOSIS

Symbiotic relationships are those in which two kinds of organisms live together consistently. All symbiotic relationships provide the potential for coevolution between the organisms involved, and in many instances the results of this coevolution are fascinating. The major kinds of symbiotic relationships include (1) **commensalism,** in which one species benefits whereas the other neither benefits nor is harmed; (2) **mutualism,** in which both participating species benefit; and (3) **parasitism,** in which one species benefits but the other is harmed. Parasitism, as mentioned in Chapter 23, can also be viewed as a form of predation, in which the organism that is parasitized does not necessarily die.

Examples of symbiosis include lichens, which are associations of certain fungi with green algae or cyanobacteria, as we shall see in more detail in Chapter 32. Another important example is the association between fungi and the roots of most kinds of plants, called mycorrhizae, in which the fungi expedite the absorption of certain nutrients by the plants, which in turn provide them with carbohydrates. Similarly, root nodules occur in legumes and certain other kinds of plants, in which bacteria occur that fix atmospheric nitrogen and make it available for their host plants and thus for the ecosystems in which they occur. Turning to marine systems, a coral reef is a highly complex symbiotic system, involving not only the coral animals but also coralline algae and other autotrophic organisms that are intermingled with the coral animals and contribute greatly to the overall productivity of the reef. More broadly, a coral reef provides the basis for the existence of an entire associated biological community, one in which symbiotic relationships are especially prominent.

Some of the most spectacular examples of symbioses are those between flowering plants and their animal visitors, including insects, birds, and bats. As we will see in Chapter 34, the characteristics of flowers have been formed in large part, during the course of their evolution, in relation to the characteristics of the mouthparts and the habits of the animals that visit them for food and, in doing so, may spread their pollen from individual to individual. At the same time, the characteristics of the animals have changed in relation to increasing specialization for obtaining food or other substances from flowers. Such symbiotic relationships, which also provide excellent examples of coevolution, can be observed all around us.

> **Symbiotic relationships are those in which two or more kinds of organisms live together in often elaborate, more or less permanent relationships.**

Commensalism

In nature, the individuals of one species are often physically attached to those of another. For example, birds nest in trees, and **epiphytes** are plants that grow on the branches of other plants. In general, the host plant is unharmed, whereas the organism that grows or nests on it benefits. Similarly, various marine animals, such as barnacles, grow on other, often actively moving sea animals and thus are carried passively from place to place (Figure 24-13). These "passengers" presumably gain more protection from predation than they would if they were fixed in one place, and they also reach new sources of food. The increased water circulation that such animals receive as their host moves around may be of great importance, particularly if the passengers are filter feeders. The gametes of the passenger are also more widely dispersed than would be the case otherwise.

The best-known examples of commensalism involve the relationships between certain small tropical fishes and sea anemones, which are marine animals that have stinging tentacles (see Chapter 39). These fishes have evolved the ability to live among the tentacles of the sea anemones, even though these tentacles would quickly paralyze other fishes that touched them (Figure 24-14). The anemone fishes feed on the detritus left from the meals of the host anemone, remaining uninjured under remarkable circumstances.

On land, an analogous relationship exists between certain birds and grazing animals, such as cattle or the rhinoceros, which may benefit by having their parasites

FIGURE 24-13
Commensalism. The barnacles that are evident on the back of this blowing gray whale, surfacing in San Ignacio Lagoon, Baja California, are carried from place to place with the whale, straining their food from water as they move about. Unlike most barnacles, which are anchored to fixed substrates, these animals have continuous access to fresh sources of the plankton on which they feed.

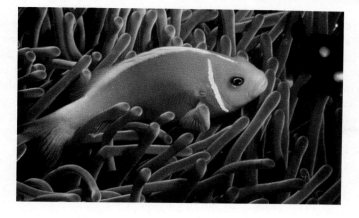

FIGURE 24-14
Symbiosis in the sea. Clown-fishes, such as this individual of *Amphiprion perideraion* in Guam, often form symbiotic associations with sea anemones, gaining protection by remaining among their tentacles and gleaning scraps from their food. Different species of anemones secrete different chemical mediators; these attract particular species of fishes and may be toxic to the fish species that occur symbiotically with other species of anemones in the same habitat. There are 26 species of clown-fishes, all of them found only in association with sea anemones; ten species of anemones are involved in such associations, so that some of the species of anemone are host to more than one species of clown-fish.

removed by birds. The birds spend most of their time clinging to the animals, picking off insects and other small bits of food, and carry out their entire life cycles in close association with the host animal. Cattle egrets, which have extended their range greatly during the past few decades, provide an example of a loosely coupled relationship of this kind.

In each of these instances, it is difficult to be certain whether the second partner receives a benefit or not, and there is no clear-cut boundary between commensalism and mutualism. For instance, it may be advantageous to the sea anemone to have particles of food removed from its tentacles; it may then be better able to catch other prey. The association of the grazing mammals and the birds, on the other hand, is quite clearly an example of mutualism. The mammal benefits by having parasites and other insects removed from its body, and the birds benefit by having a dependable source of food.

Mutualism

Examples of mutualism are of fundamental importance in determining the structure of biological communities. In the tropics, for example, leafcutter ants (Figure 24-15) are often so abundant that they can remove a quarter or more of the total leaf surface of the plants in a given area. They do not eat these leaves directly; rather, they take them to their underground nests, where they chew them up and inoculate them with the spores of particular fungi. These fungi are cultivated by the ants and brought from one specially prepared bed to another, where they grow and reproduce. In turn, the fungi constitute the primary food of the ants and their larvae. The relationship between the leafcutter ants and these fungi, fueled by material cut from the leaves of plants, is an excellent example of mutualism.

Another relationship of this kind involves ants and aphids. Aphids, also called greenflies, are small insects that suck fluids from the phloem of living plants with their piercing mouthparts. They extract a certain amount of the sucrose and other nutrients from this fluid, but much runs out in an altered form through their anus. Cer-

FIGURE 24-15
Mutualism. Leafcutter ants are enormously abundant in the tropics, and they often consume a high proportion of the available leaves at a particular spot to build their "fungus gardens."

FIGURE 24-16
Domesticating aphids. These ants *(Crematogaster)* are tending willow aphids, obtaining the "honeydew" that the aphids excrete continuously, moving them from place to place and protecting them from potential predators.

tain ants have taken advantage of this habit—in effect domesticating the aphids—by carrying the aphids to new plants where they come into contact with new sources of food and using the "honeydew" that the aphids excrete as food (Figure 24-16).

A particularly striking example of mutualism involving ants concerns certain Latin American species of the plant genus *Acacia*. In these species, the leaf parts called stipules are modified as paired, hollow thorns; consequently, these particular species are called "bull's horn acacias." The thorns are inhabited by stinging ants of the genus *Pseudomyrmex*, which do not nest anywhere else. Like all thorns that occur on plants, they serve to deter herbivores.

At the tip of the leaflets of these acacias are unique, protein-rich bodies called Beltian bodies—after Thomas Belt, a nineteenth-century British naturalist who first wrote about them based on his experiences in Nicaragua. Beltian bodies do not occur in species of *Acacia* that are not inhabited by ants and their role is clear: they serve as a primary food for the ants. In addition, the plants secrete nectar from glands near the bases of their leaves. The ants consume this nectar also, and feed it and the Beltian bodies to their larvae as well (Figure 24-17).

Apparently this association is beneficial to the ants, and one can readily see why they inhabit acacias of this group. The ants and their larvae are protected within the swollen thorns and the trees provide a ready source of a balanced diet, including the sugar-rich nectar and the protein-rich Beltian bodies. What, if anything, do the ants do for the plants? This had been the question that had fascinated observers for nearly a century until it was answered by Daniel Janzen, then a graduate student at the University of California, Berkeley, in a beautifully conceived and executed series of field experiments.

Whenever any herbivore lands on the branches or leaves of an acacia that is inhabited by such ants, the ants immediately attack and devour it. Thus the ants protect the acacias from being eaten, and the herbivore also provides additional food for the ants, which continually patrol the branches. Related species of acacias that do not have the special features of the bull's horn acacias and are not protected by ants have

A

B

C

D

FIGURE 24-17
Mutualism between ants and bull's horn acacias.
A Ants of the genus *Pseudomyrmex* live within the inflated, spinelike stipules of the bull's horn acacias, a unique group of this large genus of plants that occurs from Mexico to northern South America.
B Nectaries at the base of the leaves provide a sugar-rich fluid that the ants harvest.
C Beltian bodies are unique, protein-rich structures borne at the ends of the leaflets; the ants harvest them for food.
D As shown here, the ants cut away much of the surrounding vegetation that crowds the acacias. They also attack most kinds of herbivores that threaten the acacias.

A   B

FIGURE 24-18

A parasitic flowering plant and a close, nonparasitic relative. Both are members of the plant family Convolvulaceae, the morning glory family.
A Morning glory *(Convolvulus).*
B Dodder *(Cuscuta)* is a parasite. It has lost its chlorophyll and its leaves in the course of its evolution, and, like animals in general, is heterotrophic—unable to manufacture its own food; instead, it obtains its food from the host plants on which it grows.

bitter-tasting substances in their leaves that the bull's horn acacias lack. Evidently, these bitter-tasting substances protect the acacias in which they occur as do the ants that inhabit other species of the same genus of plants.

The ants that live in the bull's horn acacias also help their hosts to compete with other plants. The ants cut away any branches of other plants that touch the bull's horn acacia in which they are living—creating, in effect, a tunnel of light through which the acacia can grow, even in the lush deciduous forests of lowland Central America. Without the ants, as Janzen showed experimentally by poisoning the ant colonies that inhabited individual plants, the acacia is unable to compete successfully in this habitat. Finally, the ants bring organic material into their nests, and the part that they do not consume, together with their excretions, provides the acacias with an abundant source of nitrogen, which is an essential nutrient.

Parasitism

Parasitism may be regarded as a special form of predation in which the predator is much smaller than the prey and remains closely associated with it. Parasitism is harmful to the prey organism and beneficial to the parasite. The concept of parasitism seems obvious, but individual instances are often surprisingly difficult to distinguish from predation (discussed in Chapter 23) and from various other kinds of symbiosis.

Many instances of parasitism are well known; for example, vertebrates are parasitized by members of many different phyla of animals and protists. Invertebrates also have many kinds of parasites that live within their bodies. However, bacteria and viruses are often not considered parasites, even though they fit our definition precisely. Lice, which live on the bodies of vertebrates—mainly birds and mammals—are normally considered parasites, but mosquitoes are not, even though they draw food from the same birds and mammals in a similar manner, because their interaction with their host lasts for only a short time. However, mosquitoes are closely associated ecologically with the animals from which they draw blood. Mosquitoes also synchronize their diurnal and seasonal activities closely with those of their hosts, so that the interrelationship is very close.

Many fungi and some flowering plants, too, are parasitic on other plants (Figure 24-18), and a few are serious pests of crops. In Chapters 39 and 41, a number of examples of parasites will be considered in relation to the groups of organisms that they represent.

Internal parasitism is generally marked by much more extreme specialization than external parasitism, as shown by the many protist and invertebrate parasites that infect humans. The more closely the life of the parasite is linked with that of its host, the more its morphology and behavior are likely to have been modified during the course of its evolution (Figure 24-19). The same, of course, is true of symbiotic relationships of all sorts. Conditions within the body of another organism are different from those encountered outside and are apt to be much more constant in every way. Consequently, the structure of the parasite is often simplified, and unnecessary armaments and structures are lost as it evolves.

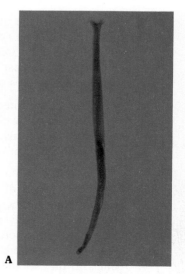

A

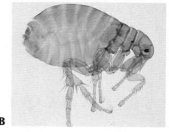

B

FIGURE 24-19
Parasitic animals.
A A male hookworm, *Ancylostoma duodenale.* These parasites of humans live in the intestine; they range in size from 8 to 10 mm for males and 10 to 13 mm for females.
B The human flea, *Pulex irritans.* Fleas are flattened from side to side and slip easily through hair; their ancestors were larger, more brightly colored, and had wings. The structural and behavioral modifications of fleas have come about in relation to a parasitic way of life.

1. Communities are repeatable sets of organisms in nature; they exist because the ecological requirements and environmental tolerances of the organisms that comprise them overlap in the areas where they occur together.

2. Even when we examine a community such as the redwood community, which is dominated by one very obvious kind of organism that also affects its habitat directly, we realize that the different organisms function individually and that they display genetic variability from place to place. Communities are continually evolving, and change dynamically through time.

3. Coevolution is the process by which different kinds of organisms adjust to one another by genetic change over long periods of time. It is a stepwise process that ultimately involves adjustment of both groups of organisms.

4. Plants are often protected from herbivores, fungi, and other agents by chemicals that they manufacture. Such chemicals, which are not part of the primary metabolism of the plant, are called secondary compounds.

5. Particular classes of secondary compounds are usually characteristic of individual plant families or groups of closely related families. The herbivores that can feed on such plants either break the secondary compounds down or store them in their bodies. If they do the latter, the chemicals may in turn protect the herbivores from their predators.

6. Warning, or aposematic, coloration makes the organisms that possess it obvious and is characteristic of organisms that are poisonous, sting, or are otherwise harmful. In contrast, cryptic coloration, or camouflage, is characteristic of organisms that are not specially protected.

7. Batesian mimicry is a situation in which a palatable or nontoxic organism resembles another kind of organism that is distasteful or toxic. Muellerian mimicry occurs when several toxic or dangerous kinds of organisms resemble one another.

8. In Batesian mimicry, the models are generally herbivores that feed on plants that contain toxic secondary compounds. The models retain such chemicals in their bodies. The mimics, in contrast, usually do not feed on such plants, but instead feed on others that are not protected in this way.

9. Symbiotic relationships are those in which two kinds of organisms live together. There are three principal sorts of symbiotic relationships: in commensalism, one benefits and the other is unaffected; in mutualism, each organism benefits; and in parasitism, one benefits and the other is harmed.

REVIEW QUESTIONS

1. What is coevolution? What are examples of the advantages of coevolution?

2. How do plants structurally defend themselves from being eaten by herbivores? How does the nutritional suitability of a plant affect its being eaten by animals?

3. What is the difference between primary and secondary plant chemical compounds? How do secondary chemical compounds protect plants from being overgrazed?

4. How are a plant's secondary chemical compounds further used by some animals?

5. How does cryptic coloration differ from aposematic coloration? How does the mechanism by which an animal circumvents plant secondary chemicals indicate whether it may exhibit aposematic or cryptic coloration?

6. What is Batesian mimicry? What purpose does it serve for its possessor?

7. What numerical relationship must exist between aposematically colored individuals and Batesian mimics? Why?

8. Of aposematic coloration, cryptic coloration, and Batesian mimicry, which term would be associated with an adult viceroy butterfly? Which term would be associated with a larval monarch butterfly? Which term would be associated with a larval viceroy butterfly?

9. What defines a symbiotic relationship? What are the three general classes of symbiosis and how do they compare to one another?

10. Classify the following symbiotic relationships as to whether they are commensalism, mutualism, or parasitism: ants on the acacia tree, dodder and host plant, sea anemones and anemone fish, ants and aphids, lice and birds, lichens, legume root nodules, fleas and humans.

 THOUGHT QUESTIONS

1. Why is it more difficult to demonstrate Muellerian mimicry than Batesian mimicry experimentally? How would you go about demonstrating each?

2. In Batesian mimicry, why are the mimics rare?

3. Some kinds of insects are more resistant to insecticides than others. Can you think of a reason why, based on your reading in this chapter?

 FOR FURTHER READING

ABRAHAMSON, W.G.: *Plant-Animal Interactions*, McGraw-Hill Book Company, New York, 1988. An excellent series of contemporary accounts of these complex systems.

AHMADJIAN, V., and S. PARACER: *Symbiosis: An Introduction to Biological Associations*, University Press of New England, 1986. A broad account of biological symbiosis.

BARRETT, S.C.H.: "Mimicry in Plants," *Scientific American*, vol. 257, pages 76-83, 1987. Mimicry in plants attracts pollinators or deters herbivores.

DAVIES, N. and M. BROOKE,: "Coevolution of the Cuckoo and its Hosts," *Scientific American*, January 1991, pages 92-98. The classic story of an evolutionary "arms race" between a parasite and its host.

DUFFY, J.E. and M.E. HAY: "Seaweed Adaptations to Herbivory," *BioScience* 40:368-385, May 1990. Chemical, structural, and morphological defenses are abundant in seaweeds.

GRIFFITHS, G., A. LEITH, and M. GREEN: "Proteins That Play Jekyll and Hyde," *New Scientist*, vol. 115, pages 59-61, 1987. The potent toxins that plants produce can kill or cure; they are under intensive investigation.

HANDEL, S. and A. BEATTIE,: "Seed Dispersal by Ants," *Scientific American*, August, 1990, pages 76-83. Many plant species induce ants to spread their seeds with special food lures and other adaptations.

HUNTER, A.E. and L.W. AARSSEN: "Plants Helping Plants," *BioScience* 38:34-40, January 1988. Outlines some interesting aspects of the effects of plants on one another in communities.

JACKSON, J.B.C., and T.P. HUGHES: "Adaptive Strategies of Coral-Reef Invertebrates," *American Scientist*, vol. 73, pages 265-274, 1985. Coral-reef environments, which are regularly disturbed by storms and predation, provide an excellent example of a complex, balanced marine community.

NAP, J. and T. BISSELING,: "Developmental Biology of a Plant-Prokaryote Symbiosis-the Legume Root Nodule," *Science*, November 1990, vol. 250, pages 948-954. The story of one of earth's most critical symbiotic relationships, nitrogen fixation in plants.

PIETSCH, T.W. and D.B. GROBECKER: "Frogfishes," *Scientific American*, June 1990, pages 96-103. Masters of aggressive mimicry, these voracious carnivores can gulp prey faster than any other vertebrate predator.

RENNIE, J.: "Living Together," *Scientific American*, January 1992, pages 120-133. An excellent overview of how co-evolution and symbiosis have shaped life today.

ROSENTHAL, G.A.: "The Chemical Defenses of Higher Plants," *Scientific American*, vol. 254, pages 94-99, 1986. Interesting account of the array of chemical defenses produced by plants.

WALKER, T.: "Butterflies and Bad Taste," *Science News*, June 1991, pages 348-349. It turns out that viceroy butterflies, the classic example of tasty creatures mimicking an unpalatable one, don't taste good after all.

Dynamics of Ecosystems

OVERVIEW

Ecosystems are complex associations of plants, animals, fungi, and microorganisms that interact with their nonliving environment in such a way as to regulate the flow of energy through them and the cycling of nutrients within them. Carbon dioxide, nitrogen gas, oxygen gas, and water are the reservoirs of the carbon, nitrogen, oxygen, and hydrogen that are used in biological processes. All of the other elements that organisms incorporate into their bodies come from the earth's rocks. Through photosynthesis, plants growing under favorable circumstances capture and lock up about 1% of the sun's energy that falls on their green parts. They may then be eaten by herbivores (primary consumers), which in turn may be eaten by secondary consumers (carnivores or decomposers), and so on. This type of sequence constitutes a food chain, and the different links in it, called trophic levels, include organisms that can transfer about 10% of the energy that exists at each level to the next level. Communities become more complex and stable through succession, but human beings are threatening this stability everywhere because of activities of their very large and rapidly growing population.

FOR REVIEW

Here are some important terms and concepts that you will encounter in this chapter. If you are not familiar with them, you should review them before proceeding.

Respiration (Chapter 8)

Nitrogen fixation (Chapter 8)

Populations (Chapter 23)

Competition (Chapter 23)

Interactions in communities (Chapter 24)

FIGURE 25-1
Ecosystems. In the Coast Ranges of California, the boundary between the evergreen shrub association known as chaparral and grassland is often sharp, as shown in this photograph taken along the western edge of the Santa Clara Valley near Morgan Hill. The plant associations shown here are distinct ecosystems, each including a characteristic set of nonliving factors with different characteristics. The oak trees represent the edge of an oak woodland community that occurs in the same region.

The ecosystem is the most complex level of biological organization. Communities are composed of the organisms present at a particular place, whereas ecosystems include the nonliving factors interacting with these organisms. In ecosystems there is a regulated transfer of energy and a controlled cycling of nutrients. The individual organisms and populations of organisms in an ecosystem act as parts of an integrated whole, adjust over time to their role in the ecosystem, and relate to one another in complex ways that we only partly understand. Despite their differences, all ecosystems regulate the flow of energy—ultimately derived from the sun—and the cycling of nutrients. The earth is a closed system with respect to the chemicals, but an open one in terms of energy. Collectively, the organisms that occur in ecosystems regulate the capture and expenditure of that energy and the cycling of those chemicals. As we will see in this chapter, all organisms, including humans, depend on the ability of a few other organisms—plants, algae, and some bacteria in the case of carbon, and certain bacteria in the case of nitrogen, for example—for the basic components of life.

As distinct functional units, different kinds of ecosystems have more or less clearly recognizable boundaries, but they also intergrade into one another, sometimes almost imperceptibly; boundaries then become arbitrary (Figure 25-1). Ecosystems also change over time and slowly become modified into new ecosystems, whose characteristics come to differ increasingly from those that preceded them. Thus the complex ecosystems of the tropical rain forests have changed gradually in adapting to the particular conditions of temperature, seasonality, and soil that typify these places. The ecosystems of the tundra have developed in a similar way, but in relation to the different environmental conditions of the far north. As the climate changes in a given place, the ecosystems that are present there change along with it, as do the individual populations within these ecosystems. By this process, the overall characteristics of the populations gradually adjust to the new conditions. Not all ecosystems are natural; we may also speak of an ecosystem in an aquarium or in a cultivated field.

BIOGEOCHEMICAL CYCLES

All of the substances that occur in organisms cycle through ecosystems. Although there are local exceptions to this generality, on a global scale the bulk of these substances is not contained within the bodies of organisms, but rather exists in the atmosphere, the water, or in rocks. Carbon (in the form of carbon dioxide), nitrogen, and oxygen primarily enter the bodies of organisms from the atmosphere, whereas phosphorus, potassium, sulfur, magnesium, calcium, sodium, iron, and cobalt, all of which are required for plant growth (see Chapter 36), come from rocks. All organisms require carbon, hydrogen, oxygen, nitrogen, phosphorus, and sulfur in relatively large quantities; the other elements are required in smaller amounts.

We speak of the cycling of materials in ecosystems because they first are incorporated from the atmosphere or from weathered rock into the bodies of organisms; they then sometimes pass from these organisms into the bodies of other organisms that feed on the primary ones, and ultimately through decomposition are returned to the nonliving world. When this occurs, the nutrients may possibly be incorporated again into the bodies of other organisms. Some examples will help to clarify the ways in which different cycles function.

The Water Cycle

The water cycle (Figure 25-2) is the most familiar of all biogeochemical cycles. All life depends directly on the presence of water, since the bodies of most organisms consist mainly of this substance. Water is the source of hydrogen ions, whose movements generate ATP in organisms, and for that reason alone it is indispensable to their functioning.

The oceans cover three fourths of the earth's surface. From their surface, water evaporates into the atmosphere, a process powered by energy from the sun. Approximately 90% of the water that reaches the atmosphere comes from plants via transpiration, a process that we will discuss in Chapter 36. Most of it falls directly into the oceans, but some falls onto the land, where it passes into surface and subsurface bodies of fresh water. Only about 2% of all the water on earth is fixed in any form—frozen, held in the soil, or incorporated into the bodies of organisms. All of the rest is free water, circulating between the atmosphere and the earth. Regardless of where this water is held temporarily, it eventually returns to the atmosphere and the oceans.

Organisms live or die on the basis of their ability to capture some of this water and incorporate it into their bodies. Plants take up water from the earth in a continuous stream; crop plants require about 1000 kilograms of water to produce one kilogram of food, and the relationships in natural communities are similar. Animals obtain water directly or from the plants or other animals they eat. The amount of free water present at a particular place often determines the nature and abundance of the living organisms present there.

Much less obvious than the surface waters, which we see in streams, lakes, and ponds, is the groundwater, which occurs in **aquifers**—permeable, saturated, underground layers of rock, sand, and gravel. In many areas, groundwater is the most important reservoir of water; it amounts to more than 96% of all fresh water in the United States. The upper, unconfined portion of the groundwater constitutes the **water table,** which flows into streams and is partly accessible to plants; the lower confined layers are generally out of reach, although they can be "mined" by humans.

Groundwater flows much more slowly than surface water, anywhere from a few millimeters to as much as a meter or so per day. In the United States, groundwater provides about 25% of the water used for all purposes and provides about 50% of the

FIGURE 25-2
The water cycle. Transpiration is the process by which water evaporates from the surface of plants.

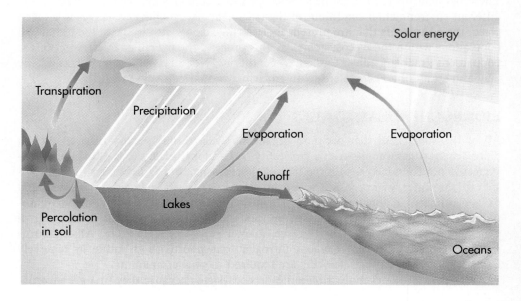

population with drinking water. Rural areas tend to depend on groundwater almost exclusively, and its use is growing at about twice the rate of surface water use. In the Great Plains of the central United States, the depletion of the Ogallala Aquifer seriously threatens the potential for agriculture; similar problems are appearing throughout the drier portions of the globe.

Because of the greater rate at which groundwater is being used, and also because it flows so slowly, the increasing chemical pollution of groundwater is a very serious problem. It is estimated that about 2% of the groundwater in the United States is already polluted, and the situation is worsening. Pesticides, herbicides, and fertilizers have become a serious problem, and another key source of groundwater pollution consists of the roughly 200,000 surface pits, ponds, and lagoons that are actively used for the disposal of chemical wastes in the United States alone. Recharging the aquifers is an important strategy for conservation, but it depends on the purity of the water there initially. Because of the large volume of water, its slow rate of turnover, and its inaccessibility, removing pollutants from aquifers is virtually impossible.

Some 96% of the fresh water in the United States consists of groundwater. This groundwater, which already provides 25% of all the water used in this country, will be used even more extensively in the future, even though problems of pollution are increasing.

The Carbon Cycle

The carbon cycle is based on carbon dioxide, which makes up only about 0.03% of the atmosphere. The worldwide synthesis of organic compounds from carbon dioxide and water results in the fixation of about 10% of the roughly 700 billion metric tons of carbon dioxide in the atmosphere each year (Figure 25-3). This enormous amount of biological activity takes place as a result of the combined activities of photosynthetic bacteria, algae, and plants. All heterotrophic organisms—including the nonphotosynthetic bacteria and protists, the fungi, the animals, and a relatively few plants, such as dodder (see Figure 24-20, *B*), that have lost the ability to photosynthesize—obtain their carbon indirectly, from the organisms that fix it. When their bodies decompose, organisms release carbon dioxide to the atmosphere again. Once there, it can be reincorporated into the bodies of other organisms.

About 10% of the estimated 700 billion metric tons of carbon dioxide in the atmosphere is fixed annually by the process of photosynthesis.

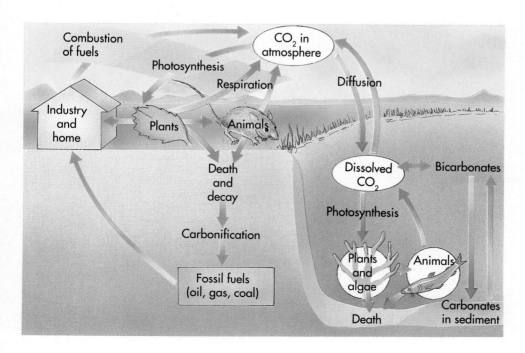

FIGURE 25-3
The carbon cycle.

Most of the organic compounds that are formed as a result of carbon dioxide fixation in the bodies of photosynthetic organisms are ultimately broken down and released back into the atmosphere or water. Certain carbon-containing compounds, such as cellulose, are more resistant to breakdown than others, but certain bacteria and fungi, as well as a few kinds of insects, are able to accomplish this feat. Some cellulose, however, accumulates as undecomposed organic matter, such as peat. The carbon in this cellulose may eventually be incorporated into fossil fuels, such as oil or coal.

In addition to the roughly 700 billion metric tons of carbon dioxide in the atmosphere, approximately 1000 billion metric tons are dissolved in the ocean; more than half of this quantity is in the upper layers, where photosynthesis takes place. The fossil fuels, primarily oil and coal, contain more than 5000 billion additional metric tons of carbon, and between 600 and 1000 billion metric tons are locked up in living organisms at any one time. In global terms, photosynthesis and respiration are approximately balanced, but the balance has been shifted recently because of our consumption of fossil fuels. The release of the carbon in coal, oil, and natural gas as carbon dioxide, a process that is proceeding rapidly as a result of combustion of fuels by humans, currently appears to be changing global climates, and may do so even more rapidly in the future, as we will discuss in more detail in Chapter 27.

The Nitrogen Cycle

Nitrogen gas constitutes 78% of the earth's atmosphere, but the total amount of fixed nitrogen in the soil, oceans, and the bodies of organisms is only about 0.03% of that figure. The nitrogen cycles between reservoirs and organisms via the **nitrogen cycle** (Figure 25-4). In this process, relatively few kinds of organisms—all of them bacteria—can convert, or fix, atmospheric nitrogen into forms that can be used for biological processes. The triple bond that links together the two atoms that make up diatomic atmospheric nitrogen (N_2) makes it a very stable molecule. In living systems the cleavage of atmospheric nitrogen is catalyzed by three proteins—ferredoxin, nitrogen reductase, and nitrogenase. This process uses ATP as a source of energy, electrons derived from photosynthesis or respiration, and a powerful reducing agent. The overall reaction can be written this way:

$$N_2 + 3H_2 \rightarrow 2NH_3$$

All living organisms depend on nitrogen fixation to synthesize proteins, nucleic acids, and other necessary nitrogen-containing compounds.

FIGURE 25-4
The nitrogen cycle.

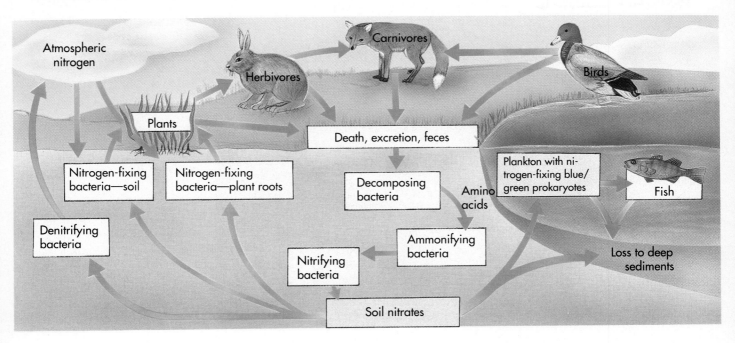

Dozens of genera of bacteria, most of them free-living, have the ability to fix atmospheric nitrogen. Among these, some are free-living and some form symbiotic relationships with the roots of legumes (plants of the pea family, Fabaceae) and other plants. Only the latter usually fix enough nitrogen to be of major significance in nitrogen production. Plants that have symbiotic associations with nitrogen-fixing bacteria can grow in soils that have such low amounts of available nitrogen that they are unsuitable for most other plants. Because of the activities of such organisms in the past, however, there now exists a large reservoir of ammonia and nitrates in ecosystems; this is the immediate source of much of the nitrogen used by organisms.

Although nitrogen gas constitutes about 78% of the earth's atmosphere, it becomes available to organisms almost entirely through the metabolic activities of a few genera of bacteria, some of which are free-living and others of which live symbiotically on the roots of legumes and other plants.

Bacteria of the genus *Rhizobium*, which inhabit nodules that they form on the roots of legumes (Figure 25-5), fix greater amounts of atmospheric nitrogen than do any other organisms. Bacteria of this genus make possible the growth of many legumes, a huge family of plants with about 18,000 species, in soils that are poor in nitrogen. Because of the presence of *Rhizobium*, legumes not only fix enough nitrogen for their own use, but also release excess fixed nitrogen into the soil, where it can be used by other plants. Nitrogen fixation by *Rhizobium* often exceeds 100 kilograms of nitrogen per hectare per year provided directly to the legumes. When legumes such as alfalfa are grown to enrich the soil, they are often plowed under so that all of the nitrogen they have fixed is available for future crops.

Specific strains of *Rhizobium* that will form nodules on the roots of different crop legumes, such as soybeans, alfalfa, and clover, are now available commercially. The individuals of *Rhizobium* enter the root hairs of their host legumes when the plants are still seedlings and move into the outer layers of the root tissue. There they stimulate repeated division of the outer cells of the root to form tumorlike nodules (see Figure 25-5).

FIGURE 25-5
Nitrogen-fixing bacteria. Bacteria of the genus *Rhizobium* form nodules (**A**) on the roots of many legumes. The bacteria (**B**) move into the roots of the legumes through cellulose tubes, which the plant forms when the bacteria are present. In the nodules, the bacteria are abundant in the individual plant cells (**C**).

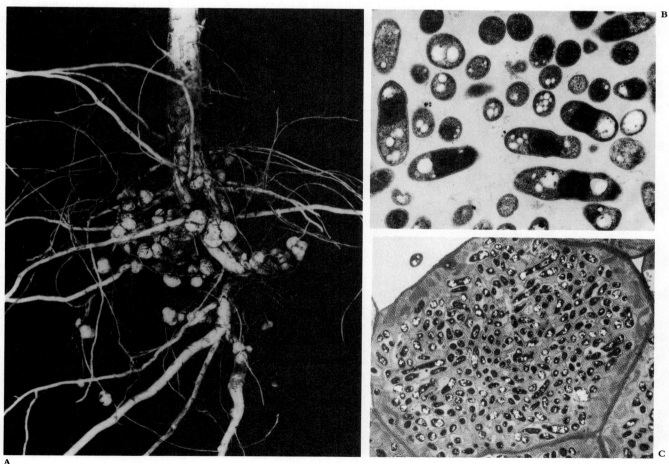

FIGURE 25-6
Alder trees and nitrogen-fixing bacteria. The alder trees, which form the dark band in the foreground, are among the first trees to colonize these relatively infertile, glaciated soils in Alaska. They can grow better than most other plants in these circumstances, because the bacteria (genus *Frankia*) that occur in nodules on their roots are able to fix nitrogen. The alders replace willows, which do not have nitrogen-fixing bacteria, and eventually will give way to poplars and then spruces.

The roots of a few other kinds of plants form associations with nitrogen-fixing bacteria of the group known as actinomycetes, which we will consider further in Chapter 30. These bacteria constitute the genus *Frankia*. Some of the plants involved are alders *(Alnus)* (Figure 25-6), sweet gale *(Myrica)*, and mountain lilac *(Ceanothus)*.

In the flooded paddies of China and Southeast Asia, rice is cultivated continuously. The nitrogen that the rice plants require is supplied by cyanobacteria of the genus *Anabaena*, which occur symbiotically between the leaves of the small floating water fern *Azolla* (Figure 25-7). In these regions, people place the *Azolla* plants out among the rice crop; this mode of biological fertilization, therefore, can be practiced only where labor costs are relatively low.

It is estimated that the activities of all free-living bacteria, including cyanobacteria, may add, on the average, about 7 kilograms of fixed nitrogen per hectare of soil per year—about the same amount of fixed nitrogen that is added with the rain, which dissolves ammonia that reaches the atmosphere from various sources. In contrast, a crop of the legume alfalfa that is plowed back into the soil may add as much as 300 to 500 kilograms of nitrogen per hectare per year. These figures clearly indicate the overwhelming importance of symbiotic associations in nitrogen fixation.

Chemoautrophic bacteria also contribute to the conversion of some ions to essential amino acids needed by animals. For example, ammonia, in the presence of water, is in equilibrium with the ammonium ion (NH_4^+). The chemoautotrophic nitrifying bacterium *Nitrosomonas* is of primary importance for the oxidation of ammonia, which is called **nitrification**. This reaction releases energy:

$$2NH_3 + 3O_2 \rightarrow 2NO_2^- + 2H + 2H_2O$$

FIGURE 25-7
Rice cultivation is aided by nitrogen-fixing cyanobacteria. The nitrogen-fixing cyanobacterium *Anabaena azollae* (**A**) lives in cavities between the leaves of the floating water fern *Azolla* (**B**), which is deliberately introduced into the rice paddies of the warmer parts of Asia. Rice (**C**), here cultivated in Sri Lanka, is the major food for well over one fourth of the human race.

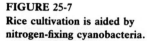

A **B** **C**

Another genus of bacteria, *Nitrobacter,* in turn oxidizes the nitrite ion (NO_2^-), which is toxic to plants, to nitrate (NO_3^-), in which form it is absorbed by plants and is directly available to them:

$$2NO_2^- + O_2 \rightarrow 2NO_3^-$$

This reaction, too, yields energy; *Nitrobacter,* like *Nitrosomonas,* is chemoautotrophic.

Inside the plant cells, the nitrate ions are reduced back to ammonium ions, an energy-requiring process, and then are transferred to amino acids and other nitrogen-containing molecules. Some amino acids are formed directly; others are formed by the transfer of the amino group ($-NH_2$) from one amino acid to another suitable molecule. Plants can produce all of the 20 amino acids they require. Most animals, however, can produce only 8; they obtain the other 12 from plants they consume directly or from animals that have eaten the plants.

Nitrogen-containing compounds, such as proteins, are decomposed rapidly by certain bacteria and fungi. These bacteria and fungi use the amino acids they obtain in this way to synthesize their own proteins and to release excess nitrogen in the form of ammonium ions (NH_4^+). This process, known as ammonification, occurs in most other kinds of organisms, as well. The ammonium ions can be converted to nitrites and nitrates by certain kinds of organisms and then be absorbed by plants.

A certain proportion of the fixed nitrogen in the soil is steadily lost. Under anaerobic conditions, nitrate is often converted to nitrogen gas (N_2) and nitrous oxide (N_2O), both of which return to the atmosphere. This process, which several genera of bacteria carry out, is called **denitrification.** In its absence, all nitrogen would eventually become fixed, converted into nitrate, and washed into the oceans. Life would thus be possible only in marine and littoral habitats. Denitrification and nitrogen fixation together constitute the mechanism for returning nitrogen from the oceans to the land.

The Oxygen Cycle

Earth is the only place in the solar system where free oxygen exists in significant quantities. This free oxygen, which constitutes nearly a fifth of our atmosphere by volume, is a product of photosynthesis, carried out over more than 3 billion years of earth history. In the process of respiration, free oxygen reacts rapidly with reduced organic material; most processes of this sort are now catalyzed by organisms. In the absence of photosynthesis, respiration would consume all organic material containing oxygen in about 50 years, but a large pool of oxygen would still remain in the atmosphere.

The Phosphorus Cycle

In all biogeochemical cycles other than those involving water, carbon, oxygen, and nitrogen, the reservoir of the nutrient exists in mineral form, rather than in the atmosphere. The **phosphorus cycle** (Figure 25-8) is presented as a representative example of all other mineral cycles because of the critical role phosphorus plays in plant nutrition worldwide.

Phosphorus is, more than any of the other required plant nutrients except nitrogen, apt to be so scarce that it limits plant growth. Phosphates—phosphorus anions—exist in the soil only in small amounts, because they are relatively insoluble and are present only in certain kinds of rocks. As phosphates weather out of soils, they are transported by rivers and streams to the oceans, where they are precipitated. They are naturally brought up again only by the uplift of lands, such as along the Pacific coast of North and South America, or by marine animals. Such animals are often consumed by sea birds, which deposit enormous amounts of guano (feces) rich in phosphorus along certain coasts; (Figure 25-9) these deposits have traditionally been used for fertilizer. Crushed phosphate-rich rocks, found in certain regions, are also used in this way, but the seas are the only inexhaustible source of phosphorus, one of the reasons that deep-seabed mining now looks so commercially attractive.

FIGURE 25-8
The phosphorus cycle.

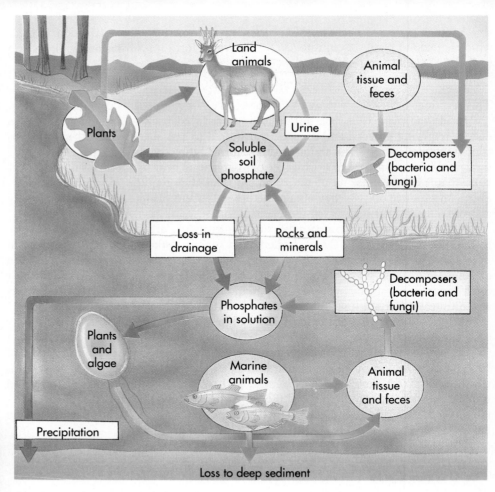

FIGURE 25-9
Living the phosphorus cycle.
These pelicans are roosting on a
small island in the Gulf of California, Mexico. Sea birds bring up
phosphorus from the deeper layers
of the sea by eating fishes and
other marine animals and depositing their remains as guano on the
rookeries.

Every year, millions of tons of phosphate are added to agricultural lands in the
belief that it becomes fixed to and enriches the soil. In general, four times as much
phosphate as a crop requires is added each year, usually in the form of **superphos-
phate,** which is soluble calcium dihydrogen phosphate, $Ca(H_2PO_4)_2$, derived by treating bones or apatite, the mineral form of calcium phosphate, with sulfuric acid. But
the enormous quantities of phosphates that are being added annually to the world's
agricultural lands are not leading to proportionate gains in crops; plants can apparently use only so much of the phosphorus that is added to the soil. Agricultural scientists are actively seeking new ways to approach the problem of phosphate supply.

Symbiotic associations occur between the roots of most plants and fungi. Called
mycorrhizae, they play a major role in absorbing phosphorus from the soil. The application of suitable fungi to agricultural soils and the fostering of conditions suitable
to their growth is one strategy that might be used to increase the supply of phosphorus available for plant growth. The recycling of animal and human wastes to recover
phosphates would also make an important contribution to the availability of phosphorus in industrial countries, where such practices are not normally employed. Human
sewage, in fact, represents a potential source of about 1 kilogram of phosphorus per
hectare of cultivated land worldwide per year. The world's rivers carry 17 million
metric tons of phosphorus to the sea every year, half from natural erosion and half
from domestic and industrial use and from sewage.

> Phosphates are relatively insoluble and are present in most soils only in small
> amounts. They often are so scarce that their absence limits plant growth.

Biogeochemical Cycles Illustrated: Recycling in a Forested Ecosystem

The overall recycling pattern of some nutrients has been revealed in impressive detail
by an ongoing series of studies conducted at the Hubbard Brook Experimental Forest

FIGURE 25-10
Gene Likens. One of the principal investigators at Hubbard Brook, Dr. Likens is shown here collecting a sample of water from a precipitation collector at Hubbard Brook.

FIGURE 25-11
Experimental weir at Hubbard Brook. All of the water is forced over the concrete, and samples of it are representative of the flow from the valley where the stream is located.

in New Hampshire. The way in which this ecosystem functions, and especially the way in which nutrients cycle within it, has been the subject of study since 1963 by Herbert Bormann of the Yale School of Forestry and Environmental Studies, Gene Likens of the New York Botanical Garden's Institute for Ecosystem Research (Figure 25-10), and their colleagues. These studies have yielded much of the information that we now have about the cycling of nutrients in forest ecosystems. They have also provided the basis for the development of much of the experimental methodology that is being applied successfully to the study of other ecosystems.

Hubbard Brook is the central stream of a large watershed that drains a region of temperate deciduous forest. For measurement of the flow of water and nutrients within the Hubbard Brook ecosystem, concrete weirs with V-shaped notches were built across six tributary streams that were selected for study (Figure 25-11). All of the water that flowed out of those valleys had to pass through the notch, since the weirs were anchored in bedrock. The precipitation that fell in the six valleys was measured, and the amounts of nutrients that were present in the water flowing in the six streams were also determined. By these methods, it was demonstrated that the undisturbed forests in this area were very efficient at retaining nutrients. The small amounts of nutrients that precipitated from the atmosphere with the rain and snow were approximately equal to the amounts of nutrients that ran out of the valleys. These quantities were very low in relation to the total amount of nutrients in the system. There was a small net loss of calcium—about 0.3% of the total calcium in the system per year—and small net gains of nitrogen and potassium.

In 1965 to 1966, the investigators felled all of the trees and shrubs in one of the six watersheds and then prevented their regrowth by spraying the area with herbicides. The effects of these activities were dramatic. The amount of water running out of that valley was increased by some 40%, indicating that water that normally would have evaporated into the atmosphere from the leaves of the trees and shrubs was now running off. For the 4-month period of June to September 1966, the runoff was actually four times higher than it had been in comparable periods during the preceding years. The amounts of nutrients running out of the system also increased greatly. For example, the loss of calcium was 10 times higher than it had been previously. Phosphorus, on the other hand, did not increase in the stream water; it apparently was locked up in the soil. A great deal of the available phosphorus may have reached deeper levels in the soil and thus become less available for plant growth.

The change in the status of nitrogen in the disturbed valley was especially striking. The undisturbed ecosystem in this valley had been accumulating nitrogen at a rate of about 2 kilograms per hectare per year, but the cut-down ecosystem *lost* it at a rate of about 120 kilograms per hectare per year! The nitrate level of the water rapidly increased to a level exceeding that judged safe for human consumption, and the stream that drained the area generated massive blooms of cyanobacteria and algae. In other words, the fertility of this logged-over valley decreased rapidly, while at the same time the danger of flooding greatly increased. This experiment is particularly instructive in the 1990s, as large areas of tropical rain forest are being destroyed to make way for cropland, a topic that will be discussed further in Chapter 27.

> **When the trees and shrubs in one of the valleys in the Hubbard Brook watershed were cut down and the area was sprayed with herbicide, the loss of nutrients such as calcium from that valley became much greater than it had been previously. Nitrogen, which had been accumulating at a rate of about 2 kilograms per hectare per year, was lost at a rate of 120 kilograms per hectare per year.**

THE FLOW OF ENERGY

An ecosystem includes autotrophs and heterotrophs. The **autotrophs,** consisting of plants, algae, and some bacteria, are able to capture light energy and manufacture their own food. To support themselves, the **heterotrophs,** including animals, fungi, most protists and bacteria, and nongreen plants, must obtain organic molecules that have been synthesized by autotrophs.

Once energy enters an ecosystem, mainly after it is captured as a result of photosynthesis, it is slowly released as metabolic processes proceed. The autotrophs that first acquire this energy provide all of the energy that heterotrophs use. The organisms that make up an ecosystem delay the release of the energy obtained from the sun back into space.

Productivity

Approximately 1% to 5% of the solar energy that falls on a plant is converted to food or other high-quality organic material. **Primary production** or **primary productivity** are terms used to describe the amount of organic matter produced from solar energy in a given area during a given period of time. **Gross primary productivity** is the total amount produced, including that used by the photosynthetic organism for the amount of organic matter that plants, algae, or photosynthetic bacteria store in excess of their own needs. Net primary productivity, therefore, is a measure of the amount of organic matter produced in a community in a given time that is available for heterotrophs. The net weight of all of the organisms living in an ecosystem, its **biomass,** increases as a result of its net production.

Some ecosystems—for example, a cornfield or a cattail swamp—have a high net primary productivity. Others, such as tropical rain forests (Figure 25-12), also have a

FIGURE 25-12
The tropical rain forest has a large biomass and is highly productive.

relatively high net primary productivity, but a rain forest has a much larger biomass than a cornfield, so the net primary productivity of rain forests is much lower in relation to its total biomass.

> Gross primary productivity occurs as a result of photosynthesis, which is carried out by green plants, algae, and some bacteria. Net primary productivity is defined as the total amount of energy fixed per unit of time minus the amount of energy expended by the metabolic activities of the *photosynthetic* organisms in the community.

In tropical forests and in marshlands, between 1500 and 3000 grams of organic material are normally produced per square meter per year. Corresponding figures for other communities are: temperate forests, 1100 to 1500 grams; dry deserts, 200 grams. For such highly productive communities as estuaries, coral reefs, and sugarcane fields, the figures may range from 10 to 25 grams per day, for comparable annual yields of from 3600 to 9100 grams. The most productive biological communities, however, appear to be those of the wave-battered intertidal zone, where for example communities of the brown alga known as sea palms (*Postelsia palmiformis*) may produce yields as high as 14,600 grams per square meter per year. How such yields are achieved is a matter currently receiving considerable attention.

Trophic Levels

Green plants, the primary producers of a terrestrial ecosystem, generally capture about 1% of the energy that falls on their leaves, converting it to food energy. In especially productive systems, this percentage may be a little higher. When these plants are consumed by other organisms, only a portion of the plant's accumulated energy is actually converted into the bodies of the organisms that consume them.

Among these consumers, several different levels may be recognized. The **primary consumers,** or herbivores, feed directly on the green plants. **Secondary consumers,** carnivores and the parasites of animals, feed in turn on the herbivores. **Decomposers** break down the organic matter accumulated in the bodies of other organisms. Another more general term that includes decomposers is **detritivores.** Detritivores are organisms that live on the refuse of an ecosystem—not only on dead organisms but also on the cast-off parts of organisms. They include large scavengers, such as crabs, vultures, and jackals, as well as decomposers (Figure 25-13).

All of these levels, and probably additional ones, are represented in any fairly complicated ecosystem. They are called **trophic levels,** from the Greek word *trophos,* which means "feeder." Organisms from each of these levels, feeding on one another, make up a series called a food chain. The length and complexity of food chains vary greatly. In real life, it is rather rare for a given kind of organism to feed on only one other kind of organism; usually, each will feed on two or more other kinds, and in turn will be fed on by several other kinds of organisms. When diagrammed, the relationship appears as a series of branching lines, rather than as one straight line; it is called a **food web** (Figure 25-14).

A certain amount of the energy ingested and retained by the organisms at a given

FIGURE 25-13
Lower levels of the food chain.
A The East African grasslands are dominated by a dense cover of grasses, with interspersed trees; these plants are primary producers, capturing energy from the sun. Grazing mammals obtain their food from the plants, and may in turn be consumed by predators, such as lions.
B This crab, *Gecarcinus quadratus,* photographed on the beach at Mazatlán, Mexico, is a detritivore, playing the same role that vultures and similar animals do in other ecosystems.
C Fungi, such as the basidiomycete whose mycelium is shown here growing through the soil in Costa Rica, are, together with the bacteria, the primary decomposers of terrestrial ecosystems.

A

B

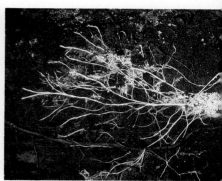

C

FIGURE 25-14

trophic level goes toward heat production. A great deal of the energy is used for digestion and work, and usually 40% or less goes toward growth and reproduction. An invertebrate typically uses about a quarter of this 40% for growth; in other words, about 10% of the food that an invertebrate eats is turned into its own body, and thus into potential food for its predators. Although the comparable figure varies from approximately 5% in carnivores to nearly 20% for herbivores, generally 10% is a good

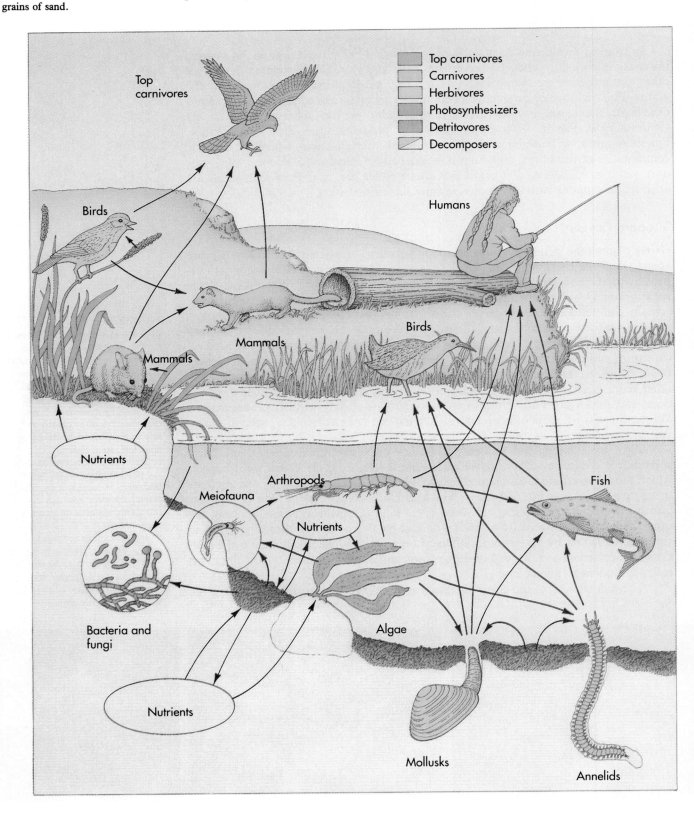

average value for the amount of organic matter that is present at each step in a food chain, or each successive trophic level, and the amount that reaches the next level.

A plant fixes about 1% of the sun's energy that falls on its green parts. The successive members of a food chain, in turn, process into their own bodies about 10% of the energy available in the organisms on which they feed.

Lamont Cole, of Cornell University, studied the flow of energy in a freshwater ecosystem in Cayuga Lake in upstate New York. He calculated that about 150 calories of each 1000 calories of potential energy that are fixed by algae and cyanobacteria are transferred into the bodies of small heterotrophs (Figure 25-15). Of these, about 30 calories are incorporated into the bodies of smelt, the principal secondary consumers of the system. If humans eat the smelt, they gain about 6 calories from each 1000 calories that originally entered the system. If, on the other hand, trout eat the smelt and humans eat the trout, humans gain only about 1.2 calories from each original 1000.

Relationships of this kind make it clear that organisms, including people, that subsist on an all-plant diet have more food available to them than do carnivores. Such considerations will become increasingly important in the future, not only for the efficient management of fisheries, but in general in the effort to maximize the yield of food for a hungry and increasingly overcrowded world.

Food chains generally consist of only three or four steps. The loss of energy at each step is so great that very little of the original energy remains in the system as usable energy after it has been incorporated successively into the bodies of organisms at four trophic levels. There are generally far more individuals at the lower trophic levels of any ecosystem than at the higher ones. Similarly, the biomass of the primary producers present in a given ecosystem is greater than that of the primary consumers, with successive trophic levels having a lower and lower biomass and correspondingly less potential energy. Larger animals characteristically are members of the higher levels; to some extent, they *must* be larger to capture enough prey to support themselves.

These relationships, if shown diagrammatically, appear as pyramids (Figure 25-16). We may therefore speak of "pyramids of biomass," "pyramids of energy," "pyramids of number," and so forth, as characteristic of ecosystems. Occasionally, the pyramids are inverted. For example, in a planktonic ecosystem—one consisting of small organisms floating in water—the turnover at the lower levels is rapid and the autotrophic organisms may reproduce rapidly, thus supporting a population of heterotrophs that is larger in biomass and more numerous than themselves. Pyramids of energy, on the other hand, cannot be inverted because of the necessary loss of energy at each step.

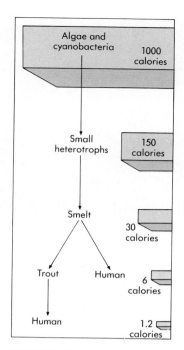

FIGURE 25-15
The food web in Cayuga Lake. In Cayuga Lake, one of New York's Finger Lakes, autotrophic plankton (algae and cyanobacteria) fix the energy of the sun, heterotrophic plankton feed on them, and both are consumed by smelt. The smelt are eaten by trout, with about a tenfold loss in fixed energy; for humans, the amount of biomass available in smelt is at least ten times greater than that available in trout, although they prefer to eat trout.

ECOLOGICAL SUCCESSION

Even when the climate of a given area remains stable year after year, ecosystems have a tendency to change from simple to complex in a process known as **succession.** This

FIGURE 25-16
Pyramids of numbers, biomass, and energy.

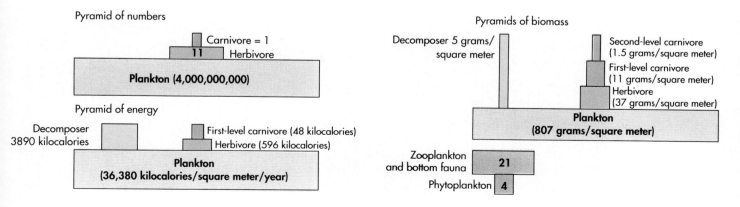

FIGURE 25-17
Succession in a pond. A pond near Tallahassee, Florida, rapidly filling with aquatic vegetation. As the process of succession proceeds, the pond will eventually be filled, and then gradually the area where it once existed will become indistinguishable from the surrounding vegetation.

FIGURE 25-18
Secondary succession. Mount St. Helens in the state of Washington erupted violently on May 18, 1980; the lateral blast devastated more than 600 square kilometers of forest and recreation lands within 15 minutes, as seen in (**A**), which shows an area near Clearwater Creek 4 months after the eruption. Five years later, succession was under way at the same spot (**B**), with shrubs such as blackberries, blueberries, and dogwoods following the first herbaceous plants that became established immediately after the blast.

process is familiar to anyone who has seen a vacant lot or cleared woods slowly but surely become occupied with larger and larger plants and more and more different kinds of them, or a pond become filled with vegetation that encroaches from the sides and gradually turns it into dry land (Figure 25-17).

Succession is continuous and worldwide in scope. If a wooded area is cleared and the clearing is left alone, plants will slowly reclaim the area. Eventually, the traces of the clearing will disappear and the whole area will again be woods. This kind of succession, which occurs in areas that have been disturbed and that were originally occupied by living organisms, is called **secondary succession.** Humans are often responsible for initiating secondary succession throughout portions of the world that they inhabit. Secondary succession may also take place after fire has burned off an area, for example, or after the eruption of a volcano (Figure 25-18).

Primary succession, in contrast to secondary succession, occurs on some bare, lifeless substrate, such as rocks, or in open water, where organisms gradually occupy the area and change its nature. Thus primary succession occurs in lakes left behind after the retreat of glaciers; another would be the processes that occur on a volcanic island that rises above the sea. Primary succession that occurs on places such as dry rocks is called **xerarch succession,** to distinguish it from the **hydrarch** primary succession that occurs in open water. On bare rocks, lichens may grow first, forming small pockets of soil and breaking down the stone. Acidic secretions from the lichens and from the plants that grow on the rocks later may help to break down the substrate and add to the accumulation of soil. Mosses may then colonize these pockets of soil, eventually followed by ferns and the seedlings of flowering plants. Over many thousands of years, or even longer, the rocks may be completely broken down, and the vegetation over an area where there was once a rock outcrop may be just like that of the surrounding grassland or forest.

In a similar example involving hydrarch succession, an **oligotrophic** lake—one poor in nutrients—may gradually, by the accumulation of organic matter, become **eutrophic**—rich in nutrients (see Figure 25-17). Plants standing along the edges of

A

B

the lake, such as cattails and rushes, and those growing submerged, such as pond-weeds, together with other organisms, may contribute to the formation of a rich organic soil. As this process continues, the pond may increasingly be filled in with terrestrial vegetation. Eventually, the area where the pond once stood, like the rock outcrop we just described, may become an indistinguishable part of the surrounding vegetation.

> **Primary succession takes place in areas that initially are devoid of life, such as dry rock faces or oligotrophic ponds. Secondary succession, in contrast, takes place in areas that have been disturbed after having been occupied by living things earlier.**

Oligotrophic ponds and bare rocks in a given region may over the very long term come to feature the same kind of vegetation as one another—the vegetation characteristic of the region as a whole. This relationship led the American ecologist F.E. Clements, at about the turn of the century, to propose the concept of **climax vegetation** (and the related term *climax community*) (Figure 25-19). However, with an increasing realization that (1) the climate keeps changing, (2) the process of succession is often very slow, and (3) the nature of a region's vegetation is being determined to a greater extent by human activities, ecologists do not consider the concept of "climax vegetation" to be as useful as they once did.

The characteristics that we will now outline are general ones that appear to hold for succession of all sorts. As ecosystems mature, there is an increase in total biomass but a decrease in net productivity. The earlier successional stages are more productive than the later ones. Agricultural systems are examples of early successional stages in which the process is intentionally not allowed to go to completion and the net productivity is high (Figure 25-20). There are many more species in mature ecosystems than in immature ones, and the number of heterotrophic species increases even more rapidly than the number of autotrophic species. This progression is related to the decreasing net productivity of increasingly mature ecosystems and to the fact that mature ecosystems have a greater ability to regulate the cycling of nutrients than do disturbed and immature ones. It appears that the plants and animals that appear in the later stages of succession may be more specialized, in general, than those that exist in the earlier stages. The late-successional species appear to fit together into more complex communities and to have much narrower ecological requirements, or niches, as we discussed in Chapter 23.

> **Communities at early successional stages have a lower total biomass, higher net productivity, fewer species, many fewer heterotrophic species, and less capacity to regulate the cycling of nutrients than do communities at later successional stages.**

In many communities, there is a constant progression in areas where trees have fallen or other local disturbances have occurred. The ways in which various kinds of organisms in the community refill the gaps, which gradually come to be occupied by mature forest, are of central importance in understanding community dynamics. Tornadoes, landslides, killing of trees over large areas by pests, and other similar natural disturbances bring about a local renewal of the communities in which they occur. Fires may release nutrients into the soil, accelerating the progress of colonization.

Some species are **fugitive species,** which occur at the earlier successional stages and disappear from the area as succession proceeds. Such species often have high reproductive rates and efficient means of dispersal. Foxglove (*Digitalis purpurea*), for example, is a fugitive species that temporarily becomes abundant in forest clearings and then disappears as the canopy cover of the trees becomes more complete. Other opportunities for fugitive species are afforded by fires; certain species may be seen only growing on burned areas. Certain fugitive species may occur only in newly formed ponds or islands and then apparently disappear from the area. Many weeds, also, are fugitive species, disappearing soon from the communities where they appear.

FIGURE 25-19
Climax community. A stage in succession toward a climax community on the west side of the San Francisco Peaks in northern Arizona. Here ponderosa pine (*Pinus ponderosa*) is replacing aspen (*Populus tremuloides*).

FIGURE 25-20
Agriculture stops succession. Modern agriculture often uses the land very intensely, as shown here in the wheatfields of eastern Washington.

The most spectacular instances of primary succession occur when a new lake is formed or an island first rises above the surface of the sea. The island of Surtsey, on the mid-Atlantic Ridge off the south coast of Iceland, rose above the sea on November 14, 1963, as a result of a volcanic eruption that began earlier on the ocean floor 120 meters below the surface. Even before the volcanic activity that created Surtsey began to subside, sea gulls were observed nesting on the warm cinders. The island continued to grow rapidly to the south, as a result of the continued accumulation of lava, until 1967. Its area was then about 280 hectares, and its maximum elevation was about 140 meters. Since then, Surtsey has been changed mainly by erosion, which has removed about 7.5 hectares annually.

The rocks on Surtsey are extremely low in phosphates, and nitrogen has only slowly been accumulating as a result of biological activity. After the island formed, bacteria and fungi soon appeared, their spores blown in from other areas. Among them were the free-living, nitrogen-fixing bacteria *Nitrosomas* and *Nitrobacter;* their activities led rapidly to the establishment of a nitrogen cycle on the island. Nitrogen-fixing and other cyanobacteria likewise became established very early, as did some of the green algae. The widespread weedy moss *Funaria hygrometrica* appeared in 1967, forming a patch by a sand bank. In 1970 the first lichens were found, and a number of additional species of mosses and lichens have since been discovered.

The first vascular plant to be found on Surtsey was a sea rocket, *Cakile edentula* (Figure 25-A, *1*), which appeared in 1965, only two years after the island emerged from the sea. This plant grows in open, sandy habitats; its fruits contain a large amount of corky tissue and float readily in seawater. Over the succeeding 20 years, an additional 19 species of vascular plants have taken hold on the island, and some of them have become quite abundant (Figure 25-A, *2* and *3*). Apparently all of these plants either drifted in from the sea or were imported inadvertently by birds.

Animals appeared on the island almost as soon as it was formed; the first fly was found in 1964. By 1976, nearly 200 species of insects and arachnids had become established. The introduction of these animals took place in several ways—by flying, being blown by the wind, drifting on flotsam in the sea, or, despite extensive precautions, as a result of the visits of scientists and other people to the island. More than 60 species of birds have been recorded, and some of them have been observed carrying seeds to Surtsey. Five of the bird species now nest on the island, further enriching the soil and increasing the complexity of the food webs.

Although the climate of Surtsey is cool, windy, and rainy, the many species of plants, animals, and microorganisms that became established there in only two decades provide striking testimony to their powers of dispersal and establishment. This small island in the North Atlantic has been colonized in the same way as the early Hawaiian Islands or the Galapagos, and it has taught us a great deal about the ways in which an extensive biota can develop rapidly under such circumstances.

FIGURE 25-A

1 Sea rocket, *Cakile edentula*, was the first vascular plant to start growth on Surtsey, but the species has not become permanently established on the island.

2 Lyme-grass, *Elymus arenarius*, the second species to grow on Surtsey, probably also became established from seeds that drifted from nearby islands.

3 Lyme-grass; northern shore-wort, *Mertensia maritima;* and sea-sandwort, *Honkenya peploides*, forming a community on the volcanic slopes of Surtsey. Sea-sandwort is the most widely distributed vascular plant on Surtsey; like the other species shown here, it is a highly prolific perennial.

1. Ecosystems are assemblages of organisms, along with the nonliving factors of their environment, that regulate the flow of energy, ultimately derived from the sun, and the cycling of nutrients.

2. Only 2% of the water on earth is fixed in any way; the rest is free. In the United States, 96% of the fresh water is groundwater. The contamination of our limited supplies of water is a serious problem.

3. About 10% of the roughly 700 billion metric tons of free carbon dioxide in the atmosphere is fixed each year through photosynthesis. An additional trillion metric tons of carbon dioxide is dissolved in the ocean, and five times that amount is locked up as coal, oil, and gas. About as much carbon exists in living organisms at any one time as there is in the atmosphere.

4. Carbon, nitrogen, and oxygen have gaseous or liquid reservoirs, as does water. All of the other nutrients, such as phosphorus, have solid reservoirs.

5. Atmospheric nitrogen is converted to ammonia by several genera of symbiotic and free-living bacteria. The ammonia, in turn, is converted to nitrites and then to nitrates by other bacteria. Nitrates are incorporated into the bodies of plants and there converted back into ammonium ions, which are used in the manufacture of many kinds of molecules in the bodies of living organisms. The breakdown of these molecules either converts them to recyclable forms or results in the release of atmospheric nitrogen.

6. Phosphorus is a key component of many biological molecules; it weathers out of soils and is transported to the world's oceans, where it tends to be lost. Phosphorus is relatively scarce in rocks; this scarcity often limits or excludes the growth of certain kinds of plants.

7. The Hubbard Brook experiments vividly illustrate the role of a forested ecosystem in regulating the cycling of nutrients. When a particular tributary valley was deforested, four times as much water and ten times as much calcium ran off as had been the case previously. Nitrogen, which had been accumulating in the area at a rate of about 2 kilograms per hectare per year, was now lost at a rate of about 120 kilograms per hectare per year.

8. Plants convert about 1% of the light energy that falls on their leaves to food energy. The herbivores that eat the plants and the other animals that eat the herbivores, constitute a set of trophic levels. At each of these levels, only about 10% of the energy fixed in the food is fixed in the body of the animal that eats that food. For this reason, food chains are always relatively short.

9. Primary succession takes place in areas that are originally bare, like rocks or open water. It is said to be either xerarch, if it occurs in dry places, or hydrarch, if it occurs in water or wet places. Secondary succession takes place in areas where the communities of organisms that existed initially have been disturbed.

10. Both types of succession lead ultimately to the formation of climax communities, whose nature is controlled primarily by the climate of the area concerned, although the human influence on many of these communities is increasing. Such communities have more total biomass, less net productivity, more species, many more heterotrophic species, and a higher capability of regulating the cycling of nutrients within them than do the earlier successional stages.

REVIEW QUESTIONS

1. What are biogeochemical cycles? What are the primary reservoirs for the chemicals in these cycles? Is there a greater amount of life-sustaining chemicals in these reservoirs or in the whole of the earth's living organisms?

2. What are the earth's repositories for nitrogen? What is the process of nitrogen fixation? What types of organisms fix nitrogen?

3. What is nitrification? Between nitrate and nitrite, which is toxic to plants and which is used by them to produce amino acids? Which organisms are more competent at producing all 20 amino acids, plants or animals?

4. How is the phosphorus cycle different from the water, carbon, nitrogen, and oxygen cycles? What are the natural sources for phosphorus?

5. What is the effect of deforestation on the water cycle and flood control? What is its effect on the nutrient cycles and overall fertility of the land?

6. What type of biological community is most productive? Rank deserts, tropical rain forests, and temperate forests in terms of primary productivity, from highest to lowest.

7. How efficient is the transfer of energy from one trophic level to the next? Which type of diet, carnivorous or herbivorous, provides more food value to any given living organism? Why? Is it therefore more efficient to feed the starving peoples of the world with corn or hamburgers?

8. What is the process of ecological succession? How does secondary succession differ from primary succession?

9. Why have scientists altered their concept of a final, climax vegetation in a given ecosystem? What types of organisms are often associated with early stages of succession?

10. How are biomass and productivity related to the successional age of an ecosystem? How is the number of species, especially heterotrophic species, related to the successional age of an ecosystem? Why?

THOUGHT QUESTIONS

1. At what successional stage would you characterize a field of wheat? What does this imply as to its stability and productivity?

2. How could you increase the net primary productivity of a desert?

3. Why does the net productivity of an ecosystem decrease as it becomes more mature?

FOR FURTHER READING

BORMANN, F.H., and G.E. LIKENS: *Pattern and Process in a Forested Ecosystem*, Springer-Verlag, New York, 1979. A fascinating and well-written description of the Hubbard Brook experiments.

FORMAN, R.T.T., and M. GORDON: *Landscape Ecology*, John Wiley & Sons, New York, 1986. This highly readable and well-illustrated volume explores the connection between the landscapes we see and the processes that create them.

FRANKLIN, J.F., H.H. SHUGART, and M.E. HARMON: "Tree Death as an Ecological Process," *BioScience*, vol. 37, pages 550-556, 1987. Outstanding article in an excellent issue of *BioScience* devoted to the life cycles and ecological role of trees.

PAUL, W.M. AND OTHERS: "The Global Carbon Cycle," *American Scientist*, vol. 78, pages 310-326. 1990. The dynamic responses of natural systems to carbon dioxide may determine the future of the earth's climate.

RICKLEFS, R.E.: *The Economy of Nature*, ed. 2, Chiron Press, Newton, Mass., 1983. This well-illustrated text is an excellent introduction to all aspects of ecology.

SANDERSON, S.L. and R. WASSERSUG: "Suspension-Feeding Vertebrates," *Scientific American*, March 1990, pages 96-101. Animals that filter their food out of the water can reap the abundance of plankton and grow in huge numbers or to enormous size.

SEBENS, K.P.: "The Ecology of the Rocky Subtidal Zone," *American Scientist*, vol. 73, pages 548-557, 1985. Succession and community structure at the edges of the sea.

SMITH, R.L.: *Ecology and Field Biology*, ed. 3, Harper & Row, Publishers, Inc., New York, 1984. A fine, field-oriented introduction to ecology and evolution.

SPENCER, C.N., B.R. McCLELLAND, and J.A. STANFORD: "Shrimp stocking, salmon collapse, and eagle displacement," *BioScience*, vol. 41, pages 14-21, 1991. Cascading interactions in the food web of a large aquatic ecosystem.

VALIELA, I.: *Marine Ecological Processes*, Springer-Verlag, New York, 1984. An outstanding introduction to life in the sea, with an emphasis on the flow of energy and the cycling of nutrients.

WARING, R.H., and W.H. SCHLESINGER: *Forest Ecosystems: Concepts and Management*, Academic Press, Orlando, Fla., 1985. This book brings together much information on ecological processes in forests, providing a good introduction to the field.

26

Atmosphere, Oceans, and Biomes

CHAPTER OUTLINE

OVERVIEW

There are great differences in climate across the face of the globe that over billions of years have resulted in the evolution of diverse terrestrial biomes and comparable associations of organisms in the sea. These are major assemblages of plants, animals, and microorganisms that occur over wide areas and have distinctive characteristics that separate them from others. The major circulation patterns of the atmosphere and the oceans, driven by the unequal distribution of heat from the sun, underlie these differences. Of all the biomes and comparable marine communities, those of the tropics, where at least two thirds of all kinds of organisms occur, are by far the richest biologically.

FOR REVIEW

Here are some important terms and concepts that you will encounter in this chapter. If you are not familiar with them, you should review them before proceeding.

Major features of evolution (Chapter 22)

Communities (Chapter 24)

Symbiosis (Chapter 24)

Biogeochemical cycles (Chapter 25)

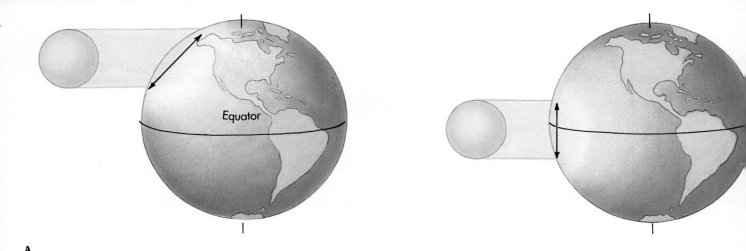

A

FIGURE 26-1
Relationships between the earth and the sun are critical in determining the nature and distribution of life on earth.

A A beam of solar energy striking the earth in middle latitudes is spread over a wider area of the earth's surface than a similar beam striking the earth near the equator.

The major biomes, which are terrestrial communities that occur over wide areas, are easily recognized by their overall appearance and characteristic climates. These biomes will be reviewed here, as will the analogous assemblages that occur in the seas. Each biome is similar in its structure and appearance wherever it occurs on earth and differs significantly from other kinds of biomes. Biomes could be classified in a number of ways; those that we use here are chosen merely as convenient means for discussing the properties of life on earth from an ecological perspective.

THE GENERAL CIRCULATION OF THE ATMOSPHERE

The distribution of biomes results from the interaction of the features of the earth itself, such as different soil types or the occurrence of mountains and valleys, with two key physical factors: (1) the amounts of heat from the sun that reach different parts of the earth and the seasonal variations in that heat; and (2) global atmospheric circulation and the resulting patterns of oceanic circulation. Together these factors determine the local climate, including the amounts and distribution of precipitation.

The Sun and Atmospheric Circulation

The earth receives an enormous quantity of heat from the sun in the form of short-wave radiation, and it radiates an equal amount of heat back to space in the form of long-wave radiation. About 10^{24} calories arrive at the upper surface of the earth's atmosphere each year, or about 1.94 calories per square centimeter per minute. About half of this energy reaches the earth's surface. The wavelengths that reach the earth's surface are not identical to those that reach the outer atmosphere; for example, most of the ultraviolet radiation is absorbed by the oxygen and ozone in the atmosphere. As we will see in Chapter 27, the depletion of the ozone layer, apparently as a result of human activities, poses serious ecological problems.

The world contains a great diversity of biomes because its climate varies so much from place to place. On a given day, Miami, Florida, and Bangor, Maine, often have very different weather. There is no mystery about this. The tropics are warmer than the temperate regions because the sun's rays arrive almost perpendicular to regions near the equator, whereas near the poles their angle of incidence spreads them out over a much greater area (Figure 26-1, *A*), providing less energy per unit area. Because the earth is a sphere, some parts of it receive more energy from the sun than others. This is responsible for many of the major climatic differences that occur over the earth's surface, and thus indirectly for much of the diversity of biomes.

The earth's annual orbit around the sun and its daily rotation on its own axis are both important in determining world climate (Figure 26-1, *B*). Because of the daily cycle, the climate at a given latitude is relatively constant; there is a constant mixing of climates and temperatures at that latitude. Because of the annual cycle, and the inclination of the earth's axis at approximately 23.5 degrees from its plane of revolution around the sun, there is a progression of seasons in all parts of the earth away

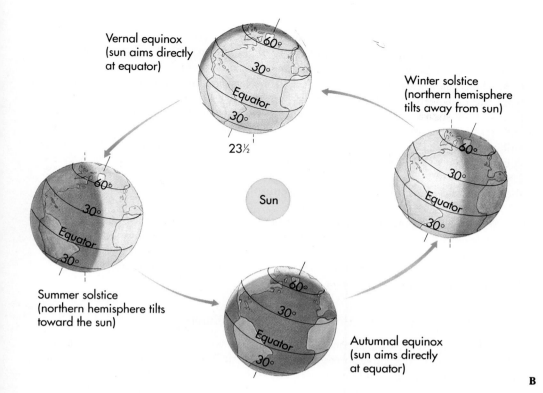

Vernal equinox
(sun aims directly
at equator)

23½

Sun

Winter solstice
(northern hemisphere
tilts away from sun)

Summer solstice
(northern hemisphere tilts
toward the sun)

Autumnal equinox
(sun aims directly
at equator)

B

FIGURE 26-1, cont'd
B The rotation of the earth around the sun has a profound effect on climate. In the northern and southern hemispheres, temperatures change in an annual cycle because the earth is slightly tilted on its axis in relation to its pathway around the sun.

from the equator. One or the other of the poles is tilted closer to the sun than the other at all times except during the spring and autumn equinoxes.

Major Circulation Patterns

Near the equator, warm air (which holds more moisture than cool air) rises and flows toward the poles (Figure 26-2). As it rises, it is cooled and loses most of its moisture; consequently, near the equator, where the air is warmest and therefore has the highest moisture content, its rising leads to the greatest amounts of precipitation anywhere. This equitorial region of rising air is one of low pressure, the doldrums, which draws air from both north and south of the equator. When the air masses that have risen reach about 30 degrees north and south latitude, the air, now cooler, sinks and becomes reheated, producing a zone of decreased precipitation. The air is still warmer than it is in the polar regions, and it continues to flow toward the poles. It rises again at about 60 degrees north and south latitude and flows back toward the equator. At this latitude there is another low-pressure area, the polar front. The air that rises here descends near the poles, producing a zone of low precipitation.

FIGURE 26-2
General patterns of atmospheric circulation.
A The pattern of air movement out from and back to the earth's surface.
B The major wind currents across the face of the earth.

A

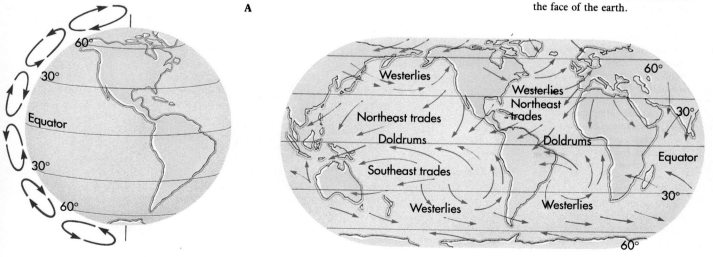

B

Related to these bands of north-south circulation are three major air currents, generated mainly by the interaction of the earth's rotation with the patterns of worldwide heat gain. Between about 30 degrees north latitude and 30 degrees south latitude, the **trade winds** blow, from the east-southeast in the southern hemisphere and from the east-northeast in the northern hemisphere. The trade winds blow all year long and are the steadiest winds found anywhere on earth; they are stronger in winter and weaker in summer. Between 30 and 60 degrees north and south latitude, the **prevailing westerlies,** often strong winds, blow from west to east and tend to dominate climatic patterns in these latitudes, particularly lands that lie along the western edges of the continents. Zones of weaker winds, blowing from east to west, occur farther north and south in their respective hemispheres.

> **Warm air rises near the equator, descends and produces arid zones at about 30 degrees north and south latitude, flows toward the poles, then rises again at about 60 degrees north and south latitude, and moves back toward the equator. Part of this air, however, moves toward the poles, where it produces zones of low precipitation.**

Atmospheric Circulation, Precipitation, and Climate

Since the moisture-holding capacity of air increases when it is warmed and decreases when it is cooled, precipitation is generally low near 30 degrees north and south latitude, where air is falling and being warmed, and relatively high near 60 degrees north and south latitude, where it is rising and being cooled. Partly as a result of these factors, all of the great deserts of the world lie near 30 degrees north or 30 degrees south latitude, and some of the great temperate forests are near 60 degrees north and south latitude. Other major deserts are formed in the interiors of the large continents; these areas have limited precipitation because of their distance from the sea, the ultimate source of most precipitation. Other deserts sometimes occur because mountain ranges intercept the moisture-laden winds from the sea (Figure 26-3). When this occurs, the moisture-holding capacity of the air decreases, resulting in increased precipitation on the windward side of the mountains—the side from which the wind is blowing. As the air descends the other side of the mountains, the leeward side, it is warmed, and its moisture-holding capacity increases, tending to block precipitation. The eastern sides of the mountains are much drier than their western sides, and the vegetation is often very different; this phenomenon is called the **rain shadow effect.**

Four relatively small areas, each located on a different continent, share a climate that resembles that of the Mediterranean region of southern Europe, North Africa, and portions of the Near East. Such a climate is found in portions of California (Figure 26-4), southwestern Oregon, and northwestern Baja California, Mexico; in central Chile; in southwestern and part of southern Australia; and in the Cape region of South Africa. In all of these areas the prevailing westerlies blow during the summer from a cool ocean onto warm land. As a result, the air's moisture-holding capacity is increased, so precipitation is completely blocked or limited during the summer. Such

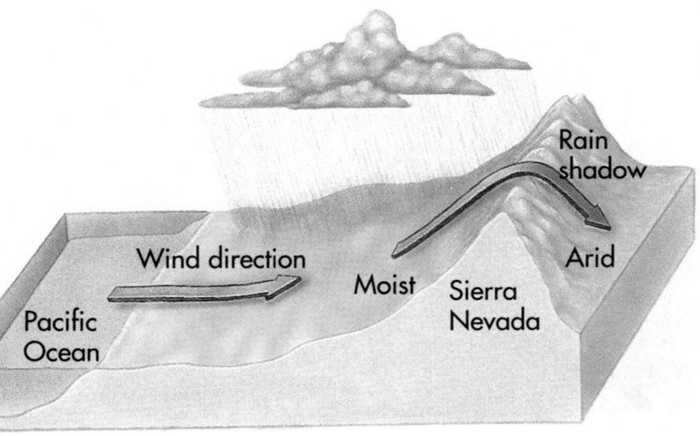

FIGURE 26-3
The rain shadow effect.
Moisture-laden winds from the Pacific Ocean rise and are cooled when they encounter the Sierra Nevada. As their moisture-holding capacity decreases, precipitation occurs, making the middle elevation of the range one of the snowiest regions on earth; it supports tall forests, including those that include the famous big trees (*Sequoiadendron giganteum*). As the air descends on the east side of the range, its moisture-holding capacity increases again, and the air picks up moisture from its surroundings rather than releasing it. As a result, desert conditions prevail on the east side of the mountains.

FIGURE 26-4
The effects of prevailing wester-lies. When winds blow across cold water onto warm land, as shown here in the summer near Point Sur, California, fog forms. The moisture-holding content of the air increases as it is warmed over the land, and precipitation is effectively blocked in the summer. This effect accounts for the dry, hot summers and cool, moist winters that occur in areas with a mediterranean climate.

climates are unusual on a world scale; in the regions where they occur, many unusual kinds of plants and animals, often local in distribution, have evolved. Because of the prevailing westerlies, the great deserts of the world (other than those in the interior of continents) and the areas of mediterranean climate lie on the western sides of the continents.

The great deserts and associated arid areas of the world, including those with mediterranean (summer-dry) climates, lie along the western sides of the continents at about 30 degrees north and south latitude. Other major deserts occur in the interiors of the large continents.

Another kind of major regional climate occurs in southern Asia. The monsoon climatic conditions that are characteristic of India and southern Asia occur during the summer months. During the winter the trade winds blow from the east-northeast off the cool land onto the warm sea. From June to October, though, when the land is heated, the direction of the air flow is reversed, and the winds blowing from the east-southeast south of India veer around to blow onto the Indian subcontinent and adjacent areas from the southwest. The duration and strength of the monsoon winds spell the difference between food sufficiency and starvation for hundreds of millions of people in this region each year. Under normal conditions, the monsoons bring heavy rains to the entire area, and the year's crop succeeds; if the rains do not come or are inadequate, the crop fails.

Patterns of Circulation in the Ocean

In the ocean, patterns of circulation are determined by the patterns of atmospheric circulation just discussed, but they are modified by the location of the land masses around which and against which the ocean currents must flow. Oceanic circulation is dominated by huge surface gyrals (Figure 26-5), which move around the subtropical zones of high pressure between approximately 30 degrees north and 30 degrees south latitude. These gyrals move clockwise in the northern hemisphere and counterclockwise in the southern hemisphere. They profoundly affect life not only in the oceans but also on coastal lands by the ways in which they redistribute heat. For example, the Gulf Stream, in the North Atlantic, swings away from North America near Cape Hatteras, North Carolina, and reaches Europe near the southern British Isles. Because of the Gulf Stream, western Europe is much warmer and thus more temperate than is eastern North America at similar latitudes. As a general principle, the western sides of continents in the temperate zones of the Northern Hemisphere are warmer than their eastern sides; the opposite is true of the Southern Hemisphere. In addition, and as we have just seen, winds passing over cold water onto warm land increase their moisture-holding capacity; precipitation is limited in such areas (see Figure 26-4).

In South America the Humboldt Current carries cold water northward up the west coast and helps make possible an abundance of marine life that supports the fisheries of Peru and northern Chile. Marine birds, which feed on these organisms, are

FIGURE 26-5

FIGURE 26-5
Ocean circulation. The circulation in the oceans moves in great surface spiral patterns called gyrals; it profoundly affects the climate on adjacent lands.

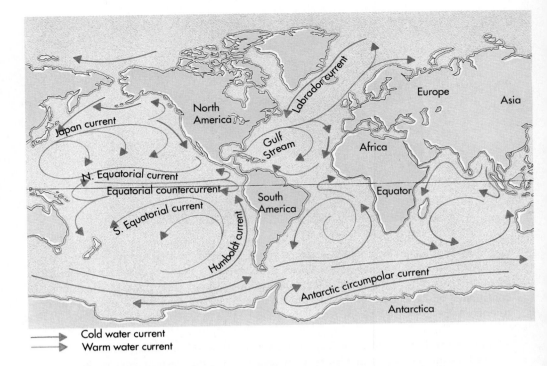

Cold water current
Warm water current

responsible for the commercially important guano deposits of these countries (see Figure 25-9). These deposits are rich in phosphorus, which is brought up from the ocean depths by the upwelling of cool water that occurs as a result of the frequent offshore wind from the generally mountainous slopes that border the Pacific Ocean. When the coastal waters are not as cool as usual, a devastating phenomenon called "El Niño" occurs (see boxed essay), which influences climatic patterns globally.

In the ocean, huge surface gyrals move around the subtropical zones of high pressure between approximately 30 degrees north and south latitude, clockwise in the northern hemisphere and counterclockwise in the southern hemisphere.

THE OCEANS

Nearly three quarters of the earth's surface is covered by ocean. The seas have an *average* depth of more than 3 kilometers, and they are, for the most part, cold and dark. Heterotrophic organisms are found at the greatest ocean depths, which reach nearly 11 kilometers in the Marianas Trench of the western Pacific Ocean, but photosynthetic organisms are confined to the upper few hundred meters of water (Figure 26-6). Organisms that live below this level obtain almost all of their food indirectly, as a result of photosynthetic activities that occur above; these activities result in organic debris that drifts downward.

Because water is much denser than air, the minerals and gases dissolved in it diffuse much more slowly. The supply of oxygen can often be critical in the ocean, since it is present in some areas only in limited quantities. In addition, the warmer the water, the less the solubility of oxygen; the amount of available oxygen is therefore extremely important in limiting the occurrence of organisms in the warmer marine regions of the globe. Carbon dioxide, in contrast, is almost never limiting in the oceans, and the distribution of minerals in the ocean is much more uniform than it is on land, where individual soils reflect faithfully the composition of the parent rocks from which they have weathered.

Despite the many new forms of animals, protists, and bacteria still being discovered in the sea and the huge biomass that occurs there, many fewer species live in the sea than on land. Probably more than 90% of all species of organisms occur on land, including all but a few species of some of the largest groups of organisms, including insects, mites, nematodes, fungi, and plants. Each of these groups has marine repre-

FIGURE 26-6
Life in the oceans. Fishes and many other kinds of animals find food and shelter among the kelp beds that occur in the coastal waters of temperate regions.

The west coast of South America is normally a highly productive area, cooled by the Humboldt Current, which sweeps up from the south and even makes it possible for penguines, which are cold-water birds, to live on the Galapagos Islands near the equator. The productivity of the surface waters depends on this current and on the cool, nutrient-rich waters that well up from the depths to replace the depleted ones at the surface. In a normal January, however, warm water flows down the coast from the tropics to southern Peru and northern Chile. The local fishermen named this current El Niño ("The Christ Child") because it occurs near Christmas and is normally benevolent. But scientists now reserve this name for a catastrophic version of the same phenomenon, one that is felt not only locally but on a global scale.

Such events, which are generally known as El Niño–Southern Oscillation or ENSO events, are now known to be of global significance in perturbing the general circulation of the atmosphere, as do episodes of enhanced cold-water upwelling. ENSOs occur every 2 to 6 years.

The term "southern oscillation" is used to describe the atmospheric changes that accompany the ocean warming of an El Niño event. The factors that lead to an ENSO are unknown, although they seem to involve major changes in atmospheric pressure systems, and there is a net movement of warm water from the western Pacific to the central and eastern Pacific. When this occurs, commercial fish stocks virtually disappear from the waters of Peru and northern Chile and the plankton drop to a twentieth of their normal abundance. The commercially valuable anchovy fisheries of Peru were essentially permanently destroyed by the 1972 ENSO event. Violent winter storms lash the coast of California, accompanied by flooding and a colder and wetter winter than normal in Florida and along the Gulf Coast. A particularly severe ENSO seems to occur about every 25 years or so. One such oscillation began in October 1986 and extended through early 1988; it was apparently correlated with unusual drought in the states of Washington and Oregon in the summer of 1987.

In the ENSO of 1982 to 1983, many birds starved to death on islands throughout the central Pacific, and a huge number of bird colonies simply gave up breeding. On Christmas Island, an important breeding site about 2000 kilometers south of Honolulu, 95% of the 14 million birds that normally nest on the island, representing 18 species, simply left, abandoning their eggs, young, and nests; by June 1983, there were only 150,000 birds on the island. Even for the Farallon Islands, which lie about 45 kilometers off San Francisco, many birds starved, and the large breeding ground was tragically disrupted. Many eggs failed to hatch, and thousands of chicks starved. On the other hand, the moist, rainy conditions that accompany an ENSO event (Figure 26-A) are favorable for crops on land; the Peruvian rice crop, for example, is strongly favored by such a climatic regimen. Land birds, such as Darwin's finch on the Galapagos Islands, breed abundantly during an ENSO, and their population sizes increased remarkably during the ENSO of 1982 to 1983.

FIGURE 26-A

1 Daphne Crater, Galapagos Islands, containing about 800 pairs of nesting blue-footed boobies in 1975, a normal year.

2 The same crater in 1983, an El Niño year, ringed with abundant vegetation but with no nesting boobies.

sentatives, but they comprise only a very small fraction of the total number of species. On land the barriers between habitats are sharper, and variations in elevation, parent rock, degree of exposure, and other factors have all been crucial to the evolution of the millions of species of terrestrial organisms; in other words, there is a greater variety of available niches. This pattern of radiation into many clearly defined habitats seems to account for the much larger number of species that live on land than in the sea. In terms of overall diversity, the pattern is quite different: of the major groups of organisms—phyla—every one occurs in the sea, whereas relatively few phyla are found on land or in freshwater habitats. Most phyla originated in the sea, but only a few of them have been successful on land, although some of these have given rise to extraordinarily large numbers of species.

Although representatives of almost every phylum occur in the sea, an estimated 90% of living species of organisms are terrestrial. This is because of the enormous evolutionary success of a few phyla on land, where the boundaries between different habitats are sharper than they are in the sea.

The marine environment consists of three major kinds of habitats: (1) the neritic zone, the zone of shallow waters along the coasts of the continents; (2) the surface layers of the open sea; and (3) the abyssal zone, the deep-water areas of the oceans.

The Neritic Zone

The neritic zone—generally considered to be that area of the ocean along the coasts of the continents and islands, and less than 300 meters below the surface—is small in area, but it is inhabited by large numbers of species as compared with other parts of the ocean. The intense and sometimes violent interaction between sea and land in this zone gives a selective advantage to well-secured organisms that can withstand being washed away by the continual beating of the waves. Part of this zone, the intertidal, or littoral, region, is exposed to the air whenever the tides recede (Figure 26-7).

Because of the way in which it affords access to the land, the intertidal zone must have been home for the ancestors of those organisms that originally colonized terrestrial habitats. Perhaps the greater complexity needed to anchor and fasten the animals and plants that dwell in this zone constituted a kind of preadaptation to life on land, where the environmental stresses are even more extensive. The organisms that live successfully in habitats that are regularly exposed to the air must also have some kind of waterproof covering or habits that protect them from the drying action of the air when the tide is out (Figure 26-8); such adaptations are of central importance for terrestrial organisms.

The world's great fisheries occur in shallow waters over continental shelves, either near the continents themselves or in the open ocean, where huge banks come

FIGURE 26-7
The intertidal zone. In the intertidal zone, as shown here along the coast south of San Francisco, California, the organisms are exposed to the pounding of the waves and periodically subject to drying out, as the tides move in and out. Tide pools often form among the rocks when the tides recede.

FIGURE 26-8
A waterproof organism of the intertidal zone. The crustaceans known as hermit crabs live within a shell, thus protecting themselves both from drying out and from predators. The mollusks shown here are periwinkles (*Littorina*), a marine snail that is found in intertidal areas throughout much of the world.

A

B

C

D

near the surface. Nutrients, derived from land, are much more abundant in coastal and other shallow regions, where upwelling from the depths occurs, than in the open ocean, accounting for their greater productivity. The preservation of these fisheries, which are a source of high-quality protein exploited throughout the world, has become a matter of growing concern. In Chesapeake Bay (Figure 26-9), where complex systems of rivers enter the ocean from heavily populated areas, the environmental stresses have become so serious that they not only threaten the continued existence of formerly highly productive fisheries, but also diminish the quality of human life in these regions. For example, the increased amounts of runoff from farms and other activities and sewage effluent in areas like Chesapeake Bay provide nutrients for greater numbers of marine organisms, which in turn use up more and more of the oxygen in the water and thus may disturb the established populations of organisms such as oysters. Such effects may be enhanced by climatic shifts, and large numbers of marine animals may die suddenly as a result.

In tropical waters, where the water temperature remains at about 21° C, coral reefs occur. These are highly productive ecosystems that can successfully concentrate nutrients, even from the relatively nutrient-poor waters that are characteristic of the tropics, and that constitute about three fourths of the surface area of the world's oceans (see Chapters 31 and 39). The coral animals, which provide most of the structure of the reefs, together with the photosynthetic coralline algae and the many other kinds of organisms that live in and around the reefs, make up one of the most complex and fascinating living systems on the earth. The unicellular organisms known as dinoflagellates that live symbiotically within the coral animals contribute a great deal of the reef's productivity, as do the other algae that grow among the coral animals. Many of the oceans of the world are highly unproductive biologically, resembling deserts in this respect.

The Surface Zone

Drifting freely or swimming in the upper, better-illuminated waters of the ocean, a diverse biological community exists, primarily consisting of microscopic organisms called plankton. Fish and other larger organisms that swim in these same waters constitute the nekton, whose members feed on organisms in the plankton and on one an-

FIGURE 26-9

Chesapeake Bay. Chesapeake Bay has more than 11,300 kilometers of shoreline and drains more than 166,000 square kilometers in one of the most densely populated and heavily industrialized areas in North America. The body of open water is about 320 kilometers long and, at some points, nearly 50 kilometers wide.

A Large metropolitan areas and shipping facilities make the Bay one of the busiest natural harbors anywhere.

B One of the most biologically productive bodies of water in the world, the Bay yielded an annual average of about 275,000 kilograms of fish in the 1960s but only a tenth as much in the 1980s. The human population of the area grew 50% during the same period.

C This grebe is coated from an oil spill off the mouth of the Potomac River.

D Uncontrolled erosion from certain agricultural practices, pesticides, and increases in nutrients block the light needed for photosynthesis and upset the delicate ecological balance on which the productivity of the Bay depends. The states that border the Bay are cooperating, with the assistance of the Environmental Protection Agency, to try to bring back its former productivity.

other. Together, the organisms that make up the plankton and the nekton provide all of the food for those that live below. Some members of the plankton, including the algae and some bacteria, are photosynthetic. Collectively, these organisms account for about 40% of all the photosynthesis that takes place on earth, and even more by some calculations. Most of the plankton occurs in the top 100 meters of the sea, the zone into which light from the surface penetrates freely. Perhaps half of the total photosynthesis in this zone is carried out by organisms less than 10 micrometers in diameter—at the lower limits of size for organisms—including cyanobacteria and the smallest algae. Such organisms are so small that their abundance and ecological importance have been unappreciated until relatively recently.

Many heterotrophic protists and animals also live in the plankton and feed directly on the photosynthetic organisms, as well as on one another. Gelatinous animals, especially jellyfish and ctenophores, which may consist of as much as 95% water, are abundant in the plankton but are relatively poorly known because their fragility makes them difficult to collect and study. The whales are the largest animals that graze on the plankton and nekton; indeed, they are the largest animals that have ever existed on earth—heavier than the largest dinosaurs known. A number of heterotrophic, free-swimming organisms, such as fishes and crustaceans, move up into the plankton at times—in some instances on a regular cycle—to feed on the other organisms there at these times.

The populations of organisms that make up the plankton are able to increase so rapidly and the turnover of nutrients in the sea is so great that the amount of productivity in these systems, although it is still low, has been seriously underestimated in the past. Even though nitrogen and phosphorus are often present in only small amounts and organisms may be relatively scarce, this productivity reflects rapid use and recycling rather than the abundance of these nutrients. The smallest organisms turn over the phosphorus much more rapidly than do the larger ones, and their role in these complex and productive ecosystems is just starting to be understood properly.

About 40% of the world's photosynthetic productivity is estimated to occur in the oceans. Of this, perhaps half is carried out by organisms less than 10 micrometers long. The turnover of nutrients in the plankton is much more rapid than in most other ecosystems, and the total amounts of nutrients are very low.

The Abyssal Zone

The area of sea floor at depths below 1000 meters is about twice that of all the land on earth. The sea floor itself is a thick blanket of mud, consisting of fine particles that settle from the overlying water and that have accumulated over millions of years. Because of high pressures (an additional atmosphere of pressure for every 10 meters of depth), cold temperatures (2° to 3° C), darkness, and lack of food, the first deep-sea biologists thought that nothing could live there. In fact, although the number of individual animals decreases with increasing depth in the sea, the number of species that live at great depth is now known to be quite high. Most of these animals are only a few millimeters in their largest dimension, although larger ones also occur in these regions. Some of the larger ones are bioluminescent (Figure 26-10) and thus are able to communicate with one another or attract their prey.

Many deep-sea animals are known from only a few samples. Today, they are sampled quantitatively by removing cubes of mud from the sea floor. Such samples have revealed a great diversity of animal species, suggesting that the deep sea might represent a reservoir of biological diversity perhaps even rivaling that of some tropical land areas.

Most of the organic matter in the plankton is recycled in the surface layers of the ocean; the animals on the bottom depend for food on the meager leftovers from organisms living kilometers overhead. The low densities and small size of most deep-sea bottom animals is at least in part a consequence of this low food supply. In 1977,

FIGURE 26-10
A deep-sea organism. The luminous spot below the eye of this deep-sea fish results from the presence of a symbiotic colony of luminous bacteria. Similar luminous signals are a common feature of deep-sea animals that move about.

FIGURE 26-11

Life in the abyssal zone. These giant beardworms are members of a small phylum of animals, the Pogonophora. They are living along warm-water vents in fissures along the Galapagos Trench in the Pacific Ocean; similar colonies occur at depths of up to 3000 meters. Water jets from these fissures at a scalding 350° C, but it soon cools to the 2° C temperature of the surrounding water. Hydrogen sulfide also emerges from these vents in abundance. Bacteria use the hydrogen sulfide as a source of energy; these bacteria in turn make possible the existence of the diverse community of animals—including these remarkable beardworms—which was discovered in 1977. These ecosystems have proved to be of extraordinary interest, since they are among the most important on earth that do not depend in any way on photosynthesis or on energy from the sun.

oceanographers diving in a research submarine were surprised to find dense clusters of large animals living on geothermal energy at 2500 meters depth. These deep-sea oases were located as a result of a discovery of equal importance to our understanding of the chemistry of the oceans. Seawater circulates through porous rock at sites where molten material from beneath the earth's crust comes close to the rocky surface of the Mid-Ocean Ridge—the worldwide feature where basalt erupts through the ocean floor, causing the phenomenon of plate tectonics.

This water is heated to temperatures in excess of 350° C and, in the process, becomes rich in reduced compounds. These compounds, such as hydrogen sulfide, provide the energy for bacterial primary production through chemosynthesis instead of photosynthesis. Mussels, clams, and large red-plumed worms in a phylum unrelated to any shallow-water invertebrates cluster around the vents (Figure 26-11). The mass of animal tissue from these animals in the few square meters around the vents is about 30 kilograms per square meter. This tissue contains bacteria living symbiotically within the animals. The animal supplies a place for the bacteria to live, and transports CO_2, H_2S, and O_2 to them for their growth; the bacteria supply the animal with organic compounds to use as food. Polychaete worms, anemones, and limpets—animals that we will discuss in later chapters—live on free-living chemosynthetic bacteria. Crabs act as scavengers and predators, and some of the fish are predators, thus completing the ecosystem—one of the few on earth not dependent on the sun's energy.

FRESH WATER

Freshwater habitats are distinct from both marine and terrestrial ones, but they are limited in area. Inland lakes cover about 1.8% of the earth's surface, and running water covers about 0.3%. All freshwater habitats are strongly connected with terrestrial ones, with marshes and swamps constituting intermediate habitats. In addition, a large amount of organic and inorganic material continuously enters bodies of fresh water from communities growing on the land nearby (Figure 26-12). Many kinds of organisms are restricted to freshwater habitats (Figure 26-13); when they occur in rivers and streams, they must be able to attach themselves in such a way as to resist or

FIGURE 26-12

A freshwater habitat. An abundant growth of algae occurs in the waters of this nutrient-rich stream in the North Coast Ranges of California in summer. As in all streams, much organic material falls or seeps into the water from the communities along the edges; this input is responsible for much of the biological productivity of the stream.

FIGURE 26-13
Freshwater organisms.
A Speckled darter (*Etheostoma stigmaeum*).
B Green frog (*Rana clamitans*).
C Freshwater snail.
D Giant waterbug with eggs on its back.
E Damselfly nymph.
F Bladderwort.

avoid the effects of current, or risk being swept away. In bodies of standing water such matters are of much less importance.

Ponds and lakes, like the ocean, have three zones in which organisms occur: a littoral zone; a limnetic zone, inhabited by plankton and other organisms that live in open water; and a profundal zone, below the limits of effective light penetration. Thermal stratification is characteristic of the larger lakes in temperate regions (Figure 26-14). Since water is densest at about 4° C, further cooling of the water as winter progresses will result in a layer of cooler, lighter water, which freezes to form a layer of ice at the surface. Below the ice, the water remains between 0° and 4° C, and plants and animals survive. In spring, as the ice melts, warmer water is formed in the surface layers of the lake, mixing with the cooler water below. This process is known as the spring overturn; in it, nutrients formerly held in the depths of the lake return to the surface.

In summer, warmer water forms a layer, known as the epilimnion, over the cooler water, called the hypolimnion (about 4° C), that lies below. There is an abrupt change in temperature, the thermocline, between these two layers. Depending on the climate of the particular area, the epilimnion may become as much as 20 meters thick during the summer. In the autumn the temperature of the epilimnion drops until it is the same as that of the hypolimnion, 4° C. When this occurs, the epilimnion and hypolimnion mix—a process called the fall overturn. Colder water reaches the surfaces

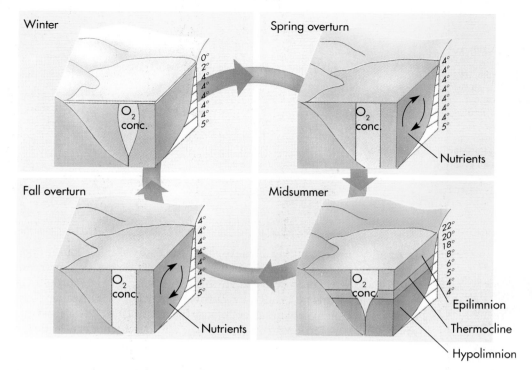

of lakes, therefore, in the spring and fall, bringing up near the surface fresh supplies of dissolved nutrients that have accumulated in them.

In temperate lakes and ponds, there is an upper layer, or epilimnion; a lower layer, or hypolimnion; and a thermocline, which is a zone of abrupt temperature change, between them. The surface waters of the lake are heated in summer but cool in autumn to 4° C, the point at which water is most dense. At that point, when their temperatures are equal, the shallower and deeper waters of the lake mix, with oxygen being carried to the depths and nutrients being brought to the surface.

Lakes can be divided into two categories, based on their production of organic matter. In eutrophic lakes there is an abundant supply of minerals and organic matter (Figure 26-15). Oxygen depletion occurs below the thermocline in the summer, and the organic material there is used up rapidly by oxygen-requiring organisms. Most of this organic material drifts down from the well-illuminated surface waters of the lake; it is highly variable in quantity, and thus in oxygen-depleting properties, from lake to lake. The lack of oxygen in the deeper waters of some lakes may have other profound effects, such as allowing the conversion of relatively harmless materials such as sulfate and nitrate into toxic materials such as hydrogen sulfide and ammonia. When the entire lake again reaches 4° C, intermixture occurs and the waters become more uniform

FIGURE 26-14
Stratification in fresh water. The pattern of stratification in a large pond or lake in temperate regions (**A**) is upset in the spring and fall overturns (**B**). Of the three layers of water shown in B at the far right, the hypolimnion consists of the densest water, at 4° C; the epilimnion consists of warmer water that is less dense; and the thermocline is the zone of abrupt change in temperature that lies between them. If you have dived into a pond in temperate regions in the summer, you have experienced the existence of these layers directly.

FIGURE 26-15
Eutrophic farm pond, with evident bloom of green algae, rapidly filling in with vegetation.

FIGURE 26-16
Oligotrophic lakes are highly susceptible to pollution. Lake Tahoe, an oligotrophic lake, lies high in the Sierra Nevada on the border between California and Nevada. The drainage of fertilizers applied to the plantings around residences, business concerns, and recreational facilities bordering the lake poses an ever-present threat to the maintenance of the deep blue color of its water.

FIGURE 26-17
The distribution of biomes. Temperate evergreen forest and chaparral are discussed under the heading "temperate deciduous forest" in the text; warm, moist evergreen forest and tropical monsoon forest are transitional between tropical evergreen rain forest and savannas; and semideserts are grouped with deserts for purposes of discussion.

in fertility throughout. In oligotrophic lakes, on the other hand, organic matter and nutrients are relatively scarce; such lakes are often deeper than eutrophic ones (Figure 26-16), and their deep water is always rich in oxygen. Oligotrophic lakes are highly susceptible to drastic change by nutrient pollution, because they have such limited quantities of many potential pollutants when undisturbed. Excess phosphorus from sources such as fertilizer runoff, sewage, and detergents can lead to harmful effects such as runaway weed growth or the promotion of "blooms" of algae that can rapidly deplete the lake's oxygen supply.

BIOMES

Biomes are climatically delineated assemblages of organisms that have a characteristic appearance and that are distributed over a wide land area. Biomes are classified in several ways, but, for our purposes, we will discuss them under seven categories: (1) tropical rain forests, (2) savannas, (3) deserts, (4) temperate grasslands, (5) temperate deciduous forests, (6) taiga, and (7) tundra. They vary remarkably from one another

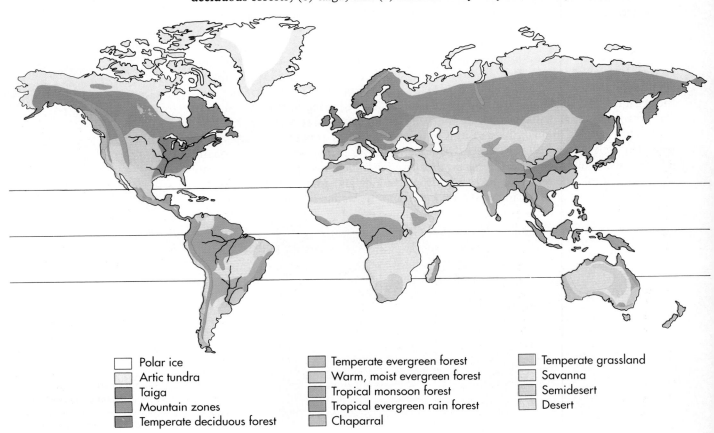

Polar ice
Artic tundra
Taiga
Mountain zones
Temperate deciduous forest

Temperate evergreen forest
Warm, moist evergreen forest
Tropical monsoon forest
Tropical evergreen rain forest
Chaparral

Temperate grassland
Savanna
Semidesert
Desert

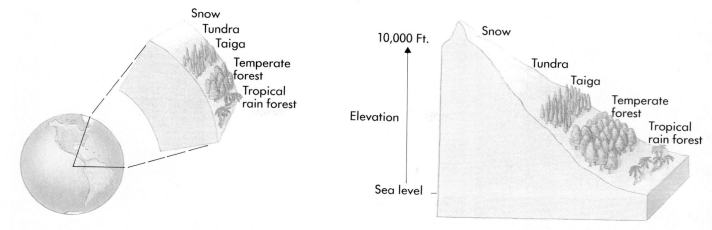

because they have evolved in regions with very different climates. The distributions of the principal biomes that we will discuss, along with those of some of the related communities, are shown in Figure 26-17.

If there were no mountains and no climatic effects caused by the irregular outlines of the continents and by different sea temperatures, each biome would form an even belt around the globe. In fact, their distribution is greatly affected by these factors, especially by elevation. Those biomes that normally occur at high latitudes follow an altitudinal gradient along mountains; thus the summits of the Rocky Mountains are covered with a vegetation type that resembles tundra, whereas other forest types that resemble taiga occur further down (Figure 26-18). It is for reasons such as these that the distributions of the biomes are so irregular. We will now examine some of the distinctive features of the individual biomes, using the seven enumerated before to organize our discussion.

FIGURE 26-18
Elevation affects the distribution of biomes. Biomes that normally occur far north and far south of the equator at sea level occur also in the tropics but at high mountain elevations. Thus on a tall mountain in southern Mexico or Guatemala one might see a sequence of biomes like the one illustrated here.

Tropical Rain Forests

The tropical rain forests (Figure 26-19) are the richest biome in terms of number of species, probably containing at least half of the species of terrestrial organisms—more than 2 million species, and possibly as much as ten times that many. In such

B   C

FIGURE 26-19
The tropical rain forest.
A The tropical rain forest is characterized by an enormous variety of trees, often with vines and epiphytes growing on them.
B A member of the pineapple family (Bromeliaceae), characteristic of the tropics of Latin America. Most of the members of this family are epiphytes. Water, held within the "tanks" formed at the base of the leaves in epiphytes such as the one shown here, provides a home for many kinds of animals, including frogs and salamanders.
C A male resplendent quetzal (*Pharomachrus mocinno*), which lives in the canopy of the forest, feeding largely on fleshy fruits.

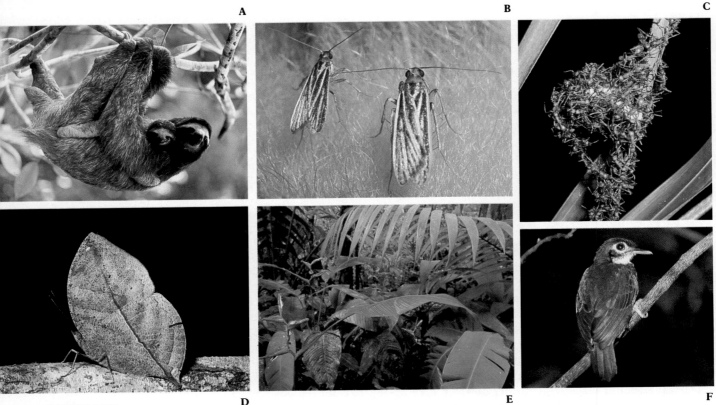

A   **B**   **C**

D   **E**   **F**

FIGURE 26-20

Ecological specialization in the tropical rain forest.

A A mother and baby three-toed sloth (*Bradypus infuscatus*). Sloths reside in the canopy of the tropical rain forest, as shown here in Panama.

B Sloth moths (*Bradypodicola*). These moths carry out their entire life cycle in the fur of the three-toed sloth, where their larvae feed on the green algae that grow luxuriantly there.

C Army ants live as huge, mobile, foraging colonies in the tropical rain forest. Here a small column is transporting a wasp larva as food.

D A dead leaf butterfly (*Kallima*) in the forests of Sumatra. When its wings are folded, this butterfly is superbly camouflaged.

E The orange and red flower clusters of this individual of *Heliconia irrasa* stand out like beacons in the Costa Rican forest. They are visited and pollinated by the hummingbirds they attract.

F Some species of birds, such as the bicolored antbird, *Gymnopithys bicolor*, shown here, make a permanent living by flying along above the columns of army ants and feeding on the insects that they flush from the foliage.

rain forests neither water nor temperature is a limiting factor, and an evergreen forest of giant trees supports a rich and diverse assemblage of plants and animals on its branches, high above the forest floor. Plants that grow on the branches of other plants are called epiphytes (see Figure 26-19, *B*) and, along with the numerous vines that occur in tropical forests (Figure 26-20, *A*), they reach the light high up in the forest canopy. Hemiepiphytes, such as strangler figs, begin their life as epiphytes, but eventually their roots reach down to the soil, and they become rooted in the forest floor. The communities that make up tropical forests are rich in species and are diverse, so that each kind of plant, animal, or microorganism is often represented in a given area by few individuals. There are seldom fewer than 40 species of tree per hectare, four or five times as many as are typical in temperate forests. In a single square mile of tropical forest in Peru or Brazil there may be 1500 or more species of butterflies—twice the total number found in the United States and Canada combined. The ways of life of tropical organisms are often specialized and highly unusual (Figure 26-20); some were discussed in Chapter 24. This pattern of narrow ecological specialization enables many species to occur together in the tropical rain forests.

The rainfall in areas where tropical rain forests occur is generally 200 to 450 centimeters per year, with little difference in its distribution from season to season. There are, of course, substantial changes with elevation. About two thirds of the soils of the tropics are acidic and deficient in phosphorus, potassium, calcium, magnesium, and other nutrients. In addition, the phosphorus in them tends to combine with iron or aluminum to form insoluble compounds that are not available to plants. Such soils also tend to have toxic levels of aluminum. Most of the roots of the trees spread out in a thin layer of soil, often no more than a few centimeters thick. These roots transfer the nutrients from the leaves and other fallen organic debris quickly and efficiently back to the trees themselves. Below the thin layer of topsoil, there is virtually no organic matter, and nutrients are scarce; most of the nutrients that exist in the system are concentrated in the trees themselves.

Tropical rain forests are highly productive, even though they exist mainly on infertile soils. Most of the nutrients are held within the plants themselves and are rapidly recycled when the plants die or when parts, such as leaves, are lost.

Tropical rain forests are widespread in South America, particularly in and around the Amazon Basin; in Africa, particularly in Central and West Africa; and in Southeast Asia. Their ecological properties are so unusual that we simply do not know, in most cases, how to cultivate them in such a way as to maintain the agricultural productivity of these areas year after year once their forest cover has been removed. Even so, these forests are being cut down at an ever-increasing rate, mainly by people living at the edge of starvation. The human population of the tropics and subtropics now constitutes more than half of the world total, and their numbers are growing rapidly. Extremely poor people constitute more than a third of the population of most of these countries. The small fields that these people clear by what are called slash-and-burn methods can generally be cultivated for only a few years, after which they are worthless, unless they are given many decades to recover. When wide areas are cleared, the forests will probably never recover. The thin soils erode rapidly, and the minerals are carried away with the trees and the crops that are harvested from the cultivated plants. Consequently, few tropical forests are likely to be left in an undisturbed condition anywhere in the world by the first part of the next century.

The destruction and disturbance of all tropical rain forests will be accompanied by the extinction of a major proportion of all plant, animal, and microorganism species on earth—perhaps one fourth—during the lifetime of many of us. Studying the plants and animals of these forests—which are the most poorly known, as well as the most numerous in terms of numbers of species on earth at present—is a matter of pressing importance from both a scientific point of view and from the point of view of improving the conditions of life for human beings. Such study would undoubtedly uncover many species of organisms of great scientific interest and of potential importance in terms of contributions to the quality of human life.

Savannas

In areas of reduced annual precipitation, or prolonged annual dry seasons, open tropical and subtropical deciduous forests give way to a kind of open grassland with scattered shrubs and trees, called savannas (Figure 26-21). The savanna biome, on a global scale, is in a sense transitional between tropical rain forest, which is evergreen, and desert. Generally 90 to 150 centimeters of rain fall each year in savannas, which generally grow on nutrient-poor soils that are often rich in aluminum, a substance that is toxic to most plants. Soil conditions apparently exert the major influence on

FIGURE 26-21
Savanna. The herds of grazing mammals that inhabit the African savannas are one of the world's greatest sights.

the formation of savannas, at least in Latin America, but water relationships are important also. There is a wider fluctuation in temperature here during the year than in the tropical rain forests, and there is seasonal drought. These factors have led to the evolution of an open landscape, often with widely spaced trees, in which large grazing mammals are sometimes characteristic, as in Africa. Periodic fires also are an important factor in the maintenance of savannas.

Some tens of thousands of years ago, toward the close of the Pleistocene Epoch, vegetation similar in appearance to that of today's African savannas was widespread in North America (see Figure 22-2). It disappeared as the climate became more and more like it is now—with even greater extremes of temperature and longer periods of seasonal drought, especially in the West. In many areas, humans seem to have slaughtered the remaining large animals that made up the extraordinary Pleistocene herds, and the vegetation evolved into the modern communities that we know today.

In savannas, the trees are usually deciduous and lose their leaves in the dry season. Savannas often gradually give way on their drier borders to thorn forest, plant communities dominated by thorny trees that are seasonally deciduous. In Southeast Asia, a similar plant community, called monsoon forest, occurs in such dry regions. Under and among the scattered trees of these communities, perennial grasses and other plants with food stored in roots or underground stems are common.

Desert

Less than 25 centimeters of annual precipitation usually falls in the world's desert areas—so low an amount that water is the predominant controlling variable for most biological processes and is also highly variable in quantity both during a given year and from year to year. In desert regions the vegetation is characteristically rather sparse (Figure 26-22). Such regions occur around 20 to 30 degrees north and south latitude, where the warm air that rises near the equator falls and precipitation is limited. Deserts are most extensive in the interiors of continents, especially in Africa (the Sahara Desert), Eurasia, and Australia. Less than 5% of North America is desert.

Because the vegetation is sparse and the skies are usually clear, deserts radiate heat rapidly at night. This leads to large daily changes in temperature, sometimes exceeding 30° C, between day and night. Summer daytime temperatures in deserts are extremely hot, frequently exceeding 40° C. Indeed, atmospheric temperatures of 58° C have been recorded both in Libya and in San Luis Potosi, Mexico—the highest that have been recorded on earth.

FIGURE 26-22
Scenes in the North American deserts.
A Flats dominated by cholla cactus (*Opuntia*) and shrubs, Borego Valley, southern California. Annuals appear seasonally in abundance on these flats.
B Fan palms (*Washingtonia filifera*), the only native species of palm in the western United States, around an oasis near Palm Springs, California. Most desert trees have deep roots, that enable them to reach water.
C Great Basin sagebrush (*Artemisia tridentata*) desert in the John Day Valley, Oregon. Cold desert of this kind is widespread in the interior western United States.

A

B

C

A

B

Annual plants often are abundant in deserts and simply bypass the unfavorable dry season in the form of seeds. After the rainfall, they germinate and grow rapidly, sometimes forming spectacular natural displays. Exhibiting a similar seasonal rhythm, animals such as fairy shrimps (see Figure 41-23, *B*) often appear and breed rapidly in temporary ponds.

The trees and shrubs that live in deserts often have deep roots that reach sources of water far below the surface of the ground. Thus trees may grow, even in regions that essentially lack precipitation, such as in the Atacama Desert of northern Chile. The woody plants that grow in deserts may be either deciduous, losing their leaves during the hot, dry seasons of the year, or evergreen, with hard, reduced leaves: the creosote bush (*Larrea*) of the deserts of North and South America is an example of an evergreen desert shrub. Near the coasts, in areas where there is cold water offshore, deserts may be foggy, and the water that the plants obtain from the fog may allow them to grow quite luxuriantly.

Succulent plants are more common in desert regions than elsewhere. Like many other kinds of plants and animals that occur in desert regions, they often have evolved a life-form that is successful under desert conditions (Figure 26-23). Such a phenomenon, in which unrelated groups of organisms evolve to become superficially similar, is called **convergent evolution.** Such succulent plants often exhibit CAM photosynthesis (see Chapter 9). Their stomata open at night and close during the day, thus conserving water while cutting down on photorespiration during the day. Succulent plants that have this system are thus able to avoid the effects of the intense heat of the desert days and still photosynthesize efficiently.

Desert animals, too, have fascinating adaptations that enable them to cope with the limited water and high temperatures of the deserts (Figure 26-24). They often limit their activity to a relatively short period of the year when water is available, or even plentiful: they resemble annual plants in this respect. Many desert vertebrates live in deep, cool, and sometimes even somewhat moist burrows; and those that are active over a greater portion of the year emerge from these burrows only at night, when temperatures are relatively cool. Some, like camels, drink large quantities of water when it is available and then safely withstand the loss of much of it. Many animals simply migrate to or through the desert, where they exploit food that may be abundant seasonally; when the food disappears, the animals move on to more favorable areas.

FIGURE 26-23

Convergent evolution in two unrelated groups of plants makes these two desert scenes resemble one another.

A Organ pipe cactus (*Lemaireocereus thurberi*) in Organ Pipe National Monument, New Mexico.
B A giant spurge (*Euphorbia*) near Windhoek, Namibia, southwestern Africa.

FIGURE 26-24

Two examples of strategies by which desert animals conserve water.

A Spadefoot toad, *Scaphiophus.* Spadefoot toads, which live in the deserts of North America, can burrow nearly a meter below the surface and remain there for as much as 9 months of each year. Under such circumstances, their metabolic rate is greatly reduced, and they depend largely on their fat reserves. When moist, cool conditions return to the desert, they emerge and breed rapidly. The young toads mature rapidly and burrow back underground, using the horny projections on their feet, which give these unusual desert toads their name.
B On the dry sand dunes of the Namib Desert in southwestern Africa, the beetle *Onymacris unguicularis* collects fog water by holding up its abdomen at the crest of a dune, thus gathering condensed water on its body.

A

B

Many arthropods combine several of these strategies with a tough exoskeleton and colors suited to the absorption or reflection of heat, depending on their habits. The cuticles and exoskeletons of all arthropods constitute an excellent preadaptation to desert conditions and are an important reason why the members of this phylum are so successful in deserts.

Desert survival depends on water conservation, achieved by structural, behavioral, or physiological adaptations. Plants and animals may restrict their activity to favorable times of the year, when water is present.

One of the means by which desert animals avoid seasonal extremes in heat and dryness in the desert is **estivation.** Estivation is a prolonged state of torpor that occurs under hot, dry conditions, in contrast to the hibernation of animals in cold climates, which is a more profound and prolonged condition of dormancy. Ground squirrels (*Spermophilus*) hibernate in the cold regions of North America, whereas the species of the same genus that occur on the deserts estivate, each thereby avoiding the extremes of the most unfavorable seasons. Estivation occurs in a number of desert rodents and some other animals, including a few birds such as the whippoorwill.

Temperate Grasslands

Temperate grasslands cover much of the interior of North America, and they are widespread in Eurasia and South America as well. Such grasslands are often highly productive when they are converted to agriculture, and many of the rich agricultural lands in the United States and southern Canada were originally occupied by prairies, another name for temperate grasslands. The roots of perennial grasses characteristically penetrate far into the soil, and grassland soils tend to be deep and fertile, the best agricultural soils in the world. Temperate grasslands differ from savannas in that they occur in regions with relatively long, cold winters, whereas savannas occur where there is a relatively cool dry season and a hot, rainy one. In addition, savannas grow on relatively infertile soils, and temperate grasslands on rich soils.

Where precipitation is relatively abundant, as in the eastern portion of the North American prairies, a kind of vegetation known as tall-grass prairie once existed (Figure 26-25). Where there is even more rainfall, as occurs in North America east of the prairie zone, forests occur. The line between tall-grass prairie and forest in the United States lies roughly along the border between Illinois (prairie) and Indiana (forest), but this boundary is not sharp. The forests of the eastern United States contain numerous local patches of prairie, sometimes called glades, in which prairie plants and animals occur, often far from their main ranges. Similarly, patches of woods may occur in prairies and often are found along streams.

FIGURE 26-25
Tall-grass prairie. Tall-grass prairie stretched over thousands of square kilometers in the interior of North America when Europeans first came to the area. This scene, in which a native sunflower (*Helianthus grosseserratus*) is prominent, is in western Illinois. Over much of its former area, tall-grass prairie has been replaced by cultivated fields.

FIGURE 26-26
Short-grass prairie with grazing bison.

Farther west in the United States and Canada, another kind of grassland, the short-grass prairie, occurs (Figure 26-26). Short-grass prairie receives less rain than tall-grass prairie and is exceptionally sensitive to disturbance through overuse. The infamous "Dust Bowl" conditions that occurred in the United States in the 1930s stemmed from the mismanagement of short-grass prairies, linked with a succession of dry years that was not abnormal for such areas.

Temperate grasslands are often populated by herds of grazing mammals, among which the North American bison is a familiar example. In this respect, temperate grasslands resemble tropical savannas. Both biomes are characterized by large quantities of perennial grasses that are highly productive when properly managed and that can support such herds. In turn, the grasslands are maintained both by the grazing and by periodic fires. In their absence, litter accumulation, which ties up nutrients and blocks seedling growth, greatly limits their net productivity.

Temperate Deciduous Forests

In areas of the Northern Hemisphere with relatively warm summers, relatively cold winters, and sufficient precipitation, temperate deciduous forest occurs (Figure 26-27). Temperate deciduous forest and related vegetation types cover large areas, including much of the eastern United States and Canada and an extensive region in Eurasia. Much of the development of civilization in the North Temperate region took place within the boundaries of the temperate deciduous forest. The existing deciduous forests are remnants of a more continuous, richer forest that stretched across the

FIGURE 26-27
Temperate deciduous forest. The leaves of the trees in the temperate deciduous forest often change color before they fall.

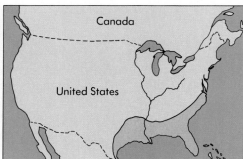

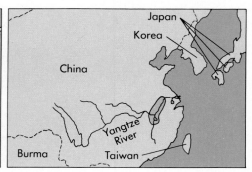

FIGURE 26-28
Distribution of alligators in North America and China. Above is an American alligator; the Chinese species has become rare and is regarded as endangered. These are the only two species of the genus *Alligator*.

northern hemisphere before the Pleistocene Epoch but that has become more restricted in distribution and in the array of species it houses as the world climate has become colder, drier, and more seasonal, especially during the past several million years. For this reason the plants and animals of the rich forests of southwestern and western China resemble those of eastern North America (Figure 26-28).

Annual precipitation in areas of temperate deciduous forest is generally from about 75 to about 250 centimeters. Precipitation is well distributed throughout the year, but water is generally unavailable during the winter because it is frozen. Where there is less precipitation, temperate deciduous forests are replaced by grassland, as in the prairies of North America or the steppes of Eurasia. Where conditions are more limiting—where restricted, for example, by intense cold—these forests may be replaced by others that consist of fewer species, as in the taiga or the predominantly coniferous forests of western North America.

> **Temperate deciduous forests cover extensive areas in North America and Eurasia; they are characterized by cold winters and warm, rainy summers.**

Many perennial herbs live in areas of temperate deciduous forest (Figure 26-29). They characteristically grow rapidly and flower early in the year, before the trees have begun to unfold their leaves. The herbs are able to grow so rapidly because of their underground storage organs, swollen roots or shoots in which the photosynthetic products of earlier seasons have been stored.

The chaparral of California—as well as related evergreen scrub associations that occur in the five areas of the world with a mediterranean, or summer-dry, climate—is historically derived from forests like the richest temperate deciduous forests that now grow in eastern Asia and eastern North America (see Figure 25-1). The regions where this kind of evergreen scrub grows are dry in summer and moist in the winter growing season. In the Mediterranean area itself, in central Chile, in the Cape region of South Africa, and perhaps to a lesser extent in southern and southwestern Australia and in California, human activities have also contributed significantly to the formation

FIGURE 26-29
Perennial herbs are common in the temperate deciduous forest. Many species of perennial herbs that grow on the floor of temperate deciduous forest, such as this pink ladyslipper orchid (*Cypripedium acaule*), flower and are visited by insects in the well-lighted conditions that exist before the trees have put out their leaves in spring.

FIGURE 26-30
Fire is a prominent ecological factor in chaparral. Following a fire in the chaparral of southern California, nutrients are released in abundance from the burned shrubs, light is plentiful, and seeds of many annual plants, rarely seen except following a fire, germinate in abundance, often producing colorful displays. Here plants of the genera *Phalecia* and *Mimulus* are conspicuous, along with a number of others.

of such plant communities. Fire is a prominent ecological factor in chaparral, and the individual kinds of plants respond to it either by sprouting from the base or seeding on the burn (Figure 26-30). Those who build houses in the chaparral expose themselves to the risk of being burned out, a perfectly natural phenomenon in the region. This is especially true since contemporary practices limit the occurrence of fires, which means that when they do occur, a great deal of organic material will probably have accumulated and the fires may then be very destructive, as seen, for example, in Yellowstone Park in the summer of 1988.

Taiga

Taiga is the northern coniferous forest, primarily spruce, hemlock, and fir, that extends across vast areas of Eurasia and North America (Figure 26-31). It is characterized by winters that are long and cold, in which the cold air contains little moisture; most of the precipitation falls during the summer. Because of the latitude where taiga occurs, the days are short in winter (as little as 6 hours) and correspondingly long in summer. During the summer, plants may grow rapidly, and crops often attain a large size in a surprisingly short time. Marshes, lakes, and ponds are common here, and they are often fringed by willows or birches. Most of the trees in the taiga tend to occur in dense stands of one or a few species. Alders, which are common, harbor nitrogen-fixing bacteria in nodules on their roots; partly for this reason, they are able to colonize raw, infertile soils such as those left behind by the recent retreat of glaciers or other ice (see Figure 25-6).

In winter, the taiga receives a deep blanket of snow. During the long, snowy season, animals must adapt to a way of life different from the one they pursue during the brief summer, or they may migrate to escape the snow. There are changes in color, in food habits, and in many other basic aspects of life. The snow protects the ground from freezing in the taiga, and thus allows the growth of forest; in the less snowy tundra biome, further north, ice lies too close to the surface of the soil for forest to grow. Beneath the thick cover of snow in the taiga there exists an active community of rodents and other animals, completely protected from most predators.

Many large mammals live in the taiga, including elk, moose, deer, and carnivores such as wolves, bear, lynx, and wolverines. Traditionally, much fur trapping has gone on in this region, which is also an important lumber-producing region. To the south, taiga grades into forests or grasslands, depending on the amount of precipitation. Northward, it gradually gives way to open tundra. Coniferous forests also occur in the mountains to the south, but these are often richer and more diverse in species than

A

B

FIGURE 26-31
Taiga.
A Taiga, here a forest dominated by spruces and alders, in Alaska.
B Moose are one of the characteristic mammals of the taiga.

FIGURE 26-32

Tundra.

A Tundra in Mt. McKinley National Park, Alaska.

B Caribou live in large herds that migrate across the tundra. Sometimes called reindeer, they have been domesticated by the Lapps in northern Scandinavia as a source of meat, milk, and hides.

C Arctic ground squirrels hibernate in burrows during the long arctic winters. They consume large quantities of plant material during the short arctic summers to build up energy reserves.

A

B

C

those of the taiga. In the early Tertiary Period, 40 to 60 million years ago, all of these conifers grew in richer, mixed forests that extended far to the north, above the Arctic Circle. As more severe winters developed over much of the region and sufficient summer rainfall disappeared from the ranges of western North America, ecosystems formed that were poorer in their representation of species, such as taiga.

Tundra

Farthest north in Eurasia, North America, and their associated islands, between the taiga and the permanent ice, occurs the open, often boggy community known as the tundra (Figure 26-32). This is an enormous biome, extremely uniform in appearance, that covers a fifth of the earth's land surface. Trees are small and are mostly confined to the margins of streams and lakes; in general, the tundra is dominated by scattered patches of grasses and sedges (grasslike plants), heathers, and lichens, with denser stands in wet places.

Annual precipitation in the tundra is low, usually less than 25 centimeters, and the water is unavailable for most of the year because it is frozen. During the brief arctic summers, water sits on frozen ground, and the surface of the tundra is often extremely boggy then. Permafrost, or permanent ice, usually exists within a meter of the surface.

> **The tundra receives little precipitation, usually less than 25 centimeters per year, but the water is often trapped near the surface by the widespread permafrost. For that reason, the tundra is often boggy.**

As in the taiga, the herbs of the tundra are perennials that grow rapidly during the brief summers, using food stored underground. Large grazing mammals, including musk-oxen, caribou, reindeer, and carnivores such as wolves, foxes, and lynx, live in the tundra, which teems with life in the short summer. Lemmings, a genus of small rodents, are animals of the tundra, and their populations rise rapidly and then crash on a long-term cycle, with important effects on the populations of the animals that prey on them.

THE FATE OF THE EARTH

Now that you have completed your survey of the biomes and the corresponding communities that occur in the sea, you have gained some appreciation of the different kinds of climate under which these areas have evolved over tens or hundreds of millions of years. In just the last several hundred years, in contrast, human populations have spread over most of the land surface of the globe. Having already converted large stretches of many biomes to agriculture, urban areas, or other uses, human populations are attacking those biomes—such as the tropical rain forests—that are less suitable for exploitation and about which we know much less. The human population doubled from 1950 to 1987, and most of that growth has occurred in the warmer portions of the globe. There are now so many of us that we must manage the whole earth as a single system, if those who follow us are to find a stable ecological situation and one in which they can live out their lives in peace and relative prosperity. In the next chapter, we will examine some of the factors that will determine our success or failure in this great enterprise.

SUMMARY

1. Warm air rises near the equator and flows toward the poles, descending at about 30 degrees north and south latitude. Since the air is falling in these regions, it is warmed, and its moisture-holding capacity is therefore increased. The great deserts of the world are formed in these latitudes.

2. The ocean comprises three major environments: the neritic zone, the surface layers, and the abyssal zone. The neritic zone, which lies along the coasts, is small in area but very productive and rich in species. The surface layers are the habitat of plankton (drifting organisms) and nekton (actively swimming ones). The productivity of this zone has been underestimated because of the very small size (less than 10 micrometers) of many of its key organisms and because of its rapid turnover of nutrients.

3. Freshwater habitats comprise only about 2.1% of the earth's surface; most of them are ponds and lakes. These possess a littoral zone, a limnetic zone, and a profundal zone. As autumn changes to winter, cooler water forms and sinks, since water is most dense at 4° C, and the water of the lake is mixed by this means. When winter changes to spring, warmer water (at 4° C) sinks, producing a similar mixing.

4. We have grouped our discussion of biomes in seven major categories. These are: (1) tropical forests, (2) savannas, (3) deserts, (4) temperate grasslands, (5) temperate deciduous forests, (6) taiga, and (7) tundra. Tropical forests contain at least half of the total species of plants, animals, and microorganisms. These are the most poorly

known organisms on earth and the most directly threatened by extinction in the near future.

5. Savannas, which were much more widespread during the Pleistocene Epoch, are highly productive ecosystems. They are often inhabited by large herds of grazing mammals and their predators, including humans.

6. Deserts are the hottest and driest habitats on earth. They are of great biological interest because of the extreme behavioral, morphological, and physiological adaptations of the plants and animals that live in them.

7. Temperate grasslands cover a large area in the interior of North America; they are also widespread in Eurasia and South America. When they receive sufficient precipitation, they are often well suited to agriculture.

8. Temperate deciduous forests once dominated huge areas of the northern hemisphere. Today, though, they are confined largely to eastern North America and parts of southern China. Areas with mediterranean (summer-dry) climates feature distinctive life-forms that are historically derived from those of the temperate deciduous forests.

9. Taiga is a vast coniferous forest that stretches across Eurasia and North America. Its winters are long and cold, but plants may grow rapidly here during the long summer days.

10. Even farther north is the tundra, which covers about 20% of the earth's land surface and consists largely of open grassland, often boggy in summer, which lies over a layer of permafrost.

REVIEW QUESTIONS

1. What is a biome? What are the two key physical factors that affect the distribution of biomes across the earth?

2. Why are the majority of great deserts located near 30 degrees north and south latitude? Is it more likely that a desert will form in the interior or at the edge of a continent? Explain. Why is the windward side of a mountain generally moister than the leeward side?

3. Is there a greater variety of species on the land or in the oceans? Why?

4. What are the three major kinds of oceanic habitats? What distinguishes each zone? What variety of life is characterized by each zone?

5. What types of organisms constitute the plankton of the ocean's surface zone? What types of organisms constitute the nekton? How important are the photosynthetic plankton to the survival of the earth? What is the level and turnover of nutrients in the surface zone?

6. What proportion of the earth's water is present in the form of fresh water? List and describe the three zones characteristic of freshwater lakes and ponds.

7. How do eutrophic lakes differ from oligotrophic lakes? Which is more susceptible to the effects of pollution? Why?

8. List the seven most prominent biomes in the world according to their apparent distance from the equator. How is this stratification imitated in changes in elevation?

9. What are the key characteristics of the tropical rain forests in terms of number of species, level of specialization, amount of rainfall, amount of nutrients in the soil, and location of nutrient concentration? Why is slash-and-burn agriculture so damaging to the tropical rain forest?

10. What are the key characteristics of the desert biome with respect to rainfall, amount of vegetation, temperature fluctuations on daily as well as seasonal scale, and tree type? What special adaptations have desert plants and animals developed?

11. What are the key characteristics of the temperate grasslands with respect to productivity, seasonal temperature, and rainfall? In what respect do they resemble the savannas?

12. What are the key characteristics of the taiga in terms of seasonal temperatures, precipitation, day length, soil nutrition, type of trees, and species variation.

THOUGHT QUESTIONS

1. How do you think the decomposition rates of organic matter would differ in a tropical rain forest, a temperate deciduous forest, a desert, and the tundra? How would these differences affect human existence in each area?

2. What kinds of biological communities would you expect to find on the windward and leeward sides of a mountain range in an area where the annual precipitation ranged between 20 and 100 centimeters per year and was distributed mainly in one rainy season? How would the height of the mountain range affect the situation?

3. Near the coast in southern California there are two major plant communities. Right along the ocean occurs the coastal sage community, which is dominated by low shrubs that often wither or lose their leaves in the summer. Higher up, on the ridges, occurs the chaparral, a community dominated by tall evergreen shrubs. Which of these communities would you think grows in the area that receives the most rainfall? Which grows on the better soils? Which is the more productive on an annual basis? Where would you expect to find the most annual plants?

FOR FURTHER READING

BARBOUR, M.G. and W.D. BILLINGS, editors: *North American Terrestrial Vegetation*, Cambridge University Press, New York, 1988. Contemporary treatment of the major vegetation types in North America.

BURGIS, M., and P. MORRIS: *The Natural History of Lakes*, Cambridge University Press, Cambridge, England, 1987. Excellent and well-written survey of lakes and their ecology.

CAUFIELD, C.: *In the Rainforest*, Alfred A. Knopf, New York, 1985. Beautifully written account of the tropical rain forest and the factors that are leading to its destruction.

FORSYTH, A.: *Portraits of the Rainforest*, Camden House Publishing, Camden East, Ontario, Canada, 1990. A magnificently illustrated and balanced account of the rainforest.

MYERS, N.: *The Primary Source: Tropical Forests and Our Future*, W.W. Norton, New York, 1984. An excellent overall account of the characteristics of tropical forests and their importance for us all.

NORSE, E.A.: *Ancient Forests of the Pacific Northwest*, The Island Press, Washington, D.C., 1990. A thorough and interesting account of the controversy surrounding the harvest of ancient forests in the Pacific Northwest of the United States.

PARFIT, M.: "The Dust Bowl," *Smithsonian*, June 1989, pages 44-57. Accentuates some of the characteristics of prairies by showing how they contributed to one of the greatest ecological disasters in North America.

PRANCE, G.T., and T.E. LOVEJOY, editors: *Amazonia*, Pergamon Press, Oxford, England, 1986. Excellent collection of essays that will give you a genuine feeling for research in the tropics.

RASMUSSON, E.M.: "El Niño and Variations in Climate," *American Scientist*, vol. 73, pages 168-177, 1985. An analysis of the way in which large-scale interactions between the ocean and the atmosphere over the tropical Pacific Ocean can dramatically affect weather patterns around the world.

27

The Future of the Biosphere

OVERVIEW

In 1990 the world population surpassed 5.3 billion individuals, having doubled in less than 40 years. At this unprecedented level, the human population is putting a nearly unbearable strain on the earth's capacity to support us all: almost one out of four people lives in extreme poverty, one of ten is severely malnourished, and other species are being driven to extinction at a rate that has not been duplicated since the end of the Cretaceous Period 65 million years ago, when the dinosaurs disappeared forever. The loss of a fifth of our topsoil since 1950 is only one of the signs that the world cannot continue to support us indefinitely if we continue to live as we are—and yet our numbers are growing by some 95 million people a year, towards a projected doubling or tripling a century from now. To improve the situation and achieve global stability, it will be necessary to use the best knowledge available from the natural and social sciences, to alter our behavior drastically, to improve our agricultural systems, and to control the pollution that we are spreading throughout the planet and the upper layers of the atmosphere. Understanding biology will be one essential requirement for intelligent citizens who wish to carry out their lives successfully in the future.

FOR REVIEW

Here are some important terms and concepts that you will encounter in this chapter. If you are not familiar with them, you should review them before proceeding.

Overpopulation (Chapter 23)
Population growth characteristics (Chapter 23)
Cycling of minerals and flow of energy (Chapter 25)
Characteristics of biomes (Chapter 26)

FIGURE 27-1
New York City by satellite in 1985.

The view in (Figure 27-1) is of the city of New York, photographed from a satellite in the spring of 1985. The wakes of ships can be seen in the harbor, wharves lining the shores of the Hudson River; the runways of J.F. Kennedy International Airport are visible at the far right, and the Verrazano Bridge can be seen in the lower center, joining Brooklyn to New Jersey where the Hudson River empties into the Atlantic Ocean. Individual buildings cannot be seen—the scale is too small for that—and from the satellite it is not obvious that the ground below teems with people. At the moment this picture was taken, millions of people within its view were talking, hundreds of thousands of cars struggled through traffic, hearts were broken, babies were born, and dead people were buried. Were our lens sharp enough, we would see frozen in time all of this and more, 15,300,000 people busy at life, a panorama of modern industrial society. All of human history has led to this photograph, and in it the future of humanity can be seen.

Our futures and those of everyone on the planet are linked to the unseen millions in this photograph, for we share the earth with them. The wisdom with which they manage their environment has deep significance for us all, for their environment is also ours. As human numbers increase, so does their impact on our common environment, posing new challenges to us all.

The earth is peppered with cities like New York, growing rapidly. More than a dozen other cities contain more than 10 million inhabitants; the three cities larger than New York City in 1990 (Mexico City, Tokyo, São Paulo) together have over 50 million inhabitants. In 1987 the human population of the earth reached—and surpassed—a significant milestone: 5 billion individuals. The central challenge to the ingenuity of the environmental scientists, politicians, and every citizen is this: there are a lot of people on our earth.

A lot of people consume a lot of food and water, use a great deal of energy and raw materials, and produce a great deal of waste. They also have the potential to solve the problems that arise in an increasingly crowded world, to do what can be done to meet their challenge. In this chapter we will study how today's living affects the environment within which all future humans must live, and the efforts being mounted to lessen adverse impact and increase potential benefits. Let us first consider human population growth in a historical context.

A GROWING POPULATION

The earliest fossils that are clearly *Homo sapiens,* our species, come from western Europe and are about 500,000 years old. Humans reached North America at least 12,000 years ago, crossing the narrow straits between Siberia and Alaska and spreading to the

FIGURE 27-2

Much of today's agriculture is still primitive. In southern India the people still thresh sorghum using primitive tools whose designs are probably thousands of years old.

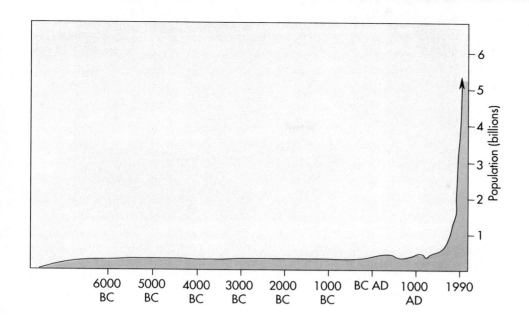

FIGURE 27-3
Growth curve of the human population.

southern tip of South America. By 10,000 years ago, there were about 5 million people on earth. With the new and much more dependable sources of food that became available as a result of agriculture (Figure 27-2), the human population began to grow more rapidly. Towns and then cities developed in the areas where agriculture was practiced and food was sufficiently abundant; such settlements were widespread by 3000 BC. At the time of Christ, there were an estimated 130 million people on earth—half the population of the United States and Canada today.

The ability to live together on a long-term basis in relatively large settlements made possible the specialization of professions in these centers. This was the necessary condition for the development of modern culture and the development of metal tools and utensils.

The dip you see at the bend of Figure 27-3 reflects the outbreak of bubonic plague in 1348, which killed nearly four fifths of the world's population. The world population recovered quickly, and by 1650 it had reached 500 million. The Renaissance in Europe, with its renewed interest in science, ultimately led to the establishment of industry in the seventeenth century and to the Industrial Revolution of the late eighteenth and early nineteenth centuries.

The Present Situation

For the past 300 years, the human birth rate (as a global average) has remained nearly constant, at about 30 births per year per 1000 people. Today it is about 27 births per year per 1000 people, but this difference is probably not significant. However, better sanitation and improved medical techniques have caused the death rate to fall steadily, to about 10 deaths per 1000 people per year as of 1990. The difference between these two figures amounts to an annual worldwide increase in human population of approximately 1.8%. Such a rate of increase may seem relatively small, but it would double the world's population in only 39 years!

The *annual* increase in population amounts to about 95 million people, a number substantially larger than the total population of either Britain, Germany, or Mexico. At this rate, more than 260,000 people are added to the world population each day, or more than 180 every minute! The world population is expected to continue to rise, to well over 6 billion people by the end of the century, and then perhaps to stabilize at 13 billion by 2090, according to United Nations estimates (Figure 27-4).

In view of the limited resources available and the necessity of our learning how to manage these resources well, the first and most necessary step toward global prosperity is to stabilize the human population. One of the surest signs of the pressure we are putting on the environment is our use of about 40% of the total global net photosyn-

FIGURE 27-4

India faces an uncertain future. At present rate of growth, India will surpass China in total population during the first half of the twenty-first century. This scene was photographed at a camp for starving refugees in New Delhi.

thetic productivity on land. Given that relationship, the rapid growth of our population poses severe problems if even present levels of relative prosperity are to be maintained. The facts virtually demand restraint in population growth: if and when we develop the technology that would allow greater numbers of people to inhabit the earth in a stable condition, it will be possible to increase our numbers to whatever level might be appropriate at that time.

> In the early 1990s, the global human population of more than 5.4 billion people was growing at a rate of approximately 1.8% annually. At that rate, it would reach about 6.3 billion people by 2000, and about 8.2 billion by 2020.

The Future Situation

By the year 2000, about 60% of the people in the world will be living in tropical or subtropical regions. An additional 20% will be living in China, and the remaining 20%—one in five—in the developed or industrialized countries: Europe, the Commonwealth of Independent States (formerly the Soviet Union), Japan, the United States, Canada, Australia, and New Zealand. Although the populations of the industrialized countries are growing at an annual rate of only about 0.5%, those of the less developed, mostly tropical countries (excluding China) are growing at an annual rate estimated in 1990 to be about 2.4%. For every person living in an industrialized country like the United States in 1950, there were two people living elsewhere; in 2020, just 70 years later, there will be five.

FIGURE 27-5

The age structure of the United States and Mexico in 1990. The rate of population growth in Mexico, already rapid, will increase considerably in the future because so much of its population will be entering into their reproductive years. The "bulge" in the population structure in the United States (BB) indicates the "baby boom" generation. Males are indicated on the left in each graph, females on the right.

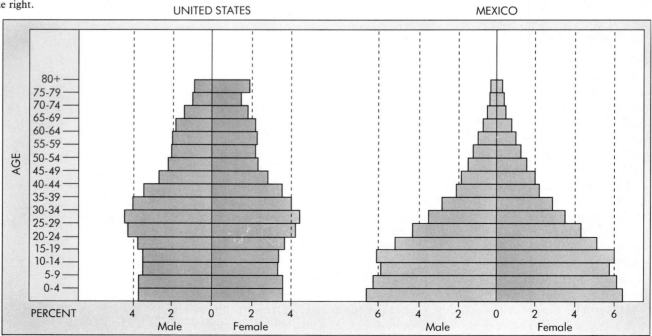

As we enter the 1990s, we note that many developing countries have begun the process of fertility decline and some—such as Thailand and Mexico—are well along the path. The onset of these declines is a significant watershed in world demographic history. But the question still remains: will declines continue smoothly or will birth rates decrease spasmodically? This is not purely academic. The course of fertility decline over the next decade will largely determine whether the world's ultimate population reaches 10 billion, 15 billion, or some other plateau. The 1990s is truly a decisive decade.

Population projections do not simply extrapolate the effect of present rates into the future. Demographers assume that as countries industrialize, their fertility will decline and that by supporting family planning, these declines can be accelerated. Such an assumption is based on the demographic experience of today's more developed countries. Projections also must make assumptions about when a country's fertility will fall to the "replacement level" of about two children per woman. At this level, couples just "replace" themselves, and thus do not increase the size of successive generations. Replacement level fertility ultimately leads to a zero population growth rate and a stationary population size.

To date, world projections have enjoyed a reasonably good track record. In 1958, the United Nations projected a total world population of

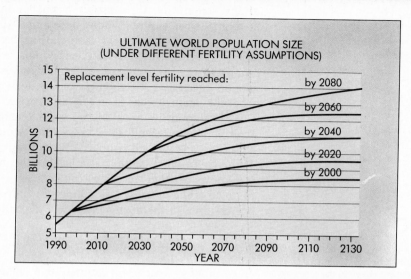

6.3 billion. Such a performance may be striking, but as further demographic changes are observed in more and more countries, predicting future size will become less certain.

The timing and place of fertility declines have a large impact on the future size of a country's population. If, for example, India resumes a period of fertility decline, ending a "stall" of many years, and reaches replacement level in 2010, its 1990 population of 853 million would grow to 1.3 billion by 2025, and the annual growth rate would fall to half the current rate of 2.1%. If India does not resume its fertility decline, the 2025 population will exceed 2 billion and will be growing at an even faster rate than today's. Figure 27-A shows the effect of dif-

ferent fertility declines on "ultimate" world population size. If replacement rates achieved as by 2000, a highly unlikely scenario, the world population would not reach 9 billion. If it is delayed to 2080, world population would be triple today's size.

There can be little doubt that the world's birth rate has entered a period of decline, but its future path will be pivotal to humanity's future numbers. What happens now is of central importance. It is highly likely that world population growth will slow to a near halt during the next century. But of its ultimate size we are far less certain.

Reprinted from *1990 World Data Sheet of the Population Reference Bureau, Inc.*

Since the people living in industrialized countries control about 85% of the world's wealth and material goods and enjoy a standard of living perhaps 20 times higher than those in many developing countries, the changing ratios of total population in each of these sectors should be a matter of special concern for everyone. To provide a few examples, the infant mortality rate in industrial countries in 1990 was 16 per thousand, whereas in developing countries (excluding China) it was 91 per thousand; the respective life expectancies at birth were 74 years versus 59 years, and even lower in those developing countries where many people are malnourished.

As you learned in Chapter 23, the age structure of a population determines how fast the population will grow. For this reason, it is essential in predicting the future growth patterns of a population to know what proportion of its individuals have not yet reached childbearing age. In industrialized countries such as the United States, about a fifth of the population is under 15 years of age; in developing countries such as Mexico, the proportion is typically about twice as high (Figure 27-5). Thus, even if the policies that most tropical and subtropical countries have established to limit their own population growth are carried out consistently for decades, the populations of these countries will continue to grow well into the twenty-first century, and the industrialized countries will constitute a smaller and smaller proportion of the world's population. For example, if India, with a mid-1990 population level of about 853 million people (with 39% under 15 years old) managed to reach a simple replacement reproductive rate by the year 2000, its population would still not stop growing until the middle of the twenty-first century. At present rates of growth, India will have a population of nearly 1.4 billion people by 2020 and will still be growing rapidly, judged from the country's age structure and current patterns of growth.

Most countries are devoting considerable attention to slowing the growth rate of their populations, and there are genuine signs of progress. It is estimated that if these efforts are maintained strongly, the world population might stabilize toward the close of the twenty-first century at a level of about 2 to 3 times its present 5.4 billion. No one knows whether the planet can support so many people indefinitely. Finding a way to do so is the greatest task facing humanity's future. The quality of life available to our children in the next century will depend to a large extent on our success.

FOOD AND POPULATION

Even though experts still estimate that enough food is produced in the world to provide an adequate diet for everyone, the distribution of this food is so unequal that large numbers of people live in hunger. In the United States, the Soviet Union, Japan, and Europe, there is, on the average, 25% to 33% *more* food available per person than the United Nations Food and Agriculture Organization (FAO) regards as the calorie intake necessary to maintain moderate physical activity. In contrast, such countries as Bangladesh, Ecuador, and Kenya had 10% to 15% *less* than this minimum available per person, and, for the most part, little cash with which to purchase more. Worldwide, 300 to 400 million people each consume less than 80% of the United Nations–recommended minimum standards for caloric intake, a diet insufficient to prevent stunted growth and serious health risks. Only the United States, Canada, Argentina, and a few European countries are consistent exporters of food.

Of the approximately 4.1 billion people living in the tropics and China in 1990, the World Bank estimated that some 1.2 billion—nearly a quarter of the total world population— were living in extreme poverty. These people cannot provide adequate food, clothing, and shelter for themselves and their families from day to day. Among the malnourished people mentioned above, UNICEF estimates that every year approximately 13 million children under the age of 5 starve to death or die of diseases complicated by malnutrition (Figure 27-6). Worse, many millions of other children exist only in a state of lethargy, their mental capacities often permanently damaged by their lack of access to adequate quantities of food.

About 1.2 billion people lived in a state of extreme poverty in the early 1990s; many of them were malnourished.

One of the most alarming trends taking place in developing countries is the massive movement to urban centers. For example, Mexico City, the largest city in the world, is plagued by smog, traffic, inadequate waste disposal, and other problems—and it will have a population of more than 30 million by the end of the century (Figure 27-7). The prospects of supplying adequate food, water, and sanitation to these people, in a country whose 1990 population (nearly 90 million) is growing at a rate at which it would double in only 29 years, are almost unimaginable. The lot of the rural poor, mainly farmers, in Mexico is even worse, and it is no wonder that nearly a quarter of all Mexicans indicated in a 1989 poll that they were either "likely" or "very likely" to move to the United States over the next 2 years.

The proportion of poor, hungry people in the less developed countries, now amounting to a third of their populations, will grow rapidly unless we are remarkably successful in developing solutions to their problems in the near future.

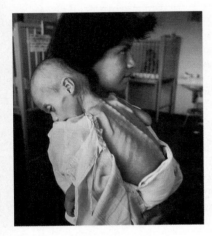

FIGURE 27-6
Starvation. Consuelo Andena, 16, holding her 1-year-old baby, Carolina, in the malnutrition ward of the Women and Children's Hospital in Tegucigalpa, Honduras, in 1983. Such tragic scenes are common throughout the tropics, where more than 40,000 babies perish every day from starvation.

THE FUTURE OF AGRICULTURE

One of the greatest and most immediate challenges facing today's world is to produce enough food to feed its expanding population. Even though it is estimated that world food production has expanded 2.6 times since 1950—more rapidly than the human population—virtually all the land that can be cultivated is already in use; 20% of the world's topsoil has been lost from agricultural lands during that period; and the human population continues to grow explosively. Much of the world is populated by large numbers of hungry people who are rapidly destroying the sustainable productivity of the lands they inhabit; at the same time, consumption in industrialized countries such as the United States is running at 20 to 30 times the rate in developing countries and is having an even greater adverse effect on the future of humanity. In the face of these massive problems, we need to consider what the prospects are for increased agricultural productivity in the future and what can we do to improve those prospects.

For a start, we are going to have to identify more kinds of crops, especially ones that will do well in the topics and subtropics. Nearly all of the major crops now grown in the world have been cultivated for hundreds or even thousands of years. Only a few, including rubber and oil palms (Figure 27-8), have entered widespread

A
B

FIGURE 27-8
Tropical tree crops.
A Oil palms (*Eleais guineensis*), are grown throughout the tropical regions of the world as a source of oil. On infertile tropical soils, trees often are much more successful than the sorts of herbaceous crops that are the mainstay of temperate agriculture.
B The roots of *Casuarina*, a fast-growing tree, are inhabited by nitrogen-fixing actinomycetes, which form nodules there. *Casuarina* is receiving increasing attention as a possible source of firewood and for watershed protection in the tropics.

With their vast and diverse arsenal of natural insecticides, plants offer a reservoir of useful products that we have barely tapped. A number of these plants have proved useful already, and many more will do so in the future, as we explore further.

One of the most fertile areas for biological exploration is the study of plant use by people who live in direct contact with natural plant communities. These people know a great deal about the plants with which they come into contact, and the screening that they and their ancestors have performed over many generations, if the results are understood and catalogued, help us recognize new, useful plants. For example, the herbal curer in the jungles of southern Surinam has a profound knowledge of uses of the plants that grow in his region, but that knowledge is being lost as civilization and modern medicine move into the area.

Some recent triumphs in the search for economically important plants are the following:

Guayule (*Parthenium argenteum*) (Figure 27-B, *1*): This widespread shrub of northern Mexico and the southwestern United States is a rich source of rubber. The yield from some cultivated strains is up to 20% rubber. Guayule is now being cultivated in more than 30 countries, and efforts are being made to bring it into even wider cultivation.

FIGURE 27-B, 1
Guayule

Periwinkle (*Catharanthus roseus*) (Figure 27-B, *2*): Periwinkle is a garden plant, now widespread in cultivation; as a native plant, it is found only in Madagascar. Two drugs, vinblastine and vincristine, which have been developed from periwinkle by Eli Lilly & Company, are effective in the treatment of certain forms of leukemia. For example, a child who develops leukemia now has a 95% chance of survival past the age of 5 if treated with these drugs. In 1950, such a child would have had only a 20% chance to live.

Jojoba (*Simmondsia chinensis*) (Figure 27-B, *3*): Despite its scientific name, which was conferred because

FIGURE 27-B, 2 .
Rosy periwinkle.

cultivation since 1800. One key feature for which nearly all of our important crops were first selected was ease of growth by relatively simple methods.

How many plants do we use at present? Just three species—rice, wheat, and corn—supply more than half of all human energy requirements; only about 150 kinds of plants are used extensively; and only about 5000 have ever been used for food. There may be tens of thousands of additional kinds of plants, among the 250,000 known species, that could be used for human food if their properties were fully explored and they were brought into cultivation. There are many uses for plants other than food, too. For example, oral contraceptives for many years were produced from Mexican yams; the muscle relaxants used in surgery came from curare, an Amazonian vine used traditionally to poison darts used in hunting; and the cure for Hodgkin's disease was developed in the early 1970s from the rosy periwinkle, a widely cultivated plant native to Madagascar.

The reasons for which we select new crops now are often different from those that appealed to our ancestors, living as they did in small groups around the foothills of the Near East or on the temperate slopes of the mountains in Mexico. Standards of cultivation have changed, and we use many products from plants—oils, drugs, and the scientist who studied the first specimens incorrectly believed that they had come from China, jojoba is native to the deserts of northwestern Mexico and southwestern United States. It is now an important crop in production in several parts of the world since the late 1970s, because the liquid waxes in its seeds have characteristics similar to those of sperm whale oil and are important for certain kinds of fine lubrication. Like guayule, jojoba can be cultivated in lands too arid for most other crops. The new availability of the liquid waxes from jojoba has helped make the hunting of sperm whales uneconomical, and these threatened animals may thus be saved from extinction.

Grain amaranths (*Amaranthus* species) (Figure 27-B, *4*): Important grain crops of the Latin American highlands in the days of the Incas and Aztecs, the grain amaranths are now minor crops in latin America; their use was suppressed because they played a role in pagan ceremonies of which the Spaniards disapproved. The grain amaranths, fast-growing plants that produce abundant grain rich in lysine—an amino acid that is rare in most plant proteins but essential for animal nutrition—are now being investigated seriously for widespread development.

Winged bean (*Psphocarpus tetragonolobus*) (Figure 27-B, *5*): The winged bean is a nitrogen-fixing tropical vine that produces highly nutritious seeds, pods, and leaves. Its tubers are eaten like potatoes, and its seeds produce large quantities of edible oil. First cultivated locally in New Guinea and Southeast Asia, the winged bean has spread since the 1970s throughout the tropics, where it holds great promise as a productive source of food, especially for small farmers.

FIGURE 27-B, 3
Jojoba.

FIGURE 27-B, 4
Grain amaranth.

FIGURE 27-B, 5
Winged bean.

other chemicals, for example—that often would not have led to their cultivation earlier. In a time when human activities threaten to drive many of the world's plants and other organisms to extinction during our lifetimes and those of our children, we need to search systematically for new and useful crops that fit the multiple needs of modern society in ways that would not have been considered earlier. We must begin more diligent efforts to find and bring into cultivation more kinds of useful plants before they are gone forever.

Only about 150 kinds of plants, out of the roughly 250,000 known, play an important role in international trade at present. Many more could be developed by a careful search for new crops.

The Prospects for More Food

In improving the world food supply, the most promising strategy is to improve the productivity of crops that are already being grown. Much of the improvement in food production must take place in the tropics and subtropics, where a rapidly growing

FIGURE 27-9
Results of the Green Revolution.
Improved strains, such as this dwarf wheat, helped make Mexico an exporter, rather than an importer, of wheat during some years in the 1960s and 1970s.

majority of the world's population already live, including most of those enduring a life of extreme poverty. These people cannot be fed by exports from the industrial nations, which contribute only about 8% of their total food at present, and where agricultural lands are already heavily exploited. During the 1950s and 1960s, the so-called Green Revolution took place as a result of the development of new, improved strains of wheat and rice. As a result of these efforts, the production of wheat in Mexico, for example, increased nearly ten-fold between 1950 and 1970, and Mexico temporarily became an exporter of wheat rather than an importer (Figure 27-9). During the same decades, food production in India was largely able to outstrip even a population growth of approximately 2.3% annually, and China became self-sufficient in food.

Despite the apparent success of the Green Revolution, the improvements were limited. The agricultural techniques that were employed require the expenditure of large amounts of energy and abundant supplies of fertilizers, pesticides, and herbicides, as well as adequate machinery. For example, in the United States it requires about 1000 times as much energy to produce the same amount of wheat that results from traditional farming methods in India. Introducing industrialized-country methods obviously increases the yield, but who will pay the bill? Energy prices are often held at artificially low levels in developing countries, so that it may actually be more expensive for a poor rural farmer to grow an equivalent amount of grain as a large-scale farmer using industrialized technology. In this respect, the introduction of Green Revolution methods in some regions has actually worsened poverty for many of the people and lessened their access to food, fuel, and other commodities on which they depend.

> **The Green Revolution has resulted in increased food production in many parts of the world through the use of improved crop strains. These strains often depend on increased inputs of fertilizers, water, pesticides, and herbicides, as well as the greater use of machinery—means that are not normally available to the poor.**

Certain nutritional problems have also arisen in connection with the Green Revolution. The overconcentration on cereal crops has tended to lower the production of other nutritionally important plants, including legumes, oilseeds, and vegetables of all kinds. The reason that legumes and cereals are often nutritionally combined is that they provide a balanced set of amino acids required by humans for proper growth. The varied strains of crops grown on small farms may also be driven out by fewer kinds of modern strains and fewer crops, which produce a better yield if large-scale inputs of chemicals and the use of machinery are possible (Figure 27-10). Despite these short-term advantages, the loss of the unique, traditional strains of crops presently cultivated by small, rural farmers throughout the world may ultimately prevent the particular crop plants from being able to grow in less favorable habitats or to withstand important diseases in the long run (Figure 27-11). **Monoculture**—the exclusive cultivation of a single crop over wide areas—is an efficient way to use certain kinds of soils, but it carries the risk of an entire crop being destroyed with the appearance of a single pest species or disease.

In improving the world food supply, the most promising strategy is to improve the productivity of crops that are already being grown. There are relatively few parts of the world where additional land can be brought into cultivation using currently available technology. Biologists play a crucial role in the improvement of crops and in the development of new ones by fully applying traditional methods of plant breeding

FIGURE 27-10
A traditional farm in Thailand. On it grow coconuts, *Leucaena* (a fast-growing, nitrogen-fixing legume that provides food and firewood), sugar palms, litchi nuts, and the nitrogen-fixing water fern *Azolla* growing in ponds. Land crabs are frequent in such farms in Thailand, and are harvested as a source of protein. Farms of this type, with their trees and mixed crops, are extremely suitable for many sites in the tropics.

A

B

C

FIGURE 27-11
Who cares if a tropical plant becomes extinct? You should.
Corn (*Zea mays*) and a wild relative.
A The genetic uniformity of hybrid corn, enhanced by selection over decades, is clearly demonstrated by this crop in Illinois, one of the most productive areas for corn in the world.
B Corn is an annual grass, but this abandoned corn field in Mexico is dominated by a wild perennial relative, *Zea diploperennis*, which was discovered in 1977. The two species cross breed to produce fertile hybrids, and *Z. diploperennis* is resistant to the seven main types of virus diseases that damage corn as a crop. Attempts have also been made to develop a new kind of perennial corn for cultivation in some of the infertile soils of subtropical areas.
C Fruiting ear of *Zea diploperennis*.
D This commercially important and scientifically interesting relative of corn might never have been found at all, because it occurs in only one small area less than one square mile wide. Logging operations were widespread in the Sierra de Manantlan, Jalisco, Mexico, where it lives, and its populations could easily have been destroyed by pressures from the increasing human populations of the area. In 1987, this mountain was declared a Biosphere Reserve, and the survival of this rare species in the world now seems assured.

D

and selection (Figure 27-12) to many important crops in the tropics and subtropics (Figure 27-13), in addition to wheat, corn, and rice. Genetic engineering techniques (see Chapter 14) will make it possible to produce plants that are resistant to specific herbicides, which can then be applied much more effectively for weed control and without damaging the crop plants. Genetic engineers are also developing new strains of plants that will grow successfully in areas where the particular crop plant could not grow before. Eventually, desirable characteristics could be introduced into important crop plants, such as the ability to fix nitrogen, carry out C_4 photosynthesis, or produce substances that deter pests and diseases. The ability to transfer genes between organisms, which became a practical technique in 1973, is of great importance in the improvement of crop plants before the twentieth century ends.

FIGURE 27-12
A greenhouse laboratory. In these experiments at the University of Arizona, Tucson, scientists are attempting to select salt-tolerant crops that will be suited for the naturally saline soils of the world, as well as for those that have become saline through continued irrigation.

FIGURE 27-13

Plants as energy factories of the future.

A Water hyacinth, despite its beautiful flowers, is generally regarded as a noxious weed, clogging waterways throughout the tropics.

B The same rapid growth that makes it such a harmful weed allows it to produce large amounts of biomass rapidly. Here it is being harvested in India as animal feed; it can readily be dried and burned as a source of energy also.

The oceans were once regarded as an inexhaustible source of food, but overexploitation of their resources is limiting the world catch from year to year and these catches are costing more in terms of energy. The mismanagement of fisheries, mainly through overfishing, local pollution, and the destruction of fish breeding and feeding grounds, has already lowered the catch of fish in the sea by about 20% from its maximum levels. The decline in the numbers of whales is a tragic and well-known example of the way fisheries have been and are being destroyed.

The development of new kinds of food, such as microorganisms cultured in nutrient solutions, should definitely be pursued. For example, the photosynthetic, nitrogen-fixing cyanobacterium *Spirulina* is being investigated in several countries as a possible commercial food source (Figure 27-14); it is a traditional food in Africa, Mexico, and other regions. *Spirulina* thrives in very alkaline water, and it has a higher protein content than soybeans; the ponds in which it grows are about 10 times more productive than wheat fields. Such protein-rich concentrates of microorganisms could provide important nutritional supplements, however, psychological barriers must be overcome to persuade people to eat such foods, and the processes required to produce these foods tend to be energy-expensive.

THE TROPICS

More than half of the world's people live in the tropics, and this percentage is increasing rapidly. For global stability, and for the sustainable management of the world ecosystem, it will be necessary to solve the problems of food production and regional stability in the areas where most people live. World trade, political and economic stability, and the future of most species of plants, animals, fungi, and microorganisms depend on our addressing these problems in an adequate fashion.

Many people in the tropics engage in **shifting agriculture**—clearing a patch of forest, growing crops for a few years, and then moving on (Figure 27-15). The fertil-

FIGURE 27-14

Spirulina, a cyanobacterium that is being grown increasingly as a source of protein.

A Individuals of *Spirulina,* each about 250 micrometers long.

B A mat of *Spirulina* on a conveyor belt in a processing plant in Mexico.

C Ponds for the production of *Spirulina* in Thailand.

FIGURE 27-15
Shifting agriculture. A family has cleared a small patch of forest in the northern Amazon Basin of Brazil; they will be able to grow crops on it for a few years and then will need to move on to another part of the original forest. Such activities can be sustained indefinitely if population levels are low enough, but that will rarely be the case in the closing years of the twentieth century.

ity of the soil has by then returned to its original, very low level—the temporary enrichment that resulted from cutting and burning the forest has been exhausted—and the cultivator must move on and clear another patch of forest. Such agricultural systems work well where human populations are relatively low, but as these numbers grow, there is little opportunity even for the cultivation of traditional crops such as manioc (tapioca, cassava) (Figure 27-16). Firewood gathering is also hastening the demise of many tropical forests; about 1.5 billion people worldwide—a third of the global population—depend on firewood as their major source of fuel and are cutting the local supplies faster than the trees can regenerate themselves.

Shifting agriculture is one major factor in the destruction of tropical forests, but a number of other factors are significant also. We can illustrate them by reference to the tropical rain forest, biologically the richest of the world's biomes. Most other kinds of tropical forest have already been largely destroyed; they tend to grow on more fertile soils, and have been exploited by humans earlier than the tropical rain forests that are now under such massive attack. In the early 1990s, it is estimated that about 6 million square kilometers of tropical rain forest still exist in a relatively undisturbed form. This area, about three quarters of the size of the United States (excluding Alaska), represents about half of the original extent of the rain forest. From it, about 160,000 square kilometers were being clear cut per year, with perhaps an equivalent amount severely disturbed by shifting cultivation, firewood gathering, and allied practices.

FIGURE 27-16
Manioc: food or fuel?
A Manioc (also called cassava; *Manihot esculenta*) is a widespread and important tropical root crop. It is often grown on marginal lands that are fertile for a few years after the clearing of the forests.
B A Brazilian woman is preparing the starchy tubers to eat.
C In Brazil, manioc is being used extensively for the production of ethyl alcohol, which is mixed with gasoline to make a fuel (gasohol) that is suitable for cars, buses, and trucks and helps to make Brazil more independent of foreign sources of oil. However, large tracts of land that might have produced food for the poor are being devoted to another purpose. Such are the trade-offs that must be made when weighing scarce resources; the decisions rarely favor the poor, however.

A   **B**   **C**

The total area of tropical rain forest destroyed—and therefore permanently removed from the world total—amounted to an area greater than the size of Indiana each year. At such a rate, all of the tropical rain forest in the world will be gone in about 30 years, but in many regions, the rate of destruction is much more rapid.

As a result of such overexploitation, experts predict, there will be little undisturbed tropical forest left anywhere in the world by early in the next century. Many areas now occupied by forest will still be tree-covered, but those trees will represent only a small percentage of those that now grow in these areas. Many species of plants, animals, fungi, and microorganisms can reproduce only under the conditions in which they live in undisturbed forest. Consequently, they are threatened with extinction, or, at the very least, exclusion from large areas. In fact, a fifth or even more of the species on earth may become extinct during the next 30 to 40 years, amounting to a million or more species. Viewing the situation in another way, several species per day are probably becoming extinct now (see Figure 23-1). Many of these species that are becoming extinct inhabit ecologically devastated islands such as St. Helena in the south Atlantic and Rodrigues (Figure 27-17) in the western Indian Ocean, or similarly devastated areas on continents, such as the Atlantic forests of Brazil or the lowlands of western Ecuador. By early next century, the rate could easily reach several species per *hour;* and it would continue to climb for at least another 50 years. Overall, this would amount to an extinction event that has been unparalleled for at least 65 million years, since the end of the age of dinosaurs. The number of species in danger of extinction during our lives is far greater than the number that became extinct at that distant time.

The prospect of such an extinction event is of major concern. Only one out of every six tropical organisms has even been given a scientific name, so many species that are about to become extinct will never have been seen by any humans. As these species disappear, so does our opportunity to learn about them, not only scientifically, but also in terms of their possible benefits for ourselves. We have an intrinsic curiosity about the plants, animals, and microorganisms with which we share this planet, and we would like to understand them better individually. Each is unique, and with its loss we lose forever the chance to use it for any purpose whatever. The fact that our entire supply of food is based primarily on 20 kinds of plants, out of the 250,000 kinds that are available, should give us pause to consider what it means to be living in a generation during which a high proportion of the remainder are being lost permanently. Many of them would surely be of great use to us if we knew about their properties.

What biologists can do in the face of this crisis is to help design intelligent plans for finding those organisms most likely to be of use and saving them from extinction. They must also participate in sound, globally based schemes to preserve as much as possible of the biological diversity of life on earth so that the options for our descen-

FIGURE 27-17
Escape from extinction.
A This palm is the last individual of its species, *Hyophorbe verschaffeltii,* known in nature. It is shown here on the island of Rodrigues in the western Indian Ocean, the same island where that famous flightless pigeon, the dodo, became extinct more than three centuries ago.
B *Hyophorbe verschaffeltii* is not likely to become extinct, because it is preserved in cultivation on the campus of the University of Mauritius, on the island of Mauritius. The palms are the second most important family of plants in the world economically, being surpassed only by the grasses.

A

B

dants may be as wide as possible. With the loss of tropical forest, and of biological communities throughout the world, we are permanently losing many opportunities not only for knowledge, but for increased prosperity—whether we realize it or not. It is biologists who must understand this message and inform their fellow citizens of its importance to them.

Viewing the consequences of uncontrolled deforestation in the tropics and subtropics in another way, tropical forests are complex, productive ecosystems that work well in the areas where they have evolved. The sad truth is that we do not know, for the most part, how to replace them with other productive ecosystems that will support humans. When we cut a forest or open a prairie in the North Temperate Zone, we provide the basis for a farm that we know can be worked for generations. In the tropics, for the most part, we simply do not know how to engage in continuous agriculture in most areas that are not now under cultivation. When we clear a tropical forest, we engage in a one-time consumption of natural resources that will not ever be available again (Figure 27-18). The complex ecosystems that have been built up over billions of years are now being dismantled, in almost complete ignorance, by the human species.

What biologists must do is to learn more about the construction of sustainable agricultural ecosystems that will meet human needs in tropical and subtropical regions. The ecological principles that we have been reviewing in the last three chapters are universal principles. The undisturbed tropical rain forest has one one of the highest rates of net primary productivity of any plant community on earth, and it is therefore logical to assume that it can be harvested for human purposes in a sustainable, intelligent way. Simply passively allowing it to be consumed is something that biologists and concerned individuals should attempt to avoid. Sound development of the tropics, an urgent matter in view of the large numbers of humans who live in tropical countries, must be based on sound biology, as well as achieving stable human population levels and alleviating the problems of poverty and widespread malnutrition. Biologists must address themselves increasingly to problems that have traditionally been the concern of agronomists, animal breeders, and other agriculturists, and must apply the results of their research to the creation of a stable global ecosystem.

A

B

FIGURE 27-18
Destroying the tropical forests.
A When tropical evergreen forests are removed, the ecological consequences can be disastrous. These fires are destroying rain forest in Brazil, which is being cleared for cattle pasture.
B The consequences of deforestation can be seen on these middle elevation slopes in Ecuador, which now support only low-grade pastures and where highly productive forest, which protected the watersheds of the area, grew in the 1970s.
C As time goes by, the consequences of tropical deforestation may become even more severe, as shown here by the extensive erosion in this area of Tanzania from which forest has been removed.

C

OUR IMPACT ON THE ENVIRONMENT

The simplest way to gain a feeling for some of the other dimensions of the problem before us is simply to scan the front pages of any newspaper or news magazine or to watch television. Although they are only a sampling, the features that are now selected by these media teach us a great deal about the scale and complexity of the challenge we face.

Nuclear Power

At 1:24 AM on April 26, 1986, one of the four reactors of the Chernobyl nuclear powerplant blew up. Located in the Ukraine 100 kilometers north of Kiev, Chernobyl was one of the largest nuclear power plants in Europe, producing a thousand megawatts of electricity, enough to light a medium-sized city. Before dawn on April 26, workers at the plant hurried to complete a series of tests of how the generator Reactor Number 4 performed during a power reduction, and took a foolish short-cut: they shut off all the safety systems. The reactors at Chernobyl were graphite reactors designed with a series of emergency systems that shut the reactors down at low power, because the core is unstable then—and the workers turned these emergency systems off. A power surge occurred during the test, and there was nothing to dampen it. Power zoomed to hundreds of times maximum, and a white-hot blast with the force of a ton of dynamite partially melted the fuel rods and heated a vast head of steam that blew the reactor apart.

The explosion and heat sent up a plume five kilometers high, carrying several tons of uranium dioxide fuel and fission products. Over 100 megacuries of radioactivity were released, making it the largest nuclear accident ever reported. By way of comparison, the Three Mile Island accident in Pennsylvania in 1979 released 17 curies, millions of times less (Figure 27-19). This cloud traveled first northwest, then southeast, spreading the radioactivity in a band across central Europe from Scandinavia to Greece. Within a 30-kilometer radius of the reactor, at least one fifth of the population, some 24,000 people, received serious radiation doses (greater than 45 rem). Thirty-one individuals died as a direct result of radiation poisoning—most of them firefighters who succeeded in preventing the fire from spreading to nearby reactors.

For the western Independent States and the rest of Europe, the radiation dose was much lower but still significant. Data indicate that radiation outside of the immediate Chernobyl area will be expected to be responsible for from 5000 to 75,000 cancer deaths, because of the large numbers of people exposed.

FIGURE 27-19
Three Mile Island nuclear power plant.

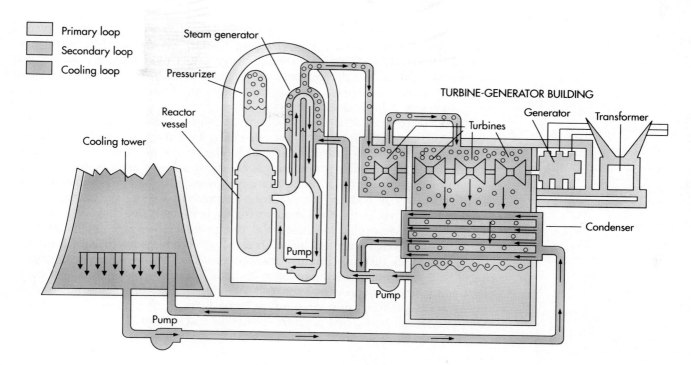

Primary loop
Secondary loop
Cooling loop

Steam generator

Pressurizer

Reactor vessel

Cooling tower

TURBINE-GENERATOR BUILDING

Generator   Transformer

Turbines

Condenser

Pump

Pump

Pump

FIGURE 27-20
Diagram of a nuclear power plant.

The Promise of Nuclear Power.   Our industrial society has grown for over 150 years on a diet of cheap energy. Until recently, much of this energy has been derived from burning fossil fuels—wood, coal, and oil. However, as these sources of fuel become increasingly scarce and the cost of locating and extracting new deposits ever more expensive, modern society is being forced to look elsewhere for energy. The great promise of nuclear power to our future is that it provides an alternative source of plentiful energy (Figure 27-20). Although nuclear power is not cheap, using current technology—power plants are expensive to build and operate—its raw material, uranium ore, is so common in the earth's crust that it is unlikely we will ever run out of it.

The burning of coal and oil to obtain energy produces two undesirable chemical by-products, sulfur and carbon dioxide. As we will see in this chapter, the sulfur emitted from burning coal is a principal cause of acid rain, whereas the CO_2 produced from the burning of all fossil fuels is a major greenhouse gas (see discussion of "global warming" below). For these reasons, we need to find replacements for fossil fuels—quite apart from the fact that they will surely run out in time anyway.

For all of its promise of plentiful energy, nuclear power presents several new problems that must be mastered before its full potential can be realized. You have met one serious challenge already in this chapter: the need to ensure safe operation of the approximately 390 nuclear reactors now in operation. A second challenge is the

A VOCABULARY OF RADIATION

radioactivity   The emission of nuclear particles and rays by unstable atoms as they decay into more stable forms. It is measured in curies; one curie equals 37 billion disintegrations per second.

dose   A measure of the amount of radiation absorbed by the body, usually stated in rem units. An acute dose of 600 rem will usually cause death within 60 days.

background radiation   Radiation from natural sources such as cosmic rays and radon gas, as well as from artificial sources such as atomic testing. In the United States, this averages about 100 millirem (one millionth of a rem) per year; for comparison, a chest radiograph exposes you to 20 millirem.

need to safely dispose of the radioactive wastes produced by nuclear power plants, and to safely decommission plants that have reached the end of their useful lives (about 25 years). By 1990 approximately 35 plants will be more than 25 years old, and not one has been safely decommissioned. A third challenge is the need to guard against terrorism and sabotage, because the technology of nuclear power generation is closely linked to that of nuclear weapons.

For these reasons, it is also important to continue to investigate and develop other alternatives to the fossil fuels, such as solar energy and wind energy, which also hold great promise when properly developed. The generation of electricity accounts for only about 15% of global warming gas emissions in the United States, however; therefore, energy conservation is far more important, and the most immediate and cost-effective way to address this problem. As much of 75% of the electricity produced in the United States and Canada currently is wasted through the use of inefficient appliances, according to scientists at the Lawrence Berkeley Laboratory. The use of highly efficient motors, lights, heaters, air conditioners, refrigerators, and other technologies that are currently available could lead to saving large amounts of energy, and greatly alleviate the problem of global warming gas emission. For example, a new, compact fluorescent light bulb uses only 20% of the amount of electricity of conventional lighting, provides equal or better lighting, lasts up to 13 times longer than incandescent bulbs, and provides substantial cost savings.

Carbon Dioxide and Global Warming

One of the major problems associated with increasing energy use is the rising concentration of carbon dioxide in the world's atmosphere. Most of this carbon dioxide comes from the burning of fossil fuels; a major proportion also comes from the destruction of forests, which releases large amounts of carbon dioxide when they are disturbed. Roughly seven times as much carbon dioxide is locked up in fossil fuels—approximately 5 trillion metric tons—as exists in the atmosphere today. Before widespread industrialization, the concentration of carbon dioxide in the atmosphere was approximately 260 to 280 parts per million (ppm). During the 25-year period starting in 1958, this concentration increased from 315 ppm to more than 340 ppm, and it is continuing to rise rapidly. A problem arises because carbon dioxide and other gases trap the longer wavelengths of infrared light, or heat, and prevent them from radiating into space. By doing so, they create what is known as **global warming,** an effect that keeps the mean global temperature about 30° C higher than it would be otherwise. Climatologists have calculated that the actual mean global temperature has increased about 1° C since 1900. These and further increases may be masked for a time by the great heat-absorbing capacity of the oceans; heat now stored in the oceans may be released over the next several decades, but the interactions involved in the overall system remain poorly understood and affected by many factors, such as the abundance of clouds and the capacity of warmer oceans to dissolve carbon dioxide.

In a recent study, the U.S. National Research Council estimated that the concentration of carbon dioxide in the atmosphere would pass 600 ppm (roughly double the current level) by the third quarter of the next century, and that level might be expected as soon as 2035. Such concentrations of carbon dioxide, if maintained indefinitely, would lead to a global surface-air warming of between 1.5° and 4.5° C. The actual increase might be considerably greater, however, because a number of trace gases—such as nitrous oxide, methane, ozone, and chlorofluorocarbons—that are also increasing rapidly in the atmosphere as a result of human activities have warming or "greenhouse" effects similar to those of carbon dioxide. All of them absorb the infrared wavelengths of light even more efficiently than carbon dioxide, and their combined effect might be even more important. For example, one of these gases, methane, has increased from 1.14 ppm in the atmosphere in 1951 to 1.68 ppm in 1986—nearly a 50% increase. Termites, now present in greater numbers because of all the forest that we are destroying, and such domestic animals as sheep, as well as rice paddies and many other human activities, add greatly to the amount of methane in the atmosphere through the substances that they excrete.

Major problems associated with climatic warming include rising sea levels; they have probably already risen 2 to 5 centimeters for this reason. If the climate becomes so warm that the polar ice caps melt, the sea level would rise by more than 150 meters, flooding the entire Atlantic coast of North America for an average distance of several hundred kilometers inland. Increased levels of carbon dioxide would cause some plants to grow more vigorously than they do at present, and would greatly change the relationships between plants and other organisms in natural communities—giving some an advantage and suppressing others. At the same time, the changing climatic patterns would make some of the best farmlands much drier than they are at present. Furthermore, there may be threshold effects—effects that lead to sudden changes in climate when certain levels are reached, and the ocean's role as a sink for carbon dioxide, for example, changes rapidly. Furthermore, studies of the air trapped inside of Antarctic ice have demonstrated a correlation between ice ages and reduced carbon dioxide levels and between higher carbon dioxide levels and warmer interglacial periods. If the climate warms as rapidly as many scientists project, the next 50 years may be marked by greatly altered weather patterns, a rising sea level, and major shifts of deserts and fertile regions. Certainly, our effects on the world's atmosphere must be understood better and considered much more seriously for our common welfare in the future.

Pollution

The river below the castle in Figure 27-21 is the Rhine, a broad ribbon of water running through the heart of Europe. From high in the Alps that separate Italy and Switzerland, the Rhine flows north across the industrial regions of Germany before reaching Holland and the sea. Judged by the sheer amount of goods that are produced and shipped on or near its shores, the Rhine is one of the world's most commercially important rivers, far exceeding the Mississippi, for example. The Rhine is also, where it crosses the mountains between Mainz and Coblenz, one of the most beautiful rivers on earth. On the first day of November, 1986, the Rhine almost died.

The blow that struck at the life of the Rhine did not at first seem so deadly. Firemen were fighting a blaze that morning in Basel, Switzerland. The fire was gutting a huge warehouse, into which the firemen poured streams of water to dampen the flames. The warehouse was that of a giant chemical company, Sandoz; in the rush to contain the fire no one thought to ask what chemicals were stored in it. By the time the fire was out, the streams of water had washed about 30 tons of mercury and pesticides into the Rhine.

Flowing down the river, the deadly wall of poison killed everything it passed. For hundreds of kilometers, the surface of the river was blanketed by dead fish. Many

FIGURE 27-21
The Rhine river.

cities that use the water of the Rhine for drinking had little time to make other arrangements. Even the plants in the river began to die. All across Germany, from Switzerland to the sea, the river reeked of rotting fish, and not one drop of the water was safe to drink or even to touch.

Six months later, Swiss and German environmental scientists monitoring the effects of the accident were able to report that the blow to the Rhine was not mortal. Enough small aquatic invertebrates and plants had survived to provide a basis for the eventual return of fish and other water life, and the river was rapidly washing out the remaining residues from the spill. A lesson difficult to ignore, the spill on the Rhine has caused the governments of Germany and Switzerland to intensify efforts to protect the river from future industrial accidents and to regulate the growth of chemical and industrial plants on its shores.

The Threat of Pollution

The pollution of the Rhine is a story that can be told countless times in different places in the industrial world, from Love Canal in New York to the James River in Virginia to Times Beach in Missouri. Nor are all of the pollutants that threaten the sustainability of life immediately toxic. Many forms of pollution arise as by-products of industry. For example, the polymers known as plastics, which we produce in abundance, break down slowly if at all in nature. Even though scientists are attempting to develop strains of bacteria that can decompose plastics, their efforts have been largely unsuccessful. Consequently, virtually all of the plastic items that have ever been produced are still with us. Collectively, they constitute a new form of pollution for which there is, as yet, no solution.

Water pollution is another serious problem that exists on a global scale. There is simply not enough water available to dispose of the diverse substances that today's enormous human population produces continuously. Despite the implementation of ever-improved methods of sewage treatment throughout the world, our lakes, streams, and groundwater are becoming increasingly polluted. For example, household detergents that contain phosphates may flow into oligotrophic lakes and lead to their eutrophication, as we discussed in Chapter 22. This leads to an overgrowth of algae and a rapid deterioration of water quality.

Widespread agriculture, carried out increasingly by modern methods, also causes large amounts of many new kinds of chemicals to be introduced into the global ecosystem. These include pesticides, herbicides, and fertilizers (Figure 27-22). Industrialized countries like the United States now attempt to monitor the side effects of these chemicals carefully. Unfortunately, however, large quantities of many toxic chemicals that were manufactured in the past still circulate in the ecosystems of these nations.

For example, the chlorinated hydrocarbons, a class of compounds that includes DDT, chlordane, lindane, and dieldrin, have all been banned for normal use in the

FIGURE 27-22
Huge quantities of pesticides and herbicides are used in connection with agriculture each year. One of the most important tasks facing agriculturists and other biologists is the development of integrated pest management and systems involving genetically engineered organisms that will avoid the necessity for such ecological excesses.

United States, where they were once widely used. These molecules break down slowly and accumulate in animal fat. Furthermore, as they pass through a food chain, they are increasingly concentrated. DDT, for example, caused serious problems by leading to the production of thin, fragile eggshells in many bird species in the United States and elsewhere until the late 1960s, when it was banned in time to save the birds from extinction. Chlorinated hydrocarbons have many other undesirable side effects, several of which are still poorly understood.

Obviously, a "back to nature" approach—one that ignores the important contributions made to our standard of living by the intelligent use of chemicals—will not allow us to care adequately for the the needs of the current world population. Even less would such a backward approach allow us to feed the additional billions of people who will join us during the next few decades. On the other hand, it is essential that we use our technology as intelligently as possible and with due regard for the protection of the productive capacity of all parts of the earth, on which we all depend.

Acid Precipitation

The four smokestacks you see in Figure 27-23 are those of the Four Corners power plant in New Mexico. This facility burns coal, sending the smoke up high into the atmosphere with these stacks, each over 65 meters tall. The smoke that the stacks belch out contains high concentrations of sulfur, which smells bad (like rotten eggs) and produces acid when it combines with the water vapor in air. The intent of those who designed the plant was to release the sulfur-rich smoke high up in the atmosphere, where the winds would disperse and dilute it. This sort of solution to the problem posed by burning high-sulfur coal was first introduced in Britain in the mid-1950s, and rapidly became popular in the United States alone. Basically they function to carry the acid produced by the fuels away from the areas where they are produced: London no longer suffers from acid fogs, but the forests and lakes of Sweden are being destroyed.

The environmental effects of this acidity are serious. The sulfur introduced into the upper atmosphere combines with water vapor to produce sulfuric acid, and when the water later falls as rain or snow, the precipitation is acid. Natural rainwater rarely has a pH lower than 5.6; in the northeastern United States, rain and snow now have a pH of about 3.8, about a hundred times as acid as the usual limit. Because the prevailing winds in the temperate latitudes (where most industries are concentrated) are westerlies, the sulfur emissions released by plants in the midwestern United States primarily return to the earth in rain and snow that falls in the eastern United States and Canada, and similar patterns occur in Europe.

Acid precipitation destroys life. Thousands of the lakes of northern Sweden and Norway no longer support fish; these lakes are now eerily clear. In the northeastern United States and eastern Canada, tens of thousands of lakes are dying biologically as

FIGURE 27-23
Four Corners power plant in New Mexico. In August 1991 the plant agreed to install "scrubbers" on its stacks, one of first benefits of new federal clean air legislation.

FIGURE 27-24

Acid precipitation kills. Twin Pond in the Adirondacks of upstate New York, one of the many lakes in the region in which the levels of acidity in the lake have killed the fishes, amphibians, and most of the other kinds of animals and plants that once occurred there.

a result of acid precipitation (Figure 27-24). At pH levels below 5.0, many fish species and other aquatic animals die, unable to reproduce under these conditions. In southern Sweden and elsewhere, groundwater is now regularly found to have a pH between 4.0 and 6.0, its acidity resulting from the acid precipitation that is slowly filtering down into the underground reservoirs, thus threatening the water supplies of future generations.

Trees also suffer. There has been enormous forest damage in the Black Forest in Germany and in the forests of the eastern United States and Canada. It has been estimated that at least 3.5 million hectares of forest in the northern hemisphere are being affected by acid precipitation (Figure 27-25), and the problem is clearly growing.

Its solution at first seems obvious: capture and remove the emissions instead of releasing them into the atmosphere. There are, however, serious difficulties in executing this solution. First, it is expensive. Reliable estimates of the cost of installing and maintaining the necessary "scrubbers" in the United States are on the order of 4 to 5 billion dollars per year. Although this is not more than one percent of the amount that will ultimately be spent to "bail out" failed savings and loan associations, our national priorities evidently do not clearly focus yet on a healthy environment. An additional difficulty is that that the pollutor and the recipient of the pollution are far from one another, and neither wants to pay so much for what they view as someone

FIGURE 27-25

Damage to trees at Camels Hump Mountain in Vermont.

A The healthy forest as it was in 1963.

B A view from the same spot 20 years later, showing many trees diseased and dying. Air pollution weakens trees and makes them more susceptible to their pests and predators.

C A closer view of some of the dying trees in this area.

A

B

C

else's problem. The Clean Air Act revisions of 1990 addressed this problem in the United States significantly for the first time.

"Acid rain" is a blanket term for the process whereby industrial pollutants such as nitric and sulfuric acids—introduced into the upper atmosphere by factory smokestacks often over 50 meters tall—are spread over wide areas by the prevailing winds and then fall to earth with the precipitation, lowering the pH of ground water and killing life.

The Ozone Hole

The swirling colors of Figure 27-26 are a view of the South Pole in 1989 as viewed from a satellite. This is a computer reconstruction in which the colors represent different concentrations of ozone (O_3), a form of oxygen gas (O_2). As you can easily see, over Antarctica there is an "ozone hole" that is about the size of the United States, within which the ozone concentration is much less than elsewhere. This ozone hole was first reported in 1985 by British environmental scientists. Reviewing the available satellite data, we now know the zone of ozone thinning appeared for the first time in 1975. The hole is not a permanent feature, but rather one that becomes evident each year for a few months at the onset of the Antarctic spring. Every September from 1975 onward, the ozone "hole" has reappeared—and each year the layer of ozone is thinner, and the hole is larger, sometimes reaching southern new Zealand, Australia, and southern South America. In 1985 the minimum ozone concentration in the hole was 30% lower than 5 years earlier.

The major cause of the ozone depletion was suggested in the early 1970s, but generally accepted slowly. Chlorofluorocarbons (CFCs) are chemicals that have been manufactured in large amounts since they were invented in the 1920s, and are largely used in cooling systems, fire extinguishers, and styroforam containers. The CFCs were percolating up through the atmosphere and reducing O_3 molecules to O_2. Although other factors have also been implicated in ozone depletion, the role of these CFCs is so predominant that worldwide agreements to phase out their production by the year 2000 have been signed. Production of CFCs and other ozone-destroying chemicals is banned in the United States after 1995. Nonetheless, large amounts of CFCs that were manufactured earlier, and which are still being manufactured, are moving slowly upward through the atmosphere, so that the problem will grow worse before the ozone layer that protects all life is stabilized once again.

The thinning of the ozone layer in the stratosphere, 25 to 40 kilometers above the surface of the earth, is a matter of serious concern. This layer protects key biological molecules, especially proteins and nucleic acids, from the harmful ultraviolet rays that

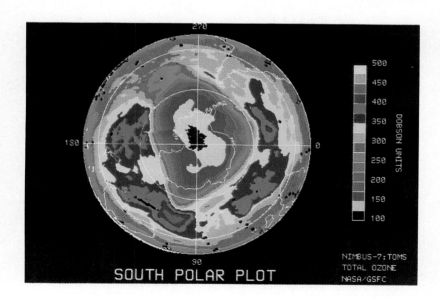

FIGURE 27-26
Satellite picture of the Antarctic ozone hole.

FIGURE 27-C
1 Acid precipitation and fog.
2 Drought.
3 Pollution.
4 Overpopulation.
5 Destruction of the rain forest.

bombard the earth continuously in huge amounts that originate from the sun. Life on land, in fact, may have become possible only when the oxygen layer was sufficiently thick that it generated enough ozone, in chemical equilibrium with the oxygen, that the surface of the earth was sufficiently shielded from these destructive rays. This factor may account for the billions of years in which all life was aquatic.

Ultraviolet radiation is a serious human health concern. Thus every 1% drop in the atmospheric ozone content is estimated to lead to a 6% increase in the incidence of skin cancers. At middle latitudes, the drop of approximately 3% that has occurred worldwide, therefore is estimated to have led to an increase of perhaps as much as 20% in skin cancers, which are one of the more lethal diseases afflicting humans. Humans are relatively resistant to increased ultraviolet radiation, but other organisms, such as the highly susceptible photosynthetic plankton species that are so important to global productivity, are apparently much more highly susceptible.

ENVIRONMENTAL SCIENCE

Environmental scientists attempt to find solutions to environmental problems, considering them in a broad context. Unlike biology or ecology, which are sciences that seek to learn general principles about how life functions, environmental science is an applied science that is dedicated to solving problems. Its basic tools are derived from the science of ecology, from geology, from meteorology (the study of climate), from the social sciences, and from the many other areas of knowledge that bear on the functioning of the environment and our management of it. Environmental science addresses the problem created by rapid human population growth: an increasing need for energy, a depletion of resources, and a growing level of pollution. These problems are both unavoidable and obvious. Environmental science focuses on solving them.

The problems faced by our severely stressed planet (Figure 27-27) are not insurmountable. A combination of scientific investigation and public action, when brought to bear effectively, can solve environmental problems that seem intractable. How is success to be achieved? Viewed simply, there are five components to solving any environmental problem:

1. **Assessment.** The first stage in addressing any environmental problem is scientific analysis, the gathering of information. Data must be collected and experiments performed to construct a "model" that describes the situation. Such a model can be used to make predictions about the future course of events.

FIGURE 27-27
Air pollution. Smog, associated with a temperature inversion, in the Los Angeles Basin.

2. **Risk analysis.** Using the results of scientific analysis as a tool, it is then possible to analyze the **environmental impact,** the potential effects of environmental intervention—what could be expected to happen if a particular course of action were followed. It is necessary to evaluate not only the potential for solving the environmental problem, but also any adverse effects that a plan of action might create.

3. **Public education.** When a clear choice can be made among alternative courses of action, the public must be informed. This involves explaining the problem in terms the public can understand, presenting the alternative actions available, and explaining the probable costs and results of the different choices.

4. **Political action.** The public, through its elected officials, makes a choice, selecting a course of action and implementing it. The choices are particularly difficult to implement when the environmental problems transcend national boundaries.

5. **Follow through.** The result of any action taken should be carefully monitored, to see whether the environmental problem is being solved and, more basically, to evaluate and improve the initial evaluation and modeling of the problem. Every environmental intervention is an experiment, and we need the knowledge gained from each one for its future applications.

WHAT BIOLOGISTS HAVE TO CONTRIBUTE

The development of appropriate solutions to the world's environmental problems must rest partly on the shoulders of politicians, economists, bankers, engineers—many different kinds of people. However, the application of basic biological principles is vital to finding permanent solutions to these problems and achieving a stable, productive world. We have reviewed a number of biological solutions to environmental problems in this chapter.

FIGURE 27-28
The future is now. The girl gazing from this page faces an uncertain future. She is a refugee, an Afghani. The whims of war have destroyed her home, her family, all that is familiar to her. Her expression carries a message about our own future. The message is that our future, like hers, is now upon us, and that the problems which humanity faces in living on an increasingly unstable, overcrowded, and polluted earth are no longer hypothetical, a dilemma for our children, but are with us today and demanding a solution.

The energy that bombards the earth comes in essentially inexhaustible amounts from the sun; living systems capture that energy and fix it in molecules that can be used by all organisms to sustain their life processes. We and all other living beings depend on the proper use of that renewable energy and of the other materials on which our civilization depends. We must come to understand better the processes involved and to develop new systems that will work efficiently in areas where they are not available now. That is our challenge for the future. It is a challenge that must be met soon. In many parts of the world, the future is happening right now (Figure 27-28).

It is clear that a scientific education has become necessary for everyone, so that we may understand the basis for our continued existance on earth and the steps that we will need to take to improve the quality of our lives. Biology should play a major part in that education and is of critical importance in improving the standard of living for our fellow humans. Biological literacy is no longer a luxury for intelligent people who want to play a constructive role in improving the world; it has become a necessity.

SUMMARY

1. When the 1990 college class started school in 1977, there were 4 billion people in the world. There are over 5.3 billion people today, and the world's population is increasing by 1.8% per year. At this rate, there will be over 8 billion people living when their children attend college, double the number living in 1977.

2. An explosively growing human population is placing considerable stress on the environment, consuming ever-increasing quantities of food and water, using a great deal of energy and raw materials, and producing enormous amounts of waste and pollution.

3. Among the many challenges to the environment posed by human activities are releases of materials into the environment that may harm it: escape of radioactive materials into the atmosphere may increase the incidence of cancer; release of pollutants into rivers may make the water unfit for human consumption; release of industrial smoke into the upper atmosphere leads to acid rain that kills forests and lakes; release of chemicals such as chlorofluorocarbons may destroy the atmosphere's ozone and so expose the world to more ultraviolet radiation; release of carbon dioxide by burning fossil fuels may increase the world's temperature and so alter weather and ocean levels.

4. Other challenges are created by our attempts to "develop" natural resources; the cutting and burning of the tropical rain forests of the world to make pasture and cropland is producing a massive wave of extinction.

5. All of these challenges to our future can and must be addressed. Today, environmental scientists and concerned citizens are actively searching for constructive solutions to these problems.

6. The application of the principles of biology to the human condition has never been more necessary than it is now. Only the full attention of society and many talented individuals to the solution of these grave problems will make it possible for our children and grandchildren to enjoy the same benefits that we now enjoy.

REVIEW QUESTIONS

1. What biological event fostered the rapid growth in human populations? How did this event affect the location in which humans lived? What major cultural event eventually took place?

2. Between the industrialized and undeveloped countries, which has the larger population of individuals under the age of 15? Of what significance is this to the population over the next few decades? How will this further affect the proportion of people and their standards of living in industrialized versus undeveloped countries?

3. What occurred during the Green Revolution of the 1950s and 1960s? What are some of its limitations? Do these limitations more strongly affect the undeveloped or industrialized countries? Explain. How much more energy is required to produce 1 calorie of wheat by these techniques as opposed to traditional methods.

4. Why is the concentration of agriculture on cereal crops at the exclusion of all others not a sound nutritional practice? What type of plant in particular nutritionally balances cereal crops? Why?

5. Define shifting agriculture.

6. Why are the agricultural techniques in the tropics destroying the tropical forests? What other human need is destroying the forests? Why? Given current rates of destruction, when will the tropical forests be gone?

7. Even if some tropical forest area does survive, how will it be different from that which exists today? What is the current estimate regarding the rate of species extinction early in the next century? What is the biological implication of such extinction, other than the obvious organismal loss?

8. What are the benefits of nuclear power? What problems must be mastered before its full potential can be realized?

9. How has the amount of carbon dioxide in the atmosphere changed since the advent of industrialization? Why? What is global warming? What effect does this have on the world as a whole?

10. What is the ozone layer? How is it formed? What are the harmful effects of decreasing the earth's ozone layer? What may be the primary cause of this damage? Are high levels of ozone deleterious as well? Explain.

11. What are the goals of environmental science? What tools do environmental scientists use? What are the five components to solving environmental problems?

THOUGHT QUESTIONS

1. Can we ever produce enough food and other materials so that population growth will not be a matter of concern? How?

2. Why don't most people consider the Green Revolution to have been an unqualified success?

3. Some have argued that attempts by the United States to promote lowering of the birth rate in the underdeveloped countries of the tropics is nothing more than economic imperialism, and that it is in the best interest of these countries that their populations grow as rapidly as possible. Give the reasons that you agree or disagree with this assessment.

FOR FURTHER READING

AUSUBEL, J.: "A Second Look at the Impacts of Climate Change," *American Scientist*, May 1991, pages 210-221. We may be worrying too much about crops and coastlines, and too little about water and wildlife.

BROWN, L., editor: *State of the World*, W.W. Norton & Co., New York, 1990. A highly recommended, easily read summary of the ecological problems faced by an overcrowded and hungry world, with an excellent chapter on suggested solutions. A new edition appears every year.

EHRLICH, P.R. and A. EHRLICH: *The Population Explosion*, Simon and Schuster, New York, 1990. A contemporary view of the global effects of human population by two of its leading students.

HEISER, C.B., Jr.: *Seed to Civilization: The Story of Food*, 2nd ed., W.H. Freeman and Co., San Francisco, 1981. A concise and outstanding account of the plants that feed us.

HOUGHTON, R.A. and G.M. WOODWELL: "Global Climatic Change," *Scientific American*, April 1989, pages 36-44. Excellent summary of recent research on global warming.

"Managing Planet Earth," *Scientific American*, entire issue, September 1989. An entire issue devoted to many aspects of the management of our earth, with a number of excellent articles.

MYERS, N.: *The Gaia Atlas of Future Worlds*, Anchor Books, Doubleday, New York, 1990. A concise and outstanding account of global environmental trends and what we can do about them.

ODUM, E.P.: *Ecology and Our Endangered Life Support Systems*, Sinauer Associates, Inc., Publishers, Sunderland, Mass., 1989. An excellent paperback by one of our leading ecologists, who draws on a sound scientific background to analyze the world's environmental problems and their solutions.

REGANOLD, J.P., R.I. PAPENDICK, and J.F. PARR: "Sustainable Agriculture," *Scientific American*, June 1990, pages 112-120. Nontraditional approaches to agriculture offer both financial and environmental rewards.

SAULL, M.: "Nitrates in Soil and Water," *New Scientist*, vol. 122, supplement, pages 1-4, 15 September 1990. The degree to which nitrate fertilizers pollute the environment depends on their behavior in soil.

TOON, O. and R. TURCO,: "Polar Stratospheric Clouds and Ozone Depletion," Scientific American, June 1991, pages 68-74. The story of how the "ozone hole" forms every spring in the high skies over Anarctica.

"The Endless Cycle," *Natural History*, entire issue, May 1990. An entire issue devoted to recycling. Unless we follow nature's example and return used goods to the production cycle, we face a bleak future.

"Trends in Ecology and Evolution," vol. 5 (9) entire issue, September 1990. This entire issue presents an excellent account of the historical and contemporary effects of global climate change on living systems.

VAUGHAN, D.A. and L.A. SITCH: "Gene Flow from the Jungle to Farmers," *BioScience*, vol. 41, pages 22-28, 1991. Charts the ways in which wild-rice genetic resources can be used for the improvement of this vital crop.

VIETMEYER, N.D.: "Lesser-Known Plants of Potential Use in Agriculture and Forestry," *Science*, vol. 232, pages 1379-1384, 1986. An excellent, brief survey of little-known but extremely useful plants that humans could use to improve food and firewood supplies in the future.

VITOUSEK, P., AND OTHERS: "Human Appropriation of the Products of Photosynthesis," *BioScience*, vol. 36, pages 268-372, 1986. Nearly 40% of terrestrial photosynthesis productivity is consumed as a result of human activities now, leaving little room for future growth.

WEINER, J.: *The Next One Hundred Years: Shaping the Future of Our Living Earth*, Bantam Books, New York, 1990. An thorough review of the ecological problems that confront us.

WEISSKOPF, M.: "Plastic Reaps a Grim Harvest in the Oceans of the World," *Smithsonian*, March 1988, pages 59-67. Human products of all kinds are causing ecological havoc, and need our attention.

"What on Earth Are We Doing," *National Wildlife*, vol. 28(20), entire issue, 1990. An excellent issue of this outstanding magazine, entirely devoted to the state of the environment, which presents many actions that you can take to help.

WHITE, R.M.: "The Great Climate Debate," *Scientific American*, July 1990, pages 36-43. Outstanding analysis of appropriate actions that we might take now in the face of impending climatic change.

WILSON, E.O., editor: *Biodiversity*, National Academy Press, Washington, D.C., 1988. A global overview of biodiversity, with articles by many of the leading experts.

WILSON, E.O.: "Threats to Biodiversity," *Scientific American*, September 1989, pages 108-116. Habitat destruction, mainly in the tropics, is driving thousands of species to extinction.

Biology of Viruses and Simple Organisms

28

The Five Kingdoms of Life

OVERVIEW

At least 10 million kinds of organisms inhabit the earth. To discuss and study these organisms, scientists name each kind and group the named species in categories. Less than one third of earth's species have been catalogued and given names. Scientists uniquely identify each kind of organism, or species, by giving it two names. Every species is assigned to a genus, a group of one or more species, and given a specific name to distinguish it from others in its genus. The genus name is written first, followed by the specific name. In turn, genera are grouped into families, families into orders, orders into classes, and classes into phyla (called divisions for plants and fungi). Lastly, phyla are grouped into kingdoms. Each of these more inclusive groups has its own distinctive properties, so that the organism's name, which indicates its position in the system, tells us a great deal about that organism.

Biologists recognize five kingdoms of organisms, falling into two distinctly different groups. The kingdom Monera includes only prokaryotic organisms: the bacteria. Of the four exclusively eukaryotic kingdoms, Fungi, Animalia, and Plantae include primarily multicellular organisms. Single-celled organisms are gathered into a catchall kingdom, Protista. Two of the most important characteristics of eukaryotes, multicellularity and sexuality, evolved among the protists.

Viruses, which are nonliving, are portions of the genomes of bacteria and eukaryotes that have broken loose and taken up a genetically independent existence of their own.

FOR REVIEW

Here are some important terms and concepts that you will encounter in this chapter. If you are not familiar with them, you should review them before proceeding.

Prokaryotic versus eukaryotic cells (Chapter 4)

Meiosis (Chapter 11)

Major features of evolution (Chapter 20)

Definition of species (Chapter 21)

Origin of species (Chapter 21)

All organisms share many biological characteristics. They are composed of one or more cells, carry out metabolism and transfer energy with ATP, and encode hereditary information in DNA. All species have evolved from simpler forms and continue to evolve. Individuals live in populations, which themselves have identifiable properties. These populations make up communities and ecosystems, which provide the overall structure of life on earth. So far, we have stressed these common themes, considering the general principles that apply to all organisms. Now we will focus on the *differences* among groups of organisms, considering the diversity of the biological world and the different properties of the individual organisms that make it up. For the rest of the text, we will examine the different kinds of life on earth, from bacteria and amoebas to blue whales and sequoia trees. Through our study, we will see that examining biological diversity is interesting, challenging, and a great deal of fun.

THE CLASSIFICATION OF ORGANISMS

Millions of different kinds of organisms exist; to talk about them and study them, it is necessary that they have names. Going back as early as we can trace, organisms have been grouped in basic units, such as oaks, cats, and horses. Eventually, these units, which were often given the same names used by the Greeks and Romans, began to be called **genera** (singular, **genus**). Starting in the Middle Ages, these names were written in Latin, the language of scholars at that time, or given a Latin form. Thus oaks were assigned to the genus *Quercus*, cats to *Felis*, and horses to *Equus*—names that the Romans applied to these groups of organisms. For genera that were not known in antiquity, new names had to be invented.

> **The names of genera were the basic point of reference in classification systems, and these names eventually came to be written in Latin.**

About 300 years ago, in the second half of the seventeenth century, scientists first became concerned with overall classification systems for organisms. The specialists who produced these systems became known as **taxonomists** or **systematists,** and their subject is called **taxonomy** or **systematics.**

The Polynomial System

Before the 1750s, scholars usually added a series of additional descriptive terms to the name of the genus when they wanted to designate a particular species. These strings of 15 or more Latin words, starting with the name of the genus, made up what came to be known as **polynomials.** Not only were polynomials cumbersome, but scholars felt free to alter them after the original description. As a result, a single species did not have a single name that was its own alone. Instead, its names were a series of different descriptive phrases that could be related to one another only by scholars. This was a burdensome and imprecise system of naming organisms. The European honeybee, for example, carried 12 names: *Apis pubescens, thorace subgriseo, abdomine fusco, pedibus posticis glabris utrinque margine ciliatis.*

The Binomial System

The Swedish biologist Carolus Linnaeus (1707-1778) first offered a solution to the century-old problem of naming organisms (Figure 28-1). Linnaeus' ambition, like that of many of his predecessors, was to catalog all of the known kinds of organisms and minerals. In the 1750s he produced several major works that, like his earlier books, employed the polynomial system. As a kind of shorthand, however, Linnaeus included a two-part name for each species—the honeybee became *Apis mellifera.*

Other species names shrunk as well. Linnaeus named the willow oak of the southern United States *Quercus phellos* (Figure 28-2), but the name he considered "correct" included four adjectives: *Quercus foliis lanceolatis integerrimis glabris* ("oak with spear-shaped, smooth leaves with absolutely no teeth along the margins"). The

FIGURE 28-1
Carolus Linnaeus (1707-1778).
This Swedish biologist devised the system of naming organisms that is still in use today.

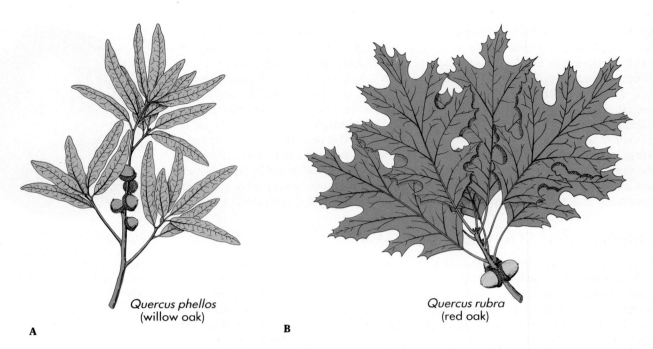

Quercus phellos
(willow oak)

A

Quercus rubra
(red oak)

B

FIGURE 28-2
Two species of oaks.
A Willow oak, *Quercus phellos.*
B Red oak, *Quercus rubra.*
Although they are clearly oaks
(*Quercus*), these two species differ
sharply in the shapes and sizes of
their leaves and in many other
features, including their overall
geographical distributions.

convenience of the short names was so great that they were immediately adopted as the standard by Linnaeus and all subsequent authors. These two-part names, or **binomials,** have continued as our standard way of identifying species. Species names may not be used without a genus, but the genus may be abbreviated to a single letter ("*Q. phellos*") if the reference is clear.

An international association of taxonomists establishes the scientific names of organisms according to a precise set of rules. The names are thus the same throughout the world and provide a uniform way of communicating about organisms. A Chinese-speaking biologist can be sure of the identity of *Quercus rubra* whether she is reading a paper written in English, Japanese, or Russian. In contrast, common names often differ greatly from place to place. A "robin" in Europe is *Erithacus rubicula*, in North America *Turdus migratorius*. The clay-colored robin of Mexico and Central America is *Turdus grayi;* the hill robin of Hawaii and Asia is *Leiothrix lutea.* Some of these "robins" are related, some are not, but the universal scientific names make their relationships clear to scientists the world over.

Even within one country, common names can confuse. The cat flea, sand flea, snow flea, and water flea have little in common besides being arthropods. Despite sharing part of their common names, each of these organisms is in a different order.

> For scientific communication, it is important to have one standard set of
> names. Linnaeus' binomial shorthand system has served the science of
> biology well for nearly 250 years.

THE TAXONOMIC HIERARCHY

Starting about 300 years ago, taxonomists began to construct a **hierarchical** system, first grouping species into genera and genera into larger, more inclusive categories known as **families** to reflect perceived relationships between the groups included. Genera of oaks (*Quercus*), beeches (*Fagus*), and chestnuts (*Castanea*) belong to the beech family, Fagaceae, because they share the characteristics of that family. Similarly, the tree squirrels (*Tamiasciurus*), Siberian and western North American chipmunks (*Eutamias*) and marmots (*Marmota*) are grouped, along with other genera, in the family Sciuridae (Figure 28-3). All have four front toes, five hind toes, a hairy tail and other features characteristic of the family. If one knows that a genus belongs to a particular family, one immediately knows a great many of its features.

The **taxonomic system** was eventually extended to include several more inclusive units. Families are grouped into **orders,** orders into **classes,** and classes into **phyla**

A

B

C

FIGURE 28-3

Genera. Three genera of the squirrel family, Sciuridae, which differ greatly in their characteristics and are adapted for different modes of life.

A Yellow-bellied marmot, *Marmota flaviventris*, a large squirrel that lives in burrows. Yellow-bellied marmots are social, tolerant, and playful; they live in harems that consist of a single territorial male together with numerous females and their offspring.

B Townsend's chipmunk, *Eutamias townsendii*, a member of a genus of diurnal, brightly colored, very active ground-dwelling squirrels.

C The red squirrel, *Tamiasciurus hudsonicus*, an agile dweller in the trees.

(singular **phylum**), which are called **divisions** among plants, fungi, and algae. Phyla and divisions are in turn grouped into **kingdoms,** the most inclusive units of classification (Figure 28-4).

> By convention, species are grouped into genera, genera into families,
> families into orders, orders into classes, and classes into phyla (or divisions).
> Phyla are the basic units within kingdoms; such a system is hierarchical.

Table 28-1 shows how the human species, the honey bee, and the red oak are placed in taxonomic categories at the seven hierarchical levels. The categories at the different levels may include many, a few, or only one smaller taxon, depending on the nature of the relationships in the particular groups involved. Thus, for example, there is only one living genus of the family Hominidae, but several living genera of Fagaceae. To someone familiar with classification or with access to the appropriate

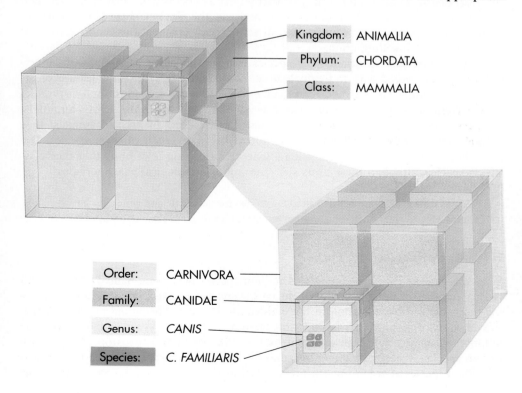

Kingdom: ANIMALIA
Phylum: CHORDATA
Class: MAMMALIA
Order: CARNIVORA
Family: CANIDAE
Genus: *CANIS*
Species: *C. FAMILIARIS*

FIGURE 28-4

The hierarchical system used in classifying organisms, in this case a dog.
Kingdom: Monera, Plantae, Fungi, Protista, ANIMALIA.
Phylum: Anthophyta, Mollusca, CHORDATA, etc.
Class: Amphibia, Reptilia, MAMMALIA, etc.
Order: Insectivora, Sirenia, CARNIVORA, etc.
Family: Mustelidae, Ursidae, CANIDAE, etc.
Genus: *Vulpes, Urocyon,* CANIS, etc.
Species: *Canis latrans, C. lupus, C. FAMILIARIS,* etc.

TABLE 28-1	SAMPLE CLASSIFICATIONS OF THREE REPRESENTATIVE ORGANISMS		
	HUMAN	HONEY BEE	RED OAK
Kingdom	Animalia	Animalia	Plantae
Phylum (division)	Chordata	Arthropoda	Anthophyta
Class	Mammalia	Insecta	Dicotyledones
Order	Primates	Hymenoptera	Fagales
Family	Hominidae	Apidae	Fagaceae
Genus	*Homo*	*Apis*	*Quercus*
Species	*Homo sapiens*	*Apis mellifera*	*Quercus rubra*

reference books, each taxon implies both a set of characteristics and a group of organisms belonging to the taxon.

Convention governs how scientific names are printed. The genus is always capitalized, the species lower case, and both are italicized, or written in distinctive print, for example, *Homo sapiens*. The scientific names of the other taxonomic units are not printed this way or underlined; they are capitalized.

What is a Species?

In Chapter 21 we reviewed the nature of species and saw that there are no absolute criteria that can be applied to the definition of this category. Individuals that belong to a given species (for example, dogs) may look very unlike one another. Nevertheless, they are generally capable of hybridizing with one another, and different forms can appear in the progeny of a single mated pair. Species remain relatively constant in their characteristics, can be distinguished from other species, and do not normally interbreed when they occur together with other species in nature. For example, dogs are not capable of interbreeding with other carnivores such as weasels or foxes, which, although they are generally similar, are members of other distinct groups. In contrast, dogs can and do form fully or partly fertile hybrids with related species such as wolves and coyotes, which are also members of the genus *Canis*. The transfer of characteristics among these species has, in some areas, changed the characteristics of both of the interbreeding units.

The criteria just mentioned for species apply primarily to those that regularly **outcross**—interbreed with individuals other than themselves. In some groups of organisms, including bacteria and many eukaryotes, **asexual reproduction**—reproduction without sex—predominates. The species of these organisms clearly cannot be characterized in the same way as the species of outcrossing plants and animals; they do not interbreed with one another, much less with individuals of other species. Despite these difficulties, biologists generally agree on the kinds of units that they classify as species, although these units share no uniform biological characteristics.

Species differ from one another in at least one characteristic and generally do not interbreed freely with one another where their ranges overlap in nature.

How Many Species are There?

Since the time of Linnaeus, about 1.4 million species have been named. This is a far greater number of organisms than Linnaeus suspected to exist when he was developing his system of classification in the eighteenth century. The actual number of species in the world, however, is undoubtedly much greater. Basing their estimates on numbers of species found in different parts of the world and the proportions of unknown species typically obtained in samples, taxonomists have concluded that at least 10 million species of organisms exist on earth. Most of these—more than 8 million spe-

FIGURE 28-5
How many specis are there? Scientists are attempting to determine the number of species in the canopy of moist tropical forest, the richest biological community on earth. In these experiments, the insects and other animals living in the canopy are killed with insecticide and then sampled as they drop onto the sheets below. By such methods, Terry L. Erwin, Jr., of the Smithsonial Institution has estimated that there may be as many as 30 million kinds of organisms in the world. Taxonomists have thus far recognized only about 1.4 million of these.

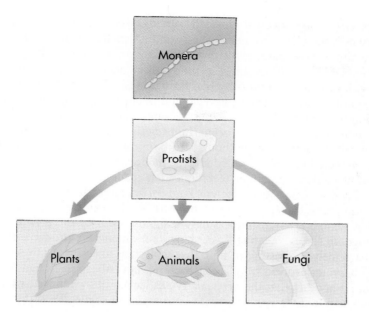

FIGURE 28-6
The five kingdoms of life.

FIGURE 28-7
Kingdoms. Kingdoms are the most inclusive level of biological classification. For obvious reasons, people have always considered plants and animals to be distinct from one another. **A,** The preying mantis, *Hymenopus coronatus*, is camouflaged by its resemblance to the orchid flowers; it will seize any insect or other small animal that comes within range. Such a pattern of activity appears completely different from that of a plant such as *Coleus*, **B.** Animals ingest their food, and most move about from place to place; plants manufacture their food through the process of photosynthesis and are stationary. These and other major differences have traditionally caused plants and animals to be regarded as distinct kingdoms, with their similarities being appreciated only in the past hundred years or so.

cies—are thought to occur in the tropics. Of that number, only 400,000 tropical species have been named so far. Furthermore, the actual number of unknown species may be much greater, since some habitats remain relatively unexplored (Figure 28-5).

THE HISTORY OF LIFE ON EARTH
The Five Kingdoms of Life

The earliest classification systems recognized only two kingdoms of living things: animals and plants. But as biologists discovered microorganisms and learned more about other organisms, the number of kingdoms increased. Systematists added kingdoms in recognition of fundamental differences discovered among organisms. Although even at the kingdom level classification remains in flux, most biologists now use the **five-kingdom system** presented here (Figure 28-6).

In this system, four of the kingdoms consist of eukaryotic organisms. The two most familiar kingdoms, **Animalia** and **Plantae,** contain only organisms that have many cells during most of their life cycle. The kingdom **Fungi** contains multicellular forms and the single-celled yeasts, which are thought to have multicellular ancestors. Fundamental differences divide the three multicellular kingdoms. They differ in their morphologies, in motility, and in their modes of nutrition. Plants are mainly stationary, but some have motile sperm; fungi have no motile cells. Animals ingest their food, plants manufacture it, and fungi digest it by means of secreted extracellular enzymes and absorb the digested food. Each of these kingdoms probably evolved from a different single-celled ancestor.

The large number of unicellular eukaryotes are arbitrarily gathered into a single kingdom, called **Protista** (see Chapter 31). Protista also includes the algae, all of which are unicellular during important parts of their life cycle and have multicellular morphologies similar to plants.

The fifth kingdom, **Monera,** contains only prokaryotic organisms—vastly different from all other living things (see Chapter 30). Within this kingdom are two vastly different kinds of prokaryotes, archaebacteria and eubacteria, and many biologists believe they should be considered separate kingdoms.

Evolution of Prokaryotes

Though not reflected directly in the kingdom system, the most fundamental distinction among living things is not between animals and plants (Figure 28-7), but between bacteria and all other organisms (Figure 28-8).

Early microscopists discovered that bacteria lack a membrane-bound nucleus and called them **prokaryotes** (*pro,* before and *karyon,* kernel or nucleus). But prokaryotes

A

B

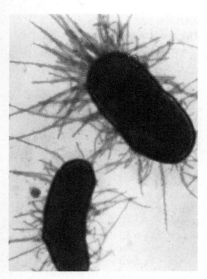

FIGURE 28-8

The oldest kingdom. Bacteria, such as these individuals of *Escherichia coli*, are the only members of the kingdom Monera. They are prokaryotes, a term that describes a pattern of cellular organization completely different from that of all other organisms. The slender projections extending from the surfaces of these bacteria are called pili.

differ from eukaryotes (*eu*, true) in many other respects as well (see Chapter 4); unlike eukaryotes, they have no membrane-bound organelles or microtubules, and their flagella are simple, compared with those of eukaryotes. Even prokaryotic chemistry differs; their DNA and RNA differ in the ratios of nucleotides, and their DNA does not bind with proteins to form complex chromosomes. The cell walls of most bacteria contain muramic acid, a chemical never found in eukaryotes. Bacteria do not undergo sexual recombination in the same sense that eukaryotic organisms do, although other forms of genetic recombination occur in bacteria. Until recently the prokaryotes, commonly called bacteria, were thought to be a relatively uniform group and all of them were assigned to a single kingdom, the monera. During the 1980s, however, the application of modern molecular techniques has made it clear that the monera kingdom in fact contains two very different kinds of bacteria, the eubacteria ("true bacteria") and the archaebacteria ("ancient bacteria"). The evidence has become so compelling that taxonomists are increasingly assigning each of these two very different groups kingdom status, creating TWO kingdoms among the prokaryotes.

It is extremely interesting to note that molecular analysis of DNA indicates that the first eukaryotes evolved from archaebacteria, not eubacteria as had been supposed in the past. It appears that the first eukaryotes lacked mitochondria, which later representatives acquired via symbiosis with eubacteria. In Figure 28-9, the establishment of such symbiotic relationships is indicated by dotted lines, while solid lines represent phylogenetic relationships. Chloroplasts are another instance in which eubacteria were acquired by eukaryotes as organelles via symbiosis—as many as six different times.

Evolution of Eukaryotes

For 2 billion years bacteria ruled the earth. No other organisms existed to eat them or compete with them, and traces of their tiny, prokaryotic cells appear in the world's oldest fossils. Eukaryotic cells appear full-fledged in the fossil record much later. Because of the lack of fossils from this **Precambrian** period, the intervening steps remain conjectural. Once eukaryotic cells appeared, however, the evolutionary picture

TABLE 28-2 CHARACTERISTICS OF THE FIVE KINGDOMS

KINGDOM	CELL TYPE	NUCLEAR ENVELOPE	MITO-CHONDRIA	CHLOROPLASTS	CELL WALL
Monera	Prokaryotic	Absent	Absent	None (photosynthetic membranes in some types)	Noncellulose (polysaccharide plus amino acids)
Protista	Eukaryotic	Present	Present or absent	Present (some forms)	Present in some forms, various types
Fungi	Eukaryotic	Present	Present or absent	Absent	Chitin and other non-cellulose polysaccharides
Plantae	Eukaryotic	Present	Present	Present	Cellulose and other polysaccharides
Animalia	Eukaryotic	Present	Present	Absent	Absent

Some biologists suggest that basal bodies and centrioles may have arisen from endosymbiotic spirochaete-like bacteria. Even today, so many bacteria and unicellular protists form symbiotic alliances that the incorporation of smaller organisms with desirable features into cells appears to be a relatively easy process.

Multicellularity. The unicellular body plan is a tremendously successful one, comprising about half of the biomass on earth. Bacteria have continued a slow evolution for billions of years, using nearly every energy source available. Modern protists are not only numerous, but they are extraordinarily diverse both in their form and in their biochemistry (Figure 28-10).

Yet a single cell has limits. The evolution of multicellularity allowed organisms to deal with their environments in novel ways. Distinct types of cells, tissues, and organs can be differentiated within the complex bodies of multicellular organisms. With such a functional division within its body, a multicellular organism can protect itself, move about, seek mates and prey, and carry out other activities on a scale and with a complexity that would have been impossible for its unicellular ancestors. With all of these advantages, it is not surprising that multicellularity has arisen independently so many times.

Multicellularity evolved several times among the protists. All of the complex differentiation that we associate with advanced life-forms depends on multicellularity, which must have been highly advantageous to have evolved independently so often.

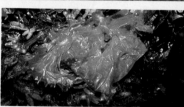

True multicellularity, in which the activities of the individual cells are coordinated and the cells themselves are in contact, occurs only in eukaryotes and is one of their major characteristics. The cell walls of bacteria occasionally adhere to one another, and bacterial cells may also be held together within a common sheath (see Chapter 30). Some bacteria form filaments, sheets, or three-dimensional aggregates, but little or no integration occurs. Such bacteria may be considered colonial, but none are truly multicellular. Many protists also form colonial aggregates, consisting of many cells with little differentiation or integration.

Other protists—for example, the red, brown, and green algae—have independently attained true but simple multicellularity (Figure 28-11). These structurally simple multicellular organisms fill an evolutionary gap between unicellular protists and the complex, multicellular animals, plants, and fungi. Multicellular green algae were almost certainly the direct ancestors of the plants (see Chapters 31 and 33), and were considered plants in previous classification schemes. However, green algae are basically aquatic and much simpler in structure than plants, and are considered protists within the five-kingdom system.

Fungi and animals arose from unicellular ancestors with different characteristics. The kingdom Protista includes several groups of animal-like protists: flagellates, ciliates, and amebas, for example. Funguslike protists, including the slime molds and water molds, are not thought to resemble ancestors of fungi.

EVOLUTIONARY TAXONOMY
Should Taxonomy Reflect History?

Taxonomy is often thought of as the grey science, old men in musty rooms arguing about names. In fact, it is one of the more lively branches of biology, often controver-

FIGURE 28-11
Multicellular protists. Three phyla of algae have become multicellular independently and differ greatly in their photosynthetic pigments and structure. **A** The brown algae (phylum Phaeophyta) are large, complex algae, which often dominate northern temperate seacoasts. **B** *Ulva* is a green alga (phylum Chlorophyta), a group that includes many unicellular organisms as well as a number that are multicellular. **C** Some red algae (phylum Rhodophyta) are shown here growing on sponges. Members of this phylum occur at greater depths than the members of either of the other phyla. **D** *Volvox* (**phylum Chlorophyta**). Individual unicellular green algae are united in *Volvox* as a hollow ball of cells.

sial and rarely dull. Perhaps the oldest and most fundamental disagreement within taxonomy concerns what its role in biology should be, what it should be trying to accomplish. There are two fundimental viewpoints, one the Linnaean approach of classifying and naming, and the other the Darwinian approach of tracing evolutionary history. Each viewpoint has led to extreme schools within taxonomy, and in practice both viewpoints have an important influence on how taxonomy is done today.

Classifying by Morphological Similarity

Ever since Linnaeus, biologists have been naming new species by carefully noting how different kinds of creatures differ from one another. The essence of this process lies in making a judgement about what differences between species are important. The shape, size, and appearance of an individual is called its morphology, and differences in morphology have always formed the backbone of taxonomy. In the 1950s some taxonomists began to use computers in order to compare a great many morphological traits at once, not only their presence or absence, but also quantitative information on size and color. This approach, called **numerical taxonomy** or **phenetics,** judges taxonomic affinities entirely on the basis of measurable similarities and differences. As many characteristics as possible are compared, in order to minimize the contribution of a few characters which might resemble one another because of parallel evolution such as discussed on pages 397-398 (**analogy**) rather than because of taxonomic affinity (**homology**). Strictly phenetic approaches are not widely used today, as they make no attempt to judge the phylogenetic affinity of organisms.

Classifying by Evolutionary Relationships

At the opposite end of the taxonomic spectrum are biologists that consider only evolutionary relatedness in assigning taxonomic affinity, ignoring their degree of morphological similarity or difference. This school of taxonomy is called **cladistics.** Cladistics (from the Greek word clados, "branch") classifies organisms according to the historical order in which branches arise along a phylogenetic tree. Only the order of branching is considered in assigning position on the phylogenetic tree.

Cladistics is ideally suited to molecular data, particularly data on DNA sequence divergence such as discussed on pages 394-395. All that is necessary is to ascertain which difference arose after a branch diverged from the evolutionary tree, called **derived characters.** Derived characters are characters which are shared by all members of a branch but not present before the branch. On the phylogenetic tree of plants, for example, vascular tissue is a derived character, shared by all vascular plants but not present in plants that evolved before them like mosses. Among the vascular plants, all the plants which possess seeds are placed after one branch on the cladogram, distinct from plants like ferns which evolved before the advent of seeds. Among the seed plants, the plants on the branch occupied by angiosperms possess flowers, while the gymnosperms do not. Further along the angiosperm branch is another branch containing a group of plants whose seedlings have two leaves, called dicots, while no other angiosperms do. A phylogenetic tree constructed in this fashion is called a **cladogram** (Figure 28-12). A cladogram shows the *order* of evolutionary descent, not the extent of divergence.

Taxonomy Today

In practice, taxonomy today utilizes information from both phenetics and cladistics. A very clear example of the conflict that can arise between *order* of divergence and *magnitude* of divergence is provided by the taxonomists' assignment of birds to a separate class, *Aves,* while retaining crocodiles in the class *Reptilia*—even though crocodiles seem more closely related to birds than to other reptiles. From a cladist's viewpoint, birds and crocodiles shared many characters (such as a four chambered heart, right), and a cladogram would group birds and crocodiles together if the traits are viewed as derived rather than having evolved independently (Figure 28-13). Tax-

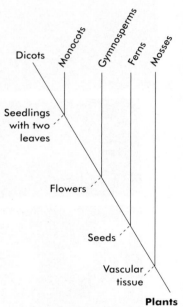

FIGURE 28-12
A plant cladogram. The derived characters connected to the cladogram branch points by dotted lines are shared by all plants above the branch point and are not present in any plants below it.

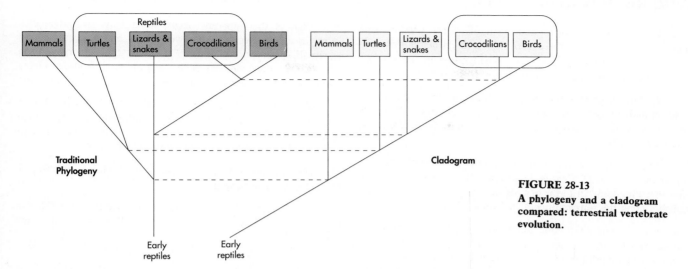

FIGURE 28-13
A phylogeny and a cladogram compared: terrestrial vertebrate evolution.

onomists today, however, assign birds to their own unique class *Aves* because birds have diverged so fundamentally since they separated from the crocodiles, evolving feathers, hollow bones, and a host of other adaptations to flight. Ultimately, the taxonomic decision to classify crocodiles with reptiles is a practical one—The extent of divergence since birds branched from crocodiles is in this case to important biologically to ignore.

SUMMARY

1. Carolus Linnaeus, an eighteenth-century Swedish scientist, developed the binomial system of naming organisms that remains in use today. Biologists give every species two names, the first of which is the genus and the second is the species. Both parts combine to uniquely identify the species.

2. In the hierarchical system of classification used to describe organisms, genera are grouped into families, families into orders, orders into classes, classes into phyla (called divisions in some groups), and phyla into kingdoms. (Many students have used this time-honored mnemonic to memorize taxon order from the top down: **K**ings **P**lay **C**hess **O**n **F**ine-**G**rained **S**and.)

3. There are at least 5 million species of plants, animals, and microorganisms in the world. Biologists have named only about one third of them so far.

4. The fundamental division among organisms is between prokaryotes, all of which lack a true nucleus, and eukaryotes, which have a true nucleus and several membrane-bound organelles. All prokaryotes are placed in the kingdom Monera.

Eukaryotes are divided into four kingdoms: Protista, Fungi, Plantae, and Animalia.

5. Many diverse phyla of unicellular eukaryotes are placed in the kingdom Protista. Virtually all protists have at least some simple multicellular representatives, and two—red algae and brown algae—are primarily multicellular.

6. Three groups of almost entirely multicellular eukaryotes are so large and differ so sharply from their ancestors among the Protista that each is classified as a different kingdom. These are Animalia, Plantae, Fungi.

7. True multicellularity and sexuality are found only among eukaryotes. Multicellularity confers a degree of protection from the environment and the ability to carry out a wider range of activities than is available to unicellular organisms. Sexuality permits extensive, orderly reproduction and genetic variation among descendents.

8. Traditional taxonomy involves a study of both the order in which groups evolve (cladistics) and the degree to which they diverge from one another.

REVIEW QUESTIONS

1. From smallest to largest and in singular form, what are the names of the groups in the current hierarchical taxonomic system. Which two are given special consideration in the way in which they are printed? What are these distinctions?

2. Is there a greater fundamental difference between plants and animals or between prokaryotes and eukaryotes? How do prokaryotes differ from eukaryotes? Into which kingdom are prokaryotes placed?

3. From which of the four eukaryotic kingdoms have the other three evolved? What is the primary characteristic that unifies that kingdom? In terms of nutrition and locomotion, distinguish among the four multicellular kingdoms.

4. Of the four eukaryotic kingdoms, which is the most diverse? What common name is given to the heterotrophic members of this kingdom? What common name is given to the autotrophic members of this kingdom? What three groups within this kingdom exhibit some multicellular forms?

5. What is the apparent origin of at least two organelles found in eukaryotes? Of the two, which one is possessed by nearly all eukaryotic organisms? To what type of organism is this organelle most similar? How many ancestors appear to have given rise to the other organelle? Why?

6. What defines whether or not a collection of individuals of a single type is truly multicellular? Did multicellularity arise once or many times in the evolutionary process? What advantages do multicellular organisms have over unicellular ones?

7. What are the three major types of life cycles in eukaryotes? Describe the major events of each.

8. Are viruses technically living or nonliving? Why? What is their most basic composition? Into which kingdom are they placed? Why?

THOUGHT QUESTIONS

1. What are some of the advantages of the binomial system of naming organisms?

2. Do you think a better system of classification would be to group all photosynthetic eukaryotes, regardless of whether or not they are single-celled, as plants, and to group all nonphotosynthetic ones as animals? Present arguments in favor of doing so and other arguments in favor of the system of classification used in this book.

3. Supposing that viruses were considered living, and emphasizing the split between prokaryotes and eukaryotes, construct a new, overall classification system, including all of the present kingdoms.

FOR FURTHER READING

GOULD, S.J.: "Fuzzy Wuzzy Was a Bear, Andy Pandy Too," *Discover*, February 1986, pages 40-48. Outstanding essay on the problem of biological relationships.

LOWENSTEIN, J.M.: "Molecular Approaches to the Identification of Species," *American Scientist*, vol. 73, pages 541-547, 1985. Discusses the application of immunological and other techniques to the problem of determining relationships.

MARGULIS, L., and D. SAGAN: *Garden of Microbial Delights*, Harcourt Brace Jovanovich, Publishers, Boston, 1988. A lively and readable guide to the viruses, bacteria, and protists.

MARGULIS, L., and D. SAGAN: *Origins of Sex*, Yale University Press, New Haven, Conn., 1986. The authors present arguments about the evolution of sex in a stimulating and beautifully illustrated book that is well worth reading.

MARGULIS, L., and K.V. SCHWARTZ: *Five Kingdoms: An Illustrated Guide to the Phyla of Life on Earth*, W.H. Freeman and Co., San Francisco, 1982. A marvelous account of the diversity of organisms, beautifully illustrated and presented in a logical fashion.

MAYR, E.: "Biological Classification: Toward a Synthesis of Opposing Methodologies," *Science*, vol. 214, pages 510-516, 1981. A thoughtful consideration of alternative methods of classifying, with emphasis on genera and species.

McDERMOTT, J: "A Biologist Whose Heresy Redraws Earth's Tree of Life," *Smithsonian*, August 1989, pages 72-80. Lynn Margulis' innovative approach to endosymbiosis has revolutionized the way we think about organisms.

O'BRIEN, S.B., AND OTHERS: "A Molecular Solution to the Riddle of the Giant Panda's Phylogeny," *Nature*, vol. 317, pages 140-144, 1985. Fascinating report of the use of multiple evidence from macromolecular structure to solve a long-standing mystery.

SIBLEY, C.G., and J.E. AHLQUIST: "Reconstructing Bird Phylogeny by Comparing DNAs," *Scientific American*, February 1986, pages 82-92. How macromolecular evidence has clarified relationships among the major groups of birds.

WOESE, C., O. KANDLER, and M. WHEELIS: "Towards a Natural System of Organism: Proposal for the Domains Archaea, Bacteria, and Eucarya," *Proc. Nat. Acad. Sci. US*, vol. 87, pages 4576-4579, 1990. Conveys much of the excitement of the new taxonomy being generated from molecular anaylsis; an alternative to six kingdoms.

29

Viruses

OVERVIEW

A virus is a fragment of another genome, often a small fragment, which if it gains entry to a cell can replicate by using that cell's machinery. Many viruses have evolved complex sets of genes that enable them to infect host cells and multiply within them. Some viruses produce serious diseases as side effects in their host organisms. Viruses are not classified as living and are therefore not considered to be organisms; they have only some of the elements necessary for self-replication. Because of their disease-producing potential, however, viruses are important biological entities. Much of what we know about molecular biology has been learned from studies of viruses. It is thought that all organisms possess viruses and that new viruses are "escaping" from bacterial and eukaryotic genomes and evolving even now.

FOR REVIEW

Here are some important terms and concepts that you will encounter in this chapter. If you are not familiar with them, you should review them before proceeding.

Definition of organism (Chapter 4)

TMV and T2 viruses (Chapter 15)

Lambda virus integration (Chapter 17)

Retroviruses and cancer (Chapter 17)

Vaccinia virus and vaccination (Chapter 18)

The simplest organisms living on earth today are bacteria, and we think that they closely resemble the first living organisms to evolve on earth. Even simpler than the bacteria, however, are the viruses. **Viruses** are strands of nucleic acid that are encased within a protein coat. Viruses do not have the ability to grow or replicate on their own. Viruses reproduce only when they enter cells and use the cellular machinery of their hosts. They are able to reproduce themselves in this way because they carry genes that are translated into proteins by the host cell's genetic machinery, leading to the production of more viruses.

Outside of its host cell, a virus is simply a fragment of nucleic acid encased in protein. The fact that individual kinds of viruses contain only a single type of nucleic acid, either DNA or RNA, is one of the major reasons that they can reproduce only within living cells. All true organisms contain both DNA and RNA, and both are essential components of their genetic machinery. Viruses also lack ribosomes, as well as all of the enzymes necessary for protein synthesis and energy production.

In light of their simple chemical nature and total inability to exist independently from other organisms, earlier theories that viruses represent a kind of halfway point between life and nonlife have now largely been abandoned. Instead, viruses are now viewed as fragments of the genomes of organisms; they could not have existed independently of preexisting organisms.

As an analogy to a virus, consider a computer whose operation is directed by a set of instructions in a program, just as a cell is directed by DNA-encoded instructions. A new program can be introduced into the computer that will cause the computer to cease what it is doing and instead devote all of its energies to making copies of the introduced program. The new program is not itself a computer, however, and cannot make copies of itself when outside the computer, lying on the desk. The introduced program, like a virus, is simply a set of instructions.

> **Viruses are fragments of DNA or RNA that have become detached from the genomes of bacteria or eukaryotes and have the ability to replicate themselves within cells. Viruses are considered to be *nonliving,* because they lack all of the necessary biochemical features that would allow them to replicate on their own; consequently, they are *not considered to be organisms.* Their genetic material consists of DNA or RNA, but not both.**

THE DISCOVERY OF VIRUSES

Depending on which genes it carries, a virus can seriously disrupt the normal functioning of the cells that it infects. Such effects occur in humans and other animals, plants, protists, and bacteria; in fact, cells that are completely free of viruses may be rare.

Diseases caused by viruses have been known and feared for thousands of years. Among the diseases that viruses cause are smallpox, chickenpox, measles, German measles (rubella), viral encephalitis, mononucleosis, mumps, shingles, influenza, colds, infectious hepatitis, yellow fever, polio, rabies, and AIDS, as well as many other diseases not as well known. In addition, viruses have been implicated in some cancers and leukemias; it has been estimated that as many as a quarter of all cases of cancer may be caused by viruses. For many autoimmune diseases, such as multiple sclerosis and rheumatoid arthritis, and for diabetes, specific viruses have been found associated with certain cases. In view of their effects, it is easy to see why the late Sir Peter Medawar, Nobel laureate in physiology of medicine, wrote: "A virus is a piece of bad news wrapped in protein." They not only cause many human diseases, but probably infect every other kind of organism as well, causing major losses in agriculture and forestry and in the productivity of natural ecosystems.

Viral diseases were studied centuries before it was understood that viruses caused them. Thousands of years ago, the Chinese practiced **variolation,** in which people were exposed to skin scabs from others who had survived smallpox infection; although such practices had a fatality rate of about 1%, they did afford a degree of protection to those who had been treated. In the 1790s, Edward Jenner, an English country doctor, observed that milkmaids often caught a mild form of "the pox," pre-

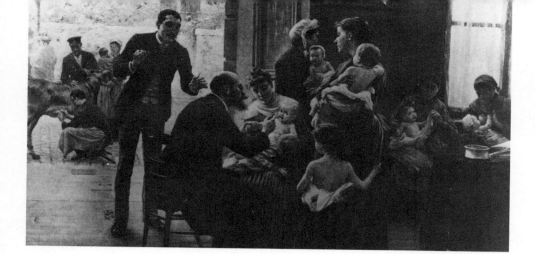

FIGURE 29-1
Vaccination. Edward Jenner inoculating patients with cowpox, and thus protecting them from smallpox, in the 1790s. The underlying principles were not understood until more than a century later.

sumably from cows, and that milkmaids who had caught cowpox rarely contracted smallpox, a very serious disease. It was as if the cowpox were protecting them from the smallpox. Using this observation, Jenner developed the process of **vaccination.** He deliberately infected people with a relatively harmless cowpox virus, causing the recipient's body to develop antibodies against cowpox, antibodies that also were effective against smallpox (Figure 29-1). Jenner learned how to carry out this process despite total ignorance of the nature of antibodies and viruses. The explanation of how vaccination works was not established until more than a hundred years later.

The earliest indirect observations of viruses, other than simple observations of their effects, were made near the end of the nineteenth century. At that time, several groups of European scientists who working independently concluded that the infectious agents associated with a plant disease known as tobacco mosaic and those associated with hoof-and-mouth disease in cattle were not bacteria. They reached this conclusion because these infectious units were not filtered out of solutions by the kinds of fine-pored porcelain filters that were routinely used to remove bacteria from various media. In 1933, Wendell Stanley of the Rockefeller Institute prepared an extract of tobacco mosaic virus and purified it. Surprisingly the purified virus precipitated in the form of crystals (Figure 29-2). Stanley was able to show by this method that viruses can better be regarded as chemical matter than as living organisms, at least in any normal sense of the word "living." The purified crystals still retained the ability to reinfect healthy tobacco plants and so clearly were the virus itself, not merely a chemical derived from it. In fact, they can be kept indefinitely in a crystalline form, and then reinfect their hosts when they are introduced into them. With Stanley's experiments, scientists began to understand the nature of viruses for the first time.

Within a few years, other scientists were able to follow up on Stanley's discovery and demonstrate that tobacco mosaic virus consisted of a protein in combination with a nucleic acid. They discovered that the particles that constituted this virus were rods about 300 nanometers long (Figure 29-3). Subsequently it was shown that tobacco

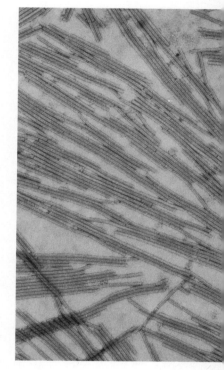

FIGURE 29-2
Tobacco mosaic virus. Electron micrograph of tobacco mosaic virus, showing virus particles in a crystalline array taken from infected tobacco leaves.

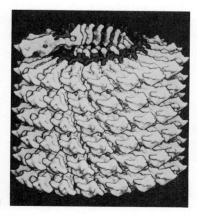

FIGURE 29-3
Computer-generated model of a portion of tobacco mosaic virus. An entire virus consists of 2130 identical protein molecules—the yellow knobs, which form a cylindrical coat around a single strand of RNA, colored red in this model. The RNA molecule is the backbone of the virus, and determines its shape; along it, identical protein molecules are packed tightly together, protecting the RNA. This model is based on radiographic analyses of the virus structure.

mosaic virus consists of RNA surrounded by a protein coat. Many plant viruses have a similar composition, but most other viruses have DNA in place of RNA. Nearly all viruses form a protein sheath, or **capsid,** around their nucleic acid core. In addition, many viruses form an **envelope** that is rich in proteins, lipids, and glycoprotein molecules around the capsid.

> **Most viruses form a protein sheath, or capsid, around their nucleic acid core, and a lipid-rich protein envelope around the capsid.**

The simple structure of viruses, the large numbers that are produced in an infection of a cell, and the fact that their genes are related to those of their host, have led many scientists to study viruses in attempts to unravel the nature of genes and how they work. For more than three decades, the history of virology has been thoroughly intertwined with those of genetics and molecular biology. In the future, it is expected that viruses will be one of the principal means by which genetic traits are experimentally carried from one organism to another, and may be used, for example, in the treatment of human genetic diseases.

THE NATURE OF VIRUSES

Viruses probably occur in every kind of organism; they usually produce no disease or other outward sign of their presence. Viruses are often highly specific in the hosts they infect and do not reproduce anywhere else. This means that there may be approximately as many kinds of viruses as there are kinds of organisms—perhaps millions of them, even though only a few thousand have been described.

Most viruses are able to transmit their nucleic acid component from one host individual to another. Having entered the host cell, some viruses "take over" the genetic machinery of the host cell and use it to produce more viruses. The cell usually **lyses** as a result of such an infection, rupturing and releasing the newly made virus particles. Viruses that cause lysis to occur in their host cells are called **virulent viruses.** Other viruses become established within the host cell as stable parts of its genome; they are called **temperate viruses.**

In general, viruses possess only those nucleotide sequences associated with the enzymes that are necessary for them to replicate their own nucleic acids, capsids, and envelopes, and thus make possible the invasion of new host cells. Although the existence of a class of infectious particles called **prions,** which consist only of proteins, has been postulated by some scientists, the available evidence now suggests that these proteins are associated with nuclear genes. The matter is still under active study, and understanding it properly may provide a key to treating Alzheimer's disease, in which prions have been suggested as the infectious agents.

Viruses are so completely integrated into the normal metabolism of their host that they are difficult to control by chemical means, although immunization is effective. Once the viruses become established, anything that inhibits them also inhibits the proper development of their host. Antibiotics, which selectively kill bacteria but not their hosts, are useless against viruses; they act on features of living organisms that viruses lack, as we will see in Chapter 30. On the other hand, viruses themselves are being used as agents of biological control against insects and are increasingly being considered as control agents for some kinds of bacteria.

> **Viral diseases are difficult to treat because a virus becomes an integral part of the cell it infects and has no metabolic traits of its own, only those of the host cell.**

THE STRUCTURE OF VIRUSES

The smallest viruses are only about 17 nanometers in diameter, the largest ones up to 1000 nanometers (1 micrometer) in their greatest dimension; they vary greatly in appearance (Figure 29-4). The largest viruses, therefore, are barely visible with a light

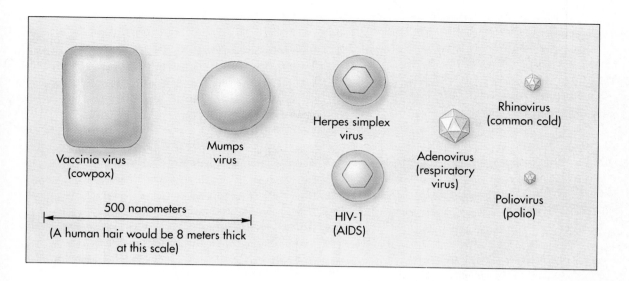

FIGURE 29-4
The diversity of viruses.

microscope. Most viruses can be detected only by using the higher resolution of an electron microscope. Viruses are directly comparable to molecules in size, a hydrogen atom being about 0.1 nanometer in diameter and a large protein molecule being several hundred nanometers in its greatest dimension.

The simplest viruses consist of a single molecule of a nucleic acid surrounded by a capsid, which is made up of one to a few different protein molecules, repeated many times (see Figure 29-3). In more complex viruses, there may be several different kinds of molecules of either DNA or RNA in each virus particle and many different kinds of proteins (Figure 29-5). Most viruses have an overall structure that is usually either **helical** or **isometric**. Helical viruses, such as the tobacco mosaic virus, have a rodlike or threadlike appearance; isometric ones have a roughly spherical shape.

The only structural pattern that has been found among the isometric viruses is the **icosahedron,** a figure with 20 equilateral triangular facets, like the adenovirus shown in Figure 29-4. The icosahedron is the basic design of the geodesic dome; it is the most efficient symmetrical arrangement that subunits can take to form an external shell with maximum internal capacity.

Most viruses are icosahedral in basic structure; an icosahedron has 20 equilateral triangular facets, each of which is characteristically subdivided in viruses. Many other viruses are helical, such as tobacco mosaic virus. Some viruses have more elaborate forms.

Some bacterial viruses, or **bacteriophages,** are among the most complex viruses (Figure 29-6). Each of them is made up of at least five separate proteins; these make up the head, the tail core, the molecules of the capsid, the base plate of the tail, and the tail fibers. A long DNA molecule is coiled within the head.

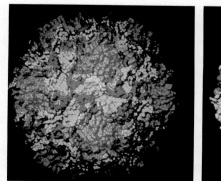
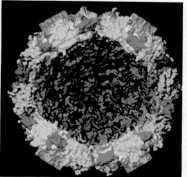

FIGURE 29-5
Viruses contain DNA or RNA as well as several proteins.
A Computer-generated diagram of the major features on the surface of poliovirus, an unenveloped virus.
B Different proteins on the surface are shown in blue, yellow, and red; another major class of protein is shown in green in the interior view. Diagrams of this sort assist scientists greatly in relating form and function. The RNA molecule of this virus is enclosed in the protein structure shown here.

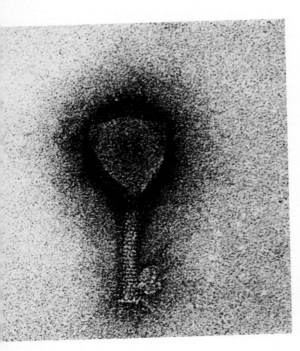

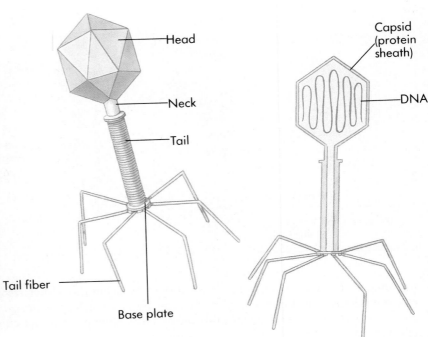

Head

Neck

Tail

Tail fiber

Base plate

Capsid (protein sheath)

DNA

FIGURE 29-6

A bacterial virus. Electron micrograph and diagram of the structure of a T4 bacteriophage.

FIGURE 29-7

Diagram of a flu virus. In this diagram, the coiled RNA has been revealed by cutting through the outer, lipid-rich envelope, with its two kinds of projecting spikes, and the inner protein capsid. The infective qualities of the flu virus are determined largely by the composition of the proteins that are present in the two different kinds of spikes.

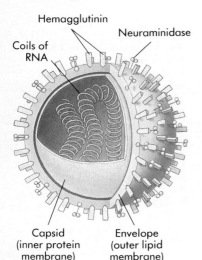

Hemagglutinin

Neuraminidase

Coils of RNA

Capsid (inner protein membrane)

Envelope (outer lipid membrane)

VIRUS REPLICATION

Before any virus can infect an animal cell, it must first bind to a specific receptor molecule in the cell membrane, probably a glycoprotein. As mentioned earlier, many but not all viruses have a lipid-rich envelope around the capsid. From the envelope of many viruses protrude spikes that may contain glycoproteins and lipids (Figure 29-7). The properties of the molecules that make up the outer covering of the virus have much to do with its adhesion to various substrates. If an envelope is not present, the properties of the capsid determine the adhesive qualities of the virus. Mammals protect themselves from viral infections by producing antibodies to envelope and capsid proteins, enabling their immune system (see Chapter 52) to identify and remove the virus particles.

Changes can occur in the base sequence in the genetic material of a virus that affect the detailed structure of its protein or glycoprotein capsid, envelope, and spikes; these changes may make that virus infectious to host organisms that had earlier been immune. The relationships may be very complex: in one kind of virus it has been shown that one of the proteins in the capsid binds to surface cell receptors; another determines the capacity for growth of the virus on mucosal surfaces, like those that line your nasal cavities; and a third is responsible for inhibiting the synthesis of proteins by the host cell. The lipids present in the envelopes of many viruses are determined by the genetic machinery of their hosts.

Among the most important characteristics of a virus is the nature of the proteins, especially the glycoproteins, that make up its capsid, envelope, and spikes. These proteins determine the infective properties of the virus.

As an example of how a viral infection proceeds, we will examine how HIV (human immunodeficiency virus), which causes AIDS, infects humans. Most other viruses follow a similar course, although the details of entry and replication differ in individual cases. The infection cycle of HIV is outlined in Figure 29-8.

When HIV is introduced into the human bloodstream, almost always sexually or through a contaminated hypodermic needle, the virus particle circulates throughout the body but does not infect most of the cells that it encounters.

When HIV encounters a certain kind of white blood cell called a T4 cell, however, it begins its assault by infecting these cells. As we shall see in Chapter 52, T4 cells are the key to the body's defenses against disease; thus it is highly significant

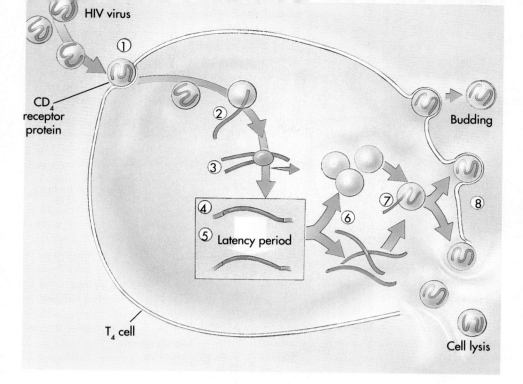

FIGURE 29-8

The HIV infection cycle.

1 The HIV virus responsible for AIDS attaches to a CD4 receptor protein on the surface of a T4 lymphocyte cell, and enters the cell by endocytosis.

2 The viral RNA is released into the cell's cytoplasm.

3 A DNA copy is made of the virus RNA.

4 The DNA copy enters the cell's chromosomal DNA.

5 After a long period (an average of at least 8 years), the virus genes initiate active transcription.

6 Both HIV RNA and HIV proteins are made.

7 Complete HIV particles are assembled.

8 Some cells are lysed, releasing free HIV, whereas others bud out HIV by exocytosis.

that they are the targets for HIV. Most other animal viruses are similarly narrow in their specificity. Poliovirus goes only to certain epithelial and nerve cells, hepatitis virus to the liver, and rabies virus to the brain. How does a virus like HIV recognize a specific kind of target cell like a T4 cell? Recall from Chapter 6 that every kind of cell in your body has a specific array of cell surface receptors, designed to bind particular hormones and growth factors. Cells also possess one or more kinds of cell surface markers that they use to identify themselves to other similar cells. HIV recognizes T4 cells because each HIV particle possesses a glycoprotein on its surface that precisely fits a specific protein (CD4) on the T4 cell surface.

After docking with a T4 cell, HIV penetrates the cell membrane. Like other animal viruses, HIV enters the cell by endocytosis, with the cell membrane folding inward to form a deep cavity around the virus particle. Once it is within the host cell, the HIV particle sheds its protective coat. This leaves a double strand of virus RNA floating in the cytoplasm, along with a virus enzyme that was also within the virus shell. This enzyme, called **reverse transcriptase,** synthesizes a double strand of DNA complementary to the virus RNA. This double-stranded DNA then inserts itself into the chromosomes of the host T4 cell, where one of two things then happens: either the copy remains **latent,** its genes becoming incorporated into the chromosomes of its host and not being transcribed into virus proteins; or the copy becomes active, taking over part of the host cell machinery and directing it to produce many copies of the virus. In the latter case, the cell eventually dies, releasing thousands of new virus particles, which infect other T4 cells (Figure 29-9).

The conditions that determine when latent HIV becomes active are not well known. Infections by other microbes seem to trigger HIV activity, perhaps because the infected T4 cells are involved in the immune response.

Plant viruses normally enter the cells of their hosts at points of injury, whereas bacteriophages shed their coats outside their host cell, injecting their nucleic acids through the host cell wall. A latent bacteriophage is called a **prophage,** and a cell in which a prophage exists is termed a **lysogenic** cell (Figure 29-10), because it may be lysed in the future. Those bacteriophage strains that have the ability to exist as prophages are, of course, temperate viruses. As discussed with HIV, temperate bacteriophages may be released subsequently and become active; while they are incorporated into a host chromosome, they are not transcribed. Many latent human viruses are known to be activated by external stimuli, such as ultraviolet radiation and some chemicals. This is precisely what happens when a fever blister develops on your lip

FIGURE 29-9

The AIDS virus.

HIV (blue) released from infected T4 cells soon spread over neighboring T4 cells, infecting them in turn. The individual viruses are very small; more than 200 million would fit on the period at the end of this sentence.

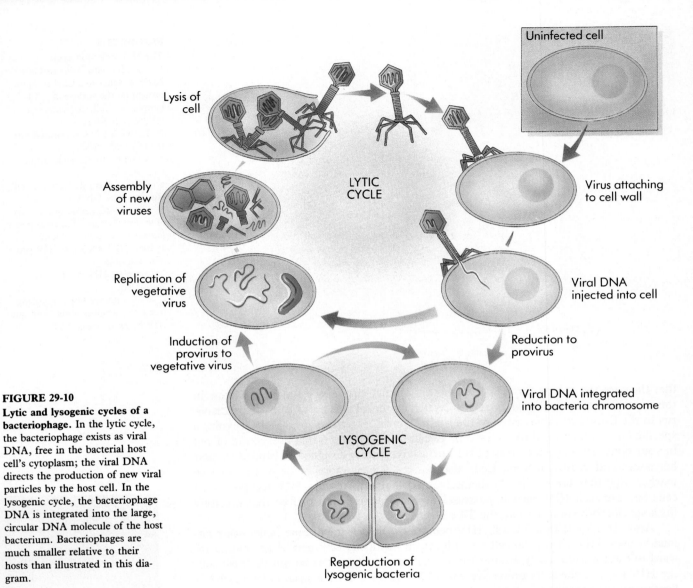

LYTIC CYCLE

Uninfected cell

Lysis of cell

Virus attaching to cell wall

Assembly of new viruses

Viral DNA injected into cell

Replication of vegetative virus

Induction of provirus to vegetative virus

Reduction to provirus

Viral DNA integrated into bacteria chromosome

LYSOGENIC CYCLE

Reproduction of lysogenic bacteria

FIGURE 29-10
Lytic and lysogenic cycles of a bacteriophage. In the lytic cycle, the bacteriophage exists as viral DNA, free in the bacterial host cell's cytoplasm; the viral DNA directs the production of new viral particles by the host cell. In the lysogenic cycle, the bacteriophage DNA is integrated into the large, circular DNA molecule of the host bacterium. Bacteriophages are much smaller relative to their hosts than illustrated in this diagram.

because of the activation of a latent herpesvirus, or when a lung cell bearing a latent retrovirus is suddenly converted into a rapidly dividing cancerous cell by a smoke-borne carcinogen, a process described in Chapter 16.

Some viruses incorporate themselves into the chromosomes of their hosts; in this form, they are said to be latent if they do not begin replicating themselves immediately.

DIVERSITY AMONG THE VIRUSES

The diversity of the viruses is great and is almost certainly related to their modes of origin. To provide a systematic idea of some of this diversity, we will discuss the viruses under eight main headings. The main characteristics of these groups are summarized in Table 29-1.

Unenveloped Plus-Strand RNA Viruses

Unenveloped plus-strand RNA viruses are called plus strands because they act directly as mRNA after infecting a host cell, attaching to the host's ribosomes and being translated. As indicated by their name, these viruses lack envelopes and consist only of a nucleic acid core surrounded by a protein capsid. They all have similar shapes

TABLE 29-1   CHARACTERISTICS OF SOME GROUPS OF VIRUSES

NAME OF GROUP	DNA OR RNA	NUMBER OF STRANDS	ENVELOPE PRESENT	EXAMPLES OF HOSTS	EXAMPLES OF HUMAN DISEASES
Unenveloped plus-strand RNA	RNA	One	No	Plants, bacteria, animals	Many colds, polio-myelitis
Enveloped plus-strand RNA	RNA	One	Yes	Arthropods, vertebrates	Some cancers, AIDS, yellow fever
Minus-strand RNA	RNA	One	Yes	Plants, animals	Flu, mumps, rabies
Viroids	RNA	One	Naked	Plants only	
Double-stranded RNA	RNA	Double	Yes	Plants and animals	Colorado tick fever
Small-genome DNA	DNA	One or double	Yes	Mainly animals	Viral hepatitis, warts
Medium- and large-genome DNA	DNA	Double	Yes	Animals	Herpes, shingles, some cancers, poxes
Bacteriophages	DNA	Double	Yes	Bacteria	

and dimensions. Some are bacteriophages; others infect plants. In both of these groups the protein coats are made up of multiple copies of a single molecule. Still others occur in the cells of animals and have coats made up of as many as four different kinds of proteins. The divided-genome plant viruses are unusual members of this group; they are called **divided-genome viruses** because they contain several different kinds of RNA that are packaged separately, but which collectively determine the characteristics of a particular virus.

In plus-strand RNA viruses the RNA acts directly as messenger RNA when it enters the host cell.

We have already discussed one of the best-known plant viruses in this group, tobacco mosaic virus (see Figures 29-2 and 29-3). Other viruses classified here cause various diseases of plants called **mosaics** or **stunt diseases,** such as alfalfa mosaic or tomato bushy stunt. The plant viruses in this group are more numerous in kinds and diverse in structure than are either bacteriophages or animal viruses with similar characteristics. The plus-strand RNA viruses that cause plant diseases vary greatly in size; some are icosahedral, but many are elongate, like the tobacco mosaic virus.

Several of the animal viruses classified here cause important diseases. Poliomyelitis, caused by poliovirus (see Figure 29-5) was a widespread, crippling disease through the first half of the twentieth century. Soon after he took office in 1933, President Franklin Delano Roosevelt, himself a victim of polio, proclaimed a "war on polio." This effort, largely funded by private contributions, was won with the development of vaccines against polio approximately 20 years later. Jonas Salk played a leading role in the discovery of these vaccines. Although poliomyelitis is now largely under control by vaccination in the industrialized countries, it remains a serious and common disease in the tropics and elsewhere in the less developed parts of the world. Hoof-and-mouth disease, which plays a major role in determining the geographical limits within which cattle can be raised successfully, is also caused by an unenveloped plus-strand RNA virus.

Colds are viral infections of the upper respiratory tract. About one third of all colds are caused by the **rhinoviruses,** which are unenveloped plus-strand RNA viruses. The colds caused by other kinds of viruses differ in their symptoms and in the seasons at which they occur most frequently. There are dozens of different strains of cold-causing rhinoviruses alone, each of them with different properties and none conferring cross-immunity to the others. More than 200 strains of viruses that cause colds have been identified, which makes the development of appropriate immunization methods very difficult, if not impossible. For about 40% of all colds, the responsible agents have not even been identified.

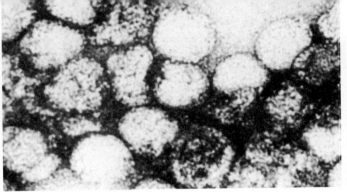

A

B

Enveloped Plus-Strand RNA Viruses

The enveloped plus-strand RNA viruses, all of which parasitize animals, are distinguished from the members of the preceding group by their lipid-rich envelopes, which determine their infective properties and make their structures more difficult to visualize. Members of this group cause many widespread diseases, such as many kinds of encephalitis, dengue, and yellow fever. They are classified as **arboviruses** (arthropod-borne viruses), together with many unrelated viruses, because they are transmitted by insects and other arthropods.

One of the most fascinating stories in the annals of disease control relates to the unraveling of the mystery of yellow fever by Walter Reed and his colleagues while the Panama Canal was being built. Yellow fever is maintained in native populations of monkeys and is spread to humans by mosquitoes of the genus *Aedes* (Figure 29-11). The disease is now largely controlled by a combination of the eradication of mosquito populations and vaccination of people who live in or visit areas where it occurs. Rubella, or German measles, is also caused by a virus of this group; it too is largely controlled by vaccination in many parts of the world.

Retroviruses, discussed earlier in connection with AIDS, are enveloped plus-strand RNA viruses. They are called retroviruses because they contain reverse transcriptase, an enzyme that catalyzes the synthesis of a double strand of DNA complementary to the RNA genome of the virus. This double-stranded DNA is integrated into the chromosomes of the host cell, from which it is copied into both mRNA (used to produce viral proteins) and the RNA copies of the viral genetic material itself. A virus of this group, HIV, was discovered in 1985 to be the cause of acquired immunodeficiency syndrome (AIDS; see Figures 29-8 and 29-9).

Minus-Strand RNA Viruses

Minus-strand RNA viruses are distinguished from plus-strand RNA viruses because they carry the RNA strand complementary to the mRNA that carries the genetic information of the virus. After a minus-strand RNA virus enters a host cell, it directs the formation of the appropriate mRNA, which then functions in the cell.

Among the members of this group are the **rhabdoviruses,** which are rod-shaped or bullet-shaped viruses about 180 nanometers long and about 70 nanometers thick, with a helical capsid and a lipid-rich envelope. Different rhabdoviruses infect humans and other vertebrates, arthropods (some rhabdoviruses are arboviruses), and plants. Rabies, which was the subject of the path-breaking discoveries of Louis Pasteur in the nineteenth century, is caused by a rhabdovirus. A second group of minus-strand RNA viruses, the paramyxoviruses, includes those that cause measles and mumps in humans, Newcastle disease in poultry, and distemper in various other animals.

Today, rabies is most often spread by small mammals, such as skunks, raccoons, and foxes; in 1984 more than 25,000 people in the United States had to get shots after being bitten by animals suspected of having rabies. Worldwide, it is estimated that some 70,000 people die of rabies each year. Fortunately it is a simple matter to vaccinate pets against rabies, and human vaccines are also being developed. The rabies virus is spread by the saliva of the host animal, often after bites, but it can also be contracted from handling a dead infected animal. An animal infected with rabies may go into a mad frenzy, often running great distances in its confusion.

Viroids

Viroids are infectious circular molecules of RNA that include from 250 to 400 nucleotides. Viroids lack capsids and have no proteins associated with them. They are functionally related to the RNA viruses that we have been discussing, but viroids are less than one-tenth the size of the next smallest known agents of infectious disease. The first viroid to be identified was the causative agent of potato spindle tuber disease (Figure 29-12). All viroids, as far as is known, infect plants: various viroids are asso-

---

FLU VIRUSES: RAPIDLY CHANGING VIRUSES THAT CAN ELUDE THE IMMUNE SYSTEM

Flu viruses are minus-strand RNA animal viruses that consist of the following major types. Type A flu viruses cause most of the serious flu epidemics in humans; they also occur in mammals and birds. Type B and also probably type C viruses are restricted to humans. Type B viruses cause relatively small outbreaks of flu, whereas type C viruses rarely cause serious health problems.

An individual flu virus resembles a ball studded with spikes (see Figure 29-7); these spikes are composed of two kinds of protein. One of these proteins, **hemagglutinin** (HA), is the substance that allows the virus to gain access to the cell interior, and the other, **neuraminidase** (NA), permits the daughter viruses to break free of the host cell once the viral replication has been completed. The "types" of flu virus (A, B, C, and so forth) are defined by how they react immunologically; their capsid proteins, which are associated with these reactions, are similar within each type and differ between the types. In contrast, the proteins that make up the spikes and the envelope are highly variable and specific, and they determine the infective characteristics of the individual strains. For example, the A-type flu viruses can be classified into 13 distinct HA subtypes and nine distinct NA subtypes. The virus that caused the Hong Kong flu epidemic of 1968 is characterized by one specific type of HA together with one specific type of NA.

Flu viruses change their genetic constitution so rapidly that vaccines against them soon become obsolete. The structures of both the HA and NA molecules are now known in detail. The HA molecule, for example, is made of three parts; it stands on the surface of the virus somewhat like a tripod and has clublike projections on top. Each of its three legs has four "hot spots," or segments of amino acids that display unusual tendencies to change readily as a result of changes in the viral RNA. These segments function as antigens against which the body's antibodies are directed. Each time the segments change, new antibodies are required to lock them up and thus neutralize the virus.

A change in one of the amino acids of the spike protein occurs as a result of point mutation in 1 of 100,000 viruses during the course of each generation. Such changes rarely alter the shape of the spike protein enough for it to escape the surveillance of the human immune system. The problem in combating the flu viruses arises through recombination. Viral genes are readily reassorted and recombined, even among fundamentally different viruses. Such reassortment can produce major shifts in the amino acids that make up the spikes. When such shifts occur, worldwide epidemics may result, since a large change in the shape of the spike protein makes it unrecognizable to human antibodies specific for the old spike configuration. Viral recombination of this kind seems to have been responsible for the three major flu epidemics that have occurred in this century: the "killer flu" of 1918, the Asian flu of 1957, and the Hong Kong flu of 1968. The flu epidemic of 1918 to 1919 resulted in 20 million deaths.

Somewhat surprisingly, it appears that wild ducks may be the major reservoirs for flu viruses. Viruses are abundant in ducks, infecting about 1 of 30 adult individuals and most juveniles. They live in the cells that line the intestinal tract, cause no disease in the ducks, and are spread readily in the feces. All of the known types of HA and NA molecules occur in ducks, so that recombination can occur readily. Since ducks fly long distances, they spread newly originated mutant flu viruses to all parts of the globe.

Flu virus researcher Robert G. Webster of St. Jude Children's Research Hospital in Memphis, Tennessee believes that the Hong Kong flu virus originated when the genetic material of a human flu virus recombined with another virus from a duck. This unfortunate event probably occurred in China, where someone infected with Asian flu was working with an infected domestic duck and contracted a second viral infection from the duck. As a result of this postulated event, 50 million people in the United States alone contracted flu, 70,000 died, and approximately 4 billion dollars were was lost in medical expenses and absences from work.

In the winter of 1985 and early 1986, a new flu virus, later named A/Taiwan, appeared in Asia. It appeared in the United States in October 1986, by which time a specific vaccine against it had been developed and distributed. The main vaccines used each year actually consist of mixtures that are active against different flu strains. Conventional vaccines offer strong immunity for several months, the likely duration of a flu season. Only a relatively small proportion of the high-risk groups for a particular strain of flu actually are immunized each year, despite the unpleasant and potentially lethal effects of the different viral strains.

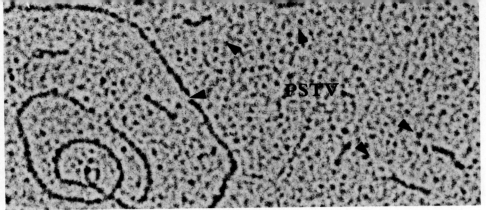

FIGURE 29-12

Viroids. Symptoms of potato spindle disease, as seen with an electron micrograph of viroids. The viroids (labeled PSTV for potato spindle tuber viroid) are much smaller than a typical virus, such as the T7 virus DNA that here looks like a rope by comparison.

ciated with about a dozen plant diseases. One of them causes cadang-cadang, a disease of coconuts that has caused millions of dollars of damage in the Philippines.

> **Viroids are naked, infectious, circular molecules of RNA that cause diseases in certain plants. They have from 250 to 400 nucleotides.**

Because viroids are circular, their RNA has no end groups. They do not seem to encode any polypeptides, so all components necessary to their replication—which involves direct RNA-to-RNA copying—must be present in the cells of the host plant. They probably produce their disease symptoms by the proliferation of viroid RNA in a normal synthetic pathway. Viroids may have originated from the introns of their hosts, which they resemble closely in base sequence. Other kinds of RNA particles, which are less well understood than viroids, have also been isolated.

Double-Stranded RNA Viruses

The **reoviruses,** which are double-stranded, icosahedral RNA viruses, infect mammals, arthropods, and plants. Some cause diseases; Colorado tick fever is an example. Double-stranded RNA was first discovered as a stable natural product in the reoviruses, which were then studied intensively. Reoviruses have a double envelope about 80 nanometers in diameter, with three kinds of proteins in the outer shell and four in the inner one. From 10 to 13 molecules of double-stranded RNA are present in each reovirus, the exact number depending on the kind of reovirus. Each virus contains a molecule of RNA-dependent RNA polymerase.

Another group of reoviruses, the **rotaviruses,** causes serious diarrhea in young children, especially during the cooler months of the year. They cause an estimated 80 deaths and 65,000 hospitalizations in the United States each year, primarily in babies less than 2 years old.

Small-Genome DNA Viruses

Many DNA viruses have small genomes; some of these viruses have single-stranded DNA, others have double-stranded DNA. Among them are the **parvoviruses,** which infect animals; they are icosahedral and about 20 nanometers in diameter. Among the parvoviruses are some of the smallest and simplest viruses. They contain a single molecule of single-stranded DNA and a coat that consists of either three or four different proteins. Recently, it has been shown that some patients with early symptoms of arthritis have antibodies to human parvovirus, showing that they had been infected. Individuals that seem to be especially susceptible are those with a specific gene, called DR4, and who have been infected with parvovirus as adults.

A second group of small-genome DNA viruses, the **papilloma viruses,** causes warts in animals, including humans. More than 50 different kinds of papilloma viruses have been identified in humans, and some appear to have consequences more serious than the production of warts. For example, cervical cancer, which kills about 7000 American women annually, appears to be associated with viruses of this group, although the evidence suggests that the viruses themselves are not sufficient to cause

One of the most intensively studied of the DNA viruses are those that cause viral hepatitis. These viruses have been difficult to study because it has not been possible to infect animals other than humans or to grow the viruses in cell culture. Not all kinds of hepatitis are caused by viruses of this group; for example, hepatitis A is caused by an RNA virus that is poorly understood, but hepatitis B (serum hepatitis) is caused by an unusual DNA virus. The causative agent of hepatitis B contains a small, circular molecule of partly, but not completely, double-stranded DNA. The viral genome encodes two kinds of proteins, a core protein and a surface protein, as well as DNA polymerase.

The amount of genetic information in the hepatitis B viral genome—359 nucleotides—is less than that of any other pathogen, except for the viroids.

The hepatitis B virus poses a serious public health problem, particularly among Asians, Africans, and male homosexuals. It often persists in carriers without causing any symptoms, but it may still be highly infectious. People infected early in life often become carriers, and it is estimated that there are about 200 million such carriers worldwide. Since most of these people are not recognized as carriers, there is a real possibility of the frequent transmission of hepatitis during immunization, blood transfusion, and similar

medical procedures. Although the proportion of carriers in Europe and the United States appears to be less than 0.1%, it may be as high as 20% in the tropics. Inoculations, medical procedures, and any kind of skin contact may lead to transmission of the virus. Not only is the hepatitis itself serious, but the virus also may play a role in causing human liver cancer, even among carriers who show no other symptoms. New vaccines against the virus have been produced by recombinant DNA techniques and are of great importance, especially for those who require frequent blood transfusions, as well as for homosexuals, prostitutes, and others who run a severe and continuing risk of infection.

cancer. Papilloma viruses can be spread sexually; the current estimate is that at least 10% of American adults are infected with papilloma virus of the genital tract. The presence of genital warts too small to be easily seen may be detected by Pap smears.

Medium-Genome and Large-Genome DNA Viruses

Herpesviruses. The **herpesviruses**, one of the major groups of large-genome, double-stranded DNA viruses, are of considerable medical importance. Herpesviruses are icosahedral and about 200 nanometers in diameter. They are common in most vertebrates and are involved in cold sores (Figure 29-13), shingles, venereal disease, mononucleosis, birth defects, and probably several kinds of cancer in humans. They have a complex set of proteins that includes more than 30 different kinds.

The five major human herpesviruses and their associated diseases are as follows:

Herpesvirus 1 (herpes simplex 1): cold sores and fever blisters

Herpesvirus 2 (herpes simplex 2): the genital virus, which causes the sexually transmitted disease

Herpesvirus 3: chickenpox and, if it becomes persistent, shingles, an acute inflammatory disease of the ganglia

Herpesvirus 4: Epstein-Barr virus, which causes mononucleosis and certain human cancers

Herpesvirus 5: cytomegaloviruses, which may cause fatal systemic infections of newborn babies

All of these different kinds of human herpesviruses are related to forms that occur widely among other mammals and birds, and there is some evidence for cross-infectivity between humans and other vertebrates. Herpesviruses of the first three groups replicate rapidly and are present in many hosts and tissues. They cause persistent infections in which the outward symptoms may be visible only during periodic outbreaks, perhaps half a dozen or so times a year. Herpesvirus 2, which is spread primarily by sexual contact, afflicts 10 to 20 million people in the United States alone. It attacks the lining of the uterus in women, sometimes damaging or destroying an embryo. A baby that passes through the birth canal of a mother with active herpes can be infected and sometimes blinded or killed; this risk can be eliminated by a cae-

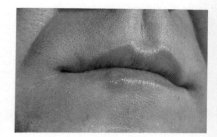

FIGURE 29-13
Herpesvirus 1. The fever blister is a familiar symptom of herpesvirus 1 infection.

sarean delivery. In addition, evidence is accumulating that herpesvirus 2 infections may be responsible for about one third of all miscarriages. Women with genital herpes are five to eight times more likely to develop cervical cancer as those who do not have the disease. Group 4 herpesviruses are generally present only in chronic infections, particularly those associated with the lymphatic system. Group 5 herpesviruses replicate slowly, mainly in fibroblasts, and tend to be host specific. Other herpesvirus are being discovered, and there is some evidence that they may eventually be implicated in other disorders, such as Hodgkin's disease and Kaposi's sarcoma.

> **The herpesviruses can be classified into five groups, and they cause for cold sores, shingles, venereal diseases, chickenpox, and certain cancers. They are large-genome, double-stranded DNA viruses.**

Epstein-Barr virus (EBV) is a widespread member of herpesvirus group 4. Many people infected with EBV display no signs of the infection. Others may develop the signs of infectious mononucleosis, also called glandular fever, and in some, infection with EBV leads to cancer—either Burkitt's lymphoma, a cancer of the lymph nodes below the jaw, or nasopharyngeal carcinoma, a tumor that develops in the space behind the nose. EBV is often isolated from the milk of women whose families have a history of breast cancer. It primarily infects lymphocytes, a type of white blood cell, but apparently replicates in skin cells. Once the lymphocytes have been infected, the virus will remain in them, and thus becomes a permanent resident in the body.

Recently, EBV has been linked with a chronic flulike condition that sometimes persists for a year or more and is accompanied by several low-grade symptoms, among them fatigue, headache, depression, sore throat, fever, aches, and pains. Whether or not this mysterious disease is actually caused by EBV has proved difficult to determine, despite the fact that it afflicts thousands of people.

Poxviruses. The poxviruses are the largest and most complex animal viruses. Often about 400 nanometers in the largest dimension, poxviruses can be seen with a light microscope. About 90% of a poxvirus is protein. Living cells contain about the same percentage; all other viruses contain much less. The viruses are somewhat brick-shaped. One of the poxviruses, smallpox virus, was the first virus actually to be seen. Another well-known poxvirus is the vaccinia virus, which causes cowpox and which Jenner used many years ago in successful inoculations against smallpox. As discussed in Chapter 18, scientists can use this virus in the production of herpes and hepatitis vaccines by splicing the genes specifying the envelopes of these viruses into a fragment of the vaccinia virus genome and inoculating mammals with the engineered vaccinia virus. Vaccinia virus is especially suitable for such procedures because of its large size—187,000 base pairs. The virus can easily accommodate an extra 25,000 base pairs, which allows engineered versions to be produced that are able to simultaneously protect against multiple pathogens.

Bacteriophages

Although double-stranded DNA bacteriophages are logically grouped with the other medium- and large-genome DNA viruses, they are so important in relation to molecular biology that we will discuss them separately here. As we mentioned earlier, many of these bacteriophages are large and complex, with relatively large amounts of DNA and proteins. Actually, bacteriophages are diverse both structurally and functionally; they are united by their occurrence in bacterial hosts rather than by other common features. Some of them have been named as members of a "T" series (T1, T2, and so forth); others have been given different kinds of names. To illustrate the diversity of these viruses, T3 and T7 bacteriophages are icosahedral and have short tails. In contrast, the so-called T-even bacteriophages (T2, T4, and T6) have an icosahedral head, a capsid that consists primarily of three proteins, a long tail, a connecting neck with a collar and long "whiskers", and a complex base plate (see Figure 29-6).

During the process of bacterial infection by T4 phage, at least one of the tail fibers of the phage—they are normally held near the phage head by the "whiskers"—contacts the lipoproteins of the host bacterial cell wall. The other tail fibers set the phage perpendicular to the surface of the bacterium and bring the base plate into contact with the cell surface. The tail contracts, and the tail tube passes through an opening that appears in the base plate, piercing the bacterial cell wall. The contents of the head, mostly DNA, are then injected into the host cytoplasm.

The T-series bacteriophages are all virulent, invariably multiplying within infected cells and eventually lysing them. Among the temperate bacteriophages is the lambda (λ) phage of *Escherichia coli*, discussed in Chapter 17. We know as much about this bacteriophage as we do about virtually any other biological particle; the complete sequence of its 48,502 bases has been determined. At least 23 proteins have been identified with the development and maturation of lambda phage, and many enzymes are involved in the integration of these viruses into the host genome.

> **Bacteriophages are a diverse group of viruses that attack bacteria. Some of them have a complex structure, such as the double-stranded DNA bacteriophages.**

VIRUSES: PARTICLES OF GENOMES

Viruses are most conveniently thought of as particles of genomes. Because viruses probably originate as fragments of bacterial and eukaryotic genomes, their great diversity is not surprising. We have learned a great deal about the structure of viruses, but there is doubtless much more to be discovered, probably including the existence of kinds of particles that are unsuspected at present. As parts of the genetic machinery of cells before they became independent, viruses are still intimately associated with these cells. The nature of viruses suggests that new forms are evolving constantly. While this process continues, we clearly have a great deal to learn about the existing viruses.

SUMMARY

1. Viruses are fragments of bacterial or eukaryotic genomes that are able to replicate within cells by using the genetic machinery of those cells. In most cases, the fragments subsequently acquire the ability to synthesize protein coats, and sometimes envelopes, around the DNA or RNA that originally split from the genome from which they were derived.

2. Viruses are not alive and are not organisms; they cannot reproduce outside of living cells, since they lack the machinery to do so by themselves.

3. Viruses are either virulent, destroying the cells in which they occur, or temperate, becoming integrated into their genomes and remaining stable there for long periods of time.

4. Viruses are basically either helical or isometric. Most isometric viruses are icosahedral. The adhesion properties of viruses are determined by those of the proteins that make up their coats and envelopes.

5. The simplest viruses use the enzymes of the host cell for both protein synthesis and gene replication; the more complex ones contain up to 200 genes and are capable of synthesizing many structural proteins and enzymes.

6. AIDS is caused by HIV (human immunodeficiency virus), which possesses a glycoprotein on its surface that precisely fits a specific protein on the T4 cell surface. Docking with a T4 cell, HIV penetrates the cell membrane, sheds its protective coat, and reproduces.

7. Among the major groups of viruses are the unenveloped plus-strand RNA viruses, which act directly as messengers after infecting the host cell; the enveloped plus-strand RNA viruses, which are similar but have an envelope; the minus-strand RNA viruses, which must synthesize mRNA in the cells they infect; viroids and similar forms, which consist of RNA alone with no associated proteins; double-stranded RNA viruses; small-genome DNA viruses, which have either single-stranded or double-stranded DNA; medium- and large-genome DNA viruses; and bacteriophages.

8. The flu viruses, minus-strand RNA viruses, resemble balls studded with spikes. The spikes are composed of two kinds of protein, hemagglutinin and neuraminidase, both of which play a role in determining the infective qualities of the virus. Recombination of the genetic material of the flu viruses appears to play the critical role in causing worldwide flu epidemics.

REVIEW QUESTIONS

1. Why are viruses not considered to be living organisms? What is the most general structure of a virus? When do viruses act like living organisms?

2. Are viruses specific or nonspecific in terms of the hosts that they infect? What could one assume, therefore, about the actual number of kinds of viruses?

3. What is a virulent virus? What are temperate viruses? Why is it so difficult to treat a viral infection of either type? How is this different from treating bacterial infections?

4. What type of microscope is generally required to visualize an actual viral particle? What, therefore, is the approximate size range of viruses? What are the two types of viral structures? What shape does each impart?

5. What component of the virus affects its adherence to various substrates, including a cell membrane? How do many mammals combat viral infections?

6. What specific type of human cell does the AIDS virus infect? What does this cell possess that other human cells do not? How does this animal virus penetrate the host cell? What viral components remain in the host cell?

7. How does a plant virus infect its host? How does a bacterial virus infect its host? What is a prophage and what name is given to the cell in which it is found?

8. What are minus-strand RNA viruses? What must be done by the host cell for these viruses to reproduce? What type of viruses are rhabdoviruses? What kinds of diseases do they cause? What kinds of diseases do paramyxoviruses cause?

9. What are viroids? What viral structural elements do they possess? What types of organisms do they infect? What is their possible origin?

THOUGHT QUESTIONS

1. What are the advantages to a virus of having a protein coat or envelope? The disadvantages?

2. In what ways might the early, self-replicating particles that gave rise to the first organisms have resembled, or differed from, viruses?

3. Did the different groups of viruses arise from one another? Describe the process by which they probably evolved.

FOR FURTHER READING

FINCHER, J.: "America's Deadly Rendezvous with the 'Spanish Lady'," *Smithsonian*, January 1989, pages 131-145. In 1918, before the end of World War I, the world found itself fighting a killer flu for which there was no cure.

GALLO, R.C.: "The AIDS Virus," *Scientific American*, January 1987, pages 47-56. Traces the steps by which the cause of AIDS was shown to be the third known human retrovirus.

GALLO, R.C.: "The First Human Retrovirus," *Scientific American*, December 1986, pages 88-98. A retrovirus discovered in 1978 causes a rare human leukemia; its discovery made possible the later recognition of the related virus that causes AIDS.

HAPGOOD, R.: "Viruses Emerge as a New Key for Unlocking Life's Mysteries," *Smithsonian*, November 1987, pages 117-127. Emphasizes the role of viral studies in learning about molecular systems in cells.

HIRSCH, M.S., and J.C. KAPLAN: "Antiviral Therapy," *Scientific American*, April 1987, pages 76-85. New antiviral drugs exploit the subtle molecular contrasts between viruses and their hosts.

HOGLE, J.M., M. CHOW, and D.J. FILMAN: "The Structure of Poliovirus," *Scientific American*, March 1987, pages 42-49. Poliovirus has become a model for investigating the molecular links between form and function.

LANGONE, J.: "Emerging Viruses," *Discover*, December 1990, pages 63-68. With environmental changes paving their way, armies of once-obsure viruses may be poised to launch an attack on unprepared humans.

OLDSTONE, M.: "Viral Alteration of Cell Function," *Scientific American*, August 1989, pages 42-48. Certain viruses interfere subtly with a cell's ability to produce specific hormones and neurotransmitters, often residing indefinately within the cell's genome.

RADETSKY, P.: "Taming the Wily Rhinovirus," *Discover*, April 1989, pages 38-43. Rhinoviruses cause over half of all "common colds." Molecular biologists are mounting a concerted attack on this annoying illness.

TIOLLAIS, P., and M. BUENDIA: "Hepatitis B Virus," *Scientific American*, April 1991. An unusual virus causes liver disease and cancer. Hundreds of millions of people are affected.

VARMUS, H.: "Reverse Transcription," *Scientific American*, September 1987, pages 56-64. Studies of viruses have led to the discovery that the conversion of RNA to DNA is widespread among organisms.

30

Bacteria

OVERVIEW

Bacteria are the most ancient, the structurally simplest, and the most abundant of organisms, and are the only ones characterized by a prokaryotic organization of their cells. Because they are so distinct, they are classified as the only members of the kingdom Monera. Bacteria have been mentioned repeatedly in this book because their key position in the evolution of life places them at the beginning of many evolutionary developments. Life on earth would not exist without bacteria, which make possible many of the essential functions of ecosystems. This chapter presents an overview of the bacteria: their structure, life history, ecology, and diversity.

FOR REVIEW

Here are some important terms and concepts that you will encounter in this chapter. If you are not familiar with them, you should review them before proceeding.

Prokaryotic cell structure (Chapter 5)

Bacterial flagella (Chapter 5)

Bacterial cell division (Chapter 10)

Bacterial metabolism and chemoautotrophy (Chapter 9)

Bacterial photosynthesis (Chapter 9)

Transduction and bacterial conjugation (Chapter 17)

Geochemical cycles (Chapter 25)

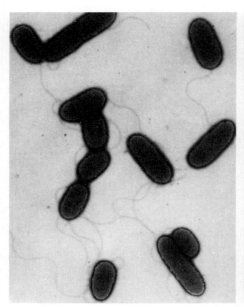

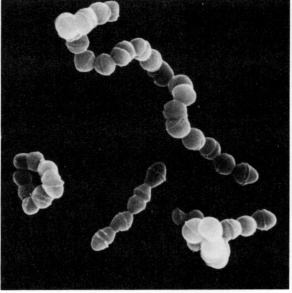

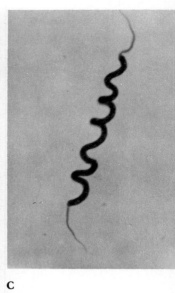

C

A   B

FIGURE 30-1

The diversity of bacteria.

A *Pseudomonas aeruginosa*, a rod-shaped, flagellated bacterium (bacillus). *Pseudomonas* includes the bacteria that cause many of the most serious plant diseases.

B *Streptococcus*. The spherical individual bacteria (cocci) adhere in chains in the members of this genus.

C *Spirillum volutans*, one of the spirilla. This large bacterium, which occurs in stagnant fresh water, has a tuft of flagella at both ends.

D *Chondromyces crocatus*, one of the gliding bacteria. The rod-shaped individuals move together, forming the composite spore-bearing structures shown here; millions of spores, which are basically individual bacteria, eventually are released from these structures.

E *Chroococcus*, a cyanobacterium in which the individuals adhere within a gelatinous capsule in groups of four.

Bacteria are the oldest, the structurally simplest, and the most abundant forms of life on earth; they are also the only organisms with prokaryotic cellular organization. Represented in the oldest rocks from which fossils have been obtained—rocks 3.0-3.5 billion years old—bacteria were abundant for well over 2 billion years before eukaryotes appeared in the world (see Figure 4-13). Bacteria are the only members of the kingdom Monera.

About 4800 different kinds of bacteria are currently recognized, but there are doubtless many thousands more awaiting proper description (Figure 30-1). The structural differences among various bacteria, even as viewed with an electron microscope, often are not great, and species are recognized largely by their metabolic characteristics. Bacteria can be characterized properly only when they are grown on a defined medium (see Chapter 14), because the characteristics of these organisms often change, depending on their growth conditions and the substances with which they are in contact.

Bacteria were largely responsible for creating the properties of the atmosphere and the soil over billions of years. They are metabolically much more diverse than are the eukaryotes, which is why they are able to exist in such an exceptionally wide range of habitats and are so important ecologically in so many of them. Many bacteria are autotrophic—both photosynthetic and chemoautotrophic—and make major contributions to the world carbon balance in terrestrial, freshwater, and marine habitats. Others are heterotrophic and play a key role in world ecology in breaking down organic compounds. Some of these heterotrophic bacteria cause major diseases of plants and animals, including humans. One of the most important roles of bacteria in the world ecosystem relates to the fact that only a few genera of bacteria—and no other organisms—have the ability to fix atmospheric nitrogen and thus make it available for use by other organisms (see Chapter 25).

> **Bacteria are the oldest and most abundant organisms on earth. The maintenance of life depends on them, since they play a vital role both in productivity and as decomposers, in cycling the substances essential to all other life-forms. They are also the organisms capable of fixing atmospheric nitrogen.**

Bacteria are very important, and becoming more so, in industrial processes. Bacteria are used now, for example, in the production of acetic acid and vinegar, various amino acids and enzymes, and especially in the fermentation of lactose into lactic acid, which coagulates milk proteins and is used in the production of almost all cheeses, yogurt, and similar products. In the production of bread and other foods, the addition of strains of bacteria with appropriate characteristics can lead to the enrich-

D

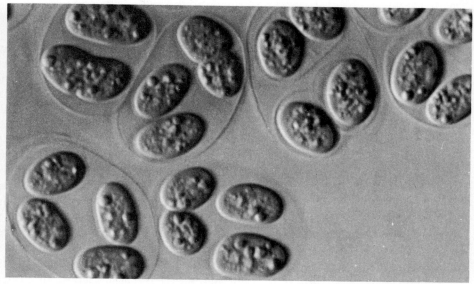

E

ment of the final product with respect to its mix of amino acids, a key factor in its nutritive value. Many products that have traditionally been manufactured by using yeasts, such as ethanol, can also be made by using bacteria; the comparative economics of these processes will determine which group of organisms is used in the future. Also, many of the most widely used antibiotics, including streptomycin, aureomycin, erythromycin, and chloromycetin, are derived from bacteria. Most antibiotics seem to be substances used by bacteria to compete with one another and with fungi in nature, allowing one species to exclude others from a favored habitat.

The application of genetic engineering methods to the production of improved strains of bacteria for commercial use holds enormous promise for the future. For example, bacteria are being investigated increasingly as nonpolluting agents to use in insect control; *Bacillus thuringiensis*, which attacks insects in nature, is particularly useful in this regard. Improved, highly specific strains of *B. thuringiensis* greatly increase its usefulness as a biological control agent. Considering the enormous metabolic diversity of the bacteria, it seems logical to assume that we have barely begun to understand and use them fully.

PROKARYOTES VERSUS EUKARYOTES

Prokaryotes, or bacteria, differ from eukaryotes in numerous features of fundamental importance. These profound differences began to be understood properly only about 50 years ago. At that time, some scientists began to realize that these differences were far greater than the traditional ones used to separate plants and animals. These differences have been discussed at several places in this book, since a consideration of the simpler features of bacteria illustrates the evolutionary origins of many of the structures and capabilities of eukaryotes. We will review the differences briefly, with cross-reference to the earlier discussion. They represent the most fundamental distinctions that separate any groups of organisms.

1. *Multicellularity*. All bacteria are fundamentally single celled. Even though in some of them the cells may adhere within a matrix and form filaments (Figures 30-1, *E*, and 30-2), especially in some cyanobacteria, their cytoplasm is not directly interconnected as often is the case in multicellular eukaryotes. The activities of a bacterial colony are less integrated and coordinated than in multicellular eukaryotes. Even though bacteria may line up end to end, they remain single cells. A primitive form of colonial organization occurs in the gliding bacteria, which move together and form spore-bearing structures (see Figure 30-1, *D*). Such approaches to multicellularity, however, are rare among the bacteria.

Heterocyst

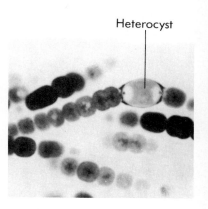

FIGURE 30-2
Filamentous bacteria. The cyanobacterium *Anabaena*, in which the individual cells adhere in filaments. The large cell is a heterocyst, a specialized cell in which nitrogen fixation occurs. Heterocysts are essentially like ordinary cells, but they form an outer bilayered envelope consisting of a polysaccharide and an inner layer of glycolipids. Within them, the membranes of the cyanobacterium are reorganized into a concentric or netlike pattern. It has recently been learned that some of the nitrogen-fixation genes are actually rearranged during heterocyst formation in cyanobacteria. These organisms exhibit one of the closest approaches to multicellularity among the bacteria.

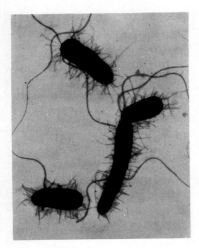

FIGURE 30-3
Pili in the common intestinal bacterium, *Escherichia coli.* **The** long strands are flagella. A greater proportion of the genetic information is known for this much-studied bacterium than for any other organism.

2. *Cell size.* Most bacteria are cells only about 1 micrometer in diameter (see Chapter 5), whereas most eukaryotic cells are well over 10 times that size.

3. *Chromosomes.* Eukaryotic cells have a membrane-bound nucleus containing chromosomes that include both nucleic acids and proteins. Bacteria do not have a nucleus, nor do they have chromosomes of the kind present in eukaryotes, in which DNA forms a structural complex with proteins. Instead, their DNA, which is circular, is free in the cytoplasm.

4. *Cell division and genetic recombination.* Cell division in eukaryotes takes place by mitosis and involves spindles made up of microtubules; cell division in bacteria takes place mainly by binary fission (see Chapter 10). True sexual reproduction is present only in eukaryotes and involves syngamy and meiosis, with an alternation of diploid and haploid forms. Despite their lack of sexual reproduction, bacteria do have mechanisms that lead to the transfer of genetic material. These mechanisms are far less regular than those of eukaryotes and do not involve the equal participation of the individuals between which the genetic material is transferred.

5. *Internal compartmentalization.* In eukaryotes the enzymes for cellular respiration are packaged in mitochondria; in bacteria the corresponding enzymes are not packaged separately but are bound to the cell membranes (see Chapters 5 and 8). The cytoplasm of bacteria, unlike that of eukaryotes, contains no internal compartments or cytoskeleton, and no organelles except ribosomes.

6. *Flagella.* Bacterial flagella (Figure 30-1, *A* and *C*, and Figure 30-3 and Chapter 5) are simple, composed of a single fiber of the protein flagellin; the flagella and cilia of eukaryotes are complex and have a 9 + 2 structure of microtubules (see Figure 5-26). Bacterial flagella also function differently from those of eukaryotes, spinning like a propellor, whereas eukaryotic flagella have a whiplike motion.

7. *Autotrophic diversity.* In photosynthetic eukaryotes, the enzymes for photosynthesis are packaged in membrane-bound organelles, the plastids. Only one kind of photosynthesis is found in eukaryotes, and it involves the release of oxygen. In photosynthetic bacteria, the enzymes for photosynthesis are bound to the cell membrane. These bacteria have several different patterns of anaerobic and aerobic photosynthesis (see Chapter 9), involving the formation of end products such as sulfur, sulfate, and oxygen. This photosynthetic diversity again underscores the metabolic diversity of the bacteria. **Chemosynthesis** is the process whereby certain bacteria obtain their energy from the oxidation of inorganic compounds and obtain their carbon from carbon dioxide; chemoautotrophs are discussed in Chapter 9. In addition, only bacteria have the ability to fix atmospheric nitrogen.

BACTERIAL STRUCTURE

Bacterial cell walls usually consist of a network of polysaccharide molecules connected by polypeptide cross-links (Figure 30-4). Many species of bacteria are gram positive; in them, this network makes up the basic structure of the bacterial cell wall, which is about 15 to 80 nanometers thick. In the more common gram-negative bacteria, large molecules of **lipopolysaccharide**—a polysaccharide chain with lipids attached to it— are deposited over this layer, forming an outer membrane. This lipopolysaccharide layer makes gram-negative bacteria resistant to many antibiotics to which gram-positive ones are susceptible. This is because antibiotics that influence cell wall synthesis do not affect gram-negative bacteria. Outside of the layers we have just mentioned, a gelatinous layer, the **capsule,** surrounds the cells of some kinds of bacteria.

Bacteria are mostly simple in form, varying mainly from straight and rod shaped (**bacilli**) or spherical (**cocci**) to long and spirally coiled (**spirilla**). Some bacteria change into stalked structures, grow long, branched filaments, or form erect structures that release **spores,** single-celled bodies that grow into new bacterial individuals (see Figure 30-1, *D*). Among the rod-shaped and spherical bacteria, some adhere end-to-end after they have divided, forming a chain.

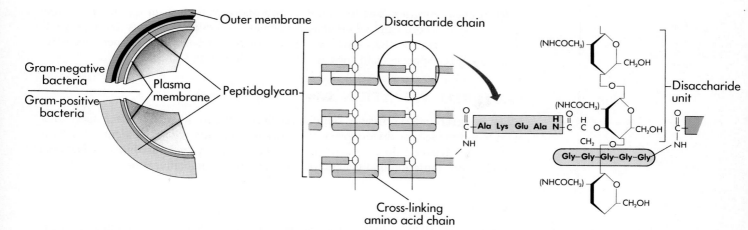

The members of one of the major photosynthetic groups, the cyanobacteria (formerly called "blue-green algae"), often form filaments (see Figure 30-2) that may aggregate into large masses up to 1 meter or more in length. In these filaments, the cells are connected to one another by their outer walls or by gelatinous sheaths, and there is a degree of coordination between them. Some of these filamentous bacteria are capable of gliding motion, often combined with rotation around a longitudinal axis. The mechanism by which they move has not yet been determined.

Bacterial cells have a simple structure. There are two structurally different kinds of cell walls, gram negative and gram positive. The cytoplasm of a bacterium contains no internal compartments or organelles and is bounded by a membrane encased within a cell wall composed of one or more layers of polysaccharide.

Many kinds of bacteria have slender, rigid, helical flagella composed of flagellin (see Figure 5-25). These flagella range from 3 to 12 micrometers in length and are very thin—only 10 to 20 nanometers thick. Flagella may be distributed all over the cells of the bacteria in which they occur, or they may be confined to one or both ends of the cell.

Pili (singular, **pilus**) are other kinds of hairlike outgrowths that occur on the cells of some bacteria (see Figure 30-3). They are shorter than bacterial flagella, up to several micrometers long, and about 7.5 to 10 nanometers thick. Pili help the bacterial cells attach to appropriate substrates. The number and arrangement of flagella and pili are useful aids in bacterial identification.

Some bacteria form thick-walled **endospores** around their chromosome and a small portion of the surrounding cytoplasm when they are exposed to drying conditions and high temperatures. These endospores (Figure 30-5) are highly resistant to environmental stress; they may germinate and form new bacterial individuals after de-

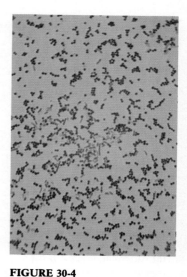

FIGURE 30-4
Bacterial cell walls and the Gram stain. Bacterial cell walls are composed of an interlocking network of disaccharide molecules cross-linked by short amino acid chains. The complex, called peptidoglycan, forms what is in actuality a single molecule woven around the outside of the cell. The peptidoglycan layer is much thicker in gram-positive bacteria than in gram-negative ones, causing them to retain gentian violet dye and thus appear purple, as shown here, in a Gram-stained smear; gram-negative bacteria do not retain the violet dye and so exhibit the background red stain.

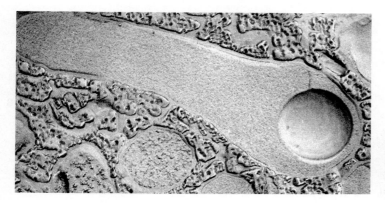

FIGURE 30-5
Endospores. The round circle at the bottom is an endospore forming within a cell of the botulism bacterium, *Clostridium botulinum*. These resistant endospores allow the bacterium to survive in improperly sterilized canned and bottled foods.

cades or even centuries. The formation of endospores is the reason that the bacterium responsible for botulism, *Clostridium botulinum*, sometimes persists in improperly sterilized cans and bottles—ones that have not been heated at a temperature sufficient to kill the spores—and then is able to multiply within them.

BACTERIAL VARIATION

When bacteria undergo fission, the two resulting cells are identical. After many divisions—some bacteria divide every few minutes—a large clone of identical cells is formed. Yet bacteria do vary genetically. Two processes lend variability to bacterial reproduction: mutation and genetic recombination during conjugation.

Mutation

In a bacterium such as *Escherichia coli* there are about 5000 genes, judging from the amount of DNA that an individual cell contains. When considering any one of these genes, we may conservatively expect one mutation to have occurred by chance among every million copies. Since 200 individual bacteria contain a total of a million genes (of 5000 kinds), we can expect one of every 200 bacteria to have a mutant characteristic (Figure 30-6). A spoonful of soil typically contains over a billion bacteria, and therefore should contain something on the order of 5 million mutant individuals!

Because bacteria multiply so rapidly, favorable mutations may also spread rapidly in a population, and the characteristics of that population will then change rapidly. For example, when adequate food and nutrients are available, a population of *Escherichia coli* growing under optimal conditions will double every 12.5 minutes, so that a new favorable mutation will soon be represented in large numbers of individuals better suited for growth than individuals that lack the mutation.

> **Because of the short generation time of bacteria, whose populations often double in number in a few minutes, mutation plays an important role in generating genetic diversity.**

The ability of bacteria to change rapidly in response to new challenges often has adverse effects on humans. For example, in the past decade or so, a number of strains of *Staphylococcus aureus* associated with serious infections in hospitalized patients have appeared, some of them with alarming frequency. Unfortunately, many of these strains have acquired resistance to penicillin; other strains have acquired the ability to produce an enzyme that breaks down the molecules of penicillin. *Staphylococcus* infec-

FIGURE 30-6

A mutant hunt in bacteria. Mutations in bacteria can be detected by the technique of replica plating, which allows the genetic characteristics of the colonies to be investigated without destroying them. The bacterial colonies, growing on a semisolid agar medium, are transferred from *A* to *B* by using a sterile velveteen disk pressed on the plate. Plate *A* has a medium that includes all of the factors necessary for the growth of the bacterium, whereas *B* has a medium that lacks some of the essential growth factors. The colonies absent in *B* were not able to grow in the deficient medium; they were mutant colonies, already present but undetected in *A*. This experiment demonstrates that these mutants were not caused by the new environment, but were selected by it, just like the mutants in Welsh bentgrass discussed in Chapter 20.

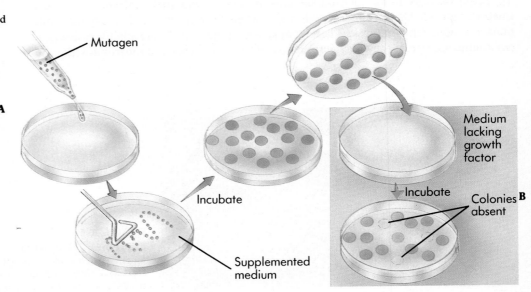

tions provide an excellent example of the way in which mutation and intensive selection can bring about rapid change in bacterial populations. Such changes occur in all bacteria and have serious medical implications, as mentioned in Chapter 17, particularly when strains of bacteria emerge that are resistant to a variety of antibiotics. In the mid-1980s the routine administration of small doses of antibiotics to livestock to improve their weight gains came into question because of the danger to humans posed by the antibiotic-resistant strains of bacteria that inevitably result from this practice. Current practices in this area are being reviewed.

Genetic Recombination

Another source of genetic variation in populations of bacteria is recombination, which we have discussed in detail in Chapter 17. Bacterial recombination occurs by the transfer of genes from one cell to another as portions of viruses, plasmids, or other DNA fragments. The rapid transfer of newly produced, antibiotic-resistant genes by plasmids has been an important factor in the appearance of the resistant strains of *Staphylococcus aureus* discussed above. An even more important example in terms of human health involves the Enterobacteriaceae, the family of bacteria to which the common intestinal bacterium, *Escherichia coli,* belongs. In this family, there are many important pathogenic bacteria, including the organisms that cause dysentery, typhoid, and other major diseases. At times, some of their genetic material is exchanged with or transferred to *E. coli;* because of its abundance in the human digestive tract, it poses a special threat if it acquires harmful traits.

In a sense, the rapid generation time of bacteria may be seen as an evolutionary strategy alternative to the complex patterns of recombination and segregation that occur in eukaryotes. Eukaryotes have a much more complex cell structure and a much slower generation time than do bacteria, and they may be simply unable to reproduce as fast. In both bacteria and eukaryotes, ample variation is present to form the raw material of evolution, but the variation is organized and presented in different ways.

BACTERIAL ECOLOGY AND METABOLIC DIVERSITY

Bacteria are the most abundant organisms in most environments, whether measured by numbers of individuals or by absolute weight. For example, in a single gram of fertile agricultural soil, there may be 2.5 billion bacteria, in addition to 400,000 fungi, 50,000 algae (photosynthetic eukaryotes other than plants), and 30,000 individual protozoa (heterotrophic protists). Another calculation has shown that in a hectare (2.47 acres) of wheat land in England, the weight of the microorganisms in the soil is approximately equal to that of 100 sheep! Bacteria may be equally abundant in the sea, where a significant proportion of the productivity depends on the activities of photosynthetic bacteria.

Bacteria occur in the widest possible range of habitats and play key ecological roles in virtually all of them. Some thrive in hot springs, for example, where the usual temperatures may range as high as 100° C; others have been recovered living beneath 430 meters of ice in Antarctica. Still other bacteria, capable of dividing only under high pressures, exist around deep-sea thermal vents, where the water is at temperatures over 100° C. These bacteria are used as food sources, directly or indirectly, by all the other organisms that live around the vents; they, in turn, obtain their energy by converting hydrogen sulfide to elemental sulfur (see Figure 26-11). Symbiotic sulfur-oxidizing bacteria also occur in clams, which then are able to grow in many unlikely habitats, including sewer outlets in the sea near Los Angeles. Similar deep-sea communities are being discovered in many additional areas. They are based on the ability of bacteria, symbiotic in bivalve mollusks such as clams and mussels, to metabolize and thus to capture energy from reduced carbon sources, including methane.

The metabolic diversity of bacteria is what allows them to play such varied ecological roles. Among the members of this group are **obligate anaerobes** (organisms that cannot grow in the presence of oxygen), **facultative anaerobes** (organisms that may function either as anaerobes or as aerobes), and **aerobes** (organisms that require

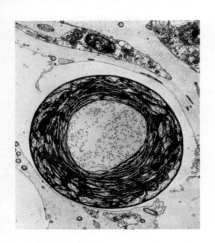

FIGURE 30-7
Electron micrograph of an individual of *Prochloron* species.

oxygen). In contrast, almost all eukaryotes are aerobes—they need oxygen at high concentrations to live—and most of the few anaerobic eukaryotes clearly have been derived from ancestors that were aerobes. There are a few anaerobic protists that appear to have descended directly from ancestral primitive eukaryotes prior to the establishment of the mitochondrial symbiosis. Bacteria vary in nearly every aspect of their metabolism, in marked contrast to eukaryotes, which are relatively uniform. All eukaryotes, for example, follow the same general patterns of respiration, photosynthesis, and synthesis of nucleic acids and proteins.

Autotrophic Bacteria

The life modes of bacteria are as diverse as their metabolisms; different groups exhibit a variety of ecological patterns, or relationships with their environments. Most bacteria are heterotrophs, obtaining their energy from organic material formed by other organisms. Less common than their heterotrophic relatives, but exceedingly important in the world ecosystem, are the autotrophic bacteria, which obtain their energy from nonorganic sources. There are two major groups of bacterial autotrophs: photosynthetic bacteria and chemoautotrophic bacteria.

Photosynthetic Bacteria. Like plants, the photosynthetic bacteria contain chlorophyll, but it is not held within plastids. One group of photosynthetic bacteria, the cyanobacteria, was discussed extensively in Chapters 4 and 22 in connection with the evolution of life on earth. The other major groups of photosynthetic bacteria are the green sulfur bacteria, the purple sulfur bacteria, and the purple nonsulfur bacteria. The different colors of these organisms are caused by their photosynthetic pigments. The photosynthetic processes carried out by these different bacteria are diverse and reflect the long evolutionary path to green plant photosynthesis. Such recently discovered organisms as *Prochloron* (Figure 30-7) and *Heliobacterium* (p. 603) are thought to represent steps in the evolution of chloroplasts in different groups of eukaryotes.

Chemoautotrophic Bacteria. In contrast to the photosynthetic bacteria, chemoautotrophic bacteria derive the energy they use for biosynthesis from the oxidation of inorganic molecules such as nitrogen, sulfur, and iron compounds, or from the oxidation of gaseous hydrogen. The metabolic chemistry of these remarkable organisms was described in Chapter 9, and some groups are discussed further on p. 607.

Heterotrophic Bacteria

Most bacteria are heterotrophs; they cannot make organic compounds from simple inorganic substances but must obtain them from other organisms. Together with the fungi, bacteria play the leading role in breaking down the organic molecules formed by biological processes. By doing so, bacteria and fungi make the nutrients in these molecules available once more for recycling. Decomposition is just as indispensable as photosynthesis to the continuation of life on earth.

Among the heterotrophic bacteria, the best represented are **saprobes,** those bacteria that obtain their nourishment from dead organic material. Many of the characteristic odors associated with soil come from substances produced by saprobic bacteria. Mutant bacteria have appeared in recent decades that are able to break down synthetic products such as nylon, herbicides, and pesticides, substances that are released by humans into the air, water, and soil in large amounts. The techniques of genetic engineering are used to produce bacteria capable of disposing of such waste materials. However, bacteria break down herbicides and pesticides so rapidly in some situations that these chemicals are much less effective than they would be otherwise.

> **Autotrophic bacteria, which are capable of manufacturing their own food, obtain their energy either from light or from the oxidation of inorganic molecules. Heterotrophic bacteria, in contrast, obtain their energy by breaking down organic compounds made by other organisms.**

Other activities of the heterotrophic bacteria may be destructive, threatening, or beneficial to humans. Some by-products of the metabolism of heterotrophic bacteria, such as antibiotics, are commercially valuable; others are harmful. For example, one of the species of the bacterial genus *Clostridium, C. botulinum* (see Figure 30-5), causes the type of food poisoning known as botulism. The bacterium produces one of the most powerful toxins known: 1 gram is enough to kill about 14 million people. As mentioned earlier, the endospores of this species are relatively resistant to boiling, although the toxins can be destroyed in this way. Botulism is rare in the United States, but another kind of food poisoning, associated with *Staphylococcus aureus* and fortunately much milder, is common. The bacterium secretes a toxin into food that causes severe reactions, including nausea with diarrhea and vomiting, within a few hours after ingesting the food. If foods are kept refrigerated, this type of food poisoning does not occur, since *Staphylococcus* is unable to grow at low temperatures. Even certain strains of the common intestinal bacterium *E. coli* can produce diarrhea by releasing toxins that indirectly cause the movement of fluid into the gut and inhibit the absorption of sodium there; they are the most frequent cause of "traveler's diarrhea" throughout the world.

Bacteria of the genus *Salmonella* may also cause gastrointestinal disease, which often develops several days after contaminated food is eaten. *Salmonella* frequently contaminates rare hamburger, and was found on 4 of 10 chickens sampled from supermarket coolers in the United States in the mid-1980s. About 36,000 cases of *Salmonella* food poisoning were reported in the United States in 1984, and 56,000 in 1985; health officials think the actual figures may be 10 or even 100 times higher. This bacterium is easily killed by cooking, which accounts for why it is rarely transmitted via chickens. Surprisingly, pet turtles *(Pseudemys scripta-elegans)* were a major source of *Salmonella* infections until their trade was banned within the United States in 1975. There are hundreds of strains of *Salmonella;* the symptoms of infection include fever, headache, diarrhea, and nausea. Most people recover from *Salmonella* infections without antibiotics. Infants, the elderly, and the infirm are at greatest risk. In addition, *Salmonella* has recently been identified as a factor in causing certain forms of arthritis. A prototype vaccine against *Salmonella*, announced in 1987, could be used to inoculate food animals and thus limit the occurrence of the bacterium.

Nitrogen-Fixing Bacteria

Both autotrophic and heterotrophic bacteria play an important role in the nitrogen cycle (see Figure 25-4). Since organic nitrogen is often in short supply, its availability is one of the chief factors limiting plant growth in specific areas—and therefore, indirectly, animal growth. Nitrogen fixation is carried out primarily by nodule-forming bacteria of the genus *Rhizobium* and the actinomycetes. In addition, free-living genera, especially *Azotobacter* and *Clostridium* (see Figure 30-5), which are both common in soils, as well as many cyanobacteria also make significant contributions to the total nitrogen fixation. Bacteria also participate in the nitrogen cycle by releasing fixed nitrogen, which occurs when they break down proteins. Some microorganisms break down the original protein molecules into their constituent peptides; many are able to break down these peptides into their constituent amino acids and in turn degrade them to release ammonium ions.

> One of the most essential links in the chain of life, the cycling of nitrogen, is carried out exclusively by the bacteria. Both heterotrophic and autotrophic genera are involved.

BACTERIA AS PLANT PATHOGENS

Many costly diseases of plants are associated with particular bacteria; almost every kind of plant is susceptible to one or more kinds of bacterial disease. The symptoms of these plant diseases vary, but they are commonly manifested as spots of various sizes on the stems, leaves, flowers, or fruits. Other common and destructive diseases of plants, including blights, soft rots, and wilts, also are associated with bacteria. Fire blight, which destroys pears and apple trees and related plants, is a well-known example of bacterial disease (Figure 30-8). Most bacteria that cause plant diseases are

FIGURE 30-8

Fire blight. Fire blight, a bacterial disease that is very destructive in apples, pears, cottoneaster, and allied plants, is caused by the bacterium *Erwinia amylovora*. Members of this genus are the most frequent and serious plant pathogens.

A Dead branch of a pear tree in the process of being killed by fire blight.

B Canker on the stem of an apple.

A   **B**

members of the group of rod-shaped bacteria known as pseudomonads (see Figure 30-1, *A*).

An outbreak in Florida of a bacterial disease known as citrus canker, first detected in August 1984, led farmers to destroy more than 4 million citrus seedlings in 4 months in an effort to halt the spread of the disease. Citrus canker, which had not been detected in Florida for more than 40 years, probably was introduced on fruits that were brought from abroad by a traveler and escaped detection at a customs station. The disease is caused by one of more than 100 distinct varieties of the pseudomonad *Xanthomonas campestris;* other varieties of this bacterium cause diseases with similar symptoms in beans, cabbages, peaches, and other plants. Citrus canker now threatens the continued existence of the Florida citrus industry, a $2.5 billion business that was weakened by the worst freezes of the century in 1983 to 1985.

BACTERIA AS HUMAN PATHOGENS

Bacteria cause many diseases in humans, including cholera, leprosy, tetanus, bacterial pneumonia, whooping cough, and diphtheria. Enormous sums are spent annually in the effort to reduce the likelihood of these infections and in limiting the other destructive activities of bacteria.

Several genera of pathogenic bacteria are of particular importance to humans. For example, members of the genus *Streptococcus* (Figure 30-1, *B*) are associated with scarlet fever, rheumatic fever, pneumonia, and other infections. Tuberculosis, an-

THE DISCOVERY OF A NEW ANTIBIOTIC

The study of biology is filled with surprises, but one of the most interesting recent ones has been the discovery of a new class of antibiotics, named **magainins,** in the skin of the African clawed frog (*Xenopus laevis*). In 1986, Micheal Zasloff (a scientist at the National Institutes of Health) noticed that the wounds of the frogs he was operating on for his biomedical experiments always healed perfectly, even though the frogs were replaced in the same aquarium water where they had been living—water filled with bacteria, fungi, and other parasites.

Investigating this phenomenon, Zasloff first found that secretions from the frog's skin, although they were toxic to humans, failed to kill bacteria and fungi. He then mixed ground skin with solvents and discovered a group of small peptides that did have the antibiotic properties—an entire chemical defense system separate from the immune system. Many organisms have evolved substances that protect them from the attack of other organisms (usually bacteria) or simply give them space in which to grow. In this case, the frogs are aquatic, and

subject to continuous attack by bacteria and fungi. In such an environment, the evolution of a protective system is easily understandable. The newly discovered magainins are now being investigated for their antibiotic properties in other circumstances and against other kinds of pathogens. In addition, magainin-like molecules are being sought in humans and other animals. In this instance, a serendipitous discovery has led to an entire promising field of investigation.

other bacterial disease, is still a leading cause of death in humans. In 1986, nearly 23,000 cases were reported in the United States, the first significant increase in at least 23 years. This increase was apparently linked with immune suppression caused by AIDS. These diseases are mostly spread through the air, as are some of those caused by the pathogenic genus *Staphylococcus*, which causes widespread infections in hospitalized patients.

Many bacterial diseases are dispersed primarily in food or water, including typhoid fever, paratyphoid fever, and bacillary dysentery. Typhus is spread among rodents and humans by insect vectors. The bacterium *Brucella abortus*, which causes the disease called brucellosis in animals, may also cause undulant fever in humans who drink milk from an infected cow. The bacteria are destroyed by pasteurization of the milk, however, and as a result of widespread pasteurization practices, undulant fever is becoming rare throughout the world.

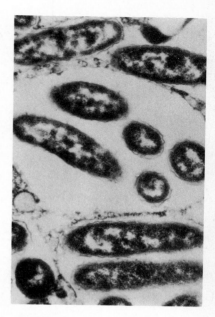

FIGURE 30-9
Legionella, **a recently discovered genus of bacteria.** Shown here is one of the bacteria that causes legionellosis.

Legionnaires' Disease

One of the more recently detected bacterial diseases of humans is legionellosis (Legionnaires' disease), which is believed to affect about 125,000 people in the United States annually. The disease develops into a severe form of pneumonia, and is fatal in about 15% to 20% of its victims if it is untreated. Legionellosis was discovered in 1976, when a mysterious lung ailment proved fatal to 35 American Legion members attending a convention in Philadelphia. This disease is caused by *Legionella*, which are small, flagellated, rod-shaped, gram-negative bacteria with pointed ends (Figure 30-9). These bacteria are common in water, preferring warm water at about 40° to 50° C, and may be abundant in whirlpool baths and similar facilities where they grow on minute algae and organic matter. They also occur under similar circumstances in airconditioning systems. In the human body they attack the monocytes, a kind of white blood cell that normally plays a major defensive role against most microorganisms. These bacteria can be destroyed with erythromycin treatment.

Sexually Transmitted Diseases

A number of important bacterial diseases are sexually transmitted. Among the most common of these diseases are gonorrhea, caused by the bacterium *Neisseria gonorrhoeae*, and syphilis, caused by *Treponema pallidum*, a spirochaete. Both of these diseases are often controlled with antibiotics, and syphilis can also be controlled with penicillin. Gonorrhea is more common and often less serious than syphilis, which can be fatal; gonorrhea infected about 500 people per 100,000 in the United States each year during the early 1980s, syphilis fewer than 15 per 100,000 per year. The rates of occurrence of both diseases are increasing. In these and other bacterial diseases, the appearance of antibiotic-resistant forms has posed great problems for the medical profession.

Infections caused by the bacterium *Chlamydia trachomatis* are more common than either syphilis or gonorrhea. *Chlamydia* are unusual bacteria that can survive only as parasites and are thus difficult to diagnose and to culture. An estimated 3 million people in the United States are infected by *Chlamydia* each year; this disease is often associated with painful and serious secondary symptoms that may result in sterility. It is estimated that as many as 50,000 women per year in the United States are involuntarily rendered infertile by chlamydial infections, and the first sign of an earlier infection may be the infertility itself. *Chlamydia* can also affect the health of infants born to infected mothers; for example, *Chlamydia* is the most frequent cause of pneumonia in newborn babies. If detected, chlamydial infections can be controlled with the antibiotic tetracycline, although it requires a full week of treatment for the cure to be complete; however tetracycline cannot be used to treat pregnant women and young children. It is estimated that chlamydial infections cost more than $1.4 billion per year, directly and indirectly, in the United States alone, yet they have only recently been recognized as a major public health problem. In 1987, it was reported that *Chlamydia* is also a common cause of arthritis in young people.

Dental Caries

One human disease that we do not usually consider bacterial in origin—dental caries (decay), which causes cavities—arises in the film on our teeth, the **dental plaque**. This film consists largely of bacterial cells surrounded by a polysaccharide matrix. Most of the bacteria in plaque are filamentous cells, classified as *Leptotrichia buccalis*, which extend out perpendicular to the surface of the tooth; many other bacterial species are also present in plaque. Tooth decay is caused by the bacteria present in the plaque, which persists especially in places that are difficult to reach with a toothbrush. Diets that are high in sugars are especially harmful to the teeth, because lactic acid bacteria (especially *Streptococcus sanguis* and *Streptococcus mutans*) ferment the sugars to lactic acid, a substance that causes the local loss of calcium from the teeth. Thus, frequent eating of sugary snacks or sucking on candy over a period of time keeps the pH level of the mouth low and actively promotes tooth decay. Bacteria will not cause decay in the absence of such a diet. Once the hard tissue has started to break down, the lysis of proteins in the matrix of the tooth enamel starts, and tooth decay begins in earnest. Fluoride makes the teeth more resistant to decay because it retards the loss of calcium. It was first realized that tooth decay was caused by bacteria when germ-free animals were raised; their teeth do not decay even if they are fed sugary diets.

Since the presence of bacteria is necessary for tooth decay to occur, it can be prevented by antibiotics or theoretically by vaccination. Strains of *Streptococcus mutans* that produce only low levels of acid are also known; attempts are under way to replace the harmful strains already present in the mouth with them, and thus to lower the frequency of tooth decay. Even if tooth decay is prevented, however, gum disease can develop, and we need to continue cleaning our teeth to prevent it. Like tooth decay, gum disease is caused by several kinds of bacteria, and may also be controlled by different antibiotics or by vaccination.

BACTERIAL DIVERSITY

Bacteria are not easily classified according to their forms, and only recently has enough been learned about their metabolic characteristics to develop a satisfactory

TABLE 30-1   CHARACTERISTICS OF SOME PHYLA OF BACTERIA

NAME OF GROUP	FORM[a]	MOTILITY[b]	METABOLISM[c]	ECOLOGICAL ROLE
Archaebacteria	R,S,C	N,F	C,P	Some digest cellulose; some produce methane; some reduce sulfur
Omnibacteria	R	N,F	H	Saprobes, pathogens, decomposers
Cyanobacteria	R,C,M	G,N	P	Carbon and nitrogen fixation
Chloroxybacteria	C	N	P	Symbiotic in sea squirts
Mycoplasmas, spiroplasmas	No wall[d]	N	H	Pathogens of plants and animals
Spirochaetes	S	F[e]	H	Decomposers and pathogens
Pseudomonads	R	F	H,C	Decomposers and plant pathogens
Actinomycetes	M,R	N	H	Soil, plants, decomposers, nitrogen fixers
Myxobacteria	R[f]	G	H	Soil, some in animals
Nitrogen-fixing aerobes	R	N,F	H	Free-living and in nodules on plant roots
Chemoautotrophs	R,C	N,F	C	Stages in nitrogen cycle; oxidize sulfur compounds; oxidize methane or methanol

[a]*R*, Rods (bacilli); *C*, cocci; *S*, spirilla; *M*, regularly forming chains or aggregations.
[b]*F*, Flaggelated; *N*, nonmotile; *G*, gliding.
[c]*H*, Heterotropic; *C*, chemosynthetic; *P*, photosynthetic.
[d]Either more or less spherical or elongate and twisted.
[e]Flagella inserted below the outer lipoprotein membrane of cell wall.
[f]The bacteria move together and form complex spore-bearing structures at certain stages in their life cycle.

overall classification comparable to that used for other organisms. Lynn Margulis and Karlene Schwartz proposed a useful classification system that divides bacteria into 16 phyla, according to their most significant features. We have organized the following discussion, which deals with most of these phyla, according to this system. Table 30-1 outlines some of the major features of the phyla that we describe.

Archaebacteria

Carl Woese and his colleagues at the University of Illinois have identified a diverse group of bacteria, which he has termed the **Archaebacteria,** that resemble one another in a number of fundamental respects, but are sharply distinct from all other bacteria in the same respects. For example, the base sequences in portions of the ribosomal RNA (rRNA) in different members of this group of bacteria are virtually identical. These sequences differ sharply from the corresponding ones found in other bacteria, as well as from those found in eukaryotes. The cell walls of all methane-producing bacteria and the related groups mentioned below lack muramic acid, which is a component of the cell walls of all other bacteria.

The Archaebacteria have distinctive membranes, unusual cell walls, and unique metabolic cofactors. The Archaebacteria are so distinct from other bacteria that many scientists treat them as a separate kingdom, with the remaining bacteria grouped as the kingdom **Eubacteria,** and this treatment is beginning to win general acceptance because the facts on which it is based are clearly established (see Figure 28-9). Among the Archaebacteria may be the oldest forms of life on earth, and both Eubacteria and eukaryotes may have diverged from them independently; certainly many Archaebacteria are capable of living in an anaerobic atmosphere rich in CO_2 and H_2. Some Archaebacteria have ribosomes that resemble those of eukaryotes, whereas others have smaller ribosomes that resemble those of Eubacteria, thus strengthening this hypothesis. At any event, the surviving Archaebacteria seem to be the remnants of ancient lines that have persisted in specific habitats where conditions continued like those that existed throughout the world when these bacteria first evolved.

Among the Archaebacteria, the methanogenic bacteria, or *methanogens*, are prominent, and include most of the known genera of this phylum (Figure 30-10). These bacteria, which we discussed in Chapter 4 in connection with the evolution of life on earth, are obligate anaerobes. They produce methane (CH_4) from carbon dioxide and hydrogen and obtain their energy from this process. They are the source of marsh

FIGURE 30-10
Examples of the remarkable diversity of Archaebacteria.
A *Pyrodictium occultum*, a sulfur-reducing bacterium that grows on sulfur vents on the sea floor off Vulcano, Italy. (×45,000.) The optimum temperature for growth by this bacterium, which forms "cobwebs," shown here, on the surface of the sulfur, is 105° C; growth does not take place below 80° C. Pyrite ("fools gold"; iron disulfide, FeS_2) was precipitated where *Pyrodictium occultum* was growing actively; this suggests that the bacterium plays a role in the formation of ores.
B Individuals of *Pyrodictium occultum*. (×75,000.)
C The "square" bacterium is a species of *Halobacterium* from a salt pond near the shore of the Red Sea. In most organisms, such a shape would have been precluded by the osmotically generated internal hydrostatic pressure, but these bacteria, which live in ponds saturated with brine, have little or no cell turgor pressure. These thin, nearly transparent bacteria derive buoyancy from their gas vacuoles and float at the surface of the brine. Square organisms are uncommon in nature.

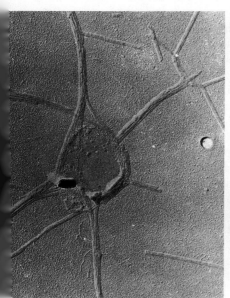

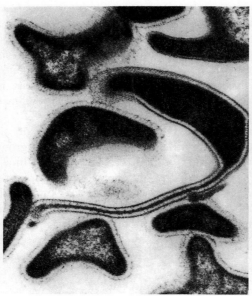

B

C

gas, which is produced in swamps and sewage-treatment plants, and of a major portion of our natural gas reserves as well. Recently it has been shown that the methanogens can reduce elemental sulfur to form hydrogen sulfide, just as do other sulfur-reducing bacteria not of this phylum. By reducing elemental sulfur, these bacteria produce their own anaerobic environment, which is important for them because they cannot function when oxygen is present. A methanogen capable of fixing atmospheric nitrogen was discovered in 1984, thus adding an unexpected dimension to the fundamental metabolic diversity among the members in the phylum.

Archaebacteria are an ancient group of prokaryotes that are very different from eubacteria and seem to be the direct ancestors of eukaryotes.

Omnibacteria

Bacteria similar to *Escherichia coli* (see Figure 30-3) constitute the phylum Omnibacteria, also called the Eubacteria (as a phylum), a phylum of rigid, rod-shaped, heterotrophic, gram-negative bacteria that includes many important pathogens. Most of these bacteria have flagella and none of them produce spores. The members of this large and diverse phylum are usually aerobic, but in the absence of oxygen they can usually continue to function, using compounds such as nitrate (NO_3^-) or other fermentative pathways as the terminal electron acceptor in respiration.

In addition to the gram-negative rods similar to *Escherichia coli*, the Omnibacteria also include a series of comma-shaped bacteria, most of which have a single terminal flagellum. Such bacteria of this form are called **vibrios.** Members of the genus *Vibrio* itself (see Figure 5-25) cause cholera, a serious and widespread disease. At least three genera of Omnibacteria are bioluminescent, including the genus *Photobacterium*, which is associated with deep-sea fishes, where the bacteria grow in special light organs (see Figure 26-10). Another group of small Omnibacteria, the rickettsias, are **obligate parasites** (organisms that can live only as parasites) within the cells of vertebrates and arthropods. Rocky Mountain spotted fever, which is spread by ticks, is an example of a human disease caused by rickettsias.

Cyanobacteria

Cyanobacteria are among the most prominent of the photosynthetic bacteria (see Figure 30-2). We have already discussed the critical role that the members of this ancient phylum have played for at least 3.5 billion years in the history of the earth and its atmosphere (see Chapters 4 and 22). The activities of cyanobacteria appear to have been decisive in bringing about the increase of free oxygen in the earth's atmosphere from far less than 1% to about 20%. This change, in turn, appears to have been crucial for the evolution of eukaryotic life based on aerobic respiration. Cyanobacteria produced the accumulation of the massive limestone deposits known as stromatolites (see Figure 22-6), which became abundant about 2.8 billion years ago but are still being formed where appropriate conditions exist. The few hundred living species of cyanobacteria are important relics of the earth's history, still of great interest and ecological importance today.

Cyanobacteria contain chlorophyll *a*, as do all photosynthetic eukaryotes. The process of photosynthesis is identical in both groups. Accessory pigments in the cyanobacteria consist of both carotenoids and blue and red, water-soluble pigments known as **phycobilins.** Carotenoids are widespread in photosynthetic organisms, both prokaryotes and eukaryotes. Phycobilins, on the other hand, are known to occur only in cyanobacteria, in the phylum of protists known as red algae (see p. 629), and in one other small phylum of eukaryotic organisms, the cryptomonads. Because the accessory pigments shared by the cyanobacteria and the chloroplasts of red algae are unique and also resemble each other in additional biochemical and structural characteristics, it seems probable that the chloroplasts of red algae have been derived directly from symbiotic cyanobacteria.

Cyanobacteria usually have a mucilaginous sheath, which is often deeply pigmented and can be yellow, brown, green, blue, red, violet, or blue-black. Colorful "blooms" may occur in polluted water as a result of the rampant growth of cyanobacteria under conditions in which these organisms flourish temporarily, such as an excess of phosphorus. The colors of such "blooms" usually result from the photosynthetic pigments of the cyanobacteria involved. The cyanobacteria form masses near the water surface because of the formation of gas vacuoles within their cells. When such population explosions occur, the results may become traumatic for the bodies of water in which they happen. Certain cyanobacteria can, under appropriate circumstances, produce toxic substances that may poison other living things. Masses of cyanobacterium are sometimes used as a source of food for humans (see Figure 27-11).

Many cyanobacteria can fix atmospheric nitrogen, and for this reason they are especially important in rice fields, where they significantly augment the supply of biologically available nitrogen. Nitrogen fixation occurs among almost all cyanobacteria within specialized cells called heterocysts (see Figure 30-2)—enlarged cells that occur in many of the filamentous members of this phylum. These cells begin to form when available nitrogen falls below a certain threshold; when nitrogen is abundant, their formation is inhibited.

Being able to photosynthesize and to fix atmospheric nitrogen, cyanobacteria are nutritionally the most independent organisms that exist. In view of this, it is not surprising that they play such a variety of ecological roles. Common in soil, they form mats that become obvious in moist conditions, often binding sand, for example. Cyanobacteria and cyanobacteria-containing lichens (see Chapter 32) are frequently found on and in rock surfaces, etching and eroding them slowly and thus providing a means by which buildings can be dated. For example, colonies on a Roman castle in the Judean hills of Israel were found to have grown at the rate of 1 millimeter per 800 years! The mats that line the surfaces of sediments in the sea are often dominated by cyanobacteria, with many other types of bacteria growing among and often under them; and modern stromatolites are growing in suitable habitats throughout the world, including the bottoms of frigid Antarctic lakes. Travertine, a porous rock often used in construction because it is light and attractive, is limestone altered substantially by cyanobacteria that have grown into the rock. Perhaps most interesting of all is the recent discovery that the bulk of our modern petroleum deposits were formed by masses of decayed cyanobacteria. No group of organisms has played such a significant role in the evolution of the modern world.

Chloroxybacteria

The phylum Chloroxybacteria includes a single genus of photosynthetic bacteria, *Prochloron*. This organism is especially interesting because it contains the same photosynthetic pigments, chlorophylls *a* and *b*, that the eukaryotic green algae and the plants do (Figure 30-11). *Prochloron* lives mainly in association with colonial ascidians (sea squirts or tunicates, phylum Chordata) along tropical and subtropical seashores (see Figure 30-7). Within the tissues of its host, *Prochloron*, like the cyanobacteria, is capable of fixing nitrogen. *Prochloron* is important to students of evolution because of its biochemical characteristics, including its possession of chlorophyll *b*. These characteristics are compatible with those postulated for the prokaryote that gave rise, by symbiosis, to the chloroplasts of green algae and thus, indirectly, to those of plants.

Recently, other pigmented bacteria have been discovered that promise, when studied further, to clarify the origin of chloroplasts more completely. One of these is *Heliobacterium chlorum*, an anaerobic, brownish, nitrogen-fixing bacterium discovered in 1983. *Heliobacterium* resembles in some respects the chloroplasts of the diatoms and brown algae, and may be related to the group that gave rise to them initially.

Mycoplasmas and Spiroplasmas

The two groups of small bacteria known as mycoplasmas and spiroplasmas have been placed by Margulis and Schwartz into a separate phylum, which they call Aphragmabacteria. These organisms differ from all other bacteria in that they lack cell walls.

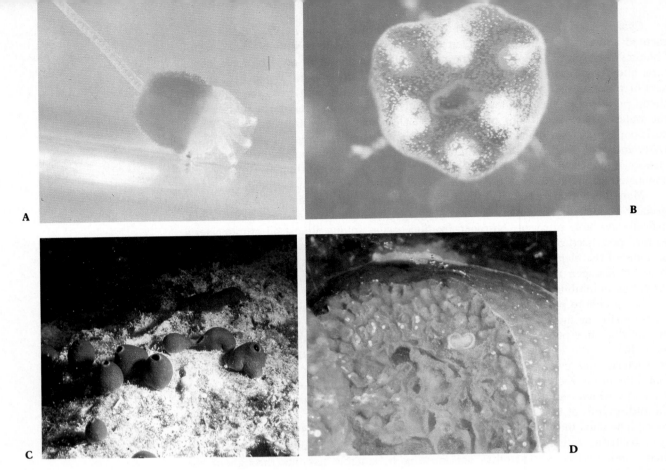

A

B

C

D

FIGURE 30-11

The colonial tunicate *Didemnum molle*, a marine animal, at Lizard Island, on the Great Barrier Reef, Australia, is home to the chloroxybacteria called *Prochloron*.

A A larva, about 2.5 millimeters long; the green material inside the animal is a colony of *Prochloron*.
B A juvenile individual, which, having settled down, has divided into six parts and is forming a colony.
C Colonies of the tunicate at a depth of about 2 meters. One of the colonies is in the process of dividing.
D A colony broken open, showing the masses of *Prochloron* within.

Their cells are bounded by a single triple-layered membrane composed of lipids (Figure 30-12). Because they lack cell walls, mycoplasmas and spiroplasmas are resistant to penicillin and other antibiotics that work by inhibiting cell wall growth. Some members of this phylum are less than two-tenths of a micrometer in diameter—tiny irregular blobs that cannot be distinguished with a light microscope.

Some mycoplasmas cause diseases in mammals and birds, and are especially important as the causative agents of certain types of pneumonia in humans and domestic mammals. Some have recently been associated with newly detected kinds of sexually transmitted diseases in humans. In addition, it is suspected that mycoplasmal infection is the most common cause of premature labor in humans. Nevertheless, they are difficult to detect, requiring up to a month to reach detectable levels in culture. When diagnosed properly, they are easily treated with antibiotics, especially tetracycline.

Spiroplasmas and other members of this phylum cause significant plant diseases, having been found to infect more than 200 plant species thus far (Figure 30-13). Among the most severe diseases are aster yellows, the stubborn and little-leaf diseases of citrus, and the lethal yellowing disease of coconuts and at least 30 other species of palms. Since it was first observed in 1971, the latter disease has killed an estimated 1.5 million coconut palms in south Florida, some 80% of the total that were originally cultivated in this area. Resistant strains of palms are now being planted to repopulate Florida. The disease is spread from tree to tree by a leafhopper, a kind of insect. It can be controlled by injecting an antibiotic, such as tetracycline, into the trees. In general, spiroplasmas are abundant in insects, which are the principal agents that spread them from plant to plant.

Spirochaetes

The spirochaetes are long spirilla in which the flagella are inserted beneath the outer lipoprotein membrane of their gram-negative outer cell wall (see Figure 30-1, *C*). Spi-

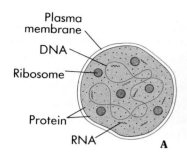

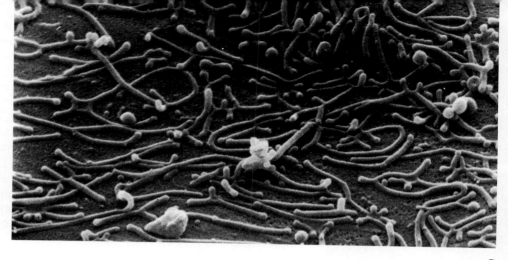

B

FIGURE 30-12
Mycoplasmas.
A Diagram of a mycoplasma, one of the smallest bacteria known.
B Scanning electron micrograph of the mycoplasma organism that causes pneumonia in humans.

rochaetes may have from 2 to more than 100 flagella in this space. They can move through thick, viscous liquids with great speed and through liquids of thinner consistency with more complex motions. Each flagellum originates near one end of the cell and extends about two thirds of the body length. Thus the flagella that originate at the different ends of the cell often overlap. The internal rotation of these flagella is thought to cause the mobility of the spirochaetes.

Spirochaetes of the genus *Treponema* are the causative agents of syphilis and of yaws, a disfiguring eye disease that is frequent in some tropical areas. Another spirochaete, *Borrelia burgdorferi,* is responsible for Lyme disease, an inflammatory ailment of humans that is named for the Connecticut village of Old Lyme (Figure 30-14). Lyme disease is spreading throughout the United States and has also been found a few times on each of the other continents. In the eastern United States, Lyme disease is spread by a tick, *Ixodes dammini.* After emerging in the spring, the juvenile stages of this tick attach themselves mainly to white-footed mice. It is during these stages that they are also most likely to attach to humans. For many victims, the first sign of infection is a rash that resembles a bull's-eye, which may appear up to a month after the tick's bite. At the end of summer, when the ticks are adults, they move to different mammals, especially (in the eastern United States) white-tailed deer, and mate on the host. The ticks that carry Lyme disease infect at least 30 species of birds in North

FIGURE 30-13
Spiroplasmas.
A Dying coconut palms. More than 1.5 million individuals have been killed since 1971 by lethal yellowing, a spiroplasma-caused disease.
B Spiroplasmas in plant tissue.

A

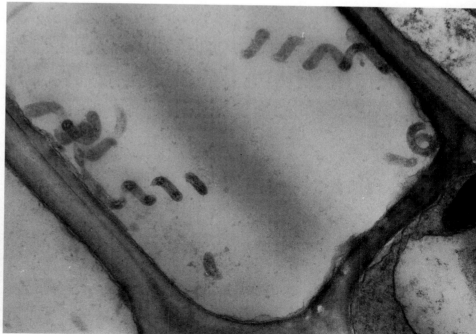

B

FIGURE 30-14
Lyme disease. A shows a liquid live culture of the spirochaete *Borrelia burgdorferi*, the bacterium that causes Lyme disease. This spirochaete is wound as a helix and shows a characteristic undulating motility. *Borrelia burgdorferi* live in the mid-gut of ticks, such as this specimen of *Ixodes ricinus* (**B**), the tick responsible for the transmission of Lyme disease in Europe.

America, in addition to mammals, and it is in this way that ticks presumably are carried over long distances. Fortunately, Lyme disease, if properly diagnosed, can be treated with penicillin or tetracycline.

Pseudomonads

Pseudomonads are straight or curved gram-negative rods that have one or several flagella at one end. Members of this phylum are found nearly everywhere in soil and water and have a broad ability to break down organic compounds of many kinds. Some pseudomonads are autotrophic, and many, especially those of the genera *Pseudomonas* and *Xanthomonas*, are serious plant pathogens. The bacterium *Pseudomonas aeruginosa* (see Figure 30-1, *A*) occurs in soil, water, and raw vegetables. Although it is usually harmless, it can form serious infections in people whose physical condition is poor. Pseudomonads of the genus *Bdellovibrio* occur in salt water and fresh water, and prey on other bacteria.

Actinomycetes

The actinomycetes are a distinctive phylum of bacteria that are sometimes mistaken for fungi because of their filamentous growth form. Members of this phylum produce spores by the division of their terminal, erect branchlets into chains of small segments that are tough and resistant to drought and other unfavorable environmental conditions (Figure 30-15). Actinomycetes of the genus *Frankia* form nitrogen-fixing nodules on the roots of a wide range of flowering plants (see Chapter 25). Dental plaque, in which the enamel-destroying bacteria grow, consists mainly of a dense mat of actinomycetes. *Mycobacterium leprae*, which causes leprosy, and *Mycobacterium tuberculosis*, the causative agent of **tuberculosis,** are actinomycetes. Tuberculosis used to be a major killer, until the advent of an arsenal of effective drugs to treat it. In 1991 new strains of *M. tuberculosis* were reported in the United States that are resistant to most of the drugs used to treat tuberculosis. Because tuberculosis is a deadly disease and *M. tuberculosis* is highly infectious, health officials are quite concerned. Many common antibiotics, including tetracycline, chloramphenicol, aureomycin, erythromycin, and neomycin, were derived originally from actinomycetes, although some of them are now synthesized commercially. Ivermectin is a potent new antiparasitic agent that has been derived from actinomycetes and is used widely in the treatment of diseases that previously have been difficult to cure. About 150 new antibiotics from actinomycetes are being discovered each year.

Myxobacteria

Myxobacteria are gliding bacteria, some of which form upright spore-bearing structures of unusual complexity for bacteria (see Figure 30-1, *D*). The members of this phylum are unicellular rods usually less than 1.5 micrometers thick but up to 5 mi-

FIGURE 30-15
Actinomycetes are filamentous. An actinomycete, *Streptomyces ambofaciens*, showing the characteristic filamentous growth habit of members of this phylum of bacteria.

crometers long. The cells are commonly embedded in a slimy mass of polysaccharides, which they excrete. These bacteria often aggregate into gliding masses, and they may form compound, spore-filled cysts that open when wet, releasing large numbers of individual gliding bacteria. Their interactions are mediated by both diffusible and contact-mediated stimuli; consequently, the myxobacteria are being considered as useful models for studying development in multicellular organisms. All myxobacteria are obligate aerobes; some break down other bacteria or protists enzymatically and then digest the contents of their cells. A few others live by oxidizing sulfides to sulfates. Most myxobacteria occur in soils.

Nitrogen-Fixing Aerobic Bacteria

The nitrogen-fixing aerobic bacteria are a phylum of great economic importance. All members of this group are gram negative and most are flagellated. *Azotobacter*, mentioned earlier, is a free-living member of this phylum. With three other recognized genera, it is a common resident in soil and water, and plays a role in converting the vast reservoir of atmospheric nitrogen into a form in which it can be used by itself and other organisms.

Clostridium, also a free-living bacterium, is a member of another phylum, the fermenting bacteria, which is not treated separately in this book. It is an obligate anaerobe, and some species are capable of nitrogen fixation. The cyanobacteria are also important in the global nitrogen cycle (see Figure 25-7), sharing with the other bacteria just mentioned the ability to fix atmospheric nitrogen. One of the most important genera of the same phylum as *Clostridium* is *Rhizobium*, which forms nodules on the roots of legumes (see Figure 25-5). *Rhizobium* plays the major role in nitrogen fixation worldwide.

Chemoautotrophic Bacteria

Only bacteria are capable of chemoautotrophy, a form of metabolism that depends on chemical sources of energy such as the reduced gases ammonia (NH_3), methane (CH_4), or hydrogen sulfide (H_2S). Chemoautotrophs do not require sunlight; in the presence of one of the chemicals just mentioned, and with nitrogenous salts, oxygen, and carbon dioxide, they can manufacture all of their own nucleic acids and proteins. The chemoautotrophs appear to be an ancient and primitive phylum of bacteria.

One group in this phylum consists of bacteria that use nitrogen compounds to gain energy. Among these organisms are *Nitrosomonas* and *Nitrobacter*, which have been discussed in relation to their role in the nitrogen cycle in Chapter 25. Other members of this phylum oxidize inorganic sulfur or iron compounds to gain energy. The sulfur bacteria live in habitats in which there are high concentrations of hydrogen sulfide or other reduced sulfur compounds; their cells contain sulfur granules. The bacteria that use ferrous iron (Fe^{++}) as an electron source live in habitats rich in iron. One of them, *Aquaspirillum magnetotacticum*, accumulates crystals of magnetite (Fe_3O_4), also known as lodestone, and uses them to orient itself when swimming (Figure 30-16). The members of a third group of chemoautotrophic bacteria use methane or methanol (CH_3OH) in the same way that the other members of the phylum use nitrogen or sulfur compounds.

SIMPLE BUT VERSATILE ORGANISMS

Bacteria are the simplest organisms, but in the diversity of their metabolism they far surpass all the rest. Although many bacteria are decomposers of one kind or another, a number are photosynthetic or chemoautotrophic; in addition, only bacteria have the ability to convert atmospheric nitrogen into a form in which it can be used by organisms. For at least 2 billion years—more than half of the history of life on earth— bacteria were the only organisms in existence. They have survived to the present in rich variety by exploiting an amazingly diverse set of habitats, some of them unchanged since the beginnings of the world as we know it. Indirectly their metabolic activities are of fundamental importance. Symbiotic bacteria also contribute directly

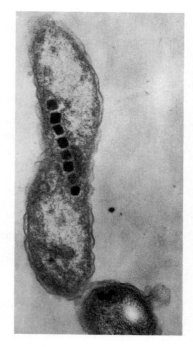

FIGURE 30-16
The magnetotactic bacterium *Aquaspirillum magnetotacticum*. The particles that form a chain of objects that are opaque to the electron beam are magnetite. With the aid of these crystals, the bacteria are able to orient themselves in a magnetic field, swimming steadily by means of their single flagellum. Since the magnetic field slopes downward toward the earth's surface, the bacteria also swim downward and thus reach the sediments at the bottom of the water in which they live.

to the functioning of all but a few eukaryotes in their role as mitochondria. Considering the high probability that all chloroplasts are symbiotic bacteria, it may be said that even photosynthesis, like chemoautotrophism and nitrogen fixation, is exclusively a property of the bacteria.

The many ways in which symbiotic relationships of eukaryotes with bacteria have contributed to the formation of evolutionary lines among the eukaryotes will be an important theme in the next chapter, as we begin to trace the evolution of eukaryotic organisms.

SUMMARY

1. The kingdom Monera is made up of the prokaryotes, or bacteria, with about 4800 species named so far. Bacteria are the oldest and simplest organisms, but they are metabolically much more diverse than all of the other forms of life on earth combined.

2. Most bacteria have cell walls; when they are present, they consist of a network of polysaccharide molecules connected by polypeptide cross-links. In the gram-negative bacteria, large molecules of lipopolysaccharide are deposited over this layer.

3. Bacteria are rod shaped (bacilli), spherical (cocci), or spiral (spirilla) in form. Bacilli or cocci may adhere in small groups or chains.

4. Bacteria reproduce asexually by binary fission. Mutation is the most important source of variability in bacteria, but they also exhibit genetic recombination mediated by bacteriophages or plasmids.

5. Bacteria and fungi break down organic compounds and thus recycle them for use by other living organisms. Different groups of bacteria fix the energy from the sun in different ways, and some bacterial groups use inorganic molecules as a source of energy.

6. Only bacteria are capable of fixing atmospheric nitrogen, thus making it available for their own metabolic activities and those of other organisms.

Some of these bacteria live in nodules on the roots of plants, whereas others are free-living, mainly in the soil. The best-known nitrogen-fixing bacterium is *Rhizobium*, which forms nodules on the roots of legumes.

7. The Archaebacteria, which consist of the methanogenic bacteria and a few related groups, differ markedly from the other bacteria and from the eukaryotes in their ribosomal sequences and in other respects. Because they are so different from other bacteria, a number of investigators have suggested that they should be considered a separate kingdom.

8. Cyanobacteria, formerly and misleadingly called "blue-green algae," are photosynthetic bacteria in which chlorophyll *a* occurs along with characteristic accessory pigments known as phycobilins. The plastids of the eukaryotic red algae are almost certainly derived from cyanobacterial ancestors. On the other hand, the plastids of the green algae, and therefore those of the plants, are more similar to the newly discovered genus *Prochloron*, which is a member of the phylum Chloroxybacteria.

9. The actinomycetes are bacteria in which the cells adhere as filaments. These bacteria are the source of a majority—about 2000 kinds—of the antibiotics discovered thus far, including erythromycin, neomycin, and tetracycline. Some actinomycetes fix nitrogen in nodules on the roots of nonleguminous plants.

REVIEW QUESTIONS

1. In what seven ways do prokaryotes differ substantially from eukaryotes?

2. What is the structure of the bacterial cell wall? How does the cell wall differ between gram-positive and gram-negative bacteria? In general, which type of bacteria is more resistant to the action of most antibiotics? Why? What is a capsule?

3. Why does mutation play such an important role in creating genetic diversity in the bacteria? How does bacterial recombination result in genetic diversity in bacteria? What is an example of the effects of such variability?

4. What chemicals are used by the chemoautotrophic bacteria? What are saprophobic bacteria? How may scientists use these organisms to combat pollution? What is a potential hazard of such research?

5. How do the Archaebacteria differ from all other bacteria? What group of bacteria is the most prominent of these ancient forms? What unique metabolism do they exhibit? What are their oxygen requirements? How do they control this themselves?

6. Why are certain animals dependent on methanogenic bacteria?

7. What is the greatest importance of the cyanobacteria? What types of photosynthetic pigments are found in this phylum? What do they indicate the about the evolution of certain eukaryotes? Why are cyanobacteria nutritionally independent organisms?

8. What one genus constitutes the phylum Chloroxybacteria? What is the greatest evolutionary significance of this organism?

9. What structure is most characteristic of the spirochaetes? What human diseases do members of this phylum cause?

10. Why are the actinomycetes a distinctive group of bacteria? What diseases are caused by members of this group? Of what enormous benefit are these organisms?

 ## THOUGHT QUESTIONS

1. Justify the assertion that life on earth could not exist without bacteria.

2. What do you think the functions of antibiotics are in the bacteria that produce them?

 ## FOR FURTHER READING

ARAL, S.O. and K.K. HOLMES: "Sexually Transmitted Diseases in the AIDS Era," *Scientific American*, February 1991, pages 62-69. Gonorrhea, syphilis, and other infections are still common and need increased attention.

BOYD, R.: *General Microbiology*, ed. 2, Mosby–Year Book, Inc. St. Louis, 1988. Excellent review of the bacteria, viruses, and some of the unicellular protists.

CHILDRESS, J.J.: "Symbiosis in the Deep Sea," *Scientific American*, May 1987, pages 114-120. Bacteria colonize the tube worms and clams at hot-water vents in the deep ocean and supply them with nutrients.

DONOGHUE, H.: "A Mouthful of Microbial Ecology," *New Scientist*, 5 February 1987, pages 61-65. Excellent and well-illustrated account of the communities of bacteria that inhabit our mouths and their effects.

FISCHETTI, V.: "Streptococcal M Protein," *Scientific American*, June 1991. The bacteria that cause strep throat and rheumatic fever depend on this cell surface protein to evade the body's defenses.

GOODFIELD, J.: *Quest for the Killers*, Birkhauser Press, Boston, 1985. An illuminating account of how several important diseases are being combated in the field by public health scientists.

HABICHT, G.S., G. BECK, and J.L.BENACH: "Lyme Disease," *Scientific American*, July 1987, pages 78-83. This rapidly spreading disease has many fascinating properties.

KIESTER, E.: "A Curiosity Turned Into the First Silver Bullet Against Death," *Smithsonian*, November 1990. A very engaging account of the discovery of penicillin.

LYNCH, J.: "Microbes are Rotting for Better Crops," *New Scientist*, 28 April 1988, pages 45-49. The communities of microorganisms that live around a plant's roots play a major role in determining its success.

McNAMARA, K.: "Survivors from the Planetary Soup," *New Scientist*, 1990, pages 50-52. Excellent account of stromatolites in the past and at the present day.

MEE, C.L., JR.: "How a Mysterious Disease Laid Low Europe's Masses," *Smithsonian*, February 1990, page 67-79. In the 1300s a third of the population of Europe died of plague brought by fleas.

READ, R.: "Bacteria with a Sticky Touch," *New Scientist*, October 1989, vol. 124, pages 38-41. A bacteria's first step when it enters the body is often to anchor itself to a cell—understanding how may make it possible to design new drugs to combat infection.

RONA, P.A.: "Metal Factories of the Deep Sea," *Natural History*, January 1988, pages 52-56. Bacteria, abundant around hot springs far below the surface of the sea, are producing lodes of copper and iron.

SAGAN, D. and L. MARGULIS: *Garden of Microbial Delights: A Practical Guide to the Subvisible World*, Harcourt Brace Jovanovich, Publishers, Boston, Mass., 1988. Excellent account of the diversity of microorganisms and the roles they play in the earth's ecosystems.

SHAPIRO, J.A.: "Bacteria as Multicellular Organisms," *Scientific American*, June 1988, pages 82-89. Explores the sophisticated temporal and spatial control systems that govern colonies of bacteria.

WOESE, C.R.: "Archaebacteria," *Scientific American*, June 1981, pages 98-122. Presents arguments that the Archaebacteria represent an independent kingdom of organisms.

31

Protists

OVERVIEW

Of the five kingdoms of organisms, the protists, a eukaryotic group, are far and away the most diverse structurally and in terms of their life cycles. Two important characteristics—sexuality and multicellularity—evolved within this group. Although the great majority of protists are unicellular, multicellularity originated a number of different times independently. The kingdom Protista contains the ancestors of each of the three kingdoms of more advanced multicellular organisms—the fungi, the plants, and the animals. Many other protists are unique single-celled forms that have not given rise to multicellular groups. Chloroplasts have originated from symbiotic bacteria in a number of these unrelated groups; their origin has involved at least three distinct groups of photosynthetic bacteria.

FOR REVIEW

Here are some important terms and concepts that you will encounter in this chapter. If you are not familiar with them, you should review them before proceeding.

Symbiotic origin of mitochondria and chloroplasts (Chapter 5)

Flagella and cellular motility (Chapter 5)

Evolution of mitosis (Chapter 10)

Sexuality and multicellularity (Chapter 28)

Life cycles of eukaryotes (Chapter 28, Figure 28-13)

Cyanobacteria, *Heliobacterium,* and *Prochloron* (Chapter 30)

Between the bacteria and the eukaryotes there is a gap in the evolutionary record, a gap that is not bridged by any living organisms. The evolution of the first eukaryotes could not have happened at once, quickly, because they differ from bacteria in so many complex ways (see Chapter 28). No organisms that are transitional between bacteria and eukaryotes survive, however. The following chapters deal with the eukaryotes. For all their incredible diversity, the various groups of eukaryotes are not nearly as different from one another as are the eukaryotes from the bacteria.

EVOLUTIONARY RELATIONSHIPS OF PROTISTS

Of the four kingdoms of eukaryotes, the most diverse kingdom is Protista, the protists (Figure 31-1). It contains the simplest of the eukaryotes as well as many groups that consist of very complex organisms. Few protists are large; most are microscopic and single celled. The multicellular animals and plants were derived independently from different groups of protists, as were the primarily multicellular fungi (Figure 31-2). Animals, plants, and fungi are considered distinct kingdoms; the remaining groups of eukaryotic organisms are assigned to the kingdom Protista.

> Among the protists are the ancestors of the primarily multicellular fungi and those of the entirely multicellular animals and plants. Except for these three kingdoms, all remaining eukaryotes, a very diverse assemblage, are considered to be protists.

Features of the Eukaryotes

Two of the most important features that evolved among the eukaryotes are sexuality and multicellularity; multicellularity occurred many times independently in different groups, as discussed in Chapter 28. The flagella and cilia of all eukaryotes, when they are present, have the complex 9 + 2 arrangement of microtubules discussed in Chapter 5. Such flagella and all of the other characteristics of eukaryotes are common to the protists. Protists are distinguished from plants, animals, and fungi because of their lack of the specialized features characteristic of those groups. For example, pro-

FIGURE 31-1
The protists. The kingdom Protista includes predominantly single-celled organisms. However, multicellularity has evolved in many lines. One of these, representative of the phylum Phaeophyta, is a gigantic kelp (**A**); a representative single celled form is *Paramecium* (phylum Ciliophora) (**B**). Still others have more unusual forms of multicellular construction, such as the multinucleate plasmodium of a plasmodial slime mold (phylum Myxomycota) (**C**). In such a plasmodium, which moves about in search of the bacteria and other organic particles that it ingests, there are many free nuclei.

A

B

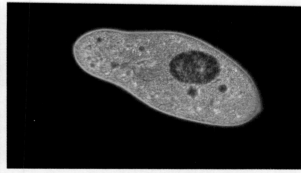

C

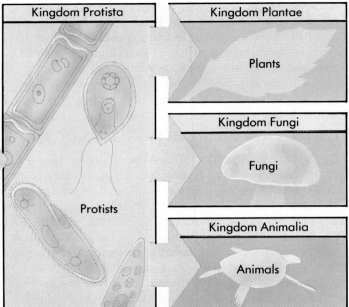

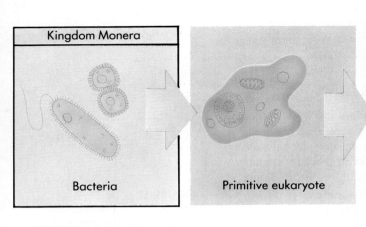

FIGURE 31-2
The evolutionary radiation of the eukaryotes. Of the five kingdoms
of living organisms, four are eukaryotes. The protist kingdom is by far
the most diverse of these, and individual protists gave rise to the other
three kingdoms. The ancestral groups of animals and plants are well
established, but no living protist has the characteristics that we believe
occurred in the ancestors of the fungi.

FIGURE 31-3
**The Protist Kingdom is a catch-
all kingdom that contains many
different phyla of unicellular organisms.** Complex unicellular organisms, such as *Vorticella*
(phylum Ciliophora)—which is
heterotrophic, feeds largely on
bacteria, and has a retractable
stalk—do not really fit the definition of either animals or plants,
and are classified as protists.

tists do not form embryos, and they do not develop the kinds of complex, multicellular sexual organs that characterize plants. Some protists have chloroplasts and manufacture their food, like the plants; others ingest their food, like the animals; and still
others absorb their food, like the fungi. The phyla of protists are, with very few exceptions, only distantly related to one another.

In the older systems of classification, the green, photosynthetic protists were considered plants, together with all protists that absorbed their food, as do the fungi.
Historically, the fungi have been included among the plants as well. Those protists
that ingested their food like the animals were considered small, very simple animals
(Figure 31-3). These schemes of classification do not reflect the complex relationships
between the different groups of protists, however, and thus they generally have been
replaced by the classification scheme used in this book. To consider some groups of
protists to be like plants, simply because they have independently acquired chloroplasts in the course of their evolution, or to "resemble" animals, because they have
not acquired chloroplasts and move, is highly misleading. It is for that reason that the
phyla of protists are considered alphabetically here.

Grouping all predominantly unicellular, eukaryotic organisms together in the
kingdom Protista allows us to compare them directly and to emphasize the similarities
and differences among them. It should be kept in mind, however, that the kingdom
Protista is a much more diverse group than any of the other three kingdoms of eukaryotic organisms.

Although those phyla of protists that do have chloroplasts and carry out photosynthesis are classified according to the botanical rules of nomenclature and therefore
should technically be called *divisions*, we will use the term *phyla* for all such groups
for clarity here. Protists that have chloroplasts are informally called **algae.** The scientists who study algae are called **phycologists,** whereas those who study heterotrophic
protists are called **protozoologists,** since these organisms collectively are called **protozoa,** a term similar to *algae* in its general, nontechnical meaning.

Two phyla of algae are primarily multicellular: the brown algae, phylum Phaeophyta, and the red algae, phylum Rhodophyta. A third phylum of algae, the green
algae, phylum Chlorophyta, includes many kinds of multicellular organisms as well as
larger numbers of unicellular ones (see Figure 28-11). Multicellularity has been
achieved separately in each of these three phyla, which have direct relationships to

different unicellular protists. The green algae include the ancestors of the plants, but we will retain them in the kingdom Protista here, since plants and green algae are so sharply distinct from each other. In addition to these three important groups, a number of other phyla of protists have also become multicellular in part during the course of their evolution.

> Three of the most important phyla of photosynthetic protists that have become wholly or partly multicellular are the green algae, brown algae, and red algae. These phyla are assigned to the kingdom Protista, even though the green algae include the ancestors of the plants.

The Role of Symbiosis

At least two organelles characteristic of all eukaryotes, including protists, are probably described from prokaryotes. Both mitochondria and chloroplasts share many features with bacteria, including a distinct kind of inner membrane; and different ribosomes, DNA, and reproductive schedules. As we reviewed in Chapter 28, mitochondria probably arose only once, early in the history of the eukaryotes—probably descending from symbiotic nonsulfur purple bacteria. Only a few eukaryotes lack mitochondria, including *Pelomyxa palustris* (Figure 31-4), an unusual amoeba-like organism that occurs on the muddy bottoms of freshwater ponds.

> Mitochondria, which probably originated from symbiotic nonsulfur purple bacteria, are absent in two unusual groups of protists. Of these, *Pelomyxa* may represent a stage of evolution before ancestral eukaryotes acquired these organelles.

In contrast to mitochondria, chloroplasts fall into three classes that differ in their biochemistry. They are as follows:

1. The chloroplasts of red algae, which contain chlorophyll *a*, carotenoids, and phycobilins. These chloroplasts were almost certainly derived from symbiotic cyanobacteria.
2. The chloroplasts of brown algae, diatoms, and dinoflagellates, with chlorophylls *a* and *c*, carotenoids, and distinctive yellowish brown pigments. The newly discovered bacterial genus *Heliobacterium* resembles these chloroplasts most closely and may be similar to the bacteria that gave rise to them originally.
3. The green chloroplasts of plants, green algae, and another, very distinctive phylum, the euglenoids (see p. 624). These chloroplasts contain chlorophylls *a* and *b* in addition to carotenoids. The bacterium *Prochloron* has essentially the same biochemical features as the chloroplasts of these groups and apparently resembles the bacteria that gave rise to them.

As mentioned in Chapter 28, there seem clearly to have been multiple symbiotic events in the evolution of chloroplasts, with the three groups of bacteria just outlined probably representing the major lines that were involved. Members of these lines seem also to have been involved in symbiotic events more than once; for example, the same kind of bacterium might have been involved in the origin of the chloroplasts of green algae and euglenoids independently of each other. These relationships are outlined in Figure 28-8.*

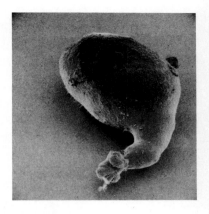

FIGURE 31-4
Pelomyxa palustris. This unique, amoeba-like protist lacks mitochondria and does not undergo mitosis. *Pelomyxa* may represent a very early stage in the evolution of eukaryotic cells. This species is the only member of the phylum Caryoblastea.

---

*Lynn Margulis of the University of Massachusetts–Amherst, believes that symbiosis has contributed even more to eukaryotic cells than mitochondria and chloroplasts. She has proposed that the mitotic apparatus is also a result of symbiotic events, and that the 9 + 2 flagella of eukaryotes were derived from spirochetes—elongated bacteria. She believes that the subsequent specialization of these spirochetes not only gave rise to flagella but also led to the evolution of centrioles and of chromosomal centromeres during the course of evolution of mitosis. The recent demonstration that DNA is present in the centrioles of the protist *Chlamydomonas* lends credibility to these hypotheses, which are being investigated further.

TABLE 31-1   THE PROTISTS

PHYLUM	COMMON NAME
PHOTOSYNTHETIC PROTISTS ("ALGAE")	
Bacillariophyta	Diatoms
Chlorophyta	Green algae
Dinoflagellata	Dinoflagellates
Euglenophyta	Euglena
Phaeophyta	Brown algae
Rhodophyta	Red algae
HETEROTROPHIC PROTISTS (MOSTLY "PROTOZOA")	
Apicomplexa	Sporozoans
Ciliophora	Ciliates
Forminifera	Forams
Rhizopoda	Amoebas
Zoomastigina	Unicellular flagellates
Acrasiomycota★	Cellular slime molds
Myxomycota★	Plasomodial slime molds
Oomycota★	Water molds

★These were regarded as fungi in the past.

PROTISTAN PHYLA

We will consider 14 phyla of protists in alphabetical order to reflect the catchall nature of the kingdom Protista, emphasizing evolutionary relationships between phyla wherever appropriate. Table 31-1 groups them artificially into photosynthesizers (algae), heterotrophs (protozoa) and absorbers (fungi-like protists). Table 31-2 summarize some of the characteristics of these 14 phyla of protists in alphabetical order.

Before studying the protistan phyla, it is particularly important to review the general material in Chapter 28, such as the characteristics of different kinds of life cycles (Figure 28-17).

TABLE 31-2   FEATURES OF 14 PHYLA OF KINGDOM PROTISTA

NAME*	NUMBER OF SPECIES	CHLOROPHYLLS	WALL/SHELL	FLAGELLA
Acrasiomycota	65	None	Cellulose (spores)	0
Bacillariophyta	11,500	$a + c$	Opaline silica	0
Chlorophyta	7000	$a + b$	Cellulose†	2
Ciliophora	8000	None	None	Many
Dinoflagellata	1000	$a + b$	Mostly cellulose plates	2; Un‡
Euglenophyta	800	$a + b$ or none	Flexible pellicle	2; Un‡
Foraminifera	Hundreds	None	Shells	0 (podia)
Myxomycota	450	None	None	0 or 2
Oomycota	475	None	Cellulose†	2
Phaeophyta	1500	$a + c$	Cellulose†	2
Rhizopoda	Hundreds	None	Cysts, tests	0
Rhodophyta	4000	a	Cellulose†	0
Sporozoa	3900	None	Spores	Variable
Zoomastigina	Thousands	None	None	Many

*Phyla whose names are printed in italics consist of multicellular organisms; Chlorophyta also includes many unicellular ones, and Rhodophyta includes a few unicellular ones. Myxomycota moves about in a mass, called a plasmodium, in which there are many nuclei.
†These phyla have cellulose as the primary constituent in the cell walls of many or all members, but other substances are present, sometimes exclusively so in some members of these phyla.
‡Un, Unequal in length.

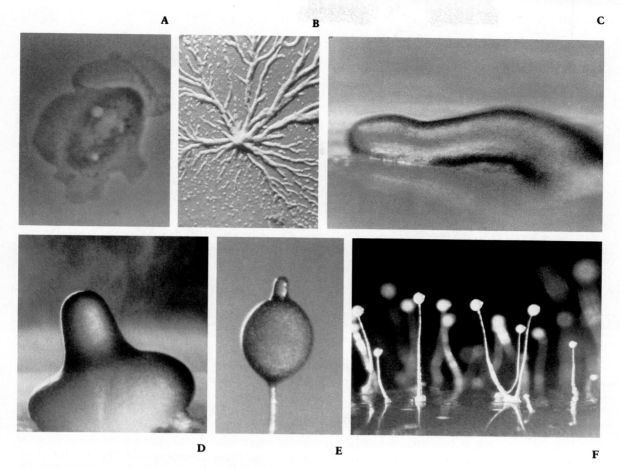

D E F

Acrasiomycota: The Cellular Slime Molds

There are about 70 species of cellular slime molds, a phylum with extraordinarily interesting features were once thought to be related to fungi, "mold" being a general term for funguslike organisms. In fact, the cellular slime molds are probably more closely related to the amoebas (phylum Rhizopoda) than to any other group, but they have many special features that mark them as distinct. Cellular slime molds are common in fresh water, in damp soil, and on rotting vegetation, especially on fallen logs. They have become one of the most important groups of organisms for studies of differentiation, because of their relatively simple developmental systems and the ease of analyzing them (Figure 31-5).

The individual organisms of this group behave as separate amoebas, moving through the soil or other substrate and ingesting bacteria and other smaller organisms. At a certain phase of their life cycle, the individual organisms aggregate and form a moving mass, the slug, that eventually transforms itself into a spore-containing mass, the **sorocarp,** in which the amoebas become encysted as spores. Some of the amoebas fuse sexually to form **macrocysts,** which have diploid nuclei; meiosis occurs in them after a short period (zygotic meiosis). Other amoebas are released directly, eventually aggregating again to form a new slug.

The development of *Dictyostelium discoideum,* a cellular slime mold, has been studied extensively because of the implications of its unusual life cycle for understanding the developmental process in general. When the individual amoebas of this species exhaust the supply of bacteria in a given area and are near starvation, they aggregate and form a compound, motile mass. The aggregation of the individual amoebas is induced by pulses of cyclic adenosine monophosphate (cAMP), which the cells begin to secrete when they are starving. In the new habitat the colony differentiates into a basal portion, a stalk, and a terminal swollen portion, within which spores differentiate. Each of these spores, if it falls into a suitably moist habitat, releases a new amoeba, which begins to feed, and the cycle is started again.

Individual amoebas of the cellular slime mold *Dictyostelium,* aggregate in response

FIGURE 31-5

Development in *Dictyostelium discoideum*, a cellular slime mold (phylum Acrasiomycota).

A Germinating spore, forming amoebas.

B The amoebas aggregate and move toward a fixed center.

C They form a multicellular slug 2 to 3 millimeters long that migrates toward light.

D The slug stops moving and begins to differentiate into a sorocarp.

E The differentiated head of a sorocarp.

F Many sorocarps together.

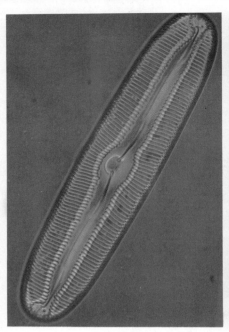

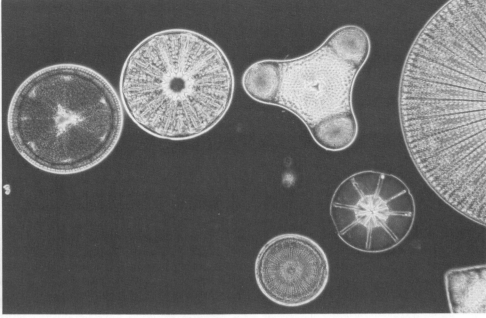

A   B

FIGURE 31-6
Diatoms (phylum Bacillario-phyta).
A A pennate (bilaterally symmetrical) diatom.
B Several different kinds of diatoms.

to cAMP released by starving cells and migrate as a mass to a new area. There they form a multicellular sorocarp, within which the spores differentiate.

Bacillariophyta: The Diatoms

Diatoms are photosynthetic, unicellular organisms with unique double shells, made of opaline silica, which are often strikingly and characteristically marked (Figure 31-6). The shells of diatoms are like small boxes with lids, one half of the shell fitting inside the other. Their chloroplasts, with chlorophylls *a* and *c*, as well as carotenoids, resemble those of the brown algae and dinoflagellates. Each of these groups probably acquired their chloroplasts independently from bacteria similar to *Heliobacterium chlorum*, and there are few other similarities between them.

There are more than 11,500 living species of diatoms, with many more known in the fossil record. The shells of fossil diatoms often form very thick deposits, which are sometimes mined commercially. The resulting "diatomaceous earth" is used as an abrasive or to add the sparkling quality to the paint used on roads, among other purposes. Living diatoms are often abundant both in the sea and in fresh water, where they are important food producers. Diatoms occur in the plankton and are attached to submerged objects in relatively shallow water. Many species are able to move by means of a secretion that is produced from a fine groove along each shell. The diatoms exude and perhaps also retract this secretion as they move.

There are two major groups of diatoms, one with radial symmetry (like a wheel) and the other with bilateral (two-sided) symmetry. Diatom shells are rigid, and the organisms reproduce asexually by separating the two halves of the shell, each half then regenerating another half shell within it. Because of this mode of reproduction, there is a tendency for the shells and consequently the individual diatoms to get smaller and smaller with a given sequence of asexual reproduction. When the resulting individuals have diminished to about 30% of their original size, one may slip out of its shell, grow to full size, and regenerate a full-sized pair of new shells.

Individual diatoms are diploid. Meiosis occurs more frequently under conditions of starvation. Among marine diatoms, some produce numerous sperm and others a single egg. If fusion occurs, the resulting zygote regenerates a full-sized individual. In some freshwater diatoms, the gametes are amoeboid, and all of them appear similar.

> **Diatoms are unicellular, photosynthetic organisms with chloroplasts that resemble those of the brown algae. They have unique double shells made of opaline silica. Diatoms are abundant both in the sea and in fresh water.**

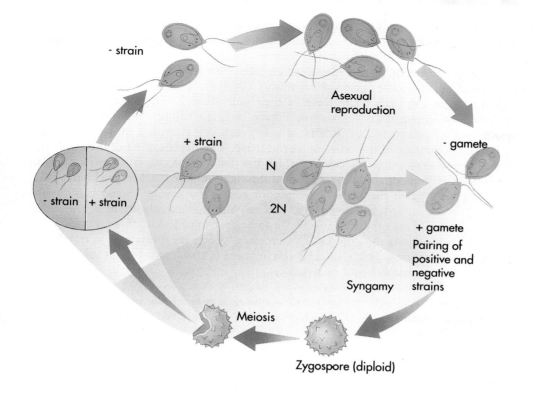

FIGURE 31-7
Life cycle of *Chlamydomonas* (phylum Chlorophyta). Individual cells of this microscopic, biflagellated alga, which are haploid, divide asexually, producing identical copies of themselves. At times, such haploid cells act as gametes—fusing, as shown at the lower right-hand side of the diagram, to produce a zygote. The zygote develops a thick, resistant wall, becoming a zygospore; this is the only diploid cell in the entire life cycle. Within this diploid zygospore, meiosis takes place, ultimately resulting in the release of four haploid individuals. Because of the segregation during meiosis, two of these individuals are what are called the + strain (shown in green), and the other two are the − strain (uncolored). Only + and − individuals are capable of mating with one another when syngamy does take place, although both may divide asexually and reproduce themselves in that way also.

Chlorophyta: The Green Algae

Green algae are of special interest, both because of their unusual diversity and because their characteristics indicate that the ancestor of the plants was a member of this group. Their chloroplasts are biochemically similar to those of the plants; they contain chlorophylls *a* and *b,* as well as carotenoids, and are similar to the bacteria *Prochloron* and *Prochlorothrix* in these respects. It seems probable that a bacterium with these characteristics originally gave rise to the chloroplasts of Chlorophyta, and independently to those of the Euglenophyta—a group that differs from Chlorophyta in nearly every other characteristic, as we will see.

Green algae are an extremely varied group of more than 7000 species, mostly aquatic but also semiterrestrial in moist places, such as on tree trunks or in soil. Many of these algae are microscopic and unicellular, but some, such as sea lettuce, *Ulva* (see Figure 28-11, *B*), are tens of centimeters across and easily visible on rocks and pilings around the coasts.

Among the unicellular green algae, *Chlamydomonas* (Figure 31-7) is a well-known genus. Individuals are microscopic (usually less than 25 micrometers long), green, and rounded and have two flagella at the anterior end. They move rapidly in water as a result of the beating of their flagella in opposite directions. Each individual has an eyespot, which contains about 100,000 molecules of rhodopsin, the same pigment that is used in the vision of vertebrates. Light received by this eyespot is used by the alga to help direct its swimming. Most individuals of *Chlamydomonas* are haploid. *Chlamydomonas* reproduces asexually by cell division and sexually by the functioning of some of the products of cell division as gametes; these gametes fuse to form a four-flagellated zygote that ultimately enters a resting phase in which the flagella disappear. Meiosis occurs at the end of this resting period and results in the production of four haploid cells.

Several lines of evolutionary specialization have been derived from organisms like *Chlamydomonas.* The first is the evolution of nonmotile, unicellular green algae. Even *Chlamydomonas* itself is capable of retracting its flagella and settling down as an immobile, unicellular organism if the ponds in which it lives dry out. Some common algae of soil and bark, such as *Chlorella,* are essentially like *Chlamydomonas* in this condition but do not have the ability to form flagella. *Chlorella* is widespread in both fresh and salt water and in soil and is only known to reproduce asexually. Recently,

Chlorella has been widely investigated as a possible food source for humans and other animals, and pilot farms have been established in Israel, the United States, Germany, and Japan.

The second major line of specialization from cells like *Chlamydomonas* concerns the formation of motile, colonial organisms. In these genera of green algae, the *Chlamydomonas*-like cells retain some of their individuality. The most highly elaborated of these organisms is *Volvox* (see Figure 28-12), a hollow sphere made up of a single layer of 500 to 60,000 individual cells, each cell provided with two flagella. Only a small number of the cells are reproductive. The flagella of all of the cells beat in such a way as to rotate the colony, which has definite anterior and posterior ends, in a clockwise direction as it moves forward through the water. The reproductive cells of *Volvox* are located mainly at the posterior end of the colony. Some may divide asexually, bulge inward, and give rise to new colonies that initially are held within the parent colony. Others produce gametes. In some species of *Volvox* there is a true "division of labor" among the different types of cells, which are specialized in relation to their ultimate function throughout the development of the organism; in these species, therefore, true multicellularity exists.

> From green algae like *Chlamydomonas*—a biflagellated, unicellular
> organism—have been derived nonmotile, unicellular algae and multicellular,
> flagellated colonies.

In addition to these two lines of specialization from *Chlamydomonas*-like cells, there are many other kinds of green algae of less certain derivation. Many filamentous genera, such as *Spirogyra*, with its ribbonlike chloroplasts (see Figure 28-10, *B*), differ substantially from the remainder of the green algae in their modes of cell division and reproduction. Some of these genera have even been placed in separate phyla. The study of the green algae, involving modern methods of electron microscopy and biochemistry, is beginning to reveal unexpected new relationships with this phylum.

Ulva, or sea lettuce (see Figure 28-11, *B*), is a genus of marine green algae that is extremely widespread. The glistening individuals of this genus, often more than 10 centimeters across, consist of undulating sheets only two cells thick. Sea lettuce attaches by protuberances of the basal cells to rocks or other substrates on which it occurs. *Ulva* has an alternation of generations (sporic meiosis; see Figure 28-13) in which the gametophytes and sporophytes resemble one another closely.

The stoneworts, a group of about 250 living species of green algae, many of them in the genera *Chara* and *Nitella*, have complex structures (Figure 31-8). Whorls of short branches arise regularly at their nodes, and the **gametangia** (singular, **gametangium**)—gamete-producing structures—are complex and multicellular. Stoneworts are often abundant in fresh to brackish water and are common as fossils.

FIGURE 31-8
Stoneworts (phylom *Chlorophyta*). In the stoneworts, as seen in this view of *Chara*, the egg-bearing structures *(right)* and the sperm-bearing ones *(left)* are multicellular and complex. Stoneworts also have a complex growth pattern, with nodal regions from which whorls of branches arise, and hollow, multinucleate internodal regions.

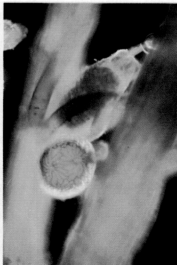

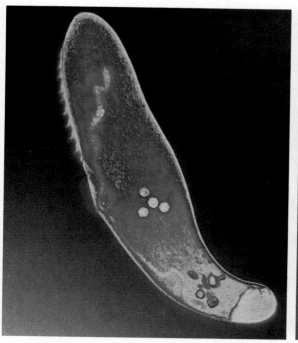

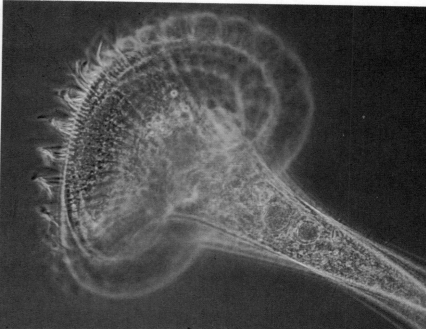

A

B

FIGURE 31-9
Ciliates (phylum Ciliophora).
A *Blepharisma japonica*, showing micronuclei (small round objects) and a macronucleus (elongated structure within cell).
B *Stentor*, a funnel-shaped ciliate, showing spirally arranged cilia.

Ciliophora: The Ciliates

As the name indicates, most members of the Ciliophora feature large numbers of cilia. These heterotrophic, unicellular protists range in size from about 10 to about 3,000 micrometers long (see Figures 31-1 and 31-9). About 8,000 species have been named. Despite their unicellularity, ciliates are extremely complex organisms, inspiring some biologists to consider them organisms without cell boundaries rather than single cells.

Their most characteristic feature, cilia, are usually arranged either in longitudinal rows or in spirals around the body of the organism (Figure 31-9, *B*). Cilia are embedded in an outer proteinaceous layer of the cell, and they beat in a coordinated fashion. In some groups the cilia have specialized locomotory and feeding functions, becoming fused into sheets, spikes and rods which may then function as mouths, paddles, teeth, or feet. The body wall of ciliates is a tough but flexible outer **pellicle** that enables the organism to squeeze through or move around many kinds of obstacles. The pellicle consists of an outer membrane with numerous fluid-filled cavities beneath it.

Ciliates form vacuoles for ingesting food and regulating their water balance (Figure 31-10). Food first enters the gullet, which in the well-known ciliate *Paramecium* is

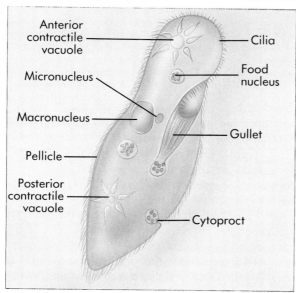

Anterior contractile vacuole

Micronucleus

Macronucleus

Pellicle

Posterior contractile vacuole

Cilia

Food nucleus

Gullet

Cytoproct

FIGURE 31-10
Diagram of *Paramecium*.

Giardia lamblia is found throughout the world, including all parts of the United States and Canada. It occurs in water, including the clear water of mountain streams and the water supplies of some cities, infecting at least 40 species of wild and domesticated animals in addition to humans. In 1984 in Philadelphia, for example, 175,000 people had to boil their drinking water for several days following the appearance of *Giardia* in the city's water system. Although most individuals exhibit no symptoms if they drink water infested with *Giardia*, many suffer nausea, cramps, bloating, vomiting, and diarrhea. Only 35 years ago, *Giardia* was thought to be harmless; today, it is estimated that at least 16 million residents of the United States alone are infected by it.

Giardia lives in the upper small intestine of its host; it occurs there in a motile form that cannot survive outside the host's body. It is spread in the feces of infected individuals in the form of dormant, football-shaped cysts—sometimes at levels as high as 300 million individuals per gram of stool. These cysts can survive at least 2 months in cool water, such as that of mountain streams. They are relatively resistant to the usual water-treatment agents such as chlorine and iodine but are killed at tem-

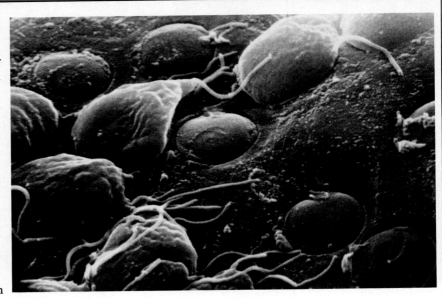

FIGURE 31-A
Giardia lamblia.

peratures greater than about 65° C. Apparently, pollution by humans seems to be the main way *Giardia* is released into stream water. There are at least three species of *Giardia* and many distinct strains; how many of them attack humans and under what circumstances are not known with certainty.

In the wilderness, good sanitation is important in preventing the spread of *Giardia*. Dogs, which readily contract and spread the dis-

ease, should not be taken into pristine wilderness areas. Drinking water should be filtered—the filter must be capable of eliminating particles as small as 1 micrometer in diameter—or boiled for at least 1 minute; water from natural streams or lakes should never be consumed directly, regardless of how clean it looks. In other regions, good sanitation methods are important to prevent not only *Giardia* but also other diseases.

lined with cilia fused into a membrane. From the gullet, the food passes into food vacuoles, where enzymes and hydrochloric acid aid in its digestion. After absorption of the digested material has been completed, the vacuole empties its waste contents in a special region of the pellicle known as the **cytoproct.** The cytoproct appears periodically when solid particles are ready to be expelled. The contractile vacuoles, which function in the regulation of water balance, periodically expand and contract as they empty their contents to the outside of the organism.

Most ciliates that have been studied have two very different types of nuclei within their cells, small **micronuclei** and larger **macronuclei.** The micronuclei, which contain apparently normal chromosomes, divide by mitosis. Macronuclei are derived from certain micronuclei immediately after fertilization by a complex series of steps. Within the macronuclei, the DNA is divided into small pieces—smaller than individual chromosomes; in one group of ciliates, these are equivalent to single genes. Ma-

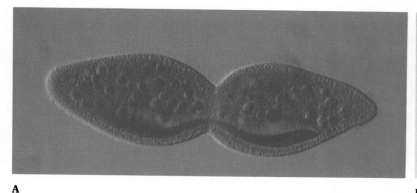

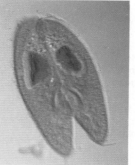

A   B

FIGURE 31-11
Reproduction among paramecia.
When a mature *Paramecium* divides (**A**), two complete individuals result. When two mature cells fuse (**B**), the process is called conjugation.

cronuclei divide by elongating and constricting and play an essential role in routine cellular functions, such as the production of mRNA to direct protein synthesis.

In addition to their characteristic cilia, most ciliates have two types of nucleus. The single macronucleus contain multiple copies of certain genes. The many micronuclei contain multi-gene chromosomes.

Ciliates usually reproduce by transverse fission of the parent cell across its short axis, thus forming two equal offspring (Figure 31-11, *A*). In this process of cell division, the mitosis of the micronuclei proceeds normally and the macronuclei divide as just described.

In *Paramecium*, the cells divide asexually for about 700 generations and then die if sexual reproduction has not occurred. Like most ciliates, however, *Paramecium* has a sexual process called **conjugation** (Figure 31-11, *B*), in which the individual cells remain attached to one another for up to several hours. Cells of two different genetically determined mating types, *odd* and *even*, are able to conjugate. Meiosis in the micronuclei of each individual produces several haploid micronuclei, and a pair of these is then exchanged through a cytoplasmic bridge that appears between the two partners.

In each conjugating individual, the new micronucleus fuses with one of the micronuclei that was already present in that individual, resulting in the production of a new diploid micronucleus in each individual. After conjugation, the macronucleus in each cell disintegrates, while the new diploid micronucleus undergoes mitosis, thus giving rise to two new identical diploid micronuclei within each individual. One of these micronuclei becomes the precursor of the future micronuclei of that cell, while the other micronucleus undergoes multiple rounds of DNA replication, becoming the new macronucleus. This kind of complete segregation of the genetic material is a unique feature of the ciliates and makes them ideal organisms for the study of certain aspects of genetics.

Progeny from a sexual division in *Paramecium* must go through about 50 asexual divisions before they are able to conjugate. When they do so, their biological clocks are restarted, and they can conjugate again. After about 600 asexual divisions, however, *Paramecium* loses the protein molecules around the gullet that enable it to recognize an appropriate mating partner. As a result, the individuals are unable to mate, and death follows about 100 generations later. The exact mechanisms by which these unusual events occur are unknown, but they involve the accumulation of a protein, which is being studied in detail now.

Conjugating ciliates exchange a pair of haploid micronuclei, which fuse in both cells to form new diploid micronuclei, while the macronucleus in each disintegrates. After their fusion, the new micronuclei divide by mitosis. One of the products of this division gives rise to more micronuclei while the other undergoes multiple rounds of DNA replication and becomes the new macronucleus.

FIGURE 31-12

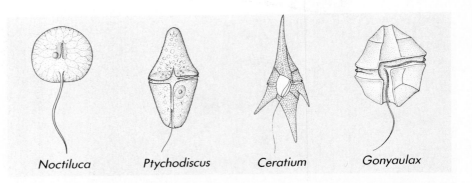

Noctiluca   Ptychodiscus   Ceratium   Gonyaulax

FIGURE 31-12

The Dinoflagellates. Noctiluca, Ptychodiscus, Ceratium, and *Gonyaulax.* *Noctiluca,* which lacks the heavy cellulosic armor characteristic of most dinoflagellates, is one of the bioluminescent organisms that cause the waves to sparkle in warm seas at certain times of year. In the other three genera, the shorter, encircling flagellum may be seen in its groove, with the longer one projecting away from the body of the dinoflagellate.

Dinoflagellata: The Dinoflagellates

The dinoflagellates consist of about 2100 known species of unicellular, photosynthetic organisms, most of which have two flagella. A majority of the dinoflagellates are marine, where they are often abundant in the plankton, but some occur in fresh water. Some of the planktonic dinoflagellates are luminous and contribute to the twinkling or flashing effects that sometimes are seen in the sea at night, especially in the tropics.

The flagella, protective coats, and biochemistry of the dinoflagellates are distinctive, and they do not appear to be directly related to any other phylum. Their flagella beat in two grooves, one encircling the body like a belt, and the other perpendicular to it. By beating in their respective grooves, these flagella cause the dinoflagellate to rotate like a top as it moves. Many dinoflagellates are clad in stiff cellulose plates, often encrusted with silica, which give them a very unusual appearance (Figure 31-12). Most have chlorophylls *a* and *c*, in addition to carotenoids, so that in the biochemistry of their chloroplasts, they resemble the diatoms and the brown algae. The members of each of these groups probably acquired their chloroplasts as a result of independent symbiotic events, although it is just possible that they might actually be related.

Some dinoflagellates are capable of ingesting other cells, and a number are colorless and heterotrophic. Other dinoflagellates occur as symbionts in many other groups of organisms, including jellyfish, sea anemones, mollusks (see Figure 40-5), and, notably, corals (Figure 31-13, *A*). When dinoflagellates grow as symbionts within other cells, they lack their characteristic cellulose plates and flagella, appearing as spherical, golden-brown globules in their host cells. In such a state they are called **zooxanthellae** (Figure 31-13, *B* to *D*). Zooxanthellae are the primary factor responsible for the productivity of corals and their ability to grow in tropical waters, which are often extremely low in nutrients. Most of the carbon that the zooxanthellae fix is translocated to the host corals, which can obtain all of the fixed carbon that they need from zooxanthellae if they are growing in well-lighted places.

> **Zooxanthellae, the symbiotic form of dinoflagellates, are primarily responsible for the productivity of coral reefs, which usually occur in waters that are poor in nutrients. The zooxanthellae live within the bodies of the coral animals.**

The poisonous and destructive "red tides" that occur frequently in coastal areas are often associated with great population explosions, or "blooms," of dinoflagellates (Figure 31-14). The pigments in the individual, microscopic cells of the dinoflagellates, or, in some cases, other organisms are responsible for the color of the water. Such red tides have a profound, detrimental effect on the fishing industry in the United States. Approximately 20 species of dinoflagellates are known to produce powerful toxins, most poorly known chemically, which inhibit the diaphragm and cause respiratory failure in many vertebrates. When the toxic dinoflagellates are abundant, fishes, birds, and marine mammals may die in large numbers. In addition, toxic dinoflagellates are accumulated by shellfish (mollusks), which strain them out of the water. The mollusks themselves are not killed, but they become poisonous to humans and other animals that consume them. Unfortunately, relatively little can be done to

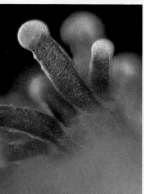

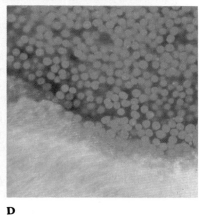

A

C

D

FIGURE 31-13
Zooxanthellae, the primary factor in the productivity of corals.
A A coral reef—a complex, highly productive community in tropical and subtropical waters—fringes each of these small islands off the coast of Java in this aerial view.
B In their symbiotic form, dinoflagellates, then known as zooxanthellae, contribute most of the productivity to the well-illuminated portions of coral reefs. Each of the tentacles of this coral animal is packed with golden-brown zooxanthellae.
C Individuals of the sea anemone *Anthopleura elegantissima*, freshly collected from the intertidal zone on the Pacific coast of the United States. The brown-green pigmentation of these animals is caused almost entirely by pigments in the zooxanthellae; animals of the same species that lack zooxanthellae are milky pink.
D Densely packed zooxanthellae in tissue of the anemone shown in **C** This photograph was taken with filters that make the zooxanthellae fluoresce red, while the tissue of the anemone fluoresces yellowish or bluish green.

reduce the effects of such outbreaks of poisonous dinoflagellates other than to keep humans from eating the contaminated seafood.

Dinoflagellates reproduce primarily by longitudinal cell division, but sexual reproduction has also been shown in more than 10 genera. Their form of mitosis is unique: the chromosomes, which are almost entirely DNA and are permanently condensed, remain within the nucleus but are distributed along the sides of channels containing bundles of microtubules that run through the nucleus.

Mitosis in dinoflagellates is unique and takes place within the nucleus. The dinoflagellates do not appear to have any close relatives.

FIGURE 31-14
Red tide. Red tides are caused by population explosions of dinoflagellates. The pigments in the dinoflagellates or, in some cases, other organisms, are responsible for the color of the water.

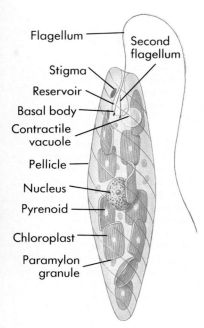

FIGURE 31-15
Diagram of an individual of the genus *Euglena* (phylum Euglenophyta). Starch forms around pyrenoids; paramylon granules are areas where food reserves are stored.

Labels (top to bottom):
Flagellum
Second flagellum
Stigma
Reservoir
Basal body
Contractile vacuole
Pellicle
Nucleus
Pyrenoid
Chloroplast
Paramylon granule

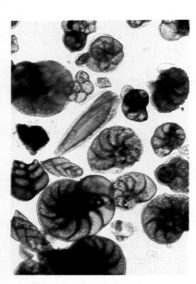

FIGURE 31-16
The calcareous tests of some representative foraminifera.

Euglenophyta: The Euglenoids

Most of the approximately 1000 known species of euglenoids live in fresh water. The members of this phylum clearly illustrate the impossibility of distinguishing "plants" from "animals" among the protists. About a third of the approximately 40 genera of euglenoids have chloroplasts and are fully autotrophic; the others lack chloroplasts, ingest their food, and are heterotrophic. These organisms are not significantly different from some groups of zoomastigotes (see p. 632), and many biologists believe that the two phyla should be merged into one.

Some of the euglenoids that have chloroplasts may also apparently lose them and thus become heterotrophic if the organisms are kept in the dark; the chloroplasts become small and nonfunctional. If they are put back in the light, they may become green within a few hours. In one species of euglenoid, the chloroplasts may be removed permanently by treatment with antibiotics or ultraviolet light. Even without such drastic treatments, the normally photosynthetic euglenoids may sometimes feed on dissolved or particulate food.

Individual euglenoids range from only 10 to 500 micrometers long and are highly variable in form. Interlocking proteinaceous strips arranged in a helical pattern form a flexible structure called the pellicle, which lies within the cell membrane of the euglenoids. Since its pellicle is flexible, a euglenoid is able to change its shape. Reproduction in this phylum occurs by mitotic cell division. The nuclear envelope remains intact throughout the process of mitosis. No sexual reproduction is known to occur in this group.

In *Euglena* (Figure 31-15), the genus for which the phylum is named, two flagella are attached at the base of a flask-shaped opening called the **reservoir,** which is located at the anterior end of the cell. The **stigma,** an organ that also occurs in the green algae (phylum Chlorophyta), is light sensitive and aids these photosynthetic organisms to move toward light. Contractile vacuoles (see Figure 6-14) collect excess water from all parts of the organism and empty it into the reservoir, which apparently functions in regulating the osmotic pressure within the organism. One of the flagella is long and has a row of very fine, short, hairlike projections along one side. A second, shorter flagellum is located within the reservoir but does not emerge from it.

Cells of *Euglena* contain numerous small chloroplasts. These chloroplasts, like those of the green algae and plants, contain chlorophylls a and *b*, together with carotenoids. Although the chloroplasts of the euglenoids differ somewhat in structure from those of the green algae, they probably had a common origin; there is clearly no direct relationship between the two phyla, however, as indicated by the fact that they differ in virtually every other characteristic.

> **Euglenoids (phylum Euglenophyta) consist of about 40 genera, about a third of which have chloroplasts similar biochemically to those of green algae and plants. Euglenoids probably acquired these chloroplasts independently, because their structure closely resembles that of certain zoomastigotes and is very different from green algae and plants.**

Foraminifera: Forams

Members of the phylum Foraminifera are heterotrophic, marine protists. They range in diameter from about 20 micrometers to several centimeters. Characteristic of the group are pore-studded shells (called **tests**) composed of organic materials usually reinforced with grains of inorganic matter (Figure 31-16). These grains may be calcium carbonate or sand or even plates from the shells of echinoderms or spicules (minute needles of calcium carbonate) from sponge skeletons. Depending on the building materials they use, foraminifera—often informally called "forams"—may have shells of very different appearance. Some of them are brilliantly colored—red, salmon, or yellow-brown.

Most foraminifera live in sand or are attached to other organisms, but two families consist of free-swimming, planktonic organisms. Their tests may be single chambered but more often are multichambered; they tend to have a spiral shape resembling that of a tiny snail. Thin cytoplasmic projections called **podia** emerge through openings in the tests (Figure 31-17); the podia are used for swimming, gathering materials for the tests, and feeding. Forams eat a wide variety of small organisms.

The life cycles of foraminifera are extremely complex, involving an alternation between haploid and diploid generations (sporic meiosis). Forams have contributed massive accumulations of their tests to the fossil record for more than 200 million years. Because of the excellent preservation of their tests and the often striking differences among them, forams are very important as geological markers. The pattern of occurrence of different forams, for example, is often used as a guide in searching for oil-bearing strata. Limestones all over the world are often rich in forams, and the White Cliffs of Dover, the famous landmark of southern England, are made up almost entirely of their tests (Figure 31-18).

The pore-studded tests, or shells, of the forams are characteristic of this group of unicellular protists. Their life cycles involve a complex alternation of generations.

Myxomycota: The Plasmodial Slime Molds

Plasmodial slime molds are a group of about 500 species. These bizarre organisms stream along as a **plasmodium,** a nonwalled, multinucleate mass of cytoplasm, which resembles a moving mass of slime (see Figure 31-1, *C*). The plasmodia may be orange, yellow, or another color. Plasmodia show a back-and-forth streaming of the cytoplasm, which is very conspicuous, especially under a microscope. They are able to pass through the mesh in cloth or simply to flow around or through other obstacles. As they move, they engulf and digest bacteria, yeasts, and other small particles of organic matter. Plasmodia contain many nuclei, but these are not separated by cell walls; nuclei undergo mitosis synchronously, with the nuclear envelope breaking down, but only at late anaphase or telophase. Centrioles are lacking in cellular slime molds. Although they have similar common names, there is no strong evidence that the plasmodial slime molds are related to the cellular slime molds; they actually differ in most features of their structure and life cycles.

When either food or moisture is in short supply, the plasmodium migrates relatively rapidly to a new area. Here it stops moving and either forms a mass in which spores differentiate or divides into a large number of small mounds, each of which

FIGURE 31-17
A representative of phylum Foraminifera. A living foram, showing the podia, thin cytoplasmic projections that extend through pores in the calcareous test, or shell, of the organism.

FIGURE 31-18
White Cliffs of Dover. The limestone that forms these cliffs is composed almost entirely of the tests of foraminifera.

A

B

C

FIGURE 31-19

Sporangia of three genera of plasmodial slime molds (phylum Myxomycota).
A *Arcyria.*
B *Fuligo.*
C Developing sporangia of *Tubifera.*

produces a single, mature sporangium, the structure in which spores are produced. These sporangia are often extremely complex in form and beautiful (Figure 31-19). These spores are either diploid or haploid, depending on the condition found in the nuclei within the plasmodium. In most species of plasmodial slime molds in which the plasmodium is diploid, meiosis occurs in the spores within 24 hours of their formation. Three of the four nuclei in each spore disintegrate, leaving each spore with a single haploid nucleus.

The spores are highly resistant to unfavorable environmental influences and may last for years if dry. When conditions are favorable, they split open and release their protoplast, the contents of the individual spore; the protoplast may be amoeboid or bear two flagella. These two stages appear to be interchangeable, and conversions in either direction occur readily. Later, after the fusion of haploid protoplasts (gametes), a usually diploid plasmodium may be reconstituted by repeated mitotic divisions.

> **The feeding phase of plasmodial slime molds consists of a multinucleate mass of protoplasm; a plasmodium can flow through a silk mesh and rejoin. If the plasmodium begins to dry out or is starving, it forms often elaborate sporangia. Meiosis occurs in the spores once they have been cleaved within the sporangium.**

Oomycota

The oomycetes comprise about 580 species, among them the water molds, white rusts, and downy mildews. All of the members of the group are either parasites or saprobes (organisms that live by feeding on dead organic matter). The cell walls of the oomycetes are composed of cellulose or polymers that resemble cellulose; these walls therefore differ remarkably from the chitin cell walls of the fungi, with which the oomycetes have at times been grouped. They also differ from fungi in that their life cycles are characterized by gametic meiosis resulting in a diploid phase. Further, mitosis in the oomycetes resembles that in most other organisms, whereas mitosis in fungi has a number of unusual features, as you will see in Chapter 32. Filamentous structures of fungi and, by convention, those of oomycetes also are called **hyphae.** Most oomycetes live in fresh or salt water or in soil, but some are plant parasites that depend on the wind to spread their spores. A few aquatic oomycetes are animal parasites.

Oomycetes are distinguished from other protists by the structure of their motile spores, or **zoospores,** which bear two unequal flagella, one of which is directed forward, the other backward. Such zoospores are produced asexually in a sporangium. Sexual reproduction in the group involves gametangia of two different kinds. The female gametangium is called an **oogonium,** and the male gametangium is called an **antheridium.** The antheridia contain numerous male nuclei, which are the functional male gametes; the oogonia contain from one to eight eggs, which are the female gametes. The flowing of the contents of an antheridium into an oogonium, which leads to the individual fusion of one or more pairs of male nuclei with eggs. This is followed by the thickening of the cell wall around the resulting zygote or zygotes. This

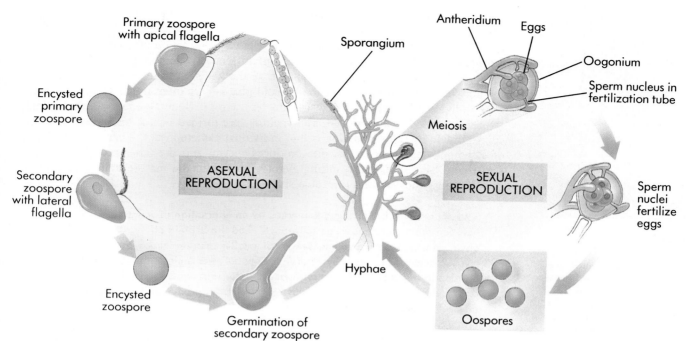

FIGURE 31-20
Life cycle of *Saprolegnia*, an oomycete. Asexual reproduction by means of flagellated zoospores is shown at left, sexual reproduction at right. Hyphae with diploid nuclei are produced by germination of both zoospores and oospores.

produces a special kind of thick-walled cell called an **oospore,** the structure that gives the phylum its name. Details from the life cycle of one of the oomycetes, *Saprolegnia,* are shown in Figure 31-20.

The aquatic oomycetes, or water molds, are common and easily cultured. Some water molds cause fish diseases, which may be seen as a kind of white fuzz on aquarium fishes. Among their terrestrial relatives are such oomycetes of great importance as plant pathogens, including *Plasmopara viticola,* which causes downy mildew of grapes, and *Phytophthora infestans,* which causes the late blight of potatoes. This oomycete was responsible for the Irish potato famine of 1845 and 1847, during which about 400,000 people starved to death or died of diseases complicated by starvation. Millions of Irish people emigrated to the United States and elsewhere as a result of this disaster.

Phaeophyta: The Brown Algae

While the kingdom Protista contains the smallest eukaryotes, it also contains some of the longest, fastest growing and most photosynthetically productive living things. The phaeophyta, or brown algae, consists of about 1,500 species of multicellular protists, almost exclusively marine. They are the most conspicuous seaweeds in many northern regions, dominating rocky shores almost everywhere in temperate North America and Eurasia. In the habitats were the larger brown algae, and in particular a group known as the **kelps** (order Laminariales), occur abundantly, they are responsible for most of the food production through photosynthesis. Many kelps are conspicuously differentiated into flattened blades, stalks, and grasping basal portions that anchor them to the rocks. The organic matter that kelp produces supports the myriad invertebrates, fishes, marine mammals and birds that live among these kelp forests (Figure 31-21).

Among the larger brown algae are genera such as *Macrocystis,* in which some individuals may reach 100 meters in length (see Figure 31-1, *A*). The flattened blades of this kelp float out on the surface of the water, while the base is anchored tens of meters below the surface. Another ecologically important member of this phylum is sargasso weed, *Sargassum,* which forms huge floating masses that dominate the vast Sargasso Sea, an area of the Atlantic Ocean northeast of the Caribbean. The stalks of the larger brown algae often exhibit a complex internal differentiation of conducting tissues analogous to that of plants. The chloroplasts of brown algae resemble those of

FIGURE 31-21
Brown algae (phylum Phaeophyta). The massive "groves" of giant kelp that occur in relatively shallow water along the coasts of the world provide food and shelter for many different kinds of organisms.

diatoms and dinoflagellates in having chlorophylls *a* and *c*. All of these chloroplasts were probably derived from a symbiotic bacterium that had features in common with *Heliobacterium*.

The life cycle of the brown algae is marked by alternation of generations between a diploid phase, or **sporophyte**, and a haploid phase, or **gametophyte**. The large individuals that we recognize, for example, as the kelps, are the sporophytes. The gametophytes—the individuals of the haploid phase—are often much smaller, filamentous individuals, perhaps a few centimeters across. Sporangia, in which haploid, swimming spores are produced after meiosis, are formed on the sporophytes. These spores divide by mitosis, giving rise to individual gametophytes. There are two kinds of gametophytes in the kelps; one produces sperm, and the other produces eggs. If sperm and eggs fuse, the resulting zygotes grow into the mature kelp sporophytes, provided that they reach a favorable site.

> The life cycle of brown algae is marked by an alternation of generations between the diploid phase, or sporophyte, and the haploid phase, or gametophyte, in which the individuals are much smaller than the sporophytes in many species. The sporophytes produce spores after meiosis, and the spores develop into gametophytes by mitosis. Gametes (eggs and sperm) are borne on two different kinds of gametophytes. They fuse, producing a zygote, the first cell of the sporophyte generation.

Some of the larger kelps form extensive beds that are harvested commercially for sodium and potassium salts, iodine, and alginates (carbohydrates that are used in the formation of gels). The possibility of using kelp, which grows continuously and produces large amounts of organic material, as an inexpensive source of fuel is also being investigated.

Rhizopoda: The Amoebas

The hundreds of species of amoebas are found throughout the world in both fresh and salt waters (Figure 31-22). They are also abundant in soil. Many kinds of amoebas are parasites of animals. Reproduction in amoebas occurs by fission, or the direct division into two cells of equal volume. Amoebas of the phylum Rhizopoda lack cell walls, flagella, meiosis, and any form of sexuality. They do undergo mitosis and have a spindle apparatus that resembles that of most other eukaryotes. *Actinosphaerium* (Figure 31-23) is an unusual kind of amoeba that belongs to another phylum, Heliozoa, which is not discussed further in this book.

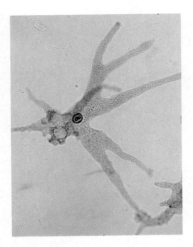

FIGURE 31-22
Amoeba proteus. The relatively large amoeba is commonly used in teaching and for research in cell biology. The projections are pseudopods; an amoeba moves by flowing into them. The nucleus of the amoeba is plainly visible.

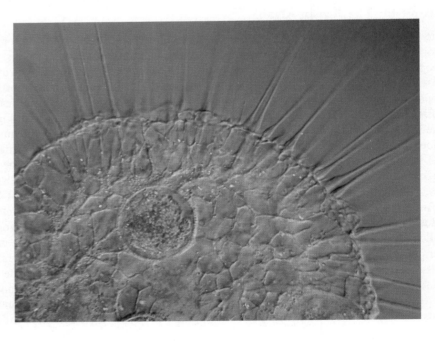

FIGURE 31-23
Actinosphaerium, an unusual amoeba of the phylum Heliozoa.

Amoebas move from place to place by means of their **pseudopods,** from the Greek words for "false" and "foot" (see Figure 5-24). The pseudopods are flowing projections of the cytoplasm that extend and pull the amoeba forward or engulf food particles. An amoeba puts a pseudopod forward and then flows into it. Actin and myosin proteins of microfilaments similar to those found in muscles are associated with these movements.

Some kinds of amoebas form resistant cysts. In parasitic species such as *Entamoeba histolytica*, which causes amoebic dysentery, the cysts enable the amoebas to resist digestion by their animal hosts, which include humans (Figure 31-24). Mitotic division takes place within the cysts, which may then ultimately break and release four, eight, or even more amoebas within the digestive tracts of their host animals. The primary infection takes place in the intestine, but it often moves into the liver and sometimes into other parts of the body. The cysts are dispersed in feces and may be transmitted from person to person in infected food and water, by flies, and by direct contact. It is estimated that up to 10 million people in the United States have infections of parasitic amoebas, and some 2 million show symptoms of the disease, ranging from abdominal discomfort with slight diarrhea to much more serious conditions. In some tropical areas, more than half of the people may be infected. The spread of amoebic dysentery can be limited by proper sanitation and hygiene.

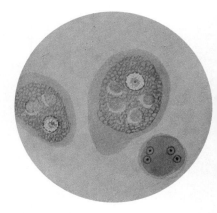

FIGURE 31-24
Entamoeba histolytica, **the cause of amoebic dysentery.** At left are two amoebas within resistant cysts.

Amoebas are unicellular, heterotrophic protists that lack cell walls, flagella, meiosis, and sexuality; they move from place to place by means of extensions called pseudopods.

Rhodophyta: The Red Algae

Along with the brown algae and the green algae, the red algae are the seaweeds that we see along shores and cast up on beaches. Their characteristic colors result from their possession of the same pigments, phycobilins, that are responsible for the colors of the cyanobacteria. Chlorophyll *a* also occurs with the phycobilins, just as it does in cyanobacteria. Undoubtedly cyanobacteria became symbiotic in the cells of the heterotrophic ancestors of the red algae and gave rise to their chloroplasts.

Almost all red algae are multicellular. The great majority of the estimated 4000 species occur in the sea, where they are common (see Figures 28-11, *C*, and 31-25). In warm waters, red algae are more common than brown algae. Phycobilins are especially efficient in absorbing the green, violet, and blue light that penetrates into the deepest waters. For this reason red algae are able to grow at greater depths than are brown algae or green algae. In 1985 a red alga was reported growing attached to rocks 268 meters below the surface of the sea in the Bahamas, a record for any photosynthetic organism.

Red algae have complex bodies made up of interwoven filaments of cells. Some, the coralline algae (see Figure 31-25), deposit calcium carbonate in their cell walls, which otherwise are mostly cellulose. In the cell walls of other red algae there is a mucilaginous outer component usually composed of sulfated polysaccharides such as agar and carrageenan, which make these algae important economically. Agar is used to make gelatin capsules, as a material for making dental impressions, and as a base for cosmetics. It is also the basis of the laboratory media on which bacteria, fungi, and other organisms are often grown. In addition, agar is used to prevent baked goods from drying out, for rapid-setting jellies, and as a temporary preservative for meat and fish in warm regions. Carrageenan is used mainly for the stabilization of emulsions such as paints, cosmetics, and dairy products such as ice cream. In addition to these uses, red algae such as *Porphyra*, called "nori," are eaten and, in Japan, are even cultivated as a food crop for human consumption.

The life cycles of red algae are complex but usually involve an alternation of generations (sporic meiosis). None of the red algae have flagella or cilia at any stage in their life cycle, and they may have descended directly from ancestors that never had them, especially since the red algae also lack centrioles. Together with the fungi, which also lack flagella and centrioles, red algae may be one of the most ancient groups of eukaryotes.

FIGURE 31-25
A coralline alga (phylum Rhodophyta). In the coralline algae, the cellulose cell walls are heavily impregnated with calcium carbonate. Other species of coralline algae contribute greatly to overall food production in coral reefs, but members of the phylum, such as the one shown here, also occur widely elsewhere.

Sporozoa: The Sporozoans

All sporozoans are nonmotile, spore-forming parasites of animals. Their spores are small, infective bodies that are transmitted from host to host. These organisms are distinguished by a unique arrangement of fibrils, microtubules, vacuoles, and other cell organelles at one end of the cell. There are about 3900 described species of this phylum; best known among them is the malarial parasite, *Plasmodium.*

Sporozoans have complex life cycles that involve both asexual and sexual phases. Sexual reproduction involves an alternation of haploid and diploid generations. Both haploid and diploid individuals can also divide rapidly by mitosis, thus producing a large number of small infective individuals. Sexual reproduction involves the fertilization of a large female gamete by a small, flagellated male gamete. The zygote that results soon becomes a thick-walled cyst called an **oocyst,** which is highly resistant to drying out and other unfavorable environmental factors. Within the oocyst, meiotic divisions produce infective haploid spores.

An alternation between different hosts often occurs in the life cycles of the sporozoans. The sporozoans of the genus *Plasmodium,* for example, are spread from person to person by mosquitoes of the genus *Anopheles;* at least 65 different species of this genus are involved. When an *Anopheles* mosquito penetrates human skin to obtain blood, it injects saliva mixed with an anticoagulant. If the mosquito is infected with *Plasmodium,* it will also inject elongated *Plasmodium* cells known as **sporozoites** (a stage of its life cycle) into the bloodstream of its victim. The parasite makes its way through the bloodstream to the liver, where it rapidly divides asexually. After this division phase, merozoites, which are the next stage of the life cycle, are formed, either reinvading other liver cells or entering the host's bloodstream. In the bloodstream, they invade the red blood cells, dividing rapidly within them and causing them to become enlarged and ultimately to rupture. This event releases toxic substances throughout the body of the host, bringing about the well-known cycle of fever and chills that is characteristic of malaria. The cycle repeats itself regularly every 48 hours, 72 hours, or longer, depending on the species of *Plasmodium* involved.

Plasmodium enters a sexual phase when some merozoites develop into **gametocytes,** cells capable of producing gametes. There are two types of gametocytes, male and female. Gametocytes are incapable of producing gametes within their human hosts and do so only when they are extracted from an infected human by a mosquito. Within the gut of the mosquito, the male and female gametocytes form sperm and eggs, respectively. Zygotes develop within the mosquito's intestinal walls and ultimately differentiate into oocysts. Within the oocysts, repeated mitotic divisions take place, producing large numbers of sporozoites. These sporozoites migrate to the salivary glands of the mosquito, from which they may be injected by the mosquito into the bloodstream of a human, thus starting the life cycle of the parasite again.

> In the life cycle of *Plasmodium,* a mosquito injects sporozoites into the bloodstream of its victim. These pass to the liver, where they divide rapidly by mitosis, eventually liberating large numbers of merozoites, which infect the red blood cells. Sexual reproduction may then occur. Eventually a mosquito may again draw some of the resulting sporozoites from the blood of an infected person, starting the cycle again.

A Possible Missing Link. *Pelomyxa palustris*—and possibly the zoomastigotes, a group of heterotrophic protists (p. 632)—may represent an early stage in the evolution of eukaryotic cells, a stage before they had acquired mitochondria and before mitosis had evolved. *Pelomyxa* lacks mitochondria and does not undergo mitosis. Its nuclei divide somewhat like those of bacteria do, by simply pinching apart into two nuclei, with new membranes forming around the daughter nuclei. The cells of Pelomyxa are much larger than bacteria and are visible to the naked eye. Although *Pelomyxa* lacks mitochondria, its cells do have two kinds of bacterial symbionts within them, and these may play the same role that mitochondria do in all other eukaryotes. This organism is so distinct that it is assigned to a phylum of its own, Caryoblastea.

Malaria, caused by infections by the sporozoan *Plasmodium*, is one of the most serious diseases in the world. About 100 million people are affected by it at any one time, and approximately 1 million of them, mostly children, die each year. Malaria kills most children under 5 years old who contract it; in areas where malaria is prevalent, most survivors more than 5 or 6 years old do not become seriously ill again from malaria infections. The symptoms, familiar throughout the tropics, include severe chills, fever, and sweating, an enlarged and tender spleen, confusion, and great thirst. Ultimately, a victim of malaria may die of anemia, kidney failure, or brain damage, or the disease may be brought under control by the person's immune system, or by drugs. As discussed in Chapter 20, some individuals are genetically resistant to malaria. Other persons develop immunity to it.

Efforts to eradicate malaria have focused on (1) the elimination of the mosquito vectors, (2) the development of drugs to poison the parasites once they have entered the human body, and (3) the development of vaccines. Regarding the first method, the widescale applications of DDT from the 1940s to the 1960s led to the elimination of the mosquito vectors in the United States, Italy, Greece, and certain areas of Latin America. For a time, the elimination of malaria worldwide appeared possible, but this hope was soon rendered impossible by the development of DDT-resistant strains of malaria-carrying mosquitoes in many regions; no less than 64 resistant strains were identified in a 1980 survey. Even though the worldwide use of DDT, long banned in the United States, nearly doubled from its 1974 level to more than 30,000 metric tons in 1984, its effectiveness in controlling mosquitoes is dropping. Further, there are serious environmental concerns about the use of this long-lasting chemical anywhere in the world. In addition to the problems with resistant strains of mosquitoes, strains of *Plasmodium* have appeared that are resistant to the drugs that have historically been used to kill them.

As a result of these problems, the number of new cases of malaria per year roughly doubled from the mid-1970s to the mid-1980s, largely because of the spread of resistant strains of the mosquito and the parasite. In many tropical regions, malaria is blocking permanent settlement. Scientists have therefore redoubled their efforts to produce an effective vaccine against the disease.

Antibodies to the parasites have been isolated and produced by genetic engineering techniques (see Chapter 18); they are starting to produce promising results.

The three different stages of the life cycle of *Plasmodium* (Figure 31-B) each produce different antigens, and they are sensitive to different antibodies. The gene encoding the sporozoite antigen was cloned in 1984, but it is not certain how effective a vaccine against sporozoites might be. When a mosquito inserts its proboscis into a human blood vessel, it injects about a thousand sporozoites. They travel to the liver within a few minutes, where they are no longer exposed to antibodies circulating in the blood. If even one sporozoite reaches the liver, it will multiply rapidly there and still cause malaria. The number of malaria parasites increases roughly eightfold every 24 hours after they enter the host's body. A compound vaccination against sporozoites, merozoites, and gametocytes would probably be the most effective preventive measure, and such a vaccine is now under development. Human trials are under way and promise eventually to control the disease, even though the complexity of the system is great.

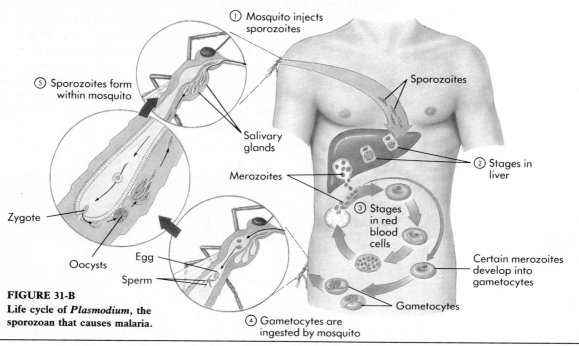

FIGURE 31-B
Life cycle of *Plasmodium*, the sporozoan that causes malaria.

① Mosquito injects sporozoites

⑤ Sporozoites form within mosquito

Salivary glands

Sporozoites

② Stages in liver

Merozoites

Zygote

③ Stages in red blood cells

Oocysts

Egg

Sperm

Certain merozoites develop into gametocytes

Gametocytes

④ Gametocytes are ingested by mosquito

FIGURE 31-26

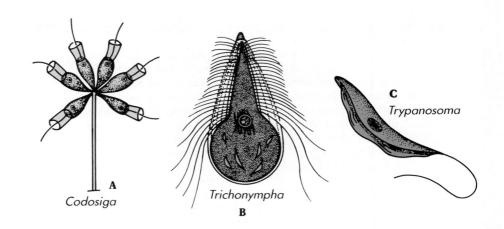

Three genera of zoomastigotes (phylum Zoomastigina), a highly diverse group.

A *Codosiga*, a colonial cho-anoflagellate that remains attached to its substrate; other colonial cho-anoflagellates swim around as a colony, resembling the green alga *Volvox* in this respect (see Figure 28-12).

B *Trichonympha*, one of the zoomastigotes that inhabit the guts of termites and wood-feeding cockroaches and digest cellulose there. *Trichonympha* has rows of flagella in its anterior regions.

C *Trypanosoma*, which causes sleeping sickness, an important tropical disease. It has a single, anterior flagellum.

Codosiga A

Trichonympha B

C
Trypanosoma

Zoomastigina: The Zoomastigotes

The zoomastigotes are unicellular, heterotrophic organisms that are highly variable in form (Figure 31-26). Each has at least one flagellum, and some species have thousands. Both free-living and parasitic organisms are included in this large phylum. Many zoomastigotes apparently reproduce only asexually, but sexual reproduction by gametes is known in some. Members of one of the included groups, the amoeboflagellates, alternate between an amoeboid stage and a flagellated one, depending on environmental conditions. The members of another group, the trypanosomes, include the genera *Trypanosoma* (Figures 31-26, *C*, and 31-27, *A*) and *Crithidia*, which are important pathogens of human beings and domestic animals. The euglenoids (see p. 624) can in many respects be viewed as a specialized group of zoomastigotes, some of which have acquired chloroplasts during the course of their evolution.

Among the diseases for which the trypanosomes are responsible are trypanosomiasis (sometimes called "sleeping sickness"), East Coast fever, and Chagas' disease, all of great importance in tropical areas. Another tropical disease, leishmaniasis, which is transmitted by sand flies, afflicts about 4 million people a year; its effects range from skin sores to deep, eroding lesions that can almost obliterate the face. The trypanosomes that cause these diseases are spread by biting insects, including tsetse flies (Figure 31-27, *B*) and assassin bugs.

A serious effort is now under way to produce a vaccine for trypanosome-caused diseases, which deprive a large portion of Africa of meat and milk from domestic cattle and thus pose a serious obstacle to the alleviation of hunger there with high-quality food. Control is especially difficult because of the unique attributes of these organisms. For example, the tsetse fly–transmitted trypanosomes have evolved an elaborate genetic mechanism for repeatedly changing the antigenic nature of their protective glycoprotein coat, thus avoiding the effects of the antibodies that their hosts produce against them. Only a single one out of some 1000 to 2000 variable antigen genes is expressed at a time; the DNA is rearranged in order to move the gene that is expressed to a position near the end of the chromosome, which is the only place it will

FIGURE 31-27

Trypanosoma, **the zoomastigote that causes sleeping sickness, among red blood cells.**

A The nuclei (dark-staining bodies), anterior flagella; and undulating, changeable shape of the trypanosomes may be seen in this photograph.

B The tsetse fly, shown here sucking blood from a human arm, can carry trypanosomes.

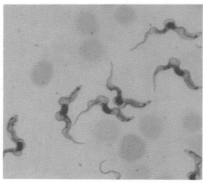

A

B

be expressed. Because these rearrangements occur during the asexual cycle of the organism, they provide a system that allows for the expression of a seemingly endless variety of different antigen genes and thus the maintenance of infectivity by the trypanosomes.

When the trypanosomes are ingested by a tsetse fly, they embark on a complicated cycle of development and multiplication, first in the fly's gut and later in its salivary glands. Recombination has been observed between different strains of trypanosomes introduced into a single fly, thus suggesting that mating and meiosis occur, even though they have not been observed directly. Although most trypanosome reproduction is asexual, this sexual cycle, reported for the first time in 1986, affords still further possibilities for recombination in these organisms.

In the guts of the flies that spread them, trypanosomes are noninfective; when they are prepared for transfer to a mammalian skin or bloodstream, they acquire a thick coat of glycoprotein antigens that protect them from the host's antibodies. When they are taken up by a fly, the trypanosomes again shed their coats. The production of vaccines against such a system is complex, but field tests of some are under way. Releasing sterilized flies to impede the reproduction of populations is another technique used to try to control them. Traps made of dark cloth and scented like cows, but poisoned with insecticides, have likewise proved effective. Research is proceeding at a rapid rate, because the presence of tsetse flies with their associated trypanosomes blocks the use of some 11 million square kilometers of potential grazing land in Africa.

Some zoomastigotes (Figure 31-26, B) occur in the guts of termites and other wood-eating insects; they possess the enzymes that allow them to digest the wood and thus make it available to their hosts. The relationship is similar to that between certain bacteria and protozoa that function in the rumens of cattle and related mammals (see Chapters 30 and 49).

Another group of zoomastigotes, the choanoflagellates (Figure 31-26, A), are certainly the group from which the sponges (phylum Porifera), and probably independently all other animals, were derived (see Chapter 39). Choanoflagellates have a single emergent flagellum surrounded by a funnel-shaped, contractile collar composed of closely placed filaments, a unique structure that is exactly matched in the sponges. These protists feed mostly on bacteria, which are pulled in a column of water into the collar and strained out of the water there.

THE MOST DIVERSE KINGDOM OF EUKARYOTES

By now, you are probably convinced of the diversity of the protists. The 14 phyla we have just discussed in this chapter consist mostly of microscopic, unicellular organisms, but there are many multicellular ones that have been derived from them. In their ways of life, the habitats in which they occur, and the details of their life cycles, the protists are extraordinarily diverse.

Among the phyla that make up this great group, we are especially interested in those that apparently gave rise to the other exclusively or predominantly multicellular evolutionary lines that we call animals, plants, and fungi. The choanoflagellates, one of the groups of zoomastigotes, appear certainly to be the ancestors of the sponges and probably independently of all other animals; the fundamental similarity between their unique structure and that of the collar cells of sponges, which we will describe in Chapter 39, is too great to be explained in any other way. Green algae, which share with plants their photosynthetic pigments, cell wall structure, and chief storage product (starch), doubtlessly include the ancestors of plants, a kingdom that achieved its distinctive features in connection with its invasion of the land. With regard to the third kingdom of predominantly multicellular organisms, the fungi, we will see in Chapter 32 that their ancestry is unclear.

Other groups of protists did not give rise to other kingdoms of organisms, but many of them, notably brown algae and red algae, have achieved multicellularity. Taken as a whole, the protists are extremely diverse and, as we learn more about the details of their structures and their lives, endlessly fascinating.

1. The kingdom Protista consists of the exclusively or predominantly unicellular phyla of eukaryotes, together with three phyla that include large numbers of multicellular organisms: the red algae, brown algae, and green algae. The phyla of protists are highly diverse.

2. Chloroplasts originated a number of times among the protists, and the process seems to have involved the members of at least three different groups of bacteria: cyanobacteria (in the red algae); *Prochloron*-like organisms (in the green algae and euglenoids); and *Heliobacterium chlorum*-like organisms (in the diatoms, dinoflagellates, and brown algae).

3. The three major multicellular groups of eukaryotes—plants, animals, and fungi—all originated from protists, but they are so large and important that they are considered distinct kingdoms. They are not related directly to one another. The plants originated from green algae; at least the sponges, and probably all animals, originated from the choanoflagellates, one of the groups of zoomastigotes. The ancestors of the fungi are unknown.

4. Cellular slime molds (phylum Acrasiomycota) exist as single cells that aggregate, migrate as a slug, and work together as a unit. Spores are formed from encysted amoebas that secrete a cellulose wall.

5. Green algae (phylum Chlorophyta) are a highly diverse group of organisms that are abundant in the sea, fresh water, and damp terrestrial habitats, such as on tree trunks and in soil. Plants were derived from a multicellular green alga.

6. There are about 8000 named species of ciliates (phylum Ciliophora); these protists have a very complex morphology with numerous cilia. They also have a life cycle that involves micronuclei, which function like the nuclei of other organisms, and macronuclei, which contain a large number of small chromatin bodies.

7. Dinoflagellates (phylum Dinoflagellata) are a major phylum of unicellular organisms that have unique chromosomes and a very unusual form of mitosis that takes place entirely within the nucleus. They are responsible for many of the poisonous red tides. In the symbiotic form known as zooxanthellae, dinoflagellates are widespread and are responsible for most of the productivity of coral reefs and of many marine organisms.

8. Euglenoids (phylum Euglenoph|yta) have chloroplasts that share the biochemical features of those found in green algae and plants; independent symbiotic events probably resulted in these similarities. With their flexible proteinaceous pellicles, euglenoids might better be considered one of the groups of zoomastigotes (phylum Zoomastigina), a group of heterotrophic, mostly unicellular protists that includes the organism responsible for sleeping sickness, rather than as an independent phylum.

9. Plasmodial slime molds (phylum Myxomycota) move about as a plasmodium, containing numerous nuclei. On drying, starving, or being subjected to other environmental cues, plasmodial slime molds form sporangia, which cleave to form spores that may undergo meiosis.

10. The oomycetes (phylum Oomycota) occur mainly in water or as plant parasites, including the organism responsible for the great Irish potato famine of the mid-nineteenth century. They have a filamentous structure and exhibit an alternation of generations.

11. Brown algae (phylum Phaeophyta) are multicellular, marine protists, some reaching 100 meters in length. The kelps—larger brown algae—contribute greatly to the productivity of the sea, especially along the coasts in relatively shallow areas. Kelps have an alternation of generations in which the gametophytes are very small and filamentous and the sporophytes are large and evident.

12. Red algae (phylum Rhodophyta), which have chlorophyll *a* and phycobilins, have more species than the other two phyla of seaweeds—about 4000—and occur at greater depths, down to more than 250 meters. They have very complex life cycles involving an alternation of generations. Their chloroplasts were almost certainly derived from symbiotic cyanobacteria. Red algae lack flagellated cells and centrioles and may be descendants of one of the most ancient lines of eukaryotes.

13. The malarial parasite, *Plasmodium*, is a member of the phylum Sporozoa. Carried by mosquitoes, it multiplies rapidly in the liver of humans and other primates and brings about the cyclical fevers characteristic of malaria by releasing toxins into the bloodstream of its host.

REVIEW QUESTIONS

1. What is the probable origin of eukaryotic mitochondria? Was this apparently a single or multiple event? Why? Do all eukaryotes possess mitochondria? Explain.

2. What is the most striking physical characteristic of the Bacillariophyta? What are they commonly called? How do these organisms reproduce? Are these individuals diploid or haploid? Biochemically, what other protists do they resemble?

3. What is the evolutionary significance of the Chlorophyta? Why is Chlamydomonas an important member of this phylum? What determines whether a collection of individuals is truly multicellular?

4. What unique characteristic differentiates the members of Ciliophora from other protists? Why are these one-celled organisms not as simple as one might expect? What is the function of two vacuoles exhibited by most members of Ciliophora?

5. What are the structural characteristics of the members of Dinoflagellata? What other phyla do they biochemically resemble? Does it appear that they are evolutionarily related to one another? Why or why not? Why is mitosis in this group unique?

6. What are red tides? What are zooxanthellae? Why are they important to coral reefs?

7. Why do the euglenoids illustrate the difficulty in assigning the designations animal-like and plantlike to the kingdom Protista? What type of reproduction is exhibited by this group?

8. What differentiates the oomycetes from the kingdom Fungi in which they were previously placed? What is the feeding strategy of this phylum? What distinguishes this group from the other protists? Why are these organisms generally considered harmful?

9. What is meant by the term *alternation of generations* as it applies to the phaeophyte life cycle? How many types of sporophytes are produced? How many types of gametophytes are produced? What types of reproductive cells are produced by the sporophytes and gametophytes? Which cells produce the zygote? Into which form does it develop?

10. What disease is caused by a member of the phylum Sporozoa and transmitted by the *Anopheles* mosquito? What stage of the organism is transmitted from the mosquito to a human? What is the stage that is actually responsible for the debilitating symptoms of malaria in humans? Why? What stages are transmitted from human infected with malaria to a mosquito? Does actual production of gametes and their fusion into zygotes occur in the mosquito or in the human host?

THOUGHT QUESTIONS

1. If plants were derived from green algae, why don't we classify green algae as plants in this book?

2. Different phyla of protists consist of organisms that have different numbers and kinds of flagella. Enumerate some of these different conditions and suggest how one may have given rise to another.

3. If mitochondria and chloroplasts originated as symbiotic bacteria, what would you suggest were the characteristics of the organism in which they became symbiotic?

4. List the evidence for and against the hypothesis that multicellularity arose only once.

FOR FURTHER READING

BOLD, H.C., and M.J. WYNNE: *Introduction to the Algae*, ed. 2, Prentice-Hall, Englewood Cliffs, N.J., 1985. Excellent overall account of the algal phyla.

FENCHEL, T.: *Ecology of Protozoa: The Biology of Free-Living Phagotrophic Protists*, Science Tech Publishers, Madison, Wis., 1987. An interesting book about protozoa from an ecological point of view.

HICKMAN, C.P., Jr., L.S. ROBERTS, and F.M. HICKMAN: *Integrated Principles of Zoology*, ed. 8, Mosby–Year Book Inc., St. Louis, 1988. The outstanding treatment of general zoology, including the heterotrophic phyla discussed here.

LEE, J.J., S.H. HUNTER, and E.C. BOVEE (editors): *An Illustrated Guide to the Protozoa*, Society of Protozoologists, Lawrence, Kan., 1985. Outstanding visual impression of the diversity of protists.

MARGULIS, L.: *Symbiosis in Cell Evolution*, W.H. Freeman and Co., San Francisco, 1980. Outstanding treatment of the origin of eukaryotic cells by serial symbiosis.

MARGULIS, L., J.O. CORLISS, M. MELKONIAN, and D.J. Chapman, eds.: *Handbook of Protoctista*, Jones & Bartlett, Publishers, Boston, 1990. The authoratative account of the groups of the kingdom Protista; indispensible to any serious consideration of these diverse and fascinating organisms.

RAPER, K.B.: *The Dictyostelids*, Princeton University Press, Princeton, N.J., 1984. An up-to-date account of these fascinating protists, written from a developmental point of view.

SAFFO, M.B.: "New Light on Seaweeds," *BioScience*, vol. 37, pages 654-664, 1987. Discusses recent studies of the role of light-harvesting pigments in depth zonation of seaweeds.

SLEIGH, M.: *Protozoa and Other Protists*, Edward Arnold, London, 1989. Concise outline of the features of the major groups, well presented and illustrated.

WATERS, A., and others: *Plasmodium falciparum* Appears to Have Arisen as a Result of Lateral Transfer Between Avian and Human Hosts," *Proceedings of the National Academy of Science USA*, April 1991, vol. 88, pages 3140-3144. Humans may have caught malaria from chickens.

WICHTERMAN, R.: *The Biology of Paramecium*, ed. 2, Plenum Press, New York, 1986. A comprehensive account of the best-studied protist.

32

Fungi

CHAPTER OUTLINE

OVERVIEW

The filaments of fungi grow through soil, wood, and other substrates. By secreting enzymes, fungi digest organic matter and absorb the products of this external digestion. Together with the bacteria, fungi are the major decomposers of the biosphere, breaking down organic molecules, including those that have been incorporated in the bodies of organisms, and making the material in them available for recycling. This process is as necessary for the continuation of life on earth as photosynthesis. Cytoplasm, including its nuclei and other organelles, streams through the bodies of fungi; their filaments either lack cross walls completely, except when reproductive structures are formed, or have only perforated ones. Thus fungi are not multicellular in the same sense that plants and animals are. They are a unique kingdom of organisms that have no evident similarities with any other group.

FOR REVIEW

Here are some important terms and concepts that you will encounter in this chapter. If you are not familiar with them, you should review them before proceeding.

Mitosis (Chapter 10)

Mutualism (Chapter 24)

Relationships among the eukaryotes (Chapter 28)

Cyanobacteria (Chapter 30)

Diversity of protists (Chapter 31)

A

B

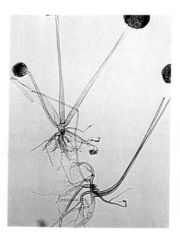

FIGURE 32-1

Representatives of the three divisions of fungi.

A A cup fungus, *Cookeina tricholoma,* a kind of ascomycete, from the rain forest of Costa Rica.

B *Amanita muscaria,* the fly agaric, a highly toxic basidiomycete. In the cup fungi, the spore-producing structures line the cup; in basidiomycetes that form mushrooms, like *Amanita,* they line the gills beneath the cap of the mushroom. All visible structures of fleshy fungi, such as the ones shown here, arise from an extensive network of filamentous hyphae that penetrates and is interwoven with the substrate on which they grow.

C *Rhizopus,* a zygomycete that grows on moist bread and other similar substrates. The spore-bearing structures are about a centimeter tall.

The fungi are a distinct kingdom of organisms, comprising about 77,000 named species, but **mycologists,** scientists who study them, believe that there may be several times more actually in existence. Although fungi have traditionally been included in the plant kingdom, they lack chlorophyll and resemble plants only in lacking mobility and growing from the ends of somewhat linear bodies (Figure 32-1). However, even these similarities prove to be misleading when they are examined closely. Some plants have motile sperm with flagella; no fungi do. Fungi are basically filamentous in their growth form (they consist of slender filaments), even though these filaments may be packed together to form complex structures such as mushrooms; plants, in contrast, are basically three dimensional. Although both plants and fungi have cell walls, they differ in their chemistry. Unlike plants, fungi obtain their food by secreting enzymes into their substrate and absorbing the materials that these enzymes make available. Fungi are an ancient group of organisms, no less than 400 million years old and possibly much older.

Many fungi are harmful because they decay, rot, and spoil many different materials as they obtain food and because they cause serious diseases of plants and animals, including humans. Other fungi, however, are extremely useful. The manufacture of both bread and beer depends on the biochemical activities of yeasts, single-celled fungi that produce abundant quantities of ethanol and carbon dioxide (Figure 32-2). Both cheese and wine achieve their delicate flavors because of the metabolic processes of certain fungi, and others make possible the manufacture of such Oriental delicacies as soy sauce and tofu. Vast industries depend on the biochemical manufacture of organic substances such as citric acid by fungi in culture, and yeasts are now being used on a large scale to produce protein for the enrichment of animal food. Many antibiotics, including the first one that was used on a wide scale, penicillin, are derived from fungi. Some fungi are used on a major scale to convert one complex organic molecule into another, such as in the synthesis of many commercially important steroids. Others are being considered increasingly as a way to clean up toxic substances in the environment. For example, at least three species of fungi have been isolated that combine selenium, accumulated at the San Luis National Wildlife Refuge in California's San Joaquin Valley (see Figure 27-23), with harmless volatile chemicals. Fungi are very important to us in both harmful and beneficial ways.

> **Fungi absorb their food after digesting it with secreted enzymes. Their mode of nutrition, combined with a filamentous growth form, makes the members of this kingdom highly distinctive.**

FIGURE 32-2

Using yeasts. Baking bread involves the metabolic activities of the yeasts, single-celled ascomycetes. Wine making also uses the metabolic properties of yeasts, and the same species, *Saccharomyces cerevisiae,* is usually employed for both processes. In baking bread, the generation of carbon dioxide is important; in making beer or wine, it is the production of alcohol.

NUTRITION AND ECOLOGY

Fungi, together with bacteria, are the principal decomposers in the biosphere, breaking down organic materials and returning the substances locked in those molecules to circulation in the ecosystem. For example, fungi break down lignin, one of the major constituents of wood, and are virtually the only organisms capable of doing so. In ways such as this, critical biological building blocks—such as compounds of carbon, nitrogen, and phosphorus—that have been incorporated into the bodies of living organisms are released and made available for other organisms.

In breaking down organic matter, some fungi attack living plants and animals, whereas others attack dead ones; both are sources of organic molecules. Fungi are often important as disease-causing organisms for both plants and animals; they are responsible for billions of dollars in agricultural losses every year. Not only are fungi the most harmful pests of living plants, but also they attack food products once they have been harvested and stored. In addition, fungi often secrete substances into the foods that they are attacking that make these foods unpalatable or even poisonous.

Two kinds of mutualistic associations between fungi and autotrophic organisms are ecologically important. **Lichens,** which are prominent nearly everywhere in the world, especially in unusually harsh habitats such as bare rocks, are symbiotic associations between fungi and either green algae or cyanobacteria. **Mycorrhizae,** specialized symbiotic associations between the roots of plants and fungi, are characteristic of about 80% of all plants. Both of these associations will be discussed further in this

---

HOW FUNGI PENETRATE LEAVES

Fungi cause billions of dollars of damage to crops annually, so the details of their interactions with plants are of special interest to mycologists. Plants manufacture many chemicals that retard or prevent fungal growth (Chapter 24). In some instances, fungi have in turn evolved the ability to break down these compounds and continue to use the plants as food. Manipulating the charactistics of such systems is one promising approach to improving the world food supply.

Recently, Harvey Hoch and his colleagues at Cornell University reported some fascinating details of the interactions between the bean rust (*Uromyces appendiculatus*) and the common bean (*Phaseolus vulgaris*). When the spores of the rust fall on bean leaves, they germinate into hyphae that seem to grow around the edge of the leaf toward the stomata on the underside—the pores in the leaves by which plants lose water vapor and allow carbon dioxide to enter. As the Cornell investigators showed, the fungus responds to the presence of ridges about 0.5 micrometer high that surround the stomata. When a hypha encounters a ridge, it balloons out into a stomatal infection structure

(Figure 32-A), growing rapidly into the stomata. In experiments with silicon wafers etched with ridges, Hoch and his associates found that the fungus usually produced infection structures at ridges 0.5 micrometer high, but no such structures when the ridges were shorter than 0.1 micrometer or taller

than 1 micrometer. Hoch and his colleagues have identified several fungal genes that are involved in the construction of infectious structures. They are now investigating the bean leaf fungus system in more detail, hoping to produce beans with hipless stomata that will not attract a fungal attack.

1

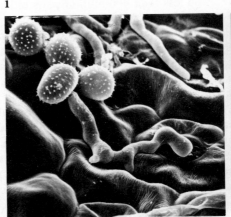

2

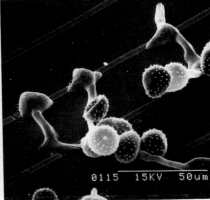

FIGURE 32-A
Electron micrograph of infection structures produced by the rust fungus *Uromyces appendiculatus* on the leaves of the common bean *Phaseolus vulgaris.*
1. A hypha growing down into a stoma on a bean leaf.
2. The fungus producing infection structures after growing across ridges of the appropriate height on a silicon wafer. Infection structures can be indentified in 1 by reference to this photograph.

chapter. In each of them the photosynthetic organisms fix atmospheric carbon dioxide and thus make available a source of organic material to the fungi, and the metabolic activities of the fungi in turn enhance the overall ability of the symbiotic association to exist in a particular habitat. In the case of mycorrhizae, the fungal partner expedites the absorption of essential nutrients such as phosphorus by the plant.

STRUCTURE

Fungi exist mainly in the form of slender filaments, barely visible with the naked eye, which are called **hyphae** (singular, *hypha*). These hyphae may be divided into cells by cross walls called **septa** (singular, *septum*). The septa rarely form a complete barrier, however, except for those separating the reproductive cells. Cytoplasm characteristically flows or streams freely throughout the hyphae, passing right through the major pores in the septa (Figure 32-3). Because of this streaming, proteins, which are synthesized throughout the hyphae, may be carried to their actively growing tips. As a result, the growth of fungal hyphae may be very rapid when food and water are abundant and the temperature is optimum.

A mass of hyphae is called a mycelium (plural, *mycelia*). This word and the term *mycologist* are both derived from the Greek word for fungus, *myketos*. The mycelium of a fungus (Figure 32-4) constitutes a system that may, in the aggregate, be many meters long. This system grows through and penetrates the substrate of the fungus, resulting in a unique relationship between a fungus and its environment. All parts of a fungus are metabolically active, continually interacting with the soil, wood, or other material in which the mycelium is growing.

In two of the three divisions of fungi, reproductive structures formed of interwoven hyphae—such as mushrooms, puffballs, and morels—are produced at certain stages of the life cycle. These structures expand rapidly because of rapid elongation of the hyphae. For this reason, mushrooms can appear suddenly in your lawn.

The cell walls of fungi are formed of polysaccharides and chitin, not cellulose like those of plants and many groups of protists. Chitin is the same material that makes up the major portion of the hard shells, or exoskeletons, of arthropods, a group of animals that includes insects and crustaceans (see Chapter 41). Chitin is far more resistant to microbial degradation than is cellulose.

> Fungi exist primarily in the form of filamentous hyphae, with incomplete division into individual cells by septa, except when reproductive organs are formed. These hyphae surround and penetrate the substrate within which the fungi are growing. The cell walls of fungi are composed of chitin. These and other unique features indicate that fungi are not closely related to any other group of organisms.

Mitosis in fungi differs from that found in other organisms: the nuclear envelope does not break down and re-form, and the spindle apparatus is formed within it. Centrioles are lacking in all fungi; instead, fungi regulate the formation of microtubules during mitosis with small, relatively amorphous structures called **spindle plaques**. This unique combination of features strongly suggests that fungi originated from some unknown group of single-celled eukaryotes with these characteristics. There is no indication of a direct evolutionary relationship between fungi and any other group of living organisms.

> The cell walls of fungi are composed of chitin; mitosis occurs within the nuclear envelope, which remains intact at all times. These and other unique features indicate that fungi are not closely related to any other group of organisms.

REPRODUCTION

All fungal nuclei except for the zygote are haploid, and there are many such haploid nuclei in the common cytoplasm of a fungal mycelium. In the sexual reproduction of fungi, hyphae of two genetically different mating types come together and fuse. In

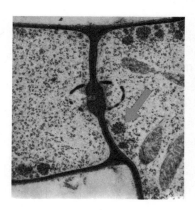

FIGURE 32-3
A septum. This transmission electron micrograph of a section through a hypha of the basidiomycete *Laetisaria arvalis*, showing the kind of septum characteristic of secondary mycelia (those mycelia formed by the fusion of hyphae of two different mating types; mycelia are masses of hyphae) in basidiomycetes. Such septa are lined with thick, barrel-shaped structures, seen here in transection (*arrows*). The septa of ascomycetes and those of the primary mycelia of basidiomycetes do not have a barrel-shaped structure associated with them.

FIGURE 32-4
Fungal mycelium growing through leaves on the forest floor in Maryland.

two of the three divisions of fungi, the genetically different nuclei that are associated in a common cytoplasm after such fusion do not fuse immediately. Instead, the two types of nuclei simply coexist for most of the life of the fungus. A fungal hypha that has nuclei within it derived from two genetically distinct individuals is called a **heterokaryotic** hypha; if all of the nuclei are genetically similar to one another, the hypha is said to be **homokaryotic**. If there are two genetically distinct nuclei within each compartment of the hyphae, they are **dikaryotic**; if each compartment has only a single nucleus, it is **monokaryotic**. A dikaryotic hypha will always be heterokaryotic; a monokaryotic one will always be homokaryotic. Dikaryotic hyphae have some of the genetic properties of diploids, because both genomes are transcribed. These distinctions are important in understanding the life cycles of the individual groups.

> Fungal hyphae in which two genetically distinct kinds of nuclei occur together are said to be heterokaryotic; those in which all of the nuclei are of the same type are homokaryotic.

The cytoplasm in fungal hyphae normally flows right through perforated septa or moves freely in their absence, but we have already mentioned one important exception to this general pattern. When reproductive structures are formed, they are cut off by complete septa that either lack perforations or have perforations that soon become blocked. Three kinds of reproductive structures occur in fungi: (1) **sporangia**, which are involved in the formation of spores; (2) **gametangia**, structures within which gametes form; and (3) **conidia**, mostly multinucleate asexual spores not produced in sporangia.

Spores, always nonmotile, are a common means of reproduction among the fungi. They may be formed as a result of either asexual or sexual processes, as we will see in our discussion of the individual divisions. When spores or conidia land in a suitable place, they germinate, giving rise to a new fungal hypha. Since they are very small, they may remain suspended in the air for long periods of time. Because of this property, fungal spores or conidia may be blown great distances from their place of origin, a factor in the extremely wide distributions of many kinds of fungi. Unfortunately, many of the fungi that cause diseases of plants and animals are spread rapidly and widely by such means. The spores and conidia of other fungi are routinely dispersed by insects and other small animals.

FUNGAL DIVISIONS

There are three divisions of fungi: Zygomycota, the **zygomycetes**; Ascomycota, the **ascomycetes**; and Basidiomycota, the **basidiomycetes**. They are called divisions instead of phyla because fungi are named according to the same rules used for plants. Several other groups that historically have been associated with fungi, such as the slime molds and water molds (phylum Oomycota; see Chapter 31), now are considered to be protists, not fungi. Oomycetes are sharply distinct from fungi in (1) their motile spores; (2) their cellulose-rich cell walls; (3) their pattern of mitosis, which does not have the unique features found in fungi; and (4) their diploid hyphae.

The three divisions of fungi are distinguished primarily by their sexual reproductive structures. In the zygomycetes, the fusion of hyphae leads directly to the formation of a zygote, which divides by meiosis when it germinates. In the other two divisions, an extensive growth of dikaryotic hyphae may lead to the formation of massive structures of interwoven hyphae within which are formed the distinctive kind of reproductive cell characteristic of that particular division. Syngamy, followed immediately by meiosis, occurs within these cells, and haploid spores are formed. On release the spores are dispersed, some of them giving rise to new hyphae.

Division Zygomycota

The zygomycetes (division Zygomycota) lack septa in their hyphae except when they form sporangia or gametangia. Zygomycetes are by far the smallest of the three divisions of fungi, with only about 600 named species. Included among them are some of

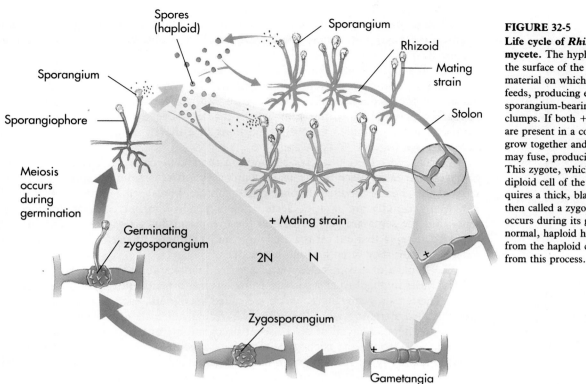

Spores (haploid)

Sporangium

Sporangium

Sporangiophore

Meiosis occurs during germination

Germinating zygosporangium

Sporangium

Rhizoid

Mating strain

Stolon

+ Mating strain

2N   N

Zygosporangium

Gametangia

FIGURE 32-5
Life cycle of *Rhizopus*, a zygomycete. The hyphae grow over the surface of the bread or other material on which the fungus feeds, producing erect, sporangium-bearing stalks in clumps. If both + and − strains are present in a colony, they may grow together and their nuclei may fuse, producing a zygote. This zygote, which is the only diploid cell of the life cycle, acquires a thick, black coat and is then called a zygospore. Meiosis occurs during its germination, and normal, haploid hyphae grow from the haploid cells that result from this process.

the more frequent bread molds (see Figure 32-1, *C*), as well as a variety of other microscopic fungi found on decaying organic material. The group is named after a characteristic feature of the life cycle of its members, the production of temporarily dormant structures called **zygospores**.

In the life cycle of the zygomycetes (Figure 32-5), sexual reproduction occurs by the fusion of gametangia, which contain numerous nuclei. The gametangia are cut off from the hyphae by complete septa. These gametangia may be formed on hyphae of different mating types or on a single hypha. If different mating types are involved, fusion between the pairs of nuclei occurs immediately. Once the haploid nuclei have fused, forming diploid zygote nuclei, the area in which the fusion has taken place develops into an often massive and elaborate zygospore. Except for the zygote nuclei, all nuclei of the zygomycetes are haploid. Meiosis occurs during the germination of the zygospore.

> **Zygomycetes form characteristic resting structures, called zygospores, around the cell in which one or more zygotes are formed. The hyphae of zygomycetes are multinucleate, with septa only where gametangia or sporangia are cut off.**

Asexual reproduction occurs much more frequently than sexual reproduction in the zygomycetes. During asexual reproduction, haploid spores are produced within more or less specialized sporangia formed on specialized hyphae. These sporangia are cut off by septa and are usually formed at the tips of erect hyphae in zygomycetes. Their spores are thus shed above the substrate, in a position where they may be picked up by the wind and blown about.

Division Ascomycota

The second division of fungi, the ascomycetes (division Ascomycota), is a very large group of about 30,000 named species, with many more being discovered each year. Among the ascomycetes are such familiar and economically important fungi as yeasts, common molds, morels, and truffles (see Figures 32-1, *A*, and 32-6). Also included in

FIGURE 32-6
An ascomycete. This morel, *Morchella esculenta*, is a delicious edible ascomycete that appears in early spring in the north-temperate woods, especially under oaks. Morels are gathered in large quantities in nature, but mushroom growers began to learn how to cultivate them commercially only in the 1980s.

this division are many of the most serious plant pathogens, including the chestnut blight, *Cryptonectria (Endothia) parasitica*, and Dutch elm disease, *Ceratocystis ulmi* (Figure 32-7).

The ascomycetes are named for their characteristic reproductive structure, the microscopic, club-shaped **ascus** (plural, **asci**). The zygote, which is the only diploid nucleus of the ascomycete life cycle (Figure 32-8), is formed within the ascus. The asci are differentiated within a structure that is made up of densely interwoven hyphae, corresponding to the visible portions of a morel or cup fungus, called the **ascocarp** (see Figures 32-1, *A*, and 32-6).

Asexual reproduction is very common in the ascomycetes, as it is among the zygomycetes. It takes place by means of conidia (singular, conidium), spores cut off by septa at the ends of modified hyphae called conidiophores (see Figures 32-14 and 32-15). Many conidia are multinucleate. The hyphae of ascomycetes are divided by septa, but the septa are perforated and the cytoplasm flows along the length of each hypha. The septa that cut off the asci and conidia are initially perforated like all other septa, but later they often become blocked.

The hyphae of ascomycetes may be either homokaryotic or heterokaryotic. The cells of these hyphae usually contain from several to many nuclei, as do the gametangia. The female gametangia, which are called **ascogonia,** each have a beaklike outgrowth called a **trichogyne.** When the **antheridium,** or male gametangium, forms, it fuses with the trichogyne of an adjacent ascogonium; nuclei from the antheridium then migrate through the trichogyne into the ascogonium and pair with nuclei of the opposite mating type. Initially, both kinds of gametangia contain a number of nuclei. Heterokaryotic hyphae then arise from the area of the fusion. Throughout such hyphae, nuclei that represent the two different original mating types occur (dikaryotic). Several nuclei, some derived from each of the parents of their mitotic products, are present within each cell of the hyphae. These hyphae are dikaryotic and heterokaryotic.

The asci are cut off by the formation of septa at the tips of the heterokaryotic hyphae. There are two haploid nuclei within each ascus, one of each of the two mating types represented in the dikaryotic hypha. Fusion of these two nuclei occurs within each ascus, forming a zygote. Each zygote divides immediately by meiosis,

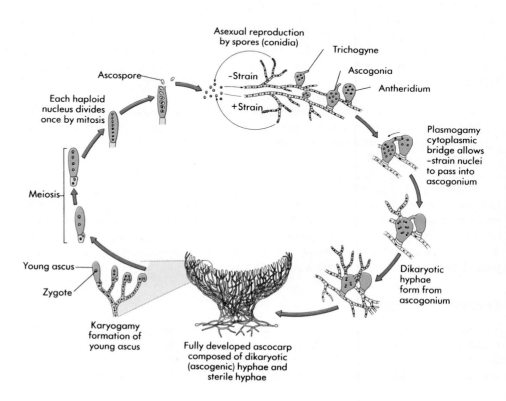

FIGURE 32-8
Life cycle of an ascomycete.

forming four haploid daughter nuclei; these usually divide again by mitosis, producing eight haploid nuclei that become walled **ascospores** (Figure 32-9). In most ascomycetes the ascus becomes highly turgid at maturity and ultimately bursts, often at a perforated area. When this occurs, the ascospores may be thrown as far as 30 centimeters, an amazing distance considering that most ascospores are only about 10 micrometers long. Aside from differences in scale, this would be equivalent to throwing a baseball (diameter, 7.5 centimeters) 1.25 kilometers—about 10 times the length of a home run!

Ascocarps (see Figures 32-1, *A,* and 32-6) are made up of tightly interwoven monokaryotic and dikaryotic hyphae. Within an ascocarp, on special fertile layers of dikaryotic hyphae, the asci are formed. The ascocarps of the cup fungi and the morels are open, with the asci lining the open cups. Other ascocarps are closed or have a small opening at the apex; the ascocarps of *Neurospora,* an important organism in genetic research, are of this latter kind.

Ascomycetes form their zygotes within a characteristic club-shaped structure, the ascus. Meiosis follows zygote formation immediately and results in the production of ascospores. Asexual reproduction by conidia, many of which are multinucleate, is common among the members of this group of fungi.

Yeasts. Yeasts, which are unicellular, are one of the most interesting and economically important groups of microscopic ascomycetes (Figure 32-10). There are about 40 genera of yeasts, with about 350 species. Most of their reproduction is asexual and takes place by cell fission or budding (the formation of a smaller cell from a larger one). Sometimes, however, whole yeast cells may fuse. One of these cells, containing two nuclei, may then function as an ascus with syngamy followed immediately by meiosis; the resulting ascospores function directly as new yeast cells.

Because they are single celled, yeasts might be considered primitive fungi; it appears certain, however, that instead they are reduced in structure and were originally derived from multicellular ancestors, most of which were ascomycetes. The word *yeast,* in fact, actually relates only to the fact that these fungi are single celled; some yeasts have been derived from each of the three classes of fungi, although ascomycetes are best represented. Even the yeasts that were derived from filamentous ascomycetes are not necessarily directly related to one another but instead seem to have been derived from different groups of ascomycetes.

The ability of yeasts to ferment carbohydrates, breaking down glucose to produce ethanol and carbon dioxide in the process, is fundamental in the production of bread, beer, and wine. Through the millennia, many different strains of yeast have been domesticated and selected for these processes. Wild yeasts—ones that occur naturally in the areas where wine is made—were important in wine making historically, but domesticated yeasts are normally used now. The most important yeast in all these processes is *Saccharomyces cerevisiae.* This yeast has been used by humans throughout recorded history. Other yeasts are important pathogens and cause diseases such as thrush and cryptococcosis; one of them, *Candida,* causes a common vaginal infection.

FIGURE 32-9
Asci from the lining of the cup of *Peziza,* a cup fungus like the one shown in Figure 32-1, *A*. Within each ascus, the zygote—the only diploid cell in the life cycle of an ascomycete—divides by meiosis; subsequently, the four haploid nuclei divide once by mitosis, and ascospores form around each of them. The fact that the spores remain lined up in order during the process makes it possible to use *Neurospora,* another ascomycete, to study crossing-over and the inheritance of metabolic traits. The ascospores of *Neurospora* can be dissected out individually from the asci and grown on media of selected composition.

FIGURE 32-10
Scanning electron micrograph of a yeast, showing the characteristic method of cell division by budding. The cells tend to hang together in chains, a feature that calls to mind the derivation of single-celled yeasts from multicellular ancestors.

Over the past few decades yeasts have become increasingly important in genetic research. They were the first eukaryotes to be manipulated extensively by the techniques of genetic engineering, and they still play the leading role as models for research in eukaryotic cells. In 1983 investigators synthesized a functional artificial chromosome in *Saccharomyces cerevisiae* by assembling the appropriate DNA molecule chemically; this has not yet been possible in any other eukaryote. With their rapid generation time and rapidly increasing pool of genetic and biochemical information, the yeasts in general and *S. cerevisiae* in particular are becoming the eukaryotic cells of choice for many types of experiments in molecular and cellular biology. Yeasts have become, in this respect, comparable to *Escherichia coli* among the bacteria and are continuing to provide significant insights into the functioning of eukaryotic systems.

Yeasts are unicellular fungi, mainly ascomycetes, which have evolved from hypha-forming ancestors; not all yeasts are directly related to one another. Long useful for baking, brewing, and wine making, yeasts are now becoming very important in genetic research; they are more easily manipulated than any other eukaryote.

Other Ascomycetes. Ascomycetes are by far the most frequent fungal partners in lichens (see p. 648). Most species of fungi that normally lack sexual reproduction likewise are ascomycetes, as judged by the features of their hyphae and conidia. Such fungi are classified as Fungi Imperfecti (see p. 647), since they cannot be placed in the proper group of ascomycetes unless the details of their sexual structures, such as their asci and ascocarps, are known.

Division Basidiomycetes

The last division of fungi, the basidiomycetes (division Basidiomycota), has about 16,000 named species. More is known about some members of this group than about any other fungi. Among the basidiomycetes are not only the mushrooms, toadstools, puffballs, jelly fungi, and shelf fungi, but also many important plant pathogens among the groups called rusts and smuts (Figure 32-11). Many mushrooms are used as food, but others are deadly poisonous. Still other species are poisonous to some people and harmless to others.

Basidiomycetes are named for their characteristic sexual reproductive structure, the **basidium**. A basidium is club shaped, like an ascus. Syngamy, or nuclear fusion, occurs within the basidium, giving rise to the zygote, the only diploid cell of the life cycle (Figure 32-12). As in all fungi, meiosis occurs immediately after the formation of the zygote. In the basidiomycetes, the four haploid products of meiosis are incorporated into **basidiospores**; in most of the members of this division, the basidiospores are borne at the end of the basidium on slender projections called sterigmata (singular, **sterigma**). In this way the structure of a basidium differs from that of an ascus, although functionally the two are identical. The septum that cuts off the young

FIGURE 32-11
Representative basidiomycetes.
A An earth star, *Geastrum saccatus*. Earth stars are a kind of puffball. Basidia form within puffballs, which may have pores, as shown here, or no external openings; they release hundreds of thousands or even millions of basidiospores when mature.
B A shelf fungus, *Trametes versicolor*, growing on a tree trunk. Basidia line the inner surface of tubes that are on the lower surface of many shelf fungi, although some have gills.
C A jelly fungus, growing in the Amazonian rain forest of Brazil. Although many familiar basidiomycetes form their basidia within mushroomlike basidiocarps, many do not do so, as these photographs illustrate.

A

B

C

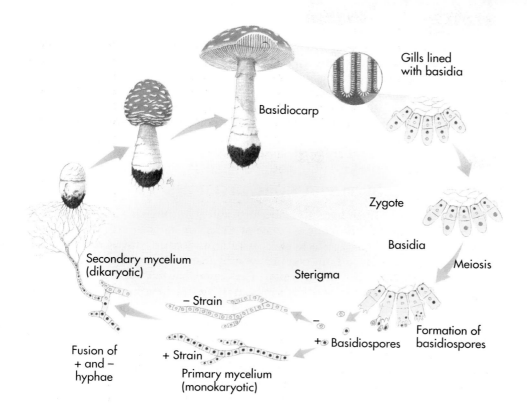

Gills lined
with basidia

Basidiocarp

Zygote

Basidia

Meiosis

Secondary mycelium
(dikaryotic)

Sterigma

– Strain

Formation of
basidiospores

Fusion of
+ and –
hyphae

+ Strain

Primary mycelium
(monokaryotic)

+ • Basidiospores

FIGURE 32-12
Life cycle of a basidiomycete.

basidium is initially perforated but often becomes blocked, as it does in the asco-mycetes.

The life cycle of a basidiomycete starts with the production of homokaryotic hyphae after spore germination. These hyphae lack septa at first. Eventually, how-ever, septa are formed between the nuclei of the monokaryotic hyphae. A basidio-mycete mycelium made up of monokaryotic hyphae is called a **primary mycelium.** Ultimately, different mating types of monokaryotic hyphae may fuse, forming a dikaryotic or **secondary mycelium.** Such a mycelium is heterokaryotic, with two nu-clei, representing the two different mating types, between each pair of septa. Most of the mycelium of basidiomycetes that occur in nature is dikaryotic, and often only dikaryotic mycelium is able to form basidiocarps. It is interesting that the haploid nu-clei occur together in this dikaryotic mycelium for such a long time before they finally fuse in the basidiocarp.

The dikaryotic hyphae of basidiomycetes have a unique pattern of cell division. They grow by the simultaneous division of the nuclei in the cell at the tip of each hypha and the progressive formation of new, perforated septa. Other cells within the mycelium that will form lateral branches divide in the same way. The **basidiocarps,** or mushrooms, are formed entirely of secondary (dikaryotic) mycelium.

Gills are radiating pleated structures that resemble the folds of an accordion; they occur on the undersurface of the cap of a mushroom. On these gills, the basidia occur in a dense layer from which vast numbers of minute spores may be released at matu-rity. It has been estimated, for example, that a mushroom with a cap that is 7.5 cen-timeters across produces as many as 40 million spores per hour! In contrast to their effective sexual reproduction, asexual reproduction is rare in most basidiomycetes.

Most of the hyphae of basidiomycetes in nature are dikaryotic, with two nuclei, one of each mating type, within each cell. These nuclei divide simultaneously, and ultimately the hyphae mass to form basidiocarps, within which basidia line the gills or pores. Meiosis immediately follows syngamy in these basidia.

Some species of basidiomycetes are commonly cultivated; for example, the button mushroom, *Agaricus campestris,* is grown in more than 70 countries, producing a crop

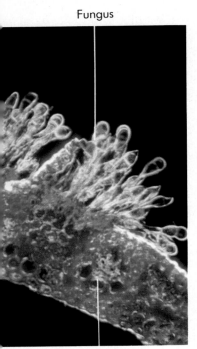

Fungus

Leaf

FIGURE 32-13
Wheat rust, *Puccinia graminis,*
one of about 7000 species of
rusts, all plant pathogens. This
species causes enormous economic
losses to wheat wherever it is
grown, and it is combated largely
by breeding resistant wheat varieties. Mutation and recombination
in wheat rust produce new virulent strains and make it necessary
to replace the existing wheat varieties constantly.

with a value of over $15 billion. An effort is being made to increase the cultivation of additional species of mushrooms, especially in Asia. Some can be grown on waste products, and many have a high protein content, as well as a delicate flavor.

Although we are more familiar with the mushrooms, another kind of basidiomycete is represented by the rusts (Figure 32-13) and smuts, which are important plant pathogens. In these fungi, basidiocarps are not formed. Instead, the basidia, which differ in structure from those of the mushrooms and other large, fleshy fungi, arise from hyphae at the surface of the host plant.

A COMPARISON OF THE FUNGAL DIVISIONS

The three divisions of fungi have a number of features in common, but they are separated by many important differences, some of which are summarized in Table 32-1. The zygomycetes have nonseptate hyphae, with septa formed only when the reproductive structures are cut off. In contrast, the hyphae of ascomycetes and basidiomycetes are divided by perforated septa. The septa that cut off the reproductive structures of the ascomycetes and basidiomycetes (asci or basidia, respectively) may become blocked and thus isolate the respective structures, but they are still perforated initially. The sexual reproductive organs of the zygomycetes are relatively unspecialized and are simply called gametangia; in contrast, the ascogonia of ascomycetes have a distinctive structure, with a beak, through which the contents of the antheridium empty. Basidiomycetes have no distinct sexual reproductive organs.

In zygomycetes the fusion of gametangia results in the production of a zygote, around which a specialized, resistant structure called a zygospore forms. Sexual fusion in the ascomycetes occurs between ascogonia and antheridia; in the basidiomycetes it occurs between normal-looking hyphae. In the ascomycetes, sexual fusion is followed by the formation of heterokaryotic hyphae in which the individual cells normally are multinucleate. In the basidiomycetes, sexual fusion is followed by the formation of dikaryotic hyphae, in which each cell has two genetically distinct nuclei.

The sexual spores of the ascomycetes are produced in complex structures called ascocarps, formed of both homokaryotic and heterokaryotic mycelia. Those of basidiomycetes are produced in similar structures, called basidiocarps, which are formed exclusively of dikaryotic, or secondary, mycelia. The characteristic asci and basidia of these divisions occur within or on the surfaces of their ascocarps or basidiocarps. In them, syngamy, meiosis, and the production of haploid spores occur. Asexual reproduction is very common in zygomycetes, in which the spores are produced in sporangia, and in ascomycetes, in which multinucleate spores called conidia are formed. Asexual reproduction is rare among the basidiomycetes, but a number of rusts and smuts form conidia.

It is generally assumed that the zygomycetes are the most primitive fungi. They lack septa except for those that are formed when reproductive structures are cut off from their hyphae. The perforated hyphae of the ascomycetes and basidiomycetes

	TABLE 32-1	CHARACTERISTICS OF THE DIVISIONS OF FUNGI		
		DIVISION		
		ZYGOMYCOTA	ASCOMYCOTA	BASIDIOMYCOTA
Number of species		665	30,000	16,000
Asexual reproduction		Sporangia	Conidia	Fragments, conidia
Sexual reproduction		Zygospores	Asci	Basidia
Septa*		Absent	Present, perforated	Present, perforated, specialized

*The reproductive structures are cut off by complete septa in all fungi; these descriptions refer to the septa in the ordinary, nonreproductive hyphae.

clearly seem to be specializations, and these groups appear to be more advanced than the zygomycetes. Both zygomycetes and ascomycetes have specialized and sometimes quite elaborate asexual spores. Most basidiomycetes have apparently lost asexual reproduction in the course of their evolution; they have a very elaborate mode of cell division, evidently representing a different line of evolution from ascomycetes; both groups may have been derived from zygomycete ancestors.

DIVISION DEUTEROMYCOTA: FUNGI IMPERFECTI

Most of the so-called Fungi Imperfecti, a group that is also called Deuteromycetes, are ascomycetes that have lost the ability to reproduce sexually. The fungi that are classified in this group, however, are simply those in which the sexual reproductive stages have not been observed; some basidiomycetes and zygomycetes are included, in addition to the well-represented ascomycetes. The division of fungi from which a particular nonsexual strain has been derived usually can be determined by the features of its hyphae and asexual reproduction. It cannot, however, be classified by the standards of that group because the classification systems are based on the features related to sexual reproduction. Especially in view of the great economic importance of many of the Fungi Imperfecti, it is necessary to classify them separately, and on the basis of their asexual structures, so that the individual species and genera can be identified (Figures 32-14 and 32-15).

There are some 17,000 described species of Fungi Imperfecti. Even though sexual reproduction is absent among Fungi Imperfecti, there is a certain amount of genetic recombination. This becomes possible when hyphae of different genetic types fuse, as sometimes happens spontaneously. Within the heterokaryotic hyphae that arise from such fusion, genetic recombination of a special kind called **parasexuality** may occur. In parasexuality, the exchange of portions of chromosomes between the genetically distinct nuclei within a common hypha takes place. Recombination of this sort also occurs in other groups of fungi and seems to be responsible for some of the production of new pathogenic strains of wheat rust (see Figure 32-13), for example.

Among the economically important genera of Fungi Imperfecti are *Penicillium* and *Aspergillus* (see Figure 32-15). Some species of *Penicillium* are sources of the well-known antibiotic penicillin, and other species of the genus give the characteristic flavors and aromas to such cheeses as Roquefort and Camembert. Species of *Aspergillus* are used for fermenting soy sauce and soy paste, processes in which certain bacteria and yeasts also play important roles. Citric acid is produced commercially with members of this genus under highly acidic conditions. In addition, the enrichment of livestock feed by the products of fermentation of other species is being investigated. Some species of both *Penicillium* and *Aspergillus* form ascocarps, but the genera are still classified primarily as Fungi Imperfecti, since the ascocarps are found only in a

A

B

FIGURE 32-14
Scanning electron micrograph of the conidiophores of Fungi Imperfecti.
A Alfalfa verticillium wilt, *Verticillium alboatrum*, an important pathogen of alfalfa, has whorled conidia. The single-celled conidia of this member of the Fungi Imperfecti are borne at the ends of the conidiophores.
B *Tolypocladium inflatum*, in which the conidia arise along the branches. This soil fungus is one of the sources of cyclosporine, a drug that suppresses immune reactions and thus assists in making human organ grafts possible; the drug was put on the market in 1979.

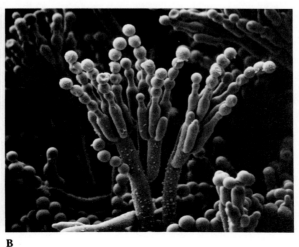

A   **B**

FIGURE 32-15
Conidiophores.
A Characteristic conidiophores of *Aspergillus*, as viewed with the scanning electron microscope.
B A colony of *Penicillium*, another very important genus of Fungi Imperfecti, growing on an orange.

few species, and usually relatively rarely even then. Most of the fungi that cause skin diseases in humans, including athlete's foot and ringworm, are also Fungi Imperfecti.

> **Fungi Imperfecti are fungi either that have lost the capacity for sexual reproduction or in which sexual reproduction has not been observed. For this reason, they cannot be classified by the standards applied to, or placed among the members of, the three divisions of fungi. Instead, they are classified by their asexual reproductive structures. The great majority of Fungi Imperfecti are clearly ascomycetes.**

Fusarium, a member of the Fungi Imperfecti group that occurs widely on food, produces highly toxic substances such as trichothecenes, which have been claimed to be agents of chemical warfare. The phenomenon of "yellow rain" in Southeast Asia, once claimed to be a means of dispersing trichothecenes, has been shown to consist of honeybee feces sprayed by swarms of bees flying at great heights. *Fusarium*, however, is very important as a dangerous agent of food spoilage.

FUNGAL ASSOCIATIONS
Lichens

Lichens (Figures 32-16 and 32-17) are symbiotic associations between a fungus and a photosynthetic partner; they provide an outstanding example of mutualism, the kind of symbiotic association in which both partners benefit. Ascomycetes (including some Fungi Imperfecti) are the fungal partners in all but about 20 of the approximately 15,000 species of lichens estimated to exist; the exceptions, mostly tropical, are basidiomycetes. Most of the visible body of a lichen consists of its fungus, but within the tissues of that fungus are found cyanobacteria or green algae or sometimes both (Figure 32-18). Specialized fungal hyphae penetrate or envelop the photosynthetic cells within them and transfer nutrients directly to the fungal partner. Biochemical "signals" given by the fungus apparently direct its cyanobacterial or green algal component to produce metabolic substances that it does not produce when growing independently of the fungus. The photosynthetic member of the association normally is held between thick layers of interwoven fungal hyphae and is not directly exposed to the light. Enough light penetrates the translucent layers of fungal hyphae to make photosynthesis possible, however. The fungi found in lichens are unable to grow normally without their photosynthetic partner. Overall, this particular symbiotic relationship might be characterized as one of controlled parasitism of the photosynthetic organism by the fungus.

FIGURE 32-16

A fruticose ("shrubby") lichen, *Ramalina menziesii*, growing on the twigs of a tree. The Spanish moss of the southern United States, although superficially similar, is actually a flowering plant, *Tillandsia*, belonging to the pineapple family.

A

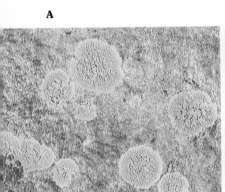

B

C

FIGURE 32-17

Lichens are found in a variety of habitats.

A Crustose (encrusting) lichens growing on a rock in California.

B A fruticose lichen, *Cladina evansii*, growing on the ground in Florida. Fruticose lichens predominate in deserts, because they are more efficient in capturing water from moist air than either of the other two morphological types shown here.

C A foliose ("leafy") lichen, *Parmotrema gardneri*, growing on the bark of a tree in the mountain forest in Panama.

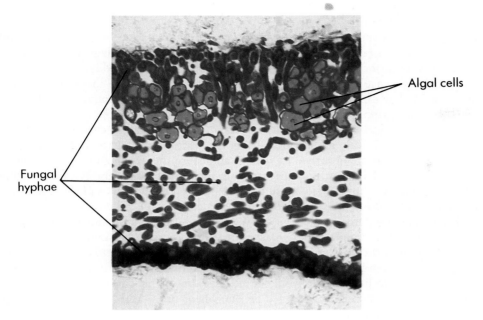

FIGURE 32-18
Section of a lichen. The section shows fungal hyphae more densely packed into a protective layer on the top, and, especially, the bottom layer of the lichen. The green cells near the upper surface of the lichen are those of a green alga. Penetrated by fungal hyphae, these cells supply carbohydrate to the fungus.

The reproduction of lichens may take place by a combination of normal sexual processes: the formation of ascospores in the fungal component and the reproduction, usually asexual, of the photosynthetic component. These two components may come together under favorable circumstances and produce a lichen body. Alternatively, lichens can simply break up into fragments, each of which contains both the fungal and photosynthetic components, and can give rise to a new individual.

The durable construction of the fungus, combined with the photosynthetic properties of its partner, has enabled lichens to invade the harshest of habitats at the tops of mountains, in the farthest northern and southern latitudes, and on dry, bare rock faces in the desert. In such harsh, exposed areas, lichens are often the first colonists, breaking down the rocks and setting the stage for the invasion of other organisms. Those lichens in which the primary or secondary photosynthetic partner is a cyanobacterium are able to fix atmospheric nitrogen, which is leached to the environment, where it can be used by other pioneering organisms.

Lichens are often strikingly colored because of the presence of pigments that probably play a role in protecting the photosynthetic partner from the destructive action of the sun's rays. These same pigments may be extracted from the lichens and used by people as natural dyes, as, for example, in the traditional method of manufacturing Harris tweed, which now, however, is colored with synthetic dyes.

Lichens are symbiotic associations between a fungus—an ascomycete in all but a very few instances—and a photosynthetic partner, which may be a green alga or a cyanobacterium or both.

Lichens are able to survive in inhospitable habitats partly by being able to dry or freeze to a condition that we might call suspended animation. Once the drought or cold has passed, the lichens recover quickly and resume their normal metabolic activities, including photosynthesis. The growth of lichens may be extremely slow in harsh environments: many relatively small ones actually appear to be thousands of years old and therefore are among the oldest living things on earth.

Lichens and Pollution

Paradoxically, lichens are extremely sensitive to pollutants in the atmosphere, and thus they can be used as bioindicators of air quality. Their sensitivity is the result of their ability to absorb substances dissolved in rain and dew readily, a property on which their existence depends but one that can also prove fatal to them in a polluted environment. For example, lichens are generally absent in and around cities because of automobile traffic and industrial activity, even though suitable substrates exist. They are acutely sensitive to sulfur dioxide and are disappearing even from national parks and other relatively remote areas being reached increasingly by industrial pollution. The pollutants are absorbed by lichens and cause the destruction of chlorophyll, decreases in photosynthesis, and alterations in membrane permeability, among other problems. Similarly, the physiological balance between the fungus and the alga or cyanobacterium is upset, and degradation or destruction of the lichen results.

Lichens are used as a means of assessing the pollution by radionuclides that occur in the vicinity of uranium mines, crashed satellites, and other similar sites, just as they were used in the days of atmospheric nuclear weapon testing to estimate the dispersal of radioactive fallout from the explosions. Similarly, the absorption of radioactive dust by arctic lichens following the nuclear disaster at Chernobyl in 1985 rendered the meat of the reindeer that fed on the lichens unsuitable for human consumption for many months.

MYCORRHIZAE

The roots of about 80% of all kinds of plants normally are involved in symbiotic relationships with certain specific kinds of fungi; it has been estimated that these fungi probably amount to 15% of the total weight of the world's plant roots. Associations of this kind are termed **mycorrhizae** (from the Greek words for "fungus" and "roots"). To a certain extent, the fungi in mycorrhizal associations replace and play the same function as the fine projections from the epidermis, or outermost cell layer, of the terminal portions of the roots—root hairs. When mycorrhizae are present, they aid in the direct transfer of phosphorus, zinc, copper, and probably other nutrients from the soil into the roots. The plant, on the other hand, supplies organic carbon to the symbiotic fungus (Figure 32-19) so that the system represents an excellent example of mutualism.

FUNGI AS CARNIVORES

It might surprise you to know that some fungi are predatory. For example, the mycelium of the edible oyster fungus, *Pleurotus ostreatus* (Figure 32-B), excretes a substance that anesthetizes tiny roundworms known as nematodes (Chapter 39) that feed on the fungus. When the worms become sluggish and inactive, the fungal hyphae envelop and penetrate their bodies and absorb their nutritious contents. Oyster fungi usually grow within living trees or on old stumps, periodically producing tiers of large, fleshy fruiting bodies. Their mycelium obtains the bulk of its glucose through the enzymatic digestion of cellulose from the wood, so that the nematodes it consumes apparently serve mainly as a source of nitrogen—a substance almost always in short supply in biological systems. Other fungi are even more active predators than *Pleurotus*, snaring, trapping or firing projectiles into nematodes, rotifers and other small animals on which they prey.

FIGURE 32-B
The oyster mushroom, *Pleurotus ostreatus*, immobilizes nematodes, which the fungus uses as a source of food.

FIGURE 32-19
**Endomycorrhizae enhance
growth in plants.** Shown here are
soybean plants without endomycorrhizae (*left*) and with different
strains of endomycorrhizae (*center*
and *right*).

There are two principal types of mycorrhizae: **endomycorrhizae,** in which the fungal hyphae penetrate the outer cells of the plant root, forming coils, swellings, and minute branches, and also extend out into the surrounding soil; and **ectomycor-rhizae,** in which the hyphae surround but do not penetrate the roots. In both kinds of mycorrhizae, the mycelium extends far out into the soil.

Endomycorrhizae (Figure 32-20) are by far the more common of these two types. The fungal component in them is a zygomycete. Only about 30 species of zygo-mycetes are known to be involved in such relationships throughout the world. These few species of zygomycetes become associated with many species of plants, perhaps more than 200,000. Endomycorrhizal fungi are being studied intensively because they are potentially capable of leading to increased crop yields with lower phosphate and energy inputs. When cultivated fields are fumigated, for example, to control nema-todes (worms that attack plants; Chapter 39), the populations of endomycorrhizal zy-gomycetes may be suppressed and need to be restored for full productivity.

Ectomycorrhizae (Figure 32-21) involve far fewer kinds of plants than do endom-ycorrhizae, perhaps only a few thousand. They are characteristic of certain groups of trees and shrubs, particularly those of temperate regions, including pines, firs, oaks, beeches, and willows. The fungal components in most ectomycorrhizae are basidio-

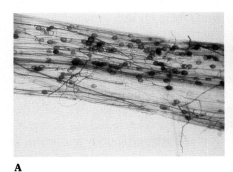

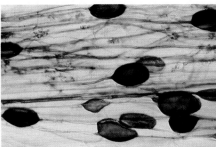

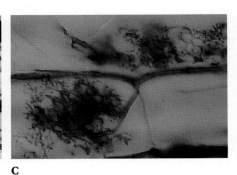

A   B   C

FIGURE 32-20
Endomycorrhizae. The zygomycete *Glomus versiforme* is shown here growing in the roots of leeks, *Allium porrum.*
A General view of the leek root, showing vesicles (sac-like structures), photographed with a dissecting microscope. The root has been squashed in fluid. (×16.)
B Vesicles in leek root. (×63.)
C Arbuscules in leek root. (×63.) Arbuscules are branching structures characteristic of young infections, with vesicles predominating later.

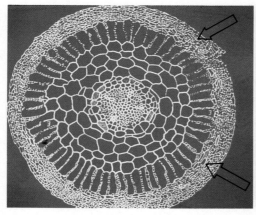 

A   B

FIGURE 32-21
Ectomycorrhizae.
A Transection of a pine root, showing thick mantle (*arrows*) of ectomycorrhizae.
B Ectomycorrhizae on roots of pines. At the left are the yellow-brown mycorrhizae formed by *Pisolithus tinctorius;* in the center are the white mycorrhizae formed by *Rhizopogon;* at the right are pine roots not associated with a fungus. The different basidiomycetes that participate in the formation of ectomycorrhizae are usually highly specific, and, as seen here, several kinds may form mycorrhizal associations with one plant species. Different combinations have different effects on the physiological characteristics of the plant and its ability to survive under different environmental conditions.

mycetes, but some are ascomycetes. At least 5000 species of fungi are involved in ectomycorrhizal relationships, and most of them are restricted to a single species of plant.

> **Mycorrhizae are symbiotic associations between plants and fungi. There are two main types: (1) endomycorrhizae, which occur in about 80% of all plants and in which the fungal partner is a zygomycete, and (2) ectomycorrhizae, which form in certain kinds of trees and shrubs and in which the fungal partner usually is a basidiomycete.**

The kinds of plants involved in ectomycorrhizal associations evidently are more resistant to drought, cold, and other harsh environmental conditions than are the plants that lack such fungal partners. This may be why they are so well represented among trees and shrubs that grow at timberline in the mountains and in the far north and south. Recently, it has been found that some kinds of ectomycorrhizal fungi provide better protection from the effects of acid precipitation to the trees with which they are associated than do others. In some way, the fungi often seem to protect the plants with which they are associated from the high accumulations of heavy metals, such as copper and zinc, that are absorbed by the plants under acid conditions.

Another type of mycorrhizal relationship involves orchids. The seeds of orchids, which are so minute as to appear dustlike, germinate in nature only in the presence of suitable fungi, members of a particular group of basidiomycetes. Apparently the minute seeds of the orchids do not have enough stored food material to allow them to grow without fungi, which supply them with carbon. Orchid growers achieve the same results with many species of orchids by growing them on agar in which a carbon source is included. Some of these, however, may germinate in a matter of days rather than months in association with the appropriate fungus. Perhaps 100 species of basidiomycetes are associated with orchids in this way.

The earliest fossil plants often have endomycorrhizal roots. Such associations, which were common during the initial period of invasion of the land by plants, may have played an important role in allowing this invasion to take place. The soils that were available at such times would have been sterile and completely lacking in organic matter. Since plants that form mycorrhizal associations are particularly successful in infertile soils now, and considering the fossil evidence, the suggestion that mycorrhizal associations were characteristic of the earliest plants seems reasonable. In addition, the most primitive vascular plants surviving today continue to depend strongly on mycorrhizae.

1. The fungi are a distinct kingdom of eukaryotic organisms characterized by their filamentous growth form, lack of chlorophyll and motile cells, chitin-rich cell walls, and external digestion of food by the secretion of enzymes. Together with the bacteria, they are the decomposers of the biosphere.

2. Fungal filaments, called hyphae, collectively make up the fungus body, which is called the mycelium. A hypha may contain genetically uniform nuclei and thus be homokaryotic, or it may contain two or more genetically different kinds of nuclei and be heterokaryotic. Mitosis in fungi occurs without centrioles and within the nuclear envelope.

3. In many fungi, the two kinds of nuclei that will eventually undergo syngamy occur together in hyphae for a long period before they fuse. Most hyphae of basidiomycetes found in nature, for example, are dikaryotic.

4. Meiosis occurs immediately after the formation of the zygote in all fungi; the zygote therefore is the only diploid nucleus of the entire life cycle in these organisms.

5. There are three divisions of fungi: Zygomycota, the zygomycetes; Ascomycota, the ascomycetes; and Basidiomycota, the basidiomycetes. Zygomycetes form septa only when gametangia or sporangia are cut off at the ends of their hyphae; otherwise, their hyphae are multinucleate. Most hyphae of ascomycetes and basidiomycetes have perforated septa through which the cytoplasm, but not necessarily the nuclei, flows freely.

6. Cells within the heterokaryotic hyphae of the ascomycetes are multinucleate; those within the heterokaryotic hyphae of the basidiomycetes are dikaryotic. Zygotes in ascomycetes form within club-shaped structures known as asci, and those in basidiomycetes form within structures known as basidia, which initially are somewhat similar. Asci and basidia are formed within or on the surfaces of often large structures like mushrooms, which are formed of interwoven hyphae.

7. Asexual reproduction in zygomycetes takes place by means of multinucleate sporangia; that in ascomycetes takes place by means of multinucleate conidia. Asexual reproduction in basidiomycetes is relatively rare, and may involve either hyphal fragmentation or the formation of conidia.

8. The yeasts are a group of unicellular fungi, mostly ascomycetes. They are commercially important because of their roles in baking and fermentation.

9. The Fungi Imperfecti are a large artificial group of fungi in which sexual reproduction does not occur or is not known; most are ascomycetes. They are classified according to the characteristics of their spore-forming structures and spores. Many are of great commercial importance.

10. Symbiotic systems involving fungi include lichens and mycorrhizae. The fungal partners in lichens are almost entirely ascomycetes, which derive their nutrients from green algae, cyanobacteria, or both.

11. Mycorrhizae are symbiotic associations between plants and fungi that are characteristic of the great majority of plants. Endomycorrhizae, which are more common, have zygomycetes as the fungal partner, whereas ectomycorrhizae have mainly basidiomycetes.

REVIEW QUESTIONS

1. What are the superficial similarities between fungi and the plant kingdom with which they were formerly classified? What are four of the more apparent differences between the two groups?

2. What are two of the known mutualistic associations that involve the fungi? What is the nonfungal partner of each association? What is the function of each partner in the associations?

3. What is a hypha? What is a mass of hyphae called? What is its function? What is a septum as it applies to the fungi? Under what circumstances are they complete? What is the composition of the fungal cell wall? Why is this composition an advantage to the fungi?

4. Which fungal nuclei are diploid? Which are haploid? What event occurs during sexual reproduction in the fungi? How does this affect the nuclei within the majority of the fungi? To what do the following terms refer and which are associated with one another: heterokaryotic, homokaryotic, dikaryotic, and monokaryotic?

5. What are the three divisions of the kingdom Fungi? What characteristic distinguishes the divisions from one another? Which of the divisions is least like the other two? Why?

6. What are the ascomycete asexual spores called? Are these spores uninucleate or multinucleate? What are the structures on

which they are borne called? Are nonreproductive hyphae of this division septate or nonseptate?

7. What are ascogonia and antheridia? From which gametangium does the trichogyne grow? What is its function? What type of hyphae arise from its product?

8. To what division do the yeasts belong? How are they different from other fungi? Is it more likely that this characteristic is primitive or degenerate? What is the yeasts' primary mode of reproduction? Why are the yeasts gaining popularity in scientific research?

9. With reference to the division Basidiomycota, what is primary mycelium? What is secondary mycelium? What is a basidiocarp? Are the basidiomycetes composed of monokaryotic or dikaryotic hyphae? Where in a common mushroom are the basidia found?

10. What are the Fungi Imperfecti? Which division seems to be best represented in this group? By what means can individuals in the class Deuteromycota be classified?

11. What are lichens? Which fungal division is best represented in the lichens? Are these species able to grow independently of this association? Why or why not?

12. What is the difference between endomycorrhizae and ectomycorrhizae? Which fungi are most prominent in endomycorrhizae?

THOUGHT QUESTIONS

1. Why do we say there are three divisions of fungi when we have mentioned five major groups: Ascomycota, Zygomycota, Basidiomycota, Fungi Imperfecti, and lichens? What are the major distinguishing characteristics of these groups? Why are only three considered divisions?

2. If you had a sample of fungal hyphae, without characteristic reproductive structures, could you determine the major group to which the organism belonged? How?

3. What is there about the way fungi live in nature that helps to make them particularly valuable in industrial processes?

FOR FURTHER READING

ANGIER, N.: "A Stupid Cell with All the Answers," *Discover,* November 1986, pages 70-83. An excellent account of the many uses of yeast for experimental biology.

BESSETTE, A., and W.J. SUNDBERG: *Mushrooms: A Quick Reference Guide to Mushrooms of North America,* Macmillan Publishing Co., New York, 1987. Color photographs illustrate the species most frequently encountered.

COOKE, W.B.: *The Fungi of Our Moldy Earth,* J. Cramer, Berlin, 1986. Description of techniques for collecting, preparing, isolating, and identifying fungi, in addition to a discussion of their ecology and classification.

HALE, M.E.: *The Biology of Lichens,* ed. 3, University Park Press, Baltimore, Md., 1983. A concise account of all aspects of the biology of lichens.

KENDRICK, B.: *The Fifth Kingdom,* Mycologue Publications, Waterloo, Ontario, 1985. Excellent text on fungi that deals with their classification and biology successfully and imaginatively.

KOSIKOWSKI, F.V.: "Cheese," *Scientific American,* May 1985, pages 88-99. A fascinating account of the ways in which more than 2000 varieties of cheese are made and of the participation of bacteria and fungi in the process.

McKNIGHT, K.H., and V.B. McKNIGHT: *A Field Guide to Mushrooms of North America,* Peterson Field Guide Series, Princeton, N.J., 1987. Excellent identification guide to the common edible and poisonous species of the United States and Canada.

NEWHOUSE, J.R.: "Chestnut Blight," *Scientific American,* July 1990, pages 106-111. Biological control is now being used in an effort to stop the ravages of this fungus.

WEBB, A.D.: "The Science of Making Wine," *American Scientist,* vol. 72, pages 360-367, 1984. Traces the ways in which an ancient practical art has become a modern science.

Biology of Plants

Diversity of Plants

OVERVIEW

Plants evolved more than 430 million years ago from multicellular green algae. By 300 million years ago, trees had evolved and formed forests, within which the diversification of vertebrates, insects, and fungi occurred. Roughly 266,000 species of plants are now living. Bryophytes and ferns require free water so that sperm can swim between the male and female sex organs; most other plants do not. Vascular plants have elaborate water- and food-conducting strands of cells, cuticles, and stomata; many of these plants are much larger than any bryophyte. Seeds evolved among the vascular plants and provide a means to protect young individuals. Flowers, which are the most obvious characteristic of angiosperms (division Anthophyta), guide the activities of insects and other pollinators so that pollen is dispersed rapidly and precisely from one flower to another of the same species, thus promoting outcrossing. Many angiosperms display other modes of pollination, including self-pollination.

FOR REVIEW

Here are some important terms and concepts that you will encounter in this chapter. If you are not familiar with them, you should review them before proceeding.

Alternation of generations (Chapter 22)

Evolutionary history of plants (Chapter 22)

Classification (Chapter 28)

Green algae (Chapter 31)

Mycorrhizae (Chapter 32)

Despite all the diversity of life on earth, only three groups are responsible for virtually all of the fixation of carbon that occurs and thus for providing the substance of all living organisms. The algae and photosynthetic bacteria carry out most marine photosynthesis, and plants are the dominant photosynthetic organisms on land. Plants are multicellular eukaryotic organisms that have cellulose-rich cell walls, chloroplasts that contain chlorophylls *a* and *b* together with carotenoids, and, as their primary carbohydrate food reserve, starch. In this chapter, we will discuss the characteristics of the divisions of plants, focusing on differences in their life cycles. In Chapters 34 to 38, we will treat specific aspects of the biology of plants in more detail.

THE EVOLUTIONARY ORIGINS OF PLANTS

Plants are almost certainly derived from an organism that, if it existed today, would be classified as multicellular green algae (division Chlorophyta). The extensive biochemical and morphological similarities between green algae and plants make this conclusion inescapable. Recall, for example, the patterns of cytokinesis discussed in Chapter 10: after mitosis is complete, animal cells pinch inward toward the center, whereas plant cells form an interior cell plate, which grows outward between the two daughter cells. Aside from plants, the only organisms that form a cell plate are a few groups of green algae.

By tradition, the major groups of plants are called *divisions*, rather than phyla, although the two units of classification are identical in rank and significance. The living plants, according to the system of classification used in this book, are grouped into 12 divisions (Figure 33-1). The members of nine of these, considered to be vascular plants, have water- and food-conducting strands, consisting of elongate, specialized cells, in their stems and roots. The other three divisions, traditionally considered "bryophytes," either lack such strands or have poorly developed ones. These three divisions are the mosses (division Bryophyta; the name has traditionally be used to include all "bryophytes," but it is here restricted to mosses); liverworts (division Hepaticophyta); and hornworts (division Anthocerotophyta). All other plants, including ferns, conifers, flowering plants, and related groups, are vascular plants.

The similarities between all plants are so great that we believe they had a single common ancestor among the green algae. This ancestor was probably semiterrestrial.

FIGURE 33-1
Representatives of four divisions of plants.
A A moss gametophyte; no sporophytes have formed here.
B A branch of Norway spruce, *Picea nigra*, in the Alps. Seeds are produced in the large cones, and pollen is produced in the smaller ones.
C Maidenhair fern, *Adiantum pedatum*.
D The white flowers of this *Bougainvillea* are surrounded by bright purple bracts—modified leaves. Only flowering plants add color to the landscape; their flowers are specialized to attract insects and other animals. In all vascular plants, including the plants here (except for the moss), the sporophyte is the conspicuous generation.

B   C   D

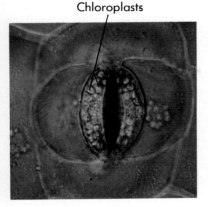

Chloroplasts

FIGURE 33-2
A stoma. The guard cells flanking the stoma, unlike the other epidermal cells, contain chloroplasts. Water passes out through the stomata, and carbon dioxide enters by the same portals.

Fossils of simple vascular plants are known from the Silurian period, some 430 million years ago, so that ancestral plants, lacking vascular tissue, must have evolved even earlier.

Plants evolved from a multicellular green alga that possessed cell walls composed predominantly of cellulose and accomplished cytokinesis by means of a cell plate.

THE GREEN INVASION OF THE LAND

Plants, fungi, and insects are the only major groups of organisms that occur almost exclusively on land; several other phyla, including chordates (see Chapter 42) and mollusks (see Chapter 40), are very well represented on land. The groups that occur almost exclusively in terrestrial and freshwater habitats, however—the plants, fungi, and insects—all probably evolved there, whereas the chordates and mollusks certainly originated in the water. Of the groups that did evolve on land, the ancestors of the plants were almost certainly the first to become terrestrial. As we have seen in the case of fungi, and will see in the next section when we discuss insects, a major evolutionary challenge in the transition from a marine to a terrestrial habitat is **desiccation**—the tendency of organisms to lose water to the air. A key factor that doubtless helped to make possible the invasion of the land by the first plants was the possession by their green algal ancestors of relatively tough and durable cell walls that resisted drying out. Cell walls of this kind also made possible the evolution of reproductive structures that could exist in dry air without damage.

Most plants also are well protected from desiccation in air by their **cuticle,** an outer covering formed from a waxy substance called **cutin.** The cuticles that cover the exposed surfaces of plants are relatively impermeable to water and so provide a key barrier to water loss. However, a completely impermeable barrier would prevent the entry of carbon dioxide; as a result, food production by photosynthesis would be inadequate. This problem is solved by passages through the cuticle and epidermis—specialized pores called **stomata** (singular, **stoma;** Figure 33-2) in the leaves and sometimes in the green portions of the stems. Stomata allow carbon dioxide to pass into leaves and oxygen to leave the plant by diffusion. Unfortunately the pores also allow water vapor to escape, necessitating its constant replacement. The cells that border stomata expand and contract, thus controlling the movement of water and carbon dioxide.

As in the case of the insects, efficient conducting systems helped the earliest vascular plants to achieve their ecological dominance of the land. Vascular plants are so named because they have **vascular tissue.** The word *vascular* comes from the Latin *vasculum,* meaning a vessel or duct, and refers, in the case of plants, to their conducting systems (Figure 33-3). Vascular tissue consists of specialized strands of elongated cells that run from near the tips of a plant's roots through its stems and into its leaves. The vascular tissue conducts both water with dissolved minerals, which comes in through the roots, and carbohydrates, which are manufactured in the green, illuminated portions of the plant. Therefore, water and nutrients reach all parts of the plant, as do the carbohydrates that provide energy for the synthesis of its different structures.

The earliest plants of which we have fossil remains were very simple in structure. Some of them, however, had efficient conducting systems (Figure 33-4), and we do not know what the very earliest plants looked like.

Vascular plants are defined primarily by their possession of a specialized conducting system, the vascular system, which involves strands of elongated, specialized cells that transport water and dissolved nutrients, and other strands that transport carbohydrate molecules. The possession of a cuticle and stomata is also more characteristic of vascular plants.

In addition to their structural features, plants developed a special kind of relationship with fungi that evidently has been of key importance in their successful oc-

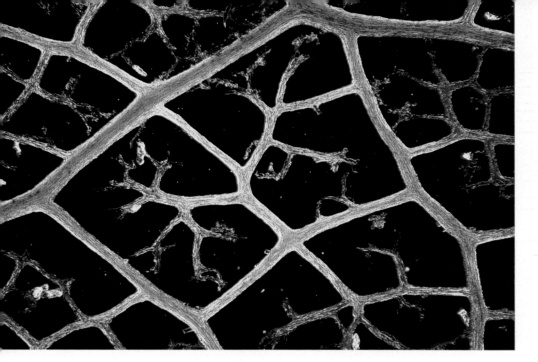

FIGURE 33-3
Vascular tissues. The veins of a vascular plant—which contain strands of specialized cells for conducting water, with its dissolved minerals and carbohydrates, respectively—run from the tips of the roots to the tips of the shoots and throughout the leaves, as shown here in this greatly enlarged photomicrograph of a cleared leaf.

cupation of terrestrial habitats. Endomycorrhizae are characteristic of some 80% of all plants and are frequently seen in fossils of the earliest plants. We presume that they probably also occurred in the common ancestor of plants. These symbiotic associations were probably critical to the ancestral plants as they adjusted to life on land. Certainly, mycorrhizae play an important role among today's plants in the assimilation of phosphorus and probably other ions, and they undoubtedly did so in the raw, unaltered soils that existed before the plants changed them. Just as lichens (see Chapter 32), which are symbiotic associations of fungi and green algae or cyanobacteria, grow in many places where neither of their component organisms could survive alone, so the early plants, with their mycorrhizae, probably fared better on terrestrial soils than would have been possible otherwise.

Many other features developed gradually and aided the evolutionary success of plants on land. For example, in the first plants there was no fundamental difference between the aboveground and underground parts. Later, roots and shoots with specialized structures evolved, each better suited to its particular environment than the others (see Chapter 35). Leaves—expanded areas of photosynthetically active tissue—evolved and diversified in relationship to the varied habitats that existed on land. Specializations in key reproductive features improved the methods by which plants protected their young and were dispersed from place to place.

The newly evolved and more specialized roots, stems, leaves, and reproductive features that evolved in plants all were important factors in the rise and overwhelming success of this group on land. There are some 266,000 species of plants now in existence. They dominate every part of the terrestrial landscape, except for the extreme deserts, polar regions, and the tops of the highest mountains.

THE PLANT LIFE CYCLE
Alternation of Generations

Understanding the different kinds of life cycles that occur among plants provides an important key to understanding their evolutionary relationships. All plants exhibit alternation of generations, which also occurs in the brown, red, and green algae (see Chapter 31). In all of them, the diploid generation, or sporophyte, alternates with the haploid generation, or gametophyte. *Sporophyte* literally means "spore-plant," and *gametophyte* means "gamete-plant"; these terms indicate the kinds of reproductive structures that the respective generations produce (Figure 33-5).

Most adult animals are diploid, and in this respect they resemble the sporophyte

Sporangia

FIGURE 33-4
***Cooksonia*, the first known vascular land plant.** Its upright, branched stems, which were no more than a few centimeters tall, terminated in sporangia, as seen here. It probably lived in moist environments such as mud flats, had a resistant cuticle and produced spores typical of vascular plants. This fossil represents a plant that occurred some 410 million years ago, in what is now New York State. *Cooksonia* belonged to a now extinct division of vascular plants and is also shown in Figure 33-13 in a reconstructed community from the early Devonian Period.

FIGURE 33-5
Generalized plant life cycle.

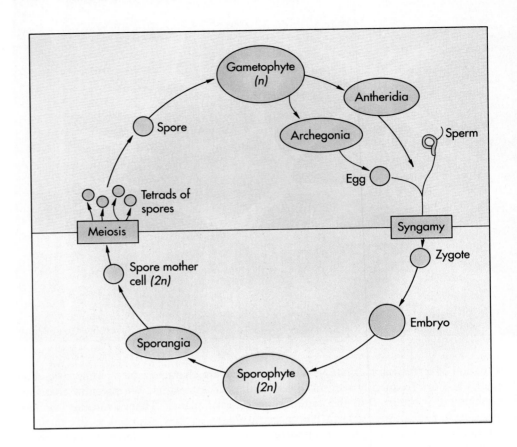

generation of a plant. Such animals, however, produce eggs and sperm, which fuse directly to form a zygote. In contrast, the sporophyte generation of a plant does not produce gametes as a result of meiosis. Instead, meiosis takes place in specialized cells, called **spore mother cells,** and results in the production of haploid **spores,** the first cells of the gametophyte generation. Spores do not fuse with one another, as gametes do; instead, they divide by mitosis, producing a multicellular haploid individual, the gametophyte (Figure 33-6).

In turn, the gametes—eggs and sperm—eventually are produced by the gametophyte, as a result of mitosis. They are haploid, like the gametophyte that produces them. When they fuse to form a zygote, the first cell of the next sporophyte generation has come into existence. The zygote grows into a sporophyte in which meiosis ultimately occurs.

Plant life cycles are marked by an alternation of generations of diploid sporophytes with haploid gametophytes. Sporophytes, as a result of meiosis, produce spores, which grow into gametophytes. Gametophytes produce gametes, as a result of mitosis.

FIGURE 33-6
Alternations of generations.
A The flattened, somewhat circular plants growing on the forest floor in New Zealand are the gametophytes of a tree fern; like most fern gametophytes, they are bisexual and photosynthetic.
B Pine gametophytes (pollen grains) are just large enough to be visible with the naked eye.

A

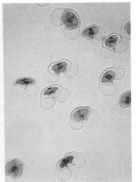

B

The basic cycle involved in alternation of generations can be summarized as zygote-sporophyte-meiosis-spores-gametophyte-gametes (eggs and sperm)-syngamy-zygote (see Figure 33-5). Such a cycle is characteristic of all plants. In some, including bryophytes and ferns, the gametophyte is green and free living; in others it is not green and is nutritionally dependent on the sporophyte. When you look at a moss or liverwort, what you see is largely gametophyte tissue; the sporophytes are usually smaller brownish or yellowish structures attached to or enclosed within the tissues of the gametophyte. In vascular plants, the gametophytes are always much smaller than the sporophytes, and, in most groups, are nutritionally dependent on them and enclosed within their tissues. What you see when you look at a vascular plant, with rare exceptions, is a sporophyte.

The Specialization of Gametophytes

Gametes of the first plants differentiated within specialized organs called **gametangia** (singular, **gametangium**), in which the eggs and sperm were surrounded by a jacket of cells. Complex multicellular gametangia of this sort are still found in all of the less-specialized members of the plant kingdom, including mosses, liverworts, ferns, and a few other groups. Eggs and sperm are formed within different kinds of gametangia. Those in which eggs are formed are called **archegonia** (singular, **archegonium**), and those in which sperm are formed are called **antheridia** (singular, **antheridium**). An archegonium produces only one egg; an antheridium produces many sperm. These structures usually look very different from one another. In some plants, including the ferns, antheridia and archegonia exist together on the same gametophyte; in other members of these groups, and commonly in mosses, the two kinds of gametangia are borne on separate gametophytes. In the more specialized vascular plants, including all but a few of the vascular plants that form seeds, the gametangia have been lost during the course of evolution, and the eggs or sperm differentiate from individual cells of the respective gametophyte. Some of the gametophytes bear only eggs, and others bear only sperm.

> **Gametophytes may produce two different kinds of gametangia: (1) antheridia, which produce sperm, and (2) archegonia, each of which produces a single egg. In bryophytes these gametangia are multicellular, but these complex structures have been lost during the evolution of the more advanced vascular plants, although they are still present in a few seed-bearing plants.**

When one kind of gametophyte bears antheridia and another kind bears archegonia, the two kinds of gametophytes may look different from one another. If they do, the gametophytes that form antheridia are called **microgametophytes,** and those that form archegonia are called **megagametophytes.** These two formidable-looking terms literally mean no more than "small gametophytes" and "large gametophytes." The terms were used originally because the two kinds of gametophytes differ greatly in size in many kinds of plants.

In nearly all plants that do have two different kinds of gametophytes, the gametophytes arise from two different kinds of spores, **microspores** and **megaspores.** Plants that produce two morphologically different kinds of spores are called **heterosporous** (Figure 33-7); those that produce only one kind of spore are **homosporous.** Under this terminology, all bryophytes are homosporous, even though in many of them morphologically identical spores may give rise to either male or female gametophytes.

Spores are formed as a result of meiosis in the sporophyte generation. Their differentiation occurs within specialized, multicellular structures called **sporangia** (singular, **sporangium**). If a plant forms both megaspores and microspores, each of these will be formed in a different kind of sporangium, called, respectively, **microsporangia** and **megasporangia.**

Having now completed our overview of plant life cycles, we will consider the major groups of plants. As we do so, we will see a progressive reduction of the gameto-

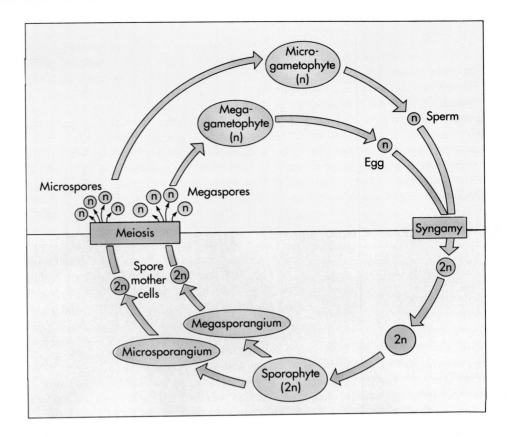

phyte from group to group, a loss of multicellular gametangia, and increasing specialization for life on the land, culminating with the remarkable structural adaptations of the flowering plants, the dominant plant group today.

MOSSES, LIVERWORTS, AND HORNWORTS

Bryophytes are especially common in relatively moist places both in the tropics and in temperate regions (Figure 33-8). In the Arctic and the Antarctic, they are the most abundant plants, boasting not only the highest number of individuals in these harsh regions, but also the highest number of species. Many mosses are able to withstand prolonged periods of drought, although they are not common in deserts; they do not appear to share with lichens the ability to absorb an adequate supply of moisture directly from the air, and thus require free water for growth. Like lichens, however, most mosses are remarkably sensitive to air pollution, and they are rarely found in abundance in or near cities or in other areas with high levels of air pollution. Also, regardless of how well or how long they can resist drought, the mosses, like ferns and certain other plants, require free water to reproduce sexually, as well as for their growth and development; they resemble the amphibians among the vertebrates in this respect. The places where they grow must be wet, at least during certain seasons.

Most mosses, liverworts, and hornworts (Figure 33-9) are small, and few exceed 2 centimeters in length. Their gametophytes are green and manufacture their own food; they are relatively large and evident (see Figures 33-1, *A*, and 33-9) as compared with the sporophytes, which obtain their food directly from the gametophytes. Some of their sporophytes are completely enclosed within gametophyte tissue; others, which are not enclosed, turn brownish or straw-colored at maturity.

The two major features that distinguish the members of these three divisions from vascular plants are:

1. The general lack of specialized vascular tissues; the sporophytes of many moss species, and the gametophytes of some, have a central strand of somewhat specialized water-conducting tissue in their stems, a condition analogous to that which occurs in vascular plant sporophytes, but food-conducting tissue has

A

B

FIGURE 33-9
Bryophytes.
A A liverwort, *Marchantia*. The sporophytes are borne within the tissues of the umbrella-shaped structures that arise from the surface of the flat, green, creeping gametophyte. These particular structures develop archegonia within their tissues; another kind of similar structure, borne on different plants of *Marchantia*, produces the antheridia.
B A hair-cup moss, *Polytrichum*. The leaves below belong to the gametophyte. Each of the yellowish brown stalks, with the capsule at its summit, is a sporophyte. Although moss sporophytes may be green and carry out a limited amount of photosynthesis when they are immature, they are soon completely dependent, in a nutritional sense, on the gametophyte.

been identified in only a few genera. Even if such tissues are present, their structures are much less complex than those found in the vascular plants.

2. The fact that the sporophytes of mosses, liverworts, and hornworts. are almost always smaller than, and always derive their food from, the gametophytes.

Most mosses have distinct leaves and stems, whereas hornworts and many liverworts do not. The gametophytes of many of the liverworts and all of the hornworts are strap shaped. Even where the stems and leaves are distinct, as in most mosses, there is no anatomical differentiation between the roots and shoots; in this respect they differ from nearly all living vascular plants. A characteristic feature of the plants in these three divisions is the presence of slender, usually colorless projections called **rhizoids,** which consist of one or a few cells. Rhizoids anchor these simple plants to their substrate, but do not play a major role in absorbing water or minerals, which often enter a bryophyte directly through its stems or leaves.

Liverworts and hornworts lack specialized vascular tissue, whereas some mosses have both water-conducting strands and a few also have food-conducting strands. In all three groups, the sporophytes photosynthesize only a limited extent, if at all, and draw their food from the gametophytes on which they are produced.

The Divisions of Non-vascular Plants

As previously mentioned, the three divisions of plants that have traditionally been grouped as "bryophytes," because of their relatively simple structure, are the liverworts (division Hepaticophyta), the hornworts (division Anthocerophyta), and mosses (division Bryophyta, a name that has customarily included the other two groups as well as mosses). The features that these three divisions have in common are generalized, primitive ones, and they therefore do not indicate that they are directly related to one another. The frequent presence of specialized water-conducting tissue in mosses, for example, probably indicates that mosses are, in effect, reduced vascular plants; in contrast, liverworts and hornworts have no such tissue. The structural simplicity of mosses may have resulted from reduction, whereas the ancestors of the other two divisions may never have been more complex than the living hornworts and liverworts.

Hepaticophyta: The Liverworts.   There are about 6500 species of liverworts. They were given their name in medieval times, when the common belief in a "Doctrine of Signatures" held that a plant that resembled a particular part of the body was probably good for treating a disease of that organ; some common liverworts have an outline that resembles a liver, and thus they were thought to be useful in treating liver

ailments. The ending -*wort* simply means "herb"—that is, not a tree or shrub—and is found in many plant names of Anglo-Saxon derivation. In some liverworts, the gametophytes (see Figure 33-9, *A*) grow flat along the ground and are relatively undifferentiated; they have a growing point at one end, where cell division takes place continuously, adding to the length of the liverwort. Other liverwort gametophytes have simple leaves, stems, and rhizoids, resembling mosses in their complexity. The sporophytes of liverworts are often unstalked and more or less spherical and usually are held within the gametophyte tissue until they shed their spores.

Anthocerophyta: The Hornworts.

The gametophytes of the approximately 100 species of hornworts resemble those of *Marchantia* and similar genera among the liverworts. The sporophytes of hornworts, however, differ remarkably from those of liverworts: they are elongated sporangia that stand up from the surface of the creeping gametophytes like horns, thus giving the English name to the members of this division of plants.

Bryophyta: The Mosses.

There may be as many as 10,000 species of mosses, small plants that are found everywhere on earth. Many superficially similar organisms are called mosses; for example, "Spanish moss" is actually a flowering plant, a relative of the pineapple. True mosses have a distinctive set of features, however, that sets them apart from other plants. Their gametophytes are almost always leafy, with small, simple leaves; and the plants may be tufted (see Figure 33-9, *B*) or creeping (see Figure 33-1, *A*). Although moss leaves are basically in three rows, spirally arranged around the stem, the leafy branches appear flattened in the creeping species. Moss sporophytes are often yellowish or brownish at maturity, bear a sporangium or capsule near their tip, and are borne individually on the gametophytes.

The flask-shaped archegonia of mosses, which are similar to those of liverworts (Figure 33-10, *A*) are found among the leaves at the ends of the stems. A single egg is produced in the swollen basal portion of each archegonium. When this egg is mature, the central cells in the neck of the archegonium disintegrate, leaving an opening at the top and a column of fluid through which the sperm, which meanwhile have been released from the nearby antheridia, swim to the egg.

Moss antheridia commonly are stalked (Figure 33-10, *B*). In them, an outer layer of cells surrounds the specialized cells that become the sperm, which are twisted and bear two flagella each. These sperm swim through drops of water from rain, dew, or other sources to the neck of the archegonium and then into it to the egg. Fertilization takes place within the swollen basal portion of the archegonium. Inside the archegonium, the zygote develops into a young sporophyte, which grows out of the archegonium and differentiates into a slender, basal stalk, or **seta,** which has a swollen capsule, or **sporangium,** at its apex.

Haploid spores are produced within the sporangium by meiosis. In mosses, the

FIGURE 33-10
Gametangia of nonvascular plants.
A Transection through archegonia of the liverwort *Marchantia*. Such archegonia are borne on the umbrella-shaped structures shown in Figure 33-9, *A*. A single egg differentiates within the lower, swollen portion of each archegonium.
B Transections through a group of moss antheridia. The small cells in each of the elongate structures will give rise to biflagellated sperm. The sperm, when liberated by the rupturing of the antheridium, will swim through free water to the mouth of the archegonium.

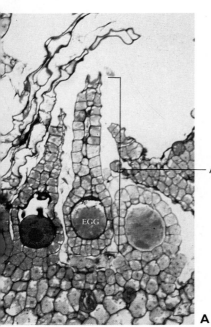

Archegonium

EGG

A

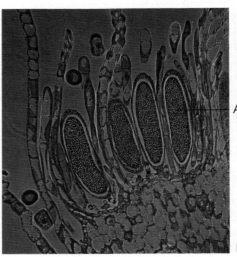

Antheridium

B

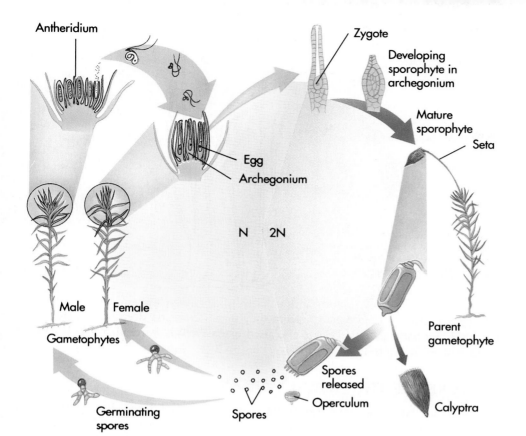

Antheridium

Zygote

Developing sporophyte in archegonium

Egg
Archegonium

Mature sporophyte

Seta

N   2N

Male   Female

Gametophytes

Parent gametophyte

Germinating spores

Spores released

Operculum

Spores

Calyptra

FIGURE 33-11
Moss life cycle. Gametophyte tissue (haploid) is shown in green, sporophyte tissue (diploid) in brown.

capsule often is covered, at least at first, with a cap, or calyptra, formed from the swollen archegonium. The spores are shed from the capsule, usually in dry weather, after a specialized lid, the operculum, drops off, usually revealing a complex set of interlocking, toothlike projections below it. The sporophyte, with its stalk, or seta, and capsule, usually contains chlorophyll and may produce some of its own food, especially when young. By the time the sporophyte is mature, however, the chlorophyll usually has disintegrated; mature moss sporophytes draw their food from the photosynthetic gametophytes, remaining attached to them.

When the spores germinate, they give rise to threadlike filaments called **protonemata** (singular, **protonema**), which resemble green algae and grow over moist surfaces. The characteristic leafy gametophytes arise from buds that form on the protonemata. A moss life cycle is summarized in Figure 33-11.

One of the kinds of mosses that is most important economically is *Sphagnum*, the peat mosses (Figure 33-12). These mosses, a highly distinctive group with more than

FIGURE 33-12
Peat mosses, *Sphagnum.*
A The glistening black sporangia emerge from the green, leafy gametophytes of *Sphagnum* on a stalk of gametophytic tissue. Unlike the sporangia of other mosses, those of *Sphagnum* have a lid that blows off explosively, releasing the spores.
B A peat bog being drained by the construction of a ditch. In many parts of the world, accumulations of peat, which may be many meters thick, are used as fuel.

A

B

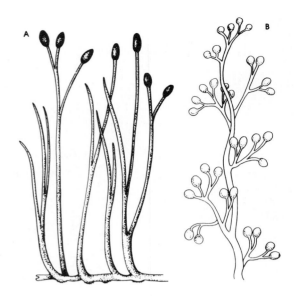

Figure 33-13

Rhyniophyta. Rhyniophyta comprises the oldest known division of vascular plants, dating to the Silurian period of 400 million years ago. In the genus *Rhynia* (**A**) the shoot is leafless and branched. *Cooksonia* (**B**) is the oldest known vascular plant.

290 species, grow in boggy places. Here they form dense and deep masses that often are dried and used as fuel. Peat is also used extensively in gardening and lawn care because of its moisture-holding capacity.

FEATURES OF THE VASCULAR PLANTS

The first vascular plants appeared no later than the early Silurian Period, some 430 million years ago. The first group for which we have a relatively complete record belonged to the extinct division Rhyniophyta, which lived some 410 million years ago. One of these plants, *Cooksonia*, is shown in Figure 33-3; it was little more than a simple branching axis with sporangia at the tips of the branches. Other ancient vascular plants evolved leaves, which first appeared as simple emergences from the stems, and more complex arrangements of sporangia. Some of these are shown in Figure 33-13.

The vascular plants (Table 33-1) are distinguished by their large, dominant, and nutritionally independent sporophytes; efficient conducting tissues (see Figure 33-4); specialized leaves, stems, and roots; cuticles and stomata (see Figure 33-2); and the fact that seeds evolved in the group (see Figure 33-14).

Heterospory, which occurs only in the vascular plants, has evolved several times independently in different groups. Recall that heterospory is a condition in which a kind of plant regularly forms two different kinds of spores, one of which gives rise to egg-producing gametophytes, and the other of which gives rise to sperm-producing gametophytes (see Figure 33-7). Seeds, in which the embryo is protected within a coat of sporophyte tissue, evolved in the advanced vascular plants, and characterize the groups known as gymnosperms and flowering plants. They represent the culmination of a major trend of evolution among the life cycles of vascular plants in which a progressive reduction in size of the gametophyte and its complete nutritional subordination, to the sporophyte, have taken place. Before illustrating this trend as it applies to particular groups of vascular plants, however, we will discuss specialization in the conducting systems of the group, a feature that has been critical to the success of the group.

> The vascular plants are characterized by their efficient conducting tissues; specialized stems, leaves, and roots; cuticles and stomata; and, in many, seeds. Seeds occur only in heterosporous plants, and heterospory occurs only in vascular plants.

Growth in the Vascular Plants

Early vascular plants were characterized by primary growth, growth that results from cell division at the tips of the stems and roots (see Chapter 35). Within the stems

TABLE 33-1　THE NINE DIVISIONS OF LIVING VASCULAR PLANTS

PSILOPHYTA

Whisk ferns. Homosporous. Motile sperm. No differentiation between root and shoot. Two genera and several species.

LYCOPHYTA

Lycopods (including club mosses and quillworts). Homosporous or heterosporous. Motile sperm. Five genera and about 1000 species.

SPHENOPHYTA

Horsetails. Homosporous. Motile sperm. One genus, 15 species.

PTEROPHYTA

Ferns. Homosporous; a very few heterosporous. Motile sperm. About 12,000 species.

ANTHOPHYTA

Flowering plants, or **angiosperms.** Heterosporous, seed-forming. Sperm not motile. Range from very diverse, tiny plants to some of the largest trees known. About 235,000 species.

CYCADOPHYTA

Cycads ("sago palms"). Heterosporous, seed-forming. Sperm flagellated and motile but carried to the vicinity of the egg by a pollen tube. Palmlike plants with sluggish secondary growth compared with that of the conifers. Ten genera, about 100 species.

GINKGOPHYTA

Ginkgo. Heterosporous, seed-forming. Sperm flagellated and motile but carried to the vicinity of the egg by a pollen tube. Deciduous tree with fanlike leaves and seeds resembling a small plum, with fleshy, ill-scented outer covering (see p. 6). One species.

GNETOPHYTA

Mormon tea (*Ephedra, Gnetum, Welwitschia*). Heterosporous, seed-forming. Sperm not motile. Shrubs, vines, three very diverse genera, about 70 species.

CONIFEROPHYTA

Conifers (including pines, spruces, firs, yews, redwoods, and others). Heterosporous, seed-forming. Sperm lack flagella. About 50 genera, with about 550 species.

---

formed as a result of primary growth were well-marked vascular bundles, which played the same conducting roles that they do in contemporary vascular plants. In the earliest vascular plants, however, there was no differentiation of the plant body into shoots and roots. The sharp separation of these two kinds of organs in a number of anatomical details is a property of nearly all modern plants. It reflects increasing specialization in relation to the demands of a terrestrial existence.

Secondary growth was an important early development in the evolution of vascular plants. It is basically an elaboration of the types of tissues that make up the primary vascular tissues. In secondary growth, cell division takes place actively in regions around the plant's periphery (see Chapter 35). In it, the conducting elements of the primary tissues are multiplied many times over as a result of the repeated divisions that occur in most plants in a cylindrical zone of dividing cells. Secondary growth makes it possible for a plant to increase in diameter. Only after the evolution of secondary growth could vascular plants become thick trunked, and therefore, tall. This evolutionary advance made possible the development of forests and, consequently, the domination of the land by plants. Judged from the fossil record, secondary growth had evolved independently in several different groups of vascular plants by the middle of the Devonian Period, some 380 million years ago. A reconstruction of a forest from about 300 million years ago, in the late Carboniferous Period, showing many primitive seed plants, is presented in Figure 22-14.

Primary growth results from cell division at the tips of stems and roots. Secondary growth results from the division of a cylinder of cells around the plant's periphery. Primary growth increases the length of stems and roots, whereas secondary growth increases their diameter.

anther   The swollen, usually lobed portion of a stamen in which pollen are formed.

antheridium   A haploid structure containing sperm cells; antheridia are found in seedless plants.

antipodal cells   A component of the 8-nucleate embryo sac in angiosperms that lie farthest away from the micropyle. Their function is unknown.

archegonium   A multicellular haploid structure in which an egg is produced.

carpel   The female reproductive part of a flower, consisting of a stigma, style and ovary.

calyx   All the sepals of a flower, collectively.

corolla   All the petals of a flower, collectively.

cotyledon   The embryonic "seed leaves" in gymnosperms and angiosperms that often contain stored food.

double fertilization   The process, unique to angiosperms, in which one sperm fuses with the egg and the other sperm fuses with the two polar nuclei.

endosperm   The triploid food supply of an angiosperm seed.

filament   The stalklike structure of a stamen that supports the anther.

flower   The reproductive structure of angiosperms; a complete flower consists of one or more carpels, and whorls of stamens, petals and sepals.

frond   The leaf of a fern.

fruit   A ripened carpel or group of carpels.

gametophyte   Any haploid portion of a plant life cycle; it is the portion which ultimately produces gametes.

gamete   A haploid reproductive cell that directly fuses with another gamete.

gametangium   A structure in which gametes are produced.

heterosporous   Refers to a plant that produces two types of spores—microspores and megaspores.

homosporous   Refers to a plant that produces only one type of spore.

integument   The outermost covering of an ovule; it will develop into the seed coat.

megasporangium   A sporangium in which are produced megaspores.

megaspore   A haploid cell that will develop into a female gametophyte in heterosporous plants.

megasporocyte   A diploid cell that will divide by meiosis producing four haploid megaspores.

Conducting Systems of the Vascular Plants

There were two types of conducting elements in the earliest plants, elements that have become characteristic of the vascular plants as a group. **Sieve elements** are soft-walled cells that conduct carbohydrates away from the areas where they are manufactured. **Tracheary elements** are hard-walled cells that transport water and dissolved minerals up from the roots. Both kinds of cells are elongated, and both occur in strands. Sieve elements are the characteristic cell types of a kind of tissue called **phloem;** tracheary elements are the characteristic cell types of a kind of tissue called **xylem.** In primary tissues, which result from primary growth, these two types of tissue often are associated with one another in the same vascular strands. These cell types, and the tissues they constitute, will be discussed in more detail in Chapter 35.

> **Water and dissolved minerals are carried in the xylem, which consists primarily of hard-walled cells called tracheary elements. Carbohydrates, in contrast, are carried in the phloem, which consists of soft-walled cells called sieve elements.**

micropyle   The opening in the integument of an ovule through which the pollen tube grows.

microsporangium   A sporangium in which are produced microspores.

microspore   A haploid cell that will develop into a male gametophyte in heterosporous plants.

microsporocyte   A diploid cell that will divide by meiosis producing four haploid microspores.

ovule   The forerunner of a seed. Initially it consists of integuments surrounding a megasporangium.

petals   A floral component consisting of modified leaflike structures that are often brightly colored to attract insect pollinators.

pollination   The transfer of a pollen grain from the microsporangium to the ovule in gymnosperms and to the stigma in angiosperms.

pollen grain   The male gametophyte in angiosperms and gymnosperms.

polar nuclei   A member of the 8-nucleate embryo sac in angiosperms. Most embryo sacs contain two polar nuclei in a large central cell.

primary endosperm nucleus   The triploid cell resulting from the fusion of a single sperm with the two polar nuclei.

protonema   The filamentous gametophyte structure that emerges from a bryophyte spore.

seed   A reproductive structure found in gymnosperms and angiosperms consisting of an embryo and a food supply surrounded by a resistant seed coat.

sepals   Usually the bottom-most whorl of modified leaflike structures in a flower.

sporangium   A structure in which spores are produced.

spore   A reproductive cell that can develop into a new individual without fusing with another cell.

sporophyte   Any diploid portion of a plant life cycle; it is the portion which ultimately produces spores.

stigma   The top-most portion of a carpel, usually containing sticky cells to which pollen can adhere.

synergid   A member of the 8-nucleate embryo sac in angiosperms. Most embryo sacs contain two synergids: the pollen tube grows into one of them.

vascular tissues   Tissues such as xylem and phloem that are specialized for conducting water, minerals, and carbohydrates throughout the plant.

Seeds

Seeds are a characteristic feature of some groups of vascular plants but not of others. In seed plants, both kinds of gametophytes have become greatly reduced in the course of their evolution, and a basically new kind of life cycle has been established. A seed is a developing sporophyte individual whose embryonic development has been temporarily arrested and is surrounded by a tough protective coat (see Figure 33-14). In seeds, all of the products of the mature megagametophyte (although they may have been altered greatly during the seed's maturation), together with the young plant of the next generation, are included together in a compact, drought-resistant package. Seeds are units by means of which plants, being rooted in the ground, are dispersed to new places; many seeds have devices, like the wings on the seeds of pines or the fruits of maples, or the plumes on the fruits of a dandelion, that help them to travel very efficiently.

The seed is a crucial adaptation to life on land because it protects the embryonic plant from drying out when it is at its most vulnerable stage. The seed coat also assists in protecting the embryo and its stored food material from being eaten by pred-

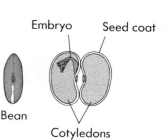

FIGURE 33-14
Diagram of a seed. The hard coat, formed of sporophytic tissue from the parent plant, protects the embryo—the dormant young plant of the next sporophyte generation—within. The first leaves, specialized in a bean for food storage, are the cotyledons. Seeds are drought resistant and readily dispersed.

ators or parasites. Most kinds of seeds have abundant food stored in them, either inside the embryo or in specialized storage tissue. This food, playing the same role as the yolk of an egg, is used as a ready source of energy by the rapidly growing young plant. The evolution of the seed clearly was a critical step in the domination of the land by plants.

SEEDLESS VASCULAR PLANTS

The earliest vascular plants lacked seeds, which are characteristic of the more advanced divisions. Members of four divisions of living vascular plants are lack seeds, with at least three others known only as fossils. Rather than starting our discussion of the seedless vascular plants with the most primitive groups, let us introduce them by outlining the features of the division that is most familiar—the ferns.

Pterophyta: The Ferns

Ferns are the most abundant group of seedless vascular plants, with about 12,000 living species. They are found throughout the world, but are much more abundant in the tropics than elsewhere. Ferns range from reduced aquatic forms less than a centimeter in diameter, such as *Azolla* (see Figure 25-7, *B*), to tree ferns that may have trunks more than 24 meters tall, with leaves up to 5 meters or more long (Figure 33-15, *A*). All but a few ferns are homosporous, with the sporophyte generation, as in all vascular plants, much larger, more conspicuous, and more complex than the gametophyte, from which it is completely independent nutritionally and in every other sense.

Let us discuss the life cycle of a representative fern in detail (Figure 33-16), beginning with the gametophyte generation. Fern gametophytes are green, photosyn-

A   B   C

FIGURE 33-15
Representatives of three of the divisions of vascular plants in which all of the living representatives are seedless.
A A tree fern in the forests of Malaysia (division Pterophyta). The ferns are by far the largest group of spore-producing vascular plants.
B The club moss *Lycopodium lucidulum* (division Lycophyta). Although superficially similar to the gametophytes of mosses, club moss plants are sporophytes. Meiosis occurs in the sporangia of the cones, ultimately resulting in the shedding of haploid spores. These spores grow into gametophytes that, in Lycopodium, bear both antheridia and archegonia.
C A horsetail, *Equisetum telmateia,* a representative of the only living genus of the division Sphenophyta. This species forms two kinds of erect stems, one green and photosynthetic, the other brownish and terminating in spore-producing cones. The spores produced by meiosis in the cones give rise to a single kind of tiny, green, nutritionally independent gametophytes.

FIGURE 33-16
Fern life cycle.

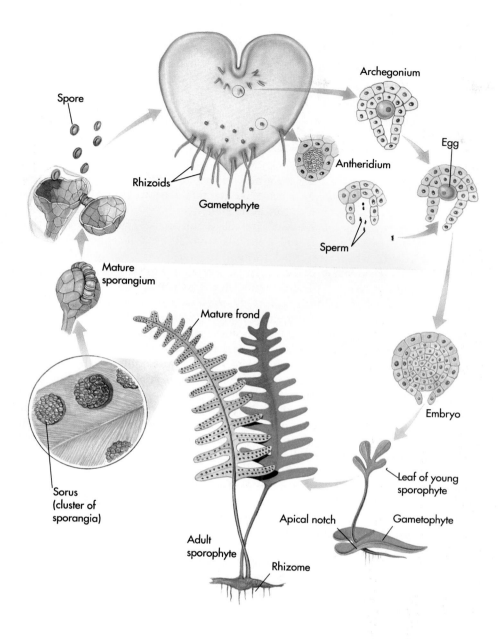

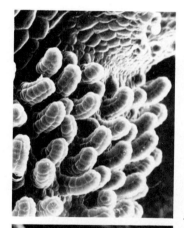

FIGURE 33-17
Scanning electron micrographs of gametangia of the fern *Anemia mexicana.*

A A mass of archegonia, not yet open, with their necks projecting above the lower surface of the gametophyte just behind its apical notch.

B An antheridium, open and showing the masses of sperm within. A sperm swims through water to the mouth of an archegonia, entering it and fertilizing the egg within.

thetic plants that live in relatively moist places and resemble the gametophytes of certain liverworts. They are flat, thin, and heart shaped, usually no more than a centimeter in length and somewhat less than that in width (see Figure 33-6, *A*). Rhizoids project from their lower surface. Archegonia form on the lower surface of the gametophytes near the **apical notch,** the region of the most rapid cell division (Figure 33-17, *A*). The antheridia are formed somewhat farther back, but also on the lower surface. When the sperm are mature, they are released from the antheridia (Figure 33-17, *B*). They require the presence of droplets of free water to make their way to the mouth of the archegonium. The sperm probably are attracted by chemicals released from the archegonium. When the sperm reach the mouth of the archegonium, they swim into the neck and eventually reach the egg in the swollen lower portion of the archegonium. When they do, a sperm fuses with the single egg, producing a zygote.

Following fertilization, the zygote begins to divide within the archegonium of the fern. Soon the growing sporophyte becomes much larger than the gametophyte and

FIGURE 33-18

Details of fern fronds.
A A "fiddlehead," or young, unfolding leaf, in a tree fern, *Sphaeropteris*. This pattern of leaf maturation is characteristic of nearly all ferns.
B The sporangia of most ferns develop on the underside of the leaves, and they are usually clustered, as shown here in a fern from the rain forests of Sumatra.

A   B

also nutritionally independent from it. When this occurs, the sporophyte anchors itself in the substrate and grows into a plant like the familiar fern plants that we see in the woods. Most ferns have more or less horizontal stems that creep along below the ground; subterranean stems of this kind are called **rhizomes.** True roots arise from these rhizomes and absorb water and nutrients from the soil in which the fern is growing. The leaves, which are called **fronds,** are borne on the rhizomes, just as leaves are borne on the stems of vascular plants in general, but they stand vertically in ferns and are more or less tufted, depending on the nature of the rhizome. In most ferns, the leaves are coiled in bud; in this stage, they are called "fiddleheads " (Figure 33-18, *A*). Sporangia ultimately differentiate on the leaves, often in clusters called sori (Figure 33-18, *B*). Meiosis takes place within these sporangia, followed by differentiation of the haploid spores, which ultimately are released from the sporangia. Only a few genera of ferns are heterosporous; one of them, incidentally, is the water fern *Azolla,* cultivated in rice paddies to encourage nitrogen fixation by its symbiotic cyanobacterium *Anabaena* (see Figure 25-7).

The fern life cycle differs from that of a moss primarily in the much greater development, independence, and dominance of the fern's sporophyte. In addition, the fern's sporophyte is much more complex than that of the moss, having vascular tissue and well-differentiated roots, stems, and leaves.

> For the most part, fern life cycles resemble those of mosses. Fern sporophytes, however, are green and become independent nutritionally. The fern plants that we see are sporophytes; fern gametophytes are only a centimeter or two across. Most ferns are homosporous.

In considering the life cycle of the fern, you should compare its features with those of a pine (division Coniferophyta, see Figure 33-21), and a flowering plant (division Anthophyta, see Figure 33-26). These three life cycles progress from simple to complex, but it should not be assumed that one of these groups has given rise to another. Nevertheless, the kind of progress that these life cycles represent must be similar to that which occurred during the evolution of these groups. For example, the ancestors of the flowering plants must have had complex gametophytes.

Turning now to the three other divisions of seedless vascular plants, we can first observe that they have many features in common with ferns. For example, all of them form antheridia and archegonia and produce free-swimming sperm that require the presence of free water for fertilization. In contrast, most seed plants have nonflagellated sperm; none form antheridia; and few form archegonia. Although a few extinct horsetails and clubmosses formed structures somewhat like seeds, none of the members of these groups have seeds that are homologous to those of contemporary seed plants.

Psilophyta: The Whisk Ferns

In the two genera of living whisk ferns, the vascular structure remains the same throughout the vegetative axis, and there are no true leaves. One genus, *Psilotum,* is widespread in the tropics and subtropics, reaching the southern states of the United

States (Figure 33-19). In *Psilotum*, the sporophyte stem forks into two branches that may continue to fork as they grow. This trait, termed **dichotomous branching,** is characteristic of many groups of primitive fossil vascular plants (see Figure 33-13). The leaflike structures that occur in the wisk ferns lack vascular tissue.

Whisk ferns are homosporous, producing haploid spores as a result of meiosis in sporangia located on their stems. The subterranean gametophytes, which are mycorrhizal, bear antheridia and archegonia.

Lycophyta: The Club Mosses

The living members of the division Lycophyta are small plants, although some fossil forms exceeded 30 meters in height (see Figure 22-14). The familiar members of this division belong to the genera *Lycopodium*, now usually subdivided into a number of distinct genera (see Figure 33-15, *B*) and *Selanginella*.

Lycopodium is homosporous, producing sporangia either at intervals along the leafy stem or clustered in conelike structures at the tips of the branches (see Figure 33-15, *B*). In other respects, the life cycle resembles that of other seedless vascular plants.

In contrast to *Lycopodium, Selanginella* is heterosporous. As generalized in Figure 33-7, *Selanginella* produces both megaspores and microspores that develop into female and male gametophytes, respectively. Archegonia are produced on the female gametophytes, and antheridia on the male gametophytes.

Sphenophyta: The Horsetails

Horsetails are characterized by their jointed stems, with whorls of scalelike leaves at the top of each joint (see Figure 33-15, *C*). Some fossil horsetails were medium-sized trees, but the living representatives, all members of the genus *Equisetum*, are much smaller, usually not much more than a meter tall. *Equisetum* occurs throughout North America and much of the rest of the world.

The sporangia of *Equisetum*, which is homosporous, are clusted in conelike structures at the tips of some of the stems. The life history resembles that of other homosporous vascular plants.

SEED PLANTS

Botanists now generally agree that seed plants were derived from a single common ancestor. There are five divisions; in one of them, the **angiosperms,** or **flowering plants,** the ovules, at the time of pollination, are completely enclosed by the tissues of the sporophytic individual on which they are borne. In the members of the other four divisions, collectively called **gymnosperms** because of this feature, the ovules are not completely enclosed by sporophytic tissue at this time. The four divisions of gymnosperms are not directly related to one another; they are simply seed plants that lack the special characteristics of angiosperms.

In all seed plants, the male and female gametophytes develop within the parent sporophyte and are completely dependent on it for nutrients and water. Pollen grains, which are actually immature male gametophytes, transport the sperm to the vicinity of the egg; they are reduced to a few cells and surrounded by a thick, protective wall. The pollen grain is shed from the parent plant and carried by wind, insects, or other animals to the female gametophyte in the process of **pollination.** The pollen grain then cracks open, or **germinates;** and the pollen tube, containing the sperm cells, grows out, transporting the sperm directly to the egg. Thus there is no need for free water in the process of pollination.

From an evolutionary and ecological perspective, the seed represents an important advance. Developing from an ovule, it protects the embryo with specialized structures derived from the parent sporophyte. Viewed in this sense, having a seed is of great benefit to a plant that grows on land; its embryos are protected from drought and, to some extent, from predators; dispersal is enhanced; and there is no immediate need for water for establishment.

A

B

FIGURE 33-19
One of the two genera of living psilophytes (division Psilophyta), the whisk-fern, *Psilotum*. Psilophytes are unique among living vascular plants in that their axes are not differentiated into roots and shoots.
A View of an entire plant, hanging over the rocks in Jamaica.
B The erect branches of another species, showing the brown sporangia.

A

B

C

FIGURE 33-20

Representatives of the four divisions of gymnosperms with living representatives. The gymnosperms and angiosperms are the seed plants.

A Longleaf pines, *Pinus elliottii*, in Florida, a representative of the largest division of gymnosperms, the Coniferophyta. The understory in this forest is dominated by palmetto palms, Sabal palmetto, which are flowering plants.
B An African cycad, *Encephalartos kosiensis*. The cycads are seed plants with fernlike leaves and sperm that swim within the pollen tubes. Seed-forming cones, like the one shown here, and pollen-forming cones are borne on separate plants.
C Maidenhair tree, *Ginkgo biloba*, the only living representative of the division Ginkgophyta, a group of plants that was abundant 200 million years ago. Among living seed plants, only the cycads and *Ginkgo* have swimming sperm.
D *Welwitschia mirabilis*, one of the three genera of Gnetales. *Welwitschia* is found in the extremely dry deserts of southwestern Africa. This plant is producing seed-bearing cones; other plants produce pollen-forming ones. In *Welwitschia*, two enormous, strap-shaped leaves grow from a circular zone of cell division that surrounds the apex of the carrot-shaped root; the cone-bearing branches also form in this zone.

D

Gymnosperms

As we have seen, the four divisions of living seed plants in which the ovules are not completely enclosed by the tissues of the sporophytic individual on which they are borne at the time of pollination—conifers, cyads, ginkgo, and gnetophyte—are called **gymnosperms** (see examples in Figure 33-20). This name combines the Greek root *sperma*, or "seed," with *gymnos*, or "naked": in other words, naked-seeded plants. In fact, the seeds of gymnosperms often become enclosed by the tissues of their parents by the time they are mature, but their ovules are indeed naked at the time of pollination.

The four divisions of gymnosperms are very diverse. For example, the cycads and ginkgo still have motile sperm (Figure 33-20), even though it is borne within a pollen tube. Gnetophyta appear to be more closely related to flowering plants than are the other gymnosperms. Some of the gymnosperms have archegonia, and others have lost them in the course of evolution, as the flowering plants have done. Archegonia are found in all other groups of vascular plants; their progressive reduction in size and eventual loss constitute a clear evolutionary trend.

The most familiar gymnosperms are the conifers, division Coniferophyta (Figure 33-20, *A*), which includes the pines, spruces, firs, hemlocks, and cypresses. The tallest living vascular plant, the giant redwood (*Sequoia sempervirens*) found in coastal California and southwestern Oregon, is a conifer. Many of the conifers have needle-like leaves, an evolutionary adaptation for retarding water loss. Thus conifers are often found growing in moderately dry regions of the world, including the northern latitudes and high on the sides of mountains.

The other three divisions of gymnosperms differ greatly from each other. The cycads, division Cycadophyta, have short stems and palmlike leaves (Figure 33-20, *B*). The nine living genera are widespread throughout the tropical and subtropical parts of the world.

The maidenhair tree, *Ginkgo biloba,* is the only living species of the division Ginkgophyta (Figure 33-20, *C*). It has fan-shaped leaves, which are shed in the autumn. Because it is resistant to air pollution, it is commonly planted along city streets.

The division Gnetophyta includes three living genera, which differ greatly from one another. As mentioned above, these genera appear to be the closely living relatives of the angiosperms, and probably share a common ancestor with that group. One of the most bizarre of all plants is *Welwitschia,* which occurs in the Namib Desert of southwestern Africa (Figure 33-20, *D*). Its two strap-shaped, leathery leaves are generated continuously from meristems at their base, splitting as they grow out over the desert sand.

The four divisions of living gymnosperms share a common ancestor with the angiosperms, or flowering plants; seeds evolved in that ancestor.

The Pine Life Cycle

The life cycle of the pine is presented here as an example of the kinds of life cycles found in conifers. It should be compared with the life cycles of ferns and flowering plants, and the points of similarity and difference considered. In the pine, the megaspores are held within ovules, as they are in all other seed plants (Figure 33-21). The megagametophytes differentiate within the protective tissue of the parent sporophyte. In the pine and other gymnosperms, the megagametophyte continues to grow and becomes the place in which food is concentrated in the seed. The seeds of pines, by virtue of this stored food, not only protect but also nourish the young embryos. From a morphological point of view, the outer layers of a seed consist of tissues derived

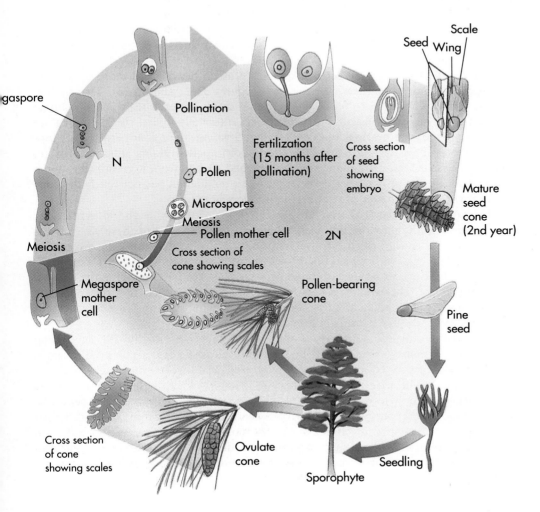

FIGURE 33-21
Pine life cycle.

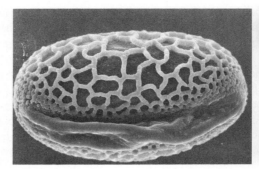 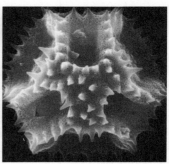

A Easter lily, *Lilium candidum*. The pollen tube emerges from a pollen grain such as this through the groove or furrow that occurs on one side of the grain.
B Pollen from a plant of the sunflower family, *Hyoseris longiloba*. There are three pores hidden among the ornamentation of this pollen grain; the pollen tube may grow out through any one of them.

from the parent sporophyte on which it occurs; inside are the derivatives of the megasporangium of that sporophyte, enclosing the derivatives of the gametophyte, and, within them, the young sporophyte of the next generation, the embryo.

To understand how the seeds are produced, we need to return to our consideration of the generalized plant life cycle. Microspores, which are produced within the microsporangia as a result of meiosis, ultimately differentiate into microgametophytes. The microgametophytes of seed plants are called pollen grains (Figure 33-22) when they are ready to be shed by their parent plant. Some of these pollen grains are transported to the vicinity of the megagametophyte, which remains held within the protective coverings of the parent sporophyte. Inside the pollen grain, the cells divide further, leading ultimately to the production of a mature microgametophyte. Among the products of this division are the cells that become the sperm.

Pine microsporangia are borne in pairs on the surface of the thin scales of the relatively delicate pollen-bearing cones (shown in the closely related genus *Picea* in Figure 33-1, *B*). When the young microgametophytes, or pollen grains, are shed, they consist of four cells. The pollen grains of pines are shed in huge quantities (see Figure 19-4), often appearing as a yellow scum on the surface of ponds, lakes, and even windshields.

The familiar seed-bearing cones of pines are much heavier and more substantial structures than the pollen-bearing cones (see Figures 33-21 and 33-23). Two ovules, and ultimately two seeds, are borne on the upper surface of each scale. Within a pine megagametophyte, cell division also takes place, ultimately giving rise to an egg. In spring, when the seed-bearing cones are small and young, their scales are slightly separated. Drops of sticky fluid, to which the airborne pollen grains adhere, form between these scales. After a pollen grain has reached such a drop, it germinates, and a slender **pollen tube** grows toward the egg. When pollination has occurred, the cone scales grow together, better protecting the developing ovules.

The pollen tube grows slowly through the tissue of the ovule and ultimately, by two further cell divisions, becomes six-celled and mature. Two of these six cells become sperm. When the pollen tube bursts, the sperm, which lack flagella, are carried passively to the egg, with which one of the sperm ultimately fuses, producing a zygote. All of these events occur within the protective covering surrounding the original megasporangium. The whole process takes place so slowly that this fusion occurs some 15 months after pollination. At any event, following syngamy, the resulting zygote starts to divide and eventually grows into an embryo.

The growth and differentiation of the embryo take place within the developing seed, where the embryo is nourished by the food-rich tissue of the expanded megagametophyte and protected by the developing seed coat, which ultimately becomes tough and hard. The seed coat and embryo are diploid, whereas the nutritious tissue of the megagametophyte is haploid. This food-rich tissue, therefore, represents the gametophytic generation that intervenes between the two sporophytic ones. Dormant within the seed, the embryo consists of a shoot-root axis and several (often eight) seed leaves. Pine seeds usually are shed from the cone during the autumn of the year that follows the cone's first appearance. These seeds are winged and flutter to the ground, germinating if conditions are favorable and then giving rise to a new tree.

FIGURE 33-23
Ovulate cone and a group of microsporangiate cones of lodgepole pine, *Pinus contorta*, in Montana.

Angiosperms

The flowering plants (division Anthophyta) differ from all other seed plants in that their ovules are enclosed within the tissue of the parent sporophyte in structures called **carpels** (Figure 33-24). The pollen grains of the flowering plants reach a specialized portion of the carpel, the **stigma,** and then germinate; thus pollination in the plants of this division is indirect. Pollen tubes that emerge from the pollen grains grow down through the tissue of the carpel, reaching the ovule and ultimately the egg. This contrasts with the gymnosperms, in which the ovules are directly exposed to the air, at least at the time of pollination. The pollen reaches them directly or falls in their vicinity. Because of their enclosed ovules and seeds, the flowering plants are called **angiosperms,** a name derived from the Greek words *angion,* or "vessel," and *sperma.* The "vessel," or carpel, ultimately matures into a fruit that encloses the mature ovules, or seeds. The fruit itself has become an important unit of dispersal in the flowering plants, as we will consider further in the next chapter.

> **Angiosperms are characterized primarily by features of their reproductive system. The unique structure known as the carpel encloses the ovules and matures into the fruit. Since the ovules are enclosed, pollination is indirect.**

The angiosperms, although they too have seeds, could not have evolved directly from any of the living divisions of gymnosperms; they differ in too many important ways. However, we would classify the ancestor of the angiosperms as a gymnosperm if we knew what it was—it must have formed seeds and had naked ovules. The form of enclosure of the seeds is what defines an angiosperm, and seeds must originally have been present—and naked—for this form of enclosure to have evolved.

In angiosperms, the life cycle is generally similar to that of pines and other gymnosperms, but there are some major differences. To illustrate these differences, it is necessary to describe the basic structure of the flower. A complete flower consists of four **whorls** (Figure 33-25), circles of similar parts arising at about the same point on an axis. Our discussion will proceed from the inward whorls outward.

The innermost whorl of a flower consists of the carpels. The carpels are closed structures, folded around the ovules (see Figure 33-24). The **stigma,** the receptive area of the carpel, is often separated from the body of the carpel by a stalk called the **style.** The swollen, lower portion of the carpel is often called the ovary. Pollination occurs in angiosperms when pollen reaches the stigma. Subsequently, the pollen tube must grow through the stigma and style into the carpel to reach the ovule.

Because carpels often have a thick lower portion, slender stalk, and terminal stigma, resembling the pestles used to grind powders in mortars, separate or fused carpels have traditionally been called **pistils.** The term *carpel* is morphologically more precise, however, since it refers to one member of this whole whether it is fused with others or not; we will therefore use *carpel* in this book. Many angiosperms may have only one carpel in each flower, but the whorl in which the carpels occur collectively is still called the **gynoecium,** a term derived from the Greek words *gynos,* which means "female," and *oikos,* or "house." Just as the mature ovules become the seeds, so the mature carpel, with its enclosed seeds and sometimes other adhering flower parts, be-

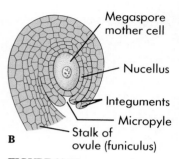

FIGURE 33-24
Diagram of a carpel.
A Shows its position in a portion of a flower, with details of an ovule.
B The ovary will mature to form a fruit, and the outer covering, or integument, of the ovule will mature to form the seed coat.

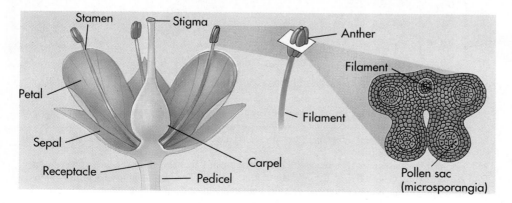

FIGURE 33-25
Diagram of an angiosperm flower.

comes the fruit. Fruits are just as characteristic of the angiosperms as are the flowers from which they develop.

The two outer whorls of the angiosperm flower together are called the **perianth.** Although not directly connected with the production of gametes, they also help to establish the distinctive character of the angiosperms. Below the androecium is the corolla, the members of which are called **petals.** Petals are often brightly colored and may perform an important role in attracting insects and other pollinators. The fourth and outermost whorl of the flower is the **calyx,** made up of **sepals.** The sepals, like the petals, may be colored and may play a role in luring pollinators, but they are more often greenish and not particularly attractive. The calyx forms the outer layer of the bud before the flower opens and appears to serve mainly a protective function.

In the course of evolution, the members of any whorl of the flower may have become fused with one another or with those of another whorl. Tubular corollas made up of fused petals are frequent, for example. Another common floral plan is one in which the calyx, corolla, and stamens are fused around the gynoecium, whose members are also fused. The arrangement of the parts in such a flower gives the impression that all the flower parts arise at the top of the ovary, which is a term used for the portion of the flower that becomes the fruit—it includes the gynoecium. Examples of several different kinds of angiosperm flowers may be seen in the next chapter.

The next whorl of the flower, located below the gynoecium on the floral axis, is the **androecium** (Greek *andros,* "male"). The individual members of the androecium are the **stamens.** In most angiosperms, a stamen consists of a slender, threadlike portion, the **filament,** and a terminal, thicker, often four-lobed **anther.** The anther contains four microsporangia, or pollen sacs (see Figure 33-25), within which the microspores are formed. The products of differentiation within these microspores are shed as pollen grains at either a two- or three-celled stage. When mature, the microgametophytes have three cells; two are sperm, and the third may play a role in pollen tube growth. The sperm of angiosperms lack flagella. If the pollen grains are shed at the two-celled stage, one of the cells divides after the germination of the pollen tube, while it is growing through the carpel toward the ovule. In such plants, the microgametophyte reaches its mature, three-celled state within the tissue of the carpel.

A mature angiosperm microgametophyte has three cells. As pollen, it may be shed at either the two- or the three-celled stage. If it is shed when it has only two cells, one further mitotic division occurs afterward.

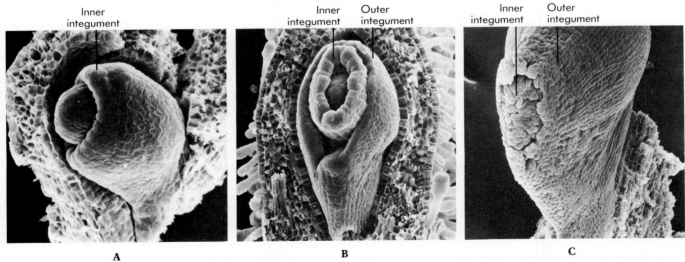

FIGURE 33-26
Stages in the formation of an ovule, shown in scanning electron micrographs of an elm (*Ulmus parviflora*).
A A young ovule. The inner integument has already formed, and the outer one is just beginning to develop around the base of the ovule.
B Later stage in maturation. The inner and outer integuments are enveloping the megasporangium.
C A mature ovule, comparable to the diagram in Figure 33-24. The integuments have grown nearly completely around the megasporangium which is bordered with the inner integument.

In many angiosperms, the pollen grains are carried from flower to flower by insects and other animals that visit the flowers for food or other rewards, or are deceived into doing so because the characteristics of the flower suggest that they may offer such rewards. The ways in which pollination occurs, and some of the floral characteristics associated with different modes of pollination, will be discussed in the next chapter. In other angiosperms, the pollen is blown about by the wind, as it is in most gymnosperms, and reaches the stigmas passively. Although in a few instances insects have been observed feeding on the pollen of gymnosperms, such visits apparently are rare. In still other angiosperms, the pollen does not reach other individuals at all; instead, it is shed directly onto the stigma of the same flower. This results in self-pollination and inbreeding.

The mature megagametophyte of an angiosperm develops from the functional megaspore derived from a single meiotic event. In other words, meiosis produces four spores, but three of them disintegrate. Meiosis occurs within the **nucellus,** a term used to describe the specific kind of megasporangium found in angiosperms. The nucellus is enfolded by specialized tissues, called the **integuments,** which are derived from the parent sporophyte. Together, these structures make up the ovule (see Figures 33-24 and 33-26), the structure that ultimately matures to form the seed.

A mature angiosperm megagametophyte is called an **embryo sac.** An embryo sac is produced by the division of the functional megaspore, which is haploid, within the tissues of the nucellus. In the kind of embryo sac that is characteristic of about 70% of the angiosperms, three successive mitotic divisions, coupled with the differentiation of cell walls around the resulting haploid nuclei, result in an embryo sac with eight nuclei and seven cells (Figure 33-27). Within the embryo sac are the following kinds of cells: the egg, flanked by two cells, at one end; three cells antipodal at the far end; and a large cell at the approximate center of the embryo sac that contains two polar nuclei. The embryo sacs of the other 30% of angiosperm species also have an egg at one end and one or more polar nuclei in their central regions, but they differ in other respects.

> In angiosperms, the mature megagametophyte is called an embryo sac. The most frequent type has eight nuclei, enclosed within seven cells. At one end is the egg, flanked by two cells; at the other end are three other cells; and in the center are two nuclei, the polar nuclei, in a single, large cell.

Fertilization in angiosperms is a unique process that is highly characteristic of this division of plants. Both of the sperm in the mature microgametophyte are functional. The first fuses with the egg, as in all organisms, forming the zygote. The second sperm fuses with the polar nuclei, forming the **primary endosperm nucleus,** which divides more rapidly than the zygote and gives rise to a nutritious tissue, the **endosperm** (Figure 33-28). If two polar nuclei fuse with the second sperm, then the primary endosperm nucleus, and consequently the endosperm, is triploid, since triploid cells have three sets of chromosomes. The process whereby both the egg and the polar nuclei are fertilized, forming the zygote and the primary endosperm nucleus, is called **double fertilization.**

Endosperm, which only angiosperms possess, is the primary nutritional tissue on which the developing embryos of these plants depend. In some angiosperms, such as the common bean or pea, the endosperm is fully used up by the time the seed is mature. In such plants, the seedling leaves often become swollen and fleshy and contain most of the stored food reserves. In other plants, such as corn, the mature seed contains abundant endosperm, which the embryo draws on as it germinates and starts to grow. In any case, endosperm plays the same role in nourishing angiosperm embryos that the haploid tissue of the megagametophyte does in gymnosperms.

> Double fertilization, a process that is unique to the angiosperms, occurs when one sperm nucleus fertilizes the egg and the second one fuses with the polar nuclei. These two events result in the formation of, respectively, the zygote and the primary endosperm nucleus. The latter divides to produce the endosperm, the nutritive tissue that occurs in the seeds of angiosperms.

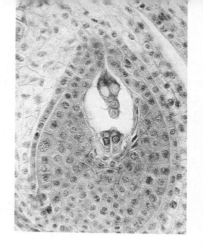

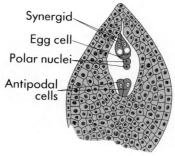

FIGURE 33-27

The embryo sac. The mature embryo sac of a water buttercup, *Ranunculus nipponicus,* showing seven of the eight nuclei present at maturity. The third antipodal is located below the plane of this section.

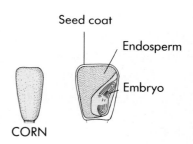

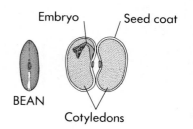

FIGURE 33-28

Endosperm in corn and a bean. The corn seed has endosperm which is still present at maturity, and the bean seed has endosperm which has been incorporated in the embryo cotyledon.

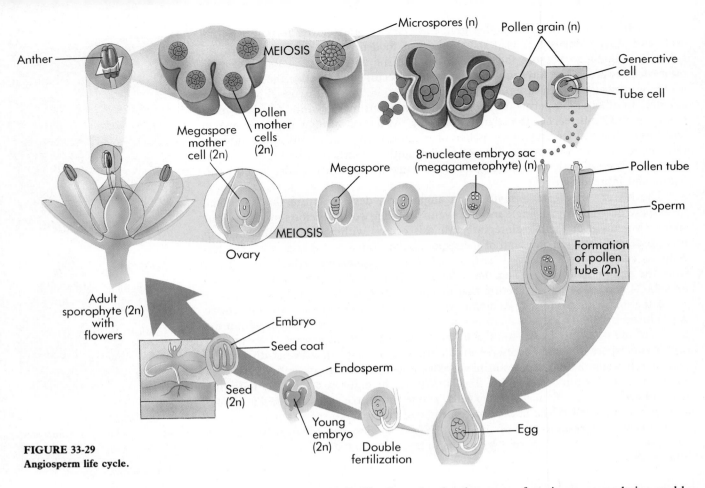

FIGURE 33-29
Angiosperm life cycle.

Apart from double fertilization, the development of angiosperm seeds is roughly the same as that of gymnosperm seeds. When the seed coats are mature and tough, the seed is shed. If it germinates in favorable conditions, a sporophyte is established. The angiosperm life cycle is summarized in Figure 33-29.

A Very Successful Group

The vascular plants, which originated in connection with their invasion of the land, have come to dominate terrestrial habitats everywhere, except for the highest mountains and the polar regions. All plants, including bryophytes, have an alternation of generations, but this alternation has become modified by the reduction of the gametophyte during the evolution of the various divisions of vascular plants.

The reduction in size and complexity of the gametophyte generation in vascular plants can be traced easily. In ferns, as in bryophytes, the gametophyte is a green, photosynthetic, independent plant that forms multicellular archegonia and antheridia. Pines and flowering plants, in contrast, have two kinds of gametophytes, the egg-producing megagametophytes and the sperm-producing microgametophytes. In the pine, the megagametophyte is a complex, multicellular structure that bears archegonia within which eggs are formed. In flowering plants, the megagametophyte is an embryo sac, most often containing eight nuclei and seven cells at maturity.

Accompanying this reduction in the size and complexity of the gametophytes has been the appearance of the seed. Seeds are highly resistant structures that are well suited to protect a plant embryo from drought and to some extent from predators. In addition, most seeds contain a supply of food for the young plant. Seeds are dispersed in many different ways, some involving animals. Flowers, which evolved among the angiosperms, are primarily a device whereby animals are induced to spread the pollen from the flowers of one individual plant to those of another. They are, in effect, a mechanism whereby plants can overcome their rooted, immobile nature and secure the benefits of wide outcrossing in promoting genetic diversity.

1. Plants are derived from an aquatic ancestor, but the evolution of their conducting tissues, cuticle, stomata, and seeds has made them progressively less dependent on water. The oldest plant fossils date from the Silurian Period, some 430 million years ago.

2. The common ancestor of the plants was a green alga. The similarity of the members of these two groups can be demonstrated by their photosynthetic pigments (chlorophylls *a* and *b*, carotenoids); chief storage product (starch); cellulose-rich cell walls (in some green algae only); and cell division by means of a cell plate (in certain green algae only).

3. The major groups of plants are the bryophytes (division Bryophyta)—mosses, liverworts, and hornworts—and the vascular plants, which make up nine other divisions. Vascular plants have two kinds of well-defined conducting strands: xylem, which is specialized to conduct water and dissolved minerals, and phloem, which is specialized to conduct the food molecules the plants manufacture.

4. All plants have an alternation of generations, in which haploid gametophytes alternate with diploid sporophytes. The spores that sporophytes form as a result of meiosis grow into gametophytes, which produce gametes—sperm and eggs—as a result of mitosis.

5. The gametophytes of bryophytes are nutritionally independent and remain green. The sporophytes of bryophytes are usually nutritionally dependent on the gametophytes and mostly are brown or straw-colored at maturity. In ferns, sporophytes and gametophytes usually are nutritionally independent; both are green. Among the gymnosperms and angiosperms, the gametophytes are nutritionally dependent on the sporophytes.

6. In all seed plants—gymnosperms and angiosperms—and in certain lycopods and a few ferns, the gametophytes are either female (megagametophytes) or male (microgametophytes). Megagametophytes produce only eggs; microgametophytes produce only sperm. These are produced, respectively, from megaspores, which are formed as a result of meiosis within megasporangia, and microspores, which are formed in a similar fashion within microsporangia.

7. In gymnosperms, the ovules are exposed directly to pollen at the time of pollination; in angiosperms, the ovules are enclosed within a carpel, and a pollen tube grows through the carpel to the ovule.

8. The nutritive tissue in gymnosperm seeds is derived from the expanded, food-rich gametophyte. In angiosperm seeds, the nutritive tissue, endosperm, is unique and is formed from a cell that results from the fusion of the polar nuclei of the embryo sac with a sperm cell.

9. The pollen of gymnosperms is usually blown about by the wind; although some angiosperms are also wind-pollinated, in many the pollen is carried from flower to flower by various insects and other animals. The ripened carpels of angiosperms grow into fruits, structures that are as characteristic of members of the division as flowers are.

REVIEW QUESTIONS

1. To what does alternation of generations refer in the plants? Define sporophyte and gametophyte. With which stage is an adult animal comparable? How are they reproductively dissimilar?

2. In what specialized sporophytic cell does meiosis take place? What type of cells do they produce? Into what do these cells grow? Does this growth involve meiosis or mitosis? What reproductive cells are in turn produced by this stage? Is this a meiotic or mitotic process? Is fusion required for the reproductive cycle to continue? Why or why not? Summarize alternation of generations, indicating structures in parentheses () and processes in brackets <>.

3. How does the appearance and nutritional status of the gametophyte differ between bryophytes and ferns and the rest of the vascular plants? How does the sporophyte of the bryophytes compare to that of the vascular plants?

4. Differentiate between microgametophytes and megagametophytes. What name is given to the haploid spores that give rise to each of these? What is a plant that produces two morphologically different kinds of spores called? What is one that produces only one type of spore called? Where are the haploid spores produced? What are they called if two different types of spores are produced?

5. How does the level of development of vascular tissue in the mosses indicate their evolution and overall relatedness to the liverworts and hornworts? Compare liverworts, hornworts, and mosses in terms of their vegetative and reproductive morphologies.

6. How do the sperm travel from the bryophyte antheridium to the archegonium? Where does the zygote develop? What characteristic spore-bearing structure is produced in the mosses? What are protonema?

7. What is a seed? Why is the seed a crucial adaptation to terrestrial life?

8. Where are the archegonia and antheridia of a fern located? Which generations of the fern are nutritionally independent? What is a rhizome? Where are the sporangia located on a fern?

What structure germinates from the spores produced in the sporangia?

9. Are the familiar large pine cones mega- or microsporangia? What tissues of the mature seed are diploid? Which tissues are haploid? What tissues does the hard seed coat surround? What part of the pine seed provides nourishment for the developing embryo?

10. In what way do the flowering plants differ from the rest of the seed plants? What is the stigma? Is fertilization in angiosperms direct or indirect? From what tissue does angiosperm fruit develop?

11. What is double fertilization? What two distinctly different tissues arise from this process? What is the genetic complement and function of each?

 ## THOUGHT QUESTIONS

1. Compare and contrast the adaptations of fungi and of plants to a terrestrial existence. Which group of organisms has been more successful, and why?

2. Why is the ability to form seeds always associated with a separation of the gametophyte generation into two distinct kinds, megagametophytes and microgametophytes? How has this relationship changed during the course of evolution?

3. Why do mosses and ferns both require free water to complete their life cycles? At what stage of the life cycle is the water required? Do angiosperms also require free water to complete their life cycles? What are the reasons for the difference, if any?

 ## FOR FURTHER READING

DYER, A.F., and J.G. DUCKETT, (editors): *The Experimental Biology of Bryophytes*, Academic Press, London, 1984. Excellent collection of papers on all aspects of the biology of bryophytes.

GIFFORD, E.M. and A.S. FOSTER: *Morphology and Evolution of Vascular Plants*, 3rd ed., W.H. Freeman and Company, New York., 1989. A classic, well-written text dealing with many aspects of vascular plants.

GENSEL, P.G., and H.N. ANDREWS: "The Evolution of Early Land Plants," *American Scientist*, vol. 75, pages 478-489, 1987. Excellent, well-illustrated account of the earliest plants.

LELLINGER, D.B.: *A Field Manual of the Ferns and Fern-Allies of the United States and Canada*, Smithsonian Institution Press, Washington, D.C., 1985. Useful account of the 406 kinds of spore-bearing vascular plants found in this area.

NIKLAS, K.J.: "Computer-Simulated Plant Evolution," *Scientific American*, March 1986, pages 78-84. A computer can be used to recreate the major trends in plant evolution and thus to understand them more completely.

NIKLAS, K.J.: "Aerodynamics of Wind Pollination," *Scientific American*, July 1987, pages 90-95. The complex aerodynamics of pollen grains and the structures on which they land exert a high degree of influence on the apparently random process of pollen dispersal.

NORSTOG, K.: "Cycads and the Origin of Insect Pollination," *American Scientist*, vol. 75, pages 270-279, 1987. Studies of one of the most ancient lines of living seed plants suggest that insect pollination was established before the origin of the angiosperms.

RICHARDSON, D.H.S.: *The Biology of Mosses*, John Wiley and Sons, Inc., New York, 1981. Excellent, concise account of the mosses.

SCAGEL, R.F., and others: *Plants: An Evolutionary Survey*, Wadsworth Publishing Co., Belmont, Calif., 1984. Comprehensive account of all groups of plants; highly recommended.

SCHOFIELD, W.B.: *Introduction to Bryology*, Macmillan Publishing Co., New York, 1985. An excellent introduction to all aspects of the group.

TANG, W.: "Insect Pollination in the Cycad *Zamia pumila* (Zamiaceae)," *American Journal of Botany*, vol. 74, pages 90-99, 1987. Reports a fine study of pollination in the only cycad found native in the United States.

TAYLOR, T.N.: *Paleobotany: An Introduction to Fossil Plant Biology*, McGraw-Hill Book Co., New York, 1981. A survey of fossil plants.

34

Flowering Plants

CHAPTER OUTLINE

OVERVIEW

The flowering plants dominate every spot on land except for the polar regions, the high mountains, and the driest deserts. Despite their overwhelming success, they are a group of relatively recent origin. Although they may be about 150 million years old as a group, the oldest definite angiosperm fossils are from about 123 million years ago; the group achieved world dominance only about 80 or 90 million years ago. Among the features that have contributed to the success of angiosperms are their unique reproductive features, which include the flower and the fruit. Flowers bring about the precise transfer of pollen by insects and other animals, a critical adaptation enabling an organism that is rooted in one place to achieve outcrossing. Pollen dispersal by wind, self-pollination, and other alternative modes of pollination have arisen repeatedly in the group. Fruits play a role of extraordinary importance in the dispersal of angiosperms from place to place. Both flowers and fruits were key elements in making possible the early success of the flowering plants.

FOR REVIEW

Here are some important terms and concepts that you will encounter in this chapter. If you are not familiar with them, you should review them before proceeding.

Hybridization and the origin of species (Chapter 21)

Origin of land plants (Chapter 22)

Evolution of the seed (Chapter 33)

Gymnosperms (Chapter 33)

Flowering plant life cycle (Chapter 33)

A

B

FIGURE 34-1

Some of the remarkable diversity of angiosperms is shown in these photographs. D and **E** are monocots; the other species shown here are dicots (see p. 685).

A Fragrant water lily, *Nymphaea odorata*, a flower with numerous, free, spirally arranged parts that intergrade with one another. The floating leaves and flowers, as well as the specialized stems and roots, are adapted to life in an aquatic environment.

B Wild geranium, *Geranium*, a perennial that occurs in woodlands. It illustrates a reduction in the number of flower parts compared with less specialized angiosperm flowers.

C An angiosperm that lacks chlorophyll, Indian pipe, *Monotropa uniflora*. This plant obtains its food from other plants, apparently transferring it by means of fungal hyphae. Its four or five petals are separate, but form a tube.

Flowering plants, division Anthophyta, are the dominant photosynthetic organisms nearly everywhere on land. Included in this great group of some 235,000 species are our familiar trees (except conifers), shrubs, and herbs, grasses, vegetables, and grains—in short, nearly all of the plants that we see on a daily basis (Figure 34-1). Virtually all of our food is derived, directly or indirectly, from the flowering plants; in fact, more than 85% of it comes from just 20 species! The remarkable evolutionary success of the angiosperms is such that this division deserves special attention.

HISTORY OF THE FLOWERING PLANTS

In view of the ecological dominance and overwhelming importance of the angiosperms, the problem of their origin and relationships to other groups has interested biologists for many years. Although seeds are shared with other groups of plants, including the four divisions of living gymnosperms, those seeds do not contain endosperm, a tissue unique to the angiosperms. In addition, flowers and fruits are unique structures, found only in angiosperms. When and how did these features originate?

Before considering possible answers to that question, it is necessary to establish that the angiosperms are descended from a single common ancestor. We have concluded that is the case because the complex common features of angiosperms' flowers and fruits are so numerous that they seem unlikely to have evolved more than once. In addition, all of the characteristic and sometimes unique features of angiosperm embryology, which were outlined in the last chapter, including double fertilization and the presence of endosperm, seem even less likely to have evolved repeatedly. In our search for the origins of the division, therefore, we are justified in seeking one group of organisms that lived at a particular time and place.

Early History

As mentioned in the last chapter, the common ancestor of the angiosperms would clearly be classified as a gymnosperm if we knew what it was. None of the four divisions of gymnosperms that has living members qualifies as the ancestor of the angiosperms, however: each is too specialized in its own way. Although the Gnetales share a number of features with angiosperms, indicating that they are the most closely related living group, each of the three surviving genera is so peculiar in its own way that none could have given rise to the flowering plants.

Certain fossil seed plants that shared key features with angiosperms existed before the end of the Jurassic Period, 144 million years ago. Even though the first unequivocal angiosperm fossils date from about 123 million years ago (fossil pollen), in the

D

E

F

FIGURE 34-1, cont'd

D An orchid from Algeria, *Ophrys tenthredinifera*. Like those of all orchids, the flowers of this species are complex, with the parts highly modified and fused. In the members of this genus and a few other orchids, the flowers resemble female insects; males of these insect species attempt to copulate with the flowers, spreading pollen from plant to plant in the process of doing so. Orchids may be the largest family of flowering plants, with an estimated 20,000 to 23,000 species—10% of the total.

E Tiger lily, *Lilium canadense*, with six, free, colored perianth parts (not differentiated into petals and sepals) and six free anthers. The three carpels are fused, forming a compound gynoecium, or pistil, the end of which is seen between the stamens.

F Scarlet gilia, *Ipomopsis aggregata;* in the flowers of this species, the sepals and petals are fused with one another, each forming a tube. The greenish calyx, formed of the fused sepals, can be seen at the base of the flower. There are five fused petals, and the stamens are fused to the inside walls of the corolla tube.

early Cretaceous Period, therefore, we may assume that the group is at least 150 million years old. Indeed, all of the divisions of plants are at least that old, and it would be strange if the angiosperms had had a much more recent origin.

As the geological record has become better known, reconstructions have been made of primitive and unusual angiosperms with no close relationships to any living families (Figure 34-2). Some living angiosperms share a number of features with these plants (Figure 34-3). The earliest fossil records of angiosperms that we have recognized, however, are pollen grains, well preserved in the rocks by virtue of their highly resistant cell walls. As additional fossil flowers and fruits are obtained and studied in detail, however, and their features become better understood, it is likely that the geological record of this great division will be pushed back several tens of millions of years.

Monocots and Dicots

There are two classes in the angiosperms, division Anthophyta: the Monocotyledones, or **monocots** (about 65,000 species), and the Dicotyledones, or **dicots** (about 170,000 species). Among the monocots are the lilies, grasses, cattails, palms, agaves, yuccas, pondweeds, orchids, and irises; the dicots include the great majority of familiar angiosperms of all kinds—almost all kinds of trees and shrubs, snapdragons, mints, peas, sunflowers, and other plants.

Monocots and dicots differ from one another in a number of features. For example, the venation in the leaves of monocots usually consists of parallel veins, and that in the leaves of dicots usually consists of netlike (reticulate) veins. In the flowers of dicots, the members of a given whorl are generally in fours or fives; in monocots, they are commonly in threes (Figure 34-4). The embryos of dicots, as the name of the class implies, generally have two seedling leaves, or **cotyledons**; those of monocots have one cotyledon. The condition in monocots has been derived from that in dicots by the suppression of one of the cotyledons. Similarly, monocots share with the most primitive dicots a very similar kind of single-pored pollen, in contrast to the three-pored and multipored pollen that is characteristic of most dicots.

> The two classes of angiosperms are monocotyledons (monocots) and dicotyledons (dicots). Monocots have one cotyledon, or seedling leaf, and usually have parallel venation in their leaves; and their flower parts are often in threes. Dicots have two cotyledons, usually netlike (reticulate) venation, and flower parts in fours or fives. There are a number of other anatomical differences between the members of these classes.

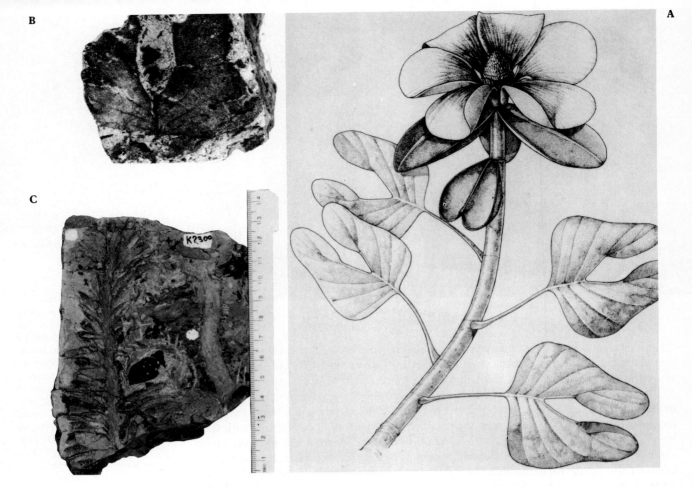

FIGURE 34-2

Reconstructions of primitive angiosperms from fossils.

A *Archaeanthus linnenbergeri*, an angiosperm discovered in the Cretaceous Period clays of Kansas by David Dilcher of Indiana University. This plant, which lived about 100 million years ago in forests dominated by ferns and gymnosperms, has no close living relatives; it is the only known member of an extinct family of angiosperms. Its flower parts are free and mostly spirally arranged, as in a magnolia or water lily flower, although the petals are reduced in number and are attached at one level on the axis, forming a whorl. A fossil of one of the unusual, deeply lobed leaves of *Archaeanthus* is shown in **B**;
C shows a fossil of the fruiting axis, which becomes elongate, like that of *Magnolia* today. Flowers of this sort may well have been visited and pollinated by beetles, like the contemporary flowers they resemble. The reconstruction was drawn by Megan Rohn.

FIGURE 34-3

Some living angiosperms share features with ancient iplants. *Drimys winteri*, of the family Winteraceae, is one of the few dicots in which the pollen has a single groove, a condition that is characteristic of the monocots. The highly distinctive pollen, of this archaic family is shed as tetrads—groups of four pollen grains. Such pollen is among the oldest angiosperm pollen fossils known. *Drimys* is also one of the very few angiosperms in which vessels are lacking; all of the conducting elements in its wood are tracheids (see Chapter 35), a very primitive condition. As shown here, the perianth parts of *Drimys* are numerous and spirally arranged, as are the stamens; the green carpels form a single whorl.

A

B

FIGURE 34-4

A typical monocot and dicot. The bright red flowers of fire-pink, *Silene virginica* (**A**), with their flower parts in fives, are typical of the dicots; those of *Trillium ozarkanum* var. *pusillum* (**B**), with flower parts in threes, are typical of the monocots. Fire-pink and trillium occur side by side in the understory of the woods throughout the temperate parts of eastern and central Canada and the United States. Fire-pink is pollinated primarily by hummingbirds. The flowers of this trillium, like those of many angiosperms, change their color after pollination, thus signaling the pollinator which flowers are still unvisited and have a supply of food.

Aside from the differences just enumerated, monocots and dicots differ fundamentally in many other ways. Here are a few examples. About a sixth of the species of dicots, but very few monocots, are **annuals** (plants that complete their entire growth cycle within a year). Monocots have underground swollen storage organs, such as bulbs, much more frequently than do dicots. In mature monocot seeds, endosperm is usually present; it is often absent in the mature seeds of dicots. We will mention more differences between monocots and dicots in the rest of this chapter and in the next one, in relation to particular features.

Monocotyledons and dicotyledons are known to have been distinct from one another from very early in the history of the division. Dicots are the more primitive of these two classes, and monocots probably were derived from early dicots before the middle of the Cretaceous Period.

Why Were the Angiosperms Successful?

When flowering plants originated, Africa and South America were still connected to one another, to Antarctica and India, and, via Antarctica, to Australia–New Zealand (see Figure 22-5). These land masses formed the great continent known as Gondwanaland. In the north, Eurasia and North America also were united, forming another supercontinent called Laurasia. The huge land mass formed by the union of South America and Africa spanned the equator and probably had a climate characterized by extreme temperatures and aridity in its interior. Similar climates occur in the interiors of the major continents at present. In the patches of drier and less favorable habitat found in the interior of Gondwanaland, much of the early evolution of angiosperms probably took place. Many of the features of the flowering plants seem to be correlated with successful growth under such arid and semiarid conditions.

Flowers effectively ensure the transfer of gametes over substantial distances and therefore promote outcrossing, for reasons that we will explore later in this chapter (Figure 34-5). The ability to transfer pollen between widely separated individuals may have been important in the early success of the angiosperms. In addition, their efficient means of fruit dispersal would have been an important element in the early success of the group. The tough, often leathery leaves of angiosperms, their efficient cuticles and stomata, and their specialized conducting elements all would have been important in survival and growth under somewhat arid conditions, just as they are in similar situations today. The many natural insecticides produced by these plants, which we discussed in Chapter 24, also would have been important for their survival. As the early angiosperms evolved, all of these features that contributed to their success became further elaborated and developed, and the group continued to evolve more and more rapidly.

The Rise to Dominance

Angiosperms became more frequent than the members of other plant groups and began to dominate terrestrial communities about 80 to 90 million years ago, during the second half of the Cretaceous Period. We can document the relative abundance of the different groups of plants by studying groups of fossils that occur at the same time

FIGURE 34-5

Scarab beetles eating pollen at the head of flowers of a relative of the sunflower (family Asteraceae) in Greece. Beetles were abundant and diverse at the time the angiosperms first evolved; they still predominate among the visitors to the flowers of many relatively primitive angiosperms—and some very advanced ones, as seen here. Beetles seem to have played a key role in the early evolution of flowers.

and place. The remains of various other divisions of plants, from more than 80 million years ago including lycopods, horsetails, ferns, and gymnosperms, are more numerous than those of the angiosperms in the strata in which they occur; afterward, and with local exceptions, angiosperms predominate.

By 80 to 90 million years ago, angiosperms were dominant in terrestrial habitats throughout the world.

At about the time that angiosperms first became the most abundant group of plants in the fossil record, there also began to appear in this record the pollen, leaves, flowers, and fruits of some of the families that still survive. Such families as the magnolias, beeches, and legumes were in existence before the end of the Cretaceous Period (65 million years ago). Among the remains of these modern families, we can even recognize a few genera that are still living. A number of the orders of insects that are particularly associated with the flowers of the angiosperms appeared or became more abundant along with the rise to dominance of this group of plants. Plants and insects have clearly played a major role in one another's patterns of evolution, and their interactions continue to be of fundamental importance in the modern world.

EVOLUTION OF THE FLOWER

As we saw in the last chapter, pollination in the angiosperms is indirect; the pollen reaches the stigma, germinates, and grows down to transport the sperm nuclei to the ovule. There, double fertilization occurs, the endosperm and embryo both grow, the seed ripens within the fruit, and the life cycle begins again. Pollen matures within the anthers and is transported, often by insects or other animals that visit the flowers for food, to another flower. The flower as a whole is attractive mainly because of the petals and, when in bud, it is protected by the sepals, which are themselves colored and attractive in some angiosperms.

Successful pollination in many angiosperms depends on the plants attracting insects and other animals regularly enough that these animals will carry pollen from one flower of that particular species to another (see Figure 34-5). Flowers that are visited regularly by animals normally provide some kind of food reward, often in the form of a liquid called nectar. Nectar is rich not only in sugars but also in amino acids and other substances. Other forms of reward that animals may find in the flowers of the angiosperms will be discussed later in this chapter in relation to particular groups of visitors.

The relationship between such animals, which are known as **pollinators,** and the flowering plants has been one of the central features of the evolution of both groups. By virtue of the evolution of such pollen-transfer systems, the flowering plants are able to disperse their gametes on a regular and more or less controlled basis, even though they, like all other plants, are anchored to their substrate. In this way, and in a certain sense, the animals perform the same functions for the flowering plants that they do for themselves when they actively search out mates.

Characteristics of Floral Evolution

Scientists have determined which kinds of angiosperms are primitive and which are specialized by correlating other kinds of features that are known to be primitive—such as those of the wood and pollen—with floral characteristics. Some of the trends are also clear simply from the nature of the features involved; for example, the fusion or reduction of flower parts. In addition, of course, the features of fossil angiosperms are taken into account.

Using these different lines of evidence, botanists have deduced that the primitive flower probably had numerous, spirally arranged sepals, petals, stamens, and carpels (see Figures 34-1, *A*, and 34-6). The differences between the petals and sepals of such a flower may not have been great, and the members of both of these floral whorls may have had similar coloration and form, grading into one another along a continuous

spiral. The members of all whorls in primitive flowers tended to be free from—not fused with—both the other members of the same whorl and the members of other floral whorls. In the flowers of more advanced groups of plants, the flower parts are arranged in definite whorls.

The first angiosperms had numerous, free, spirally arranged flower parts.

Flowers are distinguished from the ordinary leafy shoots of the same plant by being **determinate** in growth. This means that their **apical meristem**—the terminal group of dividing cells from which all shoots and roots in plants, including flowers, arise—does not continue to divide after the flower is formed during the course of its development in the young bud. In contrast, most leafy shoots are **indeterminate** and continue to grow under appropriate circumstances, progressively differentiating leaves along the way.

Flowers are determinate shoots; their apical meristem does not continue to grow after the flower is formed. Most vegetative shoots, in contrast, are indeterminate, growing continuously.

Calyx

The calyx is the outer whorl of a complete flower and consists of the sepals, which are, in effect, modified leaves that protect the flower in bud. The similarities between sepals and leaves are extensive and include their pattern of veins, as well as coloration and form. In addition, many genes are known that affect both sepals and leaves, but not petals. Sepals and leaves are two different kinds of flattened appendages that arise from the stems of angiosperms; they may have had a common evolutionary origin.

Corolla

The evolution of petals, which collectively make up the corolla, is more complex and difficult to interpret than that of sepals. Petals and stamens are similar structurally; they are affected by some genes that do not affect the carpels or sepals. In view of this relationship, the petals and stamens of most flowering plants might be homologous structures that share a common evolutionary origin. In other plants, such as water lilies (see Figure 34-1, *A*), the petals seem to have originated as modified sepals (leaves); in these plants, there are often obvious transitional structures between sepals and petals. In them, the petals were apparently derived from specialized sepals that became larger, colored, and attractive to insects during the course of their evolution.

Petals, then, appear to have had two different evolutionary origins, one—in most flowering plants—from stamens that have become flattened and leaflike in the course of evolution, and the other—in a minority of flowering plants—from sepals that have undergone the same kind of evolutionary specialization. Although it is possible that

FIGURE 34-6

A primitive flower. Species of *Magnolia*, an angiosperm genus have relatively primitive, although large, flowers. A is *M. glauca*, the southern bay magnolia; the other two photographs are of evergreen magnolia, *M. grandiflora*.

A A flower, illustrating the numerous free, spirally arranged carpels and stamens and the whorled petals, an arrangement precisely like that of the 100-million-year-old *Archaeanthus* shown in Figure 34-2. The stamens have fallen off into the bowl-shaped flower, and a longhorn beetle is eating pollen from them.

B Young flower with sepals and petals removed, showing the spirally arranged carpels (the curly structures at the top of the floral axis are their styles) and stamens (packed around the lower two thirds of the floral axis).

C Fruiting axis, again showing the numerous, spirally arranged carpels. The bright red seeds of this species emerge from the individual fruits and hang by slender threads. These seeds are attractive to and dispersed by birds, which eat their fleshy outer layers.

FIGURE 34-7
Stamens.
A A flower of strawberry, *Fragaria ananassa*, showing the many separate stamens around the central fleshy portion of the flower, where there are many short, spirally arranged carpels.
B Central column from a flower of *Hibiscus*. In *Hibiscus* and other plants of the mallow family, the stamens are also fused with the style and appear to arise from it. The stigmas are bright red in this flower.

sepals should actually be regarded as leaves that have become specialized in the course of evolution, most petals and, as we will see, all stamens and carpels, are not directly comparable to leaves.

> Petals function mainly in attracting insects and other animals to the flowers in which they occur. The petals of most kinds of plants seem to be homologous with stamens; in other plants, petals clearly evolved as enlarged, colored sepals.

Androecium

As we have seen, stamens, which collectively make up the androecium, are specialized organs that bear the microsporangia of angiosperms. In the pollen cones of gymnosperms, there are similar structures on which the microsporangia are borne. In both cases, the structures, in terms of their evolutionary history, probably evolved from systems of small branches among which the microsporangia were borne. These branch systems have progressively become reduced and modified over the course of evolutionary history. The great majority of living angiosperms have stamens whose filaments are slender—often threadlike—and whose four microsporangia are evident in a swollen portion, the anther, at the apex (Figure 34-7). Some more primitive angiosperms have stamens that are flattened and leaflike, with the sporangia protruding from the upper or lower surface.

Gynoecium

Carpels, collectively the gynoecium, are highly specialized organs that are unique in the flowering plants. Some of the more primitive flowering plants, as judged by their other features, have rather leaflike carpels. In most modern flowering plants, however, the carpels are highly specialized; the ovules occur in their swollen lower portion, the ovary, and there is a slender style, with a swollen, receptive stigma at its apex.

TRENDS OF FLORAL SPECIALIZATION

The evolution of the wide array of modern flowers can be understood largely in terms of trends involving the aggregation—grouping together—and reduction, or loss, of parts (Figure 34-8). The central axis of many flowers has been shortened in the course of their evolution. In such flowers, the whorls are close to one another. The spiral patterns that characterize the attachment of all floral parts in primitive angiosperms give way in the course of evolution to a single whorl at each level. In more advanced angiosperms, the number of parts in each of these whorls has often been reduced from many to few. The members of one or more whorls have in some evolutionary lines become fused with one another and sometimes actually joined into a tube; in other kinds of flowering plants, different whorls may be fused. Whole whorls may even be lost from the flower, which may lack sepals, petals, stamens, carpels, or various combinations of these.

Factors Promoting Outcrossing

Outcrossing, as we have stressed repeatedly, is of critical importance for the adaptation and evolution of all eukaryotic organisms. Although the flowers of most kinds of plants have both stamens and carpels, in others, one of these whorls is lacking: it has been lost during the course of evolution of the group. Flowers in which the stamens are missing or do not produce pollen are called **pistillate** flowers (they have pistils, a term discussed in the preceding chapter); those in which the carpels (pistils) are missing are called **staminate** flowers.

Staminate and pistillate flowers may occur on separate individuals in a given plant species, as for example in willows. Such plants, whose sporophytes produce either only ovules or only pollen, are called **dioecious,** from the Greek words for "two

FIGURE 34-8

Angiosperm flowers display a wealth of structural modifications.

A Baobab, *Adansonia digitata*. In the hanging flowers of this species, the numerous stamens are united into a tube around the style.

B An orchid, a species of *Laelia*, from Venezuela. In orchids, the three carpels are fused, and the flower parts appear to be attached at the top of the ovary, which is therefore called an inferior ovary. In most orchids, there is only one stamen, which is fused with the style and stigma into a single complex structure, the column. A particularly bizarre orchid is shown in Figure 34-1, *D*.

C The large, brown, ill-scented flowers of this South African milkweed, *Stapelia schinzii*, are mistaken for rotten meat by fleshflies like the one shown here. The adult flies lay eggs in the flowers, where the larvae cannot find sufficient food to complete their development; in moving from flower to flower, however, the flies effect self-pollination.

D Dutchman's pipe, *Aristolochia californica*, in which the calyx is elaborated into a bizarre structure somewhat resembling an old-fashioned pipe. Midges (small flies) enter this convoluted, balloonlike structure, are trapped, and eventually emerge, spreading pollen to another flower.

E In passion vines, *Passiflora*, the flowers are elaborate, with the five stamens united into a column around the style.

F In most members of the very large family Asteraceae, the sunflower or daisy family, which has more than 21,000 species, the flowers are crowded together into heads, with the central ones—disc flowers—radially symmetrical and usually bisexual, and the marginal ones—the ray flowers—distended into a ray and usually pistillate with carpels, but lacking stamens.

houses." In them, outcrossing is obviously complete. In other kinds of plants—oaks, birches (Figure 34-9), and ragweed (Figure 34-10) are familiar examples—the two types of flowers may be produced on the same plant; such plants are called **monoecious,** meaning "one house." In moneocious plants, the separation of pistillate and staminate flowers, which may also become mature at different times, greatly enhances the percentage of outcrossing.

Even if functional stamens and carpels are both present in each flower of a particular plant species, as is usually the case, these organs may reach maturity at different times. Plant species in which this occurs are said to be **dichogamous.** If the stamens become mature first, shedding their pollen before the stigmas are receptive, the flower is effectively staminate at that time. Once the stamens have completed shedding pollen, the stigma or stigmas may then become receptive, and the flower may

FIGURE 34-9

FIGURE 34-9
The wind-pollinated flowers of a birch, *Betula*. Birches are monoecious; their staminate flowers hang down in long, yellowish tassles, whereas their pistillate flowers, which mature into characteristic conelike structures, occur in small, reddish brown clusters.

FIGURE 34-10
Ragweed, a monecious plant.
A Flowers of great ragweed, *Ambrosia trifida*.
B Pollen of western ragweed, *Ambrosia psilostachya*. Ragweed pollen causes allergic rhinitis—more commonly known as hay fever—in millions of sufferers each year.

then be pistillate (Figure 34-11). The flower may thus achieve the same effect as if it completely lacked either functional stamens or functional carpels and may thereby significantly increase its outcrossing rate.

An even simpler way in which many flowers promote outcrossing is through the physical separation of the anthers and the stigmas. If the flower is constructed in such a way that these organs do not come into contact with one another, there will naturally be a tendency for the pollen to be transferred to the stigma of another flower, rather than to the stigma of its own flower.

Another device that occurs widely in flowering plants and increases outcrossing is **genetic self-incompatibility.** In self-incompatible plants, the pollen from a given individual does not function on the stigmas of that individual, or the embryos resulting from self-fertilization do not function. Self-incompatibility, which is widespread in flowering plants, is a mechanism that increases outcrossing even though the flowers of such plants may produce both pollen and ovules, and their stamens and stigmas may be mature at the same time.

> Plants may promote outcrossing through self-incompatibility, dioecism, monoecism, or by the separation of the different parts of the flower in space (within the flower) or of their maturation in time.

A   B

FIGURE 34-11
Dichogamy, as illustrated by the flowers of fireweed, *Epilobium angustifolium*. This species, which is strongly outcrossing, was one of the first plants in which the process of pollination was described (in the 1790s). In it, the anthers first shed pollen, then the style swings up into a similar position in the flower and its four lobes open; these flowers are functionally staminate at first, becoming pistillate about 2 days later. The flowers open progressively up the stem, so that the lowest ones are visited first. Working up the stem, the bees encounter pollen-shedding, staminate-phase flowers, become covered with pollen, and carry it passively to the lower, functionally pistillate flowers of another plant. Shown here are flowers in (**A**) the staminate phase and (**B**) the pistillate phase.

Trends in Floral Symmetry

Other evolutionary trends in floral evolution have affected the symmetry of the flower. Primitive flowers are radially symmetrical; those of many advanced groups are bilaterally symmetrical; that is, they are equally divisible into two parts along a single plane. Examples of such flowers are the orchids shown in Figures 34-1, *D*, and 34-8, *B*, as well as the Dutchman's pipe shown in Figure 34-8, *D*. Such tubular, bilaterally symmetrical flowers are especially frequent in advanced groups of flowering plants, such as among the snapdragons or peas. In these groups, they are connected with advanced and often highly precise pollination systems.

POLLINATION IN FLOWERING PLANTS
Pollination in Early Seed Plants

The early seed plants were pollinated passively, by the action of the wind. They simply shed great quantities of pollen, which were blown about and occasionally reached the vicinity of the ovules of the same species. For such a system to operate efficiently, the individuals of a given plant species must grow relatively close together. Otherwise, the chance that any pollen will arrive at the appropriate destination is very small because the average distance that pollen is carried in large quantities by the wind is actually quite short (the vast majority of the pollen travels less than 100 meters), compared with the long distances to which pollen is routinely carried very accurately by certain insects and other animals that visit flowers.

Keeping this fact in mind, one can begin to understand why many of the uniform stands of trees that dominate huge areas in North America and in Eurasia, including gymnosperms such as spruces, pines (see Figure 19-4), and hemlocks, and angiosperms such as birches and alders, are wind pollinated (see Figures 34-9 and 34-10). As mentioned earlier, all but a few living gymnosperms are wind pollinated, but insect pollination might already have been established in the ancestor of the angiosperms.

Pollination by Animals

Animals, visiting the flowers of specific kinds of angiosperms and spreading their pollen from individual to individual precisely, have clearly played an important role in the evolutionary success of the group. It now seems clear that the earliest angiosperms, and perhaps their ancestors also, were insect pollinated, and the coevolution of insects and plants has been important for both groups for well over 100 million years. Such interactions have likewise been important in bringing about increased floral specialization, as an examination of the flowers illustrated in this chapter will readily show. As the flowers become increasingly specialized, so do their relationships with particular groups of insects and other animals.

The shift from wind pollination to insect pollination among the ancestors of the angiosperms was an important element in the evolutionary success of the group.

Bees and Flowers

Among the insect-pollinated angiosperms, the most numerous are those pollinated by bees (see Figures 19-6 and 34-12). Like most insects, bees locate sources of food largely by odor at first, but they then orient themselves on the flower or group of flowers by its shape, color, and texture. Many flowers that are characteristically visited by bees are blue or yellow. Many have characteristic lines of dots or stripes that indicate the position in the respective flowers of the **nectaries,** the glands where the nectar is produced, which often are located within the throats of specialized flowers. Some bees visit flowers for nectar, primarily as a source of food for the adults and sometimes also for the larvae; most of the approximately 20,000 species of bees, however, visit flowers to obtain pollen. They use this pollen to provision the cells in which their larvae complete their development (Figure 34-13).

FIGURE 34-12
Pollination by a bumblebee. As this bumblebee, *Bombus*, squeezes into the bilaterally symmetrical, advanced flower of a member of the mint family, the stigma contacts its back and picks up any pollen that the bee may have acquired there during a visit to a previous flower.

FIGURE 34-13
An underground nest of the solitary bee, *Nomia melanderi*. The female bee that dug this burrow has provisioned it with a large ball of alfalfa pollen, on which the larval bee can be seen feeding. In order to be able to set seed, alfalfa requires visits by bees to trip its floral mechanism. Since alfalfa seed is an important crop, this species of *Nomia* and other bees are often introduced in large numbers into the neighborhood of alfalfa fields.

Only a few hundred species of bees are social or semisocial in their nesting habits. These occur in colonies like the familiar honeybee, *Apis mellifera*, or the bumblebees, *Bombus* (see Figure 34-12). Such bees produce several generations during a year and must shift their attentions to different kinds of flowers as the season progresses. Because of the size of their colonies, they also must often use the flowers of more than one kind of plant as a food source at any given time.

Except for these social and semisocial bees and about 1000 species that are parasitic in the nests of other bees, the great majority of bees—at least 18,000 species—are solitary. Solitary bees characteristically have only a single generation in the course of a year in temperate regions. Often they are active as adults for as little as a few weeks a year.

Solitary bees often use the flowers of a given group of plants exclusively, or nearly so, as sources of their larval food. The highly constant relationships of such bees with these flowers may lead to modifications, over time, in both the flowers and the bees. For example, the time of day when the flowers open may be correlated with the time when the bees appear; the mouthparts of the bees may become elongated in relation to tubular flowers; or the bees' pollen-collecting apparatus may reflect in its characteristics those of the pollen of the plants that they normally visit. When such relationships are established, they provide both an efficient mechanism of pollination for the flowers and a constant source of food for the bees that "specialize" on them.

Bees are the most frequent and characteristic pollinators of flowers. About 90% of the roughly 20,000 species of bees are solitary; morphological and physiological specializations often link these solitary bees with particular kinds of plants.

Bees first appeared at least 100 million years ago, judging from the occurrence of modern-appearing fossils 80 million years old. They later became abundant and diverse as the world climate became drier. As the abundance and diversity of bee species increased, their activities clearly exerted a powerful selective influence on the pattern of angiosperm diversification. In addition, the bees themselves have become even more diverse as their relationships with flowers have grown in complexity and specificity. Because of the great precision of these relationships, which are the most specific of any between flowering plants and their insect visitors, you might assume that bees were responsible for much of the evolutionary diversification of angiosperms. There is some truth to this statement, but there are only about 20,000 species of bees in the world, and about 235,000 species of angiosperms. Clearly, then, bees have not been responsible for the overall diversification of angiosperms, even though they have had an important influence on the evolution of many individual species.

Insects Other Than Bees

Among flower-visiting animals other than bees, a few groups are especially prominent. Flowers that are visited regularly by butterflies, for example, phlox, often have flat "landing platforms" on which the butterflies perch. They also tend to have long, slender floral tubes filled with nectar that is accessible to the long, coiled proboscis characteristic of the Lepidoptera, the order of insects that includes butterflies and moths (Figure 34-14). Flowers visited regularly by moths are often pale in color, white or yellow, like jimsonweed or evening primrose; heavily scented, thus serving to make the flowers easy to locate at night; and long tubed, like the flowers visited regularly by butterflies.

Pollination by Birds

A particularly interesting group of plants are those regularly visited and pollinated by birds, especially by hummingbirds in North and South America (Figure 34-15). Such plants must produce large amounts of nectar. If they do not, the birds, which consume a great deal of energy, will not be able to find enough food to maintain them-

FIGURE 34-14
Flower visitation by a butterfly and a moth.
A The copper butterfly *Lycaena gorgon*, visiting the flowers of a composite. The proboscis of the butterfly, which coils like a watchspring when at rest, is being thrust into one flower after another to extract the nectar.
B A hawkmoth, *Hyles lineata*, at the flowers of scarlet lobelia, *Lobelia cardinalis*. Such red flowers are frequented by hummingbirds, but hawkmoths will occasionally visit any flowers that have sufficient nectar, extracting it by means of a long proboscis, which is coiled at rest. In *Lobelia cardenalis*, the flowers first enter a staminate phase and then become pistillate; staminate-phase flowers are nectar-rich and clustered at the top of the stems.

selves or to continue visiting the flowers of that plant, for energy reasons alone. If these flowers do produce such large amounts of nectar, however, it is not advantageous for them to be visited by insects, because the insects could supply their energy requirements at perhaps a single flower and therefore would not move to another flower. Although the nectar of that flower might be very adequate for the insect, the plant would not be cross-pollinated as a result of the insect's visit. How do flowers that "specialize" on hummingbirds balance these different selective forces?

To birds, red is a very conspicuous color, just as it is to us. To insects, however, ultraviolet is a highly visible color, but red is not. Carotenoids, yellow pigments frequently found in plants, are responsible for the yellow colors of many flowers, such as sunflowers and mustard. Carotenoids reflect both in the yellow range and in the ultraviolet range, the mixture resulting in a distinctive color called "bee's purple." Such yellow flowers may then be marked in distinctive ways, normally invisible to us but highly visible to bees and other insects (Figure 34-16).

Red, in contrast, does not stand out as a distinct color to most insects. To most insects, the red upper leaves of poinsettias look just like the other leaves of the plant. Even though the red upper leaves of poinsettias are obvious to us, they do not attract insects to the small, yellowish green flowers of the same plants, even though these flowers produce abundant supplies of nectar. Consequently, insects are not apt to visit these flowers, but hummingbirds are. Thus red is a perfect color both to signal birds of the presence of abundant nectar and, at the same time, to make that nectar as inconspicuous as possible to insects.

As hummingbirds migrate northward through North America every summer, they encounter a progression of red flowers that provide an abundant source of food for them along the way. Even if they are unfamiliar with these flowers, the red color of the flowers signals the birds that they are suitable sources of food.

Hummingbird-visited flowers are also specialized in other, analogous ways. Such flowers generally are odorless, because birds normally do not have well-developed ol-

FIGURE 34-15

Hummingbirds and flowers.
A Long-tailed hermit hummingbird extracting nectar from the flowers of *Heliconia imbricata* in the forests of Costa Rica. Note the pollen on the bird's beak. Hummingbirds of this group obtain nectar primarily at long, curved flowers that more or less match the length and shape of their beak.
B Poinsettia (*Euphorbia pulcherrima*) flowers. The individual flowers each have a large nectary at one side. The clusters of yellowish flowers are made more attractive to birds because of the large red leaves that surround them.
C *Penstemon centranthifolius* in the deserts of southern California, showing how attractive the plants that regularly attract hummingbirds may be, even at a distance.

A   B

FIGURE 34-16

How a bee sees a flower. The yellow flower of *Ludwigia peruviana* (Onagraceae) photographed in normal light (**A**) and with a filter that selectively transmits ultraviolet light (**B**). The outer sections of the petals reflect both yellow and ultraviolet, a mixture of colors called "bee's purple"; the inner portions of the same petals reflect yellow only and therefore appear dark in the photograph that emphasizes ultraviolet reflections. To a bee, this makes the flower appear as if it has a conspicuous central bull's-eye.

FIGURE 34-17
Wind-pollinated flowers. Most of the parts of the flowers of grasses and other wind-pollinated plants are greenish; they are therefore not particularly attractive to insects. At this stage, the large yellow anthers, dangling on very slender filaments, are hanging out, about to shed their pollen to the wind; later these flowers will become pistillate, with long, feathery stigmas—well suited for trapping wind-blown pollen—sticking far out of them. Many grasses, like this one, are therefore dichogamous.

factory senses and do not orient to odor. Insects, as we have mentioned, often do, and odors strongly signal the presence of food to them. The nectar in hummingbird-visited flowers is often held within strong tubes or otherwise protected where it can be obtained only by the action of the beak of the bird and not, in general, by insects.

> **Insects often are attracted by the odors of flowers and do not perceive red as a distinct color. Birds are not attracted by odors and do perceive red. Bird-pollinated flowers, which must produce large quantities of nectar to support the birds that visit them and to keep them visiting, are characteristically odorless and red, with the nectar well protected.**

Wind-Pollinated Angiosperms

Many angiosperms, representing a number of different groups, have reverted to the wind pollination that was characteristic of early seed plants. Among them are such familiar plants as oaks, birches, cottonwoods, grasses, sedges, and nettles. The flowers of these plants are small, greenish, and odorless; and their corollas are reduced or absent (see Figures 34-9, 34-10, and Figure 34-17). Such flowers often are grouped together in fairly large numbers and may hang down in tassels that wave about in the wind and shed pollen freely. Many wind-pollinated plants are dioecious or monoecious, with the staminate and pistillate flowers separated on separate individuals or on a single individual. If the pollen-producing and ovule-bearing flowers are separated, it is certain that pollen released to the wind will fertilize a flower other than the one that sheds it, a strategy that greatly promotes outcrossing.

Self-Pollination

All of the modes of pollination that we have considered thus far tend to lead to outcrossing, which is as highly advantageous for plants as it is for eukaryotic organisms generally. Notwithstanding this, self-pollination is also very frequent among angiosperms. In fact, probably more than half of the angiosperms that occur in temperate regions self-pollinate regularly. Most of these have small, relatively inconspicuous flowers in which the pollen is shed directly onto the stigma, sometimes even before the bud opens (Figure 34-18). You might logically ask why there are so many self-pollinated plant species if outcrossing is just as important genetically for plants as it is for animals. There are two basic reasons for the very frequent occurrence of self-pollinated angiosperms. They are as follows:

1. Self-pollination obviously is ecologically advantageous under certain circumstances, because the plants in which it occurs do not need to be visited by animals to produce seed. As a result, self-pollinated plants can grow in areas where the kinds of insects or other animals that might visit them are absent or very scarce—as in the Arctic or at high elevations.
2. In genetic terms, self-pollination produces progenies that are more uniform than those that result from outcrossing. Such progenies may contain high proportions of individuals well adapted to particular habitats. Self-pollination initially tends to produce large numbers of ill-adapted individuals, many of which may be incapable of reaching reproductive size. Functionally, however, this kind of bottleneck in fitness may be overcome within a few generations, with no further obstacle to continued self-pollination. If the habitat to which the uniform progenies produced by self-pollination are well adapted continues to exist, it may be advantageous for the plant to continue self-pollinating indefinitely. This is the main reason that self-pollinating plant species are well represented as weeds—the habitat of weeds has been made uniform and spread all over the world by humans.

> **Self-pollinated angiosperms are frequent where there is a strong selective pressure to produce large numbers of genetically uniform individuals adapted to particular relatively uniform habitats.**

FIGURE 34-18
Self-pollinating flowers. In the genus *Epilobium*, most species are self-pollinating, like *E. ciliatum*, shown here. Note that the anthers are shedding pollen directly on the stigma. A strongly outcrossing species of the same genus is shown in Figure 34-11.

A

B

C

FIGURE 34-19

Animal-dispersed fruits.

A The spiny fruits of this burgrass, *Cenchrus incertus*, adhere readily to any passing animal, as you will know if you have stepped on them.

B The bright red berries of this honeysuckle, *Lonicera hispidula*, are highly attractive to birds, just as are red flowers. Birds may carry the seeds they contain for great distances after eating the fruits, with their sticky pulp, either on their feet or other body parts or internally.

C The fruits of figs, *Ficus carica*, are flower clusters turned inside out. They are pollinated by tiny specialized wasps, which require figs to complete their life cycle. The wasps enter through the hole in the end of the fig. When mature, figs are consumed and the seeds of the individual tiny flowers are scattered about by birds and mammals.

THE EVOLUTION OF FRUITS

Paralleling the evolution of flowers of the angiosperms, and nearly as spectacular, has been the evolution of their fruits. Aside from the many ways in which fruits can be formed, they exhibit a wide array of modes of specialization in their dispersal.

Fruits that have fleshy coverings, often shiny black or bright blue or red, normally are dispersed by birds and other vertebrates (see Figure 34-6, *C,* and Figure 34-19, *B* and *C*). Like the red flowers that we discussed in relation to pollination by birds, red fruits signal an abundant food supply. By feeding on these fruits, birds and other animals may carry seeds from place to place and thus transfer the plants from one suitable habitat to another.

Other kinds of specialized fruit dispersal have evolved many different times in the flowering plants. Fruits with hooked spines, like snakeroot, beggar-ticks, and burdock (Figure 34-19, *A*), are particularly characteristic of several genera of plants that occur in the northern deciduous woods. Such fruits are often spread from place to place by mammals, including humans. In addition, mammals like squirrels disperse and bury seeds, not always finding them again. Other fruits or seeds have wings and are blown about by the wind; the wings on the seeds of pines and those on the fruits of ashes or maples (Figure 34-20) play identical ecological roles. The dandelion pro-

FIGURE 34-20

Wind-dispersed fruits.

A False dandelion, *Pyropappus caroliniana.* The "parachutes" disperse the fruits of dandelions widely in the wind, much to the gardener's despair.

B The double fruits of maples, *Acer,* when mature, are blown considerable distances from their parent trees.

C Tumbleweed, *Salsola,* in which the whole dead plant becomes a light, windblown structure that rolls about scattering seeds.

A   **B**   **C**

A   B

FIGURE 34-21
Water-dispersed fruits.
A Mangrove, *Rhizophora mangle*. The embryos begin to grow on the parent plants and soon become established when they wash away to another silty shore.
B A fruit of the coconut, *Cocos nucifera*, sprouting on a sandy beach. One of the most useful plants for humans in the tropics, coconuts have become established even on the most distant islands by drifting in the waves.

vides a familiar example of a kind of fruit that is dispersed by the wind, and the dispersal of the seeds of plants such as milkweeds, willows, and cottonwoods is similar.

Still other fruits, like the coconut and those of certain other plants that characteristically occur on or near beaches, are regularly spread from place to place by water (Figure 34-21). Dispersal of this sort is especially important in the colonization of distant island groups, such as the Hawaiian Islands. It has been calculated that the seeds of about 175 original angiosperms must have reached Hawaii to have evolved into the roughly 970 species found there today (see Figure 19-4 for an example). Some of these seeds blew through the air, others were transported on the feathers or in the guts of birds, and still others drifted across the Pacific to Hawaii, nearly a third of them from North America. Although the distances are rarely as great as that between Hawaii and the mainland, dispersal is just as important for those mainland plant species in which the habitats are discontinuous, such as mountaintops, marshes, or north-facing cliffs. All in all, dispersal is one of the most fascinating aspects of plant biology.

> **Fruits, which are characteristic of angiosperms, are extremely diverse. The evolution of structures in particular fruits that have improved their possibilities for dispersal in some special way has produced many examples of parallel evolution.**

The examples given and illustrated before provide some idea of the kinds of evolutionary changes that have been involved in the evolution of dispersal mechanisms in fruits and seeds. You may be interested in studying the fruits and seeds produced by the plants in your area and in trying to discover the mechanisms by which they are dispersed. For some plants it is advantageous for the seeds to be widely dispersed, and for others it is advantageous to keep them near the parent plant. You may find it of interest to consider the factors that are important in determining which of these conditions applies to a particular plant species.

SUMMARY

1. The ancestor of angiosperms was a seed-bearing plant that was probably already pollinated by insects to some degree. No living group of plants has the correct combination of characteristics to be this ancestor, but seeds have originated a number of times during the history of the vascular plants.

2. Although angiosperms are probably at least 150 million years old as a group, the oldest definite fossil evidence of this division is pollen from the early Cretaceous Period, some 123 million years ago. By 80 to 90 million years ago, angiosperms were more common worldwide than other plant groups. They became abundant and diverse as drier habitats became widespread during the last 30 million years or so.

3. Among the reasons that angiosperms have been successful are their relatively drought-resistant vegetative features, including their vascular systems, cuticles, and stomata. Most important, however, are their flowers and fruits. Flowers make possible the precise transfer of pollen, and therefore, outcrossing, even when the stationary individual plants are widely separated. Fruits, with their complex adaptations, facilitate the wide dispersal of angiosperms.

4. The flowers of primitive angiosperms had numerous, separate, spirally arranged flower parts, as we know from the correlation of flowers of this kind with primitive pollen, wood, and other features. Sepals are homologous with leaves; the petals of most angiosperms appear to be homologous with stamens, although some appear to have originated from sepals; and stamens and carpels probably are modified branch systems whose spore-producing organs were incorporated into the flower during the course of evolution.

5. Outcrossing in different angiosperms is promoted by the separation of the pollen- and ovule-producing structures into different flowers, or even onto different individuals; by genetic self-incompatibility; and by the separation of the pollen and the stigmas within a given flower with respect to space or time of maturation.

6. Bees are the most frequent and constant visitors of flowers. The solitary bees, which constitute about 90% of the estimated 20,000 bee species, are the most specialized and specific visitors. They often have morphological and physiological adaptations related to their specialization in visiting the flowers of particular plant species.

7. Flowers visited regularly by birds must produce abundant nectar to provide the birds with enough energy so that they will continue to be attracted to them. If insects, most of which need less energy than birds, visit such flowers, they may not be stimulated to move from flower to flower. Many bird-visited flowers, therefore, are red, a color not attractive to most insects, and are odorless; insects often orient by odor, but birds usually do not. The nectar of bird-visited plants tends to be well protected by the structure of the flowers.

8. Fruits and seeds are highly diverse in terms of their dispersal, often displaying wings, barbs, or other structures that aid their dispersal. Means of fruit dispersal are especially important in the colonization of islands or other distant patches of suitable habitat.

REVIEW QUESTIONS

1. Why is it virtually certain that all angiosperms were derived from a single ancestor? Which of the four living groups of gymnosperms are most closely related to the angiosperms? What are the earliest known definite angiosperm fossils? How old are they?

2. What two classes comprise the angiosperms? How do the two classes structurally differ from one another? Which class derived from the other? Explain.

3. What are the characteristics of a primitive angiosperm flower? How can scientists determine whether a given flower is primitive or advanced? What other plant part does a primitive flower most resemble?

4. What flower whorl is collectively made up of petals? Which other flower parts are the petals of most flowers homologous with? In what ways? What is the origin of petals in a small number of flowers?

5. What is an androecium? Of which flower parts is it composed? What is the probable evolution of these flower parts? In what structure are the pollen grains produced?

6. What are pistillate versus staminate flowers? What is a dioecious versus a monoecious plant? What is the advantage of these diverse characteristics?

7. What kind of growth pattern is necessary for passive pollination to be successful? In relative terms, how much pollen must be produced by such kinds of plants? What is the distance such pollen is carried compared to the distance pollen is carried by active pollinators? What has been the effect of coevolution of flowering plants and animal pollinators, primarily insects?

8. How does the reproductivity and colony size of social versus solitary bees relate to the kinds of flowers that they pollinate? Is it more likely that a flower visited by a social or a solitary bee will become highly specialized toward that bee? Why? How is such coevolution advantageous to both parties?

9. What are characteristics of flowers that are pollinated by butterflies? What are characteristics of flowers that are pollinated by moths?

10. Why must flowers pollinated by birds produce greater amounts of nectar than those pollinated by most kinds of insects? How do such plants deter insects from wasting their nectar?

11. What are the primary characteristics of wind-pollinated flowers? Do most wind-pollinated plants have separate pistillate and staminate flowers, or do they have flowers with both parts functional? Why?

12. What types of fruit are dispersed by birds and other vertebrates? How are fruit that have evolved hooks dispersed? How do animals like the squirrel aid in seed dispersal? What structures aid some seeds in becoming dispersed passively, by the wind?

THOUGHT QUESTIONS

1. Why haven't we found the "missing link" between the flowering plants and their ancestors? What would you suggest we do to try to solve this riddle?

2. What role did bees play in the origin of angiosperms and in their subsequent diversification?

3. Angiosperms usually produce both pollen and ovules within a single flower, or at least on a single plant. Why don't all angiosperms simply self-pollinate?

FOR FURTHER READING

GIBSON, A.C., and P.S. NOBEL: *The Cactus Primer*, Harvard University Press, Cambridge, Mass., 1986. Outstanding account of all aspects of the biology of this fascinating plant family.

HEYWOOD, V.H. (editor): *Flowering Plants of the World*, Prentice-Hall, New York, 1985. An outstanding guide to the families of flowering plants, with excellent illustrations and diagrams.

HOBSON, G.: "How the Tomato Lost Its Taste," *New Scientist*, September 1988, vol. 119, pages 46-50. No one seems to know exactly when the flavor fled the tomato, only that it did. Scientists are searching for the lost ingredient.

JOHNSON, W.C., and C.S. ADKISSON: "Airlifting the Oaks," *Natural History*, October 1986, pages 40-46. Bluejays spread acorns widely to new habitats; a familiar example of animal dispersal of plant fruits and seeds.

JONES, S.B., Jr., and A.E. LUCHSINGER: *Plant Systematics*, ed. 2, McGraw-Hill Book Co., New York, 1986. A comprehensive account of plant systematics, including a useful review of plant families.

PROCTOR, M., and P. YEO: *The Pollination of Flowering Plants*, Taplinger Publishing Co., Inc., New York, 1973. An excellent introduction to pollination biology, clearly and interestingly presented.

RICHARDS, A.J.: *Plant Breeding Systems*, George Allen & Unwin, London, 1986. An excellent account of all aspects of the subject.

ROBACKER, D.C., MEEUSE, J.D. BASTIAAN, and E.H. ERICKSON: "Floral Aroma," *BioScience*, vol. 38, pages 390-396, 1988. An interesting discussion of the odors that plants utilize in attracting their pollinators.

STUESSY, T.F.: *Plant Taxonomy: The Systematic Evaluation of Comparative Data*. Columbia University Press, New York, 1990. An advanced but well-written text that examines much of the evidence dealing with angiosperm classification.

TYRRELL, E.Q., and R.A. TYRRELL: *Hummingbirds: Their Life and Behavior: A Photographic Study of the North American Species*, Crown Publishers, Inc., New York, 1985. A magnificently illustrated account of the most highly specialized birds that visit flowers.

35

Vascular Plant Structure

OVERVIEW

The stems and roots of vascular plants differ in structure, but both grow at their apices and consist of the same three kinds of tissues. Epidermis covers their external surfaces, vascular tissue conducts materials within them, and ground tissue performs photosynthesis and stores nutrients. One component of the vascular tissue, the xylem, conducts water and its dissolved minerals; and the other component, the phloem, distributes carbohydrates, which are manufactured in the green parts of the plant. Plants are immobile, but they adjust to their environment by growing and changing their form. Roots are distinguished from shoots by the way their branches originate—deep within the ground tissue, ultimately bursting through to the surface; in stems, branches arise on the surface in the axils of the leaves. The leaves themselves are flat organs, specialized for photosynthesis. When flowers form on an axis, the stem stops growing permanently along that axis.

FOR REVIEW

Here are some important terms and concepts that you will encounter in this chapter. If you are not familiar with them, you should review them before proceeding.

Cell structure (Chapter 5)

How cells divide (Chapter 10)

Major groups of plants (Chapter 33)

Reproduction in flowering plants (Chapter 34)

Monocots and dicots (Chapter 34)

A plant never becomes an adult in the way in which an animal does. Rather, plants simply keep growing, adding new cells, tissues, and organs at the ends of their shoots and their roots, becoming ever larger. Some large patches of prairie grasses are thought to be single individuals that have been growing in one place since the glaciers receded, more than 10,000 years ago. Trees attain great ages, and potatoes and many other crops are simply propagated over and over again as parts of a single cloned plant, producing generation after generation of genetically identical individuals.

The cells and tissues of vascular plants, and the ways in which they grow and develop, will be our major focus in this chapter. We will discuss the fundamental differences between the underground portions of the plant—the roots—and the above-ground portions—the shoots—as well as the functional and structural relationships between them. First, though, we will analyze the way in which these organs, together with their specialized cell and tissue types and their appendages, form the plant body.

Although the similarities between a cactus, an orchid, and a pine tree might not be obvious at first sight, plants have a fundamental unity of structure (Figure 35-1). This unity is reflected in the construction plan of their respective plant bodies; in the way that they grow, produce, and transport their food; and in the means by which they regulate their development.

ORGANIZATION OF THE PLANT BODY

A vascular plant is basically an axis consisting of root and shoot. The **root** penetrates the soil and absorbs water and various ions, which are crucial for plant nutrition, and

FIGURE 35-1
Diagram of a dicot plant body, showing the parts. The gray areas are zones of active elongation.

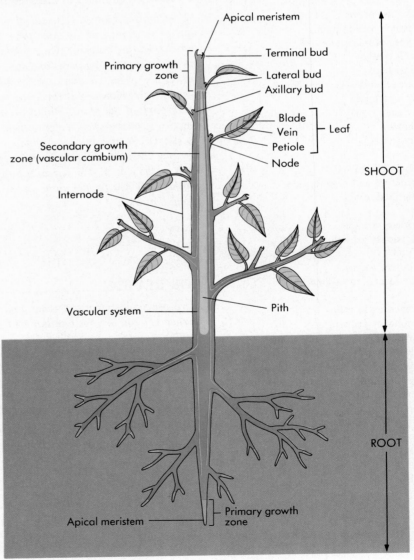

PART IX BIOLOGY OF PLANTS

it also anchors the plant. The **shoot** consists of stem and leaves. The **stem** serves as a framework for the positioning of the **leaves,** the principal places where photosynthesis takes place. The arrangement, size, and other characteristics of the leaves are of critical importance in the plant's production of food. Flowers, other reproductive organs, and ultimately fruits and seeds are formed on the shoot as well.

Tissue Types in Plants

There are three major types of tissue in plants. One is the conducting tissue, or **vascular tissue.** There are two kinds of vascular tissue: (1) **xylem,** which conducts water and dissolved minerals, and (2) **phloem,** which conducts carbohydrates, mainly sucrose, which the plant uses for food, as well as hormones, amino acids, and other substances that are necessary for plant growth. Xylem and phloem, the two kinds of vascular tissue, differ not only in function but also in structure, as we will see.

The other two major tissue types in plants are **ground tissue,** in which the vascular tissue is embedded, and **dermal tissue,** the outer protective covering of the plant. The dermal tissue often is covered with a waxy substance known as cuticle. Each of the three major tissue types—vascular tissue, ground tissue, and dermal tissue—consists of its own distinctive cell types, which are related to the functions of the tissues in which they occur. Some of the characteristic cell types will be discussed later in this chapter.

> **The three major types of tissues in plants are (1) dermal tissue, (2) ground tissue, and (3) vascular tissue.**

Types of Meristems

Primary growth in plants is initiated by the **apical meristems.** These are regions of active cell division that occur relatively close to the tips of roots and stems. The growth of these meristems results primarily in the extension of the plant body. As it elongates, it forms what is known as the primary **plant body,** which is made up of **primary tissues.** The primary plant body comprises the young, soft shoots and roots of a tree or shrub, or the entire plant in some short-lived plants.

> **An apical meristem is a region of active cell division that occurs relatively close to the tips of the roots and shoots of plants. Apical meristems are responsible for a plant's primary growth.**

Secondary growth involves the activity of **lateral meristems,** which are cylinders of meristematic tissue; the continued division of their cells results primarily in the thickening of the plant body. There are two kinds of lateral meristems: (1) the vascular cambium, which gives rise to ultimately thick accumulations of secondary xylem and also to secondary phloem, which is gradually sloughed off, and (2) the cork cambium, from which arise the outer layers of the bark in both roots and shoots. The tissues formed from the lateral meristems, comprising most of the bulk of trees and shrubs, are known collectively as the secondary plant body; and its tissues are known as secondary tissues. Figure 35-1 compares primary and secondary growth zones.

> **The primary plant body, which includes the young, soft shoots and roots, arises from the apical meristems. Once the lateral meristems begin to function, they produce the secondary plant body, which is characterized by thick accumulations of conducting tissue and the other cell types associated with it.**

PLANT CELL TYPES
Meristems

Meristems, both apical and lateral, consist of small, usually unspecialized cells, some of which divide repeatedly. When one of these cells divides, one of its two derivative cells remains in the meristem, and the other becomes part of the plant body. The cells that ultimately become part of the plant body usually divide further before they begin

A

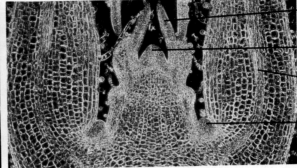

Young leaf primordium

Apical meristem

Older leaf primordium

Lateral bud

B

FIGURE 35-2

Leaf development in *Coleus*.

A The leaves of *Coleus*, which are often colorful, are a familiar sight in greenhouses and as bedding plants. They are borne in opposite pairs, alternating in direction.

B Transection of a shoot apex in *Coleus*, showing how this kind of leaf arrangement is initiated.

to differentiate and assume the characteristics of the type of cell that they will become (Figure 35-2).

The apical meristem gives rise to three types of primary meristems. These are partly differentiated tissues; in them, cell division continues to take place as they develop into the tissues of the plant body (Figure 35-3). The three primary meristems are the **protoderm**, which differentiates further into epidermis; the **procambium**, which differentiates further into primary vascular strands; and the **ground meristem**, which differentiates further into ground tissue. The basic tissue patterns are established by early activity in the region of the apical meristem, when the primary meristems begin to differentiate.

The primary meristems, which differentiate from the apical meristems, are (1) the protoderm, which becomes the epidermis, (2) the procambium, which becomes the primary vascular strands, and (3) the ground meristem, which becomes the ground tissue.

Parenchyma and Collenchyma

Parenchyma cells, which often are somewhat spherical, are the least specialized and the most common of all plant cell types (see Figures 35-5 and 35-14, for example,

FIGURE 35-3

Diagram of primary meristems in the root, showing their relation to the apical meristem.

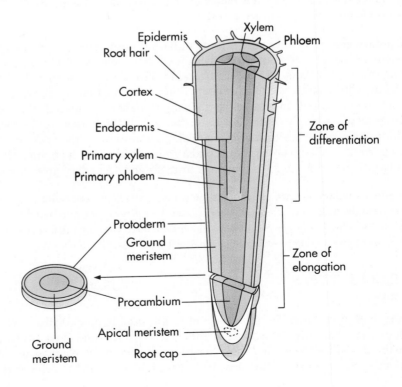

Epidermis

Root hair

Cortex

Endodermis

Primary xylem

Primary phloem

Xylem

Phloem

Zone of differentiation

Protoderm

Ground meristem

Zone of elongation

Procambium

Ground meristem

Apical meristem

Root cap

show many parenchyma cells). They form masses in leaves, stems, and roots in which secondary growth has not taken place and also in secondary tissues. Parenchyma cells, unlike some of the other cell types that we will consider, are characteristically alive at maturity, with fully functional protoplasts and a nucleus. They are capable, therefore, of further division. Most parenchyma cells have only **primary walls,** which are cell walls that are laid down while the cells are still growing. **Secondary cell walls,** in contrast, are deposited inside the primary walls of fully expanded cells.

Collenchyma cells, which also are living at maturity, form strands or continuous cylinders beneath the epidermis of stems or leaf stalks and along veins in leaves. They are usually elongated, with unevenly thickened primary walls (Figure 35-4), which are their distinguishing feature. Strands of collenchyma provide much of the support for plant organs in which secondary growth has not taken place. The "strings" of celery leaves—the part of celery that we eat is the stalk of the leaf—consist mainly of collenchyma.

> **Parenchyma cells are somewhat spherical and usually lack secondary cell walls. Collenchyma cells are elongate and have unevenly thickened cell walls; they provide much of the support for primary tissues. Both cell types are usually living at maturity.**

Sclerenchyma

Sclerenchyma cells have tough, thick secondary walls; they usually do not contain living protoplasts when they are mature. Their secondary cell walls are often impregnated with lignin, a complex polymer; cell walls containing lignin are said to be **lignified.** Lignin is deposited in the primary walls of some kinds of cells, as well as the secondary ones. It is a rigid substance and makes the walls in which it is deposited more rigid. Lignin is common in the walls of plant cells that have a supporting or mechanical function. Part of the reason that fruits become softer as they ripen has to do with the breakdown of lignin in their cell walls.

There are two types of sclerenchyma: fibers, which are long, slender cells that usually form strands, and **sclereids,** which are variable in shape but often branched. Both of these tough, thick-walled cell types serve to strengthen the tissues in which they occur. For example, linen is woven from strands of fibers that occur mainly in association with the phloem of flax. Sclereids, on the other hand, may occur singly or in groups. The gritty texture of pears is caused by the groups of sclereids that occur among the soft flesh of the fruit (Figure 35-5).

> **Sclerenchyma cells have thick, often lignified secondary walls and are nonliving at maturity. The two types of sclerenchyma are fibers, which are elongated, and sclereids, which are not.**

Xylem

Xylem is the principal water-conducting tissue of plants. It forms a continuous system running throughout the plant body and supports the plant body as well as providing for the conduction of water and dissolved minerals throughout it. Within the xylem, water passes from the roots up through the shoot in an unbroken stream. Dissolved minerals also are taken into plants through their roots, as a part of this stream of water. When the water reaches the leaves, it passes into the air as water vapor, mainly through the stomata. **Primary xylem** is derived from procambium, which in turn comes from the apical meristem of that respective root or shoot. **Secondary xylem,** on the other hand, is formed by the vascular cambium, a lateral meristem that develops later. Wood consists of accumulated secondary xylem.

The two principal types of conducting elements in the xylem are **tracheids** and **vessel elements** (Figure 35-6), both of which resemble fibers in having thick, lignified secondary walls, in being elongated, and in having no living protoplast at maturity. The continuous stream of water in a plant flows from tracheid to tracheid through pits in their secondary walls. In contrast, vessel elements have not only pits

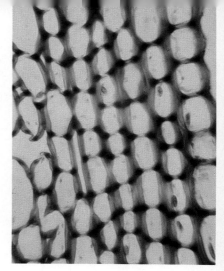

FIGURE 35-4
Collenchyma. Cross section of collenchyma cells, with thickened side walls, from a young branch of elderberry *(Sambucus)*. In other kinds of collenchyma cells, the thickened areas may occur at the corners of the cells or in other kinds of strips.

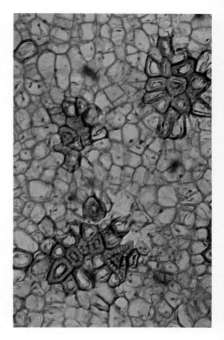

FIGURE 35-5
Sclereids. Clusters of sclereids ("stone cells"), stained red in this preparation, in the pulp of a pear. The surrounding, thin-walled cells, stained light blue, are parenchyma. Such clusters of sclereids give pears their gritty texture.

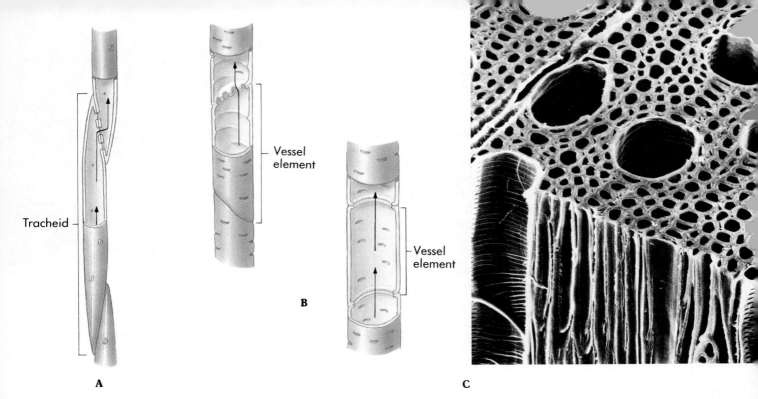

Tracheid

Vessel element

B

Vessel element

A

C

FIGURE 35-6

Comparison between vessel elements and tracheids. In tracheids (**A**) the water passes from cell to cell by means of pits, whereas in vessel elements (**B**) it moves by way of pores, which may be simple or interrupted by bars. In gymnosperm wood, tracheids both conduct water and provide support; in most kinds of angiosperms, vessels conduct the water, and fibers provide the support.

C Scanning electron micrograph of the wood of red maple, *Acer rubrum* (×350).

(on their side walls), but also definite perforations in their end walls by which they are linked together and through which water flows even more efficiently. A linked row of vessel elements forms a vessel.

Vessels apparently conduct water much more efficiently than strands of tracheids do. We conclude this partly because vessel elements have evolved from tracheids independently in several groups of plants, indicating that they have been favored strongly by natural selection. It is probably also the case that at least some kinds of fibers have evolved from tracheids (see Figure 35-6, *A*), becoming specialized for a strengthening, rather than a conducting, function. Some primitive flowering plants have only tracheids, but the great majority of angiosperms have vessel elements.

In addition to the conducting cells, xylem likewise includes fibers and parenchyma cells. The arrangements of these and other kinds of cells in the xylem make possible the identification of various plant species from their wood alone.

> The major types of conducting cells of the xylem are tracheids and vessel elements. Incorporated in the xylem, however, are also fibers and parenchyma cells.

Phloem

Phloem (Figure 35-7) is the principal food-conducting tissue in the vascular plants. Different kinds of plants have one of the two different kinds of elongate, slender conducting cells that occur in phloem: **sieve cells** and **sieve-tube members;** clusters of pores known as **sieve areas** occur on both types of cells. Such clusters are more abundant on the overlapping ends of the cells; they connect the protoplasts of adjoining sieve cells and sieve-tube members. Both of these types of cells are living, but neither has a nucleus.

In sieve-tube members, the pores in some of the sieve areas are larger than those in the others; such sieve areas are called **sieve plates.** Sieve-tube members occur end to end, forming longitudinal series called sieve tubes. Sieve cells are less specialized than sieve-tube members; the pores in all of their sieve areas are roughly the same diameter. They are the only type of food-conducting cell that most vascular plants have, except for the angiosperms, which have only sieve-tube members. In an evolutionary sense, sieve-tube members clearly are advanced over sieve cells; they are more specialized and presumably more efficient.

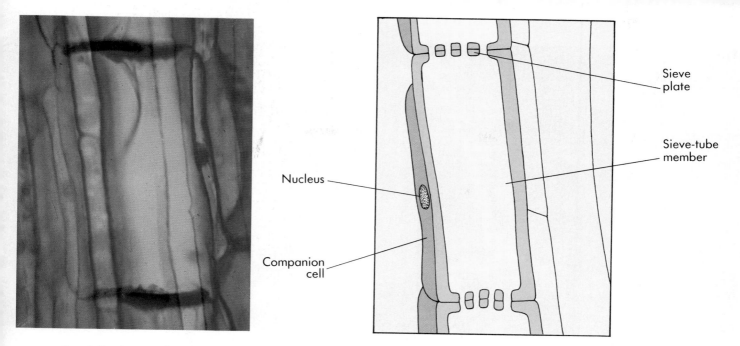

Specialized parenchyma cells known as **companion cells** occur regularly in association with sieve-tube members (see Figure 35-7). Companion cells apparently carry out some of the metabolic functions that are needed to maintain the individual sieve-tube members with which they are associated. They have all of the components of the normal parenchyma cells of the plants in which they occur, including nuclei, and their numerous pores connect their cytoplasm with that of the conducting cells. Fibers and parenchyma cells likewise occur in phloem.

The principal cell types of phloem are sieve cells or sieve-tube members, which lack a nucleus at maturity. If sieve-tube members are present, they are associated with specialized living, parenchyma cells called companion cells.

FIGURE 35-7

A sieve tube. Sieve-tube member from the phloem of squash (*Cucurbita*), connected with the cells above and below to form a sieve tube. Note the thickened end walls, which are at right angles to the sieve tube. The narrow cell with the nucleus at the left of the sieve-tube member is a companion cell.

Epidermis

Flattened epidermal cells, often coated by a thick layer of cuticle, cover all parts of the primary plant body. A number of types of specialized cells occur in the **epidermis,** including guard cells, trichomes, and **root hairs.**

Guard cells are paired, and the pairs flank a stoma, the epidermal opening through which gas exchange takes place and water is lost. Stomata (Figure 35-8) occur frequently in the epidermis of leaves, and they sometimes occur on other parts of the shoot, such as stems or fruits. The passage of oxygen and carbon dioxide into and out of the leaves, as well as the loss of water from them, takes place almost exclusively through the stomata. In most kinds of plants, stomata are more numerous on the lower surface of the leaf than on its upper surface, and in many plants they are entirely confined to the lower surface. The stomata open and shut in relation to external factors such as the supply of moisture. During periods of active photosynthesis, the stomata are open, allowing the free passage of carbon dioxide into the leaf. We will consider the mechanism that governs such movements in Chapter 36.

Trichomes are outgrowths of the epidermis that vary greatly in form in different kinds of plants. They also occur frequently on stems, leaves, and reproductive organs (Figure 35-9). A "fuzzy" or "woolly" leaf is covered with trichomes, which can be seen clearly with a microscope under low magnification. Trichomes play an important role in regulating the heat and water balance of the leaf. Some are glandular, often secreting sticky or toxic substances that may deter potential herbivores.

Near the tips of roots occur extensions of the epidermis called **root hairs.** The root hairs, which are single cells, occur in masses just behind the very ends of the roots. They keep the root in intimate contact with the particles of soil, eventually dry-

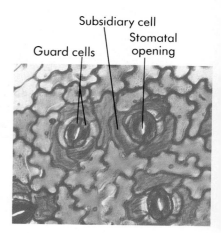

FIGURE 35-8

Epidermis of a leaf. Stomata occur frequently among the epidermal cells of this member of the aralia family (Araliaceae). The epidermis has been peeled off the leaf and stained with a red dye. Around the guard cells are specialized epidermal cells, found in association with the guard cells, called subsidiary cells.

A

B

C

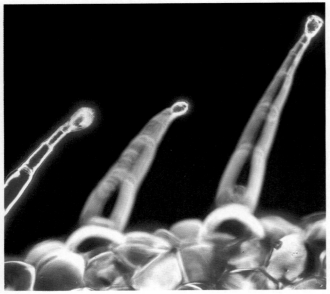
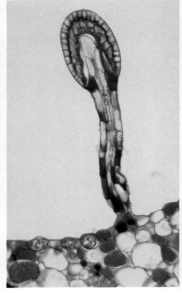
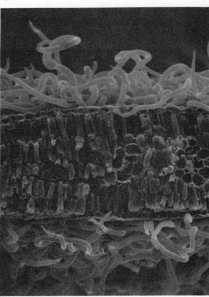

FIGURE 35-9
Trichomes.
A Three of the multicellular trichomes that cover the leaves of African violets, *Saintpaulia*. A covering of trichomes creates a layer of more humid air near the surface of a leaf, enabling the plant to conserve available supplies of water.
B A complex multicellular hair from the leaves of the sundew, *Drosera*. Such hairs secrete the enzymes that the plant uses to digest the bodies of its insect prey.
C The dense mats of trichomes on both sides of the silvery leaves of the desert composite *Encelia farinosa*, seen in this scanning electron micrograph, cause the leaf absorption of visible sunlight to drop from about 85%, as in a green leaf, to about 30%; as a consequence, the leaf temperatures are about 8° to 10° C cooler than if they were green. This lower leaf temperature reduces water loss by some 30% to 50%.

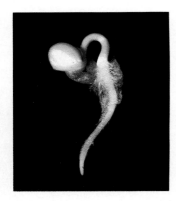

FIGURE 35-10
Root hairs on a radish seedling.

ing up as the root continues to grow. Virtually all of the absorption of water and minerals occurs by way of the root hairs in herbaceous plants (Figure 35-10). In mature woody plants, water is also absorbed in substantial quantities directly through the roots. Mycorrhizal fungi, if present, also play an important role in the process of absorption.

LEAVES

Leaves, which arise as outgrowths of the shoot apex, are the most important light-capturing organs in the majority of plants. The only exceptions are found in some plants, such as cacti, in which the stems are green and have largely taken over the function of photosynthesis. Since leaves are so crucial to a plant, such features as their arrangement, shape, size, and other aspects are highly significant. These factors differ greatly in plants that grow in different environments.

The apical meristems of stems and roots are capable of growing indefinitely under the appropriate circumstances. Leaves, however, grow by means of **marginal meristems,** which flank their thick central portions. These marginal meristems grow outward and ultimately form the blade of the leaf, while the central portion becomes the midrib. Once a leaf is fully expanded, its marginal meristems cease to function. The growth of roots and stems is indeterminate, but that of leaves, like that of flowers, is determinate.

General Features

Most leaves have a flattened portion, the **blade,** and a slender stalk, the **petiole** (Figure 35-11). In addition, there may be two leaflike organs, the **stipules,** that flank the

FIGURE 35-11

Angiosperm leaves are stunningly variable: they are the primary factor in controlling the specific ways plants capture light and regulate their water loss.

A Diverse leaves in the herb layer of a Costa Rican rain forest.

B A compound leaf: marijuana *(Cannabis sativa)*. Such a compound leaf, in which the leaflets join the petiole at one point, is said to be palmately compound. A compound leaf is associated with a single lateral bud, located where the petiole is attached to the stem.

C A simple leaf, its margin deeply lobed, from the tulip tree *(Liriodendron tulipifera)*.

D Compound leaf, from a member of the legume family in the lowland forest of Peru. Such a compound leaf, in which the leaflets are attached all along the main axis, is said to be pinnately compound.

E Many unusual arrangements of leaves occur in different kinds of plants. For example, in miner's lettuce *(Claytonia perfoliata)*, an herb of the Pacific states, two leaves are completely fused below each of the clusters of flowers, which seem, therefore, to arise from the center of a single leaf.

base of the petiole, where it joins the stem. These stipules may be microscopic or relatively large, leaflike, and conspicuous. Veins, usually consisting of both xylem and phloem, run through the leaves. In many monocots, the veins are parallel; in most dicots, the pattern is net or reticulate venation (Figure 35-12). What are called simple leaves are undivided, although they may be deeply lobed. In contrast, compound leaves consist of clearly separated **leaflets.**

In their placement on the stem, leaves follow definite patterns (Figure 35-13). In most plants they are spirally arranged, a condition that is often called **alternate.** In many others the leaves are **opposite,** occurring in pairs, and in a few they are **whorled,** with more than two leaves attached at one level on the stem.

FIGURE 35-12

Dicot and monocot leaves. The leaves of dicots, such as this African violet relative from Sri Lanka **(A),** have net, or reticulate, venation; those of monocots, like this Latin American palm **(B),** have parallel venation. Both leaves have been cleared with chemicals, and the dicot one has been stained with a red dye to make its veins show up more clearly.

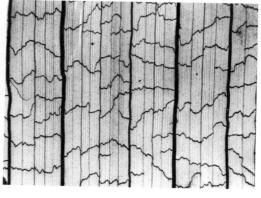

A   B

FIGURE 35-13
The three common types of leaf arrangements.

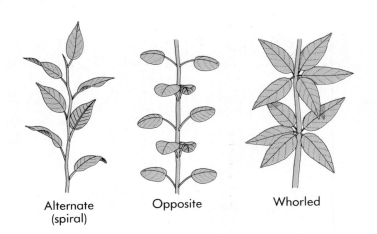

Alternate (spiral) Opposite Whorled

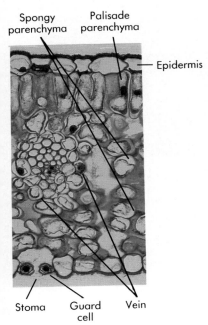

Spongy parenchyma Palisade parenchyma

— Epidermis

Stoma Guard cell Vein

FIGURE 35-14
Transection of a lily leaf. This transection shows palisade and spongy parenchyma; a vascular bundle, or vein; and the epidermis, with paired guard cells flanking the stoma that is visible on the lower surface of the leaf and the substomatal space.

Structure and Organization

A typical leaf contains masses of parenchyma, through which the vascular bundles, or veins, run (Figure 35-14). The masses of parenchyma that occur in leaves are called **mesophyll** ("middle-leaf"). Beneath the upper epidermis of a leaf, there are one or more layers of columnar parenchyma cells, which are called **palisade parenchyma.** Certain kinds of plants, including some species of *Eucalyptus*, have leaves that hang down, rather than being extended horizontally. Their palisade parenchyma is found on both sides of the leaf, and there is, in effect, no upper side. But regardless of whether palisade parenchyma is found on one or both sides of a leaf, the rest of the interior, except for the veins, consists of a tissue called **spongy parenchyma.** Spongy parenchyma is composed of parenchyma cells that are more or less spherical. Between these cells are located large intercellular spaces that function in gas exchange and the passage of water vapor from the leaves. These intercellular spaces are connected, directly or indirectly, with the stomata.

The cells of the mesophyll, especially those near the leaf surface, are packed with chloroplasts. These cells constitute the plant's primary site of photosynthesis. Water and minerals are brought from the roots to the leaves in the xylem strands of the veins. Water delivered by the veins to the leaves passes into the mesophyll cells. Some of it moves, following the secretion of sugar, into the phloem for downward transport. Commonly a large fraction of the leaf's water escapes into the intercellular spaces of the mesophyll by evaporating from the surfaces of the mesophyll cells, and then diffuses to the outside atmosphere through the open stomata. The intercellular spaces have within them a saturated atmosphere that slows down the evaporation of additional water into them from the surfaces of the cells.

> The mesophyll in a leaf consists of two kinds of parenchyma cells—palisade and spongy parenchyma. Both are packed with chloroplasts, especially when they lie near the leaf surface.

Modified Leaves

Because of their great significance in the life of the plant, leaves have become modified in various ways during the course of their evolution (Figure 35-15). The huge, soft, highly dissected leaves of a tree fern lose water much more rapidly than do the tough, scalelike leaves of a juniper. Furthermore, the arrangement of leaves on a stem, the distance between the leaves, and the height and branching pattern of the stem are closely related to the way that a particular kind of plant functions in its environment.

As mentioned before, the stems have taken over the function of photosynthesis in some plants. For example, the large, succulent stems of cacti are protected from predators by their spines, which are highly modified leaves that do not function in photosynthesis and contain no living tissue at maturity. A few genera of cacti still have leaves of normal appearance, but in most they have been lost in the course of evolution.

Tendrils are organs that assist plants in climbing to relatively well-lighted places. They are sometimes modified stems (as in grapes or ivy) and sometimes modified

A

B

C

FIGURE 35-15
Modified leaves.
A A tendril in the garden pea, *Pisum sativum*.
B Jade plant, *Crassula*, with succulent, opposite leaves and CAM photosynthesis.
C An onion, *Allium cepa*, in which the fleshy organs that make up the bulb are colorless, underground leaves modified for food storage.

leaves (as in the garden pea; see Figure 35-15, *A*). Another example of modified leaves is provided by the layers of an onion, which are actually fleshy leaves, modified for food storage and clustered around an underground stem (see Figure 35-15, *C*). The scales around a bud are modified leaves that fall off as the bud unfolds and the shoot within begins to grow. Perhaps the most spectacular modified leaves, however, are those of the carnivorous plants (see p. 740).

Succulent leaves, like those of jade plants (see Figure 35-15, *B*), hen and chickens, and ice plant, function in water storage and thus help plants to withstand drought. These and many similar plants, as well as plants such as cacti that have succulent stems, usually have a modified form of photosynthesis called CAM (Crassulacean acid metabolism) photosynthesis, which is a variant of the C_4 pathway that we discussed in Chapter 9. In plants that have CAM photosynthesis, the stomata are closed during the day, thus conserving water. Carbon dioxide enters them, through the open stomata at night, and C_4 compounds are accumulated, entering the Calvin cycle during the day. Such a process allows the fixation of much more carbon dioxide during a 24-hour cycle than would be possible otherwise. CAM photosynthesis allows the plant to obtain carbon dioxide at night, when transpiration occurs at a slower rate than during the day, and to use the carbon dioxide later. It is a highly advantageous system for succulent plants that grow in hot, arid regions.

SHOOTS
Primary Growth

Leaves first appear as **leaf primordia** (singular, **primordium**), or rudimentary young leaves, which cluster around the apical meristem, unfolding and growing as the stem itself elongates (Figure 35-16). The places on the stem at which leaves are attached

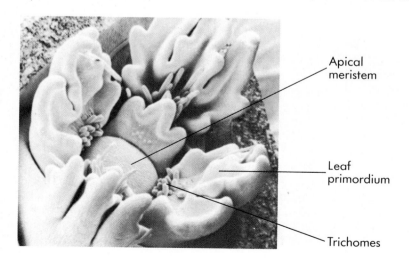

Apical meristem

Leaf primordium

Trichomes

FIGURE 35-16

A shoot apex. Scanning electron micrograph of shoot apex of silver maple, *Acer saccharinum*, showing a developing shoot during summer, the season of active growth. The apical meristem, leaf primordia, and trichomes are plainly visible at this stage.

are called **nodes,** and the portions of the stem between these points of attachment are called the **internodes.** As the leaves expand to maturity, a **bud,** which is a tiny, undeveloped side shoot, develops in the axil of each leaf (see Figure 35-1). These buds, which have their own leaves, may elongate and form lateral branches, or may remain small and dormant. As we will discuss in detail in Chapter 38, a hormone diffusing downward from the terminal bud of the shoot continuously suppresses the expansion of the lateral buds. These buds begin to expand when the terminal bud is removed. Therefore, gardeners who wish to produce bushy plants or dense hedges crop off the tops of the plants, removing their terminal buds.

Within the soft, young stems, the strands of procambium either occur as a cylinder in the outer portion of the ground meristem, as is common in dicots, or are scattered throughout it, an arrangement that is common in monocots (Figure 35-17). At the stage when only primary growth has occurred, the inner portion of the ground tissue of a stem is called the **pith,** and the outer portion is the **cortex.** The outermost parenchyma cells in the cortex frequently are packed with chloroplasts, in which case the stem is green and photosynthetically active.

FIGURE 35-17

Stems. Transections of a young stem in (**A**) a dicot, the common sunflower, *Helianthus annuus,* in which the vascular bundles are arranged around the outside of the stem; and (**B**) a monocot, asparagus, *Asparagus officinalis,* with the scattered vascular bundles characteristic of the class.

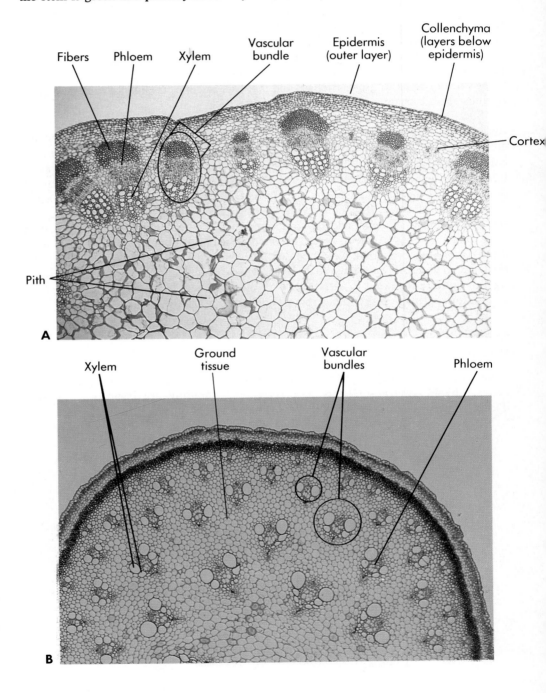

As the stem matures, the strands of procambium within it differentiate into vascular bundles, which contain both primary xylem and primary phloem (Figure 35-18). The procambial strands of the stem grow upward into the developing leaf primordia by the differentiation of additional procambial cells as the primordia expand. As a consequence of this pattern of development, one or more vascular bundles diverge from the group of strands in the stem at each node and enter the leaf or leaves attached at that point. The pattern of vascular strands in the stem of a particular kind of plant, therefore, directly reflects the arrangement of the leaves.

Secondary Growth

Secondary growth is initiated by the differentiation of the **vascular cambium,** which consists of a thin cylinder of cells that in mature woody plants is located near the area where the bark and the main stem come together. The vascular cambium differentiates from parenchyma cells within the vascular bundles of the stem, between the primary xylem and the primary phloem. The cylindrical form of the vascular cambium is completed by the differentiation of some of the parenchyma cells that lie between the bundles (Figure 35-19). The pattern just described is the predominant pattern of formation of the vascular cambium in dicots, but other patterns also occur, especially in monocots, where true secondary growth occurs only in a few groups.

> **The vascular cambium differentiates, in most dicots at least, as a cylinder of dividing cells derived both from the parenchyma of the ground tissue, or cortex, and from the region within the primary vascular bundles that lies between the xylem and the phloem.**

The vascular cambium consists of elongated, somewhat flattened cells with large vacuoles. Such cells are very different in appearance from the small, nonelongated ones that occur in the apical meristems, which have small vacuoles. Cell division in the vascular cambium produces cells that differentiate either into secondary phloem (the outer derivatives) or secondary xylem (the inner derivatives). The cells of the vascular cambium also divide laterally, allowing the cambium to increase in diameter as the stem thickens. The elongated cambial cells that divide to produce the conducting elements of the xylem and phloem, and other cambial cells similar to themselves, are called **fusiform initials.** Much smaller, more nearly isodiametric **ray initials** divide, giving rise to the rays, which are radial strands of parenchyma that function in lateral water movement through the stem.

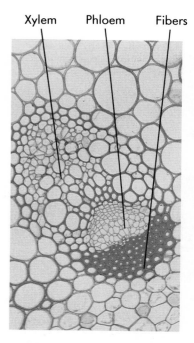

FIGURE 35-18
Vascular bundles in a buttercup. This transection of vascular bundles from a buttercup, *Ranunculus acris*, shows the xylem and phloem.

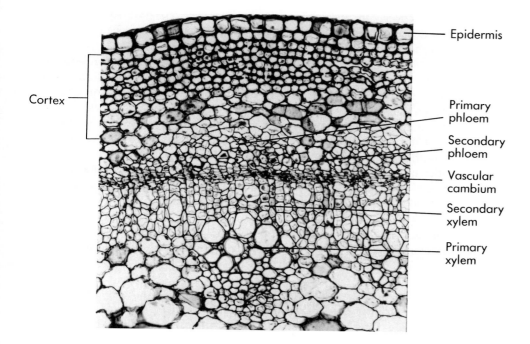

FIGURE 35-19
Early stage in differentiation of vascular cambium in an elderberry (*Sambucus canadensis*) stem. The outer part of the cortex consists of collenchyma, and the inner part of parenchyma.

FIGURE 35-20
Periderm. An early stage in the development of periderm in elderberry (*Sambucus canadensis*).

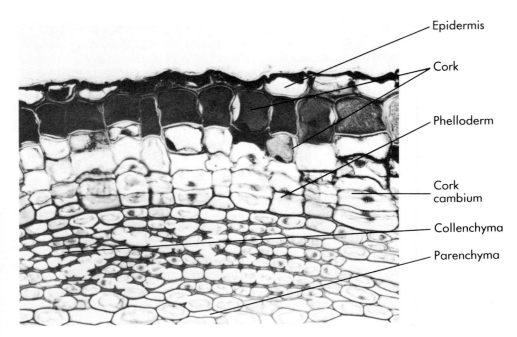

Epidermis

Cork

Phelloderm

Cork cambium

Collenchyma

Parenchyma

A

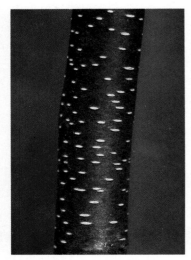

B

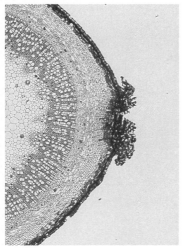

FIGURE 35-21
Lenticels.
A Lenticels, the numerous, small, pale, raised areas shown here on the bark of paper birch, *Betula papyrifera*, allow oxygen to readily diffuse into the living tissues immediately below the bark of woody plants. Highly variable in form in different species, lenticels are a valuable aid to the identification of deciduous trees and shrubs in winter.
B Transection through a lenticel in a stem of elderberry, *Sambucus canadensis*.

Because the vascular cambium consists of a thin cylinder and is indispensable for the continued growth of the stem, trees can be killed by girdling them in such a way as to interrupt the vascular cambium all the way around. This is the method often used to eliminate trees from areas where they are not wanted. The cells of xylem and phloem that are active in transporting substances within the plant are located close to the vascular cambium that gives rise to them also, making a tree all the more vulnerable to girdling its trunk. For this reason, the activities of beavers, or, more commonly, beetles that burrow through the vascular cambium and associated food-rich tissues (see Figure 35-22, *C*), threaten the lives of the trees that are affected.

While the vascular cambium is first becoming established, a second kind of lateral cambium, the cork cambium, normally also develops in the outer layers of the stem. The cork cambium usually consists of plates of dividing cells that, in effect, move deeper and deeper into the stem as they divide. The cells that the divisions of the cork cambium produce outwardly are mainly radial rows of densely packed cork cells. Their inner layers contain large amounts of a fatty substance, **suberin,** which makes the layers of cork nearly impermeable to water. Cork cells are dead at maturity. The **phelloderm,** a dense tissue composed primarily of parenchyma, may be differentiated inwardly from the cork cambium.

Together, the cork, cork cambium, and phelloderm make up the **periderm** (Figure 35-20). The periderm constitutes the outer protective covering of the mature plant's stem or root; it is formed only after a considerable amount of secondary growth has taken place. The periderm is renewed continuously by the cork cambium, formed within the secondary phloem at deeper and deeper levels, which is why cork for commercial use can be harvested from the bark of the cork oak (*Quercus suber*) without killing the trees. Gas exchange through the periderm is necessary for the metabolic activities of the living cells of the phelloderm and vascular cambium beneath. This exchange takes place through **lenticels,** areas of loosely organized cells, which often are easily identifiable on the outer surface of bark (Figure 35-21).

The periderm, a layer differentiated from the lateral meristem known as the cork cambium, consists of cork to the outside and phelloderm to the inside of a stem or root. It retards water loss from the secondary plant body.

Cork, which covers the surfaces of mature stems or roots, takes the place of the epidermis, which performs a similar function in the younger portions of the plant. **Bark** is a term used to refer to all of the tissues of a mature stem or root outside of the

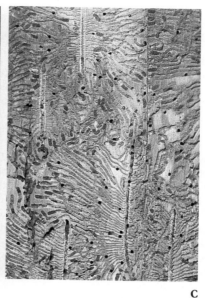

A   B   C

FIGURE 35-22
Bark.

A Bark of valley oak, *Quercus lobata*, in California. Bark is highly characteristic of individual trees and shrubs, which often can be recognized by the bark's characteristics even when the trees are leafless, as in this madrone, *Arbutus mensiesii* (**B**).
C Galleries constructed by elm bark beetles, which spread Dutch elm disease from tree to tree, in the cambium of an elm tree. Such beetles bore through the thin layer of living cells that separates bark from the trunk of a tree, thus gaining access to the carbohydrates passing through the phloem.

vascular cambium (Figure 35-22). Since the vascular cambium has the thinnest-walled cells that occur anywhere in a secondary plant body, it is the layer at which the bark breaks away from the accumulated secondary xylem. The inner layers of the bark are primarily secondary phloem, with the remains of the primary phloem crushed among them. Its outer layers consist of the periderm, and the very outermost ones are cork.

Wood

Wood is one of the most useful, economically important, and beautiful products that we obtain from plants. From an anatomical point of view, wood is accumulated secondary xylem. As the secondary xylem ages, its cells become infiltrated with gums and resins, and the wood becomes darker. For this reason, the wood located nearer the central regions of a given trunk, called heartwood, is often darker in color and denser than the wood nearer the vascular cambium. That wood, called **sapwood,** is still actively involved in transport within the plant. The proportion of sapwood to heartwood, and even whether they are distinct or not, differs widely from one kind of tree to another (Figure 35-23).

Because of the way in which it is accumulated, wood often displays rings. In temperate regions, these rings are **annual rings.** They reflect the fact that the vascular cambium divides actively when water is plentiful and temperatures are suitable for growth and ceases to divide when water is scarce and the weather is cold. In most temperate regions, growth rings form during the spring and summer. The structural basis for their appearance is the difference in density between the wood that forms early in the growing season and the wood that forms later. The abrupt discontinuity between the layers of larger cells, with proportionately thinner walls, that form early in the growing season and those that formed at the end of the previous growing season is often very evident.

Annual rings in wood generally are thicker when formed in years of plentiful rainfall and thinner when formed in dry years. For this reason, the annual rings in a tree trunk can be used not only to calculate the age of the tree, but also to learn about

A

B

FIGURE 35-23
Some features of wood.

A Annual rings in a section of pine (*Pinus*).
B The distinction between heartwood (dark central portion) and sapwood (light outer portion) is evident in this sawed-off limb of ponderosa pine (*Pinus ponderosa*) in the mountains of California.

In many Asian bamboo species, all the individuals of the species flower and set seed simultaneously. These cycles tend to occur at very long intervals, ranging from 3 years upward. Most of the cycles are between 15 and 60 years long. The most extreme example is that of the Chinese species *Phyllostachys bambusoides*, which seeded massively and throughout its range in 919, 1114, between 1716 and 1735, in 1833-1847, and in the late 1960s. The last event involved cultivated plants in widely separated areas, including England, Russia, and Alabama. *Phyllostachys bambusoides*, therefore, has a flowering cycle of about 120 years, set by an internal clock that runs more or less independently of environmental circumstances!

The bamboos that have cycles of this kind spread mainly by the production of rhizomes. When they do flower, nearly all individuals flower at once, set large quantities of seed, and die. Huge numbers of animals, including rats, pigs, and pheasants, often migrate into areas where bamboos are fruiting to feed on the seeds; humans also gather them for food. Apparently the plants put all of their energy into producing numbers of seeds huge enough for large numbers of seedlings to survive, even though most of the seeds are eaten.

A particular conservation problem that attracted widespread notice in the 1980s is the relationship between the flowering of bamboos and the survival of the giant panda, a spectacular animal that survives in a few mountain ranges in Sichuan Province of southwestern China. Poaching and habitat destruction seriously threaten the future of the panda, of which probably only about 700 individuals exist in the wild. In their scattered reserves, they are also threatened by the mass flowering of one or more of the few species of bamboo that constitute their main sources of food locally (Figure 35-A, *1*). When this occurs, the pandas often move to lower elevations, seeking their species of bamboo as alternative sources of food (Figure 35-A, *2*). Because of widespread cultivation and other human pressures, however, the pandas may not find suitable stands of bamboo and may become more subject to poaching in the settled areas.

1   2

FIGURE 35-A
1   A Chinese scientist examining a stand of the bamboo *Gelidocalamus fangianus*, the normal food of giant pandas at high elevations in the Min Mountains, which has flowered and dropped its leaves; the whole plant will soon be dead. Most of the plants of this species in the area died in 1983.
2   During the winter of 1986-1987, more than 3 years after the mass flowering, many of the giant pandas had moved to lower elevations, where they fed mainly on this bamboo, *Fargesia spathacea*. This panda is in an enclosure.

past climates. By correlating the patterns formed by the alternation of thick rings, formed in moist years, and thin ones, formed in dry years, it is possible to date pieces of wood accurately, provided that they were formed during a sequence for which the pattern of alternation has been established. Such methods have been used to date wood from European oaks back for more than 9000 years and to determine the ages of the frames used to mount old paintings.

Commercially, wood is divided into hardwoods and softwoods. **Hardwoods** are the woods of dicots, regardless of how hard or soft they actually may be; **softwoods** are the woods of conifers. Many hardwoods used commercially come from the tropics, whereas almost all softwoods come from the great forests of the north temperate zone. When they are used as building materials or for other commercial purposes, woods often are given names that are not related directly to the names of the trees from which they come. Thus, the wood called "Oregon pine" does not come from a pine at all, but from Douglas fir (*Pseudotsuga menziesii*).

Individual woods differ widely in their microscopic characteristics, including such features as the width and height of the rays, type of pits in the conducting ele-

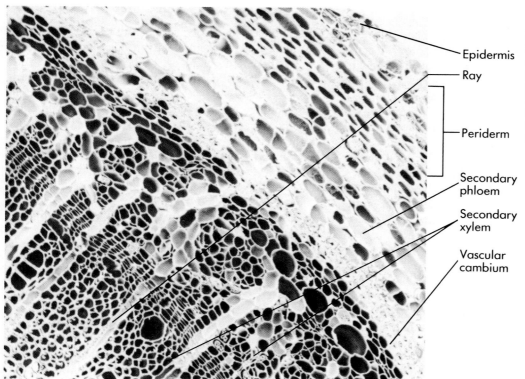

Epidermis

Ray

Periderm

Secondary
phloem

Secondary
xylem

Vascular
cambium

FIGURE 35-24
Scanning electron micrograph of
a diagonal view across the outer
portion of a trunk of silver maple, *Acer saccharinum.* The tracheids, vessels, and rays in the
secondary xylem; the vascular
cambium; phloem; and periderm
can be clearly seen.

ments, abundance of fibers, and nature of the conducting elements and other cells
(Figure 35-24). They can be identified readily by experts and often provide valuable
clues to the evolutionary position of the plants that form them.

**Wood is accumulated secondary xylem; when it is formed in regions with a
seasonal climate, wood is marked by annual rings. The woods of dicots are
called hardwoods; those of conifers are called softwoods.**

Modified Stems

Stems have been modified to perform many different functions during the course of
plant evolution. One example that we mentioned earlier concerns plants such as
grapes, Virginia creeper, and ivy, whose tendrils are modified stems (Figure 35-25).
The tendrils of other plants, including those of garden peas, are modified leaves.

Stems running underground, which are called **rhizomes,** may be important in the
vegetative propagation of plants; they often give rise to new individuals, sometimes
quite far from the parent plant. Rhizomes are primarily responsible for the rapid
spread of some of our most noxious weeds, such as nutgrass, and they also occur in

FIGURE 35-25
Unmodified, compound leaves
(A) and modified leaves—
adhesive, flattened tendrils (B)—
in Virginia creeper
(*Parthenocissus quinquefolia*).

A

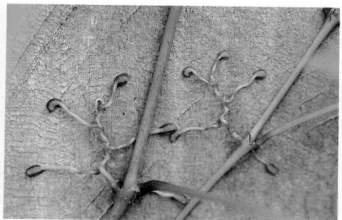

B

many other plants, including irises and violets. As we saw in the last chapter, fern fronds arise from rhizomes also. Similar kinds of stems that run horizontally aboveground are called **stolons;** they are characteristic of plants such as strawberries and Bermuda grass. The clumps of bamboo discussed in the boxed essay above also are connected by rhizomes.

Tubers are underground stems modified to store food; the most familiar tuber is that of the potato. The "eyes" of the potato are buds, each arising in the axil of a scale (modified leaf) and capable of giving rise to a new potato plant. **Corms** are thick, fleshy underground stems that are modified for food storage, as in cyclamens and gladiolus. Unlike tubers, corms are upright. Finally, bulbs are short underground stems that bear thickened, fleshy scale leaves. Onions have bulbs whose scales are modified leaves that enfold the stem (see Figure 35-15, *C*); lilies have bulbs whose scales are small, very fleshy, and easily detached. When the scales of lily bulbs do become detached, they can give rise to new plants.

The Formation of Flowers

When a flower or group of flowers forms at a vegetative shoot apex, the meristematic activity of that apex ceases, usually permanently. In other words, the shoot changes from indeterminate to determinate growth. A number of internal and external factors that we will discuss in the following chapters are involved in the switch from vegetative growth to the production of a flower. As the change occurs, the shoot apex often becomes broad and domelike. The organs of a flower develop from primordia that resemble leaf primordia; they are bumps or shelves of tissue that project from the shoot apex. During the development of these organs, they may fuse with members of the same or another whorl, thus giving rise to the complex structures that occur in mature flowers.

ROOTS

Roots have a simpler pattern of organization and development than do stems (Figure 35-26). Although different patterns exist, we will describe a kind of root that is found in many dicots. There is no pith in the center of the vascular tissue of the root in most dicots. Instead, these roots have a central column of primary xylem with radiating

FIGURE 35-26
Root structure. A root-tip in corn, *Zea mays*. This medium, longitudinal section of a root cap, apical meristem, protoderm, epidermis, and ground tissue.

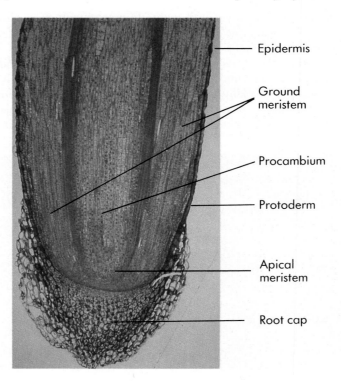

Epidermis

Ground
meristem

Procambium

Protoderm

Apical
meristem

Root cap

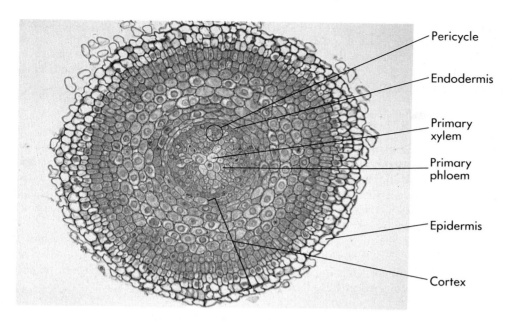

FIGURE 35-27
Cross section through a root of a buttercup, *Ranunculus californicus*. The central xylem and phloem, cortex, and epidermis are visible.

Pericycle

Endodermis

Primary xylem

Primary phloem

Epidermis

Cortex

arms. Alternating with the radiating arms of xylem are strands of primary phloem (Figure 35-27). If you drew a line from the center of a stem outward through the xylem, that line would pass through phloem also; in dicot roots this is not the case because primary xylem and primary phloem are found on alternating radii. The anatomy of the monocot root differs from that of the dicot root. As in the vascular system of the monocot shoot, the root system too is made up of bundles that are composed of both xylem and phloem, although the arrangement of these bundles usually differs markedly from that in a stem of the same species. These bundles have no particular arrangement in the root, and in cross section are seen to be scattered throughout the root.

Surrounding the vascular tissue in dicot roots are distinct regions and layers of cells. The cortex comprises the largest region (see Figure 35-27). Outside the cortex, and completely encircling it, is the epidermis. The epidermis protects the root and, through the production of root hairs, functions in the uptake of water and minerals from the soil. The innermost layer of the cortex has a special name, the **endodermis**. This layer has an important function in the physiology of the plant, for it is here that the plant determines which minerals and nutrients can enter the vascular tissue and thereby move to the shoot system. The cells of the endodermis are surrounded by a thickened waxy band called the Casparian strip. The movement of water and minerals into the root is discussed further in Chapter 36.

Just inside the endodermis is another specialized layer of cells known as the **pericycle**. From cells of the pericycle, branch or lateral roots are formed, as we will discuss further.

Most dicot roots have a central column of primary xylem with radiating arms and strands of primary phloem between these arms surrounded by a layer one or more cells thick called the pericycle. The innermost layer of the cortex, or endodermis, consists of cells surrounded by a thickened waxy band called the Casparian strip.

The apical meristem of the root divides and produces cells both inwardly—back toward the body of the plant—and outwardly. Outward cell division results in the formation of a thimblelike mass of relatively unorganized cells, the root cap, which covers and protects the root's apical meristem as it grows through the soil. The cells of the root cap are loose and slough off, facilitating the passage of the fragile meristem through the soil.

The root elongates relatively rapidly just behind its tip. Above that zone are formed abundant root hairs, which, as noted earlier, are slender projections of the

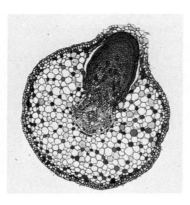

FIGURE 35-28
A lateral root. A lateral root growing out through the cortex of black willow, *Salix nigra*. The origin of lateral roots occurs beneath the surface of the main root, whereas the origin of lateral stems is superficial.

epidermis (see Figure 35-10). Virtually all of the absorption of water and minerals from the soil takes place through the root hairs. These root hairs greatly increase the surface area and therefore the absorptive powers of the root. In plants that have ectomycorrhizae (see Chapter 32), the root hairs often are greatly reduced in number; the fungal filaments take their place in promoting absorption.

Branching in Roots

Roots have a simpler structure than stems do, primarily because roots have no appendages comparable to leaves. Branching in stems occurs by means of buds, which form in leaf axils; the formation of the lateral branches of stems is superficial and does not involve the deeper layers of the stem. Branching in roots, on the other hand, is initiated well behind the root apex and deep within the tissues of the root. Branch roots are initiated by cell divisions in the pericycle.

Deep within the cortex, the lateral root primordia grow out toward the surface of the root (Figure 35-28). As they do so, they develop the characteristics of the main root apex, including a root cap. Eventually, they break through the surface of the root and become established as fully formed lateral roots.

> Branch root primordia form well beneath the surface of the root and grow out through the cortex. There are no externally formed organs in roots comparable to leaves or lateral buds.

Secondary Growth

Secondary growth in roots is similar to that in shoots. The vascular cambium is initiated from undifferentiated procambial cells between the primary xylem and the primary phloem. The areas where this cell division is initiated are connected by corresponding areas of cell division in the pericycle. Ultimately, these areas of cell division fuse, and the vascular cambium then forms a cylinder that surrounds the primary xylem. The first differentiation of cork cambium and periderm occurs in the pericycle; later cork cambia develop in patches from the parenchyma of secondary phloem.

Adventitious Roots and Shoots

Many dicots have a single taproot, with smaller lateral branches; the taproot is sometimes the only major underground structure (Figure 35-29, *A*). In some, like carrots and radishes, the taproot is fleshy and modified for food storage. The taproot that develops in monocots, on the other hand, often dies during the early growth of the

FIGURE 35-29

Two types of roots.

A Taproots in dandelion, *Taraxacum officinale*. Even a small portion of such a taproot can regenerate a new plant, which is one reason that dandelions are so difficult to eliminate from lawns and gardens.

B Prop roots in corn, *Zea mays*. Such roots, which are adventitious—they arise from stem tissue—take over the function of the main root, which is soon lost, in many monocots.

A   B

FIGURE 35-30
Clones of aspen. The contrast between the deciduous quaking aspens (*Populus tremuloides*) and the dark green, evergreen Engelmann spruces (*Picea engelmannii*) is evident in this autumn scene near Durango, Colorado. Each distinct clump of aspen is a clone, in which most of the individual trunks have originated from adventitious buds forming on roots spreading underground. Because aspen is dioecious, some of the clones consist entirely of staminate individuals, and others entirely of pistillate ones.

plant, and new roots differentiate from the tissues of the stem's lower part (Figure 35-29, *B*). These **adventitious roots**—which arise from kinds of tissue other than root tissue—take over the function of the taproots and branching root systems that are characteristic of dicots.

Similarly, **adventitious shoots** may arise from roots, often at some distance from the parent plant. Such shoots may then give rise to new plants that eventually may become fully independent of the parent plant. This method of propagation is characteristic of a number of kinds of plants, including quaking aspen (*Populus tremuloides*); thus the beautiful and conspicuous clumps shown in Figure 35-30 are clones (and therefore genetically identical)—each originated from the roots of a single individual. Most plants that spread underground, however, do so by means of rhizomes, which, as we have seen, are subterranean stems directly connected to other stem tissue.

EARLY DEVELOPMENT

The beginnings of patterns of growth and development that are characteristic of plants emerge in the early stages of embryo development, which take place within the growing seed. Within such a seed, the mature angiosperm embryo consists of an axis with either one or two cotyledons, or embryonic leaves (Figure 35-31). In general, monocot embryos have only one cotyledon, and dicot embryos have two. In dicots, the food stored initially in the endosperm may either be absorbed into the cotyledons, which may then become quite thick and fleshy, as in peas and beans, or remain in the seed at maturity. In monocots, the single cotyledon functions mainly as a food-absorbing organ, transferring food from the endosperm to the young embryo during the germination of the seed (see Figure 33-29). In either case, the food that is initially concentrated in the endosperm during the early development of the embryo is used during the germination and early establishment of the young plant. Starches, fats, and oils are converted into sugars and used to nourish and sustain the young plant before its photosynthesis begins.

In the embryo, the apical meristems differentiate early, continuing to divide throughout the life of the plant. The shoot apical meristem in a plant embryo is located at the tip of the **epicotyl,** the portion of the axis that extends above the cotyledons. The epicotyl may be short and relatively undifferentiated, or it may be longer and even include one or more embryonic leaves. The epicotyl, together with its young leaves, is called a **plumule.** The axis of the embryo below the cotyledons is called the **hypocotyl.** A distinct embryonic root at the lower end of the hypocotyl is called the radicle; it eventually develops into the primary root.

The ways in which the embryonic root and shoot emerge from the seed during germination vary widely from species to species. In many, the root emerges first and

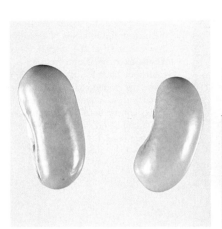

A

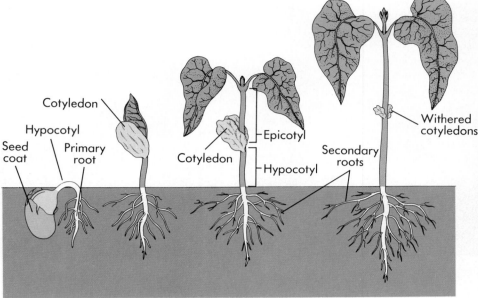

B

FIGURE 35-31
Seeds and stages of germination in a dicot, the common bean, *Phaseolus vulgaris* (A and B), and a monocot, corn, *Zea mays* (C and D).

anchors the plant in the soil before the shoot appears (see Figure 35-10). The cotyledons may be held either below or above ground, and they may or may not become green and then contribute to the nutrition of the seedling while it is becoming established. The period from the germination of the seed to the establishment of the young plant is a very critical one for the plant's survival; it is unusually susceptible to disease and drought during this period.

HOW LONG DO INDIVIDUAL PLANTS LIVE?

Once established, plants live for a highly variable period of time, depending on the species involved (Figure 35-32). Woody plants are those in which secondary growth has been extensive; herbaceous plants are those in which it is limited or absent. Herbaceous plants send up new stems above the ground every year, producing them from woody underground structures, or germinate and grow to flowering just once. Depending on the length of their life cycles, herbaceous plants may be annual, biennial, or perennial in duration. Shorter-lived plants rarely become very woody, since there is not enough time in their limited life spans for the massive accumulation of secondary tissues.

Annual plants grow, flower, and form fruits within a period of less than a year and die when the process is complete. Many crop plants, for example, are annuals, including corn, wheat, and soybeans. Annuals generally grow rapidly under favorable conditions and increase greatly in size in proportion to the availability of additional water or fertilizer. Some annuals, like sunflowers or giant ragweed, do form wood as a result of secondary growth, if they live long enough. Many, however, are entirely herbaceous.

Biennial plants, of which there are many fewer kinds than there are annuals, have life cycles that take 2 years to complete. During the first year, biennials generally form a tuft of leaves, or **rosette**. In the second year of these plants' growth, when they flower, the energy stored in the rosette and in the underground parts of the plant is used to produce flowering stems. Many crop plants, including carrots, cabbage, and beets, are biennials, but these plants generally are harvested for food during their first season, before they flower. They are grown for their leaves or roots, not for their fruits or seeds.

Perennial plants grow on from year to year and may be herbaceous, such as many woodland and prairie wildflowers, or woody, as are trees and shrubs. The majority of vascular plant species are perennials. Herbaceous perennials rarely have any secondary growth in their stems; the stems die back promptly each year after a period of

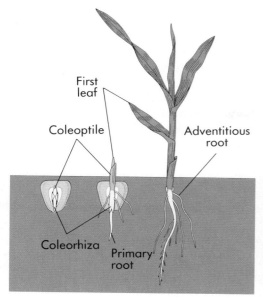

C

D

A

relatively rapid growth and food accumulation. The food is often stored in their roots, which may then become quite large in relation to their delicate aboveground stems.

Annual plants complete their whole growth cycle within a single year. Biennial plants flower only once, after two seasons of growth. Perennials, which may be either herbaceous or woody, flower repeatedly once they begin to do so and live for an indefinite period.

Perennials with woody stems above ground are either trees or shrubs. Trees are large woody plants, usually with a single stem, whereas shrubs are smaller and often have several stems. There is, however, no real dividing line between trees and shrubs, and some may grow in either form, depending on local conditions. Trees and shrubs may be either **deciduous,** with all the leaves falling at one particular time of year and the plants remaining bare for a period, or **evergreen,** with the leaves dropping throughout the year and the plants never appearing completely bare (see Figure 35-30). In north temperate regions, conifers are the most familiar evergreens; but in tropical and subtropical regions, most angiosperms are evergreen, except for those that occur where there is severe seasonal drought. Many tropical and subtropical angiosperms in regions where there is a drought may be deciduous, losing their leaves during the drought and thus conserving water.

Plants that flower only once and then die, like annuals and biennials, are termed **monocarpic** ("once-fruiting"); perennials, which flower repeatedly, are called **polycarpic** ("many-fruiting"). Century plants (*Agave* species) are well-known examples of monocarpic plants; they grow for very long periods of time and then produce massive flowering stalks. Most of the bamboos are also monocarpic, producing so many seeds in the years when they do flower that the predators which consume large numbers of them spare many. The missed seeds produce the new plants of the next generation. After they have set fruit, the entire plant dies. Many plants that are considered biennials actually do not flower until they are 3 or more years of age.

PLANT GROWTH AND DEVELOPMENT

As we have shown in this chapter, plants grow by an open system—one that keeps functioning throughout their lives and producing similar structures over and over. The apical meristems of the shoot differentiate leaves and the primary tissues of the shoot, whereas the apical meristems of the root, protected by a root cap, simply produce the primary tissues of the root. No structure in roots is comparable to leaves; if such structures did exist, they would tend to retard the growth of the root through

B

FIGURE 35-32

The age of plants. Plants live for very different lengths of time. Desert annuals like those shown in A complete their entire life span in a few weeks; whereas some trees, such as the giant redwood, *Sequoiadendron giganteum* (B), which occurs in scattered groves along the west slope of the Sierra Nevada in California, live 2000 years or more.

the soil. A major difference in the patterns of growth of roots and shoots is that each branch shoot develops from buds, whereas branch roots develop as a result of the functioning of areas of rapid cell division that lie in the pericycle, deep within the root. There is a bud at the place where each leaf joins the stem, but most of these buds simply remain as buds rather than growing into lateral branches.

The control of this seemingly simple and repetitive growth pattern is by hormones, which the plant produces internally, and by external factors; it is complex and only partly understood. The three remaining chapters of this section will be devoted to a review of our knowledge concerning the way in which external and internal factors interact to produce a mature, functioning plant.

SUMMARY

1. A plant body is basically an axis that includes two parts: root and shoot. Within the plant are three principal tissue types: dermal tissue, vascular tissue, and ground tissue. Dermal tissue covers the outside of the plant; vascular tissue conducts substances through it; and ground tissue is the matrix in which the vascular tissue is embedded.

2. Plants grow by means of their apical meristems, zones of active cell division at the ends of the roots and the shoots. The apical meristem gives rise to three types of primary meristems, partly differentiated tissues in which active cell division continues to take place. These are the protoderm, which gives rise to the epidermis; the procambium, which gives rise to the vascular tissues; and the ground meristem, which becomes the ground tissue.

3. Parenchyma cells, which are the most abundant cell type in the primary plant body (the portion of the plant that differentiates from the primary meristems), usually have only primary walls, which are laid down while the cells are still growing. They are living at maturity.

4. Collenchyma cells often form strands or continuous cylinders in the primary plant body, for which they provide the chief source of strength. They may be recognized by the uneven thickening of their primary cell walls.

5. Sclerenchyma cells, including fibers, have tough, thick walls with secondary thickening—thickening that is laid down after the cells have reached full size. Tracheids and vessel elements, the principal conducting elements of the xylem, are similar to sclerenchyma cells in having thick cell walls.

6. Carbohydrates are conducted through the plant primarily in the phloem, whose elongated conducting cells—sieve cells and sieve-tube members—are living but lack a nucleus. Associated with sieve-tube members are specialized parenchyma cells called companion cells.

7. The growth of leaves is determinate, like that of flowers; the growth of stems and roots is indeterminate.

8. Water reaches the leaves of a plant after entering it through the roots and passing upward via the xylem. Water vapor passes out of leaves by entering intercellular spaces, evaporating, and moving out through stomata.

9. Stems branch by means of buds that form externally at the point where the leaves join the stem; roots branch by forming centers where pericycle cells begin dividing. Young roots grow out through the cortex, eventually breaking through the surface of the root.

10. Secondary growth in both stems and roots takes place after the formation of lateral meristems, known as vascular cambia. These cylinders of dividing cells form xylem internally and phloem externally. As a result of their activity, the girth of a plant increases.

11. The cork cambium forms in both roots and stems during the initial stages of secondary growth and increase in diameter. It produces cork externally and phelloderm, a dense tissue composed mainly of parenchyma, internally. The cork, cork cambium, and phelloderm collectively are called the periderm. The periderm, the outermost layer of bark, is perforated in the stem by areas of loosely organized tissue called lenticels.

12. An angiosperm embryo consists of an axis with one or two cotyledons, or seedling leaves. In the embryo, the epicotyl will become the shoot, and the radicle, a portion of the hypocotyl, will become the root. Food for the developing seedling may be stored in the endosperm at maturity or in the embryo itself.

REVIEW QUESTIONS

1. Where in a plant does primary growth occur? How does primary growth change the body of a plant? In what kind of general tissue does secondary growth occur? Where specifically is this tissue found? How does it change the body of a plant?

2. What is the appearance of a parenchyma cell? Are these cells highly specialized? In what types of tissue are they found? Are they alive at maturity? What ability does this confer upon them that most other plant cells lack? How do secondary cell walls differ from primary cell walls?

3. What is the appearance of the two types of sclerenchyma cells? Are they alive at maturity? What special compound is deposited in their secondary cell walls? What property does it confer to this type of cell? In what types of tissue are they found?

4. What is the function of xylem? What is the difference between primary and secondary xylem? What are the two types of conducting cells within xylem? How do they compare to one another? What other cell types are present in xylem?

5. What is the function of phloem? What two types of conducting cells are present in phloem? What is their general appearance? Are they alive at maturity? What are sieve areas? How do the conducting cells differ from one another? Which is more evolutionarily advanced? What are companion cells? What is their function?

6. By what means do angiosperm leaves grow? Is this growth determinate or indeterminate? What are the basic components of most leaves? How do simple and compound leaves differ from one another? Name and describe the three types of growth pattern exhibited by most leaves.

7. What are leaf primordia? At which region on a stem are the leaves attached? What is the difference between lateral and terminal buds? Which produces chemicals to suppress the other? What types of cells are present in a young stem? From what tissue does leaf vascular tissue develop?

8. What are the location and function of the vascular cambium? From what tissue(s) is it derived? What type of cell is produced when the vascular cambium divides: outwardly, inwardly, or laterally? Why does a girdled tree ultimately die?

9. What is the function of cork cambium? What fatty substance is contained within these cells? What is its function? Are these cells dead or living at maturity? What three tissues make up the periderm? Does removing the periderm of a tree have the same effect as girdling it? Why or why not? What tissues comprise a tree's bark? What is the ultimate fate of the primary phloem in a typical tree?

10. How is the season growth in a tree growing in a temperate region reflected in its wood? Which season produces the larger cells? What else can be determined by examining a horizontal section through a tree? What is the difference between softwoods and hardwoods from a biological standpoint?

11. What is the appearance of a section cut horizontally through a monocot root? How does the inside of a dicot root differ from its stem? What is the appearance of the xylem and phloem in a cross section of such a root? What are the layers of a dicot root, traveling from the outside inward? What key features are associated with each layer? With which layer is the Casparian strip associated?

12. Why do roots have a simpler anatomy than do stems? How do roots form lateral branches? Does a lateral root develop a root cap before or after it breaks the outer epidermis of the main root? Does the secondary vascular growth in a root form isolated bundles of secondary xylem and phloem? Explain.

THOUGHT QUESTIONS

1. If you hammer a nail into the trunk of a tree 2 meters above the ground when the tree is 6 meters tall, how far above the ground will the nail be when the tree is 12 meters tall?

2. Can you suggest a reason why branches are not formed the same way in roots that they are in stems?

FOR FURTHER READING

ESAU, K.: *Anatomy of Seed Plants,* ed. 2, John Wiley & Sons, Inc., New York, 1977. Short but outstanding textbook on plant anatomy.

GALSTON, A.W., P.J. DAVIS, and R.L. SATTER: *The Life of the Green Plant,* ed. 3, Prentice-Hall, Inc., Englewood Cliffs, N.J., 1980. Very well-written, physiologically oriented treatment of the structure and functioning of plants.

HARDWICK, R.: "Construction Kits for Modular Plants," *New Scientist,* April 10, 1986, pages 39-42. Much attention is properly being paid to the ways in which plant growth follows repetitive patterns, and this short article provides an introduction to the topic.

HUTCHINGS, M.J., and I.K. BRADBURY: "Ecological Perspectives on Clonal Perennial Herbs," *BioScience,* 1986, vol. 36, pages 178-182. Does a clonal plant have an advantage in nature if it is functionally integrated or if the parts are independent?

RAVEN, P.H., R.F. EVERT, and S.E. EICHHORN: *Biology of Plants,* ed. 5, Worth Publishers, Inc., New York, 1991. A comprehensive treatment of general botany, emphasizing structural botany.

SANDVED, K.B., and G.T. PRANCE: *Leaves,* Crown Publishers, Inc., New York, 1984. This book provides an incredibly beautiful introduction to the diversity of leaves.

ZIMMERMANN, M.H.: *Xylem Structure and the Ascent of Sap,* Springer-Verlag New York, Inc., New York, 1983. An outstanding monograph on the way xylem is put together and functions.

36

Nutrition and Transport in Plants

OVERVIEW

The body of a plant is basically a tube embedded in the ground and extending up into the light, where expanded surfaces—the leaves—capture the sun's energy and participate in gas exchange. The warming of the leaves by sunlight increases evaporation from them, creating a suction that draws water into the plant through the roots and up the plant through the xylem to the leaves. Transport from the leaves and other photosynthetically active structures to the rest of the plant occurs through the phloem. This transport is driven by osmotic pressure; the phloem actively picks up sugars near the places where they are produced, expending ATP in the process, and unloads them where they are used. Most of the minerals critical to plant metabolism are accumulated by the roots, which expend ATP in the process. The minerals are subsequently transported in the water stream through the plant and distributed to the areas where they are used—another energy-requiring process.

FOR REVIEW

Here are some important terms and concepts that you will encounter in this chapter. If you are not familiar with them, you should review them before proceeding.

Adhesion, cohesion, and capillary movement (Chapter 2)

Osmosis (Chapter 6)

Transport of ions across membranes (Chapter 6)

Biogeochemical cycles (Chapter 25)

Xylem and phloem (Chapter 35)

Endodermis (Chapter 35)

Stomata (Chapter 35)

Structurally, a plant is essentially a tube embedded in the ground. At the base of the tube are roots, and at its top are leaves. For a plant to function, two kinds of transport processes must occur. First, the carbohydrate molecules that are produced in the leaves by photosynthesis must be carried to all of the other living cells of the plant. To accomplish this, liquid, with these carbohydrate molecules dissolved in it, must move both up and down the tube. Second, minerals and water in the ground must be taken up by the roots and ferried to the leaves and other cells of the plant. In this process, liquid must move up the tube.

As you saw in Chapter 35, plants accomplish these two processes by using chains of specialized cells; those of the phloem transport photosynthetically produced carbohydrates up and down the tube, and those of the xylem carry water and minerals upward (Figure 36-1). Perhaps you are wondering why plants use the narrow channels in xylem and phloem to transport liquids instead of large-diameter pipes like our blood vessels, in which the rapid movement of water would be possible. The answer is that we actively pump our blood, whereas plants rely on passive forces to drive the movement of liquids through their bodies. As you will see, these forces depend heavily on the existence of very narrow transport tubes.

In this chapter we will deal with the forces that move water and solutes in plants, with the nutritional requirements of plants, with the ways in which plants conserve water, and with some of the ecological consequences of these relationships. We will begin with some remarks about the soil, the substrate in which nearly all plants grow, and then consider how the water and minerals derived from the soil move into and through the plant.

THE SOIL

Soils are produced by the weathering of rocks in the earth's crust; they vary according to the composition of those rocks. The crust includes about 92 naturally occurring elements; the abundance of some of them in the crust is listed in Table 2-1. Most elements are combined into inorganic compounds called minerals; most rocks consist of several different minerals.

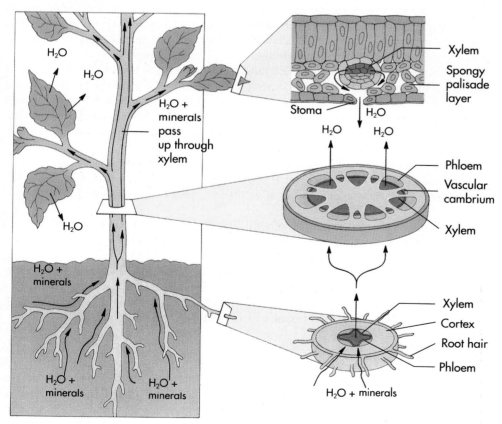

FIGURE 36-1
Diagram of flow of materials into, out of, and within a plant.

FIGURE 36-2

Soil organisms. About 5 metric tons of carbon are tied up in the organisms that are present in the soil under a hectare (0.06 square miles) of wheatland in England, an amount that approximately equals the weight of 100 sheep!

Composition of Soil

Millions of years of weathering of rocks have produced the particles from which soils are formed. All soils on earth show unmistakable signs of biological processes: these signs are both direct (shown in the structure of soils and their load of organic materials) and indirect. For example, the atmosphere itself, which plays the major role in weathering rocks on earth, has been changed profoundly over time by the effects of biological evolution. Biological effects are obvious not only in the topsoil, where organisms and organic debris are abundant and diverse (Figure 36-2), but also lower down. The ways in which this organic debris is broken down and recycled, mainly as a result of the activities of organisms living in the soil, have a great deal to do with the fertility of the soil and play a key role in controlling the cycling of substances through it. **Topsoil** is a mixture of mineral particles of varying size, living organisms, and **humus,** which consists of partly decayed organic material; most roots occur in topsoil.

> **Rocks are made up of one or more kinds of minerals. They weather to give rise to soils, which differ according to the composition of their parent rocks. The amount of organic materials in soils affects their fertility and other properties.**

Soil is made up of particles of varying size, from coarse sand, with particles that are 200 to 2000 micrometers in diameter, to clay, with particles that are less than 2 micrometers in diameter. The size of particles in fine sand and silt lies between these two extremes, and most soils are a mixture of particles of different sizes.

About half of the total soil volume is occupied by empty space, which may be filled with air or water depending on moisture conditions. Not all of the water in soil, however, is available to plants, because of the nature of water itself.

Water in Soil

You will recall from Chapter 2 that the chemical and physical behavior of water is dictated largely by its tendency to form hydrogen bonds with itself and other materials. Two aspects of the behavior of water are particularly important here. First, water is tightly bound to objects that bear electrostatic charges and to those which are able to form hydrogen bonds. Many soils contain minerals that have exactly these properties. Clays, for example, are often negatively charged, and water is tightly bound to clay particles by hydrogen bonds.

Second, the greater the surface area of the sum total of all the soil particles in a quantity of soil, the more water will adhere to them. Far more water will adhere to soils with very small particles than to soils with large ones. Thus water drains rapidly through sandy soils, which consist of relatively coarse particles. Clay soils may hold a great deal more water, but much of this water may be held so tightly that plants cannot extract it from the soil. For these reasons, as well as for those concerned with the availability of nutrients, a soil composed of a balanced mixture of coarse and fine particles—of sand, silt, and clay—is ideal for plant growth. Such soils are known as loams.

> **Water is tightly bound to objects that bear electrostatic charges and to those which form hydrogen bonds. The more soil particle surface area in a given soil, the more water it will bind. For these reasons, water runs through sandy soils and is so tightly bound to clay soils that it is often unavailable to plants.**

Some of the water that reaches a given soil will drain through it immediately, because of the force of gravity. Another fraction of the water is held in the smaller soil pores, which are generally less than about 50 micrometers in diameter. Such water is readily available to plants; it is called **capillary water.** The amount of water held in a given soil after gravity has removed the excess is called the **field capacity** of that soil. When evaporation and removal by plants have eliminated the capillary water from the soil, leaving only water that is not available to plants, that soil is said to have reached its permanent wilting point, which is defined as the moisture content of a given soil at which plants will wilt permanently unless additional water is added.

WATER MOVEMENT THROUGH PLANTS

It is not unusual for many of the leaves of a large tree to be more than 10 stories off the ground. Did you ever wonder how a tree manages to raise water so high? To understand how this happens, imagine filling a long, hollow tube with water, closing it at one end, and placing the tube, open end down, in a full bucket of water. Gravity affects the water in the tube in two ways. First, the weight of the water tends to move it down the tube. Second, gravity acts on the column of air over the bucket. Consequently, the air's weight (at sea level) exerts an amount of pressure that is defined as 1 atmosphere downward on the water in the bucket, thus forcing the water up into the tube.

These two effects counteract each other, and there is a limit to how high the tube can be and still be filled with water. That limit is set by the pressure exerted by the atmosphere. At sea level, this pressure is sufficient to raise the water in a tube to a height of about 10.4 meters. The weight of the water in a tube any taller than about 10.4 meters pulls itself down, leaving a vacuum at the upper, closed end of the tube that fills with water vapor. The water is said to **cavitate** (Figure 36-3).

Could adhesion of water to the sides of small conducting vessels such as those in the xylem or phloem create capillary movement of water up the plant? No. The capillary forces are not strong enough. Water is lifted less than 1 meter in a glass tube the diameter of a xylem element. How then does water get to the top of the tree?

The answer was suggested by Otto Renner in Germany in 1911. To understand his suggestion, we will return for a moment to our analogy of a plant as an open tube embedded in soil. Imagine that water fills the tube to a certain height, and, for simplicity, further assume that the water is held at that height by capillary action—although we will later identify the true cause as something else. Now blow across the top of the tube. The stream of relatively dry air will cause water molecules to evaporate from the surface of the water in the tube. Does the level of water in the tube fall? No. As water molecules are drawn from the top, they are replenished by new ones that are taken in from the bottom. This, in essence, is what Renner proposed actually

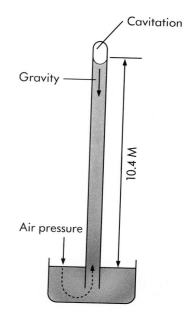

FIGURE 36-3
Cavitation. Closed tube in a bucket, with water rising 10.4 meters, and cavitation at the upper end of the tube.

happens in plants. The passage of air across leaf surfaces results in the loss of water by evaporation, creating a pull at the open upper end of the "tube," while water is pushed up from below by atmospheric pressure.

> **Water rises in a plant beyond the point at which it would be supported by atmospheric pressure (10.4 meters at sea level) because evaporation from its leaves produces a tension on the entire water column that extends all the way down to the roots.**

But what of cavitation? Why does a column of water in a tree more than 10.4 meters tall not cavitate simply because of its weight? The answer is that water has an inherent tensile strength that arises from the cohesion of its molecules, their tendency to form hydrogen bonds with one other. The tensile strength of a column of water varies inversely with the diameter of the column; that is, the smaller the diameter of the column, the greater the tensile strength. Therefore plants must have very narrow transporting vessels in order to take advantage of the tensile strength of water.

Cavitation can occur locally within individual conducting cells in the xylem because of deformation or freezing. If it does, it will result in the formation of small bubbles of air within the cell. You will recall from Chapter 35 that individual vessel elements and tracheids are connected by one or more pores at their end walls, rather than simply being open and directly connected, like the segments of a sewer pipe. The bubbles formed by cavitation are larger than the pores, so they cannot pass through them. Furthermore, the cohesive force of water is so great that the bubbles are forced into rigid spheres. These spheres are unable to change their shape and thereby deform and squeeze through the pores of the vessel element. Because vessel elements and tracheids are connected by these fine-diameter pores, any cavitation that does occur is limited to the cells where it begins, and water may continue to rise in parallel elements.

> **When cavitation occurs within a column of water in a plant, the bubbles formed are held as rigid spheres by the cohesive force of the water and are unable to pass through the pores between the vessel elements or tracheids.**

Plant biologists often discuss the forces that act on water within a plant in terms of the water's potential energy. Water high in the plant has more potential energy than water at ground level, the difference being that more energy is required to oppose gravity and hold the water up there. In general, water in a plant may possess potential energy for two reasons: (1) **pressure potential,** or pressure exerted by the atmosphere, as we discussed before; and (2) water movements driven by diffusion forces. Thus water tends to move by osmosis into a root because the root cells contain many metabolites and other solutes. For this reason the root is said to have a greater **solute potential** than the fluid outside: solutes are more concentrated in it. The algebraic sum of the pressure potential and the solute potential represents the total potential energy of the water in the plant and is called its **water potential.** In these terms water rises in a plant because it is driven by a water potential that is created by the positive pressure of the atmosphere and the negative pressure (pull) that is caused by the evaporation of water from the leaves.

> **Water rises in a plant because it is driven by a water potential that is created by the positive pressure of the atmosphere and the negative pressure caused by the evaporation of water from the leaves.**

The water potential in a plant, and thus the movement of water through the plant, depends on its osmotic absorption by the roots, the positive pressures driving its movement through the xylem, and the negative pressures created by its transpiration from the leaves and other plant surfaces. These three processes are closely linked, and, as you will see, the negative pressure generated by transpiration is largely responsible for the operation of the other two processes.

PLANT RESPONSES TO FLOODING

The continued existence and growth of plants depends on an adequate supply of water. Plants lack a closed circulation system, and only the continuous stream of water that flows through them keeps them healthy.

Even plants, however, can receive too much water. This occurs when the soil is flooded, a condition that can arise when rivers or streams overflow their banks or when rainfall is heavy, irrigation is excessive, or drainage is poor. Flooding rapidly depletes the available oxygen in the soil and blocks the normal reactions that take place in roots and make possible the transport of minerals and carbohydrates. Abnormal growth patterns may result, and the plants may ultimately "drown." Hormone levels change in flooded plants—ethylene, for example, often increases, while gibberellins and cytokinins usually decrease—and these changes may also contribute to the abnormal growth patterns. Flooding involving moving water, which brings in new supplies of oxygen, is much less harmful than flooding involving standing water, which does not; flooding that occurs when a plant is dormant is much less harmful than flooding when it is growing actively.

Physical changes that may occur in the roots as a result of oxygen deprivation may halt the flow of water through the plant, paradoxically drying out the leaves—even though the roots of the same plant may be standing in water. Because of such stresses, the stomata of flooded plants often close. In some plants the closing of the stomata maintains the turgor of the leaves.

Many plants, of course, grow in places that are often flooded naturally; they have adapted to these conditions during the course of their evolution (Figure 36-A, *1*). One of the most frequent adaptations among such plants is the formation of **aerenchyma,** loose parenchymal tissue with large air spaces in it. Aerenchyma is very prominent in water lilies and many other aquatic plants. Oxygen may be transported from the parts of plants above the water to those below by way of passages in the aerenchyma. This supply of oxygen allows oxidative respiration to take place even in the submerged portions of the plant. Some plants normally form aerenchyma, whereas others, subject to periodic flooding, can form it when necessary. In corn, ethylene, which becomes abundant under the anaerobic conditions of flooding, induces aerenchyma formation. Plants also respond to flooded conditions by forming larger lenticels, which facilitate gas exchange, and additional adventitious roots.

Plants such as mangroves (Figure 36-A, *2* and *3*), which are normally flooded with salt water, must not only provide a supply of oxygen for their submerged parts, but also control their salt balance. The salt must be excluded, actively secreted, or diluted as it enters. The arching stilt roots of mangroves are connected to long, spongy, air-filled roots that emerge above the mud. These spongy air roots have large lenticels on their abovewater portions through which oxygen enters; it is then transported to the submerged roots. In addition, the succulent leaves of mangroves contain large quantities of water, which dilutes the salt that reaches them. Many plants that grow in such conditions secrete large quantities of salt through more or less specialized glands.

1

2

3

FIGURE 36-A

Adaptations of plants to flooded conditions.
1 The "knees" of the bald cypress (*Taxodium*) are formed wherever it grows in wet conditions.
2 Red mangrove (*Rhizophora*), with its stilt roots, is a familiar sight along shores throughout the world's tropics and subtropics; this scene is in Honduras.
3 Closer view of the stilt roots of the red mangrove.

Transpiration

More than 90% of the water that is taken in by the roots of a plant is ultimately lost to the atmosphere, almost all of it from the leaves. It passes out primarily through the stomata in the form of water vapor. The process by which water leaves the plant is known as **transpiration**. First, the water passes into the pockets of air within the leaf from the walls of the spongy mesophyll cells that line the intercellular spaces (see Figure 35-14). As you saw in Chapter 35, these intercellular spaces open to the outside of the leaf by way of the stomata. The water that evaporates from the surfaces of the spongy mesophyll cells that line the intercellular spaces is continuously replenished from the tips of the veinlets in the leaves. Since the strands of xylem conduct water within the plant in an unbroken stream all the way from the roots to the leaves, when a portion of the water vapor in the intercellular spaces passes out through the stomata, the supply of water vapor in these spaces is continually renewed.

Because they are constantly losing so much water to the atmosphere and because the presence of this water is essential to their metabolic activities, growing plants depend on the continuous stream of water entering and leaving their bodies at all times. Water must always be available to their roots. Such structural features as the stomata, the cuticle, and the substomatal spaces have evolved in response to one or both of two contradictory requirements: to minimize the loss of water to the atmosphere, on the one hand, and to admit carbon dioxide into the plant, on the other. Before considering how plants resolve this problem, however, we must consider the absorption of water by the roots.

> **More than 90% of the water that enters a plant reaches the intercellular spaces and is lost through the stomata. A plant can grow only if it has access to water, externally or stored within its body.**

The Absorption of Water by Roots

Most of the water absorbed by the plant comes in through the root hairs, which collectively have an enormous surface area (see Figure 35-10). Water enters the root hairs, which are always turgid because of their greater solute potential. Mineral ions enter the plant with the water. Because their concentration in the soil solution is usually much less than within the plant, the accumulation of such ions usually requires the expenditure of energy in the form of ATP. The membranes of these cells contain a variety of ion transport channels, which actively pump specific ions into them even against large concentration gradients. These ions, many of which are plant nutrients, are then transported throughout the plant as a component of the water flowing through the xylem.

Once they are inside the roots, the water and mineral ions pass inward to the conducting elements of the xylem, either directly through the cells themselves by way of the protoplasm of adjacent cells or through and between the cell walls (Figure 36-4). When the mineral ions pass between the cells, they do so nonselectively. On their journey inward, however, they eventually reach the endodermis. When they finally do pass through this layer, they must, because of the presence of the waxy Casparian strips, go directly through the plasma membranes and protoplasts of the endodermal cells. The transport through the cells of the endodermis is selective, and the endodermis, with its unique structure, serves as a way of controlling which ions actually reach the xylem.

Under most circumstances, water with its dissolved ions is drawn into the roots indirectly because of transpiration from the leaves; the continuous water column that extends through the plant is pulled from the top. At night, when the relative humidity may approach 100%, there may be no transpiration, however. Under these circumstances, the negative pressure component of the water potential—caused by evaporation—becomes very small or nonexistent.

Active transport of ions into the roots, however, continues to take place under these circumstances. It results in an increasingly high ion concentration within the cells, which causes water to be drawn into the root hair cells by osmosis. In terms of

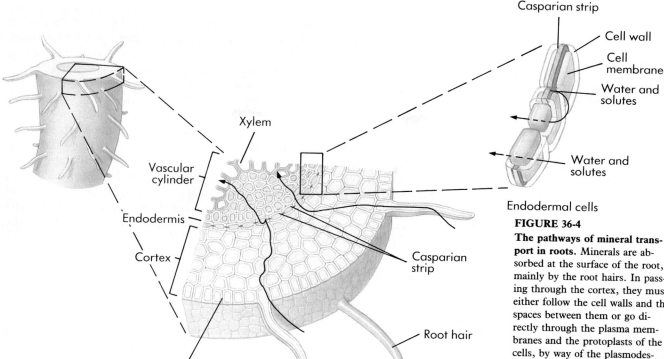

Casparian strip

Cell wall

Cell membrane

Water and solutes

Water and solutes

Endodermal cells

FIGURE 36-4

The pathways of mineral transport in roots. Minerals are absorbed at the surface of the root, mainly by the root hairs. In passing through the cortex, they must either follow the cell walls and the spaces between them or go directly through the plasma membranes and the protoplasts of the cells, by way of the plasmodesmata. When they reach the endodermis, however, their further passage through the cell walls is blocked by the Casparian strip, and they must pass through the membranes and protoplast of an endodermal cell before reaching the xylem.

Xylem

Vascular cylinder

Endodermis

Cortex

Casparian strip

Root hair

Epidermis

water potential, we say that active transport increases the solute potential of the roots. The result is the movement of water into the plant and up the xylem columns, caused by osmosis. The phenomenon is called **root pressure.** In reality, root pressure is an osmotic phenomenon that results from the lack of transpiration-driven water movement from the roots. Thus root pressure is primarily a nighttime activity that disappears when transpiration resumes the next morning.

> **Root pressure, which is active primarily at night, is caused by the continued, active accumulation of ions by the roots of a plant at times when transpiration from the leaves is very low or absent.**

Under certain circumstances, root pressure is so strong that water will ooze out of a cut plant stem for hours or even days or will be forced up into a glass tube attached to the end of the stump. When root pressure is very high, it may force water up to the leaves, from which the water may be lost in a liquid form through a process known as **guttation** (Figure 36-5). Guttation takes place not through the stomata, but through special groups of cells that are located near the ends of small veins that function only in this process. Root pressure is never sufficient to push water up great distances, and guttation thus takes place only in relatively short plants.

Some water enters plants by way of organs other than roots. Plants that grow in coastal deserts, for example, may absorb large quantities of water from fog that condenses on their branches and leaves, and epiphytes that grow on the branches of trees in the humid tropics sometimes also obtain large quantities of water in this way. The significance of such uptake in lowering water loss from the leaves needs much more extensive evaluation, however. In mosses and lichens, water absorption from moist air is clearly important for survival.

Water Movement in Plants

Because of the interactions of the forces that we have just considered, water is transported to the tops of tall trees and to the uppermost branches of vines. The tension that arises within the xylem of trees during periods of active transpiration may create such a strong pull on the conducting cells that their walls may be pulled closer to-

FIGURE 36-5

Guttation. In herbaceous plants, water passes through specialized groups of cells at the edges of the leaves; it is visible here as small droplets around the edge of the leaf of this strawberry plant (*Fragaria ananassa*).

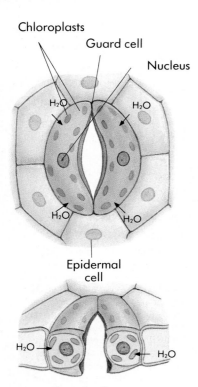

Chloroplasts

Guard cell

Nucleus

H_2O H_2O

H_2O H_2O

Epidermal
cell

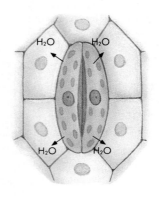

H_2O → ← H_2O

A Guard cells open

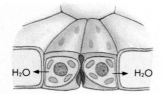

H_2O H_2O

H_2O H_2O

H_2O → ← H_2O

B Guard cells closed

FIGURE 36-6
How guard cells open and close.
A The thick inner side of each of the two guard cells that make up a stoma bow outwardly, opening the stoma, when solute pressure is high (water pressure is low) within the guard cells.
B When the solute pressure is low (water pressure is high) within the guard cells, they become flaccid and close the stoma.

gether, and the diameter of the tree trunk as a whole may be measurably less during the day than it is at night. Negative water potential occurs in columns of dead vessel elements or tracheids, as the newer and still-living xylem elements draw water laterally from them. This negative water potential created in the column of dead xylem cells becomes better established as the season progresses from spring to autumn.

Each morning, the sun illuminates and warms the leaves and small branches of a tree, and this warming leads to the increased evaporation of water from them. This increased evaporation creates a pull at the upper end of the water column, in the veinlets of the leaves; this pull is transferred to the water column in the small branches and through them to that in the main trunk, ultimately creating a force that pulls water into the roots. After the sun sets and transpiration from the leaves decreases, the water potential becomes less negative, first in the upper portions of the plant. In the movement of water through the plant, the sun is the ultimate source of potential energy. The water potential that is responsible for water movement is the product largely of the negative pressure generated by transpiration, which is driven by the warming effects of sunlight.

The Regulation of Transpiration Rate

The only way that plants can control water loss on a short-term basis is to close their stomata. Many plants can do this when they are subjected to water stress. The stomata must be open at least part of the time, however, so that CO_2, which is necessary for photosynthesis, can enter the plant. In its pattern of opening or closing its stomata, a plant must respond both to the need to conserve water and to the need to admit CO_2.

When CO_2 has entered the intercellular spaces, it must dissolve before it can enter the plant's cells. The gas dissolves mainly on the walls of the intercellular spaces below the stomata. These walls are kept moist by the continuous stream of water that reaches the leaves from the roots.

The stomata open and close because of changes in the water pressure of their guard cells (Figure 36-6). The guard cells are the only epidermal cells with chloroplasts, and they stand out for this reason and because of their distinctive shape—thicker on the side next to the stoma opening and thinner on their other sides and ends. When the guard cells are turgid, or plump and swollen with water, they become bowed in shape, as do the thick inner walls, thus opening the stoma as wide as possible.

The guard cells use ATP-powered ion transport channels through their membranes to concentrate ions actively. This concentration creates a solute potential within the guard cells that causes water to enter osmotically. As a result, these cells accumulate water and become turgid, opening the stomata. The guard cells remain turgid only so long as the active transport channels pump ions into the cells and so maintain the higher solute concentration there. Thus keeping the stomata open requires a constant expenditure of ATP. When the active transport of ions into the guard cells ceases, the higher concentration of ions within the guard cells causes ions to move by diffusion into the surrounding cells. This diffusion reduces the solute potential; water leaves the guard cells, which become flaccid, and the stomata between them close.

The ion most important in controlling the turgidity of guard cells is the potassium ion, large amounts of which are held in the cells surrounding the guard cells. The gradient of K^+ between the guard cells and these surrounding cells is capable of changing rapidly in certain circumstances, K^+ moving from one to the other. When such a change occurs, the stomata may open or close rapidly. Apparently photosynthesis in the guard cells provides an immediate source of ATP, which drives the active transport of K^+ ions into and out of the guard cells by way of a specific K^+ channel that has now been isolated and studied. In some species Cl^- ions accompany the K^+ ions into and out of the guard cells, thus maintaining electrical neutrality. In many other species, H^+ ions move in the direction opposite the K^+ ions.

When water is so scarce that the whole plant wilts, the guard cells may become limp, and the stomata may close as a result. The guard cells of many plant species, however, regularly become turgid in the morning, when photosynthesis is possible, and flaccid in the evening, regardless of the availability of water. When they are turgid, the stomata open, and CO_2 enters freely; when they are flaccid, CO_2 is largely excluded, but water loss is retarded also.

Abscisic acid, a plant hormone that we will discuss in Chapter 38, plays a primary role in allowing the K^+ ions to pass rapidly out of the guard cells and, therefore, in causing the stomata to close. This hormone is produced by leaf tissue under water stress, and it brings about a direct response from the guard cells, binding to specific receptor sites in their plasma membranes. Perhaps abscisic acid will prove useful in the conservation of water in crops grown in dry regions, provided that enough CO_2 would still be able to reach the mesophyll of the plants whose stomata had been closed by applying the hormone. At any rate, the accumulating evidence suggests that plants control the duration of their stomatal opening through the integration of several stimuli in a way that we only partly understand.

Stomata open when their guard cells become turgid. Keeping the guard cells turgid requires a constant expenditure of ATP to pump potassium ions into the guard cells; potassium ions pass out when the guard cells become flaccid and thus control the opening and closing of the stomata.

Other Factors Regulating Transpiration

Factors such as CO_2 concentration, light, and temperature can also affect stomatal opening. When CO_2 concentrations are high, the guard cells of many plant species become flaccid, and their stomata close. At such times the plant has no need to bring in additional CO_2 and conserves water by closing its guard cells. When the temperature exceeds 30° to 34° C, the stomata also close. In the dark the stomata will open at low concentrations of CO_2. In Chapter 35 we mentioned CAM photosynthesis, which occurs in some kinds of succulent plants such as cacti. In this process CO_2 is taken in at night and fixed during the day. CAM photosynthesis conserves water in dry environments where succulent plants grow.

Many mechanisms have evolved in plants by which they regulate their rate of water loss. One strategy involves dormancy at times of the year when water is in short supply. The deciduous habit, which is so common in plants that grow in areas that experience a severe drought at some season of the year, is perhaps the most familiar of these; one way in which a plant can avoid water loss is to lose its leaves. When water is locked up in ice and snow, plants experience drought and thus are often deciduous in regions with severe winters. Annual plants conserve water, in a sense, by simply not being present, except as seeds, when conditions are unfavorable.

Thick, hard leaves, often with relatively few stomata—and frequently with stomata on the lower side of the leaf only—lose water much more slowly than large, soft leaves with abundant stomata. Leaves covered with masses of woolly looking trichomes retain a more humid layer of air near their surfaces; an even more important factor in slowing down the loss of water vapor to the air is that the trichome layers hold down leaf temperatures greatly (see Figure 35-9, C). Plants in arid or semiarid habitats often have their stomata in crypts or pits in the leaf surface. Within these depressions the water content of the air may be high, therefore reducing the rate of water loss.

NUTRIENT MOVEMENT

Apparently most of the movement of ions into a plant takes place through the protoplasts of the cells rather than between their walls. Ion passage through cell membranes seems to be active and carrier mediated, although the details are not well understood. We do know, however, that the initial movement of nutrients into the roots is an active process that requires energy and that, as a result, specific ions can be maintained within the plant at very different concentrations from the soil. When roots are deprived of oxygen, they lose their ability to absorb ions, a definite indication that they require energy for this process to occur successfully. A starving plant—one from which light has been excluded—will eventually exhaust its nutrient supply and be unable to replace it.

Once the ions reach the xylem, they are distributed rapidly throughout the plant, eventually reaching all metabolically active parts. Ultimately the ions are removed from the roots and relocated to other parts of the plant, their passage taking place in the xylem, where phosphorus, potassium, nitrogen, and sometimes iron may be abundant in certain seasons. In many plants, such a pattern of ionic concentration helps to conserve these essential nutrients, which may be translocated from parts such as leaves and twigs that a plant is constantly shedding as it grows. Among the essential ions, however, calcium is not removed once it has been deposited in plant parts.

The accumulation of ions by plants is an active process that usually takes place against a concentration gradient and requires the expenditure of energy.

CARBOHYDRATE MOVEMENT

Most carbohydrates manufactured in leaves and other green parts of the plant are moved through the phloem to other parts of the plant. This process, known as translocation, is responsible for the availability of suitable carbohydrate building blocks at the actively growing regions of the plant. The carbohydrates that are concentrated in storage organs such as tubers, often in the form of starch, are also converted into transportable molecules, such as sucrose, and moved through the phloem. The pathway that sugars and other substances travel within the plant has been demonstrated precisely by using radioactive tracers. The route is fairly well understood, despite the fact that living phloem is delicate and the process of transport within it is easily disturbed.

Aphids, a group of insects that suck the sap of plants, have been valuable tools in understanding translocation. Aphids thrust their piercing mouthparts into the phloem cells of leaves and stems to obtain the abundant sugars there (Figure 36-7). When the aphids are decapitated, the liquid continues to flow from the detached mouthparts and is thus available in pure form for analysis. The liquid in the phloem contains 10% to 25% dry matter, almost all of which is sucrose. Using aphids to obtain the critical samples and radioactive tracers to mark them, it has been possible to demonstrate that the movement of substances in the phloem can be remarkably fast; rates of 50 to 100 centimeters per hour have been measured.

The **mass flow** of materials transported in the phloem occurs because of hydrostatic pressure, which develops as a result of osmosis (Figure 36-8). First, sucrose, which is produced as a result of photosynthesis, is actively loaded into the phloem tubes of the veinlets. This process increases the solute potential of the sieve tubes, so water passes into them by osmosis. An area where the sucrose is made is called a **source;** an area where it is being taken from the sieve tubes is called a **sink.** Sinks include roots, stems, growing fruits, and other regions where the sucrose solution is being unloaded. There, the solute potential of the sieve tubes is decreased as the sucrose is removed. As a result of these processes, water moves in the sieve tubes from

FIGURE 36-7

Feeding on phloem. Aphids like this individual of *Schizaphis graminarum*, shown here on the edge of a wheat leaf, feed on the food-rich contents of the phloem, which they suck out with their piercing mouthparts. The contents from the aphids can then be analyzed.

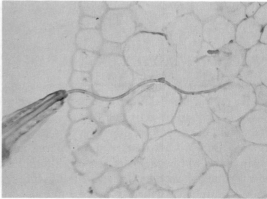

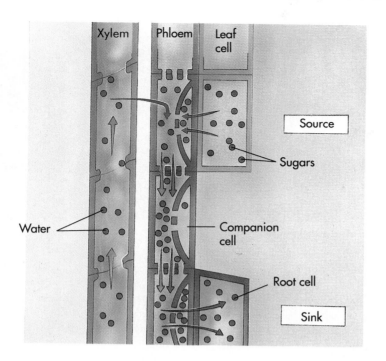

FIGURE 36-8
Diagram of mass flow. In this diagram, sucrose molecules are indicated by red dots, water molecules by blue ones. Moving from the parenchyma cells of a leaf or another part of the plant into the conducting cells of the phloem, the sucrose molecules are then transported to other parts of the plant by mass flow and unloaded where they are required.

the areas where sucrose is being taken in to those areas where it is being withdrawn, and the sucrose moves passively with it.

The transport of water, sucrose, and other substances through the sieve tubes does not require energy, but the loading and unloading of these substances into and from the sieve tubes does. This energy is supplied in the form of ATP by the companion cells, if they are present, or other parenchyma cells, if they are not, through plasmodesmata into the sieve tubes. As you will recall from Chapter 35, sieve cells and sieve-tube elements lack nuclei at maturity and are not directly involved in providing this energy. However, the manufacture of sugar in some regions of the plant and its use in others make possible the mass flow of solutes through the phloem at such rapid rates.

> **Water moves through the phloem as a result of decreased water potential in areas of active photosynthesis, where sucrose is actively being loaded into the sieve tubes, and increased water potential in those areas where sucrose is being unloaded. Energy for the loading and unloading of the sucrose and other molecules is supplied by companion cells or other parenchyma cells. However, the movement of water and dissolved nutrients within the sieve tubes is a passive process that does not require the expenditure of energy.**

PLANT NUTRIENTS

Plants require a number of inorganic nutrients. Some of these are **macronutrients,** which the plants need in relatively large amounts, and others are **micronutrients,** those required in trace amounts. There are nine macronutrients: carbon, hydrogen, and oxygen—the three elements found in all organic compounds—as well as nitrogen, potassium, calcium, phosphorus, magnesium, and sulfur. Each of these nutrients approaches or exceeds 1% of the dry weight of a healthy plant. The seven micronutrient elements that constitute from less than one to several hundred parts per million by dry weight in most plants are iron, chlorine, copper, manganese, zinc, molybdenum, and boron.

> **Plant macronutrients—carbon, hydrogen, oxygen, nitrogen, potassium, calcium, phosphorus, magnesium, and sulfur—approach or exceed 1% of a plant's dry weight, whereas the micronutrients—iron, chlorine, copper, manganese, zinc, molybdenum, and boron—are present only in trace amounts.**

A   **B**   **C**

D   **E**   **F**

FIGURE 36-9

Mineral deficiencies in plants.
Individuals of Marglobe tomatoes
(*Lycopersicon esculentum*) are grown
in hydroponic solution—water to
which nutrients have been added.
A Control (healthy) plant, grown
in a complete nutrient solution
(one that contains all of the essential minerals), with a plant showing the symptoms of calcium
deficiency. Calcium was not added
to the nutrient solution of the
plant at right, resulting in its un-
healthy, dwarfed appearance.
Without calcium, the young
shoots and roots die.
B Leaf of a healthy control plant.
C Chlorine-deficient plant, with
curled and necrotic leaves (leaves
with patches of dead tissue).
D Copper-deficient plant, with
blue-green, curled leaves. Copper
is required only in very small
amounts for normal plant growth.
E Zinc-deficient plant, with
small, necrotic leaves.
F Manganese-deficient plant,
with chlorosis (yellowing) between
the veins. This chlorosis is caused
by a drastic reduction in chlorophyll production. The agricultural
implications of deficiencies such as
these are obvious; a trained observer can determine the nutrient
deficiencies that are affecting a
plant simply by inspecting it.

The macronutrients were, in general, discovered in the last century, but the micronutrients were detected only more recently because plants need them in such small
quantities. For studying their nutritional requirements, plants are usually grown in
hydroponic culture; their roots are suspended in aerated water containing nutrients.
Most plants grow satisfactorily in hydroponic culture, and the method is occasionally
practical for commercial purposes, although expensive.

The solutions in which test plants are grown contain all of the necessary nutrients
in the right proportions but with certain known or suspected nutrients left out. The
plants are then allowed to grow and are studied for the presence of abnormal symptoms that might indicate a need for the missing element (Figure 36-9). However, the
water or vessels used often contain enough micronutrients to allow the plants to grow
normally, even though these substances were not added deliberately to the solutions.
To give an idea what small quantities of micronutrients plants may require, the standard dose of molybdenum added to seriously deficient soils in Australia amounts to
about 34 grams (about one handful) per hectare, once every 10 years!

The Roles of Plant Nutrients

There are many ways in which the six macronutrients other than carbon, hydrogen,
and oxygen and the seven micronutrients are involved in plant metabolism; some of
these are listed in Table 36-1.

Some of these functions are obvious; for example, potassium ions regulate the
turgor pressure of guard cells and, therefore, the rate at which the plant loses water
and takes in carbon dioxide. Calcium is an essential component of the middle lamellae
that are laid down between plant cell walls, and it also helps to maintain the physical
integrity of membranes. Magnesium is a part of the chlorophyll molecule. The presence of phosphorus in many key biological molecules, such as nucleic acids and ATP,
has been explored in detail in the earlier chapters of this book. Nitrogen is an essential part of amino acids and proteins, of chlorophyll, and of the nucleotides that make
up nucleic acids.

TABLE 36-1   ESSENTIAL NUTRIENTS IN PLANTS

ELEMENT	PRINCIPAL FORM IN WHICH ELEMENT IS ABSORBED	APPROXIMATE PERCENT OF DRY WEIGHT	IMPORTANT FUNCTION (EXAMPLES)
MACRONUTRIENTS			
Carbon	(CO_2)	44	Major component of organic molecules
Oxygen	(O_2, H_2O)	44	Major component of organic molecules
Hydrogen	(H_2O)	6	Major component of organic molecules
Nitrogen	(NO_3^-, NH_4^+)	1-4	Component of amino acids, proteins, nucleotides, nucleic acids, chlorophyll, coenzymes
Potassium	(K^+)	0.5-6	Component of enzymes, protein synthesis, operation of stomata
Calcium	(Ca^{++})	0.2-3.5	Component of cell walls, maintenance of membrane structure and permeability, activates some enzymes
Magnesium	(Mg^{++})	0.1-0.8	Component of chlorophyll molecule, activates many enzymes
Phosphorus	$(H_2PO_4^-, HPO_4^=)$	0.1-0.8	Component of ADP and ATP, nucleic acids, phospholipids, several coenzymes
Sulfur	$(SO_4^=)$	0.05-1	Components of some amino acids and proteins, coenzyme A
MICRONUTRIENTS (CONCENTRATIONS IN PPM)			
Chlorine	(Cl^-)	100-10,000	Osmosis and ionic balance
Iron	$(Fe^{++}$ or $Fe^{+++})$	25-300	Chlorophyll synthesis, cytochromes, nitrogenase
Manganese	(Mn^{++})	15-800	Activator of certain enzymes
Zinc	(Zn^{++})	15-100	Activator of many enzymes, active in formation of chlorophyll
Boron	$(BO_3^-$ or $B_4O_7^=)$	5-75	Possibly involved in carbohydrate transport and nucleic acid synthesis
Copper	(Cu^{++})	4-30	Activator or component of certain enzymes
Molybdenum	(MoO)	0.1-5	Nitrogen fixation, nitrate reduction

Some kinds of plants have specific nutritional requirements that are not shared by others. Silica, for example, is essential for the growth of many grasses—it helps to retard their complete destruction by herbivores (Figure 36-10)—but not for plants in general. Cobalt is necessary for the normal growth of the nitrogen-fixing bacteria associated with the nodules of legumes and is therefore an essential element for the normal growth of these plants. Nickel seems to be essential for soybeans, and its role in the nutrition of other plants should be investigated further. Sodium, although it is important in maintaining osmotic and ionic balances, is probably not essential for many plants; it does appear to be required by some desert and salt-marsh species.

In general, the elements that animals require reach them through plants, which therefore form an indispensable link between animals and the reservoirs of chemicals in nature. Some of the elements that animals require, such as iodine, come by way of plants but are not required by the plants. Iodine is very rare in soils; a shortage of iodine in the human diet can lead to the condition known as goiter. Selenium, which might be an essential element for at least some plants, can sometimes be toxic to animals, as we saw in our discussion of the San Luis National Wildlife Refuge in the San Joaquin Valley of California in Chapter 25. Plants also may concentrate minerals of economic interest, such as gold, so they are sometimes used as an assay for the presence of these minerals in a particular region.

FIGURE 36-10
Silica defends plants from grazing. These zebras graze selectively on the grass species that make up this East African savanna, depending in part on the degree to which their leaves are reinforced with silica. In general, the more silica in the leaf cells, the less likely zebras and other grazing animals are to eat that particular kind of grass.

Some plants are able to use other organisms directly as sources of nitrogen, and some require minerals, just as animals normally do. These are the carnivorous plants. Carnivorous plants often grow in acidic soils such as bogs—habitats that are not favorable for the growth of most legumes or of nitrifying bacteria. By capturing and digesting small animals directly, such plants obtain adequate supplies of nitrogen and thus are able to grow in these seemingly unfavorable environments.

The carnivorous plants have adaptations that are used to lure and trap insects and other small animals. The plants digest their prey with enzymes secreted from various kinds of glands.

The Venus flytrap (*Dionea muscipula*) (Figure 36-B, *1* and *2*), which grows in the bogs of coastal North and South Carolina, has three sensitive hairs on each side of each leaf, which, when touched, trigger the two halves of the leaf to snap together. Once enfolded by a leaf, the prey of a Venus flytrap is digested by enzymes secreted from the leaf surfaces.

Pitcher plants (Figure 36-B, *3* and *4*) attract insects by the bright, flowerlike colors within their pitcher-

1

2

shaped leaves and perhaps also by sugar-rich secretions. Once inside the pitchers, the insects may slide down into the cavity of the leaf, which is filled with water, digestive enzymes, and half-digested prey. Bladderworts, *Utricularia*, are aquatic; they sweep small animals into their bladderlike leaves by the rapid action of a springlike trapdoor and then digest these animals. In the sundews (Figure 36-B, *5*), the glandular trichomes (see Figure 35-9, *B*) secrete both sticky mucilage, which traps small animals, and digestive enzymes.

FIGURE 36-B

Carnivorous plants.

1 Venus flytrap, *Dionea muscipula*, which inhabits low boggy ground in North and South Carolina.

2 A Venus flytrap leaf has snapped together, imprisoning a fly.

3 A tropical Asian pitcher plant, *Nepenthes*. Insects enter the pitchers, which are modified leaves, seeking nectar, and are trapped and digested. Complex communities of invertebrate animals and protists inhabit the pitchers.

4 Yellow pitcher plant, *Sarracenia flava*, which grows in bogs in the southeastern United States. Its pitchers, with their yellow borders, resemble flowers and secrete a sweet-smelling nectar that aids the plants in trapping insects. The flowers of this species, with their hanging petals, are also evident in this photograph.

5 Sundew, *Drosera*. A small fly has been trapped by the glandular hairs.

3

(unnumbered image of yellow pitcher plants)

4

5

Fertilizer

In natural communities, available nutrients are recycled and made available to organisms on a continuous basis. When these communities are replaced by cultivated crops, the situation changes drastically. First, the soil is much more exposed to erosion and to the loss of nutrients, which may simply wash away. Second, large amounts of nutrients may be removed with the crops themselves or with the animals to which they are fed. For this reason, cultivated crops and garden plants must usually be supplied with additional mineral nutrients. In the tropics the soils are often especially poor in nutrients, and fertilizers significantly increase their yield.

The most important mineral nutrients that need to be added to soils are usually nitrogen (N), phosphorus (P), and potassium (K). All of these elements are needed in large quantities, and they are the ones most apt to become deficient in the soil; because they are often added in large quantities, they are also important sources of pollution in certain situations (see Chapter 25). Commercial fertilizers are normally given a "grade," which reflects the percentages of N, P, and K by dry weight that they contain; thus 6-6-6 means 6% of each element in the fertilizer. The suitable proportions are best determined in relation to the tested fertility of the soil and the requirements of the particular crop or other plant that is being grown on it. The nitrogen used in fertilizers is produced through the Haber process, in which nitrogen gas is recombined with hydrogen gas at high temperatures and pressures to yield ammonia. Nitrogen is often supplied to crops in very large amounts, since it seems to be especially important in affecting crop yield. Nutrients other than these three may also be scarce in individual kinds of soil, but such situations are rather rare and must be dealt with individually.

Organic fertilizers were used long before chemical fertilizers were understood and applied widely. Such substances as manure or remains of dead animals have traditionally been applied to crops, and plants are often plowed into the soil to increase its fertility as well. There is no basis for believing that organic fertilizers supply any element to plants that inorganic fertilizers cannot also provide. However, organic fertilizers do build up the humus content of the soil, which often enhances its water- and nutrient-retaining properties. For this reason the availability of nutrients to plants at different times of year may be improved, under certain circumstances, with organic fertilizers. Compost also serves to build up the humus content of the soil, although it may or may not be rich in the nutrients that the plants require, depending on the original source of the compost.

HOW DO PLANTS GROW?

Overall, plants are very different from animals. Their rigid cell walls lead to rigid bodies, by virtue of which they are confined to a fixed, or sessile, mode of existence. Manufacturing their own food, plants bathe themselves in a continuous stream of water that they take in through their roots and lose through their leaves. They have evolved a whole series of adaptations that fit them as well to the demands of their primarily terrestrial existence as the animals, but they do so in an entirely different way. Animals have flexible cells and often move freely, hunting food, seeking mates, and fleeing from predators. In the next chapter, we will consider the ways in which growth and development are regulated in plants, given the limitations that their structures impose.

1. Soils are formed by the weathering of rocks; soils have been deeply modified, in most cases, by the action of biological systems. Clay soils, which are composed of very fine particles, hold a great deal of water but do so very tightly, so the water is mostly unavailable for plant growth.

2. Water flows through plants in a continuous column, driven mainly by transpiration through the stomata. The plant can control water loss primarily by closing its stomata. The cohesion of water molecules and their adhesion to the walls of the very narrow cell columns through which they pass are additional important factors in maintaining the flow of water to the tops of plants.

3. Root pressure develops at night, when there is high osmotic concentration in the root's cells and the stomata are closed. In most circumstances root pressure makes only a minor contribution to the flow of water through plants.

4. Stomata open when their guard cells are turgid and bow out, thus causing the thickened inner walls of these cells to bow away from the opening. Abscisic acid, which is produced in the leaf under conditions of water stress, allows potassium ions to move rapidly out of the guard cells; these cells then lose their turgidity and the stomata close.

5. Nutrients move through plants primarily as solutes in the water column in the xylem. Their selective admission into the plant and their subsequent movement between living cells require the expenditure of energy.

6. The movement of water, with its dissolved sucrose and other substances, in the phloem does not require energy. Sucrose is loaded into the phloem near sites of synthesis, or sources, using energy supplied by the companion cells or other nearby parenchyma cells. The sucrose is unloaded in sinks, at the places where it is required. The water potential is lowered where the sucrose is loaded into the sieve tube and raised where it is unloaded.

7. The nine macronutrients, substances that are each present at concentrations of 1% or more of a plant's dry weight, are carbon, hydrogen, oxygen, nitrogen, potassium, calcium, phosphorus, magnesium, and sulfur. The seven micronutrients, each present at concentrations of from one to several hundred parts per million of dry weight, are iron, chlorine, copper, manganese, zinc, molybdenum, and boron. These elements play different roles in plant metabolism.

REVIEW QUESTIONS

1. What is capillary water? What is meant by a soil's field capacity? When is a soil said to have reached its permanent wilting point?

2. In what two ways does water held high in a plant possess potential energy? What specific term is associated with this potential energy? Why does water move into a root cell?

3. What amount of water that enters a plant leaves it via transpiration? What leaf structures help regulate this activity? Why must plants be exposed to a constant source of water? Must this source always be external? Why or why not?

4. Why are root hairs always turgid? Does the accumulation of minerals within a plant root require the expenditure of energy? Why or why not? Is mineral passage between most plant cells a selective or a nonselective process? What is an exception to this generalization? Why?

5. Under what environmental condition is water transport through the xylem reduced to near zero? How much transpiration occurs under these circumstances? Does active transport at the roots continue under these conditions? What is the result? What term is associated with this phenomenon?

6. What is guttation? Does it occur through the leaf stomata? Is it more likely to occur in very tall or very short plants? Why? Why is a tree's diameter often less during the day than it is at night? What is the ultimate source of a plant's water potential?

7. Does stomatal control require energy? Explain. Which ion is an integral part of this control mechanism? Which epidermal cells possess chloroplasts? Why? When does leaf tissue produce abscisic acid? What is its effect on leaf cells?

8. Is most nutrient movement within a plant active or passive? What two observations support this? Which nutrients can be relocated within a plant? Why? Why is calcium different from these nutrients?

9. What is translocation? In what form are plant carbohydrates generally stored? In what form are they transported? What is the driving force behind translocation?

10. Describe the movement of carbohydrates through a plant beginning with the source and ending with the sink. Is this process active or passive? How are the companion cells involved in this process?

11. What are macronutrients? List them. In what ways are they important to a plant's survival? What are micronutrients? What are the primary micronutrients? What are the most common ways in which they function?

12. Why are fertilizers generally required in cultivated environments when they are not needed in natural communities? What three mineral nutrients are most commonly found in commercial fertilizers? Are organic fertilizers nutritionally superior to chemical fertilizers? For what reason(s) may organic fertilizers be better than chemical fertilizers?

THOUGHT QUESTIONS

1. Why do gardeners often remove many of a plant's leaves after transplanting it?

2. Is a legume likely to become a carnivorous plant? Why do you think so? If you did find a carnivorous legume, what else would you expect to be true about it?

3. If you grew a plant that initially weighed 200 grams, but eventually weighed 50 kilograms, in a pot, would you expect the soil in the pot to change weight? If so, how much, and why?

FOR FURTHER READING

CRAFTS, A.S., and C.E. CRISP: *Phloem Transport in Plants*, W.H. Freeman and Company, San Francisco, 1971. Excellent, experimentally oriented treatment of the topic.

EPSTEIN, E.: *Mineral Nutrition of Plants*, John Wiley and Sons, New York, 1972. Good overview of the field.

HESLOP-HARRISON, Y.: "Carnivorous Plants," *Scientific American*, February 1978, pages 104-115. Provides experimental evidence about the way that carnivorous plants function.

KOZLOWSKI, T.T.: "Plant Responses to Flooding of Soil," *BioScience*, vol. 34, pages 162-167, 1984. A review of the metabolic, structural, and evolutionary responses of plants to excess water.

MANSFIELD, T.A., and W.J. DAVIES: "Mechanisms for Leaf Control of Gas Exchange," *BioScience*, vol. 35, pages 158-164, 1985. Excellent review of some of the factors involved in stomatal opening and the ways in which they are integrated.

MOONEY, H.A., AND OTHERS: "Plant Physiological Ecology Today," *BioScience*, vol. 37, pages 18-67, 1988. Nearly an entire issue of *BioScience* devoted to the ways in which plants cope with their environments and scientists study them. Highly recommended.

POSTGATE, J.R.: *The Fundamentals of Nitrogen Fixation*, Cambridge University Press, New York, 1982. Concise, authoritative account of biological nitrogen fixation.

SALISBURY, F.B., and C.W. ROSS: *Plant Physiology*, ed. 3, Wadsworth Publishing Co., Inc., Belmont, Calif., 1985. An outstanding textbook of plant physiology.

ZIMMERMAN, M.H.: *Xylem Structure and the Ascent of Sap*, Springer-Verlag New York, Inc., New York, 1983. A comprehensive treatment of the structure of xylem and the way in which water moves through it.

37

Plant Development

OVERVIEW

Plants, unlike animals, are always undergoing development. Their cells do not move in relation to one another during the course of development, which is a continuous process. In plants, all nucleated, living cells that have not developed a secondary wall are capable of developing into other cells; when individual cells taken from mature plants are cultured under appropriate conditions, they are capable of growing into whole plants. The embryos of most divisions of vascular plants are shed from their parents enclosed in seeds, structures that enable young plants to delay their further development until appropriate conditions occur. In this way, otherwise immobile plants increase greatly the likelihood that they will be dispersed safely and efficiently.

FOR REVIEW

Here are some important terms and concepts that you will encounter in this chapter. If you are not familiar with them, you should review them before proceeding.

Eukaryotic cell structure (Chapter 5)

Mechanism of gene action (Chapter 15)

Reproduction in plants (Chapter 33)

Plant structure (Chapter 35)

This chapter begins to consider the dynamic aspects of plant growth. It deals first with the development patterns of individual plant cells and their integration into tissues; then it considers the integration of these tissues into individuals. Although plant hormones are introduced here, more detail about them is presented in Chapter 38. The aim of this chapter is to indicate how plants grow, develop, and attain their mature form.

CONTINUOUS DEVELOPMENT: A CHARACTERISTIC OF PLANTS

A tree, like a human, consists of an intricate array of tissues. Each of these tissues differs from the others, and each stands in precise physiological and morphological relation to the others. Similarly, the control of the developmental processes that affect the organization of complex tissues is often exceedingly specific. In animals these developmental controls are largely internalized, as you will see in more detail in Chapter 55. They are exercised via cytoplasmic commands expressed as early as the initial formation of the female gamete, the egg. The responses to such controls in animals involve both the movement of cells and the irreversible commitment of cells to particular developmental paths.

Animal development proceeds as if it were unfolding according to a faithfully followed blueprint. At each stage, particular and highly specific cell movements and changes in gene expression occur—just as when the notes in a musical score are read, both the order and the timing are critical to the overall result. Such a program permits a high degree of tissue-specific specialization and physiological interaction. The final, complex arrangement of the cells and tissues in an adult animal normally is not influenced much by outside environmental changes during development, and no environmental cues are used during the process. Instead, all of the instructions that pertain to it are internal. Indeed, the very essence of animal development is internal **homeostasis,** or stability, which provides freedom from the influences of unprogrammed disturbances that might upset the delicate balance required to produce complex organized tissues. In the course of animal development, homeostasis reflects an integrated set of feedbacks that compensates for most normal environmental influences. We can upset that balance with unusual stimuli—for example, in humans, cigarette smoking or consuming alcohol during pregnancy is apt to be harmful to the development of the embryo—but for the most part, animal development is tightly controlled. For most animal tissues and organs, and for the animal as a whole, there is an end point, at which the individual animal is said to be an **adult.**

Plants, in contrast, are always undergoing development. They develop, like all other organisms, according to a genetic blueprint, but the way in which their particular blueprint is expressed is greatly influenced by external factors (Figure 37-1). The

FIGURE 37-1
The development of an individual plant is strongly affected by its environment. This wind-swept Jeffrey pine (*Pinus jeffreyi*) is growing on the slopes of Sentinel Dome in Yosemite National Park, California.

differentiation of specific tissues in plants is carried out under the direction of hormones—chemical substances produced usually in minute quantities in one part of an organism and transported to other parts on which they have specific effects. The particular expression of these hormones is mediated not only by genes, but also by external environmental factors. Much of our discussion of plant growth and development in this chapter and the next will be concerned with the patterns of production and functioning of these hormones. We also will address the ways in which plant growth responds to external environmental signals.

> **Both animals and plants develop according to a predetermined blueprint, but in animals, development reaches a conclusion—the adult individual— whereas in plants, some cells undergo development continuously.**

Plants cannot move and therefore must continue to exist in the particular microhabitat where they develop. Much of what a plant does (photosynthesis, flowering, seed dispersal, growth) is sensitive to the exact nature of local conditions. In view of this close and important relationship, it is highly adaptive for a plant to respond to, and to be modified by, the conditions in which it will continue to grow and develop. Animals, of course, respond to their environments too—think of the thick coats that many mammals grow in the winter—but the range of such responses is only a tiny fraction of what it is in plants. Animals can meet the demands of their environment or move to a new environment by means of their complex patterns of behavior; plants do not have such opportunities.

DIFFERENTIATION IN PLANTS: EXPERIMENTAL EVIDENCE

Perhaps related to the lack of a fixed developmental blueprint in plants is the fact that almost all of their differentiation is fully reversible. In living plant cells—those that retain their protoplasts at maturity—gene expression can be reactivated, leading to alternative modes of differentiation or even to a complete plant.

The process of differentiation in plants has been studied in detail throughout the twentieth century. Most of the impetus for this study has come from a single hypothesis: in 1902 the German botanist Gottlieb Haberlandt proposed that all living plant cells are **totipotent,** each of them possessing the full genetic potential of the organism. Haberlandt explained the differentiation of tissues by theorizing that only part of this potential is normally realized in any given differentiated tissue. He went on to suggest that one ought to be able to develop a whole mature plant from isolated plant cells if one could devise the correct medium to support growth. He tried repeatedly and unsuccessfully to do this himself. Indeed, for more than 50 years no one was able to do it.

> **Haberlandt suggested that all living plant cells are totipotent, each containing the ability to express the full genetic potential of that organism. His hypothesis was not confirmed for more than 50 years, however.**

Cell Culture

Before scientists could determine whether plant cells were indeed totipotent, they needed to learn how to grow them in culture. This proved to be a difficult technical problem. It was not difficult to isolate a single plant cell, but for differentiation to occur, that cell would have to divide repeatedly. The conditions promoting such rapid cell division had not been identified. It has since turned out that the substances needed for cell division diffuse out of individual cells when they are isolated in an agitated medium, and the cells then will not divide further. This problem was finally solved in several different ways in the 1950s. One solution was to place an isolated single cell within a tiny drop of culture medium onto a piece of filter paper floating on an established cell culture. In this way the single cell was isolated from other cells by the filter paper, but at the same time it could be influenced by them. Under these conditions, the isolated cell could obtain from the medium that it shared with the established cell culture any substances necessary to promote cell growth and division, and the isolated cell grew and divided rapidly, establishing the kind of mass of dividing, undifferentiated cells that is called a **callus**. This callus could then be grown indefinitely in culture.

Using such methods, it has been possible to determine the nutritional requirements of some plants and to grow them in defined media. For other plants, this has not been possible, and it is necessary to add coconut milk, the liquid endosperm that forms within coconuts, to the medium. This substance, which we will discuss in more detail in the next section, clearly contains additional factors necessary for the proper functioning of these plant cells.

SOMATIC VARIATION IN PLANTS

All plant cells are not genetically uniform, and scientists have long attempted to understand the principles governing the differences between them. In cell culture, for example, genetic changes occur; the individual plants that are recovered from such cultures often differ genetically. Some of the variation in tomatoes grown from cell culture is shown in Figure 37-A. Such variability can be used as a basis for introducing genetic changes into crop plants and thus developing new breeding lines. Obtaining variants in this way permits reduction in the time period required for the development of new varieties and permits

access to classes of genetic variants that may not appear in the course of natural, whole-plant variation.

Within a whole plant, such as a tree, there may likewise be genetic variation. An individual tree may have between 10,000 and 1 million shoot tips at maturity, depending on the species. Mutations clearly occur during the course of development of the cell lines in many individual branch tips, and these tips therefore differ genetically from one another to some extent in such features as their resistance to herbivores or fruit production. Indeed, many cultivated forms of plants were obtained as a result of genetic mutations that

affected a single branch tip, and then were cultured asexually from that tip. Do long-lived plants change genetically over the course of many years, and thus enhance their chances of survival? Do different portions of a grass or clover spreading aggressively for many years survive differentially in different microhabitats according to the genetic changes that accumulate in them? The extent of genetic variation that exists within individual plants in nature, and its adaptive significance, should clearly be examined in detail and evaluated according to the findings.

FIGURE 37-A

Genetic variation in tomatoes (*Lycopersicon esculentum*) obtained from cell cultures.

1 Progeny test from selected, self-fertilized, first-generation, regenerated plants, showing true breeding for mottled leaves.

2 Fruits of normal (*bottom*) and somatic variant (*top*) plants, the latter with higher pigment contents.

3 Chlorophyll-deficient mutants with a green stripe on the cotyledon, selected from a cell culture.

4 Normal red fruits, shown with orange and yellow ones derived in different lines from cell culture. Orange is controlled by a single dominant allele, and yellow is controlled by a single recessive one.

FIGURE 37-2

Growing a new plant from a bit of mature tissue. The technique of isolating phloem tissue from carrots *(Daucus carota)*, as carried out in the laboratory of F.C. Steward at Cornell University. The disks of tissue were grown in a flask in which the medium was constantly agitated so as to bring a fresh supply of nutrients to the masses of callus that soon formed.

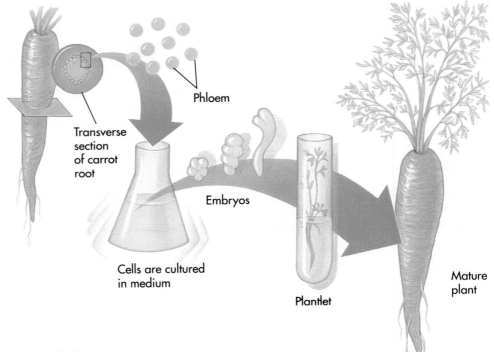

Phloem

Transverse section of carrot root

Cells are cultured in medium

Embryos

Plantlet

Mature plant

FIGURE 37-3

Tomato *(Lycopersicon esculentum)* plants regenerated from single parenchyma cells.
A Roots differentiating from masses of callus tissue in culture.
B A mass of callus tissue anchored on agar, a firm medium, with a shoot emerging.
C Regenerated tomato plant, which will mature, producing flowers and fruits, if it is transferred to soil. F.C. Steward obtained similar results in his classical experiments with carrots, which confirmed Haberlandt's hypothesis.

Tissue Culture

A second solution to the problem of culturing plant cells was devised by Cornell University plant physiologist F.C. Steward and his co-workers in 1958. His methods also depended on supplying differentiated cells with substances obtained from dividing cells. Steward isolated small bits of secondary phloem tissue from carrot root and placed them in liquid growth medium in a flask (Figure 37-2). Such growth media contained sucrose and the minerals that were essential for plant growth, as well as certain vitamins—organic molecules that the particular plant parts cannot manufacture for themselves.

In such a medium, many of the new cell clumps, each of which had originated from a single phloem cell, differentiated roots (Figure 37-3, *A*). Placed on agar, the clumps also developed shoots (Figure 37-3, *B*) and eventually grew into whole plants, thus confirming Haberlandt's hypothesis (Figure 37-3, *C*). Steward's results clearly demonstrated that the original differentiated phloem tissue still contained all of the genetic potential needed for the differentiation of whole plants. Indeed, the embryonic growth of isolated cells resembled closely that of normal zygotes. In their initial stages of development, the dividing cells differentiated into masses that resembled embryos closely. These "embryoids" had shoot and root apices from which further development proceeded, as well as distinct tissue systems that ordinary embryos have at comparable stages of development.

Steward experimentally confirmed Haberlandt's hypothesis: under appropriate circumstances, and provided that they still have a living protoplast and nucleus, individual plant cells can resume growth and differentiation.

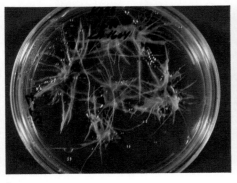

A

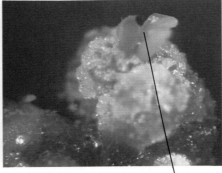

B

Emerging shoot

C

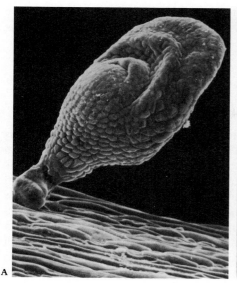
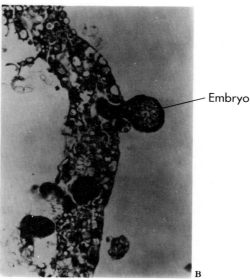

FIGURE 37-4
Direct embryogenesis from leaf mesophyll cells in orchard grass, *Dactylis glomerata*. Grass leaves grow largely from a meristem at their base, which is why, when you mow your lawn, it remains green and healthy. In these experiments, carried out by scientists from the University of Tennessee, sections of such meristems were brought into an agar culture containing the necessary nutrients, to which the chemical 3,6-dichloro-*o*-anisic acid (dicamba) had been added. Meristematic regions farther from the base of the leaf, and therefore older, produced masses of callus, but the younger ones produced typical embryos directly, as shown in the photographs.
A Scanning electron micrograph of a well-developed embryo arising directly from a grass segment.
B A very young embryo protruding from the leaf surface. Each of these embryos is connected to the leaf surface by a suspensor, a structure that plays a similar role in embryos that develop normally.

Many plants will differentiate "embryoids" and ultimately whole plants only if they form masses of callus first. In others, such as recently demonstrated in orchard grass, *Dactylis glomerata* (Figure 37-4), it has been possible to induce the formation of embryos directly from plant tissue—either leaves or flower parts—an impressive demonstration of the totipotency of plant cells.

Our general understanding of the factors that lead to the production of roots and shoots in masses of callus tissue is still limited, and the results obtained with a given treatment often vary widely from species to species. Despite these difficulties, however, more plant species are added each year to the list for which regeneration is possible. Researchers are successful mainly by "tinkering" with tissue cultures derived from individual kinds of plants; experiments now being carried out by commercial firms involved in the application of biotechnology to agriculture are making a great contribution to the available stock of information. Cotton, tomatoes, and black cherries are among the plants that recently have been regenerated in this way, assuring a supply of genetically identical individuals growing under specified experimental conditions. For woody plants, the techniques are particularly important, since they greatly shorten the life cycle of the plant and thus allow the selection for improved characteristics to proceed much more rapidly than would otherwise be possible. This area of biotechnology is a very promising one that will continue to afford satisfying careers to many scientists in the future.

Figure 37-5 illustrates the application of biotechnology to the breeding of an important forest tree of western North America, the Douglas fir (*Pseudotsuga menziesii*). Similar techniques have been applied to the breeding of many other commercially important trees, such as coconuts (*Cocos nucifera*). Coconuts, like most other palms, have only a single apical meristem and are difficult to propagate vegetatively, but scientists have recently succeeded in producing young plants from coconut floral tissue, a considerable advantage in multiplying genetically superior individuals rapidly. Similar techniques are also being applied to other slow-breeding species of commercial importance, such as elms (*Ulmus*) and redwoods (*Sequoia sempervirens*). Millions of individuals can be produced rapidly by these methods and then subjected to selection for desirable traits—a process that might otherwise take centuries. In shorter-lived plants such as strawberries and chrysanthemums, in which the production of large numbers of genetically similar individuals in a short period of time is required, **micropropagation** of this sort is the standard method of commercial propagation. Different species of plants differ greatly in their requirements for propagation, and the field remains highly experimental.

An important general conclusion that follows from these experimental results is that the differentiated plant tissue is indeed capable of *expressing* its hidden genetic complement when suitable environmental signals are provided. What is the mecha-

FIGURE 37-5
Douglas fir (*Pseudotsuga menziesii*) shoots growing in sterile culture. These shoots are the result of genetic experimentation aimed at cloning trees that have commercially valuable traits.

nism that halts the expression of genetic potential when the same kinds of cells are incorporated into normal, growing plants? As we will discuss in this chapter and the next, the expression of these genes is largely controlled by plant hormones.

Embryo Culture

Methods analogous to those discussed above have been used widely in recent years to learn about the development of plant embryos. Early efforts in this field involved dissecting out whole ovules and placing them in a suitable nutrient solution. If this is done carefully, the zygote within usually will develop into a mature embryo. When an ovule is taken from an ovary and placed in a nutrient solution by itself, it exhibits complex nutrient requirements. Once the embryo is mature, it is autotrophic; from that point on, it will grow and develop normally if it is simply kept at a favorable temperature and provided with water and oxygen. Earlier, however, when it is still heterotrophic, growth and development will not occur, even if the embryo is supplied with sucrose and the necessary mineral nutrients. In addition, young embryos will grow to maturity only if they are supplied with substances that the embryo needs to develop properly. The older an isolated embryo is, the better its chances for normal growth and survival will be. Embryo culture techniques are important in the development of new crops; they allow the development of plants with a much wider range of genotypes than would be possible if the embryos remained attached and developed normally, since the chances of survival for unusual genotypes are greater in culture.

Pollen Culture

In recent years, whole plants have been regenerated from pollen grains under certain circumstances. These plants, of course, have the same number of chromosomes as the pollen grains that give rise to them. If the sporophyte plant that produced them is diploid, then its pollen grains and the plants regenerated from them will be haploid (Figure 37-6). Such haploid sporophytes may be important in experimental plant breeding because of the direct expression of the genes that are present in that particular haploid set of chromosomes.

> **Haploid individuals can be obtained by culturing the pollen of some kinds of plants. Such haploid plants have a great potential importance for plant breeding, since all of their genes are expressed directly.**

These regeneration techniques have been applied successfully to over 200 kinds of plants since they were first developed in the mid-1960s. The stage of pollen development in the particular anther is critical in determining whether the pollen grain can be induced to develop into a new plant, but the importance of this factor seems to vary widely among species and even perhaps within a single species. The actual culture methods used, however, are similar to those used for embryo growth and for the growth of isolated plant cells or masses of tissue. The further refinement of these methods appears to hold great promise for plant breeding in the future.

FACTORS IN THE GROWTH MEDIUM
Experimental Studies

One way to learn more about the details of plant growth is to investigate the growth medium in which development takes place. Experimental results obtained throughout the twentieth century, some of which we have just reviewed, have made it abundantly clear that there are special chemical factors present in various growth media that play an important role in controlling plant development. The success of coconut milk in the regeneration of certain kinds of tissue caused it to come under investigation early. Eventually it was discovered that coconut milk is rich in reduced nitrogen compounds, such as amino acids, which are necessary for growth. In addition, coconut milk was found to contain cytokinins, growth hormones that are necessary for the differentiation of plant tissues.

FIGURE 37-6
A sporophyte. A haploid sporophyte of tobacco (*Nicotiana tabacum*) grown by culturing anthers on agar in an appropriate medium.

Additional studies have demonstrated, however, that other types of differentiated plant cells exhibit the ability to redifferentiate even when they are grown in a medium that lacks coconut milk. No one set of conditions is broadly effective, and to some extent each plant cell type is a special case. For example, Steward failed to regenerate whole potato plants, even when he used the apparently undifferentiated cells of potato tubers and exposed them to the same treatment as the carrot cells with which he had been successful earlier.

These results make it clear that plant development is not entirely molded by the environment in which the cells exist: genetically specified controls also play an important role. Control of the highly ordered progression of cell differentiation that occurs to produce a mature plant tissue is vested in the genes of the plant cells. The environment determines the ultimate outcome by triggering the expression of one particular developmental pattern from among several genetically determined possibilities.

Special chemical factors present in various growth media play an important role in controlling plant development, which occurs as a result of the interactions between genetic factors and the environment.

Regeneration in Nature

If the cues that trigger plant development come from the external environment, then they ought to be familiar stimuli that we see around us. Are everyday plants regenerating whole individuals from differentiated tissue out in the real world? Yes, indeed. This is particularly evident after injury or removal of tissue from plants. The common practice of using cuttings to produce mature plants reflects this property. At the base of a severed stem, adventitious roots may arise from mature pericycle tissue, which has divided to form a root meristem.

For reasons that are not well understood, adventitious roots seem to form easily in plants, whereas adventitious shoots form much less easily. Cuttings from plants such as beans or garden geraniums (*Pelargonium*) will form roots if they are simply left with their lower ends in water or in sand that is watered; an example is shown in Figure 37-7. In some plants, however, root formation occurs only with much difficulty, if at all. Stems that have leaves on them will form roots much more readily than those from which the leaves have been removed, apparently because the buds produce auxin, a plant hormone that stimulates root production in such situations.

In some kinds of plants, for example, succulents such as jade plants (see Figure 35-15, *B*), whole plants may be regenerated from bits of differentiated leaf tissue. The tiny plantlets that differentiate along the edge of the leaves of the familiar houseplant *Kalanchoë*—aptly called "mother-of-thousands"—provide an example of the way in which such vegetative propagation happens frequently in nature. As you saw in the last chapter, many plants may likewise differentiate whole new individuals from rhizomes, stolons, or horizontal roots. For example, colonies of aspen, *Populus tremuloides*, often consist of a single individual that has given rise to a whole genetically identical colony by producing adventitious buds on its roots (see Figure 35-30). Sometimes plants propagate by modification of their normal reproductive structures. In some century plants (species of *Agave*, for example), plantlets may form among the flowers. Seeds can also form that contain unfertilized embryos, a process that is genetically equivalent to the other kinds of asexual reproduction discussed here.

FIGURE 37-7
Adventitious roots. A leaf of African violet, *Saintpaulia*, forming adventitious roots and new plantlets at the base of the petiole.

Plant stems regularly produce adventitious roots under the appropriate circumstances; roots produce adventitious stems less frequently.

PLANT EMBRYONIC DEVELOPMENT

We will now consider the factors that control the normal development of a plant embryo, comparing its development with that of the sort of callus tissue differentiation that we have been discussing. The first stage in the development of a plant zygote is active cell division. The zygote divides repeatedly to form an organized mass of cells, the embryo. Within such an embryo, one can observe the initial differentiation of the

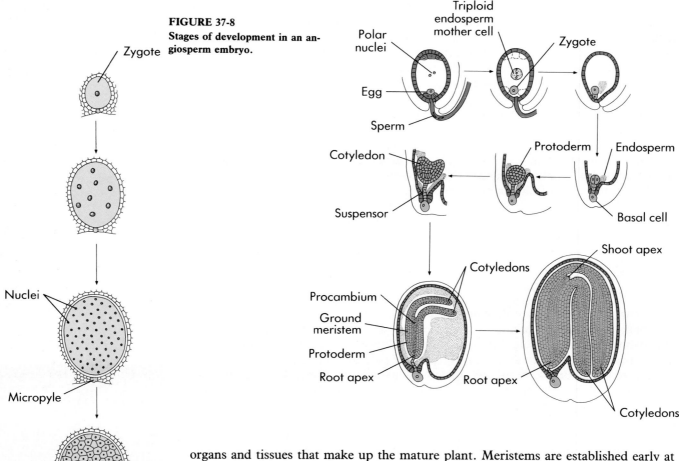

FIGURE 37-8
Stages of development in an angiosperm embryo.

Labels in figure: Zygote, Polar nuclei, Egg, Sperm, Triploid endosperm mother cell, Zygote, Protoderm, Endosperm, Cotyledon, Suspensor, Basal cell, Procambium, Ground meristem, Protoderm, Root apex, Cotyledons, Shoot apex, Root apex, Cotyledons

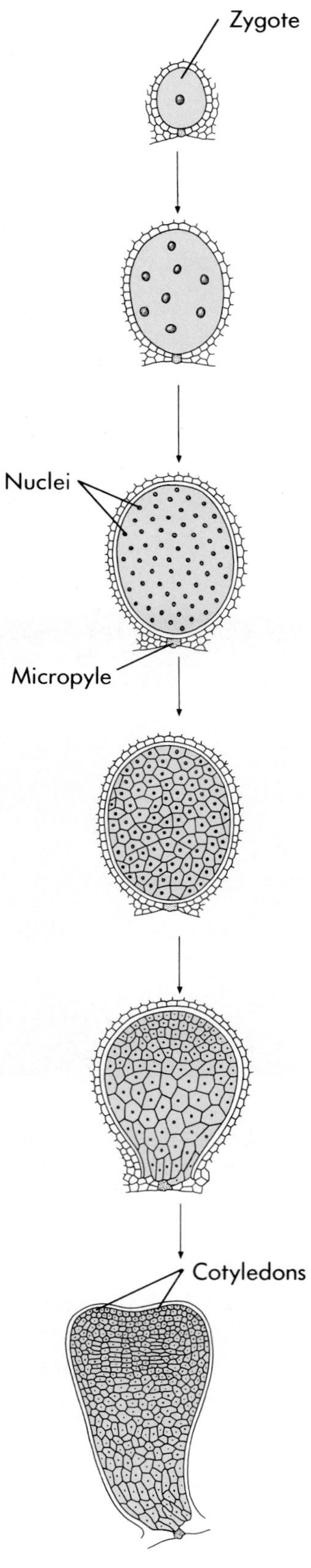

(Left column figure labels: Zygote, Nuclei, Micropyle, Cotyledons)

FIGURE 37-9
Diagrams of developing embryo in a gymnosperm.

organs and tissues that make up the mature plant. Meristems are established early at the shoot and root apices.

In angiosperms the differentiation of cell types in the embryo begins almost immediately after fertilization (Figure 37-8). The zygote first divides twice in one plane. After this division, the **basal cell,** which is the cell nearest the **micropyle**—the opening in the ovule through which the pollen tube entered—undergoes a series of further lateral divisions. As a result, it forms a narrow column of cells, the **suspensor.**

The other three cells into which the zygote first divided develop differently. After several days, they come to form a mass of about 40 ordered cells arranged in layers, like stacked bricks. By about day 5 of cell division, the principal tissue systems of the developing plant can be detected within this mass, and differentiation has begun. Soon thereafter, perhaps 6 days after the initiation of cell division, the root and shoot apical meristems can be detected as a zone of small, densely staining cells at the pole of the embryo opposite the suspensor. The shoot meristem grows upward and differentiates leaves and lateral branches; the root meristem grows downward, differentiating the various structures of the root. Both meristems continue to function throughout the life of the plant, so the shoot and root systems have a potentially unlimited pattern of growth. In addition, the lateral meristems—the vascular cambium and the cork cambium—make possible the plant's potentially unlimited increase in girth by secondary growth.

> A developing angiosperm embryo consists of a suspensor and a mass of layered cells. When about 40 of these cells are present, the axes of the shoot and root begin to become evident, and differentiation begins.

In gymnosperms (Figure 37-9) the zygote nucleus divides repeatedly after fertilization, but in many gymnosperms, no cell walls form initially between the daughter nuclei. After eight rounds of cell division, a single large embryonic cell contains about 256 nuclei. Cell walls then form, producing a mass of 256 cells of equal size. Differentiation then commences. The cells farthest from the micropyle divide faster than those nearer to it. This soon results in a marked gradient in cell size, since the rapidly dividing cells are smaller than the others. The larger cells near the micropyle develop into a suspensor. At the same time, the smaller cells on the opposite pole of the embryo, which constitute about a third of the total cell mass, give rise first to the apical meristem of the root and then to that of the shoot.

Thus, although the details are different, the same basic pattern can be seen in the embryogenesis of angiosperms and gymnosperms. A particularly important feature, and one that clearly separates plants from animals, is that cell movement does not occur in plants during the course of embryonic development: plant cells differentiate where they are formed. Their position in relation to other cells is important in determining how they differentiate; their specific course of development probably is determined in part by chemical gradients, although this has not been demonstrated fully.

The pattern of development of an animal zygote, as we will see in Chapter 55, is strongly affected by specific chemical signals that are present in the egg. In plants, however, this is not the case; instead, the pattern of embryo development in plants has been shown experimentally to be an expression of the genotype of the zygote itself. A plant embryo will develop normally even if it is removed from its ovule; this observation makes it clear that no chemical signals other than those the plant embryo received initially play a role in determining its subsequent pattern of development. An animal embryo, on the other hand, will not develop properly without chemical signals of just this kind. In plants, the function of the tissue in the egg and in the zygote that is formed when the egg is fertilized appears to be primarily a nutrient one. The developmental pattern of an individual plant is also influenced by its environment, which alters the concentrations and distributions of the hormones that direct the patterns of differentiation.

GERMINATION IN PLANTS
The Role of Seed Dormancy

Early in the development of an angiosperm embryo, a profoundly significant event occurs: the embryo simply stops developing! In many plants, the development of the embryo is arrested soon after apical meristems and the first leaves, or cotyledons, are differentiated. The integuments—the coats surrounding the embryo—develop into a relatively impermeable seed coat, which encloses the quiescent embryo, together with a source of stored food, within the seed. Seeds are important adaptively in at least three respects:

1. They permit plants to postpone development when conditions are unfavorable and to remain dormant until more advantageous conditions arise. Under marginal conditions, a plant can "afford" to have some seeds germinate, because others remain dormant.
2. By tying the reinitiation of development to environmental factors, seeds permit the course of embryo development to be synchronized with critical aspects of the plant's habitat, such as temperature and moisture.
3. Perhaps most important, the dispersal of seeds facilitates the migration and dispersal of genotypes into new habitats and also offers maximum protection to the young plant at its most vulnerable stage of development.

Once a seed coat is formed around the embryo, most of the embryo's metabolic activities cease; a mature seed contains only about 10% to 40% water. Under these conditions, the seed and the young plant within it are very stable; it is primarily the progressive and severe desiccation of the embryo and the associated reduction in metabolic activity that are responsible for its arrested growth. Germination (Figure 37-10) cannot take place until water and oxygen reach the embryo, a process that sometimes involves cracking the seed. Seeds of some plants have been known to remain viable for hundreds of years (Figure 37-11).

Specific adaptations often help to ensure that the plant will germinate only under appropriate conditions. Sometimes the seeds are held within tough fruits that themselves will not crack until, for example, they are exposed to the heat of a fire, a strategy that clearly results in the germination of a plant in an open, fire-cleared habitat;

FIGURE 37-10
Germination. The germination of a seed, such as this garden pea (*Pisum sativum*), involves the fracture of the seed coat, from which the epicotyl and hypocotyl emerge.

FIGURE 37-11
A seedling grown from seeds of lotus, *Nelumbo nucifera,* **recovered from the mud of a dry lake bed in Manchuria, northern China.** The radiocarbon age of this seed indicates that it was formed in about 1515; another seed that was germinated was estimated to be at least a century older.

nutrients will also be relatively abundant there, having been released from the plants burned up in the fire. The seeds of other plants will germinate only when inhibitory chemicals have been leached from their seed coats, thus guaranteeing their germination when sufficient water is available. Still other plants will germinate only after they pass through the intestines of birds or mammals or are regurgitated by them, which both weakens the seed coats and ensures the dispersal of the plants involved. Sometimes, under particular environmental circumstances, the seeds of plants thought to have become extinct in a particular area may germinate, and the plants may then reappear locally.

In terms of gene expression, the events underlying embryo dormancy are not complex. The encapsulation of the embryo within the seed coat denies it water and oxygen and simply causes its metabolism to run down. To reactivate metabolism, the embryo has only to be provided with water, oxygen, and a source of metabolic energy. The final release of arrested development, germination, is then cued to specific signals from the environment.

> **Seed dormancy is an important evolutionary factor in plants, ensuring their survival in unfavorable conditions and allowing them to germinate when the chances of survival for the young plants are the greatest.**

Germination

The first step in the germination of a seed occurs when it absorbs water. Because the seed is so dry at the start of germination, it takes up water with great force. Once this has occurred, metabolism within the seed resumes (Figure 37-12). Initially, metabolism may be anaerobic, but when the seed coat ruptures, oxidative metabolism takes over. At this point, it is important that oxygen be available to the developing embryo, because plants, like animals, require oxygen for active growth. Few plants produce seeds that germinate successfully underwater, although some, such as rice, have evolved a tolerance of anaerobic conditions (see boxed essay, p. 731).

A dormant seed, although it may have imbibed a full supply of water and may be respiring, synthesizing proteins and RNA, and apparently carrying on perfectly normal metabolism, may nonetheless fail to germinate without an additional signal from the environment. This signal may be light of the correct wavelengths and intensity (see Chapter 38), a series of cold days, or simply the passage of time at the appropriate temperatures for germination. The seeds of many plants will not germinate unless they have been **stratified**—held for a period of time at low temperatures. Stratification, which is also called *after-ripening*, prevents the seeds of many plants that grow in cold areas from germinating until they have passed the winter, thus protecting their seedlings from cold conditions.

Germination can occur over a wide temperature range (5° to 30° C), although it is generally optimum over a relatively narrow one, 25° to 30° C. Even under the best conditions, not all seeds will germinate. In some species, a significant fraction of seeds remain dormant, providing a genetic reservoir, or **seed pool,** of great evolutionary significance to the future of the plant population. One of the most interesting questions that has puzzled students of plant population biology is whether the genotypes of the seeds that germinate readily differ from those of the seeds that germinate later. This is a difficult question to investigate, but a highly significant one.

> **Because oxidative metabolism usually takes over soon after a plant embryo starts to grow, most seeds require oxygen for germination. Many seeds in a population may not germinate immediately, perhaps doing so when appropriate conditions occur later.**

The Mobilization of Reserves

Germination occurs when all internal and external requirements are met. Germination and early seedling growth require the mobilization of metabolic reserves stored in the starch grains of **amyloplasts** (chloroplasts that are specialized to store starch) and pro-

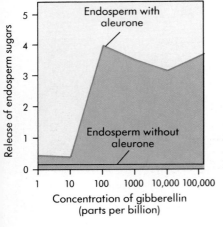

FIGURE 37-12
The release of sugar from endosperm is induced by gibberellin treatment. These data show that sugars are produced only when the aleurone layer is present. It is, in fact, the aleurone layer that is the source of the enzyme, α-amylase, which digests the starches stored in the endosperm.

tein bodies. Fats and oils also are important food reserves in some kinds of seeds. They can readily be digested during germination to produce glycerol and fatty acids, which yield energy through oxidative respiration and can also be converted to glucose. Any of these reserves may be stored in the embryo itself or in the endosperm, depending on the kind of plant.

In the cereal grains, the cotyledon is modified into an organ called the **scutellum**, a term that comes from the Latin word meaning "shield." The mobilization of the endosperm in these plants is not required during germination and does not occur then. The abundant food stored in the scutellum is mostly used up first. Later, while the seedling is becoming established, the scutellum absorbs the additional food that is stored in the endosperm, doing so by means of a two-stage process:

1. The initial mobilization of the starch in the endosperm is accomplished by hydrolases, which are secreted by the epithelial layer of the scutellum.
2. The later and more extensive mobilization of the starches in the endosperm is achieved by the secretion of amylase and other hydrolytic enzymes from the **aleurone layer,** a layer of specialized endosperm cells that lies just inside the seed coat. The synthesis and secretion of these aleurone hydrolases are controlled by a class of hormones called **gibberellins,** which are synthesized in the embryo (see Figure 37-12).

How are the genes that transcribe the enzymes involved in the mobilization of food resources activated? Experimental studies have shown that, in the endosperm of the cereal grains at least, this occurs when the gibberellins initiate a burst of messenger RNA (mRNA) and protein synthesis. It is not known whether the gibberellins act directly on the DNA or via chemical intermediates in the cytoplasm. DNA synthesis apparently does not occur during the early stages of seed germination but becomes important when the radicle, or embryonic root, has grown out of the seed coats.

During early germination and seed establishment, the vital mobilization of the food reserves stored in the embryo or the endosperm is mediated by hormones, which in at least some cases are gibberellins.

THE ROLE OF THE APICAL MERISTEMS

Once a seed has germinated, the plant's further development depends on the activities of the meristematic tissues, which themselves interact with the environment. As described earlier, the shoot and root apical meristems give rise to all of the other cells of the adult plant. It is interesting to study the way in which a plant's genetic programming interacts with environmental influences to determine its structure.

A series of simple experiments sheds light on this matter. First, one may ask whether the meristems are influenced by other parts of the plant or whether they function independently. To test this most directly, one can isolate the apical meristem so that it is free of influences from the other parts of the plant. In one experiment on the wood fern, *Dryopteris*, four deep incisions were cut into the tip of the stem, separating its apical meristem from the surrounding tissue (Figure 37-13). The isolated stem tip continued to function as a normal meristem, continuing to act by itself as an autonomous source of differentiated tissue.

The results of this experiment suggest that shoot apical meristems cut from plants ought to be able to develop normally in a medium containing the necessary nutrients, vitamins, and a source of energy. Indeed, this has proved to be the case. The apical meristem produces only shoots and leaves, however, and never roots. Complete, mature plants never develop, regardless of how long the experiment proceeds. Eventually the excised apical meristem will even cease producing additional leaves and stems.

Even transferring the apical meristem to fresh medium will not help. The apical meristems of plants produce a hormone called **auxin** that continually diffuses back from the shoot apex and suppresses the active growth of the lateral branches. Auxins act largely by promoting cell elongation, apparently by affecting the plasticity of the cell walls. For them to work properly in normal cell division, however, another group

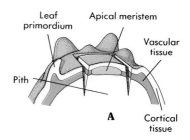

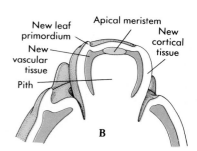

FIGURE 37-13
Surgical isolation of the apical meristem of the wood fern *Dryopteris dilatata*.
A A view of the shoot tip, showing the position of the four incisions that isolated the terminal meristem from the surrounding leaf primordia and differentiating tissues (×30).
B Longitudinal section of a surgically isolated meristem, showing the tissues that have been severed by the incisions, and the intact pith connection below the meristem (×70).

FIGURE 37-14

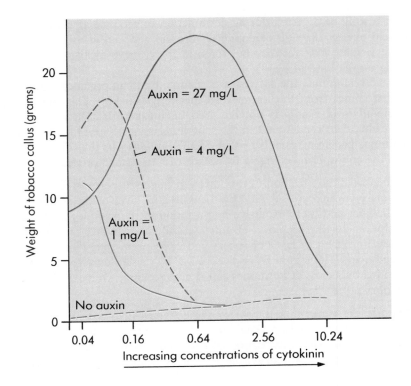

Auxin-cytokinin interactions. Cytokinin alone has little effect on the growth of tobacco callus in tissue culture. Auxin alone, regardless of the concentration used, causes the culture to grow to a weight of about 10 grams. When both hormones are present, growth is greatly increased. Note, however, that when optimum concentrations are exceeded, the growth responses cease.

of hormones, the **cytokinins,** also must be involved. Cytokinins are also necessary to cause masses of plant tissue being grown in a culture to differentiate and produce roots. When shoots are grown experimentally in a medium, they produce auxin themselves but not cytokinins, which must be added for normal growth and development (Figure 37-14). When the proper concentration of hormones is applied, complete little plants may develop.

You will encounter auxins repeatedly in your study of plant development. They have simple structures, similar to the amino acid tryptophan. The naturally occurring auxin, **indoleacetic acid** (IAA), is critically important in determining the patterns of plant growth. Some of the synthetic auxins, which act in a similar way, have wide commercial use, as shown in the next chapter.

The Autonomy of Apical Meristems

The experiments described above, taken as a whole, offer convincing evidence that apical meristems can function autonomously. Their activities are influenced by the other parts of the plant only in generalized ways, such as by supplying nutrients and hormones. The apical meristem itself is maintained as a coherent unit because its cells somehow act to inhibit the differentiation of the surrounding meristematic cells. As a result, new meristems normally do not grow out of an existing one. This relationship must reflect the presence of an inhibitor, which itself is presumably another plant hormone.

We have already noted the phenomenon of **apical dominance,** in which auxin, diffused backward from the apical meristem of the stem, inhibits the active growth of the lateral buds. The auxin may do so by stimulating the production of the hormone **ethylene** ($H_2C{=}CH_2$). Ethylene apparently is produced around the nodes of the stem and acts to inhibit differentiation of the surrounding cells, thus preventing lateral bud formation. It might also act in a similar way to maintain the integrity of the apical meristem of the stem. Cytokinins may likewise be involved in this complex system. By promoting cell division, they allow the development of lateral branches in some circumstances, even in the presence of auxin.

Leaf Primordia: Irreversibly Determined?

Finally, let us explore the degree to which the leaf primordia and buds that the apical meristem produces are irreversibly determined to be leaf primordia and buds at the time they are produced. Initially these structures appear as simple cellular outgrowths at the base of the shoot meristem. How are the major differences in leaf form among, say, a fern, a pine, and an oak determined? Are they under genetic control, or are they set by environmental demands? Since the primordia of these different kinds of leaves look similar at first, the question may be restated another way: can structures in plants be determined before they differentiate?

This has been tested directly in ferns, which tend to have large leaves with complex patterns of division (see Figures 33-1,C). The leaf primordia of ferns have been isolated in culture medium at various stages of development, and the ability of these different isolates to produce mature leaves has been evaluated. The results of this experiment have been unequivocal: when the proper nutrients, vitamins, and hormones have been present in the medium, normal fern leaves have differentiated regardless of the stage at which the leaf primordia were isolated (Figure 37-15). No additional specific factor or factors from the original plant are necessary for the primordia to develop properly.

The leaf primordium is an independent developmental subsystem. By the time it is differentiated, its ultimate "fate" has already been determined by the meristem on which it was produced. When it is cut off from that meristem, it proceeds along its own predetermined developmental path. The executive decision is made by the meristem ("What tissue will this cell be?"), and the operational decisions are left to be expressed by the tissue itself ("When will it differentiate, and into what kind of leaf?").

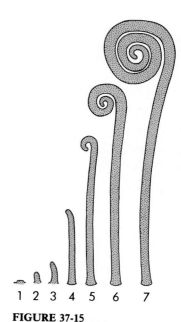

FIGURE 37-15

Fern leaf development. Development of a frond of the cinnamon fern, *Osmunda cinnamomea*, excised from the plant at the end of the third growing season and grown in sterile culture to maturity. The normal course of frond development, from its inception to the end of the fourth growing season: (*1*) soon after inception, (*2*) end of the first season, (*3*) end of the second season, (*4*) end of the third season, (*5* to *7*) developmental stages during the fourth season. The coiled form of the frond develops entirely during the fourth season.

THE VASCULAR CAMBIUM: A DIFFERENTIATED MERISTEM

The apical meristems of plants retain the ability to proliferate indefinitely. As they do so, they produce cells that are differentiated in various ways and to various degrees. Some of these cells give rise to the vascular cambium, a permanently meristematic tissue. The ways in which the vascular cambium produces different kinds of cells inwardly and outwardly apparently involve both (1) the identity of the cells that are immediately adjacent and (2) differences in physical factors, such as pressure, to which the derivatives of the vascular cambium are subjected. In this sense, the vascular cambium is partly differentiated, although permanently meristematic.

The vascular cambium is a differentiated, permanently meristematic tissue.

In this way the vascular cambium represents an intermediate stage of differentiation. A developmental commitment to produce vascular tissue, which comprises a variety of different cell types, has already been made. At the same time, the proliferation of the meristematic tissue itself continues; the subsequent commitment of individual derivative cells to xylem, phloem, or ray cells has yet to be made. It is convenient to visualize a series of discrete stages of differentiation (Figure 37-16), each of which represents an increased level of developmental commitment, with each stage functioning more or less autonomously.

The ways in which the processes we have discussed in this chapter are integrated during the course of development in the whole plant involve several kinds of hormones, some of which we have already mentioned. These hormones strongly influence one another as they affect plant growth, and the form of the plant that results from their action is also strongly influenced by the environment in which that plant is growing. These interactions, and the ways in which plant growth is regulated, will form the subject of the next chapter.

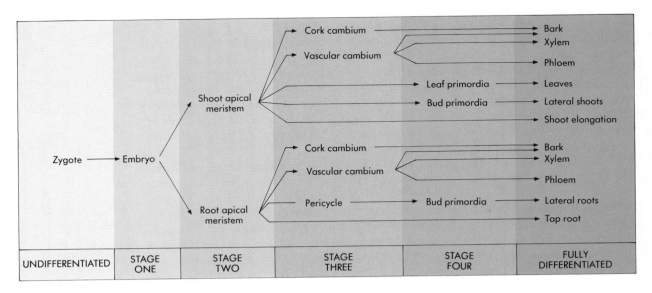

FIGURE 37-16
Stages in plant differentiation.

SUMMARY

1. Animals undergo development according to a fixed blueprint that is followed rigidly until they are mature. Plants, in contrast, develop continuously. The course of their development is mediated by hormones, which are produced as a result of interactions with the external environment.

2. Almost all of the differentiation in plants is fully reversible; plant cells are totipotent as long as they retain a living protoplast and a nucleus.

3. Whole plants can be regenerated from cultures of single cells, provided that these cells are kept in fairly close contact with others. If they are not, they usually tend to lose the factors that make it possible for them to divide.

4. Embryo development in animals involves extensive movements of cells in relation to one another, but the same process in plants consists of an orderly production of cells, rigidly bound by their cellulose-rich cell wall. The cells do not move in relation to one another in plant development, as they do in animal development. By the time about 40 cells have been produced in an angiosperm embryo, differentiation begins; the meristematic shoot and root apices are evident.

5. In most gymnosperms, cell walls do not form in the first part of the developmental process. After about 256 nuclei have been produced by succes sive mitotic divisions within a common cytoplasm, cell walls are formed and differentiation begins.

6. Seeds are held within a rigid, relatively impermeable seed coat, which may crack only when it receives appropriate environmental cues. These may include the quality and periodicity of light, cold temperatures, "ripening" (the passage of time), or mechanical or chemical cracking. All of these cues tend to concentrate seed germination at the times when the young plants are most likely to become established.

7. In the germination of seeds, the mobilization of the food reserves stored in the cotyledons and in the endosperm is critical. In the cereal grains, this process is mediated by hormones of the kind known as gibberellins, which appear to activate transcription of the loci involved in the production of amylase and other hydrolase enzymes.

8. Apical meristems will develop normally into whole plants when they are excised, provided that cytokinins, which stimulate root production and are not produced in the apical meristems or in the shoot, are added. Leaf primordia also are capable of development independent of the plant on which they are formed, starting with the point at which they are first visible as independent structures.

REVIEW QUESTIONS

1. How do the developmental control systems of animals and plants compare to one another? How is each affected by changes in the environment? Why must a plant be sensitive to the conditions of the environment in which it lives?

2. How reversible is the differentiation of most plant cells? Explain. What are the limitations of this ability? What term defines this ability?

3. How did Steward experimentally confirm the concept of plant cell totipotency? What are embryoids? Do they generally form directly from isolated plant tissue? Explain.

4. What type of structure develops from cuttings of plant stems that have been immersed in water or wet sand? Will such growths occur more readily if the leaves are left on or if they are removed? Why? What structures develop from underground horizontal roots? Is it more likely that a complete plant will regenerate from stem or from root tissue?

5. How does early development in a gymnosperm differ from that in an angiosperm? When and how does cell differentiation occur in gymnosperms?

6. In many plants, when does development of the embryo come to a temporary halt? What seed structures are present at this point to protect the arrested embryo? Why are seeds adaptively important?

7. Why may a seed showing proper respiration, protein, and nucleic acid synthesis and all other normal metabolisms still fail to germinate? What is stratification as it relates to seed germination? What purpose does it serve the seed?

8. Into what specialized structure is the cotyledon of a cereal grain modified? How does it regulate the use of stored food by the germinating seedling? What enzymes are secreted by the epithelial layer of this structure?

9. What enzymes are secreted by the layer of cells just inside the seed coat? What name is given to this layer? What hormone controls the synthesis and secretion of these enzymes? What is the direct biochemical effect of the application of this hormone in cereal grains?

10. From what meristematic tissues are all cells of the adult plant derived? What plant structures are formed in an excised apical meristem further supplied with all vital nutrients? Is it capable of producing a complete, mature plant? Why or why not?

11. Into what class of hormones is indoleacetic acid placed? Is it a natural or synthetic hormone? What are two of its most basic functions in terms of plant growth? What is apical dominance? What hormones are associated with this phenomenon? How might they work?

12. Is the development of the leaf primordium dependent upon or independent of the presence of signals from its plant? Given proper nutritional requirements will an excised leaf primordium develop into a normal leaf? Why or why not?

 THOUGHT QUESTIONS

1. Why is plant development so much more closely linked with environmental cues than animal development is?

2. If you take parenchyma cells from the stem of a plant and put them in a culture medium, what will that medium have to contain in order for the cells to develop into whole plants?

3. What experiments would you suggest for studying the differentiation of individual plant cells? How could you begin to understand the factors involved in their assumption of mature forms?

 FOR FURTHER READING

COOK, R.E.: "Clonal Plant Populations," *American Scientist*, vol. 71, pages 244-253, 1983. Excellent discussion on the role of asexual reproduction among plants in natural populations.

DALE, J.E., and F.L. MILTHORPE (editors): *The Growth and Functioning of Leaves*, Cambridge University Press, New York, 1983. An excellent series of articles on all aspects of leaf development and functioning.

MAYER, A., and A. POLJAKOFF: *The Germination of Seeds*, ed. 2, Pergamon Press, Inc., Elmsford, N.Y., 1975. Excellent and clearly written book on seed biology.

SHEPARD, J.F.: "The Regeneration of Potato Plants from Leaf-Cell Protoplasts," *Scientific American*, May 1982, pages 112-121. Provides valuable insight into the way in which experiments in this area are designed.

WAREING, P.F., and I.D.J. PHILLIPS: *Growth and Differentiation in Plants*, ed. 3, Pergamon Press, Ltd., Oxford, England, 1981. Outstanding review of plant development and differentiation.

WILKINS, M.B. (editor): *Advanced Plant Physiology*, Longman, Inc., White Plains, N.Y., 1984. This collection of essays includes excellent accounts of the areas covered in this chapter.

38

Regulation of Plant Growth

CHAPTER OUTLINE

OVERVIEW

The major classes of plant hormones—auxins, cytokinins, gibberellins, ethylene, and abscisic acid—interact in complex ways to produce a mature, growing plant. Unlike the highly specific hormones of animals, plant hormones are not produced in definite organs nor do they have definite target areas. They stimulate or inhibit growth in response to environmental cues such as light, day length, temperature, touch, and gravity and thus allow plants to respond efficiently to environmental demands by growing in specific directions, producing flowers, or displaying other responses appropriate to their survival in a particular habitat.

FOR REVIEW

Here are some important terms and concepts that you will encounter in this chapter. If you are not familiar with them, you should review them before proceeding.

Turgor pressure (Chapter 6)

Flowering plant life cycle (Chapter 33)

Apical meristems (Chapter 35)

Seed germination (Chapter 35)

Auxin and plant growth (Chapter 37)

Cytokinins and differentiation (Chapter 37)

As we emphasized in the last chapter, plants respond to their environment by growing. Unlike animals, plants cannot move from place to place to seek more favorable circumstances, except by the very slow process of growth, or by reproduction and seed or spore disposal. We also introduced plant hormones and how they are stimulated by external factors and interact with one another during plant growth. In this chapter we will consider each of the major classes of plant hormones in greater detail. We will then explore some of the ways in which external factors affect plant growth and thus the appearance of individual plants.

PLANT HORMONES

Hormones are chemical substances that are produced in small, often minute, quantities in one part of an organism and then are transported to another part of the organism, where they bring about physiological responses. The activity of hormones results from their ability to stimulate certain physiological processes and to inhibit others. How they act in a particular instance is influenced both by what the hormones themselves are and by how they affect the particular tissue that receives their message.

In animals, hormones usually are produced at definite sites, normally in organs that are solely concerned with hormone production. In plants, on the other hand, hormones are produced in tissues that are not specialized for that purpose but that carry out other, usually more obvious, functions as well. There are at least five major kinds of hormones in plants: auxin, cytokinins, gibberellins, ethylene, and abscisic acid (Table 38-1). Other kinds of plant hormones certainly exist, but they are less

TABLE 38-1   FUNCTIONS OF THE MAJOR PLANT HORMONES

	HORMONE	MAJOR FUNCTIONS	WHERE PRODUCED OR FOUND IN PLANT
	Auxin (IAA)	Stem elongation: promotion of vascular tissue growth; suppression of lateral buds	Shoot apical meristems
	Cytokinins	With auxin, stimulation of cell division and determination of course of differentiation	Produced in roots and transported from there
	Gibberellins (GA$_1$)	Stem and internode elongation; mobilization of enzymes during seed germinaton	Apical portions of roots and shoots
	Ethylene	Controls of abscission of leaves, flowers, fruits; retardation of lateral bud elongation; hastening of fruit ripening	Leaves, stems, young fruits
	Abscisic acid	Suppression of bud growth; important role in stomatal opening; promotion of leaf senescence	Mature leaves, fruits, root caps

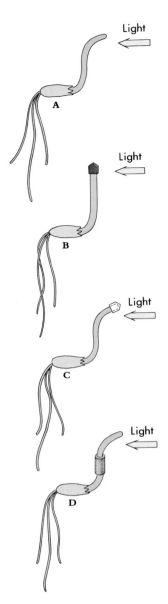

FIGURE 38-1
The Darwins' experiment. Young grass seedlings normally bend toward the light (**A**). When the tip of a seedling was covered by a lightproof collar (**B**) (but not when it was covered by a transparent one) (**C**), this bending did not occur. When the collar was placed below the tip (**D**), the characteristic light response took place. From these experiments, the Darwins concluded that, in response to light, an "influence" that caused bending was transmitted from the tip of the seedling to the area below the tip, where bending normally occurs.

well understood. The study of plant hormones, especially our attempts to understand how they produce their effects, is an active and important field of current research.

Five major kinds of plant hormones are reasonably well understood: auxins, cytokinins, gibberellins, ethylene, and abscisic acid.

Auxins

In the previous chapter, we mentioned auxin with respect to its ability to control the growth of lateral buds on the stem. Because the effects of auxin on grass seedlings are so evident, auxin was the first plant hormone to be discovered and studied more than a century ago. Related compounds that have similar modes of action are classified as auxins and are also the most important commercially. Although the rate of stem elongation in mature flowering plants is regulated largely by gibberellins, auxin regulates stem elongation in young grass seedlings and in herbs generally.

Discovery of Auxin. In his later years, the great evolutionist Charles Darwin became increasingly devoted to the study of plants. In 1881 he and his son Francis published a book called *The Power of Movement in Plants*. In this book, the Darwins reported their systematic experiments concerning the way in which growing plants bend toward light, a phenomenon known as **phototropism**. They made many observations in this field and also conducted experiments using young grass seedlings.

Charles and Francis Darwin found that the seedlings they were studying normally bent strongly toward a source of light, if the light came primarily from one side. However, if they covered the upper part of a seedling with a cylinder of metal foil, so that no light reached its tip, the shoot would not bend (Figure 38-1). The Darwins obtained this result even though the region where the bending normally occurred was still exposed. Light reached this part of the seedling directly, but bending did not occur. However, if they covered the end of the shoot with a gelatin cap, which transmitted light, the shoot would bend as if it were not covered at all.

In explaining this unexpected finding, the Darwins hypothesized that when the shoots were illuminated from one side, an "influence" arose in the uppermost part of the shoot, was transmitted downward, and caused the shoot to bend. For some 30 years, the Darwins' perceptive experiments remained the sole source of information about this interesting phenomenon. Then a series of experiments was performed by several other botanists. Some of the most significant were those conducted independently by the Danish plant physiologist Peter Boysen-Jensen and the Hungarian plant physiologist Arpad Paal, who demonstrated that the substance that caused the shoots to bend was a chemical. They showed that if one cut off the tip of a grass seedling and then replaced it but separated it from the rest of the seedling by a block of agar, the seedling would react as if there had been no change. Something evidently was passing from the tip of the seedling through the agar into the region where the bending occurred. On the basis of these observations, Paal suggested that an unknown substance, under conditions of uniform illumination or of darkness, continually moves down the grass seedlings from their tips and promotes growth on all sides. Such a pattern would not, of course, cause the shoot to bend.

The next telling results in this area were obtained by Frits Went, a Dutch plant physiologist, in the course of studies for his doctoral dissertation and were published in 1926. Carrying Paal's experiments an important step further, Went cut off the tips of grass seedlings that had been illuminated normally and set these tips on agar. He then took grass seedlings that had been grown in the dark and cut off their tips in a similar way. Finally, Went cut tiny blocks from the agar on which the tips of the light-grown seedlings had been placed and put them on the tops of the decapitated dark-grown seedlings but set off to one side (Figure 38-2). Even though these seedlings had not been exposed to the light themselves, they bent *away* from the side on which the agar blocks were placed.

Went then observed the seedlings on which untreated agar blocks had been placed. He put blocks of pure agar on the decapitated stem tips and noted either no effect or a slight bending *toward* the side where the agar blocks were placed. Finally,

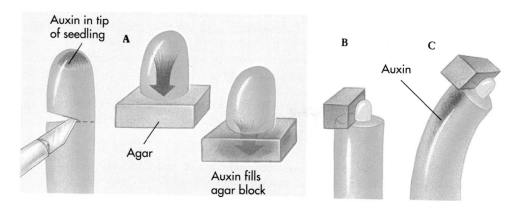

Auxin in tip
of seedling

A

Agar

Auxin fills
agar block

B

C

Auxin

FIGURE 38-2

Frits Went's experiments.

A Went removed the tips of the grass seedlings and put them on agar, a gelatinous inert substance. **B** Blocks of the agar were then put on one side of the ends of other grass seedlings from which the tips had been removed. **C** The seedlings bent away from the side on which the agar block was placed. Went concluded that the substance that he named auxin promoted the elongation of the cells and that it accumulated on the side of a grass seedling away from the light.

Went cut sections out of the lower portions of the light-grown seedlings to see whether the active principle was present in them. He placed these sections on the tips of decapitated, dark-grown grass seedlings and again observed no effect.

As a result of his experiments, Went was able to show that the substance that had flowed into the agar from the tips of the light-grown grass seedlings could make seedlings curve when they otherwise would have remained straight. In other words, it enhanced rather than retarded cell elongation. Went also showed that this chemical messenger caused the tissues on the side of the seedling into which it flowed to grow more than those on the opposite side. He named the substance that he had discovered **auxin,** from the Greek word *auxein,* which means "to increase."

Went's experiments provided a basis for understanding the responses that the Darwins had obtained some 45 years earlier. The grass seedlings bent toward the light because the auxin contents on the two sides of the shoot differed. The side of the shoot that was in the shade had more auxin, and its cells therefore elongated more than those on the lighted side, bending the plant toward the light.

Thus auxin acts to fit the form of the plant to its environment in a highly advantageous way. It is a "releaser" of growth and elongation, and the means by which the plant is able to respond to its environment; signals received from the environment influence the distribution of auxin in the plant directly. How does the environment exert this influence? It might theoretically destroy the auxin, or decrease the cell's sensitivity to auxin, or cause the auxin molecules to migrate away from the light into the shaded portion of the shoot. This last possibility has since proved to be the case.

In a simple but effective experiment, Winslow Briggs, of the Department of Plant Biology of the Carnegie Institution of Washington, inserted a thin sheet of mica vertically between the half of the shoot oriented toward the light and the half oriented away from it (Figure 38-3). He found that lateral light does not cause a plant into which such a glass barrier has been inserted to bend. When Briggs examined the illuminated plant itself, he found that the auxin levels in the light and dark sides of the barrier were equal. He concluded that auxin in normal plants migrates from the light side to the dark side in response to light. The presence of this glass barrier, by preventing the migration, blocked the light response.

FIGURE 38-3

Phototropism and auxin: diagrams of experiments originally performed by Winslow Briggs. Experiments **A** and **B,** performed in the dark, showed that splitting the tip of the seedling leaf and inserting a barrier did not significantly affect the total amount of auxin that was diffused from its tip. The amount of curvature produced, which is related to the amount of auxin produced, is shown by the numbers below the agar block. The other experiments (**C** to **F**) were performed with light coming from the right side, as indicated by the shading. A comparison of **C** and **D** with **A** and **B** shows that auxin production is not dependent on light. The slight differences in curvature shown are not significant. If a barrier was inserted in the agar block (**E**), light caused the displacement of the auxin away from the light. Finally, experiment **F** showed that it was displacement that had occurred and not different rates of auxin production on the dark and light sides, because when displacement was prevented with a barrier, auxin production was not significantly different in the two sides. The exact nature of the auxin migration process is not known, but it is thought to involve a light-sensitive pigment that perhaps alters membrane permeability to auxin. Only light of wavelengths less than 500 nanometers (blue light) can promote the lateral migration of auxin.

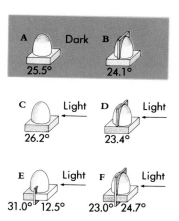

A Dark B
25.5° 24.1°

C Light D Light
26.2° 23.4°

E Light F Light
31.0° 12.5° 23.0° 24.7°

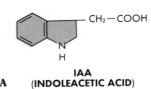

A
IAA
(INDOLEACETIC ACID)

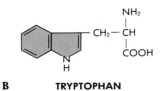

B   **TRYPTOPHAN**

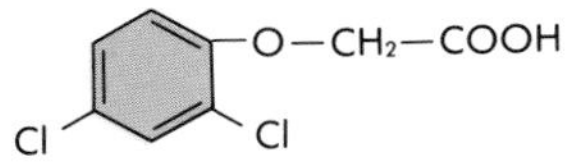

C DICHLOROPHENOXYACETIC ACID
(2, 4-D)

FIGURE 38-4
Auxins.
A Indoleacetic acid (IAA), the
only known naturally occurring
auxin.
B Tryptophan, the amino acid
from which plants synthesize IAA.
C Dichlorophenoxyacetic acid
(2,4-D), a synthetic auxin, is a
widely used herbicide.

Subsequent experiments have revealed the chemical structure of what seems to be the only naturally occurring auxin, **indoleacetic acid (IAA)** (Figure 38-4, *A*). These experiments have also yielded a great deal of information about how auxins act in plants. The term *auxin* is now used to refer both to the naturally occurring substance and to those related synthetic molecules that produce similar effects. As we saw in the last chapter, IAA is produced at the shoot apex, in the region of the apical meristem, and diffuses continuously downward, suppressing the growth of lateral buds. Auxin resembles the amino acid tryptophan (Figure 38-4, *B*) in structure, and it is in fact synthesized from tryptophan by plants.

The only known naturally occurring auxin, indoleacetic acid (IAA), is synthesized in apical meristems of shoots. Auxin diffuses down stems and suppresses the growth of lateral buds as it goes. In young grass seedlings and other herbs, it causes stems to bend toward light by migrating toward the darker side and causing the cells there to elongate.

Auxin and Plant Growth.   Auxin increases the plasticity of the plant cell wall. A more plastic wall will stretch more during active cell growth, while its protoplast is swelling. Since very low concentrations of auxin promote cell wall plasticity, the hormone must be broken down rapidly to prevent its accumulation. Plants do this by means of the enzyme **indoleacetic acid oxidase**. By controlling the levels of both IAA and IAA oxidase, plants can regulate their growth very precisely. Using methods involving monoclonal antibodies (see Chapter 18), scientists have recently been able to locate the transport sites involved in the movement of auxin downward from the shoot apex through the plant. These sites are in the plasma membranes near the basal ends of the cells in which the auxins are promoting elongation. The further application of such methods will certainly help to clarify other properties of the auxin-mediated process of cell elongation.

The speed with which auxin increases cell plasticity—grass seedlings may begin to bend within 3 minutes of its application—has made it difficult to determine the chemical basis of these reactions. It seems highly unlikely that these rapid-response effects could be caused by changes in the rates of transcription or translation of genes. To act so rapidly, auxin must affect some system that already exists. Extensive changes have been reported in the polysaccharides of plant cell walls, involving the formation and breaking of covalent bonds, in response to treatment with auxin. An increase in the concentration of H^+ ions in the wall is also very likely to occur. In addition, auxin appears to mediate the stimulation of mRNA transcription, which would lead to long-term growth changes.

Auxin also promotes the growth of vascular tissue in stems and the growth of the vascular cambium itself. It likewise increases fruit growth, an effect that has important commercial implications. Fruits normally will not develop if fertilization has not occurred and seeds are not present, but they often will do so if auxins are applied. Apparently auxins, which are present in pollen in large quantities, play a key role in making the development of a mature fruit possible. Synthetic auxins are used commercially for the same purpose.

Auxin controls various different plant responses in addition to those involved in phototropism. One of these is the suppression of lateral bud growth, an effect we have mentioned earlier. How can auxin, a growth promoter, also inhibit growth? Apparently the cells around lateral buds, in the nodal regions of the stem, produce ethylene under the influence of auxin. The ethylene in turn inhibits the growth of the lateral buds. When the terminal bud on a stem is removed and the lateral buds grow, bushy plants are produced and the number of flowers on an individual plant is increased. Similar effects may be achieved chemically or by selectively breeding individuals that have such characteristics.

Auxin stimulates the elongation of cells by making their walls more plastic. By interacting with other hormones, auxin also suppresses the growth of lateral buds and stimulates activity of the vascular cambium, which promotes an increase in girth.

Synthetic Auxins. Synthetic auxins have many uses in agriculture and horticulture. One of their most important uses is based on their prevention of **abscission,** the process by which a leaf or other organ falls from a plant. Before abscission occurs, a **separation layer** or *abscission layer* forms at the base of the organ. Such a layer consists of a thin layer of often small cells oriented at right angles to the stalk of the particular structure. The walls of the cells in the separation layer become soft and gelatinous, so that the leaf, flower, or fruit readily breaks off at that point and falls from the plant when the wind moves it.

Before a leaf, flower, or fruit is ready to fall, it becomes **senescent;** its cells begin to die, and the organ often turns yellowish or brownish. In this process the slowdown or cessation in the production of auxin by the dying leaf is critical. Leaf fall often can be retarded, therefore, by applying auxin to the leaf blade. Synthetic auxins are also used to prevent fruit drop in apples before they are ripe, to hold the berries on holly that is being prepared for shipping, to promote flowering and fruiting in pineapples, and to induce the formation of roots in cuttings.

Further, synthetic auxins are routinely used to control weeds. When they are used as herbicides, they are applied in higher concentrations than those at which IAA normally occurs in plants. One of the most important of the synthetic auxins used in this way is 2,4-dichlorophenoxyacetic acid, usually known as 2,4-D (Figure 38-4, *C*), which effectively kills weeds in lawns, selectively eliminating broad-leaved dicots. The weeds literally grow to death, their stems elongating from a few centimeters to several decimeters within days.

A molecule closely related to 2,4-D is the herbicide 2,4,5-trichlorophenoxyacetic acid, better known as 2,4,5-T, which has long been widely used to kill woody seedlings and weeds. Notorious as a component of "Agent Orange" of the Vietnam War, 2,4,5-T is easily contaminated with a by-product of its manufacture, 2,3,7,8-tetrachlorodibenzo-para-dioxin. This substance, known as **dioxin** for short, is apparently toxic to people, although its effects are poorly documented. Dioxin is the subject of great environmental concern in the United States and elsewhere.

Cytokinins

Another group of naturally occurring growth hormones in plants is the **cytokinins.** Cytokinins were discovered because of their role in promoting the differentiation of organs in masses of plant tissue growing in culture. Studies by Haberlandt around the turn of the century demonstrated the existence of a chemical that would induce the parenchyma tissue of potatoes to become meristematic. Subsequent studies have focused on the role of cytokinins in bringing about the differentiation of callus tissue.

A cytokinin is a plant hormone that, in combination with auxin, stimulates cell division in plants and determines the course of differentiation (see Table 38-1). Substances with these properties are widespread, both in bacteria and in other eukaryotes. In vascular plants, most cytokinins seem to be produced in the roots and are then transported throughout the rest of the plant. Developing fruits apparently are also important sites of cytokinin synthesis. In mosses, cytokinins cause the formation of vegetative buds on the protonemata. In all plants, cytokinins seem to have been incorporated into the regulation of growth patterns, working with other hormones.

The naturally occurring cytokinins all appear to be derivatives of the purine base adenine (Figure 38-5). Other molecules, not known to occur naturally, have similar effects but are very diverse chemically. In contrast to auxins, cytokinins *promote* the growth of lateral branches. Similarly, auxins promote the formation of lateral roots, whereas cytokinins inhibit their formation. As a consequence of these relationships, the balance between cytokinins and auxin determines, among other factors, the appearance of a mature plant. In addition, the application of cytokinins prevents the yellowing of leaves that are detached from the plant.

> **Cytokinins are plant hormones that, in combination with auxin, stimulate cell division and determine the course of differentiation. In contrast to auxins, they stimulate the growth of lateral branches. The naturally occurring cytokinins are derivatives of the purine base adenine.**

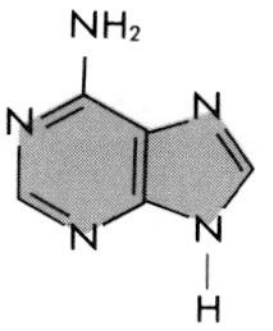

ADENINE

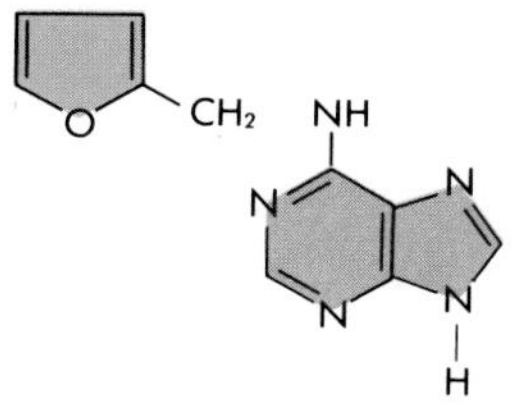

KINETIN

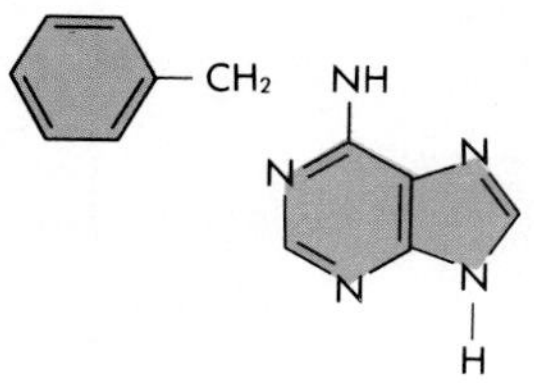

6-BENZYLAMINO PURINE (BAP)

FIGURE 38-5

Some cytokinins. Two commonly used synthetic cytokinins: kinetin and 6-benzylamino purine. Note their resemblance to the purine base adenine, also shown.

The action of cytokinins, like that of other hormones, has been studied in terms of its effects on the growth and differentiation of masses of tissue growing in defined media. Cytokinins seem to be necessary for mitosis and cell division to take place. They apparently work by influencing the synthesis or activation of proteins that are specifically required for mitosis.

Gibberellins

FIGURE 38-6
A genetically dwarf variety of tomato (*Lycopersicon esculentum*). When treated with gibberellin, the plants at right attained the normal stature.

Gibberellins are named for the fungus *Gibberella*, which produces a disease of rice causing the plants to grow abnormally tall. This "foolish seedling" disease of rice was investigated in the 1920s by the Japanese plant pathologist Kurosawa, who grew the fungus in culture and obtained a substance from the fungal filtrate itself that would mimic the "foolish seedling" disease when added to rice. This substance was isolated by Japanese chemists in 1939 and identified by British chemists in 1954. Although such chemicals were first thought to be only a curiosity, they have since turned out to belong to a large class of naturally occurring plant hormones called the **gibberellins** (see Table 38-1).

Synthesized in the apical portions of both stems and roots, gibberellins have important effects on stem elongation in plants and play the leading role in controlling this process in mature trees and shrubs. In these plants, the application of gibberellins characteristically promotes internode elongation, and this effect is enhanced if auxins are present also. Many dwarf mutants of plants are known in which normal growth and development can be restored if gibberellins are applied (Figure 38-6). It is supposed that such mutants lack naturally occurring gibberellins, at least in sufficient quantities to promote normal growth.

Gibberellins are also involved with many other aspects of plant growth. In the last chapter, for example, we examined their role in stimulating production of amylase and other hydrolytic enzymes that are mobilized during the germination and establishment of cereal seedlings. Biennial plants often can be induced by the application of gibberellins to grow out of the rosette stage and to flower (Figure 38-7). These hormones also hasten seed germination, apparently because they can substitute for the effects of cold or sometimes light requirements in this process.

> **Gibberellins are an important class of plant hormones that are produced in the apical regions of shoots and roots. They play the major role in controlling stem elongation for most plants, acting in concert with auxin and other hormones.**

According to recent analyses carried out by B.O. Phinney of the University of California, Los Angeles, probably only one kind of gibberellin, GA_1, is active in the control of shoot elongation in most plants. The natural functions of the other gibberellins that occur widely both in bacteria and in eukaryotes other than flowering plants are not known, although a number of them function like GA_1 if they are applied to flowering plants. Among the first vascular plants, GA_1 seems to have acquired the properties of a plant hormone; it is probably characteristic of all vascular plants.

Like all other plant hormones, the gibberellins produce their effects largely on the basis of their interactions with other hormones. We still have a great deal to learn about these interactions. Gibberellins are enjoying increased commercial use and doubtlessly will become even more important in the future. They are particularly significant in increasing fruit size and set and cluster size in grapes. They are used to delay the ripening of citrus fruits on the trees and to speed up the flowering of strawberries. The most important commercial use of gibberellins is to stimulate the partial digestion of starches in germinating barley during the process of brewing beer.

Ethylene

FIGURE 38-7
Effects of gibberellins. The biennial plant honesty (*Lunularia annua*) will "bolt," or flower, when it is treated with gibberellins.

Long before its important role as a plant hormone was appreciated, the simple hydrocarbon **ethylene** ($H_2C{=}CH_2$) was known to defoliate plants when it leaked from gaslights in street lamps. Ethylene is, however, a natural product of the metabolism of plants that interacts with other plant hormones in minute amounts (see Table 38-1). We have already mentioned the way in which auxin, diffusing down from the apical meristem of the stem, may stimulate the production of ethylene in the tissues around the lateral buds and thus retard their growth. Ethylene also suppresses stem and root elongation, probably for similar reasons.

However, ethylene plays other important roles in plants as well. For example, ethylene appears to be the main factor in the formation of the separation layer during abscission. Both pollinated flowers and fruits that are developing properly produce large amounts of auxin; if such amounts of auxin are present, or if, as we described earlier, auxin is applied to the flowers or fruits, abscission will not occur. Ethylene, however, which is being produced constantly in all parts of the plant, counteracts the effects of the auxin and promotes the formation of the separation layer.

Ethylene is also produced in large quantities during a certain phase of the ripening of fruits, when their respiration is proceeding at its most rapid rate. At this phase, which is called the **climacteric,** complex carbohydrates are broken down into simple sugars, chlorophylls are also broken down, cell walls become soft, and the volatile compounds associated with flavor and scent in the ripe fruits are produced. When ethylene is applied to fruits, it hastens their ripening. One of the first lines of evidence that led to the recognition of ethylene as a plant hormone was the observation that gases that came from oranges caused premature ripening in bananas. Such relationships have led to major commercial uses. For example, tomatoes are often picked green and then artificially ripened later by the application of ethylene. Ethylene is widely used to speed the ripening of lemons and oranges as well. Carbon dioxide produces effects in fruits opposite to those of ethylene, and fruits are often shipped in an atmosphere of carbon dioxide if they are not intended to ripen yet.

Ethylene, a simple gaseous hydrocarbon, is a naturally occurring plant hormone. It controls the abscission of leaves, flowers, and fruits from the plants on which they form; their abscission is counteracted by auxin. Auxin probably retards the development of lateral buds by stimulating the production of ethylene around them.

Ethylene can also be used to hasten fruit drop, when the fruits are ready to be harvested, and leaf fall, when this is desirable. It is used to induce flowering in pineapples and other members of the same plant family, Bromeliaceae, a number of which are strikingly beautiful ornamentals. Ethylene also stimulates the germination of seeds and is used commercially in weed control because of this property: the weed seedlings can be killed easily by the application of a herbicide, but their seeds, being dormant, are much more difficult to kill. A problem in the commercial use of ethylene, as one might imagine, is that it is a gas. Plant scientists are seeking chemicals that produce a steady supply of ethylene once they have been applied to a plant but that remain liquid themselves.

The ecological role of ethylene has been stressed by the results of recent studies. Ethylene production increases rapidly following exposure of a plant to toxic chemicals (such as ozone), temperature extremes, drought, or attack by pathogens or herbivores. The increased production of ethylene that occurs under such circumstances can accelerate the abscission of leaves or fruits that have been damaged by these stresses. Similarly, it now appears that the ethylene that the plants produce is responsible for the damage associated with exposure to ozone. Some recent results suggest that the production of ethylene by plants subjected to attack by herbivores or infected with diseases may be a signal for the plants to activate their defense mechanisms, including the production of molecules that are toxic to the animals or pests attacking them. Understanding these relationships thoroughly is obviously very important for agriculture and forestry.

Abscisic Acid

Abscisic acid (see Table 38-1), a naturally occurring plant hormone, appears to be synthesized mainly in mature green leaves, fruits, and root caps. The hormone was given its name because applications of it stimulate leaf senescence and abscission, but there is little evidence that it plays an important natural role in this process. When it is spotted on a green leaf, the spots turn yellow. Thus abscisic acid has the exact opposite effect on a leaf from that of the cytokinins: a yellowing leaf will remain green in an area where cytokinins are spotted.

Abscisic acid may induce the formation of winter buds—dormant buds that remain through the winter—by converting leaf primordia into bud scales, and, like ethylene, it may suppress the growth of dormant lateral buds. From what we know, it appears likely that abscisic acid suppresses the growth and elongation of buds and promotes senescence, also counteracting some of the effects of the gibberellins (which stimulate the growth and elongation of buds) and auxin (which tends to retard senescence). Abscisic acid is also important in controlling the opening and closing of stomata, as we discussed in Chapter 37.

> **Abscisic acid, which is produced chiefly in mature green leaves and in fruits, suppresses the growth of buds and plays an important role in controlling the opening and closing of stomata. It also promotes leaf senescence.**

Abscisic acid occurs in all groups of plants and apparently has been functioning as a growth-regulating substance since early in the evolution of the plant kingdom. Relatively little is known about the exact nature of its physiological and biochemical effects. These effects are so rapid, however—often taking place within a minute or two—that they must be at least partly independent of protein synthesis. Some longer-term effects of abscisic acid do involve the suppression of protein synthesis, but the way in which this occurs is poorly understood. Recently, the binding sites for abscisic acid have been demonstrated to be proteins located on the outer surface of the plasma membrane and not involved with the transport of the hormone into the cells. Like other plant hormones, abscisic acid probably will prove to have valuable commercial applications when its mode of action is better understood.

In 1986 abscisic acid was discovered in the brains of several kinds of mammals, where it is apparently manufactured. Abscisic acid may act as a hormone in the mammals, perhaps as a gene repressor—which seems to be one of its major roles in plants. Further investigations of its mode of action will be most interesting.

The Combined Action of Plant Hormones

Hormones are relatively difficult to study because they are produced in such small quantities; also, the actions of plant hormones are so thoroughly integrated with one another that they have sometimes been difficult to separate. Indeed, the most outstanding characteristic of plant hormones may be the way in which they interact to produce their effects. For example, auxins promote cell elongation but only in certain kinds of stems and under certain conditions; cytokinins promote cell division; ethylene inhibits cell division; and gibberellins promote both cell division and cell elongation. An example of the interaction between auxin and cytokinin was shown in Figure 37-14, and other evidence of hormonal interaction has been mentioned in other sections of this chapter.

All of these hormones are produced at varying rates, and all are also broken down by enzymes produced within the plant body. When it comes to interpreting a particular plant growth response, therefore, it is usually the reinforcing or contradictory actions of two or more hormones that prove to be critical. As a result, it probably will be many years before the interactions of hormones in determining the form of a mature plant will be well understood.

TROPISMS

Tropisms, or orientation in response to external stimuli, control the growth patterns of plants and thus their appearance. Plants adjust to the conditions of their environment by growth responses. Here we will consider three major classes of plant tropisms: phototropism, gravitropism, and thigmotropism.

Phototropism

We have already introduced **phototropism,** the bending of plants toward unidirectional sources of light, in our discussion of the action of auxin (see pp. 762-765). In general, stems are positively phototropic, growing toward the light, whereas roots are negatively phototropic, growing away from it. The phototropic reactions of stems clearly are of adaptive value because they allow plants to capture greater amounts of light than would otherwise be possible. They are also very important in determining how the organs of plants develop and consequently in the appearance of the plant. Individual leaves display phototropic responses, as do whole stems, and the position of these leaves is of great importance to the photosynthetic efficiency of the plant. On the other hand, the phototropic responses of roots are also of adaptive value; they cause roots to grow downward, away from light, and so to reach a supply of water and nutrients. Auxin is certainly involved in most, if not all, of the phototropic growth responses of plants.

FIGURE 38-8
Gravitropism. The stem of this fallen plant is growing straight up because it is negatively gravitropic and also because it is positively phototropic.

> Phototropisms are growth responses of plants to a unidirectional source of light. They are mostly, if not entirely, mediated by auxin and are very important in determining the form of a plant.

Gravitropism

Another familiar plant response is **gravitropism,** formerly known as *geotropism*. This kind of tropism causes stems to tend to grow upward and roots downward (Figure 38-8); both of these responses clearly are of adaptive significance. Stems that grow upward are apt to receive more light than those that do not; roots that grow downward are more apt to encounter a favorable environment than those that do not. The phenomenon is now called gravitropism because it is clearly a response to gravity and not to the earth (prefix "geo") as such.

In shoots that are placed horizontally, differences in auxin concentration soon develop between the upper and lower sides with greater concentrations on the lower side. Auxin is a powerful inhibitor of root growth, and very small concentrations induce a root to curve toward the side where the auxin concentration is greater. These differences cause the growth responses that are responsible for the shoots growing upward against the force of gravity—negative gravitropism. In roots, such gradients in hormone concentration have not been as well documented. Nevertheless, in somewhat horizontally growing roots the upper sides grow more rapidly than the lower sides, causing the root ultimately to grow downward: this phenomenon is known as positive gravitropism.

> Gravitropism, the response of a plant to gravity, generally causes shoots to grow up (negative gravitropism) and roots to grow down (positive gravitropism).

It may surprise you to learn that in tropical forests, roots often grow *up* the stems of neighboring plants, contrary to the normal phototropic and gravitropic responses of these organs. By doing so, they are able to absorb nutrients from the predictable flow of precipitation, with nutrients that the precipitation absorbs from the upper parts of the plants—a more favorable environment for them than the sterile soils in which they are rooted.

FIGURE 38-9
Thigmotropism. These tendrils coil around the stem because of their positive thigmotropism.

Thigmotropism

Still another commonly observed plant response is **thigmotropism,** a name derived from the Greek root *thigma,* meaning "touch." Thigmotropism is the response of plants to touch. The responses by which tendrils curl around and cling to stems or other objects are surprisingly rapid and clear (Figure 38-9); the ways in which twining plants, such as bindweed, coil around objects are analogous. This behavior is the result of rapid growth responses to touch. Specialized groups of cells in the epidermis appear to be concerned with thigmotropic reactions, but again, their exact mode of action is not well understood.

Thigmotropisms are growth responses of plants to touch. The way in which such stimuli are perceived is poorly understood.

The tendrils of many kinds of plants will coil around an object in the direction from which the tendril is touched, but the tendrils of other plants always coil in the same direction, regardless of which side of the tendril is touched first. Both auxin and ethylene appear to be involved in the movements of tendrils; they can induce coiling in the absence of any contact stimulus.

TURGOR MOVEMENTS

Some kinds of plant movements are based on reversible changes in the turgor pressure of specific cells, rather than on differential growth or patterns of cell enlargement or division. One of the most familiar of these reversible changes has to do with the changing leaf positions that certain plants exhibit at night and in the day (Figure 38-10). For example, the attractively spotted leaves of the prayer plant (*Maranta*) spread horizontally during the day but become more or less vertical at night.

Many other kinds of plants exhibit comparable leaf movements, associated with the bending or straightening of a jointed, multicellular structure at the base of the leaf called the **pulvinus**. Enlarged cells called **motor cells** flank the pulvinus on both sides; when those on one or the other side imbibe water differentially, the leaf moves away from the side on which the most water has been imbibed. Using similar mechanisms, the leaves of certain plants track the sun. Although the ways in which they direct their orientation are not understood, such leaves often can move rapidly (as much as 15 degrees an hour) as a result of the changes in turgor pressure of the pulvini.

The ways in which leaves orient in relation to the sun depend on the kind of plant and its habitat. For example, in annual plants that grow in the desert and must complete their life cycles while water is available and before temperatures become too high, many kinds of plants orient their leaves at right angles to the sun, which makes rapid growth possible. Plants that are able to orient their leaves in this way are from one and a third to nearly two times as efficient in assimilating carbon as those that cannot do so. In contrast, plants that grow in the hot sun and photosynthesize in the summer, such as wild lettuce, may hold their leaves parallel to the sun. Doing so keeps down the temperatures in the leaves of these plants and thus retards their loss of water but not their rates of photosynthesis. Plants that grow in habitats where it freezes every night, such as on mountains in the tropics, may fold their leaves up around their shoot apical meristem and thus protect it from freezing. Other plants may do so when dry conditions prevail.

FIGURE 38-10

Turgor pressure. Prayer plant (*Maranta*) in the day with leaflets held horizontally (**A**) and at night with leaflets held down (**B**).

Turgor movements of plants are reversible and involve changes in the turgor pressure of specific cells. They allow plants to orient their leaves and flowers in ways that allow them to photosynthesize more successfully, or to protect themselves from cold, drought, or other unfavorable environmental influences.

A   B

Especially rapid are the movements of the leaves and leaflets of the sensitive plant (*Mimosa pudica*), which fall into a folded position when they are touched (Figure 38-11). The leaves are able to do so because the motor cells on one side of their pulvini become permeable to potassium and other ions when they are touched or when they are stimulated by the electrical currents that result from being touched. When this happens, ions rapidly flow out of the cells, thus equalizing their concentrations on both sides of the plasma membrane. The cells then rapidly lose much of their water content and therefore their turgor, which results in the rapid collapse of the leaf.

A

Even more spectacular are the rapid movements of the leaves of the carnivorous Venus flytrap (*Dionaea muscipula*) when capturing their animal prey (see boxed essay, Chapter 36). These movements, however, are not caused by changes in turgor pressure; there are no pulvini in these leaves. In the lobes of the leaves are located trigger hairs, under which there are motor cells. When an insect or any other object touches two specific trigger hairs in sequence, the leaf snaps shut. The process involves irreversible cell enlargement, which is initiated by a drop in the pH of the cell walls. The leaves close most rapidly at a pH of 3 to 4, a level at which the walls of the motor cells are very flexible. When the leaves are snapping shut, the motor cells expand their own walls rapidly by releasing H^+ ions, expending energy in the form of ATP. During the 1 to 3 seconds required for the leaves to close, about 29% of the ATP in the cells is lost! The ATP apparently is used in the very fast transport of H^+ ions from the motor cells to the outer leaf cells, where the ions acidify the cell walls. This rapid transfer of H^+ ions is, in effect, an electrical signal. During closure, the outer surface of the leaf expands rapidly; as the leaf slowly opens, over a period of about 10 hours, its inner surface expands.

B

FIGURE 38-11
Sensitive plant (*Mimosa pudica*). The leaves of *Mimosa* are divided into numerous segments, or leaflets. An undisturbed leaf is shown in **A**; **B** shows the effects of striking the leaf partway down from its tip.

The movement of flowers, such as sunflowers, that "follow" the sun is controlled by structures similar to pulvini. By tracking the sun, the sunflower keeps the inside of the flower head warm, creating a favorable environment for its insect visitors, even when the temperature of the surrounding air is relatively cool. For this reason, it is possible for insects to visit the flower heads over a wider range of conditions than would otherwise be possible. The flowers of many species of angiosperms close at night, and their closing is controlled by pulvini also.

PHOTOPERIODISM

Essentially all eukaryotic organisms are affected by the cycle of night and day, and many features of plant growth and development are keyed to the changes in the proportions of light and dark in the daily 24-hour cycle. Such responses constitute **photoperiodism,** a mechanism by which organisms measure seasonal changes in relative day and night length. One of the most obvious of these photoperiodic reactions concerns the production of flowers by angiosperms.

Flowering Responses

Day length changes with the seasons; the farther from the equator one is, the greater the variation. The flowering responses of plants fall into two basic categories in relation to day length. **Short-day plants** begin to form flowers when the days become shorter than a critical length (Figure 38-12). **Long-day plants,** on the other hand, initiate flowers when the days become longer than a certain length. In both kinds of plants, it is actually the length of darkness (night) that is significant, and not the length of day, as we will see.

The critical day lengths for both long-day and short-day plants tend to fall in the 12- to 14-hour range. At middle latitudes, therefore, short-day plants bloom in late summer and autumn when the days have become short enough, whereas long-day plants bloom in spring and early summer when the days have become long enough. Many fall flowers, including chrysanthemums, goldenrods, and poinsettias; important crops, such as soybeans; and weeds, such as ragweed, are short-day plants. Conversely, many spring and early-summer flowers, including clover, irises, and hollyhocks, are long-day plants. Commercial plant growers use these responses to day length to bring plants into bloom when they are wanted for sale. Supplementary light

FIGURE 38-12

How flowering responds to day-length. The goldenrod is a short-day plant that blooms in the fall throughout the Northern hemisphere, stimulated to do so by the long nights. The iris is a long-day plant that blooms in the spring, stimulated to do so by the short nights that occur in the spring. If the long night of winter is artificially interrupted by a flash of light, the goldenrod will not bloom, and the iris will. In each case, it is the duration of uninterrupted darkness that determines when flowering will occur.

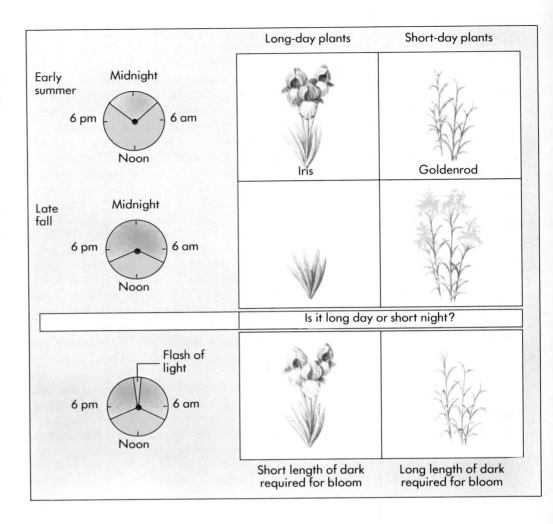

is used to "force" (as the process is termed) long-day plants, and artificial protection from light is used to bring short-day plants into bloom.

In addition to the long-day and short-day plants, a number of plants are described as **day neutral**. These plants produce flowers whenever environmental conditions are suitable, without reference to day length. Day-neutral plants include roses, snapdragons, and tomatoes.

> **Short-day plants start to form flowers when the days become shorter than a certain critical length; long-day plants form flowers when the days become longer than a certain length. Day-neutral plants do not have specific day-length requirements for flowering.**

Flowering responses to day length may be important in controlling the distribution of certain plants. For example, in the tropics, where the days may never be long enough to satisfy the requirements of certain long-day plants or short enough to satisfy those of some short-day plants, these plants may simply be unable to flower and therefore to reproduce.

The Chemical Basis of the Photoperiodic Response

Students of flowering responses made an early discovery that was to prove extremely valuable in their later studies: short-day plants require a certain amount of darkness in each 24-hour cycle to initiate flowering; however, if the period of darkness is interrupted by a brief period of light, often less than a minute, the plants will not flower.

Once this discovery had been made, scientists proceeded to determine the wavelength of light that was most effective in inhibiting flowering. This was found to be red light with a wavelength of about 660 nanometers. Curiously, however, if the plants were exposed to red light at 660 nanometers and then to far-red (longer-wavelength red) light at a wavelength of about 730 nanometers, the effect was canceled. What is the chemical basis for this strange observation?

Plants contain a pigment, **phytochrome,** which exists in two interconvertible forms, P_r and P_{fr}. In the first form, phytochrome absorbs red light; in the second, it absorbs far-red light. When a molecule of P_r absorbs a photon of red light, it is instantly converted into a molecule of P_{fr}, and when a molecule of P_{fr} absorbs a photon of far-red light, it is quickly converted into P_r. P_{fr} is biologically active; P_r is biologically inactive. In other words, when P_{fr} is present, a given biological reaction that is affected by phytochrome will occur. When most of the P_{fr} has been replaced by P_r, the reaction will not occur.

Phytochrome is a receptor of light and does not itself act directly to bring about the reactions to light. In short-day plants, however, the presence of P_{fr} leads to a biological reaction that suppresses flowering. The amount of P_{fr} steadily declines in darkness, the molecules being converted to P_r—and when the period of darkness has been long enough, the flowering response is triggered. A single flash of red light at a wavelength of about 660 nanometers, however, will convert most of the molecules of P_r to P_{fr}, and the flowering reaction will be blocked (Figure 38-13). Since in darkness most of the P_{fr} is converted to P_r within 3 to 4 hours, however, the conversion of P_r to P_{fr} cannot be the full explanation for the flowering responses of short-day plants; other factors, still not understood, must also be involved.

The existence of phytochrome was conclusively demonstrated in 1959 by Harry A. Borthwick and his collaborators at the U.S. Department of Agriculture Research Center at Beltsville, Maryland. It has since been shown that the molecule consists of two parts: a smaller one that is sensitive to light and a larger portion that is a protein. The phytochrome pigment is blue, and its light-sensitive portion is similar in structure to the phycobilins that occur in cyanobacteria and red algae. Phytochrome is present in all groups of plants and in a few genera of green algae, but not in bacteria, fungi, or protists other than those few green algae. Therefore it is provisionally as-

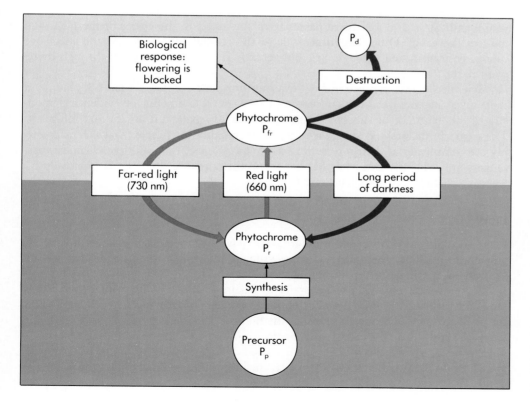

FIGURE 38-13

How phytochrome works. Phytochrome is synthesized in the P_r form from amino acids, designated P_p for precursor. When exposed to red light, P_r changes to P_{fr}, which is the active form that elicits a response in plants. P_{fr} is converted to P_r when exposed to far-red light, as well as in darkness, where it may also be destroyed. The destruction product is designated P_d.

FIGURE 38-14
Corn plants grown in light and darkness, showing the characteristics of an etiolated plant. Seedlings are naturally etiolated when they are growing through the ground, and etiolation may be considered a strategy to concentrate the energy of the plant in such a way that it will reach the surface, and light, as rapidly as possible. When it does so, normal growth commences.

sumed that phytochrome systems for measuring light evolved among some of the green algae and were present in the common ancestor of the plants.

Two forms of phytochrome, a large molecule that includes a protein, are interconverted by exposure to red light of different wavelengths. Phytochrome plays an important role in controlling the expression of flowering in short-day–sensitive plants.

Phytochrome is also involved in many other plant growth responses. For example, seed germination is inhibited by far-red light and stimulated by red light in many kinds of plants. Another example is the elongation of the shoot of an **etiolated** seedling, that is, a very slender and colorless seedling that has been kept in the dark. Such plants become normal when exposed to light, especially red light; but the effects of such exposure are canceled by far-red light, indicating a relationship similar to that observed in seed germination (Figure 38-14).

In 1983 Dina Mandoli and Winslow Briggs, working at the Department of Plant Biology at the Carnegie Institution of Washington, reported findings that further enlarged our understanding of the possible ways in which a plant could receive light and react to it. Mandoli and Briggs found that an etiolated seedling, in essence, acts like a bundle of fiber-optic strands, guiding light over distances as great as 4.5 centimeters through the interiors of dozens, even hundreds, of cells, as well as through the junctions between them. Such a system enables light directly and immediately to affect and begin to coordinate the responses of plant portions below ground. These findings open a promising avenue for research into plant growth.

The Flowering Hormone: Does It Exist?

Working with long-day and short-day plants, some investigators have gathered evidence for the existence of a flowering hormone. It has been shown that plants will not flower in response to day-length stimuli if their leaves have been removed before exposure to the light. However, the presence of a single leaf or the exposure of a single leaf to the appropriate stimuli will generally bring about flowering. If the leaf is removed immediately after exposure, the plant will not produce flowers; but if it is left on the plant for a few hours and then removed, flowering will occur normally. These results indicate that a substance passes from the leaves to the apices of the plant and induces flowering. Other experiments have shown that, unlike auxin, the substance cannot be transmitted through agar but actually requires a connection through living plant parts.

Scientists have searched for a flowering hormone for more than 50 years, but their quest has not yet been successful. A considerable amount of evidence demonstrates the existence of substances that promote flowering and others that inhibit it. These poorly understood substances appear to interact in complex ways. It is probably the complexity of their interactions as well as the fact that multiple chemical messengers are evidently involved that has made this scientifically and commercially interesting matter such a difficult one to resolve.

DORMANCY

Plants respond to their external environment largely by changes in growth rate. As you might imagine, the ability to cease growing altogether when conditions are not favorable is a critical factor in their survival.

In temperate regions, we generally associate dormancy with winter, when low temperatures and the unavailability of water because of freezing make it impossible for plants to grow. During this season, the buds of deciduous trees and shrubs remain dormant, and the apical meristems remain well protected inside enfolding scales. Perennial herbs spend the winter underground as stout stems or roots packed with stored food. Many other kinds of plants, including most annuals, pass the winter as seeds.

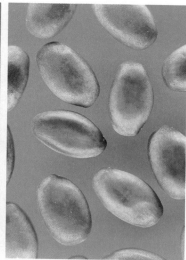

FIGURE 38-15

Palo verde *(Cercidium floridum)*, a desert tree with tough seeds that germinate only after they are cracked.

In climates that are seasonally dry, dormancy will occur primarily during the dry season, whenever in the year it falls. In dry conditions, plants stay in a dormant condition by using strategies similar to those that the plants of temperate areas rely on in winter.

Annual plants occur frequently in areas of seasonal drought. Seeds are ideal for allowing annual plants to bypass the dry season when there is insufficient water for growth. When it rains, they can germinate and the plants can grow rapidly to take advantage of the relatively short periods when water is available. Chapter 37 covered some of the mechanisms involved in breaking seed dormancy and allowing germination under favorable circumstances. These include leaching out from the seed coats chemicals that inhibit germination or cracking the seed coats mechanically, a procedure that is particularly suitable for promoting growth in seasonally dry areas. Rains will leach out the chemicals in the seed coats whenever they occur, and the hard coats of other seeds may be cracked when they are being washed down along arroyos in temporary floods (Figure 38-15).

Seeds may remain dormant for surprisingly long periods of time (see Figure 37-11). Many legumes (plants of the pea and bean family, Fabaceae—also known as Leguminosae), for example, have tough seeds that are virtually impermeable to water and oxygen; these seeds often last decades and possibly even longer without special care. They will germinate eventually, after their seed coats have been cracked and when water is available.

A period of cold is necessary before some kinds of seeds will germinate, as we mentioned in Chapter 37. The seeds of certain other plants will germinate only if they are able to obtain adequate water when the temperatures are relatively high. For this reason, certain weeds germinate and grow in the cool part of the year and others in the warm part of the year. Similarly, a period of cold is necessary before the buds of some trees and shrubs will break dormancy and develop normally. For this reason, many of the plants that normally grow in temperate regions do not do well in the tropics—even at high elevations where it may be relatively cool all year—because it does not get cold enough and because the day-length relationships are different from those that occur in temperate regions.

In dry or cold seasons, which are unfavorable for growth, adult plants may become dormant, such as by losing their leaves and forming drought-resistant winter buds. Seeds enable plants to bypass long unfavorable periods.

1. Hormones are chemical substances produced in small quantities in one part of an organism and transported to another part of the organism, where they bring about physiological responses. The tissues in which plant hormones are produced are not specialized particularly for that purpose, nor are there usually clearly defined receptor tissues or organs.

2. There are five major classes of naturally occurring plant hormones: auxins, cytokinins, gibberellins, ethylene, and abscisic acid. They often interact with one another in bringing about growth responses.

3. Auxins are produced at the tips of shoots and diffuse downward, suppressing the growth of lateral buds. In young grass seedlings and other herbs, they play a major role in promoting stem elongation. In such stems, the auxin migrates away from the light and makes the cells on the side to which they have migrated more plastic. By expanding more than the less plastic cells on the other side of the stem, the elongating cells cause the stem to bend away from themselves.

4. Cytokinins are necessary for mitosis and cell division in plants. They promote the growth of lateral buds and inhibit the formation of lateral roots.

5. Gibberellins play the major role in stem elongation in most plants. They also tend to hasten seed germination, to break dormancy in buds, and to cause the stems in the rosettes of biennials to elongate and the plants ultimately to flower.

6. Ethylene is a gas that functions as a plant hormone. Auxin may retard the growth of lateral buds because it stimulates the production of ethylene near them. Ethylene is widely used to hasten fruit ripening.

7. Used experimentally, abscisic acid promotes senescence and the abscission of plant parts, and it may induce the formation of winter buds. It also plays a key role in the opening and closing of stomata.

8. Tropisms in plants are growth responses to external stimuli. A phototropism is a response to light, gravitropism is a response to gravity, and thigmotropism is a response to touch.

9. Turgor movements are reversible but important elements in the adaptation of plants to their environments. By means of turgor movements, leaves, flowers, and other structures of plants track light and take full advantage of it.

10. The flowering responses of plants fall into two basic categories in relation to day length; in both, it is the period of darkness that is critical. Short-day plants begin to form flowers when the days become shorter than a given critical length; long-day plants do so when the days become longer than a certain length. In temperate regions, the progression of the seasons stimulates plants of both classes to flower—the short-day plants in the autumn, and the long-day plants in the spring. Many other plants are day neutral.

11. A blue pigment known as phytochrome is interconverted between two forms by red light of different wavelengths. It plays a role in determining the flowering response and in mediating certain other plant responses.

12. Dormancy is a necessary part of plant adaptation that allows a plant to bypass unfavorable seasons, such as winter, when the water may be frozen, or periods of drought. Dormancy also allows plants to survive in many areas where they would be unable to grow otherwise.

REVIEW QUESTIONS

1. In what ways are plant and animal hormones similar to one another? In what ways are plant hormones significantly different from animal hormones? What are the five major kinds of plant hormones?

2. What is the only naturally occurring auxin? From what biomolecule is it synthesized? Where in the mature plant is it produced and to where is it transported? Does this occur via the plant vascular tissue? Explain.

3. How does auxin affect the plasticity of the plant cell wall? What is the function of indoleacetic acid oxidase? What is the relative rate of speed with which auxin acts in response to light?

4. To what naturally occurring plant hormones are 2,4-D and 2,4,5-T related? Of what commercial value are they? Upon which class of angiosperms do they selectively act? How do they work?

5. What is the primary role of cytokinins in plant development? How do their actions compare to those of auxins? Where are most cytokinins produced? From what biomolecule do cytokinins appear to derive?

6. What plant hormone is usually lacking in genetically dwarfed plants? How can one determine if this is indeed true? What are three secondary effects of this hormone? What are three commercial uses of this hormone?

7. What is the role of ethylene in leaf and fruit abscission? What effect does it have upon the ripening of fruit? What common gas counteracts this effect of ethylene? What are some of the commercial applications of ethylene? Why is this compound a valuable component in weed control products?

8. Where is abscisic acid produced in a mature plant? What is the effect of spotting this hormone on a green leaf? What hormone reverses this? What are the associations of abscisic acid with leaf buds? How does abscisic acid affect stomata?

9. In general which part of a plant is positively phototropic and which part is negatively phototropic? Explain each. What is the adaptive significance of each reaction? What hormone most completely mediates this reaction?

10. A burst of which wavelength of light is most effective in inhibiting flowering? How can this effect be cancelled? What compound mediates these reactions? Explain.

11. What two plant growth responses are mediated by phytochromes? Briefly explain each. Describe the appearance of an etiolated seedling.

THOUGHT QUESTIONS

1. If day length in a particular place at a particular time of year were 10 hours, which would produce flowers: a short-day plant, a long-day plant, both, or neither? Why? Do you think there are any short-day plants in the tropics? Why?

2. How can we be certain that it is the length of the dark period, rather than the length of the light period, that actually determines whether a short-day or a long-day plant will flower? Describe experiments that have led scientists to this conclusion.

3. When poinsettias are kept inside a house after the holiday season, they rarely bloom again. Why do you think this might be, and what might you do to get them to produce flowers a second time?

FOR FURTHER READING

ALBERSHEIN, P., and A.G. DARVILL: "Oligosaccharins," *Scientific American*, September 1985, pages 58-64. Fragments of the cell wall serve as regulatory molecules and interact with hormones in ways that are just beginning to be understood.

BASKIN, J.M., and C.C. BASKIN: "The Annual Dormancy Cycle in Buried Weed Seeds: A Continuum," *BioScience*, vol. 35, pages 492-498, 1985. This interesting article shows how differences between the temperatures that different kinds of weeds require for germination determine the times when they germinate and grow.

CHAPIN, F.S., III: "Integrated Responses of Plants to Stress," *BioScience*, vol. 41, pages 29-36, 1991. All plants respond to stress of many types in basically the same way, which we are beginning to understand more completely.

EVANS, M.L., R. MOORE, and K.H. HASENSTEIN: "How Roots Respond to Gravity," *Scientific American*, December 1986, pages 112-119. Modern studies of this fascinating phenomenon.

LESHEM, Y.Y., A.H. HALEVY, and C. FRENKEL: *Processes and Control of Plant Senescence*, Elsevier Science Publishers, Amsterdam, 1986. Animals age and die, but what are the analogous processes in plants?

MANDOLI, D.F., and W.R. BRIGGS: "Fiber Optics in Plants," *American Scientist*, vol. 251, pages 90-98, 1984. A good summary of the subject.

SALISBURY, F.B., and C.W. ROSS: *Plant Physiology*, ed. 3, Wadsworth Publishing Co., Belmont, Calif., 1985. A fine account of the way plants grow and function.

SCHMIDT, W.: "Bluelight Physiology," *BioScience*, vol. 34, pages 698-704, 1984. Many plants and fungi respond in specific ways to blue light.

SISLER, E.C., and S.F. YANG: "Ethylene, the Gaseous Plant Hormone," *BioScience*, vol. 33, pages 238, 1984. A review of this important plant hormone.

TREWAVAS, A.: "How Do Plant Growth Substances Work?" *Plant, Cell, and Environment*, vol. 4, pages 203-228, 1981. An excellent, concise introduction to plant hormones.

WAREING, P.F., and I.D.J. PHILLIPS: *Growth and Differentiation in Plants*, ed. 3, Pergamon Press, Inc., Elmsford, N.Y., 1981. Excellent treatment of the subject.

X

Biology of Animals

The Primitive Invertebrates

OVERVIEW

In this chapter we begin to trace the long evolutionary history of the animals; in it we encounter the simplest members of this kingdom—sponges, jellyfish, and several kinds of worms. These animals are important ecologically, and they illustrate the advent of the major characteristics that are important in the more advanced animal phyla. These characteristics include the development of tissues and organs, the use of internal digestion, the appearance of radial and then bilateral body organization, and the appearance of internal body cavities. The evolution of the primitive invertebrates accomplished the first major organization of the animal body, an organization that was maintained and elaborated by all later forms.

FOR REVIEW

Here are some important terms and concepts that you will encounter in this chapter. If you are not familiar with them, you should review them before proceeding.

Evolutionary theory (Chapter 20)

Evolutionary history (Chapter 22)

Symbiosis (Chapter 24)

Classification (Chapter 28)

Choanoflagellates (Chapter 31)

The two most conspicuous kingdoms of organisms are the plants, which we have just considered, and the animals, which we begin to consider here. Animals are a very distinct group of organisms and seem always to have been recognized as such. Animals are the eaters of the earth, and all of them are heterotrophs. All animals depend directly or indirectly for their nourishment on plants, photosynthetic protists (algae), or autotrophic bacteria. Many animals are able to move from place to place in search of their food, which they ingest. In most of them, ingestion of food is followed by digestion in an internal cavity.

Animals, constituting millions of species, are the most abundant living things. Found in every conceivable habitat, they bewilder us with their diversity.

A

All animals are multicellular. Several decades ago a biologist would not have made that statement. In recent years the generally accepted scientific definition of the kingdom Animalia, which comprises the animals, has been changed in one respect. The unicellular, heterotrophic organisms called "Protozoa," which were at one time regarded as simple animals, are now considered to be members of the kingdom Protista, the large and diverse group that we discussed in Chapter 31.

There are many kinds of animals—at least 9 to 10 million living species, and perhaps many times more. Of these, only about 42,500 have a dorsal backbone. Those animals are the vertebrates; they will be discussed in Chapters 42 to 56. The other 99% of animal species lack such a backbone and are collectively referred to as the invertebrates. In this chapter we will consider the simplest of these (Figure 39-1), and in Chapters 40 and 41, the more complex ones. But first we will step back a bit and consider animals as a group. Despite their great diversity, they have much in common.

B

THE ANIMAL KINGDOM

Animals are extraordinarily diverse in form. They range in size from a few that are smaller than many protists to others like the truly enormous whales and giant squids. The animal kingdom includes about 35 phyla, most of which are found in the sea, with fewer in fresh water, and fewer still on the land. Two phyla, Arthropoda—a gigantic group that includes crustaceans, spiders, and insects—and Chordata—our own phylum, which includes the vertebrates—dominate animal life on the land. In addition, members of several other phyla, such as the terrestrial snails and slugs (phylum Mollusca) are abundant and diverse in terrestrial communities.

Animal cells are exceedingly diverse in structure and function; they lack rigid walls, and are usually quite flexible. Animal cells, except for those of sponges, are organized into **tissues,** groups of cells organized into a structural and functional unit. In most animals the tissues are organized into organs, complex structures that are composed of two or more kinds of tissues.

In more complex animals, such as vertebrates, specialized cells make up many kinds of distinctive tissues. These include the incredibly complex tissues that are concerned with the senses and with locomotion. The ability of animals to move—and to move more rapidly and in more complex ways than the members of other kingdoms—is perhaps their most striking characteristic, one that is directly related to the flexibility of their cells. Animals usually move by means of muscle cells, contractile cells that contain the proteins actin and myosin. The most remarkable form of animal movement, perhaps, is flying, an ability that is well developed among both vertebrates (birds, bats, and some extinct reptiles) and insects.

The kingdom Animalia consists of about 35 phyla, most of which occur in the sea. The diverse cells of animals, except for sponges, are organized into structurally and functionally distinctive groups called tissues, and the tissues of most animals in turn are organized into complex structures called organs, which are made up of two or more kinds of tissues.

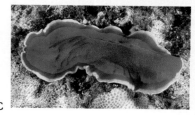

C

FIGURE 39-1

Diversity of lower invertebrates.
A Yellow tube sponge (phylum Porifera). The cells of sponges are not organized into tissues.
B Sea anemone (phylum Cnidaria). The cnidarians are radially symmetrical animals in which the tissues are not grouped into organs.
C Free-living flatworm (phylum Platyhelminthes)—an acoelomate, a member of one of the groups of animal phyla in which the members are bilaterally symmetrical.

Most animals reproduce sexually. Their eggs, which are nonmotile, are much larger than their small, usually flagellated sperm. In animals the cells formed by meiosis function directly as gametes; they do not divide by mitosis, as they do in plants and some protists, but rather fuse directly with one another. Consequently, there is no counterpart among animals of the alternation of gametophytic (haploid) and sporophytic (diploid) generations, which is characteristic of plants. With few exceptions, some of which we will discuss, animals are diploid: the gametes are the only haploid cells in their life cycles.

The complex form of a given animal develops from a zygote by a characteristic process of embryonic development. The zygote first undergoes a series of mitotic divisions and becomes a hollow ball of cells, the **blastula,** a developmental stage that occurs in all animals. In most, the blastula folds inward at one point to form a hollow sac with an opening at one end called the **blastopore.** An embryo with a blastopore is called a **gastrula.** The subsequent growth and movement of the cells of the gastrula produce the digestive system, also called the gut or intestine. The details of embryonic development differ widely from one phylum of animals to another and often provide important clues to the evolutionary relationships among the phyla. Taken as a whole, however, the pattern of embryology is characteristic of the animal kingdom.

THE CLASSIFICATION OF ANIMALS

Two subkingdoms are generally recognized within the kingdom Animalia: (1) Parazoa—animals that for the most part lack a definite symmetry and possess neither tissues nor organs—and (2) Eumetazoa—animals that have a definite shape and symmetry and, in most cases, tissues, which are organized into organs and organ systems. The subkingdom Parazoa consists primarily of the sponges, phylum Porifera. The other animals, composing about 35 phyla, belong to the subkingdom Eumetazoa.

The structure of eumetazoans is much more complex than that of sponges. For example, all eumetazoans form three distinct embryonic layers: an outer ectoderm, an inner endoderm, and an intermediate mesoderm. These layers differentiate into the tissues of the adult animal. In general, the nervous system and outer covering layers develop from the ectoderm, the muscles and skeletal elements from the mesoderm, and the intestine and digestive organs from the endoderm. No such layers are present in sponges.

An outline of the animal phyla treated in this book is presented in Table 39-1. (See also Appendix A). Some of the terms used in the table will be explained during the course of this chapter.

> The animal kingdom is traditionally divided into two subkingdoms. In the simpler subkingdom, Parazoa, the animals lack symmetry and possess neither tissues nor organs. In the more advanced subkingdom, Eumetazoa, the animals are symmetrical, and nearly all possess tissues.

TABLE 39-1   OUTLINE OF ANIMAL CLASSIFICATION: KINGDOM ANIMALIA

I. Subkingdom Parazoa: phylum Porifera
II. Subkingdom Eumetazoa
   A. Radially symmetrical animals: phyla Cnidaria and Ctenophora; probably unrelated to one another
   B. Bilaterally symmetrical animals
      (1) Acoelomates (animals that lack a body cavity): phyla Mesozoa, Platyhelminthes, Rhynchocoela
      (2) Pseudocoelomates (animals with a pseudocoel): phyla Nematoda, Rotifera, Loricifera
      (3) Coelomates (animals with a coelom)
         (a) Protostomes (coelomates in which the mouth develops from or near the blastopore during the course of embryonic development): phyla Mollusca, Annelida, Pogonophora, Onychophora, and Arthropoda
         (b) Deuterostomes (coelomates in which the anus forms from or near the blastopore): phyla Echinodermata, Chaetognatha, Hemichordata, and Chordata

---

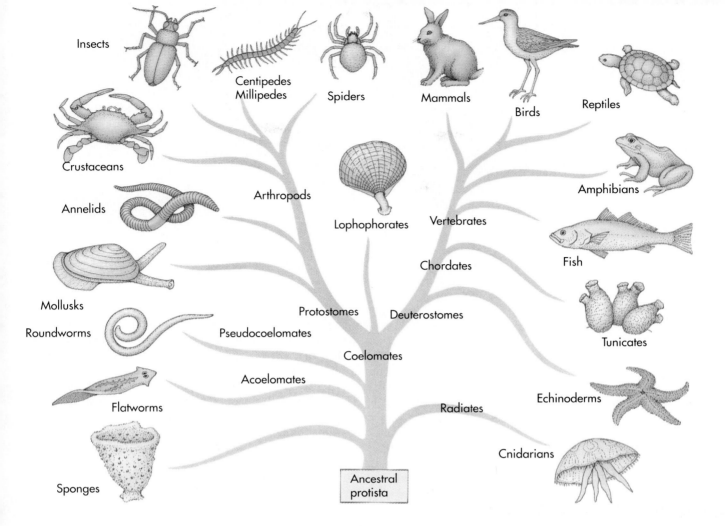

Insects

Centipedes
Millipedes

Spiders

Mammals

Birds

Reptiles

Crustaceans

Arthropods

Lophophorates

Vertebrates

Amphibians

Annelids

Chordates

Fish

Mollusks

Protostomes

Deuterostomes

Tunicates

Roundworms

Pseudocoelomates

Coelomates

Acoelomates

Echinoderms

Flatworms

Radiates

Sponges

Cnidarians

Ancestral protista

FIGURE 39-2
The phylogeny of the major groups of animals.

Sponges and Eumetazoa were thought to have had different unicellular ancestors, but it is now believed that both of these evolutionary lines arose from a choanoflagellate ancestor (phylum Zoomastigina; see Figures 31-26 and 39-2). Because sponges and Eumetazoa have so many features in common, they are grouped as a single kingdom.

The so-called "lower" or "primitive" invertebrates, which have a less complex tissue organization than the "higher" invertebrates, make up about 14 phyla. In this chapter we will focus on four of these that have been the most successful, as measured by their size, distribution, and important ecological roles. The four phyla that we will consider are the sponges (phylum Porifera), which lack any tissue organization; the radially symmetrical jellyfish, hydroids, sea anemones, and corals (phylum Cnidaria); the bilaterally symmetrical flatworms (phylum Platyhelminthes); and the nematodes (phylum Nematoda), a phylum that includes both free-living and parasitic worms.

PHYLUM PORIFERA: THE SPONGES

There are perhaps 5000 species of marine sponges (see Figures 39-1, *A*, and 39-3) and about 150 additional species that live in fresh water. In the sea, sponges are abundant at all depths. A few of the smaller ones are radially symmetrical, but most of the

FIGURE 39-3
Diversity in sponges (phylum Porifera).
A Red boring sponge (*Cliona delitrix*).
B Barrel sponge, a large sponge in which the form is somewhat organized.

A

B

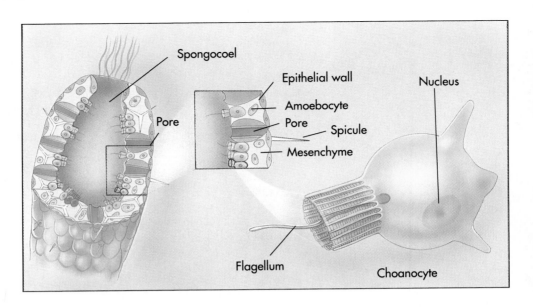

FIGURE 39-4
The basic structure of a sponge.

members of this phylum lack symmetry completely. Sponges lack specialized tissues and organs. Although some sponges are tiny, no more than a few millimeters across, some, like the loggerhead sponges, may reach 2 meters or more in diameter. Many sponges are colonial. Some are low and encrusting, while others may be erect and lobed, sometimes in complex patterns. Although larval sponges are free swimming, the adults are **sessile,** or anchored in place. Individual adults remain attached to submerged objects once they have settled down.

There is relatively little coordination among the cells of sponges. Like a plasmodial slime mold (Chapter 31), a sponge can pass through a fine silk mesh and then reaggregate on the other side. In sponges, reaggregation takes a few weeks. Although a sponge seems to be little more than a mass of cells embedded in a gelatinous matrix, these cells recognize one another with a high degree of fidelity, and are specialized for different functions of the body.

Sponges have a unique form of feeding that involves the flow of water through a system of tiny pores and canals. The water is forced through these passageways by the beating of the flagella with which they are lined. The name of the phylum, *Porifera,* refers to this system of pores.

The basic structure of a sponge can best be understood by examining the form of a young individual. When it attaches to a substrate, a young sponge grows into a vaselike shape. The walls of such a vase have three functional layers:

1. Facing into the internal cavity are specialized, flagellated cells called choanocytes, or collar cells (Figure 39-4). These cells line either the entire body cavity or, in many large and more complex sponges, specialized chambers.
2. The bodies of sponges are bounded by an outer epithelial layer, consisting of flattened cells somewhat like those that make up the epithelia, or outer layers, of other animal phyla. Some portions of this layer contract when touched or exposed to the appropriate chemical stimuli, and this contraction may cause some of the pores to close.
3. Between these two layers, sponges consist mainly of a gelatinous, protein-rich matrix called the **mesenchyme,** within which various types of amoeboid cells occur. In addition, many kinds of sponges have minute needles of calcium carbonate or silica known as spicules (Figure 39-5), fibers of a tough protein called spongin, or both, within this matrix. Spicules and spongin strengthen the bodies of the sponges in which they occur. A spongin skeleton is the model for the bathtub sponge, once the skeleton of a real animal, but now largely known from its cellulose and plastic mimics.

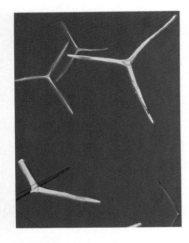

FIGURE 39-5
Spicules in a sponge, as viewed with polarized light.

Water is drawn into a sponge through very numerous small pores and is circulated internally through a system of channels. Eventually the water is forced out through an **osculum,** a specialized, larger pore.

The Choanocyte

Each choanocyte (see Figure 39-4) closely resembles a protist with a single flagellum, a similarity that reflects its evolutionary derivation. The beating of the flagella of the many choanocytes that line the body cavity draws the water in through the pores and drives it through the sponge, thus providing the means by which the sponge acquires food and oxygen and expels wastes. Each choanocyte beats independently, and the pressure they create collectively in the cavity forces water out of the osculum. In some sponges the inner wall of the body cavity is highly convoluted, increasing the surface area and therefore the number of flagella that can drive the water. In such a sponge, one cubic centimeter of tissue can propel more than 20 liters of water a day!

Sponges are unique in the animal kingdom in possessing choanocytes, special flagellated cells whose beating drives water through the body cavity.

The microscopic examination of individual choanocytes (see Figure 39-4) reveals an important substructure: the base of each flagellum is surrounded by a collar of small, hairlike projections, analogous to a picket fence. The strands that make up the collar of each choanocyte are connected to one another by delicate microfibrils. The beating flagellum of the choanocyte draws water, and any food particles in it, through the openings between these microfibrils. The flagellum then forces this water out through the open top of the collar. Any food particles that enter the collar are trapped in the mucus at the base of the flagellum. The food of a sponge is about four-fifths particulate organic matter and one-fifth small organisms. Once trapped at the base of the flagellum, this food is digested either by the choanocyte itself or by a neighboring amoeboid cell. Choanocytes are essentially identical to the choanoflagellates, a group of the protist phylum Zoomastigina (Chapter 31). Sponges seem clearly to have evolved from the choanoflagellates.

Reproduction in Sponges

As you might suspect from their ability to re-form themselves once they have passed through a silk mesh, sponges frequently reproduce themselves by simply breaking into fragments. If a sponge breaks up, the resulting fragments usually are able to reconstitute whole new individals. Sexual reproduction is also frequent in sponges, mature individuals producing eggs and sperm. Larval sponges, which may undergo their initial stages of development within their parent, have numerous external, flagellated choanocytes and are free swimming. After a short planktonic stage, they settle down on a suitable substrate, where they begin their transformation into adults by turning inside out so that their choanocytes are internal.

PHYLUM CNIDARIA: THE CNIDARIANS

The subkingdom Eumetazoa, animals that have a definite shape and symmetry and nearly always distinct tissues, is divided into two major groups. The first, which includes only two phyla, consists of radially symmetrical organisms. **Radial symmetry** is a condition in which the parts are arranged around a central axis in such a way that any plane passing through the central axis divides the organism into halves that are approximate mirror images. These two phyla are Cnidaria, or cnidarians—hydroids, jellyfish, sea anemones, and corals (see Figures 39-1, *B*, and 39-6)—and Ctenophora, the comb jellies or ctenophores. The bodies of all other eumetazoans are marked by a fundamental bilateral symmetry. Echinoderms (Chapter 42) may be radially symmetrical as adults, but are bilaterally symmetrical when young. The radially symmetrical eumetazoans—cnidarians and ctenophores—have often been considered primitive; the fact that they are the only eumetazoans in which the tissues are not organized into organs appears to be in agreement with this conclusion. Cnidarians were widespread, abundant, and diverse in Precambrian times, as much as 630 million years ago (see Figure 22-8 for a reconstruction of a Cambrian Period animal of this group). There are about 9100 living species.

Cnidarians are nearly all marine, although a few live in fresh water. These fascinating and simply constructed animals are often abundant, especially in shallow,

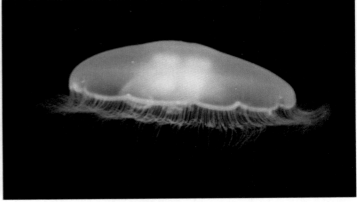

A

B

C

FIGURE 39-6
Representatives of the three classes of cnidarians (phylum Cnidaria).
A *Hydra* (class Hydrozoa).
B Jellyfish, *Aurelia aurita* (class Scyphozoa).
C Soft coral (class Anthozoa).

warm-temperature or subtropical waters. They are basically gelatinous in composition. Like all eumetazoans, they differ markedly from the sponges in organization: their bodies are made up of distinct tissues; as we just mentioned, however, cnidarians lack organs. These animals are carnivores. For the most part, they do not move from place to place, but rather capture their prey (which includes fishes, crustaceans, and many other kinds of animals) with the tentacles that ring their mouth.

There are two basically different body forms among the cnidarians, **polyps** and **medusae** (singular, **medusa**) (Figure 39-7). Polyps are cylindrical animals, usually found attached to a firm substrate, where they may be solitary or colonial. In a polyp, the mouth faces away from the substrate on which the animal is growing, and therefore often upward. Many polyps build up a chitinous or calcareous (made up of calcium carbonate) external or internal skeleton, or both. Some polyps are free floating, but most are anchored to the substrate. In contrast, most medusae are free floating and are often umbrella shaped. Their mouths usually point downward, and the tentacles hang down around them. Medusae, particularly those of the class Scyphozoa, are commonly known as jellyfish because of their thick, gelatinous **mesoglea.**

Many cnidarians occur only as polyps, while others exist only as medusae; still others alternate between these two phases during the course of their life cycles. Both phases, of course, consist of diploid individuals. Polyps may reproduce asexually by budding; if they do, they may produce either new polyps or medusae. Medusae are produced in specialized buds.

In most cnidarians, fertilized eggs give rise to free-swimming, multicellular, ciliated larvae, known as **planulae** (Figure 39-8). Planulae are common in the plankton at times and may be dispersed widely in the currents.

> Individual cnidarian species may be either medusae—floating, bell-shaped animals with the mouth directed downward— or polyps—anchored animals with the mouth directed upward. In some cnidarians, these two forms alternate during the life cycle of the organism.

A major evolutionary innovation in cnidarians, compared with sponges, is the internal extracellular digestion of food—digestion within a gut cavity, rather than within individual cells. This evolutionary modification is characteristic of all of the more advanced phyla of animals. In cnidarians, digestive enzymes (primarily proteases) are released from cells lining the walls of the cavity; these partially break down food. Unlike the process in more advanced invertebrates, however, cnidarian digestion is not completely extracellular; rather, food is fragmented extracellularly into small bits that are subsequently engulfed by the cells lining the gut—the process of phagocytosis (Chapter 6).

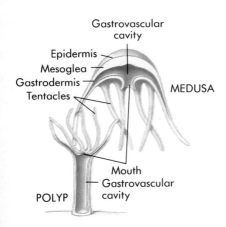

Gastrovascular
cavity

Epidermis
Mesoglea
Gastrodermis
Tentacles

MEDUSA

Mouth
Gastrovascular
cavity

POLYP

FIGURE 39-7
The two kinds of cnidarians, the medusa (above) and the polyp (below). These two phases alternate in the life cycles of many cnidarians, but a number—including the corals and sea anemones, for example—exist only as polyps.

The extracellular fragmentation that precedes phagocytosis and intracellular digestion allows cnidarians to digest animals larger than individual cells, an important improvement over the strictly intracellular digestion of sponges. Any food particles that cannot be digested are released out the same opening through which they were pulled into the animal originally.

Cnidarians are the most primitive animals that exhibit extracellular digestion.

In cnidarians, the digestive cavity has only one opening (see Figure 39-7). This digestive cavity, usually called the **gastrovascular cavity,** has sometimes been called the **coelenteron.** Based on this word, a phylum was once recognized that included both Cnidaria and Ctenophora and was named "Coelenterata." The term seemed especially appropriate, since some of the hydroids—members of one of the characteristic groups of Cnidaria—apparently consist of little more than a gut. When the phylum was divided, however, its two parts were given different names, and the more inclusive group Coelenterata is no longer used.

Nets of nerve cells coordinate the contraction of cnidarian muscles. Such nerve nets apparently have little central control in most cnidarians, although some, such as the jellyfish, display a degree of coordination. Cnidarians have no blood vessels, no respiratory system, and no specialized internal cavity, such as those that occur in more advanced animals.

On their tentacles and sometimes body surface, cnidarians bear specialized cells called **cnidocytes.** The name of the phylum Cnidaria refers to these cells, which are highly distinctive and occur in no other group of organisms. Within each cnidocyte there is a **nematocyst,** in effect a small but powerful harpoon (Figure 39-9). Each nematocyst features a coiled, threadlike tube, which is essentially an extension of the capsule. Lining the inner wall of the tube is a series of barbed spines. Cnidarians use the threadlike tube to spear their prey and then draw the harpooned prey back with the tentacle containing the cnidocyte; nematocysts may also serve a defensive purpose. To propel the harpoon, the cnidocyte uses water pressure. Before firing, the cnidocyte contains a very high concentration of ions, and its membrane is not permeable to water. Within the undischarged nematocyst, the osmotic pressure reaches about 140 atmospheres!

If a flagellum-like trigger on the cnidocyte is touched (and correct chemical stimuli are also present), the nematocyst is stimulated to discharge. Its walls become permeable to water, which rushes inside and violently pushes out the barbed filament. As the filament flies outward, it turns inside out, exposing the barbs. Nematocyst discharge is one of the fastest cellular processes in nature. The stinging sensation associ-

FIGURE 39-8
Planula larva of *Aurelia*, a jellyfish.

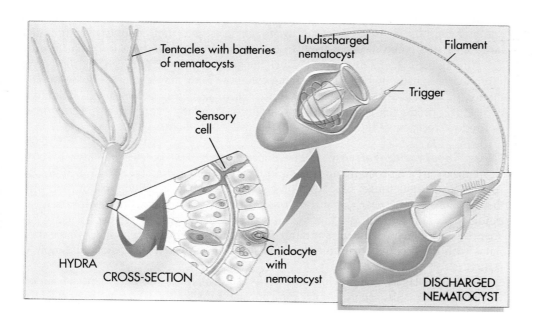

FIGURE 39-9
Basic structure of *Hydra*.

ated with the discharge of the nematocysts results from the injection of a toxic protein.

> **Cnidarians characteristically possess a specialized kind of cell called a cnidocyte. Each cnidocyte contains a nematocyst, a harpoon used to attack prey. Nematocysts are produced only by cnidarians.**

The powerful neurotoxins (nerve poisons) that are secreted by the nematocysts of the Portuguese man-of-war, a large, floating, colonial, marine cnidarian (Figure 39-10), are dangerous and occasionally even fatal to humans. A number of species of jellyfish also have stings that are exceedingly painful; the box jelly (*Chironex fleckeri*) has been responsible for the deaths of at least 60 people in northern Australia alone. If the jellyfish are abundant, they can drive swimmers out of the water.

Some nudibranchs—marine slugs of the phylum Mollusca —and flatworms are able to eat cnidarians, tentacles and all, without causing the nematocysts to discharge. These predators have the ability to retain the nematocysts within their own bodies and use them to defend themselves.

Classes of Cnidarians

There are three classes of cnidarians: Hydrozoa (hydroids), Scyphozoa (jellyfish), and Anthozoa (corals and sea anemones). Hydroid medusae are generally considered to represent the most primitive form of any cnidarian. The Scyphozoa, which are complex in overall structure and in musculature, are medusae, alternating with inconspicuous polyps during the course of their life cycles. They are generally thought to have evolved from hydroid medusae. Increasing emphasis on the polyp generation, which has developed in relation to a particular, sedentary way of life, is pronounced among many hydroids. This evolutionary trend reaches full development in the Anthozoa, all of which are polyps. Like the Scyphozoa, the Anthozoa are believed to have evolved from the Hydrozoa.

Class Hydrozoa: Hydroids. Most of the approximately 2700 species of hydroids (class Hydrozoa) have both polyp and medusa stages in their life cycle. Most of these animals are marine and colonial; an example we have already mentioned is the Portuguese man-of-war (see Figure 39-10). Some of the marine hydroids are bioluminescent.

A well-known hydroid is the abundant freshwater genus *Hydra*, which is exceptional in that it has no medusa stage and exists as a solitary polyp (see Figures 39-6, *A*, and Figure 39-9). Each polyp sits on a basal disk, by means of which it can glide around, aided by mucous secretions. It can also move by somersaulting, bending over and attaching itself to the substrate by its tentacles, and then looping over to a new location. If the polyp detaches itself from the substrate, it can float to the surface.

Among the colonial hydroids, such as the genus *Obelia* (Figure 39-11), the polyps reproduce asexually by budding, forming colonies. They may also give rise by the formation of specialized buds to medusae, in which gametes are produced. These gametes fuse, producing zygotes that develop into planulae—which in turn settle down to produce the new polyps.

Class Scyphozoa: Jellyfish. The approximately 200 species of jellyfish (class Scyphozoa) are transparent or translucent marine organisms (see Figure 39-6, *B*), some of a striking orange, blue, or pink color. These animals spend most of their time floating horizontally near the surface of the sea. In all of them, the medusa stage is dominant—much larger and more complex than the polyps. These medusae are bell shaped, with hanging tentacles around their margins. The polyp stage, which alternates with the medusa stage in most jellyfish, is small, inconspicuous, and simple in structure.

The outer layer, or epithelium, of a jellyfish contains a number of specialized epitheliomuscular cells, each of which can contract individually. Together the cells form

FIGURE 39-10
Portuguese man-of-war, *Physalia utriculus*. The Portuguese man-of-war is a colonial hydrozoan that has adopted the way of life characteristic of the jellyfish. This highly integrated colonial organism can ensnare good-sized fishes by using its painful stings and tentacles, which are sometimes over 15 meters long.

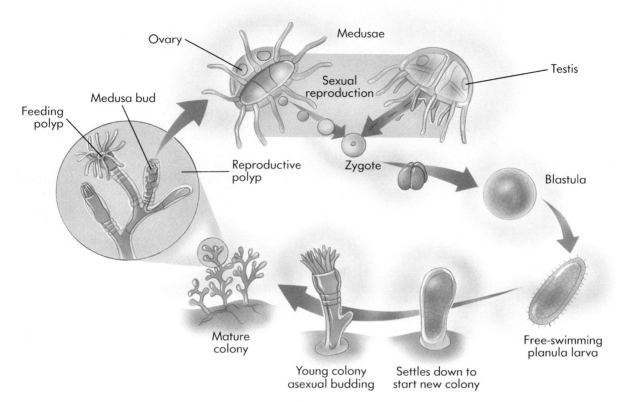

FIGURE 39-11
The life cycle of *Obelia*, a marine colonial hydroid.

a muscular ring around the margin of the bell that pulses rhythmically and propels the animal through the water. Jellyfish have separate male and female individuals. After fertilization, planulae are formed, which develop into the polyps. The polyps reproduce themselves asexually, in addition to budding off medusae. In some jellyfish that live in the open ocean, the polyp stage is suppressed, and the planulae develop directly into medusae.

Class Anthozoa: Sea Anemones and Corals. By far the largest class of cnidarians is Anthozoa, the "flower animals" (from the Greek word *anthos,* meaning "flower"). The approximately 6200 species of this group are solitary or colonial marine animals (see Figures 39-1, *B,* 39-6, *C,* and 39-12). They include the stonelike corals, the soft-bodied sea anemones, and other groups known by such fanciful names as sea pens, sea pansies, sea fans, and sea whips. All of these names reflect a plantlike body topped by a tuft or crown of hollow tentacles, built on a plan involving multiples of six. Like other cnidarians, the anthozoans use these tentacles in feeding. Nearly all members of this class that live in shallow waters harbor symbiotic algae, which supplement the nutrition of their hosts through photosynthesis (see Figure 31-13, *B*). The fertilized eggs of anthozoans usually develop into planulae that settle and develop into polyps; no medusae are formed by these animals.

Sea anemones are a large group of soft-bodied anthozoans. They are found in coastal waters all over the world and are especially abundant in the tropics. When they are touched, most sea anemones retract their tentacles into the body cavity and fold up. Sea anemones are highly muscular and relatively complex organisms, with greatly divided internal cavities. These animals range from a few millimeters to about 10 centimeters in diameter and are perhaps twice that high, but some of them are considerably larger. Many sea anemones are quite colorful.

The corals are another major group of anthozoans. Many of them secrete tough outer skeletons, or exoskeletons, and are thus stony in texture. Others, including the gorgonians, or soft corals, do not secrete such exoskeletons. Some of the hard corals participate in the formation of coral reefs, which are shallow-water limestone ridges

FIGURE 39-12
Symbiosis in corals. Corals, such as this yellow cup coral, are inhabited by zooxanthellae, the symbiotic form of dinoflagellates.

FIGURE 39-13
A comb jelly (phylum Ctenophora). Note the comblike plates
and two tentacles.

that occur in warm seas. Coral reefs are formed as a result of the accumulation of carbonates over long periods. Although the waters where these reefs develop are often nutrient poor, the coral animals themselves, particularly because of the abundant algae associated with them, some of which are symbiotic, are able to grow actively.

Coral reefs are built up in warm seas by colonial coral animals, which harbor algae and are thus able to manufacture food to support themselves.

PHYLUM CTENOPHORA: THE COMB JELLIES

Traditionally the ctenophores (phylum Ctenophora), a group of about 90 species, have been considered to be closely related to the cnidarians; recently, this assumption has been seriously questioned. The members of this small phylum, which range from spherical to ribbonlike, are known as comb jellies or sea walnuts. Ctenophores are structurally more complex than cnidarians; for example, they have anal pores, so that water and other substances pass completely through the animal. Comb jellies themselves are transparent and usually a few centimeters long; they are often abundant in the open ocean (Figure 39-13). The members of one group have two long, retractable tentactles that they use to capture their prey.

Ctenophores propel themselves through the water by means of eight comblike plates of fused cilia that beat in a coordinated fashion; they are the largest animals that use cilia for locomotion. Many ctenophores are luminescent, giving off bright flashes of light that are particularly evident in certain areas of the open ocean at night.

Ctenophores are radially symmetrical animals that propel themselves through the water by means of eight comblike plates of fused cilia. The members of one group have two long, retractable tentacles.

THE EVOLUTION OF BILATERAL SYMMETRY

Unlike a radially symmetrical animal, a bilaterally symmetrical animal (Figure 39-14) has a right and a left half that are mirror images of one another; there is a top and a bottom, better known respectively as the **dorsal** and **ventral** portions of the animal. There is also a front, or **anterior** end, and a back, or **posterior,** end, and therefore right and left sides. Bilaterally symmetrical animals constitute a major evolutionary line; this unique form of organization allowed the differential adaptation of the various parts of the body. In some of them, the adults are radially symmetrical, but the larvae are bilaterally symmetrical. Bilaterally symmetrical animals move from place to place more efficiently than radially symmetrical ones, which, in general, lead a seden-

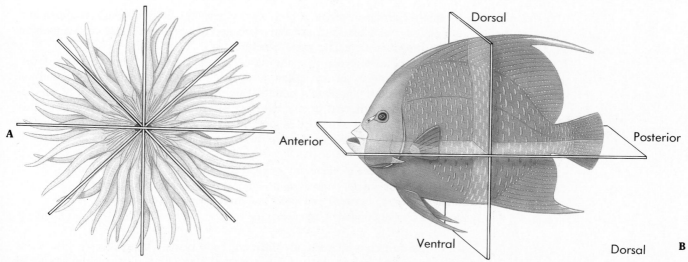

tary existence. Bilaterally symmetrical animals are, therefore, efficient in seeking food and mates and avoiding predators.

During the early evolution of bilaterally symmetrical animals, those structures that were important to the organism in monitoring its environment, and thereby capturing prey or avoiding enemies, came to be grouped at the anterior end. Other functions tended to be located farther back in the body. The number and complexity of sense organs are much greater in bilaterally symmetrical animals than they are in radially symmetrical ones.

Much of the nervous system in bilaterally symmetrical animals is in the form of major longitudinal nerve cords, which constitute the **central nervous system.** In a very early evolutionary advance, nerve cells became grouped around the anterior end of such animals. These nerve cells probably first functioned mainly to transmit impulses from the anterior sense organs to the rest of the nervous system. This trend ultimately led to the evolution of the head and the brain, as well as to the increasing dominance and specialization of these organs in the more advanced animal phyla.

Among bilaterally symmetrical animals, different parts of the body are specialized in relation to different functions. All of the higher animals are bilaterally symmetrical, although in some of them the pattern is evident only in the embryos.

FIGURE 39-14
How radial (A) and bilateral (B) symmetry differ.

Body Plans of Bilaterally Symmetrical Animals

Three basic body plans occur in bilaterally symmetrical animals (Figure 39-15):

1. Some bilaterally symmetrical animals have no body cavity, other than the digestive system. Animals with this kind of a body plan are called acoelomates.

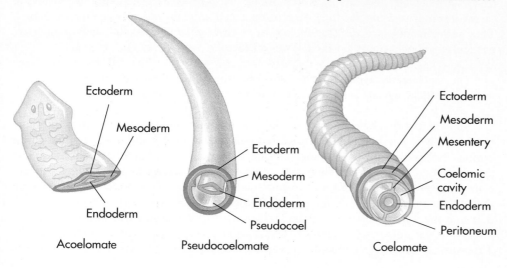

FIGURE 39-15
Three body plans for bilaterally symmetrical animals.

2. In another group of phyla, a body cavity develops between the mesoderm and the endoderm. This kind of cavity is called a pseudocoel, and the animals in which it occurs are called pseudocoelomates.

3. In the more advanced phyla, the tissues of the mesoderm open during development, leading to the formation of a particular kind of body cavity called the coelom, which is completely bounded by mesoderm. The digestive, reproductive, and other internal organs develop within or around the margins of this coelom, suspended within it by double layers of mesoderm known as mesenteries. Animals in which a coelom develops are called coelomates.

The body architecture of bilaterally symmetrical animals follows one of three patterns: (1) acoelomate, possessing no body cavity; (2) pseudocoelomate, possessing a cavity between endoderm and mesoderm; (3) coelomate, possessing a cavity bounded by mesoderm.

Each of these three fundamentally different body plans is characteristic of a series of animal phyla that belong to a distinct evolutionary line. The acoelomates and pseudocoelomates are all worms, and each of these lines comprises several phyla. Coelomates, however, are the most numerous and the most diverse of the three groups; all vertebrates and more advanced invertebrates are coelomates. The relationships between coelomates, pseudocoelomates, and acoelomates are not clear, however; acoelomates, for example, could have given rise to coelomates or been derived from them. The different phyla of pseudocoelomates might have had separate origins.

ACOELOMATES: THE SOLID WORMS

Among bilaterally symmetrical animals, those with the simplest body plan are the acoelomates; they lack any internal cavity other than the digestive tract (see Figure 39-15). We will focus our discussion of the acoelomates on the largest phylum of the group, the flatworms.

Phylum Platyhelminthes: The Flatworms

The flatworms (phylum Platyhelminthes; see Figures 39-1, *C*, and 39-16) consist of some 12,200 species. These ribbon-shaped, soft-bodied animals are so called because they are flattened dorsoventrally, from top to bottom. Flatworms are among the simplest of bilaterally symmetrical animals, but they have a definite head at the anterior

A end. Their bodies are solid: the only internal space consists of the digestive cavity.

Flatworms range in size from a millimeter or less to many meters long, as in some of the tapeworms. Most species of flatworms are parasitic; members of this phylum occur within the bodies of members of many other kinds of animals. A number of other kinds of flatworms, however, are free living, occurring in a wide variety of marine and freshwater habitats, as well as moist places on land. Free-living flatworms are carnivores and scavengers; they eat various small animals and bits of organic debris. They move from place to place by means of their ciliated epithelial cells, which are particularly concentrated on their lower surfaces.

The acoelomates, typified by the flatworms, are the most primitive bilaterally symmetrical animals and the simplest animals in which organs occur.

B

FIGURE 39-16
Flatworms (phylum Platyhelminthes).
A A marine, free-living flatworm.
B The human liver fluke,
Clonorchis sinensis.

Many flatworms have a gut with only one opening (Figure 39-17). Since they cannot feed, digest, and eliminate undigested particles of food simultaneously, flatworms cannot feed continuously, as more advanced animals can. Muscular contractions in the upper end of the gut of flatworms cause a strong sucking force by which the flatworms ingest their food and tear it into small bits. The gut is branched and extends throughout the body, functioning in both digestion and transport of food. The cells that line the gut engulf most of the food particles by phagocytosis and digest them; but, as in the cnidarians, some of these particles are partly digested extracellularly. Tapeworms, which are parasitic flatworms, lack digestive systems; they absorb their

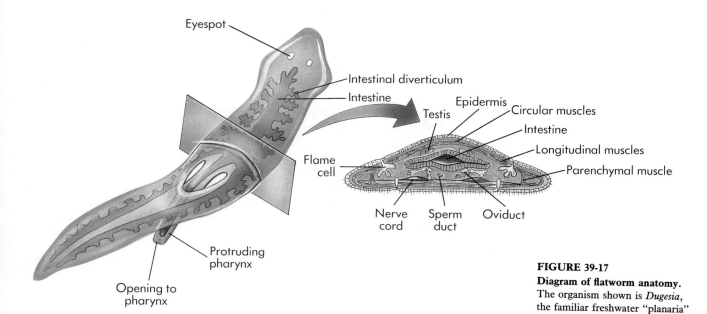

Eyespot

Intestinal diverticulum

Intestine

Testis

Epidermis

Circular muscles

Intestine

Longitudinal muscles

Parenchymal muscle

Flame cell

Nerve cord

Sperm duct

Oviduct

Protruding pharynx

Opening to pharynx

FIGURE 39-17
Diagram of flatworm anatomy.
The organism shown is *Dugesia*, the familiar freshwater "planaria" of many biology laboratories.

food directly through their body walls. They attach themselves to the intestine wall, where they are bathed by dissolved nutrients.

Unlike the cnidarians, flatworms do have an excretory system, which consists of a network of fine tubules (little tubes) that runs throughout the body. Cilia line the hollow centers of the bulblike **flame cells,** which are located on the side branches of the tubules. By doing so, the cilia move water with the substances to be excreted into a system of tubules and then to exit pores located between the epidermal cells. Flame cells were named because of the flickering movements of the tuft of cilia within them. They primarily regulate the water balance of the organism. The excretory function of the flame cells appears to be a secondary one, since a large proportion of the metabolic wastes excreted by flatworms probably diffuses directly into the gut and is eliminated through the mouth.

Flatworms lack circulatory systems, which transport oxygen and food molecules in higher organisms. Consequently all of the cells of flatworms must be within diffusion distance of oxygen and food. Flatworms have thin bodies and highly branched digestive cavities, which make such a relationship possible.

The nervous system of flatworms is very simple compared to that of the other bilaterally symmetrical animals. Some primitive flatworms have only a nerve net, like that of the cnidarians. In most of the members of this phylum, however, there are definite longitudinal nerve cords that constitute a simple central nervous system. The tiny swellings at the anterior end of these cords are, in essence, primitive brains. Although they are fairly well developed in some of the free-living flatworms, these brains exert only a limited dominance over the activities of the animals in which they occur. Between the cords there are cross-connections, so that the flatworm nervous system resembles a ladder.

Flatworms lack circulatory systems, and most of them have a gut with only one opening. They excrete wastes directly from the gut, but also by means of a network of fine tubules that has ciliated flame cells on the side branches. Their nervous systems are simple.

Free-living flatworms use sensory pits or tentacles along the sides of their heads to detect food, chemicals, or movements of the fluid in which they are moving. Once they have sensed food, they move directly toward it and begin to feed. The free-living members of this phylum also have eyespots on their heads. These are inverted, pigmented cups containing light-sensitive cells, connected with the nervous system. These eyespots enable the worms to distinguish light from dark; the worms move away from strong light.

Flatworms are far more active than are cnidarians or ctenophores. Such activity is characteristic of bilaterally symmetrical animals; in flatworms, it seems to be related to the greater concentration of sensory organs and, to a degree, of the nervous system elements in the heads of these animals. This concentration marks the beginning of an evolutionary trend of critical importance for the kingdom Animalia.

The reproductive systems of flatworms are complex. Most flatworms are hermaphroditic; in many of them, fertilization is internal. When they mate, each partner deposits sperm in the copulatory sac of the other. The sperm travel along special tubes to reach the egg. In free-living flatworms, the fertilized eggs are laid in cocoons strung in ribbons and hatch into miniature adults; in some parasitic flatworms, there is a complex succession of distinct larval forms.

Flatworms are capable of asexual regeneration to a much greater degree than are the members of most more advanced animal phyla. In some genera, when a single individual is divided into two or more parts, each part can regenerate an entire new flatworm.

Class Turbellaria: The Turbellarians.
Of the three classes of flatworms, the turbellarians (class Turbellaria) are free living. One of the most familiar is the freshwater genus *Dugesia*, the common planaria (see Figure 39-17), which is used in biology laboratory exercises. Other members of this class are widespread and often abundant in lakes, ponds, and the sea. Some of them also occur in moist places on land.

Class Trematoda: The Flukes.
There are two classes of parasitic flatworms, which live within the bodies of other animals: flukes (class Trematoda) and tapeworms (class Cestoda). In both groups the worms have epithelial layers and linings of the gut that are resistant to the digestive enzymes produced by their hosts—an im-

FIGURE 39-18
Life cycle of the human liver fluke, *Clonorchis sinensis*.

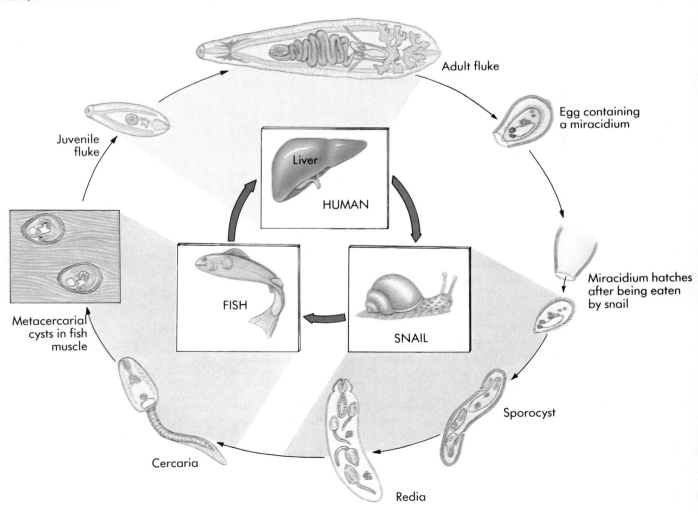

Adult fluke

Egg containing a miracidium

Juvenile fluke

Liver

HUMAN

Miracidium hatches after being eaten by snail

Metacercarial cysts in fish muscle

FISH

SNAIL

Sporocyst

Cercaria

Redia

portant feature in relation to their parasitic way of life. However, they lack certain features of the free-living flatworms, such as cilia in the adult stage, eyespots, and other sensory organs, all of which lack adaptive significance for an organism that lives within the body of another animal.

Flukes take in food through their mouth, just like their free-living relatives. There are more than 8100 named species, ranging in length from less than 1 millimeter to more than 8 centimeters. Flukes attach themselves within the bodies of their hosts by means of suckers, anchors, or hooks. Some have a life cycle that involves only one host—usually a fish; most, however, have life cycles involving two or more hosts. Their larvae almost always occur in snails, and there may be secondary intermediate hosts. The final host of such flukes, however, is almost always a vertebrate.

To human beings, one of the most important flatworms is the human liver fluke, *Clonorchis sinensis* (see Figure 39-16, *B*), which lives in the bile passages of the liver of humans, cats, dogs, and pigs. It is especially common in the Orient. The worms are 1 to 2 centimeters long and have a complex life cycle (Figure 39-18). Although they are hermaphroditic, with each individual containing a full complement of male and female sexual structures, cross-fertilization between different individuals is usual. The eggs, each of which contains a complete, ciliated first-stage larva, or **miracidium,** are passed out in the feces. If they reach water, they may be ingested by a snail, within which they transform into a sporocyst—a baglike structure with embryonic germ cells. Within the sporocysts are produced **rediae,** which are elongate, nonciliated larvae. These larvae continue growing within the snail, giving rise there to several individuals of the tadpolelike next larval stage, the **cercariae.**

The cercariae, which are produced within a snail, escape into the water, where they swim about. If they encounter a fish of the family Cyprinidae—the family that includes carp and goldfish—they bore into the muscles or under the scales, lose their tails, and transform into metacercariae within cysts in the muscle tissue. If a human being or other mammal eats raw infected fish, the cysts dissolve in the intestine, and the young flukes migrate to the bile duct, where they mature. An individual fluke may live for 15 to 30 years in the liver. In humans a heavy infestation of liver flukes may cause cirrhosis of the liver and death.

Other very important flukes are the blood flukes of genus *Schistosoma* (Figure 39-19), which afflict about 1 in 20 of the world's population, more than 300 million people throughout tropical Asia, Africa, Latin America, and the Middle East. Three species of *Schistosoma* cause the disease called schistosomiasis, or bilharzia. Some 800,000 people die each year from this disease.

The tiny yellow eggs of these worms (see Figure 39-19, *B*) leave the human body through the urine and feces. If they reach water, they hatch into miracidia. The miracidia, which are completely formed within the eggs, resemble ciliates. They must reach their freshwater snail hosts within a few hours to survive; if they do, they develop into sporocysts. These sporocysts divide asexually to give rise to daughter sporocysts, and the daughter sporocysts in turn give rise to cercariae like those of *Clonorchis.* A single infected snail may release 100,000 cercariae during its lifetime of several months; it starts to release them 25 to 40 days after it is infected. Instead of attacking fishes, the cercariae of *Schistosoma* burrow into human skin. They ultimately reach the bloodstream and are swept along to the lungs, where they remain for about 10 days. Subsequently they reenter the circulatory system and migrate to the hepatic and portal veins, which supply the liver. The worms become sexually mature and pair (see Figure 39-19, *A*) on their journey to this destination, where they mate and then migrate joined together to the veins that supply the upper intestine, the lower intestine, or the bladder, depending on the species of *Schistosoma* involved. In these sites, the adult worms live from several to more than 30 years, copulating and shedding from 300 to more than 3000 eggs per day continuously, but not reproducing themselves within the human body.

The very large numbers of eggs that these worms produce are the primary cause of the inflammation, extreme discomfort, and sometimes death associated with schistosomiasis. They cause widespread ulceration and bleeding of the intestinal and bladder walls and may also cause cirrhosis of the liver by blocking the small capillaries in that organ. Proper sanitation is important in controlling this disease, since it can con-

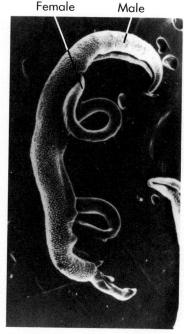

Female   Male

A

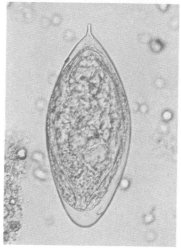

B

FIGURE 39-19
Schistosoma.
A Scanning electron micrograph of copulating male and female individuals. The thicker male has a canal that runs the length of its body, within which the female is held during insemination and egg laying. The worms mate in the human bloodstream.
B An egg containing a mature larva.

tinue only if urine or feces reach bodies of fresh water in which snails are living. It can also be controlled by eliminating the snail populations or reducing them to very low levels, since the miracidia can live only a few hours before they must reach such a host. The widespread introduction of irrigation in the tropics has contributed greatly to the spread of this serious disease, by spreading habitats for the snails that carry the worms.

Recently there has been a great deal of effort to control schistosomiasis. The worms protect themselves in part from the body's immune system by coating themselves with a variety of the host's own antigens that effectively render the worm immunologically invisible (see Chapter 52). Despite these difficulties, the search for a vaccine that would cause the host to develop antibodies to one of the antigens of the young worms before they protect themselves with host antigens is being actively pursued. The disease can also be cured with drugs once humans have been infected, by stopping larvae from reaching people, or by stopping eggs from reaching the water in human feces. All of these alternatives are being investigated.

Class Cestoda: The Tapeworms.
Tapeworms (class Cestoda) are the third class of flatworms; like flukes, they live as parasites within the bodies of other animals. In contrast to flukes, tapeworms simply hang on to the inner walls of their hosts by means of specialized terminal attachment organs and absorb food through their skins. Tapeworms lack digestive cavities as well as digestive enzymes. Thus they are extremely specialized in relation to their parasitic way of life (Figure 39-20). Most species of tapeworms occur in the intestines of vertebrates, about a dozen of them regularly in humans.

The long, flat bodies of tapeworms are divided into three zones: the **scolex,** or attachment organ; the unsegmented **neck;** and a series of repetitive segments, the proglottids. The scolex usually bears several suckers, and it may also have hooks. Each proglottid is a complete hermaphroditic unit, containing both male and female reproductive organs. Proglottids are formed continuously in an actively growing zone at the base of the neck, with the maturing ones moving farther back as new ones are formed in front of them. Ultimately the proglottids near the end of the body form mature eggs. As these eggs are fertilized, the zygotes in the very last segments begin to differentiate, and these segments become filled with embryos. These embryos, each surrounded by a shell, emerge from the proglottid through either a pore or the ruptured body wall, leave their host with the feces, and are deposited on leaves, in water, or in other places from which they may be picked up by another animal.

The beef tapeworm, *Taenia saginata* (see Figure 39-20), occurs as a juvenile primarily in the intermuscular tissue of cattle but as an adult in the intestines of human

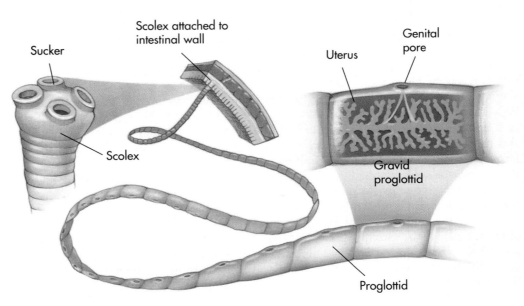

FIGURE 39-20
Structure of the beef tapeworm,
Taenia saginata.

Sucker

Scolex attached to intestinal wall

Scolex

Uterus

Genital pore

Gravid proglottid

Proglottid

FIGURE 39-21
A ribbon worm, *Lineus* (phylum
Rhynchocoela).

beings. A mature adult beef tapeworm may reach a length of 10 meters or more. These worms attach themselves to the intestinal wall of their host by a scolex with four suckers. The segments that are shed from the end of the worm pass out of the human in the feces and may crawl onto vegetation; they ultimately rupture and scatter the embryos. The embryos may remain viable for up to 5 months. If they are ingested by cattle, they burrow through the wall of the intestine and ultimately reach the muscle tissues by way of the blood or lymph vessels. About 1% of the cattle in the United States are infected, and some 20% of the beef consumed is not federally inspected; when such beef is eaten rare, infection of humans by these tapeworms is likely to occur. As a result, the beef tapeworm is a frequent parasite of humans.

Phylum Rhynchocoela: The Ribbon Worms

The members of phylum Rhynchocoela, the ribbon worms (Figure 39-21), share many characteristics with free-living flatworms but differ from them in several significant ways. These acoelomate aquatic worms, the majority of which are marine, may be thread shaped or ribbon shaped. There are about 650 species. They are characterized by their proboscis, a long muscular tube that can be thrust out quickly from a sheath to capture prey. Ribbon worms are large, often 10 to 20 centimeters and sometimes many meters in length. They are the simplest animals with a complete digestive system and also the simplest ones with a circulatory system in which the blood flows in vessels. In these ways, the ribbon worms provide early indications of important evolutionary trends that become fully developed in more advanced animals.

THE EVOLUTION OF A BODY CAVITY

The body organization of the other bilaterally symmetrical animals differs from the solid worms in an important respect: all the other bilaterally symmetrical animals possess an internal body cavity. Among them, seven phyla are characterized by their possession of a pseudocoelomate body plan (see Figure 39-15). Only one of the seven, however, the phylum Nematoda, includes a large number of species. In all pseudocoelomates, the pseudocoel serves as a hydrostatic skeleton—one that gains its rigidity from being filled with fluid under pressure. The animal's muscles can work against this "skeleton," thus making the movements of pseudocoelomates far more efficient than those of the acoelomates.

The pseudocoelomates were the first animals to possess an internal body cavity. Among many other advantages, this cavity makes the animal's body rigid, permitting resistance to muscle contraction and thus opening the way to muscle-driven body movement.

A

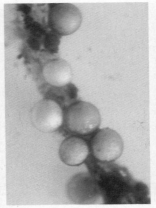
B

C

FIGURE 39-22
Many nematodes are serious pests of plants.

A Nematode damage to a field of onions; damage caused by lesion nematodes (*Pratylenchus penetrans*) in Indiana. The area at right was treated with a nematicide to kill most of the nematodes before the onions were planted; the area at left was untreated.
B Cysts of the golden nematode (*Globodera rostochiensis*) on roots of potato. Although most nematode species remain wormlike throughout their lives, the mature females of some species that feed on plants become round cysts filled with eggs, as shown here.
C Swellings on crown and roots of African violet (*Saintpaulia*), caused by root knot nematode (*Meloidogyne*).

The pseudocoelomates lack a defined circulatory system; its role is performed by the fluids that move within the pseudocoel. All pseudocoelomates have a complete, one-way digestive tract, as do all coelomates. This digestive tract acts like an assembly line, with the food being acted on in different ways in each section. First it is broken down, then absorbed, then the wastes are treated and stored, and so on. Considerable specialization has evolved in the digestive tracts of the pseudocoelomate and coelomate phyla, as we will see.

Phylum Nematoda: The Nematodes

The nematodes, eelworms, and roundworms comprise a large phylum, Nematoda, with some 12,000 recognized species; scientists estimate that the actual number might approach 100 times that many, however. The members of this phylum are ubiquitous. Nematodes are abundant and diverse in marine and freshwater habitats, and many members of this phylum are parasites of vertebrates, invertebrates, and plants (Figure 39-22). Many nematodes are microscopic animals that live in soil; it has been estimated that a spadeful of fertile soil may contain, on the average, a million nematodes (Figure 39-23).

Almost every species of plant and animal that has been studied has been found to have at least one parasitic species of nematode living in it. Many of these worms live in environments that are very hot, dry, cold, or salty; when these conditions occur seasonally, the nematodes may be able to avoid them by entering a dormant state. Some nematodes are being investigated as agents of biological control of insects and other agricultural pests and, provided that they can be produced in sufficient quantities, they may be useful for such purposes.

Nematodes are bilaterally symmetrical, cylindrical, unsegmented worms. They are covered by a flexible, thick cuticle, which is molted as they grow. Their muscles constitute a layer beneath the epidermis and extend along the length of the worm, rather than encircling its body. These longitudinal muscles pull both against the cuticle and against the pseudocoel, which forms a hydrostatic skeleton. When nematodes move, their bodies whip about from side to side.

Near the mouth of a nematode, at its anterior end, there are usually 16 raised, often hairlike, sensory organs. The mouth is often equipped with piercing organs called stylets. In the complete digestive tract of nematodes, the food first passes through the mouth as a result of the sucking action of a muscular chamber called the **pharynx**. After passing through a short corridor into the pharynx, food continues through the other portions of the digestive tract, where it is broken down and then digested. Some of the water with which the food has been mixed is reabsorbed near the end of the digestive tract, and the material that has not been digested is eliminated through the anus.

Nematodes completely lack flagella or cilia, even on the sperm cells. Their excretory system consists of cells that function as glands, or systems of canals, and does not depend on flagella or cilia for its functioning, thus differing from the excretory sys-

FIGURE 39-23
Living dirt. One square meter of ordinary garden, lawn, or forest soil teems with 2 to 4 million nematodes. Although most are similar in form, they range from about 0.2 millimeter to about 6 millimeters long. A trained nematologist (student of nematodes) must examine slide mounts with a compound microscope to determine which species are present.

tems of many animals. Reproduction in nematodes is sexual, with the sexes usually separate. Their development is simple, and the adults consist of very few cells that develop in a precise manner; for this reason, nematodes have become extremely important subjects for genetic and developmental studies. The 1-millimeter-long *Caenorhabditis elegans* matures in only three days, its body is transparent, and with only 1,000 cells, it is the only animal whose complete cellular anatomy is known. After two decades of intense international research, scientists are nearing completion of a complete physical map of its approximately 10,000 genes—the first such map for a multicellular organism.

About 50 species of nematodes, including several that are rather common in the United States, regularly parasitize human beings. Hookworms, mostly of the genus *Necator*, can be common in certain warm regions. The larvae occur in soil and burrow into skin that comes into contact with them. Their anterior end curves dorsally, suggesting a hook, which is why they are so named. Inside the host they attach to the wall of the small intestine and feed on blood. Because the worms suck much more blood from the intestinal wall than they use, their activities often result in anemia if they are abundant enough.

Pinworms, *Enterobius,* are abundant throughout the United States, where it is estimated they infect about 30% of all children and about 16% of adults. Fortunately the symptoms they cause are not severe, and the worms can easily be controlled by drugs. Adult pinworms live in the human large intestine and blind gut, laying fertilized eggs in the anal region of their hosts at night, which causes itching. The adult female worms are up to about 12 millimeters in length. The eggs, which become infective within 6 hours, are spread by scratching the anal region. Pinworms can be irritating but are not regarded as a particularly serious parasite.

The intestinal roundworm, *Ascaris,* infects approximately one of six people worldwide but is rare in most areas with modern plumbing. Various species of this genus parasitize other animals. Like pinworms, intestinal roundworms occur only in human beings, and there is no intermediate host. They live in the intestine, and their fertilized eggs are spread in feces; these eggs can remain viable for years in the soil. In relatively unsanitary conditions, the eggs may be ingested easily because of improper washing of the hands. The young worms hatch in the small intestine, bore through the wall of the gut, pass through the heart and lungs, and eventually pass out of the breathing passage. They are then swallowed and reach the intestine again. The adult females, which are up to 30 centimeters long (Figure 39-24), contain up to 30 million eggs and can lay up to 200,000 of them each day! Obviously very few of these eggs ever reach another host.

The most serious common nematode-caused disease in temperate regions is trich-

FIGURE 39-24
Structure of the large roundworm of humans, *Ascaris lumbri-
coides.*

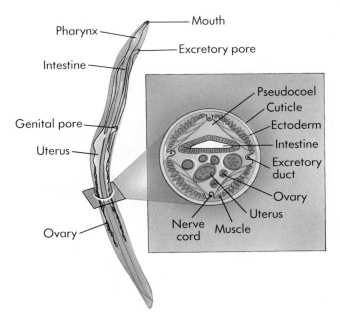

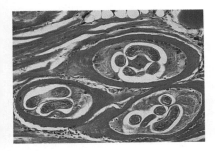

FIGURE 39-25
Cysts of *Trichinella* in pork.
These worms cause the disease
trichinosis in human beings.

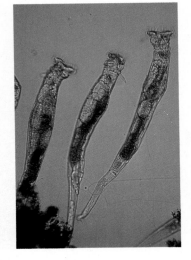

FIGURE 39-26
Rotifers (phylum Rotifera), common, small aquatic animals.

inosis, caused by worms of the genus *Trichinella*. These worms live in the small intestine of pigs, where the fertilized females burrow into the intestinal wall. Once it has
penetrated these tissues, each female produces about 1500 live young. The young enter the lymph channels and thereby reach muscle tissue throughout the body, within
which they mature and form highly resistant, calcified cysts (Figure 39-25). Infection
in human beings or other animals arises from eating undercooked or raw pork in
which the cysts of *Trichinella* are present. If the worms are abundant, a fatal infection
can result, but such infections are rare. It is thought that about 2.4% of the people in
the United States are infected with trichinosis, but only about 20 deaths have been
attributed to this disease during the past decade. The same parasite occurs in bears,
and when humans eat improperly cooked bear meat, there seems to be an even
greater chance of infection than with pork.

Other nematode-caused diseases are extremely serious in the tropics. *Filaria* and
other genera cause filariasis, which infects at least 250 million people worldwide.
These worms can be up to 10 centimeters long and live in the lymphatic system,
which they may seriously obstruct, causing potentially severe inflammation and swelling. The female worms release the larvae into the blood and lymph, from which they
may be picked up by female mosquitoes along with their blood meal. When an infected mosquito bites another animal, the larval worms may penetrate into the wound
and thus reach a new host. The condition known as elephantiasis, a grotesque swelling of the legs or other extremities, may result from extreme filariasis. In certain tropical regions, filariasis-causing nematodes are the most serious health problem, effectively blocking settlement in areas where they are abundant.

Phylum Rotifera: Rotifers

Rotifers (phylum Rotifera; Figure 39-26) are common, small, basically aquatic animals that have a crown of cilia at their heads. Rotifers are pseudocoelomates, as are
nematodes; they have several features that suggest that their ancestors may have resembled flatworms. There are about 2000 species of this phylum. Some of them live
in the soil or in the capillary water in cushions of mosses; most occur in fresh water
and are common everywhere. A very few rotifers are marine. Most rotifers are between 100 and 500 micrometers in length; they are bilaterally symmetrical and covered by an external layer of chitin. They depend on their cilia for both feeding and
locomotion, sweeping their prey into their mouths and swimming from place to place.
Rotifers are often called "wheel animals" because the cilia, when they are beating together, resemble spokes radiating from a wheel.

In 1983, the scientific world was surprised by the announcement of the discovery of a new phylum of pseudocoelomates, named Loricifera. The members of this phylum are tiny animals that grow from larvae less than 195 micrometers long into adults only about 230 micrometers long. They are members of the **meiofauna,** animals adapted to living in the spaces between grains of sand in the ocean. Previously, members of five or six other phyla had been known to live in this habitat.

The new phylum was discovered by Reinhardt Kristensen of the University of Copenhagen. He chose the name Loricifera, from Greek words meaning "girdle-bearers." The most complete samples were obtained from gravel or sand taken from the ocean floor off France, Greenland, and Florida, at depths of 25 to 30 meters. The animals cling tenaciously to the grains of sand or gravel by means of clawlike and club-shaped spines on their head. They can be shocked into releasing

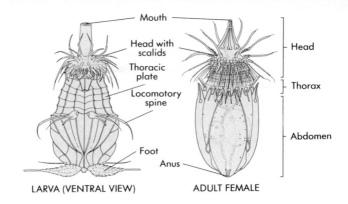

LARVA (VENTRAL VIEW)   ADULT FEMALE

their grip on the grains by flooding the samples with fresh water. They have unusual mouthparts, consisting of a unique flexible tube that can be telescopically retracted into the body. The larvae, which molt as they grow, swim by means of a pair of very unusual locomotory spines, which can propel them in any direction. These spines are attached to a kind of ball-and-socket joint. In contrast, the adults lack such appendages and may be sedentary. More and more Loricifera are turning up in samples that have been screened in the proper way, and the animals may actually be fairly common but undetected because of their unusual habits. A larva and an adult female of *Nanaloricus mysticus,* a species collected off the coast of France, are shown in Figure 39-A.

Free-living rotifers feed on bacteria, protists, and small animals. Rotifers have a muscular pharynx and complex jaws inside it, which grind rapidly. They also have flame cells like those of the flatworms, and they regulate their osmotic pressure by excreting water from these cells. Some rotifers are sessile, whereas others creep, float, or swim. The sexes of rotifers are separate. In many species, parthenogenesis (in which unfertilized eggs develop only into females) is the rule; males do not occur in these species. Each female rotifer lays between 8 and 20 large eggs during her life.

EVOLUTIONARY EXPERIMENTS IN BODY PLAN

In this chapter we have examined representatives of most of the kinds of body plans that occur in animals. We began with the sponges, very simple animals in which there are no tissues differentiated, and continued with the radially symmetrical phyla, animals that lack organs. These radially symmetrical animals, the cnidarians and ctenophores, apparently were derived originally from radially symmetrical ancestors. Even though the echinoderms, which we will meet in Chapter 42, are often radially symmetrical as adults, their larvae are bilaterally symmetrical. From this fact we are justified in concluding that the radial symmetry of adult echinoderms developed secondarily during the course of their evolution.

Once bilateral symmetry had developed among the animals, the stage was set for extensive specialization of the different parts of the body. Among the bilaterally symmetrical animals, there are three fundamentally different kinds of body plans; we call these the acoelomate, pseudocoelomate, and coelomate patterns. All of the representatives of the first two groups are worms, although some of them are very elaborate. It is among the coelomates, which we will begin to consider in the next chapter, that animal diversity reaches its full extent.

1. The animals, kingdom Animalia, comprise some 35 phyla and at least 9 to 10 million species—well over 90% of all living species. Animals are heterotrophic, multicellular organisms that ingest their food.

2. A mature animal develops from a zygote by a characteristic embryological process involving blastula and gastrula stages and complex cell movements.

3. The kingdom Animalia is divided into two subkingdoms: Parazoa, which includes only the asymmetrical phylum Porifera, and Eumetazoa. Eumetazoa is further divided into two branches: Radiata, the radially symmetrical animals and Bilateria, the bilaterally symmetrical animals.

4. The sponges (phylum Porifera) are characterized by specialized, flagellated cells called choanocytes, by a lack of symmetry in their body organization in most species, and by a lack of tissues and organs.

5. The remaining animals have radially or bilaterally symmetrical body plans and distinct tissues. The cnidarians (phylum Cnidaria) are predominantly marine, radially symmetrical animals with unique stinging cells called cnidocytes, each of which contains a specialized harpoon apparatus, or nematocyst. The comb jellies (phylum Ctenophora) are also radially symmetrical, but may not be related to the cnidarians.

6. Bilaterally symmetrical animals include three major evolutionary lines: acoelomates, which lack a body cavity; pseudocoelomates, which develop a body cavity (pseudocoel) between the mesoderm and the endoderm; and coelomates, which develop a body cavity (coelom) within the mesoderm.

7. Acoelomates are the most primitive bilaterally symmetrical animals. They lack an internal cavity, except for the digestive system, and are the simplest animals that have organs—structures made up of two or more tissues. The most prominent phylum of acoelomates, Platyhelminthes, includes the free-living flatworms and the parasitic flukes and tapeworms.

8. The ribbon worms (phylum Rhynchocoela) are similar to free-living flatworms, but have a complete digestive system and a circulatory system in which the blood flows in vessels. They are marine.

9. Pseudocoelomates, exemplified by the nematodes (phylum Nematoda), have a body cavity that develops between the mesoderm and the endoderm. The lining of this pseudocoel provides a place to which the muscles can attach, giving these worms enhanced powers of movement. Rotifers (phylum Rotifera), or wheel animals, are another phylum of pseudocoelomates.

REVIEW QUESTIONS

1. Are all animals multicellular? Why or why not? What is the primary difference between vertebrate and invertebrate animals? Which group is more populous in terms of representative species? What is the structural nature of animal cells that differentiates them from plant cells? Into what functional units are these cells organized? What are collections of these units called?

2. What are the two subkingdoms of animals? How do they differ in terms of symmetry and body organization? What embryological tissues are present in the more primitive subkingdom? Into what organ systems do the embryological tissues of the more advanced subkingdom differentiate?

3. From what unicellular ancestor did each of the animal subkingdoms evolve? What single, general characteristic divides the "lower" invertebrates from the "higher" invertebrates? What are the four major phyla of "lower" invertebrates?

4. Where is food digested in a sponge? Why is the interior of some sponges highly convoluted? What are the primary stages in sponge sexual reproduction?

5. What is the basic composition of a cnidarian body? What are the two body forms present in this group? Describe each, including the ploidy of each form. What is a planula?

6. Where does digestion occur in cnidarians? What major evolutionary innovation is associated with this system? Why is this important? By what chemical means is this digestion accomplished? What happens to indigestible particles?

7. What are the three basic body plans exhibited by bilaterally symmetrical animals? Describe the type of body cavity found in each. How are the internal organs arranged in the most complex plan? Which body plans are found in wormlike organisms? Which plan is found in all advanced invertebrates and vertebrates?

8. What body plan do members of the phylum Platyhelminthes possess? What is their general appearance? Are these animals parasitic or free living? How do they move from place to place? What type of digestive system do they possess?

9. What sensory structures are found on the heads of free-living flatworms? Why are these structures important? What is the nature of sexual reproduction in flatworms? Are they capable of asexual reproduction?

10. How do flukes ingest food? Generally, how many hosts are needed to complete an entire fluke life cycle? What are the stages of a typical fluke life cycle? Indicate in what host each stage is found.

11. What is one of the most important functions of the pseudocoel in those animals that possess one? What evolutionary advantage does this give these animals? How are the internal systems of these animals generally organized?

12. Briefly describe the digestive tract of a typical nematode. Why are nematodes structurally unique in the animal world? How are these animals frequently damaging to humans?

THOUGHT QUESTIONS

1. Outline some of the advantages of bilateral symmetry.

2. How might you investigate whether or not sponges shared a common ancestor with the other animals?

3. Parasites, especially those that require two or more hosts to complete their life cycles, often produce very large numbers of offspring. What advantage would this present to them?

FOR FURTHER READING

ALEXANDER, R.M.: *Animals,* Cambridge University Press, Cambridge, 1990. An exciting and lucid account of the ways in which the various groups of animals hae adjusted to their modes of life.

FIELD, K.G., and others: "Molecular Phylogeny of the Animal Kingdom," *Science,* 1988, vol. 239, pages 748-753. Although fairly technical, this article illustrates beautifully the way in which molecular evidence is being used to clarify the evolutionary relationships of animals.

HICKMAN, C.P., Jr., L.S. ROBERTS, and F.M. HICKMAN: *Integrated Principles of Zoology,* ed. 8, Times Mirror/Mosby College Publishing, St. Louis, 1988. An outstanding general zoology text, with thorough coverage of all groups; the standard in its field.

JACKSON, J.B.C., and T.P. HUGHES: "Adaptive Strategies of Coral-Reef Invertebrates," *American Scientist,* 1985, vol. 73, pages 265-274. Provides outstanding insight into the way in which a coral-reef ecosystem functions.

KUHLMANN, D.H.H.: *Living Coral Reefs of the World,* Arco Publishing, New York, 1985. Beautifully illustrated account of coral reefs and research concerning them.

MONTGOMERY, G.: "Worm Watching: The Case of the Suicidal Sex Cell," *Discover,* October 1987, pages 44-54. This article makes clear some of the reasons that nematodes have become important in genetic and developmental studies.

PECHENIK, J.A.: *Biology of the Invertebrates,* PWS Publishers, Boston, 1985. Clear account of all groups of invertebrates.

RICKETTS, E.F., J. CALVIN, and J.W. HEDGPETH: *Between Pacific Tides,* ed. 5, Stanford University Press, Palo Alto, Calif., 1985. An outstanding, ecologically oriented treatment of the marine organisms of the Pacific Coast of the United States.

SCHMIDT, G.D., and L.S. ROBERTS: *Foundations of Parasitology,* ed. 4, Times Mirror/Mosby College Publishing, St. Louis, 1989. General treatment of parasitology, with much information about parasitic groups of animals.

TRAGER, W.: *The Biology of Animal Parasitism,* Plenum Publishing, New York, 1986. Excellent, thoughtful account of all aspects of parasitism.

WHARTON, D.A.: *A Functional Biology of Nematodes,* Croom Helm, a division of Routledge, London, 1986. Contemporary account of an abundant and highly significant phylum of animals.

WILLMER, P.: *Invertebrate Relationships: Patterns in Animal Evolution,* Cambridge University Press, Cambridge, 1990. For a deeper look at the relationships of invertebrate groups, this book is highly recommended.

Mollusks and Annelids

OVERVIEW

*The evolution of a coelom—an internal body cavity surrounded by mesoderm—
was a major advance for the animal kingdom that facilitated the development of
complex internal organs. Among coelomate animals, there are two major evolutionary lines: the protostomes, in which the mouth develops from the blastopore,
include mollusks, annelids, and arthropods (insects, crustaceans, and related
groups), whereas the deuterostomes, in which the anus develops from the blastopore, consist primarily of echinoderms and chordates. Several small phyla of
primarily marine animals, comprising the lophophorates, are somewhat intermediate between these two major evolutionary branches in their features. They will
be treated here, along with the mollusks (the snails, bivalves, and octopuses),
phylum Mollusca. Most mollusks are unsegmented, but it is not certain whether
their ancestors were segmented or not. In the segmented worms, or annelids (marine worms, earthworms, and leeches), phylum Annelida, the body consists of a
few to many dozen similar segments, lined one behind the other like a string of
beads. Mollusks and some annelids (the polychaetes) have similar trochophore
larvae; the two phyla probably share a relatively close common ancestor.*

FOR REVIEW

*Here are some important terms and concepts that you will encounter in this
chapter. If you are not familiar with them, you should review them before proceeding.*

Major features of evolutionary history (Chapter 22)

Classification (Chapter 28)

Body plans of bilaterally symmetrical animals (Chapter 39)

A

B

C

FIGURE 40-1
Representatives of phylums Mollusca, Annelida, and Lophophorata.
A An ectoproct, representing the lophophorate phyla.
B An annelid, the Christmas-tree worm, *Spirobranchus giganteus*. This annelid is a polychaete, a member of the largest of the three annelid classes.
C A mollusk, the terrestrial snail *Allogona townsendiana*.

There are three basic kinds of body plans found in bilaterally symmetrical animals. Two of them were introduced in Chapter 39: the acoelomates, represented by the flatworms, and the pseudocoelomates, worms with a body cavity that develops between their mesoderm and endoderm. Although the body plans of both of these types have proven successful, a third way of organizing the body evolved in the common ancestor of the "higher" invertebrates and vertebrates, which together include most of the species of animals. It involves the development of a coelom, a body cavity that originates within the mesoderm.

We will begin our discussion of the coelomate animals with the mollusks and annelids, two major phyla that include such animals as clams, snails, slugs, octopuses, earthworms, and clamworms (Figure 40-1). In these phyla we can observe all of the major evolutionary advances associated with the evolution of the coelom. The lophophorate phyla, mainly marine animals of uncertain relationship, will likewise be discussed in this chapter. The remaining groups of coelomate animals will be discussed in Chapters 41 and 42.

THE ADVENT OF THE COELOM

The evolution of an internal body cavity made possible a significant advance in animal architecture. Consider for a moment the limitations of a solid body: a solid worm has no internal circulatory or digestive system, and all of its internal organs are pressed on by muscles and deformed by muscular activity. Although flatworms do have digestive systems, they are subject to problems of this kind.

An internal body cavity circumvents these limitations; its development was an important step in animal evolution. Perhaps the most important advantage of an internal body cavity is that the body's organs are located within a fluid-filled enclosure in which they can function without having to resist pressures from the surrounding muscles. In addition, the fluid that fills the cavity may act as a circulatory system, freely transporting food, water, waste, and gases throughout the body. Without the free circulation made possible by such a system, every cell of an animal must be within a short distance of oxygen, water, and all of the other substances that it requires.

The digestive system of an animal, like its circulatory system, functions much more efficiently within an open internal body cavity such as a coelom than it can when embedded in other tissues. Food can pass through the gut freely and at a rate controlled by the animal because the opening and closing of the gut does not depend on the movements of the animal. When the rate of passage of food is controlled, it can be digested much more efficiently than would be possible otherwise. Waste removal is also carried out much more efficiently under such circumstances.

In addition, the presence of a coelom allows the digestive tract, by its coiling or folding within the coelom, to be longer than the animal itself. The longer passage allows for storage organs for undigested food, longer exposure to the enzymes for more complete digestion, and even storage and final processing of food remnants. Such an arrangement allows an animal to eat a great deal when it is safe to do so and then to hide during the digestive process, thus limiting the animal's exposure to predators. The tube within the coelom architecture is also more flexible, thus allowing the animal greater freedom to move.

An internal body cavity also provides space within which the gonads (ovaries and testes) can expand, allowing the accumulation of large numbers of eggs and sperm. Such accumulation helps to make possible all of the diverse modifications of breeding strategy that characterize the more advanced phyla of animals. Furthermore, large numbers of gametes can be released when the conditions are as favorable as possible for the survival of the young animals.

Both pseudocoelomate and coelomate animals possess a fluid-filled body cavity, a great improvement in body design compared with the less advanced solid worms. So what is the functional difference between a pseudocoel and a coelom, and why has the latter kind of body cavity been so much more overwhelmingly successful in evolutionary terms? The digestive tract of a coelomate animal is composed of an endoderm lining and a mesoderm outer portion, suspended within the coelom cavity. In pseudocoelomates, the mesoderm outer portion of the digestive tract is missing, which evidently makes the differentiation of the gut into regions more difficult.

The evolutionary success of coelomates seems to reflect the repositioning of the body cavity. In coelomates, the cavity develops not between endoderm and mesoderm, but entirely within the mesoderm, and all of the internal organs and the circulatory system can now be assembled entirely by a single system of embryological controls—the mesoderm's; these controls direct their development, making it easier for complex organ systems to develop. The evolutionary specialization of the internal organs of the coelomates has far exceeded that of the pseudocoelomates. Few pseudocoelomates, for example, possess a closed circulatory system (Rhynchocoela are an exception). Many coelomates, however, have developed a closed circulatory system—a system of blood vessels derived from mesoderm, by means of which they circulate blood throughout the body.

> **The evolution of the coelom was a major improvement in animal body architecture. It permitted the development of a closed circulatory system; provided a fluid environment within which digestive, sexual, and other organs could be suspended; and facilitated muscle-driven body movement.**

The gut is suspended, along with other organ systems of the animal, in the coelom; the coelom, in turn, is surrounded by a layer of cells, the epithelium, which is entirely derived from the mesoderm. The portion of the epithelium that lines the outer wall of the coelom is called the parietal peritoneum, and the portion that lines the internal organs suspended within the cavity is called the visceral peritoneum.

In the smaller coelomates, the coelom serves as a hydrostatic system, one based on the pressure of enclosed fluids, like the brake systems in most automobiles. It then plays a role similar to that of the pseudocoel, as described in Chapter 39. Such a hydrostatic system greatly increases the efficiency of the movements of animals that possess it, compared with those of the acoelomates.

AN EMBRYONIC REVOLUTION: PROTOSTOMES AND DEUTEROSTOMES

There are two major branches of coelomate animals, representing two distinct evolutionary lines. In the first, which includes the mollusks—annelids, and arthropods, as well as some smaller phyla—the mouth (stoma) develops from or near the blastopore. This pattern of embryonic development also occurs in all noncoelomate animals. Animals whose mouth develops in this way are called **protostomes** (from the Greek

PART X   BIOLOGY OF ANIMALS

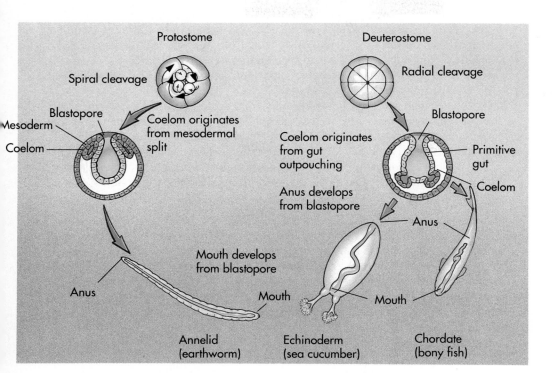

FIGURE 40-2
Protostomes and deuterostomes.
Patterns of embryonic development in coelomates and of egg
cleavage in protostomes and deuterostomes.

words *protos*, first, and *stoma*, mouth). If such an animal has a distinct anus or anal pore, it develops later in another region of the embryo. The fact that this kind of developmental pattern is so widespread in diverse phyla makes it virtually certain that it is the original one for animals as a whole and that it was characteristic of the common ancestor of all eumetazoan animals.

A second distinct pattern of embryological development occurs in the echinoderms, the chordates, and a few other small related phyla; it doubtless arose in the common ancestor of this group. In these animals, the anus forms from or near the blastopore, and the mouth forms subsequently on another part of the blastula (Figure 40-2). This group of phyla consists of animals that are called the **deuterostomes** (Greek, *deuteros*, second, and *stoma*, mouth). They are clearly related to one another by their shared pattern of embryonic development, which differs radically from the protostome pattern and was undoubtedly derived from it.

In addition to the pattern of blastopore formation, deuterostomes differ from protostomes in a number of other fundamental embryological features:

1. The progressive division of cells during embryonic growth is called cleavage. The pattern of cell cleavage relative to the embryo's polar axis determines the way in which the cells are arrayed. In nearly all protostomes, each new cell buds off at an angle oblique to the polar axis. As a result, a new cell nestles into the space between the adjacent older ones, resulting in a closely packed array. This kind of pattern is called **spiral cleavage,** because a line drawn through a sequence of dividing cells spirals outward from the polar axis (see Figure 40-2). In deuterostomes, however, the cells divide parallel to and at right angles to the polar axis. As a result, the pairs of cells that result from each division are positioned directly above and below one another; this process gives rise to a loosely packed array of cells. Such a pattern is called **radial cleavage,** because a line drawn through a sequence of dividing cells describes a radius outward from the polar axis.

2. In protostomes, the developmental fate of each cell in the embryo is fixed when that cell first appears. Even at the four-celled stage, each cell is different, and no one of them, if separated from the others, can develop into a complete animal because the chemicals that act as developmental signals have already been localized in different parts of the egg. Consequently the cleavage divisions

that occur after fertilization separate different signals into different daughter cells. In deuterostomes, on the other hand, the first cleavage divisions of the fertilized embryo result in identical daughter cells, any one of which can, if separated, develop into a complete organism. The commitment to prescribed developmental pathways occurs later.

3. In all coelomates, the coelom originates from mesoderm. In protostomes, this occurs simply and directly: the cells simply move away from one another as the coelomic cavity expands within the mesoderm. In deuterostomes, however, whole groups of cells usually move around to form new tissue associations. The coelom is normally produced by an invagination of the gut cavity lining, forming a depression that is closed over, which becomes the hollow cavity of the coelom.

As we discussed in Chapter 22, the first abundant and well-preserved animal fossils are about 630 million years old; they occur in the Ediacara series of Australia and similar formations elsewhere. Among these fossils, many represent groups of animals that no longer exist (see Figure 22-3, A). In addition, however, these ancient rocks bear evidence of the two major evolutionary lines of coelomates, the protostomes and the deuterostomes. The coelomates are certainly the most advanced evolutionary line of animals, and it is remarkable that their two major subdivisions were differentiated so early. It is clear that deuterostomes are derived from protostomes, since the protostome pattern of development is characteristic of all animals except for the four deuterostome phyla. The event, however, occurred very long ago and presumably did not involve groups of organisms that closely resemble any that are living now.

The deuterostomes—echinoderms, chordates, and two minor phyla—are an evolutionary line descended from a common ancestor in which the embryo develops in a new and different way from the protostomes. All other coelomates are protostomes. The orientation of the embryo, the pattern of its cell cleavage, the timing of the developmental commitment of its cells, and the degree of cell movement during embryo development—all of these features distinguish deuterostomes from protostomes.

In this chapter we will consider two of the three major phyla of protostomes, the mollusks and the annelids. The third major phylum, the arthropods, will be discussed in the next chapter. The giant tube worms (phylum Pogonophora) illustrated in Figure 26-11, are members of another phylum of protostomes, a small group with about 100 species. The deuterostomes, the two largest phyla of which are the echinoderms and the chordates, will be covered in Chapter 42. First, however, we will discuss a fascinating group of primarily marine animals with features that seem intermediate between those of protostomes and deuterostomes, the lophophorates.

LOPHOPHORATES

Three phyla of marine animals—(Ectoprocta; formerly Bryozoa; see Figure 40-1, A), Brachiopoda, and Phoronida—are characterized by a lophophore, a circular or U-shaped ridge around the mouth, bearing either one or two rows of ciliated, hollow tentacles (Figure 40-3). Because of this unusual feature, they are thought to be related to one another; the lophophore presumably arose in their common ancestor. The coelomic cavity of the lophophorates lies within the lophophore and its tentacles, and the anus is always located elsewhere. In these animals, the lophophore functions as a food-collection organ and as a surface for gas exchange. Lophophorates are attached to their substrate or move slowly, using the cilia of their lophophore to capture the plankton on which they feed.

The lophophorates share features with both the protostomes and the deuterostomes, as mentioned before. Cleavage is basically radial, as in deuterostomes. The formation of the coelom, however, varies; some lophophorates resemble protostomes, others deuterostomes, in this respect. In Ectoprocta, the method of coelom formation is unique, occurring only after the completion of larval life. In Phoronida,

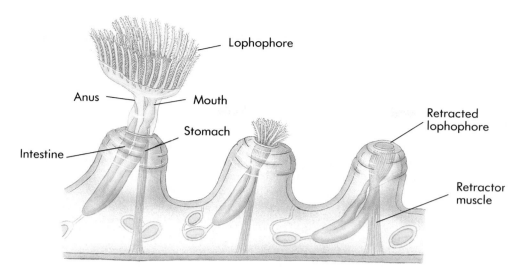

Lophophore

Anus — Mouth

Stomach

Intestine

Retracted lophophore

Retractor muscle

FIGURE 40-3

A lophophorate. A small portion of a colony of the freshwater ecto-proct *Plumatella* (phylum Ecto-procta), which grows on the underside of rocks. The individual at the left has a fully extended lophophore, the structure charac-teristic of the three lophophorate phyla. The tiny individuals of *Plumatella* disappear into their shells when they are disturbed.

the mouth forms from the blastopore, whereas in the other two phyla, it forms from the anus. Despite these great differences, the lophophore itself is a unique structure that seems to indicate that the members of these three phyla share a common ances-tor. For these reasons, their relationships continue to present a fascinating puzzle.

Phylum Phoronida: The Phoronids

Phoronids (phylum Phoronida) superficially resemble common tube worms seen on dock pilings. Each phoronid secretes a chitinous tube within which it lives out its life, and they also extend tentacles to feed and quickly withdraw them when disturbed, but there the resemblance to tube worms ends. Instead of a straight tube-within-a-tube body plan, phoronids have a U-shaped gut within a sac. Only about 10 phoronid species are known, ranging in length from a few millimeters to 30 centimeters. Some species lie buried in sand, others attached to rocks either singly or in groups. Phoron-ids develop as protostomes, with the anus developing secondarily; some species show spiral cleavage.

Phylum Ectoprocta: The Ectoprocts

Ectoprocts (phylum Ectoprocta, previously Bryozoa) look like tiny, short versions of phoronids (see Figure 40-1, *A* and 40-3). Because they are small—usually less than 0.5 millimeters long—and mostly colonial, they were called moss animals or Bryozoa. The name Ectoprocta refers to the location of the anus (proct), which is external to the lophophore. The 4000 species include both marine and freshwater forms—the only non-marine lophophorates. Most live in shallow water, but some species have been found at 18,000 feet.

Individual ectoprocts secrete a tiny chitinous chamber, or **zoecium,** attached to other members of the colony and to rocks. Individuals communicate chemically through pores between chambers. Brachiopods develop as deuterostomes, with the mouth developing secondarily; cleavage is radial.

A

Phylum Brachiopoda: The Brachiopods

Brachiopods, or lamp shells, superficially resemble clams, with two calcified shells (Figure 40-4). Brachiopod shells develop on the dorsal and ventral surfaces. Many species attach to rocks or sand by a stalk that protrudes through an opening in one shell. Within the shell lies the lophophore, which performs the same function seen in phoronids and ectoprocts when the brachiopod's shells are opened slightly.

Although only 300 species of brachiopods (phylum Brachiopoda) exist today, more than 30,000 species of this phylum are known as fossils. The genus *Lingula* has a fossil record extending back more than 500 million years, although the species that

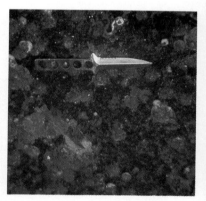

B

FIGURE 40-4

Representative brachiopods (phy-lum Brachiopoda).
A *Laqueus californicus.*
B *Crania californica.*

exist today are much younger. *Lingula* is probably the most ancient surviving genus of animals.

Brachiopods develop as protostomes, but they show radial cleavage. Furthermore, coelom development follows the enterocoelic pattern more commonly seen in deuterostomes.

PHYLUM MOLLUSCA: THE MOLLUSKS

The mollusks (phylum Mollusca) include the snails, clams, scallops, oysters, cuttlefish, octopuses, slugs, and many other familiar animals. The durable shells of mollusks are often beautiful and elegant; they have long been favorite objects for professional scientists and amateurs alike to collect, preserve, and study. The mollusks are one of the most successful of all phyla. They are the largest animal phylum, except for the arthropods, in terms of named species; there are at least 110,000, and probably at least that many more still to be discovered. Mollusks are widespread and often abundant in marine, freshwater, and terrestrial habitats.

A number of mollusks have invaded the land, including the snails and slugs that live in your garden. Terrestrial mollusks are often abundant in places that are at least seasonally moist. Some of these places, such as the crevices of desert rocks, may appear very dry, but even these habitats have at least a temporary supply of water at certain times. There are so many terrestrial mollusks, in fact, that only the arthropods have more species adapted to a terrestrial way of life. The 35,000 species of terrestrial mollusks far outnumber the roughly 20,000 species of terrestrial vertebrates.

As a group, mollusks are an important source of food for humans. Oysters, clams, scallops, mussels, octopuses, and squids are among the culinary delicacies that belong to this large phylum. The mollusks are also of economic significance to us in many other ways—for example, the production of pearls and of the shell material that is used as mother-of-pearl in jewelry and other decorative objects. Mollusks are not wholly beneficial to humans, however. For example, certain mollusks are major destructive agents of timbers that are submerged in the sea, including those used to build boats, docks, and pilings. Slugs and terrestrial snails often cause extensive damage to garden flowers, vegetables, and crops. Other mollusks serve as hosts to the intermediate stages for many serious diseases, including several carried by nematodes and flatworms, which we discussed in Chapter 39.

Most mollusks measure several centimeters in their largest dimension, and a number are minute. Some, however, reach formidable sizes. The giant squid, which is occasionally cast ashore but has rarely been seen alive, may be up to 21 meters long! Weighing up to at least 250 kilograms, the giant squid is the largest invertebrate and, along with the giant clam, the heaviest also. There are thought to be millions of giant squid in the ocean, even though they are seldom caught, and they are probably increasing in number as whales, their major competitors as predators, are decreasing. Among the bivalves, one species of giant clam, *Tridacna maxima*, may have a pair of shells as much as 1.5 meters long and may weigh as much as 270 kilograms (Figure 40-5).

FIGURE 40-5

The giant clam, *Tridacna maxima*.

A A giant clam, *Tridacna maxima*, on the bottom of the sea at Rose Atoll, American Samoa. The chocolate-brown color of the mantle is caused by the presence of symbiotic dinoflagellates (zooxanthellae), which probably contribute most of the food supply of the clam, although it remains a filter feeder like most bivalves. Some individual giant clams may be nearly 1.5 meters long and weigh up to 270 kilograms.
B Zooxanthellae in the modified hemal (blood-containing) sinuses of a giant clam. Like the corals, which also harbor zooxanthellae, the giant clams grow in nutrient-poor water and, despite that, attain very large sizes.

A

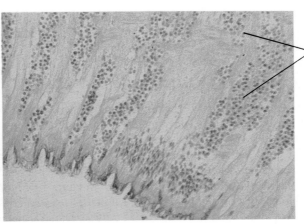

Zooxanthellae

B

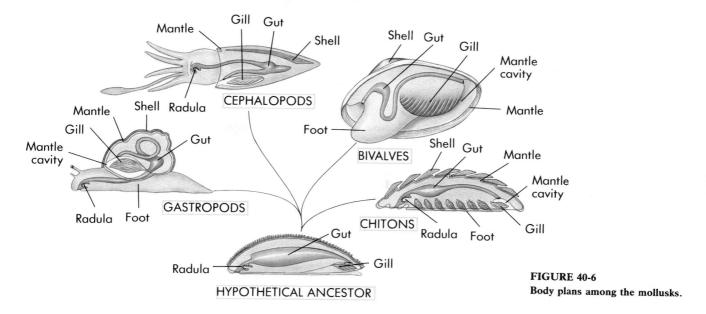

FIGURE 40-6
Body plans among the mollusks.

Body Plan of the Mollusks

In their basic body plan, mollusks have distinct bilateral symmetry (Figure 40-6). They have a **visceral mass** covered with a soft epithelium and a muscular foot that is used in locomotion. There may also be a differentiated head at the anterior end of the body. Within the visceral mass are found the organs of digestion, excretion, and reproduction. Folds, often two in number, arise from the dorsal body wall and enclose a cavity between themselves and the visceral mass; these folds comprise the **mantle**. Within the mantle cavity are the mollusk's gills or lungs. **Gills** are specialized portions of the mantle that usually consist of a system of filamentous projections, rich in blood vessels. These projections greatly increase the surface area available for gas exchange and therefore the animal's overall respiratory potential. Oxygen moves inward, carbon dioxide outward. Mollusk gills are very efficient, and many gilled mollusks extract 50% or more of the dissolved oxygen from the water that passes through the mantle cavity. Finally, in most members of this phylum, the outer surface of the mantle also secretes a protective shell.

A mollusk shell consists of a horny outer layer, rich in protein, which protects the underlying calcium-rich layers from erosion. A middle layer consists of densely packed crystals of calcium carbonate. The inner layer is pearly and increases in thickness throughout the animal's life. Pearls are occasionally produced inside the bodies of **bivalve** (two-shelled) **mollusks,** including clams and oysters, as a result of secretions of shell material around foreign bodies.

Many mollusks can withdraw for protection into their mantle cavity, which lies within the shell. In aquatic mollusks, a continuous stream of water is passed into and out of this cavity by the cilia on the gills. This water brings in oxygen and, in the case of the bivalves, food; it also carries out waste materials. When the gametes are being produced, they are carried out in the same stream. In the squids and octopuses, the mantle cavity has been modified to create the jet-propulsion system that enables the animals to move rapidly through the water.

The foot of a mollusk is muscular and may be adapted for locomotion, for attachment, for food capture (in squids and octopuses), or for various combinations of these functions. Some mollusks secrete mucus, forming a path that they glide along on their foot. In the cephalopods—squids and octopuses—the foot is divided into arms, also called tentacles. In some pelagic forms—mollusks that are perpetually free-swimming—the foot is modified into winglike projections or thin, mobile fins.

One of the most characteristic features of mollusks is the **radula,** a rasping, tonguelike organ that all members of the phylum have—except bivalves, in which it has almost certainly been lost; the bivalves obtain their food by filter feeding. The radula consists primarily of chitin, can be protruded or moved by the animal, and is

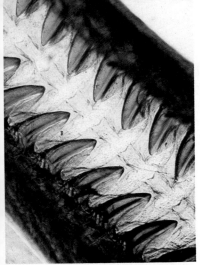

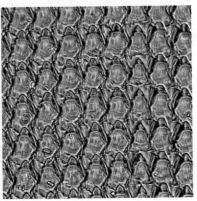

A

B

FIGURE 40-7
Structure of the radula in a snail.
A The radula consists of chiton
and is covered with rows of teeth.
B Enlargement of the rasping
teeth on a radula.

covered with rows of pointed, backward-curving teeth (Figure 40-7). It is used by some of the snails and their relatives, which are members of a group known as the gastropods, to scrape algae and other food materials off their substrates and then to convey this food to the digestive tract. Other gastropods are active predators; they use their radula to puncture their prey and extract food from it.

> Mollusks are the second-largest phylum of animals in terms of named species. They characteristically have bodies with three distinct sections: head, visceral mass, and foot. All mollusks, except the bivalves, also possess a unique rasping tongue called a radula.

The circulatory system of mollusks consists of a heart and, in most cases, an open system in which the blood circulates freely. The mollusk heart usually has three chambers, two of which collect aerated blood from the gills, while the third pumps it to the other body tissues. In the cephalopods, a closed system of vessels carries blood to and from the heart. These dexterous and rapidly moving animals also have auxiliary hearts, which increase the efficiency of their overall pumping system. In mollusks, the coelom is primarily represented by a small area around the heart.

In most coelomate animals, most of which have a closed circulatory system, the blood vessels are intimately associated with the excretory organs, making possible the direct exchange of materials between these two systems. For this reason, the excretory systems of these coelomates are much more efficient than the flame cells of the acoelomates, which pick up substances only from the body fluids. Nitrogen-rich wastes are removed from the mollusk by one or two tubular structures called **nephridia.** Mollusks were one of the earliest evolutionary lines to have developed an efficient excretory system.

A typical nephridium has an open funnel, the **nephrostome,** which is lined with cilia. From the nephrostome, a coiled tubule runs into an enlarged bladder, which in turn is connected to an excretory pore. Wastes are gathered by the nephridia from the coelom, which is located around the heart only, and discharged into the mantle cavity, from which they are expelled by the continuous pumping of the gills. As the wastes leave the mollusk, sugars, salts, water, and other materials are absorbed by the walls of the nephridia and returned to the animal's body as needed to achieve an appropriate osmotic balance. Eventually they pass into the mantle cavity, by way of which they ultimately leave the animal. In animals with a closed circulatory system, such as annelids, some mollusks, and the vertebrates, the coiled tubule of a nephridium is surrounded by a network of capillaries. Wastes are extracted from the circulatory system through these capillaries and are transferred into the nephridium, from where they are subsequently discharged, whereas salts, water, and other associated materials may also be reabsorbed from the tubule of the nephridium back into the capillaries. All coelomates except for arthropods and chordates have basically similar excretory systems.

> Mollusks were among the earliest animals to evolve an efficient excretory system, based on nephridia, small tubes that gather wastes by diffusion from the coelomic fluid.

Reproduction in Mollusks

Most mollusks have distinct male and female individuals, although a few bivalves and many gastropods are hermaphroditic, with both sexes occurring in a single individual. Even in hermaphroditic mollusks, however, cross-fertilization, rather than self-fertilization, is most common. Remarkably, some sea slugs and oysters are able to change from one sex to the other several times during a single season.

Many marine mollusks have free-swimming larvae called **trochophores** (Figure 40-8). Trochophores are distinctive in structure; they are propelled through the water by a row of cilia that encircles the middle of their body. In most marine snails and in the bivalves, there is a second free-swimming stage that follows the trochophore

FIGURE 40-8
The trochophore larvae of a mollusk. Similar larvae, as you will see, are characteristic of the annelid worms.

PART X   BIOLOGY OF ANIMALS

stage. In this **veliger** stage, the beginnings of a foot, shell, and mantle can be seen (Figure 40-9). Mollusks are dispersed from place to place largely as trochophores and veligers. The trochophores and veligers drift widely in the currents to new areas.

The Classes of Mollusks

There are seven classes of mollusks. Some of the smaller classes have greatly helped us to understand the limits of variation in the phylum and the nature of its probable ancestor. By comparative study of these animals, some scientists have concluded that the ancestral mollusk was probably a dorsoventrally flattened, unsegmented, worm-like animal that glided on its ventral surface. These scientists believe that this hypothetical animal had a moderate amount of chitinous cuticle and overlapping calcareous scales. Alternative hypotheses are certainly possible, but this view of ancestral mollusks provides one plausible way of viewing the evolution of the phylum.

A contemporary class of mollusks in which many of these characteristics still exist is the class Polyplacophora, the chitons. These marine mollusks have oval bodies with eight overlapping calcareous plates (Figure 40-10). Underneath the plates is a broad, flat foot that chitons use to creep along, surrounded by a groove or mantle cavity in which the gills are arranged. Most of the chitons are grazing herbivores that live in shallow marine habitats, but some have been found in depths of more than 7000 meters.

We will treat three classes of mollusks in some detail as representatives of the phylum: (1) Gastropoda—snails, slugs, limpets, and their relatives; (2) Bivalvia—clams, oysters, scallops, and their relatives; and (3) Cephalopoda—squids, octopuses, cuttlefishes, and nautilus.

FIGURE 40-9
Veliger stage of a mollusk.

FIGURE 40-10
A chiton, *Tonicella lineata*, class *Polyplacophora*. The shells of chitons, unlike those of other mollusks, consist of overlapping calcareous plates.

Class Gastropoda: The Snails and Slugs. About 80,000 named species of snails and slugs compose the gastropods (class Gastropoda; see Figures 40-1, *C*, and 40-11). This class is primarily a marine group that is also abundant in freshwater and terrestrial habitats. Gastropods have a single shell, which has been lost in some groups, and a body that is generally divisible into a head, foot, and visceral mass. These animals generally creep along on their foot, which may be modified for swimming. Slugs and nudibranchs (marine slugs) are clearly descended from ancestors that had a shell. The shell of most marine gastropods is closed by a plate, the **operculum,** that the animal can pull into place. Most land gastropods lack an operculum as adults.

On their head, most gastropods have a pair of tentacles, on which the eyes may be located. The tentacles have been lost, however, in some of the more advanced forms of the class. The mouth opening in gastropods may be simple, or it may be modified as a proboscis. Within the mouth cavity of many members of this class are horny jaws and a radula.

CULTIVATING MOLLUSKS AS FOOD

Because they are often so delicious and prized as food, several kinds of mollusks are encouraraged or cultivted. For example, oysters have long been cultivated in underwater beds, both for food and for their pearls.

The conch, a large marine snail that was plentiful throughout the Caribbean, has been so overfished that it is now becoming relatively difficult to obtain. Millions of kilograms per year were once harvested in the Florida Keys alone. Conchs are being propagated commercially at Providencia, in the Turks and Caicos Islands, but efforts to reestablish them in this way are just beginning.

Scallops are grown commercially in Japan, and efforts to cultivate them are under way in Western Canada also.

Edible snails have long been cultivated and harvested in France and Italy. The same snails, introduced into California in Gold Rush days, have become serious pests of gardens and commercial crops. Some 37 million dollars is spent annually within the state to control them. These snails, however, are now being cultivated by some farmers; Americans consume nearly 300 million dollars worth of snails per year. Before being canned, snails are fed a special diet to improve their taste, much like cattle are fed corn.

FIGURE 40-11
Gastropod mollusks.
A The textile cone, *Conus textilis*. The radulas of the cone shells, which are active hunters, are modified into harpoons. The animals of this genus, some of which hunt fishes, secrete powerful neurotoxins; they include some of the most dangerous marine animals.
B Hawaiian limpet, *Helcioniscus exaratus*, viewed from below.
C A slug, *Arion ater*, one of the approximately 500 species of terrestrial mollusks that lack shells.
D A nudibranch, *Hermissenda crassicornis*.

The visceral mass of the gastropods is twisted and has become asymmetrical during the course of evolution. This twisting, or **torsion**, occurs during the embryological development of the gastropod and results in a thorough rearrangement of the body. It takes place separately from the twisting of the shell and results from the fact that one side of the larva grows much more rapidly than the other. This evolutionary process, which is considerably advanced in certain members of this class, is sometimes associated with other changes. For example, many gastropods have lost their right gill and sometimes also their right nephridium.

> Gastropods typically, but not always, possess a hard shell, in which they live, and a "door" called an operculum. During gastropod development, one side of the embryo grows more rapidly than the other, producing a characteristic twisting of the visceral mass.

Gastropods display extremely varied feeding habits. Some are predaceous, others scrape algae off rocks (or aquarium glass), and others are scavengers. Many are herbivores, and some of the terrestrial ones, as we have mentioned, can be serious garden and agricultural pests. The radula of whelks is stalked and is used to bore holes in the shells of other mollusks, through which the contents are then sucked out. In cone shells (Figure 40-11, *A*), the stalked radula has been modified into a kind of harpoon, which is shot suddenly into the prey with an injection of poison. Nudibranchs (Figure 40-11, *D*) are active predators; a few species of nudibranchs protect themselves with nematocysts they obtain from the polyps they eat.

In terrestrial gastropods, an area under the mantle that is extremely rich in blood vessels serves, in effect, as a lung. This lung evolved in animals living in environments with plentiful oxygen. It absorbs oxygen more efficiently than a gill does for an animal living on land and breathing air. The lung is essentially an empty mantle cavity in the space that gills occupy in the aquatic ancestors of terrestrial mollusks. Some

A

B

FIGURE 40-12
Bivalves.
A Scallop, *Chlamys hericia*. Note the blue eyes around the margin of the body.
B File shell, *Lima scabra*, swimming.

aquatic snails were probably derived from terrestrial forms that have returned to the water, as evidenced by the fact that they still come to the surface and breathe by means of lungs.

Class Bivalvia: The Bivalves.

Members of the class Bivalvia—which includes clams, scallops, mussels, and oysters—have two shells hinged together dorsally and a wedge-shaped foot (see Figures 40-5 and 40-12). A ligament hinges the shells together and causes them to gape open. Pulling against this ligament are one or two large **adductor muscles** that can draw the shells together. The mantle secretes the shells and ligament and envelops the internal organs within the pair of shells. Additionally, the mantle is drawn out to form two siphons, one for the incoming and one for the outgoing stream of water. There is a pair of gills on each side of the visceral mass, each pair covered by a fold of the mantle. These gills consist of pairs or sheets of filaments that contain many blood vessels. Within the gills is a rather elaborate pattern of water circulation.

Bivalves do not have distinct heads or radulas, differing from gastropods in this respect (see Figure 40-6). They also lack tentacles, which are characteristic of cephalopods. However, their foot is large and muscular. It may be adapted, in different species, for creeping about, burrowing, cleansing the animal, or anchoring it in place in its burrow. Some species of clams can dig into sand or mud very rapidly by means of muscular contractions of their foot.

Most bivalves are sessile filter-feeders. They extract small organisms from the water that they filter through their mantle cavity by the ciliary action of their gills, squirting it out through a siphon. The food particles that bivalves ingest are entangled by masses of mucus secreted by glands, mainly located on the gills. On either side of the mouth is usually a pair of organs, the **palps,** which aid in handling particles of food.

Bivalves disperse from place to place largely as larvae. Most of the adults are fixed and relatively immobile or even confined to a burrowing way of life. Some adult bivalves do, however, move about readily. Many genera of scallops, for example, can move swiftly through the water by using their large adductor muscles, which are what we usually eat as "scallops," to clap their shells together. (Small circular plugs of muscle cut from skates and rays, which are cartilaginous fishes, are sometimes served as "scallops," however.) One of a scallop's shells is larger than the other, and the edge of its body is lined with tentacle-like projections. Their complex eyes can differentiate light and darkness and therefore, presumably, the shadows that potential predators, such as starfish, cast. Scallops can also detect predators by means of chemical signals, which prompts them to flee.

There are about 10,000 species of bivalves. Most of these are marine, although many species also live in fresh water. One of the more interesting freshwater groups is the pearly freshwater mussels, or naiads. There are nearly 1200 species of this group, which is distributed worldwide. One of its families, Unionidae, includes more than

FIGURE 40-13

Pearly freshwater mussels, members of the family Unionidae, from the United States. More than 50 of the estimated 500 species in North America may already have become extinct, and many of the remainder are threatened with extinction by the pollution of the waters in which they occur.

A *Epioblasma obliquata perobliqua* from the St. Joseph River in Indiana. This species was formerly widespread in the Wabash and upper Maumee River systems in Indiana. It is almost certainly extinct in the former drainage, and there are no records at all for more than 20 years, despite intensive searching in the areas where it formerly occurred.

B *Quadrila sparsa*, officially listed by the United States government as "endangered," once occurred widely in the Clinch and Holston rivers of Tennessee and in the Cumberland River in Kentucky. Reduced greatly in extent by dam construction, pollution, gravel dredging, and strip-mining, this species now occurs only in the Powell River in Tennessee, where it is endangered by silting, which occurs primarily as a result of strip-mining.

500 species that occur in the rivers and lakes of North America (Figure 40-13). Formerly, some of them were so abundant—for example, at certain places along the Mississippi River—that they supported pearl button factories. The larvae of this family of mollusks parasitize fishes; they are brooded in a special pouch before they are released, thus exhibiting a very unusual life cycle for a mollusk.

Class Cephalopoda: The Octopuses, Squids, and Nautilus.

The more than 600 species of the class Cephalopoda—the octopuses, squids, and nautilus—are the most intelligent of the invertebrates. They are active marine predators that swim, often swiftly, and compete successfully with fish. The giant squid, mentioned earlier, plays an ecological role like that of the large marine mammals, such as the killer whales, and the large, predaceous fishes. Cephalopods feed primarily on fishes, other mollusks, crustaceans, and worms. The foot has evolved into a series of tentacles equipped with suction cups, adhesive structures, or hooks that seize prey efficiently. Squids have 10 tentacles; octopuses, as indicated by their name, 8; and the nautilus, about 80 to 90 (Figure 40-14). Once the tentacles have snared the prey, it is bitten with strong, beaklike paired jaws and pulled into the mouth by the tonguelike action of the radula.

Cephalopods have highly developed nervous systems, and their brains are unique among the mollusks. Their rapid responses are made possible by a bundle of giant nerve fibers attached to the muscles of the mantle. Their eyes are very elaborate, and the retina has a structure much like that of vertebrate eyes, although there are fundamental differences; they evolved separately (see p. 955). Squid eyes can also be quite large; in fact, the eyes of a giant squid that was washed up on a beach in New Zealand in 1933 were 40 centimeters across, the largest eyes known in any animal! Many cephalopods exhibit complex patterns of behavior and a high level of intelligence. For example, octopuses can be trained easily to distinguish among classes of objects. Most members of this class have closed circulatory systems; they are the only mollusks that do have such systems.

FIGURE 40-14
Cephalopod diversity.
A An octopus. Octopuses generally move slowly along the bottom of the sea.
B A squid. Squids are active predators, competing effectively with fish for prey.
C Pearly nautilus, *Nautilus pompilius*.

The octopuses and other cephalopods are efficient and often large predators. They possess well-developed brains and are the most intelligent of the invertebrates.

A   B   C

Although they evolved from shelled ancestors, living cephalopods, except for the few species of nautilus, lack an external shell. The squids and cuttlefish retain an internal remnant of their ancestral shells, which serves as an internal stiffening support, but the octopuses have no trace of a shell. Cuttlefish "bones" are used in canary cages to provide calcium for the birds.

Like other mollusks, cephalopods take water into the mantle cavity and expel it through a siphon. The cephalopods, however, have modified this system into a means of jet propulsion. When threatened, they eject water violently and shoot themselves through the water. Squids, which have streamlined bodies and fins, have a particularly efficient jet-propulsion system; they also have lateral undulating fins running along their sides, that help propel them through the water. Octopuses can also swim, but they spend most of their time climbing over rocks and other objects on the ocean floor. The members of both groups can swim in various directions by turning their siphon. Both can also release from special sacs a dark fluid, once used for ink, that clouds the water and thus helps to disguise the direction of their escape.

In cephalopods the sexes are separate. The sperm are stored in spermatophores in a sac that opens into the mantle cavity of the male. During copulation the male uses a specialized arm, or tentacle, to transmit a **spermatophore**—a mass enclosing many sperm—from its own mantle cavity into the female's. The female then fertilizes the eggs as they leave the **oviduct,** the passageway by which they leave the female, attaching them to stones and other objects.

The pearly nautilus (see Figure 40-14, C), or chambered nautilus, comprises a single genus of five or perhaps six species, which live in the western Pacific and Indian Oceans. The kinds of chambered shells that are characteristic of the living pearly nautilus are frequent in the fossil record for the past 200 million years. For example, the ammonites are a related group that became extinct at the end of the Cretaceous Period, 65 million years ago (Figure 40-15). Fossils as much as 75 million years old are very similar to the living species. The shells of these living species are about 10 to 27 centimeters across their flattened plane. The exterior of these shells resembles porcelain, whereas the interior is pearly, hence one of the names of these animals. The animal occupies the outermost chamber of the shell. A tube leads from the inner chambers, by means of which the animal can regulate the relative proportions of gases and fluids in these chambers, and thus its density.

The 80 to 90 tentacles of a pearly nautilus are adhesive, but they lack suction cups. The outermost tentacles have thickened sheaths and, when held together, form a kind of hood that, when it is retracted, helps to protect the animal. In the center of the circle of tentacles is a strong, horny beak. The eyes are small and apparently function like a pinhole camera. The nautilus uses a specialized funnel leading from its mantle cavity to propel itself through the water. The animals are nocturnal, remaining hidden during the day; they feed on crustaceans and dead animals on and just above the ocean floor. They grow slowly, living a considerable length of time after they reach sexual maturity at about 10 to 15 years of age. Reproduction occurs year-round, with a few large eggs laid at any one time. In contrast, the octopuses and squids typically grow very quickly, reproduce once, and die. The nautilus is usually most abundant at depths of 200 to 500 meters but is sometimes found floating near the surface.

FIGURE 40-15
An ammonite from the Cretaceous Period. The ammonites, shelled cephalopods, were abundant during the Mesozoic Era, but abruptly went extinct at its end. Ammonites were apparently pelagic animals that could not compete successfully with the bony fishes that became abundant toward the close of the Mesozoic Era. Their shells are somewhat similar to those of the modern, deep-water pearly nautilus.

THE ADVENT OF SEGMENTATION

Of the major phyla of protostomes, the annelids and arthropods have segmented bodies, whereas most living mollusks do not. Just as it is efficient for workers to construct a tunnel from a series of identical prefabricated parts, so these advanced protostome coelomates are "assembled" from a serial succession of identical segments. During the animal's early development, these segments become most obvious in the mesoderm but later are reflected in the ectoderm and endoderm as well. Two advantages result from early embryonic segmentation:

1. Each segment of mesoderm may go on to develop a more or less complete set of the several organ systems. Damage to any one segment need not be fatal to the individual, since the other segments duplicate that segment's functions.

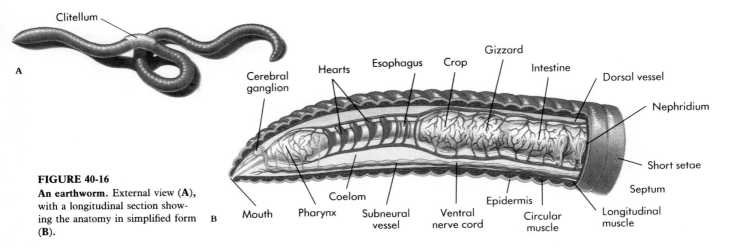

FIGURE 40-16
An earthworm. External view (**A**), with a longitudinal section showing the anatomy in simplified form (**B**).

2. Locomotion is far more effective when individual segments can move independently because the animal as a whole has more flexibility of movement.

Segmentation underlies the organization of all advanced animals. In some adult arthropods the segments are fused, making it difficult to perceive the underlying segmentation, but it is usually apparent in embryological development. Even the backbone and muscular areas of the vertebrates are segmented.

> **Segmentation is a feature of the advanced coelomate phyla, notably the annelids and the arthropods, although it is not obvious in some of them.**

PHYLUM ANNELIDA: THE ANNELIDS

The annelids (phylum Annelida), one of the major animal phyla, are segmented worms (see Figure 40-1, *B*). They can be recognized by their strongly marked segmentation and are abundant in marine, freshwater, and terrestrial habitats throughout the world. Their segments are visible externally as a series of ringlike structures running the length of the body. Internally the segments are divided from one another by partitions called **septa.** In each of the cylindrical segments of these animals the digestive and excretory organs are repeated in tandem.

Body Plan of Annelids

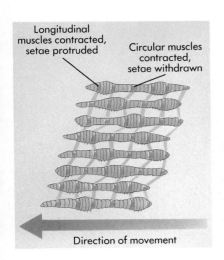

FIGURE 40-17
Locomotion in an earthworm. As the worm moves, some segments become long and thin, and others become thick and fat.

The basic body plan of the annelids (Figure 40-16) is a tube within a tube, with the internal digestive tract suspended within the coelom—a tube running from mouth to anus. The anterior segments have become modified according to the habits of the particular kind of annelid, and there is a well-developed cerebral ganglion, or brain, in one of these segments. The diverse sensory organs are mainly concentrated near the anterior end of the worm's body. Some of these are sensitive to light, and elaborate eyes with lenses and retinas have evolved in certain members of the phylum. Separate nerve centers, or ganglia, are located in each segment but are interconnected by nerve cords. These nerve cords are responsible for the coordination of the worm's activities.

Annelids use their muscles to crawl, burrow, and swim. In each segment, the muscles play against the coelomic fluid, which creates a hydrostatic skeleton that gives the segment rigidity. Because the septa isolate each segment into an individual hydrostatic skeletal unit, each is able to contract or expand autonomously. Therefore, a long body can move in ways that are often quite complex. When an earthworm crawls on a flat surface, for example, it lengthens some parts of its body while shortening others (Figure 40-17).

Each segment of an annelid typically possesses **setae,** bristles of chitin, that help to anchor the worms during locomotion or when they are in their burrows. In a crawling worm, for example, the setae are retracted (withdrawn into the animal) in the expanding segments and extended to make contact with the surface in the contracted segments, anchoring them. Because of these setae, annelids are sometimes called "bristle worms." The setae may also aid some of the small marine annelids in swimming. They are absent in the leeches, except for a single species.

The members of one class of annelids, the polychaetes, develop from free-swimming trochophore larvae similar to those of the mollusks (see Figure 40-8). The existence of such larvae in both phyla makes it extremely probable that they share a common ancestor, which may have resembled a flatworm. A primitive coelom would then have developed in this common ancestor. Clear segmentation is found in one small class of mollusks, the chitons, but their other features suggest that chitons are not particularly primitive and that the common ancestor of the mollusks was unsegmented. This indicates that the common ancestor of the annelids and mollusks was also unsegmented and that segmentation evolved within the annelids. Certainly annelids are an ancient phylum; Figure 22-9, C, shows a polychaete from the Cambrian Period, 530 million years ago. Segmentation clearly arose early in the evolution of coelomate animals.

In any case, the elaborate and obvious segmentation of the annelid body clearly represents an evolutionary specialization. Many scientists believe that segmentation is an adaptation for burrowing; it makes possible the production of strong peristaltic waves along the length of the worm, thus enhancing vigorous digging. Segmentation itself is also characteristic of the arthropods, a phylum that may share a common ancestor with the annelids. The segmentation that is characteristic of the vertebrates, on the other hand, very likely evolved independently, judging from their overall features.

Like all coelomates except for the arthropods and most mollusks, the annelids have a closed circulatory system. Blood moves through a closed circulatory system faster and more efficiently than it does through an open system. The annelids exchange oxygen and carbon dioxide with the environment through their body surfaces; they lack gills, lungs, and similar organs. Much of their oxygen supply reaches the different parts of their bodies by way of their blood vessels, however. Some of these vessels are enlarged and heavily muscular, serving as hearts in the pumping of blood. Earthworms, for example, have five pulsating blood vessels on each side that serve as hearts, helping to pump blood from the main dorsal vessel, which is their major pumping structure, to the main ventral one (see Figure 40-16). The blood of the larger annelids has respiratory pigments, such as hemoglobin, dissolved in it.

The excretory system of the annelids consists of ciliated, funnel-shaped nephridia generally similar to those of the mollusks. These nephridia—each segment has a pair—collect waste products and transport them out of the body through the coelom by way of specialized excretory tubes.

> **The annelids are characterized by serial segmentation. The body is composed of numerous similar segments, each with its own circulatory, excretory, and neural elements and each with its own array of setae.**

Classes of Annelids

The roughly 12,000 described species of annelids occur in many different habitats. They range in length from as little as 0.5 millimeter to more than 3 meters in some giant Australian earthworms and some polychaetes. There are three classes of annelids: (1) the Polychaeta, which are free-living, almost entirely marine bristleworms, comprising some 8000 species; (2) the Oligochaeta, the terrestrial earthworms and related marine and freshwater worms, with some 3100 species; and (3) the Hirudinea, the leeches, mainly freshwater predators or bloodsuckers, with about 500 species. The annelids are believed to have evolved in the sea, with the polychaetes being the primitive class. Oligochaetes seem to have evolved from polychaetes, perhaps by way of brackish water to freshwater estuaries and then to streams. Leeches share with oligochaetes an organ called a **clitellum,** which secretes a cocoon specialized to receive the eggs. It is generally agreed that leeches evolved from oligochaetes by specialization in relation to their bloodsucking habits and way of life as external parasites.

A

B

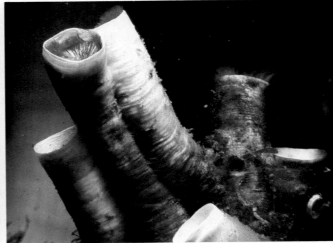

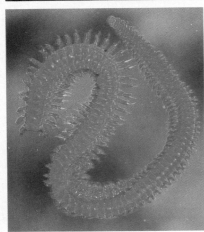

C

FIGURE 40-18
Polychaete annelids.
A Fan worm, Sabella melanostigma.
B Tube worms, *Eudistylia polymorpha,* withdrawn into their tubes.
C Shiny bristleworm, *Oenone fulgida.*

Class Polychaeta: The Polychaetes. The polychaetes (class Polychaeta) include the clamworms, plumed worms, scaleworms, lugworms, twin-fan worms, sea mice, peacock worms, and many others. These worms are often surprisingly beautiful, with unusual forms and sometimes iridescent colors (see Figures 40-1, *B,* and 40-18). Polychaetes live in such places as in burrows, under rocks, in tubes of hardened mucus that they manufacture, and inside shells. They are often a crucial part of marine food chains, being extremely abundant in certain habitats. A number of polychaetes are **commensal;** they live inside sponges, in the shells of mollusks, within echinoderms or crustaceans, and in other animals, eating the food particles left over by these organisms. A few of these worms are parasites; some are active predators.

Polychaetes have a well-developed head with specialized sense organs; they differ from other annelids in this respect. Their bodies are often highly organized into distinct regions formed by groups of segments related in function and structure. Their sense organs include eyes, which range from simple eyespots to quite large and conspicuous stalked eyes; all of these kinds of eyes primarily serve to concentrate light and indicate the direction from which it is coming.

Another distinctive characteristic of the polychaetes is the paired, fleshy, paddlelike flaps, called **parapodia,** on most of their segments. These parapodia, which often bear bristlelike setae, are used in swimming, burrowing, or crawling. They also play an important role in gas exchange, because they greatly increase the surface area of the body. In some polychaetes that live in burrows or tubes, the parapodia may feature hooks that help to anchor the worm. Slow crawling is carried out by means of the parapodia, and rapid crawling is aided by undulating motions of the body. In addition, the polychaete epidermis often includes ciliated cells, which the worms use to set up currents of water, thus aiding in respiration and food procurement.

The sexes of polychaetes are usually separate, and fertilization is often external, occurring in the water and away from both parents. Also unlike other annelids, polychaetes usually lack permanent male gonads, the sex organs in which sperm are produced. These polychaetes produce their sperm directly from cells in the lining of the coelom or in their septa. The eggs of polychaetes may be laid in gelatinous masses and attached to the substrate, or incubated in a tube or brood chamber in or on the female's body. Fertilization results in the production of ciliated, mobile trochophore larvae similar to the larvae of mollusks. The trochophores develop for long periods in the plankton before beginning to add segments and thus changing to a juvenile form that more and more closely resembles the adults.

Class Oligochaeta: The Earthworms. The earthworms (class Oligochaeta) literally eat their way through the soil, anchoring themselves by means of their setae. Earthworms suck in organic and other material by contracting their strong pharynx;

FIGURE 40-19
Earthworms mating.

everything that they ingest passes through their long, straight digestive tracts. In one region of this tract, the gizzard, muscles are concentrated; here the worm grinds up the organic material with the help of the soil particles that it also takes in. The body of an earthworm (see Figure 40-16) consists of 100 to 175 similar segments, with a mouth on the first of these and an anus on the last.

The material that passes through an earthworm is deposited outside of its burrow in the form of castings, familiar objects that look as if they had been extruded from a toothpaste tube. In this way, earthworms aerate and enrich the soil, a subject that fascinated Charles Darwin, particularly in his later years. In his book, *The Formation of Vegetable Mould through the Action of Worms,* Darwin pointed out that a worm could eat its own weight in soil every day and that the equivalent of 22 to 40 metric tons of soil per hectare pass through their intestines every year. This indicates a population of well over 16,000 worms per hectare!

In view of the underground life-style that earthworms have evolved, it is not surprising that they have no eyes. However, earthworms do have light-sensitive and touch-sensitive organs. These are not as elaborate as are the corresponding organs in polychaetes. The light-sensitive organs of the oligochaetes are concentrated in those segments near each end of the body—those regions most likely to be exposed to light. Earthworms also have sensitive moisture-detecting cells, an important feature since they, like all annelids, carry out their gas exchange, a highly moisture-sensitive process, through their skins. Earthworms have fewer setae than do the polychaetes and have no parapodia. In addition, earthworms lack the distinct head regions that characterize polychaetes.

Earthworms are hermaphroditic, another way in which they differ from most polychaetes. When they mate (Figure 40-19), their anterior ends point in opposite directions, and their ventral surfaces touch. The clitellum is obvious as a thickened band on an earthworm's body. The mucus it secretes holds the worms together during copulation. Sperm cells are released from pores in specialized segments of one partner into the sperm receptacles of the other, the process going in both directions simultaneously.

Two or three days after the worms separate, the clitellum of each worm secretes a mucous cocoon, surrounded by a protective layer of chitin. As this sheath passes over the female pores of the body—a process that takes place as the worms move—it receives eggs. As it subsequently passes along the body, it incorporates the sperm that were deposited during copulation. When the mucous sheath finally passes over the end of the worm, its ends pinch together. The sheath then encloses the fertilized eggs in a cocoon from which the young worms ultimately hatch.

Class Hirudinea: The Leeches.
Leeches (class Hirudinea) occur mostly in fresh water, although a few are marine and some tropical leeches occur in terrestrial habitats. Most leeches are 2 to 6 centimeters long, but one tropical species reaches up to 30 centimeters. Leeches are usually flattened dorsoventrally, (Figure 40-20) like flatworms. They are hermaphroditic, like the oligochaetes, and develop a clitellum dur-

FIGURE 40-20
A freshwater leech, with young leeches visible in a ventral brood pouch.

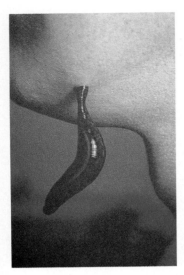

FIGURE 40-21
The medicinal leech, *Hirudo medicinalis,* **attached to a person's chin.** The large posterior sucker is used in crawling; the smaller sucker at the head encloses the mouth, which has three jaws. This anterior sucker has a diverse array of sensory structures used in feeding behavior.

ing the breeding season; cross-fertilization is obligatory. Leech eggs are enclosed in cocoons and laid on damp earth, attached to objects in the water, or occasionally attached to the host or to the parent leech itself.

The coelom of leeches is reduced and continuous throughout the body. It is not divided into individual segments like those of the polychaetes and oligochaetes. Leeches have evolved suckers at either one or both ends of the body. Those that have a sucker at both ends move by attaching first one and then the other end to the substrate and looping along. Except for one species, leeches have no setae.

Most leeches are predators or scavengers, but some have evolved the habit of sucking blood from mammals, including humans, and other vertebrates; a few suck blood from crustaceans. Many freshwater leeches live as external parasites, remaining on their hosts for long periods and sucking their blood from time to time. Some terrestrial leeches even climb trees to seek mammals and birds.

The best-known leech is the medicinal leech, *Hirudo medicinalis* (Figure 40-21), which was used for centuries to remove what was thought to be "excess" blood responsible for certain illnesses. Individuals of *Hirudo* are 10 to 12 centimeters long and can suck a great deal of blood from a person relatively rapidly. They were gathered in large numbers from the ponds of Europe for this purpose until the nineteenth century. Their mouth has chitinous teeth that rasp through the skin of the victim, and the leech secretes an anticoagulant into the wound to prevent clotting of the blood as it flows out. The powerful sucking muscles of the leech pump the blood out quickly once the hole has been opened. Bloodsucking leeches gorge themselves when they feed, and the animals can survive for months between meals.

The medicinal leech is now used as a source of anticoagulant, which is used in research into blood clotting. The animal is still widely collected by European pharmaceutical companies for this purpose—to remove excess blood after surgery and to treat atherosclerosis and thrombosis. Following surgery, venous failure often occurs, and the accumulating blood "turns off" the arterial supply of fresh blood, so that the tissue often dies. When leeches are used to remove blood from the area, new capillaries form in about a week, and the tissues remain healthy.

As a result of continued harvesting, leeches are becoming rare in certain areas. Attempts are now being made to transfer the genes coding for the anticoagulant to bacterial plasmids to facilitate commercial production, applying modern molecular biology to an age-old treatment. Leech anticoagulants also may find more widespread use. One of the anticoagulants that leeches secrete was shown in 1987 to be effective against the spread of cancer.

THE FIRST TWO LARGE PROTOSTOME PHYLA

Mollusks and annelids are two great phyla of protostome coelomate animals, both abundantly represented in the sea, but with large numbers of species also in terrestrial and freshwater habitats. The phyla originated early in the course of evolution of multicellular animals and have proliferated extensively over hundreds of millions of years subsequently. In the next chapter, we will consider the third great group of protostomes, the arthropods, a phylum that in the number of its species exceeds all others.

SUMMARY

1. The coelomates, or higher animals, are characterized by their internal body cavity, or coelom, which develops entirely within the mesoderm.

2. An internal body cavity, whether a pseudocoel or a coelom, not only makes an animal rigid but also provides an area where circulatory, digestive, sexual, and other systems can function free of the pressure of surrounding muscles.

3. A pseudocoel—which develops between the endoderm and the mesoderm—and a coelom are functionally equivalent, but the developmental processes that lead to the formation of organ systems within a coelom are much simpler because the stimuli originate within a single type of embryonic tissue.

4. The two major evolutionary lines of coelomate animals—the protostomes and the deuterostomes—were both represented among the oldest known fossils of multicellular animals, dating back some 630 million years.

5. In the protostomes, the mouth develops from or near the blastopore, and the early divisions of the embryo are spiral. At early stages of development, the fate of the individual cells is already determined, and they cannot develop individually into a whole animal.

6. In the deuterostomes, the anus develops from or near the blastopore, and the mouth forms subse

quently on another part of the blastula. The early divisions of the embryo are radial. At early stages of development, each cell of the embryo can differentiate into a whole animal.

7. The lophophorates consist of three phyla of marine animals—Phoronida, Ectoprocta, and Brachiopoda—that are characterized by a circular or U-shaped ridge, the lophophore, around the mouth. They share features both with protostomes and with deuterostomes.

8. The major phyla of protostomes are Mollusca, Annelida, and Arthropoda; the major phyla of deuterostomes are Echinodermata and Chordata.

9. The mollusks constitute the second largest phylum of animals in terms of named species. Their body plan consists of distinct parts: a visceral mass, a foot, a head, and a mantle.

10. Of the seven classes of mollusks, the gastropods (snails and slugs), bivalves, and cephalopods (octopuses, squids, and nautilus), are best known. Only a few mollusks show traces of segmentation.

11. Annelids are segmented worms, consisting of the largely marine polychaetes, the largely terrestrial earthworms, and the largely freshwater leeches. These animals have a highly repetitive body plan.

REVIEW QUESTIONS

1. What are some advantages of possessing a body cavity with respect to circulation, digestion, and reproduction? What are the advantages of a completely mesodermally enclosed cavity? Differentiate between a parietal and a visceral peritoneum.

2. What patterns of embryonic development related to cleavage and the blastopore occur in protostomic coelomates? What patterns occur in deuterostomic coelomates? Which major coelomate phyla are protostomes and which are deuterostomes? How is the early developmental fate of cells different between the two groups? How is the development of the coelom from mesodermal tissue different between them?

3. What prominent feature characterizes the lophophorate animals? What are the functions of this feature? Where is the coelomic cavity found in these animals? What are the protostomic and deuterostomic features of these animals? What are the three phyla that compose this group?

4. What is the basic body plan of a mollusk? Where is the mantle located? Why is it important in the mollusks? What occurs in the mantle cavity of aquatic mollusks?

5. What type of circulatory system is found in most mollusks? How does cephalopod circulation differ from the others? What part of the mollusk body is represented by the coelom? Describe the mollusk excretory system.

6. What types of reproductive individuals are found in the mollusks? What are trochophores? What is a veliger?

7. What are four characteristics of the bivalves? What is the function of the adductor muscles? What specializations are associated with the mantle? How do bivalves feed? Do bivalves generally disperse as larvae or adults? Explain.

8. What are two advantages of embryonic segmentation? In which two animal phyla is segmentation most notable?

9. How is the annelid nervous system specialized? How is annelid locomotion similar to that of the nematodes? How is it more advanced? What are annelid setae? What function do they serve?

10. How are annelids developmentally similar to mollusks? What is the likely evolutionary connection between these phyla? What is the relationship between the segmentation of annelids and arthropods as compared with the segmentation found in vertebrates?

11. What type of circulatory system is found in annelids? How do they exchange gases with their environment? How does the typical annelid excretory unit compare to that of mollusks?

12. How do earthworms obtain their nutrients? What sensory structures do earthworms possess? Where are earthworm parapodia located? How do these animals reproduce?

13. How are leeches similar to earthworms? How are they different? What adaptations enable leeches to suck blood?

 ## THOUGHT QUESTIONS

1. Would you expect the pattern of cell division in the early embryos of the acoelomate and pseudocoelomate animals to be spiral? Why?

2. The ancestral mollusk is thought to have had a very limited shell, consisting mainly of calcareous plates; many contemporary mollusks now seem to be in the process of losing their shells. Most mollusks, however, still have well-developed shells. What is the evolutionary advantage of having a shell? Of not having one?

3. In what ways is an earthworm more complex than a flatworm?

4. Why do we believe that the ancestor of the deuterostomes is a protostome?

 ## FOR FURTHER READING

BAVERDAM, F.: "Even for Ethereal Phantasms, It's a Dog-Eat-Dog World," *Smithsonian*, August 1989, pages 94-101. A fascinating account of the ways in which nudibranchs, marine slugs, deter predators.

DALES, R.P.: *Annelids*, ed. 2, The Hutchinson Publishing Group Ltd., London, 1967. A good overall account of this important phylum.

GOSLINE, J.M., and M.E. DEMONT: "Jet-Propelled Swimming in Squids," *Scientific American*, January 1985, pages 96-103. Describes how a squid jets through the water as rapidly as a fish for short distances by contracting radial and circular muscles in its boneless mantle wall.

HEDEEN, R.A.: *The Oyster: The Life and Lore of the Celebrated Bivalve*, Tidewater Publishers, Centreville, Md., 1986. The natural history of oysters, with emphasis on their role in Chesapeake Bay.

HESLINGA, G.A., and W.K. FITT: "The Domestication of Reef-Dwelling Clams," *BioScience*, 1987, vol. 37, pages 332-339. Farming, using land-based nurseries and natural coral reefs, may restore giant clams to importance for people in tropical Pacific and Indian Ocean areas.

HICKMAN, C.P., Jr., L.S. ROBERTS, and F.M. HICKMAN: *Integrated Principles of Zoology*, ed. 8, Mosby—Year Book Publishing Co., St. Louis, 1988. An outstanding general zoology text, with thorough coverage of all groups; the standard in its field.

KNIGHT, D.: "Nice Work for a Worm," *New Scientist*, vol. 123, pages 55-59. An interesting and concise account of some of the ecologically significant activities of earthworms.

LEE, K.E.: *Earthworms: Their Ecology and Relationships with Soils and Land Use*, Academic Press, New York, 1985. An excellent account of the biology of earthworms, a group of critical ecological importance.

LENT, C.M., and M.H. DICKINSON: "The Neurobiology of Feeding in Leeches," *Scientific American*, June 1988, pages 98-103. Illustrates beautifully the use of leeches as experimental animals.

MORTON, J.E.: *Mollusks*, ed. 5, The Hutchinson Publishing Group Ltd., London, 1979. A comprehensive account of the biology of mollusks.

RICHARDSON, J.R.: "Brachiopods," *Scientific American*, September 1986, pages 100-106. Fascinating discussion of the ecology and diversity of this ancient phylum.

WELLS, M.: "Legend of the Living Fossil," *New Scientist*, 23 October 1986, pages 36-41. Very interesting analysis of the features of the chambered nautilus—a very specialized animal with a shell that resembles that of its ancestors.

41

Arthropods

OVERVIEW

Arthropods are the most successful of all living animals in terms of their number of individuals and species, total mass, and complete occupation of terrestrial habitats. More arthropods are living on earth than all other kinds of animals put together. All arthropods have an exoskeleton that is rich in chitin and is often hard. The major evolutionary advance that led to their success was the development of jointed appendages, a development that may have occurred independently at least three times. The insects and their relatives were clearly derived from the annelids, but the origins of the other two major lines of arthropods are not clear. The nature of their respiratory, circulatory, excretory, and skeletal systems limits the size that arthropods can attain.

FOR REVIEW

Here are some important terms and concepts that you will encounter in this chapter. If you are not familiar with them, you should review them before proceeding.

Chitin (Chapter 3)

The origin of species (Chapter 21)

Evolutionary history (Chapter 22)

Protostomes-deuterostomes (Chapter 40)

With the evolution of the first annelids, many of the major innovations of animal structure had already appeared: the division of tissues into three primary types (endoderm, mesoderm, and ectoderm), bilateral symmetry, coelomic body architecture, and segmentation. An innovation yet remained, however, whose appearance originated the body plan characteristic of the most successful of all animal groups, the arthropods. This innovation was the development of jointed appendages. Such legs seem to have evolved after the ancestors of the arthropods acquired a relatively rigid, protective exoskeleton. For legs of this kind to be able to move, joints were a necessity; therefore, jointed appendages should not be taken as strong evidence for a common ancestry.

Arthropods are by far the most successful of all animals. Approximately 900,000 species—about two thirds of all the named species on earth—are members of this gigantic phylum, and it is estimated that there may be *at least* 2 million more still awaiting discovery. One scientist has recently estimated, based on the number and diversity of insects in tropical forests, that there might be as many as 30 million species of this one class alone (see Figure 28-5). If he is correct, there would be 10 times as many species of insects as of all other kinds of living animals put together! At the very least, several times as many species of arthropods exist as of all other plants, animals, and microorganisms. A hectare of lowland tropical forest is estimated to be inhabited by as many as 41,000 species of insects, on the average; many suburban gardens may have 1500 or more species of this gigantic class of organisms. The insects and other arthropods are abundant in every habitat on this planet, but they especially dominate the land, where, along with the flowering plants and the vertebrates, they determine the very structure of life. In terms of individuals, it has been estimated that approximately a billion billion (10^{18}) insects are alive at any one time!

Arthropods and especially the largest class—the insects—are of enormous economic importance and affect all aspects of human life. They compete with humans for food of every kind and cause billions of dollars of damage to crops, both before and after harvest. They are by far the most important herbivores in all terrestrial ecosystems; virtually every kind of plant is eaten by one or many species of insect. Insects also prey on, parasitize, or otherwise obtain food from most kinds of animals. The diseases that they spread cause enormous financial damage each year and strike every kind of domesticated animal and plant, as well as human beings.

On the positive side, the pollination of certain crops by insects is of key significance, as is their role in controlling other insects and weeds. A number of products, such as silk and honey, are produced directly by insects. The crustaceans, a class of primarily marine and fresh water arthropods that includes crabs, lobsters, shrimps, and related animals, are important as sources of human food. Insects and other arthropods are important elements in recycling organic matter within the soil and elsewhere, and herbivores play an especially important role in this process. At normal population levels, insects and other arthropods in the ecosystem speed the cycling of nutrients, have little effect on the amount of living plant material, and strongly influence the stock of dead and decaying organic matter. These are only a few of the ways in which arthropods are involved in human welfare.

GENERAL CHARACTERISTICS OF ARTHROPODS

The name "arthropod" comes from two Greek words, *arthros*, jointed, and *podes*, feet. We recognize the members of this phylum especially because of their jointed appendages. The numbers of these appendages are progressively reduced in the more advanced members of the phylum, and their nature differs greatly in different subgroups. Thus individual appendages may be modified into antennae, mouthparts of various kinds, or legs. Still others (the wings of certain insects, for example) are not homologous to the other appendages of arthropods; insect wings evolved separately.

Arthropod bodies are segmented like those of annelids, a phylum to which at least some of the arthropods are clearly related. The members of some classes of arthropods have many body segments. In others, the segments have become fused together into functional groups, or **tagmata** (singular, **tagma**), such as the head or thorax of an insect (Figure 41-1), by a process known as **tagmatization**, which is of cen-

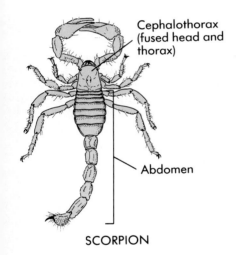

Cephalothorax (fused head and thorax)

Abdomen

SCORPION

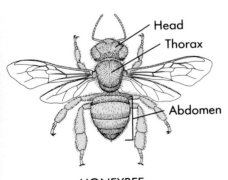

Head

Thorax

Abdomen

HONEYBEE

FIGURE 41-1
Evolution from many to few body segments among the arthropods. The scorpion and the honeybee are examples of arthropods with different numbers of body segments.

tral importance in the evolution of the arthropods. But even in these arthropods, the original segments can be distinguished during the development of the larvae. To a certain extent, the segments can also be reconstructed on the basis of careful morphological comparisons of the adult structures. All arthropods have a distinct head, sometimes fused with the thorax to form a tagma called the **cephalothorax.** This is another distinguishing feature of the phylum.

Arthropods, like annelid worms, are segmented, but in many arthropods the individual segments are fused into functional assemblies called tagmata. All arthropods have a distinct head, sometimes fused into a single unit with the thorax.

The arthropods have a rigid external skeleton, or **exoskeleton.** As an individual outgrows an exoskeleton, that exoskeleton splits open and is shed, allowing the animal to increase in size. The eggs of arthropods develop into immature forms that may bear little or no resemblance to the adults of the same species; most members of this phylum change their characteristics as they develop from stage to stage, a process called **metamorphosis** (see Figure 41-32).

The great majority of arthropod species consist of small animals—mostly about a millimeter in length—but members of the phylum range in adult size from about 80 micrometers long (some parasitic mites) to 3.6 meters across (a gigantic crab found in the sea off Japan). Some lobsters are nearly a meter in length. The largest living insects are about 33 centimeters long, but the giant dragonflies that lived during the Carboniferous Period, about 300 million years ago, had wingspans of as much as 60 centimeters!

ONYCHOPHORA

The phylum Onychophora consists of some 70 species of wormlike animals, ranging from about 1.5 centimeters to more than 15 centimeters long, with a soft, chitinous exoskeleton. Called velvet worms because of their slightly fuzzy cuticle, and walking worms because of their numerous paired, unjointed legs, onychophorans appear to be closely related to arthropods. When first discovered in 1825, they were thought to be mollusks—thin, dry slugs with legs.

Onychophora crawl amongst humid leaf litter in tropical and subtropical regions (Figure 41-A), and they also occur in some temperate regions of the southern hemisphere. They capture prey, usually insects, by shooting a gluey substance from openings on each side of the mouth. They then inject lethal saliva. The fossil record indicates that Onychophora have changed little for more than 500 million years. Early Cambrian fossils from China described in 1991 suggest that *Hallucigenia* and other odd forms discovered in the Burgess Shale are also members of the phylum.

Manton and some other taxonomists would include onycophorans in the subphylum Uniramia because they share some characteristics with arthropods: a cuticle, open circulatory system with a dorsal tubular heart, tracheae, and a hemocoel. But they also have annelid characteristics: non-jointed appendages; segmentally arranged, ciliated nephridia; and ciliated reproductive ducts. Their mouthparts resemble neither arthropods nor annelids, and the ladderlike ventral nerve cords appear more widely separated than in either phylum. Onychophoran embryology resembles that of both the Uniramia and the oligochaete annelids.

Despite the mixture of annelid and arthropod characteristics—some of which may have evolved independently in this group—most taxonomists would not call Onychophora a missing link between the segmented worms and jointed leg animals. They are probably most closely related to arthropods, but appear to represent an independent evolutionary line in which generalized features have persisted for hundreds of millions of years.

FIGURE 41-A
One of the Onychophora, *Peripatodes*, in the moist forests of Queensland, Australia.

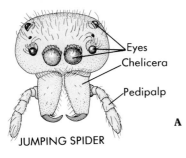

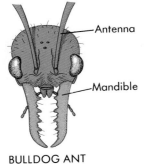

FIGURE 41-2
Chelicerates and mandibulates.
In the chelicerates, such as the jumping spider (**A**), the chelicerae are the foremost appendages of the body. Mandibulates, in contrast, such as the bulldog ant (**B**), have the antennae at the anterior end of the body.

MAJOR GROUPS OF ARTHROPODS

Many arthropods, traditionally called the **mandibulates,** have jaws, or **mandibles,** formed by the modification of one of the pairs of anterior appendages, but not the one nearest the anterior end. The mandibulates include the crustaceans, insects, centipedes, millipedes, and a few other small groups. The remaining arthropods, which include the spiders, mites, scorpions, and a few other groups, lack mandibles. These animals are called the **chelicerates.** Their mouthparts, called chelicerae (Figure 41-2), evolved from the appendages nearest the anterior end of the animal. They often take the form of pincers or fangs.

In the mandibulates, the appendages nearest the anterior end are one or more pairs of sensory antennae, and the next are the mandibles. The mandibles of the crustaceans, sometimes called the "aquatic mandibulates," are similar to their other appendages. All crustacean appendages are basically biramous, or two branched (Figure 41-3), although some of these appendages have become single branched by reduction in the course of their evolution. Mandibulates other than crustaceans have traditionally been called the "terrestrial mandibulates." They include the insects, centipedes, millipedes, and two other small classes. Another phylum, the Onychophora, also seems to be related to this group of arthropod classes (see boxed essay, p. 827). The mandibles of the terrestrial mandibulates are fundamentally similar to their other appendages, which are uniramous, or single branched. It is clear that mandibles evolved independently in the crustaceans and in the terrestrial mandibulates; the ancestors of both groups evidently lacked mandibles.

Among the chelicerates, the second pair of appendages is usually pincerlike or feelerlike, and the remaining pairs of appendages are legs. Chelicerae resemble the other appendages of the chelicerates more closely than mandibles resemble the other appendages of the mandibulates. The fundamental differences in the derivation and structure of their mouthparts indicate that the chelicerates and the mandibulates represent different evolutionary lines among the arthropods and that neither group gave rise to the other.

The late English zoologist S.M. Manton argued convincingly that the arthropods consist of three groups that evolved independently; she called these groups the subphyla Chelicerata, Biramia (now Crustacea), and Uniramia (Table 41-1). Manton concluded that the exoskeleton, which is characteristic of arthropods, evolved independently in each of these lines. Once it had done so, segmentation of both the body and

TABLE 41-1   MAJOR GROUPS OF PHYLUM ARTHROPODA
Subphylum **Chelicerata:** chelicerates
Class Arachnida: arachnids
Order Scorpiones: scorpions
Order Araneae: spiders
Order Acari: mites (including ticks and chiggers)
Order Opiliones: harvestmen or daddy longlegs
Class Merostomata: horseshoe crabs
Class Pycnogonida: sea spiders
Subphylum **Crustacea:** crustaceans
Class Crustacea: crustaceans
Order Cladocera: water fleas
Order Anostraca: fairy shrimps
Order Ostracoda: ostracods
Order Copepoda: copepods
Order Cirripedia: barnacles
Order Isopoda: isopods
Order Amphipoda: amphipods
Order Decapoda: crabs, lobsters, and shrimps
Subphylum **Uniramia:** insects, centipedes, and millipedes
Class Insecta: insects (many orders are mentioned in text)
Class Chilopoda: centipedes
Class Diplopoda: millipedes

FIGURE 41-3
Diagrams of a biramous leg in crustacean (crayfish) and of a uniramous leg in an insect.

its appendages became necessary. Otherwise the animals would have been unable to move!

Embryological evidence appears to support Manton's suggestion. For example, all Crustacea share a unique kind of larva called a **nauplius** (Figure 41-4). In contrast, the early development of the Uniramia closely resembles that of certain annelids, from which they were probably derived. In contrast, the embryology of the chelicerates differs greatly from that of both the crustaceans and the Uniramia. Despite this, other features—the structure of certain pigments, for example—suggests that the three groups of arthropods descended from a common ancestor. Additional evidence will be needed to clarify the situation further.

The phylum Arthropoda is divided into three subphyla: Chelicerata, Crustacea, and Uniramia. The chelicerates are characterized by chelicerae, mouthparts that often take the form of pincers or fangs, which evolved from the anterior appendages. Members of the other two subphyla have mandibles, originally biting jaws that also evolved from appendages, but from the second or third pair back from the anterior end. All appendages in crustaceans are fundamentally biramous, or two branched, whereas those in Uniramia are uniramous, or single branched.

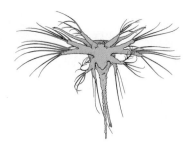

FIGURE 41-4
Although crustaceans are diverse, they have fundamentally similar larvae. The nauplius larva of a crustacean is an important unifying feature found in all members of this group.

EXTERNAL FEATURES

Aside from such features as the segmentation and appendages of arthropods, two characteristic features of the phylum deserve special discussion: the exoskeleton and the compound eye.

Exoskeleton

The bodies of all arthropods are covered by a hardened exoskeleton or cuticle, a key adaptation that facilitated the invasion of the land by the members of this group. This tough outer covering, against which the muscles work, is secreted by the epidermis and fused with it. The exoskeleton varies greatly in toughness and thickness in different arthropods (Figure 41-5); its versatility is doubtlessly one of the keys to the amazing success of the phylum. In most crustaceans, the exoskeleton is impregnated with calcium carbonate and is thus relatively inflexible. In other arthropods, the exoskeleton may be fairly soft and flexible; although in some large insects, horseshoe crabs, and other groups, it can also be thick and very hard.

Arthropods periodically undergo **ecdysis,** until they are mature; when they outgrow their exoskeletons, they form a new one underneath the old one. This process is controlled by hormones. When the new exoskeleton is complete, it becomes separated from the old one by fluid. This fluid dissolves the chitin and, if it is present, calcium carbonate, from the old exoskeleton, and the fluid increases in volume. Finally, the original exoskeleton cracks open, usually along the back, and is shed. The arthropod

A   B

FIGURE 41-5
Exoskeletons. Some arthropods have a tough exoskeleton, like this South American scarab beetle, *Dilobderus abderus* (order Coleoptera) (**A**); others have a fragile exoskeleton, like the green darner dragonfly, *Anax junius* (order Odonata) (**B**).

emerges, clothed in a new, pale, and still somewhat soft exoskeleton. As a result of the blood circulation to all parts of the body, the arthropod ultimately expands to full size; many insects and spiders take in air to assist them in this expansion. The exoskeleton subsequently hardens after exposure to the air or water in which the animal lives. The exoskeleton protects arthropods from water loss and helps to protect them from predators, parasites, and injury. While the exoskeleton is soft, however, the animal is especially vulnerable. At this stage arthropods often hide under stones, leaves, branches, or other safe places.

> **All arthropods have a rigid, chitinous exoskeleton that provides places for muscle attachment, protects the animal from predators and injury, and, most important, impedes water loss.**

Compound Eye

An important structure of many arthropods is the **compound eye** (Figure 41-6, *A*). Compound eyes are composed of many independent visual units, often thousands of them, called **ommatidia.** Each ommatidium is covered with a lens and linked to a complex of eight retinula cells and a light-sensitive central core, or **rhabdom.** Compound eyes among the insects are of two main types: **apposition eyes** and **superposition eyes.** Apposition eyes are like those of bees, in which each ommatidium acts in isolation, collecting light from one sector of the external world and throwing an inverted image of that scene on the rhabdom of that ommatidium. In such an eye, the individual ommatidia are surrounded by **pigment cells,** which keep the light that reaches each ommatidium separate. The individual images are combined in the insect's brain to form its visual image of the external world. In superposition eyes, such as those found in moths, the images from a series of ommatidia are combined on a

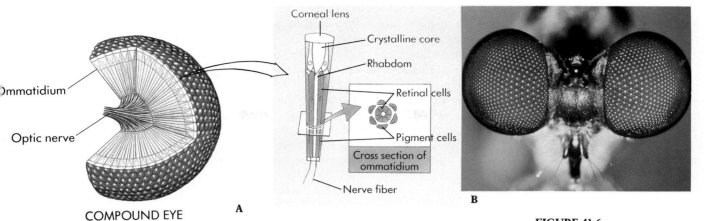

COMPOUND EYE

FIGURE 41-6
The compound eye.
A The compound eyes found in insects are complex structures.
B Three ocelli can be seen between the compound eyes of the robberfly (order Diptera).

cornea that lies at the back of the compound eye; and there are no screening pigment cells. Such an image is right-side-up.

Simple eyes, or **ocelli,** with single lenses are found in the other groups and sometimes occur together with compound eyes, as is often the case in insects (Figure 41-6, *B*). Ocelli function in distinguishing light and darkness; the ocelli of some flying insects, namely locusts and dragonflies, function as horizon detectors and are involved in the visual stabilization of their course in flight.

INTERNAL FEATURES

In the course of arthropod evolution, the coelom has become greatly reduced; in essence, it consists only of the cavities that house the reproductive organs and some glands. Arthropods completely lack cilia, both on the external surfaces of the body and on the internal organs. Like the annelids, the arthropods have a tubular gut that extends from the mouth to the anus. In the next paragraphs we will discuss the circulatory, respiratory, excretory, and nervous systems of the arthropods (Figure 41-7).

Circulatory System

The circulatory system of arthropods is open; their blood flows through cavities between the internal organs and not through closed vessels (see Figure 41-7, *B*). The principal component of an insect's circulatory system is a longitudinal vessel, which is

FIGURE 41-7
A grasshopper (order Orthoptera) illustrates the major structural features of the insects, the most numerous group of arthropods.
A External anatomy.
B Internal anatomy.

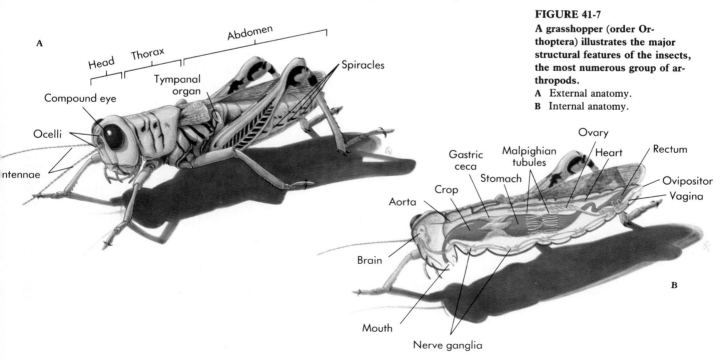

called the heart. This vessel runs near the dorsal surface of the thorax and abdomen. When it contracts, blood flows from its anterior end into the head region of the insect. From there, the blood circulates gradually through all of the body's internal spaces. Blood flows most rapidly when the insect is running, flying, or otherwise active. At such times the blood efficiently delivers nutrients to the tissues and removes wastes from them.

When an insect's heart relaxes, blood returns to it through a series of valves. These valves are located in the posterior region of the heart and allow the blood to flow inward only. Thus blood from the head and other anterior portions of the insect gradually flows through the spaces between the tissues toward the posterior end and then back through the one-way valves into the heart.

Respiratory System

Insects and other Uniramia, which are fundamentally terrestrial, depend on their respiratory rather than their circulatory system to carry oxygen to their tissues. In vertebrates, the blood moves within a closed circulatory system to all parts of the body and carries the oxygen with it. This is a much more efficient arrangement than exists in the arthropods, in which all parts of the body need to be near an air passage to obtain oxygen. As a result, the size of the arthropod body is much more limited than those of the vertebrates.

Unlike most animals, arthropods have no single major respiratory organ. The respiratory system of the terrestrial arthropods consists of small, branched, cuticle-lined air ducts called **tracheae** (Figure 41-8). These tracheae, which ultimately branch into very small **tracheoles,** are a series of tubes that transmit oxygen throughout the body. The tracheoles are in direct contact with the individual cells, and oxygen diffuses from them to other cells directly across the cell membranes. Air passes into the trachea by way of specialized openings called **spiracles,** which can, in most insects, be closed and opened by valves. The ability to prevent water loss by closing their spiracles was a key adaptation that facilitated the invasion of the land by arthropods. In many insects, especially the larger ones, the contraction of the muscles helps to increase the flow of gases into and out from the trachea. In other terrestrial arthropods, the flow of gases is essentially a passive process.

> **Arthropods have no cilia and only a limited coelom. Their circulatory system is open. A network of tubes called tracheae transmits oxygen from the outside to the organs; external tracheal openings are controlled by opening and closing spiracles.**

Many spiders and some other chelicerates have a unique system of respiration that involves **book lungs,** a series of leaflike plates within a chamber into which air is drawn and from which it is expelled by muscular contraction. Such book lungs may exist alongside tracheae, or they may function instead of tracheae, in particular groups of organisms. One small class of marine chelicerates, the horseshoe crabs, have book gills, which are analogous to book lungs but function in water. Tracheae,

FIGURE 41-8
Trachae and tracheoles. Trachae and tracheoles are connected to the exterior by specialized openings called spiracles, and carry oxygen to all parts of the insect's body.
A The tracheal system of a grasshopper.
B A portion of the tracheal system of a cockroach.

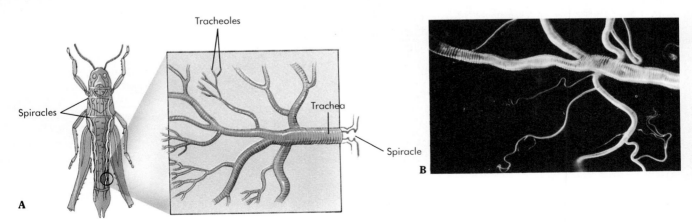

book lungs, and book gills are all structures found only in arthropods and in the phylum Onychophora (see boxed essay, p. 827), which have tracheae and are related to the insects. The crustaceans lack such structures and have gills.

Excretory System

Though there are various kinds of excretory systems in different groups of arthropods, we will focus here on the unique excretory system that evolved in terrestrial arthropods in relation to their open circulatory system. The principal structural element in such an excretory system is the **Malpighian tubules,** which are slender projections from the digestive tract. These are attached at the junction of the midgut and hindgut (see Figure 41-7, *B*). Fluid is passed through the walls of the Malpighian tubules to and from the blood in which the tubules are bathed. As this fluid passes through the tubules toward the hindgut, the nitrogenous wastes in it are precipitated as concentrated uric acid or guanine. These substances are then emptied into the hindgut and thus eliminated. Most of the water and salts in the fluid is reabsorbed by the hindgut and returned to the arthropod's body. Malpighian tubules constitute an efficient mechanism for water conservation and were another key adaptation facilitating invasion of the land by arthropods.

> **Arthropods eliminate metabolic waste by a unique system of Malpighian tubules that extend from the digestive tract into the blood. Fluid enters the tubules, waste is precipitated, and the fluid is reabsorbed, passing out through the tubule walls back into the body cavity.**

Nervous System

The central feature of the arthropod nervous system is a double chain of segmented ganglia running along the animal's ventral surface (see Figure 41-7, *B*). At the anterior end of the animal are three fused pairs of dorsal ganglia, which constitute the brain. However, much of the control of an arthropod's activities is relegated to ventral ganglia. Therefore, many functions, including eating, movement, and copulation, can be carried out even if the brain has been removed. Indeed, the brain of arthropods seems to be a control point, or inhibitor, for various actions, rather than a stimulator, as it is in vertebrates. On the other hand, the degree of coordinated response by many arthropods is high, and their movements may be very rapid.

EVOLUTIONARY HISTORY OF THE ARTHROPODS

Fossils whose features resemble those of the chelicerates existed with the earliest well-preserved multicellular animals; this group of arthropods, therefore, goes back at least 630 million years. The terrestrial chelicerates evolved from marine ancestors that had thick exoskeletons and clearly marked segmentation.

The trilobites were another important group of early arthropods, but they have been extinct for 250 million years (see Figure 22-1). Trilobites, which lived in the early seas, were the first animals whose eyes were capable of a high degree of resolution. These eyes were unusual in structure; the eye of a trilobite contained as many as 15,000 individual elements! Each of these elements provided a separate image, which the animal could integrate into a highly detailed composite. Although these eyes were compound, they were clearly not equivalent to those of the crustaceans or Uniramia; eyes clearly evolved independently in trilobites.

> **The fossil record of chelicerates goes back as far as that of any multicellular animals, about 630 million years. One group of extinct chelicerates, the trilobites, was particularly abundant more than 250 million years ago.**

Fossil crustaceans more than 500 million years old are known, but they are so similiar to modern forms that they do not help us understand the derivation of the group. The crustaceans are diverse, but many have a fundamentally similar nauplius

FIGURE 41-9
A flying insect. This African blister beetle (order Coleoptera) has spread its two tough, leathery forewings, exposing its delicate flying wings normally concealed beneath them, and taken off. More than 350,000 species of beetles—more than a third of all named animal species—have been described.

larva (see Figure 41-4) and all were clearly derived from a common ancestor. Aside from a generalized relationship with the annelids and the Uniramia, little can be said about the evolutionary relationships of these animals.

Fossils similar to Onychophora (see boxed essay, p. 827) go back some 500 million years. Because of their close relationship to Uniramia, we may assume that the evolutionary line leading to Onychophora and Uniramia had originated, probably from ancestors similar to oligochaetes, before that time. Among the Uniramia are the most diverse and abundant of all arthropods, the insects. The earliest known insect fossils are more than 300 million years old. Some groups of insects that date that far back in the fossil record still have living representatives—cockroaches and dragonflies, for example. One of the early and important evolutionary innovations among the insects was the ability to fly. Flying insects have been in existence for more than 300 million years and have achieved remarkable success (Figure 41-9). For more than 100 million years, insects were the only flying organisms in existence; even today among all the living groups of organisms, only birds, bats, and insects truly fly.

Despite their great age and enormous evolutionary success, the arthropods have not evolved intelligent forms. Large brains, a prerequisite for intelligence, can develop only within large bodies. The external skeletons of the arthropods are more massive in relation to the support that they provide than the internal ones of the vertebrates. Because of this large mass-to-support ratio, exoskeletons do not provide the structural basis for the evolution of large bodies. In addition, because of their small diameters and the need for direct connections to the outside, tracheae are not as efficient as lungs and a circulatory system in distributing oxygen throughout a large body; it would be impossible for tracheae to function in a human-sized organism. Furthermore, a closed circulatory system seems to be necessary for a large animal, judging from the fact that all large animals have systems of this sort. Nevertheless, the drought-resistant cuticles and Malpighian tubules of the arthropods are excellent adaptations for a terrestrial existence. Their versatile, highly manipulable appendages are another important element in their remarkable success.

SUBPHYLUM CHELICERATA: THE CHELICERATES

Chelicerates (subphylum Chelicerata) are a distinct evolutionary line of arthropods, one in which the appendages nearest the anterior end of the body have been modified into chelicerae. These chelicerae, which often function as fangs or pincers, had a different evolutionary origin from the mandibles that originated separately in the crustaceans and in the Uniramia, the other two subphyla of arthropods. There are three classes of chelicerates: arachnids, horseshoe crabs, and sea spiders.

Class Arachnida: The Arachnids

By far the largest of the three classes of chelicerates is the largely terrestrial class Arachnida, with some 57,000 named species—the spiders, ticks, mites, scorpions, and daddy longlegs. Arachnids such as spiders have a pair of chelicerae, a pair of pedipalps, and four pairs of walking legs. The **chelicerae** are the foremost appendages; they consist of a stout basal portion and a movable fang, which is connected with a poison gland by a duct. The **pedipalps,** the next pair of appendages, may resemble the legs, but they have one less segment. In male spiders, the pedipalps are specialized copulatory organs. In other groups of arthropods, they may be specialized in other ways; often they have a sensory function. Despite their appearance, the pedipalps of carnivorous arachnids, such as spiders, are often used for catching and handling prey. Spiders also chew with the basal portions of their pedipalps. The pedipalps are rarely used for locomotion.

Although arachnids are sometimes confused with insects, they are actually different in many structural and other features. Most arachnids are carnivorous, although the mites are largely herbivorous. Most arachnids can ingest only preliquefied food, which they often digest externally by secreting enzymes into their prey and then suck up with their muscular, pumping pharynx. Arachnids occur primarily in terrestrial habitats, where they have evolved habits that lead to the direct transfer of sperm—

thus protecting it from drying out—including the complex mating rituals of many species of spiders. In other arachnids, such as the scorpions, sperm transfer is indirect, but the sperm is held in packets and thus protected from drying out. Not all arachnids live on land; some 4000 known species of mites and one species of spider live in fresh water, and a few mites live in the sea. Arachnids breathe by means of tracheae, book lungs, or both.

Eleven orders of arachnids include living species. Of these, we will briefly discuss four: Scorpiones, the scorpions; Araneae, the spiders; Acari, the mites and ticks; and Opiliones, the harvestmen or daddy longlegs. These are the most familiar of the orders of arachnids, and our discussion of them will introduce the group.

FIGURE 41-10
The scorpion *Uroctonus mordax*. This photograph shows the characteristic pincers and segmented abdomen, ending in a stinger, raised over the animal's back. White young scorpions cluster on this individual's back.

Order Scorpiones: The Scorpions.
The scorpions (order Scorpiones) are a familiar group of arachnids whose pedipalps are modified into pincers. Scorpions use these pincers to handle their food and tear it apart (see Figures 41-1 and 41-10). The venomous stings of scorpions are used mainly to stun their prey and less commonly in self-defense. The sting is located in the terminal segment of the body, which is slender toward the end. A scorpion holds its abdomen folded forward over its body when it is moving about. The elongated, jointed abdomens of scorpions are distinctive; in most chelicerates, the abdominal segments are more or less fused together and appear as a single unit.

Scorpions are probably the most ancient group of terrestrial arthropods; they are known from as early as the Silurian Period, some 425 million years ago. The adults of this order of arachnids range in size from 1 to 18 centimeters. Respiration in living scorpions occurs by means of book lungs, but their ancestors, all of which were marine, had gills. There are some 1200 species of scorpions, all terrestrial, which occur throughout the world, although they are more common in tropical, subtropical, and desert regions. The courtship of scorpions is elaborate, with the spermatophores being fixed to a substrate by the male and then picked up subsequently by the female. The young are born alive, with 1 to 95 in a given litter.

Order Araneae: The Spiders.
There are about 35,000 named species of spiders (order Araneae) (Figure 41-11). These animals play a major role in virtually all terrestrial ecosystems, where they are particularly important as predators of insects and other small animals. Spiders hunt their prey or catch it in webs of remarkable diver-

A

B

C

FIGURE 41-11
Spiders.
A Tarantulas are large hunting spiders, powerful enough to prey on small lizards, mammals, and birds, as well as insects. They do not spin webs
B This crab spider, in a flower in Malaya, has just captured a fly.
C A wolf spider, *Phidippus audax*, in Maryland. Wolf spiders, like the tarantulas, do not spin webs but instead actively hunt their prey. Most spiders have eight pairs of simple eyes; these differ in size, as seen here.

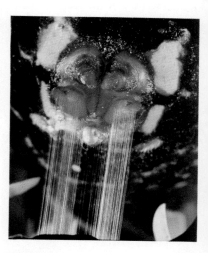

FIGURE 41-12
Spinnerets of a Peruvian orb-weaving spider in action producing fine strands of silk.

A

B

FIGURE 41-13

Two common poisonous spiders.
A The black widow spider, *Latrodectus mactans.*
B The brown recluse spider, *Loxosceles reclusa.* Both species are common throughout temperate and subtropical North America, but bites are rare in humans.

sity that they have constructed (see Figure 3-15, *D*). The silk of the webs is formed from a fluid protein that is forced out of spinnerets, modified appendages on the posterior portion of the spider's abdomen (Figure 41-12). There may be up to six pairs of these silk glands, with which different kinds of spiders produce many adaptive modifications of webs; some spiders, for example, can spin gossamer floats that allow them to drift away in the breeze to a new site. The webs and habits of individual kinds of spiders are often distinctive.

Many kinds of spiders, like the familiar wolf spiders (Figure 41-11, *C*), do not spin webs but instead hunt their prey actively. Others, the trap-door spiders, construct silk-lined burrows with lids from which they seize their prey as it passes by. One species of spider, *Argyroneta aquatica,* lives in fresh water, spending most of its time below the surface. Its body is surrounded by a bubble of air, while its legs, which are used both for underwater walking and for swimming, are not. Several other kinds of spiders walk about freely on the surface of water.

Spiders have poison glands leading through their chelicerae, which are pointed and used to bite and paralyze prey. Some members of this order, such as the black widow and brown recluse (Figure 41-13), have bites that are poisonous to humans and other large mammals.

Male spiders make a sperm web from special silk glands in the anterior portion of their abdomen. A drop of sperm deposited on the sperm web is picked up by the spider with the pedipalps. Many spiders have an elaborate courtship, following which the pedipalps fit into a special plate on the lower side of the female's abdomen, permitting the sperm to enter specific receptacles. In many species, the female eats the male once fertilization is completed. The eggs are enclosed in a silken egg sac, which is abandoned, guarded, or carried about, depending on the species of spider. Young spiders resemble adults and become mature after 3 to 15 molts, depending on the species.

Order Acari: The Mites.

The order Acari, the mites, is the largest—in terms of number of species—and most diverse of the arachnids (Figure 41-14). Although only

FIGURE 41-14

Mites.
A The tiny red flecks on this flower of creamcups, *Platystemon californicus,* are mites of the genus *Balaustium;* they have eaten all of the pollen.
B A water mite, perhaps of the genus *Hydrachna.* Note the feathery legs, modified for swimming.
C Ticks on the hide of a tapir in Peru. Many ticks spread diseases in humans and other vertebrates.

A

B

C

Chiggers (suborder Prostigmata, order Acari) are a large group of more than 150 genera and perhaps 3000 species of mites that parasitize amphibians, reptiles, birds and mammals, as well as a few kinds of invertebrates. In the first juvenile (larval) stage, most species of chiggers are ectoparasites on vertebrates. In the second juvenile (nymphal) stage and as adults, most species are free-living, carnivorous predators that prowl the soil-surface litter in search of prey such as insect eggs.

Larval chiggers do not feed on the blood of their hosts. Instead, they insert piercing mouthparts into the outer layers of the skin. They then digest some of the deeper skin tissue with salivary enzymes. The larvae remain on the skin for several days, sucking up extracellular fluid and dissolved cells. If the larvae obtain sufficient food, they drop off the host and continue their life cycle.

In temperate regions, chiggers appear during the warm months of the year. They occur throughout the United States. Most of the bites on humans in the United States are inflicted by several species of the genus *Eutrombicula* (Figure 41-*B*),

FIGURE 41-B

Larvae (1) and an adult (2) of a chigger (*Eutrombicula splendens*). The larvae are about 0.2 millimeter long, and the free-living, predaceous adult is about 1 millimeter long.

which normally parasitize lizards and snakes. Since they are so small as to be barely visible—the larvae are about 0.2 millimeter long—the presence of chiggers is usually detected only from the irritation caused by their bites. They crawl up and seek moist skin areas, and sitting or lying on chigger-infested vegetation can have painful consequences. Chiggers are especially common in areas with clay, rather than sandy, soil. Chigger bites are usually easily prevented with mosquito repellent to the skin and clothing. Bathing soon after contact with chiggers can remove them before they bite.

about 30,000 species of mites have been named, students of the group estimate that there may actually be a million or more members of this order in existence. In many regions of the world, the members of this group are poorly known; moreover, relatively few scientists study them.

Most mites are small, less than 1 millimeter long, but the adult length of different species ranges from 100 nanometers to 2 centimeters. In most mites, the cephalothorax and abdomen are fused into an unsegmented ovoid body. Respiration occurs either by means of tracheae or directly through the exoskeleton. Many mites pass through several distinct stages during their life cycle. In most, an inactive eight-legged prelarva gives rise to an active six-legged larva, which hatches from the egg and in turn produces a succession of three eight-legged stages and finally the adult males and females. In a number of different mites, however, various juvenile stages have become reproductive, and the development of those particular species then stops at that stage. The evolutionary process whereby juvenile stages become reproductive is known as paedomorphosis.

Mites are diverse, not only in their structure but also in their habits. They are found in virtually every terrestrial, freshwater, and shallow marine habitat known and feed on fungi, plants, and animals; they act as predators and as internal and external parasites of both invertebrates and vertebrates. Many plants—at least 1000 species—have pits, pores, or crypts on their leaves that are inhabited by predaceous mites, which protect the plants from herbivores.

Many mites are well known to human beings because of their irritating bites and the diseases that they transmit. Follicle mites live in the hair follicles and wax glands

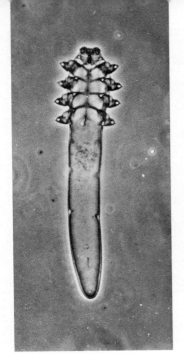

FIGURE 41-15

A follicle mite (*Desmodex folliculorum*). Extensively modified in structure in relation to their symbiotic habits, follicle mites occur in the glands at the base of the hairs of virtually every mammal (including humans), feeding on host cells and fluids. Usually a single mite occurs in each follicle, although there may be as many as three.

of the human forehead, nose, and chin, but usually cause no symptoms (Figure 41-15). Other mites cause mange in dogs and cats, often with severe consequences. A number of species of mites can cause house-dust allergy, which is aggravating to millions of people throughout the world; they are inhaled along with dust. The larvae of some species of chiggers (see boxed essay, p. 837) inflict annoying bites on humans.

Ticks, which are also members of this order, are blood-feeding ectoparasites—parasites that occur on the surface of their host—of vertebrates (see Figure 41-14, *C*). They are larger than most other mites and cause discomfort directly by sucking the blood of humans and other animals. Some of them also inject toxins into their hosts. A few ticks of this kind are known to cause paralysis in the people they bite. Ticks also carry many diseases, including some caused by viruses, bacteria, and even protozoa. The spotted fevers, of which Rocky Mountain spotted fever is a familiar example, are caused by members of a group of small omnibacteria known as rickettsiae, whereas Colorado tick fever is caused by a virus. Lyme disease is apparently caused by spirochaetes transmitted by ticks; tularemia is another example of a bacterial disease spread, in part, by ticks. Red-water fever or Texas fever is the best-known and most important tick-borne protozoan disease of cattle, horses, sheep, and dogs.

In addition to the diseases they cause in humans and other animals, mites cause extensive and often severe damage to plants. The group known as spider mites, or red spiders, are often the most serious pests of houseplants. Mites of this group also damage many crops. On the positive side, mites recently have been used as agents of biological control: certain mites attack harmful insects or other mites and have been introduced to control their numbers and therefore the harmful effects of their prey.

Order Opiliones: The Daddy Longlegs.
Another familiar group of arachnids consists of the daddy longlegs or harvestmen (order Opiliones). The members of this order are easily recognized by their oval, compact bodies and extremely long, slender legs (Figure 41-16). They respire by means of a single pair of tracheae and are unusual among the arachnids in that they engage in direct copulation. The males have a penis, and the females an ovipositor, or egg-laying organ, by means of which they deposit their eggs in cracks and crevices. Most daddy longlegs are predators of insects, other arachnids, snails, and worms, but some live on plant juices and many scavenge dead animal matter. The order includes about 5000 species. Although it occurs throughout the world, it is best represented in the tropics of Asia and South America.

Class Merostomata: The Horseshoe Crabs

A second class of chelicerates is that of the horseshoe crabs (class Merostomata). There are three genera of horseshoe crabs. One, *Limulus* (Figure 41-17), is common along the East Coast of North America. The other two genera live in the seas surrounding tropical Asia and its neighboring islands. The horseshoe crabs are an an-

FIGURE 41-16

A harvestman, or daddy long-legs.

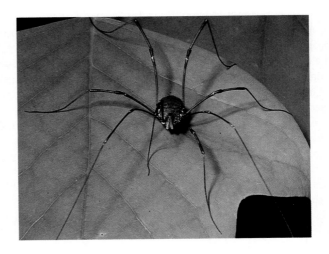

FIGURE 41-17
Diagram of a horseshoe crab,
Limulus, from below, illustrating
the principal features of this archaic animal.

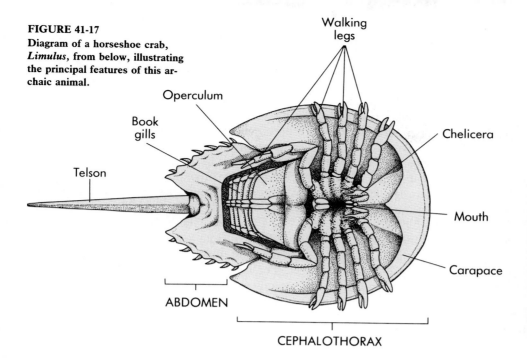

FIGURE 41-18
Horseshoe crabs, *Limulus*,
emerging from the sea to mate at
the edge of Delaware Bay, New
Jersey, in early May.

cient group; fossils virtually identical to *Limulus* date back 220 million years to the Triassic Period. Other members of the class, the eurypterans, are known from as much as 400 million years ago. Horseshoe crabs may have been derived from the trilobites, a relationship that is suggested by the appearance of their larvae.

Individuals of *Limulus* are up to 30 or even 60 centimeters long. They mature in 9 to 12 years, and have a life span of 14 to 19 years. Horseshoe crabs live in deep water, but they migrate to shallow coastal waters every spring, emerging from the sea to mate on moonlit nights when the tide is high (Figure 41-18). While they are mating, the male grasps the back of the female and holds on; the fertilization is external, with the females laying their eggs in the sand. The young horseshoe crabs that emerge from these eggs swim actively. They lack the tail spine, or telson, that is characteristic of the adults, and resemble trilobites. Despite its appearance, the telson (see Figure 41-17) is not used for defense.

Horseshoe crabs feed at night, primarily on mollusks and annelids. They swim on their backs by moving their abdominal plates. They can also walk on their five pairs of legs, which are protected, along with their chelicerae, by their shell. Each of the five pairs of book gills of the horseshoe crab is located under a pair of covers, or opercula, posterior to the legs.

Class Pycnogonida: The Sea Spiders

The third class of chelicerates is the sea spiders (class Pycnogonida). Sea spiders are relatively common, especially in coastal waters, and more than 1000 species are in the class. These animals are not observed often, however, because they are small—usually only about 1 to 3 centimeters long—and rather inconspicuous (Figure 41-19). They are found in oceans throughout the world but are most abundant in the far north and far south. Adult sea spiders are mostly external parasites or predators of other animals, including sea anemones.

Sea spiders have a sucking proboscis, with the mouth located at its end. Their abdomen is much reduced, and their body appears to consist almost entirely of the cephalothorax, with no well-defined head. There are usually four, or less commonly five or six, pairs of legs. Male sea spiders carry the eggs on their legs until they hatch, thus providing a measure of parental care for them. Sea spiders completely lack excretory and respiratory systems. They appear to carry out these functions by the direct diffusion of waste products outward through the cells and of oxygen inward through them. Sea spiders are not closely related to either of the other two classes of arachnids.

FIGURE 41-19
The sea spider *Pycnogonum
littorale* crawling over a sea
anemone.

SUBPHYLUM CRUSTACEA: THE CRUSTACEANS

The crustaceans (subphylum Crustacea) are a large group of primarily aquatic organisms, consisting of some 35,000 species of crabs, shrimps, lobsters, crayfish, barnacles, water fleas, pillbugs, and related groups. Often incredibly abundant in marine and freshwater habitats and playing a role of critical importance in virtually all aquatic ecosystems, crustaceans have been called "the insects of the water." Most crustaceans have two pairs of antennae, three pairs of chewing appendages, and various numbers of pairs of legs. All of the appendages of crustaceans, with the possible exception of the first pair of antennae, are basically biramous. One of these branches has been lost during the course of evolutionary specialization in many crustaceans, however, so that the appendages then have only a single branch. The nauplius larva stage through which crustaceans pass (see Figure 41-4) provides evidence that all of the members of this diverse group are descended from a common ancestor. In many groups, however, this nauplius stage has been lost during the course of evolution, and development to the adult form is direct. The nauplius hatches with three pairs of appendages; it metamorphoses through several other stages before reaching maturity.

Crustaceans differ from the insects but resemble the centipedes and millipedes in that they have legs on their abdomen as well as on their thorax. They are the only arthropods with two pairs of antennae. Their mandibles are thought to have originated from a pair of limbs that during the course of evolution took on a chewing function, a process that apparently occurred independently in the common ancestor of the terrestrial mandibulates. Many crustaceans have compound eyes. In addition, they have delicate tactile hairs that project from the cuticle all over the body. The larger crustaceans have feathery gills near the bases of their legs. In the smaller members of this class, gas exchange takes place directly through the thinner areas of the cuticle or the entire body.

The excretion of nitrogenous wastes in crustaceans also takes place mostly by diffusion across thin areas of cuticle, such as the gills. The osmotic and ionic composition of the crustacean's blood is regulated by a pair of glands, located in the ventral region of the head, consisting of a bladder, a tubule, and a spongy mass called the **labyrinth,** or green gland. Salts, amino acids, and some water is extracted as the waste products pass along the tubule, eventually to be excreted through the labyrinth.

Large, primarily marine crustaceans such as shrimps, lobsters, and crabs, along with their freshwater relatives the crayfish, are collectively called decapod crustaceans (Figure 41-20). The term decapod means "ten-footed." In these animals, the exoskeleton is usually reinforced with calcium carbonate. Most of their body segments are fused into a cephalothorax, which is covered by a dorsal shield, or **carapace,** which arises from the head. The crushing pincers common in many decapod crustaceans are used in obtaining food, for example, by crushing mollusk shells.

In lobsters and crayfish, appendages called **swimmerets,** which are used in reproduction and also for swimming, occur in lines along the sides of the abdomen (Figure 41-21). In addition, flattened appendages known as uropods form a kind of com-

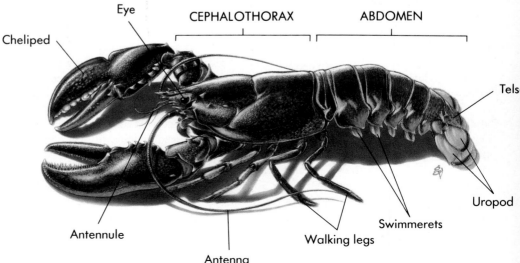

FIGURE 41-20
Decapod crustaceans.
A A freshwater crayfish, *Procambarus.*
B Montague's shrimp.
C Edible crab, *Cancer pangyrus.*
Many of the decapod crustaceans are important sources of food for humans.

FIGURE 41-21
The principal external features of a lobster, *Homarus americanus.*

Eye
Cheliped
CEPHALOTHORAX
ABDOMEN
Tels
Antennule
Antenna
Walking legs
Swimmerets
Uropod

A

B

C

pound "paddle" at the end of the abdomen. They may also have a telson, or tail spine. By snapping their abdomen, the animals propel themselves through the water rapidly and forcefully. Crabs (see Figure 41-20, *C*) differ from lobsters and crayfish in proportion; their carapace is much larger and broader and the abdomen is tucked under it. Shrimps (see Figure 41-20, *B*) and their relatives also have different proportions; their carapace is proportionately smaller than that of lobsters or crabs.

Although most crustaceans are marine, many occur in fresh water and a few have become terrestrial. These include some crabs, which may play an important role in some tropical and subtropical areas, and the pillbugs and sowbugs (Figure 41-22, *A*), which are the terrestrial members of a large, predominantly marine order of crustaceans known as the isopods (order Isopoda). Terrestrial isopods live primarily in places that are moist, at least seasonally; about half of the estimated 4500 species of the order are terrestrial. The sand fleas, or beach fleas (order Amphipoda) are another familiar group that includes many terrestrial crustaceans (Figure 41-22, *B*).

Minute crustaceans, including the larvae of larger species, are abundant in the plankton, the assemblage of small organisms suspended in the ocean's upper layers. Especially significant among them are the tiny copepods (order Copepoda; Figure 41-22, *C*), which are among the most abundant multicellular organisms on earth. Other orders of minute aquatic crustaceans include the water fleas (Cladocera) (Figure 41-23, *A*), ostracods (Ostracoda), and fairy shrimps (Anostraca). One of the fairy shrimps, the brine shrimp, is a standard food of aquarium fish (Figure 41-23, *B*).

Barnacles (order Cirripedia) (Figure 41-24; see Figure 23-12) are a group of crustaceans that are sessile as adults. Barnacles have free-swimming larvae. These ultimately attach themselves to a piling, rock, or other submerged object by their head and then stir food into their mouth with their feathery legs. Calcareous plates protect their body, and these plates are usually attached directly and solidly to the substrate.

Most crustaceans have separate sexes. Barnacles are hermaphroditic, but they generally cross-fertilize. Many different kinds of specialized copulation occur among the crustaceans, and the members of some orders carry their eggs with them, either singly or in egg pouches, until they hatch.

FIGURE 41-22

Diversity in crustaceans.

A Sowbugs, *Porcellio scaber*, representatives of the terrestrial isopods (order Isopoda).
B A marine amphipod, *Eurius* (order Amphipoda), thriving in water at −1.9° C in Antarctica. Many amphipods, known as beach fleas or sand fleas, are terrestrial or semiterrestrial in their habits.
C A copepod, member of an abundant group of marine and freshwater crustaceans (order Copepoda), most of which are a few millimeters long. Copepods are important components of the plankton.

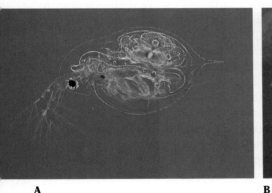

A

B

FIGURE 41-23

Crustaceans.

A Water flea (order Cladocera). Nearly all of the more than 400 species occur in fresh water. Most water fleas are about 0.2 to 3 millimeters long.
B Brine shrimp, one of the approximately 185 species of fairy shrimps (order Anostraca). The eggs of fairy shrimps can survive drying, and the members of this order occur in temporary pools or very saline ones throughout the world. Brine shrimp are commonly used to feed aquarium fishes.

FIGURE 41-24

Gooseneck barnacles, *Lepas anatifera*, feeding. These are stalked barnacles; many others lack a stalk.

A

B

FIGURE 41-25
Centipedes and millipedes. Centipedes are active predators, whereas millipedes are sedentary herbivores.
A Centipede, *Scolopendra*.
B Millipede, *Sigmoria*, in North Carolina.

SUBPHYLUM UNIRAMIA

The subphylum Uniramia is an enormous group that includes millipedes, centipedes, and insects, three distinct but clearly related classes. The wormlike Onychophora (see the boxed essay, pp. 827) are clearly related to the Uniramia. The Uniramia were certainly derived from annelids, probably ones similar to the oligochaetes, which they resemble in their embryology. All Uniramia respire by means of tracheae and excrete their waste products by means of Malpighian tubules.

Classes Diplopoda and Chilopoda: The Millipedes and Centipedes

The millipedes and centipedes both have bodies that consist of a head region followed by numerous segments that are all more or less similar and nearly all bearing paired appendages (Figure 41-25). Although the name *centipede* would imply an animal with a hundred legs and the name *millipede* one with a thousand, centipedes actually have 30 or more legs, millipedes 60 or more. Centipedes have one pair of legs on each body segment, millipedes two. Actually, each segment of a millipede is a tagma that originated during the group's evolution by the fusion of two ancestral segments. This fact explains why millipedes have twice as many legs per segment as centipedes.

In both centipedes and millipedes, fertilization is internal and takes place by direct transfer of sperm. The sexes are separate and all species lay eggs. Young millipedes usually hatch with three pairs of legs; they experience a number of growth stages, adding segments and legs as they mature, but do not change their general appearance.

The centipedes, of which some 2500 species are known, are all carnivorous and feed mainly on insects. The appendages of the first trunk segment are modified into a pair of poison fangs. The poison is often quite toxic to human beings, and many centipede bites are extremely painful, sometimes even dangerous.

In contrast, most millipedes are herbivores, feeding mainly on decaying vegetation; a few millipedes are carnivorous, like the centipedes. Many millipedes can roll their bodies into a flat coil or sphere because the dorsal area of each of their body segments is much longer than the ventral one. More than 10,000 species of millipedes have been named, but this is estimated to be no more than one sixth of the actual number of species that exists. In each segment of their body, most millipedes have a pair of complex glands that produces an ill-smelling fluid. This fluid is exuded for defensive purposes through openings along the sides of the body. The chemistry of the secretions of different millipedes has become a subject of considerable interest because of the diversity of the compounds involved and their effectiveness in protecting millipedes from attack. Some produce cyanide gas from segments near their head end. Millipedes live primarily in damp, protected places, such as under leaflitter, in rotting logs, under bark or stones, or in the soil.

Class Insecta: The Insects

The insects, class Insecta, are by far the largest group of organisms on earth, whether measured in terms of numbers of species or numbers of individuals. Insects live in every conceivable habitat on land and in fresh water, and a few have even invaded the sea. One scientist has calculated that there are probably about 200 million insects

FIGURE 41-26
Insect diversity.
A Cockroach, *Supella longipalpa* (order Blattoidea), with an egg mass, in California.
B Termite, *Macrotermes bellicosus* (order Isoptera), in Ghana. The large, sausage-shaped individual is a queen, specialized for laying eggs; most of the smaller individuals around it are nonreproductive workers, but the larger individual at the left of the queen is a reproductive male.
C Copulating grasshoppers (order Orthoptera).
D A true bug, *Edessa rufomarginata* (order Hemiptera), in Panama.
E A thorn-shaped leafhopper, *Umbonica crassicornis* (order Homoptera). Insects belonging to the related orders Hemiptera and Homoptera have sucking mouthparts.

alive at any one time for each person on earth! More than 70% of all the named animal species are insects, and the actual proportion is doubtless much higher, because millions of additional forms await detection, classification, and naming. Approximately 90,000 described species occur in the United States and Canada, and the actual number of species in this area probably approaches 125,000. In the tropics, the estimated numbers are truly amazing (see Figure 28-5). A glimpse at the enormous diversity of insects is presented in Figures 41-26 and 41-27.

External Features. Insects are primarily a terrestrial group, and most, if not all, of the aquatic insects probably had terrestrial ancestors. Most insects are relatively small, ranging in size from 0.1 millimeter to about 30 centimeters in length or wingspan. Insects have three body sections, the head, thorax, and abdomen; three pairs of legs, all attached to the thorax; and one pair of antennae (see Figure 41-7). In addition, they may have one or two pairs of wings. Most insects have compound eyes, and many have ocelli as well.

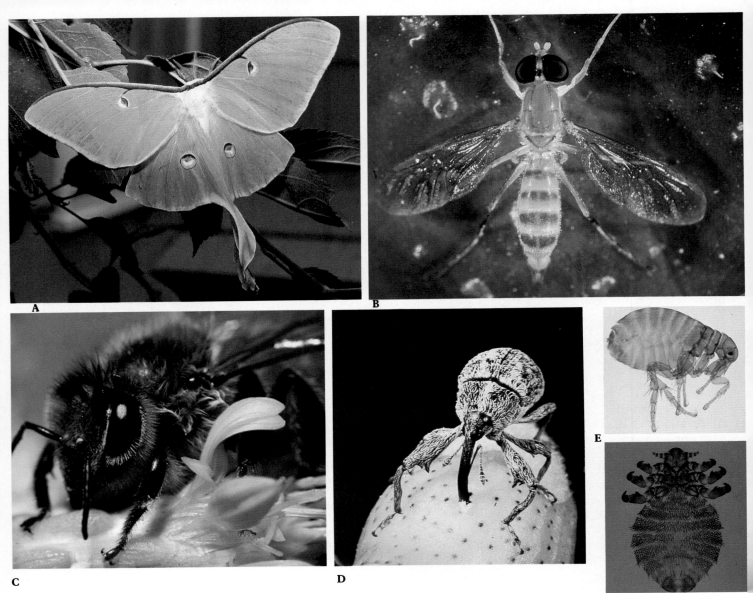

FIGURE 41-27

More examples of insect diversity.

A Luna moth, *Actias luna,* in Virginia. Luna moths and their relatives are among the most spectacular insects.

B Soldier fly, *Ptecticus trivittatus* (order Diptera), in Virginia.

C Honeybee, *Apis mellifera* (order Hymenoptera). Widely domesticated and an efficient pollinator of flowering plants.

D Boll weevil, *Anthonomus grandis.* Weevils are one of the largest groups of beetles (order Coleoptera). The boll weevil causes billions of dollars of damage to cotton crops throughout the world.

E Human flea, *Pulex irritans* (order Siphonaptera), in California. Fleas are flattened laterally, slipping easily through hair.

F Sucking louse, *Echinophthirus horridus* (order Anoplura), from a Lake Baikal seal. Flattened dorsoventrally, sucking lice have habits similar to those of fleas. Both fleas and lice have evolved from winged ancestors.

The mouthparts of insects are elaborate (Figure 41-28). They usually consist of the jaws, or mandibles, which are tough and unsegmented; a secondary pair of mouthparts, the **maxillae,** which are segmented; and the lower lip, or **labium,** which probably evolved from the fusion of another pair of maxilla-like structures. The upper lip, called the **labrum,** is of less certain origin. The hypopharynx is a short, tonguelike organ (in chewing insects) that lies between the maxillae and above the labium; the salivary glands usually open on or near the hypopharynx.

Within this basic structural framework, the mouthparts vary widely among groups of insects, mainly in relation to their feeding habits. Many orders of insects—

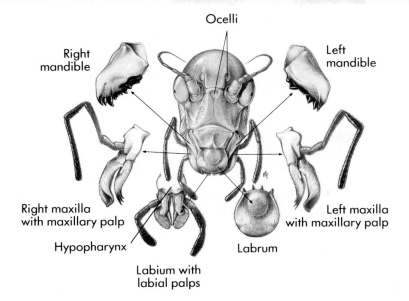

Ocelli

Right
mandible

Left
mandible

Right maxilla
with maxillary palp

Left maxilla
with maxillary palp

Hypopharynx

Labrum

Labium with
labial palps

FIGURE 41-28
**The complex chewing mouth-
parts of a grasshopper, showing
the various modified appendages.**

such as Coleoptera, the beetles; Hymenoptera, the bees, wasps, and ants; Isoptera, the termites; and Orthoptera, the grasshoppers, crickets, and their relatives (see Figure 41-28)—have chewing, or mandibulate, mouthparts. In other orders, the mouthparts may be elongated or styletlike (Figure 41-29). For example, in some flies (order Diptera), such as mosquitoes, blackflies, and horseflies, there are six piercing, fused stylets: the labrum, the mandibles, the maxillae, and the hypopharynx; the labium sheathes the stylets (see Figure 41-29, *A*). In more advanced flies, the labium may be the principal piercing organ or may be expanded into large, soft lobes through which liquid food is absorbed (see Figure 41-29, *B*). In the order Lepidoptera, the moths and butterflies, a long, coiled, sucking proboscis has evolved from the maxillae (see Figures 34-14, *A*, and 41-29, *C*); and the mandibles are absent or highly reduced in all but a few families. The proboscis of the Lepidoptera uncoils because of blood pressure and coils because of its own elasticity.

The insect thorax consists of three segments (tagmata), each of which has a pair of legs. Occasionally one or more of these pairs of legs is absent. Legs are completely absent in the larvae of certain groups—for example, in most members of the order Hymenoptera, the bees (see Figures 34-12 and 56-22), wasps, and ants—and among

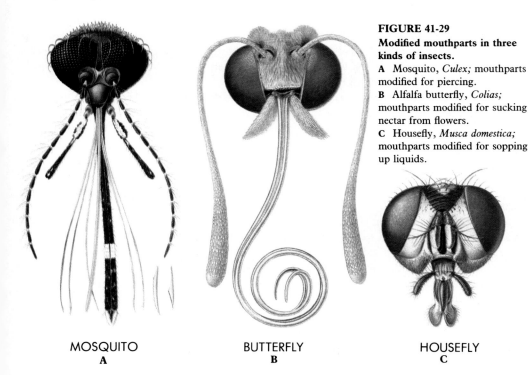

MOSQUITO
A

BUTTERFLY
B

HOUSEFLY
C

FIGURE 41-29
**Modified mouthparts in three
kinds of insects.**
A Mosquito, *Culex;* mouthparts
modified for piercing.
B Alfalfa butterfly, *Colias;*
mouthparts modified for sucking
nectar from flowers.
C Housefly, *Musca domestica;*
mouthparts modified for sopping
up liquids.

FIGURE 41-30
**Larvae ("wigglers") of a mos-
quito, *Culex pipiens*.** Most immature stages of flies resemble the
maggots (larvae) shown in Figure
41-32, but the aquatic larvae of
mosquitoes are quite active. They
breathe through tubes from the
surface of the water, as shown
here. Covering the water with a
thin film of oil, a method used
widely in mosquito control, causes
them to drown.

FIGURE 41–31
Scales on the wing of *Parnassius imperator*, a butterfly from China. Scales of this sort account for most of the colored patterns seen on the wings of butterflies and moths.

the flies, order Diptera (Figures 41-30 and 41-32). If two pairs of wings are present, they are attached to the middle and posterior segments of the thorax; if only one pair of wings is present, it is usually attached to the middle segment. The thorax is almost entirely filled with muscles that operate the legs and wings.

The wings of insects arise as saclike outgrowths of the body wall; in adult insects, they are solid except for the veins. Insect wings, therefore, are not homologous to the other appendages. Basically, insects have two pairs of wings (see Figures 41-9 and 41-27, *A*), but one of these has been lost during the course of evolution in some groups—for example, in the flies (see Figure 41-27, *B*). Most insects can fold their wings over their abdomen when they are at rest; but a few, such as the dragonflies and damselflies (order Odonata), cannot do this and keep their wings erect or outstretched at all times (see Figure 41-5, *B*).

Insect forewings may be tough and hard; if they are, they form a cover for the hindwings and usually open during flight. The beetles (Figures 41-5, *A* and 41-27, *D*) provide an example of the forewings remaining folded over the back except during flight (see Figure 41-9). The tough forewings also serve a protective function in the order Orthoptera, which includes grasshoppers (see Figures 41-7 and 41-26, *C*) and crickets. The wings of many insects are made of sheets of chitin; their strengthening veins are tubules of chitin. The moths and butterflies have wings that are covered with detachable scales, which provide most of the basis for their bright colors (Figure 41-31). In some wingless insects, such as the springtails or silverfish, wings never evolved; many others, such as the fleas (Figure 41-27, *E*) and lice (Figure 41-27, *F*), are derived from ancestral groups of insects that had wings.

All insects possess three body segments (tagmata): the head, the thorax, and the abdomen. The three pairs of legs are attached to the thorax. Fused and specialized mouthparts, ultimately derived from appendages, as well as the antennae, are located on the head. Most insects have compound eyes as well as simple eyes, or ocelli, and many have one or two pairs of wings.

Internal Organization. The internal features of insects resemble those of the other arthropods in many ways (see Figure 41-7). The digestive tract is a tube, usually somewhat coiled. It is often about the same length as the body. However, in the order Homoptera, which consists of the leafhoppers, cicadas, and related groups, and in many flies (order Diptera), it may be greatly coiled and several times longer than the body. Such long digestive tracts are generally found in insects that feed on juices rather than on protein-rich solid foods, because they offer a greater opportunity to absorb fluids and their dissolved nutrients. The digestive enzymes of the insect, also, are more dilute and thus less effective in a highly liquid medium than in a more solid one. Longer digestive tracts give these enzymes more time to work while food is passing through the insect's body. As a result, long, coiled digestive tracts often occur in those insects that have sucking mouthparts.

The anterior and posterior regions of an insect's digestive tract are lined with cuticle. Digestion takes place primarily in the stomach, or midgut; and excretion takes place through the Malpighian tubules, which arise from the junction between the midgut and hindgut and empty into the anterior end of the hindgut. Cells that line the midgut mainly secrete the digestive enzymes, although some are contributed by the salivary glands, located closer to the mouth.

The tracheae of insects extend throughout the body and permeate its different tissues (see Figure 41-8). In many winged insects, the tracheae are dilated in various parts of the body, forming air sacs. Such air sacs are surrounded by muscles and form a kind of bellows system to force air deep into the tracheal system. The spiracles, a maximum of 10 on each side of the insect, are paired and located on or between the segments along the sides of the thorax and abdomen. In most insects the spiracles can be opened by muscular action. In some parasitic and aquatic groups of insects, however, the spiracles are permanently closed. The tracheae run just below the surface of the insect, and gas exchange takes place by diffusion. Closing the spiracles at times may be important in retarding water loss. Longitudinal tubes connect the tracheae and make the outer part of the system an important route for flowing air.

The **fat body** of insects is a group of cells located in the body cavity. This structure may be quite large in relation to the size of the insect, and it serves as a food-storage reservoir, also having some of the functions of a vertebrate liver. It is often more prominent in immature insects than in adults, and it may be completely depleted when metamorphosis is finished. However, insects that do not feed as adults retain their fat bodies and live on the food stored in them throughout their adult life.

Sense Receptors. In addition to their eyes, insects have several characteristic kinds of sense receptors. These include **sensory hairs,** usually widely distributed over their bodies. The sensory hairs are linked to nerve cells and are sensitive to touch. They are particularly abundant on the antennae and legs—those parts of the insect most likely to come into contact with other objects. Similar organs are sensitive to chemicals or changes in position of the different parts of the insect. Taste organs are mostly located on the mouthparts; organs of smell, primarily on the antennae.

Sound, which is of vital importance to insects, is detected by tympanal organs in groups such as grasshoppers and crickets, cicadas, and some moths. These organs are paired structures composed of a thin membrane, the **tympanum,** associated with the tracheal air sacs (see Figure 41-7). In many other groups of insects, sound waves are detected by sensory hairs similar to those discussed previously. Male mosquitoes, for example, use thousands of sensory hairs on their antennae to detect the sounds made by the vibrating wings of female mosquitoes.

The detection of sound in insects is important not only for protection but also for communication. Many insects communicate by making sounds, most of which are quite soft, very high-pitched, or both, and thus inaudible to humans. Only a few groups of insects, especially the grasshoppers, crickets, and cicadas, make sounds that people can hear. Male crickets and longhorned grasshoppers produce sounds by rubbing their two front wings together; shorthorned grasshoppers do so by rubbing their hind legs over specialized areas on their wings. Male cicadas vibrate the membranes of air sacs located on the lower side of the most anterior abdominal segment. Other insects communicate by tapping some part of their body against an external object.

In addition to sound, nearly all insects communicate by means of chemicals or mixtures of chemicals known as **pheromones.** These compounds, which are extremely diverse in their chemical structure, are sent forth into the environment, where they are active in very small amounts and convey a variety of messages to other individuals. Such messages may include not only the attraction and recognition of members of the same species for mating, for example, but also the marking of trails for members of the same species, as in the ants. Recently, some pheromones have been used in biological control, to bait traps for pests or to disrupt mating, for example.

> **Insects possess sophisticated means of sensing their environment, including
> sensory hairs to detect touch, tympanal organs to detect sound, and
> chemoreceptors to detect chemical signals called pheromones.**

Life Histories. Most young insects hatch from fertilized eggs laid outside their mother's body. In a few insects, the eggs hatch within the mother's body. The zygote develops within the egg into a young insect, which escapes by chewing its way out or by bursting the shell in various ways. Some immature insects have specialized projections, often on the dorsal side of the head, that assist in this process.

During the course of their development into adults, young insects undergo ecdysis a number of times before they become adults and stop molting permanently. Most insects molt 4 to 8 times during the course of their development, but some may molt as many as 30 times. The stages between the molts are defined as instars. When an insect first emerges following ecdysis, it is usually pale, soft, and especially susceptible to predators. Its exoskeleton generally hardens in an hour or two; it must grow to its new size, usually by taking in air or water, during this brief period. The wings are expanded by forcing blood into their veins.

There are two principal kinds of metamorphosis in insects: **simple** and **complete** (Figure 41-32). In simple metamorphosis, the wings, if present, develop externally

FIGURE 41-32
Metamorphosis. Simple metamorphosis in a chinch bug (order Hemiptera) and complete metamorphosis in a housefly, *Musca domestica.*

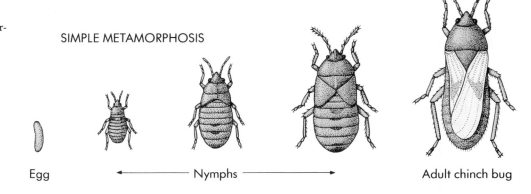

SIMPLE METAMORPHOSIS

Egg ←———— Nymphs ————→ Adult chinch bug

during the juvenile stages; ordinarily no "resting" stage exists before the last molt. In complete metamorphosis, the wings develop internally during the juvenile stages and appear externally only during the resting stage that immediately precedes the final molt (Figure 41-33). During this stage, the insect is called a pupa or chrysalis, depending on the group to which it belongs. A pupa or chrysalis does not normally move around much, although the pupae of mosquitoes do move around freely and thus provide one well-known exception to this general rule. A large amount of internal reorganization of the insect's body takes place while it is a pupa or chrysalis.

In insects with simple metamorphosis, the immature stages are often called **nymphs.** They are usually quite similar to the adults, differing mainly in their smaller size, less well-developed wings, and sometimes in their color. The most primitive orders of insects, such as the springtails and silverfish, are descended from ancestors that never had wings during their evolutionary history. In them, development is direct: no particular change occurs between the appearance of the nymphs and adults, except in size. In some other orders of insects with simple metamorphosis, such as the mayflies and dragonflies, the nymphs are aquatic and extract oxygen from the water by means of gills, which have evolved in these groups. The adult stages are terrestrial and look very different from the nymphs. In still other groups, such as the grasshoppers and their relatives, the nymphs and adults live in the same habitat. Such insects usually change gradually during their life cycle with respect to wing development, changes in body proportions, the appearance of ocelli, and other features.

More than 90% of the insects, however, including the members of all of the larg-

A

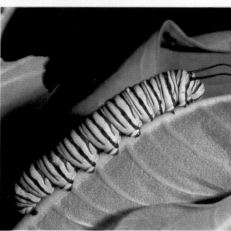

B

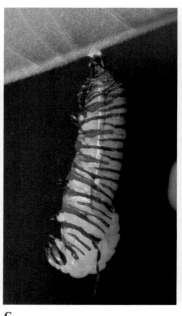

C

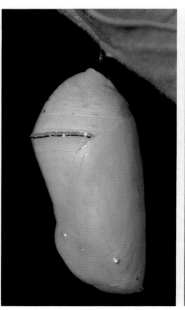

D

E

COMPLETE METAMORPHOSIS

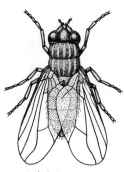

Egg   Early larva   Full-sized larva   Pupa   Adult housefly

est and most successful orders, display complete metamorphosis, in which the juvenile stages and adults often live in distinct habitats, have different habits, and are usually extremely different in form. In these insects, development is indirect. Larvae in insects are immature stages, often wormlike, which differ greatly in appearance from the adults of the same species. Larvae do not have compound eyes. They may be legless or have legs as well as sometimes having leglike appendages on the abdomen (see Figures 41-30, 41-32, and 41-33). They usually have chewing mouthparts, even in those orders in which the adults have sucking mouthparts. Chewing mouthparts are therefore clearly the primitive condition in all of these groups, and the sucking mouthparts of the adults have evolved from ancestral chewing mouthparts. When lar-

---

FIGURE 41-33

The life cycle of the monarch butterfly, *Danaus plexippus* (order Lepidoptera), illustrates complete metamorphosis.

A Egg.

B Larva (caterpillar) feeding on a leaf. The pseudopods, or false legs, that occur on the abdomen of larval butterflies and moths are not related to the true legs that occur in the adults.

C Larva preparing to shed its skin and become a chrysalis.

D Chrysalis.

E The colors of the wings of the adult butterfly can be seen through the thin outer skin of the chrysalis; the adult is nearly ready to emerge.

F to H Adult butterfly emerging.

I Adult monarch butterfly. Monarch butterflies feed on poisonous plants of the milkweed family (Asclepiadaceae), and they are protected from their predators, primarily birds, by the chemicals they obtain from these plants. Their conspicuous warning coloration advertises their toxic nature, as we saw in Chapter 24.

F   **G**   **H**   **I**

FIGURE 41-34

The hormonal control of metamorphosis in the silkworm moth, Bombyx mori. Neurosecretory cells on the surface of the brain secrete brain hormone, which in turn stimulates the prothoracic gland to produce molting hormone (ecdysone). For the caterpillars to molt, both ecdysone and juvenile hormone, produced by bodies near the brain called the *corpora allata,* must be present; but the level of juvenile hormone determines the result of a particular molt. At the late stages of metamorphosis, therefore, it is important that the corpora allata not produce large amounts of juvenile hormone.

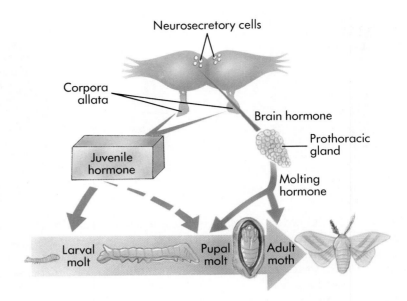

vae and adults play different ecological roles, they do not compete directly for the same resources, a fact that is clearly an advantage to the species.

Pupae do not feed and are usually, as we mentioned, relatively inactive. As pupae, insects are extremely vulnerable to predators and parasites, but they are often covered by a cocoon or some other protective structure. Groups of insects with complete metamorphosis include the moths and butterflies; beetles; bees, wasps, and ants; flies; and fleas. Fleas (order Siphonaptera) are among those orders of wingless insects that are descended from winged ancestors; their winglessness is an advantage in their specialized way of life.

Hormones control both ecdysis and metamorphosis. **Molting hormone,** or **ecdysone,** is released from a gland in the thorax when that gland has been stimulated by brain hormone, which in turn is produced by neurosecretory cells and released into the blood. The effects of the molting induced by ecdysone are determined by juvenile hormone, which is present during the immature stages but declines in quantity as the insect passes through successive molts. When the level of juvenile hormone is relatively high, the molt produces another larva; when it is lower, it produces the pupa and then the final development of the adult (Figure 41-34).

THE LARGEST PHYLUM

In this chapter we have presented an overview of the arthropods, the largest and most successful group of organisms on earth. Bewildering in their diversity, extraordinary in their numbers and ecological dominance in every habitat, the arthropods represent the culmination of one evolutionary line. We will now move on to consider the parallel line that led to the chordates and ultimately to ourselves.

SUMMARY

1. Arthropods are the most successful of all animals in terms of numbers of individuals and numbers of species, as well as in terms of ecological diversification. Like the annelids, arthropods have segmented bodies; but in arthropods, some of the segments have become fused into tagmata during the course of evolution.

2. All arthropods possess a rigid external skeleton, or exoskeleton, which provides a surface for muscular attachment, a shield against predators and diseases, and a means of conserving water.

3. Arthropods have an open circulatory system, which is not as important in conducting oxygen to the tissues as the tracheae, a system of tubes that open to the outside by way of spiracles.

4. Arthropods consist of three subphyla: chelicerates (subphylum Chelicerata); crustaceans (subphylum Crustacea); and insects, centipedes, and millipedes (subphylum Uniramia). These subphyla probably evolved from different ancestors, as suggested by the structure and derivation of their mouthparts and other appendages and by the important embryological differences among them.

5. The ancestry of the chelicerates and crustaceans is not clear, although they are generally related to the annelids and to the Uniramia. Crustaceans resemble Uniramia in their embryological features, whereas chelicerates differ from both groups.

6. Chelicerates consist of three classes, of which the Arachnida is the largest. Spiders, the best known arachnids, have a pair of chelicerae, a pair of pedipalps, and four pairs of walking legs. Spiders secrete digestive enzymes into their prey by means of the fangs on their chelicerae and then suck the contents out.

7. Crustaceans comprise some 35,000 species of crabs, shrimps, lobsters, barnacles, sowbugs, beach fleas, and many other groups. Their appendages are basically biramous, and their embryology is distinctive.

8. Uniramia, comprising the insects, centipedes, and millipedes, share a unique pattern of embryological development with the oligochaete annelids and leeches. Thus it seems clear that they evolved from ancestors similar to oligochaete annelids.

9. Insects and other Uniramia eliminate wastes by a unique system of Malpighian tubules, which extend from the digestive tract into the blood. This system is a very efficient means of water conservation, and, along with the exoskeleton and the spiracles, was a key factor in the invasion of land by the ancestors of insects.

10. Insects are the largest class of organisms, constituting at least 2 million and perhaps as many as 30 million species. They exhibit either simple metamorphosis, in which a succession of forms relatively similar to the adult progressively matures (the wings, if present, develop externally, and there is no resting phase), or complete metamorphosis, in which an often wormlike larva becomes a pupa, usually relatively sedentary, and then an adult (the wings, if present, develop internally). In complete metamorphosis, each of these stages is quite distinct from the others.

REVIEW QUESTIONS

1. Into what two groups associated with type of mouthparts are arthropods traditionally divided? Describe each group in terms of its appendages and give several examples. Which subgroup of arthropods has fundamentally biramous appendages? Which subgroup has uniramous appendages? How do biramous and uniramous appendages differ from one another?

2. What are the three arthropod subphyla? How are they evolutionarily related to each other? How is this supported by their embryology?

3. How do arthropods make their exoskeletons? Why is it valuable? What occurs during the process of ecdysis? What controls this process?

4. What is the nature of the coelom in most arthropods? What type of digestive system do they possess? What type of circulatory system do they have? Describe the direction of blood flow in arthropods. What helps to maintain this one-way flow?

5. Which arthropod organ system transports oxygen throughout the body? Is this different from vertebrates? In what way? What is the resultant evolutionary impact? Describe the respiratory system of a typical arthropod. What special systems exist in spiders, horseshoe crabs, and crustaceans?

6. What are the functional units of the terrestrial arthropod excretory system? To what other system are they connected? How are wastes processed by this system? How is water loss regulated?

7. What are the three classes of Chelicerata? Give examples of each. What is the arrangement of appendages in the largest class? What is the basic function of each set of appendages?

8. From what ancient arthropods have horseshoe crabs evolved? What type of fertilization do they exhibit? What is the general appearance of sea spiders? How do they regulate body wastes and gas exchange?

9. What is the typical arrangement of appendages in crustaceans? On which parts of the body do they possess legs? What is the function of the labyrinth, or green gland? How are decapod crustaceans different from most other crustaceans?

10. How are millipedes and centipedes similar to one another? How are they different from one another?

11. What type of digestive system do most insects possess? What digestive adaptations occur in those insects that feed on juices low in protein? Why? What respiratory adaptations have occurred in some insects?

12. What is an instar as it relates to insect metamorphosis? What are the two different kinds of metamorphosis in insects? How do they differ from one another? What are the immature forms of each type called?

 ## THOUGHT QUESTIONS

1. What are the main factors that limit the size of terrestrial arthropods? What effect has this had on the evolution of intelligence in the group?

2. What is the evidence that the annelids and the arthropods had a common ancestor? Which arthropods were involved? When would this evolutionary transition have taken place?

3. What is the evidence that the arthropods consist of groups that evolved from different ancestors at different times in the past? Why have the arthropods been treated as a single phylum if this evidence is so convincing?

 ## FOR FURTHER READING

ARNETT, R.H.: *American Insects: A Handbook of the Insects of America North of Mexico,* Van Nostrand Reinhold, New York, 1985. Synoptical account of the insects of the United States and Canada.

BORROR, D.J., D.M. DELONG, and C.A. TRIPLEHORN: *An Introduction to the Study of Insects,* ed. 5, Holt, Rinehart & Winston, New York, 1980. The standard text in the field, readable and well presented; strongly oriented to systematics and morphology.

EVANS, H.E.: *Insect Biology: A Textbook of Entomology,* Addison-Wesley Publishing Co., Reading, Mass., 1984. An outstanding modern textbook of entomology, less oriented to morphology and systematics than the text written by Borror, and others.

EVANS, H. and K. O'NEILL: "Beewolves," *Scientific American,* August 1991, pages 70–76. The females of these insects are voracious predators, each year capturing many bees to feed their young.

FOELIX, R.F.: *Biology of Spiders,* Harvard University Press, Cambridge, Mass., 1982. A readable account of all aspects of the life of spiders, ecologically oriented and well illustrated.

HADLEY, N.F.: "The Arthropod Cuticle," *Scientific American,* July 1986, pages 104-112. This complex covering accounts for much of the adaptive success of the arthropods.

HEINRICH, B.: "Thermoregulation in Winter Moths," *Scientific American,* March 1987, pages 104-111. How can certain moth species fly, feed, and mate at near-freezing temperatures?

HOLLDOBLER, B. and E.O. WILSON: *The Ants,* Harvard University Press, Cambridge, Mass., 1990. A wonderful book, filled with exciting and informative insights into this incredibly diverse group of insects.

SARGENT, W.: *The Year of the Crab: Marine Animals in Modern Medicine,* Norton, 1987. Relates how the horseshoe crab and other marine animals have become important tools in biomedical research.

SCHRAMM, F.R.: *Crustacea,* Oxford University Press, New York, 1986. A comprehensive and useful treatment of the morphology and development of the Crustacea.

SCOTT, J.A.: *The Butterflies of North America: A Natural History and Field Guide,* Stanford University Press, Stanford, Calif., 1986. An outstanding review of the butterflies of the United States and Canada, with much useful information about their biology.

SEELEY, T.D.: *Honeybee Ecology,* Princeton University Press, Princeton, N.J., 1986. Very readable account of the ecology of one of the most interesting and best-studied social insects.

WOOTTON, R.J.: "The Mechanical Design of Insect Wings," *Scientific American,* November 1990, pages 114-120. Subtle details of engineering and design reveal how insect wings are remarkably adapted to the acrobatics of flight.

VOLLRATH, F.: "Spider Webs and Silks," *Scientific American,* March 1992, pages 70-76. A delightful look at the many kinds of silk that spiders make, and how they use them.

Echinoderms and Chordates

CHAPTER OUTLINE

OVERVIEW

More than 630 million years ago, a revolution occurred within the animal kingdom, leading to the evolution of a novel pattern of embryonic development. A new line of animals appeared—the deuterostomes. Two superficially dissimilar animal phyla comprise most of this group: the echinoderms (including sea stars, sea urchins, and their relatives) and the chordates (including vertebrates). Although the echinoderms and chordates are very different in appearance, they are fundamentally similar in a number of features, which indicates that they were derived from a common ancestor. Among the chordates, vertebrates originated in the sea as jawless fishes. The bony fishes, which evolved from them, are the most plentiful vertebrates today in terms of numbers of species. Amphibians, which evolved from bony fishes, were the first vertebrate land dwellers. The first true terrestrial vertebrates, however, were the reptiles, which gave rise independently to the birds and to the mammals, including humans.

FOR REVIEW

Here are some important terms and concepts that you will encounter in this chapter. If you are not familiar with them, you should review them before proceeding.

Evolutionary theory (Chapter 20)

Vertebrate evolution (Chapter 22)

Classification (Chapter 28)

Protostomes and deuterostomes (Chapter 40)

Among the animals, two great evolutionary branches correspond to fundamental differences in the architecture of the developing embryo. In Chapters 39 to 41, we have traveled along one of these branches. Most invertebrates, including the annelids, mollusks, and arthropods, share the same pattern of embryological organization and are collectively called the protostomes. We will now consider the second major evolutionary branch, the deuterostomes.

DEUTEROSTOMES

Two outwardly dissimilar large phyla, Echinodermata and Chordata, together with two smaller phyla, have a series of key embryological features different from those of the other animal phyla (see Figure 40-2). Because it is extremely unlikely that these features evolved more than once, it is believed that these four phyla share a common ancestry. They are the members of a group called the deuterostomes, which were introduced in Chapter 40. As we saw there, deuterostomes diverged from protostome ancestors, which would not have resembled any living animals, more than 630 million years ago.

Deuterostomes, like protostomes, are coelomates. They differ fundamentally from protostomes, however, in the way in which the embryo grows. The blastopore of a deuterostome becomes the animal's anus, and the mouth develops at the other end; they have radial, rather than spiral, cleavage; daughter cells are identical for a brief period of development—each of which, if separated from the others at an early enough stage, can develop into a complete organism. Also whole groups of cells move around during embryonic development to form new tissue associations. The coelom is normally produced by an evagination of the archenteron—the main cavity within the gastrula, also called the primitive gut. The **archenteron,** which is lined with endoderm, opens to the outside via the blastopore, and eventually becomes the gut cavity.

In this chapter we will discuss the echinoderms, the two minor phyla of deuterostomes, and the simpler groups of chordates, thus encompassing the diversity of the class. We will focus our attention on the more advanced members of the phylum Chordata, the vertebrates. The concluding section of the book, which starts with Chapter 43, considers the features of the vertebrates in detail.

PHYLUM ECHINODERMATA: THE ECHINODERMS

Echinoderms (phylum Echinodermata) are an ancient group of marine animals well represented in the fossil record. The term *echinoderm* means "spine-skin," an apt description of many members of this phylum, which consists of about 6000 living species (Figure 42-1). Many of the most familiar animals seen along the seashore—the sea stars (starfish), brittle stars, sea urchins, sand dollars, and sea cucumbers—are echinoderms. All are radially symmetrical as adults. In the sea cucumbers and some other echinoderms, however, the animal's axis lies horizontally; and radial symmetry is not as obvious as it is in most other members of the phylum. Echinoderms are well represented not only in the shallow waters of the sea but also in its abyssal depths. All of them are bottom-dwellers except for a few swimming sea cucumbers. The adults range from a few millimeters to more than a meter in diameter (for one species of sea star) or in length (for a species of sea cucumber).

Basic Features of Echinoderms

Echinoderms have a delicate epidermis, containing thousands of neurosensory cells, stretched over an endoskeleton—internal skeleton—composed of either movable or fixed calcium-rich (calcite) plates called **ossicles.** The animals grow more or less continuously, but their growth slows down with age. When the plates first form, they are enclosed in living tissue. In most cases, these plates bear spines. The plates in certain portions of the body of most echinoderms are perforated. Through these perforations extend the **tube feet,** a part of the water vascular system, which is a unique feature of this phylum.

A

B

C

D   E   F

Echinoderms have a five-part body plan corresponding to the arms of a sea star or the design on the "shell" of a sand dollar. As adults, these animals have no head or brain. Their nervous systems consist of central **nerve rings** from which branches arise. The animals are capable of complex response patterns, but there is no centralization of function. Apparently, the centralization of the nervous system is not feasible in animals with radial symmetry.

The water vascular system of an echinoderm radiates from a **ring canal** that encircles the animal's esophagus. Five **radial canals,** the positions of which are determined early in the development of the embryo, extend into each of the five parts of the body and determine its basic symmetry (Figure 42-2, *A*). Water enters the water vascular system through a **madreporite,** a sievelike plate on the animal's surface, and flows to the ring canal through a tube, or stone canal, so named because of the surrounding rings of calcium carbonate. The five radial canals in turn extend out of the animal through short side branches into the hollow tube feet (Figure 42-2, *B*). In some echinoderms, each tube foot has a sucker at its end; in others, suckers are absent. At the base of each tube foot is a muscular sac, the **ampulla,** which contains fluid. When the ampulla contracts, the fluid is prevented from entering the radial canal by a one-way valve and is forced into the tube foot, thus extending it. When extended, the foot attaches itself to the substrate. Longitudinal muscles in the tube foot's wall contract, shortening it, and the animal is pulled forward to the new position as the water in the tube foot is forced back into the ampulla. By such action,

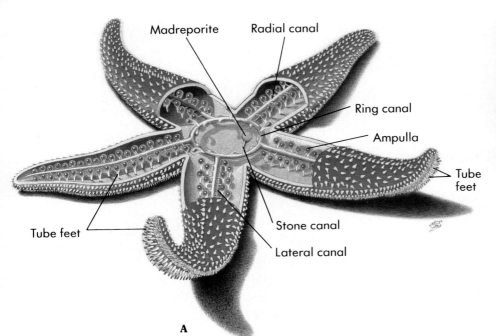

Madreporite Radial canal

Ring canal

Ampulla

Tube feet

Tube feet

Stone canal

Lateral canal

A

B

FIGURE 42-2
Structure of an echinoderm.
A The echinoderm body plan, emphasizing the water vascular system of a sea star.
B The extended tube feet of a sea star, *Ludia magnifica,* in Hawaii.

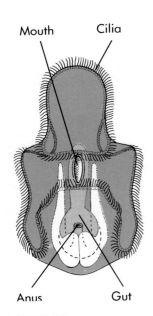

Mouth Cilia

Anus Gut

FIGURE 42-3
The free-swimming larva of an echinoderm. The bands of cilia by which the larva moves are prominent in this drawing. Such bilaterally symmetrical larvae demonstrate convincingly that the ancestors of the echinoderms were not radially symmetrical like the living members of the phylum.

repeated in many tube feet, the sea stars, sand dollars, and sea urchins slowly move along. In a sea star, any one of the five arms may go first.

> Echinoderms are radially symmetrical animals whose bodies typically have a pattern that repeats itself five times. They have characteristic calcium-rich plates called ossicles and a unique water vascular system.

In sea cucumbers (see Figure 42-1, *B*), which are shaped like their vegetable namesakes, there are usually five rows of tube feet on the body surface that are used in locomotion. Sea cucumbers also have modified tube feet around their mouth cavity; these are used in feeding. In sea lilies, which most often remain attached at a particular place in the sea, the tube feet arise from the branches of the arms, which extend from the margins of an upward-directed cup. With these tube feet, the animals take food from the surrounding water. In brittle stars (see Figure 42-1, *D*), the tube feet are pointed and specialized for feeding.

In echinoderms, the coelom, which is proportionately large, connects with a complicated system of tubes and helps to provide for circulation and respiration. In many echinoderms, respiration takes place by means of **skin gills,** which are small, fingerlike projections that occur near the spines. Waste removal also takes place through these skin gills. The digestive system is simple but usually complete, consisting of a mouth, gut, and anus.

Many echinoderms are able to regenerate lost parts, and some, especially sea stars and brittle stars, drop various parts when under attack. In a few echinoderms, asexual reproduction takes place by splitting, and the broken parts of sea stars, for example, can sometimes regenerate whole animals. Some of the smaller brittle stars, especially tropical ones, regularly reproduce by breaking into two equal parts; each half then regenerates a whole animal. Sea cucumbers, when irritated, sometimes eject a portion of their intestines by a strong muscular contraction that may send the intestinal fragments through the anus or even rupture the body wall. This is apparently a defensive mechanism.

Despite the ability of many echinoderms to break into parts and regenerate new animals from them, most reproduction in the phylum is sexual and external. The sexes in most echinoderms are separate, although usually little external difference exists between them. The fertilized eggs of echinoderms usually develop into free-swimming, bilaterally symmetrical larvae (Figure 42-3), which are different from the trochophore larvae of mollusks and annelids. These larvae form a part of the plankton until they metamorphose through a series of stages into the more sedentary adults.

Diversity of Echinoderms

There are more than 20 extinct classes of echinoderms and an additional five with living members: (1) Crinoidea, the sea lilies and feather stars; (2) Asteroidea, the sea stars or starfish; (3) Ophiuroidea, the brittle stars; (4) Echinoidea, the sea urchins and sand dollars; and (5) Holothuroidea, the sea cucumbers. It is almost certain, based on the existence of their bilaterally symmetrical, free-swimming, ciliated larvae, that adult echinoderms were originally free swimming and bilaterally symmetrical as well. During the course of their evolution, they seem to have settled down to a more sedentary life. After doing so, they developed radial symmetry, the water vascular system, and the other distinctive characteristics of the phylum as we now know it.

Class Crinoidea: The Sea Lilies and Feather Stars.

The sea lilies and feather stars, or crinoids (class Crinoidea) (see Figures 42-1, *C* and 42-4) differ from all other living echinoderms in that the mouth and anus are located on their upper surface in an open disk. They are connected by a simple gut. These animals have simple excretory and reproductive systems and an extensive water vascular system. The arms, which are the food-gathering structures of crinoids, are located around the margins of the disk. Different species of crinoids may have from 5 to more than 200 arms extending upward from their bodies, with smaller structures called pinnules branching from the arms. In all crinoids, the number of arms is initially small; species with more than 10 arms add additional arms progressively during growth. Crinoids are filter feeders, capturing the microscopic organisms on which they feed by means of the mucus that coats their tube feet, which are abundant on the animal's pinnules. Consequently, the position of an individual animal is very critical to its success in obtaining food (see Figure 42-4).

There are two basic crinoid body plans. In the sea lilies, the animal is attached to its substrate by a stalk that may be as much as a meter long in some living forms and up to 20 meters long in some of the extinct ones. If they are removed from their points of attachment to a substrate, some sea lilies can move slowly by means of their featherlike arms. Approximately all 80 living species of sea lilies are found below a depth of 100 meters in the ocean. In the second group of crinoids, the feather stars, which comprise about 520 species, the disk detaches from the stalk at an early stage of development. Adult feather stars are usually anchored directly to their substrate by clawlike structures. However, some feather stars are able to swim for short distances, and many of them can move along the substrate. Feather stars range into shallower water than do sea lilies, and only a few species of either group are found at depths greater than 500 meters. Along with the sea cucumbers, crinoids are the most abundant and conspicuous large invertebrates in the warm waters and among the coral reefs of the western Pacific Ocean. They have separate sexes, with the sex organs simply being masses of cells in special cavities of the arms and pinnules. Fertilization is usually external, with the male and female gametes shed into the water, but brooding—in which the females shelter the young individuals—occurs occasionally.

Students of echinoderms believe that when the common ancestors of this phylum first settled down to the substrate, they gave rise to sessile, sedentary, radially symmetrical animals that resembled the crinoids. Such animals were abundant in the ancient seas, and crinoids were present when the Burgess Shale (see Figure 22-9) was deposited about 530 million years ago. Some fossil crinoids measured as much as 25 meters in length, although the living ones are rarely more than a hundredth of that size. More than 6000 fossil species of this class are known, in comparison with the approximately 600 living species. The sea lilies are the only living echinoderms that are fully sessile.

Class Asteroidea: Sea Stars.

The sea stars, or starfish (see Figures 42-1, *A;* 42-2; and 42-5) (class Asteroidea), are perhaps the most familiar echinoderms. These attractive animals are among the most important predators in many marine ecosystems. Their arms are more or less prominent and sharply set off from the disk. Although many sea stars have five arms, conforming to the basic symmetry of the phylum, the members of some families have many more. The mouth of a sea star is located in the

FIGURE 42-4

Sea lilies, *Cenocrinus asterius.*
Two specimens showing a typical parabola of arms, forming a "feeding net." The water current is flowing from right to left, carrying small organisms to the stalked crinoid's arms. Prey, when captured, are passed down the arms to the central mouth. This photograph was taken at a depth of about 400 meters in the Bahamas from the Johnson-Sea-Link Submersible of the Harbor Branch Foundation, Inc.

FIGURE 42-5

Crown-of-thorns sea star, *Acanthaster planci.* This echinoderm was first reported in large numbers in 1957 near the Ryukyu Islands, between Japan and Taiwan, and rapidly spread over the western Pacific Ocean, destroying coral reefs as it went. The animals are able to move along the reef at a rate of up to 20 meters per hour. After the 1960s, the crown-of-thorns became much less abundant for unknown reasons, and the reefs began to recover. This individual, photographed on Lodestone Reef, part of the Great Barrier Reef near Townsville, Australia, has just finished feeding on the table coral (*Acropora hyacinthus*) at the top left; the sea star is an adult, about 30 centimeters in diameter.

center of its lower surface. There is striking variation in color in this group, and many species of sea stars are remarkably beautiful. They are abundant in the intertidal zone (the zone between the high and low tide marks), but they also occur at depths as great as 10,000 meters. The roughly 1500 species of sea stars occur throughout the world.

Some sea stars have an extraordinary way of feeding on bivalve mollusks. They grasp both of a bivalve's shells with their tube feet (Figure 42-6), extrude their stomach through its mouth, and push the stomach into the bivalve. A sea star can push its stomach into a bivalve through an opening as little as 0.1 millimeter wide, corresponding to the natural openings in many irregular shells! Within the mollusk, the sea star secretes its digestive enzymes and digests the soft tissues of its prey, retracting its stomach when the process is complete.

Most sea stars have separate sexes, with a pair of gonads lying in the space between each pair of arms. The eggs and sperm are shed into the water so that fertilization is external. In some species, the fertilized eggs are brooded in special cavities or simply under the animal. They mature into larvae that swim by means of conspicuous bands of cilia.

FIGURE 42-6

A sea star attacking a clam. The tube feet, each of which ends in a suction cup, are located along grooves on the underside of the arms.

Class Ophiuroidea: The Brittle Stars.

The brittle stars (see Figure 42-1, *D*), class Ophiuroidea, have slender branched arms. These arms are more sharply set off from their central disk than in many sea stars. Brittle stars move by "rowing" along the substrate—moving their arms, often in pairs or groups, from side to side. In a number of species the arms are covered with spines, which also assist in their movements. Some of the brittle stars use their arms to swim, a very unusual habit among the echinoderms.

The brittle stars feed by capturing suspended microplankton with their tube feet, long arm spines, or branching arms or by climbing over the objects on the ocean floor in search of small animals. In addition, the tube feet are important sensory organs and assist in directing food into the mouth once the animal has captured it. As implied by their English name, the arms of brittle stars detach easily, a characteristic that helps to protect the brittle stars from their predators. There are about 2000 species. In shallow water, they occur largely on hard substrates and are secretive, but in the deep ocean, they are one of the most abundant organsims. More closely related to the sea stars than to the other classes of the phylum, they are sometimes grouped with them in a single class.

Brittle stars usually have separate sexes, with the male and female gametes in most species being released into the water and fusing there. When they are mature enough, some brood their young in special cavities and release the larvae. These larvae are free-swimming by means of their conspicuous bands of cilia.

Class Echinoidea: The Sea Urchins and Sand Dollars.

The members of the class Echinoidea, the sand dollars and sea urchins (see Figure 42-1, *E* and *F*), lack distinct arms but have the same five-part body plan as all other echinoderms. Five rows of tube feet protrude through the plates of the calcareous skeleton, and there are also openings for the mouth and anus. These different openings can be seen in the globular skeletons of sea urchins and in the flat skeletons of sand dollars (Figure 42-7). Both types of skeleton, which are sometimes common along the seashore, consist of fused calcareous plates. About 950 living species constitute the class Echinoidea.

Echinoids walk by means of either their tube feet or, in the case of sea urchins, their movable spines, which are hinged to the skeleton by a joint that makes free rotation possible. Sea urchins and sand dollars move along the sea bottom, feeding on algae and small fragments of organic material. They scrape these off the substrate with the large, triangular teeth that ring their mouth. The gonads of sea urchins, often eaten raw, are considered a great delicacy by people in different parts of the world, especially along the Mediterranean and in Japan. Because of their calcareous plates, sea urchins and sand dollars are well preserved in the fossil record, where more than 5000 additional species have been described.

As with most other echinoderms, the sexes of sea urchins and sand dollars are separate. The eggs and sperm are shed separately into the water, where they fuse. Some brood their young, and others have free-swimming larvae, with bands of cilia extending onto their long, graceful arms—an unusual feature of this class.

Class Holothuroidea: The Sea Cucumbers.

Sea cucumbers (see Figure 42-1, *B*) (class Holothuroidea) differ from the preceding classes in that they are soft, slug-like organisms, often with a tough, leathery outside skin. The class consists of about 1500 species found worldwide. Except for a few forms that swim, sea cucumbers lie on their sides at the bottom of the ocean. Their mouth is located at one end and is surrounded by 8 to 30 modified tube feet called tentacles; the anus is at the other end. The tentacles around the mouth may secrete a mucous net used to capture the small planktonic organisms on which the animals feed. The tentacle is periodically wiped off within the esophagus and then brought out again, covered with a new supply of mucus. Some sea cucumbers ingest bottom sediment, as earthworms do, and extract the organic matter from it. When they do, they leave castings of excreted material similar to those deposited by earthworms.

Sea cucumbers are soft because their calcareous skeleton is reduced to widely separated microscopic plates. These interesting animals have extensive internal branch-

FIGURE 42-7
A sand dollar skeleton, showing the five similar segments into which the animal's body is divided.

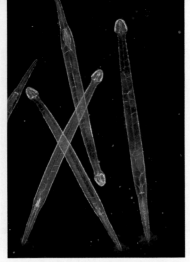

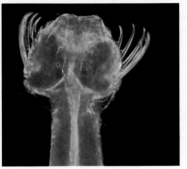

FIGURE 42-8
Arrow worms (phylum Chaetognatha).
A Several individuals of *Sagitta elegans.*
B The head of an arrow worm, *Sagitta setosa,* showing the eyes and the hooks ringing the jaws. The individual shown here is dead, having been captured by a jellyfish, *Obelia.*

A

B

ing systems called respiratory trees, which arise from the **cloaca,** or anal cavity. Water is pulled into and expelled from the respiratory tree by contractions of the cloaca; gas exchange to the various body parts takes place as this process occurs. Most kinds of sea cucumbers have tube feet on the body in addition to tentacles. If they do, these additional tube feet, which might be restricted to five radial grooves or scattered over the surface of the body, may enable the animals to move about slowly. On the other hand, sea cucumbers may simply wriggle along whether or not they have additional tube feet. Most sea cucumbers are quite sluggish, but some, especially among the deep-sea forms, swim actively. Sea cucumbers are considered a great delicacy in many parts of the world.

The sexes of sea cucumbers are separate, but some of them are hermaphroditic— that is, both male and female. Their sex organ is in the form of a single gonad consisting of one or two clusters of tubules that join together at a duct. The eggs and sperm are shed into the water and give rise to free-swimming larvae.

TWO MINOR PHYLA: THE ARROW WORMS AND ACORN WORMS
Phylum Chaetognatha: The Arrow Worms

There are roughly 70 species of arrow worms (phylum Chaetognatha) (Figure 42-8). Given their English name because of their shape, arrow worms are the most abundant primary predators in the marine plankton. Arrow worms occur all over the world, but they are most common in warm, shallow seas. In most marine planktonic environments, their numbers are exceeded only by the copepod crustaceans. They can dart rapidly forward or backward through the water to capture their prey, which includes mainly copepods but also cnidarian medusae, fish larvae, and other small animals. Arrow worms that feed on fish larvae may be harmful predators in areas where commercially important fishes breed.

Arrow worms, which range from 0.6 to 7 centimeters long as adults, are translucent and bilaterally symmetrical coelomates. Transverse septa divide arrow worms into head, trunk, and tail segments, foreshadowing the kind of specialization that occurs in chordates. The arrow worms have large eyes, powerful jaws, and a head ringed with numerous sharp movable hooks (see Figure 42-8, *B*), all key factors in their predatory role. These animals, which are linked to the other deuterostomes only in the details of their embryonic development, seem to lack close relatives. Arrow worms date back in the fossil record at least 500 million years and appear to have remained essentially unchanged throughout this long history.

Phylum Hemichordata: The Acorn Worms

The exclusively marine acorn worms consist of about 90 species comprising the phylum Hemichordata, meaning "half chordates" (Figure 42-9). In addition to those features that mark them as deuterostomes, the acorn worms share a number of features with both the echinoderms and the chordates; their ciliated larvae closely resemble those of sea stars, for example. A dorsal nerve cord is present in hemichordates, in addition to a ventral one; and in some hemichordates, part of the dorsal nerve cord is hollow, a feature otherwise found only among the chordates. Also, only in acorn worms and in chordates is the pharynx, or throat region, perforated with holes known as pharyngeal slits; these structures are found in no other animals.

The acorn worms are soft-bodied animals that live in burrows in sand or mud in the sea. Most range between about 2.5 centimeters and 2.5 meters in length. Their bodies are fleshy and contractile and consist of a proboscis, collar, and trunk. A second group of animals included in this phylum, the class Pterobranchia, consists of sessile and colonial animals less than 7 millimeters long. The members of this class have the same basic body regions and structure as the acorn worms, even though they differ so remarkably in size and appearance. Fossils similar to pterobranchs, which if confirmed as members of this class would be the oldest fossil hemichordates, occur in rocks 450 million years old. This evidence suggests that the phylum is an ancient one.

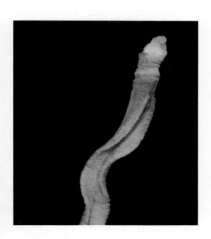

FIGURE 42-9
The anterior portion of an acorn worm, *Glossobalanus sarniensis* **(phylum Hemichordata).**

PHYLUM CHORDATA: THE CHORDATES

The chordates (phylum Chordata) are the best known and most familiar group of animals. There are some 42,500 species of chordates, a phylum that includes the birds, reptiles, amphibians, fishes, and mammals (and therefore our own species).

Characteristic Features

Three principal feature characterize the chordates (Figure 42-10):

1. A single, hollow **nerve cord,** which runs just beneath the dorsal surface of the animal.
2. A flexible rod, the **notochord,** which forms on the dorsal side of the primitive gut in the early embryo and is present at some stage of the life cycle in all chordates. The nerve cord is located just above the notochord. In adult vertebrates, as we will see, the notochord is replaced by a vertebral column that forms around it.
3. **Pharyngeal slits.** These connect the pharynx, a muscular tube that connects the mouth cavity and the esophagus with the outside. In most vertebrates, the slits do not actually connect to the outside and are better termed pharyngeal pouches.

Each of these three distinguishing features of the chordates has played an important role in the evolution of the phylum. In the more advanced vertebrates, the dorsal nerve cord becomes more and more differentiated into the brain and spinal cord. The notochord, which persists throughout the life cycle of some of the invertebrate chordates, becomes surrounded and then replaced during the embryological development of vertebrates by the vertebral column. Pharyngeal pouches are present in the embryos of all vertebrates but are lost later in the development of terrestrial vertebrates. The presence of these structures in all vertebrate embryos provides a clue to the aquatic ancestry of the group.

In addition to these three principal features, a number of other characteristics also tend to distinguish the chordates. In their body plan, chordates are more or less segmented, and distinct blocks of muscles can often be seen clearly in embryos of this phylum (Figure 42-11). Most chordates have an internal skeleton against which the muscles work; either this skeleton or the notochord makes possible the extraordinary powers of locomotion that characterize the members of this group. Finally, chordates have a tail that extends beyond the anus, at least during their embryonic development; nearly all other animals have a terminal anus. In all of these respects, chordates differ fundamentally from other animals.

Subphylum Urochordata: The Tunicates

The tunicates (subphylum Urochordata) are a group of about 1250 species of marine animals. Most of them are sessile as adults (Figure 42-12), and in these, only the larva has a notochord and a nerve cord. As adults, these animals also lack visible signs of segmentation or of a body cavity. Most species occur in shallow waters, but some are found at great depths. In some tunicates, the adults are colonial, living in masses on the ocean floor. Regardless of whether they are solitary or colonial, the tunicates obtain their food by ciliary action; their pharynx is lined with numerous cilia. These cilia beat, drawing a stream of water into the pharynx, where the microscopic particles that are eaten are removed from the current of water by mucous secretions.

The tadpolelike larvae of tunicates plainly exhibit all of the basic characteristics of chordates and mark the tunicates as having the most primitive combination of features found in any chordate (Figure 42-13). They do not feed and have a poorly developed gut. The larvae remain free swimming for no more than a few days; then they settle to the bottom and attach themselves to a suitable substrate by means of a sucker.

Tunicates change so much as they mature and adjust developmentally to a sessile, filter-feeding existence that it would be difficult to discern their evolutionary relation-

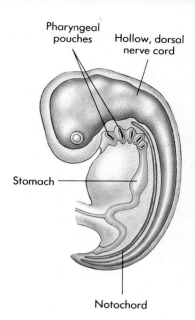

FIGURE 42-10
Some of the principal features of the chordates, as shown in a generalized embryo.

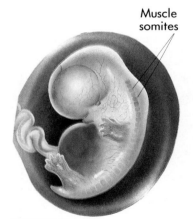

FIGURE 42-11
A human embryo. The muscle is evidently divided into segments called somites at this stage, reflecting the fundamentally segmented nature of all chordates.

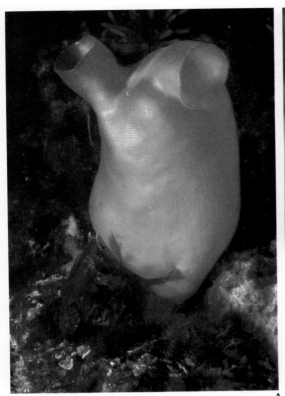

A

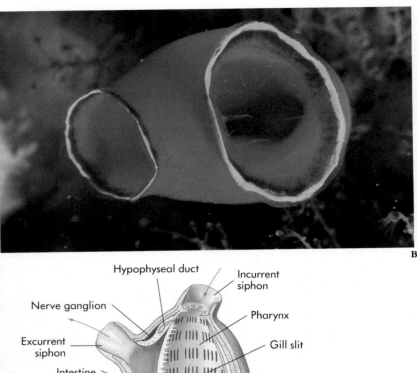

B

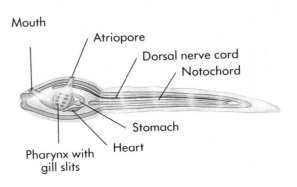

Hypophyseal duct   Incurrent siphon

Nerve ganglion

Pharynx

Excurrent siphon

Gill slit

Intestine

Tunic

Genital duct

Gonad

Stomach

Heart

C

FIGURE 42-12

Tunicates (phylum Chordata, subphylum Urochordata).

A The sea peach, *Halocynthia auranthium.*

B A beautiful blue-and-gold tunicate.

C Diagram of the structure of a tunicate.

ships by examining an adult. Many adult tunicates secrete a **tunic,** which is a tough sac composed mainly of cellulose, a substance that is frequent in the cell walls of plants and algae but is very rare in animals. The tunic surrounds the animal and gives the subphylum its name. In colonial tunicates, there may be a common sac and a common opening to the outside. Many tunicates are inhabited by photosynthetic bacteria (see Figure 30-11), which provide their principal nutrition.

In some tunicates occurring in the world's warmer regions, the tadpole-shaped larvae never change to another form; and the adults of these free-swimming species may resemble the larvae of other groups. Some of these tunicates are gelatinous animals that have several unique adaptations to their planktonic way of life, including high feeding rates and much shorter generation times than other marine planktonic herbivores.

Because it is the larval tunicate rather than the highly modified adult that shows close similarities to other chordates, it has been proposed that certain larval tunicates may have become sexually mature at some stage in their evolution and then developed

FIGURE 42-13

Diagram of the structure of a larval tunicate, showing the characteristic tadpolelike form. Larval tunicates resemble the postulated common ancestor of the chordates.

Mouth

Atriopore

Dorsal nerve cord

Notochord

Stomach

Heart

Pharynx with gill slits

directly into the ancestors of other groups. The process of paedomorphosis is one by which larvae become sexually mature and reproduce; the development of the organism then ceases at that stage. A number of unrelated groups of organisms, including salamanders, include groups that have life cycles of this sort; we pointed out earlier that it is frequent in mites.

Subphylum Cephalochordata: The Lancelets

The lancelets (subphylum Cephalochordata) (Figure 42-14) are scaleless, fishlike marine chordates a few centimeters long; they occur widely in shallow water throughout the oceans of the world. There are about 23 species of this subphylum. Lancelets were given their English name because of their similarity in appearance to a lancet—a small, two-edged surgical knife. Most of them belong to the genus *Branchiostoma*, formerly called *Amphioxus*, a name still used widely. In the lancelets, the notochord runs the entire length of the dorsal nerve cord and persists throughout the animal's life.

Lancelets spend most of their time partly buried in sandy or muddy substrates, with only their anterior ends protruding. They can, however, swim efficiently, although they seem to do so only rarely. Their muscles can easily be seen as a series of discrete blocks. Lancelets have many more pharyngeal gill slits than do fishes, which they resemble in overall shape. They lack pigment in their skin, which has only a single layer unlike the multilayered skin of all vertebrates. The lancelet body is pointed at both ends. There is no distinguishable head and no separate eyes, nose, or ears, although there are pigmented light receptors.

Lancelets feed on microscopic members of the plankton, filtering them through a current that they create by means of cilia that line the anterior end of their alimentary canal. They have an oral hood that projects beyond the mouth and bears sensory tentacles, which also ring the mouth itself. The males and females are separate, but no obvious external differences exist between them.

Biologists are not sure whether lancelets are primitive chordates that have survived with a number of ancestral features or are actually degenerate fishes. In the latter case, they would be vertebrates whose structural features have been reduced and simplified during the course of evolution so that they resemble their ancestors. The fact that lancelets feed by means of cilia and have a single-layered skin, coupled with distinctive features of their excretory systems, seem to make it unlikely that they are simply degenerate fishes. The recent discovery of fossil forms, similar to living lancelets, in rocks some 550 million years old—well before the appearance of any organisms that could reasonably be called fishes—also argues for the antiquity of this group and its importance in understanding the evolution of the vertebrates.

The remainder of this chapter will be devoted to an account of the vertebrates, with treatments of each of the major groups presented in an evolutionary context.

FIGURE 42-14
Lancelets.
A Two lancelets, *Branchiostoma lanceolatum* (phylum Chordata, subphylum Cephalochordata), partly buried in shell gravel, with their anterior ends protruding. The muscle segments are clearly visible; the square, pale yellow objects along the side of the body are gonads, indicating that these are male lancelets.
B The structure of a lancelet, showing the path that its cilia pulls the water through.

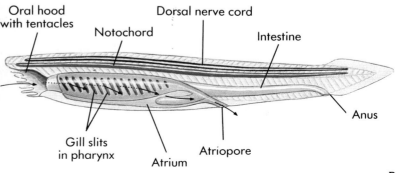

Oral hood with tentacles Notochord Dorsal nerve cord Intestine Anus Gill slits in pharynx Atrium Atriopore

A

B

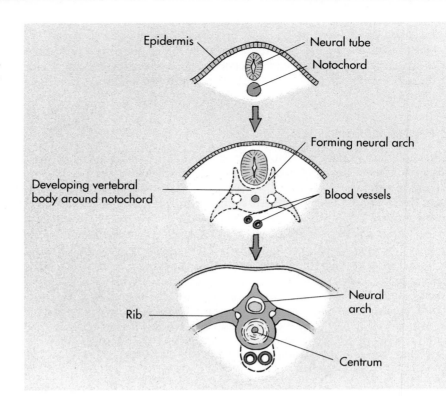

FIGURE 42-15

Embryonic development of a vertebra. During the course of evolution, or of development, the flexible notochord is surrounded and eventually replaced by a cartilaginous or bony covering, the centrum. The neural tube is protected by an arch above the centrum, and the vertebra may also have a hemal arch, which protects the major blood vessels below the centrum. The vertebral column functions as a strong, flexible rod against which the muscles pull when the animal swims or moves.

Subphylum Vertebrata: The Vertebrates

Vertebrates (subphylum Vertebrata) differ from other chordates in that they usually possess a vertebral column, for which the group is named. This column replaces the notochord to a greater or lesser extent in different members of the subphylum. The individual bony segments that make it up are called **vertebrae.** In addition, vertebrates have a distinct head with a skull (cranium) and brain; as a result, they are sometimes called the craniate chordates. The hollow dorsal nerve cord of most vertebrates is protected within a U-shaped groove formed by paired projections from the vertebral column (Figure 42-15).

Among the internal organs of the vertebrates, their livers, kidneys, and endocrine organs are characteristic of the group. The endocrine organs are ductless glands that secrete hormones that play a critical role in controlling the functions of the vertebrate body. All vertebrates have a heart and closed blood vessels. In both their circulatory and their excretory functions, vertebrates differ markedly from other animals.

The vertebrates are a diverse group, containing members adapted to life in the sea, on land, and in the air. There are seven principal classes of living vertebrates. Three of the classes are fishes that live in the water, and four are land-dwelling **tetrapods,** animals with four limbs. (The name *tetrapod* comes from two Greek words meaning "four-footed.") The classes of fishes are Agnatha, the lampreys and hagfish; Chondrichthyes, the cartilaginous fishes, sharks, skates, and rays; and Osteichthyes, the bony fishes that are dominant today. The four classes of tetrapods are Amphibia, the amphibians, including salamanders, frogs, and toads; Reptilia, the reptiles; Aves, the birds; and Mammalia, the mammals. The evolutionary relationships among these classes and some of their extinct relatives are shown in Figure 42-16.

The members of the class Agnatha differ from all other vertebrates in that they lack jaws. Partly because of this fact, they are usually taken as representative of the earliest stages of vertebrate evolution. This relationship has in the past sometimes been recognized by designating the agnathans as a distinct group of vertebrates, more different from the other classes than the other classes are from one another. Most vertebrates have a bony skeleton, although the living members of the Agnatha and Chondrichthyes have a cartilaginous one.

Vertebrates are a subphylum of chordates characterized by a vertebral column surrounding a dorsal nerve cord. Most species of vertebrates are bony fishes.

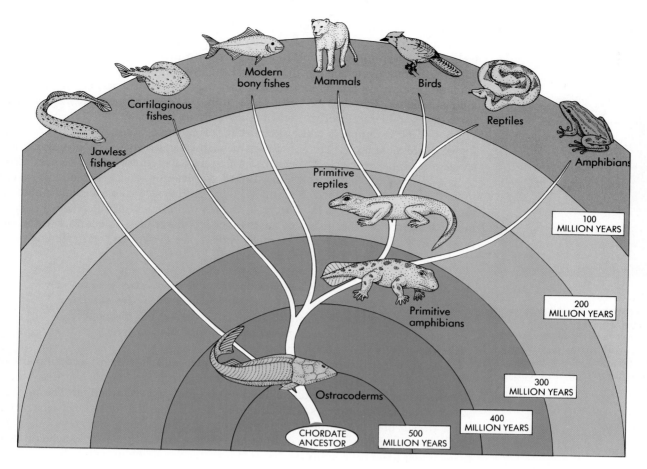

Jawless fishes
Cartilaginous fishes
Modern bony fishes
Mammals
Birds
Reptiles
Amphibians
Primitive reptiles
Primitive amphibians
Ostracoderms
CHORDATE ANCESTOR
100 MILLION YEARS
200 MILLION YEARS
300 MILLION YEARS
400 MILLION YEARS
500 MILLION YEARS

FIGURE 42-16
Vertebrate family tree. Diagram of the evolutionary relationships of the seven classes of vertebrates that include living representatives and some of their extinct relatives. Ostracoderms are an extinct group of fishes.

In the following pages, we will first review the jawless fishes, the first vertebrates, and then discuss the evolution of the Chondrichthyes and Osteichthyes (bony fishes) from them. Later, we will consider the vertebrates' invasion of the land led by the amphibians, starting about 350 million years ago. The reptiles were derived from the amphibians, and, in turn, gave rise to the two most successful terrestrial groups of vertebrates, the mammals and the birds. With the evolution of these last two classes, the vertebrates completed their conquest of the land.

Class Agnatha: The Jawless Fishes.
The agnathans, or jawless fishes, are representative of the first stages in the evolution of the vertebrates: they appeared 470 million years ago; and, for more than 100 million years, they were the only vertebrates. Many of the major groups of agnathans are extinct, the living ones—the lampreys and hagfishes—represent the two surviving lineages.

In the agnathans, the notochord persists throughout the life of the animal. Within the body of the living agnathans are portions of cartilaginous skeleton. We know from fossils, however, that their ancestors had bony skeletons comparable to those of the Osteichthyes. In contrast to the naked living lampreys and hagfishes, many of these extinct agnathans had elaborate and well-developed bony scales, although these were completely different in structure and evolutionary origin from the scales of living fishes.

The feeding habits of the living agnathans are specialized; they are parasites on other fish or are scavengers. Indeed, their success in exploiting these particular ecological niches, or ways of life, may be the reason that they alone have survived into the modern world as representatives of what was once the dominant group of vertebrates. There are about 31 living species of hagfishes and about 32 living species of lampreys. All of the other major groups of this class have been extinct for hundreds of millions of years.

One of the two groups of living agnathans, the lampreys, have round mouths that function like suction cups (Figure 42-17). Using these specialized mouths, lampreys, which are up to 1 meter long as adults, attach themselves to bony fishes. Once attached, they rasp through the skin of the fish with their tongues, which are covered with sharp spines, sucking out the blood of the fish through the hole that they make. Sometimes lampreys can be so abundant that they constitute a serious threat to commercial fisheries. Entering the Great Lakes from the sea, they have become important pests there; millions of dollars are spent annually on their control.

Spawning in lampreys takes place in the spring and is accompanied by the building of depressions in the stony bottoms of freshwater streams. The tiny transparent larvae grow into opaque, eel-like fishes, which may sometimes exceed 15 centimeters in length. They live buried in the mud, feed on plankton, and apparently move only to find richer sources of food. Eventually, these larvae mature and metamorphose into the parasitic adults. The adults occur in the sea or in brackish water, are free swimming, and feed on other fishes in the manner just described. In contrast to the lampreys, the hagfishes, which are externally similar, are scavengers, often feeding on the insides of dead fishes or large invertebrates.

Agnathans were the first vertebrates; for 60 million years they were the only vertebrates. Fossils indicate that agnathans originally had bony skeletons, but present-day agnathans, which apparently represent two distinct evolutionary lineages, have only cartilaginous skeletons and lack jaws.

Larval lampreys superficially resemble lancelets and, like them, feed on plankton. Lancelets move a stream of water through their pharynx by means of ciliary action, whereas larval lampreys, lacking cilia, do so by muscular contraction. The muscular action of the larval lampreys is much more efficient than that of lancelet cilia in forcing a stream of water through their gills; consequently, larval lampreys can extract much more food from a stream of water in a given period of time. Even when they are living under identical conditions, larval lampreys grow much more rapidly than lancelets because they obtain much more food. Primitive chordates doubtlessly fed like the lancelets, by means of cilia; and the evolution of the ability to use muscular contractions to force water through the gills represents an important evolutionary advance.

The Appearance of Jawed Fishes. Jaws first developed among vertebrates that lived 410 million years ago, toward the close of the Silurian Period. The jaws that evolved in ancient fishes during the later Silurian Period allowed them to become proficient predators, to defend themselves, and to collect food from places and in ways that had not been possible earlier. They evolved increasingly efficient fins and were then able to propel themselves through the seas rapidly and accurately and to hunt, seek mates, and avoid predators. Fishes were able to bite and chew their food instead of sucking it or filtering it like all of the more primitive chordates did. The early jawed fishes outcompeted their jawless ancestors, and agnathans have survived to the present only as parasites or scavengers, two specialized ways of life in which they have relatively few competitors. Over the same period of time, fishes have become dominant throughout the waters of the world.

The biting jaws of the vertebrates were produced by the evolutionary modification of one or more of the gill arches, originally the areas between the gill slits. These gill slits in turn developed from the pharyngeal slits or pouches that are characteristic of the chordate embryos (see Figure 42-10). In the process of jaw evolution, the gill arches moved forward in relation to the animal's body and changed in form (Figure 42-18). The progressive modification of the arches just behind the mouth produced the kind of jaw that is characteristic of modern gnathostome ("tooth-mouth") vertebrates—all vertebrates except the agnathans. The vertebrates' teeth originated through the evolutionary modification of the skin. After jaws had first evolved in the earliest gnathostomes, their descendants developed many different kinds of jaws, which vary greatly in different vertebrate classes.

Of the approximately 42,500 species of living chordates, about half are fishes. Undoubtedly many more species are awaiting discovery, especially in the fresh waters

FIGURE 42-17
Specialized mouth of a lamprey. Lampreys use their suckerlike mouths to attach themselves to the fishes on which they prey. When they have done so, they bore a hole in the fish with their teeth and feed on its blood.

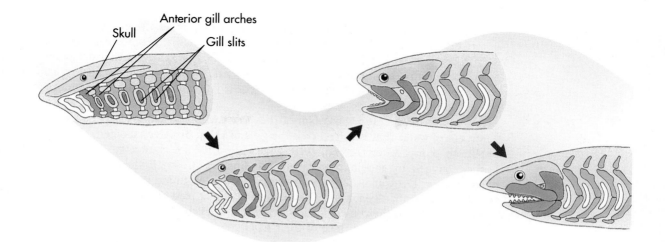

Skull

Anterior gill arches

Gill slits

of South America and Africa, which are exceptionally rich in members of this group. Fishes have efficiently and completely occupied the waters of the globe ever since their origin more than 410 million years ago. Moreover, jawed fishes, through a series of important evolutionary advances, have given rise to all of the other vertebrates.

Class Chondrichthyes: The Cartilaginous fishes.

Among the first groups of jawed fishes to evolve were members of such groups as the placoderms, bony fishes (Osteichthyes), and Chondrichthyes, the sharks, dogfish sharks, skates, and rays (Figure 42-19). These groups and other extinct lineages largely replaced agnathans throughout the world; their descendants dominate the seas and fresh water today.

Many hundreds of extinct species of class Chondrichthyes are known from the fossil record, but only about 850 are living at present. These fishes are mostly scavengers and carnivores; and the internal skeleton of the living forms is composed of cartilage, a soft, light, and elastic material. Their efficient fins and light skeletons make them efficient swimmers and predators, and they have played an important ecological role in the sea for hundreds of millions of years. In sharks, the liver may constitute up to 20% of the total weight, adding to their buoyancy.

The ancestors of sharks and other Chondrichthyes developed jaws, efficient fins, and light, cartilaginous skeletons early in their history.

The skin of Chondrichthyes is covered with small, pointed **denticles,** similar in structure to the teeth of other vertebrates. As a result, the skin of these fishes is rough like sandpaper. Their teeth, which are abundant in the fossil record, are enlarged versions of these denticles.

Sharks drive themselves through the water by sinuous motions of the whole body and by their thrashing tails. Such motion tends to drive the sharks downward, but this tendency is corrected by their two spreading **pectoral fins**—the fins that arise in the fish's thoracic region. In the skates and rays, these pectoral fins have become en-

FIGURE 42-18
Evolution of the jaw. Jaws evolved from the anterior gill arches of ancient, jawless fishes.

FIGURE 42-19
Chondrichthyes. Members of the class Chondrichthyes, which are mainly predators or scavengers, spend most of their time in graceful motion. As they move, they create a flow of water past their gills, from which they extract oxygen. Three representatives of this class are shown here.
A Blue shark.
B Diamond sting ray.
C Manta ray.

A

B

C

larged and undulate when these fishes move, which gives these animals their characteristic appearance. Their tail, which is not the principal means of locomotion that it is in the sharks, is thin and whiplike; it is sometimes armed with a poisonous spine. Skates and rays spend most of their time at or near the ocean floor, feeding mainly on invertebrates. Feeding along the ocean floor is facilitated because their mouths are located on their lower surfaces.

Many sharks (and those bony fishes that swim constantly) depend on a constant stream of water that is forced past the gills to bring dissolved oxygen. Other fishes are able to pump water through their gills while they are stationary. Fishes that obtain their oxygen by moving constantly can literally drown if they are prevented from swimming. Drowned sharks, for example, are often found trapped in the nets used to protect Australian beaches. In skates and rays that feed on or near the ocean floor and have mouths on their lower surfaces, the openings through which they draw water are located on the upper surface of their heads.

Class Osteichthyes: The Bony Fishes.

The vast majority of the more than 18,000 known species of fishes belong to the class Osteichthyes, the dominant class of fishes in the modern world. Although bones are heavier than cartilaginous skeletons, bony fishes are still buoyant because they possess a **swim bladder,** a gas-filled sac that allows them to regulate their buoyant density and so remain suspended at any depth in the water with great accuracy (Figure 42-20). Swim bladders apparently evolved from the lungs; they in turn, originated as outpocketings of the pharynx, specialized for respiration, in Devonian Period fishes. The use of such lungs to regulate buoyancy among the early bony fishes seems to have evolved secondarily. Lacking such bladders, sharks swim perpetually to keep from sinking, as well as to obtain oxygen. This single strategic difference—the ability to swim or not to swim depending on the circumstances—has apparently had much to do with the overwhelming success of the bony fishes.

Osteichthyes comprise about half of the species of living vertebrates (Figure 42-21). Chondrichthyes may originally have evolved in the sea; but bony fishes, which may have originated as an offshoot of Chondrichthyes, clearly evolved in fresh water, judged by the places where early fossils of the class are found and the characteristics of these ancient fishes. Bony fishes apparently entered the sea only after the origin of a number of distinctive evolutionary lines. They are abundant in the modern seas and

FIGURE 42-20
Diagram of a swim bladder. The bony fishes use this structure, which evolved as an outpocketing of the pharynx, to control their buoyancy in water.

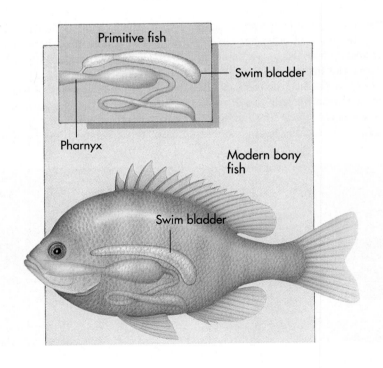

Primitive fish

Swim bladder

Pharnyx

Modern bony fish

Swim bladder

A

B

C

D

FIGURE 42-21

Osteichthyes. Only a little of the incredible diversity of the bony fishes (class Osteichthyes) is shown here. All of these species are marine.
A Puffers, which are also marine fishes, avoid being eaten by inflating themselves when attacked, thus projecting their spines outward.
B A sea horse, an example of the bizarre forms of some fishes. Sea horses move slowly and are difficult to see among the marine algae and sea grasses where they live.
C Moray eel, Rangiroa, French Polynesia. A small cleaner fish is feeding just behind the eel's mouth.
D Koran angelfish, *Pomacanthus semicircularis*, in Fiji; one of the many striking fishes that live around coral reefs in tropical seas.

in fresh water, and many kinds of fishes spend a portion of their lives in fresh water and another portion in the sea.

> **Bony fishes (class Osteichthyes) possess swim bladders, which enable them to increase their buoyancy and thus to offset the greater weight of a bony skeleton. They are the most successful of the vertebrates, accounting for more than half of all living vertebrate species.**

Bony fishes have a number of features that are peculiar to this class alone. Their scales are a covering of thin, overlapping bony plates, sometimes spiny along the edges, which provide some protection for the animal. These fishes have a highly developed **lateral line system**, which consists of organs in sunken pits connected with one another by a canal below the surface and with the surface by the openings of the pits (see Figure 46-11). This lateral line system enables fishes to detect changes of water pressure and thus the movements of predators, prey, and other objects in the water. Lateral line systems are found in the Chondrichthyes, but they are not as elaborate or as well developed as those in the Osteichthyes. Bony fishes detect sound, which travels better in water than in air, especially well and respond to it. In many kinds of fishes the swim bladders are linked to the inner ear by bony connections, thus providing an efficient sound-detecting organ.

The class Osteichthyes is divided into two subclasses: lobe-finned fishes (subclass Sarcopterygii) and ray-finned fishes (subclass Actinopterygii). There are only four living genera of lobe-finned fishes, but the group is known from fossils that date back to the very beginning of the class about 390 million years ago, in the Devonian Period. The paired fins of the lobe-finned fishes have a characteristic scaly, lobed form. These lobed fins, which differ so much in appearance from the fins of most living fishes, are thought to have given rise to the limbs of the tetrapods (Figure 42-22).

The living lobe-finned fishes include three genera of lungfishes, one each in Australia, Africa, and South America (see Figure 20-12). The African genus of lungfishes has four species, whereas the others have one each. In the lungfishes the swim bladder is paired and serves as a functional lung, which allows these remarkable organisms to breathe air. It also enables them to survive in conditions of drought, lying

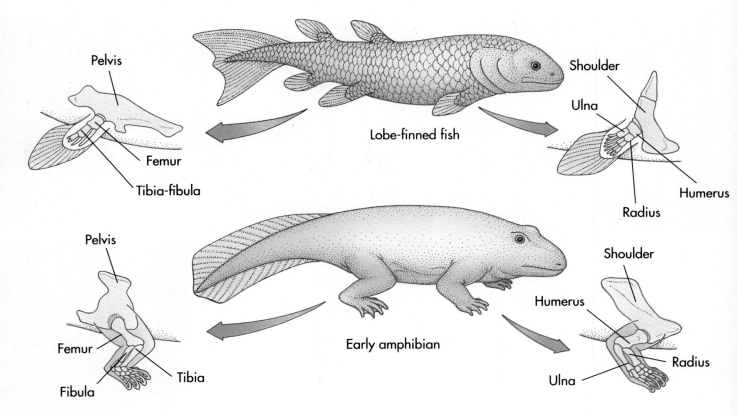

Pelvis

Femur

Tibia-fibula

Lobe-finned fish

Shoulder

Ulna

Humerus

Radius

Pelvis

Femur

Tibia

Fibula

Early amphibian

Shoulder

Humerus

Radius

Ulna

FIGURE 42-22

A comparison between the limbs of a lobe-finned fish and those of a primitive amphibian.

A A lobe-finned fish. Some of the animals of this appearance could probably move out onto land.
B A primitive amphibian. As illustrated by their skeletal structure, the legs of such an animal could clearly function on land much better than the fins of the lobe-finned fish.

dormant in burrows at the bottom of ponds that dry up seasonally. The lungfishes are specialized, and they represent a distinct evolutionary line within the lobe-finned fishes. The remaining genus of lobe-finned fishes contains one species, the coelacanth, a remarkable fish that occurs in the seas off southeastern Africa (Figure 42-23). Recent comparisons of mitochondrial DNA, reported in 1990, have shown that lungfishes are more closely related to the terrestrial tetrapods than are coelacanths. Therefore, many of the features that link the living coelacanth with terrestrial tetrapods probably resulted from convergent evolution.

Class Amphibia: The Amphibians. About 350 million years ago, the lobe-finned fishes that were to become the ancestors of all tetrapods were starting to explore a terrestrial existence at the margins of freshwater ponds or swamps (see Figure 42-22). In these animals, which ultimately became the earliest amphibians, the lungs gradually developed into the kinds of efficient air-breathing organs seen in modern tetrapods. These lungs originally had been the kinds of supplementary organs present in modern lungfishes. The evolution of strong thoracic (body) skeletal supports made efficient locomotion on land gradually possible. Such supports provided a more rigid base for the limbs, which were derived from the kinds of fins found in the lobe-finned

FIGURE 42-23

The living coelacanth, *Latimeria chalumnae.* Discovered in the western Indian Ocean in 1938, this coelacanth represents a group of fishes that had been thought to be extinct for about 70 million years. Scientists who studied living individuals in their natural habitat at depths of 100 to 200 meters observed them drifting in the current and hunting other fishes at night. Some individuals are nearly 3 meters long; they have a slender, fat-filled swim bladder. *Latimeria* is a strange animal, and its discovery was a complete surprise.

fishes. These two evolutionary advances, which led to the ability to use gaseous oxygen for respiration and to travel on land, became the basis of the success and subsequent evolutionary radiation of vertebrates on land.

The first land vertebrates were amphibians, members of a class that first became abundant about 300 million years ago during the Carboniferous Period. The early amphibians had rather fishlike bodies, short stubby legs, and lungs (see Figure 22-10); ultimately, they gave rise to all of the other tetrapods. As a group, amphibians, unlike reptiles, birds, and mammals, depend on the availability of water during their early stages of development. Many amphibians live in moist places even when they are mature. There are more than 4200 species. Members of this class were the dominant vertebrates on land for about a hundred million years, until the reptiles gradually replaced them.

The two most familiar orders of amphibians are those that possess tails—the salamanders, mud puppies, and newts, which comprise the order Caudata, with about 369 species; and those that do not have tails—the frogs, and toads, which comprise the order Anura, with about 3680 species (Figure 42-24). A third interesting order, Gymnophionia, the caecilians, comprises 168 species of legless, tropical, burrowing amphibians that have scales embedded in their skin. The first true lungs evolved in early amphibians, but these organs were relatively inefficient. Respiration also takes place through their thin, moist, glandular skins, which lack scales in both the orders Caudata and Anura, and through the lining of their mouth. The constant loss of water through the skins of amphibians is one of the reasons that these animals must remain, for the most part, in moist habitats. Amphibian larvae and the adults of those species that remain permanently in the water—certain salamanders—respire by means of gills.

Amphibians lay their eggs directly in water or in moist places; the eggs lack water-retaining external membranes and shells and tend to dry out rapidly. Anuran larvae are tadpoles, which usually live in the water, where they feed on minute algae; their adults, which are highly specialized for jumping and very different from the larvae in appearance, are carnivorous. In contrast to the anurans, young salamanders are carnivorous like the adults and look like small versions of them. Many salamanders swim efficiently and return to water to breed. Although some salamanders remain in

FIGURE 42-24

Some of the types of amphibians, representing the diversity of the class Amphibia.

Frogs and toads (order Anura):

A A tadpole, the larval form of the pickerel frog, *Rana palustris*, during metamorphosis to the adult form. The hindlegs develop at this stage, and the animal will soon become terrestrial.

B Giant toad, *Bufo marinus*.

C Red-eyed tree frog, *Agalychnis callidryas*.

Salamanders and newts (order Caudata):

D Spotted salamander, *Ambystoma maculatum*.

E Tennessee cave salamander, *Gyrinophilus palleucus*. This and other salamanders remain aquatic for their entire lifespan; permanently larval in form, they become sexually mature and breed without ever changing.

the water even as adults, most of them live in moist places, such as under stones or logs, or among the leaves of tropical bromeliads (relatives of the pineapple).

Amphibians are terrestrial, but they still depend on a moist environment; their eggs are laid in water and the development of their larvae takes place there.

Although amphibians appear primitive, they are in fact members of a successful group, one that has survived for over 300 million years. The amphibians evolved long before the dinosaurs and have, thus far, outlasted them by 65 million years.

Class Reptilia: The Reptiles.   Reptiles (class Reptilia) have a dry skin covered with scales that efficiently retards water loss. As a result, the members of this large class, consisting of nearly 6000 living species, are independent of free water, unlike their ancestors. Nonetheless, of the three major orders of reptiles that have living representatives (Figure 42-25), the Crocodilia—the crocodiles, alligators, and caimans— live in water, as do most of the members of the order Cheloni, the turtles and tortoises. Members of the third, and by far the largest order, Squamata, the lizards, snakes, and other reptiles are almost entirely terrestrial. The scales of crocodiles and turtles are replaced as they are worn away, whereas those of lizards and snakes are shed several times a year when the animals molt. Snakes evolved from lizards through the loss of their limbs and other evolutionary modifications; some lizards have also lost their limbs in relation to specialization for burrowing. There are about 2800 living species of lizards, about 2700 species of snakes, about 250 species of turtles and tortoises, and 23 species of crocodiles and alligators.

FIGURE 42-25
Representatives of the major groups of reptiles (class Reptilia).
A River crocodile, *Crocodilus acutus*. Most crocodiles resemble birds and mammals in having a four-chambered heart; all other living reptiles have a three-chambered one. Crocodiles, like birds, are related to dinosaurs, rather than to any of the other living reptiles.
B Red-bellied turtles, *Pseudemys rubriventris*. This attractive turtle occurs frequently in the northeastern United States.
C An Australian skink, *Sphenomorophus*. Some burrowing lizards lack legs, and the snakes evolved from one line of legless lizards.
D Smooth greensnake, *Opheodrys vernalis*.

Reptiles are derived from amphibians, with the first ones appearing during the age of dominance of the amphibians, the Carboniferous Period, about 300 million years ago. Reptiles became abundant over the next 50 million years, gradually replacing amphibians. Immediately following the Carboniferous Period, the Permian Period (286 to 248 million years ago) was a period of widespread glaciation and drought. The special adaptations of reptiles, including their water-resistant skin and more efficient lungs, probably gave them an advantage as arid climates became widespread. The dinosaurs, by far the best-known group of fossil organisms, appeared near the start of the Mesozoic Era, at least 225 million years ago. The reptiles dominated the land for about 180 million years, until the start of the Tertiary Period (see Figure 22-13).

Reptiles were the first truly terrestrial vertebrates. They have more efficient lungs than amphibians do, so they do not need to respire through their skin, which is dry and retards water loss efficiently.

One of the most critical adaptations of reptiles in relation to their life on land is the evolution of the **amniotic egg,** which protects the embryo from drying out, nourishes it, and enables it to develop outside of water (Figure 42-26). Amniotic eggs, characteristic of reptiles, birds, and mammals (mostly hatching, in effect, within the mother), retain their own water. They contain a large **yolk,** the primary food supply for the embryo, and abundant **albumin,** or egg white, which provides additional nutrients and water. The yolk is enclosed in a yolk sac attached to the embryo. The embryo's nitrogenous wastes are excreted into the **allantois,** a sac that grows out of the embryonic gut. These wastes are stored in it until the egg hatches, at which time they are simply left behind. Nitrogen waste in an insoluble form (uric acid) does not add to the osmotic environment of the shelled egg. Blood vessels grow out of the embryo through the membranes of the yolk sac and the allantois to the egg's surface, where they take in oxygen and release carbon dioxide. The egg is more easily permeable to these gases than to water. The **amnion,** a membrane that forms around the egg after fertilization, surrounds the developing embryo, enclosing a liquid-filled space within which the embryos pass their gilled stage. Around the embryo, amnion, yolk sac, and allantois is a membrane called the **chorion,** which controls the permeability of the egg and lies just within the shell. In most reptiles, the egg shell is leathery; unlike bird eggs, reptile eggs are somewhat permeable to water.

The key to successful invasion of the land by vertebrates, as it was by arthropods, was the elaboration of ways to avoid desiccation in dry air. In addition to a dry skin that retained water, a second pivotal innovation of reptiles was the watertight amniotic egg, which contains embryonic nutrients and is only partly permeable to gases but not to water.

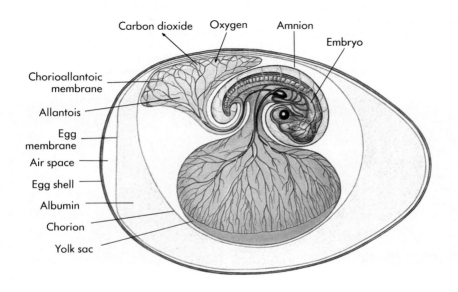

FIGURE 42-26
The water-tight egg. The amniotic egg is perhaps the most important feature that allows reptiles to live in a wide variety of terrestrial habitats.

Temperature Control in Land Animals

Reptiles, like amphibians and fishes but unlike birds and mammals, are **ectothermic** (from the Greek words *ectos*, outside, and *thermos*, heat), regulating their body temperature by taking in heat from the environment. They have evolved a wide array of behavioral mechanisms that enable them to control their temperatures with remarkable precision. Even though ectothermic animals are often called "cold-blooded," they often may actually be able to maintain body temperatures much warmer than their surroundings as a result of their behavior. Several scientists have postulated that at least some dinosaurs, like the birds and mammals, were **endothermic** (Greek *endos*, inside), capable of regulating their body temperature internally. Although endothermic animals are sometimes called "warm-blooded," the temperatures of their bodies may often be cooler than that of their surroundings, a condition that they maintain by the expenditure of energy and by appropriate behavior, as when a dog pants. Endothermic animals are also called **homeotherms**, and ectothermic ones are called **poikilotherms**; we will use both sets of terms interchangeably in this book.

Birds (class Aves) and mammals (class Mammalia) are unique among living organisms in being endothermic. Consequently, the members of these two classes are able to remain active at night, even if temperatures are cool. They are also able to live at higher elevations and farther north and south than most reptiles and amphibians. Endotherms can maintain their internal organs within a very narrow temperature range more or less independently of external conditions, an adaptation that makes maintenance and functioning of these animals more efficient than that of the ectotherms. About 80% of the calories in the food of endotherms, however, goes to maintain their body temperature; a reptile the same size as a mammal therefore can subsist on about a tenth of the amount of food. For this reason, reptiles are able to live in places, such as some extreme deserts, where food may be too scarce to support abundant bird and mammal fauna.

Class Aves: The Birds. There are approximately 9000 species of living birds (class Aves) (Figures 42-27 and 34-15). In birds, the wings are forearms, modified in the course of evolution. Birds are clothed with feathers, flexible, strong organs that can be replaced and that make up an excellent airfoil—a surface designed to obtain resistance on its surfaces from the air through which it moves—for flying (see Fig-

A

B

C

FIGURE 42-27

Aves. The birds (class Aves) are a large and successful group of about 9000 species, more than any other class of vertebrates except the bony fishes. These photographs show some of their diversity.

A Ostrich, *Struthio camelus.* The group of flightless birds that includes the ostrich (Africa), rhea (South America), emu (Australia), and kiwi (New Zealand)—all ratite birds—seems to represent an evolutionary line distinct from all other birds. They migrated between their now widely separated southern lands before these drifted apart.

B A bee-eater, *Merops apiaster*, in Tanzania. Bee-eaters nest communally, with many adult individuals participating in the care of the young.

C Northern saw-whet owl, *Aegolius acadius.* Owls, hawks, eagles, and similar predatory birds are called raptors.

D A pair of wood ducks, *Aix sponsa*, showing the marked difference between males and females.

E Tufted puffin, *Fratercula cirrhata*, one of many unique groups of sea birds.

D   E

Most vertebrates have teeth. Exceptions are the monotremes, which are specialized for eating insects, worms and small crustaceans, and the birds. Among birds, considerable variation has arisen in the hard, horny extensions of the mouth, called **beaks,** that are replacements for lips and cutting teeth. As replacements for the grinding teeth of other animals, birds have **gizzards,** which are often filled with grit that acts in place of molars to break up food. Birds such as hawks use their beaks and talons to tear food; other birds have beaks specialized for probing, prying, pounding, or crushing.

Beaks are not limited to birds. Many reptiles have beaks, including turtles; so do some fish, including the parrot fish, which uses its beak to rip away fragments of coral reef. Among the ancestors of birds—the dinosaurs—are many beaked forms, including the Triceratops.

1

2

FIGURE 42-A
1 Prairie falcon, a beak specialized for tearing.
2 Hummingbird, a beak specialized for sucking.
3 Woodpecker, a beak specialized for chiseling.
4 Cardinal, a beak specialized for crushing.
5 Roseate spoonbill, a beak specialized for stirring.

3

4

5

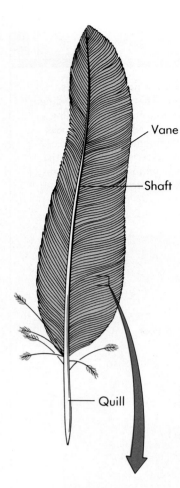

Vane

Shaft

Quill

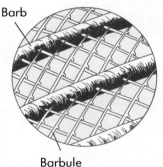

Barb

Barbule

FIGURE 42-28
A feather. The enlargement above
shows the way in which the vanes,
secondary branches, are linked
together by the microscopic barbules.

ures 3-15, *C* and 42-28). Feathers are also very important for their role in insulating birds against temperature changes. The membranes that comprise the flying surface of a bat, or of one of the extinct groups of flying reptiles, can be severely damaged by a single rip. In contrast, the feathers of birds are individually replaceable. Even after the loss of many individual feathers, a bird can keep flying. Several hundred microscopic hooks along the sides of the individual barbules of a feather attach the barbules to one another. These hooks unite the feather so that in a marvelous way, unique to birds, a feather is perfectly adapted for flight. Light, hollow bones, the replacement of scales with feathers, and the development of highly efficient lungs that supply the large amounts of oxygen necessary to sustain muscle contraction during prolonged flight all make flight in birds possible.

The earliest known fossil that may be a bird is *Protoavis* (see Figure 22-15), a bird-like reptile about the size of a chicken that was found in 225 million-year-old rocks in west Texas. Another famous fossil bird is *Archaeopteryx* (see Figure 22-16), which was discovered in 1862 and used by Darwin and his followers in support of the theory of evolution. *Archaeopteryx* had not only feathers, but also a fused collarbone ("wishbone"), a distinctive feature of birds; however, it also had archaic features such as teeth and a long, jointed tail.

Birds evolved from reptiles and, with the bats, are one of the two living vertebrate groups that have achieved a full mastery of the air.

Class Mammalia: The Mammals.   The first known mammals are from the early Mesozoic Era, about 220 million years ago. The first mammals were clearly derived from reptiles; in fact, a number of the early fossil organisms are difficult to classify satisfactorily either as reptiles or as mammals. The first mammals probably were small; fed on insects, as suggested by their teeth; and may have been nocturnal, as suggested by their large eyes. The mammal from the early Tertiary Period shown in Figure 22-20 probably represents this general sort, even though this species occurred more than 130 million years after the origin of the class. Mammals remained secondary to the reptiles, both in absolute numbers and in numbers of species, until about 65 million years ago, after the start of the Tertiary Period. Following the extinction of the dinosaurs at that time, mammals began to become abundant.

There are about 4500 species of living mammals (class Mammalia), including human beings. Like birds, mammals are homeotherms. Their skin is covered with hair during at least some stage of their life cycle. Mammals, also like birds and most crocodiles, have a four-chambered heart with complete **double circulation,** that is, separate systems for circulating oxygen-rich and oxygen-poor blood. Mammals nourish their young with milk, a nutritious substance produced in the mammary glands of the mother. Their locomotion is advanced over that of the reptiles, which in turn is advanced over that of the amphibians; the legs of mammals are positioned much farther under the body than those of reptiles and are suspended from limb girdles, which permit greater leg mobility.

Mammals nourish their young with milk and are covered with hair rather than scales or feathers.

The evolution of specialized teeth in mammals represents a major evolutionary advance. In fishes, amphibians, and reptiles, all of the teeth are essentially the same size and shape; in mammals there has been evolutionary specialization of these teeth into incisors, chisel-like teeth used for cutting; canines, for gripping and tearing; and molars, for crushing and breaking (see Figure 49-6). These evolutionary changes and the subsequent specializations that have taken place in different groups of mammals have constituted one of the major reasons for their evolutionary success.

In all but a few mammals, the young are not enclosed in eggs when they are born. The monotremes, or egg-laying mammals, consisting of the duckbilled platypus (Figure 42-29, *A*) and the two genera of echidna, or spiny anteater (one is shown in Figure 42-29, *B*), are the only exceptions to this rule. These animals occur only in

A

B

Australia and New Guinea; recently, a 110-million-year-old fossil jaw similar to that of a platypus was discovered in Australia, indicating that monotremes have inhabited that part of the world for a long time. The eggs of monotremes are similar to those of reptiles. In the echidna, the mother transfers them to a special marsupial pouch where they hatch. The young in both living genera of monotremes enter the pouch, and they are fed by milk there, just as in other mammals; the milk is produced within the pouch from specialized sweat glands. The ducts of these sweat glands are not united to open on nipples. Monotremes have a cloaca, a common channel for digestive, excretory, and reproductive products; a leathery beak or bill (they lack teeth); and the echidnas have a pouch like those of the marsupials. The ability of monotremes to regulate their own temperature is less efficient than that of the other mammals. Monotremes have no clear relationships among other groups of mammals.

No mammals other than monotremes lay eggs. The marsupials are mammals in which the young are born early in their development—sometimes as little as 8 days after fertilization—and are retained in a pouch, or marsupium. Marsupials have a cloaca, as do the monotremes. Living marsupials are found only in Australia, where they are abundant and diverse (see Figures 20-11 and 42-30), and in North and South America. Immediately before the Isthmus of Panama rose above the sea between 3.1 and 3.6 million years ago, marsupials were found only in Australia and South America. Subsequently, some of them, including the opossum (see Figure 22-6), migrated into North America. Ancient fossil marsupials, 40 to 100 million years old, are found

FIGURE 42-29
Monotremes (class Mammalia).
A Duckbilled platypus, *Ornithorhynchus anatinus*, at the edge of a stream in Australia. The "duck bill" and webbed feet of this unique mammal can readily be seen in this photograph.
B Echidna, *Tachyglossus aculeatus*.

A

B

FIGURE 42-30
Marsupials.
A Kangaroo with young in its pouch.
B Koala.

FIGURE 42-31

The placenta. The placenta is characteristic of the largest group of mammals, which are called placental mammals. It evolved from the amniotic egg. The umbilical cord evolved from the allantois. The chorion, or outermost part of the amniotic egg, forms most of the placenta itself. The placenta serves as the provisional lung, intestine, and kidney of the embryo, without ever allowing the maternal and fetal blood to mix with one another.

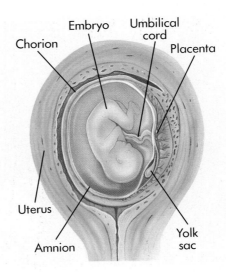

in North America. About 40 million years ago, the group became extinct in North America, where it was absent until geologically recent reintroduction. Recently a fossil marsupial was found in Antarctia, thus indicating the pathway that marsupials passed over between South America and Australia, presumably before the end of the Mesozoic Era.

Most modern mammals are placental mammals. The first organ to form during the course of their embryonic development is the placenta. The placenta is a specialized organ, held within the womb in the mother, across which she supplies the offspring with food, water, and oxygen and through which she removes wastes (Figure 42-31). Both fetal and maternal blood vessels are abundant in the **placenta,** and substances thus can be exchanged efficiently between the bloodstreams of the mother and her offspring. The fetal placenta is formed from the membranes of the chorion and allantois. The maternal side of the placenta is part of the wall of the uterus, the organ in female mammals in which the young develop. In placental mammals, unlike marsupials, the young undergo a considerable period of development before they are born.

Some mammals (monotremes) lay eggs; others (marsupials) give birth to embryos that continue their development in pouches (as do the monotremes); and still others (placental mammals) nourish their developing embryos within the body of the mother by means of a placenta, until development is almost complete.

Placental mammals are extraordinarily diverse (Figure 42-32), as indicated merely by mentioning some of the constituent orders: the insectivores (order Insectivora), including shrews, moles, and hedgehogs; the primates, including monkeys, apes, and humans; the bats (order Chiroptera); the rodents (order Rodentia), including squirrels, rats, mice, woodchucks, and marmots (see Figure 28-3); the meat eaters (order Carnivora), including bears, wolves, foxes, cats, dogs, minks, weasels (Figure 42-33), and raccoons (see Figure 51-1); elephants (order Proboscidea) and the cetaceans (order Cetacea), including whales (see Figure 23-8) and dolphins. One evolutionary line of mammals, the bats, has joined insects and birds in the only group of living animals that truly flies; another, the Cetacea, together with the Sirenia, or manatees, has reverted to an aquatic habitat, like that from which the ancestors of mammals came hundreds of millions of years ago.

In the remaining chapters of this book, we will examine in detail the biology of the vertebrates, treating in some detail their body functions, reproductive methods, and behavior. Underlying all of these is the marvelous diversity that we have reviewed in this chapter, a diversity that has unfolded during the evolutionary course of this dominant group of animals.

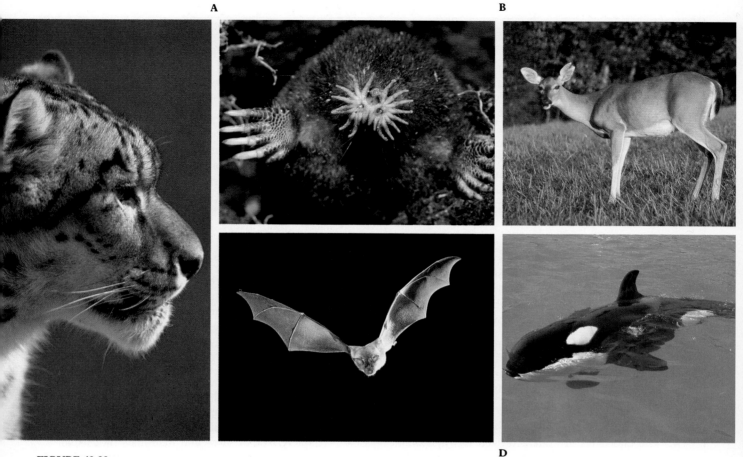

FIGURE 42-32
Placental mammals (class Mammalia).
A Snow leopard, *Unica*, a cat (order Carnivora).
B Starnosed mole, *Condylura cristata*, a burrowing insectivore (order Insectivora).
C White-tailed deer, *Odocoileus virginianus* (order Artiodactyla), abundant in eastern and central temperate North America.
D Greater horseshoe bat, *Rhinoiopus ferrum equinum*, in flight.
E Orca, *Orcinus orca*, a carnivorous whale (order Cetacea).

FIGURE 42-33
Back from the brink. The black-footed ferret, *Mustela nigripes*, is one of the most attractive weasels of North America. This animal preys on prairie dogs, which are colonial rodents that have been exterminated by poisoning over much of their former large area. As a result, black-footed ferrets became so rare that for a time they were feared to be extinct. One small population, of which this photograph was taken, was discovered by ranchers in September 1981, near Meeteetse, Wyoming. The minimum size of this population was established to include about 129 individuals in the summer of 1984, but an outbreak of canine distemper reduced the population greatly in 1985. All 18 known individuals were captured in 1986, and they had produced more than 150 individuals by 1990. Plans call for building the captive population up to at least 200 breeding pairs and then attempting to reintroduce the species in nature. Efforts are being made to develop a vaccine to the canine distemper before releasing the animals.

1. Deuterostomes differ from protostomes in several significant embryological features: their anus, not their mouth, is formed at or near the blastopore; they have radial, rather than spiral, cleavage; the fate of their embryonic cells is settled at a later stage of development; and their coelom is produced by evagination, involving extensive cell movement. The major phyla of deuterostomes are the echinoderms (phylum Echinodermata) and the chordates (phylum Chordata).

2. Echinoderms are exclusively marine deuterostomes that are radially symmetrical, at least as adults. They have a system of separate or fused calcium-rich plates and a unique water vascular system that includes tube feet, by means of which some echinoderms move.

3. The five classes of echinoderms with living members are Crinoidea, the sea lilies and feather stars; Asteroidea, the sea stars or starfish; Ophiuroidea, the brittle stars; Echinoidea, the sea urchins and sand dollars; and Holothuroidea, the sea cucumbers.

4. The chordates are characterized by their single, hollow dorsal nerve cord; by the presence, at least early in their development, of a dorsal, cartilaginous rod, the notochord; and by pharyngeal slits, at least during their embryological development.

5. The approximately 45,000 species of chordates are classified into two small, exclusively marine subphyla, the tunicates and the lancelets, which seem to represent ancient evolutionary offshoots from the group. They are also classified into one large subphylum, the vertebrates, whose members include the fishes as well as four classes that have radiated extensively in terrestrial habitats.

6. Vertebrates are a subphylum of the phylum Chordata. The members of this subphylum differ from other chordates in that they usually possess a vertebral column, a distinct and well-differentiated head, and a bony skeleton.

7. Members of the class Agnatha differ from the other vertebrates in that they lack jaws. Once abundant and diverse, now they are represented only by the lampreys and hagfishes, which are parasitic on other fishes.

8. The two classes of fishes other than Agnatha consist of animals that have jaws, as do the members of the other four classes of vertebrates. Jawed fishes constitute more than half of the estimated 42,500 species of vertebrates and are dominant in fresh and salt waters everywhere. Of the two classes of jawed fishes, the Chondrichthyes, or cartilaginous fishes, consist of about 850 species of sharks, rays, and skates; the Osteichthyes, or bony fishes, comprise about 18,000 species.

9. The first land vertebrates were the amphibians, one of the four classes of tetrapods (four-footed vertebrates). Amphibians are quite dependent on water and lay their eggs in moist places. In many species the larvae live in water; in some species the adults do, too. The amphibians in turn gave rise to the reptiles.

10. The reptiles were the first vertebrates that were fully adapted to terrestrial habitats. Amniotic eggs, which evolved in this group but are also characteristic of birds and mammals (not usually deposited externally), represent a significant adaptation to the dry conditions that are widespread on land. Reptilian scales, too, represent an efficient adaptation to drought.

11. The birds and mammals were derived from the reptiles and are now among the dominant groups of animals on land. The members of these two classes have independently become homeothermic (endothermic), capable of regulating their own body temperatures; all other living animals are poikilothermic (ectothermic), their body temperatures set by external influences.

12. The living mammals are divided into three major groups: (1) the monotremes, or egg-laying mammals, consisting only of the echindas and the duckbilled platypus; (2) the marsupials, in which the young are born at a very early stage of development and complete their early development in a pouch, or marsupium (as do the young of monotremes); and (3) the placental mammals, which lack pouches and suckle their young.

REVIEW QUESTIONS

1. What type of symmetry and body plan is exhibited by adult echinoderms? What is the composition and location of their skeleton? How centralized is their nervous system? What is the nature of their water vascular system?

2. How are tube feet specialized for functions other than locomotion? How do echinoderms respire? How developed is their digestive system? What is the value of an echinoderm's ability to regenerate lost body parts? How do most members of this phylum reproduce? What type of larva do they possess?

3. What is the basic body plan of a sea star? Where is the mouth of a sea star located? How are brittle stars visually different from sea stars?

4. How do sea cucumbers superficially differ from other echinoderms? How are some of their tube feet specially modified? How do they feed? What is the extent of their skeleton? What is the function of their unique respiratory tree? How is their reproduction different from other echinoderms?

5. What are the three primary characteristics of the chordates? What are the three subphyla of the chordates? Give an example of each.

6. What is the relationship between the notochord and the vertebral column in the vertebrates? How is this latter structure associated with the nerve cord? What are the seven classes of vertebrates? Give examples of each.

7. What is one advantage of possessing jaws? From what existing structures did jaws evolve? What is the evolutionary derivation of teeth as found in jawed animals? To what animals does the term gnathostome belong?

8. What is the primary disadvantage of a bony skeleton compared to one made of cartilage? What special structure do most bony fish have to counter this? What is the evolutionary derivation of this structure? What is the function of the lateral line system in fishes?

9. How are amphibians different from other land dwellers? What is the primary differentiating characteristic of each order of amphibians? How do most amphibians respire?

10. How does the skin of reptiles differ from that of most amphibians? What type of egg is first exhibited by the reptiles? What are its evolutionary advantages? How does the embryo obtain nutrients and excrete wastes while contained within the egg? How does the developing embryo respire?

11. Which vertebrates are ectothermic and which are endothermic? How do these processes differ from one another? What are the advantages and disadvantages of endothermy compared to ectothermy?

12. How are birds different from other tetrapods? Why are they more efficient flyers than are reptiles or mammals? From what reptilian structure are feathers derived?

THOUGHT QUESTIONS

1. How are the characteristics of the three subphyla of chordates related to the way of life for each group?

2. Why is it believed that echinoderms and chordates, which are so dissimilar, are members of the same evolutionary line?

3. List some of the advantages that the early birds, in which flight was not nearly as efficient as it is in most of their modern

descendants, might have had as a result of their feathered wings.

4. What limits the ability of amphibians to occupy the full range of terrestrial habitats, and allows other terrestrial vertebrates to occur in them successfully?

FOR FURTHER READING

BELLAIRS, A., and R. CARRINGTON: *The Life of Reptiles*, Universe Books, New York, 1970. A good overall introduction.

BIERI, R. and E.V. THUESEN: "The Strange Worm *Bethybelos*," *American Scientist*, vol. 78, pages 542-549, 1990. The discovery of a single specimen of an unusual arrow worm suggests a new way of looking at the evolution of animal nervous systems.

BILLETT, D.: "The Rise and Rise of the Sea Cucumber," *New Scientist*, 20 March 1986, pages 48-51. Sea cucumbers often swim, and there are many more of them than once thought; beautiful illustrations.

CARROLL, R.L.: *Vertebrate Paleontology and Evolution*, W.H. Freeman, New York, 1987. Outstanding review of what the fossil record tells us about the evolution of the vertebrates.

DUELLMAN, W.E., and L. TREUB: *Biology of Amphibians*, McGraw-Hill Book Company, New York, 1986. Excellent account of all aspects of this large and varied class of vertebrates.

EHRLICH, P.R., D.S. DOBKIN, and D. WHEYE: *The Birders Handbook*, Simon & Schuster Inc., New York, 1988. A field guide to the natural history of North American birds; it will greatly deepen your insight into the lives of these familiar animals.

EISENBERG, J.F.: *The Mammalian Radiations: An Analysis of Trends in Evolution, Adaptation, and Behavior*, University of Chicago Press, Chicago, Ill., 1981. A marvelous, biologically oriented account of the mammals.

FEDER, M.E., and W.W. BURGGREN: "Skin Breathing in Vertebrates," *Scientific American*, November 1985, pages 126-142. Breathing through skin either supplements or replaces breathing through lungs or gills in some vertebrates.

GILL, F.B.: *Ornithology*, W.H. Freeman and Company, New York, 1990. Excellent overall account of the structure, biology, and other features of the birds.

HANKEN, J.: "Development and Evolution in Amphibians," *American Scientist*, vol. 77, pages 336-343, 1989. The evolution of morphological diversity in amphibians has been achieved by modifications in development.

HILDEBRAND, M., D.M. BRAMBLE, K.F. LIEM, and D.B. WAKE: *Functional Vertebrate Morphology*, Belknap Press, Harvard University, Cambridge, Mass., 1985. Balanced account of vertebrate morphology, presented in an evolutionary context.

KONISHI, M., AND OTHERS: "Contributions of Bird Studies to Biology," *Science*, vol. 246, October 1989, pages 465-472. A broad-ranging review of avian research, by outstanding avian biologists.

LAWRENCE, J.: *A Functional Biology of Echinoderms*, Johns Hopkins University Press, Baltimore, Md., 1987. A comprehensive, modern perspective on echinoderm biology, clearly written and well illustrated.

MOYLE, P.B., and J.J. CECH, JR.: *An Introduction to Ichthyology*, Prentice-Hall, Inc., Englewood Cliffs, N.J., 1982. This excellent book treats all aspects of the biology and diversity of fishes.

POWERS, D.: "Fish as Model Systems," *Science*, vol. 246, October 1989, pages 352-358. A detailed look at current research on fish, covering a broad array of areas and focusing on new approaches. Recommended.

RISMILLER, P.D. and R.S. SEYMOUR: "The Echidna," *Scientific American*, February 1991, pages 96-103. The secrets of the natural history and reproductive behavior of this spiny mammal of Australia and New Guinea are now being explored.

SEIDENSTICKER, J. with S. LUMPKIN: "Playing Possum is Serious Business for Our Only Marsupial," *Smithsonian*, November 1989, pages 109-119. How do features of the basic biology of opossums explain their spectacular expansion in range in North Amercia?

TUTTLE, M.D.: *America's Neighborhood Bats*, University of Texas Press, Austin, 1988. An engaging account of how to understand and learn to live with bats, abundant but seldom-seen mammals.

TUTTLE, R.: "Apes of the World," *American Scientist*, vol.78, March 1990, pages 115-125. A survey of our closest living relatives reveals rich prospects for further study and an urgent need for conservation.

WILLSON, M.F.: *Vertebrate Natural History*, Saunders Publishing Company, Philadelphia, Penn., 1983. A rich account of the diversity of vertebrates and their activities.

Vertebrate Biology

43

Organization of the Vertebrate Body

OVERVIEW

The vertebrate body is composed of several hundred kinds of cells. These cells make up the four major tissue types: epithelium, connective tissue, muscle, and nerve. Each of these tissue types can be recognized by its structure, function, and origin. The organs of the body are usually composed of several different tissue types grouped together in a functional unit. Organ systems are composed of individual organs working together. All of these elements act in concert to maintain homeostasis.

FOR REVIEW

Here are some important terms and concepts that you will encounter in this chapter. If you are not familiar with them, you should review them before proceeding.

Eukaryotic cell structure (Chapter 5)

The evolution of vertebrates (Chapter 42)

Basic structure of chordates (Chapter 42)

When they think of animals, most people think of the animals they see at the zoo—growling lions, long-necked giraffes, armor-plated rhinoceroses, seals that swim, and birds that fly and sing. Monkeys look at us, puzzling out why we are looking at them; snakes slither, squirrels dart about, and peacocks strut. All of these animals are vertebrates, and although they may seem very different from each other, they are in fact very much alike "under the skin." All vertebrates share the same basic body plan, with the same sorts of organs operating in much the same way. In this chapter we will begin a detailed consideration of the biology of vertebrates, and of the often intricate and fascinating structure of their bodies. We will focus on human biology, both because the architecture of the human body provides a good focal point for discussing the structure and functioning of vertebrate bodies in general, and because human biology is of particular importance to all of us. We each want to know how our body works, and why it functions the way it does.

Not all vertebrates are the same, of course, and some of the differences are important. A fish doesn't breathe the same way you do, for example. So we will not limit ourselves to humans, but rather describe the human animal in the context of vertebrate diversity. The differences reflect our evolutionary history, and offer important lessons in why we function the way we do.

THE HUMAN ANIMAL

An incredible machine of great beauty, the human body has the same general body architecture as do all vertebrates. It includes a long tube that travels from one end to the other, from mouth to anus; this tube is suspended within an internal body cavity, the coelom that was mentioned in Chapter 42. In humans the coelom is divided into two parts: (1) the **thoracic cavity,** which contains the heart and lungs; and (2) the **abdominal cavity,** which contains the stomach, intestines, and liver (Figure 43-1). The body of all vertebrates is supported by an internal scaffold or **skeleton** made up of jointed bones that grow as the body grows. A bony **skull** surrounds the brain, and a column of hollow bones, the **vertebrae,** surrounds the dorsal nerve cord, or **spinal cord.**

How the Body is Organized

There are four levels of organization in the vertebrate body: (1) cells, (2) tissues, (3) organs, and (4) organ systems. Like all multicellular animals, the bodies of vertebrates are composed of different cell types. The bodies of adult vertebrates contain between 50 and several hundred different kinds of cells, depending upon which vertebrate is being considered and how finely one differentiates between cell types. Groups of cells similar in structure and function are organized into **tissues.** Blood cells are one kind of connective tissue, and bone another.

Organs are body structures composed of several different tissues grouped together into a structural and functional unit. Your heart, for example, is an organ. It contains cardiac muscle tissue, wrapped in connective tissue, embedded with nerves and blood vessels. All of these tissues work together to pump blood through your body. An **organ system** is a group of organs that function together to carry out the principal activities of the body. For example, the digestive organ system is composed of individual organs concerned with breaking up food (teeth), passage of food to the stomach (esophagus), storage of food (stomach), digestion and absorption of food (intestine), and expulsion of solid residue (rectum). The human body contains 11 principal organ systems (Table 43-1).

TISSUES

Early in the development of any vertebrate, the growing mass of cells differentiates into three fundamental embryonic tissues: **endoderm, mesoderm,** and **ectoderm.** These three kinds of embryonic tissue in turn differentiate into the scores of different cell types that are characteristic of the vertebrate body. In adult vertebrates, there are

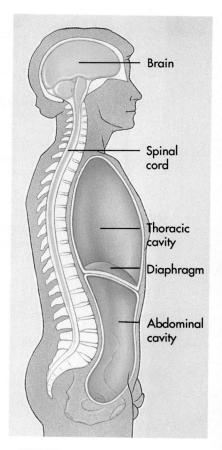

FIGURE 43-1
Architecture of the human body. Humans, like all vertebrates, have a dorsal central nervous system, consisting of a spinal cord and brain enclosed in vertebrae and the skull. In mammals, a muscular diaphragm divides the coelom into the thoracic cavity and the abdominal cavity.

TABLE 43-1   THE MAJOR VERTEBRATE ORGAN SYSTEMS

SYSTEM	FUNCTIONS	COMPONENTS	DETAILED TREATMENT
Circulatory	Transports cells and material throughout the body	Heart, blood vessels, blood, lymph, and lymph structures	Chapter 51
Digestive	Captures soluble nutrients from ingested food	Mouth, esophagus, stomach, intestines, liver, and pancreas	Chapter 49
Endocrine	Coordinates and integrates the activities of the body	Pituitary, adrenal, thyroid, and other ductless glands	Chapter 47
Urinary	Removes metabolic wastes from the bloodstream	Kidney, bladder, and associated ducts	Chapter 53
Immune	Removes foreign bodies from the bloodstream	Lymphocytes, macrophages, and antibodies	Chapter 52
Integumentary	Covers the body and protects it	Skin, hair, nails, and sweat glands	Chapter 43
Muscular	Produces body movement	Skeletal muscle, cardiac muscle, and smooth muscle	Chapter 47
Nervous	Receives stimuli, integrates information, and directs the body	Nerves, sense organs, brain, and spinal cord	Chapters 44, 45, 46
Reproductive	Carries out reproduction	Testes, ovaries, and associated reproductive structures	Chapter 54
Respiratory	Captures oxygen and exchanges gases	Lungs, trachea, and other air passageways	Chapter 50
Skeletal	Protects the body and provides support for locomotion and movement	Bones, cartilage, and ligaments	Chapters 43, 48

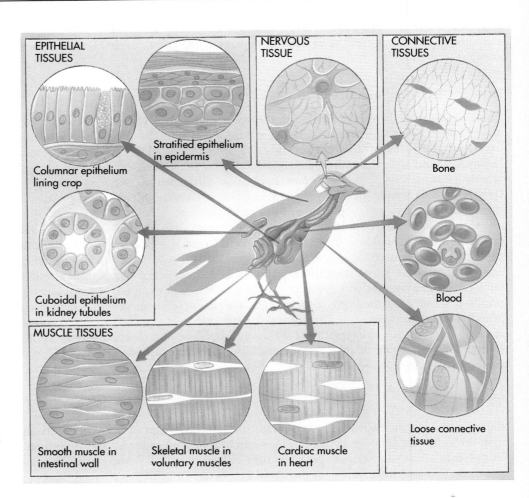

FIGURE 43-2
Vertebrate tissue types. Epithelial tissues are indicated by *blue* arrows; connective tissues by *green* arrows; muscle tissue by *red* arrows; and nervous tissue by a *yellow* arrow.

generally considered to be four principal kinds of tissues: **epithelium, connective tissue, muscle,** and **nerve** (Figure 43-2).

EPITHELIUM IS PROTECTIVE TISSUE

Epithelial cells are the guards and protectors of the body. They cover its surface and determine which substances enter it and which do not. The organization of the vertebrate body is fundamentally tubular, with ectodermal cells covering the outside (skin), endodermal cells lining the hollow inner core (gut), and mesodermal cells lining the body cavity (coelom). All of these kinds of **epithelial** or "skin" cells are broadly similar in form and function, and they are collectively called the **epithelium.** The epithelial layers of the body function in four different ways:

1. They protect the tissues beneath them from dehydration and mechanical damage, a particularly important function in land-dwelling vertebrates.
2. They provide a selectively permeable barrier that can facilitate or impede the passage of materials into the tissue beneath. Because epithelium covers or lines all of the body's surfaces, every substance that enters or leaves the body must cross an epithelial layer.
3. They provide sensory surfaces. Many sensory nerves end in epithelial cell layers.
4. They secrete materials. Most secretory glands are derived from invaginations of layers of epithelial cells that occur during development.

Layers of epithelial tissue are usually only one or a few cell layers thick. Individual epithelial cells typically possess a small amount of cytoplasm and have a relatively low metabolic rate. Few blood vessels pass through the epithelium; instead, the circulation of nutrients, gases, and wastes in epithelial tissue occurs via diffusion from the capillaries of underlying tissue.

Epithelial tissues possess remarkable regenerative powers. The cells of epithelial layers are constantly being replaced throughout the life of the organism. The liver, for example, which is a gland formed of epithelial tissue, can readily regenerate substantial portions of tissue that have been surgically removed from it.

There are three general classes of epithelial tissue: **simple epithelium, stratified epithelium,** and **glands.** These are further subdivided, based on the shape of the cells composing the epithelium, into squamous, cuboidal, and columnar.

Simple Epithelium

Simple epithelial tissues are a single cell layer thick (Figure 43-3). The three kinds of simple epithelium come in three shapes, which occur in very different kinds of membranes. **Simple squamous** epithelial cells have an irregular, flattened shape with tapered edges. The membranes that line the lungs and the major cavities of the body (enclosing the heart, lungs, and intestines) are composed of simple squamous cells. These cells, like many other epithelial cells, are relatively inactive metabolically; their

FIGURE 43-3

Types of epithelial tissue, based on shape.

A Squamous epithelium lines the artery seen here. The nuclei are characteristically flattened. The round cells above the epithelium are blood cells in the hollow interior of the artery.

B Cuboidal epithelium forms the walls of these kidney tubules, seen in cross section.

C Columnar epithelium forms the outer cell layer of this human intestine. Interspersed among the epithelial cells are goblet cells, which secrete mucus.

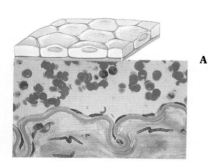

Simple squamous

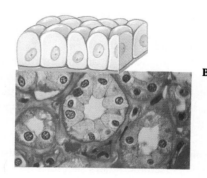

Simple cuboidal

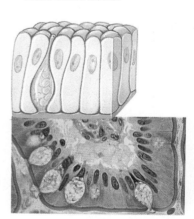

Columnar

presence is generally associated with the passive movement of water and electrolytes into and out of the body. **Simple cuboidal** and **columnar** epithelial cells have a fuller, less flat appearance, possess many mitochondria and extensive endoplasmic reticulum, and are much more active metabolically than the squamous epithelial cells. The respiratory tract and the ducts of the testes are lined with epithelium composed of cuboidal and columnar cells. Individual cells within these epithelial layers often bear cilia or have secretory functions.

Stratified Epithelium

Stratified epithelial tissues are several cell layers thick. The most abundant and prominent kind of stratified epithelial tissue is that found in the epidermis, or skin. It is **stratified squamous epithelium,** because the upper layer is squamous. The cornea of the eye is also composed of stratified epithelium in which the individual cells have become highly specialized. A characteristic property of the cells of stratified epithelium is that they are able to lay down **keratin,** a very strong fibrous protein. The deposition of keratin usually occurs as a response to stress. The callouses that occur on the palms of the hands of manual laborers are composed of keratin. Hair, which is derived from skin, is also composed of keratin.

Glands

The vertebrate body contains two major gland systems, both of which are composed of stratified columnar epithelia. These body glands produce sweat, milk, saliva, digestive enzymes, hormones, and a diverse catalogue of other substances. In one gland system, the **exocrine glands,** the connection of the gland to the epithelial sheet from which it invaginated is maintained as a duct. The liver and salivary glands are exocrine glands. The other major gland system, **endocrine glands,** are ductless glands. They arise during development as a tubular invagination of a sheet of epithelial cells; however, the connection with the epithelium is lost during the course of development. The endocrine glands, therefore, are not connected to the epithelium in the adult animal. Many of the body's most important hormones are produced by endocrine glands, the subject of Chapter 48.

> **Epithelium makes up the skin of the body and the lining of the respiratory and digestive tracts. Much of the internal lining of respiratory and digestive tracts is simple epithelium, only one cell thick, whereas external skin is composed of stratified epithelium that is many cell layers thick.**

CONNECTIVE TISSUE SUPPORTS THE BODY

The cells of connective tissue provide the body with its structural building blocks and also with its most potent defenses. Unlike epithelial cells, the cells that make up connective tissue are not stacked tightly; instead, they are spaced well apart from one another (Figure 43-4). A connective tissue is a supporting tissue composed of living cells embedded within a nonliving matrix that varies in consistency from fluid, such as blood plasma, to the semicrystalline structure of bone. Connective tissue cells are derived from the mesoderm and fall into three categories: **defensive, structural,** and **sequestering**. Cells of defensive connective tissue, those which are involved in the defense of the body, are found in blood plasma and roam the circulatory system as mobile hunters of invading bacteria and foreign substances. Structural connective tissue cells are static, secreting proteins into the empty spaces between individual cells and thus creating a loose fibrous matrix. Sequestering connective tissue cells act as storehouses, accumulating and sequestering specific substances such as fat, the skin pigment melanin, or hemoglobin.

> **Connective tissues are derived from mesoderm. Some are mobile hunters of invading bacteria, others secrete a protein matrix to provide structural support, and still others act as storage sites for fat or hemoglobin.**

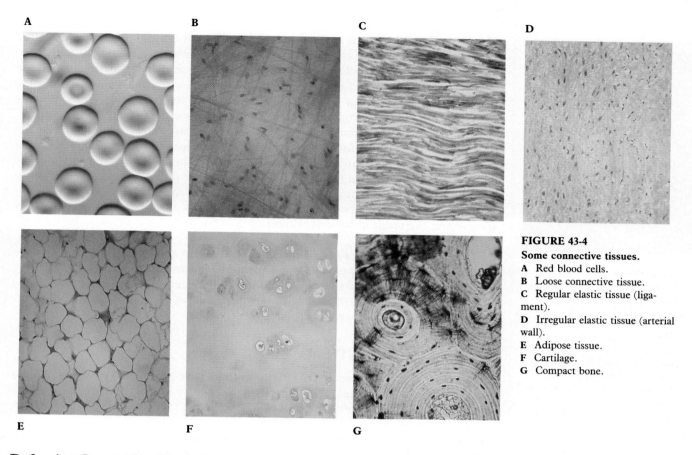

FIGURE 43-4
Some connective tissues.
A Red blood cells.
B Loose connective tissue.
C Regular elastic tissue (ligament).
D Irregular elastic tissue (arterial wall).
E Adipose tissue.
F Cartilage.
G Compact bone.

Defensive Connective Tissues

The three principal cell types that make up defensive connective tissues are **macrophages, lymphocytes,** and **mast cells.** All of these are dispersed; they are relatively small, rounded cells that move individually within the circulating fluids of the body (Figure 43-5).

Macrophages. The first or principal cell type of defensive connective tissue is the macrophage. Macrophages are abundant in the bloodstream and also in the fibrous mesh of many tissues. They usually are mobile, but sometimes they are attached to fibers. The key property of these cells is that they are phagocytic: they are able to engulf and digest cellular debris and invading bacteria. Macrophages provide the key phagocytic element in the body's two defenses against invasive elements. As elements of the reticuloendothelial system, they patrol the capillaries of the circulatory system, engulfing and digesting any bacteria that they encounter, and disposing of any damaged tissue. As elements of the immune system, macrophages phagocytize any particles that are coated with antibodies that they encounter in the capillaries. Antibodies are special proteins that bind to molecules that are not normally present in the body. For this reason, the binding of an antibody to a particle acts as a flag that alerts the macrophages to the foreign nature of the tagged molecule.

Lymphocytes. The second type of defensive connective tissue is the lymphocyte. Lymphocytes are cells that circulate in the blood. They play a key role in the body's defense against infection, and normally are not numerous except when an organism is sick. The kind called *B cells* provide a good example of the complex roles lymphocytes play in protecting the body. When a B cell encounters a foreign substance (called an **antigen**), it enlarges and begins to divide and synthesize an antibody protein capable of binding to that particular antigen. Each mature lymphocyte cell, called a **plasma cell,** possesses within its nucleus a unique version of the antibody-encoding gene, different from the version of almost any other lymphocyte. How antibodies function in the body's defense against disease is described in Chapter 52.

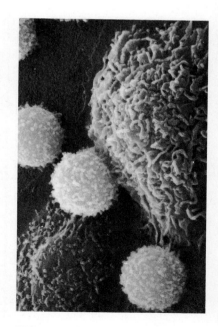

FIGURE 43-5
Lymphocytes and macrophages. The lymphocytes are small and spherical; the macrophages are larger and more irregular in form.

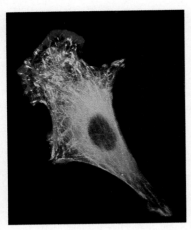

FIGURE 43-6
A complex web of protein fibers runs through the cytoplasm of this fibroblast. The different proteins are made visible by fluorescence microscopy and then combined photographically. Actin appears blue here, microtubules appear green and intermediate fibers appear red (see Chapter 5).

Mast cells. The third type of defensive connective tissue is the mast cell. Mast cells are not numerous. Their key property is that they synthesize the molecules that are involved in the body's **inflammation response** to physical injury or trauma. These molecules dilate blood vessels, bringing in more oxygen and nutrients, and diluting any toxins. The cytoplasm of mast cells is filled with many vesicles. These vesicles contain histamine and serotonin, molecules that produce the inflammation response, as well as heparin, a molecule that prevents blood clotting. The mast cells release the contents of their vesicles into the bloodstream in response to mechanical or chemical injury. The molecules that they release act to enlarge the capillaries and so speed the healing process. Interestingly, mast cells also disgorge their contents in response to very high levels of individual antibodies. Antigens that the body has encountered frequently in the past thus evoke this response, which is called **hypersensitivity.** Most allergies are the result of hypersensitivity.

> The vertebrate body uses connective tissues to defend itself against foreign infection. Lymphocytes produce proteins called antibodies, which recognize and bind to foreign bacteria and other elements, after which macrophages engulf and digest the invader. Mast cells enlarge the blood vessels in response to trauma, speeding the healing process.

Structural Connective Tissues

The three principal components of structural connective tissue are **fibroblasts** (Figure 43-6), **cartilage,** and **bone.** They are distinguished principally by the nature of the matrix laid down between their individual cells.

Fibroblasts. Of all connective tissue cells, fibroblasts are the most common within vertebrate bodies. They are flat, irregular, branching cells that secrete structurally strong proteins into the matrix between cells. The most commonly secreted protein is **collagen** (Figure 43-7), which is the most abundant protein in the bodies of vertebrates. One quarter of all animal protein is collagen.

At least one third of the amino acids of collagen are glycine. Because glycine is a very small amino acid, chains of collagen are able to wind intimately around one another, forming a triple helix. The side chains of the proline amino acids form cross-links that lock the three strands together. Collagen fibers are made up of many individual collagen strands arrayed side by side and connected with one another by additional cross-links. The strength of the fiber is determined by the nature of its cross-links. Fibroblasts are active in wound healing. They multiply rapidly in wound tissue, forming granulated fibrous scar tissue that possesses a collagen matrix.

The way in which collagen fibers are laid down is of particular interest. The loose fibrous matrix of the vertebrate body, tying different tissues together, is composed of fibroblast-secreted collagen. The deformation of preexisting collagen fibers by stress exposes charged lysine side chains that otherwise would be buried in the fibers. As a result of the exposure of these lysine side chains, a very minute electric current, called a **piezoelectric current,** is generated along the deformed fiber. Fibroblasts are able to sense these minute currents, which polarize the surfaces of the cells and thus determine their lines of movement. As a result, the new collagen fibers are laid down along the lines of stress, helping to strengthen the area and heal breaks.

Collagen is not the only fibrous tissue produced by fibroblasts. **Reticulin** is a fine-branching fiber that forms the framework of many glands, such as the spleen and the lymph nodes, and also makes up the junctions between many tissues. **Elastin** is fibrous tissue that is the principal component of the lung. Elastin cross-links are longer than those of collagen, giving great elasticity to fibers of elastin. When elastin fibers are stretched, they return promptly to their former length after being released.

Cartilage. Cartilage is a specialized connective tissue in which the collagen matrix between cells is formed at positions of mechanical stress. In cartilage, the fibers are laid down along the lines of stress in long parallel arrays. The result of this process is

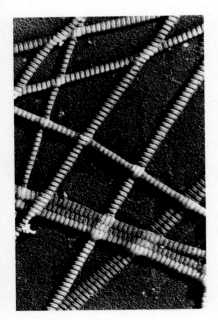

FIGURE 43-7
Collagen fibers. Each fiber is composed of many individual collagen strands and can be very strong.

a firm and flexible tissue that does not stretch, that is far tougher than the defensive or sequestering connective tissues, and that has great tensile strength. Cartilage cushions bone joints and makes up the entire skeletal system of the modern agnathans and Chondrichthyes (see Chapter 42), having replaced the body skeletons that were characteristic of the ancestors of these vertebrate groups.

Tendons, which bind muscles to bone, and ligaments, which bind together the bones that meet in joints, such as the knee, ankle, and elbow, are other connective tissues containing collagen fibers. A tendon's tensile strength is half that of steel—a tendon 1 cm across will support a ton.

Bone. Bone is a special form of cartilage in which the collagen fibers are coated with a calcium phosphate salt. The great advantage of bone over the chitin of invertebrates as a structural material is that bone is strong without being brittle (Figure 43-8). The structure of bone and how it is formed is discussed in Chapter 48.

Sequestering Connective Tissues

The third general class of connective tissue is composed of cells that specialize in accumulating and transporting particular molecules. Sequestering connective tissues include the fat cells of adipose tissue, as well as pigment-containing cells. The most important cells of this tissue class are red blood cells.

Blood cells are classified according to their appearance (Figure 43-9), either as **erythrocytes** (red blood cells) or as **leukocytes** (white blood cells). We have described white blood cells before; they include the macrophages and lymphocytes that defend the body. The role of red blood cells is very different. They act as mobile transport units, picking up and delivering gases.

Erythrocytes are the most common blood cells; there are about 5 billion in every milliliter of blood. They act as the transporters of oxygen in the vertebrate body. During their maturation in mammals they lose their nucleus and mitochondria, and their endoplasmic reticulum dissolves. As a result of these processes, mammalian erythrocytes are relatively inactive metabolically, but they are not empty. Large amounts of the iron-containing protein **hemoglobin** are produced within the erythrocytes and remain in the mature cell. Hemoglobin is the principal carrier of oxygen in vertebrates, as it is in many other groups of animals. Each erythrocyte contains about 300 million molecules of hemoglobin.

The fluid intercellular matrix, or **plasma,** in which the erythrocytes move is both the banquet table and the refuse heap of the vertebrate body. Practically every substance used by cells is found in plasma. These include the sugars, lipids, and amino acids, which are the fuel of the body, as well as the products of metabolism. The plasma also contains inorganic salts, such as calcium used to form bone; fibrinogen from the liver, which helps blood clot; albumin, which gives the blood its viscosity; and antibody proteins produced by lymphocytes. Every substance secreted or discarded by cells, such as urea, is also present in the plasma.

> **Many connective tissues are specialized for accumulating particular classes of molecules, such as fat or pigment. Blood cells called erythrocytes accumulate the oxygen-carrying protein hemoglobin.**

MUSCLE TISSUE PROVIDES FOR MOVEMENT

Muscle cells are the workhorses of the vertebrate body. The distinguishing characteristic of muscle cells, the one that makes them unique, is the relative abundance and organization of actin and myosin microfilaments within them. These microfilaments are present as a fine network in all eukaryotic cells, but they are far more common in muscle cells. In muscle cells, the actin microfilaments are bunched together with thicker filaments of myosin into many thousands of strands called **myofibrils.** The myofibrils shorten when the actin and myosin filaments slide past each other. The shortening of these myofibrils can cause the muscle cell to change shape. Because

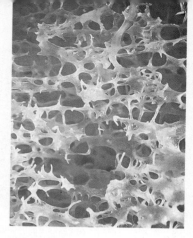

FIGURE 43-8
While we think of bone as hard and solid, the interiors of many bones are composed of a delicate and surprising latticework. Bone, like most of the many tissues present in your body, is a dynamic structure, constantly renewing itself.

FIGURE 43-9
Blood cells. White blood cells, or leukocytes, are roughly spherical and have irregular surfaces with numerous extending pili. Red blood cells, or erythrocytes, are flattened spheres, typically with a depressed center.

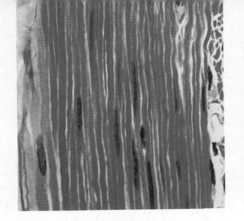
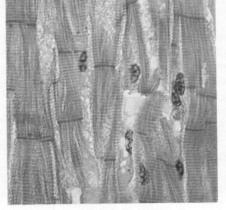
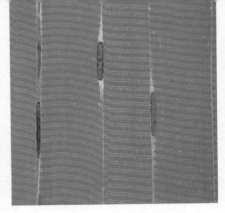

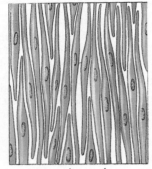

A   Smooth muscle

B   Cardiac muscle

C   Skeletal muscle

FIGURE 43-10
Types of muscle.
A Smooth muscle cells are long and spindle shaped, with a single nucleus.
B Cardiac muscle cells are organized into long branching chains that interconnect, forming a lattice.
C Skeletal or striated muscle is formed by the fusion of several muscle cells, end to end, to form a long fiber with many nuclei.

there are so many filaments all aligned in parallel in muscle cells, a considerable force is generated when all of the myofibrils contract at the same time. Vertebrates possess three different kinds of muscle cells (Figure 43-10): **smooth muscle, striated** or **skeletal muscle,** and **cardiac muscle.** Muscles are the subject of Chapter 47.

Smooth Muscle

Smooth muscle was the earliest form of muscle to evolve, and it is found throughout the animal kingdom. Smooth muscle cells are long and spindle shaped, each cell containing a single nucleus. The interiors of smooth muscle cells are packed with actin-myosin myofibrils, but the individual myofibrils of a cell are not aligned with respect to one another into organized arrays. Smooth muscle tissue is organized into sheets of cells. In some tissues the muscle cells contract only when they are stimulated by a nerve or hormone, and then all of the cells contract together as a unit. Examples of this are the muscles found lining the walls of many vertebrate blood vessels and those that make up the iris of the vertebrate eye. In other smooth muscle tissue, such as that found in the wall of the gut, the individual cells may contract spontaneously, leading to a slow, steady contraction of the tissue. All smooth muscle contraction is involuntary—unlike your skeletal muscle, you cannot will your smooth muscle to contract.

Skeletal Muscle

Skeletal muscles are the muscles of the skeleton and are also called striated muscles. Each striated muscle is a tissue made up of numerous individual muscle cells acting in concert. These striated muscle cells represent a distinct improvement in muscle cell organization, as compared with that found in the smooth muscle cells. Imagine a large raft being towed upstream by many small canoes, each canoe bound to the raft by its own towline. This is analogous to the contraction of smooth muscle, each smooth muscle cell participating individually in the contraction of the muscle. Now imagine placing all the rowers in one galley so that they are able to row in concert, pulling the raft far more effectively. This is analogous to the contraction of striated muscle, in which numerous muscle cells pool their resources.

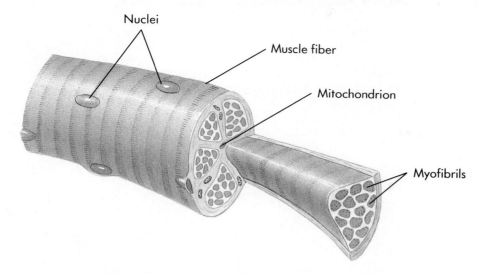

FIGURE 43-11
Diagram of a skeletal muscle
fiber.

Striated muscle cells are produced during development by the fusion of several cells, end to end, to form a very long **fiber** (Figure 43-11). A single fiber may run the entire length of a vertebrate muscle. Each cell, or **muscle fiber,** still contains all of the original nuclei, pushed out to the periphery of the cytoplasm by a central cable made up of 4 to 20 myofibrils. The cytoplasm in striated muscle is given a special name, the **sarcoplasm.** The myofibrils that run down the center of a muscle cell are highly organized to promote simultaneous contraction.

Cardiac Muscle

The hearts of vertebrates are composed of striated muscle fibers arranged very differently from the fibers of skeletal muscle. Instead of very long multinucleate cells running the length of the muscle, heart muscle is composed of chains of single cells, each with its own nucleus. These chains of cells are organized into fibers that branch and interconnect, forming a latticework. This lattice structure is critical to how heart muscle functions. As we will describe in Chapter 51, heart contraction is initiated by the opening of transmembrane channels that admit ions into muscle cells, altering the charge of their membranes. This change in membrane charge is called **depolarization.** When two cardiac muscle fibers touch one another, their membranes make an electrical junction. As a result, the electrical depolarization of any one fiber initiates a wave of contraction throughout the heart, with the wave of depolarization rapidly passing from one fiber to another across these junctions. Thus a mass of heart muscle tends to contract all at once rather than a bit at a time.

Muscle cells are rich in actin and myosin, which form microfilaments that are capable of contraction. Many muscle cells contracting in concert can exert considerable force.

NERVE TISSUE CONDUCTS SIGNALS RAPIDLY

The fourth major class of vertebrate tissue is nervous tissue. It is composed of two kinds of cells: (1) **neurons,** which are specialized for the transmission of nerve impulses; and (2) **supporting cells,** which are specialized to assist in the propagation of the nerve impulse and to provide nutrients to the neuron. Neurons can be further subdivided into **sensory** cells, signal transmitting (**motor**) cells, and information processing cells. Neurons carry impulses rapidly from one organ to another.

Neurons

Neurons (Figure 43-12) are cells specialized to conduct an electric current. The nature of nerve impulses, described in more detail in Chapter 44, can vary, but all nerve impulses are electrical in nature.

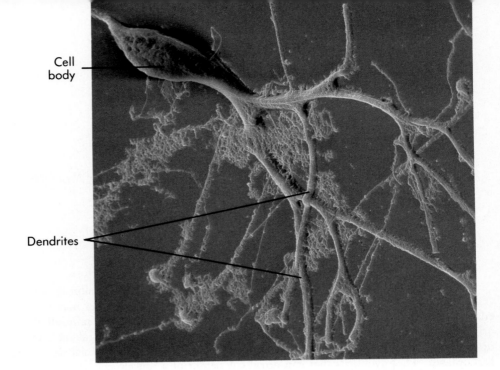

Cell body

Dendrites

FIGURE 43-12

A human neuron. The cell body is at the upper left, with the axon extending up out of view. The branching network of fibers extending down from the cell body is made up of dendrites, which carry signals to the cell body.

The cell body of a neuron (Figure 43-13) contains the cell nucleus. From the cell body project two kinds of cytoplasmic elements that carry out the transmission functions of the neuron. The first group of elements consists of the **dendrites,** which are threadlike protrusions that act as antenna for the reception of nerve impulses from other cells or sensory systems. The second kind of cytoplasmic element that projects from the cell body of a neuron is the **axon.** An axon is a long tubular extension of the cell that provides a cable for the transmission of the nerve impulse away from the cell body. Because axons can be quite long, some nerve cells can be very long indeed. A single neuron that innervates the muscles in your thumb, for example, may have its cell body in the spinal cord and possess an axon that extends all the way across your shoulder and down your arm to your thumb. Single neuron cells over a meter in length are common.

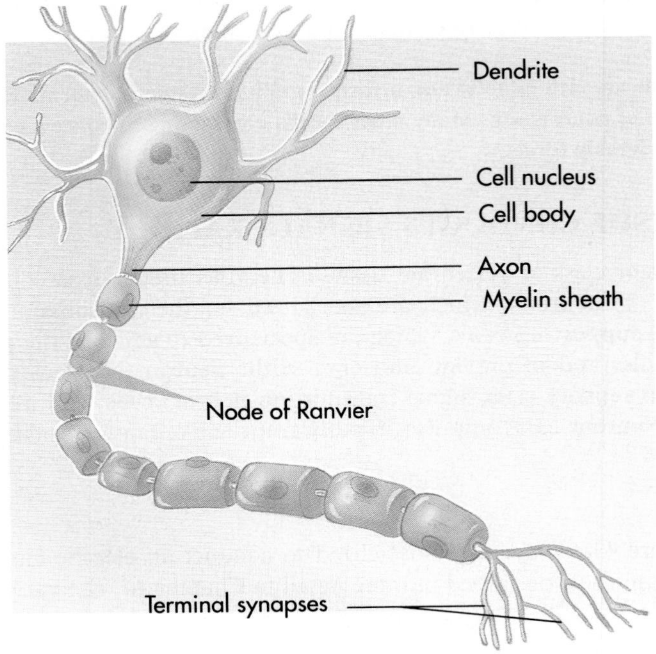

Dendrite

Cell nucleus
Cell body

Axon
Myelin sheath

Node of Ranvier

Terminal synapses

FIGURE 43-13

Idealized structure of a vertebrate neuron. Many dendrites lead to the cell body, from which a single long axon extends. In some neurons, specialized for rapid signal conduction, the axons are encased at intervals within myelin sheaths. At its far end, an axon may terminate at one cell, or branch to several cells, or connect to several locations on one cell.

Nerves

The **nerves** of the vertebrate body, which appear as fine white threads when they are viewed with the naked eye, are actually composed of clusters of axons and dendrites. Like a telephone trunk cable, they include large numbers of independent communications channels—bundles of hundreds of axons and dendrites, each connecting a different nerve cell with a different muscle fiber or sensory receptor. In addition, the nerve contains numerous supporting cells bunched around the axons called glial cells. In the brain and spinal cord, which together make up the central nervous system, these supporting cells are called neuroglia. The supporting cells associated with projecting axons and all other nerve cells, which make up the peripheral nervous system, are called **Schwann cells.**

Neurons are cells that are specialized to conduct electrical signals. Their membranes are rich in ion-transporting transmembrane channels that establish an electrical charge on the membrane surface. Opening some of these channels removes the charge, an effect that tends to open nearby channels and so to propagate the charge reversal along the surface of the nerve as a nerve impulse.

ORGANS: DIFFERENT TISSUES WORKING TOGETHER

The four major classes of vertebrate tissues that we have discussed in this chapter (Table 43-2) are the building blocks of the vertebrate body. Each organ of the body is a structure that carries out a specific function and is composed of these tissues, arrayed and assembled in various ways. A muscle, for example, is composed of muscle tissue wrapped in connective tissue and connected with nervous tissue. Different combinations of tissues are found in different organs.

The many organs that, working together, carry out the principal activities of the body are referred to as **organ systems.** Much of the biology of vertebrates is concerned with the functioning of organ systems. The 11 major organ systems of the human body all work together. The remainder of the text is devoted to describing them in more detail.

HOMEOSTASIS

As the animal body has evolved, specialization has increased, and nowhere is this more evident than in the many cell types that make up the body of a vertebrate. Each is a sophisticated machine, finely tuned to carry out a precise role within the body. Such specialization of cell function is possible only when extracellular conditions are kept within narrow limits. Temperature, pH, the concentration of glucose and oxygen—all these factors must stay very constant for cells to function efficiently and relate to one another properly.

Homeostasis Is Basic to Body Function

We call the maintenance of constant extracellular conditions within the body **homeostasis.** Your body employs many mechanisms to ensure that its internal conditions do not vary outside of a narrow range. There are many familiar examples.

Regulating Levels of Glucose in the Blood.   Stability of the internal environment is essential for homeostasis. However, the environment within our bodies is continuously being exposed to conditions that, if not regulated, would act to change it. When you eat a large dinner, for example, its digestion introduces a large amount of glucose to your body in a short time. What prevents the level of glucose in your blood from quickly rising to a high level? When glucose levels within the blood exceed normal values, the excess glucose is absorbed by liver cells, which convert it to a storage form, glycogen. When glucose levels in the blood fall below the normal range, the liver breaks down its glycogen to add more glucose to the bloodstream. Thus glucose levels in the fluid surrounding the body's cells change very little over the course

TABLE 43-2   TYPES OF VERTEBRATE TISSUE

TISSUE	TYPICAL LOCATION	TISSUE FUNCTION	CHARACTERISTIC CELL TYPES
EPITHELIAL			
Simple epithelium			
Squamous	Lining of lungs, capillary walls, and blood vessels	Cells very thin; provides a thin layer across which diffusion can readily occur	Epithelial cells
Cuboidal	Lining of some glands and kidney tubules; covering of ovaries	Cells rich in specific transport channels; functions in secretion and specific absorption	Gland cells
Columnar	Surface lining of stomach, intestines, and parts of respiratory tract	Thicker cell layer; provides protection and functions in secretion and absorption	Epithelial cells
Stratified epithelium			
Squamous	Outer layer of skin; lining of mouth	Tough layer of cells; provides protection	Epithelial cells
Columnar	Lining of parts of respiratory tract	Functions in secretion of mucus; dense with cilia that aid in movement of mucus; provides protection	Gland cells; ciliated epithelial cells
CONNECTIVE			
Defensive			
White blood cells	Circulatory system	Functions as highway of immune system and stabilizer of body temperature	Macrophages; lymphocytes; mast cells
STRUCTURAL			
Connective tissue proper			
Loose	Beneath skin and other epithelia	Support; provides a fluid reservoir for epithelium	Fibroblasts
Dense	Tendons; sheath around muscles; kidney; liver; dermis of skin	Provides flexible, strong connections	Fibroblasts
Elastic	Ligaments; large arteries; lung tissue; skin	Enables tissues to expand and then return to normal size	Fibroblasts
Adipose	Fat beneath skin and on surface of heart and other internal organs	Provides insulation, food storage, and support of breasts and kidneys	Fat cells

of a day, even though the body's intake of glucose many be concentrated within a short period. This is the essence of glucose homeostasis.

Regulating Body Temperature.   When the temperature of your blood exceeds 98.6° F (37° C), neurons in the brain detect the temperature change. These neurons provide input to the hypothalamus, which responds by triggering mechanisms for dissipating heat (sweating, dilation of blood vessels, and so on). There are also two sorts of temperature-sensitive nerve endings in your skin, one sensitive to low temperatures, the other to high temperatures. In each case the change in temperature affects ion channels in the membrane of the nerve ending, initiating a nerve impulse in a way we discuss in the next chapter. These signals are also relayed to the control center of a specific part of the brain called the hypothalamus.

Homeostasis is Maintained by Feedback Loops

To maintain homeostasis, the body must constantly monitor itself and act to correct any deviation. We call such a process of surveillance and response a **feedback loop.**

TISSUE	TYPICAL LOCATION	TISSUE FUNCTION	CHARACTERISTIC CELL TYPES
Cartilage	Spinal disks; knees and other joints; ear; nose; tracheal rings	Provides flexible support; functions in shock absorption and reduction of friction on load-bearing surfaces	Chondrocytes
Bone	Most of skeleton	Protects internal organs; provides rigid support for muscle attachment	Osteocytes
SEQUESTERING			
Red blood cells	In plasma	Transports oxygen	Red blood cells
Adipose tissue	Beneath skin	Stores fat	Fat cells
MUSCLE			
Smooth	Walls of blood vessels, stomach, and intestines	Powers rhythmical contractions not under conscious control but commanded by central nervous system	Smooth muscle cells
Cardiac	Walls of heart	Highly interconnected cells; promotes rapid spread of signals initiating contraction	Heart muscle cells
Skeletal	Voluntary muscles of the body	Powers walking, lifting, talking, and all other voluntary movement	Fibers
NERVE			
Neurons			
Sensory cells	Eyes; ears; surface of skin	Receives information about body's condition and about exterior world	Rods and cones; muscle stretch receptors
Signal-transmitting cells	Nerves	Transmits signals	Neurons
Information-processing cells	Brain and spinal cord	Integrates information	Interneurons

Homeostasis, in which changes in the body's condition are detected and reversed, involves **negative feedback** mechanisms, in which a disturbance triggers processes that reduce the disturbance. Instances in which a disturbance is accentuated are called **positive feedback** and are important components of the nerve impulse, labor and delivery, and resistance to disease.

Negative Feedback. Negative feedback prevents the departure of a controlled variable such as pH or temperature from its normal, or ideal value. The ideal level of a controllled variable is defined as its **setpoint.** Negative feedback is well illustrated by the familiar example of an individual driving a car. The controlled variable is the actual position of the car in its lane; the setpoint is the center of the lane. The eyes of the driver serve as **sensors;** the driver's brain constantly compares the car's position with the center of the lane. Deviations from the setpoint are called **error signals.** They occur as a result of disturbing factors (**perturbations**) such as bumps or curves in the road, and are opposed by a system of **effectors** that includes the muscles of the driver and the car's steering system.

In the negative feedback control systems of the body (Figure 43-14), the actual

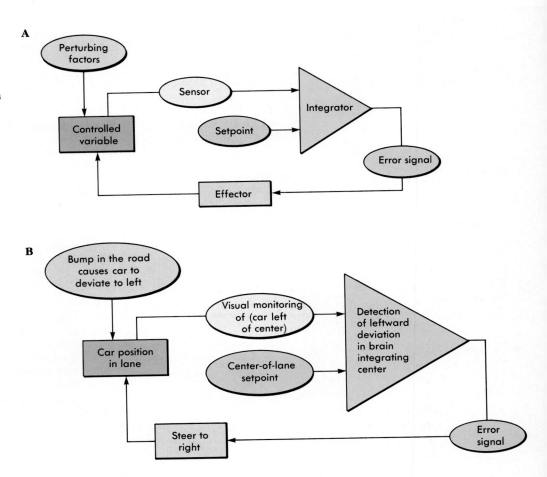

FIGURE 43-14
Negative feedback.
A A generalized diagram of a negative feedback loop.
B Application of the negative feedback loop to the analogy of a car and driver.

value of a controlled variable (such as body temperature or blood pressure) is continuously compared with its setpoint value. Changes in the controlled variable initiate responses that tend to oppose the change and restore the variable to its setpoint—thus the term *negative* is used.

The setpoint in physiological systems may be changed from time to time. For example, body temperature is regulated at a lower value during sleep and at a higher value during fever. Most physiological control systems need to monitor several variables because of the complex interactions between different organ systems. The cardiovascular control center, for example, receives information on arterial blood pressure, blood volume, and oxygen and carbon dioxide content.

Positive Feedback. Negative feedback systems stabilize variables near their setpoints because the response of the effector minimizes the error signal. Most physiological feedback is negative, but there are exceptions. Positive feedback refers to the condition where a change in the controlled variable causes the effector to drive the controlled variable even further away from the initial value. Systems in which there is positive feedback are highly unstable, analogous to a spark that ignites an explosion. One example of positive feedback occurs in the female reproductive cycle, in which the level of a particular pituitary hormone (LH) in the plasma controls ovulation. When LH is released by the pituitary it acts back on the pituitary to increase the level of production and release of LH. The result is a rapid rise of blood levels of LH, which, in turn, triggers ovulation. Positive feedback is also useful in expulsive processes such as the contractions of the uterus in childbirth (Figure 43-15). Here a positive feedback loop increases uterine contractions in response to the pressure of the baby's head on the cervix. In both examples, the amount of positive feedback is carefully regulated by other, negative feedback systems. For example, expulsion of the fetus reduces contractions of the uterus.

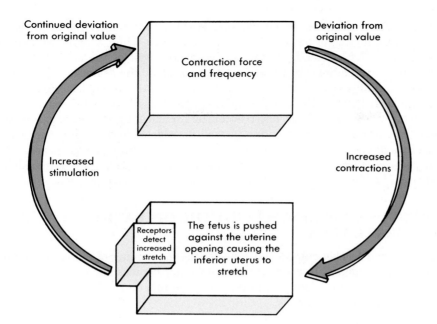

FIGURE 43-15
An example of positive feedback in childbirth.

Homeostasis in Action. The monitoring of body functions is centered in the brain, much of it in a small, marble-sized region called the **hypothalamus.** Using sensors scattered throughout your body, it monitors temperature, pH, and many other factors, and when a disturbance is detected, it issues orders to combat it. Sometimes these orders alter body functions, such as how fast you breathe; at other times they call for the production of particular hormones.

Regulating Blood pH and Salt Balance. The regulation of your blood's pH and salt balance provides an example of how a negative feedback loop works. The kidneys play a very important role in regulating the composition of vertebrate blood and thus the internal chemical environment of the body. By selectively removing substances from the blood, the kidneys maintain rigorous control over the concentrations of ions, such as H^+, Na^+, K^+, Cl^-, Mg^{++}, Ca^{++}, and HCO_3^-, all of which are maintained within narrow boundaries. This serves to maintain the proper ion balances for nerve conduction (blocked by even a small excess of Mg) and muscle contraction (a small increase in K^+ levels in the blood causes the heart to stop).

When you drink a lot of pure water, you dilute the salts in your blood. That is why athletes drink Gatorade, a properly balanced mixture of water and salts similar to that in plasma. To keep the concentration of ions in your blood within proper limits after your body absorbs a lot of water, the hypothalamus employs simple feedback loops. When uptake of water is excessive, receptors in the hypothalamus detect the resulting decrease in blood salt concentration; the hypothalamus responds by lessening its output of a hormone called **antidiuretic hormone (ADH).** A decrease in ADH inhibits the reabsorption of water by the kidneys and therefore reduces water intake that triggered the response.

Other feedback loops regulate the levels of specific ions. Thus, when levels of sodium ion in the blood rise, as happens when you eat highly salted food, cells in the kidneys detect the increase and cause a decrease in the production of the hormone **aldosterone.** Aldosterone stimulates the kidneys to absorb sodium; a decrease in aldosterone lessens the reabsorption of sodium by the kidneys and thus lowers the concentration of sodium in the blood, compensating for the excessive sodium intake that triggered the response.

Regulating Blood Pressure. The brain maintains a constant blood pressure by another feedback loop that adjusts the rate at which your heart beats, the force of contraction of the heart, and the diameter of some blood vessels to compensate for changes in blood pressure. Blood pressure is measured at special sites in the major arteries where the wall of the artery is thin and contains a highly branched network of nerve endings. When blood pressure increases, the wall of the artery is stretched, causing the nerve endings in the wall to send signals to the brainstem. The brain responds by issuing signals that slow down the heart's pacemaker, lowering the rate at which the heart beats and dilating peripheral blood vessels (vasodilation).

When you exercise, your heart becomes stretched by the extra blood flowing through it, and nerve endings within its wall fire signals to the brainstem, which responds by stimulating the heart's pacemaker and thus increasing your heartbeat. Similarly, your heartbeat is increased by the brainstem if levels of carbon dioxide in the blood rise, indicating that the body's cells probably need more oxygen.

HOMEOSTASIS IS CENTRAL TO THE OPERATION OF THE VERTEBRATE BODY

In the following 11 chapters we will consider the functioning of the vertebrate body's organ systems in detail. In every case we encounter homeostatic mechanisms that act to coordinate these functions within the narrow limits dictated by the complexity of these highly specialized animals.

SUMMARY

1. The four basic types of tissue are epithelium, connective tissue, muscle, and nerve.

2. Epithelium covers the surfaces of the body. The lining of the major body cavities is composed of simple epithelium, whereas the exterior skin is composed of stratified epithelium. The major gland systems are also derived from epithelium.

3. The defensive connective tissues contain macrophages, which engulf foreign bacteria and antibody-coated cells or particles; lymphocytes, each of which produces a single unique antibody; and mast cells, which release healing chemicals at sites of trauma.

4. The structural connective tissues include cartilage, bone, and fibroblasts, which secrete protein fibers.

5. The sequestering connective tissues include pigment-producing cells, fat cells, and red blood cells.

6. Muscle contraction provides the force for mechanical movement of the body. There are three kinds of muscle: smooth muscle, which is organized into sheets and contracts spontaneously; striated or skeletal muscle, which is organized into trunks of long fibers and contracts when stimulated by a nerve; and cardiac muscle, in which the fibers are interconnected and may initiate contraction spontaneously.

7. Nerve cells provide the body with a means of rapid communication. There are two kinds of nerve cells: neurons and supporting cells. Neurons are specialized for the conduction of electrical impulses.

8. Organs are body structures composed of several different tissues grouped together into a structural and functional unit.

9. An organ system is a group of organs that function together to carry out the principal activities of the body.

10. The organs of the body employ feedback loops to maintain constant extracellular conditions around the cells, a condition called homeostasis.

REVIEW QUESTIONS

1. What is a tissue? What are the four principal kinds? Into what kinds of structures are different types of tissues collected? How are these structures further organized into functional units?

2. What are the four primary functions of epithelial cells? Into what three general classes are epithelial tissues divided?

3. What is the definition of a connective tissue? Into what three categories are connective tissues grouped?

4. What are the three structural connective tissues? How are they different from one another? Explain.

5. From what embryonic tissue is muscle derived? What two protein microfilaments are abundant in this tissue? What are the three categories of muscle cells?

6. What type of muscle cells are associated with the vertebrate skeleton? Are these muscles under voluntary or involuntary control? What is the morphology of this type of muscle? What is a consequence of the great degree of organization within this tissue?

7. What types of cells are found within nerve tissue? What is the function of each?

8. What is the composition of a typical nerve fiber visible with the naked eye? What type of supporting cells are found in the brain and spinal column? What are the supporting cells of the peripheral nervous system?

9. What is homeostasis? What is a negative feedback loop? A positive feedback loop? Give an example of each type of feedback loop.

THOUGHT QUESTIONS

1. Among long-distance runners and committed joggers, the long bones of the legs often develop "stress fractures," numerous fine cracks running parallel to one another along the lines of stress. In most instances, stress fractures occur when runners push themselves much farther than they are accustomed to running. Runners who train by gradually increasing the distance they run rarely if ever develop stress fractures. What protects this second kind of runner?

2. Land was successfully invaded four times—by plants, fungi, arthropods, and vertebrates. Each of these four groups evolved a characteristic hard substance to lend mechanical support, since bodies are far less buoyant in air than in water. Describe and compare these four substances, discussing their advantages and disadvantages. Can you imagine any other substance that would have been superior to any of these? What about plastic?

3. What do you think would happen if there were a delay between the detection of an error signal in a negative feedback system and the response of the controller?

FOR FURTHER READING

CAPLAN, A.: "Cartilage," *Scientific American*, October 1984, pages 84-97. An interesting account of the many roles played by cartilage in the vertebrate body.

CURREY, J.: *The Mechanical Adaptations of Bones*, Princeton University Press, Princeton, N.J., 1984. A functional analysis of why different bones are structured the way they are, with an unusually well-integrated evolutionary perspective.

EYRE, D.R.: "Collagen: Molecular Diversity in the Body's Protein Scaffold," *Science*, vol. 207, pages 1315-1322, 1980. A good description of how collagen is formed and how it is utilized by the vertebrate body.

GORDON, M.: *Animal Physiology: Principles and Adaptations*, MacMillan and Company, New York, N.Y., 1982. Summary of the general physiological principles involved in the design of the major organ systems.

HOUK, J.: "Control Strategies in Physiological Systems," *FASEB Journal*, vol. 2, February 1988, p. 97. An article written for a very general audience that discusses electrical and chemical messages and control systems from both a cellular and an integrative perspective.

KENT, G.: *Comparative Anatomy of the Vertebrates*, ed. 7, Times Mirror/Mosby College Publishing Co., St. Louis, 1992. A very readable introduction to the comparative anatomy of the vertebrates.

NATIONAL GEOGRAPHIC SOCIETY: *The Incredible Machine*, National Geographic Society, Washington, D.C., 1986. A series of outstanding articles on the human body, focusing on its major organ systems. Beautifully illustrated and fun to read.

ROSENFELD, A.: "There's More to Skin Than Meets the Eye," *Smithsonian*, May 1988, pages 159-180. An entertaining and informative account of the many tasks carried out by the body's largest organ.

SEELEY, R., T. STEPHENS, and P. TATE: *Anatomy and Physiology*, ed. 2, Mosby–Year Book Publishing Co., St. Louis, 1992. A comprehensive introduction to human anatomy and physiology that is visually pleasing and highly student oriented.

44

Neurons

OVERVIEW

All animals except sponges possess special cells, called neurons, which transmit information. Just as electrical pulses pass down a phone line, so waves of electrical disturbance passing over the surfaces of neurons carry information from one part of the body to another. Like the dots and dashes of Morse code, all nerve impulses are the same, differing only in their frequency and their point of origin.

FOR REVIEW

Here are some important terms and concepts that you will encounter in this chapter. If you are not familiar with them, you should review them before proceeding.

Sodium/potassium pump (Chapter 6)

Ion channels (Chapter 6)

Neuron (Chapter 43)

Depolarization (Chapter 43)

There are several ways in which one cell of your body can communicate with another. One simple way is by direct contact, with an open channel between the two cells that permits the passage of ions and small molecules. This method is similar to face-to-face interactions between adjacent people who run a city—a slow and uncertain method. In the body, communication of this kind is provided by gap junctions, discussed in Chapter 6. It provides a ready means of communication between adjacent cells, but it does not provide rapid and efficient communication between distant tissues.

It would be better, in terms of distant communication, if the city manager sent a letter instructing various persons what to do. The body organizes various tissues in just this way, sending chemical instructions to these tissues. The instructions are in the form of **hormones,** small chemical molecules that act as messengers within the body. The hormone "letters" are secreted by glands into the bloodstream and then carried around through the body by the circulatory system. Like a letter, each hormone has an address, a chemical shape that only the target tissue will recognize and act on. However, this kind of command system can be too slow, just as the mail may sometimes be when you are waiting for an important letter.

If the message to be delivered to the leg muscles of your body is "Contract quickly, we are being pursued by a leopard," a quicker means of communication than hormones is desirable. In a city, a person in an emergency does not mail a letter, but rather uses the telephone to call for help. That, in effect, is just what the vertebrate body does. All complex animals possess specialized cells called **neurons** (Figure 44-1), which as you will recall from Chapter 43 are cells specialized for signal transmission. Neurons maintain an electrical charge on their outer surface by actively pumping certain ions out across their membranes. Electrical signals pass down the length of a neuron much as electrical impulses pass down a phone line.

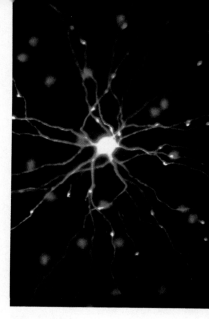

FIGURE 44-1

The neuron. A neuron in the retina of the eye injected with a fluorescent dye. The long unbranched dendrites of this neuron are readily apparent.

THE NEURON

The body contains many different kinds of neurons. Some neurons are tiny and have only a few projections, others are bushy and have more projections, and still others have extensions that are meters long (Figure 44-2). Despite their varied appearances,

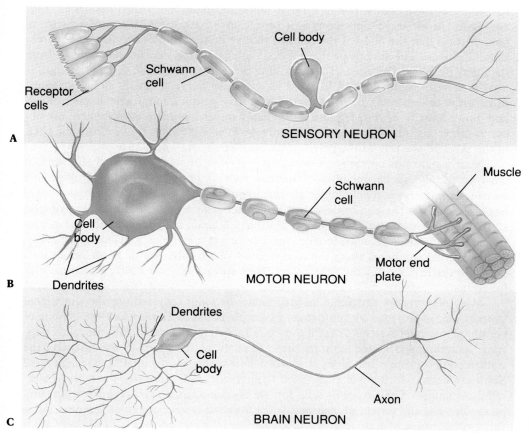

A SENSORY NEURON
Receptor cells
Schwann cell
Cell body

B MOTOR NEURON
Cell body
Dendrites
Schwann cell
Muscle
Motor end plate

C BRAIN NEURON
Dendrites
Cell body
Axon

FIGURE 44-2

Types of vertebrate neurons.

A Sensory neurons (which carry signals from sense organs to the brain) typically have dendrites only in specific receptor cells.
B The axons of many motor neurons (which carry commands from the brain to muscles and glands) are encased at intervals by Schwann cells, as are some of the axons of sensory neurons.
C Neurons within the brain often possess extensive, highly branched dendrites.

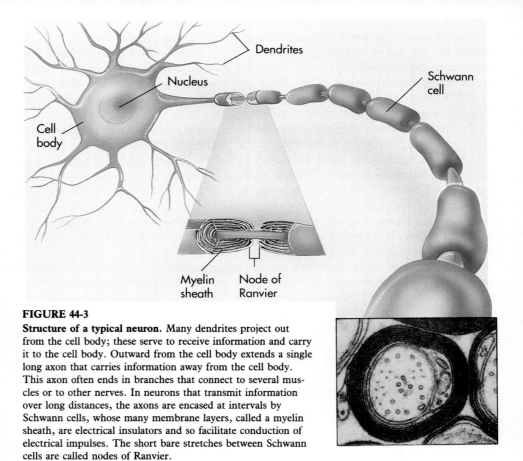

Dendrites

Schwann cell

Nucleus

Cell body

Myelin sheath   Node of Ranvier

FIGURE 44-3

Structure of a typical neuron. Many dendrites project out from the cell body; these serve to receive information and carry it to the cell body. Outward from the cell body extends a single long axon that carries information away from the cell body. This axon often ends in branches that connect to several muscles or to other nerves. In neurons that transmit information over long distances, the axons are encased at intervals by Schwann cells, whose many membrane layers, called a myelin sheath, are electrical insulators and so facilitate conduction of electrical impulses. The short bare stretches between Schwann cells are called nodes of Ranvier.

FIGURE 44-4

Although most vertebrate cells are small, some nerve cells are quite large. From each toe of this giraffe, a single nerve axon extends all the way up to the pelvis.

all neurons have the same functional architecture. Extending from the body of all but the simplest nerve cells is one or more cytoplasmic extensions called **dendrites** (Figure 44-3). Most nerve cells possess a profusion of dendrites, many of which are branched, so that the body of the cell can receive inputs from many different sources simultaneously. The surface of the cell body integrates the information arriving from the many different dendrites. If the resulting membrane excitation is great enough, it travels outward from the cell body as an electrical nerve impulse, along an **axon.** Most nerve cells possess a single axon, which may be quite long. The axons controlling motor activity in your legs are more than a meter long, and even longer ones occur in larger mammals. In a giraffe (Figure 44-4) a single axon travels from the back of the skull all the way to the pelvis, a distance of about 3 meters.

Neurons are not normally in direct contact with one another. At the far end of most axons, the axonal membrane contains packets of specialized chemicals called **neurotransmitters;** acetylcholine is an example of a neurotransmitter. When a nerve impulse arrives at the end of the axon, these chemicals are released into the **synaptic cleft,** the space between two adjacent neurons. When a sufficient number of neurotransmitter molecules reach the receiving neuron, it undergoes a depolarizing event, and the signal continues along the next neuron. Where the tips of axons innervate a skeletal muscle fiber, on the other hand, they spread out and form **neuromuscular junctions.**

Most neurons are unable to survive alone for long; they require the nutritional support that is provided by companion **neuroglia cells.** More than half the volume of vertebrate nervous systems is composed of supporting neuroglia cells. In many neurons, including the motor neurons that extend from the brain to the muscles, the transmission of impulses along the very long axons is facilitated by neuroglial **Schwann cells,** which envelop the axon at intervals (see Figure 44-3) and act as electrical insulators. The Schwann cells form a **myelin sheath** of fatty material around many, but not all, vertebrate neurons. Such a sheath is interrupted at frequent intervals by the **nodes of Ranvier,** where the axon is in direct contact with the surrounding intercellular fluid. An axon and its associated Schwann cells, or cells with similar properties, form a myelinated fiber. Bundles of nonmyelinated and myelinated neurons, which in cross section (Figure 44-5) look somewhat like telephone cables, are called **nerves**.

Nerve cells specialized for electrical signal transmission are called neurons. Typically a signal is received by a dendrite branch and passes to the cell body, where its influence is merged with that of other incoming signals; the resulting signal is passed outward along a single long axon. When the signal reaches the end of the axon, it is transmitted chemically to another neuron or a target cell.

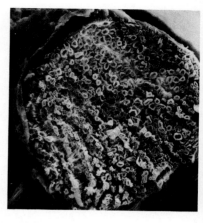

FIGURE 44-5
The nerve. A nerve is a bundle of axons bound together by connective tissue. In this cross section of a bullfrog nerve magnified 1600 times, many myelinated neuron fibers are visible, each looking in cross section something like a Cheerio.

A VOCABULARY OF NERVE TRANSMISSION

action potential   A single nerve impulse; a transient all-or-none reversal of the electric potential across a neuron membrane; because it can activate nearby voltage-sensitive channels, an action potential propagates along a nerve cell.

axon   A single long process extending out from a neuron that conducts impulses away from the neuron cell body.

chemically gated ion channel   A transmembrane pathway for a particular ion that is opened or closed by a chemical such as a neurotransmitter.

dendrite   A process extending from a neuron, typically branched, that conducts impulses inward toward the cell body; neurons may have many dendrites.

depolarization   The movement of ions across a cell membrane that wipes out locally an electrical potential difference.

excitatory synapse   A synapse in which the receptor protein is a chemically gated sodium channel; binding of a neurotransmitter opens the channel and initiates an excitatory electrical potential that increases the ease with which the membrane can be depolarized.

inhibitory synapse   A synapse in which the receptor protein is a chemically gated potassium or chloride channel; binding of a neurotransmitter opens the channel and produces an inhibitory electrical potential that reduces the ability of the membrane to depolarize.

nerve   A bundle of axons with accompanying supportive cells, held together by connective tissue.

neuron   A nerve cell specialized for signal transmission.

neurotransmitter   A chemical released at an axon tip that travels across the synapse and binds a specific receptor protein in the membrane on the far side.

potential difference   A difference in electrical charge on two sides of a membrane caused by an unequal distribution of ions.

refractory period   The recovery period after membrane depolarization during which the membrane is unable to respond to additional stimulation.

resting potential   The charge difference that exists across a neuron's membrane at rest (about 70 millivolts).

synapse   A junction between a neuron and another neuron or a muscle cell; the two cells do not touch, neurotransmitters crossing the narrow space between them.

threshold potential   The minimum change in membrane potential necessary to produce an action potential.

voltage-gated channel   A transmembrane pathway for an ion that is opened or closed by a change in the voltage, or charge differences, across the cell membrane.

THE NERVE IMPULSE

How does the nervous system work? The answer lies in the fact that the nerve membrane contains gated ion channels. They are called gated because they can open or close. Some channels only allow Na^+ ions to pass through them, whereas others are selective for K^+. Whenever ion channels open or close, there is a change in the movement of ions across the nerve membrane. This current flow then affects the charge separation across the cell, that is, its voltage. Information passes along the nerves of the vertebrate body as electrical currents and associated voltage changes, and the key to understanding how the nervous system works is understanding these nerve impulses.

Membrane Potential

When a salt crystal (such as NaCl) is placed in water it separates into Na^+ and Cl^- ions, forming an **electrolyte solution.** Positively-charged ions are termed **cations,** while negatively-charged ions are called **anions.** The cytoplasm of cells and the extracellular fluid are electrolytes. Electrolyte solutions must be neutral as a whole but small differences in the distribution of positive and negative charge in the immediate vicinity of the cell membrane can occur under two conditions: (1) when there is a **concentration gradient** for a particular ion such as K^+, or (2) when there is a difference in the relative **permeability** of the cell membrane to two different ions (usually Na^+ or K^+). Such charge differences produce electrical forces. **Voltage,** or **membrane potential,** is a measure of the electrical driving force and, like the voltage in a car battery, the membrane potential causes ions to move. Movement of electrical charge is called a **current.** Current is increased when the voltage is made larger or when ions pass across a membrane more easily. Concentration gradients represent stored energy that can do electrical work as well as chemical work.

When nerve or muscle cells are not producing electrical responses there is still a voltage difference across their membranes, termed the **resting potential.** The resting potential can be measured by inserting a fine glass capillary called a **micropipette** into the cell and connecting it to a voltmeter. The resting membrane potential is typically about -70 to -80 mV with the inside of the cell negative to the outside. These cells are said to be **polarized.** When the interior of the cell becomes less negative, we refer to the voltage shift as a **depolarization.** Voltage shifts in the opposite direction (the inside becomes more negative) are called **hyperpolarizations.**

Ion Channels

The cell membrane is a thin film of lipid studded with proteins. The only way for ions to cross the cell membrane is by way of tiny pores called **ion channels** (Figure 44-6). An ion channel is a tube (not much larger than the ions themselves) through the

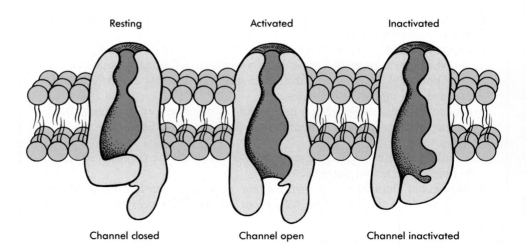

Resting   Activated   Inactivated

Channel closed   Channel open   Channel inactivated

FIGURE 44-6
A hypothetical model of the Na^+ channel illustrating the three states: resting, activated and inactivated.

membrane formed by membrane proteins. At any instant of time ion channels can only be either open or closed.

Ion channels can be described in two ways: (1) by noting the kinds of ions that can pass through them and how easily, termed the channel's **permeability;** and (2) by describing the factors that control the opening of a channel, termed **gating.** Some channels are sensitive to the voltage across a membrane. These are termed **voltage-gated** and they are responsible for the nerve impulse. Other **stimulus-gated** channels are opened by chemicals, or by physical stimuli such as pressure. Stimulus-gated channels are responsible for the transmission of nerve impulses from one cell to another and the conversion of sensory inputs into a neural code.

The Resting Potential: Result of a K^+ Concentration Gradient

The resting potential of a cell develops because: (1) there is a concentration gradient for K^+ (it is higher internally than externally), and (2) cells are more permeable to K^+ than to other cations like Na^+, or to anions such as Cl^-. Outside the cell the numbers of K^+ and Cl^- ions are equal. Inside the cell there will be more K^+ ions than Cl^- ions because within the cell there are negatively charged proteins which are impermeable.

There is a direct relationship between the magnitude of the concentration gradient for K^+ and the voltage. Internal K^+ tends to leave the cell because of its outwardly directed concentration gradient. Each K^+ ion that crosses carries a positive charge to the outside and leaves a negative charge behind in the form of an unpaired Cl^- ion. Before very many K^+ ions have crossed, the electrical force exerted by the separated charges becomes sufficient to retard K^+s leaving the cell and to favor K^+ entry, so the net movement of K^+ is zero. At this point, the system has come into **equilibrium.** The electrical gradient, called the **equilibrium potential,** exactly balances the force applied to K^+ by the concentration gradient (Figure 44-7). The equilibrium potential for K^+ is usually slightly more internally negative than the resting potential.

The Na^+ concentration is low inside cells and high outside, so there is an inwardly-directed concentration gradient for Na^+. The equilibrium potential for Na^+ is therefore of opposite sign, with the cytoplasm positive relative to the extracellular fluid. If a membrane were to become highly permeable to Na^+ (for example, due to the opening of gated Na^+ ion channels), the membrane potential would become internally positive (Table 44-1). Another way of saying this is that the Na^+ concentration gradient represents a force favoring depolarization, while the K^+ concentration gradient favors the internally-negative resting potential. The fact that the resting potential is much nearer to the equilibrium potential for K^+ than the equilibrium potential for Na^+ is because resting cells are much more permeable to K^+ than Na^+. It is only when the Na^+ permeability increases during a nerve impulse (discussed next) that the Na^+ concentration gradient dominates.

Role of the Na^+-K^+ Pump.
All cells are slightly leaky. Cells expend energy to make the Na^+ in the extracellular fluid act as if it were impermeable, and to compensate for any loss of K^+. The **Na^+-K^+ pump** is an enzyme in the membrane that returns 3 Na^+ ions to the extracellular fluid in exchange for taking up 2 K^+ ions from

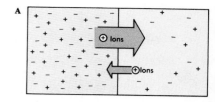

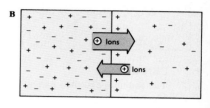

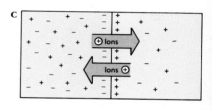

FIGURE 44-7

The role of potassium ions. The diagram illustrates the development of electrochemical equilibrium in a system in which the membrane is permeable only to the positive ion (K^+, indicated simply as + signs) of the salt in solution.

A Initially, the concentration gradient and lack of opposing charge result in a large net flow of K^+ out of the cell.

B As K^+ ions leave there is a charge separation. This charge separation causes the outward flow to decrease while the opposite flow increases driven by the electrical attraction of the charge on the membrane.

C At equilibrium, the electrical and concentration forces are equal and opposite, and there is no net flow of K^+ across the membrane. So few ions have moved relative to the total number in solution, that the change in the concentrations of the two compartments has been negligible.

TABLE 44-1	**THE IONIC COMPOSITION OF CYTOPLASM AND EXTRACELLULAR FLUID (CONCENTRATIONS IN mMOLES/LITER)**			
	CYTOPLASM	EXTRACELLULAR FLUID	RATIO	EQUILIBRIUM POTENTIAL
Na^+	15	150	10:1	+60 mV
K^+	150	5	1:30	−90 mV
Cl^-	7	110	15:1	−70 mV

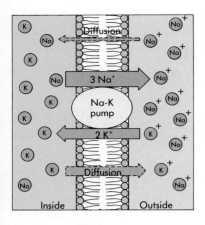

FIGURE 44-8

The sodium-potassium pump.
The Na-K pump transports Na$^+$
ions to the outside of the cell, creating a high exterior Na$^+$ concentration.

FIGURE 44-9

The patch clamp technique. By
pressing the fine tip of a micropipette against the enzymatically
cleaned surface of a cell and applying gently suction, researchers
can place a seal around a small
patch of the cell membrane and
the ion channels it contains. An
experimenter can then stimulate
the trapped channels from within
the pipette and measure their behavior.

the extracellular fluid (Figure 44-8). In this process, ATP is split to provide the energy needed to move both Na$^+$ and K$^+$ against their concentration gradients. Thus,
although cell membranes are actually slightly permeable to Na$^+$, and become more so
during nerve impulses, any excess Na$^+$ entering is continuously being removed from
the cytoplasm as rapidly as it leaks in.

It is important to realize that the role of the Na$^+$-K$^+$ pump is to maintain normal
concentration gradients across the cell membrane. It has no direct role in the generation of nerve impulses or in cell-to-cell transmission. The resting potential of a cell
can be altered in only two ways: (1) the permeability to a particular ion may change,
or (2) the concentration gradient for an ion may change. Ion concentration gradients
do not normally change for two reasons. The activity of the Na$^+$-K$^+$ pump balances
any net gain or loss of ions by the cell, while homeostatic mechanisms control the
plasma concentrations of Na$^+$ and K$^+$. In other words, under normal conditions, the
relative permeabilities of the cell membrane to Na$^+$ and K$^+$ are the only factors that
determine the voltage across it.

The behavior of Na$^+$, K$^+$ and other ion channels is studied by isolating individual channels within the mouth of a very narrow pipette, and then stimulating the
channel via the fluid within the pipette. This technique, called **patch clamping** (Figure 44-9), won its developers the Nobel Prize in 1991.

Initiating a Nerve Impulse

To appreciate what occurs during an action potential, it is important to first understand how neurons are stimulated. To stimulate a neuron it is necessary to temporarily displace the membrane potential from the resting potential in the direction of
depolarization. In **sensory receptors** and at **synapses** the membrane potential is initially displaced by a physical or chemical stimulus, whereas in nerve and muscle there
is electrical activation of one part of the membrane by an adjacent part. Once displaced, the voltage may or may not return to the resting potential, depending on which
ion channels are opened. In sensory receptors and at synapses, restoration of the membrane potential to the resting potential always occurs, while in nerve and muscle fibers,
restoration occurs only if the initial depolarization fails to initiate a nerve impulse.

If the membrane potential is depolarized, the driving forces for Na$^+$ and K$^+$
change (Figure 44-9, *A*). The driving force for K$^+$ increases, while the driving force
for Na$^+$ decreases. If nothing else happens (no ion channels are affected), there will

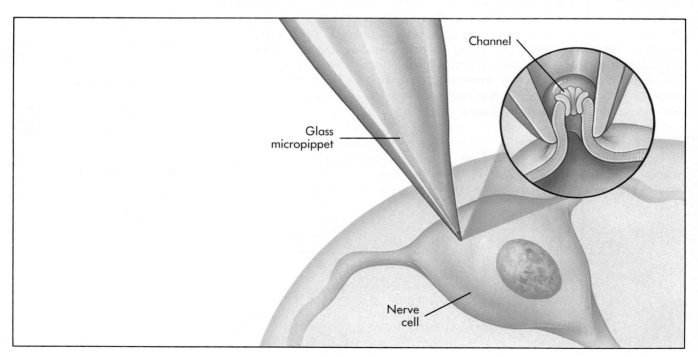

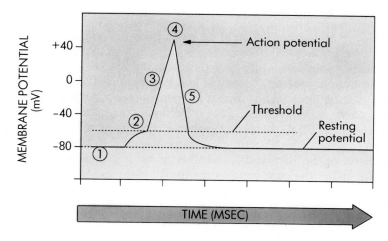

FIGURE 44-10

The action potential. Changes in membrane permeability at selected points during the action potential are shown for a patch of membrane. At the resting potential (1) some K^+ channels are open. At threshold (2) enough Na^+ channels have opened so that the influx of Na^+ exceeds the exit of K^+, causing more Na^+ channels to open (3) until all have opened (4). During the falling phase of the action potential voltage-gated K^+ channels are opening while Na^+ channels are becoming inactivated (5).

be an increased net flow of K^+ out of the cell, and a decreased net flow of Na^+ into the cell. The resulting loss of positive charge will restore the membrane potential to the resting potential.

The Action Potential

In nerve and muscle cells a depolarization opens voltage-gated Na^+ channels. The driving force for Na^+ entry is very large (Figure 44-9, *B*), so the opening of even a few Na^+ channels allows a large increase in Na^+ entry, which will further depolarize the cell and cause more of the voltage-gated Na^+ channels to open. These newly opened channels continue to depolarize the membrane, opening still more channels until all the Na^+ channels are open. At this point the polarity of the cell reverses, with the interior becoming positive relative to the exterior. This is referred to as an **action potential** or simply a nerve impulse (Figure 44-10).

An action potential either occurs or it doesn't, depending on the magnitude of the stimulus that is applied. **Threshold** is defined as the value of the membrane potential sufficient to open enough Na^+ channels so that an action potential will occur. Except for heart muscle and some smooth muscles, action potentials are short, lasting a few milliseconds. Action potentials are standard size responses that travel the length of the axon. During an action potential, the membrane potential goes from its resting, inside-negative value to an inside-positive value before returning to the resting potential.

What causes the membrane potential to return to the resting potential after an action potential? There are two factors. The most important is that after a Na^+ channel opens, it closes spontaneously, even though the membrane potential is still depolarized. This process, called **Na^+ inactivation,** is about ten times slower than Na^+ channel opening. In addition to Na^+ inactivation, depolarization opens voltage-gated K^+ channels, increasing the K^+ permeability above that in the resting cell. Opening voltage-gated K^+ channels favors more rapid repolarization. Voltage-gated K^+ channels activate at about the same rate that Na^+ channels inactivate (see Figure 44-10).

Once they have inactivated, voltage-gated Na^+ channels need time to recover. As long as the vast majority of the Na^+ channels are inactivated, no amount of stimulation can open them. The **absolute refractory period** is the period of time during which it is impossible to initiate another action potential. The absolute refractory period is followed by a somewhat longer **relative refractory period** in which a second action potential can be initiated, but a stimulus must be stronger to initiate an action potential. The length of the relative refractory period is determined by how long it takes for most of the Na^+ channels to recover from inactivation, and how long voltage-gated K^+ channels remain open.

> Action potentials are explained on the basis of the behavior of voltage-gated protein channels in the excitable membranes: Na^+ channels are activated rapidly by depolarization and inactivate spontaneously; K^+ channels are activated more slowly by depolarization and do not inactivate.

Propagation of Action Potentials

Under normal conditions, initiation of an action potential occurs in sensory receptors or at points where cell-to-cell transmission occurs. In both cases there is a depolarization that exceeds threshold. Once an action potential has been initiated at one point on a nerve or muscle cell, the quantity of ions flowing across the membrane (the electric current) will be more than sufficient to open voltage-gated Na^+ channels in adjacent regions of the cell membrane. Thus the action potential will be propagated along the membrane with each little bit of membrane stimulating the next (Figure 44-11).

The velocity with which nerve impulses propagate is determined by two factors: (1) the **diameter** of the fiber, and (2) whether or not it is **myelinated.** The larger the diameter, the higher the **conduction velocity.** In many axons of the vertebrate nervous system, insulation against current leakage is provided by the supporting cells referred to at the beginning of the chapter, either Schwann cells or oligodendrocytes, depending on where in the nervous system they are located. Each myelin sheath is composed of many layers of phospholipid cell membrane. The myelin sheaths are interrupted periodically along the length of the axons to form small gaps 1-2 microns wide called **nodes of Ranvier.**

The effect of the myelin sheaths is to dramatically increase conduction velocity by eliminating the need to excite each small area of membrane. Instead, the action potential jumps from one node to another. This is called **saltatory conduction** (from the Latin *saltare,* to jump) (Figure 44-12). The relative advantages of size and myelination can be seen in the following example. An unmyelinated nerve 10 microns in diameter would conduct action potentials at a velocity of about 0.5 meters per second. If axons like these were present in the reflex pathway that withdraws a limb from a painful stimulus, the response would have a lag time of about 4 seconds. If a 10 micron axon is myelinated, the conduction velocity increases to about 50 meters per second and the lag time would be decreased to 40 milliseconds. Myelination is used selectively in the nervous system for those pathways where speed is essential. This includes sensory nerve fibers from pressure receptors in the skin, and motor nerves that run to skeletal muscles in the arms and legs where conduction velocities can exceed 100 meters per second (Table 44-2).

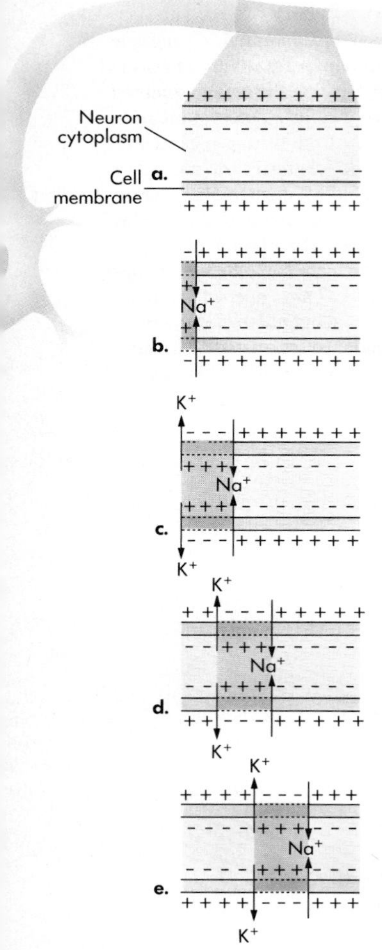

FIGURE 44-11
Transmission of a nerve impulse.

(Left margin labels: Neuron cytoplasm, Cell membrane, a., Na^+, b., K^+, Na^+, c., K^+, K^+, Na^+, d., K^+, K^+, Na^+, e., K^+)

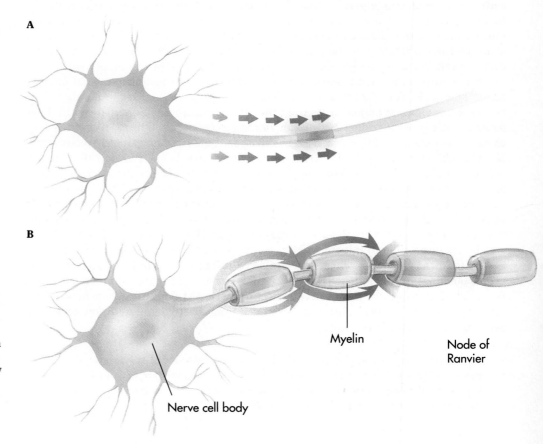

A

B

Myelin

Node of Ranvier

Nerve cell body

FIGURE 44-12

Saltatory conduction.

A In an unmyelinated fiber, each portion of the membrane becomes depolarized in turn, like a row of falling dominos.

B The signal moves faster along a myelinated fiber because the wave of depolarization jumps electrically from node to node without ever depolarizing the insulated membrane segments between nodes.

PART XI   VERTEBRATE BIOLOGY

| TABLE 44-2 | CONDUCTION VELOCITIES OF SOME NERVE FIBERS | | |

	NERVE DIAMETER	MYELIN	CONDUCTION VELOCITY
Squid giant axon	500 microns	No	25 meters per second
Large motor nerve to leg muscle	20 microns	Yes	120 meters per second
Nerve from skin pressure receptor	10 microns	Yes	50 meters per second
Nerve from temperature receptor in skin	5 microns	Yes	20 meters per second
Motor nerves to internal organs	1 micron	No	2 meters per second

TRANSFERRING INFORMATION FROM NERVE TO TISSUE

An action potential passing down an axon eventually reaches the end of the axon. That end is often branched; it may be associated either with several dendrites of other nerve cells or with sites on muscle or secretory cells. These junctions of nerves with other cells are called **synapses.** When the tip of a vertebrate axon is examined carefully, it becomes apparent that it does not actually make contact with the target cell it approaches. There is a narrow intercellular gap, 10 to 20 nanometers (nanometer = one billionth of a meter) across, separating the axon tip and the target cell (Figure 44-13). This gap is called a **synaptic cleft.** Synaptic clefts are characteristic of all vertebrate nerve junctions, except for some of the synapses in the brain.

When a nerve signal arrives at a synaptic cleft, it passes across the gap chemically. The membrane on the axonal side of the synaptic cleft is called the **presynaptic membrane.** When a wave of depolarization reaches the presynaptic membrane, it stimulates the release of neurotransmitter chemicals from vesicles at the tip into the cleft. These chemicals rapidly pass to the other side of the gap. Once there, they combine with receptor molecules in the membrane of the target cell, which is called the **postsynaptic membrane.** By doing so, they cause ion channels to open. These channels are referred to as **chemically gated.**

A synapse is a junction between an axon tip and another cell, almost always including a narrow gap which the impulse cannot bridge. Passage of the impulse across the gap is by chemical signal from the axon.

FIGURE 44-13
A synaptic cleft between two neurons.

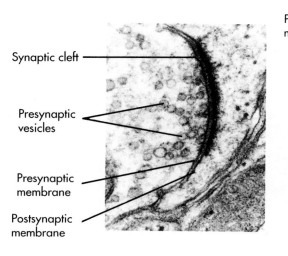

Synaptic cleft

Presynaptic vesicles

Presynaptic membrane

Postsynaptic membrane

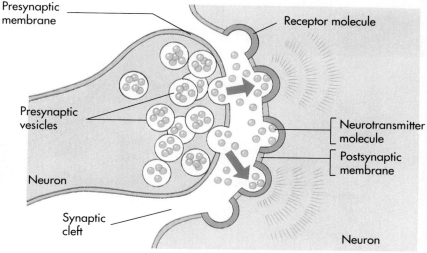

Presynaptic membrane

Receptor molecule

Presynaptic vesicles

Neurotransmitter molecule

Postsynaptic membrane

Neuron

Synaptic cleft

Neuron

The great advantage of a chemical junction such as this, compared with a direct electrical contact, is that the nature of the chemical transmitter can be different in different junctions, permitting different kinds of responses. Over 60 different chemicals have been identified that either act as specific neurotransmitters or that modify the activity of neurotransmitters. The events that occur within the synaptic cleft when a nerve signal arrives depend very much on the identity of the particular neurotransmitter chemical that is released into the cleft. To understand what happens, we will first look at the junction between a nerve and a muscle cell, where the situation is a simple one, and then consider nerve-nerve junctions, where the situation is more complex.

Neuromuscular Junctions

In synapses with skeletal muscle cells, called **neuromuscular junctions** (Figure 44-14), the neurotransmitter is **acetylcholine.** Passing across the gap, the acetylcholine molecules bind to receptors in the postsynaptic membrane, opening chemically gated Na^+ ion channels, which are different from voltage-gated channels that produce action potentials. During the millisecond that the channels are open, some 10^4 ions flow inward (Figure 44-15). This ion flow depolarizes the postsynaptic muscle cell membrane, which contains voltage-gated Na^+ channels. In this way, acetylcholine initiates a wave of depolarization that passes down the muscle cell. This wave of depolarization permits the entry of calcium ions, and so triggering muscle contraction. How depolarization accomplishes muscle contraction will be discussed in Chapter 47.

At a neuromuscular junction, acetylcholine released from an axon tip depolarizes the muscle cell membrane, permitting the entry of calcium ions, which trigger muscle contraction.

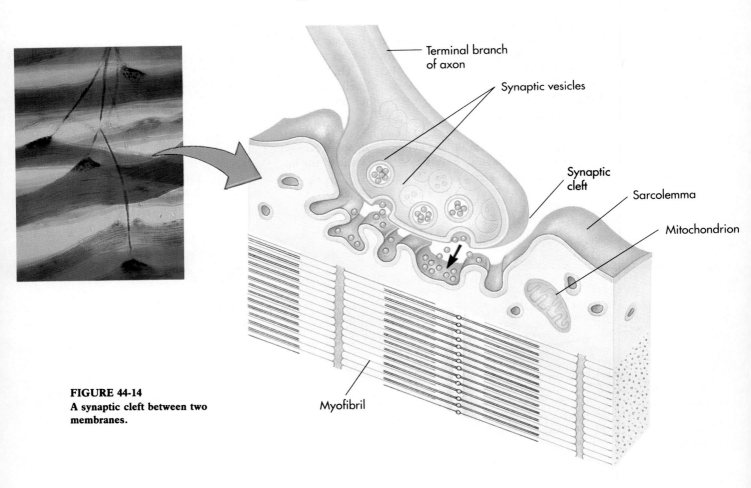

FIGURE 44-14
A synaptic cleft between two membranes.

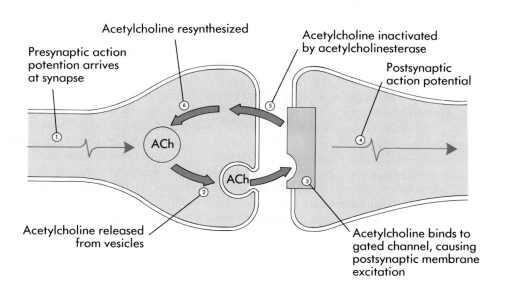

Presynaptic action potention arrives at synapse

Acetylcholine resynthesized

Acetylcholine inactivated by acetylcholinesterase

Postsynaptic action potential

ACh

Acetylcholine released from vesicles

Acetylcholine binds to gated channel, causing postsynaptic membrane excitation

FIGURE 44-15
The sequence of events in synaptic transmission.

FISH STORIES: MARINE TOXINS AND EXCITABLE SODIUM CHANNELS

There are a variety of toxins produced by marine organisms that interact with ion channels in membranes. Six different families of toxins have been identified, but only two groups commonly enter the human food chain. They are among the most potent toxins known, with lethal doses as low as 10 micrograms per kilogram of body weight.

Tetrodotoxin (TTX) is a poison found in the sex organs and liver of marine puffer fish, as well as in the blue-ringed octopus and some salamanders. TTX blocks voltage-gated sodium channels in nerve and muscle fibers, eliminating nerve impulses and paralyzing muscles. The earliest reference to tetrodotoxin appeared in the first Chinese pharmacopoeia (200 BC), where it was recommended as an effective means of arresting convulsions. A detailed account of the symptoms of TTX poisoning appears in the log of Captain Cook's second circumnavigation of the globe in 1670. Mild cases involve limb numbness, flushing of the skin, muscle weakness, and a tingling sensation in the mouth and tongue. Larger doses produce severe disturbances of the heart rhythm. Lethal doses paralyze the diaphragm, leading to asphyxiation. In Japan the puffer fish, called fugu, is regarded as a delicacy. Restaurants that serve

fugu are strictly licensed, but consumption of fugu accounts for some 200 to 400 cases of fugu poisoning annually.

Saxitoxin (STX) is structurally similar to TTX and is identical in its effects. STX is found in single-celled marine protozoa called dinoflagellates. Saxitoxin-containing dinoflagellates "bloom" at certain times of the year. The concentration of organisms sometimes reaches 20 million per liter, giving the water a reddish tint referred to as a "red tide." The dinoflagellates are consumed by clams, mussels, and scallops that then become contaminated. These shellfish are not harvested at some times of the year to avoid accidental poisoning.

Some dinoflagellates in tropical waters contain the marine toxin Ciguatoxin (CTX). These dinoflagellates are ingested by small fish which, in turn, are eaten by larger fish. Levels of CTX gradually accumulate, particularly in the largest fish which have, presumably, eaten the largest number of small fish. The presence of significant amounts of CTX is the reason barracuda cannot be legally imported into this country. However, small amounts are sometimes seen in pacific red snapper and grouper. It was recently estimated that there are about

10,000 cases of CTX poisoning annually. Unlike TTX and STX, which block Na^+ channels, CTX holds them open by interacting with a different binding site. The result is usually mild convulsions and intestinal and cardiovascular disturbances. High doses, however, cause the block of nerve and muscle action potentials and can be lethal.

With the increasing interest in seafood as a healthy alternative to beef, there has been more attention to seafood inspection. Very recently, specific antibodies to various marine toxins have been developed to aid in identifying the very low levels of these toxins in contaminated seafood. Marine toxins have been used to show that sodium channels are physically distinct from other membrane channels such as K^+ channels, and have helped in characterizing the molecular architecture of the Na^+ channel and in separating channel molecules. Because of their specificity for ion channels, many marine toxins may be of interest as pharmacological agents under certain conditions. For example, some marine toxins have antitumor activity, while others suppress the immune response.

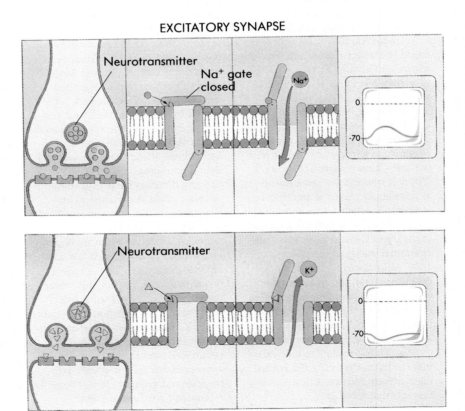

TABUN

SARIN

PARATHION

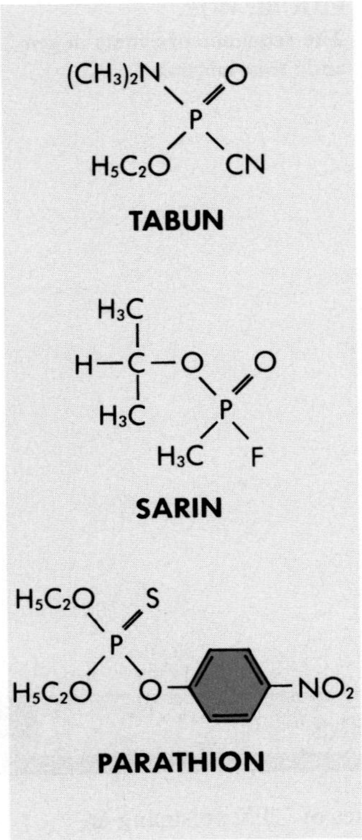

FIGURE 44-16
Some neurotoxins.

For a neuromuscular synapse to transmit more than one impulse, it is necessary to destroy the residual neurotransmitter remaining in the synaptic cleft after the previous impulse. If the residual neurotransmitter is not destroyed the postsynaptic membrane simply remains depolarized. This removal of leftover acetylcholine is accomplished by an enzyme, **acetylcholinesterase**, which is present in the synaptic cleft. Acetylcholinesterase is one of the fastest-acting enzymes in the vertebrate body, cleaving one acetylcholine molecule every 40 microseconds. The rapid removal of neurotransmitter by acetylcholinesterase permits as many as 1000 impulses per second to be transmitted across the neuromuscular junction. Many organic phosphate compounds, such as the nerve gases tabun and sarin and the agricultural insecticide parathion, are potent inhibitors of acetylcholinesterase (Figure 44-16). Because they produce continuous neuromuscular stimulation such compounds can be lethal to vertebrates. Breathing, for example, requires rhythmical muscular contraction.

Neural Synapses

When the axon connection is with another nerve cell rather than with a muscle, a synaptic event has more possible outcomes. Vertebrate nervous systems utilize dozens of different kinds of neurotransmitters, each with specific receptors on postsynaptic membranes.

In an **excitatory synapse,** the receptor protein is a sodium channel that is closed at rest. On binding a neurotransmitter which it recognizes (acetylcholine is a typical example), the sodium channel opens, and allows movement of small ions. Because the concentration gradient for Na^+ dominates, the result is primarily the entry of Na^+, resulting in a depolarization referred to as an excitatory postsynaptic potential.

In an **inhibitory synapse** the receptor protein is a chemically gated potassium channel (or in some cases a chloride channel). Binding of its neurotransmitter (**gammaaminobutyric acid, or GABA,** is a typical example) opens the channel, leading to the exit of positively charged potassium ions and a more negative interior. The result

EXCITATORY SYNAPSE

INHIBITORY SYNAPSE

FIGURE 44-17
Different kinds of synapses use different ion gates. Excitatory synapses open Na^+ gates, while inhibitory synapses open K^+ gates.

FIGURE 44-18

Integration of nerve impulses takes place on the neuron cell body. The synapses made by some axons are inhibitory, tending to counteract depolarization of the cell body membrane; these are indicated in red. The synapses made by other axons are stimulatory, tending to depolarize the cell body membrane; these are indicated in blue. The summed influences of all these inputs determines whether the axonal membrane will be sufficiently depolarized to initiate a propagating nerve impulse.

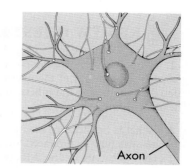

Axon

is termed an inhibitory postsynaptic potential, and this voltage shift reduces the ability of the postsynaptic membrane to depolarize (Figure 44-17).

An individual nerve cell can possess both kinds of synaptic connections to other nerve cells. When signals from both excitatory and inhibitory synapses reach the body of the neuron, the depolarizing effects (which cause less internal negative charge) and the stabilizing ones (which cause more internal negative charge) interact with one another (Figure 44-18). The result is a process of **integration** in which the various excitatory and inhibitory electrical effects tend to cancel or reinforce each other.

In the nervous system, decisions are made by many thousands or millions of neurons acting in concert, each receiving many inputs. For example, a single motor neuron in the spinal cord may have as many as 50,000 synapses on it.

The integration of neural information occurs within individual neurons as the result of the summed effect on a neuron's membrane of the electrical effects of different synapses. Some of these synapses facilitate depolarization, whereas others inhibit it.

SUMMARY

1. A neuron is a cell with an excitable membrane. It is specialized for the transmission of waves of depolarization along the membranes of processes extending out from the cell body.

2. The three structural elements common to all neurons are a cell body, dendrites that carry signals toward the cell body, and axons that carry signals away from the cell body.

3. An excitable membrane is created when the sodium-potassium pump actively pumps sodium ions out across the membrane and potassium ions in. Because the membrane is only slightly permeable to sodium ions or negatively charged anions but is permeable to potassium ions, a net negative charge within the cell results.

4. A nerve impulse is initiated by depolarization of an excitable membrane, that is, by the opening of sodium ion channels.

5. A nerve impulse propagates because the opening of some sodium ion channels facilitates the opening of other adjacent channels, causing a wave of depolarization to travel down the membrane of the nerve cell. All impulses traversing a given neuron have the same amplitude, differing only in the frequency of impulses.

6. When a nerve impulse reaches the far end of a nerve cell, the axon tip, it depolarizes the membrane at the tip, causing the release of chemical transmitter molecules from the tip.

7. These chemicals pass across a synaptic cleft and interact with channels in the membrane of another neuron or of muscle cells. They either open Na^+ channels and depolarize the postsynaptic membranes, an "excitatory" response, or open K^+ channels and hyperpolarize the postsynaptic membranes, an "inhibitory" response.

8. The integration of nerve signals occurs on the cell body membranes of individual neurons, which receive both depolarizing waves from dendrites associated with excitatory synapses and hyperpolarizing waves from dendrites associated with inhibitory synapses. These waves tend to cancel each other, the final amount of depolarization depending on the mix of the signals received.

REVIEW QUESTIONS

1. Why do most neurons possess a profusion of dendrites? What is the function of the neuron's cell body? In what direction does an impulse travel between a cell body and its axon and dendrites?

2. What is the primary function of the neuroglia cells? What special structure is produced by Schwann cells? What is its function? What are nodes of Ranvier?

3. What stimulates a nerve fiber to initiate an electrical impulse? What effect does this stimulus have on the neuron membrane?

4. How does a wave of depolarization spread along a nerve fiber? What is the refractory period?

5. Can a nerve be stimulated to conduct a partial impulse? What is the threshold value of an impulse? What is the action potential of a neuron? Do all neurons possess the same action potential? Why or why not?

6. What name is given to the region that separates one Schwann cell from another? How do these regions propagate the electrical impulse along the myelinated neurons? Why is this mode of conduction advantageous to a neuron? What is this type of conduction called?

7. What is a synapse? What term is associated with the axon side of the synapse? What term is associated with the target cell side of the synapse? How does the nerve impulse cross the synapse?

8. Why is chemical synaptic transmission in neurons superior to stimulation? In relative terms, how many neurotransmitters have been identified?

9. What is the function of acetylcholinesterase at the neuromuscular junction? Why is this necessary? What is the result of interfering with the function of this enzyme?

10. If a neuron acts to inhibit synaptic transmission, would you expect it to open or close channels? Which ones? Why?

THOUGHT QUESTIONS

1. Why do most synapses contain gaps across which an electrical impulse cannot pass, when a direct physical connection would enable the uninhibited passage of the impulse?

2. Ouabain is a drug that poisons the Na^+ ion pump. What would be the immediate effect of the application of ouabain on a nerve fiber? Would the long-term effect be different? Why?

3. Why can a nerve impulse jump from node to node in saltatory conduction, but not across a synaptic cleft?

4. The venoms of most scorpions and many insecticides act by slowing the inactivation of Na^+ channels. What effects would this have on the duration of an action potential? Would its amplitude be changed significantly? What would happen to the refractory period?

5. A plant cell has a high internal concentration of K^+, Na^+, and Cl^- compared to the surrounding fresh water in which it lives. The resting potential is internally negative. What ion(s) could be most permeable? How could you determine which?

FOR FURTHER READING

CATTERALL, W.: "Structure and Function of Voltage-Sensitive Ion Channels," *Science*, vol. 242, October 1988, pages 50-60. The principle subunits of Na^+, Ca^{++}, and K^+ channels are members of a single gene family.

COTMAN, C.W. and IVERSON, L.L.: "Excitatory Amino Acids in the Brain—Focus on NMDA Receptors." *Trends in Neurosciences* Volume 10, p. 263, 1987. Discusses the relationship between dopamine receptors and drugs used to treat psychological disorders.

DUNANT, Y., and ISRAEL, M.: "The Release of Acetylcholine," *Scientific American*, April 1985, pages 58-66. A challenge to the accepted theory that acetylcholine is emitted by synaptic vesicles.

GOTTLIEB, D.: "GABAergic Neurons," *Scientific American*, February 1988, pages 82-89. An account of how inhibitory synapses help to shape the neural networks that underlie behavior.

HOFFMAN, M.: "A New Role for Gases—Neurotransmission," *Science*, vol. 252, June 1991, pages 1788-1789. The remarkable discovery that nitric oxide carries nerve impulses is revolutionizing ideas of nerve transmission.

KUFFLER, S.W., and J.G. NICHOLLS: *From Neuron to Brain: A*

Cellular Approach to the Function of the Nervous System, ed. 2, Sinauer Associates, Inc., Sunderland, Mass., 1984. A superb overview of the mechanisms of nerve excitation and transmission.

LEVITAN, I. and L. KACZMAREK: *The Neuron: Cellular Molecular Biology*, Oxford University Press, New York, 1991. A superb new text covering all aspects of neuron structure and function.

MARX, J.: "Marijuana Receptor Gene Cloned," *Science*, vol. 249, August 1990, pages 624-626. The recent isolation of the gene encoding the marijuana receptor may lead to a better understanding of how the brain controls pain.

NEHER, E. and B. SAKMANN: "The Patch Clamp Technique," *Scientific American*, March 1992, pages 44-51. A technique for isolating and studying individual ion channels in membranes, described by the two scientists who received the 1991 Nobel Prize for discovering it.

SHEPHERD, G.M.: *Neurobiology*, ed. 2, Oxford University Press. New York, 1988. Available in paperback, it is one of the most readable texts in the field. Contains a particularly good discussion of the higher functions of the nervous system such as perception, learning and memory.

45

The Nervous System

OVERVIEW

Networks of neurons control the operation of the vertebrate body by transmitting information from muscles, organs, and sensory systems to the brain, where the information is integrated and used to determine the nature of signals sent to the muscles and glands. The brains of all vertebrates are organized along similar lines, with increasing specialization of the forebrain seen as vertebrates evolved.

FOR REVIEW

Here are some important terms and concepts that you will encounter in this chapter. If you are not familiar with them, you should review them before proceeding.

Primitive invertebrates (Chapter 39)
Depolarization (Chapter 44)
Neuron (Chapters 43 and 44)

FIGURE 45-1
Singing well takes practice. This baby coyote is greeting the approaching evening. His howling is not as impressive as his dad's—a good performance takes practice. His brain is learning by repetition to control the vocal cords properly.

Vertebrates, like all animals except sponges, use a network of neurons to gather information about the body's condition and about the external environment, to process and integrate that information, and to issue commands to the body's muscles and glands (Figure 45-1). This network of neurons, the **nervous system,** is the subject of this chapter; its basic architecture is similar throughout the animal kingdom.

ORGANIZATION OF THE VERTEBRATE NERVOUS SYSTEM

All nervous systems can be said to have one underlying mechanism and three basic elements. The underlying mechanism is the nerve impulse; the three basic elements are (1) a central processing region or **brain,** (2) nerves that bring information to the brain, and (3) nerves that transmit commands from the brain.

The vertebrate nervous system consists of two contrasting functional groups, the **central nervous system** and the **peripheral nervous system** (Figure 45-2). The central nervous system is composed of the brain and the **spinal cord;** it is the site of information processing within the nervous system (Table 45-1).

The spinal cord is composed of a central region containing nerve cell bodies (gray matter) surrounded by ascending (sensory) and descending (motor) tracts consisting of bundles of axons (white matter). Ascending tracts carry sensory information to the brain, whereas descending tracts carry impulses from the motor areas of the brain. The tracts ultimately influence the neurons in the spinal cord controlling the muscles of the body. These neurons are located in the ventral (lower) part of the gray matter of the spinal cord.

The peripheral nervous system includes all the nerve pathways of the body outside of the brain and spinal cord. These pathways are divided into two groups: the **sensory** or **afferent** pathways, which transmit information to the central nervous system; and the **motor** or **efferent** pathways, which transmit commands from it. The motor pathways are in turn partitioned into the **voluntary** ("somatic") **nervous system,** which relays commands to skeletal muscles, and the **autonomic** ("involuntary") **nervous system,** which stimulates the glands and other muscles of the body. In addition, there is the **neuroendocrine system,** which is a network of endocrine glands whose hormone production is controlled by commands from the central nervous system. The central nervous system issues commands through these three different systems.

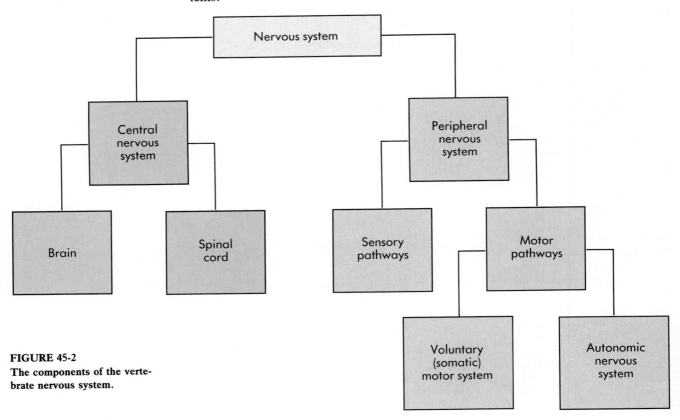

FIGURE 45-2
The components of the vertebrate nervous system.

TABLE 45-1   SUBDIVISIONS OF THE CENTRAL NERVOUS SYSTEM

MAJOR SUBDIVISION	COMPONENTS	FUNCTION
Spinal cord		Spinal reflexes; relay sensory information
Medulla		Sensory afferent nuclei; reticular activating system; visceral control centers
Pons		Reticular activating system; visceral control centers
Midbrain		Similar to pons
Cerebellum		Coordination of movements; balance
Diencephalon	Thalamus	Relay station for ascending sensory and descending motor neurons; control of visceral function
	Hypothalamus	Visceral function; neuroendocrine control
Telencephalon	Basal ganglia	Motor control
	Red nucleus	Motor control
	Corpus callosum	Connects the two hemispheres
	Hippocampus (limbic system)	Memory; emotion
	Cortex	Higher functions

These pathways are composed of individual nerve fibers, axons, and long dendrites, all of which are bundled together like the strands of a telephone cable. Within the central nervous system these bundles of nerve fibers are called **tracts,** whereas in the peripheral nervous system they are called **nerves.** The cell bodies from which pathways extend are often clustered into groups called **nuclei** (if they are within the central nervous system) and **ganglia** (if they are in the peripheral nervous system).

In this chapter, we will begin by considering the brain and central nervous system, then we will consider the sensory systems that convey information to the brain, and we will end with a brief look at the motor pathways and their mode of action.

THE EVOLUTION OF NERVOUS SYSTEMS

Sponges are the only major phylum of multicellular animals that lack nerves. If you prick a sponge, the nearby surface contracts slowly. The protoplasm of each individual cell conducts the impulse, which fades within a few millimeters. No messages dart from one part of the sponge body to others, as they do in all other multicellular animals.

The Simplest Nervous Systems: Reflex Arcs

The simplest nervous systems occur among the cnidarians (Figure 45-3, *A*), which were discussed in Chapter 39. In these animals all neurons are similar, each having fibers of approximately equal length. Cnidarian neurons are linked to one another in a

FIGURE 45-3

Evolution of the nervous system.
The evolution of nervous systems in invertebrates has involved a progressive elaboration of organized nerve cords and the centralization of complex responses in the front end of the nerve cord as seen in these four invertebrates.
A Hydra.
B Planaria.
C Earthworm.
D Grasshopper.

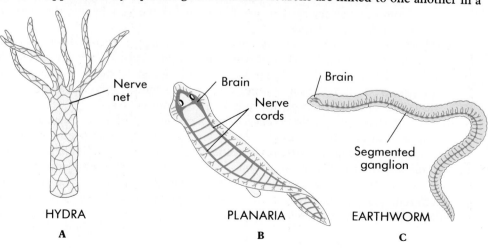

HYDRA

A

PLANARIA

B

EARTHWORM

C

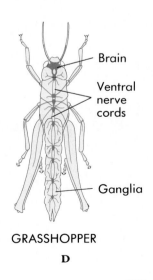

GRASSHOPPER

D

web, or **nerve net,** which is dispersed throughout the epithelium of the body. The conduction of signals through the nerve net is slow, but a stimulus anywhere can eventually spread throughout the whole net.

A nerve impulse in the cnidarians passes through the nerve net, eventually reaching the body muscles and causing them to contract. The motion that results is called a **reflex** because it is an automatic consequence of the nerve stimulation. The simple chain of events is called a **reflex arc,** in this case sensory cell → neuron → muscle. There is no **associative activity** in which other neurons can influence the outcome, no control of complex actions, and little or no coordination. The nerve net of the cnidarians possesses only the barest essentials of nervous reaction.

In the simplest reflex arcs a sensory neuron synapses directly with a motor neuron. In the human body only the "knee jerk" and a few other reflexes are so simple. Reflex arcs in complex animals more typically involve a relay of information from a sensory neuron through one or more **interneurons** to a motor neuron. The functioning of these reflex arcs may be modulated by the interneurons.

> **A reflex is a simple behavior produced by a neural information-response**
> **circuit called a reflex arc.**

The Advent of More Complex Nervous Systems: Associative Activities

Simple reflex arcs are not associative: the passage of a signal along an arc is not influenced by the level of activity of other neurons. The signal depends only on whether or not the arc is stimulated. The first associative activity in nervous systems is seen in the free-living flatworms, phylum Platyhelminthes (Figure 45-3, *B*). Running down the body of these flatworms are two nerve cords; peripheral nerves extend outward from the nerve cords to the muscles of the body. The two nerve cords converge at the front end of the body, forming an enlarged mass of nervous tissue that also contains associative neurons with synapses connecting neurons to one another in various ways. This primitive "brain" is a rudimentary central nervous system. It permits far more complex control of muscular responses than was possible in the cnidarians, because sensory input may induce any of a variety of responses, depending on the associative activity that occurs in the central nervous system.

The Evolutionary Path to the Vertebrates

All of the subsequent evolutionary changes in nervous systems can be viewed as a series of elaborations of the characteristics that are already present in the flatworms. Five principal evolutionary trends can be identified, each becoming progressively more pronounced as nervous systems evolved greater degrees of complexity. Vertebrate nervous systems, which are the most complex to have evolved, represent the farthest extension of each of these five trends:

1. *The elaboration of more sophisticated sensory mechanisms.* As such mechanisms evolved, they provided the nervous system with better information. By far the most complex sensory systems are found among the vertebrates; they are described in the next chapter.

2. *The differentiation of the nerve network into central and peripheral systems.* As nervous systems evolved, most cell bodies of neurons came to be concentrated in one or a few nerve cords (Figure 45-3, *C*) or in masses of nerve cell bodies that are part of, or located near, a nerve cord. Such a central nervous system is connected to all other parts of the body by peripheral nerves.

3. *The differentiation of afferent and efferent nerve fibers.* Individual neurons carry impulses in one direction only. As nervous systems increased in complexity, neurons operating in particular directions became specialized. Those carrying impulses toward the central nervous system are called **afferent** nerves, whereas those conducting impulses away from the central nervous system are called **efferent** nerves. In general, the sensory neurons are afferent, whereas the nerves to glands or muscles are efferent.

4. *The increased complexity of association.* Within the central nervous system, the afferent and efferent neurons are linked by associative interneurons. The path between sensory receptor and motor effector becomes more complex, the axons of interneurons are often branched, with processes that lead to several alternative cells. The particular branch that carries the impulse is determined by the degree of membrane polarization in that branch, which in turn is influenced by synapses with many other neurons: some of these synapses are excitatory and some are inhibitory. Thus, as central nervous systems with more numerous interneurons evolved, more different kinds of activities within the central nervous system were able to influence the destination of a particular impulse.

5. *The elaboration of the brain.* The central coordination of complex responses came to be increasingly localized in the front end of the nerve cord (Figure 45-3, *D*). As this region evolved, it came to contain a progressively larger number of associative interneurons and to develop tracts, which are highways within the brain that connect associative elements.

As nervous systems became more complex, there was a progressive increase in associative activity. It is important to note that no matter how complex the associations are that form within a brain, all neuron impulses are identical because of the "all-or-nothing" nature of neuronal transmission. The difference between a strong signal and a weak one is solely in terms of the frequency of impulses and the number of neurons carrying the signal. All the information that a brain has to work with is the source, frequency, and number of signals that it receives.

EVOLUTION OF THE VERTEBRATE BRAIN

The brains in primitive chordates are little more than swellings at the end of the nerve cord. In the ancestors of the vertebrates, such primitive brains served primarily as sensory centers, receiving messages from eyes and other sensory receptors. They also received information from sensors embedded within muscles and used this information to modulate neuromuscular commands. Far more complex brains and associated neuron-rich nerve trunks, collectively called central nervous systems, developed both in vertebrates (Figure 45-4) and, independently, in cephalopods and arthropods.

Basic Organization of the Vertebrate Brain

The earliest vertebrates had far more complex brains than their ancestors. Casts of the interior braincases of fossil agnathans, fishes that swam 500 million years ago, have revealed a lot about the early evolutionary stages of the vertebrate brain. Although they were very small, these brains already had the three principal divisions that characterize the brains of all contemporary vertebrates: (1) the **hindbrain,** or rhombencephalon; (2) the **midbrain,** or mesencephalon; and (3) the **forebrain,** or prosencephalon.

The hindbrain is the principal component of these early brains (Figure 45-5) as it still is in fishes today. Composed of the cerebellum, the pons, and the medulla, the hindbrain may be considered an extension of the spinal cord devoted primarily to coordinating motor reflexes. Tracts composed of large numbers of nerve fibers run like cables up and down the spinal cord to the hindbrain. The hindbrain in turn integrates the many afferent signals from the muscles and determines the pattern of efferent response.

Much of this coordination is carried on within a small extension of the hindbrain called the **cerebellum** ("little cerebrum"). In the advanced vertebrates the cerebellum plays an increasingly important role as a coordinating center and is correspondingly larger than it is in the fishes. In all vertebrates the cerebellum processes data on the current position and movement of each limb, the state of relaxation or contraction of the muscles involved, and the general position of the body and its relation to the outside world. These data are gathered in the cerebellum and synthesized, and the resulting orders are issued to efferent pathways.

FIGURE 45-4
The human nervous system.

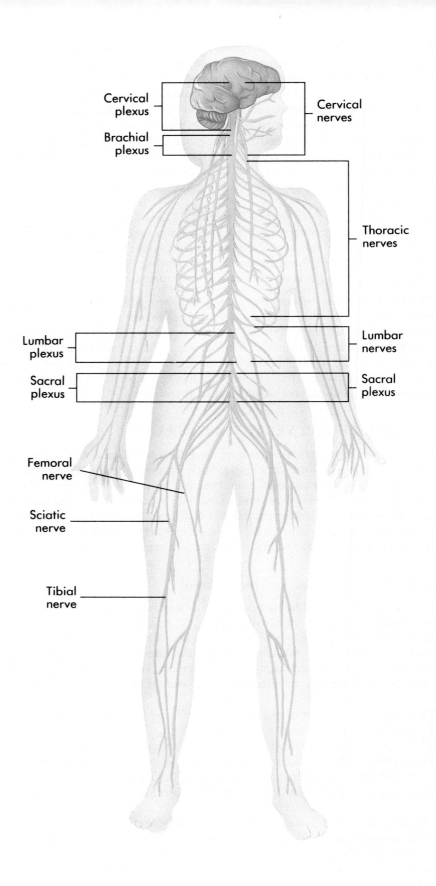

Cervical
plexus

Cervical
nerves

Brachial
plexus

Thoracic
nerves

Lumbar
plexus

Lumbar
nerves

Sacral
plexus

Sacral
plexus

Femoral
nerve

Sciatic
nerve

Tibial
nerve

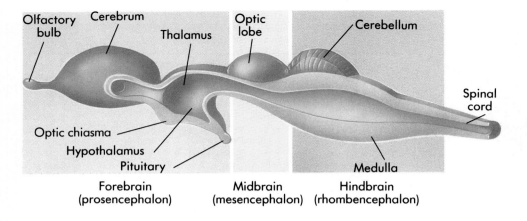

Olfactory bulb | Cerebrum | Thalamus | Optic lobe | Cerebellum

Optic chiasma | Hypothalamus | Pituitary

Spinal cord

Medulla

Forebrain (prosencephalon) | Midbrain (mesencephalon) | Hindbrain (rhombencephalon)

FIGURE 45-5
The basic organization of the vertebrate brain can be seen in the brains of primitive fishes. These brains are divided into the same regions that can be seen in differing proportions in all vertebrate brains: the hindbrain, which is the largest portion of the brain in fishes; the midbrain, which in fishes is a small zone devoted to processing visual information; and the forebrain, which in fishes is devoted primarily to processing olfactory (smell) information. In the brains of terrestrial vertebrates, the forebrain plays a far more dominant role than it does in fishes.

In fishes the remainder of the brain is devoted to the reception and processing of sensory information. The second major division of the brain, the midbrain, is composed primarily of the optic lobes, which receive and process visual information. The third major division, the forebrain, is devoted to processing olfactory (smell) information.

The brains of fishes continue growing throughout their lives. This is in marked contrast to the brains of the more advanced classes of vertebrates, which complete their development by infancy and, with a few minor exceptions, form no new neurons thereafter.

> **In early vertebrates the principal component of the brain was the hindbrain, which is devoted largely to coordinating motor reflexes.**

The Dominant Forebrain

Starting with the amphibians, and much more prominently in the reptiles, a pattern arises that becomes a dominant evolutionary trend in the further development of the vertebrate brain (Figure 45-6): the processing of sensory information becomes increasingly centered in the forebrain.

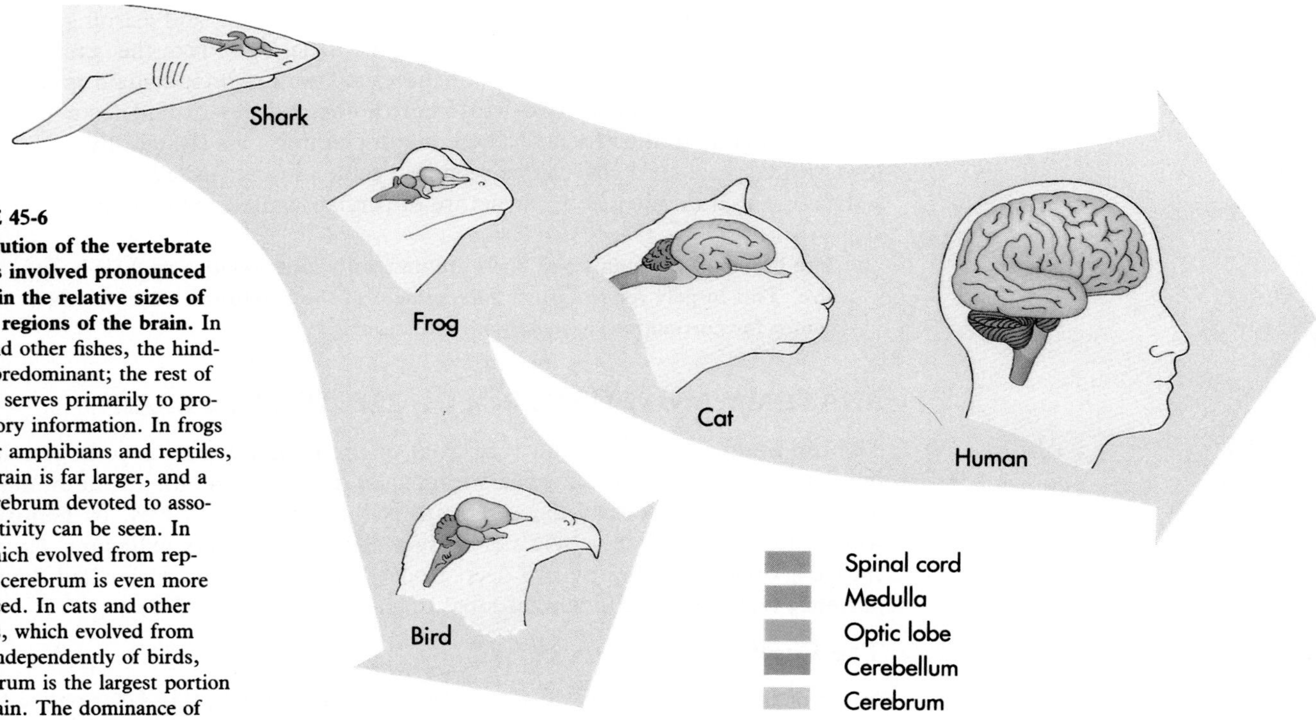

Shark

FIGURE 45-6
The evolution of the vertebrate brain has involved pronounced changes in the relative sizes of different regions of the brain. In sharks and other fishes, the hindbrain is predominant; the rest of the brain serves primarily to process sensory information. In frogs and other amphibians and reptiles, the forebrain is far larger, and a larger cerebrum devoted to associative activity can be seen. In birds, which evolved from reptiles, the cerebrum is even more pronounced. In cats and other mammals, which evolved from reptiles independently of birds, the cerebrum is the largest portion of the brain. The dominance of the cerebrum is greatest in humans, where it envelops much of the rest of the brain.

Frog

Cat

Human

Bird

Spinal cord
Medulla
Optic lobe
Cerebellum
Cerebrum

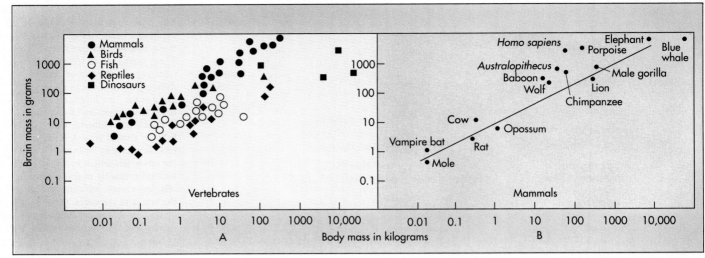

FIGURE 45-7

Brain mass versus body mass.
Among most vertebrates, brain weight is a relatively constant proportion of body weight, so that a plot of brain mass versus body mass gives a straight line.
A However, the proportion of brain mass to body mass is much greater in birds than in reptiles, and even greater in mammals.
B Among mammals, humans have the greatest brain mass per unit of body mass (that is, the farthest perpendicular distance from the plotted line). In second place are the porpoises.

The forebrain in reptiles, amphibians, birds, and mammals is composed of two elements that have distinct functions. The **diencephalon** (Greek *dia,* between) is devoted to the integration of sensory information. The **telencephalon,** or "end brain" (Greek *telos,* end), is located at the front and is devoted largely to associative activity.

> **In land-dwelling vertebrates the processing of information is increasingly centered in the forebrain.**

The Expansion of the Cerebrum

In examining the ratio of brain mass to body mass among the vertebrates (Figure 45-7), one sees a remarkable discontinuity between fishes and reptiles, on the one hand, and birds and mammals, on the other. Mammals have brains that are particularly large relative to their body mass; this is especially true of humans and porpoises. Your brain weighs about 3 pounds. The increase in brain size in the mammals largely reflects a great enlargement of the **cerebrum,** the dominant part of the mammalian brain.

The cerebrum is the center for correlation, association, and learning in the mammalian brain. It receives sensory data from the thalamus. From the cerebrum, motor fibers extend to the motor columns of the spinal cord, which contain groups of neurons servicing different muscles. Efferent motor neurons extend from these motor columns directly to the muscles. These motor columns pass straight through the brain toward the spinal cord and can be seen as a cable of fibers called the pyramidal tract, a structure that dominates the structure of primate brains particularly.

> **The brains of mammals and birds are unusually large relative to their body size. This largely reflects great enlargement of the cerebrum, which is the center for correlation, association, and learning.**

ANATOMY AND FUNCTION OF THE HUMAN BRAIN

The cerebrum, which is at the very front of the human brain, is so large relative to the rest of the brain that it appears to envelop it (Figure 45-8). In the brains of humans and other mammals the cerebrum is split into two halves, or **hemispheres,** which are connected only by a nerve tract called the **corpus callosum.** Each hemisphere is divided further by two deep grooves into four lobes, designated the **frontal, parietal, temporal,** and **occipital** lobes of the brain (Figure 45-9).

The Cerebral Cortex

Much of the neural activity of the cerebrum occurs within a thin layer only a few millimeters thick on its outer surface. This layer, called the **cerebral cortex,** is densely

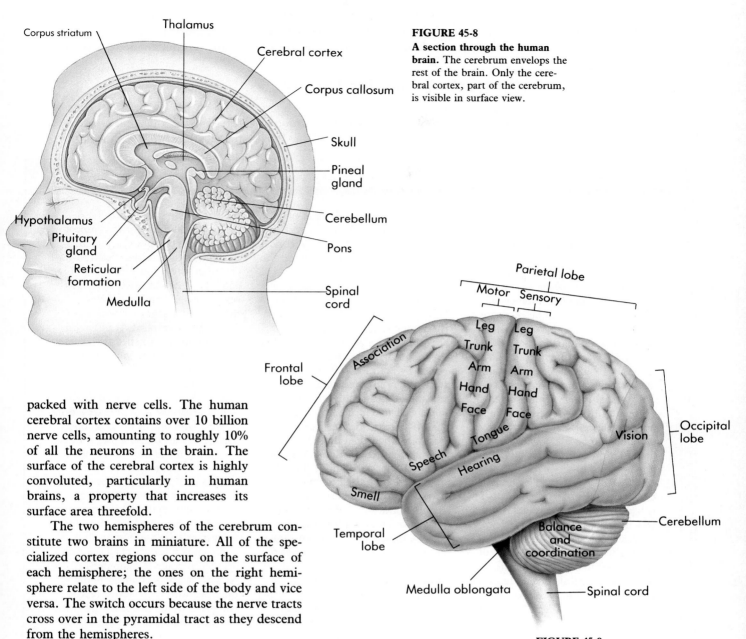

FIGURE 45-8
A section through the human brain. The cerebrum envelops the rest of the brain. Only the cerebral cortex, part of the cerebrum, is visible in surface view.

Labels (Figure 45-8): Corpus striatum, Thalamus, Cerebral cortex, Corpus callosum, Skull, Pineal gland, Cerebellum, Pons, Spinal cord, Hypothalamus, Pituitary gland, Reticular formation, Medulla

Labels (Figure 45-9): Parietal lobe, Motor, Sensory, Leg, Trunk, Arm, Hand, Face, Leg, Trunk, Arm, Hand, Face, Association, Frontal lobe, Tongue, Speech, Hearing, Vision, Occipital lobe, Smell, Temporal lobe, Balance and coordination, Cerebellum, Medulla oblongata, Spinal cord

FIGURE 45-9
The major functional regions of the human brain.

packed with nerve cells. The human cerebral cortex contains over 10 billion nerve cells, amounting to roughly 10% of all the neurons in the brain. The surface of the cerebral cortex is highly convoluted, particularly in human brains, a property that increases its surface area threefold.

The two hemispheres of the cerebrum constitute two brains in miniature. All of the specialized cortex regions occur on the surface of each hemisphere; the ones on the right hemisphere relate to the left side of the body and vice versa. The switch occurs because the nerve tracts cross over in the pyramidal tract as they descend from the hemispheres.

Internal Communication: The Corpus Callosum

Communication between neurons in different parts of each cerebral hemisphere is provided by fibers that course from one part of each hemisphere to another through the interior of the hemisphere, called the **corpus striatum.** The two hemispheres communicate with one another by way of the **corpus callosum** (see Figure 45-8), a thick bundle of fibers (white matter) that connects the two hemispheres with one another. If the corpus callosum is damaged, the two halves of the brain cannot effectively communicate with one another. Such "split-brain" individuals show a variety of abnormalities of speech and perception.

Motor Control: The Corpus Striatum

The most important elements of gray matter in the corpus striatum are several large groups of neurons in nuclei referred to collectively as the **basal ganglia.** This designation is an exception to the general convention that refers to collections of neurons within the CNS as nuclei. The basal ganglia, together with the cerebellum and the

red nucleus in the brainstem, are involved in generating the patterns of activity in motor neurons, which drive complex behavior. The corpus striatum is interconnected with the cerebellum which, as we noted earlier, helps coordinate motor activity.

Sensory Integration: The Thalamus

The thalamus (see Figure 45-8) is a primary site of sensory integration in the brain. Auditory, optical, and other information is relayed to the thalamus, passing from there to the outer surface of the brain. For example, information about posture, de-

---

IMAGING THE BRAIN

There are now several ways to examine the soft tissues of the body, including the brain, without needing to inject a contrast medium into an individual or perform surgery. They are referred to as *noninvasive* and include (1) **computed tomography** (CT scans), (2) **magnetic resonance imaging** (MRI), and **positron emission tomography** (PET) scans.

In a CT scan differences in how the brain tissues absorb x rays are used to construct a three-dimensional image. Unlike older techniques, it is not necessary in inject a contrast material into the individual to be examined, and the dose of x rays needed for a CT scan is very much lower (Figure 45 A, *1*).

The MRI is a procedure in which an externally applied magnetic field is used to detect differences in the way hydrogen nuclei (protons, ^1H) of water molecules in the tissues of the brain vibrate. The vibration of these water molecules will differ when the chemical surroundings differ. In addition to hydrogen, there now exist MRI methods for detecting the vibration of other nuclei of abundant, naturally occurring atoms. This includes fluorine (^{19}F), sodium (^{23}Na), phosphorus (^{31}P) and nitrogen (^{15}N). For example, it is possible to distinguish where ATP is

most abundant by looking at phosphorus resonance. In this way it has become possible to learn a great deal about metabolic processes within the brain (Figure 45-A, *2*).

CT and MRI scans provide different types of information about the location of tumors, sites of hemorrhage following injury, and other abnormalities. For example, bleeding following a concussion is revealed by both the MRI and CT scans. More information about the site and extent of intracranial bleeding is available in the MRI because the MRI technique has greater contrast resolution than the CT scan and bony structures do not obscure the MRI images. Both CT and MRI scans have been used to non-invasively locate tumors and other abnormalities in other parts of the body. For many applications, MRI may be the diagnostic tool of choice in the future because it does not require that a patient receive x-rays. However, if information about the presence of bony structures in relation to the site of injury or abnormality is needed, then other diagnostic tools may be more effective than the MRI.

The PET scan is based on the fact small amounts of the naturally occurring isotopes of several elements

spontaneously emit positrons. A positron is like an electron, except that is has a positive charge rather than a negative charge. Positron-emitting isotopes include ^{13}N, ^{15}O, and ^{11}C. Like the MRI, the PET scan involves forming a three-dimensional image of where these molecules are located. It differs in that compounds labelled with these isotopes are given to an individual before the PET scan is performed. For example, if you want to know where CNS-active drugs bind, all that is necessary is to prepare a drug labelled with one of these isotopes. The PET scan is also a useful way of measuring blood flow in the brain, glucose utilization, and oxygen consumption. It has great potential for diagnosing psychiatric disorders, brain tumors, epilepsy, and the degenerative changes characteristic of Alzheimer's disease (Figure 45-A, *3*).

A common feature of all these imaging techniques is computer-assisted reconstruction of an image of the brain. However, CT scans, MRIs, and PETs would not be feasible without the advances in computer speed and memory capacity that have developed over the past two decades.

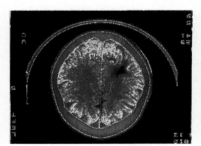

FIGURE 45-A, 1
CT scan.

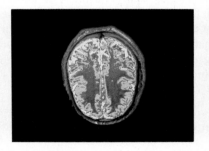

FIGURE 45-A, 2
MRI scan.

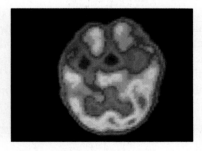

FIGURE 45-A, 3
PET scan.

---

rived from sensory receptors within the muscles, and information about orientation, derived from sensors within the ear, passes from the cerebellum of the hindbrain to the thalamus. The thalamus then processes the information and channels it to the appropriate motor center on the outer surface of the brain.

Integrating Visceral Responses: The Hypothalamus

The **hypothalamus** integrates the visceral activities. It controls body temperature, influences respiration and heartbeat, and also directs the secretions of the brain's master hormone-producing gland, the pituitary gland. The hypothalamus is linked by a network of neurons to the cerebral cortex, the brain's outermost layer. This network, together with the hypothalamus, is the so-called **limbic system** and is discussed next.

Emotion: The Limbic System

The **hippocampus** and **amygdala** are the major components of the limbic system. The limbic system is an evolutionarily old group of linked structures deep within the brain responsible for emotional responses. The hippocampus is also believed to be important to the formation and recall of memories, a topic we will discuss later.

Consciousness and Attention: The Reticular Activating System

The brainstem contains a net-like collection of neurons, referred to as the **reticular formation.** One part of the reticular formation, called the **reticular activating system,** controls consciousness and alertness. All of the sensory systems have fibers that feed into this system, which "wiretaps" all of the incoming and outgoing communications channels of the brain. In doing so, the reticular activating system monitors information concerning the incoming stimuli and identifies important ones. When the reticular system has been stimulated to arousal, it increases the activity level of many parts of the brain. Neural pathways from the brainstem reticular formation to the cortex and other brain regions are depressed by anesthetics and barbituates. In addition to its role in regulating the wakeful and sleep states, discussed next, the reticular activating system plays an important role in cardiovascular, pulmonary, and digestive system regulation.

ACTIVITIES OF THE BRAIN
Sleep and the Electroencephalogram

The reticular activating system controls both sleep and the waking state. Sleep is not the loss of consciousness. It is an active multistate process. The reticular system is periodically repressed by the release of the neurotransmitter serotonin within the brain. Serotonin causes the level of brain activity to fall, bringing on sleep. It is easier to sleep in a dark room than in a lighted one because there are fewer incoming stimuli to trigger the reticular system. In a relaxed individual whose eyes are shut, the pattern of brain wave activity, called an **electroencephalogram (EEG),** consists of large slow waves that occur at a frequency of 8 to 13 Hertz; these are referred to as *alpha waves.* In an alert subject whose eyes are open, electrical activity of the cortex is more desynchronized because multiple sensory inputs are being received, processed, and translated into motor activities.

The first change seen in the EEG with the onset of drowsiness is slowing and a reduction in the overall amplitude of the waves. There are several stages of **slow-wave** sleep but, in general, slow-wave sleep involves decreases in arousability, skeletal muscle tone, heart rate, blood pressure, and respiratory frequency. Another phase of the sleep cycle appears paradoxical. During the **REM phase** of sleep the EEG resembles that seen in a relaxed awake individual, and the heart rate, blood pressure, and respirations are all increased. Individuals in REM sleep are difficult to arouse and are more likely to awaken spontaneously. Dreaming occurs during REM stage sleep. The rapid eye movements that occur during REM sleep are similar to the tracking movements made by the eyes during waking, suggesting that dreamers "watch" their dreams.

FIGURE 45-10

Motor and sensory regions of the cerebral cortex. Each region of the cerebral cortex is associated with a different part of the human body, as indicated in this stylized map.

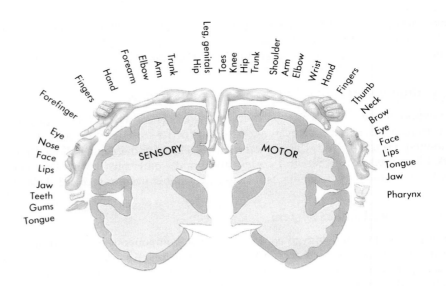

Higher Functions

In land-dwelling vertebrates the processing of information is increasingly centered in the forebrain. By examining the effect of injuries to particular sites on the cerebrum, it has been possible to plot roughly the location of various activities on the cerebral cortex. Each location is referred to as a specialized cortex. There are three general kinds: **motor, sensory,** and **associative** (see Figure 45-9).

Sensory and Motor Areas of the Cerebral Cortex. The **primary motor cortex** straddles the rearmost portion of the frontal lobe. Each point on its surface is associated with the movement of a different part of the body. Right behind the motor cortex, on the leading edge of the parietal lobe, lies the **primary somatosensory cortex.** Each point on the surface of the somatosensory cortex receives inputs from sensory receptors in a different part of the body, such as the pressure sensors of the fingertips or the taste receptors on the tongue (Figure 45-10). The **auditory cortex** lies within the temporal lobe; different surface regions of this cortex correspond to different sound frequencies. The **visual cortex** lies on the occipital lobe, with different sites corresponding to different positions on the retina, equivalent to particular points in the visual fields of the eyes.

Association Areas of the Cerebral Cortex. Only a small portion of the total surface of the cerebral cortex is occupied by the motor and sensory cortexes. The remainder of the cerebral cortex is referred to as **associative cortex.** This appears to be the site of higher mental activities. The associative cortex represents a far greater portion of the total cortex in primates than it does in any other mammals and reaches its greater extent in human beings. In a mouse, for example, 95% of the surface of the cerebral cortex is occupied by motor and sensory areas. In humans, only 5% of the surface is devoted to motor and sensory functions; the remainder is associative cortex.

Language Areas in the Brain. Although the two hemispheres of the brain seem structurally similar, the two hemispheres are responsible for different activities. The best studied specialization is language. The hemisphere in which language ability resides has been called the **dominant hemisphere.** It is almost always located on the left hemisphere of right-handed people; in about one third of left-handed people the right hemisphere is dominant. There are two language areas in the dominant hemisphere: (1) **Wernicke's area,** and (2) **Broca's area.** Wernicke's area, located in the parietal lobe between the primary auditory and visual areas, is important for interpretation of language and for formulation of thoughts into speech. Broca's area, found near the part of the motor cortex controlling the face, is responsible for generation of the patterns of motor output that result in meaningful speech.

Specializations of the Right Hemisphere.   A well-studied specialization of the right cortex is the facial recognition area. Damage to the occipital lobe eliminates the capacity to recall faces. Reading, writing, and oral comprehension remain normal and patients with this disability can still recognize their acquaintances by their voices. Damage to other parts of the right hemisphere may lead to an inability to appreciate spatial relationships and may impair musical activities such as singing. The right hemisphere is important for consolidation of memories of non-verbal experiences. The two hemispheres handle information differently. The dominant hemisphere is adept at sequential reasoning, like that needed to formulate a sentence. The non-dominant hemisphere is adept at spatial reasoning, like that needed to assemble a puzzle or draw a picture.

Memory and Learning

One of the great mysteries of the brain is the basis of memory and learning. If portions of the brain are removed, particularly the temporal lobes, memory is impaired but not lost; there is no one part of the brain in which memory appears to reside. Investigators who have tried to probe the physical mechanisms underlying memory often have felt that they were grasping at a shadow. An understanding of these mechanisms has continued to elude them.

However, researchers have learned a little about the physical basis of memory. The first stage, short-term memory, is transient, lasting only a few moments. Such memories can readily be removed from the brain by application of an electrical shock. When this is done, short-term memories are wiped from the circuits, but the long-term memories are preserved. This result suggests that short-term memories are stored electrically in the form of short-term neural excitation. Long-term memory, in contrast, appears to involve structural changes in the neural connections within the brain.

Neuroscientists now believe that changes in synaptic efficacy (see Chapter 44) are the basis of long-term memory. The response of postsynaptic receptors to a pattern of inputs could be altered, or new receptors could be synthesized. Specific cortical sites cannot be identified for particular memories, because relatively extensive cortical damage does not selectively remove memories. Many memories persist in spite of the damage,and ability to access them gradually recovers with time. Regions of the temporal lobes, the hippocampus and the amygdala (a part of the limbic system), are involved in both short-term memory and memory consolidation. Damage to them affects the ability to process recent events into long-term memories.

PERIPHERAL NERVOUS SYSTEM

Integration of the vertebrate body's many activities is the primary function of the central nervous system. All control of the bodily functions, voluntary and involuntary, is vested in this system. It directs voluntary and involuntary functions in different ways. Voluntary functions are movements of skeletal (striated) muscle that are directed by somatic motor pathways from the brain and spinal cord. These are the pathways that coordinate your fingers when you grasp a pencil, spin your body when you dance, and put one foot ahead of the other when you walk. The direction of these muscular movements by the central nervous system is in large measure subject to conscious control by the associative cortex.

Many of the functions directed by the central nervous system are not subject to conscious control, however; these are referred to as involuntary or **autonomic** (Greek *auto*, self + *nomos*, law—self-controlling). The motor pathways that carry commands from the central nervous system to regulate the glands and nonskeletal muscles of the body are collectively called the **autonomic nervous system.** The autonomic nervous system takes your temperature, monitors your blood pressure, and sees to it that your food is properly digested. The body's internal physiological condition is regulated by the autonomic nervous system (Table 45-2). Most physiological conditions are maintained within relatively narrow bounds, a condition referred to as **homeostasis.**

TABLE 45-2   COMPARISON OF SOMATIC AND AUTONOMIC MOTOR SYSTEMS

CHARACTERISTIC	SOMATIC	AUTONOMIC
Effectors	Skeletal muscle	Cardiac muscle Smooth muscle Gastrointestinal tract Blood vessels Airways Exocrine glands
Effect of motor nerves	Excitation	Excitation or inhibition
Innervation of effector cells	Always single	Typically dual
Number of neurons in path to effector	One	Two
Transmitter	Acetylcholine	Acetylcholine Norepinephrine

Both the voluntary and the involuntary nervous systems are motor pathways. In both cases, sensory input from receptors such as those described in the next chapter provides the central nervous system with information. The central nervous system processes this information and reaches decisions about the appropriate responses. The central nervous system then issues a series of commands to bring these responses about. These commands travel to the body's muscles and organs along the motor pathways of the voluntary and involuntary nervous systems.

The central nervous system employs three separate motor command systems (Figure 45-11). These systems differ in their speed of expression, duration of response, and narrowness of application. They are (1) the neuromuscular control of striated muscles by the motor neurons of the voluntary nervous system, (2) the neurovisceral control of cardiac and smooth muscles by the motor neurons of the autonomic nervous system, and (3) the neuroendocrine control of hormone-producing glands (the subject of Chapter 47) by the hypothalamus.

NEUROMUSCULAR CONTROL

The voluntary nervous system directs the striated muscles of the body and regulates their contraction by means of the stretch receptors described earlier. When a muscle extends or contracts, receptors in the muscle spindle are depolarized, initiating a

FIGURE 45-11
The central nervous system issues commands via three different systems. *(1)* The voluntary nervous system is a network of nerves that extends to the striated muscles. These are the nerves that carry the commands to arm muscles when you lift your arm.
(2) The involuntary or autonomic nervous system is a network of nerves that extend to cardiac and smooth muscles and some glands. These are the nerves that carry the commands to the smooth muscles encasing your intestines, for example, causing these muscles to undergo regular peristaltic contraction and to move food through your digestive tract.

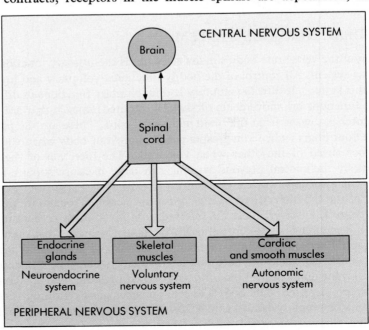

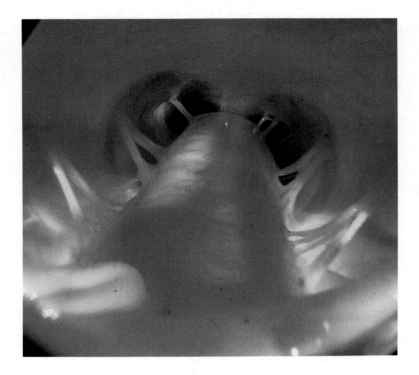

FIGURE 45-12
A view down the human spinal cord. Pairs of spinal nerves can be seen extending out from it. It is along these nerves that the central nervous system, consisting of the brain and spinal cord, communicates with the body.

nerve impulse that travels along afferent nerve fibers to a cell body in the dorsal root ganglion outside the spinal cord (Figure 45-12). Fibers from the dorsal root ganglion make synaptic contact with neurons in the middle of the spinal cord, usually termed **interneurons.** These interneurons in turn affect the motor neurons controlling the muscles.

Reflex Arcs

In animals of relatively simple construction, a nerve impulse passes through the system of neurons, eventually reaching the body muscles and causing them to contract. The motion that results is a **reflex** (see p. 931).

The simplest reflex arcs are **monosynaptic reflex arcs,** in which the afferent nerve cell makes synaptic contacts directly with a motor neuron in the spinal cord whose axon travels directly back to the muscle. A monosynaptic reflex arc is relatively stable but can be influenced by motor efferents from higher levels of the CNS. All the voluntary muscles of your body possess such monosynaptic reflex arcs, although usually in conjunction with other more complex feedback pathways. It is through these more complex paths that voluntary control is established.

In the few cases where the monosynaptic reflex arc is the only feedback loop present, its function can be clearly seen. The well-known "knee jerk" is one such reflex (Figure 45-13). If the patellar ligament just below the kneecap is struck lightly by the edge of your hand or by a doctor's rubber hammer, the sudden pull that results stretches the muscles of the upper leg, which are attached to the ligament. Stretch receptors in these muscles immediately send an impulse along afferent nerve fibers to the spinal cord, where these fibers synapse directly with motor neurons that extend back to upper leg muscles, stimulating them to contract and the leg to jerk upward. Such reflexes play an important role in maintaining posture. Most reflexes are more complex than the knee jerk. Reflexes are important components of many movements (Figure 45-14).

FIGURE 45-13
The knee-jerk reflex.

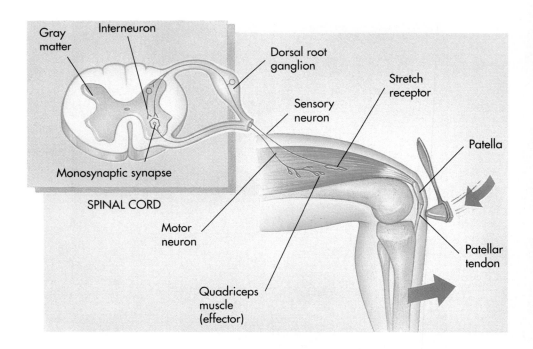

FIGURE 45-14
Yawn. There are some reflexive
behaviors that are common to almost all mammals. Do you know
any mammal that doesn't yawn?

Antagonistic Control of Skeletal Muscles

The movements of limbs and organs are sensitively controlled through the use of two opposing sets of muscles to move every limb and organ (Figure 45-15). One such set of muscles is called **flexors,** the other **extensors.** Muscles can only pull (when they contract); they cannot push. To move a limb in one direction, one set of muscles pulls it by contracting while the other set of muscles relaxes. To reverse the direction, the first set of muscles relaxes while the second contracts, pulling the limb back. All voluntary motor movements of the vertebrate body, and all of its reflex movements, follow this model.

Feedback information is used to coordinate the flexor-extensor muscle pair. The stretching of one muscle, in response to an increase in load on it, stimulates stretch receptors, sending afferent signals to the spinal cord. In the spinal cord, an interneuron inhibits the motor neuron that extends to the antagonistic (opposing) muscle, causing it to relax, while it excites the motor neuron that travels to the contracting muscle, causing it to contract. If the stretching continues, more motor neurons leading to the muscle are excited, increasing contraction until the load is balanced. In this way, the muscle is maintained at a constant length despite variations in the load it bears (Figure 45-16). This is why we are able to hold a heavy weight in our hands.

> The voluntary muscles of the vertebrate body are organized in antagonistic pairs of flexors and extensors. Stretch receptors within them report to the central nervous system on their state of contraction.

Some weights are too heavy to lift for long, as any aspiring weight lifter can tell you. If a contracted muscle is stretched very forcefully, members of another set of sensory receptors called **tendon organs,** embedded in the protein fibers (tendons) that link muscle to bone, are deformed. The tendon organs initiate impulses that travel to interneurons within the spinal cord that inhibit the motor neurons driving contraction of the muscle. The muscle relaxes and ceases to resist the load. This reflex protects the muscle from rupture by excessive loads.

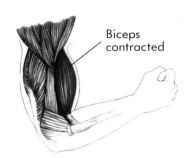

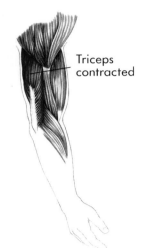

FIGURE 45-15
Antagonistic control of muscles. To raise your arm, one set of muscles, the flexors, contracts and another set, the extensors, relaxes; to lower your arm, the extensors contract and the flexors relax.

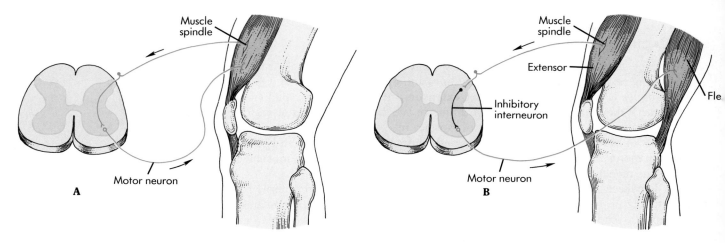

FIGURE 45-16

Monosyaptic reflexes.

A The stretch reflex arc runs from muscle spindle stretch receptors in the stretched muscle to motor neurons that excite the muscle.

B Reciprocal inhibition in the stretch reflex involves connections among afferent neurons, interneurons, and motor neurons that simultaneously excite motor neurons innervating the stretched muscle (and its synergists, not shown) and inhibit ongoing activity in motor neurons innervating the muscles' antagonists.

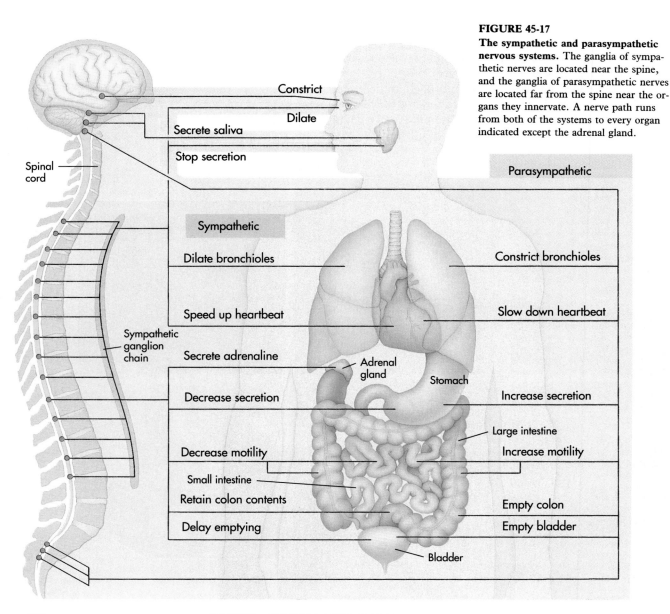

FIGURE 45-17

The sympathetic and parasympathetic nervous systems. The ganglia of sympathetic nerves are located near the spine, and the ganglia of parasympathetic nerves are located far from the spine near the organs they innervate. A nerve path runs from both of the systems to every organ indicated except the adrenal gland.

NEUROVISCERAL CONTROL

The glands, smooth muscle, and cardiac muscle of the body (the "viscera") respond to another network of motor neurons, the autonomic nervous system. This system plays a major role in fine-tuning the body's internal environment. Like the voluntary nervous system, the autonomic nervous system achieves its control by both feedback loops and antagonistic controls.

The Autonomic Nervous System

Control of the body's glands and involuntary muscles is carried out by the autonomic nervous system, composed of two elements usually acting antagonistically (in opposition to each other) (Figure 45-17):

1. The **parasympathetic nervous system** is made up of a network of long efferent central nervous system axons that synapse with organ-associated ganglia in the immediate vicinity of an organ and of short efferent neurons extending from the ganglia to the organ.
2. The **sympathetic nervous system** is made up of a network of short efferent central nervous system axons that extend to ganglia located near the spine and of long efferent neurons extending from the ganglia directly to each target organ.

Control of the Autonomic Nervous System

An autonomic nerve signal crosses two synapses in traveling from the central nervous system out to its target organ, whether it travels out along sympathetic or parasympathetic nerves. The first synapse is in the ganglion, between the axon of a neuron ex-

	TABLE 45-3 AUTONOMIC INNERVATION OF TARGET TISSUES	
ORGAN	**SYMPATHETIC STIMULATION**	**PARASYMPATHETIC STIMULATION**
EYE		
Pupil	Dilated	Constricted
GLANDS		
Salivary	Vasoconstriction Slight secretion	Vasodilation Copious secretion
Gastric	Inhibition of secretion	Stimulation of secretion
Pancreas	Inhibition of secretion	Stimulation of secretion
Sweat	Sweating	None
GASTROINTESTINAL TRACT		
Sphincters	Increased tone	Decreased tone
Wall	Decreased tone	Increased motility
GALL BLADDER	Relaxed	Contracted
BLADDER		
Muscle	Relaxed	Contracted
Sphincter	Contracted	Relaxed
HEART		
Muscle	Increased rate	Slowed rate
LUNGS	Bronchi dilated	Bronchi constricted
BLOOD VESSELS	Constricted	None

tending from the central nervous system and the dendrites of the autonomic neuron's cell body; the second synapse is between the autonomic neuron's axon and the target organ. The neurotransmitter within the ganglion is acetylcholine for both sympathetic and parasympathetic nerves. However, the neurotransmitter between the terminal autonomic neuron axon and the target organ is different in the two antagonistic autonomic nervous systems. In the parasympathetic system, the neurotransmitter at the terminal synapse is acetylcholine, just as it is in the ganglion. In the sympathetic system, the neurotransmitter at the terminal synapse is either **adrenaline** or **noradrenaline,** both of which have an effect *opposite* to that of acetylcholine. Thus, depending on which of the two nerve paths is selected by the central nervous system, an arriving signal will either stimulate or inhibit the organ (Table 45-3).

Most glands (except the adrenal gland), smooth muscles, and cardiac muscle controlled by the autonomic nervous system have inputs from *both* the sympathetic and the parasympathetic systems. Usually these two systems antagonize one another, but either may be excitatory. Thus, depending on which of the two components of the autonomic nervous system is selected by the CNS, an arriving signal will either stimulate or inhibit the organ.

NEUROTRANSMITTERS IN THE CENTRAL NERVOUS SYSTEM

As you learned in Chapter 44, neurotransmitters relay electrical signals between neurons and between neurons and muscle cells. Over 50 distinct chemicals are used as neurotransmitters in the nervous system. **Acetylcholine** was the first discovered. Besides its excitatory effects at the neuromuscular junction and its role in the autonomic nervous system, a lack of acetylcholine has been suggested as a cause of Alzheimer's disease. Both acetylcholine and norepinephrine are **monoamines,** a class of compounds which also includes **dopamine** and **serotonin.** Dopamine is involved in emotional behavior, with excesses of dopamine leading to certain types of schizophrenia. Drugs that activate dopamine receptors, such as the amphetamines, cause various psychotic reactions. Most antipsychotic drugs, on the other hand, block the action of dopamine. Dopamine is also important in the motor systems of the body. Lack of dopamine causes the shaking and wild movements seen in Parkinson's disease. Serotonin is involved in the control of sleep, and also seems to regulate a variety of emotional responses. Lack of serotonin is thought to be one cause of clinical depression. Many of the antidepressants currently available block the reuptake of serotonin into the presynaptic terminal, thus increasing its effectiveness. The drug LSD specifically blocks serotonin receptors in a region of the brainstem known as the raphe nuclei.

Another important class of transmitters are the amino acids: **glutamate, glycine,** and **gamma-aminobutyric acid** (GABA). Glutamate is the main excitatory transmitter in the central nervous system, while glycine is one of the major inhibitory transmitters. Excess glutamate has been linked to degenerative diseases, such as Huntington's chorea. GABA is also inhibitory, and a deficit of GABA is responsible for certain types of anxiety. Many of the anti-anxiety drugs such as diazepam (Valium) increase GABA levels by augmenting its release from presynaptic terminals. Recently, the gas nitric oxide has been shown to act as a neurotransmitter.

Learning and memory involves changes in the activity of the nervous system that can persist for minutes, hours or days. Some of these long-term changes appear to be mediated by a family of transmitter chemicals, sometimes termed **neuromodulators.** Most neuromodulators are small proteins called **neuropeptides.** In many cases both a neurotransmitter and a neuromodulator are stored in the same nerve ending and released together, termed *corelease.* The difference between neuromodulators and transmitters lies in the time scale involved. Neurotransmitters cause rapid, brief effects on a postsynaptic cell. Neuromodulators produce changes that are slower and more lasting. The effects of neuromodulators may be either postsynaptic or presynaptic, and usually are mediated by intracellular second messengers. Some neuromodulators were first identified as hormones secreted by the brain including **vasopressin** ("antidiuretic hormone"), **somatostatin** and **oxytocin.** All of these are now believed to function as neuromodulators.

An excellent example of the importance of neuromodulators is the perception of pain. Nerve fibers from pain receptors enter the spinal cord, where they synapse with interneurons that relay this information to the brain. The transmitter released by pain-sensitive sensory neurons is a peptide called *Substance P.* The intensity with which pain is perceived depends strongly on the effects of peptide neuromodulators called **enkephalins** and **endorphins.** Enkephalins are released by nerve fibers descending from the brain and act to inhibit the passage of pain information to the brain. Endorphins are released by neurons in the brainstem and also block the perception of pain. Both the enkephalins and the endorphins are termed **endogenous opiates** because they activate the same receptors that are activated by opium and its derivatives. Morphine and heroin have an analgesic (pain-reducing) effect because they are similar enough in chemical structure to bind to the receptors normally utilized by enkephalins and endorphins.

Each gland, smooth muscle, and cardiac muscle constantly receives stimulatory signals through one nerve and inhibitory signals by way of the other nerve. The CNS controls activity in each case by varying the ratio of the two signals.

The glands and involuntary muscles of the body are usually innervated by two antagonistic sets of efferent nerves.

Thus an organ receiving nerves from both visceral nervous systems will be subjected to the effects of two opposing neurotransmitters. If the sympathetic nerve ending excites a particular organ, the parasympathetic synapse usually inhibits it. For example, the sympathetic system speeds up the heart and slows down digestion, whereas the parasympathetic system slows down the heart and speeds up digestion. In general, the two opposing systems are organized so that the parasympathetic system stimulates the activity of normal body functions—for example, the churning of the stomach, the contractions of the intestine, and the secretions of the salivary glands. The sympathetic system, on the other hand, generally mobilizes the body for greater activity, as in increased respiration or a faster heartbeat.

THE IMPORTANCE OF THE PERIPHERAL NERVOUS SYSTEM

In this chapter, after considering how individual neurons function, we have focused on the central nervous system—the brain and spinal cord. It is within the central nervous system that all associative and most integrative activities occur, and it is natural that our attention should first be directed there. The central nervous system does not function in isolation, however. It functions as an information-processing center and central command post, a manager of the body's functions; like any manager it requires both information on which to act and a means of carrying out commands. The central nervous system obtains information from the afferent nerves of the sensory nervous system, the subject of the next chapter, and issues commands through the motor nerves of the somatic and autonomic nervous systems. These other elements make up the peripheral nervous system. The sensory, integrative, and command functions of nervous systems evolved in concert, and the system functions as a unified whole.

 ## SUMMARY

1. The evolution of the nervous system has involved the extensive elaboration of networks of associative interneurons in the brain.

2. The hindbrain is the principal component of the brain in fishes. Even in humans, the hindbrain governs many of the body's most basic activities.

3. In the more advanced vertebrates, associative activity is increasingly centered in the forebrain. The midbrain serves as a conduit, linking the lower brain to the forebrain.

4. The skeletal muscles are directed by the motor neurons of the central nervous system. The activity of the central nervous system in directing the skeletal muscles is modulated by stretch receptors, which are embedded in the muscles linked to afferent fibers returning to the central nervous system. The motor and afferent fibers constitute a feedback control loop.

5. Most skeletal muscles are organized in antagonistic pairs, one pulling in one direction, the other in the opposite direction; muscles do not push.

6. Smooth muscles, cardiac muscles, and glands are directed by antagonistic command nerve pairs of the autonomic nervous system, one of which stimulates while the other inhibits. In general, the parasympathetic nerves stimulate the activity of normal internal body functions and inhibit alarm responses, and the sympathetic nerves do the reverse.

REVIEW QUESTIONS

1. Into what functional groups is the vertebrate nervous system divided? What structures are a part of each group? How are the nerve pathways functionally divided? How do the somatic and autonomic motor pathways differ from one another?

2. What is the definition of a reflex? What units of the nervous systems are involved with a typical vertebrate reflex arc?

3. What determines which interneuronal axon branch will carry a particular impulse? Are all nerve impulses within the brain identical? If so, how does the brain differentiate among the signals that it receives?

4. What three phyla have complex nervous systems with well-developed central nervous systems? What are the three principal divisions that characterize all contemporary vertebrate brains? Which of these divisions is most prominent in primitive vertebrates, including fishes?

5. What region of the brain is most dominant in advanced vertebrates? Into what elements is it divided? How do they differ in terms of function?

6. What nerve tract connects the right and left hemispheres of the cerebrum? Into what four lobes is each hemisphere divided? What is the location of the cerebral cortex? What are its major regions based on activity of each region?

7. How many neurons are involved in a monosynaptic reflex arc? What are they and where is their synapse located? What are the functions of such reflex arcs?

8. How is an interneuron-mediated reflex arc more complicated than a monosynaptic arc? How many neurons are associated with this type of reflex? How does it allow for greater control of muscular contraction?

9. How does the system that controls the viscera exert control over its target organs? How are the two networks of neurons anatomically different?

10. What is the chemical transmitter at the ganglionic synapse of both antagonistic visceral control networks? What is the chemical transmitter at the target organ synapse of each of these networks? How are the transmitters at the target organ synapse different?

THOUGHT QUESTIONS

1. When the brain is starved for oxygen even briefly, it dies. When the body is starved for energy, it begins to metabolize its own tissues, channeling the products preferentially to the brain. This behavior points out the importance to the brain of ongoing oxidative respiratory metabolism. Why is active oxidative respiration so important to the continued well-being of the brain's nerve cells?

2. You cannot go for very long without sleeping. Do you think fish sleep? How about earthworms? Why do you think sleep has evolved? Would we not be better off if we never had to sleep? Discuss.

3. A monkey will pick up a chair and move it under a shelf so that he can climb up and get food stored on the shelf. Is the monkey "thinking"? Do you imagine that dogs think? Sharks? Or has thinking evolved only in humans?

FOR FURTHER READING

ALKON, D.: "Memory Storage and Neural Systems," *Scientific American*, July 1989, pages 42-50. Changes in the molecular and electrical properties of nerve cells take place during learning.

ALLPORT, S.: *Explorers of the Black Box: The Search for the Cellular Basis of Memory*, W.W. Norton & Company, New York, 1986. A vivid account of the pioneering studies of Eric Kandel and others in their efforts to demonstrate how we remember. Easy to read, this book shows scientists in action, gathering data and disputing among themselves about what the data mean.

BLACK, I.B. AND OTHERS: "Biochemistry of Information Storage in the Nervous System," *Science*, vol. 236, June 1987, page 1263. Discusses short- and long-term changes in synaptic efficacy and the appearance of specific intracellular proteins correlated with learning and memory in invertebrates model systems.

ECCLES, J.: *Evolution of the Brain*, Routledge, London, 1989. A deeply stimulating and thoughtful look at how our brains differ from those of apes, by a Nobel-prize winning neurobiologist.

FINE, A.: "Transplantation in the Central Nervous System," *Scientific American*, August 1986, pages 52-59. Exciting new work involving the transplanting of embryonic nerve tissue promises to revolutionize treatment of traumatic brain damage. A controversial approach that serves as a good example of how science and social values can come into conflict.

FREEMAN, W.: "The Physiology of Perception," *Scientific American*, February 1991, pages 78-85. It now seems that the chaotic collective activity of millions of neurons is essential for rapid recognition of patterns.

GAZZANGIA, M.: "Organization of the Human Brain," *Science*, vol. 245, September 1989, pages 947-952. The human brain is organized in modules with different functional roles.

GOLDSTEIN, G., and A.L. BETZ: "The Blood-Brain Barrier," *Scientific American*, September 1986, pages 74-83. A detailed description of the structure of brain capillaries, and how their special properties enable them to carefully regulate what passes between blood and brain.

HARVEY, P. and J. KREBS: "Comparing Brains," *Science,* vol. 249, July 1990, pages 140-145. How evolution promoted larger brains in some animals than others is explored by a famous neurobiologist.

HOLLOWAY, M.: "Rx for Addiction," *Scientific American,* March 1991, pages 94-104. An up-to-date look at the molecular mechanisms underlying drug addiction.

KALIL, R.: "Synapse Formation in the Developing Brain," *Scientific American,* December 1989, pages 76-85. Formation of the proper connections requires that developing neurons be used.

KUFFLER, S.W., and J.G. NICHOLLS: *From Neuron to Brain: A Cellular Approach to the Function of the Nervous System,* ed. 2, Sinauer Associates, Inc., Sunderland, Mass., 1984. A superb overview of the mechanisms of nerve excitation and transmission.

LONG, M.: "What Is This Thing Called Sleep?" *National Geographic,* December 1987, pages 787-821. A very interesting and entertaining look at the many mysteries about sleep that still baffle investigators.

MELZACK, R.: "The Tragedy of Needless Pain," *Scientific American,* February 1990, pages 27-34. Morphine administered to control pain may not be addictive.

MISHKIN, M., and T. APPENZELLER: "The Anatomy of Memory," *Scientific American,* June 1987, pages 80-89. A description of how deep structures in the brain interact with outer brain layers to transform sensory stimuli into memories.

MONTGOMERY, G.: "A Brain Reborn," *Discover,* June 1990, pages 48-53. A canary's brain can grow new nerve cells.

MUSTO, D.: "Opium, Cocaine, and Marijuana in American History," *Scientific American,* July 1991, pages 40-47. A brief history of drug use by the general public over the last 200 years.

SELKOE, D.: "Amyloid Protein and Alzheimer's Disease," *Scientific American,* November 1991, pages 68-78. New understanding of how this protein fragment accumulates in the brain may provide a key to developing a treatment for this tragic disorder.

SPECTOR, R., and C. JOHANSON: "The Mammalian Choroid Plexus," *Scientific American,* November 1989, pages 68-74. The gatekeeper of the brain, it regulates what passes out of the bloodstream and into the cerebrospinal fluid.

VECA, A. and DREISBACH, J.H.: "Classical Neurotransmitters and Their Significance Within the Nervous System," *Journal. Chemical Education,* vol. 65, page 108, 1988. Written for nonneuroscientists, provides an up-to-date survey of the role of acetylcholine, norepinephrine, and amino acid transmitters.

VELLUTINO, F.: "Dyslexia," *Scientific American,* March 1987, pages 34-41. A new look at an old problem, mirror writing. The problem is not visual reception, but rather linguistic deficiency, which can be remedied by proper instruction.

WEISSMANN, G.: "Aspirin," *Scientific American,* January 1991, pages 84-90. A fascinating look at the many effects of this drug on the nervous system.

WINSON, J.: "The Meaning of Dreams," *Scientific American,* November 1990, pages 86-96. Dreams to reflect a fundamental aspect of how mammals process their memories, this challenging article argues.

WINTER, P., and J. MILLER: "Anesthesiology," *Scientific American,* April 1985, pages 124-131. A clear description of how certain drugs act to block consciousness.

ZIVIN, J. and D. CHOI: "Stroke Therapy," *Scientific American,* July 1991, pages 56-63. Human trials have begun on several promising treatments for limiting brain damage during and after strokes.

Sensory Systems

OVERVIEW

All sensory information is acquired through the depolarization of sensory nerve endings. From a knowledge of which neurons are sending signals and how often they are doing so, the brain builds a picture. Various specialized sensory receptors cause the depolarization of sensory nerve endings and the firing of neurons in response to particular aspects of the body's internal or external environment— chemical stimuli, mechanical deformation, or electromagnetic stimulation.

FOR REVIEW

Here are some important terms and concepts that you will encounter in this chapter. If you are not familiar with them, you should review them before proceeding.

Carotene (Chapter 9)

Depolarization (Chapter 44)

Sensory neuron (Chapter 45)

Imagine floating on a still surface of water, your eyes closed, in a quiet place where no air is stirring. It would be hard not to sleep. The reticular system of your brain, experiencing little or no stimulation, would not arouse your brain to consciousness. You could not long tolerate this peace, however. Without any stimulation, you would become disoriented and awaken; if you did not find a frame of reference for yourself, you would eventually go insane. The human brain cannot endure sensory deprivation for long—its proper functioning depends on continuous sensory input.

All the input from sensory neurons to the central nervous system arrives in the same form, as nerve impulses carried on afferent sensory neurons. To the brain, every arriving nerve impulse is identical to every other one. The information that the brain derives from sensory input is based on the frequency with which these impulses arrive and on the neuron that transmits the input. A sunset, a symphony, searing pain—to the brain they are all the same, differing only in the source of the impulse and its frequency. Thus if the auditory nerve is artificially stimulated, the central nervous system perceives the stimulation as a noise. If the optic nerve is artificially stimulated in exactly the same manner and degree, the stimulation is perceived as a flash of light. To understand sensory input to the nervous system, then, we must examine the sources of sensory signals, as well as the factors that influence the frequency with which these sources send nerve impulses to the brain.

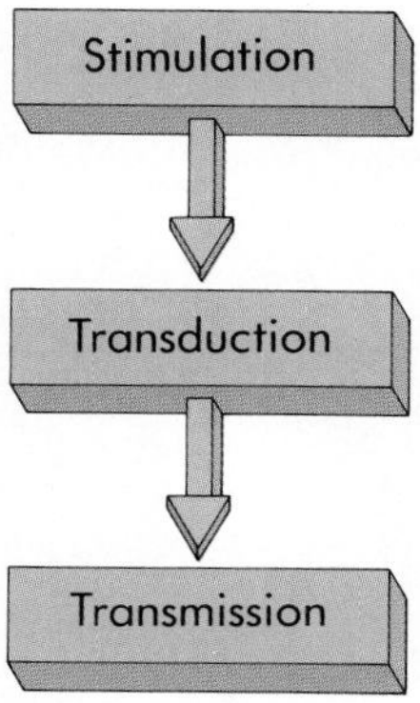

FIGURE 46-1
The path of sensory information.

THE NATURE OF NEUROSENSORY COMMUNICATION

The path of sensory information to the central nervous system is a simple and ancient one, composed of three elements (Figure 46-1):

1. *Stimulation.* A physical stimulus impinges on a neuron or an accessory structure, called a sensory receptor.
2. *Transduction.* The sensory receptor initiates the opening or closing of ion channels in a sensory neuron.
3. *Transmission.* The sensory neuron conducts an action potential along an afferent pathway to the central nervous system.

All sensory receptors are able to initiate nerve impulses by opening or closing ion channels within sensory neuron membranes. They differ from one another with respect to the nature of the environmental input that triggers this event. Many sorts of receptors have evolved among the vertebrates, with each receptor sensitive to a different aspect of the environment. Broadly speaking, we can recognize three classes of environmental stimuli (Table 46-1). Receptors capable of responding to these stimuli, singly or in concert, constitute the sensory repertory of vertebrates.

Sensing the Exterior World

The sensing of the exterior world is called **exteroception** (Figure 46-2, *A*). The simplest sensory receptors are **free nerve endings** that respond to direct physical stimulation, to temperature, to chemicals like oxygen diffusing into the nerve cell, or to a bending or stretching of the nerve cell membrane. Other sensory receptors are more

TABLE 46-1	CLASSES OF ENVIRONMENTAL STIMULI	
MECHANICAL FORCES	CHEMICALS	ELECTROMAGNETIC ENERGY
Pressure	Taste	Light
Gravity	Smell	Heat
Inertia	Humidity	Electricity
Sound		Magnetism
Touch		
Vibration		

complex, involving the association of the receptor with connective tissues or the presence of more than one cell between the stimulus and the afferent nerve ending.

The simplest sensory receptors are free nerve endings that respond to direct physical stimuli such as mechanical distortion, change in temperature, or concentrations of chemicals.

Sensing the nature of objects in the external environment presents a complex problem. Imagine an object some distance away. The information that any sensory receptor can obtain about that object is limited by the nature of the stimulus sensed by that receptor and by the medium, either air or water, through which the stimulus must move to reach the receptor. There are three levels of information that different sensory systems can provide:

1. *Attention.* Some sensory systems provide only enough information to determine that the object is present, saying little or nothing about where it is located.
2. *Location.* Other sensory systems provide information about the direction of the object. They permit the organism to locate the object by moving toward it.
3. *Imaging.* Still other sensory systems provide information concerning distance and direction. By doing so, they enable the central nervous system to construct a three-dimensional image of the object and its surroundings.

Almost all of the exterior senses of vertebrates evolved in water before vertebrates invaded the land. Consequently, many senses of terrestrial vertebrates emphasize stimuli that travel well in water. Such senses use receptors that have been carried over from the sea to the air virtually unchanged. Hearing, for example, converts an airborne stimulus to a waterborne one, using receptors similar to the ones that originally evolved in the water. A few vertebrate sensory systems that function well in the water, such as the electric organs of fish, can not function in the air and are not found among terrestrial vertebrates. On the other hand, some land-dwellers have sensory systems, such as infrared vision, that could not function in the sea.

The four primary senses use different classes of receptors. **Taste** and **smell** use chemical receptors (**chemoreceptors**); **hearing** uses mechanical receptors (mechanoreceptors), and **vision** uses electromagnetic **photoreceptors.**

Sensing the Body's Condition

Traditionally, the sensing of information that relates to the body itself, its internal condition and position, is known as **interoception,** or inner perception (Figure 46-2, *B* and Table 46-2). Interoception involves detecting changes in blood chemistry via **chemoreceptors,** blood pressure changes via **baroreceptors,** pain (**nociceptors**), and mechanical displacements (**mechanoreceptors**). Other interoceptors called **proprioceptors** detect muscle length and tension, and limb position. Many of the receptors that monitor body functions are simpler than those that monitor the external environment, more like what primitive sensory receptors were like than are any other present-day receptors.

We will consider the different types of receptors according to the type of stimulus they are specialized to respond to. Some will be interoceptors, while others are exteroceptors. We will pay particular attention to touch, hearing, and vision, because these are the senses we most often use. At the end of the chapter we will examine a few sensory systems unique to non-human vertebrates. First, however, we must learn how all sensory receptors transform the stimuli reaching them into a neural code.

THE BASIS OF SENSORY TRANSDUCTION

Sensory receptors differ from one another with respect to the nature of the environmental input that they respond to. Many sorts of receptors have evolved among the vertebrates, with each receptor sensitive to a different aspect of the environment. Receptors responding to these stimuli, singly or in concert, constitute the sensory repertoire of vertebrates.

A

B

FIGURE 46-2

By tradition, the senses are grouped into two classes: exteroception and interoception.

A One class senses the external environment. This leaf frog *(Phyllomedusa tarsius),* photographed in the Amazon rain forest of Ecuador, learns about the world around it by using eyes to perceive patterns of light, just as you are doing in reading this page. Sensing of this sort is exteroception.

B Dancers are able to maintain their proper stance by using their internal sense of balance. Sensing of this sort is interoception.

TABLE 46-2 SENSORY TRANSDUCTION AMONG THE VERTEBRATES

STIMULUS	RECEPTOR	LOCATION	STRUCTURE	TRANSDUCTION PROCESS
INTEROCEPTION				
Temperature	Heat receptors and cold receptors	Skin, hypothalamus	Simple nerve ending	Temperature of change alters activity of ion channels in membrane
Touch	Meissner's corpuscles Merkel cells	Surface of skin	Nerve ending within elastic capsule	Rapid or extended change in pressure deforms nerve
Vibration	Pacinian corpuscles	Deep within skin	Nerve ending within elastic capsule	Severe change in pressure deforms nerve
Pain	Nociceptors	Body surfaces	Simple nerve endings	Changes in pressure or temperature open membrane channels
Muscle contraction	Stretch receptors	Within muscles	Spiral nerve endings wrapped around muscle spindle	Stretch of spindle deforms nerve
Blood pressure	Baroreceptors	Arterial branches	Nerve extends over thin part of arterial wall	Stretch at arterial wall deforms nerve
EXTEROCEPTION				
Gravity	Statocysts	Outer chambers of inner ear	Pebble and cilia	Pebble presses against cilia
Motion	Cupula	Semicircular canals of inner ear	Collection of cilia	Deformation of cilia by fluid movement
	Lateral-line organ	Within grooves on body surface of fish	Collection of cilia	Deformation of cilia by fluid movement
Taste	Taste bud cells	Mouth and skin of fish	Chemoreceptors	Binding to specific receptors in membrane
Smell	Olfactory neurons	Nasal passage	Chemoreceptors	Binding to specific receptors in membrane
Hearing	Organ of Corti	Cochlea of inner ear	Cilia between membranes	Deformation of membrane by sound waves in fluid
Vision	Rod and cone cells	Retina of eye	Array of photosensitive pigment	Light initiates process that closes ion channels
Heat	Pit organ	Face of snake	Temperature receptors in two chambers	Temperature of surface and interior chambers compared
Electricity	Ampullae of Lorenzini	Within skin of fish	Closed vesicles with asymmetrical ion channel distribution	Electric field alters ion distribution on membranes
Magnetism	Unknown	Unknown	Unknown	Deflection at magnetic field initiates nerve impulse?

Receptor Potentials

Cells respond to stimuli because they possess **stimulus-sensitive receptor proteins** in their membranes which are either ion channels themselves or are coupled to ion channels (Figure 46-3). Applying the appropriate stimulus to the receptor proteins can result in either opening or closing of ion channels, depending on the sensory system involved. Opening or closing ion channels results in permeability changes and thus voltage changes. The voltage changes that occur in sensory receptors upon stimulation are referred to as **generator potentials** or **receptor potentials** (Figure 46-4). In most sensory systems (photoreceptors are the major exception), the stimulus-gated ion channels in sensory receptors allow both Na^+ and K^+ ions to pass through them.

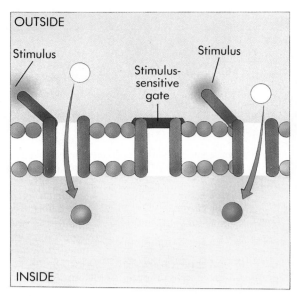

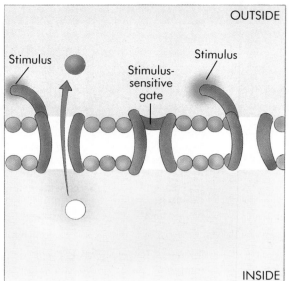

FIGURE 46-3
Conceptual diagram of sensory receptor proteins which are ion channels. In this illustration the stimulus causes the gates on Na^+ (left) or K^+ channels (right) to open.

Since the resting potential is normally close to the K^+ equilibrium potential, the dominant effect will be an influx of Na^+, resulting in **depolarization** of the nerve ending. If the depolarization reaches threshold a series of action potentials will be initiated in the afferent nerve fiber.

Encoding What? and Where?

During development, each sensory (afferent) neuron sends an axon out to a particular place within the body or on its surface. The peripheral end of the sensory neuron selectively responds to a single type of stimulus, called its **modality.** Modality describes WHAT? happened. We can recognize three classes of environmental stimuli (see Table 46-1), each with several distinct modalities such as pressure, vibration, sound, or taste.

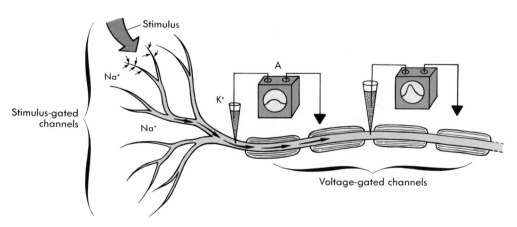

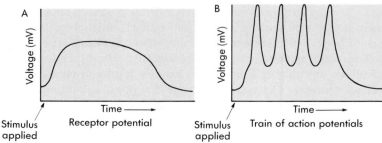

FIGURE 46-4
Events in sensory transduction.
A Depolarization of a free nerve ending leads to a receptor potential that spreads by local current flow to the axon.
B Action potentials that arise in the axon in response to a sufficiently large receptor potential.

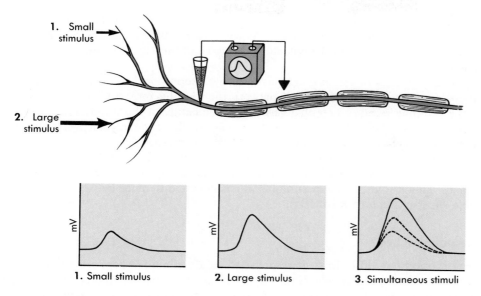

FIGURE 46-5
Spatial summation. Spatial summation occurs in the input segment of a sensory receptor. Two simultaneous stimuli to adjacent regions of the receptor result in a larger receptor potential than either one by itself.

The particular place in a sensory system where a stimulus of adequate intensity can result in a response in a particular afferent neuron is called its **receptive field.** For instance, the endings of a touch receptor in the skin will cover an approximately circular area on the skin, its receptive field. The position of a photoreceptor cell in the retina determines what part of the visual field it will cover. This is the receptive field of the photoreceptor. In the auditory system, receptive fields are sound frequencies (tones). You can think of a set of labeled lines between regions of the central nervous system specialized for processing different modalities (for example, the visual cortex) and the location of the corresponding sensory receptors (on the retina). Afferent nerve fibers enter the central nervous system in a highly orderly way, so that the central regions literally become living **maps** of the location of receptive fields. The nature of the maps differ for the different sensory systems, but each provides the answer to the question WHERE?

Encoding How Much?

The intensity of stimuli (HOW MUCH?) is coded in two ways: (1) as the frequency of action potentials in the afferent nerve fibers from a receptor, and (2) by the number of receptors activated. Receptor potentials vary in size and can *summate* with one another. There are two types of summation: **spatial summation** (Figure 46-5) and **temporal summation** (Figure 46-6). Spatial summation involves the addition of inputs

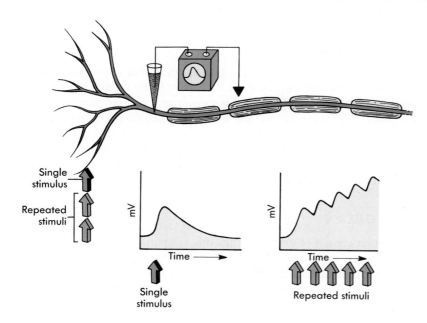

FIGURE 46-6
Temporal summation in the input segment of a sensory receptor. The receptor potential generated at a single input site stimulated once is compared to the receptor potential resulting from repeated stimulation.

arising from more than one site on the receptor ending. Temporal summation is the result of rapidly repeated stimulation. As with spatial summation, the effect is an increase in the magnitude of the receptor potential and thus the frequency of action potentials.

The frequency of action potentials is always proportional to the amplitude of the receptor potential. The relationship between the stimulus intensity and the amplitude of the resulting receptor potential varies with the nature of the stimulus a sensory receptor detects. There are two basic patterns. In the first the receptor potential is proportional to the stimulus intensity, termed **linear receptors**. The second category are **logarithmic receptors**. The logarithmic response pattern of visual and auditory receptors results in action potential frequencies that change by only seven-fold in response to a 10 million-fold change in stimulus intensity (the difference between a barely audible whisper and a rock concert).

Encoding When?

Sensory receptors differ in their response to steady stimuli. There are two classes of receptors. **Tonic**, or **non-adapting**, sensory receptors continues to produce action potentials at the same rate no matter how long the stimulus lasts. They monitor stimuli such as pain, temperature, limb position, blood pressure, and blood oxygen content. The second category gives a burst of action potentials at the onset of stimulation, but as the stimulus continues, the rate of action potentials diminishes and ultimately they cease. These are termed **phasic** or **rapidly adapting** sensory receptors. Phasic receptors are specialized to register novel stimuli or changes in stimulus intensity, and are found in systems in which change is especially significant. Examples are phasic receptors in skin, which respond only to initial contacts, and muscle receptors, which detect the speed of muscle length changes.

One of the functions of the central nervous system is to filter out information that is distracting or unchanging. Some of this filtration occurs in the sensory receptors themselves (phasic touch receptors, for example, cease to react to the clothing you are wearing). In addition, the sensitivity of many receptors is influenced by efferent fibers from the brain. The brain attends selectively to some sensations, especially those that are novel or those that experience has shown to be important. This ability to focus attention is important for higher brain functions such as learning.

> **Sensory information reaches the central nervous system along labeled lines from receptors specific for single sensory modalities. The intensity of stimuli is encoded by the frequency of action potentials in afferent nerves. The initial event in intensity coding is a graded potential change (usually a depolarization), termed a generator or receptor potential, that results from activation of stimulus-sensitive ion channels.**

SENSING TEMPERATURE

There are two populations of nerve endings in the skin that are sensitive to changes in temperature (**thermoreceptors**). One set is stimulated by a lowering of temperature, termed **cold receptors**. The other type responds to a raising of temperature, called **warm receptors**. Thermoreceptors are also found within the hypothalamus, where they monitor the temperature of the circulating blood and thus provide the central nervous system with reliable information on changes in the body's internal (core) temperature. Body temperature regulation will be discussed in Chapter 51 and also in Chapter 52 when we discuss fever and the immune response.

SENSING FORCES

Mechanoreceptors consist of nerve endings that contain specific ion channels sensitive to mechanical force applied to the membrane. These channels open in response to mechanical distortion, initiating a depolarizing receptor potential causing the afferent nerve to fire a series of action potentials.

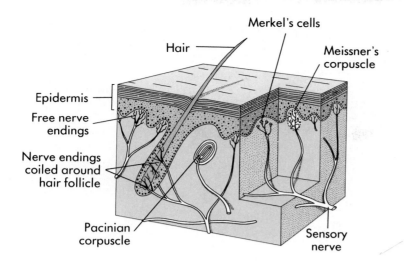

FIGURE 46-7
Sensory receptors in human skin.

Touch

Several types of mechanically-sensitive receptors are present in the skin, some in the **epidermis,** some in the **dermis** and others in the underlying **subcutaneous tissue.** Morphologically specialized receptors respond to what we think of as **fine touch** and are most dense on areas such as the fingertips and face. They are used to localize cutaneous stimuli very precisely and can be phasic or tonic. Phasic receptors include the **hair follicle receptors** and **Meissner's corpuscles.** Meissner's corpuscles are present on the body surfaces that do not contain hair—the fingers, palms, and nipples, for example. There are two types of tonic receptors. **Touch dome endings (Merkel's discs)** are located near the surface of the skin, while **Ruffini endings** are in the dermis (Figure 46-7). Merkel's discs measure the duration of pressure and the extent to which it is applied. Your brain learns to "tune out" signals from many touch receptors after a period of time, and the receptors cease to fire as rapidly. That is why you don't feel the chair you're sitting on.

> Mechanical stimuli initiate nerve impulses by deforming the sensory neuron membrane. Such deformation opens ion channels in the membrane and initiates depolarization.

Vibration

Deep below the skin of vertebrates in the subcutaneous tissue lie vibration-sensitive receptors called **Pacinian corpuscles.** Each Pacinian corpuscle consists of an afferent nerve terminal, surrounded by a capsule made up of alternating layers of connective tissue cells and extracellular fluid (see Figure 46-7). When sustained pressure is applied to the corpuscle, the elastic capsule absorbs much of it. Only the rapid onset and removal of pressure, vibration, causes the nerve ending to fire.

Pain

A stimulus that causes or is about to cause tissue damage is perceived as pain. Such a stimulus elicits reflexive withdrawal of a body segment from a source of stimulus and changes in heartbeat and blood pressure. The receptors that produce these effects are called **nociceptors.** They consist of free nerve endings located within tissue, usually near surfaces where damage is most likely to occur. Different nociceptors have different thresholds. Some respond only to actual tissue damage, and others fire when subjected to pressure or temperature that has not yet damaged tissue.

Pulling

Buried deep within the muscles of all vertebrates, except the bony fishes, are specialized muscle cells called **muscle spindles.** Wrapped around each of these cells is the

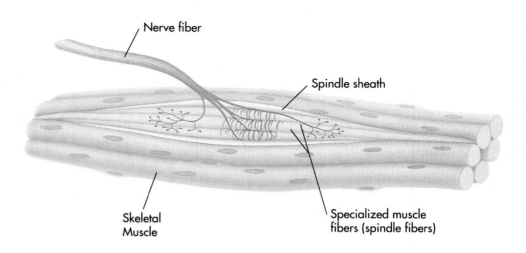

FIGURE 46-8

A stretch receptor embedded within skeletal muscle. Stretching of the muscle elongates the spindle fiber, which deforms the nerve endings, causing them to fire and send a nerve impulse out along the nerve fiber.

Nerve fiber

Spindle sheath

Skeletal Muscle

Specialized muscle fibers (spindle fibers)

end of an afferent neuron. Muscle spindles are **stretch receptors** (Figure 46-8), a type of **proprioreceptor.** When a muscle is stretched, the spindle elongates, stretching the spiral nerve endings in it and repeatedly stimulating the afferent nerve to fire. When the muscle contracts, the tension on the fiber lessens and the stretch receptors cease firing. The frequency of stretch receptor discharges in the afferent nerve fiber measures the muscle length at any given moment. Other stretch receptors, called **Golgi tendon organs,** monitor the tension produced by muscles in the tendons that attach the muscles to the skeleton. The central nervous system uses this information to control total muscle force and movements.

Blood pressure

The brain is supplied with blood by two major arteries, the carotid arteries, which branch high up in the neck at a cleft called the **carotid sinus.** In the carotid sinus, the wall of the artery is thinner than usual. Within this wall is a highly branched network of nerve endings that act as **baroreceptors,** or blood pressure receptors. When the blood pressure increases, the arterial wall balloons out where it is thinnest, in the region of the baroreceptor. This increases the rate of firing of the afferent neuron. A fall in blood pressure causes the wall to move inward and lowers the rate of neuron firing. The frequency of impulses arriving from the baroreceptor afferents provides the central nervous system with a continuous measure of blood pressure.

Gravity

All vertebrates possess gravity receptors known as **statocysts.** To illustrate how they work, imagine a pencil standing in a glass. No matter which way you tip the glass, the pencil will roll along the rim, applying pressure to the lip of the glass. If you wish to know the direction in which the glass is tipped, you need only to inquire where on the rim the pressure is being applied. The body uses receptors of this sort to sense its position in space with gravity as a reference point.

The receptors that obtain the information our brain uses to perceive balance are located in a series of hollow chambers within the inner ear. The gravity receptors are called the **saccule** and the **utricle** (Figure 46-9). In each of these receptors a gelatinous matrix containing crystals of calcium carbonate, called a **statolith,** rests on a bed of ciliated sensory receptor cells, termed **hair cells.** Those cilia directly beneath the statolith are bent by its weight. Each bent cilium exerts pressure on the membrane of the sensory cell from which it arises just as a pencil exerts pressure on the lip of a glass. Any shift in a statolith's position results in different ciliated sensory cells being activated, depending on the direction of the shift. In this way, the brain always knows the orientation of the receptor relative to "up."

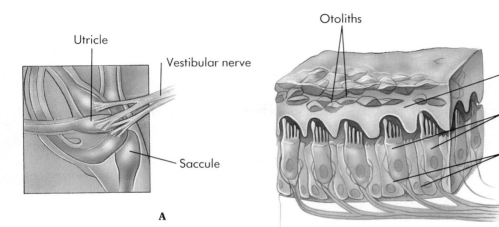

FIGURE 46-9

The structure of the macula.
A The position of the macula in relation to the cochlea and semicircular canals.
B Enlargement of a section of the macula showing the otoliths imbedded in the gelatinous matrix that covers the hair cell receptors.

Equilibrium

Vertebrates sense motion in a way similar to the way in which they detect vertical position. In this case, however, the sensory receptor is designed so that the fluid movements produced by a motion deflect the cilia on sensory cells in a direction opposite to that of the motion. Within the inner ear, past the saccule and utricle, are three fluid-filled **semicircular canals** (Figure 46-10). Each canal is oriented in a different plane at right angles to the other two, so that motion in any direction can be detected. Protruding into the canals in the **ampulla** are sensory hair cells, which are connected to afferent nerves. From the ends of these sensory cells extend a series of short cilia, with one long cilium located at one side of each cilia bundle.

The pressure of the long cilium against the shorter ones initiates the depolarization of the cell membrane, which triggers a nerve impulse. The cilia protrude into a gelatinous material that occludes the semicircular canal, called the **cupula** (the Latin word for a little cup). When the cupula is bent by the movement of liquid within the semicircular canal, the cilia bend in the direction of the fluid movement. Because of inertia, the direction of their bending is opposite to the direction in which the body is moving. When the cilia bend, they either increase or decrease the frequency of nerve firing. Because the three canals are oriented in all three planes, movement in any plane is sensed by at least one of the canals. Complex movements are analyzed by comparing the sensory input from each canal.

Lateral Line Organs

Fishes perceive their movement relative to the water surrounding them in the same way that you perceive motion when you shake your head. Fishes having cupula receptors similar to those in your semicircular canals. In fishes, however, such sensory cells

FIGURE 46-10
Equilibrium.
A The position of the semicircular canals in relation to the rest of the inner ear.
B Greater enlargement of a section of the ampulla showing how hair cell stereocilia insert into the cupula.

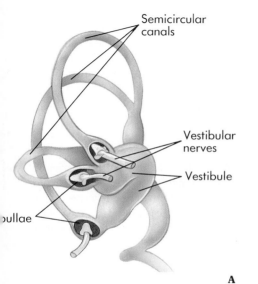

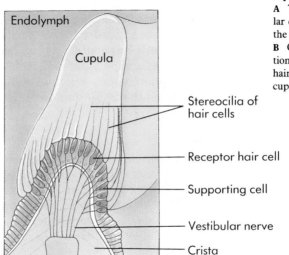

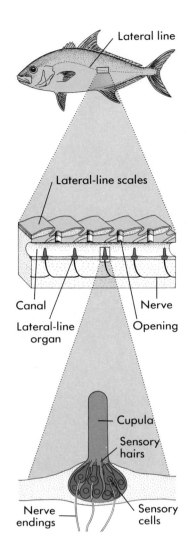

Lateral line

Lateral-line scales

Canal

Lateral-line
organ

Nerve

Opening

Cupula

Sensory
hairs

Nerve
endings

Sensory
cells

FIGURE 46-11

The lateral line system. The lateral line is a system of motion-sensing receptors very similar in structure to the cupulae of the semicircular canals that humans use to sense motion. A series of these receptors project into a canal beneath the surface of the skin. The canal is open to the exterior; it runs the length of the fish's body. Movement of water past the fish forces water through the canal, deflecting the cupulae and initiating nerve impulses.

are located on the surface of the body. The cupulae of fishes extend out into grooves that run laterally along the head and sides of the body and make up the **lateral-line** organs discussed in Chapter 42 (Figure 46-11). Within the lateral-line organs of a fish, the cilia of adjacent cells are oriented so that some are stimulated by movement of water in one direction and others by movement in the opposite direction. Nerve impulses from the lateral-line receptors permit the fish to assess its rate of movement through water. Lateral-line organs sense movement of the water around the fish's body as pressure waves against its lateral line. This is how a trout orients with its head upstream.

Lateral-line organs enable a fish to detect motionless objects at a distance by the movement of water reflected off the objects. In a very real sense, this form of perception is the fish's equivalent of hearing. It does not differ in its mechanism from what happens within your ears when you listen to a symphony, a pattern of pressure waves in the air around you. To hear, terrestrial vertebrates use hair cell receptors within the ear that are thought to have evolved through the modification of lateral-line receptors.

Complex mechanical receptors can respond to pressure, to gravity, or to angular acceleration. In each case, the receptors employ mechanical devices such as weights or levers to convert the information to a mechanical stimulus, which deforms and so depolarizes a sensory membrane.

SENSING CHEMICALS

Embedded within the membranes of afferent nerve endings or of sensory cells associated with afferent neurons are specific chemical receptors that induce depolarization when they are bound by particular molecules. There are two types of chemical sensory systems that use different receptors and process information at different locations in the brain: (1) **taste,** in which the receptors are specialized sensory cells, and (2) **smell,** or **olfaction,** where the receptors are neurons.

Taste

Taste and smell receptors of fishes are the most sensitive vertebrate chemoreceptors known. The taste receptors, or **taste buds,** are not located in the mouth like those of terrestrial vertebrates but rather are scattered over the surface of the fish's body. These taste buds are exquisitely sensitive to amino acids. A catfish, for example, can distinguish between two different amino acids at a concentration of less than 100 micrograms per liter of water (that's 1 gram per 10,000 liters!). The ability to taste the surrounding water in this way is very important to bottom-feeding fishes, enabling them to sense the presence of food in an often murky environment.

One group of chemoreceptors in human beings is also composed of specialized taste buds, although a human's taste buds are not nearly as sensitive. In all terrestrial vertebrates, the taste buds are located in the mouth (Figure 46-12). Each taste bud is associated with an afferent neuron. Humans have four kinds of taste buds, each of which responds to a broad range of chemicals. The stimuli to which the different kinds of taste buds respond are **salty, sweet, sour,** and **bitter.** Our complex perception of taste is composed of different combinations of impulses from these four kinds. The chemoreceptors are concentrated on different parts of the tongue, with sweet and salty on the front, sour on the sides, and bitter at the back.

Smell

The chemoreceptors that human beings use to smell are located in the upper portion of the nasal passage (Figure 46-13). These olfactory chemoreceptors are neurons whose cell bodies are embedded in the nasal epithelium. From the cell bodies, dendrites extend to the surface of the epithelium and project sensory cilia into the surface mucous layer. A terrestrial vertebrate uses the sense of smell in much the same way that a fish uses the sense of taste—to sense the chemical environment around it. Be-

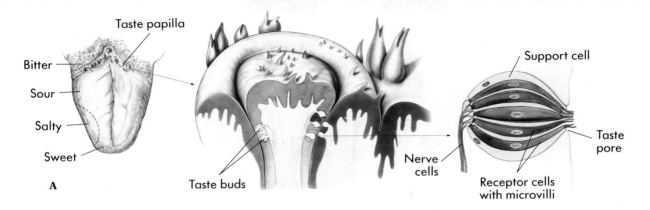

Taste papilla

Bitter
Sour
Salty
Sweet

A

Support cell

Nerve cells

Taste pore

Receptor cells with microvilli

Taste buds

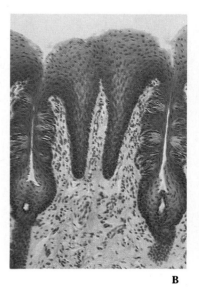

B

cause terrestrial vertebrates live surrounded by air rather than water, their sense of smell has become specialized to detect airborne particles.

External chemical stimuli are sensed by specific receptors that are components of sensory neuron membranes (smell) or of the membranes of associated receptor cells (taste). These receptors depolarize in response to the binding of specific chemicals.

The information that chemoreceptors provide to the central nervous system is sketchy at best. The stimulus arrives slowly, and the medium through which the chemical must diffuse is itself moving constantly. Chemoreception functions primarily to alert the central nervous system to the presence of a particular chemical. Some specific chemical cues, the pheromones that we discussed in Chapter 41, are detectable by chemoreceptors in trace amounts. They are employed by animals to attract potential mates, for example.

Blood Chemistry

Embedded within the walls of your arteries, at several locations in the circulatory system, are receptors called **carotid bodies.** The carotid bodies provide one of the sensory inputs that your body uses to regulate its rate of respiration. When oxygen levels fall below normal limits (**hypoxia**) this information is conveyed by these **peripheral chemoreceptors** to the central nervous system, which reacts by increasing the respiration rate. The nerve endings sense the level of oxygen in the arterial blood by changes in the rate of oxygen diffusion into the neuron. The peripheral chemoreceptors also detect the pH (acidity) of the plasma. A second set of chemoreceptors, called **central chemoreceptors,** are located within the brain and detect CO_2. You will learn more about these chemoreceptors in Chapters 50 and 51.

FIGURE 46-12
Taste.
A Human beings have four kinds of taste buds (bitter, sour, salty, sweet), located on different regions of the tongue. Groups of taste buds are typically organized in sensory projections called papillae.
B Individual taste buds are bulb-shaped collections of chemical receptor cells that open out into the mouth through a pore.

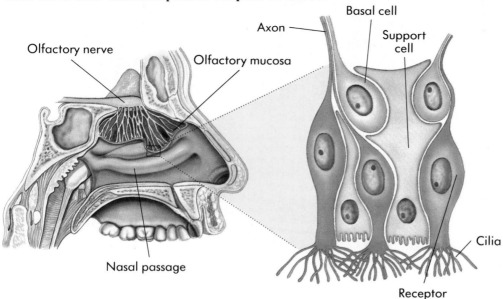

Olfactory nerve

Axon

Basal cell

Support cell

Olfactory mucosa

Nasal passage

Cilia

Receptor cell

FIGURE 46-13
Smell. Humans smell by using olfactory receptor cells located in the lining of the nasal passage. The receptor cells are neurons. Axons from these sensory neurons project back through the olfactory nerve directly to the brain.

951

tors also detect the pH (acidity) of the plasma. A second set of chemoreceptors, called **central chemoreceptors,** are located within the brain and detect CO_2. You will learn more about these chemoreceptors in Chapters 50 and 51.

HEARING

Just as fishes detect vibration in water by means of their lateral-line receptors, so terrestrial vertebrates detect vibration in air by means of mechanical receptors within the ear. These receptors, as we noted earlier, evolved from lateral-line organs. Hearing actually works less well in air than in water because water transmits pressure waves better. Despite this, hearing is widely used by terrestrial vertebrates to monitor their environments and particularly to detect possible sources of danger. Auditory stimuli travel farther and more quickly than chemical ones, and auditory receptors provide better directional information than do chemoreceptors. Auditory stimuli alone, however, provide little information about distance.

Structure of the Ear

The evolutionary modification of the lateral-line receptors of fishes into a terrestrial vertebrate hearing organ that functions in air, the **ear,** has involved ingenious solutions to a serious mechanical problem. Sound waves are not easily transmitted from air to water, so to use the same receptor, it is necessary to amplify the sound. The ears of terrestrial vertebrates achieve this amplification in two ways.

In the ears of mammals, sound waves beat against a large membrane of the outer ear called the **tympanic membrane,** or **eardrum** (Figure 46-14), causing corresponding vibrations in three small bones, called **ossicles:** the hammer, anvil, and stirrup. These bones act as a lever system, increasing the force of the vibration. The third in line of these levers, the stirrup, pushes against another membrane, the oval window. Because the oval window is smaller than the tympanic membrane, vibration against it produces more force per unit area. The oval window is the door to the inner ear, where hearing actually takes place. The fluid-filled chamber of the inner ear is shaped like a tightly coiled snail shell and is called the **cochlea** from the Latin name for snail.

The chamber in which all of these events occur is called the **middle ear.** It is connected to the throat by a tube called the **Eustachian tube** in such a way that there is no difference in air pressure between the middle ear and the outer ear. The familiar "ear popping" that is associated with the landing in an aircraft or with the rapid de-

FIGURE 46-14

Structure of the human ear. Sound waves passing through the ear canal beat on the tympanic membrane, pushing a set of three small bones, or ossicles (hammer, anvil, and stirrup) against an inner membrane called the oval window. This sets up a wave motion in the fluid filling the cochlea that travels through the vestibular and tympanic canals. Where the sound wave beats against the sides of the canals, the tectorial membrane is pushed against the basilar membrane, bending hair cells and firing associated neurons.

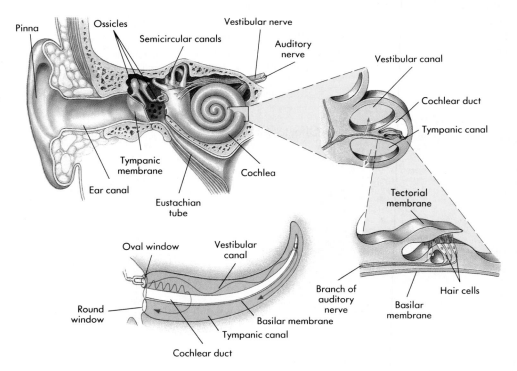

scent of an elevator in a tall building is a result of pressure equalization between these two sides of the eardrum.

The Auditory Receptors

The auditory receptors, the same hair cells we previously encountered as lateral-line receptors in fishes, are located on the **basilar membrane,** which bisects the cochlea. The cilia of these hair cells project into a covering gelatinous structure, the **tectorial membrane.** This arrangement, in which the receptors are sandwiched between two membranes, is sometimes called the **organ of Corti.** Bending of the basilar membrane as it vibrates causes the hairs of the receptor cells pressed against the tectorial membrane to bend. Deflections of the cilia in one direction depolarizes hair cells, while movements in the other hyperpolarize them. Like presynaptic nerve terminals, hair cells contain transmitter-filled vesicles. Depolarization increases transmitter release and elevates afferent activity, whereas hyperpolarization inhibits release, decreasing afferent activity. Hair cells are alternately excited and inhibited in proportion to the vertical displacement of the basilar membrane.

Frequency Localization in the Cochlea

The ability to analyze the frequency components of sounds depends on the principle of **resonance.** A tuning fork vibrates at a characteristic **resonant frequency,** as do the strings of musical instruments. In stringed musical instruments, the resonant frequency of a string depends on the length of the string and its tautness. A string on a cello or bass has a lower frequency than a string on a violin, but a musician can raise the frequency of the note played by a string by tightening it or by using the fingerboard to shorten the portion of the string free to vibrate. Induction of vibration in one object by the vibration of another is called **sympathetic resonance.** Sympathetic resonance occurs if a violin note is played near a piano. Thus, a middle C (512 Hz) on a violin causes the corresponding piano string to vibrate.

The basilar membrane consists of elastic fibers of varying length and stiffness imbedded in a jellylike material. It resembles a piano into which someone has poured a large quantity of jello. At the base of the cochlea, the fibers of the basilar membrane are short and stiff. At the apex of the cochlea the fibers are five times longer (the basilar membrane is wider) and 100 times more flexible. Like a piano, the resonant frequency of the basilar membrane is highest at the base and lowest at the apex.

Because the elastic fibers in the basilar membrane are imbedded in a gelatinous matrix, they are loosely coupled to one another. When a wave of sound energy enters the cochlea from the oval window, it initiates a traveling up-and-down motion of the basilar membrane. However, this wave imparts most of its energy to that part of the

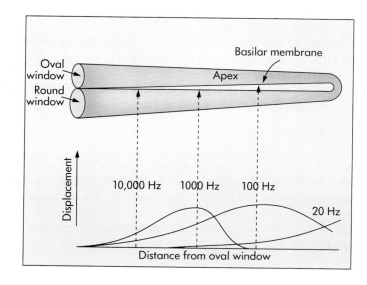

FIGURE 46-15

Frequency localization in the cochlea. The cochlea is shown unwound at the top, while the resonance of the basilar membrane in response to different sound frequencies (tones) is illustrated below. The parts of the basilar membrane nearest the oval window resonate preferentially to sounds of high frequency; at increasing distances from the oval widow progressively lower resonant frequencies are encountered.

basilar membrane that has resonant frequencies near the particular frequency of the sound wave, resulting in a maximum deflection of the basilar membrane at this point (Figure 46-15).

Changes in Auditory Sensitivity

Hair cells are innervated by efferent fibers from the central nervous system. Increases in activity in the efferent fibers can make hair cells less sensitive. This example of central control of receptor sensitivity has been demonstrated to increase the ability of individuals to concentrate on one part of the auditory range (for example, a conversation) in the midst of background noise, which is effectively "tuned out" by the efferent control. Our ability to hear depends on the flexibility of the basilar membrane, a flexibility that changes as we grow older. We are not able to hear low-pitched sounds, below 20 vibrations, or cycles, per second, although some other vertebrates can. As children, human beings can hear high-pitched sounds, up to 20,000 cycles per second, but this ability decays progressively throughout middle age. Other vertebrates can hear sounds at far higher frequencies that these. Dogs, for example, readily detect sounds at 40,000 cycles per second. Thus dogs can hear a high-pitched dog whistle when it seems silent to a human listener.

> **The tympanic membrane and middle ear bones transmit sound energy to the cochlea. The sensitivity of this transmission is regulated by a small muscle in the middle ear. The resonance properties of the cochlea translate sound energy into displacement in such a way that specific hair cells are stimulated according to the frequency of the sound.**

SONAR

Since there are two ears on opposite sides of the head, the information provided by hearing can be used by the central nervous system to determine direction with some precision. Sound sources vary in strength, however, and sounds are attenuated (weakened) to various degrees by the presence of objects in the environment. For these reasons, auditory sensors do not provide a reliable measure of distance.

A few groups of mammals that live and obtain their food in dark environments have circumvented the limitations of darkness. A bat flying in a completely dark room easily avoids objects that are placed in its path—even a wire less than a millimeter in diameter (Figure 46-16). Shrews use a similar form of "lightless vision" beneath the ground, as do whales and dolphins beneath the sea. All of these mammals perceive distance and depth by means of **sonar.** They emit sounds and then determine the time that it takes these sounds to reach an object and return to the animal. A bat, for example, emits clicks that last from 2 to 3 milliseconds and are repeated several hundred times per second. The three-dimensional imaging achieved with such an auditory sonar system is quite sophisticated.

Being able to "see in the dark" has opened a new ecological niche to bats, one largely closed to birds because birds must rely on vision. There are no truly nocturnal birds. Even owls rely on vision to hunt, and do not fly on dark nights. Because bats have solved the problem of being active and efficient in the dark, they are one of the most numerous and widespread of all orders of mammals.

VISION

Because light travels in a straight line and arrives virtually instantaneously, visual information can be used to determine both the direction and the distance of an object. No other stimulus provides as much detailed information. All of the receptors that we have described up to this point have been either chemical or mechanical ones. None of them respond to the electromagnetic energy provided by photons of light. Vision, the perception of light, is carried out in vertebrates by a specialized sensory apparatus called an **eye.**

FIGURE 46-16
Sonar. This bat is emitting high-frequency "chirps" as it flies. The bat listens for the sound's reflection against the moth and by timing how long it takes for a sound to return, it can effectively catch its prey even in total darkness.

In considering the evolution of major groups of organisms, we sometimes find that the differences among them are so great that it is difficult to trace the lines of descent. Especially complex features, such as the eye, appear so unlike any possible earlier structure from which they may have been derived that the question of their origin greatly puzzled early students of evolution.

When we consider eyes in more detail, however, we find that they have in fact evolved many times, as shown by the fundamental differences between the eyes found in different groups of organisms. Some eyes consist of a single photoreceptor cell, and others are complex, like those of the vertebrates, with focusing lenses and color sensitivity. Zoologists, analyzing these differences, have calculated that eyes have evolved independently, and through intermediate stages, at least 38 different times in various groups of animals (Figure 46-A).

In other words, not all of the structures we call eyes are homologous with one another. By analyzing and understanding their similarities and differences, biologists can more precisely interpret the relationships among the group in which they occur and more accurately understand the outlines of their evolution. Thus the difficult problem of how eyes evolved, when it is solved, assists us in interpreting broad evolutionary patterns of great interest.

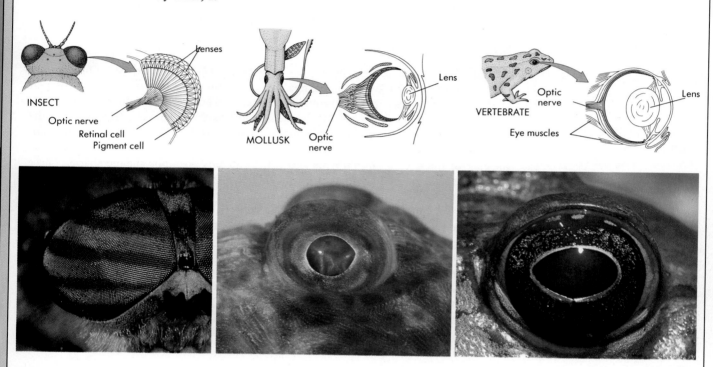

FIGURE 46-A
Although they are superficially similar, the eyes that occur in three different phyla of animals differ greatly in structure and are not homologous with one another. Each has evolved separately and, despite the apparent structural complexity, has done so from simpler structures.

The Evolution of the Eye

The evolution of photoreceptors like those involved in the human visual system is only one step in the development of a true visual sense. Less advanced animals, whose photoreceptors are clustered together in an **eyespot,** can perceive light but cannot "see." The eyespot, however, can be used to perceive the direction from which light is arriving. True image-forming eyes probably evolved from such comparatively simple structures. Eyes may have evolved independently numerous times among different groups of animals. The members of four phyla—annelids, mollusks, arthropods, and vertebrates—have each evolved well-developed image-forming eyes. Interestingly, all of them use the same visual pigment, suggesting that not many alternative pigments are able to play this role.

FIGURE 46-17

Structure of the human eye.
Light passes through the transparent cornea and is focused by the lens on the rear surface of the eye, the retina, at a particular location called the fovea. The retina is rich in rods and cones.

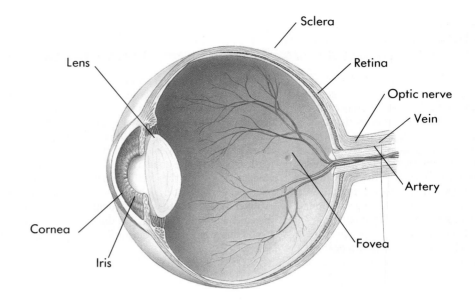

FIGURE 46-18

Focusing the human eye. Contraction of ciliary muscles pulls on suspensor ligaments and changes the shape of the lens, which alters its point of focus forward or backward. In nearsighted people the ciliary muscles place the point of focus in front of the fovea rather than on it. The problem can be corrected with glasses or contact lenses, which extend the focal point back to where it should be. In farsighted people, the ciliary muscles make the opposite error, placing the point of focus behind the retina. Corrective lenses can shorten the focal point.

Structure of the Vertebrate Eye

Vertebrate eyes are lens-focused eyes (Figure 46-17). In them, light first passes through a transparent layer, the **cornea,** which begins to focus the light onto the rear of the eye. Light then passes through the *lens,* a structure that completes the focusing. The lens is a fat disk, somewhat resembling a flattened balloon. In mammals, the lens is attached by suspending ligaments to **ciliary muscles.** When these muscles contract, they change the shape of the lens (Figure 46-18) and thus the point of focus on the rear of the eye. In amphibians and fishes the lens does not change shape. These animal instead focus their images by moving the lens in and out, thus operating in exactly the same way that a camera does. In all vertebrates, the amount of light entering the eye is controlled by a shutter, called the **iris,** between the cornea and the lens. The iris reduces the size of the transparent zone, or **pupil,** of the eye through which the light passes.

An array of receptor cells, the **rods** and **cones,** lines the back of the eye. Along with several types of interneurons, the photoreceptors form the **retina** (Figure 46-19). The retina of humans contains about 3 million cones, most of them located in the central region of the retina called the **fovea,** and approximately 1 billion rods. Rod cells are responsible for black-and-white vision, and cone cells function in color vision. The eye forms its sharpest image in the central fovea region of the retina, a region composed almost entirely of cone cells.

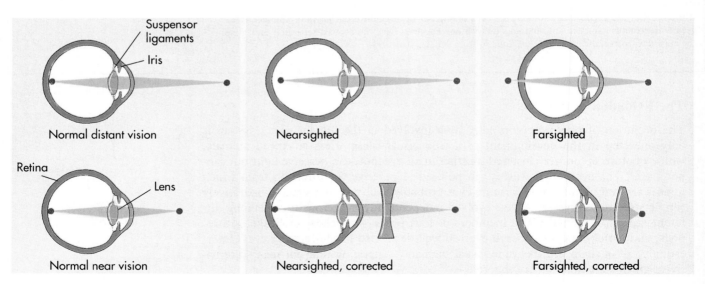

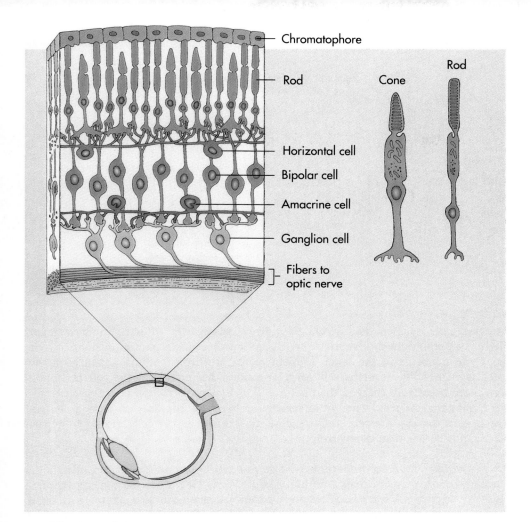

Chromatophore

Rod

Horizontal cell

Bipolar cell

Amacrine cell

Ganglion cell

Fibers to optic nerve

Cone

Rod

FIGURE 46-19
Structure of the retina. Note that the rods and cones are at the rear of the retina, not the front. Light passes through several layers of ganglion and bipolar cells before it reaches the rods and cones.

The lens of the vertebrate eye is constructed to filter out short-wavelength light. This solves a difficult optical problem: any uniform lens refracts short wavelengths more than it does longer ones, a phenomenon known as chromatic aberration. Consequently, these short wavelengths cannot be brought into focus simultaneously with longer wavelengths. Unable to focus the short wavelengths, the vertebrate eye eliminates them. Insects, whose eyes do not focus light, perceive these lower, ultraviolet wavelengths quite well and often use them to locate food or mates.

Photoreceptors

Photoreceptors detect single photons of light. How does a photon-detecting sensory receptor work? The primary event of vision is the absorption of a photon of light by a pigment. The visual pigment is called **cis-retinal** (Figure 46-20), the cleavage product of carotene, a photosynthetic pigment in plants. In photosynthesis, a pigment needs only to be able to absorb light and to donate an electron. In vision, on the other hand, the visual pigment must undergo a conformational change large enough to alter the shape of the proteins associated with it. In vertebrates, the cis-retinal pigment is coupled to a protein called **opsin** to form **rhodopsin**. In vertebrate eyes, the visual pigment is located in the tips of rod and cone cells (Figure 46-21).

Rod Cells. Rod cells are composed of two basic elements: (1) an **outer segment** densely packed with rhodopsin on the surface of about 1000 flattened disks, stacked one on top of the other, and (2) an **inner segment** rich in mitochondria and connected to the outer one only by a narrow passageway. The inner segments contain numerous vesicles filled with neurotransmitters.

The membrane of the outer segment of a rod cell has many open sodium channels. Because there is more sodium outside the cell than within it, sodium ions flow

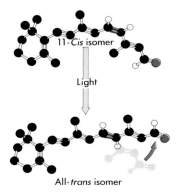

11-*Cis* isomer

Light

All-*trans* isomer

FIGURE 46-20
Absorption of light. When light is absorbed by the 11-*cis* isomer of the visual pigment retinal, the pigment undergoes a change in shape: the linear end of the molecule (right end) rotates about a double bond, indicated here in color. The new isomer is referred to as all-*trans* retinal. This change in the pigment's shape induces in turn a change in the shape of the protein opsin to which the pigment is bound, initiating a chain of events that leads to the generation of a nerve impulse.

FIGURE 46-21

Rods and cones. The broad tubular cell diagrammed on the right is a rod, the shorter tapered cell next to it a cone. Although not obvious from the electron micrograph, the pigment-containing outer segments of these cells are separated from the rest of the cells by a partition through which there is only a narrow passage, the "connecting column."

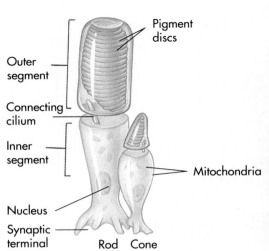
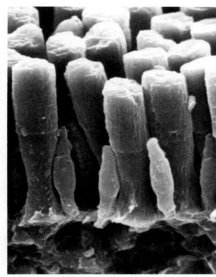

into this outer segment and across the narrow passageway into the inner segment where they continuously depolarize it. Reception of a single photon of light by one rhodopsin molecule in the outer segment closes numerous sodium channels in the outer segment. The result is a hyperpolarization, the interior of the rod becoming even more negatively charged than before. This hyperpolarization is conveyed to bipolar and ganglion cells, where it alters the frequency of impulses in the optic nerve. The stimulation of a rod cell, unlike that of most other sensory cells, results in hyperpolarization rather than depolarization.

Amplification in Photoreceptors.

The rod cell uses a single photon of light to block the entry of more than a million sodium ions. How does the pigment accomplish this prodigious feat? Each rhodopsin complex that absorbs a photon of light is changed in shape, being transformed into an enzyme that activates several hundred molecules of a protein called transducin. Each of these activated transducin molecules activates several hundred molecules of another enzyme, called phosphodiesterase, and each phosphodiesterase molecule when activated generates cyclic guanine monophosphate (cGMP) molecules that close sodium channels at a rate of about 1000 per second.

> **The photoreceptors of rod cells thus greatly amplify the original sensory stimulus by employing a cascade of interactions. Each photon that is absorbed by rhodopsin leads to the hydrolysis of more than 105 molecules of cyclic GMP!**

Cone Cells and Color Vision.

Color vision is achieved by cone cells in a way that is similar to the action of rod cells. Cone cells have the same general internal architecture as rod cells and are stimulated by light in the same way. Cone cells are shorter than rod cells (see Figure 46-21) and, as their name suggests, are shaped like inverted ice cream cones. There are three kinds of cone cells, each of which possesses an opsin molecule with a distinctive amino acid sequence and thus a different shape. These differences shift the absorption maximum of cis-retinal from the 500 nanometers characteristic of rod cells to 455 nanometers (blue-absorbing), 530 nanometers (green-absorbing), or 625 nanometers (red-absorbing), as shown in Figure 46-22.

Organization of the Retina

Each foveal cone cell makes a one-to-one connection with a special kind of neuron called a **bipolar cell.** Each of the bipolar cells is connected in turn to an individual visual **ganglion cell,** whose axon is part of the optic nerve. The bipolar cells receive the hyperpolarizing stimulus from the cone cells, but transmit a depolarization stimulus to the ganglion cell. The optic nerve transmits visual impulses directly to the

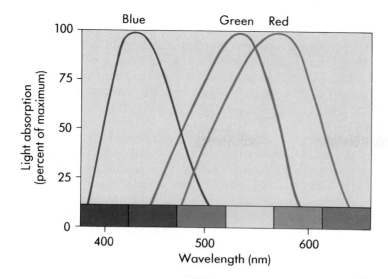

brain. The frequency of pulses transmitted by any one receptor provides information about light intensity. The pattern of firing among the different foveal cones provides a point-to-point image. Different cone cells provide color information.

The relationship of receptors to bipolar cells to ganglion cells is 1:1:1 within the fovea. Outside the fovea, the output of many receptor cells is channeled to single bipolar cells, many of which converge in turn on individual ganglion cells. In peripheral (outer edge) portions of the retina, more than 125 receptor cells feed stimuli to each ganglion axon in the optic nerve. In this peripheral region, many additional neurons cross-connect the ganglion cells with one another and carry out extensive processing of visual information. Thus peripheral vision is less acute. It has been said that we use the periphery of the eye as a detector and the fovea as an inspector.

Binocular Vision

Most vertebrates have two eyes, one located on each side of the head. When both of them are trained on the same object, the image that each sees is slightly different because each eye views the object from a different angle. This slight displacement of images, an affect called parallax, permits sensitive depth perception. By comparing the differences between the images provided by the two eyes with the physical distance to particular objects, vertebrates learn to interpret different degrees of disparity between the two images as representing different distances. This is called **stereoscopic vision.** We are not born with the ability to perceive distance; we learn it. Most predators have their eyes facing forward to maximize the field of overlap in which stereoscopic vision occurs (Figure 46-23). Prey by contrast tend to have eyes located to the sides of the head enlarging the receptive field.

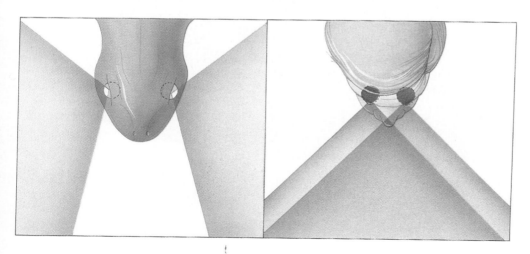

FIGURE 46-22
Color vision. The absorption spectrum of *cis*-retinal is shifted in cone cells from the 500 nanometers characteristic of rod cells. The amount of the shift determines what color the cone absorbs: a shift down to 455 nanometers yields blue absorption, a shift up to 530 nanometers yields green absorption, and a shift farther up to 625 nanometers yields red absorption. There is no reason in principle why additional kinds of cones could not be generated by other alterations in *cis*-retinal that absorb light of other wavelengths, such as yellow, but they are not needed. By comparing the relative intensities of the signals from the three cones, the brain can calculate the intensity of yellow light.

FIGURE 46-23
Stereoscopic vision. Vertebrates have eyes located toward the front of the head, so that the two fields of vision overlap. When eyes are located on the sides of the head, the two vision fields do not overlap and stereoscopic vision does not occur.

Other Environmental Senses in Vertebrates

Vision is the primary sense used by all vertebrates that live in a light-filled environment, but the wavelength and intensity of visible light are by no means the only stimuli that are available to vertebrates for assessing this environment. We will now consider some examples of other ways in which vertebrates function in this area of sensory perception.

Heat. Electromagnetic radiation with wavelengths longer than those of visible light is not detected by visual receptors because such radiation has low quantum energy. Radiation from this far red portion of the spectrum is what we normally think of as radiant heat. Heat is an extremely poor environmental stimulus in water because the water has a high thermal capacity and readily absorbs heat, causing any thermal stimulus that is introduced into a body of water to decay very rapidly. Air, in contrast to water, has a low thermal capacity, and heat in air is therefore a potentially useful environmental stimulus.

The potential of using heat as a source of information about the environment, however, has not been tapped by most groups of terrestrial vertebrates. Snakes are an exception. The pit vipers, for example, possess a pair of heat-detecting **pit organs,** one located on each side of the head between the eye and the nostril (Figure 46-24). A snake that has such organs can locate and strike a motionless warm animal in total darkness by perceiving the thermal radiation emanating from the body of the prey. A pit organ has an outer and an inner chamber, separated by a membrane. The organ apparently operates by comparing the temperatures of the two chambers. The nature of the pit organ's thermal receptor is not known; it probably consists of temperature-sensitive receptors in the neurons innervating the two chambers. The two pit organs appear to provide stereoscopic information, in much the same way that two eyes do. Indeed, the information that is transmitted from the pit organs is processed by the visual center of the snake brain.

Electricity. Air does not readily conduct an electric current, and electricity is thus a poor environmental stimulus for terrestrial animals. Water, on the other hand, conducts electricity readily. A number of different groups of fishes apparently use weak electrical discharges to locate their prey, constructing a three-dimensional image of their environment even in murky water. Although electricity is inferior to light as a stimulus—it has a limited range—it can provide these fishes with very detailed information.

The electrical discharges in these fishes are produced by special organs that consist of modified muscle. We can get some idea of how these organs work by studying an unusual adaptation: in certain fishes, such as electric eels and some rays, electric discharges are used to create a shock. In these animals, each electrical organ is com-

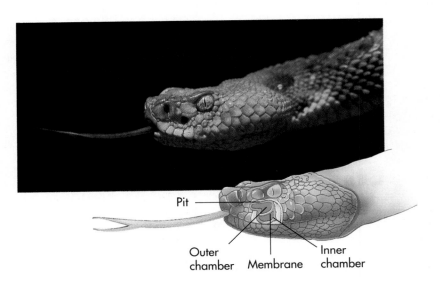

FIGURE 46-24

"Seeing" heat. The depression between the nostril and the eye of this rattlesnake opens into the pit organ. In the cutaway portion of the diagram you can see that the organ is composed of two chambers separated by a membrane.

A duckbilled platypus is able to hunt by detecting the tiny electrical signals—a few hundred microvolts—passing between nerves and muscles in the tail of a fish or shrimp. Platypuses hunt at night, when vision is of little use. When hungry, they circle the bottom of a pool, sweeping their bills from side to side. What they are looking for is electrical activity. In the front of their bills are tiny electroreceptors, nerve endings of filaments that derive from the fibers that form the trigeminal nerve; this nerve also carries information about touch between the bill and the brain. Experimental stimulation of a platypus' bill with a current as small as 300 microvolts produces a focus of activity in the cortex of the brain, next to the spot excited by touching the bill. Using these receptors, a platypus can detect the activity of a prey individual's nervous system. It is the electricity itself the platypus

detects, not a smell or other clue. If you hide a tiny pen flashlight battery in their pool along with a shrimp, they home in on the battery, not the shrimp.

While a variety of taxonomically primitive fish, like the catfish, and several amphibians are known to be able to detect electrical signals, the "electricity detectors" of the platypus are unlike those of any other animal, suggesting that the idiosyncratic platypus evolved its electrical sense quite independently.

posed of numerous columns of disk-shaped cells stacked one on top of the other. Such a column may contain from 5000 to 10,000 of these **electroplates,** and there may be more than 50 columns. One surface of each disk has nerve endings; the other does not. At rest, the individual disks carry a positive charge on both their upper and lower surfaces, actively transporting positive ions out from the cell. The electrical discharge is initiated by the neurons, which depolarize the surface with which they synapse. As a result of this discharge, there is a transient voltage difference across each disk amounting to about 150 millivolts. Because the electroplates are arranged in series, the amount of voltage adds up. An electric eel can produce 500 volts in a single discharge!

Most electric fishes use far weaker discharges to survey their surroundings. The electric catfish (Figure 46-25), for example, discharges electricity continuously—some 300 discharges per second—producing an electrical current in the water around it. The presence of any object with higher or lower conductivity distorts the field by changing the lines of flow of the current. The receptors that detect the changes in current, such as the ones called the **ampullae of Lorenzini,** are very sensitive, being able to respond to extremely weak currents. These receptors are located in chambers that lie beneath the surface of the skin, which are connected to the surrounding water by long canals.

FIGURE 46-25
Using electricity to "see." The electric catfish, *Malapterus electricus,* surveys its surroundings by producing a continuous electrical current in the water around it and monitoring the lines of current flow in the water. It can detect objects that disturb the lines of current flow, and so "see" in very murky water.

Magnetism. Eels, sharks, and many birds appear to navigate along the magnetic fields of the earth. Even some bacteria use such forces to orient themselves (see Figure 30-16). Birds that are kept in blind cages, with no visual cues to guide them, will peck and attempt to move in the direction in which they would normally migrate. They will not do so, however, if the cage is shielded from magnetic fields by steel. Indeed, if the magnetic field of a blind cage is deflected 120 degrees clockwise by an artificial magnet, a bird that normally orients to the north will orient east-southeast.

There have been many speculations about the nature of the magnetic receptor in these vertebrates, but it is still very poorly understood.

AN OVERVIEW OF SENSORY SYSTEMS

The sensory systems of vertebrates inform the brain about the condition of the body and also provide it with a detailed picture of its surroundings. As we have seen, sensory systems have evolved that utilize a broad variety of cues. Most of the sensory systems of vertebrates evolved early, while vertebrates were still confined to the sea. Some, such as sensors that detect electrical fields, are not used by land vertebrates, presumably because air is a very poor conductor of electrical currents. Others, such as sensors that respond to liquid pressure waves on the lateral line, are used by land vertebrates to detect pressure waves in air—that is, to hear. Because the terrestrial environment is so physically different from the sea, it has presented some new sensory opportunities. Air, for example, transmits heat much better than water does, and some terrestrial vertebrates have evolved heat sensors that provide quite detailed three-dimensional images.

No one vertebrate has a perfect repertoire of sensory systems able to finely discriminate all potential cues. Rather, vertebrates have evolved sensory systems that meet particular evolutionary challenges. Thus some hunting mammals such as dogs have evolved highly capable olfactory systems, but primates have not. It is of some interest to note, for example, the existence of detailed sensory information in our own environment that we could potentially use if we possessed the proper sensory receptors. We cannot use thermal radiation, as snakes do, or ultrasound, as bats do, to construct a three-dimensional image. Only by using a computer to convert these sensory cues to differences in light emission are we able to construct a visual analogue.

SUMMARY

1. Sensory receptors respond to three classes of stimuli: mechanical, chemical, and electromagnetic. No matter what type of stimulus is involved, a response usually takes the same form—the depolarization of a sensory neuron membrane.

2. Many of the body's internal receptors are simple receptors in which a nerve ending becomes depolarized as response to a chemically induced or temperature-induced opening of ion channels in the nerve membrane.

3. Muscle contraction, blood pressure, and touch are sensed by simple mechanical receptors, in which deformation of a nerve membrane by stretching or distortion opens ion channels and thus initiates depolarization.

4. Gravity, equilibrium, and lateral-line hearing are sensed by complex mechanical receptors, which use a mechanical device (a stone or a lever) to convert the sensory information into an electrical stimulus.

5. The receptors that vertebrates use to sense their external environments all evolved in water, with the exception of the heat vision that pit vipers possess.

6. The most sensitive chemical receptors are the taste and smell receptors of fishes, which can discriminate the presence of a few molecules of specific substances. Human taste and smell receptors are much less acute.

7. Hearing in terrestrial vertebrates uses an evolutionary modification of the fish lateral-line receptor, in which the airborne sound waves are amplified and directed at a fluid-containing chamber within the ear. Mechanical receptors in the chamber are then deformed by the waterborne sound waves.

8. The vertebrate eye is designed like a lens-focused camera. The fovea at the center of the human retina transmits a point-to-point image to the brain; the edges of the retina transmit information on movement and boundaries rather than a full picture.

9. Vision, like photosynthesis, uses a pigment as a primary photoreceptor. The reception of a single photon is greatly amplified by a cascade arrangement of sequential reactions that results in a spreading chain reaction of effects.

10. Other stimuli sensed by vertebrate receptors include heat, electricity, and magnetism.

REVIEW QUESTIONS

1. Define interoception and exteroception. What type of sensory receptor is simplest? Do vertebrates possess a single kind of receptor for sensing all changes in temperature? Explain.

2. Where are carotid bodies located? What changes in blood chemistry do they detect? What physiological response do they regulate?

3. What organs in the human ear detect changes in equilibrium? Why are there three of them? How does this receptor work?

4. Is the sense of hearing more acute in water or in air? Why? What changes did this necessitate in terms of the evolution of the mammalian ear?

5. Describe the mechanical means by which sound waves are transmitted and amplified through the middle ear. In which structure of the ear does the sense of hearing actually take place? What causes the recognition of individual frequencies of sound?

6. How is the chemistry of a cone cell different from a rod cell? What are the primary colors that are sensed by human cone cells? Why are vertebrates incapable of seeing ultraviolet light? Why are insects capable of seeing ultraviolet light?

7. What structure focuses the light rays entering the vertebrate eye? Does focusing occur in a similar manner in all vertebrates? Explain. What structure controls the amount of light entering the eye?

8. Are rods and cones evenly distributed over the entire surface of the retina? Why or why not? At which point on the retina is a point-to-point image formed? What is the nature of the neuronal connections in this region compared with the rest of the retina? What type of visual perception occurs in this latter region?

9. What is the advantage of having two eyes versus having a single eye? How does the position of the eyes on the head differ between predators and prey? What perceptual differences result from this difference in positioning?

10. What highly developed sense in fish has not evolved in terrestrial animals? Why?

THOUGHT QUESTIONS

1. Most of us have sensed at one time or another an oncoming storm by detecting the increase in humidity in the air. What sort of receptors detect humidity? Why do you suppose that hot days seem so much hotter when it is humid?

2. The heat-detecting pit receptor of snakes is a very effective means of "seeing" at night. It is the same sort of sensory system employed by soldiers in "snooperscopes" and in heat-seeking missiles. Why do you suppose other night-active vertebrates, such as bats, have not evolved this sort of sensory system?

3. In zero gravity, how would you expect a statocyst to behave? What would you expect the subjective impression of motion to be? Would the semicircular canals detect angular acceleration equally well at zero gravity?

4. Why haven't owls developed sonar sensory system like bats?

FOR FURTHER READING

BAHILL, A.T. and T. LaRITZ: "Why Can't Batters Keep Their Eyes On The Ball?" *American Scientist*, vol. 72, page 249, 1984. Describes the reflexes that control eye movement and the tracking of moving objects.

BARLOW, R.: "What the Brain Tells the Eye," *Scientific American*, April 1990, pages 90-95. The brain may exercise substantial control over just what the eye can detect.

BLACKMORE, C.: *The Mechanics of the Mind*, Oxford University Press, New York, 1977. A very good explanation of how the mind integrates sensory information.

BORG, E. and S.A. COUNTER: "The Middle-Ear Muscles," *Scientific American*, August 1989, pages 74-80. The muscles behind the ear drum act like acoustic shock absorbers, dampening the vibration of ear bones when sounds are loud.

FINKE, R.: "Mental Imagery and the Visual System," *Scientific American*, May 1986, pages 84-92. Explains the theory that the way in which visual information is perceived and carried to the brain has a strong influence on mental imagery.

GIBBONS, B.: "The Intimate Sense of Smell," *National Geographic*, September 1986, pages 324-361. A detailed account of the human sense of smell; interesting and well illustrated.

GILBERT, A., and C. WYSOCKI: "Smell: The Survey Results," *National Geographic*, October 1987, pages 514-525. A delightful and interesting account of how well we smell, based on a national survey. Among many surprising findings: women smell more accurately than men.

GRIFFITHS, M.: "The Platypus," *Scientific American*, May 1988, pages 84-91. Receptors in its bill detect the weak electric fields generated by the muscle activity of the animals on which it feeds.

HUDSPETH, A.J.: "The Hair Cells of the Inner Ear," *Scientific American*, January 1983, pages 54-64. An explanation of the mechanism of hearing.

HUDSPETH, A.: "How the Ear's Works Work," *Nature*, vol. 341, October 1989, pages 397-404. A review of how sensory receptors in the ear activate ion channels to initiate nerve impulses.

KORETZ, J., and G. HANDELMAN: "How the Human Eye Focuses," *Scientific American*, July 1988, pages 92-99. A detailed look at how ciliary muscles focus the lens of the eye, that explains why as people age they gradually loose the ability to focus on nearby objects.

MASLAND, R.: "The Functional Architecture of the Retina," *Scientific American*, December 1986, pages 102-111. An up-to-date description of the interconnections of the many cells beneath the retina.

MILLER, J.A.: "A Matter of Taste," *BioScience*, vol. 40(2), 1990, pages 78-82. Innovative methods and materials have made possible new insights into the way our sense of taste works.

NATHANS, J.: "The Genes for Color Vision," *Scientific American*, February 1989, pages 42-49. An interesting account of how genetic engineers isolated the genes encoding the color-detecting proteins.

POGGIO, T. and C. KOCH: "Synapses That Compute Motion," *Scientific American*, May 1987, page 46. Describes how certain cells in the visual system become differentially sensitive to moving objects.

RENOUF, D.: "Sensory Function in the Harbor Seal," *Scientific American*, April 1989, pages 90-95. The seal's sensory machinery must work on both land and sea, so both hearing and vision have evolved special adaptations.

SCHMIDT-NIELSEN, K.: *Animal Physiology: Adaptation and Environment*, ed. 2, Cambridge University Press, New York, 1984. A brilliant comparative treatment of vertebrate physiology, with an excellent section on sensory systems.

SCHNAPF, J., and D. BAYLOR: "How Photoreceptor Cells Respond to Light," *Scientific American*, April 1987, pages 40-47. A detailed account of how an individual photoreceptor in the eye is able to detect a single photon of light.

STRYER, L.: "The Molecules of Visual Excitation," *Scientific American*, July 1987, pages 42-50. A lucid description of the cascade that a photon initiates in a rod cell.

SUGA, N.: "Biosonar and Neural Computation in Bats," *Scientific American*, June 1990, pages 60-68. How the bat brain is organized to extract information from biosonar signals. Highly recommended.

47

Hormones

CHAPTER OUTLINE

OVERVIEW

In vertebrates, the central nervous system coordinates and regulates the diverse physiological activities of the body by using chemical signals called hormones to effect long-term changes. Some hormones circulate throughout the body, while others act only locally. Many of these hormones are secreted by ductless glands called endocrine glands. The body's hormones maintain physiological conditions within narrow bounds, and most of them function similarly in all vertebrates.

FOR REVIEW

Here are some important terms and concepts that you will encounter in this chapter. If you are not familiar with them, you should review them before proceeding.

Membrane transport (Chapter 6)

Operation of neurons (Chapter 44)

Central nervous system (Chapter 45)

FIGURE 47-1
Our bodies are complex machines whose activities must be coordinated for integration. The nervous control system and the hormonal control system provide this integration.

The tissues and organs of an adult mammal orchestrate a symphony of activities, all of them regulated so as to avoid conflict and to maximize interaction. This is why we think of ourselves as an *organism,* rather than as a smoothly functioning collection of organs. Integration of the vertebrate body's many activities is the primary function of the central nervous system. All control of the bodily functions, voluntary and involuntary, is vested in this system (Figure 47-1).

The central nervous system employs three separate motor command systems. These systems differ in their speed of expression, duration of response, and narrowness of application. They are the autonomic nervous system discussed in Chapter 45, the voluntary nervous system discussed in Chapters 45 and 48, and the **neuroendocrine system,** which will be discussed here.

NEUROENDOCRINE CONTROL

The effective regulation of many of the body's functions requires not only the constant modulation and integration of its internal activities, but also the ability to induce longer-term changes in levels of activity, such as initiating the production of milk by the mammary glands of nursing mothers or carrying out the sexual maturation that occurs when a young boy or girl goes through puberty. In the vertebrate body these changes are achieved by chemical signals rather than by nervous ones. These signals are hormones that the circulatory system delivers to the target organ (Figure 47-2). As we have noted earlier, **hormones** are regulating chemicals that are made at one place in the body and exert their influence at another. They are not them-

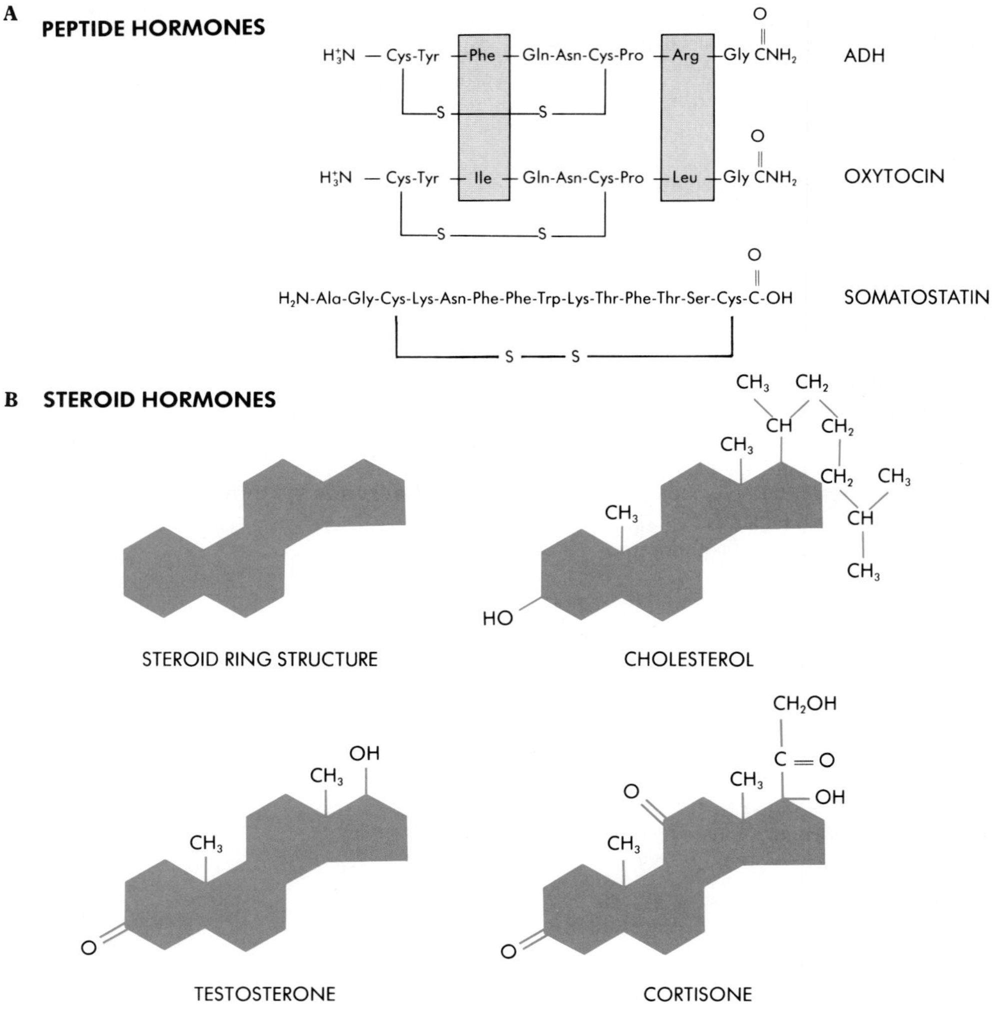

A   PEPTIDE HORMONES

H_3^+N — Cys-Tyr — Phe — Gln-Asn-Cys-Pro — Arg — Gly $\overset{O}{\overset{\|}{C}}NH_2$ ADH

H_3^+N — Cys-Tyr — Ile — Gln-Asn-Cys-Pro — Leu — Gly $\overset{O}{\overset{\|}{C}}NH_2$ OXYTOCIN

H_2N-Ala-Gly-Cys-Lys-Asn-Phe-Phe-Trp-Lys-Thr-Phe-Thr-Ser-Cys-$\overset{O}{\overset{\|}{C}}$-OH   SOMATOSTATIN

B   STEROID HORMONES

STEROID RING STRUCTURE

CHOLESTEROL

TESTOSTERONE

CORTISONE

FIGURE 47-2
Chemical structure of some important hormones.
A Peptide hormones. Antidiuretic hormone (ADH) regulates water loss by the kidneys; oxytocin, which is very similar in structure to ADH, acts on the mammary glands to stimulate milk ejection. The differences between the two molecules are highlighted. Somatostatin inhibits the secretion of growth hormone.
B Steroid hormones. Many steroid hormones are similar in structure to the blood lipid cholesterol. Testosterone stimulates development of the male genital tract; cortisone promotes the breakdown of muscle proteins in metabolism.

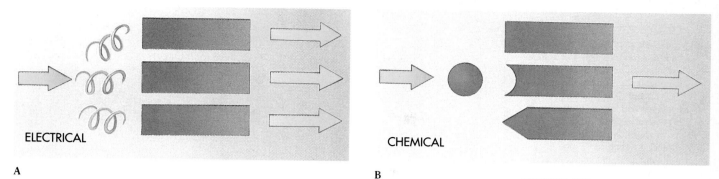

ELECTRICAL

A

CHEMICAL

B

FIGURE 47-3

A key advantage of chemical signals is specificity. An electrical signal might excite many adjacent neurons to fire (**A**), while a chemical signal can be targeted to a specific neuron (**B**).

selves enzymes, but act by regulating pre-existing processes. When the hormones reach their destination, they are recognized by specific receptors that only their target glands or tissues possess. Because most hormones circulating in the bloodstream are produced by ductless endocrine glands (Figure 47-3), whose activity is under the direct control of the nervous system.

THE IMPORTANCE OF CHEMICAL MESSENGERS

To control the vertebrate body, the central nervous system employs a battery of specific molecules called hormones as signals. Why use a chemical signal rather than an electrical one? For electrical signals to be effective they must be transmitted to individual cells. This would require many nerves and would be wasteful when the desired result is to affect the metabolic or other activity of a group of tissues, organs, or organ systems. The advantage of chemical molecules over electrical signals as messengers within the body is two-fold. First, chemical molecules can spread to all tissues via the blood. Second, each kind of hormone molecule has a unique shape unlike any other, much as every human face is unique. The mechanism (receptor) coupling a hormone to intracellular processes can also be different in different cells. The human body uses various kinds of signals. In some cases messengers are differently shaped molecules. In other cases the same messenger activates receptors that trigger diverse intercellular events (see Figure 47-3).

The Role of Receptors

How does the body recognize a molecule with a particular shape? It does so by designing a template that exactly matches the shape of a potential signal molecule—a glove to fit the hand. Using such a template, the body can recognize a signal molecule with exquisite precision, selecting one individual molecule from billions of others. These marvelous templates are the **receptor proteins** you encountered in Chapter 6. The AIDS virus, for example, infects certain cells of the immune system, and not the cells of the lungs or foot, because the immune cells possess a particular cell surface receptor that the virus recognizes.

In Chapters 44 and 45 you saw how receptor proteins play a critical role in the nervous system as the targets of neurotransmitters. As you recall, nerve cells have highly specific cell surface receptors embedded in their membranes; each receptor is tuned to respond to a different neurotransmitter molecule. How the neuron responds to stimulation depends on which of its receptors encounter its particular neurotransmitter. *The great advantage of a molecular messenger is that it can be directed at a particular protein receptor on its "target cells" that recognizes only this molecule, ignoring all other molecules.* In each case the operating principle the body uses is the same: only cells whose membranes contain an appropriate receptor protein will respond to a molecular message.

> Chemical communication within the vertebrate body involves two elements, a molecular signal and a protein receptor on target cells. The system is highly specific because each protein receptor has a shape that only its particular signal molecule fits.

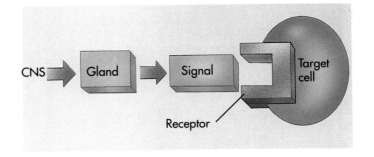

FIGURE 47-4

FIGURE 47-4
The path of hormonal communication in vertebrates. The central nervous system (CNS) issues a command that reaches the appropriate gland. The gland secretes a hormone that acts as a signal to a particular receptor, or target. After the hormone binds to the receptor, the receptor responds by changing its shape, which triggers a change in cell activity.

How Molecular Signals are Sent

The path of communication within the vertebrate body can be visualized as a series of simple steps (Figure 47-4):

1. *Issuing the command.* The central nervous system is the most important regulator of the body's activities, although other organs can also independently control the release of chemical messengers. For many hormones an area of the brain called the hypothalamus controls the release of chemical messengers from the pituitary gland. The pituitary gland is discussed later.
2. *Transporting the signal.* Hormones may act on an adjacent cell, be carried throughout the body by the bloodstream, or even pass to a different organism.
3. *Hitting the target.* When a hormone encounters a cell with a matching receptor, called a target cell, the hormone binds to that receptor.
4. *Having an effect.* When the hormone binds it, its receptor protein responds by changing shape, which triggers a change in cell activity.

Factories for Making Molecular Messengers: Endocrine Glands

There are three classes of molecular messengers: messengers that work only within cells, called second messengers; neurotransmitters released by axons; and stable molecular messengers released from endocrine glands called **hormones.** A hormone is a chemical messenger, often a steroid or peptide, that is stable enough to be transported in active form far from where it is produced and that typically acts at a distant site. Hormones thus are a very different class of molecular messenger than neurotransmitters, which tend to have effects that last only for a short time. Hormones are designed for more lasting effect.

Neurohormones are secreted by nerve cells. The most important neurohormones travel through the bloodstream from their release sites in the brain to target cells in closely associated endocrine glands. Some hormones are secreted from ductless glands that may in turn respond to different hormones, including some neurohormones. Ductless glands are called **endocrine glands.** Endocrine glands need to be clearly distinguished from glands with ducts, called **exocrine glands,** which secrete sweat, milk, digestive enzymes, and other material from the body. Endocrine glands are the hormone-producing factories of the body. By churning out large amounts of hormones, endocrine glands act to amplify greatly the initial neurohormone signal issued by the central nervous system (Figure 47-5).

FIGURE 47-5
Endocrine and exocrine glands. Endocrine glands do not have ducts and produce hormones, the body's chemical messengers. Exocrine glands do have ducts and secrete sweat, milk, digestive enzymes, and other materials.

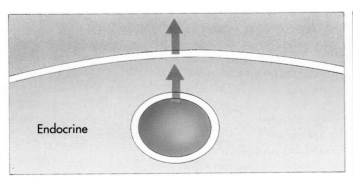

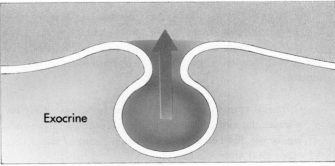

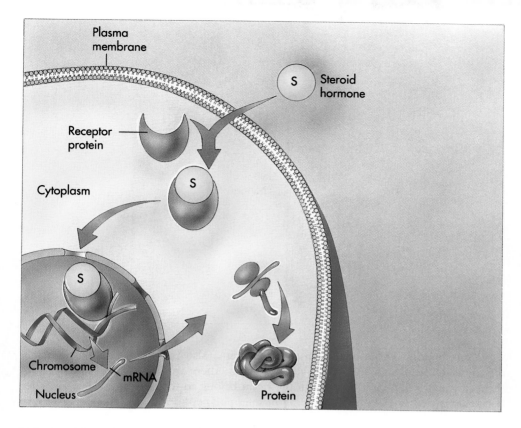

FIGURE 47-6
How steroid hormones work.
Steroid hormones are lipid soluble and thus readily pass through the plasma membrane of cells into the cytoplasm. There they bind to specific receptor proteins, forming a complex that enters the nucleus and binds to specific regulatory sites on chromosomes. The binding initiates transcription of the genes regulated by the site and thus results in the production of specific proteins.

HOW HORMONES WORK

Neurohormones and endocrine hormones act in one of two fundamental ways: either they enter the target cell or they do not.

Steroid Hormones Enter Cells

Some protein receptors designed to recognize hormones are located in the cytoplasm of the target cell. The hormones in these cases are lipid-soluble molecules, typically steroids, that pass across the cell membrane and bind to receptors within the cytoplasm (Figure 47-6). This complex of receptor and hormone then binds to the DNA in the nucleus and causes a change in the pattern of gene activity. It is the change in gene activity that is responsible for the effect of the hormone. The steroids that weight lifters and other athletes sometimes use turn on genes and thus trick their muscle cells into added growth.

All **steroid hormones** are derived from cholesterol, a complex molecule composed of three six-membered carbon rings and one five-membered carbon ring; it resembles a fragment of chain-link fence. The hormones that promote the development of the secondary sexual characteristics are steroids. They include cortisone and testosterone, as well as the hormones estrogen and progesterone, which are used in oral contraceptives.

> **Steroid hormones enter a target cell, bind to a cytoplasmic receptor, and penetrate the nucleus, where they initiate the transcription of some genes while repressing the transcription of others.**

Peptide Hormones Do Not Enter Cells

Other hormone receptors are embedded within the cell membrane, with their recognition region directed outward from the cell surface (Figure 47-7). Hormones, typically peptides, bind to these receptors on the cell surface. This binding then triggers events within the cell cytoplasm, usually through intermediates known as second messengers.

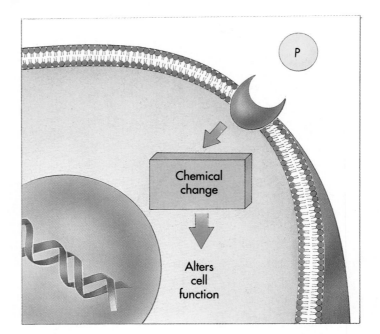

FIGURE 47-7

How peptide hormones work.
Peptide hormones (P) bind to receptors on the cell surface. This binding triggers chemical changes within the cytoplasm. These changes alter cell function, usually through the agency of chemicals known as second messengers.

Some **peptide hormones,** such as epinephrine (also called adrenaline), are small molecules derived from the amino acid tyrosine; others are short polypeptide chains. Most hormones that circulate within the brain belong to this second class. Still other hormones are large proteins, consisting of long polypeptide chains, such as insulin.

> **Peptide hormones do not enter their target cells. Instead they interact with a receptor on the cell surface and initiate a chain of events within the cell by increasing the levels of second messengers.**

A Peptide Hormone in Action: How Insulin Works

How does the binding of a peptide hormone to the *surface* of a cell produce changes within? The peptide hormone insulin provides a well-studied example of how peptide hormones achieve their effect within target cells (Figure 47-8). Most vertebrate cells have receptors for insulin in their membranes, the number ranging from fewer than 100 to more than 100,000 in some liver cells. The receptors are glycoproteins. One part protrudes from the surface of the cell and binds insulin, whereas another spans the membrane and extends into the cytoplasm.

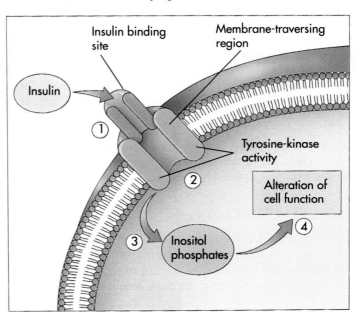

FIGURE 47-8
How insulin works.

Stage one: activation by insulin. The binding of insulin causes a change in the shape of the insulin receptor so that within the cell, phosphate groups are actively added to the tyrosine amino acid side groups of proteins. This phosphorylation of proteins activates the second stage of the insulin response.

Stage two: amplification of signal. As a result of insulin receptor–induced enzyme activity, small molecules of cyclic AMP are formed from ATP. Cyclic AMP activates a variety of enzymes in the cell that stimulate the uptake of glucose from the blood and the formation of glycogen, causing levels of glucose in the blood to fall. In addition to cyclic AMP, insulin binding also promotes the production of **inositol phosphates,** which are cleaved from the plasma membrane. A single insulin molecule binding to its receptor releases many mediator molecules into the cell, and each mediator activates many enzyme molecules in an expanding cascade of response.

Second Messengers

Insulin mediators are one example of **second messengers,** intermediary compounds that couple extracellular signals to intracellular processes and also amplify a hormonal signal. Second messengers set into action a variety of events within the affected cell, depending on the enzymatic profile of the cell. In the early 1960s Earl Sutherland described the first second messenger, a cyclic form of adenosine monophosphate, **cyclic AMP (cAMP).** When hormones such as epinephrine bind receptors on liver cells (Figure 47-9), the receptor changes shape and binds a cell protein called **G protein,** causing it in turn to bind the nucleotide GTP and activate another membrane protein, **adenylate cyclase.** The result of these complex interactions is the production of large amounts of cAMP by the activated adenylate cyclase. *The cAMP is the amplified form of the epinephrine hormonal message.* Another common second messenger is inositol phosphate.

A variety of peptide hormones use cAMP as a second messenger. The cAMP has different effects in various target cells because different enzymes are present in different target cells and tissues. In muscle cells, cAMP is induced by epinephrine and activates the enzyme protein kinase A, which in turn activates the enzyme that breaks down glycogen into glucose. In cells of the ovary, cAMP is induced by a luteinizing hormone and stimulates the cells to produce a specific follicle cell enzyme.

Second messengers are cytoplasmic signal molecules produced in response to a peptide hormone; often they greatly amplify the original signal.

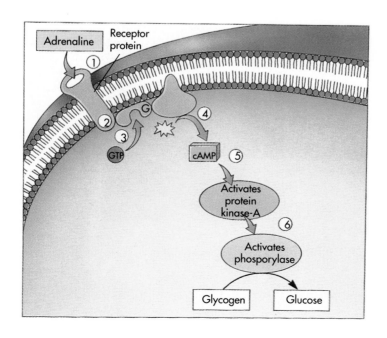

FIGURE 47-9

Second messengers.

1 Peptide hormones bind to specific receptor proteins on the surface of target cells.

2 The hormone enters the cytoplasm.

3 The receptor changes shape and binds to a protein called G.

4 The receptor then binds to GTP and activates a membrane protein called adenyl cyclase. Adenyl cyclase produces cAMP.

5 cAMP activates protein kinase A.

6 Protein kinase A activates phosphorylase, which breaks glycogen down into glucose.

FIGURE 47-10
The human endocrine system.

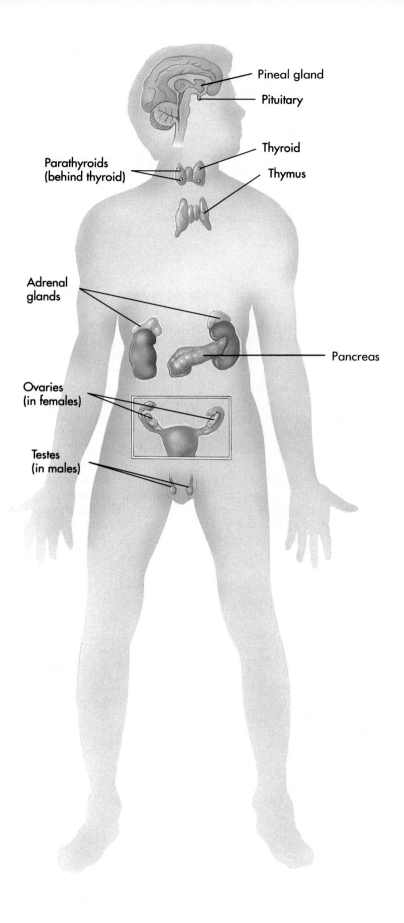

Pineal gland

Pituitary

Parathyroids
(behind thyroid)

Thyroid

Thymus

Adrenal
glands

Pancreas

Ovaries
(in females)

Testes
(in males)

Thyroxine is Unique: It Goes Directly into the Nucleus

One important endocrine hormone is neither steroid nor peptide. The hormone thyroxine, produced by the thyroid gland, is a modified form of the amino acid tyrosine and contains four iodine atoms. Thyroxine acts on target cells in a unique way. It diffuses directly into the cell cytoplasm and then into the cell nucleus, where it interacts with a protein receptor attached to DNA to initiate production of particular growth-promoting messenger RNAs. Thyroxine is the only hormone known to go directly into a target cell nucleus without first binding to receptors on the cell membrane or in the cytoplasm.

THE MAJOR ENDOCRINE GLANDS AND THEIR HORMONES

Vertebrates have about a dozen major endocrine glands that together make up the **endocrine system** (Figure 47-10). In this section we briefly examine the principal endocrine glands and the hormones they produce. These principal endocrine glands and their hormones are summarized in Table 47-1.

The Pituitary

The pituitary, located in a bony recess in the brain below the hypothalamus, is the site of production of nine major hormones. Because many of these hormones act principally to influence other endocrine glands (Figure 47-11), it was fashionable until relatively recently to regard the pituitary as a "master gland" orchestrating the endocrine system. In fact, as shown here, that role is reserved for the hypothalamus because the hypothalamus controls the pituitary. The pituitary remains, however, one of the most important and interesting of the endocrine glands.

FIGURE 47-11
The role of the pituitary. Interactions between the anterior lobe and the posterior lobe of the pituitary lobe of the pituitary (two distinct glands), and various organs of the human body are shown in this diagram.

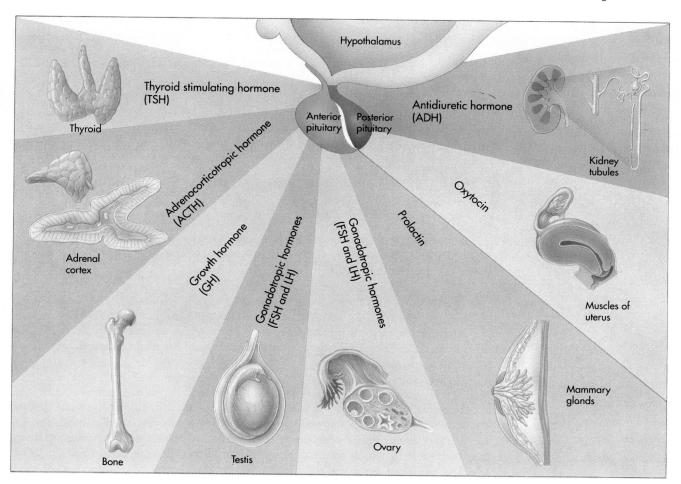

TABLE 47-1 PRINCIPAL ENDOCRINE GLANDS AND THEIR HORMONES

ENDOCRINE GLAND AND HORMONE	TARGET TISSUE	PRINCIPAL ACTIONS	CHEMICAL NATURE
Posterior Lobe of Pituitary			
Oxytocin	Uterus	Stimulates contraction of uterus	Peptide (9 amino acids)
	Mammary glands	Stimulates ejection of milk	
ADH	Kidneys	Stimulates reabsorption of water; conserves water	Peptide (9 amino acids)
Anterior Lobe of Pituitary			
Somatotropin (growth hormone, GH)	General	Stimulates growth by promoting protein synthesis and breakdown of fatty acids	Protein
Prolactin (PRL)	Mammary glands	Stimulates milk production	Protein
Thyroid-stimulating hormone (TSH)	Thyroid gland	Stimulates secretion of thyroid hormones	Glycoprotein
Adrenocorticotropic hormone (ACTH)	Adrenal Cortex	Stimulates secretion of adrenal cortical hormones	Polypeptide
Follicle-stimulating hormone (FSH)	Gonads	Stimulates ovarian follicle spermatogenesis	Glycoprotein
Luteinizing hormone (LH)	Gonads	Stimulates ovulation and corpus luteum formation in females; stimulates secretion of testosterone in males	Glycoprotein
Thyroid Gland			
Thyroid hormone (thyroxine)	General	Stimulates metabolic rate; essential to normal growth and development	Iodinated amino acid
Calcitonin	Bone	Lowers blood calcium level by inhibiting loss of calcium from bone	Polypeptide (32 amino acids)
Parathyroid Glands			
Parathyroid hormone	Bone, kidneys, digestive tract	Increases blood calcium level by stimulating bone breakdown; stimulates calcium reabsorption in kidneys; activates vitamin D	Polypeptide (34 amino acids)
Adrenal Medulla			
Adrenaline and noradrenaline (epinephrine and norepinephrine)	Skeletal muscle, cardiac muscle, blood vessels	Initiates stress responses; increases heart rate, blood pressure, metabolic rate; dilates blood vessels; mobilizes fate; raises blood sugar level	Amino acid derivatives

The pituitary is actually two glands. The back *posterior* end regulates water conservation, milk letdown, and uterine contraction in women; the front *anterior* end regulates other endocrine glands.

The Posterior Pituitary. The role of the posterior pituitary first became evident in 1912, when a remarkable medical case was reported: a man who had been shot in the head developed a surprising disorder—he began to urinate every 30 minutes or so, unceasingly. The bullet had lodged in the pituitary gland, and subsequent research demonstrated that removal of the pituitary produces these unusual symptoms. Pituitary extracts were shown to contain a substance that makes the kidneys conserve water, and eventually in the early 1950s the peptide hormone **vasopressin** (also called antidiuretic hormone, ADH) was isolated. Vasopressin is the hormone that regulates the kidneys' retention of water (Figure 47-12). When vasopressin is missing, the kidneys cannot retain water, which is why the bullet led to excessive urination (and why excessive alcohol, which inhibits vasopressin secretion, has the same effect).

The posterior pituitary also produces a second hormone of very similar struc-

ENDOCRINE GLAND AND HORMONE	TARGET TISSUE	PRINCIPAL ACTIONS	CHEMICAL NATURE
Adrenal Cortex			
Aldosterone	Kidney tubules	Maintains proper balance of sodium and potassium ions	Steroid
Cortisol	General	Adaption to long-term stress; raises blood glucose level; mobilizes fat	Steroid
Islets of Langerhans			
Insulin	General	Lowers blood glucose; increases storage of glycogen in liver	Polypetide (51 amino acids)
Glucagon	Liver, adipose tissue	Raises blood glucose level; stimulates breakdown of glycogen in liver	Polypeptide (29 amino acids)
Ovary			
Estrogen	General Female reproductive structures	Stimulates development of secondary sex characteristics in females and growth of sex organs at puberty; prompts monthly preparation of uterus for pregnancy	Steroid
Progesterone	Uterus Breasts	Completes preparation of uterus for pregnancy Stimulates development	Steroid
Testis			
Testosterone	General	Stimulates development of secondary sex characteristics in males and growth spurt at puberty	Steroid
	Male reproductive structures	Stimulates development of sex organs; stimulates spermatogenesis	Steroid
Pineal Gland			
Melatonin	Gonads, pigment cells	Function not well understood; influences pigmentation in some vertebrates; may control biorhythms in some animals; may help control onset of puberty in humans	Amino acid derivative

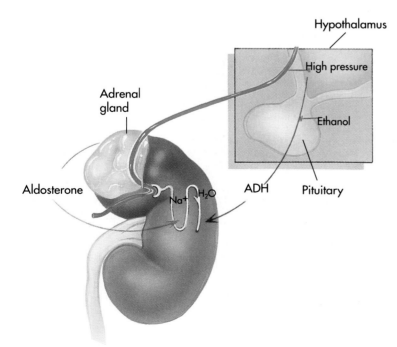

FIGURE 47-12

How hormones control water chemistry. Control of water and salt balance within the kidney is centered in the hypothalamus. The hypothalamus produces antidiuretic hormones (ADH), which renders the collecting ducts of the kidneys freely permeable to water and so maximizes water retention. If too much water retention leads to high blood pressure, pressure-sensitive receptors in the hypothalamus detect this and cause the production of ADH to be shut down. If the level of sodium in the blood falls, the adrenal gland initiates production of the hormone aldosterone, which stimulates salt reabsorption by the renal tubules of the kidney. Feedback loops are noted by the colored lines with the double bar.

ture—both are short peptides composed of nine amino acids—and very different function, called **oxytocin**. Oxytocin initiates milk release because sensory receptors in the nipples send messages to the hypothalamus, causing oxytocin release. Oxytocin stimulates the contraction of the muscles around the ducts into which mammary glands secrete milk. Oxytocin also stimulates uterine contractions in women during childbirth, which is why the uterus of a nursing mother returns to normal size after its extension during pregnancy more quickly than does the uterus of a mother who does not nurse her baby.

Vasopressin and oxytocin are synthesized inside nerve cells within the hypothalamus. The two hormones are transported down nerve cell axons from the hypothalamus to synapses located within the posterior pituitary, where they are stored in the axon terminals. The hormones are released into the bloodstream when a nerve impulse reaches the terminals from the hypothalamus.

> **The posterior lobe of the pituitary, a distinct gland, is connected with the hypothalamus by neural connections. It secretes the important hormones vasopressin and oxytocin.**

The Anterior Pituitary.   The key role of the anterior pituitary first became understood in 1909, when a 38-year-old South Dakota farmer was cured of the growth disorder acromegaly by the surgical removal of a pituitary tumor. Acromegaly is a form of giantism in which the jaw begins to protrude and features thicken (Figure 47-13). It turned out that giantism is almost always associated with pituitary tumors. Robert Wadlow, born in Alton, Illinois, in 1928, grew to a height of 8 feet, 11 inches and weighed 475 pounds before he died from infection at age 22—the tallest human being ever recorded (Figure 47-14). He was still growing the year he died. Skull x-rays showed he had a pituitary tumor. So did the 8-foot, 2-inch Irish giant Charles Byrne, born in 1761; his skeleton has been preserved in the Royal College of Surgeons, London, and shows the effects of a large pituitary tumor.

FIGURE 47-13
Acromegaly. This ancient carving of the Egyptian pharaoh Akhenaton, who ruled from 1379-1362 BC, exhibits many characteristics of acromegaly; if so, it is the oldest known case.

FIGURE 47-14
The Alton giant. This photograph, taken when Robert Wadlow of Alton, Illinois, was 13 years old, shows him towering over his father and a 9-year-old brother. Born at a normal size, he developed a growth-hormone-secreting pituitary tumor as a young child and never stopped growing during his 22 years of life.

Why did removal of the pituitary tumor cure the South Dakota farmer? Pituitary tumors produce giants because the tumor cells produce prodigious amounts of a growth-promoting hormone. This **growth hormone (GH),** a peptide of 191 amino acids, is normally produced in only minute amounts by the anterior pituitary gland and usually only during periods of body growth, such as infancy and puberty.

We now know that the anterior pituitary gland of vertebrates produces seven major peptide hormones, each controlled by a particular releasing factor secreted from cells in the hypothalamus:

Thyroid-stimulating hormone (TSH). TSH stimulates the thyroid gland to produce thyroid hormone, which in turn stimulates oxidative respiration.

Luteinizing hormone (LH). LH plays an important role in the female menstrual cycle (see Chapter 54). It also stimulates the male gonads to produce testosterone, which initiates and maintains the development of male secondary sexual characteristics, those external features not involved in reproduction.

Follicle-stimulating hormone (FSH). FSH is significant in the female menstrual cycle (see Chapter 54). In males, it stimulates certain cells in the testes to produce a hormone that regulates the development of sperm.

Adrenocorticotropic hormone (ACTH). ACTH stimulates the adrenal cortex to produce corticosteroid hormones. Some of these hormones regulate the production of glucose from fat; others regulate the balance of sodium and potassium ions in the blood; and still others contribute to the development of the male secondary sexual characteristics.

Somatotropin, or growth hormone (GH). GH stimulates the growth of muscle and bone throughout the body.

Prolactin (PRL). PRL stimulates the breasts to produce milk.

Melatonin, or *melanocyte-stimulating hormone (MSH).* In reptiles and amphibians, MSH stimulates color changes in the epidermis. This hormone has no known function in mammals.

Seven major peptide hormones are secreted by the anterior lobe of the pituitary. Very similar in structure, they have a wide variety of functions.

In addition to their endocrine functions, these and many other hormones have been shown also to be associated with particular populations of cells within the central nervous system. This is an area of very active research; biologists are attempting to uncover the as yet unknown role of these hormones in the central nervous system. Whatever their function there, it is clear that the same hormone may play different roles in different parts of the body because cells have different enzymatic profiles. Hormones are signals used in different tissues for different reasons, just as raising your hand is a signal that has different meanings in different contexts: in class it indicates you have a question; in a football game it signals a "fair catch"; and when a policeman does it on a street, it means "stop." The evolving use of hormones by vertebrates has been a conservative process. Rather than produce a new hormone for every use, vertebrates have often adapted a hormone already at hand for a new use, when this new use does not cause confusion.

The Thyroid: A Metabolic Thermostat

The thyroid gland is shaped like a shield (Greek *thyros,* meaning "shield"). It lies just below the Adam's apple in the front of the neck. It makes several hormones, the two most important of which are **thyroxine,** which increases metabolic rate and promotes growth, and **calcitonin,** which stimulates calcium uptake.

Without adequate thyroxine (also called thyroid hormone), growth is retarded. Children with underactive thyroid glands are not able to carry out carbohydrate breakdown and protein synthesis at normal rates, a condition called **cretinism,** which results in stunted growth. Mental retardation is also seen because thyroxine is needed for normal development of the CNS. Adults with too little of this hormone also have slowed metabolism, affecting their mental performance.

FIGURE 47-15
A goiter. This condition is caused by a lack of iodine in the diet.

As mentioned earlier, the hormone thyroxine is made from the amino acid tyrosine, with four iodine atoms added to it. If the iodine concentration in a person's diet is too low, the thyroid cannot make adequate amounts of thyroxine and will grow larger in a futile attempt to manufacture more of the hormone. The greatly enlarged thyroid gland that results is called a **goiter** (Figure 47-15). This need for iodine in your diet is why iodine is added to table salt.

The Parathyroids: Builders of Bones

The parathyroid glands are four small glands attached to the thyroid. Small and unobtrusive, they were ignored by researchers until well into this century. The first suggestions that the parathyroids produce a hormone came from experiments in which they were removed from dogs: the concentration of calcium in the dogs' blood plummeted to less than half the normal level. However, if an extract of parathyroid gland was administered, calcium levels returned to normal. If an excess was administered, calcium levels became *too* high, and the bones of the dogs literally were dismantled by the extract. It was clear that the parathyroid glands were producing a hormone that acted on calcium uptake into, and release from, bone.

The hormone produced by the parathyroids is **parathyroid hormone (PTH)**. It is one of only two hormones in your body that are absolutely essential for survival (the other, discussed in the next section, is aldosterone, produced by the adrenal glands). PTH acts to regulate levels of calcium in your blood. Calcium ions are the key actors in vertebrate muscle contraction; by altering calcium release, nerve impulses cause muscles to contract. You cannot live without the muscles that pump your heart and drive your body, and these muscles cannot function if calcium levels are not kept within narrow limits. Calcium is also important for normal nerve activity.

PTH acts as a fail-safe to make sure calcium levels never fall too low. PTH is released into the bloodstream, where it travels to the bones and acts on the osteoclast cells within bones, stimulating them to dismantle bone tissue and release calcium into the bloodstream. PTH also acts on the kidneys to resorb calcium ions from the urine and leads to the activation of vitamin D, necessary for calcium absorption by the intestine. A diet deficient in vitamin D leads to poor bone formation, a condition called

FIGURE 47-16

Maintenance of proper calcium levels in the blood is managed by the thyroid. If levels of calcium in the blood become too high, calcitonin stimulates calcium uptake in bone, thus lowering the calcium blood levels.

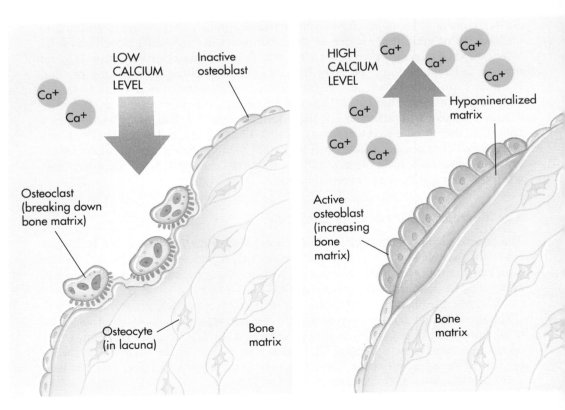

rickets. PTH is synthesized by the parathyroids in response to falling levels of calcium ions in the blood; the body essentially sacrifices bone to keep calcium levels within the narrow limits necessary for proper functioning of muscle and nerve.

The thyroid also plays a role in maintaining proper calcium levels (Figure 47-16). If levels of calcium in your blood become too high, the thyroid hormone calcitonin, mentioned earlier, stimulates calcium deposition in bone, thus lowering calcium levels in the blood.

The Adrenals: Two Glands in One

There are two adrenal glands, one located just above each kidney. Each adrenal gland is composed of two parts: an inner core called the **medulla,** which produces the peptide hormones epinephrine and norepinephrine, and an outer layer called the **cortex,** which produces the steroid hormones **cortisol** and **aldosterone.**

Adrenal Medulla: Emergency Warning Siren.   The medulla releases epinephrine and norepinephrine in times of stress. Epinephrine and norepinephrine act on the body as emergency signals that stimulate rapid deployment of body fuel. The "alarm" response throughout the body is identical to the individual effects achieved by the sympathetic nervous system but is longer lasting. Among the effects of these hormones are an accelerated heartbeat, increased blood pressure, higher levels of blood sugar, dilated blood vessels, and increased blood flow to the heart and lungs. These hormones thus can be thought of as extensions of the sympathetic nervous system.

Adrenal Cortex: Maintaining the Proper Amount of Salt.   Cortisol (also called hydrocortisone) acts on many different cells in the body to maintain nutritional well-being. It stimulates carbohydrate metabolism and acts to reduce inflammation. Derivatives of this hormone, such as prednisone, have widespread medical use as anti-inflammatory agents. The ability of many cortisol-derived steroids to stimulate muscle growth has also led to the abuse of so-called anabolic steroids by athletes (Figure 47-17).

Aldosterone acts primarily at the kidney to promote the uptake of sodium and other salts from the urine. Sodium ions play critical roles in nerve conduction and many other body functions. Their concentration also has a critical influence on blood pressure. Without aldosterone, sodium ions are not retrieved from body fluids and are lost in the urine. The loss of salt in the blood causes water to leave the bloodstream and enter cells; thus blood pressure falls. Aldosterone also acts in the opposite way to

FIGURE 47-17
Many athletes have succumbed to the abuse of steroids. Although steroids stimulate muscle growth, their abuse carries many serious health risks.

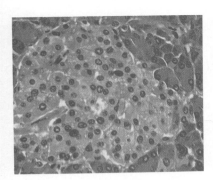

FIGURE 47-18
Islets of Langerhans. Glucagon and insulin are produced by clumps of cells within the pancreas called islets of Langerhans, which are stained dark in this preparation.

promote the export of potassium out of the body, stimulating the kidneys to secrete potassium ions into the urine. When aldosterone levels are too low, potassium levels in the blood may rise to dangerous levels. Aldosterone is, with PTH, one of the two endocrine hormones essential for survival. Removal of the adrenal glands is invariably fatal.

The Pancreas: The Body's Dietitian

The pancreas gland is located behind the stomach and is connected to the front end of the small intestine by a small tube. It secretes a variety of digestive enzymes into the gut through this tube, and for a long time was thought to be solely an exocrine gland. In 1869, however, a German medical student named Paul Langerhans described some unusual clusters of cells scattered throughout the pancreas (Figure 47-18). At the end of the last century doctors had begun to notice that patients with injuries to the pancreas often develop **diabetes mellitus,** a common and serious disorder in which the affected individuals develop elevated levels of glucose in the blood. They are unable to take up glucose from the blood, even though the level of blood glucose is high. In Type I diabetes mellitus, insulin secretion is abnormally low. In Type II diabetes mellitus, there is an abnormally low number of insulin receptors on the target tissue, but the level of insulin is normal in the blood. Such individuals literally starve; they lose weight and may eventually suffer brain damage and even death if their condition is untreated. In 1893 it was suggested that the clusters of cells in the pancreas, which came to be called **islets of Langerhans,** produced something that prevented diabetes mellitus.

The substance, which we now know to be the peptide hormone **insulin,** was not isolated until 1922, when two young doctors working in a Toronto hospital succeeded where many others had not. On January 11, 1922, they injected an extract purified from beef pancreas glands into a 13-year-old boy, a diabetic whose weight had fallen to 65 pounds and who was not expected to survive. The hospital record note gives no indication of the historic importance of the trial, only stating "15 cc of MacLeod's serum. 7½ cc into each buttock." With this single injection, the glucose level in the boy's blood fell 25%. A more potent extract soon brought levels down to near normal. This was the first instance of successful insulin therapy. Today cases of diabetes can be treated by supplying insulin to the affected individual daily. Others are treated by a combination of exercise and diet. Active research on the possibility of transplanting islets of Langerhans holds much promise of a lasting treatment for diabetes.

> **Diabetes is a condition in which individuals are unable to obtain glucose from their blood because they either lack insulin (Type I) or have an abnormally low number of insulin receptors (Type II). It is a serious disease and can be fatal if untreated.**

We now know that the islets of Langerhans in the pancreas produce *two* hormones that interact to govern the levels of glucose in the blood (Figure 47-19), insulin and **glucagon.** Insulin is a storage hormone, designed to put away nutrients for leaner times. It promotes the accumulation of glycogen in the liver and triglycerides in fat cells. When you eat, **beta cells** in the islets of Langerhans secrete insulin, storing away glucose to be used later. Later, when body activity causes the level of glucose in the blood to fall as it is used up as fuel, other cells in the islets of Langerhans called **alpha cells** secrete glucagon, which causes liver cells to release stored glucose and fat cells to break down triglycerides. The two hormones thus work together to keep glucose levels in the blood within narrow bounds.

Other Endocrine Glands

The ovary and testes are important endocrine glands, producing sex hormones (estrogen, progesterone, and testosterone), which are described in detail in Chapter 54. Surprisingly, the gut is also a major endocrine gland, secreting hormones that regu-

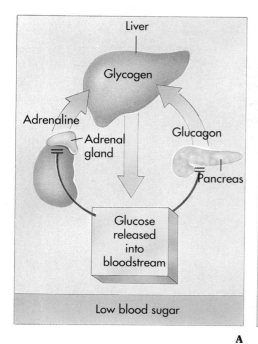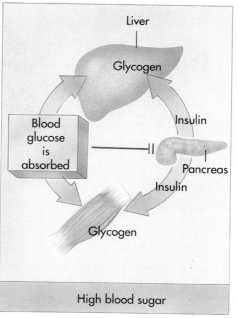

FIGURE 47-19

Hormonal control of blood glucose levels.

A When blood glucose levels are low, cells within the pancreas release the hormone glucagon into the bloodstream; and other cells within the adrenal gland, situated on top of the kidneys, release the hormone adrenaline into the bloodstream. When they reach the liver, these two hormones both act to increase the liver's breakdown of glycogen to glucose.

B When blood glucose levels are high, other cells within the pancreas produce the hormone insulin, which stimulates the liver and muscles to convert blood glucose into glycogen. The glucose level in the blood determines the levels of insulin and glucagon in the blood via feedback loops to the pancreas and adrenal gland.

Low blood sugar — A

High blood sugar — B

late the release of secretions, which in turn play a key role in food digestion; they are discussed in Chapter 49. The only other major endocrine gland is the **pineal gland,** which sits in the center of the brain. It is small, about the size of a pea, and is shaped like a pine cone (hence its name). It is the last endocrine gland discovered in the human body, and its function is not known. It was discovered in 1958 that the pineal gland secretes a hormone, **melatonin,** a derivative of the amino acid tryptophan. In hamsters, melatonin regulates reproductive biology, and in frogs it influences pigmentation, but its function in humans is not well understood.

The pineal gland in reptiles is located closer to the surface and is called the "third eye" because it is structurally similar to the retina and responds directly to light. In humans the pineal gland in not connected to the central nervous system directly, but the gland *is* connected, via the sympathetic nervous system, to the eyes. Melatonin seems to be released by the human pineal gland as a response to darkness, in a cyclical biological rhythm keyed to daylight. Perhaps the pineal gland is involved in establishing daily biorhythms. It has also been implicated in mood disorders such as winter depression, also called SADS (seasonal affective disorder syndrome), and in a variety of other roles concerning sexual development.

HOW THE BRAIN CONTROLS THE ENDOCRINE SYSTEM

The 12 major endocrine glands (Table 47-1) do not function independently; each is not a separate kingdom. Their activities are coordinated at two levels: (1) the six anterior pituitary hormones regulate many of the activities of the other endocrine glands, and (2) the pituitary gland is controlled by the CNS via the hypothalamus.

Releasing Hormones: The Brain's Chemical Messengers

How the pituitary gland is regulated by the brain was until recently one of the great mysteries of medicine. It is suspended by a short stalk to the hypothalamus (Figure 47-20), which is part of the diencephalon region of the forebrain, located at the base of the brain. Within the hypothalamus, information about the body's many internal functions (called interoceptive information) is processed and regulatory commands are issued. Some of these commands involve functions such as the regulation of body temperature, the intake of food and water, reproductive behavior, and response to

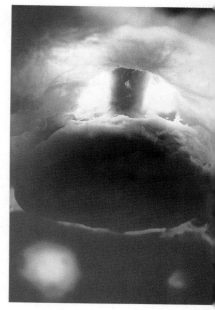

FIGURE 47-20

The pituitary gland hangs by a short stalk from the hypothalamus. The pituitary regulates the hormone production of many of the body's endocrine glands. Here it is enlarged 15 times.

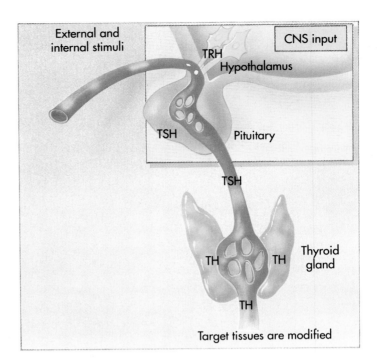

FIGURE 47-21
The release of thyroid hormone (TH) from the thyroid gland is the end result of a long series of events mediated by the central nervous system (CNS).

pain and emotion. These commands are issued to the pituitary gland, which in turn sends chemical signals to the various hormone-producing glands of the body. For this reason, injury to the hypothalamus causes a decrease in the anterior pituitary's production of hormones. The hypothalamus thus regulates the body through a chain of command in the same way that a general gives orders to a chief of staff, who relays them to lower-ranking commanders.

The hypothalamus processes information about the body's internal functions and issues regulatory commands that direct the pituitary gland to send further chemical signals to the various hormone-producing glands of the body.

No nerves connect the anterior pituitary gland to the hypothalamus or to any other part of the brain. How are the commands sent from the hypothalamus to the pituitary? In the 1930s a network of tiny blood vessels was discovered that spans the short distance between the hypothalamus and anterior pituitary; researchers wondered if perhaps the hypothalamus employed chemical messengers. The experimental difficulties in isolating such hypothalamic hormones were great because very little of them are present in any one brain. After concentrating the hypothalamus glands from 1 million pigs, however, the first of these hormones, a short peptide called **thyrotropin-releasing hormone (TRH),** was isolated in 1969 (Figure 47-21). The release of TRH from the hypothalamus triggers secretion of thyrotropin from the anterior pituitary.

Six other hypothalamic regulatory hormones have since been isolated, which together govern all the hormones secreted by the anterior pituitary. For each releasing hormone secreted by the hypothalamus, a corresponding hormone is synthesized by the anterior pituitary. When the pituitary gland receives a releasing hormone from the hypothalamus, the anterior lobe responds by secreting the corresponding pituitary hormone.

The anterior pituitary is connected to the hypothalamus by special blood vessels only a few millimeters long. Through these vessels passes a group of hormones produced in the hypothalamus called releasing hormones, which command the anterior pituitary to initiate the production (release) of specific hormones in distant endocrine glands.

How the Hypothalamus Regulates Hormone Production

Control over the production of hormones produced by the anterior pituitary gland is exercised in two ways.

CNS Control. The production of the hormones GH, PRL, and MSH is controlled by both releasing and inhibitory signals produced by the hypothalamus. Consider the growth hormone somatotropin (GH) (Figure 47-22). The releasing signal for GH is the growth-hormone–releasing hormone (GHRH) produced by the hypothalamus; GHRH stimulates the anterior pituitary to produce somatotropin. The inhibiting signal, which is also produced at the same time by the hypothalamus, is somatostatin. Somatostatin inhibits the anterior pituitary from producing somatotropin. The hypothalamus thus regulates growth by mediating the relative rates of production of GHRH and somatostatin. In a similar way the hypothalamus regulates the production of PRL and MSH. Thus the release and inhibition of hormones is controlled by the release of *other* hormones.

Feedback Control. The levels of all other hormones produced by the anterior pituitary are controlled by negative feedback from the target glands. For example, when LH stimulates the gonads to release testosterone into the bloodstream, that testosterone in turn inhibits the hypothalamus. The hypothalamus then ceases to transmit LH-releasing hormone to the pituitary.

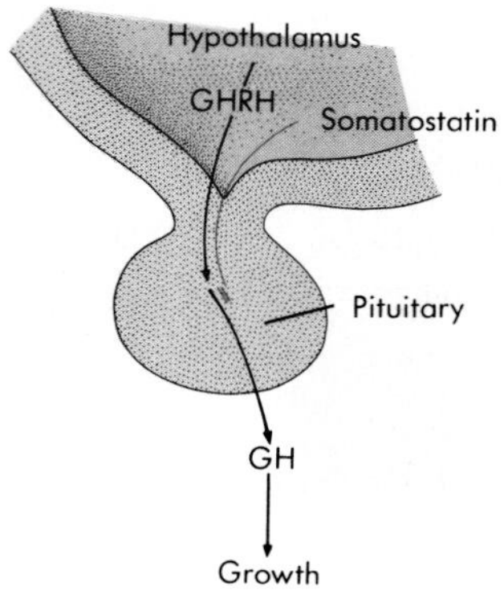

NONENDOCRINE HORMONES
Neuropeptides

In 1974 Swedish researchers first reported the existence in the brain of **neurohormones.** They isolated two small peptides called **enkephalins** (Figure 47-23), only five amino acid units long, which acted as powerful narcotics. The enkephalins appear to play a role in integrating afferent impulses from the pain receptors. A second type of peptide hormone has since been found that is larger: the brain hormones called **endorphins** are polypeptides 32 amino acid units long. Endorphins appear to regulate emotional responses in the brain. Morphine has such a potent analgesic (pain-relieving) effect on the CNS because it mimics the effects of the endorphins.

More than 20 peptides have now been identified as acting as neurotransmitters in the brain, and some investigators believe the number will eventually exceed 100. Their study is causing a revolution in how scientists view the brain's internal activity. The role of chemical messengers within the brain is one of the most active areas of biological research.

FIGURE 47-22
Levels of the growth-promoting hormone (GH) are regulated by both stimulatory and inhibitory signals. The production of GH by the pituitary is stimulated by a hypothalamic-releasing hormone, GHRH, and inhibited by another pituitary hormone, somatostatin. If levels of GHRH become too high, the excess somatostatin that is induced in the pituitary shuts down some of the production of GH (indicated by the double bar), so that effective levels of GH circulating in the blood do not fluctuate.

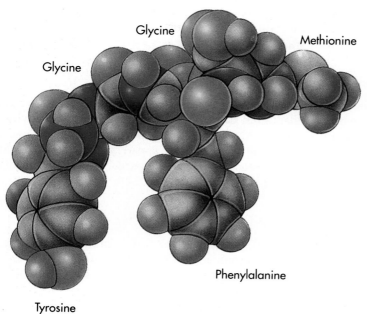

FIGURE 47-23
Molecular structure of enkephalin. A potent narcotic, enkephalin is a peptide composed of a linear chain of five amino acids.

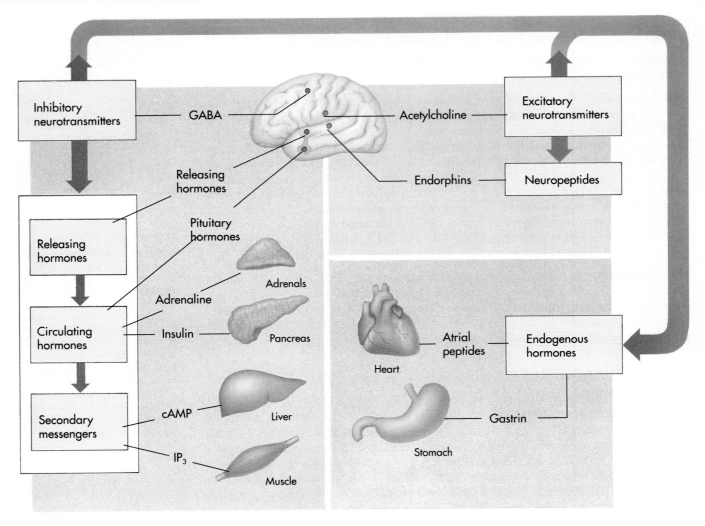

FIGURE 47-24
An overview of neuroendocrine control.

Prostaglandins

Prostaglandins are modified lipids produced from membrane phospholipids by virtually all cells. They do not circulate in the blood, but rather accumulate in regions of tissue disturbance or injury. They stimulate smooth muscle contraction and expansion and contraction of blood vessels. Aspirin relieves headache pain because it inhibits prostaglandin production; overproduction of prostaglandins swells cerebral blood vessels so that their walls press against nerve tracts in the brain, causing pain.

Atrial Peptides

Another nonendocrine hormone of considerable interest is a small peptide manufactured in the heart, called **atrial natriuretic hormone (ANH),** that circulates throughout the body. Receptors for ANH have been identified in cells of blood vessels, kidneys, and adrenal glands. The atrial peptides apparently help regulate blood pressure and volume. Because they reduce blood volume, atrial peptides are being investigated as a potential treatment for high blood pressure.

THE BODY'S MESSENGERS

Hormonal communication is essential to the body's proper functioning. The many kinds of hormones we have encountered in this chapter form a network of communication (Figure 47-24) that reaches to all the tissues of the body. The hormones are all ultimately under the control of the central nervous system, of which they are a chemical extension. You will encounter hormones repeatedly in the following chapters. They are one of the principal integrating elements of the vertebrate body.

1. Chemical messengers are highly specific, affecting only particular target cells. This exquisite sensitivity is possible because each messenger is keyed to a particular receptor protein present in the membranes of target cells and not in other cells.

2. Hormones are stable chemical messengers that act on cells located far from where the hormone is produced.

3. Most hormones are either peptides that interact with receptors on the target cell surface, thus activating enzymes within, or steroids, which enter into cells and alter their transcription patterns.

4. The posterior lobe of the pituitary, a distinct gland, secretes hormones such as ADH into the bloodstream. This gland is linked directly to the hypothalamus by neural connections.

5. The anterior lobe of the pituitary, another distinct gland, is connected to the hypothalamus by special blood vessels a few millimeters long. It secretes seven principal kinds of hormones, each corresponding to a specific releasing hormone from the hypothalamus.

6. The brain maintains long-term control over physiological processes by synthesizing releasing hormones in the hypothalamus. These hormones direct the synthesis of specific circulating hormones by the pituitary gland. Pituitary hormones travel out into the body and initiate the synthesis of particular hormones in target tissues.

7. Most hormones circulating in the bloodstream are produced by endocrine glands, whose activity is under the direct control of the nervous system.

REVIEW QUESTIONS

1. What is the definition of a hormone? Why do only certain organs or tissues respond to the presence of a particular hormone? Why is the endocrine system considered a chemical extension of the nervous system?

2. What hormones are secreted by the posterior pituitary gland? What function does each serve? Where are these hormones actually produced? How are these hormones transported to the region from which they are released?

3. What general category of hormones is secreted by the hypothalamus? What function does this class of hormones serve? How do these hormones reach the anterior pituitary? What relationship exists between the hormones produced by the hypothalamus and those produced by the anterior pituitary?

4. What are the seven principal hormones produced by the anterior pituitary? What function does each serve?

5. What two hormones are produced by the adrenal medulla? What nonhormonal function do they serve? How is their effect as hormones related to this function? How do they affect changes in heartbeat and blood flow? How does the hypothalamus regulate their production?

6. Why is tyrosine important to one class of peptide hormones? What is the primary function of endorphins? What is the basic chemical composition of insulin? How do peptide hormones effect changes in their target cells?

7. From what chemical compound are all steroid hormones derived? What are four examples of steroid hormones? In general, how do steroid hormones effect changes in their target cells?

8. What hormones are produced when the body's blood glucose levels drop below normal? How do these hormones act to return the level to normal? What hormone is produced when the body's blood glucose levels become elevated? How does this hormone act to return the level to normal?

9. In general terms, what is diabetes? What is the ultimate hormonal deficiency in these diseases? How does this affect an individual's ability to use glucose? How is this problem treated?

THOUGHT QUESTIONS

1. Why do you suppose the brain goes to the trouble of synthesizing releasing hormones, rather than simply directing the production of the pituitary hormones immediately?

2. Why do you suppose steroid hormones do not employ second messengers when so many peptide hormones do?

3. Two different organs, A and B, are sensitive to a particular hormone. Both organs have identical receptors and receptor activation produces the same intracellular second messenger. The intracellular effects of the hormone on A and B are completely opposite. Explain.

FOR FURTHER READING

AMERICAN COLLEGE OF SPORTS MEDICINE: "The Use of Anabolic-Androgenic Steroids in Sports," *Sports Medicine Bulletin*, vol. 19, 1984, page 13. Summarizes the incidence of anabolic steroid abuse and evidence for long-term health risks.

ATKINSON, M., and N. MacLAREN: "What Causes Diabetes?" *Scientific American*, July 1990, pages 62-71. For insulin-dependent diabetic patients, the answer is an autoimmune ambush of the body's insulin-producing cells. Why the attack begins and persists is now becoming clear.

BAULIEU, E. and P. KELLY: *Hormones*, Hermann Press, Paris, 1990. A comprehensive text on hormones, with chapters written by individual experts.

BERRIDGE, M. and R. IRVINE: "Inositol Phosphates and Cell Signalling," *Nature*, vol. 341, September 1989, pages 197-205. A second messenger mobilizes calcium and may stimulate entry of calcium into muscle cells.

BERRIDGE, M.: "The Molecular Basis of Communication Within the Cell," *Scientific American*, October 1985, pages 142-152. An up-to-date account of what is known about second messengers in the cell.

BLOOM, F.E.: "Neuropeptides," *Scientific American*, October 1981, pages 148-168. An account of advances in the study of endorphins and other brain hormones.

CANTIN, M., and J. GENEST: "The Heart as an Endocrine Gland," *Scientific American*, February 1986, pages 76-81. A good example of how the body self-regulates its activities; in addition to pumping blood, the heart secretes a hormone that fine-tunes the control of blood pressure.

CRAPO, L.: *Hormones, the Messengers of Life*, W.H. Freeman and Co., New York, 1985. A brief, readable account of the field of endocrinology, up-to-date and enjoyable.

DAVIS, J.: *Endorphins: New Waves in Brain Chemistry*, Doubleday & Company, Inc., New York, 1984. A popular account of research on brain hormones.

GOLDE, D., and J. GASSON: "Hormones That Stimulate the Growth of Blood Cells," *Scientific American*, July 1988, pages 62-70. There are three basic kinds of white blood cells, and which kind a stem cell develops into is under hormonal control.

"Growing Up With the Endocrine System," *Current Health*, vol. 1, February, 1987, page 3. Discussion of the structure and function of the endocrine system, including neuroendocrine mechanisms.

RASSMUSSEN, H.: "The Cycling of Calcium as an Intercellular Messenger," *Scientific American*, October 1989, pages 66-73. Calcium is involved in a variety of prolonged responses within cells.

ROSEN, O.: "After Insulin Binds," *Science*, September 1987, pages 1452-1458. A review of how the insulin receptor works—it is much like growth factor receptors and some cancer-inducing genes.

48

Locomotion

OVERVIEW

One of the obvious differences between animals and plants is that most animals move about from one place to another, and plants do not. Whether it is swimming, flying, or walking, movement almost always requires that an individual shift the position of a limb against a resistance. This is done by contracting muscles attached to the limb. In arthropods, the muscles are attached to a somewhat brittle outer skeleton, which must become thicker as the animals become larger to perform its function as an anchor point. In vertebrates, on the other hand, the muscles are attached to bones, and their contraction moves one bone relative to another within a flexible skin that stretches to accommodate the movement.

FOR REVIEW

Here are some important terms and concepts that you will encounter in this chapter. If you are not familiar with them, you should review them before proceeding.

Actin (Chapter 5)

ATP (Chapter 7)

Bone (Chapter 43)

Striated muscle (Chapter 43)

Neuromuscular junction (Chapter 44)

FIGURE 48-1
A female lion springs into motion, attempting to seize a gazelle. The integration of nerves and muscles and the flexibility of the skin combine to make this leap possible.

Some animals remain in one place all their adult lives, rooted like plants. The barnacles that encrust submerged rocks and the bottoms of ships are immobile, for example. They are an exception, however, to a very general rule, which is that animals move about from place to place, often with great speed (Figure 48-1). Many animals are creatures of constant activity, and few stay still for long. Of the three multicellular kingdoms that evolved from protists, only animals explore their environment in this active way; plants and fungi move only by growing, or as the passive passengers of wind and water. To move, all animals use the same basic mechanism, the contraction of muscles. In this chapter we will examine how animals use muscles to achieve movement, with our focus on vertebrates.

THE MECHANICAL PROBLEMS POSED BY MOVEMENT

If you've ever tried to lift a large boulder, you are familiar with the basic problem posed by movement, which is that gravity tends to hold objects in one place. The reason that you cannot toss a boulder with your little finger is that gravity is pulling down on the boulder far harder than your finger can push up. To move the boulder, you have to lift with a greater force than gravity is exerting. All motion must meet this simple requirement.

Organisms use the chemical energy of ATP to supply that force. By splitting an ATP molecule into ADP and P_i (inorganic phosphate), 7.3 kcal of energy per mole (the atomic weight of a substance—in this case ATP—expressed in grams) is made available to do the work of movement. Protists use this energy to wave cilia and so sweep themselves from place to place. Multicellular organisms apply this energy to compress the length of structural elements within certain cells called muscle cells, causing the cells to shorten. When a lot of muscle cells shorten all at once, they can exert a great deal of force.

BONE: WHAT VERTEBRATE SKELETONS ARE MADE OF

Bone is a special form of connective tissue in which collagen fibers (see Chapter 43) are coated with a calcium phosphate salt. The great advantage of bone as a structural material is that it is strong without being brittle. To understand the properties of bone, first consider those of fiberglass. Fiberglass is composed of glass fibers embedded in epoxy glue. The individual fibers are rigid, giving great strength, but they are brittle. The epoxy component of fiberglass, on the other hand, is flexible but weak. The composite, fiberglass, is both rigid and strong. When a fiber breaks because of stress and a crack starts to form, the crack runs into glue before it reaches another fiber. The glue distorts and reduces the concentration of the stress, and the adjacent fibers consequently are not exposed to the same high stress. In effect, the glue acts to spread the stress over many fibers.

The construction of bone is similar to that of fiberglass because the collection of the fibrils in bone runs in various directions. Small, needle-shaped crystals of a calcium-containing mineral, hydroxyapatite, surround and impregnate collagen fibrils of bone. The fibrils are placed parallel to the axes of long bones and also parallel to the curved ends of bones in joints. As a result of the way these fibers are placed, no crack can penetrate far into bone without encountering a hard mass of hydroxyapatite crystals embedded in a collagenous matrix. Bone is more rigid than collagen. On the other hand, bone is more flexible and resistant to fracture than is hydroxyapatite—or chitin.

Bone is a dynamic, living tissue that is constantly being reconstructed throughout the life of an individual. New bone is formed by cells called **osteoblasts,** which secrete the collagen fibers on which calcium is later deposited. Bone is laid down in thin, concentric layers called **lamellae,** like so many layers of paint on an old pipe. The lamellae are laid down as a series of tubes around narrow channels called **Haversian canals,** which run parallel to the length of the bone. The Haversian canals are interconnected and contain blood vessels and nerve cells. The blood vessels provide a lifeline to living bone-forming cells, whereas the nerves control the diameter of the blood vessels and thus the flow through them. When bone is first formed in the embryo, osteoblasts use the cartilage skeleton as a template for bone formation. Later, new bone is formed along lines of stress. Mature osteoblasts become trapped by the bone that they lay down, and are then called **osteocytes.** Flat bones, such as the sternum, originate from sheets of dense connective tissue as the osteoblasts do their work.

Bone is formed in two stages: first, collagen is laid down in a matrix of fibrils along lines of stress, and then calcium minerals impregnate the fibrils. These minerals provide rigidity, whereas the collagen provides flexibility.

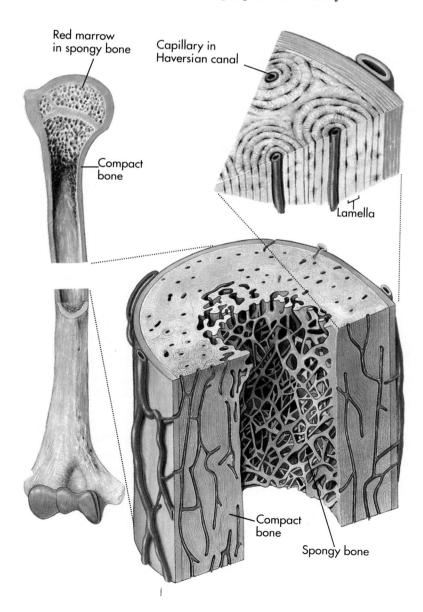

Red marrow in spongy bone

Capillary in Haversian canal

Compact bone

Lamella

Compact bone

Spongy bone

FIGURE 48-2
The organization of compact bone, shown at three levels of detail. Some parts of bone are dense and compact, giving the bone strength. Other parts are spongy, with a more open lattice; it is here that most red blood cells are formed.

The bones of the vertebrate skeleton are composed of two structural elements. The ends and interiors of long bones are composed of an open lattice of bone called **spongy bone tissue,** or **marrow.** Within this lattice framework, most of the body's red blood cells are formed. Surrounding the spongy bone tissue at the core of bones are concentric layers of **compact bone tissue** (Figure 48-2) in which the collagen fibrils are laid down in a pattern that is far denser than the marrow. The compact bone tissue gives the bone the strength to withstand mechanical stress.

JOINTS: HOW BONES OF THE SKELETON ARE ATTACHED TO ONE ANOTHER

The mechanical movement of the body occurs when bones move relative to one another. These interactions occur at **joints,** where one bone meets another. There are three kinds of joints (Figure 48-3):

Sutures are nearly immobile joints connected by a thin layer of connective tissue. The cranial bones of your skull are joined by sutures. In a fetus, these bones are not fully formed, and there are open areas of connective tissue between the bones ("soft spots," or fontanels) that allow the bones to slide over one another slightly as the fetus makes its journey down the birth canal during childbirth. Later, bone replaces this connective tissue.

Cartilaginous joints are slightly movable joints bridged by cartilage. The vertebral bones of your spine are separated by pads of cartilage called intervertebral disks. The disks allow some movement while acting as efficient shock absorbers.

Freely movable joints are swinging joints bridged by pads of cartilage and held together by bands of connective tissue called **ligaments.** The pads of cartilage that tip the two bones do not touch. Synovial membranes filled with fluid occur where tendons glide over bones, and this fluid lubricates the gliding of the bones across one another. **Rheumatoid arthritis** is a degenerative and very painful disorder in which the immune system attacks the bursa membrane; white blood cells degrade the cartilage and other connective tissue, and bone is deposited in the joint.

THE SKELETON

FIGURE 48-3
Three types of joints: fibrous, cartilaginous, and synovial.

For a muscle to produce movement, it must direct its force against another object. Some soft-bodied invertebrates like slugs, which live on land but have no shell or in-

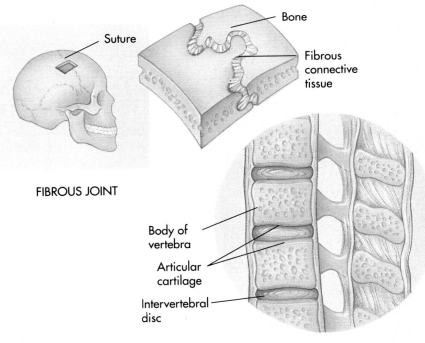

Suture

Bone

Fibrous connective tissue

FIBROUS JOINT

Body of vertebra

Articular cartilage

Intervertebral disc

CARTILAGINOUS JOINT

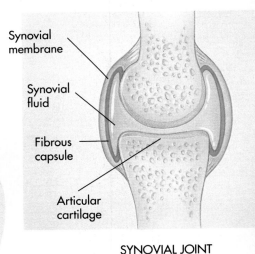

Synovial membrane

Synovial fluid

Fibrous capsule

Articular cartilage

SYNOVIAL JOINT

ternal skeleton, move by attaching themselves to the material over which they move. Other organisms use different types of skeletons against which the force of their muscles are directed.

Hydroskeletons. Most soft-bodied invertebrates, which have neither an internal nor an external skeleton, solve the problem of what to direct muscle force against by using the relative incompressibility of the water within their bodies as a kind of skeleton. They simply direct the force of their muscles against the water. This sort of skeleton is called a **hydroskeleton.** Earthworms have hydroskeletons, and so do jellyfish. In the case of jellyfish, the action of the muscles expelling the water produces an opposite reaction, causing the animal to move.

Exoskeletons. Most animals, however, are able to move because the opposite ends of their muscles are attached to hard part of their bodies, so that, for example, muscle contraction results in the beating of a wing or the lifting of a leg (Figure 48-4). When the hard body parts to which the muscles are attached is a shell that encases or surrounds the body, the shell is called an **exoskeleton.** Arthropods, for example, have muscles that are attached to a rigid, chitinous exoskeleton, enabling them to swim, to walk, and to fly (Figure 48-5). As long as an individual is small enough, this is an effective strategy. However, the rules of mechanics require that the exoskeleton must be much thicker to bear the pull of the muscles in large insects than in small ones. In an insect the size of a human being, the exoskeleton would need to be so thick that the animal could hardly move. This relationship puts real limits on the size of insects.

Endoskeletons. In vertebrates, muscles are attached to an internal scaffold of bone, an **endoskeleton,** which is both rigid and flexible and yet able to bear far more weight than chitin. Instead of a rigid exterior skeleton, vertebrates have a soft, flex-

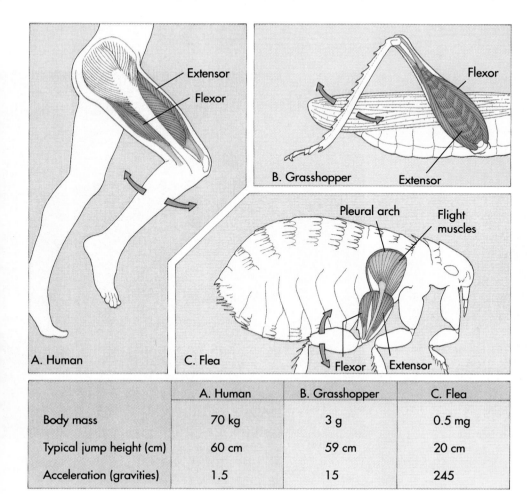

	A. Human	B. Grasshopper	C. Flea
Body mass	70 kg	3 g	0.5 mg
Typical jump height (cm)	60 cm	59 cm	20 cm
Acceleration (gravities)	1.5	15	245

FIGURE 48-4
An example of the action of muscles in a flea, a grasshopper, and a human being. Despite differing by several orders of magnitude in size, all three jump to similar heights. In all three, antagonistic muscle pairs control the movement of the legs. Muscles can exert force only by becoming shorter. All three animals shown here have double sets of muscles that work in opposite directions. In the human (**A**), the muscles are attached to the bones, whereas in the flea and the grasshopper (**B** and **C**), the muscles are attached to the inside of the skeleton. In each case, the flexor muscles move the lower leg closer to the body and the extensor muscles move it away. Because the flea muscle is more massive for its size, it is capable of producing much greater acceleration.

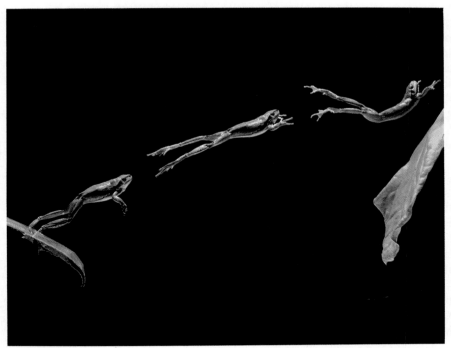

LEAPING

WALKING

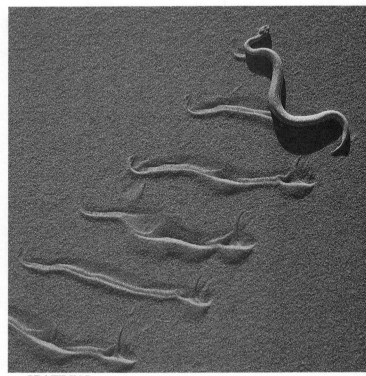

CRAWLING

FLYING

CLIMBING

RUNNING

SWIMMING

ALONG FOR THE RIDE

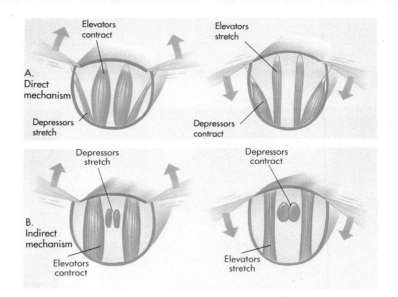

FIGURE 48-5
Muscles in insects. A. In most insects, as in birds, the raising and lowering of wings is achieved by the alternate contraction of extensor muscles (elevators) and flexor muscles (depressors). **B.** Four insect orders (including flies, mosquitoes, wasps, bees, and beetles), however, beat their wings at frequencies from 100 to more than 1000 times per second, faster than nerves can carry successive impulses! In these insects, the flight muscles are not attached to the wings at all, but rather to the wall of the thorax, which is distorted in and out by their contraction. The reason that these muscles can beat so fast is that the contraction of one set stretches the other, initiating its contraction in turn. Much less energy is expended in such beating than in that of the wings of birds or bats.

ible exterior, which stretches to accommodate the movement of their bodies (Figure 48-5). Whenever you bend you arm, the skin covering the joint of your elbow stretches; if it didn't, it would tear. The flexibility of the skin is a necessary component of vertebrate movement, which otherwise is determined largely by muscles and their attachments to bones. Vertebrate skin is described in detail in Chapter 52.

THE HUMAN SKELETON

The endoskeleton of humans is made up of 206 individual bones (Figure 48-6). These can be grouped according to their functions. The 80 bones of the **axial skeleton** support the main body axis, and the 126 bones of the **appendicular skeleton** support the arms and legs (Figure 48-7). Interestingly, the motor control systems of the body have eveolved to control the muscles of the axial skeleton (postural muscle) and the appendages (manipulatory muscle) more or less independently.

The Axial Skeleton. The axial skeleton is made up of the skull, backbone, and rib cage. Of the skull's 28 bones, 8 form the cranium that encases the brain; the rest are facial bones and middle-ear bones. The skull also contains the hyoid bone. It is suspended at the back of the jaw by muscles and a form of connective tissue called a ligament and supports the base of the tongue.

The skull is attached to the anterior (upper) end of the backbone, which is also called the **spine** or **vertebral column.** The spine has 33 vertebrae, stacked one on another to provide a flexible column that surrounds and protects the spinal cord. Curving forward from the vertebrae are 12 pairs of ribs, which are attached at the front to the breastbone, or sternum, forming a protective cage around the heart and lungs.

The Appendicular Skeleton. The 126 bones of the appendicular skeleton are attached to the axial skeleton at the shoulders and hips. The shoulder or **pectoral girdle** is composed of two large, flat shoulder blades (scapulae), each connected to the breastbone by a slender, curved collarbone (clavicle). The arms are attached to the pectoral girdle. Each arm and hand contains 32 bones. The clavicle is the most frequently broken bone of the body; if you fall on an outstretched arm, a large component of the force is transmitted to the clavicle.

The **pelvic girdle** forms a bowl that provides strong connections for the legs, which carry the weight of the body. Each leg and foot contains a total of 30 bones.

Tendons Connect Bone to Muscle

Muscles are attached to bones by straps of dense collagenous connective tissue called **tendons.** Bones pivot about joints becuase of where the tendons are attached to them. Each muscle acts to pull on a specific bone. One end of the muscle, the **origin,** at-

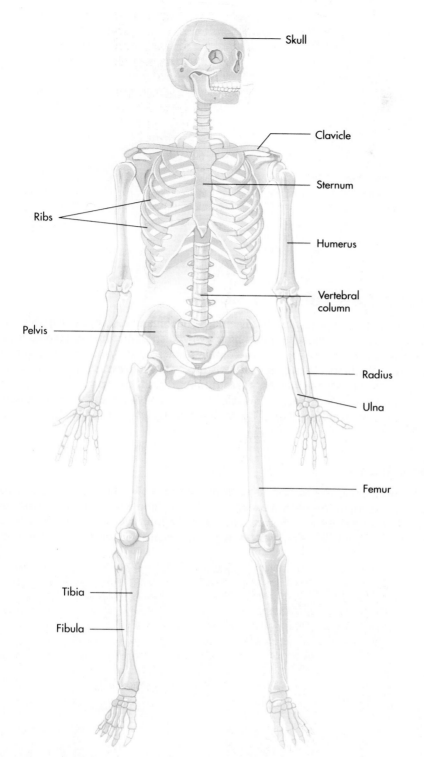

Skull

Clavicle

Sternum

Ribs

Humerus

Vertebral
column

Pelvis

Radius

Ulna

Femur

Tibia

Fibula

FIGURE 48-6
The human muscular system.

FIGURE 48-7
Axial and appendicular skeletons. The axial skeleton is shown in red, and the appendicular skeleton is shown in yellow.

taches to a bone that remains stationary during a contraction. This provides an object against which the muscle can pull. The other end of the muscle, the **insertion,** is attached to a bone that moves if the muscle contracts.

In the freely movable joints of vertebrates, muscles are attached in opposing pairs, flexors and extensors (Figure 48-8). When the flexor muscle of your leg contracts, the lower leg is moved closer to the thigh. When the extensor muscle of your leg contracts, the lower leg is moved in the opposing direction, further away. To use the body's energy efficiently, it is necessary to control antagonistic pairs of muscles. It would be futile if, when you lifted an object, the triceps contracted at the same time as the biceps. Your arm would not move.

FIGURE 48-8
Flexor and extensor muscles.

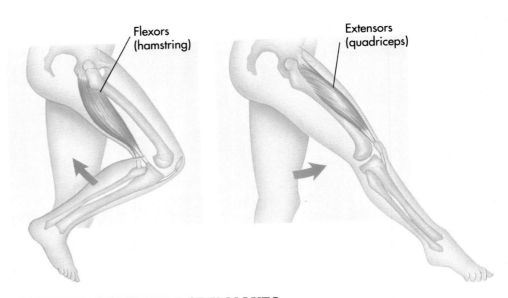

Flexors
(hamstring)

Extensors
(quadriceps)

MUSCLE: HOW THE BODY MOVES

Even though the cells of almost all eukaryotic organisms appear capable of shape changes, many multicellular animals have evolved specialized cells devoted almost exclusively to this purpose. These cells contain numerous filaments of the proteins actin and myosin. Such specialized animal cells are called **muscle cells.** As we mentioned in Chapter 43, vertebrates possess three different kinds of muscle cells: smooth muscle, skeletal muscle, and cardiac muscle. Of these, skeletal and cardiac muscle show clear patterns of striations caused by the geometric arrangement of the actin and myosin filaments.

Smooth Muscle

Smooth muscle was the earliest form of muscle to evolve, and it is found throughout the animal kingdom. **Smooth muscle** cells are long and spindle shaped, with each cell containing a single nucleus (Figure 48-9, *A*). The interiors of smooth muscle cells are packed with myofibrils composed of actin and myosin filaments, but the individual

FIGURE 48-9

Types of muscle.
A Smooth muscle cells are long and spindle shaped, with a single nucleus.
B Cardiac muscle cells, such as these from a human heart, also contain a single nucleus and are organized into long branching chains that interconnect, forming a lattice.
C Skeletal or striated muscle cells are formed by the fusion of several muscle cells, end to end, to form a long fiber with many nuclei.

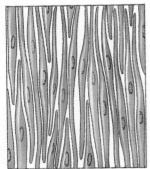
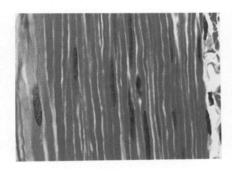

A   Smooth muscle

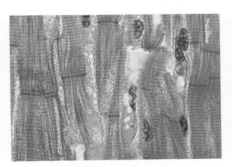
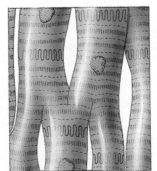

B   Cardiac muscle

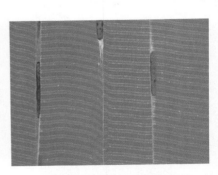

C   Skeletal muscle

FIGURE 48-10
The human muscular system.

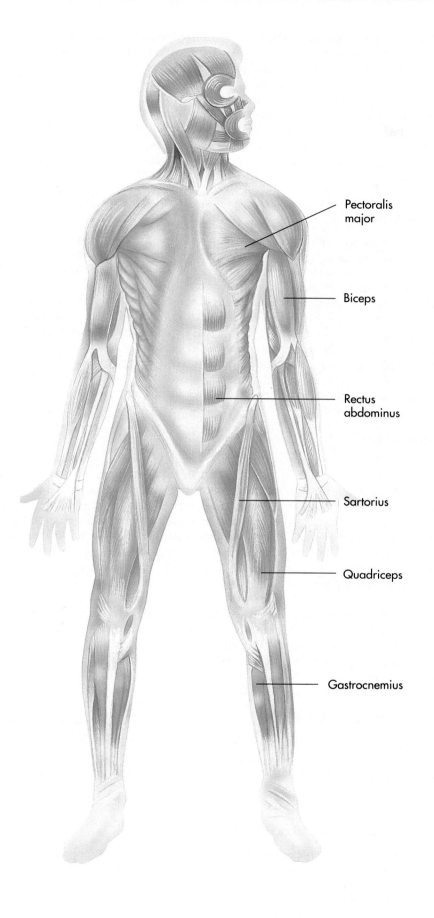

Pectoralis
major

Biceps

Rectus
abdominus

Sartorius

Quadriceps

Gastrocnemius

myofibrils of a cell are not aligned into organized assemblies as they are in skeletal and cardiac muscle. Smooth muscle tissue is organized into sheets of cells. In some tissues smooth muscle cells contract only when they are stimulated by a nerve or hormone. Examples are the muscles found lining the walls of many vertebrate blood vessels and those which make up the iris of the vertebrate eye. In other smooth muscle tissue, such as that found in the wall of the gut, the individual cells contract spontaneously, leading to a slow, steady contraction of the tissue. Smooth muscles are also classified as unitary or multiunit. In unitary smooth muscle, cells are electrically coupled to one another so the muscle contracts as a unit. In multiunit smooth muscle, individual cells are not coupled and must be activated separately.

Skeletal Muscle

Skeletal muscles are the muscles associated with the skeleton (Figure 48-10). They, along with cardiac muscle, are called **striated muscles** because they are marked with obvious lines (Latin *striae;* see Figure 48-9, *C*). Skeletal muscle cells are produced during development by the fusion of several cells at their ends to form a very long fiber. Each muscle cell or muscle fiber still contains all the original nuclei pushed out to the periphery of the cytoplasm. Each skeletal muscle is a tissue made up of numerous individual muscle cells that act as a unit. These skeletal muscle cells are specialized for rapid contractions and large forces. In contrast, smooth muscles are specialized for slow, maintained contraction with minimal energy utilization. Each smooth muscle cell participates individually in the contraction of the muscle. In skeletal muscle, numerous muscle cells pool their resources during contraction.

Cardiac Muscle

The vertebrate heart is composed of striated muscle fibers arranged very differently from the fibers of skeletal muscle. Instead of very long multinucleate cells running the length of the muscle, heart muscle is composed of chains of single cells, each with its own nucleus. Each cell is coupled to its neighbors electrically by gap junctions. These chains of cells are organized into fibers that branch and interconnect, forming a latticework (see Figure 48-9, *B*). This lattice structure is critical to the way heart muscle functions. Heart contraction is initiated at one location by the opening of transmembrane channels that admit ions into the muscle cells there, altering the voltage difference across their membranes. This change in membrane charge is a depolarization, similar to that discussed in Chapter 44. When two cardiac muscle fibers touch one another, their membranes make an electrical junction, and the electrical depolarization of the initial fiber initiates a wave of contraction throughout the heart. This wave of depolarization rapidly passes from one fiber to another across these junctions. For this reason the heart contracts in an orderly fashion as a unit so as to propel the blood throughout the body.

HOW MUSCLES WORK

Muscle cells move by expanding and contracting portions of their surfaces. This ability to alter surface relationships arises from dynamic changes in their cytoskeletons. The key elements driving these changes are the tiny cables within muscle cells called myofilaments.

The Structure of Myofilaments

Far too fine to see with the naked eye, individual myofilaments are only 6 nanometers thick. Myofilaments are long chains of the proteins **actin** and **myosin.** There are two additional proteins, **troponin,** and **tropomyosin,** that are important in initiating contractions in skeletal muscle; we discuss them more later. The organization of actin and myosin is such that muscles are arranged into repeating units called **sarcomeres.**

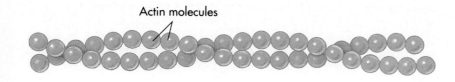
Actin molecules

FIGURE 48-11
An actin filament.

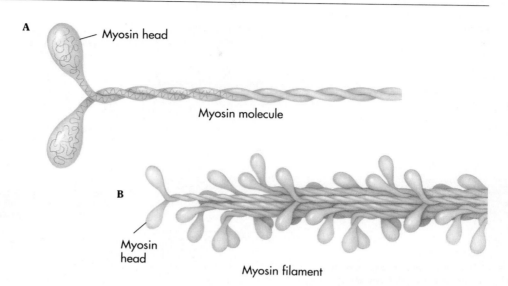

A — Myosin head

Myosin molecule

B

Myosin head

Myosin filament

FIGURE 48-12
The structure of myosin.
A Myosin molecule. Each myosin molecule is a coil of two chains wrapped around one another; at the end of each chain is a globular region referred to as the "head."
B Myosin filament. Myosin molecules are usually combined into filaments, which are cables of myosin from which the heads protrude at regular intervals.

1. *Actin*. Actin microfilaments are one of the two major components of myofilaments. The individual actin proteins are the size of a small enzyme. Actin molecules polymerize to form thin filaments (Figure 48-11). The filaments consist of two strings of actin proteins wrapped around one another, like two strands of pearls loosely wound together. The result is a long thin helical filament with a diameter of about 6 nanometers.

2. *Myosin*. The other major constituent of myofilaments, myosin, is a protein molecule more than ten times longer than an individual actin molecule. Myosin has an unusual shape: one end of the molecule consists of a very long rod, whereas the other end consists of a double-headed globular region. In electron micrographs a myosin molecule looks like a two-headed snake. Like actin, myosin spontaneously forms into filaments (Figure 48-12). The heads of the myosin molecule are capable or interacting with actin to form what are known as crossbridges.

How Myofilaments Contract

The contraction of myofilaments occurs when the heads of the myosin filaments change their orientation with respect to their rodlike backbone. The myosin heads tilt in a way similar to flexing the hand at the wrist. As a result of this tilt, the actin filaments are pulled so as to slide past the myosin filaments. In effect, myosin "walks" step by step along actin (Figure 48-13). Each step uses a molecule of ATP.

How does microfilament sliding lead to cell movement? Myofilaments convert the sliding of fibers into motion by anchoring the actin, the ends of which are bound to the anchoring protein alpha-actinin (Figure 48-14), sometimes visible in micrographs and referred to as the "Z line." Alpha-actinin is found widely distributed on the interior surfaces of the plasma membranes of eukaryotic cells. Because the actin is not free to move with respect to the alpha-actinin to which it is bound, the zones between alpha-actinin anchors shorten when the microfilaments slide past myosin. Because the myofilament may be attached at both of its ends to membrane, its overall shortening moves the membranes to which the myofilament is attached.

Myofilaments are composed of a highly ordered complex of actin and myosin.

FIGURE 48-13

The mechanisms of myofilament contraction. Myosin moves along actin (from left to right in this diagram) by first binding to it and then hunching forward as the result of a change in the shape of the myosin head. The splitting of ATP recocks the mechanism, returning the myosin head to its extended position.

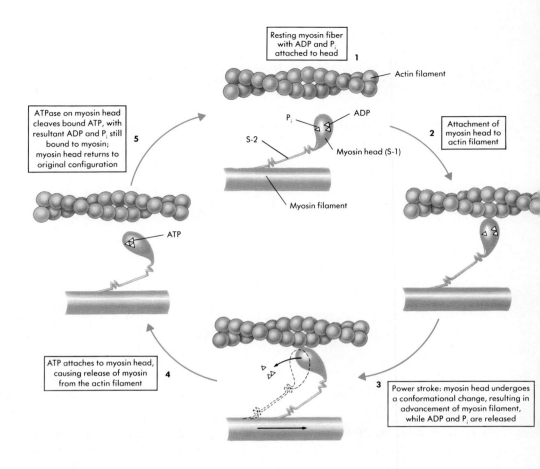

Resting myosin fiber with ADP and P$_i$ attached to head **1**

Actin filament

ATPase on myosin head cleaves bound ATP, with resultant ADP and P$_i$ still bound to myosin; myosin head returns to original configuration **5**

P$_i$ ADP

S-2

Myosin head (S-1)

Attachment of myosin head to actin filament **2**

Myosin filament

ATP

ATP attaches to myosin head, causing release of myosin from the actin filament **4**

3 Power stroke: myosin head undergoes a conformational change, resulting in advancement of myosin filament, while ADP and P$_i$ are released

FIGURE 48-14

The interaction of actin microfilaments and myosin filaments in vertebrate muscle. The heads on the two ends of the myosin filament are oriented in opposite directions, so that as the right-hand end of the myosin molecules "walks" along the actin filaments going right and so pulling them and the attached alpha-actinin (a component that in muscles is called the Z line) leftward toward the center, the left-hand end of the same myosin molecule "walks" in a leftward direction, pulling its actin filaments and their attached Z line rightward toward the center. The result is that both Z lines move toward the center—contraction.

The secret of muscle contraction lies in the way in which the actin and myosin fibers are combined. They *interdigitate*. The arrangement is diagrammed in Figure 48-14, with a myosin filament interposed between two pairs of actin filaments, with the myosin heads (crossbridges) jutting out toward the actin filaments on each side.

The contraction of vertebrate muscles and many other kinds of cell movement in eukaryotes result from the movements of myofilaments within cells. The myofilaments are composed of long parallel fibers of actin cross-connected by myosin. Their movement results from an ATP-driven conformational change in myosin.

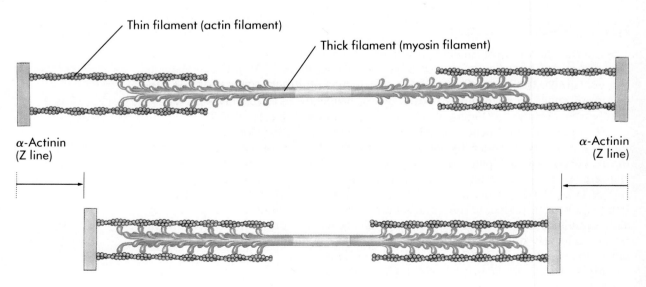

Thin filament (actin filament)

Thick filament (myosin filament)

α-Actinin (Z line)

α-Actinin (Z line)

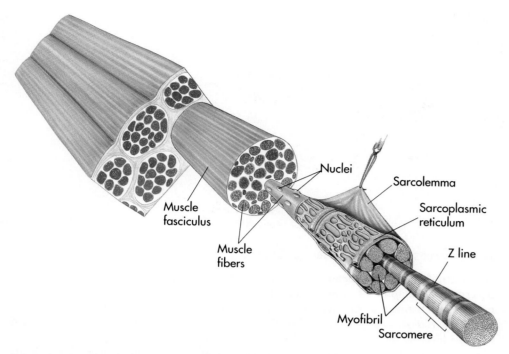

Labels on figure: Nuclei, Sarcolemma, Sarcoplasmic reticulum, Z line, Muscle fasciculus, Muscle fibers, Myofibril, Sarcomere

How Striated Muscle Contracts

Striated muscle cells are produced during development by the fusion of several cells, end to end, to form a very long fiber. A single fiber will typically run the entire length of a vertebrate muscle. Each cell, or muscle fiber, still contains all of the original nuclei, pushed out to the periphery of the cytoplasm by a central cable of 4 to 20 myofibrils. The cytoplasm in striated muscle is given a special name, the **sarcoplasm**. The myofibrils which run down the center of a muscle cell are highly organized to promote simultaneous contraction (Figure 48-15):

1. A myofibril is made up of a long chain of contracting units called **sarcomeres,** lined up like the cars on a train.
2. Each sarcomere is composed of interdigitating filaments of actin and myosin. One collection of actin filaments is attached to the front and another to the back of the sarcomere, to plates referred to as Z lines (see Figure 48-14). These front and back assemblies of actin filaments are not long enough to reach each other in the center of the sarcomere. They are joined to one another by interdigitating myosin filaments.
3. The sarcomere contracts when the heads of the myosin filaments change their shape. Since these heads are in contact with the actin filaments, the effect of their contraction is to pull the myosin along the actin.
4. The orientation of myosin is such that it moves along actin towards the Z line. Because both ends of the myosin filaments move in this manner simultaneously, the effect is to pull the two Z lines together, contracting the sarcomere.
5. Simultaneous contraction of all of the sarcomeres of a myofibril results in an abrupt and forceful shortening of the myofibril. All of the myofibrils of a muscle fiber usually also contract in concert at the same time, producing a very strong contraction in the length of the muscle fiber cell.

The sarcomeres are lined up, in register with one another, all along the length of the stacked myofibrils. This gives striated muscle, as viewed with the light microscope, the distinctive pattern of bands or striations (see Figure 43-10, *C*) that gives it its name.

Muscle cells are rich in actin and myosin, which form myofilaments that are capable of contraction. Many muscle cells contracting in concert can exert considerable force.

FIGURE 48-15
The organization of striated muscle.

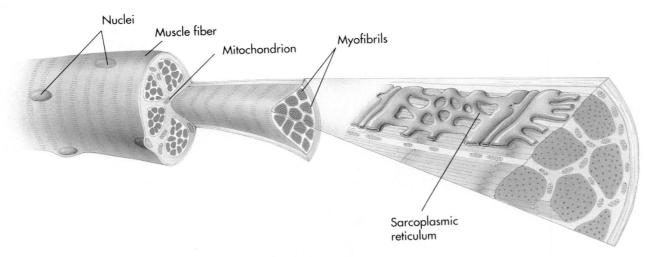

Nuclei
Muscle fiber
Mitochondrion
Myofibrils
Sarcoplasmic
reticulum

How Nerves Signal Muscles to Contract

FIGURE 48-16
The sarcoplasmic reticulum is a system of membranes that wraps around individual myofibrils of a muscle fiber. The transverse tubules are continuous with the plasma membrane of the muscle fiber and relay the wave of depolarization initiated at a neuromuscular junction to individual myofibrils.

In vertebrate striated skeletal muscle, contraction is initiated by a nerve impulse. As described in Chapter 44, a nerve impulse is a wave of depolarization moving down a nerve cell, channels in the neuron membrane opening and ions crossing into the cell as the wave moves along. The end of the nerve fiber that initiates muscle contraction is embedded in the surface of the muscle fiber, forming a neuromuscular junction. When a wave of depolarization reaches the end of a particular neuron where it attaches to a muscle—the motor end plate—the arrival of the wave of depolarization causes the neuron's membrane tip to release the neurotransmitter acetylcholine into the junction. The acetylcholine passes across to the muscle membrane and causes the opening of calcium ion channels in that membrane.

How does the depolarization of the nerve fiber membrane cause the contraction of the muscle fibers? The endoplasmic reticulum of a striated muscle cell, which is called the **sarcoplasmic reticulum,** wraps around each myofibril like a sleeve (Figure 48-16). As a result, the entire length of every myofibril is very close to the intracellular space bounded by the sarcoplasmic reticulum membrane system. Within the sarcoplasmic reticulum are embedded numerous ion channels that permit calcium to pass through them. In resting muscle, calcium ions are actively pumped into the sarcoplasmic reticulum by an ATP-driven calcium pump, thus concentrating all the calcium ions of that cell within the spaces of the sarcoplasmic reticulum. The depolarization of the muscle fiber membrane opens calcium channels in the sarcoplasmic reticular membrane* and thus causes the release of this concentrated calcium into the cytoplasm (sarcoplasm). This calcium acts as a trigger to initiate contraction of the myofibril. It does this in the following way (Figure 48-17):

1. In resting muscle, myosin filaments are not free to interact with actin, because the sites on actin where the myosin heads must make contact are covered by the protein tropomyosin. Another protein, troponin, binds to both actin and tropomyosin.
2. Troponin molecules are also able to bind calcium ions, and when they do, the troponin molecules change their shape. As a result of this change in shape, tropomyosin is repositioned to a new location, where it does not cover the myosin binding sites on actin. Only when this repositioning has occurred can contraction take place. The myosin heads are now free to form crossbridges with actin and, with ATP expenditure, move along the actin in a stepwise fashion to shorten the myofibril.

Thus the release of calcium by the nerve's stimulation of the sarcoplasmic reticulum acts as a "calcium-activated switch," triggering the contraction of the myofibril.

---

*It is not known how the signal passes from the outer cell membrane to the sarcoplasmic reticular membrane, some 50 nanometers away. This is a long distance in molecular terms. Some investigators believe that a chemical messenger such as inositol triphosphate links the two membranes, whereas others believe calcium itself passes across. The matter is under active investigation.

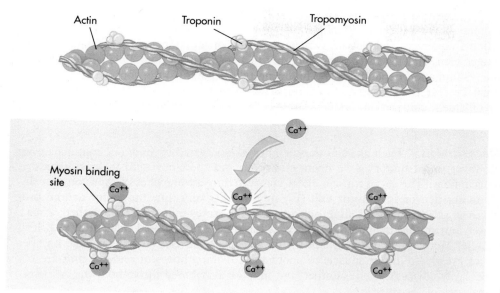

FIGURE 48-17
How calcium controls muscle contraction.
A When the muscle is at rest, a long filament composed of the molecule tropomyosin blocks the myosin-binding sites of the actin molecule. Without actin's ability to form links with myosin at these sites, muscle contraction cannot occur.
B When calcium ions bind to another protein, troponin, the resulting complex displaces the filament of tropomyosin, exposing the myosin-binding sites of actin, cross-links form between actin and myosin, and contraction occurs.

Contraction of Whole Muscles

The total amount of force developed by skeletal muscles depends on two factors: (1) *summation* of contractions of individual muscle fibers, and (2) increases in the number of fibers active in a given muscle, a process referred to as *recruitment*. In muscle contraction there may or may not be an actual shortening of the muscle.

Summation. Summation is the result of repetitive activation of the motor neuron innervating a given muscle fiber. If muscle action potentials occur very infrequently, the Ca^{++} concentration returns to rest levels before the arrival of the next action potential. When two or more muscle action potentials occur in rapid succession, not all of the Ca^{++} realeased in response to the first activation can be removed before more Ca^{++} is released. As a result, the contractions summate. As the stimulation rate is

---

A VOCABULARY OF MOVEMENT

actin One of the two major proteins that make up myofilaments (the other is myosin).

bone Hard, flexible material that forms the vertebrate endoskeleton; a fiberglass-like material composed of calcium phosphate salts embedded within a matrix of collagen.

cartilage A strong, flexible connective tissue in the skeletons of vertebrates; forms much of the skeleton of embryos, but is largely replaced by bone in adults of most species.

ligament A band or sheet of connective tissue that links bone to bone.

motor endplate The point where a neuron attaches to a muscle; a neuromuscular synapse.

myofilaments A contractile microfilament within muscle cells composed largely of actin and myosin; sometimes called myofibrils.

myosin One of the two protein components of myofilments (the other is actin).

sarcomere The fundamental unit of contraction in skeletal muscle; the repeating bands of actin and myosin that appear between two Z lines.

sarcoplasmic reticulum The endoplasmic reticulum of a muscle cell; a sleeve of membrane that wraps around each myofilament.

tendon A strap of collagenous connective tissue that attaches muscle to bone.

increased, the Ca^{++} concentration increases and the total force rises. At high stimulation frequencies individual "twitches"—firing of individual muscle fibers—are no longer seen, and the force rises smoothly to a maximum value. This is called a **tetanus.** Summation depends on the following two factors: (1) the contractile machinery must be able to respond to maintained high levels of internal Ca^{++} with continual myofilament contraction, and (2) the muscle action potential must not last as long as the twitch.

Recruitment. Each skeletal muscle fiber is innervated by only one motor neuron. However, most motor nerve fibers innervate, and therefore control, more than one muscle fiber. The set of muscle fibers innervated by all branches of the axon of a single motor neuron is a **motor unit** (Figure 48-18). Every time the motor neuron produces an action potential, all muscle fibers in the motor unit contract together. Thus a motor unit, rather than a muscle fiber, is the smallest functional element of a skeletal muscle. Motor units differ in size within any given muscle, and the average number varies greatly from one muscle to another. Individual fibers in a motor unit are not necessarily adjacent to one another, but may be distributed throughout the cross-sectional area of the muscle.

In addition to summation, the total force exerted by a skeletal muscle is regulated by selective activation of its motor units, termed **recruitment.** The motor units activated at the lowest levels of effort are those with the smallest number of muscle fibers. The initial increments to the total force generated by a muscle are therefore relatively small. If greater force is required, larger and larger motor units are recruited and the force increments become larger. This results in smooth increases in force.

Types of Contractions. When a muscle is unable to shorten the generation of force is called an **isometric** (constant length) contraction. Here the effect of crossbridge formation is to increase the **tension** or tautness of the muscle. Force generation without shortening occurs when one tries to lift an immovable object. When a muscle shortens under a constant load, the contraction is referred to as **isotonic** (constant tension). A contraction can change from isometric to isotonic and vice versa. For instance, when a weight is lifted from the ground, the contraction is isotonic as long as the weight is moving. If a weight is held at arm's length, the force exerted by the muscles is equal to the weight and muscle length is not changing, so the contraction is isometric.

Maximum tension is generated in muscles at their normal resting length. At this length the overlap between thick and thin filaments is such that all myosin heads may form crossbridges. In the rowing analogy, all the oarsmen are the boat. When a muscle is stretched by more than a few percent of rest length, the filament overlap is decreased. Some myosin heads can no longer form crossbridges, so that the maximum force decreases. Everyday experience tells us that heavier weights cannot be lifted as rapidly as light loads. The velocity of shortening of a muscle decreases when the load increases, becoming zero when the load equals the maximum force the muscle can exert.

Muscle Energy Consumption. In a contracting muscle the active crossbridges require large amounts of ATP. Since each crossbridge cycle requires the hydrolysis of one ATP, the energy requirements of active muscle are greatly influenced by the rate of crossbridge cycling (in the rowing analogy this would be the number of strokes per minute). The rate of energy use is higher when muscles shorten than when force is exerted at constant muscle length because crossbridges cycle more rapidly when filaments are sliding past one another.

Both glycolysis and oxidative phosphorylation produce ATP (see Chapter 8, p. 174). Glycolysis is rapid, but less efficient, and results in the production of lactic acid as an end-product. Oxidative phosphorylation produces a lot of ATP, but a constant oxygen supply is required. In rapidly contracting muscle the oxygen supply becomes inadequate, so muscle metabolism switches to glycolysis using the glycogen stored in muscles as a source of glucose.

FIGURE 48-18
A motor unit. The way in which the branches of this neuron axon connect with a skeletal muscle is shown. The thick central portion is the body of the axon, with small branches each ending in a motor end plate, where the nerve and muscle join to form a neuromuscular junction.

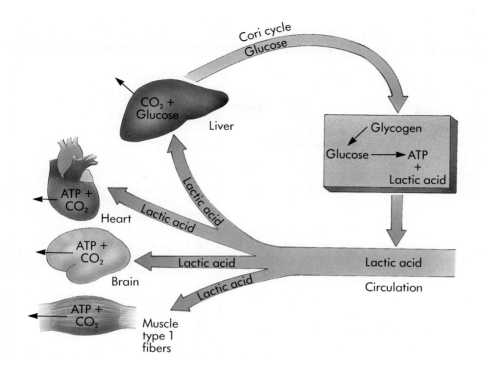

FIGURE 48-19
Metabolic fate of lactic acid produced during exercise. The Cori cycle represents one possible fate; other tissues that respire oxidatively during exercise, such as the heart and kidney can convert lactic acid to pyruvate and metabolize it to CO_2 by way of the Krebs cycle.

The Oxygen Debt. At the end of strenuous exercise, oxygen comsumption does not fall to its resting level immediately. The extra oxygen consumption after exercise is referred to as **oxygen debt.** Some of the oxygen debt is the result of accumulated lactic acid which must be metabolized to CO_2 by oxidative phosphorylation. The process by which the lactic acid is converted to glucose and returned to the muscle is the **Cori cycle** (Figure 48-19). During heavy exercise lactic acid is released by muscle cells into the bloodstream. Some lactic acid is taken up by the liver and resynthesized to glucose. The resulting glucose may be taken up by muscles that have depleted their glycogen reserves. The oxygen used by the liver to convert lactic acid to glucose is the oxygen debt.

Muscle Fatigue. **Fatigue** refers to the use-dependent decrease in the ability of a muscle to generate force. Fatigue is mostly the result of operation under anaerobic conditions. High-intensity exercise causes an increase in the lactic acid in the muscle and blood. The resulting acidic conditions alter the activity of glycolytic enzymes and interferes with the crossbridge cycle. Fatigue also involves the depletion of the glycogen stores of the liver and muscle. When this occurs the body must begin to utilize fat as the sole energy source. Energy production from fat occurs at about half the rate of energy production from glucose, so the depletion of glycogen stores is marked by a substantial decrease in muscle performance. Distance runners refer to this as "hitting the wall".

> **In isotonic contractions muscle tension is constant and shortening results**
> **from force development. In isometric contractions muscle length is constant**
> **and sarcomere shortening is accompanied by an increase in muscle tension.**

Cardiac Muscle

The molecular mechanism of force generation is the same in cardiac and skeletal muscle. However, cardiac muscle cells are shorter and, more importantly, are linked together electrically. When one cardiac muscle cell is excited, the action potential spreads throughout the entire muscle. Thus cardiac muscle contracts as a unit. Action potentials in cardiac muscle last much longer than in skeletal muscle fibers so that individual contractions of cardiac muscle cannot summate. The advantage of the pro-

The various muscles of the body have different functions. Finger and eye muscles must contract rapidly and they fatigue with continued use. The muscles that control posture contract slowly and do not fatigue. These functional differences can occur because there are three distinct muscle **fiber types.** Fiber types differ in two basic ways: (1) the ATPase activity of the myosin in their thick filaments differs, (2) they use different metabolic pathways for energy, either oxidative phosphorylation or glycolysis.

The three muscle fiber types are: **slow oxidative, fast oxidative,** and **fast glycolytic.** The speed of contraction is related to the kind of myosin a muscle fiber contains, called **myosin isozymes.** Fast oxidative and fast-glycolytic fibers contain myosin with a high ATPase activity. Slow-oxidative fibers have myosin with low ATPase activity. The higher the rate of ATP hydrolysis, the faster the crossbridges can cycle and the more rapidly a muscle can shorten. The price of faster shortening is a higher rate of ATP use.

Slow-oxidative muscle fibers are found in the greatest numbers in postural muscles, such as the long muscles in the back. These fibers have a high mitochondrial density and rich capillary blood supply. They are red in color because they contain an oxygen carrier known as **myoglobin.** Myoglobin provides a reserve oxygen supply. The two types of fast muscle are distinguished from one another by their relative ease of fatigue. Fast-oxidative fibers resist fatigue. Fast-glycolytic fibers rely on anaerobic glycolysis and fatigue quickly. Fast-glycolytic fibers have large stores of glycogen but little myoglobin and they appear pale.

Muscle fibers that are used frequently respond by increasing their performance. The changes that occur differ for different fiber types. Oxidative fibers respond with increases in the myoglobin content, number of mitochondria, and the number of capillaries serving the muscle. These changes involve little change in the mass of the muscle. Glycolytic fibers respond by increasing the number of myofibrils and the size of the glycogen stores. These changes necessitate an increase in fiber diameter and thus an increase in muscle mass, termed **hypertrophy.**

Individual muscles are mixtures of glycolytic and oxidative fibers, but for each muscle the mix corresponds to functions normally assumed by that muscle. For example, the muscles of the neck and those attached to the spine are predominantly oxidative fibers, whereas those of the arms are predominantly glycolytic. In each individual, the numbers of fast and slow fibers in each skeletal muscle are constant after maturity. Thus both the total number of fibers and the relative numbers of fast and slow fibers are more strongly affected by heredity than by sex or training.

The proportion of the three fiber types present in a particular muscle is not altered by training. However, selective use of fibers of a single type results in hypertrophy of those fibers. Development of endurance is favored by exercise that calls for sustained effort, such as distance swimming, running, or skiing. The result is an increase in the circulatory supply, mitochondria, and oxidative enzymes and myoglobin in all fibers.

The situation is quite different for brief, high-intensity training such as weight lifting. The increase in muscle mass experienced by bodybuilders is the result of two changes: (1) an increase in the diameters of glycolytic fibers, and (2) the addition of collagen and other connective tissue required to sustain the passive tension of heavy loads. The rate and amount of tension developed during training exercises is the most important factor for increasing contractile protein incorporation into muscle. The short-term demands of maximum force development do not cause the circulatory and metabolic changes associated with endurance training. The effects of both types of training are rapidly reversed if the regimen is not maintained, called **disuse atrophy.**

longed action potential is that it prevents tetanic contractions that would interfere with the heart's pumping cycle of contraction and relaxation.

Smooth Muscle

Smooth muscle surrounds hollow internal organs including the gastrointestinal tract and all blood vessels except capillaries. Smooth muscle contains actin and myosin, but these proteins are not organized into sarcomeres. Parallel arrangements of thick and thin filaments cross diagonally from one side of the cell to the other. Myosin molecules are either attached to structures called **dense bodies,** the functional equivalents of Z lines, or to the muscle membrane. Most smooth muscles have 10 to 15 actin filaments per myosin filament compared to the 3 thin filaments per thick filament of striated muscle.

Smooth muscle cells do not have the sarcoplasmic reticulum seen in striated muscle cells. Contraction in smooth muscle cells requires the entry of Ca^{++} from the extracellular space. This is often caused by autonomic neurotransmitter activation of membrane Ca^{++} channels. Inside the smooth muscle cell the Ca^{++} combines with a protein called **calmodulin.** The Ca^{++} calmodulin complex activates the enzyme **myosin light chain kinase** (MLCK). MLCK then phosphorylates the myosin heads, and this phosphorylation exposes a binding site for actin. Since initiation of contraction in smooth muscle involves a change in the thick filament, smooth muscle differs fundamentally from striated muscle, where initiation of contraction involves primarily the thin filament.

A LOOK BACK

In the previous chapters we have discussed the nerves that initiate contraction of most muscles, the nervous system of which they are an integral part, and the sensory receptors that provide information to the nervous system. The nervous system responds to this information by issuing commands to muscles and glands. The movement of vertebrates in large measure reflects the interaction of these three components: sensory input of information, association within a central nervous system, and motor nerve commands to muscles. The charging of a rhinoceros, the delicate pirouette of a ballerina, the blinking of your eye, all share this same pattern of input and response.

SUMMARY

1. In vertebrates, movement results from the contraction of muscles anchored to bones. When the limb containing the bone is pulled to a new position, the skin stretches to accommodate the change.

2. Muscle contraction provides the force for mechanical movement of the body.

3. Mechanical movement also requires something against which the force can act—a skeleton. Vertebrate skeletons are internal and made of bone; many other animals have external skeletons.

4. Striated muscle is organized into trunks of long fibers and contracts only when stimulated by a nerve.

5. Muscle cells contract as a result of sliding of myofilaments. The myofilaments are composed of the proteins actin and myosin, together with three other proteins, which together control the contraction.

6. In a myofilament the myosin is located between adjacent actin filaments. Changes in the shape of the ends of the myosin molecule, driven by the splitting of ATP molecules, cause the myosin molecule to move along the actin, producing contraction of the muscle.

7. In the vertebrate striated skeletal muscle, contraction is initiated by a nerve impulse. Acetylcholine passes across the neuromuscular junction from the nerve to the muscle, causing the muscle to contract.

8. Calcium acts to reposition molecules blocking actin, freeing myofibrils to interact.

REVIEW QUESTIONS

1. What form of chemical energy do living organisms use to counter the force of gravity? How much energy is released by splitting one mole of these molecules? To what use is this energy put in a muscle cell?

2. To what are arthropod muscles attached? How does this relationship limit the size of highly mobile arthropods? To what organs are vertebrate muscles attached? Why is this system superior to that of the arthropods?

3. In what way are muscle cells specialized for movement? What particular name is given to the microfilaments that occur in muscle cells? Of what two proteins are myofilaments primarily composed? What additional three proteins are associated with muscle? Which of these are found exclusively in skeletal muscle?

4. Briefly describe how the actin and myosin filaments slide past one another. At which step is a molecule of ATP used?

5. How does the sliding of actin past myosin result in cell movement? Where in a eukaryotic cell is alpha-actinin located?

6. What special name is given to the cytoplasm of striated muscle cells? What is the cellular derivation of an individual striated muscle fiber? What subunits comprise a single myofibril? What is the anatomy of a single one of these subunits?

7. What initiates contraction in a vertebrate skeletal muscle? What chemical is released at the junction of this regulator and the muscle? In general terms, how does this chemical cause the muscle fiber to contract?

8. In a muscle fiber, what constitutes the sarcoplasmic reticulum? How does it control the location of the calcium ions within the muscle? What is the result when it is depolarized?

9. How does tropomyosin interact with the elements of a resting muscle? Where are troponin molecules located? How does troponin interact with the elements of a resting muscle?

10. How do calcium ions interact with troponin molecules? What effect does this have on the tropomyosin trigger?

 ## THOUGHT QUESTIONS

1. Myofilaments can contract forcefully, pulling membranes attached to the two ends toward one another. Myofilaments cannot expand, however, pushing membranes attached to the two ends of a myofilament apart. Why is it that myofilaments can pull but not push?

2. Why do you suppose that Ca^{++} ions are employed to initiate muscle contration, rather that Na^+ or K^+?

 ## FOR FURTHER READING

ALEN, M. AND OTHERS: "Response of Serum Hormones to Androgen Administration in Power Athletes," *Medical* Science Sports Exercise, vol 17, 1985, page 354. Explores the long-term changes in the endocrine system in athletes taking large doses of anabolic steroids to improve performance.

ALEXANDER, R.M.: "How Dinosaurs Ran," *Scientific American*, April 1991, pages 130–136. Taking the approach of a structural engineer, the author argues that many giant dinosaurs were formidable running machines.

CARAFOLI, E., and J. PENNISTON: "The Calcium Signal," *Scientific American*, November 1985, pages 70-80. The release of calcium ion is the only known way in which the electricity of the nervous system is able to produce changes in the body. Nerves regulate all muscle contractions and hormone secretions by controlling the level of Ca^{++} ions.

COHEN, C.: "The Protein Switch of Muscle Contraction," *Scientific American*, November 1975, pages 36-45. How proteins associated with myofilaments interact with Ca^{++} ions to trigger contraction.

GORDON, K.: "Adaptative Nature of Skeleton Design," *BioScience*, December 1989, pages 784-790. Plasticity allows changes in strength and locomotion.

GOSLOW, G., AND OTHERS: "Bird Flight—Insights and Complications," *BioScience*, February 1990, pages 108-115. How flapping flight evolved is still a puzzle, although key questions have been answered.

HILDEBRAND, M.: "The Mechanics of Horse Legs," *American Scientist*, November 1987, pages 594-601. A delightful analysis of the horse leg as a lever system, from the perspective of physics.

LAI, F.A., AND OTHERS: "Purification and Reconstitution of the Calcium Release Channel from Skeletal Muscle," *Nature*, vol. 331, January 28, 1988, pages 315-319. Report of a very important research result, the isolation of the Ca^{++} channel from neuromuscular junctions.

RASMUSSEN, H., Y. TAKUWA and S. PARK: "Protein Kinase C In The Regulation Of Smooth Muscle Contraction," *FASEB Journal*, vol. 1, 1987, page 177. Comprehensive discussion of the mechanisms of smooth muscle contraction and excitation-contraction coupling in smooth muscle.

SCHMIDT-NIELSEN, K.: *Animal Physiology: Adaptation and Environment*, ed. 3, Cambridge University Press, New York, 1983. Chapter 11 presents an outstanding treatment of muscles and bones from an evolutionary perspective.

SMITH, D.: "The Flight Muscles of Insects," *Scientific American*, June 1965, pages 76-89. Some insects flap their wings more rapidly than nerves can carry successive impulses.

WEEKS, O.: "Vertebrate Skeletal Muscle: Power Source for Locomotion," *BioScience*, December 1989, pages 791-799. The organization of muscle fibers is highly adaptable.

WEXLER, M.: "Capturing the Beauty of Flight," *National Wildlife*, August 1990, pages 4-9. Ingenious photography captures some stunning images of birds frozen in flight.

Fueling Body Activities: Digestion

OVERVIEW

Digestion is the conversion of parts of organisms into small molecules that can easily be used as food by cells: sugars, lipids, and amino acids. In many simple animals digestion is intracellular; in vertebrates it is extracellular, food being digested as it passes through a long, one-way digestive tract. Food is first pulverized in the mouth and then degraded into molecular fragments in the stomach. The digestion of fragments to simple molecules is completed in the small intestine. The small molecules that are the product of digestion are then absorbed into the body through the walls of the small intestine, and the residual solids are concentrated in the large intestine and eliminated.

FOR REVIEW

Here are some important terms and concepts that you will encounter in this chapter. If you are not familiar with them, you should review them before proceeding.

Acid (Chapter 2)

Glycogen (Chapter 3)

Lysosome (Chapter 5)

Phagocytosis (Chapter 6)

Enzyme action (Chapter 7)

Oxidative respiration (Chapter 8)

The vertebrate body is a complex colony of many cells. Like a city, it contains many individuals that carry out specialized functions. It has its own police (macrophages), its own construction workers (fibroblasts), and its own telephone company (the nervous system). The many individual cells of the vertebrate body, like the people in a city, need to be fed with food that is trucked in from elsewhere. Among the cells of the vertebrate body there are no farmers. No vertebrate contains photosynthetic cells. All of the cells of the body are nourished with food that the body obtains outside itself and transports to the individual cells. Many of the major organ systems of the vertebrate body are involved in this acquisition of energy. The digestive system acquires organic foodstuffs; the respiratory system acquires the oxygen necessary to metabolize the food; the circulatory system transports both food and oxygen to the individual cells of the body; and the excretory system rids the body of wastes produced by metabolism. In this chapter we will consider the first of these activities, digestion.

THE NATURE OF DIGESTION

Animals are thermodynamic machines, expending energy to maintain order throughout their bodies. Sources of energy are therefore essential for survival. Animals obtain the metabolic energy needed for growth and activity by degrading the chemical bonds of organic molecules. In Chapters 7 and 8 we considered these degradation processes in detail. They include the breakdown of sugar molecules in glycolysis and the oxidation of pyruvate in the citric acid cycle. What these processes have in common is that they act on simple molecules: on proteins, lipids, sugars, and fragments of these molecules such as amino acids and triglycerides. Heterotrophs, then, live by degrading simple organic molecules. But few organisms contain significant concentrations of free sugars and amino acids. Instead, the simple molecules are incorporated into long chains, into starches, fats, and proteins. Thus, eating another organism (Figure 49-1) does not in itself provide a source of metabolic energy to a heterotroph. It is first necessary to degrade that organism's macromolecules into the simple compounds from which they were built. This process is called **digestion.**

THE EVOLUTION OF DIGESTIVE SYSTEMS
Phagocytosis

The first problem confronting an organism that attempts to digest part or all of another organism is where to do it. Fungi solve this problem by secreting their digestive enzymes onto food and then absorbing the products of enzyme digestion back into their hyphae. The members of most of the phyla of protists have taken different approaches. The majority of them are motile, not sitting for long periods in one spot digesting, as fungi do. In general, protists absorb food particles into their bodies and digest them there.

Protists avoid digesting themselves with their own digestive enzymes by enclosing their food particles within a membrane-bound vesicle called a **food vacuole** and digesting them there. This food vacuole is the "stomach" of the protists. The digestive enzymes of protists are compartmentalized in lysosomes when they are first synthesized. During digestion these lysosomes fuse with the membrane of a food vacuole, and the digestive enzymes pass into the food vacuole. This system serves to expose the food particle to the enzymes while protecting the cell from them. The sugars, amino acids, and other metabolites that are produced in the food vacuole by the action of the digestive enzymes are absorbed across the vacuole membrane into the cytoplasm of the protist by a variety of transport channels. Fats, which are soluble in other lipids, pass freely across the membrane.

A Digestive Cavity

The simple feeding strategy of phagocytosis was greatly improved on by various groups of multicellular animals (Figure 49-2). In the sponges, the multicellular animals with the simplest structure, each cell resembles an individual zoomastigote, and

FIGURE 49-1
Eating is something we all do. Some of us prefer hamburgers; others, like this chipmunk, enjoy nuts. Most of the substance of the nuts in its cheeks will become part of the chipmunk's body or be burned to supply it with energy. Eventually, the residue will be discarded. All the food that each of us consumes suffers the same fate, being converted to body tissue, energy, and refuse.

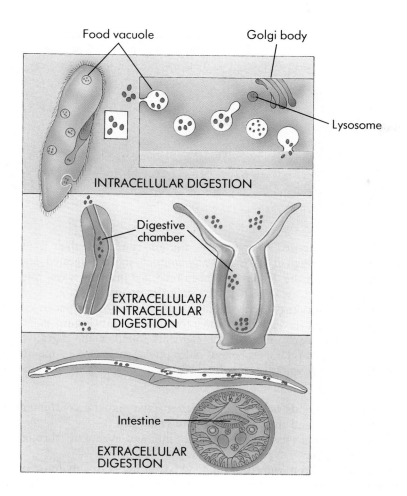

Food vacuole   Golgi body

Lysosome

INTRACELLULAR DIGESTION

Digestive chamber

EXTRACELLULAR/ INTRACELLULAR DIGESTION

Intestine

EXTRACELLULAR DIGESTION

FIGURE 49-2

Types of digestion.
A The most ancient animals used intracellular digestion, which limited the size of the food particle to a fraction of the animal's size.
B Among the cnidarians we first see a digestive cavity, in which food particles are broken down extracellularly, outside of cells; digestion within the cavity is only preliminary, however; final digestion is carried out within the cells that line the cavity.
C The roundworms are the oldest organisms that practice true extracellular digestion, in which food passes through a "digestive tube" in one direction; digestive products are absorbed into the organisms during passage, and the residual matter is eliminated from the far end of the tube.

the process of digestion is what would be expected—phagocytosis by each cell. All other multicellular animals, however, have a digestive cavity that is specialized and efficient.

In the cnidarians, the digestive cavity is a blind sac with only one opening—an opening that serves as both mouth and anus. Within this sac the first stage of digestion occurs. Large food particles are broken down into smaller ones, which are then phagocytized by the individual cells of the cnidarian. Further digestion takes place within the food vacuoles that are formed in the individual cells. The advent of a **digestive cavity** was a great modification in heterotrophy: the animals that possessed such a sac were able, for the first time, to digest other organisms larger than an individual cell.

> **The first great evolutionary change in digestion was the advent of a digestive cavity, which for the first time permitted animals to digest particles larger than a cell.**

During their evolution, the ancestors of the flatworms (phylum Platyhelminthes) developed a **pharynx,** which controls the entry of external materials into the gut. The digestive cavity of the flatworms branches far more than that of the cnidarians, although functionally it is similar. Again, digestion within the cavity is only preliminary; once the food particles have been broken into fragments, they are engulfed by cells in the lining of the cavity and then digested intracellularly.

Extracellular Digestion

The first true **extracellular digestion** among animals occurred with the evolution of the roundworms, or nematodes (phylum Nematoda). The members of this group

FIGURE 49-3
The digestive system of an
annelid.

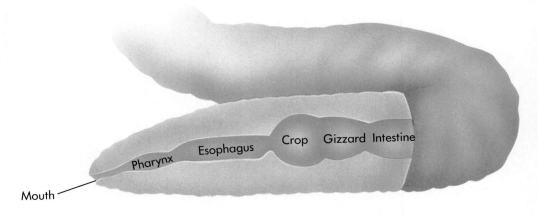

have a tubular gut composed of endoderm; the gut runs from their mouth to their anus. Food moves through it on a one-way journey, being digested and absorbed along the way. Although the details have been modified in many ways, this same general strategy is employed by all of the more complex animal phyla.

Much of the evolution of extracellular digestion has been concerned with how food is processed during its one-way journey through the gut (Figure 49-3). The principal elements evolved early; most of the elements found in the digestive system of vertebrates can be seen in the members of older phyla, such as the annelids. In earthworms, which are representative annelids, digestion proceeds in the following four stages:

1. *Acquisition.* Annelids have a muscularized pharynx, the contraction of which forces food into and through the nonmuscular gut.
2. *Storage.* From the pharynx, food passes into and eventually through a storage sac called a **crop.**
3. *Fragmentation.* From the crop, the food passes into a thickly muscled chamber, the **gizzard.** Here it is pulverized by churning, often together with grit or small stones that the animal has eaten.
4. *Extracellular digestion.* From the gizzard, food passes into a long **intestine.** Glands that secrete hydrolytic enzymes into the intestinal fluid are located in the epithelial lining of the intestine. These enzymes digest the food particle as it passes through the intestine. The sugars, amino acids, and other end products of digestion are absorbed across the intestinal wall.

In other groups of advanced invertebrates, such as mollusks and arthropods, the tubular gut is partitioned into three functional zones. The first zone combines the several functions of the anterior portion of the earthworm digestive tract, carrying out acquisition, storage, and fragmentation of food. The second zone is devoted exclusively to extracellular digestion. The third zone is devoted to the absorption of the end products of digestion. Among arthropods, for example, the mouth and foregut act to mix and prepare the food. The midgut in arthropods is the primary digestive organ; it functions in the absorption of the nutrients. Later, water is reabsorbed in the hindgut. This segmentation of the digestive tract into three functional zones is a pattern that is characteristic of all of the more advanced groups of animals.

> **The more advanced invertebrates exhibit two fundamental advances in digestion, as compared with groups such as the cnidarians and flatworms: (1) extracellular digestion—the secretion by cells of digestive enzymes, coupled to the absorption by cells of the products of digestion; and (2) the one-way flow of food particles through the zone of digestion.**

Among nonvertebrate chordates such as the lancelets, which obtain both food and oxygen from a stream of water passing through the mouth, the pharynx acts as a filtering device. It passes food particles through an esophagus into the digestive tract, while diverting the water into another channel. A pouch that is derived from a section

TABLE 49-1   DIGESTIVE ENZYMES

LOCATION	ENZYME	SUBSTRATE	DIGESTION PRODUCT
Salivary gland	Amylase	Starch Glycogen	Disaccharides
Stomach	Pepsin	Proteins	Short peptides
Small intestine	Peptidases	Short peptides	Amino acids
	Nucleases	DNA, RNA	Sugars, nucleic acid bases
	Lactase Maltase Sucrase	Disaccharides	Glucose, monosaccharides
Pancreas	Lipase	Triglycerides	Fatty acids, glycerol
	Trypsin Chymotrypsin	Proteins	Peptides
	DNase	DNA	Nucleotides
	RNase	RNA	Nucleotides

of midgut in the lancelets has become specialized for the secretion of digestive enzymes. This pouch is the forerunner of the **pancreas** of vertebrates, a more specialized organ that has exactly the same function.

In humans, digestion is carried out in two ways: by hydrochloric acid (HCl), which indiscriminately breaks up large proteins into smaller pieces, and by a variety of highly specific enzymes (Table 49-1). Enzymes that break up proteins into amino acids are called **proteases;** enzymes that break up starches and other carbohydrates into sugars are called **amylases;** and enzymes that break up lipids and fats into small segments are called **lipases.** There are several kinds of proteases, amylases, and lipases, as well as a DNase and an RNase which break up DNA and RNA. Most digestive enzymes cannot tolerate high acid concentrations, so the vertebrate digestive process is carried out in two phases: acid digestion takes place first, in the stomach, and then the food moves to the small intestine, where the acid is neutralized and a variety of digestive enzymes continues the digestive process. In addition to partial degradation, gastric HCl is also an important barrier protecting against invading bacteria and viruses.

In the stomach, proteins are partially denatured by acids; the fragments are then cleaved into individual amino acids by a variety of enzymes in the stomach and small intestine. Also in the small intestine, starches are digested by amylases and fats are digested by lipases.

Organization of Vertebrate Digestive Systems

The general organization of the digestive tract is the same in all vertebrates, although different elements are emphasized in different groups. In all vertebrates, acid digestion of proteins takes place in the stomach, after which food passes to the upper part of the small intestine, called the **duodenum,** where a battery of digestive enzymes continues the digestive process. The products of digestion then pass across the wall of the small intestine into the bloodstream. Figure 49-4 illustrates the organization of the human digestive system, which is typical of the kinds of digestive systems found in vertebrates.

Specializations among the digestive systems of different kinds of vertebrates reflect differences in the way these animals live. The initial components of the gastrointestinal tract are the mouth and pharynx. The pharynx is the gateway to the esophagus. Fishes have a large pharynx with gill slits, not unlike that of the lancelets, whereas air-breathing vertebrates have a greatly reduced pharynx. Adult amphibians, which are carnivores, have a short intestine; the food they ingest is readily digested,

FIGURE 49-4
The human digestive system.

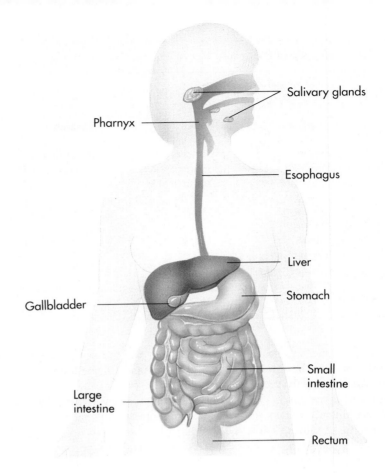

including the soluble carbohydrate glycogen. Many birds, in contrast, subsist on plant material. The primary structural component of plants is cellulose, a rigid carbohydrate that resists digestion. Birds have a convoluted small intestine, by means of which they prolong the process of digestion and aid absorption of digestion products. Many animals have teeth, and chewing (mastication) breaks up food into small particles and mixes food with fluid secretions. Birds, which lack teeth to masticate their food, break up food in their stomachs, which have two chambers. In one of these chambers—the gizzard—small pebbles that are ingested by the bird are churned together with the food by muscular action; this churning serves to grind up the seeds and other hard plant material into smaller chunks before their digestion in the second chamber of the stomach. Mammals that digest grass and other vegetation often have stomachs with multiple chambers where bacteria aid the digestion of cellulose.

WHERE IT ALL BEGINS: THE MOUTH

The food of every vertebrate is taken in through its mouth, which in all groups, except for the birds, usually contains **teeth** (Figure 49-5). Vertebrates capture their food in many different ways, from biting it off of other animals to grazing on plants. As their different methods of feeding suggest, the teeth of different kinds of vertebrates are specialized in many different ways. The teeth of carnivorous mammals, for example, are pointed and lack flat grinding surfaces. Such pointed teeth are adaptations to cutting and shearing, and characterize meat eaters. Carnivores must capture and kill their prey but have little need to chew it, since digestive enzymes can act directly on animal cells. Recall how a cat or dog gulps down food. By contrast, grass-eating herbivores, such as cows and horses, have very large and flat teeth, with complex ridges that are well suited to grinding. Broad, flat teeth characterize herbivores, which must pulverize the cellulose cell wall of plant tissue before digesting it.

Humans are omnivores, eating both plant and animal food regularly. As a result, our combination of teeth is structurally intermediate between those of carnivores and

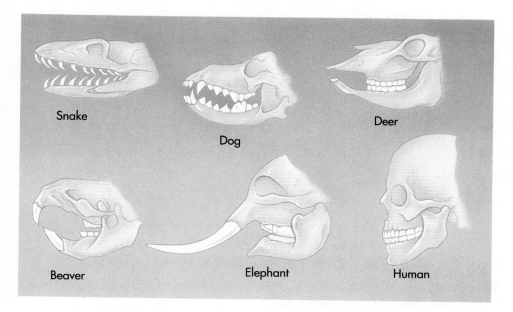

FIGURE 49-5
Vertebrate teeth fit many functions. The teeth of vertebrates are specialized for particular tasks. In the snake the teeth slope backward, to aid in retention of prey during swallowing. In carnivores such as the dog, "canine" teeth specialized for ripping food predominate. In herbivores such as deer, grinding teeth predominate. In the beaver the foreteeth are specialized as incisors—chisels. In the elephant, two of the upper front teeth are specialized as weapons. Humans as omnivores have a relatively broad-function mouth containing canine, chisel, and grinding teeth.

those of herbivores. Viewed simply, humans are carnivores in the front of the mouth and herbivores in the back (Figure 49-6). The four front teeth in the upper and lower jaws are **incisors**—sharp, chisel shaped, and used for biting. On each side of the incisors, on both jaws, are sharp, pointed teeth called **canines,** which are used in tearing food. Behind each canine, on each side of the mouth and on both top and bottom jaws, are two **premolars** and three **molars,** both of which have flattened, ridged surfaces for grinding and crushing food. Humans have only 20 teeth when they are babies. These "baby teeth" are lost later in childhood, and are replaced then by the 32 adult teeth.

Within the mouth, the tongue of vertebrates mixes the food with a mucous solution, the **saliva.** In humans, saliva is secreted into the mouth by three pairs of **salivary glands,** which empty through the mucosal lining of the mouth. The saliva moistens and lubricates the food so that it is swallowed more readily and does not abrade the tissue it passes on its way down through the esophagus. The saliva also contains the hydrolytic enzyme **amylase,** which initiates the breakdown of starch and other complex polysaccharides into smaller fragments. This action by amylase is the first of the many digestive processes that occur as the food passes through the digestive tract.

The secretions of the salivary glands are controlled by the nervous system, which maintains a constant secretion of about half a milliliter per minute in humans. This process goes on every minute, all of the time, and always keeps the mouth moist. The presence of food in the mouth acts as a stimulus to increase the rate of secretion by the salivary glands. Chemoreceptors in the mouth send a signal to the brain stem, which responds by stimulating the salivary glands. The most potent stimuli are acid solutions: lemon juice can increase the rate of salivation eightfold. Actually, even thinking about food acts as an effective stimulus in some vertebrates. The sight, sound, or smell of food is an effective stimulus to salivation in dogs, although it has only a slight effect in humans.

> Vertebrate teeth serve to shred animal tissue and to grind plant material. Saliva secreted into the mouth moistens the food, aiding its journey into the digestive system, and begins the enzyme-catalyzed process of degradation.

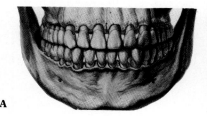

A

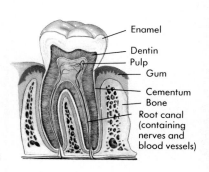

Enamel
Dentin
Pulp
Gum
Cementum
Bone
Root canal
(containing nerves and blood vessels)

B

FIGURE 49-6
Humans are carnivores in the front of their mouths and herbivores in the back.
A The front six teeth on the upper and lower jaws are canines and incisors. The remaining teeth, running along the sides of the mouth, are grinders, called premolars and molars.
B Each tooth is alive, with a central pulp containing both nerves and blood vessels. The actual chewing surface is hard enamel, layered over the softer dentin that forms the body of the tooth.

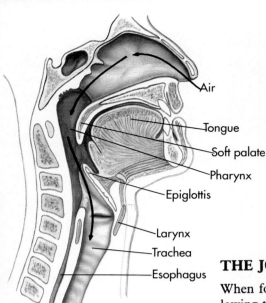

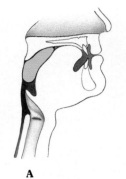

A

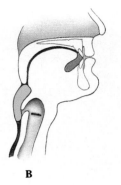

B

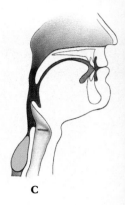

C

Air
Tongue
Soft palate
Pharynx
Epiglottis
Larynx
Trachea
Esophagus

FIGURE 49-7

How humans swallow. As food passes back past the rear of the mouth (**A**), it presses the soft palate against the back wall of the pharynx, sealing off the nasal passage; as the food passes on down, a flap of tissue called the epiglottis folds down (**B**), sealing the respiratory passage; after the food enters the esophagus, the soft palate relaxes and the epiglottis is raised (**C**), opening the respiratory passage between the nasal cavity and the trachea.

THE JOURNEY OF FOOD TO THE STOMACH

When food passes beyond the teeth to the back of the mouth in a mammal, the following three things happen (Figure 49-7):

1. The **palate** (which is soft in mammals) elevates, pushing against the back wall of the pharynx. This seals off the nasal cavity and prevents any food from entering it.
2. Pressure against the pharynx stimulates receptors within its walls to send nerve signals to the swallowing center in the brain stem.
3. The swallowing center sends out nerve signals that keep food from getting into the respiratory tract, which branches off just below the pharynx. These signals both inhibit respiration and seal the windpipe (trachea) by raising the **larynx** and thus closing the passage between the larynx and the pharynx, called the **glottis.** A flap of tissue, called the **epiglottis,** folds back over the opening, providing a seal.

After passing the tracheal opening, the food enters a tube called the **esophagus,** which connects the pharynx to the stomach. No further digestion takes place in the esophagus. Its role is that of an escalator, moving food down toward the stomach. In adult humans the esophagus is about 25 centimeters long, and its lower end opens into the stomach proper. The upper portion of the esophagus is enveloped in skeletal muscle, and the lower two-thirds is enveloped in smooth muscle. Successive waves of contraction of these muscles, which are stimulated by the swallowing center, move

FIGURE 49-8

The upper digestive tract of a human being. Food enters the stomach from the esophagus. The epithelial walls of the stomach are dotted with gastric pits, which contain glands that secrete acid and digestive enzymes. The entrance to the small intestine—the duodenum—is controlled by a band of muscle called the pyloric sphincter. The surface of the duodenum is covered with microscopic villi which greatly increase its surface area and so aid in absorption of nutrients.

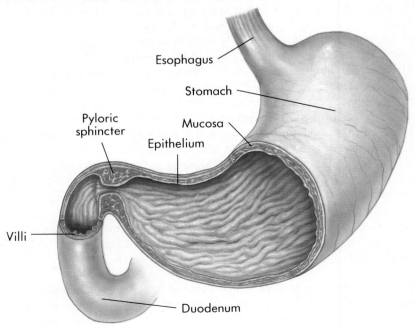

Esophagus
Stomach
Pyloric sphincter
Mucosa
Epithelium
Villi
Duodenum

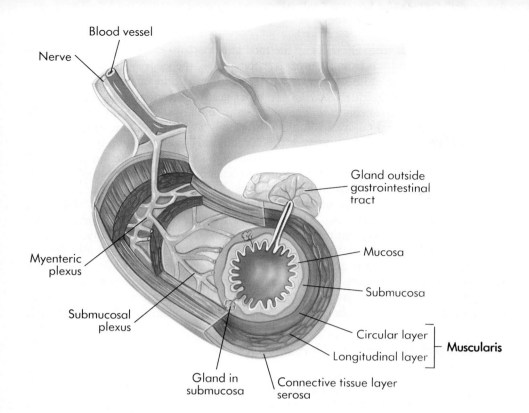

FIGURE 49-9
Organization of the digestive tube.

Blood vessel

Nerve

Gland outside gastrointestinal tract

Myenteric plexus

Submucosal plexus

Gland in submucosa

Mucosa

Submucosa

Circular layer
Longitudinal layer — **Muscularis**

Connective tissue layer serosa

food down through the esophagus to the stomach. These rhythmic sequences of waves of muscular contraction in the walls of a tube are called **peristalsis.** Because the movement of food through the esophagus is primarily caused by these peristaltic contractions, humans can swallow even if they are upside down.

The exit of food from the esophagus to the stomach is controlled by a muscular constriction, or **sphincter,** which opens in response to the pressure exerted by the food. When this sphincter is contracted, it prevents the food in the stomach from moving back up the esophagus.

PRELIMINARY DIGESTION: THE STOMACH

The stomach (Figure 49-8) is a saclike portion of the digestive tract. In it, the digestive process is organized; the stomach collects ingested food, hydrolyzes it into small molecular fragments, and feeds it in a controlled fashion into the primary digestive organ, the small intestine. The interior of the stomach, like that of the rest of the digestive tract, is continuous with the outside of the body. The epithelium overlies a deep layer of connective tissue (Figure 49-9), called **mucosa,** below which is located a complex array of muscles, blood vessels, and nerves.

How the Stomach Digests Food

The epithelium of the stomach is the source of the "digestive juices," which actually are a complex mixture of substances that continue the breakdown of food that begins in the saliva. The upper epithelial surface of the stomach is dotted with deep depressions called **gastric pits** (Figure 49-10). In them the epithelial membrane is invaginated inward, forming exocrine glands within the mucosa. These exocrine glands contain two kinds of secreting cells, **parietal cells,** which secrete hydrochloric acid (HCl), and **chief cells,** which secrete the protein pepsinogen. In the stomach, HCl cleaves a terminal fragment from the secreted pepsinogen, converting it into the protein-hydrolyzing enzyme pepsin.

Many of the epithelial cells that line the stomach are specialized for the secretion of **mucus.** This mucus, which is produced in large quantities, lubricates the stomach wall and facilitates the movement of food within the stomach. It also protects the cells

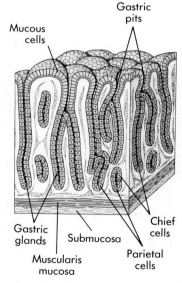

Gastric pits

Mucous cells

Gastric glands

Muscularis mucosa

Submucosa

Chief cells

Parietal cells

FIGURE 49-10

Gastric pits. Gastric pits are deep invaginations of the stomach epithelium down into the underlying mucosa, at the base of which parietal and chief cells secrete hydrochloric acid and a protein that is cleaved into the enzyme pepsin.

of the stomach wall from abrasion by the food and, perhaps most important, protects the walls of the stomach from its own digestive juices, the gastric fluid.

Food within the stomach is attacked by both acid and enzymes. The human stomach secretes about 2 liters of HCl and other gastric secretions every day, creating a very concentrated acid solution. This solution is actually about 150 millimolar HCl, and thus 3 million times more acidic than the blood. The hydrochloric acid breaks up connective tissue. It does this because the very low pH values, between 1.5 and 2.5, that are created by the HCl change the ionization of carboxyl and amino side groups of proteins. This process causes the folded proteins of connective tissue to open out and disrupts their associations with one another.

The action of the acid disintegrates the food into molecular fragments, and thus is an essential prelude to digestion. The acid itself does not carry out any further digestive activity. It has a very limited ability to break proteins down into amino acids, or carbohydrates into their constituent sugars, and it does not attack fats at all. The further digestion of protein is accomplished by the enzyme **pepsin,** which cleaves proteins into short polypeptides. Because the fragments produced by pepsin are large and frequently charged, they cannot pass across the epithelial membrane. Except for water, some vitamins, and alcohol, no absorption takes place through the stomach wall.

The Stomach Secretes the Hormone Gastrin

It is important that a stomach not produce *too* much acid. If it did, it would be impossible for the body to neutralize the acid later in the small intestine, a step that is essential for the terminal stages of digestion. The stomach controls the production of acid by means of digestive hormones (Table 49-2) produced by endocrine cells that are scattered throughout its epithelial layer. One of the principal digestive hormones, called **gastrin,** regulates the synthesis of hydrochloric acid by the parietal cells of the gastric pits, permitting such synthesis to occur only when the pH of the stomach contents is higher than about 1.5 (Figure 49-11).

Some stomachs greatly overproduce gastrin, which results in excessive acid production. The excessive acid may attack the walls of the small intestine, burning holes through the wall. These holes are called **duodenal ulcers.** The contents of the small intestine are not normally acidic, and this organ is much less able to withstand the disruptive actions of stomach acids than is the wall of the stomach. For this reason, over 90% of all ulcers are duodenal, although other ulcers do sometimes occur in the stomach.

TABLE 49-2	HORMONES OF DIGESTION				
HORMONE	CLASS	SOURCE	STIMULUS	ACTION	NOTE
Gastrin	Polypeptide	Pyloric portion of stomach	Entry of food into stomach	Secretion of HCl	Unusual in that it acts on same organ that secretes it
Cholecysto-kinin	Polypeptide	Duodenum	Arrival of food in small intestine	Stimulates gall bladder contraction, and so the release of bile salts into intestine. Stimulates secretion of digestive enzymes by pancreas	CCK bears a striking structural resemblance to gastrin
Secretin	Polypeptide	Duodenum	HCl in duodenum	Stimulates pancreas to secrete bicarbonate, which neutralizes stomach acid	The first hormone to be discovered (1902)

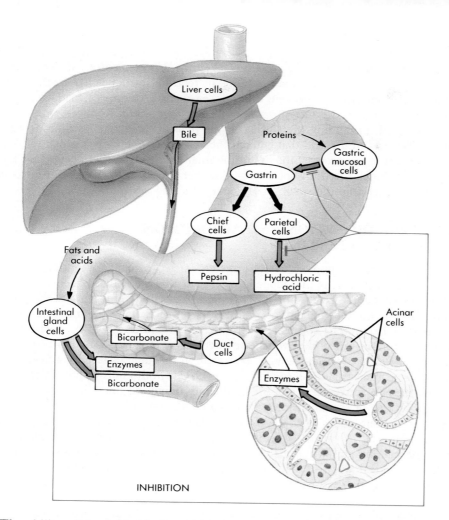

FIGURE 49-11
Regulation of digestive enzyme production. The production of pepsin and HCl by the endothelial glands of the stomach, and of bile by the liver, is regulated by the hormone gastrin. The production of gastrin, and of pancreatic digestive enzymes, is regulated by intestinal gland cells, which also directly regulate production of pepsin, HCl, and bile. In this way, levels of HCl production in the stomach are increased when stomach pH rises or duodenum pH falls below critical values.

The ability of the acid solution within the stomach to break down cells is also a very effective means of killing bacteria that are ingested with the food. Most bacteria are simply torn apart by the strong acid. The few that survive their harrowing journey through the stomach and make it to the intestine intact are able to grow and multiply there, particularly in the large intestine. There are enough survivors so that most vertebrates harbor thriving colonies of bacteria within their intestines, with the result that bacteria are a major component of feces. As we will discuss later, bacteria that live within the digestive tract of cows and other ruminants play a key role in the ability of these mammals to digest cellulose.

In the stomach, concentrated acid breaks up connective tissue and protein into molecular fragments, which are further digested by pepsin into short polypeptides. Carbohydrates and fats are not digested in the stomach.

The inner surface of the stomach is highly convoluted. For this reason it can fold up when empty and open out like an expanding balloon as it fills with food. There is, of course, a limit to how much food a stomach can hold. The human stomach has a volume of about 50 milliliters when empty; when full, it may have a volume 50 times larger, from 2 to 4 liters. Carnivores that engage in sporadic gorging as an important survival strategy possess stomachs that are able to distend much more than our stomachs can.

Leaving the Stomach

The digestive tract exits from the stomach at another muscular constriction, the **pyloric sphincter** (see Figure 49-8). The pyloric sphincter is the gate to the small intestine, the organ within which the final stages of digestion occur. The pyloric sphincter, therefore, is the traffic light of the digestive system. The capacity of the small intestine is limited, and its digestive processes take time. Consequently, efficient digestion requires that only relatively small portions of food be introduced from the stomach into the small intestine at any one time. When a small volume of food, which is by now highly acidic, passes into the small intestine, the acid introduced with the food acts as a signal, prompting the closing of the pyloric sphincter. As time passes, the food is digested and the acid that entered the small intestine with it is neutralized. At a certain point in the process, the pH of the small intestine reaches a level that signals the pyloric sphincter to open once again. Another small portion of food is introduced from the stomach into the small intestine, and the process continues.

TERMINAL DIGESTION AND ABSORPTION: THE SMALL INTESTINE

The small intestine is approximately 6 meters long. The first 25 centimeters, about 4% of the total length, is the duodenum. It is at this point that pancreatic enzymes and bile enter the intestine, initiating digestion. Absorption of water and the products of digestion by the bloodstream occurs in the later sections of the intestine, the jejunum and the ileum. The epithelial wall of the small intestine is covered with fine fingerlike projections called **villi,** which are microscopic (Figure 49-12). In turn, each of the epithelial cells covering the villi is covered on its outer surface by a field of cytoplasmic projections called **microvilli** (Figure 49-13). Both kinds of projections greatly increase the absorptive surface of the epithelium lining the small intestine. The average surface area of the small intestine of an adult human being is about 300 square meters. The membranes of the epithelial cells contain some enzymes that complete digestion, as well as carrier systems that actively transport sugars and amino acids across the membrane; fatty acids cross passively by diffusion. After absorption from the lumen, end-products of digestion enter capillaries (sugars and amino acids) in the villi and enter lymph vessels known as lacteals (triglycerides and fatty acids).

Most digestion occurs in the first 25 centimeters of the 6-meter length of the small intestine, in a zone called the duodenum. The rest of the small intestine is devoted to the absorption of the products of digestion.

FIGURE 49-12
The small intestine.
A Cross section of the small intestine.
B Microvilli, shown in a scanning electron micrograph, are very densely clustered, giving the small intestine an enormous surface area, which is very important in efficient absorption of the digestion products.

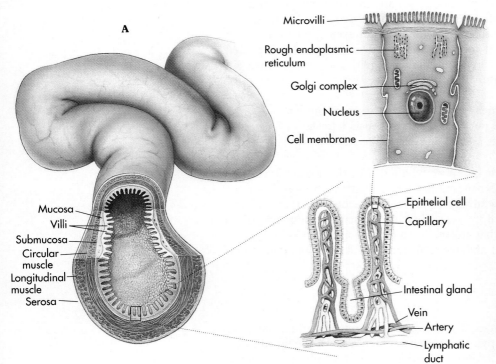

A

Microvilli
Rough endoplasmic reticulum
Golgi complex
Nucleus
Cell membrane
Epithelial cell
Capillary
Intestinal gland
Vein
Artery
Lymphatic duct

Mucosa
Villi
Submucosa
Circular muscle
Longitudinal muscle
Serosa

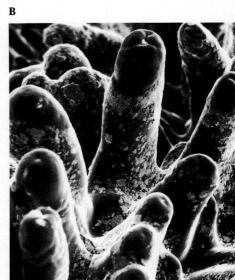

B

The amount of material passing through the small intestine is startlingly large. An average human consumes about 800 grams of solid food and 1200 milliliters of water each day, for a total volume of about 2 liters. To this amount is added about 1.5 liters of fluid from the salivary glands, 2 liters from the gastric secretions of the stomach, 1.5 liters from the pancreas, 0.5 liter from the liver, and 1.5 liters of intestinal secretions. The total adds up to a remarkable 9 liters. However, although the flux is great, the *net* passage is small. Almost all of these fluids and solids are reabsorbed during their passage through the intestines, with about 8.5 liters passing across the walls of the small intestine and an additional 350 milliliters through the wall of the large intestine. Of the 800 grams of solid and 9 liters of liquid that enter the digestive tract in 1 day, only about 50 grams of solid and 100 milliliters of liquid leave the body as feces. The normal fluid absorption efficiency of the digestive tract thus approaches 99%, which is very high indeed.

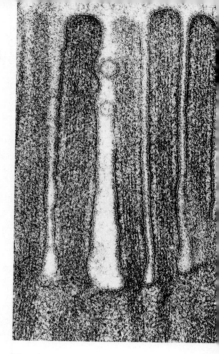

FIGURE 49-13
Intestinal microvilli.

ORGANS THAT SUPPORT THE DIGESTIVE SYSTEM
The Pancreas: Making Digestive Enzymes

The **pancreas** is a large gland situated near the junction of the stomach and the small intestine. It is one of the body's major exocrine glands, secreting a host of different enzymes that act in the duodenum to break down carbohydrates, proteins, and fats. These enzymes include proteases for breaking down proteins, lipases for digesting fats, and enzymes that break down carbohydrates. The pancreas also functions as an endocrine gland.

The pancreas has two types of exocrine cells. The first type functions exactly opposite to the parietal cells of the gastric pits. Instead of secreting an acid, HCl, these specialized cells secrete a base, **bicarbonate.** The alkaline bicarbonate that is secreted by the pancreas is critical to successful digestion, since most of the enzymes secreted by the pancreas will not work in acid solution. The introduction of bicarbonate into the duodenum neutralizes the acid derived from the stomach and thus permits the digestive enzymes to function. Since acid is secreted in the stomach, and bicarbonate is secreted in the intestine, there is no net effect of digestion on the body's acid-base balance.

The pancreas has yet a third function critical to metabolism. As we noted in Chapter 47, the islets of Langerhans, distributed throughout the exocrine regions of the pancreas, function as endocrine glands, producing the hormones that act in the liver and elsewhere to regulate the level of sugar in the blood.

The Liver

Because fats are insoluble in water, they tend to enter the small intestine as small globules that are not attacked readily by the enzymes secreted by the pancreas. Before fats can be digested by pancreatic lipases, they must be made soluble. This process is carried out by a collection of detergent molecules secreted by a second gland, the **liver.** The liver is the body's principal metabolic factory, turning foodstuffs arriving from the digestive tract in the bloodstream into substances that are utilized by the different cells of the body. It is the largest internal organ of the body. In an adult human being the liver weighs about 1.5 kilograms and is the size of a football.

The liver carries out a wide variety of metabolic functions, many of which we discuss in later chapters. It supplies quick energy, metabolizes alcohol, makes proteins, stores vitamins and minerals, regulates blood clotting, regulates the production of cholesterol, and detoxifies poisons. It also produces the detergent molecules already mentioned, known as **bile salts** and secretes them through a duct into the duodenum. Bile acts as a superdetergent. It combines with fats to form microscopic droplets called micelles in a process known as emulsification. The components of bile include bile salts, cholesterol, and the phospholipid lecithin. All these combine to render fat soluble. Human beings concentrate and store bile manufactured in the liver in the **gallbladder.** When chyme enters the small intestine, a hormone known as cholecystokinin (CCK) stimulates contraction of the gallbladder to release bile into the duodenum.

How the Liver Regulates Blood Glucose Levels.

It is important that the blood of vertebrates maintain a relatively constant composition, since the different tissues of the body need specific compounds as energy sources. Brain cells, for example, can store very little glucose and lack the enzymes to convert fat or amino acids into glucose. Brain cells are thus very sensitive to the level of available glucose. They are totally dependent on blood plasma for glucose and cease to function if the level of glucose in the blood falls much below normal values.

Maintaining a constant level of metabolites in the blood requires active control by the body's organs. A moment's reflection shows why. Most vertebrates eat sporadically. In the United States, most people eat three meals a day. Food enters the digestive system at intervals separated by long periods of fasting. Much of the food is digested relatively quickly, with the metabolites, such as glucose and amino acids, passing through the lining of the small intestine into the bloodstream. Without active control of the levels of these and other metabolites, they would suddenly become much more abundant in the blood right after a meal and then fall rapidly during a period of starvation, when the metabolites are being removed from the bloodstream by metabolizing cells and are not being replenished.

Control over metabolite levels in the blood is achieved in a very logical way: by establishing a reservoir, or metabolic bank, in the liver. One of the liver's most important functions is regulation of the blood's metabolite levels. This regulation is achieved by means of a special shunt in the circulatory system. Blood returning from the stomach and small intestine flows into the portal vein, which carries it not to the heart but to the liver (Figure 49-14). Within the liver, blood passes through a network of fine passages called sinuses. Only after flowing through the liver is the blood collected into the hepatic vein and delivered to the vena cava and the heart.

When excessive amounts of glucose are present in the blood passing through the liver, a situation that occurs soon after a meal, the liver converts the excess glucose into the starchlike glucose polymer, glycogen, stimulated by the hormone insulin. The liver stores this glycogen. Some glycogen is also stored in the muscles, where it is easily available to fuel muscle contraction. When blood glucose levels decrease, as they do in a period of fasting or between meals, the glucose deficit in the blood plasma is made up from the glycogen reservoir, stimulated by the hormone glucagon.

FIGURE 49-14
Hepatic-portal circulation. Blood from the stomach and small intestine, rich with the metabolites of digestion, is collected into the portal vein, which carries it to the liver. After flowing through the sinuses of the liver, the liver-processed blood is re-collected into the hepatic vein, which carries it back toward the heart.

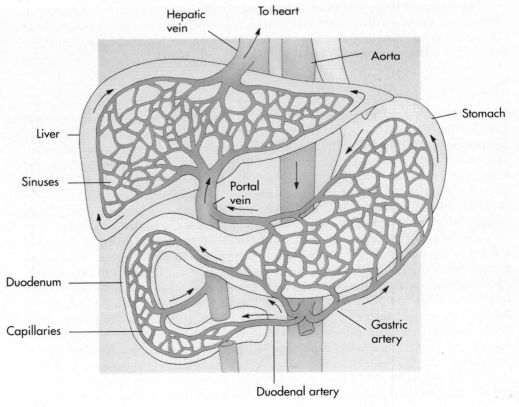

The human liver stores enough glycogen to supply glucose to the bloodstream for about 10 hours of fasting. If fasting continues, the liver begins to convert other molecules, such as amino acids, into glucose to maintain the glucose level in the blood. The liver, the body's metabolic reservoir, thus acts much like a bank, making deposits and withdrawals in the "currency" of glucose molecules.

In addition to its many other roles, the liver acts to regulate the level of blood glucose, maintaining it within narrow bounds.

Also like a bank, the liver exchanges currencies, converting other molecules, such as amino acids and fats, to glucose and subsequently glycogen for storage. Excess amino acids that may be present in the blood are converted to glucose by liver enzymes, and the glucose is stored as glycogen. The first step in this conversion is the removal of the amino group (NH_3) from the amino acid, a process called deamination. Unlike plants, animals cannot reuse the nitrogen from these amino groups, and must excrete it as nitrogenous waste. The product of amino acid deamination, ammonia (NH_4), forms a complex with carbon dioxide to form urea. In some vertebrates, amino acids are not deaminated directly, but instead are converted to uric acid. The uric acid is released by the liver into the bloodstream, where the kidneys subsequently remove it.

The liver has a limited storage capacity. When its glycogen reservoir is full, it continues to remove excess glucose molecules from the blood by converting them to fat, which is stored elsewhere in the body. In humans, for example, long periods of overeating and the resulting chronic oversupply of glucose frequently result in the deposition of fat around the stomach or on the hips.

CONCENTRATION OF SOLIDS: THE LARGE INTESTINE

The large intestine, or **colon,** is much shorter than the small intestine, occupying approximately the last meter of the intestinal tract. No digestion takes place within the large intestine, and only about 4% of the absorption of fluids by the body occurs there. The large intestine is not convoluted, lying instead in three relatively straight segments, and its inner surface does not possess villi. Consequently, the large intestine has less than one-thirtieth the absorptive surface area of the small intestine. Although sodium, vitamin K, and some other products of bacterial metabolism are absorbed across its wall, the primary function of the large intestine is to act as a refuse dump. Within it, undigested material, primarily bacterial fragments and cellulose, is compacted and stored. Many bacteria live and actively divide within the large intestine; the excess bacteria are incorporated into the refuse material. Bacterial fermentation produces gas within the colon at a rate of about 500 milliliters per day. This rate increases greatly after the consumption of beans or other vegetable matter because the passage of undigested plant material into the large intestine provides material for fermentation.

The large intestine serves primarily to compact the solid refuse remaining after digestion of food, thus facilitating its elimination.

Like all good things, digestion eventually comes to an end. In this case the end is a short extension of the large intestine called the **rectum.** Compacted solids within the colon, called **feces,** pass into the rectum as a result of the peristaltic contractions of the muscles encasing the large intestine. From the rectum, the solid material passes out the **anus** through two anal sphincters. The first of these is composed of smooth muscle; it opens involuntarily in response to a pressure-generated nerve signal from the rectum. The second sphincter, in contrast, is composed of striated muscle. It is subject to voluntary control from the brain, thus permitting a conscious decision to delay defecation.

In all vertebrates except placental mammals, the reproductive and urinary tracts exit with the rectum into a common cavity called the **cloaca.** In placental mammals

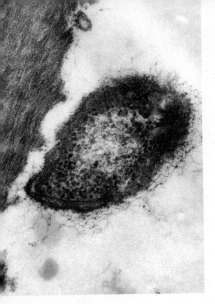

FIGURE 49-15

A bacterium, *Bacteroides succinogenes*, digesting a cellulose fiber. Its approach is not unlike that employed by the fungi: the cell adheres to the fiber by means of a loose carbohydrate mesh, and secretes bacterial enzymes such as cellulase and xylulase onto the fiber. These enzymes then digest the fiber in that immediate vicinity.

FIGURE 49-16

Digestive tract of a ruminant. In a cow, the grass and other plants that an individual eats pass first into the rumen, the first of four chambers, where they are partially digested; after passing through a second chamber, the reticulum, food may be regurgitated for rechewing before passing through the rear chambers.

(but not in monotremes and marsupials), the rectum is a nonabsorptive cavity that acts solely to control the exit of material from the large intestine. The reproductive and urinary tracts of placental mammals do not join the rectum; instead they make their own separate exits from the body.

SYMBIOSIS WITHIN THE DIGESTIVE SYSTEMS OF VERTEBRATES

Vertebrates lack the enzymes that are necessary to digest some kinds of potentially useful foodstuffs. In certain kinds of vertebrates, bacteria and protozoans living within the digestive tract possess the enzymes necessary to convert these materials into substances that the host vertebrate can digest. The relationship is mutually beneficial and provides an excellent example of symbiosis.

Although bacterial digestion within the gut plays a relatively small role in human metabolism, it is an essential element in the metabolism of many other vertebrates. Herbivorous animals are one example. No vertebrate produces the enzyme **cellulase,** and therefore no vertebrate can digest cellulose (the chief carbohydrate component of plants) without outside help. Herbivorous animals achieve the digestion of cellulose by using bacteria that live within their digestive tracts to produce this necessary enzyme (Figure 49-15). Termites, cockroaches, and a few other groups of insects operate in the same way, though they utilize protozoans rather than bacteria. Silverfish, which are members of a primitive order of insects, seem to have the ability, extremely unusual among animals, to produce cellulase themselves.

In horses, rodents, and lagomorphs (an order that includes mammals such as rabbits and hares), the digestion of cellulose by bacteria takes place in a large pouch of the intestine called the **cecum.** In cows and related mammals, the digesting pouch is called the **rumen;** it is located in front of the stomach instead of in the intestinal region. This forward location proves to be an important difference, because it allows a cow to regurgitate and rechew the contents of the rumen ("chewing the cud") (Figure 49-16). This process leads to the far more efficient digestion of cellulose in mammals such as cows, which have a rumen, than in those which lack such an organ, such as horses.

Rabbits have evolved a bizarre but effective way to digest cellulose, a way that achieves a degree of efficiency similar to that of ruminate digestion, despite the fact that a rabbit's cecum is positioned behind the stomach, which precludes regurgitation and redigestion in it. Rabbits do this by eating their feces, thus passing their food through the digestive tract for a second time. The second passage provides the rabbit with many of the important products of bacterial metabolism; rabbits cannot remain healthy if they are prevented from eating their feces and thus gaining the opportunity of digesting more of the cellulose in them.

Cellulose is not the only indigestible plant product that can be used by vertebrates as a food source because of the digestive activities of intestinal bacteria. Wax, a substance that is indigestible by almost all animals, is digested by symbiotic bacteria that live within the gut of the honey guides, African birds that eat the wax in bee nests.

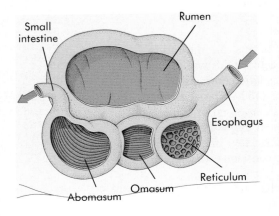

Another example of the way in which intestinal bacteria function in the metabolism of their host animals is provided by the synthesis of vitamin K. All mammals rely on intestinal bacteria to synthesize this vitamin, which is necessary for the coagulation of blood. Birds, which lack the necessary bacteria, must consume the necessary quantities of vitamin K in their food. In humans, prolonged treatment with antibiotics greatly reduces the populations of bacteria in the intestine; under such circumstances, it may be necessary to provide supplementary vitamin K.

Much of the food value of plants is tied up in cellulose, which no vertebrate is able to digest unaided. Many vertebrates, however, harbor colonies of microorganisms in their digestive tract that do have the ability to digest cellulose and thus make it available as a source of food for their hosts. Intestinal microorganisms often also excrete molecules important to the well-being of their vertebrate hosts.

NUTRITION

The ingestion of food by vertebrates serves two ends: it provides a source of energy, and it also provides raw materials that an animal is not able to manufacture for itself (Figure 49-17). The liver maintains a very constant level of glucose in the blood, and stores several hours' reserve of glucose in the form of glycogen. Any intake of food in excess of that required to maintain the glycogen reserve results in one of two consequences. Either the excess glucose is metabolized by the muscles and other cells of the body, or it is converted to fat and stored within fat cells. Thus we have the simple equation:

$$FOOD - EXERCISE = FAT$$

In wealthy countries, such as those of North America and Europe, the obesity that results from chronic overeating and from imbalanced diets high in carbohydrates is a

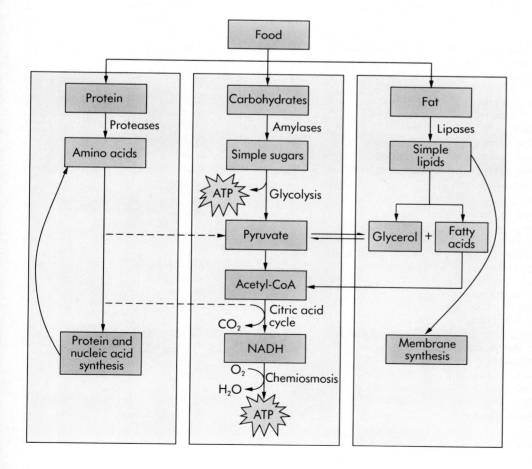

FIGURE 49-17

The fate of food. Most food can be broadly considered to consist of carbohydrate, fat, or protein. During digestion, protein is broken down into amino acids, carbohydrate is broken down into simple sugars, and fats are absorbed often without degradation. The resulting sugars and lipids may be considered members of a common pool of metabolites, as each may be converted to the other. Amino acids form a distinct metabolic pool—although there is some interconversion of amino acids and sugars, the amount is relatively small. In general, amino acids are channeled into synthesis of nitrogen-containing molecules such as proteins, hormones, and nucleotides, and the carbohydrates and fats are used to store energy and assemble structural components of the cell.

significant human health problem. In the United States, about 30% of middle-aged women and 15% of middle-aged men are classified as overweight. In other words, they weigh at least 20% more than the average weight for their height (Figure 49-18). Being overweight is strongly correlated with coronary heart disease and many other disorders.

Over the course of their evolution, many vertebrates have lost the ability to synthesize different substances that nevertheless continue to play critical roles in their metabolism. They simply no longer have the genes that encode the enzymes necessary to produce these substances. Perhaps substances that these vertebrates required were dependably present in the food they ate, so that mutations rendering these enzymes nonfunctional were not disadvantageous, and came to predominate.

TABLE 49-3   MAJOR VITAMINS

VITAMIN	FUNCTION	DIETARY SOURCE	RECOMMENDED DAILY ALLOWANCE (milligrams)	DEFICIENCY SYMPTOMS	SOLUBILITY
Vitamin A (retinol)	Used in making visual pigments, maintenance of epithelial tissues	Green vegetables, milk products, liver	1	Night blindness, flaky skin	Fat
B-complex vitamins					
B_1	Coenzyme in CO_2 removal during cellular respiration	Meat, grains, legumes	1.5	Beriberi, weakening of heart, edema	Water
B_2 (riboflavin)	Part of coenzymes FAD and FMN, which play metabolic roles	In many different kinds of foods	1.8	Inflammation and breakdown of skin, eye irritation	Water
B_3 (niacin)	Part of coenzymes NAD^+ and $NADP^+$	Liver, lean meats, grains	20	Pellagra, inflammation of nerves, mental disorders	Water
B_5 (pantothenic acid)	Part of coenzyme-A, a key connection between carbohydrate and fat metabolism	In many different kinds of foods	5 to 10	Rare: fatigue, loss of coordination	Water
B_6 (pyridoxine)	Coenzyme in many phases of amino acid metabolism	Cereals, vegetables, meats	2	Anemia, convulsions, irritability	Water
B_{12} (cyanocobalamin)	Coenzyme in the production of nucleic acids	Red meats, dairy products	0.003	Pernicious anemia	Water
Biotin	Coenzyme in fat synthesis and amino acid metabolism	Meat, vegetables	Minute	Rare: depression, nausea	Water
Folic acid	Coenzyme in amino acid and nucleic acid metabolism	Green vegetables	0.4	Anemia, diarrhea	Water
Vitamin C	Important in forming collagen, cement of bone, teeth, connective tissue of blood vessels; may help maintain resistance to infection	Fruit, green leafy vegetables	45	Scurvy, breakdown of skin, blood vessels	Water
Vitamin D (calciferol)	Increases absorption of calcium and promotes bone formation	Dairy products, cod liver oil	0.01	Rickets, bone deformities	Fat
Vitamin E (tocopherol)	Protects fatty acids and cell membranes from oxidation	Margarine, seeds, green leafy vegetables	15	Rare	Fat
Vitamin K	Essential to blood clotting	Green leafy vegetables	0.03	Severe bleeding	Fat

Substances that an animal cannot manufacture for itself but which are necessary for its health must be obtained in other ways—in its diet. Such essential organic substances utilized by organisms in trace amounts are called **vitamins** (Table 49-3). Humans, apes, monkeys, and guinea pigs, for example, have lost the ability to synthesize ascorbic acid (vitamin C) and will develop the disease scurvy—a condition characterized by weakness, spongy gums, and bleeding of the skin and mucous membranes, which can ultimately prove fatal—if vitamin C is not supplied in sufficient quantities in their diets. All other mammals, as far as is known, are able to synthesize ascorbic acid. Vitamin K, as we just mentioned, is obtained by mammals from their symbiotic intestinal bacteria, but must be consumed by birds with their food. Humans require at least 13 different vitamins.

A vitamin is an organic substance that is required in minute quantities by an organism for growth and activity, but which the organism cannot synthesize.

Some of the substances that vertebrates are not able to synthesize are required in more than trace amounts. Many vertebrates, for example, require one or more of the 20 amino acids that are necessary to synthesize proteins. Humans are unable to synthesize 8 of these 20 amino acids: lysine, tryptophan, threonine, methionine, phenylalanine, leucine, isoleucine, and valine. These amino acids, called **essential amino acids,** must be obtained by humans from proteins (Figure 49-19) in the food they eat. All vertebrates have lost the ability to synthesize certain polyunsaturated fats that provide backbones for fatty acid synthesis. Some essential substances that vertebrates do synthesize for themselves cannot be manufactured by the members of other animal groups. Some carnivorous insects, for example, are unable to synthesize the cholesterol that is required for synthesis of steroid hormones; they must therefore obtain

OBESITY STARTS HERE*			
*FOR ADULTS OF MEDIUM BUILD BETWEEN THE AGES OF 25 AND 59, INCLUDING CLOTHES AND ALLOWING FOR 1" HEELS. (BASED ON THE 1983 METROPOLITAN LIFE TABLES.)			
MEN		WOMEN	
HEIGHT	WEIGHT	HEIGHT	WEIGHT
5'6"	174	5'2"	150
5'7"	178	5'3"	154
5'8"	181	5'4"	157
5'9"	185	5'5"	161
5'10"	188	5'6"	164
5'11"	192	5'7"	168
6'0"	196	5'8"	172
6'1"	200	5'9"	175
6'2"	205	5'10"	179

FIGURE 49-18
Obesity is usually characterized as the state of being more than 20% heavier than the average person of the same sex and height. For a variety of heights, these are the weight values at which obesity begins in Americans, using average 1983 weights.

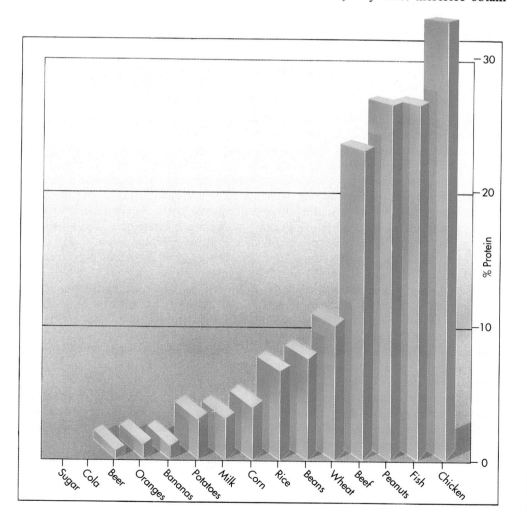

FIGURE 49-19
The protein content of a variety of common foods.

In the United States, serious eating disorders have become much more common since the mid-1970s. The most frequent of these are **anorexia nervosa,** a condition in which the afflicted person literally starves himself or herself, and **bulimia,** a condition in which individuals gorge themselves and then cause themselves to vomit, so that their weight stays constant. For reasons that we do not understand, 90% to 95% of those suffering from these eating disorders are female, and researchers estimate that 2% to 5% of the adolescent girls and young women in the United States suffer from eating disorders.

Those who suffer from anorexia are typically shy, well-behaved young women who feel embarrassed about their bodies. Eventually, they often begin to show signs of severe malnutrition and abnormally low temperatures and pulse rates. Although malfunction of the pituitary gland may produce similar symptoms, anorexia nervosa is now understood to be a psychiatric disturbance that has severe psychic and psychological consequences. Its medical consequences may be severe enough to result in death.

Bulimia is usually accompanied by severe feelings of guilt, depression, and a feeling of anxiety and helplessness. The eating binges that are characteristic of bulimia often involve large quantities of carbohydrate-rich junk food. The accompanying medical consequences result from the frequent purging, which reduces the body's supply of potassium. Potassium plays an important role in regulating body fluids; its loss may lead to muscle weakness or paralysis, an irregular heartbeat, and kidney disease. Like anorexia nervosa, bulimia often leads to decreased resistance to infection.

Although both anorexia nervosa and bulimia are now believed to be psychiatric disorders, scientists are continuing to search actively for physiological explanations for these serious conditions. Changes in the levels of certain hormones are associated with these conditions, but whether these changes cause or result from the conditions has not been demonstrated clearly. In treating these conditions, it is essential that the shame that is often associated with them be recognized as a problem, and that professional help be sought promptly. Both a psychiatrist or a psychologist and a physician should be involved, and the family as a whole must play a strong, supportive role for the treatment to be effective. Even with effective treatment, both conditions may persist for years; it is therefore essential that they be understood. Parents, teachers, and others should be alert to the existence of anorexia and bulimia and should provide the kind of education about food and diet that, coupled with a supportive environment, will lead to the early detection and cure of these eating disorders if they do occur.

cholesterol in their diet. Other insects have the ability to convert other steroids, which they consume with their food, into cholesterol.

In addition to energy and those organic compounds that cannot be synthesized, the food that an animal consumes must also supply **essential minerals** such as calcium and phosphorus, and it must include a wide variety of **trace elements,** which are minerals that are required in very small amounts (see Table 2-1). Among the trace elements are iodine (a component of thyroid hormone), cobalt (a component of vitamin B_{12}), zinc and molybdenum (components of enzymes), manganese, and selenium. All of these, with the possible exception of selenium, are also essential for plant growth. Animals that require particular trace elements obtain them either directly from plants or from animals that have eaten the plants.

Interestingly, one essential characteristic of food is simply bulk, its content of undigested fiber. The large intestine of humans, for example, has evolved as an organ adapted to process food that has a relatively high fiber content. Diets that are low in fiber, which are common in the United States, result in a slower passage of food through the colon than is desirable. This low dietary fiber content is thought to be associated with the level of colon cancer in the United States, which is among the highest in the world.

The vertebrate digestive systems that we have described in this chapter are much larger and more complex than those of the invertebrates. In vertebrate digestive systems, the working surfaces are convoluted and therefore much more extensive, and the different portions of the system are specialized for particular functions in the digestive process. Nutritional specialization in mammals has likewise become complex, and has led to many of the problems that humans face in selecting an appropriate diet. In the next chapter we shall consider the way in which vertebrates acquire oxygen, another critical component of their systems for energy use and release.

1. Digestion is the rendering of parts of organisms into small molecules which can be metabolized: amino acids, simple fats, and sugars.

2. Primitive digestion was carried out within cells, in food vacuoles that isolated the digestive process from the rest of the cell. The more advanced invertebrates evolved extracellular digestion, in which digestive enzymes are secreted into a digestive cavity and the products of digestion are absorbed from that cavity.

3. The digestive tract of vertebrates is one way. The initial portion leads from a mouth and pharynx through an esophagus to a stomach. In some vertebrates, food is further macerated in a gizzard, and in some birds it is stored in a crop. In most mammals, food is stored in the stomach, and its digestion begins there.

4. The stomach juices are concentrated acid, in which the protein-digesting enzyme pepsin is active.

5. Food passes from the stomach to the small intestine, where the pH is neutralized and a variety of enzymes, many synthesized in the pancreas, act to complete digestion. Most digestion occurs in the first 25 centimeters of the small intestine, in a zone called the duodenum.

6. The products of digestion are absorbed across the walls of the small intestine, which possess numerous villi and so achieve a very great surface area. Amino acids and sugars are transported by specific transmembrane channels, whereas simple fats, which are lipid soluble, pass readily across the membranes of the villi.

7. Glucose and other metabolic products of digestion do not enter the general circulation directly, but instead flow to the liver. The liver removes and stores any excess metabolic products and maintains blood glucose levels within narrow bounds.

8. The large intestine has little digestive or absorptive activity; it functions principally to compact the refuse that is left over from digestion for easier elimination.

9. Vertebrates lack the enzymes necessary to digest cellulose. Ruminants, such as cows and sheep, are able to digest grass and other plants by maintaining colonies of cellulase-producing bacteria in their gastrointestinal tracts, much as termites maintain symbiotic protozoa in their guts that are able to digest the cellulose in wood.

10. Vertebrates also lack the essential organic substances required to synthesize many necessary compounds and must obtain these organic substances, called vitamins, from their diet. A number of trace elements also must be present in the diet.

REVIEW QUESTIONS

1. What are the two basic agents involved in the process of digestion? On what biomolecules do proteases, amylases, and lipases act? Into what end products are these biomolecules degraded? Why is it necessary to carry out the process of digestion in two separate stages?

2. In which group of animals does true extracellular digestion first occur? How is the digestive cavity of these animals different from that of cnidarians and flatworms? What two general activities occur along this system?

3. What is the general organization of the vertebrate digestive tract in terms of function and locale? What digestive tract specializations in particular vertebrate groups reflect the environment in which a group lives or the type of food it eats?

4. What class of vertebrates lacks teeth? How does tooth structure vary among carnivores, herbivores, and omnivores? What are the four types of teeth present in adult humans? What is the function of each?

5. Food passing beyond a mammal's back teeth signals what three reflexes? How is food normally prevented from reentering the mouth once it is in the esophagus? What two kinds of muscle surround the esophagus? What term describes the rhythmical contractions of these muscles?

6. Why is it important to control the amount of acid secreted in the stomach? What is the nature of this control? What physical condition results from failure of this system?

7. What chemicals does the pancreas produce and why are they important to the digestive process? What is the function of the liver in respect to digestion occurring in the small intestine? What is the function of the gallbladder?

8. What portion of the small intestine is actively involved in digestion? What is the primary function of the remainder of this organ? How does this region appear when viewed through a microscope? What is the physiological importance of these structures? How are the three products of digestion specifically absorbed?

9. What is the importance of bacterial symbiosis in herbivorous vertebrates? How does this compare to the digestive symbionts in various insects? Which group of animals are physiologically better able to utilize plants as a food source, horses or cows? Why? What is the value of human bacterial symbionts?

10. What is the formal definition of a vitamin? What common vitamin is synthesized by most vertebrates except for man, apes, monkeys, and guinea pigs? What are essential amino acids? How do they differ from vitamins?

THOUGHT QUESTIONS

1. One of the great evolutionary changes in the process of digestion was the advent of extracellular digestion. Do fungi carry out extracellular digestion? What advantage does a vertebrate digestive system have over the digestive process of a fungus?

2. Many birds possess crops, although few mammals do. Suggest a reason for this difference.

3. Humans obtain vitamin K from symbiotic bacteria living in their gastrointestinal tract. Many bacteria also produce ascorbic acid, vitamin C. Can you suggest a reason why people have not evolved a symbiotic relationship with bacteria that would result in their obtaining bacterial vitamin C?

4. Many proteolytic enzymes are synthesized in the pancreas and released into the duodenum. Since this is the case, why do mammalian digestive systems go to the trouble of producing pepsin in the stomach? What does pepsin do that the others do not?

FOR FURTHER READING

ABRAHAM, S., and D. LLEWELLYN-JONES: *Eating Disorders— The Facts*, Oxford University Press, Oxford, 1984. Covers the physiological and psychological factors leading to obesity, anorexia, and bulimia.

ALTSCHUL, A.M.: *Weight Control: A Guide For Counselors and Therapists*, Prager, New York, 1987. Practical manual for an integrated program of weight reduction and control.

COHEN, L.A.: "Diet and Cancer," *Scientific American*, vol. 257, November, 1987, page 42. Summary of evidence for an association between high fat, low fiber diets and certain cancers. Raises the question of whether the human gastrointestinal tract evolved to handle a very different set of foodstuffs than those currently consumed.

DeGABRIELE, R.: "The Physiology of the Koala," *Scientific American*, July 1980, pages 110-117. These Australian marsupials are adapted to a highly specific and unusual diet.

DIESENDORF, M.: "The Mystery of Declining Tooth Decay," *Nature*, July 1986, pages 125-129. Large reductions in tooth decay are being reported in unfluoridated and fluoridated countries—does fluoridation do any good? This article argues no.

FURNESS, J.B., and M. COSTA: *The Enteric Nervous System*, Churchill-Livingstone, Edinburgh, 1987. Complete description of neural control of gastrointestinal function.

HARRIS, A.R.: "Why do More Women get Ulcers?" *Chatelaine*, vol. 61, March 1988, page 32. Discusses the link between ulcers in women and risk factors such as smoking, alcohol, and stress.

LOGUE, A.: *The Psychology of Eating and Drinking*, W.H. Freeman and Co., New York, 1986. Why people eat the way they do. A very good treatment of anorexia and obesity.

MOOG, F.: "The Lining of the Small Intestine," *Scientific American*, November 1981, pages 154-176. A clear description of the most important absorptive surface in the human body.

PENRY, D. and P. JUMARS: "Modeling Animal Guts as Chemical Reactors," *American Naturalist*, January 1987, pages 69-96. Different animals have quite different kinds of digestive tracts, and this article argues that the reasons for the differences are evolutionary.

WARDLAW, G.M., and P.M. INSEL: *Perspectives in Nutrition*, Mosby–Year Book, Inc., St. Louis, 1990. Provides a comprehensive, up-to-date treatment of all aspects of nutrition.

WINICK, M.: *Control of Appetite: Current Concepts in Nutrition*, John Wiley & Sons, Inc., New York, 1988. Describes appetite regulation by control centers in the brain and receptors in the gastrointestinal tract. Discusses that the CNS pathways involved in satiety involve endogenous opiates and mediate responses to stress and reward.

50

Respiration

OVERVIEW

Oxygen enters the bodies of animals by diffusion into water. The evolution of respiratory mechanisms among the vertebrates has favored changes that maximize the rate of this diffusion. The most efficient aquatic mechanism to evolve is the gill of bony fishes; the most efficient aerial respiration mechanism is the two-cycle lung of birds. Both of these structures achieve high efficiency by arranging their capillaries so that blood flows in a direction counter to that of water or air. The respiratory mechanisms of other terrestrial vertebrates have evolved in various ways, which although less efficient than those of the fishes and birds, adapt them well to their terrestrial habitat. In the vertebrates, hemoglobin plays a critical role in the transport of oxygen and carbon dioxide in the respiration process.

FOR REVIEW

Here are some important terms and concepts that you will encounter in this chapter. If you are not familiar with them, you should review them before proceeding.

Chemistry of carbon dioxide (Chapter 2)

Diffusion (Chapter 6)

Oxidative respiration (Chapter 8)

Adaptation of vertebrates to terrestrial living (Chapter 42)

Red blood cells (Chapter 43)

FIGURE 50-1

Most of us think of respiration as breathing. This boy is blowing his horn with air from his lungs. Respiration also refers to the metabolic processes which utilize oxygen and generate carbon dioxide.

No vertebrate is capable of photosynthesis. Vertebrates are heterotrophs, and they obtain carbon compounds by consuming other organisms and then burning these carbon compounds to obtain energy. The biochemical mechanism of this **oxidative metabolism** was discussed in Chapter 8. Basically, heterotrophs obtain their energy by oxidizing carbon compounds. They first remove electrons from them and then channel these electrons through a series of proton-pumping ports in the membrane to generate ATP. Finally, they donate the electrons to oxygen gas (O_2), forming water. In the process of stripping electrons from organic carbon compounds, individual carbon atoms are cleaved from the organic molecules and released as carbon dioxide (CO_2). The process thus uses up oxygen and generates carbon dioxide and water.

The water given off by this process is called **metabolic water** to emphasize its source. In some desert vertebrates the generation of metabolic water provides most of the water that they require to live. In considering the oxidative metabolism of most vertebrates, however, metabolic water is usually ignored, because vertebrates have plentiful supplies of water available to them. In these vertebrates, the metabolic water produced is simply diluted into the much larger volume of the body's internal water.

For most vertebrates, then, metabolic water can be ignored and oxidative metabolism may be viewed as a process that utilizes oxygen and produces carbon dioxide (Figure 50-1). This final balance sheet determines one of the principal physiological challenges facing all animals—how to obtain oxygen and dispose of carbon dioxide. The uptake of oxygen and the release of carbon dioxide together are called **respiration.**

WHERE OUR OXYGEN COMES FROM: THE COMPOSITION OF AIR

The oxygen gas molecules that are the raw material of respiration have—each one of them—been produced by photosynthesis. The oxygen in the earth's air is the product of photosynthesis. Initially, photosynthetic organisms in the oceans released oxygen gas molecules into the water which diffused into the atmosphere; later, terrestrial plants began to release additional oxygen directly into the air. Both processes are still going on. The present atmosphere, which we call **air,** is rich in oxygen.

Dry air has a very constant composition: 78.09% nitrogen, 20.95% oxygen, 0.93% argon and the other noble gases, and 0.03% carbon dioxide. Physiologists group the noble gases present in air (they constitute less than 1% of the total volume) with nitrogen, because like nitrogen they are inert in most physiological processes. Convection currents cause air to have a constant composition to an altitude of at least 100 kilometers, although there is much less air present at high altitudes (Figure 50-2).

The amount of air present at a given altitude is usually expressed in terms that depend on its weight. Imagine a column of air standing on the ground and extending up into the sky as far as the atmosphere goes. This column has a lot of gas molecules in it, and they all experience the force of gravity. For this reason the column, which you might consider "as light as air," actually weighs a lot. People and other terrestrial animals are not usually aware of this weight, since land-dwelling animals evolved under its influence and possess bodies that are structurally able to withstand the pressure that is exerted by the weight of the air.

How much does all this air weigh? It weighs enough to push down on one end of a U-shaped column of mercury sufficiently to raise the other end of the column 760 millimeters at sea level under a set of specified, standard conditions (Figure 50-3). An apparatus that measures air pressure in this way is called a barometer; 760 mm Hg (millimeters of mercury) is therefore the **barometric pressure** of the air at sea level, on the average. This pressure is also defined as **one atmosphere** of pressure.

Within the column of air, each separate gas exerts a pressure as if it existed alone. The total air pressure is made up of the component **partial pressures** of each of the individual gases. Thus at sea level the total of 760 millimeters air pressure is composed of:

$760 \times 79.02\% = 600.6$ millimeters of nitrogen
$760 \times 20.95\% = 159.2$ millimeters of oxygen
$760 \times 0.03\% = 0.2$ millimeter of carbon dioxide

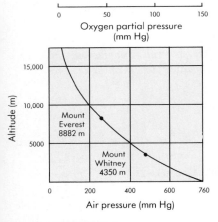

FIGURE 50-2

The relationship between air pressure and altitude above sea level. At the high altitudes characteristic of mountaintops, air pressure is much less than at sea level. At the top of Mount Everest, the world's highest mountain, there is only one-third the air present at sea level.

At altitudes above 6000 meters, humans do not survive long. The air still contains 20.95% oxygen, but there is much less air and thus less total oxygen available. The atmospheric pressure at such a height is about 380 millimeters (see Figure 50-2), so that the partial pressure of oxygen (abbreviated P_{O_2}) is only:

$$380 \times 20.95 = 80 \text{ millimeters of oxygen}$$

This figure is only half of the amount of oxygen that is available at sea level.

The Diffusion of Oxygen into Water

Molecules of oxygen and other gases are constantly diffusing into the earth's atmosphere from the oceans and diffusing back into the oceans from the air in a dynamic exchange. As in any diffusion process, the net movement is in the direction of least concentration. Thus the number of oxygen gas molecules that will diffuse from the air into a volume of water is directly proportional to the amount of oxygen that is present in that air, and inversely proportional to the amount of oxygen already present in the water. Oxygen will diffuse into the volume of water until an equilibrium is reached, an event that occurs when the partial pressure of the oxygen in the water is the same as the partial pressure of the oxygen in the air.

The Passage of Oxygen into Cells

Life evolved in an aqueous environment, and in a real sense terrestrial organisms are still prisoners of water. The cell membrane of a terrestrial organism presents no barrier to the entry of oxygen gas molecules directly from air. Like water, oxygen and carbon dioxide molecules can diffuse freely across the phospholipid layer of the cell membrane, passing through small imperfections in it. The membrane of a cell, however, cannot maintain its integrity without the presence of water surrounding it, because it is the hydrophobic interaction with water that causes the lipid bilayer to form. For this reason no terrestrial organism obtains gases across a dry cell membrane. Instead organisms import and export gases as dissolved components within the water that passes freely across cell membranes.

Like passengers on a train, oxygen and carbon dioxide gas molecules are transported into and out of cells by water. How many molecules enter or leave depends on how many are carried there by water, in which they are dissolved—in other words, on the concentrations of these gases in the water. Any factor that affects these concentrations will thus be an important one in respiration. How many molecules of a particular gas will be present in water depends on the following four factors:

1. *Composition of the air.* Molecules from the air enter and leave the water by diffusion. Recall that diffusion is the net movement of molecules to areas of lower concentration; it occurs as a result of random molecular motion. The rate of diffusion depends on molecular concentration. The more oxygen in the air, for example, the more oxygen molecules that are available to enter the water.

2. *Solubility.* Some gases are more soluble in water than others.

3. *Temperature.* Random molecular motion increases with increasing temperature, with the result that diffusion proceeds more rapidly at higher temperatures. Because molecules diffuse out of liquids as well as into them, higher temperatures decrease the solubility of gases in liquids. Oxygen is only half as soluble in water at 30° C as it is at 0° C (Figure 50-4).

4. *Solute concentration.* Salts reduce the solubility of gases. Thus the solubility of oxygen in seawater at 15° C is only 82% of what it is in fresh water.

Air enters cells by diffusion into water. Air contains 21% oxygen, but the rate of diffusion of oxygen from air into water, and thus into cells, depends on the concentration of air, which is less at higher altitudes.

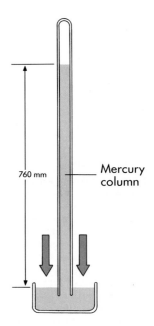

FIGURE 50-3

A simple mercury barometer. The weight of air pressing down on the surface of the mercury in the open dish pushes the mercury down into the dish and up the tube. The greater the air pressure pushing down on the mercury surface, the farther up the tube the mercury will be forced. At sea level, air pressure will cause a standard column of mercury to rise 760 millimeters.

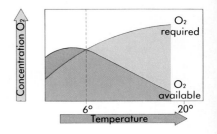

FIGURE 50-4

As temperature increases, so do metabolic demands for oxygen, but the solubility of oxygen in water falls. This is why frogs can dig into mud and survive all winter, but will drown if held underwater in the summer.

Some gases are more soluble in water than others, even when they are present at the same partial pressures. In dry air at sea level, at 15° C, and a partial pressure of one atmosphere (760 millimeters), 16.9 milliliters of nitrogen, 34.1 milliliters of oxygen, and 1019 milliliters of carbon dioxide will dissolve in a liter of water. Carbon dioxide is 30 times more soluble than oxygen.

THE EVOLUTION OF RESPIRATION

The capture of oxygen and the discharge of carbon dioxide by organisms depend on the diffusion of these gases into water. In vertebrates, the gases diffuse into the aqueous layer covering the epithelial cells that line the respiratory system of the animal. The diffusion process is in all cases passive, driven only by the difference in oxygen concentration between the interior of the organism and the exterior environment. The process is described by a relationship known as **Fick's Law of Diffusion:**

$$R = D \times A \times \frac{\Delta p}{d}$$

where

R = the rate of diffusion. In this case, the speed of oxygen or carbon dioxide movement is the rate of diffusion.

D = the diffusion constant, whose value depends on the material through which the diffusion is occurring.

A = the area across which diffusion takes place.

Δp = the difference in partial pressures, or concentration difference. In this case, the difference is the partial pressure of oxygen or carbon dioxide in the environment minus that in the organism.

d = the distance a molecule must travel to get to an area of lower concentration.

FIGURE 50-5
Gas exchange in animals may take place in a variety of ways.
A Gases diffuse directly into single-celled organisms.
B Amphibians and many other multicellular organisms respire across their skin.
C Echinoderms have protruding papulae, which provide an increased respiratory surface.
D Insects respire through spiracles, openings in their cuticle.
E The gills of fishes provide a very large respiratory surface and countercurrent exchange.
F Mammalian lungs provide a large respiratory surface, but do not permit countercurrent exchange.

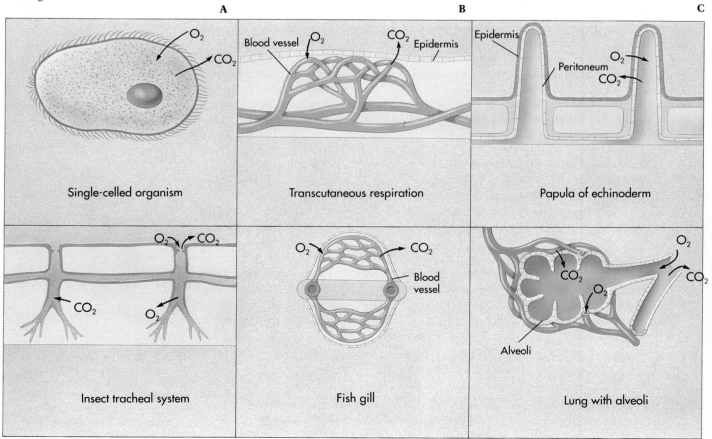

PART XI   VERTEBRATE BIOLOGY

Major changes in the mechanism of respiration have occurred during the evolution of animals (Figure 50-5). In general these changes have tended to optimize the rate of diffusion, R. By inspecting Fick's Law you can see that natural selection can optimize the rate of diffusion by favoring changes that (1) increase A (surface area); (2) decrease d (distance); and (3) increase Δp (concentration difference). The evolution of respiratory systems has involved changes of all three sorts.

Simple Diffusion

Oxygen diffuses slowly. The levels of oxygen required by oxidative metabolism in most organisms cannot be obtained by diffusion alone over distances greater than about 0.5 millimeter. This factor severely limits the size of organisms that obtain their oxygen entirely by diffusion from the environment to the metabolizing cytoplasm. Protozoa are small enough that diffusion distance presents no problem, but as the size of an organism increases, the problem soon becomes significant. An organism's mass increases as the cube of the diameter, whereas its surface area increases only as the square. In other words, as the size of an organism increases, an increasingly greater proportion of the volume of that organism is farther and farther away from the surface. Unless metabolism is greatly slowed down, as it is, for example, in the dilute cytoplasm of jellyfish, an increase in organism size must be accompanied by changes that facilitate the diffusion of gases into that organism.

Creating a Water Current

Most of the more primitive multicellular phyla possess no special respiratory organs. The sponges (phylum Porifera), cnidarians (phylum Cnidaria), many flatworms (phylum Platyhelminthes) and roundworms (phylum Nematoda), and some annelids (phylum Annelida) all obtain their oxygen by diffusion directly from surrounding water. How do they overcome the limits imposed by diffusion? They increase the term Δp, the difference in concentration, in the Fick equation. In a number of different ways, many of which involve beating cilia, these organisms create a water current, by means of which they continuously replace the water over the diffusion surface. Because of this continuous replenishment with water containing fresh oxygen, Δp *does not decrease as diffusion proceeds*. Although each oxygen gas molecule that passes into the organism has been removed from the surrounding volume of water, the exterior oxygen concentration does not fall, as a new volume of water is constantly replacing the depleted one. The result is a higher realized value of R, the rate of diffusion.

Increasing the Diffusion Surface

All of the more advanced invertebrates (mollusks, arthropods, echinoderms), as well as the vertebrates, possess special respiratory organs that increase the surface area available for diffusion and provide intimate contact with the internal fluid, which is usually circulated throughout the body. These special organs thus increase the rate of diffusion in the Fick equation by increasing A and decreasing d. These organs are of two kinds: (1) those that facilitate exchange with water, and (2) those that facilitate exchange with air. As a rough rule of thumb, aquatic respiratory organs increase the diffusion surface by extensions of tissue, called **gills,** that project from the body out into the water. Atmospheric respiratory organs, on the other hand, involve invaginations into the body.

Among the simplest of the water-exchange respiratory organs is the **external gill,** a highly convoluted outfolding membrane system with a very large surface area exposed to water. The great increase in diffusion surface provided by gills enables organisms living in water to extract far more oxygen from it than would be possible from their body surface alone.

Perhaps the simplest of the air-exchange organs are the **tracheae** of arthropods (Chapter 41). Tracheae are extensive series of passages connecting the surface of the animal to all portions of its body. Oxygen diffuses from these passages directly to the cells, without the intervention of a circulatory system. Piping air directly to the cells in this manner works very well in organisms such as insects, which have small bodies relative to those of vertebrates. Air must move only a relatively short distance within their bodies, but this relationship severely limits the potential body size of such organisms.

Enclosing the Respiratory Organ

External gills provide a greatly increased diffusion surface (A), but they do have one great disadvantage. It is very difficult to circulate water constantly past the diffusion surface to maintain a high Δp. Many fish larvae and amphibian larvae, as well as developmentally arrested (**neotenic**) amphibian larvae that remain permanently aquatic, such as the axolotl (Figure 50-6), circulate water past the respiratory surface of external gills by physically moving the gill through the water. This is not a very effective solution to the problem, especially since the highly branched gills offer significant resistance to the movement.

Among other organisms, specialized **branchial chambers** evolved, which provide a means of pumping water past the gills. Mollusks, for example, have an internal cavity called the **mantle cavity,** which opens to the outside and contains the gill. The contraction of the muscular walls of the mantle cavity causes the chamber to draw water in and then to expel it. In crustaceans the branchial chamber lies between the bulk of the body and the hard outer shell of the animal. This chamber contains gills and opens to the surface underneath a limb. Movement of the limb serves to bail out the branchial chamber, drawing water through it and thus creating currents over the gills.

> Aquatic animals have optimized the rate of diffusion of oxygen by increasing the diffusion surface with gills and by actively moving water past the diffusion surface.

THE GILL AS AN AQUEOUS RESPIRATORY MACHINE

By far the most successful branchial chamber evolved among the bony fishes. In the members of this group, water passes through the mouth into two **opercular cavities,** which are situated on each side of the head behind the mouth. From these cavities the water passes out of the body. The gills are situated between the mouth and the entrance to each of the opercular cavities, separating the cavities from the mouth just as curtains would do.

Many fishes that swim continuously, such as tuna, have practically immobile gill covers over the opercular cavities. These fishes swim with their mouths partly open, forcing water constantly over the gills, a process that amounts to a form of **ram ventilation.** Most bony fishes, however, have flexible gill covers that permit a pumping action (Figure 50-7). When the fish swallows water, the water is forced past the gills during the process of inhalation. By expanding the sides of the opercular cavities during exhalation, a negative pressure (suction) is created, which draws water from the

FIGURE 50-6
The gills of the axolotl are external. The axolotl sweeps them through the water in order to move water over the diffusion surface.

FIGURE 50-7
How a fish breathes. The gills are suspended between the mouth cavity and the outside, under a hard cover called the operculum, which when pressed down seals off the opening. Breathing occurs in two stages.
A When the oral valve of the mouth is open, closing the operculum increases the volume of the mouth cavity so that water is drawn in.
B When the oral valve is closed, opening the operculum decreases the volume of the mouth cavity, forcing water out past the gills to the outside.

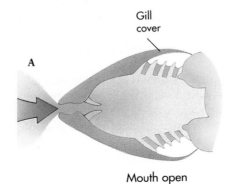

A

Gill cover

Mouth open

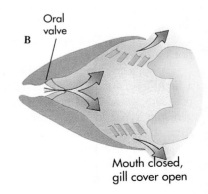

B

Oral valve

Mouth closed, gill cover open

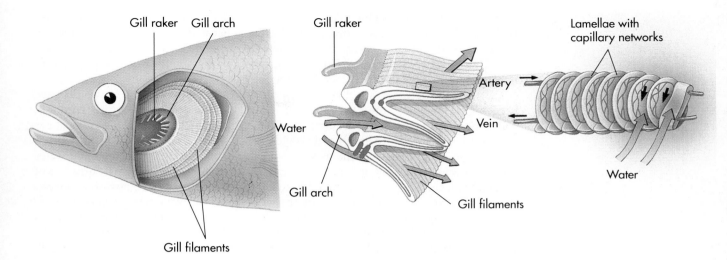

Gill raker   Gill arch

Gill raker

Lamellae with capillary networks

Water

Artery

Vein

Gill arch

Gill filaments

Water

Gill filaments

FIGURE 50-8
Structure of a fish gill.

mouth past the gills and through the opercular cavities. Thus there is an uninterrupted one-way flow of water past the gills all through the inhalation cycle. This continuous movement of water over the gills maintains Δp at a high value, since the layer of water immediately over the gills is constantly being replaced as oxygen is taken from it.

In addition to maintaining a high value of Δp by a continuous flow of water, the gills of fishes are constructed in such a way that they actually increase the value of Δp. Each gill is composed of two rows of **gill filaments,** thin membranous plates stacked one on top of the other and projecting out into the flow of water (Figure 50-8). Each filament, in turn, carries rows of thin disklike lamellae arrayed parallel to the direction of water movement. Water flows past these lamellae from front to back. Within each lamella the blood circulation is arranged so that blood is carried in the direction opposite the movement of the water, from the back of the lamella to the front.

Because the water flowing over the lamellae and the blood flowing within lamellae run in opposite directions (Figure 50-9), Δp is maximized. At the back of the gill the least oxygenated blood meets the least oxygenated water and is able to remove oxygen from it. By the time the blood reaches the front of the gill it has acquired a lot of

FIGURE 50-8
Structure of a fish gill. Water passes from the gill arch out over the filaments (from left to right in the diagram). Water always passes the lamellae in the same direction—which is opposite to the direction the blood circulates across the lamellae. The success of the gill's operation critically depends on this opposite orientation of blood and water flow.

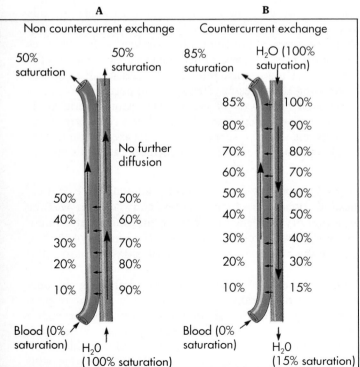

A
Non countercurrent exchange

50% saturation   50% saturation

No further diffusion

50%   50%
40%   60%
30%   70%
20%   80%
10%   90%

Blood (0% saturation)   H₂0 (100% saturation)

B
Countercurrent exchange

85% saturation   H₂O (100% saturation)

85%   100%
80%   90%
70%   80%
60%   70%
50%   60%
40%   50%
30%   40%
20%   30%
10%   15%

Blood (0% saturation)   H₂0 (15% saturation)

FIGURE 50-9
Countercurrent exchange.
A When blood and water flow in the same direction, dissolved oxygen gas can diffuse from the water into the blood rapidly at first, because of the large concentration difference (0% in blood versus 100% in water), but the difference decreases as more oxygen diffuses from water into blood, until finally the concentrations of oxygen in water and blood are equal. At this point there is no concentration difference to drive any further diffusion. In this example, blood can obtain no more than 50% dissolved oxygen in this fashion.
B When blood and water flow in opposite directions, the initial concentration difference between water and blood is not as great (0% in blood versus 15% in water) but is sufficient for diffusion to occur from water to blood. As more oxygen diffuses into the blood, raising its oxygen concentration, it encounters water with high and higher oxygen concentrations; at every point, the oxygen concentration is higher in the water, so that diffusion continues. In this example, blood obtains 85% dissolved oxygen.

oxygen, but the blood is able to acquire still more oxygen by diffusion from the water entering the gill. The reason that it can do this is because the new water is richer in oxygen than the water that has already flowed past the gills and lost some of its oxygen. This kind of **countercurrent flow** ensures a continuous gradient of concentration, and diffusion continues to occur all along the gill.

If the flow of water and blood had been in the same direction, Δp would have been high initially, as the oxygen-free blood met the new water that was entering. The concentration difference would have fallen rapidly, however, as the water lost oxygen to the blood. Since the blood oxygen concentration rises as the water oxygen falls, much of the oxygen in the water would remain there when the blood and water concentrations become equal, and diffusion would cease before maximal diffusion of oxygen from the water into the blood is achieved. In countercurrent flow, in contrast, the blood oxygen level encountered by the water becomes lower and lower as the level of oxygen in the water falls. The result is that in countercurrent flow the blood can attain oxygen concentrations as high as those that exist in the water entering the gills. Fish gills are the best Fick machines that occur among organisms. They are able to maximize the rate of diffusion in an oxygen-poor medium, obtaining up to 85% of the available oxygen.

> The gill is the most efficient of all respiratory organs. This great efficiency
> derives from the countercurrent flow of water past the blood vessels of the gills.

FROM AQUATIC TO ATMOSPHERIC BREATHING: THE LUNG

Water is relatively poor in dissolved oxygen; it contains only 5 to 10 milliliters of oxygen per liter. Air, in contrast, is rich in oxygen, containing about 210 milliliters of oxygen per liter. Not surprisingly, many members of otherwise aquatic groups utilize atmospheric air as a source of oxygen; these include many mollusks, crustaceans, and fishes.

When organisms first became fully terrestrial, air became the source of their oxygen. An entirely new respiratory apparatus evolved, one that was based on internal passages rather than on gills. Why were gills not maintained in terrestrial organisms, in view of the fact that they are such superb oxygen-capturing mechanisms? Gills were lost for two principal reasons:

1. *Air is less buoyant than water.* Because the fine membranous lamellae of gills lack structural strength, they must be supported by water to avoid collapsing on one another. A fish out of water, although awash in oxygen, soon suffocates because its gills collapse into a mass of tissue. This collapse greatly reduces the diffusion surface of the fish. Unlike gills, internal air passages can remain open, the body itself providing the necessary structural support.
2. *Water diffuses into air through the process of evaporation.* Atmospheric air is rarely saturated with water vapor, except immediately after a rainstorm. Consequently, terrestrial organisms that live surrounded by air are constantly losing water to the atmosphere. Gills would have provided an enormous surface area for water loss.

Two main systems of internal respiratory exchange evolved among terrestrial organisms. One was the tracheae of insects, mentioned earlier, and the other was the **lung.** Both systems sacrifice respiratory efficiency in order to maximize water retention. Insects prevent excessive water loss by closing the external openings of the tracheae whenever possible. They do this whenever body carbon dioxide levels are below a certain point.

Lungs, in contrast, minimize the effects of desiccation by eliminating the one-way flow of air that was such an effective means of increasing Δp in aquatic respiratory systems. In most terrestrial vertebrates, the air moves into the lung through a tubular passage and then back out again by way of the same passage. When each breath is completed, the lung still contains a volume of air, the so-called **residual volume.** In adult humans this volume is about 1200 milliliters. Each inhalation adds

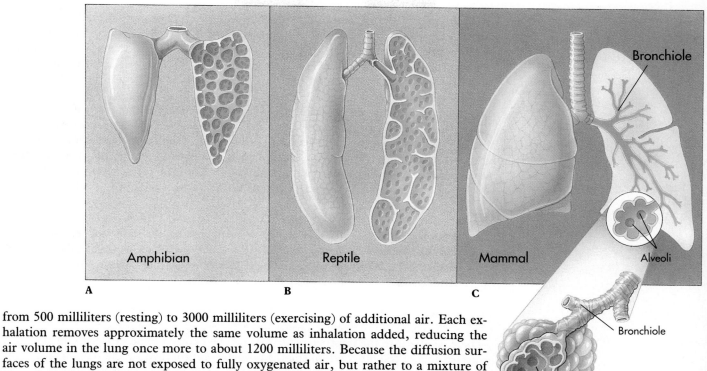

FIGURE 50-10
Evolution of the vertebrate lung.

from 500 milliliters (resting) to 3000 milliliters (exercising) of additional air. Each exhalation removes approximately the same volume as inhalation added, reducing the air volume in the lung once more to about 1200 milliliters. Because the diffusion surfaces of the lungs are not exposed to fully oxygenated air, but rather to a mixture of fresh and partly depleted air, the Δp is far from maximal, and the respiratory efficiency of lungs is much less than that of gills. Oxygen capture is lessened by this two-way flow of air—but so is water loss.

Amphibians

There is so much more oxygen in air than in water that low respiratory efficiency does not appear to have presented a critical problem to early land-dwellers; the amphibian lung is hardly more than a sac with a convoluted internal membrane (Figure 50-10, A). This internal sac is connected by a windpipe or **trachea** to the rear of the mouth chamber or oral cavity, and the opening is controlled by a valve, the **glottis.** A series of passages called **sinuses** connect the oral cavity to the nose, which opens to the outside, enabling the animal to breathe with its mouth closed. Although the inner membrane surface of the lung is convoluted, the surface area of amphibian lungs available for diffusion is still not large. Amphibians obtain much of their oxygen through their moist skin.

Reptiles

Reptiles are far more active than amphibians, and they have significantly greater metabolic demands for oxygen. The early reptiles could not rely on their skins for respiration; living fully on land, they are "watertight," avoiding desiccation by means of a dry scaly skin. Again, as in aquatic organisms, the respiratory apparatus has changed in ways that tend to optimize respiratory efficiency. The lungs of reptiles possess many small chambers within their surface called **alveoli,** which are clustered together like grapes (Figure 50-10, B). Each cluster of alveoli is connected to the main air sac in the lung by a short passageway called a **bronchiole.** Air within the lung enters the alveoli, where all gas exchange with the circulatory system takes place. The alveoli greatly increase the diffusion surface of the lung.

Mammals

The metabolic demands for oxygen became even greater with the evolution of mammals, which, unlike reptiles and amphibians, maintain a constant body temperature by heating their bodies metabolically. In the lungs of mammals the bronchioles branch, each branch connecting to many clusters of alveoli (Figure 50-10, *C*). Humans have about 300 million alveoli in each of their two lungs. The branching of the bronchioles and the increase in number of alveoli combine to increase yet again the total diffusion surface area of the lung. In humans the total surface area devoted to diffusion can be as much as 80 square meters, an area about 42 times the surface area of the body.

Some mammals are far more active than others, but the more active ones do not have proportionally larger lungs. Lung mass is proportional to body mass in all mammals. In active mammals, however, the individual alveoli are smaller and more numerous, increasing the diffusion surface (*A* in the equation on p. 1034). In addition, the epithelial layer of cells separating the alveoli from the bloodstream in such active mammals is thinner, a factor reducing the diffusion distance (*d* in the equation on p. 1034).

> When amphibians evolved lungs, one-way flow through the respiratory organ was abandoned in favor of a saclike lung. Increases in efficiency among the reptiles and mammals have been achieved by increases in the lung's internal surface area.

Birds

There is a limit to the improvements that can be realized by increases in the diffusion surface area of the lung, a limit that is probably approached by the more active mammals. With the advent of birds, flying introduced respiratory demands that exceed the capacity of a saclike lung. Unlike bats, whose flight involves considerable gliding, many birds rapidly beat their wings for prolonged periods during flight. Such rapid wing beating uses up a lot of energy quickly, because it depends on the frequent contraction of wing muscles. Flying birds must carry out intensive oxidative respiration within their cells to replenish the ATP expended by contracting flight muscles—and thus require a great deal of oxygen, more oxygen than a saclike lung, even one with a large surface area such as a mammalian lung, is capable of delivering. The lungs of birds cope with the demands of flight by employing a new respiratory mechanism, one that produces a significant improvement in respiratory efficiency.

An avian lung works like a two-cycle pump (Figure 50-11). When a bird inhales air, the air passes directly to a series of nondiffusing chambers called the **posterior air sacs.** When the bird exhales, the air flows into the lung. On the following inhalation, the air passes from the lung to a second series of air sacs, the **anterior air sacs.** Finally, on the second exhalation, the air flows from the anterior air sac out of the body. What is the advantage of this complicated passage? It creates a unidirectional flow of air through the lungs!

FIGURE 50-11
How a bird breathes.
A The respiratory system of a bird is composed of anterior air sacs, lungs, and posterior air sacs.
B Breathing occurs in two cycles.
Cycle 1: Air is drawn from the trachea into the posterior air sacs and then is exhaled through the lungs.
Cycle 2: Air is drawn from the lungs into the anterior air sacs and then is exhaled through the trachea. Passage of air through the lungs is always in the same direction, from posterior to anterior (right to left in this diagram). Because blood circulates in the lung from anterior to posterior, the lung achieves a type of countercurrent flow and thus is very efficient at picking up oxygen from the air.

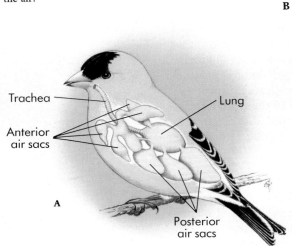

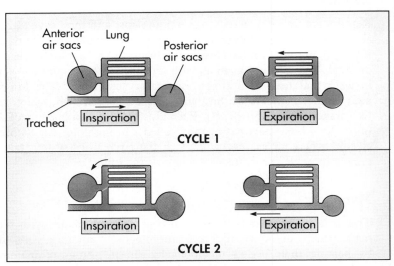

Air flows through the lung system of birds in one direction only, from posterior to anterior. There is no dead volume, as there is in the mammalian lung. For this reason, the air passing across the diffusion surfaces of the avian lung is always fully oxygenated. Unidirectional flow also permits a second major improvement in respiratory efficiency. Just as in the gills of fish, the flow of blood past the avian lung runs in a direction different from that of air flow in the lung (Figure 50-12). However, the form of flow seen in the avian lung is different from that seen in fish gills.

In fish gills, flow of blood and water are in opposite directions, 180° apart, while in bird lungs the latticework of capillaries is arranged *across* the air flow, at a 90° angle. This modified countercurrent flow mechanism, called **cross-current flow,** is not as efficient at extracting O_2 as a 180° arrangement would be, but the oxygenated blood leaving the lung can still contain more O_2 than exhaled air, a capacity not achievable by mammalian lungs. Because of the high efficiency of cross-current flow in gathering oxygen from the air into the blood, a sparrow has no trouble breathing at an altitude of 6000 meters, whereas a mouse, which has approximately the same body mass, blood with the same affinity for oxygen, and a similar high metabolic rate, cannot respire successfully at such an elevation. In terms of the Fick equation, one would say that in the kind of lungs that birds possess, Δp is greatly increased. Just as fish gills are the most efficient aquatic respiratory machines, so avian lungs are the most efficient atmospheric ones. Both achieve high efficiency by utilizing forms of countercurrent flow.

> **The most efficient atmospheric respiratory organs are the lungs of birds, which are arranged to permit a one-way flow of air without significant water loss. This arrangement permits the establishment of a type of countercurrent flow of blood, the key to high efficiency.**

THE STRUCTURE AND MECHANICS OF THE RESPIRATORY SYSTEM

Thus we see that humans are by no means the most efficient breathers among terrestrial vertebrates. Understanding how humans breathe is, however, of considerable practical interest.

Humans possess a pair of lungs located in the chest, or thoracic, cavity. The two lungs hang free within the cavity, being connected to the rest of the body only at the one position where the lung's blood vessels and air tube enter (Figure 50-13). This air

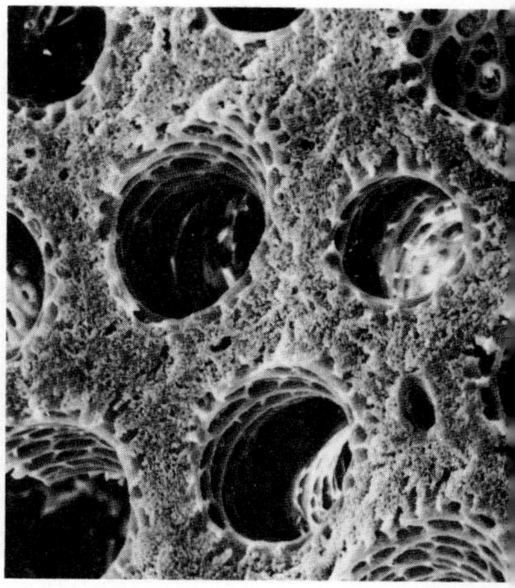

FIGURE 50-12
Cross section of a lung of a domestic chicken, magnified 75 times. Air travels through the tunnels, called parabronchi, while blood circulates in the opposite direction within the fine lattice.

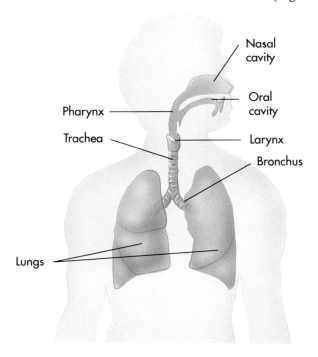

FIGURE 50-13
The human respiratory system.

tube is called a **bronchus.** It connects each lung to a long tube, the **trachea,** which passes upward in the body, past the voice box, or **larynx,** and opens into the rear of the mouth. Air normally enters through the nostrils, which are lined with hairs that filter out dust and other particles. As the air passes through the nasal cavity, an extensive array of cilia on its epithelial lining further cleans the air and moistens it (Figure 50-14). The air then passes through the back of the mouth, crossing the path of food as it enters first the larynx and then the trachea. From there it passes down through the bronchus and the **bronchioles,** and into the alveoli of the lungs.

The human respiratory apparatus is simple in structure, functioning as a one-cycle pump. The thoracic cavity is bounded on its sides by ribs, which are capable of flexing, and on the bottom by a thin layer of muscle, the **diaphragm,** which separates the thoracic cavity from the abdominal cavity. Each lung is covered by a very thin, smooth membrane called the **pleural membrane.** A second pleural membrane marks the interior boundary of the thoracic cavity, dividing it into two halves; each lung is thus suspended in its own cavity. Within the cavity the weight of the lungs is supported by water, the **intrapleural fluid.**

The intrapleural fluid not only supports the lungs, but also plays another important role by permitting an even application of pressure to all parts of the lung. You can visualize the pleural membranes as a system of two balloons of different sizes, one nested inside the other (Figure 50-15), with the space between them completely filled with fluid and the inner balloon opening out to the atmosphere. Two forces act on this inner balloon: air pressure pushing it outward and water pressure pushing it inward.

The active pumping of air in and out through the lungs is called breathing. During inhalation (Figure 50-16), the walls of the chest cavity expand. The rib cage moves upward and outward, and the diaphragm moves downward by stretching taut. In effect, we have enlarged the outer balloon by pulling it in all directions. This expansion of the fluid space causes the fluid pressure to decrease to a level less than that of the internal air pressure within the lung (the inner balloon). As a result, the wall of the lung is pushed out. As the lung expands, its internal air pressure decreases, and air moves in from the atmosphere. During **exhalation,** the ribs and diaphragm return to their original resting position. In doing so, they exert pressure on the fluid. This pressure is transmitted uniformly by the fluid over the entire surface of the lungs, forcing air from the inner cavity back out to the atmosphere.

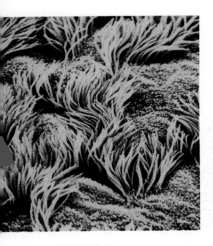

FIGURE 50-14
Respiratory cilia such as these line the trachea.

FIGURE 50-15

A simple experiment that shows how you breathe. In the jar is a balloon (**A**). When the diaphragm is pulled down, as shown in **B,** the balloon expands; when it is relaxed (**C**), the balloon contracts. In the same way, air is taken into your lungs when your diaphragm pulls down, expanding the volume of your lung cavity. When your diaphragm pushes back, the volume decreases and air is expelled.

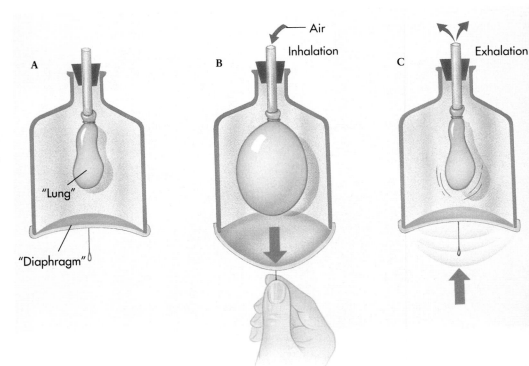

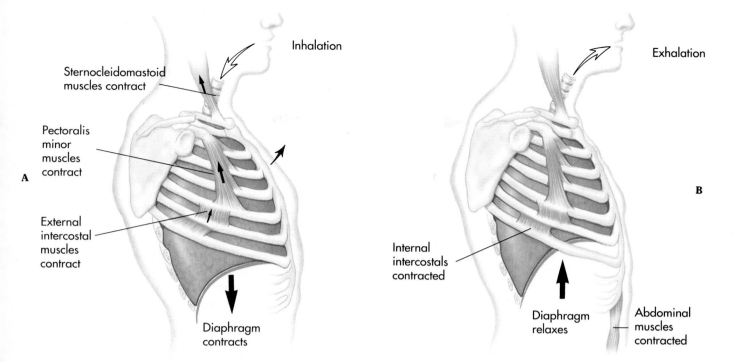

Inhalation

Sternocleidomastoid muscles contract

Pectoralis minor muscles contract

External intercostal muscles contract

Diaphragm contracts

A

Exhalation

Internal intercostals contracted

Diaphragm relaxes

Abdominal muscles contracted

B

Alveoli.   The respiratory bronchioles eventually terminate in the exchange zone of the lung, the **alveoli.** The basic functional unit of the lung is an **acinus,** a structure that resembles a miniature bunch of grapes. There are about 300 million alveoli in each lung, giving it a sponge-like consistency. Diffusion is favored by large surface area and impaired by long path length. The distance between the thin-walled alveoli and the pulmonary capillaries is only 0.5 to 1.5 microns. The surface area for gas exchange of alveoli in the lung is 60 to 80 square meters, about the same as a tennis court! In all there are about 30 billion pulmonary capillaries, or about 100 capillaries per alveolus. Alveoli can be visualized as tiny air bubbles whose entire surface is bathed by the blood flowing through the lung.

Role of Bronchiolar Smooth Muscle.   Within the lung, each primary bronchus branches extensively. These branchings result ultimately in about 150,000 **terminal bronchioles.** Terminal bronchioles are surrounded by smooth muscle cells which can adjust the diameter of these passageways. Parasympathetic activity constricts bronchiolar smooth muscle. Sympathetic activity relaxes bronchiolar smooth muscle, which dilates the airways, for example during exercise. In some people substances called **allergens** trigger **histamine release** which, in turn, constricts bronchiolar smooth muscle. This decreases ventilation. In severe allergic attacks the airway resistance may rise by twenty-fold and the effect on ventilation may be life-threatening.

Air Flow in the Lung

Certain terms are used to describe the volume changes of the lung during breathing. The volume of air alternately inspired and expired in a single breath is the **tidal volume** (typically 500 milliliters at rest), while the volume of air that remains in the lung after a normal resting expiration is the **functional residual capacity** (FRC). The extra volume of air above the FRC that can be inspired using a maximal effort is the **inspiratory reserve volume.** The extra volume that can be forced out of the lung following a normal expiration is called the **expiratory reserve volume.** The sum of the tidal volume, and the inspiratory and expiratory reserve volumes equals the maximum amount of air that can be moved into and out of the lungs by muscular effort. This is called the **vital capacity.** The average vital capacity is 4.6 liters in young men and 3.1 liters in young women. The air remaining in the lung after a maximal expiration is the **residual volume,** and is typically about 1200 milliliters. The residual volume cannot

FIGURE 50-16
How a human breathes.
A Inhalation. The diaphragm and walls of the chest cavity expand, increasing its volume. As a result of the larger volume, air is sucked in through the trachea.
B Exhalation. The diaphragm and chest walls return to their normal positions, reducing the volume of the chest cavity and forcing air outward through the trachea.

be determined using a spirometer. A normal adult male value for total lung capacity is 5800 milliliters.

The **respiratory rate** is the number of breaths per unit time. The total amount of air entering and leaving the respiratory system per minute, the **respiratory minute volume,** is the product of the tidal volume and the respiratory rate in breaths per minute. A normal respiratory minute volume is around 5 liters per minute, but minute volumes of as much as 130 liters per minute can be attained by young adult males during strenuous exercise. Movement of gas between the atmosphere and the blood occurs in two stages: **bulk flow** and **diffusion.** In the network of branching bronchi and bronchioles, there is bulk flow of air. Bulk flow is a much more effective mechanism than diffusion for transporting substances over long distances. Between the bronchioles and the alveoli, gas exchange occurs by diffusion.

Not all the air inhaled and exhaled enters the alveoli where it can be exposed to the blood. Air in the trachea, bronchi, and bronchioles does not contact the pulmonary capillaries, and these areas are referred to as the **anatomical dead space** of the lung. A fraction of the respiratory minute volume is wasted because it only ventilates the dead space. The amount of fresh air that reaches the alveoli with each breath, the **alveolar minute volume,** is equal to the tidal volume minus the dead space volume. The anatomic dead space volume of a healthy 70 kilogram man is about 150 milliliters. Thus for a single breath of 500 milliliters, only 350 milliliters of new air reaches the alveoli. Ventilation is compromised by increases in the dead space. For example, when swimmers breathe through a snorkel tube, tidal volume must be increased by the volume of the additional dead space if alveolar ventilation is to remain at its normal value.

How the Lungs Work

The human respiratory apparatus is simple in structure, functioning as a one-cycle pump. During inspiration (Figure 50-16), the walls of the chest cavity expand. The rib cage moves upward and outward, and the diaphragm moves downward. In effect, we have enlarged the outer balloon of Figure 50-16 by pulling it in all directions. The intrapleural fluid, like all fluids, is incompressible and thus its volume cannot change. The expansion of the outer balloon—the chest cavity—causes an equal expansion of the inner balloon—the lung. The lungs and chest cavity are tightly coupled by the intrapleural fluid, just as two microscope slides stick together if there is a thin film of water between them. The interplay of forces in the lung–chest wall system is like a tug of war between the chest wall and the lungs in which the intrapleural fluid is an inelastic rope.

When the breathing muscles are relaxed the inward pull of the lung is matched exactly by the outward spring of the chest wall and diaphragm. This is the condition between the end of an expiration and the beginning of the next inspiration in a quietly breathing person. As a result of the inward pull of the lung and outward spring of the chest wall, the intrapleural pressure is less than atmospheric pressure.

The intrapleural pressure becomes more negative on inspiration because, as the chest volume is increased, the lungs are stretched and the inward pull of the lungs increases. The increase in lung volume decreases the alveolar pressure, creating a pressure gradient between alveoli and atmosphere that drives air into the alveoli. When the respiratory muscles relax, the lung and chest wall system "springs back" to its resting position. While the volume of the lung is dropping back to its original value, the alveolar pressure rises above atmospheric pressure, driving air from alveoli through the airway into the atmosphere. This is called a **passive expiration** because it does not require active muscular effort. The volume of the lung at the end of a passive expiration is the functional residual capacity.

> **The active pumping of air in and out through the lungs is called breathing. The lungs are filled by suction—the expansion of the chest cavity draws air into the lungs—and the return of the ribs to their resting position drives air from the lungs.**

There is an attractive force between water molecules near an air-water interface. This force, called **surface tension,** makes water droplets bead on a nonabsorbent surface like a newly waxed car or nonstick pan and lets us make soap bubbles. The alveoli resemble soap bubbles, a sphere of air covered by a fluid, the blood. It turns out that surface tension is responsible for about 70% of the inward pull of the lung. The actual tension in the alveolar walls is much lower than would be expected if the inner surface of the alveoli was a layer of pure water. The reason is that the water on the inner surface of the alveoli contains phospholipid molecules called **lung surfactant.** Surfactant is produced by a special category of epithelial cells called **type II alveolar cells.**

Like all phospholipids, lung surfactant molecules have a hydrophilic (water-loving) end and a hydrophobic (water-hating) end. The surfactant molecules in the lung act like the detergents in soap bubbles and greatly reduce the surface tension. This reduces the force required to expand the lungs (or to expand a soap bubble) by a factor of three to four, thus greatly reducing the muscular effort required to breathe. It also turns out that these surfactant molecules compensate for the ten-

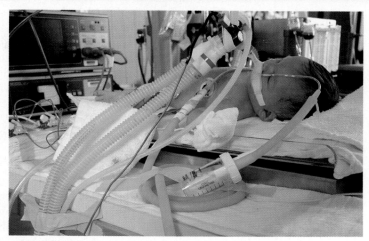

FIGURE 50-A

dency for small alveoli to collapse and large alveoli to expand. This effect can be understood by visualizing the surfactant molecules as closer together in the smaller alveolus than in the larger alveolus.

Infants born prematurely may have **respiratory distress syndrome** or **hyaline membrane disease.** This condition is caused by inadequate quantities of lung surfactant. Lung surfactant begins to be synthesized about the thirty-second week of gestation; it requires the hormone cortisol and is completed only a few weeks before full-term delivery. Infants with respiratory distress syndrome must make strenuous efforts

to breathe, and in extreme cases this increased effort leads to exhaustion and death.

Infants with respiratory distress syndrome are supported by mechanical ventilation with the hope that surfactant synthesis will occur given enough time. It is now possible for women threatened with premature delivery to be given a cortisol injection that will stimulate the synthesis of surfactant in the fetus. A recent technique, still undergoing clinical trials is the administration of **artificial surfactant** to premature infants. If successful, this would solve one of the major risk factors premature infants face.

Forced Inspiration and Expiration

Inspiration during quiet breathing is mainly the result of contraction of the diaphragm, with minimal expansion of the rib cage. Larger volumes of air can be inhaled by vigorous contraction of the **external intercostal muscles.** Contraction of external intercostal muscles lifts the rib cage outward and upward, at the same time the diaphragm moves downward. This greatly increases the volume of the thorax. Forceful expiration in vigorous breathing is the result of contraction of the **internal intercostal muscles** along with muscles of the abdominal wall. The effect of contracting these muscles is to increase the alveolar pressures attained during expiration. The changes in intra-alveolar pressure are reflected in changes of intrapleural pressure, but intrapleural pressure is always less than atmospheric pressure no matter whether breathing is quiet or forceful.

GAS TRANSPORT AND EXCHANGE

When oxygen has diffused from the air into the moist cells lining the inner surface of the lung, its journey has just begun. Passing from these cells into the bloodstream, the oxygen is carried throughout the body by the circulatory system to be described in the next chapter. It has been estimated that it would take a molecule of oxygen 3 years to be transported from your lung to your toe if the transport depended only on diffusion, unassisted by a circulatory system.

Oxygen moves within the circulatory system on carrier proteins that bind dissolved molecules of oxygen. This binding occurs in the capillaries surrounding the alveoli of the lungs. The carrier proteins subsequently release their oxygen molecules to metabolizing cells at distant locations in the body.

The carrier protein that is used by almost all vertebrates is **hemoglobin.** Hemoglobin is a protein composed of four polypeptide subunits; each of the four polypeptides is combined with an iron ion in such a way that oxygen can be bound reversibly to the ion. Hemoglobin is synthesized by **erythrocytes,** or red blood cells, and remains within these cells, which circulate in the bloodstream like ships bearing cargo.

Hemoglobin is an ancient protein, one that is also used as a carrier of oxygen in the circulatory systems of annelids, mollusks, and even some protozoans. Many invertebrates, however, also employ a second carrier protein, **hemocyanin.** Hemocyanin uses copper to bind oxygen in the same way that hemoglobin uses iron. Hemocyanin never occurs within blood cells as hemoglobin does, but it exists free in the circulating fluid (called **hemolymph**) of invertebrates. When hemocyanin combines with oxygen, it gives the hemolymph a blue color; when hemoglobin combines with oxygen, it gives blood a bright red color.

Oxygen Transport

In humans, only a small amount (5%) of oxygen gas is transported from lungs to body tissues while dissolved in blood plasma. The remaining 95% is bound to hemoglobin within red blood cells. The higher the P_{O_2} in the air within the lungs, the more oxygen will dissolve in the blood and combine with hemoglobin. The relationship is not a simple proportion, however. If the P_{O_2} is halved, the oxygen content is reduced only by about 25%. This facilitates oxygen uptake from the lungs into the blood and also aids unloading of oxygen from the blood to the tissues. Hemoglobin molecules act like little oxygen sponges, soaking oxygen up within red blood cells and causing more to diffuse in from the blood plasma. At the P_{O_2} encountered in the blood supply of the lung, most hemoglobin molecules are saturated with oxygen. In the tissues the P_{O_2} is much lower so that hemoglobin gives up its bound oxygen.

In tissue, the presence of carbon dioxide (CO_2) causes the hemoglobin molecule to assume a different shape, one that gives up its oxygen more easily (Figure 50-17). This augments the unloading of oxygen from hemoglobin. The effect of CO_2 on oxygen binding is called the Bohr effect. It is of real importance, since CO_2 is produced by the tissues at the site of cell metabolism. For this reason the blood unloads oxygen more readily to those tissues undergoing metabolism and generating CO_2.

FIGURE 50-17

Oxygen-hemoglobin dissociation curves for human hemoglobin. Hemoglobin is saturated with oxygen in blood leaving the lungs, **A,** but little of this oxygen remains after circulation through the tissues. In tissue such as exercising muscle, **B,** the concentration of oxygen, measured as pressure in millimeters of mercury, is low, about 30 millimeters. At such low pressures, oxygen tends to dissociate from hemoglobin, with the hemoglobin molecules giving up about 60% of the oxygen molecules they carry. In the lungs, where oxygen concentrations are much higher (typically 110 millimeters), oxygen molecules do not tend to dissociate from hemoglobin; hemoglobin molecules are fully saturated with all the oxygen they can carry.

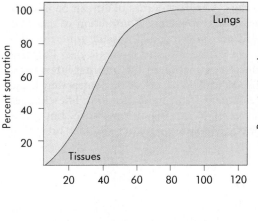

The shape of hemoglobin is also changed by molecules that bind to it, and this can also influence the ease of oxygen unloading. One of these is **2,3-diphosphoglycerate (DPG),** which shifts the dissociation curve even further to the right than carbon dioxide, greatly facilitating oxygen unloading. DPG binds to hemoglobin in many rapidly metabolizing tissues. The levels of DPG increase within a few days in the bodies of humans who stay at high altitudes (Figure 50-18). Because the partial pressure of oxygen is lower at high altitudes, it is necessary that the dissociation curve be shifted to the right for tissues to receive adequate amounts of oxygen.

Oxygen and carbon dioxide are not the only gases that can bind to hemoglobin. **Carbon monoxide (CO)** binds to hemoglobin 300 times more readily than does oxygen. Because it binds so strongly, it does not readily dissociate and so prevents the CO-bound hemoglobin from acting as an oxygen carrier. As a result, even a small amount of carbon monoxide in the air can lead to respiratory failure; as little as 0.1% of carbon monoxide in the air is dangerous.

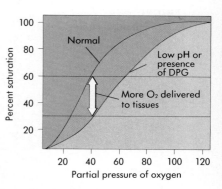

FIGURE 50-18
Lower blood pH or DPG acts to shift the oxygen dissociation curve of hemoglobin to the right, facilitating oxygen unloading.

The Exchange Process

When oxygen-rich air enters the alveoli of the lungs, it encounters oxygen-depleted blood in the capillaries surrounding the alveoli. The oxygen diffuses down the concentration gradient into the blood plasma. Within the plasma, it diffuses into the red blood cells, where it is bound by hemoglobin, one oxygen molecule to each of the four subunits of each hemoglobin protein. The binding of oxygen molecules to hemoglobin effectively removes them from solution, so that the concentration of free oxygen in the red blood cells remains low, facilitating the additional diffusion inward of more oxygen from the plasma and lungs. This soaking up of oxygen by hemoglobin enables whole blood to carry 50 times as much oxygen as the plasma alone could absorb by diffusion.

When the oxygen-loaded red blood cell reaches a capillary in tissue where the carbon dioxide concentration is high as a result of oxidative respiration carried out by the tissue cells, the hemoglobin molecules react to the higher carbon dioxide concentration by unloading their oxygen—the Bohr effect. This dissociation greatly elevates the free oxygen concentration within the plasma of the capillary, and the oxygen rapidly diffuses outward into the surrounding tissue, where the concentration of oxygen is much lower.

CO_2 Transport

At the same time that the red blood cells are unloading oxygen, they are also absorbing CO_2 from the tissue. Perhaps one fifth of the CO_2 that the blood absorbs is bound to hemoglobin. Another 8% is simply dissolved in plasma. The remaining 72% of the CO_2 diffuses from the plasma into the cytoplasm of red blood cells. There an enzyme, **carbonic anhydrase,** catalyzes the combination of CO_2 with water to form carbonic acid (H_2CO_3), which dissociates into bicarbonate (HCO_3^- and hydrogen (H^+) ions. This process removes large amounts of CO_2 from the blood plasma, facilitating the diffusion of more CO_2 into it from the surrounding tissue. The facilitation is critical to CO_2 removal, since the difference in CO_2 concentration between blood and tissue is not large (only 5%).

The red blood cells carry their cargo of bicarbonate ions back to the lungs. The lower CO_2 concentration in the air inside the lungs causes the carbonic anhydrase reaction to proceed in the reverse direction, releasing gaseous CO_2, which diffuses outward from the blood into the alveoli. With the next exhalation, this CO_2 leaves the body. The fifth of the CO_2 bound to hemoglobin also leaves because hemoglobin has a greater affinity for oxygen than for CO_2 at low CO_2 concentrations. The diffusion of CO_2 outward from the red blood cells causes the hemoglobin within these cells to release its bound CO_2 and take up oxygen instead. The red blood cells, with their newly bound oxygen, then start the next respiratory journey (Figure 50-19).

FIGURE 50-19
The respiratory journey.

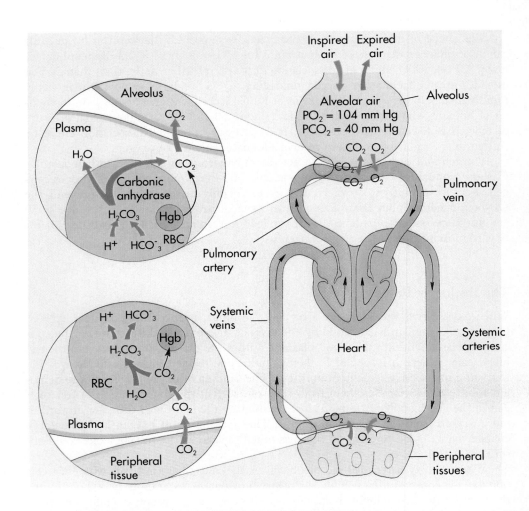

HOW THE BRAIN CONTROLS BREATHING

Each breath is initiated by a respiratory control center in the brain, which sends the nerve signals to the diaphragm and, for deeper breathing, to the intercostal muscles. These contractions regularly expand the chest cavity. Other neurons then act to inhibit the stimulation of these muscles so that they relax and the body exhales. These signals are not subject to voluntary control; we cannot consciously stop breathing. You breathe more slowly when you sleep, more rapidly when you run. You need less oxygen when asleep because many of your muscles are not being used, and more oxygen when you run because you are using muscles intensively.

The body regulates this change in rate of breathing by a simple feedback mechanism, a respiratory center in the brainstem. Chemoreceptors in the brain (central chemoreceptors) detect changes in CO_2 levels and transmit nerve impulses to the respiratory center, which sends appropriate signals to the muscles of the diaphragm and rib cage. There is also a second set of chemoreceptors in the aorta and carotid artery. These are primarily sensitive to pH and greatly reduced amounts of oxygen. When the CO_2 level rises, the respiratory center increases the rate of breathing. Thus, when you run and rapid muscle contraction is adding increased amounts of CO_2 to your blood, specific brain neurons detect the rise in CO_2, leading to more rapid and deeper breathing.

The walls of brain capillaries are not freely permeable to ions such as H^+ and HCO_3^-. This is the **blood-brain barrier** and it isolates the cerebrospinal fluid (CSF)

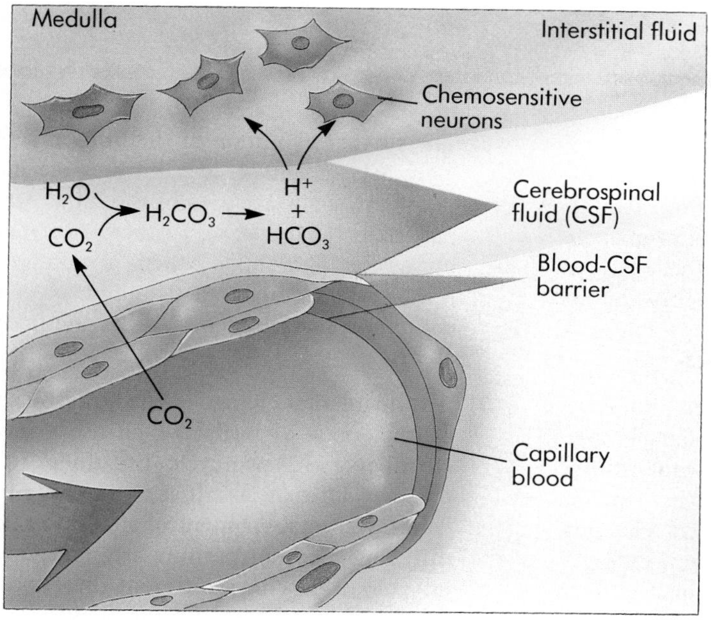

FIGURE 50-20
The blood-brain barrier.

from the arterial blood. While H^+ and HCO_3^- cannot cross the blood-brain barrier, dissolved CO_2 can cross readily. The central respiratory chemoreceptors do not detect plasma P_{CO_2} directly, but respond to the pH of the CSF. However, any increase in the arterial P_{CO_2} results in an immediate and equivalent increase in the P_{CO_2} of the CSF. Once in the CSF, the CO_2 reacts with water to form carbonic acid (H_2CO_3), which dissociates to produce H^+ and HCO_3^-. Hydrogen and HCO_3^- ions formed within the CSF are trapped there because they cannot cross the blood-brain barrier into the plasma (Figure 50-20).

Compared with the blood, CSF is weakly buffered because it contains little protein. The low buffer capacity of the CSF makes the pH of CSF sensitive to changes in plasma P_{CO_2}. When the arterial P_{CO_2} increases, the P_{CO_2} of the CSF increases proportionately. Reaction of CO_2 with H_2O in the CSF decreases its pH. The response of the central chemoreceptors to this change in pH increases respiration. In the case of CO_2, the ventilatory response is nearly linearly related to the change in arterial P_{CO_2}.

Under normal conditions, the input from the central chemoreceptors provides the major component of respiratory drive. These receptors monitor the pH of CSF, which is determined over the time scale of minutes to hours by the arterial P_{CO_2}.

LOOKING AHEAD TO CIRCULATION

In Chapter 51 we consider in more detail the journey of the red blood cells. Traveling between the lungs, where they acquire oxygen and release CO_2, and respiring tissues, where they release oxygen and acquire CO_2, the body's red blood cells traverse a complex highway that passes to all parts of the body. The red blood cells are not the only traffic on this highway. Just as the roads and sidewalks transport all the commerce of a city, so the circulatory system of the vertebrate body transports all the material that moves from one part of the body to another.

SUMMARY

1. All animals obtain oxygen by its diffusion into water. The rate of this diffusion depends on temperature and on atmospheric pressure. Air at high altitudes has the same percentage of oxygen as does air at sea level, but because there is less air at high elevations, the rate of diffusion is slower.

2. The evolution of respiratory mechanisms among animals has tended to favor changes that improve the rate of diffusion. These include changes that decrease the length of the path over which diffusion occurs, changes that increase the surface area over which diffusion occurs, and changes that maximize the difference in oxygen concentration between environment and tissue.

3. The most efficient aquatic respiratory organ is the gill of bony fishes. Fishes take water in through their mouths, move it past their gills, and pass it out of their bodies. This one-way flow is the secret to high respiratory efficiency, since it permits fishes to establish a countercurrent flow of blood; blood vessels are located within the gills in such a way that blood flows in a direction opposite that of water.

4. Gills will not work in air, since air is not buoyant enough to support their fine latticework of passages. That is why fishes drown in air, even though there is much more oxygen available in air than there is in the water in which they normally live.

5. The first successful atmospheric respiratory organ was the lung of amphibians, which was a simple sac with two-way flow in and out. It was not efficient, and gaseous diffusion across its small internal surface area has been supplemented, over the course of evolution, by diffusion across the moist skin.

6. The further evolution of the lung in reptiles and mammals has involved a progressive increase in the internal surface area of the lung, achieved by partitioning the inner surface into increasingly numerous chambers called alveoli. A human lung possesses about 300 million alveoli, with a combined surface area 42 times the body surface area.

7. A fundamental change in the atmospheric lung is characteristic of birds, which use a series of air chambers and two-cycle breathing to effect a one-way flow of air. Just as in the gill of bony fishes, the establishment of a one-way flow of the diffusing medium permits arrangement of capillary blood flow in a different direction from the flow of air or water, creating a superior diffusion mechanism.

8. Terrestrial vertebrates breathe by expanding and contracting the cavity within which the lungs hang. These actions expand the lungs, sucking air inward, and then compress the air, forcing it out of the lungs.

9. In the respiration of terrestrial vertebrates, hemoglobin within the red blood cells binds the oxygen, which diffuses across the lung capillaries into the blood cells to the respiring tissues of the body.

10. In respiring tissues, the partial pressure of oxygen is much lower than it is in the blood and the partial pressure of carbon dioxide is higher as a result of the consumption of oxygen and generation of carbon dioxide by respiring cells. Hemoglobin responds to the higher carbon dioxide concentration by unloading its oxygen, which diffuses out of the red blood cells into the tissue. The carbon dioxide is absorbed into blood and carried to the lungs, where it is discharged.

REVIEW QUESTIONS

1. As briefly as possible, describe Fick's Law of Diffusion as related to vertebrate respiration in nonmathematical terms. Which of these variables have animals altered to optimize the rate of diffusion to and evolve more efficient respiratory systems? Which variable are we as yet incapable of altering?

2. What are the simplest aquatic and terrestrial respiratory organs? What are the disadvantages associated with an external gill? Why are gills associated with branchial chambers more efficient?

3. What is the nature of the countercurrent flow associated with respiration in fishes? What is the result in terms of Δp of such a countercurrent flow? What would be the result in terms of net diffusion lacking this unique system?

4. What are the two major kinds respiratory systems found in terrestrial animals? How does each succeed in minimizing the excessive water loss?

5. Why do mammals require a more efficient respiratory system than reptiles? How do they achieve this greater efficiency? How does the ratio of lung mass to body mass in mammals change with increasing metabolic activity? Explain.

6. Why in general do flying birds have a greater respiratory need than do flying mammals? How is respiration in birds similar to that in fishes in terms of air flow? How is it similar in terms of its associated blood flow?

7. Describe the flow of air into a human lung, indicating all major structures along the way, with the assumption that the mouth remains closed at all times. Why is this a one-cycle system? How are the lungs connected to and supported within the thoracic cavity?

8. What is the carrier molecule used to transport oxygen through the blood of vertebrates? Upon what metal ion is it based? In what cells is this molecule contained and synthesized? What carrier molecule is present in most invertebrates? How does it differ from that of vertebrates? What is the color of each carrier molecule when combined with oxygen?

9. What data are indicated by an oxygen-hemoglobin dissociation curve? What is the Bohr effect as it relates to hemoglobin? Of what advantage is this to metabolically active tissues? How well does carbon monoxide bind to hemoglobin? What is the medical significance of this?

10. When and how does the blood unload its supply of oxygen? How is carbon dioxide carried by the blood? By what means does the blood increase the concentration of carbon dioxide it can absorb? How is carbon dioxide unloaded at the lungs?

 THOUGHT QUESTIONS

1. Often people who appear to have drowned can be revived, in some cases after being underwater for as long as half an hour. In every case of full recovery after extended submergence, however, the person had been submerged in very cold water. Is this observation consistent with the fact that oxygen is twice as soluble in water at 0° C as it is at 30° C?

2. Can you think of a reason why a respiratory system has not evolved in which oxygen is actively transported across respiratory membranes, in place of the passive process of diffusion across these membranes that is universally employed?

3. If by accident your pleural membrane were punctured, would you be able to breathe?

 FOR FURTHER READING

DAVSON, H., K. WELCH, and M.B. SEGAL: *The Physiology And Pathophysiology Of The Cerebrospinal Fluid*, page 453, Churchill-Livingstone, Edinburgh, 1987. Describes the role of the blood-brain barrier in the regulation of the partial pressure of carbon dioxide in the arterial blood.

DICKERSON, R.E., and I. GEIS: *Hemoglobin: Structure, Function, Evolution, and Pathology*, Benjamin-Cummings Publishing Co., Menlo Park, Calif., 1983. A comprehensive treatment of all aspects of hemoglobin, from the genes that encode it to the way in which the protein functions in the vertebrate respiratory system.

FEDER, M., and W.W. BURGGREN: "Skin Breathing in Vertebrates," *Scientific American*, November 1985, pages 126-142. Many terrestrial vertebrates breathe through their skin. Indeed, some amphibians have dispensed with lungs altogether.

PALLOT, D.J., editor: *Control Of Respiration*, Croon Helm, London, 1983. Comprehensive textbook on this subject.

PERUTZ, M.F.: "Hemoglobin Structure and Respiratory Transport," *Scientific American*, December 1978, pages 92-125. An account of how hemoglobin changes its shape to facilitate oxygen binding and unloading, by the man who won a Nobel Prize for unraveling the structure of hemoglobin.

RALOFF, J.: "New Clues to Smog's Effect on Lungs," *Science News*, vol. 132, August 8, 1987, page 86. Describes how the lung's defense mechanisms can be damaged by environmental contamination.

RANDALL, D.J., W.W. BURGGREN, A.P. FARRELL, and M.S. HASWELL: *The Evolution of Air Breathing in Vertebrates*, Cambridge University Press, Cambridge, England, 1981. An account of the physiological changes in respiration that occurred during the evolution of the terrestrial vertebrates.

SCHMIDT-NIELSEN, K.: "How Birds Breathe," *Scientific American*, December 1971, pages 73-79. A fascinating account of the discovery of unidirectional flow in avian lungs, by a great comparative physiologist.

ZAPOL, W.: "Diving Adaptations of the Weddell Seal," *Scientific American*, June 1987, pages 100-105. Seals can hold their breath longer than most mammals. This delightful account of Antarctic seals explains how they can manage this remarkable feat.

51

Circulation

OVERVIEW

In vertebrates, circulatory systems are like highways, over which red blood cells carry oxygen to the tissues and remove carbon dioxide. The fluid of the blood also transports glucose and amino acids to the cells and carries away nitrogenous wastes. Blood composition is kept constant by the liver and kidneys, which monitor and adjust metabolite and ion levels in the blood. The key to circulation in vertebrates is the organ that pumps the blood through the system, the heart. The heart has evolved in concert with the respiratory system of vertebrates. The double-pump architecture of the heart in mammals and birds plays an important role in their ability to maintain a constant body temperature higher than their surroundings.

FOR REVIEW

Here are some important terms and concepts that you will encounter in this chapter. If you are not familiar with them, you should review them before proceeding.

Erythrocytes (Chapter 43)

Cardiac muscle (Chapter 43)

Depolarization (Chapter 44)

How hemoglobin carries oxygen (Chapter 50)

Of the many tissues of the vertebrate body, few have the emotive impact of blood. In movies and in real life the sight of blood connotes violence, injury, and bodily harm. In literature, blood symbolizes the life force, which is not a bad analogy in real life either. This chapter is about blood, its circulation through the body (Figure 51-1), and the many ways in which it affects the body's functions. Although vertebrates possess many other organ systems that are necessary for life, it is the activities of the blood that bind them together into a functioning whole.

THE EVOLUTION OF CIRCULATORY SYSTEMS

The capture of nutrients and gases from the environment is one of the essential tasks that all living organisms must carry out. When a bacterium has transported nutrients across its cell membrane into its cytoplasm, its job is essentially over. The molecules it has brought in are able to move throughout the entire bacterium, a single cell, by diffusion. Among the single-celled protists and some of the simpler multicellular animals, the movement of nutrients occurs in a similar way. Some portion of every cell of the organism is exposed to the external liquid environment (Figure 51-2, A), enabling the individual cells to capture materials from that environment and to release materials directly into it.

As multicellular animals have become larger, however, with layers of cells stacked on one another, their more interior cells have experienced greater difficulty in exchanging materials with the environment by simple diffusion. This dilemma was solved with the advent of the body cavity (Figure 51-2, B), which can be seen in one of its first manifestations among the roundworms, phylum Nematoda (see Chapter 39). Fluid within this body cavity constitutes a primitive kind of circulatory system, one that permits materials to pass from one cell to another without leaving the organism. In organisms with a body cavity, one cell is able to take up material from the environment, and another distant cell can receive and use that material. This transport of material from one place to another within an organism by passage through an internal fluid is called **circulation.**

A circulatory system may be either open or closed. In a **closed system** the fluid is enclosed within blood vessels and so is separated from the rest of the body's fluids and does not mix freely with them. Materials pass into and out of the circulating fluid by diffusion. In an **open system,** by contrast, there is no distinction between circulating fluid and body fluid generally—the circulating fluid *is* the body fluid.

FIGURE 51-1
All vertebrates have circulatory systems. Yours is not very different from those of these baby raccoons.

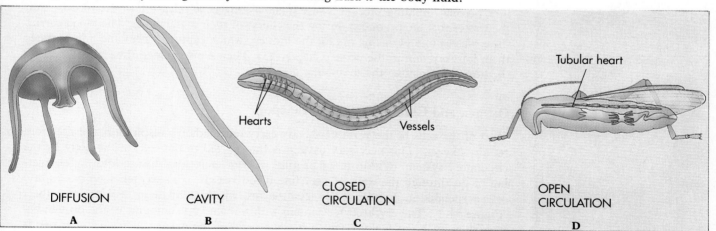

FIGURE 51-2
Evolution of circulatory systems.
A Simple diffusion; a jellyfish. Without any movement of the organism, diffusion reaches all parts of the body.
B Circulation in a body cavity; a nematode. As the worm moves, its contracting muscles push fluid back and forth through the body cavity.
C A closed circulatory system; an annelid. The series of hearts in each segment push blood out through a system of vessels to the body's tissues, then back via a second vessel system.
D An open circulatory system; an insect. The tubular heart of insects pumps blood out to the tissues of the body, from which it perfuses back rather than traveling in vessels.

Either an open or a closed circulatory system will work, and both function successfully in different phyla of animals. Annelids (see Chapter 40), for example, have a closed circulatory system (Figure 51-2, C) in which two major tubes or vessels extend the length of the animal, one on top (dorsal) and the other on the bottom (ventral), with branches extending from each tube out to the muscles, skin, and digestive system. Movement of fluid within this network of vessels is promoted by the contraction of the muscles surrounding the main dorsal vessels and by five conduits connecting the dorsal and ventral tubes. All vertebrates have closed circulatory systems.

Arthropods, by contrast, evolved an open circulatory system (Figure 51-2, D). In this system, a muscled tube within the central body cavity forces the cavity fluid out into the body through a network of interior channels and spaces. The fluid then flows back into the central cavity.

A great advantage of closed circulatory systems is that they permit regulation of fluid flow by means of muscle-driven changes in the diameters of the vessels. In other words, with a closed circulatory system the different parts of the body can maintain different circulation rates.

THE FUNCTION OF CIRCULATORY SYSTEMS IN VERTEBRATES

The closed circulatory system of vertebrates is like a roadway connecting the various muscles and organs of the body with one another. It serves four principal functions.

Nutrient and Waste Transport

The food molecules that fuel cell metabolism are transported to the cells of the body by the circulatory system. The products of digestion pass into the bloodstream largely through the wall of the small intestine, diffusing into a fine net of blood vessels below the mucosa. Metabolites, principally sugars and amino acids, are first carried from the intestines to the liver. In the liver some of these metabolites are converted to glucose, which is released into the bloodstream. Others, such as the essential amino acids and vitamins, pass through the liver unchanged. Still others, excess molecules that are sources of energy, are converted to storage compounds and accumulated for later use.

From the liver, blood carries the dissolved glucose and other metabolites to all body cells. In these cells glucose is used as an energy source, and the other metabolites become building blocks. These metabolizing cells release into the bloodstream the wastes that are produced during the course of their metabolism. The blood carries these wastes to a cleansing organ, the kidney, which captures them and concentrates them for excretion in the urine (Chapter 53). The cleansed blood then passes back to the heart, completing the **metabolic circuit**.

Oxygen and Carbon Dioxide Transport

Most of the cells of the vertebrate body carry out oxidative respiration and therefore require oxygen. This necessary oxygen is transported to the cells of the body by the circulatory system. Within lungs or gills, oxygen molecules diffuse into the circulating blood through the walls of very fine blood vessels. This oxygen soon passes into cells suspended in the blood, the erythrocytes, which circulate in very large numbers (Figure 51-3). The erythrocytes contain within them large amounts of the protein hemoglobin, each molecule of which binds several of the oxygen molecules. Oxygen is thus well packaged: hemoglobin is the primary carrier; hemoglobin is in turn carried by erythrocytes; and erythrocytes are carried within the circulating blood. From the lungs, the blood carries the carrier-bound oxygen to all of the metabolizing cells of the body, where the oxygen is released from the blood and is used to carry out oxidative respiration. The metabolizing cells release the end product of this metabolism, carbon dioxide, into the bloodstream. There some of it is bound by the hemoglobin carrier molecules (this is discussed in detail in Chapter 50). The blood then returns to the lungs, where the release of carbon dioxide and the capture of new oxygen complete the **respiratory circuit**.

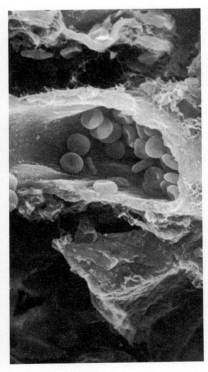

FIGURE 51-3
A blood vessel. This ruptured tube is a blood vessel, an element of the human circulatory system. It is full of red blood cells, which move through these blood vessels transporting oxygen and carbon dioxide from one place to another in the body.

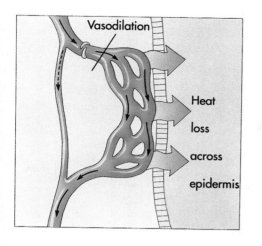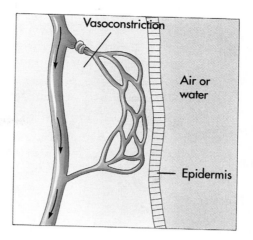

FIGURE 51-4
Regulation of heat loss. The amount of heat lost at the body's surface can be regulated by controlling the flow of blood to the surface. Constriction of the surface blood vessels limits flow and lessens heat loss; dilation of them increases blood flow and thus increases heat loss.

Temperature Maintenance

Mammals and birds maintain a constant body temperature by expending metabolic energy. They differ in this respect from most other organisms, which typically do not use metabolism to maintain a constant internal temperature. In the vertebrates, regardless of how a given body temperature is attained, the circulating blood distributes the heat more or less uniformly throughout the body. This heat circulation is accomplished first by means of a network of fine blood vessels, which passes immediately beneath the external surfaces of the body. These surfaces are in contact with the environment. The blood that circulates within this network absorbs heat from a hot environment and releases heat to a cold one. The further circulation of this heated or cooled blood to the interior of the animal's body tends to adjust the internal temperature of that animal in the direction of its external environment. At the same time the circulation of blood back out from the body's interior to its extremities completes the **thermal circuit.** Many vertebrates conserve heat by short-circuiting this pattern of circulation. They do so by constricting the blood vessels beneath the surface extremities of their body (Figure 51-4) so that little blood flows through them.

Human infants have a very limited capacity to regulate body temperature by either sweating or shivering, but they do have a mechanism for producing heat that is lost after the first year or so of life (although this mechanism is present throughout life in many mammals). **Brown fat** is a metabolically-active tissue that is located on the human infant in the neck and chest. The location of brown fat allows it to channel heat to the thorax and to blood in the arteries that serve the brain. Its color is due to the many mitochondria present in the fat cells. Activation of brown fat by sympathetic inputs elicits heat production through pathways that allow the process of oxidative metabolism to be largely uncoupled from phosphorylation. The result is that stored lipids are catabolized but very little ATP is produced, so the sole purpose served by these fat deposits is to rapidly generate heat.

Hormone Circulation

The metabolic and other activities of the body's many organs are coordinated both by direct nerve signals and by hormones. Once hormones have been released into the blood, they are transported within it throughout the body. They soon reach the target tissues capable of responding to them. Most hormones persist in the bloodstream for only a short time, since they are continuously destroyed by the enzymes present in the body. As a result of this turnover, the appearance of hormones within the bloodstream is a dependable signal for the initiation of whatever process the hormones affect.

> The closed circulatory system in vertebrates transports oxygen, nutrients, and metabolites to cells, transports carbon dioxide and metabolic wastes away from metabolizing cells, helps to minimize temperature differences, and transmits hormonal regulatory signals to the cells of the body.

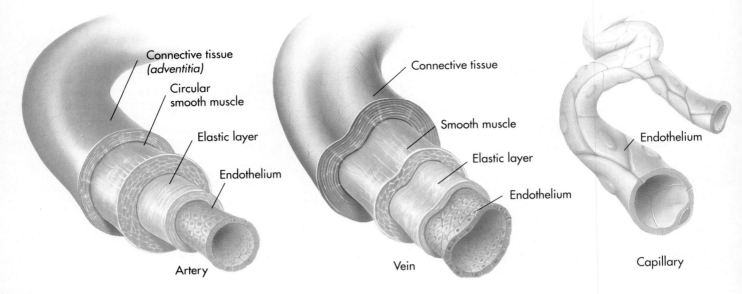

FIGURE 51-5
The structure of blood vessels.

THE CARDIOVASCULAR SYSTEM

The closed circulatory system of vertebrates is composed of three elements: (1) the **heart,** a muscular pump that we will consider later in this chapter; (2) the blood vessels (Figure 51-5), a network of tubes that passes through the body; and (3) the blood, which circulates within these vessels. The plumbing of the closed circuit, the heart and vessels, is known collectively as the **cardiovascular system.** Blood moves within this cardiovascular system, leaving the heart through vessels known as **arteries.** From the arteries the blood passes into a larger network of **arterioles,** or smaller arteries. From these it is eventually forced through the **capillaries,** a fine latticework of very narrow tubes, which get their name from the Latin word *capillus,* "a hair." It is while passing through these capillaries that the blood exchanges gases and metabolites with the cells of the body. After traversing the capillaries, the blood passes into the **venules** or small veins. The venules lead to a network of larger **veins,** which collect the circulated blood and carry it back to the heart.

Because the vertebrate cardiovascular system is closed, it has certain structural requirements that open circulatory systems do not. To illustrate this relationship, we will consider the properties of one very familiar kind of open "vascular" system, a garden hose. If you turn the water on, a pulse of water passes down the hose, emerging a few moments later out the end as a stream of water. If you turn the water on harder, the water just moves through the hose faster. Now close the nozzle at the end of the hose. You have now converted the hose into a closed vascular system. If you turn the water on even harder, will more water move into the hose? Where is it going to go? The answer is that no more water will move into the hose, because the walls of the hose cannot expand to accommodate a larger volume. They are strong, but they are not elastic.

The heart of a vertebrate faces the same plumbing problem as the closed garden hose: it must push a pulse of fluid through a closed system of vessels that meets a resistance at one end, a network of small capillaries. The capillaries have a much smaller diameter than the other blood vessels of the body. To understand why reduction in tube diameter leads to increased resistance to flow, we must recall a little physics. When a fluid, such as blood, flows through a horizontal tube, it meets a frictional resistance that is inversely proportional to the fourth power of the radius of the tube (Figure 51-6). For example, when the radius of a tube is reduced by half, the resistance to flow is 16 times greater. Now you can see why flow through the capillaries creates resistance. Blood leaves the human heart through the **aorta,** a tube that has a radius of about 1 centimeter, but when it reaches the capillaries, it passes through vessels with an average radius of only 8 micrometers, a reduction in radius of some 1250 times!

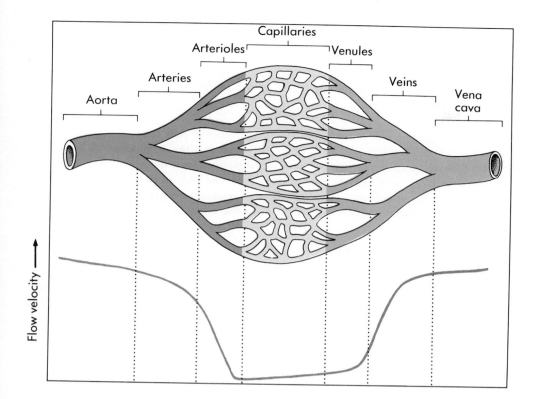

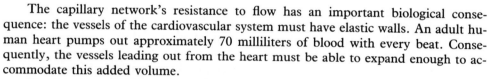

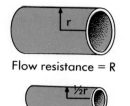
Flow resistance = R

Flow resistance = 16R

FIGURE 51-6
The effect of vessel diameter on resistance and flow velocity. The narrower a tube, the greater the resistance is, inversely proportional to the fourth power of the radius of the tube.

The capillary network's resistance to flow has an important biological consequence: the vessels of the cardiovascular system must have elastic walls. An adult human heart pumps out approximately 70 milliliters of blood with every beat. Consequently, the vessels leading out from the heart must be able to expand enough to accommodate this added volume.

Because the vascular system of vertebrates is closed, every part must have the same overall flow rate (about 5 liters per minute in adult humans). Flow rate is measured as the volume of blood pumping out through the system per unit of time. Flow rate is not the same as velocity, which is the distance that blood moves in a unit of time. Thus even though capillaries collectively have the same flow rate as arteries, the velocity of flow in individual capillaries is much less than in an artery, because there are so many capillaries that the total cross sectional area is far greater than in the arteries (see Figure 51-6).

Arteries: Highways from the Heart

Arteries are the vessels that carry blood away from the heart. Artery walls are made up of four layers of tissue. The innermost one is composed of a thin layer of **endothelial cells.** Surrounding these cells is a thick layer of **elastic fibers** and another of **smooth muscle,** which in turn are encased within an envelope of **connective tissue.** Because this sheath is elastic, the artery is able to expand its volume considerably in response to a pulse of hydrostatic pressure, much as a tubular balloon might. The steady contraction of the muscle layer strengthens the wall of the vessel against overexpansion.

Arterioles: Little Arteries

Arterioles differ from arteries simply in that they are smaller in diameter. The muscle layer that surrounds them can be relaxed under the influence of hormones and metabolites. When this happens, the blood flow can be increased, an advantage during times of high metabolic activity. Conversely, most arterioles are in contact with many nerve fibers. When stimulated, these nerves cause the muscular lining of the arteriole to contract and thus constrict the diameter of the arteriole. Such contraction limits

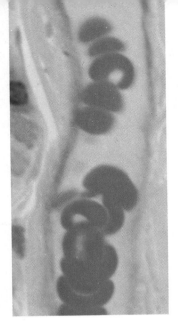

FIGURE 51-7

The red blood cells in this capillary are passing along in single file. Many capillaries are even narrower than those shown here, in the bladder of a monkey. However, red blood cells will even pass through capillaries narrower than their own diameter, pushed along by the pressure generated by a pumping heart.

the flow of blood to the extremities during periods of low temperature or stress. You turn pale when you are scared because contraction of this kind constricts the arterioles in your skin. When you blush or flush from embarrassment or because you are overheated, the opposite effect occurs: the nerve fibers connected to muscles surrounding the arterioles are inhibited, relaxing the smooth muscle and causing the arterioles in the skin to dilate.

The major vessels of the circulatory system are tubes of cells encased within three sheaths: (1) a layer of elastic fibers, which renders the diameter of the vessel elastic to accommodate pulses of blood pumped from the heart; (2) a layer of muscle serviced by nerves, which permits the body to control the diameter of the vessel and strengthens the wall against overexpansion; and (3) a layer of connective tissue, which protects the vessel.

Capillaries: Where Exchange Takes Place

Capillaries have the simplest structure of any element in the cardiovascular system. They are little more than tubes one cell thick and on the average about 1 millimeter long; they connect the arterioles with the venules. The internal diameter of a capillary is, on the average, about 8 micrometers. Surprisingly, this is little more than the diameter of a red blood cell (5 to 7 micrometers). However, red blood cells squeeze through these fine tubes without difficulty (Figure 51-7). The intimate contact between the walls of a capillary and the red blood cells facilitates the diffusion of gases and metabolites across the walls of the red blood cell and those of the capillary, as well as those of the surrounding cells. *No cell of the body is more than 100 micrometers from a capillary.* At any one moment, about 5% of your blood is in your capillaries. Some capillaries, called **through-flow channels,** connect arterioles and small veins directly (Figure 51-8). From these channels, loops of true capillaries leave and return. It is through these loops that almost all exchange between the blood and the cells of the remainder of the body occurs. The entry to each loop is guarded by a ring of muscle called a **precapillary sphincter,** which, when closed, blocks flow through the capillary. Such restriction of entry to the capillaries in surface tissue is one of several means of limiting heat loss from an animal's body during periods of cold (Figure 51-9).

FIGURE 51-8

The capillary network connects arteries with veins. The most direct connection is via through-flow channels that connect arterioles directly to venules. Branching from these through-flow channels is a network of finer channels, the capillary network. Most of the exchange between body and red blood cells occurs while they are in this capillary network. Entrance to the capillary network is controlled by bands of muscle called precapillary sphincters at the entrance to each capillary. When a sphincter is contracted it closes off the capillary. By contracting these sphincters, the body can limit the amount of blood in the capillary network of a particular tissue, and thus control the rate of exchange in that tissue.

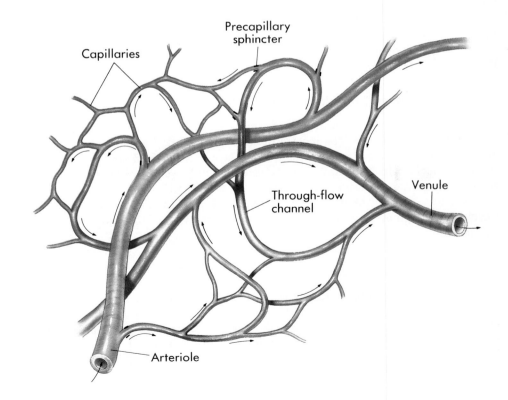

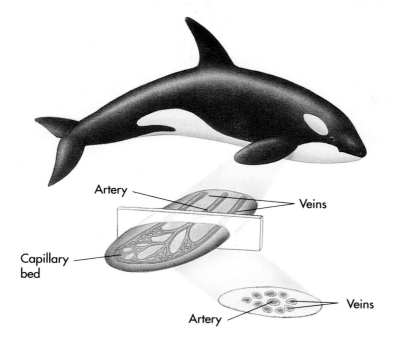

FIGURE 51-9
Many marine mammals, such as this killer whale, limit heat loss in cold water by countercurrent flow. Heat is removed from the arterial blood by nearby veins before the blood circulates to the body's surface.

The entire body is permeated with a fine mesh of these capillaries, a network that amounts to several thousand kilometers in overall length. If all of the capillaries in your body were laid end to end, they would extend across the United States. Although individual capillaries have high resistance to flow because of their small diameters, the cross-sectional area of the extensive capillary network is far greater than that of the arteries leading to it, so that the total resistance to blood flow is actually lower in the capillaries.

Veins and Venules: Returning Blood to the Heart

Veins are the passages through which blood is transported back to the heart, and contain one-way valves that permit flow in this direction only. Veins do not have to accommodate the pulsing pressures that arteries do, because much of the force of the heartbeat is attenuated by the high resistance and great cross-sectional area of the capillary network. The walls of veins, although similar in structure to arteries, have much thinner layers of muscle and elastic fiber (Figure 51-10). An empty artery is still a hollow tube, like a pipe, but when a vein is empty, its walls collapse like an empty balloon.

FIGURE 51-10
Veins and arteries. The vein (*left*) has the same general structure as an artery (*right*), but much thinner layers of muscle and elastic fiber. An artery will retain its shape when empty, but a vein will collapse.

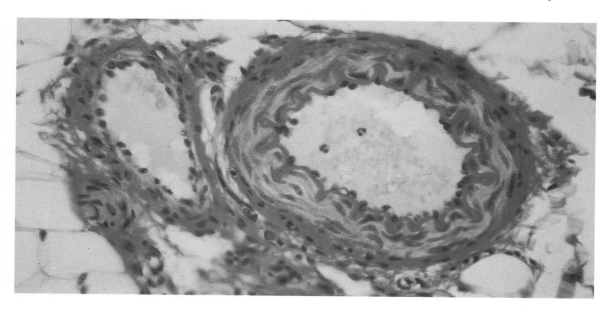

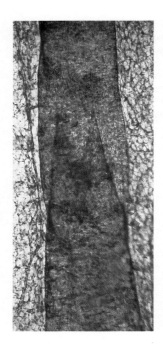

FIGURE 51-11
A lymphatic vessel valve, magnified 25 times. Flow from bottom to top is not retarded since such flow tends to force open the inner cone; flow from top to bottom is prevented because such flow tends to force the inner cone closed.

The internal passageway of veins is often quite large. The diameter of the largest vein in the human body, the **vena cava,** which leads into the heart, is fully 3 centimeters. One reason that veins are so much larger than arteries is that it is advantageous to minimize any resistance as blood flows out of an organ, so that there is no backpressure into it. A second reason is related to the lower pressure of blood flowing within veins back toward the heart—it is advantageous to minimize any further resistance to flow. Recall, as we noted above, that resistance to flow varies as the inverse of the fourth power of the radius of a tube—a larger tube presents much less resistance to flow. Veins are large because larger veins present less resistance to the flow of blood back to the heart. Unlike arterial blood, the blood in veins is far from the heart and does not get pushed hard when the heart beats. The contraction of muscles helps push the blood through veins, which have valves to prevent backflow. People pass out from standing in one place too long because blood pools in the legs and muscles do not push it along.

The Lymphatic System: Recovering Lost Fluid

The cardiovascular system is considered a closed system because all of its tubes are connected with one another and none are simply open ended. In another sense, however, the system is open—to diffusion through the walls of the capillaries. This process is a necessary part of the functioning of the circulatory system, but it poses difficulties in maintaining the integrity of that system.

Difficulties arise because diffusion from the capillaries is accompanied by the loss of large quantities of liquid from the cardiovascular system. When blood passes through the capillaries, it loses more water to the body than it reabsorbs from it. In a human being, about 3 liters of fluid leave the cardiovascular system in this way each day, a quantity amounting to more than half the body's total supply of about 5.6 liters of blood. To counteract the effects of this process, the body uses a second *open* circulatory system called the **lymphatic system.** The elements of the lymphatic system gather liquid from the body and return it to cardiovascular circulation. Open-ended lymph capillaries gather up fluids and carry them through a series of progressively larger vessels to two large lymphatic vessels, which resemble veins. The lymphatic vessels contain a series of one-way valves like those of veins (Figure 51-11), which permit movement in only one direction. These two lymphatic vessels drain into veins through one-way valves. No heart pumps fluid through the lymphatic system (Figure 51-12); instead, like veins, fluid is driven through lymphatic vessels when these vessels are squeezed by the movements of the body's muscles.

> **Much of the water within blood plasma diffuses out during passage through the capillaries. This fluid is collected by an open circulatory system, the lymphatic system, and returned to the bloodstream.**

BLOOD

About 8% of the body mass of most vertebrates is taken up by the blood circulating through their bodies. This blood is composed of a fluid plasma and several different kinds of cells that circulate within that fluid.

Blood Plasma: The Blood's Fluid

Blood plasma is a complex solution of three different components dissolved in water:

1. *Metabolites and wastes.* If the circulatory system is thought of as the "highway" of the vertebrate body, the blood contains the "traffic" passing on that highway. Dissolved within the plasma are all of the metabolites, vitamins, hormones, and wastes that circulate among the cells of the body.
2. *Salts and ions.* Like the water of the seas in which life arose, plasma is a dilute salt solution. The chief plasma ions are sodium, chloride, and bicarbonate ions. In addition, there are trace amounts of other salts, such as calcium and

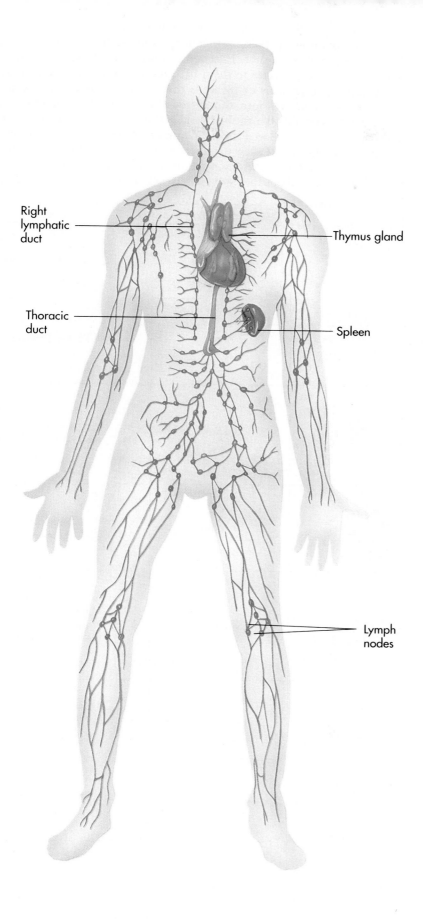

FIGURE 51-12
The human lymphatic system.

Right
lymphatic
duct

Thymus gland

Thoracic
duct

Spleen

Lymph
nodes

magnesium, as well as of metallic ions, including copper, potassium, and zinc. The composition of the plasma, therefore, is not unlike that of seawater.

3. *Proteins.* Blood is 90% water. Passing by all the cells of the body, blood would soon lose most of its water to them by osmosis if it did not contain as high a concentration of proteins as the cells that it passed. The water would move from the blood vessels into the surrounding cells, diluting their contents. Blood plasma contains the antibody and globulin proteins that are active in the immune system, as well as a small amount of **fibrinogen,** a protein that plays a role in blood clotting. Taken together, however, these proteins make up less than half of the amount of protein that is necessary to balance the protein content of the other cells of the body. The rest consists of a protein called **serum**

FIGURE 51-13
Types of blood cells.

Blood cell	Life span in blood	Function
Erythrocyte	120 days	O_2 and CO_2 transport
Neutrophil 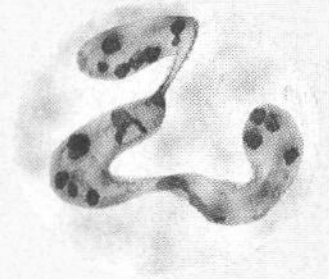	7 hours	Immune defenses
Eosinophil 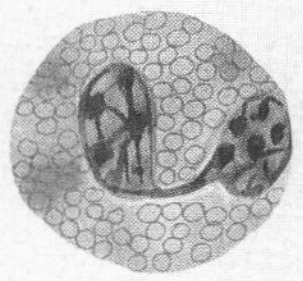	Unknown	Defense against parasites
Basophil 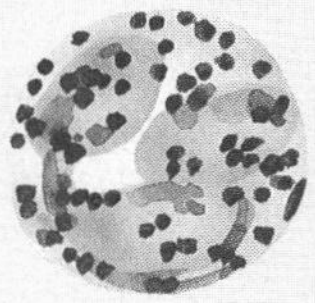	Unknown	Inflammatory response
Monocyte 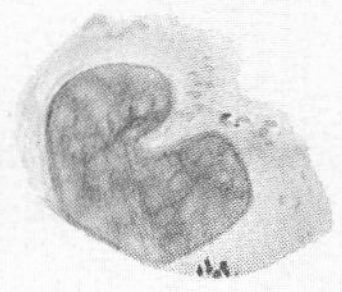	3 days	Immune surveillance (precursor of tissue macrophage)
B-lymphocyte 	Unknown	Antibody production (precursor of plasma cells)
T-lymphocyte 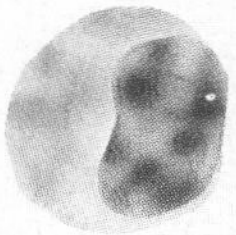	Unknown	Cellular immune response
Platelets 	7-8 days	Blood clotting

albumin, which circulates in the blood as an osmotic counterforce. Human blood contains 46 grams of serum albumin per liter. Protein deficiency diseases such as kwashiorkor produce swelling of the body because the body's cells take up water from the albumin-deficient blood.

Blood contains metabolites, wastes, and a variety of ions and salts. Blood also contains high concentrations of the protein serum albumin, which functions to keep the blood plasma in osmotic equilibrium with the cells of the body.

Blood Cells: Cells that Circulate Through the Body

The fraction of the total volume of the blood that is occupied by red blood cells is referred to as the blood's **hematocrit.** In humans, red blood cells typically occupy about 45% of the blood's volume. Figure 51-13 shows the different types of blood cells.

Erythrocytes. Each milliliter of blood contains about 5 billion **erythrocytes,** or red blood cells. Each erythrocyte is a flat disk with a central depression (Figure 51-14), something like a doughnut with a hole that does not go all the way through. Attached to the outer membranes of the erythrocytes is a collection of polysaccharides, which determines an individual's blood group (see Figure 52-10). Almost the entire interior of each cell is packed with hemoglobin.

Erythrocytes in mammals lose their nuclei and protein-synthesizing machinery when they mature. Because they lack a nucleus, these cells are unable to repair themselves, and they therefore have a rather short life; any one erythrocyte lives for only about 4 months. As erythrocytes age, they are removed from the bloodstream by macrophages, which ingest them and break them down. Balancing this loss, new erythrocytes are constantly being synthesized and released into the blood by cells within the soft interior marrow of bones. Each erythrocyte develops from a bone marrow cell, accumulating hemoglobin and gradually losing its nucleus, a process referred to as **erythropoiesis.** When the oxygen levels in the blood fall below normal values, increased amounts of the hormone erythropoietin are secreted into the blood by the kidney and other organs. This hormone stimulates the bone marrow to produce more erythrocytes, thus increasing the oxygen-carrying capacity of the blood.

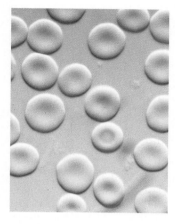

FIGURE 51-14
Human erythrocytes, magnified 1000 times. Human erythrocytes lack nuclei, which gives them a characteristic collapsed appearance, a bit like a pillow on which someone has sat.

Leukocytes Defend the Body. Less than 1% of the cells in human blood are leukocytes, or white blood cells; there are 1 or 2 leukocytes for every 1000 red blood cells. Leukocytes are larger than red blood cells; they contain no hemoglobin and are essentially colorless. There are several kinds of leukocytes, each with a different function. All these functions, however, are related to the defense of the body against invading microorganisms and other foreign substances, as you will see in Chapter 52. Leukocytes are not confined to the bloodstream; they also migrate out into the interstitial fluid.

If you prick your skin, some of the injured cells release chemicals that cause the capillaries in the vicinity to expand. The resulting increase in blood flow is one component of the **inflammatory response** making the wound look red and feel warm. In inflammation, **granulocytes,** which are circulating leukocytes, push out through the walls of the distended capillaries to the site of the injury. The granulocytes are classified into three groups by their staining properties. About 50% to 70% of them are **neutrophils,** which begin to stick to the interior walls of the blood vessels at the site of the injury. They then form projections that enable them to push their way into the infected tissues, where they engulf microorganisms and other foreign particles. **Basophils,** a second kind of leukocyte, contain granules that rupture and release chemicals that enhance the inflammatory response; they are important in causing allergic responses. The function of the third kind of leukocyte, the **eosinophils,** is not clear, but they may play a role in defending against parasitic infections.

Another kind of circulating leukocyte, the **monocyte,** is also attracted to the sites of inflammation, where they are converted into **macrophages**—enlarged, amoeba-like cells that entrap microorganisms and particles of foreign matter. They usually arrive

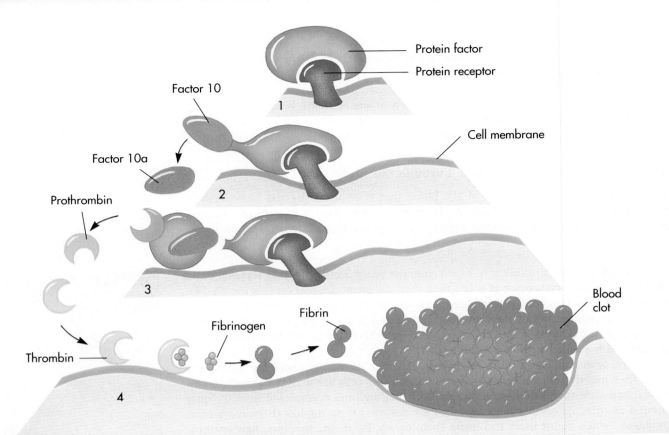

Protein factor
Protein receptor
Factor 10
Factor 10a
Prothrombin
Cell membrane
Thrombin
Fibrinogen
Fibrin
Blood clot

FIGURE 51-15

How does blood clot?

Step 1 A blood clot is initiated by the binding of a membrane receptor by a circulating protein factor.
Step 2 The complex of protein receptor and protein factor then binds with factor 10, changing it to the active form, factor 10a.
Step 3 Each molecule of 10a binds to another protein factor, and this complex catalyzes the conversion of prothrombin to thrombin.
Step 4 Thrombin interacts with fibrinogen, promoting its conversion to fibrin. Finally fibrin traps blood cells that seal the wound.

at the site of inflammation after the neutrophils, and they are important in the immune response because they play a key role in antibody production (see Chapter 52).

Platelets Help Blood to Clot. Certain large cells within the bone marrow, called **megakaryocytes,** regularly pinch off bits of their cytoplasm. These cell fragments, called platelets, contain no nuclei; they enter the bloodstream, where they play an important role in controlling blood clotting. A blood clot is a seal of a ruptured blood vessel. The ruptured vessel seals itself by generating a matrix of long fibers and trapped cells that fills the gap from components present in the plasma. In a clot the gluey substance is a protein called fibrin (derived from fibrinogen), which sticks platelets together to form a tight, strong seal.

Recently scientists have discovered that the fibrin that forms blood clots is generated in a spreading cascade of molecular events (Figure 51-15). The clotting process is initiated by injury to blood vessel cells attracting platelets, which releases a protein factor that starts the cascade. At each stage in the cascade that follows, proteins from cells and blood combine in fast-rising waves, involving many more molecules in the progressive steps of the process. Billions of molecules of fibrin can be formed from a single clot-initiating event.

THE EVOLUTION OF THE VERTEBRATE HEART

Any closed circulatory system requires both a system of passageways through which fluid can circulate and a pump to force the fluid through them. In the circulation of blood, the pump is the heart. The evolution of the heart among the vertebrates reflects two great transitions in their history. The first of these was the shift from filter feeding to active prey capture, an event that occurred at the dawn of the vertebrates. The second was the invasion of the land. Active predation by the first fishes entailed a greatly improved respiration system, the gill (see Chapter 50). The transition to land involved the evolution of a different type of breathing apparatus (the lung), a significant lessening of pressure from sea to air, and the development of intrinsic control of body temperature. All of these changes had major influences on the evolution of the heart.

The Early Chordate Heart: A Peristaltic Pump

The chordates that were ancestral to the vertebrates are thought to have had simple tubular hearts, not unlike those now seen in lancelets. The heart was little more than a specialized zone of the ventral artery, more heavily muscled than the rest of the arteries, which beat in simple peristaltic waves. **Peristaltic pumps** such as these hearts are not very efficient—blood is pushed in *both* directions as the heart contracts peristaltically. The pumping action results because the as-yet-uncontracted portion of the blood vessel has a larger diameter than the contracted portion and thus less resistance to flow.

The Fish Heart: A One-Cycle Chamber Pump

The development of gills by fishes necessitated a more efficient pump to force blood through the fine capillary network, and in fishes we see the evolution of a true chamber-pump heart. The fish heart can be considered a tube that has four chambers arrayed one after the other (Figure 51-16). The first two chambers (**sinus venosus and atrium**) are collection chambers; the second two (**ventricle and conus arteriosus**) are pumping chambers.

As might be expected from the nature of the hearts of the early chordates from which the fishes evolved, the sequence of the heartbeat is the peristaltic sequence, starting at the rear and moving to the front. The first of the four chambers to contract is the sinus venosus, then the atrium, then the ventricle, and finally the conus arteriosus. Despite shifts in the relative positions of the chambers in the vertebrates that evolved later, this heartbeat sequence is maintained and unchanged in all vertebrates.

The fish heart is admirably suited to the gill respiratory apparatus and represents one of the major evolutionary innovations of the vertebrates. Perhaps its greatest advantage is that the blood it delivers to the tissues of the body is fully aerated (oxygenated). Thus blood is pumped directly to and through the gills, where it becomes fully aerated; from the gills it flows through a network of arteries to the rest of the body and then returns to the heart through veins (Figure 51-17, *A*). This arrangement has

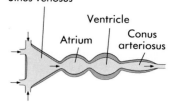

FIGURE 51-16
Schematic diagram of the fish heart.
Sinus venosus. Because water pressure pushes in on the fish's blood through the veins and thus opposes the movement of blood through the veins back to the heart, efficient operation requires that the fish heart contribute the least possible resistance to venous flow, which is minimized by a large collection chamber.
Atrium. To deliver blood to the pump quickly in increments of suitable volume, a vestibule about the size of the ventricle is required to receive blood from the sinus venosus.
Ventricle. To provide the energy to move the blood through the gills, a thick-walled pumping chamber is required.
Conus arteriosus. To smooth the pulsations and add still more thrust, blood leaving the ventricle passes through a second elongated chamber.

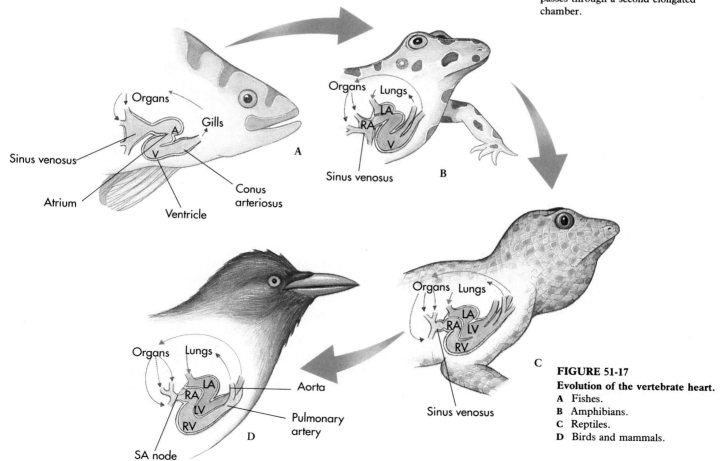

FIGURE 51-17
Evolution of the vertebrate heart.
A Fishes.
B Amphibians.
C Reptiles.
D Birds and mammals.

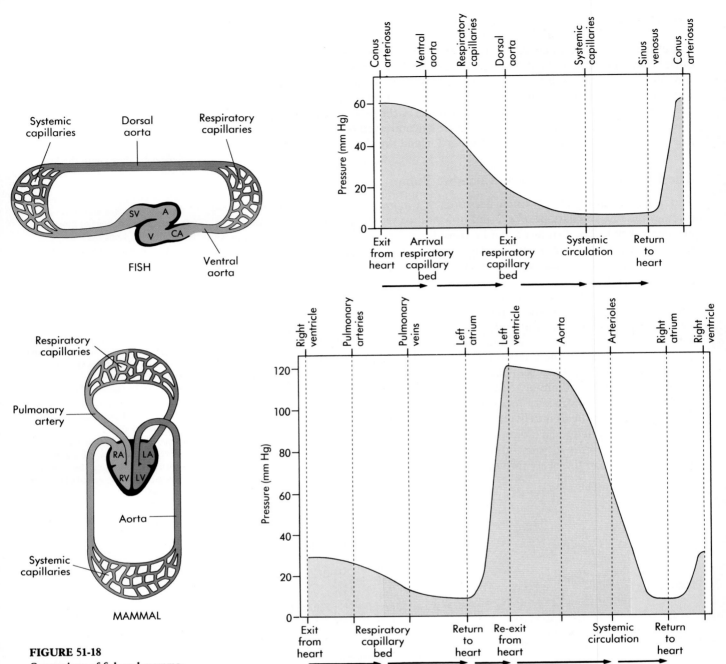

FIGURE 51-18

Comparison of fish and mammalian circulatory systems. The blood pressure in the fish circulatory system drops precipitously after passage of the blood through the gills, because of the great resistance their narrow passages present to flow. Because the blood flows from the gills directly to the rest of the body, circulation is sluggish at the low pressure that remains. In mammals, the blood is repumped after leaving the lungs; thus the resistance of the capillaries in the lungs does not affect the force of the heart's contraction, and circulation to the rest of the body remains vigorous.

one great limitation, however. After passing through the fine network of capillaries in the gills, the flow of blood has lost much of the force contributed by the contraction of the heart (Figure 51-18), so the circulation from the gills to the rest of the body is sluggish. This means that oxygen cannot be delivered to body muscles at a high rate. Fishes have never evolved a means to overcome this limitation.

Amphibian and Reptile Hearts: The Advent of Pulmonary Circulation

The advent of gaseous respiration in the lungs involved a major change in the pattern of circulation. Ultimately this evolutionary change enabled vertebrates to overcome the inherent limitations of the circulatory system that evolved in fishes. The change consisted of the development of an additional pair of major veins. After blood is pumped through the fine network of capillaries in the lungs, it is not dispersed to the tissues of the body but is instead returned to the heart through large veins called **pulmonary veins** for repumping. The development of these new veins led to a great improvement in the performance of the circulatory system, since the blood being pumped to other tissues of the body could be pumped at a much higher pressure than if it were not returned to the heart at this stage. The disadvantage of pulmonary veins is that the aerated blood from the lungs is mixed in the heart with the nonaerated blood that is constantly being returned to the heart from the rest of the body. Consequently the heart pumps out a mixture of aerated and nonaerated blood rather than fully aerated blood. Many of the subsequent changes in the evolution of the heart in birds and mammals have been modifications that minimize this disadvantage.

The amphibian heart is altered from the arrangement seen in fishes in two ways (Figure 51-17, *B*) that tend to lessen the effect of the mixing of aerated and nonaerated blood:

1. The atrium is divided into two chambers—a **right atrium,** which receives nonaerated blood from the sinus venosus for circulation to the lungs, and a **left atrium,** which receives aerated blood from the lungs through the pulmonary vein for circulation through the body.
2. The conus arteriosus is partially separated by a **septum,** or dividing wall, which directs aerated blood into the aorta and directs nonaerated blood into the pulmonary arteries, which lead to the lungs. The aorta leads to the body's network of arteries.

These two changes tend to separate the blood circulation of the body into two separate paths: (1) the **pulmonary circulation,** in which blood travels from the heart to and from the lungs; and (2) the **systemic circulation,** in which blood travels from the heart to and from the rest of the body. However, the separation of the pulmonary and systemic circulations is imperfect. A significant amount of mixing of aerated and nonaerated blood still occurs in the heart. It is no accident that amphibians are usually sluggish.

Among reptiles, further modifications occurred that reduced the amount of mixing in the heart (Figure 51-17, *C*):

1. A septum partially subdivides the pumping chamber of the heart, the ventricle. The presence of this septum results in a far more effective separation of aerated and nonaerated blood within the heart. In one order of reptiles, the crocodiles, the separation is complete.
2. The conus arteriosus is absent. It has become fully subdivided, forming the trunks of the large arteries leaving the heart, which thus become directly connected to the ventricle.

Because the reptilian heart achieves a better separation of aerated and nonaerated blood, it represents a distinct increase in efficiency over the amphibian heart. The separation within the ventricle is not complete, however, a factor that still limits the overall efficiency of the circulatory system, some of the nonaerated blood still mixing in the ventricle with aerated blood.

Mammal and Bird Hearts: A True Two-Cycle Pump

Only a relatively slight alteration is seen in the hearts of mammals, of birds, and of crocodiles (Figure 51-17, *D*), although the change, which occurred independently in these three lines of evolution,* has great consequences: the septum within the ventricle is closed, dividing the pumping chamber into two parts.

The closure of the ventricular septum in mammals and birds created for the first time the **double circulatory system** toward which evolutionary pressures had been modifying the heart since the advent of the pulmonary vein. These hearts have the advantage that is conferred by repumping the blood after its passage through the lungs, without the disadvantage of mixing aerated and nonaerated blood. The blood that is pumped by the heart of a mammal or bird into the systemic arterial system is fully aerated.

---

*Birds and crocodiles are on the same evolutionary line, and both have developed four-chambered hearts. In crocodilians, the complete ventricular septum developed after the split between crocodiles and alligators.

In mammals and birds the four-chambered heart acts as a double pump, the left side of the heart pumping aerated blood to the general body circulation and the right side pumping nonaerated venous blood to the lungs. Note that the four chambers that occur in a bird or mammalian heart evolved from only two of the chambers of the four-chambered fish heart: the atrium and the ventricle. The great increase in efficiency that is realized by the double circulatory system in mammals and birds is thought to have been important in the evolution of endothermy in them. More efficient circulation is necessary to support the great increase in metabolic rate that is required to generate body heat internally. Also, blood is the carrier of heat within the body, and an efficient circulatory system is required to distribute heat evenly throughout the body.

The separation of the blood flow into two circuits also has a second favorable result. Because the overall circulatory system is closed, in each full passage through the system the same volume of blood has to move through the smaller lung circulation path as through the much more extensive body circulation path. This means that the blood must move through the lungs very much faster than through the rest of the body. This more rapid circulation is not accomplished by higher pressure in the pulmonary circuit. Instead the blood vessels in the lung are larger in diameter than those in the rest of the body and offer less resistance to flow. The favorable result is that the rapid flow of blood through the lungs greatly increases the efficiency with which oxygen is captured by the bloodstream.

The Pacemaker: Preservation of the Sinus Venosus

The sinus venosus, a major chamber in the fish heart, is reduced in size in amphibians and is further reduced in reptiles. In mammals and birds the sinus venosus is no longer evident as a separate chamber. The disappearance of the sinus venosus is not really complete, however. Although the chamber is gone, some tissue remains, and it has a very important function. Throughout the evolutionary history of the vertebrate heart, the sinus venosus has had two functions, one as a collection chamber and the other as the site of origin of the heartbeat. This second function is indispensable, and mammals have retained the excitatory tissue of the sinus venosus in the wall of the right atrium—where it was previously located in the fish heart, near the point where the veins now empty directly into the atrium. This tissue is called the **sinoatrial node (SA node)**. In mammals it is the point of origin of each heartbeat—the **pacemaker.**

The hearts of mammals and birds evolved independently from the central two chambers of the four-chambered fish heart, each central chamber becoming divided into two chambers separated by a septum. A vestige of the initial chamber of the fish heart remains as the sinoatrial node, or pacemaker.

Increase in Pumping Capacity

During the course of vertebrate evolution, hearts have gotten bigger, and thus are able to pump harder. Within any one class of vertebrate, the size of the heart is a constant fraction of the body mass. In mammals, for example, the heart is very close to 0.6% of the total body mass, whether that body mass is 0.01 kilogram or 100 kilograms (Figure 51-19). However, the transition to more complex hearts during the evolution of the vertebrates has involved an increase in the size of the heart relative to that of the rest of the body in different classes, as shown in Table 51-1.

THE HUMAN HEART

The human heart, like that of all mammals and birds, is really two separate pumping systems operating together within a single sac. One of these pumps blood to the lungs, while the other pumps the blood that circulates to the rest of the body. A cross section through the heart shows its organization clearly (Figure 51-20). The left side has two connected chambers, and so does the right. However, even though the right and left sides of the heart pump in concert with one another, they are not connected.

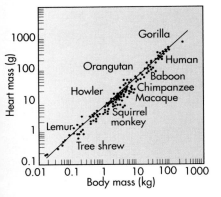

FIGURE 51-19
Heart size is constant. The heart is about 0.6% of the total body mass of any mammal, regardless of its size. This is true of a mouse and equally true of an elephant.

TABLE 51-1	HEART MASS
CLASS	HEART MASS (KG) / BODY MASS (KG)
Fishes	0.20%
Amphibians	0.46%
Reptiles	0.51%
Mammals	0.60%
Birds	0.82%

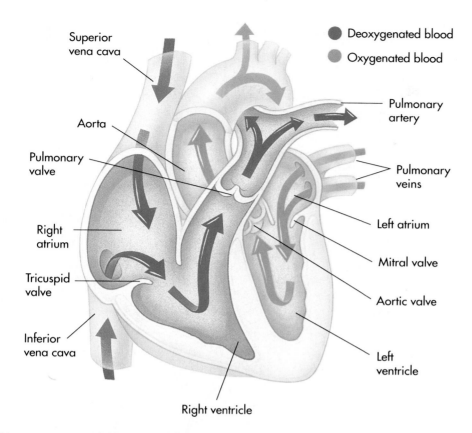

FIGURE 51-20

The path of blood through the human heart. Entering the right atrium by way of the superior vena cava, the blood passes into the right ventricle and then through the pulmonary valve to the pulmonary artery and the lungs. Oxygenated blood from the lungs then returns to the heart by way of the pulmonary veins, entering the left atrium and then the left ventricle, from which it enters the general circulatory system of the body by way of the aorta.

Labels in figure:
- Superior vena cava
- Aorta
- Pulmonary valve
- Right atrium
- Tricuspid valve
- Inferior vena cava
- Right ventricle
- Deoxygenated blood
- Oxygenated blood
- Pulmonary artery
- Pulmonary veins
- Left atrium
- Mitral valve
- Aortic valve
- Left ventricle

Circulation Through the Heart

Let us follow the journey of blood through the human heart, starting with the entry into the heart of oxygenated blood from the lungs. Oxygenated blood from the lungs enters the left side of the heart, emptying into the **left atrium** through the **pulmonary veins,** which open directly into the atrium. From the atrium, blood flows through an opening into the connecting chamber, the **left ventricle.** Most of this flow, roughly 80%, occurs while the heart is relaxed. When the heart starts to contract, the atrium contracts first, pushing the remaining 20% of its blood into the ventricle.

After a slight delay, the ventricle contracts. The walls of the ventricle are far more muscular than those of the atrium, and as a result this contraction is much stronger. It forces most of the blood out of the ventricle in a single strong pulse. The blood is prevented from going back into the atrium by a large one-way valve, the **bicuspid** or **mitral valve,** whose flaps are pushed shut as the ventricle contracts. Strong fibers that prevent the flaps from moving too far when closing are attached to their edges. If the flaps did move too far, they would project out into the atrium. The fibers that prevent this operate in much the same way as a rope might if tied from the steering wheel of a car to the driver's door handle, with a bit of slack—the door can be opened only as far as the slack in the rope permits.

Prevented from reentering the atrium, the blood within the ventricle takes the only other passage out of the contracting left ventricle. It moves through a second opening that leads into a large vessel called the **aorta.** The aorta is also bounded by a one-way valve, the **aortic valve.** Unlike the mitral valve, the aortic valve is oriented to permit the *outward* flow of the blood. Once this outward flow has occurred, the aortic valve closes, thus preventing the reentry of blood from the aorta into the heart.

The aorta and all of the other blood vessels that carry blood away from the heart are arteries (Figure 51-21). Many of these arteries branch from the aorta, carrying oxygen-rich blood to all parts of the body. The first to branch are the **coronary arteries,** which carry fresh oxygenated blood to the heart itself; the muscles of the heart do not obtain their supply of blood from within the heart.

The blood that flows into the arterial system, after delivering its cargo of oxygen to the cells of the body, eventually returns to the heart. In doing so, it passes through

FIGURE 51-21
The human circulatory system.

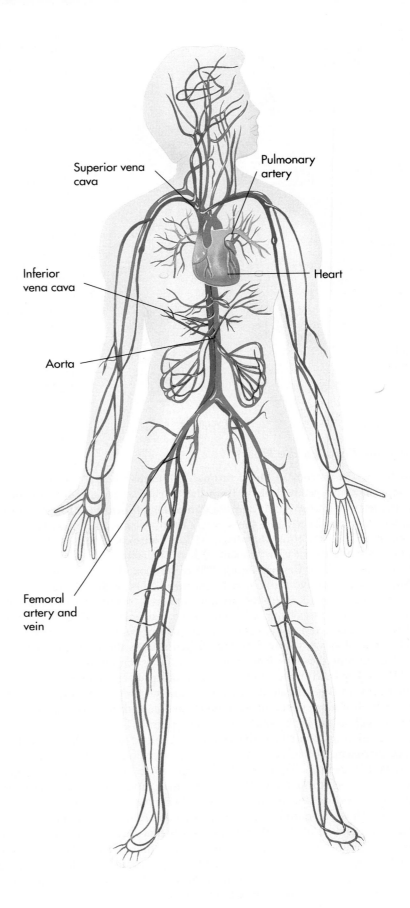

Superior vena
cava

Pulmonary
artery

Inferior
vena cava

Heart

Aorta

Femoral
artery and
vein

a series of veins, eventually entering the right side of the heart. Two large veins collect blood from the systemic circulation. The **superior vena cava** drains the upper body, the **inferior vena cava** the lower body. These veins empty oxygen-depleted blood into the right atrium. The right side of the heart is similar in organization to the left side. Blood passes from the right atrium into the right ventricle through a one-way valve, the tricuspid valve. It passes out of the contracting right ventricle through a second valve, the pulmonary valve, into the pulmonary arteries, which carry the oxygen-depleted blood to the lungs. The blood then returns from the lungs to the left side of the heart with a new cargo of oxygen, which is pumped to the rest of the body.

How the Heart Contracts

The contraction of the heart consists of a carefully orchestrated series of muscle contractions. Contraction is initiated by the sinoatrial node (Figure 51-22), the small cluster of excitatory cardiac muscle cells derived from the sinus venosus that is embedded in the upper wall of the right atrium. The cells of the SA node act as a pacemaker for the rest of the heart, their membranes spontaneously depolarizing with a regular rhythm that determines the rhythm of the heart's beating. Each depolarization initiated within this pacemaker region passes quickly from one cardiac muscle cell to another in a wave that envelops both the left and right atria almost instantaneously.

> **The contraction of the heart is initiated by the periodic, spontaneous depolarization of cells of the SA node; the resulting wave of depolarization passes over both the left and the right atria and causes their cells to contract.**

The wave of depolarization does not immediately spread to the ventricles, however. Almost 0.1 second passes before the lower half of the heart starts to contract. The reason for the delay is that the atria of the heart are separated from the ventricles by connective tissue, and connective tissue cannot propagate depolarization. The depolarization would not pass to the ventricles at all except for a slender connection of cardiac muscle cells known as the **atrioventricular node (AV node)**. These cells have a small diameter and thus propagate the depolarization slowly, causing the delay just noted. This delay permits the atria to finish emptying their contents into the corresponding ventricles before those ventricles start to contract.

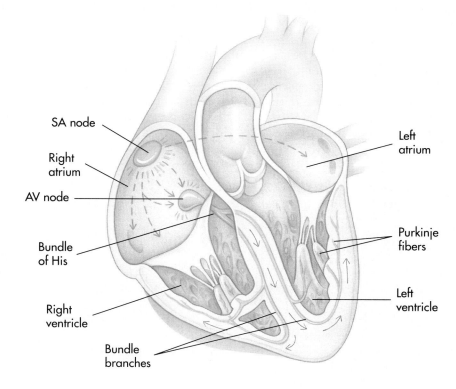

FIGURE 51-22
How the heart contracts. Contraction of the human heart is initiated by a wave of depolarization that begins at the SA node, the evolutionary vestige of the sinus venosus. After passing over the right and left atria and causing their contraction, the wave of depolarization reaches the AV node, from which it passes to the ventricles. The depolarization is conducted rapidly over the surface of the ventricles by a set of fibers called Purkinje fibers, which make up the bundle of His and branch throughout the ventricle.

From the AV node, the wave of depolarization is conducted rapidly over both ventricles by a special network of fibers called the **bundle of His.** Its passage triggers the almost simultaneous contraction of all the cells of the right and left ventricles.

> **The passage of the wave of depolarization over the narrow AV node delays the depolarization of the ventricles for a fraction of a second. When the ventricles receive the wave of depolarization, they conduct it so rapidly over their surface that all of the cells of the two ventricles contract almost simultaneously.**

Monitoring the Heart's Performance

As you can see, the heartbeat is not simply a squeeze-release, squeeze-release cycle but rather is like a little play in which a series of events occurs in a predictable order. You may watch the play in several ways, depending on the events that you are observing. The simplest way to monitor heartbeat is to listen to the heart at work. The first sound you hear, a low-pitched *lub,* is the closing of the mitral and tricuspid valves at the start of ventricular contraction. A little later you hear a higher-pitched *dub,* the closing of the pulmonary and aortic valves at the end of ventricular contraction. If the valves are not closing fully, or if they open too narrowly, turbulence is created within the heart. This turbulence can be heard as a **heart murmur.** It often sounds like sloshing.

A second way to examine the events of the heartbeat is to monitor the blood pressure. During the first part of the heartbeat the atria are filling and contracting. At this time the pressure in the arteries leading from the left side of the heart out to the tissues of the body decreases slightly as the blood moves out of the arteries, through the vascular system, and into the atria. This period is referred to as the **diastolic period.** During the contraction of the left ventricle a pulse of blood is forced into the systemic arterial system, immediately raising the blood pressure within these vessels. This pushing period, which ends with the closing of the aortic valve, is referred to as the **systolic period.** Blood pressure values are measured in millimeters of mercury (mm Hg); like atmospheric pressure, they reflect the height to which a column of mercury would be raised in a tube by an equivalent pressure. Normal blood pressure values are 70 to 90 mm Hg diastolic and 110 to 130 mm Hg systolic. When the inner walls of the arteries accumulate fats, as they do in the condition known as **atherosclerosis,** the diameters of the passageways are narrowed. If this occurs, the blood pressure is elevated.

A third way to monitor the progress of events during a heartbeat is to measure the waves of depolarization. Because the human body consists primarily of water, it conducts electrical currents rather well. A wave of membrane depolarization passing over the surface of the heart generates a minute electric current that passes in a wave throughout the body. The magnitude of this electrical pulse is tiny, but it can be detected with sensors placed on the skin. A recording made of these impulses (Figure 51-23) is called an **electrocardiogram.**

FIGURE 51-23
An electrocardiogram.

Atrial   Ventricular   Ventricular
excitation   excitation   repolarization

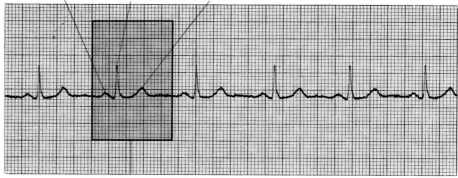

ONE HEARTBEAT

Cardiovascular diseases are the leading cause of death in the United States; more than 42 million people have some form of cardiovascular disease. Heart attacks are the main cause of cardiovascular deaths in the United States, accounting for about a fifth of all deaths. They result from an insufficient supply of blood reaching an area of heart muscle, leading to death of some cells. Heart attacks may be caused by a blood clot forming somewhere in the vessels and blocking the passage of blood through those vessels. They may also result if a vessel is blocked sufficiently by atherosclerosis. Recovery from a heart attack is possible if the segment of the heart tissue damaged was small enough that the other blood vessels in the heart can enlarge their capacity and resupply the damaged tissues. **Angina pectoris,** which literally means "chest pain," occurs for reasons similar to those which cause heart attacks, but it is not as severe. The pain may occur in the heart and often also in the left arm and shoulder. It is a warning sign that the blood supply to the heart is inadequate but not enough to cause cell death.

Strokes are caused by an interference with the blood supply to the brain. They often occur when a blood vessel bursts in the brain, and they may be associated with a thrombus, or coagulation (clotting) of blood cells or other elements in one of the vessels. Such a thrombus may be caused by cancer or other diseases. The effects of a stroke depend on how severe the damage is and where in the brain the stroke occurs.

Atherosclerosis is a buildup within the arteries (Figure 51-A, *1, 2,* and *3*). Atherosclerosis contributes both to heart attacks and to strokes. The accumulation within the arteries of fatty materials, of abnormal amounts of smooth muscle cells, of deposits of cholesterol or fibrin, or of cellular debris of various kinds can all impair the arteries' proper functioning. When this condition is severe, the arteries can no longer expand and contract properly, and the blood moves through them with difficulty. The accumulation of cholesterol is thought to be the prime contributor to atherosclerosis, and diets low in cholesterol and unsaturated fats are now prescribed to help prevent this condition.

Arteriosclerosis, or hardening of the arteries, occurs when calcium is deposited in arterial walls. It tends to occur when atherosclerosis is severe. Not only is flow through such arteries restricted, but they also lack the ability to expand as normal arteries do to accommodate the volume of blood pumped out by the heart. This forces the heart to work harder.

1
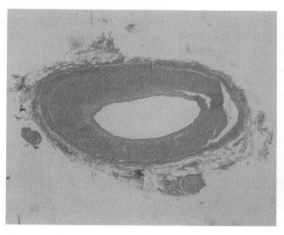

2
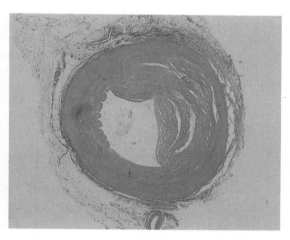

3
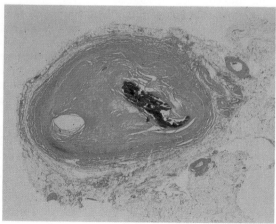

FIGURE 51-A
The path to a heart attack.
1 The coronary artery shows only minor blockage.
2 The artery exhibits severe atherosclerosis-much of the passage is blocked by buildup on the interior walls of the artery.
3 The coronary artery is essentially completely blocked.

In a normal heartbeat, three successive electrical pulses are recorded. First, there is an atrial excitation, caused by the depolarization that is associated with atrial contraction. A tenth of a second later there is a much stronger ventricular excitation, reflecting both the depolarization of the ventricles and the repolarization accompanying relaxation of the atria. Finally, perhaps two-tenths of a second later, there is a third pulse, the repolarization accompanying the relaxation of the ventricles.

Cardiac Output

Cardiac output is the rate at which the heart beats multiplied by the volume of blood delivered with each heartbeat. In a normal resting person, cardiac output will be 72 beats per minute $\times$ 0.07 liter per beat $\simeq$ 5 liters per minute.

When a person exercises heavily, the oxygen in the bloodstream is used more quickly and becomes depleted, while CO_2 levels increase. When the central nervous system detects a rise in CO_2 levels in the blood circulating to itself, it sends out signals through the nerves of the sympathetic nervous system to the SA node, stimulating these cells to depolarize at a more rapid rate. The hormone adrenaline may also be released into the bloodstream in response to nerve signals to the adrenal glands that produce it. Adrenaline also increases the rate of the heartbeat. As a result of these two stimuli, the heart beats faster—as much as twice as fast as normal. Its ventricles are squeezed much more tightly than normal. Since the ventricles never empty completely, a stronger contraction will deliver more blood. The combined result of these two changes is a greatly increased cardiac output: 140 beats per minute $\times$ 0.21 liter per beat $\simeq$ 30 liters per minute.

REGULATION OF BLOOD PRESSURE

Changes in the tension exerted by the smooth muscle cells in the arterioles controls the distribution of blood flow. There are two separate factors that control arteriolar smooth muscle tension: (1) extrinsic control by the autonomic nervous system, and (2) intrinsic control, or **autoregulation,** by chemical factors produced in the immediate vicinity of the blood vessels. Generally speaking, autoregulation allows each organ to receive the blood it needs for its activities, as long as there is enough available. Autonomic control steps in when there are competing demands, shifting the available blood flow to the sites where it is needed.

HYPERTENSION AND ITS CONSEQUENCES

The central nervous system regulates blood pressure at normal levels in several different ways. A reflex involving pressure receptors in the arteries regulates blood pressure over periods of seconds to minutes. Blood pressure regulation over periods of hours to days involves neural and hormonal mechanisms that regulate the volume of the blood through effects on water and salt reabsorption by the kidneys. When this long-term regulation breaks down, the result may be a chronic elevation of blood pressure, referred to as hypertension.

In one form of hypertension, called high-renin hypertension, there are abnormally high levels of the hormone renin, released by cells in the kidney. Renin causes salt to be reabsorbed, accompanied by water, increasing the blood volume and therefore blood pressure. A second form of hypertension, known as essential hypertension, is not caused by elevated levels of renin. Although it is not fully understood, one possible explanation for this disorder is a decrease in the release of a newly discovered hormone released from cells in the wall of the right atrium, called atrial natriuretic hormone (ANH).

Whenever the blood pressure is chronically elevated, there is an increased chance that blood vessels will rupture. When this occurs in the brain, the result is a stroke that may seriously damage this delicate structure. Hypertension also increases the work needed by the heart to pump blood, requiring an increased oxygen supply. As a result, the heart may fail. Hypertension is also associated with an increased accumulation of fat deposits on the walls of the coronary arteries (atherosclerosis), which ultimately leads to impaired blood flow to regions of the heart. Finally, hypertension damages the nephrons of the kidneys, often leading to a vicious cycle of further increases in salt and water retention and therefore in blood pressure.

Most arterioles are innervated by sympathetic nerve fibers which usually exhibit some degree of **resting tone.** An increase in sympathetic input to most vessels increases smooth muscle tension. The effect is a decrease in radius, or **vasoconstriction,** resulting in a decrease in blood flow to the region controlled by these arterioles. A decrease in resting sympathetic tone allows the smooth muscle to relax, called **vasodilation,** resulting in increased flow. A good example of the effect of autonomic control is seen in the classical "fight-or flight" reaction. Blood is redirected from organs such as the gastrointestinal tract and skin, to the skeletal muscles.

The Baroreceptor Reflex

Receptors sensitive to systemic arterial blood pressure, termed **baroreceptors,** are located in the walls of the carotid artery and in the wall of the aortic arch. Afferents from the baroreceptors reach the **cardiovascular control center** of the medulla. When blood pressure falls, the firing rates of the baroreceptors decreases. The cardiovascular control center reacts by increasing sympathetic activity and reducing parasympathetic activity. These changes increase the rate and force of contraction of the heart. The effect of increased sympathetic input to the blood vessels is arteriolar constriction. Both changes restore blood pressure to normal and maintain cardiac output (Figure 51-24).

Standing without fainting is possible because of the rapid response of the baroreceptor reflex. Standing up adds a gravitational component to the arterial and venous pressures of the lower body and reduces the pressures in the part of the body above

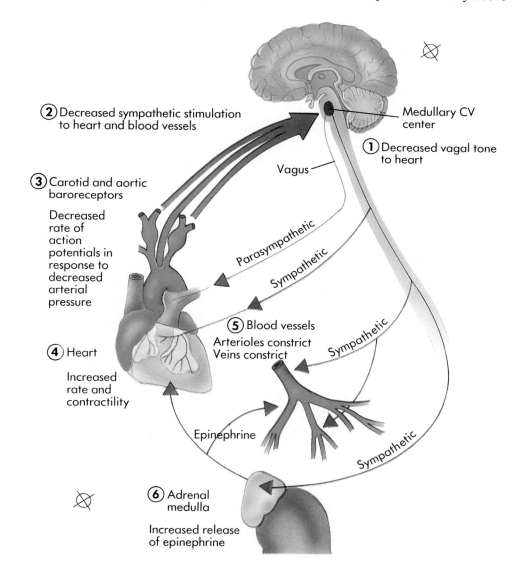

(2) Decreased sympathetic stimulation to heart and blood vessels

(3) Carotid and aortic baroreceptors

Decreased rate of action potentials in response to decreased arterial pressure

(4) Heart

Increased rate and contractility

Vagus

Parasympathetic

Sympathetic

(5) Blood vessels
Arterioles constrict
Veins constrict

Sympathetic

Epinephrine

Sympathetic

(6) Adrenal medulla

Increased release of epinephrine

Medullary CV center

(1) Decreased vagal tone to heart

FIGURE 51-24
Schematic of the baroreceptor reflex response to a decrease in mean arterial pressure. The net effect of the sympathetic inputs to heart and blood vessels is to return mean arterial pressure to a value close to the setpoint value for the medullary integrating center.

the heart (where the baroreceptors are located). The increased venous pressure in the lower body increases the volume of blood in lower veins, reducing the pressure in the veins entering the right heart. If these changes were unopposed, the drop in venous pressure and venous return would decrease cardiac output and fainting would result from inadequate blood flow to the brain. The baroreceptor reflex responds to standing by rapidly increasing the heart rate and constricting the arterioles, thus maintaining arterial blood pressure at its normal value.

Blood loss, termed **hemorrhage,** may reduce venous pressure to a point that cardiac output becomes dangerously low. The baroreceptor reflex is the first line of defense against hemorrhage. The general arteriolar constriction caused by the baroreceptor reflex does not affect the arterioles of the heart and brain since in these organs autoregulation is the major determinant of flow resistance. In effect, perfusion of the skin, skeletal muscle, and abdominal organs is sacrificed to maintain cardiac and cerebral blood flow.

Volume Receptors

Central venous pressure, the pressure in the right atrium, is directly proportional to blood volume. Volume regulation requires the cooperation of three homeostatic systems: (1) the **antidiuretic hormone** (ADH) system, (2) the **atrial natriuretic hormone** (ANH) system, and (3) the **renin-angiotensin-aldosterone** (RAA) system (Figure 51-25).

Antidiuretic hormone is secreted by the posterior pituitary in response to a decrease in activity from **stretch receptors** in the wall of the right atrium that functionally act as blood volume detectors. Decreases in central venous pressure increase the rate of secretion of ADH, reducing the rate of fluid loss in urine. Atrial walls contain

FIGURE 51-25
Regulating blood volume. Summary of three hormone systems that regulate blood volume in response to changes in atrial pressure (antidiuretic hormone-ADH, and atrial natriuretic hormone-ANH) or renal perfusion (renin-angiotensin-aldosterone). A significant loss of extracellular fluid results in increased secretion of all three hormones; their net effect is to decrease renal excretion of Na^+.

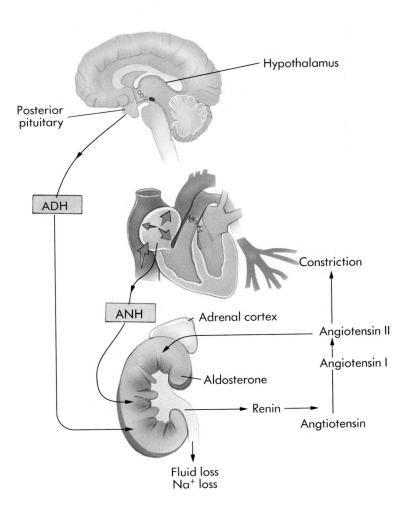

endocrine cells that secrete ANH, a peptide. Secretion of ANH decreases when central venous pressure decreases. Decreased ANH also limits the loss of plasma solutes and water in the urine.

Secretion of aldosterone by the adrenal cortex is controlled by a regulatory cascade that originates in the kidney. Endocrine cells of the kidney increase their secretion of *renin* in response to decreased blood pressure in the kidney. In the plasma, renin initiates the conversion of a plasma protein into several **angiotensins**. Angiotensins I and II stimulate contraction of vascular smooth muscle, increasing peripheral resistance. Angiotensin III stimulates aldosterone secretion. Aldosterone decreases the rate of Na^+ loss in urine, increasing total body Na^+. The ultimate effect of an increase in total Na^+ is an increase in the blood volume, elevating the blood pressure.

THE CENTRAL IMPORTANCE OF CIRCULATION

The evolution of multicellular organisms has depended critically on the ability of organisms to circulate nutrients and other materials to the various cells of the body, and to carry metabolic wastes away from them. The digestive processes described in Chapter 49, by which vertebrates obtain metabolizable foodstuffs, and the respiratory processes described in Chapter 50, by which vertebrates obtain the oxygen necessary for aerobic metabolism, both depend critically on the transport of food and oxygen to cells and the removal of the end products of their metabolism. Vertebrates carefully regulate the operation of their circulatory systems; by doing so, they are able to integrate their body activities.

SUMMARY

1. Unlike the open circulatory system of insects, the circulatory system of vertebrates is closed, permitting better control of circulation rates.

2. The general flow of blood circulation through the body is a circuit starting from the heart, which pumps blood out by way of muscled arteries to the capillary networks that permeate the tissues of the body; the blood returns to the heart from these capillaries by way of the veins. The water that is lost by diffusion during transit is recovered by the lymphatic system.

3. The circulatory fluid is blood, which contains cells rich in hemoglobin that transport oxygen to metabolizing cells and transport carbon dioxide away from them. The plasma of the circulating blood contains the proteins and ions necessary to maintain the blood's osmotic equilibrium with the surrounding tissues.

4. The heart of mammals and birds is a double pump, pushing both pulmonary (lung) circulation and systemic (general body) circulation. Because the blood is repumped after it passes through the narrow passages of the lung, the systemic circulation of mammals and birds is much more vigorous than that of the fishes. Because the two circula-

tions are kept separate within the heart, the systemic circulation receives only fully aerated blood, a great improvement in efficiency over the heart of amphibians and reptiles.

5. The four-chambered mammalian or bird heart evolved from two of the chambers of the four-chambered fish heart, by the creation of septa (dividing walls) within the two central chambers. The other two chambers of the fish heart were gradually lost, although the pacemaker cells of the sinus venosus have been retained in their original location.

6. The contraction of the heart is initiated at the SA node, or pacemaker, as a periodic spontaneous depolarization of these cells. The wave of depolarization spreads across the surface of the two atrial chambers, causing all of these cells to contract.

7. The passage of the wave of depolarization to the ventricles is briefly delayed by insulating the two segments of the heart from one another. Only a narrow channel of cardiac muscle cells connects the atria and the ventricles. The delay in the passage of the wave of depolarization permits the atria to empty completely into the ventricles before ventricular contraction occurs.

REVIEW QUESTIONS

1. What are the three elements of the closed circulatory system of vertebrates? What name is given to the plumbing elements of this system? In very general terms, trace the flow of blood through this plumbing. At which point does gas and metabolite exchange between the blood and tissues occur?

2. What is the relationship of flow resistance to blood vessel diameter? Which vessel exhibits a greater flow resistance, one that is large or one that is small? Where is the greatest amount of flow resistance in a vertebrate circulatory system? Why is it necessary for the walls of such a system to be elastic?

3. What is the basic structure of a capillary? How does the diameter of a typical capillary compare to the size of a red blood cell? What is the advantage of this? What is the approximate maximal distance between any one body cell and its nearest capillary? What is the function of a precapillary sphincter?

4. How does the structure of a vein differ from that of an artery? Why don't the veins need to accommodate the same pumping pressures as the arteries? Compare the appearance of an empty artery and an empty vein. Why are the diameters of veins larger than the diameters of corresponding arteries?

5. What is the function of the lymphatic system as it directly relates to the cardiovascular system? Is the lymphatic system a closed or an open system? What drives the flow of fluid within this system? What is the direction of flow in this system?

6. What are the three components of blood plasma? What amount of blood is water? What prevents it from being lost by osmosis to the general body tissues? What are the technical names of the three general categories of cells that are present in the blood? Define hematocrit.

7. In which vertebrate is the first chamber heart seen? What is the sequence of chambers in this type of heart? Which are collecting and which are pumping chambers? What is the major advantage of this pumping system? What is its major disadvantage?

8. What are the two functions of the sinus venosus in primitive vertebrates? What part of this structure remains evident in higher vertebrates? What is its function? How does the size of the heart relative to total body mass compare within a class of vertebrates? How does this value compare among classes of vertebrates?

9. Briefly trace the flow of blood through the human heart. Which is a stronger contraction: that of an atrium or that of a ventricle? Why? What is the general function of the various heart and large artery valves?

10. What is the sequence of muscular contractions in a single beat of the human heart? Why is there a delay in the transfer of the signal from the atria to the ventricles?

THOUGHT QUESTIONS

1. Why have mammals not improved the efficiency of their circulation by evolving hearts larger than 0.6% of their body mass?

2. Instead of evolving an entire second open circulatory system, the lymphatic system, to collect water lost from the blood plasma during passage through the capillaries, why haven't vertebrates simply increased the level of serum albumin in their blood?

3. Starving animals often exhibit swollen bodies rather than emaciated ones, in early stages of their deprivation. Why?

4. The hearts of the more advanced vertebrates pump blood entirely by pushing action. Why do you suppose hearts have not evolved that act like suction pumps, drawing blood into the heart as it expands, rather than pushing it out as the heart contracts?

5. Mean blood pressure continuously declines from the aorta to the right atrium. Where does the largest pressure drop occur? Why?

6. The diameter of a capillary is smaller than that of an arteriole, yet collectively the capillaries have a lower flow resistance than the arterioles. Explain.

FOR FURTHER READING

BROOKS, S.M.: *Basic Facts of Body Water and Ions*, ed. 3, Springer-Verlag, Inc., New York, 1973. A clear account of the principles of fluid and electrolyte balance.

CANTIN, M., and J. GENEST: "The Heart as an Endocrine Gland," *Scientific American*, February 1986, pages 76-81. An account of the discovery of atrial peptides, and interesting speculation on how they work in regulating blood pressure and volume.

COSSINS, A.R., and K. BOWLER: *Temperature Biology of Animals*, Chapman and Hall, New York, 1987. An excellent synthesis of general principles of temperature biology suitable for the nonspecialist.

DOOLITTLE, R.F.: "Fibrinogen and Fibrin," *Scientific American*, December 1981, pages 126-135. How the structure of the fibrin proteins determines the course of clot formation.

EISENBERG, M., AND OTHERS: "Sudden Cardiac Death," *Scientific American*, May 1986, pages 37-93. An account of the role of rapid medical intervention in coping with serious heart attacks.

GOLDE, D.W., and J.C. GASSON: "Hormones That Stimulate the Growth of Blood Cells" *Scientific American*, July 1988, page 62. Shows how specific hormones cause the progenitor cells in the bone marrow to differentiate into red blood cells or one of the many varieties of white blood cells. Soon it may be possible to treat diseases by calling up an extra supply of the cells needed.

GOLDSTEIN, G., and A.L. BETZ: "The Blood-Brain Barrier," *Scientific American*, September 1986, pages 74-83. A description of brain capillaries and the special ways in which they guard what enters and leaves the brain.

LAWN, R.M., and G.A. VEHAR: "The Molecular Genetics of Hemophilia," *Scientific American*, March 1986, page 48. Describes inherited defects in clotting factors responsible for hemophilia.

LEVINE, R. and S. WARDLAW: "A New Technique for Examining Blood," *American Scientist*, vol. 76, December 1988, pages 592-598. When a blood sample is spun by centrifugation, the paper-thin layer of white blood cells that lies above the red blood cells can help the physician diagnose diseases ranging from appendicitis to malaria.

LILLYWHITE, H.: "Snakes, Blood Circulation, and Gravity," *Scientific American*, December 1988, pages 92-98. Did you ever wonder why when a anake climbs a tree that all the blood doesn't rush to its tail?

RALOFF, J.: "Do You Know Your HDL?" *Science News*, vol. 136, September 1989, pages 171-173. A very clear account of recent research indicating that "good cholesterol" can be a strong predictor of coronary risk.

ROBINSON, T., S. FACTOR, and E. SONNEBLINK: "The Heart As a Suction Pump," *Scientific American*, June 1986, pages 84-91. Between birth and death our hearts beat millions of times. These authors argue that the heart is aided greatly in this Herculean task by a very clever trick: contraction compresses elastic elements within the heart muscles, which then bounce back to expand the ventricles.

ROWELL, L.B.: *Human Circulation Regulation During Physical Stress*, Oxford University Press, New York, 1986. Comprehensive coverage of cardiovascular responses to exercise.

SMITH, H.M.: *Evolution of Chordate Structure*, Holt, Rinehart & Winston, New York, 1960. A comprehensive and thoughtful treatise on the comparative anatomy of vertebrates, with a strong evolutionary perspective. A classic.

STOREY, K., and J. STOREY: "Frozen and Alive," *Scientific American*, December 1990, pages 92-97. Some animals survive the winter by freezing solid, using special proteins and antifreeze to avoid ice damage.

UTERMAN, G.: "The Mysteries of Lipoprotein (a)," *Science*, vol. 246, November 1989, pages 904-910. Lipoprotein (a), when present in high concentrations in the bloodstream, signals the likelihood of heart attack and stroke.

VINES, G.: "Diet, Drugs, and Heart Disease," *New Scientist*, vol. 121, February 1989, pages 44-49. There now seems little question that high levels of cholesterol in the blood give people heart disease. Now the question is what to do about it.

ZUCKER, M.B.: "The Functioning of the Blood Platelets," *Scientific American*, June 1980, pages 86-103. A description of the many roles of platelets in human health, with emphasis on their role in blood clotting.

52

The Immune System

CHAPTER OUTLINE

OVERVIEW

Most human diseases result from microbial infections—invasions of the body by viruses, bacteria, fungi, or protists. To defend against this onslaught and against cancer, vertebrates possess a sophisticated screening system called the immune system, which continually checks the bloodstream for the presence of any foreign cells or molecules. When an infection is detected, the invading microbes are attacked and destroyed. Without such a defense against infection, no vertebrate can survive for long. That is why infectious diseases that destroy our immune system, such as AIDS, are so very dangerous.

FOR REVIEW

Here are some important terms and concepts that you will encounter in this chapter. If you are not familiar with them, you should review them before proceeding.

Phagocytosis (Chapter 6)

Recombination (Chapter 17)

Macrophage (Chapter 43)

Lymphocyte (Chapter 43)

The lymphatic system (Chapter 51)

When you think of how animals defend themselves, it is natural to think of armor, of dinosaurs covered like tanks with heavy plates, of turtles and clams and armadillos. However, armor offers no protection against the most dangerous enemies that vertebrates face—their distant relatives, the microbes. Every vertebrate body offers a feast in nutrients for single-celled creatures too tiny for you to see with the naked eye, as well as a warm, sheltered environment in which they can grow and reproduce. Like Europeans first discovering the New World, a microbe entering a vertebrate body is faced with a rich ecosystem ripe for plundering. We live in a world awash with microbes, and no vertebrate body could long withstand their onslaught unprotected. We survive because we have evolved a variety of very effective defenses against this constant attack. These defenses are the subject of this chapter. As we review them, it is important to keep in mind that our defenses are far from perfect—microbial infection is still a major cause of death among humans. Some 22 million Americans and Europeans died of flu within 18 months in 1918 to 1919 (Figure 52-1). More than 3 million people will die of malaria *this year*. Attempts to improve our defenses against infection are among the most active areas of scientific research today.

FIGURE 52-1
The flu epidemic of 1918 killed 22 million people in 18 months. With 25 million Americans infected during the influenza epidemic, it was hard to provide care for everyone. The Red Cross often worked around the clock.

THE BODY'S DEFENSES

Your body is defended from infection the same way that knights defended medieval cities. There are walls and moats to make secret entry difficult, roaming patrols that attack strangers, and sentries that challenge anyone wandering about and call patrols if a proper ID is not presented.

Walls and moats. The outermost defense of the vertebrate body is the **skin** and the mucous membranes. In some vertebrates like rhinoceroses the skin is very thick and tough. In all vertebrates it offers a surprisingly efficient barrier to penetration by microbes. The lungs also have important barriers that protect their delicate alveoli from invasion.

Roaming patrols (nonspecific defenses). The initial response of the vertebrate body to infection is a battery of **nonspecific defenses,** including chemicals and cells that kill microbes. These defenses act very rapidly after the onset of infection.

Sentries (specific defenses). The vertebrate body also employs two types of cells that scan the surfaces of every cell in the body. Together they are called **the immune system.** One kind of cell aggressively attacks and kills any cell identified as foreign, whereas the other type marks the foreign cell or virus for elimination by the roaming patrols.

SKIN

Skin is the outermost layer of the vertebrate body, and provides its first defense against invasion by microbes. It also serves to keep the body watertight so that it does not lose excessive water to the air by evaporation. Skin is the largest organ of the vertebrate body. In an adult human, 15% of the total weight is skin. Many other specialized cells are crammed in among skin cells; one square centimeter of human skin, about what a dime covers, contains 200 nerve endings, 10 hairs and muscles, 100 sweat glands, 15 oil glands, 3 blood vessels, 12 heat-sensing organs, 2 cold-sensing organs, and 25 pressure-sensing organs.

Vertebrate skin is composed of three layers (Figure 52-2): an outer **epidermis,** and lower **dermis,** and an underlying layer of **subcutaneous tissue.**

Epidermis is the "Bark" of the Vertebrate Body

The epidermis of skin is from 10 to 30 cells thick, about as thick as this page. The outer layer, called the **stratum corneum,** is the one you see when you look at your

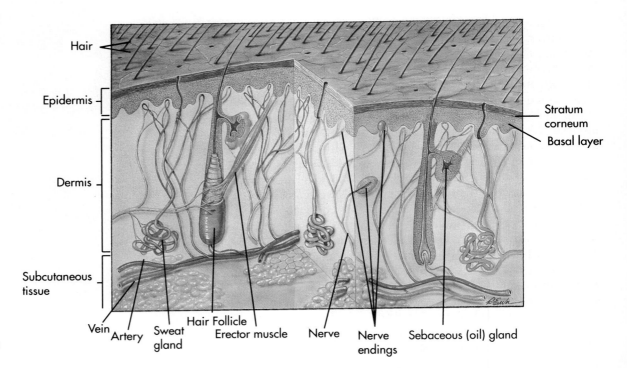

Hair
Epidermis
Dermis
Subcutaneous
tissue

Stratum
corneum
Basal layer

Vein Artery Sweat Hair Follicle Nerve Nerve Sebaceous (oil) gland
gland Erector muscle endings

FIGURE 52-2
Human skin.

arm or face. Cells from this layer are continuously subjected to damage. They are abraded, injured, and worn by friction and stress during the body's many activities. They also lose moisture and dry out. The body deals with this damage not by repairing cells but by replacing them. Cells from the stratum corneum are shed continuously. They are replaced by new cells produced deep within the epidermis. The cells of the innermost layer of the epidermis, called the **stratum basal** layer, are among the most actively dividing cells of the vertebrate body. New cells formed there migrate upward, and as they move they form keratin protein, which makes the skin tough. Each cell eventually arrives at the outer surface and takes its turn in the stratum corneum, ready to be shed and replaced by a newer cell. A cell normally lives in the stratum corneum for about a month. **Psoriasis,** familiar to some 4 million Americans as persistent dandruff, is a chronic skin disorder in which new cells reach the epidermal surface every 3 or 4 days, about 8 times faster than normal.

The Lower Skin Layers Provide Support and Insulation

The dermis of skin is from 15 to 40 times thicker than the epidermis. The thick dermis provides structural support for the epidermis and a matrix for the many nerve endings, muscles, and specialized cells residing within skin. The wrinkling that occurs as we grow older takes place here (Figure 52-3). A fine network of blood vessels passes through it. The leather used to manufacture belts is derived from very thick animal dermis.

The layer of subcutaneous tissue below the dermis is composed primarily of fat-rich cells. They act as shock absorbers and provide insulation, which conserves body heat. This tissue varies greatly in thickness in different parts of the body. The eyelids have none of it, whereas the buttocks and thighs may have a lot of it. The subcutaneous tissue of the skin on the soles of your feet may be a quarter-inch thick or more.

The Battle is on the Surface

The skin not only defends the body by providing a nearly impenetrable barrier, but also reinforces this defense with chemical weapons on the surface. The oil and sweat glands within the human epidermis, for example, lower the pH at the skin's surface to 3-5, an acid level that inhibits the growth of many microorganisms. Sweat also contains the enzyme lysozyme, which attacks and digests the cell walls of many bacteria.

FIGURE 52-3
For the most part, skin ages gradually, but on the face, the changes are more dramatic. At 19 this woman's face appears youthful and smooth. Forty years later, the production of skin oil is much less; her face appears less smooth and elastic and begins to exhibit wrinkles.

Two other routes of entry must also be guarded:

The digestive tract. Many bacteria are present in the food that we eat. Most of these are killed by saliva, which also contains lysozyme; by the strong digestive acids present in the stomach; and by protein-digesting enzymes in the intestine.

The respiratory tract. Ciliated epithelial cells in the nasal cavity entrap many bacteria before they can enter the airway. Many microbes are present in the air we breathe. The cells lining the smaller bronchi and bronchioles secrete a layer of sticky mucus that traps microorganisms before they can reach the warm, moist lungs, ideal breeding grounds for microbes. Cilia on the cells lining these passages continually sweep the mucus upward, where it can be swallowed, carrying potential invaders out of the lungs like bound prisoners to be destroyed by gastric HCl.

The surface defenses of the vertebrate body are very effective, but they are occasionally breached. Invaders now and then enter our bodies. When these invaders reach deeper tissue, a second line of defense comes into play, the vertebrate body's nonspecific defenses.

The surface defenses of the body consist of the skin and mucous membranes, which eliminate many invading organisms before they can enter the body tissues.

NONSPECIFIC DEFENSES

The vertebrate body uses a host of nonspecific cellular and chemical devices to defend itself. They all have one property in common: they respond to *any* microbial infection. They do not pause to inquire as to the identity of the invader, but swing into action immediately. Of the many nonspecific defenses, the four that are of most importance are (1) cells that ingest invading microbes; (2) antimicrobial proteins that kill pathogens; (3) the inflammatory response, which speeds defending cells to the point of infection; and (4) the temperature response, which elevates body temperature to slow the growth of invading bacteria.

Cells That Kill Invading Microbes

Perhaps the most important of the vertebrate body's nonspecific defenses are cells that attack invading microbes. These cells patrol the bloodstream and await invaders within tissues. There are three basic kinds: macrophages, neutrophils, and natural killer cells. Each of these cells kills invading organisms differently.

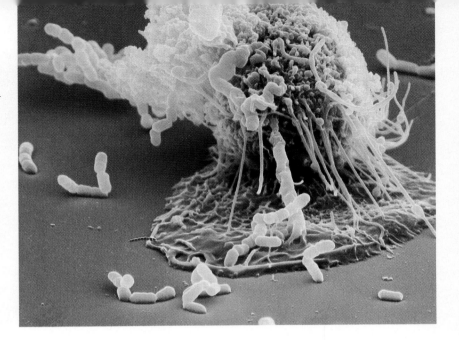

FIGURE 52-4

A macrophage in action. In this scanning electron micrograph a macrophage is "fishing" with long, sticky cytoplasmic extensions. Bacterial cells unfortunate enough to come in contact with the extensions are drawn back to the macrophage and engulfed.

Foot soldiers. **Macrophages** ("big eaters") kill bacteria one at a time by ingesting them, much as an amoeba ingests a food particle. Flowing cytoplasmic extensions stick to the invading bacterium (Figure 52-4) and pull it inside the macrophage by means of endocytosis. Once inside the macrophage, the bacterium is killed very efficiently: the membrane-lined vacuole containing the bacterium is fused with a lysosome. This activates lysosomal enzymes that liberate large quantities of oxygen free radicals that literally rip the bacteria apart chemically. Although some macrophages are fixed within particular organs, including lungs, liver sinusoids, spleen, and brain, most of the body's macrophages patrol the byways of the body, circulating in the blood, lymph, and interstitial fluid between cells.

Kamikazes. **Neutrophils** are white blood cells that ingest bacteria in the same way macrophages do. Both these defending cells are **phagocytes,** cells that kill invading cells by engulfing them. Neutrophils, however, are kamikazes; they also release chemicals (identical to household bleach) to "neutralize" the entire area, killing any other bacteria in the neighborhood and themselves in the process. Macrophages kill only one invading cell at a time but live to keep on doing it.

Internal security patrol. **Natural killer cells** do not kill invading microbes, but rather the cells that are infected by them. They are particularly effective at detecting and attacking body cells that have been infected with viruses. Natural killer cells are not phagocytes; rather they kill by attacking and puncturing the membrane of the target cell (Figure 52-5). Creation of the hole allows water to rush into the target cell, which swells and bursts (Figure 52-6). Natural killer cells are also able to detect cancer cells, which they kill before the cancer cells have a chance to develop into a tumor (Figure 52-7). The vigilant surveillance by natural killer cells is one of the body's most potent defenses against cancer.

How do these three kinds of cells distinguish *self* (the body's own cells) from *not self* (foreign cells)? It is very important not to unleash the destructive power of these cells on the body's own cells. When the body's defenders do turn on the body itself, in **autoimmune diseases,** the results can be fatal. The patrolling cells do not attack their own body because all the cells of their body contain a surface protein marker that identifies them. This protein, called a **MHC (major histocompatibility complex) marker,** acts like a "dogtag"—each person has a different version, although all the cells within a particular person contain the same one. The cells of the immune system simply ignore any cells having "self" MHC markers alone; they are called "nonspecific" because they will attack *any* cell that lacks the "self" MHC marker. In autoimmune diseases, antibodies are formed against "self" cells because the immune system fails to distinguish between foreign and host tissue.

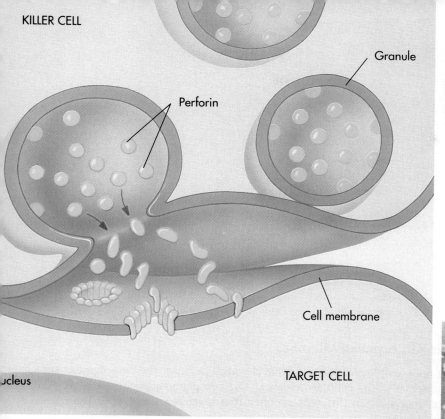

KILLER CELL

Perforin

Granule

Cell membrane

TARGET CELL

ucleus

FIGURE 52-5
How cytotoxic T cells kill target cells. The initial event is the tight binding of the killer T cell to the target cell. Binding initiates a chain of events within the T cell in which granules loaded with perforin molecules move to the outer cell membrane and disgorge their contents into the intercellular space over the target. The perforin molecules insert into the membrane like staves of a barrel to form a pore that admits water and ruptures the cell.

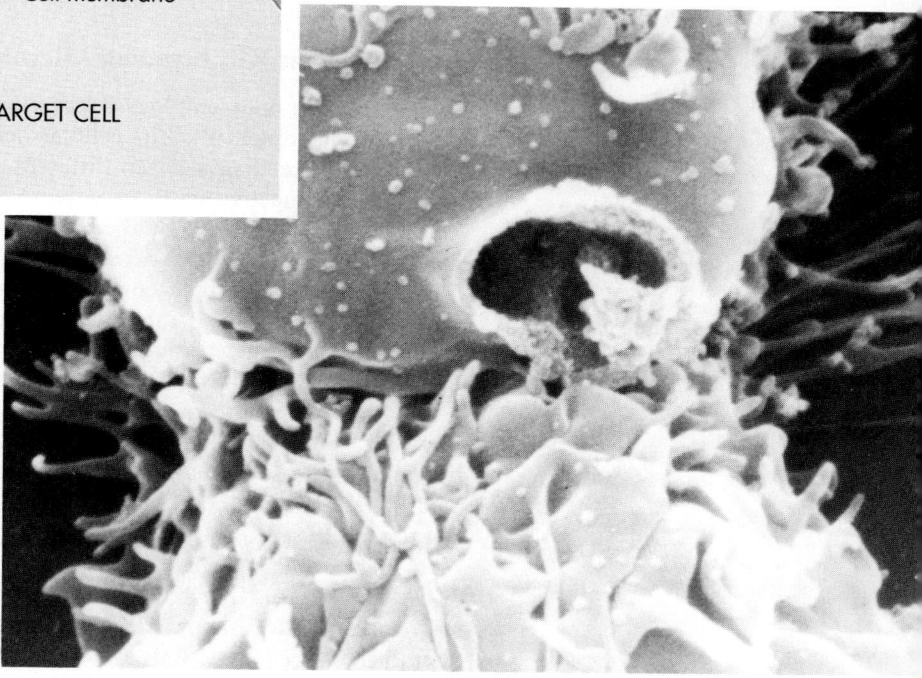

FIGURE 52-6
Death of a tumor cell. A natural killer cell has attacked this cancer cell, punching a hole in its cell membrane. Water has rushed in, making it balloon out. Soon it will burst.

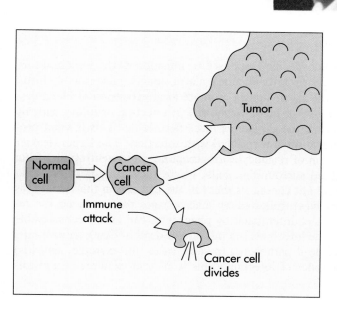

Normal cell

Cancer cell

Immune attack

Tumor

Cancer cell divides

FIGURE 52-7
The immune system protects against cancer. Every year many cells in your body become cancerous, but you do not develop cancer because your immune system detects and destroys them. AIDS patients, who lack a functional immune defense, often die of cancer.

FIGURE 52-8
**How complement creates a hole
in a cell.** As the diagram shows,
the complement proteins form a
complex transmembrane channel
resembling the perforin-lined lesion in cytotoxic T cells.

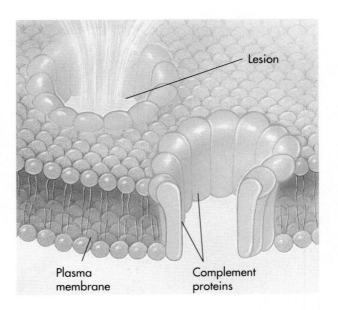

Lesion

Plasma
membrane

Complement
proteins

Proteins That Kill Invading Microbes

The vertebrate body employs a very effective chemical defense, the **complement system,** that "complements" its cellular defenses. The complement system consists of a
battery of more than a dozen different proteins that normally circulate in the blood
plasma in an inactive state. Its defensive activity is triggered by the cell walls of bacteria and fungi (and also by the binding of antibodies to invading microbes, discussed
later in this chapter). On detection of a bacterial cell wall, these proteins interact to
form a **membrane attack complex (MAC)** like that produced by natural killer cells.
The MAC inserts itself into the pathogen's cell membrane, forming a hole (Figure
52-8). Like a dagger through the heart, this wound is fatal to the invading cell—water pulled in by osmosis causes it to swell and burst.

Proteins of the complement system also act to amplify the effect of other body
defenses. Some amplify the inflammatory response (discussed next) by stimulating
histamine release, others attract phagocytes to the site of infection, and still others
coat invading microbes, roughening the microbe's surfaces so that macrophages may
more readily stick to them.

Another class of proteins that play a key role in body defenses are **interferons.**
Released by virus-infected cells, they diffuse out to other cells and inhibit the ability
of viruses to infect them. As you will see later in this chapter, their principal role is to
sound the alert for the immune system.

The Inflammatory Response

One of the most generalized nonspecific responses to infection is the **inflammatory
response.** Infected or injured cells release chemical alarm signals, most notably histamines and prostaglandins. These alarm signals promote local expansion of blood vessels, which both increases the flow of blood to the site of infection or injury and, by
stretching their thin walls, makes the capillaries more permeable. This is what produces the redness and swelling so often associated with infection. The larger, leakier
capillaries promote the migration of phagocytes (macrophages and neutrophils) from
the blood to the interstitial fluid surrounding cells, where they can engulf bacteria.
Neutrophils arrive first, spilling out chemicals that kill the bacteria in the vicinity (as
well as tissue cells and themselves), followed by macrophages that clean up the remains of all the dead cells. This counterattack by phagocytes can take a considerable
toll; the pus associated with some infections is a mixture of dead or dying neutrophils,
broken down tissue cells, and dead pathogens. In some cases (for example, arthritis)
inflammation occurs in the absence of infection. This is another example of a misdirected immune response.

The Temperature Response

When macrophages encounter invading microbes, they release chemical substances called **pyrogens** (Greek *pyr*, fire), which pass through the bloodstream to the brain. When they reach the cluster of neurons in the hypothalamus that serves as the body's thermostat, they act to boost the body's temperature several degrees above the normal value of 37° C (98.6° F). The higher-than-normal temperature that results is called a **fever.** Fever contributes to the body's defense by stimulating phagocytosis, by inhibiting microbial growth, and by causing the body to reduce blood levels of iron, which bacteria need in large amounts to grow. Very high fevers, however, are dangerous because excessive heat may inactivate critical enzymes. In general, temperatures greater than 103° F are considered dangerous; those greater than 105° F are often fatal.

The nonspecific defenses, both chemical and cellular, provide the vertebrate body with a sophisticated defense against microbial infection. Only occasionally do bacterial or viruses overwhelm them. When this happens, they face yet a third line of defense, more difficult to evade than any they have encountered. It is the immune system, the most elaborate of the body's defenses. Unlike other defenses, the immune system remembers previous encounters with potential invaders, and if they reappear, the immune system is ready for them.

SPECIFIC DEFENSES: THE IMMUNE SYSTEM

Few of us pass through childhood without being infected by a microbe. Measles, chickenpox, mumps—these are rites of passage, childhood illnesses that most of us experience before our teens. They are diseases of childhood because most of us suffer through them as children and *never catch them again*. Once you have had measles, you are immune. The mechanism that provides you with this immunity to such childhood diseases is the **immune system,** your body's most powerful means of resisting infection. Your immune system is the backbone of your health, protecting you not only from measles, but also from many other far more serious diseases. It is only in the last few years, as the result of an explosion of interest and knowledge, that biologists have begun to get a clear idea of how our immune system works.

DISCOVERY OF THE IMMUNE RESPONSE

In 1796 an English country doctor named Edward Jenner carried out an experiment that marks the beginning of the study of immunology. Smallpox was a common and deadly disease in those days, and only those who had previously had the disease and survived it were immune from the infection—except, Jenner observed, milkmaids. Milkmaids who had caught another, much milder form of "the pox," called cowpox (presumably from cows), rarely caught smallpox. It was as if they had already had the more serious disease. Jenner set out to test the idea that cowpox conferred protection against smallpox. He deliberately infected people with material that induced cowpox (Figure 52-9), causing them to catch this mild illness—and many of them became immune to smallpox, just as he had predicted.

Jenner's work demonstrated that it is possible for the human body to protect itself against disease very effectively, when it is able to make suitable preparations. We now know that smallpox is caused by a virus called variola, and that cowpox is caused by a different, although similar, virus. Jenner's patients who were injected with cowpox virus mounted a defense against the cowpox infection, a defense that was also effective against a later infection of the similar smallpox virus. Jenner's procedure of injecting a harmless microbe in order to confer resistance to a dangerous one is called **vaccination.** Modern attempts to develop resistance to malaria, herpes, and other diseases often involve a virus, vaccinia, related to the cowpox virus Jenner used.

A long time passed before people learned how one microbe can confer resistance to another. A key step was taken more than a half century after Jenner, when the famous French scientist Louis Pasteur showed that immunity was not created by the injected material, but rather invoked by it. Pasteur was studying fowl cholera, a serious disease in chickens that we now know to be caused by a bacterium. From dis-

FIGURE 52-9
This famous painting shows Edward Jenner inoculating patients with cowpox in the 1790s and thus protecting them from smallpox. The underlying principles of vaccination were not understood until more than a century later.

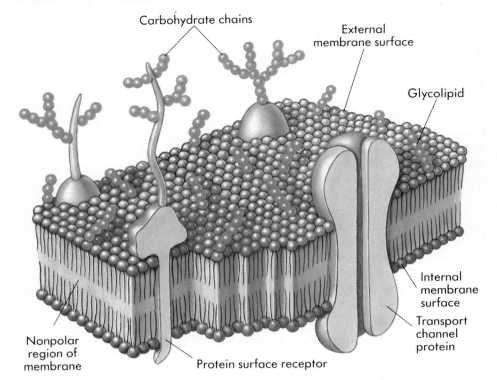

Carbohydrate chains

External
membrane surface

Glycolipid

Internal
membrane
surface

Transport
channel
protein

Nonpolar
region of
membrane

Protein surface receptor

FIGURE 52-10

The outer surface of a cell is not smooth, but rather a tangle of glycolipid and protein embedded within the membrane. Glycolipid molecules often serve as highly specific cell surface markers that identify specific cell types. The two major kinds of transmembrane protein are transport channels and receptors. Transport channels import ions, sugars, and other molecules into the cell. Receptors bind hormones, growth factors, neurotransmitters, and, in the case of immune receptors, other proteins.

eased chickens Pasteur could isolate a culture of bacteria that would elicit the disease if injected into other healthy birds. One day Pasteur accidentally left his bacterial culture out on a shelf at the end of the day and went on vacation. Two weeks later he returned and injected this culture into healthy birds. The culture had been weakened by its exposure; the injected birds became only slightly ill and then recovered. Surprisingly, however, the vaccinated birds could not then be infected with fowl cholera. They stayed healthy even if injected with massive doses of active fowl cholera bacteria, whereas control chickens receiving the same injections all died. Clearly something about the bacteria could elicit immunity, if only the bacteria did not kill the bird first.

We now know what that "something" was: molecules protruding from the surface of the bacterial cells. Every cell has on its surface a tangle of proteins, carbohydrates, and lipids (Figure 52-10), and it was the presence of foreign molecules on the surface of the cholera bacteria to which the chickens were responding. These bacterial cell surface molecules are different from any of the bird's own. "Not-self" molecules such as these are called **antigens.** Chickens injected with heat-weakened fowl cholera bacteria are immune to subsequent infection because the bacterial antigens cause the chickens to produce proteins called **antibodies.** These antibodies are able to recognize any future fowl cholera invaders and prevent them from causing disease. The production of antibodies directed against a specific antigen is one example of what we now call an **immune response.** The production of antibodies by the initial antigen is called the **primary immune response.** The second introduction of the same antigen results in an amplified production of antibodies called the **secondary immune response.** These immune responses are but one component of a complex system of recognition and defense that we call the immune system.

THE CELLS OF THE IMMUNE SYSTEM

Our immune system is not localized to one place in the body, nor is it controlled by any central organ such as the brain. Rather, it is composed of a host of individual cells, an army of defenders that rushes to the infection site to combat invading bacteria and viruses. These cells, called white blood cells, arise in the bone marrow and circulate in blood and lymph. Of the 100 trillion cells in an adult human, two in every hundred, or 2 trillion (2×10^{12}), are white blood cells. Although not bound together, the body's white blood cells exchange information and act in concert as a functional, integrated system. They are found not only in blood and lymph, but also in lymph nodes, spleen, liver, thymus, and bone marrow (Figure 52-11).

FIGURE 52-11
The human immune system.

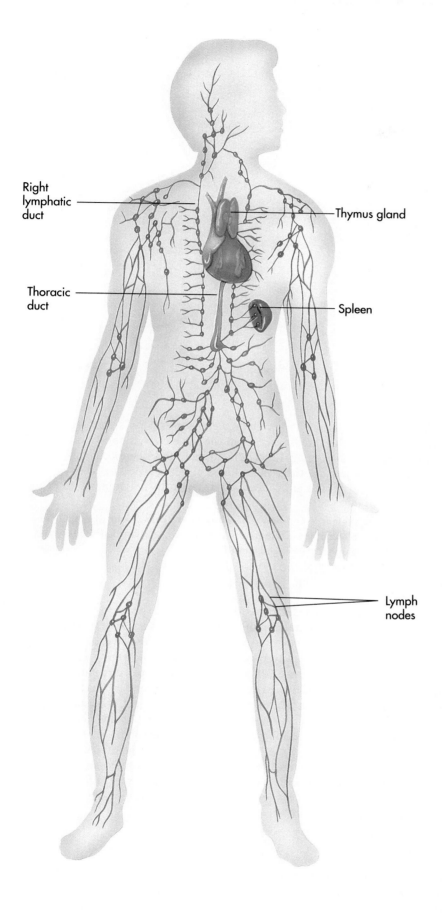

Right
lymphatic
duct

Thoracic
duct

Thymus gland

Spleen

Lymph
nodes

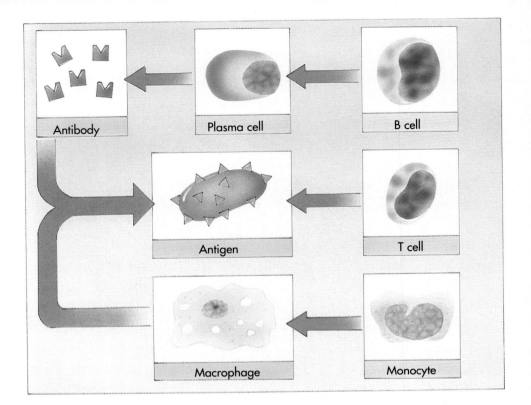

FIGURE 52-12
Players in the immune response.
Three different kinds of cells mount attacks against the antigen-studded foreign cell in the center of the diagram. As described in the text, each line of attack actually involves the cooperative action of many cell types.

White blood cells are larger than the red blood cells that ferry oxygen to the body's tissues, but they are formed from the same cells in bone marrow, called **hemopoietic stem cells** (Figure 52-12). Unlike mature red blood cells, all white blood cells have a nucleus. There are three main kinds of white blood cells: phagocytes, T cells, and B cells. T and B cells are collectively referred to as lymphocytes (Table 52-1).

> The immune system is composed of white blood cells. Four principal classes are involved; phagocytes (including macrophages), natural killer cells, and two kinds of lymphocyte (T cells and B cells).

Phagocytes

A **phagocyte** is a cell that destroys other cells by phagocytosis—it extends its membranes around other cells and engulfs them, much as an amoeba ingests a food particle. In humans some phagocytes are associated with the liver, spleen, lymph nodes, and other tissue, where they trap cellular debris and infecting bacteria and then remove them from circulation. Other mobile phagocytes circulate in the blood and lymph. Prominent among this roving army of defenders are **macrophages,** large, irregularly shaped cells that act as the body's scavengers. Constantly patrolling the body, macrophages engulf and consume anything that is not normal, including cell debris, dust particles in the lungs, and invading microbes. When a macrophage encounters a bacterium or a virus particle, the macrophage attacks, embracing the invader with its membranes and consuming it. When things are quiet, only a small number of macrophages circulate in the body's bloodstream and lymphatic system. In response to infection, however, precursors of macrophages called **monocytes** develop into mature macrophages in large numbers.

T Cells

Like macrophages and B cells, **T cells** also arise from stem cells in the bone marrow. Then, however, they migrate to the thymus, a small, gray gland that sits just above the heart (hence the designation "T"). There they develop the ability to identify invading bacteria and viruses by the foreign molecules (antigens) exposed on the invad-

TABLE 52-1 CELLS OF THE IMMUNE SYSTEM

CELL TYPE	FUNCTION
Helper T cells	Commander of the immune responses, the helper T cell detects infection and sounds the alarm, initiating both T cell and B cell responses
Inducer T cells	Not involved in the immediate response to infection, these cells mediate the maturation of T cells that are involved
Cytotoxic T cells	Recruited by helper T cells, these are the foot soldiers of the immune response, detecting and killing bacteria and infected body cells
Suppressor T cells	These cells dampen the activity of T and B cells, scaling back the defense after the infection has been checked
B cells	Precursors of plasma cells, these cells are specialized to recognize particular foreign antigens
Plasma cells	Biochemical factories, these cells are devoted to the production of antibody directed against a particular foreign antigen
Mast cells	Initiators of the inflammatory response (see p. 1101), which aids the arrival of white blood cells at a site of infection
Monocytes	Precursors of macrophages
Macrophages	The body's first line of defense, they also serve as antigen-presenting cells to B cells; later they engulf antibody-covered cells
Killer cells	These lymphocytes recognize and kill foreign cells: natural killer (NK) cells detect and kill a broad range of foreign cells; killer (K) cells attack only antibody-coated cells

A

ing microbe's surface. Tens of millions of different T cells are made, each specializing in recognizing one particular foreign antigen. No invader can escape being recognized by at least a few T cells. There are four principal kinds of T cells:

1. *Helper T cells,* which initiate the immune response
2. *Cytotoxic T cells,* which lyse cells that have been infected by viruses
3. *Inducer T cells,* which oversee development of T cells in the thymus
4. *Suppressor T cells,* which terminate the immune response

B Cells

Unlike T lymphocytes, **B cells** do not travel to the thymus; they complete their maturation in the bone marrow. From there they are released to circulate in the bloodstream and lymph (B cells were originally characterized in an immune gland of chickens called the bursa; hence the designation "B"). Individual B cells, like T cells, are specialized to recognize particular foreign antigens. When a B cell encounters the foreign antigen to which it is targeted, it begins to rapidly divide; and its progeny differentiate into miniature factories called **plasma cells** (Figure 52-13), each producing an antibody protein that sticks like a flag to the foreign antigen wherever it occurs in the body, marking it for destruction. The immunity that Pasteur observed resulted from such antibodies and from the continued presence of the B cells that produce them.

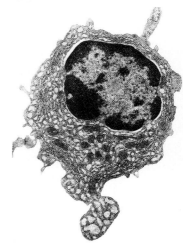

B

FIGURE 52-13
Electron micrographs, at the same magnification, of a mature T cell (A) and a plasma B cell (B). The plasma cell is filled with extensive endoplasmic reticulum on which antibody molecules are being synthesized. In contrast, the activated T cell has relatively little endoplasmic reticulum.

> The immune system is composed of white blood cells. There are three principal classes: phagocytes, including macrophages, and two kinds of lymphocytes (T cells and B cells).

THE STRATEGY OF IMMUNE SURVEILLANCE

All organisms possess mechanisms that protect them from the onslaught of other organisms. Bacteria protect themselves from viral invasion by means of restriction endonucleases, enzymes that degrade any foreign DNA lacking the specific pattern of

DNA methylation characteristic of that bacterium. Multicellular organisms face a more difficult problem in defending themselves, since animal and plant viruses often are taken up whole by host cells, rather than being injected into cells as naked DNA, as is the case with infecting bacterial viruses. Invertebrates solve this problem by marking the surfaces of their cells with special proteins that serve as "self" labels. Amoeboid phagocytic cells attack and engulf any foreign cells not identified by such labels as the invertebrate's own.

Invertebrates thus employ a *negative* test to recognize foreign tissue: cells lacking the invertebrate's specific cell surface protein are attacked and destroyed. This provides invertebrates with a very effective surveillance system, although it has one great weakness: any bacterium or virus with a surface protein resembling the invertebrate "self" marker will not be recognized as foreign. An invertebrate has no defense against such a "copycat" invader.

Vertebrates employ a multilevel defense against infection by microbes. The first line of defense is nonspecific. The skin and mucous membranes block entry of bacteria into the body. Mucus and saliva also contain the enzyme lysozyme, which digests bacterial cell walls.

If an infecting microbe gains entrance to the vertebrate body, a second line of defense comes into play, a *negative* test similar to that employed by invertebrates, but one that cannot be foiled by "copycat" foreign cells. The surface of every cell of a vertebrate individual possesses marker proteins called **major histocompatibility complex proteins (MHC proteins)**, which inform phagocytic cells that the cell is a vertebrate one; these marker proteins are different in each individual. The genes encoding the MHC proteins are highly polymorphic, so that very few individuals in a population possess the same set of alleles. A virus or bacterium mimicking one individual's MHC markers may successfully invade that individual, but not another. Only when individuals of a population are *not* polymorphic for the MHC gene are they at risk. This sometimes happens in zoo populations of endangered species, when the few individuals must be mated to one another, producing inbred populations with little MHC variation.

Vertebrates employ a third, very powerful defense strategy that relies on a *positive* test to identify foreign cells, as well as to detect cancer cells. This defense identifies molecules characteristic of invading microbes or cancer cells, molecules not present on the surfaces of normal vertebrate cells. More AIDS victims, whose immune systems are destroyed, die of cancer than any other cause. The test for "not-self" is carried out by white blood cells, which possess cell surface receptor proteins on their surfaces. It is these receptors that detect foreign molecules. The white blood cells examine the surface molecules of other cells at random. When a white blood cell encounters a cell with molecules on its surface that will bind to blood cell receptors, the white blood cell marks that cell for destruction.

IMMUNE RECEPTORS

A **cell surface receptor** is a protein that extends across a cell membrane and whose protruding end is able to bind to specific hormones or other "signal" molecules on the surface of the cell. In many cases, the binding alters the shape of the other end of the receptor protein, which protrudes into the interior of the cell. The change in the receptor protein's shape induces enzyme activity or opens an ion channel in the membrane. In this way the binding of a signal molecule to the cell surface is able to induce changes within the cell. All known neurotransmitters, protein hormones, and growth factors (such as epidermal growth factor, illustrated in Figure 16-12) bind to specific receptor proteins.

> **Cells perceive chemical signals such as hormones and growth factors by means of transmembrane proteins called receptors, which protrude from their outer surface. The binding of a signal molecule to a receptor protein on the cell surface causes an alteration in the receptor's shape that initiates a change within the cell.**

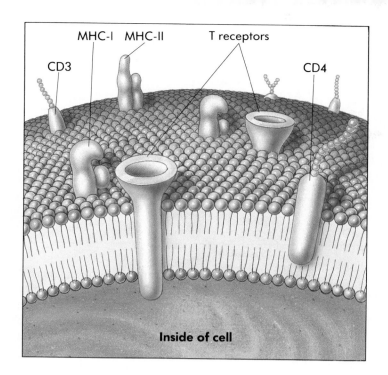

FIGURE 52-14
Cell surface proteins play critical roles in the immune system. Illustrated here is the cell surface of a kind of lymphocyte called a T_4 lymphocyte, within which are embedded three types of proteins: (1) glycoproteins, which serve as markers to identify the cell as being of a particular type; the glycoprotein labeled CD3 identifies the cell as a T cell, and the CD4 marker is characteristic of a particular kind of T cell; (2) MHC proteins, which serve to identify the cell as belonging to that individual; this cell exhibits both MHC-I and MHC-II; and (3) immune receptor proteins, which serve to identify foreign molecules; this cell utilizes receptors called T receptors. Different white blood cells exhibit different collections of these types of molecules.

The cell surfaces of white blood cells present to the outside world a variety of receptors and other proteins that play roles in the immune response. There are three general types of proteins (Figure 52-14):

1. Cell identity markers, glycoproteins that identify a cell as being of a particular type. Human T lymphocytes, for example, possess a marker designated **CD3.**
2. MHC proteins that serve as "self" labels, identifying a cell as belonging to a specific individual. These MHC proteins are recognized by immune receptors, which use them to distinguish between "self" and "not-self" cells. There are two classes of MHC protein, designated **MHC-I** and **MHC-II.**
3. Immune receptor proteins that serve to identify foreign molecules—both molecules characteristic of the cell surfaces of viruses and bacteria, and molecules characteristic of cancer cells. These receptors are unusual among vertebrate proteins in being encoded by genes assembled by somatic rearrangement. There are two classes of these receptors, called **B receptors** and **T receptors** after the cells that exhibit them. Any protein that binds to one of these receptors is recognized as an antigen, a foreign molecule. By having a very large library of "not-self" receptors, a vertebrate can correctly identify many foreign antigens and mark for destruction those cells bearing the foreign antigen.

The three principal kinds of white blood cells differ in the particular cell surface proteins they possess (Table 52-2):

1. B cells possess a "not-self" receptor called a B receptor, assembled by a unique process of DNA shuffling called **somatic rearrangement** and designed to bind a specific foreign antigen. When a B receptor binds such an antigen, it signals the B cell to secrete large amounts of the receptor. This circulating form of the B receptor, not bound to a membrane, is called an antibody.
2. T cells possess a different "not-self" receptor protein called the T receptor. Like the B receptor, the T receptor is assembled by somatic rearrangement and is designed to bind foreign antigens. There are four kinds of T cells, each with a different function in the immune response. Helper T cells and inducer T cells exhibit on their surfaces a cell-type marker protein called **CD4,** and are referred to as T_4 cells. Cytotoxic T cells and suppressor T cells exhibit instead a cell-type surface marker designated **CD8,** and are called T_8 cells.
3. Macrophages do not possess a somatically rearranged "not-self" receptor, but have the two sorts of "self-identifying" MHC proteins mentioned earlier.

TABLE 52-2   KEY CELL SURFACE PROTEINS OF THE IMMUNE SYSTEM

CELL TYPE	IMMUNE RECEPTORS		MHC PROTEINS	
	T RECEPTOR	B RECEPTOR	MHC-I	MHC-II
B cells	−	+	+	+

T$_4$ cells
  Helper
  Inducer
	+	−	+	+

T$_8$ cells
  Suppressor
  Cytotoxic
	+	−	+	−

Macrophages	−	−	+	+

White blood cells use the two MHC proteins in very different ways. One class of MHC, MHC-I, is present not only on macrophages but also on every other nucleated cell of the body. The MHC-II protein is a less widespread cell surface marker that is present only on macrophages, B cells, and T$_4$ cells. These three cell types bind together in one form of the immune response, and their MHC-II markers permit them to recognize one another.

The MHC surface markers of humans are specified by genes called **HLA genes,** some of which are highly variable from one person to the next. If a person is homozygous for the three MHC-I encoding genes, then all of the many MHC-I markers on the cells of that individual are identical in amino acid sequence. A heterozygous individual will have a few different kinds of marker, all present on each cell. It is rare, however, that the MHC-I (or MHC-II) proteins of two different individuals will have the same amino acid sequence.

> **Lymphocytes possess receptors that specialize in binding foreign antigens. The receptor characteristic of T cells is called a T receptor, that of B cells is called a B receptor. Antibodies are circulating forms of B receptors. B cells also possess the "self"-characterizing proteins MHC-I and MHC-II on their surfaces. All T cells possess MHC-I, but only T$_4$ cells among T cells possess MHC-II.**

THE ARCHITECTURE OF THE IMMUNE DEFENSE

The vertebrate immune system is part of a multilayered defense (Figure 52-15) that employs both immediate rapid responses to infection and several levels of long-lasting immune protection.

Immediate Response

Macrophages provide the first line of defense, with an immediate attack upon infected body cells. Macrophages engulf any circulating viruses or bacteria that have not yet infected cells; and they sound the alarm by activating helper T cells, which initiate the immune response.

Immune Response

Helper T cells respond to the macrophage alarm by simultaneously activating two parallel immune defenses, one involving T cells and the other involving B cells. T cells mount an effective defense against virus infection by recognizing and destroying infected body cells. This cuts off the virus infection before it can spread. B cells provide an alternative defense, effective against cancer cells, virus-infected body cells, and invading bacteria. The antibodies produced as part of this defense also indirectly provide long-term protection against renewal of the infection.

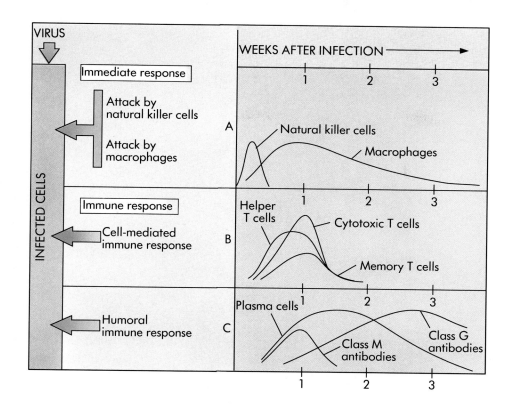

FIGURE 52-15

WEEKS AFTER INFECTION

VIRUS

INFECTED CELLS

Immediate response

Attack by natural killer cells

Attack by macrophages

A

Natural killer cells

Macrophages

Immune response

Cell-mediated immune response

B

Helper T cells

Cytotoxic T cells

Memory T cells

Humoral immune response

C

Plasma cells

Class M antibodies

Class G antibodies

FIGURE 52-15

Time course of the immune response. A virus infection is illustrated here, although a bacterial infection produces a similar chain of events. The immediate response of natural killer cells and macrophages (**A**) occurs within hours of infection and peaks in 1 or 2 days. This is followed by mobilization of helper T cells, which simultaneously initiates two parallel immune responses: cell mediated (**B**) and humoral (**C**). These two responses both peak within a few weeks, but antibody produced in the humoral response may persist in the bloodstream far longer.

The vertebrate body's defense against a measles infection provides a good example of how the immune system is organized to marshal a defense against infection.

The First Line of Defense. When a measles virus infects a vertebrate cell, the infected cell responds by secreting interferons, which circulate in the bloodstream and stimulate circulating defensive cells to activity (Figure 52-16). At the onset of the infection, these defensive cells, which include macrophages as well as natural killer (NK) cells, recognize body cells that have become infected with viruses and attack them. Natural killer cells pierce holes in the cells they attack, whereas macrophages ingest their quarry. When a macrophage consumes a virus-infected cell, it displays the antigens of the infecting virus on its own surface like a victory banner. This initial defense by macrophages and natural killer cells peaks within a day or two of the infection (see Figure 52-15, *A*).

Sounding the Alarm. At the onset of a measles infection, these encounters between viruses and macrophages provide an initial defense on the one hand, and on the other hand, serve to initiate the body's main line of defense, the immune response. Macrophages activate the immune system in two ways (see Figure 52-16):

FIGURE 52-16

Elements of the immediate response. The attacks by natural killer cells and macrophages are parallel in time, but the macrophage attack leads to further activation of the immune system.

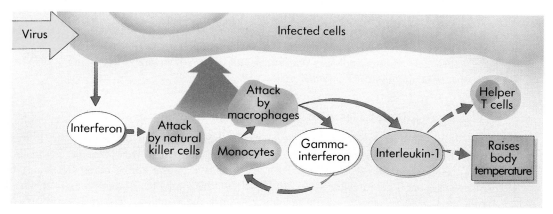

Virus

Infected cells

Interferon

Attack by natural killer cells

Attack by macrophages

Monocytes

Gamma-interferon

Interleukin-1

Helper T cells

Raises body temperature

FIGURE 52-17

How an infection causes fever.
At the site of infection *(1)*, macrophages release interleukin-1,
which passes through the bloodstream *(2)* to the brain. There it
stimulates the hypothalamus *(3)*,
the body's thermostat, triggering
it to set a higher temperature. To
raise the body's temperature to
the new setting, the brain sends
nerve impulses *(4)* to muscles,
ordering them to contract; as a
result, the body shivers, producing heat. Other nerve impulses *(5)*
order blood vessels near the skin
to constrict, minimizing heat loss.
The higher temperature aids the
immune response and inhibits the
growth of invading microorganisms. When the immune system
begins to make headway against
the infection, reducing the numbers of invading microbes, production of interleukin-1 by
macrophages stops. The fever
"breaks," and body temperature
soon falls to normal values.

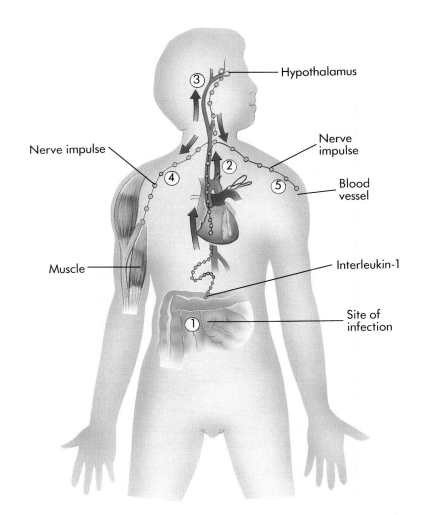

1. Macrophages respond to a virus encounter by secreting soluble proteins known
   as **monokines** (the name refers to the fact that the secreting cell is a form of
   monocyte), which signal the onset of the immune response. Among these
   monokines are **gamma-interferon,** which activates other monocytes to mature
   into macrophages, and **interleukin-1,** which activates T cells that have been
   stimulated by virus-infected cells and prepares them to proliferate. It is inter-
   leukin-1 that is responsible for the onset of fever often associated with infection (Figure 52-17).
2. Macrophages enzymatically degrade the protein envelope of an engulfed virus
   and display the resulting virus envelope fragments on the macrophage cell surface. Macrophages thus prepare the virus surface antigens for recognition by
   the immune system. Because of this behavior, macrophages are sometimes referred to as "antigen-presenting cells."

**The vertebrate body's first defense against infection is a patrolling army of
killer cells and macrophages that attack and destroy invading viruses and
bacteria and eliminate infected cells.**

The Main Line of Defense: Activating the Immune System

The immediate response is not adequate to eliminate many infections, but it buys
time for the immune system to respond. The key element in this response is the class
of T cells called helper T cells, which are sensitive to the alarm being broadcast by
macrophages. The interleukin-1 produced by macrophages activates any helper T
cells that have recognized a foreign antigen on a macrophage. When T receptors on

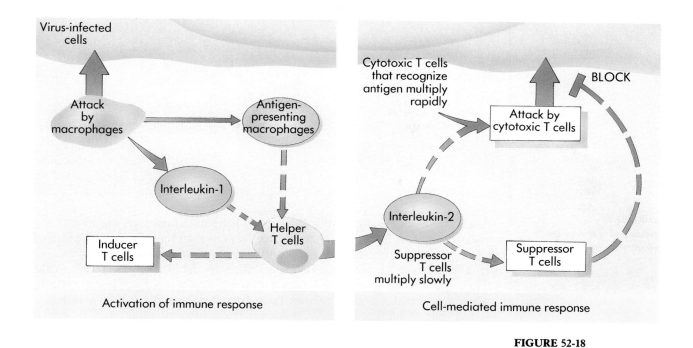

FIGURE 52-18
Elements and stages of the cell-mediated immune response.

the surfaces of the activated helper T cells encounter viral antigens presented by macrophages, the helper T cells simultaneously initiate two different but parallel immune responses: (1) the **cell-mediated response,** and (2) the **humoral response.**

The Cell-Mediated Immune Response.
The activation of helper T cells by interleukin-1 unleashes a chain of events known as the **cell-mediated immune response,** in which special **cytotoxic T cells** (*cytotoxic* means "cell poisoning") recognize and destroy infected body cells (Figure 52-18). Their mechanism of killing is the same as that of natural killer cells—they puncture the membranes of target cells. Helper T cells initiate the response, activating both cytotoxic T cells and other elements of the system. Cell-mediated immunity is essential for the destruction of host cells that have been infected by viruses or have become abnormal (for example, in some cancers).

Proliferation. When a helper T cell has been activated by interleukin-1, it also produces soluble factors; in this case, they are known as **lymphokines** because they are secreted by lymphocytes. Most important among these is **interleukin-2,** also called T cell growth factor. Interleukin-2 is the key that unleashes the proliferation of T cells. Under its influence, any T cells whose T receptors have bound to virus antigen (either exposed on macrophages, as in the case of helper T cells, or exposed on the surface of target cells, as in the case of cytotoxic T cells) begin to divide, forming large clones of T cells capable of recognizing that antigen. A virus infection usually induces proliferation of all three circulating T cell types: cytotoxic killer cells, helper T cells, and suppressor T cells.

Activation. A second lymphokine secreted by activated helper T cells is **macrophage migration inhibition factor,** which attracts macrophages to the site of infection and inhibits their migration away from it.

Induction. Helper T cells activate inducer T cells in the thymus, which trigger the maturation of immature lymphocytes into mature T cells.

Attack. The T receptors present on the surface of cytotoxic T cells recognize virus-infected body cells, binding simultaneously to the virus envelope attached to the cell surface and to the MHC-I antigen, present on all nucleated cells of the individual (Figure 52-19). The T cell proceeds to disrupt the cell membrane of the virus-infected cell to which it has bound, lysing the cell. The cytotoxic response peaks about a week after infection (see Figure 52-15, *B*).

Even though vertebrates did not evolve the immune system as a defense against tissue transplants, T cells will also attack any foreign version of MHC-I as if it sig-

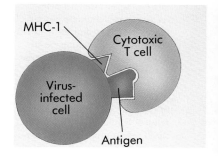

FIGURE 52-19
Cytotoxic T cells attack infected body cells. They recognize that a cell is virus-infected by the presence of virus antigens on the cell surface and recognize "self" by the presence of MHC-I. By binding only infected body cells, cytotoxic T cells avoid binding to free virus particles, which they cannot kill.

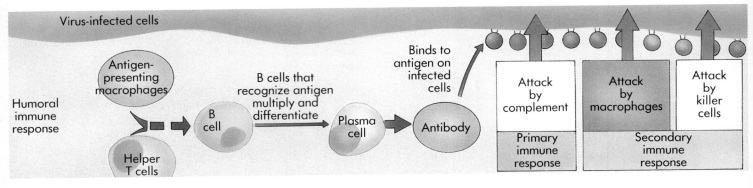

FIGURE 52-20
Elements and stages of the humoral immune response.

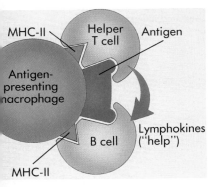

FIGURE 52-21

The humoral immune response. The humoral immune response generally involves the participation of three immune cell types: B cells, helper T cells, and antigen-presenting macrophages. B receptors on the surface of B cells recognize and bind one portion of the antigen. The antigen is usually presented by a macrophage (although naked virus and bacterial antigens can also be effective). T receptors on the surface of helper T cells bind to a different portion of the antigen. The T cell then releases lymphokines that induce the B cell to proliferate and mature into a plasma cell dedicated to production of the antigen-specific antibody. Binding of the three cell types together in the humoral immune response requires mutual recognition, provided by MHC-II proteins that identify cells of the humoral immune response system.

naled a virus-infected cell. For this reason vertebrate immune systems attack transplanted tissue. It is this sort of attack that leads to graft rejection in tissue transplants. Cyclosporine suppresses graft rejection by inactivating cytotoxic T cells.

Suppression. Suppressor T cells block the response of cytotoxic T cells to antigen. The population of suppressor T cells multiplies more slowly than do cytotoxic T cells. Their low initial numbers prevent them from blocking the cytotoxic attack on infected cells. After 1 to 2 weeks, however, the number of suppressor T cells rises to the point where they are able to shut down the cytotoxic T cell response (see Figure 52-15, *B*).

Memory. After suppression, a population of T cells persists, probably for the life of the individual. Referred to as memory cells, these helper and cytotoxic T cells provide an accelerated response to any later encounter with the virus antigen.

T cells carry out the cell-mediated immune response, in which cytotoxic T cells recognize and destroy infected body cells. Helper T cells initiate the response, activating both cytotoxic T cells and other elements of the immune system.

Interestingly, some studies suggest that macrophages and T cells may be able to respond to chemical messengers created by brain cells. Small neuropeptides such as endorphins and enkephalins may create a link between mind and body.

The Humoral Immune Response. When a helper T cell is stimulated to respond to a foreign antigen, it not only activates the T-cell–mediated immune response as just described, but also simultaneously activates a second, longer-range defense, called the **humoral** or **antibody immune response** (Figure 52-20). The key player in this stage of defense against infection is another kind of lymphocyte, the B cell. B cells recognize invading pathogens much as T cells do, but unlike T cells they do not attack the pathogen directly; rather they mark them for destruction by the nonspecific body defenses. The humoral immune response is specialized to destroy invading bacteria and viruses and inactivate foreign molecules that otherwise would be toxic.

Proliferation. Each B cell has on its surface about 100,000 B receptors. At the onset of a viral infection such as measles (and similarly in a bacterial infection), these B receptors bind to virus antigens, either to free viruses or to viral antigens displayed by macrophages. These antigen-bound B cells are detected by helper T cells, which bind simultaneously to the attached antigen and to MHC-II proteins on the B cell surface (Figure 52-21). When this happens, the helper T cells release lymphokines that induce the B cell to proliferate.

Differentiation and secretion. After about 5 days and eight cell divisions, a large clone of cells called **plasma cells** has been produced from each B cell that was stimulated by antigen to proliferate. All but a few of the proliferating B cells then stop reproducing and dedicate all of their resources to producing more copies of the B receptor protein that responded to the antigen. The receptor proteins are secreted as circulating antibodies, also called immunoglobulins. The secreting plasma cells live only a few days but secrete a great deal of antibody during that time. One cell will typically secrete more than 2000 molecules per second. Antibodies constitute about 20% by weight of the total protein in blood plasma.

Two main classes of antibody are secreted, called **class M** and **class G** (Figure 52-22; see also Figure 52-15, *C*): the first antibodies secreted are class M antibodies. This response peaks after 1 week. Plasma cells then shift to producing class G antibodies. The production of G antibody peaks after approximately 3 weeks. The two classes of antibody, M and G, induce different modes of antigen attack.

Attack. Antibodies do not destroy a virus or bacterium directly but rather mark it for destruction by one of three mechanisms:

1. *Complement.* Class M (and to a lesser degree class G) antibodies, when attached to a cell, activate a group of about 20 proteins collectively called **complement.** These proteins assemble into doughnut-shaped channels that pierce the membrane of the antibody-coated cell (Figure 52-23). Water is drawn into the cell osmotically through the resulting hole, causing the cell to swell and burst.
2. *Macrophages.* When a macrophage encounters a virus or cell coated with class G antibodies, it ingests and consumes it (Figure 52-24).
3. *K cells.* Killer (K) cells are cells resembling natural killer cells, except that K cells possess receptors that recognize antibody-coated cells. When a K cell encounters an antibody-coated cell, it binds and kills the cell.

Suppression. As in the case of the cell-mediated immune response, the antibody response is shut down after several weeks by suppressor T cells.

Memory. Some members of the clone of proliferating B cells do not go on to differentiate into plasma cells but rather persist as circulating lymphocytes, memory B cells. As in the case of memory T cells, these provide an accelerated secondary response to any later encounter with the stimulating antigen.

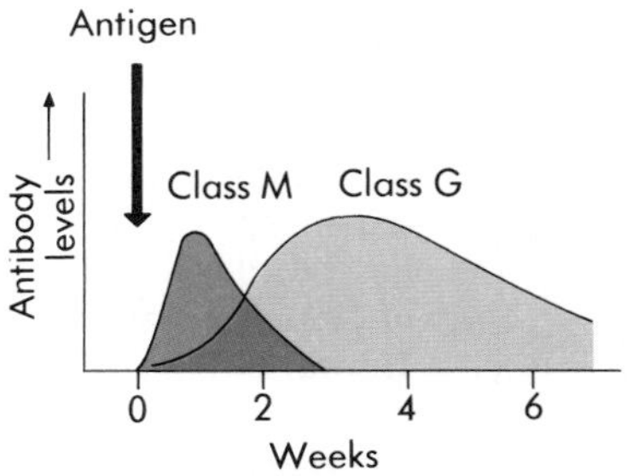

> **In the antibody defense, B cells recognize foreign antigens and, if activated by helper T cells, proceed to produce large quantities of antibody molecules directed against the antigen. The antibodies bind to any antigen they encounter and mark for destruction cells or viruses bearing the antigen.**

HOW DO ANTIBODIES RECOGNIZE ANTIGENS?

The cell surface receptors of lymphocytes are able to recognize specific antigens with great precision. Even proteins that differ by as little as one amino acid can often be discriminated, with a receptor recognizing one form and not the other. This high degree of precision is a necessary property of the immune system, since without it the identification of foreign antigens would not be possible in many cases; the differences between "self" and "not-self" (foreign) molecules can be very subtle.

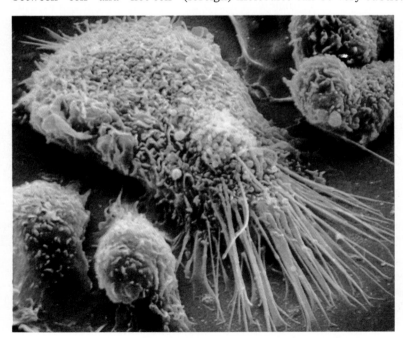

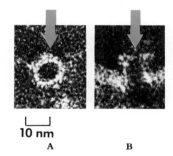

IMMUNE RESPONSE

FIGURE 52-22
Class M and class G antibodies. The first antibodies produced in the humoral immune response are class M antibodies. These antibodies are very effective at activating the complement system. This initial wave of class M antibody peaks after about 1 week and is followed by a far more extended production of antibodies, this time class G, during the secondary immune response.

FIGURE 52-23
Electron micrographs of complement lesions in the plasma membrane of a cell. The view of the lesion in **A** is from above, and the lesion in **B** is seen in cross section, as an apparent transmembrane channel. The micrographs are stained negatively, so that the complement proteins appear white and the channel appears black.

FIGURE 52-24
A macrophage ingesting cells.

Malignant melanoma is a particularly lethal form of skin cancer. Tumors grow very rapidly, quickly subverting the body's defenses. In August of 1990, the National Institutes of Health (NIH) gave its approval to use gene therapy to treat this cancer, one of the first instances in which humans will be the direct targets of genetic engineering.

The human cell that will be the target of this pioneering effort is a special kind of white blood cell called a tumor-infiltrating lymphocyte (TIL). This cell is part of the body's cancer surveillance system. It will normally seek out and attack a cancerous tumor but is not strong enough by itself to control the tumor. In previous years, NIH researchers had tried removing TIL cells from patients with malignant melanoma, culturing billions of them in test tubes, and then returning the cells to the patients' bloodstreams—but even in greatly increased numbers the cells were not strong enough to cure the cancer, although about half the patients improved.

Here is how the gene therapy will be done. Researchers will remove TIL cells from a patient with malignant melanoma and insert into each cell's chromosomes a gene that will command the cell to produce a protein called tumor necrosis factor (TNF). This protein kills tumor cells by blocking them from establishing a blood supply. The cells will then be returned to the patient's bloodstream to seek out and invade the malignant melanoma tumors. As each genetically altered TIL cell finds and enters a tumor, it will be able to attack using a much stronger weapon, TNF, in effect becoming a factory that makes the tumor-killing protein inside the tumor itself.

This first attempt at human gene therapy is experimental, and the degree to which it will be successful in treating malignant melanoma will not be known for a year or more. Because the approach shows great promise, it has medical researchers very excited. The chairman of the Recombinant DNA Advisory Committee that approved the therapy called the advance historic: "What we're doing today is adding gene therapy to vaccines, antibiotics, and radiation in the medical arsenal. Medicine has been waiting thousands of years for this."

How do immune receptors perform such fine distinctions? Biologists have learned the answer to this question by studying the amino acid sequences and three-dimensional structures of the receptors. The best known of the immune receptors is the B receptor, because its free form—antibody—is secreted in large amounts and therefore is far easier to study than membrane-bound receptors such as the T receptor. Because of the importance structure plays in immune receptor recognition, we will present here a fairly detailed discussion of the structure of the key immune molecules: antibodies (circulating B receptors), T receptors, and MHC proteins.

Antibody Structure

Antibody molecules (also called **immunoglobulins**) consist of four polypeptide chains. There are two identical short strands, called **light chains,** and two identical long strands, called **heavy chains.** The amino acid sequences of the two kinds of chains suggest that they evolved from a single ancestral sequence of about 110 amino acids. Modern light chains contain two of these 110 basic amino acid units, or domains; and heavy chains contain three or four of them. The four chains are held together by disulfide ($=S—S=$) bonds, forming a Y-shaped molecule (Figure 52-25).

From studying the amino acid sequences of different antibody molecules, it has become clear that the specificity of antibodies resides in the two arms of the Y, and the stem determines what role the antibody plays in the immune response. The terminal half of each of the two arms of the Y is where most of the variation in sequence between antibodies of different specificity is found. Within this variable region three small **hypervariable segments** come together at the end of each arm to form a cleft that acts as the binding site for the antigen. Both arms always have exactly the same cleft. The specificity of the antibody molecule for antigen depends on the precise shape of these clefts. An antigen fits into one of the clefts like a hand fits into a glove. Changes in the amino acid sequence of an antibody can alter the shape of its clefts; and by doing so, change the specific antigen that can bind to that antibody, just as changing the size of a glove will alter which hand can fit it.

> **An antibody molecule recognizes a specific antigen because it possesses two clefts, or depressions, into which an antigen can fit, much as a substrate fits into an enzyme's active site. Changes in the amino acid sequence at the position of these clefts alter their shape, and thus the antigen that can fit into them.**

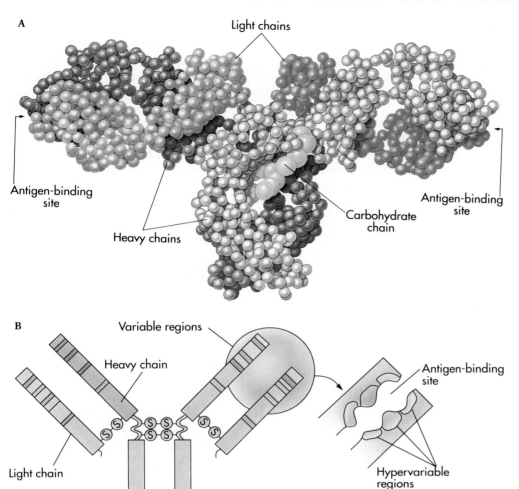

A

Light chains

Antigen-binding site

Heavy chains

Carbohydrate chain

Antigen-binding site

B

Variable regions

Heavy chain

Light chain

S S S S S S

Antigen-binding site

Hypervariable regions

FIGURE 52-25
The structure of an antibody molecule.
A A molecular model of an antibody molecule. Each amino acid is represented by a small sphere. The heavy chains are colored blue; the light chains are red. The four chains wind about one another to form a Y shape, with two identical antigen-binding sites at the arms of the Y and a tail region that serves to direct the antibody to a particular portion of the immune response.
B A schematic drawing of an antibody molecule. Each molecule is composed of two identical light (L) chains and two identical heavy (H) chains. Carbohydrate is sometimes complexed to the heavy chain. The antigen-binding sites are formed by a complex of both H and L chains, but the tail region is formed by H chains alone.

Antibodies with the same variable regions will have identical clefts and therefore recognize the same antigen, but they may differ from one another in the stem portion of the antibody molecule, which affects how the antibody functions in the immune response. The stem is formed by the constant regions of the heavy chains. In mammals there are five different classes of heavy chain: IgM, IgG, IgA, IgD, and IgE.

M chains are utilized in the first kind of antibody to be secreted by B cells, early in a humoral immune response. M antibodies produced in the primary immune response are very effective at activating the complement system. The production of antibodies with M chains peaks in about a week, after which B cells switch to using other classes of heavy chain when assembling their antibody molecules.

G chains are the major class of heavy chain employed by B cells in assembling antibodies after the first week, or during a secondary immune response. G chains are very effective at eliciting macrophage attack, although they can also initiate the complement reaction.

E chain antibodies, produced in much lower quantities than G chains during the secondary response, bind to mast cells (see p. 1091). The heavy chain stems of the E antibody molecules insert into receptors within the mast cell membrane, in effect creating B receptors on the mast cell surface. When such a mast cell encounters the specific antigen that elicited the antibody, it initiates an **inflammatory response** by releasing chemicals called **histamines.** These histamines cause nearby blood vessels to dilate, enabling lymphocytes, macrophages, and complement to reach the site where the mast cell has encountered antigen more easily.

The other two classes of antibody do not play such central roles in the immune response. The A chains are utilized in antibodies present in secretions such as milk and are thought to provide immune protection to nursing children, whose own immune systems have not yet fully developed. In secretions of mucous membranes, A chain antibodies are an important defense against polio, flu, colds, and other infections. Like E antibodies, very few D chain antibodies are produced, and their function is unknown.

The antibodies produced in the first week of an infection possess a class M heavy chain. These are directed at the complement system. After the first week, and in secondary responses, the class of heavy chain is switched to class G, and these new antibodies are very efficient at eliciting macrophage attack.

Structure of the T Receptor

T receptors perform a more complicated recognition process than B receptors, recognizing both positive (antigen) and negative (MHC) signals simultaneously. The structure of the T receptor (Figure 52-26, A) resembles one arm of a B receptor (Figure 52-26, B), with two chains referred to as alpha and beta. The amino acid sequences of the alpha and beta chains resemble those of B receptor light and heavy chains (although they are by no means identical); and like the chains of the B receptor, they exhibit constant and variable regions.

Structure of MHC Proteins

The simpler of the two MHC proteins is MHC-I (Figure 52-26, C), a single polypeptide chain. MHC-I proteins are very common on the surface of nucleated cells, where they can make up as much as 1% of the plasma membrane protein. Each copy is found associated with a second, smaller protein called **beta-microglobin.** Both beta-microglobin and a segment of MHC-I have an amino acid sequence similar to the repeated domain present in the light and heavy chains of an antibody molecule.

FIGURE 52-26

Structures of immune receptors (A and B) and of MHC proteins (C and D). All four classes of molecules have similar molecular structures and share similar sequences of amino acids. In each case, the molecules are characterized by units, or domains, of about 100 amino acids (characterized as loops) joined by S—S covalent bonds. Unlike the other three molecules, MHC-I protein (C) has only one chain, which is associated noncovalently with a small protein called beta-2 microglobulin (whose amino acid sequence also resembles the immune domain). The diagrams are schematic and are not drawn to scale; they emphasize the substantial similarities of these molecules.

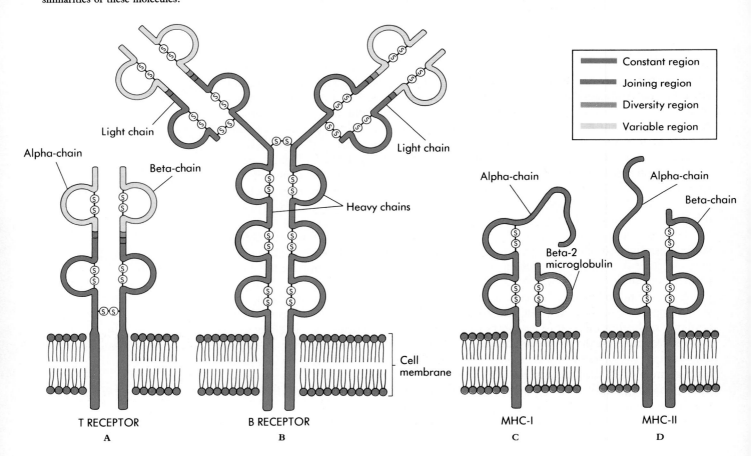

In humans three copies of the gene encoding MHC-I are arrayed side-by-side in a complex called the **human leukocyte-associated antigen (HLA) complex.** The HLA genes are the most polymorphic human genes known, with as many as 50 alleles each. Very few pairs of humans have the same combination of HLA alleles.

The other MHC protein, MHC-II (Figure 52-26, *D*), is also encoded in humans within the HLA complex. Like the T receptor, the MHC-II protein is composed of two polypeptide chains, and it also possesses antibody-like domains.

HOW CAN THE IMMUNE SYSTEM RESPOND TO SO MANY DIFFERENT FOREIGN ANTIGENS?

The vertebrate immune response is capable of recognizing as foreign practically any "not-self" molecule presented to it—literally millions of different antigens. It is estimated that a human is able to make between 10^6 and 10^9 different antibody molecules, although how this is done has long been a puzzle. Research has demonstrated that there are only a few hundred receptor-encoding genes, not millions, on vertebrate chromosomes. What process, then, is responsible for generating the great diversity of receptor and receptor-derived antibody? Two ideas have been proposed:

1. The **instructional theory** proposes that the antigen elicits the appropriate receptor, like a shopper ordering a custom-made suit.
2. The **clonal selection theory** proposes that millions of different kinds of stem cells are indeed in the bone marrow; and that an antigen causes those few encoding an appropriate receptor to proliferate, creating a clone of descendants expressing the appropriate receptor.

We now know the clonal selection theory to be correct. Within our bone marrow the stem cells destined to form B cells and T cells express an incredible diversity of receptor-encoding genes. Each cell encodes only one form of B and T receptor, but every cell is different from practically every other one.

How do vertebrates generate millions of different stem cells, each producing a unique receptor, when their chromosomes encode only a few hundred copies of such genes? They do it by **somatic rearrangement.** Immune receptor genes do not exist as single sequences of nucleotides, like the genes encoding all other proteins; rather they are first *assembled* by stitching together three or four DNA segments. Each segment, corresponding to a region of the receptor molecule, is encoded at a different site on the chromosome. These chromosomal sites are composed of a cluster of similar sequences (Figure 52-27), each sequence varying from the others in its cluster by small degrees. When an antibody is assembled (Figure 52-28), one sequence is selected at random from each cluster, and the DNA sequences selected from the various clusters are brought together by DNA recombination to form a composite gene. The process is not unlike your going into a large department store and choosing at random one coat, one shirt or blouse, one pair of pants, one pair of socks, and one pair of shoes; few people would come out of the store wearing the same outfit.

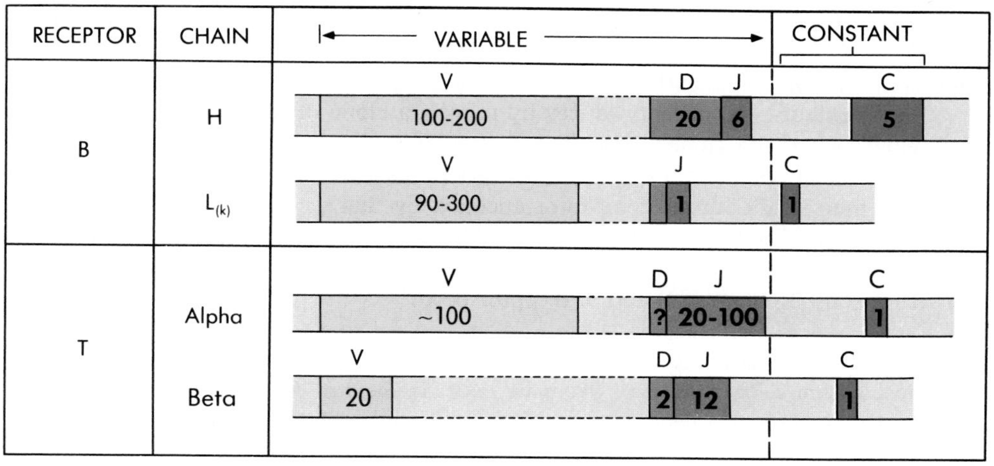

RECEPTOR	CHAIN	VARIABLE			CONSTANT
B	H	V 100-200	D 20	J 6	C 5
	L_(k)	V 90-300		J 1	C 1
T	Alpha	V ~100	D ?	J 20-100	C 1
	Beta	V 20	D 2	J 12	C 1

FIGURE 52-27

The immune receptor response library. The gene specifying the variable region of an antibody protein is assembled from DNA segments. Each segment typically exists in several copies and, in some cases, hundreds. Thus for the B receptor heavy chain there are several hundred copies of the "variable" or V segment, 20 copies of the "diversity" or D segment, and 6 copies of the "joining" or J segment. One specific heavy chain gene is produced in a particular stem cell by a maturation process that selects one copy each of V, D, and J at random and joins them to one copy of the constant region (C). A single gene encoding a specific light chain is similarly assembled at random from clusters of V and J segments, and the heavy and light chains are joined to form a complete B or T receptor. Millions of different combinations are possible.

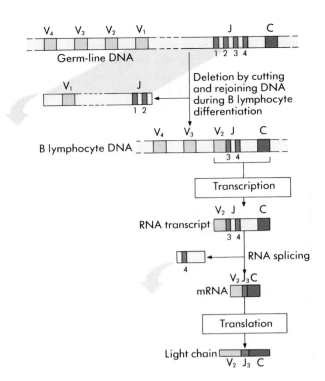

FIGURE 52-28

How a B receptor light chain is assembled. Before rearrangement, the four J segments are separated from each other and from the C segment by short introns and from the V region by a much longer stretch of DNA. During development of a particular stem cell, particular V gene and J gene segments are chosen at random (V_2 and J_3 in this case) by the deletion of the intervening DNA; this places the V and J segments precisely next to each other. The "extra" J_4 gene and intron sequences are transcribed together with V_2-J_3 and are then removed from the RNA transcript by RNA splicing enzymes.

Many different composite receptors are possible. The variable end of a human B receptor heavy chain, for example, is assembled from sequences selected at random from each of three clusters (see Figure 52-27), referred to as V (variable), D (diversity), and J (joining). As many as 200 genes are in the V cluster, about 20 genes in the D cluster, and 4 to 6 genes in the J cluster. Thus there are $200 \times 20 \times 4 = 16,000$ different possible combinations of V + D + J, leading to 16,000 different possible heavy chains.

The light chains are similarly assembled, from about 300 V genes and 4 J genes (there is no D cluster in this case), so that there are $300 \times 4 = 1200$ different possible light chains.

Two other processes generate even more additional sequences: (1) the segments are not bound together precisely, but often are joined one or two nucleotides off register, shifting the reading frame during gene translation and so generating a totally different sequence; (2) changes in amino acid sequence are generated by random mistakes in base pairing during successive replications of the clone of cells, a process called somatic mutation. It has been estimated that these two processes increase diversity by a factor of from 10 (light chain) to 100 (heavy chain).

Because a cell may end up with any heavy chain gene and any light chain gene during its maturation, the total number of different antibodies possible is staggering: HEAVY (16,000 combinations) × LIGHT (1200 combinations) = 19 million different possible antibodies. If one also takes into account the changes induced by alterations of reading frame and by somatic mutation, the total approaches 200 million!

Every mature stem cell produces by mitosis a clone of descendant lymphocytes, and each cell of the clone carries the particular rearranged gene assembled earlier, when that stem cell was undergoing maturation. As a result, all of the cells of a clone produce the specific immune receptor encoded by that stem cell, and no other. Because an adult vertebrate contains many millions of stem cells and because each stem cell undergoes the maturation process independently, cells specializing in millions of different combinations of B and T receptor occur.

Both T receptors and B receptors are encoded by genes that are assembled during stem cell maturation by somatic rearrangement of the DNA. Because each component is selected at random from many possibilities, a vast array of different T receptors and B receptors is produced.

THE NATURE OF IMMUNITY

When a particular B cell is stimulated by an invading microbe to begin dividing, producing a clone of proliferating cells with the same antibody, all these identical cells do not go on to become plasma cells. Instead many persist as circulating lymphocytes called memory cells. These provide an accelerated response to any later encounter with the stimulating antigen, because there are now many cells that can respond to it, rather than a few.

The Primary and Secondary Immune Response

The first time a particular kind of pathogen invades the vertebrate body, there are only a few B cells that by chance may have the antibody that can recognize it. The immune response that this first encounter sets off is called a **primary response.** It takes several days for these few cells to form a clone of cells that will produce antibody. The *next time* the body is invaded by the same pathogen, however, the immune system is ready; as a result of the first infection, there are now a small army of B cells that can recognize that pathogen—the memory cells. Remember, only some of the clones of dividing B cells became plasma cells in the first response; all the others are still there, a host of memory cells patrolling the bloodstream. Because each of these memory cells is well along the road to becoming a plasma cell, the **secondary immune response** is swifter, and because there are so many more of them, the response is much stronger (Figure 52-29). With each succeeding encounter, the bank of memory cells carrying that antibody becomes larger, so that the immune response grows even quicker and stronger.

Memory cells can survive for several decades, which is why most of us rarely contract mumps, chickenpox, or measles a second time once we have had them. Memory cells are also why vaccination against measles, polio, and smallpox are effective against these diseases. The microbes causing these childhood diseases have a surface that changes little from year to year, so the same antibody is effective decades later. Other diseases like flu are caused by microbes whose surface-specifying genes mutate rapidly; thus new strains appear every year or so that are not recognized by memory cells from previous infections. That is why immunity to flu lasts only a few years—the memory cells persist, but the antigen continually shifts to new forms. The ability of flu to defeat our immune defenses by constantly changing its surface is but one of several strategies pathogens have employed to defeat the vertebrate immune system.

WHY ARE THERE FEW ANTIBODIES DIRECTED AGAINST "SELF"?

How is the immune system able to distinguish foreign molecules from "self" molecules? The immune system of an embryo is able to respond to both foreign and "self" molecules, but the developing individual learns not to respond to "self" as it matures. Indeed, if a foreign tissue is introduced into embryonic animals before the immune sytem has developed, the mature animals do not recognize that tissue as foreign and will accept grafts of the donor tissue. This form of immune acceptance is referred to as **acquired immunological tolerance.**

The fact that a mature animal's immune system does not respond to its own tissue is called **natural immunological tolerance.** Both forms of tolerance have the same basis: elimination or suppression of particular clones of lymphocytes. During normal stem cell maturation, lymphocyte clones that would have been expected to respond to a particular "self" antigen are eliminated. Any that survive are specifically suppressed by suppressor T cells. Thus the only clones that survive this phase of development are those that are *not* directed against "self," that is, those directed against foreign antigens.

> **We are not attacked by our own antibodies because the cells that would have produced those antibodies are identified early in development and are removed or suppressed.**

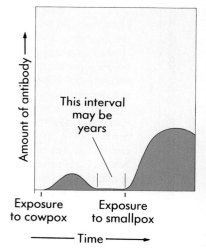

FIGURE 52-29

Immunity. Immunity to smallpox in Jenner's patients occurred because their inoculation with cowpox stimulated their bodies to produce antibodies that would recognize either cowpox or smallpox. Some of these antibodies, and the cells that produce them, remain in the bloodstream for a long time. A second exposure, this time to smallpox, stimulates the body to produce large amounts of the antibody much more rapidly than before.

The maintenance of self-tolerance requires the constant presence of the "self" antigen. If a particular antigen is removed, tolerance is lost within a few months; and the immune system regains its ability to respond with specific T and B receptors to the antigen.

Natural tolerance sometimes breaks down spontaneously, causing one or both classes of immune receptors to recognize their own tissue antigens. These losses of immune tolerance are called **autoimmune diseases.** Myasthenia gravis, for example, is an autoimmune disease in which individuals produce antibody directed against ace-

A CLOSER LOOK AT ALLERGY

Though the human immune system provides very effective protection against viruses, bacteria, and parasites, sometimes it does its job too well, mounting a major defense against a harmless antigen. Such immune responses are called allergic reactions. Hay fever, the sensitivity that many people exhibit to proteins released from plant pollen (particularly ragweed pollen), is a familiar example of an allergy. In response to as little as 20 pollen grains per cubic meter, a sensitive person's immune system will swiftly mount a defense. Many other people are sensitive to proteins released from the feces of a minute house-dust mite called *Dermatophagoides* (Figure 52-A), which lives in the house dust present on mattresses and pillows and consumes the dead skin scales that all of us shed in large quantities daily. Many people sensitive to feather pillows are in reality allergic to the mites that are residents of the feathers.

What makes an allergic reaction uncomfortable, and sometimes dangerous, is the involvement of class E antibodies. Antibodies with class E heavy chains are typically attached to mast cells. The binding of antigen to these antibodies initiates an inflammatory response: histamines and other powerful chemicals called **mediators** are released from the mast cells, causing dilation of blood vessels and a host of other physiological changes. Sneezing, runny nose, fever—all the symptoms of hay fever—result. In some instances when the body possesses substantial amounts of class E antibody directed against an antigen, allergic reactions can be far more dangerous than hay fever, resulting in anaphylactic shock in which swelling makes breathing difficult.

Not all antigens are **allergens,** initiators of strong immune re-

sponses. Nettle pollen, for example, is as abundant in the air as ragweed pollen, but few people are allergic to it. And not all people develop allergies; the sensitivity seems to run in families. It seems that allergies require both a particular kind of antigen and a high level of class E antibody: the antigen must be able to bind simultaneously to two adjacent E antibodies on the surface of the mast cell in order to trigger the mast cell's inflammatory response, and only certain antigens are able to do this. The class E antibodies must be produced in large enough amounts that many mast cells will have antibody molecules spaced close to one another, rather than the few such cells typical of a normal immune response, and only certain people churn out these high levels of E antibody. It is this combination of appropriate antigen on the one hand and inappropriately high levels of particular class E antibodies on the

other hand that produces the allergic response.

Hay fever and other allergies are often treated by injecting sufferers with extracts of the antigen, a process called **desensitization.** Allergy shots work best for pollen allergies and for allergy to the venom of bee and wasp stings; they are not effective again for food or drug allergies. The strategy of desensitization is to produce high levels of class G antibody in the bloodstream, so that when a particular antigen is encountered it will be mopped up by the G antibodies before encountering E antibodies on mast cells. Actually, there seems to be little correlation between levels of circulating G antibody and successful desensitization, and it is not clear why the procedure works as well as it does. A more ideal therapy would be to lower the amounts of class E antibody produced during the immune response, an approach that is being actively investigated.

FIGURE 52-A
The house-dust mite *Dermatophagoides*.

tylcholine receptors on their own skeletal muscle cells. The binding of antibody to the muscle cell receptors prevents the muscle receptors from responding to acetylcholine, so that arriving nerve impulses do not fire the muscle normally. Such patients can die from an inability of the chest muscles to carry out breathing. Rheumatoid arthritis and systemic lupus erythematosus (SLE) are other autoimmune diseases.

DEFEAT OF THE IMMUNE SYSTEM

All mammals and birds possess an immune system, most of them similar to the human one we have described in this chapter. During the evolutionary history of the vertebrates, several microbes have developed strategies, some of them quite successful, for defeating vertebrate immune defenses. We will discuss two. As you might expect, both strategies are responsible for very serious diseases.

Antigen Shifting

One excellent way of defeating the vertebrate immune sysem is to vary the nature of surface antigens. We are all familiar with periodic outbreaks of new strains of influenza, or "flu" (p. 583). The virus responsible for flu defeats the human immune system by undergoing frequent recombination within the gene encoding its major surface antigen. Every few years a major new strain of flu is generated with a rearranged surface antigen that human antibodies directed to previous versions do not recognize. The new strain of flu sweeps through the population. While immunity to this new version of the flu antigen eventually builds up, a lot of people may get sick. For example, over 20 million people died in the flu epidemic of 1918-1919.

An even more effective antigen-shifting strategy is practiced by trypanosomes, the protists responsible for sleeping sickness (p. 632). Trypanosomes possess several thousand different versions of the gene encoding their surface protein, but the cluster of coat genes has no promoter and so is not transcribed. The necessary promoter is located within a transposable element that jumps at random from one position to another within the cluster, transcribing a different surface protein gene with every move. Because such moves occur in at least one cell of an infective trypanosome population every few weeks (Figure 52-30), the human immune system is unable to mount an effective defense against infection. By the time an appreciable amount of antibody has been generated that recognizes one form of trypanosome surface receptor, another form is already present in the trypanosome population that survives immunological counterattack and renews the infection cycle. People with sleeping sickness rarely rid themselves of the infection.

T Cell Destruction: AIDS

A second successful strategy for defeating the vertebrate immune system has been to attack directly the immune mechanism itself. If you had to design such an attack, based on what you have learned in this chapter about the mechanism of the immune response, where would you direct it? Perhaps the most sensitive target would be the T_4 class of T cells. Helper T_4 cells are the key to the entire immune response (Figure 52-31), responsible for inducing proliferation of both T cells and B cells, whereas the maturation of all T cells (including helper T cells) requires cooperation of inducer T_4

FIGURE 52-30
Trypanosomes in the process of switching their coat antigen. Both panels are photographs of the same microscopic field. The trypanosomes in **A** express a particular surface antigen; antibody directed against that antigen has been bound by the trypanosomes, causing them to fluoresce red. The trypanosomes in **B** express a different surface antigen, to which a green-fluorescing antibody has been bound. The individual in the lower left is in the process of switching its coat antigen and so is therefore labeled with both antibodies.

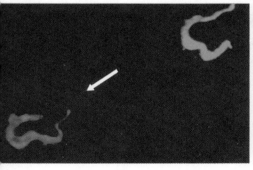

A

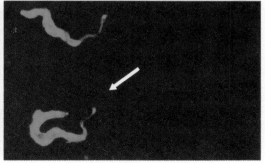

B

Only a small portion of an antigen molecule actually fits into an antibody's recognition site. In the case of protein antigens, for example, this portion (sometimes called a **determinant**) is typically in the size range of two to six amino acids. Most antigens are much larger than this, and as a result different portions of the antigen molecule can fit into different antibody sites. A typical antigen will thus elicit many different antibodies, each fitting to a different portion of the antigen surface. The antibody response is said to be **polyclonal.**

Antibodies offer great promise in medicine and research, because they recognize biological molecules with exquisite precision. However, it is often critical that biological tools be specific in order to be useful, just as a letter must have a specific address to reach its destination—and a polyclonal response presents a whole phone book of potential addresses. To find one address, an investigator needs instead an antibody that is directed against only one determinant (a **monoclonal antibody**). In 1984 Cesar Milstein of England and George Kohler of Switzerland were awarded the Nobel Prize for learning how to engineer an antibody response that is monoclonal. They devised an easy procedure for isolating a single clone of plasma cells, all producing the same antibody molecule.

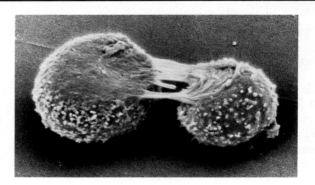

FIGURE 52-B, 1
A hybridoma dividing.

What Milstein and Kohler did was to mix plasma cells that were producing antibody with cancer cells, malignant lymphocytes called **myelomas.** Neither plasma cells nor myeloma cells lived long when Milstein and Kohler mixed them together: plasma cells normally live only a few cell generations anyway, and the cancer cells they employed were mutant ones, unable to survive without a certain metabolite the cells could no longer synthesize. Some cells did live in the mixture, however, growing and dividing—these were a new kind of cell produced by the fusion of a plasma cell with a myeloma cell (Figure 52-B, 1). These cell hybrids, called "hybridomas" (Figure 52-B, 2), utilized the genes of the plasma cell to make the metabolite necessary for growth, and were directed by the genes of the myeloma to ceaselessly grow and divide, as cancer cells. What made the experiment of profound importance was that the hybridoma cells continued to produce the antibody in which that plasma cell had specialized. Isolating single hybridoma cells from the mixture, Milstein and Kohler obtained rapidly growing cell lines that could be maintained in culture indefinitely, every cell of which was producing the same antibody molecule—monoclonal antibodies.

Monoclonal antibodies have proved to be of great importance to industry, because they can be used to purify specific molecules from complex mixtures. Interferon, present in only trace amounts in tissue extracts, was first purified in this

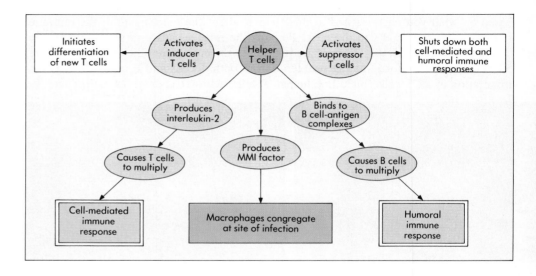

FIGURE 52-31
The many roles of the helper T cell.

way. Monoclonal antibodies have also revolutionized many aspects of biological research. It has proved possible, for example, to generate monoclonal antibodies directed against each of the many proteins that stud a cell's surface, and so to learn a great deal about cell surface receptors. The T receptor that plays such an important role in this chapter was first isolated in 1984 by employing a monoclonal antibody. In medicine, monoclonal antibodies offer great promise as vehicles for delivering specific therapies. There is an intensive search under way, for example, for antigens that occur only, or predominantly, on cancer cells, against which radioactive monoclonal antibodies could be targeted to selectively kill cancer cells.

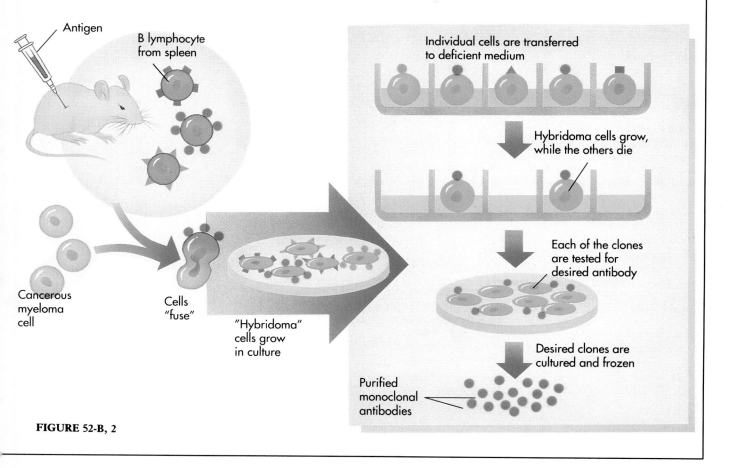

FIGURE 52-B, 2

cells. Without helper T_4 cells and inducer T_4 cells, the immune system is cut off at the knees, unable to mount a response to *any* foreign antigen.

AIDS is a deadly disease for just this reason. The AIDS retrovirus, called **human immunodeficiency virus** or **HIV,** mounts a direct attack on macrophages and T_4 lymphocytes (inducer T cells and helper T cells). Although most retroviruses are not specific, the HIV virus targets macrophages and T_4 cells, because the virus recognizes characteristic CD4 surface antigens associated with these cells.

HIV-infected T_4 cells are altered in ways that inhibit a successful immune response:

1. HIV-infected cells die but only after releasing progeny viruses (Figure 52-32) that infect other T_4 cells, until the entire population of T_4 cells is destroyed. In a normal individual, T_4 cells make up 60% to 80% of circulating T cells; in AIDS patients T_4 cells often become too rare to detect (Figure 52-33).
2. The HIV virus causes infected T_4 cells to secrete a soluble suppressing factor that blocks other T cells from responding to antigen.
3. The HIV virus may block transcription of MHC genes, hindering the recognition and destruction of infected cells and thus protecting infected T cells from any remaining vestiges of the immune system.

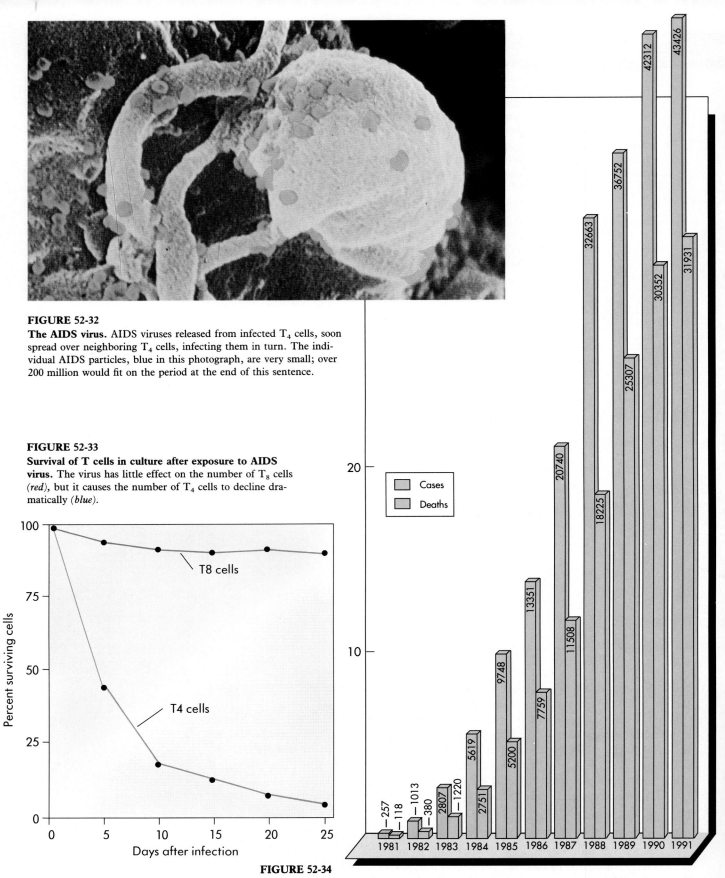

FIGURE 52-32

The AIDS virus. AIDS viruses released from infected T₄ cells, soon spread over neighboring T₄ cells, infecting them in turn. The individual AIDS particles, blue in this photograph, are very small; over 200 million would fit on the period at the end of this sentence.

FIGURE 52-33

Survival of T cells in culture after exposure to AIDS virus. The virus has little effect on the number of T₈ cells (*red*), but it causes the number of T₄ cells to decline dramatically (*blue*).

FIGURE 52-34

The AIDS epidemic in the United States. The U.S. Centers for Disease Control (CDC) reports that the end of 1991 there were 206,392 AIDS cases in the United States and 133,232 deaths. Over 1 milli other individuals are thought to have been infected with the HIV virus. The 100,000th AIDS case wa reported in August 1989, 8 years into the epidemic; the next 100,000 cases accumulated in just 26 months. In 1992 the CDC announced an expanded clinical definition of AIDS that does not rely on c tracting a particular syndrome of opportunistic infections; now, persons will be classified as having A if they are HIV-positive and have a T-cell count below 200 (a normal T-cell count is 800 to 1000). Th new broader definition is expected to add 165,000 additional AIDS cases.

The combined effect of these responses to HIV infection is to wipe out the human immune defense. An effective immune response, either a T-cell–mediated cellular immune response or a B-cell–mediated antibody response, is impossible without inducer T cells to mature the lymphocytes or helper T cells to initiate the response. With no defense against infection, any of a variety of otherwise commonplace infections proves fatal. With no ability to recognize and destroy cancer cells when they arise, death by cancer becomes far more likely. Indeed, AIDS was first recognized as a disease because of a cluster of cases of a usually rare form of cancer.

AIDS destroys the ability of the immune system to mount a defense against any infection. The HIV virus attacks and destroys T$_4$ cells, without which no immune response can be initiated.

Although the HIV virus became a human disease vector only recently, possibly transmitted to humans from African green monkeys in central Africa, it is already clear that AIDS is one of the most serious diseases in human history. The fatality rate of AIDS is 100%—no patient exhibiting the symptoms of AIDS has ever been known to survive more than a few years (Figure 52-34). The disease is *not* highly infectious; it is transmitted from one individual to another during the transfer of internal body fluids, typically in semen and in blood during transfusions. Not all individuals exposed to AIDS (as judged by antibodies in their blood directed against HIV virus) have yet acquired the disease. However, most will die within 2 years of onset of the symptoms unless additional strategies for the treatment of AIDS are discovered first.

Efforts to develop a vaccine against AIDS continue, both by splicing portions of the HIV surface protein gene into vaccinia virus (Figure 52-35) and by attempting to develop a harmless strain of HIV. (In one such attempt, for example, a gene called *tat* necessary for transcription of virus genes is deleted from the HIV virus.) These approaches, while promising, have not yet proved successful and are limited by the fact that different strains of HIV virus seem to possess different surface antigens. Like flu, AIDS indulges in some form of antigen shifting. Drugs that inhibit *tat* or block the operation of enzymes critical to synthesis of viral RNA are also being investigated.

FIGURE 52-35
How scientists are attempting to construct a vaccine for AIDS. Of the several genes of the human immunodeficiency virus (HIV), one is selected that encodes a surface feature of the virus. All of the other HIV genes are discarded. This one gene is not in itself harmful to humans; it is simply the shape of one of the HIV surface proteins. This one gene, or a fragment of it, is inserted into the DNA of a harmless vaccinia cowpox virus, resulting in the vaccinia virus eventually producing a protein coat with surface features specified by the HIV gene. Humans infected with the altered vaccinia virus do not become ill (the vaccinia virus is harmless), but they do develop antibodies directed against the infecting virus surface. Because the surface contains HIV proteins, the new antibodies would serve to protect the infected person against any subsequent exposure to the HIV virus.

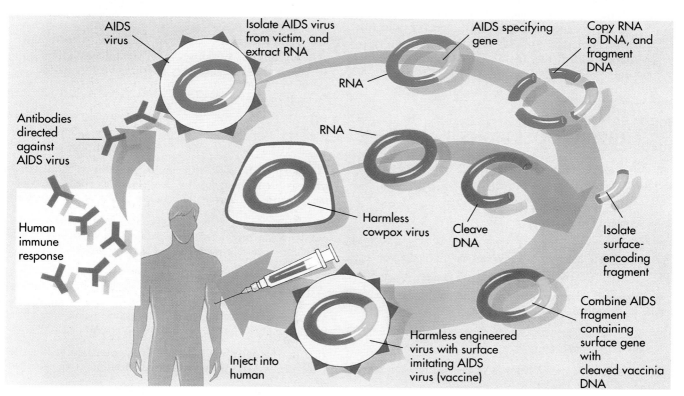

AIDS virus

Isolate AIDS virus from victim, and extract RNA

AIDS specifying gene

Copy RNA to DNA, and fragment DNA

RNA

RNA

Antibodies directed against AIDS virus

Human immune response

Harmless cowpox virus

Cleave DNA

Isolate surface-encoding fragment

Combine AIDS fragment containing surface gene with cleaved vaccinia DNA

Inject into human

Harmless engineered virus with surface imitating AIDS virus (vaccine)

SUMMARY

1. Although immunity was discovered almost 200 years ago, we have only recently learned that resistance to disease is achieved by populations of white blood cells collectively called the immune system.

2. Four types of white blood cells are involved in the immune system: T cells, B cells (collectively called lymphocytes), macrophages and other phagocytes, and killer cells.

3. Vertebrates employ two screening procedures to detect invading microbes. First, they mark all of their own cells with distinctive cell surface proteins called MHC proteins. Any cell with a different version of MHC is recognized as foreign. Second, vertebrate lymphocytes possess on their cell surfaces immune receptors that are able to bind to specific foreign molecules.

4. The immune receptors present on the surface of lymphocytes are unique among proteins in that the genes encoding them are assembled by somatic rearrangement. The two principal immune receptors are T receptors (present on T cells) and B receptors (present on B cells). Antibodies are secreted, circulating forms of B receptors.

5. The two different MHC proteins serve to channel antigen toward different populations of T cells: MHC-I is present on the surface of all nucleated cells, and antigen associated with it recruits cytotoxic T cells. MHC-II is present only on B cells and on helper and inducer T cells (together referred to as T_4 cells), and antigen associated with it recruits T_4 cells to initiate the immune response.

6. The immune response is initiated by those T_4 cells called helper T cells. Helper T cells, when stimulated by macrophages and antigen, simultaneously activate two parallel responses: the cell-mediated immune response, in which cytotoxic T cells attack infected body cells, and the humoral immune response, a longer-range defense in which B cells secrete a free form of B receptor called antibody that binds circulating antigen and marks cells or viruses bearing it for destruction.

7. The specificity of T and B receptors for particular antigens reflects the three-dimensional shape of a cleft in the molecule. Slight changes in the amino acid sequence alter the shape of the cleft and thus alter the identity of molecules able to fit into it.

8. Vertebrates can recognize many different antigens as foreign because the bone marrow of vertebrates contains many different stem cells, and each matures independently. During maturation, a stem cell assembles the two genes encoding its particular T and B receptors by splicing together component parts, randomly selecting each part from a large library of possibilities.

9. Vertebrates do not produce T or B receptors targeted against their own tissues because the stem cells that would have produced these antibodies are destroyed or suppressed early in development.

10. AIDS destroys the ability of humans to mount an immune response, by attacking and killing T_4 cells that are necessary to initiate the response.

REVIEW QUESTIONS

1. What are the four primary types of T cells and how is each unique?

2. How do invertebrates recognize foreign tissue? What is a disadvantage of this system?

3. How is the vertebrate negative defense system superior to the invertebrate defense? What is the nature of the vertebrate positive defense system?

4. What cell surface proteins are found on the three kinds of white blood cells?

5. What are the primary events of the cell-mediated immune response? Briefly explain each.

6. What are the primary events of the humoral immune response? Briefly explain each.

7. What is the basic structure of an antibody molecule? In which portion of the molecule does specificity reside? How does the structure of a T receptor compare to that of a B receptor?

8. How is an organism able to produce such a great variety of receptors and receptor-derived antibodies? What process provides for bone marrow stem cell variety? Briefly explain this process.

9. How does embryonic immune tissue respond to self and foreign molecules? What is the process of acquired immunological tolerance? What is natural immunological tolerance? What is the immunological basis for both processes?

10. In what three ways does AIDS interfere with the immune response? Is AIDS the direct cause of death in its victims? Why or why not?

THOUGHT QUESTIONS

1. Why do you suppose the human immune system encodes only a few hundred V genes and a few D and J genes, when much more diversity could be generated by encoding 1000 copies of each?

2. The African green monkeys from which the AIDS virus is thought to have arisen do not suffer from AIDS. How do you imagine they have escaped this?

3. Some breeding programs in zoos have deliberately sought to obtain highly inbred lines with individuals as genetically alike as possible, and other programs have unavoidably produced such highly inbred lines because of a shortage of wild-captured individuals. Such inbred individuals possess perfectly functional immune systems and yet are sometimes subject to disastrous outbreaks of disease. Why?

FOR FURTHER READING

ADA, G.L. and G. NOSSAL: "The Clonal Selection Theory," *Scientific American*, August 1987, p. 62. Lucid description of one of the most important concepts in immunology.

COHEN, I.: "The Self, the World, and Autoimmunity," *Scientific American*, April 1988, pages 52-60. An up-to-date description of disorders in which the immune system attacks normal healthy tissue.

DONELSON, J., and M. TURNER: "How the Trypanosome Changes Its Coat," *Scientific American*, February 1985, pages 44-51. An easily understood discussion of how parasites responsible for sleeping sickness use transposition to avoid immune surveillance.

ESSEX, M. and P.J. KANKI: "The Origins of the AIDS Virus," *Scientific American*, October, 1987, p. 64. A discussion of how AIDS-related HIV viruses interact with human beings and monkeys and how some HIV viruses seem to have evolved toward disease-free coexistence with their hosts.

GALLO, R., and L. MONTAGNIER: "AIDS in 1988," *Scientific American*, October 1988, pages 40-51. The two investigators who discovered the cause of AIDS assess where AIDS research stands in 1988.

HASELTINE, W.A. and F. WONG-STAHL: "The Molecular Biology of the AIDS Virus," *Scientific American*, October, 1988, p. 52. Shows how just three viral genes direct the machinery of a cell infected with the AIDS virus.

JARET, P.: "Our Immune System: The Wars Within," *National Geographic*, June 1986, pages 702-736. A very readable account of current progress in the study of the human immune system, with striking photographs by Lennart Nilsson.

LAWRENCE, J.: "The Immune System in AIDS," *Scientific American*, December 1985, pages 84-93. An excellent overview of how T cells function in the immune response and how AIDS thwarts that response.

LEDER, P.: "The Genetics of Antibody Diversity," *Scientific American*, May 1982, pages 102-116. An account of how a few hundred genes are shuffled to make millions of antibody combinations, by the man who first worked it out.

LERNER, R., and A. TRAMONTANO: "Catalytic Antibodies," *Scientific American*, March 1988, pages 58-70. A remarkable advance that promises to couple the catalytic power of enzymes to the incredible specificity of antibodies.

MARCHOLONIS, J. and S. SCHLUTER: "Origins of Immunoglobulins and Immune Recognition Molecules," *BioScience*, vol. 40, pages 738-768, 1990. Some immune recognition molecules emerged early in evolution, whereas others are late-comers that occur only in particular phyla.

MARRACK, P., and J. KAPPLER: "The T Cell and Its Receptor," *Scientific American*, February 1986, pages 36-45. An account of the discovery of the T cell receptor, and what is known so far of its structure and function.

MILSTEIN, C.: "Monoclonal Antibodies," *Scientific American*, October 1980, pages 66-74. The fusion of particular antibody-producing cells with cancer cells produces clones of cells that secrete antibody directed at only one antigen.

ROITT, I., J. BROSTOFF, and D. MALE: *Immunology*, The C.V. Mosby Co., St. Louis, 1985. A good introductory text with a medical slant and excellent illustrations.

SMITH, K.A.: "Interleukin-2," *Scientific American*, March 1990, pages 50-57. The first hormone of the immune system to be recognized, it helps the body to mount a defense against microorganisms by triggering the multiplication of only those cells which attack an invader.

TONEGAWA, S.: "The Molecules of the Immune System," *Scientific American*, October 1985, pages 122-131. An account of what is currently known of the structure of the B and T receptors.

VON BOEHMER, H. and KISIELOW, P.: "How the Immune System Learns About Self," *Scientific American*, October 1991, pages 74-81. Within the thymus, antibody cell directed against "self" are removed—we now know how.

WEBER, J., and R. WEISS: "HIV Infection: The Cellular Picture," *Scientific American*, October 1988, pages 100-109. What we know now about how AIDS viruses identify and enter certain immune-system cells.

YOUNG, J.D. and Z.A. COHN: "How Killer Cells Kill," *Scientific American*, January 1988, p. 38. Discusses the way killer T cells recognize and destroy their target cells.

53

Kidneys and Water Balance

CHAPTER OUTLINE

Osmoregulation
    The problems faced by osmoregulators
    How osmoregulation is achieved
The organization of the vertebrate kidney
    Filtration
    Reabsorption
    Excretion
The evolution of kidneys among the
    vertebrates
    Freshwater fishes
    Marine fishes
    Sharks
    Amphibians and reptiles
    Mammals and birds
How the mammalian kidney works
Excretion of nitrogenous wastes
The kidney as a regulatory organ
    Regulation of kidney function

OVERVIEW

Vertebrates live in salt water, in fresh water, and on land; each of these environments poses different problems for balancing water retention with proper salt concentration. Vertebrates conserve or excrete water, depending on the environment in which they live, by the regulation of the passage of water through their excretory system.

FOR REVIEW

Here are some important terms and concepts that you will encounter in this chapter. If you are not familiar with them, you should review them before proceeding.

Sodium chloride (Chapter 2)

Membrane transport (Chapter 6)

Hormones (Chapter 48)

Nitrogenous wastes (Chapter 49)

Countercurrent exchange (Chapter 50)

The first vertebrates evolved in water, and the physiology of all vertebrates still reflects this origin (Figure 53-1). Approximately two-thirds of every vertebrate's body is water. If the amount of water in the body of a vertebrate falls much lower than this, the animal will die. In this chapter we discuss the various strategies animals employ to keep from gaining or losing too much water. As we shall see, these strategies are closely tied to how vertebrates exploit the varied environments where they occur.

FIGURE 53-1
The kangaroo rat, *Dipodomys panamintensis*, shown here, has very efficient kidneys that can concentrate urine to a high degree by reabsorbing water. As a result, it avoids losing any more moisture than necessary. Since kangaroo rats live in dry or desert habitats, this feature is extremely important to them.

OSMOREGULATION

Plasma membranes are freely permeable to water, but are impermeable to salts and ions except through special channels. This property of differential permeability forms the basis for many life processes, including the nerve conduction that we discussed in the earlier chapters. If the concentration of salts and ions dissolved in the water surrounding a vertebrate's body were the same as that within the body, the differential permeability of the cells would present no problem; there would be no tendency for water to leave or enter the body. In other words, the osmotic pressure of the body fluids would be the same as that of the surroundings. Most marine invertebrates are **osmoconformers**—while they regulate the concentrations of individual ions such as magnesium ion, they maintain the total ionic concentration of their body fluids (osmolality) at the same level as that of the medium in which they are living, and they change the osmolality of their body fluids when the ionic concentration of the medium changes.

The Problems Faced by Osmoregulators

Among the vertebrates, by contrast, only sharks are osmoconformers, and they are only imperfect ones. All other vertebrates are **osmoregulators** (Figure 53-2). An osmoregulator maintains an internal solute concentration that does not vary, regardless of the environment in which the vertebrate lives. The maintenance of a constant internal solute concentration has permitted vertebrates to evolve complex patterns of internal metabolism. Maintaining this concentration, however, does require constant regulation of the animal's internal water level.

Freshwater vertebrates must maintain much higher salt concentrations in their bodies than of those in the water surrounding them. In other words, they are hyperosmotic relative to their environment, and water tends to enter their bodies. They must, therefore, exclude water to prevent self-dilution.

Marine vertebrates have only about one-third the osmotic concentration of the surrounding seawater in their bodies. They are therefore said to be hypoosmotic relative to their environments; water tends to leave their body. These animals must retain water to prevent dehydration.

On land where they are surrounded by air, the bodies of vertebrates have a higher concentration of water than the air surrounding them does. They therefore tend to

FIGURE 53-2
The concentration of ions is roughly similar in the bodies of different classes of vertebrates. Sharks hold the concentration of solutes in their blood at about the level in seawater, or at a slightly higher level, by adding urea to their bloodstream. In contrast, the body fluids of marine bony fishes contain a far lower ion concentration than the one characteristic of their seawater environment; they are said to be hypoosmotic with respect to seawater. Consequently, such fishes tend to lose water by osmosis and must struggle to retain as much water as possible. Freshwater fishes have the opposite problem: their body fluids contain far higher ion concentrations than the water in which they live. Such fluids are said to be hyperosmotic with respect to the surrounding water. Terrestrial vertebrates have ion concentrations not unlike those of the fishes from which they evolved.

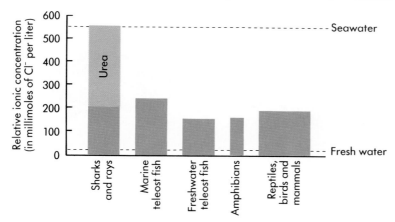

lose water to the air by evaporation. The amphibians, which live on land only part of the time, face this situation to some degree. This situation is faced by all reptiles, birds, and mammals, which must conserve water to prevent dehydration.

How Osmoregulation Is Achieved

Animals have evolved a variety of mechanisms to cope with these problems of water balance, all of them based in one way or another on the animal's excretory system. In many animals the removal of water or salts is coupled with the removal of metabolic wastes from the body. Simple organisms, such as many protists and sponges, employ contractile vacuoles for this purpose. Many freshwater invertebrates employ **nephrid organs,** in which water and waste pass from the body across a membrane into a collecting organ, from which they are ultimately expelled to the outside through a pore. The membrane acts as a filter, retaining proteins and sugars within the body while permitting water and dissolved waste products to leave.

Insects use a similar filtration system, with a significant improvement that helps them guard against water loss. The excretory organs in insects are the **Malpighian tubules** (Figure 53-3). Malpighian tubules are tubular extensions of the digestive tract that branch off before the hindgut. Potassium ions are secreted into the tubules, causing body water and organic wastes to flow into them from the body's circulatory system because of the osmotic gradient. Blood cells and protein molecules are too large to pass across the membrane into the Malpighian tubules, so that the circulatory system is not short-circuited. Because the system of tubules empties into the hindgut, however, the water and potassium can be reabsorbed by the hindgut, and only small molecules and waste products are excreted. Malpighian tubules provide a very efficient means of water conservation.

Like the insects, the vertebrates use a strategy that couples water balance and salt concentration with waste excretion. Instead of relying on the secretion of salts into the excretory organ to establish an osmotic gradient, however, vertebrates rely on pressure-driven filtration. Whereas insects use an osmotic gradient to *pull* the blood through the filter, the vertebrates *push* the blood through the filter, using the higher blood pressure that a closed circulatory system makes possible. Fluids are forced through a membrane that retains both proteins and large molecules within the blood vessel, but allows the small molecules to exit. Water is then reabsorbed from the filtrate as it passes through a long tube.

Like the osmotically collected waste fluid of insects, the fluid filtered in vertebrates contains many small molecules that are of value to the organism, such as glucose, amino acids, and vitamins. Vertebrates have evolved a means of selectively reabsorbing these valuable small molecules without absorbing the waste molecules that are also dissolved in the filtered waste fluid, urine. Selective reabsorption gives the vertebrates great flexibility, since the membranes of different groups of animals can evolve different active transport channels and thus the ability to reabsorb different

FIGURE 53-3

Malpighian tubules. The Malpighian tubules of insects are extensions of the hindgut that collect fluid from the body's circulatory system. Water is later reabsorbed, while wastes in the fluid are eliminated.

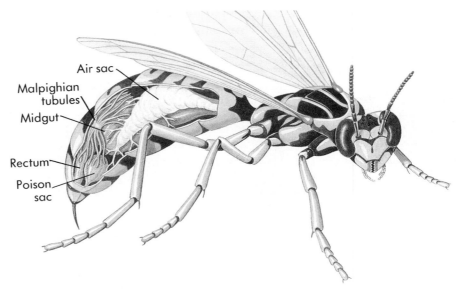

Air sac
Malpighian tubules
Midgut
Rectum
Poison sac

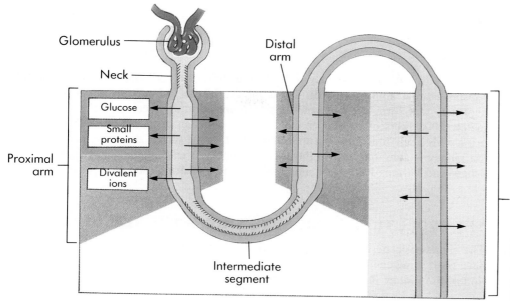

FIGURE 53-4

The basic organization of the vertebrate nephron. The nephron tube of the freshwater fish is a basic design that has been retained in the kidneys of marine fishes and terrestrial vertebrates, which evolved later. Sugars, small proteins, and ions with two or more ionic bonds (divalent ions), such as Ca^{++} and PO_4^-, are recovered in the proximal arm; ions with a single ionic bond (monovalent ions), such as Na^+ and Cl^-, are recovered in the distal arm; and water is recovered in the collection duct.

molecules. This flexibility is a key factor underlying the ability of different vertebrates to function in many diverse environments. They can reabsorb small molecules that are especially valuable in their particular habitat and not reabsorb wastes. Among the vertebrates, the apparatus that performs filtration and reabsorption is the **kidney.** It can function with modifications in fresh water, in the sea, and on land.

THE ORGANIZATION OF THE VERTEBRATE KIDNEY

The kidney is a complex structure of repeating elements called **nephrons,** each of which has a tubular and a vascular component. The outer layer of the kidney is called the cortex, and the inner region, the medulla. From the medulla a series of converging tubes leads to the ureter and, ultimately, to the bladder. All vertebrate kidneys carry out three functions:

1. *Filtration,* in which blood is passed through a filter that retains blood cells and proteins but allows passage of water and small molecules such as amino acids, glucose, and salts.
2. *Reabsorption,* in which desirable ions and metabolites are recaptured from the filtrate, leaving metabolic wastes such as urea and water behind for later elimination. Organisms have expended a great deal of energy to obtain these molecules, and they need to be reclaimed.
3. *Excretion,* in which the kidney first secretes and then excretes K^+, H^+, NH_4^+, and certain drugs and foreign organic materials.

Filtration

The filtration device of the vertebrate kidney consists of a large number of individual tubular filtration-reabsorption devices called nephrons (Figure 53-4). At the front end of each nephron tube is a filtration apparatus called a **Bowman's capsule.** In each, an arteriole enters and splits into a fine network of vessels called a **glomerulus** (Figure 53-5). It is the walls of these capillaries that act as a filtration device. Blood pressure forces fluid through the capillary walls, which are differentially permeable. These walls withhold the proteins and other large molecules, while passing water and small molecules such as glucose, ions, and ammonia, the primary nitrogenous waste product of metabolism.

Reabsorption

At the back of each nephron tube is a reabsorption device that operates like the mammalian small intestine, discussed in Chapter 49. The fluid that passes out of the

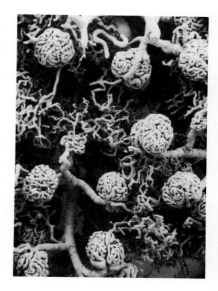

FIGURE 53-5

Bowman's capsules. The spherical structures in this micrograph are Bowman's capsules. In each, a fine network of capillaries, the glomerulus, is connected to a nephron tube by an arteriole.

capillaries of each glomerulus, called the **glomerular filtrate,** enters the tube portion of that nephron within which reabsorption takes place.

Excretion

A major function of the kidney is the elimination of a variety of potentially harmful substances that animals eat, drink, or inhale. In this process the roughly 2 million nephrons that form the bulk of the two human kidneys receive a flow of approximately 2000 liters of blood per day. The mammalian kidney is able to concentrate urine with a salt concentration well above that of the blood. As a part of the process, the kidney is able to build up the concentration of such materials as H^+, K^+, NH_4^+, drugs, and various foreign organic materials in the urine, and thus to excrete them from the body.

THE EVOLUTION OF KIDNEYS AMONG THE VERTEBRATES

The same basic design has been retained in all vertebrate kidneys, although there have been some changes. In the following sections we discuss the structure and function of the kidney in the different major groups of vertebrates.

Freshwater Fishes

Kidneys are thought to have evolved first among the freshwater fishes. A freshwater fish drinks little and produces large amounts of urine. Because the body fluid of a freshwater fish is hyperosmotic as compared with the water in which the fish lives, water is not reabsorbed in its nephrons. The excess water that enters its body passes instead through the nephron tubes to the bladder, from which it is eliminated as urine. Within the urine is not only the excess water but also all the small molecules that were not reabsorbed while passing through the nephron tubes. Notable among these molecules is ammonia, the principal metabolic waste product of nitrogen metabolism (present in solution as ammonium ion, NH_4^+). Ammonia in higher concentrations is toxic, but because the urine contains so much water, the concentration of ammonia is low enough not to harm the fish.

Marine Fishes

Although most groups of animals clearly seem to have evolved first in the sea, marine bony fishes probably evolved from freshwater ancestors, as was mentioned in Chapter 19. In making the transition to the sea, they faced a significant new problem of water balance, because their body fluids are hypoosmotic with respect to the water that surrounds them. For this reason, water tends to leave their bodies, in which the fluids are less concentrated osmotically than is seawater. To compensate, marine fishes drink a lot of water, excrete salts instead of reabsorbing them, and reabsorb water. This places radically different demands on their kidneys than those faced by freshwater fishes, thus turning the tables on an organ that had originally evolved to eliminate water and reabsorb salts. As a result, the kidneys of marine fishes have evolved important differences from their freshwater relatives. For example, they possess active ion transport channels that conduct ions with two or three charges—molecules that are particularly abundant in seawater—*out* of the body and *into* the tube. In the sea, the water that the fish drinks is rich in ions such as Ca^{++}, Mg^{++}, $SO_4^=$, and $PO_4^=$, all of which must be excreted. Because of the functioning of these ions transport channels, the direction of movement is reversed compared with that found in the kidneys of their freshwater ancestors.

Sharks

Except for one species of shark found in Lake Nicaragua, all sharks, rays, and their relatives live in the sea (Figure 53-6). Some of these members of the class Chondrichthyes have solved the osmotic problem posed by their environment in a different way

FIGURE 53-6
A great white shark. Although the seawater in which it is swimming contains a far higher concentration of ions than does the shark's body, the shark avoids losing water osmotically by maintaining such a high concentration of urea that its body fluids are osmotically similar to the sea around it.

than have the bony fishes. Instead of actively pumping ions out of their bodies through their kidneys, these fishes use their kidneys to reabsorb the metabolic waste product urea, creating and maintaining urea concentrations in their blood 100 times as high as those that occur among the mammals. As a result, the sharks and their relatives become isotonic with the surrounding sea. They have evolved enzymes and tissues that tolerate these high concentrations of urea. Because they are isotonic with the water in which they swim, they avoid the problem of water loss that other marine fishes face. Since sharks do not need to drink large amounts of seawater, their kidneys do not have to remove large amounts of divalent ions from their bodies.

Amphibians and Reptiles

The first terrestrial vertebrates were the amphibians; the amphibian kidney is identical to that of the freshwater fishes. This is perhaps not surprising, since amphibians spend a significant portion of their time in fresh water; and when on land, they generally stay in wet places.

Reptiles, on the other hand, live in diverse habitats, many of them very dry. Reptiles living mainly in fresh water, like some of the crocodilians, occupy a habitat similar to that of the freshwater fishes and amphibians and have similar kidneys. Marine reptiles, which consist of some crocodilians, turtles, sea snakes, and a few lizards, possess kidneys similar to those of their freshwater relatives. They eliminate excess salts not by kidney excretion but by means of salt glands located near the nose or the eye.

The terrestrial reptiles reabsorb much of the water in kidney filtrate before it leaves the kidneys. This urine, however, cannot become any more concentrated than the blood plasma; otherwise, the body water would simply flow into the urine while it was in the kidneys.

Ammonia is toxic and must be excreted in a dilute form. This would cause excessive fluid loss in a terrestrial organism. The solution is to convert ammonia to a less toxic substance. In the relatively concentrated urine of most reptiles, therefore, nitrogenous waste is no longer excreted in the form of ammonia, but either as the solid urea or as uric acid. These metabolic conversions take place in the liver.

Mammals and Birds

Your body possesses two kidneys, each about the size of a small fist, located in the lower back region (Figure 53-7). These kidneys represent a great increase in efficiency compared with those of reptiles. Mammalian, and to a lesser extent avian, kidneys can remove far more water from the glomerular filtrate than can the kidneys of reptiles and amphibians. Human urine may be as much as 4.2 times as concentrated as blood plasma. Some desert mammals achieve even greater efficiency: a camel's urine

FIGURE 53-7
The urinary system of the human
female.

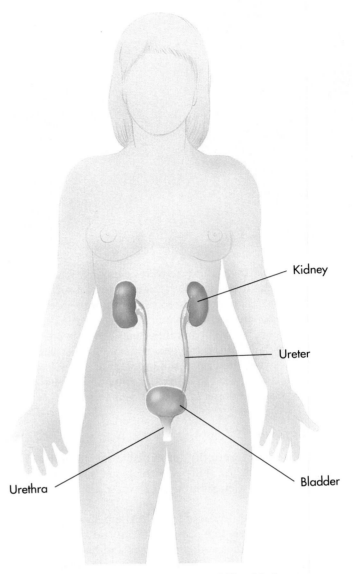

is eight times as concentrated as its plasma, a gerbil's, 14 times as concentrated; and the pocket mouse *Perognathus*, which lives in deserts, has urine 22 times as concentrated as its blood plasma. Mammals and birds achieve this remarkable degree of water conservation by using a simple but superbly designed mechanism: they greatly increase the local salt concentration in the tissue through which the nephron tube passes and then use this osmotic gradient to draw the water out of the tube.

> **The basic design of the kidney of the freshwater fish has been maintained in other vertebrates, with changes in the orientation of ion channels and in the positioning of the nephron segments.**

HOW THE MAMMALIAN KIDNEY WORKS

Remarkably, mammals and birds have brought about this major improvement in efficiency by a very simple change—they bend the nephron tube. A single mammalian kidney contains about a million nephrons. Each of these is composed of a glomerulus, which is connected to a nephron tube, called a **renal tubule** in mammals. Between its proximal and distal segments, each renal tubule is folded into a hairpin loop called the **loop of Henle** (Figure 53-8).

The kidney uses the hairpin loop of Henle to set up a countercurrent flow. Just as in the gills of a fish discussed in Chapter 50 (but with water instead of oxygen being

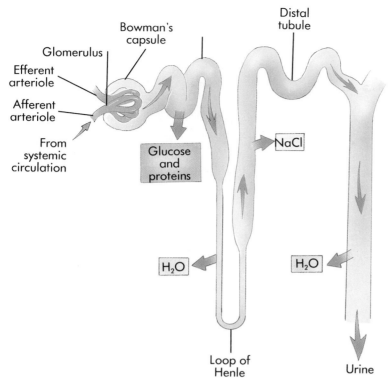

FIGURE 53-8

Organization of a mammalian renal tubule. The glomerulus is enclosed within a filtration device called Bowman's capsule. Blood pressure forces liquid through the glomerulus and into the proximal segment of the tubule, where glucose and small proteins are reabsorbed from the filtrate. The filtrate then passes through a double-loop arrangement consisting of the loop of Henle and the collecting duct, which act to remove water from the filtrate. The water is then collected by blood vessels and transported out of the kidney to the systemic (body) circulation.

reabsorbed) countercurrent flow enables water reabsorption to occur with high efficiency. In general, the longer the hairpin loop, the more water can be reabsorbed. Animals such as desert rodents that have highly concentrated urine have exceptionally long loops of Henle. The countercurrent process involves the passage of two solutes across the membrane of the loop: salt (NaCl) and urea, a waste product of nitrogen metabolism. It has long been known that animals fed high-protein diets, yielding large amounts of urea as waste products, can concentrate their urine better than animals excreting lower amounts of urea, a clue that urea plays a pivotal role in kidney function. The countercurrent process can be considered as occurring in five stages (Figure 53-9):

1. Filtrate from the glomerulus passes down the descending arm of the loop. The walls of this portion of the tubule are impermeable to either salt or urea but are freely permeable to water. Because (for reasons we will later discuss) the surrounding tissue has a high osmotic concentration of urea, water passes out of the descending arm by osmosis, leaving behind a more concentrated filtrate.

2. At the turn of the loop, the walls of the tubule become permeable to salt but much less permeable to water. As the concentrated filtrate passes up the ascending arm, salt diffuses into the surrounding tissue (the surrounding tissue, though it has lots of urea, does not contain as much salt as the concentrated filtrate). This makes salt more concentrated at the bottom of the loop.

3. Higher in the ascending arm, the walls of the tubule contain active-transport channels that pump out even more salt. This active removal of salt from the ascending loop encourages even more water to diffuse outward from the filtrate. Left behind in the filtrate is the urea that initially passed through the glomerulus as nitrogenous waste; eventually the urea concentration becomes very high in the tubule.

4. The tubule empties into a collecting duct that passes back through the tissue; unlike the tubule, the lower portions of the collecting duct are permeable to urea. During this final passage, the concentrated urea in the filtrate diffuses out into the surrounding tissue, which has a lower urea concentration. A high urea concentration in the tissue results, which is what caused water to move out of the filtrate by osmosis when it first passed down the descending arm.

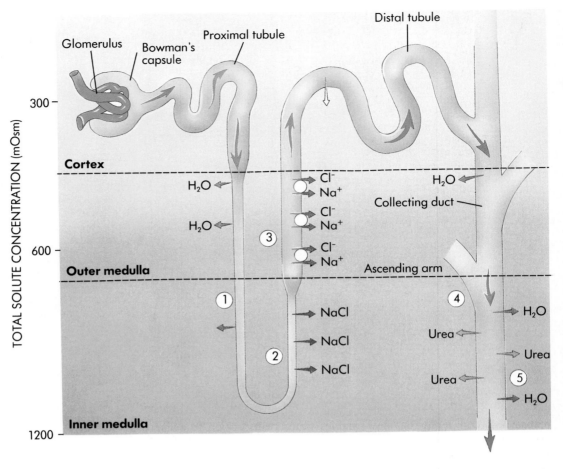

FIGURE 53-9
The flow of materials in the human kidney.

5. As the filtrate passes down the collecting duct, even more water passes outward by osmosis, since the osmotic concentration of the surrounding tissue reflects both the urea diffusing out from the duct *and* the salt diffusing out from the ascending arm. This sum is greater than the osmotic concentration of urea in the filtrate (the salt has already been removed).

In effect, the interior of the kidney is divided into two zones (Figure 53-10). The outer portion of the kidney, called the **cortex,** contains the upper portion of the loop, including the upper ascending arm where reabsorption of salt from the filtrate by active transport occurs. The inner portion of the kidney, called the **medulla,** contains both the lower portion of the loop and the bottom of the collecting duct, which is permeable to urea.

FIGURE 53-10
Structure of the human kidney.
The cortex has an osmotic concentration like that of the rest of the body. The outer medulla has a somewhat higher osmotic concentration, primarily salt. The osmotic concentration within the inner medulla becomes progressively higher at greater depths; this part of the kidney maintains substantial concentrations of urea, which is a major component of its high osmotic concentration.

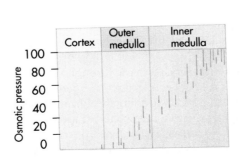

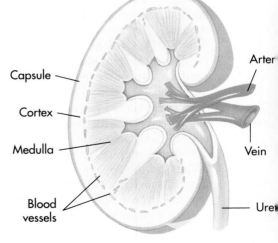

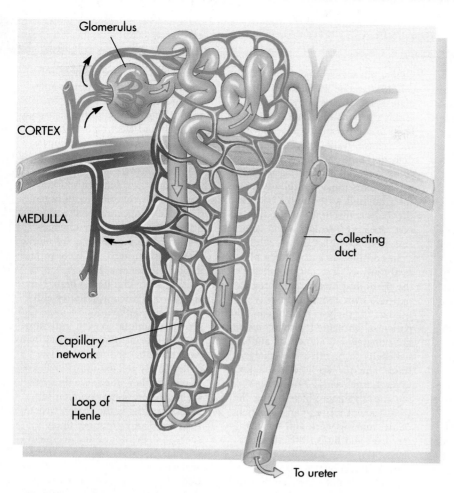

Glomerulus

CORTEX

MEDULLA

Collecting duct

Capillary network

Loop of Henle

To ureter

FIGURE 53-11
How the blood supply collects water from the kidney.

The active reabsorption of salt in the cortex of the kidney drives the process. This salt reabsorption from the filtrate in one arm of the loop establishes a gradient of salt concentration, with concentrations higher in the medulla at the bottom of the loop. It is this high salt concentration that raises the total tissue osmotic concentration so high that water passes by osmosis out of the collecting duct. Just as important is the active reabsorption of salt in the cortex concentrating the renal filtrate with respect to urea. This high urea concentration in the filtrate causes urea to diffuse outward into surrounding tissue in the only zone where it is able to do so—the lower collecting duct—creating a high urea concentration in the medulla. This high concentration causes water to diffuse out from the filtrate in the initial descending arm. The water is then collected by blood vessels in the kidney (Figure 53-11), which carry it into the body's systemic circulation.

> **The mammalian kidney achieves a high degree of water reabsorption by using the salts and urea in the glomerular filtrate to increase the osmotic concentration of the kidney tissue. This facilitates the movement of water from the filtrate out into surrounding tissue, where it is collected by blood vessels impermeable to the high urea concentration but permeable to water.**

EXCRETION OF NITROGENOUS WASTES

All animals contain proteins and nucleic acids, which are both nitrogen-containing molecules. When a heterotroph consumes another animal, the proteins and nucleic acids are broken down. In Chapters 48 and 49 we discussed the functioning of the liver. In the liver, enzymes break down amino acids by removing the amino group (NH_2), and combine it with H^+ ions to form ammonia (NH_3); the remainder of the amino acid is then converted to sugar or lipid. Ammonia is quite toxic to all cells, and so it is necessary to dispose of the ammonia produced during the breakdown of pro-

There are many ways kidneys fail. The most common diseases affecting the kidneys are acute or chronic infection, long-standing diabetes, untreated prolonged high blood pressure, and damage by misdirected reactions of the body's own immune system (autoimmune kidney disease). When kidneys stop working toxic waste materials such as urea and H^+ ions accumulate in the plasma. In addition, electrolyte levels rapidly depart from their normal physiological values. If the kidneys fail there are only two treatment options: dialysis and kidney transplant.

Dialysis is a method of removing toxic wastes from the blood, effectively substituting a machine for the kidney's homeostatic mechanisms. There are two ways dialysis is performed. In **hemodialysis** tubes called catheters are surgically inserted into an artery and a vein, usually on the lower arm. These catheters are equipped with valves. Every few days the affected individual goes to a clinic where these catheters are connected to a dialysis machine. Collected blood passes from the patient's artery into the dialysis machine and then back into a vein. Inside the dialysis machine the blood passes through a disposable unit consisting of many hollow fibers surrounded by a thin cellulose acetate membrane. This dialysis membrane allows waste materials and ions that have accumulated in the plasma to pass down their concentration gradients into a dialyzing fluid that has the same composition as normal plasma. Dialysis patients must carefully manage their salt and water intake because the dialysis machine, unlike the kidney, does not regulate blood volume and total body Na^+.

In **continuous ambulatory peritoneal dialysis,** a recent modification, the peritoneal membrane lining the patient's own abdominal cavity is used as the dialysis membrane. The patient is implanted with an abdominal catheter that allows the abdominal cavity to be filled with dialysis fluid. The dialysis fluid is changed several times daily. This procedure has the advantages of reducing the frequency of visits to an outpatient clinic, making work and travel easier. Diet and fluid restrictions are fewer, but great care is needed to prevent infections.

Dialysis is not a permanent solution. Not only is dialysis expensive but also it does not fully substitute for a normal kidney. A single healthy kidney can meet the entire homeostatic needs of the body.

Therefore a more permanent solution to renal failure is transplantation of a kidney from a healthy donor. Because the kidney can regulate fluid and electrolyte homeostasis without any neural innervation, a kidney transplant cures the recipient of renal disease.

The only problem with kidney transplant is common to all organ transplants: rejection of the transplanted kidney by the recipient's immune system. All cells are marked with "self-markers" on their surfaces to prevent the attack of one's own cells by their immune system. These are called **histocompatibility antigens.** The combination of these antigens is as unique to each individual as a fingerprint. Only identical twins have the same self-markers. The more closely related individuals are to one another, the more likely they are to possess some common self-antigens. This is the reason that tissue transplants are more likely to succeed if the donor and recipient are matched with respect to these antigens. To reduce chances of rejection, the recipient is treated with drugs that suppress the immune system. However, such drugs increase the risk of infections and may be toxic to the liver and bone marrow.

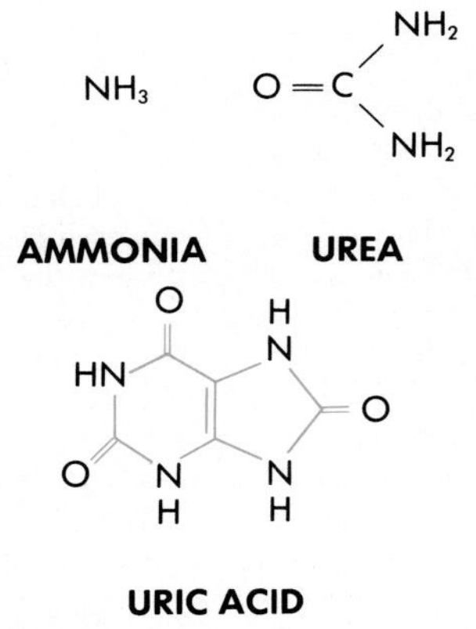

AMMONIA   UREA

URIC ACID

FIGURE 53-12
Wastes. When amino acids are metabolized, the immediate by-product is ammonia, which is quite toxic. Mammals convert ammonia to urea, which is less toxic. Animals that lay eggs convert it instead to uric acid, which is insoluble.

tein, nucleic acids, and other nitrogen-containing molecules. Because even very low concentrations of ammonia can kill cells, it is necessary to transport it in very dilute solution.

The need to transport and excrete ammonia in very dilute solution has posed a major evolutionary challenge in situations where water conservation is important. In freshwater fishes, who have if anything too much water, dilution presents no problem, and toxic ammonia is flushed out in highly diluted form. In salt water fishes and terrestrial animals, however, water is precious and it is not practical to excrete highly diluted ammonia. Three general solutions to this problem have been adopted:

1. *Flushing.* Both freshwater and salt water fishes carry out protein breakdown in the gills, so that little ammonia actually enters the body; it is simply carried away in the water passing across the gills.

2. *Detoxification.* Most mammals and many other land animals convert ammonia to **urea,** which is far less toxic and so can be transported and excreted at far higher concentrations. The urea is carried by the bloodstream to the kidneys, where it is excreted as a principal component of urine.

3. *Insolubilization.* Birds and terrestrial reptiles face a special problem: their eggs are encased within shells, and so metabolic wastes build up as the embryo grows within the egg. The solution is to convert ammonia to **uric acid,** which is insoluble (Figure 53-12). The process, while lengthy and requiring consider-

able energy, produces a compound that crystallizes and precipitates as it becomes more concentrated. Adult birds and terrestrial reptiles carry out the same process, excreting the final result as a semi-solid paste called guano.

The metabolic breakdown of proteins produces ammonia as a by-product. Because ammonia is very toxic, animals have devised a variety of strategies for removing it.

THE KIDNEY AS A REGULATORY ORGAN

The kidney plays a very important role in regulating the composition of the blood, and so the internal chemical environment of the body. By selectively removing substances from the blood, it is able to maintain rigorous control over the concentrations of ions and other chemicals. Thus, while almost all amino acids are retained by the kidney, almost half of the urea in entering blood is eliminated. In normal individuals no glucose is eliminated, but in diabetics excess glucose is excreted. Concentrations of ions such as H^+, Na^+, K^+, Cl^-, Mg^{++}, Ca^{++}, and HCO_3^- are all maintained within narrow boundaries. This serves to maintain the blood's pH at a constant value, as well as maintain the proper ion balances for nerve conduction (blocked, for example, by even a small excess of Mg^{++}) and muscle contraction (a small increase in K^+ levels in the blood causes the heart to stop).

The kidney is as much a regulatory organ as it is an excretory organ.

Regulation of Kidney Function

Like most organs concerned with **homeostasis**—the maintenance of constant physiological conditions within the body—the operation of the kidney is regulated by the central nervous system; it uses the voluntary, autonomic, and hormonal controls we have discussed in previous chapters.

Why is regulating the operation of the kidney necessary? The proper operation of the body's many organ systems requires that the osmotic concentration of the blood be maintained within narrow bounds. For this reason, it is not always desirable for your body to retain the same amount of water. If you have consumed an unusually large amount of water, for example, the balance of ions in the blood can only be preserved if you retain less of the water than you would otherwise.

Control of water and salt balance is centered in the hypothalamus. The hypothalamus produces the hormone vasopressin (antidiuretic hormone), which renders the collecting ducts of the kidneys freely permeable to urea, which passes out into the interstitial fluid of the deep medulla and thus maximizes water retention by causing water to pass out of the descending loop into the surrounding tissue. When the up-

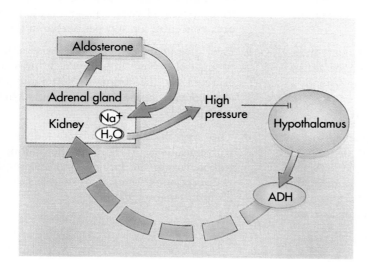

FIGURE 53-13
Control of water and salt balance within the kidney is centered in the hypothalamus. The hypothalamus produces antidiuretic hormone (ADH), which renders the collecting ducts of the kidneys freely permeable to urea and so maximizes water retention. If too much water retention leads to high blood pressure, pressure-sensitive receptors in the hypothalamus detect this and cause the production of ADH to be shut down. If the level of sodium in the blood falls, the adrenal gland initiates production of the hormone aldosterone, which stimulates salt reabsorption by the renal tubules of the kidney. (Feedback loops are noted by the colored lines with the double bar.)

take of water is excessive, blood pressure rises and pressure-sensitive neurons in the atria detect this rise. In addition, osmoreceptors in the hypothalamus detect a decrease in osmolarity and decrease the hypothalamus output of vasopressin (Figure 52-13). A decrease in levels of vasopressin renders the collecting ducts of the renal tubules less permeable to water, lowering the osmotic concentration within the medulla, and in that way inhibits the reabsorption of water and increases urine volume. Alcohol consumption is one factor that suppresses vasopressin secretion.

Another example of the central nervous system regulating the kidney concerns the salt balance in your body (see Figure 53-13). The amount of salt in your diet can vary considerably, and yet it is important to many physiological processes that salt levels in your blood not vary widely. When levels of sodium ion in the blood fall, the adrenal gland increases its production of the hormone **aldosterone,** a steroid hormone that stimulates active sodium ion reabsorption across the walls of the ascending arms of your kidneys' renal tubules. In this way, it decreases the amount of sodium that is lost in the urine. In the total absence of this hormone, humans may excrete up to 25 grams of salt a day.

SUMMARY

1. Animals that live in fresh water tend to gain water from their surroundings, whereas those that live in salt water or on land tend to lose water.

2. Insects solve the problem of dehydration by conserving water. In particular, they avoid excreting water along with body wastes. They pump ions into Malpighian tubules, so that body fluids are drawn in by osmosis. These body fluids contain the wastes created by metabolism. The wall of the tubule acts as a filter, passing wastes but retaining proteins and blood cells. Insects then reabsorb the water and useful metabolites back out, leaving the wastes behind to be excreted.

3. Instead of sucking body liquid through a filter as insects do, vertebrates push it through. They are able to do this because they possess a closed circulation system that operates under considerable pressure.

4. The vertebrate kidney is composed of many individual units called nephrons, each made up of two segments: the first, a filter and the second, a reabsorption tube.

5. Freshwater fishes do not reabsorb water; their bodies already gain too much by direct diffusion from the water. Marine fishes excrete the salts in the water they drink and reabsorb water. Some marine vertebrates, notably sharks, maintain high body levels of urea so that they are isotonic with the sea and do not tend to gain or lose water.

6. Amphibians and reptiles have kidneys much like those of freshwater fishes. Birds and mammals achieve much greater water conservation by bending the nephron tube, producing what is known as the loop of Henle. This creates a countercurrent flow, which greatly increases the efficiency of water reabsorption.

7. The longer the loop of Henle, the greater the osmotic concentration that can be achieved and the more water that can be reclaimed from the urine.

8. The mammalian kidney achieves a high degree of water reabsorption by using the salts and urea in the glomerular filtrate to increase the osmotic concentration of the kidney tissue. This facilitates the movement of the water from the filtrate out into the surrounding tissue, where it is collected by blood vessels impermeable to the high urea concentration but permeable to the water.

9. Most mammals convert ammonia, a toxic byproduct of protein breakdown, to urea, which is excreted by the kidneys. Vertebrates that lay eggs convert ammonia to uric acid, which is insoluble.

10. The hypothalamus regulates kidney function with hormones. ADH raises the osmotic concentration within the kidney, and so promotes the reabsorption of water. Aldosterone stimulates sodium ion reabsorption.

REVIEW QUESTIONS

1. What is the salt concentration of a freshwater vertebrate compared to its environment? Does water therefore tend to enter or exit its body? What must it do to maintain proper body water levels?

2. What is the salt concentration of a marine vertebrate compared to its environment? Does water therefore tend to enter or exit its body? What must it do to maintain proper body water levels? What special considerations regarding water do terrestrial vertebrates face?

3. How does the vertebrate excretory system compare to the insect system? What is selective reabsorption as it relates to the vertebrate excretory system? What kind of flexibility does this confer to the vertebrates?

4. In what class of animals did the kidney first evolve? What are the three basic functions of this kidney? Explain each.

5. What is a nephron? Where is the Bowman's capsule located? What is the glomerulus? Where is it located? How does filtration occur between these two structures?

6. What is the glomerular filtrate? What portion of the nephron is associated with reabsorption? What occurs in each of the five divisions of the nephron?

7. How does the efficiency of water reabsorption in a bird or mammal compare to that in a terrestrial reptile? Why? What is the anatomical basis for this advancement?

8. What physiological process occurs in the loop of Henle? Describe the five stages of this process.

9. How is the water that diffuses into the medulla of the kidney returned to the vascular system?

10. How does the hypothalamus detect changes in blood solute concentration? What hypothalamic hormone regulates kidney function? How is this accomplished? What is the result of decreasing the levels of this hormone in terms of kidney physiology and urine output?

THOUGHT QUESTIONS

1. In the mammalian kidney, water is reabsorbed from the filtrate across the wall of the collecting duct into the salty tissue near the bottom of the loops of Henle; blood vessels then take the water away. Why doesn't the blood in these vessels become very salty?

2. If you are lost in the desert with a case of liquor and are desperately thirsty, should you drink the liquor? Explain your answer.

3. In the Bowman capsule, fluid is forced out of the capillaries by blood pressure. Why does this not happen in the gills of freshwater fish?

FOR FURTHER READING

BEAUCHAMP, G.K.: "The Human Preference for Excess Salt," *American Scientist*, Volume 75, p. 27, 1987. Discusses salt appetite and the possible relationship of salt consumption to hypertension.

BEEUWKES, R.: "Renal Countercurrent Mechanisms, or How to Get Something for (Almost) Nothing." In C.R. Taylor and others (editors): *A Companion to Animal Physiology*, Cambridge University Press, New York, 1982. A clearly presented summary of current ideas about how the kidney works, with an account of the evidence that led to the rejection of the countercurrent multiplier hypothesis.

HEATWOLE, H.: "Adaptations of Marine Snakes," *American Scientist*, vol. 66, 1978, pages 594-604. Several groups of snakes are able to live in the sea by clever adaptations that modify salt and water balance.

MARSHALL, E.: "Testing Urine for Drugs," *Science*, vol. 241, July 1988, pages 150-152. Urinalysis now provides fast and accurate tests for the prescence of many drugs.

SMITH, H.W.: *From Fish to Philosopher*, Little, Brown & Co., Boston, 1953. A broad and well-written account of the evolution of the vertebrate kidney.

VANDER, A., J. SHERMAN, and D. LUCIANO: *Human Physiology: The Mechanisms of Body Function*, ed. 4, McGraw-Hill Book Co., New York, 1985. This basic physiology text contains a very good chapter on human body fluids and the physiology of the kidney.

54

Sex and Reproduction

CHAPTER OUTLINE

OVERVIEW

Almost all vertebrates reproduce sexually. Sex evolved in the sea, long before the vertebrates, and its modification for organisms living on land has entailed evolutionary innovations to avoid drying out. The solutions vertebrates adopted to the problem of water loss during reproduction include the amniotic egg, which first evolved among the reptiles. Amniotic eggs occur in most terrestrial vertebrates, including the birds and a few primitive mammals. The placental mammals have adopted a different solution, nourishing their developing young within the mother's body. Sex in humans plays an important role in pair bonding as well as in reproduction. A variety of procedures have been developed that permit people to engage in sex while blocking reproduction.

FOR REVIEW

Here are some important terms and concepts that you will encounter in this chapter. If you are not familiar with them, you should review them before proceeding.

Meiosis (Chapter 11)

Evolution of sexual reproduction (Chapter 11)

Adaptation (Chapter 20)

Amniotic egg (Chapter 42)

Mammals (Chapter 42)

Hormones (Chapter 48)

On any dark night you can hear biology happening. The cry of a cat in heat, insects chirping outside the window, frogs croaking in swamps, wolves howling in a frozen northern scene—all are the sounds of evolution's essential act, reproduction. Nor are we humans immune. Few subjects pervade our everyday thinking more than sex; few urges are more insistent. They are no accident, these strong feelings, no trap laid by the Devil to ensnare the innocent in sin. The frog in its swamp knows no sin, only an urgent desire to reproduce itself, a desire that has been patterned within it by a long history of evolution. It is a pattern that we share. For almost all of us, the reproduction of our families spontaneously elicits a sense of rightness and fulfillment (Figure 54-1). It is hard not to return the smile of an infant, not to feel warmed by it and by the look of wonder and delight on the parents' faces. This chapter deals with sex and reproduction among the vertebrates, of which humans are one kind.

FIGURE 54-1
This baby girl is looking for her mommy. It is difficult not to want to pick her up and comfort her.

SEX EVOLVED IN THE SEA

Sexual reproduction first evolved among marine organisms, long before the advent of the vertebrates. The eggs of most marine fishes are produced in batches by the females; when they are ripe, the eggs are simply released into the water. Seawater itself is not a hostile environment for gametes or for young organisms. Fertilization is achieved by the male's release of sperm into the water containing the eggs.

The effective union of free gametes in the sea by **external fertilization** poses a significant problem for all marine organisms. Eggs and sperm become diluted rapidly in seawater, so their release by females and males must be almost simultaneous if successful fertilization is to occur. Because of this necessity, most marine vertebrates restrict the release of their eggs and sperm to a few brief and well-defined periods. There are few seasonal cues in the ocean that organisms can use as signals, but one that is all-pervasive is the cycle of the moon. Each month, the moon approaches closer to the earth at one time; when it does, its increased gravitational attraction causes somewhat higher tides. Many different kinds of marine organisms sense the changes in water pressure that accompany the tides, and much of the reproduction that takes place in the sea is entrained to the lunar cycle. Some marine vertebrates reproduce once a year; others reproduce once every month, timing the development, production, and joint release of their male and female gametes by the lunar clock.

The invasion of the land by creatures from the sea meant that these organisms faced for the first time the danger of desiccation, or drying out. This problem was all the more severe for their small and vulnerable reproductive cells. Obviously the gametes could not simply be released near one another on land; they would soon dry up and perish. There are many possible solutions to this problem. We have already considered the adaptations of the plants, the fungi, and the arthropods to an existence on dry land, including the ways in which the transmittal of their gametes has been modified for successful union under terrestrial conditions. We now consider the vertebrates.

VERTEBRATE SEX AND REPRODUCTION: FOUR STRATEGIES

The five major groups of vertebrates have evolved reproductive strategies that are quite different and that range from external fertilization to the bearing of live young. In large measure these differences reflect different approaches to protecting gametes from desiccation, since four of the five major groups of the vertebrate subphylum are terrestrial.

Fishes

Some vertebrates, of which the bony fishes are the most abundant, have remained aquatic. All fishes reproduce in the water and in essentially the same way as all other aquatic animals. Fertilization in most species is external, the eggs containing only enough yolk material to sustain the developing zygote for a short time. After the initial dowry of yolk has been exhausted, the growing individual must seek its food from the waters around it. Many thousands of eggs are fertilized in an individual mating,

Human beings, like most vertebrates, have two sexes and reproduce exclusively by the union of the male's sperm with the female's egg. A few vertebrates, however, violate this rule.

One of the first cases of unusual modes of reproduction among vertebrates was reported by the Russian biologist Ilya Darevsky in 1958. He observed that some populations of small lizards of the genus *Lacerta* were exclusively female, and suggested that lizards could lay eggs that were viable even if they were not fertilized—virgin birth, or parthenogenesis, reproduction in the absence of sperm. Further work has shown parthenogenesis to occur among populations of other lizard genera, one of which is illustrated in Figure 11-10. Such reproduction is asexual, with offspring being clones—genetically identical replicas of their mothers.

Numerous fish genera contain species in which individuals can change their sex. Among coral reef fishes, for example, both **protogyny** (a change from female to male) and **protandry** (a change from male to female) occur. In fishes that practice protogyny (Figure 54-A, *1*), the sex change appears to be under social control. These fishes commonly live in large groups, or schools, where

successful reproduction is typically limited to one or a few large dominant males. If these large males are removed, the largest female rapidly changes sex, becoming a dominant male.

Deep-sea fishes commonly practice

yet another reproductive variant that is unusual among the vertebrates: they are **hermaphrodites**—both male and female at the same time. Among these fishes, two individuals fertilize each other's eggs (Figure 54-A, *2*)!

1  2

FIGURE 54-A
1 The bluehead wrasse, *Thalassoma bifasciatium*, is protogynous—females sometimes turn into males. Here a large male or sex-changed female is pair mating with a female, typically much smaller.
2 The hamlet bass (genus *Hypoplectrus*) is a deep-sea fish that is simultaneously male and female—a hermaphrodite. In the course of a single pair mating, one fish may alternate sexual roles as many as four times, by turns offering eggs to be fertilized and fertilizing its partner's eggs. Here the fish acting as male curves around the motionless partner, fertilizing the upward-floating eggs.

but few of the resulting zygotes survive the rigors of their aquatic environment and grow to maturity. Some of them succumb to microbial infection, many others to predation. The development of the fertilized eggs is speedy, and the young that survive achieve maturity rapidly.

Although most fish fertilize their eggs externally, fertilization is internal in a few groups of fishes, with the male injecting sperm into the female. Biologists distinguish three kinds of internal fertilization:

1. When the fertilized eggs are laid outside the mother's body to complete their development, the practice is called **ovipary.** Most fishes are oviparous.
2. When the eggs are retained within the mother and complete their development there, with fully developed young eventually hatching from the eggs and being released, the practice is called **ovovivipary.** Mollies, guppies, and mosquito fish are ovoviviparous.
3. Some fishes are even **viviparous,** the undeveloped young hatching within their mother and obtaining further nourishment from her as their development proceeds (Figure 54-2). The young are released fully developed at birth. Hammerhead and blue sharks are among the truly viviparous fishes.

The usual pattern in fishes, however, is external fertilization and development of the eggs.

FIGURE 54-2
Viviparous fishes carry live, mobile young within their bodies while these young complete their development. They are then released as small but competent adults. Here a lemon shark has just given birth to a young shark, which is still attached by the umbilical cord.

Amphibians

A second group of vertebrates, the amphibians (exemplified by the frogs and toads), have invaded the land without fully adapting to the terrestrial environment. The life cycle of the amphibians is still tied to the presence of free water. Among most amphibians, fertilization is still external, just as it is among the fishes and other aquatic animals. Many female amphibians lay their eggs in a puddle or a pond of water. Among the frogs and toads, the male grasps the female and discharges fluid containing the sperm onto the eggs as they are released (Figure 54-3).

In most amphibians, as in most fishes, large numbers of eggs become mature, are released from the ovary, and are fertilized together. It was only with the advent of reptiles, which enclose fertilized eggs within an amniotic membrane and devote considerable metabolic resources to each egg, that fewer eggs are produced at one time.

The development time of the amphibians is much longer than that of the fishes, but amphibian eggs do not include a significantly greater amount of yolk. Instead, the process of development consists of two distinct components, a larval and an adult stage, in a way reminiscent of some of the life cycles found among the insects.

The development of the aquatic larval stage of the amphibians is rapid, utilizing yolk supplied from the egg. The larvae of most amphibians then function, often for a considerable period of time, as independent food-gathering machines. They scavenge nutrients from their environments and often grow rapidly. Tadpoles, which are the larvae of frogs and toads, grow in a matter of weeks from creatures no bigger than the tip of a pencil into individuals as big as goldfish. When an individual larva has grown to a sufficient size, it undergoes a developmental transition, or **metamorphosis,** into the terrestrial adult form.

> The invasion of land by vertebrates was initially tentative, with the amphibians maintaining the external fertilization and modes of reproduction that are characteristic of most of the fishes, the group from which they evolved.

FIGURE 54-3
The eggs of frogs are fertilized externally. When frogs mate, as these two are doing, the clasp of the male induces the female to release a large mass of mature eggs, over which the male discharges his sperm.

Reptiles and Birds

The reptiles were the first group of vertebrates to abandon the marine habitat completely. Their eggs are fertilized internally within the mother before they are laid, with the male introducing his sperm directly into her body (Figure 54-4). By this means, fertilization still occurs in a nondesiccating environment, even though the adult animals are fully terrestrial. Most vertebrates that fertilize internally utilize a tube, the **penis,** to inject sperm into the female. Encased in erectile tissue, the penis can become quite rigid and penetrate far into the female reproductive tract.

Many reptiles are oviparous, the eggs being laid and then abandoned to their developmental destiny; others are viviparous or ovoviviparous, forming eggs that hatch within the body of the mother. The young are then born alive from the mother's body.

Most birds (swans are an exception) lack a penis and achieve internal fertilization by the simple expedient of the male slapping sperm against the reproductive opening of the female. This kind of mating occurs more quickly than that of most reptiles. All birds are oviparous, the young emerging from their eggs outside of the mother's body. Birds encase their eggs in a harder shell than their reptilian ancestors. They

FIGURE 54-4
The introduction of sperm by the male into the female's body is called copulation. Reptiles such as these turtles were the first terrestrial vertebrates to develop this form of reproduction, which is particularly suited to a terrestrial environment.

also hasten the development of the embryo within each egg by warming the eggs with their bodies. The young that hatch from the eggs of most bird species are not able to survive unaided, since their development is still incomplete. The young birds, therefore, are fed and nurtured by their parents, and they grow to maturity gradually.

The shelled eggs of reptiles and birds constitute one of their most important adaptations to life on land (see Chapter 42), since these eggs can be laid in dry places. Such eggs, called **amniotic eggs,** develop a system of internal cell layers which enclose the space where the embryo will develop, forming a fluid-filled "amniotic cavity." The outermost layer, the **chorion,** lines the shell membrane of the egg. The inner cellular layer, the **amnion,** forms a sac around the embryo. Each egg is provided with a large amount of yolk and is encased within a leathery cover. The zygote develops within the egg, eventually achieving the form of a miniature adult before it completely depletes its supply of yolk and leaves the egg to face its fate as an adult.

> **The first major evolutionary change in reproductive biology among land vertebrates was that of the reptiles and their descendants, the birds. Although both groups are still oviparous like fishes, they practice internal fertilization, with the zygote encased within a watertight egg.**

Mammals

The most primitive mammals, the monotremes, are oviparous like the reptiles from which they evolved. The living monotremes consist solely of the duckbilled platypus (see Figure 42-29) and the echidna. No other mammals lay eggs.

The young of viviparous mammals are nourished and protected by their mother, an outstanding characteristic of the members of this class. Mammals other than monotremes have approached the problem of nourishing their young in two ways:

1. Marsupials nourish their young with milk, giving birth to embryos at a very early stage of development. The tiny animals continue their development nursing in pouches on the mother's body, eventually emerging when they are able to function sufficiently well.
2. The placental mammals retain their young for a much longer period within the body of the mother. To nourish them, they have evolved a specialized, massive network of blood vessels called a **placenta,** through which nutrients are channeled to the embryo from the mother's blood.

> **The second and third major evolutionary changes in reproductive biology among land vertebrates were those of the marsupials and placental mammals. These groups nourish their young within a pouch or inside the body until they have reached a fairly advanced stage of development.**

FIGURE 54-5
The human reproductive system.

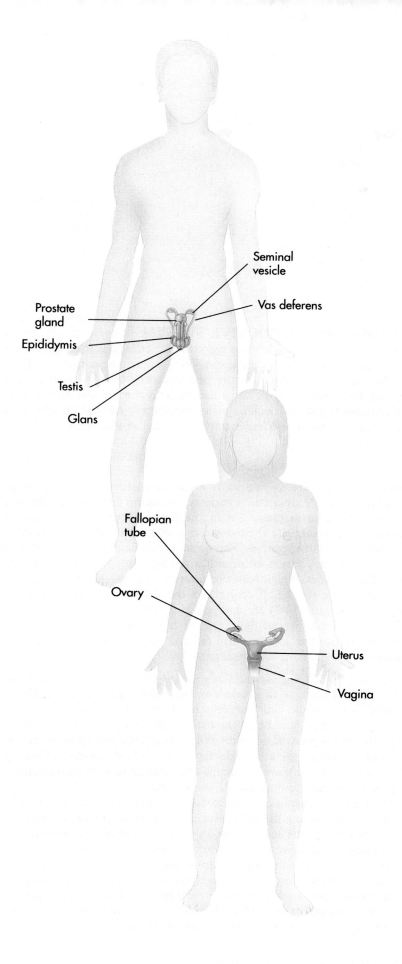

Seminal vesicle

Vas deferens

Prostate gland

Epididymis

Testis

Glans

Fallopian tube

Ovary

Uterus

Vagina

FIGURE 54-6
Organization of the male reproductive organs.

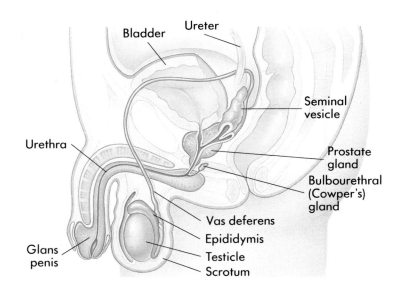

THE HUMAN REPRODUCTIVE SYSTEM

Like those of all vertebrates, human reproductive cells consist of egg cells, which are richly endowed with nutrients, and sperm cells, where nutrients are nearly absent. As we will describe in more detail in the next chapter, the egg is fertilized within the female, and the zygote develops into a mature fetus there. Most reproductive activity, then, is centered in the female. The male contributes little but his sperm. The reproductive systems of males and females are illustrated in Figure 54-5.

Males

The human male gamete, or **sperm,** is highly specialized for its role as a carrier of genetic information. Produced after meiosis, the sperm cells have 23 chromosomes instead of the 46 found in most cells of the human body. Unlike these other cells, sperm cells do not complete their development successfully at 37° C (98.6° F), the normal human body temperature. The sperm-producing organs, or **testes,** move during the course of fetal development out of the body proper and into a sac called the **scrotum.** The scrotum, which hangs between the legs of the male, maintains the testes at a temperature about 3° C cooler than the rest of the body (Figure 54-6).

The testes (Figure 54-7) are composed of several hundred compartments, each packed with large numbers of tightly coiled tubes called **seminiferous tubules.** The tubes themselves are the sites of sperm cell synthesis, or **spermatogenesis** (Figure 54-8), a process that occurs in two phases:

1. *Meiosis.* Packed against the basement membrane of the seminiferous tubules are diploid cells called **spermatogonia,** which are constantly dividing mitotically. Some of the daughter cells move inward toward the interior of the tubule, the lumen, and begin the path of differentiation that will eventually lead to the production of a sperm cell. These cells are **primary spermatocytes.** They undergo meiosis, producing four haploid gametes called **spermatids.** Each of these spermatids contains one member of each chromosome pair (an X or a Y sex chromosome and 22 autosomes).

2. *Development.* The undifferentiated spermatids then undergo a process of development, producing mature sperm cells, or **spermatozoa.** Spermatozoa are relatively simple cells and contain only a cell membrane, a compact nucleus, and a vesicle called an **acrosome,** which is derived from the Golgi body. The acrosome is located at the leading tip of the sperm cell. It contains enzymes that aid in the penetration of the protective layers surrounding the egg. Each spermatozoan also has a propulsive mechanism consisting of a flagellum, which propels the cell; a centriole, which acts as a basal body for the flagellum; and mitochondria, which generate the necessary metabolic power.

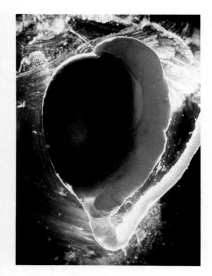

FIGURE 54-7
Human testes. The testicle is the darker sphere in the center of the photograph; within it sperm are formed. Cupped beneath the testicle is the epididymis, a highly coiled passageway within which sperm complete their maturation. Extending away from the epididymis is a long tube, the vas deferens, in which mature sperm are stored.

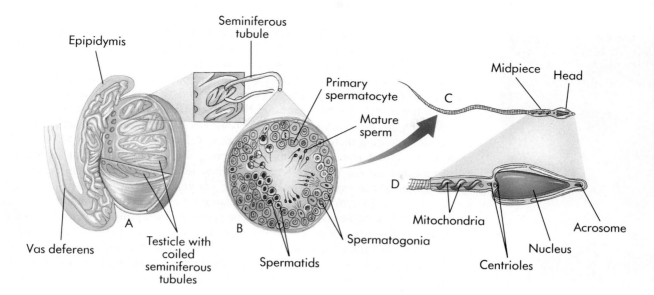

FIGURE 54-8

The interior of the testis, site of spermatogenesis. Within the seminiferous tubules of the testis (**A**), cells called spermatogonia (**B**) develop into sperm, passing through the spermatocyte and spermatid stages. Each sperm (**C**) possesses a long tail coupled to a head (**D**), which contains a haploid nucleus.

The full process of sperm development, from spermatogonia to spermatozoa, takes about 2 months. Sperm production is triggered by the hormones FSH and LH, which also regulate the secretion by the testes of the hormone testosterone, responsible for the development of male sexual characteristics. The number of sperm produced is truly incredible. A typical adult male produces several hundred million sperm each day of his life. Those which are not ejaculated from the body are reabsorbed, in a continual cycle of renewal.

After the sperm cells complete their differentiation within the testes, they are delivered to a long coiled tube called the **epididymis,** where they are stored and mature further. The sperm cells are not motile when they arrive in the epididymis, and they must remain there for at least 18 hours before their motility develops. From the epididymis the sperm are delivered to another long tube, the **vas deferens,** where they await delivery. When they are delivered during intercourse, the sperm travel through a tube from the vas deferens to the **urethra,** where the reproductive and urinary tracts join, emptying through the penis.

The penis is an external tube composed of three cylinders of spongy tissue (Figure 54-9). In cross section the arteries and veins can be seen along the dorsal surface, beneath which two of the cylinders sit side by side. Below the pair of cylinders is a third cylinder, which contains in its center the urethra, through which both sperm

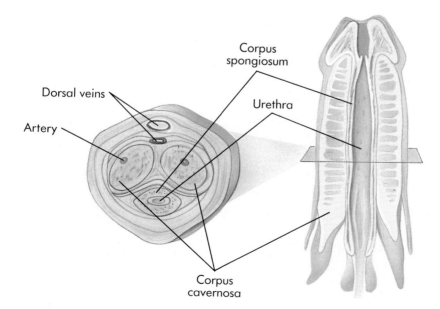

FIGURE 54-9
A penis in cross section.

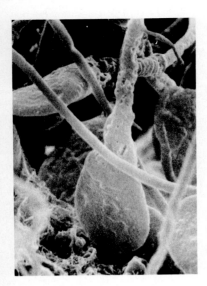

FIGURE 54-10
Human sperm. Only the heads and a portion of the long slender tails of these sperm are shown in this scanning electron micrograph.

(during intercourse) and urine (during liquid excretion) pass. The spongy tissue that makes up the three cylinders is riddled with small spaces between its cells. When nerve impulses from the central nervous system cause the dilation of the arterioles leading into this tissue, blood collects within the spaces. This causes the tissue to become distended and the penis to become erect and rigid. Continued stimulation by the central nervous system is required for this erection to be maintained.

Erection can be achieved without any physical stimulation of the penis. Fantasy is a common initiating factor. However, the physical stimulation of the penis usually is required for any delivery of sperm to take place. Prolonged stimulation of the penis, as by repeated thrusts into the female's vagina, leads first to the mobilization of the sperm. In this process muscles encircling the vas deferens contract, moving the sperm into the urethra. Eventually the prolonged stimulation leads to the violent contraction of the muscles at the base of the penis. The result is **ejaculation,** the ejection of about 5 milliliters of semen out of the penis. **Semen** is a mixture of sperm and several liquids collectively called **seminal fluid.** Within this small volume are several hundred million sperm. The odds against any one of them successfully completing the long journey to the egg and fertilizing it are extraordinarily high. Successful fertilization requires a high sperm count (Figure 54-10); males with less than 20 million sperm per milliliter are generally considered sterile.

> An adult male produces sperm continuously, several hundred million each day of his life. The sperm are stored and then delivered during sexual intercourse.

Females

We will now consider the reproductive system of females. Fertilization requires more than insemination. There must be a mature egg to fertilize. Eggs are produced within the ovaries of females (Figure 54-11). The ovaries are compact masses of cells, 2 to 3 centimeters long, located within the abdominal cavity. Eggs develop from cells called **oocytes,** which are located in follicles in the outer layer of the ovary. Unlike males, whose gamete-producing spermatogonia are constantly dividing, females have at birth all of the oocytes that they will ever produce. At each cycle of ovulation one or a few of these oocytes initiate development; the others remain in a developmental holding pattern. This long maintenance period is one reason that developmental abnormalities crop up with increasing frequency in pregnancies of women who are over 35 years old. The oocytes are continually exposed to chromosomal mutation throughout life, and after 35 years the odds of a harmful mutation having occurred become appreciable.

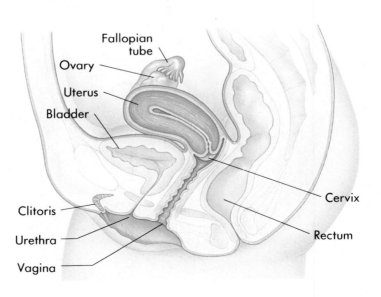

FIGURE 54-11
Organization of the female reproductive organs.

At birth a female's ovaries contain some 2 million oocytes, all of which have initiated the first meiotic division. At this stage they are called **primary oocytes.** Meiosis is arrested, however, in prophase of the first meiotic division. The development of very few oocytes ever proceeds further. With the onset of puberty, the female matures sexually. At this time the release of FSH initiates the resumption of the first meiotic division in a few oocytes, but a single one soon becomes dominant and the others regress (Figure 54-12). Approximately every 28 days thereafter, another oocyte matures. It is rare for more than about 400 of the approximately 2 million oocytes with which a female was born to mature during her lifetime. When they do mature, the egg cells are called **ova,** the Latin word for eggs.

Unlike male gametogenesis, the process of meiosis in the oocytes does not result in the production of four haploid gametes. Instead, a single haploid ovum is produced, the other meiotic products being discarded as **polar bodies.** The process of meiosis is stop-and-go rather than continuous (Figure 54-13):

1. *Developmental arrest.* The first phase of meiosis in female gametogenesis is arrested in prophase of meiosis I, leaving the gamete-forming cell as a primary oocyte.
2. *Ovulation.* The second phase of meiosis completes the first meiotic division. It is triggered by hormonal changes associated with ovulation. At the end of meiosis I the nuclear envelope disintegrates and the chromosomes move to the surface of the cell. There a bulge in the cell surface forms a pocket into which one set of chromosomes settles. This pocket is pinched off, forming a polar body and leaving the other set of chromosomes within the egg.
3. *Fertilization.* The second meiotic division does not occur until after fertilization. It results in the production of a mature ovum, which is haploid, and another polar body. The polar bodies eventually disintegrate, all reproductive investment being devoted to the single ovum. This very large cell is over 100 micrometers in diameter and rich in metabolites and biosynthetic machinery.

When the ovum is released from the follicle at ovulation, it is swept by the beating of cilia into the **oviducts,** called **fallopian tubes** in humans, which lead away from the ovaries (Figure 54-14). Smooth muscles lining the oviducts contract rhythmically;

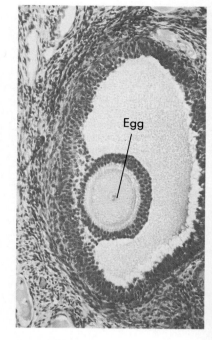

FIGURE 54-12
A mature egg within an ovarian follicle of a cat.

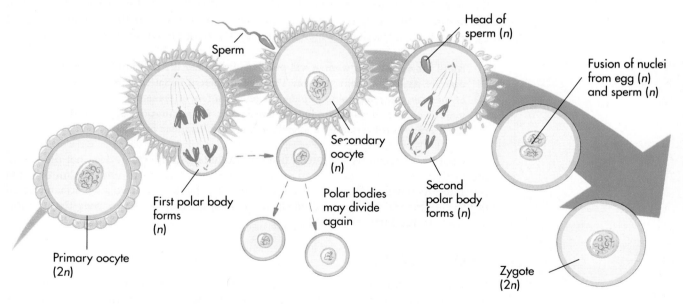

FIGURE 54-13
The meiotic events of oogenesis. A primary oocyte is diploid. In its maturation the first meiotic division is completed, and one division product is eliminated as a polar body. The other product, the secondary oocyte, is released during ovulation. The second meiotic division does not occur until after fertilization and results in production of a second polar body and a single haploid egg nucleus. Fusion of the haploid egg nucleus with a haploid sperm nucleus produces a diploid zygote, from which an embryo subsequently forms.

FIGURE 54-14

The journey of an egg. Produced within a follicle and released at ovulation, an egg is swept up into a fallopian tube and carried down by waves of contraction of the tube walls. Fertilization occurs within the tube, by sperm journeying upward. Several mitotic divisions occur while the fertilized egg continues its journey down the fallopian tube, so that by the time it enters the uterus, it is a hollow sphere of cells, a blastula-stage embryo. The blastula implants itself within the wall of the uterus, where it continues its development.

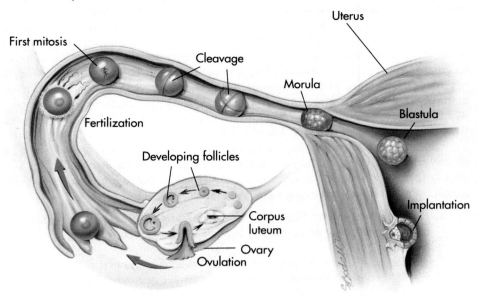

these rhythmic contractions, called peristaltic contractions, move the egg down the tube to the uterus. The journey is a slow one, taking about 3 days to complete. If the egg is unfertilized, it is dead when it reaches the uterus. It can only live approximately 24 hours unless it has been fertilized. For this reason the sperm cannot simply lie in wait within the uterus. Any sperm that is to fertilize an egg successfully must make its way up the oviducts, a long passage that few survive.

Sperm are deposited within the **vagina,** a muscular tube about 7 centimeters long, which leads to the mouth of the **uterus.** The opening to the uterus is bounded by a muscular sphincter called the **cervix.** The uterus is a hollow pear-shaped organ about the size of a small fist. Its inner wall, the **endometrium,** has two layers. The outer of these layers is shed during menstruation, while the one beneath it generates the outer layer. Sperm entering the uterus are carried upward by waves of motion in the uterus walls, enter the oviduct, and swim upward against the current generated by peristaltic contractions, which is carrying the ovum downward toward the uterus.

All of the eggs that a woman will produce during her life develop from cells that are already present at birth. Their development is halted at prophase I of meiosis, and, beginning at puberty, one or a few of these cells resume meiosis each 28 days to produce a mature egg. On maturation, eggs travel to the uterus. Fertilization by a sperm, if it occurs, happens en route.

Chapter 55 recounts the fate of a successfully fertilized egg. It does not die within the oviduct, as do its unfertilized sisters; rather, it proceeds to undergo mitosis and continues its journey. When it reaches the uterus, the new embryo soon attaches itself to the endometrial lining and thus starts the long developmental journey that eventually leads to the birth of a child.

SEXUAL CYCLES

For efficient reproduction, mating or the release of sperm must occur as soon as the mature egg or eggs become available. Otherwise, few of the sperm will survive long enough to achieve successful fertilization. Among those vertebrates that practice internal fertilization, the females typically signal the successful development and release of an egg, ovulation, by the release of chemical signals called pheromones. Female dogs signal their reproductive readiness in this manner, and the male is able to detect their pheromones in the air, even at very low concentrations. In rabbits and many other mammals, the female does not actually release a mature egg until mating has occurred. For them, the physical stimulus of copulation causes the pituitary gland to release a signal that triggers ovulation.

When a female vertebrate does not possess a mature egg, she will often reject the sexual advances of males. Most female mammals are sexually receptive, or "in heat," for only a few short periods each year. The period in which the animal is "in heat" is called **estrus;** the periods of estrus correspond to ovulation events during a periodic cycle, the **estrous cycle.** In general, small mammals have many estrous cycles in rapid succession, whereas larger ones have fewer cycles spaced farther apart.

Humans and some apes provide the sole exception to the rule of cycles of receptivity. Female humans exhibit no estrous cycle, and are sexually receptive throughout the reproductive cycle. The human reproductive cycle of egg production and release, called the **menstrual cycle,** takes on average about 28 days. One cycle follows another, continuously, and a female human may mate at any time during a cycle. Successful fertilization, however, is possible only during a period of 3 to 4 days starting at ovulation.

SEX HORMONES

A lot of signaling goes on during sex. Not only is it necessary for a viviparous female to signal the male when a mature egg has been released at the beginning of an estrous cycle, but it is also necessary that many different processes within both the female and within the male be coordinated during the process of gametogenesis and reproduction. The delayed sexual development that is common in mammals, for example, entails a nonsexual juvenile period, after which changes occur that produce sexual maturity. These changes occur in many parts of the body, a process that requires the simultaneous coordination of further development in many different kinds of tissues. The production of gametes is another closely orchestrated process, involving a series of carefully timed developmental events. Successful fertilization initiates yet another developmental "program," in which the female body prepares itself for the many changes of pregnancy.

All of this signaling is carried out by the hypothalamus, the portion of the brain specialized to process information about the body's internal functions. The hypothalamus regulates many of the basic activities of the body, dictating the responses to pain and emotion, coordinating the digestion of food, and maintaining a constant body temperature in mammals and birds, as well as coordinating reproductive events. The control of body functions by the hypothalamus involves regulating organs and glands that are situated at distant locations in the body. To accomplish this control, command signals must pass from the hypothalamus to the organs, and feedback signals must pass from that organ back to the hypothalamus.

The signals by which the hypothalamus regulates reproduction are not carried by nerves. Although a nerve-conducted signal would be very fast and specific, many of the desired changes are long-term ones, which would require a continuous nerve signal to run for days or months. The reproductive signals are thus of a slower, longer-lasting variety. They are hormones, which the bloodstream carries to the various organs of the body. Many reproductive hormones are complex carbon-ring lipids called steroids (Figure 54-15); others are peptides.

The body uses sex hormones as signals to control various body functions in the reproductive cycle.

Reproduction is only one of the many functions that the vertebrate body controls with such chemical signals from the brain. Vertebrate hormones and their production were discussed in detail in Chapter 47. Here we need only note that like any complex organization, the vertebrate brain is hierarchical. The hypothalamus initiates a reproductive change by issuing an order to a subordinate, a hormone-producing region of the brain called the pituitary gland, with which it is in intimate contact. These orders are themselves hormones, although they do not travel far.

When the pituitary gland receives a "releasing hormone" from the hypothalamus,

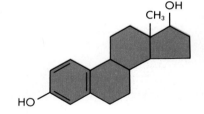

ESTROGEN

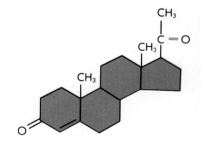

PROGESTERONE

FIGURE 54-15
The steroid sex hormones estrogen and progesterone. Estrogen prepares and maintains the uterine lining for pregnancy; progesterone stimulates thickening of the uterine lining during pregnancy.

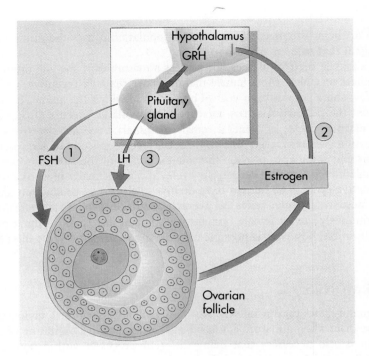

FIGURE 54-16

Mammalian egg maturation is under hormonal control.

1 Gonadotrophic releasing hormone (GRH) causes the pituitary to produce the follicle-stimulating hormone (FSH) and release it into the blood circulation; when FSH reaches the ovaries, it initiates final egg development; **2** FSH also causes the ovaries to produce the hormone estrogen; rising levels of estrogen in the blood cause the pituitary to shut down its production of FSH and instead produce the luteinizing hormone (LH); **3** When this LH circulates back to the ovaries, it inhibits estrogen production and initiates ovulation.

it initiates the production of the particular pituitary hormone that is called for by the releasing hormone. This pituitary hormone is released into the bloodstream, where it acts at a distance to bring about the change commanded by the hypothalamus. For example, the maturation of eggs, which begins the estrous cycle, is initiated when the hypothalamus produces a releasing hormone called gonadotropin-releasing hormone (GRH) (Figure 54-16). This hormone passes to the nearby pituitary, where it initiates the production of **follicle-stimulating hormone (FSH)** by the pituitary. FSH passes from the pituitary into the bloodstream, eventually reaching the ovaries. There, FSH is bound by receptors on the surface of the follicles, capsules in the ovary within which the eggs are developing. The binding of FSH initiates the final development and maturation of the eggs. Other tissues of the body also possess FSH receptors and respond to the hormone by producing pheromones and undergoing other physiological changes associated with the initiation of the estrous cycle. A summary of the actions of reproductive hormones is presented in Table 54-1.

TABLE 54-1	REPRODUCTIVE HORMONES
MALE	
Follicle-stimulating hormone (FSH)	Stimulates spermatogenesis
Luteinizing hormone (LH)	Stimulates secretion of testosterone
Testosterone	Stimulates development and maintenance of male secondary sexual characteristics
FEMALE	
Follicle-stimulating hormone (FSH)	Stimulates growth of ovarian follicle; stimulates secretion of estrogen
Luteinizing hormone (LH)	Stimulates conversion of ovarian follicles into corpus luteum; inhibits estrogen production
Estrogen	Stimulates development and maintenance of female secondary sexual characteristics; prompts monthly preparation of uterus for pregnancy
Progesterone	Completes preparation of uterus for pregnancy; helps maintain female secondary sexual characteristics
Oxytocin	Stimulates contraction of uterus; initiates milk release
Prolactin	Stimulates milk production

THE HUMAN REPRODUCTIVE CYCLE

The reproductive cycle of female mammals, including that of humans, is composed of two distinct phases—the **follicular phase** and the **luteal phase**.

Triggering the Maturation of an Egg

The first, or follicular, phase of the reproductive cycle is marked by the hormonally controlled development of eggs within the ovary. As we have described, the pituitary initiates the cycle by secreting FSH, which binds to receptors on the surface of the follicles, initiating the final development and maturation of the egg. Normally, only a few eggs at any one time have developed far enough to respond immediately to FSH. FSH levels are reduced before other eggs reach maturity so that in every cycle only a few eggs ripen.

The reduction of the FSH levels is achieved by a feedback command to the pituitary. In addition to initiating final egg development, FSH also triggers the ovary's production of the female sex hormone **estrogen**. Rising estrogen levels in the bloodstream feed back to the pituitary and cut off the further production of FSH. In this way, only the few eggs that are already developed far enough for their maturation to be initiated by FSH are taken into the final stage of development. The rise in estrogen level and the maturation of one or more eggs complete the follicular phase of the estrous cycle.

Preparing the Body for Fertilization

The second, or luteal, phase of the cycle follows smoothly from the first. The pituitary responds to estrogen by secreting a second hormone, called **luteinizing hormone (LH)**, which is carried in the bloodstream to the developing follicle. LH inhibits estrogen production and causes the wall of the mature follicle to burst. The egg within the follicle is released into one of the fallopian tubes, which extend from the ovary to the uterus. This process is called **ovulation**. Meanwhile, the ruptured follicle repairs itself, filling in and becoming yellowish. In this condition it is called the **corpus luteum**, which is Latin for "yellow body." The corpus luteum soon begins to secrete a hormone, **progesterone**, which initiates the many physiological changes associated with pregnancy and inhibits development of other follicles. The body is preparing for fertilization.

During the follicular stage, estrogen has caused the uterine lining to build up. The progesterone produced during the luteal phase induces the tissues of the lining to start secreting, while estrogen continues to increase the lining's thickness. The corpus luteum continues its production of progesterone after fertilization. If, however, fertilization does not occur soon after ovulation, then production of progesterone slows and eventually ceases, marking the end of the luteal phase.

> The reproductive cycle of mammals is composed of two alternating phases. During the follicular phase some of the eggs within the ovary complete their development. During the following luteal phase the mature eggs are released into the fallopian tubes, a process called ovulation. If fertilization does not occur, ovulation is followed by a new follicular phase, the start of another cycle.

In the absence of estrogen and progesterone the pituitary can again initiate production of FSH, thus starting another estrous cycle. In humans the next cycle follows immediately after the end of the preceding one. A cycle usually occurs every 28 days, or a little more frequently than once a month, although this varies in individual cases. The Latin word for month is *mens*, which is why the reproductive cycle in humans is called the menstrual cycle, or monthly cycle (Figure 54-17). In humans and some other primates, the hormone progesterone has among its many effects a thickening of the walls of the uterus in preparation for the implantation of the developing embryo. When fertilization does not occur, the lowering levels of progesterone cause this

FIGURE 54-17
The human menstrual cycle. The growth and thickening of the uterine (endometrial) lining is governed by levels of the hormone progesterone; menstruation, the sloughing off of this blood-rich tissue, is initiated by lower levels of progesterone.

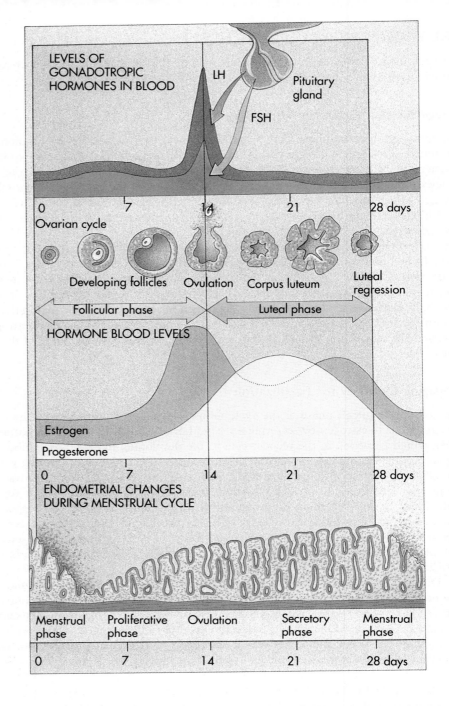

thickened layer of blood-rich tissue to be sloughed off, a process that results in the bleeding associated with menstruation. Menstruation, or "having a period," usually occurs about midway between successive ovulations, or roughly once a month, although its timing varies widely, even for individuals.

Two other hormones are important in the female reproductive system. Prolactin is needed for the production of milk and is secreted by the anterior pituitary. Oxytocin, secreted by the posterior pituitary, causes milk release. In combination with uterine prostaglandins, oxytocin initiates labor and delivery.

THE PHYSIOLOGY OF HUMAN INTERCOURSE

Few physical activities are more pleasurable to humans than sexual intercourse. The sex drive is one of the strongest drives directing human behavior, and, as such, it is circumscribed by many rules and customs. Sexual intercourse acts as a channel for

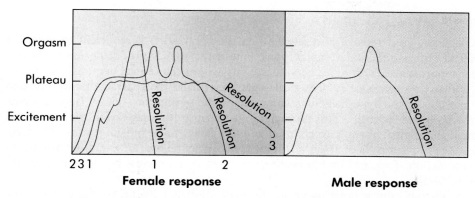

FIGURE 54-18
The human orgasmic response.
Among females, the response is
highly variable. It may be typified
by one of the three patterns illustrated here. Among males, the
response does not vary as much as
it does among females, with a single intense response peak.

the strongest of human emotions such as love, tenderness, and personal commitment.
Few subjects are at the same time more private and of more general interest. Here we
will limit ourselves to a very narrow aspect of sexual behavior, its immediate physiological effects. The emotional consequences are no less real, but they are beyond the
scope of this book.

Until relatively recently, the physiology of human sexual activity was largely unknown. Perhaps because of the prevalence of strong social taboos against the open
discussion of sexual matters, research on the subject was not being carried out, and
detailed information was lacking. Each of us learned from anecdote, from what our
parents or friends told us, and eventually from experience. Largely through the pioneering efforts of William Masters and Virginia Johnson in the last 25 years and an
army of researchers who have followed them, this gap in the generally available information about the biological nature of our sexual lives has now largely been filled.

The sexual act is referred to by a variety of names, including intercourse, copulation, and coitus, as well as a host of more informal ones. It is common to partition the
physiological events that accompany intercourse into four periods, although the division is somewhat arbitrary, with no clear divisions between the four phases. The four
periods are **excitement, plateau, orgasm,** and **resolution** (Figure 54-18).

Excitement

The sexual response is initiated by the nervous system. In both males and females,
commands from the brain increase the heartbeat, blood pressure, and rate of breathing. These changes are very similar to the ones that the brain induces in response to
alarm. Other changes increase the diameter of blood vessels, leading to increased circulation. In some women these changes may produce a reddening of the skin around
the face, breasts, and genitals (the sex flush). The nipples commonly harden and become more sensitive. In the genital area this increased circulation leads to the vasocongestion in the males's penis that produces erection. Similar swelling occurs in the
clitoris of the female, a small knob of tissue composed of a shaft and glans much like
the male penis but without the urethra running through it. The female experiences
changes that prepare the vagina for sexual intercourse. The increased circulation leads
to swelling and parting of the lips of tissue, or **labia,** which cover the opening to the
vagina; the vaginal walls become moist; and the muscles encasing the vagina relax.

Plateau

The penetration of the vagina by the thrusting penis results in the repeated stimulation of nerve endings both in the tip of the penis and in the clitoris. The clitoris,
which is now swollen, becomes very sensitive and withdraws up into a sheath or
"hood." Once it has withdrawn, the stimulation of the clitoris is indirect, with the
thrusting movements of the penis rubbing the clitoral hood against the clitoris. The
nervous stimulation produced by the repeated movements of the penis within the vagina elicits a continuous sympathetic nervous system response, greatly intensifying
the physiological changes initiated in the excitement phase. In the female, pelvic
thrusts may begin, while in the male the penis reaches its greatest length and rigidity.

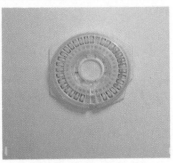

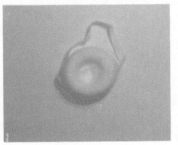

FIGURE 54-19

Shown here are five of the common means used to achieve birth control.

A Condom.

B Foams.

C Diaphragm and spermicidal jelly.

D Oral contraceptives.

E Vaginal sponge.

Orgasm

A

The climax of intercourse is reached when the stimulation is sufficient to initiate a series of reflexive muscular contractions. The nerve signals producing these contractions are associated with other nervous activity within the central nervous system, activity that we experience as intense pleasure. In females the contractions are initiated by impulses in the hypothalamus, which causes the pituitary to release large amounts of the hormone oxytocin. This hormone in turn causes the muscles in the uterus and around the vaginal opening to contract and the cervix to be pulled upward. Contractions occur at intervals of about a second. There may be one or several intense peaks of contractions (orgasm) or the peaks may be more numerous but less intense (see Figure 54-18).

B

Analogous contractions occur in the male, initiated by nerve signals from the brain. These signals first cause emission, in which the rhythmic peristaltic contraction of the vas deferens and of the prostate gland, where the vas deferens and urethra join, causes the sperm and seminal fluid to move to a collecting zone of the urethra, located at the base of the penis. Shortly after, nerve signals from the brain induce violent contractions of the muscles at the base of the penis, resulting in the ejaculation of the collected semen through the penis. As in the female orgasm, the contractions are spaced about a second apart, although in the male they continue for a few seconds only. The orgasmic contractions in the male, however, are almost invariably restricted to a single intense wave of contractions.

Resolution

C

After ejaculation, males rapidly lose their erection and enter a refractory period lasting 20 minutes or longer, in which sexual arousal is difficult to achieve and ejaculation is almost impossible. By contrast, many women can be aroused again almost immediately. After intercourse, the bodies of both men and women return slowly, over a period of several minutes, to their normal physiological state.

CONTRACEPTION AND BIRTH CONTROL

In most vertebrates, sexual intercourse is associated solely with reproduction. Reflexive behavior that is deeply ingrained in the female limits sexual receptivity to those periods of the sexual cycle when she is fertile. In humans sexual behavior serves a second important function—the reinforcement of pair bonding, the emotional relationship between two individuals living together. The evolution of strong pair bonding is certainly not unique to humans, but it was probably a necessary precondition for the subsequent evolution of our increased mental capacity. The associative activities that make up human "thinking" are largely based on learning, and learning takes time. Human children are very vulnerable during the extended period of learning that follows their birth, and they require parental nurturing. Perhaps for this reason human pair bonding is a continuous process, not restricted to short periods coinciding with ovulation. Among all of the vertebrates, only in human females and in a few species of apes has the characteristic of sexual receptivity throughout their reproductive cycle evolved and come to play a role in pair bonding.

D

E

Not all human couples want to initiate a pregnancy every time they have sexual intercourse, yet sexual intercourse may be a necessary and important part of their emotional lives together. Among some religious groups this problem does not arise or is not recognized, since members of these groups believe that sexual intercourse has only a reproductive function and thus should be limited to situations in which pregnancy is acceptable—among married couples wishing to have children. Most couples, however, do not limit sexual relations to procreation, and among them unwanted pregnancy presents a real problem. The solution to this dilemma is to find a way to avoid reproduction without avoiding sexual intercourse; this approach is commonly called **birth control** (Table 54-2).

Several different approaches are commonly taken to achieve birth control. These methods differ from one another in their effectiveness and in their acceptability to different couples. Some of them are shown in Figure 54-19.

TABLE 54-2 METHODS OF BIRTH CONTROL*

DEVICE	ACTION	FAILURE RATES	ADVANTAGES	DISADVANTAGES
Intrauterine devices (Paragard Copper T380A, Progestasert)	Small plastic or metal devices placed in the uterus that somehow prevent fertilization or implantation; some contain copper, others release hormones	1-5	Convenient, highly effective, need to be replaced infrequently	Can cause excess menstrual bleeding and pain; danger of perforation, infection, and expulsion; not recommended for those who are childless or not monogamous, risk of pelvic inflammatory disease or infertility; dangerous in pregnancy
Oral contraceptives	Hormones, either in combination or progestin only, that primarily prevent release of egg	1-5, depending on type	Convenient and highly effective; provide significant non-contraceptive health benefits, such as protection against ovarian and endometrial cancers	Pills must be taken regularly; possible minor side-effects, which new formulations have reduced; not for women with cardiovascular risks, mostly those over 35 who smoke
Condom	Thin rubber sheath for penis that collects semen	3-15	Easy to use, effective, and inexpensive; protects against some sexually transmitted diseases	Requires male cooperation; may diminish spontaneity; may deteriorate on the shelf
Diaphragm	Soft rubber cup that covers entrance to uterus, prevents sperm from reaching egg, and holds spermicide	4-25	No dangerous side-effects; reliable if used properly; provides some protection against sexually transmitted diseases and cervical cancer	Requires careful fitting; some inconvenience associated with insertion and removal; may be dislodged during sex
Cervical cap	Miniature diaphragm that covers cervix closely, prevents sperm from reaching egg, and holds spermicide	Probably comparable to diaphragm	No dangerous side-effects; fairly effective; can remain in place longer than diaphragm	Problems with fitting and insertion; comes in limited number of sizes
Foams, creams, jellies, vaginal suppositories	Chemical spermicides inserted in vagina before intercourse that also prevent sperm from entering uterus	10-25	Can be used by anyone who is not allergic; protect against some sexually transmitted diseases; no known side effects	Relatively unreliable, sometimes messy; must be used 5 to 10 minutes before each act of intercourse
Sponge	Acts as sperm barrier and releases spermicide	15-30	Safe; easy to insert; provides some protection against sexually transmitted diseases; can be left in place for 24 hours; needs no fitting	Relatively unreliable; comes in only one size; some sensitivity and removal problems; cannot be used during menstruation
Implant (Norplant)	Capsules surgically implanted under skin that slowly release a hormone that blocks release of eggs	0.3	Very safe, convenient, and effective; very long-lasting (5 years); may have non-reproductive health benefits like those of oral contraceptives	Irregular or absent periods; necessity of minor surgical procedure to insert and remove
Injectable contraceptive (Depo-Provera) Unavailable in U.S.	Injection every 3 months of a hormone slowly released from the muscle that prevents ovulation	1	Convenient and highly effective; no serious side-effects other than occasional heavy menstrual bleeding	Animal studies suggest that it may cause cancer, though new studies of women are mostly encouraging

Source: American College of Obstetricians and Gynecologists: Benefits, Risks, and Effectiveness of Contraception, Washington, D.C., ACOG.
*Approximate effectiveness of these reversible methods of birth control is measured in pregnancies per 100 actual users per year.

Abstinence

The simplest and most reliable way to avoid pregnancy is not to have sex at all. Of all methods of birth control, this is the most certain—and the most limiting, since it denies a couple the emotional support of a sexual relationship.

A variant of this approach is to avoid sexual relations only on the two days preceding and following ovulation, since this is the only period during which successful fertilization is likely to occur. The rest of the sexual cycle is relatively "safe" for intercourse. This approach, called the **rhythm method,** is satisfactory in principle but difficult in application, since ovulation is not easy to predict and may occur unexpectedly. The effectiveness of the rhythm method is low; the failure rate is estimated to be 13% to 21% (13 to 21 pregnancies per 100 women practicing the rhythm method per year).

Another variant of this approach is to have only incomplete sex—the penis is withdrawn before ejaculation, a procedure known as **coitus interruptus.** This requires considerable willpower and often destroys the emotional bonding of intercourse. It also is not as reliable as it might seem. The penis can secrete prematurely released sperm within its lubricating fluid, and a second sexual act may transfer sperm ejaculated earlier. The failure rate of this approach is estimated at 9% to 25%, which is no better than that of the rhythm method.

Sperm Blockage

If sperm are not delivered to the uterus, fertilization cannot occur. One way to prevent the delivery of sperm is to encase the penis within a thin rubber bag, or **condom.** Many males do not favor the use of condoms, since they tend to decrease the sensory pleasure of the male. In principle this method is easy to apply and foolproof, but in practice it proves to be less effective than one might expect, with a failure rate of from 3% to 15%. Nevertheless, it is the most commonly employed form of birth control in the United States. Condoms are also beginning to see widespread use as a means of avoiding AIDS. Over a billion condoms were sold last year.

A second way to prevent the entry of sperm into the uterus is to place a cover over the cervix. The cover may be a relatively tight-fitting **cervical cap,** which is worn for days at a time, or a rubber dome called a **diaphragm,** which is inserted immediately before intercourse. Because the dimensions of individual cervices vary, a cervical cap or diaphragm must be fitted by a physician. Failure rates average from 4% to 25% for diaphragms, perhaps because of the propensity to insert them carelessly when in a hurry. Failure rates for cervical caps are somewhat lower.

Sperm Destruction

A third general approach to birth control is to remove or destroy the sperm after ejaculation. This can be achieved in principle by washing out the vagina immediately after intercourse, before the sperm have a chance to travel up into the uterus. Such a procedure is called by the French name for "wash," **douche.** This method turns out to be difficult to apply well, since it involves a rapid dash to the bathroom immediately after ejaculation and a very thorough washing. The failure rate is as high as 40%.

Sperm delivered to the vagina can be destroyed there with spermicidal **jellies, sponges,** or **foams.** These generally require application immediately before intercourse. The failure rate varies widely, from 10% to 25%.

Prevention of Egg Maturation

Since about 1960 a widespread form of birth control in the United States has been the daily ingestion of hormones, or **birth control pills.** These pills contain estrogen and progesterone, either taken together in the same pill or in separate pills taken sequentially. In the normal sexual cycle of a female, these hormones act to shut down the production of the pituitary hormones FSH and LH. The artificial maintenance of

As most college students now know, AIDS is a serious disease that is rapidly becoming common in our country and around the world. The disease, first reported in 1981, is transmitted by a virus and is almost always fatal. 206,392 individuals in the United States had contracted AIDS by the end of 1991, and 133,232 of them have died. Nearly 70 Americans will die each day of AIDS this year. The World Health Organization estimates there are now 10 million cases of AIDS world wide. Few if any of these patients will recover.

The name **AIDS** is shorthand for **acquired** (transmitted from another infected individual) **immune deficiency** (a breakdown of the body's ability to defend itself against disease) **syndrome** (a spectrum of symptoms). The disease is fatal because no one can survive for long without an immune system to defend against viral and bacterial infections and to ward off cancer. The virus also infects the central nervous system, often leading to serious mental disorders.

AIDS is not the only fatal disease to threaten humans, nor is it the most contagious. What makes AIDS an unusually serious threat is that the virus which causes the disease does not have its effect immediately upon infection. Recently infected people usually show no symptoms of the disease at all. Only much later—typically 5 years—does the virus begin to multiply and attack the immune system. During these years, however, the infected person is an unknowing carrier, able to transmit the virus to others. It is the large reservoir of undiagnosed, infected individuals that casts such a shadow over our future. Current estimates of the number of adults in major cities who are infected with the virus in the United States approach 1%, suggesting both a staggering load of future suffering that will be difficult to avoid and a greater danger that the infection will spread further. It would be folly to assume that college campuses will escape infection. Every student should face that fact squarely.

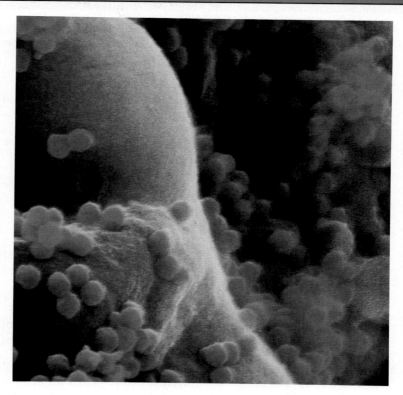

FIGURE 54-B, *1*
AIDS-causing viruses infecting a cell of the immune system.

What causes AIDS? The virus that causes AIDS is called **human immunodeficiency virus (HIV).** It is a fragile virus that does not survive outside body cells (Figure 54-B, *1*). It is present in the body fluids of infected individuals (notably in blood and in semen and vaginal fluid). The HIV virus is transmitted from one individual to another when body fluid is transferred from an infected individual. It is *not* transmitted in the air or by casual contact. You cannot catch AIDS from a bathroom seat, from a hot tub shared with an AIDS victim, from kissing an HIV carrier, or by being bitten by a mosquito that bit an AIDS victim. In studies of 619 households of AIDS victims, not one family member contracted the HIV virus. The *only* way you can become infected with HIV is to come into contact with the body fluids of an infected person. Among college students, there are two important routes of infection.

1. *Sexual intercourse.* Both semen and vaginal fluid of infected individuals have high levels of HIV virus. This means that vaginal, anal, and oral sex with an infected individual can all transmit the virus successfully—and to either sex. Absorption across the vaginal wall or into the penis offers a ready means for the virus to enter the body; the tiny tears produced during anal sex may facilitate entry even better. The microscopic abrasions that everybody has in their mouth from eating and chewing are a third easy means of entry, making oral sex also dangerous.

The only safe way to have sex with an infected individual is to use a condom, and use it correctly. When used correctly, condoms offer the best available protection from infection. It is important, however, that they be

Continued.

used properly. For some 10% of couples using condoms for birth control the woman becomes pregnant, almost always as a result of careless use. This suggests that it is not wise to mix alcohol or drugs with sexual encounters; they may cloud your judgment and lead you to do things you wouldn't do with a clearer head—such as forgetting to use a condom or using it carelessly. It is a mistake that could cost you your life.

If you are involved in a sexual relationship and don't want to use condoms, you can lessen the uncertainty (and any danger to yourself or to future partners) by sharing an AIDS antibody test with your sexual partner, well in advance of sexual intercourse. These tests, described below, detect the presence of the virus.

2. *Drug use.* A needle used more than once by an infected individual typically harbors large quantities of HIV virus, both in the fluid that remains behind in the needle and in the body of the hypodermic syringe. Anyone else who re-uses the needle will become infected. The use of intravenous drugs is itself dangerous—both illegal and life-threatening—but if you engage in such folly, do not compound the damage by employing a used needle or syringe.

Who is at risk? You are. AIDS is commonly perceived as a disease of homosexual men, because the disease first appeared in this country among the gay community. Because homosexuals tend to confine their sexual interactions to one another, the disease initially spread among homosexuals without entering the larger heterosexual community. That initial segregation appears to be ending: while only 4% of the AIDS cases diagnosed in 1986 were heterosexual non-drug users, the incidence of the HIV virus among heterosexuals is now expanding. Estimates of how fast vary widely. The Public

Health Service estimates that 2 million people in the United States now harbor the virus. In Africa, where the virus first infected humans several years before it spread to this country, the epidemic has proceeded farther than in the United States, even though homosexuality is rare there. In Africa, sexual transmission of AIDS is almost exclusively heterosexual and occurs in *both* directions, female to male as well as male to female. In some central African countries, as many as 10% of the adult individuals are thought to carry the HIV virus.

The AIDS antibody test. Can a person find out if he or she has been infected with the HIV virus? Yes, easily. A simple test identifies infected individuals by detecting antibodies in their blood directed against HIV virus. Such antibodies are detectable only in the bodies of infected individuals. They are the remnants of the body's attempt to ward off the HIV infection.

How an AIDS test works. The standard test for detecting the presence of antibody directed against HIV virus is called the ELISA antibody test. In the test, about 5 ml of blood (roughly the amount that would fit in a paper drinking straw) is drawn and checked to see whether any antibodies are present that will interact with bits of HIV attached to the surface of a plastic dish (Figure 54-B, *2*). If such antibodies are present, the test is said to be positive.

If the result of the ELISA test is negative, no further tests are required, since there are almost no false negatives unless the test is given within 6 weeks after infection, before the body has begun to produce antibodies. False positives are also rare, but if the ELISA test result is positive, it is repeated. Two positive ELISA's are the signal for a more elaborate test, called a Western blot, which although it is far more expensive is very accurate.

How to get an AIDS test. The Red Cross offers an inexpensive walk-in

AIDS antibody test at a variety of locations in most metropolitan areas. The Centers for Disease Control also maintains a national AIDS hotline that you can contact for information and advice by calling (800) 342-AIDS.

Confidentiality. At the Red Cross, the testing is completely confidential. The test is coded with a number that you select—the test is not associated with your name—and you telephone a few days later for the result of the test identified with that number. Maintaining the confidentiality of a positive result is very difficult, however, even though the test itself is completely confidential. When an individual does have a positive AIDS test, it is necessary to see a doctor frequently in order to monitor possibilities of disease progression. All of this has to be documented in the physician's office, along with billing information. Because these files may be accessed when you sign a medical release form or pay bills with insurance, it is impossible to maintain confidentiality over a long period.

The only way to survive AIDS is not to contract it. The only way (short of contaminated needles) that any student will contract AIDS is by having unprotected sex with someone who has the virus. But the 5-year lag prevents anyone from knowing who is infected and who is not, and the disease continues to spread. Although widespread antibody testing on college campuses has been proposed by some, in order to identify infected individuals and so dampen the spread of the virus while it is still confined to a relatively few individuals, the proposal is controversial, and the associated dangers of invasion of privacy are a real concern. In the absence of such wide scale on-campus testing, any student you have sex with might be a carrier and not know it. To avoid becoming part of the epidemic, you have to accept the responsibility of protecting your health.

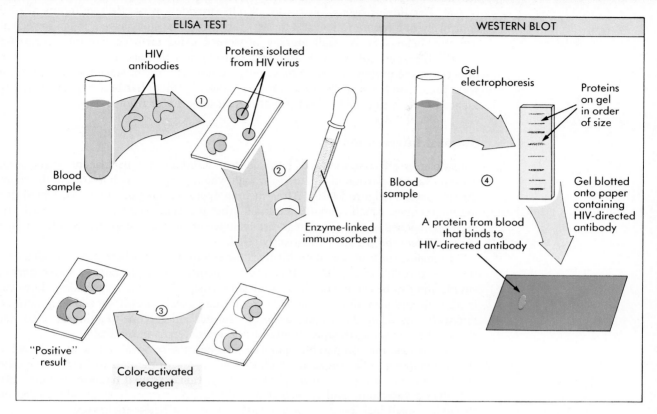

FIGURE 54-B, *2*

How an "AIDS test" works. AIDS tests are of two sorts: an inexpensive and rapid test that misses very little, but occasionally yields "false positives," and a more definitive test that is more laborious.

Step 1 The standard test for detecting the presence of antibody directed against the HIV virus (and thus identifying infected individuals) is called ELISA. The letters stand for **E**nzyme **L**inked **I**mmuno-**S**orbent **A**ssay. In the test, about 5 ml of blood is drawn, roughly the amount that would fit in a paper drinking straw. This blood sample is exposed to a plastic surface to which HIV virus proteins are chemically bound. Any HIV-specific antibodies present in the blood bind to the viral proteins on the plastic surface.

Step 2 After the blood sample is washed from the plastic plate, the plate is exposed to a solution containing a special reagent, the "enzyme linked immunosorbent." Basically, it is an antibody directed against other antibody molecules, to which a color reagent is attached. This reagent will bind to antibodies on the plastic plate only if antibody molecules are already bound there—that is, only if the blood possesses antibodies directed against HIV virus proteins.

Step 3 To detect a positive result, the plate is then treated with an enzyme that reacts with the color reagent to produce a color change. Running the full test takes about 3 hours. If the result of the ELISA test is negative, no further tests are required, since there are almost no false negatives unless the test is given within a few weeks after infection, before the body has begun to produce antibodies. False positive ELISAs are the signal for a more precise test, the Western blot.

Step 4 In the Western blot, the proteins in the blood sample are separated according to size by being induced to migrate through a gel in response to an electric field (bigger proteins migrate more slowly). The gel is then blotted onto paper containing antibodies directed against the HIV virus. Because the size of each HIV virus protein is known, the binding of any protein in the blood sample to the HIV-directed antibody on the paper—in just the place on the gel to which an HIV protein would have migrated—is considered to be a conclusively positive test.

high levels of estrogen and progesterone in a woman's bloodstream fools the body into acting as if ovulation had already occurred, when in fact it has not: the ovarian follicles do not ripen in the absence of FSH, and ovulation does not occur in the absence of LH. For these reasons, birth control pills provide a very effective means of birth control, with a failure rate of 0% to 10%. Implanted capsules have a much lower failure rate, below 1%. A small number of women using birth control pills experience undesirable side effects, such as blood clotting and nausea. The long-term consequences of the prolonged use of these pills are not yet known, since they have been in widespread use for only 25 years. To date, however, there has been no conclusive evidence of any serious side effects for the great majority of women.

Surgical Intervention

A completely effective means of birth control, although it is permanent, is the surgical removal of a portion of the tube through which gametes are delivered to the reproductive organs (Figure 54-20). After such an operation no gametes are delivered during intercourse, which is unaffected in all other respects. The failure rate of such surgical approaches is zero. The great disadvantage of surgical intervention is that it generally renders the person permanently sterile.

In males, such an operation involves the removal of a portion of the vas deferens, the tube through which sperm travel to the penis. This procedure, a **vasectomy**, is simple and can be carried out in a physician's office. Reversing a vasectomy, however, is difficult and sometimes impossible, requiring general anesthesia and long hours of microsurgery (using the magnification of a surgical microscope to guide the reconnection of the many microscopic tubules of the vas deferens individually).

In females, the comparable operation involves the removal of a section of each of the two fallopian tubes through which the egg travels to the uterus. Since these tubes are located within the abdomen, the operation, called a **tubal ligation**, is more difficult than a vasectomy and is even more difficult to reverse. It is usually carried out through a small incision made in the vaginal wall just below the cervix.

The surgical removal of the entire uterus, an operation that is not uncommon but is usually performed later in life for medical reasons other than birth control, is called a **hysterectomy**.

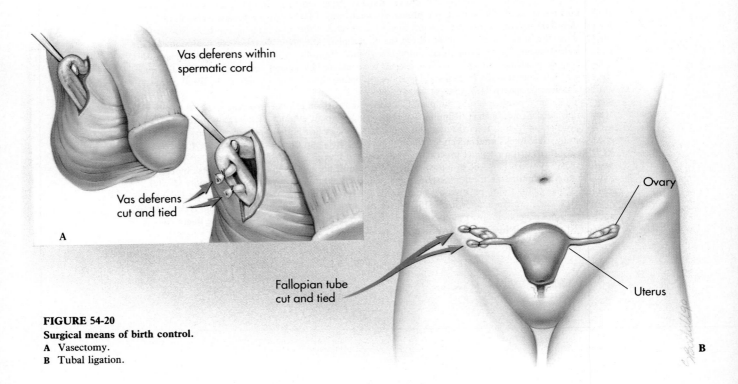

FIGURE 54-20
Surgical means of birth control.
A Vasectomy.
B Tubal ligation.

Prevention of Embryo Implantation

The insertion of a coil or other irregularly shaped object into the uterus is an effective means of birth control, since the irritation in the uterus prevents the implantation of the descending embryo within the uterine wall. Such **intrauterine devices** (IUDs) have a failure rate of only 1% to 5%. Their high degree of effectiveness probably reflects their convenience; once they are inserted, they can be forgotten. The great disadvantage of this method is that almost a third of the women attempting to use IUDs cannot; the devices cause them cramps, pain, and sometimes bleeding.

Another effective way to prevent embryo implantation is the use of the "morning after" pill, which contains 50 times the dose of estrogen present in birth control pills. The failure rate is 1% to 10%, but many women are uneasy about taking such high hormone doses.

Abortion

Reproduction can be prevented after fertilization if the embryo is **aborted** (removed) before its development and birth. During the first trimester this can be accomplished by **vacuum suction** or by **dilation and curettage,** in which the cervix is dilated and the uterine wall is scraped with a spoon-shaped surgical knife called a curette. Chemical methods are also being developed that cause abortion early in the first trimester, apparently with complete safety. One such drug, **RU 486,** is already in use in France and China, although not yet approved for use in the United States. Administration of RU 486 followed by prostaglandins that induce uterine contractions is almost 100% effective when taken within 49 days of the patient's last menstrual period.

In the second trimester the embryo can be removed by injecting a 20% saline solution into the uterus, which induces labor and delivery of the fetus. In general, the more advanced the pregnancy, the more difficult and dangerous the abortion is to the woman.

As a method of birth control, abortion takes a great emotional toll, both on the woman undergoing the abortion and often on others who know and care for her. Abortion also presents serious moral problems for some. Many people believe that the fetus is a living person from the time of conception and that abortion is simply murder. There are many countries in which abortions are defined as a crime. In the United States a fetus is not legally considered a person until birth, and abortions are permitted by law during the first two trimesters. They are illegal in the third trimester, however, except when the mother's life is endangered. The Supreme Court ruling on legalized abortion in the first two trimesters is a relatively recent one, and it is still the subject of intense controversy. The abortion rate in the United States (and most other developed countries) exceeds the birth rate.

Among methods of birth control, the rhythm method, coitus interruptus, and douches are relatively ineffective. Condoms and diaphragms are effective when used correctly, but mistakes are not uncommon. Birth control pills and IUDs are very effective. Vasectomies and tubal ligations are completely effective, although usually permanent.

AN OVERVIEW OF VERTEBRATE REPRODUCTION

Vertebrates carry out reproduction in many different ways, reflecting the evolutionary course of their invasion of the land. These differences are reflected in whether fertilization is external or internal, in whether zygotes begin life as eggs or develop internally, and in the size of broods and the frequency with which they are produced. Nor does the story end here. The course of development of the zygote is influenced in many ways by the reproductive strategy of the organism. A zygote nourished by the yolk of an egg develops differently from one nourished by its mother's blood supply. In the next chapter you will consider these differences in developmental processes in detail.

1. Sexual reproduction evolved in the sea. Among most fishes and the amphibians, fertilization is external, with gametes being released into water. Amphibians and the great majority of fishes are oviparous, the young being nurtured by the egg rather than by the mother.

2. The vertebrates' successful invasion of the land involved major changes in reproductive strategy. The first change was the watertight amniotic egg of the reptiles, which could survive in dry places. Internal fertilization, which is also characteristic of the reptiles, is another important survival strategy on land. Birds, as well as a primitive group of mammals called monotremes, have the same kinds of eggs as reptiles do.

3. The second important change in reproductive adaptation to life on land was that adopted by the marsupials. In them, fertilization is internal. The young are born within a few weeks of fertilization, but then they are nourished and protected during the course of their further development within a pouch.

4. A third change was adopted by placental mammals. Their young are nourished within the mother's body by means of a large, complex structure, the placenta, which exchanges material from her bloodstream with that of her offspring.

5. Reproduction in mammals is regulated by reproductive hormones typically produced by the pituitary on commands from the hypothalamus.

6. The reproductive cycle of mammals, the estrous cycle, is composed of two phases: (1) a follicular phase, in which some eggs in the ovary are hormonally signaled to complete their development; and (2) a luteal phase, in which one or more mature eggs are released into the oviduct, a process called ovulation. A complete menstrual cycle in a human female takes about 28 days.

7. The male gametes, or sperm, of mammals are produced within the testes. In human males hundreds of millions of sperm are produced each day. Sperm mature, become motile, and are stored in the epididymis, and are also stored in the vas deferens. Stimulation of the penis causes it to become distended and erect and causes the sperm to be delivered from the vas deferens to the urethra at the base of the penis. Further stimulation causes violent muscle contractions, which ejaculate the sperm from the penis.

8. At birth, female mammals contain all of the gametes, or oocytes, that they will ever have. A human female has approximately 2 million oocytes, arrested in meiotic prophase. At each ovulation the first meiotic division of one or a few eggs is completed. The second meiotic division does not occur until after fertilization.

9. Fertilization occurs within the oviducts. The journey of an egg to the uterus takes 3 days, and it is viable for only 24 hours unless it is fertilized. Consequently, only those eggs which have been reached within 24 hours by sperm swimming up the oviducts from the uterus can be fertilized successfully. Fertilized eggs continue their journey down the oviducts and attach to the lining of the uterus, where their development proceeds.

10. Human intercourse has four physiological periods: excitement, plateau, orgasm, and resolution. Orgasm in women is variable and may be prolonged. Orgasm in men is uniformly abrupt; it coincides with the ejaculation of sperm.

11. Humans practice a variety of birth control procedures; men using condoms and women using birth control pills are the most common. Intrauterine devices (IUDs) and surgical procedures that block the delivery of gametes are becoming increasingly common. Birth control is not always effective, and some couples terminate unwanted pregnancies. In the United States, as in other developed countries, the abortion rate exceeds the birthrate.

REVIEW QUESTIONS

1. What type of sexual reproduction is most common in marine organisms? What major problem is associated with this process? What additional problem was encountered with the invasion of the land?

2. How is reptilian and avian fertilization different from that of amphibians and most fishes? How is reptile reproduction different from bird reproduction? How is development in reptiles different from that in birds?

3. How are monotremes reproductively different from all other mammals? How do marsupials and placental mammals provide nourishment to their developing young?

4. What are the two phases of the female reproductive cycle in mammals? Briefly describe what occurs in each phase. During which phase does ovulation take place?

5. What event (or lack thereof) causes the end of an individual cycle? What stimulates the onset of another cycle? What is the reason for menstruation in some primates, including humans? How is this event hormonally mediated?

6. Into what structure do mature sperm travel? What further developmental process occurs here? Where are mature sperm stored? What are the major structures of the penis? What fluid(s) passes through the urethra?

7. What are the three stages of oogenesis? Briefly describe what occurs in each stage.

8. Into what structure does an ovulated secondary oocyte proceed? How does it travel toward the uterus? At what point in the female reproductive tract must the egg and sperm meet to ensure fertilization? Why?

9. What are the four periods in the physiology of human intercourse? What events occur in both the female and male during the first period? What events specifically occur in the male at this point? What events specifically occur in the female?

10. What are several means of birth control that use sperm-blocking devices? Why are these devices often not as effective as one might expect? What birth control methods rely on destroying deposited sperm? How do birth control pills affect birth control?

THOUGHT QUESTIONS

1. Perhaps the most controversial issue to arise in our consideration of sex and reproduction is abortion. List arguments for and against abortion. Under what conditions would you permit abortions? Forbid them? Do you think the disadvantages of abortion are in any way counterbalanced by the advantages it provides to underdeveloped countries with very high birthrates and swelling populations? Should the United States promote or oppose dissemination of information about abortion in underdeveloped countries?

2. Some fishes and many reptiles are viviparous. Birds, however, even though they evolved from reptiles, never employ this means of protecting their eggs. Can you think of a reason why?

3. Relatively few kinds of animals have both male and female sex organs in the same animal, whereas most plants do. Propose an explanation for this.

4. Explain why animals that develop by means of parthenogenesis are usually female.

FOR FURTHER READING

ARAL, S., and K. HOLMES: "Sexually Transmitted Diseases in the AIDS Era," *Scientific American*, February 1991, pages 52-59. Gonorrhea, syphilis, and other infections still exact a terrible toll.

BELL, G.: *The Masterpiece of Nature: The Evolution and Genetics of Sexuality*, University of California Press, Berkeley, 1982. A somewhat advanced but very rewarding look at the various theories of why sexual reproduction evolved the way it did.

BROMWICH, P., and T. PARSONS : *Contraception—The Facts*, Oxford University Press, Oxford, 1984. Practical guide to methods of contraception.

FRISCH, R.: "Fatness and Fertility," *Scientific American*, March 1988, pages 88-95. Dieting and exercise can lead to infertility. The author argues that fat tissue exerts a regulatory effect on female reproductive ability.

HAPGOD, F.: *Why Males Exist: An Inquiry into the Evolution of Sex*, William Morrow & Co., New York, 1979. An accurate but informal treatment that serves as a good introduction to the issues involved in this area.

HRDY, S.: "Daughters or Sons: Can Parents Influence the Sex of Their Offspring?" *Natural History*, April 1988, pages 64-83. New research says that many creatures, including humans, can influence the odds.

LAGERCRANTZ, H., and A. SLOTKIN: "The 'Stress' of Being Born," *Scientific American*, April 1986, pages 100-107. Passage through the narrow birth canal triggers the release of hormones important to the newborn's future survival.

LEISHMAN, K.: "Heterosexuals and AIDS," *The Atlantic*, February 1987, pages 39-58. A chilling account of the difficulty of modifying sexual behavior, despite the knowledge of the dangers associated with AIDS.

MARGULIS, L., and D. SAGAN: *Origins of Sex: Three Billion Years of Genetic Recombination*, Yale University Press, New Haven, Conn. 1987. A very provocative argument about the origins of sex, focusing on the life histories of primitive eukaryotes. Controversial and stimulating.

MAYNARD-SMITH, J.: *The Evolution of Sex*, Cambridge University Press, New York, 1978. An important viewpoint on the origin of sex, which clearly outlines the issues currently being argued by researchers.

SEGAL, S.J.: "The Physiology of Human Reproduction," *Scientific American*, September 1975, pages 52-62. An overview of the way in which hormones and the nervous system interact to regulate the sexual cycle.

SHORT, R.V.: "Breast feeding," *Scientific American*, April 1984, 35. Practical discussion of the advantages of breast feeding in infant nutrition and development.

WASSARMAN, P.M.: "Fertilization in Mammals," *Scientific American*, December 1988, p. 78. Describes the cellular mechanisms of capacitation, sperm penetration and fertilization.

"What Science Knows About AIDS," a single-topic issue of *Scientific American*, October 1988. Ten articles by leading scientists that present a comprehensive view of what is known about AIDS as this book goes to press.

Development

OVERVIEW

Vertebrate embryonic development may be divided artificially into several stages, although in reality these stages are parts of a continuous, dynamic process. Development starts with the union of egg and sperm to form a diploid zygote, followed by a series of rapid "cleavage" divisions, the pattern of which is influenced by the presence of yolk. The ball of cells that results from cleavage then differentiates into a structure resembling an indented tennis ball and possessing the three primary tissues. Although the details differ, this process occurs in all deuterostomes. The development of the specific tissues of the body follows. In chordates the notochord and dorsal nerve tube are among the first tissues to develop. Vertebrates go on to elaborate the neural crest, whose cells migrate to other positions in the developing embryo and ultimately give rise to most of the distinctive features of vertebrates. The developmental process takes longer in mammals than in other vertebrates; it lasts about 266 days in human beings.

FOR REVIEW

Here are some important terms and concepts that you will encounter in this chapter. If you are not familiar with them, you should review them before proceeding.

Terrestrial reproductive strategies (Chapter 22)

Ectoderm, mesoderm, endoderm (Chapter 39)

Coelom (Chapter 39)

Radial and spiral cleavage (Chapter 40)

Chordates (Chapter 42)

The amniotic egg (Chapters 22, 42)

Amnion and chorion (Chapter 54)

Sex is almost universal among the vertebrates. Reproduction in all but a few members of this subphylum involves two haploid gametes, which unite to form a single diploid cell called a zygote. This zygote grows by a process of cell division and differentiation into a complex multicellular animal, composed of many different tissues and organs (Figure 55-1). The process of development comprises the events that occur after the union of the two haploid gametes. Although some of its details differ from group to group, the process of development is fundamentally the same in all vertebrates.

In vertebrates, development occurs in six stages, outlined in Table 55-1. In this chapter we will discuss each of these stages in turn, consider the mechanisms governing developmental changes, and conclude with a detailed description of the events that occur during the course of human development.

INITIAL STAGE OF REPRODUCTION: FERTILIZATION

In vertebrates, as in all sexual animals, the first step in reproduction is the union of male and female gametes, a process called **fertilization.** Fertilization consists of three stages: (1) penetration, (2) activation, and (3) fusion. The male gametes of vertebrates, like those of other animals, are small, motile sperm. Each sperm is shaped like a comet, with a head containing a haploid nucleus and a long tail. Sperm are among the smallest cells in the body. The female gametes, called eggs or oocytes, are large cells. In many vertebrates the eggs contain significant amounts of yolk.

Penetration

In fishes and amphibians, fertilization is typically external, whereas in all other vertebrates it occurs internally. Internal fertilization is achieved when a mature egg is released into a body cavity, into which sperm can be introduced. There the egg can be fertilized by one of the many sperm introduced into the female reproductive tract during mating. The actively swimming sperm migrate up the oviduct until they encounter a mature egg.

Like a traveling princess, the mammalian egg is surrounded by a great deal of baggage (Figure 55-2). The egg cell itself is encased within an outer membrane called the **zona pellucida,** which is in turn surrounded by a protective layer of follicle cells. The first sperm to work its way through this barrier adheres to the egg membrane by

FIGURE 55-1
Development is the process that determines life's form and substance. A human fetus at 18 weeks is not yet halfway through the 38 weeks—about 9 months—it will spend within its mother, but already has developed many distinct behaviors, such as the sucking reflex that is so important to survival after birth.

TABLE 55-1	STAGES OF DEVELOPMENT
1. Fertilization	The male and female gametes form a zygote.
2. Cleavage	The zygote rapidly divides into many cells, with no overall increase in size. These divisions set the stage for development, since different cells receive different portions of the egg cytoplasm and hence different regulatory signals.
3. Gastrulation	The cells of the zygote move, forming three cell layers. These layers are the primary cell types: ectoderm, mesoderm, and endoderm.
4. Neurulation	In all chordates the first organ to form is the notochord, followed by formation of the dorsal nerve cord.
5. Neural crest formation	The first uniquely vertebrate event is the formation of the neural crest. From it develop many of the uniquely vertebrate structures.
6. Organogenesis	The three primary cell types then proceed to combine in various ways to produce the organs of the body.

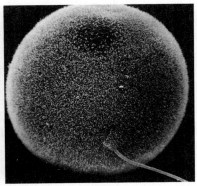
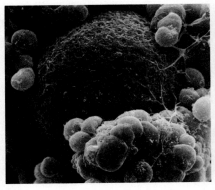

FIGURE 55-2

Mammalian reproductive cells.
The large cell on the left is a human ovum, surrounded by numerous nutritional cells. At the right, a hamster ovum has been penetrated by a solitary sperm cell.

the tip of the sperm cell head, the acrosome. From its acrosome, the sperm releases enzymes that cause the plasma membranes of the sperm and egg cell to fuse. Egg cytoplasm bulges out at this point, engulfing the head of the sperm and permitting the sperm nucleus to enter the cytoplasm of the egg (Figure 55-3).

Activation

Entry of the sperm nucleus into the egg has three effects:

1. The series of events initiated by sperm penetration is collectively called **egg activation.** In frogs, reptiles, and birds, many sperm may gain entry to the egg cell, but only the first one that enters is successful in fertilizing it. In mammals, by contrast, the penetration of the first sperm initiates changes in the egg cell membrane that prevent the entry of other sperm.

2. In the development of a mammalian egg, a polar body—one of the two products of the first meiotic division—is extruded from the egg cell as soon as that division has been completed. The second meiotic division does not occur until the first sperm has penetrated into the egg cell. Upon sperm penetration, the chromosomes in the egg nucleus complete meiosis, producing two egg nuclei. One of these two newly formed nuclei is extruded from the egg cell as a second polar body, leaving a single haploid egg nucleus within the egg.

3. A third effect of sperm penetration in vertebrates is the rearrangement of the egg cytoplasm. Around the point of sperm entry a series of cytoplasmic move-

FIGURE 55-3

Sperm penetration of a sea urchin egg.

A The stages of penetration.
B An electron micrograph of penetration (×50,000).

A

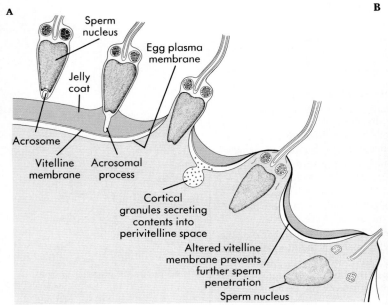

Sperm nucleus

Egg plasma membrane

Jelly coat

Acrosome

Vitelline membrane

Acrosomal process

Cortical granules secreting contents into perivitelline space

Altered vitelline membrane prevents further sperm penetration

Sperm nucleus

B

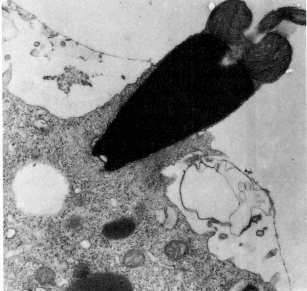

Half a million American women cannot have children because their oviducts are blocked or because they have no oviducts. Even though they have normal ovaries, produce normal eggs, and have a normal uterus, there is simply no place for their eggs to become fertilized, no highway for sperm to reach the uterus. In 1978 a child was born to such a mother. An egg was removed from the mother's ovary and fertilized on a glass plate with the father's sperm. The resulting zygote was allowed to grow and divide for 2 days, and then the embryo was placed in the mother's uterus. It embedded itself in the uterine wall, just as if it had arrived through her oviducts, and proceeded to develop. Two hundred sixty-four days later, Louise Brown was born, the world's first "test-tube baby."

The process that gave rise to Louise Brown is called "in vitro fertilization" (from the Latin word *vitro*, glass).

The procedure has now become a standard medical procedure performed in hundreds of clinics around the world. Many clinics handle more than 250 patients a year. Not all attempts are successful—indeed, less than a quarter of fertilization attempts "take." Still, thousands of babies now live that without this technique would not have been born.

In vitro fertilization has given rise to an entirely new realm of ethical questions. There is nothing in the procedure that requires the egg and sperm to be donated by a married couple, for example. What if the egg is donated by another woman, who later demands "her" child? Nor is there anything in the procedure that requires the uterus to be that of the egg donor. What if another woman carries the baby to term, and later demands "her" child? It is not clear what the answers to these ethical questions should be—the situations have never arisen before—but we

are going to have to decide what the answers are going to be, because they will continue to arise with increasing frequency.

Other avenues of in vitro fertilization present even more troubling ethical issues. An in vitro fertilized embryo need not be reimplanted immediately—the embryo can be frozen, stored, and used later by an entirely different couple. Should it be legal to charge a fee for this process—can embryos be sold? Or how about this: work with cattle has led to in vitro fertilization procedures that allow the fertilized egg to be dissected after several cleavage divisions in such a manner that all four or eight division products go on to form embryos and develop normally. All four or eight individuals are genetically identical clones. What if a couple wants to use this approach to make identical twins—should humans be cloned?

---

ments is initiated within the egg. These movements ultimately establish the bilateral symmetry of the developing organism. In frogs, for example, sperm penetration causes an outer pigmented cap of egg cytoplasm to rotate toward the point of entry, uncovering a **gray crescent** of interior cytoplasm opposite the point of penetration (Figure 55-4). The position of the gray crescent determines the orientation of initial cell division. A line drawn between the point of sperm entry and the gray crescent would bisect the right and left halves of the future adult.

In some vertebrates it is possible to activate an egg without the entry of a sperm, simply by pricking the egg membrane. If the development of any egg is stimulated in this way, the egg may go on to develop parthenogenetically. A few kinds of amphibians, fishes, and reptiles rely entirely on parthenogenetic reproduction in nature, as we mentioned in Chapter 54.

Fusion

The third stage of fertilization is the fusion of the entering sperm nucleus with the haploid nucleus of the egg to form a diploid zygotic nucleus. This fusion is triggered by the activation of the egg. If a sperm nucleus is introduced by microinjection without activation of the egg, fusion of the two nuclei will not take place. The nature of the signals that are exchanged between the two nuclei, or sent from one to the other, is not known.

The three stages of fertilization are penetration, activation, and fusion. Penetration initiates a complex series of developmental events, including major movements of cytoplasm, which eventually lead to the fusion of the egg and sperm nuclei.

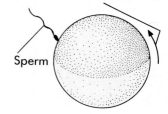

Movement of pigment opposite sperm entry

Sperm

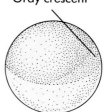

Gray crescent

FIGURE 55-4
Gray crescent formation in frogs.
Appearance of the gray crescent opposite the point of penetration.

FIGURE 55-5

Three kinds of eggs.
A In the primitive vertebrate *Amphioxus* the organization of the egg is simple, with a central nucleus surrounded by yolk.
B In a frog egg there is much more yolk, and the nucleus is displaced toward one pole.
C Bird eggs are complexly organized, with the nucleus astride the surface of a large central yolk like a spot painted on a balloon.

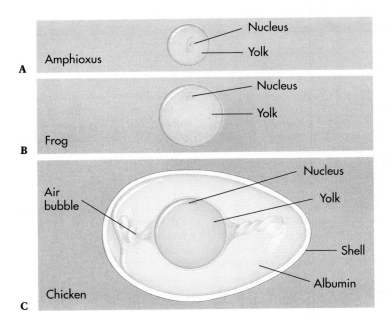

SETTING THE STAGE FOR DEVELOPMENT: CELL CLEAVAGE

The second major event in vertebrate reproduction is the rapid division of the zygote into a larger and larger number of smaller and smaller cells. This period of division, called **cleavage,** is not accompanied by any increase in the overall size of the embryo. The resulting tightly packed mass of about 32 cells is called a **morula,** and each individual cell in the morula is referred to as a **blastomere.** Blastomeres are by no means equivalent to one another. Different blastomeres contain different components of the egg cytoplasm; these components dictate different developmental fates for the cells in which they are present. The cells of the morula continue to divide without an overall increase in size, each cell secreting a fluid into the center of the cell mass. Eventually a hollow ball of 500 to 2000 cells, called a **blastula,** is formed. This hollow ball of cells surrounds a fluid-filled cavity called the **blastocoel.**

Cell Cleavage Patterns

The pattern of cleavage division is greatly influenced by the presence of the yolk (Figure 55-5). As we discussed in the last chapter, vertebrates have embraced a variety of reproductive strategies involving different patterns of yolk utilization.

Primitive Aquatic Vertebrates. When eggs contain little or no yolk, cleavage occurs throughout the whole egg (Figure 55-6). This pattern is called **holoblastic cleavage.** Holoblastic cleavage was characteristic of the ancestors of the vertebrates and is still seen in groups such as the lancelets and agnathans. It results in the formation of a symmetrical blastula, composed of cells of approximately equal size.

Amphibians and Advanced Fishes. The eggs of bony fishes and frogs contain much more yolk in one hemisphere than in the other. Because yolk-rich cells divide much more slowly than those which are poor in yolk, holoblastic cleavage of these eggs results in a very asymmetrical blastula (Figure 55-7), with large cells containing a lot of yolk at one pole and a concentrated mass of small cells containing very little yolk at the other.

Reptiles and Birds. Some eggs are composed almost entirely of yolk, with a small amount of cytoplasm concentrated at one pole. Such eggs are typical of reptiles, birds, and some fishes. In such yolk-rich eggs, cleavage occurs only in the tiny disk of polar cytoplasm, called the **blastodisc,** which lies astride the large ball of yolk mate-

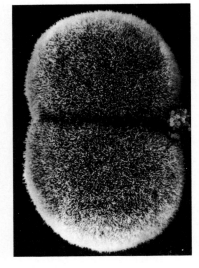

FIGURE 55-6
Holoblastic cleavage. Holoblastic cleavage is symmetrical, dividing an egg into equal portions. The egg dividing here is that of a mouse. Holoblastic cleavage was typical of the ancestors of the vertebrates.

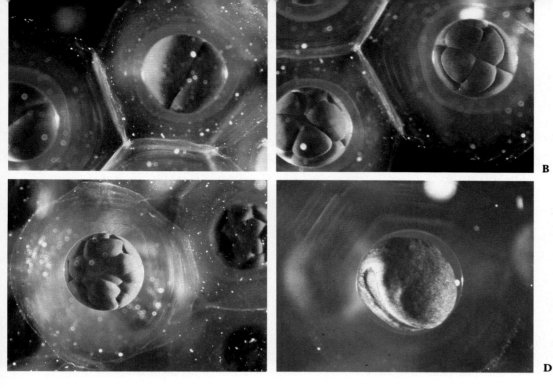

B

D

FIGURE 55-7
Cleavage divisions producing a frog embryo. The initial divisions (**A**) are, in this case, on the side of the zygote facing you, producing a cluster of cells on this side of the embryo (**B**), which soon expands to become a compact mass of cells (**C**). This mass eventually invaginates into the interior of the embryo (**D**), forming a gastrula-stage embryo.

rial (Figure 55-8). This type of cleavage pattern is called **meroblastic** cleavage. The resulting embryo is not spherical, but rather has the form of a hollow cap perched on the yolk.

Mammals. Mammalian eggs are in many ways similar to the reptilian eggs from which they evolved, except that they contain very little yolk. Because there is no mass of yolk to impede cleavage in mammalian eggs, the cleavage of the developing zygote is holoblastic. Such cleavage forms a ball of cells surrounding a blastocoel. In mammalian eggs, however, an inner cell mass is concentrated at one pole (Figure 55-9). This interior plate of cells is analogous to the blastodisc of reptiles and goes on to form the developing embryo. The outer sphere of cells is called a **trophoblast;** it is analogous to the cells that form the membrane that functions as a watertight covering and conserves water within the reptilian egg. These cells have changed during the course of mammalian evolution to carry out a very different function. The trophoblast

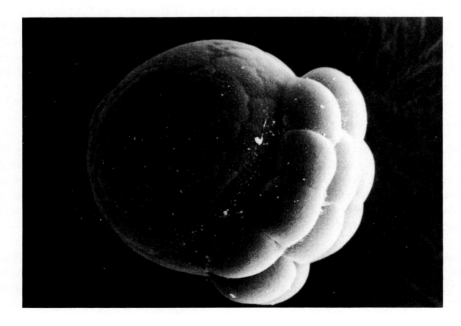

FIGURE 55-8
Meroblastic cleavage. Meroblastic cleavage is asymmetrical, with only a portion of a fertilized egg actively dividing to form a cell mass. The dividing cells of this fish embryo are magnified 400 times.

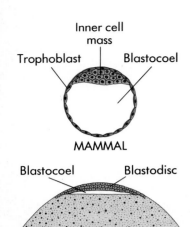

FIGURE 55-9
The eggs of mammals and birds are more similar than they seem. A mammalian blastula is composed of a sphere of cells, the trophoblast, surrounding a cavity, the blastocoel, and an inner cell mass. An avian (bird) blastula is a disk rather than a sphere, a blastodisc resting astride a large yolk mass; in bird eggs the blastocoel is a cavity between the blastodisc and the yolk.

develops into a complex series of membranes, the placenta, which connects the developing embryo to the mother's blood supply.

> A series of rapid cell divisions called cleavage divisions initiates development in the zygote. This produces a ball of cells called a blastula. The evolution of the amniotic egg in reptiles caused an alteration in the pattern of cleavage, related to the presence of a yolk. This kind of cleavage pattern is carried on by mammals, a reflection of their ancestry.

The Blastula

Viewed from the outside, the blastula looks like a simple ball of cells all resembling one another. The apparent close similarity of these cells is misleading, however. In fact, they differ from one another in three essential respects:

1. Each cell contains a different portion of cytoplasm derived from the egg.
2. Some cells are larger than others, containing more yolk and dividing more slowly.
3. Each cell is in contact with a different set of neighboring cells.

Egg cells contain many substances that act as genetic signals during the early stages of zygote development. These signal substances are not distributed uniformly within the egg cytoplasm, however. Instead, they are clustered at specific sites within the egg. The location of each site is genetically determined by information encoded on the mother's chromosomes. When the egg is activated during fertilization, its cytoplasm reorients itself with respect to the site of sperm entry. During the cleavage divisions that follow, the signal substances within this cytoplasm are partitioned into different daughter cells. The signals endow the different daughter cells with distinct developmental instructions. The egg is therefore prepatterned, since the pattern of its cytoplasm determines the future orientation of the different embryonic cells.

THE ONSET OF DEVELOPMENTAL CHANGE: GASTRULATION

The first visible results of prepatterning and of the cell orientation within the blastula can be seen immediately after the completion of the cleavage divisions. Certain groups of cells *move* inward from the surface of the sphere in a carefully orchestrated migration called **gastrulation**. Gastrulation forms the primary tissues ectoderm, endoderm, and mesoderm, and moves endoderm and mesoderm to the interior to form the gut and coelom.

How can cells move within a cell mass? Cell shape can readily be changed by microfilament contraction. Apparently the migrating cells creep over the stationary ones by means of a series of microfilament contractions. The migrating cells move as a single mass because they adhere to one another. How do cells "know" to which other cell to adhere? Within an adhering cell, genes have been expressed that bring about the synthesis of particular polysaccharides on the cell surface, which adhere to similar polysaccharides on the surfaces of the other adhering cells.

Alternative Patterns of Gastrulation

During gastrulation about half of the blastula's cells move into the interior of the hollow ball of cells. By doing so, they form a structure that looks something like an indented tennis ball. Just as the pattern of cleavage divisions in different groups of vertebrates depends heavily on the amount and distribution of yolk in the egg, so the pattern of gastrulation varies among the vertebrates, depending on the shape of the blastulas produced by the earlier cleavage divisions.

Aquatic Vertebrates. In fishes and other aquatic vertebrates with asymmetrical yolk distribution in their eggs, the blastula produced by the cleavage divisions has two distinct poles, one richer in yolk than the other. The hemisphere of the blastula that comprises cells rich in yolk is called the **vegetal pole**; the opposite hemisphere,

TABLE 55-2	DEVELOPMENTAL FATES OF THE PRIMARY TISSUES
Ectoderm	Skin, central nervous system, sense organs, neural crest
Mesoderm	Skeleton, muscles, blood vessels, heart, gonads
Endoderm	Digestive tract, lungs, many glands

comprising cells that are relatively poor in yolk, is called the **animal pole.** In primitive chordates such as lancelets, the surface bulges inward, invaginating into the blastocoel cavity. Eventually the inward-moving wall of cells pushes up against the opposite side of the blastula, and then it ceases to move. The resulting two-layered, cup-shaped embryo is the **gastrula.** The hollow crater resulting from the invagination is called the **archenteron,** and it becomes the progenitor of the gut. The opening of the archenteron, the future anus of the lancelets, is the **blastopore.**

Gastrulation in the lancelets produces an embryo with two cell layers, an outer ectoderm and an inner endoderm. A third cell layer, the mesoderm, forms soon afterward between these two layers from pouches pinched off the endoderm. The formation of these three primary cell types sets the stage for all subsequent tissue and organ differentiation, since the descendants of each cell type are destined to have very different developmental fates (Table 55-2).

In the blastula of amphibians and those aquatic vertebrates with asymmetrical yolk distribution, the yolk-laden cells of the vegetal pole are fewer and far larger than the yolk-free cells of the animal pole. Because of this cell distribution, invaginating the blastula at the vegetal pole is mechanically not feasible. Instead, a layer of cells from the animal pole folds down over the yolk-rich cells and then invaginates inward (Figure 55-10). The site where invagination begins is called the **dorsal lip.** As in the lancelets, the invaginating cell layer eventually eliminates the blastocoel cavity, its cells pressing against the inner surface of the opposite side of the embryo. In amphib-

FIGURE 55-10
Frog gastrulation.
A A layer of cells from the animal pole folds down over the yolk cells, forming the dorsal lip.
B The dorsal lip zone then invaginates into the hollow interior, or blastocoel, eventually pressing against the far wall. The three principal tissues (ectoderm, mesoderm, endoderm) become distinguished here.
C The inward movement of the dorsal lip creates a new internal cavity, the archenteron, which opens to the outside through the plug of yolk remaining at the point of invagination.
D The neural plate later forms from ectoderm and folds down over the surface of the embryo, **(E)** before moving to the interior.

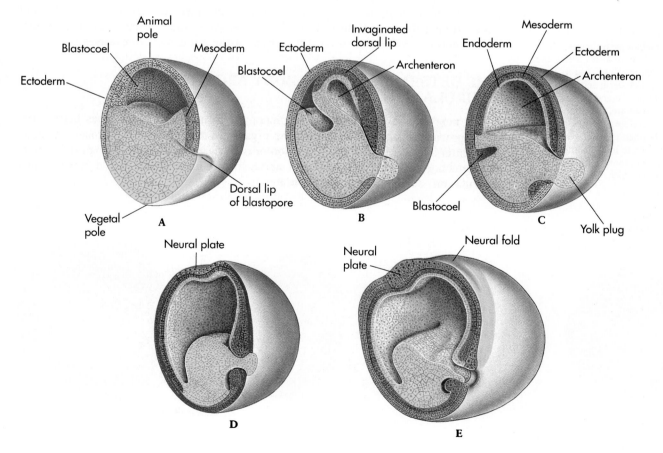

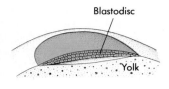

Blastodisc

Yolk

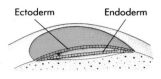

Ectoderm   Endoderm

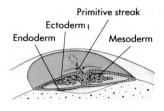

Primitive streak

Ectoderm

Endoderm   Mesoderm

FIGURE 55-11

Gastrulation of the chick blasto-disc. The upper layer of the blastodisc differentiates into ectoderm, the lower layer into endoderm. Among the cells that migrate into the interior through the dorsal primitive streak are future mesodermal cells.

FIGURE 55-12

Mammalian gastrulation. The amniotic cavity forms within the inner cell mass (**A**), and its base, layers of ectoderm and endoderm differentiate (**B** and **C**), as in the chick blastodisc. A primitive streak develops, through which cells destined to become mesoderm migrate into the interior (**D**), again reminiscent of gastrulation in the chick.

ians, as in fishes, the cavity produced by the invagination is called the archenteron, and its opening is called the blastopore. In this case the blastopore is filled with yolk-rich cells, the **yolk plug.** The outer layer of cells in the gastrula, which is formed as a result of these cell movements, is the ectoderm; and the inner layer is the endoderm. Cells from the dorsal lip migrate between these two cell layers, forming the mesoderm layer.

Reptiles, Birds, and Mammals. In the blastodisc of a chick the developing embryo is not shaped like a sphere. Instead, it is a hollow cap of cells situated over the animal pole of the large yolk mass. Despite this seemingly great difference between the developing embryos of birds, reptiles, and mammals on the one hand, and amphibians on the other hand, the pattern of establishment of the three primary cell layers is basically similar in all four groups.

No yolk separates the two sides of the blastodisc in reptiles, birds, and mammals (Figure 55-11). Consequently, the lower cell layer is able to differentiate without cell movement into endoderm and the upper layer into ectoderm. Just after this differentiation, the mesoderm layer arises by the invagination of cells from the upper layer inward, along the edges of a furrow that appears at the longitudinal midline of the embryo. The site of this invagination, which is analogous to an elongate blastopore, appears as a slit on the gastrula's surface. Because of its appearance, it is called the **primitive streak.** Gastrulation occurs at the site of formation of a primitive streak in the reptiles and their descendants, the birds and mammals (Figure 55-12).

> The many cells of the blastula gain unequal portions of egg cytoplasm during cleavage. This asymmetry results in the activation of different genes and a repositioning of cells with respect to one another, which establishes the three primary cell types: ectoderm, mesoderm, and endoderm.

The events of gastrulation determine the basic developmental pattern of the vertebrate embryo. By the end of gastrulation, the distribution of cells into the three primary tissues has been completed. Although the position of the yolk mass dictates changes in the details of gastrulation, the end result of the process is fundamentally the same in all deuterostomes: the ectoderm is destined to form the epidermis and neural tissue; the mesoderm to form the connective tissue, muscle, and vascular elements; and the endoderm to form the lining of the gut and its derivatives.

THE DETERMINATION OF BODY ARCHITECTURE: NEURULATION

In the next step in vertebrate development the three primary cell types begin their development into the body's tissues and organs. In all chordates the process of tissue differentiation begins with the formation of two characteristic morphological features, the notochord and the hollow dorsal nerve cord. This stage in development, called **neurulation,** occurs only in the chordates.

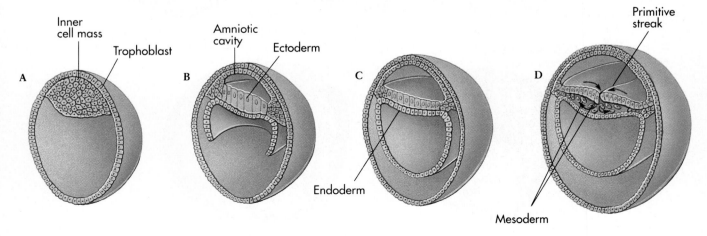

Inner cell mass

Trophoblast

A

Amniotic cavity

Ectoderm

B

C

Endoderm

Primitive streak

D

Mesoderm

The first of these two structures to form is the **notochord**. It is first visible soon after gastrulation is complete, forming from mesoderm tissue along the midline of the embryo, below its dorsal surface.

After the notochord has been laid down, the dorsal nerve cord forms. The region of the ectoderm that is located above the notochord later differentiates into the spinal cord and brain. The process is illustrated in Figure 55-13. First, a layer of ectodermal cells situated above the notochord invaginates inward, forming a long groove—called the **neural groove**—along the long axis of the embryo. The edges of this groove then move toward each other and fuse, creating a long hollow tube, the **neural tube,** which runs beneath the surface of the embryo's back.

> **The key developmental event that marks the evolution of the chordates is neurulation, the elaboration of a notochord and a dorsal nerve cord.**

While the neural tube is forming from ectoderm, the rest of the basic architecture of the body is being determined rapidly by changes in the mesoderm. On either side of the developing notochord, segmented blocks of tissue form. Ultimately, these blocks, or **somites,** give rise to the muscles, the vertebrae, and the connective tissue. As the process of development continues, more somites are formed progressively. Many of the significant glands of the body, including the kidneys, adrenal glands, and gonads, develop within another strip of mesoderm that runs alongside the somites. The remainder of the mesoderm layer moves out and around the inner endoderm layer of cells and eventually surrounds it entirely. As a result of this movement, the mesoderm forms a hollow tube within the ectoderm. The space within this tube is the coelom (Chapter 39); it contains the endoderm layers that ultimately form the lining of the stomach and gut.

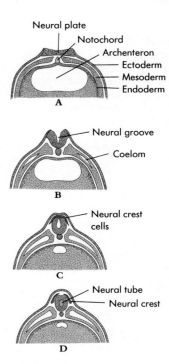

FIGURE 55-13
Neural tube formation. The neural tube forms above the notochord (**A**) when cells of the neural plate fold together to form the neural groove (**B**), which eventually closes (**C**) to form a hollow tube (**D**). As this is happening, some of the cells from the dorsal margin of the neural tube differentiate into the neural crest, which is characteristic of vertebrates.

EVOLUTIONARY ORIGIN OF THE VERTEBRATES: THE NEURAL CREST

Neurulation occurs in all chordates; the process is much the same in a lancelet as it is in a human being. The next stage in the development of a vertebrate, however, is unique to that group and is largely responsible for the characteristic body architecture of its members. Just before the neural groove closes over to form the neural tube, its edges develop a special strip of cells, the **neural crest,** which becomes incorporated into the roof of the neural tube (see Figure 55-13, *C* and *D*). Subsequently, the cells of the neural crest shift laterally to the sides of the developing embryo. The appearance of the neural crest was a key event in the evolution of the vertebrates, because neural crest cells, migrating to different parts of the embryo, ultimately develop into the structures characteristic of the vertebrate body.

The various cells of the neural crest strip develop very differently, depending on their location. At the anterior end of the embryo they merge with the anterior portion of the brain, the forebrain. Nearby clusters of ectoderm cells associated with the neural crest cells thicken into the placodes, structures that subsequently develop into parts of the sense organs of the head. The neural crest and associated placodes consist of two lateral strips, a pattern responsible for the fact that the sense organs of vertebrates come in pairs.

The remaining cells of the neural crest, located behind the anterior ones, have a very different developmental fate. These cells migrate away from the nerve tube to other locations in the head and trunk where they form connections between the nerve tube and the surrounding tissues. At these new locations they dictate the development of a series of different structures, discussed below, that are particularly characteristic of the vertebrates. This migration of neural crest cells is unique in that it is not simply a change in the relative position of the cells, such as is seen in gastrulation. Instead, the migrating neural crest cells actually pass through other tissues.

Structures Derived from the Neural Crest

Many of the differences between vertebrates and the more primitive chordates from which they evolved are related to the products of the neural crest. The characteristic

FIGURE 55-14
The gill chamber. The stiff bars of the gill chamber (in red) are readily visible in a fish embryo—and in the embryos of the major vertebrate groups that evolved from fishes.

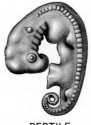
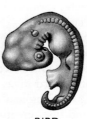
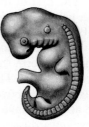

FISH   REPTILE   BIRD   HUMAN

structures that are derived from neural crest cells or from placodes induced by the anterior neural crest include the following body elements.

The Gill Chamber. Primitive chordates such as lancelets were filter-feeders, using the rapid beating of cilia to draw water into their bodies through slits in their pharynx. These pharyngeal slits have become elaborated greatly in the course of vertebrate evolution, forming the gill chamber, a structure that provides a greatly improved means of respiration (Figure 55-14). The evolution of the gill chamber was certainly a key event in making possible the transition from filter-feeding to active predation.

In the development of the gill chamber, some of the neural crest cells form cartilaginous bars between the embryonic pharyngeal slits. Other neural crest cells induce portions of the mesoderm to form muscles along the cartilage. Still others form nerves between the nerve cord and these muscles. A major blood vessel, called an aortic arch, passes through each of the bars. Lined by still more neural crest cells, these bars, with their internal blood supply, become highly branched and form the gills.

Because the stiff bars of the gill chamber can be bent inward by powerful muscles controlled by nerves, the whole structure is a very efficient pump, which serves to drive water past the gills. The gills themselves act as highly efficient oxygen exchangers, greatly increasing the respiratory capacity of the organism. The structure and operation of the gill chamber were described in more detail in Chapter 50.

Elaboration of the Nervous System. Some neural crest cells migrate downward toward the notochord and form sensory ganglia. Others become specialized as Schwann cells, which insulate nerve fibers, thus permitting the rapid conduction of nerve impulses. Most vertebrate sensory neurons are derived from neural crest cells. Other neural crest cells form the adrenal medulla, the key element of the sympathetic nervous system. The adrenal medulla produces adrenaline at times of stress or danger, allowing the animal to respond. Together, these structures make possible a great improvement in the ability of the nervous system to respond to sensory information precisely and quickly.

Skull and Sensory Organs. A variety of sense organs develop from the placodes, which are formed from the cells of the anterior neural crest. Included among them are the olfactory (smell) and lateral line (primitive hearing) organs, which were discussed in Chapter 46. The teeth develop from neural crest cells, as do the cranial bones that protect the brain, which is the enlarged anterior end of the nerve cord.

The Role of the Neural Crest in Vertebrate Evolution

The adaptations just described, and still others that are associated with neural crest cells, are thought to have played a key role in the evolution of the vertebrates (Figure 55-15). The primitive chordates were initially slow-moving, filter-feeding animals with relatively low metabolic rates. Adaptations derived from neural crest allowed the vertebrates to assume a very different ecological role. The vertebrates became fast-swimming predators with much higher metabolic rates. This metabolic pattern allowed for a much increased level of activity than was possible for the vertebrate ancestors. Other evolutionary changes associated with the derivatives of the neural crest

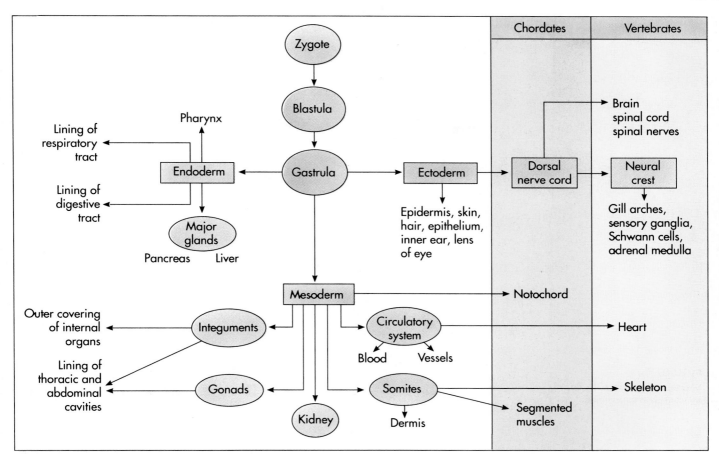

FIGURE 55-15
Derivation of the major tissue types. The key role of the neural crest is evident from the many characteristically vertebrate features that derive from it.

made possible improved detection of prey, a better ability to orient spatially during the capture of prey, and a greatly improved ability to respond quickly to sensory information. The evolution of the neural crest cells and of the structures derived from them was therefore a crucial step in the evolution of the vertebrates.

The appearance of the neural crest in the developing embryo marked the beginning of the first truly vertebrate phase of development, since many of the structures that are characteristic of vertebrates are derived directly or indirectly from neural crest cells.

HOW CELLS COMMUNICATE DURING DEVELOPMENT

In the process of vertebrate development (Figure 55-16), the relative position of particular cell layers determines, to a large extent, the organs that develop from them. By now, you may have wondered how these cell layers know where they are. For example, when cells of the ectoderm situated above the developing notochord give rise to the neural groove, how do these cells know they are above the notochord?

The solution to this puzzle is one of the outstanding accomplishments of experimental embryology, the study of how embryos form. The great German biologist Hans Spemann and his student Hilde Mangold solved it early in this century. In their investigation they removed cells from the dorsal lip of an amphibian blastula (Figure 55-17) and transplanted them to a different location on another blastula. (The dorsal lip region of amphibian blastulas develops from the gray crescent zone and is the site of origin of those mesoderm cells that later produce the notochord.) The new location corresponded to that of the animal's future belly. What happened? The embryo developed *two* notochords, a normal dorsal one and a second one along its belly!

By using genetically different donor and host blastulas, Spemann and Mangold were able to show that the notochord produced by transplanting dorsal lip cells

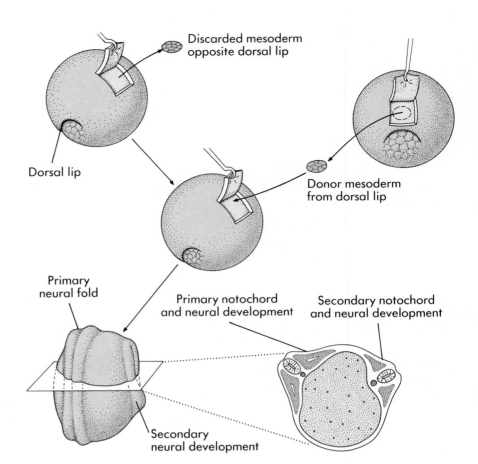

Discarded mesoderm
opposite dorsal lip

Dorsal lip

Donor mesoderm
from dorsal lip

Primary
neural fold

Primary notochord
and neural development

Secondary notochord
and neural development

Secondary
neural development

FIGURE 55-16
Spemann and Mangold's dorsal
lip transplant experiment.

contained host cells as well as transplanted ones. The transplanted dorsal lip cells had acted as **organizers** of notochord development. As such, these cells stimulated a developmental program in the belly cells of the embryos in which they were transplanted: the development of a notochord. The belly cells clearly contained this developmental program but would not have expressed it in the normal course of their development. The transplantation of the dorsal lip cells caused them to do so. These cells had indeed induced the ectoderm cells of the belly to form a notochord. This phenomenon as a whole is known as **induction.**

> **Induction is the determination of one tissue's course of development by another tissue.**

The process of induction that Spemann discovered appears to be the basic mode of development in vertebrates. Inductions between the three primary tissue types—ectoderm, mesoderm, and endoderm—are referred to as **primary inductions.** Inductions between tissues that have already been differentiated are called **secondary inductions.** The differentiation of the central nervous system during neurulation by the interaction of dorsal ectoderm and dorsal mesoderm to form the neural tube is an example of primary induction. In contrast, the differentiation of the lens of the vertebrate eye from ectoderm by interaction with tissue from the central nervous system is an example of secondary induction.

The eye develops as an extension of the forebrain, a stalk that grows outward until it comes in contact with the epidermis (Figure 55-17). At a point directly above the growing stalk, a layer of the epidermis pinches off, forming a transparent lens. When the optic stalks of the two eyes have just started to project from the brain and the lenses have not yet formed, one of the budding stalks can be removed and transplanted to a region underneath a different epidermis, such as that of the belly. When Spemann performed this critical experiment, a lens still formed, this time from belly epidermis cells in the region above where the budding stalk had been transplanted.

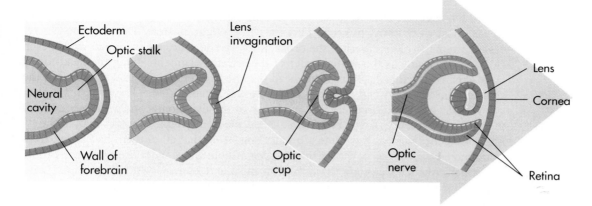

The chemical nature of the induction process is not known in detail. If one interposes a nonporous barrier, such as a layer of cellophane, between the inducer and the target tissue, no induction takes place. A porous filter, in contrast, does permit induction to occur. It is thought that the inducer cells produce a protein factor that binds to the cells of the target tissue, stimulating mitosis in them and initiating changes in gene expression.

THE NATURE OF DEVELOPMENTAL DECISIONS

All of the cells of the body, with the exception of a few specialized ones that have lost their nuclei, have an entire complement of genetic information. Despite the fact that all of its cells are genetically identical, an adult vertebrate contains hundreds of cell types, each expressing various aspects of the total genetic information for that individual. What factors determine which genes are to be expressed in a particular cell and which are not to be? In a liver cell, what mechanism keeps the genetic information that specifies nerve cell characteristics turned off? Does the differentiation of that particular cell into a liver cell entail the physical loss of the information specifying other cell types? No, it does not—but cells progressively lose the capacity to *express* ever-larger portions of their genomes. *Development is a process of progressive restriction of gene expression.*

Some cells become **determined** quite early during development. For example, all of the egg cells of a human female are set aside very early in the life of the embryo, yet some of these cells will not achieve differentiation into functional oocytes for more than 40 years. To a large degree, a cell's location in the developing embryo determines its fate. By changing a cell's location, an experimenter can alter its developmental destiny. However, this is only true up to a certain point in the cell's development. At some stage, every cell's ultimate fate becomes fixed and irreversible, a process referred to as **commitment.**

When a cell is "determined," it is possible to predict its developmental fate; when a cell is "committed," that developmental fate cannot be altered. Determination often occurs very early in development, commitment somewhat later.

The Homeobox

The molecular mechanisms of determination and commitment are not well understood, but remarkable progress has been made in the 1980s by researchers studying embryonic development in *Drosophila*. Sixty-six years ago a remarkable *Drosophila* mutation was described in which the mutant fly, called bithorax, had its third body segment converted into a second body segment—giving it a double set of wings. As more such segment-altering genes (called **homoeotic genes**) were discovered in *Drosophila*, most were found to group into one of two tight clusters. Detailed analysis revealed a remarkable regularity: the order in which the genes are arranged within each cluster is the same as the spatial order of the affected segments along the head-to-tail body axis: the further back from the head the affected region was, the further into the cluster from the right end the homeotic gene was located.

In 1983 a crucial discovery was made: all of the fly's homoeotic genes were found to include a sequence of 180 base pairs, called the **homeobox**. The fact that all the fly homeotic genes possess this same sequence suggested that all of them had arisen from a common ancestral gene. Using the *Drosophila* homeobox sequence as a gene probe, the same sequence was soon discovered in mice and humans—mammals don't have the same kind of segmentation that flies do, but we know that the mammal homeobox genes also specify structure formation: in 1991 researchers demonstrated that mutations of individual mouse homeobox genes after specific regions in the mouse embryo such as the head or neck where the genes are active in development. Homeobox genes have now been found even in such simple animals as sea urchins and jellyfish, suggesting that this method of controlling development evolved very early in life's history. Somewhat modified homeobox sequences have been reported in plants.

The fruitfly has about 10 homeobox genes, located in two clusters on one chromosome. Mice and humans have at least 40 of the genes grouped mainly in four clusters of about 10 genes each, which are located on different chromosomes (a few others are present elsewhere). What do homeobox genes do? Proteins encoded by homeobox genes are transcription factors—they bind to DNA and regulate when genes are transcribed. Each homeobox protein is composed of a variable region, which determines a protein's specific regulatory activity, linked to a 60 amino acid sequence encoded by the homeobox sequence. This sequence twists into four alpha-helixes, one of which recognizes and binds to a specific DNA sequence in the target genes. Homeobox proteins appear to be master switches in development, and are the subject of intense study today.

The way in which positional information is used in development seems to have been carefully conserved during evolution. All four mammalian clusters are organized just as is the fly cluster, the genes lined up in the same order within the clusters although 500 million years have elapsed since the insects and mammals have diverged. Just as in the fruit fly, the genes expressed are at the 3' or righthand end of the cluster, affecting the front (anterior) end of the developing body; as genes progressively more leftward in the cluster are expressed, zones further back on the embryo are affected. The first-expressed 3' gene is the one that most closely resembles in cnidarian homeobox genes, suggesting that the original homeobox genes were involved in regulating the formation of anterior structures in the metazoan (multicellular animal) body plan. The genes needed for the formation of more posterior body structures presumably evolved later, by duplication of existing homeotic genes, to yield the cluster of 10 genes seen in insects. During the course of vertebrate evolution the whole cluster duplicated at least twice as species became more complicated, ultimately producing the four clusters seen in mammals today.

Positional information is encoded by a cluster of genes that regulate when parts of the developmental program are transcribed.

THE COURSE OF HUMAN DEVELOPMENT

The development of the human embryo shows its evolutionary origins. If we did not have an evolutionary perspective, we would be unable to account for the fact that human development proceeds in much the same way as development in the chick. In both embryonic chickens and embryonic human beings, the blastodisc is flattened. In a chick egg, for example, the blastodisc is pressed against a yolk mass; in a human embryo the blastodisc is similarly flat, despite the absence of a yolk mass. In human blastodiscs a primitive streak forms and gives rise to the three primary cell types, just as it does in the chick blastodisc.

Human development takes much longer than chicken development, an average of 266 days from fertilization to birth, the familiar "9 months of pregnancy." What may not be so readily apparent, however, is how very early the critical stages of development outlined in this chapter occur during the course of human pregnancy.

First Trimester

The First Month. In the first week after fertilization occurs, the fertilized egg undergoes cleavage divisions. The first of these divisions occurs about 30 hours after the fusion of the egg and the sperm, and the second, 30 hours later. Cell divisions continue until a blastodisc forms within a ball of cells. During this period the embryo continues the journey that the egg initiated down the mother's oviduct. From 3 to 6 days later the embryo reaches the uterus, attaches to the uterine lining, or endometrium, and penetrates into the tissue of the lining. The embryo begins to grow rapidly, initiating the formation of membranes. One of these membranes, the amnion, will enclose the developing embryo, while another, the chorion, will interact with uterine tissue to form the placenta responsible for nourishing the growing embryo.

In the second week after fertilization, gastrulation takes place. The primitive streak can be seen on the surface of the embryo, and the three primary tissue types are differentiated. Around the developing embryo the placenta starts to form from the chorion.

In the third week neurulation occurs. This stage is marked by the formation of the neural tube along the dorsal axis of the embryo, as well as by the appearance of the first somites, from which the muscles, vertebrae, and connective tissue develop. By the end of the week, over a dozen somites are evident, and the blood vessels and gut have begun to develop. At this point the embryo is about 2 millimeters long.

In the fourth week organogenesis (the formation of body organs) occurs (Figure 55-18, *A*). The eyes form. The tubular heart develops its four chambers and begins to pulsate, its rhythmical beating stopping only with death. At 70 beats per minute, the little heart is destined to beat more than 2.5 billion times during a lifetime of 70 years. Over 30 pairs of somites are visible by the end of the fourth week, and the arm and leg buds have begun to form. The embryo more than doubles in length during this week, to about 5 millimeters.

All of the major organs of the body have begun their formation by the end of the fourth week of development. Although the developmental scenario is now far advanced, many women are not aware that they are pregnant at this stage.

Early pregnancy is a very critical time in development, since the proper course of events can be interrupted easily. In the 1960s, for example, many pregnant women took the tranquilizer thalidomide to minimize discomforts associated with early pregnancy. Unfortunately, this drug had not been adequately tested. It interferes with fetus limb bud development, and its widespread use resulted in many deformed babies. Also during the first and second months of pregnancy, a mother's contracting rubella (German measles) can upset organogenesis in the developing embryo. Most spontaneous abortions occur in this period.

The Second Month. Morphogenesis (the formation of shape) takes place during the second month (Figure 55-18, *B*). The miniature limbs of the embryo assume their adult shapes. The arms, legs, knees, elbows, fingers, and toes can all be seen—as well as a short bony tail! The bones of the embryonic tail, an evolutionary reminder of our past, later fuse to form the coccyx. Within the body cavity, the major organs, including the liver, pancreas, and gall bladder, become evident. By the end of the second month, the embryo has grown to about 25 millimeters in length, weighs perhaps a gram, and begins to look distinctly human.

The Third Month. The nervous system and sense organs develop during the third month (Figure 55-18, *C*). By the end of the month, the arms and legs begin to move. The embryo begins to show facial expressions and carries out primitive reflexes such

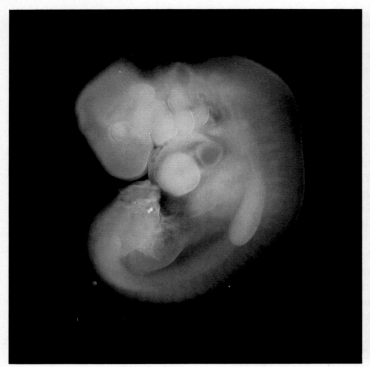

A

B

FIGURE 55-18
The developing human.
A 4 weeks.
B 7 weeks.
C 3 months.
D 4 months.

as the startle reflex and sucking. By the end of the third month, all of the major organs of the body have been established. Development of the embryo is essentially complete. From this point on, the developing human being is referred to as a fetus rather than an embryo. What remains is essentially growth.

Second Trimester

In the fourth and fifth months of pregnancy (Figure 55-18, *D*), the fetus grows to about 175 millimeters in length, with a body weight of about 225 grams. Bone enlargement occurs actively during the fourth month. During the fifth month the head and body become covered with fine hair. This downy body hair, called **lanugo,** is another evolutionary relict but is lost later in development. By the end of the fourth month, the mother can feel the baby kicking. By the end of the fifth month, she can hear its rapid heartbeat with a stethoscope. In the sixth month growth begins in earnest. By the end of that month, the baby weighs 0.6 kilogram (about 1 1/2 pounds) and is over 0.3 meter (1 foot) long; but most of its prebirth growth is still to come. The baby cannot yet survive outside the uterus without special medical intervention.

Placental Development.   As the embryo continues to grow the fetus develops a circulatory system, and the trophoblast enlarges to form a *placenta*. In the placenta the fetal blood is brought in close proximity to the maternal blood. However, the fetal and maternal bloods do not mix. The **maternal-fetal** barrier protects the placenta and fetus from a maternal immune response. Two **umbilical arteries** branch from the fetal descending aorta and pass into the umbilical cord that connects the fetus with the placenta. A single **umbilical vein** returns blood from the placenta to the fetus.

By the beginning of the second trimester, the placenta reaches its full development and secretes estrogen and progesterone. During the second and third trimesters the plasma levels of estrogen and progesterone rise continuously, reaching a peak at the time of labor. In addition to estrogen and progesterone, the placenta secretes a hormone called *human placental lactogen (HPL)*. Placental lactogen is similar to growth hormone and prolactin. It stimulates breast development in preparation for lactation, supports fetal bone growth, and alters maternal metabolism by substituting lipids for glucose for energy. The placenta also secretes antidiuretic hormone, aldosterone, and renin.

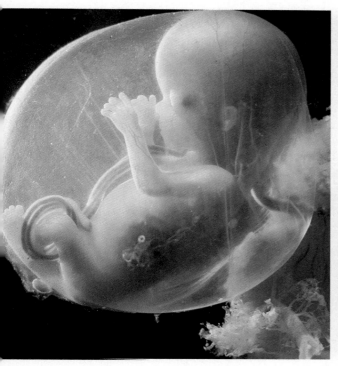

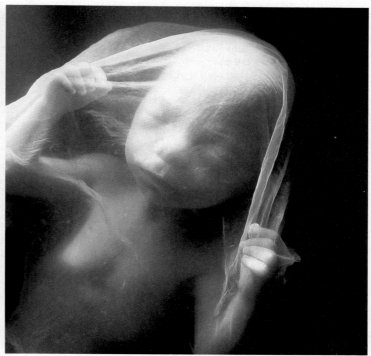

C

D

Third Trimester

The third trimester is predominantly a period of growth, rather than one of development. In the seventh, eighth, and ninth months of pregnancy the weight of the fetus doubles several times. This increase in bulk is not the only kind of growth that occurs, however. Most of the major nerve tracts in the brain, as well as many new brain cells, are formed during this period. The mother's bloodstream fuels all of this growth by the nutrients it provides. Within the placenta these nutrients pass into the fetal blood supply (Figure 55-19). The undernourishment of the fetus by a malnourished mother can adversely affect this growth and result in severe retardation of the infant. Retardation resulting from fetal malnourishment is a severe problem in many underdeveloped countries where poverty is common.

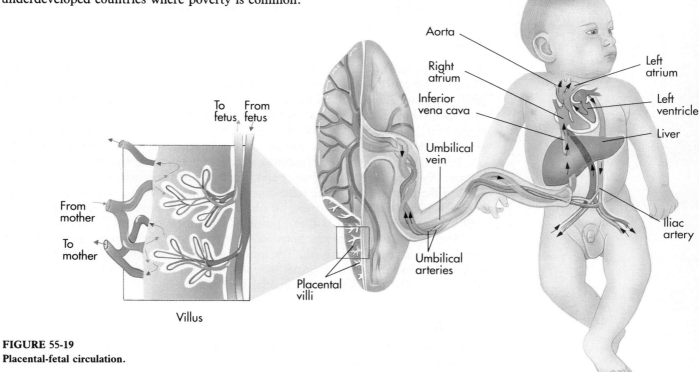

FIGURE 55-19
Placental-fetal circulation.

FIGURE 55-20
Position of the fetus just before
birth. A developing fetus is a major addition to a woman's anatomy. The stomach and intestines
are pushed far up, and there is
often considerable discomfort
from pressure on the lower back.
In a natural delivery the fetus exits through the vagina, which
must dilate (expand) considerably
to permit passage.

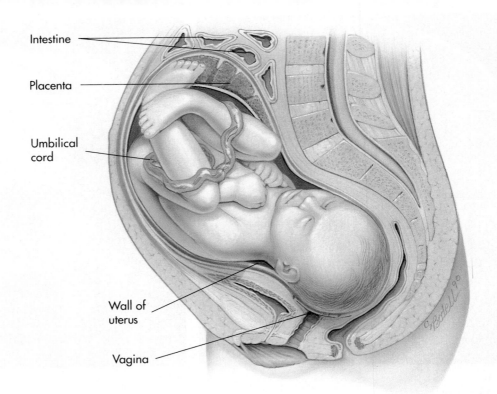

Intestine

Placenta

Umbilical
cord

Wall of
uterus

Vagina

By the end of the third trimester, the neurological growth of the fetus is far from complete, and, in fact, continues long after birth. By this time, however, the fetus is able to exist on its own. Why doesn't development continue within the uterus until neurological development is complete? Because physical growth would continue as well, and the fetus is probably as large as it can be for safe delivery through the pelvis (Figure 55-20). Birth takes place as soon as the probability of survival is high. For better or worse, the infant is then on its own, a person.

**The critical stages of human development take place quite early. All the ma-
jor organs of the body have been established by the end of the third month.
The following 6 months are essentially a period of growth.**

Birth

Birth takes place at the end of the third trimester, some 266 days after fertilization. Changing hormone levels in the developing fetus initiate the process. First, the fetal pituitary gland secretes the hormone ACTH, which stimulates the fetal adrenal glands to release steroid hormones that induce maternal placental cells to manufacture prostaglandins that help induce powerful uterine muscle contractions. Acting in parallel, the pressure of the baby's head against the cervix sends nerve impulses to the mother's brain that trigger the hypothalamus to release the hormone oxytocin from the pituitary. Working together, oxytocin and prostaglandins stimulate contractions in the walls of the uterus, forcing the fetus downward. Initially, only a few contractions occur each hour, the rate increasing to one every 2 to 3 minutes. Eventually, strong contractions, aided by the mother's pushing, expel the fetus, now a newborn baby.

After birth, uterine contractions continue and expel the placenta and associated membranes, called the afterbirth. The umbilical cord is still attached to the baby, and to free the newborn a doctor or midwife ties and cuts it; blood clots in the cord and contraction of its muscles prevent excessive bleeding.

Nursing

In the third trimester, the mammary glands begin to produce and store small amounts of a yellowish fluid called **colostrum,** which is both nutritious and rich in maternal

FIGURE 55-21
Nursing is an important bonding time between mother and child. It also provides the child with the protection of its mother's immune system while its own develops.

antibodies. When a newborn begins to nurse, this is its first food. Because the placenta is gone, there is no placental progesterone or estrogen to inhibit the production of prolactin; the infant's instinctive sucking on the mother's nipples (Figure 55-21) sends signals to the mother's brain that initiate the production of the hormones prolactin and oxytocin. Prolactin stimulates the production of true milk, and oxytocin stimulates milk secretion. This is why new mothers often experience cramps while nursing—the oxytocin induces uterine muscle contraction, just as it did during labor and delivery.

Milk production occurs in the alveoli of mammary glands. Milk from the alveoli is secreted into a series of **alveolar ducts,** surrounded by smooth muscle, that eventually lead to the nipple. Although the plasma levels of prolactin are high at the end of pregnancy, milk production does not begin because it is inhibited by estrogen and progesterone. In effect, during pregnancy the breast is primed for milk production but not activated. The drop in estrogen and progesterone after delivery allows milk production to begin. Milk production is sustained by continued prolactin secretion as long as suckling continues.

Many mothers nurse for a year or longer. During this period, important pair bonding occurs between the mother and child. When nursing stops, the accumulation of milk sends signals to the brain that cause prolactin secretion to stop, and milk production to end. Any woman who has nursed a child can resume nursing later; an infant's sucking on her nipples reinitiates prolactin secretion and milk production.

POSTNATAL DEVELOPMENT

Growth continues rapidly after birth. Babies typically double their birth weight within 2 months. Different organs grow at different rates, however. The reason that adult body proportions are different from infant ones is that different parts of the body grow at different rates or cease growing at different times. The head, for example, which is disproportionately large in infants, grows more slowly in infants than does the rest of the body. Such a pattern of growth, in which different components grow at different rates, is referred to as **allometric growth.**

In most mammals, brain growth is entirely a fetal phenomenon. In chimpanzees, for example, the growth of the brain and the cerebral portion of the skull rapidly decelerate after birth, while the bones of the jaw continue to grow. The skull of an adult chimpanzee, therefore, looks very different from that of a fetal chimpanzee. In human beings, on the other hand, the brain and cerebral skull continue to grow at the same rate after birth as before. During gestation and after birth the developing human brain generates neurons (nerve cells) at an average rate estimated at more than 250,000 per minute; it is not until about 6 months after birth that this astonishing production of new neurons ceases permanently. Because both brain and jaw continue to grow, the jaw-skull proportions do not change after birth; and the skull of an adult human being looks very similar to that of a human fetus. It is primarily for this reason that a young human fetus seems so incredibly adultlike.

In the last 13 chapters, you have learned what an adult vertebrate body is like and how it functions. You have examined the various tissues of the vertebrate body and then considered how vertebrates eat, breathe, and sense the world around them. All the complexity you have encountered, the beauty and cleverness of design, are established during the developmental process we have briefly considered in this chapter. Development is not simply the beginning of life—it is the process that determines life's form and substance.

SUMMARY

1. Fertilization is the union of an egg and a sperm to form a zygote. Fertilization is external in fishes and amphibians, internal in all more advanced vertebrates. The three stages of fertilization are (1) penetration, in which the sperm cell moves past the cells surrounding the egg and penetrates the egg membrane; (2) activation, in which a series of cytoplasmic movements is initiated by penetration; and (3) fusion, in which the sperm and egg nuclei fuse.

2. Cleavage is the rapid division of the newly formed zygote into a mass of perhaps a thousand cells, without any increase in overall size. Because the egg is structured with respect to the location of developmentally important regulating signals, the future embryo becomes structured by the cleavage divisions. These divisions, in effect, partition the egg cytoplasm into small portions that contain different regulatory elements.

3. Gastrulation is the mechanical movement of portions of the blastula, forming the three basic tissues: ectoderm, endoderm, and mesoderm. In eggs that lack a yolk the movement is one of simple invagination. When a yolk is present, the movement of the cells is affected by it. In amphibians the cell layers move down and around the yolk. In reptiles, birds, and mammals, cells establish the three primary cell types as an upper layer (ectoderm), a lower layer (endoderm), and a layer that invaginates inward from the upper layer (mesoderm).

4. Neurulation in chordates is the formation of the first secondary tissues, particularly the notochord and the dorsal nerve cord, from the primary tissues.

5. The formation of the neural crest is the first development event unique to vertebrates. Most of the distinctive structures associated with vertebrates are derived from cells of the neural crest.

6. Cells influence one another during development by a process of induction. In this process, substances that exist on the surface of one cell induce other cells to divide.

7. At some point during animal development the ultimate developmental fate of cells becomes fixed and unalterable. The cells are then said to be committed, even though they may not exhibit any of the characteristics they will eventually assume.

8. Most of the critical events in the development of a human occur in the first month. Cleavage occurs during the first week, gastrulation during the second week, neurulation during the third week, and organ formation during the fourth week.

9. The second and third months of the first trimester are devoted to morphogenesis and to the elaboration of the nervous system and sensory organs. By the end of this period, the development of the embryo is essentially complete.

10. The last 6 months of human pregnancy are essentially a period of growth, devoted to increase in size and to formation of nerve tract within the brain. Most of the weight of a fetus is added in the final 3 months of pregnancy.

REVIEW QUESTIONS

1. What are the three stages of fertilization? How is an individual egg cell protected while traveling through the oviduct? What initial role does the acrosome play in fertilization?

2. Describe the cleavage patterns in primitive vertebrates, advanced fishes/amphibians, reptiles/birds, and mammals. Consider the name associated with the pattern, the amount of yolk present in the egg, and the resultant blastula.

3. What is the general appearance of the blastula of aquatic vertebrates with asymmetrical yolk distribution? How does gastrulation alter this appearance? What part of this gastrula is the archenteron? What is the opening of this structure called? What is the future fate of this opening?

4. Why is gastrulation in amphibians different from that in fish? How does this change the cell migration pattern in amphibians? Where is the dorsal lip of the gastrula located? What is the yolk plug? How does mesoderm form?

5. What stage of development occurs only in chordates? What two unique structures form as a result of this process? From what type of tissue do they arise? What are the basic events of this process? What processes occur in the mesodermal tissue at this time?

6. Which stage of development occurs only in vertebrates? From which tissue does the primary structure of this stage form? Why is this stage special? What is the fate of this tissue? How is this stage tied to the evolution of the vertebrate animal?

7. What is the process of induction as it relates to vertebrate development? What is the difference between primary and sec-ondary induction? Give an example of each. What is the chemical nature of induction?

8. What developmental processes occur within the first month of the first trimester of human pregnancy? What are the functions of the amnion and chorion in human development? When do gastrulation, neurulation, and organogenesis occur? Why is this stage of pregnancy critical?

9. What developmental processes occur during the second trimester of human pregnancy? What is lanugo? Would a fetus born late in this stage be able to survive?

10. What occurs during the third trimester of pregnancy? What neurological growth occurs at this time? Is neurological growth complete at birth? Explain. Why is proper nourishment of the mother imperative at this time of pregnancy?

THOUGHT QUESTIONS

1. In reptiles and birds the fetus is basically masculine, and fetal estrogen hormones are necessary to induce the development of female characteristics. In mammals the reverse is true, the fetus being basically female, with fetal hormones acting to induce the development of male characteristics. Can you suggest a reason why the pattern that occurs in reptiles and birds would not work in mammals?

2. Female armadillos always give birth to four offspring of the same sex. Can you suggest a mechanism that would account for this?

FOR FURTHER READING

BOGIN, B.: "The Evolution of Human Childhood," *Bio Science*, January 1990, pages 16-25. Among mammals, only the human species has childhood as a step in the life cycle.

BEARDSLEY, T.: "Smart Genes," *Scientific American*, August 1991, pages 86-95. A very readable account of recent work on the molecular control of development.

CERAMI, A., H. VLASSARA, and M. BROWMLER: "Glucose and Aging," *Scientific American*, May 1987, pages 90-96. An argument that glucose contributes to age-associated declines in tissue functioning by permanently altering some proteins.

COOKE, J.: "The Early Embryo and the Formation of Body Pattern," *American Scientist*, February 1988, pages 35-42. A few common control mechanisms appear to underlie early developmental events in most animals.

DALE, B.: *Fertilization in Animals*, Edward Arnold, Publishers, London, 1983. A brief text devoted entirely to the process of animal fertilization, with good illustrations and up-to-date discussions of physical mechanisms.

DeROBERTIS, E., and others: "Homeobox Genes and the Vertebrate Body Plan," *Scientific American*, July 1990, pages 46-52. These genes determine the shape of the body by subdividing the embryo along the tail-to-head axis into groups of cells that eventually become limbs and other structures.

DETHIER, V.: "The Magic of Metamorphosis: Nature's Own Sleight of Hand," *Smithsonian*, May 1986, pages 122-131. A delightful account of genomes that encode two organisms. Fun to read and beautifully illustrated.

GANS, C., and R.G. NORTHCUTT: "Neural Crest and the Origin of Vertebrates: A New Head," *Science*, vol. 220, pages 268-274, 1983.

NILSSON, L., and J. LINDBERG: *Behold Man*, Little, Brown & Co., Boston, 1974. A wonderful collection of color photographs of the developing human fetus.

RUGH, R., and L.B. SHETTLES: *From Conception to Birth: The Drama of Life's Beginnings*, Harper & Row, Publishers, New York, 1971. A detailed treatment of human development, from fertilization to birth, with excellent photographs.

SAUNDERS, J.: *Developmental Biology*, Macmillan Publishing Co., New York, 1982. A very clear undergraduate text, easily understood by students with little background. A particularly strong point of this text is its emphasis on how experiments have established the basic information.

TRINKAUS, J.: *Cells into Organs: The Forces That Shape The Embryo*, ed. 2, Prentice-Hall, Inc., Englewood Cliffs, N.J., 1984. An advanced but easily understood discussion of the physical mechanisms underlying developmental change.

ULMANN, A., and others: "RU 486," *Scientific American*, June 1990, pages 42-48. A controversial drug now widely used in France to terminate unwanted pregnancies.

Animal Behavior

OVERVIEW

Animals use information in the environment to find food or a place to live, avoid predators, and locate mates. This information is coded in the form of visual, acoustical, and chemical stimuli. Behavior is crucial to survival and reproduction and is a mechanism of response to the environment. The responses an animal shows to environmental stimuli are mediated by its nervous system and physiological condition. Genetics and learning may both influence behavior; instinct and experience often interact in complex ways during development to form behavior.

FOR REVIEW

Here are some important terms and concepts that you will encounter in this chapter. If you are not familiar with them, you should review them before proceeding.

Neurons and interneurons (Chapter 44)

Memory and learning (Chapter 45)

Reproductive isolation (Chapter 21)

Sensing the environment (Chapter 46)

Hormonal control of physiological processes (Chapter 47)

During the past two decades the study of animal behavior has emerged as an important and diversified science that often bridges different disciplines within biology. Evolution, ecology, physiology, genetics, and psychology all provide natural and logical linkages with the study of behavior.

Research in animal behavior has made major contributions to understanding the organization of the nervous system, child development, and human communication as well as the process of speciation, community organization, and the mechanism of natural selection itself. The study of the behavior of non-human animals has been applied, often controversially, to human social behavior and has changed the way we perceive ourselves.

Behavior can be defined as the way an organism responds to a stimulus in its environment; the stimulus might be as simple as the odor of food. In this sense, a bacterial cell "behaves" by moving toward higher concentrations of sugar. This pattern of movement is a very simple behavior, but it is one of a variety of simple responses that are suited to the life of bacteria and allow these organisms to live and reproduce. During the course of the evolution of multicellular animals, organisms occupied different environments and faced diverse problems that affected survival and reproduction. The nervous system and behavior concomitantly became more complex. The behavior shown by any animal is appropriate to its lifestyle; more elaborate forms of behavior evolved to meet the demands of the environment. Peripheral and central nervous systems perceive and process information provided by stimuli in the environment and trigger an adaptive motor response, which we see as a pattern of behavior.

When we observe animal behavior, we can explain it in two different ways. First, we might ask *how* it all works, that is, how the animal's senses, nerve networks, or internal state provide a physiological basis for the behavior. In this way, we would be asking a question of **proximate causation**. To analyze the proximate cause, or mechanism of behavior, hormone levels might be measured or the firing patterns of nerve cells recorded. We could also ask *why* the behavior evolved; that is, what was its adaptive value. This is a question concerning **ultimate causation**. To study the ultimate or evolutionary cause of a behavior we would measure how it influenced the animal's survival or reproductive success. Thus a male songbird may sing during the breeding season because his level of testosterone, a steroid sex hormone, is high enough to stimulate enough hormone receptors in the brain to trigger the production of song; this is the proximate cause of birdsong. But the male sings to defend a territory from other males and to attract a female to reproduce; this is the ultimate, or evolutionary, explanation for the male's vocalization.

In this chapter we will consider the mechanisms by which an animal responds to its environment, as well as the way in which behavior develops in an individual. We will discuss different approaches to the study of behavior, some focusing on instinct, some emphasizing learning, still others combining the elements of both approaches. The picture that will emerge is one of behavioral biology as a diverse science that draws strongly from allied disciplines such as neurobiology, physiology, and psychology. Today, the overarching theme of the study of behavior is evolution. Although the evolution and ecology of behavior will be the focus of Chapter 57, in this chapter we also examine how the environment has influenced the neurobiological, hormonal, and developmental processes that mediate animal behavior.

APPROACHES TO THE STUDY OF BEHAVIOR

The study of behavior has had a long history of controversy. One theme of the controversy concerns the importance of genetics and learning. Do genes determine behavior, or does learning and experience shape behavior? Is behavior, therefore, the result of **nature** (instinct) or **nurture** (experience)? Although in the past it has been argued as an "either/or" situation, many studies have shown that both instinct and experience play significant roles, often interacting in complex ways to produce the final behavioral product. The scientific study of instinct and learning, as well as their relationship, has led to the growth of several disciplines such as ethology, behavioral genetics, behavioral neuroscience, and psychology.

FIGURE 56-1

The founding fathers of ethology. Karl von Frisch, Konrad Lorenz, and Niko Tinbergen were the pioneers of ethology. In 1973 they received the Nobel Prize for their pathbreaking contributions to behavioral science. von Frisch led the study of honeybee communication and sensory biology. Lorenz focused on social development (imprinting) and the natural history of aggression. Tinbergen examined the functional significance of behavior and was the first behavioral ecologist.

Ethology

Ethology is the study of the natural history of behavior. Because of their training as zoologists and evolutionary biologists, fields that emphasize the study of animal behavior under natural conditions, ethologists (Figure 56-1) believed that behavior was largely instinctive, or innate, and resulted from the programming of behavior by natural selection. Ethologists emphasized that because behavior was often **stereotyped** (appearing in the same form in different individuals of a species), it was based on programmed neural circuits. These circuits are structured from genetic blueprints, and cause animals to show a relatively complete behavior the first time it is produced.

For example, geese incubate their eggs in a nest. If a goose notices an egg that has been knocked out of the nest accidentally, the goose will extend its neck towards the egg, get up, and roll the egg back into the nest with a side-to-side motion of its neck while the egg is tucked beneath the bill. Even if the egg is removed during retrieval, the goose completes the behavior, as if it was driven by a program which was released by the initial sight of the egg outside of the nest. According to ethologists, egg retrieval behavior is triggered by detecting a **sign stimulus,** the egg out of the nest; a component of the goose's nervous system, the **innate releasing mechanism,** provides the neural instructions for the motor program, or **fixed action pattern** (Figure 56-2). More generally, the sign stimulus is the "signal" in the environment that triggers the behavior. The innate releasing mechanism refers to the sensory mechanism detecting the "signal," and the fixed action pattern in the stereotyped act. Similarly, a frog unfolds its long, sticky tongue at the sight of an insect, and a male stickleback fish will attack another male showing a bright red underside. Such responses

FIGURE 56-2

Lizard prey capture. The complex series of muscular contractions this chameleon uses to capture an insect represents a fixed action pattern.

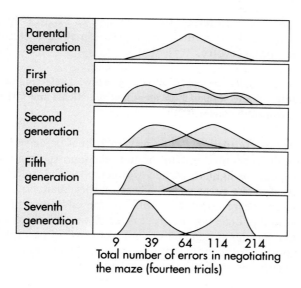

FIGURE 56-3
The genetics of learning. Tryon was able to select among rats for ability to learn to run a maze, demonstrating that this ability is directly influenced by genes. He tested a large group of rats, selected the few that ran the maze in the shortest time, and let them breed with one another; he then tested their progeny and again selected those with the quickest maze-running times for breeding. By seven generations he succeeded in halving the average time an inexperienced rat required to negotiate the maze. Parallel selection for slow running time was also successful—it more than doubled the average running time.

certainly appear to be programmed and instinctive, but what evidence supports the ethological view that behavior has an underlying genetic and neural basis?

Behavioral Genetics. In a famous experiment carried out in the 1940s, Robert Tryon studied the ability of rats to find their way through a maze with many blind alleys and only one exit, where a reward of food awaited (Figure 56-3). It took a while, as false avenues were tried and rejected, but eventually some individuals learned to zip right through the maze to the food, making few incorrect turns. Other rats never seemed to learn the correct path. Tryon bred the "maze-bright" rats to one another, establishing a colony from the fast learners, and similarly established a second "maze-dull" colony breeding the slowest learning rats to each other. He then tested the offspring in each colony to see how quickly they learned the maze. The offspring of maze-bright rats learned even more quickly than their parents had, while the offspring of maze-dull parents were even poorer at maze learning. After repeating this procedure over several generations, Tryon was able to produce two behaviorally distinct types of rat with very different maze-learning ability. Clearly the ability to learn the maze was to some degree hereditary, governed by genes that were being passed from parent to offspring. And the genes are specific to this behavior, rather than being general ones that influence many behaviors—the abilities of these two groups of rats to perform other behavioral tasks such as running a completely different kind of maze did not differ.

Tryon's research is an example of how a study can illustrate the genetic component of a behavior. Under natural conditions, rats with an ability to learn quickly may have had some advantage that led to increased survival and reproductive success.

Studies of hybrids have also provided support for the hypothesis of a genetic basis of behavior. William Dilger of Cornell University studied the genetic basis of nest material–carrying behavior in lovebirds by producing in the laboratory a hybrid of two species that differ in the way twigs, paper, and other material used to build a nest are carried. *Agapornis personata* holds nest material in the beak whereas *Agapornis roseicollis* carries material tucked under its flank feathers (Figure 56-4). The hybrid offspring of the two species carry nest material in a way that seems intermediate between that of the parents: they repeatedly shift nest material from the bill to the flank feathers. Other hybrid studies conducted with cricket and treefrog courtship song similarly illustrate the intermediate nature of the behavior of hybrids.

The understanding of behavioral genetics has also benefited from the use of mutants to identify the alleles that code for particular behavior defects. One such male fruit fly behavioral mutant does not disengage from a female to end a mating; others are unable to learn that a female has already been mated and is therefore unreceptive to any additional attempt to mate. And it has recently been shown that the rhythm of

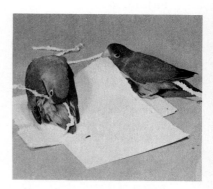

FIGURE 56-4
Genetics of lovebird behavior. The tendency to carry nest material—such as these paper strips—in the rump feathers is inherited.

a courtship song may be transferred from one species of fruit fly to another via a DNA sequence. This "molecular ethology" approach may be useful in understanding how closely-related species evolve.

> The genetic basis of behavior can be shown by artificial selection and hybridization studies. Some studies have identified the alleles and DNA sequences which control behavior.

The Neural Basis of Behavior.
The way an animal perceives its world depends upon the structure of its sensory system. Sound reception is often limited to frequencies and the perception of light to certain wavelengths. The sense of smell decodes chemical information in the air or water. These sensory channels of **audition, vision, and olfaction** govern the response to stimuli in the environment. Research on the neurobiology of stimulus detection has revealed the way in which animal senses and other aspects of the nervous system are organized to govern perception and behavior, thus providing support for the ethologist's concept of the innate releasing mechanism. The study of the neural basis of behavior is called **neuroethology.**

Nerve cells are often specialized for the detection of certain stimuli. Frogs and toads prey on insects by flipping out a sticky tongue. To do this, a toad must be able to "track" prey in the environment; this is accomplished with light-sensitive nerve cells located in the retina of each eye. These retinal cells enable the toad to identify prey, such as insects or worms. As an insect moves in front of a toad, its image enters the lens of the eye and is focused on the retina. The image thus moves over groups of retinal receptor cells that fire nerve impulses and relay information via ganglion cells (which process information from several retinal cells) to the brain (Figure 56-5).

A toad attempts to strike its tongue at and capture what it sees depending upon the size of the object on which it has focused and whether the image is moving horizontally. Thus, the innate releasing mechanism which responds to the sign stimuli provided by an insect and turns on the fixed action pattern of flipping out the tongue at the "target" may be thought of as a group of specialized "insect detector" cells in the toad's retina.

Some animals have been extremely useful as "models" for research on the neural basis of behavior because they have relatively simple nervous systems and show behaviors that are easy to record. A sea hare has a brain that features a small number of very large individual nerve cells (Figure 56-6). A simple, yet important, behavior shown by the sea hare is its **escape response,** which is made up of a series of contractions and relaxations of muscles that cause its body to flex dorsally and ventrally and thus move. The response occurs when the sea hare detects the presence of one of its

FIGURE 56-5
Prey detectors in the eye of the toad. Light-sensitive cells (rods and cones) embedded in the retina monitor the object motion. Ganglion cells receive input from bipolar cells, which in turn process signals from a group of light-receptor cells. Each ganglion cell has a receptive field. It monitors receptor activity in a portion of the retina's surface. Large objects cast an image on more of the surface of the retina than does the movement of small objects such as insects. Large objects inhibit ganglion cells from relaying impulses to the optic nerve, whereas small objects cause the ganglion cells to fire action potentials via the optic nerve to the brain.

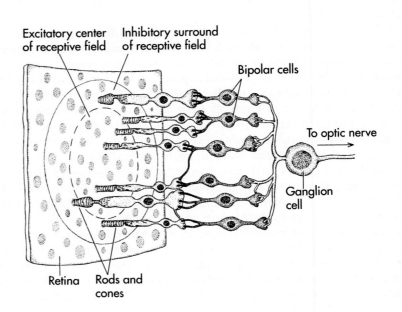

Excitatory center of receptive field

Inhibitory surround of receptive field

Bipolar cells

To optic nerve

Ganglion cell

Retina

Rods and cones

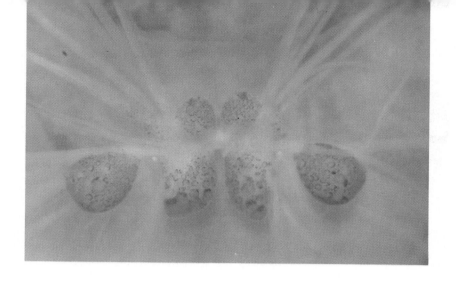

FIGURE 56-6
Small brain, large nerve cells.
The brain of this sea hare has relatively few large neurons, making this animal easy to study for neurobiologists.

predators, a sea star. The motor patterns that comprise the escape behavior result from impulses fired from three groups of neurons in the brain. One group of cells, the dorsal flexion neurons, stimulate the muscles which cause the animal to flex dorsally; a second group of cells, the ventral flexion neurons, likewise have a neural output regulating ventral flexion. A third group of neurons inhibits the triggering of action potentials in opposing groups of flexion neurons. So while the dorsal flexion neurons are firing, the ventral flexion neurons are inhibited from firing.

The neurobiology of escape behavior has also been examined in the cockroach. In this case, **mechanoreceptors,** sensory hairs located on appendages at the end of the abdomen, detect microcurrents of air and fire nerve impulses that are transmitted rapidly along a giant interneuron to a thoracic ganglion where they cause motor neurons to fire to initiate locomotion. The response time is extremely fast—it takes only about 60 milliseconds from the time the air movement is perceived to initiate the escape behavior.

Studies of the nerve networks that are the basis of response to stimuli support the concept of the innate nature of behavior. In the examples of the escape behaviors described above, it is clear that components of the nervous system evolved in response to ecological pressures like predation and the need to capture food. Research on the neuroethology of simple behaviors also helps us understand the neural units from which more complex nervous systems and behavior may be assembled.

Psychology and the Study of Animal Behavior

In direct contrast to the instinct theory of behavior that was developed by ethologists, other students of animal behavior focused almost exclusively on learning as the major element that shapes behavior. These **comparative psychologists** were not interested in naturalistic studies or evolutionary theory, and worked primarily in laboratory settings on rats. The main contribution of comparative psychology was to identify the ways by which animals learn.

Learning is defined as a modification of behavior that arises as a result of experience rather than as a result of maturation. There are generally considered to be categories of learning, or ways in which learning occurs. The simplest type of learning is **nonassociative,** that is, it does not require the animal to form an association between two stimuli, or between a stimulus and a response. Learned behaviors that require such "pairing" within the central nervous system are termed **associative.**

One form of nonassociative learning is **habituation,** which can be defined as a decrease in response to a repeated stimulus having no positive or negative consequences (that is, there is no *reinforcement*). Habituation is learning *not* to respond to a stimulus. Learning to ignore unimportant stimuli is a critical ability to an animal confronting a barrage of stimuli in a complex environment. In many cases, the stimulus

FIGURE 56-7
Learning what is edible. Associative learning is involved in predator/prey interactions. **(A)** A naive toad is offered a bumblebee as food. **(B)** It is stung, and **(C)** subsequently avoids feeding on bumblebees or any insects resembling their black and yellow coloration. The toad has associated the appearance of the insect with pain, and modifies its behavior.

evokes a strong response when it is first encountered, but then the magnitude of response gradually declines after repeated exposure. Young birds see many types of objects moving overhead; at first sight, they may crouch down and remain still in response to these stimuli. Some of these objects—like falling leaves or members of their own species flying by—are seen very frequently, and have no positive or negative consequence to the nestling. Over time, the young bird may stop responding; it habituates to the stimuli. **Sensitization** is basically the opposite of habituation. It is a process which causes an animal to show an increased response to a stimulus.

Associative learning is more complex than habituation or sensitization. Thinking is associative (Figure 56-7), as are other cognitive behaviors. Associative learning is the alteration of behavior by experience leading to the formation of an association between two stimuli or between a stimulus and a response. The mechanisms by which associative learning occurs can be grouped according to the way stimuli become associated. Behavior is thus modified, or **conditioned,** through the association.

Types of Associative Behavior.   Two major types of associative learning are recognized; these are **classical conditioning** and **operant conditioning.**

The repeated presentation of a stimulus in association with a response can cause the brain to form an association between them, even if the stimulus and the response have never been associated before. Classical conditioning is also called **Pavlovian conditioning,** after the Russian psychologist Ivan Pavlov, who first described this type of learning. Pavlov presented meat powder, an *unconditioned stimulus,* to a dog, and noted that the dog responded by salivating, an *unconditioned response.* If a second unrelated stimulus, such as the ringing of a bell, was presented at the same time that the meat powder was presented, over repeated trials the dog would salivate in response to the sound of the bell *alone.* The response of salivation was thus conditioned; the sound of the bell is the **conditioned stimulus.** The dog had learned to associate the unrelated sound stimulus with the meat powder stimulus. Early experimenters believed that *any* stimulus could be linked in this way to any response, although as you will see below, we now know this is not true.

FIGURE 56-8
Skinner box. The rat rapidly learns that pressing the lever results in the appearance of a food pellet. This kind of learning, trial and error with a reward for success, can also work for far more complex tasks.

In classical conditioning, learning does not influence whether or not the reinforcing stimulus is received. In **operant conditioning,** by contrast, the reward follows only after the animal shows the correct behavioral response, so the animal must make the proper association before it receives the reinforcing stimulus, or reward. The American psychologist B.F. Skinner studied such conditioning in rats by placing them in an apparatus that came to be called a "Skinner box" (Figure 56-8). Once inside, the rat would explore the box; occasionally, it would accidentally press a lever, and a pellet of food would appear. At first, the rat would ignore the lever and continue to move about, but soon it learned to press the lever to obtain food. When it was hungry, it would spend all of its time pushing the bar. This sort of trial-and-error learning is of major importance to most vertebrates. Comparative psychologists used to believe that animals could be conditioned to perform any learnable behavior in response to any stimulus by operant conditioning, but as in the case of classical conditioning, this is not so. Today, instinct is credited with the major role in guiding what type of information can be learned, that is, what types of stimuli may be paired through association.

> Habituation is a simple form of learning in which there is no association between a stimulus and a response. Associative learning involves either classical conditioning or operant conditioning. In associative learning an animal learns to pair a certain response with a certain stimulus.

Instincts and Learning

Roughly 20 years ago, most behavioral biologists came to believe that behavior has both genetic and learned components and the polarization of the schools of ethology and psychology drew to an end. It has become clear, for example, that certain types of learning are not always possible. Some animals have innate predispositions towards forming certain associations. Such **learning preparedness** means that what an animal can learn is guided genetically. For example, if rats are offered food pellets at the same time they are exposed to x-rays (which later produces nausea), the rat will remember the taste of the food but not the size of the pellet. And if rats are given electric shock (which is immediately painful) during a feeding they will remember the size of the pellet, but not its taste.

Similarly, pigeons associate colors with food but cannot make associations between sounds and food; they *can* associate sounds and danger—but cannot associate colors with danger. And honey bees learn to associate a reward of food with some colors more rapidly than others. The sorts of associations that are possible are genetically determined. That is, conditioning is possible only within boundaries set by instinct. Animals are innately programmed to learn some things more readily than oth-

We do not know a great deal about the neural basis of learning, although intriguing glimpses of what the mechanisms may be are emerging from studies of simple systems. Some of the most interesting results have been obtained by Eric Kandel and associates with a marine slug, *Aplysia*, (Figure 56-A). *Aplysia* possesses only a few large neurons, a property that makes its reactions simple enough to study in detail. Like many mollusks, this animal has a single large gill, which is protected by a sheet of tissue called the mantle. The mantle ends in a spout called the siphon. A gentle tap on the mantle or siphon of *Aplysia* will cause the animal's gill to withdraw into its central cavity. This response involves only two neurons, a sensory neuron and a motor neuron with which it synapses directly.

Aplysia can learn to modify this withdrawal response, and its learning exhibits habituation, sensitization, and classical conditioning.

Habituation. If the mantle is gently prodded several times, *Aplysia* will learn to ignore it.

Sensitization. An electric shock applied to its tail will render *Aplysia* more sensitive to prodding; when the mantle is prodded very gently after a shock, so gently that the prod would not previously have elicited a response, *Aplysia* withdraws its gill abruptly.

Classical conditioning. When the tail of *Aplysia* is shocked a short time after a gentle tap on the siphon, the gill is withdrawn rapidly. After about 15 exposures to this pair of stimuli, a gentle tap alone elicits the strong withdrawal.

The habituation of gill withdrawal in *Aplysia* results from the overloading of the Ca^{++} ion channels in the mantle's sensory neuron synapses. When a wave of depolarization reaches the presynaptic membrane, it opens Ca^{++} channels. The inrushing calcium causes the synaptic vesicles to release neurotransmitter into the synaptic cleft. The re-

FIGURE 56-A
Aplysia.

peated stimulation of the same nerve uses up all available neurotransmitter more quickly than it can be supplied. Thus repeated stimulation leads to the gradual loss of function of these calcium channels, with the result that less neurotransmitter is released.

The sensitization of *Aplysia* gill withdrawal involves an interneuron. Stimulation of the tail sensory nerve cell stimulates this interneuron, which synapses with the mantle's sensory neuron, to release the inhibitory neurotransmitter serotonin. The serotonin blocks the K^+ channels of the mantle's sensory neuron. When a weak action potential arrives along this sensory neuron, the blockage of the K^+ channels prolongs the action potential and thus allows the calcium channels of its presynaptic membrane to stay open longer, releasing more neurotransmitter into the synaptic cleft and thereby facilitating gill withdrawal.

Conditioning of gill withdrawal in *Aplysia* occurs by a form of enhanced sensitization. If a sensory neuron has just "fired," its synapse is more efficient. Apparently the residual Ca^{++} from the preceding release of neurotransmitter lowers the level of Ca^{++} that is required to achieve a new release. In *Aplysia* conditioning the greater activity of

the mantle's sensory nerve-motor synapse (sensitization) that is elicited by the repeated shocks facilitates the passage of subsequent impulses across the synapse, even in the absence of the shocks. Conditioning in this case appears to be only a special form of sensitization.

We do not know whether all learning processes depend on the same simple repertory of habituation and sensitization that is responsible for learning in *Aplysia*. One idea that is rapidly gaining acceptance is that the circuits of the vertebrate brain are fully wired at birth or soon thereafter and that no new connections are added by the growing brain. Instead, sensory experience changes the strength of the various synapses; behaviors capable of conditioning or other learning could thus be modified. Such modification might occur by means of the selective facilitation of the kind we see in *Aplysia*.

Aplysia teaches us that learning need not involve the laying down of new circuits in the brain. In simple animals, learning involves modification of the strength of preexisting synapses, strengthening some (sensitization) and weakening others (habituation). In humans, however, it is thought that some new circuits are established.

ers. Instinct determines the boundaries of learning. Innate programs have evolved because they underscore adaptive responses. Rats, which forage at night and have a highly developed sense of smell, are better able to identify dangerous food by its odor than size or color. The seed that a pigeon eats may have a distinctive color that it can see, but it makes no sound that a pigeon can hear. Evolution has biased behavior with instincts which make adaptive response more likely. Today, the study of learning has expanded toward its neurobiological and molecular basis (unnumbered box 56-1) or its ecological significance.

The Development of Behavior

Thus far in this chapter we have discussed the influence of genes and the role of learning in shaping behavior. For many years researchers in behavior were divided on the importance of instinct and experience. This dichotomy of approaches to the study of behavior formed the basis of the **nature versus nurture controversy** which reflected the polarization of researchers who believed behavior was either innate or learned. Of course, neither of these extreme positions is entirely correct, except in the simplest cases. Today, behavioral biologists believe that instinct and experience interact during development, each contributing to the formation of behavior.

Experience begins very early in life, even before an animal is born. If pregnant female rats are stressed, their offspring will be less active than those of control (unstressed) females. And in humans, the infants of stressed mothers grow more slowly and show higher activity levels. After birth, a number of important experiences are crucial in the development of behavior.

As an animal matures after birth, it forms attachments to other individuals and develops preferences. **Imprinting** is a process involved in the formation of social bonds. It is sometimes considered to be a special form of learning. In **filial imprinting** social attachments are formed between parents and offspring. For example, a few hours after hatching young birds will begin to follow their mother; their following response results in a bond between mother and young. Actually the young bird will follow the first object it sees after hatching and direct its social behavior toward that object. Lorenz raised geese from eggs, and offered himself as a model for imprinting. The goslings treated him as if he was their parent, and followed him dutifully (Figure 56-9). But young birds will follow *any* object and imprint on it. Black boxes, flashing lights, and watering cans are all effective imprinting objects (Figure 56-10). Imprinting is characterized by a **sensitive phase,** or a period of time after hatching or birth during which social bonds can form. During the sensitive phase, which in ducks occurs during the first 43 hours after hatching, there is also a **critical period** at roughly 13 to 16 hours when the following response, and thus the intensity of attachment formation is the greatest.

Several studies have shown that social interactions between parents and offspring occurring during a sensitive phase of development are crucial to the normal development of behavior. Harry Harlow, a psychologist, gave orphaned rhesus monkey infants the opportunity to form social attachments with two "surrogate" mothers, one of which was a cloth model, the other a wire model (Figure 56-11). The infants chose

FIGURE 56-9
An unlikely parent. The eager goslings following Konrad Lorenz think he is their mother. He is the first "animal" they saw when they hatched, and they have used him as a model for imprinting.

A   B

FIGURE 56-10
How imprinting is studied. Although the first object seen by a duckling is usually its mother, any object can be used to study imprinting. The device in (A) rotates a black box in an arena; it is followed by ducklings. In (B), ducklings have imprinted to a white sphere.

FIGURE 56-11
A young rhesus monkey clings to a terry-cloth substitute.

to spend time with the cloth "mother," even if the wire mother provided food. Texture and warmth are key qualities of a mother that promote infant social attachment. If infants are deprived of normal social contact, their development is abnormal. The greater the degree of deprivation, the greater the abnormality in social behavior as juveniles and adults. Studies on orphaned human infants suggest that a constant "mother figure" is required for normal growth and psychological development.

Recent research has shown that there is a biological need for the stimulation that occurs during parent/offspring interactions early in life. Female rats lick their pups after birth; this stimulation inhibits the release of an endorphin that can block normal growth. Pups receiving normal tactile stimulation also have more receptors in the brain for glucocorticoid hormones, are more tolerant of stress, and their brain cells have greater longevity. And premature human infants who are massaged gain weight rapidly. These studies indicate that the need for normal social interactions is based in the brain. Touch and other aspects of contact between parents and offspring are important to the development of behavior.

Interactions which occur during sensitive phases of imprinting are critical to normal behavioral development. Social attachments are formed, and physical contact is needed for psychological well-being and growth.

Recognition of members of one's own species also occurs during development. **Sexual imprinting** is a process in which an individual learns the identifying characteristics of a species. This information is acquired early in life and is used when an animal is sexually mature and courts a mate. In **cross-fostering** studies, individuals of one species are raised by parents of another species to determine if species recognition is innate or learned. In most species of birds these studies have shown that, when sexually mature, the fostered bird will try to mate with members of its foster species. Try as they will, these ardent suitors are ignored by the objects of their sexual attention.

Behavior develops as a result of interaction of instinct and experience. The work of Peter Marler and his colleagues on how white-crowned sparrows acquire their courtship song provides an excellent example. The song is sung by mature males and is characteristic only of the white-crowned sparrow species. By rearing male birds in soundproof incubators provided with a speaker and a microphone, Marler could completely control what a bird heard as it matured and could record on tape the song it produced as an adult. He found that birds that heard no song at all during growth sang a poorly developed song as adults. When played the song of the song sparrow, a related bird species, males also sang a poorly developed song as adults. But when both the white-crowned and song sparrow songs were played with developed, mature males, they sang a fully-developed, white-crowned sparrow song. If a young bird is surgically deafened after it is played a white-crowned sparrow song, it also sings a poorly developed song as an adult (Figure 56-12). This research suggests that birds have a **genetic template**, or instinctive program, to guide learning the appropriate

FIGURE 56-12
Song development in birds.
Sonograms of songs produced by white-crowned sparrow males exposed to their own species' song during development (A) or which heard no song during rearing (B). This illustrates that the genetic program itself is insufficient to produce a normal song.

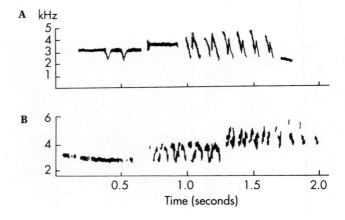

song. During a critical period the template accepts the correct song as a model, and the bird "practices" listening to himself sing, matching what he hears to the model his template has accepted. Song acquisition is based on learning, but only the song of the *correct* species can be learned. The genetic template for learning is *selective*.

Although this model of song development stood as an unchallenged paradigm for many years, recent research has shown that white-crowned sparrow males *can* learn another species' song under the right conditions. If a live strawberry finch male is placed in a cage next to his "social tutor," the white-crowned sparrow male will learn to sing the strawberry finch's song! These results are really not inconsistent with the idea of a selective template, but they do show that social stimuli are more effective than a tape recorded song in "overriding" the innate program that guides song development. In other bird species males have no opportunity to hear the song of their avian species. How does song development occur in these species? For example, cuckoos are brood parasites; females lay their eggs in the nest of another species of bird and young are reared by unsuspecting "foster" parents that sing their own species-specific songs (Figure 56-13). It appears that males of brood parasite species instinctively "know" their own species' song. Since male brood parasites would most likely hear the song of their host species during development, it is adaptive for such incorrect stimuli to be ignored. And because they hear no adult males of their own species singing, correct song models are not available. Natural selection has programmed the male with a genetically-guided song.

FIGURE 56-13
Brood parasite. Cuckoos are birds which parasitize other species of birds by laying eggs in their nests. In this case, a meadow pipit is feeding a cuckoo. Since the cuckoo is being raised by a different species, it has no opportunity to learn its song; cuckoo songs are innate.

The Physiology of Behavior

Psychologists criticized ethology's emphasis on instinct because it ignored the study of internal factors that control behavior. If asked why does a male bird defend a territory and sing only during the breeding season, ethologists would answer that birds sang when they were in the right *motivational state* or *mood*, and had the appropriate *drive*. But what do these terms mean? They are simply "black-box" concepts that give a name to some internal control mechanism that remains unknown.

Today the internal control of behavior is understood from the study of physiology. One focus has been on the physiological control of reproductive behavior. Studies on lizards, birds, rats, and other animals have shown that hormones play an important role in the control of behavior and provide a chemical basis for motivation. Animals show reproductive behaviors such as courtship only during the breeding season. They are able to monitor changes in day length to trigger a series of events that involve the release of hormones from the endocrine glands the **hypothalamus,** the **pituitary,** and the **ovaries** and **testes.** Ultimately the steroid hormones **estrogen** and **testosterone** released from the ovaries and testes travel to the brain and cause animals to show behaviors associated with reproduction. Bird song and territorial behavior depend upon the level of testosterone in the male, and the receptivity of females to male courtship depends upon estrogen.

Research in the physiology of reproductive behavior has shown that these are important interactions between hormones, behavior, and stimuli in both the physical and social environment of an individual. Daniel Lehrman's work on reproduction in ring doves provides an excellent example of how these different factors operate (Figure 56-14). Male courtship behavior (a "bow-coo" display and vocalization) is dependent upon the release of androgens, a group of hormones that includes testosterone. The male's display behavior causes the release of *follicle-stimulating hormone* in the female ring dove; this hormone stimulates the growth of the ovaries. The developing follicles (eggs) in the ovaries then release estrogen, which in turn affects others reproductive tissue. Nest construction follows after one or two days. The nest itself is a stimulus for additional hormone release, causing the secretion of *progesterone* in the female. After egg laying, this hormone turns on incubation behavior in both sexes. Feeding occurs once eggs hatch, and this behavior is also hormonally controlled.

The research of Lehrman and his colleagues paved the way for many additional studies in **behavioral endocrinology,** the study of the hormonal regulation of behavior. These studies demonstrate the interactive effects of the physical environment

FIGURE 56-14

Cycle of reproduction. Reproduction behavior in the ring dove involves a sequence of behaviors regulated by hormones: (1) courtship and copulation, (2) nest building, (3) egg laying, (4) incubation, and (5) feeding crop milk to the young squabbles after they hatch.

(temperature, day length) as well as the social environment (the presence of a nest, the courtship display of a mate) on the hormonal condition of an animal. For example, male *Anolis* lizards begin courtship after a seasonal rise in temperature. The male's courtship is needed for ovarian growth. Whether or not a female is receptive to a male's courtship depends on her level of estrogen.

Reproductive behavior is regulated by a complex series of instructions involving the physical environment, the behavior of a mate, and the release of hormones. The endocrine system and the brain orchestrate sequences of behaviors necessary for mating and the raising of offspring.

Hormones are therefore a proximate cause of behavior. To control reproductive behavior, they are released at the time of the year during which conditions for the growth of young are favorable. Environmental stimuli that trigger hormone release and thus behavior include the courtship activities of males as well as changes in the physical environment such as temperature and day length.

Behavioral Rhythms

Many animals exhibit behaviors that vary in a regular fashion. Geese migrate south in the fall, birds sing in the early morning, bats fly at night rather than in daylight hours while we humans sleep at night and are active in the daytime. Some behaviors are timed to occur in concert with lunar or tidal cycles (Figure 56-15). Why do regular repeating patterns of behavior occur, and what determines when they occur? The study of questions like these has revealed that rhythmic animal behaviors are based on both **exogenous** (external) timers and **endogenous** (internal) rhythms.

Much of the study of endogenous rhythms has focused on behaviors that seem keyed to a daily cycle, such as sleeping. Many of these behaviors have a strong endogenous component, as if they were driven by a **biological clock.** In the absence of any cues from the environment, the behaviors continue on a regular cycle. Such rhythms are termed **free-running.** Endogenous rhythms of about 24 hours, which occur even in the absence of external cues, are called **circadian** ("about a day") **rhythms.** Almost all fruit fly pupae hatch in the early morning, for example, even if kept in total darkness throughout their week-long development—they keep track of time with an internal clock whose pattern is determined by a single gene.

The rhythms of most biological clocks do not exactly match that of the environment, so an exogenous cue is required to keep the behavior properly in time with the changing real-world environment. Because the duration of each individual cycle of the behavior deviates slightly from 24 hours, an individual kept under constant conditions gradually drifts out of phase with the outside world. Exposure to an environmental cue resets the clock. Light is the most common cue for resetting circadian rhythms.

The most obvious circadian rhythm in humans is the sleep/activity cycle. In controlled experiments, humans have lived for months underground in apartments where all light is artificial and there are no external cues of any kind; left to set their own

FIGURE 56-15

Tidal rhythm. Oysters open and close their shells in rhythm to the tidal cycle. The shells open for feeding when the tide is in, and close at low tide.

schedules, most people adopt daily activity patterns (one period of activity plus one period of sleep) of about 25 hours, although there is considerable variation. Some individuals exhibited 50 hour clocks, active for as long as 36 hours each period! In the real world, the day/night cycle resets the free-running clock every day to a cycle of 24 hours. How do animals keep the time? Where is the biological clock? In some insects, time-keeping appears to be based on hormones and neurosecretions, and the clock seems to be in the optic lobes of the brain. In mammals, including humans, the biological clock lies in a specific region of the hypothalamus, the **suprachiasmatic nuclei.** This portion of the hypothalamus binds **melatonin,** an amino acid secreted in cyclic fashion by the **pineal gland** in the brain. Since the pineal gland is located just beneath the skull and is receptive to light, this organ monitors **photoperiod,** or seasonal changes in day length. Melatonin is secreted by the pineal in proportion to day length; more melatonin is secreted under conditions of short day length, since melatonin is released only at night. This clock mechanism is timed by photoperiod and is responsible for timing seasonal activities such as reproduction.

> **Circadian rhythms are endogenous cycles of about 24 hours that occur in the absence of external clues.**

Many important biological rhythms occur on cycles longer than 24 hours. Annual cycles of breeding, hibernation, and migration are examples of behaviors that occur on a yearly cycle, so-called **circannual behaviors.** These behaviors seem to be largely timed by hormonal and other physiological changes that are keyed to exogenous factors such as day length. The degree to which endogenous biological clocks underlay circannual rhythms is not known—it is very difficult to perform constant-environment experiments of several years' duration.

The physiological mechanism of endogenous biological clocks is unknown, although there has been a great deal of study and speculation, most of it centered on regularly-occurring molecular interactions. The mechanism of the biological clock remains one of the most tantalizing puzzles in biology today.

ANIMAL COMMUNICATION

The sign stimuli we discussed earlier in the chapter are used by predators to locate prey or are used by prey to avoid predators. However, other stimuli are **social releasers,** or signals produced by one individual to communicate with another individual, usually of the same species. In this case, it is adaptive for both the sender and receiver of the signal to exchange information, perhaps concerning the readiness to mate, the location of a food source, or the presence of a threat. Communication may occur through a number of sensory channels; signals may be visual, acoustical, chemical, tactile, or electric. Much of the research in animal behavior concerns analyzing the nature of the signal, determining how it is perceived, and identifying the ecological role it plays.

Courtship

Depending on their reproductive condition, animals produce signals to communicate with potential mates, and with other members of their own sex. Courtship usually consists of a series of behaviors that involve within-and between sex interactions. A **stimulus/response chain** often occurs, in which a behavior released by some action of a partner, in turn releases another behavior.

In the stickleback, males defend the nests they build on the bottom of a pond or stream against other sticklebacks. A male residing in a nest will recognize a *conspecific* male (that is, a male of his own species) and attack. Niko Tinbergen studied the social releasers responsible for this behavior by making simple clay models (Figure 56-16). Shape was unimportant; the degree of resemblance to a fish was of little consequence as long as the model had a red underside. This is how a male recognizes another male—a potential competitor—of his own species. A male responds to a female by

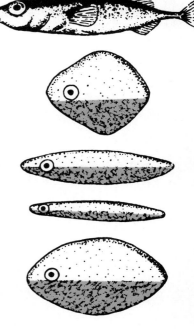

FIGURE 56-16
Dummy fish. Clay models used to study territorial behavior in male sticklebacks. Any dummy with a red underside, like the bottom four models, will elicit an attack response.

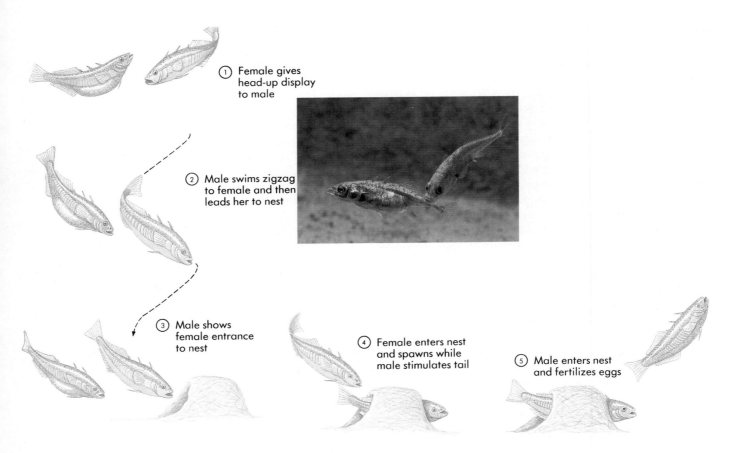

① Female gives head-up display to male

② Male swims zigzag to female and then leads her to nest

③ Male shows female entrance to nest

④ Female enters nest and spawns while male stimulates tail

⑤ Male enters nest and fertilizes eggs

FIGURE 56-17

A stimulus/response chain. Stickleback courtship involves a sequence of behaviors leading to the fertilization of eggs.

showing a zig-zag courtship "dance" (Figure 56-17). A stickleback male recognizes a female by her abdomen swollen with eggs. This was also demonstrated by using a series of clay models.

Courtship signals are often **species-specific**—they limit communication to members of the same species and thus play a key role in reproductive isolation. The flash patterns of fireflies (which are actually beetles) code for species identity; females recognize males of their own species by the number of flashes in a pattern (Figure 56-18). Males recognize the female of their species by her flash response. Thus there is a series of reciprocal responses presenting a continuous "check" on the species identity of a potential mate.

Visual courtship displays sometimes have more than one component. The male *Anolis* lizard extends and retracts his fleshy and often colorful dewlap while perched on a branch in his territory (Figure 56-19). The display thus involves movement (the extension of the dewlap as well as a series of lizard "push-ups") and color. To which component of the display does the female respond? The dewlap color can be experimentally altered with ink, so a red dewlap could be made to appear blue. Results show that for some species dewlap color is unimportant; that is, a female will be successfully courted by a male with an atypically-colored dewlap. But for other species, color may be a species-identifying cue.

Chemical signals mediate interactions between males and females. Even the human egg produces a chemical attractant to communicate with sperm! Chemical sex attractants, or **pheromones,** have been described in many species. Female silk moths, *Bombyx mori,* produce a sex pheromone in a gland associated with the reproductive system. In the 1950s a chemist extracted just 15 milligrams of the sex attractant from 500,000 female silk moths. Think of how biologically active this molecule must be! The attractant is an alcohol, named *bombykol* after the species from which it was isolated. Neurophysiological studies have shown that **chemoreceptors** which are nerve

FIGURE 56-18
Firefly fireworks. The bioluminescent displays of these lampyrid beetles are species-specific and serve as behavioral reproductive isolating mechanisms. Each number represents the flash pattern of a male of a different species.

FIGURE 56-19
Dewlap display of a male *Anolis* lizard. Under hormonal stimulation, males extend the fleshy, colorful dewlap to court females. This behavior also stimulates hormone release and egg-laying in the female.

cells designed to detect chemicals, are abundant on the male's antennae and are tuned to perceive only one form of the bombykol molecule. They can detect extraordinarily small quantities of pheremone: The perception of only 200 molecules will trigger a response by male silk moths.

Many insects, amphibians, and birds produce species specific acoustical signals to attract mates (Figure 56-20). Bullfrog males call to females by inflating and discharging air from the vocal sac, located beneath the lower jaw. The ear of the female is neurally wired to distinguish the male's call from that of other frogs which may occur in the same habitat and be mating at the same time of year. Bird songs are complex sounds composed of notes and phrases used by males to advertise their presence and to attract females. In many species of birds males vary their songs and these variations identify *particular* males in a population. Bird song thus specifies one *individual* male as well a given species.

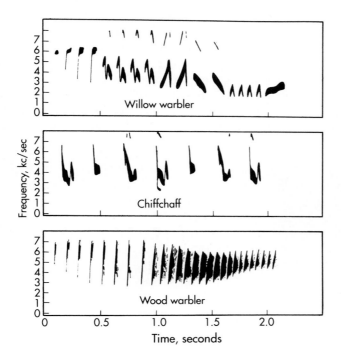

FIGURE 56-20
Singing a different tune. These sonograms illustrate how birds announce their species' names. At one time, these three species were considered to be one.

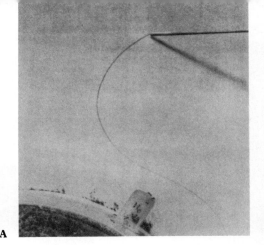 

A   **B**

FIGURE 56-21

The chemical control of fire ant foraging. Trail pheromones, produced in an accessory gland to the fire ant's stinger, organize cooperative foraging. An artificial trail has been drawn out from the nest (A); it is soon followed by ants (B).

Communication in Social Groups

Many insects, fish, birds, and mammals live in social groups in which information is communicated between group members. For example, in mammalian societies some individuals serve as "guards" and keep watch for predators. When danger occurs, an **alarm call** is given; group members respond by absconding and seeking shelter. Social insects such as ants and honey bees produce chemicals called **alarm pheromones,** which trigger attack behavior. Ants deposit **trail pheromones** between the nest and a food source to induce cooperation during foraging (Figure 56-21). Honey bees have an extremely complex **dance language** behavior that directs nestmates to rich nectar sources (see box on page 1193).

Language

Some primates have a "vocabulary" that allows individuals to communicate the identity of specific predators. African vervet monkeys, for example, have vocalizations which code for the presence of eagles, leopards, and snakes (Figure 56-22). And chimps and gorillas can learn to recognize and use a large number of symbols, and use them to communicate abstract concepts. But there is no non-human animal whose communication rivals the complexity of human language.

Language develops at an early age in humans. Human infants are programmed to recognize the consonant sounds characteristic of human speech (including those not present in the particular language they learn) while ignoring a world full of other sounds; they learn by trial and error (the "babbling" phase) how to make these sounds. "Babbling" appears to be neurally programmed: even deaf children go through a babbling phase using sign language.

Although human languages appear on the surface to be very different, in fact they share many basic structural similarities. Researchers believe these similarities in the basic structure of human languages reflect how our brains handle abstract information, a genetically determined characteristic all humans share. The differences be-

FIGURE 56-22

Primate semantics. Predation and the evolution of vervet monkey "language." (A) Predators, like this leopard, attack and feed on vervet monkeys. Different alarm calls (B) are given when leopards, eagles, and snakes are sighted by troupe members. Each distinctive call elicits a different escape behavior.

A   **B**

 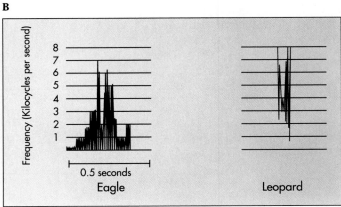

THE DANCE LANGUAGE OF THE HONEYBEE

The European honeybee, *Apis mellifera*, is extremely important as a pollinator of plant species. A bee hive is actually a society of 30,000 to 40,000 individuals whose behaviors are integrated into a complex unit. Worker bees may forage for miles from the hive, collecting nectar and pollen from a variety of plants, switching between species and populations according to how energetically rewarding their food is. A colony adjusts its food harvesting effort through the reconnaissance behavior of scout bees which sample available food sources. The nectar and pollen sources used by bees are clumped and potentially rich in food quality and quantity. Each clump offers much more food than can be transported to the hive by a single bee. A honeybee colony is able to exploit such rich resources through the behavior of scout bees which locate a resource patch and communicate its location to nest mates. The life's work of Nobel laureate Karl von Frisch was to unravel the details of this communication system which conveys information about the location of new food sources through a *dance language*.

To direct bees in the hive to a food source both the direction and distance of the patch must be communicated. After a successful scout bee returns to the nest she performs a remarkable behavior pattern called a *waggle dance* on a vertical comb. The movement resembles a figure-of-8; it is called a waggle dance because during one portion of the dance, the straight run, the dancing bee vibrates, or waggles, her abdomen while producing bursts of sound. She may stop periodically to give her hivemates a sample of the nectar she has carried with her in her crop, a storage organ. A dancing bee is followed closely throughout her performance; bees which were followers of the dance soon appear as foragers at the new food source.

Do the bees use the waggle dance to locate the food? If so, how is the information on the location of the food communicated? von Frisch and his colleagues found that the scout bee indicates the direction of the food source by transposing the visual angle between the food source and

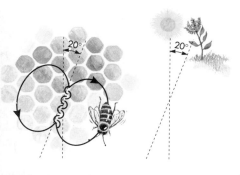

FIGURE 56-B

The waggle dance of honey bees.

1 A scout bee dances on a comb in the hive.
2 The angle between the food source and the nest is represented by the dancing bee as the angle between the "straight run" of the dance and food source. In (*1*), an angle of 45 degrees to the left of the sun is indicated. In (*2*), food is directly toward the sun, and in (*3*), food is 90 degrees to the right of the sun.
3,4 The "dance" of the robot bee is programmed by computer.

the nest in reference to the sun to the same angle given by the straight run of the dance relative to gravity on the vertical comb in the hive. Distance is given by the tempo, or degree of vigor, of the dance.

In spite of the many ingenious experiments carried out by von Frisch and his students, another scientist, Adrian Wenner, from the University of California, did not believe that the dance language communicated anything, and he challenged von Frisch's results. Wenner believed that flower odor was the key cue in allowing recruited bees to arrive at a new food source. A heated controversy ensued as each group of researchers published articles supporting their position.

The "dance language controversy" was resolved in the mid-1970s by the creative research of James L. Gould, who devised an experiment in which

a scout bee gave incorrect information to hivemates. Gould could thus cause a scout bee to tell nestmates of a given location—not where the scout had been but rather a place where automated traps would capture the "fooled" bees.

Gould's approach to solve the problem of the bee dance was convincing, but for many years researchers wanted to extend the study of the dance language by building a robot bee whose dance could be completely controlled. Several attempts failed, but recently a robot bee proved successful. The robot bee's dance is programmed by a computer and perfectly reproduces the typical dance—the robot even stops to give food samples! The use of the robot bee has allowed researchers to determine precisely which cues are used to direct hivemates to food sources.

tween English, French, Japanese, and Swahili are learned—but any human can learn them. All human languages (roughly 3,000) draw from the same set of 40 consonant sounds (English uses two dozen of them), and every normal human baby can distinguish between all 40 of them.

Learning a particular language is something most humans do while young. Children who have not heard certain consonant sounds as infants can only rarely distinguish or produce them as adults. That is why Americans never master the throaty French "r," whereas speakers of French typically replace the English "th" with "z," and why native Japanese substitute "l" for the unfamiliar English "r." Children quickly and effortlessly learn a vocabulary of thousands of words. This phase of rapid learning seems to be programmed. It is followed by a stage of simple sentence structure, which although grammatically incorrect, can convey information. Rules of grammar are next learned; this comprises the final stage of language learning.

Although language is the primary channel of human communication much evidence exists to suggest that odor and other non-verbal signals ("body language") may also be important. In an animal as socially complex and intelligent as a human, it is difficult to sort out the relative contribution of the composite signals (that is, multichannel) humans produce.

The study of animal communication involves studies of the specificity of a signal, information content, and the methods used to produce and receive it. Communication plays an important ecological role in maintaining genetic isolation between species.

ORIENTATION AND MIGRATION

Animals may travel to and from a nest to feed, or move regularly from one place to another. To do this they must orient themselves by tracking stimuli in the environment.

Movement toward or away from some stimulus is called a **taxis.** The crowding of flying insects about outdoor lights is a familiar example, insects being attracted to the light. They are said to be *positively phototactic.* Other insects, such as the common cockroach, avoid light (are negatively phototactic). Other stimuli may be used as orienting cues. Trout orient in a stream so as to face upstream, against the current. Not all responses involve such a specific orientation. In some cases an individual simply becomes more active under certain conditions. If an animal moves randomly but is active under poor conditions and quiet under favorable ones, then it will tend to stay in favorable areas. These changes in activity level are dependent on stimulus intensity and are called **kineses.**

Long-range two-way movements are called **migrations.** In animals, many migrations are tied to a circannual clock, occurring once a year. Ducks and geese migrate down flyways from Canada across the United States each fall and return each spring. Monarch butterflies migrate from the eastern United States to Mexico, a journey of over 3,000 kilometers that takes from two to five generations of butterflies to complete. Perhaps the longest migration is that of the golden plover, which flies from Arctic breeding grounds to wintering areas in southeastern South America, a distance of some 13,000 kilometers.

Migration patterns may be instinctive. When colonies of bobolinks became established in the western United States, far from their normal range in the Midwest and East, these birds did not migrate directly to their winter range in South America; rather they migrated east to their ancestral range and then south along the old flyway (Figure 56-23). The old pattern was not changed, but rather new pattern was added.

Biologists have studied migration with great interest, and we now have a good idea of how these feats of navigation are achieved. Birds and other animals navigate by looking at the sun and the stars. The indigo bunting, for example, which flies during the day and uses the sun as a guide, compensates for the movement of the sun in the sky as the day progresses by reference to the north star, which does not move in

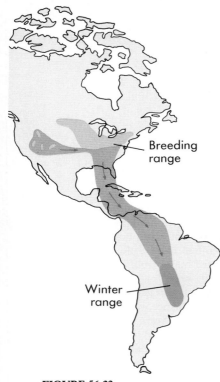

FIGURE 56-23
Birds on the move. The migratory path of California bobolinks. These birds came recently to the far West from their more established range in the Midwest. When they migrate to South America in the winter, they do not fly directly, but rather fly to the Midwest first and then use the ancestral flyway.

Breeding range

Winter range

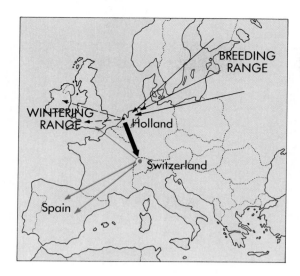

FIGURE 56-24
Migratory behavior of starlings. Navigational abilities of inexperienced birds differ from those of adults who have made the migratory journey before. Starlings were captured in Holland, halfway along their fall migratory route from Baltic breeding grounds to wintering grounds in England, and these birds were transported to Switzerland and released. Experienced older birds compensated for the displacement and flew toward the normal wintering grounds (blue arrow). Inexperienced young birds kept on flying in the same direction, on a course that took them toward Spain (red arrows). Inexperienced birds fly by compass; experienced birds learn true navigation.

the sky. Other birds, such as starlings, compensate for the sun's apparent movement by use of an internal clock. If captive birds are shown an experimental sun in a fixed position, they will change their orientation to it at a constant rate of about 15 degrees per hour.

It is important to note the distinction between **orientation** (the ability to follow a bearing) and **navigation** (the ability to set or adjust a bearing, and then follow it)—"compass" versus "map and compass." Orientation mechanisms such as those of the indigo bunting and the starling are well understood (Figure 56-24).

Many migrating bird species appear to use a compass. They have the ability to detect the earth's magnetic field, and orient themselves with respect to it. If such birds are studied in a closed indoor cage, they will attempt to move in the correct geographical direction, even though there are no visible external cues. However, if an investigator places a powerful electromagnet near the cage to change the magnetic field, he or she can alter the direction the birds attempt to move at will. Little is known about the sensory receptors that birds employ to detect magnetic fields. Magnetite, a magnetized iron ore, has been found in the heads of some birds, but the exact nature of the receptor is unknown.

Although bird migration has been well studied, we know relatively little about how other migrating animals navigate. Green sea turtles migrate from Brazil halfway across the Atlantic Ocean to Ascension Island—how do they find this tiny island, half an ocean away? How do the young that hatch on Ascension Island know how to find Brazil, thousands of miles away over the open sea? How do they find their way back, as adults, when they breed, perhaps 30 more years later? Current studies suggest wave action is an important cue.

ANIMAL AWARENESS

It is likely that all of us could tell an anecdotal story about the behavior of a pet cat or dog that would seem to suggest that the animal had a degree of reasoning ability or was capable of thinking. But remember, the study of animal behavior is a *science*, and results must be replicable and free from alternative explanations that would cause us to reject the hypothesis of thinking. For many decades students of animal behavior flatly rejected the notion that animals can think; in fact, the behaviorist Lloyd Morgan stated that we should never assume any behavior represents conscious thought if there is any other explanation that precludes the assumption of consciousness.

It is indeed difficult to know if animals can think. But the prevailing approach has been to treat animals as though they respond to the environment through reflex-like behaviors. Nevertheless, serious attention has recently been given to the question of animal awareness. The central question is: Do animals show **cognitive behavior**—that is, do they process information and respond in a manner that suggests thinking (Figure 56-25)?

A

B

FIGURE 56-25
Animal thinking?
A This chimpanzee is stripping the leaves from a twig, which it will then use to probe a termite nest. This strongly suggests that the chimpanzee is consciously planning ahead, with full knowledge of what it intends to do.
B This sea otter is using the rock as a tool to break open a clam, bashing the clam on the rock "anvil." Often a sea otter will keep a favorite rock for a long time, suggesting that it has a clear idea of what it is going to use the rock for. Behaviors such as these suggest that animals have cognitive abilities.

What kinds of behavior suggest cognition? Some birds in urban areas remove the foil caps from milk bottles to get at the cream beneath, and this behavior is known to have spread within a population to other birds. Japanese macaques learned to wash potatoes and float grain to separate it from sand. Chimpanzees pull the leaves off of a tree branch and use the stick to gather termites by probing the nest entrance of a colony. And as we discussed earlier in the chapter, vervet monkeys have a vocabulary that identified specific predators.

Currently there are relatively few experiments that test the thinking ability of non-human animals. Some of these studies indicate animals may give false information (that is, "lie") to other individuals if it is to their advantage. And studies on the dance language of the honeybee provide enticing preliminary data suggesting that bees have *cognitive maps*, or an image of their environment on which the spatial coordinates given by the dance are projected and evaluated as being plausible locations of food. Much of this type of research on animal awareness is in its infancy, but it is sure to grow and be controversial. In any case, there is nothing to be gained by a dogmatic denial of the *possibility* of animal consciousness.

SUMMARY

1. Behavior is an adaptive response to stimuli in the environment. An animal's sensory system monitors the environment and has specialized neural elements to detect and process environmental information.

2. Behavior is both instinctive and controlled by genes and is learned through experience. Genes are thought to limit the extent to which behavior can be modified and the types of associations that can be made.

3. The simplest forms of learning involve sensitization and habituation. More complex associative learning may also occur in this way, by the strengthening and weakening of existing synapses, although learning may also involve the formation of entirely new synapses.

4. An animal's internal state influences when and how a response will occur. Hormones cause an animal's behavior and perception of stimuli to change in a way that facilitates reproduction.

5. Animals communicate by producing visual, acoustical, chemical, and electric signals. These signals are involved in mating, food finding, predator defense, and other social situations.

6. Animals use the position of the sun and stars and other stimuli to orient during daily activities and to navigate during long-range migrations.

REVIEW QUESTIONS

1. What types of behaviors are categorized as instincts? Define learning. To what degree is learning associated with instinctive behaviors?

2. How does evolution limit the extent of learning?

3. How are animal behavioral rhythms controlled? What type of control is stronger in behaviors keyed to a daily cycle? What maintains such behaviors within the synchrony of the outside world? What are circadian rhythms?

4. What is the nature/nurture controversy? How does Marler's work on song learning in white-crowned sparrows illustrate that behavior is shaped from both instinct and learning?

5. What types of studies demonstrate that genes affect behavior?

6. What is ethology? How did ethologists view the regulation of behavior? Have their concepts been supported by research?

7. What is filial imprinting? What is sexual imprinting? Why do some young animals imprint on objects like a moving box?

8. How do hormones coordinate reproductive behavior? How does the environment regulate the release of hormones?

9. What is the definition of a taxis? What are kineses? How is migration different from either of these? What determines migration patterns? How do patterns change as the range of the migrating animal changes? What cues do migrating birds use to orient themselves during their migrations?

10. What roles do communication signals play in reproductive isolation? How are visual, acoustical, and chemical signals species-specific?

THOUGHT QUESTIONS

1. Why can it be said that some instincts are learned?

2. Some species of birds store seeds in caches. The seed hoardes—as many as 9000—are recovered later. What type of learning might be involved? How might the structure of the brain of such a bird be adapted to this feeding habit?

3. A species of bird is found on an island, where it is the only bird species present. Do you think that song learning would occur in the same manner as it does in the white-crowned species?

FOR FURTHER READING

CREWS, D.: "The Annotated Anole: Studies on the Control of Lizard Reproduction," *American Scientist* vol. 65, pp. 428-434, 1977. A fascinating account of how seasonal and social changes trigger the hormonal events that lead to reproduction.

GOULD, J., and P. MARLER: "Learning by Instinct," *Scientific American*, January 1987, pages 74-85. A clear and interesting account of the relative roles of instinct and learning in behavior. The authors, prominent behaviorists, argue that learning is often limited or controlled by instinct.

GRIFFIN, D.: "Animal Thinking," *American Scientist*, vol. 72, pages 456-463, 1984. An exciting discussion of the possibility that animals have conscious awareness. This article describes many examples of what appears to be conscious thinking by animals.

GWINNER, E.: "Internal Rhythms in Bird Migration," *Scientific American*, April 1986, pages 84-92. A brilliant analysis of how birds know when to migrate—and when to stop.

HUBER, F., and J. THORSON: "Cricket Auditory Communication," *Scientific American*, December 1985, pages 60-68. An unusually clear example of how nervous system activity underlies animal behavior.

MILLER, G., and P. GILDEA: "How Children Learn Words," *Scientific American*, September 1987, pages 94-99. Children learn from context, and the authors suggest that interactive videos may aid the process.

ROSE, K.: *The Body in Time*, New York, Wiley & Sons, 1988. A fascinating journey through the human clock—how our body keeps time.

WILLIAMS, A.O.D.: "Giant Brain Cells in Mollusks," *Scientific American*, vol. 244, pages 68-75. A good introduction to neural aspects of behavior and the use of the sea hare as a model system.

WURSIG, B.: "The Behavior of Baleen Whales," *Scientific American*, April 1988, pages 102-107. Whales are mammals that graze in the sea. They appear to have kept many of the behaviors of their ancestors, who grazed on land 55 million years ago.

Behavioral Ecology

OVERVIEW

Behavior serves as an adaptation to environmental conditions, and natural selection has played an important role in shaping the behavior patterns of animals. Behavior is crucial to survival and reproduction because it functions in biologically important processes such as feeding, mate selection, predator avoidance, and defense of territory. Social behaviors play a role in achieving reproductive success; animal societies represent group adaptations to ecological circumstances. Cooperation between individuals in animal societies is the central issue in the study of the evolution and ecology of social behavior. In many species, natural selection has favored the evolution of cooperative behavior, or altruism, among close relatives.

FOR REVIEW

Here are some important terms and concepts that you will encounter in this chapter. If you are not familiar with them, you should review them before proceeding.

Natural selection (Chapter 19)

Adaptation (Chapter 20)

Animal behavior (Chapter 56)

THE ADAPTIVENESS OF BEHAVIOR

In an important essay Nobel laureate Niko Tinbergen outlined the different types of questions biologists can ask about animal behavior. In essence, he divided approaches to the study of behavior into the study of its development, physiological basis, and evolution. One type of evolutionary analysis pioneered by Tinbergen himself was the study of the **survival value,** or function, of behavior. That is, Tinbergen asked questions that concerned how behavior allowed an animal to stay alive, or keep its offspring alive. For example, after gull nestlings hatched, he observed that the parents picked up the egg shells and dropped them away from the nest. To understand *why* this behavior occurred, he distributed gull eggs (which are camouflaged due to their color) throughout the area in which the birds were nesting (Figure 57-1). Next to some of the eggs he placed broken eggshells. As a control for this experimental procedure, other eggs were left alone without eggshells. He then recorded which eggs were found more readily by predators (namely, crows). Because the crows could use the white interior of a broken shell as a cue, crows ate more eggs that were nearby eggshells. The eggshell removal behavior was *adaptive:* it reduced predation and thus increased the survival of offspring.

Tinbergen is considered to have fathered this functional approach to the study of behavior; he is credited with founding the field of **behavioral ecology,** which is the study of how natural selection shapes behavior. In other words, behavioral ecology is the study of the **adaptive significance** of behavior, or how behavior may increase survival and reproduction. In the early history of the science of behavioral ecology, research dealt with how individuals of a given species selected the right habitat in which to live or how animals avoided predation or became more effective as predators themselves. More recently, the focus of research has been on the contribution made by behavior to an animal's reproductive success, or **fitness.** To study how behavior is related to fitness is to study the process of adaptation itself.

Natural selection acts on behavioral differences having underlying genetic components; behaviors that favor higher reproductive success become more prevalent in a population over evolutionary time. Behavioral ecologists focus on behavioral differences among individuals that lead to reproductive advantage. Measuring fitness and demonstrating its correlation with behavior is thus critical to test hypotheses on the evolution and adaptiveness of behavior. In place of measuring reproductive success directly, other factors associated with reproduction, such as the rate at which an animal acquires energy, may be measured.

FIGURE 57-1
The adaptive value of egg coloration. Niko Tinbergen painted chicken eggs to resemble the mottled brown camouflage appearance of gull eggs. The eggs were used to test the hypothesis that camouflaged eggs are more difficult for predators to find, and thus increase the survival of young.

FORAGING BEHAVIOR

Feeding is one of the most basic of all animal behaviors. Because all animals are heterotrophs, all of them must eat to survive, and many complex behaviors have evolved that influence what an animal eats, and how it obtains its food. These behaviors are collectively called **foraging behaviors.**

Some animals, like lions, actively hunt prey. Others are sit-and-wait predators that ambush prey or set snares, like an ant-lion's pit or a spider's web, into which prey stumble or fly (Figure 57-2).

FIGURE 57-2

Insect snare. The ant lion larva traps its prey, ants, by digging a deep pit in the sand, **A,** and hiding itself, **B,** at the bottom. Larger pits trap larger ants. **C,** The adult is a delicate-winged insect.

A   B   C

FIGURE 57-3
Foraging behavior in oyster-catchers. An oystercatcher forages by stabbing the sand in search of buried mussels.

If you consider the range of food items taken, animals can be divided into two broad groups: some animals are **specialists,** feeding on only one kind of food, while others are **generalists,** feeding on many different kinds. Some species of ants, for example, eat only spider eggs. Oystercatchers are shore birds that feed only on mussels. They have highly specialized foraging behavior (Figure 57-3), stabbing repeatedly into the mollusk to open its shells. Generalists, like species of insects that eat the leaves of a wide variety of plant species, represent the opposite feeding habit. It appears that generalists are not as efficient as specialists at feeding on any one type of food—but they can take advantage of collecting more than one kind.

Because foraging behavior is of obvious importance to growth and reproduction, one would expect it to be the target of strong evolutionary pressures. One aspect of the evolution of foraging behavior has been of importance to students of behavior: Does natural selection favor animals that are more *efficient* foragers, animals that maximize energy intake over energy expenditure? We would certainly expect evolution to favor efficiency—but does it? To investigate this question, biologists have studied situations in which foraging behavior involves a variety of tradeoffs. For many predators, food items come in a variety of sizes. Larger food items may contain more food energy but are harder to capture and may be less abundant. By measuring the food value (that is, the caloric or energy content) of different prey items and the energetic costs of pursuing and handling prey of different sizes, it is possible to calculate the **net energy** gained by feeding on each size prey.

The net energy gained would simply be the gross energy value of a food item (measurable in calories or Joules) minus the energetic costs of pursuit and handling (measured in the same units of energy). Because it is possible to measure the energetic benefits and costs of feeding on different items, foraging ecologists are able to predict which prey would be the *best* to take into the animals' diet. The prediction is easy to make: animals should feed on prey that maximize their energy intake, that is, the amount of energy gained per unit of foraging time. Such predictions are made by **optimal foraging theory,** which assumes that natural selection has designed foraging behavior to be as energetically efficient as possible (Figure 57-4).

Behavioral ecologists interested in optimality approaches attempt to predict aspects of foraging behavior such as where an animal should search for food, and how long it should stay in one area before moving on to another. Studies have shown that animals cannot always behave to maximize energy intake; sometimes different foods that supply important nutrients must be taken. And predation often forces an animal to compromise its energy intake in favor of lowering its risk of being eaten itself (Figure 57-5).

Natural selection has favored the evolution of foraging behaviors that maximize the amount of energy gained per unit time spent foraging. Animals which acquire energy efficiently during foraging will increase their fitness by having more energy available for reproduction.

FIGURE 57-4
Optimal diet. The shore crab selects a diet of energetically profitable prey. The curve describes the energy gain derived from feeding on different size mussels. The bar histogram shows the number of mussels of different sizes in the diet. Shore crabs most often feed on mussels which provide the most energy.

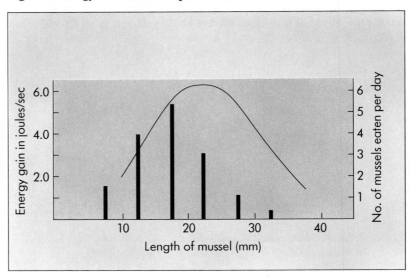

SOCIAL BEHAVIOR

While searching for food, looking for a place to nest, finding a mate, or rearing young animals often engage in interactions with other individuals of the same species, and different species as well. Although these interactions may be either competitive or co-operative, they are collectively referred to as **social behavior.** Social interactions are important in defending territory and competing for mates. Natural selection has also favored cooperation among individuals, sometimes leading to the formation of groups or highly-organized societies. Self-sacrificing behavior, or altruism, is an important and fascinating feature of the social life of animals.

Territorial Behavior

Animals often move over a large area, or **home range,** during the daily course of activity. The home ranges of several individuals may overlap in time or in space. But in many animal species, each individual *defends* a portion of its home range and uses it *exclusively.* **Territoriality** is a form of behavior in which individual members of a species will exclusively use an area holding some limiting resource, such as a foraging ground, food resources, or a group of potential mates (Figure 57-6). The critical aspect of territorial behavior is **defense** against intrusion by other individuals. Territories are defended by advertising through displays that they are occupied and by overt aggression. A bird sings from its perch within a territory to prevent a take-over by a neighboring bird. If an intruder persists, it will be attacked. But singing is energetically expensive and attacks can lead to injury. Moreover, advertisement through song or visual display can reveal one's position to a predator. Why bear these costs and take such risks?

Over the past two decades it has become increasingly clear that an *economic* approach is insightful in studying the evolution of territoriality. As there are energetic costs to defending a territory, there are also energetic benefits. Studies of nectar-feeding birds like hummingbirds and sunbirds make this point clear. A bird benefits from having exclusive use of a patch of flowers because it can efficiently harvest the nectar they produce. But to maintain exclusive use, the flowers must be actively defended. The benefits of exclusive use of such a resource outweigh the costs of defense only under certain conditions (Figure 57-7). For example, if flowers are very scarce, the nectar they yield does not have enough calories to balance the number of calories used

FIGURE 57-5
Foraging and predator avoidance. A meerkat sentinel on duty. Meerkats, *Suricata suricata,* are a species of highly social mongoose living in the semiarid sands of the Kalahari Desert. This meerkat is taking his turn to act as a lookout for predators; under the security of his vigilance the other group members can focus their attention on foraging.

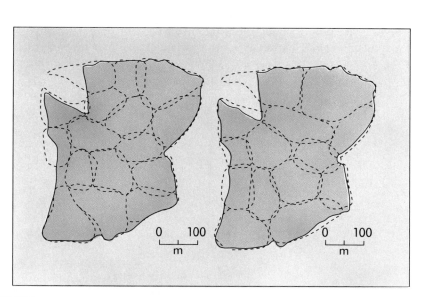

FIGURE 57-6

Competition for space. Territory size in birds is adjusted according to the number of competitors. When six pairs of great tits (*Parus major*) were removed from their territories (the shaded areas in the left figure), their space was taken over by other birds in the area and by four new pairs (right figure).

FIGURE 57-7

The benefit of territoriality. Sunbirds increase nectar availability by defending flowers.

up in defense. In this case, it is not advantageous to be territorial. Similarly, if flowers are very abundant, a bird can efficiently meet its daily energy requirements *without* behaving territorially. Defending abundant resources isn't worth it from an energetic standpoint. But for intermediate levels of flowers availability, the benefits do outweigh the costs and territorial behavior will be favored.

> The economic approach has been very useful in examining the evolution and ecology of territoriality and other behaviors, such as foraging. It is assumed that animals which gain more energy from a behavior than they expend will have an advantage in survival and reproduction over animals which behave in less efficient ways. Thus estimating the energetic benefits and costs of a behavior is one way the adaptiveness of behavior can be determined.

The Ecology of Reproduction

During the breeding season, animals make several important "decisions" concerning who to mate with, how many mates to have, and how much time and energy to devote to rearing offspring. Males and females usually differ in their **reproductive strategies,** or sets of behaviors which have evolved to maximize reproductive success. Mate choice, the number of mates taken during a breeding season, and parental care are all aspects of a reproductive strategy. The reproductive behaviors shown by males and females have in part evolved in response to ecology, which in this case refers to how food resources, nest sites, or members of the opposite sex are spatially distributed in the environment.

Darwin was the first to observe that females do not simply mate with the first male encountered, but seem to somehow "evaluate" a male's quality and then decide whether or not to mate (Figure 57-8). **Mate choice** occurs because individuals that select superior quality mates will leave behind more offspring. The benefits of choosing a mate may lie in acquiring "good genes" for offspring or resources such as favorable nest sites and energetically valuable food. But which sex should show mate choice?

Parental Investment and Mate Choice. Robert Trivers proposed that patterns of mate choice could be understood by comparing the **parental investment** made by males and females. Parental investment refers to the contribution made by each sex to rearing offspring; it is, in effect, an estimate of the energy expended by males and females in offspring care (Figure 57-9). Although one may think of parental care as the nursing or protection of young following birth, parental care has many forms. For example a fundamental difference in male and female parental investment is seen in the relative size of gametes: eggs are much larger than sperm (the human egg is

FIGURE 57-8

Feathered display. The Australian male superb lyrebird shows off before a female.

FIGURE 57-9

The young of primates and birds require intense parental care by both parents. This young lady, not yet 2 years old, still has much to learn before she can wear Daddy's hat.

FIGURE 57-10
Sexual dimorphism in deer. Stags have a rack of antlers used in defending a harem of females from other males. Antlers are absent in females.

195,000 times the size of a sperm!!). And eggs contain proteins and lipids in the yolk that provide the developing embryo with nutrition. Sperm, on the other hand, are little more than motile DNA.

In some groups of animals, like mammals, females bear additional costs associated with reproduction. Females are responsible for gestation and lactation; only females are able to carry out these important parenting functions. Many studies have shown that parental care is energetically costly to females. In other species, males make an important contribution to parental care by transferring nutrients used in egg growth to the female or by defending and feeding young.

Trivers predicted that the sex having the higher parental investment, or cost of reproduction, should be the sex that shows mate choice. It is *usually* the female but it *can* be the male. It all depends on the relative cost of parental care.

Reproductive Competition and Sexual Selection.

Mate choice has been described in many invertebrate and vertebrate species. Often it is the case that females mate with the largest male. Large size may enable a male to compete with other males. Stag deer lock antlers while trying to gain access to females for mating. Older males with large, multi-point antlers most often win such disputes, and successfully defend a larger group of females, mate with many females, and have high reproductive success (Figure 57-10). Likewise, male elephant seals are able to aggressively defend territories used by females for breeding; a relatively small number of males in the breeding population mate and sire the majority of the next generation's offspring (Figure 57-11). Such **reproductive competition,** or competitive interactions over access to mates, occurs in many species.

FIGURE 57-11
Intrasexual competition. Male elephant seals compete aggressively for breeding territories. Four percent of the males in the population sire 85% of the offspring of the next generation.

Males also compete in ways that do not involve aggression. Male birds of paradise display their elaborate feathers, and accompany their mating ritual with complex vocalizations—all to impress the females, who stand by and seem to scrutinize the strange antics of their suitors. As Darwin so eloquently phrased in Victorian prose, the key question in the study of mate choice is, "Is it not possible that the female prefers the most gallant, beautiful, or melodious male?" Recent research has shown that females *do* select males showing the brightest or longest plumage or producing the most complex songs. But how did such diverse and exquisite male ornamentation evolve?

Darwin thought that long, bright feathers or complex vocalizations posed a problem for survival for males (Figure 57-12). Even if such characteristics yielded an advantage in mating, they would make a male more conspicuous to a predator and thus place him at risk. Since natural selection favored traits permitting survival, he thought a different process was involved in the evolution of male ornamentation. He called this process **sexual selection.** Sexual selection occurs when individuals of one

FIGURE 57-12

The benefits and costs of vocalizing. The male tungara frog, *Physalaemus pustulosus* (**A**). Calls are used to attract females, but also attract predatory bats (**B**). Calls of greater complexity are seen from top to bottom in **C**. Females prefer more complex calls, but so do bats. Males compromise by giving simpler calls, thus lowering reproductive success, but also decreasing risk of predation.

A

B

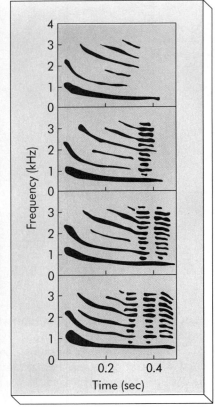

C

sex compete for mates. It involves interactions between members of one sex, or **intrasexual selection** ("the power to conquer other males in battle," as Darwin put it) and **intersexual selection,** or in effect the choice of a mate ("the power to charm"). Sexual selection thus leads to the evolution of structures like antlers or ram's horns used in combat with other males or ornamentation like long tail feathers used to try to "persuade" females to mate (Figure 57-13). These traits are called **secondary sexual characteristics.** It is understandable that antlers would evolve if they aided in combat. It is not as evident why bright color, large flashy tails, and complex songs should evolve.

The famous population geneticist R.A. Fisher proposed that once mate choice began it could lead to the exaggeration of a trait if the trait was coupled to a preference for it in the opposite sex. So if there was an initial preference by females for males having a longer-than-average tail, tail length would then continue to increase as long as females used tail length to choose a mate. Such "runaway selection" would stop

FIGURE 57-13

Products of sexual selection. Attracting mates with long feathers is common in bird species such as the African paradise whydah (**A**), which shows pronounced sexual dimorphism. Bower bird males (**B**) try to impress females by adorning their "nests" with rare flowers, or in the case of this spotted bowerbird, a toothbrush. Males of some species of bowerbirds steal long, colorful feathers of bird-of-paradise males to use as ornaments.

A

B

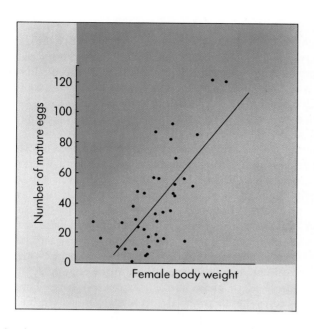

when natural selection acted as a "brake"; at some point, long tails would reduce locomotion or enhance predation. Other theories suggest that secondary sexual characteristics are true indicators of a mate's quality; they are "advertisement" of superior genes which will be passed on to the female's offspring. Brightly colored feathers, for example, may indicate that a male is resistant to parasites. Indeed, when males of some species are parasitized, their coloration becomes less vibrant and they are less able to acquire mates.

The Benefits of Mate Choice.

What does a male or female gain by selecting a mate? Behavioral ecologists recognize two types of potential benefits: individuals may get "good genes," which might promote the survival and fitness of young, or nutritional materials or resources necessary for reproduction. Fruit fly females that are allowed to select a mate have increased survivorship of larvae compared to females that have had their mates chosen for them at random. Large spadefoot toads are normally chosen by females; the sperm from a large male will produce tadpoles that mature rapidly, and thus may be able to metamorphose before an ephemeral habitat such as a small pond dries up. Both of these examples suggest that genes coding for traits affecting survival are passed from males to females. In other cases, females benefit by selecting a male defending a territory holding high quality resources, like many nectar-producing flowers or nesting sites that favor the survival of young.

Males may show mate choice as well, if their parental investment is high. Mormon cricket males transfer a protein-containing spermatophore to females during mating. Almost 30% of a male's body weight is made up by the spermatophore, which provides nutrition to a female. Females of this species compete with each other for access to males, and males choose females based on the female's body weight. Since heavier females have more eggs, males that choose a larger females will leave more offspring (Figure 57-14).

> Natural selection has favored the evolution of behaviors which maximize the reproductive success of males and females. By evaluating and selecting a mate with superior qualities, the growth, and survival, and chance of reproduction of offspring can be increased.

Mating Systems.

Individuals may mate with only one individual during the breeding season and form a long-lasting pair bond or may have more than one mate. **Mating systems** such as **monogamy** (a male mates with one female), **polygyny** (a male mates with more than one female), and **polyandry** (a female mates with more than one male) are other aspects of male and female reproductive strategy (Figure 57-

FIGURE 57-15
Pink flamingos. These birds are monogamous and show little size and plumage differences.

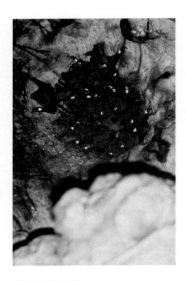

FIGURE 57-16
Female defense polygyny in bats.
The male at the left is guarding a group of females.

15). These patterns of reproduction have also evolved to maximize male and female fitness. Much research has shown that mating systems have a strong ecological component. For example, a male may defend a territory which holds nest sites or food sources necessary for a female to reproduce, and the territory might have resources sufficient for more than one female. If males differ in the quality of territories they hold, a female will do its best to mate with a male in a high quality territory. Such a male may already have a mate, but it is still to the advantage of a female to breed with a mated male in a high-quality territory than with an unmated male in a low-quality territory. Thus polygyny may evolve.

Decisions about mating are also constrained by the needs of offspring. If the presence of both parents is necessary to successfully rear young, then monogamy may be favored. This is generally the case in birds, in which over 90% of all species form monogamous pair bonds. The type of mating system that may evolve therefore depends not only on ecology but on phylogeny, or a species' history. For example, males have two options in the course of evolution—provide care for offspring or desert a mate to search for more mates and thus more opportunities for reproduction. But whether or not it is possible for a male to desert a mate trying to rear young depends upon the requirement for male assistance in feeding or defending offspring. In some species, offspring are **altricial** and need extensive care. Thus the need for two parents will influence the tendency to desert a mate to seek other matings. In other species, young are **precocial** and require little care. This rapid development may decrease the need for male parental care.

The timing of female reproduction will also determine if individuals of one sex can have access to more than one mate. If females become sexually receptive throughout the entire breeding season such that no females overlap in their readiness to mate, then males will be unable to secure many mates at any one time. So phylogeny may determine if males can take advantage of any ecological opportunities, such as defendable resources, which might favor polygyny (Figure 57-16).

Mating systems represent reproductive adaptations to ecological conditions. The need for parental care and the ability of both sexes to provide it as well as the timing of female reproduction are important influences on the evolution of monogamy, polygyny, and polyandry.

The Evolution of Animal Societies

Society is a common word; we tend to think of it in terms of the groups in which humans live and the cultural environment of man. But species of organisms as diverse as the slime mold, cnidarians, insects, fish, birds, prairie dogs, lions, whales, and chimpanzees exist in social groups. To encompass many types of social phenomena we can broadly define a society as *a group of organisms of the same species that are organized in a cooperative manner*. Why have individuals in some species given up a solitary existence to become a member of a group? Recent research has focused on the advantages and disadvantages of group living, and much attention has been given to the evolution of the trait which defines social life: cooperation.

Sociobiology: The Biological Basis of Social Behavior. In the early 1970s, E.O. Wilson of Harvard University, Richard Alexander of the University of Michigan, and Robert Trivers now at the University of California initiated what has become a major movement within biology, the attempt to study and understand animal social behavior as a biological process, a process with a genetic basis that is shaped by evolution.

An evolutionary view of social behavior predicts that the behavioral characteristics of animals are suited to their mode of living—that is, in the sense of Darwin, are adaptive. This is the approach of behavioral ecology. The study of the biological basis of social behavior and of the organization of animal societies is called **sociobiology.** Sociobiology has provided many insights into the origin of social behavior. We will then consider one of the most contentious topics in the history of biology: can evolutionary theory tell us anything about human nature?

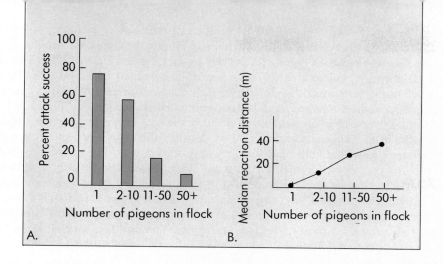

FIGURE 57-17
Flocking behavior decreases predation. As flock size increases, hawks have a lower success rate of capturing a pigeon (**A**). When more birds are present, hawks can be detected at greater distances, thus allowing more time for birds to abscond (**B**).

Group Living. Living as a member of a group is actually selfish behavior. A bird that joins a flock does so because it may have greater protection from predators. In fact, as flock size increases, the risk of predation decreases because there are more individuals present to vigilantly scan the environment (Figure 57-17). A member of a flock may also increase its feeding rate if it can acquire information from other flock members about the location of new, rich food sources. But there are costs to group living. For example, parasites and disease are more easily spread within groups, and the advantages of increased group size may be balanced by the disadvantages of parasitism. Cliff swallows offer an excellent illustration of this point. In larger groups, each bird has an increased feeding rate, but they also suffer loss of young due to blood-sucking, ectoparasitic insects called "swallow bugs" (Figure 57-18).

The Evolution of Altruism. You may think of **altruism** as a heroic human behavior such as jumping into a river to save a drowning person. But altruism, or self-sacrificing behavior, is not limited to humans and occurs in extreme forms in other animals. In many species it is an important aspect of cooperation. A honey bee worker has a barbed sting that remains in the skin of a vertebrate. As the bee flies off, the anchored sting causes the worker to eviscerate itself and thus commit suicide. Vampire bats may share blood meals at the roost. One of the most important aspects of altruism, however, concerns assisting another individual in reproducing. In some species, non-reproductive or physiologically sterile individuals exist to help other group members rear offspring. How can such behavior evolve?

One of the great misconceptions about social behavior is that altruism has evolved because the act benefits the group as a whole, or even an entire species. This **group selection** argument has been used (incorrectly!) to explain how animals regulate the size of their populations. For example, in some bird species males display and compete with each other on the mating grounds to try to secure centrally located territories. Some males are able to hold territories, others are not. Females choose males with territories as mates, and males unable to hold territories never mate. V.C. Wynne-Edwards, an animal ecologist, explained that non-territorial males were sacrificing their own reproduction to limit population growth. In other words, some males did not reproduce because it was "good for the species." That is to say that too large a population might exhaust limited resources with the result that the whole group or species might go extinct.

There is a very important flaw in this group selection explanation of the evolution of altruism. How could the trait of altruism be passed from generation to generation if males that have the trait never leave any offspring? There is simply no Darwinian logic to this argument! We must look for other evolutionary explanations of altruistic behavior.

Reciprocity. Robert Trivers proposed that individuals may form "partnerships" in which continuous exchanges of altruistic acts occur. Altruistic acts are mutually recip-

A

B

FIGURE 57-18
Communal nests of the cliff swallow (A). The benefits of colonial nesting are decreased in large colonies by blood-sucking swallow bugs (**B**), which parasitize nestling birds.

FIGURE 57-19
A vampire bat roost.

rocated, hence Trivers' theory is called **reciprocal altruism.** It is important to note that this theory requires that the individuals of an altruistic pair are unrelated, that is, they share no genes in common. Studies of alliances among male baboons show that a male usually solicits the help of a particular unrelated male and that "favorite partners" occur. In the evolution of altruism through reciprocity, "cheaters" (non-reciprocators) are discriminated against and are cut-off from receiving future aid. According to Trivers, cheating should not occur if the cost of not reciprocating exceeds the benefits of receiving future aid. That is, the altruistic act is relatively inexpensive, and the small gain to a "cheater" for not reciprocating would be far outweighed by the potential cost of not receiving future aid. Vampire bats roost in hollow trees in groups of 8 to 12 individuals (Figure 57-19). Because these bats have a high metabolism, individuals that have not fed recently may die. Bats that have found a host imbibe a great deal of blood; giving up a small amount has no great energetic cost to the donor, and can keep a roost-mate from starvation. Vampire bats tend to share blood with past reciprocators. If an individual fails to give blood to a bat from which it had received blood in the past, then it will be "cut off" (that is, discriminated against) from future blood sharing.

Kin Selection. The most influential theory of the origin of altruism was presented by sociobiologist William D. Hamilton in 1964. This theory is perhaps best introduced by quoting a passing remark made in a pub in 1932 by the great population geneticist J.B.S. Haldane. Haldane had said that he would willingly lay down his life for *two brothers* or *eight first cousins*. This would "make sense," evolutionarily speaking because each brother and Haldane share half of their genes in common—Haldane's brothers each had a 50% chance of receiving a given allele that Haldane obtained, and so forth for all other alleles. Consequently, it is statistically true that two of his brothers would carry as many of Haldane's particular combination of alleles to the next generation as Haldane himself would. Similarly, Haldane and a first cousin would share an eighth of their alleles: their sibling parents would each share half of their alleles, and each of their children would receive half of these, of which half on the average would be in common—$0.5 \times 0.5 \times 0.5 = 0.125$, or one eighth. Eight first cousins would therefore pass on as many of those genes to the next generation as Haldane himself would. Hamilton saw Haldane's point clearly: *evolution will favor any strategy that increases the net flow of a combination of genes to the next generation.*

Altruism has costs and benefits. Hamilton showed that by directing aid toward kin, or close genetic relatives, the reduction in an altruist's (personal) fitness may be outweighed by the increased reproductive success of relatives. From an evolutionary perspective, selection will favor the behavior that maximizes the propagation of alleles. If giving up one's own reproduction to help relatives reproduce accomplishes this, then even sterility can be favored by natural selection. Selection acting to favor the propagation of genes by directing altruism toward relatives is called **kin selection.**

In his theory of the evolution of altruism by kin selection, Hamilton developed the concept of **inclusive fitness** to describe the sum of genes propagated by personal reproduction and the effect of help on reproduction by relatives. It is important to note that inclusive fitness does not simply result from adding the number of genes passed on directly via an individual's own offspring and the number of genes passed on via relatives other than offspring. Inclusive fitness is rather the sum of the number of genes directly passed on in an individual's offspring and those genes passed on indirectly by kin (other than offspring) whose existence results from the benefit of the individual's altruism. Simply stated, Hamilton noted that fitness has both a *personal* and *kin-selected* component.

> **The theory of kin selection proposed by W.D. Hamilton predicts that altruism is likely to be directed toward close relatives. The closer the degree of relatedness, the greater the potential genetic payoff in inclusive fitness. This is described by "Hamilton's rule," or the expression b/c > 1/r, in which b and c are the benefits and costs of the altruistic act, respectively, and r is the coefficient of relatedness,** which describes the proportion of genes shared through common descent between individuals.

One exception to the general rule that vertebrate societies are not organized into insect-like societies is the naked mole rat, a small hairless rodent that lives in East Africa. Adult naked mole rats are 8 to 13 centimeters long and weight up to about 60 grams, and they look like a pink sausage with teeth (Figure 57-A). Unlike other kinds of mole rats, which live alone or in small family groups, naked mole rats form large underground colonies with a far-ranging system of tunnels and a central nesting area. It is not unusual for a colony to contain 80 or more animals.

Naked mole rats live by eating large, underground roots and tubers, which they locate by tunneling constantly. Naked mole rats tunnel in teams. Each mole rat has protruding front teeth, which make it look something like a pocket-sized walrus; it uses these teeth to chisel away the earth from the blind face at the end of a tunnel. When the leading mole rat has loosened a pile of earth, it pushes the pile between its feet and then scuttles backward through the tunnel, moving the pile of earth with its legs. When this animal finally reaches the opening, it gives the pile to another animal, which kicks the dirt out of the tunnel. Then, free of its pile of dirt, the tunneler returns to the end of the tunnel to dig again, crawling on tip-toe over the backs of a long train of other tunnelers that are moving backwards with their own dirt piles. And like insect societies, there is a division of labor among colony members, so some mole rats may be tunnelers whereas others perform different tasks.

Naked mole rat colonies are unusual vertebrate societies not only because they are large and well-organized. They also have a breeding structure that one might normally associate with truly social insects like ants and termites: they show reproductive division of labor. All of the breeding is done by a single female or "queen," who has one or two male consorts. The worker group, composed of both sexes, keeps the tunnel clear and forages for food. This is very unusual—few mammals surrender breeding rights without contest or challenge. If the queen is removed from the colony, however, conflict occurs among the workers: one individual attacks another, and all of them compete to become part of the new hierarchy in reproduction. When another female becomes dominant and starts breeding, things settle down once again and discord disappears.

Recent DNA fingerprinting studies have shown that colony members may share 80% of their genes due to inbreeding. Kin selection and ecological conditions may have been important in the evolution of this eusocial mammalian society.

FIGURE 57-A
A naked mole rat.

INSECT SOCIETIES

We have already mentioned the striking forms altruism can take in the social insects. The evolution of the honeybee's suicidal sting and the sterility of worker bees was indeed an enigma to Darwin, who considered the insect societies to present a fatal blow to his theory. Before we try to solve Darwin's problem, it is useful to refresh our memories as to how natural selection works. Evolution acts on *individuals*, not on populations—selection favors the genes borne by those individuals that leave the most offspring. But what constitutes an individual is an insect society?

In the **eusocial,** or truly social insects (some bees and wasps, all ants, and all termites), natural selection has acted on the *colony;* the society itself is the individual and the unit acted upon by evolution. Each colony is made up of reproductive and sterile members (Figure 57-20). A honeybee hive, for example, has a single queen who is the sole egg-layer, and tens of thousands of her offspring, who are female workers having generally non-functional ovaries. The sterility of workers is altruistic: during the course of evolution these offspring gave up their personal reproduction to help their mother rear more of their sisters.

Hamilton explained the origin of this trait with the theory of kin selection. He notes that due to an unusual system of sex determination called **haplodiploidy** that is present in bees, wasps, and ants, workers share a very high proportion of genes, theoretically as many as 75%. This is because males are haploid, having only one set of chromosomes, whereas females are diploid. Because of close genetic relatedness, *workers propagate more genes by giving up their own reproduction to assist their mother in rearing their sisters, some of which will start new colonies and thus reproduce.* This is how workers maximize their inclusive fitness.

The colony is the evolving entity in insect societies. Because workers share large fractions of genes, they propagate more genes by not directly reproducing but helping their mother reproduce.

Social insect colonies are composed of highly integrated groups of individuals, called **castes,** each of which performs a set of tasks. Their specialization and organization is so remarkable that insect societies as a whole exhibit many of the properties of an individual organism and are sometimes considered a **superorganism.** A human body, for example, relies on millions of individual cells that are specialized to perform many different tasks. In a similar way, a beehive or ant colony is a cohesively organized group of individuals, in which certain individuals perform specialized tasks on which the survival and reproduction of the entire colony depends. Only one component of a human body, the gonads, is responsible for reproduction. In a similar way, only one component of the beehive or ants' nest, the queen, is involved in the reproduction of that colony. All the cells of a human body are related to one another by descent from one fertilized zygote. Similarly, all of the members of a nest or hive are descended from an individual queen.

A honey bee colony may have up to 50,000 sterile females and a single female queen who lays all the eggs. The queen maintains her dominance in the hive by secreting a pheromone, called "queen substance," that suppresses development of the ovaries in other females, turning them into sterile workers. Drones (male bees) are produced in a hive only for purposes of mating.

When the colony grows larger in the spring, some members do not receive a sufficient quantity of queen substance, and the colony begins preparations for swarming. Workers make several new queen cells, in which new queens begin to develop.

During swarming, scout workers look for a new nest site and communicate its location to the colony; the old queen and a swarm of female workers then move to the new hive. Left behind, the new queen emerges, kills the other candidate queens, flies out to mate, and returns to assume "rule" of the hive. So conflict, as well as cooperation, characterizes sociality in honey bees.

The life-styles of social insects can be unusual and remarkable, none more so than the leafcutter ants. Leafcutters live in colonies of up to several million individuals, growing crops of fungi beneath the ground. Their moundlike nests are underground

The corpse moves slowly. It sinks deeper and deeper into the earth. While it is being embalmed, mating takes place and the young are reared.

Satanic ritual? No. It is the theme of reproduction and parental care in a remarkable group of insects known as burying beetles.

Also called gravedigger or undertaker beetles, these insects of the genus *Nicrophorus* use the carcasses of small mammals such as mice or birds as the food source for their developing offspring. Among the intriguing aspects of their biology, male and female burying beetles are monogamous and cooperate in rearing young. Although monogamy and cooperative parental care are common in birds, male parental care is unusual in most animals and is exceptional in insects. Recent investigations into the ecology of reproduction in burying beetles by Michelle Pellissier Scott of the University of New Hampshire and her colleagues at Boston University have yielded extensive details on why males cooperate with females and provide care to their offspring.

After an animal dies, it begins to emit chemicals signalling its location. The antennae of burying beetles are densely packed with receptors that detect these odors and allow males and females to find the corpse. If more than one male and female arrive at the carcass, fighting occurs between males and between females until only one individual—invariably the largest—of each sex remains.

The victorious male and female then move the corpse to a suitable location and begin to dig beneath it (Figure 57-B, *1*). As it is buried, they shave its hair or clip its feathers, roll it into a ball, and treat it with secretions to preserve it (Figure 57-B, *2*). The very act of burying the corpse triggers the growth of eggs in the female, and as a brood chamber is constructed around the corpse (now roughly four inches underground) mating takes place and eggs are layed nearby.

When larvae hatch, they assemble at the top of the corpse (which by now bears more resemblance to a meatball than to a mouse or bird) and beg for food from both their parents. Males and females respond to the irresistible stroking of their mouthparts by regurgitating food to the hungry offspring (Figure 57-B, *3*). The larvae grow quickly, and soon turn the carcass into bones through their own direct consumption. When maturation is complete, the larvae disperse and metamorphose into pupae. Females remain with their young for a significantly longer amount of time than males. Does maternal and paternal care differ?

Observations of male and female brood care show that both sexes perform essentially the same behavior. Females feed the young somewhat more than do males and males guard the brood more than females but their repetoires are very similar. Brood care behavior is flexible: if one parent is removed, the other will increase its activity to compensate for the loss of a mate.

What is the adaptive significance of male parental care? To answer this question, an experiment was carried out in which the male was removed, leaving the female to raise offspring as a single parent. In a control group, both parents remained. To test the effect of mate absence, reproductive success was measured by counting and weighing mature larvae as well as recording the number of adults that enclosed from pupae. Curiously, females successfully reared *more* young when the male was absent! This may be due to male's feeding on the carcass, reducing the amount of food available to larvae. It may also be that in an attempt to produce larger offspring, both parents practice larvicide, eating some of their own young to make more food available to a smaller number of larvae.

Although single female parenting would seem to maximize fitness, males do have a significant function in guarding the brood. Broods guarded by two parents are less likely to be taken over by other males and females of the same species that will invade the brood chamber and kill the young to gain the precious resource provided by the carcass. So, competition for reproduction may occur even after carrion has been buried.

It may seem like a great leap from burying beetles to vertebrates, but these insects can serve as a model system in which nearly all of the factors influencing the evolution of parental care can be studied. Research on parental care in an insect can thus serve to test general theories of parental care and aid in understanding its evolution in other species.

Different species of burying beetles illustrate variations of the basic theme of reproduction described above. One species, the American burying beetle, is on the Federal Endangered Species List. Studies of the ecology of reproduction in these fascinating insects may reveal important information about why populations of some species decline in numbers, and thus provide a useful example for conservation biologists as well as behavioral ecologists.

FIGURE 57-B

1 A pair of beetles prepares a mouse for burial.

2 The male and female remove the fur and roll the corpse into a ball. The larvae are on top.

3 A burying beetle feeds a larva regurgitated food.

FIGURE 57-21
Leafcutter ants cutting out sections of a leaf to carry back to their nest. Behavioral differences among workers of the leafcutter ant are associated with worker size. These workers, which have a head width of roughly 2.4 mm, cut leaves. Larger workers defend the nest and the smallest individuals tend the fungus gardens.

"cities" covering more than a hundred square yards, with hundreds of entrances and chambers as deep as 16 feet underground. Long lines of workers travel daily between the nest to a tree or bush, cut its leaves into small pieces, and carry the pieces back to the colony. Smaller worker ants chew the leaf fragments into a mulch which they spread like a carpet in underground chambers, and still smaller workers implant fungal hyphae. Soon a luxuriant garden of fungi is growing. Nurse ants carry the larvae of the nest around to graze on choice spots. Other workers weed out undesirable kinds of fungi. The **division of labor** among workers is related to worker size (Figure 57-21).

VERTEBRATE SOCIETIES

In contrast to the highly structured and integrated insect societies and their remarkable forms of altruism, vertebrates form far less rigidly organized social groups. This seems paradoxical because vertebrates have larger brains and are capable of more complex behavior, yet they show generally lower degrees of altruism. Why? Apparently the low degree of altruism is due to the lower amount of gene sharing among group members (maximally 50% in nearly all vertebrates). Nevertheless, many complex vertebrate social systems exist in which individuals show both reciprocity and kin-selected altruism. But vertebrate societies are also characterized by a greater degree of conflict and aggression among group members (although conflict does at times occur in insect societies). In vertebrates conflict generally centers around access to food resources and mates. Features of vertebrate societies that have been studied in detail in several species are cooperative breeding and alarm calling.

In some species of birds like the African pied kingfisher and the Florida scrubjay, cooperative breeding systems have evolved. For example, a pair of scrubjays may have other birds present in their territory that serve as **helpers at the nest** (Figure 57-22), which assist in feeding the pair's offspring, keeping watch for predators, and

FIGURE 57-22
A scrubjay family. Helpers cooperate with parents to rear young.

TABLE 57-1	HELPERS.		
	AVERAGE NUMBER OF OFFSPRING REARED WITH:		
	NO HELPERS		HELPERS
Inexperienced pairs	1.24		2.20
Experienced pairs	1.80		2.38

Florida Scrubjay helpers increase the reproductive success of breeding pairs whether or not those pairs are experienced breeders. Usually 1-2 birds serve as helpers

defending the territory. Helpers are fully capable of breeding on their own, but remain as non-reproductive altruists for a period of time. Nests with helpers have more offspring than those that do not (Table 57-1). Helpers are most often the fledged offspring of the pair they assist, so the situation resembles that of a family. The evolution of this type of cooperative breeding in birds has been explained by using the inclusive fitness concept we discussed earlier.

From the previous example it seems that some aspects of vertebrate behavior are self-sacrificing and therefore present a puzzle to evolutionists. Particularly puzzling is the fact that vertebrates are often organized into social groups in such a way that the activities of certain individuals benefit the group at the potential expense of those individuals themselves. The meerkat in Figure 57-5, for example, is maintaining a lookout on a termite mound deep in the Kalahari desert of Southern Africa, exposed to predators and the full glare of the sun. In doing this, the lookout draws attention to itself and thus exposes itself to greater danger than if it were not a sentry. Behaviors such as this seem to be contrary to an individual's self interest.

Individuals in some vertebrate species, such as the meerkat, not only serve as sentries for the group by keeping watch, they may also give an **alarm call** to communicate the presence of a predator when one is sighted. An alarm call causes individuals to seek shelter. Why would an individual place itself in jeopardy by revealing its location by calling? *Who benefits from hearing an alarm call?*

Paul Sherman of Cornell University has answered this question through years of field observations of alarm calling in Belding's ground squirrel. Alarm calls are given when a predator like a coyote or badger is spotted. Such predators may attack a calling squirrel, so giving a signal places the caller at risk. The social unit of a ground squirrel colony is female-based; the group tends to be composed of a female, and her daughters, sisters, aunts, and nieces. Males in the colony are not genetically related to these females. By marking all squirrels in a colony with an individual dye pattern on their fur (using Lady Clariol hair color!) and recording which individuals gave calls and the social circumstances of calling, Sherman found females with relatives living nearby were more likely to give alarm calls than females without kin nearby. Males tend to call much less frequently. Alarm calling therefore seems to represent **nepotism,** that is, it favors relatives. Again, the kin selection concept can explain the evolution of alarm calling, even when many other competing hypotheses of the origin of this trait are evaluated.

Vertebrate societies, like insect societies, have a particular type of organization (Figure 57-23). A social group of vertebrates has a certain size, stability of members, and may vary in the number of breeding males and females and type of mating system. Sociobiologists have learned that the way in which a group is organized is related to the species' ecology. The environmental features that influence the evolution of a particular type of social organization are often food type and predation. The study of the influence of the environment on social organization is called **socioecology.**

African weaver birds provide an excellent example. There are roughly 90 species of these finch-like birds which construct nests from vegetation. These species can be divided according to the type of social group they form. One group of species is found in the forest and builds camouflaged, solitary nests (Figure 57-24). Males and females

A

B

C

FIGURE 57-23
Social diversity among primates. Prosimian bush babies (**A**) are solitary; Golden-lion marmosets (**B**) form family groups, and gorillas (**C**) exist in troops headed by a dominant silver-backed male

are monogamous; they forage for insects to feed their young. The second group nests in colonies in trees on the savanna (Figure 57-24). These species feed in flocks on seeds and are polygynous. The feeding and nesting ecology is correlated with the type of breeding relationships. In the forest, insects are hard to find and both parents must cooperate in feeding the young. The camouflaged nest does not call the attention of predators to their brood. On the open savanna, building a hidden nest is not an option. Rather, savanna dwelling weaver birds are protected from predators by nesting in spiny trees, which are not very abundant. This shortage of safe nest sites means that birds must nest together in colonies. Because seeds occur abundantly, a female can acquire all the food needed to rear young without a male's help. The male, free from the duties of parenting, spends his time competing with other males for the best nest sites in the tree and courting many females, and a polygynous mating system is favored.

Similar studies of social organization have been conducted on African ungulates such as gazelle, impala, and water buffalo. In this group, sociality is correlated with diet: the more general the feeding habits, the larger the group. Social organization is also associated with feeding habits in primates.

FIGURE 57-24
Weaver bird socioecology. Monogamous pairs of weaver birds build camouflaged nests in the forest (**A**); polygynous, savanna-dwelling species form colonial nests (**B**).

Social behavior in vertebrates is often characterized by kin selected altruism. Altruistic behavior is involved in cooperative breeding in birds and alarm calling in mammals. The organization of a vertebrate society represents an adaptive response to ecological conditions.

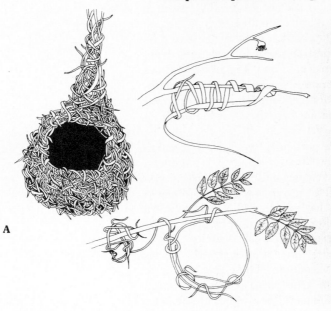

A

B

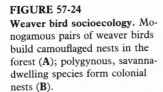

HUMAN SOCIOBIOLOGY

Sociobiology is a comparative science. The same theory is applied to study the origin of social behavior in very different species. As we saw in our discussion of insect and vertebrate societies, altruism in bees, birds, and mammals could be explained with the same concept of kin selection. We can all agree that social behavior has a biological basis, and that there is only one unifying theory of evolution to explain it. But are humans just another social animal whose behavior can be explained with Darwinian concepts and fully understood?

As a social species, humans have an unparalleled complexity. Indeed we are the only species with the intelligence to contemplate the social behavior of other animals. Intelligence, albeit unique, is just one human trait. If an ethologist took an inventory of human behavior, he or she would record both kin selected altruism and reciprocity and other elaborate social contracts, extensive parental care, conflicts between parents and offspring, and violence and warfare. A variety of mating systems like monogamy, polygyny and polyandry would be described, along with a number of sexual behaviors like adultery and homosexuality. Behaviors like adoption that appear to defy evolutionary explanation would also be a part of the ethologist's catalog. And remember, this incredible variation in behavior all occurs *in one species* and any trait can change within any *individual*. Are these behaviors rooted in our biology?

During the course of human evolution and the emergence of civilization, two processes led to adaptive change. One is **biological evolution.** We have a primate heritage, reflected in the extensive sharing of genetic material between man and our closest relatives, chimpanzees. Our upright posture, bipedal locomotion, power and precision hand grips are adaptations whose origins are traceable through our primate ancestors. Kin-selected and reciprocal altruism can also be seen in non-human primates, as well as other shared traits like aggression and different types of mating systems. In non-human primates we can demonstrate through careful study that these social traits are adaptive. We may *speculate*, based on various lines of evidence, that similar traits evolved in early humans. If individuals showing certain behaviors had an advantage in reproduction over other individuals showing alternative behaviors, and these social traits had a genetic basis, then the alleles for the expression of these traits are now part of the human genome and may influence our behavior.

The second process that underscored the emergence of civilization and led to adaptive change is **cultural evolution.** Cultural evolution refers to the transfer across generations of information necessary for survival. It is a nongenetic mode of adaptations, although some scientists feel genes influence culture. Many adaptations—the use of tools, the formation of cooperative hunting groups, the construction of shelters, and marriage practices—do not follow Mendelian rules of inheritance and are passed from generation to generation by **tradition.** To anthropologists interested in the origin of human behavior, a culture is as real a way of conveying adaptations across generations as a gene. Human cultures are also extraordinarily diverse. The way in which children are socialized among Trobriand Islanders, Pygmies, and Yanomamo Indians are very different. Again we must remember that this fantastic variation occurs within one species, and that individual behavior is very flexible.

Given this great flexibility, how can biological components of human behavior be identified? One way is to study behaviors that are cross-cultural. In spite of cultural variation, there are some traits that characterize all human societies. For example, all cultures have an incest taboo forbidding marriages between close relatives. Incestuous matings lead to a greater chance of exposing disorders such as mental retardation and hemophilia. Natural selection may have acted to create that behavioral disposition which has become a cultural norm to avoid serious biological problems. Genes that were responsible for guiding this behavior were fixed in human populations because of their adaptive effects.

Although human mating systems vary, polygyny is found to be the most common when all cultures are surveyed. Most mammalian species are polygynous; the human pattern seems to reflect our mammalian evolutionary heritage. Nonverbal communication patterns like smiling and raising the hand as a greeting also occur in many cultures. Perhaps these behaviors represent a common human heritage.

A significant number of biologists and social scientists vigorously resist any attempt to explain human behavior in evolutionary terms. In E.O. Wilson's 1975 landmark treatise *Sociobiology: The New Synthesis*, one chapter was devoted to human behavior. It was denounced vehemently by critics as an attempt to encourage thinking about human behavior in evolutionary terms. In the past, evolutionary theory has been abused to support racist, fascist, and sexist politics. Distorting Darwin's ideas, Herbert Spencer referred to starvation and poverty as "cleansing agents" of society. And the rise of Nazism in Germany was encouraged by biologists who used evolutionary ideas to explain the superiority of the caucasian race. It is clear that biology has not been free from politics. The critics of human sociobiology believe that there is still a potential for abuse. Moreover, if we consider human behavior to be the product of evolution and that our actions are influenced by genes and are at least in part "hardwired," doesn't this suggest that unpleasant aspects of human behavior such as aggression and violence cannot easily be modified? And doesn't such a view affect how we perceive the prospects for positive social change?

Darwinian theory can provide us with an overarching evolutionary *perspective* on human nature, one which stresses the unity of the human species and the origin of some important, adaptive behaviors. However, human behavior is affected *by* innate as well as learned components and many human activities (such as, art, music, religion, and seemingly biologically important behaviors such as those dealing with reproduction,) are strongly influenced by culture and are not easy to study as adaptation. It therefore appears unlikely that evolutionary biology will give us any *resolution*, or detailed explanations of the fine structure of human nature. The study of the biology of human behavior will always be provocative.

SUMMARY

1. Many behaviors are important ecologically and serve as adaptations. Foraging and territorial behaviors have evolved because they allow animals to efficiently use resources.

2. Male and female animals show different reproductive behaviors which maximize fitness. These differences relate to the extent to which each sex provides care for offspring. Usually males are competitive and females show mate choice because females have higher reproductive costs. Mating systems are related to a species' ecology.

3. Many animals show altruistic or self-sacrificing behavior. Altruism may evolve through reciprocity or be directed toward genetic relatives. Cooperative behavior often increases an individual's inclusive fitness.

4. Social groups form because it is advantageous to the individual to do so. The benefits of living in a group, such as enhanced feeding, are often balanced by the costs of disease and parasitism. Animal societies are characterized by cooperation and conflict. The organization of a society is related to the ecology of a species.

5. Human behavior is extremely rich and varied, and may result from both biology and culture. Evolutionary theory can give us limited but important insight into human nature, but such an approach to the study of human behavior may have political consequences.

REVIEW QUESTIONS

1. What is behavioral ecology? How do behavioral ecologists measure the adaptiveness of behavior?

2. How do animals forage? What is the difference between a specialist and a generalist feeder? How has natural selection acted to favor increased foraging efficiency?

3. What is a territory? Why are animals territorial?

4. What is parental investment? How do male and female reproductive costs differ?

5. Why are some animals selective in choosing a mate? What are the advantages?

6. What are the advantages and disadvantages of living in a group? How might group size be optimized?

7. Define monogamy, polygyny, and polyandry. What is the polygyny threshold?

8. What is group selection? Why is the group selection argument, as applied to the evolution of altruism, flawed?

9. What is sexual selection? How does it differ from natural selection?

10. Define altruistic behavior. To what does inclusive fitness refer? How is this related to kin selection? When can kin selection lead to the evolution of altruistic behavior? What is reciprocal altruism?

THOUGHT QUESTIONS

1. A recent study of Yanomamo Indians has shown that high-ranking men have more wives, more children, and commit murder more often than other Yanomamo men. Does this mean that there is a gene for violence in humans?

2. There are many speculations about the evolutionary origin of human behavior but it is usually not possible to rigorously test such hypotheses and critically evaluate alternative, non-evolutionary explanations. Does this mean that the evolutionary study of human behavior is not scientific?

3. Swallows often hunt in groups, whereas hawks and other predatory birds usually are solitary hunters. Can you suggest an explanation for this difference?

4. Can you suggest an evolutionary reason why in most vertebrate species polygyny is common but polyandry is rare?

FOR FURTHER READING

BORGIA, G.: "Sexual Selection in Bower Birds," *Scientific American*, June 1986, pages 92-100. Behavior among birds is often unusual, but none is more so than that of the Australasian bower birds, in which the females choose their mates depending on how well they adorn their bowers or ritualized nests.

DAWKINS, R.: *The Selfish Gene*, Oxford University Press, New York, 1976. An accurate and entertaining account of the sociobiologist's view of behavior.

HALLIDAY, T.: *Sexual Strategy*, University of Chicago Press, 1982. A colorful, broad-based treatment of animals reproduction.

HOLLDOBLER, B., and E.O. WILSON,: *The Ants*, Harvard University Press, Cambridge, MA, 1990. A Pulitzer Prize winning tour-de-force of the evolution, ecology, and social biology of the most diverse groups of social species.

HRDY, S.: *The Woman that Never Evolved*, Harvard University Press, Cambridge, MA, 1983. An evolutionary perspective on what it means to be female.

JOLLY, A.: "The Evolution of Primate Behavior," *American Scientist*, vol. 73, pages 230-239, 1985. A fascinating survey of behavior among primates that indicates a progressive development of intelligence, rather than a sudden, full-blown appearance when humans evolved.

KREBS, J., and N.B. DAVIES: *An Introduction to Behavioral Ecology*, Sinauer, Sunderland, MA, and edition, 1991. Perhaps the best source for entering the study of the adaptiveness of behavior.

RYAN, M.J.: "Signals, species, and sexual selection," *American Scientist*, vol 78 (1), pages 46-52, 1990. An excellent review of current ideas on sexual selection, including studies of the tungara frog, *Physalaemus pustulosus*.

Appendix A

Classification of Organisms

The classification used in this book is that explained in Chapter 28. It recognizes a separate kingdom, Monera, for the bacteria (prokaryotes) and divides the eukaryotes into four kingdoms: the diverse and predominantly unicellular Protista, and three large, characteristic multicellular groups derived from them: Fungi, Plantae, and Animalia.

All of the phyla or divisions (equivalent terms used for different groups of organisms) discussed in this book are outlined in this appendix, together with the classes that we have mentioned. Viruses, which are considered nonliving, are not included in this appendix, but are treated in Chapter 29.

KINGDOM MONERA

Bacteria; the prokaryotes. Single-celled organisms, sometimes forming filaments or other forms of colonies. Bacteria lack a membrane-bound nucleus and chromosomes, sexual recombination, and internal compartmentalization of the cells; their flagella are simple, composed of a single fiber of protein. They are much more diverse metabolically than the eukaryotes. Their reproduction is predominantly asexual. About 4800 species are currently recognized, but that is probably only a small fraction of the actual number.

Phylum Methanocreatrices

Methane-producing bacteria; bacteria that form methane (CH_4) by reducing CO_2 and oxidizing hydrogen (H_2). Methane-producing bacteria are either nonmotile or flagellated rods, cocci, or spirilla. They are linked with other bacteria that reduce sulfur, fix nitrogen, or are photosynthetic by their unique rRNA base sequences, which differ sharply from those of all other bacteria; a separate kingdom, Archaebacteria, has been proposed for this group.

Phylum Omnibacteria

Mainly rigid, rod-shaped, heterotrophic, gram-negative bacteria; many important pathogens are included in this phylum, as well as the ubiquitous *Escherichia coli*. Another group, the vibrios, consists of comma-shaped bacteria with a single terminal flagellum; still another comprises the rickettsias, very small bacteria that are obligate parasites.

Phylum Cyanobacteria

Photosynthetic bacteria with chlorophyll *a* and phycobilins as accessory pigments. Cyanobacteria, formerly called "blue-green algae," have been prominent for at least 3 billion years and have played a major role in raising the earth's oxygen to its present levels. Many can fix atmospheric nitrogen in specialized cells called heterocysts. Some cyanobacteria form complex filaments or other colonies. A few hundred living species.

Phylum Chloroxybacteria

Prochloron, a single genus of nonmotile, marine cocci, symbiotic in colonial ascidians; contains chlorophylls *a* and *b*.

Phylum Aphragmabacteria

Mycoplasmas and spiroplasmas; nonmotile, heterotrophic, very small bacteria that lack cell walls. Some members of this phylum are less than 0.2 micrometer in diameter, the smallest cellular organisms. They cause important diseases of plants and animals.

Phylum Spirochaetae

Spirochetes; tightly coiled spirilla in which the flagella are located beneath the outer lipoprotein membrane of their gram-negative outer cell wall; there are 2 to more than 100 flagella, which propel the spirochetes through liquid.

Pseudomonads

Pseudomonads are straight or curved gram-negative rods that bear one to several flagella at one end; some are autotrophic, many are serious plant pathogens, and some prey on other bacteria.

Phylum Actinobacteria

The actinomycetes have a filamentous growth habit; they produce spores by the division of erect, terminal branches that divide into chains of small segments. Important pathogens, causing leprosy and tuberculosis, among other diseases. *Frankia* forms nitrogen-fixing nodules on the roots of certain plants. Actinomycetes are the source of more than half of the antibiotics in use today.

Phylum Myxobacteria

The gliding bacteria; unicellular rods less than 1.5 micrometers thick but up to 5 micrometers long. Myxobacteria, which are structurally the most complex prokaryotes, often aggregate into gliding masses and may form complex spore-bearing structures; chemotrophs or heterotrophs.

Nitrogen-fixing Aerobic Bacteria

Nitrogen-fixing aerobic bacteria are gram negative and mostly flagellated. Some genera are free living, but *Rhizobium*, the key genus in nitrogen fixation worldwide, forms nodules on the roots of legumes.

Chemoautotrophic Bacteria

Chemoautotrophic bacteria use the reduced gases ammonia, methane, and hydrogen sulfide as sources of energy; they are rods or cocci, some of them flagellated. Certain chemoautotrophic bacteria make possible specific steps in the nitrogen cycle, including the conversion of nitrites to nitrates.

KINGDOM PROTISTA

Eukaryotic organisms, including many evolutionary lines of primarily single-celled organisms. Eukaryotes have a membrane-bound nucleus and chromosomes, sexual recombination, and extensive internal compartmentalization of the cells; their flagella are complex, with 9 + 2 internal organization. Some protists have pseudopods, and some are sessile (immobile). They are diverse metabolically, but much less so than bacteria; protists are heterotrophic or autotrophic, and may capture prey, absorb their food, or photosynthesize. Reproduction in protists is either sexual, involving meiosis and syngamy, or asexual. These are general features of all eukaryotes; Fungi, Animalia, and Plantae differ in their specific characteristics. Although some protists would technically be regarded as belonging to divisions, we shall consider them all phyla here for simplicity, since the two categories are equivalent.

Phylum Caryoblastea

One species of amoebalike organism, *Pelomyxa palustris*, which lacks mitosis, mitochondria, and chloroplasts, but does have two kinds of bacterial symbionts. *Pelomyxa*, which occurs on the muddy bottoms of freshwater ponds, probably represents a very early stage in the evolution of eukaryotes.

Phylum Dinoflagellata

Dinoflagellates; unicellular, photosynthetic organisms, most of which are clad in stiff, cellulose plates and have two unequal flagella that beat in grooves encircling the body at right angles. Many are symbiotic in other organisms, and are then known as zooxanthellae. Chlorophylls *a* and *c*. About 1000 species.

Phylum Rhizopoda

Amoebas; heterotrophic, unicellular organisms that move from place to place by cellular extensions called pseudopods and reproduce only asexually, by fission. Hundreds of species.

Phylum Sporozoa

Sporozoans; unicellular, heterotrophic, nonmotile, spore-forming parasites of animals. Sporozoans have complex life cycles that involve both sexual and asexual phases, the former with an alteration of haploid and diploid generations. About 3900 species.

Phylum Acrasiomycota

Cellular slime molds; unicellular, amoebalike, heterotrophic organisms that aggregate in masses at certain stages of their life cycle and form compound sporangia. Meiosis occurs within cysts in these sporangia. About 70 species.

Phylum Myxomycota

Plasmodial slime molds; heterotrophic organisms that move from place to place as a multicellular, gelatinous mass, which forms sporangia at times. Diploid spores are produced within the sporangium, and meiosis occurs within them at the time of germination. About 450 species.

Phylum Zoomastigina

Zoomastigotes; a highly diverse phylum of mostly unicellular, heterotrophic, flagellated, free-living or parasitic protists (from one flagellum to thousands). Sexual reproduction rare. Thousands of species.

Phylum Euglenophyta

Euglenoids; mostly unicellular, photosynthetic or heterotrophic protists with two unequal flagella. Chlorophylls *a* and *b* in photosynthetic forms. Euglenoids closely resemble certain zoomastigotes. About 1000 species.

Phylum Phaeophyta

Brown algae; multicellular, photosynthetic, mostly marine protists with chlorophylls *a* and *c* and an abundant carotenoid (fucoxanthin) that colors the organisms brownish. About 1500 species.

Phylum Bacillariophyta

Diatoms; mostly unicellular, photosynthetic organisms with chlorophylls *a* and *c* and fucoxanthin. Diatoms have a unique double shell made of opaline silica, resembling a small box with a lid. Sexual reproduction is apparently rare; flagella absent. About 11,500 living species.

Phylum Foraminifera

Heterotrophic marine organisms; characteristic feature is a pore-studded shell, or test, some of them brilliantly colored.

Phylum Chlorophyta

Green algae; a large and diverse phylum of unicellular or multicellular, mostly aquatic organisms with chlorophylls *a* and *b* and carotenoids, and with starch accumulated within the plastids (as it is also in plants) as the food storage product. Many green algae have two equal whiplash flagella; others are nonflagellated. Some form cell plates like those of plants; many have cellulose prominent in their cell walls. About 7000 species.

Phylum Ciliophora

Ciliates; diverse, mostly unicellular, heterotrophic protists, characteristically with large numbers of cilia. Outer layer of cell proteinaceous, flexible. Most ciliates have a complex sexual process known as conjugation. About 8000 species.

Phylum Oomycota

Oomycetes; water molds, white rusts, and downy mildews. Aquatic or terrestrial unicellular or multicellular parasites or saprobes that absorb dead organic matter; cellulose is the primary constituent of the cell wall in many oomycetes. About 580 species.

Phylum Rhodophyta

Red algae; mostly marine, mostly multicellular protists with chloroplasts containing chlorophyll *a* and, like the cyanobacteria that presumably gave rise to their plastids, phycobilins; no flagellated cells present. About 4000 species.

KINGDOM FUNGI

Filamentous, multinucleate, heterotrophic eukaryotes with cell walls rich in chitin; no flagellated cells present. Mitosis in fungi takes place within the nuclei, the nuclear envelope never breaking down. The filaments of fungi grow through the substrate, secreting enzymes and digesting the products of their activity. Septa between the nuclei in the hyphae normally complete only when sexual or asexual reproductive structures are being cut off. Asexual reproduction frequent in some groups. The nuclei of fungi are haploid, with the zygote the only diploid stage in the life cycle. About 77,000 named species.

Division Zygomycota

Zygomycetes; bread molds and other microscopic fungi that occur on decaying organic matter. Zygomycetes are the fungal partners in endomycorrhizae. Hyphae aseptate except when forming sporangia or gametangia. Sexual reproduction by equal gametangia containing numerous nuclei; after fusion, zygotes are formed within a zygospore, a structure that often forms a characteristic thick wall. Meiosis occurs during the germination of the zygospore. About 600 species.

Division Ascomycota

Ascomycetes; yeasts, molds, many important plant pathogens, morels, cup fungi, and truffles. Hyphae divided by incomplete septa except when asci, the structures characteristic of sexual reproduction, are formed. Dikaryotic hyphae form after appropriate fusions of monokaryotic ones, and eventually differentiate asci, which are often club shaped, within an ascocarp. Syngamy within each ascus produces the zygote, which divides immediately by meiosis to produce the ascospores. About 30,000 named species.

Division Basidiomycota

Basidiomycetes; mushrooms, toadstools, bracket and shelf fungi, rusts, and smuts. Most ectomycorrhizae involve basidiomycetes; a few involve ascomycetes. Hyphae and life cycle similar to that of ascomycetes but differing in important details. Basidiospores elevated from the basidium on threadlike projections called sterigmata; basidiocarps may be large and elaborate or absent, depending on the group. About 16,000 named species.

Fungi Imperfecti

An artificial group of about 17,000 species. Most are ascomycetes for which the sexual reproductive structures are not known; this may be because the stages are rare or do not occur, or because the fungus is poorly known. In either situation, the classification must be made according to the details that are known, which necessitates the recognition of the Fungi Imperfecti.

Lichens

Lichens are symbiotic associations between an ascomycete (a few basidiomycetes are involved also) and either a green alga or a cyanobacterium; the fungus provides protection and structure, and the photosynthetic organism enclosed in it manufactures carbohydrates. There are at least 15,000 species of lichens.

KINGDOM PLANTAE

Multicellular, photosynthetic, primarily terrestrial eukaryotes derived from the green algae (phylum Chlorophyta) and, like them, containing chlorophylls *a* and *b*, together with carotenoids, in chloroplasts and storing starch in chloroplasts. The cell walls of plants have a cellulose matrix and sometimes become lignified; cell division is by means of a cell plate that

forms across the mitotic spindle. The vascular plants have an elaborate system of conducting cells, consisting of xylem (in which water and minerals are transported) and phloem (in which carbohydrates are transported); the mosses have a reduced vascular system, something the liverworts and hornworts—which may not be directly related to the mosses—lack. Plants have a waxy cuticle that helps them to retain water, and most have stomata, flanked by specialized guard cells, which allow water to escape and carbon dioxide to reach the chloroplast-containing cells within their leaves and stems. All plants have an alternation of generations with reduced gametophytes and multicellular gametangia. About 270,000 species.

Division Bryophyta

Mosses, hornworts, and liverworts, three groups of simple plants that may not be directly related, comprise the bryophytes. Bryophytes have green, photosynthetic gametophytes and usually brownish or yellowish sporophytes with little or no chlorophyll. Bryophytes have multicellular gametangia and biflagellated sperm. About 10,000 species.

Class Hepaticopsida

Liverworts; some "leafy," and some thallose, strap-shaped plants that creep along the soil or other substrate on which they grow. Sporophytes are usually enclosed within gametophyte tissue. About 6500 species.

Class Antheroceratopsida

Hornworts; thallose plants with green, creeping gametophytes. The photosynthetic sporophytes arise from the upper surface of the gametophyte. About 100 species.

Class Muscopsida

The mosses; usually brownish sporophytes, with complex sporangia, arise from the green, leafy gametophytes. Both gametophytes and sporophytes usually contain reduced vascular tissue, and mosses may be, in effect, reduced vascular plants. About 10,000 species.

Division Psilophyta

Whisk ferns. Vascular plants (like those in all the remaining plant divisions) with well-developed vascular tissue, consisting of xylem and phloem. Gametophytes complex, with multicellular gametangia. Homosporous. Motile sperm. No differentiation between root and shoot. Two genera and several species.

Division Lycophyta

Lycopods (including club mosses and quillworts). Vascular plants. Homosporous (*Lycopodium*) or heterosporous (*Selaginella, Isoetes*). Gametophytes complex, with multicellular gametangia. Motile sperm. Five genera and about 1000 species.

Division Sphenophyta

Horsetails. Vascular plants with characteristic ribbed, jointed stems. Homosporous. Gametophytes complex, with multicellular gametangia. Motile sperm. One genus (*Equisetum*), 15 species.

Division Pterophyta

Ferns. Vascular plants, often with characteristic divided, feathery leaves (fronds). Mostly homosporous, a few genera heterosporous. Gametophytes complex, with multicellular gametangia. Motile sperm. About 12,000 species.

Division Coniferophyta

Conifers. Vascular plants, mostly trees, some shrubs, often with hard or leathery leaves. Heterosporous and seed-forming; the seeds are naked at the time of fertilization (therefore, the conifers are gymnosperms). Archegonia multicellular, antheridia lacking; sperm immotile, carried to the vicinity of the egg by the pollen tube. Gametophytes reduced, held within the ovule (megagametophytes) or pollen grain (microgametophytes). About 50 genera, with about 550 species.

Division Cycadophyta

Cycads; palmlike, heterosporous gymnosperms with large, pinnate leaves and terminal cones. Archegonia multicellular, antheridia lacking; sperm flagellated but carried to the vicinity of the egg by the pollen tube. Gametophytes greatly reduced. Ten genera, about 100 species.

Division Ginkgophyta

One species, the ginkgo or maidenhair tree; a tall, deciduous tree with fan-shaped leaves that have open dichotomous venation. Heterosporous gymnosperm with fleshy ovules. Details of gametes, gametangia, and gametophytes same as in cycads.

Division Gnetophyta

Genetophytes, a very diverse group of three genera of gymnosperms. Some genetophytes have archegonia, others do not; all lack antheridia. Sperm immotile; gametophytes reduced. About 70 species.

Division Anthophyta

Flowering plants, or angiosperms, the dominant group of plants. Angiosperms lack antheridia and archegonia, and have very reduced gametophytes consisting of a few cells; their sperm are immotile, carried to the vicinity of the ovule by the pollen tube. Angiosperms are characterized by their flowers, unique structures that may have as many as four whorls of appendages, and fruits, which enclose the seeds at maturity. Fertilization is indirect, and the ovules are enclosed at the time of pollination. Fertilization is double, one sperm fusing with the egg to produce a zygote, and other fusing with the polar nuclei to form the primary endosperm nucleus, which gives rise to the endosperm—a distinctive tissue that nourishes the developing embryo. About 235,000 species.

Class Monocotyledons

The monocots; angiosperms with one seedling leaf (cotyledon). Monocots usually have parallel venation and flower parts in threes or multiples of threes. The vascular bundles in the stem are scattered, and true secondary growth is present in only a few genera. About 65,000 species.

Class Dicotyledons

The dicots; angiosperms with two seedling leaves (cotyledons). Dicots usually have reticulate (netlike) venation and flower parts in fours or fives, or multiples of fours or fives. The vascular bundles in the stem usually form a ring, and true secondary growth is characteristic. About 160,000 species.

KINGDOM ANIMALIA

Animals are multicellular eukaryotes that characteristically ingest their food. Their cells are usually flexible; in all of the approximately 35 phyla except sponges, these cells are organized into structural and functional units called tissues, which in turn make up organs in most animals. In animals, the cells move extensively during the development of the embryos; the blastula, a hollow ball of cells, forms early in this process and is characteristic of the group. Most animals reproduce sexually; their nonmotile eggs are much larger than their small, flagellated sperm. The gametes fuse directly to produce a zygote and do not divide by mitosis as in plants. More than a million species of animals have been described, and at least several times that many await discovery.

SUBKINGDOM PARAZOA

Animals that mostly lack definite symmetry, and possess neither tissues nor organs.

Phylum Porifera

Sponges. Larval sponges are free swimming, whereas the adults are sessile. Sponges feed by circulating water through pores and canals lined with flagellum-bearing choanocytes. The structure of these choanocytes indicates that sponges evolved from certain zoomastigotes. Many sponges have a skeleton composed of spicules (calcium carbonate or silica) or spongin (a protein), or both. About 10,000 marine species and 150 that live in fresh water.

SUBKINGDOM EUMETAZOA

Animals with definite symmetry, either radial or bilateral, and which have definite tissues and usually organs.

Phylum Cnidaria

Corals, jellyfish, hydras. Radially symmetrical animals that mostly have distinct tissues; two basically different body forms: polyps and medusae. Digestive cavity (gastrovascular cavity, or coelenteron) has only one opening. Cnidocysts, specialized stinging cells, occur only in the members of this phylum. Nearly all marine, a few in fresh water. About 10,100 species.

Class Hydrozoa

Hydroids; most have both polyp and medusa stages in their life cycle, with the polyp stage dominant. Most hydroids are marine and colonial. About 2700 species.

Class Scyphozoa

Jellyfish, medusa stage dominant, with the polyp stage small, inconspicuous, and simple in structure. About 200 species.

Class Anthozoa

Corals, sea anemones, and their relatives; only polyps occur in the members of this class, and they may be colonial or solitary. About 6200 species.

Phylum Ctenophora

Comb jellies and sea walnuts; translucent, gelatinous, free-swimming, spherical or ribbonlike, radially symmetrical animals with a gastrovascular cavity and an anal pore. Specialized tentacles with colloblasts in one group. About 90 species, now usually thought not to be closely related to the cnidarians.

Phylum Platyhelminthes

Flatworms; bilaterally symmetrical acoelomates. Flatworms are the simplest animals that have organs; they lack a circulatory system. Flatworm guts have only one opening; they are mostly hermaphroditic, with complex reproductive systems. About 13,000 species.

Class Turbellaria

Free-living flatworms; they move from place to place by means of their ciliated epidermis, and have eyespots, which are absent in their parasitic relatives. Abundant in fresh water and the sea.

Class Trematoda

Flukes; parasitic flatworms with a digestive tract, and often complex life cycles that involve two or more hosts.

Class Cestoda

Tapeworms; parasitic flatworms that lack a digestive tract and absorb food through their body walls; complex life cycles.

Phylum Rhynchocoela

Ribbon worms; bilaterally symmetrical, large, ribbon-shaped or thread-shaped aquatic, acoelomate worms, mainly marine, with a long, extensible proboscis and complete digestive system. About 650 species.

Phylum Nematoda

Nematodes, eelworms, and roundworms; ubiquitous, bilaterally symmetrical, cylindrical, unsegmented, pseudocoelomate worms, including many important parasites of plants and animals. More than 12,000 described species, but the actual number is probably 500,000 or more species.

Phylum Rotifera

Rotifers; small, wormlike or spherical, bilaterally symmetrical, pseudocoelomate animals with a crown of cilia. Most rotifers occur in fresh water, with a few in soil and a few in the sea. About 2000 species.

Phylum Loricifera

Minute, bilaterally symmetrical, pseudocoelomate animals that live in the spaces between grains of sand in the sea; mouthparts consist of a unique flexible tube. Several species; the discovery of this phylum was announced in 1983.

Phylum Bryozoa

Bryozoans; small aquatic, mostly marine animals with a true coelom and a lophophore, a tentacle-bearing organ that rings the mouth; mostly colonial, and often plantlike. About 4000 species.

Phylum Brachiopoda

Lamp shells; marine animals, abundant during the Paleozoic Era but now represented by only about 300 living species. Brachiopods superficially resemble clams, but have distinct ventral and dorsal shells; like bryozoans, they have a lophophore.

Phylum Phoronida

Phoronid worms; tube-dwelling, marine worms in which as many as 500 tentacles may ring the lophophore. Some species are solitary, but most live in dense populations; they range from 1 millimeter to about 50 centimeters long. About a dozen species.

Phylum Mollusca

Mollusks; bilaterally symmetrical, protostome coelomate animals that occur in marine, freshwater, and terrestrial habitats. Mollusks have a visceral mass and a muscular foot that is used in locomotion; many also have a head. Many mollusks form a shell, and all except for the bivalves have a radula, a rasping, tonguelike organ used for scraping, drilling, or capturing prey; the circulatory system consists of a heart and, usually, an open system through which the blood circulates freely. At least 110,000 species.

Class Polyplacophora

Chitons; marine mollusks with eight overlapping calcareous dorsal plates embedded in the mantle. The chitons are grazing herbivores. About 600 species.

Class Gastropoda

Snails and slugs; most gastropods have a spiral shell and a head with one or two pairs of tentacles. Slugs are snails that have lost their shell during the course of evolution. Of the approximately 80,000 species, nearly half are terrestrial.

Class Bivalvia

Bivalves—clams, scallops, oysters, mussels, and related mollusks—that have two shells, hinged together, and a wedge-shaped foot; bivalves lack distinct heads and radulas, and are usually sessile filter-feeders that disperse from place to place largely as larvae. About 10,000 species.

Class Cephalopoda

Octopuses, squids, and nautilus; active, intelligent marine predators in which the foot has evolved into a series of tentacles—8 in octopuses, 10 in squids, and 80 to 90 in the nautilus. The cephalopods have two horny jaws, highly developed eyes, and complex, efficient nervous systems. The shell is internal (squids), external (nautilus), or absent (octopuses). More than 600 species.

Phylum Annelida

Annelids; segmented, bilaterally symmetrical, protostome coelomates; the segments are divided internally by septa. Cerebral ganglion (brain) well developed; circulatory system closed; digestive tract complete, one-way. About 12,000 species.

Class Polychaeta

Polychetes; clamworms, plumed worms, scaleworms, and their relatives. Mainly marine worms with a distinct head and specialized sense organs, including eyes. Fleshy, paddlelike flaps called parapodia, used in locomotion, occur on most segments. The trochophore larvae are free swimming. About 8000 species.

Class Oligochaeta

Earthworms; terrestrial, freshwater, and marine annelids with fewer setae than the polychetes and no parapodia. Earthworms lack a distinct head, and their sense organs are not as specialized as those of the polychetes. About 3100 species.

Class Hirundinea

Leeches; dorsiventrally flattened external parasites, predators, or scavengers, with suckers at one or both ends of the body. About 300 species.

Phylum Arthropoda

Arthropods; bilaterally symmetrical protostome coelomates with a segmented body, chitinous exoskeleton, complete digestive tract, dorsal brain and paired nerve cord, and jointed appendages. Arthropods are the largest phylum of animals, with nearly a million species described and many more to be found.

Subphylum Chelicerata
Class Arachnida

Arachnids; spiders, mites, ticks, and scorpions. Largely terrestrial, carnivorous, air-breathing animals with chelicerae (pincers or fangs), pedipalps (usually sensory, but often used for catching and handling prey), and four pairs of walking legs. About 57,000 species.

Class Merostomata

Horseshoe crabs. Marine chelicerates with five pairs of walking legs, compound eyes, and book gills. Three genera and four species.

Class Pycnogonida

Sea spiders; small marine chelicerates that are mostly external parasites or predators on other animals. Sea spiders have a sucking proboscis and four to six pairs of legs. More than 1000 species.

Subphylum Crustacea
Class Crustacea

Crustaceans; lobsters, crayfish, shrimps, crabs, and many others. Mainly aquatic mandibulate animals with biramous (fundamentally two-branched) appendages; many crustaceans have a distinctive, free-swimming nauplius larva. Crustaceans have two pairs of antennae and legs on both their thorax and abdomen; many have compound eyes. About 35,000 species.

Subphylum Uniramia
Class Chilopoda

Centipedes; active, mandibulate predators with a head region followed by 15 to 177 body segments, each with a pair of legs. About 2500 species.

Class Diplopoda

Millipedes; mostly herbivorous, mandibulate animals with a head region followed by 20 to 200 segments, each with two pairs of legs. More than 10,000 named species.

Class Insecta

Insects; by far the largest group of organisms. Mostly terrestrial mandibulates, with a body divided into three regions: head, thorax, and abdomen. Insects have a complex series of mouthparts, which are highly specialized in different orders; they have compound eyes and one pair of antennae on the head. The six legs are located on the thorax, as are the wings, if present. Insects breathe by means of tracheae and excrete their wastes by a system of Malpighian tubules. Many have complex metamorphosis. More than 750,000 described species, but several times that many doubtless await discovery.

Phylum Pogonophora

Giant tube worms, or beard worms; sessile deep-sea worms that live within chitin tubes they secrete on bottom sediments or decaying wood. These worms range from 10 cm to about 90 cm long, and are usually about 1 millimeter thick; they have long, beardlike tentacles behind the head. About 100 species.

Phylum Onychophora

Wormlike protostome animals with a chitinous exoskeleton and tracheae; embryo development identical to that of the annelids and Uniramia. Onychophora appear to link annelids with Uniramia; the 70 living species appear to be evolutionary relicts.

Phylum Echinodermata

Echinoderms; sea stars, brittle stars, sand dollars, sea cucumbers, and sea urchins. Complex deuterostome, coelomate, marine animals that are more or less radially symmetrical as adults; calcareous plates called ossicles are more or less abundant in the epidermis. Water vascular system extends through perforated plates as tube feet, and is a specialized feature of the phylum. About 6000 living species.

Class Crinoidea

Crinoids; sea lilies and feather stars. Filter-feeding echinoderms in which the mouth and anus are located in a disk on the upper surface, 5 to more than 200 feathery arms are located around the margins of this disk. About 600 species.

Class Asteroidea

Sea stars; star-shaped echinoderms, with the five to many arms set off more or less sharply from the disk. The mouth and anus are directed downward, and the animals move by means of rows of tube feet on the arms. About 1500 species.

Class Ophiuroidea

Brittle stars; star-shaped echinoderms with very slender, long, often spiny, highly flexible arms that are sharply set off from the disk. About 2000 species.

Class Echinoidea

Sea urchins and sand dollars; echinoderms that lack distinct arms and have a rigid external covering. Echinoids walk by means of their tube feet or jointed spines. About 950 species.

Class Holothuroidea

Sea cucumbers; soft, sluglike echinoderms that lie on their sides. About 1500 species.

Phylum Chaetognatha

Arrow worms; rapid-swimming, translucent, bilaterally symmetrical, marine, deuterostome, coelomate worms with large eyes, powerful jaws, and a head ringed with numerous sharp, movable hooks. About 70 species.

Phylum Hemichordata

Acorn worms; soft-bodied, bilaterally symmetrical, burrowing, marine, deuterostome, coelomate worms with three-segmented bodies (proboscis, collar, and trunk). About 90 species.

Phylum Chordata

Chordates; bilaterally symmetrical, deuterostome, coelomate animals that at some stage of their development have a notochord, pharyngeal slits, a hollow nerve cord on their dorsal side, and a tail. The best-known group of animals; about 45,000 species.

Subphylum Urochordata

Tunicates; larval tunicates are free-swimming and have a notochord and nerve cord, structures that are absent in the sessile, usually saclike adults, which also lack any sign of segmentation or of a body cavity. Tunicates are marine chordates that obtain their food by ciliary action. About 1250 species.

Subphylum Cephalochordata

Lancelets; marine fishlike animals with a permanent notochord, a nerve cord, a pharynx with gill slits, and no internal skeleton. Lancelets are filter-feeders that draw water through their pharyngeal slits by ciliary action. About 23 species.

Subphylum Vertebrata

Vertebrates; notochord replaced by cartilage or bone, forming a segmented vertebral column (the backbone). Vertebrates have a distinct head with a skull and brain. Their hollow dorsal nerve cord is usually protected within a U-shaped groove formed by paired projections from the vertebral column. About 43,700 species.

Class Agnatha

Agnathans; lampreys and hagfishes. Naked, eel-like, jawless fishes with a cartilaginous skeleton; the agnathans lack scales, bones, and fins. Parasites on other fishes, or scavengers. About 63 species.

Class Chondrichthyes

Sharks, skates, and rays. Almost entirely marine fishes with a cartilaginous skeleton, efficient fins, complex copulatory organs, and small, pointed scales (denticles); but lacking air bladders. About 850 species.

Class Osteichthyes

Bony fishes; the members of this class, which are abundant both in the sea and in fresh water, have bony skeletons, efficient fins, scales, and usually air bladders (by means of which they regulate their density and therefore their level in the water). More than 18,000 species.

Class Amphibia

Salamanders, frogs, and toads. Tetrapod, egg-laying, ectothermic vertebrates that lack scales; amphibians respire with gills as larvae and with lungs as adults. They have incomplete double circulation. Amphibians were the first terrestrial vertebrates; they still depend on a moist environment for at least a portion of their life cycles. About 2800 species.

Class Reptilia

The reptiles; lizards, snakes, turtles, and crocodiles. Tetrapod, ectothermic vertebrates with an amniotic egg; reptiles have lungs and are covered with scales; most are fully terrestrial, although some are aquatic. Reptiles have incomplete double circulation. The four legs are absent in snakes and some lizards. Nearly 6000 species.

Class Aves

Birds. Tetrapod, endothermic vertebrates in which the forelimbs are modified into wings; most are capable of flight, and all lay amniotic eggs. Birds have lungs and are fully terrestrial, although some live in water. They have complete double circulation; feathers are a characteristic feature of the class. About 9000 species.

Class Mammalia

Mammals. Tetrapod, endothermic vertebrates with complete double circulation and usually hairy skins. The forelimbs are modified into wings in bats, and all four limbs are modified into flippers in some aquatic mammals. Monotremes lay eggs; marsupials retain their young in a marsupium, or pouch, for a prolonged period; and the great majority of mammals— the placental mammals—nourish their young in the womb by means of a specialized structure, the placenta, modified from the amniotic egg characteristic of their ancestors. About 4500 species.

Appendix B

Answers to Review Questions

CHAPTER 1

1. A hypothesis is an educated guess based on careful observation. Deductive reasoning is used to formulate hypotheses, and inductive reasoning is used to make predictions from hypotheses.
2. A hypothesis becomes a theory through continued testing over long periods of time. Some examples are Einstein's theory of relativity, Mendel's theories about heredity, and Newton's theory of gravity.
3. Darwin's evidence included a progressive change in fossils, distinctive geographical associations, and island species that showed a relatedness to one another and to the species of the nearest continent.
4. Discoveries that supported and directed Darwin's theory of evolution included new calculations on the age of the earth and Malthus' ideas on population growth.
5. Natural selection may be defined as nature working to limit population growth, keeping it level rather than allowing it to grow geometrically. This ensures that those traits best suited to survival are passed on to successive generations.
6. Darwin's theory of natural selection states that organisms reproduce excessively, that those better able to survive will reproduce more, that the ability to survive is passed on to offspring, and that the favored characteristics will thus increase in the population.
7. Since Darwin scientists have found an expanded fossil record, more accurate estimation of the age of the earth, the mechanisms of heredity, an examination of comparative anatomy, and advances in molecular biology.
8. Homologous structures support evolution by showing that the function of structures change over time but the structures have the same evolutionary origin.

CHAPTER 2

1. The three subatomic particles are the proton (+), the neutron (o), and the electron (−). The proton and neutron are found in the nucleus of the atom.
2. The number of protons equals the atomic number. The number of protons plus the number of neutrons equals the atomic mass. Electrons have virtually no mass.
3. Isotopes have the same number of protons and thus the same atomic number, but they have different numbers of neutrons and hence different masses. They exhibit the same chemical properties because their number of electrons is the same.
4. Ions have different numbers of electrons and therefore different atomic charges and different chemical properties.
5. The electron.
6. An ionic bond refers to the transfer of an electron between elements and the subsequent attraction of oppositely charged ions (NaCl). A covalent bond refers to the sharing of electrons between atoms (CH_4).
7. Reduction is the addition of an electron to a molecule; oxidation is the removal of an electron from a molecule.
8. The six life-giving properties are cohesion/adhesion, heat storage, powerful solvent, organization of nonpolar molecules, formation of hydrogen bonds, and ionization. The most important factor is the fact that water is polar.
9. pH is equal to $-\log[H^+]$. The pH range for acids is 1 to 7; that for bases is 7 to 14. In a molar concentration of 1/10,000,000,000 hydrogen ions the pH is 10; in one of 0.001 hydrogen ions per liter, the pH is 3.
10. A buffer is an acid/base pair with sufficient H^+ ions to donate them to or remove them from a solution.

CHAPTER 3

1. The four macromolecules are carbohydrates, lipids, proteins, and nucleic acids. The elements of each are as follows: carbohydrate = C, H, O; lipids = C, H, some O; proteins = C, H, O, N; nucleic acids = C, H, O, N, P.
2. Dehydration synthesis connects two molecules with the removal of water; hydrolysis breaks two molecules apart with the addition of water.
3. The three most common monosaccharides are glucose, fructose, and galactose. The structural isomers are glucose and fructose; the stereoisomers are glucose and galactose.
4. Unlike starch, cellulose and chitin have bonds that are indigestible by most organisms. Chitin has nitrogen attached to each glucose unit.
5. Three fatty acids attached to a three-carbon glycerol backbone.
6. In saturated fats all fatty acid carbons have the maximum number of hydrogen atoms attached; these are generally solid in nature. Unsaturated fats have a double bond between one or more successive carbon atoms, have fewer hydrogen atoms, and are liquid. A polyunsaturated fat has more than one double bond in the fatty acid carbon chain.
7. Fats are more efficient because of their high concentration of energy storing C-H bonds. Fats = 9 kcal/g, carbohydrates = 4 kcal/g.
8. Amino acids are the general subunits that compose proteins. They have a central carbon with attached H, NH_2, COOH, and R group.
9. The group most affected is the tertiary structure. The primary structure is a chain of amino acids coded by DNA, the secondary is a folding or coiling controlled by hydrogen bonding among repeated elements, the tertiary is a configuration controlled by the R groups, and the quaternary is the combining of polypeptide subunits.
10. A nucleic acid is a five-carbon sugar, a phosphate group, and organic nitrogen containing purine or a pyrimidine base.
11. DNA is a deoxyribose sugar, has ATCG, is double stranded, and is the template. RNA is a ribose sugar, has AUCG, is usually single stranded, and is the copy.

CHAPTER 4

1. The atmosphere of early earth contained N_2, CO_2, H_2O, H_2S, NH_3, CH_4, and maybe H_2. O_2 was absent. Then the atmosphere was reducing, now it is oxidizing.
2. These experiments took compounds assumed present in the early atmosphere, bombarded them with electricity, and observed the formation of simple organic molecules—formic acid, urea, and the amino acids glycine and alanine.
3. Cellular organization, growth and metabolism, reproduction, and heredity are necessary. Most living things exhibit movement, sensitivity, and complexity.
4. Coacervates are protein/lipid aggregations, have a quasi-lipid bilayer membrane, accumulate more organic materials inside themselves, and divide by "budding." They are not alive because they possess no known genetic material.
5. Supporting evidence includes the fact that rocks containing the earliest fossils have organic material in them that has more ^{12}C than is found in the atmosphere and that only living organisms are known to selectively concentrate ^{12}C. They are most closely related to cyanobacteria.
6. The atmosphere acquired plentiful amounts of oxygen 2.5 billion years ago. Before that date iron was present as soluble ferrous compounds in the oceans, and after that time iron was oxidized and exists as insoluble ferric oxides.
7. The hydrogen source is H_2O; the carbon source is CO_2. The byproduct is O_2. Their ancestors produced all of the oxygen presently in our atmosphere.
8. Autotrophs are self-feeding and produce their own food, usually but not always via photosynthesis. Heterotrophs are "other feeding"

and get their food from, or directly eat, autotrophs.

9. The first eukaryotes known are fossils 1.3 to 1.5 billion years old. Eukaryotes differ from earlier prokaryotes in that they have a nucleus.

10. It is likely that there is some form of life elsewhere because by the sheer mathematical probability that with so many stars and so many planets, life should have evolved somewhere out there.

CHAPTER 5

1. (1) All organisms are composed of one or more cells. (2) Cells are the smallest living things. (3) Cells arise only from the division of other pre-existing cells.

2. There is an optimal surface-to-volume ratio; a big cell does not have enough surface area to interact properly for all its volume.

3. The plasma membrane is a lipid bilayer. The protein classes embedded in it are channels, receptors, and markers.

4. Endoplasmic reticulum is a bilayer lipid membrane. It compartmentalizes the cell interior and channels transport of molecules throughout the cell. Rough ER has ribosomes and synthesizes proteins; smooth ER does not have ribosomes and is involved in lipid synthesis and detoxification.

5. Microbodies are membrane-bound vesicles and contain oxidative enzymes for converting lipids to carbohydrates and destroying peroxides.

6. The nuclear membrane is bilayered. Proteins move in and RNA and RNA-protein molecules move out.

7. The nucleolus is the region of active RNA synthesis. It is located near the synthesizing chromosomes. It is a transient structure.

8. Golgi bodies collect, package, and distribute cell-synthesized molecules. They are closely associated with the ER.

9. Mitochondria and chloroplasts are involved in cellular energetics. Both are found in plants, and only mitochondria are found in animals. They are derived from ancient bacteria through endosymbiosis.

10. A mitochondrion has an outer membrane, an outer compartment, an inner folded membrane or cristae, and an inner matrix. It functions in oxidative metabolism.

11. Chloroplasts are the site of photosynthesis. Their functional unit is the thylakoid.

12. The cytoskeleton is composed of actin protein filaments, tubulin microtubules, and intermediate fibers. The first two are changeable and the last is stable.

13. The cytoskeleton determines the cell's shape, provides a scaffold to

hold the cell's enzymes, and assists in cell movement.

14. The term "9 + 2 flagella" means a circle of 9 pairs of microtubules surrounding 2 central microtubules. They are characteristic of eukaryotes. Cilia and flagella have the same structure but cilia are shorter and more numerous.

CHAPTER 6

1. Biological membranes are composed of phospholipid. It has two fatty acids and an organic phosphorylated alcohol; it also has both polar and nonpolar ends. Typical fats have three fatty acids and are nonpolar.

2. The polar heads like water and face it, forming hydrogen bonds. The nonpolar fatty acid tails move away from water and are sheltered from it on the other side. This is a desirable quality in waterproofing because water can't easily move through the solid sheet of nonpolar tails.

3. The four components and their functions are as follows: lipid bilayer foundation—is a flexible impermeable matrix; transmembrane proteins—provide channels to allow molecule passage; supporting fiber network—reinforces membrane shape; exterior glycolipids—act as identification markers.

4. (1) Controls passage of water; (2) controls passage of bulk materials; (3) selectively transports special molecules; (4) acts as point of information reception; (5) is the expression of cell identity; (6) has physical connections with other cells; and (7) exhibits surface enzyme activity.

5. Diffusion is the net movement of molecules from a region of high concentration to one of low concentration resulting in uniform distribution. The driving force behind diffusion is random spontaneous molecular motion.

6. Hypertonic—solution solutes greater than cell solutes; hypotonic—solution solutes less than cell solutes; isotonic—solution solutes equal to cell solutes.

7. Osmosis is a special case describing only the movement of water across a membrane. Water passage into a cell immersed in a hypotonic solution is limited by the hydrostatic/osmotic pressure equilibrium when a cell wall is present, or by the bursting of the cell membrane.

8. Endocytosis describes the bulk passage of materials into the cell; exocytosis is the opposite effect. Phagocytosis involves an organism or a fragment of organic material, whereas pinocytosis involves a liquid with dissolved molecules.

9. Selective permeability allows the passage of only certain molecules

across a membrane. It controls what goes in and what comes out of a cell.

10. In receptor-mediated endocytosis a coated pit in the membrane contains a protein receptor. When triggered by the presence of a specific molecule, the pit closes over and traps the molecule within a vesicle.

11. Facilitated diffusion refers to the transport of molecules across a membrane by a carrier protein in the direction of lowest concentration. It is specific, is passive, and may become saturated.

12. Active transport requires the expenditure of energy, is independent of solute concentration, and may maintain a concentration counter to that which would normally occur based on diffusion alone.

13. As stated in the legend to Figure 9-13, three molecules of Na^+ bind to transmembrane protein with a change in shape. The newly shaped molecule binds with one ATP. The terminal phosphate remains bound, and ADP is released into the interior. The new shape allows passage of Na^+ molecules across the membrane, and Na^+ molecules dissociate to the outside. Two K^+ molecules bind to the molecule and the shape changes again, with phosphate released. K^+ molecules cross the membrane and are released to the inside. Three Na^+ molecules are moved outside, two K^+ molecules are moved inside, and one ATP molecule is used.

14. The coupled channel is related to the Na-K pump. Amino acids and sugars are transported against a concentration gradient by coupling their passage with Na^+ as it reenters the cell via facilitated diffusion.

15. The proton pump actively transports hydrogen out of the cell, creating a proton gradient with more protons outside than inside the cell. Therefore hydrogen wants to diffuse back in and can only do it via special channels coupled to the production of ATP. This process is also called chemiosmosis. It occurs within the cell.

16. A cell surface receptor is an information-transmitting protein extending across a cell membrane. The outer portion has a shape specific for a certain molecule. When encountered it binds, changing the shape of the inner portion and eliciting a specific reaction.

17. Cell surface markers are protein or glycolipid molecules anchored in the plasma membrane that identify all cells of a particular organism as self and differentiate them from foreign invaders. They also serve to differentiate various

types of organs and tissues within an individual and are primarily involved in development.

18. Animal cells have adhering junctions (desmosomes), organizing junctions (tight junctions), and communicating junctions (gap junctions). Plant cells communicate via holes in the cell wall called plasmodesmata where the plasma membranes of two cells come into direct physical contact.

CHAPTER 7

1. The two forms of energy are kinetic and potential. Kinetic energy is actively engaged in work; potential energy is stored energy.

2. Energy can be changed from one form to another (potential to kinetic) but can't be lost or gained. Heat is the energy of random molecular motion.

3. All objects in the universe tend to become more disordered. The second law states that entropy increases, where entropy is energy so random and dissipated that it can't be used for work.

4. Oxidation occurs when a molecule loses an electron; reduction is when a molecule gains an electron. These two must occur together because every electron that is lost by one molecule must be gained by another. Oxygen is one of the most common electron acceptors.

5. Atoms store potential energy via orbiting electrons. The higher energy level of the electron is maintained as it is removed from one molecule in oxidation and gained by another in reduction. The element referred to is hydrogen.

6. In exergonic reactions products contain less energy or more disorder than the reactants, release energy, and proceed spontaneously. In endergonic reactions, products contain more free energy or more disorder than reactants and require an input of energy to proceed.

7. Activation energy is the energy required to get spontaneous reactions going. It keeps spontaneous reactions under some control to prevent reactions from immediately going to the maximum. If there is high activation energy, there is a slow reaction rate; with a low activation energy, there is a fast reaction rate.

8. Alteration of activation energy can be accomplished when chemical bonds are stressed, making them more easily broken and lowering the activation energy. The compound is a catalyst (enzyme). The final proportion is not changed; the catalyst only changes the speed of the reaction, and the reaction proceeds at the same rate forward and backward.

9. The three factors are temperature, pH, and cofactors. They alter the

three-dimensional shape of the enzyme.

10. The chief energy molecule is ATP. Its three main components are as follows: (1) a five-carbon ribose sugar, (2) adenine, and (3) a triphosphate group. The actual energy carrier is the high-energy bond between phosphate groups. The two molecules formed are ADP and P_i. This liberates 7.3 kcal of energy.

11. The carbon-hydrogen covalent bond.

12. The final reactions because pathways evolve backward, adding new steps at the beginning.

CHAPTER 8

1. The two ways to generate ATP are substrate level phosphorylation, which involves the synthesis of ATP from ADP and P_i driven by another strongly exergonic reaction, and chemiosmotic generation, which involves the transmembrane channel pumping protons out of the cell; then as they pass back in via diffusion, ATP forms from ADP and P_i. Substrate level phosphorylation evolved first. Chemiosmosis yields the most ATP.

2. Glycolysis occurs in the cytoplasm of the cell. When oxygen is absent only substrate level phosphorylation occurs; when oxygen is present both substrate level phosphorylation and chemiosmosis occur.

3. Glucose starts glycolysis. Pyruvate remains. There are four ATPs produced. The net production of ATP is two because two ATPs are used to mobilize glucose. Two additional ATPs are produced with oxygen present.

4. There are only small amounts of NAD^+ in each cell, and when these are all converted to NADH, glycolysis stops.

5. The two mechanisms are oxidative respiration, which requires oxygen and is aerobic, and fermentation, which does not require oxygen and is anaerobic.

6. There are many different nonoxygen electron acceptors used in fermentation, such as organic acids and alcohol. There is no additional energy production, but the processes recycle the NADH to NAD^+ so that the reaction can continue producing its small amount of ATP per glucose molecule.

7. The end products are alcohol and carbon dioxide; alcohol is toxic.

8. When oxygen can't get to the muscles fast enough, aerobic respiration stops, NAD^+ runs out, ATP production stops, and lactic acid fermentation allows the regeneration of NAD^+ from NADH.

9. In NADH from glycolysis, two ATP are exchanged because one ATP is used to transport the NADH into the mitochondrion. In NADH from the citric acid cycle, three ATP result. In $FADH_2$, two ATP are produced because it enters the electron transport chain at a later point.

10. -7.3 Kcal/mole $\times$ 36 ATP/-686 Kcal/mole of glucose = 263 Kcal/mole in ATP/686 Kcal/mole in glucose = 38% available energy harvested; 18 times more efficient.

CHAPTER 9

1. The photon is the basic unit. The short wavelength blue is most energetic; the long wavelength red is least.

2. Green is not absorbed at all by the plant, so green light is reflected back to our eyes.

3. (1) Chemiosmotic generation of ATP from the sun's electrons is the light reaction; (2) ATP from this reaction drives the conversion of CO_2 to organic molecules and is the dark reaction; (3) pigment rejuvenation. All three occur in the presence of light, and all but the light reaction occur in the absence of light.

4. The overall equation is $6CO_2 + 12H_2O \rightarrow C_6H_{12}O_6 + 6H_2O + 6O_2$. The final oxygen comes from the lysis of water.

5. P_{700} captures energy and passes it to ferredoxin. Electrons are not passed but rather just the high energy possessed by each electron is passed from compound to compound.

6. Light causes ejection of the high-energy electron and it travels through the photosynthetic unit, and drives the proton pump to generate ATP; the electron is then returned to the pigment. The ejected electron is high-energy and the returned one is not, so it is not a precise cycle. Two electrons yield one ATP.

7. The photocenter works as follows: a photon is absorbed by P_{680} of PS II, light excites an electron, the high-energy electron drives the proton pump to make ATP, the electron returns to its normal energy state and is passed to PS I where another photon reexcites it, and it is channeled to ferredoxin, causing $NADP^+$ to become NADPH. The PS II system generates ATP, and the PS I system NADPH.

8. It is an accessory pigment, absorbing photons that chlorophyll *a* can't.

9. In the dark reaction CO_2 is fixed, forming carbohydrates; ATP provides energy for the conversion of RuMP to RuBP; and two NADPH and two ATP are used to make one fructose-6-phosphate from 2,3-phosphoglycerates. Another name for this reaction is the Calvin cycle.

10. RuBP is carboxylated by the addition of CO_2 into two molecules of 3PG. G3P is used in the regeneration of RuBP.

11. The light reaction occurs in the thylakoid membrane; the dark reaction occurs within the stroma.

12. Photorespiration is the oxidation of ribulose 1,5-bisphosphate, releasing CO_2 without producing ATP or NADPH. The advantage of C_4 photosynthesis is the fact that bundle sheath cells concentrate CO_2 and inhibit photorespiration. C_4 photosynthesis requires use of ATP to concentrate CO_2 in bundle sheath cells, so it is not as energy efficient as C_3 photosynthesis.

CHAPTER 10

1. Eukaryotic chromosomes are 40% DNA, some RNA, and 60% chromatin protein. DNA is the site of RNA synthesis. A nucleosome is a group of 8 arginine/lysine histone polypeptides. DNA duplex wraps around the histone core of the nucleosome. Supercoiling is further coiling of the DNA/nucleosome string.

2. Heterochromatin is the portions of the chromosome that are always condensed, thus the genes are never expressed. Euchromatin is the chromosomal material condensed only during cell division; at other times the genes are expressible.

3. A karyotype is the particular array of chromosomes in an organism. Differentiation is accomplished by comparing length of arms, size, staining properties, and location of arm constrictions.

4. A normal human has 23 kinds of chromosomes. The pairs are called homologues. There are 46 individual chromosomes; 92 chromatids, after replication. The number of chromosomes is best determined by counting centromeres.

5. G_1 is growth; S is DNA synthesis; G_2 is organelle replication, chromosome condensation, and microtubule synthesis; M is mitosis; and C is cytokinesis. Cell division is associated with M and C. G_1 is generally the lengthiest.

6. (1) Interphase = G_1, S, G_2; chromosomes are not condensed. (2) Prophase = chromosome condensation. (3) Metaphase = chromosomes line up on central plane. (4) Anaphase = physical separation of sister chromatids. (5) Telophase = reformation of nuclei at poles of cell.

7. A pre-S chromosome has one chromatid, and a post-S has two. An individual chromosome has chromatids joined by a constriction called a centromere. Sister chromatids are identical.

8. The events are DNA begins to coil and the microtubular machinery begins to assemble. Animals have centrioles; plants and fungi do not.

9. At the beginning of prophase rRNA synthesis stops. At this point the nucleolus disappears.

10. Metaphase begins when chromosome pairs align along a central plane of the cell, the metaphase plate. It ends with the splitting of the chromosomes into sister chromatids.

11. Anaphase is characterized by movement of the chromatids to opposite poles. The two microtubular movements are (1) sliding of adjacent pole to pole microtubules which pushes poles apart as the cell elongates and (2) microtubules attached to centrioles shorten by actual loss of tubulin subunits pulling chromatids toward the organizing center at each pole.

12. The spindle disassembles, the nuclear envelope reforms around a group of chromatids at each pole, chromatids begin to uncoil to permit gene expression, and rRNA begins to be synthesized, making the nucleolus reappear.

13. In animals the cell pinches in two via a constricting belt of microfilaments around the cleavage furrow. In plants the cell wall prevents constriction, membrane components are assembled perpendicular to the orientation of the spindle apparatus, this cell plate grows outward to the edge of the cell, and cellulose is then laid down.

CHAPTER 11

1. Meiosis is a process of nuclear division in which the number of chromosomes is halved during gamete formation. Both meiosis and syngamy are necessary because meiosis halves the chromosome number and syngamy restores it.

2. Examples of asexual reproduction include binary fission in bacteria, unstressed protist reproduction, and animal budding. Asexual progeny are identical to the parent and one another whereas sexual progeny are different from the parents and from one another.

3. Human somatic cells (general body cells) are diploid, as are gamete-producing cells.

4. (1) In meiosis the homologous chromosomes pair, exchange genetic material during crossing-over, and separate in the first division. (2) Sister chromatids do not further replicate between divisions, and they separate in the second division. (3) In meiosis there is crossing-over and in mitosis there is not.

5. (1) Sister chromatids are joined by the centromere and (2) homologues are held together at crossing-over points of the synaptonemal complex.

6. The DNA duplex of each homologue is aligned side-by-side, chi-

asmata form between the DNA strands of homologues, and they exchange portions of DNA between one another.

7. Microtubules form the spindle but attach to only one face of each centromere at the kinetochone. The kinetochone of each homologue attaches to the pole toward its outer side, and the sides of the centromeres facing one another are not attached to a spindle fiber from either pole. In mitosis each centromere region attaches to spindle fibers from both poles.

8. The chromosomes move to the poles in a random fashion, and some paternal and some maternal chromosomes move to each pole without regard to the movement of all other chromosomes.

9. During anaphase I spindles shorten, breaking chiasmata apart and pulling centromeres with related chromosomes to poles. These two stages differ as follows: (1) in meiosis the centromere does not split, (2) the sister chromatids remain together, and (3) the sister chromatids are not identical to each other as a result of crossing-over.

10. Parthenogenesis is the development of an adult organism from an unfertilized egg.

11. It developed as a means to repair double-strand damage to DNA.

12. Sexual reproduction results in *genetic variation*. The three events that strengthen this consequence are crossing-over in meiosis prophase I, meiotic independent assortment, and fertilization/syngamy.

CHAPTER 12

1. The hybridization showed that breeding between species produced fertile offspring and that traits were masked through one generation and not blended as expected. The progeny were all different from one another, exhibiting different forms of a trait.

2. Mendel quantified his research by keeping accurate numerical records of his results.

3. (1) He used previous researchers' success to direct his work, (2) he picked seven clearly distinguishable traits to examine closely, (3) peas are small, are easy to grow, and have short generation times, and (4) peas are self-fertile but can be cross-fertilized easily as well.

4. (1) He allowed several generations of self-fertilization to ensure that the strains were true-breeding. (2) He crossed plants with two different forms of only one trait, thus forming hybrids. (3) He allowed self-fertilization of the hybrids for several generations, counting the number of offspring of each type.

5. (1) The appearance of progeny

was not blended. (2) For each pair of traits, one was not expressed in the F_1 generation. (3) The pairs of traits segregated among the progency. (4) Pairs of traits exhibited a 3 : 1 ratio of dominant to recessive.

6. (1) Traits are not passed directly from parent to offspring. (2) Each individual contains two factors for each hereditary trait. (3) Each factor exhibits alternative forms called alleles. (4) The two alleles of any trait in no way influence one another. (5) The presence of a factor in an individual does not guarantee that it will be expressed in that individual.

7. Genotype is the totality of alleles that an individual contains. Phenotype is the physical appearance of an individual. Homozygous describes the situation when the two alleles in a genotype are the same and the organism is true-breeding. Heterozygous describes the situation when the two alleles in the genotype are different and the organism is not true-breeding.

8. It is crossed with a homozygous recessive, because if the unknown is homozygous dominant, all the progeny will be purple Ww; if it is heterozygous, half of the progeny will be purple Ww and half will be white ww. Therefore the two results are clearly different from one another.

9. The law of segregation: Alternative alleles segregate from one another in gamete formation and remain distinct; they do not blend, and each gamete has an equal chance to possess either form of the allele.

10. (1) Reproduction involves the union of egg and sperm. (2) Chromosomes segregate in meiosis similar to Mendel's alleles. (3) Gametes possess one copy of homologous chromosomes, and diploid individuals have two copies, as is true in Mendel's model. (4) In meiosis homologous pairs orient on the metaphase plate independently of all others.

11. Sex-linkage experiments with fruit flies involving white/red eye color associated with the sex chromosomes. In these experiments the white color trait was on the X chromosome and there was no functional alternate on the Y chromosome.

12. The genes are so far apart on the chromosome that crossing-over between them is very likely; therefore they virtually display independent assortment.

13. (1) Multiple alleles: more than two alternatives for a trait. (2) Gene interaction: several genes act jointly to elicit a specific trait. (3) Continuous variation: gradation in variance of a group of individuals. (4) Pleiotropy: gene has more than one effect on an individual's phe-

notype. (5) Incomplete dominance: allele not fully dominant or recessive. (6) Environmental effect: the expression of allele altered by the environment.

CHAPTER 13

1. Monosomic individuals exhibit the loss of one autosome; those who are trisomic have an extra autosome. Monosomy is always lethal.

2. Primary nondisjunction is the failure of chromosomes to separate from one another in meiosis. Down syndrome is usually caused by primary nondisjunction. Women who are more than 35 years old are more prone to produce these gametes because gametes in women are held in prophase I from birth to whenever the egg is released to develop further (ovulation), and the chance for damage increases if this holding pattern is lengthy. In men, sperm are produced continuously from puberty and are not stored for long periods.

3. In a normal female the sex chromosomes are XX; in a normal male, XY. The Y chromosome possesses fewer active alleles. The genes that exist on this chromosome express some male characteristics.

4. The genotype of a person with Klinefelter's syndrome is XXY, which makes him genetically male, since the presence of the Y chromosome determines maleness. The individual appears female.

5. The genotype of a person with Turner's syndrome is XO, which makes her genetically female since there is no Y chromosome. The individual appears female.

6. A group = AA or AO; B group = BB or BO; AB group = AB; and O group = OO. The sugars present are A, galactosamine; B, galactose; AB, galactosamine and galactose; and O, none.

7. The physiological defects are thick mucus in the lungs, liver, and pancreas. Biochemically, the transmembrane channels transporting chloride ions do not function properly so that water does not pass from the blood into the passageways.

8. Hemophilia is caused by a defect in the mechanism for blood clotting. Types VIII and IX are sex linked and genes are on the X chromosome, whereas the other forms have protein clotting genes on the autosomes. An otherwise normal human male cannot be a carrier since the genes are on the X and, with no correcting genes on the Y, all males with the affected X will exhibit hemophilia.

9. Huntington's disease is dominant. Its frequency is maintained because the physiological effects are not evident until relatively late in

life, generally after some offspring have already been produced.

CHAPTER 14

1. The survivors were the ones injected with nonvirulent, noncoated bacteria or heat-killed, coated bacteria. The dead mice were injected with virulent, coated bacteria or with a combination of live, coatless bacteria and dead, coated bacteria. His conclusion was that the information specifying the coat from heat-killed bacteria was transferred to the live, coatless bacteria.

2. The primary component was DNA as evidenced by the fact that it was digested by DNAse. Most scientists still thought that proteins were the hereditary material.

3. No, there are not equal amounts. This indicates that DNA is not a simple polymer of repeating ACTG units. The proportions are equal amounts of adenine and thymine and equal amounts of cytosine and guanine.

4. The three-dimensional shape of DNA was shown to be a double helical spiral, where the nitrogenous bases pointed inward toward one another. A large purine was always paired with a small pyrimidine, so the diameter remained the same. The amount of A equals the amount of T, and the amount of G equals the amount of C; therefore A always bonds to T and G always bonds to C.

5. This is important because the sequence of one helix determines the sequence of the other helix, so that if one is ATTGCAT, the other can only be TAACGTA. DNA unzips and each strand acts as a template, forming a new strand along its now naked self. The term given to this form of replication is semiconservative.

6. No, one strand grows inward from the end to the fork and is replicated from the 3′ end. The other strand grows in the 3′ direction, but since it runs in the 5′ direction, it replicates in small batches, working backward.

7. 3′ to 5′ = DNA polymerase; 5′ to 3′ = DNA polymerase still is used for the replication of that strand. The growing 3′ strand replication is *continuous synthesis;* the growing 5′ strand is *discontinuous synthesis.*

8. In bacteria, the duplex is nicked at one site, and a strand on one or both sides is displaced, creating one or two replication forks. These proceed around the circle until complete. Eukaryotic DNA has numerous replication forks along each chromosome, working in discrete units of 10,000 to 1,000,000 base pairs in length.

9. The intuitive hypothesis of Garrod was that information encoded on

DNA specifies particular enzymes. The organism used to support this was *Neurospora*. They changed the organism by exposing spores to x-rays, inducing metabolic mutations.

10. Initial growth on a complete medium was to ensure the growth of all forms of the organisms, regardless of whether or not each could manufacture all of its own metabolites. Minimal medium was used for the progeny to isolate various metabolic deficient mutants.

CHAPTER 15

1. Ribosomal RNA (rRNA) functions in ribosomes to provide the site where a polypeptide is assembled. Transfer RNA (tRNA) transports amino acids to ribosomes to build a polypeptide. Messenger RNA (mRNA) passes from the nucleus to the cytoplasm to be a blueprint for protein synthesis, which is produced from chromosomal DNA information.

2. Translation requires rRNA, mRNA, and tRNA. The rRNA molecule in the ribosome binds to the mRNA molecule. Then the ribosome proceeds down mRNA in increments of three nucleotides. For each three, a new amino acid is added to the growing polypeptide chain from a tRNA until a stop signal is reached and the complex is disassembled.

3. The sequences are read in a simple sequence with no separating or silent nucleotides. Crick and his colleagues chemically deleted a nucleotide to change the reading frame, then determined what further deletions were required to restore the proper reading frame. They found that the reading of the code remained nonsense when one or two additional deletions were made, but three deletions corrected the reading frame.

4. Artificial RNA molecules of a single sequence, i.e., UUU, were added to a mixture of RNA and protein from ruptured cells, and scientists observed the type of protein that resulted.

5. A codon is the three-nucleotide portion of the mRNA. An anticodon is the complementary three-nucleotide sequence on the tRNA. To confer specificity, there are 20 different activating enzymes, one for each common amino acid, which recognize a particular tRNA anticodon sequence and matching amino acid.

6. The first step is formation of the initiation complex. In this process met-tRNA binds to a small ribosomal subunit that has been positioned there by initiation factors and guided to mRNA. A large ribosomal subunit binds to the complex. This is critical to ensure the correct reading frame for the rest of the protein.

7. In prokaryotes the entire gene codes for an entire protein, whereas in eukaryotes genes are much larger than the proteins they produce and contain introns, long intervening sequences cut out of eukaryotic mRNA transcripts before translation.

8. Since the regulatory protein-binding site is located between the RNA polymerase-binding site and the gene being transcribed, when the regulatory protein is on its binding site it blocks the movement of the polymerase to the gene.

9. The shape of the regulatory protein can be altered by the binding of small molecules to its own binding sites, disabling its ability to bind or enabling it to bind to a new location.

10. The *lac* operon CAP site is the regulatory site where activation proteins bind, facilitating unwinding of the DNA duplex and enabling polymerase to bind to the promoter. CAP stimulates transcription when the cell's energy levels are low. It is incapacitated otherwise because when glucose levels are high, metabolites bind to CAP so it does not recognize the CAP site and the gene is not transcribed.

CHAPTER 16

1. A free radical is an atom with unpaired electrons. Most free radicals come from water. They are damaging because they can tear DNA apart like shrapnel.

2. A pyrimidine dimer is a cross link between adjacent pyrimidines of the DNA strand. It harms a cell by blocking DNA replication. To repair this damage the cell cleaves the bond between pyrimidines or excises the entire dimer and fills in the gap using the other strand as a template. If the dimer portion is not repaired, the problem area is filled in later, but this is very prone to error.

3. A slipped mispairing is when chromosomes pair and sequences misalign, looping out a portion of one strand. Possible deletion of several hundred nucleotides from one of the DNA strands may result. Some of the deletions are frameshifts, in which the code is misread so that the protein made is useless.

4. Most mutations are detrimental because they are random changes rather than a result of natural selection which tends to favor and establish the most fit situation. Whether they are major or minor consequences is determined by the function of the altered gene and the identity of the cell in which the damage occurred.

5. A plant cell mutation will more likely affect future progeny because plant cell differentiation occurs late in development; therefore most cells of a plant are able to develop into a new adult plant. In animals the differentiation occurs at an early stage and an individual body cell is unlikely to develop into an adult animal. As a result the mutational damage is only passed on to animal progeny when it occurs in gamete-producing tissue.

6. Cancer causes uncontrolled, invasive growth. A tumor is a mass of cancer cells. Connective tissue forms a sarcoma, whereas epithelial tissue forms a carcinoma. Cancer cells that have spread are termed metastases.

7. The protein is an enzyme that phosphorylates tyrosines. It signals the initiation of cell division.

8. Scientists labeled *src* with radioactivity and permitted it to bind the complementary sequences of the chicken genome; then they examined whether the genome became radioactive. As a result radioactive *src* was bound to the RSV insertion site as well as another site. This indicated that the *src* gene was not initially a viral gene, but a normal chicken growth gene picked up sometime in the past that had now escaped control by normal regulatory mechanisms.

9. "Transfection" involves three steps: (1) isolation of nuclear DNA from human tumor cells, (2) cleavage of this DNA into random fragments using restriction endonucleases, and (3) individually testing fragments for the ability to induce cancer. This research determined that sometimes only a single gene was necessary to cause cancer, and this gene was one of the previously identified *onc* genes.

10. Anything that reduces the exposure to mutagens.

CHAPTER 17

1. Genetic recombination is the change in the chromosomal position of a gene or a fragment of a gene. The three kinds are (1) reciprocal recombination, wherein two chromosomes trade segments; (2) gene transfer, wherein one chromosome donotes a piece to another; and (3) chromosome assortment in meiosis. All occur in eukaryotes, but only reciprocal recombination occurs in prokaryotes. The most primitive is reciprocal recombination because it is the only one present in primitive prokaryotes.

2. A plasmid is a small auxiliary chromosome. It can insert on a main chromosome where a nucleotide sequence occurs that matches one on the plasmid DNA. A transposon is a small gene fragment that can move from place to place at random. Plasmids are found in bacteria; transposons are found in bacteria and eukaryotes.

3. Bacteria can pass plasmids from one cell to another by the formation of a pilus, which constitutes a conjugation bridge through which a replicated copy of the F plasmid passes to the new cell. Insertion of the bacterial F plasmid in the main chromosome at the time of transfer results in the entire bacterial genome being copied to the other cell. This is called bacterial conjugation.

4. Insertional inactivation is the loss of activity in a gene as a result of the insertion of a mobile element within that gene.

5. Crossing-over occurs in meiosis prophase I. Variation will more likely result from mutation exchange in crossing-over.

6. A mismatched pair is a region along the length of the paired chromosomes where nucleotides of one homologue are not complementary with those of the other. If during replication enzymes detect that this has occurred, they excise one of the mismatched regions and correct it to match the other. This is called gene conversion.

7. Unequal crossing-over occurs when two chromosomes exchange segments of unequal length. It usually occurs when more than one copy of a gene exists on a chromosome and the homologue lines up at the wrong site. The result is that one gamete's chromosome has hundreds of copies of a particular gene, while the other has very few if any.

8. Satellite DNA, transposable elements, tandem clusters, multigene families, dispersed pseudogenes, and single copy genes.

9. Satellite DNA is short nucleotide sequences that are repeated several million times. It is usually clustered around the centromere or near the ends of chromosomes. It is usually highly condensed, tightly coiled, and untranscribed. Its function is most likely structural.

10. Tandem clusters encode products that the cell requires in large amounts. This occurs as the cell rapidly transcribes all of the numerous copies simultaneously. Two examples are rRNA and the spacer sequence.

11. The genes in multigene families are those that are groups of related but distinctly different genes. They differ from tandem clusters in that they contain fewer genes and the genes are much more different from one another. These families are thought to have evolved from a single ancestral gene that altered and multiplied through a series of unequal crossing-overs.

12. Pseudogenes are silent copies of a

gene that have been inactivated by mutation. These differ when they are dispersed from when they are in multigene families in that they are distantly located from their original location in the multigene family cluster.

13. Protection comes from the fact that specific chromosomal proteins regulate the degree of coiling in isolated regions, decreasing the accessibility of those genes to RNA polymerase. A translocation occurs when part of a chromosome becomes part of another chromosome. An inversion is reversal in the orientation of a portion of a chromosome.

CHAPTER 18

1. Bacterial DNA is modified so that it is not recognized by its own virus-degrading enzymes. These enzymes are methylases, and they recognize the same sites as the endonucleases, but they add CH_3 groups to the nucleotides so they are not recognized by the endonucleases.

2. The sequences are 4 to 6 nucleotides long and are symmetrical. The nucleotides at one end of the recognition site are complementary to those at the other end, so the two strands have the same nucleotide sequence running in opposite directions.

3. A chimeric genome is artificially made and consists of DNA from distinctly different organisms. To make one, cut up a bacterial genome with a given restriction endonuclease so that both ends of the fragment have the same sticky ends and form a plasmid. Then use this same enzyme to cut up DNA from another organism and mix the two together. The DNA from the second organism should insert into the bacterial plasmid. This is called recombinant DNA.

4. Use bacteria incapable of surviving outside of the laboratory.

5. Cleavage, recombination, cloning, and screening.

6. During recombination, placing fragments of DNA into a plasmid or viral vehicle enables the fragment to be replicated as part of the plasmid or viral genome, a necessary step. Additionally there must be the elimination of vehicles not containing fragments.

7. Cloning in genetic engineering means reproduction of the given cells that contain plasmids with the desired fragment. This identifies certain clones that possess only the specific desired fragment.

8. Bacteria possess the characteristic of *antibiotic resistance*. To select against bacteria lacking the appropriate vehicles, include an antibiotic resistance gene in the fragment and culture the bacteria on a medium containing that antibiotic. Therefore only the bacteria

with the resistance gene, and therefore the desired fragment, will be able to grow.

9. In agriculture genetic engineering is used to manipulate genes associated with herbicide resistance, virus resistance, insect tolerance, and nitrogen fixation. The primary vehicle used is the T_i plasmid of *Agrobacterium*. This vehicle is effective for broadleaf plants.

10. Currently, genes specifying the protein-polysaccharide coat of herpes and hepatitis viruses are spliced in the vaccinia virus to produce new human vaccines. Recombinant virus dictates the production of both vaccinia proteins and the herpes of hepatitis so that individuals produce antibodies against the herpes of hepatitis coat. The recombinant virus is otherwise harmless because it does not contain the harmful herpes or hepatitis genes but rather only those of the harmless coat.

CHAPTER 19

1. Macroevolution refers to changes in populations of plants and animals so that new species develop from old ones. Microevolution refers to changes in allele frequencies within a species.

2. Adaptation results from the possession of features that promote the chance of an organism's survival and reproduction. The natural selection of Darwin's concept of evolution is the process, and adaptation is the result.

3. $(p + q)^2 = p^2 + 2pq + q^2$ where p^2 is the frequency of individuals homozygous for the more common allele, q^2 is the frequency of individuals homozygous for the less common allele, and $2pq$ is the frequency of heterozygous individuals.

4. $p = 54/100 = 0.54$ $q = 1 - p = 0.46$. Therefore: heterozygous: $2pq = 2(0.54 \times 0.46) = 0.4968 = 5/10$; homozygous dominant: $p^2 = 0.54 \times 0.54 = 0.29 = 3/10$; homozygous recessive: $q^2 = 0.46 \times 0.46 = 0.21 = 2/10$; CHECK: $5/10 + 3/10 + 2/10 = 10/10$.

5. Migration, mutation, genetic drift, nonrandom mating, and selection.

6. Genetic drift is loss of an allele from a population. This is more likely to occur if a population is small because the loss of a single breeding individual will greatly affect gene frequencies.

7. The only alleles the subsequent population can build on are those few present in the population originators.

8. It increases the proportion of homozygotes.

9. Artificial selection involves an individual choosing individuals to be the parents based on promoting certain desired traits. In natural selection, as environmental

conditions change, certain individuals do not survive to reproduce; the survivors reproduce and pass the characteristics that enabled them to survive on to their progeny.

10. Few traits are completely independent of others, and their interactions increase with successive matings; selection can only act on traits in which the homozygotes and heterozygotes are clearly distinguishable.

CHAPTER 20

1. Sickle cell anemia, peppered moths, lead tolerance.

2. Homozygous recessives die because of sickling; homozygous dominants die because of malaria; and heterozygotes survive to reproduce. In addition, women may be conferred with additional fertility.

3. Soot-darkened trees and pollution killed light-colored lichens, so light-colored moths stood out on the dark background and were eaten. Therefore fewer light-colored moths survived to reproduce and a greater number of dark-colored moths reproduced and passed on their dark color genes.

4. The fossil record, molecular record, homology, development, vestigial structures, parallel adaptation, and patterns of distribution.

5. In Darwin's day they age-dated rocks by the position of various strata and could only give ages in relation to one another. Now radioactive dating is based on the half-life decay of certain radioisotopes, a definite quantity.

6. More closely related species have a greater number of DNA sequences in common and more distant relations have a greater number of differences, based on a progressive accumulation of DNA change.

7. Vestigial structures resemble those of presumed ancestors, but serve no known function. They are important in evolution because they indicate a common ancestry based on structure independent of function.

8. Parallel adaptation is the presence of similar looking and functioning organisms in widely separated areas. Examples in Figure 20-11.

9. The gradualists believe that all change was gradual, with many transitional forms. The punctuated view is that little change occurred over long periods of time with occasional sudden large evolutionary changes.

10. Scientific creationism is not truly scientific because it is based on beliefs rather than observations and it does not infer its principles from observations.

CHAPTER 21

1. Species refers to groups that differ in one or more characteristics and do not naturally intergrade to any great extent.

2. A niche is the distinctive role an organism plays in nature, habitat, behavior, feeding, etc.

3. They adapt according to environmental demands and changes.

4. Forces that differ greatly produce rapid divergence. If forces are similar, there is slow divergence.

5. Geographical, ecological, temporal, behavioral, and mechanical isolations. Prezygotic mechanisms prevent gamete fusion; postzygotic mechanisms involve embryo developmental failure or abnormalities.

6. When the hybrids are sterile, are nearly sterile, or are less well adapted to survive in the environment.

7. Adaptive radiation refers to groups of closely related species which evolved from a common ancestor. An example would be Darwin's finches or Hawaiian *Drosophila*.

8. Behavioral isolation.

9. Varied habitats in close proximity to one another and rapid changes in climate.

10. Polyploidy is an amount of genetic material greater than 2N. Parthenogenesis is asexual reproduction wherein the egg develops directly into the embryo or wherein somatic cells fuse to produce the embryo.

CHAPTER 22

1. Fossil records are primarily found in sedimentary rocks because only these rocks are capable of "burying" an organism in a relatively intact state. The most complete fossil record is found in the seabed.

2. Mere lack of fossilization/ sedimentation at that time or location, or they support punctuated equilibria.

3. The earth is 4.5 billion years old. The rocks are 3.8 billion years old. The discrepancy exists because the actual oldest rocks have disappeared, probably associated with continental drift.

4. Continental drift assumes that the continents "float" on the surface of the planet and change position with time. Plate tectonics explains how this movement occurs.

5. Most non-plant life-forms evolved during this time, a period of intense adaptive radiation.

6. Plants, arthropods, and fungi were the first to successfully colonize the land about 410 million years ago. Special features that were developed were multicellularity and efficient mechanisms to conserve and internally transport water.

7. The first terrestrial vertebrates were amphibians; this occurred about 360 million years ago. They evolved into reptiles, which then evolved into birds and mammals.

8. There have been 5 mass extinctions, 4 in the Paleozoic era and 1 in the Mesozoic. The most drastic was the fourth one in the Paleozoic era about 248 million years ago. The most famous one was the last one in the Mesozoic era, which was the extinction of the dinosaurs.

9. The hominids include humans and their direct ancestors. They display bipedal locomotion and enlarged cranial capacity.

10. *Homo habilis* lived 2 million to 1.5 million years ago; *H. erectus* lived 1.7 million to 250,000 years ago; *H. sapiens* lived perhaps 500,000 years ago and lives to the present.

CHAPTER 23

1. A small population is more likely to become extinct because random events are more likely to adversely affect a population of only a few individuals. Inbreeding leads to direct loss of genetic vigor and loss of variability affects the ability to adjust to changing conditions, therefore increasing the possibility of extinction.

2. The three types of dispersion are random spacing, even spacing, and clumping. The most frequent in nature is clumping because specific environmental conditions are neither randomly nor evenly distributed, animals often congregate for various reasons, and the young of a species are likely to be near the parents.

3. The biotic potential of a population is the rate at which a given population will increase with no limits placed on it. The actual rate of population increase is the difference between birth rate and death rate per individuals per time. Two factors that affect it are emigration and immigration. An exponential capacity for growth means that the rate of increase remains constant while the actual increase in numbers accelerates rapidly as the population size grows.

4. Density-dependent factors are those in which the resources are in short supply, causing the individuals to compete more intensely as the population grows. Density-independent factors are those caused by factors that operate regardless of population size (i.e., weather and physical disruption).

5. Partway up the sigmoid growth curve reflects a rapid growth phase, and yields will therefore be better longer, extending the period of rapid growth. Harvesting small populations results in overharvesting, destroys long-term

productivity, and increases the likelihood of extinction.

6. A K strategist exhibits a sigmoid growth curve, and the limiting factor is K, the carrying capacity. These types of organisms reproduce late and have long generation times. They have few, large offspring and provide lengthy parental care; their offspring mature slowly.

7. An r strategist exhibits an exponential growth curve, and the limiting factor is r, the intrinsic rate of increase. These types reproduce early and have short generation times. They have many, small offspring and provide little parental care; their offspring mature rapidly.

8. Survivorship is the percentage of an original population that survives to a given age. The three types are I—large proportion of individuals reach their physiologically determined maximum age and the greatest mortality is in the aged; II—mortality constant through all ages; and III—mortality especially high in the young stages and declines with age. Examples are as follows: I, humans; II, hydra; III, oysters.

9. Demography is the statistical study of populations, which involves predicting the ways in which populations will change in the future. The two factors taken into account in demographic studies are age distribution of the population and changing population size through time. The characteristics of a stable population are births plus immigration equals deaths plus emigration, and population size and age structure remain constant.

10. The terms niche and habitat are not synonymous. The habitat is part of the niche, but only the physical part; an organism's niche additionally includes behavior, seasonal factors, and daily patterns. The theoretical niche is the potential niche if there were no competitors present, and the actual niche is what the organism occupies under natural circumstances.

11. Under these circumstances the niche widens. If the species become different enough to have different niches, there is no exclusion; if they are not different enough, additional selection will occur or one will become excluded.

CHAPTER 24

1. Coevolution is long-term mutual evolutionary adjustment in the characteristics of organisms with relation to one another. Examples include the following: fruit provides food for animals which then increase seed dispersal of the plants; pollen or nectar provides

animal food and causes flower pollination.

2. Defense structures include thorns, spines, prickles, sticky plant hairs, and high levels of silica. Nutritionally, if a plant lacks several of the 12 essential amino acids, it is less likely to be useful food.

3. Primary chemical compounds are regularly formed as components of various metabolic pathways; secondary compounds are chemicals not involved in metabolic pathways. Overgrazing is prevented by these secondary compounds because they may be unpalatable or toxic to metabolic processes or development.

4. Those that are tolerant of the compounds often retain them in their bodies, making themselves as inedible to their predators as the plants are to intolerant herbivores.

5. Cryptic coloration provides camouflage, hiding the animal from predators. Aposematic coloration is highly visible and indicates the presence of some defensive mechanism. An animal that stores a plant's secondary chemicals will generally be poisonous and brightly colored, whereas an animal that degrades the compounds generally will not be poisonous and may be camouflaged.

6. Batesian mimicry describes the situation when an unprotected species evolves to resemble a chemically protected species. The purpose for the possessor is that it will not be preyed upon by those predators familiar with the model.

7. There must be a much smaller number of mimics than aposematics. If there are the same number of or more mimics, the predators will have a good chance of getting an unprotected individual rather than a protected individual and will not learn avoidance of the model.

8. The adult viceroy butterfly exhibits Batesian mimicry; the larval monarch butterfly exhibits aposematic coloration; and the larval viceroy butterfly exhibits cryptic coloration.

9. Symbiotic relationships involve two or more kinds of organisms living together in a more or less permanent situation. The classes of symbiosis are (1) commensalism, in which one organism benefits while the other neither benefits nor is harmed; (2) mutualism, in which both organisms benefit; and (3) parasitism, in which one organism benefits while the other is harmed.

10. Ants on the acacia tree—mutualism; dodder and host plant—parasitism; sea anemones and anemone fish—commensalism; ants and aphids—mutualism; lice and birds—

parasitism; lichens—mutualism; legume root nodules—mutualism; and fleas and humans—parasitism.

CHAPTER 25

1. Biogeochemical cycles are geological cycles that involve the biologically controlled cycling of chemicals. Chemicals are generally stored in the atmosphere, water, and rocks in these cycles. There is much more of the life-sustaining chemicals in these reservoirs than in organisms.

2. The earth holds nitrogen as atmospheric nitrogen gas, fixed nitrogen in the soil, and nitrogenous compounds in organisms. The process of fixation involves converting nitrogen gas to ammonia and is performed by bacteria, especially those associated with legumes, and certain cyanobacteria.

3. Nitrification is the oxidation of ammonia to nitrate. Nitrites are toxic, and nitrates are valuable. Plants produce all 20 amino acids, and animals only produce 8.

4. The reservoir of phosphorus exists in mineral form rather than in the atmosphere, as in the other cycles. The natural sources for phosphorus are soil (small amounts), isolated rock outcroppings, ocean sediments, guano, and bone meal.

5. Deforestation produces fewer plants, which results in greater water runoff and increased land damage due to flooding. With regard to the nutrient cycles and overall fertility of the land, with increased water runoff, greater amounts of nutrients are lost from the immediate environment and fertility is decreased.

6. The wave-battered intertidal zone is the most productive biological community. Tropical rainforests are more productive than temperate forests, which are more productive than the desert.

7. On the average, 10% of the energy is transferred from one trophic level to the next. Herbivorous diets provide the greatest food value to living organisms because they involve eating the primary producer on the first trophic level rather than a primary or secondary consumer, which is a higher trophic level; therefore less energy is lost going from one trophic level to the next. The more efficient method would be to feed them corn.

8. The process of ecological succession involves a change in an ecosystem from simple to complex over time. Secondary succession occurs when large areas previously inhabited by living organisms are grossly disturbed, whereas primary succession occurs on a bare, lifeless substrate.

9. Three factors have produced this altered concept: (1) climate is con-

tinually changing; (2) succession is a slow process; and (3) the nature a region's vegetation is greatly affected by human activities. The organisms often associated with early stages of succession are those that exhibit symbiotic relationships: lichens = algae + fungi, legumes = plant + bacteria, and mycorrhizae = plant + fungi.

10. Biomass and productivity are related to the successional age of an ecosystem in that older ecosystems have greater biomass, but lower net productivity. There are more species in mature ecosystems, and especially many more heterotrophic species in mature ecosystems. This results because mature ecosystems better regulate their nutrient cycles and generally have more specialized organisms.

CHAPTER 26

1. A biome is an assemblage of organisms that has a characteristic appearance and that occurs over a wide terrestrial geographical area. The two key physical factors that effect the distribution across the earth of biomes are (1) amounts of heat that reach the earth and its seasonal variation and (2) global atmospheric and resulting oceanic circulation patterns.

2. Most deserts are located at these latitudes because here the air is falling and being warmed; since warm air holds more moisture than cool air, there is less precipitation. It is more likely that a desert will form at the interior of a continent due to the distance from the sea. The windward side of a mountain generally is moister because as the air rises on this side, it is cooled and loses its moisture-holding capacity, causing precipitation. On the lee side the air descends, is warmed, and holds more moisture, blocking precipitation.

3. There is a wider variety on the land as a result of the sharper barriers between terrestrial habitats; ocean variation is less distinct and there is less variety of available niches, resulting in less evolutionary diversity.

4. The three major oceanic habitats are (1) the neritic zone—shallow zones along the continental coasts with a large number of species; (2) the surface zone—upper layers of the open sea with diverse species; and (3) the abyssal zone—deep-water areas shown to have more diversity than was thought as sampling techniques have improved.

5. Plankton are microscopic organisms, algae, and cyanobacteria. Nekton are fish and other larger organisms that feed on the nekton. The photosynthetic plankton are important because 40% or more of the earth's photosynthesis occurs in them. The level is low and there is a rapid turnover of nutrients in the surface zone; most of the nutrients are tied up in the organisms.

6. Almost 2% of the earth's water is fresh. The three zones are (1) littoral zone—shallow water along the edges; (2) limnetic zone—upper layer of open water; and (3) profundal zone—water below the limit of light penetration.

7. Eutrophic lakes are abundant in minerals and organic; the action of decomposers makes the hypolimnion stagnant during the summer. Oligotrophic lakes are characterized by scarce organic matter and nutrients; they are generally deep with an abundant supply of oxygen in the hypolimnion. Oligotrophic lakes are more susceptible to the effects of pollution because the excess phosphorus from fertilizers increases productivity, which ultimately depletes the lake of oxygen.

8. The seven most prominent biomes are tropical rainforests, savannas, deserts, temperate grasslands, temperate deciduous forests, taiga, and tundra. The stratification is imitated in changes in elevation at a given latitude in that lower elevations are more similar to the equator and the highest elevations are more similar to the poles.

9. The key characteristics are great variety of species, highly specialized; tremendous rainfall; poor soil nutrition; and most nutrients in living beings, especially the trees. Slash-and-burn agriculture is damaging because burning destroys the nutrients rather than putting them back in the soil.

10. The key characteristics are extremely low precipitation, sparse vegetation, large daily fluctuations, extreme heat in summer, and deciduous and evergreen vegetation. The adaptations necessary include (1) succulent plants store water and exhibit CAM photosynthesis and (2) animals may limit activity to moist periods and are nocturnal; they are also able to store large quantities of water in their tissues.

11. The key characteristics are extremely rich soil; long, cold winters; and relatively abundant rainfall. They resemble savannas in that they have large quantities of perennial grasses which support herds of large grazing mammals.

12. The key characteristics are long, cold winters; short summers; winters that are very dry; greatest precipitation in summer; short days in winter and long ones in summer; infertile soils; coniferous trees; and limited variability of species.

CHAPTER 27

1. The development of reliable agriculture fostered the rapid growth in human populations. This event fostered development of permanent settlements and cities. The major cultural event that eventually took place was the specialization of professions.

2. Undeveloped countries have the larger population of individuals under the age of 15. The significance of this is that they have a greater number of potentially reproductive individuals who can increase their population at an even faster rate. As a further effect, people in developed countries will be a much smaller proportion of the whole population, while still holding the majority of the world's wealth.

3. During the Green Revolution improved strains of wheat, rice, and corn were developed. Some of the limitations of these were the fact that the agricultural techniques required to raise these crops depend on large amounts of energy and the use of herbicides, pesticides, fertilizers, and machinery. These various techniques are expensive and cannot be afforded by poor, undeveloped countries. It takes 1000 times as many calories of energy using these techniques as opposed to traditional methods.

4. This is not a sound principle because cereal crops do not produce all the required amino acids. A balanced type of plant is the legumes; they produce essential amino acids that cereals do not.

5. Shifting agriculture is clearing and cultivating a patch of forest, growing crops for a few years and moving, allowing the patch to return to its previous state. It is also termed slash-and-burn cultivation. This method succeeds with relatively small populations because groups can cultivate an area and move on to another, etc., as long as there are few people in a large expanse of land.

6. The agricultural techniques used in the tropics are destroying the tropical forests because the forests are not being given sufficient time to recover. Another human need that is contributing to the destruction is gathering wood for fuel, and wood is being gathered faster than the trees can regenerate themselves. At current rates, the forests will be gone in 30 years.

7. Whatever tropical forest survives will have significantly less diversity and be more homogeneous. The rate of species extinction estimated for the next century is several species an hour. The biological implication is that plants and animals that are lost may have had some value in terms of human usefulness.

8. The chief benefit of nuclear power is that it uses a plentiful resource. The chief drawbacks are: (1) it is *not* cheap, when the costs of plant construction are included. (2) Nuclear safety is not guarenteed with current reactor design—and what if a nuclear-powered country like France were attacked with conventional weapons in a war? (3) Spread of nuclear weapons material and technology will create a more dangerous world. (4) Nuclear waste disposal problems have not been solved, nor has a single nuclear power plant yet been de- commissioned successfully.

9. The amount of CO_2 has increased since the advent of industrialization because of the burning of fossil fuels and the destruction of forests. The greenhouse effect is the result of CO_2 and other gases in the atmosphere preventing heat from radiating into space. This produces global climatic warming; rising sea levels; increased growth of some kinds of plants, which alters community relationships; and the drying out of the best farmlands, thus decreasing their productivity.

10. Ozone is O_3 and it is formed as light stimulates the dissociation of O_2 by ultraviolet radiation. Decreasing the earth's ozone layer results in an increased rate of human skin cancers and damage to the ocean's phytoplankton and other photosynthetic organisms. The primary cause is probably manufactured chemicals, especially chlorofluorocarbons. High levels of ozone are also deleterious because they lead to damage to human lungs and to citrus plants.

11. The goals at environmental science are operational ones—to solve the problems created by modern society's impact on the environment. Eliminating pollution, conserving biodiversity, protecting groundwater and topsoil—these are but a few specific goals. The tools environmental scientists use are Ecology, Geology, Political Science, and Economics; environmental problems typically impact a broad range of scientific and social issues. The five components to solving environmental are assessment, risk analysis, public education, political action, and follow-through.

CHAPTER 28

1. The current hierarchical taxonomic system names are as follows: species, genus, family, order, class, phylum, and kingdom. Genus and species are either underlined or italicized, and the species name is not capitalized.

2. Prokaryotes and eukaryotes exhibit a greater fundamental difference. Prokaryotes lack membrane-bound organelles and microtubules, their DNA is not associated with proteins, their sexual recombination is different, and their cell walls contain muramic acid. Prokaryotes are in the kingdom Monera.

3. The other three (Fungi, Plantae, and Animalia) have evolved from Protista; most individuals in this kingdom are unicellular but there is no common characteristic—they are not plants, animals, *and* fungi. In terms of nutrition and locomotion, Animalia ingest food and are highly motile; Plantae manufacture food and are generally stationary; and Fungi digest food extracellularly and absorb it and are wholly nonmotile.

4. Of the eukaryotic kingdoms, the Protista are the most diverse; their

common names are protozoa for the heterotrophic members and algae for the autotrophic members. Multicellular forms are exhibited by the red, brown, and green algae.

5. Mitochondria and chloroplasts are apparently of symbiotic origin, and mitochondria are possessed by nearly all eukaryotic organisms. This organelle is most closely related to nonsulfur purple bacteria. Three ancestors appear to have given rise to the other organelle because there are three distinct biochemical categories of chloroplasts.

6. Multicellular organisms possess a significant degree of coordination and integration among the individual cells of the group. Multicellularity arose many times in the evolutionary process; its advantages are the ability to carry out activities, including selfprotection, movement, and search for food and mates, with a complexity not possible by unicellular organisms.

7. Stress prompts sexual reproduction in many unicellular protists. The three major life cycles in eukaryotes are (1) zygotic meiosis, in which the zygote is the only diploid cell and, on forming it, the cell immediately undergoes meiosis; (2) gametic meiosis, in which the gametes are the only haploid cells, and two fuse and form the diploid zygote that grows to adulthood; and (3) sporic meiosis, in which there is a regular alternation of generations between a multicellular haploid phase and a multicellular diploid phase with the diploid phase producing spores that grow into a haploid phase, which then produces gametes, and two of these then fuse to form the diploid zygote, the first cell of the multicellular diploid phase.

8. Technically, viruses are nonliving because, although they are capable of replication, they cannot exist on their own and must direct the machinery of their host cell to manufacture more of their viral components. Their most basic composition is that they are fragments of a eukaryotic or prokaryotic genome with the capacity to organize a protein coat around themselves. Viruses belong to no kingdom because they are not alive.

CHAPTER 29

1. Viruses are not considered to be living organisms because they cannot grow or replicate on their own. The most general structure of a virus is a nucleic acid encased in a protein coat. They act like living organisms when they infect a living host cell and use the cell's machinery to reproduce themselves.

2. Viruses are generally highly specific; therefore one can assume that there are as many viruses as there are kinds of living organisms.

3. A virulent virus is one that takes over the cell's machinery, makes more viruses, and ruptures or lyses the cell, releasing the new viral particles. Temperate viruses infect a host cell and become established in the cell's own genome. It is difficult to treat a viral infection of either type because the virus becomes so well integrated with the host genome that inhibiting the virus also inhibits the host cells. In bacterial infections, on the other hand, one can generally kill prokaryotic bacteria with little effect on the eukaryotic host cells.

4. Generally an electron microscope is required to see an actual viral particle, so the approximate size range is 17 to 1000 nanometers. The two types of structures are helical (rodlike shape) and isometric (somewhat spherical shape called an icosahedron).

5. Adherence is affected by the glycoprotein and lipid spikes on the capsid or envelope. Mammals combat viral infections by producing antibodies against the envelope and capsid proteins, enabling the immune system to identify and remove the virus particles.

6. The AIDS virus infects the T4 white blood cell. The cell possesses a CD4 marker that matches a glycoprotein on the virus. This animal virus penetrates the host cell via endocytosis. Although the coat is shed, the protein RNA and the enzyme reverse transcriptase stay in the cytoplasm.

7. A plant virus infects its host by entering at a point of injury; bacterial viruses inject their nucleic acid into the bacterium through the cell wall. A prophage is a latent bacteriophage, and it is found in a bacterial cell called a lysogenic cell because it may be lysed in the future.

8. Arboviruses are viruses transmitted by insects and other arthropods; they cause encephalitis, dengue, and yellow fever. Retroviruses are enveloped plusstrand viruses that possess reverse transcriptase; they cause cancer and AIDS. Rhabdoviruses are rod- or bullet-shaped minus-stranded RNA viruses; they cause rabies and other diseases of vertebrates, arthropods and plants. Paramyxoviruses cause measles, mumps, Newcastle disease in poultry, and distemper in animals.

9. Viroids are infectious circles of RNA molecules that do not possess capsids or proteins associated with them. They infect exclusively plants. Their possible origin is host introns.

CHAPTER 30

1. (1) Multicellularity—bacteria are not; (2) cell size—bacteria are very small; (3) chromosomes—bacteria lack a nucleus and DNA is not complexed with proteins; (4) cell division and genetic recombination—bacteria do not have true sexual reproduction, but do have means to transfer genetic material; (5) internal compartmentalization—bacteria lack membrane-bound organelles; (6) flagella—bacterial flagella are composed of single fibers of flagellin; (7) autotrophic diversity—bacteria have several different kinds of aerobic and anaerobic photosynthesis with a variety of end products, including sulfur, sulfates, and oxygen; other bacteria are chemosynthesizers metabolizing various inorganic and organic compounds.

2. The bacterial cell wall is a network of polysaccharide molecules crosslinked by polypeptides. Gram-positive bacteria have a plain polypeptide-linked polysaccharide wall, whereas gram-negative bacteria have an additional layer of large lipopolysaccharide molecules deposited over a plain layer. Gram-negative bacteria are generally more resistant to most antibiotics because of the nature of the cell wall. A capsule is a gelatinous layer that surrounds some bacteria.

3. Mutation is important because with rapid generation time, populations can double within several minutes, allowing a favorable mutation to be represented in large numbers quickly. Bacterial recombination results in genetic diversity because genes are transferred from one bacterium to another via viruses, plasmids, or other DNA fragments. An example of such variability is the development of antibiotic resistance.

4. Chemoautotrophic bacteria oxidize nitrogen, sulfur, iron compounds, and gaseous hydrogen. Saprophobic bacteria obtain nourishment from dead organic material. Scientists use such organisms to combat pollution because they already have natural mutants that can degrade nylon, herbicides, and pesticides, and genetic engineering can hopefully improve them. A potential hazard is that using bacteria to degrade pesticides and herbicides could decrease their agricultural effectiveness.

5. The archaebacteria differ in their base sequences in rRNA and the fact that their cell walls lack muramic acid. The most prominent group of these is the methanogens, which produce methane from carbon dioxide and hydrogen. They are obligate anaerobes and control this themselves by reducing elemental sulfur to hydrogen sulfide, creating an aerobic environment.

6. They are present in the rumens of various cellulose-digesting animals, and the bacteria digest the cellulose and the animals digest them.

7. The cyanobacteria are responsible for the increase in the earth's free oxygen, enabling evolution of aerobic eukaryotes. The photosynthetic pigments found are chlorophyll a, carotenoids, and phycobilins. Certain eukaryotes—red algae—have the same pigments present in the chloroplasts, indicating their derivation from cyanobacteria. The cyanobacteria are nutritionally independent because they photosynthesize and fix their own nitrogen.

8. *Prochloron* is the only genus in the phylum Chloroxybacteria; evolutionarily, this organism has chlorophyll a, b, and other biochemical characteristics, indicating its relationship to green algae and higher plant chloroplasts.

9. Spirochetes' characteristic structure is of flagella inserted beneath the outer lipoprotein membrane of the gram-negative outer cell wall; members of this phylum cause syphilis, yaws, and Lyme disease.

10. The actinomycetes are distinctive because they are filamentous like fungi. They cause tuberculosis, leprosy and dental plaque, but they benefit man because they also produce the majority of the known antibiotics.

CHAPTER 31

1. Eukaryotic mitochondria are probably derived from symbiotic nonsulfur purple bacteria, apparently in a single event because all mitochondria are the same. All eukaryotes but *Pelomyxa* and a few other groups possess mitochondria.

2. Bacillariophyta have double-boxlike opaline silica shells commonly called diatoms. Each shell regenerates another half, smaller than itself, then as a certain minimum is reached, the diatom slips its shell and grows to full size. These individuals are diploid and resemble brown algae and dinoflagellates biochemically.

3. The Chlorophyta include the ancestors of the plants and are therefore significant in evolution. *Chlamydonomas* is an important member of this phylum because many other simple green algae resemble it, but may lack flagella; other colonial green algae like *Volvox* appear similar to colonies of *Chlamydomonas*-like cells. A collection of individuals is truly multicellular if there is a division of labor among the cells so that certain of them perform very specialized functions.

4. The members of Ciliophora have

cilia arranged in longitudinal rows or spirals around the body, unique among the protists. They are not so simple as evidence by the fact that they have modified their cilia for various specialized locomotive and feeding functions and possess complex digestive systems. The two vacuoles are (1) a food vacuole, which digests food particles, and (2) a contractile vacuole, which regulates water balance.

5. In Dinoflagellata two flagella beat in grooves in the armor-like cellulose plates. Biochemically they resemble diatoms and brown algae, but it does not appear that they are evolutionarily related to each other because they appear to have only their chloroplasts in common, which were probably derived from separate symbiotic events. Mitosis is unique in this group because it takes place solely within the nucleus and chromosomes include very small amounts of histones.

6. Red tides are population explosions of dinoflagellates, which produce toxins harmful to vertebrates but not to the shellfish that concentrate them. Zooxanthellae are a symbiotic form of dinoflagellate found in jellyfish, sea anemones, and mollusks. They are important to coral reefs because they are the primary factor responsible for the productivity of the reefs, which are otherwise very poor in nutrients.

7. Some euglenoids that are otherwise identical have chloroplasts while others do not, and those possessing chloroplasts may lose them temporarily if grown in the dark, showing the difficulty of assigning them to any group. Only mitosis is exhibited by this group, but the nuclear membrane remains intact during the entire process.

8. Fungi have chitinous cell walls, whereas oomycetes have cellulose walls. They are saprophobes or parasites and are distinguished from other protists because they have zoospores with unequal flagella. They are considered harmful because they cause disease in aquatic plants and animals and caused the potato blight responsible for the infamous Irish potato famine.

9. Alternation of generations in this context means that the life cycle alternates between a diploid sporophyte phase and a haploid gametophyte phase. One type of sporocyte (haploid, motile spores) and two types of gametophytes (haploid eggs and haploid sperm) are produced. Sperm and eggs fuse to produce the zygote, and the diploid zygote becomes the sporophyte.

10. Malaria is the disease produced. It is transmitted from the mosquito to the human in the sporozoite form. The symptoms are produced by the merozoite form; the sporozoites invade red blood cells, divide within them, and cause them to rupture, releasing toxic substances. Gametocytes are transmitted from human to mosquito, and the gametes and fusion into zygotes occur in the mosquito.

CHAPTER 32

1. Fungi and plants both lack mobility and grow from somewhat linear bodies. They differ in that fungi lack chlorophyll and lack motile flagellated sperm, are basically a filamentous growth form, secrete enzymes into a substrate and absorb nutrients, and have a different cell wall chemistry.

2. Two mutalistic associations are lichens (with algae) and mycorrhizae (with plant roots). In these associations, the photosynthetic organism fixes carbon dioxide and fungus enhances the ability of the association to survive in the particular environment.

3. Hyphae are slender filaments. A mass of them is a mycelium. The mycelium penetrates the substrate in which they are growing to obtain nutrients. As it applies to fungi, a septum is a cross wall that may divide the hyphae into cells. The septum is complete when it separates the reproductive structures from the rest of the mycelium. The fungal wall is composed of polysaccharide and chitin, which is an advantage because it renders the wall more resistant to degradation bacteria.

4. The fungal nuclei that are diploid are those of the zygote. All the rest are haploid. The event that occurs during sexual reproduction is that the hyphae of two genetically different mating types fuse. This affects the nuclei because the nuclei do not fuse but coexist as separate entities. Heterokaryotic and dikaryotic hyphae contain two genetically distinct types of nuclei; homokaryotic and monokaryotic hyphae contain nuclei that are all genetically similar.

5. The three divisions of the kingdom Fungi are Zygomycota, Ascomycota, and Basidiomycota. They are differentiated by their sexual reproductive structures. Zygomycota are different because they have fusion of hyphae, which leads to zygote formation, and meiosis occurs only on spore germination. The Ascomycota and Basidiomycota have extensive dikaryotic hyphae where nuclear fusion does not take place except in structures that immediately undergo meiosis, and they produce haploid spores.

6. These asexual spores are called conidia and are multinucleate. They are carried by conidiophores, and the nonreproductive hyphae of this division are incompletely septate.

7. Ascogonia and antheridia are female and male gametangia. The trichogyne grows from the ascogonia; its function is to allow migration of nuclei from the antheridium to the ascogonium. The hyphae that arise are heterokaryotic with paired nuclei.

8. The yeasts belong mostly to the Ascomycota; they differ from other fungi because they are unicellular. It is more likely that this characteristic is degenerate, derived from multicellular ancestors. Yeasts' primary mode of reproduction is by cell fission or budding. They are becoming more popular in research because they are the first eukaryotes used in genetic engineering and are the *E. coli* of the eukaryotic world.

9. Primary mycelium in this context is mycelium made up of only monokaryotic hyphae. Secondary mycelia are hyphae containing nuclei of both mating types, dikaryotic mycelium. A basidiocarp is the structure on which basidia are found. The basidiomycetes are generally only dikaryotic and are usually found on the surface of the gills of a common mushroom.

10. The Fungi Imperfecti are those fungi in which sexual reproduction has not been observed; the best represented of this group is the Ascomycota. Individuals in the class Deuteromycota are classified by examination of hyphal structure and similarities in asexual reproduction.

11. Lichens are symbiotic associations between a green alga and/or cyanobacterium. The best represented of this group is the Ascomycota. These species are not able to grow independently; the fungus directs the photosynthetic component to produce certain special metabolic substances.

12. Endomycorrhizae hyphae penetrate the outer cells of the plant root, whereas ectomycorrhizae surround but do not penetrate the roots. The most prominent fungi are the Zygomycota.

CHAPTER 33

1. In plants, alternation of generations refers to alternation of a diploid generation with a haploid generation. Sporophytes are diploid, spore-producing reproductive structures; gametophytes are haploid, gamete-producing reproductive structures. The sporophyte is comparable to an adult animal, although it differs in that adult animals produce sperm and egg (gametes), whereas a sporophyte plant does not directly produce gametes.

2. Meiosis takes place in the spore mother cell; these cells produce haploid spores which grow into gametophytes. This growth involves mitosis. Gametes (egg and sperm) are in turn produced through mitosis. Fusion is required for the cycle to continue because the egg and sperm unite to form the diploid zygote. The sequence of alternation of generations is as follows: (zygote)—(sporophyte)[meiosis]—(spores)—(gametophyte) [mitosis](gametes = egg and sperm) [syngamy]—(zygote).

3. The gametophyte is green and free living in bryophytes and ferns, and nongreen and nutritionally dependent on the sporophyte in the rest of the vascular plants. The sporophyte is brownish or yellowish tissue enclosed in and dependent on the bryophyte gametophyte, whereas in the vascular plants the sporophyte is the free-living photosynthetic stage.

4. Microgametophytes produce antheridia, and megagametophytes produce archegonia. The haploid spores that give rise to these are microspores and megaspores, respectively. A plant that produces the spores referred to is heterosporous. The plant that produces only one kind is homosporous. The haploid spores are produced in the sporangium; if two different types are produced, they are called microsporangium and megasporangium.

5. Mosses have rudimentary vascular tissue and may be evolutionary degenerate vascular plants; the liverworts and hornworts have no vascular tissue and are probably very primitive. Liverworts: strap-shaped or leafy, unstalked, spherical sporophyte; hornworts: strap-shaped, stalked, horn-shaped sporophyte; mosses: leafy, stalked sporophyte, with capsule at tip.

6. Sperm travel by water. The zygote develops within the archegonium. The characteristic structure is a capsule or sporangium. Protonema are photosynthetic, thread like filaments.

7. A seed is a developing sporophyte in a state of arrested embryonic development surrounded by a protective coat. The seed is crucial because it protects the embryo from drying out or being eaten and provides a source of energy for the growing plant.

8. The archegonia and antheridia are located on the lower surface of the gametophyte. Both the gametophyte and sporophyte are nutritionally independent. A rhizome is the horizontal stem of the fern. Sporangia are located on the frond or leaf. More gametophytes germinate from the spores produced in the sporangia.

9. The large pine cones are megasporangia. The seed coat and embryo

are diploid; the megagametophyte is haploid. The hard seed coat surrounds the embryo and megagametophyte. The nourishment is supplied by the megagametophyte.

10. Flowering plants differ because they contain ovules enclosed within carpels, the parent sporophytic tissue. A stigma is a specialized part of the carpel. Fertilization is indirect. The angiosperm fruit develops from the carpel.

11. Double fertilization occurs when one sperm nucleus fertilizes the egg and the other fuses with one of the polar nuclei. The zygote and endosperm arise from this; the zygote is diploid and grows into the embryo; the endosperm is triploid and supplies nutritive tissue for the embryo.

CHAPTER 34

1. All angiosperms are believed to be derived from a single ancestor because of the great number of common features, embryology, double fertilization, and endosperm; these make it unlikely that such features evolved in the same manner more than once. Gnetales are most closely related to the angiosperms. The earliest known definite angiosperm fossils are pollen grains, which are 123 million years old.

2. The two classes are the monocotyledons and dicotyledons. They differ in the following ways: Monocots: parallel venation in leaves, flower parts in threes, embryos have one cotyledon seedling leaf. Dicots: netlike venation in leaves, flower parts in fours or fives, embryos have two cotyledon seedling leaves. Monocots are derived from dicots by the suppression of one of the cotyledons. This is evident from the fact that monocots and primitive dicots have single-pored pollen, whereas more advanced dicot pollen is three pored.

3. A primitive angiosperm flower has numerous, free, spirally arranged sepals, petals, stamens, and carpels, with little visual differentiation between petals and sepals. Scientists can determine whether a flower is primitive or advanced by comparing the wood and pollen of the plant to those of known primitive plants, observing fusion and/or reduction in flower parts, and comparing the flower to fossil flowers. A primitive flower most resembles a leafy shoot.

4. The corolla is collectively made up of petals. The petals of most flowers are homologous with stamens in that both are affected by genes that do not affect either carpels or sepals. In a small number of flowers, petals evolved from sepals, which enlarged and developed color.

5. An androecium is the male reproductive structure, and it is composed of the stamens, which have probably evolved from the systems of small branches among which the microsporangia were borne. The anther produces pollen grains.

6. Pistillate flowers have only pistils and sterile or no stamens; staminate flowers have only stamens and no carpels. A dioecious sporophyte produces either ovules or pollen, whereas a monoecious sporophyte produces both ovules and pollen. These diverse characteristics promote outcrossing and reduce inbreeding.

7. For passive pollination to be successful, plants must grow in large uniform stands and great quantities of pollen must be produced. Such pollen is carried only short distances compared to active pollinators' pollen. The effect of coevolution has been increased floral specialization, and close relationships between flowering plants and the animals that pollinate them.

8. Social bees have large colonies and must visit many types of flowers to get enough food; they produce several generations over a summer and must vary their flower choices over time as well; solitary bees have only a single brood to raise and use only a single type of flower. It is more likely that this flower will be visited by a solitary bee because it visits only a single type of flower; thus they can evolve together. Coevolution is advantageous to both parties as follows: plants get an efficient, reliable source of pollination and the bee gets a constant source of food.

9. Flowers pollinated by butterflies are those with flat landing platforms and long, slender floral tubes filled with nectar. Those pollinated by moths are pale colors, usually white or yellow, heavily scented, open at night, and have long floral tubes.

10. Plants pollinated by birds must produce more nectar because birds use lots more energy than insects. Deterring factors are red flowers, which attract birds but are not especially noticeable to insects; odorless flowers because birds have strong visual, but weak, olfactory senses; and well-protected nectaries that are available only to a strong beak.

11. Wind-pollinated flowers are small, greenish, and odorless; have reduced or absent corollas; and hang downward in tassels that wave in the wind. These plants have separate flowers, often even on separate plants. This promotes outcrossing.

12. The fruit dispersed by birds and other vertebrates are ones with fleshy shiny black, blue, or red coverings. Fruit with evolved hooks are dispersed by catching on the fur or hair of animals. Some animals aid in seed dispersal by burying the seeds for winter food and not finding all of them. Seed structures that aid seed dispersal are wings or tufts of cottonlike fuzz.

CHAPTER 35

1. Primary growth occurs in the apical meristems of roots and stems. The change produced is elongation of the plant body. Secondary growth occurs in the lateral meristems. This tissue is found in the vascular cambium and cork cambium. The change produced by secondary growth is thickening of the plant body.

2. A parenchyma cell is spherical with functioning cytoplasm, organelles, and nucleus with only primary walls. These cells are not highly specialized and are found ubiquitously in leaves, stems, roots, and primary and secondary tissues. They are alive at maturity, which confers on them the capability to divide further. Secondary cell walls are deposited inside the primary walls of fully expanded cells.

3. The two types of sclerenchyma cells have tough, thick secondary cell walls, fibers that are elongated, and sclerids that are variable and often branched. They are not alive at maturity. The special compound is lignin, which confers the property of ridigity. They are found in supportive or mechanical tissue.

4. Xylem conduct water. Primary xylem are derived from procambium, apical meristem, and secondary xylem are derived from vascular cambium, lateral meristem. The conducting cells are tracheids and vessel elements. These cells are both elongated and fiberlike and not alive at maturity. Tracheids have pits in the secondary walls, and vessels have pits on the side walls and perforations in the end walls. Other cell types that are in xylem are fibers and parenchyma.

5. Phloem conduct food and contain sieve cells and sieve-tube members. They are slender and elongated and lack nuclei. They are alive at maturity. Sieve areas are pores that connect protoplasts of adjoining sieve cells or sieve-tube members to one another. Sieve cells are less specialized; the pores in sieve areas are all the same size; sieve-tube members occur end to end, forming sieve tubes; and the larger pores in sieve areas are called sieve plates. Sieve-tube members are more evolutionarily advanced. Companion cells are specialized parenchyma cells associated with sieve-tube members; they carry out metabolic functions for the sieve-tube members.

6. Angiosperm leaves grow by their marginal meristems. This growth is determinate. Most leaves have a blade, stalklike petiole, stipules that may be present at the base of the petiole, and veins comprising vascular tissue. Simple leaves are undivided, and compound leaves consist of clearly divided leaflets. The three types of growth patterns of leaves are as follows: (1) alternate—leaves spirally arranged along the length of the stem; (2) opposite—leaves occur in pairs along the stem; and (3) whorled—more than two leaves are attached at any one level on a stem.

7. Leaf primordia are rudimentary young leaves surrounding apical meristem. The leaves are attached at nodes. Lateral buds form in the axil of the leaf, and terminal buds are at the end of a branch. The terminal buds produce a chemical to suppress growth of the lateral buds. The cells present in a young stem are (1) inner portion of ground tissue—pith; (2) outer portion—cortex; and (3) outermost cortical parenchyma—photosynthetic cells. Leaf vascular tissue develops from the procambium of the stem.

8. The vascular cambium is in a cylinder around the stem between the main stem and the bark, and it initiates secondary growth. It is derived from both ground tissue and cells between the xylem and phloem of the primary vascular bundles. As the stem grows in diameter, phloem, xylem, and more cambium are produced. A girdled tree eventually dies because the vascular cambium is removed so water and nutrients cannot be transported.

9. Cork cambium produces cork cells. Suberin is contained in these cells. This substance makes cork impermeable to water. These cells are dead at maturity. The periderm is composed of the cork, cork cambium, and phelloderm. Removing the periderm does not equal girdling the tree because no vascular tissue is removed and of itself this will not kill the tree. The tissue of bark includes all of the tissues in a mature root or stem outside of the vascular cambium. The primary phloem ultimately becomes crushed among the layers of secondary phloem in the inner layers of bark.

10. Annual rings are formed that show season growth. Larger cells are produced in the spring. Rings also tell climatic conditions, with rings wider in moist years and thinner in dry years. Hardwoods are dicots, and softwoods are conifers.

11. A section through a monocot root

has bundles of xylem and phloem scattered throughout the section. Inside a dicot root there is no pith. The xylem and phloem of a dicot root are as follows: central column of primary xylem with radiating arms, alternating with strands of primary phloem. The layers of a dicot root are epidermis (protection and root hairs), cortex (storage), endodermis (determines passage of water and minerals into vascular system), and pericycle (formation of lateral root). The Casperian strip is associated with the endodermis.

12. Roots are simpler than stems because they lack appendages analogous to the leaves. Roots form lateral branches that are initiated deep within the root tissues through cell divisions in the pericycle. A lateral root develops a root cap before it breaks the outer epidermis of the main root. Secondary vascular growth does not form isolated bundles of secondary xylem and phloem; it forms a cylinder of xylem on the inside and phloem on the outside, just as stems do.

CHAPTER 36

1. Capillary water is the water held in smaller soil pores that remains in soil and is available to plants. A field capacity is the amount of water held in a given soil after gravity has removed the excess. Wilting point refers to the state when the only water left in a soil is unavailable to plants, so that they will wilt permanently unless water is added.

2. Water held high in a plant possesses potential energy as follows: (a) pressure potential exerted by the atmosphere, and (b) water movement into roots driven by diffusion forces. This is termed water potential. Water moves into a root cell by osmosis, which causes the water to move from a soil with low solute concentration into the root with a greater solute concentration.

3. Of the water that enters a plant 90% leaves it via transpiration. The leaf structures that help regulate this are the stomata, cuticle, and substomatal spaces. Plants must be exposed to a constant source of water because so much is lost that it must be replenished for metabolic activities to continue. This source need not always be external; it can be water stored by the plant.

4. Root hairs are turgid as a result of their greater solute potential. Energy is expended to accumulate minerals in a plant root because their concentration is much greater in the root than in soil so ions must be pumped in at a cost of ATP and against a concentration gradient. Mineral passage is

nonselective, except passage through the endodermis. This results because of the presence of waxy Casperian strips which cause minerals to pass selectively through endodermis plasma membrane.

5. Water transport is reduced to near zero when the relative humidity is 100%, i.e., at night. Almost no transpiration occurs under these circumstances. Active transport does continue and causes higher concentrations of ions, which pulls even more water in by osmosis, so that water moves upward in the xylem. The term for this is root pressure.

6. Guttation occurs when water droplets are forced out of special cells at the ends of small veins. It is more likely in very short plants because root pressure can't push water up very great distances. Trees have a smaller diameter in the day because transpiration suction is so great that it causes the xylem walls to pull together. The ultimate source of a plant's water potential is the sun.

7. Stomatal control requires energy; ATP is used to pump ions into guard cells, water follows passively, and cells stay turgid. Potassium is an integral part of this control mechanism. Guard cells possess chloroplasts to provide an immediate source of ATP to regulate their opening. Leaf tissue produces abscisic acid during water stress. It causes potassium to pass rapidly out of the guard cells, and the stomata close.

8. Most nutrient movement is active. This is supported by (1) the fact that roots deprived of oxygen lose the ability to absorb nutrients from the soil, and (2) light-starved plants will exhaust the nutrient supply and not be able to replace it. The nutrients that can be relocated are phosphorus, potassium, nitrogen, and sometimes iron; this serves to retain them instead of losing them when the leaves drop off. Once calcium is deposited, it cannot be removed.

9. Translocation is the movement of sugars from where they are produced to where they are used; they are stored as starch and transported as sucrose. The driving force behind translocation is hydrostatic pressure.

10. Carbohydrates move through a plant as follows: sucrose is produced at the source, which is then actively loaded into the veinlet sieve tubes, which increases their solute concentration, then water flows into the tubes via osmosis, sucrose is unloaded at the sink, water follows it and moves toward the sink, where it is pulled out, and sucrose follows it. The process has an active source and sink and is passive in between. The

companion cells provide ATP to load and unload sucrose.

11. Macronutrients are inorganic nutrients that plants require in large amounts; they include carbon, hydrogen, oxygen, nitrogen, potassium, calcium, phosphorus, magnesium, and sulfur. They are important because they are components of organic molecules, enzyme activity, the maintenance of cell walls and membranes, and energy production. Micronutrients are inorganic nutrients that plants require in very small amounts; the primary ones are iron, chlorine, copper, manganese, zinc, molybdenum, and boron. Their most common functions are as enzyme activators and in the formation of chlorophyll.

12. Fertilizers are required because cultivation increases erosion and the loss of nutrients, because large amounts of nutrients are taken away with the crops, and because soils may be naturally poor. Natural communities recycle all available nutrients so that they are used over and over. The three mineral nutrients most commonly found in commercial fertilizers are nitrogen, phosphorus, and potassium. Organic fertilizers may be superior because they help build up the humus content of the soil, enhancing water and nutrient-retaining properties.

CHAPTER 37

1. Animal control systems are internalized cell movements and changes in gene expression, which allow for a high degree of tissue-specific specialization. Plant systems are external influences that alter the expression of the genetic blueprint and the production of plant hormones. Animal systems are affected little by environmental changes because homeostatic processes shield the developing animal from these influences. Plant systems are almost exclusively under the control of the environment in their development because it cannot move away from the environment so it must meet the demands of that environment.

2. Most plant cells are very reversible in the aspect of differentiation because gene expression can be reactivated, leading to alternate differentiation or growth of the whole plant from an isolated group of cells. The limitation is that the cell must retain the protoplast at maturity. This is termed totipotent.

3. Stewart excised secondary phloem cells, and when they were cultured in liquid nutrient medium they grew roots, whereas when they were placed on solid medium, they grew shoots and developed into complete plants. Embroids are developing isolated

cells that resemble true embryos in having comparable distinct tissue systems, shoot and root apices. They do not generally form from isolated plant tissue because the cells first revert to callus.

4. Adventitious roots develop from cuttings in water or wet sand; these growths will occur more readily if the leaves are left on. In this way buds produce auxin, a hormone that stimulates root production. The structures that develop from underground horizontal roots are adventitious stems. It is more likely that a complete plant will regenerate from stem tissue.

5. In a gymnosperm the nucleus divides but no cell walls form between nuclei until 256 have formed, making 256 equally sized cells. Cell differentiation occurs at this 256-cell stage. The larger cells near the micropyle develop into suspensor and smaller rapidly dividing cells become root and shoot apices.

6. Development of the embryo temporarily halts after differentiation of apical meristems and cotyledons. The seed structures present are integuments or seed coat. Seeds are adaptively important because (1) they allow development to be postponed until environmental conditions are advantageous, (2) they permit development to be synchronized with environmental conditions, and (3) they allow for wider dispersal of plants and protect the embryo at its most vulnerable stage.

7. A seed may fail to germinate because it needs additional environmental signals, such as a certain wavelength or intensity of light or a period of cold. As it relates to seed germination, stratification is a seed held for a period of time at low temperature; this prevents the seed from accidentally germinating in the middle of winter.

8. The cotyledon is modified into the scutellum. The seedling uses the food reserves in the scutellum first, then the scutellum absorbs food from the endosperm to pass on to the seedling. The enzymes secreted are hydrolases.

9. The enzymes secreted are amylase and hydrolases, and they come from the aleurone layer. The controlling hormone is gibberellins, and it initiates a burst of mRNA and protein synthesis.

10. All cells of the adult plant are derived from the root and shoot apical meristems. The plant structures formed are shoots and leaves. These are not capable of producing a complete mature plant because they don't form roots because no cytokinins are produced.

11. Indoleacetic acid is an auxin, and

is a natural hormone whose two basic functions in plants are to promote cell elongation and suppress growth of lateral branches. Apical dominance is the inhibition of growth of lateral branches. The associated hormones are ethylene, which inhibits the differentiation of cells around nodes, thus preventing lateral bud formation, and cytokinins, which may override auxins by promoting cell division.

12. The development of the leaf primordium is independent of the presence of signals from its plant. An excised leaf primordium will therefore develop into a normal leaf because when the primordium is cut from the meristem, it proceeds along its own predetermined developmental path, so that its developmental fate is determined by the time it has differentiated from the apical meristem.

CHAPTER 38

1. Both plant and animal hormones are chemical substances that are produced in small quantities in one region and transported to another region where they result in physiological responses. Plant hormones are not produced as definite sites solely involved with the production of that hormone, whereas those of animals are. The five major plant hormones are auxin, cytokinins, gibberellins, ethylene, and abscisic acid.

2. The only naturally occurring auxin is indoleacetic acid (IAA); it is synthesized from tryptophan. It is produced in the shoot apex and transported downward, not by vascular tissue, but by an unknown mode of transport.

3. Auxin increases the plasticity of the plant cell wall, allowing greater stretching during active cell growth. Indoleacetic acid oxidase degrades IAA, preventing its accumulation. Auxin acts rapidly in response to light.

4. The substances are related to the auxins and IAA. Commercially they are used as weed killers and are of the class of dicots; they act upon dicots to grow themselves to death.

5. Cytokinins stimulate cell division and regulate differentiation. Cytokinins promote lateral bud formation and inhibit lateral root formation, which are opposite to the affects of auxins. Cytokinins are produced by roots and developing fruit, derived from the biomolecule adenine.

6. Gibberellins is usually lacking in genetically dwarfed plants. If it is, the application of gibberellins will restore normal growth. The three secondary effects of this hormone are (1) to stimulate the production of seed amylase, (2) to induce biennial plants to flower, and (3) to hasten seed germination, substi-

tuting for the effects of cold or light. The three commercial uses are (1) to increase fruit and cluster size in grapes, (2) to delay ripening of citrus fruits and speed the flowering of strawberries, and (3) to stimulate the digestion of starches in brewing.

7. Ethylene is the main factor in forming the separation layer between the stem and leaf. It hastens fruit ripening. The common gas that counteracts this effect is carbon dioxide. Commercially, ethylene hastens fruit drop, induces flowering in the pineapple family, and stimulates seed germination. In weed control products, ethylene stimulates the germination of weed seeds, and weeds can then be killed.

8. Abscisic acid is produced in mature green leaves and in fruit and root caps. If spotted on a green leaf, this hormone makes the leaf under the spot turn yellow; cytokinins reverse this effect. Abscisic acid induces the formation of winter buds, suppresses the growth and elongation of buds, and promotes senescence. It also regulates the opening and closing of stomata.

9. The stems are positively phototropic and grow toward the light. This allows the leaves to capture a greater amount of light energy. The roots are negatively phototropic and grow away from the light. This causes the roots to grow downward toward a potential supply of water and nutrients. Auxin most completely mediates this action.

10. A burst of red light at 660 nanometers inhibits flowering but can be cancelled by exposure to far-red light of 730 nanometers. The compound that mediates these reactions is phytochrome pigment. One form absorbs red light, while the other form absorbs far-red light. Therefore a flash of red light will convert phytochrome to the far-red sensitive form that blocks flowering and a second flash of far-red light converts phytochrome to the red-sensitive form that does not interfere with flowering.

11. The two plant growth responses mediated by phytochromes are (1) seed germination inhibited by far-red and stimulated by red, and (2) shoot of an etiolated seedling becoming normal after exposure to red light, but not far-red light. An etiolated seedling is a slender, colorless shoot.

CHAPTER 39

1. All animals are multicellular; unicellular heterotrophs are protists. Vertebrates have dorsal backbones, whereas invertebrates do not. Invertebrates (99%) are most numerous. Animal cells lack rigid

walls, and are usually quite flexible. The functional units are tissues, and the collection of these units are organs.

2. The two subkingdoms of animals are Parazoa, which are asymmetrical and have no tissues or organs, and Eumetazoa, which are symmetrical and have tissues organized into organs and organ systems. No embryological tissues are present in the more primitive subkingdom. The embryological tissues of the more advanced subkingdom are the ectoderm (nervous system and outer coverings), the mesoderm (skeleton and muscles), and the endoderm (digestive system).

3. Both of the animal subkingdoms evolved from choanoflagellates. The general characteristic that divides the lower and higher invertebrates is that the lower have less complex tissue organization. The four major phyla of the lower invertebrates are Porifera, Cnidaria, Platyhelminthes, and Nematoda.

4. In a sponge, food particles stick to microfibrils connecting collar strands, which serve as a food sieve. It is digested directly by collar cells or by amoeboid cells. The highly convoluted interior increases the surface area that is exposed to water, and the number of choanocytes. The primary stages in sponge reproduction are as follows: initial development occurs within the body of the adult; when released, the larvae have many external flagellated choanocytes that are free swimming. After the planktonic stage, they settle onto a suitable substrate, and turn inside out so that the choanocytes of the adult are on the inside.

5. A cnidarian body is a gelatinous matrix with distinct tissue differentiation. The two body forms present are the polyp, a cylindrical diploid form that is attached to a substrate and is solitary or colonial, whose mouth points upward, and who may build an external skeleton diploid; and the medusa, an umbrella shaped, free-floating organism, whose mouth points downward. A planula is a free-swimming, multicellular, ciliated cnidarian larva derived from fusion of gametes.

6. Digestion occurs within the gastrovascular cavity in cnidarians. The majority evolutionary innovation is the advent of internal extracellular digestion, which allows the animal to digest organisms larger than individual cells. The chemical means of this digestion are enzymes released into the gastrovascular cavity that break the food into bits that are then phagocytized by the cells lining the gut. The indigestible particles go out

through the same opening that they entered.

7. The three basic body plans of bilaterally symmetrical animals are acoelomates, wherein no body cavity other than the digestive tract exists; pseudocoelomates, wherein there is a cavity between the mesoderm and the endoderm; and the coelomates, wherein there is a cavity completely bounded by mesoderm. The internal organs are suspended in coelom by mesodermal double-layer mesenteries. In wormlike organisms are found acoelomate and pseudocoelomate body plans; in advanced invertebrates and vertebrates are found coelomate only.

8. Platyhelminthes' body plan is acoelomate. They are dorsoventrally flattened and have a distinct head at the anterior end. Most are parasitic, although some are free living. They move through ciliated epithelial cells on their dorsal surface. The digestive system is with the gut branched, with only one opening. Food is partly digested extracellularly, and phagocytic cells line the gut. Some parasites absorb food through the body wall.

9. The sensory structures are chemical-sensing tentacles and light-sensitive eyespots. These structures orient the animal to food and away from the light. With regard to sexual reproduction, the individuals are hermaphroditic, and each deposits sperm in the copulatory sac of the mate, laid in cocoons in the free-living forms. The parasitic forms may have several larval forms. Some can regenerate whole individuals when divided.

10. Flukes ingest food through the mouth. Two or more hosts are needed. The stages of the life cycle are (1) egg hatches into miracidium, (2) when digested by a snail it transforms into a sporocyst, (3) from which rediae hatch, (4) these grow into cercariae, (4) these attach to fish and transform into metacercariae encysting in muscles, and (5) a mammal eats the fish and cysts dissolve, releasing the young flukes, which then lay eggs.

11. The pseudocoel serves as a hydrostatic skeleton that muscles can work against. Therefore movements are far more efficient. The internal systems of these animals are generally complete oneway digestive tracts.

12. The digestive tract of a typical nematode has sensory organs at the anterior end and a mouth equipped with piercing stylets. The muscular pharynx sucks food inside the mouth, and food then continues through the digestive tract and is eliminated through the anus. The nematodes are

structurally unique because they completely lack cilia and flagella, even in the sperm and excretory organs. They are damaging to humans because many are parasites in the digestive tract, causing anemias, damaging tissues, and obstructing circulatory vessels.

CHAPTER 40

1. Among the advantages of possessing a body cavity are the fact that body organs within a fluid-filled cavity can function without pressure from surrounding muscles; fluid can act as a circulatory system, increasing the distance that any cell can be away from the exterior; the opening and closing of the gut is not altered by muscle movements; and the cavity provides space into which the gonads can expand. In a completely mesodermally enclosed cavity, the mesoderm can direct development of the more complex internal systems. The parietal peritoneum is epithelium that lines the outer wall of the coelom. The visceral peritoneum lines internal organs suspended within the coelom.
2. The pattern of embryonic development related to cleavage and the blastopore that occurs in protostomic coelomates is the spiral cleavage, wherein the mouth develops at or near the blastopore and the anus develops elsewhere. Radial cleavage, wherein the anus forms from or near the blastopore and the mouth forms elsewhere, is the pattern in deuterostomic coelomates. The major coelomate phyla that are protostomes are mollusks, annelids, and arthropods. Those that are deuterostomes are echinoderms and chordates. These two groups differ in early developmental fate as follows: protostomes—fate sealed at cleavage; separated cells cannot survive; deuterostomes—fate set at later date; early cells can be separated, with each developing into a normal individual. The development of the coelom differs as follows: protostomes—coelom origin simple and direct; deuterostomes—whole groups of cells move around, forming new tissue associations.
3. The prominent feature of lophophore animals is the lophophore, a circular or U-shaped ridge around the mouth, with one or two rows of ciliated hollow tentacles. This functions in food collection and as a gas exchange organ. The coelomic cavity is within the lophophore and tentacles. The protostomic and deuterostomic features are radial cleavage like deuterostomes and, in two phyla, the blastopore becomes a mouth (protostomic) and, in the third, it becomes the anus (deuterostomic). The three phyla

in this group are brachiopods, bryozoans, and phoronida.
4. The basic body plan of a mollusk is a visceral mass with a soft epithelium and a muscular foot, and in the mass is contained digestive, excretory, and reproductive organs; this animal may have a differentiated head at the anterior end. The mantle is outside the visceral mass. This allows the gills to develop from its tissues, and the mantle may secrete the protective shell. In the mantle cavity of aquatic mollusks the water passes through, bringing in oxygen and food and carrying out wastes and gametes. This has developed into the jet propulsion system of squid and octopus.
5. The circulatory system of most mollusks is an open system with a three-chambered heart. Cephalopod circulation differs from others because it has a closed system of vessels to carry the blood to and from the heart. The coelom represents the area around the heart. The mollusk excretory system is as follows: two tubular structures called nephridia; the opening is a funnel-like nephrostome that is lined with cilia that collect fluid from the coelom; and a coiled tubule leads to the bladder which is connected to an excretory pore.
6. The reproductive individuals in the mollusks are generally male and female individuals. Trochophores are distinct mollusk larvae with a row of cilia around the middle of the body. A veliger is a second larval stage of marine snails and bivalves with a distinct foot, shell, and mantle.
7. The four characteristics of the bivalves are (1) two hinged shells, (2) a wedge-shaped foot, (3) lack of radula, and (4) tentacles. The adductor muscles open and close the shells. The mantle secretes the shell and forms the ingoing and outgoing siphons. Bivalves feed as the cilia on the gills bring in water, and food particles become entangled in the mucus secreted by glands; then the palps direct the mass into the mouth. Bivalves disperse as larvae because adults are relatively fixed and immobile.
8. The two advantages of embryonic segmentation are (1) each segment contains a more or less complete set of organs so the loss of one segment is not necessarily fatal, since there is functional duplication in other csegments, and (2) more effective locomotion is possible due to the independent movement of segments and increased flexibility. The two phyla in which segmentation is most notable are the annelids and arthropods.
9. The annelid nervous system has diverse sensory organs that are concentrated on the anterior end,

cerebral ganglion, and a primitive brain that is found in the anterior segment with separate interconnected ganglia in each segment. Annelid and nematode locomotion is similar because muscles move against a hydrostatic skeleton. Annelid locomotion is more advanced because each segment can move independently. Annelid setae are bristles of chitin that serve as anchors in certain crawling movements and may aid in swimming.
10. Annelids are similar developmentally to mollusks because both possess free-swimming trochophore larvae. It is likely that mollusks and annelids are derived from a common unsegmented ancestor. The relationship between segmentation in annelids and arthropods and that in vertebrates is that annelids and arthropods may share a same segmented ancestor whereas vertebrate segmentation evolved independently.
11. The circulatory system of annelids is a closed system with five hearts. They exchange gases with the environment directly through the skin—they lack gills or lungs. The typical annelid excretory unit exists as numerous structurally similar nephridia, one pair per segment, that collect waste products and transport them out of the body through the coelom by means of specialized excretory tubes.
12. Earthworms obtain nutrients by sucking soil into their mouth via contraction of the pharynx; then muscles in the gizzard grind up the organic material and the food moves through a long, straight digestive tract, with undigested material deposited as castings. Earthworm sensory structures sense light, touch, and moisture. Earthworms do not have parapodia. Reproduction is as follows: Earthworms are hermaphroditic. They orient head to tail when mating, held together by a mucous band produced by the clitellum. Each exchanges sperm with the other. After separation the clitellum secretes a cocoon, and as it passes the female openings, it receives the eggs. It then picks up the other worm's sperm from the sperm receptacles. Young worms eventually hatch from the cocoon.
13. Leeches and earthworms both possess a clitellum (during breeding season) and are both hermaphroditic with cross-fertilization. Leeches differ in that they have a reduced coelom, they are not divided into individual segments, they have suckers at one or both ends of the body, they generally have no setae, and many are external parasites. The adaptations that allow leeches to suck blood are a mouth with chitinous teeth that rasp through skin, an anti-

coagulant that prevents the wound from clotting, and powerful sucking muscles that suck up the blood.

CHAPTER 41

1. The two groups are chelicerates, with mouthparts formed from the most anterior pair of appendages and the second set pincer or feeler like, with the rest legs (examples: spiders, mites, scorpions), and mandibulates, with the most anterior appendages sensory antennae and jaws formed by the second or third set (examples: crustaceans, insects, millipedes, and centipedes). The subgroup with fundamentally biramous appendages is the crustaceans, specifically aquatic mandibulates. The subgroup with uniramous appendages is the terrestrial mandibulates. Biramous appendages are two-branched, and uniramous are one-branched.
2. The three Arthropod a subphyla are Chelicerata, Crustacea (Biramia), and Uniramia. They all evolved independently, in terms of exoskeletons first, then body segmentation and jointed appendages. This is supported by the embryology as follows: Biramia have unique nauplius larvae, early Uniramia embryology resembles annelids, and chelicerates are different from the others.
3. Arthropods make their exoskeletons through a substance that is secreted by and fused with the epidermis. The exoskeleton provides a surface for the muscles to work against, protects from predators and injury, and reduces water loss. During ecdysis a new exoskeleton is grown under the old one, with fluid separating them; then the outer one is shed and the body is expanded by blood circulation and air intake, and the soft new skeleton hardens as it is exposed to air or water. The control for this process involves hormones.
4. In most arthropods the coelom is reduced to cavities that house the reproductive organs and some glands. The arthropod digestive system is a tubular gut that extends from the mouth to the anus. Their circulatory system is an open one, with a dorsal, longitudinal heart. The blood flows from anterior end to the head, through internal body spaces toward the posterior end and back in the dorsal vessel. This one-way flow is maintained by valves in the posterior region of the heart.
5. The respiratory system transports oxygen throughout the arthropod's body. This is different from vertebrates in that vertebrates use the circulatory system. The resultant impact is a limitation of the size of the organism, with each

cell necessarily within diffusion distance of a respiratory structure in the arthropod. The respiratory system of a typical arthropod is numerous, small, branched cuticle-lined air ducts called tracheoles with an opening to the outside via spiracles that can be closed to reduce water loss. The special systems are as follows: spiders—book lungs; horseshoe crabs—book gills, crustaceans—gills.

6. The functional units of the terrestrial arthropod excretory system are the Malpighian tubules. They are projections from the digestive tract, between the midgut and the hindgut. The wastes are processed as follows: fluid is absorbed from the blood through the walls of the Malpighian tubules, nitrogenous wastes are concentrated, then they are emptied into the hindgut and eliminated. Water loss is regulated because water and salts are reabsorbed by the hindgut and returned to the circulation.

7. The three classes of Chelicerata are Arachnids (spiders), Merostomata (horseshoe crabs), and Pycnogonida (sea spiders). The arrangement of appendages in the largest class, arachnids, is a pair of chelicerae, a pair of pedipalps, and four pairs of walking legs. These appendages function as follows: chelicerae—fangs attached to poison glands; pedipalps—specialized for copulation in male spiders, but also serve sensory and feeding functions; walking legs—locomotion.

8. The ancient arthropods that are ancestors to horseshoe crabs are trilobites. Horseshoe crabs exhibit external fertilization. Sea spiders are very small, with a reduced abdomen, a body that is mostly cephalothorax, no well-defined head, and four to six pairs of legs. They regulate body wastes and gas exchange by direct diffusion across the body surface.

9. Crustaceans have two pairs of antennae, three pairs of chewing appendages, and various numbers of walking legs. The legs are on the abdomen as well as the thorax. The labyrinth functions to regulate the osmotic condition of the body primarily, and secondarily to excrete salts, amino acids, and water. Decapod crustaceans are different because they have an exoskeleton reinforced with calcium carbonate and a cephalothorax covered by a shieldlike carapace.

10. Millipedes and centipedes are both members of Uniramia, with bodies that consist of a head region followed by numerous segments with paired appendages. However, centipedes have one pair of legs per segment and are carnivores. Millipedes have two pairs of legs per segment (actually two segments fused to form one) and are herbivores.

11. Most insects possess a tubular digestive tract that is somewhat coiled. Insects that feed on juices low in protein have a greatly coiled tract to provide greater opportunity to absorb nutrients and to allow their digestive enzymes, which are weaker, more length over which to work. Respiratory adaptations in insects include (1) tracheae enlarged to form air sacs surrounded by muscles that help force air deep into the body, (2) permanently closed spiracles, and (3) gases that cross between the trachea and exoskeleton via diffusion.

12. An instar in this context refers to stages between molts. The two different kinds of metamorphosis are simple—if wings are present, they develop externally during juvenile stages with no resting stage before the last molt into adulthood (immature stages—nymphs); and complete—wings develop internally during the juvenile stages and appear only during the resting stage prior to the adult (immature stages—larvae; resting stage—pupa or chrysalis).

CHAPTER 42

1. The symmetry is radial and there is a five-part body plan in adult echinoderms. The skeleton is calcium-rich plates called ossicles that make up the internal skeleton covered by epidermis beset with numerous spines. Their nervous system is not centralized, with no head or brain, but rather the system is central nerve rings that branch. The water vascular system is a ring canal with five radial canals so that water enters through the madreporite, flows to the ring canal through the stone canal, and ultimately into tube feet. They extend and contract their tube feet through the contraction of muscular ampullae at the base of each tube foot, which forces fluid into the tube foot, causing it to extend; then when the muscles contract, fluid is forced back into the ampulla.

2. In sea cucumbers, sea lilies, and brittle stars the tube feet are specialized for feeding. Echinoderms respire via skin gills, which are fingerlike projections of epidermis that occur near spines. The digestive system is complete with mouth, gut, and anus. The ability to regenerate lost body parts means that they can asexually reproduce from parts, and sea cucumbers in particular can eject intestinal parts as a defensive mechanism. Reproduction is usually sexual, with the sexes separate and externally indistinguishable; there is external fertilization although some species brood eggs in cavities or underneath bodies. The larvae are bilaterally symmetrical.

3. A sea star has five arms sharply set off from the disk. The mouth of a sea star is in the center of its lower surface. Brittle stars have arms that are more slender, branched and set off from the disk even more sharply.

4. Sea cucumbers are soft bodies, with leathery skin, and lie on their sides so that the radial symmetry is less evident. Their tube feet form tentacles that surround the mouth. To eat, the tube feet exude mucus that captures small organisms, then the tentacles are brought to the mouth and the food organisms are wiped off within the esophagus. Their skeleton is widely separated microscopic plates. The respiratory tree has internal branches from the cloaca that function in gas exchange as water is pulled in and out. Reproductively, some are hermaphroditic.

5. The three primary characteristics are single, hollow dorsal nerve cord; dorsal notochord; and pharyngeal gill slits. The three subphyla are Urochordata (tunicates), Cephalochordata (lancets), and Vertebrata (vertebrates).

6. In the vertebrates, bony vertebrae replace the notochord. The nerve cord is protected within a U-shaped groove in the vertebrae. The seven classes of vertebrates are as follows: Agnatha (lampreys, hagfishes), Chondrichthyes (sharks, rays, skates), Osteichthyes (bony fish), Amphibia (frogs, toads, salamanders), Reptilia (reptiles), Aves (birds), Mammalia (mammals).

7. An advantage to possessing jaws is that it encourages active predation; it also improves the ability to collect food over sucking or filtering it. Jaws are a modification of one or more gill arches. Teeth are modified from skin. Gnathostome refers to all vertebrates except the agnathans.

8. The bony skeleton weighs more so that fish tend to sink with it. An air bladder provides buoyancy to counter the sinking. This bladder developed from a lunglike outpocketing of the pharynx. The lateral line system is the ability to detect changes in the pressure of water and sense movements of objects.

9. Amphibians differ in that they depend on the availability of free water during the early stages of development and to combat water loss through the skin. Salamanders and newts have tails; frogs and toads lack tails as adults; and caecilians lack legs. Most amphibians respire through inefficient lungs, across the skin and lining of the mouth.

10. Reptile skin is dry and scaly and retards water loss; amphibian skin is moist and lacks scales. The egg is an amniotic egg, which protects the embryo from drying out, nourishes it, and enables the development of the embryo independent of free water. Nutrients are in the yolk, and wastes are excreted into the allantois and stored until hatching. Respiration is through the shell and membranes, which are permeable to oxygen and carbon dioxide.

11. The ectothermic or poikilothermic vertebrates are fish, amphibians, and reptiles; the endothermic or homeothermic vertebrates are birds and mammals. Ectotherms regulate body temperature by taking heat from the environment, whereas endotherms regulate the body temperature internally, independent of environmental temperature. Endotherms are more active at night and in cool temperatures and are generally more efficient. Ectotherms subsist with minimal food needs and can live where endotherms cannot.

12. Birds differ in that their forelegs have evolved into wings and their bodies are covered with feathers. They have light, hollow bones; and they have highly efficient lungs. Feathers are derived from scales.

CHAPTER 43

1. A tissue is a group of structurally and functionally similar cells. The four principal kinds are epithelium, connective tissue, muscle, and nerve. A collection of tissues is an organ. These organs are further grouped into organ systems to carry out the principal activities of the body.

2. The four primary functions of epithelial cells are (1) protect tissue from dehydration and mechanical damage, (2) provide a selectively permeable barrier, (3) provide sensory surfaces, and (4) secrete materials. The three general classes are simple epithelium, stratified epithelium, and glands.

3. A connective tissue is a supporting tissue composed of living cells embedded in a nonliving matrix. The three categories are defensive, structural, and sequestering.

4. The three structural connective tissues are fibroblasts, cartilage, and bone. They differ in the nature of the matrix cells laid down, as follows: fibroblasts—secrete collagen in loose irregular mats; cartilage—secretes collagen in parallel arrays; bone—secretes collagen fibers coated with calcium salts.

5. Muscle is derived from mesoderm. Actin and myosin are abundant in muscle. The three categories of muscle cells are smooth, striated, and cardiac.

6. Striated muscle cells are associated with the vertebrate skeleton. These muscles are under voluntary control. The morphology of this type of muscle is cells fused end-to-end to form long fibers, many myofibrils thick, each with many nuclei at the periphery of the fiber. This organization allows simultaneous contraction of fibers.

7. The cells in nerve tissue are neurons, which transmit nerve impulses, and supporting cells, which assist in the propagation of the nerve impulse and provide nutrients to neurons.

8. A typical nerve fiber is bundles of axons, dendrites, and supporting cells. Glial cells are found in the brain and spinal column. The supporting cells are Schwann cells.

9. Homeostasis is the maintenance of constant extracellular conditions within the body. A negative feedback loop prevents the departure of a controlled variable such as pH or temperature from its normal, or ideal, value. Changes in the controlled variable trigger responses that oppose the change and restore the variable to its ideal level. An example of negative feedback is the regulation of body temperature. Positive feedback refers to a condition in which a change in a controlled variable actually causes the body to drive the controlled variable even further away from the initial value. An example of positive feedback is the contractions of the uterus during childbirth.

CHAPTER 44

1. Most neurons possess a profusion of dendrites to receive simultaneous input from several different sources. A neuron's cell body integrates information received from many different dendrites. The impulse travels from dendrite to cell body to axon.

2. Neuroglia cells provide nutritional support to neurons. Schwann cells produce the myelin sheath, which provides insulation. The nodes of Ranvier are breaks in the myelin sheath where an axon is in direct contact with the intercellular fluid.

3. A nerve fiber is stimulated by pressure, chemical activity, etc., applied at some site on the neuron membrane. The effect is to change membrane proteins so that sodium is freely admitted into the cell.

4. A wave of depolarization spreads by localized current at the site of the stimulus that causes nearby closed sodium ion channels to open, allowing sodium in there as well, which causes other channels to open, and so on, which passes the depolarization along the length of the fiber. The refractory period

is the time required to restore the original ion concentrations, during which another series of depolarization events cannot occur.

5. A nerve cannot conduct a partial impulse; it is an all-or-nothing process. The threshold value of an impulse is the amount of stimulation required to open enough sodium channels to initiate depolarization. The action potential is the shift of ions and consequent shift in electrical charge; not all neurons possess the same action potential. It is dependent upon the density of the sodium ion channels.

6. The region between Schwann cells is the node of Ranvier. These nodes are in direct contact with the intercellular fluid, and the action potential jumps from node to node. This type of conduction (called saltatory conduction) is advantageous because it travels faster than an impulse passed in continuous fashion, and requires less metabolism than depolarizing and repolarizing an entire length of neuron.

7. A synapse is the junction between an axon and another cell. The axon side is the presynaptic membrane and the target cell side is the postsynaptic membrane. The nerve impulse crosses the synapse by the release of neurotransmitter chemicals from the vesicles at the presynaptic membrane, which combine with receptor molecules on the postsynaptic membrane.

8. Synaptic conduction is superior because the nature of the chemical transmitter can be different in different junctions, allowing for various types of responses. Many neurotransmitters have been identified, in fact, over 60.

9. Acetylcholinesterase cleaves acetylcholine from the postsynaptic membrane receptors. This is needed to allow passage of the next impulse. Interference with this enzyme results in continuous neuromuscular transmission.

10. Inhibition can occur when K^+ channels open, allowing positive K^+ ions to leave the cell, therefore making the intracellular potential more negative. Inhibition can also occur if Cl^- channels are opened, allowing negative Cl^- ions to enter the cell.

CHAPTER 45

1. The functional groups are the central and peripheral nervous systems. The central system includes the brain and spinal cord; the peripheral, all nerve pathways. The nerve pathways are the afferent, which transmits information to the central system, and the efferent, which transmits commands from the central nervous system. Somatic motor pathways are under voluntary control and transmit

commands to skeletal muscles; autonomic pathways are under involuntary control and transmit commands to glands and involuntary muscle systems.

2. A reflex is a response that is an automatic consequence of nerve stimulation, exerts no control over complex actions, and shows little coordination. This involves a sensory neuron, a motor neuron, and possibly an interneuron.

3. The determining factor is the degree of membrane polarization in that branch as influenced by excitatory and inhibitory synapses with other neurons. All nerve impulses within the brain are considered identical; the brain differentiates through source of signal, number of signals received, and frequency of signal.

4. The three phyla are chordates, arthropods, and cephalopods. The three principal divisions are the hindbrain (rhombencephalon), midbrain (mesencephalon), and forebrain (prosencephalon). The most prominent in primitive vertebrates is the hindbrain.

5. The dominant part in advanced vertebrates is the forebrain. Its elements are the diencephalon, which functions in the integration of sensory information, and the telencephalon, which functions in associative activity.

6. The connecting nerve tract is the corpus callosum. The four lobes are the frontal, parietal, temporal, and occipital. The cerebral cortex is located in the thin layer of tissue just under the surface of the cerebrum. Its major regions are motor, sensory, and associative cortexes.

7. There are two neurons involved; they are the afferent sensory neuron and the efferent motor neuron. They synapse in the spinal cord. Reflex arcs provide immediate reaction to physical damage and maintain posture.

8. An interneuron-mediated reflex arc is more complicated because an afferent neuron synapses with an interneuron in the spinal cord which then synapses with an efferent neuron that extends back to the source of the afferent neuron. There are three neurons involved. It allows for processing by the central nervous system because interneuron may receive signals from other neurons.

9. Control of the viscera is through the parasympathetic and sympathetic systems. These two networks differ as follows: Parasympathetic—long efferent central nervous system neurons, organ-associated ganglia, short efferent neurons from ganglia to target organ; sympathetic—short efferent central nervous system axons, ganglia near the spine, long efferent neurons from the ganglia to target organs.

10. The chemical transmitter is acetylcholine at the ganglionic synapse of both antagonistic visceral control networks. At the target organ synapse it is acetylcholine for the parasympathetic and adrenaline or noradrenaline for the sympathetic. These transmitters are different at the target organ in that they cause opposite reactions in the target organ. The parasympathetic system stimulates normal body activities, whereas the sympathetic system suspends normal activities and mobilizes the body for greater activity.

CHAPTER 46

1. Interoception is sensing information that relates to the body itself, its internal condition and position. Exteroception is sensing of the exterior. The simplest sensory receptor is one that depolarizes in response to direct physical stimulation. Vertebrates do not possess a single kind of receptor for all changes in temperature; one senses an increase in temperature and another senses a decrease in temperature.

2. The carotid bodies are embedded within the artery walls at several locations. They detect changes in oxygen and carbon dioxide concentrations and regulate respiratory rate.

3. The semicircular canals detect changes in angular acceleration. There are three to detect changes in all three dimensions perpendicular to one another. This works as long cilia of sensory cells move against short cilia in the same direction as fluid moves, altering firing of neurons of short cilia.

4. Hearing is more acute in water because water transmits pressure waves better. This necessitated the evolution of mechanisms to amplify sound waves.

5. Sound waves vibrate the tympanic membrane; the hammer, anvil, and stirrup vibrate correspondingly; and the stirrup vibrates the oval window. The sense of hearing actually takes place in the basilar membrane of the cochlea. Recognition of individual frequencies is accomplished as the basilar membrane, which is differentially flexible, vibrates in a portion according to the frequency of the sound wave, stimulating hair cells in that region.

6. The cone cell has *cis*-retinal complexes with one of three kinds of proteins, making molecules capable of absorbing photons of three different wavelengths. The three primary colors are blue, green, and red. Vertebrates cannot see ultraviolet light because very short wavelengths are filtered out to prevent a chromatic aberration that occurs when the eye focuses short and long wavelengths at the

same time. Insects don't focus light.

7. The lens focuses light rays. Amphibians and fishes use muscles to move the lens position relative to the retina, whereas other vertebrates use muscles to alter the shape of the lens. The iris controls the amount of light entering the eye.

8. Rods and cones are not evenly distributed because most cones are at the central region of the retina, especially at the point of best focus, the fovea. A point-to-point image is formed at foveal cone cells. At this point each cone cell is connected to a single visual ganglion cell with an axon in the optic nerve; outside the fovea many receptor cells are connected to a single ganglion cell. This latter region provides information about movement and boundaries.

9. Two eyes provide images at slightly different angles, providing sensitive depth perception. Predators have eyes facing forward on the head, whereas prey have eyes located to the sides of the head. Predators have better stereoscopic vision, whereas prey have better vision over a broad receptive field.

10. Fish have developed a sense of electrical perception, but terrestrial animals have not because air does not readily conduct electricity but water does.

CHAPTER 47

1. A hormone is a regulating chemical made at one place in the body that exerts influence at another part. Only certain organs or tissues respond because only they have the proper receptors. The endocrine system is considered a chemical extension of the nervous system because the activity of the endocrine glands is under direct nervous system control.

2. The posterior pituitary gland secretes antidiuretic hormone (ADH), or vasopressin, which regulates water reabsorption in the kidneys and intestines, and oxytocin, which contracts muscles around the ducts of the mammary glands, initiating milk release, and causes uterine contractions during childbirth. These hormones are actually produced in the hypothalamus and are transported via nerve axons connecting the hypothalamus and the posterior pituitary gland.

3. Generally the hormones secreted by the hypothalamus are releasing hormones; they initiate the production of specific anterior pituitary hormones. They reach the anterior pituitary via the very short blood vessels that are between the hypothalamus and the anterior pituitary. The relationship is that each hypothalamic releasing hormone triggers the

production of a corresponding pituitary hormone.

4. The seven principal hormones produced by the anterior pituitary are thyroid-stimulating hormone (TSH), which stimulates the thyroid to produce thyroxine; luteinizing hormone (LH), which plays a role in the female menstrual cycle as well as stimulating male gonads to produce testosterone; follicle-stimulating hormone (FSH), which has a role in the female menstrual cycle and stimulates testicular cells to produce the hormone regulating sperm development; adrenocorticotropic hormone (ACTH), which stimulates the adrenal cortex to produce corticosteroid hormones and regulates glucose production from fat, regulates blood sodium and potassium balance, and contributes to male sex characteristics; somatotropin or growth hormone (GH), which stimulates the growth of bone and muscle; prolactin (PRL), which stimulates milk production; and melanocyte-stimulating hormone (MSH), which stimulates epidermal color changes in reptiles and amphibians.

5. Noradrenaline and adrenaline are produced by the adrenal medulla; they serve as neurotransmitters in the sympathetic nervous system, which is related to their effect as hormones because it is similar to nervous function but longer lasting. The changes in heartbeat and blood flow are the typical alarm response—accelerated heartbeat, increased blood pressure, and blood flow to the heart and lungs, and increased blood sugar. The hypothalamus does not regulate their production.

6. One class of peptide hormones is derived from tyrosine. The primary function of endorphins is to regulate emotional responses in the brain. Insulin is a protein. Peptide hormones interact with the receptor on the target cell surface, initiating a chain of events.

7. All steroid hormones are derived from cholesterol. Four examples of steroids are cortisone, testosterone, estrogen, and progesterone. Steroid hormones enter the nucleus of the target cell, alter gene expression, and thereby initiate or repress transcription.

8. Glucagon and adrenaline are produced when glucose levels drop, and they act within the liver to increase the conversion of glycogen to glucose. Insulin is produced when glucose levels become elevated, and it mediates the uptake of glucose by the liver cells, stimulating formation of glycogen in the liver.

9. Diabetes is a diverse group of diseases in which affected individuals develop elevated blood glucose levels. The ultimate hormonal de-

ficiency is lack of insulin. This causes an individual to be unable to turn glucose into glycogen to be stored; it is treated by maintaining a careful balance between glucose intake and the addition of insulin.

CHAPTER 48

1. ATP is used to counter the force of gravity. Splitting one mole of ATP releases 7.3 kcal. This energy is used to compress the length of structural elements within the cell, causing it to shorten.

2. Arthropod muscles are attached to a rigid chitinous exoskeleton. Chitin is brittle, so areas where muscles are attached are thicker than others, making larger arthropods require thicker exoskeletons which reduce mobility. Vertebrate muscles are attached to bone, a superior system because bone is rigid and flexible and can support more weight than chitin. The skeleton is internal, and the soft flexible exterior can stretch to accommodate movement.

3. Muscle cells contain more actin microfilaments than other cells. The name given to the microfilaments in muscle cells is myofilaments. The two proteins are actin and myosin; the three additional proteins are alpha-actinin, tropomyosin, and troponin. Those found exclusively in muscle are tropomyosin and troponin.

4. The resting myosin head attaches to the actin filament at the angle between the S-1 and S-2 units, which is 90°; a power stroke reduces the angle between S-1 and S-2, resulting in the advancement of myosin filament relative to the actin filament. The globular head detaches from the actin filament and the head returns to the previous 90° angle configuration and starts the cycle again. ATP is used at the power stroke step, when the angle between S-1 and S-2 is reduced.

5. Actin is anchored to alpha-actinin, so when the zones between anchor points shorten with filament sliding, movement occurs. Alpha-actinin is on the interior surfaces of plasma membrane.

6. Sarcoplasm is the name of the cytoplasm of striated muscle cells. Cells fuse, forming long fibers, and all original nuclei are retained and pushed to the periphery of the fiber. The subunits are sarcomeres. A sarcomere is made of interdigitating filaments of actin and myosin, with the actin connected to the front and back alpha-actinin Z lines and a space in the middle that is broached by the myosin filament.

7. Contraction is initiated by a nerve impulse. The chemical released is acetylcholine. It depolarizes the

muscle membrane which opens calcium channels in the muscle membrane.

8. The sarcoplasmic reticulum is the endoplasmic reticulum that wraps around each myofibril. Calcium is concentrated in the sarcoplasmic reticulum in resting muscle. When it is depolarized, the sarcoplasmic reticulum releases its concentration of calcium into the sarcoplasm.

9. The myosin heads are covered with tropomyosin, which also winds around actin at regular intervals. Troponin molecules are located along the tropomyosin. Troponin holds tropomyosin in position on the actin.

10. Calcium ions bind to troponin molecules, changing their shape. This repositions the tropomyosin trigger to a location where it does not interfere with myosin interaction.

CHAPTER 49

1. The two agents are hydrochloric acid and enzymes. The biomolecules are proteins converted to amino acids, starches converted to simple sugars, and fats converted into small lipid segments. The two separate stages are needed because most enzymes cannot tolerate highly acidic conditions.

2. Roundworms or nematodes have true extracellular digestion first. Their digestive cavity is a oneway tube with a separate mouth and anus. The general activities occurring in this system are digestion and absorption.

3. The general organization of the vertebrate digestive tract involves acid digestion in the stomach, enzyme digestion in the duodenum, and absorption in the small intestine. The specializations are as follows: fish possess an extensive pharynx, carnivorous adult amphibians possess a short gut, and herbivorous birds have a convoluted gut to prolong digestion and a gizzard to fragment the seeds and hard plant material.

4. Birds lack teeth. Carnivores have pointed teeth and lack flat grinding surfaces. Herbivores have large, flat teeth to pulverize the cellulose plant wall. Omnivores have mixed tooth surfaces to accommodate both kinds of food—carnivore-like teeth in the front and herbivore-like teeth in the back. The four types of teeth in adult humans are incisors for biting, canines for tearing food, and premolars and molars, both for grinding and crushing.

5. The three reflexes are (1) elevation of the palate, which seals off the nasal cavity; (2) pressure against the pharynx, which initiates swallowing; and (3) nervous signals as a result of swallowing center stimulation, which inhibits ventilation,

raises the larynx, and folds the epiglottis, forcing food into the esophagus and not the trachea. Food is prevented from reentering the mouth because of a muscular sphincter at the top of the esophagus. The two types of muscles are skeletal in the upper third of the esophagus and smooth in the lower two-thirds. The rhythmic contractions of these muscles are termed peristalsis.

6. Controlling the amount of acid enables proper neutralization later in the small intestine. This control is produced by the hormone gastrin, which is produced in the stomach; it regulates the production of HCl only when the pH of the stomach exceeds 1.5. If this system fails, duodenal ulcers result.

7. Secretory cells secrete enzymes which digest carbohydrates, proteins, and fats; and bicarbonate which neutralize stomach acid, which enables enzymes to operate at a neutral pH. The liver produces bile salts, cholesterol, and lecithin to solubilize fats. The gallbladder stores bile.

8. The duodenum is the portion of the small intestine actively involved in digestion. The remainder of this organ absorbs digestive products and water. Through a microscope this region appears to possess projections; these microvilli increase the absorptive surface. The three products of digestion are absorbed as follows: sugars and amino acids are absorbed via carrier systems into the blood; fatty acids diffuse into the lymphatic circulation.

9. Herbivorous vertebrates need bacteria to digest cellulose. Symbionts have protists instead of bacteria to do the same thing. Cows are physiologically better able to use plants as food because they digest cellulose in the rumen, at the front end of the digestive tract, and can digest the materials a second time; horses degrade cellulose in the cecum, at the end of the line where it isn't redigested. Human bacterial symbionts provide a source of vitamin K.

10. A vitamin is formally defined as an essential organic substance used in small amounts that must be consumed in the diet because the organism is incapable of synthesizing it. Vitamin C is synthesized by most vertebrates except for those listed. The essential amino acids are those which must be ingested. Amino acids are required in large amounts, whereas vitamins are only needed in trace amounts.

CHAPTER 50

1. Fick's Law of Diffusion is diffusion rate equals the diffusion constant of the respiratory membrane times the surface area of the respiratory organ times the difference in O_2 concentration of the environment versus the inside of the organ divided by the distance the O_2 travels. Animals have increased the surface area, decreased the distance, and increased the differences in concentration. We are as yet incapable of altering the diffusion constant of the membrane.

2. The simplest aquatic respiratory organ is the external gill, and the simplest terrestrial one is the insect trachea. The disadvantages associated with an external gill are that the water must constantly circulate over the gills, which requires lots of energy to move the highly branched organs through the water. Gills associated with branchial chambers are on the inside of the body, and the various structures formed to move water over the gill surface meet with less resistance.

3. In the countercurrent flow, blood circulates in the direction opposite the flow of the water across the gills. In terms of Δp, at the front of the gill the most oxygenated water meets the most oxygenated blood, optimizing diffusion, and at the back of the gill, the least oxygenated water meets the least oxygenated blood, again optimizing diffusion, so the whole gill extracts O_2 from the water. If blood and water flowed in the same direction, the difference in O_2 concentration between blood and water would initially be high, resulting in significant diffusion, but as the blood becomes more oxygenated, water becomes less oxygenated and diffusion can no longer occur; thus only a short portion of the gill would work well.

4. The two major kinds of respiratory systems in terrestrial animals are the insect trachea and lungs. In trachea the external openings close whenever the body CO_2 levels fall below a certain level; in lungs air flows in and out through the same passage, reducing both O_2 diffusion and water loss.

5. Mammals have a higher body metabolism since they use metabolic heat to maintain a constant body temperature. This greater efficiency is achieved by their greater surface area and more numerous alveoli. The ratio of lung mass to body mass doesn't change; the alveoli get smaller but more numerous to increase the surface area, and the epithelial layer separating the alveoli and blood gets thinner, decreasing diffusion distance.

6. Flying mammals glide, whereas birds more actively beat their wings. In both birds and fishes air flow is one way. In terms of asso-ciated blood flow there is a countercurrent flow of blood versus water or air.

7. Air flow in the human lung is as follows: air enters the nostrils, goes into the nasal cavity, and then goes through the back of the mouth to the larynx and down the trachea. Next it goes into one of the branching bronchi, into the bronchioles, and into the alveoli. This is a one-cycle system in that air completes one cycle with only a single inhalation and exhalation. The lungs are connected at the junction of the lung and the bronchus and supported by the water tension of the interpleural fluid between the two pleural membranes.

8. The carrier molecule is hemoglobin, and it is based on iron. It is contained and synthesized in erythrocytes or red blood cells. Most invertebrates have hemocyanin as their carrier molecule, which is based on copper ion and not contained in cells, but rather is free in the blood fluid. Hemoglobin is red and hemocyanin is blue.

9. The oxygen-hemoglobin dissociation curve indicates the amount of O_2 that combines with hemoglobin plotted against the partial pressure of O_2. The Bohr effect occurs when a high CO_2 concentration lowers blood pH, causing hemoglobin to dissociate from O_2 at a greater rate than normal. Metabolically this means that blood unloads more O_2 where it is especially needed. Carbon monoxide binds to hemoglobin 300 times more easily than oxygen and dissociates much less readily. Medically, a small amount can irreversibly prevent O_2 from getting to cells.

10. Blood unloads its oxygen via the Bohr effect at capillaries in tissues undergoing metabolism because O_2 concentration is then greater in the plasma than in the tissue, so the O_2 diffuses inward. One-fifth of the carbon dioxide is carried bound to hemoglobin, and the rest diffuses into the cytoplasm of red blood cells. Increased absorption of CO_2 is achieved in red blood cells as CO_2 is catalyzed by carbonic anhydrase to dissociate into bicarbonate and H^+ ions. CO_2 is unloaded at the lungs because lower CO_2 levels cause a reverse reaction to occur; at low CO_2 levels hemoglobin has a greater affinity for O_2 than CO_2, causing the cells to give up the CO_2 for new O_2.

CHAPTER 51

1. The three elements of the closed circulatory system are heart, blood vessels, and blood. The plumbing elements are referred to as the cardiovascular system. The flow of blood is as follows: blood leaves the heart by way of the arteries, then goes through the arterioles to the capillaries; it begins to return by way of the venules, then the veins, and finally back into the heart. Gas and metabolite exchange occurs in the capillaries.

2. Flow resistance is inversely proportional to the fourth power of the tube radius. Therefore small vessels exhibit greater flow resistance. The greatest amount of resistance is at the capillaries. The walls must be elastic because the vessels out of the heart must be able to expand to accommodate the blood volume until resistance allows the blood to flow through the capillaries.

3. A capillary is a tube one cell layer thick. The diameter of a capillary is just slightly larger than a red blood cell. The advantage of this is that it facilitates diffusion of gases and metabolites across the red blood cells and capillary to the surrounding tissues. The maximal distance is 100 micrometers. The precapillary sphincter closes off the flow to that capillary.

4. Veins have one-way valves to prevent backflow, and the muscle and elastic fiber layers are thinner. By the time the blood reaches the veins, the force of the heartbeat has been attenuated by the high resistance and great area of the capillary network. When empty, an artery is a semirigid, hollow tube; a vein is like a collapsed balloon. Veins have larger diameters to minimize the flow resistance in vessels exiting organs and thereby prevent backflow of blood into the organ as well as to minimize resistance of blood flow on its way back to the heart.

5. The lymphatic system gathers liquid from body tissues as it seeps out at the capillaries and returns the fluid to the cardiovascular system. It is an open system. The flow of fluid is driven by the pressure created when vessels are squeezed by the movement of body muscles. The direction of flow is one-way into the venous blood flow.

6. The three components of blood plasma are metabolites and wastes; salts and ions; and proteins. About 90% of blood is water. It is not lost by osmosis because the blood proteins, particularly albumin, act as an osmotic counterforce. The technical names are erythrocytes, leukocytes, and platelets. Hematocrit is that fraction of the total volume of blood that is occupied by red blood cells.

7. The fish is the vertebrate with the first chambered heart. The sequence of chambers in this type of heart are sinus venosus, atrium, ventricle, and conus arteriosus. The sinus venosus and atrium are

collecting chambers, and the ventricle and conus arteriosus are pumping chambers. The major advantage of this heart is that blood is pumped directly to the gills, thus all blood delivered to the body is fully oxygenated. The major disadvantage of this heart is that the flow of blood through the gill capillary bed reduces its force, so that the flow through the body is generally sluggish.

8. The sinus venosus serves as a collection chamber and a pacemaker in primitive vertebrates. The excitatory tissue of the SA node in the wall of the right atrium is retained in higher vertebrates and serves a pacemaker function. It is the point of origin for each heartbeat. The size of the heart relative to total body mass is a constant value within a class. This value compares as follows: fish, smallest ratio, then amphibian, reptile, mammal, and finally the largest ratio in the bird.

9. Blood flow is as follows: Superior body—vena cava—right atrium—right ventricle—pulmonary artery—lungs—pulmonary veins—left atrium—left ventricle—aorta—body. The stronger contraction is in the ventricle because it is more muscular. The various valves function to prevent backflow of blood during strong contractions.

10. Muscular contractions are initiated by the periodic, spontaneous contraction of the SA node, then a wave of depolarization passes over the left and right atria and on to the atrioventricular node, and finally to the left and right ventricles through the bundle of His. The delay results because the ventricle is separated from the atria by connective tissue, which cannot propagate a depolarization wave.

CHAPTER 52

1. The T cell types are as follows: (1) helper—initiate the immune response; (2) cytotoxic—lyse virus-infected cells; (3) inducer—oversee the development of T cells; and (4) suppressor—terminate the immune response.

2. Vertebrates recognize foreign tissue because all cells are marked with self-marker protein and amoeboid phagocytes engulf any cells that are without the label, i.e., invaders. The disadvantage of this is that phagocytes will not engulf copycat invading cells with a protein marker that resembles the self-marker.

3. The vertebrate negative defense system is superior because self-marker proteins are unique to each individual (called a major histocompatibility complex [MHC] protein). The positive defense system identifies invaders by

their non-self molecules, which match cell surface receptors of the white blood cells, and are marked for destruction.

4. (1) B cells—not-self B receptor is designed by somatic rearrangement to bind to specific antigens and signal the cell to produce large amounts or receptor (antibody) into the blood. (2) T cells—not-self T receptor is designed by somatic rearrangement to bind a specific antigen. Helper and inducer cells have a CD4 marker; cytotoxic and suppressor cells have CD8 markers. (3) Macrophages—there are two types of MHC self-marker proteins; MHC-I is also present on all nucleated cells and MHC-II is present only on macrophages, B cells, and T_4 (helper and inducer) T cells.

5. The primary events of the cell-mediated immune response are as follows: proliferation—lymphokines cause stimulated T cells to multiply, increasing the number of cells capable of recognizing antigens; activation—macrophage migration inhibition factor attracts the macrophage to the site of infection; induction—triggers the maturation of immature thymus lymphocytes into T cells; attack—cytotoxic cells bind to and disrupt cell membrane of the virus-infected cell, lysing it; suppression—suppressor cells multiply more slowly than cytotoxic cells but eventually outnumber them and shut down their activity; and memory—generation of a population of memory T cells that can later provide an accelerated response to a later infection of the same antigen.

6. The primary events of the humoral immune response are as follows: proliferation—B receptors bind to antigens recognized by helper T cells, which further bind antigen and MHC-II on B cells, releasing lymphokines and causing B cells to multiply; differentiation and secretion—B plasma cells secrete M (primary) and G (secondary) antibodies; attack—M antibodies activate complement, which bursts antibody-coated cells, macrophages ingest G-coated cells, and killer cells attack antibody-coated cells; suppression—response shut down by suppressor T cells; and memory—B cells that do not become plasma cells circulate as memory B cells to provide an accelerated response.

7. The basic structure of an antibody molecule is four polypeptide chains: two identical, short light chains, and two identical, long heavy chains held together by disulfide bonds, that form a Y-shape. Specificity resides in the terminal half of each arm of the

Y. A T receptor resembles one arm of the antibody molecule, with alpha and beta chains, with both constant and variable regions.

8. The great variety comes from the millions of different kinds of stem cells, so that each antigen stimulates a few with the proper receptors to proliferate. The process is referred to a somatic rearrangement. It proceeds as follows: immune receptor genes do not exist as unique entities but are composites assembled by stitching three or four DNA segments together, with each segment chosen at random from a cluster of similar sequences.

9. Embryonic immune tissue responds to both as non-self until it learns to recognize self. Acquired immunological tolerance refers to the process in which foreign tissue introduced in the embryo is not recognized as non-self and grafts of that tissue are then not rejected by the adult. Natural immunological tolerance refers to the lack of an immunological response against one's own tissues. The immunological basis for these processes is the elimination or suppression of those particular clones of lymphocytes.

10. AIDS interferes with the immune response by three means: (1) AIDS-infected cells release progeny viruses that infect other T_4 cells, killing them and releasing more viruses; (2) AIDS virus causes T_4 cells to secrete a factor that blocks other T cell responses to the viral antigen; and (3) AIDS virus blocks transcription of the MHC genes, hindering the recognition and destruction of infected cells. Usually death results from other infections or cancers, not directly from AIDS.

CHAPTER 53

1. A freshwater vertebrate has a higher concentration of salt than the environment (is hyperosmotic), and therefore water tends to enter its body. To maintain proper body water levels, it must excrete water to prevent self-dilution.

2. A marine vertebrate has a lower salt concentration than its environment by about 1/3 (is hypoosmotic), so water tends to exit its body. It retains water to prevent dehydration. Terrestrial vertebrates face the added complication of water loss due to evaporation into the air.

3. The insect system uses an osmotic gradient to pull blood through the filter, and the vertebrate system has a pressure-driven filtration system to push blood through the filter. Selective reabsorption is the ability to reabsorb small molecules without absorbing waste mole-

cules. The flexibility of this ability is that the body can reabsorb various molecules depending on the membrane; the kidney can work under all conditions.

4. The kidney first evolved in freshwater fishes. (1) filtration—water and small molecules pass through and cells and proteins remain; (2) reabsorption—desirable ions and metabolites are recaptured from the filtrate; and (3) excretion—wastes, excess ions, drugs, and foreign organic materials are secreted into the tubules and excreted from the body.

5. A nephron is the filtration-reabsorption device in the kidney. Bowman's capsule is located at the front of the nephron. The glomerulus is a fine network of capillaries located within Bowman's capsule. Filtration occurs as blood pressure forces fluid through the capillary walls, retaining proteins, cells, and large molecules and allowing water and small molecules to pass through to the interior of the capsule.

6. The glomerular filtrate is fluid that enters the nephron. The tubules of the nephron are associated with reabsorption. The actions of each of the five divisions of the nephron are as follows: (1) beating cilia speed flow of the filtrate and increase the amount of fluid that passes through the kidney; (2) proximal segments first actively absorb glucose and small proteins, then, second, actively reabsorb Mg^{++}, Ca^{++}, and SO_4; (3) intermediate segment is a secondary pump with beating cilia; (4) distal segment has channels that reabsorb Na^+, K^+, and Cl^-; and (5) ends of nephrons connect, forming ducts that conduct remaining filtrate to the urinary bladder.

7. Birds and mammals are much more efficient because they increase the local salt concentration in the tissue surrounding the nephron, using the concentration gradient to draw additional water out of the tube. The anatomical basis for this is the fact that the nephron is bent in a hairpin turn.

8. In the loop of Henle there is a countercurrent flow that increases the efficiency of water reabsorption, as follows: (1) filtrate passes into the descending arm of the loop, a tubule permeable only to water, which passes outward; (2) at the turn of the loop and the beginning of the ascending arm, the tubule becomes permeable to salt and less permeable to water, so that salt passes by diffusion into the surrounding tissues, with the highest concentration at the hairpin turn; (3) the higher concentration in the ascending loop causes the tubule to actively pump salt out of itself and into tissue,

causing water to also diffuse out here and through the distal tubule; the filtrate here is concentrated urea; (4) the tubule empties into the surrounding tissue and is the driving force behind water diffusing out of the descending tubule; and (5) the filtrate in the collecting duct becomes more concentrated as water diffuses out driven by the osmotic concentration of both salt from the ascending arm and urea from the collecting tubule.

9. The blood vessel distal to the glomerulus is in close proximity to the loop of Henle, so water is reabsorbed by blood vessels that are permeable to water but impermeable to urea.

10. The hypothalamus possesses osmoreceptors to detect any decrease in solute concentration. Antidiuretic hormone (ADH) regulates kidney function. It makes collecting ducts permeable to urea, which further allows for the reabsorption of water. Decreasing the levels of ADH means that ducts are not permeable to urea, which lowers the osmotic concentration in the medulla so that less water diffuses outward and is reabsorbed, and urine volume is increased.

CHAPTER 54

1. The sexual reproduction most common in marine organisms involves releasing both eggs and sperm into water where fertilization can occur. The major problem is that the gametes are diluted by water and union is not ensured without the controls associated with the timing of gamete release. The terrestrial environment was hostile to gametes and zygotes, neither of which was resistant to drying out.

2. Reptilian and avian fertilization is accomplished internally. Most male reptiles have penises, whereas few male birds do; all birds are oviparous, whereas reptiles may be ovoviviparous or viviparous, and the bird egg shell is thicker than that of the reptiles. Newly hatched reptiles generally are fully developed and don't require parental care; most birds are hatched relatively underdeveloped and do require parenting.

3. Monotremes are oviparous. Marsupials have embryos born very underdeveloped and continue development in the pouch; in placentals all development takes place in the mother's body and internal nourishment is provided by means of a network of blood vessels called the placenta.

4. The two phases are (1) follicular (hypothalamus releases GRH, which causes the pituitary to release FSH, which causes the eggs to mature and triggers estrogen

production in the ovary, which feeds back to the hypothalamus to stop FSH production; estrogen also causes the uterine lining to proliferate) and (2) luteal (pituitary responds to estrogen and produces LH that acts at the ovary to inhibit estrogen production and cause mature follicle to burst and release the egg into the fallopian tubes; follicle becomes the corpus luteum and secretes progesterone, which stimulates the further development of the uterine lining. Ovulation marks the beginning of the luteal phase.)

5. Lack of fertilization after ovulation causes the cessation of progesterone production and ends an individual cycle. A new one starts when there is a lack of estrogen and progesterone. Menstruation results from the thickening of the uterus in preparation for the embryo implantation, which must be sloughed off when there is no implantation. Hormonally this layer is lost by a process initiated by a decrease in progesterone levels.

6. Sperm travel through the epididymis, where they become motile. They are stored there and in the vas deferens. The major structures of the penis are two cylinders of spongy tissue with an artery running through each and a third spongy cylinder with the urethra running through it. The fluids that pass through the urethra are semen containing sperm and urine.

7. The three stages of oogenesis are as follows: (1) developmental arrest—primary oocytes are suspended in prophase I of meiosis; (2) ovulation—hormonal changes stimulate the completion of meiosis I, resulting in secondary oocyte and the first polar body; and (3) fertilization—initiates the completion of meiosis, producing ovum and the second polar body.

8. The ovulated secondary oocyte proceeds into the fallopian tube and travels toward the uterus by peristaltic contractions of the fallopian tubes. The egg and sperm must meet in the fallopian tube because it takes 3 days to travel down the fallopian tube and an unfertilized egg only lives 1 day.

9. The four periods are excitement, plateau, orgasm, and resolution. In both female and male during excitation there is increased heartbeat, blood pressure, respiration, and vasodilation, as well as increased sensitivity in the nipples and genital tissues. The male experiences vasocongestion in the penis, resulting in erection. The female experiences increased circulation with swelling in the labia and clitoris, and the vaginal walls become moist and its muscles relax.

10. Sperm-blocking methods include

condoms, cervical cap, and diaphragm. Failure of these methods is attributable to improper use/application. Destroying deposited sperm is the focus of spermicidal jellies, foams, and sponges. Birth control pills prevent the maturation of the egg by providing hormonal signs that ovulation has already taken place.

CHAPTER 55

1. The three stages of fertilization are penetration, activation, and fusion. The egg cell is protected by the zona pellucida and protective layer of follicle cells. The acrosome releases enzymes that cause fusion of the sperm and egg plasma membranes.

2. The cleavage patterns are as follows: primitive—holoblastic, little yolk, symmetrical blastula, cells all the same size; fish/amphibians—holoblastic, lots of yolk, asymmetrical cleavage, large yolk-containing cells at one end, small cells with less yolk at the other; reptiles/birds—meroblastic, egg mostly yolk, cleavage occurs only through blastodisc, blastoderm not spherical, but hollow cap atop yolk; mammals—holoblastic, little yolk, blastula mass concentrated at one end, analogous to blastodisc, remaining region is trophoblast, which forms placental membranes.

3. The blastula is two distinct hemispheres, an animal pole and a yolk-rich vegetal pole. Gastrulation causes the animal hemisphere to bulge inward, forming a cup-shaped embryo. The hollow crater resulting from the invagination is the archenteron. The opening is the blastopore. This develops into the anus.

4. Gastrulation in amphibians differs in that it is not mechanically possible to invaginate at the vegetal pole due to the few, large, yolk-laden cells. This changes the cell migration pattern so that the layer of animal pole cell folds over the yolk cells and invaginates inward. The dorsal lip of the gastrula is the region at which animal cells invaginate. The yolk plug is a region of the blastopore that is filled with yolk-rich cells. The mesoderm forms from cells of the dorsal lip that migrate between the ectoderm and endoderm.

5. Only chordates have neurulation. The two unique structures are the notochord and the hollow dorsal nerve cord. They arise from mesoderm and ectoderm, respectively. The basic events of this process are as follows: a layer of cells above the notochord invaginates inward, forming the neural groove; edges of the groove move toward each other and fuse, forming the neural tube; the blocks of mesodermal somites develop,

eventually to become muscles, vertebrae, and connective tissue; then glandular tissues develop, and the hollow tube within the ectoderm forms the coelom, which becomes the gut.

6. Only vertebrates experience neural crest formation. The primary structure of this process forms from the edges of the neural groove prior to neural tube formation. This stage is special because cells of this crest migrate throughout the body, forming structures that are characteristic of vertebrates. This tissue develops into forms that merge with the forebrain, becoming gill arches, Schwann cells, the adrenal medulla, sense organs, and the skull. This is tied to the evolution of the vertebrate animal because it allowed for the evolution of fast-swimming predators, with increased activity, who were better able to detect and spatially orient to their prey.

7. Induction is the determination of the course of development of one tissue by another tissue. Primary induction involves inductions between endoderm, mesoderm, and ectoderm and includes interactions between the dorsal ectoderm and mesoderm to form the neural tube. Secondary induction involves inductions between already differentiated tissues and includes differentiation of the lens from ectoderm by the interaction with the CNS tissue. The details of the chemical nature of induction are unknown, but a porous filter permits induction, a nonporous filter inhibits it, and inducer cells produce a factor that stimulates other cells to divide, initiating changes in gene expression.

8. During the first month of pregnancy, cleavage occurs while the cells are in the oviduct, the blastodisc is formed within the trophoblast, and there is implantation after 6 days, which initiates growth of the developmental membranes. The amnion encloses the developing embryo; the chorion interacts with the uterus to form the placenta. Gastrulation occurs in the second week; neurulation in the third week; and organogenesis in the fourth week. This entire period is critical because the stage is set for the entire development of the organism, chemicals can have a great effect of development and can upset organogenesis, and this is the period when spontaneous abortions can be caused by chemicals or other events.

9. During the second trimester there is continued growth and development, bone enlargement, the development of body hair, and the development of a heartbeat. Lanugo is body hair, a relict of evolution. Survival would not be possi-

ble without special medical intervention.

10. During the third trimester primarily growth occurs, with most development completed. The neurological growth includes formation of the major nerve tracts and new brain cells. Neurological growth is not complete at birth; however, remaining in utero longer to further neurological development would also result in an increase in bulk and the baby would grow too large to pass through the birth canal. Maternal nourishment is vital because the fetus needs nourishment to grow properly; malnourished fetuses are likely to be retarded in both growth and neurological development.

CHAPTER 56

1. Instincts are highly programmed behaviors. Learning is defined as the creation of long-term changes in behavior due to experience. Instincts rely relatively little on learning.

2. Evolution biases learning in an adaptive way, through learning preparedness. Some stimuli are learned more rapidly than others, and some associations are made more quickly.

3. Animal behavior rhythms are controlled by exogenous timers and endogenous rhythms. The stronger control in behaviors keyed to a daily cycle is endogenous rhythms. These behaviors are maintained by exposure to an environmental cue such as light, which is an exogenous timer. Circadian rhythms are endogenous rhythms of about 24-hour cycles even in the absence of external cues.

4. The nature/nurture controversy involved whether behavior was instinctive or learned. Marler's work on bird song development showed that both instinct and learning make important contributions in shaping behavior.

5. Studies of hybrids show that the behavior of a hybrid offspring is

intermediary between that of the parents. Behavioral traits like learning ability can also be selected from the laboratory. Studies of mutants are also useful in identifying the genetics of behavior.

6. Ethology is the naturalistic study of behavior. Ethologists believed key stimuli in the environment trigger a fixed action pattern, which was controlled by the innate releasing mechanism. Neuroethologists have provided much support for this mechanistic view.

7. Filial imprinting involves the formation of parent/offspring bonds. Sexual imprinting concerns how early experience affects an animal's ability to identify mates of its own species. A moving box is just as good an imprinting stimulus as a real parent because the first object seen by a young animal is, in fact, its parent.

8. Hormones coordinate reproductive behavior by serving on the internal chemical messengers that signal when an animal should reproduce. Environmental stimuli, like photoperiod and temperature are perceived by the sensory system. For example, the pineal gland perceives changes in daylength an secretes melatonin, which triggers hormonal release by the hypothalamus.

9. A taxis is movement toward or away from a stimulus; kineses are changes in activity levels due to a stimulus. These differ from migration in that migrations are long-range movements. Migration patterns are genetically determined. Patterns do not change, but new information is added to the old. Migrating birds use the sun, the stars, and the detection of the earth's magnetic field to orient themselves on migration.

10. Communication signals are species-specific, thus limiting sexual communication to members of one species. Firefly flash patterns, bird song, and some insect sex pheromones are species-specific.

1. Behavioral ecology is the study of the adaptive significance of behavior. Behavioral ecologists measure the adaptiveness of behavior by estimating fitness (reproductive success). Fitness can be estimated from the effect behavior has on survival, or some measure of reproduction such as the number of matings an individual has, the benefits and costs are of the behavior, or the number of offspring produced.

2. Animals may be sit-and-wait predators that ambush prey, or they may actively search for food. A generalist has a broad range of food types in its diet, a specialist only one or a few. Natural selection favors foraging behavior that maximize the nt energy gained from feeding on a given food item.

3. A territory is an area which is used exclusively by one animal or social group and is actively defended against intrusion by others. Animals are territorial when the benefits of exclusive use of resources in a territory outweigh the costs of defense and risk of injury or predation.

4. Parental investment is any energy expenditure which increases the fitness of offspring. Female reproductive costs are usually greater than those of males because of the cost of producing eggs and rearing offspring.

5. Mate choice can be adaptive if individuals in a breeding population differ in the quality of their genes or resources they hold. The advantages of mate choice include selecting parasite-free individuals (a trait which may be heritable) or acquiring food or nest sites needed for reproduction.

6. The advantages of group living include enhanced predator defense and improved food location. The disadvantages are increased incidence of disease or parasitism. Group size may be optimized when any benefit of group living

is maximized relative to a constraint, or cost, or social life.

7. Monogamy is a mating system in which a male and female form a pair bond during a breeding season. Polygyny occurs when a male mates with more than one female, and polyandry occurs when a female mates with more than one male. The polygyny threshold is the difference in territory quality that would give a female greater fitness by mating with a mate male in a higher quality territory than mating with an unmated male in a lower quality territory.

8. Group selection involves the survival or extinction of whole populations, or species. Applied to the evolution of altruism, group selection theory states that self-sacrificing behavior evolves because it benefits the whole group or species. This argument is incorrect because behavior favoring altruism cannot evolve if individuals showing that behavior never have offspring.

9. Sexual selection is the process by which individuals compete for access to mates; it involves competition among members of one sex for mates, and mate choice. Natural selection concerns the evolution of traits favoring survival, sexual selection concerns the evolution of traits that favor mating success.

10. Altruistic behavior is that which is not in the individual's self-interest. Inclusive fitness refers to the total number of one's genes that are passed on to the next generation. Kin selection maximizes one's inclusive fitness if by sacrificing one's self for one's relatives his total gene contribution to the next generation increases. Kin selection can lead to altruistic behavior when the individuals in a society are related to one another. Reciprocal altruism is when individuals perform an act of altruism in expectation that they will receive similar treatment from others.

Appendix C

Answers to Genetics Problems

CHAPTER 12 MENDELIAN GENETICS PROBLEMS

1. He only chose to study two pure-breeding varieties of any given trait; he could have chosen to study more. In any given cross, the maximum number of alleles that he could have observed is four, if the two parents were each heterozygous for a different pair of alleles.

2. Alleles segregate in meiosis, and the products of that segregation are contained *within* a pod. Each pea is a gamete. In this diagram, the segregation is incorrectly shown as being *between* pods, each pod shown as uniformly *wrinkled* or *round*.

3. The probability of getting two genes on the same chromosome is $\frac{1}{2}^{23}$.

4. Somewhere in your herd you have cows and bulls that are not homozygous for the dominant gene "polled." Since you have many cows and probably only one or some small number of bulls, it would make sense to concentrate on the bulls. If you have only homozygous "polled" bulls, you could never produce a horned offspring regardless of the genotype of the mother. The most expedient thing to do would be or keep track of the matings and the phenotype of the offsprings resulting from these matings and render ineffective any bull found to produce horned offspring.

5. It would not be possible on the basis of the information presented to substantiate a claim of infidelity. You do not know if the woolly trait is the result of a single gene product, or even if the trait is dominant or recessive. Assuming for the moment that it was the effect of a single dominant allele *W*, the man could still be a heterozygote for the gene and, when mated to a recessive homozygous female, would expect to produce woolly headed offspring only one-half the time.

6. ½

7. Albinism, *a*, is a recessive gene. If heterozygotes mated you would have the following:

	A	a
A	AA	Aa
a	Aa	aa

Clearly one-fourth would be expected to be albinos.

8. The best thing to do would be to mate Dingleberry to several dames homozygous for the recessive gene that causes the brittle bones. Half of the offspring would be expected to have brittle bones if Dingleberry were a heterozygous carrier of the disease gene. Although you could never be 100% certain Dingleberry was not a carrier, you could reduce the probability to a reasonable level.

9. Your mating of *DDWw* and *Ddww* individuals would look like the following:

	Dw	Dw	dw	dw
DW	DDWw	DDWw	DdWw	DdWw
Dw	DDww	DDww	Ddww	Ddww
DW	DDWw	DDWw	DdWw	DdWw
Dw	DDww	DDww	Ddww	Ddww

Long-wing, red-eyed individuals would result from eight of the possible 16 combinations, and dumpy, white-eyed individuals would never be produced.

10. Breed Oscar to Heidi. If half of the offspring are white eyed, then Oscar is a heterozygote.

11. Both parents carry at least one of the recessive genes. Since it is recessive, the trait is not manifest until they produce an offspring who is homozygous.

12. To solve this problem, let's first look at the second cross, where the individuals were crossed with the homozygous recessive sepia flies *se/se*. In one case, all the flies were red eyed:

		Unknown genotype	
		Se	Se
Sepia	Se	Se/se	Se/se
	se	Se/se	Se/se

The only way to have all red-eyed flies when bred to homozygous sepia flies is to mate the sepia fly with a homozygous red-eyed fly. In the other case, half of the offspring were black eyed and the other half red eyed:

		Unknown genotype	
		Se	se
Sepia	se	Se/se	se/se
	se	Se/se	se/se

The unknown genotype in this case must have been *Se/se*, since this is the only mating that will produce the proper ratio of sepia-eyed flies to red-eyed flies. Since the ratio of this unknown genotype and the one previously determined was 1:1, we must deduce the genotype of the original flies, which when mated, will produce a 1:1 ratio of *Se/se* to *Se/Se* flies.

		Unknown original 1	
		Se	Se
Original 2	Se	Se/Se	Se/Se
	se	Se/se	Se/se

You can see from this diagram that if one of the original flies was homozygous for red eyes and the other was a heterozygous individual, the proper ratio of heterozygous and homozygous offspring would be obtained.

13. a. It could have originated as a mutation in his germ cell line. b. Since their son Alex was a hemophiliac, the disease almost certainly originated with Alexandra, Nicholas II would have contributed only a silent Y chromosome to Alex's genome. There is 50% chance that Anastasia was a carrier.

CHAPTER 13 HUMAN GENETIC PROBLEMS

1. b—autosomal dominant.
2. Sex-linked recessive.
3. Dominant.
4.

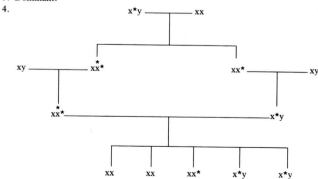

5. No. $I^A I^O \times I^B I^O \rightarrow I^O I^O$

6.

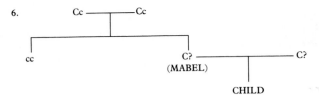

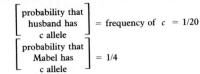

c is a rare allele, and Mabel's child can develop cystic fibrosis only if her husband also carries the allele:

$$\left[\begin{array}{c}\text{probability that}\\ \text{husband has}\\ c\text{ allele}\end{array}\right] = \text{frequency of } c = 1/20$$

$$\left[\begin{array}{c}\text{probability that}\\ \text{Mabel has}\\ c\text{ allele}\end{array}\right] = 1/4$$

7. 45(44 autosomes + one X).

8. AO

9. Dominant

10. Dominant

11.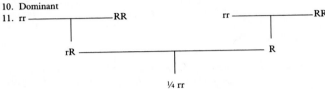

The chances are 25%.

12. Let a = albino. The genotype of the father is Aa.

13.

parents	baby
O and O	O
A and B	AB
B and B	B
AB and O	A

14. It would be very difficult to establish negligence, especially if the child is a boy. If the child's maternal grandmother was a carrier, there's a 50:50 chance that the mother is a carrier. If the mother is a carrier, there's a 50:50 chance that a boy will express the disease. One would be suspicious about the causes of *Duchenne muscular dystrophy* in a girl. In the first place, symptomatic males may not live to reproductive age, and it would be rare for a symptomatic male to be virile. A symptomatic girl from a nonsymptomatic father could result from a genetic anomaly in the father, or from a mutation unrelated to radiation. The case for mutation on the job is clearly stronger for the girl, but very weak in either case.

Glossary

abdomenal cavity The posterior portion of the body. (p. 885)

abortion The removal or induced elimination of the embryo from the uterus before the third trimester of pregnancy. (p. 1151)

abscisic acid (ABA) (L. *ab*, away, off + *scissio*, dividing) A plant hormone with a variety of inhibitory effects; brings about dormancy in buds, maintains dormancy in seeds, effects stomatal closing, and is involved in the geotropism of roots; also known as the "stress hormone." (p. 767)

abscission (L. *ab*, away, off + *scissio*, dividing) In vascular plants, the dropping of leaves, flowers, fruits, or stems at the end of a growing season, as the result of formation of a layer of specialized cells (the abscission zone) and the action of a hormone (ethylene). (p. 765)

absorption (L. *absorbere*, to swallow down) The movement of water and of substances dissolved in water into a cell, tissue, or organism. (p. 732)

accessory pigment In plants, a secondary pigment that captures light energy and transfers it to chlorophyll *a*. (p. 182)

acetylcholine The most important of the numerous chemical neurotransmitters responsible for the passing of nerve impulses across synaptic junctions; the neurotransmitter in neuromuscular (nerve-muscle) junctions. (p. 912)

acid A proton donor; a substance that dissociates in water, releasing hydrogen ions (H^+) and so causing a relative increase in the concentration of H^+ ions; having a pH in solution of less than 7; the opposite of "base." (p. 35)

acoelomate (Gr. *a*, not + *koiloma*, cavity) Without a coelom, as in flatworms and proboscis worms. (p. 792)

acquired immune deficiency syndrome (AIDS) An infectious and usually fatal human disease caused by a retrovirus, HIV (human immunodeficiency vi-

rus), which attacks T cells. The virus multiplies within and kills individual T cells, releasing thousands of progeny that infect and kill other T cells, until no T cells remain, leaving the affected individual helpless in the face of microbial infections because his or her immune system is now incapable of marshaling a defense against them. *See* T cell. (p. 1147)

actin (Gr. *actis*, a ray) One of the two major proteins that make up vertebrate muscle (the other is myosin). (p. 99)

action potential A transient, all-or-none reversal of the electric potential across a membrane; its strength depends only on characteristics of the membrane (in nerves, the diameter) and is independent of the strength of the stimulus that triggers it; in neurons, an action potential initiates transmission of a nerve impulse. (p. 909)

activation energy The energy that must be possessed by a molecule in order for it to undergo a specific chemical reaction. (p. 140)

active site The region of an enzyme surface to which a specific set of substrates binds, lowering the activation energy required for a particular chemical reaction and so facilitating it. (p. 142)

active transport The pumping of individual ions or other molecules across a cellular membrane from a region of lower concentration to one of higher concentration (that is, against a concentration gradient); because the ion or molecule is moved in a direction other than the one in which simple diffusion would take it, this transport process requires energy, which is typically supplied by the expenditure of ATP. (p. 122)

adaptation (L. *adaptare*, to fit) A peculiarity of structure, physiology, or behavior that promotes the likelihood of an organism's survival and repro-

duction in a particular environment. (p. 371)

adaptive radiation The evolution of several divergent forms from a primitive and unspecialized ancestor. (p. 416)

adaptive significance The extent to which a behavior may increase survival and rate of reproduction. (p. 1199)

adenine (Gr. *aden*, gland) A purine base; a nitrogen-containing ring compound present in nucleic acids (DNA and RNA), in adenosine phosphates (ADP and ATP), and in nucleotide coenzymes. (p. 57)

adenosine diphosphate (ADP) A nucleotide consisting of adenine, ribose sugar, and two phosphate groups; formed by the removal of one phosphate from an ATP molecule; the formation of ADP from ATP releases usable metabolic energy. (p. 147)

adenosine monophosphate (AMP) A nucleotide consisting of adenine, ribose sugar, and one phosphate group; can be formed by the removal of two phosphate groups from an ATP molecule, a reaction that releases energy and is often found at irreversible steps of metabolic pathways. The formation of AMP from ATP often renders a reaction irreversible because the two-phosphate ("pyrophosphate") reaction product is readily cleaved into individual phosphate units that are not easily rejoined. (p. 69)

adenosine triphosphate (ATP) A nucleotide consisting of adenine, ribose sugar, and three phosphate groups. ATP is the energy currency of cell metabolism in all organisms. ATP is formed from ADP + P in an enzymatic reaction that traps chemical energy released by metabolic processes or light energy captured in photosynthesis. Upon hydrolysis, ATP loses one phosphate and one hydrogen to become adenosine disphophate (ADP), releasing energy in the process. (p. 124)

adrenaline (L. *ad*, to + *renalis*, pertaining to kidneys) A hormone produced by the medulla of the adrenal gland. Adrenaline is responsible for the physiological changes associated with the alarm response; it increases the concentration of sugar in the blood, raises blood pressure and heart rate, and increases muscular power and resistance to fatigue; also a neurotransmitter across synaptic junctions. Sometimes called "epinephrine." (p. 936)

adventitious (L. *adventicius*, not properly belonging to) Referring to a structure arising from an unusual place, such as stems from roots or roots from stems. (p. 721)

aerobic (Gr. *aer*, air + *bios*, life) Requiring free oxygen; any biological process that can occur in the presence of gaseous oxygen (O_2). (p. 95)

afferent (L. *ad*, to + *ferre*, to carry) Adjective meaning leading inward or bearing toward some organ, for example, nerves conducting impulses toward the brain, or blood vessels carrying blood toward the heart; opposite of "efferent." (p. 918)

AIDS *See* acquired immune deficiency syndrome

aldosterone (Gr. *aldainein*, to nourish + *stereos*, solid) A hormone produced by the adrenal cortex that regulates the concentration of ions in the blood by stimulating the reabsorption of sodium and the excretion of potassium by the kidney. (p. 899)

alga, pl. **algae** A unicellular or simple multicellular photosynthetic organism lacking multicellular sex organs; the blue-green algae, or cyanobacteria, are photosynthetic bacteria. (p. 612)

allantois (Gr. *allas*, sausage + *eidos*, form) A membrane of the amniotic egg that functions in respiration and excretion in birds and reptiles and plays an important role in the development of the placenta in most mammals. (p. 873)

allele (Gr. *allelon*, of one another) One of two or more alternative states of a gene. (p. 239)

allele frequency The relative proportion of a particular allele among the chromosomes carried by individuals of a population. Not equivalent to "gene frequency," although the two are sometimes confused. (p. 375)

allergy An inappropriately severe response by IgE antibodies to a harmless antigen, one that does not invoke a response in most people. (p. 1106)

allometry (Gr. *allos*, other + *metros*, measure) Relative growth of a part in relation to the whole organism. (p. 1173)

allosteric interaction (Gr. *allos*, other + *stereos*, shape) A change in the shape of a protein resulting from the binding to the protein of a nonsubstrate molecule, called an effector. The new shape typically has different properties, becoming activated or inhibited. Cells control the activities of enzymes, transcription regulators, and other proteins by modulating the cellular concentration of particular allosteric effectors. (p. 144)

alpha helix (Gr. *alpha*, first + L. *helix*, spiral) A right-handed coil; in DNA, the usual form of biological DNA molecules (the one originally proposed in 1961 by Watson and Crick), although left-handed forms (called "Z DNA") also occur; in proteins, a secondary structure composed of a regular right-handed coil. (p. 54)

alternation of generations A reproductive cycle in which a haploid ($1n$) phase, the gametophyte, gives rise to gametes which, after fusion to form a zygote, germinate to produce a diploid ($2n$) phase, the sporophyte. Spores produced by meiotic division from the sporophyte give rise to new gametophytes, completing the cycle. (p. 659)

altruism Self-sacrifice for the benefit of others; in formal terms, a behavior that increases the fitness of the recipient while reducing the fitness of the altruistic individual. (p. 1207)

alveolus, pl. **alveoli** (L. a small cavity) One of the many small, thin-walled air sacs within the lungs in which the bronchioles terminate. (p. 1039)

amino acids (Gr. *Ammon*, referring to the Egyptian sun god, near whose temple ammonium salts were first prepared from camel dung) Nitrogen-containing organic molecules. Amino acids are the units, or "building blocks," from which protein molecules are assembled; all amino acids have the same underlying structure, H_2N—RCH—$COOH$, where R stands for a side group that is different for each kind of amino acid; 20 different amino acids combine to make proteins. (p. 51)

amniocentesis (Gr. *amnion*, membrane around the fetus + *centes*, puncture) Examination of a fetus indirectly, by the carrying out of tests on cell cultures grown from fetal cells obtained from a sample of the amniotic fluid surrounding the developing embryo; a procedure often carried out in pregnant women over the age of 35, since the examination of fetal chromosomes readily reveals Down syndrome if it is present, permitting the choice of a therapeutic abortion. (p. 272)

amnion (Gr., membrane around the fetus) The innermost of the extraembryonic membranes; the amnion forms a fluid-filled sac around the embryo in amniotic eggs. (p. 873)

amniotic egg An egg that is isolated and protected from the environment by a more or less impervious shell during the period of its development and that is completely self-sufficient, requiring only oxygen from the outside. (p. 873)

amylase (L. *amylum*, starch + *-ase*, suffix meaning enzyme) An enzyme that breaks down (hydrolyzes) starch (amylose) into smaller units. (p. 1013)

anabolism (Gr. *ana*, up + *bolein*, to throw) The biosynthetic or constructive part of metabolism; those chemical reactions involved in biosynthesis; opposite of "catabolism." (p. 163)

anaerobic (Gr. *an*, without + *aer*, air + *bios*, life) Any process that can occur without oxygen, such as anaerobic fermentation or H_2S photosynthesis; anaerobic organisms can live without free oxygen; obligate anaerobes cannot live in the presence of oxygen; facultative anaerobes can live with or without oxygen. (p. 70)

analogous (Gr. *analogos*, proportionate) Structures that are similar in function but different in evolutionary origin, such as the wing of a bat and the wing of a butterfly. (p. 14)

anaphase In mitosis and meiosis II, the stage initiated by the separation of sister chromatids, during which the daughter chromosomes move to opposite poles of the cell. In meiosis II, marked by separation of replicated homologous chromosomes. (p. 213)

anatomy The study of the internal structure of organisms; opposed to morphology, the study of the external structure of organisms. (p. 13)

androecium (Gr. *andros*, man + *oilos*, house) The floral whorl that comprises the stamens. (p. 678)

aneuploidy (Gr. *an*, without + *eu*, good + *ploid*, multiple of) An organism whose cells have lost or gained a chromosome; cells that have a chromosome number of $2n - 1$ (one fewer than normal) or $2n + 1$ (one extra chromosome). Down syndrome, which results from an extra copy of human chromosome 21, is an example of aneuploidy in humans. (p. 348)

anhydrase (Gr. *an*, not + *hydor*, water + *-ase*, enzyme suffix) An enzyme that catalyzes a reaction that involves the removal of water from a substrate molecule. Carbonic anhydrase, for example, catalyzes the removal of water from carbonic acid, producing carbon dioxide. (p. 141)

anion (Gr. *anienae*, to go up) A negatively charged ion. (p. 121)

annual (Gr. *annus*, year) A plant that completes its life cycle (from seed germination to seed production) and dies within a single growing season. (p. 687)

anterior (L. *ante*, before) Located before or toward the front; in animals, the head end of an organism. (p. 790)

anther (Gr. *anthos*, flower) In angiosperm flowers, the pollen-bearing portion of a stamen. (p. 678)

antheridium, pl. **antheridia** A sperm-producing organ; antheridia occur in some plants and protists. (p. 642)

anthropoid (Gr. *anthropos*, human) Closely related to humans; a higher primate (monkeys, apes, and humans). (p. 440)

antibiotic (Gr. *anti*, against + *biotikos*, pertaining to life) An organic molecule that is produced by a microorganism and kills or retards the growth of other microorganisms. (p. 591)

antibody (Gr. *anti*, against) A protein called an immunoglobulin that is produced by lymphocytes in response to a foreign substance (antigen) and released into the bloodstream. (p. 1088)

anticodon The three-nucleotide sequence at the end of a transfer RNA molecule that is complementary to, and base pairs with, an amino-acid specifying codon in messenger RNA. (p. 306)

antidiuretic hormone (ADH) (Gr. *anti*, against + *diurgos*, thoroughly wet + *hormaein*, to excite) A peptide hormone produced by the hypothalamus that promotes the reabsorption of water from the nephrons of the kidneys; also called vasopressin. Drinking excessive alcohol increases urine output because alcohol inhibits ADH. (p. 899)

antigen (Gr. *anti*, against + *genos*, origin) A foreign substance, usually a protein or polysaccharide, that stimulates one or more lymphocyte cells to begin to proliferate and secrete specific antibodies that bind to the foreign substances, labeling it as foreign and destined for destruction. (p. 889)

anus The terminal opening of the gut; the solid residues of digestion are eliminated through the anus. (p. 1023)

aorta (Gr. *aeirein*, to lift) The major artery of vertebrate systemic blood circulation; in mammals, carries oxygenated blood away from the heart to all regions of the body except the lungs. (p. 1056)

apical dominance (L. *apex*, top) In plants, the influence of a terminal bud in suppressing the growth of lateral buds by producing hormones. (p. 756)

apical meristem (L. *apex*, top + Gr. *meristos*, divided) In vascular plants, the growing point at the tip of the root or stem. (p. 689)

aposematic (Gr. *sema*, sign) A warning; usually refers to particular colors or structures that signal possession of defense systems by their bearer. (p. 477)

archegonium, pl. **archegonia** (Gr. *archegonos*, first of a race) In bryophytes and some vascular plants, the multicellular egg-producing organ. (p. 661)

archenteron (Gr. *arche*, beginning + *enteron*, gut) The principal cavity of a vertebrate embryo in the gastrula stage; lined with endoderm, it opens to the outside and represents the future digestive cavity. (p. 854)

artery A thick-walled, elastic tube, lined with muscle, that carries blood from the heart to the tissues. (p. 1056)

artificial selection The breeding of selected organisms for the purpose of producing descendants with desired traits. (p. 11)

ascospore A fungal spore produced within an ascus. (p. 643)

ascus pl. **asci** (Gr. *askos*, wineskin, bladder) A specialized cell, characteristic of the ascomycetes, in which two haploid nuclei fuse to produce a zygote that divides immediately by meiosis; at maturity, an ascus contains ascospores. (p. 642)

asexual Without distinct sexual organs; an organism that reproduces without forming gametes—that is, without sex. Asexual reproduction, therefore, does not involve sex. (p. 564)

atmospheric pressure (Gr. *atmos*, vapor + *sphaira*, globe) The weight of the earth's atmosphere over a unit area of the earth's surface; measured with a mercury barometer at sea level, this corresponds to the pressure required to lift a column of mercury 760 millimeters. (p. 730)

atom (Gr. *atomos*, indivisible) The smallest particle into which a chemical element can be divided and still retain the properties characteristic of the element; consists of a central core or nucleus composed of protons and neutrons, encircled by one or more electrons that move around the nucleus in characteristic orbits whose distance from the nucleus depends on their energy. (p. 21)

atomic mass The weight of a representative atom of an element (estimated as the average weight of all its isotopes) relative to the weight of an atom of the most common isotope of carbon (which is by convention assigned the integral value of 12); the atomic mass of an atom is approximately equal to the number of protons plus neutrons in its nucleus. (p. 21)

atomic number The number of protons in the nucleus of an atom; in an atom that does not bear an electric charge (i.e., one that is not an ion), the atomic number is also equal to the number of electrons. (p. 21)

atomic weight *See* atomic mass.

ATP *See* adenosine triphosphate.

atrium (L., vestibule or courtyard) An antechamber; in the heart, a thin-walled chamber that receives venous blood and passes it on to the thick-walled ventricle; in the ear, the tympanic cavity. (p. 1065)

autonomic nervous system (Gr. *autos*, self + *nomos*, law) The involuntary neurons and ganglia of the peripheral nervous system of vertebrates; regulates the heart, glands, visceral organs, and smooth muscle; subdivided into the sympathetic and parasympathetic divisions, whose effects oppose one another (one stimulating, the other inhibiting). (p. 918)

autosome (Gr. *autos*, self + *soma*, body) Any eukaryotic chromosome that is not a sex chromosome; autosomes are present in the same number and kind in both males and females of the species. (p. 259)

autotroph (Gr. *autos*, self + *trophos*, feeder) An organism able to build all the complex organic molecules that it requires as its own food source, using only simple inorganic compounds. Plants, algae, and some bacteria are autotrophs. Contrasts with *heterotroph*. (p. 154)

auxin (Gr. *auxein*, to increase) A plant hormone that controls cell elongation, among other effects. (p. 755)

axon (Gr., axle) A process extending out from a neuron that conducts impulses away from the cell body. (p. 894)

B cell A type of lymphocyte which, when confronted with a suitable antigen, is capable of secreting a specific antibody protein; each individual lymphocyte B cell is capable of secreting only one form of antibody, but few if any B lymphocytes secrete the same form of antibody. (p. 1094)

bacillus, pl. **bacilli** (L. *baculum*, rod) A rod-shaped bacterium. (p. 592)

bacteriophage (Gr. *bakterion*, little rod + *phagein*, to eat) A virus that infects bacterial cells; also called a *phage*. (p. 284)

bark In plants, all tissues outside the vascular cambium in a woody stem. (p. 714)

basal body A self-reproducing cylinder-shaped cytoplasmic organelle composed of nine triplets of microtubules from which the flagella or cilia arise; identical in structure to the centriole, which is involved in mitosis and meiosis in most protists and animals and in some plants. (p. 98)

base A substance that dissociates in water, causing a decrease in the relative number of hydrogen ions (H^+), often by producing hydroxide ions (OH^-), which combine with hydrogen ions to form water and thus reduce the concentration of free hydrogen ions; having a pH greater than 7.0; the opposite of acid. (p. 36)

basidiospore A spore of the basidiomycetes, produced within and borne on a basidium after nuclear fusion and meiosis. (p. 644)

basidium, pl. **basidia** (L., a little pedestal) A specialized reproductive cell of the basidiomycetes, often club shaped, in which nuclear fusion and meiosis occur. (p. 644)

behavior A coordinated neuromotor response to changes in external or internal conditions; a product of the integration of sensory, neural, and hormonal factors. (p. 1176)

behavioral ecology The study of how natural selection shapes behavior. (p. 1198)

biennial A plant that normally requires two growing seasons to complete its life cycle. Biennials flower in the second year of their lives. (p. 722)

bilateral symmetry (L. *bi*, two + *lateris*, side; Gr. *symmetria*, symmetry) A body form in which the right and left halves of an organism are approximate mirror images of each other. (p. 616)

bile salts A solution of organic salts that is secreted by the vertebrate liver and temporarily stored in the gallbladder; emulsifies fats in the small intestine. (p. 1021)

binary fission (L. *binarius*, consisting of two things or parts + *fissus*, split) Asexual reproduction by division of one cell or body into two equal, or nearly equal, parts. (p. 203)

binomial system (L. *bi*, twice, two + Gr. *nomos*, usage, law) A system of nomenclature in which the name of a species consists of two parts, the first of which designates the genus. (p. 562)

biogeochemical cycle (Gr. *bios*, life + *geo*, earth + *chemeia*, alchemy; *kyklos*, circle, wheel) The cyclic path of an inorganic substance, such as carbon or nitrogen, through an ecosystem. (p. 487)

bioluminescence Light production by living organisms; in bioluminescent organisms, proteins called luciferins, in the presence of oxygen, are converted by an enzyme called luciferase to oxyluciferins, with the liberation of light. (p. 514)

biomass (Gr. *bios*, life + *maza*, lump or mass) The weight of all the living organisms in a given population, area, or other unit being measured. (p. 496)

biome One of the major terrestrial ecosystems, characterized by climatic and soil conditions; the largest ecological unit. (p. 451)

biosphere (Gr. *bios*, life + *sphaira*, globe) That part of earth containing living organisms. (p. 531)

biotic Pertaining to life. (p. 453)

blade The broad, expanded part of a leaf; also called the lamina. (p. 708)

blastocoel (Gr. *blastos*, a sprout + *koilos*, hollow) The central cavity of the blastula stage of vertebrate embryos. (p. 1158)

blastodisc (Gr. *blastos*, sprout + *discos*, a round plate) In the development of birds, a disklike area on the surface of a large, yolky egg that undergoes cleavage and gives rise to the embryo. (p. 1158)

blastomere (Gr. *blastos*, sprout + *meros*, part) One of the cells of a blastula; a product of the early cleavage divisions in the development of a vertebrate embryo. (p. 1158)

blastopore (Gr. *blastos*, sprout + *poros*, a path or passage) In vertebrate development, the opening that connects the archenteron cavity of a gastrula stage embryo with the outside; represents the future mouth in some animals (protostomes), the future anus in others (deuterostomes). (p. 782)

blastula (Gr., a little sprout) In vertebrates, an early embryonic stage consisting of a hollow, fluid-filled ball of cells one layer thick; a vertebrate embryo after cleavage and before gastrulation. (p. 782)

blood group In humans, the type of cell surface antigens present on the red blood cells of an individual; genetically determined, alternative alleles yield different surface antigens. When two different blood types are mixed, the cell surfaces often interact, leading to agglutination. One genetic locus encodes the ABO blood group, another the Rh blood group, and still others encode other surface antigens. (p. 265)

blood pressure In vertebrates, the hydrostatic or fluid pressure within the circulatory system, generated by heart contractions; usually measured at large arteries of the systemic circulation, such as in an arm. (p. 1072)

Bowman's capsule In the vertebrate kidney, the bulbous unit of the nephron, which surrounds the glomerulus. The kidney works by forced filtration, blood pressure driving blood plasma from the glomerular capillaries into the Bowman's capsule, after which it passes through the nephron, where most water and ions are reabsorbed into the bloodstream and the residue is excreted as urine. (p. 1117)

brainstem The most posterior portion of the vertebrate brain; includes the medulla, pons, and midbrain. (p. 1048)

branchial (Gr. *branchia*, gills) Referring to gills. (p. 1036)

bronchus, pl. **bronchi** (Gr. *bronchos*, windpipe) One of a pair of respiratory tubes branching from the lower end of the trachea (windpipe) into either lung; the respiratory path further subdivides into progressively finer passageways, the bronchioles, culminating in the alveoli. (p. 1042)

bud An asexually produced outgrowth that develops into a new individual. In plants, an embryonic shoot, often protected by young leaves; buds may give rise to branch shoots. (p. 712)

buffer Any substance or chemical compound that tends to keep pH levels constant when acids or bases are added; a combination of a weak acid and a weak base that counteracts sudden swings in pH by combining with hydrogen ions (H^+) when the H^+ concentration rises, and by releasing H^+ ions when the H^+ concentration falls. (p. 36)

bundle sheath cell In plants, a layer or layers of cells surrounding a vascular bundle, especially characteristic of C_4 plants. (p. 195)

C_4 pathway A form of photosynthesis that avoids photorespiration in hot climates by increasing the internal CO_2 concentration of cells within which the dark reactions of photosynthesis occur. C_4 plants initially fix carbon dioxide to a compound known as phosphoenolpyruvate (PEP) to yield oxaloacetate, a four-carbon compound—hence the name C_4. Also known as the Hatch-Slack pathway. (p. 195)

callus (L. *callos*, hard skin) Undifferentiated tissue; a term used in tissue culture, grafting, and wound healing. (p. 747)

Calvin cycle The dark reactions of photosynthesis; a series of enzymatically mediated photosynthetic reactions in which carbon dioxide is bound to ribulose 1,5-bisphosphate and then reduced to 3-phosphoglyceraldehyde, and the carbon-dioxide-accepting ribulose 1,5-bisphosphate is regenerated. For every six molecules of carbon dioxide entering the cycle, a net gain of two molecules of glyceraldehyde 3-phosphate results. Also called the Calvin-Benson cycle. (p. 192)

calyx (Gr. *kalyx*, a husk, cup) The sepals collectively; the outermost flower whorl. (p. 678)

CAM *See* crassulacean acid metabolism.

cambium, pl. **cambia** In vascular plants, embryonic tissue zones (meristems) that run parallel to the sides of roots and stems; consists of the vascular cambium and the cork cambium. (p. 713)

cancer Unrestrained invasive cell growth; a tumor or cell mass resulting from uncontrollable cell division; a result of mutational damage that destroys the normal control of cell division. A cancer cell is said to be "malignant," and in many (although by no means all)

tissues, malignant cells can migrate to other body regions (metastasize), where they form secondary tumors. (p. 1092)

capillary (L. *capillaris*, hairlike) The smallest of the blood vessels; the very thin walls of capillaries are permeable to many molecules, and exchanges between blood and the tissues occur across them; the vessels that connect arteries with veins. (p. 1056)

capillary action The movement of a liquid along a surface as a result of the combined effects of cohesion and adhesion. (p. 32)

carapace (Fr. from Sp. *carapacho*, shell) Shieldlike plate covering the cephalothorax of decapod crustaceans; the dorsal part of the shell of a turtle. (p. 840)

carbohydrate (L. *carbo*, charcoal + *hydro*, water) An organic compound consisting of a chain or ring of carbon atoms to which hydrogen and oxygen atoms are attached in a ratio of approximately 2:1; a compound of carbon, hydrogen, and oxygen having the generalized formula $(CH_2O)''$; carbohydrates include sugars, starch, glycogen, and cellulose (p. 42)

carbon cycle The worldwide circulation and reutilization of carbon atoms. (p. 489)

carbon fixation The conversion of CO_2 into organic compounds during photosynthesis; the first stage of the dark reactions of photosynthesis, in which carbon dioxide from the air is combined with ribulose 1,5-bisphosphate. (p. 185)

carbonic anhydrase *See* anhydrase.

carcinogen Any agent capable of inducing cancer; because cancer occurs as a result of mutation, most carcinogens are also potent mutagens. Programs that screen chemicals to detect potential carcinogens employ tests which measure the propensity of a chemical to cause mutations. (p. 326)

cardiovascular system The blood circulatory system and the heart that pumps it; collectively, the blood, heart, and blood vessels. (p. 1056)

carnivore (L. *carnis*, flesh + *voro*, to devour) Any organism that eats animals; a meat eater, as opposed to a plant eater, or herbivore. (p. 1015)

carpel (Gr. *karpos*, fruit) A leaflike organ in angiosperms that encloses one or more ovules; one of the members of the gynoecium. (p. 677)

cartilage (L. *cartilago*, gristle) A connective tissue in skeletons of vertebrates. Cartilage forms much of the skeleton of em-

bryos, very young vertebrates, and some adult vertebrates, such as sharks and their relatives. In most adult vertebrates much of it is converted into bone. (p. 890)

Casparian strip (Robert Caspary, a German botanist) In plants, a thickened, waxy strip that extends around and seals the walls of endodermal root cells, thus restricting the diffusion of solutes across the endodermis into the vascular tissues of the root. (p. 719)

catabolism (Gr. *ketabole*, throwing down) In a cell, those metabolic reactions that result in the breakdown of complex molecules into simpler compounds, often with the release of energy. (p. 163)

catalyst (Gr. *kata*, down + *lysis*, a loosening) A substance that accelerates the rate of a chemical reaction by lowering the activation energy, without being used up in the reaction; enzymes are the catalysts of cells. (p. 140)

cation (Gr. *katienai*, to go down) A positively charged ion. (p. 906)

cecum, also **caecum** (L. *caecus*, blind) In vertebrates, a blind pouch at the beginning of the large intestine; any similar pouch. (p. 1024)

cell (L. *cella*, a chamber or small room) The structural unit of organisms; the smallest unit that can be considered living; a cell consists of cytoplasm encased within a membrane. (p. 80)

cell cycle The repeating sequence of growth and division through which cells pass each generation. (p. 207)

cell division The division of a cell and its contents into two roughly equal parts. Bacteria divide by fission, eukaryotes by mitosis. (p. 213)

cell membrane The outermost membrane of the cell; also called the plasma membrane. (p. 79)

cell plate The structure that forms at the equator of the spindle during early telophase in the dividing cells of plants and a few green algae. The cell plate becomes the middle lamella during the course of development. (p. 215)

cell wall The rigid, outermost layer of the cells of plants, some protists, and most bacteria; the cell wall surrounds the cell (plasma) membrane. (p. 85)

cellular respiration The metabolic harvesting of energy by oxidation; carried out by the citric acid cycle and oxidative phosphorylation, in which considerable energy is extracted from

sugar molecule fragments left over from glycolysis. (p. 139)

cellulase An enzyme that hydrolyzes cellulose. (p. 1024)

cellulose (L. *cellula*, a little cell) The chief constituent of the cell wall in all green plants, some algae, and a few other organisms. Cellulose is an insoluble complex carbohydrate $(C_6H_{10}O_5)''$ formed of microfibrils of glucose molecules. (p. 46)

central nervous system That portion of the nervous system where most association occurs; in vertebrates, it is composed of the brain and spinal cord; in invertebrates it usually consists of one or more cords of nervous tissue, together with their associated ganglia. (p. 791)

centriole (Gr. *kentron*, center of a circle + L. *olus*, little one) A cytoplasmic organelle located outside the nuclear membrane, identical in structure to a basal body; found in animal cells and in the flagellated cells of other groups. The centriole divides and organizes spindle fibers during mitosis and meiosis. (p. 209)

centromere (Gr. *kentron*, center + *meros*, a part) Condensed region on eukaryotic chromosomes where sister chromatids are attached to one another after replication; composed of highly repeated DNA sequences (satellite DNA). *See* kinetochore. (p. 206)

cephalothorax (Gr. *kephale*, head + *thorax*) A body division found in many arachnids and crustaceans, in which the head is fused with some or all of the thoracic segments. (p. 827)

cerebellum (L., little brain) The hindbrain region of the vertebrate brain which lies above the medulla (brainstem) and behind the forebrain; it integrates information about body position and motion, coordinates muscular activities, and maintains equilibrium. (p. 921)

cerebral cortex The thin surface layer of neurons and glial cells covering the cerebrum; well developed only in mammals, and particularly prominent in humans. The cerebral cortex is the seat of conscious sensations and voluntary muscular activity. (p. 924)

cerebrum (L., brain) The portion of the vertebrate brain (the forebrain) that occupies the upper part of the skull, consisting of two cerebral hemispheres united by the corpus callosum. The cerebrum, which consists of paired hemispheres occupying the upper part of the skull overlying the thalamus, hypothalamus, and pituitary, is the primary association center of the brain. It coordinates and processes sensory input and coordinates motor responses. (p. 924)

chelicera, pl. **chelicerae** (Gr. *chele*, claw + *keras*, horn) The first pair of appendages in horseshoe crabs, sea spiders, and arachnids—the chelicerates, a group of arthropods. Chelicerae usually take the form of pincers or fangs. (p. 828)

chemical reaction The making or breaking of chemical bonds between atoms or molecules. (p. 138)

chemiosmosis The mechanism by which ATP is generated in mitochondria and chloroplasts. Energetic electrons excited by light (chloroplasts) or extracted by oxidation in the citric acid cycle (mitochondria) are used to drive proton pumps, creating a proton concentration gradient. When protons subsequently flow back across the membrane, it is through channels that couple their passage to the synthesis of ATP. (p. 126)

chemoautotroph An autotrophic bacterium that uses chemical energy released by specific inorganic reactions to power its life processes, including the synthesis of organic molecules. (p. 191)

chemoreceptor A sensory cell or organ that responds to the presence of a specific chemical stimulus by initiating a nerve impulse; includes taste and smell receptors. (p. 942)

chiasma, pl. **chiasmata** (Gr., a cross) During meiosis, the region of contact between homologous chromatids where crossing-over has occurred during synapsis; a chiasma has the appearance of the letter X. (p. 222)

chitin (Gr. *chiton*, tunic) A tough, resistant, nitrogen-containing polysaccharide that forms the cell walls of certain fungi, the exoskeleton of arthropods, and the epidermal cuticle of other surface structures of certain other invertebrates. (p. 47)

chlorophyll (Gr. *chloros*, green + *phyllon*, leaf) The green steroid pigment of plant cells, which is the receptor of light energy in photosynthesis; chlorophyll is also found in some protists and bacteria. (p. 49)

chloroplast (Gr. *chloros*, green + *plastos*, molded) A cell-like organelle present in algae and plants that contains chlorophyll (and usually other pigments) and carries out photosynthesis. (p. 80)

choanocyte A type of flagellated cell that lines the interior surface of a sponge; essentially

identical to choanoflagellates, a group of the protist phylum Zoomastigina, from which sponges seem to have evolved. (p. 785)

chordate (L. *chorda*, chord, string) A member of the animal phylum Chordata, of which all members possess a notochord, dorsal nerve cord, pharyngeal gill pouches, and a tail, at least at some stage of the life cycle. (p. 861)

chorion (Gr., skin) The outer member of the double membrane that surrounds the embryo of reptiles, birds, and mammals; in placental mammals, it contributes to the structure of the placenta. (p. 1132)

chromatid (Gr. *chroma*, color + L. *-id*, daughters of) One of the two daughter strands of a duplicated chromosome which are joined by a single centromere; separates and becomes a daughter chromosome at anaphase of mitosis or anaphase of the second meiotic division. (p. 209)

chromatin (Gr. *chroma*, color) The complex of DNA and proteins of which eukaryotic chromosomes are composed. (p. 205)

chromosome (Gr. *chroma*, color + *soma*, body) The vehicle by which hereditary information is physically transmitted from one generation to the next; the organelle that carries the genes. In bacteria, the chromosomes consist of a single naked circle of DNA; in eukaryotes, they consist of a single linear DNA molecule and associated proteins. (p. 86)

cilium, pl. **cilia** (L., eyelash) A short, hairlike flagellum, especially used to refer to such structures when they are numerous. Cilia may be used in locomotion; in animals, they aid the movement of substances across surfaces. (p. 103)

circadian rhythms (L. *circa*, about + *dies*, day) Regular rhythms of growth or activity that occur on an approximately 24-hour cycle. (p. 1188)

cisterna, pl. **cisternae** (L., a reservoir) In cells, a flattened or saclike space between membranes of the endoplasmic reticulum or Golgi body. (p. 94)

citric acid cycle The cyclic series of reactions in which pyruvate, the product of glycolysis, enters the cycle to form citric acid and is then oxidized to carbon dioxide. Also called the Krebs cycle (after its discoverer) and the tricarboxylic acid (TCA) cycle (citric acid possesses three carboxyl groups). (p. 157)

class A taxonomic category between phyla (divisions) and orders. A class contains one or more orders, and belongs to a particular phylum or division. (p. 562)

cleavage In vertebrates, a rapid series of successive cell divisions of a fertilized egg, forming a hollow sphere of cells, the blastula. (p. 161)

climax community In ecology, the final stage in a successional series; the climax stage is determined primarily by the climate and soil type of the area. (p. 501)

cloaca (L. sewer) In some animals, the common exit chamber from the digestive, reproductive, and urinary systems; in others, the cloaca may also serve as a respiratory duct. (p. 1023)

clone (Gr. *klon*, twig) A line of cells, all of which have arisen from the same single cell by mitotic division; one of a population of individuals derived by asexual reproduction from a single ancestor; one of a population of genetically identical individuals. (p. 409)

cloning Producing a cell line or culture all of whose members contain identical copies of a particular nucleotide sequence; an essential element in genetic engineering, cloning is usually carried out by inserting the desired gene into a virus or plasmid, infecting a cell culture with the hybrid virus, and selecting for culture a cell that has taken up the gene. (p. 351)

cnidocyte (Gr. *knide*, nettle + *kytos*, vessel) Specialized cell that holds a nematocyst; cnidocytes are characteristic of cnidarians. (p. 787)

coacervate (L. *coacervatus*, heaped up) A spherical aggregation of lipid molecules in water, held together by hydrophobic forces. (p. 67)

coccus, pl. **cocci** (Gr. *kokkos*, a berry) A spherical bacterium. (p. 592)

cochlea (Gr. *kochlios*, a snail) In terrestrial vertebrates, a tubular cavity of the inner ear containing the essential organs of hearing; occurs in crocodiles, birds, and mammals; spirally coiled in mammals. (p. 952)

co-dominance In genetics, a situation in which the effects of both alleles at a particular locus are apparent in the phenotype of the heterozygote. (p. 265)

codon (L., code) The basic unit ("letter") of the genetic code; a sequence of three adjacent nucleotides in DNA or mRNA that code for one amino acid,

or for polypeptide chain termination. See anticodon. (p. 304)

coelenteron (Gr. *koilos*, a hollow + *enteron*, gut) A digestive cavity with only one opening, characteristic of the cnidarians and ctenophores. (p. 787)

coelom (Gr. *koilos*, a hollow) A body cavity formed between layers of mesoderm and in which the digestive tract and other internal organs are suspended. (p. 1163)

coenzyme (L. *co-*, together + Gr. *en*, in + *zyme*, leaven) A nonprotein organic molecule that plays an accessory role in enzyme-catalyzed processes, often by acting as a donor or acceptor of electrons. NAD⁺, FAD, and coenzyme A are common coenzymes. (p. 146)

coevolution (L. *co-*, together + *e-*, out + *volvere*, to fill) The simultaneous development of adaptations in two or more populations, species, or other categories that interact so closely that each is a strong selective force on the other. (p. 438)

cofactor One or more nonprotein components required by enzymes in order to function; many cofactors are metal ions, others are coenzymes. See coenzyme. (p. 145)

cohesion (L. *cohaerer*, to stick together) The mutual attraction (coherence) of molecules of the same substance. (p. 32)

collagen (Gr. *kolla*, glue + *genos*, descent) A tough, fibrous protein occurring in vertebrates as the chief constituent of bones, tendons, and other connective tissue. Collagen also occurs in nonvertebrates—for example, in the cuticle of nematodes. (p. 890)

collenchyma (Gr. *kolla*, glue + *en*, in + *chymein*, to pour) In plants, a supporting tissue composed of collenchyma cells; often found in regions of primary growth in stems and in some leaves. Collenchyma cells are elongated living cells with irregularly thickened primary cell walls. (p. 704)

colony A group of organisms living together in close association. (p. 569)

commensalism (L. *cum*, together with + *mensa*, table) A relationship in which one individual lives close to or on another and benefits, and the host is unaffected; a kind of symbiosis. (p. 480)

community (L. *communitas*, community, fellowship) All of the organisms inhabiting a common environment and interacting with one another. (p. 451)

companion cell In the phloem of flowering plants, a specialized kind of elongate parenchyma cell associated with a sieve-tube member. (p. 707)

competition Interaction between members of the same population or of two or more populations in order to obtain a mutually required resource available in limited supply; in competition, one species interferes with another enough to keep it from gaining access to the resource. (p. 460)

competitive exclusion The hypothesis that two species with identical ecological requirements cannot exist in the same locality indefinitely, and that the more efficient of the two in utilizing the available scarce resources will exclude the other; also known as Gause's principle, after the Russian biologist G.F. Gause. (p. 461)

complement A group of 11 proteins that act in concert to destroy foreign cells during the immune response. (p. 1099)

compound (L. *componere*, to put together) A molecule composed of two or more kinds of atoms in definite ratios, held together by chemical bonds. (p. 26)

compound eye In arthropods, a complex eye composed of many separate elements, each element composed of light-sensitive cells and a lens that can form an image; all of the individual images are merged in the arthropod central nervous system to form a detailed, three-dimensional image. (p. 830)

compound leaf A leaf whose blade is divided into several distinct leaflets. (p. 709)

concentration gradient The concentration difference of a substance as a function of distance; in a cell, a greater concentration of its molecules in one region than in another. (p. 906)

condensation reaction *See* dehydration reaction. (p. 170)

conditioning A form of learning in which a behavioral response becomes associated (by means of a reinforcing stimulus) with a new stimulus not previously capable of invoking the response. (p. 1182)

cone (1) In plants, the reproductive structure of a conifer. (p. 676) (2) In vertebrates, a type of light-sensitive neuron in the retina, concerned with the perception of color and with the most acute discrimination of detail. (p. 956)

conjugation (L. *conjugare*, to yoke together) Temporary union of two unicellular organisms, during which genetic material is transferred from one cell to the other; occurs in bacteria, pro-

tists, and certain algae and fungi. (p. 339)

conjugation bridge A cytoplasmic tube through which genetic material passes during conjugation. (p. 339)

connective tissues A collection of vertebrate tissues derived from the mesoderm. Some kinds of connective tissue cells (lymphocytes and macrophages) are mobile hunters of invading bacteria, others secrete a matrix (as in cartilage and bone) that provides the body with structural support, and still others provide sites of fat storage (adipose tissue) or a site for hemoglobin (red blood cells). (p. 887)

consumer In ecology, a heterotroph that derives its energy from living or freshly killed organisms or parts thereof. Primary consumers are herbivores; higher-level consumers are carnivores. (p. 497)

continuous variation Variation in traits to which many different genes make a contribution; in such traits, a gradation of small differences is observed; often exhibiting a "normal" or bell-shaped distribution. (p. 251)

contractile vacuole In protists and some animals, a clear fluid-filled cell vacuole that takes up water from within the cell and then contracts, releasing it to the outside through a pore in a cyclical manner. Contractile vacuoles function primarily in osmoregulation and excretion. (p. 118)

convergent evolution The independent development of similar structures in organisms that are not directly related; often found in organisms living in similar environments. (p. 523)

copulation (Fr., from L. *copulare*, to couple) Sexual union of opposite sexes, in which sperm are introduced into the body of the female to facilitate internal fertilization. (p. 1132)

copy DNA (cDNA) DNA "copied" from an mRNA molecule, using the viral enzyme reverse transcriptase. (p. 357)

cork (L. *cortex*, bark) A secondary tissue produced by a cork cambium; made up of flattened cells, nonliving at maturity, with walls infiltrated with suberin, a waxy or fatty material that is resistant to the passage of gases and water vapor; the outer part of the periderm. (p. 714)

cork cambium The lateral meristem that forms the periderm, producing cork (phellem) toward the surface (outside) of the plant and phelloderm toward the inside; common in

stems and roots of gymnosperms and dicots. (p. 714)

cornea (L. *corneus*, horny) The transparent outer layer of the vertebrate eye. (p. 956)

corolla (L. *cornea*, crown) The petals, collectively; usually the conspicuously colored flower whorl. Petals may be free or fused with one another or with the members of other floral whorls. (p. 689)

corpus luteum (L., yellowish body) A structure that develops from a ruptured follicle in the ovary after ovulation; the corpus luteum secretes estrogens and progesterone, which maintain the uterus during pregnancy. (p. 1141)

cortex (L., bark) The outer layer of a structure; in animals, the outer, as opposed to the inner, part of an organ, as in the adrenal, kidney, and cerebral cortexes (p. 925); in vascular plants, the primary ground tissue of a stem or root, bounded externally by the epidermis and internally by the central cylinder of vascular tissue. (p. 712)

cotyledon (Gr. *kotyledon*, a cup-shaped hollow) Seed leaf; cotyledons generally store food in dicotyledons and absorb it in monocotyledons. The food is used during the course of seed germination. (p. 685)

countercurrent exchange In organisms, the passage of heat or of molecules (such as oxygen, water, or sodium ions) from one circulation path to another moving in the opposite direction; because the flow of the two paths is in opposite directions, there is always a concentration difference between the two channels, facilitating transfer. (p. 1037)

coupled reactions In cells, the linking of an endergonic (energy-requiring) reaction to an exergonic (energy-releasing) reaction. In coupled reactions, the decrease in free energy associated with the exergonic reaction is typically greater than the increase in free energy associated with the endergonic reaction, so that the net change in free energy is negative and the two-reaction process proceeds—even though the first step would not have proceeded alone. In metabolism, coupling reactions to ATP hydrolysis (a highly exergonic reaction) is a common strategy employed to drive endergonic reactions. (p. 154)

covalent bond (L. *co-*, together + *valere*, to be strong) A chemical bond formed between atoms as a result of the sharing of one or more pairs of electrons. (p. 27)

crassulacean acid metabolism (CAM metabolism) A variant of the C_4 pathway; phosphoenolpyruvate fixes CO_2 in C_4 compounds at night and then, during the day, the fixed CO_2 is transferred to the ribulose bisphosphate of the Calvin cycle within the same cell. CAM metabolism is characteristic of such succulent plants as cacti. (p. 196)

crista, pl. **cristae** (L., crest) In mitochondria, the enfoldings of the inner mitochondrial membrane, which form a series of "shelves" containing the electron-transport chains involved in ATP formation. (p. 96)

crossing-over In meiosis, the exchange of corresponding chromatid segments between homologous chromosomes; responsible for genetic recombination between homologous chromosomes. (p. 221)

cuticle (L. *cuticula*, little skin) A waxy or fatty, noncellular layer on the outer wall of epidermal cells. The cuticle is formed of a substance called cutin. (p. 658)

cyclic AMP (cAMP) A form of adenosine monophosphate (AMP) in which the atoms of the phosphate group form a ring; found in almost all organisms. cAMP functions as an intracellular hormone that regulates a diverse array of metabolic activities. (p. 316)

cyclic photophosphorylation In photosynthesis, the path taken by a light-excited electron from a photosystem through a membrane-embedded transport chain to proton pumps, then back to the photosystem; more generally, the chemiosmotic synthesis of ATP in photosynthesis. (p. 186)

cytochrome (Gr. *kytos*, hollow vessel + *chroma*, color) Any of several iron-containing protein pigments that serve as electron carriers in transport chains of photosynthesis and cellular respiration. (p. 392)

cytokinesis (Gr. *kytos*, hollow vessel + *kinesis*, movement) Division of the cytoplasm of a cell after nuclear division. (p. 207)

cytokinin (Gr. *kytos*, hollow vessel + *kinesis*, motion) A class of plant hormones that promote cell division, among other effects. (p. 756)

cytoplasm (Gr. *kytos*, hollow vessel + *plasma*, anything molded) The living matter within a cell, excluding the nucleus; the protoplasm. (p. 79)

cytosine A pyrimidine; one of the four nitrogen-containing bases in the nucleic acids DNA and RNA; in a DNA duplex molecule, cytosine pairs with guanine, forming three hydrogen bonds. (p. 57)

cytoskeleton A network of protein microfilaments and microtubules within the cytoplasm of a eukaryotic cell that maintains the shape of the cell, anchors its organelles, and is involved in animal cell motility. (p. 88)

deciduous (L. *decidere*, to fall off) In vascular plants, shedding all the leaves at a certain season. (p. 723)

decomposers Organisms (bacteria, fungi, heterotrophic protists) that break down organic material into smaller molecules, which are then recirculated. (p. 497)

dehydration reaction (L. *co-*, together + *densare*, to make dense) A type of chemical reaction in which two molecules join to form one larger molecule, simultaneously splitting out a molecule of water. One molecule is stripped of a hydrogen atom and another is stripped of a hydroxyl group (—OH), resulting in the joining of the two molecules. The H^+ and —OH released may combine to form a water molecule. The biosynthetic reactions in which monomers (e.g., monosaccharides, amino acids) are joined to form polymers (e.g., polysaccharides, polypeptides) are condensation reactions. (p. 41)

demography (Gr. *demos*, people + *graphein*, to draw) The properties of the rate of growth and the age structure of populations. (p. 459)

denaturation The loss of the native configuration of a protein or nucleic acid as a result of excessive heat, extremes of pH, chemical modification, or changes in solvent ionic strength or polarity that disrupt hydrophobic interactions; usually accompanied by loss of biological activity. (p. 55)

dendrite (Gr. *dendron*, tree) A process extending from the cell body of a neuron, typically branched, that conducts impulses inward toward the cell body; although they may be long, most dendrites are short, and a single neuron may possess many of them. (p. 894)

denitrification The conversion of nitrate to gaseous nitrogen; carried out by certain soil bacteria. (p. 493)

deoxyribonucleic acid (DNA) The genetic material of all organisms; composed of two complementary chains of nucleotides wound in a double helix; local unwinding of the helix by disruption of hydrogen bonds between strands permits RNA polymerase molecules to transcribe mRNA copies of genes, and permits DNA polymerase molecules to replicate copies of the duplex molecule. *See* alpha helix. (p. 56)

dermis (Gr. *derma*, skin) The inner, sensitive mesodermal layer of skin, beneath the epidermis; corium. (p. 947)

determinate growth Growth of limited duration, as is characteristic of floral meristems and of leaves. (p. 689)

detritivores (L. *detritus*, worn down + *vorare*, to devour) Organisms that live on dead organic matter; included are large scavengers, smaller animals such as earthworms and some insects, and decomposers (fungi and bacteria). (p. 497)

deuterostome (Gr. *deuteros*, second + *stoma*, mouth) An animal in whose embryonic development the anus forms at or near the blastopore and the mouth forms secondarily elsewhere. Deuterostomes are also characterized by radial cleavage during the earliest stages of development and by enterocoelous formation of the coelom. *See* protostome. (p. 807)

diaphragm (Gr. *diaphrassein*, to barricade) (1) In mammals, a sheet of muscle tissue that separates the abdominal and thoracic cavities and functions in breathing. (p. 1042) (2) A contraceptive device used to block the entrance to the uterus temporarily and thus prevent sperm from entering during sexual intercourse. (p. 1146)

dicot Short for dicotyledon; a class of flowering plants generally characterized as having two cotyledons, net-veined leaves, and flower parts usually in fours or fives. (p. 685)

differentiation A developmental process by which a relatively unspecialized cell undergoes a progressive change to a more specialized form or function; differentiation in plants may be reversed under suitable conditions, but it is rarely reversible in animals. (p. 746)

diffusion (L. *diffundere*, to pour out) The net movement of dissolved molecules or other particles from a region where they are more concentrated to a region where they are less concentrated, as a result of the random movement of individual molecules; the process tends to distribute molecules uniformly. (p. 115)

digestion (L. *digestio*, separating out, dividing) The breakdown of complex, usually insoluble foods into molecules that can be absorbed into cells and there degraded to yield energy and the raw materials for synthetic processes. (p. 1009)

dihybrid (Gr. *dis*, twice + L. *hibrida*, mixed offspring) An individual heterozygous at two different loci; for example, A/a B/b. (p. 243)

dikaryotic (Gr. *di*, two + *karyon*, kernel) In fungi, having pairs of nuclei within each cells. (p. 640)

dioecious (Gr. *di*, two + *oikos*, house) Having the male and female elements on different individuals. (p. 690)

diploid (Gr. *diploos*, double + *eidos*, form) Having two sets of chromosomes ($2n$); in animals, twice the number characteristic of gametes; in plants, the chromosome number characteristic of the sporophyte generation; in contrast to haploid ($1n$). (p. 207)

disaccharide A carbohydrate formed of two simple sugar molecules bonded covalently; sucrose is a disaccharide composed of two glucose molecules linked together. (p. 44)

division A major taxonomic group; kingdoms are divided into divisions (or phyla, which are equivalent), and divisions are divided into classes. (p. 563)

DNA *See* deoxyribonucleic acid.

dormancy (L. *dormire*, to sleep) A period during which growth ceases and is resumed only if certain requirements, as of temperature or day length, have been fulfilled. (p. 735)

dorsal (L. *dorsum*, the back) Toward the back, or upper surface; opposite of ventral. (p. 790)

double fertilization The fusion of the egg and sperm (resulting in a $2n$ fertilized egg, the zygote) and the simultaneous fusion of the second male gamete with the polar nuclei (resulting in a primary endosperm nucleus, which is often triploid, $3n$); a unique characteristic of all angiosperms. (p. 679)

Down syndrome A congenital syndrome, whose manifestations include mental retardation, caused by the cells in a person's body having an extra copy of a segment of chromosome 21; also called trisomy 21. (p. 260)

duodenum (L. *duodeni*, 12 each—from its length, about 12 fingers' breadth) In vertebrates, the upper portion of the small intestine; the principal site of food digestion, where carbohydrates, proteins, and fats are broken down into sugars, amino acids, and fatty acids. (p. 1013)

ecdysis (Gr. *ekdysis*, stripping off) Shedding of outer cuticular layer; molting, as in insects or crustaceans. (p. 829)

ecdysone (Gr. *ekdysis*, stripping off) Molting hormone of arthropods, which stimulates growth and ecdysis. (p. 850)

ecology (Gr. *oikos*, house + *logos*, word) The study of the interactions of organisms with one another and with their physical environment. (p. 18)

ecosystem (Gr. *oikos*, house + *systema*, that which is put together) A major interacting system that involves both organisms and their nonliving environment. (p. 451)

ecotype (Gr. *oikos*, house + L. *typus*, image) A locally adapted variant of an organism; differing genetically from other ecotypes. (p. 409)

ectoderm (Gr. *ecto*, outside + *derma*, skin) One of the three embryonic germ layers of early vertebrate embryos; ectoderm gives rise to the outer epithelium of the body (skin, hair, nails) and to the nerve tissue, including the sense organs, brain, and spinal cord. (p. 885)

ectotherm (Gr. *ectos*, outside + *therme*, heat) An organism, such as a reptile, that regulates its body temperature by taking in heat from the environment or giving it off to the environment; contrasts with endotherm. *See* poikilotherm. (p. 874)

effector (L. *efficere*, to bring to pass) In animals, an organ, tissue, or cell that becomes active in response to stimulation; within cells, a small molecule that induces allosteric changes in the shapes of proteins. *See* allosteric interaction. (p. 897)

efferent (L. *ex*, out of + *ferre*, to bear) Leading or conveying away from some origin—for example, nerve impulses conducted away from the brain, or blood conveyed away from the heart; contrasts with afferent. (p. 918)

egg A female gamete; nonmotile and usually containing abundant cytoplasm; often larger than a male gamete. (p. 1158)

electrolyte A substance that dissociates into ions in aqueous solution. (p. 906)

electron A subatomic particle with a negative electric charge equal in magnitude to the positive charge of the proton but with a much smaller mass; electrons orbit the atom's positively charged nucleus and determine its chemical properties. *See* atom. (p. 21)

electron transport chain The passage of energetic electrons through a series of membrane-associated electron-carrier molecules to proton pumps embedded within mitochondrial or chloroplast membranes; as the electrons arrive at the proton pumping channel, their energy drives the transport of protons out across the membrane, leading to the chemiosmotic synthesis of ATP. *See* chemiosmosis. (p. 172)

electrophoresis A technique in which different molecules can be separated and identified by their rate of movement in an electric field. (p. 372)

element A substance composed only of atoms of the same atomic number, which cannot be decomposed by ordinary chemical means; one of more than 100 distinct natural or synthetic types of matter that, singly or in combination, compose all materials of the universe. (p. 25)

embryo (Gr. *en*, in + *bryein*, to swell) The early developmental stage of an organism produced from a fertilized egg; in plants, a young sporophyte, before its initial period of rapid growth; in animals, a young organism before it emerges from the egg, or from the body of its mother; in humans, refers to the first 2 months of intrauterine life. (p. 861)

embryo sac The female gametophyte of angiosperms, generally an eight-nucleate, seven-celled structure; the seven cells are the egg cell, two synergids and three antipodals (each with a single nucleus), and the central cell (with two nuclei). (p. 679)

endergonic (Gr. *endon*, within + *ergon*, work) Describing a chemical reaction that requires energy; energy from an outside source must be added before the reaction proceeds; a thermodynamically "uphill" process; opposite of exergonic. (p. 139)

endocrine gland (Gr. *endon*, within + *krinein*, to separate) Ductless gland that secretes hormones into the extracellular spaces, from which they diffuse into the circulatory system; in vertebrates, includes the pituitary, sex glands, adrenal, thyroid, and others. (p. 888)

endocytosis (Gr. *endon*, within + *kytos*, hollow vessel) The uptake of material into cells by inclusion within an invagination of the plasma membrane; the material becomes trapped within a vacuole when the edges of the invagination fuse together. If solid material is included, the uptake is called phagocytosis; if dissolved material, it is called pinocytosis. (p. 579)

endoderm (Gr. *endon*, within + *derma*, skin) One of the three embryonic germ layers of early vertebrate embryos, destined to give rise to the epithelium that lines certain internal structures, such as most of the digestive tract and its outgrowths, most of the respiratory tract, and the urinary bladder, liver, pancreas, and some endocrine glands. (p. 885)

endodermis (Gr. *endon*, within + *derma*, skin) In vascular plants, a layer of cells forming the innermost layer of the cortex in roots and some stems. The endodermis is characterized by a Casparian strip within radial and transverse walls. (p. 719)

endomembrane system A dynamic system of membranes and the organelles formed from them within a eukaryotic cell. (p. 88)

endometrium (Gr. *endon*, within + *metrios*, of the womb) The lining of the uterus in mammals; thickens in response to secretion of estrogens and progesterone and is sloughed off in menstration. (p. 1138)

endoplasmic reticulum (Gr. *endon*, within + *plasma*, from cytoplasm; L. *reticulum*, network) An extensive system of membranes present in most eukaryotic cells, dividing the cytoplasm into compartments and channels; those portions containing a dense array of ribosomes are called "rough ER," and other portions with fewer ribosomes are called "smooth ER." (p. 79)

endorphin One of a group of small neuropeptides produced by the vertebrate brain; like morphine, endorphins modulate pain perception; they are also implicated in many other functions. (p. 983)

endosperm (Gr. *endon*, within + *sperma*, seed) A storage tissue characteristic of the seeds of angiosperms, which develops from the union of a male nucleus and the polar nuclei of the embryo sac. The endosperm is digested by the growing sporophyte either before the maturation of the seed or during its germination. (p. 679)

endothermic (Gr. *endon*, within + *therme*, heat) An organism that regulates its body temperature internally through metabolic processes, as do birds and mammals; contrasts with ectotherm. *See* also homeotherm. (p. 874)

energy Capacity to do work. (p. 135)

entropy (Gr. *en*, in + *tropos*, change in manner) A measure of the randomness or disorder of a system; a measure of how much energy in a system has become so dispersed (usually as evenly distributed heat) that it is no longer available to do work. (p. 137)

enzyme (Gr. *enzymes*, leavened, from *en*, in + *zyme*, leaven) A protein that is capable of speeding up specific chemical reactions by lowering the required activation energy, but is unaltered itself in the process; a biological catalyst. (p. 42)

epidermis (Gr. *epi*, on or over + *derma*, skin) The outermost layers of cells; in plants, the exterior primary tissue of leaves, young stems, and roots; in vertebrates, the nonvascular external layer of skin, or ectodermal origin; in invertebrates, a single layer of ectodermal epithelium. (p. 707)

epididymis (Gr. *epi*, on + *didymos*, testicle) A sperm storage vessel; a coiled part of the sperm duct that lies near the testis. (p. 1135)

epinephrine *See* adrenaline.

epiphyte (Gr. *epi*, on + *phyton*, plant) A plant that grows on another organism but is not parasitic on it. (p. 480)

epistasis (Gr. *epi*, on + *stasis*, a standing still) Interaction between two nonallelic genes in which one of them modifies the phenotypic expression of the other; the masking or prevention of the expression of one gene by another gene at another locus. (p. 251)

epithelium (Gr. *epi*, on + *thele*, nipple) In animals, a type of tissue that covers an exposed surface or lines a tube or cavity. (p. 887)

equilibrium (L. *aequus*, equal + *libra*, balance) A stable condition; a system in which no further net change is occurring; the point at which a chemical reaction proceeds as rapidly in the reverse direction as it does in the forward direction, so that there is no further net change in the concentrations of products or reactants. (p. 907)

erythrocyte (Gr. *erythros*, red + *kytos*, hollow vessel) Red blood cell, the carrier of hemoglobin. In mammals, erythrocytes lose their nuclei, whereas in other vertebrates the nuclei are retained. (p. 891)

estrogens (Gr. *oestros*, frenzy + *genos*, origin) A group of steroid hormones, such as estradiol, that affect female secondary sex characteristics, estrus, and the human menstrual cycle. (p. 1187)

estrus (L. *oestrus*, frenzy) The period of maximum female sexual receptivity, associated with ovulation of the egg; being "in heat." (p. 1139)

ethology (Gr. *ethos*, habit or custom + *logos*, discourse) The study of patterns of animal behavior in nature. (p. 1178)

ethylene A simple hydrocarbon that is a plant hormone involved in the ripening of fruit; $H_2C{=}CH_2$. (p. 756)

etiolation (Fr. *etioler*, to blanch) A condition that develops when plants are grown with insufficient or no light; it involves increased stem elongation, poor leaf development, and lack of chlorophyll. (p. 774)

euchromatin (Gr. *eu*, good + *chroma*, color) That portion of eukaryotic chromosomes that is transcribed into mRNA; contains active genes. (p. 206)

eukaryote (Gr. *eu*, good + *karyon*, kernel) A cell characterized by membrane-bound organelles, most notably the nucleus, and one that possesses chromosomes whose DNA is associated with proteins; an organism composed of such cells; contrasts with prokaryote. (p. 70)

eutrophication Process whereby a body of fresh water becomes enriched with nutrients, increases in productivity, and accumulates organic debris. (p. 500)

evolution (L. *evolvere*, to unfold) Genetic change in a population of organisms; in general, evolution leads to progressive change from simple to complex. Darwin proposed that natural selection was the mechanism of evolution. (p. 15)

exergonic (L. *ex*, out + Gr. *ergon*, work) An energy-yielding process or chemical reaction; energy is released from the reactants, so that the products contain less chemical potential energy than the reactants; a "downhill" process that will proceed spontaneously. (p. 139)

exocrine glands (Gr. *ex*, out of + *krinein*, to separate) A type of gland that releases its secretion through a duct, such as digestive glands and sweat glands; contrasts with endocrine. (p. 888)

exocytosis (Gr. *ex*, out of + *kytos*, vessel) A type of bulk transport out of cells; cytoplasmic particles are encased within membranes, forming a vacuole that is transported to the cell surface; there, the vacuole membrane fuses with the cell membrane, discharging the vacuole's contents to the outside. (p. 120)

exon (Gr. *exo*, outside) A segment of DNA that is both transcribed into RNA and translated into protein, specifying the amino acid sequence of part of a polypeptide; contrasts with intron. Exons are characteristic of eukaryotes. (p. 311)

exoskeleton (Gr. *exo*, outside + *skeletos*, hand) An external skeleton, as in arthropods. (p. 827)

exteroception (L. *exter*, outward, + *capere*, to take) The condition when a sense organ is excited by stimuli from the external world. (p. 941)

F₁ (first filial generation) The offspring resulting from a cross; the parents of the cross are referred to as the parental generation. (p. 237)

F₂ (second filial generation) The offspring resulting from a cross between members of the F₁ generation; if these F₂ offspring were to mate and produce progeny, the progeny would be the F₃ generation, and so forth. (p. 237)

facilitated diffusion Carrier-assisted diffusion; the transport of molecules across a cellular membrane through specific channels (carrier molecules embedded in the membrane) from a region of high concentration to a region of low concentration; the process is driven by the concentration difference and does not require energy. The chief difference from free diffusion is that the membrane is impermeable to the molecule except for passage through the carrier channels, contrasts with active transport, which is in the direction of higher concentration and requires energy. (p. 121)

family A taxonomic group made up of one or more genera; one of the subdivisions of an order. The ending of family names in animals and heterotrophic protists is *-idae;* in all other organisms it is *-aceae.* (p. 562)

fat A molecule composed of glycerol and three fatty acid molecules; the proportion of oxygen to carbon is much less in fats than it is in carbohydrates; fats in the liquid state are called oils. (p. 47)

fatty acid A long hydrocarbon chain ending with a —COOH group; fatty acids are components of fats, oils, phospholipids, and waxes. (p. 47)

feedback inhibition Control mechanism whereby an increase in the concentration of some molecule inhibits the synthesis of that molecule; more generally, the regulation of the level of any factor sensitive to its own magnitude; important in the regulation of enzyme and hormone levels, ion concentrations, temperature, and many other factors. (p. 150)

fermentation (L. *fermentum*, ferment) The enzyme-catalyzed extraction of energy from organic compounds without the involvement of oxygen; the enzymatic conversion, without oxygen, of carbohydrates to alcohols, acids, and carbon dioxide; the conversion of pyruvate to ethanol or lactic acid. (p. 164)

ferredoxin One of a class of electron-transferring proteins characterized by high iron content; some are involved in photosynthetic photophosphorylation. (p. 186)

fertilization The fusion of two haploid gamete nuclei to form a diploid zygote nucleus. (p. 219)

fetus (L., pregnant) An unborn or unhatched vertebrate that has passed through the earliest development stages; in humans, a developing individual is a fetus from about the second month of gestation until birth. (p. 1155)

filtration The filtering of blood in the kidney; blood plasma is forced, under pressure, out of the glomerular capillaries into Bowman's capsule, through which it enters the renal tubule; the filtrate contains water and ions (which are recovered) and metabolic wastes (which are eliminated as urine), but not red blood cells or large proteins, which are too large to pass through the glomerular capillary wall. (p. 1117)

fission (L., a splitting) Asexual reproduction by a division of the cell or body into two or more parts of roughly equal size. *See* asexual, clone. (p. 203)

fitness The genetic contribution of an individual to succeeding generations, relative to the contributions of other individuals in the population. (p. 383)

flagellum, pl. **flagella** (L. *flagellum*, whip) A fine, long, threadlike organelle protruding from the surface of a cell; in bacteria, a single protein fiber, capable of rotary motion, that propels the cell through the water; in eukaryotes, an array of microtubules with a characteristic internal 9 + 2 microtubule structure, capable of vibratory but not rotary motion; used in locomotion and feeding; common in protists and motile gametes. A cilium is a small flagellum. (p. 79)

flower The reproductive structure of an angiosperm, which contains at least one stamen or one carpel, and may contain both kinds of structures and sepals and petals as well. (p. 677)

food chain A portion of a food web. A sequence of prey species and the predators that consume them. (p. 158)

food web The food relationships within a community. A diagram of who eats whom. (p. 497)

fossil fuels The altered remains of once-living organisms that are burned to release energy. Examples are coal, oil, and natural gas. (p. 490)

fovea (L., a small pit) A small depression in the center of the retina with a high concentration of cones; the area of sharpest vision. (p. 956)

free energy Energy available to do work. (p. 137)

free energy change The total change in usable energy that results from a chemical reaction or other process; equal to the change in total energy (the heat content of enthalpy) minus the change in unavailable energy (the disorder or entropy times temperature). *See* entropy. (p. 138)

frond The leaf of a fern; any large, divided leaf. (p. 672)

fruit In angiosperms, a mature, ripened ovary (or group of ovaries), containing the seeds; also applied informally to the reproductive structures of some other kinds of organisms. (p. 678)

gametangium, pl. **gametangia** (Gr. *gamein*, to marry + L. *tangere*, to touch) A cell or organ in which gametes are formed. (p. 618)

gamete (Gr., wife) A haploid reproductive cell; upon fertilization, its nucleus fuses with that of another gamete of the opposite sex; the resulting diploid cell (zygote) may develop into a new diploid individual, or, in some protists and fungi, may undergo meiosis to form haploid somatic cells. (p. 207)

gametophyte In plants, the haploid (1n), gamete-producing generation, which alternates with the diploid (2n) sporophyte. (p. 628)

ganglion, pl. **ganglia** (Gr., a swelling) An aggregation of nerve cell bodies; in invertebrates, ganglia are the integrative centers; in vertebrates, the term is restricted to aggregations of nerve cell bodies located outside the central nervous system. (p. 919)

gap junction A junction between adjacent animal cells that allows the passage of materials between the cells; a system of pipes joining the cytoplasms of two adjacent cells. (p. 130)

gastrula (Gr., little stomach) In vertebrates, the embryonic stage in which the blastula with its single layer of cells turns into a three-layered embryo made up of ectoderm, mesoderm, and endoderm, surrounding a cavity (archenteron) with one opening (blastopore). (p. 782)

gene (Gr. *genos*, birth, race) The basic unit of heredity; a sequence of DNA nucleotides on a chromosome that encodes a protein, tRNA, or rRNA molecule, or regulates the transcription of such a sequence. (p. 239)

gene conversion Alteration of one homologous chromosome by the cell's error-detection and repair system to make it resemble the other homologue. (p. 342)

gene frequency The relative occurrence of a particular allele in a population. (p. 372)

genetic code The "language" of the genes, dictating the correspondence between nucleotide sequence in DNA and amino acid sequence in proteins; a series of 64 different three-nucleotide sequences, or triplets (codons); except for three "stop" signals, each codon corresponds to one of the 20 amino acids. (p. 305)

genetic drift Random fluctuation in allele frequencies over time by chance. (p. 375)

genome (Gr. *genos*, offspring + L. *oma*, abstract group) The total genetic constitution of an organism; in bacteria, all of the genes in the main circular chromosome or in associated plasmids; in a eukaryote, all the genes in a haploid set of chromosomes. (p. 203)

genotropism *See* gravitropism.

genotype (Gr. *genos*, offspring + *typos*, form) The total set of genes present in the cells of an organism, as contrasted with the phenotype, which is the realized expression of these genes; often also used to refer to the genetic constitution underlying a single trait or set of traits. (p. 239)

genus, pl. **genera** (L., race) A taxonomic group that includes species; families are divided into genera. (p. 561)

germination (L. *germinare*, to sprout) The resumption of growth and development by a spore or seed. (p. 721)

germline cells Gametes, or cells that give rise directly to gametes. (p. 325)

gibberellins (*Gibberella*, a genus of fungi) A group of plant growth hormones, the best-known effect of which is on the elongation of plant stems. (p. 755)

gill A respiratory organ of aquatic animals, usually a thin-walled projection from some part of the external body surface, endowed with a rich capillary bed and having a large surface area; in basidiomycete fungi, the plates on the underside of the cap. (p. 811)

gland (L. *glans*, acorn) A cluster of secretory cells; in animals, typically an organ composed of modified epithelial cells specialized to produce one or more secretions. (p. 887)

glomerulus (L., a little ball) In the vertebrate kidney, a cluster of capillaries enclosed by Bowman's capsule; also a small spongy mass of tissue in the proboscis of hemichordates, presumed to have an excretory function. Also, a concentration of nerve fibers situated in the olfactory bulb. *See* Bowman's capsule, filtration. (p. 1117)

glucagon (Gr. *glukus*, sweet + *ago*, to lead toward) A vertebrate hormone produced in the pancreas that acts to initiate the breakdown of glycogen to glucose subunits and so raise the concentration of blood sugar. (p. 980)

glucose A common six-carbon sugar ($C_6H_2O_6$); the most common monosaccharide in most organisms. (p. 43)

glycerol Three-carbon molecule with three hydroxyl groups attached; combines with fatty acids to form fat or oil. (p. 47)

glycogen (Gr. *glykys*, sweet + *gen*, of a kind) Animal starch; a complex branched polysaccharide that serves as a food reserve in animals, bacteria, and fungi; can be broken down readily into glucose subunits. (p. 1022)

glycolysis (Gr. *glykys*, sweet + *lyein*, to loosen) The anaerobic breakdown of glucose; the enzyme-catalyzed breakdown of glucose to two molecules of pyruvate with the net production of two molecules of ATP. (p. 156)

glyoxysome A small cellular organelle or microbody containing enzymes necessary for the conversion of fats into carbohydrates; glyoxysomes play an important role during seed germination in plants. (p. 94)

Golgi body (after Camillo Golgi, Italian histologist) An organelle present in many eukaryotic cells; consisting of flat, disk-shaped sacs, tubules, and vesicles, it functions as a collecting and packaging center for substances that the cell manufactures for export; also called dictyosome in plants. The terms "Golgi apparatus" and "Golgi complex" are used to refer collectively to all of the Golgi bodies of a given cell. (p. 93)

grana, sing. **granum** (L., grain or seed) In chloroplasts, stacks of membrane-bound disks (thylakoids); the thylakoids contain the chlorophylls and carotenoids and are the sites of the light reactions of photosynthesis. (p. 97)

gravitropism (L. *gravis*, heavy + *tropes*, turning) Growth response to gravity in plants; formerly called geotropism. (p. 769)

ground meristem (Gr. *meristos*, divisible) The primary meristem, or meristematic tissue, that gives rise to the plant body (except for the epidermis and vascular tissues). (p. 704)

guanine (Sp. from Quechua, *huanu*, dung) A purine base found in DNA and RNA; its name derives from the fact that it occurs in high concentration as a white crystalline base, $C_5H_5N_5O$, in guano and other animal excrements. (p. 57)

guard cells Pairs of specialized epidermal cells that surround a stoma; when the guard cells are turgid, the stoma is open, and when they are flaccid, it is closed. (p. 707)

guttation (L. *gutta*, a drop) The exudation of liquid water from leaves due to root pressure. (p. 732)

gymnosperm (Gr. *gymnos*, naked + *sperma*, seed) A seed plant with seeds not enclosed in an ovary; the conifers are the most familiar group. (p. 673)

gynoecium (Gr. *gyne*, woman + *oikos*, house) The aggregate of carpels in the flower of a seed plant. (p. 677)

habitat (L. *habitare*, to inhabit) The environment of an organism; the place where it is usually found. (p. 466)

habituation (L. *habitus*, condition) A form of learning; a diminishing response to a repeated stimulus; the ignoring of an often-repeated stimulus. (p. 1181)

haploid (Gr. *haploos*, single + *ploion*, vessel) Having only one set of chromosomes (*n*), in contrast to diploid (*2n*); characteristic of eukaryotic gametes, of gametophytes in plants, and of some protists and fungi. (p. 207)

Hardy-Weinberg equilibrium A mathematical description of the fact that the relative frequencies of two or more alleles in a population do not change because of Mendelian segregation; allele and genotype frequencies remain constant in a random-mating population in the absence of inbreeding, selection, or other evolutionary forces; usually stated: if the frequency of allele *a* is *p* and the frequency of allele *b* is *q*, then the genotype frequencies after one generation of random mating will always be $p^2(a) + 2pq(ab) + q^2(b)$. (p. 373)

heart In animals, a muscular organ that pumps the blood through a closed circulatory system. (p. 1056)

hemoglobin (Gr. *haima*, blood + L. *globus*, a ball) A globular protein in vertebrate red blood cells and in the plasma of many invertebrates that carries oxygen and carbon dioxide; an essential part of each molecule is an iron-containing heme group, which both binds O_2 and CO_2 and gives blood its red color. (p. 267)

hemophilia (Gr. *haima*, blood + *philios*, friendly) A group of hereditary diseases characterized by failure of the blood to clot and consequent excessive bleeding from even minor wounds; a mutation in a gene encoding one of the protein factors involved in blood clotting. (p. 268)

herb (L. *herba*, grass) A nonwoody seed plant with a relatively short-lived aerial portion. A herb is said to be *herbaceous*. (p. 526)

herbivore (L. *herba*, grass + *vorare*, to devour) Any organism subsisting on plants. Adj., *herbivorous*. (p. 158)

heredity (L. *heredis*, heir) The transmission of characteristics from parent to offspring through the gametes. (p. 67)

hermaphrodite (Gr. *hermaphroditos*, containing both sexes; from Greek mythology, Hermaphroditos, son of Hermes and Aphrodite) An organism with both male and female functional reproductive organs; many plants and some animals such as deepsea fishes are hermaphroditic; hermaphrodites may or may not be self-fertilizing. (p. 1130)

heterochromatin (Gr. *heteros*, different + *chroma*, color) The portion of eukaryotic chromosomes that is not transcribed into RNA; stains intensely in histological preparations; characteristic of centromeres. (p. 206)

heterocyst (Gr. *heteros*, different + *kystis*, bladder) A large, transparent, thick-walled cell that forms under appropriate conditions in the filaments of certain cyanobacteria; nitrogen fixation takes place in heterocysts. (p. 191)

heterokaryotic (Gr. *heteros*, other + *karyon*, kernel) In fungi, having two or more genetically distinct types of nuclei within the same mycelium. (p. 640)

heterosporous (Gr. *heteros*, other + *sporos*, seed) In vascular plants, having spores of two kinds, namely, microspores and megaspores. (p. 661)

heterotroph (Gr. *heteros*, other + *trophos*, feeder) An organism that cannot derive energy from photosynthesis or inorganic chemicals, and so must feed on other plants and animals, obtaining chemical energy by degrading their organic molecules; animals, fungi, and many unicellular organisms are heterotrophs; *see also* autotroph. (p. 154)

heterozygote (Gr. *heteros*, other + *zygotos*, a pair) A diploid individual that carries two different alleles on homologous chromosomes at one or more genetic loci. Adj., *heterozygous*. Opposite of homozygote. (p. 239)

hibernation (L. *hiberna*, winter) In mammals, the passing of winter in a torpid state in which the body temperature drops nearly to freezing and the metabolism drops close to zero. (p. 524)

histamine A chemical secreted by connective tissue cells that stimulates dilation of blood vessels and increases their permeability. It is responsible for many of the symptoms of inflammation and allergy. (p. 1101)

histone (Gr. *histos*, tissue) A group of relatively small, very basic polypeptides, rich in arginine and lysine. An essential component of eukaryotic chromosomes, histones form the core of nucleosomes around which DNA is wrapped in the first stage of chromosome condensation. (p. 206)

HIV *See* human immunodeficiency virus.

holoblastic cleavage (Gr. *holos*, whole + *blastos*, germ) Process in vertebrate embryos in which the cleavage divisions all occur at the same rate, yielding a uniform cell size in the blastula; found in mammals, lancelets, and many aquatic invertebrates that have eggs with a small amount of yolk. (p. 1158)

homeostasis (Gr. *homeos*, similar + *stasis*, standing) The maintaining of a relatively stable internal physiological environment in an organism, or steady-state equilibrium in a population or ecosystem; usually involves some form of feedback self-regulation. (p. 745)

homeotherm (Gr. *homoios*, same or similar + *therme*, heat) An organism, such as a bird or mammal, capable of maintaining a stable body temperature independent of the environmental temperature; at lower environmental temperatures this involves the metabolic generation of heat at considerable expense in terms of ATP utilized;

"warm blooded." Contrasts with poikilotherm. *See* endotherm. (p. 874)

hominid (L. *homo*, man) Any primate in the human family, Hominidae. *Homo sapiens* is the only living representative; *Australopithecus* is an extinct genus. (p. 441)

hominoid (L. *homo*, man) Collectively, hominids and apes; together with the monkeys, hominoids constitute the anthropoid primates. (p. 441)

homokaryotic (Gr. *homos*, same + *karyon*, kernel) In fungi, having nuclei with the same genetic makeup within a mycelium. (p. 640)

homologous chromosome (Gr. *homologia*, agreement) In diploid cells, one chromosome of a pair that carry equivalent genes; chromosomes that associate in pairs in the first stage of meiosis; also called "homologues." (p. 221)

homology (Gr. *homologia*, agreement) A condition in which the similarity between two structures or functions is indicative of a common evolutionary origin. Adj., *homologous*. (p. 396)

homosporous (Gr. *homos*, same or similar + *sporos*, seed) In some plants, production of only one type of spore rather than differentiated types. Compare heterosporous. (p. 661)

homozygous Being a homozygote; the term is usually applied to one or more specific loci, as in "homozygous with respect to the *w* locus" (that is, the genotype is *w/w*). (p. 239)

hormone (Gr. *hormaein*, to excite) A chemical messenger; a molecule, usually a peptide or steroid, that is produced in one part of an organism and triggers a specific cellular reaction in target tissues and organs some distance away. (p. 903)

human immunodeficiency virus The virus responsible for AIDS, a deadly disease that destroys the human immune system. HIV is a retrovirus (its genetic material is RNA) that is thought to have been introduced to humans from African green monkeys. (p. 1109)

hybrid (L. *hybrida*, the offspring of a tame sow and a wild boar) Offspring of two different varieties or of two different species; alternatively, offspring of two parents that differ in one or more heritable characteristics. (p. 237)

hybridization The mating of unlike parents. (p. 234)

hybridoma (contraction of *hybrid* + *myeloma*) A fast-growing cell line produced by fusing a cancer cell (myeloma) to some other cell, such as an antibody-producing cell. *See* monoclonal antibody. (p. 1108)

hydrocarbon (Gr. *hydor*, water + L. *carbo*, charcoal) An organic compound consisting only of carbon and hydrogen atoms. (p. 550)

hydrogen bond A weak and very directional molecular attraction involving hydrogen atoms; produced by the interaction of the partial positive charge of a polar hydrogen atom (typically, hydrogen atoms covalently linked to oxygen or nitrogen, which more strongly attract the shared electron and so render the hydrogen nucleus partially positive) with the partial negative charge of another polar atom (typically an oxygen or nitrogen atom without a covalently bound hydrogen). *See* polar. (p. 31)

hydroid The polyp form of a cnidarian, as distinguished from the medusa form. Any cnidarian of the class Hydrozoa, order Hydroida. (p. 788)

hydrolysis reaction (Gr. *hydoe*, water + *lysis*, loosening) Splitting of one molecule into two parts by addition of H$^+$ and OH$^=$ ions, derived from water. *See opposite terms* condensation reaction, dehydration reaction. (p. 42)

hydrophilic (Gr. *hydor*, water + *philios*, friendly) Having an affinity for water; applied to polar molecules, which readily form hydrogen bonds with water and so readily dissolve in water (are water soluble). *See* polar molecule. (p. 34)

hydrophobic (Gr. *hydor*, water + *phobos*, hating) Repelled by water; refers to nonpolar molecules, which do not form hydrogen bonds with water and so are not soluble in water. (p. 34)

hydrophobic interaction (L. *hydrophobos*, hated by water) The propensity for water molecules to exclude nonpolar molecules, as oil is excluded from water; water tends to form the maximum number of hydrogen bonds, and more hydrogen bonds are possible if nonpolar molecules (which do not form hydrogen bonds) are not present to interfere with the hydrogen bonds between water molecules. Hydrophobic interactions are responsible for much of the three-dimensional structure of proteins, which is why the addition of nonpolar solvents denatures (unfolds) proteins. (p. 34)

hypertonic (Gr. *hyper*, above + *tonos*, tension) Refers to a solution that contains a higher concentration of solute particles; water moves across a semipermeable membrane into hypertonic solution. (p. 117)

hypha, pl. **hyphae** (Gr. *hyphe*, web) A filament of a fungus or oomycete; collectively, the hyphae comprise the mycelium. (p. 626)

hypothalamus (Gr. *hypo*, under + *thalamos*, inner room) A region of the vertebrate brain just below the cerebral hemispheres, under the thalamus; a center of the autonomic nervous system, responsible for the integration and correlation of many neural and endocrine functions. (p. 899)

hypothesis (Gr. *hypo*, under + *tithenai*, to put) A guess as to what might be; a postulated explanation of a phenomenon consistent with available information; no hypothesis is ever *proved* to be correct—all hypotheses are provisional, working ideas that are accepted for the time being but may be rejected in the future if not consistent with data generated by further experiments; a hypothesis that survives many tests and is very unlikely to be discarded is referred to as a "theory." (p. 6)

hypotonic (Gr. *hypo*, under + *tonos*, tension) Refers to a solution that contains a lower concentration of solute particles; water moves across a semipermeable membrane out of a hypotonic solution. (p. 117)

IAA *See* indoleacetic acid.

immune response In vertebrates, a defensive reaction of the body to invasion by a foreign substance or organism; the invader is recognized as foreign or "not-self" by antibodies produced by B cell lymphocytes, and then is eliminated by macrophages. T cells are responsible for protecting an individual from its own antibodies (cell-mediated immunity). *See* antibody, B cell. (p. 1088)

immunoglobulin (L. *immunis*, free + *globus*, globe) An antibody. (p. 1100)

inbreeding The breeding of genetically related plants or animals. In plants, inbreeding results from self-pollination; in animals, inbreeding results from matings between relatives; inbreeding tends to increase homozygosity. (p. 377)

incomplete dominance The ability of two alleles to produce a heterozygous phenotype that is different from either homozygous phenotype. (p. 251)

independent assortment Mendel's second law; the principle that segregation of alternative alleles at one locus into gametes is independent of the segregation of alleles at other loci; only true for gene loci located on different chromosomes, or so far apart on one chromosome that crossing-over is very frequent between the loci. *See* Mendel's second law. (p. 223)

indeterminate growth In plants, unrestricted or unlimited growth, as with a vegetative apical meristem that produces an unrestricted number of lateral organs indefinitely. (p. 689)

indoleacetic acid (IAA) A naturally occurring auxin, one of the plant hormones. (p. 756)

induction Processes by which one group of embryonic cells affects an adjacent group, thereby inducing those cells to differentiate in a manner they otherwise would not have. (p. 1167)

inflammation (L. *inflammare*, from *flamma*, flame) The mobilization of body defenses against foreign substances and infectious agents, and the repair of damage from such agents; involves phagocytosis by macrophages and is often accompanied by an increase in the local temperature. (p. 890)

innate (L. *innatus*, inborn) Describing a characteristic based partly or wholly on inherited gene differences. (p. 1178)

inositol phosphates Small molecules cleaved from the plasma membrane that mediate the cellular responses to insulin. (p. 971)

instinct (L. *instinctus*, impelled) Stereotyped, predictable, genetically programmed behavior. Learning may or may not be involved. (p. 1183)

insulin A peptide hormone produced by the vertebrate pancreas which acts to promote glycogen formation and thus to lower the concentration of sugar in the blood. (p. 980)

integration, neural The summation of the depolarizing and repolarizing effects contributed by all excitatory and inhibitory synapses acting on a neuron. (p. 915)

integument (L. *integumentum*, covering) In plants, the outermost layer or layers of tissue enveloping the nucellus of the ovule; develops into the seed coat. (p. 679)

interferon In vertebrates, a protein produced in virus-infected cells that inhibits viral multiplication. (p. 1086)

interneuron Neuron that transmits nerve impulses from one neuron to another within the central nervous system; an individual may receive impulses from and transmit impulses to many different neurons. (p. 920)

internode In plants, the region of a stem between two successive nodes. (p. 712)

interphase The period between two mitotic or meiotic divisions in which a cell grows and its DNA replicates; includes G1, S, and G2 phases. (p. 209)

intron (L. *intra*, within) Portion of mRNA as transcribed from eukaryotic DNA that is removed by enzymes before the mature mRNA is transplanted into protein. These untranscribed regions comprise the bulk of most eukaryotic genes; typically, the transcribed portion of the gene exists as numerous short segments called "exons" that are scattered in no particular order within a much longer stretch of nontranscribed DNA. Those segments of the background nontranscribed DNA that fall between two exons are called introns. *See* exon. (p. 311)

invagination (L. *in*, in + *vagina*, sheath) The local infolding of a layer of tissue, especially in animal embryos, so as to form a depression or pocket opening to the outside. (p. 1161)

inversion (L. *invertere*, to turn upside down) A reversal in order of a segment of a chromosome; also, to turn inside out, as in embryogenesis of sponges or discharge of a nematocyst. (p. 347)

ion Any atom or molecule containing an unequal number of electrons and protons and therefore carrying a net positive or net negative charge; gain of an extra electron produces a positively charged cation, and loss of an electron produces a negatively charged anion. (p. 22)

ionic bond A chemical bond formed as a result of the mutual attraction of ions of opposite charge; ionic bonds are nondirectional and form between one ion and nearly all ions of opposite charge. (p. 26)

isomer (Gr. *isos*, equal + *meros*, part) One of a group of molecules identical in atomic composition but differing in structural arrangement, for example, glucose and fructose. (p. 324)

isotonic (Gr. *isos*, equal + *tonos*, tension) Refers to two solutions that have equal concentrations of solute particles; if two isotonic solutions are separated by a semipermeable membrane, there will be no net flow of water across the membrane. (p. 117)

isotope (Gr. *isos*, equal + *topos*, place) An alternative form of a chemical element; differs from other atoms of the same element in the number of neutrons in the nucleus; isotopes thus differ in atomic mass. All isotopes have the same chemical behavior, as all contain the same number of protons and electrons. Some isotopes are unstable and emit radiation. (p. 21)

karyotype (Gr. *karyon*, kernel + *typos*, stamp or print) The morphology of the chromosomes of an organism as viewed with a light microscope. (p. 206)

keratin (Gr. *kera*, horn + *in*, suffix used for proteins) A tough, fibrous protein formed in epidermal tissues and modified into skin, feathers, hair, and hard structures such as horns and nails. (p. 100)

kidney In vertebrates, the organ that filters the blood to remove nitrogenous wastes and regulates the balance of water and solutes in blood plasma. *See* Bowman's capsule, filtration, glomerulus. (p. 1117)

kin selection Selection favoring relatives; an increase in the frequency of related individuals (kin) in a population, leading to an increase in the relative frequency in the population of those alleles shared by members of the kin group. (p. 1208)

kinetic energy Energy of motion. (p. 135)

kinesis (Gr. *kinesis*, motion) changes in activity level in an animal that are dependent on stimulus intensity. (see **kinetic energy**) (p. 1194)

kinetochore (Gr. *kinetikos*, putting in motion + *choros*, chorus) Disk-shaped protein structure within the centromere to which the spindle fibers attach during mitosis or meiosis. *See* centromere. (p. 209)

kingdom The chief taxonomic category, for example, Monera or Plantae. In this book we recognize five kingdoms. (p. 563)

Krebs cycle Another name for the citric acid cycle; also called the tricarboxylic acid (TCA) cycle. (p. 157)

labrum (L., a lip) The upper lip of insects and crustaceans situated above or in front of the mandibles. (p. 844)

lamella (L., a little plate) A thin, platelike structure; in chloroplasts, a layer of chlorophyll-containing membranes; in bivalve mollusks, one of the two plates forming a gill; in vertebrates, one of the thin layers of bone laid concentrically around an osteon (Haversian) canal. (p. 989)

larva, pl. **larvae** (L., a ghost) Immature form of an animal that is quite different from the adult and undergoes metamorphosis in reaching the adult form; examples are caterpillars and tadpoles. (p. 1131)

larynx The voice box; a cartilaginous organ that lies between the pharynx and trachea and is responsible for sound production in vertebrates. (p. 1016)

lateral meristems (L. *latus*, side + Gr. *meristos*, divided) In vascular plants, the meristems that give rise to secondary tissue; the vascular cambium and cork cambium. (p. 703)

leaf primordium (L. *primordium*, beginning) A lateral outgrowth from the apical meristem that will eventually become a leaf. (p. 711)

learning The modification of behavior by experience. (p. 929)

lenticels (L. *lenticella*, a small window) Spongy areas in the cork surfaces of stem, roots, and other plant parts that allow interchange of gases between internal tissues and the atmosphere through the periderm. (p. 714)

leucoplast (Gr. *leukos*, white + *plasein*, to form) In plant cells, a colorless plastid in which starch grains are stored; usually found in cells not exposed to light, such as roots and internal stem tissue. (p. 98)

leukocyte (Gr. *leukos*, a white + *kytos*, hollow vessel) A white blood cell; a diverse array of nonhemoglobin-containing blood cells, including phagocytic macrophages and antibody-producing lymphocytes. (p. 891)

lichen Symbiotic association between a fungus and a photosynthetic organism such as a green alga or cyanobacteria. (p. 638)

life cycle The sequence of phases in the growth and development of an organism, from zygote formation to gamete formation. (p. 570)

light reactions First stage of photosynthesis; the absorption by a pigment of a photon of light from sunlight, the transport of the resulting excited electron to a proton pump to drive the chemiosmotic synthesis of ATP or the reduction of NADP$^+$, and the return of the electron to the pigment. (p. 184)

lignin A stiffening substance that is the most abundant polymer in plant cell walls after cellulose. (p. 638)

linked Lack of independent segregation; the tendency for two or more genes to segregate together in a cross owing to the fact that they are located on the same chromosome. (p. 249)

lipase (Gr. *lipos*, fat + *-ase*, enzyme suffix) An enzyme that catalyzes the hydrolysis of fats. (p. 1013)

lipid (Gr. *lipos*, fat) a nonpolar organic molecule; fatlike; one of a large variety of nonpolar hydrophobic molecules that are insoluble in water (which is polar) but that dissolve readily in nonpolar organic solvents; includes fats, oils, waxes, steroids, phospholipids, prostaglandins, and carotenes. (p. 47)

locus, pl. **loci** (L., place) The location of a gene on a chromosome; more precisely, the position of a particular transcription unit on a chromosome. (p. 239)

loop of Henle (after F.G.J. Henle, German pathologist) In the kidney of birds and mammals, a hairpin-shaped portion of the renal tubule in which water and salt are reabsorbed from the glomerular filtrate by diffusion. (p. 1120)

lymphatic system In animals, an open vascular system that reclaims water which has entered interstitial regions from the bloodstream (lymph); also transports fat from small intestine to the bloodstream; consists of lymph capillaries, which begin blindly in the tissues and lead to a network of progressively larger vessels that empty into the vena cava; also includes the lymph nodes, spleen, thymus, and tonsils. (p. 1060)

lymphocyte (L. *lympha*, water + Gr. *kytos*, hollow vessel) A type of white blood cell. Lymphocytes are responsible for the immune response; there are two principal classes—B cells (which differentiate into antibody-producing plasma cells) and T cells (which interact directly with the foreign invader and are responsible for cell-mediated immunity). (p. 889)

lymphokines A regulatory molecule that is secreted by lymphocytes. In the immune response, lymphokines secreted by helper T cells unleash the cell-mediated immune response. (p. 1097)

lysis (Gr., a loosening) Disintegration of a cell by rupture of its cell membrane. (p. 576)

lysogenic (Gr. *lysis*, a loosening + *genos*, race or descent) Bacteria carrying a silent virus integrated into the chromosome as if it were a bacterial gene; such bacteria carry the seeds of their own destruction (are "lysogenic") because at some future time the virus may initiate active reproduction, causing lysis of the bacterial cells. (p. 579)

lysosome (Gr. *lysis*, a loosening + *soma*, body) A membrane-bound cell organelle containing hydrolytic (digestive) enzymes that are released when the lysosome ruptures; important in recycling worn-out mitochondria and other cellular debris. (p. 95)

macromolecule (Gr. *makros*, large + L. *moleculus*, a little mass) An extremely large molecule; a molecule of very high molecular weight; refers specifically to proteins, nucleic acids, polysaccharides, and complexes of these. (p. 40)

macronutrients (Gr. *makros*, large + L. *nutrire*, to nourish) Inorganic chemical elements required in large amounts for plant growth, such as nitrogen, potassium, calcium, phosphorus, magnesium, and sulfur. (p. 737)

macrophage A large phagocytic cell that is able to engulf and digest cellular debris and invading bacteria. (p. 889)

major histocompatibility complex (MHC) A large group of cell surface antigens unique to each individual; may trigger T cell responses leading to rejection of a tissue or organ transplant. (p. 1084)

Malpighian tubules (Marcello Malpighi, Italian anatomist, 1628-1694) Blind tubules opening into the hindgut of terrestrial arthropods; they function as excretory organs. (p. 833)

mandibles (L. *mandibula*, jaw) In crustacèans, insects, and myriapods, the appendages immediately posterior to the antennae; used to seize, hold, bite, or chew food. (p. 828)

mantle The soft, outermost layer of the body wall in mollusks; the mantle secretes the shell. (p. 811)

mass In chemistry, the total number of protons and neutrons in the nucleus of an atom. Approximately equal to the atomic weight. (p. 22)

matrix (L. *mater*, mother) In mitochondria, the solution in the interior of a mitochondrion, surrounding the cristae; contains the enzymes and other molecules involved in oxidative respiration; more generally, the intercellular substance of a tissue, or that part of a tissue within which an organ or process is embedded. (p. 96)

mechanoreceptors A sensory cell or organ that responds to mechanical stimuli associated with gravity and pressure; includes hearing, touch, and balance. (p. 942)

medulla (L., marrow) (1) The inner portion of an organ, in contrast to the cortex or outer portion, as in the kidney or adrenal gland. (p. 979) (2) Also, the most posterior region of the vertebrate brain, the hindbrain. (p. 1075)

medusa (Gr. mythology, a female monster with snake-entwined hair) A jellyfish, or the free-swimming stage in the life cycle of cnidarians in general; medusae are free swimming and bell or umbrella shaped. (p. 786)

megagametophyte (Gr. *megas*, large + *gamos*, marriage + *phyton*, plant) In heterosporous plants, the female gametophyte; located within the ovule of seed plants. (p. 661)

megasporangium, pl. **megasporangia** A sporangium in which megaspores are produced in heterosporous plants. (p. 661)

megaspore (Gr. *megas*, large + *sporos*, seed) In heterosporous plants, a haploid (1*n*) spore that develops into a female gametophyte; megaspores are usually larger than microspores. (p. 661)

meiosis (Gr. *meioun*, to make smaller) Reduction division; the two successive nuclear divisions in which a single diploid (2*n*) cell forms four haploid (1*n*) nuclei, halving the chromosome number; segregation, crossing-over, and reassortment all occur during meiosis; in animals, meiosis usually occurs in the last two divisions in the formation of the mature egg or sperm; in plants, spores—which divide by mitosis—are produced as a result of meiosis. Compare with mitosis. (p. 219)

Mendel's first law The law of allele segregation; the factors specifying a pair of alternative characteristics (alleles) are separate, and only one may be carried in a particular gamete; gametes combine randomly in forming progeny. Chromosomes had not been observed in Mendel's time, and meiosis was unknown. Modern form: alleles segregate as chromosomes do. (p. 242)

Mendel's second law The law of independent assortment: The inheritance of alternative characteristics (alleles) of one trait is independent of the simultaneous inheritance of other traits—different traits (genes) assort independently. Mendel never tested a pair of traits that were located close together on a chromosome (although two of the seven genes he studied in fact were), and so never discovered that only genes that are not close together assort independently. Modern form:

unlinked genes assort independently. (p. 244)

menstrual cycle (L. *mens*, month) Periodic sloughing off of the blood-enriched lining of the uterus when pregnancy does not occur. The menstrual cycle in primates is the cycle of hormone-regulated changes in the condition of the uterine lining, which is marked by the periodic discharge of blood and disintegrated uterine lining through the vagina (menstruation). (p. 1139)

meristem (Gr. *merizein*, to divide) Undifferentiated plant tissue from which new cells arise. (p. 703)

meroblastic (Gr. *meros*, part + *blastos*, germ) In vertebrates, a pattern of embryonic cleavage divisions in fertilized eggs having a large amount of yolk at the vegetal pole; cleavage divisions occur more rapidly at the animal pole and in extreme cases are restricted to a small area at the animal pole. (p. 1159)

mesenchyme (Gr. *mesos*, middle + *enchyma*, infusion) Embryonic connective tissue; irregular or amoebocytic cells, often embedded in a gelatinous matrix. (p. 784)

mesoderm (Gr. *mesos*, middle + *derma*, skin) One of the three embryonic germ layers that form in the gastrula; gives rise to muscle, bone and other connective tissue, the peritoneum, the circulatory system, and most of the excretory and reproductive systems. (p. 885)

mesoglea (Gr. *mesos*, middle + *glia*, glue) The layer of jellylike or cement material between the epidermis and gastrodermis in cnidarians; may also be used to refer to the jellylike matrix between the epithelial layers in sponges. (p. 786)

mesophyll (Gr. *mesos*, middle + *phyllon*, leaf) The photosynthetic parenchyma of a leaf, located within the epidermis. The vascular strands (veins) run through the mesophyll. (p. 710)

messenger RNA (mRNA) The RNA transcribed from structural genes; RNA molecules complementary to one strand of DNA, which are translated by the ribosomes into protein. (p. 302)

metabolism (Gr. *metabole*, change) The sum of all chemical processes occurring within a living cell or organism; includes carbon fixation (photosynthesis), digestion (catabolism), extraction of chemical energy (respiration), and synthesis of organic molecules (anabolism). (p. 67)

metamorphosis (Gr. *meta*, after + *morphe*, form + *osis*, state of) Process in which there is a marked change in form during postembryonic development, for example, tadpole to frog or larval insect to adult. (p. 827)

metaphase (Gr. *meta*, middle + *phasis*, form) The stage of mitosis or meiosis during which microtubules become organized into a spindle and the chromosomes come to lie in the spindle's equatorial plane. (p. 212)

microbody A cellular organelle bounded by a single membrane and containing a variety of enzymes; generally derived from endoplasmic reticulum; includes peroxisomes and glyoxysomes. (p. 94)

microgametophyte (Gr. *mikros*, small + *gamos*, marriage + *phyton*, plant) In heterosporous plants, the male gametophyte. (p. 661)

micronutrient (Gr. *mikros*, small + L. *nutrire*, to nourish) A mineral required in only minute amounts for plant growth, such as iron, chlorine, copper, manganese, zinc, molybdenum, and boron. (p. 737)

micropyle In the ovules of seed plants, an opening in the integuments through which the pollen tube usually enters. (p. 752)

microspore (Gr. *mikros*, small + *sporos*, seed) In plants, a spore that develops into a male gametophyte; in seed plants, it develops into a pollen grain. (p. 661)

microtubule (Gr. *mikros*, small + L. *tubulus*, little pipe) In eukaryotic cells, a long, hollow protein cylinder, about 25 nanometers in diameter, composed of the protein tubulin. Microtubules influence cell shape, move the chromosomes in cell division, and provide the functional internal structure of cilia and flagella. (p. 98)

microvillus (Gr. *mikros*, small + L. *villus*, shaggy hair) Cytoplasmic projection from epithelial cells, usually containing microtubules; microvilli greatly increase the surface area of the small intestine. (p. 1020)

middle lamella The layer of intercellular material, rich in pectic compounds, that cements together the primary walls of adjacent plant cells. (p. 215)

mimicry (Gr. *mimos*, mime) The resemblance in form, color, or behavior of certain organisms (mimics) to other more powerful or more protected ones (models), which results in the mimics being protected in some way. (p. 478)

mineral A naturally occurring element or inorganic compound. (p. 1028)

mitochondrion, pl. **mitochondria** (Gr. *mitos*, thread + *chondrion*, small grain) The site of oxidative respiration in eukaryotes; a bacterium-like organelle found within the cells of all but one species of eukaryote. Mitochondria contain the enzymes catalyzing the citric acid cycle, which generates electrons that drive proton pumps within the mitochondrial membrane and so fosters the chemiosmotic synthesis of ATP. Almost all of the ATP of nonphotosynthetic eukaryotic cells is produced in mitochondria. (p. 79)

mitosis (Gr. *mitos*, thread) Somatic cell division; nuclear division in which the duplicated chromosomes separate to form two genetically identical daughter nuclei; usually accompanied by cytokinesis, producing two daughter cells. After mitosis, the chromosome number in each daughter nucleus is the same as it was in the original dividing cell. Mitosis is the basis of reproduction of single-celled eukaryotes, and of the physical growth of multicellular eukaryotes. (p. 207)

molecule (L. *moliculus*, a small mass) A collection of two or more atoms held together by chemical bonds; the smallest unit of a compound that displays the properties of the compound. (p. 26)

molting Shedding of all or part of an organisms's outer covering; in arthropods, periodic shedding of the exoskeleton to permit an increase in size. (p. 847)

monoclonal antibody An antibody of a single type that is produced by genetically identical plasma cells (clones); a monoclonal antibody is typically produced from a cell culture derived from the fusion product of a cancer cell and an antibody-producing cell. *See* hybridoma. (p. 1108)

monocot Short for monocotyledon; a flowering plant in which the embryos have only one cotyledon, the floral parts are generally in threes, and the leaves typically are parallel-veined. Compare dicot. (p. 685)

monocyte (Gr. *monos*, single + *kytos*, hollow vessel) A type of leukocyte that becomes a phogocytic cell (macrophage) after moving into tissues. (p. 1063)

monoecious (Gr. *monos*, single + *oecos*, house) A plant in which the staminate and pistillate flowers are separate, but borne on the same individual. (p. 691)

monosaccharide (Gr. *monos*, one + L. *saccharum*, sugar). A simple sugar that cannot be decomposed into smaller sugar molecules; the most common are five-carbon pentoses (such as ribose) and six-carbon hexoses (such as glucose). (p. 42)

morphogenesis (Gr. *morphe*, form + *genesis*, origin) The development of form; construction of the architectural features of organisms: the formation and differentiation of tissues and organs. (p. 1169)

morphology (Gr. *morphe*, form + *logos*, discourse) The study of form and its development; includes cytology (the study of cell structure, histology (the study of tissue structure), and external anatomy (the study of gross structure). (p. 471)

morula (L., a little mulberry) Solid ball of cells in the early stage of embryonic development. (p. 1158)

motor neuron Neuron that transmits nerve impulses from the central nervous system to an effector, which is typically a muscle or a gland; an efferent neuron. (p. 903)

mRNA *See* messenger RNA.

multigene family A collection of related genes on a chromosome; a family of genes duplicated by unequal crossing-over whose members have since diverged in nucleotide sequence but are clearly related to one another. The majority of eukaryotic genes appear to be members of multigene families. (p. 345)

muscle fiber Muscle cell; a long, cylindrical, multinucleated cell containing numerous myofibrils, which is capable of contraction when stimulated. (p. 893)

mutagen (L. *mutare*, to change + Gr. *genaio*, to produce) An agent that induces changes in DNA (mutations): includes physical agents that damage DNA and chemicals that alter one or more DNA bases, link them together, or delete one or more of them. An X ray is an example of a physical mutagen that breaks both strands of a DNA molecule; cigarette tar is an example of a chemical that produces cancer-inducing mutations in lung cells. (p. 322)

mutant (L. *mutare*, to change) A mutated gene; alternatively, an organism carrying a gene that has undergone a mutation. (p. 246)

mutation A permanent change in a cell's DNA; includes changes in nucleotide sequence, alteration of gene position, gene loss or duplication, and insertion of foreign sequences. (p. 266)

mutualism (L. *mutuus*, lent, borrowed) The living together of two or more organisms in a symbiotic association in which both members benefit. (p. 480)

mycelium (Gr. *mykes*, fungus) In fungi or oomycetes, a mass of hyphae. (p. 639)

mycology The study of fungi. One who studies fungi is called a *mycologist*. (p. 637)

mycorrhiza, pl. **mycorrhizae** (Gr. *mykes*, fungus + *rhiza*, root) A symbiotic association between fungi and the roots of a plant. (p. 494)

myelin sheath (Gr. *myelinos*, full of marrow) A fatty layer surrounding the long axons of motor neurons in the peripheral nervous system of vertebrates; made up of the membranes of Schwann cells. (p. 904)

myofibril (Gr. *myos*, muscle + L. *fibrilla*, little fiber) A contractile microfilament within muscle, composed of myosin and actin. (p. 891)

myosin (Gr. *mys*, muscle + *in*, belonging to) One of the two protein components of vertebrate muscle (the other is actin). (p. 998)

NAD (nicotinamide adenine dinucleotide) A coenzyme that functions as an electron acceptor in many of the oxidation reactions of respiration. NAD$^+$ is the oxidized form of NAD; NADH is the reduced form. (p. 56)

NADP (nicotinamide adenine dinucleotide phosphate) A coenzyme that functions as an electron donor in many of the reduction reactions of biosynthesis. NADP$^+$ is the oxidized form, and NADPH$_2$, the reduced form of NADP. (p. 188)

natural selection The differential reproduction of genotypes; caused by factors in the environment; leads to evolutionary change. (p. 11)

nectar (Gr. *nektar*, the drink of the gods) A sugary fluid that attracts insects to plants. Nectar is produced in structures called nectaries. (p. 688)

negative feedback A homeostatic control mechanism whereby an increase in some substance or activity inhibits the process leading to the increase; also known as feedback inhibition. (p. 897)

nematocyst (Gr. *nema*, thread + *kystos*, bladder) In cnidarians, a specialized cellular capsule containing a tiny barb with a poisonous, paralyzing substance, which can be discharged against predator or prey. (p. 787)

neotony (Gr. *neos*, new + *teinein*, to extend) The attainment of sexual maturity in the larval condition. Also, the retention of larval characters into adulthood. (p. 1036)

nephridium, pl. **nephridia** (Gr. *nephros*, kidney) In invertebrates such as earthworms, a tubular excretory structure. (p. 812)

nephron (Gr. *nephros*, kidney) Functional unit of the vertebrate kidney; one of numerous tubules (a human kidney contains about 1 million) involved in filtration and selective reabsorption of blood; each nephron consists of a Bowman's capsule, an enclosed glomerulus, and a long attached tubule; in humans, called a renal tubule. (p. 1117)

nerve A group or bundle of nerve fibers (axons) with accompanying neurological cells, held together by connective tissue; located in the peripheral nervous system (a bundle of nerve fibers within the central nervous system is known as a tract). (p. 887)

nerve impulse An action potential; a rapid, transient, self-propagating reversal in electric potential that travels along the membrane of a neuron. (p. 906)

nerve net In some invertebrates, neurons dispersed through epithelium yet functionally linked to sensory cells, to each other, and to muscle tissue. Permits diffuse response to stimuli; unlike nervous system, has little orientation to information flow (the neurons can carry signals in either direction). (p. 920)

nervous system All the nerve cells of an animal. In humans, the nervous system consists of sensory receptors, efferent sensory nerves that carry sensory information to the brain, interneurons within the brain and spinal cord, and afferent motor neurons that carry commands to muscles and glands. (p. 918)

neural crest A special strip of cells that develops just before the neural groove closes over to form the neural tube; appearance of the neural crest was a key event in the evolution of vertebrates, because these cells ultimately develop into the structures characteristic of the vertebrate body. (p. 1163)

neuroglia (Gr. *neuron*, nerve + *glia*, glue) Nonconducting nerve cells that are intimately associated with neurons and appear to provide nutritional support; in vertebrates they represent at least half of the volume of the nervous system, yet the function of most is not well understood. (p. 904)

neuron (Gr., nerve) A nerve cell specialized for signal transmission; includes cell body, dendrites, and axon. (p. 21)

neurotransmitter (Gr. *neuron*, nerve + L. *trans*, across + *mittere*, to send) A chemical released at the axon terminal of a neuron that travels across the synaptic cleft, binds a specific receptor on the far side, and, depending on the nature of the receptor, depolarizes or hyperpolarizes a second neuron or a muscle or gland cell. (p. 904)

neutron (L. *neuter*, neither) An uncharged subatomic particle of about the same size and mass as a proton but having no electric charge. (p. 893)

niche The role played by a particular species in its environment. (p. 405)

nicotinamide adenine dinucleotide *See* NAD.

nicotinamide adenine dinucleotide phosphate *See* NADP.

nitrification The oxidation of ammonia or ammonium to nitrites, carried out by certain soil bacteria. (p. 492)

nitrogen cycle Worldwide circulation and reutilization of nitrogen atoms, chiefly caused by metabolic processes of living organisms. Plants take up inorganic nitrogen and convert it into organic compounds (chiefly proteins), which are assimilated into the bodies of one or more animals; excretion and bacterial and fungal action on dead organisms return nitrogen atoms to the inorganic state. (p. 490)

nitrogen fixation The incorporation of atmospheric nitrogen into nitrogen compounds, a process that can be carried out only by certain microorganisms. (p. 190)

node (L. *nodus*, knot) The part of a plant stem where one or more leaves are attached; *see* internode. (p. 712)

noradrenaline A hormone produced by the medulla of the adrenal gland which increases the concentration of sugar in the blood, raises blood pressure and heart rate, and increases muscular power and resistance to fatigue; also one of the principal neurotransmitters of the parasympathetic nervous system; sometimes called *norepinephrine*. (p. 936)

notochord (Gr. *noto*, back + L. *chorda*, cord) In chordates, a dorsal rod of cartilage that runs the length of the body and forms the primitive axial skeleton in the embryos of all chordates; in most adult chordates the notochord is replaced by a vertebral column that forms around (but not from) the notochord. (p. 861)

nucellus (L. *nucella*, a small nut) Tissue composing the chief part of the young ovule, in which the embryo sac develops; equivalent to a megasporangium. (p. 679)

nuclear membrane (L. *nucleus*, a kernel) The double membrane that surrounds the nucleus within a eukaryotic cell; the outer membrane is often continuous with endoplasmic reticulum. Also called the nuclear envelope. (p. 79)

nucleic acid A nucleotide polymer; a long chain of nucleotides; chief types are deoxyribonucleic acid (DNA), which is double stranded, and ribonucleic acid (RNA), which is typically single stranded. (p. 56)

nucleolar organizer region A special area on certain chromosomes associated with the formation of the nucleolus. (p. 345)

nucleolus (L., a small nucleus) In eukaryotes, the site of rRNA synthesis; a spherical body composed chiefly of rRNA in the process of being transcribed from multiple copies of rRNA genes. (p. 92)

nucleosome (L. *nucleus*, kernel + *soma*, body) The fundamental packaging unit of eukaryotic chromosomes; a complex of DNA and histone proteins in which one-and-three quarters turns of the double-helical DNA are wound around eight molecules of histone. Chromatin is composed of long strings of nucleosomes, like beads on a string. (p. 206)

nucleotide A single unit of nucleic acid, composed of a phosphate, a five-carbon sugar (either ribose or deoxyribose), and a purine or a pyrimidine. (p. 287)

nucleus In atoms, the central core, containing positively charged protons and (in all but hydrogen) electrically neutral neutrons. In eukaryotic cells, the membranous organelle that houses the chromosomal DNA; in the central nervous system, a cluster of nerve cell bodies. (p. 70)

nymph (L. *nympha*, nymph, bride) An immature stage (following hatching) of a hemimetabolous insect that lacks a pupal stage. (p. 848)

obligate anaerobe An organism that is metabolically active only in the absence of oxygen. (p. 595)

ocellus, pl. **ocelli** (L., little eye) A simple light receptor common among invertebrates. (p. 831)

olfactory (L. *olfacere*, to smell) Pertaining to smell. (p. 950)

ommatidium, pl. **ommatidia** (Gr., little eye) The visual unit in the compound eye of arthropods; contains light-sensitive cells and a lens able to form an image. (p. 830)

oncogene (Gr. *oncos*, cancer + *genos*, birth) A cancer-causing gene; a mutant form of a growth-regulating gene that is inappropriately "on," causing unrestrained cell growth and division. Cancer appears to result only when several such controls have been abrogated. (p. 328)

ontogeny (Gr. *ontos*, being + *geneia*, act of being born, from *genes*, born) The course of development of an individual from egg to senescence. (p. 1168)

oocyte (Gr., *oion*, egg + *kytos*, vessel) A cell that gives rise to an ovum by meiosis. (p. 1136)

operator A site of gene regulation; a sequence of nucleotides overlapping the promoter site and recognized by a repressor protein. Binding of the repressor prevents binding of the polymerase to the promoter site and so blocks transcription of the structural gene. (p. 315)

operon (L. *operis*, work) A cluster of adjacent structural genes transcribed as a unit into a single mRNA molecule; transcription of all the genes of an operon is regulated coordinately by controlling binding of RNA polymerase to the single promoter site, typically via an adjacent and overlapping operator site. A common mode of gene organization in bacteria, but rare in eukaryotes. (p. 315)

orbital (L. *orbis*, circle) The volume of space surrounding the atomic nucleus in which an electron will be found most of the time. (p. 23)

order A category of classification above the level of family and below that of class; orders are composed of one or more families. (p. 562)

organ (Gr. *organon*, tool) A body structure composed of several different tissues grouped together in a structural and functional unit. (p. 885)

organelle (Gr. *organella*, little tool) Specialized part of a cell; literally, a small organ analogous to the organs of multicellular animals. (p. 79)

organic Pertaining to living organisms in general, to compounds formed by living organisms, and to the chemistry of compounds containing carbon. (p. 40)

organism Any individual living creature, either unicellular or multicellular. (p. 362)

osmoregulation Maintenance of constant internal salt and water concentrations in an organism; the active regulation of internal osmotic pressure. (p. 1115)

osmosis (Gr. *osmos*, act of pushing, thrust) The diffusion of water across a selectively permeable membrane (a membrane that permits the free passage of water but prevents or retards the passage of a solute). In the absence of differences in pressure or volume, the net movement of water is from the side containing a lower concentration of solute to the side containing a higher concentration. (p. 117)

osmotic pressure The potential pressure developed by a solution separated from pure water by a differentially permeable membrane. Measured as the pressure required to stop the osmotic movement of water into a solution, it is an index of the solute concentration of the solution; the higher the solute concentration, the greater the osmotic potential of the solution. *Osmotic potential* is a synonym. (p. 117)

osteoblast (Gr. *osteon*, bone + *blastos*, bud) A bone-forming cell. (p. 989)

outcrossing Breeding with individuals other than oneself or one's close relatives. (p. 378)

ovary (L. *ovum*, egg) (1) In animals, the organ in which eggs are produced. (p. 1136) (2) In flowering plants, the enlarged basal portion of a carpel, which contains the ovule(s); the ovary matures to become the fruit. (p. 1187)

oviduct (L. *ovum*, egg + *ductus*, duct) In vertebrates, the passageway through which ova (eggs) travel from the ovary to the uterus. (p. 817)

ovipary (L. *ovum*, egg + *parere*, to bring forth) Reproduction in which unfertilized eggs are released by the female; fertilization and development of offspring occur outside the maternal body. Adj., *oviparous*. (p. 1130)

ovovivipary (L. *ovum*, egg + *vivere*, to live + *parere*, to bring forth) Reproduction in which fertilized eggs develop within the maternal body without obtaining additional nourishment from it, and hatch there or immediately after laying. Adj., *ovoviviparous*. (p. 1130)

ovulation In animals, the release of an egg or eggs from the ovary. (p. 1141)

ovule (L. *ovulum*, a little egg) A structure in seed plants that contains the female gametophyte and is surrounded by the nucellus and one or two integuments; when mature, an ovule becomes a seed. (p. 678)

ovum, pl. **ova** (L., egg) The egg cell; female gamete. (p. 1137)

oxidation (Fr. *oxider*, to oxidize) Loss of an electron by an atom or molecule. In metabolism, often associated with a gain of oxygen or loss of hydrogen. Oxidation (loss of an electron) and reduction (gain of an electron) take place simultaneously, because an electron that is lost by one atom is accepted by another. Oxidation-reduction reactions are an important means of energy transfer within living systems. (p. 24)

oxidative phosphorylation Using electrons derived from oxidative repiration to make ATP; the energetic electrons are delivered by NADH to a chain of carriers within the mitochondrial membrane that deliver them to proton pumps, driving the chemiosmotic synthesis of ATP. *See* chemiosmosis. (p. 1004)

pacemaker A patch of excitatory tissue in the vertebrate heart that initiates the heartbeat; the evolutionary remnant of the first chamber of the fish heart, it is called the sinoatrial node and is located in the same position, where the superior vena cava enters the right atrium. Mechanical pacemakers, inserted near the heart of certain human patients, have the same function. (p. 1068)

palisade parenchyma (L. *palus*, stake) In plant leaves, the columnar, chloroplast-containing parenchyma cells of the mesophyll. Also called *palisade cells*. (p. 710)

pancreas (Gr. *pan*, all + *kreas*, meat, flesh) In vertebrates, the principal digestive gland; a small gland located between the stomach and the duodenum which produces digestive enzymes and the hormones insulin and glucagon. (p. 1013)

parapodia (Gr. *para*, beside + *pous*, foot) One of the paired lateral processes on each side of most segments in polychete annelids; variously modified for locomotion, respiration, or feeding. (p. 820)

parasexuality In certain fungi, the fusion and segregation of heterokaryotic haploid nuclei to produce recombinant nuclei. (p. 647)

parasitism (Gr. *para*, beside + *sitos*, food) A living arrangement in which an organism lives on or in an organism of a different species and derives nutrients from it. It may be regarded as a form of predation. (p. 480)

parasympathetic nervous system (Gr. *para*, beside + *syn*, with + *pathos*, feeling) In vertebrates, one of two subdivisions of the autonomic nervous system (the other is the sympathetic). The two subdivisions operate antagonistically, the parasympathetic system stimulating resting activities such as digestion and restoring the body to normal after emergencies by inhibiting alarm functions initiated by the sympathetic system. (p. 934)

parenchyma (Gr. *para*, beside + *en*, in + *chein*, to pour) A plant tissue composed of *parenchyma* cells; such cells are living, thin walled, and randomly arranged, and have large vacuoles; usually photosynthetic or storage tissue. (p. 704)

parthenogenesis (Gr. *parthenos*, virgin + *genesis*, birth) The development of an egg without fertilization, as in aphids, bees, ants, and some lizards. (p. 226)

pedigree A family tree showing lines of descent; tool for studying how a human genetic trait is inherited. (p. 263)

penis In reptiles and mammals, the male reproductive organ through which sperm are delivered into the female reproductive tract during sexual intercourse. (p. 1131)

peptide Two or more amino acids linked by peptide bonds. (p. 50)

peptide bond (Gr. *peptein*, to soften, digest) The type of bond that links amino acids together in proteins; formed by removing an OH from the carboxy (—COOH) group of one amino acid and an H from the amino (—NH₂) group of another to form an amide group CO=NH=. (p. 53)

perennial (L. *per*, through + *annus*, a year) A plant that lives for more than a year and produces flowers on more than one occasion. (p. 722)

perianth (Gr. *peri*, around + *anthos*, flower) In flowering plants, the petals and sepals taken together. (p. 678)

pericycle (Gr. *peri*, around + *kykos*, circle) In vascular plants, one or more cell layers surrounding the vascular tissues of the root, bounded externally by the endodermis and internally by the phloem. (p. 719)

periderm (Gr. *peri*, around + *derma*, skin) Outer protective tissue in vascular plants that is produced by the cork cambium and functionally replaces epidermis when it is destroyed during secondary growth; the periderm includes the cork, cork cambium, and phelloderm. (p. 714)

peripheral nervous system (Gr. *peripherein*, to carry around) All of the neurons and nerve fibers outside the central nervous system, including motor neurons, sensory neurons, and the autonomic nervous system. (p. 918)

peristalsis (Gr. *peristaltikos*, compressing around) Pumping by waves of contraction; in animals, a series of alternating contracting and relaxing muscle movements along the length of a tube such as the oviduct or alimentary canal that tend to force material such as an egg cell or food through the tube. (p. 1017)

permeability (L. *permeare*, to pass through) Refers to a membrane through which a specified molecule, ion, or other solute can pass freely. (p. 906)

peroxisome A microbody that plays an important role in glycolic acid metabolism associated with photosynthesis; the site of photorespiration. (p. 94)

petal A flower part, usually conspicuously colored; one of the units of the corolla. (p. 678)

petiole (L. *petiolus*, a little foot) The stalk of a leaf. (p. 708)

pH A measure of the relative concentration of hydrogen ions in a solution; equal to the negative logarithm of the hydrogen ion concentration. pH values range from 0 to 14; the lower the value, the more hydrogen ions it contains (the more acidic it is); pH 7.0 is neutral, less than 7.0 is acidic, more than 7.0 is alkaline. (p. 143)

phage *See* bacteriophage.

phagocyte (Gr. *phagein*, to eat + *kytos*, hollow vessel) Any cell that engulfs and devours microorganisms or other particles. (p. 1084)

phagocytosis (Gr., cell-eating) Endocytosis of solid particles; the cell membrane folds inward around the particle (which may be another cell) and then discharges the contents into the cell interior as a vacuole; characteristic of protists, amoebas, the digestive cells of some invertebrates, and vertebrate white blood cells. (p. 119)

pharynx (Gr., gullet) In vertebrates, a muscular tube that connects the mouth cavity and the esophagus; it serves as the gateway to the digestive tract and to the windpipe (trachea). (p. 798)

phenotype (Gr. *phainein*, to show + *typos*, stamp or print) The realized expression of the genotype; the physical appearance or functional expression of a trait; the result of the biological activity of proteins or RNA molecules transcribed from the DNA. (p. 239)

pheromone (Gr. *pherein*, to carry + *hormonos*, exciting, stirring up) Chemical substance released by the exocrine glands of one organism that influences the behavior or physiological processes of another organism of the same species. Some pheromones serve as sex attractants, as trail markers, and as alarm signals. (p. 847)

phloem (Gr. *phloos*, bark) In vascular plants, a food-conducting tissue basically composed of sieve elements, various kinds of parenchyma cells, fibers, and sclereids. (p. 668)

phosphate group —PO₄; a chemical group commonly involved in high-energy bonds. (p. 287)

phosphodiester bond The type of bond that links nucleotides in a nucleic acid; formed when the phosphate group of one nucleotide binds to the hydroxyl group of another. (p. 56)

phospholipid A phosphorylated lipid; similar in structure to a fat, but only two fatty acids are attached to the glycerol backbone, with the third space linked to a phosphorylated molecule. Phospholipid molecules have a polar hydrophilic "head" end (contributed by the phosphate group) and a nonpolar hydrophobic "tail" end (contributed by the fatty acids); they orient spontaneously in water to form bimolecular membranes in which the nonpolar tails of the molecules are oriented inward toward one another, away from the polar water environment. Phospholipids are the foundation for all cell membranes. *See* lipid, fatty acid. (p. 108)

phosphorylation (Gr. *phosphoros*, bringing light) A reaction in which a phosphate group is added to a compound; phosphorylation often results in the formation of a high-energy bond (as in the formation of ATP from ADP and inorganic phosphate) and also plays a key role in preventing cellular loss of metabolites by diffusion. (By phosphorylating glucose and other sugars, for example, the cell prevents them from diffusing outward to where the concentration is lower because the sugar's phosphate group is repelled by the interior of the cell membrane.) (p. 171)

photon (Gr. *photos*, light) The elementary particle of electromagnetic energy; light. (p. 180)

photoperiodism (Gr. *photos*, light + *periodos*, a period) The tendency of biological reactions to respond to the duration and timing of day and night; a mechanism for measuring seasonal time. (p. 771)

photophosphorylation (Gr. *photos*, light + *phosphoros*, bringing light) The formation of ATP in the chloroplast during photosynthesis. (p. 186)

photoreceptor (Gr. *photos*, light) A light-sensitive sensory cell. (p. 942)

photorespiration The light-dependent production of glycolic acid in chloroplasts and its subsequent oxidation in peroxisomes; this process tends to short-circuit photosynthesis, and becomes progressively more of a drain to plants at higher temperatures. C₄ and CAM photosynthesis are evolutionary responses to this dilemma. (p. 194)

photosynthesis (Gr. *photos*, light + *syn*, together + *tithenai*, to place) The utilization of light energy to create chemical bonds; the synthesis of organic compounds from carbon dioxide and water, using chemical energy (ATP) and reducing power (NADPH) generated by photosynthesis. (p. 192)

photosystem A functional light-trapping unit; an organized collection of chlorophyll and other pigment molecules embedded in the thylakoids of chloroplasts which trap photon energy and channel it in the form of energetic electrons to the thylakoid membrane. (p. 188)

phototropism (Gr. *photos*, light + *trope*, turning) In plants, a growth response to a light stimulus. (p. 762)

phycobilins A group of water-soluble accessory pigments, including phycocyanins and phycoerythrins, which occur in red algae and cyanobacteria. (p. 602)

phycologist (Gr. *phykos*, seaweed) The scientist who studies algae. (p. 612)

phylogeny (Gr. *phylon*, race, tribe) The evolutionary relationships among any group of organisms. (p. 1168)

phylum, pl. **phyla** (Gr. *phylon*, race, tribe) A major category, between kingdom and class, of taxonomic classifications. *Division* is an equivalent term used in all groups except animals and heterotrophic protists. (p. 562)

phytochrome (Gr. *phyton*, plant + *chroma*, color) A plant pigment that is associated with the absorption of life; photoreceptor for red to far-red light; involved in a number of timing processes, such as flowering, dormancy, leaf formation, and seed germination. (p. 773)

pigment (L. *pigmentum*, paint) A molecule that absorbs light. (p. 181)

pinocytosis (Gr. *pinein*, to drink + *kytos*, hollow vessel + *osis*, condition) The taking up of fluid by endocytosis; refers to cells. (p. 119)

pistil (L. *pistillum*, pestle) Central organ of flowers, typically consisting of ovary, style, and stigma; a pistil may consist of one or more fused carpels, and is more technically and better known as the gynoecium. A flower with carpels but no functional stamens is called *pistillate*. (p. 677)

pith The ground tissue occupying the center of the stem or root within the vascular cylinder; usually consists of parenchyma. (p. 712)

pituitary (L. *pituita*, phlegm) Perhaps the most important of the endocrine glands in vertebrates; under the hormonal control of the hypothalamus, which directs it via releasing hormones; the anterior lobe secretes tropic hormones, growth hormone, and prolactin; the posterior lobe stores and releases oxytocin and ADH produced by the hypothalamus. (p. 1187)

placenta, pl. **placentae** (L., a flat cake) (1) In flowering plants, the part of the ovary wall to which the ovules or seeds are attached. (2) In mammals, a tissue formed in part from the inner lining of the uterus and in part from other membranes, through which the embryo (later the fetus) is nourished while in the uterus and wastes are carried away. (p. 878)

plankton (Gr. *planktos*, wandering) Free-floating, mostly microscopic, aquatic organisms. (p. 513)

planula (L. *planulus*, a little wanderer) The ciliated, free swimming type of larva formed by many cnidarians. (p. 786)

plasma (Gr., form) The fluid of vertebrate flood; contains dissolved salts, metabolic wastes, hormones, and a variety of proteins, including antibodies and albumin; blood minus the blood cells. (p. 891)

plasma cell An antibody-producing cell resulting from the multiplication and differentiation of a B lymphocyte that has interacted with an antigen; a mature plasma cell can produce from 3000 to 30,000 antibody molecules per second. (p. 1091)

plasma membrane The membrane surrounding the cytoplasm of an animal cell; the outermost membrane of a cell; consists of a single bilayer of membrane; also called cell membrane and plasmalemma. (p. 79)

plasmid (Gr. *plasma*, a form or mold) A small fragment of extrachromosomal DNA, usually circular, that replicates independently of the main chromosome, although it may have been derived from it. Plasmids make up about 5% of the DNA of many bacteria, but are rare in eukaryotic cells. (p. 337)

plasmodesma, pl. **plasmodesmata** (Gr. *plassein*, to mold + *desmos*, a bond) Minute cytoplasmic connections that connect adjacent cells; in plants they extend through pores in the cell walls. (p. 130)

plasmodium (Gr. *plasma*, a form, mold + *eidos*, form) Stage in the life cycle of myxomycetes (plasmodial slime molds); a multinucleate mass of protoplasm surrounded by a membrane. (p. 625)

plastid (Gr. *plastos*, formed or molded) An organelle in the cells of photosynthetic eukaryotes (plants and algae) that is the site of photosynthesis and, in plants and green algae, of starch storage; plastids are bounded by a double membrane. (p. 98)

platelet (Gr. dim. of *plattus*, flat) In mammals, a fragment of a white blood cell that circulates in the blood and functions in the formation of blood clots at sites of injury. (p. 1064)

pleiotropic (Gr. *pleros*, more + *trope*, a turning) Describing a gene that produces more than one phenotypic effect. (p. 251)

pleural membrane (Gr., side, rib) In vertebrate animals, the membrane that lines each half of the thorax and covers the lungs. (p. 1042)

poikilotherm (Gr. *poikilos*, changeable + *therme*, heat) An animal with a body temperature that fluctuates with that of the environment; cold blooded. *See* ectotherm. (p. 874)

point mutation An alteration of one nucleotide in a chromosomal DNA molecules; includes addition, deletion, or substitution of individual nucleotides. (p. 321)

polar body Minute, nonfunctioning cell produced during the meiotic divisions leading to gamete formation in vertebrates; contains a nucleus but very little cytoplasm. (p. 1137)

polar molecule A molecule with positively and negatively charged ends; one portion of a polar molecule attracts electrons more strongly (is more electronegative) than another portion, with the result that the electron-rich portion carries a partial negative charge contributed by the electron excess,

and the electron-poor portion carries a partial positive charge because of the electron deficit. (p. 31)

pollen grains (L., fine dust) Collectively referred to as term pollen. In seed plants, each pollen grain contains an immature male gametophyte enclosed within a protective outer covering; pollen grains may be two-celled or three-celled when shed. (p. 676)

pollen tube A tube formed after germination of the pollen grain; carries the male gametes into the ovule. (p. 676)

pollination The transfer of pollen from an anther to a stigma. (p. 673)

polygamy (Gr. *polyps*, many + *gamos*, marriate) Condition of having more than one mate at one time. (p. 1205)

polymer (Gr. *polus*, many + *meris*, part) A molecule composed of many similar or identical molecular subunits. Starch is a polymer of glucose. (p. 40)

polymorphism (Gr. *polys*, many + *morphe*, form) The presence in a population of more than one allele of a gene at a frequency greater than that of newly arising mutations; operationally, a population in which the most common allele at a locus has a frequency of less than 99%. (p. 373)

polyp (Fr. *polype*, octopus, from L. *polypus*, many footed) The vase-shaped, sessile, sedentary stage in the life cycle of cnidarians. (p. 786)

polypeptide (Gr. *polys*, many + *peptein*, to digest) A molecule consisting of many joined amino acids; not as complex as a protein. (p. 54)

polyploid (Gr. *polys*, many + *ploin*, vessel) An organism, tissue, or cell with more than two complete sets of homologous chromosomes. (p. 348)

polysaccharide (Gr. *polys*, many + *sakcharon*, sugar, from Latin *sarkara*, gravel, sugar) A sugar polymer; a carbohydrate composed of many monosaccharide sugar subunits linked together in a long chain; examples are glycogen, starch, and cellulose. (p. 45)

population (L. *populus*, the people) Any group of individuals, usually of a single species, occupying a given area at the same time. (p. 451)

posterior Of or pertaining to the rear end. In humans, the "rump" is often referred to as the posterior. (p. 790)

potential energy Energy that is not being used, but could be; energy in a potentially usable form; often called "energy of position." (p. 136)

prey (L. *prehendere*, to grasp, seize) An organism eaten by another organism. (p. 462)

primary endosperm nucleus In flowering plants, the result of the fusion of a sperm nucleus and the (usually) two polar nuclei. (p. 679)

primary growth In vascular plants, growth originating in the apical meristems of shoots and roots, as contrasted with secondary growth; results in an increase in length. (p. 703)

primary structure of a protein The amino acid sequence of a protein. (p. 54)

primary wall In plants, the wall layer deposited during the period of cell expansion. (p. 705)

primitive streak (L. *primus*, first) In the early embryos of birds, reptiles, and mammals, a dorsal, longitudinal strip of ectoderm and mesoderm that is equivalent to the blastopore in other forms. (p. 1162)

primordium, pl. primordia (L. *primus*, first + *ordiri*, to begin to weave) A cell or organ in its earliest stage of differentiation. (p. 711)

procambium (L. *pro*, before + *cambiare*, to exchange) In vascular plants, a primary meristematic tissue that gives rise to primary vascular tissues. (p. 704)

productivity A measure of the rate at which energy is assimilated by an organism or group of organisms. (p. 496)

progesterone (L. *progerere*, to carry forth or out + *steiras*, barren) In mammals, a steroid hormone secreted by the corpus luteum that prepares the uterus for implantation of the fertilized egg (ovum) and maintains the uterus during pregnancy. (p. 1141)

prokaryote (Gr. *pro*, before + *karyon*, kernel) A bacterium; a cell lacking a membrane-bound nucleus or membrane-bound organelles. Prokaryotic cells are more primitive than eukaryotic cells, which evolved from them. (p. 70)

promoter A specific nucleotide sequence on a chromosome to which RNA polymerase attaches to initiate transcription of mRNA from a gene. (p. 315)

prophage Noninfectious bacteriophage units linked with the bacterial chromosome that multiply with the growing and dividing bacteria but do not bring about lysis of the bacteria. Prophage is a stage in the life cycle of a temperate phage. *See* lysogenic. (p. 579)

prophase (Gr. *pro*, before + *phasis*, form) An early stage in nuclear division, characterized by the formation of a microtubule spindle along the future axis of division, the shortening and thickening of the chromosomes, and their movement toward the equator of the spindle (the "metaphase plate"). (p. 209)

proprioceptor (L. *proprius*, one's own) In vertebrates, a sensory receptor that senses the body's position and movements; located deep within the tissues, especially muscles, tendons, and joints. (p. 942)

prostaglandins (Gr. *prostas*, a porch or vestibule + L. *glans*, acorn) A group of modified fatty acids, originally discovered in semen, that function as chemical messengers and have powerful effects on smooth muscle, nerves, circulation, and reproductive organs; synthesized in most, possibly all, cells of the body. (p. 984)

prostate gland (Gr. *prostas*, a porch or vestibule) In male mammals, a mass of glandular tissue at the base of the urethra that secretes an alkaline fluid which has a stimulating effect on the sperm as they are released. (p. 1144)

protease (Gr. *proteios*, primary + *ase*, enzyme ending) An enzyme that digests proteins by breaking peptide bonds. Also called peptidases. (p. 1013)

protein (Gr. *proteios*, primary) A chain of amino acids joined by peptide bonds; a protein typically contains over 100 amino acids and may be composed of more than one polypeptide. (p. 50)

protist (Gr. *protos*, first) A member of the kingdom Protista, which includes the unicellular eukaryotic organisms and some multicellular lines derived from them. (p. 610)

protoderm (Gr. *protos*, first + *derma*, skin) A primary meristematic tissue in vascular plants that gives rise to epidermis. (p. 704)

proton A subatomic, or elementary, particle with a single positive charge equal in magnitude to the charge of an electron and a mass of 1, very close to that of a neutron; the nucleus of a hydrogen atom is composed of a single proton. (p. 21)

proton pump A protein channel in a membrane of the cell that expends energy to transport protons against a concentration gradient; involved in the chemiosmotic generation of ATP. (p. 126)

protostome (Gr. *protos*, first + *stoma*, mouth) An animal in whose embryonic development the mouth forms at or near the blastopore. Protostomes are also characterized by spiral cleavage during the earliest stages of development and by schizocoelous formation of the coelom. (p. 806)

pseudocoel, or pseudocoelom (Gr. *pseudos*, false + *koiloma*, cavity) A body cavity not lined with peritoneum and not a part of the blood or digestive systems, embryonically derived from the blastocoel. (p. 792)

pseudogene (Gr. *pseudos*, false + *genos*, birth) A silent gene; a copy of a gene that is not transcribed. (p. 346)

pseudopod, or pseudopodium (Gr. *pseudes*, false + *pous*, foot) "False foot"; a nonpermanent cytoplasmic extension of the cell body. (p. 629)

pulmonary circulation In terrestrial vertebrates, the pathway of blood circulation leading to and from the lungs. (p. 1067)

punctuated equilibrium A model of the mechanism of evolutionary change which proposes that long periods of little or no change are punctuated by periods of rapid evolution. (p. 399)

pupa (L., girl, doll) A developmental stage of some insects in which the organism is nonfeeding, immotile, and sometimes encapsulated or in a cocoon; the pupal stage occurs between the larval and adult phases. In butterflies, it is called a chrysalis. (p. 848)

purine (Gr. *purinos*, fiery, sparkling) The larger of the two general kinds of nucleotide base found in DNA and RNA; a nitrogenous base with a double-ring structure, such as adenine or guanine. (p. 57)

pyrimidine (alt. of pyridine, from Gr. *pyr*, fire) The smaller of the two general kinds of nucleotide base found in DNA and RNA; a nitrogenous base with a single-ring structure, such as cytosine, thymine, or uracil. (p. 57)

pyruvate The three-carbon compound that is the end product of glycolysis and the starting material of the citric acid cycle. (p. 160)

quaternary structure of a protein The level of aggregation of a globular protein molecule that consists of two or more polypeptide chains. (p. 55)

radial cleavage In animals, a type of embryological development in which early cleavage planes are symmetrical to the polar axis, each blastomere of one tier lying directly above the corresponding blastomere of the next layer; indeterminate cleavage. (p. 807)

radial symmetry (L. *radius*, a spoke of a wheel + Gr. *summetros*, symmetry) The regular arrangement of parts around a central axis such that any plane passing through the central axis divides the organism into halves that are approximate mirror images. (p. 785)

radioactive isotope An unstable isotope of an element that decays or disintegrates spontaneously, emitting radiation. (p. 22)

radula (L., scraper) Rasping tongue found in most mollusks. (p. 811)

recessive allele (L. *recedere*, to recede) An allele whose phenotypic effect is masked in the heterozygote by that of another, dominant allele; heterozygotes, although they contain a copy of the recessive allele, are phenotypically indistinguishable from dominant homozygotes. (p. 380)

reciprocal altruism Performance of an altruistic act with the expectation that the favor will be returned. A key and very controversial assumption of many theories dealing with the evolution of social behavior. *See* altruism. (p. 1208)

recombinant DNA Fragments of DNA from two different species, such as a bacterium and mammal, spliced together in the laboratory into a single molecule; a key component of genetic engineering technology, in which, for example, a gene that makes a certain bacterium resistant to a chemical weed killer is transferred to a crop plant. (p. 354)

recombination The formation of new gene combinations; in bacteria, it is accomplished by the transfer of genes into cells, often in association with viruses; in eukaryotes, it is accomplished by reassortment of chromosomes during meiosis, and by crossing-over. (p. 336)

reduction (L. *reductio*, a bringing back; originally "bringing back" a metal from its oxide) The gain of an electron by an atom; takes place simultaneously with oxidation (loss of an electron by an atom), because an electron that is lost by one atom is accepted by another. (p. 24)

reflex (L. *reflectere*, to bend back) In the nervous system, an "automatic" response to a stimulus; a motor response subject to little associative modification; among the simplest neural pathways, involving only a sensory neuron, sometimes (but not always) an interneuron, and one or more motor neurons. (p. 920)

releasing hormone A peptide hormone produced by the hypothalamus that stimulates or inhibits the secretion of specific hormones by the anterior pituitary. (p. 981)

renal (L. *renes*, kidneys) Pertaining to the kidney. (p. 1120)

replication (L. *replicatio*, a folding back) The production of a second "daughter" molecule of DNA exactly like the first "parent" molecule; the parent molecule is used as a template. (p. 203)

repressor (L. *reprimere*, to press back, keep back) A protein that regulates DNA transcription by preventing RNA polymerase from attaching to the promoter and transcribing the structural gene. *See* operator. (p. 313)

respiration (L. *respirare*, to breathe) The utilization of oxygen; in terrestrial vertebrates, the inhalation of oxygen and the exhalation of carbon dioxide; in cells, the oxidation (electron removal) of food molecules, particularly pyruvate in the citric acid cycle, to obtain energy. (p. 1032)

resting membrane potential The charge difference (difference in electric potential) that exists across a neuron at rest (about 70 millivolts). (p. 906)

restriction endonuclease An enzyme that cleaves a DNA duplex molecule at a particular base sequence ("restriction enzyme"). (p. 351)

reticular (L., *reticulum*, small net) Resembling a net in appearance or structure. (p. 927)

retina (L., a small net) The photosensitive layer of the vertebrate eye; contains several layers of neurons and light receptors (rods and cones); receives the image formed by the lens and transmits it to the brain via the optic nerve. (p. 956)

retrovirus (L. *retro*, turning back) An RNA virus; a virus whose genetic material is RNA. When a retrovirus enters a cell, the cell's machinery "reads" the virus RNA, which contains a gene encoding an enzyme—reverse transcriptase—that then transcribes the virus RNA into duplex DNA, which the cell's machinery replicates as if it were its own; the DNA copies of many retroviruses appear able to enter eukaryotic chromosomes, where they act much like transposons. Retroviruses are associated with many human diseases, including cancer and AIDS. (p. 326)

reverse transcriptase An enzyme that transcribes RNA into DNA; found only in association with retroviruses. (p. 579)

rhizoid (Gr. *rhiza*, root) Slender, rootlike anchoring structure in some fungi, algae, and plant gametophytes. (p. 663)

rhizome (Gr. *rhizoma*, mass of roots) In vascular plants, a usually more or less horizontal underground stem; may be enlarged for storage, or may function in vegetative reproduction. (p. 672)

ribonucleic acid (RNA) A class of nucleic acids characterized by the presence of the sugar ribose (DNA contains deoxyribose instead) and the pyrimidine uracil (DNA contains thymine instead); includes mRNA, tRNA, and rRNA. (p. 58)

ribose A five-carbon sugar. (p. 147)

ribosomal RNA (rRNA) A class of RNA molecules found, together with characteristic proteins, in ribosomes; transcribed from the DNA of the nucleolus. (p. 92)

ribosome The molecular machine that carries out protein synthesis; the most complicated aggregation of proteins in a cell, also containing three different rRNA molecules; may be free in the cytoplasm or in eukaryotes sometimes attached to the membranes of the endoplasmic reticulum. (p. 92)

RNA *See* ribonucleic acid.

RNA polymerase An enzyme that catalyzes the assembly of an mRNA molecule, the sequence of which is complementary to a DNA molecule used as a template. *See* transcription. (p. 302)

rod Light-sensitive nerve cell found in the vertebrate retina; sensitive to very dim light; responsible for "night vision." (p. 956)

root The usually descending axis of a plant, normally below ground, which anchors the plant and serves as the major point of entry for water and minerals. (p. 702)

root hairs In vascular plants, tubular outgrowths of the epidermal cells of the root just back of its apex; most water enters through the root hairs. (p. 707)

root pressure The vascular plants, the pressure that develops in roots as the result of osmosis, which causes guttation of water from leaves and exudation from cut stumps. (p. 733)

salt Ionic substance containing neither H^+ nor OH^-. (p. 27)

sarcomere (Gr. *sarx*, flesh + *meris*, part of) Fundamental unit of contraction in skeletal muscle; repeating bands of actin and myosin that appear between two Z lines. (p. 1001)

sarcoplasm (Gr. *sarx*, flesh + *plasma*, mold) The cytoplasm of a vertebrate striated muscle cell; clear, semifluid substance between fibrils of muscle tissue. (p. 893)

satellite DNA A nontranscribed region of the chromosome with a distinctive base composition; a short nucleotide sequence repeated tandemly many thousands of times. (p. 344)

sclereid (Gr. *skleros*, hard) In vascular plants, a sclerenchyma cell with a thick, lignified, secondary wall having many pits; not elongate like a fiber, the other principal kind of sclerenchyma cell. (p. 705)

sclerenchyma cell (Gr. *skleros*, hard + *en*, in + *chymein*, to pour) A cell of variable form and size with more or less thick, often lignified, secondary walls; may or may not be living at maturity; includes fibers and sclereids. Collectively, sclerenchyma cells may make up a kind of tissue called sclerenchyma. (p. 705)

scrotum (L., bag) The pouch that contains the testes in most mammals. (p. 1134)

secondary cell wall In plants, the innermost layer of the cell wall, formed in certain cells after cell elongation has ceased; secondary walls have a highly organized microfibrillar structure, and are often impregnated with lignin. (p. 705)

secondary growth In vascular plants, an increase in stem and root diameter made possible by cell division of the lateral meristems. Secondary growth produces the secondary plant body. (p. 703)

secondary sex characteristics External differences between male and female animals; not directly involved in reproduction. (p. 1204)

secondary structure of a protein The twisting or folding of a polypeptide chain; results from the formation of hydrogen bonds between different amino acid side groups of a chain; the most common structures that form are a single-stranded helix, an extended sheet, or a cable containing three (as in collagen) or more strands. (p. 55)

seed A structure that develops from the mature ovule of a seed plant; seeds generally consist of seed coat, embryo, and a food reserve. (p. 669)

seed coat The outer layer of a seed, developed from the integuments of the ovule. (p. 753)

segregation of alleles *See* Mendel's first law.

selectively permeable membrane (L. *seligere*, to gather apart + *permeare*, to go through) A membrane that permits passage of water and some solutes but blocks passage of one or more solutes. Same as differentially permeable. Used to be referred to as "semipermeable." (p. 120)

self-fertilization The union of egg and sperm produced by a single hermaphroditic organism. (p. 236)

self-pollination The transfer of pollen from another to stigma in the same flower or to another flower of the same plant, leading to self-fertilization. (p. 696)

semen (L., seed) In reptiles and mammals, sperm-bearing fluid expelled from the penis during male orgasm; contains millions of sperm in each milliliter of seminal fluid, an alkaline, fructose-containing liquid that nourishes the sperm cells suspended in it. (p. 1136)

semipermeable membrane *See* selectively permeable membrane.

sensory neuron A neuron that transmits nerve impulses from a sensory receptor to the central nervous system or central ganglion; an afferent neuron. (p. 893)

sensory receptor A cell, tissue, or organ that responds to internal or external stimuli by initiating a nerve impulse. (p. 908)

sepal (L. *sepalum*, a covering) A member of the outermost floral whorl of a flowering plant; collectively, the sepals constitute the calyx. (p. 678)

septate (L. *septum*, fence) Divided by cross-walls (septa) into cells or compartments. (p. 1067)

septum, pl. **septa** (L., fence) A wall between two cavities. (p. 639)

serotonin (L. *serum*, serum) 5-Hydroxytryptamine; a phenolic amine found in many animal tissues that has profound but poorly understood metabolic and vascular effects, particularly on the central nervous system. (p. 927)

sessile (L. *sessilis*, of or fit for sitting, low dwarfed) Attached; not free to move about. (p. 784) In vascular plants, a leaflacking a petiole is said to be sessile.

sex chromosomes Chromosomes that are different in the two sexes and that are involved in sex determination. All other chromosomes are called autosomes. *See* autosome. (p. 259)

sex-linked characteristic A genetic characteristic, such as color blindness in humans or white eye in fruit flies, that is determined by a gene located on a sex chromosome and that therefore shows a different pattern of inheritance in males than in females. (p. 246)

sexual reproduction The fusion of gametes followed by meiosis and recombination at some point in the life cycle. (p. 219)

sexual selection A type of differential reproduction that results from variable success in obtaining mates. More successful individuals may possess traits that are preferred by the opposite sex and/or traits that allow them to compete more successfully with other individuals of their same sex for mates. (p. 419)

shoot In vascular plants, the aboveground portions, such as the stem and leaves. (p. 703)

sieve cell In the phloem (food-conducting tissue) of vascular plants, a long, slender sieve element with relatively unspecialized sieve areas and with tapering end walls that lack sieve plates; found in all vascular plants except angiosperms, which have sieve-tube members. (p. 706)

sieve tube In the phloem of angiosperms, a series of sieve-tube members arranged end-to-end and interconnected by sieve plates. (p. 706)

sieve-tube member One of the component cells of a sieve tube. (p. 706)

sinoatrial node *See* pacemaker.

sinus (L., curve) A cavity or space in tissues or in bone. (p. 1039)

smooth muscle Nonstriated muscle; lines the walls of internal organs and arteries and is under involuntary control. (p. 892)

sociobiology The study of the biological basis of social behavior and of the organization of animal societies. (p. 1206)

sodium-potassium pump Transmembrane channels engaged in the active (ATP-driven) transport of sodium ions, exchanging them for potassium ions; maintains the resting membrane potential of neurons. (p. 123)

solute A molecule dissolved in some solution. As a general rule, solutes dissolve only in solutions of similar polarity—for example, glucose (polar) dissolves in (forms hydrogen bonds with) water (also polar), but not in vegetable oil (nonpolar). (p. 115)

solution A homogeneous mixture of the molecules of two or more substances; the substance present in the greatest amount (usually a liquid) is called the solvent, and the substances present in lesser amounts are called solutes. (p. 115)

solvent Medium in which one or more solutes is dissolved. *See* solution. (p. 115)

somatic cells (Gr. *soma*, body) The differentiated cells composing body tissues of multicellular plants and animals; all body cells except those giving rise to gametes. (p. 325)

somatic nervous system (Gr. *soma*, body) In vertebrates, the neurons of the peripheral nervous system that control skeletal muscle; the "voluntary" system, as contrasted with the "involuntary," or autonomic, nervous system. (p. 930)

somite (Gr. *soma*, body) One of the blocks, or segments, of tissue into which the mesoderm is divided during differentiation of the vertebrate embryo. (p. 1163)

speciation (L. *species*, kind) The process by which new species are formed during the course of evolution. (p. 408)

species, pl. **species** (L., kind, sort) A kind of organism; species are designated by binomial names written in italics. (p. 405)

specific heat The amount of heat (in calories) required to raise the temperature of 1 gram of a substance 1° C. The specific heat of water is 1 calorie per gram. (p. 33)

sperm (Gr. *sperma*, seed) A mature male gamete, usually motile and smaller than the female gamete. (p. 1134)

spermatid (Gr. *sperma*, seed) In animals, each of four haploid (1*n*) cells that result from the meiotic divisions of a spermatocyte; each spermatid differentiates into a sperm cell. (p. 1134)

spermatocytes (Gr. *sperma*, seed + *kytos*, vessel) In animals, the diploid (2*n*) cells formed by the enlargement of the spermatogonia; they give rise by meiotic division to the spermatids. (p. 1134)

spermatogenesis (Gr. *sperma*, seed + *genesis*, origin) In animals, the process by which spermatogonia develop into sperm. (p. 1134)

spermatogonia (Gr. *sperma*, seed + *gonos*, a child) In animals, the unspecialized diploid (2*n*) cells on the walls of the testes that, by meiotic division, become spermatocytes, then spermatids, then sperm cells. (p. 1134)

spermatozoon, pl. **spermatozoa** (Gr. *sperma*, seed + *zoos*, living) A sperm cell. (p. 1134)

sphincter (Gr. *sphinkter*, band, from *sphingein*, to bind tight) In vertebrate animals, a ring-shaped muscle capable of closing a tubular opening by constriction (such as the one between stomach and small intestine, or between anus and exterior). (p. 1017)

spindle apparatus The motive assembly that carries out the separation of chromosomes during cell division; composed of microtubules and assembled during prophase at the equator of the dividing cell. (p. 209)

spindle fibers A group of microtubules that together make up the spindle apparatus. (p. 209)

spiracle (L. *spiraculum*, from *spirare*, to breathe) External opening of a trachea in arthropods. (p. 832)

spiral cleavage In animals, a type of early embryonic cleavage in which cleavage planes are diagonal to the polar axis and unequal cells are produced by the alternate clockwise and counterclockwise cleavage around the axis of polarity; determinate cleavage. (p. 807)

spirillum, pl. **spirilla** (L. *spira*, coil) A long coiled or spiral bacterium. (p. 592)

spongy parenchyma A leaf tissue composed of loosely arranged, chloroplast-bearing cells. *See* palisade parenchyma. (p. 710)

sporangium, pl. **sporangia** (Gr. *spora*, seed + *angeion*, a vessel) A structure in which spores are produced. (p. 640)

spore A haploid reproductive cell, usually unicellular, capable of developing into an adult without fusion with another cell. Spores result from meiosis, as do gametes, but gametes fuse immediately to produce a new diploid cell. (p. 592)

sporophyte (Gr. *spora*, seed + *phyton*, plant) The spore-producing, diploid (2*n*) phase in the life cycle of a plant having alternation of generations. (p. 628)

stamen (L., thread) The organ of a flower that produces the pollen; usually consists of anther and filament; collectively, the stamens make up the androecium. (p. 678)

starch (Mid. Eng. *sterchen*, to stiffen) An insoluble polymer of glucose; the chief food storage substance of plants; typically composed of 1000 or more glucose units. (p. 45)

statocyst (Gr. *statos*, standing + *kystis*, sac) A sensory receptor sensitive to gravity and motion; consists of a vesicle containing granules of sand (statoliths) or some other material that stimulates surrounding tufts of cilia when the organism moves. (p. 948)

statolith (Gr. *statos*, standing + *lithos*, stone) Small calcareous body resting on tufts of cilia in the statocyst. (p. 948)

stem The aboveground axis of vascular plants; stems are sometimes below ground (as in rhizomes and corms). (p. 711)

stereoscopic vision (Gr. *stereos*, solid + *optikos*, pertaining to the eye) Ability to perceive a single, three-dimensional image from the simultaneous but slightly divergent two-dimensional images delivered to the brain by each eye. (p. 959)

steroid (Gr. *stereos*, solid + L. *ol*, from *oleum*, oil) One of a group of lipids having a molecular skeleton of four fused carbon rings and, often, a hydrocarbon tail. Cholesterol, sex hormones, and the hormones of the adrenal cortex are steroids. (p. 969)

stigma (Gr., mark, tattoo mark) (1) In angiosperm flowers, the region of a carpel that serves as a receptive surface for pollen grains. (2) Light-sensitive eye spot of some algae. (p. 677)

stimulus (L. goad, incentive) Any internal or external change or signal that can be detected by an organism; every stimulus is some form of energy change (for example, a change in heat, sound wave, chemical, or light energy). (p. 943)

stolon (L. *stolo*, shoot) (1) A stem that grows horizontally along the ground surface and may form adventitious roots, such as runners of the strawberry plant. (2) A functionally similar structure that occurs in some colonial cnidarians and ascidians. (p. 718)

stoma, pl. **stomata** (Gr., mouth) In plants, a minute opening bordered by guard cells in the epidermis of leaves and stems; water passes out of a plant mainly through the stomata, and CO_2 passes in chiefly by the same pathway. (p. 193)

striated muscle (L., from *striare*, to groove) Skeletal voluntary muscle and cardiac muscle. The name derives from its striped appearance, which reflects the arrangement of contractile elements. (p. 998)

style (Gr. *stylos*, column) In flowers, the slender column of tissue which arises from the top of the ovary and through which the pollen tube grows. (p. 677)

subspecies A subdivision of a species, often a geographically distinct race. (p. 407)

substrate (L. *substratus*, strewn under) The foundation to which an organism is attached; a molecule upon which an enzyme acts. (p. 142)

substrate level phosphorylation Formation of ATP that takes place via a coupled reaction during glycolysis. (p. 155)

succession In ecology, the slow, orderly progression of changes in community composition that takes place through time. *Primary succession* occurs in nature over long periods of time; *sec-*

ondary succession occurs when a climax community has been disturbed. (p. 499)

sugar Any monosaccharide or disaccharide. (p. 42)

surface tension A tautness of the surface of a liquid, caused by the cohesion of the molecules of liquid. Water has an extremely high surface tension. *See* cohesion. (p. 1045)

symbiosis (Gr. *syn*, together with + *bios*, life) The living together in close association of two or more dissimilar organisms; includes parasitism (in which the association is harmful to one of the organisms), commensalism (in which it is beneficial to one, of no significance to the other), and mutualism (in which the association is advantageous to both). (p. 480)

sympathetic nervous system A subdivision of the autonomic nervous system of vertebrates that functions as an alarm response; increases heartbeat and dilates blood vessels while putting the body's everyday functions, such as digestion, on hold; usually operates antagonistically with parasympathetic nerves; in times of stress, danger, or excitement, it mobilizes the body for rapid response. (p. 935)

synapse (Gr. *synapsis*, a union) A junction between a neuron and another neuron or muscle cell; the two cells do not touch, the gap being bridged by neurotransmitter molecules. (p. 908)

synapsis (Gr., contact, union) The point-by-point alignment (pairing) of homologous chromosomes that occurs before the first meiotic division; crossing-over occurs during synapsis. (p. 222)

syngamy (Gr. *syn*, together with + *gamos*, marriage) The process by which two haploid cells fuse to form a diploid zygote; fertilization. (p. 219)

synthesis (Gr. *syntheke*, a putting together) The formation of a more complex molecule from simpler ones. (p. 190)

systematics (Gr. *systema*, that which is put together) Scientific study of the kinds and diversity of organisms and of the relationships between them; equivalent to *taxonomy*. (p. 561)

systemic circulation The circulation path of blood leading to and from all body parts except the lungs. (p. 1067)

T cell A type of lymphocyte involved in cell-mediated immunity and interactions with B cells; the "T" refers to the fact that T cells are produced in the thymus; also called a T lymphocyte. (p. 1094)

tagma pl. **tagmata** (Gr., arrangement, order, row) A compound body section of an arthropod resulting from embryonic fusion of two or more segments; for example, head, thorax, abdomen. The process of fusion is called *tagmatization* or *tagmosis*. (p. 842)

tandem clusters Multiple copies of the same gene lying side by side in series. (p. 345)

taxis, pl. **taxes** (Gr., arrangement) An orientation movement by a (usually) simple organism in response to an environmental stimulus. (p. 1194)

taxonomy (Gr. *taxis*, arrangement + *nomos*, law) The science of the classifications of organisms; equivalent to *systematics*. (p. 561)

telencephalon (Gr. *telos*, end + *encephalon*, brain) The most anterior portion of the brain, including the cerebrum and associated structures. (p. 924)

telophase (Gr. *telos*, end + *phasis*, form) The last stage of the nuclear division of mitosis and meiosis, during which the chromosomes become reorganized into two nuclei. (p. 214)

template A pattern guiding the formation of a negative or complementary image; with reference to DNA duplication, each strand serves as a template on which a complementary strand is assembled. (p. 1186)

tendril (L. *tendere*, to extend) A slender coiling structure, usually a modified leaf or stem, that aids in the support of a plant. (p. 710)

territory (L. *territorium*, from *terra*, earth) An area or space occupied and defended by an individual or a group of animals. (p. 1201)

tertiary structure of a protein The three-dimensional shape of a protein; primarily the result of hydrophobic interactions of amino acid side groups and, to a lesser extent, of hydrogen bonds between them; forms spontaneously. (p. 55)

testcross A mating between a phenotypically dominant individual of unknown genotype (it could in principle be either homozygous or heterozygous) and a homozygous recessive "tester," done to determine whether the phenotypically dominant individual is homozygous or heterozygous for the relevant gene. *See* backcross. (p. 241)

testis, pl. **testes** (L., *witness*) In mammals, the sperm-producing organ; also the source of male sex hormone. (p. 1134)

testosterone (Gr. *testis*, testicle + *steiras*, barren) The male sex

hormone; a steroid hormone secreted by the testes of mammals that stimulates the development and maintenance of male sex characteristics and the production of sperm. (p. 1187)

tetrapods (Gr. *tetras*, four + *pous*, foot) Four-footed (actually, four-limbed) vertebrates; the group includes amphibians, reptiles, birds, and mammals. (p. 864)

thalamus (Gr. *thalamos*, chamber) That part of the vertebrate forebrain just posterior to the cerebrum; governs the flow of information from all other parts of the nervous system to the cerebrum. (p. 926)

theory (Gr. *theorein*, to look at) A well-tested hypothesis, one unlikely to be rejected by future tests. (p. 6)

thermodynamics (Gr. *therme*, heat + *dynamis*, power) The study of transformations of energy, using heat as the most convenient form of measurement of energy. The first law of thermodynamics states that the total energy of the universe remains constant. The second law of thermodynamics states that the entropy, or degree of disorder, tends to increase. (p. 136)

thigmotropism In plants, unequal growth in some structure that comes about as a result of physical contact with an object. The unequal growth that causes a vine to curl around a fence post is an example. (p. 769)

thorax (Gr., a breastplate) (1) In vertebrates, that portion of the trunk containing the heart and lungs. (p. 885) (2) In crustaceans and insects, the fused, leg-bearing segments between head and abdomen.

threshold value In neurons, the minimum change in membrane potential necessary to produce an action potential. (p. 909)

thylakoid (Gr. *thylakos*, sac + *-oides*, like) A saclike membranous structure in cyanobacteria and the chloroplasts of eukaryotic organisms; in chloroplasts, stacks of thylakoids form the grana; chlorophyll is found in the thylakoids. (p. 97)

thymine A pyrimidine occurring in DNA but not in RNA; *see also* uracil. (p. 57)

tight junction Region of actual fusion of cell membranes between two adjacent animal cells that prevents materials from leaking through the tissue; for example, intestinal epithelial cells are surrounded by tight junctions. (p. 129)

tissue (L. *texere*, to weave) A group of similar cells organized into a structural and functional unit. (p. 128)

trachea, pl. **tracheae** (L., windpipe) A tube for breathing; in terrestrial vertebrates, the windpipe that carries air between the larynx and bronchi (which leads to the lungs); in insects and some other terrestrial arthropods, a system of chitin-lined air ducts. (p. 832)

tracheid (Gr. *tracheia*, rough) In vascular plants, an elongated, thick-walled conducting and supporting cell of xylem. Tracheids, which are dead when functional, have tapering ends and pitted walls without perforations. Compare with *vessel element*. (p. 705)

transcription (L. *trans*, across + *scribere*, to write) The enzyme-catalyzed assembly of an RNA molecule complementary to a strand of DNA; may be mRNA (for protein-encoding genes), tRNA, or rRNA. (p. 302)

transcription unit The portion of a gene that is actually transcribed into mRNA; a sequence of DNA that begins with a promoter and encompasses the leader region (where the ribosome binds to mRNA) and one or more structural genes. A transcription unit does not include the regulatory portions of a gene such as operators, enhancers, or repressor-encoding sequences. (p. 315)

transduction The transfer of genes from one organism to another, with a virus used as a vector (carrier); "piggyback" recombination; occurs naturally among bacteria, when viruses mistakenly incorporate bacterial DNA into themselves and so transfer it to other cells they infect. (p. 942)

transfection The transformation of eukaryotic cells in culture. (p. 329)

transfer RNA (tRNA) (L. *trans*, across + *ferre*, to bear or carry) A class of small RNAs (about 80 nucleotides) with two functional sites; to one site an "activating enzyme" adds a specific amino acid, the other site carries the nucleotide triplet (anticodon) for that amino acid. Each type of tRNA transfers a specific amino acid to a growing polypeptide chain as specified by the nucleotide sequence of the messenger RNA being translated. (p. 301)

transforming principle (L. *trans*, across + *formare*, to shape) The transfer of naked DNA from one organism to another; also called simply "gene transfer." First observed as uptake of DNA fragments among pneumococcal bacteria, transformation is now routinely carried out in fruit flies and other eukaryotes by microinjection of eggs, with transposons often used as vectors. (p. 284)

translation (L. *trans*, across + *latus*, that which is carried) The assembly of a protein on the ribosomes, using mRNA to direct the order of amino acids. (p. 303)

translocation (L. *trans*, across + *locare*, to put or place) (1) In plants, the long-distance transport of soluble food molecules (mostly sucrose), which occurs primarily in the sieve tubes of phloem tissue. (2) In genetics, the interchange of chromosome segments between nonhomologous chromosomes. (p. 310)

transpiration (L. *trans*, across + *spirare*, to breathe) The loss of water vapor by plant parts; most transpiration occurs through the stomata. (p. 732)

transposition Type of genetic recombination in which transposable elements move from one site in the DNA to another. (p. 321)

transposon (L. *transponere*, to change the position of) A DNA sequence carrying one or more genes and flanked by insertion sequences that confer the ability to move from one DNA molecule to another; an element capable of transposition, the changing of chromosomal location. (p. 337)

tricarboxylic acid cycle or **TCA cycle** *See* Krebs cycle.

triglycerides A group of lipids formed when three fatty acids replace the three hydrogen atoms in the hydroxyl groups of glycerol. (p. 48)

trophic level (Gr. *trophos*, feeder) A step in the movement of energy through an ecosystem. (p. 497)

trophoblast (Gr. *trephein*, to nourish + *blastos*, germ) In vertebrate embryos, the outer ectodermal layer of the blastodermic vesicle; in mammals it is part of the chorion and attaches to the uterine wall. (p. 1159)

tropism (Gr. *trope*, a turning) A response to an external stimulus. (p. 768)

tropomyosin (Gr. *tropos*, turn + *myos*, muscle) Low-molecular-weight protein surrounding the actin filaments of striated muscle. (p. 998)

troponin Complex of globular proteins positioned at intervals along the actin filament of skeletal muscle; thought to serve as a calcium-dependent "switch" in muscle contraction. (p. 998)

tubulin (L. *tubulus*, small tube + *in*, belonging to) Globular protein subunit forming the hollow cylinder of microtubules. (p. 99)

turgor (L. *turgere*, to swell) The pressure exerted on the inside of a plant cell wall by the fluid contents of the cell; the interior of the cell is hypertonic in relation to the fluids surrounding it and so gains water by osmosis. (p. 770)

turgor pressure (L. *turgor*, a swelling) The pressure within a cell resulting from the movement of water into the cell. A cell with high turgor pressure is said to be *turgid*. *See* osmotic pressure. (p. 118)

umbilical (L. *umbilicus*, navel) In humans, refers to the navel, or umbilical cord. (p. 1170)

unequal crossing-over A crossover between two identical sequences lying at different locations on homologous chromosomes. (p. 342)

uracil A pyrimidine found in RNA but not in DNA; *see also* thymine. (p. 58)

urea (Gr. *ouron*, urine) An organic molecule formed in the vertebrate liver; the principal form of disposal of nitrogenous wastes by mammals. (p. 1124)

urethra (Gr. from *ourein*, to urinate) The tube carrying urine from the bladder to the exterior of mammals. (p. 1135)

uric acid Insoluble nitrogenous waste product produced largely by reptiles, birds, and insects. (p. 1124)

urine (Gr. *ouron*, urine) The liquid waste filtered from the blood by the kidney and stored in the bladder pending elimination through the urethra. (p. 1118)

uterus (L., womb) In mammals, a chamber in which the developing embryo is contained and nurtured during pregnancy. (p. 1138)

vacuole (L. *vacuus*, empty) A space or cavity within the cytoplasm of a cell; typically filled with a watery fluid, the cell sap; part of the lysosomal compartment of the cell. (p. 79)

vagina Female accessory reproductive organ that receives sperm from the male penis; forms part of the birth canal, and acts as a channel to the exterior for menstrual flow. (p. 1138)

variety A group of plants of less than species rank; some botanists view the term as equivalent to subspecies in animals, whereas others view it as a taxon of lower rank. (p. 407)

vas deferens (L. *vas*, a vessel + *deferre*, to carry down) In mammals, the tube carrying sperm from the testes to the urethra. (p. 1135)

vascular bundle In vascular plants, a strand of tissue containing primary xylem and primary phloem (and procambium if still present) and frequently enclosed by a bundle sheath of parenchyma or fibers. (p. 713)

vascular cambium In vascular plants, a cylindrical sheath of meristematic cells, the division of which produces secondary phloem outwardly and secondary xylem inwardly; the activity of the vascular cambium increases stem or root diameter. (p. 713)

vascular tissue (L. *vasculum*, a small vessel) Containing or concerning vessels that conduct fluid. (p. 658)

vector (L., a bearer, carrier, from *vehere*, to carry) In genetics, any virus or plasmid DNA into which a gene is integrated and subsequently transferred into a cell; in disease, any agent that carries and transmits pathogenic microorganisms from one host to another host. (p. 354)

vein (L. *vena*, a blood vessel) (1) In plants, a vascular bundle forming a part of the framework of the conducting and supporting tissue of a stem or leaf. (2) In animals, a blood vessel carrying blood from the tissues to the heart. (p. 1056)

ventral (L. *venter*, belly) Pertaining to the undersurface of an animal that moves on all fours, and to the front surface of an animal that holds its body erect. (p. 790)

ventricle (L. *ventriculus*, the stomach—i.e., the belly of the heart) A muscular chamber of the heart that receives blood from an atrium and pumps blood out to either the lungs or the body tissues. (p. 1065)

vertebrate An animal having a backbone made of bony segments called vertebrae. (p. 864)

vesicle (L. *vesicula*, a little bladder) A small, intracellular, membrane-bound sac in which various substances are transported or stored. (p. 79)

vessel (L. *vas*, a vessel) A tubelike element in the xylem of angiosperms, composed of dead cells (vessel elements) arranged end to end. Its function is to conduct water and minerals from the soil. (p. 705)

vessel element In vascular plants, a typically elongated cell, dead at maturity, which conducts water and solutes in the xylem; vessel elements make up vessels. *Tracheids* are less specialized conducting cells. (p. 705)

villus, pl. **villi** (L., a tuft of hair) In vertebrates, one of the minute, fingerlike projections lining the small intestine that serve to increase the absorptive surface area of the intestine. (p. 1020)

viroids A single-stranded RNA virus with no coat. In plants, an infectious agent responsible for diseases. (p. 583)

visceral mass (L., internal organs) Internal organs in the body cavity of an animal. (p. 811)

vitamin (L. *vita*, life + *amine*, of chemical origin) An organic substance that cannot be synthesized by a particular organism but is required in small amounts for normal metabolic function; must be supplied in the diet or synthesized by bacteria living in the intestine. (p. 1027)

vivipary (L. *vivus*, alive + *parere*, to bring forth) Reproduction in which eggs develop within the mother's body, with her nutritional aid; characteristic of mammals, many reptiles, and some fishes; offspring are born as juveniles. Adj., *viviparous*. (p. 1130)

water cycle Worldwide circulation of water molecules, powered by the sun. (p. 488)

water potential The potential energy of water molecules. Regardless of the reason (e.g., gravity, pressure, concentration of solute particles) for the water potential, water moves from a region where water potential is greater to a region where water potential is lower. (p. 730)

whorl A circle of leaves or of flower parts borne at a single node. (p. 677)

wild type In genetics, the phenotype or genotype that is characteristic of the majority of individuals of a species in a natural environment. (p. 249)

wood Accumulated secondary xylem. (p. 715)

xylem (Gr. *xylon*, wood) In vascular plants, a specialized tissue, composed primarily of elongate, thick-walled conducting cells, which transports water and solutes through the plant body. (p. 668)

yolk The stored food in egg cells; it nourishes the embryo. (p. 873)

zoospore A motile spore. (p. 626)

zygote (Gr. *zygotos*, paired together) The diploid ($2n$) cell resulting from the fusion of male and female gametes (fertilization). A zygote may either develop into a diploid individual by mitotic divisions or undergo meiosis to form haploid (n) individuals that divide mitotically to form a population of cells. (p. 751)

Illustration Acknowledgement

7-9 Bill Ober
7-10 Nadine Sokol after Bill Ober
7-11 Pagecrafters/Nadine Sokol
7-12 Bill Ober
7-13 Bill Ober
7-14 Bill Ober
7-15 Kevin Somerville after Bill Ober
7-16 Nadine Sokol after Bill Ober
7-17 Pagecrafters/Nadine Sokol
7-18 Barbara Cousins

CHAPTER 8
8-1 Nadine Sokol
8-2 Nadine Sokol
8-3 Kevin Somerville after Bill Ober
8-4 Barbara Cousins
8-5 Kevin Somerville
8-6 Pagecrafters/Nadine Sokol
8-7 Grant Heilman/Grant Heilman Photography, Inc.
8-8 Pagecrafters/Nadine Sokol
8-9 Pagecrafters/Nadine Sokol
8-10 Photo by Dr. Lester Reed; Illustration by Pagecrafters and Nadine Sokol after Bill Ober
8-11 Kevin Somerville after Bill Ober
8-12 Pagecrafters/Nadine Sokol
8-13 Bill Ober
8-14 Kevin Somerville after Bill Ober
8-15 Nadine Sokol after Bill Ober
8-16 Pagecrafters/Nadine Sokol after Bill Ober
8-17 Pagecrafters/Nadine Sokol
8-18 Pagecrafters/Nadine Sokol
8-A 1-5, E.S. Ross
8-B Bill Ober

CHAPTER 9
9-1 Kjell Sandved
9-2 Bill Ober
9-3 Pagecrafters/Nadine Sokol after Bill Ober
9-4 Bill Ober
9-5 Raychel Ciemma
9-6 Raychel Ciemma
9-7 Raychel Ciemma
9-8 Raychel Ciemma
9-9 Bill Ober
9-10 A, Sherman Thomson/Visuals Unlimited; B, David Dennis/Tom Stack & Assoc.
9-11 Pagecrafters/Nadine Sokol
9-12 Pagecrafters/Nadine Sokol
9-13 Manfred Kage/Peter Arnold, Inc.
9-14 Bill Ober
9-15 A, Dr. Lewis K. Shumway; B, Bill Ober
9-16 Kevin Somerville
9-17 Raychel Ciemma after Bill Ober
9-18 Bill Ober
9-19 Pagecrafters/Nadine Sokol
9-20 Richard Gross/Biological Photography
9-21 Pagecrafters/Nadine Sokol

CHAPTER 10
10-1 Zig Leszczynski/Animals Animals
10-2 Bill Ober
10-3 Lee D. Simon/Photo Researchers
10-4 Raychel Ciemma
10-5 Photo by Olins and Olins/Biological Photo Service; Illustration by Raychel Ciemma
10-6 Photo Researchers/Science Photo Library
10-7 Lili Robins
10-8 Photo by Biophoto Associates/Photo Researchers; Illustration by Nadine Sokol

10-9 Photos from Jan de Mey, Janssen Pharmaceutica, *Scientific American*, vol. 243, August 1980, page 76; Andrew S. Bajer, University of Oregon; Illustrations by Bill Ober
10-10 Photo by Andrew Bajer; Illustration by Barbara Cousins
10-11 David M. Phillips/Visuals Unlimited
10-12 Illustration by Nadine Sokol after Bill Ober; Photo by B.A. Palevitz & E.H. Newcomb, BPS/Tom Stack & Assoc.

CHAPTER 11
11-1 Science VU-L. Maziarski/Visuals Unlimited
11-2 Barbara Cousins
11-3 John D. Cunningham/Visuals Unlimited
11-4 Barbara Cousins
11-5 Diedrich von Wettstein: Reproduced with permission from Annual Review of Genetics, vol. 6, 1972 by Annual Review, Inc.
11-6 Kevin Somerville
11-7 Bill Ober
11-8 Bill Ober
11-9 Photo by C.A. Hasenkampf/ BPS; Illustrations by Barbara Cousins
11-10 Charles J. Cole, American Museum of Natural History
11-11 Raychel Ciemma
11-12 Sheldon Wolff & Judy Bodycote, University of California-San Francisco

CHAPTER 12
12-1 Alan Carey/The Image Works
12-2 A, Dan Coffey/The Image Bank; B, Gisela Caspersen/The Image Bank; C, Hank DeLespinasse/The Image Bank;
12-3 A-B, Fred Bruemmer
12-4 Science VU-NIH/Visuals Unlimited
12-5 Richard Gross/Biological Photography
12-6 Courtesy of V. Orel, Mendelianum Musei Moraviae, Brno
12-7 R.W. Van Norman
12-8 A, John D. Cunningham/Visuals Unlimited; B, Bill Ober
12-9 Nadine Sokol
12-10 Pagecrafters/Nadine Sokol after Bill Ober
12-11 Courtesy of V. Orel, Mendelianum Musei Moraviae, Brno
12-12 Globe Photos
12-13 Nadine Sokol
12-14 Nadine Sokol
12-15 Nadine Sokol after Bill Ober
12-16 Nadine Sokol after Bill Ober
12-17 A-B, Carolina Biological Supply
12-18 Raychel Ciemma
12-19 Pagecrafters/Nadine Sokol
12-20 Bill Ober
12-21 Bill Ober
12-22 Pagecrafters/Nadine Sokol
12-23 Photo from Albert & Blakeslee, "Corn and Man," *Journal of Heredity*, vol. 5, pg. 511, 1914; Illustration by Pagecrafters/Nadine Sokol
12-24 Nadine Sokol
Genetics problems illustrations by Pagecrafters/Nadine Sokol

CHAPTER 13
13-1 The Bettman Archive
13-2 A, Loris McGavran/Denver Children's Hospital; B, Richard Hutchings/Photo Researchers
13-3 Pagecrafters/Nadine Sokol
13-4 Raychel Ciemma after Bill Ober
13-5 Photo by Field Museum of Natural History, Neg #118, Chicago; Illustration by Pagecrafters/Nadine Sokol
13-6 Pagecrafters/Nadine Sokol
13-7 Photo by Dr. Frank B. Sloop, Jr.; Illustration by Pagecrafters/Nadine Sokol
13-8 Ingalls & Salerno, *Maternal & Child Health Nursing*, ed. 6, Mosby-Year Book, 1987
13-9 Murayama 1981/BPS
13-10 Pagecrafters/Nadine Sokol
13-11 Pagecrafters/Nadine Sokol
13-12 Pagecrafters/Nadine Sokol
13-13 Photo by Michael Ochs Archives/Venice, CA; Illustration by Pagecrafters/Nadine Sokol
13-14 Kevin Somerville after Bill Ober
13-15 R. Winborn, RDMS, Washington University, St. Louis, MO
13-16 Pagecrafters/Nadine Sokol
13-17 Nadine Sokol
13-18 Nadine Sokol
13-19 Nadine Sokol
Genetics problems by Pagecrafters/Nadine Sokol

CHAPTER 14
14-1 Eastcott/Momatiuk/The Image Bank
14-2 Biological Photo Service
14-3 Bill Ober
14-4 Anne Crossway/Calgene, Inc.
14-5 Bill Ober
14-6 Bill Ober
14-7 Photo A by A.K. Kleinschmidt; Photo B-C, by Lee D. Simon/Photo Researchers; Illustration by Kevin Somerville after Bill Ober
14-8 Barbara Cousins
14-9 Pagecrafters/Nadine Sokol
14-10 Bill Ober
14-11 Pagecrafters/Nadine Sokol
14-12 From J.D. Watson's The Double Helix, Antheneum, New York, 1968
14-13 A.C. Barrington Brown from J.D. Watson, The Double Helix, Atheneum, New York, 1968
14-14 B, Bill Ober
14-15 Bill Ober
14-16 George Klatt, Nadine Sokol after Bill Ober
14-17 Nadine Sokol after Bill Ober
14-18 Nadine Sokol after Bill Ober
14-19 Barbara Cousins
14-20 Ulrich Laemmli
14-21 A, David R. Wolstenholme; B, Bill Ober
14-22 Barbara Cousins
14-23 Barbara Cousins/Raychel Ciemma after Bill Ober
14-24 Bill Ober

CHAPTER 15
15-1 Barbara Cousins
15-2 Pagecrafters/Nadine Sokol
15-3 Kevin Somerville after Bill Ober
15-4 Nadine Sokol after Bill Ober
15-5 Nadine Sokol
15-6 Bill Ober
15-7 A, O.L. Miller, Jr., Barbara A.

Hamkalo, C.A. Thomas, Jr. 1970. "Visualization of Bacterial Genes in Action" *Science* 169:392–395; B, Bill Ober
15-8 Nadine Sokol after Bill Ober
15-9 Nadine Sokol after Bill Ober
15-10 Nadine Sokol after Bill Ober
15-11 Nadine Sokol after Bill Ober
15-12 Oscar Miller
15-13 Nadine Sokol after Bill Ober
15-14 Bill Ober
15-15 Claus Pelling
15-16 Photo by Jack Griffith; Illustration by Nadine Sokol
15-17 Raychel Ciemma after Bill Ober
15-18 Bill Ober
15-19 Raychel Ciemma after Bill Ober
15-20 Nadine Sokol
15-A 1-2, Bill Ober; Photo by Bert O'Malley

CHAPTER 16
16-1 David Scharf/Discover 1987, Family Media, Inc.
16-2 Nadine Sokol after Bill Ober
16-3 Bill Ober
16-4 Ackerman and del Regato, Cancer, by J.A. del Regato, H. Spjut and J.D. Cox, Mosby-Year Book, St. Louis, MO 1985
16-5 Bill Ober
16-6 Bill Ober
16-7 Custom Medical Stock
16-8 Raychel Ciemma
16-9 A, Nadine Sokol; B-C, Lili Robins
16-10 Barbara Cousins after Bill Ober
16-11 Raychel Ciemma after Bill Ober
16-12 Barbara Cousins after Bill Ober
16-13 Bill Ober
16-14 Lilli Robins
16-15 Bill Ober
16-16 Lili Robins
16-A American Cancer Society
16-B Pagecrafters/Nadine Sokol
16-C Bill Ober

CHAPTER 17
17-1 Nik Kleinberg/The Picture Group, Inc.
17-2 Bill Ober
17-3 Bill Ober
17-4 Bill Ober
17-5 Bill Ober
17-6 Photo by S. Dohen & J. Shapiro, *Scientific American*, February 1980, pages 40-49; Illustration by Bill Ober
17-7 Bill Ober
17-8 Bill Ober
17-9 Bill Ober
17-10 M.L. Pardue, Chromosomes Today, vol. 3, pg. 47-52; Donald Brown, Proceedings of the National Academy of Sciences vol. 68, pg. 3175, 1975
17-11 Bill Ober
17-12 Bill Ober
17-13 Bill Ober
17-14 Bill Ober

CHAPTER 18
18-1 R.L. Brinster
18-2 Barbara Cousins
18-3 Barbara Cousins after Bill Ober
18-4 Stanley N. Cohen, Stanford University
18-5 Barbara Cousins after Bill Ober
18-6 Photo from BioProducts,

54-7 Lennart Nilsson, *Behold Man,* Little Brown and Co.

54-8 Nadine Sokol after Bill Ober

54-9 Nadine Sokol after Bill Ober

54-10 Lennart Nilsson, Behold Man, Little, Brown & Co.

54-11 Barbara Cousins (Rhonda has this as Ed Reschke)

54-12 Ed Reschke

54-13 Barbara Cousins

54-14 Scott Bodell after Bill Ober

54-16 Nadine Sokol

54-17 Kevin Somerville

54-18 Pagecrafters/Nadine Sokol

54-19 A-E, Trent Stephens

54-20 Bill Ober

54-A Steven G. Hoffman

54-B 1, Lennart Nilsson, Boehringer Ingelheim International GmbH, THE INCREDIBLE MACHINE, National Geographic Society.; 2, Bill Ober

CHAPTER 55

55-1 Lennart Nilsson, *A Child is Born,* Dell Publishing Company

55-2 A-B, David M. Phillips/Visuals Unlimited

55-3 A, Bill Ober

55-4 Bill Ober

55-5 Barbara Cousins after Bill Ober

55-6 David M. Phillips/Visuals Unlimited

55-7 A-D, Carolina Biologica Supply Co.

55-8 David M. Phillips/Visuals Unlimited

55-9 Bill Ober

55-10 Kevin Somerville after Bill Ober

55-11 Bill Ober

55-12 Kevin Somerville after Bill Ober

55-13 Kevin Somerville after Bill Ober

55-14 Bill Ober

55-15 Nadine Sokol

55-16 Bill Ober

55-17 A, Barbara Cousins after Bill Ober; B, Courtesy of Norman K. Wessels & Kathryn W. Tosney from Tissue Interactions & Development, Benjamin Cummings, Menlo Park, Calif. 1977

55-18 A-C, Lennart Nilsson, A CHILD IS BORN, Dell Publishing Company; D, Lennart Nilsson, BEHOLD MAN, Little, Brown & Company

55-19 Nadine Sokol after Bill Ober

55-20 Scott Bodell after Bill Ober

55-21 Claire Parry/The Image Works

CHAPTER 56

56-1 Bettman Archive

56-2 Stephen Dalton/Photo Researchers

56-3 Nadine Sokol/Pagecrafters

56-4 Jim Grier: *Biology of Animal Behavior,* Mosby-Year Book, St. Louis, MO, 1984

56-5 John Alcock: *Animal Behavior,* 4th Ed., Sinauer, Sunderland, MA 1988

56-6 A.O.D. Willows

56-7 A-C, Lee Boltin Picture Library

56-8 Omikron/Photo Researchers

56-9 Thomas McAvoy, LIFE Magazine, Time Inc.

56-10 *Grzimek's Encyclopedia of Ethology,* Van Nostrand Reinhold Co.

56-11 Harlow Primate Laboratory, University of Wisconsin

56-12 Redrawn from John Alcock, *Animal Behavior,* 4th ed., Sinauer, Sunderland, MA 1988 by Pagecrafters

56-13 Jennifer Owne, *Feeding Strategy,* University of Chicago Press, 1980

56-14 Elizabeth Rhone Rudder

56-15 B. Miller/Biological Photo Service

56-16 Niko Tinbergen, *The Study of Instinct,* Oxford University Press, 1951, Oxford England

56-17 Oxford Scientific Films/Animals Animals

56-18 A. Thornall and J. Alcock: *Insect Mating Systems,* Sinauer, Sunderland, MA 1984

56-19 Michael Fogden/Animals Animals

56-20 P. Marler and W. Hamilton: *Mechanisms of Animal Behavior,* John Wiley & Sons, New York, NY, 1984

56-21 Scientific American, May 1963

56-22 A, Stan Osolinski/Oxford Scientific Films; B, *Animal Behavior,* 36 (2), pg. 484, 1988

56-23 Pagecrafters

56-24 Bill Ober

56-25 Linda Koebner/Bruce Coleman, Inc. (chimpanzee); Jeff Foott/Bruce Coleman, Inc. (sea otter)

56-26 Bill Ober

56-A Cleveland Hickman

56-B 1, Dr. Mark Moffett, National Geographic Society; 2, Elizabeth Rhone Rudder; 3-4, Dr. Mark Moffett, National Geographic Society

CHAPTER 57

57-1 Nina Leen, LIFE Magazine/Time Warner, Inc.

57-2 A-B, E.S. Ross

57-3 Joe McDonald/Tom Stack & Assoc.

57-4 Pagecrafters

57-5 Dr. David Macdonald

57-6 Pagecrafters

57-7 K. Preston-Mafham/Animals Animals

57-8 J.R. Brownlie/Bruce Coleman Limited

57-9 Terry G. Murphy/Animals Animals

57-10 William J. Weber/Visuals Unlimited

57-11 Dale & Marian Zimmerman/Animals Animals

57-12 A, Jim Traniello; B, Dr. Merlin Tuttle, Bat Conservatory International; C, Pagecrafters

57-13 A-B, from Halliday, *Sexual Strategy,* University of Chicago Press, 1980

57-14 Pagecrafters

57-15 Zig Leszczynski/Animals Animals

57-16 Gary McCracken, University of Tennessee, Knoxville

57-17 Pagecrafters

57-18 A, Tom Mangelsen; B, Art Gingert/Wildlands Photography

57-19 from Owen, *Feeding Strategy,* University of Chicago Press, 1980

57-20 E.S. Ross

57-21 Paulette Brunner/Tom Stack & Assoc.

57-22 E.O. Wilson: *Sociobiology: The New Synthesis,* Harvard University Press, Cambridge, MA 1975

57-23 A, Anthony Bannister/Animals Animals/Oxford Scientific Films; B-C, Napier, *The Natural History of Primates*/MIT Press

57-24 A, from T. Halliday, *Sexual Strategy,* University of Chicago Press, 1980; B, Jim Grier: *Biology of Animal Behavior,* Mosby-Year Book, St. Louis, MO 1984

57-A Christopher Springman

57-B 1-3, Dr. Mark Moffett

Index

Family trees
    human, possible, 443
    vertebrate, 865
Fan palms, 522
Fan worm, 820
Farallon Islands, 511
Fargesia spathacea, 716
Farm animals, genetic engineering
    in, 361-362
Fast-glycolytic muscle fiber, 1006
Fast-oxidative muscle fiber, 1006
Fat(s), 47-49
    brown, 1055
    cellular respiration of, 175-177
    digestion of, and liver, 1021
    saturated and unsaturated, 48
Fat body in insects, 847
Fatigue, muscle, 1005
Fatty acids, 47-48
    chains of, 108-109
    oxidation of, 175-176
Feather stars, 855, 857
Feathers, 876
    brightly colored, 1204-1205
Feces, 1023
Feedback
    of endocrine hormones, 983
    in muscle control, 933
Feedback inhibition, 150, 177
Feedback loops, 149
    in homeostasis, 896-897
Feeding
    of anthozoans, 789
    of brittle stars, 859
    of Darwin's finches, 417
    in flatworms, 792-793
    in sponges, 784
Felis, 561
Female reproductive system,
    1136-1138
Fermentation, 164-166
Ferns, 670-672
Ferredoxin, 186
Ferret, 879
Fertility
    declines in, 535
    sickle cell anemia and, 387
Fertilization, 219, 1155-1157
    double, 679
    external, 1129
    human, 1137
Fertilizer, 741
    in water pollution, 550
Fetal malnourishment, 1171
Fetus, human, 1155
    ultrasound appearance of, 272
Fever
    infection causing, 1096
    role of, in immunity, 1087
Fever blisters, 585
Fibers
    in foods, 1028
    muscle, interdigitation of, 1000
    in plants, 705
    spindle, 209
Fibrin, 51, 1064
Fibrinogen, 1062
Fibroblasts, 890
Fibrous proteins, 56
Fick's law of diffusion, 1034-1035
Ficus carica, 697
Fiddleheads, 672
Field capacity of soil, 729
Fight or flight reaction, 1075
Figs, 697
Filament, 678
Filaria, 800
Filariasis, 800
File shell, 815
Filial generations, 237, 240-241
Filial imprinting, 1185
Filtration by kidney, 1117
Finches, Darwin's, 10, 416-418, 466,
    511
    patterns of distribution of, 397
Fine touch receptors, 947

Fins, pectoral, 867
Fire in chaparral, 526-527
Fire ants, foraging in, 1192
Fire blight, 598
Fireflies, flash patterns of, 1190-1191
Fire-pink, 687
Fireweed, 692
Firewood gathering, 543
First filial (F_1) generation, 237, 240
First law of thermodynamics, 136
Fish(es)
    bony, 868-870
    brain in, 921, 923
    cell cleavage in, 1158
    breathing in, 1036
    electricity in, 960-961
    freshwater kidneys of, 1118
    heart of, 1065-1066
    jawed, 866-867
    jawless, 864, 865-867
    lateral line organs of, 949-950
    lobe-finned, 869-870
    marine, kidneys of, 1118
    ray-finned, 869-870
    reproduction in, 1129-1130
    and sea anemones, commensalism
        in, 480
    sense of taste of, 950
    viviparity in, 1131
Fisher, R.A., 1204
Fisheries, 512-513
    anchovy, 511
    mismanagement of, 542
Fission, binary, 203-204
Fitness, 383, 1199
    inclusive, 1208
Five kingdoms of life, 560-572
Fixed action pattern in behavior,
    1178
Flagellates, reproduction in, 226
Flagellum(a), 79, 101-104, 622, 785
    of bacteria, 101, 102, 592, 593
Flame cells, 793
Flamingos, 1205
Flatfish, 477
Flatworms, 781, 792-797
    water current in, 1035
Flavine adenine dinucleotide (FAD),
    168
Flea, 483
Fleming, Walter, 205
Flexors, 933
Flies, 845
Flocking behavior, 1207
Flooding, response of plants to, 731
Florida scrubjay, 1212-1213
Flour beetle, 461, 466
Flow rate, 1057
Flower(s), 677-678, 680, 687
    of dicots, 685
    evolution of, 688-690
    formation of, 718
    pistillate and staminate, 690
    primitive, 688-689
    in species isolation, 412
Flower fly, 479
Flowering
    bamboo, and panda starvation, 716
    ethylene in, 767
    photoperiodism in, 771-772
Flowering hormone, 774
Flowering plants, 438, 677-700
    definition of, 677
    development of embryo in, 752
    evolution of flower in, 688-690
    evolution of fruits in, 697-698
    history of, 684-688
    pollination in, 693-696
    specialization trends in, 690-693
Flu
    epidemic of 1918, 1081, 1107
    immunity to, 1105
    viruses in, 578, 583
Fluid, lipid bilayer as, 110
Flukes, 794-796
Fluoride and dental caries, 600

Flying, 876
    metabolic demands of, 1040
Flying insects, 834
Foams, spermicidal, 1145, 1146
Focusing human eye, 956
Fog, absorption of water from, 733
Foliose lichens, 648
Follicle-stimulating hormone (FSH),
    974, 977, 1135, 1137,
    1140, 1141, 1187
Follicular phase of reproductive
    cycle, 1141-1142
Food
    fungi in, 637
    human, mollusks as, 810
    and population, 536
    prospects for more, 539-542
Food chain, 497
    metabolic efficiency and, 158-159
Food poisoning, 597
Food vacuole, 1010
Food web, 497-498
Foot of mollusk, 811
Foraging behavior, 1199-1200
Foraminifera, 439, 624-625
Foraminiferans, 108, 624
Forams, 108, 624
Forcer inspiration, 1045
Forebrain, 921, 923-924, 981
Foreign cells, distinguishing from
    self, in immune system,
    1084, 1105-1107
Foreign species, 455
Forest(s)
    biogeochemical cycles in, 494-496
    coniferous, 528
    monsoon, 522
    primary productivity in, 497
    redwood, 470
    temperate deciduous, 525-527
    tropical rain, 496-497, 519-521
        destruction of, 543-545
Fork, replication, 292-294
Fossil(s), 424-427
    of angiosperms, 686
    of bacteria, 71
    bird, 876
    brachiopod, 809
    of crinoids, 857
    crustacean, 833-834
    dating of, 426-427
    of earliest plants, 659
    as evidence of macroevolution,
        390-391, 392
    living, 70
    of protostomes and deuterostomes,
        808
    similar to lancelets, 863
    in testing theory of evolution, 13
Fossil fuels, 490
    and atmosphere, 547, 548
Fossilization, preservation problems
    in, 425-426
Founder principle, 377
Fovea, 956
Fowl cholera, 1087-1088
Foxglove, 501
Fraenkel-Conrat, Heinz, 286
Fragaria ananassa, 690, 733
Frameshift mutation, 324-325, 342
Frankia, 492, 606
Franklin, Rosalind, 289
Fratercula cirrhata, 874
FRC; *see* Functional residual capacity
Free energy, 137-138
Free nerve endings, 941-942
Free radical, 322
Free-running rhythms, 1188
Frequency in population genetics,
    373-375
Fresh water, 515-518
    habitats of, 515-518
    stratification in, 517
Freshwater fishes, kidneys of,
    1118
Freshwater mussels, 815

Freshwater vertebrates,
    osmoregulation in, 1115
Frog(s), 516, 871
    chemical defenses of, 477
    egg of, 1158
    embryo of, 1159
    gastrulation in, 1161
    gray crescent in, 1157
    species of, 413
    tadpoles of, 1131
    vocalizing of, 1204
Fronds, fern, 672
Fruit(s)
    auxin and, 764
    dispersal of, 687, 697
        by animals, 697
    evolution of, 697-698
    as ovary, 678
    ripening of, ethylene in, 767
Fruticose lichens, 648
FSH; *see* Follicle-stimulating
    hormone
Fuels, fossil, 490
Fugitive species, 501
Fugu, 913
Fuligo, 626
Funaria hygrometrica, 502
Functional groups, 40, 41
Functional residual capacity (FRC),
    1043
Fundamental niche, 466
Fungus(i), 494, 495, 636-654
    associations with, 648-650
    as carnivores, 650
    divisions of, 640-646
        comparison of, 646-647
    in food chain, 497
    "gardens" of, and leafcutter ants,
        481
    Imperfecti, 647-648
    mycorrhizae, 650-652
    number of chromosomes in, 205
    nutrition and ecology of, 638-639
    reproduction in, 639-640
    structure of, 639
    success of, on land, 433
Fur trapping, 527
Fusarium, 648
Fusiform initials, 713
Fusion in fertilization, 1157

G

G chains, antibody, 1100
G protein, 971
G_1 phase, 207
G_2 phase, 207
G3P; *see* Glyceraldehyde-3-phosphate
GA_1, 766
GABA; *see* Gammaaminobutyric acid
Galapagos Islands, 8, 10-11, 416-418,
    511
    finches of, patterns of distribution
        of, 397
    tortoises of, 10
Gallbladder, 1021
Gametangium(a), 618, 646, 661
    of ferns, 671
    of fungi, 640, 641
    of nonvascular plants, 664
Gametes
    prevention of fusion of, as barrier
        to hybridization, 413
    plant, 660
Gametic meiosis, 570
Gametocytes, 630, 631
Gametogenesis, female human, 1137
Gametophytes, 628, 659-660
    fern, 670-671
    specialization of, 661-662
Gamma rays, 181
Gammaaminobutyric acid (GABA),
    914-915, 936
Gamma-interferon, 1096
Gancer-causing genes, 260
Ganglia
    basal, 925-926

Updated Version
BIOLOGY
Third Edition

List of Changes

page xv Corrected page references; changed "photocenter" to "photosystem."

page xvii Corrected page references.

page xxi Corrected contents entries.

page 33 Text correction; fixed folio.

page 56 Text correction ("itself" to "themselves" in second paragraph).

page 87 Correction to numbered item 4.

page 104 Correction in Table 5.2.

page 120 Text correction ("carriers" boldfaced in third paragraph).

page 121 Text correction ("The" lowercase in third paragraph).

page 122 Text correction (deleted one paragraph under "Active Transport").

page 185 Text corrections ("photosystem" and "Reaction Center Chlorophyll" boldfaced).

page 186 Text corrections (first two sentences clarified).

page 188 Text corrections("chlorophyll *a* " changed to "chlorophyll").

page 235 Figure 12.8 legend revised.

page 242 Text correction (new paragraph started with "Mendel's model").

page 278 Changed "white" to "Caucasian."

page 282 Art correction.

page 304 Art label correction.

page 310 Art label correction.

page 355 Text correction ("vectors" boldfaced).

page 356 Art label corrections.

page 357 Text corrections ("98 degrees Celsius" to "98° C," "60 degrees Celsius" to "60° C," "enzume" to "enzyme," "2 to the twentieth power" to "2²⁰", "JURRASIC PARK" to "Jurrassic Park").

page 422 Corrected alphabetizing of readings.

page 427 Art label correction.

page 445 Text correction ("support to "supports").

page 563 Corrected leader line in Figure 28.4.

page 565 Corrected paragraph heading.

page 567 Figure 28.9 legend revised.

page 568 Corrected paragraph heading.

page 569 Corrected paragraph heading.

page 570 Text corrections ("ORDER" to *order,* "MAGNITUDE" to "magnitude").

page 571 Figure 28.13 legend rewritten.

page 664 Text correction, deleted incorrect figure reference.

page 678 Corrected figure reference in Figure 33.26 legend.

page 680 Art label correction.

page 783 Art label correction in Figure 39.2.

page 812 Text correction ("whereas" to "where").

page 852 Added "Vollrath" to list of readings.

page 857 Text correction ("sarfish" to "starfish").

page 862 Corrected Figure 42.12 legend.

page 863 Text correction ("musles" to "muscles").

page 867 Text correction ("cartilaganeous" to "cartilaginous").

page 874 Text correction ("mannals" to "mammals," "cones" to "ones").

page 906 Art label correction.

page 908 Text correction ("patch clamping" boldfaced).

page 925 Text correction("haves" to "halves").

page 931 Text correction ("a" added before reflex in second paragraph).

page 944 Art label correction.

page 981 Figure 47.19 legend corrected.

page 991 Text corrections (art reference corrected).

page 994 Text and art corrections ("be cuase" to "because").

page 1011 Art label corrections.

page 1046 Text corrections (first line in first paragraph under "Oxygen Transport" clarified).

page 1067 Footnote correction.

page 1068 Text added before last paragraph head.

page 1076 Text correction ("constrition" to "constricting").

page 1086 Text correction ("monocytes" to "neutrophils").

page 1117 Text corrections ("excludes" to "allow passage of," "secretion" to "excretion").

page 1123 Art correction.

page 1137 Text corrections (fertilization clarified).

page 1139 Text corrections (fertilization clarified).

page 1140 Table 54.1 corrected.

page 1141 Text correction ("estrogen" boldfaced).

page 1161 Text corrections ("animal hemisphere" to "surface," "symmetrical" to "asymmetrical").

page 1182 Text correction ("associated" to "associate").

page 1188 Text correction (third paragraph clarified).

page 1190 Text correction ("but" to "But").

page 1191 Figure 56.18 legend corrected.

page 1193 Order of box photos corrected.

page 1200 Text corrections ("comes" to "come," "measure" to "measured").

page 1203 Text correction ("Reproduction" to "Reproductive").

page 1208 Text correction ("not" added before "receiving" in tenth text line).

THE METRIC SYSTEM

Metric Length

1 meter (the unit)	×	10	=	dekameter (10 m)
		100	=	hectometer (10^2 m)
		1000	=	kilometer (10^3 m)
		1,000,000	=	megameter (10^6 m)

1 meter	÷	10	=	decimeter (10^{-1} m)
		100	=	centimeter (10^{-2} m)
		1000	=	millimeter (10^{-3} m)
		1,000,000	=	micrometer (10^{-6} m)
		1,000,000,000	=	nanometer (10^{-9} m)
		1,000,000,000,000	=	picometer (10^{-12} m)
		10,000,000,000	=	Angstrom (Å) (10^{-10} m) (an older unit of measurement)

Metric Weights or Masses

1 gram (the unit)	×	1000	=	kilogram

1 gram	÷	1000	=	milligram (mg) (10^{-3} g)
		1,000,000	=	microgram (μg) (10^{-6} g)
		1,000,000,000	=	nanogram (ng) (10^{-9} g)
		1,000,000,000,000	=	picogram (pg) (10^{-12} g)